Selected Diode Model Parameters

Parameter	Default Value	Description
Saturation current	10^{-14} A	I_S from Eq. (4.1)
Emission coefficient	1	η from Eq. (4.1)
Parasitic resistance	0 Ω	Series lead and bulk resistance
Zero-bias *pn* capacitance	0 F	C_{jo} from Eq. (6.79)
pn junction potential	1 V	V_j, as in Eq. (6.80)
pn grading coefficient	0.5	0.5 = abrupt 0.33 = graded
Forward-bias-depletion capacitance coefficient	0.5	C_T saturates for $v_R < -FC \times VJ$
Transit time	0 s	τ
Reverse breakdown voltage	∞ V	Magnitude of BV
Reverse breakdown current	10^{-10} A	Reverse current at BV
Band-gap potential	1.11 eV	0.67 = Schottky 1.11 = silicon
I_S temperature coefficient	3	Temperature coefficient for n_i^2
Flicker noise coefficient	0	
Flicker noise exponent	1	

Selected SPICE MOSFET Model Parameters

Name	Parameter	Default Value	Description
LEVEL	Model index	1	Shichman–Hodges
VTO	Zero-bias threshold voltage	0 V	V_t, as in Eq. (7.5)
KP	Transconductance parameter	2×10^{-5} A/V^2	$K_m = K_n$, as in Eq. (7.5)
LAMBDA	Channel-length modulation	0 V^{-1}	λ, as in Eq. (7.16)
GAMMA	Bulk threshold parameter	$0\sqrt{V}$	γ
RD	Drain ohmic resistance	0 Ω	r_d
RS	Source ohmic resistance	0 Ω	r_s
TOX	Oxide thickness	10^{-5} cm	
NSUB	Substrate doping	0 cm^{-3}	
UO	Surface mobility	600 cm^2/V.s	

Selected SPICE BJT Model Parameters

Name	Parameter	Default Value	Description
IS	Transport saturation current	10^{-16} A	I_S from Eq. (8.83)
NF	Forward I emission coefficient	1	η
BF	Ideal maximum forward β_F	100	β in Eq. (8.6)
IKF	Corner for high-current β_F rolloff	∞ A	
VAF	Forward Early voltage	∞ V	V_A
BR	Ideal maximum reverse β_R	1	As in Eq. (D.3)
NR	Reverse I emission coefficient	1	
IKR	Corner for high-current β_R rolloff	∞ A	
RB	Base ohmic resistance	0 Ω	Base spreading resistance r_b
RC	Collector resistance	0 Ω	Collector lead and bulk resistance
RE	Emitter resistance	0 Ω	Emitter lead resistance
TF	Forward transit time	0 s	$\tau_F = 1/2\pi f_T$
TR	Reverse transit time	0 s	
CJE	Zero-bias B-E capacitance	0 F	C_{je0}, transition capacitance, as in Eq. (8.159)
VJE	B-E junction potential	0.75 V	V_{je} for base–emitter junction, as in Eq. (8.159)
MJE	B-E junction grading coefficient	0.33	0.5 = abrupt 0.33 = graded
FC	Forward-bias-depletion capacitance coefficient	0.5	C_{je} saturates for $v_{BE} > \mathrm{FC} \times \mathrm{VJE}$
CJC	Zero-bias B-C capacitance	0 F	$C_{\mu 0}$, as in Eq. (8.161)
VJC	B-C junction potential	0.75 V	V_{jc} for base–collector junction
MJC	B-C junction grading coefficient	0.33	0.5 = abrupt 0.33 = graded
CJS	Zero-bias collector-substrate C	0 F	C_{CS0}, as in Eq. (8.162)
VJS	C-S junction potential	0.75 V	V_{js} for collector-substrate junction, as in Eq. (8.162)
MJS	C-S junction grading coefficient	0	0.5 if needed
EG	Band-gap potential	1.11 eV	Silicon
XTI	I_S temperature coefficient	3	Temperature coefficient for I_S
KF	Flicker noise coefficient	0	
AF	Flicker noise exponent	1	

Physical Constants

Quantity	Symbol	Value
Boltzmann's constant	k	1.38066×10^{-23} J/K
Elementary charge	q	1.60218×10^{-19} C
Electron rest mass	m_e	9.1095×10^{-31} kg
Electron volt	eV	1 eV = 1.60218×10^{-19} J
Permittivity of free space	ϵ_o	8.85418×10^{-14} F/cm
		8.854×10^{-12} F/m
Planck's constant	h	6.62617×10^{-34} J·s
Speed of light in vacuum	c	3.0×10^{10} cm/s = 3×10^{8} m/s
Thermal voltage at 300 K	kT/q	0.0259 V

Conversion Factors and Prefixes

Unit	Symbol	Conversion
Angular frequency	ω	$\omega = 2\pi f$ rps (radians/s)
Frequency	f	$f = \omega/2\pi$ Hz (cycles/s)
Micron, micrometer	µm	10^{-6} m = 10^{-4} cm = 10^{-3} mm
Mil	mil	10^{-3} in. = 25.4 µm
Temperature, Celsius	°C	°C − K − 273.15°
femto	f	$\times 10^{-15}$
pico	p	$\times 10^{-12}$
nano	n	$\times 10^{-9}$
micro	µ	$\times 10^{-6}$
milli	m	$\times 10^{-3}$
kilo	k	$\times 10^{3}$
mega	M	$\times 10^{6}$
giga	G	$\times 10^{9}$
tera	T	$\times 10^{12}$

Useful Properties of Crystalline Silicon, Germanium, and Gallium Arsenide

Property (at 300 K)	Silicon	Germanium	Gallium Arsenide (GaAs)
Atomic number	14	32	Gallium: 31 Arsenic: 33
Atoms/cm^3	5×10^{22}	4.41×10^{22}	4.42×10^{22}
Relative dielectric constant, ϵ_r	11.8	15.8	13.1
Energy gap, E_g (eV)	1.12	0.66	1.42
Electron mobility, μ_n ($cm^2/V \cdot s$)	1500	3900	8500
Hole mobility, μ_p ($cm^2/V \cdot s$)	480	1900	400
Nominal $n_i = p_i$ (number/cm^3)	1.5×10^{10}	2.4×10^{13}	1.8×10^{6}
Melting point, °C	1420	936	1238

Key Typical BJT Parameters at 300 K (Room Temperature)

Parameter	npn	pnp (Substrate)	pnp (Lateral)
β_F	200	50	30
β_R	2	4	3
V_A (volts)	150	50	50
V_J (volts)	0.7	0.55	0.55
I_S (amperes)	2×10^{-15}	10^{-14}	2×10^{-15}
r_b (ohms)	200	100	300

SPICE Models

Device	Device Name	Model	Order of Node Numbers
Diode	D	D(IS = I_s N = n)	Anode, cathode
npn BJT	Q	NPN(BF = β IS = I_S)	Collector, base, emitter
pnp BJT	Q	PNP(BF = β IS = I_S)	Collector, base, emitter
n-channel JFET	J	NJF(VTO = V_P BETA = K)	Drain, gate, source
p-channel JFET	J	PJF(VTO = $-V_P$ BETA = $\lvert K \rvert$)	Drain, gate, source
n-channel MOSFET	M	NMOS(VTO = V_t KP = $2K$)	Drain, gate, source, substrate
p-channel MOSFET	M	PMOS(VTO = V_t KP = $2\lvert K \rvert$)	Drain, gate, source, substrate

Two-port amplifier models

Voltage amplifier	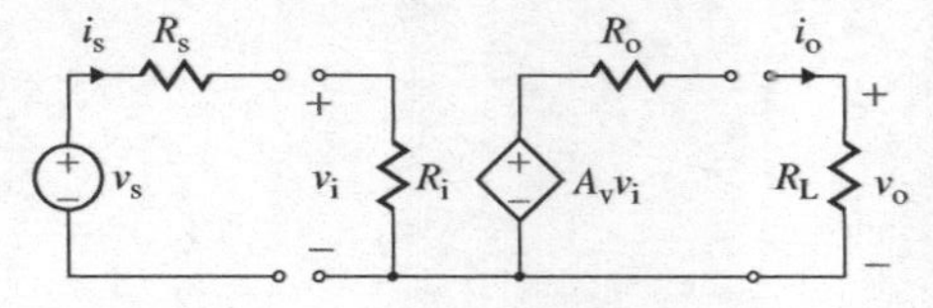 i_s R_s v_s + v_i − R_i $A_v v_i$ R_o i_o + R_L v_o −
Current amplifier	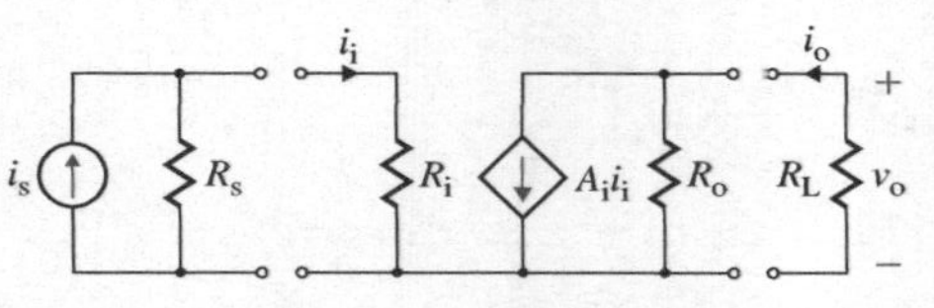 i_s R_s i_i R_i $A_i i_i$ R_o i_o + R_L v_o −
Transconductance amplifier	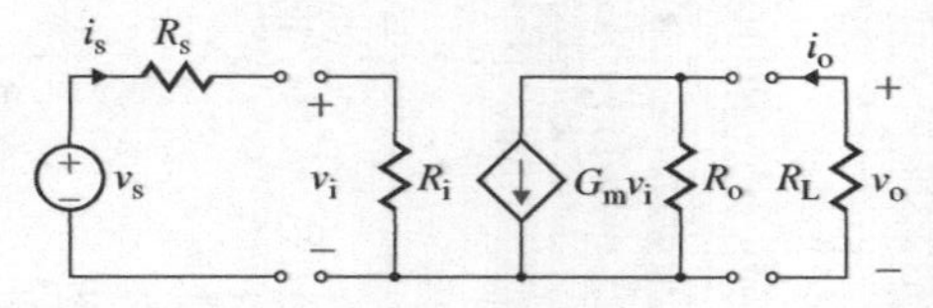 i_s R_s v_s + v_i − R_i $G_m v_i$ R_o i_o + R_L v_o −
Transresistance amplifier	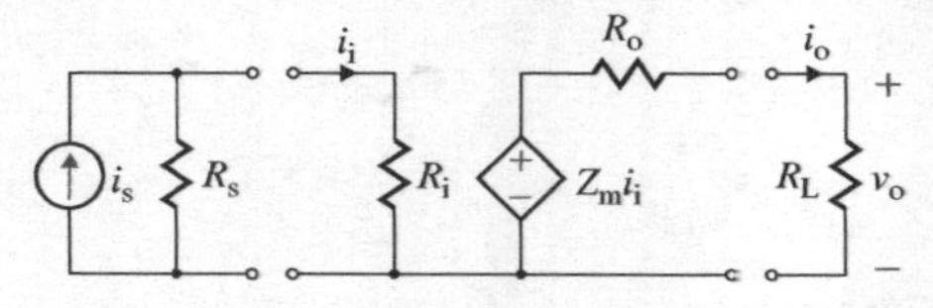 i_s R_s i_i R_i $Z_m i_i$ R_o i_o + R_L v_o −

Single-stage amplifier configurations

Amplifier Type	Transistor Type			
	NMOS	PMOS	*npn*	*pnp*
Common-source/ common-emitter (CS/CE)	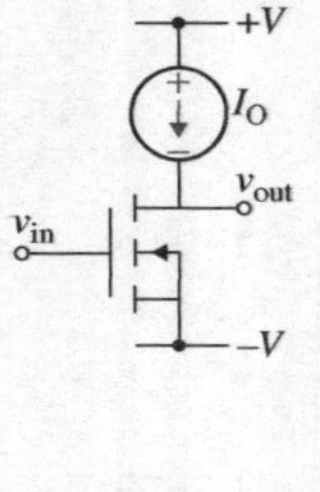 $+V$ I_O v_{out} v_{in} $-V$	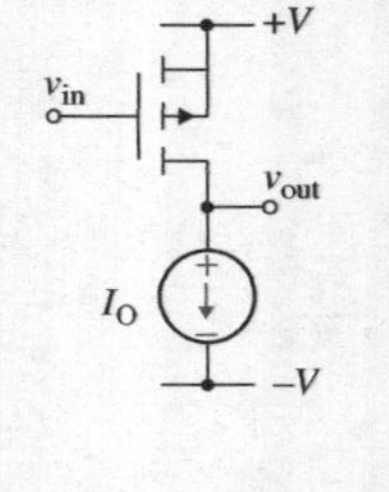 $+V$ v_{in} v_{out} I_O $-V$	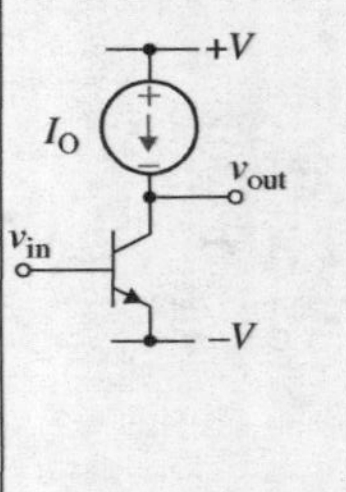 $+V$ I_O v_{out} v_{in} $-V$	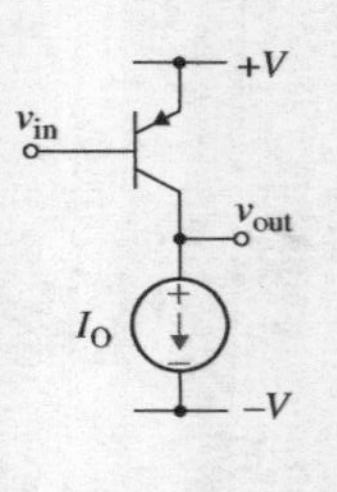 $+V$ v_{in} v_{out} I_O $-V$
Common-drain/ common-collector (CD/CC)	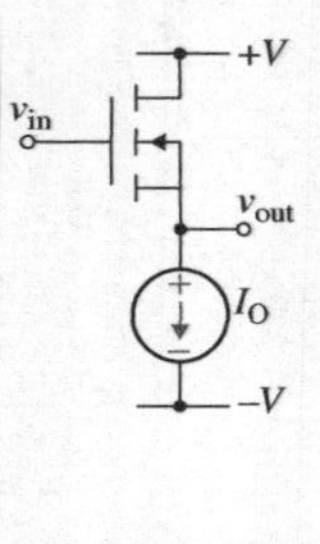 $+V$ v_{in} v_{out} I_O $-V$	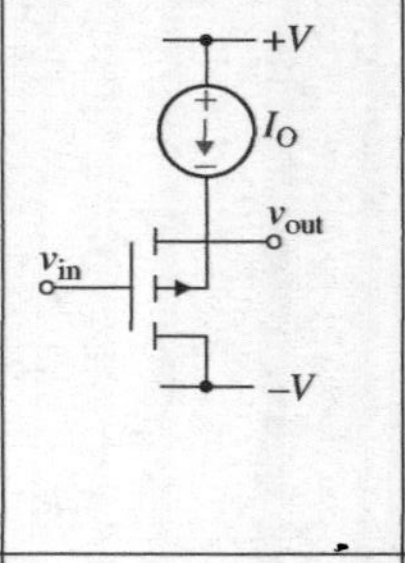 $+V$ I_O v_{out} v_{in} $-V$	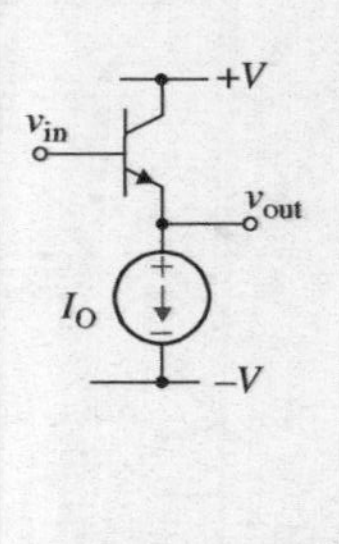 $+V$ v_{in} v_{out} I_O $-V$	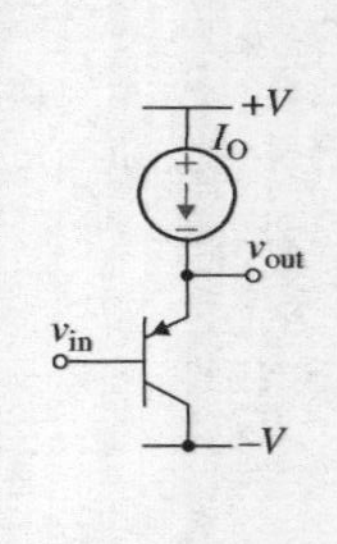 $+V$ I_O v_{out} v_{in} $-V$
Common-gate/ common-base (CG/CB)	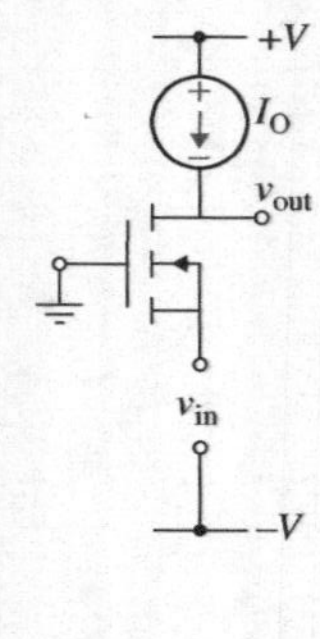 $+V$ I_O v_{out} v_{in} $-V$	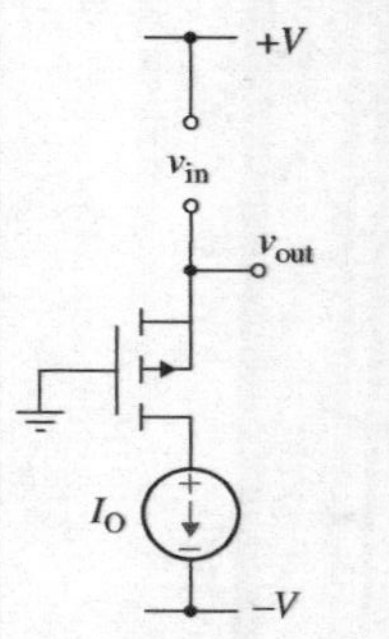 $+V$ v_{in} v_{out} I_O $-V$	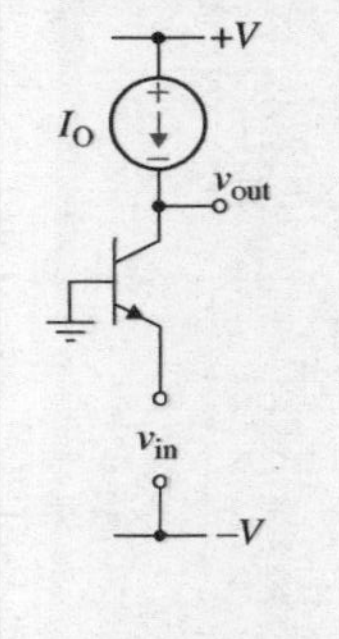 $+V$ I_O v_{out} v_{in} $-V$	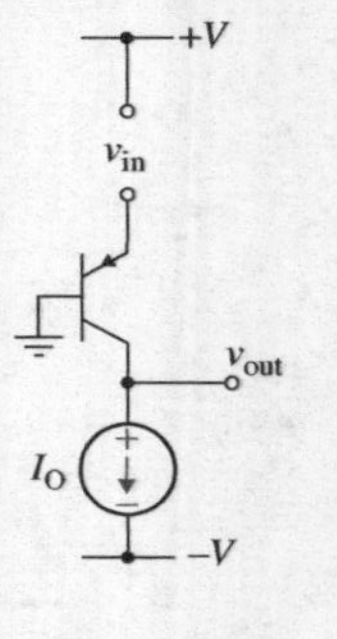 $+V$ v_{in} v_{out} I_O $-V$

Microelectronic Circuits
Analysis and Design

Second Edition

Muhammad H. Rashid
University of West Florida

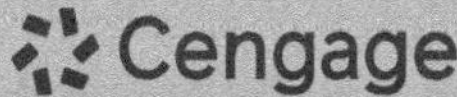

Australia • Brazil • Canada • India • Mexico • Singapore • United Kingdom • United States

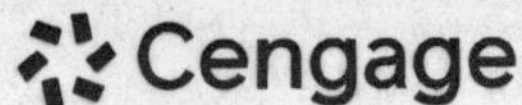

Microelectronic Circuits: Analysis and Design, 2nd Edition

Muhammad H. Rashid

For product information and technology assistance, contact us at **asia.infoindia@cengage.com**

For permission to use material from this text or product, submit all requests online at **www.cengage.com/permissions**

ISBN-13: 978-81-315-1683-6
ISBN-10: 81-315-1683-0

Cengage Learning India Pvt. Ltd.
T-303, 3rd Floor
DAV Complex, Plot 3
Mayur Vihar Phase 1
Delhi 110091

Cengage is a leading provider of customized learning solutions with employees residing in nearly 40 different countries and sales in more than 125 countries around the world. Find your local representative at: **www.cengage.com.**

Cengage is committed to creating inclusive and diverse products, including perspectives that represent the diversity of people across all backgrounds, identities, opinions, and experiences. If you have any feedback related to this, please contact us at **support.cengage.com.**

To learn more about Cengage products, platforms and services, visit **www.cengage.co.in.**

Disclaimer:
This book contains link to resources which are a part of the imported edition. The same may not be available with the Indian edition. The buyer needs to purchase a copy of the imported edition to have full access to those resources. In no event the publisher shall be liable for any limitation or performance due to unavailability of these resources.

Printed by:
Saurabh Printers Pvt. Ltd.
67A-68, Ecotech Ext-I, Greater Noida,
Tel-0120-2341180

Printed in India
Print Number: 04 Print Year: 2025

To my parents,
my wife, Fatema,
my children, Faeza, Farzana, and Hasan

CONTENTS

Chapter 11 Power Amplifiers

Chapter 12 Active Filters

Chapter 13 Oscillators

Chapter 16 Integrated Analog Circuits and Applications

Chapter 16 Integrated Analog Circuits and Applications

PREFACE

Semiconductor devices and integrated circuits (ICs) are the backbone of modern technology, and thus the study of electronics—which deals with their characteristics and applications—is an integral part of the undergraduate curriculum for students majoring in electrical, electronics, or computer engineering. Traditionally, the basic course in electronics has been a one-year (two-semester) course at most universities and colleges. However, with the emergence of new technologies and university-wide general education requirements, electrical engineering departments are under pressure to reduce basic electronics to a one-semester course. This book can be used for a one-semester course as well as a two-semester course. The only prerequisite is a course in basic circuit analysis. A one-semester course would cover Chapters 1 through 8, in which the basic techniques for analyzing electronic circuits are introduced using ICs as examples. In a two-semester course, the second semester would focus on detailed analysis of devices and circuits within the ICs and their applications.

The objectives of this book are:

- To develop an understanding of the characteristics of semiconductor devices and commonly used ICs
- To develop skills in analysis and design of both analog and digital circuits
- To introduce students to the various elements of the engineering design process, including formulation of specifications, analysis of alternative solutions, synthesis, decision-making, iterations, consideration of cost factors, simulation, and tolerance issues

Approach

This book adopts a top-down approach to the study of electronics, rather than the traditional bottom-up approach. In the classical bottom-up approach, the characteristics of semiconductor devices and ICs are studied first, and then the applications of ICs are introduced. Such an approach generally requires a year of instruction, as it is necessary to cover all the essential materials in order to give students an overall knowledge of electronic circuits and systems. In the top-down approach used here, the ideal characteristics of IC packages are introduced to establish the design and analytical techniques, and then the characteristics and operation of devices and circuits within the ICs are studied to understand the imperfections and limitations of IC packages. This approach has the advantage of allowing the instructor to cover only the basic techniques and circuits in the first semester, without going into detail on discrete devices. If the curriculum allows, the course can continue in the second semester with detailed analysis of discrete devices and their applications.

In practice, the lectures and laboratory experiments run concurrently. If students' experimental results differ from the ideal characteristics because of the practical limitations of IC packages, students may become concerned. This concern may be addressed by a brief explanation of the causes of discrepancies. The experimental results, however, will not differ significantly from the theoretically obtained results.

Current ABET (Accreditation Board of Engineering and Technology) criteria and other engineering criteria under the Washington Accord (http://www.washingtonaccord.org/) require the integration of design and computer usage throughout the curriculum. After students have satisfied other ABET and accreditation requirements in math, basic science, engineering science, general education electives, and free electives, they find that not many courses are available to satisfy the design requirements. The lack of opportunities for design credits in engineering curricula is a common concern. Electronics is generally the first electrical engineering course well suited to the integration of design components and computer usage. This book is structured to permit design content to constitute at least 50% of the course, and it integrates computer usage through PSpice. Many design examples use PSpice to verify the design requirements, and the numerous computer-aided design examples illustrate the usefulness of personal computers as design tools, especially in cases in which design variables are subjected to component tolerances and variations.

New to This Edition

The second edition offers a reorganized order of chapters with the required material augmented and the nonessential topics abridged. The key changes to this edition are summarized below:

- All new chapter on MOSFETs and amplifiers
- All new chapter on semiconductors and *pn* junctions
- Fully revised chapter on BJTs
- More emphasis on MOSFETs and active biasing techniques to allow students to move easily on to differential amplifiers and ICs
- Extensive revision of power amplifiers to include MOSFET circuits with class C, D, and E amplifiers
- Integrated PSpice/OrCAD examples for both analysis and design verifications
- Developed Mathcad files for calculations of worked-out examples so that students can try similar problems and explore the effects of design parameters

Content and Organization

After an introduction to the design process in Chapter 1, the book may be divided into six parts:

I. Chapters 2 and 3 on characteristics of amplifiers and their frequency responses
II. Chapters 4 and 5 on diodes and applications
III. Chapters 6 to 8 and 11 on semiconductor fundamentals, transistors, and amplifiers
IV. Chapters 10, 12, and 13 on characteristics and analyses of electronic circuits
V. Chapter 15 on digital logic gates
VI. Chapters 9, 14, and 16 on integrated circuits and applications

A review of basic circuit analysis and an introduction to PSpice are included in the appendices.

Modern semiconductor technology has evolved to such an extent that many analog and digital circuits are available in the form of integrated circuit (IC) packages. Manufacturers of these packages provide application notes that can be used to implement circuit functions. Knowledge of the characteristics and operation of devices within the IC packages is essential, however, to understand the limitations of these ICs when they are interfaced as building blocks in circuit designs. Such knowledge also serves as the basis for developing future generations of IC packages. Although the trend in IC technology suggests that discrete circuit design may disappear entirely in the future, transistor amplifiers (in large-scale or very-large-scale integrated forms) will continue to be the building blocks of ICs. Thus, semiconductor fundamentals and transistor amplifiers are covered in Chapters 6 to 8, after the general types and specifications of amplifiers have been introduced in Chapter 2. Because diodes are the building blocks of many electronic circuits, and because the techniques for the analysis of diodes are similar to those for transistor amplifiers, diodes and their applications are addressed in detail in Chapters 4 and 5.

Pedagogy and Supplements

The pedagogical approach of the first edition has been enhanced and augmented in this edition. Mathematical derivations are kept to a minimum by using approximate circuit models of operational amplifiers, transistors, and diodes. The significance of these approximations is established by computer-aided analysis using PSpice. Important circuits are analyzed in worked-out examples in order to introduce the basic techniques and emphasize the effects of parameter variations. At the end of each chapter, review questions and problems test students' learning of the concepts developed in the chapter. The student learning outcomes (SLOs) are listed at the beginning of each chapter. Symbols and their meanings have been uniquely identified at the beginning of each chapter to serve as a quick reference to the students. Every chapter opens with an introduction that puts the content of the chapter in perspective of the field of microelectronics. Solved examples carry captions that identify the objective of the example. Notes interspersed through the text provide a link to other chapters and serve to guide students against common misconceptions and mistakes. Key points of most of the sections are summarized in a box in addition to an end-of-chapter summary. A list of references is included at the end of each chapter for those interested in further reading. End-of-chapter exercises are divided into Review Questions and Problems. Design problems and PSpice problems are identified by relevant symbols.

Student support from Cengage Learning is available on the book's student website www.cengage.com/engineering/rashid. This website contains tools that are designed to help the student learn about electronics more effectively. It includes electronic copies of all the PSpice schematics printed in this book, and Mathcad files for all worked-out examples in the book, which can be downloaded and allow students to work their own problems.

The student version PSpice schematics and/or OrCAD capture software can be obtained or downloaded from:

Cadence Design Systems, Inc.
2655 Seely Avenue
San Jose, CA 95134, USA
Websites: http://www.cadence.com
http://www.orcad.com
http://www.ema-eda.com

Support for Instructors

A solutions manual (in both print and electronic forms) and slides of the figures in this book are available on request from Cengage Learning through the Global Engineering website www.cengage.com/engineering.

Teaching plans and suggested course outlines for one- and two-semester courses using this book are included just after this preface.

Acknowledgments

Thanks are due to the editorial team at Cengage Learning, Chris Carson, Chris Shortt, Hilda Gowans, Swati Meherishi, and Yumnam Ojen Singh for their guidance and support.

I would also like to thank the following reviewers for their comments and suggestions on the first and the second editions:

Dr. Ezzat G. Bakhoum
University of West Florida

Dr. William T. Baumann
Virginia Polytechnic Institute and State University

Dr. Paul J. Benkeser
Georgia Institute of Technology

Dr. Alok K. Berry
George Mason University

Dr. Michael A. Bridgwood
Clemson University

Dr. Nadeem N. Bunni
Clarkson University

Dr. Wai-Kai Chen
University of Illinois at Chicago

Dr. Shirshak K. Dhali
Southern Illinois University

Dr. Constantine Hatziadoniu
Southern Illinois University

Dr. Bruce P. Johnson
University of Nevada-Reno

Dr. Frank Kornbaum
South Dakota State University

Dr. Oguz Kucur
Gebze Institute of Technology, Turkey

Dr. John A. McNeill
Worcester Polytechnic Institute

Dr. Bahram Nabet
Drexel University

Dr. Hemanshu R. Pota
Australian Defense Force Academy

Dr. Jack R. Smith
University of Florida

Dr. Robert D. Strattan
University of Tulsa

Finally, thanks to my family for their patience while I was occupied with this and other projects.

Any comments and suggestions regarding this book are welcome. They should be sent to the author at mrashidfl@gmail.com.

Muhammad H. Rashid
Web: http://uwf.edu/mrashid

TEACHING PLANS AND SUGGESTED COURSE OUTLINES

As with any comprehensive microelectronics textbook, this text has more material than can be covered in two single-semester courses. Instructors are often lost on what topics to cover in two semesters of 16 weeks each.

The book covers diodes after op-amp circuits so that the complete coverage of op-amp circuits including nonlinear circuits cannot be included in the same chapter. However, if nonlinear op-amp circuits are not to be covered in the course, then the op-amp circuits can be covered at the beginning, after Chapter 2. Most of the materials in Chapter 2 on introduction to amplifiers and in Chapter 5 on applications of diodes can, however, be skipped in a first course. Some approaches to typical first and second electronics courses are delineated below.

First Electronics Course

This course usually covers (a) characteristics and models of amplifiers and their frequency responses; (b) IC op-amps and their applications; (c) physical operation, characteristics, and modeling of diodes, which form the basis for understanding small-signal operation and modeling of transistors; (d) the operation, characteristics, modeling, and biasing of transistors; (e) the fundamentals of active sources and differential amplifiers, which are generally used in IC amplifiers; and (f) understanding of frequency responses of electronic circuits. These can be covered by one of the following two approaches. The suggested sequences of course topics are shown in Tables 1 and 2, respectively.

TABLE 1 Suggested topics for first electronics course—Approach A

Number of Weeks	Topics	Chapters	Sections
1	Introduction to Electronics and Design	1	1.3–1.9
1	Introduction to Amplifiers and Frequency Responses	2	2.1–2.7
2	Introduction to Op-Amps	3	3.1–3.4, 3.5.1–3.5.6
2	Diodes	4	4.1–4.7
1	Applications of Diodes	5	3.1–3.3
1	Semiconductors and *pn* Junctions	6	6.1–6.4
3	MOSFETs and Amplifiers	7	7.1–7.9
3	BJTs and Amplifiers	8	8.1–8.9
1	Differential Amplifiers	9	9.1–9.5
1	Exams		

TABLE 2 Suggested topics for first electronics course—Approach B

Number of Weeks	Topics	Chapters	Sections
1	Introduction to Electronics and Design	1	1.3–1.9
2	Diodes	4	4.1–4.7
1	Applications of Diodes	5	3.1–3.3
1	Semiconductors and *pn* Junctions	6	6.1–6.4
3	MOSFETs and Amplifiers	7	7.1–7.9
3	BJTs and Amplifiers	8	8.1–8.9
1	Introduction to Amplifiers and Frequency Responses	2	2.1–2.7
2	Introduction to Op-Amps	3	3.1–3.4, 3.5.1–3.5.6
1	Differential Amplifiers	9	9.1–9.5
1	Exams		

Approach A: Op-amps are covered before diodes in which the course is not expected to cover nonlinear op-amp circuits (using diodes). Since op-amps are the building blocks of many electronic circuits, the analyses of simple op-amp circuits are often covered in the first Basic Circuit Analysis course, which is generally a prerequisite for the electronics course. This approach has the advantage of continuity with the circuits course and is more of a systems-based approach. This approach may be viewed as a top-down approach.

Approach B: Op-amps are covered after diodes, so that students can work on nonlinear op-amp circuits (using diodes) as design projects. This has the advantage of logical progression from the devices (diodes and transistors) to op-amp amplifiers.

Second Electronics Course

This course covers the characteristics and applications of amplifiers. The course usually covers (a) the frequency response of amplifiers; (b) introduction to active filters; (c) feedback amplifiers; (d) oscillators; (e) differential amplifiers with active current sources; (f) power amplifies; (g) op-amps; and (h) IC applications. The sequence of course topics is shown in Table 3.

TABLE 3 Suggested topics for second electronics course

Number of Weeks	Topics	Chapters	Sections
1	Frequency Response of Amplifiers	3, 7, 8	2.7, 7.13, 8.15
1	Differential Amplifiers	9	9.1–9.5
2	Feedback Amplifiers	10	10.1–10.8, 10.14
2	Power Amplifiers	11	11.1–11.9
2	Active Filters	12	12.1–12.9, 12.14
2	Oscillators	13	13.1–13.7
2	Introduction to Digital Electronics	15	15.1–15.4, 15.7–15.8
1	Operational Amplifiers	14	14.1–14. 4
2	IC Applications	16	16.1, 16.2, 16.5–6.8
1	Exams		

ABOUT THE AUTHOR

Muhammad H. Rashid is currently Professor (and past Director from 1997 to 2007) of Electrical and Computer Engineering at the University of West Florida. Dr. Rashid received his B.Sc. degree in Electrical Engineering from Bangladesh University of Engineering and Technology, and M.Sc. and Ph.D. degrees from the University of Birmingham in the UK. Previously, he worked as Professor of Electrical Engineering and the Chair of the Engineering Department at Indiana University-Purdue University at Fort Wayne. He has also served as Visiting Assistant Professor of Electrical Engineering at the University of Connecticut, Associate Professor of Electrical Engineering at Concordia University (Montreal, Canada), Professor of Electrical Engineering at Purdue University Calumet, Visiting Professor of Electrical Engineering at King Fahd University of Petroleum and Minerals (Saudi Arabia), design and development engineer with Brush Electrical Machines Ltd. (England, UK), Research Engineer with Lucas Group Research Centre (England), and Lecturer and Head of Control Engineering Department at the Higher Institute of Electronics (Malta).

Dr. Rashid is actively involved in teaching, researching, and lecturing, especially in the area of power electronics. He has published 16 books and more than 130 technical papers. His books are adopted as textbooks all over the world. His books have been translated into several world languages, including Spanish, Portuguese, Indonesian, Korean, and Persian.

Dr. Rashid was a registered Professional Engineer in the Province of Ontario (Canada), and a registered Chartered Engineer (UK). He is a Fellow of the Institution of Electrical Engineers (IEE, UK) and a Fellow of the Institute of Electrical and Electronics Engineers (IEEE, USA). He was elected as an IEEE Fellow with the citation "Leadership in power electronics education and contributions to the analysis and design methodologies of solid-state power converters." Dr. Rashid is the recipient of the 1991 Outstanding Engineer Award from the IEEE. He received the 2002 IEEE Educational Activity Board (EAB) Meritorious Achievement Award in Continuing Education with the following citation "For contributions to the design and delivery of continuing education in power electronics and computer-aided simulation." He is the recipient of the 2008 IEEE Undergraduate Teaching Award with the citation "For his distinguished leadership and dedication to quality undergraduate electrical engineering education, motivating students and publication of outstanding textbooks."

Dr. Rashid was an ABET program evaluator for electrical engineering from 1995 to 2000 and an engineering evaluator for the Southern Association of Colleges and Schools (SACS, USA). He has been elected as an IEEE Industry Applications Society (IAS) Distinguished Lecturer and Speaker. He is the Series Editors of *Power Electronics and Applications* and *Nanotechnology and Applications* with the CRC Press. He serves as the Editorial Advisor of *Electric Power and Energy* with Elsevier Publishing. He lectures and conducts workshops on outcome-based education (OBE) and its implementations including assessments.

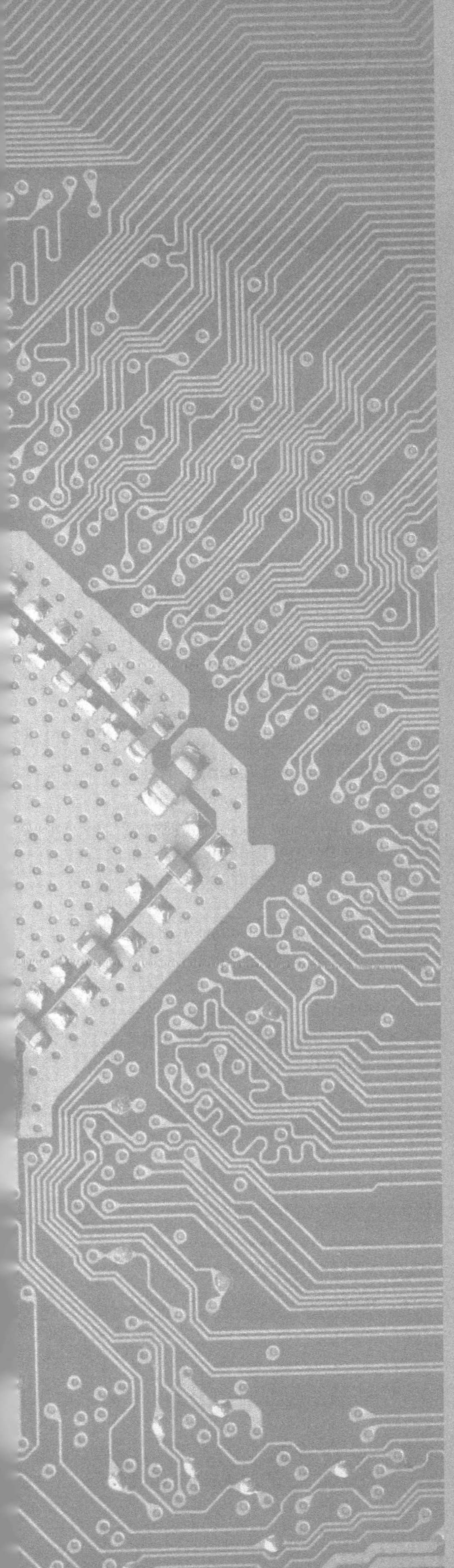

CHAPTER 1

INTRODUCTION TO ELECTRONICS AND DESIGN

Learning Outcomes

After completing this chapter, students should be able to do the following:

- Describe the historical development of electronics.
- List electronic systems and their classifications.
- List the types of electronic amplifiers.
- Describe what constitutes engineering design.
- Describe the design process of electronic circuits and systems.
- List some electronic devices and describe their basic input and output characteristics.

Symbols and Their Meanings

Symbol	Meaning
A_V, A_v	DC and small-signal voltage gains
BW, A_{PB}	Bandwidth and pass-band voltage gain
f_H, f_L	High and low cutoff frequencies
t_d, t_r, t_f, t_{on}, t_{off}	Delay, rise, fall, on, and off times
T_s	Sampling time
T, f	Period and frequency of a signal
$v_I(t)$, $v_o(t)$	Instantaneous input and output voltages
V_i, V_o	rms (root mean square) input and output voltages

1.1 Introduction

We encounter electronics in our daily life in the form of telephones, radios, televisions, audio equipments, home appliances, computers, and equipments for industrial control and automation. Electronics have become the stimuli for and an integral part of modern technological growth and development. The field of *electronics* deals with the design and applications of electronic devices. This chapter serves as an introduction to electronics.

1.2 History of Electronics

The age of electronics began with the invention of the first amplifying device, *the triode vacuum tube*, by Fleming in 1904. This invention was followed by the development of the *solid-state point-contact diode* (silicon) by Pickard in 1906, the first *radio circuits* from diodes and triodes between 1907 and 1927, the *super heterodyne receiver* by Armstrong in 1920, demonstration of *television* in 1925, the *field-effect device* by Lilienfield in 1925, *frequency modulation* (FM) by Armstrong in 1933, and *radar* in 1940.

The first electronics revolution began in 1947 with the invention of the *silicon transistor* by Bardeen, Bratain, and Shockley at Bell Telephone Laboratories. Most of today's advanced electronic technologies are traceable to that one invention. This revolution was followed by the first demonstration of *color television* in 1950 and the invention of the *unipolar field-effect transistor* by Shockley in 1952.

The next breakthrough came in 1956, when Bell Laboratories developed the *pnpn triggering transistor*, also known as a *thyristor* or a *silicon-controlled rectifier* (SCR). The second electronics revolution began with the development of a commercial thyristor by General Electric Company in 1958. That was the beginning of a new era for applications of electronics in power processing or conditioning, called *power electronics.* Since then, many different types of power semiconductor devices and conversion techniques have been developed.

The first *integrated circuit* (IC) was developed in 1958 simultaneously by Kilby at Texas Instruments and Noyce and Moore at Fairchild Semiconductor, marking the beginning of a new phase in the microelectronics revolution. This invention was followed by development of the first commercial IC *operational amplifier*, the μA709, by Fairchild Semiconductor in 1968; the 4004 microprocessor by Intel in 1971; the 8-bit microprocessor by Intel in 1972; and the gigabit memory chip by Intel in 1995. The progression from vacuum tubes to microelectronics is shown in Fig. 1.1. Integrated circuit development

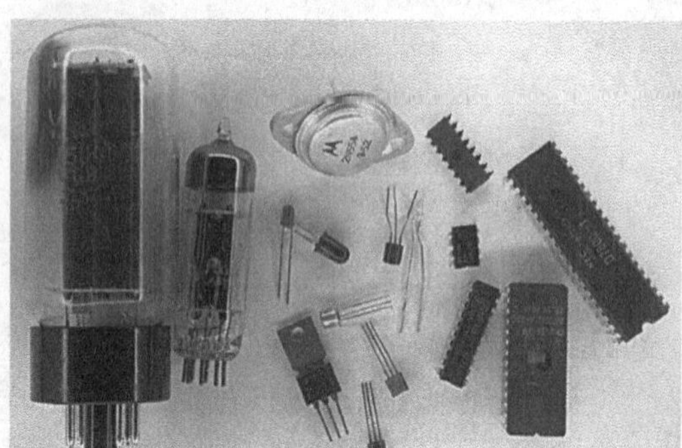

FIGURE 1.1 Progression from vacuum tubes to microelectronics

TABLE 1.1 Levels of integration

Date	Degree of Integration	Number of Components per Chip
1950s	Discrete components	1 to 2
1960s	Small-scale integration (SSI)	Fewer than 10^2
1966	Medium-scale integration (MSI)	From 10^2 to 10^3
1969	Large-scale integration (LSI)	From 10^3 to 10^4
1975	Very-large-scale integration (VLSI)	From 10^4 to 10^9
1990s	Ultra-large-scale integration (ULSI)	More than 10^9

continues today in an effort to achieve higher-density chips with lower power dissipation; historical levels of integration in circuits are shown in Table 1.1.

The degree of device integration continues to follow Moore's law, which is an observation made by Gordon E. Moore that the number of transistors inside an IC could be doubled every 24 months at a density that also minimizes the cost of a transistor [1]. Figure 1.2(a) shows the growth in the number of transistors on ICs over the years. Figure 1.2(b) shows the generations of microelectronics technology [2].

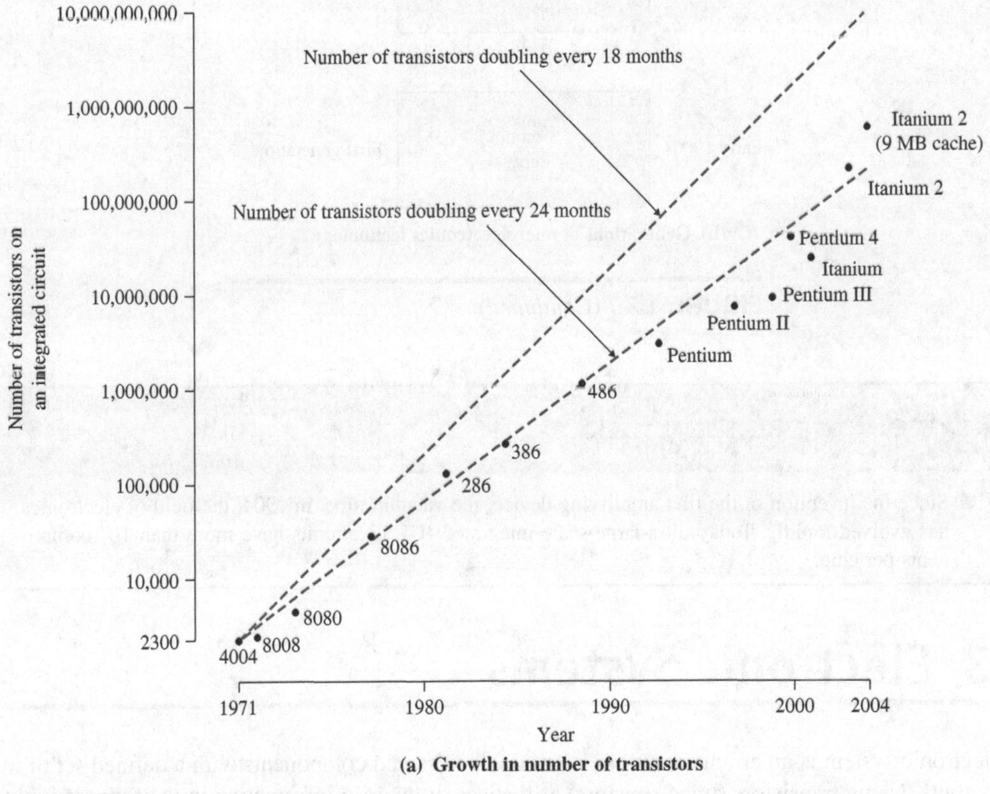

(a) Growth in number of transistors

FIGURE 1.2 Growth in the number of transistors in an integrated circuit (http://commons.wikimedia.org/wiki/File:Moore_Law_diagram_(2004).jpg) and generations of microelectronic technology (*Continued*)

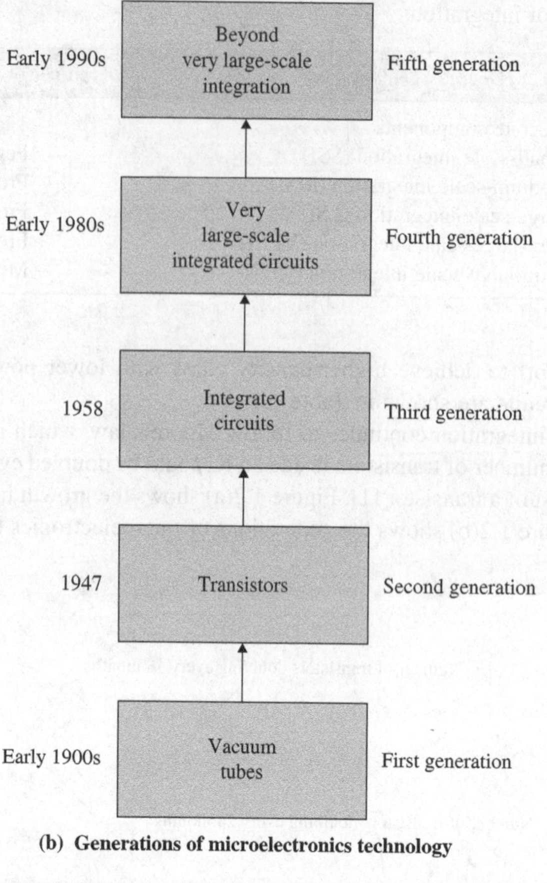

(b) Generations of microelectronics technology

FIGURE 1.2 *(Continued)*

KEY POINT OF SECTION 1.2

- Since the invention of the first amplifying device, the vacuum tube, in 1904, the field of electronics has evolved rapidly. Today ultra-large-scale integrated (ULSI) circuits have more than 10^9 components per chip.

1.3 Electronic Systems

An electronic system is an arrangement of electronic devices and components with a defined set of inputs and outputs. Using transistors (trans-resistors) as devices, it takes in information in the form of input signals (or simply inputs), performs operations on them, and then produces output signals (or outputs). Electronic systems may be categorized according to the type of application, such as communication system, medical electronics, instrumentation, control system, or computer system.

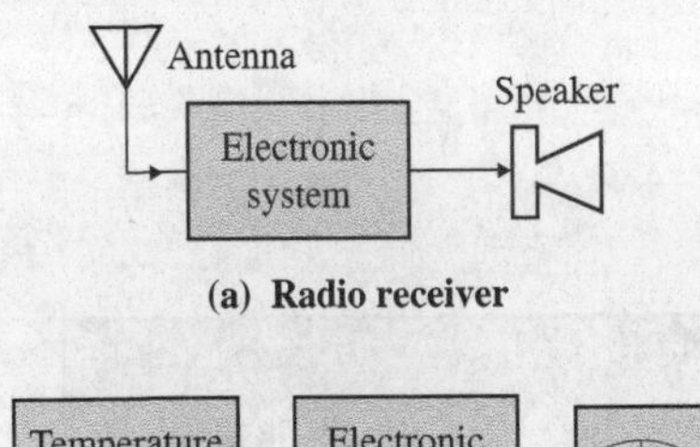

(a) Radio receiver

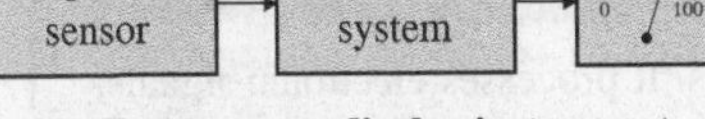

(b) Temperature display instrument

FIGURE 1.3 Examples of electronic systems

A block diagram of an FM radio receiver is shown in Fig. 1.3(a). The antenna acts as the sensor. The input signal from the antenna is small, usually in the microvolt range; its amplitude and power level are amplified by the electronic system before the signal is fed into the speaker. A block diagram of a temperature display instrument is shown in Fig. 1.3(b). The output drives the display instrument. The temperature sensor produces a small voltage, usually in millivolts per unit temperature rise above 0°C (e.g., 1 mV/°C). Both systems take an input from a sensor, process it, and produce an output to drive an actuator.

An electronic system must communicate with input and output devices. In general, the inputs and outputs are in the form of electrical signals. The input signals may be derived from the measurement of physical qualities such as temperature or liquid level, and the outputs may be used to vary other physical qualities such as those of display and heating elements. Electronic systems often use *sensors* to sense external input qualities and *actuators* to control external output qualities. Sensors and actuators are often called *transducers.* The loudspeaker is an example of a transducer that converts an electronic signal into sound.

1.3.1 Sensors

There are many types of sensors, including the following:

- Thermistors and thermocouples to measure temperature
- Phototransistors and photodiodes to measure light
- Strain gauges and piezoelectric materials to measure force
- Potentiometers, inductive sensors, and absolute position encoders to measure displacement
- Tachogenerators, accelerometers, and Doppler effect sensors to measure motion
- Microphones to measure sound
- Anemometer to measure the wind speed

1.3.2 Actuators

Actuators produce a nonelectrical output from an electrical signal. There are many types of actuators, including the following:

- Resistive heaters to produce heat
- Light-emitting diodes (LEDs) and light dimmers to control the amount of light
- Solenoids to produce force

- Meters to indicate displacement
- Electric motors to produce motion or speed
- Speakers and ultrasonic transducers to produce sound

KEY POINTS OF SECTION 1.3

- An electronic system consists of electronic devices and components. It processes electronic signals, acting as an interface between sensors on the input side and as actuators on the output side.
- Sensors convert physical qualities to electrical signals, whereas actuators convert electrical signals to physical qualities. Sensors and actuators are often called *transducers*.

1.4 Electronic Signals and Notation

Electronic signals can be categorized into two types: analog and digital. An analog signal has a continuous range of amplitudes over time, as shown in Fig. 1.4(a). Figure 1.4(b) is the sampled form of the input signal in Fig. 1.4(a). A digital signal assumes only discrete voltage values over time, as shown in Fig. 1.4(c). A digital signal has only two values, representing binary logic state 1 (for high level) and binary logic state 0 (for low level). To accommodate variations in component values, temperature, and noise (or extraneous

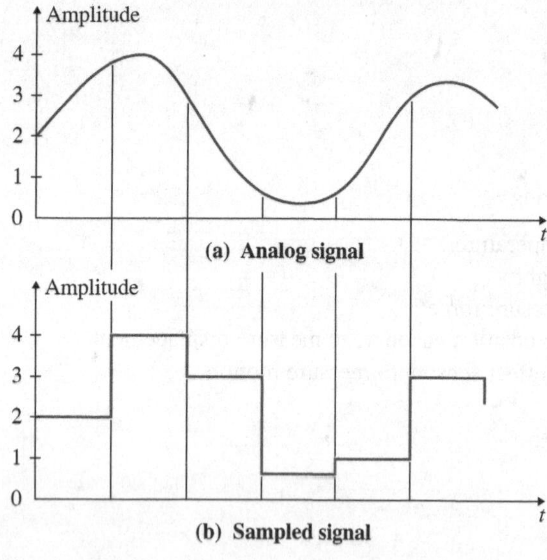

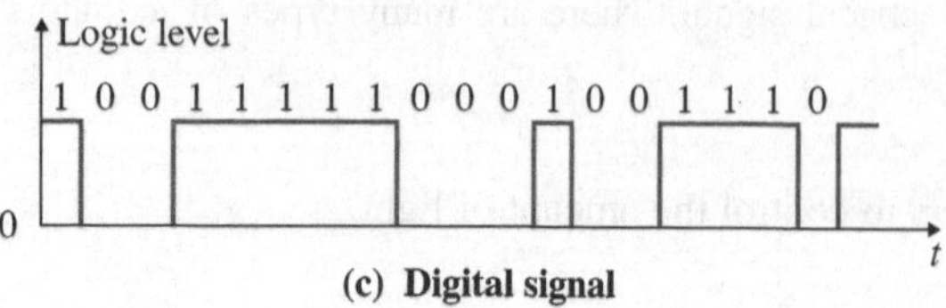

FIGURE 1.4 Types of electronic signals

signals), logic state 1 is usually assigned to any voltage between 2 V and 5 V. Logic state 0 may be assigned to any voltage between 0 and 0.8 V.

The output signal of a sensor is usually of the analog type, and actuators often require analog input to produce the desired output. An analog signal can be converted to digital form and vice versa. The electronic circuits that perform these conversions are called *analog-to-digital* (A/D) and *digital-to-analog* (D/A) *converters.*

1.4.1 Analog-to-Digital Converters

An A/D converter converts an analog signal to digital form and provides an interface between analog and digital signals. Consider the analog input voltage shown in Fig. 1.5(a). The input signal is sampled at periodic intervals determined by the *sampling time* T_S, and an n-bit binary number ($b_1 b_2 \ldots b_n$) is assigned to each sample, as shown in Fig. 1.5(b) for $n = 3$. The n-bit binary number is a binary fraction

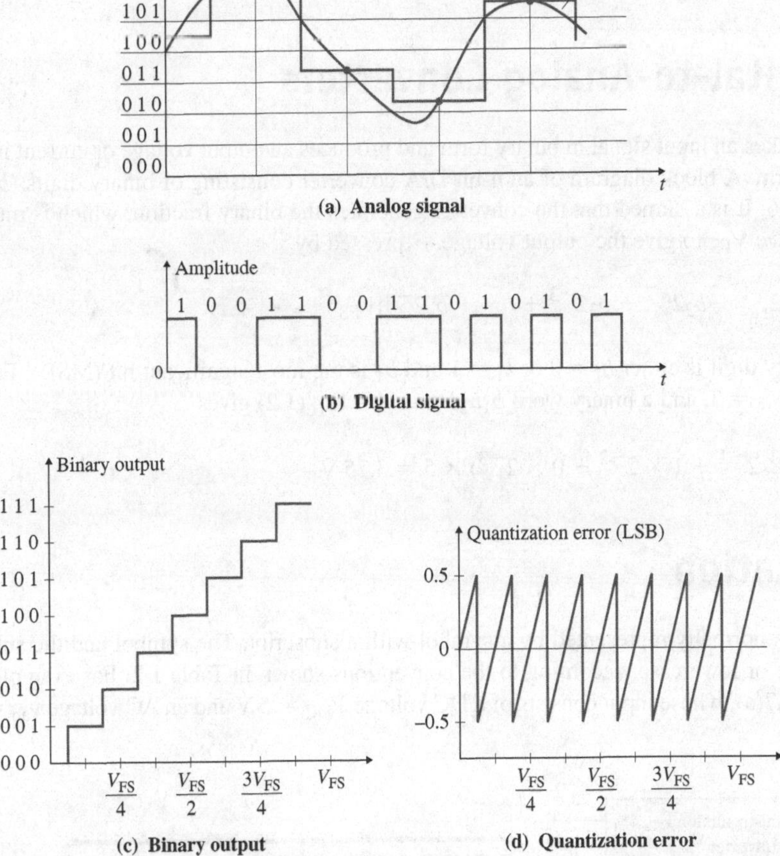

FIGURE 1.5 Analog-to-digital conversion

that represents the ratio between the unknown input voltage v_I and the full-scale voltage V_{FS} of the converter. For $n = 3$, each binary fraction is $V_{FS}/2^n = V_{FS}/8$. The output voltage of a 3-bit A/D converter is shown in Fig. 1.5(c).

The input–output relation shown in Fig. 1.5(c) indicates that as the input voltage increases from 0 to full-scale voltage, the binary output steps up from 000 to 111. However, the binary number remains constant for an input voltage range of $V_{FS}/2^n$ ($=V_{FS}/8$ for $n = 3$), which is equal to one least significant bit (LSB) of the A/D converter. Thus as the input voltage increases, the binary output will give first a negative error and then a positive error, as shown in Fig. 1.5(d). This error, called the *quantization error*, can be reduced by increasing the number of bits n. Thus, the quantization error may be defined as the smallest voltage that can change the LSB of the binary output from 0 to 1. The quantization error is also called the *resolution* of the converter, and it can be found from

$$V_{LSB} = V_{error} = \frac{V_{FS}}{2^n} \tag{1.1}$$

where V_{FS} is the full-scale voltage of the converter. For example, V_{LSB} for an 8-bit converter of $V_{FS} = 5$ V is

$$V_{LSB} = \frac{V_{FS}}{2^n} = \frac{5}{2^8} = 19.53 \text{ mV} \approx 20 \text{ mV}$$

1.4.2 Digital-to-Analog Converters

A D/A converter takes an input signal in binary form and produces an output voltage or current in an analog (or continuous) form. A block diagram of an n-bit D/A converter consisting of binary digits ($b_1b_2 \ldots b_n$) is shown in Fig. 1.6. It is assumed that the converter generates the binary fraction, which is multiplied by the full-scale voltage V_{FS} to give the output voltage, expressed by

$$V_O = (b_1 2^{-1} + b_2 2^{-2} + b_3 2^{-3} + \cdots + b_n 2^{-n})V_{FS} \tag{1.2}$$

where the ith binary digit is either $b_i = 0$ or $b_i = 1$ and b_1 is the most significant bit (MSB). For example, for $V_{FS} = 5$ V, $n = 3$, and a binary word $b_1b_2b_3 = 110$, Eq. (1.2) gives

$$V_O = (1 \times 2^{-1} + 1 \times 2^{-2} + 0 \times 2^{-3}) \times 5 = 3.75 \text{ V}$$

1.4.3 Notation

An analog signal is normally represented by a symbol with a subscript. The symbol and the subscript can be either uppercase or lowercase, according to the conventions shown in Table 1.2. For example, consider the circuit in Fig. 1.7(a), whose input consists of a DC voltage $V_{DC} = 5$ V and an AC voltage $v_{ab} = 2 \sin \omega t$.

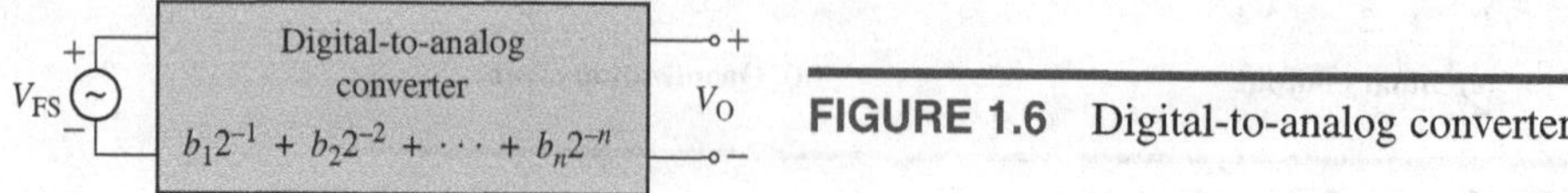

FIGURE 1.6 Digital-to-analog converter

TABLE 1.2 Definition of symbols and subscripts

Definition	Quantity	Subscript	Example
DC value of the signal	Uppercase	Uppercase	V_D
AC value of the signal	Lowercase	Lowercase	v_d
Total instantaneous value of the signal (DC and AC)	Lowercase	Uppercase	v_D
Complex variable, phasor, or rms value of the signal	Uppercase	Lowercase	V_d

The instantaneous voltages are shown in Fig. 1.7(b). The definitions of voltage and current symbols are as follows:

1. V_{DC} **and** I_{DC} **are DC values: uppercase variables and uppercase subscripts.**

$$V_{DC} = 5\text{ V}$$

$$I_{DC} = \frac{V_{DC}}{R_L} = 5\text{ mA}$$

2. v_{ab} **and** i_a **are instantaneous AC values: lowercase variables and lowercase subscripts.**

$$v_{ab} = 2 \sin \omega t$$

$$i_a = 2 \sin \omega t \text{ mA} \quad (\text{for } R_L = 1\text{ k}\Omega)$$

3. v_{AB} **and** i_A **are total instantaneous values: lowercase variables and uppercase subscripts.**

$$v_{AB} = V_{DC} + v_{ab} = 5 + 2 \sin \omega t$$

$$i_A = I_{DC} + i_a = 5\text{ mA} + 2 \sin \omega t \text{ mA} \quad (\text{for } R_L = 1\text{ k}\Omega)$$

4. V_{ab} **and** I_a **are total rms values: uppercase variables and lowercase subscripts.**

$$V_{ab} = \sqrt{5^2 + \left(\frac{2}{\sqrt{2}}\right)^2} = 5.20\text{ V}$$

$$I_a = \sqrt{5^2 + \left(\frac{2}{\sqrt{2}}\right)^2} = 5.20\text{ mA} \quad (\sqrt{2} \text{ factor is used to convert the peak value to a rms value})$$

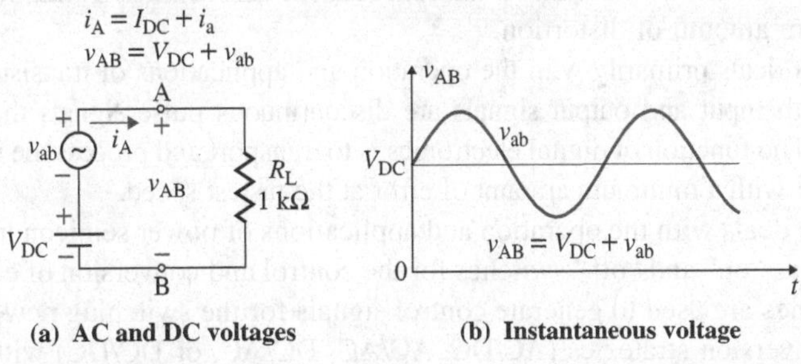

FIGURE 1.7 Notation for electronic signals

- There are two types of electronic signals: analog and digital. An analog signal can be converted to digital form and vice versa.
- A lowercase symbol is used to represent an instantaneous quantity, and an uppercase symbol is used for DC and rms values. A lowercase subscript is used to represent instantaneous AC and rms quantities, and an uppercase subscript is used for the total value, which includes both AC and DC quantities.

1.5 Classifications of Electronic Systems

The form of signal processing carried out by an electronic system depends on the nature of the input signals, the output requirements of the actuators, and the overall functional requirement. However, certain functions are common to a large number of systems. These include amplification, addition and subtraction of signals, integration and differentiation of signals, and filtering. Some systems require a sequence of operations such as counting, timing, setting, resetting, and decision making. Also, it may be necessary to generate sinusoidal or other signals within a system.

Electronic systems find applications in automobiles, home entertainment, office and communication equipments, and medicines, among other areas, and help us maintain our high-tech lifestyles. Electronic systems are often classified according to the type of application:

- Automobile electronics
- Communication electronics
- Consumer electronics
- Industrial electronics
- Instrumentation electronics
- Mechatronics
- Medical electronics
- Office electronics

The field of electronics is divided into three distinct areas, depending on the type of signals and processing required by the electronic systems.

Analog electronics deals primarily with the operation and applications of transistors as amplifying devices. The input and output signals take on a continuous range of amplitude values over time. The function of analog electronics is to transport and process the information contained in an analog input signal with a minimum amount of distortion.

Digital electronics deals primarily with the operation and applications of transistors as "on" and "off" switching devices. Both input and output signals are discontinuous pulse signals that occur at uniformly spaced points in time. The function of digital electronics is to transport and process the information contained in a digital input signal with a minimum amount of error at the fastest speed.

Power electronics deals with the operation and applications of power semiconductor devices, including power transistors, as "on" and "off" switches for the control and conversion of electric power. Analog and/or digital electronics are used to generate control signals for the switching power devices in order to obtain the desired conversion strategies (AC/DC, AC/AC, DC/AC, or DC/DC) with the maximum conversion efficiency and the minimum amount of waveform distortion. The input to a power electronic system is a DC or an AC power supply voltage (or current). Power electronics is primarily concerned with power content and quality rather than the information contained in a signal. For example, a power

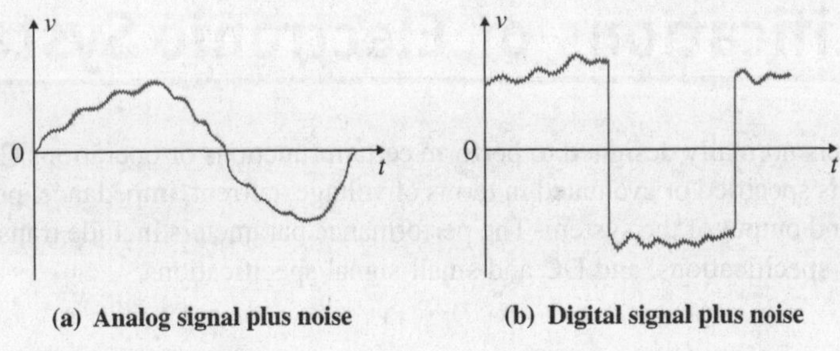

FIGURE 1.8 Effects of noise on analog and digital signals

electronic circuit can provide a stable DC power supply, say 12 V to an analog system and 5 V to a digital system, from an AC supply of 120 V at 60 Hz.

Microelectronics has given us the ability to generate and process control signals at an incredible speed. Power electronics has given us the ability to shape and control large amounts of power with a high efficiency—between 94% and 99%. Many potential applications of power electronics are now arising from the marriage of power electronics—the muscle—with microelectronics—the brain. Also, power electronics has emerged as a distinct discipline and is revolutionizing the concept of power processing and conditioning for industrial power control and automation.

Many electronic systems use both analog and digital techniques. Each method of implementation has advantages and disadvantages, summarized in the following list:

- Noise is usually present in electronic circuits. It is defined as the extraneous signal that arises from the thermal agitation of electrons in a resistor, the inductive or capacitive coupling of signals from other systems, or other sources. Noise is added directly to analog signals and hence affects the signals, as shown in Fig. 1.8(a). Thus noise is amplified by the subsequent amplification stages. Since digital signals have only two levels (high or low), noise will not affect the digital output, shown in Fig. 1.8(b), and can effectively be removed from digital signals.
- An analog circuit requires fewer individual components than a digital circuit to perform a given function. However, an analog circuit often requires large capacitors or inductors that cannot be manufactured in ICs.
- A digital circuit tends to be easier to implement than an analog circuit in ICs, although it can be more complex than an analog circuit. Digital circuits, however, generally offer much higher quality and speed of signal processing.
- Analog systems are designed to perform specific functions or operations, whereas digital systems are adaptable to a variety of tasks or uses.
- Signals from sensors and to actuators in electronic systems are generally analog. If an input signal has a low magnitude and must be processed at very high frequencies, then the analog technique is required. For optimal performance and design, both analog and digital approaches are often used.

KEY POINT OF SECTION 1.5

- Electronics can be classified into three areas: analog, digital, and power electronics. The classification is based primarily on the type of signal processing. Electronic systems are often classified according to the type of application such as medical electronics and consumer electronics.

1.6 Specifications of Electronic Systems

An electronic system is normally designed to perform certain functions or operations. The performance of an electronic system is specified or evaluated in terms of voltage, current, impedance, power, time, and frequency at the input and output of the system. The performance parameters include transient specifications, distortion, frequency specifications, and DC and small-signal specifications.

1.6.1 Transient Specifications

Transient specifications refer to the output signal of a circuit generated in response to a specified input signal, usually a repetitive pulse signal, as shown in Fig. 1.9(a). The output signal usually goes through a delay time t_d, rise time t_r, on time t_{on}, fall time t_f, and off time t_{off} in every cycle, as shown in Fig. 1.9(b). Depending on the damping factor of the circuit, the response may exhibit an overshoot before settling into the steady-state condition, as shown by the dashed curve in Fig. 1.9(b). The times associated with an output signal are defined as follows:

- *Delay time* t_d is the time before the circuit can respond to any input signal.
- *Rise time* t_r is the time required for the output to rise from 10% to 90% of its final (high) value.
- *On time* t_{on} is the time during which the circuit is fully turned on and is functioning in its normal mode.
- *Fall time* t_f is the time required for the output to decrease from 90% to 10% of its initial (high) value.
- *Off time* t_{off} is the time during which the circuit is completely off, not operating.

Thus, the switching period T is

$$T \approx t_d + t_r + t_{on} + t_f + t_{off} \tag{1.3}$$

and the switching frequency is $f = 1/T$. These times limit the maximum switching speed f_{max} of a circuit. For example, the maximum switching frequency of a circuit with $t_d = 1\ \mu s$ and $t_r = t_f = 2\ \mu s$ is

$$f_{max} = \frac{1}{(t_d + t_r + t_f)} = \frac{1}{5\ \mu s} = 200\ \text{kHz}$$

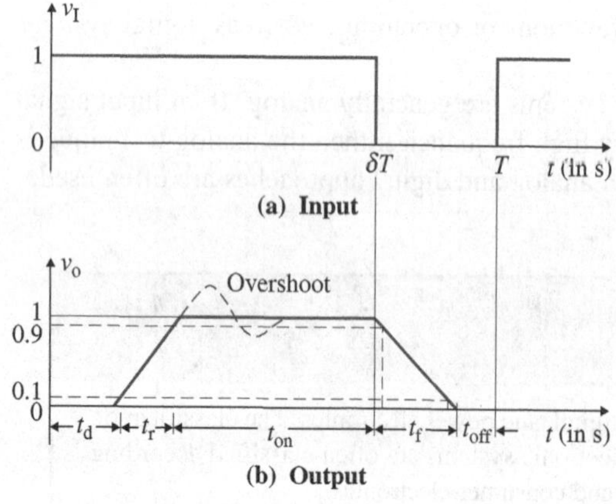

FIGURE 1.9 Pulse response of a circuit

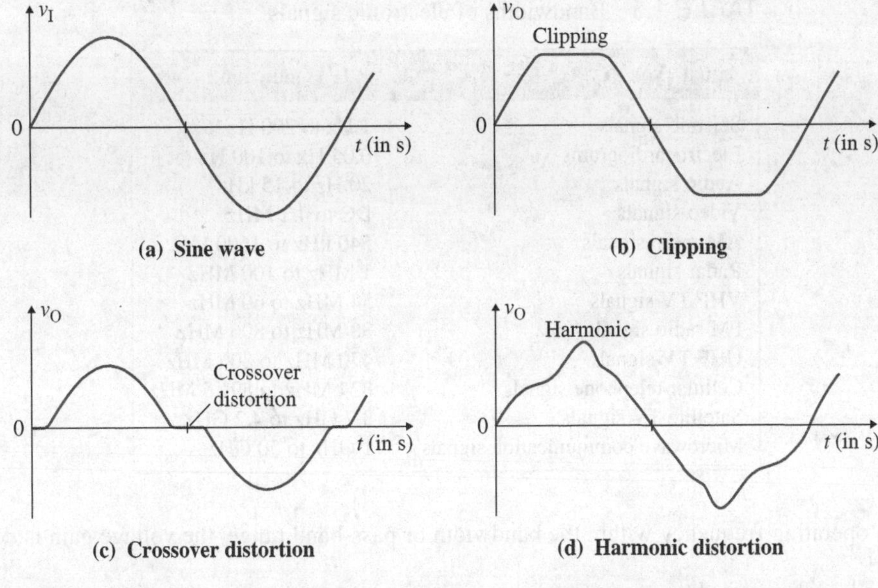

FIGURE 1.10 Some examples of distortion

1.6.2 Distortion

While passing through different stages within an electronic system, a signal often gets distorted. Distortion may take many forms and can alter the shape, amplitude, frequency, or phase of a signal. Some examples of distortion are shown in Fig. 1.10: part (b) shows clipping of the original sine wave in part (a) due to the power supply limit, part (c) shows crossover distortion due to ineffectiveness of the circuit near zero crossing, and part (d) shows harmonic distortion due to nonlinear characteristics of electronic devices. A sinusoidal input signal of a specified frequency is usually applied to the input of a circuit, and then the fundamental and harmonic components of the output signal are measured. The amount of distortion is specified as the *total harmonic distortion* (THD), which is the ratio of the rms value of the harmonic component to the rms value of the fundamental component (at the frequency of the sinusoidal input). The THD should be as low as possible.

1.6.3 Frequency Specifications

The range of signal frequencies of electronic signals varies widely, depending on the application, as shown in Table 1.3. The frequency specifications refer to the plot of the output signal as a function of the input signal frequency. A typical plot for a system such as the one in Fig. 1.11(a) is shown in Fig. 1.11(b). For frequencies less than f_L and greater than f_H, the output is attenuated. But for frequencies between f_L and f_H, the output remains almost constant. The frequency range from f_L to f_H is called the *bandwidth* BW of the circuit. That is, BW $= f_H - f_L$. A system with a bandwidth like the one shown in Fig. 1.11(b) is said to have a band-pass characteristic. If $f_L = 0$, the system is said to have a low-pass characteristic. If $f_H = \infty$, the system is said to have a high-pass characteristic.

TABLE 1.3 Bandwidths of electronic signals

Signal Type	Bandwidth
Seismic signals	1 Hz to 200 Hz
Electrocardiograms	0.05 Hz to 100 Hz
Audio signals	20 Hz to 15 kHz
Video signals	DC to 4.2 MHz
AM radio signals	540 kHz to 1600 kHz
Radar signals	1 MHz to 100 MHz
VHF TV signals	54 MHz to 60 MHz
FM radio signals	88 MHz to 806 MHz
UHF TV signals	470 MHz to 806 MHz
Cellular telephone signals	824 MHz to 891.5 MHz
Satellite TV signals	3.7 GHz to 4.2 GHz
Microwave communication signals	1 GHz to 50 GHz

For an operating frequency within the bandwidth or pass-band range, the voltage gain is defined as

$$A_{PB} = \frac{V_o}{V_i} \tag{1.4}$$

where V_i and V_o are the rms values of the input and output voltages, respectively. The input impedance is defined as

$$Z_i = \frac{V_i}{I_i} \tag{1.5}$$

where I_i is the rms value of the input current of the circuit. Z_i is often referred to as the small-signal input resistance R_i because the output is almost independent of the frequency in the midband range. Ideally, R_i should tend to infinity. Thevenin's equivalent resistance seen from the output side is specified as the output impedance Z_o or the output resistance R_o, which should ideally be zero.

1.6.4 DC and Small-Signal Specifications

The DC and small-signal specifications include the DC power supply V_{CC}, DC biasing currents (required to activate and operate internal transistors), and power dissipation P_D (power requirement from the DC power supply). The voltage gain (the ratio of the output voltage v_O to the input voltage v_I) is

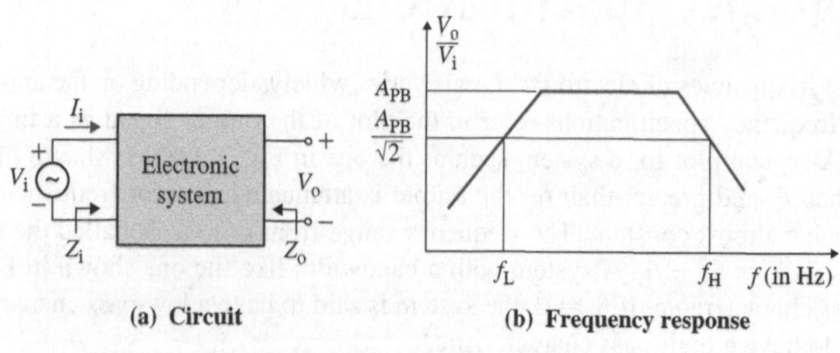

FIGURE 1.11 Typical frequency characteristic

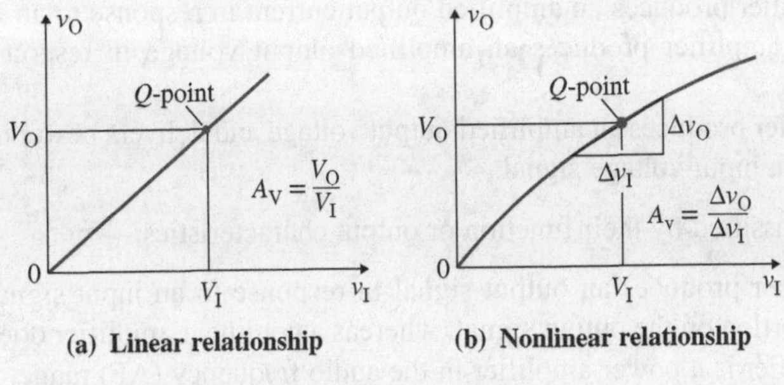

FIGURE 1.12 Large-signal and small-signal characteristics

often specified. If the v_O–v_I relationship is linear, as shown in Fig. 1.12(a), and the circuit operates at a quiescent point Q, the voltage gain is given by

$$A_V = \frac{v_O}{v_I} = \frac{V_O}{V_I} \tag{1.6}$$

A_V is often called the *large-signal voltage gain.* The characteristic plot of transistors is generally nonlinear, as shown in Fig. 1.12(b), and the circuit is operated at a quiescent operating point, the Q-point. The input signal is made to vary over a small range so that the v_O–v_I relation is essentially linear. The voltage gain is then referred to as the *small-signal gain* A_v, expressed by

$$A_v = \left.\frac{\Delta v_O}{\Delta v_I}\right|_{\text{at } Q\text{-point}} \tag{1.7}$$

Electronic circuits, especially amplifiers, are normally operated over a practically linear range of the characteristic. For an operating frequency within the BW of the circuit, $A_v \equiv A_{PB}$, where A_{PB} is the pass-band or midfrequency gain of the amplifier.

KEY POINT OF SECTION 1.6

- The parameters that describe the performance of electronic circuits and systems usually include transient specifications, distortion, frequency specifications, and large- and small-signal specifications.

1.7 Types of Amplifiers

There are many types of amplifiers, which can be classified according to the type of signal amplification, the function, the type of interstage coupling, the frequency range, and the type of load.

Signal amplification types are classified by the types of input and output signals:

1. A voltage amplifier produces an amplified output voltage in response to an input voltage signal.
2. A transconductance amplifier produces an amplified output current in response to an input voltage signal.

3. A current amplifier produces an amplified output current in response to an input current signal.
4. An impedance amplifier produces an amplified output voltage in response to an input current signal.
5. A power amplifier produces an amplified output voltage and delivers power to a low resistance load in response to an input voltage signal.

Functional types are classified by their function or output characteristics:

1. A linear amplifier produces an output signal in response to an input signal without introducing significant distortion on the output signal, whereas a nonlinear amplifier does introduce distortion.
2. An audio amplifier is a power amplifier in the audio frequency (AF) range.
3. An operation amplifier performs some mathematical functions for instruments and for signal processing.
4. A wideband amplifier amplifies an input signal over a wide range of frequencies to boost signal levels, whereas a narrowband amplifier amplifies a signal over a specific narrow range of frequencies.
5. A radio frequency (RF) amplifier amplifies a signal for use over the RF range.
6. A servo amplifier uses a feedback loop to control the output at a desired level.

Interstage coupling types are classified by the coupling method of the signal at the input, at the output, or between stages:

1. An *RC*-coupled amplifier uses a network of resistors and capacitors to connect it to the following and preceding amplifier stages.
2. An *LC*-coupled amplifier uses a network of inductors and capacitors to connect it to the following and preceding amplifier stages.
3. A transformer-coupled amplifier uses transformers to match impedances to the load side and input side.
4. A direct-coupled amplifier uses no interstage elements, and each stage is connected directly to the following and preceding amplifier stages.

Frequency types are classified in accordance to the frequency range:

1. A DC amplifier is capable of amplifying signals from zero frequency (DC) and above.
2. An AF amplifier is capable of amplifying signals from 20 Hz to 20 kHz.
3. A video amplifier (VA) is capable of amplifying signals up to a few hundred megahertz (<10 MHz for TV).
4. An ultra-high-frequency (UHF) amplifier is capable of amplifying signals up to a few gigahertz.

Load types are classified in accordance to the type of load:

1. An audio amplifier has an audio type of load.
2. A video amplifier has a video type of load.
3. A tuned amplifier amplifies a single RF or band of frequencies.

KEY POINT OF SECTION 1.7

- Amplifiers can be classified according to the type of signal amplification, the function, the type of interstage coupling, the frequency range, and the type of load.

1.8 Design of Electronic Systems

Engineering systems are becoming increasingly complex. Thus, it is highly desirable that engineers have the skills needed to analyze, synthesize, and design complex systems. A design transforms specifications into circuits that satisfy those specifications. Designing a system is a challenging task involving many variables. One can use different approaches to implement the same specifications, and hence many decisions must be made in implementing the specifications.

In practical design work, the most challenging tasks are attacked first, and then the simple tasks are tackled. That way, if an acceptable solution cannot be found to the difficult problems, time and money are not wasted on solving easier problems. Thus, the engineering design process follows a hierarchy in which systems are designed first through functional block diagrams, after which circuits and then devices are designed. This approach is the opposite of what is normally taught in academic courses. The system-level design is conceptualized and expressed in terms of functional blocks and system integration [3]. The major steps in the design process, shown in Fig. 1.13, are as follows:

1. General product description
2. Definition of specifications/requirements

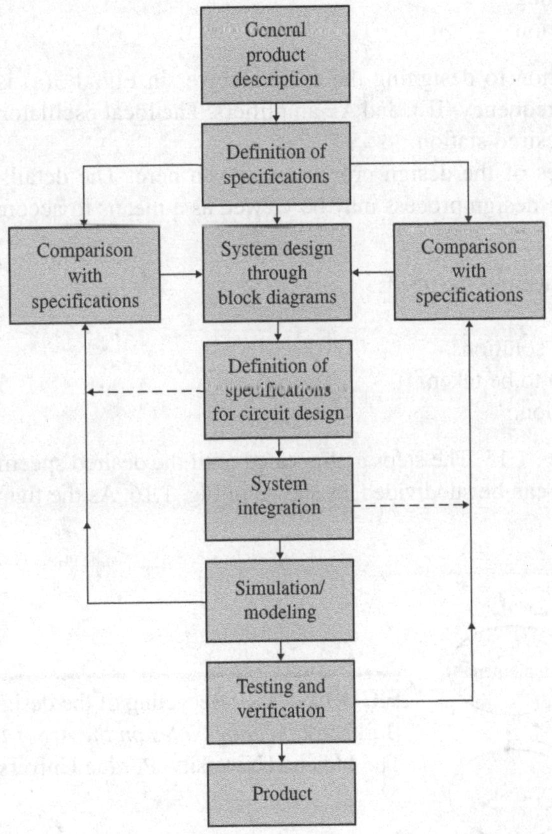

FIGURE 1.13 System-level design process

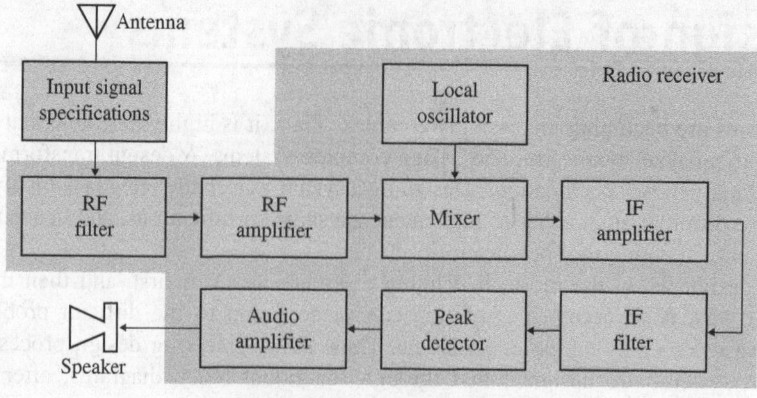

FIGURE 1.14 System-level block diagram of radio receiver

3. System design through functional block diagrams
4. Definition of specifications of functional blocks for circuit-level synthesis and implementation
5. System integration
6. Simulation or modeling
7. Testing and verification

The system-level solution to designing the radio receiver in Fig. 1.3(a) is shown in Fig. 1.14. It includes RF, intermediate frequency (IF), and AF amplifiers. The local oscillator tunes the radio receiver to receive the signal of a desired station.

Only the broad outlines of the design process are given here. The details depend on the type of system being designed. The design process may be viewed as a means to accomplish the following [4]:

1. Identify needs.
2. Generate ideas for meeting the needs.
3. Refine the ideas.
4. Analyze all possible solutions.
5. Decide on the action to be taken.
6. Implement the decision.

These steps are shown in Fig. 1.15. The steps are repeated until the desired specifications have been satisfied. Each of these six steps can be subdivided, as shown in Fig. 1.16. As the figure suggests, engineering

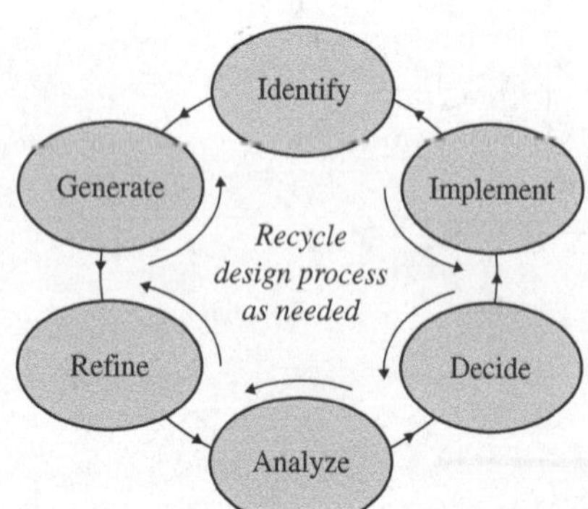

FIGURE 1.15 Recycling of the design process (John Burkhardt, *Lecture Notes on the Art of Design*. Fort Wayne, IN: The Indiana University–Purdue University Fort Wayne, 1996)

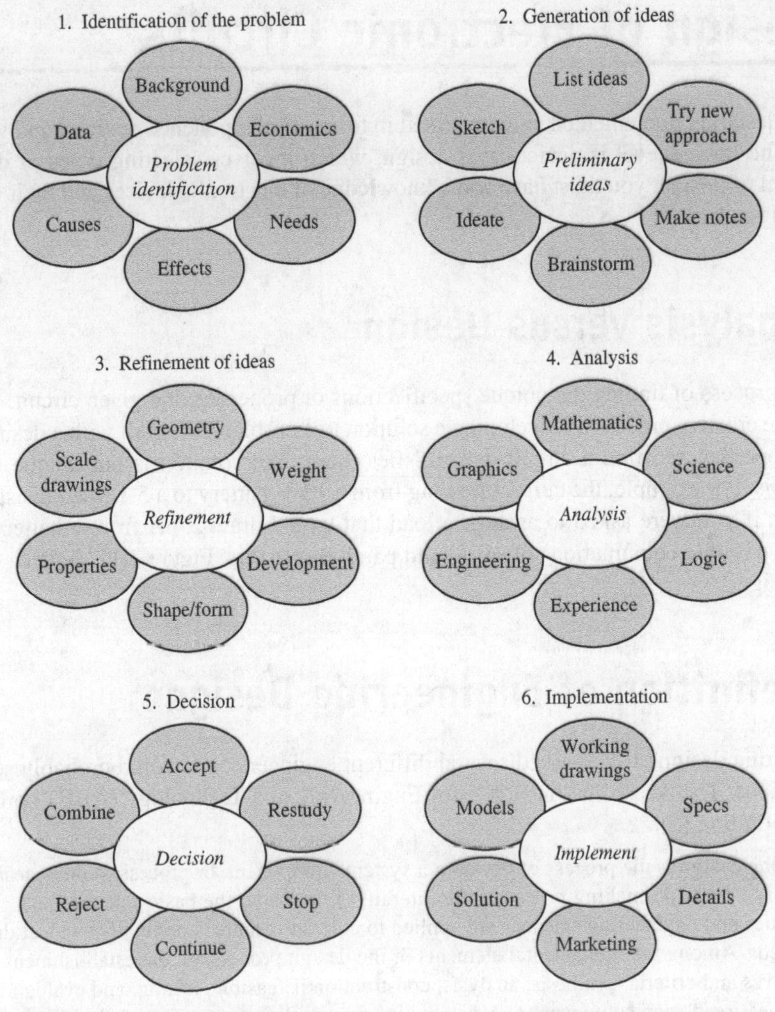

FIGURE 1.16 Elements of the design process (John Burkhardt, *Lecture Notes on the Art of Design.* Fort Wayne, IN: The Indiana University–Purdue University Fort Wayne, 1996)

design involves many disciplines, and a design engineer must be able to function in a multidisciplinary team and communicate effectively with other team members.

KEY POINT OF SECTION 1.8

- In practical design work, the most challenging tasks are attacked before the simple tasks. Thus the engineering design process follows a hierarchy in which systems are designed first through functional block diagrams, after which circuits, and then devices are designed.

1.9 Design of Electronic Circuits

A circuit-level design is implemented and expressed in terms of components, devices, and voltage–current relationships. The lowest level is device-level design, which involves selecting types of devices. Before starting this level of design, you must have some knowledge of electronic devices and their characteristics, parameters, and models.

1.9.1 Analysis versus Design

Analysis is the process of finding the unique specifications or properties of a given circuit. Design, on the other hand, is the creative process of developing a solution to a problem. We start with a desired set of specifications or properties and find a circuit that satisfies them. The solution is not unique, and finding it requires synthesis. For example, the current flowing from a 12-V battery to a 5-Ω load resistance is simply 2.4 A. However, if you were asked to arrange a load that would draw 2.4 A from a battery of 12 V, you could use many possible combinations of series and parallel resistors. Figure 1.17 shows a comparison of analysis and design.

1.9.2 Definition of Engineering Design

What is engineering design? If you asked several different engineers, you would probably set with several different definitions. The Accreditation Board for Engineering and Technology (ABET) provides the following broad definition [5]:

> Engineering design is the process of devising a system, component, or process to meet desired needs. It is a decision-making process (often iterative), in which the basic sciences and mathematics and engineering sciences are applied to convert resources optimally to meet these stated needs. Among the fundamental elements of the design process are the establishment of objectives and criteria, synthesis, analysis, construction if feasible, testing, and evaluation. The engineering design component of a curriculum must include most of the following features: development of student creativity, use of open-ended problems, development and use of design theory and methodology, formulation of design problem statements and specifications, consideration of alternative solutions, feasibility considerations, production processes, concurrent engineering design, and detailed system descriptions. Further, it is essential to incorporate appropriate engineering standards and include multiple realistic constraints such as economic, environmental, social, political, ethical, health and safety, manufacturability, and sustainability.

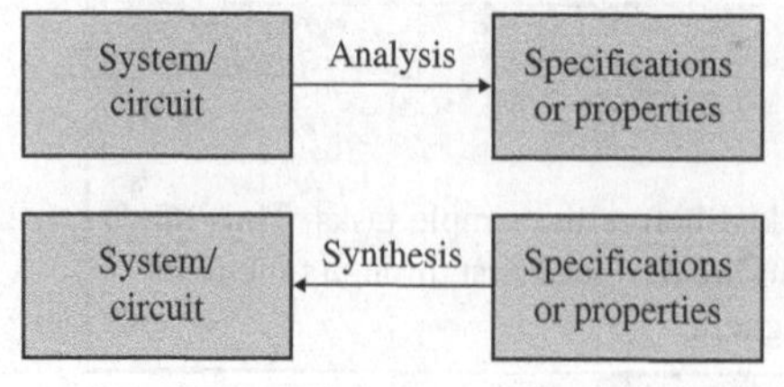

FIGURE 1.17 Analysis versus design

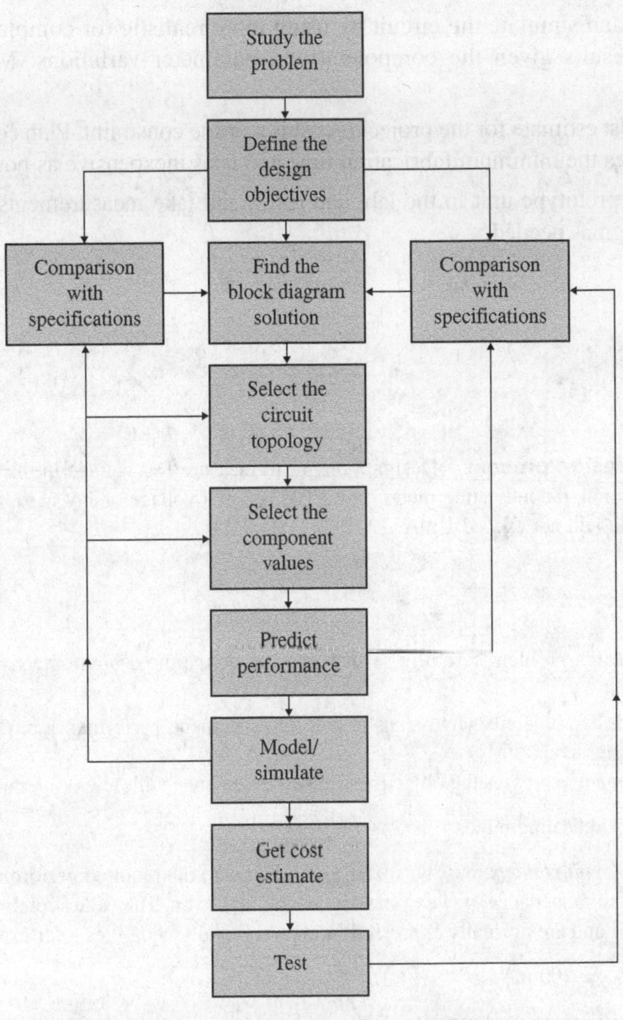

FIGURE 1.18 Circuit-level design process

1.9.3 The Circuit-Level Design Process

The major steps in the circuit-level design process, shown in Fig. 1.18, are as follows:

Step 1. Study the design problem.

Step 2. Define the design objectives—that is, establish the design's performance requirements.

Step 3. Establish the design strategy, and find the functional block diagram solution.

Step 4. Select the circuit topology or configuration after evaluating alternative solutions.

Step 5. Select the component values and devices. Analysis and synthesis may be required to find the component values. Use simple device models to simplify the analytical derivations.

Step 6. Evaluate your design, and predict its performance. Modify your design values if necessary.

Step 7. Model and simulate the circuit by using more realistic (or complex) device models. Get the worst-case results given the component and parameter variations. Modify your design as needed.

Step 8. Get a cost estimate for the project if cost is a prime constraint. Plan component layout so that the project requires the minimum fabrication time and is as inexpensive as possible.

Step 9. Build a prototype unit in the lab, and test it and take measurements to verify your design. Modify your design as needed.

EXAMPLE 1.1

D **Carrying out the design process** Design a circuit to measure DC voltage in the range from 0 to 20 V. For a full-scale deflection, the indicating meter draws 100 μA at a voltage of 1 V across it. The current drawn from the DC supply should not exceed 1 mA.

SOLUTION

Step 1. Study the design problem carefully so that you can define the design objectives succinctly in engineering terms.

Step 2. Define the design objectives by means of a design statement, performance requirements, design constraints, and design criteria.

The *design statement* expresses the objective in a single sentence with few or no numbers—for example,

Design of a DC indicating meter

The *performance requirements* must be specific and related to the required performance characteristics in terms of voltage, current, impedance, power, time, frequency, and so on. The values refer to the input and output terminals of the circuit and are normally expressed in mathematical inequalities—for example,

Meter current $I_M \leq 100\ \mu A$

The *design constraints* are the limitations imposed by the system-level design process—for example,

DC supply voltage $V_{DC} = 0$ to 20 V

DC supply current $I_{DC} \leq 1$ mA

Meter voltage $V_M = 1$ V

The designer, in general, has no flexibility to modify these constraints.

The *design criteria* are the criteria for judging the quality of a design and may include factors such as accuracy, cost, reliability, efficiency, response time, bandwidth, and power dissipation—for example,

- The excess of I_M over 100 μA should be a minimum, say 5%:

  $\Delta I_M \leq 5\ \mu A$

- The value of I_{DC} under 1 mA should be a maximum, say within 15%:

  $\Delta I_{DC} \leq 150\ \mu A$

- The cost must be kept to a minimum.

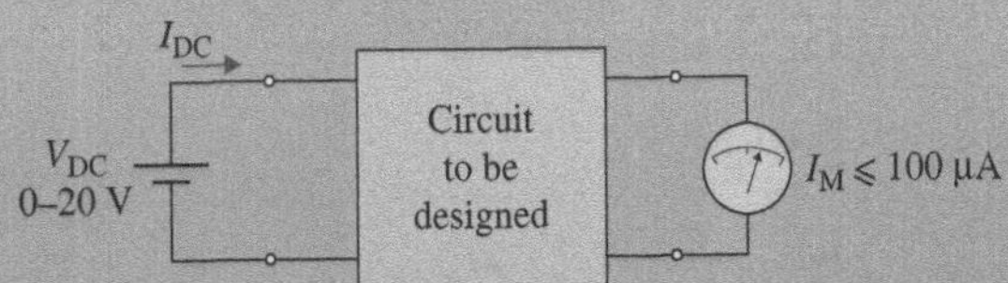

FIGURE 1.19 Block diagram solution of the design for Example 1.1

Step 3. Establish the design strategy, and find the functional block diagram solution. This solution is shown in Fig. 1.19.

Step 4. Select the circuit configuration after evaluating alternative solutions. Many different circuit configurations (such as using zener diodes) could perform the function of providing the meter current at a specific voltage. For this example we will use a simple circuit that utilizes the principle of voltage division. This circuit is shown in Fig. 1.20. Note that this is not a unique arrangement. We could remove R_2 and still satisfy the specifications.

Step 5. Select the component values after analyzing the circuit. The meter can be represented by a resistance R_M:

$$R_M = \frac{V_M}{I_M} = \frac{1\text{ V}}{100\ \mu\text{A}} = 10\text{ k}\Omega$$

The value of resistance R_1 can be found from

$$R_1 \geq \frac{(V_{DC} - V_M)}{I_{DC}} = \frac{(20 - 1)}{1\text{ mA}} = 19\text{ k}\Omega$$

To keep the cost down, we will use a 5% carbon resistor. From the tables of available resistors in Appendix E, we find that 20 kΩ is the nearest higher value for a 5% carbon resistor. That is,

$$R_1 = 20\text{ k}\Omega \pm 5\%$$

From the voltage divider rule, V_M is related to V_{DC} by

$$V_M = \frac{R_M \| R_2}{R_1 + (R_M \| R_2)} V_{DC} \tag{1.8}$$

which, for $V_M = 1$ V, $V_{DC} = 20$ V, $R_1 = 20$ kΩ, and $R_M = 10$ kΩ, gives $R_2 = 1.18$ kΩ. We find that 1.2 kΩ is the nearest higher value for a 5% carbon resistor. That is, $R_2 = 1.2\text{ k}\Omega \pm 5\%$.

Step 6. Evaluate your design, and predict its performance. Using Eq. (1.8), we get

$$V_M = \frac{R_M \| R_2}{R_1 + (R_M \| R_2)} V_{DC} = 1.01695\text{ V}$$

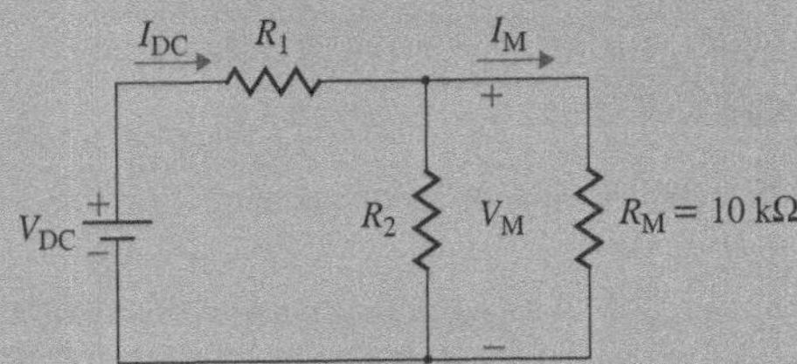

FIGURE 1.20 Proposed circuit configuration for Example 1.1

which in turn gives $I_M = 1.01695/10\ \text{k}\Omega = 102\ \mu\text{A}$, which falls within the specified criteria.

$$I_{DC} = \frac{(V_{DC} - V_M)}{R_1} = \frac{(20 - 1.01695)}{20\ \text{k}\Omega} = 949\ \mu\text{A}$$

This also falls within the specified criteria. The power rating of R_1 is

$$P_{R1} = (V_{DC} - V_M) \times I_{DC} = (20 - 1.01695) \times 949\ \mu\text{A} = 18\ \text{mW}$$

We choose the lowest power rating, 1/8 W. The power rating of R_2 is

$$P_{R2} = \frac{V_M^2}{R_2} = \frac{1.01695^2}{1.2\ \text{k}\Omega} = 862\ \mu\text{W}$$

We choose the next larger power rating, 1/8 W.

NOTE: At this step we may need to modify our design because we did not consider the effects of resistor tolerance on performance.

Step 7. Simulate the circuit. We will use PSpice to find the meter current. The complete circuit for PSpice simulation is shown in Fig. 1.21. R_{break} is the model name for resistors to assign tolerance to resistors. Refer to Appendix A, "Introduction to OrCAD," and to Rashid [6].

The PSpice plot of the meter current $I(R_m)$ versus the DC supply voltage V_{DC} is shown in Fig. 1.22. The design meets the specifications under nominal values, but under worst-case conditions the design falls short of specifications. Fine-tuning and several modifications will be required to find the final solution. This is generally true for open-ended design problems. We could try changing the value of R_1 or R_2 so that the specifications are met under worst-case conditions. Of course we could meet the specifications easily, but at a higher cost, if we chose resistors with 1% tolerance. The second cursor of the first plot is used to measure the current of the second plot.

Step 8. Get a cost estimate. From the resistor cost tables provided by suppliers, we find

R_1: 20 kΩ, 1/8 W $0.20 (approx.)

R_2: 1.2 kΩ, 1/8 W $0.15 (approx.)

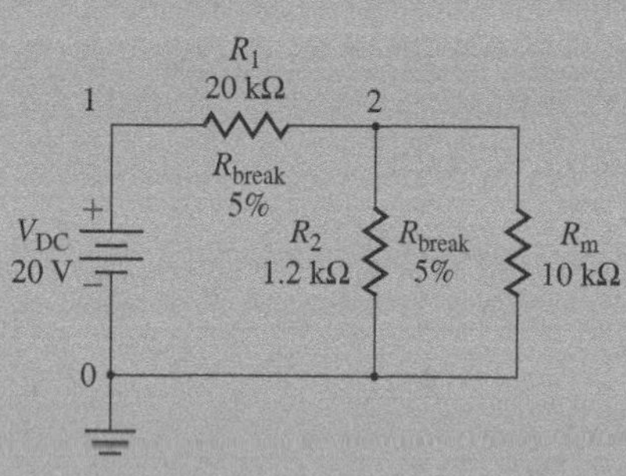

FIGURE 1.21 Proposed circuit for PSpice simulation

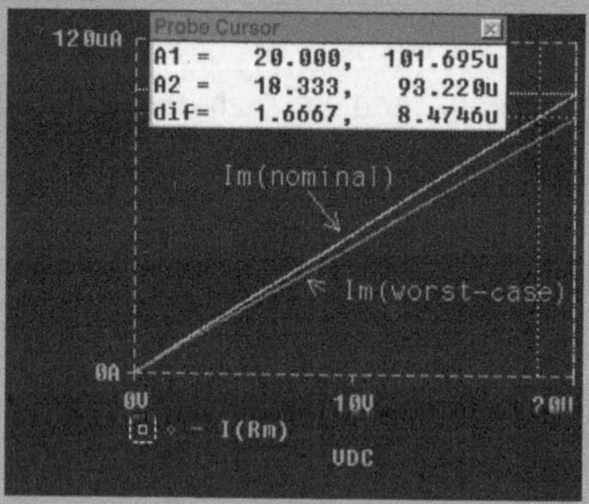

FIGURE 1.22 PSpice plots of meter current

which gives a total component cost of $0.35. Note that this estimate does not take into account costs usually associated with production, manufacturing, or company overhead.

1.9.4 Benefits of Studying from a Design Perspective

On a concrete level, the use of the design process helps you translate complex design tasks into simple circuits in a systematic way. Also, it helps you learn to establish procedures for tackling the system blocks, circuits, or subcircuits and to find, use, and integrate information from various sources, such as manufacturers' data sheets, modeling, and simulations.

On a more abstract level, the design requires decision making on trade-offs and alternative solutions and challenges you to select the best answer from a large number of acceptable ones. Thus, it strengthens decision-making skills and develops judgment, as well as fosters self-confidence and expertise in applying theory to solve real problems. Also, it lets you solve problems in your own unique way; thus, it motivates and develops creativity as well as critical thinking skills. Creativity is important to the design process, which requires you to go beyond what you learn in classroom lectures.

Because it integrates different topics in electronics as well as material from other courses on basic circuits, physics, mathematics, simulation and modeling, and laboratory techniques, a design perspective emphasizes that broad general knowledge is essential for engineering design. The design at the system level requires knowledge of system methodologies such as analysis and top-down design and system characteristics such as safety and reliability. It also requires communication skills to prepare reports and present data and management skills to coordinate product development and discover why designs fail to meet performance specifications.

1.9.5 Types of Design Projects

Engineering design involves open-ended problems whose objectives are only partially defined. Problem definition and identification of constraints are needed to achieve a satisfactory solution. Often considerable ingenuity is required to find an acceptable solution from alternative pathways, and generally several iterations are needed to reach the solution. Also, it is necessary to verify the solution by simulation and/or testing to ensure that the design objectives have been satisfied. The design projects that appear in this text vary in complexity; the time required to complete them varies from 1 h to 1 month. The design projects can be classified into four categories, depending on the time required: short design projects, mini design projects, medium design projects, and large design projects. *Electrical Engineering Design Compendium* [7] is an excellent source of other open-ended problems.

Short Design Projects

Short design projects can be completed in 1 or 2 h. Some of the problems at the end of each chapter fall into this category, including the following:

1. Defining the specifications of rectifier circuits (Chapter 5)
2. Defining the specifications of amplifiers (Chapter 2)
3. Design of transistor biasing circuits (Chapters 7 and 8)
4. Design of simple operational amplifier (op-amp) circuits (Chapter 3)
5. Design for offset minimization of op-amp circuits (Chapter 14)
6. Design of simple current source biasing circuits (Chapter 9)

Mini Design Projects

Mini design projects can be completed in approximately 1 week. Following are some of the end-of-chapter problems in the text that fall into this category:

1. Design of half-wave and full-wave rectifiers with an output filter (Chapter 5)
2. Design of diode wave-shaping circuits (Chapter 5)
3. Design of op-amp differentiators or integrators (Chapter 3)
4. Design of power amplifiers (Chapter 11)
5. Design of Schmitt trigger circuits (Chapter 16)
6. Design of sample and hold circuits (Chapter 16)
7. Design of timing circuits (Chapter 16)

Medium Design Projects

Medium design projects can be completed in approximately 2 to 3 weeks. Following are some of the end-of-chapter problems in the text that fall into this category:

1. Design of single-stage transistor amplifiers (Chapters 7 and 8)
2. Design of active filters (Chapter 12)
3. Design of instrumentation amplifiers (Chapter 3)
4. Design of transistor amplifiers to meet frequency specifications (Chapters 7 and 8)
5. Design of single-stage feedback amplifiers (Chapter 10)
6. Design of oscillators (Chapter 13)
7. Design of active current sources (Chapter 9)
8. Design of differential amplifiers with current source biasing (Chapter 9)
9. Design of electronic circuits using A/D and D/A converters (Chapter 15)
10. Design of electronic circuits using ICs of phase-lock loop (PLL) and voltage-controlled oscillators (VCO) (Chapter 16)

Large Design Projects

Large design projects can be completed in approximately 4 to 5 weeks. Following are some of the end-of-chapter problems in the text that fall into this category:

1. Design of multistage amplifiers (Chapters 7 and 8)
2. Design of higher-order active filters (Chapter 12)
3. Design of power amplifiers with current source biasing (Chapter 11)
4. Design of operational amplifiers (Chapter 14)
5. Design of multistage feedback amplifiers (Chapter 10)
6. Design of logic gates (Chapter 15)

1.9.6 Design Report

It is recommended that in your design reports you do the following:

- Give the complete design, including the ratings and values of each component.
- Justify the use of a particular circuit topology.

- Verify your design objectives by simulating your circuit using PSpice/SPICE/OrCAD or Electronics Workbench. Include a worst-case analysis (with 10% tolerances for all passive components, unless specified).
- Give a cost estimate. The project should be as inexpensive as possible.

A suggested format for design reports is as follows:

1. Title page (including your name, the course number, and the date)
   Include the following statement on the cover sheet:
   I/we certify that this assignment is the result of my/our own efforts.
   Signature(s): Date:
2. Design objectives and specifications
3. Design steps (including the circuit topology)
4. Design modifications
5. Computer simulation and design verification
6. Components and costs
7. Flowchart of the design process
8. Costs versus reliability and safety considerations
9. Conclusions

KEY POINTS OF SECTION 1.9

- Design is the creative process of developing a solution to an open-ended problem. Use of the engineering design process has many benefits. It develops creativity as well as critical thinking skills, and it promotes decision making and develops judgment.
- Following the design process helps in translating complex design tasks into simple circuits in a systematic way.

1.10 Electronic Devices

Electronic devices constitute the heart of electronics. The many types of devices can be classified into three categories: semiconductor diodes, bipolar junction transistors (BJTs), and field-effect transistors (FETs). All are nonlinear devices.

1.10.1 Semiconductor Diodes

A diode is a two-terminal semiconductor device. It offers a low resistance in the forward direction and a high resistance in the reverse direction. Thus a diode permits easy current flow in only one direction. The symbol for a diode is shown in Fig. 1.23(a). The arrow indicates the direction of current flow. If the anode–cathode voltage v_D is greater than zero, a diode is like a short circuit; if the voltage v_D is less than zero, the diode is like an open circuit. Thus, a diode is a logic device and can be represented by a controlled switch, as shown in Fig. 1.23(b). We will study the characteristic and modeling of diodes in Chapter 4 and the applications of diodes in Chapter 5.

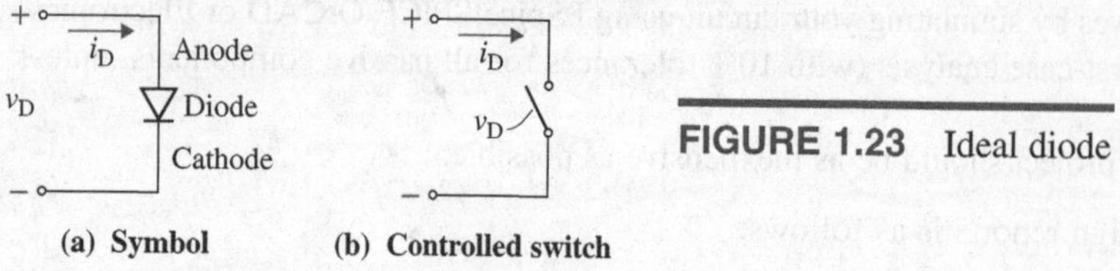

FIGURE 1.23 Ideal diode

1.10.2 Bipolar Junction Transistors

Bipolar junction transistors (BJTs), developed in the 1950s, are the oldest devices for amplification of signals. There are two types of transistors: *npn* and *pnp*. Their symbols are shown in Fig. 1.24[(a) and (b)]. A BJT has three terminals: the *emitter* (E), the *base* (B), and the *collector* (C). The arrowhead on the emitter identifies the transistor as an *npn* or a *pnp* transistor. Voltages V_{BE} and V_{CC} are required to activate and bias the transistors appropriately in their normal operating modes.

A BJT is a current-dependent device, and its collector (output) current i_C depends on the base current i_B, as shown in Fig. 1.24(c). The base emitter behaves like a diode and can be represented by a diode. Thus, a small change in the base current i_b causes an amplified change in the collector current i_c. That is,

$$i_c = \beta_F i_b \tag{1.9}$$

where β_F is called the *forward current gain* of the transistor. The small-signal model of a BJT in response to a small input base current i_b is shown in Fig. 1.24(d). In Chapter 8, we will study the characteristics and modeling of bipolar transistors.

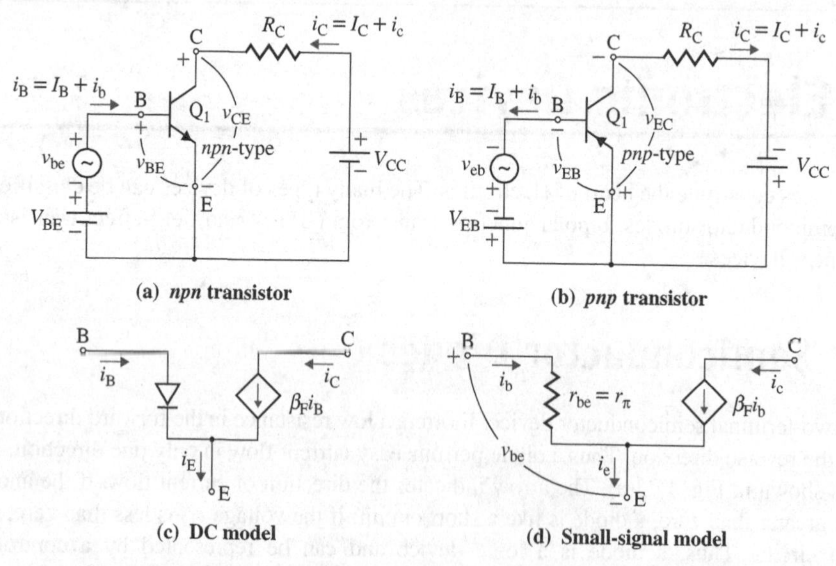

FIGURE 1.24 Bipolar junction transistor

1.10.3 Field-Effect Transistors

Field-effect transistors (FETs) are the next generation of transistors after BJTs. An FET has three terminals: the *drain* (D), the *gate* (G), and the *source* (S). The output current of an FET depends on an electric field that depends on a gate control voltage. An FET operates as a voltage-dependent device. That is, the drain (output) current depends on the input gate voltage. There are three types of FETs: enhancement metal oxide semiconductor field-effect transistors (enhancement MOSFETs), depletion metal oxide semiconductor field-effect transistors (depletion MOSFETs), and junction field-effect transistors (JFETs). In Chapter 7, we will study the characteristics and modeling of FETs.

Enhancement MOSFETs

There are two types of enhancement MOSFETs: *n*-channel and *p*-channel. Their symbols are shown in Fig. 1.25[(a) and (b)]. The arrowhead on the substrate indicates the type, either *p* or *n*. The substrate B is normally connected to the source terminal. A channel is induced under the influence of electric field action. As shown by the break in the lines from the drain to the source, there is no physical channel between the drain and the source. Voltages $v_{GS} = V_{GS}$ and V_{DD} are required to activate and bias the FETs appropriately in their normal operating modes. The gate current i_G is very small, practically zero. The drain (output) current i_D depends on the gate–source voltage v_{GS}, as shown in Fig. 1.25(c), and it is given by

$$i_D = K_p(v_{GS} - V_t)^2 \quad \text{for } |v_{GS}| \geq |V_t| \tag{1.10}$$

where K_p is the MOSFET constant, in ampere per square volts V_t is the threshold voltage of the MOSFET, in volts. v_{GS} and V_t are positive for *n*-type enhancement MOSFETs and negative for *p*-type enhancement MOSFETs. The value of v_{GS} must exceed the value of V_t for any drain current to flow. That is,

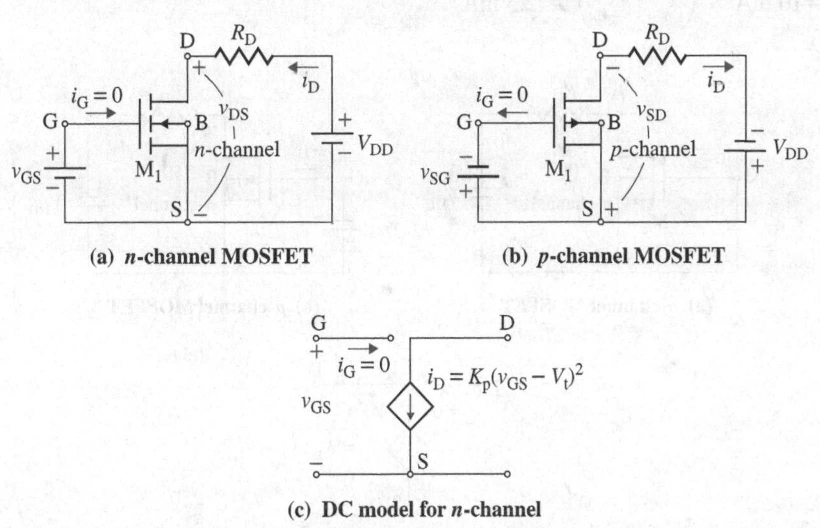

FIGURE 1.25 Enhancement MOSFETs

$|v_{GS}| > |V_t|$. For example, if $K_p = 2\ \text{mA/V}^2$, $V_t = 1.5\ \text{V}$ (for an n-channel MOSFET), and $v_{GS} = 3\ \text{V}$, Eq. (1.10) gives

$$i_D = \frac{2\ \text{mA}}{\text{V}^2} \times (3 - 1.5)^2\ \text{V}^2 = 4.5\ \text{mA}$$

Depletion MOSFETs

There are two types of depletion MOSFETs: n-channel and p-channel. Their symbols are shown in Fig. 1.26[(a) and (b)]. As shown by the continuous lines from the drain to the source, there is a physical channel between the drain and the source. However, the channel can be enhanced or depleted by the influence of electric field action. Voltages v_{GS} $(=V_{GS})$ and V_{DD} bias the FETs appropriately in their normal operating modes. The gate current i_G is practically zero. The drain (output) current i_D depends on the gate–source voltage v_{GS}, as shown in Fig. 1.26(c), and it is given by

$$i_D = K_p(v_{GS} - V_p)^2 \tag{1.11}$$

$$= I_{DSS}\left(1 - \frac{v_{GS}}{V_p}\right)^2 \tag{1.12}$$

where K_p = MOSFET constant, A/V^2

$I_{DSS} = K_p / V_p^2$, the steady-state drain current at $v_{GS} = 0$, A

V_p = pinch-off voltage of the MOSFET, V

V_p is the voltage at which the drain–source channel is effectively pinched off and no drain current flows. V_p is negative for n-type depletion MOSFETs and positive for p-type depletion MOSFETs. v_{GS} can be either positive or negative, but its magnitude cannot exceed $|V_p|$. That is, $v_{GS} > V_p$ for n-channel and $v_{GS} > V_p$ for p-channel. For example, if $I_{DSS} = 10$ mA, $v_{GS} = -2$ V and $V_p = -4$ V, Eq. (1.12) gives

$$i_D = 10\ \text{mA} \times \left(1 - \frac{-2}{-4}\right)^2 = 2.5\ \text{mA}$$

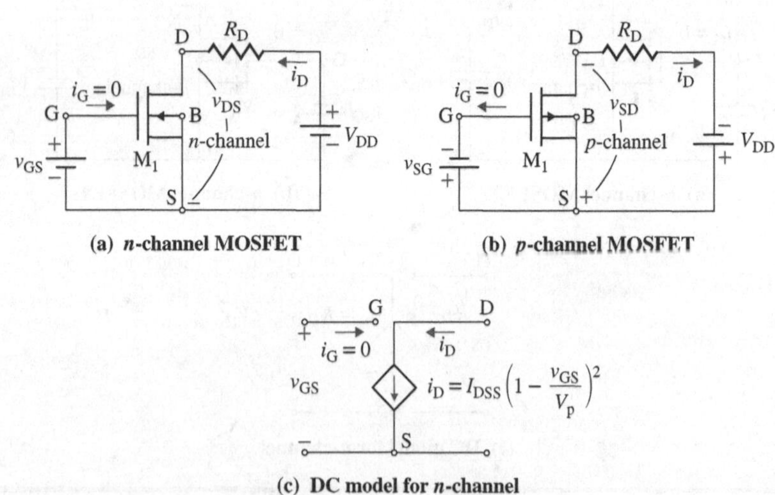

FIGURE 1.26 Depletion MOSFETs

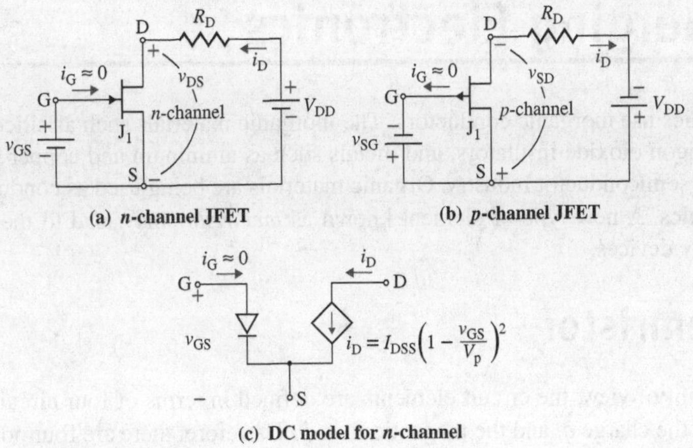

(a) *n*-channel JFET (b) *p*-channel JFET

(c) DC model for *n*-channel

FIGURE 1.27 JFETs

Junction FETs

Depletion MOSFETs can perform the same functions as JFETs [8] and have advantages over JFETs. As a result, JFETs are being phased out and are not covered further in this text. There are two types of JFETs: *n*-channel and *p*-channel. Their symbols are shown in Fig. 1.27[(a) and (b)]. The gate–source junction is reverse biased so that it behaves as a reverse-biased diode. As shown by the continuous lines from the drain to the source, there is a physical channel between the drain and the source. The drain current is controlled by the influence of electric field action. Voltages v_{GS} ($=V_{GS}$) and V_{DD} bias the JFETs appropriately in their normal operating modes. There is a small gate current i_G on the order of microamperes. The drain (output) current i_D depends on the gate–source voltage v_{GS}, as shown in Fig. 1.27(c), and it is given by

$$i_D = I_{DSS}\left(1 - \frac{v_{GS}}{V_p}\right)^2 \quad \text{for } |v_{GS}| \le |V_p| \tag{1.13}$$

where I_{DSS} is drain current at $v_{GS} = 0$, in amperes and V_p is pinch-off voltage of the JFET, in volts. V_p is the voltage at which the drain–source channel is effectively pinched off and no drain current flows. V_p is negative for *n*-type JFETs and positive for *p*-type JFETs. v_{GS} is negative for *n*-type JFETs and positive for *p*-type JFETs, but its magnitude cannot exceed V_p. For example, if $I_{DSS} = 20$ mA, $V_p = -3$ V (for an *n*-channel JFET), and $v_{GS} = -1.5$ V, Eq. (1.13) gives

$$i_D = 20 \text{ mA} \times (1 - 1.5/3)^2 = 5 \text{ mA}$$

KEY POINTS OF SECTION 1.10

- Electronic devices constitute the heart of electronics. There are three categories of devices: semiconductor diodes, bipolar junction transistors (BJTs), and field-effect transistors (FETs).
- A diode acts as a switch, which may be on or off depending on the voltage across its terminals. A BJT is a current-dependent device that can be operated as a switch or as an amplifying device. An FET is a voltage-dependent device that can be operated as a switch or as an amplifying device.

1.11 Emerging Electronics

Traditional electronics use inorganic conductors. The inorganic materials such as silicon, gallium arsenide semiconductors, silicon dioxide insulators, and metals such as aluminum and copper have been the backbone of the modern semiconductor industry. Organic materials are being used as conductors in organic and biomedical electronics. A new type of element known as *memristor* may lead to the invention of a new class of high-density devices.

1.11.1 Memristor

From the circuit point of view, the circuit elements are defined in terms of four circuit variables: the current i, the voltage v, the charge q, and the magnetic flux ϕ. Therefore, there are four possible combinations of these variables: i and v, v and q, q and ϕ, and ϕ and i. Three of these relationships describe the three basic circuit elements: the resistance R, the inductance L, and the capacitance C. The fourth relationship between q and ϕ was discovered and described by Chua [9] as the fourth circuit element, known as *memristor*. Thus, the four circuit elements are related to the circuit variables by

$$\text{Resistance: } R(t) = \frac{dv}{di} \tag{1.14}$$

$$\text{Inductance: } L(i) = \frac{d\phi}{di} \tag{1.15}$$

$$\text{Capacitance: } C(q) = \frac{dq}{dv} \tag{1.16}$$

$$\text{Memristance: } M(q) = \frac{d\phi}{dq} \tag{1.17}$$

If the charge q and the flux ϕ vary with the time t, the current i and the voltage v are given by

$$\text{Current: } i(t) = \frac{dq}{dt} \tag{1.18}$$

$$\text{Voltage: } v(t) = \frac{d\phi}{dt} \tag{1.19}$$

Therefore, if the charge q and the flux ϕ vary, the ratio of $v(t)$ and $i(t)$ gives the memristance $M(q) = v(t)/i(t) = d\phi/dq$. The memristor is a circuit element in which the magnetic flux ϕ is a function of the accumulated charge; its symbol is shown in Fig. 1.28(a). The memristance is the rate of change of flux

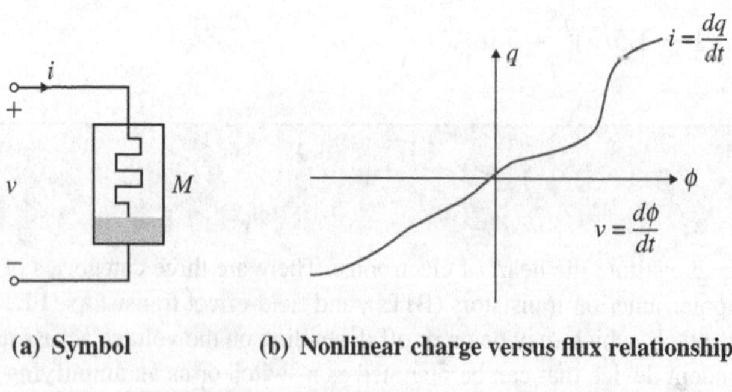

(a) Symbol (b) Nonlinear charge versus flux relationship

FIGURE 1.28 Memristor

with the charge as shown in Fig. 1.28(b). Since $M(q)$ has the unit of resistance, it is the incremental memristance whose value at any time t_0 depends on the time integral of the memristor current from $t = -\infty$ to $t = t_0$.

Thus, a memristor behaves like an ordinary resistor at any given instant of time t_0, but its resistance depends on the history of its current or its accumulated charge that has flowed through the device in the same way that the voltage of capacitors does. The name *memristor* is a portmanteau of *memory resistor*. A memristor (one for which M varies) is a nonlinear resistor. A linear memristor (one for which M is constant) would be similar to an ordinary linear resistor (one for which R is constant), with $M = R$. It is important to note that unlike a capacitor that stores charge, a memristor does not use magnetic flux to achieve its nonlinear characteristic; rather it achieves the characteristic of a resistor whose value depends on the history of the current using the chemical mechanism of charge changes.

A memristor is capable of "remembering" how much electrical charge most recently passed through it in a particular direction, which is reversible by changing the direction of current [10]. The memristor has unique analog properties that may lead to the invention of a new class of high-density devices having the characteristics of low heat generation. These include applications in high-density nonvolatile digital memory [10, 11, 12], programmable logic, signal processing, neural networks, and control systems.

1.11.2 Organic Electronics

There has been a growing research effort in "organic electronics" to improve the semiconducting, conducting, and light-emitting properties of *organics* (polymers, oligomers) and *hybrids* (organic–inorganic composites) [13, 14]. Organic electronics, also called *plastic electronics*, use organic material that has semiconductor properties [15]. A *semiconductor* is any compound whose electrical conductivity is between that of typical metals and that of insulating compounds. Carbon-based conductive polymers and molecules, which are similar to the molecules of living cells, are used as conductors exhibiting semiconductor properties. Single-molecule organic semiconductors are of two types: short-chain (oligomers) and long-chain (polymers). Like inorganic semiconductors, organic semiconductors can be doped to obtain an electron donor molecule or an electron acceptor molecule.

Conductive polymers are lighter, more flexible, and less expensive than inorganic conductors. They are finding new applications as smart windows and electronic papers, and they are expected to play an important role in the emerging area of molecular computers. Conductive polymers, however, have a higher resistance and therefore conduct electricity poorly and inefficiently compared to inorganic conductors. The organic materials, which are used as either sacrificial stencils (photoresists) or passive insulators, do not take any active role in the electronic functioning of a device. They neither conduct current to act as switches or wires nor emit light. Organic electronics fall broadly into two categories:

1. Organic thin-film transistors (OTFTs) [16]: These offer unique advantages over inorganic transistors and are cheaper to produce. They find applications such as a flat screen that can be rolled up and put into a jacket pocket. They can be constructed on flexible surfaces, such as plastic film, making organic circuits ideal for portable and mobile devices. However, organic transistors have a big disadvantage of much higher energy consumption. OTFTs also have low charge mobilities, which limit their operation at high switching frequencies.
2. Organic light-emitting diodes (OLEDs): These are breaking into the consumer electronics market and replacing the small liquid crystal displays (LCDs) [17] found in music players, cameras, and mobile phones [18]. The production of flexible OLED screens will allow roll-up screens for easy transport of unique, large-screen personal digital assistants (PDAs) [19] and flexible and impact-resistant display units.

Here are some advantages of organic electronics:

- Organic electronics have photoconductive properties, which makes them suitable in applications such as liquid crystal displays, light-emitting diodes, solar cells, and even superconductors.
- They are lighter, more flexible, and cheaper to manufacture than inorganic electronics—for example, in making radio frequency identification (RFID) tags.
- Environment-polluting chemicals are not needed in manufacturing organic circuits because the circuits can be printed from organic "ink." Silicon components, in contrast, require a long, tedious manufacturing process that requires dangerous chemicals such as arsenic, lead, and mercury, along with millions of gallons of water and solvent.

The disadvantages of organic electronics include the following:

- Their major disadvantage is their relatively higher energy consumption.
- The mobility of their electrons is much lower than that of silicon. This affects the speed at which the transistors or other devices can switch on and off.

After the disadvantages are overcome, organic electronics are expected to take over a major share of the silicon-based component marketplace.

1.11.3 Bioelectronics

Electronic principles find applications in many biomedical devices. Bioelectronics use biomaterials as conductors and apply the principle of intermolecular electron transfer in physiological processes. Biocompatible materials are used to construct artificial organs, rehabilitation devices, or prostheses and replace natural body tissues. Materials that are used and adapted for medical applications are essentially biomaterials. Therefore, a *biomaterial* is any material, natural or human-made, that makes up all or part of a living structure or biomedical device that performs, augments, or replaces a natural function. Applications include but are not limited to cartilage replacement, bone plates, organ replacement, blood vessel prostheses, skin repair devices, and contact lenses.

There are many different types of bioelectronics, depending on the types of applications. These include sensors that can distribute medications and supplements according to the user's needs, nanowires capable of sending information and power, biofuel cells capable of producing reasonable power, solar energy, bioelectronic noses, medical devices, prostheses, and electronic biosensors. An implant biosensor can restore partial sight.

Nanowires [20] could yield optical chips and sensors for biological and chemical molecules. The wires are about a thousandth the width of a human hair and function with minimal signal loss. Bioelectronic principles include concepts for curing diseases, DNA-based nanofabrication, fabricating nanoelectronic devices using self-assembling protein molecules, cell restoration, organ prostheses, and nanoelectronic circuitry.

Some applications and terminology of bioelectronics fall broadly into 1 of 12 categories:

1. *Biotechnology* [21] uses microorganisms, such as bacteria or yeasts, or biological substances, such as enzymes, to perform specific industrial or manufacturing processes.
2. *Nanotechnology* [22] refers to a field of applied science concerning the control of matter and construction of devices on the atomic and molecular scale. Nanofabrication based on DNA architecture can synthesize and program reliably self-organizing building blocks for any desired structure—such as metals, semiconductor nanoparticles, carbon nanotubes, and biomolecules.

3. *Nanomedicine* is a specific application of nanotechnology. Nanomedicine is being used for drug delivery: Medicines are delivered by basic nanites that can identify particular cells that need a medicine and release the drug into the cells. This greatly increases the efficiency of the medication. Nanotechnology is also currently used to augment the quality of MRI scans by providing greater contrast with nanoparticle agents. Nanites can help identify certain cancers and can sometimes destroy the cancers at a cellular level. Nanoscopic biosensors can be injected to provide medical personnel more complete information about patients [22].
4. *Biosensors* [23–26] are devices that detect, record, and transmit information about a physiological change or process. They use biological materials to monitor the presence of various chemicals in a substance. A biosensor consists of three elements: the sensitive biological element, the transducer or the detector element, and the user interface. Blood glucose sensors are the most widespread example of a commercial biosensor. Other instances of biosensor development include electronic noses [24] and eyes [27].
5. *Medical telesensors* [28] are fingertip chips that measure and transmit data about human body temperature, blood pressure, oxygen level, and pulse rate. This type of medical telesensor can report measurements of vital functions to remote recorders.
6. *Microcantilevers* can measure the presence of substances by nonoptical methods and are alternatives to optical fiber. They can act as physical, chemical, or biological sensors by detecting changes in cantilever bending or vibrational frequency. Microcantilevers are a million times smaller than the molecules they detect; but molecules adsorbed on a microcantilever cause vibrational frequency changes. Viscosity, density, and flow rate can also be measured through these changes in vibrational frequency.
7. *Bioreporters* are a type of biosensor based on detection of light emitted by a specially engineered microorganism that is involved in bioremediation. The light originates from a particular protein that has been installed in certain bacteria by modern molecular genetic methods. This type of biosensor can find application within the fields of pollution cleanup and detection and reporting of biodegradation.
8. *Microchip implants* [23] in animal or human brains are linked to a communication device on a computer. This process, called brain–computer interface, is linked directly to the neural part of the brain. The computer can either send or receive signals from the microchip—but not both. The computer can send signals to the brain to help control and improve eyesight, hearing, or movement. The microchip can also send signals to a computer system to warn of potential health problems of an animal or human.
9. *Biofuel cells* [23] are microorganisms when consumed by a substrate such as sugar in an aerobic condition, that is, a solution involving air, usually oxygen, produce carbon dioxide and water. However, when oxygen is not present, they produce electrons, thus providing current for power generation. Therefore, a biofuel cell uses living organisms to produce electricity. Two types are microbial fuel cells and enzymatic biofuel cells. Microbial fuel cells use living microorganisms. Enzymatic biofuel cells use enzymes derived from organisms.
10. *Microbial fuel cells* [23, 29] convert chemical energy to electrical energy by the catalytic reaction of microorganisms. They have a number of potential uses. The most obvious is harvesting the electricity produced; virtually any organic material could be used to "feed" the fuel cells. Microorganisms known as *electricigens* can efficiently convert organic wastes, renewable biomass, and even mud into electricity and harmless by-products. This capability offers the potential for using microbes (or their components) to generate electricity at low cost while transforming industrial, domestic, and farm wastes.
11. *Nanobiotechnology* refers to the use of nanotechnology to further the goals of biotechnology.
12. *Nanofabrication* refers to any technique used to create objects or mechanisms on the scale of nanotechnology.

References

1. "Plot of Moore's Law: Number of Transistors Integrated on an Integrated Circut." http://commons.wikimedia.org/wiki/File:Moore_Law_diagram_(2004).jpg. Accessed April 2009.
2. US Congress, Office of Technology Assessment, "Microelectronics Research and Development—A Background Paper," OTA-B P-C IT-40. Washington, DC: US Government Printing Office, March 1986.
3. B. B. Blanchard and W. J. Fabrycky, *Systems Engineering and Analysis*. Englewood Cliffs, NJ: Prentice Hall, 2006.
4. J. Burkhardt, *Lecture Notes on the Art of Design*. Fort Wayne, IN: The Indiana University–Purdue University Fort Wayne, 1996.
5. *Criteria for Accrediting Programs in Engineering in the United States*. Baltimore, MD: Engineering Accreditation Commission, ABET Inc., 2008.
6. M. H. Rashid, *Introduction to PSpice Using OrCAD for Circuits and Electronics*. Upper Saddle River, NJ: Prentice Hall, 2004.
7. R. L. McConnell, W. L. Cooley, and N. T. Middleton, *Electrical Engineering Design Compendium*. Reading, MA: Addison-Wesley, 1993.
8. M. N. Horenstein, *Design Concepts for Engineers*. Upper Saddle River, NJ: Prentice Hall, 2002.
9. L. O. Chua, "Memristor—The Missing Circuit Element." *IEEE Transactions on Circuit Theory*, Vol. 18, No. 5 (September 1971): 507–519.
10. P. Marks, "Engineers Find 'Missing Link' of Electronics." *New Scientist*, April 30, 2008.
11. D. B. Strukov, G. S. Snider, Stewart, R. Duncan, and S. R. Williams, "The Missing Memristor Found." *Nature*, Vol. 453, No. 7191 (May 2008): 80–83.
12. J. Fildes, "Getting More from Moore's Law." BBC, November 13, 2007. http://technology.newscientist.com/article/dn13812-engineers-find-missing-link-of-electronics.html. Accessed April 2009.
13. H. Klauk, *Organic Electronics: Materials, Manufacturing, and Applications*. New York: Wiley, August 2006.
14. J. M. Shaw and P. F. Seidler, "Organic Electronics: Introduction." *IBM Journal of Research and Development*, Vol. 45, No. 1 (January 2001): 3–10.
15. "Complementary Circuits Made from Organic Materials." *Nanowerk*, March 2007. www.nanowerk.com/news/newsid=1556.php. Accessed April 2009.
16. C. D. Dimitrakopoulos and D. J. Mascaro, "Organic Thin-Film Transistors: A Review of Recent Advances." *IBM Journal of Research and Development*, Vol. 45, No. 1 (January 2001): 11–27.
17. J. R. Sheats, H. Antoniadis, M. Hueschen, W. Leonard, J. Miller, R. Moon, D. Roitman, and A. Stocking, "Organic Electroluminescent Devices." *Science*, Vol. 273 (August 1996): 884.
18. OLED-info.com, "Introduction." OLED-info.com, June 2006. www.oled-info.com/introduction. Accessed April 2009.
19. "Organic Light-Emitting Diode." Wikipedia, April 2007. http://en.wikipedia.org/wiki/Organic_light-emitting_diode. Accessed April 2009.
20. www.fas.harvard.edu/home/news_and_events/releases/nanowire_12172003.html. Accessed April 2009.
21. www.imec.be/wwwinter/community/en/JobsOfTheFuture/biotechnologist.shtml. Accessed April 2009.
22. www.nanotech-now.com/news.cgi?story_id=10441. Accessed April 2009.
23. R. M. Allen and H. P. Bennetto, "Microbial Fuel Cells—Electricity Production from Carbohydrates." *Biotechnology and Applied Biochemistry*, Vol. 39/40 (1993): 27–40.
24. www.rsc.org/Publishing/ChemTech/Volume/2006/8/bioelectronic_nose.asp. Accessed April 2009.
25. G. Orellana, M. C. Moreno-Bondi, and J. Alonso, *Springer Series on Chemical Sensors and Biosensors*. New York: Springer Science & Business Media, 2005.
26. "Biosensors." Wikipedia, April 2008. http://en.wikipedia.org/wiki/Biosensor. Accessed April 2009.

27. www.engadget.com/2008/03/12/bio-electronic-implant-seeks-to-restore-partial-sight/. Accessed April 2009.
28. T. Sienko, *Molecular Computing*. Cambridge, MA: MIT Press, 2003.
29. D. R. Bond et al., "Electrode-Reducing Microorganisms That Harvest Energy from Marine Sediments." *Science*, Vol. 295, No. 5554 (January 2002): 483–485.

Problems

1.1 Design a circuit to measure a DC voltage in the range from 0 to 400 V. For a full-scale deflection, the indicating meter draws 100 μA at a voltage of 1 V (DC) across it. The accuracy should be better than 2%. The input current from the source should be less than 1 mA.

1.2 Design a circuit to supply 6 V to a load from a DC supply of 24 V. The load current should be 5 A. The accuracy should be better than 5%.

1.3 Design a circuit to provide 60 W at 50 V to a resistive lamp from an AC supply of 120 V ± 10% at 60 Hz. The circuit should be energy efficient—that is, it should consume the least power possible.

1.4 Design a circuit to charge a 2-μF capacitor from a DC supply of 24 V. The charging current should be limited to 1 mA. The accuracy should be better than 5%.

1.5 The input and output voltages of an amplifier are

$$v_i = 5 \sin (1000\,\pi t + 30°)\text{ mV} \quad \text{and} \quad v_o = 400 \sin (1000\,\pi t + 90°)\text{ mV}$$

Find the magnitude and phase of the voltage gain of the amplifier.

1.6 An n-channel MOSFET has $K_p = 20$ mA/V^2 and $V_t = 1.5$ V. If the gate–source voltage is $v_{GS} = 3$ V, find the small-signal transconductance g_m of the MOSFET.

$$g_m = \frac{di_D}{dv_{GS}}$$

1.7 An n-channel JFET has $I_{DSS} = 20$ mA and $V_p = 23$ V. If the gate–source voltage is $v_{GS} = -1.5$ V, find the small-signal transconductance g_m of the JFET. Assume the devices are in saturation.

$$g_m = \frac{di_D}{dv_{GS}}$$

1.8 The base current of a bipolar transistor is $i_B = 2\,(1 + \sin 2000\,\pi t)$ mA and the current gain of the transistor is $\beta_F = 100$. What are I_B, i_b, I_C, and i_c?

1.9 For a bipolar transistor, the collector–emitter voltages are $V_{CE} = 6$ V and $v_{ce} = -100 \sin (2000\,\pi t)$ mV, and the base–emitter voltages are $V_{BE} = 0.7$ V and $v_{be} = 1 \sin (2000\,\pi t)$ mV.

a. What are the expressions for v_{CE} and v_{BE}?

b. Find the small-signal voltage gain A_v.

1.10 For an FET, the drain–source voltages are $V_{DS} = 6$ V and $v_{ds} = -50 \sin (1000\,\pi t)$ mV, and the gate–source voltages are $V_{GS} = 3$ V and $v_{gs} = 2 \sin (1000\pi t)$ mV.

a. What are the expressions for v_{DS} and v_{GS}?

b. Find the small-signal voltage gain A_v.

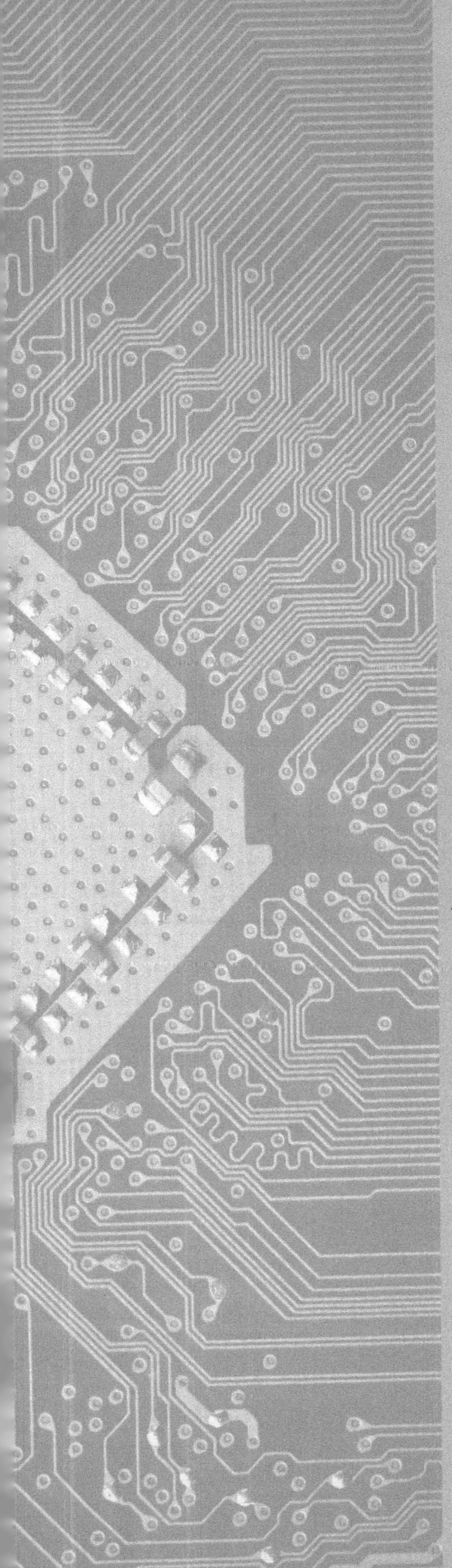

CHAPTER 2

INTRODUCTION TO AMPLIFIERS AND FREQUENCY RESPONSE

Learning Outcomes

After completing this chapter, students should be able to do the following:

- Describe the characteristics, types, circuit models, and applications of amplifiers.
- Determine the design specifications of an amplifier in order to meet the input and output requirements of an application.
- Describe the frequency behavior of amplifiers.
- Apply the frequency response techniques to determine the low and high cutoff frequencies of amplifiers.
- Describe the external characteristics and circuit models of op-amps.

Symbols and Their Meanings

Symbol	Meaning
A_v, G_m, Z_m, A_i	Voltage, transconductance, transimpedance, and current gain
A_{vo}, G_{mo}, Z_{mo}, A_{is}	Open-circuit voltage, transconductance, transimpedance, and current gain

Symbol	Meaning
BW, A_{PB}	Bandwidth and pass-band voltage gain
CMRR	Common-mode rejection ratio
f_H, f_L	High and low cutoff frequencies
i_s, i_i, i_o	Small-signal source and input and output currents
R_i, R_o	Input and output resistances
R_{CI}, τ_{CI}	Equivalent resistance of an *RC* circuit and the *RC* time constant
$v_s(t)$, $v_i(t)$, $v_o(t)$	Instantaneous input signal and input and output voltages
V_s, V_o	DC input signal and output voltages

2.1 Introduction

The output signals from transducers are weak (in the range of microvolts [μV] or millivolts [mV]) and possess a very small amount of energy. These signals are generally too small to be processed reliably to perform any useful function. Signal processing is much easier if the magnitude of the signal is large (in the range of volts). Amplifiers are used in almost every electronic system to increase the strength of a weak signal. An amplifier consists of one or more amplifying devices. The complexity of an amplifier depends on the number of amplifying devices. To analyze a complex circuit consisting of a number of amplifiers, a model representing the terminal behavior of each amplifier is often necessary.

2.2 Amplifier Characteristics

An amplifier may be considered as a two-port network with an input port and an output port. It is represented by the circuit symbol shown in Fig. 2.1, which indicates the direction of signal flow from the input side to the output side. Normally, one of the input terminals is connected to one of the output terminals to form a *common ground.* The output voltage (or current) is related to the input voltage (or current) by a *gain parameter.* If the output signal is directly proportional to the input signal such that the output is an exact replica of the input signal, the amplifier is said to be a *linear amplifier.* If there is any change in the output waveform, it is considered to have *distortion,* which is undesirable. The amplifier is then said to

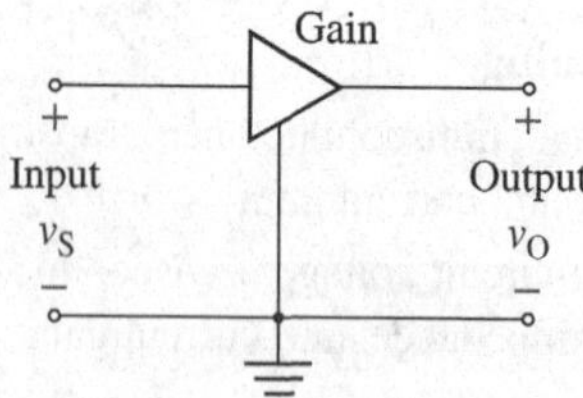

FIGURE 2.1 Symbol for an amplifier

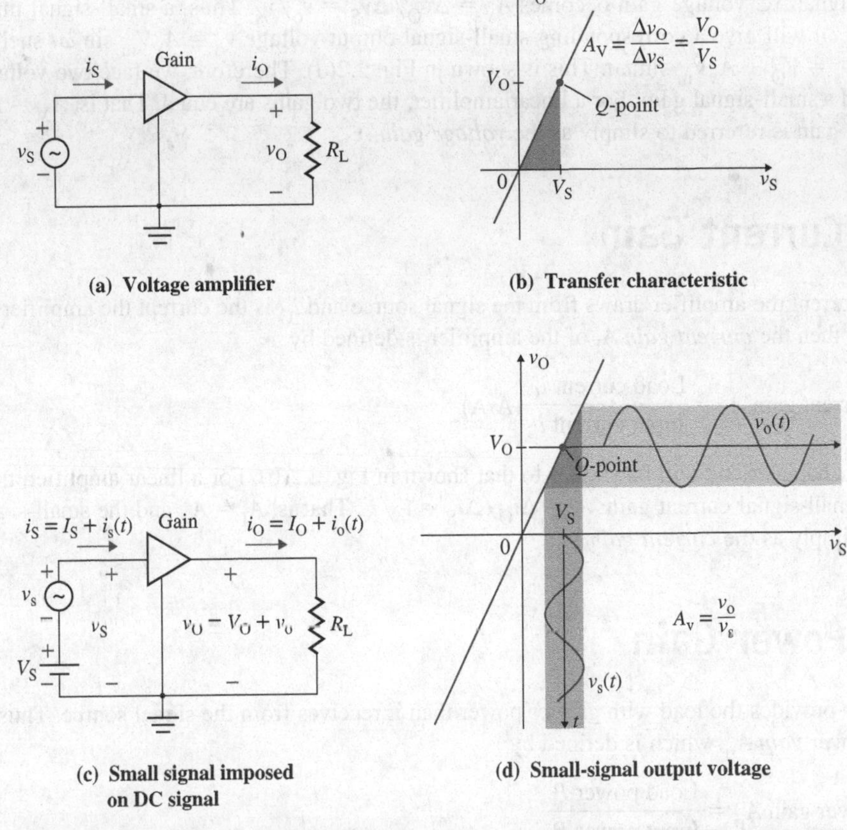

FIGURE 2.2 Voltage amplifier

be a *nonlinear amplifier.* The characteristics of amplifiers are defined by a number of parameters, which are described in the sections that follow.

2.2.1 Voltage Gain

If the input signal voltage to a linear amplifier is v_S, the amplifier will provide an output voltage v_O, which will be a magnified facsimile of v_S. This situation is shown in Fig. 2.2(a) for an amplifier with a load resistance of R_L. The *voltage gain* A_V of the amplifier is defined by

$$\text{Voltage gain } A_V = \frac{\text{Output voltage } v_O}{\text{Input voltage } v_S} (\text{V/V}) \tag{2.1}$$

The transfer characteristic, shown in Fig. 2.2(b), will be a straight line with a slope of A_V. Thus, if we apply a DC input signal of $v_S = V_S$, the DC output voltage will be $v_O = V_O = A_V V_S$ and the amplifier will operate at point Q. The DC voltage gain then becomes $A_V = V_O/V_S$. However, if we superimpose a small sinusoidal signal $v_s = V_m \sin \omega t$ on V_S, as shown in Fig. 2.2(c), the output voltage becomes $v_O = V_O + v_o$.

The small-signal AC voltage gain becomes $A_v = \Delta v_O / \Delta v_S = v_o / v_s$. Thus, a small-signal input voltage $v_s = V_m \sin \omega t$ will give a corresponding small-signal output voltage $v_o = A_v V_m \sin \omega t$ such that $v_O = A_V V_S + A_v v_s = V_O + A_v V_m \sin \omega t$. This is shown in Fig. 2.2(d). Therefore, we face two voltage gains: a DC gain and a small-signal gain. For a linear amplifier, the two gains are equal. That is, $A_V = A_v$, and the small-signal gain is referred to simply as the *voltage gain.*

2.2.2 Current Gain

If i_S is the current the amplifier draws from the signal source and i_O is the current the amplifier delivers to the load R_L, then the *current gain* A_I of the amplifier is defined by

$$\text{Current gain } A_I = \frac{\text{Load current } i_O}{\text{Input current } i_S} \text{(A/A)} \tag{2.2}$$

The transfer characteristic will be similar to that shown in Fig. 2.2(b). For a linear amplifier, the DC gain equals the small-signal current gain: $A_i = \Delta i_O / \Delta i_S = i_o / i_s$. That is, $A_I = A_i$, and the small-signal gain is referred to simply as the *current gain.*

2.2.3 Power Gain

An amplifier provides the load with greater power than it receives from the signal source. Thus, an amplifier has a *power gain* A_p, which is defined by

$$\text{Power gain } A_p = \frac{\text{Load power } P_L}{\text{Input power } P_i} \tag{2.3}$$

$$= \frac{v_o i_o}{v_s i_s} \text{(W/W)} \tag{2.4}$$

After substitution of $A_v = v_o / v_s$ and $A_i = i_o / i_s$, Eq. (2.4) can be written as

$$A_p = A_v A_i \tag{2.5}$$

Thus, the power gain is the product of the voltage gain and the current gain.

2.2.4 Logarithmic Gain

The gains of amplifiers can be expressed either as dimensionless quantities or with units (V/V for a voltage gain, A/A for a current gain, or W/W for a power gain). Their values are usually very large and extend over several orders of magnitude. It is not convenient to plot such large numbers against other parameters. Gains are normally expressed in terms of logarithms, as follows:

$$\text{Power gain in decibels (dB)} = 10 \log A_p = 10 \log_{10}\left(\frac{P_L}{P_i}\right) = 10 \log_{10}\left(\frac{v_o^2/R_L}{v_s^2/R_i}\right)$$

$$= 20 \log_{10}\left(\frac{v_o}{v_s}\right) + 10 \log_{10}\left(\frac{R_i}{R_L}\right)$$

where R_i is the input resistance of the amplifier, which is seen as a load by the signal source v_S. The term $20 \log_{10} (v_o/v_s)$ is referred to as the voltage gain of the amplifier in decibels. That is,

$$\text{Voltage gain in dB} = 20 \log |A_v| \quad \text{for } R_i = R_L$$

The power gain can also be expressed in terms of the input and output current:

$$\text{Power gain in dB} = 10 \log A_p = 10 \log_{10}\left(\frac{P_L}{P_i}\right) = 10 \log_{10}\left(\frac{i_o^2 R_L}{i_s^2 R_i}\right)$$

$$= 20 \log_{10}\left(\frac{i_o}{i_s}\right) + 10 \log_{10}\left(\frac{R_L}{R_i}\right)$$

The term $20 \log_{10} (i_o/i_s)$ is referred to as the current gain of the amplifier in decibels. That is,

$$\text{Current gain in dB} = 20 \log |A_i|$$

If $R_i = R_L$, the power gain in decibels is equal to the voltage and current gains in decibels. That is,

$$\text{Power gain in dB} = \text{Voltage gain in dB} = \text{Current gain in dB}$$

Some amplifiers, such as operational amplifiers (op-amps), have a very high voltage gain, which is quoted in decibels. For example, rather than writing $A_v = 10^5$ V/V, it is common to write 100 dB, which equals $20 \log 10^5$.

▶ NOTES

1. If there is a phase difference of 180° between the input and output voltages (or currents), the voltage gain A_v (or current gain A_i) will be negative. Therefore, the absolute value of A_v (or A_i) must be used for calculating the gain in decibels. However, the power gain A_p is always positive.
2. If the absolute value of the voltage (or current) gain is less than 1, the output is said to be *attenuated* rather than amplified, and the gain in decibels will be negative.

2.2.5 Input and Output Resistances

Input resistance R_i is a measure of the current drawn by the amplifier. It is a ratio of the input signal voltage to the input current:

$$R_i = \frac{v_S}{i_S} \tag{2.6}$$

Output resistance R_o is the internal resistance seen from the output terminals of an amplifier—that is, Thevenin's equivalent resistance.

2.2.6 Amplifier Saturation

An amplifier needs a DC power supply (or supplies) so that an operating Q-point can be established, as shown in Fig. 2.2(b), that allows variation in the output signal in response to a small change in the input signal. The DC supply (or supplies) provides the power delivered to the load, as well as any power that is dissipated as heat within the amplifier itself. An amplifier with two power supplies, V_{CC} and V_{EE}, is shown in

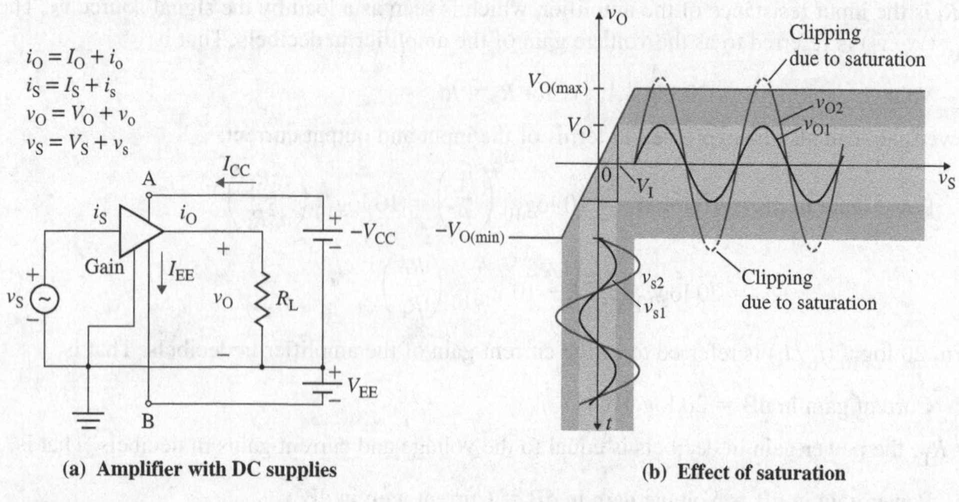

FIGURE 2.3 Amplifier power supplies and saturation

Fig. 2.3(a). I_{CC} and I_{EE} are the currents drawn from the DC supplies V_{CC} and V_{EE}, respectively. Terminal A is connected to the positive side of the DC source V_{CC}, and terminal B is connected to the negative side of the DC source V_{EE}. The output voltage of the amplifier cannot exceed the positive saturation limit $V_{O(max)}$ and cannot decrease below the negative saturation limit $-V_{O(min)}$.

Each of the two saturation limits is usually within 1 V or 2 V of the corresponding power supply. This fact is a consequence of the internal circuity of the amplifiers and the nonlinear behavior of the amplifying devices. Therefore, to avoid distortion of the output voltage as shown in Fig. 2.3(b), the input voltage must be kept within the range defined by

$$\frac{-V_{O(min)}}{A_V} \leq v_S \leq \frac{V_{O(max)}}{A_V} \tag{2.7}$$

As long as the amplifier operates with the saturation limits, the voltage gain can normally be assumed to be linear. The power delivered by the DC supplies will be

$$P_{dc} = V_{CC}I_{CC} + V_{EE}I_{EE} \tag{2.8}$$

and the power delivered P_i by the input signal will be small compared to P_{dc}. Therefore, the efficiency η of an amplifier is defined by

$$\text{Amplifier efficiency } \eta = \frac{\text{Load power } P_L}{\text{Power delivered by DC supplies } P_{dc}} \tag{2.9}$$

The efficiency of an amplifier ranges from 25% to 80%, depending on the type of amplifier. For amplifiers with a very low input signal (millivolts or microvolts), the voltage gain rather than the efficiency is the prime consideration. On the other hand, for power amplifiers (covered in Chapter 11), efficiency is the major consideration because the amplifier should supply the maximum power to the load (such as the speakers of an audio amplifier).

EXAMPLE 2.1

Finding amplifier parameters The measured small-signal values of the linear amplifier in Fig. 2.3(a) are $v_s = 20 \sin 400t$ (mV), $i_s = 1 \sin 400t$ (μA), $v_o = 7.5 \sin 400t$ (V), and $R_L = 0.5$ kΩ. The DC values are $V_{CC} = V_{EE} = 12$ V and $I_{CC} = I_{EE} = 10$ mA. Find **(a)** the values of amplifier parameters A_v, A_i, A_p, and R_i; **(b)** the power delivered by DC supplies P_{dc} and the power efficiency η; and **(c)** the maximum value of the input voltage so that the amplifier operates within the saturation limits.

SOLUTION

$v_{s(peak)} = 20$ mV, $v_{o(peak)} = 7.5$ V, and $i_{s(peak)} = 1$ μA.

(a) The load current is

$$i_o = \frac{v_o}{R_L} = \frac{7.5 \sin 400t \text{ (V)}}{0.5 \text{ k}\Omega} = 15 \times 10^{-3} \sin 400t = 15 \sin 400t \text{ (mA)}$$

The voltage gain is

$$A_v = \frac{v_{o(peak)}}{v_{s(peak)}} = \frac{7.5 \text{ V}}{20 \text{ mV}} = 375 \text{ V/V} \qquad [\text{or } 20 \log (375) = 51.48 \text{ dB}]$$

The current gain is

$$A_i = \frac{i_{o(peak)}}{i_{s(peak)}} = \frac{15 \text{ mA}}{1 \text{ μA}} = 15 \text{ kA/A} \qquad [\text{or } 20 \log (15 \text{ k}) = 83.52 \text{ dB}]$$

The power gain is

$$A_p = A_v A_i = 375 \times 15 \text{ k} = 5625 \text{ kW/W} \qquad [\text{or } 10 \log (5625 \text{ k}) = 67.5 \text{ dB}]$$

The input resistance is

$$R_i = \frac{v_{s(peak)}}{i_{s(peak)}} = \frac{20 \text{ mV}}{1 \text{ μA}} = 20 \text{ k}\Omega$$

(b) The power delivered by the DC supplies is

$$P_{dc} = V_{CC} I_{CC} + V_{EE} I_{EE} = 2 \times 12 \text{ V} \times 10 \text{ mA} = 240 \text{ mW}$$

The load power is

$$P_L = \left(\frac{v_{o(peak)}}{\sqrt{2}}\right)\left(\frac{i_{o(peak)}}{\sqrt{2}}\right) = \left(\frac{7.5 \text{ V}}{\sqrt{2}}\right)\left(\frac{15 \text{ mA}}{\sqrt{2}}\right) = 56.25 \text{ mW}$$

By using $\sqrt{2}$ factor for converting a peak value to a rms value, the input power is

$$P_i = \left(\frac{v_{s(peak)}}{\sqrt{2}}\right)\left(\frac{i_{s(peak)}}{\sqrt{2}}\right) = \left(\frac{20 \text{ mV}}{\sqrt{2}}\right)\left(\frac{1 \text{ μA}}{\sqrt{2}}\right) = 10 \text{ mW}$$

The power efficiency is

$$\eta = \frac{P_L}{P_{dc} + P_i} = \frac{56.25 \text{ mW}}{250 \text{ mW}} = 22.5\%$$

(c) Since

$$A_v v_{s(max)} = v_{O(max)} = V_{CC} = V_{EE} \quad \text{or} \quad v_{S(max)} = \frac{12\text{ V}}{375} = 32\text{ mV}$$

the limit of the maximum input voltage is $0 \le v_{S(max)} \le 32$ mV, and the limit of the minimum input voltage is $v_{S(min)} = -v_{S(max)} = -32$ mV.

2.2.7 Amplifier Nonlinearity

Practical amplifiers exhibit a nonlinear characteristic, which is caused by nonlinear devices such as transistors (discussed in Chapters 7 and 8). For the amplifier shown in Fig. 2.4(a) with one DC supply, its nonlinear characteristic is shown in Fig. 2.4(b). Fortunately there is a region in the midrange of the output voltage where the gain remains almost constant. If the amplifier can be made to operate in this region, a small variation in the input voltage will cause an almost linear variation in the output voltage, and the gain will remain approximately constant. This goal is accomplished by biasing the amplifier to operate at a quiescent point, generally called the *Q-point,* having a DC input voltage V_S and a corresponding DC output voltage V_O. If a small instantaneous input voltage $v_s(t) = V_m \sin \omega t$ is superimposed on the DC input voltage V_S, as shown in Fig. 2.4(b), the total instantaneous input voltage becomes

$$v_S(t) = V_S + v_s(t) = V_S + V_m \sin \omega t$$

which will cause the operating point to move up and down along the transfer characteristic around the Q-point. This movement will cause a corresponding time-varying output voltage

$$v_O(t) = V_O + v_o(t)$$

If $v_s(t)$ is sufficiently small, then $v_o(t)$ will be directly proportional to $v_s(t)$; so

$$v_o(t) = A_v v_s(t) = A_v V_m \sin \omega t$$

where A_v is the slope of the transfer characteristic at the Q-point. That is,

$$A_v = \left. \frac{dv_O}{dv_S} \right|_{\text{at } Q\text{-point}} \tag{2.10}$$

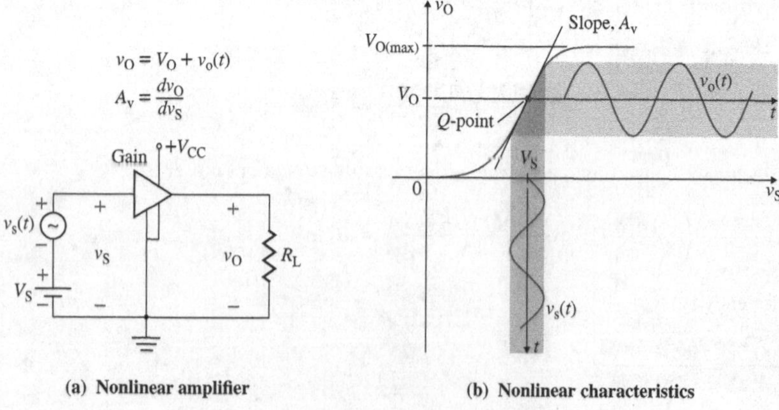

(a) Nonlinear amplifier

(b) Nonlinear characteristics

FIGURE 2.4 Amplifier nonlinearity

Therefore, as long as the input signal is kept sufficiently small, the amplifier will exhibit an almost linear characteristic. However, increasing the magnitude of the input signal is expected to cause distortion of the output voltage and may even cause saturation. A_v is known as the *small-signal voltage gain* (or simply the *voltage gain*) of the amplifier; it should not be confused with the *DC gain*, which is defined by

$$A_{dc} = A_V = \left.\frac{v_O}{v_S}\right|_{\text{at } Q\text{-point}} = \frac{V_O}{V_S} \tag{2.11}$$

Thus, we can conclude that the analysis and the design of a nonlinear amplifier involve two signals: a DC signal and an AC signal. However, the characteristics of an amplifier are described by its behavior in response to a small AC input signal.

▶ **NOTE** In practical amplifiers, the *Q*-point is set internally and the amplifiers operate from a small input signal. The input signal v_s is superimposed on the *Q*-point (which consists of V_O and V_S) to produce a small-signal output voltage v_o.

EXAMPLE 2.2

Finding the limiting parameters of a nonlinear amplifier The measured values of the nonlinear amplifier in Fig. 2.4(a) are $v_O = 4.3$ V at $v_S = 18$ mV, $v_O = 5$ V at $v_S = 20$ mV, and $v_O = 5.8$ V at $v_S = 22$ mV. The DC supply voltage is $V_{CC} = 9$ V, and the saturation limits are $2\text{ V} \le v_O \le 8\text{ V}$.

(a) Determine the small-signal voltage gain A_v.

(b) Determine the DC voltage gain A_{dc}.

(c) Determine the limits of input voltage v_S.

SOLUTION

Let $v_O = 5$ V at $v_S = 20$ mV be the *Q*-point. Then

$$\Delta v_O = v_O(\text{at } v_S = 22\text{ mV}) - v_O(\text{at } v_S = 18\text{ mV}) = 5.8\text{ V} - 4.3\text{ V} = 1.5\text{ V}$$

$$\Delta v_S = v_S(\text{at } v_O = 5.8\text{ V}) - v_S(\text{at } v_O = 4.3\text{V}) = 22\text{ mV} - 18\text{ mV} = 4\text{ mV}$$

(a) The small-signal voltage gain is

$$A_v = \frac{\Delta v_O}{\Delta v_S} = \frac{1.5\text{ V}}{4\text{ mV}} = 375\text{ V/V} \qquad (\text{or } 51.48\text{ dB})$$

(b) The DC voltage gain is

$$A_{dc} = A_V = \frac{v_O}{v_S} = \frac{5\text{ V}}{20\text{ mV}} = 250\text{ V/V} \qquad (\text{or } 47.96\text{ dB})$$

(c) The limits of input voltage v_S are

$$\frac{-(v_O - v_{O(\min)})}{A_v} \le v_S - 20\text{ mV} \le \frac{(v_{O(\max)} - v_O)}{A_v}$$

That is, $-(5-2)/A_V \le v_S - 20\text{ mV} \le (8-5)/A_V$, or $-8\text{ mV} \le v_S - 20\text{ mV} \le 8\text{ mV}$, which gives $12\text{ mV} \le v_S \le 28\text{ mV}$.

2.2.8 Rise Time

The *rise time* is the time required for the output voltage to rise from 10% to 90% of the steady-state value. If the voltage gain of the amplifier is assumed to be unity, the output voltage due to a step input voltage V_S can be expressed as

$$v_O = V_S(1 - e^{-t/\tau}) \tag{2.12}$$

where τ is the time constant due to the internal resistance and capacitance of the amplifier. From Eq. (B.40) in Appendix B, the time rise is related to the time constant τ by

$$t_r = 2.2\tau \tag{2.13}$$

The typical value of rise time is 0.3 μs for the μA741 op-amp. Note that a linear operation was assumed in deriving Eq. (2.13) and the effect of slew rate (discussed in Sec. 2.2.9) was ignored.

2.2.9 Slew Rate

The *slew rate* (SR) is the maximum rate of rise of the output voltage per unit time, and it is measured in volts per microsecond. If a sharp step input voltage is applied to an amplifier, the output will not rise as quickly as the input because the internal capacitors require time to charge to the output voltage level. SR is a measure of how quickly the output of an amplifier can change in response to a change of input frequency. The slew rate depends on the voltage gain, but it is normally specified at unity gain. SR for the LF411 op-amp is 10 V/μs, whereas it is 0.5 V/μs for the μA741C op-amp. The output response due to a step input is shown in Fig. 2.5(a). The output, which follows the slew rate of the op-amp, will be distorted because the op-amp output cannot rise as fast as the input voltage.

With a unity-gain amplifier, the rate of rise of the output voltage for a step signal V_S can be found from Eq. (2.12) to be

$$\frac{dv_O}{dt} = \frac{V_S}{\tau} e^{-t/\tau} \tag{2.14}$$

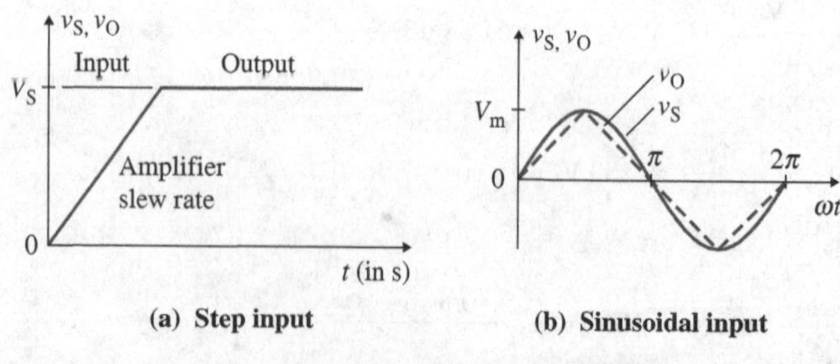

FIGURE 2.5 Effect of slew rate on amplifier response

which becomes maximum at $t = 0$ and gives the slew rate SR as

$$\mathrm{SR} = \left.\frac{dv_{\mathrm{O}}}{dt}\right|_{t=0} = \frac{V_{\mathrm{S}}}{\tau}$$

The 3-dB frequency can be related to the time constant τ or to the rise time t_{r} by

$$f = \frac{1}{2\pi\tau} = \frac{2.2}{2\pi t_{\mathrm{r}}} = \frac{0.35}{t_{\mathrm{r}}} \tag{2.15}$$

Thus, the frequency response is inversely proportional to the rise time t_{r}. The input signal frequency f_{s} should be less than the maximum op-amp frequency; otherwise the output voltage will be distorted. For example, if the rise time of an input signal is $t_{\mathrm{r}} = 0.1$ μs, its corresponding input frequency is $f_{\mathrm{s}} = 0.35/0.1$ μs $= 3.5$ MHz, and the output voltage will be distorted in an op-amp unity-gain bandwidth of $f_{\mathrm{bw}} = 1$ MHz.

Substituting τ from Eq. (2.13) and t_{r} from Eq. (2.15) gives

$$\mathrm{SR} = \frac{2.2V_{\mathrm{S}}}{t_{\mathrm{r}}} = \frac{2.2V_{\mathrm{S}}f}{0.35} = 6.286V_{\mathrm{S}}f \tag{2.16}$$

For a sinusoidal input voltage with a unity gain and without limiting by the slew rate, the output voltage becomes

$$v_{\mathrm{O}} = V_{\mathrm{m}} \sin \omega t$$

$$\frac{dv_{\mathrm{O}}}{dt} = \omega V_{\mathrm{m}} \cos \omega t$$

which becomes maximum at $\omega t = 0$, and SR is given by

$$\mathrm{SR} = \left.\frac{dv_{\mathrm{O}}}{dt}\right|_{t=0} = \omega V_{\mathrm{m}} = 2\pi f V_{\mathrm{m}} \tag{2.17}$$

which gives the maximum frequency $f_{\mathrm{s(max)}}$ of the sinusoidal input voltage as

$$f_{\mathrm{s(max)}} = \frac{\mathrm{SR}}{2\pi V_{\mathrm{m}}} \tag{2.18}$$

Slew rate can introduce a significant error if the rate of change of the input voltage is more than the SR of the amplifier. Note that the rate of change of the input voltage rather than the change indicates how fast the input can rise. For example, if the rate of change of a sinusoidal input voltage is very high compared to the SR of the amplifier, the output will be highly distorted and will tend to have a triangular waveform. This situation is shown in Fig. 2.5(b) for a sinusoidal input voltage.

EXAMPLE 2.3

Finding the effect of slew rate on the input frequency The slew rate of a unity-gain amplifier is SR = 0.7 V/μs. The frequency of the input signal is f_s = 300 kHz. Calculate **(a)** the peak sinusoidal input voltage V_m that will give an output without any distortion and **(b)** the maximum input frequency $f_{s(max)}$ that will avoid distortion if the input has a peak sinusoidal voltage of V_m = 5 V.

SOLUTION

SR = 0.7 V/μs = 0.7×10^6 V/s, $f = f_s$ = 300 kHz.

(a) Using Eq. (2.17), we find that the peak value of input voltage is

$$V_m = \frac{SR}{2\pi f_s} = \frac{0.7 \times 10^6}{2\pi \times 300 \times 10^3} = 371.4 \text{ mV}$$

(b) From Eq. (2.18), the maximum frequency $f_{s(max)}$ becomes

$$f_{s(max)} = \frac{SR}{2\pi V_m} = \frac{0.7 \times 10^6}{2\pi \times 5} = 22.28 \text{ kHz}$$

KEY POINTS OF SECTION 2.2

- The performance of an amplifier is described by its voltage gain, current gain, power gain, input resistance, and output resistance.
- The gains of an amplifier have high magnitudes and are quoted generally in decibels (dB).
- The power gain is very large because the signal power is very low. The DC power supply (or supplies) provides the load power.
- A DC power supply (or supplies) is needed to establish a Q-point. The small-signal source is then superimposed on the DC input so that the operating point can move up and down around the Q-point, and a magnified replica of the signal source is obtained on the output. As long as the signal source is sufficiently small, a nonlinear amplifier exhibits an almost linear characteristic.
- The DC power supply (or supplies) sets the saturation limit(s) of an amplifier.
- There are two types of gain: a DC gain and a small-signal AC gain. The small-signal gain is normally quoted as the gain of the amplifier.

2.3 Amplifier Types

The input signal to an amplifier can be either a voltage source or a current source. The output of an amplifier can be either a voltage source or a current source. Therefore, there are four possible input and output combinations: *v-v*, *i-i*, *v-i*, and *i-v*. Based on the input and output relationships, amplifiers can be classified into four types: voltage amplifiers, current amplifiers, transconductance amplifiers, and transimpedance amplifiers.

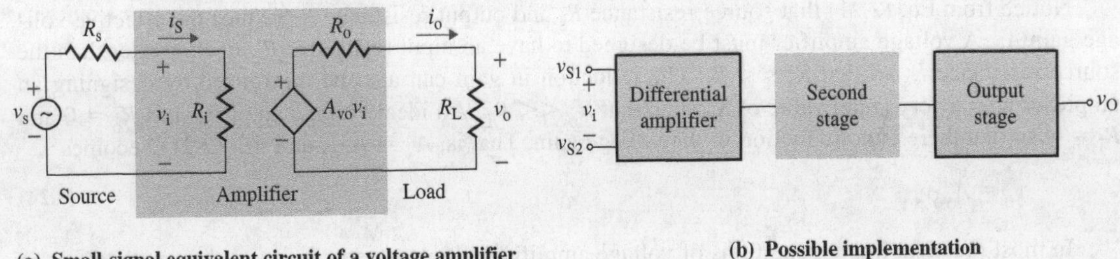

(a) Small-signal equivalent circuit of a voltage amplifier

(b) Possible implementation

FIGURE 2.6 Voltage amplifier

2.3.1 Voltage Amplifiers

An amplifier whose output voltage is proportional to its input voltage is known as a *voltage amplifier.* The input signal is a voltage source, and the output of the amplifier is also a voltage source. Such an amplifier is referred to as a voltage-controlled voltage source (VCVS); an example is shown in Fig. 2.6(a). The amplifier is connected between a voltage source v_s and a load resistance R_L. R_s is the source resistance. A_{vo} is the voltage gain with load resistance R_L disconnected, and it is known as the *open-circuit voltage gain.* R_o is the output resistance of the amplifier.

The output voltage of a voltage amplifier can be obtained by using the voltage divider rule:

$$v_o = i_o R_L = A_{vo} v_i \frac{R_L}{R_L + R_o} \tag{2.19}$$

From the voltage divider rule, the input voltage v_i to the amplifier is related to the signal voltage v_s by

$$v_i = \frac{R_i}{R_i + R_s} v_s \tag{2.20}$$

Substituting v_i from Eq. (2.20) into Eq. (2.19), we get the *effective voltage gain* A_v, which is defined as the ratio of v_o to v_s. That is,

$$A_v = \frac{v_o}{v_s} = \frac{v_o}{v_i} \times \frac{v_i}{v_s} = \frac{A_{vo} R_i R_L}{(R_i + R_s)(R_L + R_o)} = \frac{A_{vo}}{(1 + R_s/R_i)(1 + R_o/R_L)} \tag{2.21}$$

The current gain A_i, which is defined as the ratio of the output current i_o to the input current i_s, is given by

$$A_i = \frac{i_o}{i_s} = \frac{A_{vo} v_i}{R_L + R_o} \times \frac{1}{v_i/R_i} = \frac{A_{vo} R_i}{R_L + R_o} \tag{2.22}$$

The power gain will be the product of the voltage gain and the current. That is,

$$A_p = A_v A_i \tag{2.23}$$

Notice from Eq. (2.21) that source resistance R_s and output resistance R_o reduce the effective voltage gain A_v. A voltage amplifier must be designed to have an input resistance R_i much greater than the source resistance R_s so that $R_s << R_i$. The reduction in gain can also be minimized by designing an amplifier with a very small value of R_o such that $R_o << R_L$. An ideal voltage amplifier has $R_o = 0$ and $R_i = \infty$ so that there is no reduction in the voltage gain. That is, $A_v = A_{vo}$, and Eq. (2.21) becomes

$$v_o = A_{vo} v_s \tag{2.24}$$

In most practical implementations of voltage amplifiers in integrated circuits, a differential input is desirable from the viewpoint of performance. The implementation of a VCVS normally begins with a differential input to give a high input resistance and then has an output stage to give a low output resistance. This arrangement is shown in Fig. 2.6(b). If sufficient gain is available from the differential amplifier, only the output stage is required to give a low output resistance. If more gain is required, a second stage will be needed. The specifications of the VCVS and the judgment of the circuit designer will be the major factors in the choice of the implementation.

EXAMPLE 2.4

D

Determining the design specifications of a voltage amplifier A voltage amplifier is required to amplify the output signal from a communication receiver that produces a voltage signal of $v_s = 20$ mV with an internal resistance of $R_s = 1.5$ kΩ. The load resistance is $R_L = 15$ kΩ. The desired output voltage is $v_o \geq 10$ V. The amplifier must not draw more than 1 μA from the receiver. The variation in output voltage when the load is disconnected should be less than 0.5%. Determine the design specifications of the voltage amplifier.

SOLUTION

Since the input current is $i_s \leq 1$ μA, the input resistance of the amplifier can be found from

$$R_s + R_i = \frac{v_s}{i_s} \geq \frac{20\text{ mV}}{1\ \mu\text{A}} = 20\text{ k}\Omega$$

which gives

$$R_i \geq 20\text{ k}\Omega - R_s = 20\text{ k}\Omega - 1.5\text{ k}\Omega = 18.5\text{ k}\Omega$$

The variation in output voltage, which depends on the ratio R_o/R_L, can be found from

$$\frac{\Delta v_o}{v_o} = \frac{R_o}{R_L + R_o} \tag{2.25}$$

which, for $\Delta v_o/v_o \leq 0.5\%$ and $R_L = 15$ kΩ, gives $R_o \leq 75$ Ω. The desired effective voltage gain is $A_v = v_o/v_s \geq 10\text{ V}/20\text{ mV} = 500$ V/V (or 53.98 dB). The open-circuit voltage gain can be found from Eq. (2.21):

$$500 \leq \frac{A_{vo}}{(1 + R_s/R_i)(1 + R_o/R_L)} = \frac{A_{vo}}{(1 + 1.5\text{ k}/18.5\text{ k})(1 + 75/15\text{ k})}$$

or $A_{vo} \geq 543$ V/V (or 54.7 dB)

The amplifier specifications are $R_i \geq 18.5$ kΩ, $R_o \leq 75$ Ω, and $A_{vo} \geq 543$ V/V (or 54.7 dB).

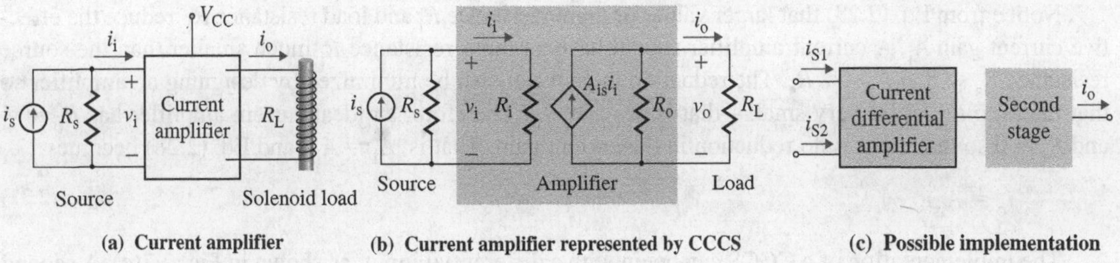

FIGURE 2.7 Current amplifier

2.3.2 Current Amplifiers

An amplifier whose output current is proportional to its input current is called a *current amplifier.* Its input is a current source, as shown in Fig. 2.7(a), with a load resistance R_L. A current amplifier is represented by a current-controlled current source (CCCS), as shown in Fig. 2.7(b). A_{is} is called the *short-circuit current gain* (or simply the *current gain*) with output terminals shorted. R_i is the input resistance, and R_o is the output resistance. A current amplifier is normally used to provide a modest voltage gain but a substantial current gain so that it draws little power from the signal source and delivers a large amount of power to the load. Such an amplifier is often known as a *power amplifier.*

The output current i_o of the amplifier can be obtained by using the current divider rule:

$$i_o = A_{is} i_i \frac{R_o}{R_o + R_L} \tag{2.26}$$

The input current i_i of the amplifier is related to the signal source current i_s by

$$i_i = \frac{R_s}{R_s + R_i} i_s \tag{2.27}$$

Substituting i_i from Eq. (2.27) into Eq. (2.26), we get the *effective current gain* A_i, which is defined as the ratio of i_o to i_s. That is,

$$A_i = \frac{i_o}{i_s} = \frac{i_o}{i_i} \times \frac{i_i}{i_s} = \frac{A_{is} R_s R_o}{(R_s + R_i)(R_o + R_L)} = \frac{A_{is}}{(1 + R_i/R_s)(1 + R_L/R_o)} \tag{2.28}$$

The voltage gain A_v, which is defined as the ratio of the output voltage v_o to the input voltage v_s, is given by

$$A_v = \frac{v_o}{v_s} = \frac{i_o R_L}{i_s R_s} = A_i \frac{R_L}{R_s} \tag{2.29}$$

The power gain is the product of the voltage gain and the current gain. That is,

$$A_p = A_v A_i \tag{2.30}$$

Notice from Eq. (2.28) that larger values of input resistance R_i and load resistance R_L reduce the effective current gain A_i. A current amplifier should have an input resistance R_i much smaller than the source resistance R_s so that $R_i << R_s$. The reduction in gain can also be minimized by designing an amplifier so that the ratio R_L/R_o is very small—that is, $R_o >> R_L$. Therefore, an ideal current amplifier has $R_o = \infty$ and $R_i = 0$ so that there is no reduction in the current gain. That is, $A_i = A_{is}$, and Eq. (2.28) becomes

$$i_o = A_{is} i_s \tag{2.31}$$

The implementation of a CCCS can begin with a differential input, as shown in Fig. 2.7(c). A second stage will be necessary because the current differential amplifier will have a low current gain. An output stage may be necessary to give a high output resistance.

EXAMPLE 2.5

D

Determining the design specifications of a current amplifier A current amplifier is required to amplify the output signal from a transducer that produces a constant current of $i_s = 1$ mA at an internal resistance varying from $R_s = 1.5$ kΩ to $R_s = 10$ kΩ. The desired output current is $i_o = 0.5$ A at a load resistance varying from $R_L = 10$ Ω to $R_L = 120$ Ω. The variation in output current should be kept within ±3%. Determine the design specifications of the current amplifier.

SOLUTION

Since the variation in output current should be kept within ±3%, the variation in the effective current gain A_i should also be limited to ±3%. According to Eq. (2.28), the variation in A_i will be contributed by A_{is}, R_s, and R_L. Let us assume that each of them contributes equally to the variation—that is, each contributes ±1%. The nominal short-circuit current gain is $A_{is} = i_o/i_s = 0.5\text{ A}/1\text{ mA} = 500$ A/A. Thus, the value of R_o that will keep the variation in positive current gain within 1% for variation in R_L from 10 Ω to 120 Ω can be found approximately from

$$0.99\frac{R_o}{R_o + 10} = \frac{R_o}{R_o + 120}$$

which gives $R_o \geq 10.88$ kΩ when solved for R_o. Similarly, the value of R_i that will keep the variation in current gain within 1% for variation in R_s from 1.5 kΩ to 10 kΩ can be found approximately from

$$0.99\frac{10\text{ k}}{10\text{ k} + R_i} = \frac{1.5\text{ k}}{1.5\text{ k} + R_i}$$

which gives $R_i \leq 17.86$ Ω when solved for R_i. Thus, the amplifier specifications are $A_{is} = 500$ A/A ± 1%, $R_o \geq 10.88$ kΩ, and $R_i \leq 17.86$ Ω.

The exact change in ΔA_i can be found from

$$\frac{\Delta A_i}{A_i} = \frac{\Delta A_{is}}{A_{is}} + \frac{1}{1 + R_{s1}/R_i} \times \frac{\Delta R_{s1}}{R_{s1}} - \frac{1}{1 + R_o/R_{L1}} \times \frac{\Delta R_L}{R_{L1}}$$

$$= 1\% + \frac{1}{1 + 1.5\text{ k}/17.86} \times \frac{1\%}{1.5\text{ k}} - \frac{1}{1 + 10.88\text{ k}/10} \times \frac{1\%}{10} = 1\%$$

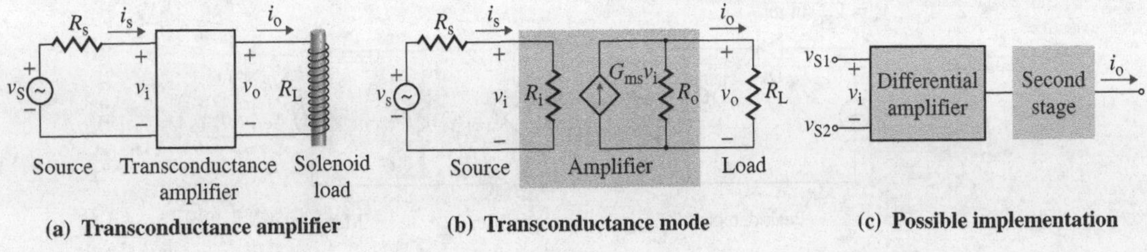

FIGURE 2.8 Transconductance amplifier

2.3.3 Transconductance Amplifiers

An amplifier that receives a voltage signal as input and provides a current signal as output is called a *transconductance amplifier;* an example is shown in Fig. 2.8(a). It can be represented by a voltage-controlled current source (VCCS), as shown in Fig. 2.8(b). The amplifier is connected between a voltage source v_s and a load resistance R_L. Gain parameter G_{ms}, which is the ratio of the short-circuit output current to the input voltage, is called the *short-circuit transconductance.* From the current divider rule, the output current i_o is

$$i_o = G_{ms} v_i \frac{R_o}{R_o + R_L} \tag{2.32}$$

The input voltage v_i of the amplifier is related to source voltage v_s by

$$v_i = \frac{R_i}{R_i + R_s} v_s \tag{2.33}$$

Substituting v_i from Eq. (2.33) into Eq. (2.32) gives the effective transconductance gain G_m as

$$G_m = \frac{i_o}{v_s} = \frac{G_{ms} R_o R_i}{(R_o + R_L)(R_i + R_s)} = \frac{G_{ms}}{(1 + R_L/R_o)(1 + R_s/R_i)} \tag{2.34}$$

The effective voltage gain A_v can be found from

$$A_v = \frac{v_o}{v_s} = \frac{i_o R_L}{v_s} = \frac{v_o}{v_i} \times \frac{v_i}{v_s} = \frac{G_{ms} R_o R_L R_i}{(R_o + R_L)(R_i + R_s)} = \frac{G_{ms} R_L}{(1 + R_L/R_o)(1 + R_s/R_i)} \tag{2.35}$$

Notice from Eq. (2.34) that the source resistance R_s and the load resistance R_L reduce the effective transconductance gain G_m. A transconductance amplifier should have a high input resistance R_i so that $R_i >> R_s$ and a very high output resistance R_o so that $R_o >> R_L$. Therefore, an ideal transconductance amplifier has $R_o = \infty$ and $R_i = \infty$ so that there is no reduction in the voltage gain. That is, $G_m = G_{ms}$, and Eq. (2.34) becomes

$$i_o = G_{ms} v_s \tag{2.36}$$

The implementation of a VCCS can begin with a differential input, as shown in Fig. 2.8(c). Since the output resistance of a differential amplifier is reasonably high, one differential stage should be adequate. If more gain is needed, however, a second stage can be added.

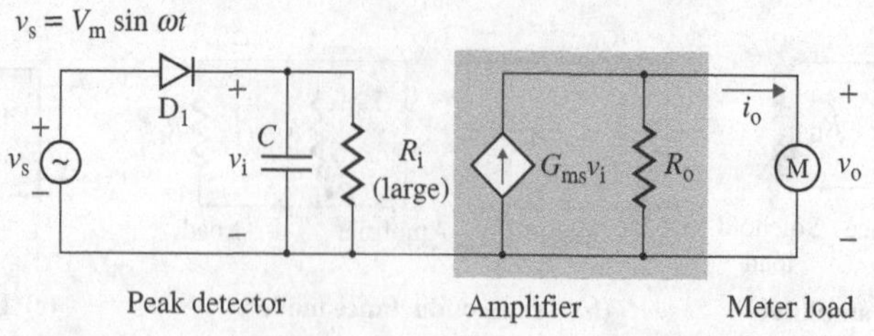

FIGURE 2.9 Impedance matching between two circuits

A transconductance amplifier can be used to eliminate interaction between two circuits, as shown in Fig. 2.9. The amplifier is connected between the meter and the peak detector. The amplifier should offer a very high resistance to the detector; at the same time, the meter current will be proportional to the peak voltage. The capacitor will continuously monitor the peak value V_m of the input signal. This peak value is indicated by the meter, whose reading depends on the current flowing through it. This technique is often used in electronic circuits to isolate two circuits from each other.

EXAMPLE 2.6

D

Determining the design specifications of a transconductance amplifier A transconductance amplifier is needed to record the peak voltage of the circuit in Fig. 2.9. The output recorder needs 10 mA for a reading of 1 cm, and it should read 10 cm ± 2% for a peak input voltage of 100 V. The internal resistance of the recorder varies from $R_L = 100\ \Omega$ to $R_L = 500\ \Omega$. The frequency of the input voltage is $f_s = 1$ kHz.

(a) Determine the value of capacitance C.

(b) Determine the design specifications of the transconductance amplifier.

SOLUTION

(a) The capacitor C will charge to the peak input voltage when the diode conducts, and it will discharge through the amplifier when the diode is off. Let us assume that the discharging time constant $\tau\ (=CR_i)$ is related to the input frequency by $\tau = 10/f_s$. For $f_s = 1$ kHz, $CR_i = 10/(1\text{ kHz}) = 10$ ms. Let us choose $C = 0.01\ \mu$F. Then $R_i = 10\text{ ms}/0.01\ \mu\text{F} = 1\ \text{M}\Omega$.

(b) Since the output variation should be kept within ±2%, the variation in the effective transconductance G_m should also be limited to ±2%. According to Eq. (2.34), the variation in G_m will be contributed by G_{ms} and R_L. Let us assume that each of them contributes equally to the variation—that is, each contributes ±1%. Note that there is no source resistance: $R_s = 0$. The nominal transconductance gain is

$$G_{ms} = \frac{i_o}{v_s} = \left(\frac{10\text{ cm}}{100\text{ V}}\right)\left(\frac{10\text{ mA}}{1\text{ cm}}\right) = 1\text{ mA/V} \pm 1\%$$

Thus, the value of R_o that will keep the gain variation within 1% for variation in R_L from 100 Ω to 500 Ω can be found from

$$0.99\frac{R_o}{R_o + 100} = \frac{R_o}{R_o + 500}$$

which gives $R_o \geq 39.5\ \text{k}\Omega$ when solved for R_o. Thus, the amplifier specifications are $G_{ms} = 1\text{ mA/V} \pm 1\%$, $R_i \approx 1\ \text{M}\Omega$, and $R_o \geq 39.5\ \text{k}\Omega$.

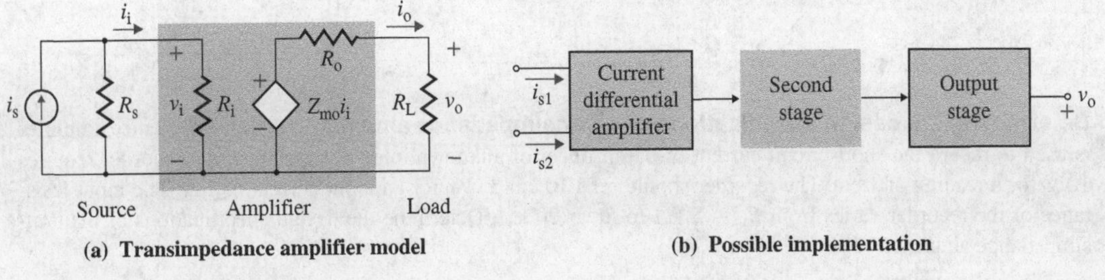

FIGURE 2.10 Transimpedance amplifier

2.3.4 Transimpedance Amplifiers

The input signal to a transimpedance amplifier is a current source, and its output is a voltage source. Such an amplifier can be represented as a current-controlled voltage source (CCVS), as shown in Fig. 2.10(a). The gain parameter Z_{mo} is the ratio of the open-circuit output voltage to the input current, and it is called the *open-circuit transimpedance* (or simply the *transimpedance*).

The output voltage v_o is related to i_i by

$$v_o = \frac{Z_{mo} i_i R_L}{R_L + R_o} \tag{2.37}$$

The input current i_i of the amplifier is related to i_s as follows:

$$i_i = \frac{R_s}{R_s + R_i} i_s \tag{2.38}$$

Substituting i_i from Eq. (2.38) into Eq. (2.37) gives the effective transimpedance Z_m:

$$Z_m = \frac{v_o}{i_s} = \frac{Z_{mo} R_L R_s}{(R_L + R_o)(R_s + R_i)} = \frac{Z_{mo}}{(1 + R_o/R_L)(1 + R_i/R_s)} \tag{2.39}$$

The effective voltage gain A_v is given by

$$A_v = \frac{v_o}{v_s} = \frac{i_o R_L}{i_s R_s} = \frac{Z_{mo} R_L}{(R_s + R_i)(R_L + R_o)} \tag{2.40}$$

A transimpedance amplifier must have an input resistance R_i much smaller than the source resistance R_s and an output resistance R_o much smaller than the load resistance R_L. An ideal transimpedance amplifier has $R_i = 0$ and $R_o = 0$. That is,

$$v_o = Z_{mo} i_s \tag{2.41}$$

The implementation of a CCVS can begin with a current differential input, as shown in Fig. 2.10(b). If the output stage has a high input resistance and a low output resistance and the gain is adequate, a second stage may not be necessary.

EXAMPLE 2.7

D

Determining the design specifications of a transimpedance amplifier A transimpedance amplifier is used to record the short-circuit current of a transducer of unknown internal resistance; the recorder requires 10 V for a reading of 1 cm. The recorder should read 10 cm ± 2% for an input current of 1 A. The input resistance of the recorder varies from $R_L = 5\ \text{k}\Omega$ to $R_L = 20\ \text{k}\Omega$. Determine the design specifications of the transimpedance amplifier.

SOLUTION

Since the output variation should be kept within ±2%, the variation of the effective transimpedance Z_m should also be limited to ±2%. According to Eq. (2.39), the variation of Z_m will be contributed by Z_{mo} and R_L. Let us assume that each of them contributes equally to the variation—that is, each contributes ±1%. Since the source resistance is unknown, we will assume that the input resistance is very small, tending to zero (say, $R_i = 10\ \Omega$).

The nominal transimpedance gain is

$$Z_{mo} = \frac{v_o}{i_i} = \left(\frac{10\ \text{V}}{1\ \text{cm}}\right)\left(\frac{1\ \text{A}}{10\ \text{cm}}\right) = 100\ \text{V/A} \pm 1\%$$

Thus, the value of R_o that will keep the gain variation within 1% for variation in R_L from 5 kΩ to 20 kΩ can be found from

$$0.99\frac{20\ \text{k}}{20\ \text{k} + R_o} = \frac{5\ \text{k}}{5\ \text{k} + R_o}$$

which gives $R_o \leq 67.6\ \Omega$. Therefore, the amplifier specifications are $Z_{mo} = 100\ \text{V/A} \pm 1\%$, $R_o \leq 67.6\ \Omega$, and $R_i \leq 10\ \Omega$.

KEY POINTS OF SECTION 2.3

- Amplifiers can be classified into four types: voltage, current, transconductance, and transimpedance. Their characteristics are summarized in Table 2.1.
- Amplifiers are used in such applications as capacitance multiplication, creating negative resistance, and inductance simulation.
- Establishing the design specifications of an amplifier requires identifying the gain, the input resistance, and the output resistance.

TABLE 2.1 Characteristics of ideal amplifiers

Amplifier Type	Gain	Input Resistance R_i	Output Resistance R_o
Voltage	A_{vo} (V/V)	∞	0
Current	A_{is} (A/A)	0	∞
Transconductance	G_{ms} (A/V)	∞	∞
Transimpedance	Z_{mo} (V/A)	0	0

2.4 Cascaded Amplifiers

Normally one amplifier alone cannot meet the specifications for gain, input resistance, and output resistance. To satisfy the specifications, two or more amplifiers are often cascaded. Any combination of the four types of amplifiers can be used. As illustrations, we will discuss cascaded voltage amplifiers and cascaded current amplifiers.

2.4.1 Cascaded Voltage Amplifiers

Voltage amplifiers are cascaded to increase the overall voltage gain. Consider three cascaded voltage amplifiers, as shown in Fig. 2.11(a). The overall open-circuit voltage gain A_{vo} of the cascaded amplifiers can be found from

$$A_{vo} = \frac{v_o}{v_{i1}} = \frac{v_{i2}}{v_{i1}} \times \frac{v_{i3}}{v_{i2}} \times \frac{v_o}{v_{i3}} \tag{2.42}$$

If A_{vo1}, A_{vo2}, and A_{vo3} are the voltage gains of stages 1, 2, and 3, respectively, such that $v_{i2} = A_{vo1}v_{i1}$, $v_{i3} = A_{vo2}v_{i2}$, and $v_o = A_{vo3}v_{i3}$, then Eq. (2.42) becomes

$$A_{vo} = A_{vo1}A_{vo2}A_{vo3} \tag{2.43}$$

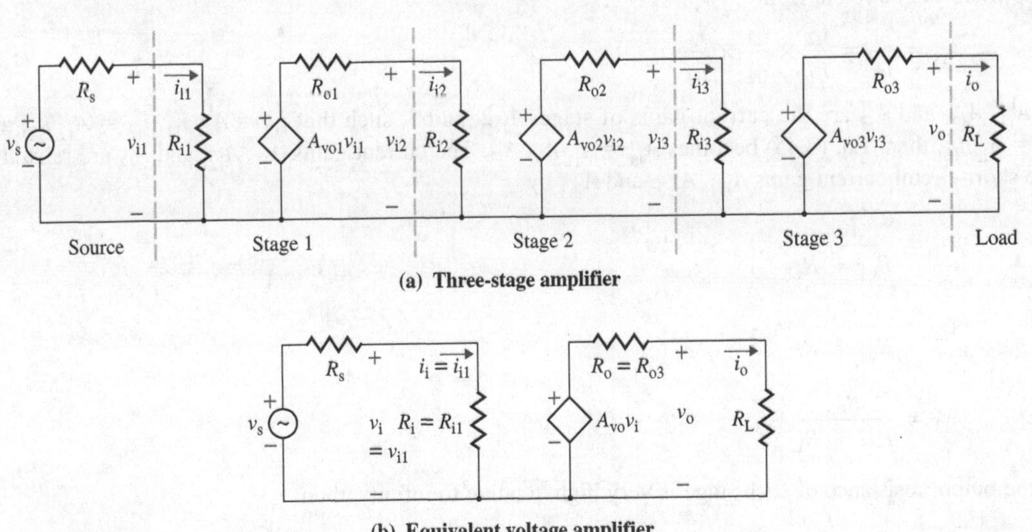

FIGURE 2.11 Cascaded voltage amplifiers

which indicates that the overall open-circuit voltage gain is the product of the individual gain of each stage. The voltage gains A_{v1}, A_{v2}, and A_{v3} are related to the no-load voltage gains A_{vo1}, A_{vo2}, and A_{vo3} by

$$A_{v1} = \frac{R_{i2}}{R_{o1} + R_{i2}} A_{vo1}$$

$$A_{v2} = \frac{R_{i3}}{R_{o2} + R_{i3}} A_{vo2}$$

$$A_{v3} = \frac{R_{i3}}{R_{o3} + R_{L}} A_{vo3}$$

If the output resistance of each stage is negligible so that

$$R_{o1} = R_{o2} = R_{o3} \approx 0$$

then the voltage gain of each stage becomes the same as its open-circuit voltage gain. That is,

$$v_{i2} = A_{vo1} v_{i1} \qquad v_{i3} = A_{vo2} v_{i2} \qquad v_{o} = A_{vo3} v_{i3}$$

The overall open-circuit voltage gain A_{vo} in Eq. (2.43) is then given by

$$A_{vo} = A_{vo1} A_{vo2} A_{vo3} \tag{2.44}$$

Therefore, the three voltage amplifiers can be represented by an equivalent single voltage amplifier with a voltage gain of A_{vo}, $R_o = R_{o3}$, and $R_i = R_{i1}$, as shown in Fig. 2.11(b).

2.4.2 Cascaded Current Amplifiers

Current amplifiers can be connected to increase the effective current gain. Consider three cascaded current amplifiers, as shown in Fig. 2.12(a). The overall short-circuit current gain A_{is} of the cascaded current amplifiers can be found from

$$A_{is} = \frac{i_o}{i_{i1}} = \frac{i_{i2}}{i_{i1}} \times \frac{i_{i3}}{i_{i2}} \times \frac{i_o}{i_{i3}} \tag{2.45}$$

If A_{i1}, A_{i2}, and A_{i3} are the current gains of stages 1, 2, and 3, such that $i_{i2} = A_{i1}\, i_{i1}$, $i_{i3} = A_{i2}\, i_{i2}$, and $i_o = A_{i3}\, i_{i3}$, then Eq. (2.45) becomes $A_{is} = A_{i1}\, A_{i2}\, A_{i3}$. The current gains A_{i1}, A_{i2}, and A_{i3} are related to the short-circuit current gains A_{is1}, A_{is2}, and A_{is3} by

$$A_{i1} = \frac{R_{o1}}{R_{o1} + R_{i2}} A_{is1}$$

$$A_{i2} = \frac{R_{o2}}{R_{o2} + R_{i3}} A_{is2}$$

$$A_{i3} = \frac{R_{o3}}{R_{o3} + R_{L}} A_{is3}$$

If the output resistance of each stage is very high, tending to infinity, then

$$R_{o1} = R_{o2} = R_{o3} = \infty$$

and $\quad i_{i2} = A_{is1} i_{i1} \qquad i_{i3} = A_{is2} i_{i2} \qquad i_{o} = A_{is3} i_{i3}$

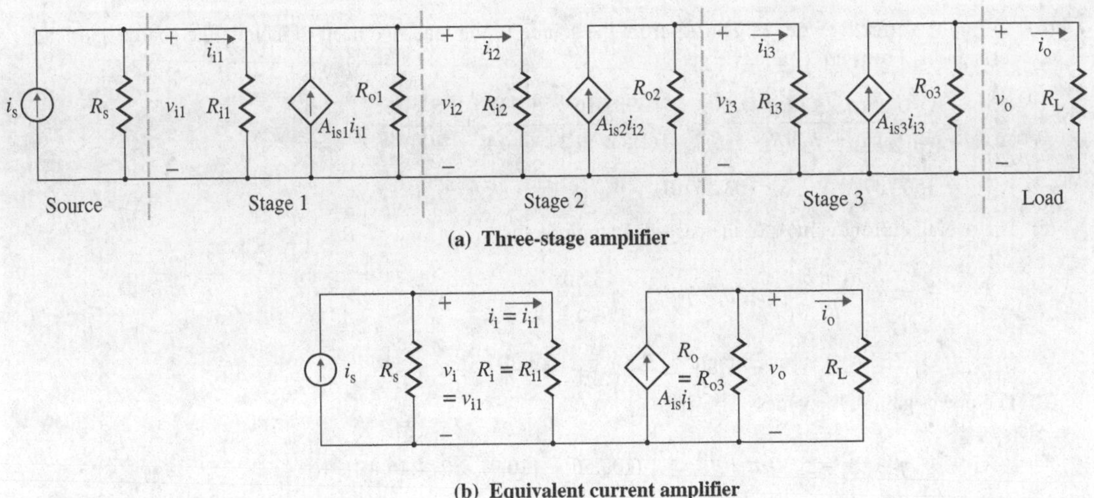

FIGURE 2.12 Cascaded current amplifiers

Then Eq. (2.45) becomes

$$A_{is} = A_{is1}A_{is2}A_{is3} \tag{2.46}$$

which indicates that the overall short-circuit gain is the product of the individual gain of each stage. Therefore, the three current amplifiers can be represented by an equivalent single current amplifier with a current gain of A_{is}, as shown in Fig. 2.12(b).

EXAMPLE 2.8

Finding the parameters of cascaded voltage amplifiers The parameters of the cascaded voltage amplifiers in Fig. 2.11(a) are $R_s = 2\ \text{k}\Omega$, $R_o = R_{o1} = R_{o2} = R_{o3} = 200\ \Omega$, $R_i = R_{i1} = R_{i2} = R_{i3} = R_L = 1.5\ \text{k}\Omega$, and $A_{vo1} = A_{vo2} = A_{vo3} = 80$. Calculate **(a)** the overall open-circuit voltage gain $A_{vo} = v_o/v_i$, **(b)** the effective voltage gain $A_v = v_o/v_s$, **(c)** the overall current gain $A_i = i_o/i_s$, and **(d)** the power gain $A_p = P_L/P_i$.

SOLUTION

(a) Using Eq. (2.19), we can calculate the voltage gain of stage 1 and stage 2 as follows:

$$A_{v1} = A_{v2} = \frac{A_{vo1}R_{i2}}{(R_{i2} + R_{o1})} = \frac{80 \times 1.5\ \text{k}}{(1.5\ \text{k} + 200)} = 70.588\ \text{V/V}$$

From Eq. (2.43), the overall open-circuit voltage gain of the cascaded amplifiers is

$$A_{vo} = \frac{v_o}{v_{i1}} = A_{v1}A_{v2}A_{vo3} = (70.588)^2 \times 80$$

$$= 398{,}616\ \text{V/V} \quad [\text{or } 20 \times \log(398{,}616) = 112.01\ \text{dB}]$$

(b) To find the effective voltage gain A_v from the source to the load, we need to include the source and load resistances. From Eq. (2.21), we get

$$A_v = \frac{v_o}{v_s} = \frac{A_{vo}R_iR_L}{(R_i + R_s)(R_L + R_o)} = \frac{398{,}616 \times 1.5\text{ k} \times 1.5\text{ k}}{(1.5\text{ k} + 2\text{ k})(1.5\text{ k} + 200)}$$

$$= 150{,}737\text{ V/V} \quad (\text{or } 103.56\text{ dB})$$

(c) The overall current gain A_i of the cascaded amplifiers is

$$A_i = \frac{i_o}{i_s} = A_v\frac{R_s + R_i}{R_L} = 150{,}737\text{ V/V} \times \frac{3.5\text{ k}}{1.5\text{ k}}$$

$$= 351{,}720\text{ A/A} \quad [\text{or } 20\log(351{,}720) = 110.9\text{ dB}]$$

(d) The power gain A_p becomes

$$A_p = \frac{P_L}{P_i} = A_vA_i = 5.30 \times 10^{10} \quad [\text{or } (103.56 + 110.9) = 214.46\text{ dB}]$$

KEY POINTS OF SECTION 2.4

- Amplifiers are often cascaded to satisfy the requirements for gain, input resistance, and output resistance.
- The overall short-circuit gain of cascaded amplifiers is the product of the individual gains of the various stages.

2.5 Frequency Response of Amplifiers

So far, we have assumed that there are no reactive elements in an amplifier and that the gain of an amplifier remains constant at all frequencies. However, the gain of practical amplifiers is frequency dependent, and even the input and output impedances of amplifiers vary with the frequency. If ω is the frequency of the input signal in radians per second, the output sinusoid $V_o(\omega)$ can have a different amplitude and phase than the input sinusoid $V_i(\omega)$. The voltage gain $A_v(\omega) = V_o(\omega)/V_i(\omega)$ will have a magnitude and phase angle. If a sine-wave signal with a specific frequency is applied at the input of an amplifier, the output should be a sinusoid of the same frequency. The frequency response of an amplifier refers to the amplitude of the output sinusoid and its phase relative to the input sinusoid (see Appendix B).

An amplifier is operated at a DC Q-point and is subjected to two types of signals: AC signals and DC signals. Often several amplifiers are cascaded by *coupling capacitors,* as shown in Fig. 2.13, so that the AC signal from the source can flow from one stage to the next stage while the DC signal is blocked. As a result, the DC biasing voltages of the amplifiers do not affect the signal source, adjacent stages, or the load. Such cascaded amplifiers are called *capacitive-* (or AC-) *coupled amplifiers.* However, amplifiers in integrated circuits are connected directly, as shown in Fig. 2.14, because capacitors cannot be fabricated in integrated form; such amplifiers are called *direct-* (or DC-) *coupled amplifiers.*

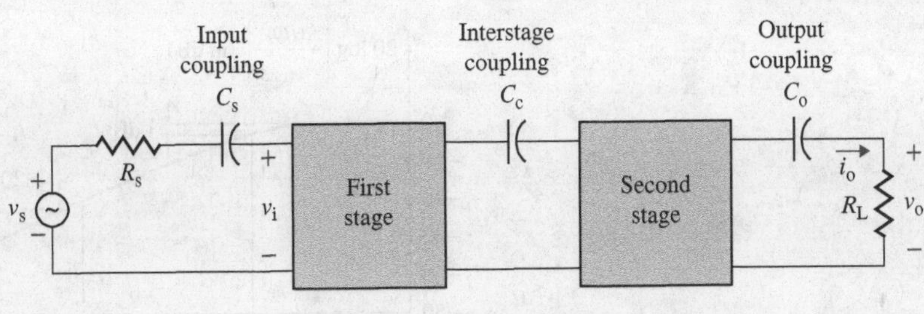

FIGURE 2.13 Capacitive-coupled amplifiers

At low frequencies, coupling capacitors, which are on the order of 10 μF, offer high reactance on the order of 1 kΩ and attenuate the signal source. At high frequencies, these capacitors have reactance on the order of 1 Ω and thus essentially short-circuit. Therefore, AC-coupled amplifiers will pass signals of high frequencies only.

There are no coupling capacitors in DC-coupled amplifiers. However, the presence of small capacitors on the order of 1 pF is due to the internal capacitances of the amplifying devices and also due to stray wiring capacitance between the signal-carrying conductors and the ground. The frequency response of an amplifier depends on the type of coupling. An amplifier can exhibit one of three frequency characteristics: low-pass, high-pass, or band-pass.

2.5.1 Low-Pass Characteristic

Consider the transconductance amplifier shown in Fig. 2.15(a). C_2, which is connected across the load R_L, could be the output capacitance of the amplifier or the stray capacitance between the output terminal and the ground. C_2 forms a parallel path to the signal flowing from the amplifier to the load R_L. The output voltage in Laplace's domain is

$$V_o(s) = -G_{ms}V_i(s)\left(R_L \parallel \frac{1}{sC_2}\right) = -G_{ms}R_L\frac{1}{1 + sC_2R_L}V_i(s) \tag{2.47}$$

Using the voltage divider rule, we get $V_i(s) = V_sR_i/(R_s + R_i)$, which, after substitution in Eq. (2.47), gives the voltage gain as

$$A_v(s) = \frac{V_o(s)}{V_s(S)} = -\frac{G_{ms}R_LR_i}{(R_s + R_i)(1 + sC_2R_L)} \tag{2.48}$$

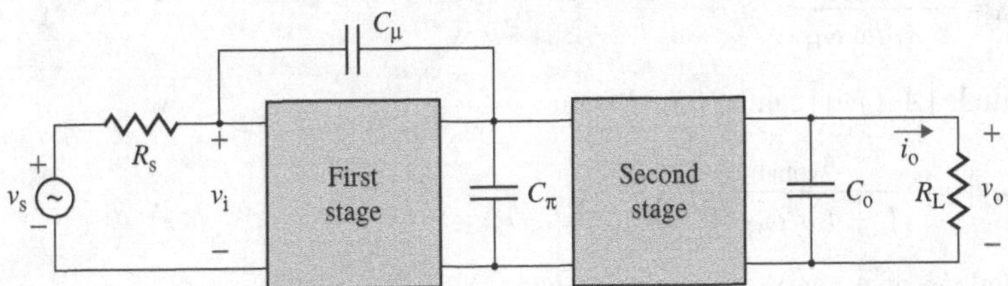

FIGURE 2.14 Direct-coupled amplifiers

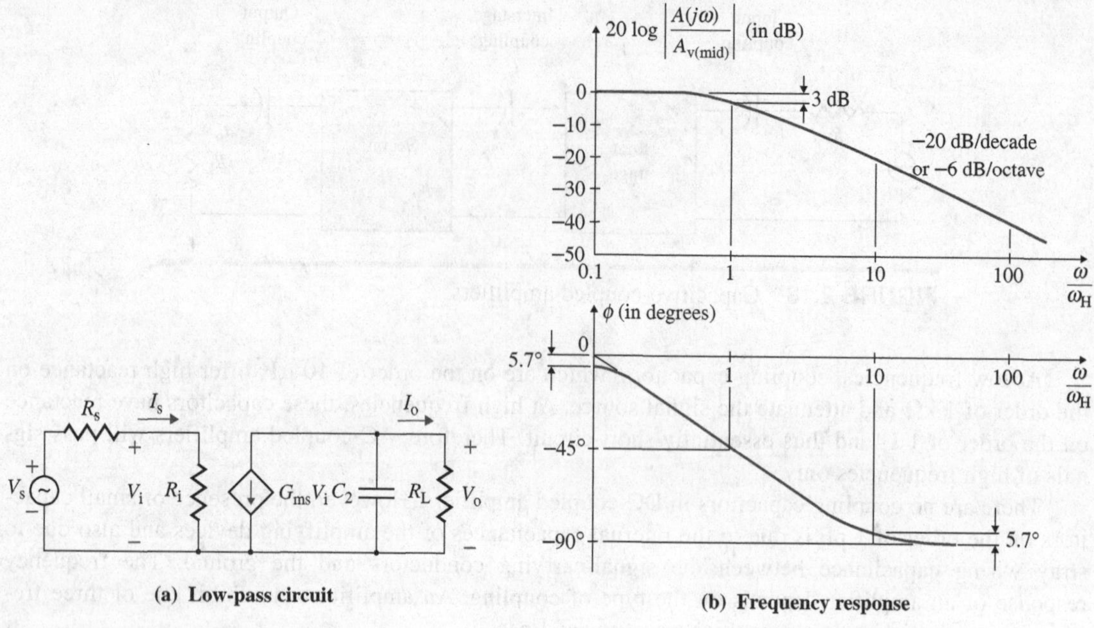

FIGURE 2.15 Low-pass amplifier

Equation (2.48) can be written in general form as

$$A_v(s) = \frac{A_{v(mid)}}{1 + s\tau_2} = \frac{A_{v(mid)}}{1 + s/\omega_H} \tag{2.49}$$

where

$$A_{v(mid)} = -\frac{G_{ms}R_LR_i}{R_s + R_i} \tag{2.50}$$

$$\tau_2 = C_2R_L \tag{2.51}$$

$$\omega_H = \frac{1}{\tau_2} = \frac{1}{C_2R_L} \tag{2.52}$$

In the frequency domain, $s = j\omega$ and Eq. (2.49) becomes

$$A_v(j\omega) = \frac{A_{v(mid)}}{1 + j\omega/\omega_H} \tag{2.53}$$

Thus, the magnitude $|A_v(j\omega)|$ can be found from

$$|A_v(j\omega)| = \frac{A_{v(mid)}}{[1 + (\omega/\omega_H)^2]^{1/2}} \tag{2.54}$$

and the phase angle ϕ of $A_v(j\omega)$ is given by

$$\phi = -\tan^{-1}\left(\frac{\omega}{\omega_H}\right) \tag{2.55}$$

For $\omega << \omega_H$, let us assume that $A_{v(mid)} = 1$. That is,

$$|A_v(j\omega)| \approx A_{v(mid)} = 1$$

$$20 \log_{10}|A_v(j\omega)| \approx 0$$

$$\phi = 0$$

Therefore, at a low frequency, the magnitude plot of $A_v(j\omega)$ is approximately a straight horizontal line at 0 dB. For $\omega >> \omega_H$,

$$|A_v(j\omega)| \approx \left(\frac{\omega_H}{\omega}\right)$$

$$20 \log_{10}|A_v(j\omega)| = 20 \log_{10}\left(\frac{\omega_H}{\omega}\right)$$

$$\phi \approx -\frac{\pi}{2} \quad \text{(90 degrees)}$$

For $\omega = \omega_H$,

$$|A_v(j\omega)| = \frac{1}{\sqrt{2}}$$

$$20 \log_{10}|A_v(j\omega)| = 20 \log_{10}\left(\frac{1}{\sqrt{2}}\right) = -3 \text{ dB}$$

$$\phi = -\frac{\pi}{4} \quad \text{(45 degrees)}$$

Let us consider a high-frequency $\omega = \omega_1$ such that $\omega_1 >> \omega_H$. The magnitude is $20 \log_{10} (\omega_H/\omega_1)$ at $\omega = \omega_1$. At $\omega = 10\omega_1$, the magnitude is $20 \log_{10} (\omega_H/10\omega_1)$. The change in magnitude becomes

$$20 \log_{10}\left(\frac{\omega_1}{10\omega_H}\right) - 20 \log_{10}\left(\frac{\omega_H}{\omega_1}\right) = 20 \log_{10}\left(\frac{1}{10}\right) = -20 \text{ dB}$$

If the frequency is doubled so that $\omega = 2\omega_1$, the change in magnitude becomes

$$20 \log_{10}\left(\frac{\omega_1}{2\omega_H}\right) - 20 \log_{10}\left(\frac{\omega_H}{\omega_1}\right) = 20 \log_{10}\left(\frac{1}{2}\right) = -6 \text{ dB}$$

The frequency response is shown in Fig. 2.15(b). If the frequency is doubled, the increase on the frequency axis is called an *octave increase.* If the frequency is increased by a factor of 10, the increase is called a *decade increase.* For a decade increase in frequency, the magnitude changes by -20 dB and the magnitude plot is a straight line with a slope of -20 dB/decade (or -6 dB/octave). The magnitude curve is therefore defined by two straight-line asymptotes, which meet at the corner frequency ω_H. The difference between the actual magnitude curve and the asymptotic curve is largest at the break frequency. The error can be found by substituting ω for ω_H. That is, $|A_v(j\omega)| = 1/\sqrt{2}$ and $20 \log_{10} (1/\sqrt{2}) = -3$ dB. This error is symmetrical with respect to the *break* (or *corner*) *frequency,* which is defined as the frequency at which the magnitude of the gain falls to 70.7% of the constant gain. The break frequency is also known as the *3-dB* (or *cutoff* or *half-power*) *frequency.* The voltage gain will fall as the frequency increases beyond ω_H.

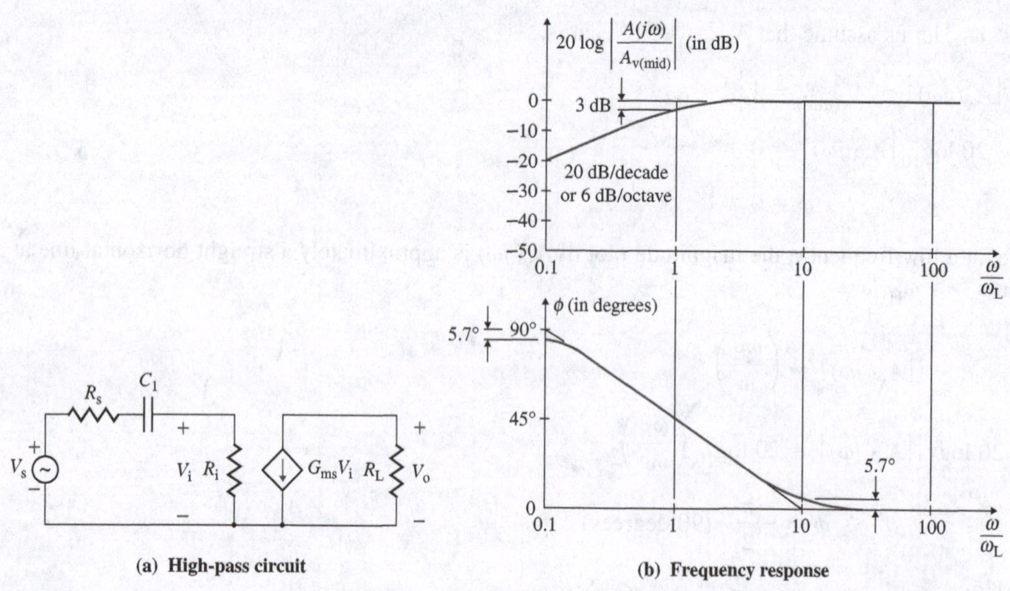

FIGURE 2.16 High-pass amplifier

For frequencies $\omega << \omega_H$, the gain will be almost independent of frequency. An amplifier with this type of response is known as a *low-pass amplifier.* $A_{v(mid)}$ is the pass-band or midband gain. The *bandwidth* (BW) of an amplifier is defined as the range of frequencies over which the gain remains within 3 dB (29.3%) of constant gain $A_{v(mid)}$. That is, BW = ω_H. Amplifiers for video signals are generally DC coupled, and the frequencies vary from 0 (DC) to 4.5 MHz.

2.5.2 High-Pass Characteristic

Consider the transconductance amplifier shown in Fig. 2.16(a). C_1 is the isolating capacitor between the signal source and the amplifier. The output voltage in Laplace's domain is

$$V_o(s) = -G_{ms} R_L V_i(s) \tag{2.56}$$

From the voltage divider rule, the voltage $V_i(s)$ is related to $V_s(s)$ by

$$V_i(s) = \frac{R_i}{R_s + R_i + 1/sC_1} V_s(s) = \frac{sC_1 R_i}{1 + sC_1(R_s + R_i)} V_s(s) \tag{2.57}$$

Substituting $V_i(s)$ from Eq. (2.57) into Eq. (2.56) gives the voltage gain:

$$A_v(s) = \frac{V_o(s)}{V_s(s)} = \frac{-G_{ms} R_L R_i}{R_s + R_i} \times \frac{sC_1(R_s + R_i)}{1 + sC_1(R_s + R_i)} \tag{2.58}$$

Equation (2.58) can be written in general form as

$$A_v(s) = \frac{A_{v(mid)} s\tau_1}{1 + s\tau_1} = \frac{A_{v(mid)} s}{s + 1/\tau_1} = \frac{A_{v(mid)} s}{s + \omega_L} \tag{2.59}$$

where $$A_{v(mid)} = \frac{G_{ms}R_L R_i}{R_s + R_i} \tag{2.60}$$

$$\tau_1 = C_1(R_s + R_i) \tag{2.61}$$

$$\omega_L = \frac{1}{\tau_1} = \frac{1}{[C_1(R_s + R_i)]} \tag{2.62}$$

In the frequency domain, $s = j\omega$ and Eq. (2.59) becomes

$$A_v(j\omega) = \frac{A_{v(mid)} j\omega}{j\omega + \omega_L} \tag{2.63}$$

Thus, the magnitude $|A_v(j\omega)|$ can be found from

$$|A_v(j\omega)| = \frac{A_{v(mid)}\,\omega}{[\omega^2 + \omega_L^2]^{1/2}} \tag{2.64}$$

and the phase angle ϕ of $A_v(j\omega)$ is given by

$$\phi = 90^\circ - \tan^{-1}(\omega/\omega_L) \tag{2.65}$$

Let us assume that $A_{v(mid)} = 1$. For $\omega << \omega_L$,

$$|A_v(j\omega)| = \frac{\omega}{\omega_L}$$

$$20 \log_{10}|A_v(j\omega)| = 20 \log_{10}\left(\frac{\omega}{\omega_L}\right)$$

$$\phi = \frac{\pi}{2} \quad \text{(90 degrees)}$$

Therefore, for a decade increase in frequency, the magnitude changes by +20 dB. The magnitude plot of $A_v(j\omega)$ is a straight line with a slope of +20 dB/decade (or +6 dB/octave). For $\omega >> \omega_L$,

$$|A_v(j\omega)| = A_{v(mid)} = 1$$

$$20 \log_{10}|A_v(j\omega)| = 0$$

$$\phi \approx 0$$

Therefore, at a high frequency, the magnitude plot is a straight horizontal line at 0 dB. At $\omega = \omega_L$,

$$|A_v(j\omega)| = \frac{1}{\sqrt{2}}$$

$$20 \log_{10}\left(\frac{1}{\sqrt{2}}\right) = -3 \text{ dB}$$

$$\phi = \frac{\pi}{4} \quad \text{(45 degrees)}$$

The frequency response is shown in Fig. 2.16(b). This circuit passes only the high-frequency signal, and the amplitude is low at a low frequency. The voltage gain will vary with the frequency for $\omega << \omega_L$. For $\omega >> \omega_L$, the gain will be almost independent of frequency. This type of amplifier is known as a *high-pass amplifier*. ω_L is known as the *break* (*corner, cutoff, 3-dB*, or *half-power*) *frequency*, and $A_{v(mid)}$ is the pass-band or midband gain. Note that for sufficiently high frequencies, the high-pass characteristic of practical amplifiers will tend to attenuate because of the internal capacitances of the amplifying devices.

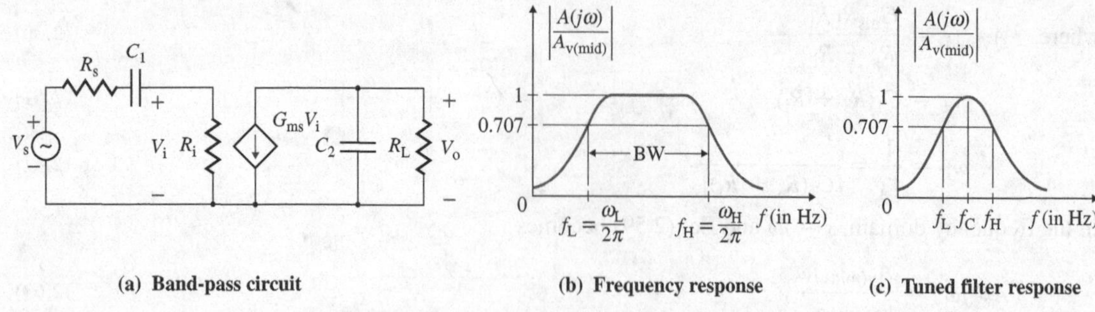

FIGURE 2.17 Band-pass amplifier

2.5.3 Band-Pass Characteristic

A capacitive-coupled amplifier will have both coupling capacitors and device capacitors (or stray capacitors). Let us connect both C_1 and C_2, as shown in Fig. 2.17(a). The circuit will exhibit a band-pass characteristic. Substituting $V_i(s)$ from Eq. (2.57) into Eq. (2.47) gives the voltage gain as

$$A_v(s) = \frac{V_o(s)}{V_s(s)} = \frac{-G_{ms}R_LR_i}{R_s + R_i} \times \frac{sC_1(R_s + R_i)}{1 + sC_1(R_s + R_i)} \times \frac{1}{1 + sC_2R_L} \tag{2.66}$$

which can be written in general form as

$$A_v(s) = \frac{A_{v(mid)}s}{(s + \omega_L)(1 + s/\omega_H)} \tag{2.67}$$

In the frequency domain, $s = j\omega$ and Eq. (2.67) becomes

$$A_v(j\omega) = \frac{A_{v(mid)}j\omega}{(j\omega + \omega_L)(1 + j\omega/\omega_H)} \tag{2.68}$$

Thus, the magnitude $|A_v(j\omega)|$ can be found from

$$|A_v(j\omega)| = \frac{A_{v(mid)}\omega}{[\omega^2 + \omega_L^2]^{1/2}[1 + (\omega/\omega_H)^2]^{1/2}} \tag{2.69}$$

and the phase angle ϕ of $A_v(j\omega)$ is given by

$$\phi = 90° - \tan^{-1}(\omega/\omega_L) - \tan^{-1}(\omega/\omega_H) \tag{2.70}$$

Thus, the voltage gain will remain almost constant if $\omega_L < \omega < \omega_H$. The frequency behavior is shown in Fig. 2.17(b). This is a band-pass circuit, and $A_{v(mid)}$ is the midfrequency (or pass-band) gain. The *bandwidth* (BW), which is the range of frequencies over which the gain remains within 3 dB (29.3%) of constant gain $A_{v(mid)}$, is thus the difference between the cutoff frequencies. That is, BW $= \omega_H - \omega_L$. Note that $A_{v(mid)}$ is not the DC gain because under DC conditions capacitor C_1 will be open-circuited and there will be no output voltage. Audio amplifiers are generally AC coupled because the frequency range of audio signals is 20 Hz to 15 kHz. The audio signal source and the loudspeakers are isolated by coupling capacitors.

If the bandwidth of a band-pass amplifier is shortened so that the gain peaks around a particular frequency (called the *center frequency*) and falls off on both sides of this frequency, as shown in Fig. 2.17(c), the amplifier is called a *tuned amplifier*. Such an amplifier is generally used in the front end of radio and TV receivers. The center frequency f_C of a tuned amplifier can be adjusted to coincide with the frequency of a desired channel so that the signals of that particular channel can be received and signals of other channels are attenuated or filtered out.

2.5.4 Gain and Bandwidth Relation

Using $\omega = 2\pi f$, we can write the voltage gain of a low-pass amplifier as

$$A_v(j\omega) = \frac{A_{v(mid)}}{1 + jf/f_H} \tag{2.71}$$

where f_H is the break (or 3-dB) frequency in hertz.

For $f >> f_H$, Eq. (2.71) is reduced to

$$A_v(j\omega) = \frac{A_{v(mid)}}{jf/f_H} = \frac{A_{v(mid)}f_H}{jf} \tag{2.72}$$

The magnitude of this gain becomes unity (or 0 dB) at frequency $f = f_{bw}$. That is,

$$f_{bw} = A_{v(mid)}f_H \tag{2.73}$$

where f_{bw} is called the *unity-gain bandwidth.* Bandwidth (BW) is often quoted as the frequency range over which the voltage gain $|A(j\omega)|$ is unity. The unity-gain bandwidth of a band-pass amplifier becomes $A_{v(mid)}(f_H - f_L)$. It is important to note that according to Eq. (2.73), the gain–bandwidth product of an amplifier remains constant.

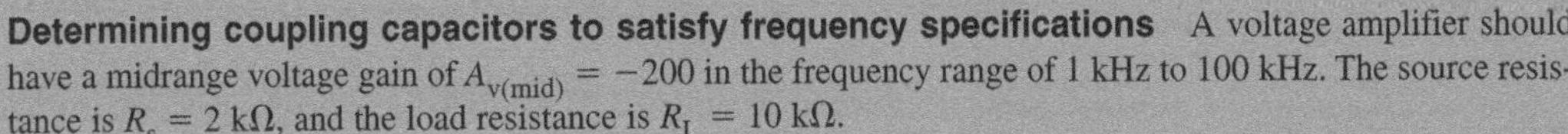

D **Determining coupling capacitors to satisfy frequency specifications** A voltage amplifier should have a midrange voltage gain of $A_{v(mid)} = -200$ in the frequency range of 1 kHz to 100 kHz. The source resistance is $R_s = 2\text{ k}\Omega$, and the load resistance is $R_L = 10\text{ k}\Omega$.

(a) Determine the specifications of the amplifier and the values for coupling capacitor C_1 and shunt capacitor C_2 shown in Fig. 2.17(a).

(b) Use PSpice/SPICE to verify your design by plotting the frequency response $|A_v(j\omega)|$ against frequency.

SOLUTION

(a) Let us choose a transconductance amplifier of $R_i = 1\text{ M}\Omega$ and $R_o = \infty$. From Eq. (2.60), we can find the value of G_{ms} that will give $A_{v(mid)} = -200$. That is,

$$G_{ms} = \frac{A_{v(mid)}(R_s + R_i)}{R_L R_i} = \frac{-200 \times (2\text{ k}\Omega + 1\text{ M}\Omega)}{10\text{ k}\Omega \times 1\text{ M}\Omega} = -20.04\text{ mA/V}$$

For $f_H = 100$ kHz, Eq. (2.52) gives the required value of C_2 as

$$C_2 = \frac{1}{(R_L\omega_H)} = \frac{1}{(2\pi f_H R_L)} = \frac{1}{(2\pi \times 100\text{ kHz} \times 10\text{ k}\Omega)} = 159.15\text{ pF}$$

For $f_L = 1$ kHz, Eq. (2.62) gives the required value of C_1 as

$$C_1 = \frac{1}{[(R_s + R_i)\omega_L]} = \frac{1}{[2\pi f_L(R_s + R_i)]} = \frac{1}{[2\pi \times 1\text{ kHz} \times (2\text{ k}\Omega + 1\text{ M}\Omega)]} = 158.84\text{ pF}$$

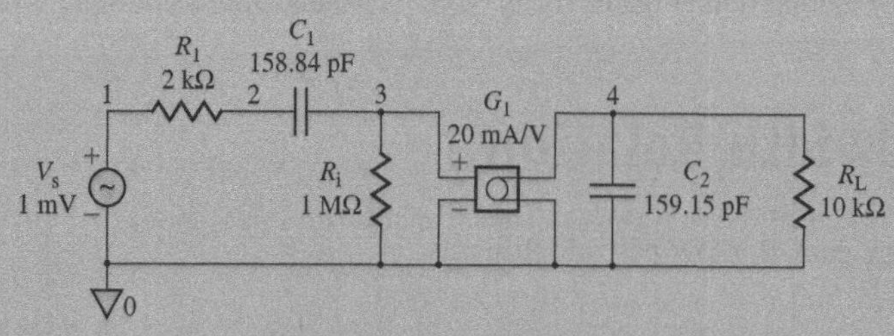

FIGURE 2.18 Circuit for PSpice simulation

(b) The circuit for PSpice simulation is shown in Fig. 2.18.

The PSpice plot of the frequency response is shown in Fig. 2.19, which gives $A_{v(mid)} = 198$ (expected value is 200), $f_L = 984$ Hz (expected value is 1 kHz), and $f_H = 102$ kHz (expected value is 100 kHz).

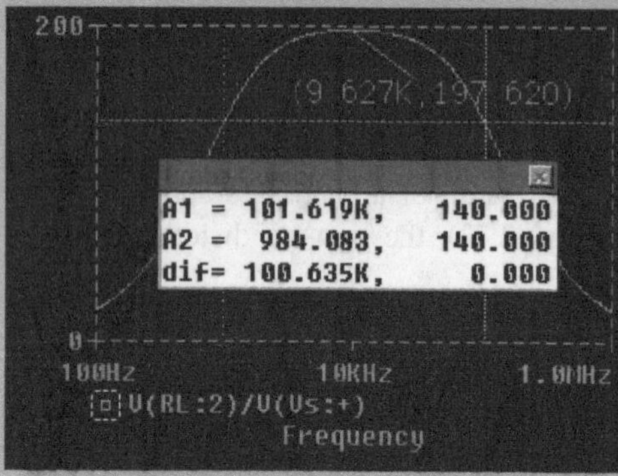

FIGURE 2.19 PSpice plot of frequency response for Example 2.9

KEY POINTS OF SECTION 2.5

- The gain of practical amplifiers is frequency dependent. The frequency response of an amplifier refers to the amplitude and phase of the output sinusoid relative to the input sinusoid. The frequency response is an important specification of an amplifier.
- Video amplifiers operate in the frequency range from 0 (DC) to 4.5 MHz and use direct coupling. That is, there are no coupling capacitors. However, the presence of small capacitors is due to the internal capacitances of the amplifying devices and also to stray wiring capacitance. These capacitors form a parallel path with the AC signal and therefore pass signals of low frequencies only.
- Audio amplifiers, which operate in the frequency range from 20 Hz to 15 kHz, use coupling capacitors so that the AC signal can flow from one stage to the next stage and the DC signals are blocked. These capacitors form a series path with the AC signal and therefore pass signals of high frequencies only. The upper frequency is limited by the device and/or stray capacitances.
- Frequency response also known as a Bode plot, which is a plot of the magnitude and the phase against the frequency, can describe the frequency characteristic and stability of an amplifier.
- Depending on the frequency response, an amplifier falls into one of three categories: low-pass, high-pass, or band-pass.
- The coupling capacitances of an amplifier normally determine the low break frequencies, whereas internal capacitances determine the high break frequencies.

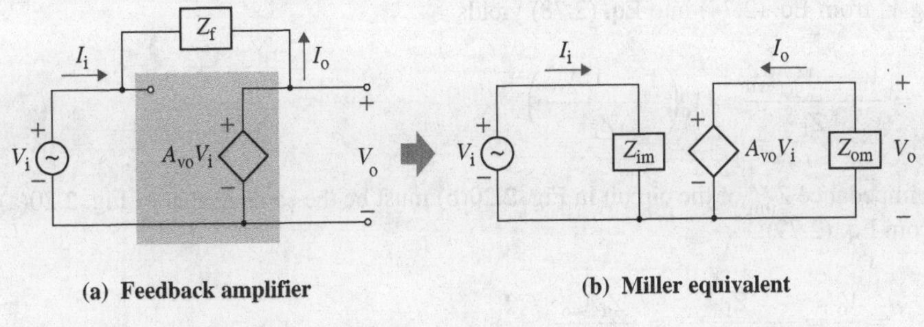

FIGURE 2.20 Circuits illustrating Miller's theorem

2.6 Miller's Theorem

An impedance known as *feedback impedance* is often connected across the input and output sides of an amplifier. Miller's theorem [1] simplifies the analysis of feedback amplifiers. The theorem states that if an impedance is connected between the input side and the output side of a voltage amplifier, this impedance can be replaced by two equivalent impedances—one connected across the input and the other connected across the output terminals. Figure 2.20 shows the relationship between the amplifier and its equivalent circuit. If we choose the appropriate values of impedances Z_{im} and Z_{om}, the two circuits in Fig. 2.20[(a) and (b)] can be made identical. In Sec. 2.7.3 we will apply Miller's theorem to find the frequency response of amplifiers.

If A_{vo} is the open-circuit voltage gain of the amplifier, the output voltage V_o is related to the input voltage V_i by

$$V_o = A_{vo}V_i \tag{2.74}$$

The input current I_i of the amplifier in Fig. 2.20(a) is given by

$$I_i = \frac{V_i - V_o}{Z_f} \tag{2.75}$$

Substituting V_o from Eq. (2.74) into Eq. (2.75) yields

$$I_i = \frac{V_i - A_{vo}V_i}{Z_f} = V_i\left(\frac{1 - A_{vo}}{Z_f}\right) \tag{2.76}$$

The input impedance Z_i of the circuit in Fig. 2.20(b) must be the same as that of Fig. 2.20(a), and it can be found from Eq. (2.76):

$$Z_{im} = \frac{V_i}{I_i} = \frac{Z_f}{1 - A_{vo}} \tag{2.77}$$

The output current I_o of the circuit in Fig. 2.20(a) is given by

$$I_o = \frac{V_o - V_i}{Z_f} \tag{2.78}$$

Substituting V_i from Eq. (2.74) into Eq. (2.78) yields

$$I_o = \frac{V_o - V_o/A_{vo}}{Z_f} = V_o\left(\frac{1 - 1/A_{vo}}{Z_f}\right) \quad (2.79)$$

The output impedance Z_{om} of the circuit in Fig. 2.20(b) must be the same as that of Fig. 2.20(a), and it can be found from Eq. (2.79):

$$Z_{om} = \frac{V_o}{I_o} = \frac{Z_f}{1 - 1/A_{vo}} = \frac{Z_f A_{vo}}{A_{vo} - 1} \quad (2.80)$$

▶ **NOTES**

1. Equations (2.77) and (2.80) are derived with the assumption that the voltage amplifier is an ideal one and that the open-circuit voltage gain A_{vo} can be found without connecting the impedance Z_f. That is, the input impedance R_i of the amplifier in Fig. 2.20(a) is very high, tending to infinity, and the output resistance R_o is very small, tending to zero. They have no effect on the analysis. Z_{im} and Z_{om} are called the *Miller impedances.*
2. The Miller theorem is applicable provided the amplifier has no independent source. The open-circuit voltage gain A_{vo} of the amplifier must be negative so that $(1 - A_{vo})$ is a positive quantity. Otherwise Z_{im} will have a negative value.
3. If a capacitor is connected between the input and output terminals of an amplifier with a negative voltage gain, this capacitor has a dominant effect and lowers the high break frequency significantly.

KEY POINT OF SECTION 2.6

■ According to Miller's theorem if an impedance is connected between the input side and the output side of a voltage amplifier, this impedance can be replaced by two equivalent impedances—one connected across the input and the other connected across the output terminal.

2.7 Frequency Response Methods

An amplifier generally receives a small AC signal from the input side, then amplifies the signal and delivers it to the output side. The amplifier requires DC supplies to operate the internal devices such as transistors. The internal DC voltages and DC currents within the amplifiers are subjected to variations. The amplifiers are often connected to the input signal source and the load resistor through coupling capacitors that effectively block low-frequency signals. The internal transistors [2, 3] have small capacitances that limit the maximum useful frequency of the amplifier.

A typical arrangement is shown in Fig. 2.21(a). Coupling capacitors C_1 and C_2, which have much higher values (typically on the order of 10 μF) than internal capacitances, are in series with the signal flow and set the low-frequency limit of the amplifier. Let us assume that the amplifier can be modeled by an equivalent circuit consisting of R_i, R_o, C_i, C_o, and g_m. This is shown in Fig. 2.21(b). A typical frequency

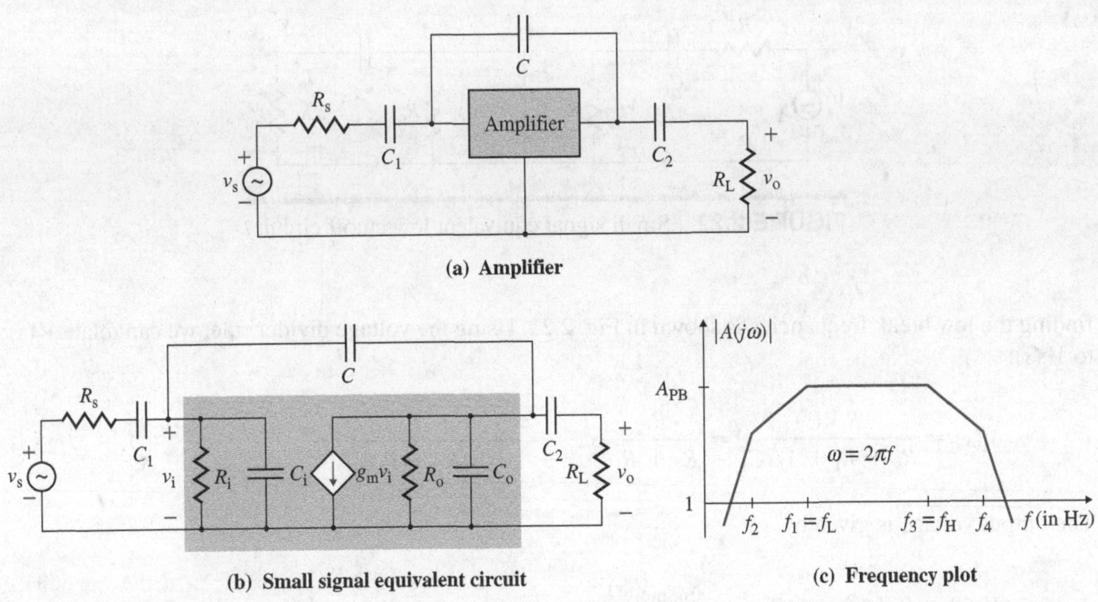

FIGURE 2.21 AC-coupled amplifier

plot (magnitude versus frequency) is shown in Fig. 2.21(c). f_L is the dominant low cutoff frequency, f_H is the dominant high cutoff frequency, and A_{PB} is the pass-band voltage gain. Thus, the performance of amplifiers depends on the input signal frequency, and the design specifications usually quote the voltage gain at a specified bandwidth.

Since there are five capacitors in Fig. 2.21(b), the denominator of the transfer function $A(s)$ will be a fifth-order polynomial in s. Finding the exact cutoff frequencies requires the calculation of five polynomial roots. Because derivation of the voltage transfer function $A(s)$ (similar to Eq. [2.48]) for the circuit in Fig. 2.21(b) is a tedious task, the analysis is normally carried out on a computer. However, the analysis can be simplified by assuming that f_L and f_H are separated by at least one decade so that f_L does not affect f_H. Then the low and high cutoff frequencies can be found separately. Thus, we can use the following steps to determine the complete frequency response of an amplifier:

1. Find the small-signal AC equivalent circuit of the amplifier as shown in Fig. 2.21(b).
2. Find the low break frequency or frequencies due to the coupling capacitors.
3. Find the high break frequency or frequencies due to the internal capacitors.
4. Find the pass-band gain of the amplifier.

2.7.1 Low-Frequency Transfer Function Method

At low frequencies (usually less than f_L = 1.5 kHz), the internal capacitors, which are typically on the range of 1 pF to 10 pF, have reactance on the order of 10 MΩ and are essentially open-circuited. We will assume that the internal capacitances are small so that the capacitors are effectively open-circuited. Thus, the low-frequency behavior is determined mostly by the coupling capacitors. The equivalent circuit for

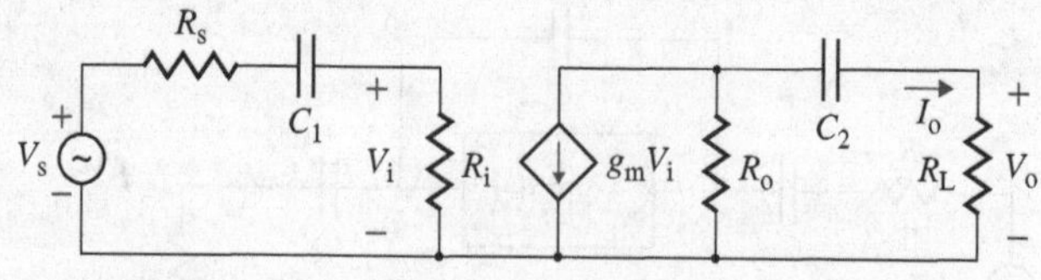

FIGURE 2.22 Small signal equivalent low cutoff circuit

finding the low break frequencies is shown in Fig. 2.22. Using the voltage divider rule, we can relate $V_i(s)$ to $V_s(s)$:

$$V_i(s) = \frac{R_i V_s(s)}{R_s + R_i + 1/sC_1} = \frac{R_i}{R_s + R_i} \times \frac{s}{s + 1/[C_1(R_s + R_i)]} V_s(s) \tag{2.81}$$

The output voltage is given by

$$V_o(s) = R_L I_o(s) = -R_L \frac{R_o g_m V_i(s)}{R_o + R_L + 1/sC_2}$$

$$= -\frac{R_L R_o g_m}{R_o + R_L} \times \frac{s}{s + 1/[C_2(R_o + R_L)]} V_i(s) \tag{2.82}$$

Substituting $V_s(s)$ from Eq. (2.81) into Eq. (2.82) and simplifying, we get the voltage transfer function at low frequencies. That is,

$$A(s) = \frac{V_o(s)}{V_s(s)}$$

$$= -\frac{R_i R_L R_o g_m}{(R_s + R_i)(R_o + R_L)} \times \frac{s}{s + 1/[C_1(R_s + R_i)]} \times \frac{s}{s + 1/[C_2(R_o + R_L)]}$$

which gives the low break frequencies and high-pass gain as

$$f_{C1} = \frac{1}{2\pi C_1(R_s + R_i)} \tag{2.83}$$

$$f_{C2} = \frac{1}{2\pi C_2(R_o + R_L)} \tag{2.84}$$

$$A_{PB} = -\frac{R_i R_L R_o g_m}{(R_s + R_i)(R_o + R_L)} \tag{2.85}$$

We can notice from Eqs. (2.83) and (2.84) that the Thevenin's equivalent resistances for C_1 and C_2 are $R_{C1} = (R_s + R_i)$ and $R_{C2} = (R_o + R_L)$, respectively. The corresponding time constants are $\tau_{C1} = C_1 R_{C1}$ and $\tau_{C2} = C_2 R_{C2}$. Either f_{C1} or f_{C2} will be the dominant low cutoff (or 3-dB) frequency f_L. For a voltage amplifier, the input resistance R_i is normally much higher than the output resistance R_o, so $f_{C2} > f_{C1}$ and $f_{C2} = f_L$(low cut-off frequency). The steps in setting the low cutoff (or 3-dB) frequency are as follows:

Step 1. Set the low 3-dB frequency f_L with the capacitor that has the lowest resistance.

Step 2. Keep the other frequencies sufficiently lower than f_L so that interactions are minimal. Separating the first break frequency f_{L1} from the second break frequency f_{L2} by a decade is generally adequate, as long as the other frequencies are kept lower than f_{L2} by means of the following relations:

$$f_{L1} = f_L$$
$$f_{L2} = f_L/10$$
$$f_{L3} = f_L/20$$
$$f_{L4} = f_L/20$$

That is, $f_{L1} = f_L$ for Thevenin's equivalent resistance R_{C1}, $f_{L2} = f_L/10$ for Thevenin's equivalent resistance R_{C2}, and $f_{L3} = f_L/20$ for Thevenin's equivalent resistance R_{C3}, where $R_{C1} < R_{C2} < R_{C3}$. Since we are interested only in keeping other frequencies far away from the cutoff frequency f_L and since a wider separation would require a higher capacitor value, it is not necessary to keep a separation of one decade between subsequent frequencies.

2.7.2 High-Frequency Transfer Function Method

At high frequencies greater than $f_H > 15$ kHz, the bypass and coupling capacitors, which are on the order of 10 μF, have reactances on the order of 1 Ω and are essentially short-circuited. Let us assume that the coupling capacitances are large so that the capacitors are effectively short-circuited. Thus, the high-frequency behavior is determined solely by the internal capacitors of the amplifier. The equivalent circuit for determining the high break frequencies is shown in Fig. 2.23. Capacitance C, between the input and output terminals of the amplifier, can be replaced by Miller's equivalent capacitances. Therefore, we can find the frequency response by s-domain analysis or by Miller's capacitor method.

As we did for the low break frequencies, we will derive the transfer function for high break frequencies. Applying Kirchhoff's current law at nodes 1 and 2, we get the following equations in Laplace's domain of s:

$$\frac{V_s - V_i}{R_s} = \frac{V_i}{R_i} + V_i C_i s + (V_i - V_o)Cs \tag{2.86}$$

$$g_m V_i + \frac{V_o}{R_o} + \frac{V_o}{R_L} + V_o C_o s + (V_o - V_i)Cs = 0 \tag{2.87}$$

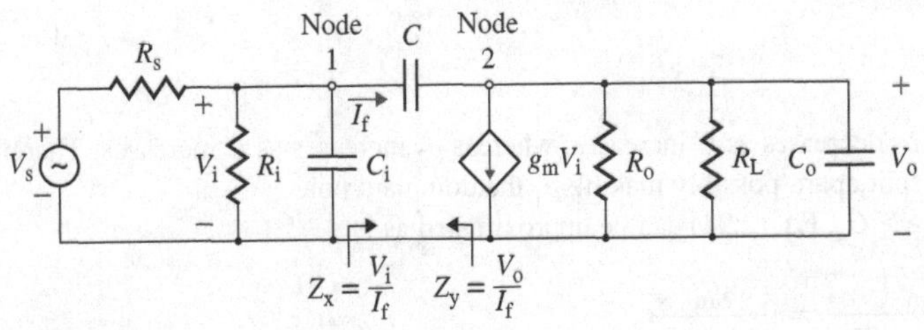

FIGURE 2.23 Small signal equivalent high cutoff circuit

Equations (2.86) and (2.87) can be solved to give the voltage transfer function

$$\frac{V_o(s)}{V_s(s)} = \frac{-(g_m - Cs)R_1R_2/R_s}{1 + s[R_1(C_i + C) + R_2(C_o + C) + g_mCR_1R_2] + s^2R_1R_2(C_iC_o + C_iC + C_oC)} \tag{2.88}$$

where $R_1 = (R_s \| R_i)$ and $R_2 = (R_o \| R_L)$. The denominator of Eq. (2.88) has two poles. If p_1 and p_2 are the two poles, the denominator can be written as

$$D(s) = \left(1 + \frac{s}{p_1}\right)\left(1 + \frac{s}{p_2}\right) = 1 + s\left(\frac{1}{p_1} + \frac{1}{p_2}\right) + \frac{s^2}{p_1p_2} \tag{2.89}$$

If the poles are widely separated, which is generally the case, and p_1 is assumed to be the dominant pole, while p_2 is assumed large, then Eq. (2.89) can be approximated by

$$D(s) = 1 + \frac{s}{p_1} + \frac{s^2}{p_1p_2} \tag{2.90}$$

Equating the coefficients of s in Eq. (2.88) to those in Eq. (2.90) yields

$$p_1 = \frac{1}{R_1(C_i + C) + R_2(C_o + C) + g_mCR_1R_2} \tag{2.91}$$

Equating the coefficients of s^2 in Eq. (2.88) to those in Eq. (2.90) yields

$$p_2 = \frac{R_1(C_i + C) + R_2(C_o + C) + g_mCR_1R_2}{R_1R_2(C_iC_o + C_iC + C_oC)} \tag{2.92}$$

In practice, the value of C is higher than that of C_i and C_o, and Eqs. (2.91) and (2.92) can be simplified further as follows:

$$p_1 \approx \frac{1}{g_mCR_1R_2} \tag{2.93}$$

$$p_2 = \frac{g_mC}{C_iC_o + C_iC + C_oC} \tag{2.94}$$

Here p_1 and p_2 correspond to the break frequencies ω_1 and ω_2 in frequency domain.

▶ **NOTES**

1. The dominant pole p_1 decreases as C increases, whereas p_2 increases as C increases. Therefore, increasing C causes the poles to split apart, possibly making p_1 the dominant pole.
2. If $C >> C_i$ and $C >> C_o$, Eq. (2.94) can be approximated as

$$p_2 \approx \frac{g_mC}{C(C_i + C_o)} = \frac{g_m}{C_i + C_o} \tag{2.95}$$

3. If there is no feedback capacitance ($C = 0$), Eq. (2.88) gives the following poles:

$$p_1 = \frac{1}{C_i R_1} \tag{2.96}$$

$$p_2 = \frac{1}{C_o R_2} \tag{2.97}$$

2.7.3 Miller's Capacitor Method

Assuming that the current through capacitor C in Fig. 2.23 is very small compared to the voltage-dependent current source $g_m V_i$, the output voltage in Laplace's domain is

$$V_o(s) = -g_m V_i(s)(R_o \| R_L)$$

The current $I_f(s)$ flowing through C (from the left side to the right side) is given by

$$I_f(s) = sC[V_i(s) - V_o(s)] = sC[V_i(s) + g_m V_i(s)(R_o \| R_L)]$$

$$= sC[1 + g_m(R_o \| R_L)]V_i(s) = sC_m V_i(s)$$

where $\quad C_m = C[1 + g_m(R_o \| R_L)] \quad$ (2.98)

The current $-I_f(s)$ flowing through C (from the right side to the left side) is given by

$$-I_f(s) = sC[V_o(s) - V_i(s)] = sC\left[V_o(s) + \frac{V_o(s)}{g_m(R_o \| R_L)}\right]$$

$$= sC\left[1 + \frac{1}{g_m(R_o \| R_L)}\right]V_o(s) = sC_n V_o(s)$$

where $\quad C_n = C\left[1 + \frac{1}{g_m(R_o \| R_L)}\right] \quad$ (2.99)

Thus, capacitor C, which is connected between the input and output terminals of a high-gain amplifier with 180° phase reversal, can be replaced by a shunt capacitor C_m on the input side and another capacitor C_n on the output side. This arrangement is shown in Fig. 2.24. The value of C is seen on the input side as a multiplying factor almost equal to the voltage gain $g_m(R_o \| R_L)$. This effect, known as *Miller's effect*, is

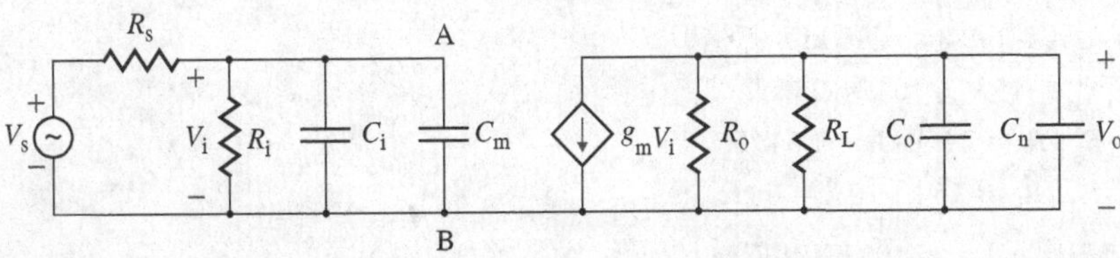

FIGURE 2.24 Miller's equivalent high cutoff circuit

dominant in amplifiers with a high voltage gain and a phase reversal. Therefore, the high-frequency poles can be found from

$$f_{H1} = \frac{1}{2\pi(C_i + C_m)(R_s \| R_i)} \tag{2.100}$$

$$f_{H2} = \frac{1}{2\pi(C_o + C_n)(R_o \| R_L)} \tag{2.101}$$

Since $A_v = -g_m(R_o \| R_L)$ is the voltage gain, we can rewrite Eqs. (2.98) and (2.99) as

$$C_m = C(1 - A_v) \tag{2.102}$$

$$C_n = C\left(1 - \frac{1}{A_v}\right) \tag{2.103}$$

Therefore, if the voltage A_v is negative, then $C_m > C$ and there is a capacitance multiplication. Thus, the effective capacitance C_m between nodes 1 and 2 (in Fig. 2.23) can be increased by a voltage amplifier. That is, a small capacitance connected between the input and output terminals of a voltage amplifier will have a much larger effective capacitance between the input terminals A and B (in Fig. 2.24). C_m is called the *Miller capacitance.* The Miller capacitance plays an important role in designing the high-frequency response of amplifiers and in the design of active filters. For example, f_{H1} in Eq. (2.100), which is contributed mainly by C_m, will be much lower than f_{H2}, and thus f_{H1} will set the dominant high frequency.

EXAMPLE 2.10

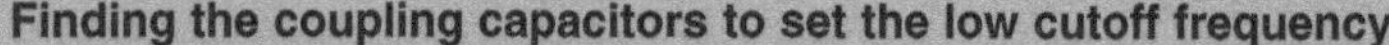

Finding the coupling capacitors to set the low cutoff frequency

(a) The amplifier in Fig. 2.21(a) has $g_m = 50$ mA/V, $R_s = 2$ kΩ, $R_i = 8$ kΩ, $R_o = 15$ kΩ, $R_L = 10$ kΩ, $C_i = 5$ pF, and $C_o = 1$ pF. Calculate the coupling capacitances C_1 and C_2 in order to set the low 3-dB frequency at $f_L = 1.5$ kHz, the pass-band gain (A_{PB}), and the feedback capacitance C so that the frequency of the dominant pole is $f_H = 100$ kHz.

(b) Use Miller's method to find the high cutoff frequencies.

(c) Use PSpice/SPICE to plot the voltage gain against the frequency.

SOLUTION

We have

$$R_s + R_i = 2\text{ k} + 8\text{ k} = 10\text{ k}\Omega$$
$$R_o + R_L = 15\text{ k} + 10\text{ k} = 25\text{ k}\Omega$$
$$R_1 = R_s \| R_i = 2\text{ k} \| 8\text{ k} = 1.6\text{ k}\Omega$$

and $R_2 = R_o \| R_L = 15\text{ k} \| 10\text{ k} = 6\text{ k}\Omega$

(a) $f_L = 1.5$ kHz and $f_H = 100$ kHz. Since

$$(R_s + R_i) = 10\text{ k}\Omega < (R_o + R_L) = 25\text{ k}\Omega$$

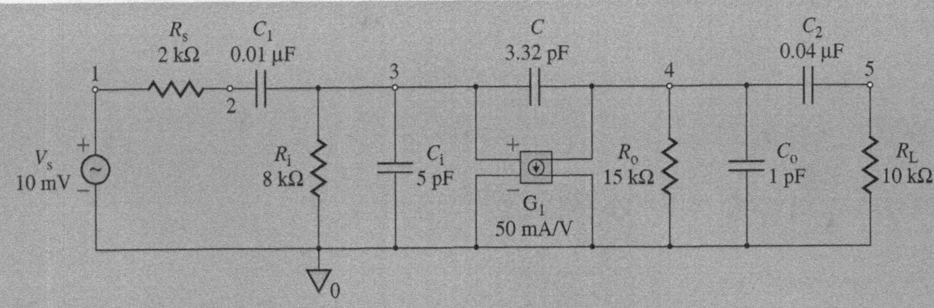

FIGURE 2.25 PSpice simulation circuit

let us set f_{C1} equal to the low 3-dB frequency. That is, $f_{C1} = f_L = 1.5$ kHz. The capacitance can be found from Eq. (2.83):

$$C_1 = \frac{1}{2\pi f_L R_s + R_i)} = \frac{1}{2\pi \times 1.5\text{ k} \times (2\text{ k} + 8\text{ k})} = 0.01\ \mu\text{F}$$

Let $f_{C2} = f_{C1}/10 = 1.5\text{ k}/10 = 150$ Hz. We get C_2 from Eq. (2.84):

$$C_2 = \frac{1}{2\pi f_{C2}(R_o + R_L)} = \frac{1}{2\pi \times 150 \times (15\text{ k} + 10\text{ k})} = 0.04\ \mu\text{F}$$

The pass-band gain is

$$A_{PB} = -\frac{R_i R_L R_o g_m}{(R_s + R_i)(R_o + R_L)} = -\frac{8\text{ k} \times 10\text{ k} \times 15\text{ k} \times 50\text{ mA/V}}{(2\text{ k} + 8\text{ k})(15\text{ k} + 10\text{ k})} = -240$$

From Eq. (2.93), we get the capacitance C for the dominant pole:

$$C \approx \frac{1}{2\pi f_H g_m R_1 R_2} = \frac{1}{2\pi \times 100\text{ k} \times 50\text{ mA/V} \times 1.6\text{k} \times 6\text{ k}} = 3.32\text{ pF}$$

From Eq. (2.91), we get

$$f_{H1} = \frac{10^{12}}{2\pi[1.6\text{ k} \times (5 + 3.32) + 6\text{ k} \times (1 + 3.32) + 50\text{ mA/V} \times 3.32 \times 1.6\text{ k} \times 6\text{ k}]}$$

$$= 97.47\text{ kHz}$$

From Eq. (2.92), we get

$$f_{H2} = \frac{[1.6\text{ k} \times (5 + 3.32) + 6\text{ k} \times (1 + 3.32) + 50\text{ mA/V} \times 3.32 \times 1.6\text{ k} \times 6\text{ k}] \times 10^{12}}{2\pi\,[1.6\text{ k} \times 6\text{ k} \times (5 \times 1 + 5 \times 3.32 + 1 \times 3.32)]}$$

$$= 1.09\text{ GHz}$$

(b) From Eq. (2.98),

$$C_m = C[1 + g_m(R_o \| R_L)] = (3.32 \text{ pF})(1 + 50 \text{ mA/V} \times 6 \text{ k}) = 999.3 \text{ pF}$$

From Eq. (2.99),

$$C_n = C\left[1 + \frac{1}{g_m(R_o \| R_L)}\right] = (3.32 \text{ pF})\left(1 + \frac{1}{300}\right) = 3.33 \text{ pF}$$

From Eqs. (2.100) and (2.101), we get

$$f_{H1} = \frac{1}{2\pi(C_1 + C_m)(R_s \| R_i)} = \frac{1}{2\pi \times (5 \text{ pF} + 999.3 \text{ pF}) \times 1.6 \text{ k}} = 99.05 \text{ kHz}$$

$$f_{H2} = \frac{1}{2\pi(C_o + C_n)(R_o \| R_L)} = \frac{1}{2\pi \times (1 \text{ pF} + 3.33 \text{ pF}) \times 6 \text{ k}} = 6.13 \text{ MHz}$$

Thus, Miller's capacitor method gives $f_{H1} = 99.05$ kHz, compared to 97.47 kHz calculated by s-domain analysis. However, $f_{H2} = 6.13$ MHz, compared to 1.09 GHz. The error is due to the fact that Miller's method does not take into account the effect of pole splitting.

(c) The circuit for PSpice simulation is shown in Fig. 2.25. Let us assume an input voltage of $v_s = 10$ mV.

The results of the simulation are shown in Fig. 2.26, which gives $A_{mid} = 221.7$ (expected value is 240), $f_L = 1.376$ kHz (expected value is 1.5 kHz) at $|A(j\omega)| = 0.707 \times 221.7 = 156.7$, and $f_H = 107.23$ kHz (expected value is 100 kHz) at $|A(j\omega)| = 0.707 \times 221.7 = 156.7$.

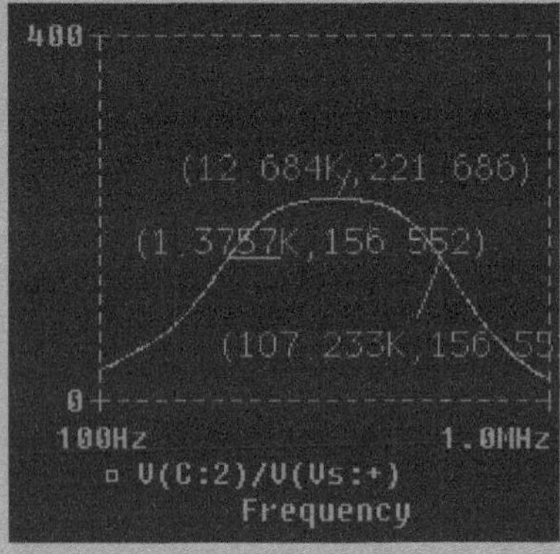

FIGURE 2.26 PSpice frequency response for Example 2.10

2.7.4 Low-Frequency Short-Circuit Method

As we have done in Sec. 2.7.1, we can determine the low cutoff frequencies and the pass-band voltage gain from the low-frequency voltage transfer function $A(s)$. In many cases, the analysis becomes laborious and it is not a simple matter to find $A(s)$. In such cases, an approximate value of the low 3-dB break frequency f_L can be found by the *short-circuit method.*

Let us assume that the voltage gain of an amplifier has two low break frequencies. Then, applying Eq. (2.59) for the high-pass characteristic with two break frequencies, we get

$$A(j\omega) = \frac{A_{v(high)}}{(1 + \omega_{L1}/s)(1 + \omega_{L2}/s)} \tag{2.104}$$

where $A_{v(high)} = A_{PB}$ is the high-frequency gain and ω_{L1} and ω_{L2} are the two break frequencies. At the low 3-dB frequency, the denominator of Eq. (2.104) should be such that $A_{PB} = 1/\sqrt{2} = 0.707$. That is,

$$\left|\left(1 + \frac{\omega_{L1}}{j\omega}\right)\left(1 + \frac{\omega_{L2}}{j\omega}\right)\right| = \sqrt{2}$$

or $$\left|1 - j\frac{\omega_{L1} + \omega_{L2}}{\omega} - \frac{\omega_{L1}\omega_{L2}}{\omega^2}\right| = \sqrt{2}$$

If $\omega > \sqrt{\omega_{L1}\omega_{L2}}$, the product term can be neglected. The imaginary term will be unity when

$$\omega_L = \omega = \omega_{L1} + \omega_{L2} = \frac{1}{\tau_{C1}} + \frac{1}{\tau_{C2}} \tag{2.105}$$

where ω_L is the effective low 3-dB frequency and is the sum of the reciprocals of the time constants τ_{C1} and τ_{C2}. For a circuit with multiple capacitors, the time constant τ_{Ck} for the kth capacitor is found by considering one capacitor at a time while setting the other capacitors to ∞ (or effectively short-circuiting them). This method assumes that only one capacitor contributes to the voltage gain. Thus, the low 3-dB frequency is determined from the effective time constant of all capacitors. That is,

$$f_L = \frac{1}{2\pi}\sum_{k=1}^{n}\frac{1}{\tau_{Ck}} = \frac{1}{2\pi}\sum_{k=1}^{n}\frac{1}{C_k R_{Ck}} \tag{2.106}$$

where τ_{Ck} is the time constant due to the kth capacitor only and R_{Ck} is Thevenin's equivalent resistance presented to C_k. One cutoff frequency will push the next higher frequency toward the right and thereby influence the effective cutoff frequency of the amplifier. If one of the break frequencies is larger than the other frequencies by a factor of 5 to 10, f_L can be approximated by the highest frequency—say f_{C1}. If $f_L \approx f_{C1}$, the error introduced will usually be less than 10%. Otherwise the error could be as high as 20%.

Let us apply this method to the circuit in Fig. 2.21(b). We will consider the effect of C_1 only; C_2 is short-circuited, as shown in Fig. 2.27(a). Thevenin's equivalent resistance presented to C_1 is

$$R_{C1} = R_s + R_i$$

Thus, the break frequency due to C_1 only is

$$f_{C1} = \frac{1}{2\pi R_{C1}C_1} = \frac{1}{2\pi(R_s + R_i)C_1} \tag{2.107}$$

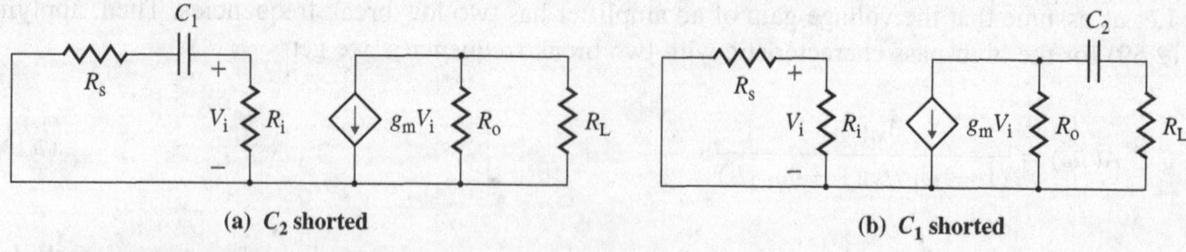

FIGURE 2.27 Small signal equivalent circuits for the short-circuit method

The equivalent circuit, with C_1 considered to be short-circuited, is shown in Fig. 2.27(b). Thevenin's equivalent resistance presented to C_2 is given by

$$R_{C2} = R_o + R_L$$

The break frequency due to C_2 only is given by

$$f_{C2} = \frac{1}{2\pi R_{C2}C_2} = \frac{1}{2\pi(R_o + R_L)C_2} \tag{2.108}$$

Therefore, the effective 3-dB frequency can be found from

$$f_L = f_{C1} + f_{C2} \tag{2.109}$$

In general, one of the low break frequencies is set to the desired 3-dB frequency f_L and the other frequencies are made much lower, normally separated by a decade. That is, if $f_L \approx f_{C1}$, then $f_{C2} = f_L/10$. The steps in setting the low 3-dB frequency are as follows:

Step 1. Draw the equivalent circuit with all but one capacitor shorted.

Step 2. Find Thevenin's equivalent resistance for each capacitor.

Step 3. Set the low 3-dB frequency f_L with the capacitor that has the lowest resistance. This will give the smallest capacitor value.

Step 4. Keep the other frequencies sufficiently lower than f_L so that interactions are minimal. That is, if $f_{C1} = f_L$ for Thevenin's equivalent resistance R_{C1}, $f_{C2} = f_L/10$ for Thevenin's equivalent resistance R_{C2}, and $f_{C3} = f_L/20$ for Thevenin's equivalent resistance R_{C3}, where $R_{C1} < R_{C2} < R_{C3}$.

2.7.5 High-Frequency Zero-Value Method

Let us assume that the voltage gain of an amplifier has two high break frequencies. Then, applying Eq. (2.49) for $s = j\omega$ and for the low-pass characteristic with two break frequencies, we get

$$A(j\omega) = \frac{A_{v(\text{low})}}{(1 + j\omega/\omega_{H1})(1 + j\omega/\omega_{H2})} \tag{2.110}$$

where $A_{v(low)} = A_{PB}$ is the low-frequency gain and ω_{H1} and ω_{H2} are the two high break frequencies. At the high 3-dB frequency, the denominator of Eq. (2.110) should be such that $A_{PB} = 1/\sqrt{2} = 0.707$. That is,

$$\left|\left(1+\frac{j\omega}{\omega_{H1}}\right)\left(1+\frac{j\omega}{\omega_{H2}}\right)\right| = \sqrt{2}$$

or

$$\left|1-\left(\frac{\omega}{\omega_{H1}}\right)\left(\frac{\omega}{\omega_{H2}}\right)+j\omega\left(\frac{1}{\omega_{H1}}+\frac{1}{\omega_{H2}}\right)\right| = \sqrt{2}$$

If $\omega < \sqrt{\omega_{H1}\omega_{H2}}$, the product term can be neglected. The imaginary term will be unity when

$$\frac{1}{\omega_H} = \frac{1}{\omega} = \frac{1}{\omega_{H1}} + \frac{1}{\omega_{H2}} = \tau_{C1} + \tau_{C2} \tag{2.111}$$

where ω_H is the effective high 3-dB frequency and is the reciprocal of the sum of the time constants τ_{C1} and τ_{C2}. For a circuit with multiple capacitors, the time constant τ_{Cj} for the jth capacitor is found by considering one capacitor at a time while setting the other capacitors to zero (or effectively open-circuiting them). Thus, the high 3-dB frequency is determined from the effective time constant of all capacitors. That is,

$$f_H = \frac{1}{2\pi\sum_{j=1}^{n}\tau_{Cj}} = \frac{1}{2\pi\sum_{j=1}^{n}C_jR_{Cj}} \tag{2.112}$$

where τ_{Cj} is the time constant due to the jth capacitor only and R_{Cj} is Thevenin's equivalent resistance presented to C_j.

Let us apply this method to the circuit in Fig. 2.23. The equivalent circuit, with C and C_o open-circuited, is shown in Fig. 2.28(a). The resistance seen by C_i is given by

$$R_{Ci} = (R_s \| R_i)$$

The equivalent circuit, with C and C_i open-circuited, is shown in Fig. 2.28(b). The resistance faced by C_o is given by

$$R_{Co} = (R_o \| R_L)$$

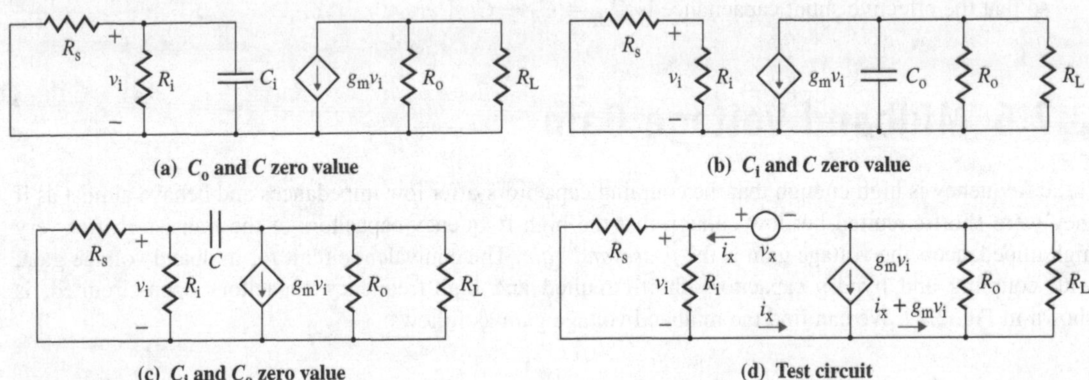

FIGURE 2.28 Small signal equivalent circuits for zero-value method

The equivalent circuit, with C_i and C_o open-circuited, is shown in Fig. 2.28(c). Let us replace C with voltage source v_x, as shown in Fig. 2.28(d). Then, applying Kirchhoff's voltage law (KVL), we get

$$v_x = v_i + (R_o \| R_L)(i_x + g_m v_i) = (R_s \| R_i)i_x + (R_o \| R_L)[i_x + g_m i_x(R_s \| R_i)]$$

$$= (R_o \| R_L) + (R_s \| R_i)[1 + g_m(R_o \| R_L)]i_x$$

which gives Thevenin's equivalent resistance seen by C as

$$R_{Cc} = \frac{v_x}{i_x} = (R_o \| R_L) + (R_s \| R_i)[1 + g_m(R_o \| R_L)]$$

$$= R_{L(eff)} + R_{i(eff)}(1 + g_m R_{L(eff)}) \tag{2.113}$$

where $R_{i(eff)} = (R_s \| R_i)$ and $R_{L(eff)} = (R_o \| R_L)$. Thus, the high 3-dB frequency f_H is given by

$$f_H = \frac{1}{2\pi(R_{Ci}C_i + R_{Co}C_o + R_{Cc}C)} \tag{2.114}$$

The steps in applying the zero-value method are as follows:

Step 1. Determine Thevenin's resistance seen by each capacitor acting alone while the other capacitors are open-circuited.

Step 2. Calculate the time constant due to each capacitor.

Step 3. Add all the time constants to find the effective time constant:

$$\tau_H = \tau_{H1} + \tau_{H2} + \cdots + \tau_{Hi}$$

Step 4. Find the high 3-dB frequency from Eq. (2.112).

Step 5. To set the high 3-dB frequency to a desired value, add an extra capacitor C_x in parallel with C so that the effective shunt capacitance is $C_{eff} = C_x + C$.

2.7.6 Midband Voltage Gain

If the frequency is high enough that the coupling capacitors offer low impedances and behave almost as if they were short-circuited but low enough that the high-frequency capacitors of the transistor offer very high impedances, the voltage gain is the *pass-band gain*. The equivalent circuit for midband voltage gain, with coupling and bypass capacitors short-circuited and high-frequency capacitors open-circuited, is shown in Fig. 2.29. We can find the midband voltage gain as follows:

$$A_{PB} = \frac{v_o}{v_s} = -g_m(R_o \| R_L)\frac{R_i}{R_s + R_i} \tag{2.115}$$

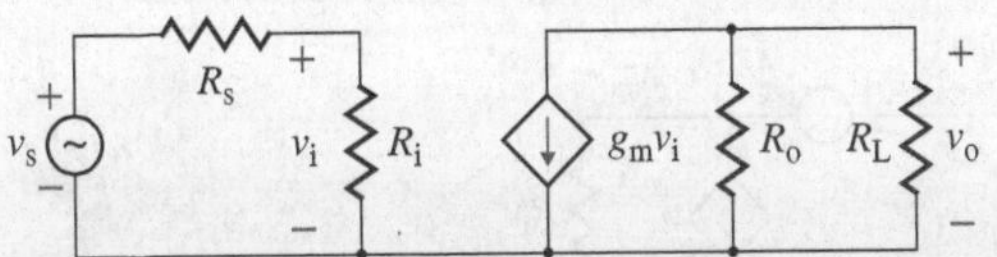

FIGURE 2.29 Equivalent circuit for determining the midband gain

EXAMPLE 2.11

Finding the 3-dB break frequencies by the short-circuit and zero-value methods The amplifier in Fig. 2.21(a) has $g_m = 50$ mA/V, $R_s = 2$ kΩ, $R_i = 8$ kΩ, $R_o = 15$ kΩ, $R_L = 10$ kΩ, $C_i = 5$ pF, and $C_o = 1$ pF. Use the following capacitor values: $C_1 = 0.01$ μF, $C_2 = 0.04$ μF, and $C = 3.32$ pF.

(a) Use the short-circuit method to find the low 3-dB frequency f_L.

(b) Use the zero-value method to find the high 3-dB frequency f_H.

(c) Find the pass-band voltage gain A_{PB}.

SOLUTION

(a) $R_{C1} = R_s + R_i = 2\text{ k} + 8\text{ k} = 10\text{ k}\Omega$, $R_{C2} = R_o + R_L = 15\text{ k} + 10\text{ k} = 25\text{ k}\Omega$

From Eq. (2.107), we get $f_{C1} = 1/(2\pi C_1 R_{C1}) = 1/(2\pi \times 0.01\ \mu\text{F} \times 10\text{ k}\Omega) = 1.592$ kHz.
From Eq. (2.108), we get $f_{C2} = 1/(2\pi C_2 R_{C2}) = 1/(2\pi \times 0.04\ \mu\text{F} \times 25\text{ k}\Omega) = 159.2$ kHz.
From Eq. (2.109), we get $f_L = f_{C1} + f_{C2} = 1.592\text{ k} + 159.2 = 1.751$ kHz.

(b) $R_{Ci} = R_s \| R_i = 2\text{ k} \| 8\text{ k} = 1.6\text{ k}\Omega$

$R_{Co} = R_o \| R_L = 15\text{ k} \| 10\text{ k} = 6\text{ k}\Omega$

Using Eq. (2.113),

$$R_{Cc} = R_o \| R_L + (R_s \| R_i)[1 + g_m (R_o \| R_L)] = 6\text{ k}\Omega + 1.6\text{ k}\Omega \times (1 + 50 \times 10^{-3} \times 6\text{ k}\Omega) = 487.6\text{ k}\Omega$$

From Eq. (2.114), we get

$$f_H = \frac{1}{[2\pi (R_{Ci}C_i + R_{Co}C_o + R_{Cc}C)]}$$

$$= \frac{1}{[2\pi \times (1.6\text{ k} \times 5 \times 10^{-12} + 6\text{ k} \times 1 \times 10^{-12} + 487.6\text{ k} \times 3.32 \times 10^{-12})]}$$

$$= 97.47\text{ kHz}$$

(c) From Eq. (2.115), we get the pass-band voltage gain

$$A_{PB} = -g_m (R_o \| R_L)\left[\frac{R_i}{(R_s + R_i)}\right] = -50 \times 10^{-3} \times [6\text{ k} \times 8\text{ k}/(2\text{ k} + 8\text{ k})] = -240\text{ V/V}$$

NOTE: The short-circuit method gives the low 3-dB frequency $f_L = 1.751$ kHz (designed for 1.5 kHz), and the zero-value method gives the high 3-dB frequency $f_H = 97.47$ kHz (designed for 100 kHz). In Example 2.10, Miller's capacitor method gives $f_H = 99.05$ kHz compared to 97.47 kHz from *s*-domain analysis.

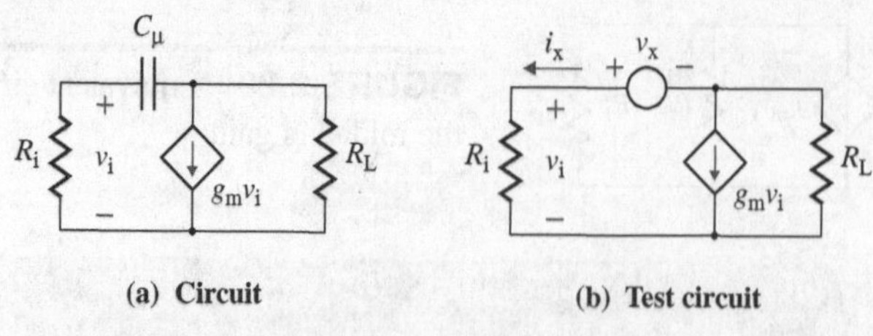

FIGURE 2.30 Capacitor with a voltage-controlled current source

2.7.7 Multistage Amplifiers

Multistage amplifiers are often used to meet voltage gain, frequency range, input impedance, and/or output impedance requirements. In this section we will apply the short-circuit and zero-value methods to determine the cutoff frequencies of multistage amplifiers. Some equations will be similar to those used in the preceding sections because the equivalent circuits for the amplifiers are similar to those encountered previously.

When a capacitor C_μ is connected between the input and the output of an amplifier, it greatly influences the high 3-dB frequency. It is often necessary to find the time constant for C_μ, and we will derive a generalized equation. Let us consider the circuit of Fig. 2.30(a). If the capacitor C_μ is replaced by a voltage source v_x, the equivalent circuit is shown in Fig. 2.30(b). Using KVL, we get

$$v_x = R_i i_x + R_L(i_x + g_m v_i) = R_i i_x + R_L(i_x + g_m R_i i_x)$$
$$= [R_L + R_i(1 + g_m R_L)]i_x$$

which gives Thevenin's equivalent resistance faced by C_μ as

$$R_{eq} = \frac{v_x}{i_x} = R_L + R_i(1 + g_m R_L) \tag{2.116}$$

KEY POINTS OF SECTION 2.7

- The method of s-domain analysis can be used to determine the transfer function and the frequency characteristics of an amplifier. However, the analysis can be laborious, especially for a circuit with more than three capacitors.
- Miller's capacitance method is a quick but approximate method for determining the high cutoff frequency.
- The short-circuit method gives the low 3-dB break frequency, and the zero-value method gives the high 3-dB break frequency. These are simple but effective methods for determining the break frequencies of amplifiers.
- In the short-circuit method, the time constant τ_{Ck} for the kth capacitor is determined by considering one capacitor at a time while setting other capacitors to ∞ (or effectively short-circuiting them). The effective low 3-dB frequency is the sum of the reciprocal of individual time constants.
- In the zero-value method, the time constant τ_{Cj} for the jth capacitor is determined by considering one capacitor at a time while setting the other capacitors to zero (or effectively open-circuiting them). The effective high 3-dB frequency is the reciprocal of the sum of the individual time constants.

2.8 PSpice/SPICE Amplifier Models

Amplifiers can be modeled in PSpice/SPICE as linear controlled sources [4]. However, these models will not exhibit the nonlinear characteristics expected in practical amplifiers. The PSpice/SPICE results must be interpreted in relation to the practical limits of a particular type of amplifier. For a current-controlled source, a dummy voltage source of 0 (say, $V_x = 0$) is inserted to monitor the controlling current, which gives the output current or voltage. The controlling current is assumed to flow from the positive node of V_x, through the voltage source V_x, to the negative node of V_x.

2.8.1 Voltage Amplifier

A voltage amplifier can be modeled as a voltage-controlled voltage source (VCVS). The symbol for a VCVS, as shown in Fig. 2.31(a), is *E*. The linear form is

```
E<name> N+ N- NC+ NC- <(voltage gain) value>
```

N+ and N− are the positive and negative output nodes, respectively. NC+ and NC− are the positive and negative nodes, respectively, of the controlling voltage.

2.8.2 Current Amplifier

A current amplifier can be modeled as a current-controlled current source (CCCS). The symbol of a CCCS, as shown in Fig. 2.31(b), is *F*. The linear form is

```
F<name> N+ N- VX <(current gain) value>
```

N+ and N− are the positive and negative nodes, respectively, of the output (current) source.

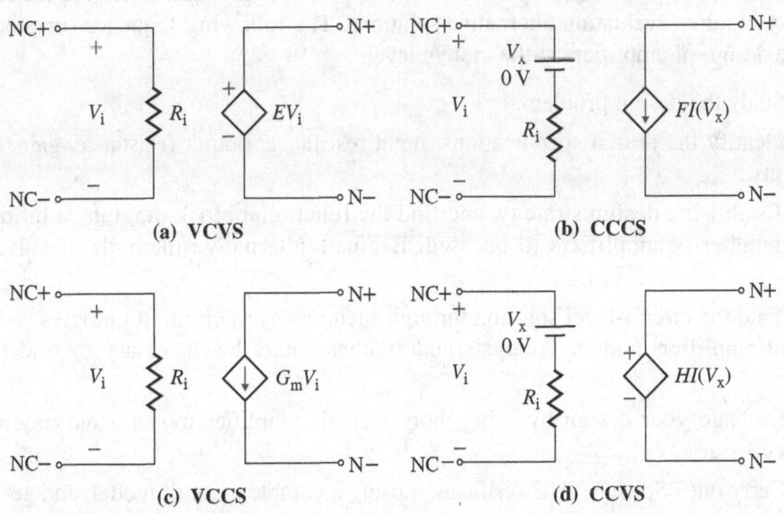

FIGURE 2.31 Dependent sources

2.8.3 Transconductance Amplifier

A transconductance amplifier can be modeled as a voltage-controlled current source (VCCS). The symbol of a VCCS, as shown in Fig. 2.31(c), is *G*. The linear form is

```
G<name> N+ N- NC+ NC- <(transconductance) value>
```

N+ and N− are the positive and negative output nodes, respectively. NC+ and NC− are the positive and negative nodes, respectively, of the controlling voltage.

2.8.4 Transimpedance Amplifier

A transimpedance amplifier can be modeled as a current-controlled voltage source (CCVS). The symbol of a CCVS, as shown in Fig. 2.31(d), is *H*. The linear form is

```
H<name> N+ N- VX <(transimpedance) value>
```

N+ and N− are the positive and negative nodes, respectively, of the output (voltage) source.

2.9 Amplifier Design

So far we have regarded amplifiers as parts of a system. Several amplifiers may be cascaded to meet some design specifications. However, viewed from the input and output sides, cascaded amplifiers may be represented by a single equivalent amplifier. That is, an amplifier may consist of one or more internal amplifiers. At this stage of the course, amplifier design will be at the system level rather than at the level of the internal components of an amplifier itself, which we will cover in Chapters 7 and 8. This chapter has illustrated a number of design examples relating to each topic area. The circuit topology was given, and the design task was mainly to find the component values. Often a designer has to choose the circuit topology, which generally requires evaluating alternative solutions. The following sequence (or process) is recommended for the design of amplifiers at the system level:

Step 1. Study the design problem.

Step 2. Identify the design specifications: input resistance, output resistance, gain, and bandwidth requirements.

Step 3. Establish a design strategy, and find the functional block diagram solution. Identify the type and number of amplifiers to be used. Evaluate alternative methods of solving the design problem.

Step 4. Find the circuit-level solution through such means as circuit topologies and hand analysis using ideal amplifier models. Analysis and synthesis may be necessary to find the component values.

Step 5. Evaluate your design by using more realistic amplifier models, and modify your design values, if necessary.

Step 6. Carry out PSpice/SPICE verification using a complex circuit model, and get the worst-case results given your components and parameter variations. Modify your design, if needed.

Step 7. Get a cost estimate of the project, and have a plan for component layout so that the project requires the minimum fabrication time and is least expensive.

Step 8. Build a prototype unit in the lab, and take measurements to verify your design. Modify your design, if needed.

EXAMPLE 2.12

D

Illustration of design steps Two signals are coming from two different transducers: $v_1 = 180$ mV to 200 mV with $R_{s1} = 2$ kΩ and $v_2 = 150$ mV to 170 mV with $R_{s2} = 2$ kΩ. Amplify the differential voltage so that the output voltage is $v_o = 200(v_1 - v_2)$. The gain variation should be less than ±3%. The load resistance is $R_L = 5$ kΩ. Determine the specifications of the amplifier.

SOLUTION

Step 1. Study the design problem. $v_1 = 180$ mV to 200 mV with $R_{s1} = 2$ kΩ, and $v_2 = 150$ mV to 170 mV with $R_{s2} = 2$ kΩ.

Step 2. Identify the design specifications. $A_v = 200 \pm 3\%$, $R_L = 5$ kΩ, and there is no bandwidth limit.

Step 3. Establish a design strategy, and find the functional block diagram solution. Since the input side will have two voltage signals whose difference is to be amplified, we need a voltage differential amplifier at the input stage. The output of this stage could be either voltage or current, which will be amplified by a gain stage, shown in Fig. 2.32(a).

Step 4. Find the circuit-level solution. We will use two identical transconductance amplifiers to give differential gain because it allows us to add (or subtract) two currents at a node. We will also use a transresistance amplifier at the output side to give the desired voltage gain and a low output resistance. This arrangement is shown in Fig. 2.32(b). Assuming ideal amplifiers of $G_{ms1} = G_{ms2} = G_{ms}$, the output voltage is given by

$$v_o = (G_{ms1}v_1 - G_{ms2}v_2)Z_{mo} = Z_{mo}G_{ms}(v_1 - v_2)$$

which gives $A_{vo} = Z_{mo}G_{ms}$. Assuming $G_{ms} = 20$ mA/V, we get

$$Z_{mo} = \frac{A_{vo}}{G_{ms}} = \frac{200 \text{ V}}{20 \text{ mA}} = 10 \text{ kV/A}$$

Step 5. Evaluate your design. Let us take practical amplifiers with input and output resistances as shown in Fig. 2.32(c). Using Eq. (2.35), we can find the effective voltage gain A_v from

$$A_v = \frac{R_{i1}}{R_{i1} + R_{s1}} \times \frac{R_{oe}}{R_{oe} + R_{i2}} \times \frac{R_L}{R_L + R_{o2}} Z_{mo}G_{ms} \; (R_{oe} = R_{o1} \,\|\, R'_{o1})$$

Since A_v will vary with variations in R_{i1}, R'_{i1}, R_{o2}, Z_{mo}, and G_{ms}, let us allow ±0.5% variation for each of them so that the overall variation is limited to ±2.5%. Assume R_{s1}, R_{s2}, and R_L do not vary.

$$\frac{R_{i1}}{R_{i1} + R_{s1}} = \frac{R_{i1}}{R_{i1} + 2\text{ k}} = 0.995$$

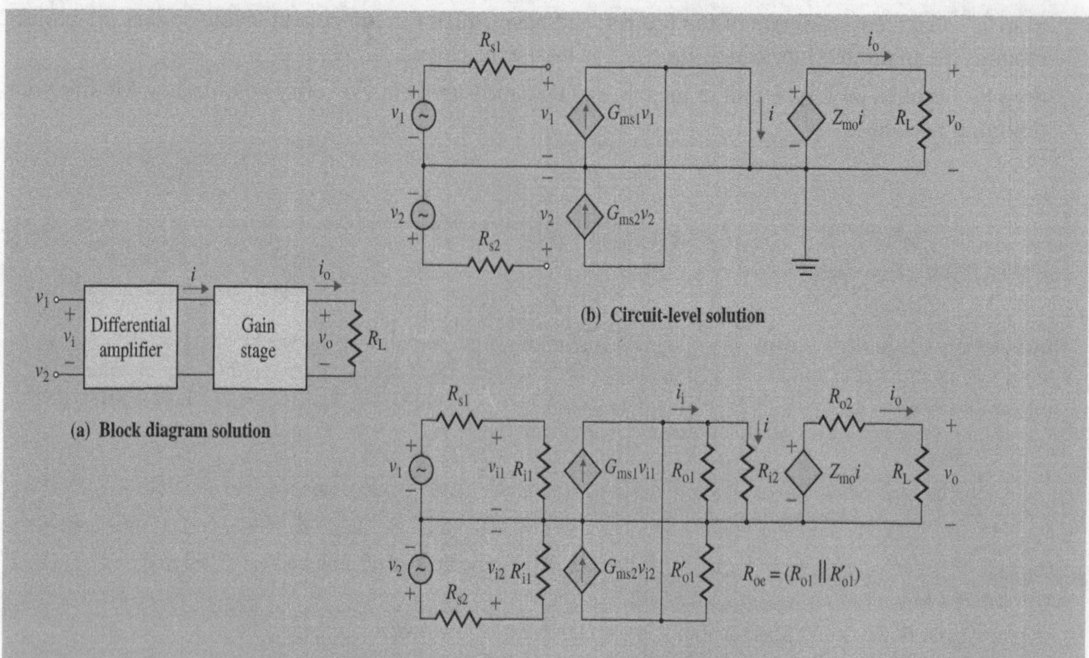

FIGURE 2.32 Amplifier design stages for Example 2.12

which gives $R_{i1} = R'_{i1} \geq 398\ \text{k}\Omega$.

$$\frac{R_L}{R_L + R_{o2}} = \frac{5\text{ k}}{5\text{ k} + R_{o2}} = 0.995$$

which gives $R_{o2} \leq 25.1\ \Omega$.

Let us assume that $R_{o1} = R'_{o1} \geq 200\ \text{k}\Omega$. Since $R_{oe} = (R_{o1} \| R'_{o1}) = 100\ \text{k}\Omega$,

$$\frac{R'_{o1}}{R'_{o1} + R_{i2}} = \frac{100\text{ k}}{100\text{ k} + R_{i2}} = 0.995$$

which gives $R_{i2} \leq 502\ \Omega$.

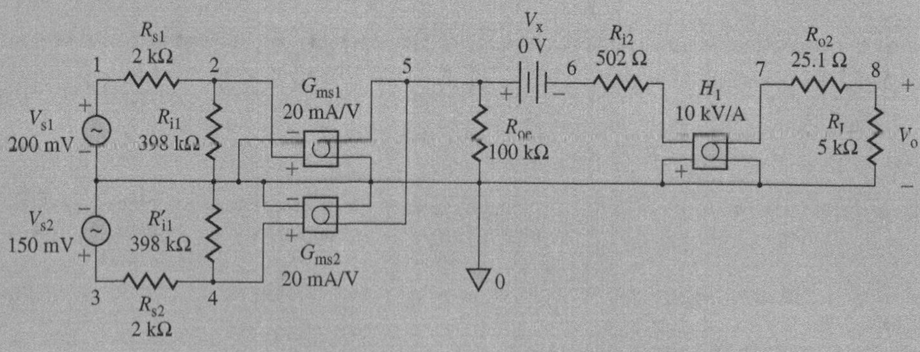

FIGURE 2.33 Circuit for PSpice simulation for Example 2.12

Step 6. Use PSpice/SPICE verification. The results of PSpice simulation for the circuit shown in Fig. 2.33 are as follows:

NODE	VOLTAGE	NODE	VOLTAGE	NODE	VOLTAGE	NODE	VOLTAGE
(1)	.2000	(2)	.1990	(3)	.1500	(4)	.1493
(5)	.4970	(6)	.4970	(7)	9.9003	(8)	9.8508

The output voltage is $v_o = 9.8796$ V, and $A_v = 9.8508$ V/(200 mV − 150 mV) = 197.02.

Step 7. Get a cost estimate.

Two identical transconductance amplifiers for the differential stage: $G_{ms} = 20$ mA/V ± 0.5%, $R_{i1} \geq 398$ kΩ, and $R_{o1} \geq 200$ kΩ. Estimated cost is $1.50.

One transresistance amplifier for the gain stage: $Z_{mo} = 10$ kV/A ± 0.5%, $R_{i2} \leq 502$ Ω, and $R_{o2} \leq 25.1$ Ω. Estimated cost is $1.

Two DC power supplies: $V_{CC} = V_{EE} = 12$ V.

Summary

Amplifiers are normally specified in terms of gain, input resistance, and output resistance. An amplifier can be classified as one of four types: a voltage amplifier, a current amplifier, a transconductance amplifier, or a transimpedance amplifier. The gain relationships of various amplifiers can be related to each other. In addition to amplifying signals, amplifiers can serve as building blocks for other applications, such as impedance matching, negative resistance simulation, inductance simulation, and capacitance multiplication. Cascaded amplifiers are often used to increase the overall gain.

Amplifiers use transistors as amplifying devices. Transistors have internal capacitances and also coupling capacitors for isolating the signal source and the load from DC signals. The gain of practical amplifiers varies with the frequency of the signal source, and amplifiers can be classified based on their frequency response as low-pass or band-pass.

Because it uses coupling, bypass, and transistor capacitors, an amplifier operates within a frequency range called a bandwidth. There are three types of frequency characteristics: low-pass, high-pass, and band-pass. An amplifier normally exhibits a band pass characteristic. Analysis or design of an amplifier requires computer-aided methods because of the complexity of the circuits and the frequency-dependent parameters involving complex numbers.

In general, a capacitor that forms a series circuit with the input signal limits the low cutoff frequency, whereas a capacitor that forms a parallel circuit limits the upper cutoff frequency. Analysis of low break frequencies can be simplified by the short-circuit method, in which the time constant due to one capacitor is determined by assuming that the other capacitors are effectively short-circuited. This method can be extended to the analysis of multistage amplifiers. The dominant low cutoff (or 3-dB) frequency can be set to one of the low break frequencies. In that case, if one of the cutoff frequencies is less than the other frequencies by a factor of 5 to 10, the error introduced by this method is usually less than 10%. Otherwise the error could be as high as 20%.

At high frequencies, any capacitor that is connected between the input and output terminals dominates the frequency response as a result of Miller's multiplication effect. Miller's capacitor method, which can be applied to determine the approximate value of the high cutoff frequency, gives a value higher than the actual one. The zero-value method, which assumes that only one capacitor contributes to the circuit response and other capacitors have a value of zero, calculates the high 3-dB cutoff frequency from the effective time constant of all capacitors and gives a conservative estimate of the frequency.

References

1. W. H. Hyatt, Jr., and G. W. Neudeck, *Electronic Circuit Analysis and Design*. Boston, MA: Houghton Mifflin, 1984.
2. P. E. Gray and C. L. Searle, *Electronic Principles*. New York: Wiley, 1969.
3. P. R. Gray and R. G. Meyer, *Analysis and Design of Analog Integrated Circuits*. New York: Wiley, 2001.
4. M. H. Rashid, *Introduction to SPICE Using OrCAD for Circuits and Electronics*. Englewood Cliffs, NJ: Prentice Hall, 2004, Chapters 8 and 9.

Review Questions

1. What are the parameters of an amplifier?
2. What is the purpose of DC biasing of an amplifier?
3. What is the effect of rise time on the frequency response of an amplifier?
4. What is a slew rate?
5. What is the slew rate of a step input voltage?
6. What are the four types of amplifiers?
7. What is the circuit model of a voltage amplifier?
8. What is the open-circuit voltage gain of a voltage amplifier?
9. What is the effect of source resistance on the effective voltage gain of a voltage amplifier?
10. What is an ideal voltage amplifier?
11. What is the circuit model of a current amplifier?
12. What is the short-circuit current gain of a current amplifier?
13. What is the effect of source resistance on the effective current gain of a current amplifier?
14. What is an ideal current amplifier?
15. What is the circuit model of a transconductance amplifier?
16. What is the short-circuit transconductance of a transconductance amplifier?
17. What is the effect of source resistance on the overall voltage gain of a transconductance amplifier?
18. What is the open-circuit transimpedance of an amplifier?
19. What is the effect of source resistance on the effective current gain of a transimpedance amplifier?
20. What is an ideal transimpedance amplifier?
21. What is the effect on the overall gain of cascading amplifiers?
22. What is the principle of negative resistance simulation (see Prob. 2.18)?
23. What is a gyrator (see Prob. 2.21)?
24. What is the frequency response of an amplifier?
25. What is a low-pass amplifier?
26. What is a high-pass amplifier?
27. What is a band-pass amplifier?
28. Which capacitors contribute to the low cutoff frequency of amplifiers?
29. What is the short-circuit method?
30. What are the advantages and disadvantages of the short-circuit method?
31. Which capacitors contribute to the high cutoff frequency of amplifiers?
32. What is Miller's capacitor method?
33. What are the advantages and disadvantages of Miller's capacitor method?

34. What is the zero-value method?
35. What are the advantages and disadvantages of the zero-value method?
36. What are the steps involved in applying the zero-value method?

Problems

The symbol **D** indicates that a problem is a design problem. The symbol **P** indicates that you can check the solution to a problem using PSpice/SPICE or Electronics Workbench.

2.2 Amplifier Characteristics

2.1 The measured small-signal values of the linear amplifier shown in Fig. 2.3(a) are as follows: $v_i = 50 \times 10^{-3} \sin 1000\pi t$, $i_i = 1 \times 10^{-6} \sin 1000\pi t$, $v_o = 6.5 \sin 1000\pi t$, and $R_L = 5\ \text{k}\Omega$. The DC values are $V_{CC} = V_{EE} = 15$ V and $I_{CC} = I_{EE} = 15$ mA. Find **(a)** the values of amplifier parameters A_v, A_i, A_p, and R_i; **(b)** the power delivered by the DC supplies P_{dc} and the power efficiency η; and **(c)** the maximum value of the input voltage so that the amplifier operates within the saturation limits.

2.2 The measured values of the nonlinear amplifier in Fig. 2.4(a) are $v_o = 5.3$ V at $v_I = 21$ mV, $v_O = 5.5$ V at $v_I = 24$ mV, and $v_O = 5.8$ V at $v_I = 27$ mV. The DC supply voltage is $V_{CC} = 12$ V, and the saturation limits are $2\text{ V} \le v_O \le 11$ V.

a. Determine the small-signal voltage gain A_v.

b. Determine the DC voltage gain A_{dc}.

c. Determine the limits of input voltage v_I.

2.3 Determine the power gain A_p of the amplifier for the measured values.

a. $v_O = 2$ V, $v_I = 1$ mV, $R_i = 100\ \text{k}\Omega$, and $R_L = 10\ \text{k}\Omega$.

b. $i_O = 100$ mA, $i_I = 1$ mA, $R_i = 100\ \Omega$, and $R_L = 1\ \text{k}\Omega$.

2.3 Amplifier Types

2.4 The voltage amplifier shown in Fig. 2.6(a) has an open-circuit voltage gain of $A_{vo} = 150$, an input resistance of $R_i = 1.8\ \text{k}\Omega$, and an output resistance of $R_o = 50\ \Omega$. It drives a load of $R_L = 4.7\ \text{k}\Omega$. The signal source voltage is $v_s = 100$ mV with a source resistance $R_s = 200\ \Omega$.

a. Calculate the effective voltage gain $A_v = v_o/v_s$, the current gain $A_i = i_o/i_i$, and the power gain $A_p = P_L/P_i$.

b. Use PSpice/SPICE to check your results in part (a).

2.5 For the amplifier in Prob. 2.4, what should the load resistance R_L be for maximum power transfer to the load? Calculate the maximum output (or load) power $P_{L(max)}$.

2.6 When a load resistance of $R_L = 1.5\ \text{k}\Omega$ is connected to the output of a voltage amplifier, the output voltage drops by 15%. What is the output resistance R_o of the amplifier?

2.7 The voltage amplifier shown in Fig. 2.6(a) has an open-circuit voltage gain of $A_{vo} = 200$, an input resistance of $R_i = 100\ \text{k}\Omega$, and an output resistance of $R_o = 20\ \Omega$. The signal source voltage is $v_s = 50$ mV, the source resistance is $R_s = 1.5\ \text{k}\Omega$, and the load resistance is $R_L = 22\ \Omega$. Calculate **(a)** the output voltage v_o, **(b)** the output power P_L, **(c)** the effective voltage gain $A_v = v_o/v_s$, **(d)** the current gain $A_i = i_o/i_s$, and **(e)** the power gain $A_p = P_L/P_i$.

2.8 **D** An amplifier is required to amplify the output signal from a transducer that produces a voltage signal of $v_s = 10$ mV with an internal resistance of $R_s = 2.5\ \text{k}\Omega$. The load resistance is $R_L = 2\ \text{k}\Omega$ to $10\ \text{k}\Omega$. The desired output voltage is $v_o = 5$ V. The amplifier must not draw more than 1 μA from the transducer.

The variation in output voltage when the load is disconnected should be less than 0.5%. Determine the design specifications of the amplifier.

2.9 An amplifier is required to give a voltage gain of $A_v = 100 \pm 1.5\%$. The source resistance is $R_s = 500\ \Omega$ to 5 kΩ, and the load resistance is $R_L = 5$ kΩ to 20 kΩ. Determine the design specifications of the amplifier.

D

2.10 A resistance R is connected between the input and output terminals of a voltage amplifier, as shown in Fig. 2.6. The input voltage signal is $v_s = 20$ mV with an internal resistance of $R_s = 1.5$ kΩ.

D

a. Derive an expression for the input resistance $R_x = v_i / i_s$.

b. Calculate R_x and i_s for $R_i = 50$ kΩ, $R_o = 75\ \Omega$, $A_{vo} = 2$, and $R = 10$ kΩ.

c. Design an amplifier circuit that will simulate a negative resistance so that the input current drawn from the source is $|i_s| \leq 2.5\ \mu$A.

2.11 A capacitor C is connected between the input and output terminals of a voltage amplifier, as shown in Fig. P2.11(a). The peak input voltage is $V_{s(peak)} = 20$ mV with an internal resistance of $R_s = 1.5$ kΩ, and the signal frequency is $f_s = 100$ Hz.

a. Derive an expression for the input impedance $Z_x = V_i / I_s$.

b. Assuming an ideal amplifier, as shown in Fig. P2.11(b), calculate Z_x and I_s for $C = 0.01\ \mu$F and $A_{vo} = -100$. That is, $R_i = \infty$ and $R_o = 0$.

FIGURE P2.11

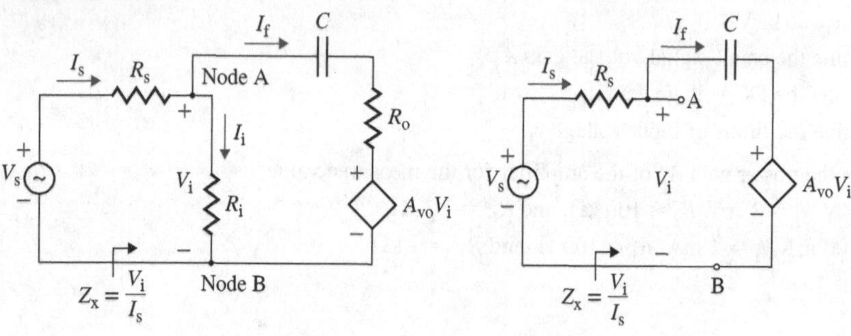

(a) Voltage amplifier **(b) Ideal voltage amplifier**

2.12 Design a voltage amplifier circuit that will simulate a negative resistance of $R = -5$ kΩ (see Prob. 2.10).

2.13 The current amplifier shown in Fig. 2.7(b) has a short-circuit current gain of $A_{is} = 200$, an input resistance of $R_i = 150\ \Omega$, and an output resistance of $R_o = 2.5$ kΩ. The load resistance is $R_L = 100\ \Omega$. The input source current is $i_s = 4$ mA with a source resistance of $R_s = 47$ kΩ.

a. Calculate the current gain $A_i = i_o / i_s$, the voltage gain $A_v = v_o / v_s$, and the power gain $A_p = P_L / P_i$.

b. Use PSpice/SPICE to check your results in part (a).

2.14 The current amplifier shown in Fig. 2.7(b) has a short-circuit current gain of $A_{is} = 100$, an input resistance of $R_i = 50\ \Omega$, an output resistance of $R_o = 22$ kΩ, and a load resistance of $R_L = 150\ \Omega$. The input source current is $i_s = 50$ mA with a source resistance of $R_s = 100$ kΩ. Calculate the output current i_o.

2.15 The current amplifier shown in Fig. 2.7(b) has a source current of $i_s = 5\ \mu$A, a source resistance of $R_s = 100$ kΩ, and an input resistance of $R_i = 50\ \Omega$. The short-circuit output current is $i_o = 100$ mA for $R_L = 0$, and the open-circuit output voltage is $v_o = 12$ V for $R_L = \infty$. The load resistance is $R_L = 2.7$ kΩ. Calculate **(a)** the voltage gain $A_v = v_o / v_s$, **(b)** the current gain $A_i = i_o / i_s$, and **(c)** the power gain $A_p = P_L / P_i$.

2.16 D An amplifier is required to amplify the output signal from a transducer that produces a constant current of $i_s = 100\ \mu A$ at an internal resistance varying from $R_s = 10\ k\Omega$ to $100\ k\Omega$. The desired output current is $i_o = 20\ mA$ at a load resistance varying from $R_L = 20\ \Omega$ to $500\ \Omega$. The variation in output current should be kept within $\pm 3\%$. Determine the design specifications of the amplifier.

2.17 D An amplifier is required to give a current gain of $A_i = 50 \pm 1.5\%$. The source resistance is $R_s = 100\ k\Omega$, and the load resistance is $R_L = 100\ \Omega$. Determine the design specifications of the amplifier.

2.18 D A resistance R is connected to a current amplifier, as shown in Fig. P2.18.

a. Derive an expression for the input resistance $R_x = v_i/i_i$.

b. Design an amplifier circuit that will simulate a negative resistance of $R_x = -10\ k\Omega$.

FIGURE P2.18

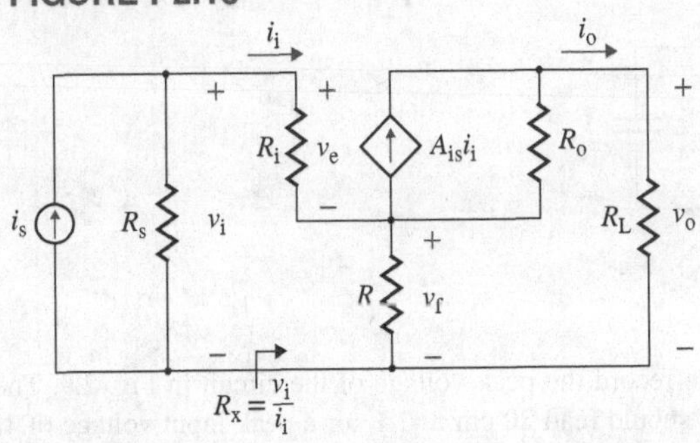

2.19 D Suppose the resistance R of the current amplifier in Fig. P2.18 is replaced by an impedance Z consisting of R, C, and $-R$. This arrangement is shown in Fig. P2.19 for $A_{is} = 2$, converting the current source to a voltage source, V_s.

a. Derive an expression for the input impedance $Z_x(s) = V_i(s)/I_i(s)$, where s is Laplace's operator. Note that $-R$ can be generated by another current amplifier such as the one shown in Fig. P2.18 with $A_{is} = 2$.

b. Design an amplifier circuit that will simulate an inductance of $L_e = 10$ mH.

c. Use PSpice/SPICE to calculate the input impedance Z_x for frequencies from 1 kHz to 5 kHz with a linear increment of 1 kHz. Use a PSpice/SPICE F-type dependent source (see Sec. 2.8).

FIGURE P2.19

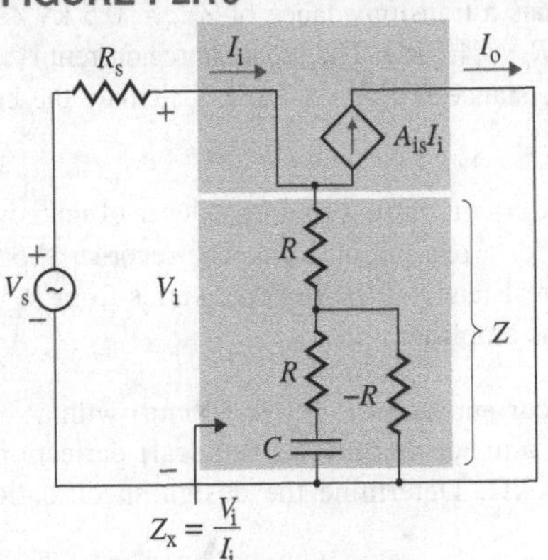

2.20 **a.** Design an amplifier circuit that will simulate an inductance of $L_e = 50$ mH (see Prob. 2.19).
D **b.** Use PSpice/SPICE to verify your design.

2.21 The gyrator circuit shown in Fig. P2.19 has a capacitance of $C = 100$ pF. Determine the value of resistance R that will give an effective inductance of $L_e = 15$ mH.

2.22 Two ideal transconductance amplifiers are connected back to back, as shown in Fig. P2.22.

a. Find the relation between the input voltage and the input currents, and find the input impedance $Z_i = V_i/I_i$.

b. If $v_s = 1 \sin(2000\pi t)$, $C = 0.1$ μF, and $G_{m1} = G_{m2} = 3$ mA/V, use PSpice/SPICE to plot the transient response of the output voltage $v_o(t)$ for a time interval from 0 to 1.5 ms with an increment of 15 μs.

FIGURE P2.22

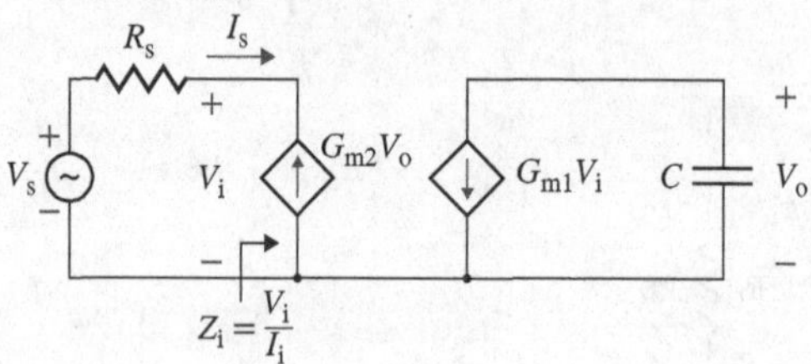

2.23 **D** A transconductance amplifier is needed to record the peak voltage of the circuit in Fig. 2.9. The recorder needs 5 mA for a reading of 1 cm, and it should read 20 cm ± 2% for a peak input voltage of 170 V. The input resistance of the recorder varies from $R_L = 20\ \Omega$ to $500\ \Omega$. The frequency of the input voltage is $f_s = 60$ kHz.

a. Determine the value of capacitance C.

b. Determine the design specifications of the transconductance amplifier.

2.24 **D** An amplifier is required to give a transconductance gain of $Z_m = 20$ mA/V ± 2%. The source resistance is $R_s = 1$ kΩ, and the load resistance is $R_L = 200\ \Omega$. Determine the design specifications of the amplifier.

2.25 **D** An amplifier is used to measure a DC voltage signal $v_s = 0$ to 10 V with a source resistance of $R_s = 2$ kΩ to 5 kΩ. The output of the amplifier is a meter that gives a full-scale deflection at a current of $i_o = 100$ mA and whose resistance is $R_m = 20\ \Omega$ to $100\ \Omega$. Determine the design specifications of the amplifier.

2.26 The transimpedance amplifier shown in Fig. 2.10(a) has a transimpedance of $Z_{mo} = 0.5$ kV/A, an input resistance of $R_i = 1.5$ kΩ, and an output resistance of $R_o = 4.7$ kΩ. The input source current is $i_s = 50$ mA with a source resistance of $R_s = 10$ kΩ. The load resistance is $R_L = 4.7$ kΩ. Calculate the current gain $A_i = i_o/i_s$ and the voltage gain $A_v = v_o/v_s$.

2.27 **D** A transimpedance amplifier is used to record the short-circuit current of a transducer of unknown internal resistance; its output is a recorder that requires 10 V for a reading of 2 cm. The recorder should read 20 cm ± 2% for an input current of 100 mA. The input resistance of the recorder varies from $R_L = 2$ kΩ to $R_L = 10$ kΩ. Determine the design specifications of the amplifier.

2.28 **D** A transimpedance amplifier is used to measure a DC current signal $i_s = 0$ to 500 mA with a source resistance of $R_s = 100$ kΩ. The output of the amplifier is a meter that gives a full-scale deflection at a voltage of $v_o = 5$ V and whose resistance is $R_m = 20$ kΩ. Determine the design specifications of the amplifier.

2.29 The parameters of the voltage amplifier in Fig. 2.6(a) are $v_s = 100$ mV, $R_s = 2$ kΩ, $A_{vo} = 250$, $R_i = 50$ kΩ, $R_o = 1$ kΩ, and $R_L = 10$ kΩ. Calculate the values of the equivalent current, transconductance, and transimpedance amplifiers.

2.30 The parameters of the transconductance amplifier in Fig. 2.8(b) are $v_s = 100$ mV, $R_s = 2$ kΩ, $G_{ms} = 20$ mA/V, $R_i = 100$ kΩ, $R_o = 2$ kΩ, and $R_L = 200$ Ω. Calculate the values of the equivalent voltage, current, and transimpedance amplifiers.

2.31 The slew rate of a unity-gain amplifier is SR = 0.5 V/μs, and the rise time is 0.3 μs. What is the maximum value $V_{S(max)}$ of a step input voltage?

2.32 The slew rate of a unity-gain amplifier is SR = 0.5 V/μs. The input frequency is $f_s = 100$ kHz. Calculate the maximum voltage $V_{s(max)}$ of a sinusoidal input voltage.

2.33 The slew rate of a unity-gain amplifier is SR = 0.5 V/μs. The input is a sinusoidal peak voltage $V_m = 10$ V. Determine the maximum input frequency $f_{s(max)}$ that will avoid distortion.

2.4 *Cascaded Amplifiers*

2.34 The parameters of the cascaded voltage amplifiers in Fig. 2.11(a) are $R_s = 200$ kΩ, $R_{o1} = R_{o2} = R_{o3} = 100$ Ω, $R_{i1} = R_{i2} = R_{i3} = R_L = 2.5$ kΩ, and $A_{vo1} = A_{vo2} = A_{vo3} = 50$.

a. Calculate the overall open-circuit voltage gain $A_{vo} = v_o/v_i$, the effective voltage gain $A_v = v_o/v_s$, the overall current gain $A_i = i_o/i_{i1}$, and the power gain $A_p = P_L/P_i$.

b. Use PSpice/SPICE to check your results in part (a).

2.35 The parameters of the cascaded voltage amplifiers in Fig. 2.11(a) are $R_s = 200$ kΩ, $R_{o1} = R_{o2} = 100$ Ω, $R_{o3} = 300$ Ω, $R_{i1} = R_{i2} = R_{i3} = 2.5$ kΩ, $R_L = 1.5$ kΩ, and $A_{vo1} = A_{vo2} = A_{vo3} = 80$.

a. Calculate the overall voltage gain $A_{vo} = v_o/v_s$, the overall current gain $A_i = i_o/i_{i1}$, and the power gain $A_p = P_L/P_i$.

b. Use PSpice/SPICE to check your results in part (a).

2.36 The parameters of the cascaded current amplifiers in Fig. 2.12(a) are $R_s = 20$ kΩ, $R_{o1} = R_{o2} = R_{o3} = 4.7$ kΩ, $R_{i1} = R_{i2} = R_{i3} = R_L = 100$ Ω, and $A_{is1} = A_{is2} = A_{is3} = 100$.

a. Calculate the effective current gain $A_i = i_o/i_s$, the overall voltage gain $A_v = v_o/v_s$, and the power gain $A_p = P_L/P_i$.

b. Use PSpice/SPICE to check your results in part (a).

2.37 One transconductance amplifier is cascaded with a transimpedance amplifier, as shown in Fig. P2.37. The parameters are $R_s = 5$ kΩ, $R_{i1} = 50$ kΩ, $R_{o1} = 200$ Ω, $Z_{mo} = 10$ kV/A, $R_{i2} = 1$ MΩ, $R_{o2} = 100$ kΩ, $R_L = 1$ kΩ, and $G_{ms} = 20$ mA/V.

a. Calculate the overall open-circuit voltage gain $A_{vo} = v_o/v_i$, the effective voltage gain $A_v = v_o/v_s$, the overall current gain $A_i = i_o/i_i$, and the power gain $A_p = P_L/P_i$.

b. Use PSpice/SPICE to check your results in part (a).

FIGURE P2.37

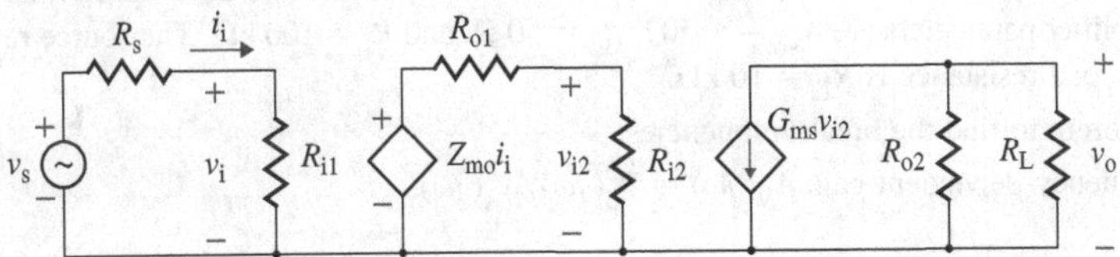

2.5 Frequency Response of Amplifiers

2.38 The voltage amplifier shown in Fig. 2.17(a) is required to have a midrange voltage gain of $A_{v(mid)} = -50$ in the frequency range of 10 kHz to 50 kHz. The source resistance is $R_s = 1\ \text{k}\Omega$, and the load resistance is $R_L = 5\ \text{k}\Omega$.

D

a. Determine the specifications of the amplifier and the values for coupling capacitor C_1 and shunt capacitor C_2.

b. Use PSpice/SPICE to verify your design by plotting the frequency response $|A_v(j\omega)|$ against frequency.

2.39 The voltage gain of an amplifier is given by

$$A_v(j\omega) = \frac{100(10 + j\omega)}{(100 + j\omega)(10^4 + j\omega)}$$

Calculate **(a)** the cutoff frequencies f_L and f_H, **(b)** the bandwidth BW $= f_H - f_L$, and **(c)** the passband gain in decibels.

2.40 The voltage gain of an amplifier is given by

$$A_v(j\omega) = \frac{200}{1 + j\omega/100}$$

Calculate **(a)** the bandwidth BW frequency if $|A_v(j\omega)| = 100$ and **(b)** the bandwidth BW frequency if $|A_v(j\omega)| = 50$.

2.41 A low-pass transconductance amplifier is shown in Fig. P2.41. The circuit parameters are $C = 0.1\ \mu\text{F}$, $R_s = 5\ \text{k}\Omega$, $G_{ms} = 20\ \text{mA/V}$, $R_i = 500\ \text{k}\Omega$, and $R_o = 50\ \text{k}\Omega$. Calculate the unity-gain bandwidth $f_{bw} = A_{v(mid)} f_H$ for **(a)** $R_L = 1\ \text{k}\Omega$ and **(b)** $R_L = 10\ \text{k}\Omega$.

FIGURE P2.41

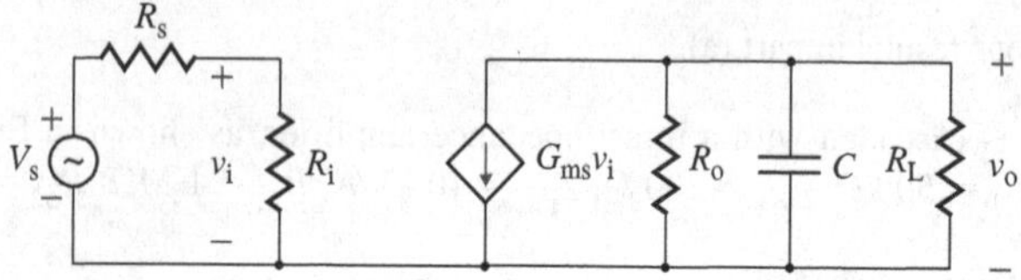

2.6 Miller's Theorem

2.42 A capacitor of $C = 0.01\ \mu\text{F}$ is connected across the input and output sides of an amplifier, as shown in Fig. P2.42. The amplifier parameters are $A_{vo} = -502$, $R_o = 50\ \Omega$, and $R_i = 100\ \text{k}\Omega$. The source resistance is $R_s = 2\ \text{k}\Omega$, and the load resistance is $R_L = 10\ \text{k}\Omega$.

a. Use Miller's theorem to find the break frequencies.

b. Express the frequency-dependent gain $A_v(j\omega) = V_o(j\omega)/V_s(j\omega)$.

FIGURE P2.42

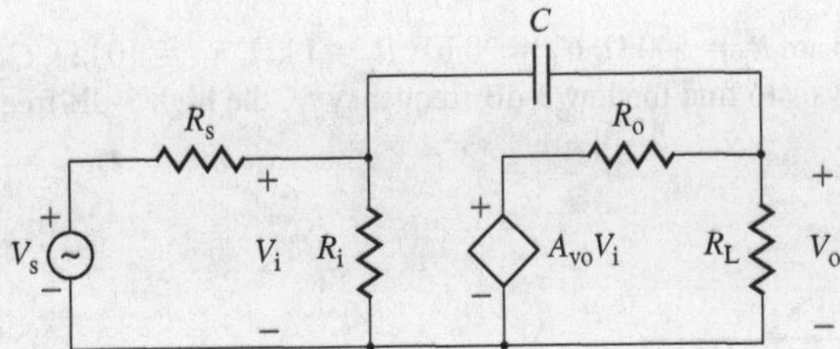

2.43 A capacitor of $C = 10$ nF is connected across the input and output sides of an amplifier, as shown in Fig. P2.42. The amplifier parameters are $A_{vo} = -1000$, $R_o = 100\ \Omega$, and $R_i = 200\ \text{k}\Omega$. The source resistance is $R_s = 5\ \text{k}\Omega$, and the load resistance is $R_L = 5\ \text{k}\Omega$.

a. Use Miller's theorem to find the break frequencies.

b. Express the frequency-dependent gain $A_v(j\omega) = V_o(j\omega)/V_s(j\omega)$.

2.44 A capacitor of $C = 0.1\ \mu\text{F}$ is connected across the input and output sides of an amplifier, as shown in Fig. P2.44(a). Determine the equivalent Miller capacitance C_x seen by the source, as shown in Fig. P2.44(b), for **(a)** $A_{vo} = -200$ and **(b)** $A_{vo} = -1$.

FIGURE P2.44

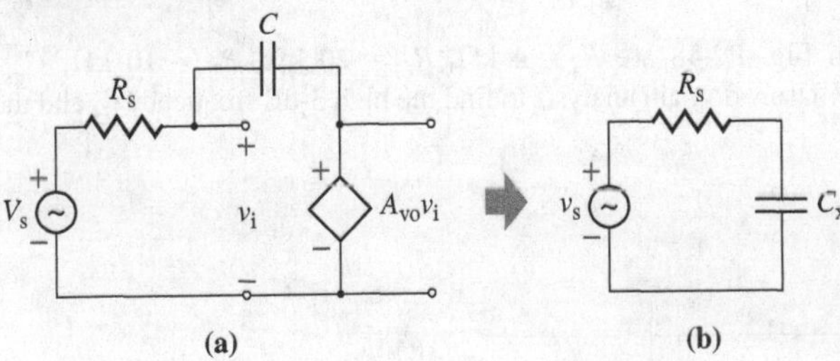

2.45 A resistance R_F is connected across the input and output sides of an amplifier, as shown in Fig. P2.45. The circuit parameters are $R_s = 1\ \text{k}\Omega$, $A_{vo} = -2 \times 10^5$, $R_i = 2\ \text{M}\Omega$, $R_o = 75\ \Omega$, $R_F = 20\ \text{k}\Omega$, and $R_L = 5\ \text{k}\Omega$.

a. Use Miller's theorem to find the effective voltage gain $A_v = v_o/v_s$.

b. Use PSpice/SPICE to check your results in part (a).

FIGURE P2.45

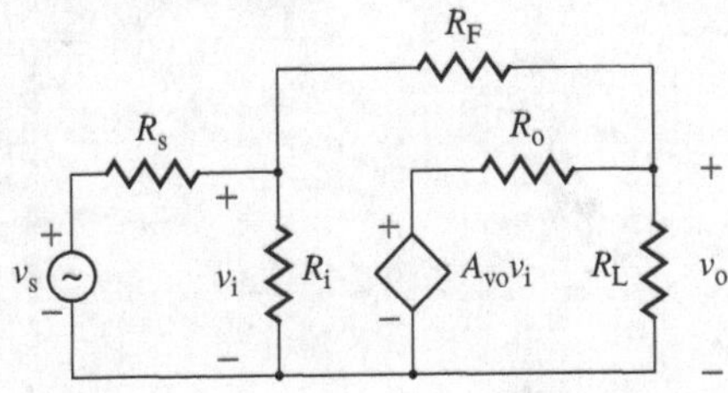

2.7 Frequency Response Methods

2.46 The parameters of the circuit in Fig. P2.46 are $R_s = 500\ \Omega$, $C_1 = 20\ \mu F$, $R_i = 1\ k\Omega$, $R_L = 10\ k\Omega$, $C_2 = 10\ pF$, and $g_m = 15\ mA/V$. Use *s*-domain analysis to find the low 3-dB frequency f_L, the high 3-dB frequency f_H, and the midband gain $A_{v(mid)} = A_{PB}$.

FIGURE P2.46

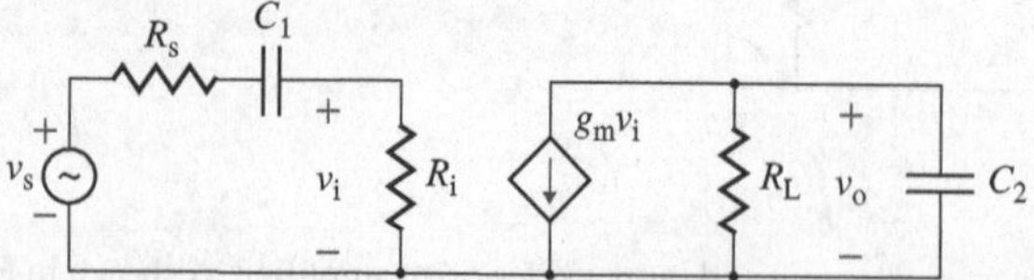

2.47 The parameters of the circuit in Fig. P2.47 are $R_s = 1\ k\Omega$, $C_1 = 10\ \mu F$, $C_i = 20\ pF$, $R_i = 25\ k\Omega$, $R_L = 10\ k\Omega$, $R_o = 10\ k\Omega$, $C_o = 10\ pF$, and $g_m = 15\ mA/V$. Use *s*-domain analysis to find the low 3-dB frequency f_L, the high 3-dB frequency f_H, and the midband gain $A_{v(mid)} = A_{PB}$.

FIGURE P2.47

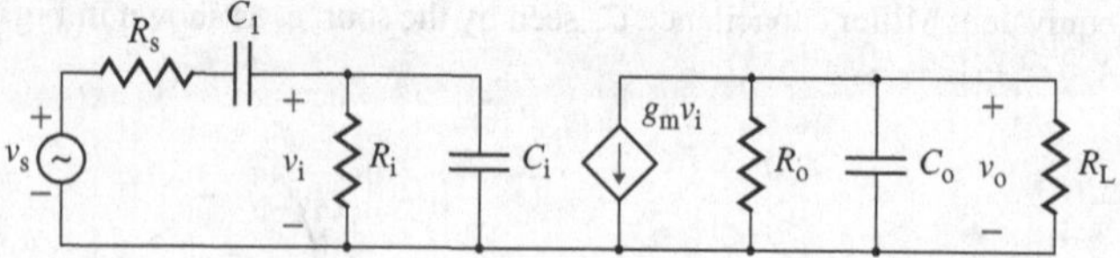

2.48 The parameters of the circuit in Fig. P2.48 are $R_s = 4\ k\Omega$, $R_G = 20\ k\Omega$, $R_L = 10\ k\Omega$, $C_{gs} = 10\ pF$, $C_{gd} = 20\ pF$, and $g_m = 10\ mA/V$. Use *s*-domain analysis to find the high 3-dB frequency f_H and the low-pass gain $A_{v(low)} = A_{PB}$.

FIGURE P2.48

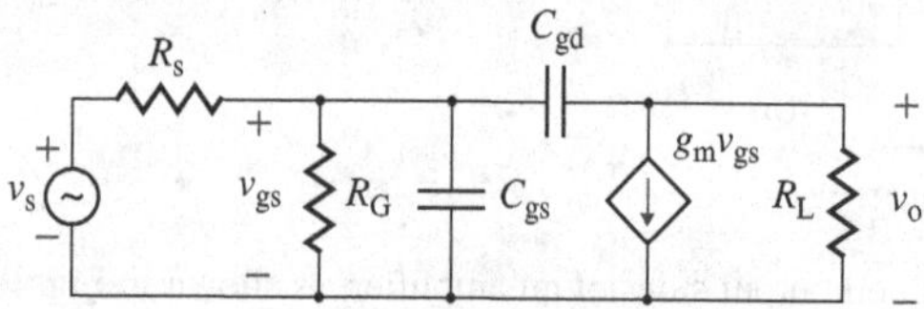

2.49 An amplifier circuit is shown in Fig. P2.49. Use the zero-value method to find the high 3-dB frequency f_H and the low-pass gain $A_{v(low)} = A_{PB}$.

FIGURE P2.49

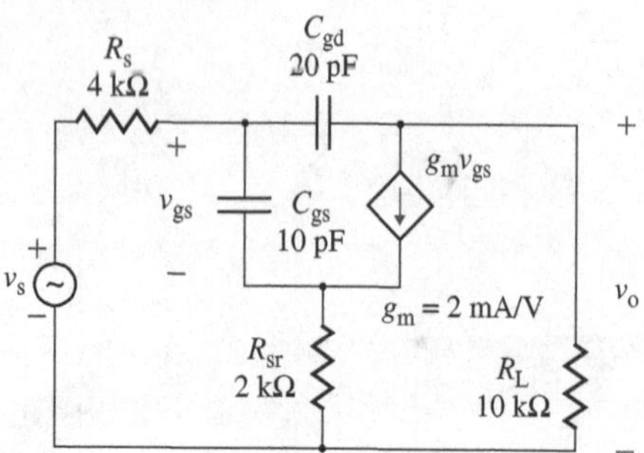

2.50 Repeat Prob. 2.49 for $R_{sr} = 0$.

2.51 An amplifier circuit is shown in Fig. P2.51. Use the short-circuit and zero-value methods to find the low
P 3-dB frequency f_L, the high 3-dB frequency f_H, and the midband gain $A_{v(mid)} = A_{PB}$.

FIGURE P2.51

2.52 Repeat Prob. 2.51 for $R_E = 0$.

2.53 An amplifier circuit is shown in Fig. P2.53. Use the short-circuit and zero-value methods to find the low
P 3-dB frequency f_L, the high 3-dB frequency f_H, and the midband gain $A_{v(mid)} = A_{PB}$.

FIGURE P2.53

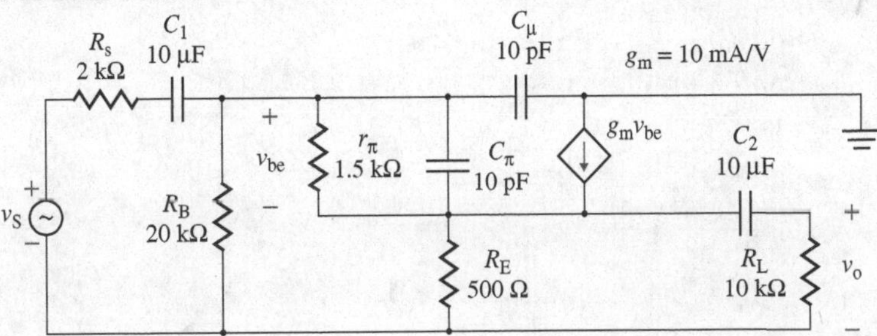

2.54 An amplifier circuit is shown in Fig. P2.54. Use the short-circuit and zero-value methods to find the low
P 3-dB frequency f_L, the high 3-dB frequency f_H, and the midband gain $A_{v(mid)} = A_{PB}$.

FIGURE P2.54

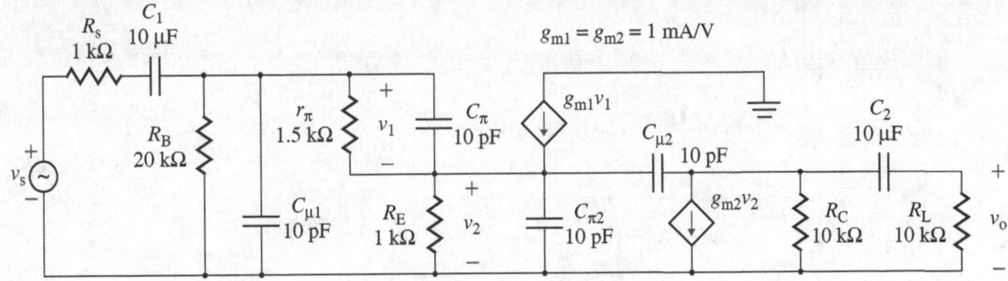

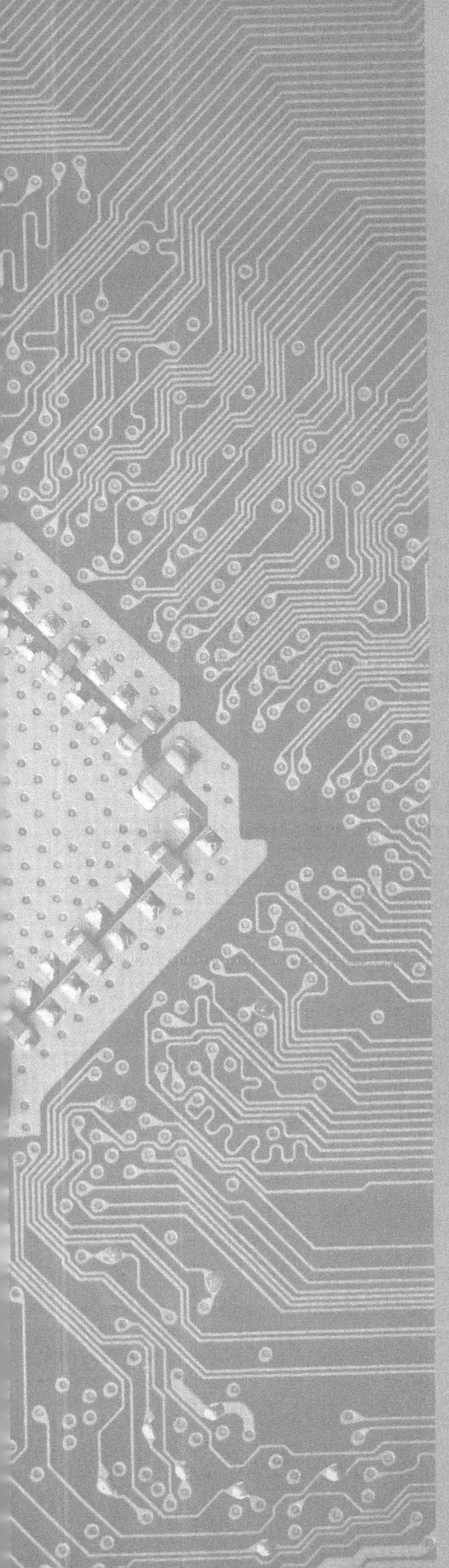

CHAPTER 3

INTRODUCTION TO OPERATIONAL AMPLIFIERS AND APPLICATIONS

Learning Outcomes

After completing this chapter, students should be able to do the following:

- Describe the external characteristics and circuit models of op-amps.
- Analyze op-amp circuits to derive their input and output relationships.
- Calculate the effect of a finite op-amp gain on the overall voltage gain of op-amp circuits.
- Determine the bandwidth of op-amp circuits.
- Design op-amp circuits to meet certain input and output specifications.
- List a few examples of op-amp applications in signal conditioning.

Symbols and Their Meanings

Symbol	**Meaning**
A_o, A_f	Op-amp open-loop voltage gain and overall (closed-loop) voltage gain of an op-amp circuit
BW, A_{PB}	Bandwidth and pass-band voltage gain
CMRR	Common-mode rejection ratio
f_H, f_L	High and low cutoff or break frequencies

Symbol	Meaning
R_i, R_o	Input and output resistances
v_p (or v_+), v_n (or v_-)	Voltages at the noninverting and inverting terminals of an op-amp
$v_S(t)$, $v_O(t)$	Instantaneous input signal and output voltages
$v_s(t)$, $v_o(t)$	Small-signal input signal and output voltages
V_S, V_O	DC input signal and output voltages
V_{CC}, V_{EE}	Positive and negative DC supply voltages

3.1 Introduction

The operational amplifier (or op-amp) is a high-gain, direct-coupled amplifier consisting of multiple stages: an input stage to provide a high input resistance with a certain amount of voltage gain, a middle stage to provide a high voltage gain, and an output stage to provide a low output resistance. It operates with a differential voltage between two input terminals, and it is a complete, integrated-circuit, prepackaged amplifier. An op-amp, often referred to as a linear (or analog) integrated circuit (IC), is a popular and versatile integrated circuit. It serves as a building block for many electronic circuits. For most applications, knowledge of the terminal characteristics of op-amps is all you need to design op-amp circuits. However, for some applications requiring precision, internal knowledge of op-amps is necessary.

3.2 Characteristics of Ideal Op-Amps

The symbol for an op-amp is shown in Fig. 3.1. An op-amp has at least five terminals. Terminal 2 is called the "inverting input" because the output that results from the input at this terminal will be inverted. Terminal 3 is called the "noninverting input" because the output that results from the input at this terminal will have the same polarity as the input. Terminal 4 is for negative DC supply $-V_{EE}$. Terminal 6 is the output terminal. Terminal 7 is for positive DC supply V_{CC}.

Instead of using two DC power supplies, we can generate V_{CC} and V_{EE} from a single power supply V_{DC}, as shown in Fig. 3.2(a). The value of R should be high enough (usually $R \geq 10\ \text{k}\Omega$) that it does not draw much current from the DC supply V_{DC}. Capacitors are used for decoupling (bypass) of the DC power supply, and the value of C is typically in the range of 0.01 μF to 10 μF. Instead of two resistors, a potentiometer can be used to ensure that $V_{CC} = V_{EE}$, as shown in Fig. 3.2(b). Diodes D_1 and D_2 (see Chapter 4)

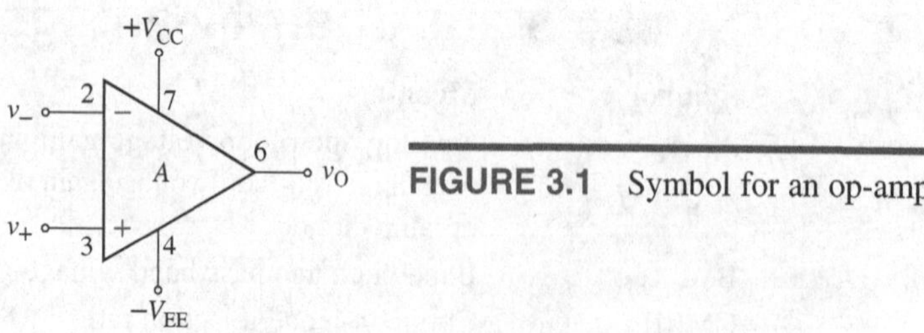

FIGURE 3.1 Symbol for an op-amp

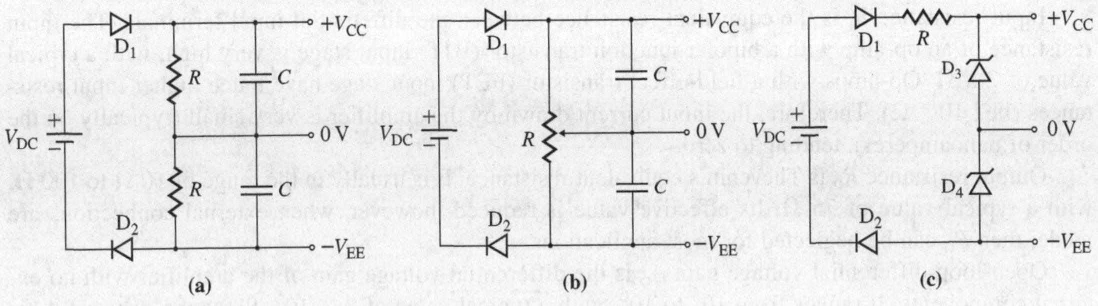

FIGURE 3.2 Arrangements for positive and negative supply voltages

prevent any reverse current flow; they are often used to protect the op-amp in case the positive and negative terminals of the supply voltage V_{DC} are reversed accidentally. Also, two zener diodes (see Chapter 4) can be used to obtain symmetrical supply voltages, as shown in Fig. 3.2(c). The value of R should be low enough to force the zener diodes to operate in the zener or avalanche mode (see Sec. 4.7). Note that these circuits will not work if the DC supply comes with a ground.

3.2.1 Op-Amp Circuit Model

The output voltage of an op-amp is directly proportional to the small-signal differential (or difference) input voltage. Thus, an op-amp can be modeled as a voltage-dependent voltage source; its equivalent circuit is shown in Fig. 3.3(a). The output voltage v_O is given by

$$v_O = A_o v_d = A_o(v_p - v_n) \tag{3.1}$$

where A_o = small-signal open-loop voltage gain
v_d = small-signal differential (or difference) input voltage
v_n = small-signal voltage at the inverting terminal with respect to the ground
v_p = small-signal voltage at the noninverting terminal with respect to the ground

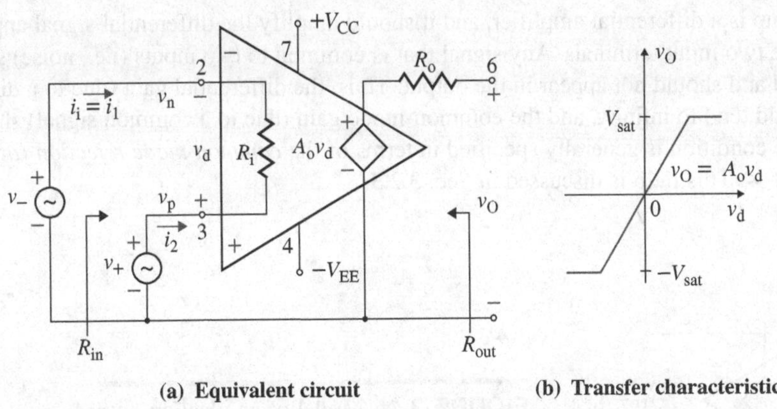

(a) Equivalent circuit **(b) Transfer characteristic**

FIGURE 3.3 Equivalent circuit of an op-amp

Input resistance R_i is the equivalent resistance between the differential input terminals. The input resistance of an op-amp with a bipolar junction transistor (BJT) input stage is very high, with a typical value of 2 MΩ. Op-amps with a field-effect transistor (FET) input stage have much higher input resistances (i.e., 10^{12} Ω). Therefore, the input current drawn by the amplifier is very small (typically on the order of nanoamperes), tending to zero.

Output resistance R_o is Thevenin's equivalent resistance. It is usually in the range of 10 Ω to 100 Ω, with a typical value of 75 Ω. Its effective value is reduced, however, when external connections are made; then R_o can be neglected for most applications.

Open-loop differential voltage gain A_o is the differential voltage gain of the amplifier with no external components. It ranges from 10^4 to 10^6, with a typical value of 2×10^5. Since the value of A_o is very large, v_d becomes very small (typically on the order of microvolts), tending to zero. The transfer characteristic (v_O versus v_d) is shown in Fig. 3.3(b). In reality, the output voltage cannot exceed the positive or negative saturation voltage $\pm V_{sat}$ of the op-amp, which is set by supply voltages V_{CC} and V_{EE}, respectively. The saturation voltage is usually 1 V lower than the supply voltage V_{CC} or V_{EE}. Thus, the output voltage will be directly proportional to the differential input voltage v_d only until it reaches the saturation voltage; thereafter the output voltage remains constant. The gain of practical op-amps is also frequency dependent. Note that the model in Fig. 3.3(a) does not take into account the saturation effect and assumes that gain A_o remains constant for all frequencies.

The analysis and design of circuits employing op-amps can be greatly simplified if the op-amps in the circuit are assumed to be ideal. Such an assumption allows you to approximate the behavior of the op-amp circuit and to obtain the approximate values of circuit components that will satisfy some design specifications. Although the characteristics of practical op-amps differ from the ideal characteristics, the errors introduced by deviations from the ideal conditions are acceptable in most applications. A complex op-amp model is used in applications requiring precise results. The circuit model of an ideal op-amp is shown in Fig. 3.4; its characteristics are as follows:

- The open-loop voltage gain is infinite: $A_o = \infty$.
- The input resistance is infinite: $R_i = \infty$.
- The amplifier draws no current: $i_i = 0$.
- The output resistance is negligible: $R_o = 0$.
- The gain A_o remains constant and is not a function of frequency.
- The output voltage does not change with changes in power supplies. This condition is generally specified in terms of the *power supply sensitivity* (PSS): PSS = 0.
- An op-amp is a differential amplifier, and it should amplify the differential signal appearing between the two input terminals. Any signal that is common to two inputs (i.e., noise) should not be amplified and should not appear in the output. Thus, the differential gain (due to a differential signal) should tend to infinity, and the common-mode gain (due to a common signal) should tend to zero. The condition is generally specified in terms of the *common-mode rejection ratio* (CMRR): CMRR = ∞. This ratio is discussed in Sec. 3.2.3.

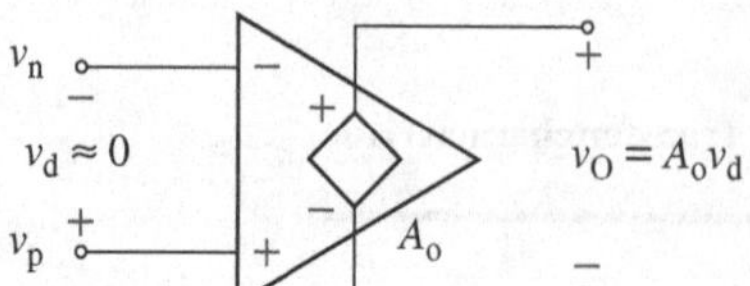

FIGURE 3.4 Model of an ideal op-amp

EXAMPLE 3.1

Finding the differential input voltage and input current of an op-amp The op-amp of Fig. 3.3(a) has an open-loop gain of $A_o = 2 \times 10^5$. The input resistance is $R_i = 0.6\ \text{M}\Omega$. The DC supply voltages are $V_{CC} = 12$ V and $-V_{EE} = -12$ V. Assume that $V_{sat} = 11$ V.

(a) What value of v_d will saturate the amplifier?

(b) What are the values of the corresponding input current i_i?

SOLUTION

(a) $v_d = \pm V_{sat}/A_o = \pm 11/(2 \times 10^5) = \pm 55\ \mu\text{V}$

(b) $i_i = -v_d/R_i = \mp 55\ \mu\text{V}/0.6\ \text{M}\Omega = \mp 0.1\ \text{nA}$

EXAMPLE 3.2

Finding the maximum output voltage of an op-amp The op-amp of Fig. 3.3(a) has $A_o = 2 \times 10^5$, $R_i = 2\ \text{M}\Omega$, $R_o = 75\ \Omega$, $V_{CC} = 12$ V, and $-V_{EE} = -12$ V. The maximum possible output voltage swing is ± 11 V. If $v_- = 100\ \mu$V and $v_+ = 25\ \mu$V, determine the output voltage v_O.

SOLUTION

From Eq. (3.1),

$$v_O = A_o(v_+ - v_-) = 2 \times 10^5 \times (25 - 100) \times 10^{-6} = -15\ \text{V}$$

Because of the saturation, the output voltage cannot exceed the maximum voltage limit of -11 V, and therefore $v_O = -11$ V.

3.2.2 Op-Amp Frequency Response

The differential voltage gain of an op-amp has the highest value at DC or low frequencies. The gain decreases with frequency. A typical frequency response is shown in Fig. 3.5. The gain falls uniformly with a slope of -20 dB/decade. This uniform slope is maintained by internal design in internally compensated op-amps. The voltage gain of an internally compensated op-amp at frequency f can usually be expressed as

$$A_o(j\omega) = \frac{A_o}{1 + j\omega/\omega_b} = \frac{A_o}{1 + jf/f_b} \tag{3.2}$$

where A_o is DC gain, typically 2×10^5 and f_b is break (or 3-dB) frequency in hertz.

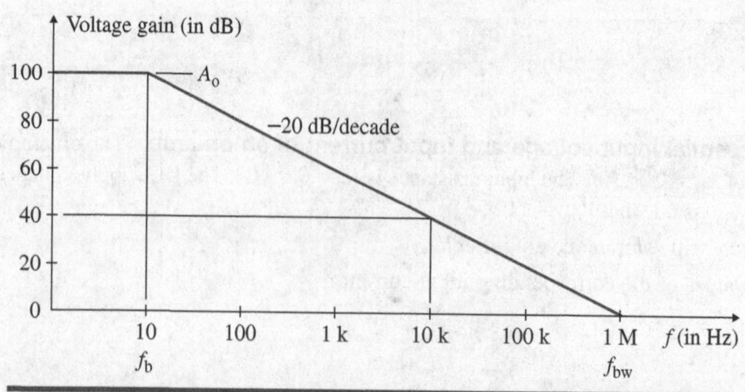

FIGURE 3.5 Voltage gain of an internally compensated op-amp

For $f >> f_b$ and $(f/f_b) >> 1$, Eq. (3.2) is reduced to

$$A_o(j\omega) = \frac{A_o}{if/f_b} = \frac{A_o f_b}{if} \tag{3.3}$$

The magnitude of this gain becomes unity (or 0) at frequency $f = f_{bw}$. That is,

$$f_{bw} = A_o f_b \tag{3.4}$$

where f_{bw} is called the *unity-gain bandwidth*. The typical value of f_{bw} for the LF411 op-amp is 4 MHz. The 3-dB frequency can be related to time constant τ or to rise time t_r by

$$f = \frac{1}{2\pi\tau} = \frac{2.2}{2\pi t_r} = \frac{0.35}{t_r} \tag{3.5}$$

Thus, the frequency response is inversely proportional to the rise time t_r. The input signal frequency f_s should be less than the maximum op-amp frequency; otherwise the output voltage will be distorted. For example, if the rise time of an input signal is $t_r = 0.1$ μs, its corresponding input frequency is $f_s = 0.35/0.1$ μs = 3.5 MHz, and the output voltage will be distorted in an op-amp of $f_{bw} = 1$ MHz.

3.2.3 Common-Mode Rejection Ratio

Because an op-amp is a differential amplifier, it should amplify the differential voltage between the input terminals. Any signal (i.e., noise) that appears simultaneously at both inputs should not be amplified. Let v_1 and v_2 be the input voltages at the noninverting and inverting terminals, respectively, as shown in Fig. 3.6(a). As shown in Fig. 3.6(b), these voltages can be resolved into two components: differential voltage v_d and common-mode voltage v_c.

Let us define the differential voltage v_d as

$$v_d = v_2 - v_1 \tag{3.6}$$

and the common-mode voltage v_c as

$$v_c = \frac{v_1 + v_2}{2} \tag{3.7}$$

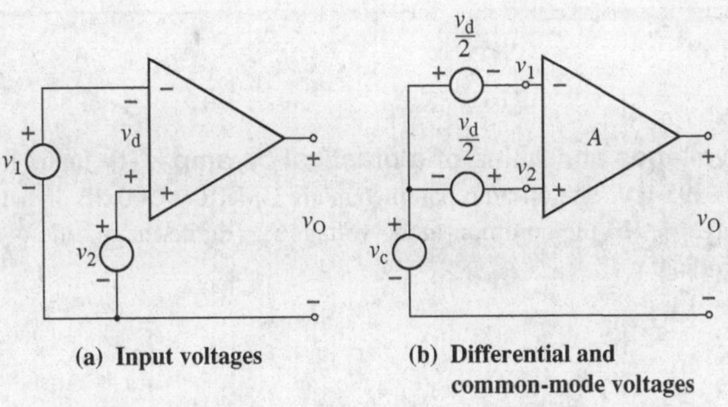

FIGURE 3.6 Op-amp with differential and common-mode inputs

Then, the two input voltages can be expressed as

$$v_2 = v_c + \frac{v_d}{2} \tag{3.8}$$

$$v_1 = v_c - \frac{v_d}{2} \tag{3.9}$$

Let A_1 be the voltage gain with an input at the inverting terminal and the noninverting terminal grounded. Let A_2 be the voltage gain with an input at the noninverting terminal and the inverting terminal grounded. We can obtain the output voltage of the op-amp by applying the superposition theorem. That is,

$$v_O = A_1 v_1 + A_2 v_2 \tag{3.10}$$

Substituting v_1 from Eq. (3.9) and v_2 from Eq. (3.8) into Eq. (3.10) yields

$$v_O = A_1\left(v_c - \frac{v_d}{2}\right) + A_2\left(v_c + \frac{v_d}{2}\right)$$

$$= \left(\frac{A_2 - A_1}{2}\right)v_d + (A_2 + A_1)v_c$$

$$= A_d v_d + A_c v_c \tag{3.11}$$

$$= A_d\left(v_d + \frac{A_c}{A_d} v_c\right) \tag{3.12}$$

where $A_d = (A_2 - A_1)/2$ is differential voltage gain and $A_c = (A_2 + A_1)$ is common-mode voltage gain.

According to Eq. (3.12), the output voltage depends on the common-mode voltage v_c and the differential voltage v_d. Since A_1 is negative and A_2 is positive, $A_d > A_c$. If A_d can be made much greater than A_c, $v_O \approx A_d v_d$ and the output voltage will be almost independent of the common-mode signal v_c. The ability of an op-amp to reject the common-mode signal is defined by a performance criterion called the *common-mode rejection ratio* (CMRR), which is defined as the magnitude of the ratio of the voltage gains. That is,

$$\text{CMRR} = \left|\frac{A_d}{A_c}\right| \tag{3.13}$$

$$= 20 \log\left|\frac{A_d}{A_c}\right| \quad \text{(dB)} \tag{3.14}$$

The value of CMRR should ideally be infinity; a typical value is 100 dB for the LF411 op-amp.

EXAMPLE 3.3

Finding the output voltages and gains of a practical op-amp The input voltages of an op-amp are $v_2 = 1005\ \mu V$ and $v_1 = 995\ \mu V$. The op-amp parameters are CMRR = 100 dB and $A_d = 2 \times 10^5$. Determine **(a)** the differential voltage v_d, **(b)** the common-mode voltage v_c, **(c)** the magnitude of the common-mode gain A_c, and **(d)** the output voltage v_O.

SOLUTION

From Eq. (3.14),

$$20 \log (\text{CMRR}) = 100 \text{ dB}$$

or $$\log (\text{CMRR}) = \frac{100}{20} = 5$$

which gives CMRR = $|A_d/A_c| = 10^5$.

(a) The differential voltage v_d is

$$v_d = v_2 - v_1 = 1005\ \mu V - 995\ \mu V = 10\ \mu V$$

(b) From Eq. (3.7), the common-mode voltage is

$$v_c = \frac{(v_1 + v_2)}{2} = \frac{(1005\ \mu V + 995\ \mu V)}{2} = 1000\ \mu V$$

(c) From Eq. (3.13),

$$\left|\frac{A_d}{A_c}\right| = 10^5$$

or $$|A_c| = \frac{|A_d|}{10^5} = \frac{2 \times 10^5}{10^5} = 2$$

(d) From Eq. (3.11), the output voltage v_o becomes

$$v_o = A_d v_d + A_c v_c$$

$$= 2 \times 10^5 \times 10\ \mu V \pm 2 \times 1000\ \mu V = 2 \pm 0.002 = 2.002 \text{ V or } 1.998 \text{ V}$$

NOTE: In the absence of the common-mode signal, $v_c = 0$ and

$$v_o = A_d v_d = 2 \times 10^5 \times 10 \times 10^6 = 2 \text{ V}$$

v_c is 100 times v_d, but the CMRR introduces only a 0.1% error in the output voltage. Therefore, the effect of the common-mode signal can be neglected. An ideal op-amp will have CMRR = ∞ so that $v_o = A_d v_d$.

KEY POINTS OF SECTION 3.2

- An op-amp is a direct-coupled differential amplifier. It has a high gain (typically 2×10^5), a high input resistance (typically 1 MΩ), and a low output resistance (typically 50 Ω).
- An ideal op-amp has the characteristics of infinite gain, infinite input resistance, and zero output resistance.
- An op-amp requires DC power supplies, and the maximum output voltage swing is limited to the DC supply voltages.

3.3 Op-Amp PSpice/SPICE Models

There are many types of op-amps, as we will see in Chapter 14. An op-amp can be simulated from its internal circuit arrangement. The internal structure of op-amps is very complex, however, and differs from one model to another. For example, the μA741 type of general-purpose op-amp consists of 24 transistors. It is too complex for the student version of the PSpice circuit simulation software to analyze; however, a macromodel, which is a simplified version of the op-amp and requires only two transistors, is quite accurate for many applications, and can be simulated as a subcircuit or a library file [1, 2]. Some manufacturers of op-amps supply macromodels of their products [3]. The student version of PSpice has a library called NOM.LIB, which contains models of three common types of op-amps: μA741, LM324, and LF411. The parameters of the three op-amps for the circuit model in Fig. 3.3(a) are as follows:

- The μA741 op-amp is a general-purpose op-amp with a BJT input stage. It is capable of producing output voltages of ±14 V with DC power supply voltages of ±15 V. The parameters are $R_i = 2$ MΩ, $R_o = 75$ Ω, $A_o = 2 \times 10^5$, break frequency $f_b = 10$ Hz, and unity-gain bandwidth $f_{bw} = 1$ MHz.
- The LF411 op-amp is a general-purpose op-amp with an FET input stage. It is capable of producing output voltages of ±13.5 V with DC power supply voltages of ±15 V. The parameters are $R_i = 10^{12}$ Ω, $R_o = 50$ Ω, $A_o = 2 \times 10^5$, break frequency $f_b = 20$ Hz, and unity-gain bandwidth $f_{bw} = 4$ MHz.
- The LM324 op-amp has a BJT input stage and is used with a single DC power supply voltage. It can produce output voltages in the range from approximately 20 mV to 13.5 V with a DC supply voltage of +15 V. The parameters are $R_i = 2$ MΩ, $R_o = 50$ Ω, $A_o = 2 \times 10^5$, break frequency $f_b = 4$ kHz, and unity-gain bandwidth $f_{bw} = 1$ MHz.

The professional version of PSpice supports library files for many devices. It is advisable to check the name of the current library file by listing the files of the PSpice programs.

If the PSpice/SPICE model of an op-amp is not available, it is possible to represent the op-amp by simple models that give reasonable results, especially for determining the approximate design values of op-amp circuits. PSpice/SPICE models can be classified into three types: DC linear models, AC linear models, and nonlinear macromodels. Taking the μA741 op-amp as an example, we will develop simple PSpice/SPICE models of these three types.

3.3.1 DC Linear Model

An op-amp may be modeled as a voltage-controlled voltage source, as shown in Fig. 3.7. Two zener diodes (see Sec. 4.7) are connected back to back in order to limit the output swing to the saturation voltages (say, between −14 V and +14 V). This simple model, which assumes that the voltage gain is independent of

FIGURE 3.7 DC linear model

the frequency, is suitable only for DC or low-frequency applications. The list of the PSpice/SPICE subcircuit UA741_DC for Fig. 3.7 is shown here:

```
* Subcircuit definition for UA741_DC
.SUBCKT UA741_DC      1   2   3   4
*   Subcircuit name  Vi+ Vi- Vo+ Vo-
RI   1  2  2MEG                ; Input resistance
RO   5  3  75                  ; Output resistance
EA   5  4  1   2  2E+5         ; Voltage-controlled voltage source
D1   3  6  DMOD                ; Zener diode with model DMOD
D2   4  6  DMOD                ; Zener diode with model DMOD
.MODEL DMOD D (BV=14V)         ; Ideal zener model with a zener voltage of 14 V
.ENDS UA741_DC                 ; End of subcircuit definition
```

In PSpice/SPICE, the name of a subcircuit must begin with X. For example, the calling statement for the amplifier A1, which uses the subcircuit UA741_DC, is as follows:

```
XA1        7         8         9         10        UA741_DC
*          Vi+       Vi-       Vo+       Vo-       Subcircuit name
```

This subcircuit definition UA741_DC can be inserted into the circuit file. Alternatively, it can reside in a user-defined file, say USER.LIB in C drive, in which case the circuit file must contain the following statement:

```
C:USER.LIB ; Library file name must include the drive and directory location
```

3.3.2 AC Linear Model

The frequency response of internally frequency-compensated op-amps can be approximated by a single break frequency, as shown in Fig. 3.8(a). This characteristic can be modeled by the circuit of Fig. 3.8(b), which is a frequency-dependent model of an op-amp. The dependent sources have a common node, 4. Without this common node, PSpice/SPICE will give an error message because there will be no DC path from the nodes of the dependent current source to the ground. The common node could be either with the input stage or with the output stage. The time constant $\tau = R_1C_1$ gives the break frequency f_b. If an op-amp has more than one break frequency, it can be represented by using as many capacitors as there are breaks. R_i and R_o are the input and output resistances, respectively. A_o is the open-circuit DC voltage gain. Two zener diodes are connected back to back in order to limit the output swing to the saturation voltages (say, between −14 V and +14 V).

The no-load output voltage can be expressed in Laplace's domain as

$$V_o(s) = A_oV_2(s) = \frac{A_oR_1I_1}{1 + R_1C_1s} = \frac{A_oV_d}{1 + R_1C_1s} \tag{3.15}$$

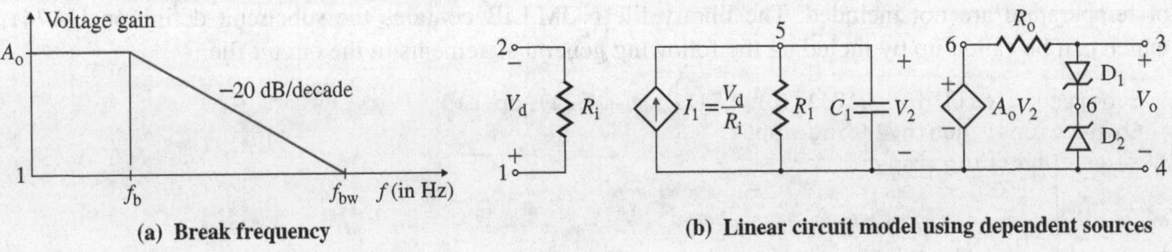

FIGURE 3.8 AC linear model with a single break frequency

Substituting $s = j\omega = j2\pi f$ into Eq. (3.15) gives

$$A_o(j\omega) = \frac{A_oV_d}{1 + j2\pi fR_1C_1} = \frac{A_oV_d}{1 + jf/f_b} \tag{3.16}$$

which gives the frequency-dependent open-loop voltage gain of an op-amp with a single break frequency as

$$A_o(j\omega) = \frac{V_o}{V_d} = \frac{A_o}{1 + jf/f_b} \tag{3.17}$$

where $f_b = 1/(2\pi R_1C_1)$ is break frequency in hertz and A_o is large-signal (or DC) voltage gain of the op-amp.

For μA741 op-amps, $f_b = 10$ Hz, $A_o = 2 \times 10^5$, $R_i = 2$ MΩ, and $R_o = 75$ Ω. If we let $R_1 = 10$ kΩ (used as a typical value), $C_1 = 1/(2\pi \times 10 \times 10 \times 10^3) = 1.5619$ μF.

Note that we could also choose a different value of R_1. The list of the PSpice/SPICE subcircuit UA741_AC for Fig. 3.8(b) is shown here:

```
* Subcircuit definition for UA741_AC
.SUBCKT UA741_AC  1   2   3   4
* Subcircuit name Vi+ Vi- Vo+ Vo-
RI  1  2  2MEG              ; Input resistance
RO  6  3  75                ; Output resistance
GB  4  5  1  2  0.1M        ; Voltage-controlled current source
R1  5  4  10K
C1  5  4  1.5619UF
EA  6  4  5  4  2E+5        ; Voltage-controlled voltage source
D1  3  7  DMOD              ; Zener diode with model DMOD
D2  4  7  DMOD              ; Zener diode with model DMOD
.MODEL DMOD D (BV=14V)      ; Ideal zener model with a zener voltage of 14 V
.ENDS UA741_AC              ; End of subcircuit definition
```

3.3.3 Nonlinear Macromodel

The subcircuit definitions of op-amp macromodels are described by a set of .MODEL statements. The macromodels are normally simulated at room temperature and contain nominal values. The effects of temperature are not included. The library file NOM.LIB contains the subcircuit definition UA741, which can be called up by including the following general statements in the circuit file:

```
* Subcircuit call for UA741 (or LF411 or LM324) op-amp
* Connections: noninverting input
*    |  Inverting input
*    |  |
*    |  |  Positive power supply
*    |  |  |  Negative power supply
*    |  |  |  |  Output
*    |  |  |  |  |        Subcircuit name
XA1  1  2  4  5  6  UA741 (or LF411 or LM324)  ; Subcircuit calling must begin with X
*   Vi+ Vi- Vp+ Vp- Vout                        ; Calling UA741 for amplifier A1
.LIB NOM.LIB                                    ; Calling library file NOM.LIB
```

▶ **NOTE** With PSpice/OrCAD schematics, there is no need for developing subcircuit definitions of an op-amp macromodel. PSpice or OrCAD automatically creates the macromodel from the op-amp schematic.

KEY POINTS OF SECTION 3.3

- An op-amp can be represented in PSpice/SPICE by one of three models: (a) a DC linear model, which is simple but suitable only for low frequencies (typically less than 20 Hz); (b) an AC linear model, which is simple and frequency dependent; (c) a nonlinear macromodel, which is more complex.
- The student version of PSpice limits the number of active devices and nodes, allowing only one macromodel in a circuit. Thus, the choice of a model depends on the complexity of the circuit; the preferred model is the macromodel, followed by the AC model and then the DC model.

3.4 Analysis of Ideal Op-Amp Circuits

In Eq. (3.1) there are three possible conditions for the output voltage v_O: (a) if $v_n = 0$, v_O will be positive ($v_O = A_o v_p$); (b) if $v_p = 0$, v_O will be negative ($v_O = -A_o v_n$); or (c) if both v_p and v_n are present, $v_O = A_o(v_p - v_n)$. Therefore, depending on the conditions of the input voltages, op-amp circuits can be classified into three basic configurations: noninverting amplifiers, inverting amplifiers, or differential (or difference) amplifiers (see Sec. 3.5.3).

3.4.1 Noninverting Amplifiers

The configuration of a noninverting amplifier is shown in Fig. 3.9(a). The input voltage v_S is connected to the noninverting terminal. The voltage v_x, which is proportional to the output voltage, is connected via R_1 and R_F to the inverting terminal. Using Kirchhoff's voltage law (KVL), we get

$$v_S = v_x + v_d$$

The differential voltage v_d, given by

$$v_d = v_S - v_x$$

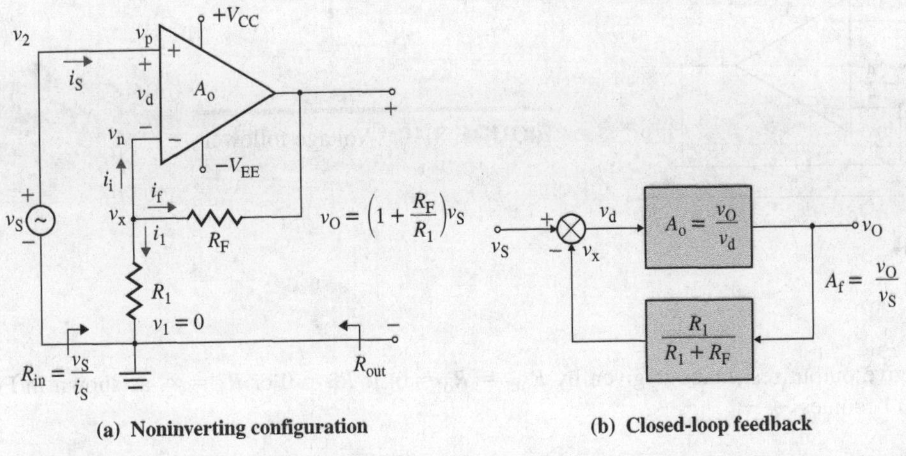

FIGURE 3.9 Noninverting amplifier

is then amplified by the op-amp, whose output is then fed back to the inverting terminal. Thus, this is a feedback circuit; the block diagram is shown in Fig. 3.9(b). We will cover feedback in Chapter 10.

Let us assume an ideal op-amp. That is, $v_d = 0$, $i_S = 0$, and $A_o \approx \infty$. The voltage v_x at the inverting terminal is

$$v_x = v_S - v_d \approx v_S$$

Using Kirchhoff's current law (KCL) at the inverting terminal, we get

$$i_1 + i_f + i_i = 0$$

Since the current i_i drawn by an ideal op-amp is zero, $i_1 = -i_f$. That is,

$$\frac{v_x}{R_1} = -\frac{v_x - v_O}{R_F} \quad \text{or} \quad \frac{v_S}{R_1} = -\frac{v_S - v_O}{R_F}$$

which, after simplification, yields

$$v_O = \left(1 + \frac{R_F}{R_1}\right)v_S$$

giving the closed-loop voltage gain A_f as

$$A_f = \frac{v_O}{v_S} = 1 + \frac{R_F}{R_1} \tag{3.18}$$

Since the current drawn by the amplifier is zero, the effective input resistance of the amplifier is very high, tending to infinity:

$$R_{in} = \frac{v_S}{i_S} = \infty$$

▶ **NOTE** v_d and i_s tend to be close to zero due to the large op-amp gain A_o, not zero. Otherwise, the circuit will not work as expected.

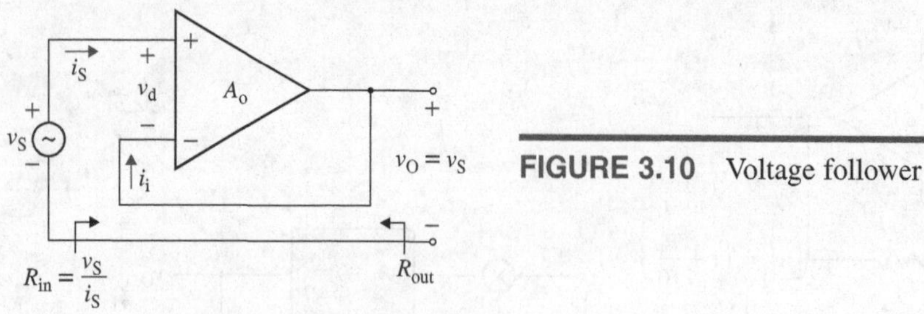

FIGURE 3.10 Voltage follower

The effective output resistance is given by $R_{out} = R_o \approx 0$. If $R_F = 0$ or $R_1 = \infty$, as shown in Fig. 3.10, Eq. (3.18) becomes

$$A_f = 1 \tag{3.19}$$

That is, the output voltage equals the input voltage: $v_O = v_S$. The circuit of Fig. 3.10 is commonly referred to as a *voltage follower* because its output voltage follows the input voltage. It has the inherent characteristics of a high input impedance (or resistance, typically 10^{10} Ω) and a low output impedance (or resistance, typically 50 mΩ). The exact values can be found by applying the feedback analysis techniques discussed in Chapter 10. A voltage follower is commonly used as the *buffer stage* between a low impedance load and a source requiring a high impedance load.

CMRR of a Noninverting Amplifier

If CMRR_o is the CMRR of the op-amp, we can find out the CMRR of the noninverting amplifier from the following derivations. Let v_{id} and v_{icm} denote the difference and common-mode signals of the amplifier, respectively. Considering two input signals v_1 and v_2 to the noninverting amplifier as shown in Fig. 3.9(a), we have $v_{id} = (v_2 - v_1) = v_2 = v_s$ and $v_{icm} = (v_2 + v_1)/2 = v_2/2 = v_S/2$ since $v_1 = 0$.

The voltage v_p at the noninverting terminal is $v_p = v_2$, and the voltage v_n at the inverting terminal can be expressed in terms of v_O by

$$v_n = \frac{R_1}{R_1 + R_F} v_O$$

Similarly, let v_{do} and v_{cmo} denote the difference and common-mode signals at the op-amp input. Then, we have $v_{do} = (v_p - v_n)$ and $v_{cmo} = (v_p + v_n)/2$. Thus, the output voltage v_o can be expressed in terms of the op-amp differential gain A_d and common-mode gain A_{cm} as

$$v_o = A_d v_{do} + A_{cm} v_{cmo} = A_d(v_p - v_n) + \frac{A_{cm}(v_p + v_n)}{2}$$

After substituting for v_p ($v_S = 2v_{icm} = v_{id}$) and v_n using the relationships in terms of v_{id} and v_{icm}, we can find the following expression for v_O:

$$v_o = A_d(v_p - v_n) + \frac{A_{cm}}{2}(2v_{icm} + v_n) = A_d\left(v_{id} - v_o\frac{R_1}{R_1 + R_F}\right) + A_{cm}\left(v_{icm} + v_O\frac{R_1/2}{R_1 + R_F}\right) \tag{3.20}$$

This can be solved for v_O after we collect all the terms containing v_o in terms of v_{id} and v_{cm}:

$$v_o = \frac{A_d(1 + R_F/R_1)v_{id} + A_{cm}(1 + R_F/R_1)v_{icm}}{1 + R_F/R_1 + A_d - A_{cm}/2} \tag{3.21}$$

Since $A_d >> A_{cm}$ and $A_d >> R_F/R_1 >> 1$, the denominator of Eq. (3.21) is approximately equal to A_d, and v_o in Eq. (3.21) can be approximated to

$$v_o \approx \frac{A_d}{A_d}\left(1 + \frac{R_F}{R_1}\right)v_{id} + \frac{A_{cm}}{A_d}\left(1 + \frac{R_F}{R_1}\right)v_{icm} = \left(1 + \frac{R_F}{R_1}\right)v_{id} + \frac{1}{\text{CMRR}_o}\left(1 + \frac{R_F}{R_1}\right)v_{icm} \tag{3.22}$$

where $\text{CMRR}_o = A_d/A_{cm}$ is the CMRR of the op-amp. From Eq. (3.22), we can find the differential voltage gain $A_{d\text{-amp}}$ and the common-mode gain $A_{cm\text{-amp}}$ of the noninverting amplifier:

$$A_{d\text{-amp}} = 1 + \frac{R_F}{R_1} \tag{3.23}$$

$$A_{cm\text{-amp}} = \frac{1}{\text{CMRR}_o}\left(1 + \frac{R_F}{R_1}\right) \tag{3.24}$$

Therefore, we can find the CMRR of the noninverting amplifier, which is the same as that of the op-amp:

$$\text{CMRR}_{amp} = \frac{A_{d\text{-amp}}}{A_{cm\text{-amp}}} = \text{CMRR}_o \tag{3.25}$$

▶ **NOTES**

1. The current i_S flowing into an op-amp and the differential voltage v_d are very small, tending to zero. Thus, the inverting terminal is at a ground potential with respect to the noninverting terminal, and it is said to be at the *virtual short*.
2. A_o is the open-loop voltage gain of the op-amp, whereas A_f is the closed-loop voltage gain of the op-amp circuit (or amplifier) and is dependent only on external components.
3. A noninverting amplifier can be designed to give a specified gain A_f simply by choosing the appropriate ratio R_F/R_1. A small value of R_1 will load the amplifier and cause it to draw appreciable current, and a large value of R_F will increase the noise generated in the resistor. As a guide, all resistances in op-amp circuits should be between 1 kΩ and 10 MΩ.
4. Designing a noninverting voltage amplifier is very simple: Given gain A_f, choose R_1 and then find R_F.

D **Designing a noninverting op-amp circuit** Design a noninverting amplifier as shown in Fig. 3.9(a) to provide a closed-loop voltage gain of $A_f = 80$. The input voltage is $v_S = 200$ mV with a source resistance of $R_s = 500\ \Omega$. Find the value of output voltage v_O. The DC supply voltages are $V_{CC} = V_{EE} = 12$ V.

SOLUTION

Choose a suitable value of R_1: Let $R_1 = 5\ \text{k}\Omega$. Find the value of R_F from Eq. (3.18). Since $A_f = 80 = 1 + R_F/R_1$,

$$\frac{R_F}{R_1} = 79$$

and $\quad R_F = 79 \times 5 = 395\ \text{k}\Omega$

Find the output voltage v_O from Eq. (3.18):

$$v_O = A_f v_S = 80 \times 200 \times 10^{-3} = 16\ \text{V}$$

which exceeds the maximum DC supply voltage $V_{CC} = 12$ V. Thus, the output voltage will be $v_O = V_{CC} = 12$ V.

NOTE: R_s is in series with the op-amp input resistance R_i, which is very large in comparison to R_s. Therefore, R_s will not affect the closed-loop gain A_f.

EXAMPLE 3.5

Finding the voltage gain of a noninverting op-amp circuit For the noninverting amplifier in Fig. 3.9(a), the input voltage is $v_S = 100$ mV with a source resistance of $R_s = 500\ \Omega$. The circuit parameters are $R_F = 395\ \text{k}\Omega$, $R_1 = 5\ \text{k}\Omega$, and $A_o = 2 \times 10^5$. Calculate **(a)** the closed-loop gain A_f, **(b)** the output voltage v_O, and **(c)** the errors in the output voltage v_O and the gain A_f if A_o tends to infinity.

SOLUTION

Since the current drawn by the op-amp is zero, $i_1 = -i_f$. That is,

$$\frac{v_x}{R_1} = -\frac{v_x - v_O}{R_F}$$

which gives

$$v_x = \frac{R_1}{R_1 + R_F} v_O \tag{3.26}$$

The output voltage v_O is

$$v_O = A_o(v_S - v_x) = A_o v_d \tag{3.27}$$

The input voltage at the noninverting terminal is the sum of v_x and v_d. That is,

$$v_S = v_x + v_d$$

which, after substitution of v_x from Eq. (3.26) and v_d from Eq. (3.27), becomes

$$v_S = \frac{R_1 v_O}{R_1 + R_F} + \frac{v_O}{A_o} = v_O\left(\frac{R_1}{R_1 + R_F} + \frac{1}{A_o}\right)$$

Thus, the closed-loop voltage gain A_f is given by

$$A_f = \frac{v_O}{v_S} = \frac{A_o(R_1 + R_F)}{A_oR_1 + R_1 + R_F} = \frac{1 + R_F/R_1}{1 + (1 + R_F/R_1)/A_o} = \frac{1 + R_F/R_1}{1 + x} \qquad (3.28)$$

where $$x = \frac{1}{A_o}\left(1 + \frac{R_F}{R_1}\right) \qquad (3.29)$$

For a small value of x, which is usually the case, $(1 + x)^{-1} \approx 1 - x$, and Eq. (3.28) can be approximated by

$$A_f = \left(1 + \frac{R_F}{R_1}\right)(1 - x) \qquad (3.30)$$

Therefore, the error introduced for a finite value of gain A_o is x.

(a) From Eq. (3.29),

$$x = \left(\frac{1 + 395\text{ k}/5\text{ k}}{2 \times 10^5\text{ k}}\right) = 40 \times 10^{-5} = 40 \times 10^{-3}\%$$

From Eq. (3.28),

$$A_f = \left(\frac{1 + 395/5}{1 + 40 \times 10^{-5}}\right) = 79.968$$

(b) The output voltage v_O is

$$v_O = A_f v_S = 79.968 \times 100 \times 10^{-3} = 7.9968\text{ V}$$

(c) From Eq. (3.30), the error in the output voltage v_O is

$$\Delta v_O = -x\left(1 + \frac{R_F}{R_1}\right) = -40 \times 10^{-5} \times 80 = -32\text{ mV}, \quad \text{or } -0.04\%$$

The error in the gain A_f is

$$\Delta A_f = -x = -40 \times 10^{-5} = -0.04\%$$

NOTE: To minimize the dependence of the closed-loop gain A_f on the open-loop gain A_o, the value of x should be made very small. That is,

$$A_o >> \left(1 + \frac{R_F}{R_1}\right) \qquad (3.31)$$

This condition is often satisfied by making A_o at least 10 times larger than $(1 + R_F/R_1)$. That is,

$$\left(1 + \frac{R_F}{R_1}\right) \leq 0.1A_o \qquad (3.32)$$

EXAMPLE 3.6

Finding the parameters of a practical noninverting op-amp circuit The noninverting amplifier in Fig. 3.9(a) has $R_1 = 10\ \text{k}\Omega$ and $R_F = 10\ \text{k}\Omega$. The op-amp parameters are $A_o = 2 \times 10^5$, $f_b = 10$ Hz, $R_o = 75\ \Omega$, and $R_i = 2\ \text{M}\Omega$. The frequency of the input signal is $f_s = 10$ kHz. Determine **(a)** the unity-gain bandwidth f_{bw} of the op-amp, **(b)** the closed-loop voltage gain A_f, and **(c)** the closed-loop break frequency f_c of the op-amp circuit.

SOLUTION

Using Eq. (3.28), we find the frequency-dependent voltage gain of the noninverting amplifier to be

$$A_f(j\omega) = \frac{1 + R_F/R_1}{1 + (1 + R_F/R_1)/A_o(j\omega)}$$

Substituting the frequency-dependent gain $A(j\omega)$ from Eq. (3.2), we get

$$A_f(j\omega) = \frac{1 + R_F/R_1}{1 + (1 + R_F/R_1)/A_o + jf(1 + R_F/R_1)/(A_o f_b)} \tag{3.33}$$

since $\omega = 2\pi f$. If we assume that $(1 + R_F/R_1) << A_o$, which is generally the case, and substitute $f_{bw} = A_o f_b$, Eq. (3.33) becomes

$$A_f(j\omega) = \frac{1 + R_F/R_1}{1 + jf(1 + R_F/R_1)/f_{bw}} \tag{3.34}$$

which gives the closed-loop break (or 3-dB) frequency as

$$f_c = \frac{f_{bw}}{1 + R_F/R_1} = \frac{f_{bw} R_1}{R_1 + R_F} = \beta f_{bw} = \beta A_o f_b \tag{3.35}$$

where $\beta = R_1/(R_1 + R_F)$ is called the *feedback ratio,* or the *feedback factor.* This should not be confused with the current gain β_F of bipolar transistors in Chapters 1 and 8.

The closed-loop DC gain is $(1 + R_F/R_1)$ (as expected), and this gain also falls off at a rate of -20 dB decade after a break frequency of $f_c = \beta f_{bw}$.

For $R_1 = 10\ \text{k}\Omega$, $R_F = 10\ \text{k}\Omega$, $A_o = 2 \times 10^5$, $f_b = 10$ Hz, and $f = f_s = 10$ kHz,

$$\frac{R_F}{R_1} = \frac{10\ \text{k}\Omega}{10\ \text{k}\Omega} = 1$$

$$\beta = \frac{R_1}{(R_1 + R_F)} = \frac{10\ \text{k}\Omega}{(10\ \text{k}\Omega + 10\ \text{k}\Omega)} = 0.5$$

which is small compared to $A_o = 2 \times 10^5$.

(a) From Eq. (3.4),

$$f_{bw} = A_o f_b = 2 \times 10^5 \times 10 = 2\ \text{MHz}$$

(b) From Eq. (3.34), we get

$$A_f(j\omega) = \frac{1 + 1}{1 + jf(1 + 1)/(f_{bw})} = \frac{2}{1 + j(10 \times 10^3) \times 2/(2 \times 10^6)}$$

$$= \frac{2}{1 + j0.01} = 1.9999\ \angle -0.57^\circ$$

(c) From Eq. (3.35), we get

$$f_c = \beta A_o f_b = 0.5 \times 2 \times 10^5 \times 10 = 1\ \text{MHz}$$

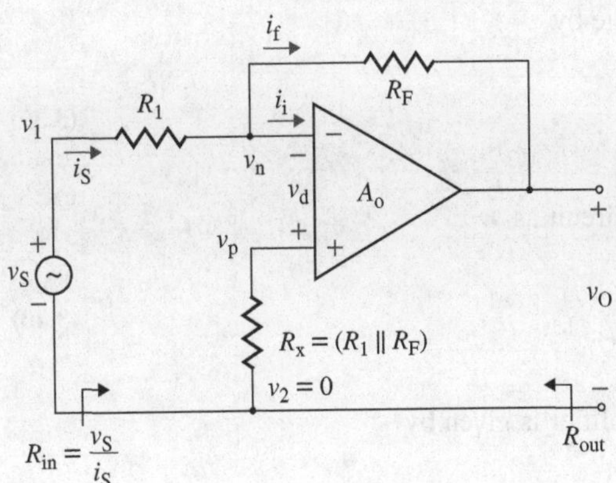

FIGURE 3.11 Inverting amplifier

3.4.2 Inverting Amplifiers

Another common configuration is the inverting voltage amplifier, as shown in Fig. 3.11. R_F is used to feed the output voltage back to the inverting terminal of the op-amp. Using Kirchhoff's voltage law, we have

$$v_S = R_1 i_S - v_d \tag{3.36}$$

$$-v_d = R_F i_f + v_O \tag{3.37}$$

▶ **NOTE** The circuit does not require the resistance R_x for the normal operation as an inverting amplifier. However, R_x establishes a nonzero voltage at the noninverting terminal, and any voltage signals generated due to the op-amp offset parameters will appear at both the inverting and noninverting terminals. By making $R_x = R_1 \parallel R_F$, which is equal to Thevenin's equivalent resistance looking at the inverting terminal, we minimize or eliminate the effect of the offset voltage on the difference voltage v_d.

Using Kirchhoff's current law at the inverting terminal, we get

$$i_S = i_f + i_i \tag{3.38}$$

For an ideal op-amp, $v_d \approx 0$ and $i_i \approx 0$. That is, Eq. (3.36) becomes

$$v_S = R_1 i_S$$

which gives

$$i_S = \frac{v_S}{R_1}$$

Also, $R_F i_f + v_O = 0$, which gives the feedback current as

$$i_f = -\frac{v_O}{R_F}$$

For $i_i \approx 0$, Eq. (3.38) becomes

$$i_S = i_f \quad \text{or} \quad \frac{v_S}{R_1} = -\frac{v_O}{R_F}$$

Therefore, the output voltage is related to the input voltage by

$$v_O = -\left(\frac{R_F}{R_1}\right)v_S \tag{3.39}$$

which gives the closed-loop voltage gain of the op-amp circuit as

$$A_f = \frac{v_O}{v_S} = -\frac{R_F}{R_1} \tag{3.40}$$

Since $v_d \approx 0$, the effective input resistance R_{in} of the amplifier is given by

$$R_{in} = \frac{v_S}{i_S} = \frac{v_S}{(v_S + v_d)/R_1} \approx R_1$$

The effective output resistance is given by $R_{out} = R_o \approx 0$.

CMRR of an Inverting Amplifier

Similar to the CMRR expression in Eq. (3.25) for a noninverting amplifier, we can derive the CMRR of an inverting amplifier. Considering two input signals v_1 and v_2, $v_{id} = v_2 - v_1 = -v_1$ and $v_{icm} = (v_2 + v_1)/2 = v_1/2$ since $v_2 = 0$.

The voltage v_p at the noninverting terminal is $v_p = 0$, and the voltage v_n at the inverting terminal can be expressed in terms of v_O by

$$v_n = v_O + \frac{R_F}{R_1 + R_F}(v_1 - v_O) = v_O\frac{R_1}{R_1 + R_F} + v_1\frac{R_F}{R_1 + R_F}$$

Similar to the noninverting op-amp, let v_{do} and v_{cmo} denote the difference and common-mode signals at the op-amp input. Then, we have $v_{do} = (v_p - v_n)$ and $v_{cmo} = (v_p + v_n)/2$. Thus, the output voltage v_O can be expressed in terms of the op-amp differential gain A_d and common-mode gain A_{cm} as

$$v_O = A_d v_{do} + A_{cm} v_{cmo} = A_d(v_p - v_n) + \frac{A_{cm}(v_p + v_n)}{2}$$

After substituting for v_p $(= 0)$ and v_n using the relationships in terms of $v_{id} = -v_1$ and $v_{cm} = v_1/2$, we can find the following expression for v_o:

$$v_O = A_d(0 - v_n) + \frac{A_{cm}}{2}(0 + v_n) = \left(v_{id}\frac{R_F}{R_1 + R_F} - v_O\frac{R_1}{R_1 + R_F}\right)$$

$$+ A_{cm}\left(v_{icm}\frac{R_F}{R_1 + R_F} + v_O\frac{R_1/2}{R_1 + R_F}\right) \tag{3.41}$$

We can solve this for v_o after collecting all the terms containing v_o in terms of v_{id} and v_{cm}:

$$v_O = \frac{A_d R_F/R_1 v_{id} + A_{cm} R_F/R_1 v_{icm}}{1 + R_F/R_1 + A_d - A_{cm}/2} \tag{3.42}$$

Since $A_d >> A_{cm}$ and $A_d >> R_F/R_1 >> 1$, the denominator of Eq. (3.42) is approximately equal to A_d, and v_o in Eq. (3.42) can be approximated to

$$v_O \approx \frac{A_d}{A_d}\frac{R_F}{R_1} v_{id} + \frac{A_{cm}}{A_d}\frac{R_F}{R_1} v_{icm} = \frac{R_F}{R_1} v_{id} + \frac{1}{\text{CMRR}_o}\frac{R_F}{R_1} v_{icm} \tag{3.43}$$

where $\text{CMRR}_o = A_d/A_{cm}$ is the CMRR of the op-amp. From Eq. (3.43), we can find the differential voltage gain $A_{d\text{-amp}}$ and common-mode gain $A_{cm\text{-amp}}$ of the noninverting amplifier:

$$A_{d\text{-amp}} = \frac{R_F}{R_1} \tag{3.44}$$

$$A_{cm\text{-amp}} = \frac{1}{\text{CMRR}_o}\frac{R_F}{R_1} \tag{3.45}$$

Therefore, we can find the CMRR of the noninverting amplifier, which is the same as that of the op-amp:

$$\text{CMRR}_{amp} = \frac{A_{d\text{-amp}}}{A_{cm\text{-amp}}} = \text{CMRR}_o \tag{3.46}$$

▶ **NOTES**

1. The negative sign in Eq. (3.39) signifies that the output voltage is out of phase with respect to the input voltage by 180° (in the case of an AC input) or of opposite polarity (in the case of a DC input).
2. The current i_i flowing into the op-amp is very small, tending to zero, and the voltage v_d at the inverting terminal is also very small, tending to zero. Although the inverting terminal is not the ground point, this terminal is said to be a virtual short.
3. We can design an inverting amplifier to give a specified gain simply by choosing the appropriate ratio R_F/R_1. A small value of R_1 will load the input source, and a large value of R_F will increase the noise generated in the resistor. As a guide, all resistances in op-amp circuits should be between 1 kΩ and 10 MΩ.
4. If $R_1 = R_F$, Eq. (3.40) gives $A_f = -1$ and $v_O = -v_S$. The circuit then behaves as a *unity-gain inverter* (or simply an *inverter*).
5. Designing an inverting voltage amplifier is straightforward: Given R_{in} and gain A_f, find R_1 and then find R_F.

EXAMPLE 3.7

Designing an inverting op-amp circuit to limit the input current A transducer produces an input-signal voltage of $v_S = 100$ mV with an internal resistance of $R_s = 2$ kΩ. Design the inverting op-amp amplifier of Fig. 3.11 by determining the values of R_1, R_F, and R_x. The output voltage should be $v_O = -8$ V. The current drawn from the transducer should not be more than 10 μA. Assume an ideal op-amp and $V_{CC} = V_{EE} = 15$ V.

SOLUTION

$R_s = 2\ \text{k}\Omega$, $v_S = 100$ mV, and $v_O = -8$ V. The source resistance R_s (not shown in Fig. 3.11) is in series with R_1. Let

$$R_1' = R_1 + R_s$$

$$A_f = \frac{v_O}{v_S} = -\frac{8}{(100 \times 10^{-3})} = -80$$

From Eq. (3.40),

$$-80 = -\frac{R_F}{R_1'} = -\frac{R_F}{(R_1 + R_s)}$$

The maximum input current is

$$i_{S(\max)} = 10\ \mu\text{A}$$

The minimum input resistance is

$$R_{\text{in(min)}} = \frac{v_S}{i_{S(\max)}} = \frac{100\ \text{mV}}{10\ \mu\text{A}} = 10\ \text{k}\Omega$$

Thus, $R_1' = R_1 + R_s = R_{\text{in(min)}} = 10\ \text{k}\Omega$

Thus, $R_F = 80(R_1 + R_s) = 80R_1' = 80 \times 10\ \text{k}\Omega = 800\ \text{k}\Omega$

Thus, $R_1 = R_1' - R_s = 10\ \text{k} - 2\ \text{k} = 8\ \text{k}\Omega$

$R_x = R_1 + R_s = 10\ \text{k}\Omega$

EXAMPLE 3.8

Finding the voltage gain of an inverting op-amp circuit The parameters of the op-amp circuit in Fig. 3.11 are $R_F = 800\ \text{k}\Omega$, $R_1 = 10\ \text{k}\Omega$, and $A_o = 2 \times 10^5$. Calculate **(a)** the closed-loop gain $A_f = v_O/v_S$, **(b)** the output voltage v_O, and **(c)** the errors in the output voltage v_O and the gain A_f if A_o tends to infinity. Assume that source resistance $R_s = 0$ and $V_S = 100$ mV.

SOLUTION

$R_1 = 10\ \text{k}\Omega$, $R_F = 800\ \text{k}\Omega$, $R_F/R_1 = 80$, $A_o = 2 \times 10^5$, and $v_S = 100$ mV. From Fig. 3.11, $v_O = A_o v_d$ or $v_d = v_O/A_o$. The input current i_S through R_1 can be found from

$$i_S = \frac{v_S + v_d}{R_1} = \frac{v_S + v_O/A_o}{R_1} \tag{3.47}$$

From Fig. 3.11, the output voltage is given by

$$v_O = -v_d - i_f R_F = -v_d - i_S R_F \quad (\text{since } i_f \approx i_S)$$

$$= -\frac{v_O}{A_o} - \frac{v_S + v_O/A_o}{R_1} R_F$$

which, after simplification, gives the closed-loop voltage gain A_f as

$$A_f = \frac{v_O}{v_S} = -\frac{R_F/R_1}{1 + (1 + R_F/R_1)/A_o} = -\frac{R_F}{R_1(1 + x)} \tag{3.48}$$

where $$x = \frac{1}{A_o}\left(1 + \frac{R_F}{R_1}\right) \tag{3.49}$$

For a small value of x, which is usually the case, $(1 + x)^{-1} \approx 1 - x$, and Eq. (3.48) can be approximated by

$$A_f = -\frac{R_F}{R_1}(1 - x) \tag{3.50}$$

Therefore, the error introduced for a finite value of gain A_o is x.

(a) From Eq. (3.49),

$$x = \left(\frac{1 + 80}{2 \times 10^5}\right) = 40.5 \times 10^{-5} = 40.5 \times 10^{-3}\%$$

From Eq. (3.48),

$$A_f = \frac{-80}{(1 + 40.5 \times 10^{-5})} = -79.9676$$

(b) The output voltage v_O is

$$v_O = A_f v_S = -79.9676 \times 100 \times 10^{-3} = -7.99676 \text{ V}$$

(c) From Eq. (3.50), the error in the output voltage v_O is

$$\Delta v_O = \frac{x R_F}{R_1} = 40.5 \times 10^{-5} \times 80 = 32.4 \text{ mV}, \text{ or } 0.0405\%$$

The error in the gain A_f is

$$\Delta A_f = x = 40.5 \times 10^{-5} = 0.0405\%$$

EXAMPLE 3.9

Finding the parameters of a practical inverting op-amp circuit

(a) An inverting amplifier has $R_1 = 10\ \text{k}\Omega$ and $R_F = 800\ \text{k}\Omega$. The op-amp parameters are $A_o = 2 \times 10^5$, $f_b = 10$ Hz, $R_o = 75\ \Omega$, and $R_i = 2\ \text{M}\Omega$. The frequency of the input signal is $f_s = 10$ kHz. Determine the unity-gain bandwidth f_{bw}, the closed-loop voltage gain A_f, and the closed-loop break frequency f_c of the op-amp.

(b) Use PSpice/SPICE to plot the closed-loop frequency response of the voltage gain. Assume $v_s = 0.1$ V (AC), and use the linear AC model.

SOLUTION

(a) Using Eq. (3.48), we find the frequency-dependent voltage gain of the inverting amplifier to be

$$A_f(j\omega) = \frac{-R_F/R_1}{1 + (1 + R_F/R_1)/A_o(j\omega)}$$

Substituting the frequency-dependent gain $A_o(j\omega)$ from Eq. (3.2), we get

$$A_f(j\omega) = \frac{-R_F/R_1}{1 + (1 + R_F/R_1)/A_o + jf(1 + R_F/R_1)/(A_o f_b)} \tag{3.51}$$

since $\omega = 2\pi f$. If we assume that $(1 + R_F/R_1) << A_o$, which is generally the case, and substitute $f_{bw} = A_o f_b$, Eq. (3.51) becomes

$$A_f(j\omega) = \frac{-R_F/R_1}{1 + jf(1 + R_F/R_1)/f_{bw}} \tag{3.52}$$

which gives the closed-loop break (or 3-dB) frequency as

$$f_c = \frac{f_{bw}}{1 + R_F/R_1} = \frac{f_{bw} R_1}{R_1 + R_F} = \beta f_{bw} = \beta A_o f_b \tag{3.53}$$

where $\beta = R_1/(R_1 + R_F)$ is called the *feedback ratio* or the *feedback factor*. (It should not be confused with the current gain β_F of a bipolar transistor.) Notice from Eq. (3.52) that the DC gain is $-R_F/R_1$ (as expected), and it falls off at a rate of -20 dB/decade after a break frequency of $f_c = \beta f_{bw}$.

For $R_1 = 10\ \text{k}\Omega$, $R_F = 800\ \text{k}\Omega$, $A_o = 2 \times 10^5$, $f_b = 10$ Hz, and $f = f_s = 10$ kHz,

$$\frac{R_F}{R_1} = \frac{800\ \text{k}\Omega}{10\ \text{k}\Omega} = 80$$

$$\beta = \frac{R_1}{(R_1 + R_F)} = \frac{10\ \text{k}\Omega}{(10\ \text{k}\Omega + 800\ \text{k}\Omega)} = 12.346 \times 10^{-3}$$

which is small compared to $A_o = 2 \times 10^5$.

From Eq. (3.4),

$$f_{bw} = A_o f_b = 2 \times 10^5 \times 10 = 2\ \text{MHz}$$

Substituting the values in Eq. (3.52), we get the voltage gain at $f = f_s = 10$ kHz as

$$A_f(j\omega) = \frac{-80}{1 + j2\pi f(1 + 80)/2\pi f_{bw}} = \frac{-80}{1 + j(10 \times 10^3) \times 81/(2 \times 10^6)}$$

$$= \frac{-80}{1 + j0.405} = -74.15 \angle -22°$$

Thus, the magnitude of the closed-loop voltage gain at $f_s = 10$ kHz is -74.15. If the input is a sinusoidal signal, the output voltage will be phase shifted by $(180° - 22°) = 158°$.

From Eq. (3.53),

$$f_c = \beta f_{bw} = 12.346 \times 10^{-3} \times 2 \times 10^6 = 24.69 \text{ kHz}$$

(b) The inverting amplifier for PSpice simulation is shown in Fig. 3.12(a).

The frequency response, which is shown in Fig. 3.13, gives the low-frequency gain $A_{f(dc)} = 79.95$ and $A_f = 61.67$ at $f_s = 10$ kHz. At $A_f = 56.48$ (estimated value is $0.707 \times 79.95 = 56.53$), $f_c = 12.198$ kHz. The calculated values are $f_c = 24.69$ kHz and $A_f = 74.15$ (at $f_s = 10$ kHz). However, this simulation was done using the nonlinear macromodel of UA741. If we run the simulation with the linear op-amp model shown in Fig. 3.12(b), we get the low-frequency gain $A_{f(dc)} = 79.93$ and $A_f = 74.07$ at $f_s = 10$ kHz. At $A_f = 56.45$ (estimated value is $0.707 \times 79.93 = 56.51$), $f_c = 24.5$ kHz. The calculated values are $f_c = 24.69$ kHz and $A_f = 74.15$ (at $f_s = 10$ kHz).

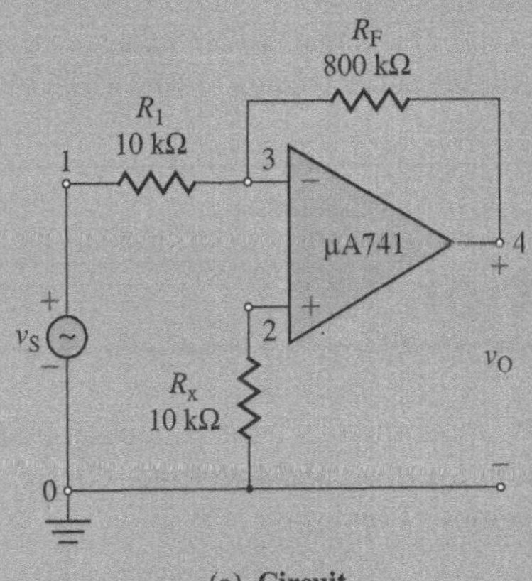

(a) Circuit

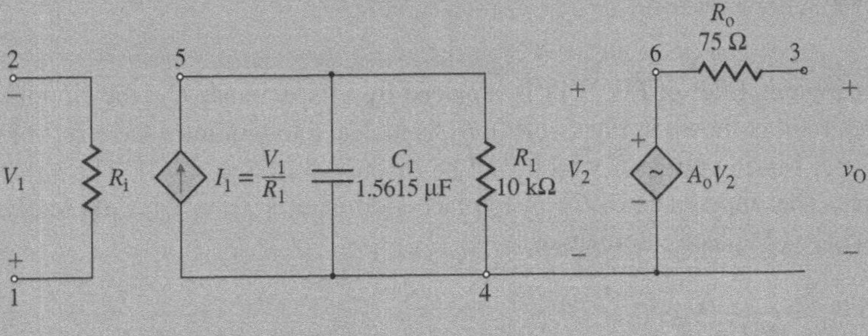

(b) Op-amp model

FIGURE 3.12 Inverting amplifier for PSpice simulation

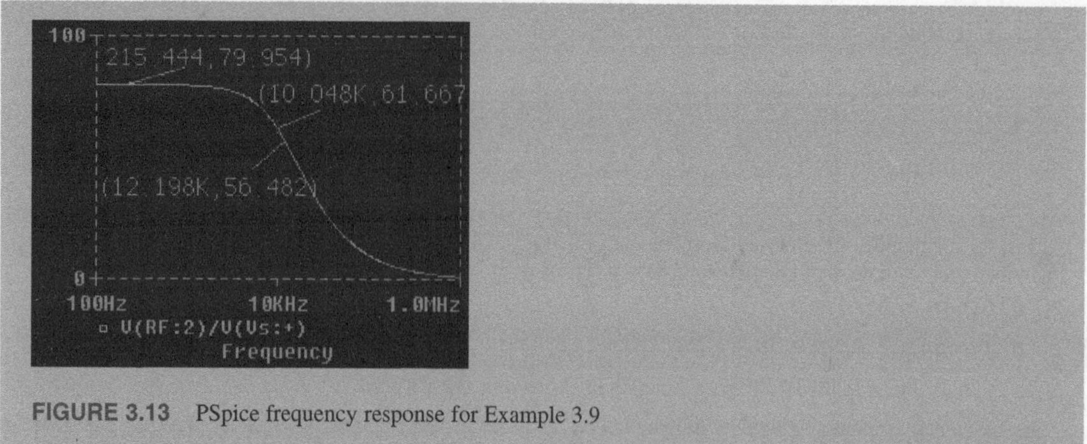

FIGURE 3.13 PSpice frequency response for Example 3.9

KEY POINTS OF SECTION 3.4

- The output voltage of an op-amp circuit is almost independent of the op-amp parameters; it depends largely on the external circuit elements.
- We can simplify the analysis of an op-amp circuit by assuming that the voltage across the op-amp terminals and the current drawn by the op-amp are very small, tending to zero. The error due to these assumptions generally is less than 0.1%.

3.5 Op-Amp Applications

The applications of op-amps are endless, and there are numerous books about op-amps [4–6, 7–10]. Most of the applications are derived from the basic noninverting and inverting, configurations described in Sec. 3.4. In this section we discuss several applications of op-amps.

3.5.1 Integrators

If the resistance R_F in the inverting amplifier of Fig. 3.11 is replaced by a capacitance C_F, the circuit will operate as an integrator. Such a circuit is shown in Fig. 3.14(a). R_x is included to minimize the effect of op-amp imperfections (i.e., the input biasing current, which will be discussed in Sec. 14.3). The value of R_x should be made equal to R_1. The impedance of C_F in Laplace's domain is $Z_F = 1/(sC_F)$. Applying Eq. (3.39) gives the output voltage in Laplace's domain as

$$V_o(s) = -\left(\frac{Z_F}{Z_1}\right)V_s(s) = -\frac{1}{sR_1C_F}\,V_s(s) \tag{3.54}$$

from which the output voltage in the time domain becomes

$$v_O(t) = -\frac{1}{R_1C_F}\int_0^t v_S\,dt - v_C(t=0) \tag{3.55}$$

where $v_C(t = 0) = V_{co}$ represents the initial capacitor voltage. That is, the output voltage is given by the integral of the input voltage v_S. Equation (3.55) can also be derived from a circuit analysis similar to that discussed in Sec. 3.4.2. That is,

$$i_f = i_S = \frac{v_S + v_d}{R_1} = \frac{v_S}{R_1} \tag{3.56}$$

since the op-amp input current is zero. Therefore, the output voltage, which is the negative of the capacitor voltage, is given by

$$v_O(t) = -v_C(t) = -\frac{1}{C_F}\int_0^t i_S\, dt - v_C(t = 0) \tag{3.57}$$

Substituting $i_S = v_S/R_1$ from Eq. (3.56) into Eq. (3.57), we can obtain Eq. (3.55). Time constant $\tau_i = R_1C_F$ for Fig. 3.14(a) is known as the *integration time constant*. If the input is a constant current $i_S = I_S$, then Eq. (3.57) gives

$$v_O(t) = -v_C(t) = -\frac{I_S t}{C_F} - v_C(t = 0) = -\frac{Q}{C_F} - v_C(t = 0) \tag{3.58}$$

That is, the output voltage is the integral of the input current I_S and is proportional to the input charge Q. Thus, the circuit in Fig. 3.14(a) can also be used as a current integrator, or charge amplifier. The plot of the output voltage for a pulse input is shown in Fig. 3.14(b). Due to the half-wave symmetry of the input voltage v_S, the output voltage v_O will also be half-wave symmetrical. That is, the first zero crossing of the output voltage will be at $t = T/4$, where T is the period of the input voltage. The area $+A$ under the input voltage during the interval ($T/2$ to $T/4$) will be equal and opposite to the area $-B$ during the interval ($3T/4$ to $T/2$) such that the output waveform crosses the zero axis at every $T/2$ interval.

At lower frequencies, the impedance Z_F of C_F will increase, and less signal will be fed back to the inverting terminal of the op-amp. Thus, the output voltage will increase. At higher frequencies, the impedance Z_F will decrease, causing more signal to be fed back to the inverting terminal. Thus, the output voltage will decrease. Therefore, an integrator circuit behaves like a low-pass filter. The magnitude plot of the voltage gain $V_o(j\omega)/V_s(j\omega)$ in Eq. (3.54) will have a low-pass characteristic with a zero break frequency, as shown in Fig. 3.14(c).

For the case in which the input signal is a constant DC voltage, Eq. (3.55) simplifies to

$$v_O(t) = -\left(\frac{V_S}{R_1C_F}\right)t - V_{co} \tag{3.59}$$

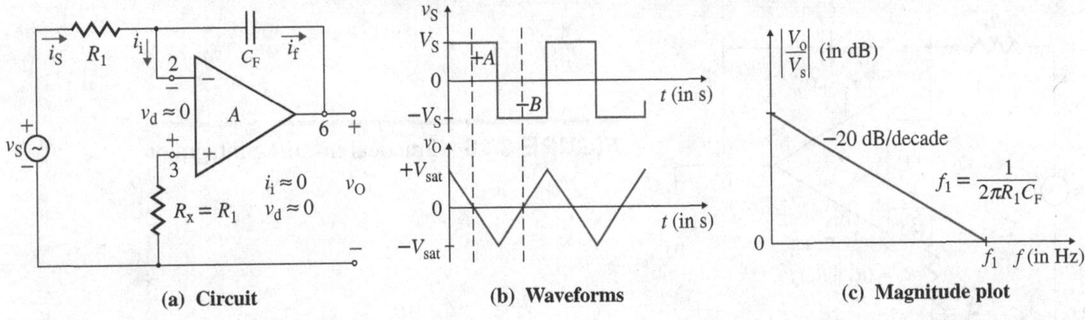

FIGURE 3.14 Integrator circuit

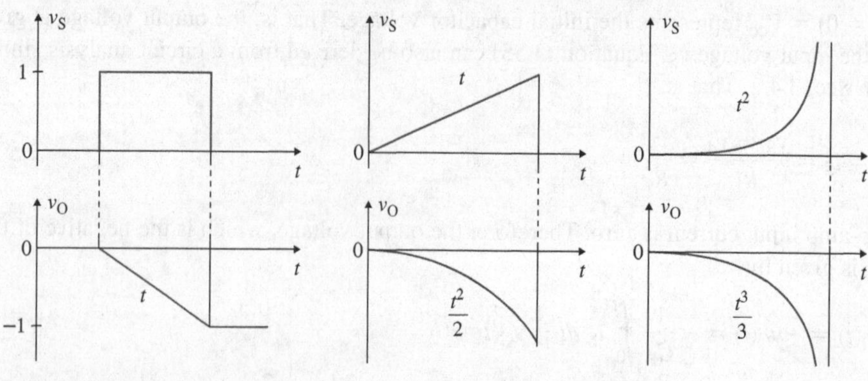

FIGURE 3.15 Typical input and output signals of an integrator

Typical plots of some input signals and the resulting output signals are shown in Fig. 3.15.

In practice, as a result of its imperfections (e.g., drift, input offset current), an op-amp produces an output voltage even if the input signal is zero ($v_S = 0$), and the capacitor will be charged by the small but finite current through it. The capacitor prevents any DC signal from feeding back from the output terminal to the input side of the op-amp. As a result, the capacitor will be charged continuously, and the output voltage will build up until the op-amp saturates. A resistor with a large value of R_F is normally connected in parallel with the capacitor of capacitance C_F, as shown in Fig. 3.16. R_F provides the DC feedback and overcomes this saturation problem. Time constant τ_F (= R_FC_F) must be larger than the period T_s (= $1/f_s$) of the input signal. A ratio of 10 to 1 is generally adequate; that is, $\tau_F = 10T_s$. For Fig. 3.16, the feedback impedance is

$$Z_F = R_F \| \left(\frac{1}{sC_F}\right) = \frac{R_F}{(1 + sR_FC_F)}$$

and Eq. (3.39) gives the output voltage in Laplace's domain as

$$V_o(s) = -\frac{R_F/R_1}{1 + sR_FC_F}V_s(s) \tag{3.60}$$

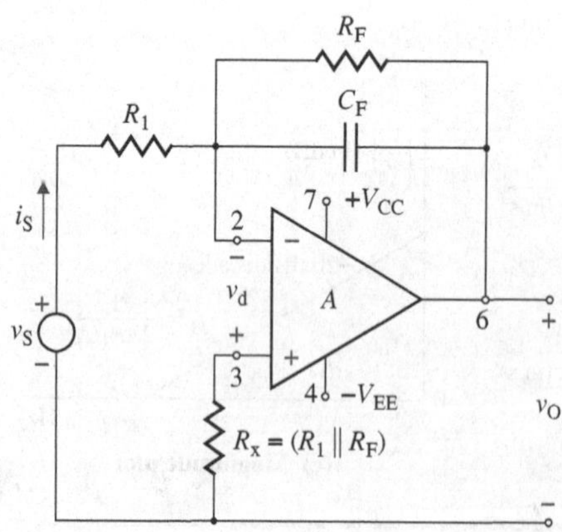

FIGURE 3.16 Practical inverting integrator

For a step input voltage of $v_S = V_S$, $V_S(s) = V_S/s$, and Eq. (3.60) can be simplified to give the output voltage in the time domain as

$$v_O(t) = -V_S \frac{R_F}{R_1}(1 - e^{-t/R_F C_F}) \tag{3.61}$$

For $t \leq 0.1 R_F C_F$, Eq. (3.61) can be approximated by

$$v_O(t) = -V_S \frac{R_F}{R_1}\left(\frac{t}{R_F C_F}\right) = -\left(\frac{V_S}{R_1 C_F}\right)t \tag{3.62}$$

which is the time integral of the input voltage. Therefore, the analysis and the input–output relation of the integrator in Fig. 3.14 can be applied to the one in Fig. 3.16, provided $\tau_F \geq 10T$.

EXAMPLE 3.10

D

Designing an op-amp integrator

(a) Design an integrator of the form shown in Fig. 3.16. The frequency of the input signal is $f_s = 500$ Hz. The voltage gain should be unity at a frequency of $f_1 = 1590$ Hz. That is, the unity-gain bandwidth is $f_{bw} = 1590$ Hz.

(b) The integrator in part (a) has $V_{CC} = 12$ V, $-V_{EE} = -12$ V, and maximum voltage swing $= \pm 10$ V. The initial capacitor voltage is $V_{co} = 0$. Draw the waveform of the output voltage for the input voltage shown in Fig. 3.17.

(c) Use PSpice/SPICE to plot the output voltage for the input voltage in part (b).

SOLUTION

(a) The steps in completing the design are as follows:

Step 1. Choose a suitable value of C_F: Let $C_F = 0.1$ μF.

Step 2. Calculate the time constant required to satisfy the unity-gain frequency requirement:

$$\tau_i = \frac{1}{2\pi f_1} = \frac{1}{2\pi \times 1590 \text{ Hz}} = 100 \text{ μs}$$

Step 3. Calculate the value of R_1 from τ_i:

$$R_1 = \frac{\tau_i}{C_F} = \frac{100 \text{ μs}}{0.1 \text{ μF}} = 1 \text{ kΩ}$$

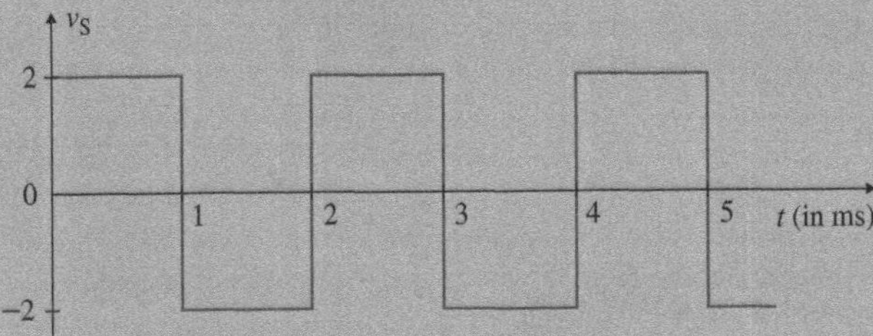

FIGURE 3.17 Input voltage for Example 3.10

Step 4. Choose time constant $\tau_F = 10T_s = 10/f_s$:

$$\tau_F = \frac{10}{500\ \text{Hz}} = 20\ \text{ms}$$

Step 5. Calculate the value of R_F from τ_F:

$$R_F = \frac{\tau_F}{C_F} = \frac{20\ \text{ms}}{0.1\ \mu\text{F}} = 200\ \text{k}\Omega$$

(b) $V_{sat} = \pm 10$ V, and $\tau_i = R_1 C_F = 1 \times 10^3 \times 0.1 \times 10^{-6} = 0.1$ ms. Since $\tau_F >> \tau_i$, the effect of τ_F can be neglected.

Under the steady-state condition after the initial transient, the capacitor voltage at $t = 0$ will be $V_{co} = -10$ V (negative of the output voltage). If we assume an initial capacitor voltage of $V_{co} = 0$, then we must start the integration at $T/4$.

For $0 \leq t \leq 1$ ms: From Eq. (3.55), the output voltage is given by

$$v_O = 10 - \frac{1}{R_1 C_F}\int_0^t 2\,dt = 10 - 2 \times 10{,}000t$$

where t is in milliseconds. At $t = 1$ ms, $v_O = -10$ V, which is more than the saturation voltage and thus is not possible. The time required for the output voltage to reach the saturation voltage of -10 V is $t_1 = 10/(2 \times 10{,}000) = 0.5$ ms. For 0.5 ms $\leq t \leq 1$ ms, the capacitor voltage is $V_{co} = 10$ V.

For 1 ms $\leq t \leq 2$ ms: From Eq. (3.55), the output voltage is given by

$$v_O = -10 + \frac{1}{R_1 C_F}\int_0^{t-1} 2\,dt = -10 + 2 \times 10{,}000(t-1)$$

where t is in milliseconds. At $t = 2$ ms, $v_O = 10$ V, and the capacitor voltage is $V_{co} = -10$ V.

For 2 ms $\leq t \leq 3$ ms: From Eq. (3.55), the output voltage is given by

$$v_O = 10 - \frac{1}{R_1 C_F}\int_0^{t-2} 2\,dt = 10 - 2 \times 10{,}000(t-2)$$

where t is in milliseconds. At $t = 3$ milliseconds, $v_O = -10$ V, and the capacitor voltage is $V_{co} = 10$ V.

For 3 ms $\leq t \leq 4$ ms: From Eq. (3.55), the output voltage is given by

$$v_O = -10 + \frac{1}{R_1 C_F}\int_0^{t-3} 2\,dt = -10 + 2 \times 10{,}000(t-3)$$

where t is in milliseconds. At $t = 4$ ms, $v_O = 10$ V, and the capacitor voltage is $V_{co} = -10$ V.

The waveforms for input and output voltages are shown in Fig. 3.18.

(c) The integrator for PSpice simulation is shown in Fig. 3.19. The plot of the output voltage $v_O \equiv$ V(CF:2) is shown in Fig. 3.20. Under the steady-state condition for the time interval of 10 ms to 14 ms after the initial transient. Note that the input signal voltage is delayed by 0.5 ms for PSpice simulation, and the PSpice model parameters of VS are

```
V1=-2V V2=2V TD=0.5ms TR=1ns TF=1ns PW=1ms PER=2ms
```

The output voltage waveform is close to the expected values: $V_{o(max)} = 10.103$ V (expected 10 V) and $V_{o(min)} = -10.238$ V. If we plot the output voltage, starting from 0, we can see the transient behavior, and it will take a number of cycles before the waveform reaches its steady-state condition

NOTE: While running the PSpice simulation, you must select Use Initial Condition in the setup; otherwise the output plot will differ from what is shown in Fig. 3.20.

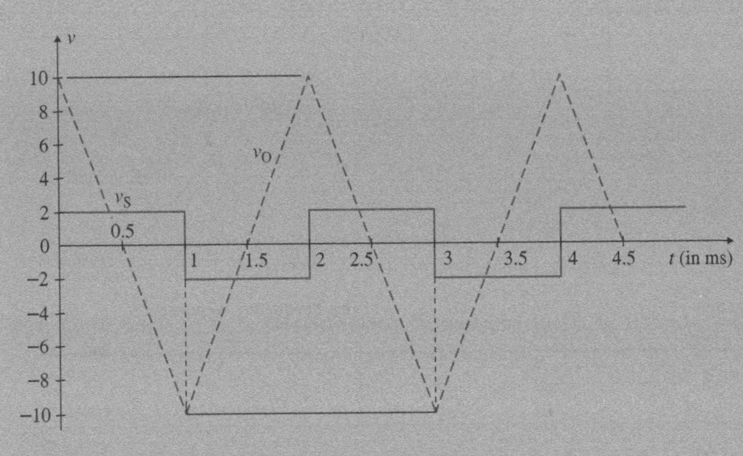

FIGURE 3.18 Waveforms for Example 3.10

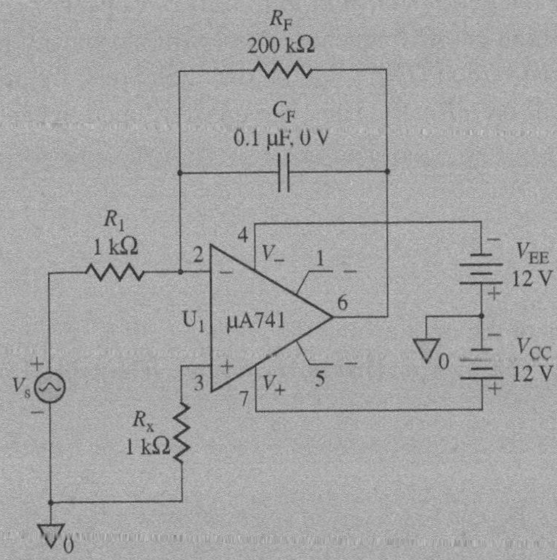

FIGURE 3.19 Integrator circuit for PSpice simulation

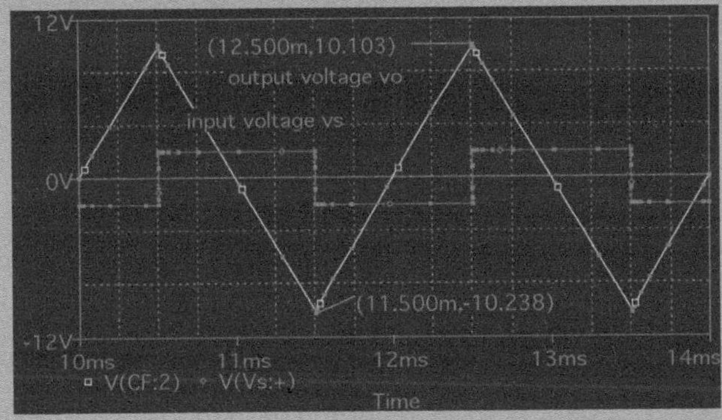

FIGURE 3.20 PSpice plots for Example 3.10

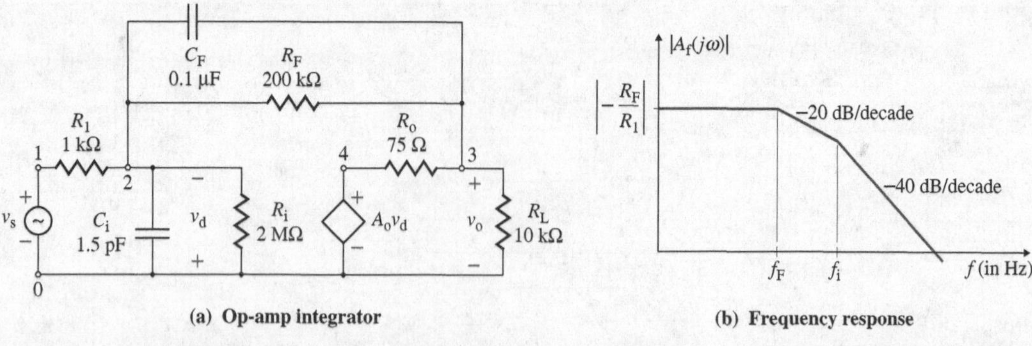

FIGURE 3.21 Op-amp integrator circuit

Frequency Response of Op-Amp Integrators

Replacing the op-amp by its equivalent circuit gives the integrator shown in Fig. 3.21(a). Capacitor C_i is the input capacitor of the op-amp, and it influences the high cutoff frequency. There will not be any low cutoff frequency, and the circuit will behave as a high-pass circuit. There will be two high break frequencies: ω_i for C_i and ω_F for C_F. The low-frequency gain will be $-R_F/R_1$. Thus, the transfer function can be expressed as

$$A_f(s) = \frac{-R_F/R_1}{(1 + s/\omega_i)(1 + s/\omega_F)} \tag{3.63}$$

Since C_F is connected between the input and output sides of the op-amp and the voltage gain is very high, C_F will dominate the high cutoff frequency $f_H = f_F = 1/(2\pi\, C_F R_F)$. That is, $\omega_i >> \omega_F$. The typical frequency response is shown in Fig. 3.21(b).

EXAMPLE 3.11

Finding the 3-dB frequency of an integrator using Miller's theorem The integrator of Fig. 3.14(a) has $C_F = 0.001\ \mu\text{F}$ and $R_1 = 1\ \text{k}\Omega$. The open-loop gain of the op-amp is $A_o = 2 \times 10^5$. Use Miller's theorem (discussed in Sec. 2.6) to find the 3-dB frequency of the integrator.

SOLUTION

Miller's theorem can be applied to replace the feedback capacitance C_F by an equivalent input capacitance C_x and an output capacitance C_y, as shown in Fig. 3.22. With the open-loop gain $A_o = A_{vo}$ and the capacitive impedance $Z_F = 1/(j2\pi f C_F)$, we can apply Eqs. (2.102) and (2.103) to find the Miller capacitances:

$$C_x = C_F(1 + A_o) = 0.001\ \mu\text{F} \times (1 + 2 \times 10^5) = 200.001\ \mu\text{F}$$

$$C_y = C_F(1 + 1/A_o) \approx C_F = 0.001\ \mu\text{F}$$

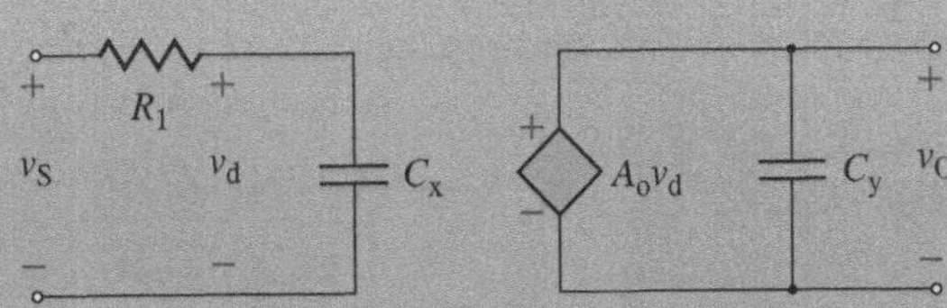

FIGURE 3.22 Equivalent circuit for Example 3.11

The output voltage in Laplace's domain is

$$V_o(s) = A_o V_d(s)$$

$$V_d(s) = \frac{1/(sC_x)}{R_i + 1/(sC_x)} V_s(s) = \frac{V_s(s)}{1 + R_1 C_x s}$$

The transfer function between the input and output voltages is given by

$$A(s) = \frac{V_o(s)}{V_s(s)} = \frac{A_o}{1 + R_1 C_x s}$$

Therefore, the 3-dB frequency is

$$\omega_b = \frac{1}{(R_i C_x)} = \frac{1}{(1 \times 10^3 \times 200.001 \times 10^{-6})} = 5 \text{ rad/s} \quad \text{or} \quad f_b = \omega_b/2\pi = 0.7958 \text{ Hz}$$

EXAMPLE 3.12

Finding the frequency response of an op-amp integrator The op-amp integrator in Fig. 3.21(a) has $R_1 = 1 \text{ k}\Omega$, $R_F = 200 \text{ k}\Omega$, $C_F = 0.1 \text{ }\mu\text{F}$, $C_i = 1.5 \text{ pF}$, $R_i = 2 \text{ M}\Omega$, $R_o = 75 \text{ }\Omega$, and open-loop voltage gain $A_o = 2 \times 10^5$.

(a) Calculate the low-frequency (or DC) voltage gain $A_{PB} = v_o/v_S$.

(b) Use the zero-value method to calculate the high 3-dB frequency f_H.

(c) Use PSpice/SPICE to plot the frequency response.

SOLUTION

(a) Assuming all capacitors are open-circuited, the low-frequency pass-band voltage gain can be found as follows:

$$A_{PB} \approx -\frac{R_F}{R_1} = -\frac{200 \text{ k}\Omega}{1 \text{ k}\Omega} = -200$$

(b) The high-frequency equivalent circuits of the op-amp integrator are shown in Fig. 3.23. If C_F is open-circuited and the voltage-controlled voltage source is converted to a voltage-controlled current source, we get the circuit in Fig. 3.23(a). The transconductance g_m is given by

$$g_m = \frac{A_o}{R_o} \tag{3.64}$$

$$= \frac{2 \times 10^5}{75} = 2.67\ \text{kA/V}$$

Applying a test voltage v_x and KVL, we get

$$v_x = R_F i_x + (R_o \| R_L)(i_x - g_m v_x)$$

which gives the resistance R_x as

$$R_x = \frac{v_x}{i_x} = \frac{R_F + R_o \| R_L}{1 + g_m(R_o \| R_L)} \tag{3.65}$$

$$= \frac{200\ \text{k}\Omega + 75\ \Omega \| 10\ \text{k}\Omega}{1 + 2.67 \times 10^3 \text{A/V} \times (75\Omega \| 10\ \text{k}\Omega)} = 1\ \Omega$$

Thevenin's equivalent resistance presented to C_i is

$$R_{Ci} = R_1 \| R_i \| R_x = 1\ \text{k}\Omega \| 2\ \text{M}\Omega \| 1\ \Omega \approx 1\ \Omega$$

and $$f_{Ci} = \frac{\omega_i}{2\pi} = \frac{1}{[2\pi (C_i R_{Ci})]} = \frac{1}{[2\pi \times (1.5\ \text{pF} \times 1)]} = 106.1 \times 10^9\ \text{Hz}$$

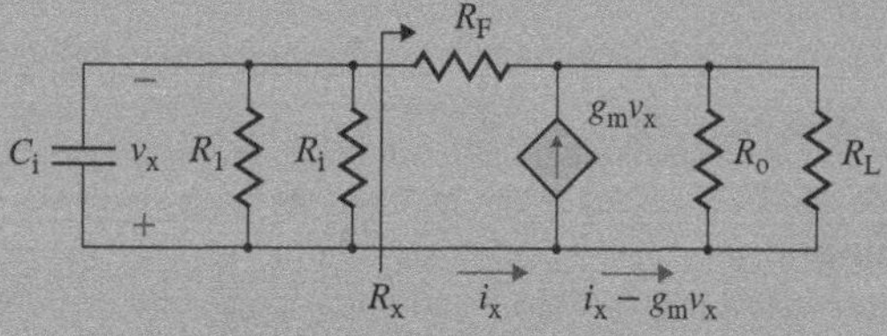

(a) C_F open-circuited

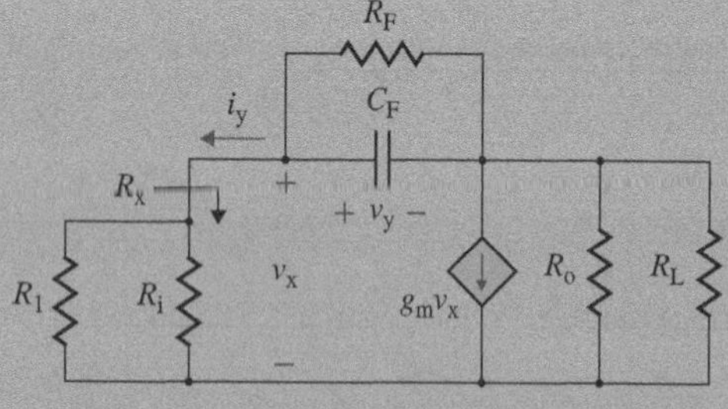

(b) C_i open-circuited

FIGURE 3.23 High-frequency equivalent circuits for op-amp integrator

If we assume C_i is open-circuited, the equivalent circuit is shown in Fig. 3.23(b). Applying a test voltage v_y and using Eq. (2.116), we can find the equivalent resistance $R_y = v_y/i_y$. That is,

$$R_y = \frac{v_y}{i_y} = R_o \| R_L + (R_1 \| R_i)[1 + g_m(R_o \| R_L)] \tag{3.66}$$

$$= 75\ \Omega \| 10\ \text{k}\Omega + (1\ \text{k}\Omega \| 2\ \text{M}\Omega)[1 + 2.667 \times 10^3\ \text{A/V} \times (75\ \Omega \| 10\ \text{k}\Omega)] = 198.4\ \text{M}\Omega$$

Thevenin's equivalent resistance presented to C_F is

$$R_{CF} = R_F \| R_y = 200\ \text{k}\Omega \| 198.4\ \text{M}\Omega = 199.8\ \text{k}\Omega$$

Thus, using Eq. (2.112), high 3-dB frequency f_H is

$$f_H = \frac{1}{2\pi\,(R_{Ci}C_i + R_{CF}C_F)} = \frac{1}{2\pi\,(1 \times 1.5\ \text{pF} + 199.8\ \text{k}\Omega \times 0.1\ \mu\text{F})} = 7.97\ \text{Hz}$$

which is dominated by C_F as expected and can be approximated by

$$f_H = \frac{\omega_F}{2\pi} = \frac{1}{[2\pi(C_F R_{CF})]} = \frac{1}{[2\pi \times (0.1\ \mu\text{F} \times 199.8\ \text{k}\Omega)]} = 7.97\ \text{Hz}$$

There really was no need to find the value of R_y, which is usually very large for an op-amp circuit, and $R_{CF} \approx R_F$.

(c) Node numbers are assigned to the AC equivalent circuit of Fig. 3.21(a) for PSpice simulation. The PSpice plot of the frequency response is shown in Fig. 3.24, which gives the midfrequency gain as $|A_{low}| = 198.16$. The high 3-dB frequency is approximately $f_H = 7.96$ Hz. The expected values are $f_H = 7.97$ Hz and $A_{PB} = -200$.

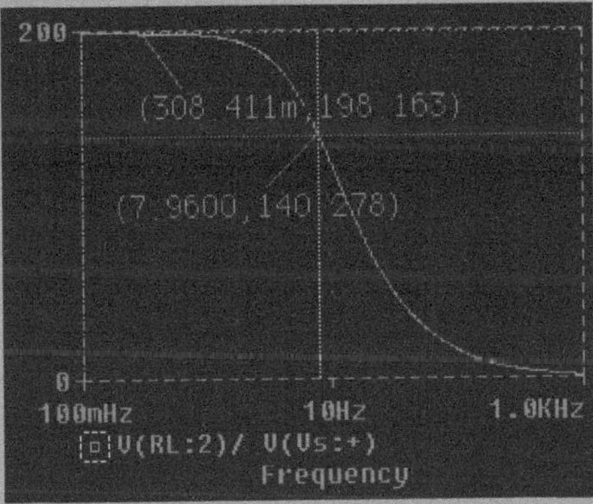

FIGURE 3.24 Frequency response for Example 3.12

3.5.2 Differentiators

If the resistance R_1 in the inverting amplifier of Fig. 3.11 is replaced by a capacitance C_1, as shown in Fig. 3.25(a), the circuit will operate as a differentiator. The value of R_x should be made equal to R_F. The impedance of C_1 in Laplace's transform is $Z_1 = 1/(sC_1)$. Using Eq. (3.39), we can find the output voltage in Laplace's domain as

$$V_o(s) = -\left(\frac{R_F}{Z_1}\right)V_s(s) = -sR_FC_1V_s(s) \tag{3.67}$$

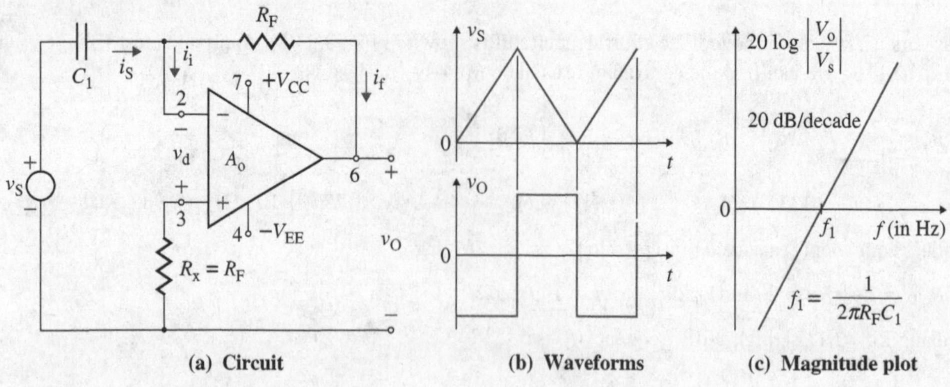

(a) Circuit (b) Waveforms (c) Magnitude plot

FIGURE 3.25 Differentiator circuit

which gives the output voltage in the time domain as

$$v_O = -R_F C_1 \frac{dv_S}{dt} \tag{3.68}$$

This equation can also be derived from a circuit analysis similar to that discussed in Sec. 3.4.2. That is,

$$i_S = i_f = C_1 \frac{dv_S}{dt} \tag{3.69}$$

$$v_O = -R_F i_f = -R_F i_S \tag{3.70}$$

Substituting i_S from Eq. (3.69) into Eq. (3.70) gives Eq. (3.68). Time constant $\tau_d = R_F C_1$ in Fig. 3.25(a) is known as the *differentiator time constant*. The output voltage in response to a triangular wave is shown in Fig. 3.25.

A differentiator circuit is useful in producing sharp trigger pulses to drive other circuits. When the frequency is increased, the impedance Z_1 of C_1 decreases and the output voltage increases. Therefore, a differentiator circuit behaves like a high-pass network. The magnitude plot of the voltage gain $V_o(j\omega)/V_s(j\omega)$ in Eq. (3.67) has a high-pass characteristic with an infinite break frequency, as shown in Fig. 3.25(c). Typical plots of some input signals and the resulting output signals are shown in Fig. 3.26.

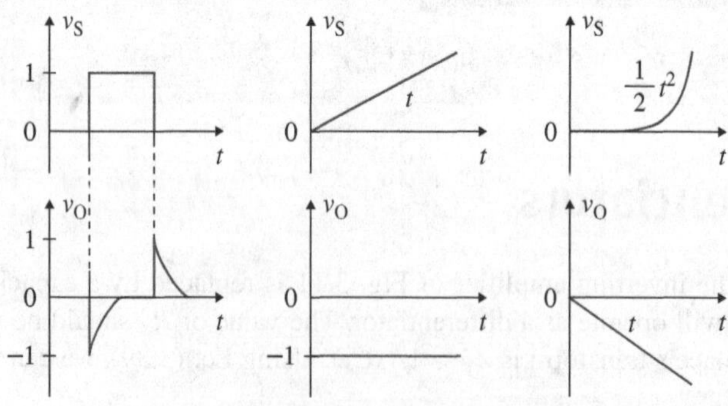

FIGURE 3.26 Typical input and output signals of a differentiator

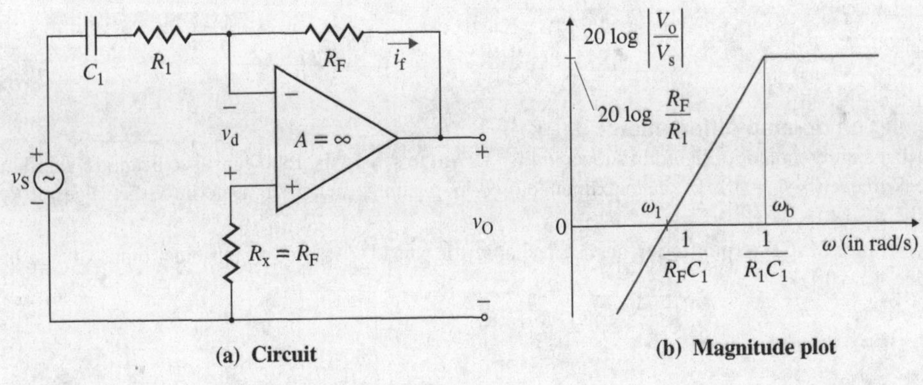

FIGURE 3.27 Practical inverting differentiator

If there is any sharp change in the input voltage $v_S(t)$ due to noise or picked-up interference, there will be amplified spikes at the output, and the circuit will behave like a noise magnifier. Thus, this type of differentiating circuit is not often used. A modified circuit that is often utilized as a differentiator is shown in Fig. 3.27(a), in which a small resistance R_1 ($<R_F$) is connected in series with C_1 to limit the gain at high frequencies. However, this arrangement also limits the high-frequency range, as shown in the magnitude plot in Fig. 3.27(b).

The impedance Z_1 for R_1 and C_1 in Laplace's domain is

$$Z_1 = R_1 + \frac{1}{sC_1} = \frac{1 + sR_1C_1}{sC_1}$$

Using Eq. (3.67), we have for the transfer function of the circuit in Fig. 3.27(a)

$$A_f(s) = \frac{V_o(s)}{V_s(s)} = -\frac{R_F}{Z_1} = -\frac{R_FC_1s}{1 + sR_1C_1} \tag{3.71}$$

For $s = j\omega$,

$$A_f(j\omega) = -\frac{R_FC_1j\omega}{1 + j\omega R_1C_1} \tag{3.72}$$

The magnitude of Eq. (3.72) is given by

$$|A_f(j\omega)| = \frac{R_FC_1\omega}{[1 + (\omega R_1C_1)^2]^{1/2}} \tag{3.73}$$

Therefore, the break frequency is $\omega_b = 1/(R_1C_1)$. For frequencies greater than ω_b, $(\omega R_1C_1)^2 >> 1$, and Eq. (3.73) reduces to

$$|A_f(j\omega)| = \frac{R_F}{R_1} \tag{3.74}$$

EXAMPLE 3.13

D

Designing an op-amp differentiator

(a) Design a differentiator of the form shown in Fig. 3.27(a) to satisfy the following specifications: gain-limiting break frequency $f_b = 1$ kHz, and maximum closed-loop gain $A_{f(max)} = 10$. Determine the values of R_1, R_F, and C_1.

(b) Use PSpice/SPICE to plot the frequency response for part (a). Assume a sinusoidal input voltage of rms value $V_s = 0.1$ V.

SOLUTION

$A_{f(max)} = 10$, and $f_b = 1$ kHz.

(a) The steps in completing the design are as follows:

Step 1. Choose a suitable value for capacitance C_1: Let $C_1 = 0.1\ \mu$F.

Step 2. Calculate the value of R_1 from the break frequency f_b:

$$f_b = \frac{1}{(2\pi R_1 C_1)}$$

$$1\ \text{kHz} = \frac{1}{(2\pi R_1 \times 0.1 \times 10^{-6})}$$

$$R_1 = 1592\ \Omega$$

Step 3. Calculate the value of R_F from Eq. (3.74):

$$A_{f(max)} = \frac{R_F}{R_1}$$

$$R_F = 1592\ A_{f(max)} = 1592 \times 10 = 15.92\ \text{k}\Omega$$

(b) The differentiator circuit for PSpice simulation is shown in Fig. 3.28. The plot of the frequency response for the output voltage is shown in Fig. 3.29, which gives $A_{f(max)} = 9.995$ (expected value is $100 \times 0.1 = 10$).

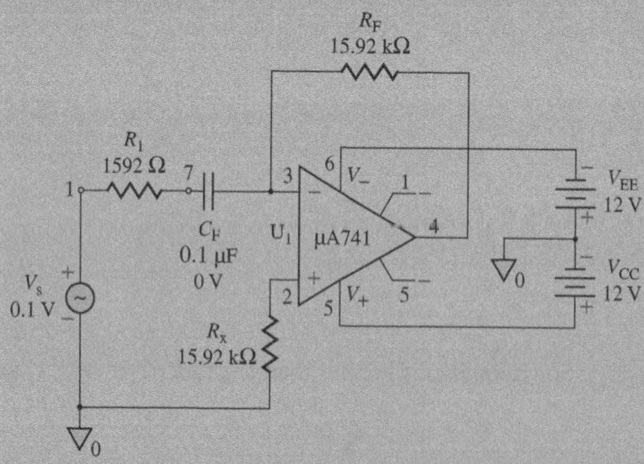

FIGURE 3.28 Differentiator circuit for PSpice simulation

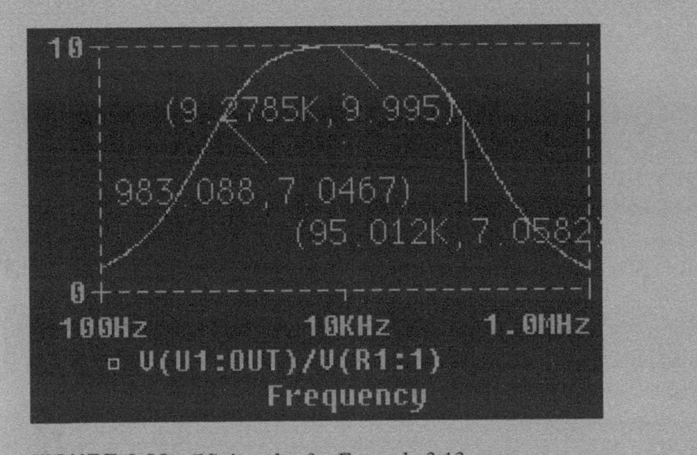

FIGURE 3.29 PSpice plot for Example 3.13

The break frequency f_b (at $A_f = 9.995 \times 0.707 = 7.07$) is 983 Hz (expected value is 1 kHz). The upper frequency limit (i.e., 95 kHz) is due to the internal frequency behavior of the op-amp.

Frequency Response of Op-Amp Differentiators

Replacing the op-amp by its equivalent circuit gives the differentiator shown in Fig. 3.30(a). The addition of capacitor C_1 to the integrator in Fig. 3.21(a) sets a low cutoff frequency ω_L. Thus, the transfer function can be expressed as

$$A_f(s) = \frac{-(R_F/R_1)s}{(s + \omega_L)(1 + s/\omega_i)(1 + s/\omega_F)} \tag{3.75}$$

C_F will dominate the high cutoff frequency f_H. That is, $\omega_L < \omega_H << \omega_i$. The typical frequency response is shown in Fig. 3.30(b). The output will increase with frequency until $f = f_L = 1/(2\pi C_1 R_1)$ and then remain constant between f_L and f_H. The circuit can be made to operate effectively until f_L only, after which we can let the gain fall by making $f_L = f_H$, as shown in Fig. 3.30(b) by the light-colored line.

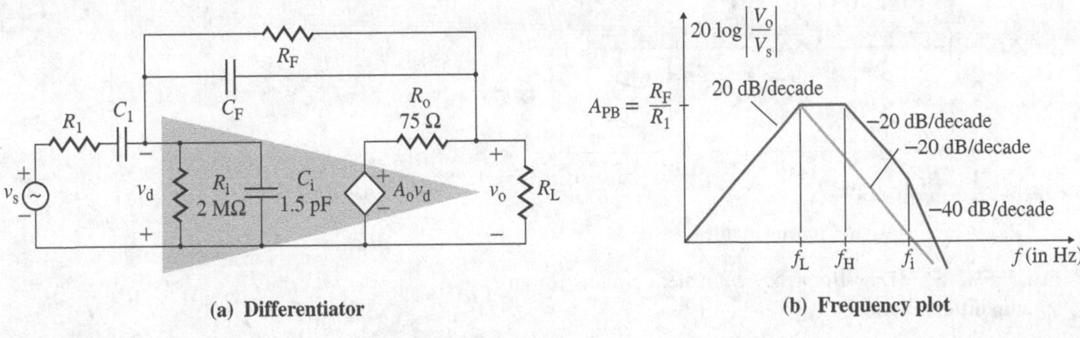

FIGURE 3.30 Op-amp differentiator circuit

EXAMPLE 3.14

D **Designing a differentiator circuit to give a specified frequency response** Design a differentiator circuit as shown in Fig. 3.30(a) to give **(a)** $f_L = 1$ kHz and $f_H = 5$ kHz and **(b)** $f_L = f_H = 5$ kHz. The pass-band gain is $A_{PB} = -20$. The op-amp parameters are $C_i = 1.5$ pF, $R_i = 2$ MΩ, $R_o = 75$ Ω, and open-loop voltage gain $A_o = 2 \times 10^5$.

SOLUTION

If we assume that C_1 is short-circuited and the other capacitors are open-circuited, the pass-band voltage gain is given by $A_{PB} \approx -R_F/R_1$. If we let $R_1 = 5$ kΩ,

$$R_F = |A_{PB}|R_1 = 20 \times 5 = 100 \text{ k}\Omega$$

(a) We will first consider $f_L = 1$ kHz and $f_H = 5$ kHz. The low-frequency equivalent circuit is shown in Fig. 3.31. We can see from Eq. (3.65) that the effective resistance R_x due to R_F is very small because the voltage gain A_o (i.e., g_m [$=A_o/R_o$]) is very large and the op-amp input voltage v_d is very small. Thus, Thevenin's equivalent resistance seen by C_1 becomes

$$R_{C1} = R_1 + (R_i \| R_x) \approx R_1$$

and Thevenin's equivalent resistance seen by C_F becomes

$$R_{CF} = R_F \| R_y \approx R_F$$

The low 3-dB frequency is given by

$$f_L = \frac{1}{2\pi R_{C1} C_1} \tag{3.76}$$

which gives

$$C_1 = \frac{1}{2\pi R_1 f_L} = \frac{1}{2\pi \times 5\text{ k} \times 1 \text{ kHz}} = 31.83 \text{ nF}$$

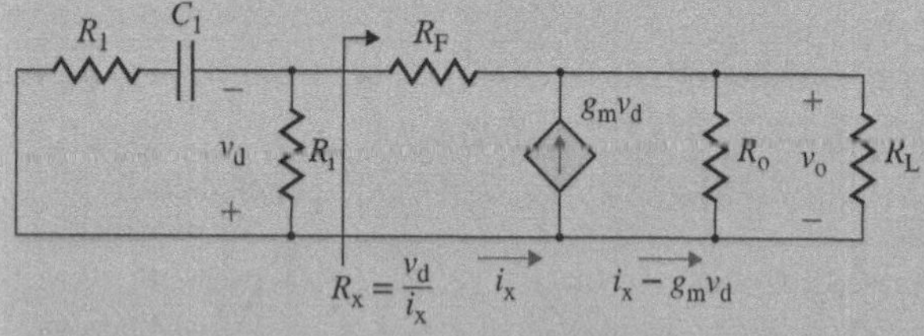

FIGURE 3.31 Low-frequency equivalent circuits for an op-amp differentiator

The high 3-dB frequency is given by

$$f_H = \frac{1}{2\pi R_F C_F} \tag{3.77}$$

which gives

$$C_F = \frac{1}{2\pi R_F f_H} = \frac{1}{2\pi \times 100\text{ k} \times 5\text{ k} \times} = 318.3\text{ pF}$$

(b) For $f_L = f_H = 5$ kHz, $C_F = 318.3$ pF, and Eq. (3.76) gives

$$C_1 = \frac{1}{2\pi R_1 f_L} = \frac{1}{2\pi \times 5\text{ k} \times 5\text{ k} \times} = 6.37\text{ pF}$$

3.5.3 Differential Amplifiers

In the differential amplifier configuration, shown in Fig. 3.32, two input voltages (v_a and v_b) are applied—one to the noninverting terminal and another to the inverting terminal. Resistances R_a and R_x are used to step down the voltage applied to the noninverting terminal. Let us apply the superposition theorem to find the output voltage v_O. That is, we will find the output voltage v_{oa}, which is due to the input voltage v_a only, and then we will find the output voltage v_{ob}, which is due to v_b only. The output voltage will be the sum of v_{oa} and v_{ob}.

The voltage v_p can be related to the input voltage v_a by

$$v_p = \frac{R_x}{R_x + R_a} v_a \tag{3.78}$$

Applying Eqs. (3.18) and (3.78) gives the output voltage v_{oa}, which is due to the input at the noninverting terminal, as

$$v_{oa} = \left(1 + \frac{R_F}{R_1}\right) v_p = \left(1 + \frac{R_F}{R_1}\right)\left(\frac{R_x}{R_x + R_a}\right) v_a \tag{3.79}$$

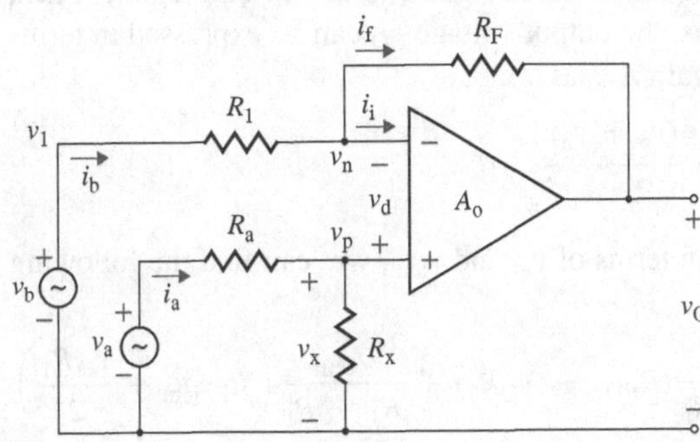

FIGURE 3.32 Differential amplifier

Applying Eq. (3.39) gives the output voltage v_{ob}, which is due to the input at the inverting terminal, as

$$v_{ob} = -\frac{R_F}{R_1}v_b \tag{3.80}$$

Therefore, the resultant output voltage is given by

$$v_O = v_{ob} + v_{oa} = -\frac{R_F}{R_1}v_b + \left(1 + \frac{R_F}{R_1}\right)\left(\frac{R_x}{R_x + R_a}\right)v_a \tag{3.81}$$

which, for $R_a = R_1$ and $R_F = R_x$, becomes

$$v_O = (v_a - v_b)\frac{R_F}{R_1} \tag{3.82}$$

Thus, the circuit in Fig. 3.32 can operate as a differential voltage amplifier with a closed-loop voltage gain of $A_f = R_F/R_1$. For example, if $v_a = 3$ V, $v_b = 5$ V, $R_a = R_1 = 12$ kΩ, and $R_F = R_x = 24$ kΩ, then Eq. (3.82) gives

$$v_O = \frac{(3 - 5) \times 24\text{ k}\Omega}{12\text{ k}\Omega} = -4\text{ V}$$

If all the resistances have the same values (i.e., $R_a = R_1 = R_F = R_x$), Eq. (3.82) is reduced to

$$v_O = v_a - v_b \tag{3.83}$$

in which case the circuit will operate as a difference amplifier. For example, if $v_a = 3$ V, $v_b = 5$ V, and $R_a = R_F = R_1 = R_x = 20$ kΩ, then Eq. (3.83) gives

$$v_O = v_a - v_b = 3 - 5 = -2\text{ V}$$

CMRR of a Differential Amplifier

Similar to the CMRR expression in Eq. (3.46) for an inverting amplifier, we can derive the CMRR of a differential amplifier. Considering two input signals v_b and v_a, we have $v_{id} = v_a - v_b$ and $v_{icm} = (v_a + v_b)/2$. The voltage v_p at the noninverting terminal is

$$v_p = \frac{R_x}{R_a + R_x}v_a = \frac{R_F}{R_1 + R_F}v_a \quad \text{for} \quad R_a = R_1 \text{ and } R_x = R_F$$

and the voltage v_n at the inverting terminal can be expressed in terms of v_O by

$$v_n = v_O + \frac{R_F}{R_1 + R_F}(v_b - v_O) = v_O\frac{R_1}{R_1 + R_F} + v_b\frac{R_F}{R_1 + R_F}$$

Similarly, let v_{do} and v_{cmo} denote the differential and common-mode signals at the op-amp input. Then, we have $v_{do} = (v_p - v_n)$ and $v_{cmo} = (v_p + v_n)/2$. Thus, the output voltage v_O can be expressed in terms of the op-amp differential gain A_d and common-mode gain A_{cm} as

$$v_O = A_d v_{do} + A_{cm}v_{cmo} = A_d(v_p - v_n) + \frac{A_{cm}(v_p + v_n)}{2}$$

After substituting for v_p and v_n using the relationships in terms of v_{id} and v_{icm}, we can find the following expression for v_O:

$$v_O = A_d(v_p - v_n) + \frac{A_{cm}}{2}(v_p + v_n) = \frac{A_d}{R_1 + R_F}(R_F v_{id} - v_O R_1) + \frac{A_{cm}}{R_1 + R_F}\left(R_F v_{icm} + \frac{v_O R_1}{2}\right)$$

This can be solved for v_O after we collect all the terms containing v_O in terms of v_{id} and v_{cm}:

$$v_O = \frac{A_d R_F/R_1 v_d + A_{cm} R_F/R_1 v_{icm}}{1 + R_F/R_1 + A_d - A_{cm}/2} \tag{3.84}$$

Since $A_d >> A_{cm}$ and $A_d >> R_F/R_1 >> 1$, the denominator of Eq. (3.84) is approximately equal to A_d, and v_O in Eq. (3.84) can be approximated to

$$v_O \approx \frac{A_d}{A_d}\frac{R_F}{R_1} v_{id} + \frac{A_{cm}}{A_c}\frac{R_F}{R_1} v_{icm} = \frac{R_F}{R_1} v_{id} + \frac{1}{\text{CMRR}_o}\frac{R_F}{R_1} v_{icm} \tag{3.85}$$

where $\text{CMRR}_o = A_d/A_{cm}$ is the CMRR of the op-amp. From Eq. (3.85), we can find the differential voltage gain $A_{d\text{-amp}}$ and the common-mode gain $A_{cm\text{-amp}}$ of the difference amplifier:

$$A_{d\text{-amp}} = \frac{R_F}{R_1}$$
$$A_{cm\text{-amp}} = \frac{1}{\text{CMRR}_o}\frac{R_F}{R_1} \tag{3.86}$$

Therefore, we can find the CMRR of the difference amplifier, which is the same as that of the op-amp:

$$\text{CMRR}_{amp} = \frac{A_{d\text{-amp}}}{A_{cm\text{-amp}}} = \text{CMRR}_o \tag{3.87}$$

Since the common-mode signal can be orders of magnitude higher than the differential signal, the common-mode component of Eq. (3.85) can be significant. Therefore, the basic differential amplifier shown in Fig. 3.32 suffers from two disadvantages: a low input resistance and an insufficient common-mode rejection, because R_F/R_1 has in general a high value.

3.5.4 Instrumentation Amplifiers

An instrumentation amplifier is a dedicated differential amplifier with an extremely high input impedance. Its gain can be precisely set by a single resistance. It has a high common-mode rejection capability (i.e., it can reject a signal that is common to both terminals but amplify a differential signal), and this feature is useful for receiving small signals buried in large common-mode offsets or noise. Therefore, instrumentation amplifiers are commonly used as signal conditioners of low-level (often DC) signals in large amounts of noise. The circuit diagram of an instrumentation amplifier is shown in Fig. 3.33. The amplifier consists of two stages. The first stage is the differential stage. Each input signal (v_{S1} or v_{S2}) is applied directly to the noninverting terminal of its op-amp in order to provide the very high input impedance. The second stage is a difference amplifier, which gives a low output impedance and can also allow voltage gain.

The voltage drop between the input terminals of an op-amp is very small, tending to zero: $v_{d1} = v_{d2} = 0$. Thus, the voltage drop across the middle resistor R_g of the potential divider is

$$v_{rg} = v_{S1} - v_{S2}$$

which gives the current i_{rg} through R_g as

$$i_{rg} = \frac{v_{rg}}{R_g} = \frac{v_{S1} - v_{S2}}{R_g}$$

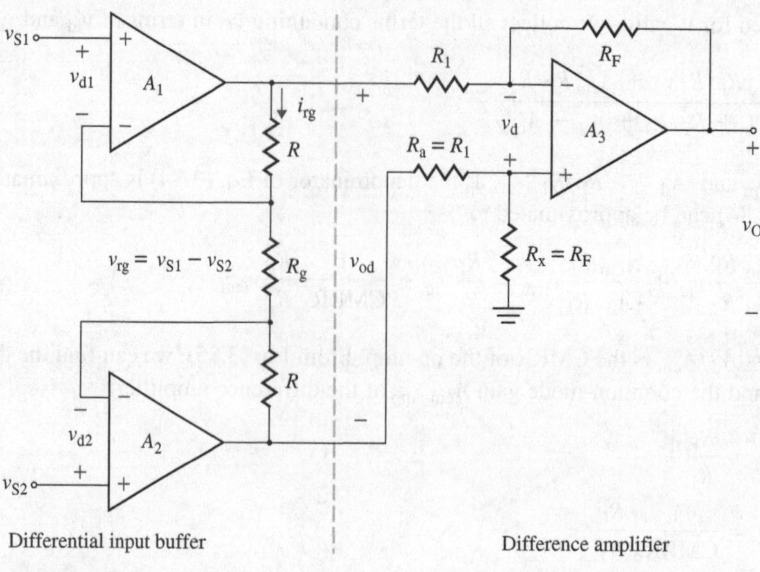

FIGURE 3.33 Instrumentation amplifier

This current flows through all three of the resistors because the currents flowing into the input terminals of the op-amps are practically zero. Therefore, the output voltage of the differential stage becomes

$$v_{od} = i_{rg}(R_g + 2R) = \frac{v_{S1} - v_{S2}}{R_g}(R_g + 2R) = (v_{S1} - v_{S2})\left(1 + \frac{2R}{R_g}\right).$$

Using Eq. (3.82), we can calculate the output voltage v_O as

$$v_O = -\,v_{od}\frac{R_F}{R_1} = -\,(v_{S1} - v_{S2})\left(1 + \frac{2R}{R_g}\right)\left(\frac{R_F}{R_1}\right) \tag{3.88}$$

which is the output of the instrumentation amplifier. This gain is normally varied by R_g. If the gain variation is not desired, then R_g can be removed and the differential amplifier can be made with two unity-gain voltage followers. This arrangement is shown in Fig. 3.34 by making $R_g = \infty$.

CMRR of an Instrumentation Amplifier

We can find the overall differential gain of an instrumentation amplifier from Eq. (3.88) as

$$A_{\text{d-amp}} = \left(1 + \frac{2R}{R_g}\right)\frac{R_F}{R_1} \tag{3.89}$$

It can easily be shown that the common-mode voltage gain of the first stage is unity. Applying, therefore, the common-mode gain of the second stage from Eq. (3.86), we can find the common-mode gain of the amplifier:

$$A_{\text{cm-amp}} = \frac{1}{\text{CMRR}_o}\frac{R_F}{R_1} \tag{3.90}$$

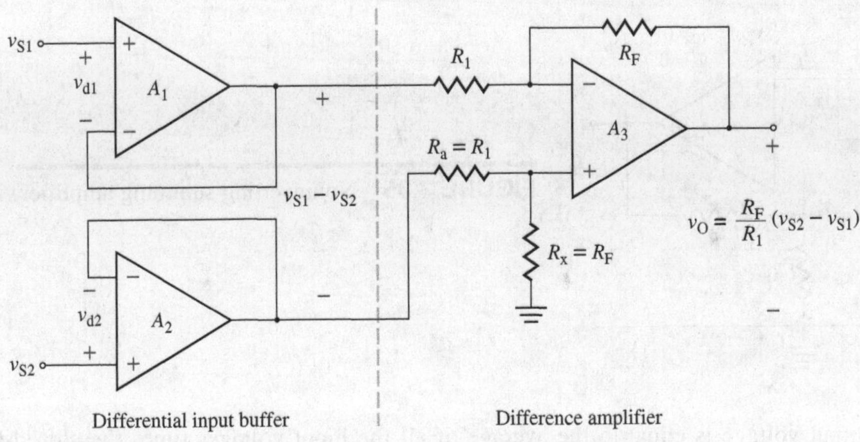

FIGURE 3.34 Instrumentation amplifier with fixed gain

Thus, the ratio of Eqs. (3.89) and (3.90) gives the CMRR of the instrumentation amplifier:

$$\text{CMRR}_{\text{amp}} = \left(1 + \frac{2R}{R_g}\right)\text{CMRR}_o \tag{3.91}$$

Therefore, the CMRR of the instrumentation amplifier is greater than that of the op-amps by a factor of $(1 + 2R/R_g)$, which can be large. For example, if we make $R_F = R_1$, there is a large multiplying factor of $(1 + 2R/R_g)$ for the differential voltage gain, but not for the common-mode gain.

3.5.5 Noninverting Summing Amplifiers

The basic noninverting amplifier in Fig. 3.9 can be operated as a summing amplifier. A noninverting summing amplifier with three inputs is shown in Fig. 3.35. Summing amplifiers are commonly employed in analog computing. By the superposition theorem, the voltage v_p at the noninverting terminal is

$$\begin{aligned} v_p &= \frac{R_b \| R_c}{R_a + R_b \| R_c} v_a + \frac{R_a \| R_c}{R_b + R_a \| R_c} v_b + \frac{R_a \| R_b}{R_c + R_a \| R_b} v_c \\ &= \frac{R_A}{R_a} v_a + \frac{R_A}{R_b} v_b + \frac{R_A}{R_c} v_c \end{aligned} \tag{3.92}$$

where $R_A = (R_a \,\|\, R_b \,\|\, R_c)$ (3.93)

Applying Eq. (3.18) for the noninverting amplifier and Eq. (3.92) gives the output voltage:

$$v_O = \left(1 + \frac{R_F}{R_B}\right) v_p = \left(1 + \frac{R_F}{R_B}\right)\left(\frac{R_A}{R_a} v_a + \frac{R_A}{R_b} v_b + \frac{R_A}{R_c} v_c\right) \tag{3.94}$$

For $R_a = R_b = R_c = R$, Eq. (3.93) gives $R_A = R/3$, and Eq. (3.94) becomes

$$v_O = \left(1 + \frac{R_F}{R_B}\right)\left(\frac{v_a + v_b + v_c}{3}\right) \tag{3.95}$$

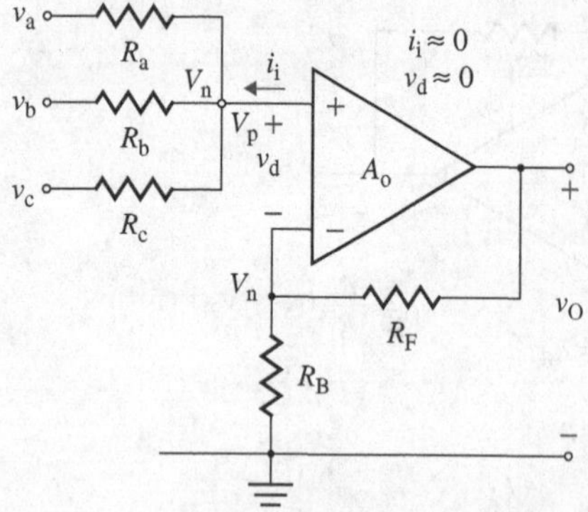

FIGURE 3.35 Noninverting summing amplifier

Thus, the output voltage is equal to the average of all the input voltages times the closed-loop gain $(1 + R_F/R_B)$ of the circuit. If the circuit is operated as a unity follower with $R_F = 0$ and $R_B = \infty$, the output voltage will equal the average of all the input voltages. That is,

$$v_O = \frac{v_a + v_b + v_c}{3} \tag{3.96}$$

If the closed-loop gain $(1 + R_F/R_B)$ is made equal to the number of inputs, the output voltage becomes equal to the sum of all the input voltages. That is, for three inputs, $n = 3$, and $(1 + R_F/R_B) = n = 3$. Then, Eq. (3.95) becomes

$$v_O = v_a + v_b + v_c \tag{3.97}$$

3.5.6 Inverting Summing Amplifiers

The basic inverting amplifier in Fig. 3.11 can be operated as an inverting summing amplifier. An inverting summing amplifier with three inputs is shown in Fig. 3.36. Depending on the values of the feedback resistance R_F and the input resistances R_1, R_2, and R_3, the circuit can be operated as a *summing amplifier*, a *scaling amplifier*, or an *averaging amplifier.* Since the output voltage is inverted, another inverter may be required, depending on the desired polarity of the output voltage.

The value of R_x should equal the parallel combination of R_1, R_2, R_3, and R_F. That is,

$$R_x = (R_1 \,\|\, R_2 \,\|\, R_3 \,\|\, R_F) \tag{3.98}$$

For an ideal op-amp, $v_d \approx 0$. Using Ohm's law, we get

$$i_1 = \frac{v_1}{R_1}, \quad i_2 = \frac{v_2}{R_2}, \quad i_3 = \frac{v_3}{R_3}, \quad i_f = -\frac{v_O}{R_F}$$

Since the current flowing into the op-amp is zero ($i_i = 0$),

$$i_1 + i_2 + i_3 = i_f$$

or

$$\frac{v_1}{R_1} + \frac{v_2}{R_2} + \frac{v_3}{R_3} = -\frac{v_O}{R_F} \tag{3.99}$$

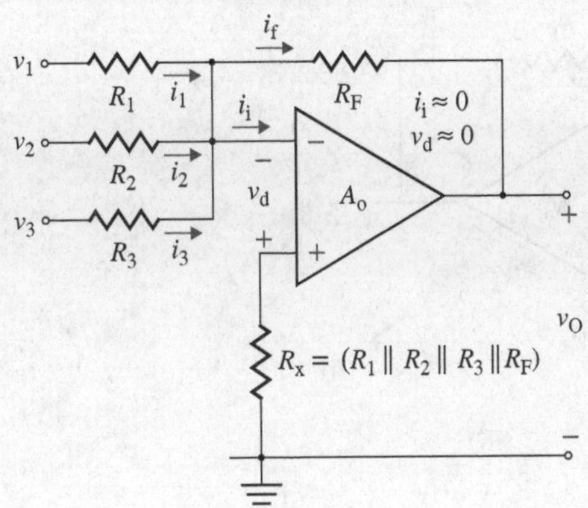

FIGURE 3.36 Inverting summing amplifier

which gives the output voltage as

$$v_O = -\left(\frac{R_F}{R_1}v_1 + \frac{R_F}{R_2}v_2 + \frac{R_F}{R_3}v_3\right) \tag{3.100}$$

Thus, v_O is a weighted sum of the input voltages, and this circuit is also called a *weighted*, or *scaling*, *summer*. If $R_1 = R_2 = R_3 = R_F = R$, Eq. (3.100) is reduced to

$$v_O = -(v_1 + v_2 + v_3) \tag{3.101}$$

and the circuit becomes a summing amplifier. If $R_1 = R_2 = R_3 = nR_F$, where n is the number of input signals, the circuit operates as an averaging amplifier. For three inputs, $n = 3$, and Eq. (3.100) becomes

$$v_O = -\frac{v_1 + v_2 + v_3}{3} \tag{3.102}$$

3.5.7 Addition–Subtraction Amplifiers

The functions of noninverting and inverting summing amplifiers can be implemented by only one op-amp, as shown in Fig. 3.37, in order to give output voltage of the form

$$v_O = A_1v_a + A_2v_b + A_3v_c - B_1v_1 - B_2v_2 - B_3v_3$$

where A_1, A_2, A_3, B_1, B_2, and B_3 are the gain constants. The resistances R_x and R_y are included to make the configuration more general. Applying Eqs. (3.94) and (3.100) gives an expression for the resultant output voltage:

$$v_O = \left(1 + \frac{R_F}{R_B}\right)\left(\frac{R_A}{R_a}v_a + \frac{R_A}{R_b}v_b + \frac{R_A}{R_c}v_c\right) - \left(\frac{R_F}{R_1}v_1 + \frac{R_F}{R_2}v_2 + \frac{R_F}{R_3}v_3\right) \tag{3.103}$$

where $R_A = (R_a \| R_b \| R_c \| R_x)$ (3.104)

$$R_B = (R_1 \| R_2 \| R_3 \| R_y) \tag{3.105}$$

To minimize the effects of offset biasing currents on the output of op-amps (discussed further in Sec. 14.3), Thevenin's equivalent resistance looking from the noninverting terminal is normally made equal to that looking from the inverting terminal. That is,

$$(R_B \| R_F) = R_A \tag{3.106}$$

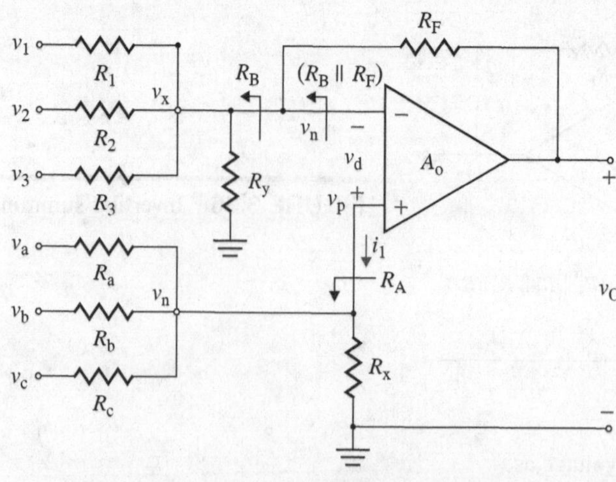

FIGURE 3.37 Addition–subtraction amplifier

or
$$\frac{R_B R_F}{R_B + R_F} = R_A \tag{3.106}$$

Using this condition, we can simplify the term $(1 + R_F/R_B)R_A$:

$$\left(1 + \frac{R_F}{R_B}\right)R_A = \left(1 + \frac{R_F}{R_B}\right)\left(\frac{R_B R_F}{R_B + R_F}\right) = R_F$$

Substituting this relation into Eq. (3.103) yields

$$v_O = \left(\frac{R_F}{R_a}v_a + \frac{R_F}{R_b}v_b + \frac{R_F}{R_c}v_c\right) - \left(\frac{R_F}{R_1}v_1 + \frac{R_F}{R_2}v_2 + \frac{R_F}{R_3}v_3\right) \tag{3.107}$$

which has the general form

$$v_O = A_1 v_a + A_2 v_b + A_3 v_c - B_1 v_1 - B_2 v_2 - B_3 v_3$$

Equation (3.107) is valid only if the condition of Eq. (3.106) is satisfied. For known values of gain constants As and Bs, the resistance values can be determined. Difficulty arises, however, in determining values of R_x and R_y that meet the criteria of Eq. (3.106). A technique proposed by W. P. Vrbancis [11] can be applied to determine the values of R_x and R_y. If details and proof of this technique are omitted, the design procedures can be simplified to the following steps:

Step 1. Add all the positive coefficients: $A = A_1 + A_2 + A_3$.

Step 2. Add all the negative coefficients: $B = B_1 + B_2 + B_3$.

Step 3. Define a parameter $C = A - B - 1$.

Step 4. Depending on the value of C, determine the values of R_x and R_y:

a. If $C > 0$, $R_x = \infty$ and $R_y = R_F/C$.

b. If $C < 0$, $R_x = -R_F/C$ and $R_y = \infty$.

c. If $C = 0$, $R_x = \infty$ and $R_y = \infty$.

Step 5. Choose a suitable value of R_F, and find the values of the other components. R_F is normally chosen to meet one of the following constraints:

a. If the equivalent resistance R_A is to be set to a particular value, R_F can be found from the relation $R_F = MR_A$, where M is the largest value of A, or $(B + 1)$.

b. If the minimum value of any resistances is to be limited to R_{min}, R_F can be found from the relation $R_F = NR_{min}$, where N is the largest value of $A_1, A_2, A_3, B_1, B_2, B_3$, or C.

(If it is not necessary to meet any of these conditions, we can complete the design by choosing a suitable value of R_F.)

Step 6. If the value of any resistor is too high or too low, we can multiply all the resistances by a constant without affecting the output voltage or the condition of Eq. (3.106).

EXAMPLE 3.15

D **Designing a summing op-amp circuit for a certain resistance R_A** Design an inverting and a noninverting summing amplifier of the configuration shown in Fig. 3.37 to give an output voltage of the form

$$v_O = 4v_a + 6v_b + 3v_c - 7v_1 - v_2 - 5v_3$$

The equivalent resistance R_A is to be set to 15 kΩ.

SOLUTION

The coefficients are $A_1 = 4$, $A_2 = 6$, $A_3 = 3$, $B_1 = 7$, $B_2 = 1$, and $B_3 = 5$. Let us follow the design steps just described.

Step 1. $A = 4 + 6 + 3 = 13$.

Step 2. $B = 7 + 1 + 5 = 13$.

Step 3. $C = A - B - 1 = 13 - 13 - 1 = -1$.

Step 4. Since $C < 0$, $R_x = -R_F/C = R_F$ and $R_y = \infty$.

Step 5. The design can be completed by choosing a value of R_F. For the given value of $R_A = 15$ kΩ, $R_F = MR_A$. In this case, $M = B + 1 = 13 + 1 = 14$. Thus, the values are as follows:

$$R_F = R_y = MR_A = 14 \times 15 = 210 \text{ k}\Omega$$

$$R_a = \frac{R_F}{A_1} = \frac{210 \text{ k}}{4} = 52.5 \text{ k}\Omega$$

$$R_b = \frac{R_F}{A_2} = \frac{210 \text{ k}}{6} = 35 \text{ k}\Omega$$

$$R_c = \frac{R_F}{A_3} = \frac{210 \text{ k}}{3} = 70 \text{ k}\Omega$$

$$R_x = -\frac{R_F}{C} = 210 \text{ k}\Omega$$

$$R_1 = \frac{R_F}{B_1} = \frac{210 \text{ k}}{7} = 30 \text{ k}\Omega$$

$$R_2 = \frac{R_F}{B_2} = \frac{210\text{ k}}{1} = 210\text{ k}\Omega$$

$$R_3 = \frac{R_F}{B_3} = \frac{210\text{ k}}{5} = 42\text{ k}\Omega$$

$$R_y = \infty$$

Check:
From Eq. (3.104),

$$R_A = (52.5\text{ k}\Omega \,\|\, 35\text{ k}\Omega \,\|\, 70\text{ k}\Omega \,\|\, 210\text{ k}\Omega) = 15\text{ k}\Omega$$

From Eq. (3.105),

$$R_B = (30\text{ k}\Omega \,\|\, 210\text{ k}\Omega \,\|\, 42\text{ k}\Omega) = 16.15\text{ k}\Omega$$

From Eq. (3.106),

$$R_B \,\|\, R_F = (16.15\text{ k}\Omega \,\|\, 210\text{ k}\Omega) = 15\text{ k}\Omega$$

Thus, the condition of $R_A = (R_B \,\|\, R_F)$ is satisfied.

EXAMPLE 3.16

D **Designing a summing op-amp circuit for a minimum resistance R_{min}** Design an inverting and a noninverting summing amplifier of the configuration shown in Fig. 3.37 to give an output voltage of the form

$$v_O = 8v_a + 6v_b + 3v_c - 7v_1 - v_2 - 5v_3$$

The minimum value of any resistance is to be set to $R_{min} = 15\text{ k}\Omega$.

SOLUTION

The coefficients are $A_1 = 8$, $A_2 = 6$, $A_3 = 3$, $B_1 = 7$, $B_2 = 1$, and $B_3 = 5$. Let us follow the design steps described earlier.

Step 1. $A = 8 + 6 + 3 = 17$.

Step 2. $B = 7 + 1 + 5 = 13$.

Step 3. $C = A - B - 1 = 17 - 13 - 1 = 3$.

Step 4. Since $C > 0$, $R_x = \infty$ and $R_y = R_F/C = R_F/3$.

Step 5. The design can be completed by choosing a value of R_F. For the given value of $R_{min} = 15\text{ k}\Omega$, $R_F = NR_{min}$, where N is the largest value of A_1, A_2, A_3, B_1, B_2, B_3, or C. In this case, $N = 8$. Thus, the values are as follows:

$$R_F = NR_{min} = 8 \times 15\text{ k} = 120\text{ k}\Omega$$

$$R_a = \frac{R_F}{A_1} = \frac{120\text{ k}}{8} = 15\text{ k}\Omega$$

$$R_b = \frac{R_F}{A_2} = \frac{120\text{ k}}{6} = 20\text{ k}\Omega$$

$$R_c = \frac{R_F}{A_3} = \frac{120\text{ k}}{3} = 40\text{ k}\Omega$$

$$R_x = \infty$$

$$R_1 = \frac{R_F}{B_1} = \frac{120\text{ k}}{7} = 17.14\text{ k}\Omega$$

$$R_2 = \frac{R_F}{B_2} = \frac{120\text{ k}}{1} = 120\text{ k}\Omega$$

$$R_3 = \frac{R_F}{B_3} = \frac{120\text{ k}}{5} = 24\text{ k}\Omega$$

$$R_y = \frac{R_F}{C} = \frac{120\text{ k}}{3} = 40\text{ k}\Omega$$

Check:
From Eq. (3.104),

$$R_A = (15\text{ k}\Omega \parallel 20\text{ k}\Omega \parallel 40\text{ k}\Omega) = 7.06\text{ k}\Omega$$

From Eq. (3.105),

$$R_B = (17.14\text{ k}\Omega \parallel 120\text{ k}\Omega \parallel 24\text{ k}\Omega \parallel 40\text{ k}\Omega) = 7.5\text{ k}\Omega$$

From Eq. (3.106),

$$R_B \parallel R_F = (7.5\text{ k}\Omega \parallel 120\text{ k}\Omega) = 7.06\text{ k}\Omega$$

Thus, the condition of $R_A = (R_B \parallel R_F)$ is satisfied.

3.5.8 Optocoupler Drivers

Optocouplers, also known as *optical isolators,* are generally used to transfer electrical signals from one part of a system to another without direct electrical connection. They find many applications in instrumentation for electrical power engineering, where direct electrical connections between low-level signals and high-current power lines must be avoided, and in medical electronics, where direct connections between patients and electrical power systems must be avoided.

An optocoupler consists of a light-emitting diode (LED), which emits light when forward current is applied, and a photodiode, which converts light to electrical current proportional to the incident light. The light power produced by an LED is directly proportional to the current through the diode. However, the output power is a nonlinear function of the diode voltage. Therefore, an optocoupler is supplied by a current source.

An optocoupler drive circuit is shown in Fig. 3.38. This circuit is a modification of the inverting op-amp shown in Fig. 3.11. Since the current flowing through the op-amp is very small, tending to zero, $i_S = i_f$. Thus, the voltage across R_2 is

$$v_O = -R_F i_f = -R_F i_S$$

The load current i_O is given by

$$i_O = i_f - i_1 = i_S - \frac{v_O}{R_2} = i_S + \frac{R_F i_S}{R_2} = \left(1 + \frac{R_F}{R_2}\right) i_S \tag{3.108}$$

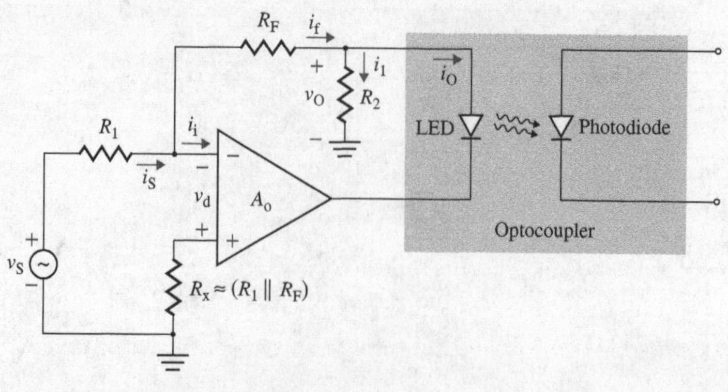

FIGURE 3.38 Optocoupler drive circuit

Therefore, the circuit operates as a current amplifier. The LED acting as the load does not determine the load current i_O. Only the multiplier R_F/R_2 determines the load current. Substituting $i_S \approx v_S/R_1$ gives the output current as a function of the input voltage. That is,

$$i_O = \left(1 + \frac{R_F}{R_2}\right)\left(\frac{1}{R_1}\right)v_S \tag{3.109}$$

The circuit then operates as a transconductance amplifier (or voltage–current converter).

3.5.9 Photodetectors

A photodiode produces a current that is a linear function of the light intensity; this current is normally measured as incident optical power density D_P. The ratio of the output current to the incident optical power density is called the *current responsitivity*. This current can be measured by an inverting op-amp of the type shown in Fig. 3.11, which is a current–voltage converter. The output voltage depends on the input current. From Eq. (3.37) for $v_d = 0$, we get

$$v_O = -R_F i_f = -R_F i_S$$

A simple light-sensing circuit consisting of a photodiode and an inverting op-amp is shown in Fig. 3.39. The anode terminal of the diode can be connected to either the ground or a negative voltage. However, a reverse-biasing voltage will reduce the diode junction capacitance, which in turn decreases the frequency (or transient) response time of the circuit.

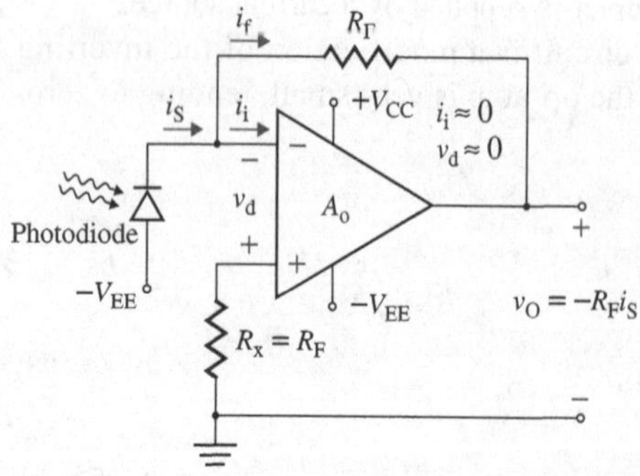

FIGURE 3.39 Photodetector circuit

EXAMPLE 3.17

D

Designing an op-amp photodetector circuit Design a photodetector circuit of the form shown in Fig. 3.39 to give an output voltage of $v_O = -200$ mV at an incident power density of $D_P = 500$ nW/cm^2. The current responsivity of the photodiode is $D_i = 1$ A/W, and the active area is $a = 40$ mm^2.

SOLUTION

The power produced by the photodiode is

$$P = D_P a = (500 \text{ nW/cm}^2) \times 40 \text{ mm}^2 = 200 \text{ nW}$$

Therefore, the current produced by the diode is

$$i_S = PD_i = 1 \text{ A/W} \times 200 \text{ nW} = 200 \text{ nA}$$

The output voltage is $v_O = -R_F i_S$, which, for $i_S = 200$ nA and $v_O = -200$ mV, gives

$$R_F = -\frac{v_O}{i_S} = \frac{200 \text{ mV}}{200 \text{ nA}} = 1 \text{ M}\Omega$$

3.5.10 Voltage–Current Converters

If the input signal is a voltage source and it is transmitted to a remote load, the load current will depend on the series resistance between the input signal and the load. Even a small drop across the series resistance could significantly change the percentage error of the load voltage. Any changes in the load resistance due to wear and tear or temperature will contribute to the error. The simplest type of voltage–current converter, shown in Fig. 3.40(a), is a modification of the basic noninverting amplifier shown in Fig. 3.9(a). The current through the resistor R_1 is given by

$$i_O = i_1 = \frac{v_S - v_d}{R_1} = \frac{v_S}{R_1} \tag{3.110}$$

Thus, the output current i_O through the load resistance R depends only on v_S and R_1, not on R. For a fixed value of R_1, i_O is directly proportional to v_S. Note that none of the load terminals in Fig. 3.40(a) is connected to the ground. That is, the load is floating. The advantage of this arrangement is that no common-mode signal (i.e., noise) will appear across the load.

Op-amps are primarily voltage amplifiers; their current-carrying capability is very limited. Many applications (such as indicators and actuators) require regulated variable current, which is beyond the op-amp's capability. The circuit shown in Fig. 3.40(b) can provide the load current i_L proportional to the input voltage v_S. The output of the op-amp forces the base current through transistor Q_1, resulting in a proportional collector current through Q_1, the load R_L, and R_1. The load current i_L can be controlled by varying either the input voltage or the value of R_1. The value of the base resistance R must be sufficiently large to protect the base–emitter junction of Q_1 and to limit the output current of the op-amp. Also, the DC supply voltage $V_{CC} \geq R_L i_L$ ($\approx R_L i_1 = v_S R_L / R_1$). The load resistance R_L is floating. Thus, the circuit cannot be used with a grounded load.

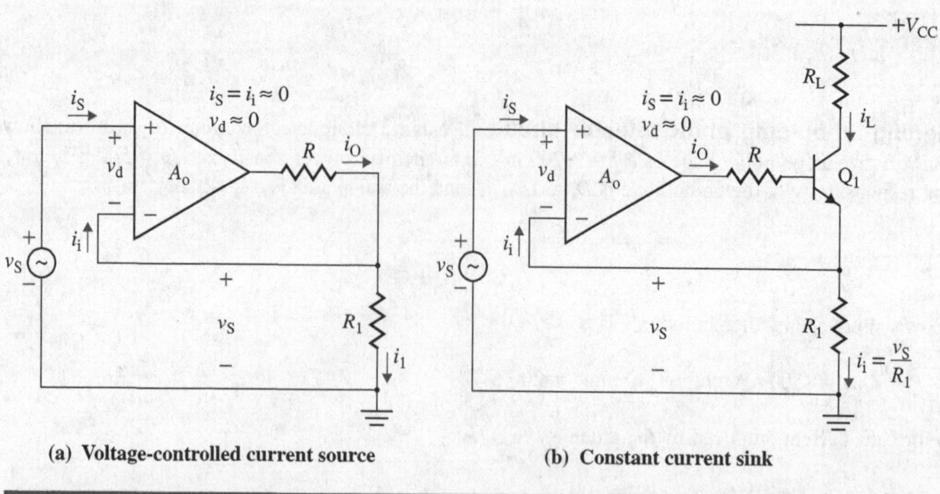

FIGURE 3.40 Voltage–current converter

3.5.11 DC Voltmeters

The voltage–current converter in Fig. 3.40(a), which consists of a noninverting amplifier, can be used as a DC voltmeter, as shown in Fig. 3.41. Since all signals are DC quantities, we will use uppercase symbols. A moving coil meter with an internal resistance of R_m is connected in the feedback path. For an ideal op-amp, $v_d \approx 0$; the meter current is given by

$$I_M = I_1 = \frac{V_x}{R_1} = \frac{V_S - v_d}{R_1} = \frac{V_S}{R_1} \tag{3.111}$$

which gives the relation between the input voltage and the meter current as

$$V_S = R_1 I_M \tag{3.112}$$

Thus, the input voltage V_S can be measured from the deflection of the meter, which is proportional to I_M. If the full-scale deflection current of the moving coil is $I_{M(max)} = 100\ \mu A$ and $R_1 = 2\ M\Omega$, the full-scale reading will be $V_{S(max)} = R_1 I_{M(max)} = 2\ M\Omega \times 100\ \mu A = 200$ V.

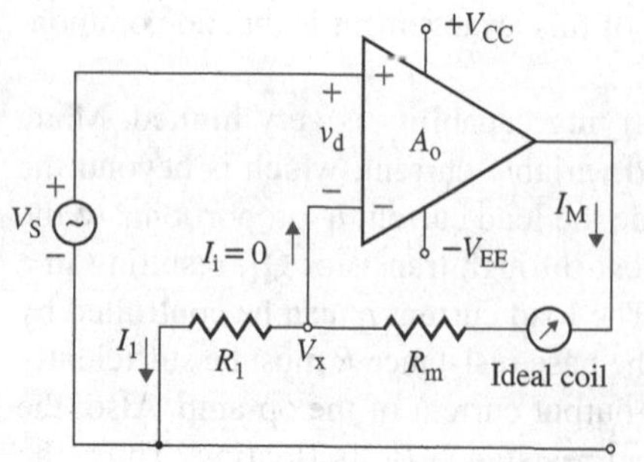

FIGURE 3.41 DC voltmeter

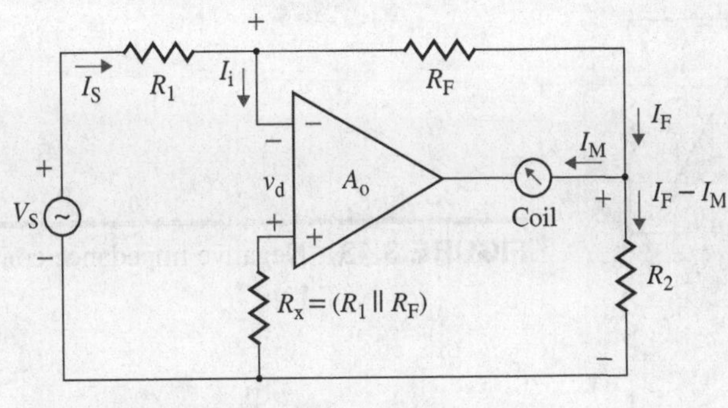

FIGURE 3.42 DC millivoltmeter

3.5.12 DC Millivoltmeters

The inverting amplifier in Fig. 3.11 can be operated as a DC millivoltmeter, as shown in Fig. 3.42. This circuit is similar to the optocoupler drive in Fig. 3.38, except that the LED is shorted, and we expect similar equations. As before, we will use uppercase symbols for DC quantities. For an ideal amplifier, $v_d = 0$ and $I_i = 0$. The current through R_1, which is the same as that through R_F, is

$$I_S = I_F = \frac{V_S}{R_1} \tag{3.113}$$

Applying Kirchhoff's voltage law around the loop formed by op-amp inputs R_F and R_2 yields

$$-v_d = R_F I_F + R_2(I_F - I_M) \quad \text{or} \quad 0 = R_F I_F + R_2(I_F - I_M)$$

from which we can find the meter current I_M:

$$I_M = \frac{R_F + R_2}{R_2} I_F = \left(1 + \frac{R_F}{R_2}\right) I_F = \left(1 + \frac{R_F}{R_2}\right)\frac{V_S}{R_1} \tag{3.114}$$

This equation is the same as Eq. (3.109) for the optocoupler in Fig. 3.38. If $R_F >> R_2$, which is usually the case, Eq. (3.114) can be approximated by

$$I_M \approx \frac{R_F}{R_1}\left(\frac{1}{R_2}\right) V_S \tag{3.115}$$

from which we can find the input voltage V_S in terms of the meter current I_M:

$$V_S = \frac{R_1 R_2}{R_F} I_M \tag{3.116}$$

$$= R_2 I_M \quad \text{for } R_1 = R_F$$

If $R_1 = R_F = 150\ \text{k}\Omega$, $R_2 = 1\ \text{k}\Omega$, and the full-scale deflection current of the moving coil is $I_{M(max)} = 100\ \mu\text{A}$, the full-scale reading will be $V_{S(max)} = R_2 I_{M(max)} = 1\ \text{k}\Omega \times 100\ \mu\text{A} = 100\ \text{mV}$.

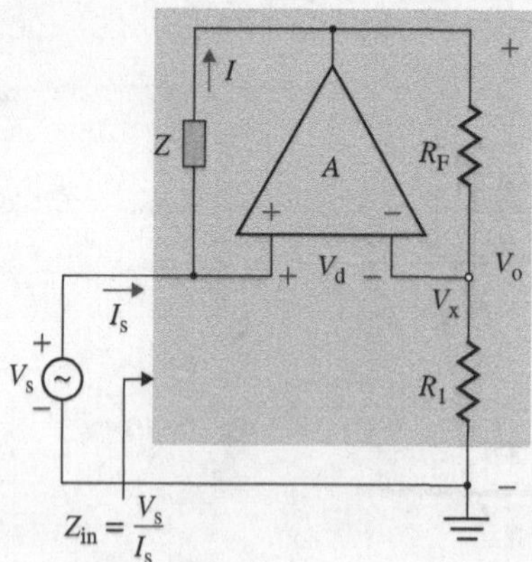

FIGURE 3.43 Negative impedance converter

3.5.13 Negative Impedance Converters

Some applications (e.g., oscillators, which we will study in Chapter 13) require the characteristic of negative resistance (or impedance) to compensate for any undesirable resistance (or impedance). The op-amp circuit shown in Fig. 3.43 can be employed to obtain this characteristic. Since the circuit has an impedance Z, all voltages and currents will have a magnitude and a phase angle. All quantities are expressed in rms values, and we will use uppercase symbols.

Since $v_d \approx 0$,

$$V_s = V_x + V_d = V_x$$

Applying Eq. (3.18) for the noninverting amplifier, we get the rms output voltage:

$$V_o = \left(1 + \frac{R_F}{R_1}\right)V_s$$

Since the current drawn by the op-amp is zero, the current I flowing through the impedance Z is the same as the input current I_s. That is,

$$I = I_s = \frac{V_s - V_o}{Z} = \frac{1}{Z}\left(V_s - V_s - \frac{R_F}{R_1}V_s\right) = -\frac{R_F}{ZR_1}V_s \tag{3.117}$$

which gives the input impedance Z_{in} as

$$Z_{in} = \frac{V_s}{I_s} = -Z\left(\frac{R_1}{R_F}\right) \tag{3.118}$$

If Z is replaced by a resistance R, then $Z = R$. The circuit will behave as a negative resistance, and Eq. (3.118) becomes

$$Z_{in} = R_{in} = -R\left(\frac{R_1}{R_F}\right) \tag{3.119}$$

Thus, the ratio R_1/R_F acts as a multiplying factor for R. If $R_1 = R_F = R$, Eq. (3.119) becomes

$$R_{in} = -R \tag{3.120}$$

For example, if $R_1 = R_F = R = 10\ \text{k}\Omega$, the circuit in Fig. 3.43 will behave as a resistance of $R_{in} = -10\ \text{k}\Omega$.

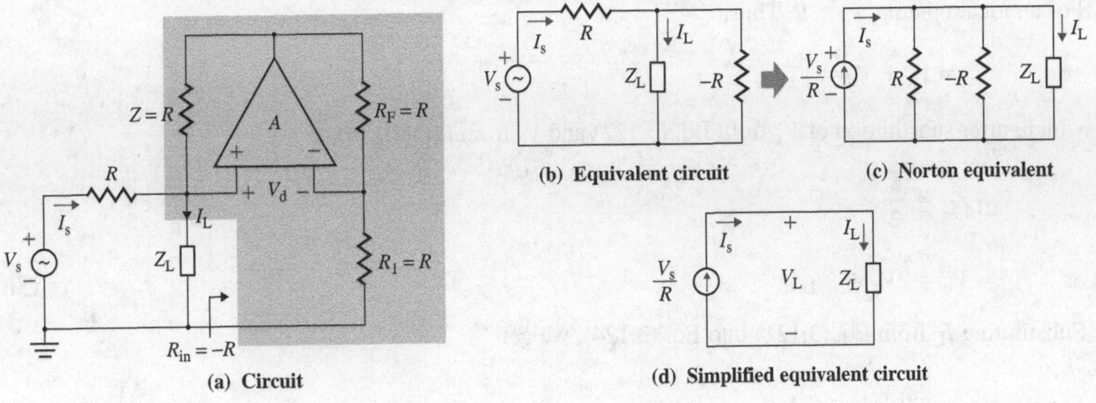

FIGURE 3.44 Constant current source

3.5.14 Constant Current Sources

It is often necessary to generate a constant current source from a voltage source. The circuit of Fig. 3.43 can be modified to convert a voltage source to a current source, as shown in Fig. 3.44(a). One side of the load Z_L is connected to the ground. If $R_1 = R_F$ and $Z = R$, the input resistance becomes $R_{in} = -R$. The circuit inside the shaded area can be replaced by $-R$; the equivalent circuit is shown in Fig. 3.44(b). The voltage source V_s can be replaced by its Norton equivalent, as shown in Fig. 3.44(c). Since the parallel combination of R and $-R$ is infinite, or an open circuit, Fig. 3.44(c) can be reduced to Fig. 3.44(d). The current flowing into load impedance Z_L is simply

$$I_L = I_s = \frac{V_s}{R} \tag{3.121}$$

Thus, the load current I_L is directly proportional to the input voltage V_s and is independent of the load impedance Z_L. To simplify the design, we can choose $R_1 = R_F = R$.

3.5.15 Noninverting Integrators

The integrators in Figs. 3.14(a) and 3.16(a) invert the polarity of the input signal and thus require an additional unity-gain inverter to get a signal of the same polarity. The circuit of Fig. 3.44(a) can operate as a noninverting integrator if the impedance Z_L is replaced by a capacitor, as shown in Fig. 3.45(a). That is,

$$R_1 = R_F = R \quad \text{and} \quad Z_L = X_c = \frac{1}{(j\omega C)}$$

Since $I_i \approx 0$, the voltage at the inverting terminal is given by

$$V_x = \frac{R_1}{R_1 + R_F} V_o = \frac{R}{R + R} V_o = \frac{V_o}{2} \tag{3.122}$$

The voltage across the capacitor is given by

$$V_c = I_L Z_L \tag{3.123}$$

For an ideal op-amp, $v_d \approx 0$. Thus,

$$V_c = V_x + V_d = V_x$$

which, after substitution of V_x from Eq. (3.122) and V_c from Eq. (3.123), gives

$$I_L Z_L = \frac{V_o}{2}$$

or $$V_o = 2V_c = 2I_L Z_L \tag{3.124}$$

Substituting I_L from Eq. (3.121) into Eq. (3.124), we get

$$V_o = \frac{2Z_L V_s}{R} = \frac{2V_s}{j\omega CR} \tag{3.125}$$

which, if converted into the time domain, gives the output voltage as

$$v_O(t) = \frac{2}{CR}\int v_S(t)\,dt + 2V_{co} \tag{3.126}$$

where V_{co} is the initial capacitor voltage at the beginning of integration. The charging of the capacitor can be represented by an equivalent circuit, as shown in Fig. 3.45(b). Thus, the capacitor voltage v_C can be found directly from Fig. 3.45(b) as follows:

$$v_C(t) = \frac{1}{C}\int i_S(t)\,dt + V_{co} = \frac{1}{RC}\int v_S\,dt + V_{co} \tag{3.127}$$

Thus, $v_O(t) = 2v_x(t) = 2v_C(t)$.

▶ **NOTE** Since one terminal of the capacitor C is grounded, the capacitor can be charged easily to a desired initial condition at the beginning of integration.

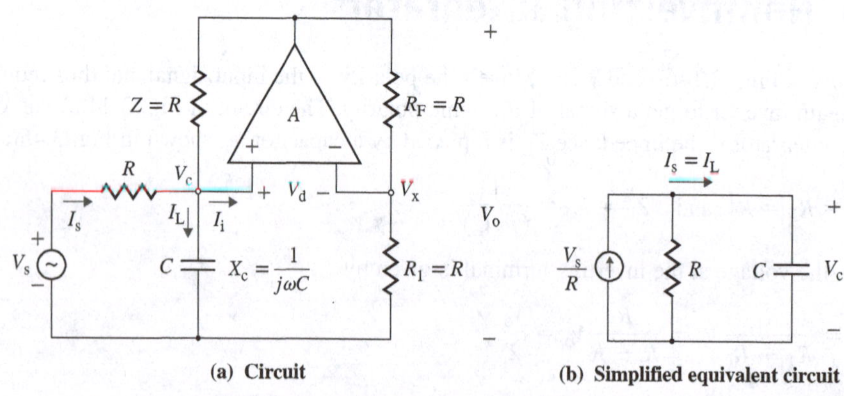

FIGURE 3.45 Noninverting integrator

3.5.16 Inductance Simulators

An op-amp circuit can be used to simulate the characteristic of an inductor. Such an op-amp circuit is shown in Fig. 3.46(a). It consists of two op-amps. The part of the circuit within the shaded area is identical to the negative impedance converter of Fig. 3.43; we can apply Eq. (3.118) to replace it with an equivalent impedance, provided we substitute $Z = R_3$, $R_1 = R_4$, and $R_F \equiv Z_C = 1/(j\omega C)$. Thus, the equivalent impedance is given by

$$Z_L = \frac{V_1}{I_1} = -R_3\left(\frac{R_4}{Z_C}\right) \tag{3.128}$$

If the circuit within the shaded area is replaced by Z_L, the resultant circuit also becomes a negative impedance converter, as shown in Fig. 3.46(b). Applying Eq. (3.118) gives the input impedance of the circuit:

$$Z_{in} = \frac{V_s}{I_s} = -R_1\left(\frac{Z_L}{R_2}\right) \tag{3.129}$$

Substituting Z_L from Eq. (3.128) into Eq. (3.129) yields

$$Z_{in} = \frac{V_s}{I_s} = -R_1\left(\frac{1}{R_2}\right)(-R_3)\left(\frac{R_4}{Z_C}\right)$$

$$= j\omega C\frac{R_1R_3R_4}{R_2} = j\omega L_e \tag{3.130}$$

where L_e is the effective inductance given by

$$L_e = \frac{R_1R_3R_4}{R_2}C \tag{3.131}$$

Therefore, by choosing the values of R_1, R_2, R_3, R_4, and C, we can simulate the desired value of inductance L_e.

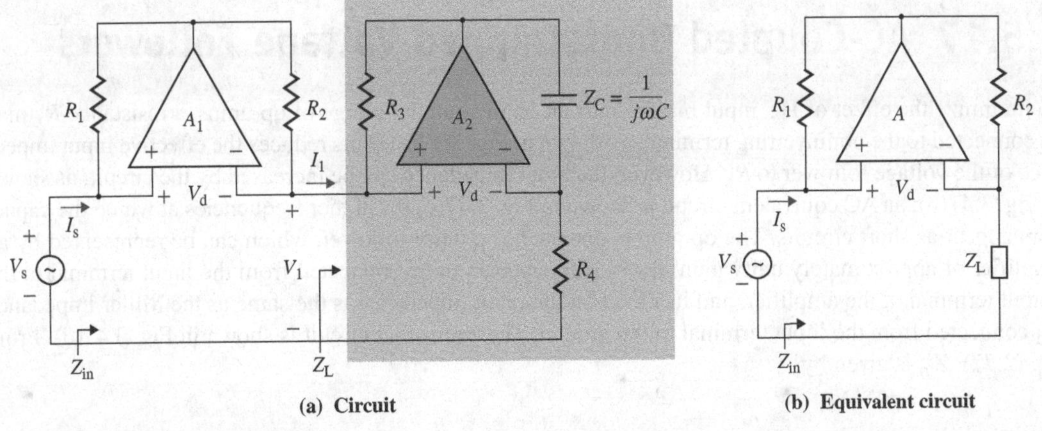

(a) Circuit

(b) Equivalent circuit

FIGURE 3.46 Inductance simulator

▶ **NOTES**

1. From this theoretical development it might appear that inductance simulators could be used in many applications as a replacement for bulky physical inductors. Because of the physical limitations of op-amps, however, inductance simulators suffer from many drawbacks and do not find many practical applications.
2. The op-amp nonlinearities begin limiting the behavior of the inductance simulator at appallingly low frequencies (even less than 20 Hz), and the inductor does not reduce the current at high frequencies as expected.
3. Inductors are commonly used in electrical power applications for storing magnetic energy. A simulated inductor cannot be used to store energy in a magnetic field, so it cannot be used in electrical power circuits (i.e., as a power filter).

EXAMPLE 3.18

D **Designing an op-amp inductance simulator** Determine the values required for the components in Fig. 3.46(a) in order to simulate an inductor of $L = 1$ mH.

SOLUTION

Let $R_3 = R_4 = 100$ kΩ and $C = 10$ pF. From Eq. (3.131), we get

$$\frac{R_2}{R_1} = \frac{R_3 R_4 C}{L_e} = \frac{100 \times 10^3 \times 100 \times 10^3 \times 10 \times 10^{-12}}{(1 \times 10^{-3})} = 100$$

If $R_1 = 5$ kΩ, then $R_2 = 100 \times 5 = 500$ kΩ.

▶ NOTE: To use Eq. (3.131), the designer needs to know the values of five quantities to find the value of L_e. The designer has to assume four values, and there is no unique solution to this design problem.

3.5.17 AC-Coupled Bootstrapped Voltage Followers

To minimize the effect of DC input biasing current on the output voltage of op-amps, a resistance R_x may be connected to the noninverting terminal, as shown in Fig. 3.47(a). This reduces the effective input impedance of the voltage follower to R_x. However, the input impedance can be increased by the circuit, as shown in Fig. 3.47(b); an AC equivalent circuit is shown in Fig. 3.47(c) for higher frequencies at which the capacitors appear as short circuits. The op-amp is operated as a unity follower, which can be represented by an amplifier of approximately unity gain: $A_v \approx 1$. R_F appears to be connected from the input terminal to the output terminal of the amplifier, and its effect on the input impedance is the same as the Miller impedance Z_{in} connected from the input terminal to the ground. The equivalent circuit is shown in Fig. 3.47(d). From Eq. (2.77), Z_{in} is given by

$$Z_{in} = \frac{V_s}{I_s} = \frac{R_F}{1 - A_v} \qquad (3.132)$$

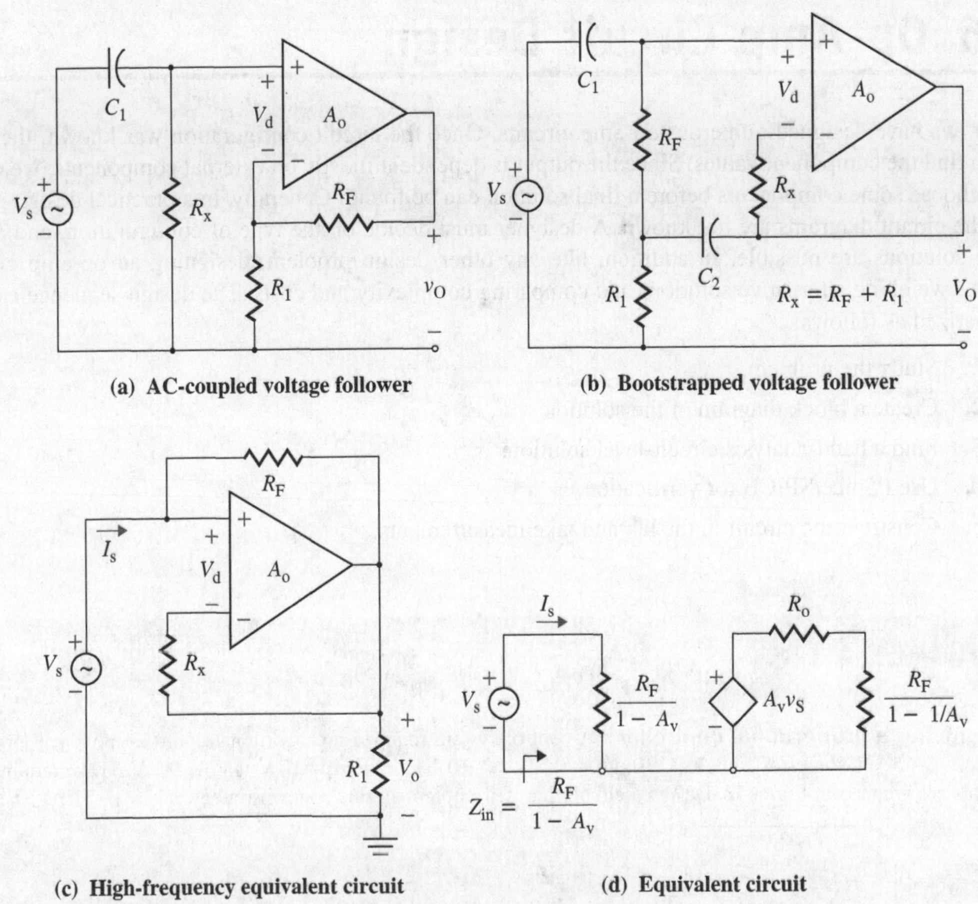

FIGURE 3.47 AC-coupled bootstrapped voltage follower

which, for $A_v \approx 1$, yields $Z_{in} = \infty$. Since the amplifier gain is unity, the output voltage equals the input voltage and there is no voltage drop across R_F. Therefore, no current flows through R_F, and the input impedance is very high—ideally, infinity. Notice that the voltage at the end of R_F in Fig. 3.47(c) is "pulled up" to the value of the input voltage, thereby offering infinite input impedance. Because of this "bootstrap" characteristic, the circuit is known as a *bootstrapped amplifier*.

KEY POINT OF SECTION 3.5

- The three basic op-amp configurations—inverting, noninverting, and differential—can be applied to perform various signal-processing functions *such as integrators, differentiators, inductance simulators, meters, limiters, detectors, comparators, and precision rectifiers.*

3.6 Op-Amp Circuit Design

So far, we have designed numerous op-amp circuits. Once the circuit configuration was known, the task was to find the component values. Since the output is dependent mostly on external components, we often must choose some components before a final solution can be found. Generally, in a practical design problem, the circuit diagrams are not known. A designer must decide on the type of configuration, and alternative solutions are possible. In addition, like any other design problem, designing an op-amp circuit requires weighing alternative solutions and comparing complexity and costs. The design sequence can be summarized as follows:

Step 1. Study the problem.
Step 2. Create a block diagram of the solution.
Step 3. Find a hand-analysis circuit-level solution.
Step 4. Use PSpice/SPICE for verification.
Step 5. Construct the circuit in the lab and take measurements.

EXAMPLE 3.19

D **Designing a proportional controller** A control system requires a proportional controller that will produce $v_O = 5$ V if the error signal $v_e = 0$, $v_O = 0$ if $v_e \leq -0.1$ V, and $v_O = 10$ V if $v_e \geq 0.1$ V. These requirements are graphed in Fig. 3.48. Design a circuit that will implement this control strategy.

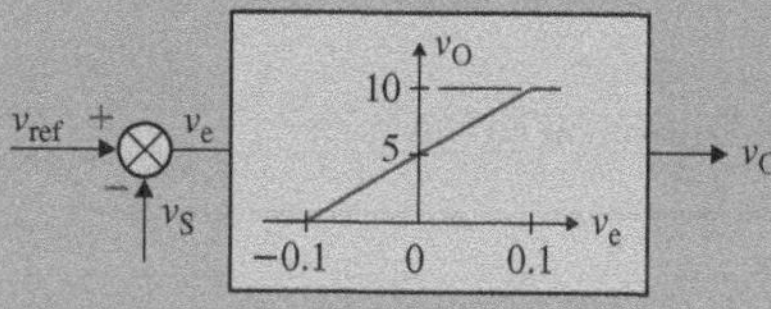

FIGURE 3.48 Proportional controller

SOLUTION

Step 1. Study the problem. The output voltage is related to the error voltage by

$$v_O = 50v_e + 5 = 50(V_{ref} - v_S) + 5 = 50V_{ref} - 50v_S + 5$$

Step 2. Create a block diagram of the solution. The problem requires a summing amplifier, as shown in Fig. 3.49(a). Since the signal v_S is expected to be positive, we also need an inverter.

Step 3. Devise a hand-analysis circuit-level solution. The inverting summing amplifier and the circuit implementation are shown in Fig. 3.49(b). Let $R_1 = R_2 = 10$ kΩ, $R_F = 50R_1 = 500$ kΩ, and $R_3 = R_F = 500$ kΩ. Choose $V_{CC} = 12$ V. Since the maximum output voltage is 10 V, there is no need for a voltage-limiting circuit.

Step 4. Use PSpice/SPICE for verification. You are encouraged to plot v_O against v_S for $v_S = 4.6$ V to 5.4 V in increments of 0.01. Invoke DC sweep with the following statement: v_{ref} is set to −5 V.

```
.DC VS 4.6 5.4 0.01
```

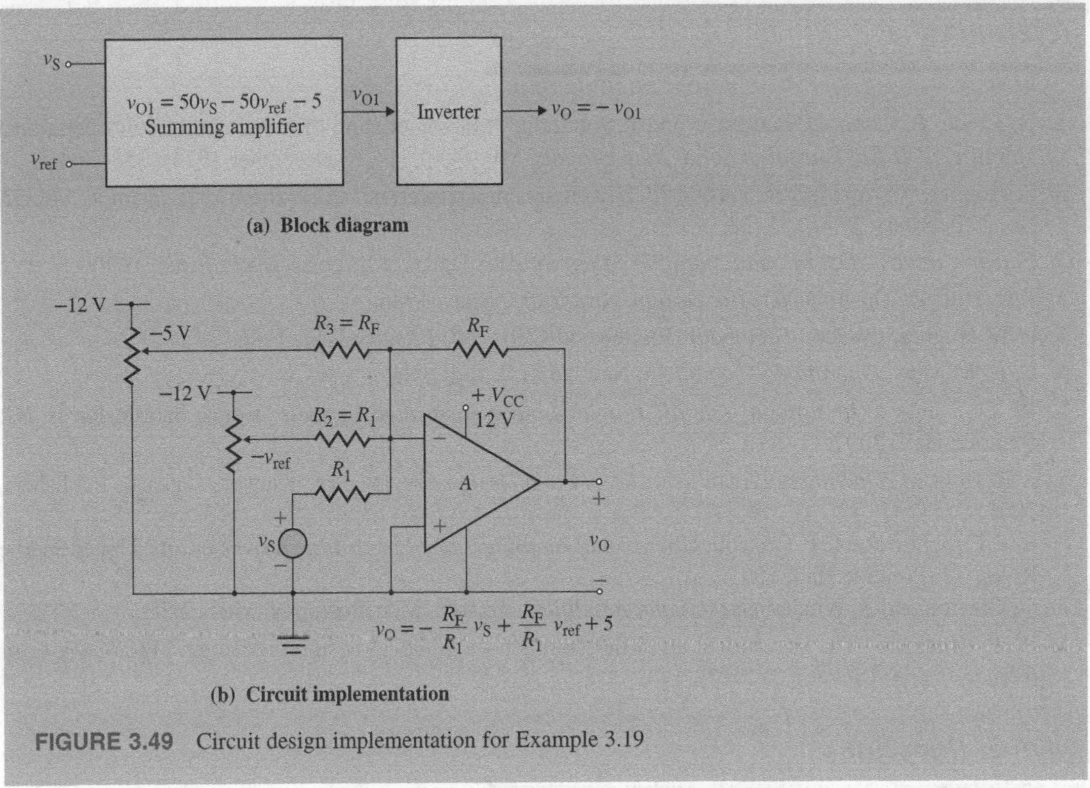

FIGURE 3.49 Circuit design implementation for Example 3.19

Summary

An op-amp is a high-gain differential amplifier that can perform various functions in electronic circuits. Op-amps are normally used with a feedback circuit, and the output voltage becomes almost independent of the op-amp parameters. The basic configurations of op-amp amplifiers can be used in many applications such as integrators, differentiators, inductance simulators, meters, limiters, detectors, comparators, and precision rectifiers.

The analysis of an op-amp circuit can be simplified by assuming ideal characteristics. An ideal op-amp has a very high voltage gain, a very high input resistance, a very low output resistance, and a negligible input current. The characteristics of practical op-amps differ from the ideal characteristics, but analyses based on the ideal conditions are valid for many applications and provide the starting point for practical circuit design. Although the DC model of op-amps can be used to analyze complex op-amp circuits, it does not take into account the frequency dependence and op-amp nonlinearities. If the op-amp is operated at frequencies higher than the op-amp break frequency, the effect of frequency dependence should be evaluated.

The op-amp macromodel gives better accuracy. However, the student version of PSpice allows simulation of an amplifier with only one op-amp. If the limit is reached, then the use of the AC model is recommended. The DC model should be the last choice unless the input signal is DC.

References

1. G. Boyle, B. Cohn, D. Pederson, and J. Solomon, "Macromodeling of integrated circuit operational amplifiers." *IEEE Journal of Solid-State Circuits*, Vol. SC-9, No. 6 (December 1974): 353–364.
2. S. Progozy, "Novel applications of SPICE in engineering education." *IEEE Trans. on Education*, Vol. 32, No. 1 (February 1990): 35–38.
3. *Linear Circuits—Operational Amplifier Macromodels*. Dallas, TX: Texas Instruments, 1990.
4. J. R. Hufault, *Op-Amp Network Design*. New York: Wiley, 1986.
5. F. W. Hughes, *Op-Amp Handbook*. Englewood Cliffs, NJ: Prentice Hall, 1986.
6. C. F. Wojslow, *Operational Amplifiers*. New York: Wiley, 1986.
7. W. D. Stanley, *Operational Amplifiers with Linear Integrated Circuits*. Upper Saddle River, NJ: Prentice Hall, 2002.
8. J. H. Huijsing, *Operational Amplifiers: Theory and Design*. Boston, MA: Kluwer Academic Publishers, 2001.
9. R. F. Coughlin and F. F. Driscoll, *Operational Amplifiers and Linear Integrated Circuits*. Upper Saddle River, NJ: Prentice Hall, 2001.
10. G. Clayton and S. Winder, *Operational Amplifiers*. Oxford, MA: Boston Newnes, 2003.
11. W. P. Vrbancis, "The operational amplifier summer—a practical design procedure." *WESCON Conference Record* (Session 2, 1982): 1–4.

Review Questions

1. What are the characteristics of an ideal op-amp?
2. What is the minimum number of terminals in an op-amp?
3. What is the typical open-loop voltage gain of an op-amp?
4. What is the typical input resistance of an op-amp?
5. What are the saturation voltages of an op-amp?
6. What is the purpose of supply voltages in an op-amp?
7. What is the PSS of an op-amp?
8. What is the CMRR of an op-amp?
9. What is the typical value of the output resistance of an op-amp?
10. Ideally, what should be the differential voltage gain of an op-amp?
11. Ideally, what should be the common-mode voltage gain of an op-amp?
12. What is the unity-gain bandwidth of an op-amp?
13. What is the effect of rise time on the frequency response of an op-amp?
14. What is the difference between a closed-loop gain and an open-loop gain?
15. What is the virtual ground of an op-amp?
16. What is the integration time constant?
17. What is the frequency response of an integrator?
18. What is the differentiator gain constant?
19. What are the problems of a differentiator?

20. What is the frequency response of a differentiator?
21. What is a voltage follower?
22. What are the advantages of a voltage follower?
23. What is the significance of negative resistance?
24. What is a weighted summing amplifier?

Problems

The symbol **D** indicates that a problem is a design problem. The symbol **P** indicates that you can check the solution to a problem using PSpice/SPICE.

3.2 Characteristics of Ideal Op-Amps

3.1 The op-amp in Fig. 3.3(a) has an open-loop gain of $A_o = 2 \times 10^5$. The input resistance is $R_i = 2\ \text{M}\Omega$. The DC supply voltages are $V_{CC} = 15$ V and $-V_{EE} = -15$ V.

a. What value of v_d will saturate the amplifier?

b. What is the value of op-amp input current i_i?

3.2 The op-amp shown in Fig. P3.2 is used as a noninverting amplifier. The values are $A_o = 10^5$, $V_{CC} = 12$ V, and $-V_{EE} = -12$ V. If $v_S = 50\ \mu$V, determine the output voltage v_O.

FIGURE P3.2

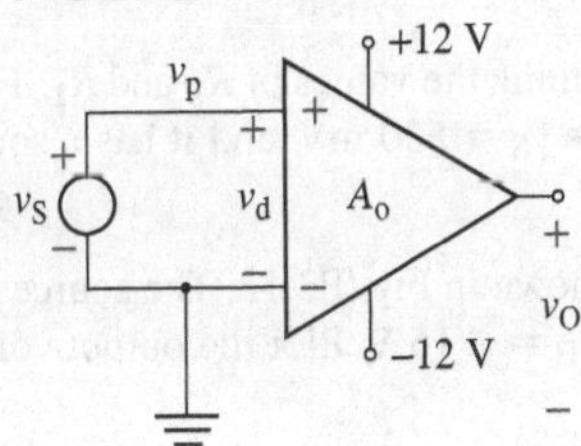

3.3 The op-amp shown in Fig. P3.3 is used as an inverting amplifier. The op-amp parameters are $A_o = 10^5$, $V_{CC} = 12$ V, and $-V_{EE} = -12$ V. If $v_S = 10\ \mu$V, determine the output voltage v_O.

FIGURE P3.3

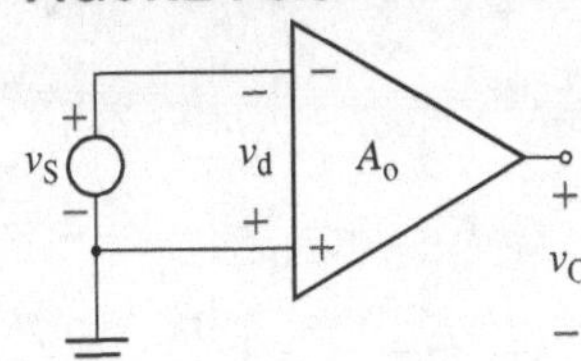

3.4 The op-amp in Fig. 3.3(a) has the following specifications: $A_o = 2 \times 10^5$, $R_i = 2\ \text{M}\Omega$, $R_o = 75\ \Omega$, $V_{CC} = 15$ V, $-V_{EE} = -15$ V, and maximum output voltage swing $= \pm 14$ V. If $v_+ = 0$ and $v_- = 2 \sin 377t$, plot the instantaneous output voltage v_O.

3.5 The op-amp in Fig. 3.3(a) has the following specifications: $A_o = 2 \times 10^5$, $R_i = 2\ \text{M}\Omega$, $R_o = 75\ \Omega$, $V_{CC} = 15$ V, $-V_{EE} = -15$ V, and maximum output voltage swing $= \pm 14$ V. If $v_+ = 75\ \mu$V and $v_- = -25\ \mu$V, determine the output voltage v_O.

3.6 The input voltages of an op-amp are $v_1 = 100\ \mu V$ and $v_2 = 60\ \mu V$. The op-amp parameters are CMRR = 90 dB and $A_d = A_o = 2 \times 10^5$. Determine **(a)** the differential voltage v_d, **(b)** the common-mode voltage v_c, **(c)** the magnitude of the common-mode gain A_c, and **(d)** the output voltage v_O.

3.3 *Op-Amp PSpice/SPICE Models*

3.7 P Develop PSpice/SPICE subcircuits for the DC model (in Fig. 3.7) and the AC model (in Fig. 3.8) for the LF411 op-amp. The parameters are $R_i = 10^{12}\ \Omega$, $R_o = 50\ \Omega$, $A_o = 2 \times 10^5$, break frequency $f_b = 20$ Hz, and unity-gain bandwidth $f_{bw} = 4$ MHz. Assume DC power supply voltages of ± 15 V.

3.8 P Develop PSpice/SPICE subcircuits for the DC model (in Fig. 3.7) and the AC model (in Fig. 3.8) for the LM324 op-amp. The parameters are $R_i = 2\ M\Omega$, $R_o = 50\ \Omega$, $A_o = 2 \times 10^5$, break frequency $f_b = 4$ kHz, and unity-gain bandwidth $f_{bw} = 1$ MHz. Assume a DC supply voltage of $+15$ V.

3.4 *Analysis of Ideal Op-Amp Circuits*

3.9 D P Design a noninverting amplifier as shown in Fig. 3.9(a) to provide a closed-loop voltage gain of $A_f = 100$. The input voltage is $v_S = 100$ mV with a source resistance of $R_s = 1\ k\Omega$. Find the value of output voltage v_O. The DC supply voltages are given by $V_{CC} = V_{EE} = 15$ V. Assume an ideal op-amp.

3.10 With the design values in Prob. 3.9, find the output voltage v_O, the input resistance $R_{in} = v_S/i_S$, and the output resistance R_{out} under the following conditions:
a. $A_o = 25 \times 10^3$, $R_i = 10^{12}\ \Omega$, and $R_o = 50\ \Omega$.
b. $A_o = 5 \times 10^5$, $R_i = 10^{12}\ \Omega$, and $R_o = 50\ \Omega$.
c. Use PSpice/SPICE to verify your results in parts (a) and (b).

3.11 Design a noninverting amplifier as shown in Fig. 3.9(a) by determining the values of R_F and R_1. The closed-loop gain should be $A_f = 10$. The input voltage to the amplifier is $v_S = 500$ mV, and it has a source resistance of 200 Ω. What is the value of output voltage v_O?

3.12 The input voltage to the noninverting amplifier in Fig. 3.9(a) is shown in Fig. P3.12. The source resistance R_s is negligible, $R_F = 20\ k\Omega$, $R_1 = 5\ k\Omega$, $V_{CC} = 15$ V, and $-V_{EE} = -15$ V. Plot the output voltage v_O if $R_F = 20\ k\Omega$ and $R_1 = 5\ k\Omega$.

FIGURE P3.12

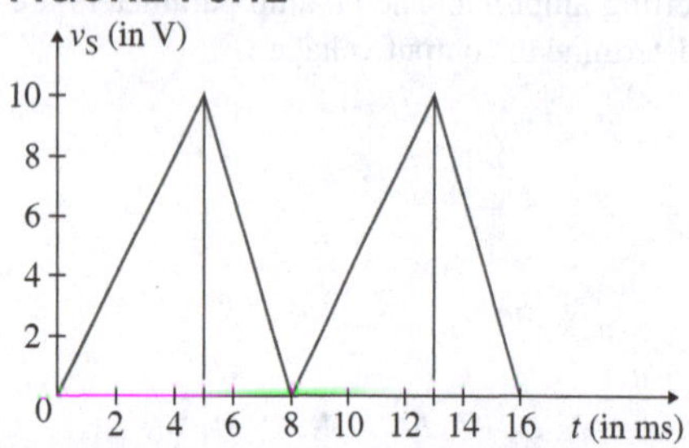

3.13 The noninverting op-amp amplifier in Fig. 3.9(a) has an open-loop gain of $A_o = 5 \times 10^3$, $R_1 = 10\ k\Omega$, and $R_F = 30\ k\Omega$. Calculate **(a)** the closed-loop voltage gain A_f, **(b)** the output voltage v_O, and **(c)** the error in output voltage if A_o is assumed to be infinite.

3.14 The input voltage to the noninverting amplifier in Fig. 3.9(a) is $v_S = 10 \sin(2000\pi t)$. The source resistance R_s is negligible. If $R_F = 20\ k\Omega$, $R_1 = 5\ k\Omega$, $V_{CC} = 15$ V, and $-V_{EE} = -15$ V, plot the output voltage v_O.

3.15 A voltage follower is shown in Fig. P3.15. The op-amp parameters are $A_o = 5 \times 10^5$, $R_o = 75\ \Omega$, and $R_i = 2\ \text{M}\Omega$. The input voltage is $v_S = 5$ V, and $R_s = 10$ kΩ. Find the output voltage v_O, the input resistance $R_{in} = v_S/i_S$, and the output resistance R_{out}.

FIGURE P3.15

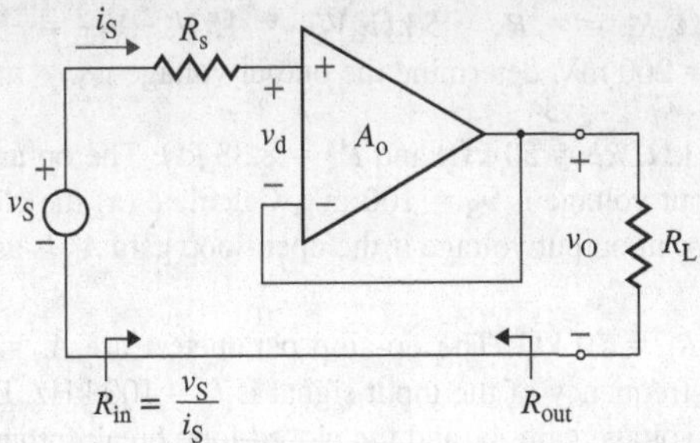

3.16 a. A noninverting amplifier has $R_1 = 15$ kΩ and $R_F = 50$ kΩ. The op-amp parameters are $A_o = 2 \times 10^5$, $f_b = 10$ Hz, $R_o = 75\ \Omega$, and $R_i = 2\ \text{M}\Omega$. The frequency of the input signal is $f_s = 100$ kHz. Determine the unity-gain bandwidth f_{bw}, the closed-loop voltage gain A_f, and the closed-loop break frequency f_c of the op-amp circuit.

b. Use PSpice/SPICE to plot the closed-loop frequency response of the voltage gain. Assume $v_s = 0.1$ V (AC), and use the linear AC model.

3.17 Repeat Prob. 3.16 for $R_1 = R_F = 15$ kΩ.

3.18 Two noninverting op-amps are cascaded as shown in Fig. P3.18. The unity-gain bandwidth of the op-amps is f_u=1 MHz, and the slew rate is SR = 6 V/μs.

a. If R_1=20 kΩ, R_2=200 kΩ, R_3=180 kΩ, R_4=50 kΩ, R_5=500 kΩ, and R_6=45 kΩ, determine the voltage gain $A_f = v_O/v_1$.

b. Determine the maximum frequency f_{max} of the input signal v_1 if the amplitude of the output voltage is limited to 10 V.

c. Use PSpice to verify your results.

FIGURE P3.18

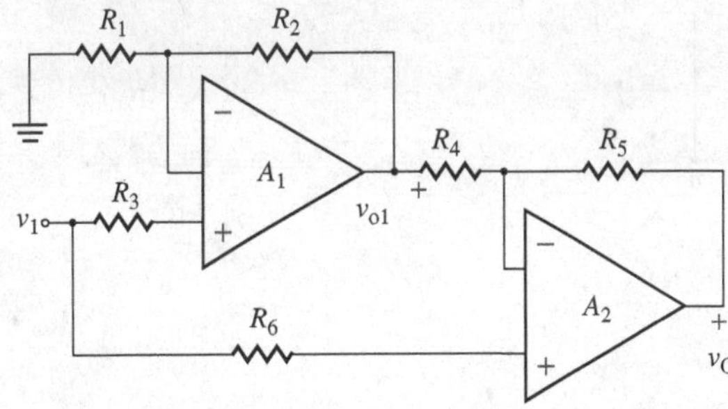

3.19 A transducer produces a voltage signal of $v_S = 50$ mV and has an internal resistance of $R_s = 5$ kΩ. Design the inverting op-amp amplifier of Fig. 3.11 by determining the values of R_1, R_F, and R_x. The output voltage should be $v_O = -5$ V. The current drawn from the transducer should not be more than 20 μA. Assume an ideal op-amp and $V_{CC} = V_{EE} = 12$ V.

3.20 With the design values in Prob. 3.19, find the value of output voltage v_O, the input resistance $R_{in} = v_S/i_S$, and the output resistance R_{out} under the following conditions:

P

a. $A_o = 25 \times 10^3$, $R_i = 10^{12}\ \Omega$, and $R_o = 50\ \Omega$.

b. $A_o = 5 \times 10^5$, $R_i = 10^{12}\ \Omega$, and $R_o = 50\ \Omega$.

c. Use PSpice/SPICE to verify your results in parts (a) and (b).

3.21 The inverting amplifier in Fig. 3.11 has $R_1 = 5\ \text{k}\Omega$, $R_F = \infty$, $R_x = 5\ \text{k}\Omega$, $V_{CC} = 15$ V, $-V_{EE} = -15$ V, and maximum output voltage swing $= \pm 14$ V. If $v_S = 200$ mV, determine the output voltage v_O.

3.22 The inverting amplifier in Fig. 3.11 has $R_1 = 10\ \text{k}\Omega$, $R_F = 50\ \text{k}\Omega$, and $R_x = 8.33\ \text{k}\Omega$. The op-amp has an open-loop voltage gain of $A_o = 2 \times 10^5$. The input voltage is $v_S = 100$ mV. Calculate **(a)** the closed-loop gain A_f, **(b)** the output voltage v_O, and **(c)** the error in output voltage if the open-loop gain A_o is assumed to be infinite.

3.23 **a.** An inverting amplifier has $R_1 = 15\ \text{k}\Omega$ and $R_F = 50\ \text{k}\Omega$. The op-amp parameters are $A_o = 2 \times 10^5$, $f_b = 10$ Hz, $R_o = 75\ \Omega$, and $R_i = 2\ \text{M}\Omega$. The frequency of the input signal is $f_s = 100$ kHz. Determine the unity-gain bandwidth f_{bw}, the closed-loop voltage gain A_f, and the closed-loop break frequency f_c of the op-amp circuit.

P

b. Use PSpice/SPICE to plot the closed-loop frequency response of the voltage gain. Assume $v_s = 0.1$ V (AC), and use the linear AC model.

3.24 Repeat Prob. 3.23 for $R_1 = R_F = 15\ \text{k}\Omega$.

3.25 The inverting amplifier shown in Fig. P3.25 has $R_1 = 50\ \text{k}\Omega$, and $R_2 = R_3 = 20\ \text{k}\Omega$. The unity-gain bandwidth of the op-amps is $f_u = 1$ MHz, and the slew rate is SR $= 6$ V/μs.

a. Determine the value of R_x that will give a voltage gain of $A_f = v_O/v_1 = -10$ V/V.

b. Determine the maximum frequency f_{max} of the input signal v_1 if the amplitude of the output voltage is limited to 10 V.

c. Use PSpice to verify your results.

FIGURE P3.25

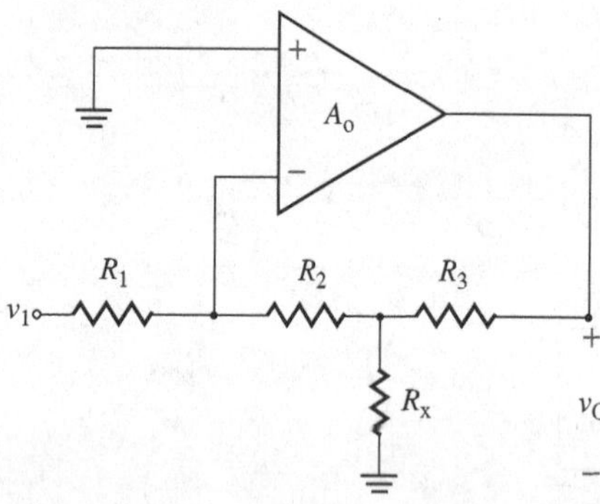

3.26 Two inverting op-amps are cascaded as shown in Fig. P3.26. The unity-gain bandwidth of the op-amps is $f_u = 1$ MHz, and the slew rate is SR $= 6$ V/μs.

a. If $R_1 = 20\ \text{k}\Omega$, $R_2 = 100\ \text{k}\Omega$, $R_3 = 150\ \text{k}\Omega$, $R_4 = 20\ \text{k}\Omega$, and $R_5 = 160\ \text{k}\Omega$, determine the voltage gain $A_f = v_O/v_1$.

b. Use PSpice to verify your results.

FIGURE P3.26

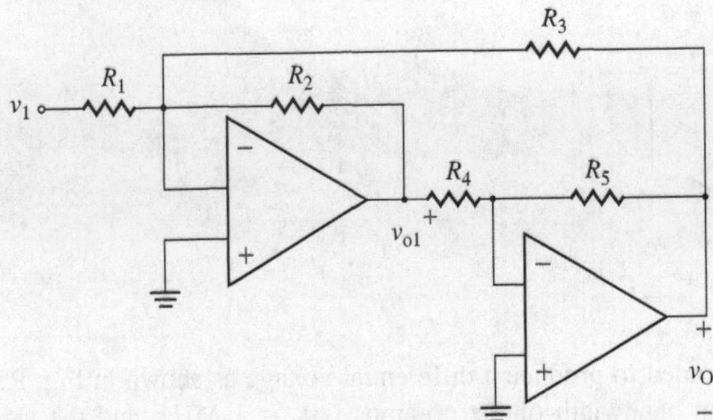

3.27 (D) Two transducers produce voltage signals of $v_b = 200$ mV and $v_a = 220$ mV. Design a differential amplifier as shown in Fig. 3.32 to produce an output voltage $|v_O| = 5$ V. Assume an ideal op-amp and $V_{CC} = V_{EE} = 12$ V.

3.28 (P) With the design values in Prob. 3.27, find the value of output voltage v_O under the following conditions of the op-amp:

a. $A_o = 25 \times 10^3$ and $R_i = 10^{12}\ \Omega$.

b. $A_o = 5 \times 10^5$ and $R_i = 10^{12}\ \Omega$.

c. Use PSpice/SPICE to verify your results in parts (a) and (b).

3.29 (D) **a.** Design a differential amplifier as shown in Fig. 3.32 to give a differential voltage gain of $|A_f| = 200$. The input voltages are $v_b = 70$ mV and $v_a = 50$ mV. Assume an ideal op-amp and $V_{CC} = V_{EE} = 12$ V.

b. Calculate the error in output voltage if the open-loop gain is $A_o = 5 \times 10^5$.

3.30 The values of the differential amplifier in Fig. 3.32 are $A_o = 5 \times 10^5$, $R_1 = 5$ kΩ, $R_F = 50$ kΩ, $R_a = 2$ kΩ, and $R_x = 20$ kΩ. The input voltages are $v_b = 5$ mV and $v_a = -15$ mV. Find the output voltage v_O.

3.31 The differential amplifier shown in Fig. P3.31 has $R_1 = 100$ kΩ, $R_2 = 150$ kΩ, $R_3 = 50$ kΩ, and $R_4 = R_5 = R_6 = R_7 = 50$ kΩ. The unity-gain bandwidth of the op-amps is $f_u = 1$ MHz, and the slew rate is SR = 6 V/μs.

a. Determine the output voltage in terms of v_1 and v_2.

b. Use PSpice to verify your results.

FIGURE P3.31

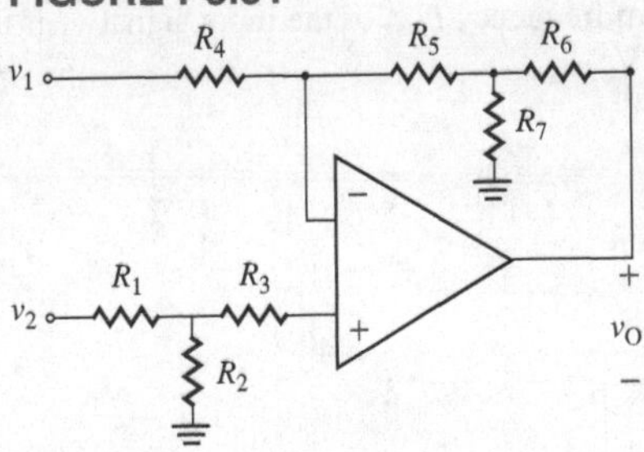

3.32 The differential amplifier shown in Fig. P3.32 has $R_1 = 40$ kΩ, $R_2 = 80$ kΩ, $R_3 = 50$ kΩ, $R_4 = 100$ kΩ, and $R_F = 500$ kΩ. The unity-gain bandwidth of the op-amps is $f_u = 1$ MHz, and the slew rate is SR = 6 V/μs.

a. Determine the voltage gain $A_f = v_O/v_i$.

b. Use PSpice to verify your results.

FIGURE P3.32

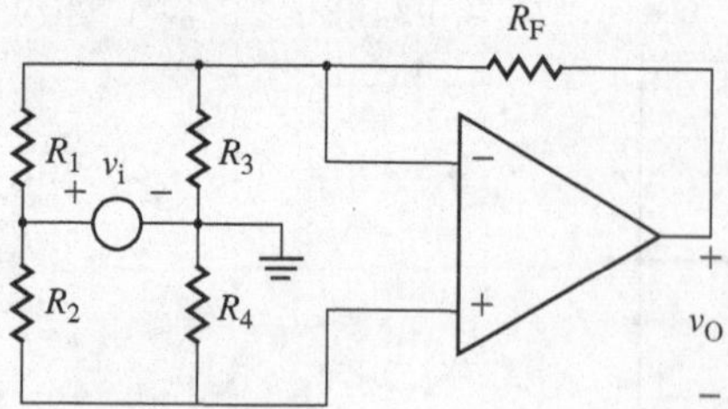

3.33 Two noninverting op-amps can be cascaded to produce a differential voltage as shown in Fig. P3.33 such that $v_O = a_1 v_1 - b_1 v_2$. The unity-gain bandwidth of the op-amps is $f_u = 1$ MHz, and the slew rate is SR = 6 V/μs.

a. If $R_1 = R_5 = 10$ kΩ, $R_2 = R_4 = 500$ kΩ, and $R_3 = R_6 = 9.8$ kΩ, determine the differential voltage gains a_1 and a_2.

b. Using the values in part (a), determine the maximum frequency f_{max} of the input signal v_1 if the amplitude of the output voltage due to v_1 is limited to 10 V.

c. Use PSpice to verify your results.

FIGURE P3.33

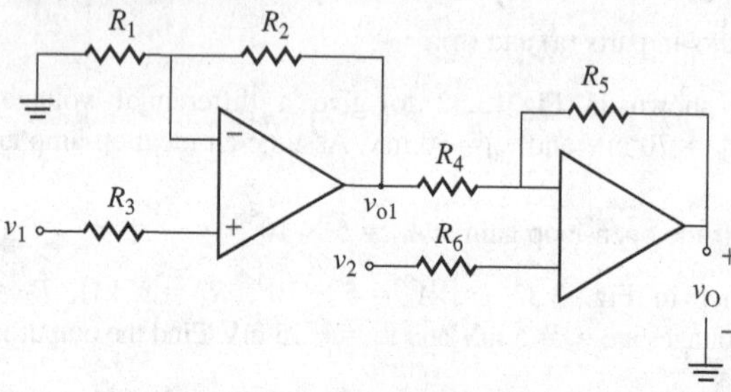

3.34 Two noninverting op-amps can be cascaded to produce a differential voltage as shown in Fig. P3.34 such that $v_O = a_1 v_1 - b_1 v_2$. The unity-gain bandwidth of the op-amps is $f_u = 1$ MHz, and the slew rate is SR = 6 V/μs.

a. If $R_1 = R_5 = 10$ kΩ, $R_2 = R_4 = 500$ kΩ, and $R_3 = R_6 = 9.8$ kΩ, determine the differential voltage gains a_1 and a_2.

b. Using the values in part (a), determine the maximum frequency f_{max} of the input signal v_1 if the amplitude of the output voltage due to v_1 is limited to 10 V.

c. Use PSpice to verify your results.

FIGURE P3.34

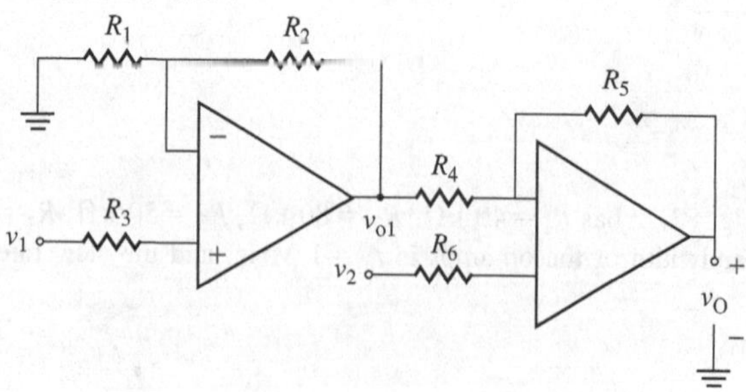

3.35 The amplifier shown in Fig. P3.35 can produce a differential voltage such that $v_O = a_1v_1 - a_2v_2$. The unity-gain bandwidth of the op-amps is $f_u = 1$ MHz, and the slew rate is SR = 6 V/μs.

a. If $R_1 = R_3 = 10$ kΩ and $R_2 = R_4 = 1$ MΩ, determine the differential voltage gains a_1 and a_2.

b. Using the values in part (a), determine the maximum frequency f_{max} of the input signal v_1 if the amplitude of the output voltage due to v_1 is limited to 10 V.

c. Use PSpice to verify your results.

FIGURE P3.35

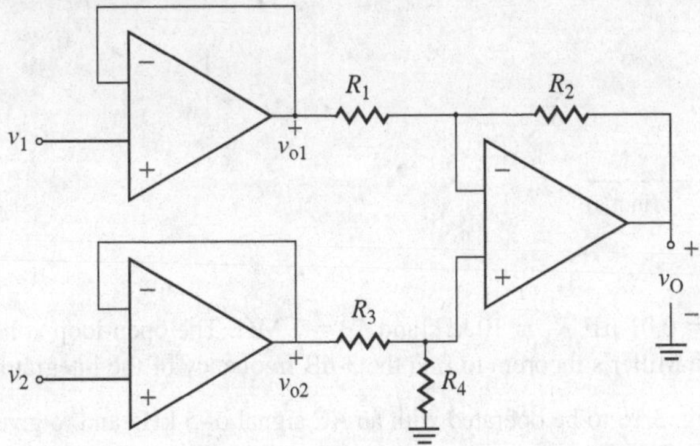

3.36 The amplifier shown in Fig. P3.36 can produce a differential voltage such that $v_O = a_1v_1 - a_2v_2$. The unity-gain bandwidth of the op-amps is $f_u = 1$ MHz, and the slew rate is SR = 6 V/μs. Derive an expression of the output voltage if $R_2 = R_3$, $R_1 = R_4$, and $R_F = R_5$.

FIGURE P3.36

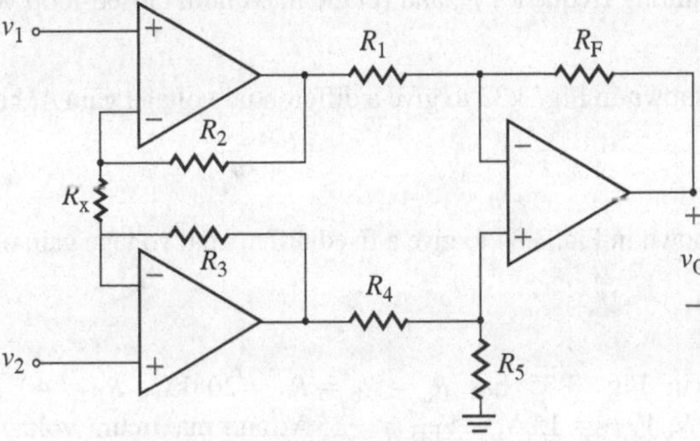

3.5 *Op-Amp Applications*

3.37 The integrator in Fig. 3.16 has $V_{CC} = 15$ V, $-V_{EE} = -15$ V, maximum voltage swing = ±14 V, $C_F = 0.01$ μF, $R_1 = 1$ kΩ, and $R_F = 1$ MΩ. The initial capacitor voltage is $V_{co} = 0$. Draw the waveform for the output voltage if the input voltage is described by

$$v_S = \begin{cases} 1\text{ V} & \text{for } 0 \le t < 1\text{ ms} \\ -1\text{ V} & \text{for } 1 \le t < 2\text{ ms} \\ 1\text{ V} & \text{for } 2 \le t < 3\text{ ms} \\ -1\text{ V} & \text{for } 3 \le t < 4\text{ ms} \end{cases}$$

3.38 The integrator in Fig. 3.16 has $V_{CC} = 15$ V, $-V_{EE} = -15$ V, maximum voltage swing $= \pm 14$ V, $C_F = 0.1$ μF, $R_1 = 10$ kΩ, and $R_F = 1$ MΩ. The initial capacitor voltage is $V_{co} = 0$. Draw the waveform of the output voltage for the input voltage shown in Fig. P3.38.

FIGURE P3.38

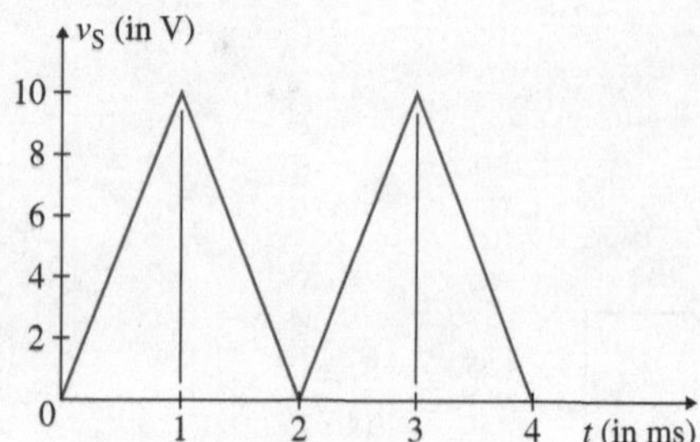

3.39 The integrator in Fig. 3.16 has $C_F = 0.01$ μF, $R_1 = 10$ kΩ, and $R_F = 1$ MΩ. The open-loop voltage gain of the op-amp is $A_o = 5 \times 10^5$. Use Miller's theorem to find the 3-dB frequency of the integrator.

3.40 D P Design an integrator as shown in Fig. 3.16 to be operated with an AC signal of 5 kHz and to give a closed-loop voltage gain of $A_f = 10$ at $\omega = 1$ rad/s.

3.41 D P **a.** Design a differentiator as shown in Fig. 3.27(a) to satisfy the following specifications: maximum voltage gain of $A_{f(max)} = 20$ and gain-limiting frequency $f_b = 10$ kHz. Determine the values of R_1, R_F, and C_1.
b. Use PSpice/SPICE to check your results by plotting the frequency response in part (a).

3.42 The differentiator in Fig. 3.27(a) has $R_1 = 2$ kΩ, $R_F = 10$ kΩ, and $C_1 = 0.01$ μF. Determine **(a)** the differentiator time constant τ_d, **(b)** the gain-limiting frequency f_b, and **(c)** the maximum closed-loop voltage gain $A_{f(max)}$.

3.43 D P Design an instrumentation amplifier as shown in Fig. 3.33 to give a differential voltage gain A_f between 500 and 1000.

3.44 D P Design an instrumentation amplifier as shown in Fig. 3.34 to give a fixed differential voltage gain of $A_f = 750$.

3.45 The noninverting summing amplifier in Fig. 3.35 has $R_a = R_b = R_c = 20$ kΩ, $R_F = 40$ kΩ, $R_B = 20$ kΩ, $v_a = 2$ V, $v_b = -3$ V, $v_c = -2$ V, $V_{CC} = 15$ V, $-V_{EE} = -15$ V, and maximum voltage swing $= \pm 14$ V. Determine the output voltage v_O.

3.46 The inverting summing amplifier in Fig. 3.36 has $R_1 = R_2 = R_3 = 20$ kΩ, $R_F = 40$ kΩ, $R_x = 5.71$ kΩ, $v_1 = 2$ V, $v_2 = -3$ V, $v_3 = -2$ V, $V_{CC} = 15$ V, $-V_{EE} = -15$ V, and maximum voltage swing $= \pm 14$ V. Determine the output voltage v_O.

3.47 D Design an add–subtract summing amplifier as shown in Fig. 3.37 to give an output voltage of the form $v_O = 5v_a + 7v_b + 3v_c - 2v_1 - v_2 - 6v_3$. The equivalent resistance R_A should be set to 20 kΩ.

3.48 D Design an add–subtract summing amplifier as shown in Fig. 3.37 to give an output voltage of the form $v_O = 5v_a + 9v_b + 3v_c - 8v_1 - 2v_2 - 6v_3$. The minimum value of any resistance should be $R_{min} = 20$ kΩ.

3.49 D P Design an optocoupler drive circuit as shown in Fig. 3.38 to produce a drive current of 500 mA from a signal voltage of 10 mV.

3.50 D P Design a photodetector circuit as shown in Fig. 3.39 to give an output voltage of 1 V at an incident power density of $D_P = 1\ \mu W/cm^2$. The current responsitivity of the photodiode is $D_i = 1$ A/W, and the active area is $a = 40\ mm^2$.

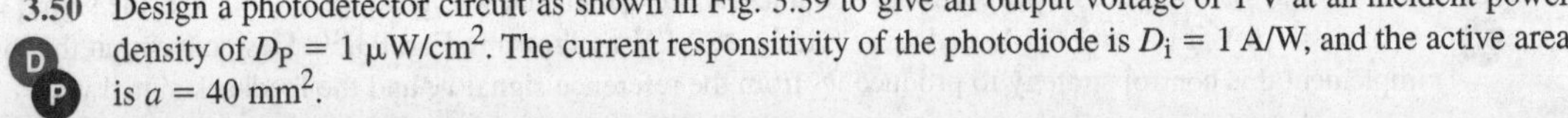

3.51 The voltage-to-current converter circuit in Fig. 3.40(a) has $R_1 = R = 10\ k\Omega$ and $v_S = 200$ mV. Determine the load current i_O.

3.52 The full-scale current of the moving coil for the DC voltmeter in Fig. 3.41 is $I_M = 200\ \mu A$. Determine the value of R_1 to give a full-scale reading of $V_S = 300$ V.

3.53 D P Design a DC millivoltmeter as shown in Fig. 3.42. The full-scale current of the moving coil is $I_M = 0.5\ \mu A$. Determine the values of R_1, R_F, and R_2 to give a full-scale voltage reading of $V_S = 200$ mV.

3.54 D P Design a negative impedance converter as shown in Fig. 3.43 by determining the component values such that the input resistance will be $Z_{in} = R_{in} = -15\ k\Omega$.

3.55 D P **a.** The noninverting integrator in Fig. 3.45(a) has $V_{CC} = 15$ V, $-V_{EE} = -15$ V, maximum voltage swing $= \pm 14$ V, $C = 0.01\ \mu F$, and $R_1 = R_F = R = 1\ M\Omega$. The initial capacitor voltage is $V_{co} = 0$. Draw the waveform for the output voltage if the input is a step voltage described by

$$v_S = 1 \text{ V} \quad \text{for } t \geq 0$$

b. Use PSpice/SPICE to plot the output voltage in part (a).

3.56 D P Design an inductance simulator as shown in Fig. 3.46 by determining the values of components. The inductance should be $L_e = 2$ mH.

3.6 Op-Amp Circuit Design

3.57 D A control system requires a proportional controller that will produce $v_O = 5$ V if the error signal $v_e = 0$, $v_O = 10$ V if $v_e \leq -0.1$ V, and $v_O = 0$ if $v_e = 0.1$ V. These requirements are graphed in Fig. P3.57. Design a circuit that will implement this control strategy to produce v_O from v_S and v_{ref}.

FIGURE P3.57

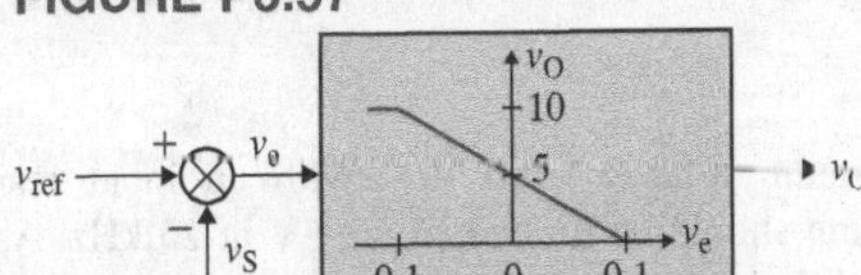

3.58 D A triggering circuit requires short pulses v_O of approximately 10 V magnitude and pulse width of $t_w = 200\ \mu s$, as shown in Fig. P3.58. Design a circuit that will generate triggering pulses. (There is no unique solution.)

FIGURE P3.58

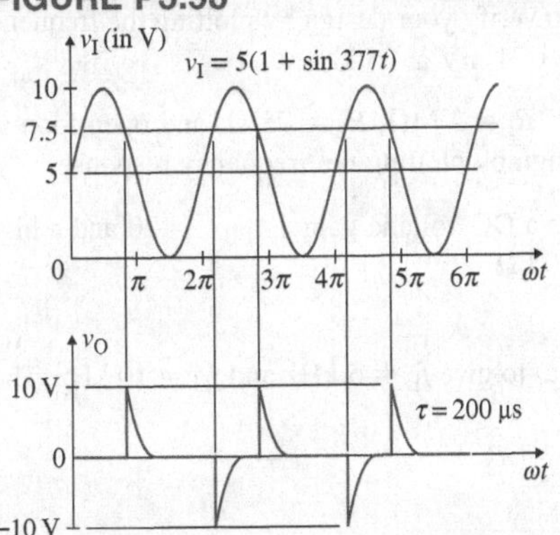

3.59 D A control system requires a proportional and integral controller that will produce $v_O = 5$ V if the signal $v_{e1} = 0$, $v_O = 10$ V if $v_{e1} \leq -0.1$ V, and $v_O = 0$ if $v_{e1} \geq 0.1$ V, as shown in Fig. P3.59. Design a circuit that will implement this control strategy to produce v_O from the reference signal v_S and the feedback signal v_{ref}.

FIGURE P3.59

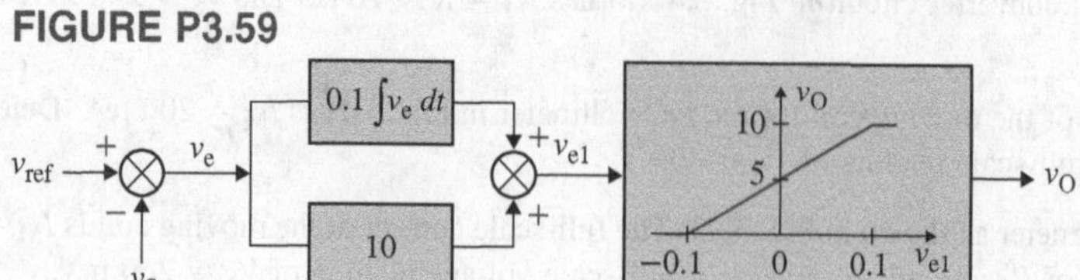

3.60 D The inverting amplifier shown in Fig. P3.60 can give high voltage gain and requires a narrow range of resistor values. The output voltage should be $v_O = 12$ V for $v_S \leq -0.05$ V and $v_O = -12$ V if $v_S \geq 0.05$ V. Design a circuit that will implement this control strategy.

FIGURE P3.60

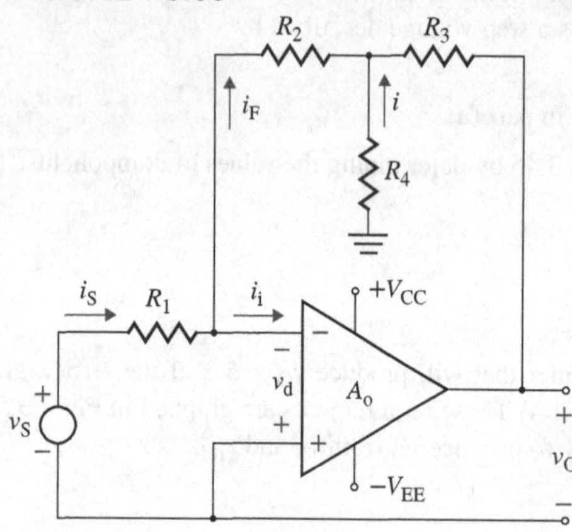

3.61 Design an op-amp circuit to obtain a voltage gain of $A_f = 100$ V/V with an input resistance of $R_i \geq 25$ kΩ. The peak-to-peak output voltage swing should be limited to ±11 V at 25 kHz. Assume DC supply voltages of ±12 V. Use PSpice to verify your design by plotting the frequency response and the transient response with an input signal of 1 mV at 25 kHz.

3.62 Design an op-amp differential amplifier circuit to obtain a voltage gain of $A_f = 5$ kV/V with an input resistance of $R_i \geq 500$ kΩ. The peak-to-peak output voltage swing should be limited to ±11 V at 25 kHz. Assume DC supply voltages of ±12 V. Use PSpice to verify your design by plotting the frequency response and the transient response with a differential voltage of 1 mV at 25 kHz.

For Probs. 3.63 to 3.66, the op-amp has $C_i = 1.5$ pF, $R_i = 2$ MΩ, $R_o = 75$ Ω, and open-loop voltage gain $A_o = 2 \times 10^5$. Use PSpice/SPICE to check your design by plotting the frequency response.

3.63 D P Design an integrator as shown in Fig. 3.21(a) to give a DC voltage gain $|A_{PB}| = 20$ and a high 3-dB frequency $f_H = 1$ kHz. Assume $R_1 = 1$ kΩ and $R_L = 20$ kΩ.

3.64 D P Design a differentiator circuit as shown in Fig. 3.30(a) to give $f_L = 5$ kHz and $f_H = 10$ kHz. The pass-band gain is $|A_{PB}| = 20$.

3.65 P An amplifier circuit is shown in Fig. P3.65. Use the short-circuit and zero-value methods to find the low 3-dB frequency f_L, the high 3-dB frequency f_H, and the pass-band gain A_{PB}.

FIGURE P3.65

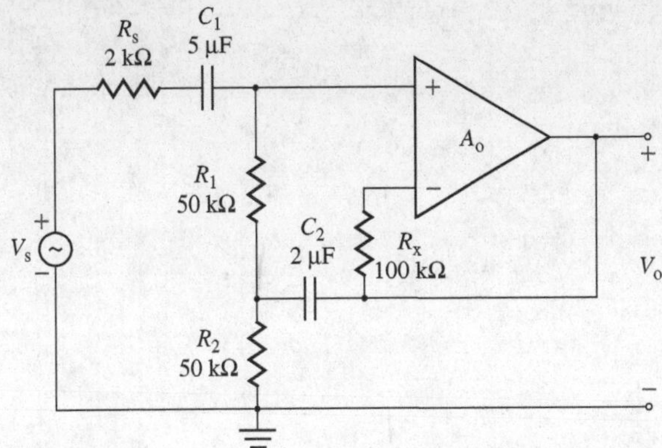

3.66 P An amplifier circuit is shown in Fig. P3.66. Use the short-circuit and zero-value methods to find the low 3-dB frequency f_L, the high 3-dB frequency f_H, and the pass-band gain A_{PB}.

FIGURE P3.66

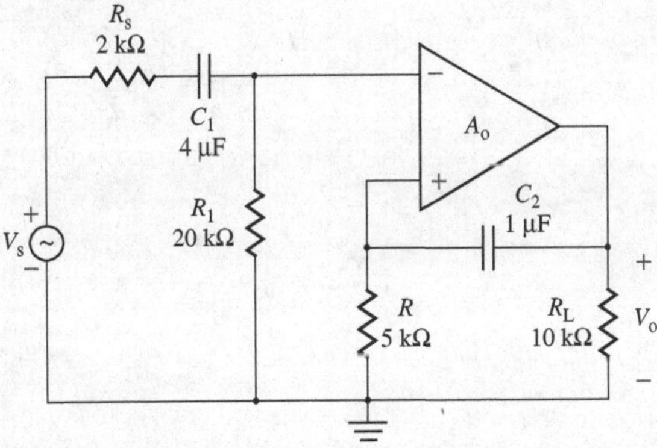

CHAPTER 4

SEMICONDUCTOR DIODES

Learning Outcomes

After completing this chapter, students should be able to do the following:

- Describe the ideal and practical characteristics of semiconductor diodes.
- Determine the circuit models of a diode and apply them for analyzing diode circuits.
- Determine the DC and small-signal performances of simple diode circuits.
- Explain the characteristics of zener diodes and their applications as voltage regulators.

Symbols and Their Meanings

Symbol	Meaning
i_D, i_d, I_D	Instantaneous DC and AC and quiescent DC diode currents
$v_O(t)$, $v_o(t)$	Instantaneous DC and AC output voltages
$V_{o(av)}$, $V_{o(rms)}$	Average and rms output voltages
$V_{r(pp)}$, $V_{r(p)}$	Peak-to-peak and peak ripple output voltages
v_D, v_d, V_D	Instantaneous DC and AC and quiescent DC diode voltages
V_Z, I_{ZT}	DC zener voltage and current
V_{ZK}, I_{ZK}	DC zener knee voltage and current
$v_Z(t)$, $i_Z(t)$	Instantaneous zener voltage and current

Symbol	Meaning
$v_L(t)$, $i_L(t)$	Instantaneous zener voltage and current
r_d, R_D	Small-signal AC and DC diode resistances
V_{ZO}, R_Z	Threshold zener voltage and small-signal zener resistance
T_J, T_A	Junction and ambient temperature

4.1 Introduction

A diode is a two-terminal semiconductor device. It offers a low resistance on the order of milliohms in one direction and a high resistance on the order of gigaohms in the other direction. Thus a diode permits an easy current flow in only one direction. A diode is the simplest electronic device, and it is the basic building block for many electronic circuits and systems. In this chapter, we will discuss the characteristics of diodes and their models through analysis of a diode circuit.

A diode exhibits a nonlinear relation between the voltage across its terminals and the current through it. However, analysis of a diode can be greatly simplified with the assumption of an ideal characteristic. The results of this simplified analysis are useful in understanding the operation of diode circuits and are acceptable in many practical cases, especially at the initial stage of design and analysis. If more accurate results are required, linear circuit models representing the nonlinear characteristic of diodes can be used. These models are commonly used in evaluating the performance of diode circuits. If better accuracy is required, however, computer-aided modeling and simulation are normally used.

4.2 Ideal Diodes

The symbol for a semiconductor diode is shown in Fig. 4.1(a). Its two terminals are the anode and the cathode. If the anode voltage is held positive with respect to the cathode terminal, the diode conducts and offers a small forward resistance. The diode is then said to be *forward biased,* and it behaves as a short circuit, as shown in Fig. 4.1(b). If the anode voltage is kept negative with respect to the cathode terminal, the diode offers a high resistance. The diode is then said to be *reverse biased,* and it behaves as an open circuit, as shown in Fig. 4.1(c). Thus, an ideal diode will offer zero resistance and zero voltage drop in the forward direction. In the reverse direction, it will offer infinite resistance and allow zero current.

An ideal diode behaves as a short circuit in the forward region of conduction ($v_D = 0$) and as an open circuit in the reverse region of nonconduction ($i_D = 0$). The *v-i* characteristic of an ideal diode is shown in Fig. 4.1(d). Because the forward voltage tends to be greater than zero, the forward current through the diode tends to be infinite. In practice, however, a diode is connected to other circuit elements, such

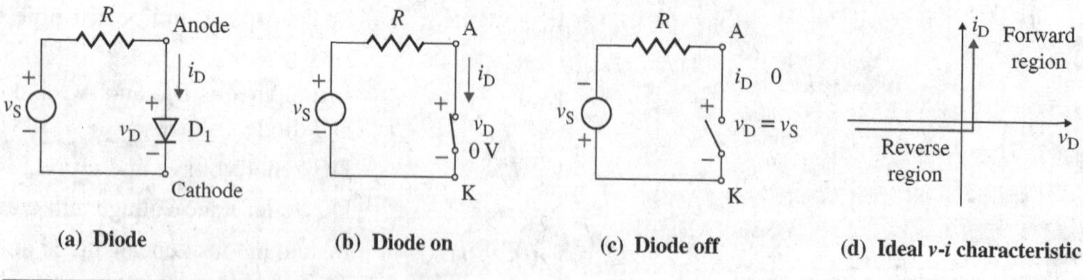

FIGURE 4.1 Characteristics of an ideal diode

as resistances, and its forward current is limited to a known value, which will depend on the values of the circuit elements.

EXAMPLE 4.1

Application as a diode OR logic function A diode circuit that can generate an OR logic function is shown in Fig. 4.2. A positive logic convention denotes logic 0 for 0 V and logic 1 for a positive voltage, typically 5 V. Show the truth table that illustrates the logic output.

SOLUTION

If both inputs have 0 V (i.e., $V_A = 0$ and $V_B = 0$), both diodes will be off, and the output V_C will be 0 (or logic 0) only. If either V_A or V_B (or both) is high (+5 V), the corresponding diode (D_1 or D_2 or both) will conduct, and the output voltage will be high at $V_C = 5$ V. As we will see later, a real diode has a finite voltage drop of approximately 0.7 V, and the output voltage will be approximately $5 - 0.7 = 4.3$ V (or logic 1). The truth table that illustrates the logic functions is shown in Table 4.1. We can define the logic level at any desired value. That is, for example, we can say greater than 3 V for logic 1, and less than 1 V for logic 0.

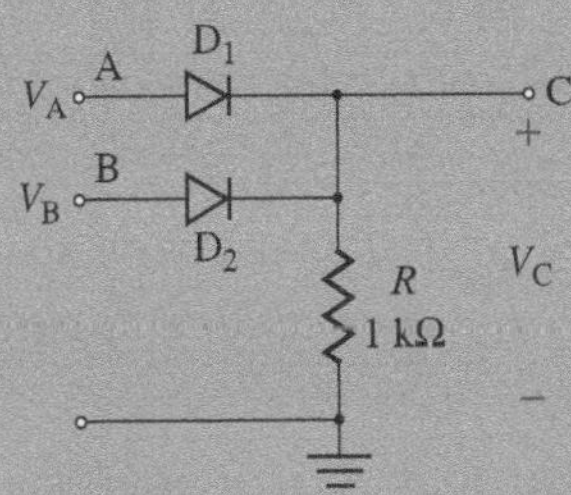

FIGURE 4.2 Diode OR logic circuit

TABLE 4.1 Truth table for Example 4.1

Voltages			Logic Levels		
V_A	V_B	V_C	A	B	C
0 (V)	0 (V)	0 (V)	0	0	0
0 (V)	5 (V)	4.3 (V)	0	1	1
5 (V)	0 (V)	4.3 (V)	1	0	1
5 (V)	5 (V)	4.3 (V)	1	1	1

EXAMPLE 4.2

Application as a diode AND logic function A diode circuit that can generate an AND logic function is shown in Fig. 4.3. A positive–logic convention denotes logic 0 for 0 V and logic 1 for a positive voltage, typically 5 V. Show the truth table that illustrates the logic output.

SOLUTION

If input V_A or V_B (or both) is 0, the corresponding diode (D_1 or D_2 or both) will conduct, and the output voltage will be 0. In practice, a diode has a finite voltage drop of approximately 0.7 V, and the output voltage will be approximately 0.7 V (or logic 0). If both inputs are high (i.e., $V_A = 5$ V and $V_B = 5$ V), both diodes will be reverse biased (off), and the output voltage will be high at $V_C = 5$ V. The output will be logic 1. The truth table for an AND logic gate is shown in Table 4.2.

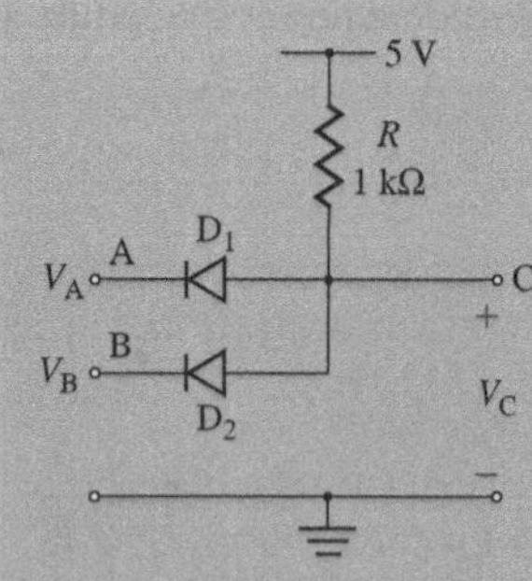

FIGURE 4.3 Diode AND logic circuit

TABLE 4.2 Truth table for Example 4.2

Voltages			Logic Levels		
V_A	V_B	V_C	A	B	C
0 (V)	0 (V)	0.7 (V)	0	0	0
0 (V)	5 (V)	0.7 (V)	0	1	0
5 (V)	0 (V)	0.7 (V)	1	0	0
5 (V)	5 (V)	5 (V)	1	1	1

▶ **NOTE** Although it is possible to use diodes to perform logic functions, diode logic circuits are slow and thus are rarely used in practice. We will see in Chapter 15 that the performance of many logic families is far superior. These examples, however, illustrate the "on" and "off" behaviors and conditions of the diodes.

EXAMPLE 4.3

Application as a diode rectifier The input voltage of the diode circuit shown in Fig. 4.4 is $v_S = v_s = V_m \sin \omega t$. The input voltage has zero DC component—that is, $V_S = 0$ and $v_S = V_S + v_s = v_s$. Draw the waveforms of the output voltage v_O and the diode voltage v_D.

SOLUTION

During the interval $0 \le \omega t \le \pi$, the voltage across the diode will be positive, and the diode will behave as a short circuit. This is shown in Fig. 4.5(a). Thus, the output voltage v_O will be the same as the input voltage v_S, and the diode voltage v_D will be zero. That is,

$$v_O = v_S \quad \text{for } 0 \le \omega t \le \pi$$
$$v_D = 0$$

During the interval $\pi \le \omega t \le 2\pi$, the voltage across the diode will be negative, and the diode will be an open circuit, as shown in Fig. 4.5(b). Thus, the output voltage v_O will be zero, and the diode voltage v_D will be the

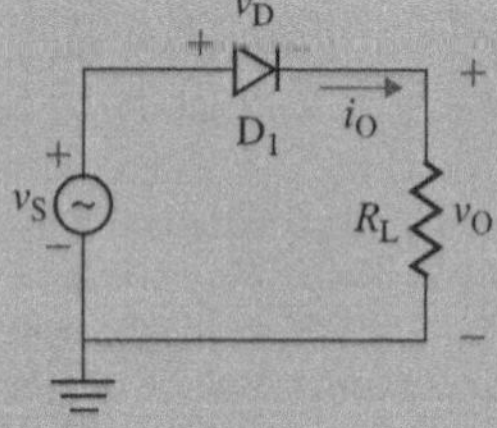

FIGURE 4.4 Diode circuit for Example 4.3

same as the input voltage v_S. That is,

$$v_O = 0 \quad \text{for } \pi \leq \omega t \leq 2\pi$$
$$v_D = v_S$$

The waveforms of the input voltage v_S, the output voltage v_O, and the diode voltage v_D are shown in Fig. 4.5(c).

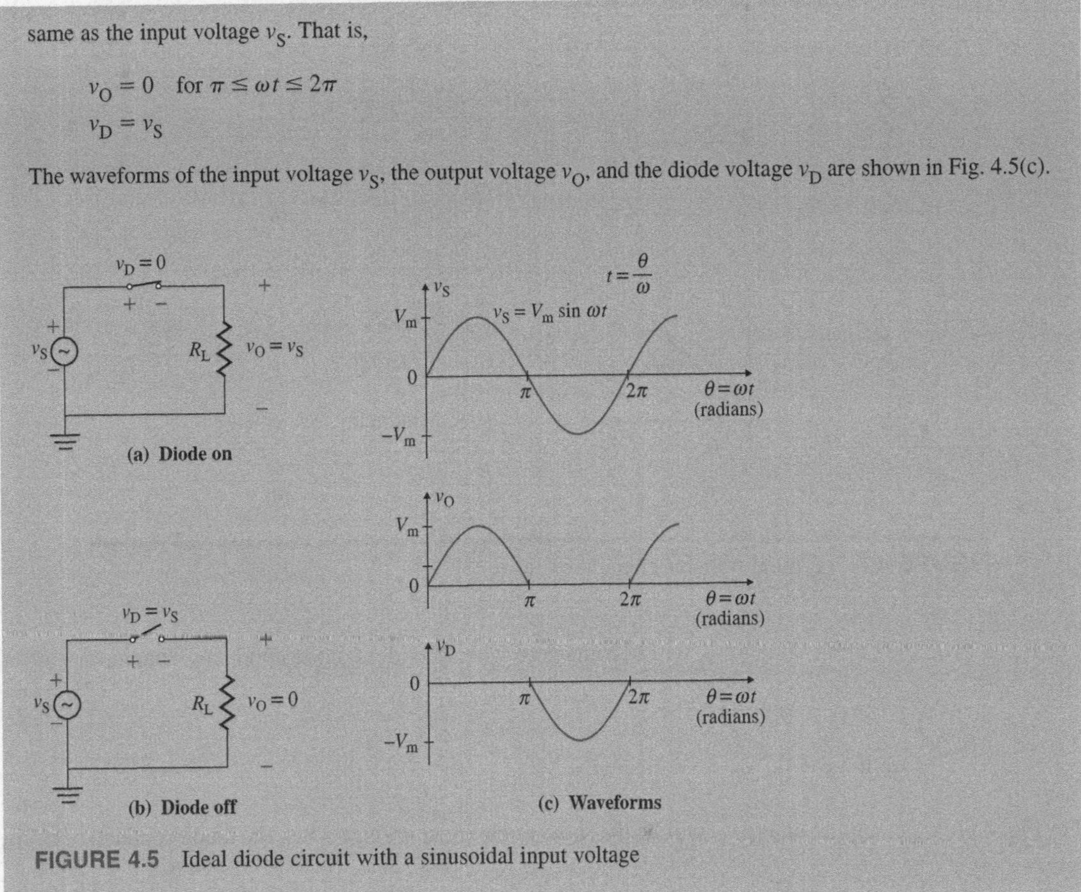

FIGURE 4.5 Ideal diode circuit with a sinusoidal input voltage

4.3 Transfer Characteristics of Diode Circuits

The output voltage of a diode circuit depends on whether the diode is on or off. If the input voltage changes with time, as illustrated in Example 4.3, the output voltage is based on the on or off status of the diode(s). The *transfer characteristic* of a circuit is the relationship between the output voltage and the input voltage. It shows the manner in which the output voltage varies with the input voltage and is independent of the input waveform. Therefore, once the transfer characteristic is known, the output waveform can be determined directly for any given input waveform. The transfer characteristic is useful in describing the behavior of a circuit.

The output voltages of the circuits in Fig. 4.6 can be described as follows. For Fig. 4.6(a), the output voltage v_O will be the same as the input voltage when the ideal diode conducts. When the diode is off, the output voltage will be zero. That is,

$$v_O = \begin{cases} v_S & \text{if } v_S > 0 \\ 0 & \text{if } v_S \leq 0 \end{cases}$$

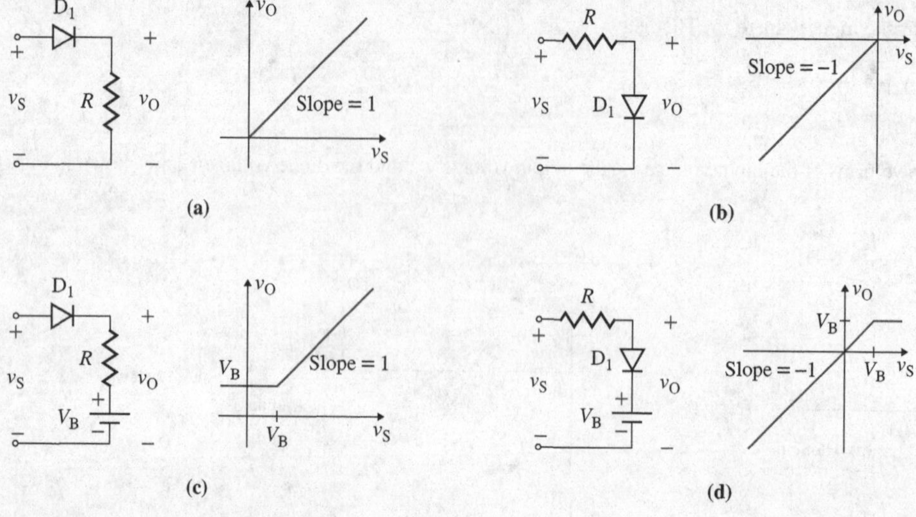

FIGURE 4.6 Typical transfer characteristics

For Fig. 4.6(b), the output voltage v_O will become zero when the ideal diode conducts. That is,

$$v_O = \begin{cases} 0 & \text{if } v_S > 0 \\ v_S & \text{if } v_S \le 0 \end{cases}$$

For Fig. 4.6(c), the output voltage v_O will be the same as the input voltage when the diode conducts. That is,

$$v_O = \begin{cases} v_S & \text{if } v_S > V_B \\ V_B & \text{if } v_S \le V_B \end{cases}$$

For Fig. 4.6(d), the output voltage v_O will be clamped to V_B (i.e., it will remain fixed at V_B) when the diode conducts. When the diode is off, the output voltage will be the same as the input voltage. Otherwise, it will be V_B. That is,

$$v_O = \begin{cases} V_B & \text{if } v_S > V_B \\ v_S & \text{if } v_S \le V_B \end{cases}$$

Typical transfer characteristics are also shown in Fig. 4.6.

KEY POINT OF SECTION 4.3

- The transfer characteristic relates the output voltage to the input voltage and does not depend on the magnitude and waveform of the input voltage.

4.4 Practical Diodes

The characteristic of a practical diode that distinguishes it from an ideal one is that the practical diode experiences a finite voltage drop when it conducts. This drop is typically in the range of 0.5 V to 0.7 V. If the input voltage to a diode circuit is high enough, this small drop can be ignored. The voltage drop may, however, cause a significant error in electronic circuits, and the diode characteristic should be taken into account in evaluating the performance of diode circuits. To understand the internal characteristics of a practical diode [1], we need to understand its physical operation, which is covered in Chapter 6.

4.4.1 Characteristic of Practical Diodes

The voltage-versus-current (*v-i*) characteristic of a practical diode is shown in Fig. 4.7. This characteristic, which can be well approximated by an equation known as the *Shockley diode equation* [2–4], is given by

$$i_D = I_S(e^{v_D/nV_T} - 1) \tag{4.1}$$

where i_D = current through the diode, in A
v_D = diode voltage with the anode positive with respect to the cathode, in V
I_S = leakage (or reverse saturation) current, typically in the range of 10^{-6} A to 10^{-15} A
n = empirical constant known as the *emission coefficient* or the *ideality factor,* whose value varies from 1 to 2

The emission coefficient n depends on the material and the physical construction of the diode. For germanium diodes, n is considered to be 1. For silicon diodes, the predicted value of n is 2 at very small or large currents; but for most practical silicon diodes, the value of n falls in the range of 1.1 to 1.8.

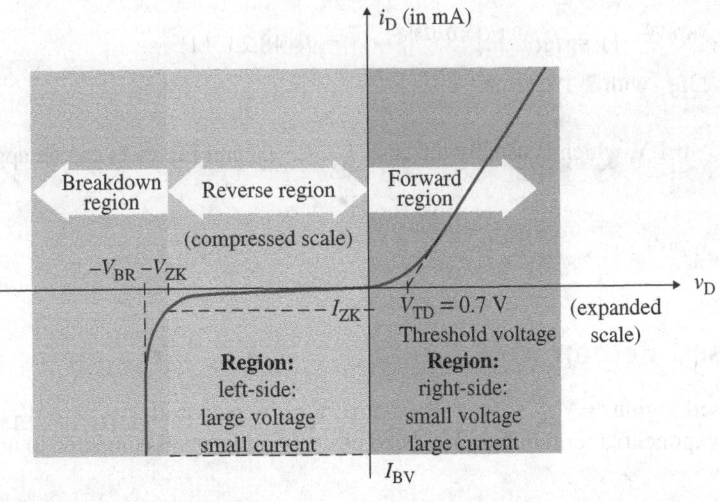

FIGURE 4.7 Voltage-versus-current characteristic of practical diode

V_T in Eq. (4.1) is a constant called the *thermal voltage,* and it is given by

$$V_T = \frac{kT_K}{q} \tag{4.2}$$

where q = electron charge = 1.6022×10^{-19} coulomb (C)
T_K = absolute temperature in kelvins = $273 + T_{Celsius}$
k = Boltzmann's constant = 1.3806×10^{-23} J per kelvin

At a junction temperature of 25°C, Eq. (4.2) gives the value of V_T as

$$V_T = \frac{kT_K}{q} = \frac{(1.3806 \times 10^{-23})(273 + 25)}{1.6022 \times 10^{-19}} = \frac{T_K}{11,605.1} \approx 25.8 \text{ mV}$$

At a specific temperature, the leakage current I_S will remain constant for a given diode. For small-signal (or low-power) diodes, the typical value of I_S is 10^{-9} A. We can divide the diode characteristic of Fig. 4.7 into three regions, as follows:

Forward-biased region, where $v_D > 0$
Reverse-biased region, where $v_D < 0$
Breakdown region, where $-V_{ZK} > v_D > 0$

Forward-Biased Region

In the forward-biased region, $v_D > 0$. The diode current i_D will be very small if the diode voltage v_D is less than a specific value V_{TD}, known as the *threshold voltage* or the *cut-in voltage* or the *turn-on voltage* (typically 0.7 V). The diode conducts fully if v_D is higher than V_{TD}. Thus, the threshold voltage is the voltage at which a forward-biased diode begins to conduct fully.

Assume that a small forward voltage of $v_D = 0.1$ V is applied to a diode of $n = 1$. At room temperature, $V_T = 25.8$ mV. From Eq. (4.1), we can find the diode current i_D as

$$\begin{aligned} i_D &= I_S(e^{v_D/nV_T} - 1) = I_S(e^{0.1/(1\times 0.0258)} - 1) = I_S(48.23 - 1) \\ &\approx 48.23 I_S \quad \text{with } 2.1\% \text{ error} \end{aligned}$$

Therefore, for $v_D > 0.1$ V, which is usually the case, $i_D >> I_S$, and Eq. (4.1) can be approximated within 2.1% error by

$$i_D = I_S(e^{v_D/nV_T} - 1) \approx I_S e^{v_D/nV_T} \tag{4.3}$$

Reverse-Biased Region

In the reverse-biased region, $-V_{ZK} < v_D < 0$. That is, v_D is negative. If $|v_D| >> V_T$, which occurs for $v_D < -0.1$ V, the exponential term in Eq. (4.1) becomes negligibly small compared to unity, and the diode current i_D becomes

$$i_D = I_S(e^{-|v_D|/nV_T} - 1) \approx -I_S \tag{4.4}$$

which indicates that the diode current i_D remains constant in the reverse direction and is equal to I_S in magnitude.

Breakdown Region

In the breakdown region, the reverse voltage is high—usually greater than 100 V. If the magnitude of the reverse voltage exceeds a specified voltage known as the *breakdown voltage* V_{BR}, the corresponding reverse current I_{BV} increases rapidly for a small change in reverse voltage beyond V_{BR}. Operation in the breakdown region will not be destructive to the diode provided the power dissipation ($P_D = v_D i_D$) is kept within the safe level specified in the manufacturer's data sheet. It is often necessary, however, to limit the reverse current in the breakdown region so that the power dissipation falls within a permissible range.

4.4.2 Determination of Diode Constants

Diode constants I_S and n can be determined either from experimentally measured *v-i* data or from the *v-i* characteristic. There are a number of steps to be followed. Taking the natural (base *e*) logarithm of both sides of Eq. (4.3), we get

$$\ln i_D = \ln I_S + \frac{v_D}{nV_T}$$

which, after simplification, gives the diode voltage v_D as

$$v_D = nV_T \ln\left(\frac{i_D}{I_S}\right) \tag{4.5}$$

If we convert the natural log of base *e* to the logarithm of base 10, Eq. (4.5) becomes

$$v_D = 2.3nV_T \log\left(\frac{i_D}{I_S}\right) \tag{4.6}$$

which indicates that the diode voltage v_D is a nonlinear function of the diode current i_D. If V_{D1} is the diode voltage corresponding to diode current I_{D1}, Eq. (4.5) gives

$$V_{D1} = nV_T \ln\left(\frac{I_{D1}}{I_S}\right) \tag{4.7}$$

Similarly, if V_{D2} is the diode voltage corresponding to the diode current I_{D2}, we get

$$V_{D2} = nV_T \ln\left(\frac{I_{D2}}{I_S}\right) \tag{4.8}$$

Therefore, the difference in diode voltages can be expressed by

$$V_{D2} - V_{D1} = nV_T \ln\left(\frac{I_{D2}}{I_S}\right) - nV_T \ln\left(\frac{I_{D1}}{I_S}\right) = nV_T \ln\left(\frac{I_{D2}}{I_{D1}}\right) \tag{4.9}$$

which can be converted to the logarithm of base 10 as

$$V_{D2} - V_{D1} = 2.3nV_T \log\left(\frac{I_{D2}}{I_{D1}}\right) \tag{4.10}$$

This shows that for a decade (i.e., a factor of 10) change in diode current $I_{D2} = 10I_{D1}$, the diode voltage will change by $2.3nV_T$. Thus, Eq. (4.6) can be written as

$$v_D = 2.3nV_T \log i_D - 2.3nV_T \log I_S \tag{4.11}$$

If this equation is plotted on a semilog scale with v_D on the vertical linear axis and i_D on the horizontal log axis, the characteristic will be a straight line with a slope of $+2.3nV_T$ per decade of current, and its equation will have the form of a standard straight-line equation—that is,

$$y = mx - c$$

where $c = 2.3nV_T \log I_S$ and $m = 2.3nV_T$ per decade of current. The plot of Eq. (4.11) is shown in Fig. 4.8. With v_D in the linear scale and i_D in the log scale.

Thus, based on the experimental results from an unknown diode, the *v-i* characteristic can be plotted on a semilog scale. The values of I_S and n can be calculated as follows:

Step 1. Plot v_D against i_D on a semilog scale, as shown in Fig. 4.8 with v_D in the linear scale and i_D in the log scale.

Step 2. Find the slope m per decade of current change on the v_D-axis.

Step 3. Find the emission coefficient n for the known value of slope m—that is,

$$n = \frac{m}{2.3V_T} = \frac{m}{2.3 \times 0.0258}$$

Step 4. Find the intercept c on the v_D-axis.

Step 5. Find the value of I_S from

$$2.3nV_T \log I_S = c$$

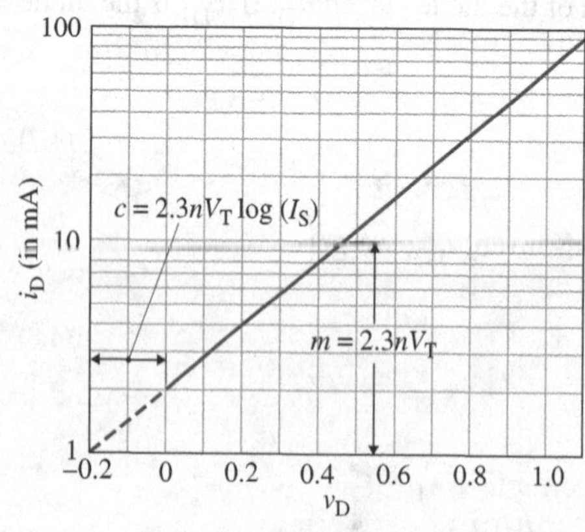

FIGURE 4.8 Diode *v-i* characteristic plotted on a semilog scale

Once the values of I_S and n have been determined, the diode voltage v_D can be expressed explicitly as a function of the diode current i_D, as in Eq. (4.5).

EXAMPLE 4.4

Finding diode constants The measured values of a diode at a junction temperature of 25°C are given by

$$V_D = \begin{cases} 0.5\text{ V} & \text{at } I_D = 5\ \mu\text{A} \\ 0.6\text{ V} & \text{at } I_D = 100\ \mu\text{A} \end{cases}$$

Determine (a) the emission coefficient n and (b) the leakage current I_S.

SOLUTION

$V_{D1} = 0.5$ V at $I_{D1} = 5\ \mu$A, and $V_{D2} = 0.6$ V at $I_{D2} = 100\ \mu$A. At 25°C, $V_T = 25.8$ mV.

(a) From Eq. (4.9),

$$V_{D2} - V_{D1} = nV_T \ln\left(\frac{I_{D2}}{I_{D1}}\right) \quad \text{or} \quad 0.6 - 0.5 = nV_T \ln\left(\frac{100\ \mu\text{A}}{5\ \mu\text{A}}\right)$$

which gives $nV_T = 0.03338$, and $n = 0.03338/V_T = 0.03338/(25.8 \times 10^{-3}) = 1.294$.

(b) From Eq. (4.5),

$$V_{D1} = nV_T \ln\left(\frac{I_{D1}}{I_S}\right) \quad \text{or} \quad 0.5 = 0.03338 \ln\left(\frac{5 \times 10^{-6}}{I_S}\right)$$

which gives $I_S = 1.56193 \times 10^{-12}$ A.

4.4.3 Temperature Effects

The leakage current I_S depends on the junction temperature T_j (in Celsius) and increases at the rate of approximately +7.2% per degree Celsius for silicon and germanium diodes [1, 5]. Thus, by adding the increments for each degree rise in the junction temperature up to 10°C, we get

$$\begin{aligned} I_S(T_j = 10) = I_S[1 &+ 0.072 + (0.072 + 0.072^2) + (0.072^2 + 0.072^3) \\ &+ (0.072^3 + 0.072^4) + (0.072^4 + 0.072^5) + (0.072^5 + 0.072^6) \\ &+ (0.072^6 + 0.072^7) + (0.072^7 + 0.072^8) + (0.072^8 + 0.072^9) \\ &+ (0.072^9 + 0.072^{10})] \\ \approx 2I_S & \end{aligned}$$

That is, I_S approximately doubles for every 10°C increase in temperature and can be related to any temperature change by

$$I_S(T_j) = I_S(T_o) \times 2^{(T_j - T_o)/10} = I_S(T_o) \times 2^{0.1(T_j - T_o)} \quad \textbf{(4.12)}$$

where $I_S(T_o)$ is the leakage current at temperature T_o. Substituting $V_T = kT_K/q$ in Eq. (4.5) gives the temperature dependence of the forward diode voltage. That is,

$$v_D = \frac{nk(273 + T_j)}{q} \ln\left(\frac{i_D}{I_S}\right) \quad \textbf{(4.13)}$$

which, after differentiation of v_D with respect to T_j, gives

$$\frac{\partial v_D}{\partial T_j} = \frac{nk}{q} \ln\left(\frac{i_D}{I_S}\right) - \frac{nk(273 + T_j)}{qI_S}\frac{dI_S}{dT_j} = \frac{v_D}{273 + T_j} - \frac{nV_T}{I_S}\frac{dI_S}{dT_j} \quad \textbf{(4.14)}$$

which decreases with the temperature T_j for a constant v_D. At a given diode current i_D, the diode voltage v_D decreases with the temperature. The temperature dependence of the forward diode characteristic is shown in Fig. 4.9.

The threshold voltage V_{TD} also depends on the temperature T_j. As the temperature increases, V_{TD} decreases, and vice versa. V_{TD}, which has an approximately linear relationship to temperature T_j, is given by

$$V_{TD}(T_j) = V_{TD}(T_o) + K_{TC}(T_j - T_o) \quad \textbf{(4.15)}$$

where T_o = junction temperature at 25°C
T_j = new junction temperature, in °C
$V_{TD}(T_o)$ = threshold voltage at junction temperature T_o, which is 0.7 V for a silicon diode, 0.3 V for a germanium diode, and 0.3 V for a Schottky diode (discussed in Sec. 6.6)
$V_{TD}(T_j)$ = threshold voltage at new junction temperature T_j
K_{TC} = temperature coefficient, in V/°C, which is −2.5 mV/°C for a germanium diode, −2 mV/°C for a silicon diode, and −1.5 mV/°C for a Schottky diode

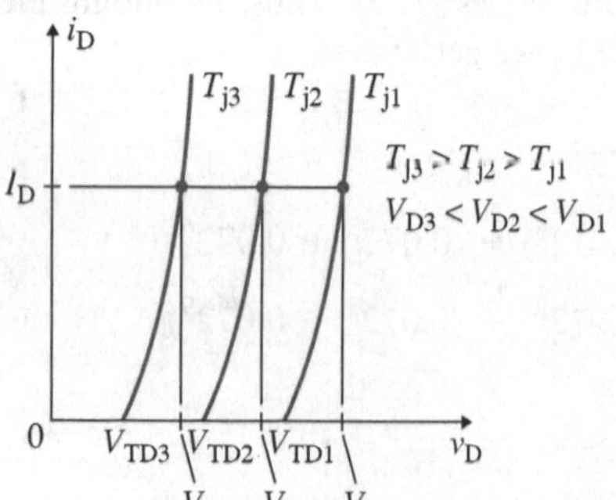

FIGURE 4.9 Temperature dependence of diode current

EXAMPLE 4.5

Finding the temperature dependence of threshold voltage The threshold voltage V_{TD} of a silicon diode is 0.7 V at 25°C. Find the threshold voltage V_{TD} at **(a)** $T_j = 100°C$ and **(b)** $T_j = -100°C$.

SOLUTION

At $T_o = 25°C$, $V_{TD}(T_o) = 0.7$ V. The temperature coefficient for silicon is $K_{TC} = -2$ mV/°C.

(a) At $T_j = 100°C$, from Eq. (4.15),

$$V_{TD}(T_j) = V_{TD}(T_o) + K_{TC}(T_j - T_o)$$

$$= 0.7 - 2 \times 10^{-3} \times (100 - 25) = 0.55 \text{ V}$$

(b) At $T_j = -100°C$, from Eq. (4.15),

$$V_{TD}(T_j) = V_{TD}(T_o) + K_{TC}(T_j - T_o)$$

$$= 0.7 - 2 \times 10^{-3} \times (-100 - 25) = 0.95 \text{ V}$$

Thus, a change in the temperature can significantly change the value of V_{TD}.

EXAMPLE 4.6

Finding the temperature dependence of diode current The leakage current of a silicon diode is $I_S = 10^{-9}$ A at 25°C, and the emission coefficient is $n = 2$. The operating junction temperature is $T_j = 60°C$. Determine **(a)** the leakage current I_S and **(b)** the diode current i_D at $v_D = 0.8$ V.

SOLUTION

$I_S = 10^{-9}$ A at $T_o = 25°C$, $T_j = 60°C$, and $v_D = 0.8$ V.

(a) From Eq. (4.12), the value of I_S at $T_j = 60°C$ is

$$I_S(T_j = 60) = I_S(T_o)2^{0.1(T_j - T_o)} = 10^{-9} \times 2^{0.1\times(60-25)} = 11.31 \times 10^{-9} \text{ A}$$

(b) At $T_K = 273 + 60 = 333$ K, Eq. (4.2) gives

$$V_T = \frac{kT_K}{q} = \frac{1.3806 \times 10^{-23} \times (273 + 60)}{1.6022 \times 10^{-19}} = 28.69 \text{ mV}$$

From Eq. (4.3), we can find the diode current i_D:

$$i_D \approx I_S e^{v_D/nV_T} = 11.31 \times 10^{-9} \times e^{0.8/(2\times 0.02869)} = 12.84 \text{ mA}$$

KEY POINTS OF SECTION 4.4

- A practical diode exhibits a nonlinear *v-i* characteristic, which can be represented by the Shockley diode equation.
- The *v-i* characteristic curve of a diode can be divided into three regions: the forward-biased region, the reverse-biased region, and the breakdown region. A diode is normally operated in either the forward- or the reverse-biased region.
- Diode constants I_S and n can be determined by plotting the *v-i* characteristic of a diode on a semi-log scale.
- The leakage current I_S increases at the rate of approximately +7.2% per degree Celsius for silicon and germanium diodes.
- Both the diode voltage v_D and the threshold voltage V_{TD} decrease with temperature.

4.5 Analysis of Practical Diode Circuits

A diode is used as a part of an electronic circuit, and the diode current i_D becomes dependent on other circuit elements. A simple diode circuit is shown in Fig. 4.10. Applying Kirchhoff's voltage law (KVL), we can express the source voltage V_S and the diode current i_D by

$$V_S = v_D + R_L i_D$$

which gives the diode current i_D as

$$i_D = \frac{V_S - v_D}{R_L} \tag{4.16}$$

Since the diode will be forward biased, the diode current i_D is related to the diode voltage v_D by the Shockley diode equation,

$$i_D = I_S(e^{v_D/nV_T} - 1) \tag{4.17}$$

which shows that i_D depends on v_D, which in turn depends on i_D. Thus, Eqs. (4.16) and (4.17) can be solved for v_D and i_D by any of the following methods: graphical method, approximate method, or iterative method.

4.5.1 Graphical Method

Let us assume that v_D is positive. Then Eq. (4.17) represents the diode characteristic in the forward direction. Equation (4.16) is the equation of a straight line with a slope of $-1/R_L$ and represents the load characteristic known as the *load line.* If Eqs. (4.16) and (4.17) are plotted on the same graph, as shown in Fig. 4.11,

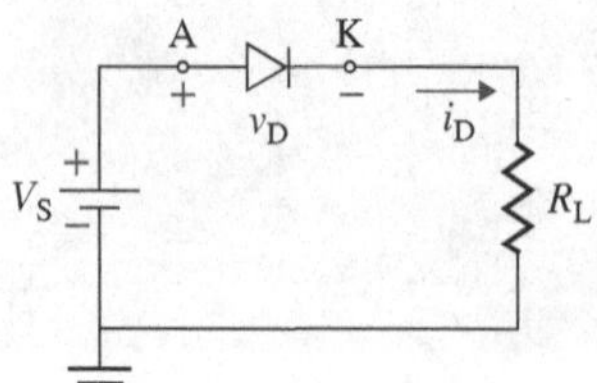

FIGURE 4.10 Simple diode circuit

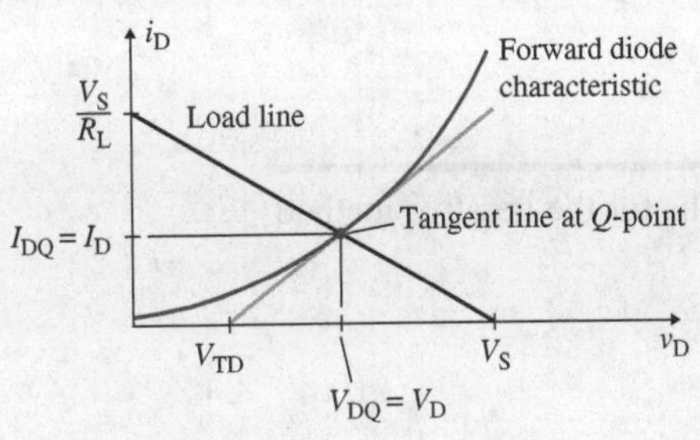

FIGURE 4.11 Graphical method of analysis

the diode characteristic will intersect the load line at a point Q, which is the *operating point* (or *quiescent point*) of the diode. The coordinates of this *Q-point* give the *quiescent diode voltage* V_{DQ} (or simply V_D) and the *quiescent diode current* I_{DQ} (or simply I_D). This graphical approach [6] is not a convenient method of analysis, and thus it is rarely used in the analysis of diode circuits. However, it helps us understand the concept of Q-point and the mechanism of diode circuit analysis.

4.5.2 Approximate Method

To solve Eqs. (4.16) and (4.17) by the approximate method, we assume the diode to have a constant voltage drop equal to the threshold voltage V_{TD}. That is, $v_D = V_{TD}$, and the diode characteristic is approximated as a vertical line, as shown in Fig. 4.12. The threshold voltage V_{TD} of small-signal diodes lies in the range of 0.5 V to 1.0 V. The diode drop for silicon diodes is approximately $v_D = V_{TD} = 0.7$ V, and that for germanium diodes is $v_D = V_{TD} = 0.3$ V. Using the approximate value of v_D, we can find the diode current i_D from Eq. (4.16) as follows:

$$i_D = \frac{V_S - v_D}{R_L} = \frac{V_S - 0.7 \text{ (or 0.3 for germanium)}}{R_L}$$

As an example, let $V_S = 10$ V, $v_D = V_{TD} = 0.7$ V, and $R_L = 1$ kΩ. Then the operating Q-point current I_D becomes $I_D = i_D = (10 - 0.7)\text{ V}/(1\text{ k}\Omega) = 9.3$ mA.

This method gives an approximate solution and does not take into account the nonlinear characteristic described by Eq. (4.17). This approximation is adequate, however, for many applications and is useful as a starting point for a circuit design.

▶ **NOTE** v_D and i_D are the variable quantities, whereas V_D and I_D are their fixed values, respectively.

4.5.3 Iterative Method

The iterative method uses an iterative solution to find the values of i_D and v_D from the load line of Eq. (4.16) and the nonlinear diode characteristic of Eq. (4.17). First a small value of v_D is assumed and Eq. (4.16) is used to find an approximate value of i_D, which is then used to calculate a better

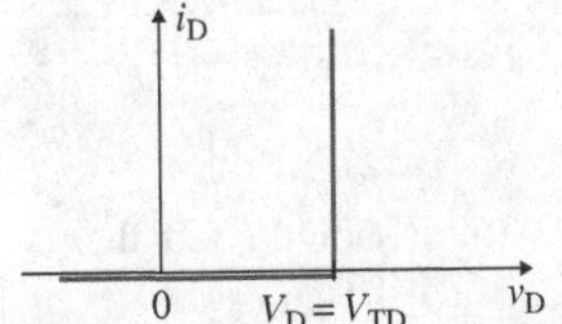

FIGURE 4.12 Approximate diode characteristic

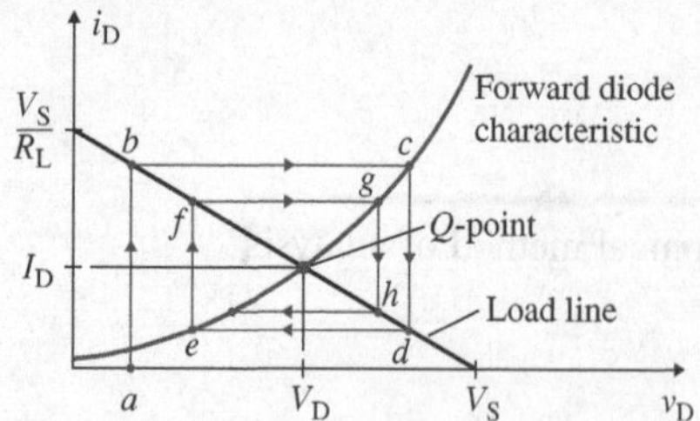

FIGURE 4.13 Paths for the iterative method

approximation of diode voltage v_D from Eq. (4.17). This completes one iteration; the iterations continue until the desired accuracy has been obtained. The steps can be described as follows:

Step 1. Start with an arbitrary point *a*, as shown in Fig. 4.13, and assume a fixed value of v_D (say 0.7 V) at a specified value of i_D.
Step 2. Find point *b* by calculating the value of i_D from the load characteristic described by Eq. (4.16).
Step 3. Find point *c* by calculating a modified value of v_D from the diode characteristic described by Eq. (4.17) or Eq. (4.9). This completes one iteration.
Step 4. Find point *d* by calculating the value of i_D from the load characteristic described by Eq. (4.16).
Step 5. Find point *e* by calculating a modified value of v_D from the diode characteristic described by Eq. (4.17). This completes two iterations.
Step 6. Find point *f* by calculating the value of i_D from the load characteristic described by Eq. (4.16).
Step 7. Find point *g* by calculating a modified value of v_D from the diode characteristic described by Eq. (4.17) or Eq. (4.9). This completes three iterations.

This process is continued until the values of i_D and v_D converge to within the range of desired accuracy.

4.5.4 Mathematical Method

Equating i_D in Eq. (4.16) with that of Eq. (4.17), we get the following relationship:

$$i_D = \frac{(V_S - v_D/R_L)}{R_L} = I_S\left(e^{v_D/\eta V_T} - 1\right) \simeq I_S e^{v_D/\eta V_T} \tag{4.18}$$

This can be solved for the diode voltage v_D by using computer software such as MATHCAD or MATLAB if the values of η, R_L, V_T, and I_S are known. Once the value of v_D is found, the value of i_D can be determined from Eq. (4.16) or Eq. (4.17).

EXAMPLE 4.7

Finding the *Q*-point of a diode circuit The diode circuit shown in Fig. 4.10 has $R_L = 1\text{ k}\Omega$ and $V_S = 10$ V. The emission coefficient is $n = 1.84$, the thermal voltage is $V_T = 25.8$ mV, and the leakage current is $I_S = 2.682 \times 10^{-9}$ A. Calculate the *Q*-point (or the operating point) V_D and I_D by **(a)** the approximate method, **(b)** the iterative method with three iterations, and **(c)** the mathematical method. Assume a default value of $v_D = 0.61$ V as the initial guess.

SOLUTION

If the initial guess of v_D is not specified, we can use the default value of $v_D = 0.70$ V. But it should affect the final results as long as the initial guess value is reasonable around 0.7 V.

(a) $V_D = v_D = 0.61$ V. From Eq. (4.16), $i_D = (V_S - v_D)/R_L = (10 - 0.61)/1\text{ k}\Omega = 9.39$ mA.

(b) $R_L = 1\text{ k}\Omega$, $n = 1.84$, $V_T = 25.8$ mV, and $v_D = 0.61$ V at $i_D = 1$ mA.

Iteration 1: Assume $v_D = 0.61$ V. From Eq. (4.16),

$$i_D = \frac{V_S - v_D}{R_L} = \frac{(10 - 0.61)\text{ V}}{1\text{ k}\Omega} = 9.39\text{ mA}$$

From Eq. (4.5), the new value of v_D is

$$v_{D(new)} = nV_T \ln\left(\frac{i_D}{I_S}\right)$$

$$= 1.84 \times 0.0258 \ln\left(\frac{9.39\text{ mA}}{2.682 \times 10^{-9}}\right) = 0.7153\text{ V}$$

Iteration 2: Assume the values of v_D from the previous iteration. That is, set $v_D = v_{D(new)} = 0.7153$ V. From Eq. (4.16),

$$i_{D(new)} = \frac{V_S - v_D}{R_L} = \frac{(10 - 0.7153)\text{ V}}{1\text{ k}\Omega} = 9.2847\text{ mA}$$

From Eq. (4.5), the new value of v_D is

$$v_{D(new)} = nV_T \ln\left(\frac{i_{D(new)}}{I_S}\right)$$

$$= 1.84 \times 0.0258 \ln\left(\frac{9.2847}{2.682 \times 10^{-9}}\right) = 0.7148\text{ V}$$

Iteration 3: Assume the values of v_D from the previous iteration. That is, set $v_D = v_{D(new)} = 0.7148$ V. From Eq. (4.16),

$$i_{D(new)} = \frac{V_S - v_D}{R_L} = \frac{(10 - 0.7148)\text{ V}}{1\text{ k}\Omega} = 9.285\text{ mA}$$

From Eq. (4.5), the new value of v_D is

$$v_{D(new)} = nV_T \ln\left(\frac{i_{D(new)}}{I_S}\right)$$

$$= 1.84 \times 0.0258 \ln\left(\frac{9.285}{2.682 \times 10^{-9}}\right) = 0.7158\text{ V}$$

Therefore, after three iterations, $V_D = v_{D(new)} = 0.7158$ V and $I_D = i_{D(new)} = 9.285$ mA. Note that the results of iteration 3 do not differ significantly from those of iteration 2. In fact, there was no need for iteration 3.

(c) Substituting for the given values in Eq. (4.18), $(V_S - v_D)/R_L \simeq I_S e^{v_D/\eta V_T}$, we get $(10 - v_D)/\text{k}\Omega = 2.682 \times 10^{-9} \times e^{v_D/(1.84 \times 25.8 \times 10^{-3})}$ which, after solving by MATHCAD software function, gives $V_D = 0.7148$ V and

$$I_D = I_S e^{-V_D/\eta V_T} = (2.682 \times 10^{-9})e^{-0.7148/(1.84 \times 25.8 \times 10^{-3})} = 9.2845\text{ mA}$$

NOTE: Four-digit answers were used to control computational errors and the number of iterations needed to reach the solution. In reality, resistors will have tolerances, and such accuracy may not be necessary.

TABLE 4.3 The values of V_D and I_D obtained by different methods

Method	V_D (V)	I_D (mA)
Approximate	0.7	9.39
Iterative	0.7148	9.2852
Mathematical	0.7148	9.2845

KEY POINTS OF SECTION 4.5

- The analysis of a diode circuit involves solving a nonlinear diode equation.
- The graphical method is rarely used.
- The approximate method gives a quick answer but approximate values. The mathematical solution by MATHCAD gives quick but accurate results. In the absence of any computer-aided solution, the iterative method also gives accurate results. Comparisons of the values obtained by these three methods for Example 4.7 are shown in Table 4.3.

4.6 Modeling of Practical Diodes

In practice, multiple diodes are used in a circuit. Therefore, diode circuits become complex, and analysis by the graphical or iterative method becomes time-consuming and laborious. To simplify the analysis and design of diode circuits, we can represent a diode by one of the following models: constant-drop DC model, piecewise linear DC model, low-frequency AC model, high-frequency AC model, or SPICE diode model.

4.6.1 Constant-Drop DC Model

The constant-drop DC model assumes that a conducting diode has a voltage drop v_D that remains almost constant and is independent of the diode current. Therefore, the diode characteristic becomes a vertical line at the threshold voltage; that is, $v_D = V_{TD}$. The Q-point is determined by adding the load line to the approximate diode characteristic, as shown in Fig. 4.14(a). The diode voltage v_D is expressed by

$$v_D = \begin{cases} V_{TD} & \text{for } v_D \geq V_{TD} \\ 0 & \text{for } v_D < V_{TD} \end{cases}$$

The circuit model is shown in Fig. 4.14(b). The typical value of V_{TD} is 0.7 V for silicon diodes and 0.3 V for germanium diodes. With this model, the diode current i_D can be determined from

$$i_D = \frac{V_S - V_{TD}}{R_L} \tag{4.19}$$

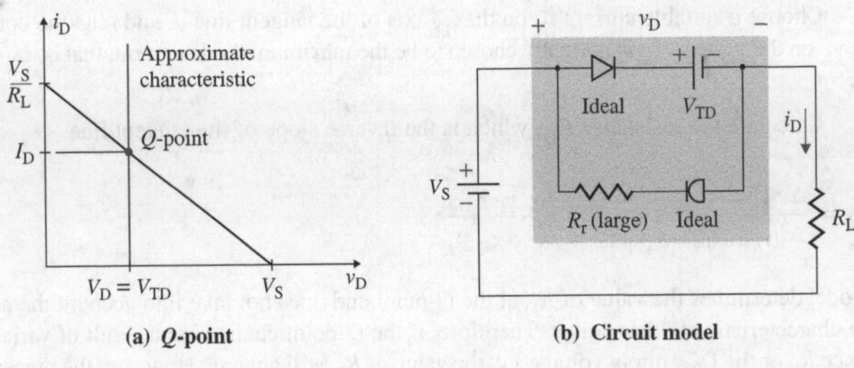

FIGURE 4.14 Constant-drop DC model

4.6.2 Piecewise Linear DC Model

The voltage drop across a practical diode increases with its current. The diode characteristic can be represented approximately by a fixed voltage drop V_{TD} and a straight line, as shown in Fig. 4.15(a). The straight line *a* takes into account the current dependence of the voltage drop, and it represents a fixed resistance R_D, which remains constant. The line *a* can pass through at most two points; it is usually drawn tangent to the diode characteristic at the estimated *Q*-point. This model represents the diode characteristic approximately by two piecewise parts: a fixed part and a current-dependent part. A piecewise linear representation of the diode is shown in Fig. 4.15(b). The steps for determining the model parameters are as follows:

Step 1. Draw a line tangent to the current-dependent part of the forward diode characteristic at the estimated *Q*-point. A best-fit line through the current-dependent part is generally acceptable.

Step 2. Use the intercept on the v_D-axis as the fixed drop V_{TD}.

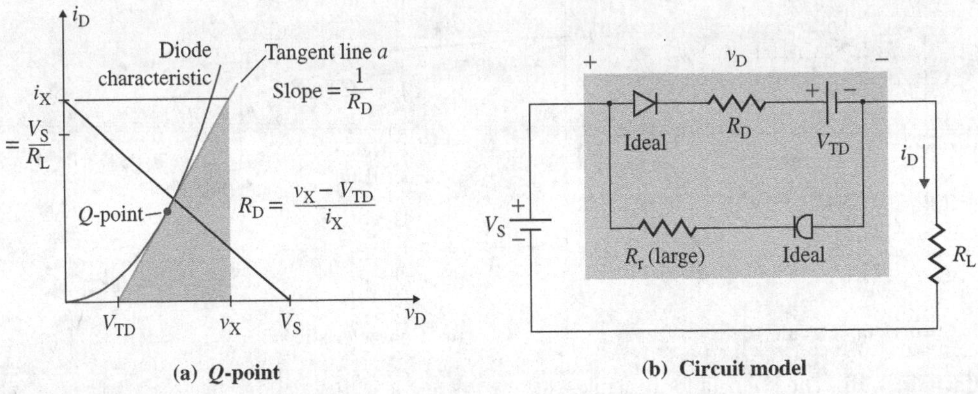

FIGURE 4.15 Piecewise linear DC model

Step 3. Choose a suitable current i_X on the i_D-axis of the tangent line a, and read the corresponding voltage v_X on the v_D-axis. i_X is normally chosen to be the maximum diode current; that is, $i_X = i_{D(max)} = V_S/R_L$.

Step 4. Calculate the resistance R_D, which is the inverse slope of the tangent line.

$$R_D = \left.\frac{\Delta v_D}{\Delta i_D}\right|_{\text{at estimated } Q\text{-point}} = \frac{v_X - V_{TD}}{i_X} \tag{4.20}$$

This model determines the value of R_D at the Q-point and does not take into account the actual shape of the diode characteristic at other points. Therefore, if the Q-point changes as a result of variations in the load resistance R_L or the DC supply voltage V_S, the value of R_D will change. However, the piecewise model is quite satisfactory for most applications. Using this model and applying KVL, we find that the diode current i_D in Fig. 4.15(b) is given by

$$V_S = V_{TD} + R_D i_D + R_L i_D \tag{4.21}$$

which gives the diode current i_D as

$$i_D = \frac{V_S - V_{TD}}{R_D + R_L} \tag{4.22}$$

EXAMPLE 4.8

Finding the *Q*-point of a diode circuit and diode model parameters by two different methods

The diode circuit shown in Fig. 4.16(a) has $V_S = 10$ V and $R_L = 1$ kΩ. The diode characteristic is shown in Fig. 4.16(b). Determine the diode voltage v_D, the diode current i_D, and the load voltage v_O by using **(a)** the piecewise linear DC model and **(b)** the constant-drop DC model.

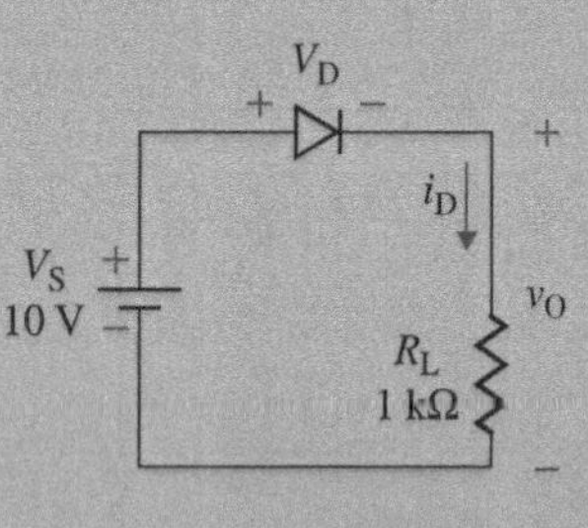

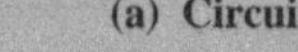

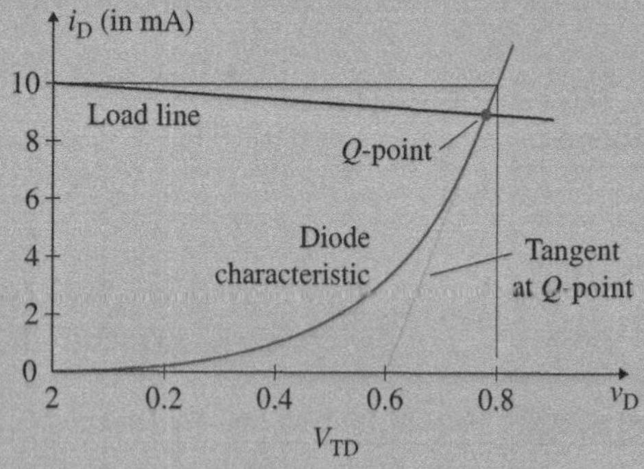

FIGURE 4.16 Diode circuit for Example 4.8

SOLUTION

$V_S = 10$ V and $R_L = 1$ kΩ. Thus, $i_{D(max)} = \dfrac{V_S}{R_L} = \dfrac{10 \text{ V}}{1 \text{ k}\Omega} = 10$ mA.

(a) If we follow the steps described in Sec. 4.6.2, the tangent or best-fit line gives $V_{TD} = 0.6$ V and $v_X = 0.8$ V at $i_X = i_{D(max)} = 10$ mA. From Eq. (4.20), the resistance R_D of the current-dependent part is

$$R_D = \frac{v_X - V_{TD}}{i_X} = \frac{(0.8 - 0.6) \text{ V}}{(10 \text{ mA})} = 20 \ \Omega$$

From Fig. 4.15(b), the diode current is

$$i_D = \frac{V_S - V_{TD}}{R_L + R_D} = \frac{(10 - 0.6) \text{ V}}{1 \text{ k}\Omega + 20\Omega} = 9.22 \text{ mA}$$

From Fig. 4.15(b), the diode voltage is

$$v_D = V_{TD} + R_D i_D = 0.6 + 20 \times 9.22 \times 10^{-3} = 0.784 \text{ V}$$

Thus, the load voltage becomes

$$v_O = V_S - V_D = 10 - 0.784 = 9.216 \text{ V}$$

NOTE: The diode mode parameters are $V_{TD} = 0.6$ V and $R_D = 20$ Ω.

(b) Using Eq. (4.19) for the constant-drop DC model of Fig. 4.14(b), we get the diode current

$$i_D = \frac{V_S - V_{TD}}{R_L} = \frac{(10 - 0.6) \text{ V}}{1 \text{ k}\Omega} = 9.4 \text{ mA}$$

The load voltage is

$$v_O = V_S - V_{TD} = 10 - 0.6 = 9.4 \text{ V}$$

for an error of (9.4 − 9.216)/9.4 = 1.96% compared to the piecewise linear model.

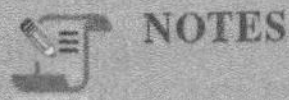

NOTES:

1. The diode mode parameters are $V_{TD} = 0.6$ V and $R_D = 0$ Ω.
2. If the supply voltage V_S is much greater than the diode voltage drop v_D, the constant-drop DC model will give acceptable results. If the diode voltage v_D is comparable to the supply voltage V_S, the piecewise linear DC model, which gives better results, is generally acceptable in most applications.

4.6.3 Low-Frequency Small-Signal Model

In electronic circuits, a DC supply normally sets the DC operating point of electronic devices including diodes, and an AC signal is usually then superimposed on the operating point. Thus, the operating point, which consists of both a DC component and an AC signal, will vary with the magnitude of the AC signal.

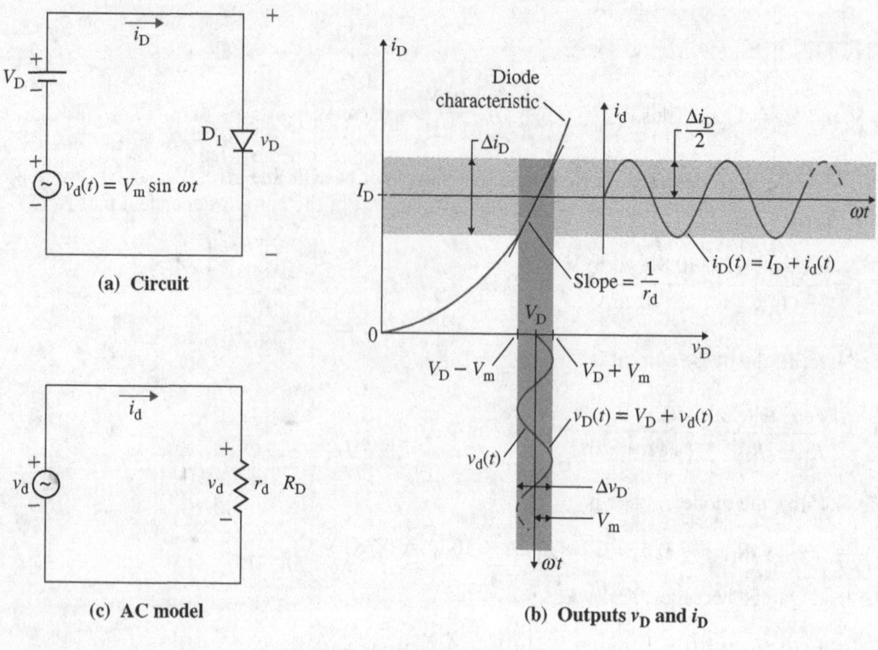

FIGURE 4.17 Low-frequency AC model

Since the i_D-v_D characteristic of a diode is nonlinear, the diode current i_D will also vary nonlinearly with the AC signal voltage. The magnitude of the AC signal is generally small, however, so the operating point changes by only a small amount. Thus, the slope of the characteristic (Δi_D versus Δv_D) can be approximated linearly. Under this condition, we can represent the diode as a resistance in order to determine the response of the circuit to this small AC signal. That is, the nonlinear diode characteristic can be linearized at the operating point. A small-signal model [7] is widely used for the analysis and design of electronic circuits in order to obtain their small-signal behavior.

Figure 4.17(a) shows a diode circuit with a DC source V_D, which sets the operating point at Q, defined by coordinates V_D and I_D. If a small-amplitude sinusoidal voltage v_d is superimposed on V_D, the operating point will vary with the time-varying AC signal v_d. Therefore, if the diode voltage varies between $(V_D + V_m)$ and $(V_D - V_m)$, the corresponding diode current will vary between $(I_D + \Delta i_D/2)$ and $(I_D - \Delta i_D/2)$. This is illustrated in Fig. 4.17(b), in which the change in the AC diode current i_d is assumed to be approximately sinusoidal in response to a sinusoidal voltage v_d. However, the diode characteristic is nonlinear and the diode current will be slightly distorted.

Under small-signal conditions, the diode characteristic around the Q-point can be approximated by a straight line and modeled by a resistance called the *dynamic resistance* or *AC resistance or small-signal resistance* r_d, which is defined by

$$\frac{1}{r_d} = g_d = \left.\frac{\Delta i_D}{\Delta v_D}\right|_{\text{at } Q\text{-point}} \quad (4.23)$$

where g_d is the *small-signal diode transconductance* and depends on the slope of the diode characteristic at the operating point. Since r_d is determined from the slope of the diode characteristic at the Q-point, its value should be the *same* as R_D of Fig. 4.15(b) if the tangent line is drawn accurately at the DC operating point.

Determining r_d by Differentiating

If the operating point (V_D, I_D) is known for a given diode characteristic and a load line, we can determine the value of r_d (= R_D) directly by considering a change in diode voltage around the operating point. If Δv_D and Δi_D are small, tending to zero, Eq. (4.23) becomes

$$g_d = \frac{1}{r_d} = \left.\frac{di_D}{dv_D}\right|_{\text{at } Q\text{-point}} \tag{4.24}$$

If $v_D > 0.1$ V, which is usually the case when the diode is operated in the forward direction, then the diode current i_D is related to the diode voltage v_D by

$$i_D = I_S(e^{v_D/nV_T} - 1) \approx I_S e^{v_D/nV_T} \tag{4.25}$$

Substituting i_D from Eq. (4.25) into Eq. (4.24) and differentiating i_D with respect to v_D gives

$$g_d = \frac{1}{r_d} = \left.\frac{di_D}{dv_D}\right|_{\text{at } Q\text{-point}} = I_S \frac{1}{nV_T} e^{v_D/nV_T} = \frac{i_D + I_S}{nV_T} \tag{4.26}$$

which gives the AC resistance ($r_d = R_D$) at the operating point (V_D, I_D). That is,

$$r_d = R_D = \frac{1}{g_d} = \frac{nV_T}{i_D + I_S} \approx \frac{nV_T}{I_D} \quad \text{since } i_D = I_D, \text{ and } I_D >> I_S \tag{4.27}$$

$$\approx \frac{0.0258}{I_D} \quad \text{at 25°C and for } n = 1 \tag{4.28}$$

Notice from Eq. (4.27) that the determination of the AC resistance requires the determination of the diode current i_D at the *Q*-point.

Determining r_d by Taylor Series Expansion

Equation (4.27) can also be derived by Taylor series expansion. The instantaneous diode voltage v_D is the sum of V_D and v_d. That is,

$$v_D = V_D + v_d \tag{4.29}$$

Substituting $v_D = V_D + v_d$ into Eq. (4.25) gives the instantaneous diode current i_D:

$$i_D \approx I_S e^{(V_D+v_d)/nV_T} = I_S e^{V_D/nV_T} e^{v_d/nV_T}$$

$$= I_D e^{v_d/nV_T} \text{ since } I_D = I_S e^{V_D/nV_T} \tag{4.30}$$

If the amplitude of the sinusoidal voltage v_d is very small compared to nV_T, so that $v_d << nV_T$, we can use the relation $e^x \approx 1 + x$. Equation (4.30) can be expanded in Taylor series with the first two terms:

$$i_D \approx I_D\left(1 + \frac{v_d}{nV_T}\right) = I_D + i_d(t) \tag{4.31}$$

Thus, the instantaneous diode current i_D has two components: a DC component I_D and a small-signal AC component i_d. This is a mathematical derivation of the principle of superposition introduced in Appendix B. From Eq. (4.31), the AC diode current i_d is defined by

$$i_d = \frac{v_d}{nV_T} I_D \tag{4.32}$$

which gives the small-signal AC resistance r_d as

$$r_d = \frac{v_d}{i_d} = \frac{nV_T}{I_D}$$

which is the same as Eq. (4.27). The value of r_d should ideally be the same as the value of R_D if the tangent line is drawn accurately at the DC operating point.

The small-signal AC model of a diode is shown in Fig. 4.17(c). This model is known as the *low-frequency small-signal AC model.* It does not take into account the frequency dependency of the diode.

▶ NOTE

1. The AC resistance r_d takes into account the shape of the curve and represents the slope of the characteristic at the Q-point. If the Q-point changes, the value of r_d will also change.
2. R_D is determined from the slope of the diode characteristic at an estimated Q-point, whereas r_d is determined from the Shockley diode equation. If r_d and R_D are determined from the two methods, their values should be the same, although there may be a small but generally negligible difference.
3. We will see in Chapters 7 and 8 that the concept of small-signal resistance r_d in Eq. (4.28) can be applied to model the small-signal behavior of transistors.

EXAMPLE 4.9

Small-signal analysis of a diode circuit The diode circuit shown in Fig. 4.18 has $V_S = 10$ V, $V_m = 50$ mV, and $R_L = 1$ kΩ. Use the Q-point found in Example 4.7 by mathematical method to determine the instantaneous diode voltage v_D. Assume an emission coefficient of $n = 1.84$.

SOLUTION

$V_T = 25.8$ mV, $n = 1.84$, $V_S = 10$ V, and $R_L = 1$ kΩ. The iterations of the Q-point analysis in Example 4.7 gave $V_D = 0.7148$ V and $I_D = 9.284$ mA. Using Eq. (4.27), we can find the AC resistance r_d from

$$r_d = \frac{nV_T}{I_D} = \frac{1.84 \times 25.8 \times 10^{-3}}{(9.284 \times 10^{-3})} = 5.11\ \Omega$$

The AC equivalent circuit is shown in Fig. 4.19. From the voltage divider rule, the AC diode voltage v_d is given by

$$v_d = \frac{r_d}{r_d + R_L} V_m \sin \omega t \tag{4.33}$$

$$= \frac{5.11}{5.11 + 1\text{ k}\Omega} 50 \times 10^{-3} \sin \omega t = 0.2542 \times 10^{-3} \sin \omega t$$

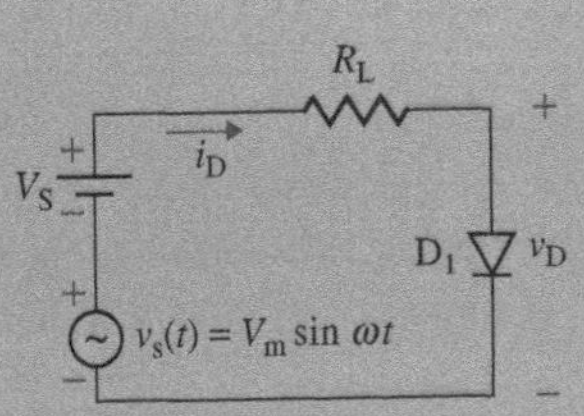

FIGURE 4.18 Diode circuit for Example 4.9

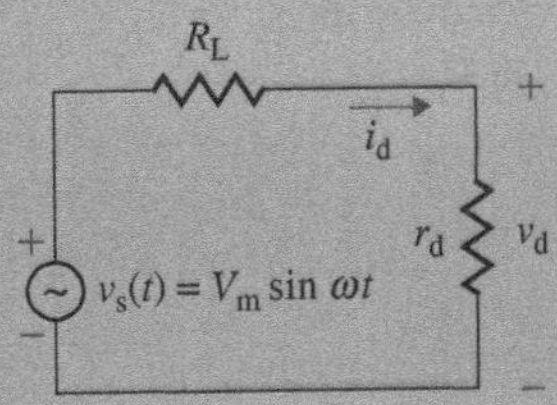

FIGURE 4.19 AC equivalent diode circuit

Therefore, the instantaneous diode voltage v_D is the sum of V_D and v_d. That is,

$$v_D = v_D + v_d$$

$$= 0.7158 + 0.2542 \times 10^{-3} \sin \omega t \text{ V}$$

EXAMPLE 4.10

Finding the *Q*-point of a diode circuit and the diode model parameters from tabular data The diode circuit shown in Fig. 4.10 has $V_S = 15$ V and $R_L = 250\ \Omega$. The diode forward characteristic, which can be obtained either from practical measurement or from the manufacturer's data sheet, is given by the following table:

i_D (mA)	0	10	20	30	40	50	60	70
v_D (V)	0.5	0.87	0.98	1.058	1.115	1.173	1.212	1.25

Determine **(a)** the *Q*-point (V_D, I_D), **(b)** the parameters (V_{TD} and R_D) of the piecewise linear DC model, and **(c)** the small-signal AC resistance r_d. Assume that the emission coefficient is $n = 1$ and that $V_T = 25.8$ mV.

SOLUTION

$V_S = 15$ V and $R_L = 250\ \Omega$.

(a) From Eq. (4.16), the load line is described by

$$i_D = \frac{(V_S - v_D)}{R_L}$$

The diode characteristic is defined by a table of data. The *Q*-point can be determined from the load line and the data table by an iterative method, as discussed in Sec. 4.5.3.

Iteration 1: Assume $v_D = 0.7$ V. From Eq. (4.16),

$$i_D = \frac{(V_S - v_D)}{R_L} = \frac{(15 - 0.7)}{250} = 57.2 \text{ mA}$$

which lies between 50 mA and 60 mA in the table. Thus, we can see from the table of data that the new value of diode drop $v_{D(new)}$ lies between 1.173 V and 1.212 V. Let us assume that the diode voltage $v_D(k)$ corresponds to the diode current $i_D(k)$ and the diode voltage $v_D(k+1)$ corresponds to the diode current $i_D(k+1)$. This is shown in Fig. 4.20. If i_D lies between $i_D(k)$ and $i_D(k+1)$, then the corresponding value of v_D will lie between $v_D(k)$ and $v_D(k+1)$. Thus, $v_{D(new)}$ can be found approximately by linear interpolation from

$$v_{D(new)} = v_D(k) + \frac{v_D(k+1) - v_D(k)}{i_D(k+1) - i_D(k)}[i_D - i_D(k)] \tag{4.34}$$

$$= 1.173 + \frac{1.212 - 1.173}{60 \text{ mA} - 50 \text{ mA}}(57.2 \text{ mA} - 50 \text{ mA}) = 1.201 \text{ V}$$

Iteration 2: Use the value of v_D from the previous iteration; that is, set $v_D = v_{D(new)} = 1.201$ V. From Eq. (4.16),

$$i_D = i_{D(new)} = \frac{(15 - 1.201)}{250} = 55.2 \text{ mA}$$

From Eq. (4.34), the new value of v_D is

$$v_{D(new)} = 1.173 + \frac{1.212 - 1.173}{60 \text{ m} - 50 \text{ m}}(55.2 \text{ m} - 50 \text{ m}) = 1.193 \text{ V}$$

This process is repeated until a stable Q-point is found. After two iterations, we have $V_D = v_{D(new)} = 1.193$ V and $I_D = i_{D(new)} = 55.2$ mA.

(b) Since R_D is the slope of the tangent at the Q-point, we get

$$R_D = \left.\frac{\Delta v_D}{\Delta i_D}\right|_{\text{at } Q\text{-point}} = \frac{v_D - v_D(k)}{i_D - i_D(k)} = \frac{v_D(k+1) - v_D}{i_D(k+1) - i_D}$$

$$= \frac{1.193 - 1.173}{(55.2 - 50) \times 10^{-3}} = 3.9\,\Omega \tag{4.35}$$

The diode threshold voltage is

$$V_{TD} = V_D - R_D I_D = 1.193 - 3.9 \times 55.2 \text{ mV} = 0.98 \text{ V}$$

(c) From Eq. (4.27), the small-signal AC resistance r_d is

$$r_d = \frac{nV_T}{I_D} = \frac{1 \times 25.8 \text{ mV}}{(55.2 \times 10^{-3})} = 0.5\ \Omega$$

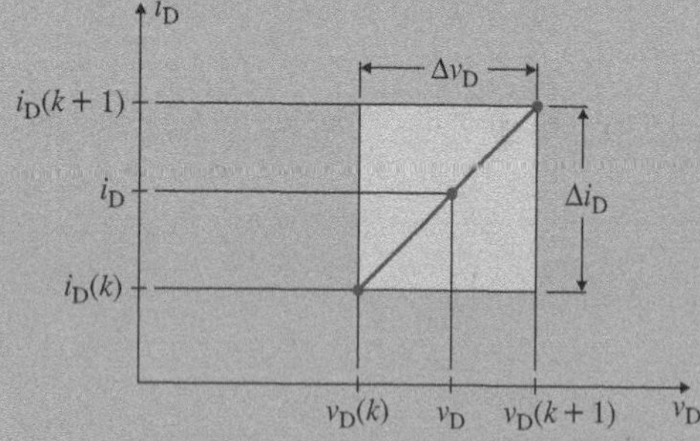

FIGURE 4.20 Linear interpolation for diode voltage

NOTE: The difference between r_d and R_D is due to the fact that a diode follows the Shockley diode equation, whereas the values in the data table are quoted without regard to any relationship. The correlation will depend on how closely the tabular data match with the Shockley equation [Eq. (4.1)].

4.6.4 PSpice/SPICE Diode Model

PSpice/SPICE uses a voltage-dependent current source, as shown in Fig. 4.21(a). r_s is the series resistance, known as the *bulk* (or *parasitic*) *resistance.* It is due to the resistance of the semiconductor and is dependent on the amount of doping. It should be noted that Fig. 4.21(a) is a nonlinear diode model, whereas the constant-drop DC model, the piecewise linear DC model, and the low-frequency AC model are linear or piecewise linear models.

At first PSpice/SPICE finds the DC biasing point and then calculates the parameter of the small-signal model shown in Fig. 4.21(b). C_j is a nonlinear function of the diode voltage v_D, and its value equals $C_j = dq_j/dv_D$, where q_j is the depletion layer charge. (The junction capacitances and the high-frequency model are discussed in Sec. 6.8.) PSpice/SPICE generates the small-signal parameters from the operating point and adjusts the values of r_d and C_j for the forward or reverse condition. The diode characteristic can be described in PSpice/SPICE in either a model statement or a tabular representation.

Model Statement

The PSpice/SPICE model statement of a diode has the general form

```
.MODEL DNAME D (P1=A1 P2=A2 P3=A3 .......PN=AN)
```

where DNAME is the model name, which can begin with any character but is normally limited to eight characters. D is the type symbol for diodes. P1, P2, . . . and A1, A2, . . . are the model parameters and their values, respectively. The model parameters can be found in the PSpice/SPICE library file or can be determined from the data sheet [8, 9]. For example, a typical statement for diode D1N4148 is as follows:

```
.MODEL D1N4148 D(IS=2.682N N=1.836 RS=.5664 IKF=44.17M XTI=3 EG=1.11
+ CJO=4P M=.3333 VJ=.5 FC=.5 ISR=1.565N NR=2 BV=100 IBV=100U TT=11.54N)
```

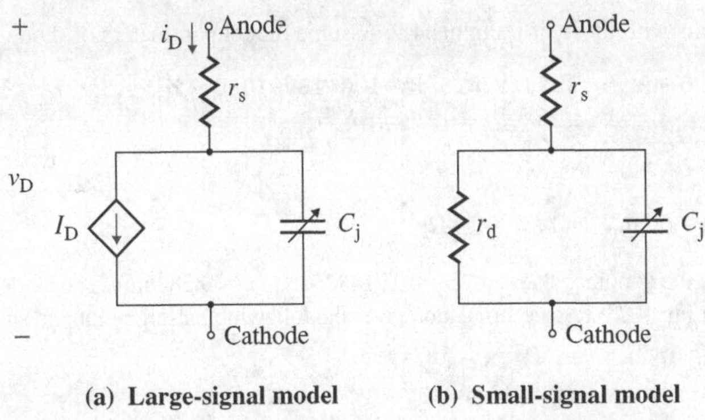

FIGURE 4.21 PSpice/SPICE diode model

FIGURE 4.22 Diode TABLE representation

Tabular Representation

The TABLE representation is available only in PSpice. It allows the *v*-*i* characteristic to be described, and it has the general form

```
E<name> N+ N- TABLE {<expression>} = <<(input) value>, <(output) value>>
```

E<name> is the name of a voltage-controlled voltage source, and N+ and N− are the positive and negative nodes of the voltage source, respectively. The keyword TABLE indicates that the relation is described by a table of data. The table consists of pairs of values: <(input) value> and <(output) value>. The first value in a pair is the input, and the second value is the corresponding output. The <expression> is the input value and is used to find the corresponding output from the lookup table. If an input value falls between two entries, the output is found by linear interpolation. If the input falls outside the table's range, the output is assumed to remain constant at the value corresponding to the smallest or the largest input.

The diode characteristic is represented by a current-controlled voltage source—say ED. That is, the diode is replaced by a voltage source of ED in series with a dummy voltage source VX of 0. VX acts as an ammeter and measures the diode current. This is shown in Fig. 4.22. ED is related to I_D [i.e., I(VX)] by a table. The PSpice representation for the diode characteristic in Example 4.10 is shown here:

```
VX 2 3 DC 0V                                        ; measures the diode current ID
ED 1 2 TABLE {I(VX)} = (0, 0.5) (10m, 0.87) (20m, 0.98) (30m, 1.058)
+ (40m, 1.115) (50m, 1.173) (60m, 1.212) (70m, 1.25) (80m, 1.5) (300m, 3.0)
```

EXAMPLE 4.11

PSpice/SPICE diode model and analysis The diode circuit shown in Fig. 4.18 has $V_S = 10$ V, $V_m = 50$ mV at 1 kHz, $R_L = 1$ kΩ, and $V_T = 25.8$ mV. Assume an emission coefficient of $n = 1.84$.

(a) Use PSpice/SPICE to generate the *Q*-point and the small-signal parameters and to plot the instantaneous output voltage $v_O = v_D$.

(b) Compare the results with those of Example 4.9. Assume model parameters of diode D1N4148:

```
IS=2.682N CJO=4P M=.3333 VJ=.5 BV=100 IBV=100U TT=11.54N
```

SOLUTION

$V_S = 10$ V, $V_m = 50$ mV, and $R_L = 1$ kΩ.

(a) From Example 4.7, the *Q*-point values are $V_D = 0.7148$ V and $I_D = 9.284$ mA. The diode circuit for PSpice simulation is shown in Fig. 4.23. PSpice simulation gives the following biasing point and small-signal parameters:

ID	9.28E-03	$I_D = 9.28$ mA
VD	7.18E-01	$V_D = 718$ mV
REQ	5.53E+00	$r_d = 5.53$ Ω
CAP	2.10E-09	$C_j = 2.1$ nF

The PSpice plot of the transient response is shown in Fig. 4.24, which gives $V_D = 718.35$ mV and $v_{d(peak)} = v_{o(peak)} = 600.52\ \mu V/2 = 300.3\ \mu V$. Thus.

$$v_d = 300.3 \times 10^{-6} \sin \omega t \quad \text{and} \quad v_{d(peak)} = 300.3\ \mu V$$

(b) Example 4.9 gives $V_D = 0.7148$ V, $I_D = 9.284$ mA, $r_d = 5.11\ \Omega$, $v_d = 254.2 \times 10^{-6} \sin \omega t$, and $v_{d(peak)} = 254.2\ \mu V$, which agree closely with the PSpice results.

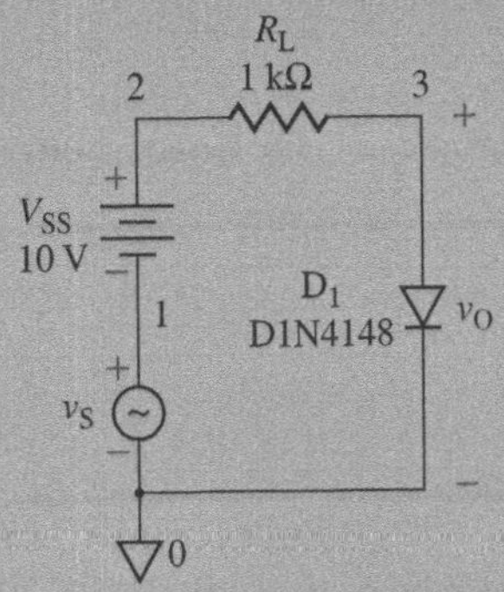

FIGURE 4.23 Diode circuit for PSpice simulation

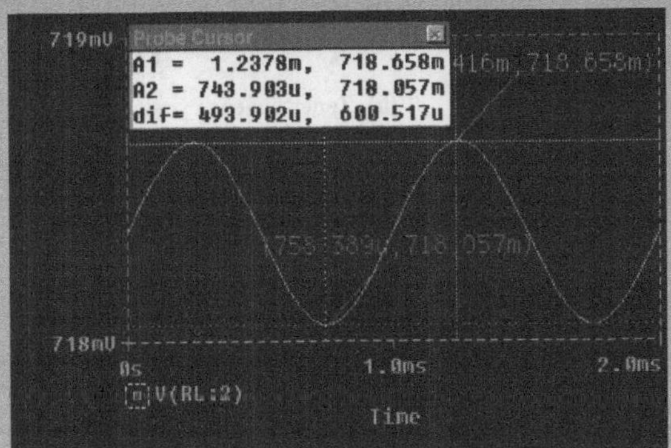

FIGURE 4.24 PSpice plot for Example 4.11

KEY POINTS OF SECTION 4.6

- The constant-drop DC model assumes a fixed voltage drop of the diode. It gives a quick but approximate result. It is best suited for finding the approximate behavior of a circuit, especially at the initial design stage.
- The piecewise linear DC model breaks the nonlinear diode characteristic into two parts: a fixed DC voltage and a current-dependent voltage drop across a fixed resistance. The resistance is determined by drawing a best-fit line through the estimated Q-point on the current-dependent part. This model is commonly used for the analysis of diode circuits, and it gives reasonable results for most applications.
- The low-frequency AC model represents the behavior of a diode in response to a variation of the Q-point caused by a small signal. It is modeled by a small-signal resistance drawn as a tangent at the Q-point and is dependent on the diode current. The resistance can be approximated by that of the piecewise linear model. Thus this model can be regarded as an extension of the piecewise linear DC model.
- The high-frequency AC model represents the frequency response of the diode by incorporating two junction capacitances (diffusion and depletion layer) into the low-frequency AC model. The depletion layer capacitance is dependent on the diode reverse voltage. But the diffusion capacitance is directly proportional to the diode current and is present only in the forward direction.
- PSpice/SPICE generates a complex but accurate model. However, it is necessary to define the PSpice/SPICE model parameters, which can be obtained from the PSpice/SPICE library or from the manufacturer. These parameters can also be determined from the diode characteristic.

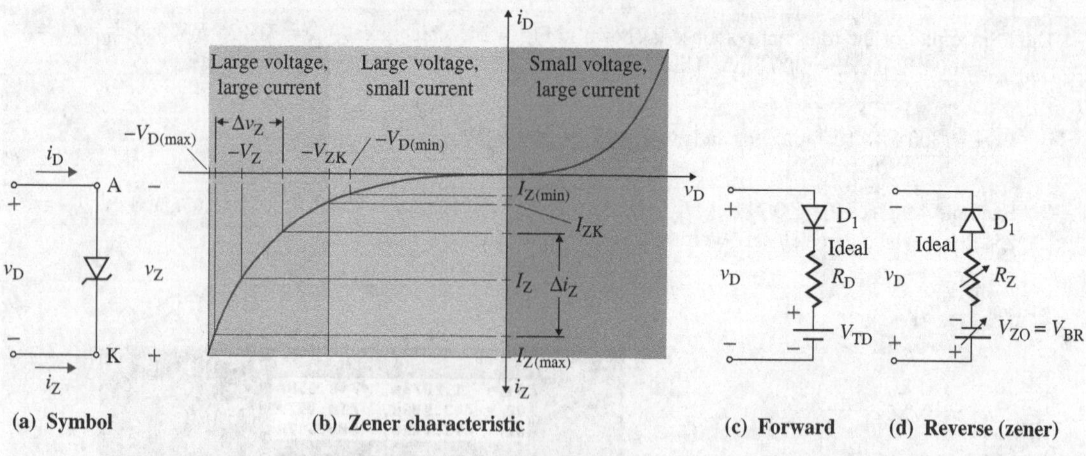

FIGURE 4.25 Characteristic of zener diodes

4.7 Zener Diodes

If the reverse voltage of a diode exceeds a specific voltage called the *breakdown voltage,* the diode will operate in the breakdown region. In this region the reverse diode current increases rapidly. The diode voltage remains almost constant and is practically independent of the diode current. However, operation in the breakdown region will not be destructive if the diode current is limited to a safe value by an external circuitry so that the power dissipation within the diode is within permissible limits specified by the manufacturer and the diode does not overheat.

A diode especially designed to have a steep characteristic in the breakdown region is called a *zener diode.* The symbol for a zener diode is shown in Fig. 4.25(a), and its v-i characteristic appears in Fig. 4.25(b). V_{ZK} is the knee voltage, and I_{ZK} is its corresponding current. A zener diode is specified by its breakdown voltage, called the *zener voltage* (or *reference voltage*) V_Z, at a specified test current $I_Z = I_{ZT}$. $I_{Z(max)}$ is the maximum current that the zener diode can withstand and still remain within permissible limits for power dissipation. $I_{Z(min)}$ is the minimum current, slightly below the knee of the characteristic curve, at which the diode exhibits the reverse breakdown. That is, $I_{Z(min)} \approx I_{ZK}$.

The forward and reverse characteristics of a zener diode are represented by an arrow symbol. The arrow points toward the positive current i_D. In the forward direction, the zener diode behaves like a normal diode; its equivalent circuit is shown in Fig. 4.25(c). In the reverse direction, it offers a very high resistance, acting like a normal reverse-biased diode if $|v_D| < V_Z$ and like a low-resistance diode if $|v_D| > V_Z$. For example, let us consider a zener diode with a nominal voltage $V_Z = 5\text{ V} \pm 2\text{ V}$. For $3\text{ V} < |v_D| < 5\text{ V}$ in the reverse direction, the diode will normally exhibit a zener effect. For $5\text{ V} < V_Z < 7\text{ V}$, the breakdown could be due to the zener effect, the avalanche effect, or a combination of the two.

The reverse (zener) characteristic of Fig. 4.25(b) can be approximated by a piecewise linear model with a fixed voltage V_{ZO} and an ideal diode in series with resistance R_Z. The equivalent circuit of the zener action is shown in Fig. 4.25(d) for $|v_D| > V_Z$. R_Z depends on the inverse slope of the zener characteristic and is defined as

$$R_Z = \left.\frac{\Delta v_Z}{i_Z}\right|_{\text{at } v_Z} = \left.\frac{\Delta v_D}{\Delta i_D}\right|_{\text{for } v_D < 0 \text{ and } i_D < 0} \tag{4.36}$$

R_Z is also called the *zener resistance.* The value of R_Z remains almost constant over a wide range of the zener characteristic. However, its value changes rapidly in the vicinity of the knee point. Thus, a zener diode should be operated away from the knee point. The typical value of R_Z is a few tens of ohms, but it increases with current i_D. At the knee point of the zener characteristic, R_Z has a high value, typically 3 kΩ. The zener current i_Z ($= -i_D$) can be related to V_{ZO} and R_Z by

$$v_Z = v_{ZO} + R_Z i_Z \tag{4.37}$$

4.7.1 Zener Regulator

A zener diode may be regarded as offering a variable resistance whose value changes with the current so that the voltage drop across the terminals remains constant. Therefore, it is also known as a *voltage reference diode.* The value of R_Z is very small. Thus, the zener voltage v_Z is almost independent of the reverse diode current $i_D = -i_Z$. Because of the constant voltage characteristic in the breakdown region, a zener diode can be employed as a *voltage regulator.* A *regulator* maintains an almost constant output voltage even though the DC supply voltage and the load current may vary over a wide range. A zener voltage regulator is shown in Fig. 4.26(a). A zener voltage regulator is also known as a *shunt regulator* because the zener diode is connected in shunt (or parallel) with the load R_L. The value of current-limiting resistance R_s should be such that the diode can operate in the breakdown region over the entire range of input voltages v_S and variations of the load current i_L.

If the zener diode is replaced by its piecewise linear model with V_{ZO} and R_Z, the equivalent circuit shown in Fig. 4.26(b) is created. If the supply voltage v_S varies, then the zener current i_Z will vary because of the presence of R_Z, thereby causing a variation of the output voltage. This variation of the output voltage is defined by a factor called the *line regulation,* which is related to R_s and R_Z:

$$\text{Line regulation} = \frac{\Delta v_O}{\Delta v_S} = \frac{R_Z}{R_Z + R_s} \tag{4.38}$$

If the load current i_L increases, then the zener current i_Z will decrease because of the presence of R_Z, thereby causing a decrease of the output voltage. This variation of the output voltage is defined by a factor called the *load regulation,* which is related to R_s and R_Z:

$$\text{Load regulation} = \frac{\Delta v_O}{\Delta i_L} = -(R_Z \| R_s) \tag{4.39}$$

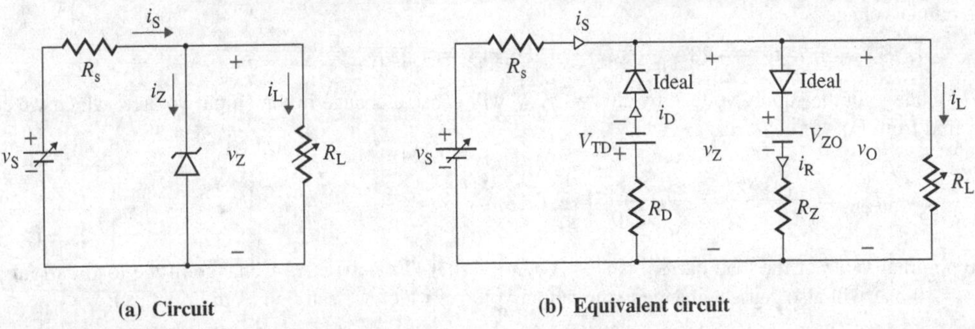

FIGURE 4.26 Zener shunt regulator

Any change in the zener voltage V_{ZO} will increase the output voltage. The variation of the output voltage is defined by a factor called the *zener regulation,* which is related to R_s and R_Z:

$$\text{Zener regulation} = \frac{\Delta v_O}{\Delta V_{ZO}} = \frac{R_s}{R_Z + R_s}$$

Thus, applying the superposition theorem, we can find the effective output voltage v_O of the regulator in Fig. 4.26(b) as follows:

$$v_O = \frac{\Delta v_O}{\Delta V_{ZO}}\Delta V_{ZO} + \frac{\Delta v_O}{\Delta v_S}\Delta v_S + \frac{\Delta v_O}{\Delta i_L}\Delta i_L$$

$$= \frac{R_s}{R_Z + R_s}\Delta V_{ZO} + \frac{R_Z}{R_Z + R_s}\Delta v_S - (R_Z \| R_s)\Delta i_L \quad \textbf{(4.40)}$$

EXAMPLE 4.12

D

Design of a zener regulator The parameters of the zener diode for the voltage regulator circuit of Fig. 4.26(a) are $V_Z = 4.7$ V at test current $I_{ZT} = 53$ mA, $R_Z = 8\ \Omega$, and $R_{ZK} = 500\ \Omega$ at $I_{ZK} = 1$ mA. The supply voltage is $v_S = V_S = 12 \pm 2$ V, and $R_s = 220\ \Omega$.

(a) Find the nominal value of the output voltage v_O under no-load condition $R_L = \infty$.

(b) Find the maximum and minimum values of the output voltage for a load resistance of $R_L = 470\ \Omega$.

(c) Find the nominal value of the output voltage v_O for a load resistance of $R_L = 100\ \Omega$.

(d) Find the minimum value of R_L for which the zener diode operates in the breakdown region.

SOLUTION

Using Eq. (4.37), we have

$$V_{ZO} = V_Z - R_Z I_{ZT} = 4.7\text{ V} - 8\ \Omega \times 53\text{ mA} = 4.28\text{ V}$$

(a) For $R_L = \infty$, the zener current is

$$i_Z = \frac{V_S - V_{ZO}}{R_Z + R_s} = \frac{12 - 4.28}{8 + 220} = 33.86\text{ mA}$$

The output voltage is

$$v_O = V_{ZO} + R_Z i_Z = 4.28\text{ V} + 8\ \Omega \times 33.86\text{ mA} = 4.55\text{ V}$$

(b) A change in the supply voltage by $\Delta v_S = \pm 2$ V will cause a change in the output voltage, which we can find from Eq. (4.38):

$$\Delta v_{O(\text{supply})} = \frac{\Delta v_S R_Z}{R_Z + R_s} = \frac{\pm 2 \times 8}{8 + 220} = \pm 70.18\text{ mV}$$

The nominal value of the load current is $i_L = V_Z/R_L = 4.7/470 = 10$ mA. A change in the load current by $\Delta i_L = 10$ mA will also cause a change in the output voltage, which we can find from Eq. (4.39):

$$\Delta v_{O(\text{load})} = -(R_Z \| R_s)\Delta i_L = -(8\ \Omega \| 220\ \Omega) \times 10\text{ mA} = -77.19\text{ mV}$$

Therefore, the maximum and minimum values of the output voltage can be found from

$$v_{O(max)} = 4.55\text{ V} + 70.18\text{ mV} - 77.19\text{ mV} = 4.54\text{ V}$$

$$v_{O(min)} = 4 - 70.18\text{ mV} - 77.19\text{ mV} = 4.47\text{ V}$$

(c) The nominal value of the load current is $i_L = V_Z/R_L = 4.7/100 = 47$ mA, which is not possible because the maximum current that can flow through R_Z is only 33.86 mA. Thus, the zener diode will be off, and the output voltage will be the voltage across R_L. That is,

$$v_O = \frac{R_L}{R_L + R_s} V_S = \frac{100}{100 + 220} \times 12 = 3.75\text{ V}$$

(d) For the zener diode to be operated in the breakdown region, allowing only I_{ZK} to flow, the maximum current that can flow through R_L is given by (assuming $I_{ZK} = I_Z$ at $v_Z = V_{ZO}$)

$$i_{L(max)} = \frac{V_{S(min)} - V_{ZO}}{R_s} - I_{ZK} \tag{4.41}$$

$$= \frac{(10 - 4.28)\text{ V}}{220\ \Omega} - 1\text{ mA} = 25\text{ mA}$$

Therefore, the minimum value of R_L that guarantees operation in the breakdown region is given by

$$R_{L(min)} \geq \frac{V_{ZO}}{i_{L(max)}} \tag{4.42}$$

$$\geq \frac{4.28\text{ V}}{25\text{ mA}} = 171.2\ \Omega$$

4.7.2 Design of a Zener Regulator

If i_Z is the zener current and i_L is the load current, the value of resistance R_s can be found from

$$R_s = \frac{V_S - V_{ZO} - R_Z i_Z}{i_Z + i_L} \quad \text{for } v_S = V_S \tag{4.43}$$

To ensure that the zener diode operates in the breakdown region under the worst-case conditions, the regulator must be designed to do the following:

1. To ensure that the zener current will exceed $i_{Z(min)}$ when the supply voltage is minimum $V_{S(min)}$ and the load current is maximum $i_{L(max)}$. Applying Eq. (4.43), we can find R_s from

$$R_s = \frac{V_{S(min)} - (V_{ZO} + R_Z i_{Z(min)})}{i_{Z(min)} + i_{L(max)}} \tag{4.44}$$

2. To ensure that the zener current will not exceed $i_{Z(max)}$ when the supply voltage is maximum $V_{S(max)}$ and the load current is minimum $i_{L(min)}$. Using Eq. (4.43), we can find R_s from

$$R_s = \frac{V_{S(max)} - (V_{ZO} + R_Z i_{Z(min)})}{i_{Z(max)} + i_{L(min)}} \tag{4.45}$$

Equating R_s in Eq. (4.44) to R_s in Eq. (4.45), we get the relationship of the maximum zener current in terms of the variations in V_S and i_L. That is,

$$(V_{S(min)} - V_{ZO} - R_Z i_{Z(min)})(i_{Z(max)} + i_{L(min)}) = (V_{S(max)} - V_{ZO} - R_Z i_{Z(max)})(i_{Z(min)} + i_{L(max)}) \tag{4.46}$$

As a rule of thumb, the minimum zener current $i_{Z(min)}$ is normally limited to 10% of the maximum zener current $i_{Z(max)}$ to ensure operation in the breakdown region. That is,

$$i_{Z(min)} = 0.1 \times i_{Z(max)} \tag{4.47}$$

EXAMPLE 4.13

D

Design of a zener regulator The parameters of a 6.3-V zener diode for the voltage regulator circuit of Fig. 4.26(a) are $V_Z = 6.3$ V at $I_{ZT} = 40$ mA and $R_Z = 2\ \Omega$. The supply voltage $v_S = V_S$ can vary between 12 V and 18 V. The minimum load current is 0 mA. The minimum zener diode current $i_{Z(min)}$ is 1 mA. The power dissipation $P_{Z(max)}$ of the zener diode must not exceed 750 mW at 25°C. Determine **(a)** the maximum permissible value of the zener current $i_{Z(max)}$, **(b)** the value of R_s that limits the zener current $i_{Z(max)}$ to the value determined in part (a), **(c)** the power rating P_R of R_s, and **(d)** the maximum load current $i_{L(max)}$.

SOLUTION

$V_Z = 6.3$ V at $i_{ZT} = 40$ mA, $i_{L(min)} = 0$ and $i_{Z(min)} = 1$ mA. Using Eq. (4.37), we have

$$V_{ZO} = V_Z - R_Z I_{ZT} = 6.3 - 2 \times 40\text{ mA} = 6.22\text{ V}$$

(a) The maximum power dissipation $P_{Z(max)}$ of a zener diode is

$$P_{Z(max)} = i_{Z(max)} V_Z = 0.75\text{ W}$$

or $$i_{Z(max)} = \frac{P_{Z(max)}}{V_Z} = \frac{0.75}{6.3} = 119\text{ mA}$$

(b) The zener current i_Z becomes maximum when the supply voltage is maximum and the load current is minimum—that is, $V_{S(max)} = 18$ V, $i_{L(min)} = 0$, and $i_{Z(max)} = 119$ mA. From Eq. (4.45),

$$R_s = \frac{V_{S(max)} - V_{ZO} - R_Z i_{Z(max)}}{i_{Z(max)} + i_{L(min)}} = \frac{18\text{ V} - 6.22\text{ V} - 2\ \Omega \times 119\text{ mA}}{119\text{ mA} + 0} = 96.96\ \Omega$$

(c) The power rating P_R of R_s is

$$P_R = (i_{Z(max)} + i_{L(min)})(V_{S(max)} - V_{ZO} - R_Z i_{Z(max)})$$
$$= 119\text{ mA} \times (18\text{ V} - 6.22\text{ V} - 2\ \Omega \times 119\text{ mA}) = 1.373\text{ W}$$

The worst-case power rating of R_s will occur when the load is shorted. That is,

$$P_{R(max)} = \frac{V_{S(max)}^2}{R_s} = \frac{18^2}{96.99} = 3.34\text{ W}$$

(d) i_L must be maintained at the maximum when V_S is minimum and i_Z is minimum—that is, $V_{S(min)} = 12$ V and $i_{Z(min)} = 1$ mA. From Eq. (4.44), we get

$$I_{L(max)} = \frac{V_{S(min)} - V_{ZO} - R_Z i_{Z(min)}}{R_s} - i_{Z(min)} = \frac{12\text{ V} - 6.22\text{ V} - 2\ \Omega \times 1\text{ mA}}{96.99\ \Omega} - 1\text{ mA}$$

$$= 58.57\text{ mA}$$

EXAMPLE 4.14

D

Design of a zener regulator and PSpice/SPICE verification The parameters of the zener diode for the voltage regulator in Fig. 4.26(a) are $V_Z = 4.7$ V at $I_{ZT} = 20$ mA, $R_Z = 19\ \Omega$, $I_{ZK} = 1$ mA, and $P_{Z(max)} = 400$ mW at 4.7 V. The supply voltage $v_S = V_S$ varies from 20 V to 30 V, and the load current i_L changes from 5 mA to 50 mA.

(a) Determine the value of resistance R_s and its power rating.
(b) Use PSpice/SPICE to check your results by plotting the output voltage v_O against the supply voltage v_S. Assume PSpice model parameters of zener diode D1N750:

```
IS=880.5E-18 N=1 CJO=175P VJ=.75 BV=4.7 IBV=20.245M
```

SOLUTION

$V_Z = 4.7$ V, $P_{Z(max)} = 400$ mW, $i_{L(min)} = 5$ mA, $i_{L(max)} = 50$ mA, $V_{S(min)} = 20$ V, and $V_{S(max)} = 30$ V. Using Eq. (4.37), we have

$$V_{ZO} = V_Z - R_Z I_{ZT} = 4.7 - 19 \times 20\text{ mA} = 4.32\text{ V}$$

Also, $i_{Z(max)} = \dfrac{P_{Z(max)}}{V_Z} = \dfrac{400\text{ mW}}{4.7\text{ V}} = 85.1$ mA (from specifications)

(a) Since the minimum value of the zener current is not specified, we can assume for all practical purposes that

$$i_{Z(min)} = 0.1 \times i_{Z(max)} = 0.1 \times 85.1\text{ mA} = 8.51\text{ mA}$$

From Eq. (4.44) and Fig. 4.26(b), we can find the value of R_s:

$$R_s = \frac{V_{S(min)} - V_{ZO} - R_Z i_{Z(min)}}{i_{Z(min)} + i_{L(max)}} = \frac{20\text{ V} - 4.32\text{ V} - 19\ \Omega \times 8.51\text{ mA}}{8.51\text{ mA} + 50\text{ mA}} = 265\ \Omega$$

From Eq. (4.45), we can find

$$R_s(i_{Z(max)} + i_{L(min)}) = V_{S(max)} - V_{ZO} - R_Z i_{Z(max)}$$

which can be solved to find the actual value of the maximum zener current $i_{Z(max)}$:

$$i_{Z(max)} = \frac{V_{S(max)} - V_{ZO} - R_s i_{L(min)}}{R_s + R_Z} = \frac{30\text{ V} - 4.32\text{ V} - 265\ \Omega \times 5\text{ mA}}{(265 + 19)\ \Omega} = 85.76\text{ mA}$$

The power rating P_R of R_s is

$$P_R \approx (i_{Z(max)} + i_{L(min)})(V_{S(max)} - V_Z)$$

$$= (85.76 \text{ mA} + 5 \text{ mA})(30 \text{ V} - 4.7 \text{ V}) = 2.3 \text{ W}$$

The worst-case power rating will be

$$P_{R(max)} = \frac{V_{S(max)}^2}{R_s} = \frac{30^2}{265} = 3.4 \text{ W}$$

From $i_{L(min)} = 5$ mA and $i_{L(max)} = 50$ mA, we find that the corresponding maximum and minimum values of the load resistance are

$$R_{L(max)} = \frac{V_Z}{i_{L(min)}} = \frac{4.7 \text{ V}}{5 \text{ mA}} = 940 \ \Omega$$

$$R_{L(min)} = \frac{V_Z}{i_{L(max)}} = \frac{4.7 \text{ V}}{50 \text{ mA}} = 94 \ \Omega$$

The zener voltage regulator for the PSpice simulation is shown in Fig. 4.27. The zener diode is normally modeled by setting the diode parameter $BV \cong V_Z$ in the PSpice model.

(b) The PSpice plot of the output voltage v_O against supply voltage v_S is shown in Fig. 4.28. The zener action begins at an output voltage of $v_O = 4.74$ V, which is close to the expected value of 4.7 V.

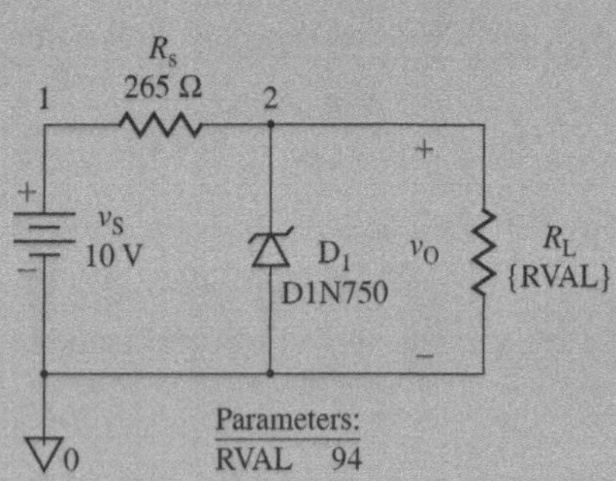

FIGURE 4.27 Zener voltage regulator for PSpice simulation

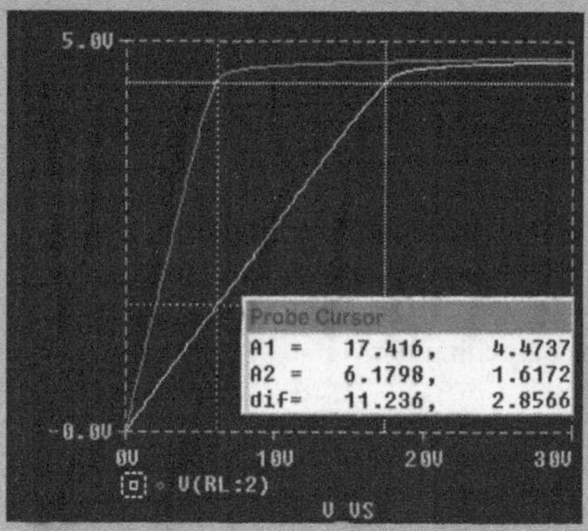

FIGURE 4.28 PSpice plots for Example 4.14

4.7.3 Zener Limiters

The zener characteristic shown in Fig. 4.25(b) can be approximated by the piecewise linear characteristic shown in Fig. 4.29(a). In the forward direction, a zener diode behaves like a normal diode, and it can be represented by a piecewise linear model with voltage V_{TD} and resistance R_D. The model of a zener diode

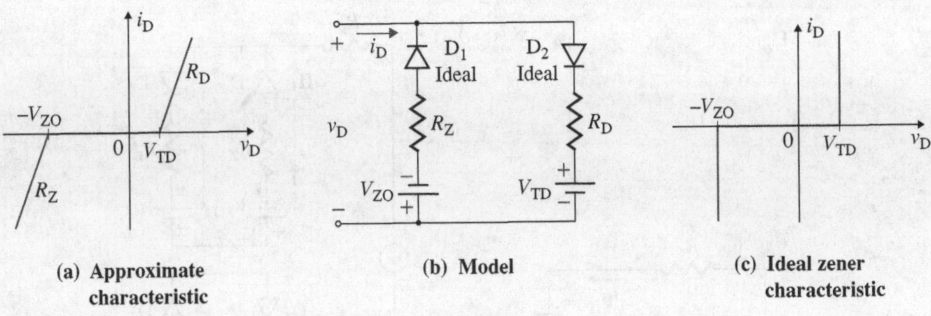

FIGURE 4.29 Piecewise linear model of zener diodes

in the forward and reverse directions is shown in Fig. 4.29(b). The current through a zener diode can be expressed as follows:

$$i_D = \begin{cases} 0 & \text{for } -V_{ZO} < v_D < V_{TD} \\ \dfrac{v_D}{R_D} - \dfrac{V_{TD}}{R_D} & \text{for } v_D \geq V_{TD} \\ \dfrac{v_D}{R_Z} + \dfrac{V_{ZO}}{R_Z} & \text{for } v_D \leq -V_{ZO} \end{cases}$$

The values of R_Z and R_D are very small, typically 20 Ω, and can be neglected for most analysis. The characteristic of Fig. 4.29(b) can be represented by the ideal zener characteristic shown in Fig. 4.29(c). Thus, a zener diode forms a natural limiter. By replacing the zener diode by its ideal characteristic (i.e., neglecting R_D and R_Z), we can simplify the circuit model of Fig. 4.29(b) to the circuit shown in Fig. 4.30(a). For a positive supply voltage $v_S \geq V_{TD}$, the output voltage v_O will be limited to V_{TD}. However, a negative input supply $v_S \leq -V_{ZO}$ will limit the output voltage v_O to $-V_{ZO}$. The approximate transfer characteristic of a zener limiter is shown in Fig. 4.30(b). This is an unsymmetrical limiter.

A symmetrical limiter can be obtained by connecting two zener diodes in series such that one diode opposes the other, as shown in Fig. 4.31(a). By replacing each zener diode by its model, shown in Fig. 4.29(b), we can create the equivalent circuit of a zener limiter, as shown in Fig. 4.31(b). If $v_S >$ $(V_{TD} + V_{ZO})$, ideal diodes D_2 and D_3 behave as short circuits and can be replaced by an equivalent

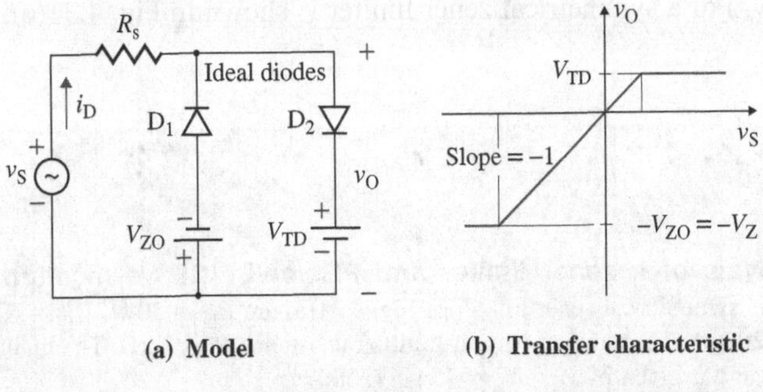

FIGURE 4.30 Unsymmetrical limiter

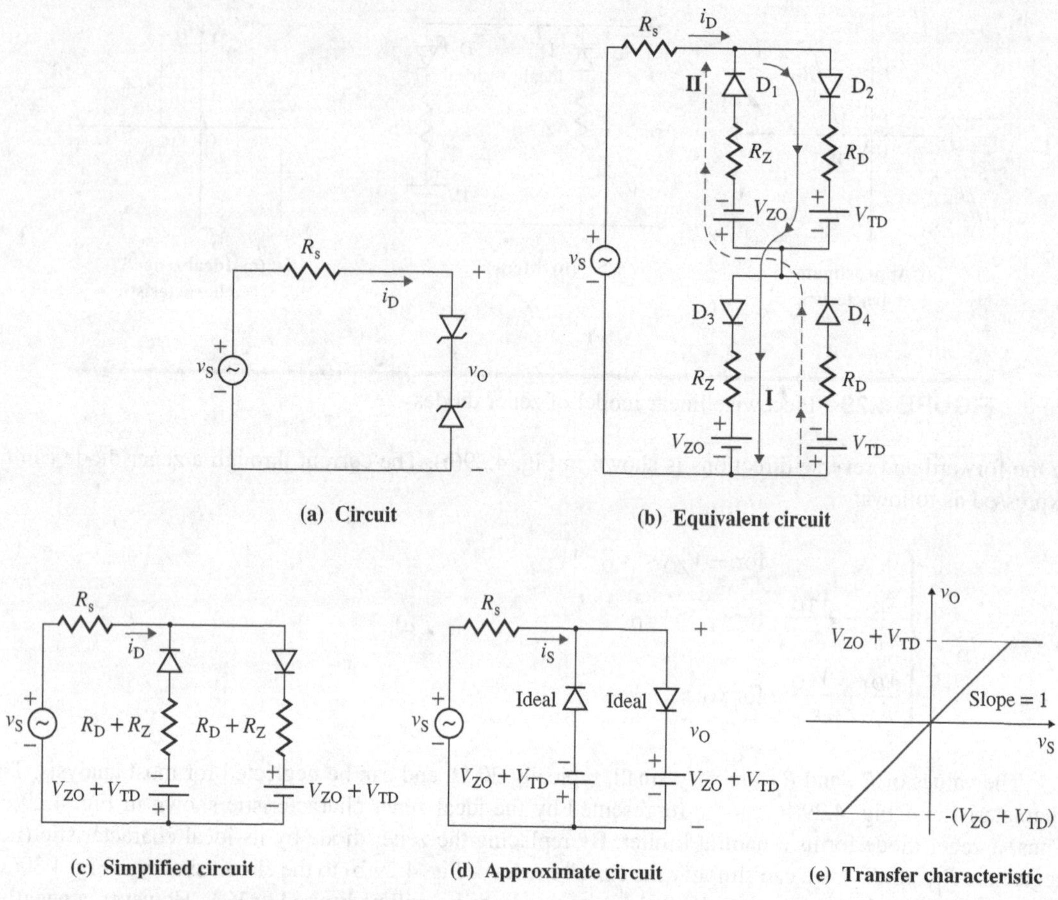

FIGURE 4.31 Symmetrical zener limiter

single diode in series with a voltage $V_{TD} + V_Z$ and a resistance $R_D + R_Z$. Similarly, when $v_S < -(V_{TD} + V_{ZO})$, diodes D_1 and D_4 can be replaced by a diode in series with a voltage $V_{TD} + V_{ZO}$ and a resistance $R_D + R_Z$. This arrangement is shown in Fig. 4.31(c). If we assume an ideal zener diode such that the values of R_D and R_Z are negligible, Fig. 4.31(c) can be reduced to Fig. 4.31(d). The transfer characteristic (v_O versus v_S) of a symmetrical zener limiter is shown in Fig. 4.31(e).

Small-signal analysis of a zener limiter and PSpice/SPICE verification The parameters of the zener diodes in the symmetrical zener limiter of Fig. 4.31(a) are $R_D = 50\ \Omega$, $V_{TD} = 0.7$ V, $R_Z = 20\ \Omega$, and $V_Z = 4.7$ V at $I_{ZT} = 20$ mA. The value of current-limiting resistance R_s is 1 kΩ. The input voltage to the limiter is AC rather than DC and is given by $v_S = v_s = 15 \sin(2000\pi t)$.

(a) Determine the instantaneous output voltage v_O and the peak zener diode current $I_{p(diode)}$.

(b) Use PSpice/SPICE to plot the instantaneous output voltage v_O. Assume PSpice/SPICE model parameters of zener diode D1N750:

```
IS=880.5E-18 N=1 CJO=175P VJ=.75 BV=4.7 IBV=20.245M
```

SOLUTION

(a) $R_D = 50\ \Omega$, $V_{TD} = 0.7$ V, $R_Z = 20\ \Omega$, $V_Z = 4.7$ V, $R_s = 1$ kΩ, and $v_S = 15 \sin(2000\pi t)$. Using Eq. (4.37), we have

$$V_{ZO} = V_Z - R_Z I_{ZT} = 4.7\text{ V} - 20\ \Omega \times 20\text{ mA} = 4.3\text{ V}$$

There are four possible intervals, depending on the value of v_S.
If $15 \sin 2000\pi t = V_{ZO} + V_{TD} = 5$, then

$$2000\pi t = \sin^{-1}\left(\frac{5}{15}\right)$$
$$= 0.34\text{ rad}$$

Interval 1: This interval is valid for $0 \le v_S \le (V_{ZO} + V_{TD})$.

$$i_D = 0$$
$$v_O = v_s = 15 \sin(2000\pi t) \quad \text{for } 0 \le 2000\pi t \le 0.34 \text{ and } (\pi - 0.34) \le 2000\pi t \le \pi$$

Interval 2: This interval is valid for $v_s \ge (V_{ZO} + V_{TD})$. From Fig. 4.31(c), we can find the instantaneous diode current i_D:

$$i_D = \frac{v_S}{R_s + R_D + R_Z} - \frac{V_{ZO} + V_{TD}}{R_s + R_D + R_Z} \tag{4.48}$$

$$= \frac{15 \sin(2000\pi t)}{1\text{ k}\Omega + 50\ \Omega + 20\ \Omega} - \frac{(4.3 + 0.7)\text{ V}}{1\text{ k}\Omega + 50\ \Omega + 20\ \Omega} = [14.02 \sin(2000\pi t) - 4.67]\text{ mA}$$

The instantaneous output voltage v_O is given by

$$v_O = V_{ZO} + V_{TD} + (R_D + R_Z)i_D \tag{4.49}$$

Substituting for i_D, we get

$$v_O = (4.3 + 0.7) + (50 + 20) \times [14.02 \sin(2000\pi t) - 4.67] \times 10^{-3}$$
$$= 4.67 + 0.981 \sin(2000\pi t) \quad \text{for } 0.34 \le 2000\pi t \le (\pi - 0.34)$$

Interval 3: This interval is valid for $0 \ge v_S \ge -(V_{ZO} + V_{TD})$.

$$i_D = 0$$
$$v_O = v_S = -15 \sin(2000\pi t) \quad \text{for } -0.34 \le 2000\pi t \le 0 \text{ and } -\pi \le 2000\pi t \le (-\pi + 0.34)$$

Interval 4: This interval is valid for $v_S \le -(V_{ZO} + V_{TD})$.

$$i_D = -[14.02 \sin(2000\pi t) - 4.67] \text{ mA}$$

$$v_O = -4.67 - 0.981 \sin(2000\pi t) \quad \text{for } (-\pi + 0.34) \le 2000\pi t \le -0.34$$

The peak diode current $i_{p(diode)}$ occurs at $2000\pi t = \pi/2$. That is,

$$i_{p(diode)} = \left[14.02 \sin\left(\frac{\pi}{2}\right) - 4.67\right] \text{mA} = 14.02 \text{ mA} - 4.67 \text{ mA} = 9.35 \text{ mA}$$

(b) The symmetrical zener limiter for PSpice simulation is shown in Fig. 4.32. The PSpice plot of instantaneous output voltage v_O is shown in Fig. 4.33, which gives +5.435 V, compared to the expected value of 4.67 + 0.981 = 5.65 V (from the expression of v_O for the interval 2).

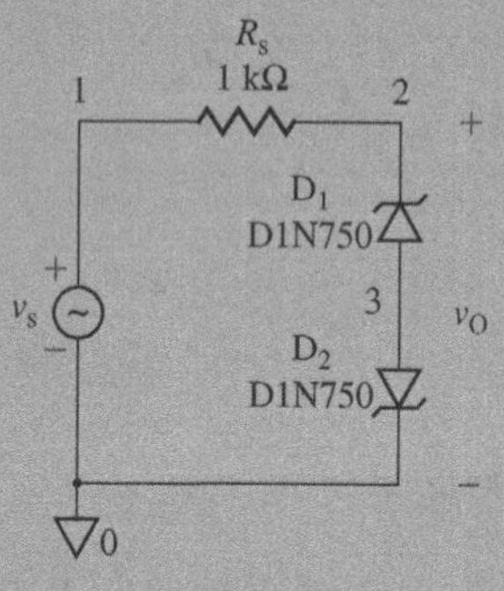

FIGURE 4.32 Symmetrical zener limiter for PSpice simulation

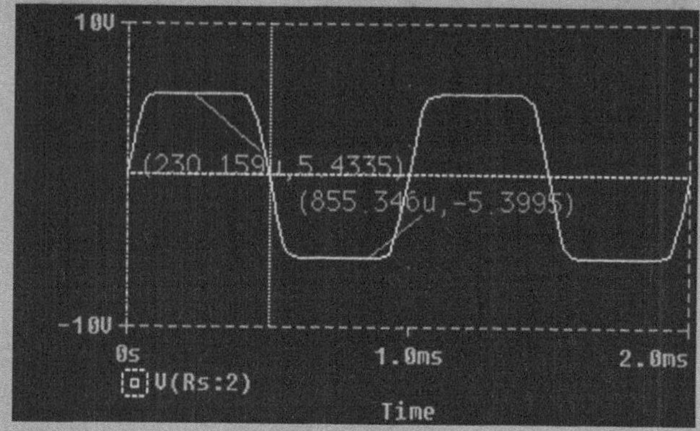

FIGURE 4.33 PSpice plots for Example 4.15

4.7.4 Temperature Effects on Zener Diodes

Any change in junction temperature generally changes the zener zoltage V_Z. The temperature coefficient is approximately +2 mV/°C, which is the same as but opposite that of a forward-biased diode. However, if a zener diode is connected in series with a forward-biased diode, as shown in Fig. 4.34, the temperature coefficients of the two diodes tend to cancel each other. This cancellation greatly reduces the overall temperature coefficients, and the effect of temperature changes is minimized.

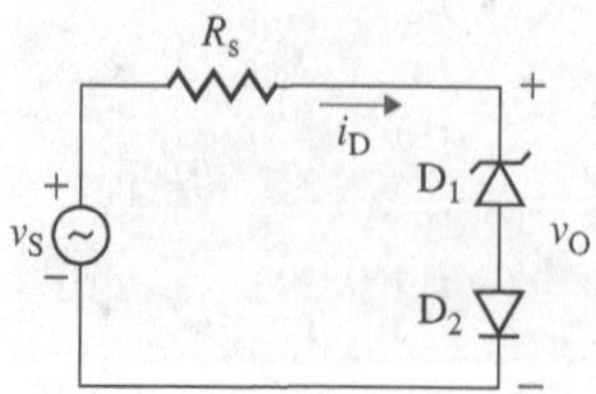

FIGURE 4.34 Zener diode in series with a forward-biased diode

EXAMPLE 4.16

Finding the temperature effect of a zener regulator with PSpice/SPICE The zener voltage of the regulator in Fig. 4.26(a) is $V_Z = 4.7$ V. The current-limiting resistance R_s is 1 kΩ, and the load resistance R_L is very large, tending to infinity. The supply voltage v_S varies from 0 to 20 V. Use PSpice/SPICE to plot the output voltage v_O against the input voltage v_S for junction temperatures $T_j = 25°C$ and $T_j = 100°C$. Assume PSpice/SPICE model parameters of zener diode D1N750:

```
IS=880.5E-18 N=1 CJO=175P VJ=.75 BV=4.7 IBV=20.245M
```

SOLUTION

The zener diode regulator for PSpice simulation is shown in Fig. 4.35. The PSpice plots of the output voltage v_O against the supply voltage v_S are shown in Fig. 4.36, which shows that the junction temperature affects the zener voltage slightly. For example, at $v_S = 5$ V, $v_O = 4.2336$ V at 25°C and 4.2285 V at 100°C.

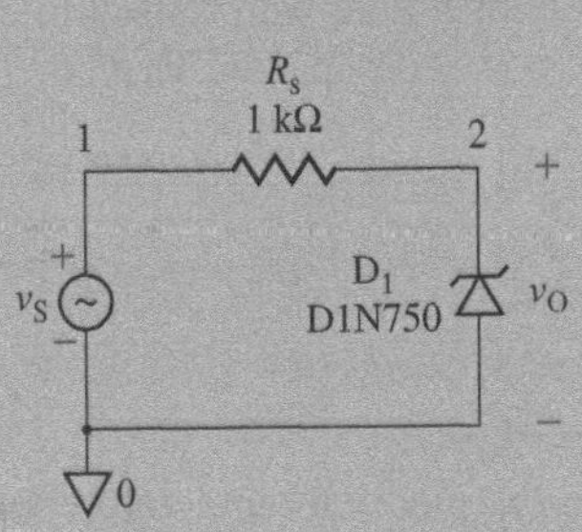

FIGURE 4.35 Zener diode regulator for PSpice simulation

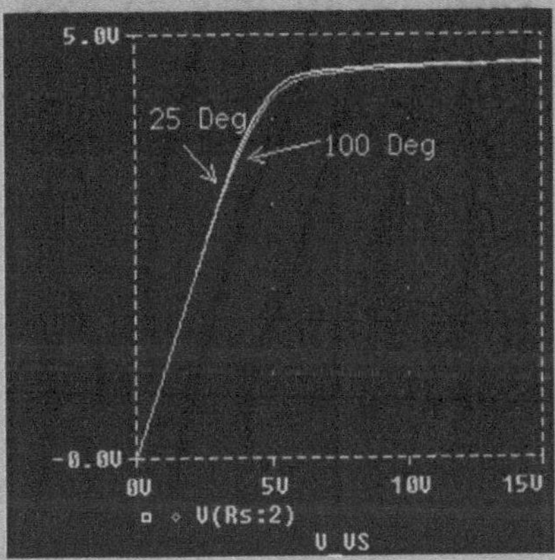

FIGURE 4.36 PSpice plots for Example 4.16

KEY POINTS OF SECTION 4.7

- A zener diode behaves like a normal diode in the forward direction. In the reverse direction, it maintains an almost constant voltage under varied load conditions if its voltage is greater than the zener voltage.
- A practical zener diode has a finite zener resistance, and the zener voltage will vary slightly with the zener current.
- Any change in junction temperature generally causes a change in the zener voltage.

4.8 Light-Emitting Diodes

A light-emitting diode (LED) is a special type of semiconductor diode that emits light when it is forward biased. The light intensity is approximately proportional to the forward diode current i_D. Light-emitting diodes are normally used in low-cost applications such as calculators, cameras, appliances, and automobile instrument panels.

4.9 Power Rating

Under normal operation, the junction temperature of a diode will rise as a result of power dissipation. Semiconductor materials have low melting points. The junction temperature, which is specified by the manufacturer, is normally limited to a safe value in the range of 150°C–200°C for silicon diodes and in the range of 60–110°C for germanium diodes. The power dissipation of a diode can be found from

$$P_D = I_D V_D \tag{4.50}$$

The power dissipation of a small-signal diode is low (on the order of milliwatts), and the junction temperature does not normally rise above the maximum permissible value specified by the manufacturer. However, power diodes are normally mounted on a heat sink. The function of the heat sink is to dissipate heat on the ambient (i.e., the material surrounding the device) in order to keep the junction temperature of power diodes below the maximum permissible value. The steady-state rise in the junction temperature with respect to the ambient temperature has been found, by experiment, to be proportional to the power dissipation. That is,

$$\Delta T = T_j - T_a = \theta_{ja} P_D \tag{4.51}$$

where T_j = junction temperature, in °C
T_a = ambient temperature, in °C
θ_{ja} = thermal resistance from junction to ambient, in °C/W

If the power dissipation P_D exceeds the maximum permissible value, the junction temperature will rise above the maximum allowable temperature. Excessive power dissipation can damage a diode. The permissible junction power dissipation P_D can be found by rearranging Eq. (4.51) to give

$$P_D = -\frac{T_a}{\theta_{ja}} + \frac{T_j}{\theta_{ja}} \tag{4.52}$$

which indicates that the permissible power dissipation will increase if the ambient temperature T_a can be reduced below the normal temperature of 25°C. However, in practice, power dissipation is limited to the value that corresponds to the permissible diode current. This limiting power P_{Dm} corresponds to the value of P_D at T_j = 25°C and is specified by the manufacturer. Thus,

$$P_D = \begin{cases} P_{Dm}\ (\text{at } T_j = 25°\text{C}) & \text{for } T_a \le 25°\text{C} \\ -\dfrac{T_a}{\theta_{ja}} + \dfrac{T_{jm}}{\theta_{ja}} & \text{for } T_a > 25°\text{C} \end{cases} \tag{4.53}$$

where T_{jm} = maximum junction temperature.

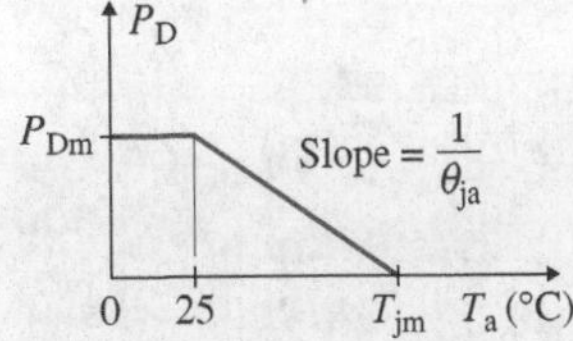

FIGURE 4.37 Power dissipation–temperature derating curve

As the junction temperature increases, the permissible power dissipation is reduced. A power derating curve is given by the manufacturer. The derating curve indicates the required adjustment in the power as the junction temperature increases above a specified temperature. A typical power dissipation–temperature derating curve is shown in Fig. 4.37. In the absence of such a characteristic, the values of T_{jm} and $P_{Dm}(T_a = 25°C)$ are usually provided.

KEY POINTS OF SECTION 4.9

- Power and temperature ratings are important parameters of a diode, and they are related to each other.
- The maximum power rating of a diode is specified at an ambient temperature. The diode must be derated if the operating ambient temperature is above this specified value. The diode can handle higher power if the operating ambient temperature is below the specified temperature.

EXAMPLE 4.17

Finding the power dissipation of a diode A diode is operated at a Q-point of $V_D = 0.7$ V and $I_D = 1$ A. The diode parameters are $P_D = 1$ W at $T_a = 50°C$ and $P_{derating} = 6.67$ mW/°C. The ambient temperature is $T_a = 25°C$, and the maximum permissible junction temperature is $T_{jm} = 200°C$. Calculate **(a)** the junction temperature T_j, **(b)** the maximum permissible junction dissipation P_{Dm}, and **(c)** the permissible junction dissipation P_D at an ambient temperature of $T_a = 75°C$.

SOLUTION

From Eq. (4.50), the junction power dissipation at the Q-point is

$$P_D = I_D V_D = 1\text{ A} \times 0.7\text{ V} = 0.7\text{ W}$$

From Eq. (4.51), the thermal resistance from junction to ambient is

$$\theta_{ja} = \frac{\Delta T}{P_D} = \frac{T_{jm} - T_a}{P_D} = \frac{200 - 50}{1} = 150°\text{C/W}$$

(a) From Eq. (4.51), the junction temperature at the Q-point is

$$T_j = T_a + \theta_{ja} P_D = 25 + 150 \times 0.7 = 130°\text{C}$$

(b) From Eq. (4.53),

$$P_{Dm}(T_a = 25°C) = \frac{T_{jm} - T_a}{\theta_{ja}} = \frac{200°C - 25°C}{150°C/W} = 1.17\text{ W}$$

(c) For $T_a = 75°C$,

$$P_{Dm}(T_a = 75°C) = \frac{200°C - 75°C}{150°C/W} = 833\text{ mW}$$

4.10 Diode Data Sheets

Diode ratings specify the current, voltage, and power-handling capabilities. This information is supplied by the manufacturer in data (or specifications) sheets. Typical data sheets for general-purpose diodes of types 1N4001 through 1N4007 are shown in Fig. 4.38. The important parameters of diodes of type 1N4001 are as follows:

1. Type of device with generic number or manufacturer's part number: 1N4001.
2. Peak inverse voltage (or peak repetitive reverse voltage) PIV = V_{RRM} = 50 V.
3. Operating and storage junction temperature range $T_j = -65°C$ to $+175°C$.
4. Maximum reverse current I_R (at DC rated reverse voltage) at PIV (50 V) = 10 μA at $T_j = 25°C$ and 50 μA at $T_j = 100°C$.
5. Maximum instantaneous forward voltage drop $v_D = v_F = 1.1$ V at $T_j = 25°C$.
6. Average rectified forward current $I_{F(AV)} = 1$ A at $T_a = 75°C$.
7. Repetitive peak current I_{FRM} is not quoted for 1N4001.
8. Nonrepetitive peak surge current $I_{FSM} = 30$ A for one cycle.
9. Average forward voltage drop $V_{F(AV)} = V_D = 0.8$ V.
10. DC power dissipation $P_D = V_{F(AV)} I_{F(AV)}$ (not quoted for 1N4001).

Typical data sheets for zener diodes of types 1N4728A through 1N4764A are shown in Fig. 4.39. The important parameters of zener diodes of type 1N4732 are as follows:

1. Type of device with generic number or manufacturer's part number: 1N4732.
2. Nominal zener voltage (avalanche breakdown voltage) $V_Z = 4.7$ V.
3. Operating and storage junction temperature range $T_j = -65°C$ to $+200°C$.
4. Zener test current $I_{ZT} = 53$ mA.
5. Zener impedance $Z_{ZT} = 8\ \Omega$.
6. Knee current $I_{ZK} = 1$ mA.
7. Nonrepetitive peak surge current $I_{FSM} = 970$ A for one cycle.
8. DC power dissipation $P_D = 1$ W at $T_a = 50°C$.
9. Power derating curve: Above 50°C, P_D is derated by 6.67 mW/°C.

▶ **NOTE** To allow a safety margin, designers should ensure that the operating values of voltage, current, and power dissipation are at least 20% to 30% less than the published maximum ratings. For military applications, the derating could be up to 50%.

1N4001 thru 1N4007

ON Semiconductor®

http://onsemi.com

Axial Lead Standard Recovery Rectifiers

This data sheet provides information on subminiature size, axial lead mounted rectifiers for general–purpose low–power applications.

LEAD MOUNTED RECTIFIERS
50–1000 VOLTS
DIFFUSED JUNCTION

CASE 59–10
AXIAL LEAD
PLASTIC

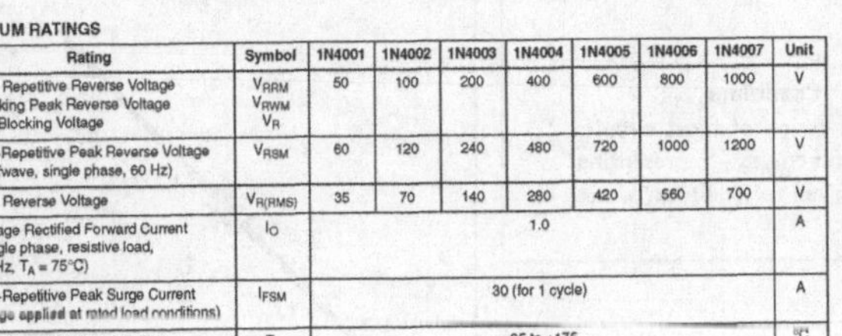

MAXIMUM RATINGS

Rating	Symbol	1N4001	1N4002	1N4003	1N4004	1N4005	1N4006	1N4007	Unit
†Peak Repetitive Reverse Voltage Working Peak Reverse Voltage DC Blocking Voltage	V_{RRM} V_{RWM} V_R	50	100	200	400	600	800	1000	V
†Non–Repetitive Peak Reverse Voltage (halfwave, single phase, 60 Hz)	V_{RSM}	60	120	240	480	720	1000	1200	V
†RMS Reverse Voltage	$V_{R(RMS)}$	35	70	140	280	420	560	700	V
†Average Rectified Forward Current (single phase, resistive load, 60 Hz, $T_A = 75°C$)	I_O	1.0							A
†Non–Repetitive Peak Surge Current (surge applied at rated load conditions)	I_{FSM}	30 (for 1 cycle)							A
Operating and Storage Junction Temperature Range	T_J T_{stg}	–65 to +175							°C

Maximum ratings are those values beyond which device damage can occur. Maximum ratings applied to the device are individual stress limit values (not normal operating conditions) and are not valid simultaneously. If these limits are exceeded, device functional operation is not implied, damage may occur and reliability may be affected.

ELECTRICAL CHARACTERISTICS†

Rating	Symbol	Typ	Max	Unit
Maximum Instantaneous Forward Voltage Drop, ($i_F = 1.0$ Amp, $T_J = 25°C$)	v_F	0.93	1.1	V
Maximum Full–Cycle Average Forward Voltage Drop, ($I_O = 1.0$ Amp, $T_L = 75°C$, 1 inch leads)	$V_{F(AV)}$	–	0.8	V
Maximum Reverse Current (rated DC voltage) ($T_J = 25°C$) ($T_J = 100°C$)	I_R	 0.05 1.0	 10 50	μA
Maximum Full–Cycle Average Reverse Current, ($I_O = 1.0$ Amp, $T_L = 75°C$, 1 inch leads)	$I_{R(AV)}$	–	30	μA

†Indicates JEDEC Registered Data

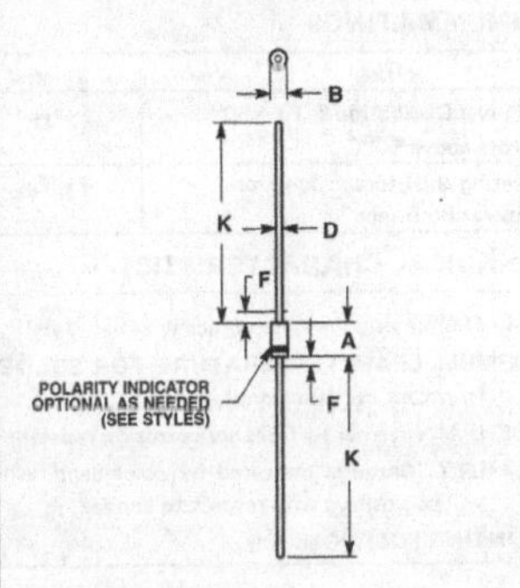

NOTES:
1. DIMENSIONING AND TOLERANCING PER ANSI Y14.5M, 1982.
2. CONTROLLING DIMENSION: INCH.
3. ALL RULES AND NOTES ASSOCIATED WITH JEDEC DO–41 OUTLINE SHALL APPLY
4. POLARITY DENOTED BY CATHODE BAND.
5. LEAD DIAMETER NOT CONTROLLED WITHIN F DIMENSION.

DIM	INCHES MIN	INCHES MAX	MILLIMETERS MIN	MILLIMETERS MAX
A	0.161	0.205	4.10	5.20
B	0.079	0.106	2.00	2.70
D	0.028	0.034	0.71	0.86
F	---	0.050	---	1.27
K	1.000	---	25.40	---

Mechanical Characteristics

- Case: Epoxy, Molded
- Weight: 0.4 gram (approximately)
- Finish: All External Surfaces Corrosion Resistant and Terminal Leads are Readily Solderable
- Lead and Mounting Surface Temperature for Soldering Purposes: 260°C Max. for 10 Seconds, 1/16 in. from case
- Polarity: Cathode Indicated by Polarity Band

FIGURE 4.38 Data sheet for diodes (Copyright of ON Semiconductor. Used by permission.)

1N4728,A thru 1N4764,A

Designers Data Sheet

ONE WATT HERMETICALLY SEALED GLASS SILICON ZENER DIODES

- Complete Voltage Range – 3.3 to 100 Volts
- DO-41 Package – Smaller than Conventional DO-7 Package
- Double Slug Type Construction
- Metallurgically Bonded Construction
- Nitride Passivated Die

Designer's Data for "Worst Case" Conditions

The Designers Data sheets permit the design of most circuits entirely from the information presented. Limit curves – representing boundaries on device characteristics – are given to facilitate "worst case" design.

1.0 WATT

ZENER REGULATOR DIODES

3.3–100 VOLTS

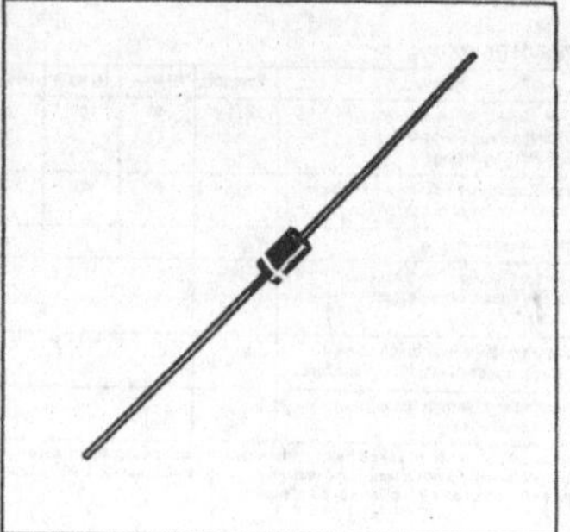

***MAXIMUM RATINGS**

Rating	Symbol	Value	Unit
DC Power Dissipation @ $T_A = 50°C$ Derate above 50°C	P_D	1.0 6.67	Watt mW/°C
Operating and Storage Junction Temperature Range	T_J, T_{stg}	-65 to +200	°C

MECHANICAL CHARACTERISTICS

CASE: Double slug type, hermetically sealed glass

MAXIMUM LEAD TEMPERATURE FOR SOLDERING PURPOSES: 230°C, 1/16" from case for 10 seconds

FINISH: All external surfaces are corrosion resistant with readily solderable leads.

POLARITY: Cathode indicated by color band. When operated in zener mode, cathode will be positive with respect to anode.

MOUNTING POSITION: Any

FIGURE 1 – POWER TEMPERATURE DERATING CURVE

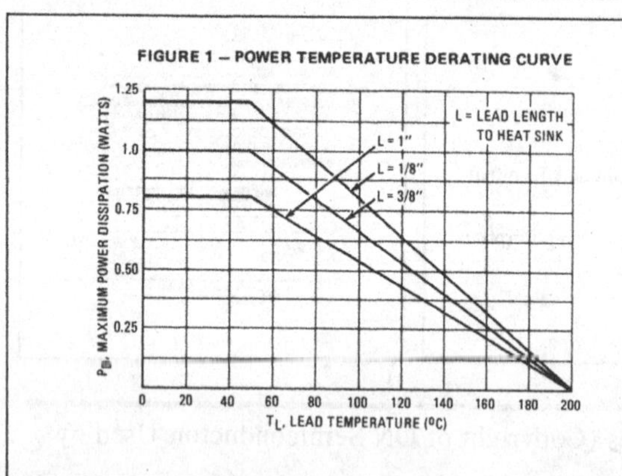

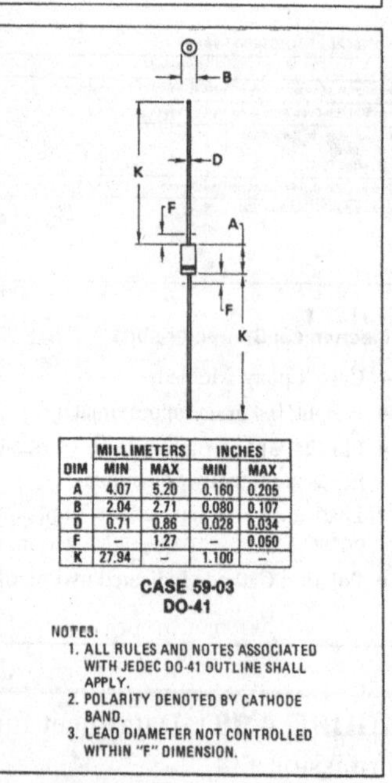

DIM	MILLIMETERS MIN	MILLIMETERS MAX	INCHES MIN	INCHES MAX
A	4.07	5.20	0.160	0.205
B	2.04	2.71	0.080	0.107
D	0.71	0.86	0.028	0.034
F	–	1.27	–	0.050
K	27.94	–	1.100	–

CASE 59-03 DO-41

NOTES:
1. ALL RULES AND NOTES ASSOCIATED WITH JEDEC DO-41 OUTLINE SHALL APPLY.
2. POLARITY DENOTED BY CATHODE BAND.
3. LEAD DIAMETER NOT CONTROLLED WITHIN "F" DIMENSION.

*Indicates JEDEC Registered Data

FIGURE 4.39 Data sheet for zener diodes (Copyright of Motorola. Used by permission.)
Updated information on the product can be found at www.onsemi.com

1N4728, A thru 1N4764, A

***ELECTRICAL CHARACTERISTICS** (T_A = 25°C unless otherwise noted) V_F = 1.2 V max, I_F = 200 mA for all types.

JEDEC Type No. (Note 1)	Nominal Zener Voltage V_Z @ I_{ZT} Volts (Notes 2 and 3)	Test Current I_{ZT} mA	Maximum Zener Impedance (Note 4)			Leakage Current		Surge Current @ T_A = 25°C i_r – mA (Note 5)
			Z_{ZT} @ I_{ZT} Ohms	Z_{ZK} @ I_{ZK} Ohms	I_{ZK} mA	I_R μA Max	V_R Volts	
1N4728	3.3	76	10	400	1.0	100	1.0	1380
1N4729	3.6	69	10	400	1.0	100	1.0	1260
1N4730	3.9	64	9.0	400	1.0	50	1.0	1190
1N4731	4.3	58	9.0	400	1.0	10	1.0	1070
1N4732	4.7	53	8.0	500	1.0	10	1.0	970
1N4733	5.1	49	7.0	550	1.0	10	1.0	890
1N4734	5.6	45	5.0	600	1.0	10	2.0	810
1N4735	6.2	41	2.0	700	1.0	10	3.0	730
1N4736	6.8	37	3.5	700	1.0	10	4.0	660
1N4737	7.5	34	4.0	700	0.5	10	5.0	605
1N4738	8.2	31	4.5	700	0.5	10	6.0	550
1N4739	9.1	28	5.0	700	0.5	10	7.0	500
1N4740	10	25	7.0	700	0.25	10	7.6	454
1N4741	11	23	8.0	700	0.25	5.0	8.4	414
1N4742	12	21	9.0	700	0.25	5.0	9.1	380
1N4743	13	19	10	700	0.25	5.0	9.9	344
1N4744	15	17	14	700	0.25	5.0	11.4	304
1N4745	16	15.5	16	700	0.25	5.0	12.2	285
1N4746	18	14	20	750	0.25	5.0	13.7	250
1N4747	20	12.5	22	750	0.25	5.0	15.2	225
1N4748	22	11.5	23	750	0.25	5.0	16.7	205
1N4749	24	10.5	25	750	0.25	5.0	18.2	190
1N4750	27	9.5	35	750	0.25	5.0	20.6	170
1N4751	30	8.5	40	1000	0.25	5.0	22.8	150
1N4752	33	7.5	45	1000	0.25	5.0	25.1	135
1N4753	36	7.0	50	1000	0.25	5.0	27.4	125
1N4754	39	6.5	60	1000	0.25	5.0	29.7	115
1N4755	43	6.0	70	1500	0.25	5.0	32.7	110
1N4756	47	5.5	80	1500	0.25	5.0	35.8	95
1N4757	51	5.0	95	1500	0.25	5.0	38.8	90
1N4758	56	4.5	110	2000	0.25	5.0	42.6	80
1N4759	62	4.0	125	2000	0.25	5.0	47.1	70
1N4760	68	3.7	150	2000	0.25	5.0	51.7	65
1N4761	75	3.3	175	2000	0.25	5.0	56.0	60
1N4762	82	3.0	200	3000	0.25	5.0	62.2	55
1N4763	91	2.8	250	3000	0.25	5.0	69.2	50
1N4764	100	2.5	350	3000	0.25	5.0	76.0	45

*Indicates JEDEC Registered Data.

NOTE 1 – Tolerance and Type Number Designation. The JEDEC type numbers listed have a standard tolerance on the nominal zener voltage of ±10%. A standard tolerance of ±5% on individual units is also available and is indicated by suffixing "A" to the standard type number.

NOTE 2 – Specials Available Include:

A. Nominal zener voltages between the voltages shown and tighter voltage tolerances,

B. Matched sets.

For detailed information on price, availability, and delivery, contact your nearest Motorola representative.

NOTE 3 – Zener Voltage (V_Z) Measurement. Motorola guarantees the zener voltage when measured at 90 seconds while maintaining the lead temperature (T_L) at 30°C ± 1°C, 3/8" from the diode body.

NOTE 4 – Zener Impedance (Z_Z) Derivation. The zener impedance is derived from the 60 cycle ac voltage, which results when an ac current having an rms value equal to 10% of the dc zener current (I_{ZT} or I_{ZK}) is superimposed on I_{ZT} or I_{ZK}.

NOTE 5 – Surge Current (i_r) Non-Repetitive. The rating listed in the electrical characteristics table is maximum peak, non-repetitive, reverse surge current of 1/2 square wave or equivalent sine wave pulse of 1/120 second duration superimposed on the test current, I_{ZT}, per JEDEC registration; however, actual device capability is as described in Figure 5.

APPLICATION NOTE

Since the actual voltage available from a given zener diode is temperature dependent, it is necessary to determine junction temperature under any set of operating conditions in order to calculate its value. The following procedure is recommended:

Lead Temperature, T_L, should be determined from

$$T_L = \theta_{LA} P_D + T_A$$

θ_{LA} is the lead-to-ambient thermal resistance (°C/W) and P_D is the power dissipation. The value for θ_{LA} will vary and depends on the device mounting method. θ_{LA} is generally 30 to 40°C/W for the various clips and tie points in common use and for printed circuit board wiring.

The temperature of the lead can also be measured using a thermocouple placed on the lead as close as possible to the tie point. The thermal mass connected to the tie point is normally large enough so that it will not significantly respond to heat surges generated in the diode as a result of pulsed operation once steady-state conditions are achieved. Using the measured value of T_L, the junction temperature may be determined by:

$$T_J = T_L + \Delta T_{JL}.$$

ΔT_{JL} is the increase in junction temperature above the lead temperature and may be found as follows:

$$\Delta T_{JL} = \theta_{JL} P_D$$

θ_{JL} may be determined from Figure 3 for dc power conditions. For worst-case design, using expected limits of I_Z, limits of P_D and the extremes of $T_J(\Delta T_J)$ may be estimated. Changes in voltage, V_Z, can then be found from:

$$\Delta V = \theta_{VZ} \Delta T_J$$

θ_{VZ}, the zener voltage temperature coefficient, is found from Figure 2.

Under high power-pulse operation, the zener voltage will vary with time and may also be affected significantly by the zener resistance. For best regulation, keep current excursions as low as possible.

Surge limitations are given in Figure 5. They are lower than would be expected by considering only junction temperature, as current crowding effects cause temperatures to be extremely high in small spots resulting in device degradation should the limits of Figure 5 be exceeded.

FIGURE 4.39 *Continued*

Summary

A diode is a two-terminal semiconductor device. It offers a very low resistance in the forward direction and a very high resistance in the reverse direction. The analysis of diode circuits can be simplified by assuming an ideal diode model in which the resistance in the forward-biased condition is zero and the resistance in the reverse direction is very large, tending to infinity.

A practical diode exhibits a nonlinear characteristic, analysis of which requires a graphical or iterative method. In order to linearize the diode characteristic to apply linear circuit laws, a practical diode is normally represented by (a) a constant DC drop V_{TD}, (b) a piecewise linear DC model, (c) a small-signal AC resistance r_d, or (d) a high-frequency AC model.

In a zener diode, the reverse breakdown is controlled, and the zener voltage is the reverse breakdown voltage. The diode characteristic depends on the operating temperature, and the leakage current almost doubles for every 10°C increase in the junction temperature.

References

1. C. G. Fonstad, *Microelectronic Devices and Circuits.* New York, NY: McGraw-Hill, 1994.
2. A. S. Sedra and K. C. Smith, *Microelectronic Circuits.* New York, NY: Oxford University Press, 2004.
3. D. A. Neamen, *Electronic Circuit Analysis and Design.* Boston, MA: Irwin Publishing, 2001.
4. D. A. Neamen, *Microelectronics: Circuit Analysis and Design.* New York, NY: McGraw-Hill, 2007.
5. M. S. Ghausi, *Electronic Devices and Circuits: Discrete and Integrated.* New York, NY: Holt, Rinehart, and Winston, 1985.
6. A. R. Hambley, *Electronics—A Top-Down Approach to Computer-Aided Circuit Design.* New York, NY: Macmillan Publishing, 1994.
7. P. R. Gray, P. Hurst, S. H. Lewis, and R. G. Meyer, *Analysis and Design of Integrated Circuits.* New York, NY: Wiley, 2001.
8. M. H. Rashid, *Introduction to SPICE Using or CAD for Circuits and Electronics Using PSpice.* Englewood Cliffs, NJ: Prentice Hall, 1995.
9. M. H. Rashid, *Electronics Circuit Design Using Electronics Workbench.* Boston, MA: PWS Publishing, 1998.

Review Questions

1. What is a diode?
2. What is the characteristic of an ideal diode?
3. What is a rectifier?
4. What is doping?
5. What is the depletion region of a diode?
6. What are the forward and reverse characteristics of a practical diode?
7. What is the forward-biased region of a diode?
8. What is the reverse-biased region of a diode?
9. What is the breakdown region of a diode?

10. What is the effect of junction temperature on the diode characteristic?
11. What are the three methods for analyzing diode circuits?
12. What is the low-frequency AC model of a diode?
13. What is the AC resistance of a diode?
14. What is the high-frequency AC model of a diode?
15. What is the PSpice/SPICE model of a diode?
16. What is a zener diode?
17. What is zener voltage?
18. What is a shunt regulator?
19. What is zener resistance?
20. What is the bulk resistance of a diode?

Problems

The symbol **D** indicates that a problem is a design problem.

4.2 *Ideal Diodes*

4.1 The diode circuit shown in Fig. P4.1 has $R = 30\ \text{k}\Omega$ and $V_{DD} = 10$ V. Determine the voltage v_O and the current i_O if **(a)** $v_S = 5$ V and **(b)** $v_S = 12$ V. Assume a diode drop of $V_D = 0.7$ V.

FIGURE P4.1

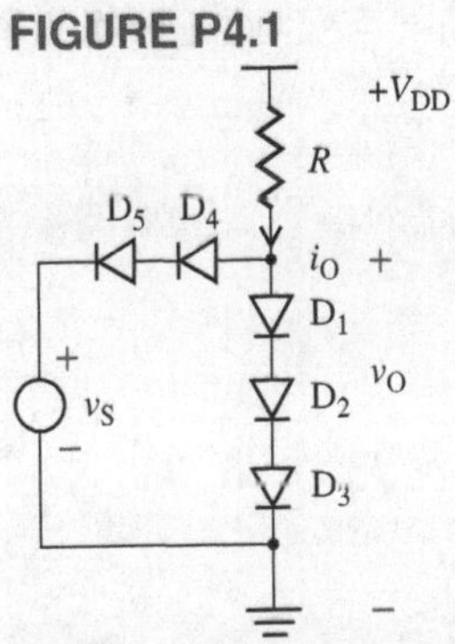

4.2 The diode circuit shown in Fig. P4.2 has $R_1 = 30\ \text{k}\Omega$, $R_2 = 10\ \text{k}\Omega$, $R_3 = 80\ \text{k}\Omega$, and $R_4 = 20\ \text{k}\Omega$. Determine the voltage v_O and the current i_O. Assume a diode drop of $V_D = 0.7$ V and $V_1 = 10$ V.

FIGURE P4.2

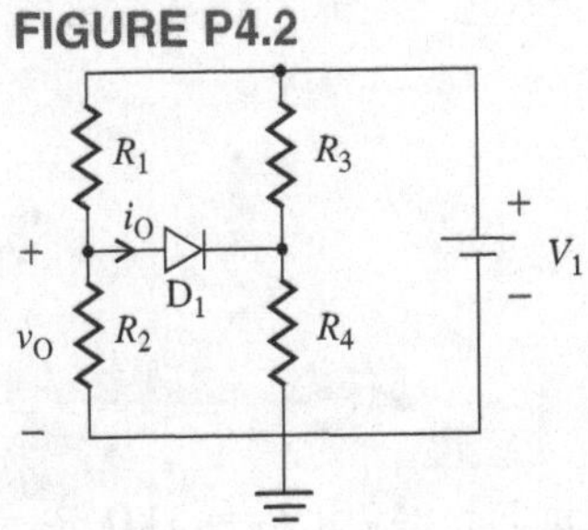

4.3 The diode circuit shown in Fig. P4.3 has $R_1 = 30\ \text{k}\Omega, R_2 = 10\ \text{k}\Omega, R_3 = 80\ \text{k}\Omega$, and $R_4 = 20\ \text{k}\Omega$. Determine the voltage v_O and the current i_O. Assume a diode drop of $V_D = 0.7$ V and $V_1 = 15$ V.

FIGURE P4.3

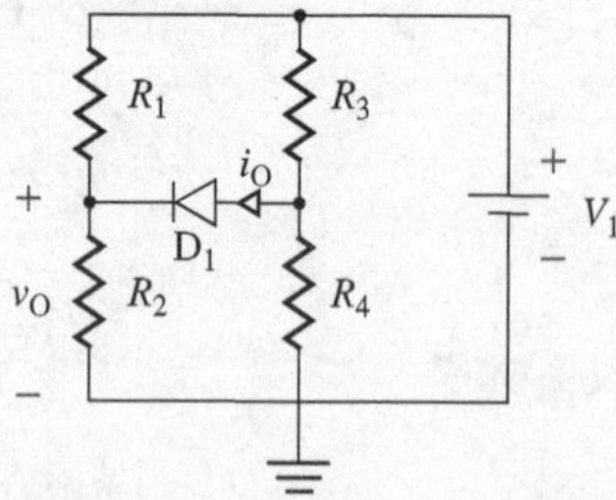

4.4 The diode circuit shown in Fig. P4.4 has $R_1 = 1\ \text{k}\Omega, R_2 = 2\ \text{k}\Omega, V_1 = 12$ V, and $V_2 = 15$ V. Determine the diode currents i_1 and i_2. Assume a diode drop of $V_D = 0.7$ V.

FIGURE P4.4

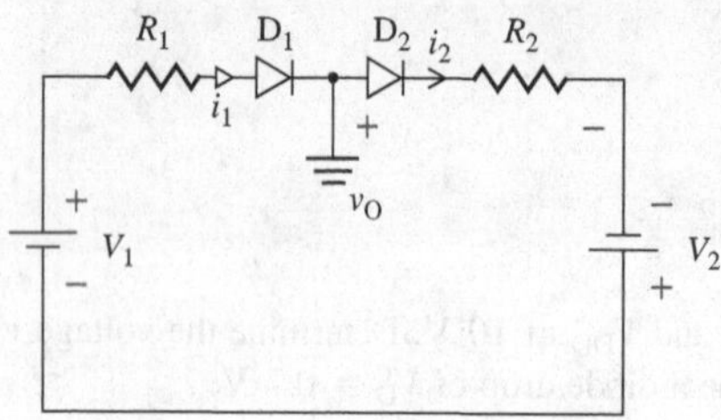

4.5 The diode circuit shown in Fig. P4.5 has $R_1 = 5\ \text{k}\Omega, R_2 = 15\ \text{k}\Omega, E_1 = 5$ V, and $v_S = 15$ V. Determine the voltage v_O and the current i_O. Assume a diode drop of $V_D = 0.7$ V.

FIGURE P4.5

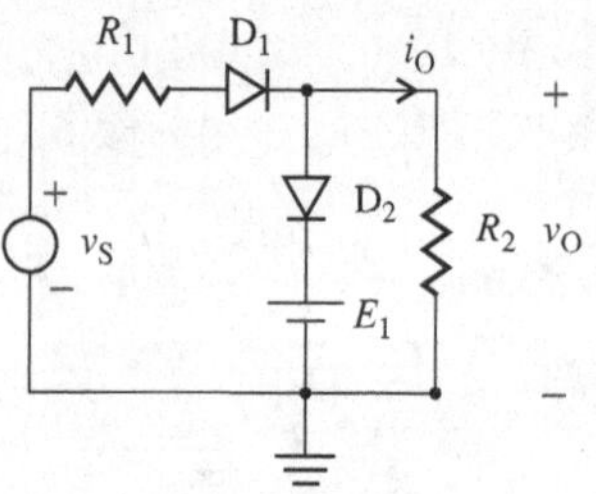

4.6 The diode circuit shown in Fig. P4.6 has $R_1 = 5\ \text{k}\Omega, R_2 = 15\ \text{k}\Omega, E_1 = 10$ V, $E_2 = 15$ V, and $v_S = 15$ V. Determine the voltage v_O and the current i_O. Assume a diode drop of $V_D = 0.7$ V.

FIGURE P4.6

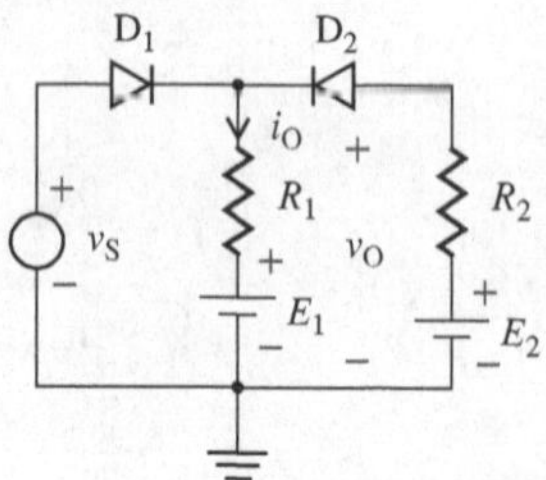

4.7 Find the voltage v_O and the current i_O of the diode circuit in Fig. P4.7 if $R_1 = 5\ \text{k}\Omega, R_2 = 10\ \text{k}\Omega$, $V_1 = 5$ V, and $V_2 = 10$ V. Assume diode voltage drop $V_D = 0.7$ V.

FIGURE P4.7

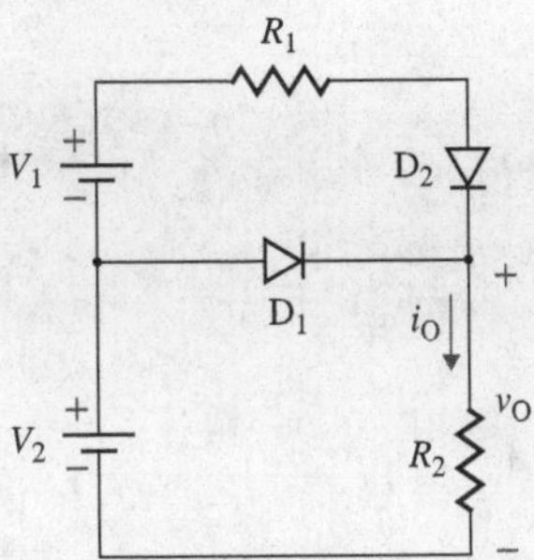

4.8 Find the voltage v_O and the current i_O of the diode circuit in Fig. P4.8 if $R_1 = 1\ \text{k}\Omega, R_2 = 5\ \text{k}\Omega$, $V_1 = 10$ V, and $V_2 = 5$ V. Assume diode voltage drop $V_D = 0.7$ V.

FIGURE P4.8

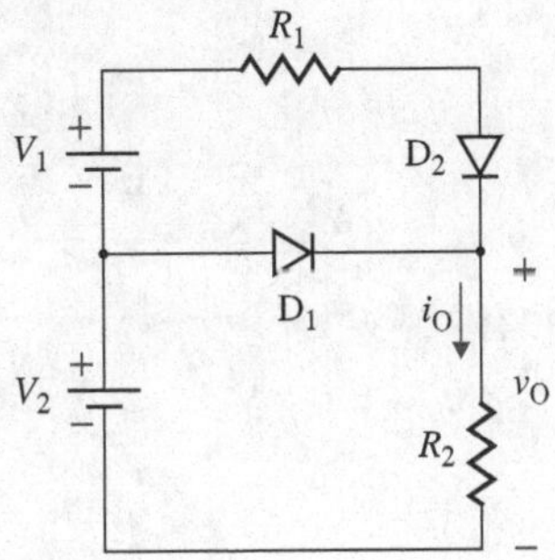

4.9 Find the voltage v_O and the current i_O of the diode circuit in Fig. P4.9 if $R_1 = 1\ \text{k}\Omega$ and $V_1 = 5$ V. Assume diode voltage drop $V_D = 0.7$ V.

FIGURE P4.9

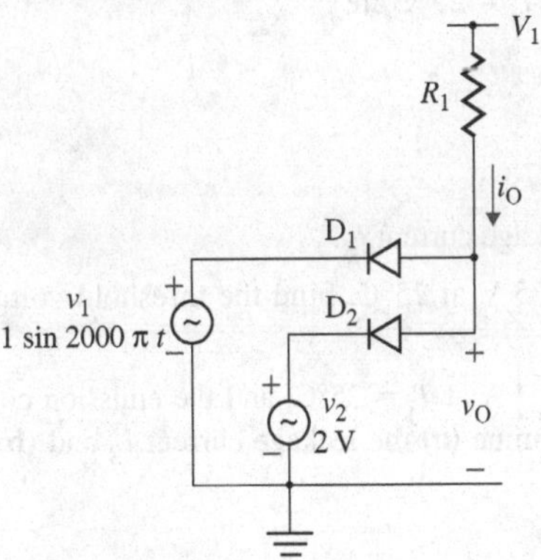

4.3 *Transfer Characteristics of Diode Circuits*

4.10 Plot the transfer characteristic (v_O versus v_S) of the diode circuit in Fig. P4.10 if the input voltage v_S is varied from 0 to 10 V in increments of 2 V. Assume $R_1 = 5\ \text{k}\Omega, R_2 = 1\ \text{k}\Omega$, $V_1 = 10$ to 5 V, and a diode voltage drop $V_D = 0.7$ V.

FIGURE P4.10

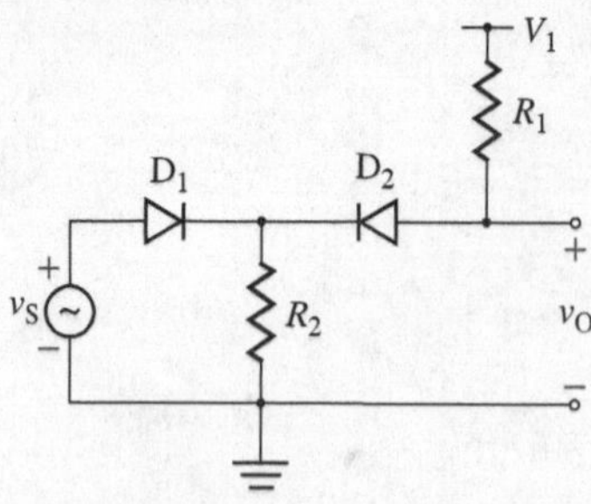

4.11 Plot the transfer characteristic (v_O versus v_S) of the diode circuit in Fig. P4.11 if the input voltage v_S is varied from −10 V to 10 V in increments of 2 V. Assume $R_1 = 5\ \text{k}\Omega$, $R_2 = 1\ \text{k}\Omega$, $V_1 = 5$ V, $V_2 = 5$ V, and a diode voltage drop $V_D = 0.7$ V.

FIGURE P4.11

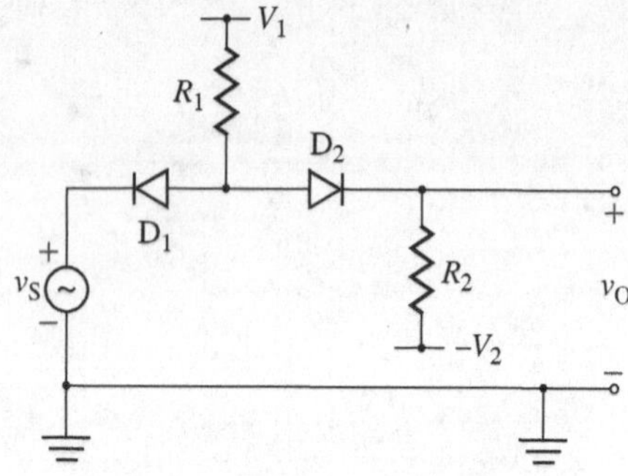

4.4 *Practical Diodes*

4.12 The measured values of a diode at junction temperature $T_j = 25°\text{C}$ are

$$V_D = \begin{cases} 0.65\ \text{V} & \text{at } I_D = 10\ \mu\text{A} \\ 0.8\ \text{V} & \text{at } I_D = 1\ \text{mA} \end{cases}$$

Determine **(a)** the emission coefficient n and **(b)** the leakage current I_S.

4.13 The threshold voltage of a silicon diode is $V_{TD} = 0.75$ V at 25°C. Find the threshold voltage V_{TD} at **(a)** $T_j = 125°\text{C}$ and **(b)** $T_j = -150°\text{C}$.

4.14 The leakage current of a silicon diode is $I_S = 5 \times 10^{-14}$ A at $T_j = 25°\text{C}$, and the emission coefficient is $n = 1.8$. The junction temperature is $T_j = 90°\text{C}$. Determine **(a)** the leakage current I_S and **(b)** the diode current i_D at a diode voltage of $v_D = 0.9$ V.

4.5 *Analysis of Practical Diode Circuits*

4.15 The diode circuit shown in Fig. 4.10 has $R_L = 4\ \text{k}\Omega$ and $V_S = 15$ V. The emission coefficient is $n = 1.8$. Use the iterative method to calculate the Q-point (or operating point), whose coordinates are V_D and I_D. Assume an approximate diode drop of $v_D = 0.75$ V at $i_D = 0.1$ mA. Assume a junction temperature of 25°C. Use three iterations only.

4.16 Repeat Prob. 4.15 using the approximate method with $V_D = 0.75$ V.

4.17 The diode circuit shown in Fig. 4.10 has $R_L = 1$ kΩ and $V_S = 10$ V. The diode characteristic is described by

$$i_D = Kv_D^2 = 5 \times 10^{-4} v_D^2 \quad (i_D \text{ in amps and } v_D \text{ in volts})$$

Determine the values of V_D and I_D at the Q-point (or operating point) by using **(a)** the iterative method and **(b)** the mathematical method.

4.6 *Modeling of Practical Diodes*

4.18 The diode circuit shown in Fig. 4.16(a) has $V_S = 15$ V and $R_L = 2.5$ kΩ. The diode characteristic is shown in Fig. 4.16(b). Determine the diode voltage v_D, the diode current i_D, and the load voltage v_O by using **(a)** the piecewise linear DC model and **(b)** the constant-drop DC model.

4.19 The diode circuit shown in Fig. 4.18(a) has $V_S = 12$ V, $V_m = 150$ mV, and $R_L = 5$ kΩ. Assume emission coefficient $n = 1.8$, diode voltage drop $v_D = 0.75$ V at $i_D = 0.5$ mA, and $V_T = 25.8$ mV at a junction temperature of 25°C. Determine **(a)** the Q-point (V_D, I_D), **(b)** the parameters (V_{TD}, R_D) of the piecewise linear DC model, and **(c)** the instantaneous diode voltage v_D.

4.20 The diode circuit shown in Fig. 4.18(a) has $V_S = 12$ V, $V_m = 150$ mV, $R_L = 5$ kΩ, and $V_T = 25.8$ mV. Assume an emission coefficient of $n = 2$ and the diode saturation current $I_S = 2.682 \times 10^{-9}$. Use PSpice/SPICE to **(a)** calculate the Q-point and small-signal parameters and **(b)** plot the instantaneous output voltage $v_O = v_D$. Assume PSpice/SPICE model parameters of diode D1N4148:

```
IS=2.682N CJO=4P M=.3333 VJ=.5 BV=100 IBV=100U TT=11.54N N=1.8
```

4.21 The diode circuit shown in Fig. 4.10 has $V_S = 18$ V and $R_L = 1.5$ kΩ. The diode forward characteristic, which can be obtained from practical measurements, can be represented by the following data:

i_D (mA)	0	15	30	45	60	75	90	105
v_D (V)	0.5	0.87	0.98	1.058	1.115	1.173	1.212	1.25

Determine **(a)** the Q-point (V_D, I_D), **(b)** the small-signal DC resistance R_D and threshold voltage V_{TD}, and **(c)** the small-signal AC resistance r_d. Assume $n = 1$ and $V_T = 25.8$ mV.

4.22 The diode circuit shown in Fig. 4.10 has $R_L = 1$ kΩ and $V_S = 10$ V. The diode characteristic is described by

$$i_D = Kv_D^2 = 5 \times 10^{-4} v_D^2 \quad (i_D \text{ in amps and } v_D \text{ in volts})$$

Determine **(a)** the diode voltage V_D, **(b)** the diode current I_D, and **(c)** the load voltage V_O.

4.23 The characteristic of the diode in Fig. P4.23 is described by

$$i_D = Kv_D^2 = 5 \times 10^{-4} v_D^2 \quad (i_D \text{ in amps and } v_D \text{ in volts})$$

Determine **(a)** the values of V_D and I_D at the Q-point (or operating point), **(b)** the small-signal ac resistance r_d, and **(c)** the rms output voltage $V_{o(rms)}$. Assume that the capacitor C offers a negligible impedance at the operating frequency.

FIGURE P4.23

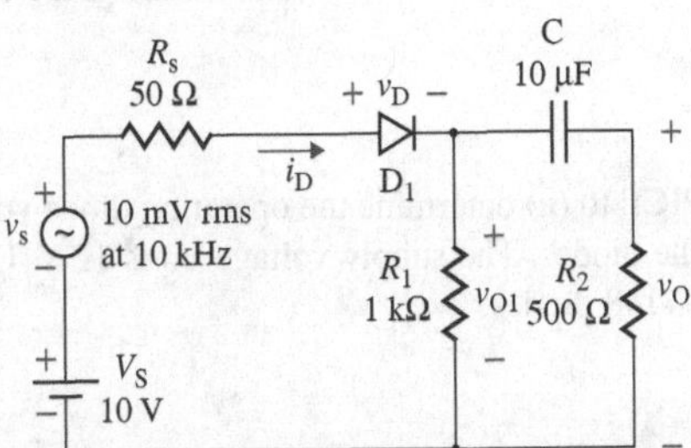

4.24 The characteristic of the diode circuit in Fig. P2.23 follows the Shockley diode equation with a leakage current of $I_S = 2.682 \times 10^{-9}$ A at 25°C and an emission coefficient of $n = 1.8$. Use PSpice/SPICE to **(a)** calculate the *Q*-point and the small-signal parameters and **(b)** plot the instantaneous output voltage $v_O = v_D$. Assume these PSpice/SPICE model parameters:

```
IS=2.682N M=.3333 VJ=.5 BV=100 IBV=100U TT=11.54N CJO=10PF N=1.8
```

4.25 A diode circuit is shown in Fig. P4.25. The diode characteristic is given by

$$i_D = 5 \times 10^{-2} v_D^2 \quad (i_D \text{ in amps and } v_D \text{ in volts})$$

Determine **(a)** the values of V_D and I_D at the *Q*-point (or operating point), **(b)** the small-signal AC resistance r_d, **(c)** the threshold voltage V_{TD}, and **(d)** the rms output voltage $V_{o(rms)}$.

FIGURE P4.25

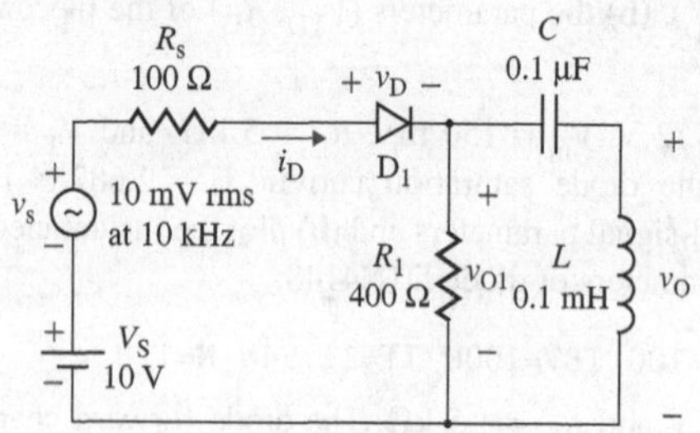

4.26 The characteristic of the diode in Fig. P4.25 follows the Shockley diode equation with a leakage current of $I_S = 2.682 \times 10^{-9}$ A at 25°C and an emission coefficient of $n = 1.8$. Use PSpice/SPICE to **(a)** calculate the *Q*-point and the small-signal parameters and **(b)** plot the instantaneous output voltage v_O. Assume these PSpice/SPICE model parameters:

```
IS=2.682N M=.3333 VJ=.5 BV=100V IBV=100U TT=11.54N  CJO=10PF N=1.8
```

4.27 A diode circuit is shown in Fig. P4.27. Use PSpice/SPICE to **(a)** determine the operating diode voltages and currents and **(b)** find the small-signal parameters of the diodes. The supply voltage V_S is 12 V. Use default values for the PSpice/SPICE model parameters of 1N4148 diodes.

FIGURE P4.27

4.28 A diode circuit is shown in Fig. P4.28. Use PSpice/SPICE to **(a)** determine the operating diode voltages and currents and **(b)** find the small-signal parameters of the diodes. The supply voltage V_S is 12 V. Use default values for the PSpice/SPICE model parameters of 1N4148 diodes.

FIGURE P4.28

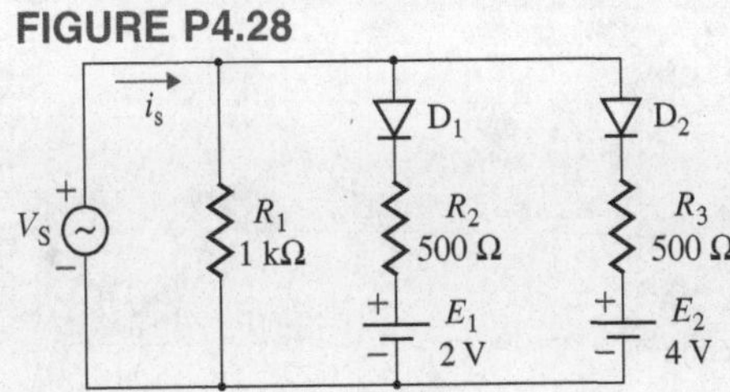

4.7 *Zener Diodes*

4.29 The parameters of the zener diode for the voltage regulator circuit of Fig. 4.26(a) are $V_Z = 6.8$ V at $I_{ZT} = 37$ mA, $R_Z = 3.5\ \Omega$, and $R_{ZK} = 700\ \Omega$ at $I_{ZK} = 1$ mA. The supply voltage is $V_S = 15 \pm 3$ V, and $R_s = 500\ \Omega$.

a. Find the nominal value of the output voltage v_O under no-load condition $R_L = \infty$.

b. Find the maximum and minimum values of the output voltage for a load resistance of $R_L = 570\ \Omega$.

c. Find the nominal value of the output voltage v_O for a load resistance of $R_L = 100\ \Omega$.

d. Find the minimum value of R_L for which the zener diode operates in the breakdown region.

4.30 D The parameters of the zener diode for the voltage regulator circuit of Fig. 4.26(a) are $V_Z = 7.5$ V at $I_{ZT} = 34$ mA, $R_Z = 5\ \Omega$, and $I_{ZK} = 0.5$ mA. The supply voltage v_S varies between 10 V and 24 V. The minimum load current i_L is 0. The minimum zener diode current $i_{Z(min)}$ is 1 mA. The maximum power dissipation $P_{Z(max)}$ of the zener diode must not exceed 1 W at 25°C. Determine **(a)** the maximum permissible value of the zener current $i_{Z(max)}$, **(b)** the value of R_s that limits the zener current $I_{Z(max)}$ to the value determined in part (a), **(c)** the power rating P_R of R_s, and **(d)** the maximum load current $I_{L(max)}$.

4.31 D The parameters of the zener diode for the voltage regulator in Fig. 4.26(a) are $V_Z = 5.1$ V at $I_{ZT} = 49$ mA, $R_Z = 7\ \Omega$, and $I_{ZK} = 1$ mA. The supply voltage v_S varies from 12 V to 18 V, and the load current i_L changes from 0 to 20 mA.

a. Determine the value of resistance R_s and its power rating.

b. Use PSpice/SPICE to check your results by plotting the output voltage v_O against the supply voltage v_S. Assume these PSpice/SPICE model parameters:

```
IS=2.682N CJO=4P M=.3333 VJ=.5 BV=5.1 IBV=49M TT=11.54N N=1.8
```

4.32 The zener diode for the regulator circuit in Fig. 4.26(a) has $V_Z = 6.2$ V at $I_{ZT} = 41$ mA, $R_Z = 2\ \Omega$, and $I_{ZK} = 1$ mA. The supply voltage v_S varies from 12 V to 18 V, and the load current i_L changes from 0 to 10 mA. Determine the minimum zener current $i_{Z(min)}$ and the maximum zener current $i_{Z(max)}$ of the diode and its maximum power rating $P_{Z(max)}$. Assume $R_s = 270\ \Omega$.

4.33 The parameters of the zener diodes in the symmetrical zener limiter of Fig. 4.31(a) are $R_D = 150\ \Omega$, $V_{TD} = 0.9$ V, $R_Z = 5\ \Omega$, and $V_Z = 6.8$ V at $I_{ZT} = 20$ mA. The value of current-limiting resistance R_s is 1.5 kΩ. The supply voltage to the limiter is AC and is given by $v_S = v_s = 20 \sin(2000\pi t)$ V.

a. Determine the instantaneous output voltage v_O and the peak diode current $I_{p(diode)}$.

b. Use PSpice/SPICE to plot the instantaneous output voltage v_O. Assume these PSpice/SPICE model parameters:

```
IS=2.682N CJO=4P M=.3333 VJ=.5 BV=6.8V IBV=20M TT=11.54N N=1
```

4.34 D A DC voltmeter is constructed using a DC meter, as shown in Fig. P4.34. The full-scale deflection of the meter is 150 μA, and the internal resistance R_m of the meter is 100 Ω. The zener voltage V_Z is 10 V, and the zener resistance R_Z is negligible. The voltmeter is required to measure 220 V at a full-scale deflection.

FIGURE P4.34

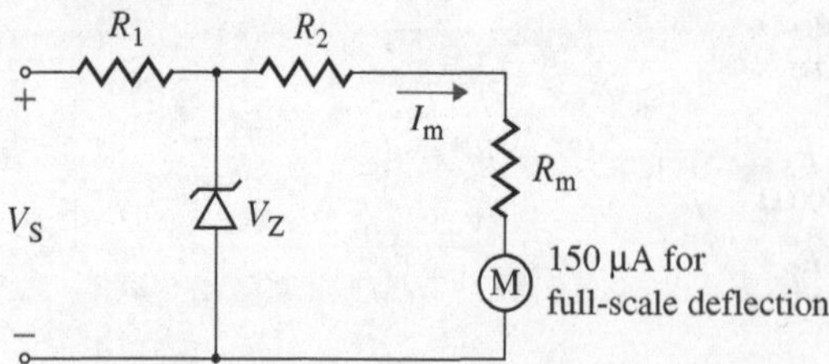

a. Design the voltmeter by determining the values of R_1 and R_2.

b. Use PSpice/SPICE to check your design by plotting the meter current I_m against the supply voltage V_S. Assume these PSpice/SPICE model parameters:

```
IS=2.682N CJO=4P M=.3333 VJ=.5 BV=10V IBV=20M TT=11.54N N=1
```

4.35 The zener voltage of the unsymmetrical regulator in Fig. 4.26(a) is $V_Z = 6.3$ V at $I_{ZT} = 20$ mA. The current-limiting resistance R_s is 1.5 kΩ, and the load resistance R_L is very large, tending to infinity. The supply voltage v_S varies from 0 to 30 V. Use PSpice/SPICE to plot the output voltage v_O against the input voltage v_S for $T_j = 25°C$ and $T_j = 150°C$. Assume these PSpice/SPICE model parameters:

```
IS=2.682N CJO=4P M=.3333 VJ=.5 BV=6.3V IBV=20M TT=11.54N N=1
```

4.36 The zener voltage of the symmetrical regulator in Fig. 4.31(a) is $V_Z = 6.3$ V at $I_{ZT} = 20$ mA. The current-limiting resistance R_s is 1.5 kΩ, and the load resistance R_L is very large, tending to infinity. The supply voltage v_S varies from 0 to 30 V. Use PSpice/SPICE to plot the output voltage v_O against the input voltage v_S for $T_j = 25°C$ and $T_j = 150°C$. Assume these PSpice/SPICE model parameters:

```
IS=2.682N CJO=4P M=.3333 VJ=.5 BV=6.3V IBV=20M TT=11.54N N=1
```

4.37 Two zener diodes are connected as shown in Fig. P4.37. The diode current in the forward direction is described by

$$i_D = I_S(e^{v_D/V_T} - 1)$$

where $V_T = 0.026$ and $I_S = 5 \times 10^{-15}$ A. The supply voltage v_S is 7.5 V. The zener voltage V_Z of each diode is 6.7 V, and the zener resistance R_Z is negligible. The forward voltage drop V_{TD} of each diode is 0.8 V. Determine **(a)** the expression for each diode voltage v_{D1} and v_{D2}, **(b)** the operating diode voltages V_{D1} and V_{D2}, and **(c)** the diode current I_D.

FIGURE P4.37

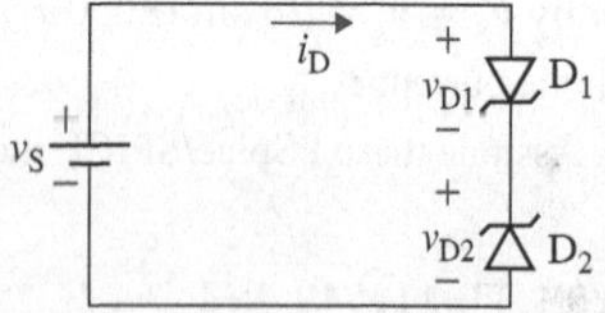

4.38 A zener regulator is shown in Fig. P4.38. Use PSpice/SPICE to plot the transfer characteristic between v_O and v_S. v_S varies from −18 V to 18 V in increments of 0.5 V. The PSpice/SPICE model parameters of the zener diodes are

```
IS=2.682N CJO=4P M=.3333 VJ=.5 BV=6.5V IBV=20M TT=11.54N N=1
```

FIGURE P4.38

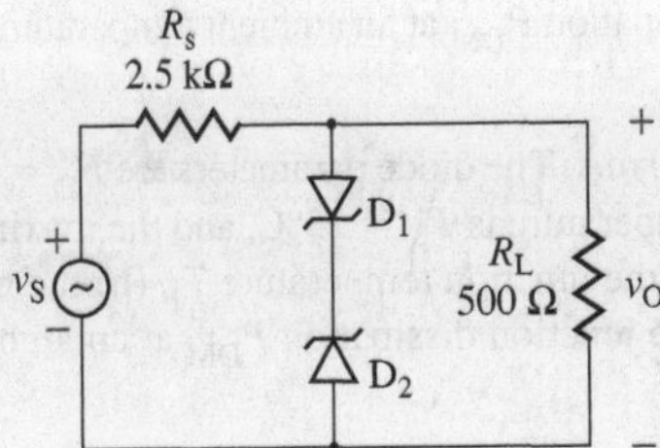

4.39 The zener regulator shown in Fig. P4.39 has $R_1 = 10\ \text{k}\Omega$, $R_2 = 100\ \text{k}\Omega$, and $R_3 = 100\ \text{k}\Omega$. Determine the voltage v_O and the power ratings of all elements if $v_S = 15$ V. The zener parameters are $V_{Z1} = 7$ V, $R_{Z1} = 0$, and $V_{Z2} = 5$ V, $R_{Z2} = 0$.

FIGURE P4.39

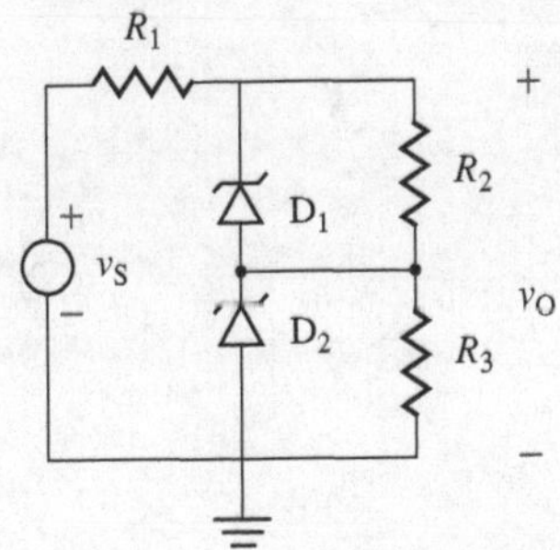

4.40 The zener regulator shown in Fig. P4.40 has $R_1 = 1\ \text{k}\Omega$, $R_2 = 5\ \text{k}\Omega$, and $R_3 = 10\ \text{k}\Omega$. Determine the output voltage v_O and its ripple voltage if the supply voltage v_S varies from 20 V to 30 V. The zener parameters are $V_{Z1} = 12$ V, $R_{Z1} = 40\ \Omega$, and $V_{Z2} = 7.5$ V, $R_{Z2} = 25\ \Omega$.

FIGURE P4.40

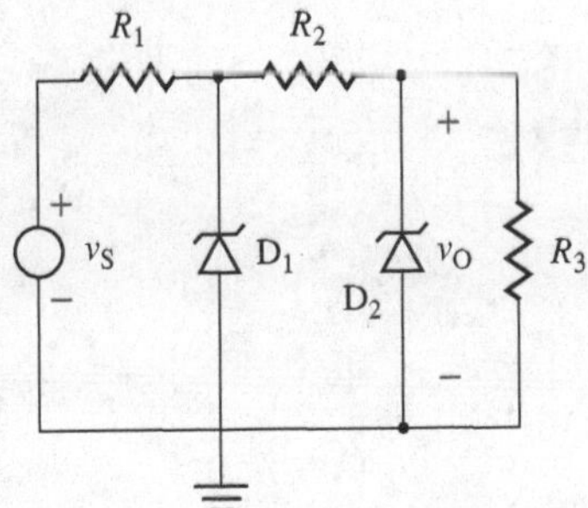

4.8–4.9 *Light-Emitting Diodes and Power Ratings*

4.41 Design an LED circuit so that the diode current I_D is 1 mA. Assume an emission coefficient of $n = 2$, a leakage current $I_S = 10^{-10}$ A, and $V_T = 25.8$ mV at a junction temperature of 25°C.

4.42 A diode is operated at a Q-point of $V_D = 0.7$ V and $I_D = 1$ A. The diode parameters are $P_D = 1.5$ W at $T_A =$ 50°C and $P_{derating} = 6.67$ mW/0°C. The ambient temperature is $T_A = 25$°C, and the maximum permissible junction temperature is T_{JM} = 250°C. Calculate **(a)** the junction temperature T_J, **(b)** the maximum permissible junction dissipation P_{DM}, and **(c)** the permissible junction dissipation P_{DM} at an ambient temperature of T_A = 65°C.

4.43 A diode is operated at a Q-point of V_D = 0.625 V and I_D = 100 mA. The diode parameters are P_D = 200 mW at T_A = 50°C and $P_{derating}$ = 6.67 mW/0°C. The ambient temperature is T_A = 25°C, and the maximum permissible junction temperature is T_{JM} = 200°C. Calculate **(a)** the junction temperature T_J, **(b)** the maximum permissible junction dissipation P_{DM}, and **(c)** the permissible junction dissipation P_{DM} at an ambient temperature of T_A = 75°C.

4.44 A diode is operated at a Q-point of V_D = 0.75 V and I_D = 200 mA. The diode parameters are P_D = 500 mW at T_A = 50°C and $P_{derating}$ = 6.67 mW/0°C. The ambient temperature is T_A = 25°C, and the maximum permissible junction temperature is T_{JM} = 250°C. Calculate **(a)** the junction temperature T_J, **(b)** the maximum permissible junction dissipation P_{DM}, and **(c)** the permissible junction dissipation P_{DM} at an ambient temperature of T_A = 55°C.

CHAPTER 5

APPLICATIONS OF DIODES

Learning Outcomes

After completing this chapter, students should be able to do the following:

- Select a diode rectifier to meet DC output voltage requirements.
- Design diode rectifiers to produce a DC supply voltage from an AC supply.
- Calculate the values of filter components in order to limit the ripple content on the DC output to a specified value.
- List some applications of diodes in wave-shaping of signals.
- Describe the diode applications as clippers, clampers, voltage multipliers, and transfer function synthesis.

Symbols and Their Meanings

Symbol	Meaning
$I_{D(av)}$, $I_{D(rms)}$	Average and rms diode currents
$I_{o(av)}$, $I_{o(rms)}$	Average and rms output currents
I_p, I_s	rms primary and secondary currents of an input transformer
$P_{o(dc)}$, $P_{o(ac)}$	Average and AC output powers
RF, PF	Output ripple factor and power factor
$v_D(t)$, $i_D(t)$, V_D	Instantaneous diode voltage, diode current, and DC diode voltage drop
$v_s(t)$, $v_O(t)$, $v_r(t)$	Instantaneous input supply, output, and ripple voltages

Symbol	Meaning
f, f_r	Frequency of the input signal and output ripple voltage
V_m, $V_{o(av)}$, $V_{o(rms)}$	Peak, average, and rms output voltages
$V_{r(pp)}$, $V_{r(p)}$, $V_{r(rms)}$	Peak-to-peak, peak, and rms output ripple voltages
V_p, V_s, n	rms primary voltage, secondary voltage, and transformer turns ratio

5.1 Introduction

We saw in Chapter 4 that a diode offers a very low resistance in one direction and a very high resistance in the other direction, thus permitting an easy current flow in only one direction. This chapter will illustrate the applications of diodes in wave-shaping circuits. For the sake of simplicity, we will assume ideal diodes—that is, diodes in which the voltage drop across the diode is zero rather than the typical value of 0.7 V.

5.2 Diode Rectifier

The most common applications of diodes are as rectifiers. A rectifier that converts an AC voltage to a unidirectional voltage is used as a DC power supply for many electronic circuits, such as those in radios, calculators, and stereo amplifiers. A rectifier is also called an *AC–DC converter*. Rectifiers can be classified on the basis of AC input supply into two types: single-phase rectifiers, in which the AC input voltage is a single-phase source, and three-phase rectifiers, in which the AC input voltage is a three-phase source [1]. Three-phase rectifiers, which are normally used in high-power applications, are outside the scope of this book. The following single-phase rectifiers are commonly used in electronic circuits: single-phase half-wave rectifiers, single-phase full-wave center-tapped rectifiers, and single-phase full-wave bridge rectifiers. For simplicity, we will assume ideal diodes in the following analysis and derivations; that is, the DC voltage drop across a diode is zero rather than a typical value of $V_D = 0.7$ V.

5.2.1 Single-Phase Half-Wave Rectifiers

The circuit diagram of a single-phase half-wave rectifier is shown in Fig. 5.1(a). Let us consider a sinusoidal input voltage $v_S = v_s = V_m \sin \omega t$, where $\omega = 2\pi ft$ and f is the frequency of the input voltage. Thus, there is no DC component on the input voltage; that is, $V_S = 0$ and $v_S = V_S + v_s = v_s$. Since v_S is positive from $\omega t = 0$ to π and negative from $\omega t = \pi$ to 2π, the operation of the rectifier can be divided into two intervals: interval 1 and interval 2.

Interval 1 is the interval $0 \leq \omega t \leq \pi$ during the positive half-cycle of the input voltage. Diode D_1 conducts and behaves like a short circuit, as shown in Fig. 5.1(b). The input voltage appears across the load resistance R_L. That is, the output voltage becomes

$$v_O = V_m \sin \omega t \quad \text{for } 0 \leq \omega t \leq \pi$$

If we include the DC diode drop V_D (= 0.7 V), the peak output voltage V_m will be reduced to $(V_m - V_D)$ and the instantaneous output voltage will become

$$v_O = (V_m - V_D) \sin \omega t \quad \text{for } 0 \leq \omega t \leq \pi$$

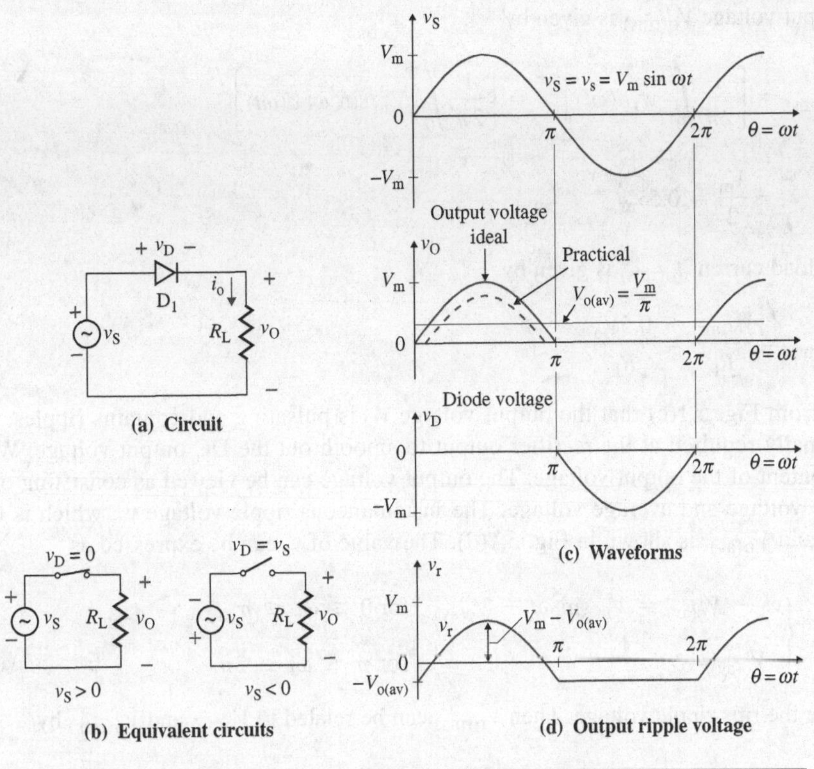

FIGURE 5.1 Single-phase half-wave rectifier

Interval 2 is the interval $\pi \leq \omega t \leq 2\pi$ during the negative half-cycle of the input voltage. Diode D_1 is reverse biased and behaves like an open circuit, as shown in Fig. 5.1(b). The output voltage v_O becomes zero. That is,

$$v_O = 0 \quad \text{for } \pi \leq \omega t \leq 2\pi$$

The waveforms of the input voltage, the output voltage, and the diode voltage are shown in Fig. 5.1(c). The output voltage will be reduced due to the diode drop of approximately 0.7 V as shown by the dotted lines. When diode D_1 conducts, its voltage becomes zero. When the diode is reverse biased, the diode current becomes zero and the diode has to withstand the input voltage. The peak inverse voltage (PIV) the diode must withstand is equal to the peak input voltage V_m. The voltage on the anode side of the diode is AC, whereas on the cathode side it is DC. That is, the diode converts AC voltage to DC. The average output voltage $V_{o(av)}$ is found using the following equation:

$$V_{o(av)} = \frac{1}{2\pi}\int_0^{\pi} v_O \, d(\omega t) = \frac{1}{2\pi}\int_0^{\pi} V_m \sin \omega t \, d(\omega t) = \frac{V_m}{\pi} = 0.318V_m \quad (5.1)$$

Therefore, the average load current $I_{o(av)}$ for a resistive load can be found from

$$I_{o(av)} = \frac{V_{o(av)}}{R_L} = \frac{V_m}{\pi R_L} = \frac{0.318V_m}{R_L} \quad (5.2)$$

The rms output voltage $V_{o(rms)}$ is given by

$$V_{o(rms)} = \left[\frac{1}{2\pi}\int_0^{\pi} v_O^2\, d(\omega t)\right]^{1/2} = \left[\frac{1}{2\pi}\int_0^{\pi} V_m^2 \sin^2\omega t\, d(\omega t)\right]^{1/2} \tag{5.3}$$

$$= \frac{V_m}{2} = 0.5V_m$$

and the rms load current $I_{o(rms)}$ is given by

$$I_{o(rms)} = \frac{V_{o(rms)}}{R_L} = \frac{0.5V_m}{R_L} \tag{5.4}$$

Notice from Fig. 5.1(c) that the output voltage v_O is pulsating and contains ripples. In practice, a filter is normally required at the rectifier output to smooth out the DC output voltage. We often know the ripple content of the output voltage. The output voltage can be viewed as consisting of two components: ripple voltage and average voltage. The instantaneous ripple voltage v_r, which is the difference between v_O and $V_{o(av)}$, is shown in Fig. 5.1(d). The value of v_r can be expressed as

$$v_r = \begin{cases} v_S - V_{o(av)} = V_m \sin\omega t - V_{o(av)} & \text{for } 0 \le \omega t \le \pi \\ -V_{o(av)} & \text{for } \pi \le \omega t \le 2\pi \end{cases} \tag{5.5}$$

Let $V_{r(rms)}$ be the rms ripple voltage. Then $V_{r(rms)}$ can be related to $V_{o(av)}$ and $V_{o(rms)}$ by

$$V_{r(rms)}^2 + V_{o(av)}^2 = V_{o(rms)}^2$$

or

$$V_{r(rms)}^2 = V_{o(rms)}^2 - V_{o(av)}^2 \tag{5.6}$$

Substituting V_m from Eq. (5.1) into Eq. (5.3), we get $V_{o(rms)} = \pi V_{o(av)}/2$, which is then applied to Eq. (5.6) to give $V_{r(rms)}$:

$$V_{r(rms)} = \left[\frac{\pi^2}{4}V_{o(av)}^2 - V_{o(av)}^2\right]^{1/2} = V_{o(av)}\left[\frac{\pi^2}{4} - 1\right]^{1/2} = 1.21V_{o(av)} \tag{5.7}$$

The ripple content of the output voltage is measured by a factor known as the *ripple factor* (RF), which is defined by

$$\text{RF} = \frac{V_{r(rms)}}{V_{o(av)}} = \frac{1.21V_{o(av)}}{V_{o(av)}} = 1.21 \text{ or } 121\% \tag{5.8}$$

▶ **NOTE** This numerical value of RF = 121% is valid only for the single-phase half-wave rectifier.

The *AC output power* $P_{o(ac)}$ is the average power and is defined as

$$P_{o(ac)} = \frac{1}{2\pi}\int_0^{2\pi} i_O^2 R_L\, d(\omega t) = I_{o(rms)}^2 R_L = V_{o(rms)}I_{o(rms)} = \left(\frac{V_m}{2}\right)^2 \frac{1}{R_L} \tag{5.9}$$

which will be the same as the input power P_{in} if we assume there is no power loss in the rectifier. That is, the input power is given by

$$P_{in} = P_{o(ac)} = \left(\frac{V_m}{2}\right)^2 \frac{1}{R_L} \tag{5.10}$$

The *DC output power* $P_{o(dc)}$ is defined by

$$P_{o(dc)} = V_{o(av)} I_{o(av)} = \frac{V_m^2}{\pi^2 R_L} \tag{5.11}$$

It is generally smaller than $P_{o(ac)}$ because the rms values are larger than the average (DC) values. The effectiveness of a rectifier in delivering DC output power is generally measured by the *rectification efficiency* η_R, which is defined as

$$\eta_R = \frac{P_{o(dc)}}{P_{o(ac)}} = \frac{V_{o(av)} I_{o(av)}}{V_{o(rms)} I_{o(rms)}} = \frac{(V_m/\pi)^2/R_L}{(V_m/2)^2/R_L} = \frac{4}{\pi^2} = 40.5\% \tag{5.12}$$

If we assume there is no power loss in the rectifier, then the input power factor (PF), which is a measure of the power drawn from the input power supply, is related to the input power (P_{in}) by

$$V_s I_s \text{PF} = P_{in} = P_{o(ac)} \tag{5.13}$$

This gives the input power factor as given by

$$\text{PF} = \frac{P_{o(ac)}}{V_s I_s} = \frac{(V_m/2)^2/R_L}{(V_m/\sqrt{2})(V_m/2R_L)} = \frac{\sqrt{2}}{2} = 0.707 \tag{5.14}$$

where V_s and I_s are the rms input supply voltage and the input supply current, respectively.

▶ **NOTE** These numerical values of $\eta_R = 40.5\%$ and PF $= 0.707$ are valid only for the single-phase half-wave rectifier.

Rectifiers are generally supplied through a transformer from a fixed AC input voltage of 120 V (rms) in order to satisfy the output voltage requirement. This arrangement is shown in Fig. 5.2(a). Let us assume an ideal transformer. Then the primary rms voltage V_p is related to the secondary rms voltage V_s by the *turns ratio n*, as follows:

$$\frac{V_p}{V_s} = \frac{N_p}{N_s} = n \tag{5.15}$$

where N_p is the number of turns of the primary winding and N_s is the number of turns of the secondary winding.

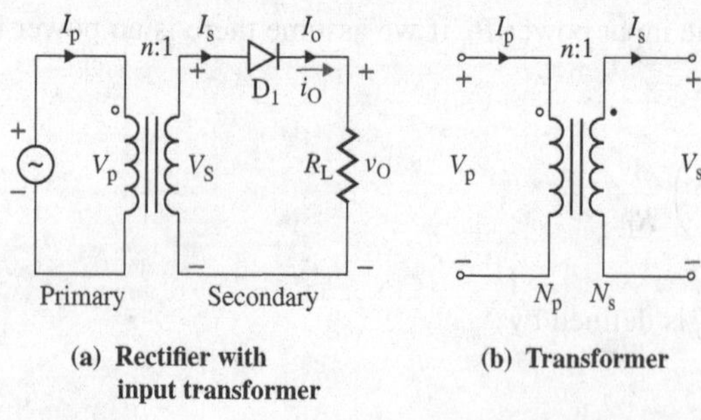

FIGURE 5.2 Half-wave rectifier with an input side transformer

Assuming there is no power loss in the transformer, the input (primary) side power must equal the output (secondary) side power. That is,

$$V_p I_p = V_s I_s \tag{5.16}$$

which, after we use the relationship between the voltages on the primary side and the secondary side $V_p = nV_s$, gives the relationship between the primary side current to the secondary side current as

$$I_s = nI_p \tag{5.17}$$

▶ **NOTES**

1. If the rectifier is connected to a battery charger, $P_{o(dc)}$ is the useful power transferred to the battery. Since $P_{o(ac)}$ is greater than $P_{o(dc)}$, $P_{loss} = P_{o(ac)} - P_{o(dc)}$ will be responsible for heating the battery. For a resistive load, however, the AC power $P_{o(ac)}$ becomes the average output power and will produce the effective heat.
2. The average current through the input side of an ideal transformer will be $I_{o(av)}/n$. A transformer is normally designed to operate from a sinusoidal AC source so that the magnetic core of the transformer is set and reset in every cycle. The unidirectional DC current flow through the transformer may cause the transformer core to saturate. Therefore, this circuit is suitable only for very low-power applications, typically tens of watts.
3. Unless noted otherwise, the AC input voltage is always specified in rms values, so $V_m = \sqrt{2}V_s$.

EXAMPLE 5.1

Finding the performance parameters of a single-phase half-wave rectifier The single-phase half-wave rectifier of Fig. 5.2(a) is supplied from a 120-V, 60-Hz source through the step-down transformer of Fig. 5.2(b) with turns ratio $n = 10:1$. The load resistance R_L is 5 Ω. Determine **(a)** the average output voltage $V_{o(av)}$, **(b)** the average load current $I_{o(av)}$, **(c)** the rms load voltage $V_{o(rms)}$, **(d)** the rms load current $I_{o(rms)}$, **(e)** the ripple factor RF of the output voltage, **(f)** the rms ripple voltage $V_{r(rms)}$, **(g)** the average diode current $I_{D(av)}$, **(h)** the rms diode current $I_{D(rms)}$, **(i)** the peak inverse voltage PIV of the diode, **(j)** the average output power $P_{o(ac)}$, **(k)** the DC output power $P_{o(dc)}$, **(l)** the frequency f_r of the output ripple voltage, and **(m)** the input power factor PF.

SOLUTION

The primary transformer voltage is $V_p = 120$ V. From Eq. (5.15), the secondary transformer voltage is $V_s = V_p/n = 120/10 = 12$ V. The peak input voltage of the rectifier is

$$V_m = \sqrt{2}V_s = \sqrt{2} \times 12 = 16.97 \text{ V}$$

(a) From Eq. (5.1),

$$V_{o(av)} = 0.318V_m = 0.318 \times 16.97 = 5.4 \text{ V}$$

(b) From Eq. (5.2),

$$I_{o(av)} = \frac{V_{o(av)}}{R_L} = \frac{5.4}{5} = 1.08 \text{ A}$$

(c) From Eq. (5.3),

$$V_{o(rms)} = 0.5V_m = 0.5 \times 16.97 = 8.49 \text{ V}$$

(d) From Eq. (5.4),

$$I_{o(rms)} = \frac{V_{o(rms)}}{R_L} = \frac{8.49}{5} = 1.7 \text{ A}$$

(e) From Eq. (5.8), RF = 1.21, or 121%.

(f) From Eq. (5.8),

$$V_{r(rms)} = \text{RF} \times V_{o(av)} = 1.21 \times 5.4 = 6.53 \text{ V}$$

(g) The average diode current $I_{D(av)}$ will be the same as that of the load. That is, $I_{D(av)} = I_{o(av)} = 1.08$ A.

(h) The rms diode current $I_{D(rms)}$ will be the same as that of the load. That is, $I_{D(rms)} = I_{o(rms)} = 1.7$ A.

(i) PIV = V_m = 16.97 V.

(j) From Eq. (5.9),

$$P_{o(ac)} = I^2_{o(rms)}R_L = (1.7)^2 \times 5 = 14.45 \text{ W}$$

(k) From Eq. (5.11),

$$P_{o(dc)} = V_{o(av)}I_{o(av)} = 5.4 \times 1.08 = 5.83 \text{ W}$$

(l) Notice from Fig. 5.1(d) that the frequency of the output ripple voltage is the same as the input frequency, $f_r = f = 60$ Hz.

(m) $V_s = 12$ V, $I_s = I_{o(rms)} = 1.7$ A, and $P_{in} = P_{o(ac)} = 14.45$ W

From Eq. (5.14),

$$\text{PF} = \frac{P_{o(ac)}}{V_s I_s} = \frac{14.45}{12 \times 1.7} = 0.7071$$

EXAMPLE 5.2

Fourier components of the output voltage of a single-phase half-wave rectifier The single-phase half-wave rectifier of Fig. 5.1(a) is connected to a source of $V_s = 120$ V, 60 Hz. Express the instantaneous output voltage $v_O(t)$ by a Fourier series. Assume ideal diodes with zero voltage drops.

SOLUTION

The output voltage v_O can be described by

$$v_O = \begin{cases} V_m \sin \omega t & \text{for } 0 \le \omega t \le \pi \\ 0 & \text{for } \pi \le \omega t \le 2\pi \end{cases}$$

which can be expressed by a Fourier series as

$$v_O(\theta) = V_{o(av)} + \sum_{n=1,2,\ldots}^{\infty} (a_n \sin n\theta + b_n \cos n\theta) \quad \text{where } \theta = \omega t = 2\pi f t \tag{5.18}$$

$$V_{o(av)} = \frac{1}{2\pi}\int_0^{2\pi} v_O \, d\theta = \frac{1}{2\pi}\left[\int_0^{\pi} V_m \sin\theta \, d\theta + \int_{\pi}^{2\pi} 0 \, d\theta\right] = \frac{V_m}{\pi}$$

$$a_n = \frac{1}{\pi}\int_0^{2\pi} v_O \sin n\theta \, d\theta = \frac{1}{\pi}\int_0^{\pi} V_m \sin\theta \sin n\theta \, d\theta$$

$$= \begin{cases} \dfrac{V_m}{2} & \text{for } n = 1 \\ 0 & \text{for } n = 2, 3, 4, 5, \ldots, \infty \end{cases}$$

$$b_n = \frac{1}{\pi}\int_0^{2\pi} v_O \cos n\theta \, d\theta = \frac{1}{\pi}\int_0^{\pi} V_m \sin\theta \cos n\theta \, d\theta$$

$$= \begin{cases} \dfrac{-2V_m}{\pi}\left(\dfrac{1}{n^2 - 1}\right) & \text{for } n = 2, 4, 6, \ldots, \infty \\ 0 & \text{for } n = 1, 3, 5, \ldots, \infty \end{cases}$$

When the values of a_n and b_n are inserted into Eq. (5.18), the expression for the instantaneous output voltage v_O becomes

$$v_O(t) = \frac{V_m}{\pi} + \frac{V_m}{2}\sin\omega t - \frac{2V_m}{3\pi}\cos 2\omega t - \frac{2V_m}{15\pi}\cos 4\omega t - \frac{2V_m}{35\pi}\cos 6\omega t - \cdots \tag{5.19}$$

$$- \frac{2V_m}{(2n-1)(2n+1)\pi}\cos 2n\omega t \quad \text{for } n = 1, 3, 5, \ldots, \infty$$

where $V_m = \sqrt{2} \times 120 = 169.7$ V and $\omega = 2\pi \times 60 = 377$ rad/s.

NOTE: Equation (5.19) contains sine and cosine components, which are known as *harmonics*. Except for the sine term, only the even harmonics are present, and their magnitudes decrease with the order of the harmonic frequency.

EXAMPLE 5.3

D **Application of the single-phase rectifier as a battery charger** A single-phase rectifier can be employed as a battery charger, as shown in Fig. 5.3(a). The battery capacity is 100 Wh, and the battery voltage is $E = 12$ V. The average charging current should be $I_{o(av)} = 5$ A. The primary AC input voltage is $V_p = 120$ V (rms), 60 Hz, and the transformer has a turns ratio of $n = 2:1$.

(a) Calculate the angle δ over which the diode conducts, the current-limiting resistance R, the power rating P_R of R, the charging time h in hours, the rectification efficiency η_R, and the peak inverse voltage PIV of the diode.

(b) Use PSpice/SPICE to plot $P_{o(ac)}$ and $P_{o(dc)}$ as a function of time. Assume model parameters of diode D1N4148:

```
IS=2.682N CJO=4P M=.3333 VJ=.5 BV=100 IBV=100U TT=11.54N
```

SOLUTION

(a) If the secondary input voltage is $v_S > E$, diode D_1 will conduct. The angle θ_1 at which the diode starts conducting can be found from the condition

$$V_m \sin \theta_1 = E$$

or

$$\theta_1 = \sin^{-1}\left(\frac{E}{V_m}\right) \tag{5.20}$$

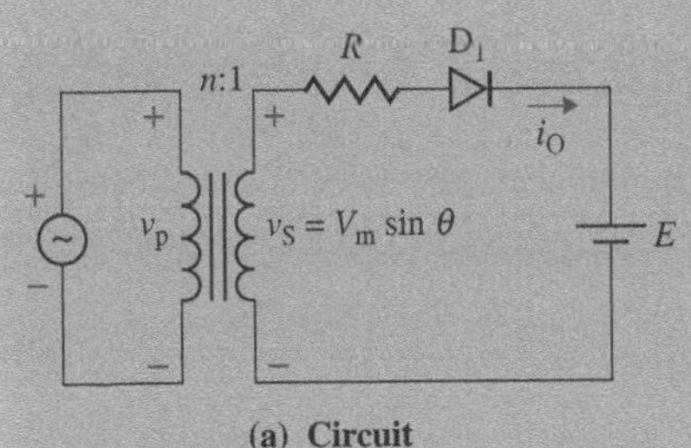

(a) Circuit

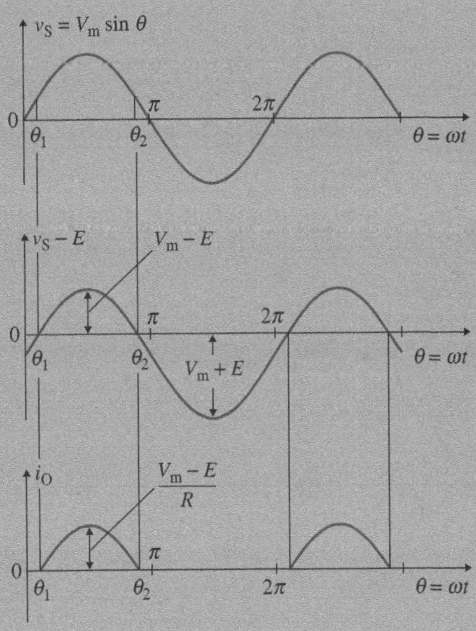

(b) Waveforms

FIGURE 5.3 Battery charger

Diode D_1 will be turned off when $v_S \le E$ at

$$\theta_2 = \pi - \theta_1$$

The charging current i_O, which is shown in Fig. 5.3(b), can be found from

$$i_O = \frac{v_S - E}{R} = \frac{V_m \sin\theta - E}{R} \quad \text{for } \theta_1 \le \theta \le \theta_2 \tag{5.21}$$

Since $V_s = V_p/2 = 120/2 = 60$ V,

$$V_m = \sqrt{2}V_s = \sqrt{2} \times 60 = 84.85 \text{ V}$$

From Eq. (5.20), $\theta_1 = \sin^{-1}(12/84.85) = 8.13°$, or 0.1419 rad. Thus,

$$\theta_2 = 180 - 8.13 = 171.87°$$

The interval over which the diode will conduct is called the *conduction angle* and is given by

$$\delta = \theta_2 - \theta_1 = 171.87 - 8.13 = 163.74°$$

The average charging current $I_{D(av)}$ is

$$I_{o(av)} = \frac{1}{2\pi}\int_{\theta_1}^{\theta_2 = \pi - \theta_1} \frac{V_m \sin\theta - E}{R}\, d\theta$$

$$= \frac{1}{2\pi R}(2V_m \cos\theta_1 + 2E\theta_1 - \pi E) \tag{5.22}$$

which gives the limiting resistance R as

$$R = \frac{1}{2\pi I_{o(av)}}(2V_m \cos\theta_1 + 2E\theta_1 - \pi E)$$

$$= \frac{1}{2\pi \times 5}(2 \times 84.85 \cos 8.13° + 2 \times 12 \times 0.1419 - \pi \times 12) = 4.26\ \Omega$$

The rms battery current $I_{o(rms)}$ is

$$I_{o(rms)}^2 = \frac{1}{2\pi}\int_{\theta_1}^{\theta_2 = \pi - \theta_1} \frac{(V_m \sin\theta - E)^2}{R^2}\, d\theta$$

$$= \frac{1}{2\pi R^2}\left[\left(\frac{V_m^2}{2} + E^2\right)(\pi - 2\theta_1) + \frac{V_m^2}{2}\sin 2\theta_1 - 4V_m E\cos\theta_1\right] \tag{5.23}$$

$$= 67.31 \text{ A}^2$$

which gives $I_{o(rms)} = \sqrt{67.31} = 8.2$ A. The power rating of R is

$$P_R = I_{o(rms)}^2 R = 8.2^2 \times 4.26 = 286.4 \text{ W}$$

The power delivered to the battery $P_{o(dc)}$ is

$$P_{o(dc)} = EI_{o(av)} = 12 \times 5 = 60 \text{ W}$$

For 100 Wh,

$$hP_{o(dc)} = 100$$

or
$$h = \frac{100}{P_{o(dc)}} = \frac{100}{60} = 1.667 \text{ h}$$

The rectification efficiency η_R is

$$\eta_R = \frac{\text{Power delivered to the battery}}{\text{Total input power}} = \frac{P_{o(dc)}}{P_{o(dc)} + P_R} = \frac{60}{60 + 286.4} = 17.32\%$$

The peak inverse voltage PIV of the diode is

$$\text{PIV} = V_m + E = 84.85 + 12 = 96.85 \text{ V} \tag{5.24}$$

(b) The battery charger circuit for PSpice simulation is shown in Fig. 5.4. Since inductance is proportional to the square of the number of turns, the primary and the secondary leakage inductances of the input transformer are selected with a ratio of 2^2 (or 4) to 1. That is, $L_1 = 40$ mH and $L_2 = 10$ mH for a linear transformer.

The PSpice plots of $I_{o(rms)}$, $P_{o(dc)}$, and $P_{o(rms)}$ are shown in Fig. 5.5, which gives $I_{o(rms)} \approx 7.3$ A, $P_{o(dc)} \approx 53.5$ W, and $P_{o(rms)} = 86.7$ W. The value of $I_{o(rms)}$ is equal to the rms current through resistance R—that is, I(R). These plots reach their steady-state values after a transient interval of approximately 80 ms.

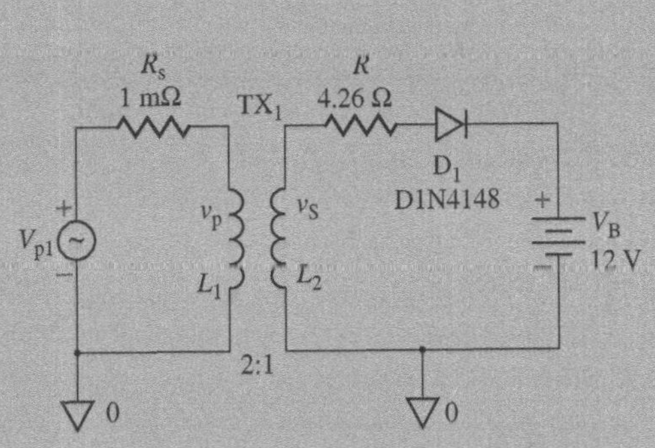

FIGURE 5.4 Battery charger circuit for PSpice simulation

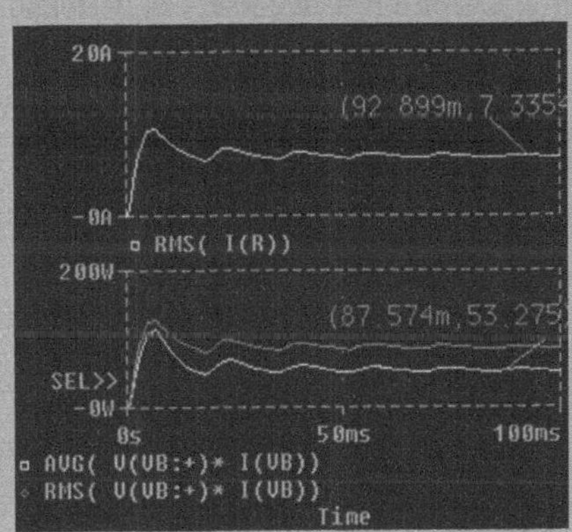

FIGURE 5.5 PSpice plots for Example 5.3

5.2.2 Single-Phase Full-Wave Center-Tapped Rectifier

For a half-wave rectifier, the average (or DC) voltage is only 0.318 V_m. A full-wave rectifier has double this output voltage, and it can be constructed by combining two half-wave rectifiers, as shown in Fig. 5.6(a). Since v_S is positive from $\omega t = 0$ to π and negative from $\omega t = \pi$ to 2π, the operation of the rectifier can be divided into two intervals: interval 1 and interval 2.

Interval 1 is the interval $0 \le \omega t \le \pi$ during the positive half-cycle of the input voltage. Diode D_2 is reverse biased and behaves like an open circuit, as shown in Fig. 5.6(b). The peak inverse voltage PIV of diode D_2 is $2V_m$. Diode D_1 conducts and behaves like a short circuit. The half-secondary voltage $v_S = V_m \sin \omega t$ appears across the load resistance R_L. That is, the output voltage becomes

$$v_O = V_m \sin \omega t \quad \text{for } 0 \le \omega t \le \pi$$

Interval 2 is the interval $\pi \le \omega t \le 2\pi$ during the negative half-cycle of the input voltage. Diode D_1 is reverse biased and behaves like an open circuit, as shown in Fig. 5.6(c). The peak inverse voltage PIV of diode D_1 is also $2V_m$. Diode D_2 conducts and behaves like a short circuit. The negative of the half-secondary voltage $v_S = V_m \sin \omega t$ appears across the load resistance R_L. That is, the output voltage becomes

$$v_O = -V_m \sin \omega t \quad \text{for } \pi \le \omega t \le 2\pi$$

The instantaneous output voltage v_O during interval 2 is identical to that for interval 1. The waveforms for the input and output voltages are shown in Fig. 5.6(d). Now we need to find the average voltage and the ripple content. Similar to that of the half-wave rectifier, the output voltage of a full-wave rectifier can be viewed as consisting of two components: ripple voltage and average voltage. The instantaneous ripple voltage v_r, which is the difference between v_O and $V_{o(av)}$, is shown in Fig. 5.6(e). The average output voltage $V_{o(av)}$ with two identical positive pulses can be found from the following equation:

$$V_{o(av)} = \frac{2}{2\pi}\int_0^{\pi} v_O \, d(\omega t) = \frac{2}{2\pi}\int_0^{\pi} V_m \sin \omega t \, d(\omega t) = \frac{2V_m}{\pi} \simeq 0.636V_m \tag{5.25}$$

It is twice the average output voltage of a half-wave rectifier, $V_{o(av)} = 0.318V_m$. Therefore, the average load current $I_{o(av)}$ for a resistive load can be found from Eq. (5.25):

$$I_{o(av)} = \frac{V_{o(av)}}{R_L} = \frac{2V_m}{\pi R_L} = \frac{0.636V_m}{R_L} \tag{5.26}$$

The rms output voltage $V_{o(rms)}$ is given by

$$V_{o(rms)} = \left[\frac{2}{2\pi}\int_0^{\pi} v_O^2 \, d(\omega t)\right]^{1/2} = \left[\frac{2}{2\pi}\int_0^{\pi} V_m^2 \sin^2\omega t \; d(\omega t)\right]^{1/2} \tag{5.27}$$
$$= \frac{V_m}{\sqrt{2}} = 0.707V_m$$

compared to $V_{o(rms)} = 0.5V_m$ for a half-wave rectifier. Therefore, the rms load current $I_{o(rms)}$ is given by

$$I_{o(rms)} = \frac{V_{o(rms)}}{R_L} = \frac{0.707V_m}{R_L} \tag{5.28}$$

To find the ripple factor, we have to find the amount of ripple content. The instantaneous ripple voltage v_r, which is shown in Fig. 5.6(e), can be expressed as

$$v_r = \begin{cases} v_S - V_{o(av)} = V_m \sin \omega t - V_{o(av)} & \text{for } 0 < \omega t < \pi \\ -V_m \sin \omega t - V_{o(av)} & \text{for } \pi \le \omega t \le 2\pi \end{cases}$$

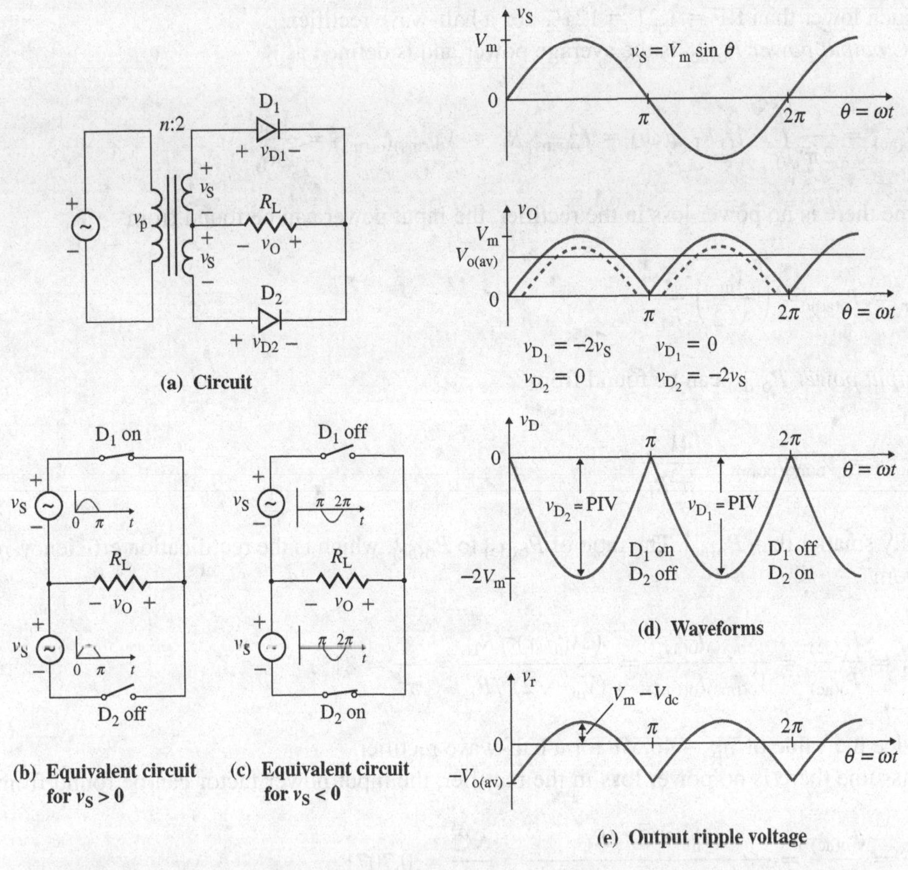

FIGURE 5.6 Full-wave rectifier with a center-tapped transformer

Let $V_{r(rms)}$ be the rms ripple voltage. Then $V_{r(rms)}$ can be related to $V_{o(av)}$ and $V_{o(rms)}$ by the mean square values. That is,

$$V_{r(rms)}^2 + V_{o(av)}^2 = V_{o(rms)}^2$$

or

$$V_{r(rms)}^2 = V_{o(rms)}^2 - V_{o(av)}^2 \tag{5.29}$$

Substituting V_m from Eq. (5.25) into Eq. (5.27), we get $V_{o(rms)} = \pi V_{o(av)}/(2\sqrt{2})$, which, when substituted into Eq. (5.29), gives

$$V_{r(rms)} = \left[\frac{\pi^2}{8}V_{o(av)}^2 - V_{o(av)}^2\right]^{1/2} = V_{o(av)}\left[\frac{\pi^2}{8} - 1\right]^{1/2} = 0.483V_{o(av)} \tag{5.30}$$

which is much less than $V_{r(rms)} = 1.21V_{o(av)}$ for a half-wave rectifier.

The ripple factor RF of the output voltage, which is a measure of the ripple content, can be found from

$$\text{RF} = \frac{V_{r(rms)}}{V_{o(av)}} = \frac{0.483V_{o(av)}}{V_{o(av)}} = 0.483, \text{ or } 48.3\% \tag{5.31}$$

which is much lower than RF = 1.21 = 121% for a half-wave rectifier.

The *AC output power* $P_{o(ac)}$ is the average power and is defined as

$$P_{o(ac)} = \frac{1}{2\pi}\int_0^{2\pi} i_O^2 R_L \, d(\omega t) = I_{o(rms)}^2 R_L = V_{o(rms)} I_{o(rms)} = \frac{V_m^2}{2R_L} \tag{5.32}$$

If we assume there is no power loss in the rectifier, the input power can be found from

$$P_{in} = P_{o(ac)} = \left(\frac{V_m}{\sqrt{2}}\right)^2 \frac{1}{R_L} \tag{5.33}$$

The *DC output power* $P_{o(dc)}$ can be found from

$$P_{o(dc)} = V_{o(av)} I_{o(av)} = \frac{4V_m^2}{\pi^2 R_L} \tag{5.34}$$

It is generally smaller than $P_{o(ac)}$. The ratio of $P_{o(dc)}$ to $P_{o(ac)}$, which is the rectification efficiency η_R, can be found from

$$\eta_R = \frac{P_{o(dc)}}{P_{o(ac)}} = \frac{V_{o(av)} I_{o(av)}}{V_{o(rms)} I_{o(rms)}} = \frac{(2V_m/\pi)^2/R_L}{(V_m/\sqrt{2})^2/R_L} = \frac{8}{\pi^2} = 81\% \tag{5.35}$$

which is twice the value of $\eta_R = 40.5\%$ for a half-wave rectifier.

If we assume there is no power loss in the rectifier, the input power factor can be found from

$$\text{PF} = \frac{P_{o(ac)}}{2V_s I_s} = \frac{(V_m/\sqrt{2})^2/R_L}{2 \times (V_m/\sqrt{2})(V_m/2R_L)} = \frac{\sqrt{2}}{2} = 0.7071 \tag{5.36}$$

▶ **NOTE** This numerical value of $\eta_R = 81\%$ is valid only for the single-phase full-wave rectifier.

The peak inverse voltage PIV of the diodes is $2V_m$. A full-wave rectifier develops twice the average output voltage of a half-wave rectifier for the same peak secondary voltage; however, it requires a center-tapped transformer. This circuit is suitable for low-power applications only—typically tens of watts.

EXAMPLE 5.4

Finding the performance parameters of a single-phase full-wave rectifier The single-phase full-wave center-tapped rectifier of Fig. 5.6(a) is supplied from a 120-V, 60-Hz source through a step-down center-tapped transformer with turns ratio $n = 10:2$. The load resistance R_L is 5 Ω. Determine **(a)** the average output voltage $V_{o(av)}$, **(b)** the average load current $I_{o(av)}$, **(c)** the rms load voltage $V_{o(rms)}$, **(d)** the rms load current $I_{o(rms)}$, **(e)** the ripple factor RF of the output voltage, **(f)** the rms ripple voltage $V_{r(rms)}$, **(g)** the average diode current $I_{D(av)}$, **(h)** the rms diode current $I_{D(rms)}$, **(i)** the peak inverse voltage PIV of the diodes, **(j)** the average output power $P_{o(ac)}$, **(k)** the DC output power $P_{o(dc)}$, **(l)** the frequency f_r of the output ripple voltage, and **(m)** the input power factor PF.

SOLUTION

The rms voltage of the transformer primary is $V_p = 120$ V. From Eq. (5.15) the rms voltage of the transformer secondary is $2V_s = 2V_p/n = 120 \times 2/10 = 24$ V. The rms voltage of the transformer half-secondary is $V_s = 24/2 = 12$ V. The peak voltage of each half-secondary is

$$V_m = \sqrt{2} \times 12 = 16.97 \text{ V}$$

(a) From Eq. (5.25),

$$V_{o(av)} = 0.636V_m = 0.636 \times 16.97 = 10.8 \text{ V}$$

(b) From Eq. (5.26),

$$I_{o(av)} = \frac{V_{o(av)}}{R_L} = \frac{10.8}{5} = 2.16 \text{ A}$$

(c) From Eq. (5.27),

$$V_{o(rms)} = 0.707V_m = 0.707 \times 16.97 = 12 \text{ V}$$

(d) From Eq. (5.28),

$$I_{o(rms)} = \frac{V_{o(rms)}}{R_L} = \frac{12}{5} = 2.4 \text{ A}$$

(e) From Eq. (5.31), RF = 0.483, or 48.3%.

(f) From Eq. (5.30),

$$V_{r(rms)} = 0.483V_{o(av)} = 0.483 \times 10.8 = 5.22 \text{ V}$$

(g) Since the average load current is supplied by two diodes, the average diode current $I_{D(av)}$ will be one-half of the load current. That is, $I_{D(av)} = I_{o(av)}/2 = 2.16/2 = 1.08$ A.

(h) Since the load current is shared by two diodes, the rms load current $I_{o(rms)}$ will be $\sqrt{2}$ times the rms diode current. That is, $I_{D(rms)} = I_{o(rms)}/\sqrt{2} = 2.4/\sqrt{2} = 1.7$ A.

(i) PIV $= 2V_m = 2 \times 16.97 = 33.94$ V.

(j) From Eq. (5.32),

$$P_{o(ac)} = I^2_{o(rms)}R_L = (2.4)^2 \times 5 = 28.8 \text{ W}$$

(k) From Eq. (5.34),

$$P_{o(dc)} = V_{o(av)}I_{o(av)} = 10.8 \times 2.16 = 23.33 \text{ W}$$

(l) The output voltage contains two pulses per cycle of the input voltage. That is, $f_r = 2f = 2 \times 60 = 120$ Hz.

(m) $V_s = 12$ V, $I_s = I_{o(rms)}/\sqrt{2} = 1.7$A, and $P_{in} = P_{o(ac)} = 28.8$ W

From Eq. (5.36),

$$\text{PF} = \frac{P_{in}}{2V_sI_s} = \frac{28.8}{2 \times 12 \times 1.7} = 0.7071$$

EXAMPLE 5.5

Fourier components of the output voltage of a single-phase full-wave rectifier The single-phase full-wave rectifier of Fig. 5.7 is supplied from a 120-V, 60-Hz source through a step-down center-tapped transformer with a turns ratio of $n = 10:2$.

(a) Express the instantaneous output voltage $v_O(t)$ by a Fourier series.

(b) Use PSpice/SPICE to calculate the harmonic components of the output voltage. Assume default diode parameters.

NOTE: The voltage-controlled voltage source representation of the input transformer will give only the correct input and output voltage waveforms, but not the correct value of the input current. To get the actual input current, we should consider the power balances such that the primary volt-amp is equal to the secondary volt-amp, $V_p I_p = 2V_s I_s$. This will require connecting two back-to-back current-controlled current sources (not shown) across the primary side [2].

SOLUTION

(a) $2V_s = 120 \times 2/10 = 24$ V, and $V_s = 12$ V. $V_m = \sqrt{2}V_s = \sqrt{2} \times 12 = 16.97$ V. The output voltage v_O can be described by

$$v_O = \begin{cases} V_m \sin \omega t & \text{for } 0 \le \omega t \le \pi \\ -V_m \sin \omega t & \text{for } \pi \le \omega t \le 2\pi \end{cases}$$

which can be expressed by a Fourier series as

$$v_O(\theta) = V_{o(av)} + \sum_{n=1,2,\ldots}^{\infty} (a_n \sin n\theta + b_n \cos n\theta) \quad \text{where } \theta = \omega t = 2\pi f t = 377t$$

$$V_{o(av)} = \frac{1}{2\pi}\int_0^{2\pi} v_O \, d\theta = \frac{2}{2\pi}\int_0^{\pi} V_m \sin\theta \, d\theta = \frac{2V_m}{\pi}$$

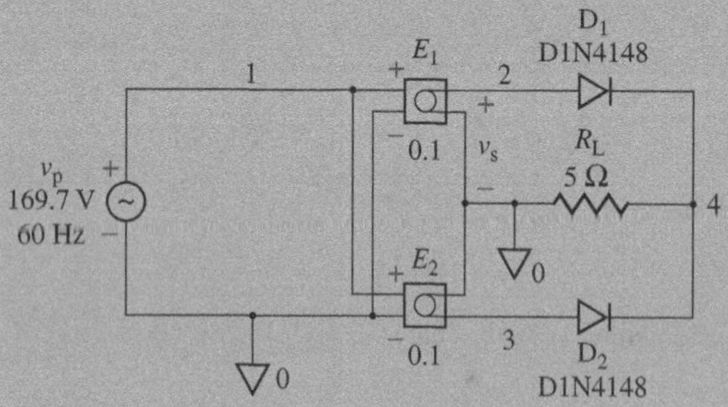

FIGURE 5.7 Single-phase full-wave rectifier circuit for PSpice simulation

$$a_n = \frac{1}{\pi}\int_0^{2\pi} v_O \sin n\theta \, d\theta$$

$$= \frac{1}{\pi}\left(\int_0^{\pi} V_m \sin\theta \sin n\theta \, d\theta + \int_{\pi}^{2\pi} -V_m \sin\theta \sin n\theta \, d\theta\right) = 0$$

$$b_n = \frac{1}{\pi}\int_0^{2\pi} v_O \cos n\theta \, d\theta = \frac{2}{\pi}\int_0^{\pi} V_m \sin\theta \cos n\theta \, d\theta$$

$$= \frac{4V_m}{\pi} \sum_{n=2,4,\ldots}^{\infty} \frac{1}{(n-1)(n+1)} \qquad \text{for } n = 2, 4, 6, \ldots, \infty$$

When the values of a_n and b_n are inserted into Eq. (5.18), the expression for the instantaneous output voltage v_O becomes

$$v_O(t) = \frac{2V_m}{\pi} - \frac{4V_m}{3\pi}\cos 2\omega t - \frac{4V_m}{15\pi}\cos 4\omega t - \frac{4V_m}{35\pi}\cos 6\omega t - \cdots$$

$$- \frac{4V_m}{(2n-1)(2n+1)}\cos 2n\omega t \qquad \text{for n} = 1, 2, 3, \ldots, \infty \tag{5.37}$$

where $V_m = \sqrt{2} \times 120 = 16.97$ V and $\omega = 2\pi \times 60 = 377$ rad/s.

Equation (5.25) gives $V_{o(av)} = 2V_m/\pi = 2 \times 16.97/\pi = 10.8$ V. From Eq. (5.37), we can find the peak magnitudes of harmonic components are

$$V_{2(peak)} = \frac{4V_m}{3\pi} = 4 \times \frac{16.97}{3\pi} = 7.2 \text{ V}$$

$$V_{4(peak)} = \frac{4V_m}{15\,\pi} = 4 \times \frac{16.97}{15\pi} = 1.44 \text{ V}$$

$$V_{6(peak)} = \frac{4V_m}{35\pi} = 4 \times \frac{16.97}{35\pi} = 0.617 \text{ V}$$

$$V_{8(peak)} = \frac{4V_m}{63\pi} = 4 \times \frac{16.97}{63\pi} = 0.343 \text{ V}$$

Note that the output voltage v_O contains only even harmonics, and the second harmonic is the dominant one at a ripple frequency of $f_r = 2f = 120$ Hz.

(b) The single-phase full-wave center-tapped rectifier circuit for PSpice simulation is shown in Fig. 5.7. The center-tapped transformer is modeled by a voltage-controlled voltage source.

The PSpice results of Fourier analysis are as follows. The hand-calculated values are shown in parentheses on the right.

FOURIER COMPONENTS OF TRANSIENT RESPONSE V(RL:2)
DC COMPONENT= 8.757443 $V_{o(dc)}$=8.757 V (10.8 V)

HARMONIC NO	FREQUENCY (HZ)	FOURIER COMPONENT	NORMALIZED COMPONENT	PHASE (DEG)	NORMALIZED PHASE (DEG)	
1	6.000E+01	2.727E-02	1.000E+00	-8.789E+01	0.000E+00	
2	1.200E+02	**6.312E+00**	2.315E+02	-9.018E+01	-2.289E+00	(7.2 V)
3	1.800E+02	2.705E-02	9.922E-01	-8.373E+01	4.161E+00	
4	2.400E+02	**1.199E+00**	4.398E+01	-9.065E+01	-2.763E+00	(1.44 V)
5	3.000E+02	2.658E-02	9.750E-01	-7.975E+01	8.142E+00	
6	3.600E+02	**4.806E-01**	1.763E+01	-9.177E+01	-3.882E+00	(0.617 V)
7	4.200E+02	2.576E-02	9.448E-01	-7.603E+01	1.186E+01	
8	4.800E+02	**2.478E-01**	9.089E+00	-9.391E+01	-6.015E+00	(0.343 V)
9	5.400E+02	2.447E-02	8.973E-01	-7.254E+01	1.535E+01	

TOTAL HARMONIC DISTORTION= 2.364960E+04 PERCENT

NOTE: The calculated values do not take into account the diode voltage drops, whereas the PSpice simulation assumes a real diode characteristic. This accounts for the differences between the PSpice and the hand-calculated values.

5.2.3 Single-Phase Full-Wave Bridge Rectifier

A single-phase full-wave bridge rectifier is shown in Fig. 5.8(a). It requires four diodes. The advantages of this rectifier are that it requires no transformer in the input side and the PIV rating of the diodes is V_m. The disadvantages are that it does not provide electrical isolation and it requires more diodes than the center-tapped version. However, an input transformer is normally used to satisfy the output voltage requirement. Since v_S is positive from $\omega t = 0$ to π and negative from $\omega t = \pi$ to 2π, the circuit operation can be divided into two intervals: interval 1 and interval 2.

Interval 1 is the interval $0 \le \omega t \le \pi$ during the positive half-cycle of the input voltage v_S. Diodes D_3 and D_4 are reverse biased, as shown in Fig. 5.8(b). The peak inverse voltage PIV of diodes D_3 and D_4 is V_m. Diodes D_1 and D_2 conduct and behaves like short circuits. The input voltage $v_S = V_m \sin \omega t$ appears across the load resistance R_L. That is, the output voltage becomes

$$v_O = V_m \sin \omega t \quad \text{for } 0 \le \omega t \le \pi$$

Interval 2 is the interval $\pi \le \omega t \le 2\pi$ during the negative half-cycle of the input voltage v_S. Diodes D_1 and D_2 are reverse biased, as shown in Fig. 5.8(c). The peak inverse voltage PIV of diodes D_1 and D_2 is V_m. Diodes D_3 and D_4 conduct and behave like short circuits. The negative of voltage $v_S = V_m \sin \omega t$ appears across the load resistance R_L. That is, the output voltage becomes

$$v_O = -V_m \sin \omega t \quad \text{for } \pi \le \omega t \le 2\pi$$

The waveforms for the input and output voltages are shown in Fig. 5.8(d). The output voltage will be reduced due to DC diode drop as shown by the dotted lines. The output ripple voltage is shown in Fig. 5.8(e). The equations that were derived earlier for a single-phase full-wave center-tapped transformer are also valid for the bridge rectifier.

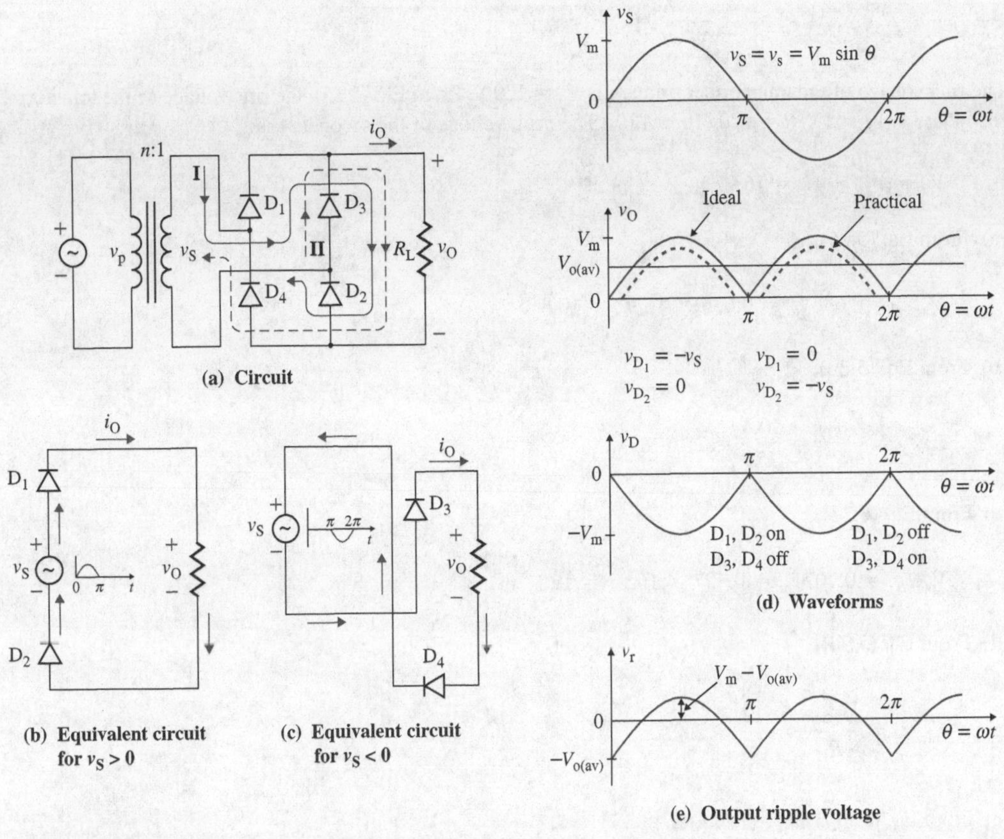

FIGURE 5.8 Single-phase full-wave bridge rectifier

If we assume there is no power loss in the rectifier, the input power factor can be found from

$$\text{PF} = \frac{P_{o(ac)}}{V_s I_s} = \frac{(V_m/\sqrt{2})^2/R_L}{(V_m/\sqrt{2})[V_m/(\sqrt{2}\,R_L)]} = 1.0 \tag{5.38}$$

EXAMPLE 5.6

Performance parameters of a single-phase full-wave bridge rectifier The single-phase full-wave bridge rectifier of Fig. 5.8(a) is supplied from a 120-V, 60-Hz source through a transformer with turns ratio $n = 10:1$. The load resistance R_L is 5 Ω. Determine **(a)** the average output voltage $V_{o(av)}$, **(b)** the average load current $I_{o(av)}$, **(c)** the rms load voltage $V_{o(rms)}$, **(d)** the rms load current $I_{o(rms)}$, **(e)** the ripple factor RF of the output voltage, **(f)** the rms ripple voltage $V_{r(rms)}$, **(g)** the average diode current $I_{D(av)}$, **(h)** the rms diode current $I_{D(rms)}$, **(i)** the peak inverse voltage PIV of the diode, **(j)** the average (or AC) output power $P_{o(ac)}$, **(k)** the DC output power $P_{o(dc)}$, **(l)** the frequency f_r of the output ripple voltage, and **(m)** the input power factor PF.

SOLUTION

The rms voltage of the transformer primary is $V_p = 120$ V. From Eq. (5.15), the rms voltage of the transformer secondary is $V_s = V_p/n = 120/10 = 12$ V. The peak voltage of the secondary is

$$V_m = \sqrt{2} \times 12 = 16.97 \text{ V}$$

(a) From Eq. (5.25),

$$V_{o(av)} = 0.636V_m = 0.636 \times 16.97 = 10.8 \text{ V}$$

(b) From Eq. (5.26),

$$I_{o(av)} = \frac{V_{o(av)}}{R_L} = \frac{10.8}{5} = 2.16 \text{ A}$$

(c) From Eq. (5.27),

$$V_{o(rms)} = 0.707V_m = 0.707 \times 16.97 = 12 \text{ V}$$

(d) From Eq. (5.28),

$$I_{o(rms)} = \frac{V_{o(rms)}}{R_L} = \frac{12}{5} = 2.4 \text{ A}$$

(e) From Eq. (5.31), RF = 0.483, or 48.3%.

(f) From Eq. (5.30),

$$V_{r(rms)} = 0.483\,V_{o(av)} = 0.483 \times 10.8 = 5.22 \text{ V}$$

(g) The load current flows through one of the top diodes (D_1 or D_3), the load, and then one of the bottom diodes (D_2 or D_4). Thus, the same current flows through two diodes, which are conducting. The time-average diode current $I_{D(av)}$ will be one-half of the load current. That is, $I_{D(av)} = I_{o(av)}/2 = 2.16/2 = 1.08$ A.

(h) The rms diode current $I_{D(rms)}$ will be $1/\sqrt{2}$ times the rms load current. That is, $I_{D(rms)} = I_{o(rms)}/\sqrt{2} = 2.4/\sqrt{2} = 1.7$ A

(i) PIV = V_m = 16.97 V

(j) From Eq. (5.32),

$$P_{o(ac)} = I^2_{o(rms)}R_L = (2.4)^2 \times 5 = 28.8 \text{ W}$$

(k) From Eq. (5.34),

$$P_{o(dc)} = V_{o(av)}I_{o(av)} = 10.8 \times 2.16 = 23.33 \text{ W}$$

(l) $f_r = 2f = 2 \times 60 = 120$ Hz

(m) $V_s = 12$ V, $I_s = I_{o(rms)} = 2.4$ A, and $P_{in} = P_{o(ac)} = 28.8$ W

From Eq. (5.38),

$$PF = \left(\frac{P_{in}}{V_s \times I_s}\right) = \left(\frac{28.8}{12 \times 2.4}\right) = 1.0$$

NOTE: The results of Examples 5.4 and 5.6 are identical, except that the PIV of a bridge rectifier is PIV = V_m = 16.97 V whereas the PIV of a center-tapped rectifier is PIV = $2V_m$ = 33.94 V for the same $V_{o(av)}$ = 10.8 V. The bridge rectifier has the best power factor.

EXAMPLE 5.7

Transfer (output versus input) characteristic of a single-phase bridge rectifier A single-phase bridge rectifier is shown in Fig. 5.9. The load resistance R_L is 4.5 kΩ. The source resistance R_s is 500 Ω.

(a) Determine the transfer characteristic (v_O versus v_S) of the rectifier.

(b) Use PSpice/SPICE to plot the transfer characteristic for $v_S = -10$ V to 10 V. Assume model parameters of diode D1N4148:

```
IS=2.682N CJO=4P M=.3333 VJ=.5 BV=100 IBV=100U TT=11.54N
```

SOLUTION

(a) $R_L = 4.5$ kΩ and $R_s = 500$ Ω. When the input voltage v_S is positive, only diodes D_1 and D_2 conduct. The output voltage v_O can be obtained by applying the voltage divider rule. That is,

$$v_O = \frac{v_S R_L}{R_L + R_s} = \frac{v_S(4.5\text{ k}\Omega)}{4.5\text{ k}\Omega + 500\ \Omega} = 0.9v_S \quad \text{for } v_S > 0$$

If the input voltage v_S is negative, only diodes D_3 and D_4 conduct. The output voltage v_O can be obtained from

$$v_O = \frac{-v_S R_L}{R_L + R_s} = \frac{-v_S(4.5\text{ k}\Omega)}{4.5\text{ k}\Omega + 500\ \Omega} = -0.9v_S \quad \text{for } v_S < 0$$

The transfer characteristic is shown in Fig. 5.10(a).

(b) The PSpice plot of v_O against v_S is shown in Fig. 5.10(b). The dead zone around 0 (between 0.82 V and −0.671 V) is due to the voltage drops across the diodes.

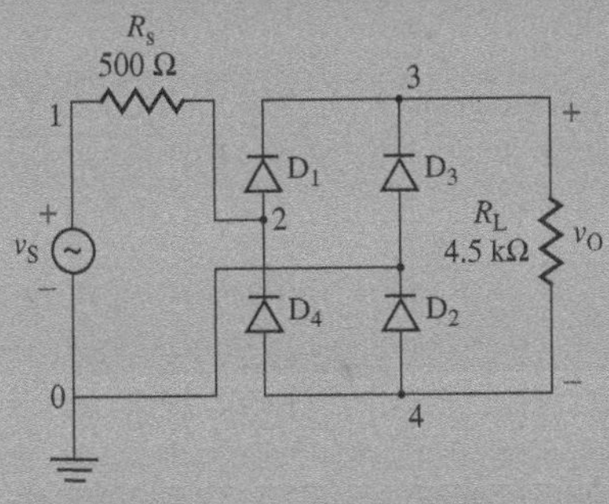

FIGURE 5.9 Single-phase bridge rectifier circuit for PSpice simulation

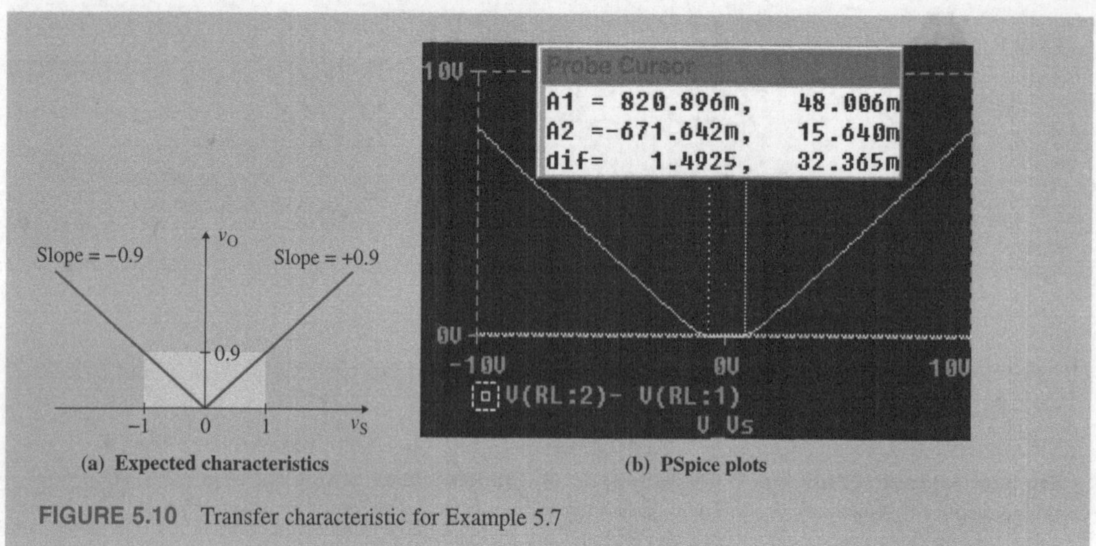

(a) Expected characteristics

(b) PSpice plots

FIGURE 5.10 Transfer characteristic for Example 5.7

EXAMPLE 5.8

Application of a single-phase bridge rectifier as an AC voltmeter An AC voltmeter is constructed by using a DC meter and a bridge rectifier, as shown in Fig. 5.11(a). The meter has a resistance of $R_m = 100\ \Omega$, and its average current is $I_m = 100$ mA for a full-scale deflection. The current-limiting resistance is $R_s = 1\ k\Omega$.

(a) Determine the rms value of the AC input voltage V_s that will give a full-scale deflection if the input voltage v_S is sinusoidal.

(b) If this meter is used to measure the rms value of an input voltage with a triangular waveform, as shown in Fig. 5.11(b), calculate the necessary correction factor K to be applied to the meter reading.

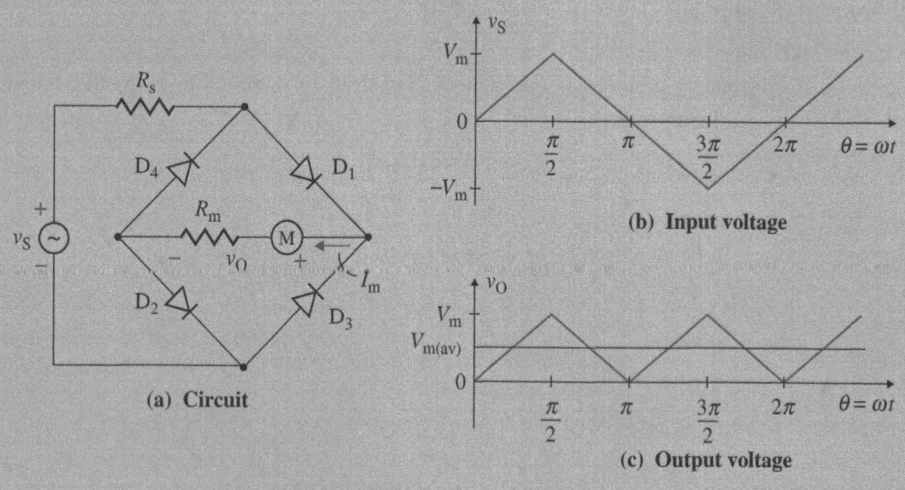

(a) Circuit

(b) Input voltage

(c) Output voltage

FIGURE 5.11 AC voltmeter

SOLUTION

$R_m = 100\ \Omega$, $R_s = 1\ \text{k}\Omega$, and $I_m = 100$ mA.

(a) The peak value V_m of a sinusoidal voltage is related to its rms value V_s by $V_m = \sqrt{2}V_s$. The average meter voltage $V_{m(av)}$ can be found by applying the voltage divider rule between resistances R_s and R_m:

$$V_{m(av)} = \frac{R_m}{R_s + R_m} V_{o(av)}$$

where $V_{s(rms)}$ is the rms value of v_S.

Using $V_{o(av)} = 2V_m/\pi$ from Eq. (5.25), we can find the average meter current $I_{m(av)}$ from

$$I_{m(av)} = \frac{V_{o(av)}}{R_s + R_m} = \frac{1}{R_s + R_m} \times \frac{2V_m}{\pi} = \frac{2\sqrt{2}\,V_s}{\pi\,(R_s + R_m)} \tag{5.39}$$

The meter reading θ_1, which is proportional to the average meter current $I_{m(av)}$, must measure the rms input voltage. That is,

$$\theta_1 = K_1 I_{m(av)} = V_s \tag{5.40}$$

where K_1 is a meter scale factor. Substituting $I_{m(av)}$ from Eq. (5.39), we get K_1:

$$K_1 \frac{2\sqrt{2}V_s}{\pi\,(R_s + R_m)} = V_s$$

which gives the constant K_1 as

$$K_1 = \frac{\pi\,(R_s + R_m)}{2\sqrt{2}} \tag{5.41}$$

$$= \frac{\pi\,(1 \times 10^3 + 100)}{2\sqrt{2}} = 1221.8\ \text{V/A}$$

Therefore, using Eq. (5.40), we can find the rms input voltage V_s that will give the full-scale deflection:

$$V_s = K_1 I_{m(av)} = 1221.8 \times 100 \times 10^{-3} = 122.2\ \text{V}$$

(b) If a triangular waveform v_S with a peak value of V_m is applied to the bridge rectifier, the output voltage v_O is as shown in Fig. 5.11(c). The rms input voltage V_s of the triangular voltage with four identical triangular areas can be found from

$$V_s = \left[\frac{4}{2\pi}\int_0^{\pi/2}\left(\frac{V_m}{\pi/2}\,\theta\right)^2 d\theta\right]^{1/2} = \frac{V_m}{\sqrt{3}} \quad \text{(after the integration is completed)} \tag{5.42}$$

The average output voltage $V_{o(av)}$ with four identical triangular areas can be found from

$$V_{o(av)} = \frac{4}{2\pi}\int_0^{\pi/2} \frac{V_m}{\pi/2}\,\theta\, d\theta = \frac{V_m}{2} \tag{5.43}$$

Substituting $V_{o(av)} = V_m/2$, we can find the average meter current $I_{m(av)}$ from

$$I_{m(av)} = \frac{V_{o(av)}}{R_s + R_m} = \frac{V_m/2}{R_s + R_m} = \frac{V_m}{2(R_s + R_m)} \tag{5.44}$$

The meter reading θ_1 must measure the rms input voltage. Substituting K_1 from Eq. (5.41) and $I_{m(av)}$ from Eq. (5.44), we get

$$\theta_1 = K_1 I_{m(av)} = \frac{\pi(R_s + R_m)}{2\sqrt{2}} \times \frac{V_m}{2(R_s + R_m)} = \frac{\pi V_m}{4\sqrt{2}} \tag{5.45}$$

But Eq. (5.42) showed that the rms value is $V_s = V_m/\sqrt{3}$. Letting K be the correction factor, we have

$$V_s = K\theta_1 = \frac{V_m}{\sqrt{3}}$$

which, after substitution for θ_1 from Eq. (5.45), gives the value of correction factor K as

$$K = \frac{V_m}{\sqrt{3}\theta_1} = \frac{V_m}{\sqrt{3}} \times \frac{4\sqrt{2}}{\pi V_m} = \frac{4\sqrt{2}}{\pi\sqrt{3}} = 1.0396 \tag{5.46}$$

Therefore, the meter will read KV_s (for sine wave) $= 1.0396 \times 122.2 = 127.04$ V at a full-scale deflection with the triangular waveform.

KEY POINTS OF SECTION 5.2

- Diodes can be used for rectification—that is, for converting AC voltage to DC voltage.
- The output voltage of a diode rectifier has harmonic content, which is measured by the harmonic factor RF.
- A half-wave rectifier has more harmonic content than a full-wave rectifier. However, it is simple and is generally used for low-power output on the order of 10 W. The center-tapped rectifier and the bridge rectifier are normally used for output in the ranges of 100 W and 1 kW, respectively.
- An input transformer is normally used to isolate the load from the supply and also to step the voltage up (or down).

5.3 Output Filters for Rectifiers

In Eqs. (5.19) and (5.37), the rectifier output voltage $v_O(t)$ has a DC component (V_m/π or $2V_m/\pi$) and other cosine components at various frequencies. The magnitudes of the cosine components are called the *harmonics.* The output should ideally be pure DC; these harmonics are undesirable. Filters are normally used to smooth out the output voltage. Since the input supply to these filters is DC, they are known as *DC filters.* Three types of DC filters are normally used: *L* filters, *C* filters, and *LC* filters. *L* filters and *LC* filters are generally used for high-power applications, such as DC power supplies. In integrated circuits, *C* filters are usually used.

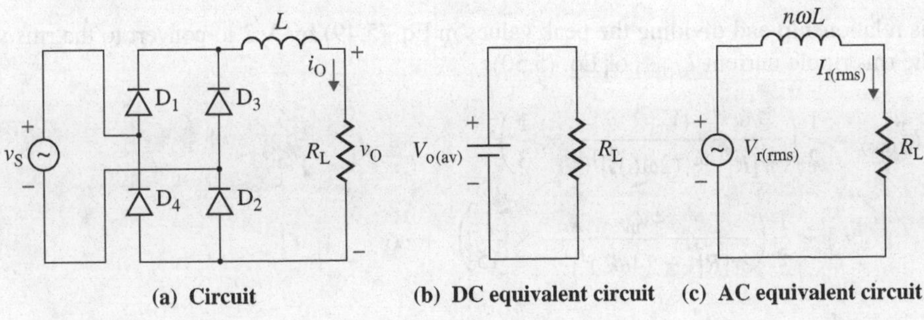

FIGURE 5.12 Single-phase bridge rectifier with an L filter

5.3.1 *L* Filters

An inductor, which is an energy storage element, tries to maintain a constant current through the load so that the variation in the output voltage is low. Let us assume that an inductor with zero internal resistance is connected in series with the load resistance R_L of a bridge rectifier. This arrangement is shown in Fig. 5.12(a). At the ripple frequencies, the inductance offers a high impedance and the load current ripple is reduced. The equivalent circuits for the DC and harmonic components are shown in Fig. 5.12[(b) and (c)], respectively. The load impedance is given by

$$Z = R_L + j(n\omega L) = \sqrt{R_L^2 + (n\omega L)^2}\angle\phi_n \tag{5.47}$$

$$\text{where} \quad \phi_n = \tan^{-1}\left(\frac{n\omega L}{R_L}\right) \tag{5.48}$$

Dividing the frequency-dependent components of the output voltage v_O in Eq. (5.37) by the impedance Z of Eq. (5.47) gives the instantaneous load current i_O:

$$i_O(t) = I_{o(av)} - \frac{4V_m}{\pi}\left[\sum_{n=2,4,6} \frac{1}{(n-1)(n+1)} \frac{\cos n\omega t - \phi_n}{\sqrt{R_L^2 + (n\omega L)^2}}\right] \tag{5.49}$$

where $I_{o(av)}$ is obtained by dividing $V_{o(av)}$ by the load resistance R_L. That is,

$$I_{o(av)} = \frac{V_{o(av)}}{R_L} = \frac{2V_m}{\pi R_L}$$

Let us consider the first two harmonic components only, ignoring the higher-order ones. Let $I_{o2(rms)}$ and $I_{o4(rms)}$ be the rms currents of the second and fourth harmonic components, respectively. Since these currents are in rms values, the resultant rms ripple current $I_{r(rms)}$ can be found by adding the mean square values of $I_{o2(rms)}$ and $I_{o4(rms)}$. That is,

$$I_{r(rms)}^2 = I_{o2(rms)}^2 + I_{o4(rms)}^2 \tag{5.50}$$

Using this relationship and dividing the peak values in Eq. (5.49) by $\sqrt{2}$ to convert to the rms values, we can get the rms ripple current $I_{r(rms)}$ of Eq. (5.50):

$$I^2_{r(rms)} = \frac{1}{2}\left(\frac{4V_m}{\pi[R_L^2 + (2\omega L)^2]^{1/2}} \times \frac{1}{3}\right)^2 + \frac{1}{2}\left(\frac{4V_m}{\pi[R_L^2 + (4\omega L)^2]^{1/2}} \times \frac{1}{15}\right)^2 + \cdots \tag{5.51}$$

EXAMPLE 5.9

D **Designing an output *L* filter** The single-phase bridge rectifier of Fig. 5.12(a) is directly supplied from a 120-V, 60-Hz source without any input transformer. The average output voltage is $V_{o(av)} = 158$ V. The load resistance is $R_L = 500\ \Omega$.

(a) Design an L filter so that the rms ripple current $I_{r(rms)}$ is limited to less than 5% of $I_{o(av)}$. Assume that the second harmonic $I_{o2(rms)}$ is the dominant one and that the effects of higher-order harmonics are negligible.

(b) Use PSpice/SPICE to check your design by plotting the output current. Use diode default parameters of 1N4148 diodes.

SOLUTION

(a) Since $V_m = \sqrt{2}V_s = \sqrt{2} \times 120 = 169.7$ V,

$$I_{o(av)} = \frac{V_{o(av)}}{R_L} = \frac{158}{500} = 316 \text{ mA}$$

$$I_{r(rms)} = 5\% \text{ of } I_{o(av)} = 0.05 \times 316 \text{ mA} = 15.8 \text{ mA}$$

Assume that the ripple current is approximately sinusoidal. Then, the peak ripple current is $\sqrt{2}$ times the value of $I_{r(rms)}$. That is,

$$I_{r(peak)} = \sqrt{2} \times I_{r(rms)} = \sqrt{2} \times 15.8 \text{ mA} = 22.34 \text{ mA}$$

The peak-to-peak ripple current $I_{r(pp)}$ is twice the value of $I_{r(peak)}$. Thus,

$$I_{r(pp)} = 2 \times I_{r(peak)} = 2 \times 22.34 \text{ mA} = 44.69 \text{ mA}$$

Let us consider only the lowest-order harmonic—that is, $n = 2$. Equation (5.51) yields

$$I_{r(rms)} \approx I_{o2(rms)} = \frac{4V_m}{\sqrt{2}\pi[R_L^2 + (2\omega L)^2]^{1/2}} \times \frac{1}{3}$$

The ripple factor RF_i of the output current is given by

$$RF_i = \frac{I_{r(rms)}}{I_{o(av)}} \approx \frac{I_{o2(rms)}}{I_{o(av)}} = \frac{4V_m}{\sqrt{2}\pi\,[R_L^2 + (2\omega L)^2]^{1/2}} \times \frac{1}{3} \times \frac{\pi R_L}{2V_m} \tag{5.52}$$

$$= \frac{4/(\sqrt{2} \times 3 \times 2)}{\sqrt{1 + (2\omega L/R_L)^2}} = \frac{0.4714}{\sqrt{1 + (2\omega L/R_L)^2}}$$

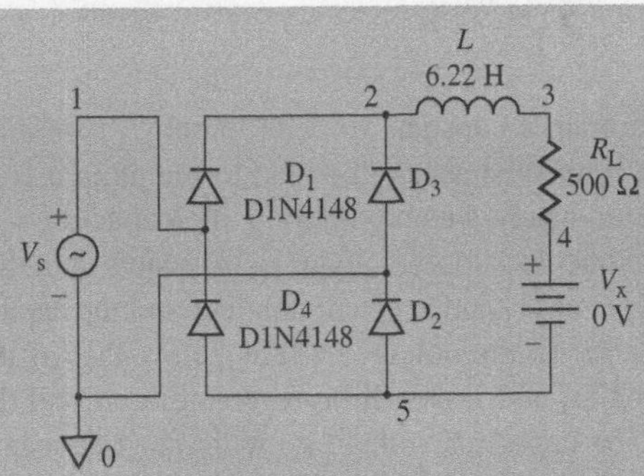

FIGURE 5.13 Bridge rectifier circuit with L filter for PSpice simulation

which can be solved to find the value of L for the known values of $R_L = 500\ \Omega$, $f = 60$ Hz, and $RF_i = 5\% = 0.05$. That is,

$$0.05 = \frac{0.4714}{\sqrt{1 + (2\omega L/R_L)^2}}$$

$$0.4714^2 = (0.05)^2 \times \left[1 + \left(\frac{2 \times 2 \times 60 \times \pi L}{500}\right)^2\right]$$

$$L = 6.22 \text{ H}$$

(b) The bridge rectifier circuit with an L filter for PSpice simulation is shown in Fig. 5.13. PSpice allows us to find the current through resistors, I(RL). It is not necessary to have a fictitious voltage source VX = 0.

The PSpice plot of load current i_O, shown in Fig. 5.14, gives the peak-to-peak ripple current as $I_{r(pp)} = 166.67 - 150.82 = 15.85$ mA, compared to the calculated value of $I_{r(pp)} = 22.34$ mA. The difference between the values is a result of neglecting the higher-order harmonics in determining the value of L and also the fact that PSpice uses real diodes rather than ideal ones with zero forward resistance. With ideal diodes, PSpice would give 27.9 mA. The DC current from PSpice is $I_{o(av)} \approx (166.61 + 150.82)/2 = 158.72$ mA, which is below the calculated value of 316 mA. This is caused by the fact that the effect of inductor L was not included in the calculated values.

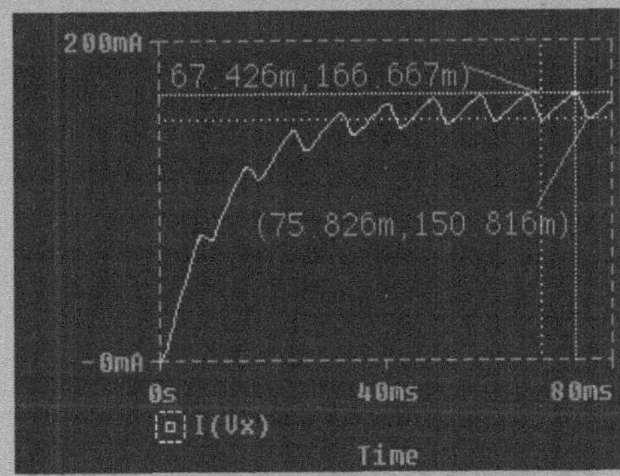

FIGURE 5.14 PSpice plot for Example 5.9

5.3.2 *C* Filters

A capacitor is also an energy storage element; it tries to maintain a constant voltage, thereby preventing any change in voltage across the load. A capacitor C can be connected across the load to maintain a continuous output voltage v_O, as shown in Fig. 5.15(a) [3]. Under steady-state conditions, the capacitor will have a finite initial voltage. When the magnitude of the instantaneous supply voltage v_S is greater than that of the instantaneous capacitor voltage v_C, the diodes (D_1 and D_2 or D_3 and D_4) will conduct and the capacitor will be charged from the supply. However, if the magnitude of the voltage v_S falls below that of the instantaneous capacitor voltage v_C, the diodes (D_1 and D_2 or D_3 and D_4) will be reverse biased and the capacitor C will discharge through the load resistance R_L. The capacitor voltage v_C will vary between a minimum value $V_{o(min)}$ and a maximum value $V_{o(max)}$. The waveforms of the output voltage v_O and ripple voltage v_r are shown in Fig. 5.15(b). If *f* is the supply frequency, the period of the input voltage is $T = 1/f$.

For a single-phase half-wave rectifier, the period of the output ripple voltage is the same as the period *T* of the supply voltage. However, for a single-phase full-wave rectifier, the period of the output ripple voltage is $T/2$. The output operation can be divided into two intervals: interval 1 for charging and interval 2 for discharging.

The equivalent circuit during charging is shown in Fig. 5.15(c). The capacitor charges almost instantaneously to the supply voltage v_S. The capacitor C will be charged approximately to the peak supply voltage V_m, so $v_C(\omega t = \pi/2) = V_m$. Figure 5.15(d) shows the equivalent circuit during discharging. The capacitor discharges exponentially through R_L. When one of the diode pairs is conducting, the

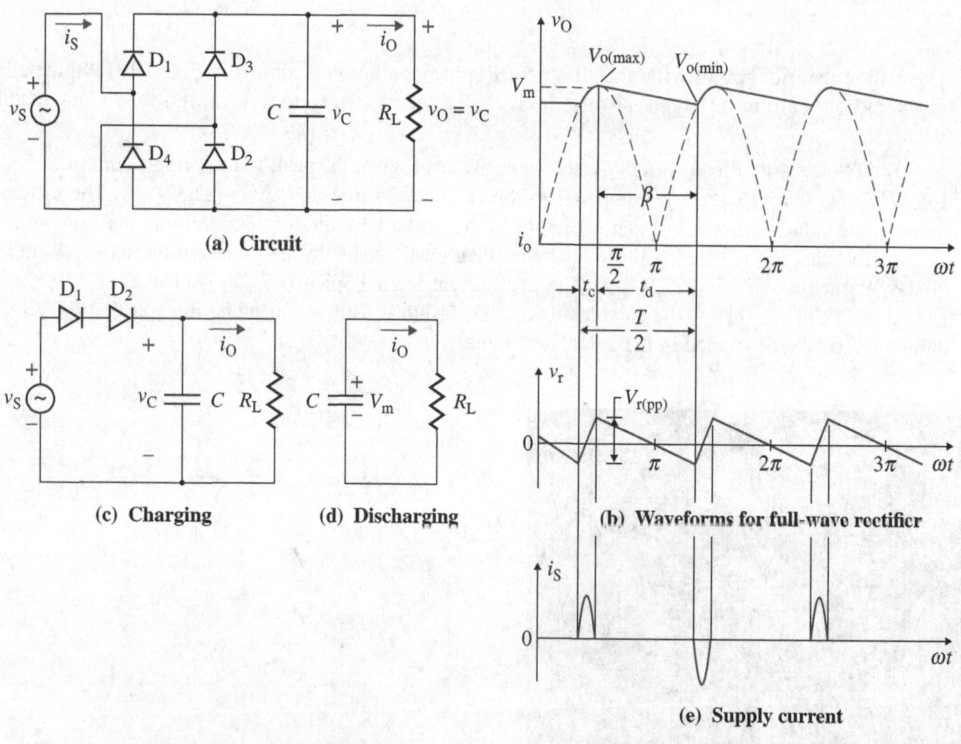

FIGURE 5.15 Bridge rectifier with a *C* filter

TABLE 5.1 Terms for measuring output ripple voltage

Definition of Terms	Relationship
The peak value of output voltage	$V_{o(max)} = V_m$
The peak-to-peak output ripple voltage	$V_{r(pp)} = V_{o(max)} - V_{o(min)} = V_m - V_{o(min)}$
The ripple factor of the output voltage	$RF = \frac{V_{r(pp)}}{V_m} = \frac{V_m - V_{o(min)}}{V_m} = 1 - \frac{V_{o(min)}}{V_m}$
The minimum value of output voltage	$V_{o(min)} = V_m(1 - RF)$

capacitor C draws a pulse of charging current from the AC supply, as shown in Fig. 5.15(e). As a result, the rectifier generates harmonic currents into the AC supply. For high-power applications, an input filter is normally required to reduce the amount of harmonic injection into the AC supply. Thus, a rectifier with a C filter is used only for low-power applications [4].

The output ripple voltage, which is the difference between maximum voltage $V_{o(max)}$ and the minimum voltage $V_{o(min)}$, can be specified in different ways, as shown in Table 5.1.

During the charging interval, under steady-state conditions the capacitor charges from $V_{o(min)}$ to V_m. Let us assume that at an angle α (rad/s), the positive input voltage is equal to the minimum capacitor voltage $V_{o(min)}$ at the end of the capacitor discharge. As the input voltage rises sinusoidally from zero to V_m, at the first cycle the angle α can be determined from

$$V_{o(min)} = V_m \sin(\alpha) \quad \text{or} \quad \alpha = \sin^{-1}\left(\frac{V_{o(min)}}{V_m}\right) \tag{5.53}$$

By redefining the time origin ($\omega t = 0$) at $\pi/2$, as the beginning of interval 1, we can deduce the discharging current from

$$\frac{1}{C}\int i_O \, dt - v_C(t = 0) + R_L i_O = 0$$

which, with an initial condition of $v_C(\omega t = 0) = V_m$, gives

$$i_O = \frac{V_m}{R_L} e^{-t/R_L C} \quad \text{for } 0 \le t \le t_d$$

The instantaneous output (or capacitor) voltage v_O during the discharging period can be found from

$$v_O(t) = R_L i_O = V_m e^{-t/R_L C} \tag{5.54}$$

From Fig. 5.15(b), we can find the discharging time t_d or the discharging angle β (rad/s) as

$$\omega t_d = \beta = \frac{\pi}{2} + \alpha \quad \text{for a full-wave rectifier} \tag{5.55a}$$

$$= \frac{3\pi}{2} + \alpha \quad \text{for a half-wave rectifier} \tag{5.55b}$$

At $t = t_d$, $v_O(t)$ in Eq. (5.54) becomes equal to $V_{o(min)}$, and we can relate t_d to $V_{o(min)}$ by

$$v_O(t = t_d) = V_{o(min)} = V_m \, e^{-t_d/R_L C} \tag{5.56}$$

which gives the discharging time t_d as

$$t_d = R_L C \ln\left(\frac{V_m}{V_{o(min)}}\right) \tag{5.57}$$

Equating t_d in Eq. (5.57) to t_d in Eq. (5.55), we get

$$\omega R_L C \ln\left(\frac{V_m}{V_{o(min)}}\right) = \frac{\pi}{2} + \alpha = \frac{\pi}{2} + \sin^{-1}\left(\frac{V_{o(min)}}{V_m}\right) \quad \text{for a full-wave rectifier} \tag{5.58a}$$

$$= \frac{3\pi}{2} + \alpha = \frac{3\pi}{2} + \sin^{-1}\left(\frac{V_{o(min)}}{V_m}\right) \quad \text{for a half-wave rectifier} \tag{5.58b}$$

Therefore, the filter capacitor C can be found from

$$C = \frac{\pi/2 + \sin^{-1}(V_{o(min)}/V_m)}{\omega R_L \ln(V_m/V_{o(min)})} \quad \text{for a full-wave rectifier} \tag{5.59a}$$

$$= \frac{3\pi/2 + \sin^{-1}(V_{o(min)}/V_m)}{\omega R_L \ln(V_m/V_{o(min)})} \quad \text{for a half-wave rectifier} \tag{5.59b}$$

Redefining the time origin ($\omega t = 0$) at $\pi/2$ when the discharging interval begins, we can find the average output voltage $V_{o(av)}$ from

$$V_{o(av)} = \frac{V_m}{\pi}\left[\int_0^{\beta} e^{-\omega t/R_L C}\, d(\omega t) + \int_{\pi}^{\beta} \cos(\omega t)\, d(\omega t)\right]$$

$$= \frac{V_m}{\pi}[\omega R_L C(1 - e^{-\beta/\omega R_L C}) + \sin\beta] \quad \text{for a full-wave rectifier} \tag{5.60}$$

$$V_{o(av)} = \frac{V_m}{2\pi}\left[\int_0^{\beta} e^{-\omega t/R_L C}\, d(\omega t) + \int_{2\pi}^{\beta} \cos(\omega t)\, d(\omega t)\right]$$

$$= \frac{V_m}{2\pi}[\omega R_L C(1 - e^{-\beta/\omega R_L C}) + \sin\beta] \quad \text{for a half-wave rectifier} \tag{5.61}$$

The equations just given for C in Eq. (5.59) and $V_{o(av)}$ in Eq. (5.60) are nonlinear. We can derive simple explicit expressions for the ripple voltage in terms of the capacitor value if we make the following assumptions:

- t_c is the charging time of the capacitor C.
- t_d is the discharging time of the capacitor C.

If we assume that the charging time t_c is small compared to the discharging time t_d (i.e., $t_d >> t_c$, which is generally the case), we can relate t_c and t_d to the period T of the input supply as

$$t_d = \frac{T}{2} - t_c \simeq \frac{T}{2} = \frac{1}{2}f \quad \text{for a full-wave rectifier} \tag{5.62a}$$

$$= T - t_c \simeq T = \frac{1}{f} \quad \text{for a half-wave rectifier} \tag{5.62b}$$

Using Taylor series expansion of $e^{-x} = 1 - x$ for small values of $x << 1$, we can simplify Eq. (5.56) to

$$V_{o(min)} = V_m\, e^{-t_d/R_L C} = V_m\left(1 - \frac{t_d}{R_L C}\right) \tag{5.63}$$

This gives the peak-to-peak ripple voltage $V_{r(pp)}$ as

$$V_{r(pp)} = V_m - V_{o(min)} = V_m \frac{t_d}{R_L C} = \frac{V_m}{2fR_L C} \quad \text{for a full-wave rectifier} \tag{5.64a}$$

$$= \frac{V_m}{fR_L C} \quad \text{for a half-wave rectifier} \tag{5.64b}$$

Equations (5.64a and 5.64b) can be used to find the value of capacitor C with reasonable accuracy for most practical purposes as long as the ripple factor is within 10%. We can observe from Eq. [5.64(a) and (b)] that the ripple voltage depends inversely on the supply frequency f, the filter capacitance C, and the load resistance R_L. For the same amount of voltage ripple, the full-wave rectifier will require half the capacitance C due to having double the output ripple frequency $2f$ as compared to the half-wave rectifier.

If we assume that the output voltage decreases linearly from $V_{o(max)}(=V_m)$ to $V_{o(min)}$ during the discharging interval, the average output voltage can be found approximately from

$$V_{o(av)} = \frac{V_m + V_{o(min)}}{2} = \frac{1}{2}\left[V_m + V_m\left(1 - \frac{t_d}{R_L C}\right)\right] \tag{5.65}$$

After we substitute for t_d in Eq. (5.65), this becomes

$$V_{o(av)} = \frac{1}{2}\left[V_m + V_m\left(1 - \frac{1}{2R_L fC}\right)\right] = \frac{V_m}{2}\left[2 - \frac{1}{2R_L fC}\right] \quad \text{for a full-wave rectifier} \tag{5.66a}$$

$$= \frac{1}{2}\left[V_m + V_m\left(1 - \frac{1}{R_L fC}\right)\right] = \frac{V_m}{2}\left[2 - \frac{1}{R_L fC}\right] \quad \text{for a half-wave rectifier} \tag{5.66b}$$

The ripple factor RF can be found from

$$\text{RF} = \frac{V_{r(pp)}/2}{V_{o(av)}} = \frac{1}{4R_L fC - 1} \quad \text{for a full-wave rectifier} \tag{5.67a}$$

$$= \frac{1}{2R_L fC - 1} \quad \text{for a half-wave rectifier} \tag{5.67b}$$

The peak input voltage V_m is generally fixed by the supply, whereas we can vary the minimum voltage $V_{o(min)}$ from almost zero to V_m by varying the values of C, f, and R_L. Therefore, it is possible to design for an average output voltage $V_{o(av)}$ in the range from $V_m/2$ to V_m. We can find the value of capacitor C to meet either a specific value of the minimum voltage $V_{o(min)}$ or the average output voltage $V_{o(av)}$ so that $V_{o(min)} = (2\,V_{o(av)} - V_m)$.

EXAMPLE 5.10

D **Designing an output C filter** The single-phase full-wave bridge rectifier of Fig. 5.15(a) is supplied directly from a 120-V, 60-Hz source without any input transformer. The load resistance is $R_L = 500\ \Omega$.

(a) Design a C filter so that the peak-to-peak ripple voltage $V_{r(pp)}$ is within 10% of V_m.

(b) With the value of C found in part (a) calculate the actual output voltage $V_{o(av)}$, and the capacitor voltage if the load resistance R_L is disconnected.

(c) Use PSpice/SPICE to check the design by plotting the instantaneous output voltage v_O. Use default diode parameters of 1N4148 type.

SOLUTION

(a) $V_m = \sqrt{2}V_s = \sqrt{2} \times 120 = 169.7$ V

The peak-to-peak ripple voltage is

$$V_{r(pp)} = 10\% \text{ of } V_m = 0.1 \times 169.7 = 16.97 \text{ V}$$

The minimum output voltage is

$$V_{o(min)} = V_m - V_{r(pp)} = 170 - 16.97 = 152.74 \text{ V}$$

From Eq. (5.53), we get the angle α as

$$\alpha = \sin^{-1}\left(\frac{V_{o(min)}}{V_m}\right) = \sin^{-1}\left(\frac{152.74}{1169.7}\right) = 1.12 \text{ rad, or } 64.16^\circ$$

From Eq. (5.55a), we get the discharge β as

$$\beta = \frac{\pi}{2} + \alpha = \frac{\pi}{2} + 1.12 = 2.698 \text{ rad, or } 154.16^\circ$$

From Eq. (5.59a), we get the filter capacitor C:

$$C = \frac{\pi/2 + \alpha}{2\pi f R_L \ln(V_m/V_{o(min)})} = \frac{\pi/2 + 1.12}{2\pi \times 86 \times 500 \times \ln(169.7/152.74)} = 135.48\ \mu\text{F}$$

(b) From Eq. (5.60), we get the average output voltage:

$$V_{o(av)} = \frac{V_m}{\pi}[2\pi\ fR_L C(1 - e^{-\beta/\omega R_L C}) + \sin\beta] = 161.49 \text{ V}$$

The approximate Eq. (5.64a) gives $C = 166.7\ \mu$F, and Eq. (5.65) gives $V_{o(av)} = (V_m + V_{o(min)})/2 = 161.2$ V.

If the load resistance R_L is disconnected, the capacitor will charge to the peak input voltage V_m. Therefore, the average output voltage with no load is

$$V_{o(no\text{-}load)} = V_m = 169.7 \text{ V}$$

The average output voltage $V_{o(av)}$ will change from 169.7 V to 158.49 V if the load is connected. This change in voltage is normally specified by a factor known as the *voltage regulation,* which is defined as

$$\text{Voltage regulation} = \frac{V_{o(no\text{-}load)} - V_{o(load)}}{V_{o(load)}} = \frac{V_{o(load)} - V_{o(av)}}{V_{o(av)}} \tag{5.68}$$

$$= \frac{169.7 - 161.49}{161.49} = 5.08\%$$

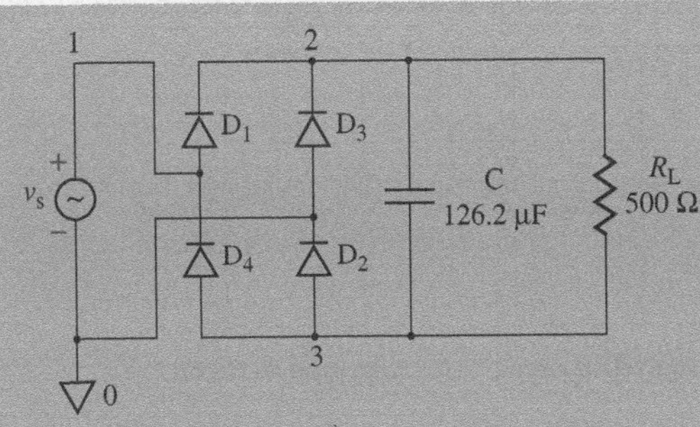

FIGURE 5.16 Single-phase bridge rectifier circuit with a C filter for PSpice simulation

(c) The single-phase bridge rectifier circuit with a C filter for PSpice simulation is shown in Fig. 5.16.

The PSpice plot of v_O, shown in Fig. 5.17 (which was obtained by using the PSpice model of diode IN4148), gives the peak-to-peak ripple voltage as $V_{r(pp)} = 4.85$ V (15.23 V with ideal diodes), compared to the calculated value of 22.34 V. The average output voltage is $V_{o(av)} = (98.98 + 94.1)/2 = 96.54$ V. The error results from neglecting the voltage drops of the diodes in hand calculations. The value of v_O reaches a steady state after a transient interval of approximately 40 ms. If we run the simulation using the ideal diode model, we get $V_{r(pp)} = 17.6$ V and $V_{o(av)} = 159.2$ V. This difference is caused by the finite resistance of the PSpice diode model during the charging interval of capacitor C.

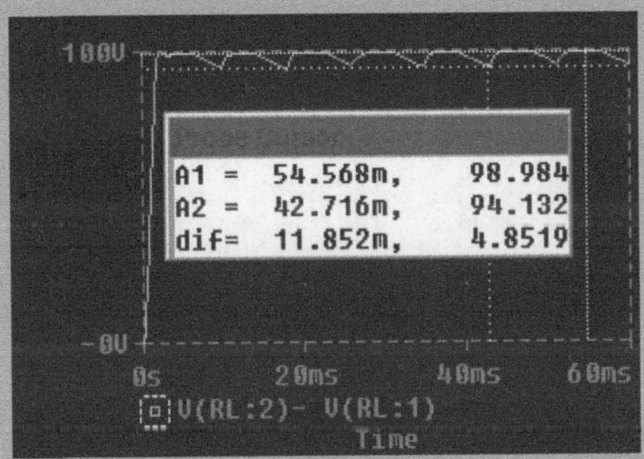

FIGURE 5.17 PSpice plot of output voltage for Example 5.10

5.3.3 *LC* Filters

An LC filter, which opposes any change in either the voltage or the current, reduces the harmonics more effectively than an L filter or a C filter. A rectifier with an LC filter is shown in Fig. 5.18(a). The equivalent circuit for harmonics is shown in Fig. 5.18(b) where V_{rn} is the nth harmonic component of the rms ripple voltage.

To make it easier for the nth harmonic ripple current to pass through the filter capacitor C rather than through the load resistance R_L, the load impedance Z_L $(= R_L)$ must be greater than that of the capacitor. That is,

$$R_L >> \frac{1}{n\omega C}$$

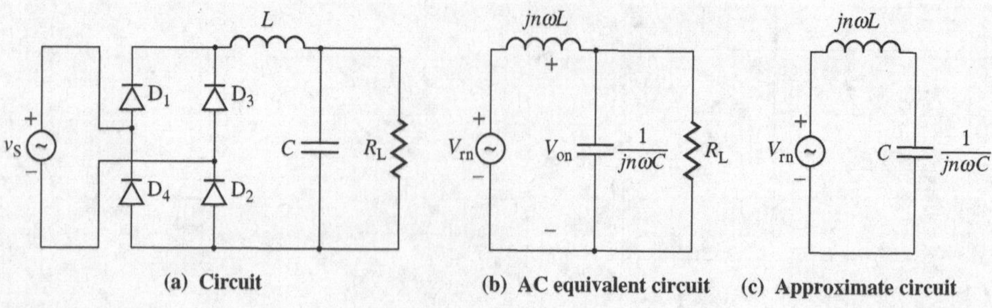

FIGURE 5.18 Rectifier with an *LC* filter

This condition is generally satisfied by choosing a ratio of 1:10. That is,

$$R_L = \frac{10}{n\omega C} \tag{5.69}$$

Under this condition, R_L can be neglected and the effect of the load resistance R_L will be negligible. Thus, Fig. 5.18(b) is reduced to Fig. 5.18(c). Using the voltage divider rule, we can find the rms value of the *n*th harmonic voltage component appearing after filtering on the output from

$$V_{rn(rms)} = \left| \frac{-j/(n\omega C)}{(j\omega L) - j/(n\omega C)} \right| V_{on(rms)} = \frac{1}{|1-(n\omega)^2 LC|} V_{on(rms)} \tag{5.70}$$

where $V_{on(rms)}$ is the rms *n*th harmonic voltage of Eq. (5.19) or Eq. (5.37). If the higher-order harmonics are neglected and the second harmonic becomes the dominant one, $V_{o2(rms)}$ becomes the output ripple voltage of the rectifier, and Eq. (5.70) can be written as

$$V_{r(rms)} = V_{r2(rms)} = \frac{1}{|1-(n\omega)^2 LC|} V_{o2(rms)} \tag{5.71}$$

With the value of *C* from Eq. (5.69), the value of *L* can be computed for a specified value of $V_{r(rms)}$.

EXAMPLE 5.11

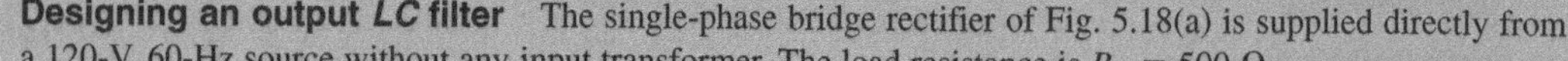

D **Designing an output *LC* filter** The single-phase bridge rectifier of Fig. 5.18(a) is supplied directly from a 120-V, 60-Hz source without any input transformer. The load resistance is $R_L = 500\ \Omega$.

(a) Design an *LC* filter so that the rms ripple voltage $V_{r(rms)}$ is within 5% of $V_{o(av)}$.

(b) Use PSpice/SPICE to check your design by plotting the instantaneous output voltage v_O. Use default diode parameters.

SOLUTION

(a) $f = 60$ Hz, $\omega = 2\pi f = 377$ rad/s, $R_L = 500\ \Omega$, and RF = 5% = 0.05.

$$V_m = \sqrt{2}V_s = \sqrt{2} \times 120 = 169.7 \text{ V}$$

$$V_{o(av)} = \frac{2V_m}{\pi} = \frac{2 \times 169.7}{\pi} = 108.03 \text{ V}$$

$$V_{r(rms)} = 5\% \text{ of } V_{o(av)} = 0.05 \times 108.03 = 5.4 \text{ V}$$

Assume that the ripple voltage is approximately sinusoidal. Then, the peak ripple voltage is given by

$$V_{r(peak)} = \sqrt{2} \times V_{r(rms)} = \sqrt{2} \times 5.4 = 7.64 \text{ V}$$

The peak-to-peak ripple voltage $V_{r(pp)}$ is

$$V_{r(pp)} = 2 \times V_{r(peak)} = 2 \times 7.64 = 15.28 \text{ V}$$

Let us consider only the dominant harmonic—that is, the second harmonic. From the second term in Eq. (5.37), the rms value of the second harmonic is

$$V_{o2(rms)} = \frac{4V_m}{3\sqrt{2}\pi}$$

NOTE: $\sqrt{2}$ converts the peak value to a rms value.

For $n = 2$, the value of C can be found from Eq. (5.69) as follows:

$$C = \frac{10}{n\omega R_L} = \frac{10}{2 \times 377 \times 500} = 26.53 \ \mu\text{F}$$

Using Eqs. (5.71) and (5.25), we can find the ripple factor RF of the output voltage from

$$\text{RF} = \frac{V_{r(rms)}}{V_{o(av)}} = \frac{V_{o2(rms)}}{|1 - (n\omega)^2LC|} \times \frac{\pi}{2V_m} = \frac{1}{|1 - (n\omega)^2LC|} \times \frac{4V_m}{3\sqrt{2}\pi} \times \frac{\pi}{2V_m}$$

$$= \frac{\sqrt{2}/3}{|1 - (n\omega)^2LC|}$$

which can be solved for L:

$$L = \frac{1}{(n\omega)^2C}\left[\frac{\sqrt{2}}{3\text{RF}} - 1\right] = \frac{1}{(2 \times 377)^2 \times 26.53 \times 10^{-6}}\left[\frac{\sqrt{2}}{3 \times 0.05} - 1\right] = 0.56 \text{ H}$$

(b) The single-phase bridge rectifier circuit with an LC filter for PSpice simulation is shown in Fig. 5.19. The PSpice plot of v_O, shown in Fig. 5.20, gives the peak-to-peak ripple voltage as $V_{r(pp)} = 14.99$ V, compared to the calculated value of 15.28 V. There is an error of 1.39 V, which can arise from various factors such as neglecting the higher-order harmonics, not considering the loading effect of R_L, and assuming an ideal diode with zero voltage drop. Thus, the design values should be revised until the desired specifications are satisfied.

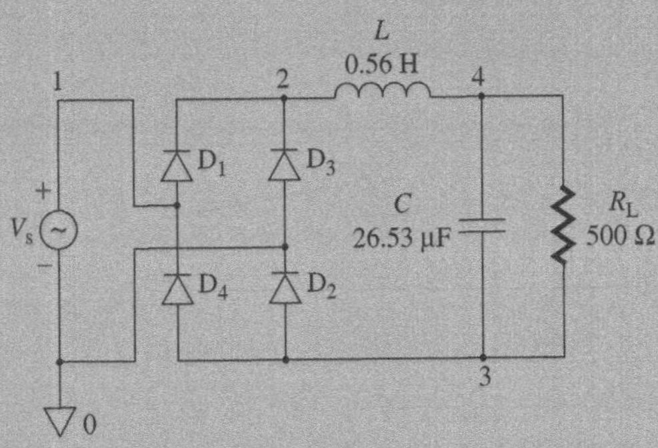

FIGURE 5.19 Single-phase bridge rectifier circuit with an LC filter for PSpice simulation

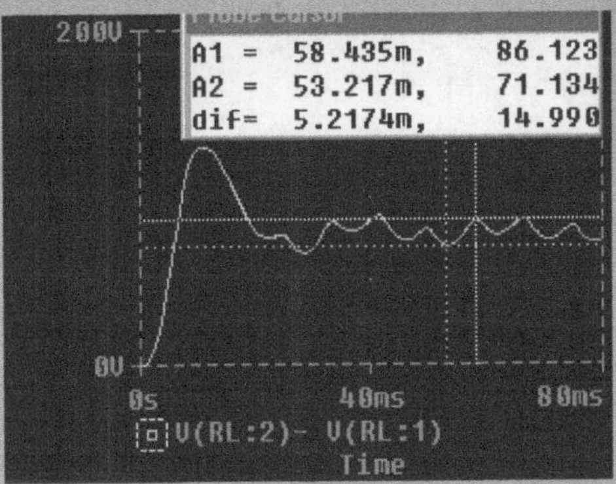

FIGURE 5.20 PSpice plot for Example 5.11

KEY POINTS OF SECTION 5.3

- The output voltage of a diode rectifier has harmonic content, and filters are normally used to smooth out the ripples.
- A *C* filter connected across the load is the simplest and the most commonly used filter. It maintains a reasonably constant DC output voltage.
- An *L* filter connected in series with the load tries to maintain a constant DC load current.
- An *LC* filter combines the features of both *C* and *L* filters. It is more effective in filtering the ripple contents from the output voltage.

5.4 Diode Peak Detectors and Demodulators

The half-wave rectifier shown in Fig. 5.21(a) can be employed as a peak signal detector. Let us consider a sinusoidal input voltage, $v_S = V_m \sin \omega t$. During the first quarter-cycle, the input voltage will rise, the capacitor C will be charged almost instantaneously to the input voltage, and the capacitor (or output) voltage v_O will follow the input voltage v_S until the instantaneous v_S reaches V_m at time $t = \pi/2\omega$. When the input voltage v_S tries to decrease, diode D_1 will be reverse biased and the capacitor C will discharge through resistance R. If we define the time $t = t_1$ when C is charged to V_m, the output (or capacitor) voltage v_O, which falls exponentially, takes the form

$$v_O(t) = V_m e^{-(t-t_1)/RC} \quad \text{for } t_1 \leq t \leq (t_1 + t_2) \tag{5.72}$$

The waveform of the output voltage is shown in Fig. 5.21(b). If the time constant $\tau = RC$ is too small, the capacitor will discharge its voltage very quickly and will not maintain its voltage close to V_m. The output voltage will be discontinuous and will not be a true representation of the peak input signal. On the other hand, if the time constant τ is too large, the output voltage will not change rapidly with a change in the peak value V_m of the input voltage. If the time constant τ is properly selected, the output voltage should approximately represent the peak input signal, within a reasonable error.

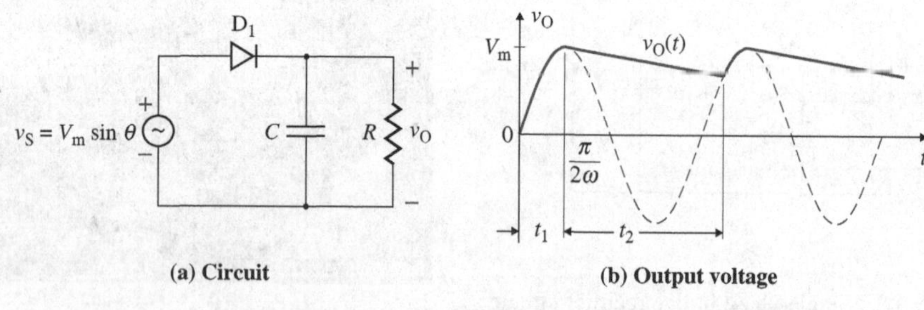

FIGURE 5.21 Peak detector

A peak detector can be used as a demodulator to detect the audio signal in an *amplitude modulated* (AM) radio signal. Amplitude modulation is a method of translating a low-frequency signal into a high-frequency one. The AM waveform can be described by

$$v_S(t) = V_m[1 + M \sin(2\pi f_m t)] \sin(2\pi f_c t) \tag{5.73}$$

where f_c = carrier frequency, in Hz
f_m = modulating frequency, in Hz
M = modulation index, whose value varies between 0 and 1
V_m = peak modulating voltage

The term $V_m[1 + M \sin(2\pi f_m t)]$ represents the envelope of the modulated waveform. Its slope (or rate of change) S is given by

$$S = \frac{d}{dt}[V_m + MV_m \sin(2\pi f_m t)] = M2\pi f_m V_m \cos(2\pi f_m t) \tag{5.74}$$

The waveform of a modulated signal is shown in Fig. 5.22(a). Since the demodulator gives the peak value, the corresponding output of the peak detector is shown in Fig. 5.22(b). A low-pass filter can be used to smooth the demodulated signals. With a proper choice of time constant $\tau = RC$, the output will trace each peak of the modulating signal. If the time constant is too large, the output will not be able to change fast enough and the audio signal will be distorted. If the time constant is too small, there will be too much "ripple" superimposed on the modulating signal.

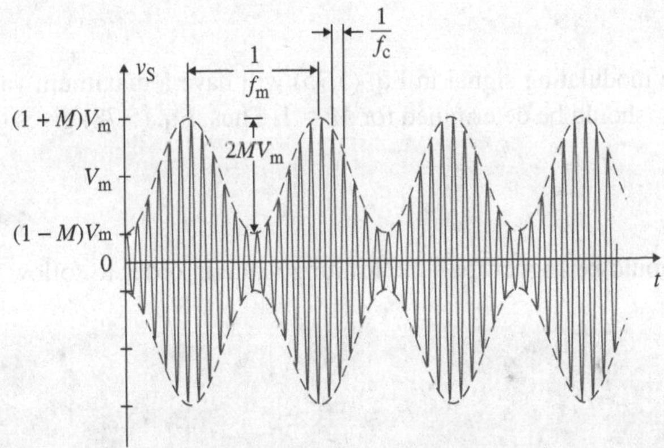

(a) Input voltage to demodulator

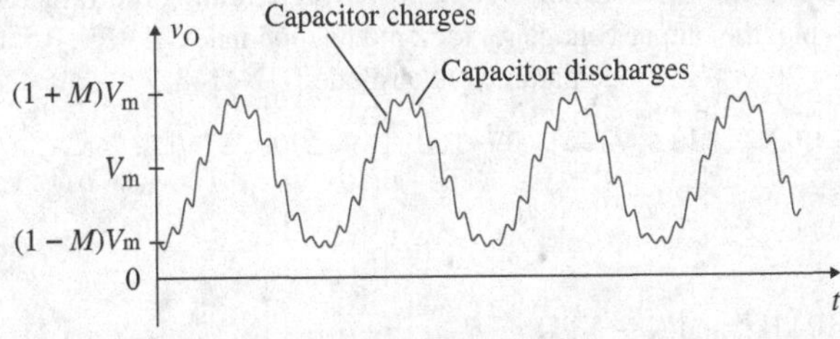

(b) Output voltage of demodulator

FIGURE 5.22 Amplitude modulated waveform

The slope S in Eq. (5.74) will be maximum at $\theta = 2\pi f_m t = 0$ or π. Therefore, the peak slope (or rate of change) S_m is given by

$$S_m = \pm M 2\pi f_m V_m \tag{5.75}$$

From Eq. (5.72), we can find the peak slope S_D of the detector as

$$S_D = \left.\frac{dv_O}{dt}\right|_{t=t_1} = -V_m \frac{1}{RC} e^{-(t-t_1)/RC}\bigg|_{t=t_1} = -\frac{V_m}{RC} \tag{5.76}$$

For the detector to cope with a rapid change in the peak input voltage, the magnitude of the slope S_D of the detector must be greater than that of the modulating signal. That is,

$$|S_D| \geq |S_m|$$

Substituting $S_m = -2\pi M f_m V_m$ from Eq. (5.75) under the falling slope condition and S_D from Eq. (5.76), we get

$$\left|-\frac{V_m}{RC}\right| \geq |-2\pi\, M f_m V_m| \tag{5.77}$$

which gives the desired value of capacitance C as

$$C \geq \frac{1}{2\pi f_m M R} \tag{5.78}$$

The peak slope S_m of the modulating signal in Eq. (5.75) will have a maximum value if $M = 1$. Therefore, the value of capacitance C should be determined for $M = 1$. Thus, Eq. (5.78) gives the limiting value of C as

$$C \geq \frac{1}{2\pi f_m R} \tag{5.79}$$

The design value of C should be higher than the limiting value in order to follow the peaks.

EXAMPLE 5.12

D **Designing a demodulator circuit** The carrier frequency f_c of a radio signal is 100 kHz, and the modulating frequency f_m is 10 kHz. The load resistance R of the detector is 5 kΩ.

(a) Design a demodulator for the waveform of Fig. 5.22(a) by determining the value of capacitance C.

(b) Use PSpice/SPICE to plot the output voltage v_O for a modulation index of $M = 0.5$ and 1.0. The peak modulating voltage is $V_m = 20$ V. Use diode parameters of diode D1N4148:

```
IS=2.682N CJO=4P M=.3333 VJ=.5 BV=100 IBV=100U TT=11.54N
```

SOLUTION

(a) $f_c = 100$ kHz, $f_m = 10$ kHz, and $R = 5$ kΩ.

From Eq. (5.79), $C = \dfrac{1}{2\pi f_m R} = \dfrac{1}{2\pi \times 10 \times 10^3 \times 5 \times 10^3} = 1605$ nF

(b) PSpice allows sine functions only, so we need to convert the cosine term into a sine term. Using the trigonometric relationship

$$\sin A \sin B = \frac{1}{2}[\cos(A - B) - \cos(A + B)]$$

we can expand Eq. (5.73) to

$$\begin{aligned} v_S(t) &= V_m \sin(2\pi f_c t) + MV_m \sin(2\pi f_m t)\sin(2\pi f_c t) \\ &= V_m \sin(2\pi f_c t) + \frac{MV_m}{2}\cos[2\pi(f_c - f_m)t] - \frac{MV_m}{2}\cos[2\pi(f_c + f_m)t] \\ &= V_m \sin(2\pi f_c t) + \frac{MV_m}{2}\sin[2\pi(f_c - f_m)t + 90°] - \frac{MV_m}{2}\sin[2\pi(f_c + f_m)t + 90°] \end{aligned} \tag{5.80}$$

For $M = 0.5$, $MV_m/2 = 0.5 \times 20/2 = 5$ V

$$f_1 = f_c - f_m = 100 \text{ kHz} - 10 \text{ kHz} = 90 \text{ kHz}$$

$$f_2 = f_c + f_m = 100 \text{ kHz} + 10 \text{ kHz} = 110 \text{ kHz}$$

The demodulator circuit for PSpice simulation is shown in Fig. 5.23. The PSpice plot of v_O, shown in Fig. 5.24, gives the peak value of output voltage as $V_{o(peak)} = 29.1$ V, compared to the calculated value of $(1 + m)V_m = (1 + 0.5) \times 20 \text{ V} = 30$ V.

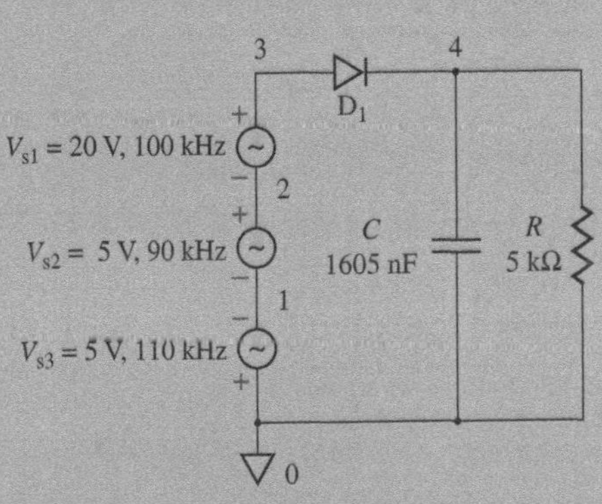

FIGURE 5.23 Demodulator circuit for PSpice simulation

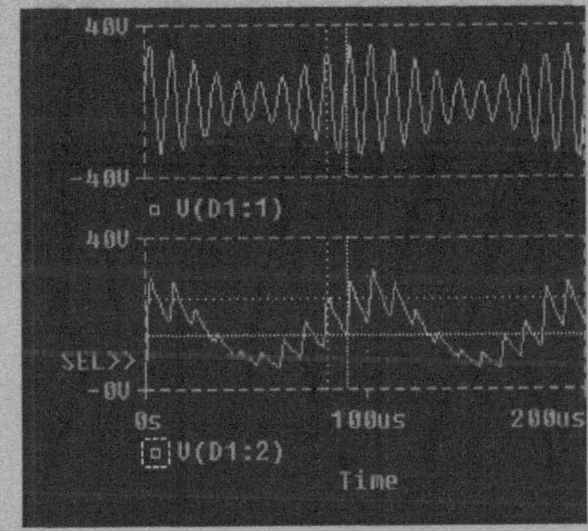

FIGURE 5.24 PSpice plot for Example 5.12

KEY POINTS OF SECTION 5.4

- A diode can charge a capacitor to the peak value of the input voltage and thus can be used as a peak detector.
- A peak detector can be used as a demodulator to detect the audio signal in an amplitude modulated (AM) radio signal.

5.5 Diode Clippers

A clipper is a limiting circuit; it is basically an extension of the half-wave rectifier. The output of a clipper circuit looks as if a portion of the output signal was cut off (clipped). Although the input voltage can have any waveform, we will assume that the input voltage is sinusoidal, $v_S = V_m \sin \omega t$, in order to describe the output voltage. Clippers can be classified into two types: parallel clippers and series clippers. The diode can be connected either in series or in parallel with the load.

5.5.1 Parallel Clippers

A clipper in which the diode is connected across the output terminals is known as a *parallel clipper* because the diode will be in parallel (or shunt) with the load. In a shunt connection, elements are connected in parallel such that each element carries a different current. Some examples of parallel clipper circuits and their corresponding output waveforms are shown in Fig. 5.25. The resistance R limits the diode current when the diode conducts. In determining the output waveform of a clipper, it is important to keep in mind that a diode will conduct only if the anode voltage is higher than the cathode voltage.

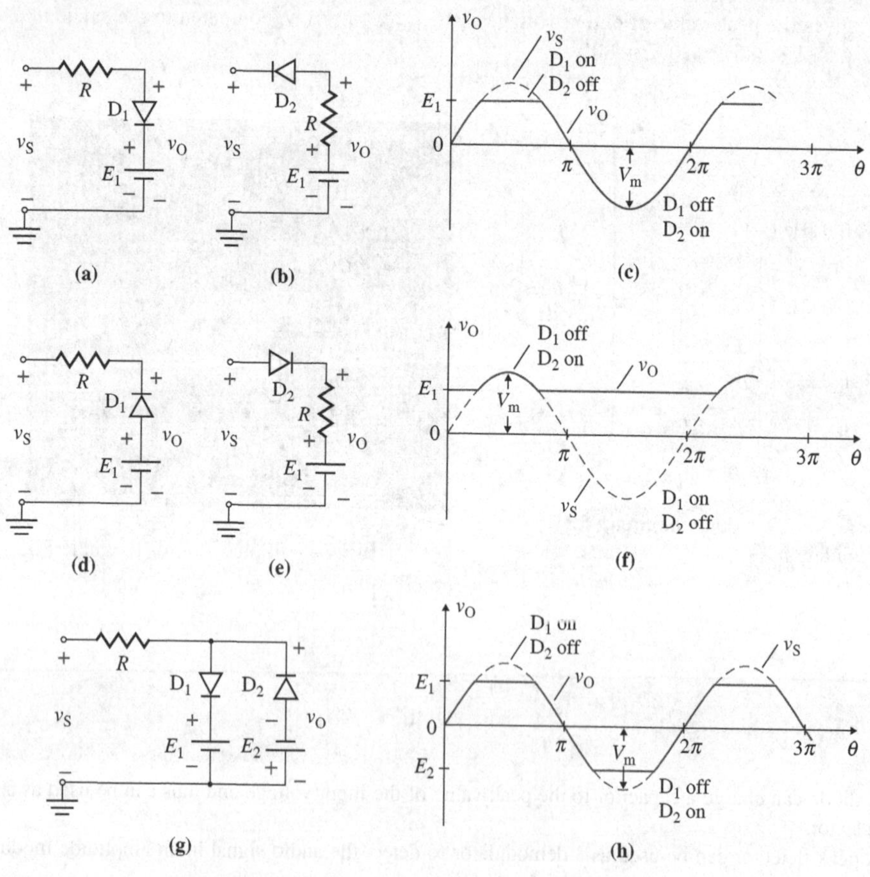

FIGURE 5.25 Diode parallel clipper circuits

When diode D_1 in Fig. 5.25(a) is off, the instantaneous output voltage v_O equals the instantaneous input voltage v_S. Diode D_1 will conduct for the portion of the positive half-cycle during which the instantaneous input voltage v_S is higher than the battery voltage E_1. On the other hand, diode D_2 in Fig. 5.25(b) will conduct when the input voltage is less than the battery voltage E_1. Although the output waveforms of these two circuits are identical, as shown in Fig. 5.25(c), diode D_2 in Fig. 5.25(b) remains on for a longer time than diode D_1 in Fig. 5.25(a). For this reason, the clipper of Fig. 5.25(a) is preferable to that of Fig. 5.25(b).

Diode D_1 in Fig. 5.25(d) will conduct most of the time and be off for the portion of the positive half-cycle during which the instantaneous input voltage v_S is higher than the battery voltage E_1 whereas diode D_2 will remain on for a short time. The output waveforms for the clippers of Fig. 5.25[(d) and (e)] are identical, as shown in Fig. 5.25(f).

The circuits of Fig. 5.25[(a) and (d)] (with E_1 reversed and renamed as E_2) can be combined to form a two-level clipper, as shown in Fig. 5.25(g). The positive and negative voltages are limited to E_1 and E_2, respectively, as shown in Fig. 5.25(h). One battery terminal of the clippers in Fig. 5.25 is common to the ground.

5.5.2 Series Clippers

A clipper in which the diode forms a series circuit with the output terminals is known as a *series clipper.* The current-limiting resistance R can be used as a load, as shown in Fig. 5.26(a). If the direction of the battery is reversed, the negative part of the sine wave is clipped as shown in Fig. 5.26(b). If the direction of

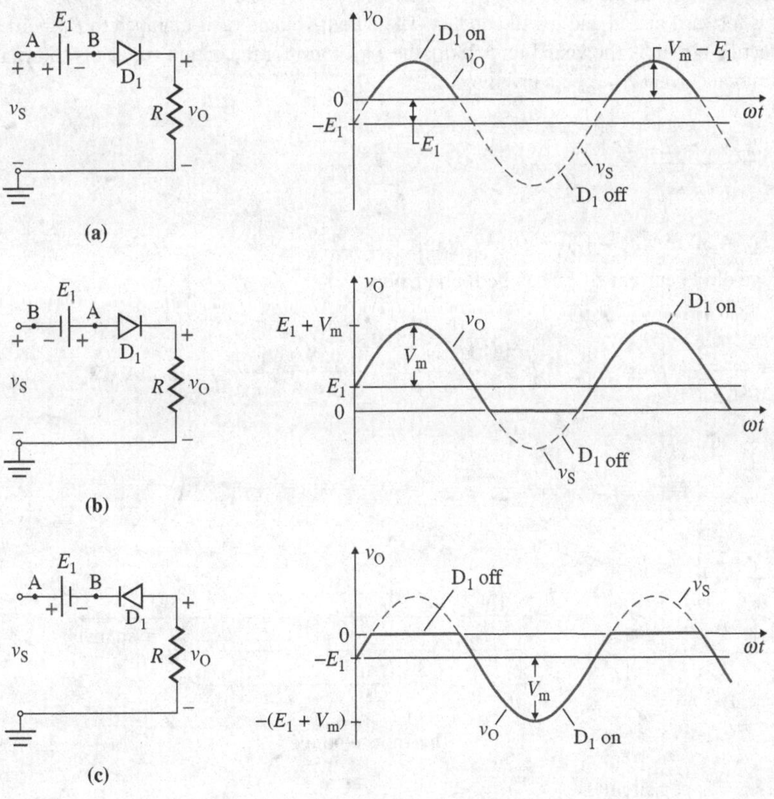

FIGURE 5.26 Diode series clipper circuits

the diode is reversed, the clipping becomes the opposite of that in Fig. 5.26(a); this situation is shown in Fig. 5.26(c). The potential difference between terminals A and B of the battery must be E_1. But terminal B cannot be at zero or ground potential. Therefore, these circuits require an isolated DC voltage (or battery) of E_1. Note that the zero level of the output voltage v_O is different from that of the input voltage v_S and is shifted by an amount equal to E_1.

EXAMPLE 5.13

D **Designing a clipper circuit** The clipper circuit shown in Fig. 5.27(a) is supplied from the input voltage shown in Fig. 5.27(b). The battery voltage is $E_1 = 10$ V. The peak diode current $I_{D(peak)}$ is to be limited to 30 mA. Determine **(a)** the value of resistance R, **(b)** the average diode current $I_{D(av)}$ and the rms diode current $I_{D(rms)}$, and **(c)** the power rating P_R of the resistance R.

SOLUTION

$I_{D(peak)} = 30$ mA, and $E_1 = 10$ V. Imagine a line at $E_1 = 10$ V on the plot of v_S in Fig. 5.27(b).

(a) During the period $0 \le t \le t_1$, the input voltage v_S is 20 V. Diode D_1 is reverse biased, and the output voltage v_O becomes the same as the input voltage v_S. That is, $v_O = v_S = 20$ V. During the period $t_1 \le t \le (t_1 + t_2)$, diode D_1 is forward biased and it will conduct. The output voltage v_O is clamped to $E_1 = 10$ V. The equivalent conducting circuit is shown in Fig. 5.28(a); the waveform for the output voltage is shown in Fig. 5.28(b). The peak diode current $I_{D(peak)}$ is given by

$$I_{D(peak)} = \frac{V_m + E_1}{R} = \frac{20 + 10}{R}$$

For $I_{D(peak)} = 30$ mA, $R = (20 + 10)$ V/30 mA $= 1$ kΩ.

(b) The average diode current $I_{D(av)}$ can be found from

$$I_{D(av)} = \frac{1}{t_1 + t_2}\int_{t_1}^{t_1+t_2} I_{D(peak)}\,dt = \frac{I_{D(peak)}t_2}{t_1 + t_2} = \frac{30\text{ mA} \times 6\text{ ms}}{4\text{ ms} + 6\text{ ms}} = 18\text{ mA}$$

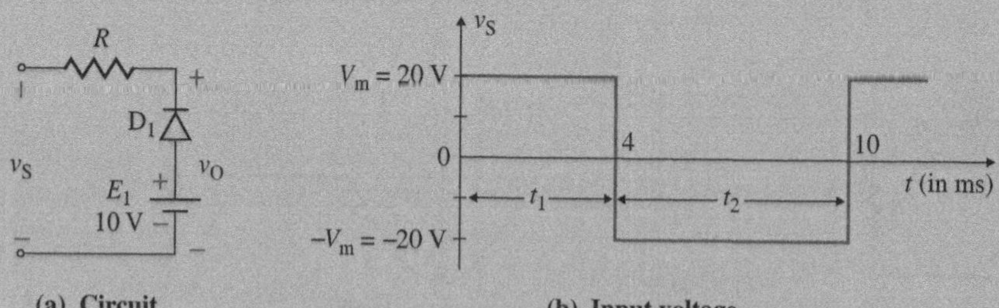

(a) Circuit

(b) Input voltage

FIGURE 5.27 Clipper circuit

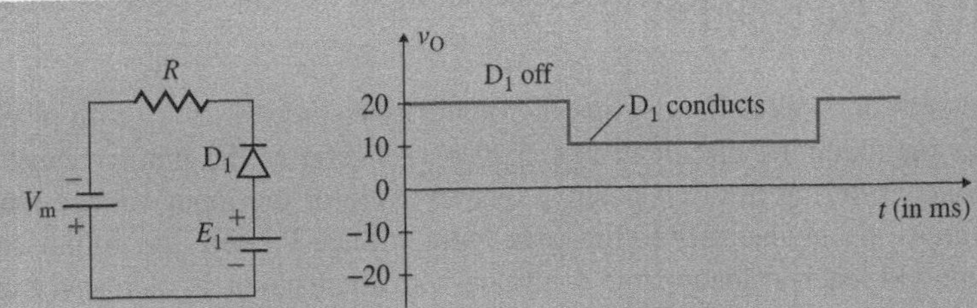

(a) **Equivalent conducting circuit** (b) **Output voltage**

FIGURE 5.28 Equivalent circuit and waveforms for Example 5.13

The rms diode current $I_{D(rms)}$ can be found from

$$I_{D(rms)} = \left[\frac{1}{t_1 + t_2}\int_{t_1}^{t_1+t_2} I_{D(peak)}^2\,dt\right]^{1/2} = I_{D(peak)}\left[\frac{t_2}{t_1 + t_2}\right]^{1/2}$$

$$= 30\text{ mA}\sqrt{\frac{6\text{ mA}}{10\text{ mA}}} = 23.24\text{ mA}$$

(c) Then

$$P_R = I_{D(rms)}^2 R = (23.24 \times 10^{-3}\text{ A})^2 \times 1\text{ k}\Omega = 0.54\text{ W}$$

KEY POINTS OF SECTION 5.5

- A diode clipper can cut off a portion of its output voltage.
- If the diode forms a series circuit with the load, it is called a series clipper. If the diode forms a parallel circuit with the load, it is called a parallel clipper.
- The output voltage of a clipper can be determined as follows:

Step 1. Draw a clockwise loop to determine the polarity of the battery. If the positive terminal of the battery is encountered first, then E_1 is positive. If the negative terminal is encountered first, then E_1 is negative.

Step 2. Draw a line at $\pm E_1$ on the plot of the input voltage.

Step 3. Find out when the diode will conduct. Then clip the appropriate portion of the input voltage, depending on the state of the diode (on or off), in order to obtain the output voltage v_O.

Step 4. Draw the final output voltage.

5.6 Diode Clamping Circuits

A clamping circuit simply shifts the output waveform to a different DC level. Thus, it is often known as a *level shifter.* The shapes of the input and output waveforms are identical; only the DC level is shifted. The input voltage can have any shape. However, we will assume that the input voltage is sinusoidal, $v_S = V_m \sin \omega t$. Clampers can be classified into two types: fixed-shift clampers and variable-shift clampers.

5.6.1 Fixed-Shift Clampers

As shown in Fig. 5.29, a fixed-shift clamper shifts the output voltage by an amount $\pm V_m$ with respect to the zero level. Let us consider the clamping circuit in Fig. 5.29(a). As soon as the input voltage v_S is switched on, diode D_1 will conduct during the first positive quarter-cycle of the input voltage, and the capacitor C will be charged almost instantaneously to the peak input voltage V_m. But the output voltage will be zero, $v_O \approx 0$. The circuit will reach a steady-state condition with a voltage of V_m across the capacitor C, as depicted in Fig. 5.29(a). Therefore, after the first quarter-cycle, the capacitor voltage will be $v_C = V_m$, and the output voltage v_O will become

$$v_O = v_S - v_C = v_S - V_m = V_m \sin \omega t - V_m = V_m (\sin \omega t - 1) \quad \text{for } \omega t \geq \frac{\pi}{2}$$

as shown in Fig. 5.29(b).

Let us assume that the input voltage v_S falls below the initial peak voltage of V_m (say, 20 V) to a new peak value of V_{m1} (say, 10 V). This situation is shown in Fig. 5.29(c). The diode voltage is now $v_O = v_S - v_C = 10 \sin \omega t - 20$, which is negative for all ωt, and the diode becomes reverse biased.

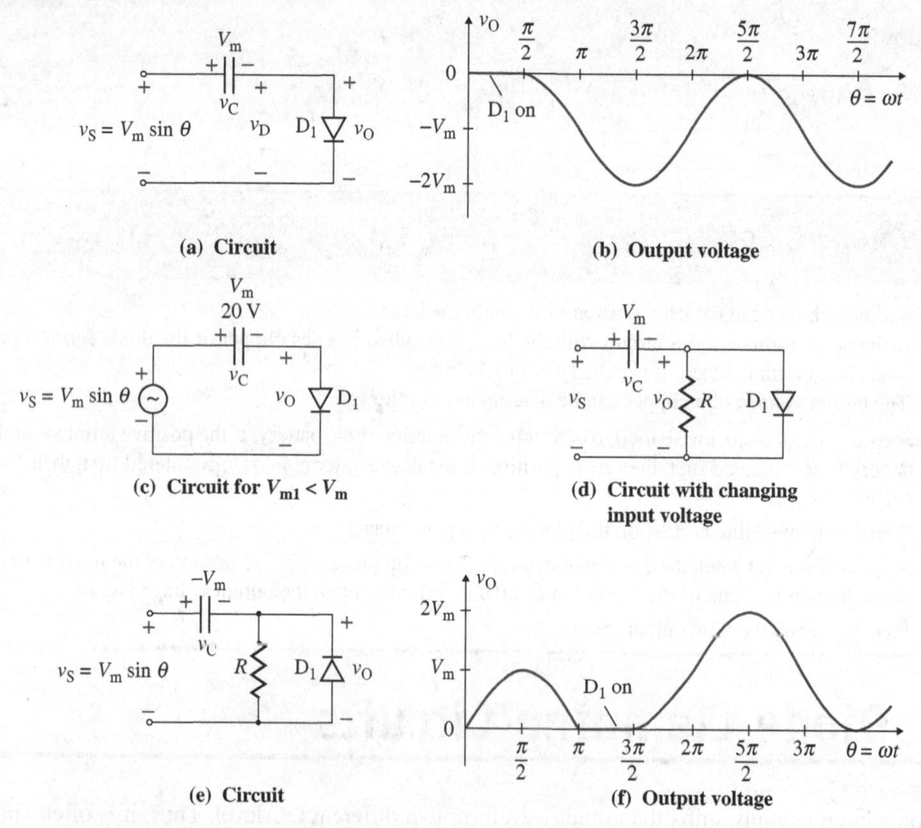

FIGURE 5.29 Fixed-shift clamping circuit

The capacitor voltage cannot adjust to the new value V_{m1} because diode D_1 is now reverse biased and there is no discharge path for the capacitor. The output voltage will be $v_O = V_m - V_{m1} \sin \omega t$, instead of $v_O = V_{m1}(\sin \omega t - 1)$ as expected. To allow the capacitor voltage to adjust to the change in the peak input voltage, a resistance R is connected across diode D_1, as shown in Fig. 5.29(d). If the input voltage then falls to a new peak, the capacitor C can discharge slowly through the resistance R. Similarly, if the input voltage is increased to a new peak, the capacitor C can charge through the resistance R. However, the voltage across the capacitor must remain fairly constant during the whole period. The values of R and C must be chosen such that the time constant $\tau = RC$ is large enough to ensure that the capacitor voltage does not change significantly within one period T of the input voltage. This condition is generally satisfied by making the time constant τ equal to 10 times the period T. That is, $\tau = 10T$.

If the direction of diode D_1 is reversed, as shown in Fig. 5.29(e), the diode will be reverse biased during the first positive half-cycle of the input voltage, and the output voltage will be equal to the input voltage, $v_O = v_S$. Diode D_1 will conduct during the first negative half-cycle of the input voltage. The capacitor C will be charged almost instantaneously to the negative peak input voltage $-V_m$, and the output voltage will become zero, $v_O = 0$. This process is completed during the first cycle, and the circuit reaches a steady-state condition with an input voltage of $-V_m$ across the capacitor C. After the first cycle, the capacitor voltage remains constant at $v_C = -V_m$. The output voltage v_O under steady-state conditions becomes

$$v_O = v_S - v_C = v_S - (-V_m) = V_m \sin \omega t + V_m = V_m (\sin \omega t + 1) \quad \text{for } \omega t \geq \frac{3\pi}{2}$$

as shown in Fig. 5.29(f). Therefore, switching the direction of the diode makes the output inverted with a phase shift of π.

▶ **NOTE** If we ignore the initial transient interval, which is required to charge the capacitor for normal operation, the output waveform of the clamping circuit in Fig. 5.29(f) becomes positive with respect to that in Fig. 5.29(b). That is, one shifts the input signal in the positive direction and the other shifts it in the negative direction.

5.6.2 Variable-Shift Clampers

The output voltage v_O can be shifted to a predefined value by introducing a battery voltage E_1. This type of clamper shown in Fig. 5.30 shifts the output voltage by an amount $\pm V_m \pm E_1$ with respect to the zero level. Consider the clamping circuit in Fig. 5.30(a). The capacitor C will be charged to $v_C = V_m - E_1$ during the first positive quarter-cycle of the input voltage, and the instantaneous output voltage v_O under steady-state conditions becomes

$$v_O = v_S - v_C = V_m \sin \omega t - (V_m - E_1) = V_m \sin \omega t - V_m + E_1 \quad \text{for } \omega t \geq \frac{\pi}{2}$$

The capacitor C in Fig. 5.30(b) will be charged to $v_C = -(V_m + E_1)$ during the first negative quarter-cycle of the input voltage. There will be an instantaneous charging to $-E_1$ at $t = 0$. Thus, the instantaneous output voltage v_O under steady-state conditions becomes

$$v_O = v_S - v_C = v_m \sin \omega t + (V_m + E_1) = V_m \sin \omega t + V_m + E_1 \quad \text{for } \omega t \geq \frac{3\pi}{2}$$

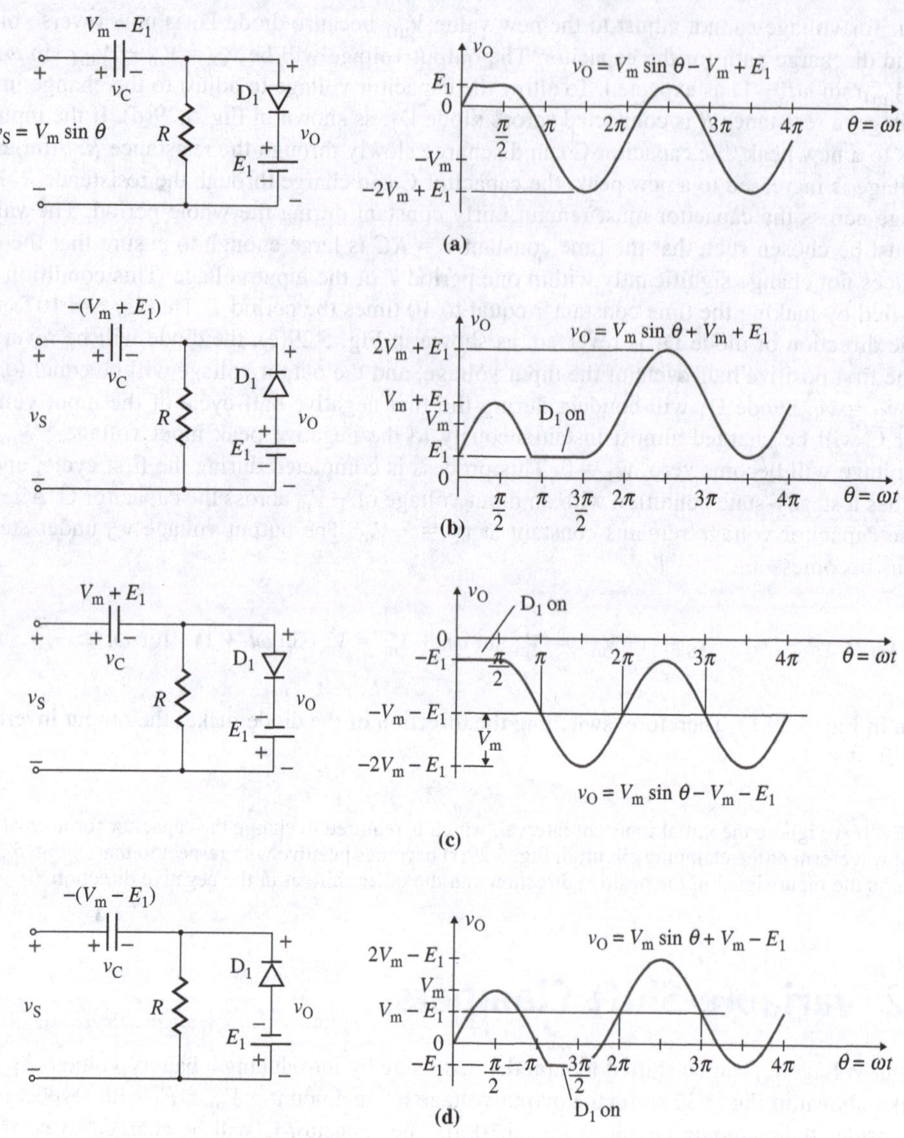

FIGURE 5.30 Variable-shift clamping circuits

The capacitor C in Fig. 5.30(c) will be charged to $v_C = (V_m + E_1)$ during the first positive quarter-cycle of the input voltage. The instantaneous output voltage v_O under steady-state conditions becomes

$$v_O = v_S - v_C = V_m \sin \omega t - (V_m + E_1) = V_m \sin \omega t - V_m - E_1 \quad \text{for } \omega t \geq \frac{\pi}{2}$$

The capacitor C in Fig. 5.30(d) is charged to $v_C = -(V_m - E_1)$ during the first negative quarter-cycle of the input voltage. The instantaneous output voltage v_O under steady-state conditions becomes

$$v_O = v_S - v_C = V_m \sin \omega t + (V_m - E_1) = V_m \sin \omega t + V_m - E_1 \quad \text{for } \omega t \geq \frac{3\pi}{2}$$

EXAMPLE 5.14

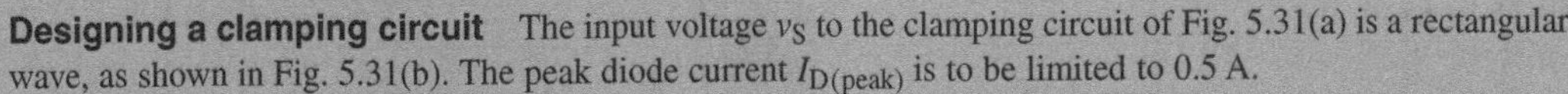

D **Designing a clamping circuit** The input voltage v_S to the clamping circuit of Fig. 5.31(a) is a rectangular wave, as shown in Fig. 5.31(b). The peak diode current $I_{D(peak)}$ is to be limited to 0.5 A.

(a) Design the clamping circuit by determining the peak inverse voltage (PIV) of the diode and the values of R_S, R, and C.

(b) Use PSpice/SPICE to plot the output voltage v_O. Use diode parameters of diode D1N4148:

```
IS=2.682N CJO=4P M=.3333 VJ=.5 BV=100 IBV=100U TT=11.54N
```

SOLUTION

(a) $I_{D(peak)} = 0.5$ A. The period T of the input waveform is $T = t_1 + t_2 = 6 \text{ ms} + 8 \text{ ms} = 14$ ms.

$$\text{PIV} = -v_S + v_C - E_1 = 10 + 25 - 5 = 30 \text{ V} \qquad \text{for } 6 \text{ ms} \le t \le 14 \text{ ms}$$

For $0 \le t \le 6$ ms, $v_O = -E_1 = -5$ V and for 6 ms $\le t \le$ 14 ms, $v_O = -25 - 10 = -35$ V. The waveform of the output voltage v_o is shown in Fig. 5.31(c).

The peak diode current $I_{D(peak)}$ is given by

$$I_{D(peak)} = \frac{20 + E_1}{R_s}$$

or

$$R_s = \frac{20 + E_1}{I_{D(peak)}} = \frac{20 + 5}{0.5} = 50 \ \Omega$$

Let $\tau = (R + R_s)C = 10T = 10 \times 14 \text{ ms} = 140$ ms. Choose a suitable value of C. Let $C = 0.1$ μF. Then

$$R + R_s = \frac{\tau}{C} = \frac{140 \times 10^{-3}}{0.1 \times 10^{-6}} = 1.4 \text{ M}\Omega$$

which gives $R = 1.4 \text{ M}\Omega - R_s = 1.4 \text{ M}\Omega - 50 \ \Omega \approx 1.4 \text{ M}\Omega$.

(b) The clamping circuit for PSpice simulation is shown in Fig. 5.32. The PSpice plot of v_O, shown in Fig. 5.33, gives the peak-to-peak output voltage as $V_{o(pp)} = 29.99$ V, compared to the calculated value of 30 V.

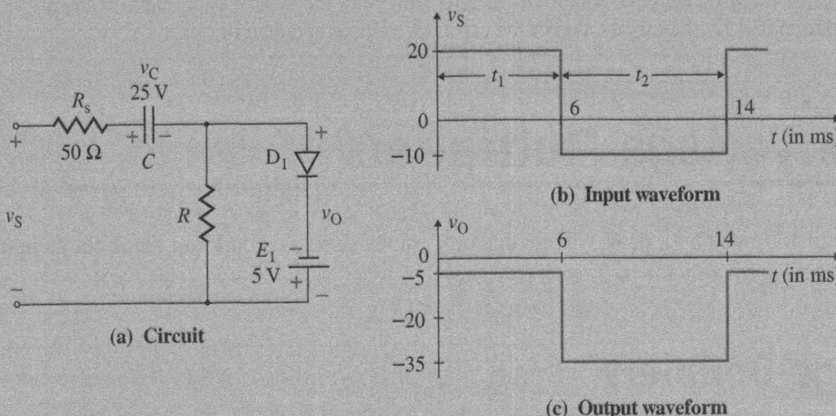

FIGURE 5.31 Circuit for Example 5.14

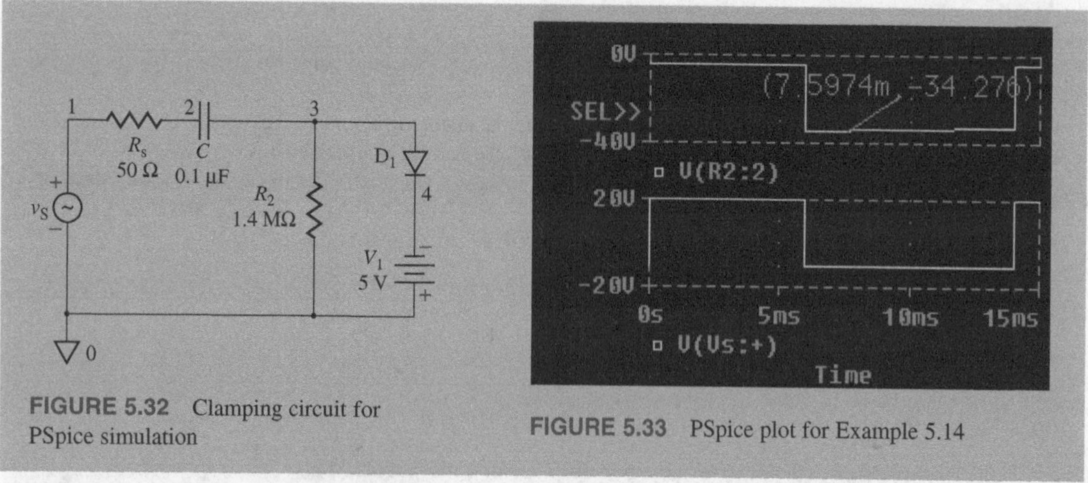

FIGURE 5.32 Clamping circuit for PSpice simulation

FIGURE 5.33 PSpice plot for Example 5.14

KEY POINTS OF SECTION 5.6

- A clamping circuit can shift the output waveform to a different DC level by either a fixed or a variable amount with respect to the zero level.
- A capacitor is initially charged through the diode to the peak input voltage during the positive or negative half-cycle of the input voltage. After the completion of the initial charging process, the capacitor voltage is in series with the input voltage. Thus, the output voltage becomes the algebraic sum of the input voltage and the capacitor voltage. That is, the capacitor voltage is added to (or subtracted from) the input voltage to produce the output voltage.
- The output voltage of a clamping circuit can be determined as follows:

Step 1. Start with the time interval of the input voltage so that the diode is forward biased. Then determine the magnitude and direction of the initial capacitor voltage $V_c = \pm V_m \pm E_1$.

Step 2. Add (or subtract) this capacitor voltage from the instantaneous input voltage v_S to obtain the instantaneous output voltage v_O.

Step 3. Then draw the instantaneous output voltage. To draw only the steady-state output voltage, just shift the input voltage by the initial value of the capacitor voltage obtained in step 1.

5.7 Diode Voltage Multipliers

A diode clamping circuit followed by a peak voltage detector can be used as a building block for stepping up the peak input voltage V_m by a factor of 2, 3, 4, or more.

5.7.1 Voltage Doublers

A half-wave voltage doubler circuit, shown in Fig. 5.34(a), uses a clamping circuit and a peak detector. Let us consider a sinusoidal input voltage of $v_S = V_m \sin \omega t$. The circuit operation can be divided into four intervals: interval 1, interval 2, interval 3, and interval 4.

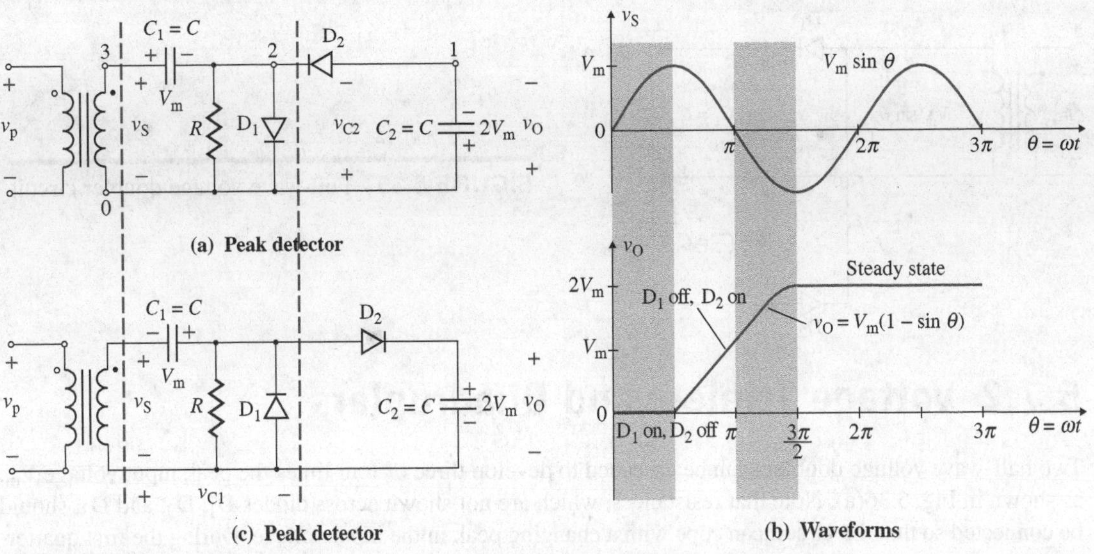

FIGURE 5.34 Half-wave voltage doubler circuit

Interval 1 is the interval $0 \leq \omega t \leq \pi/2$. As soon as the input voltage is switched on, diode D_1 will conduct, but diode D_2 will be reverse biased. The output voltage is $v_O = 0$. The capacitor C_1 will be charged during the first quarter-cycle to V_m (at $\omega t = \pi/2$) with the polarities shown.

Interval 2 is the interval $\pi/2 \leq \omega t \leq \pi$. D_1 will be off, and D_2 will be on. If the value of R is large enough that $RC > 1/f$, where f = supply frequency, then capacitor C_1 will not have time to discharge through R and the voltage on capacitor C_1 will remain approximately at V_m.

Interval 3 is the interval $\pi \leq \omega t \leq 3\pi/2$. The polarity of the input voltage is negative. Diode D_1 will be off, and diode D_2 will conduct. The output voltage v_O, which will be the same as the voltage across capacitor C_2, will become $v_O = v_{C1} - v_S = V_m - V_m \sin \omega t$. At $\omega t = 3\pi/2$, the output voltage will become $2V_m$ and the capacitor C_2 will be charged to $2V_m$.

Interval 4 is the interval $3\pi/2 \leq \omega t \leq 2\pi$. Diodes D_1 and D_2 will be off. The voltage on capacitor C_1 will be $v_{C1} = V_m$, and that on capacitor C_2 will be $v_{C2} = 2V_m$. However, we have assumed that capacitor C_1 acts as the voltage source of V_m and contributes to charging C_2. In fact, C_1 and C_2 form a series circuit and share $2V_m$, so the voltage on capacitor C_2 will be less than $2V_m$. It will take a couple of cycles before the steady-state condition is reached.

The waveforms for instantaneous input and output voltages are shown in Fig. 5.34(b). If the directions of the diodes are reversed, as shown in Fig. 5.34(c), the polarities of the output voltage will also be reversed. If a load resistance R_L is connected across capacitor C_2, the output voltage will fall when D_1 is off and will rise when D_2 is on. More time will be required to reach the steady-state condition.

Figure 5.35 shows a full-wave voltage doubler circuit. During the first quarter-cycle, v_S is positive, diode D_1 will conduct, and diode D_2 will be reverse biased, thereby causing the capacitor C_1 to be charged to $v_{C1} = V_m$ with polarities as shown. During the third quarter-cycle, v_S is negative, diode D_1 is reverse biased, and diode D_2 will conduct. Thus capacitor C_2 will be charged to $v_{C2} = V_m$ with polarities as shown. The steady-state output voltage after a complete cycle will be $v_O = 2V_m$. If a load resistance R_L is connected across the output, the effective capacitance seen by the load is $C = (C_1 \parallel C_2)$, which will be less than C_2 for the half-wave doubler circuit of Fig. 5.34(a). A lower value of effective capacitance indicates poorer filtering than that provided by a single capacitor filter. The peak inverse voltage PIV of the diodes in Figs. 5.34 and 5.35 will be $2V_m$.

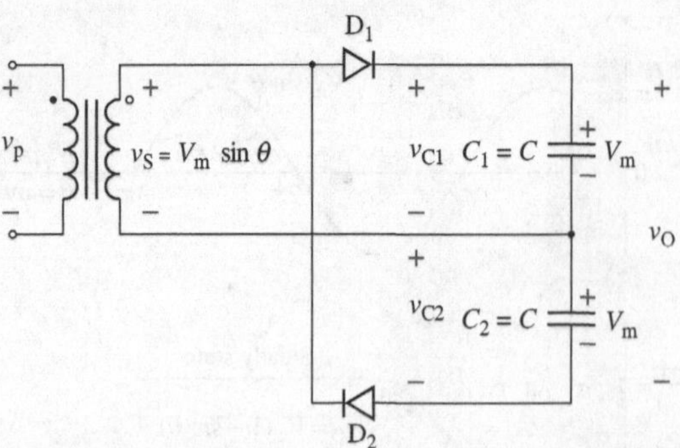

FIGURE 5.35 Full-wave voltage doubler circuit

5.7.2 Voltage Triplers and Quadruplers

Two half-wave voltage doublers can be cascaded to develop three or four times the peak input voltage V_m, as shown in Fig. 5.36(a). Note that resistances, which are not shown across diodes D_1, D_2, and D_3, should be connected so that the circuit can cope with a changing peak in the input voltage. During the first quarter-cycle ($0 \le \omega t \le \pi/2$) of input voltage v_S, capacitor C_1 will be charged to V_m through D_1. During the third quarter-cycle ($\pi \le \omega t \le 3\pi/2$), capacitor C_2 will be charged to $2V_m$ through C_1 and D_2. During the fifth quarter-cycle ($2\pi \le \omega t \le 5\pi/2$), capacitor C_3 will be charged to $2V_m$ through C_1, C_2, and D_3. During the seventh quarter-cycle ($3\pi \le \omega t \le 7\pi/2$), capacitor C_4 will be charged to $2V_m$ through C_1, C_2, C_3, and D_4. Depending on the output connections, the steady-state output voltage can be V_m, $2V_m$, $3V_m$, or $4V_m$. The instantaneous output voltages across various terminals are shown in Fig. 5.36(b) (e.g., $v_{O1} = v_{C1}$, $v_{O2} = v_{C2}$, $v_{O3} = v_{C3}$, and $v_{O4} = v_{C4}$). If additional sections of diode and capacitor are used, each capacitor will be charged to $2V_m$. The peak inverse voltage PIV of each diode is $2V_m$. It will take a couple of cycles before the steady-state conditions are reached.

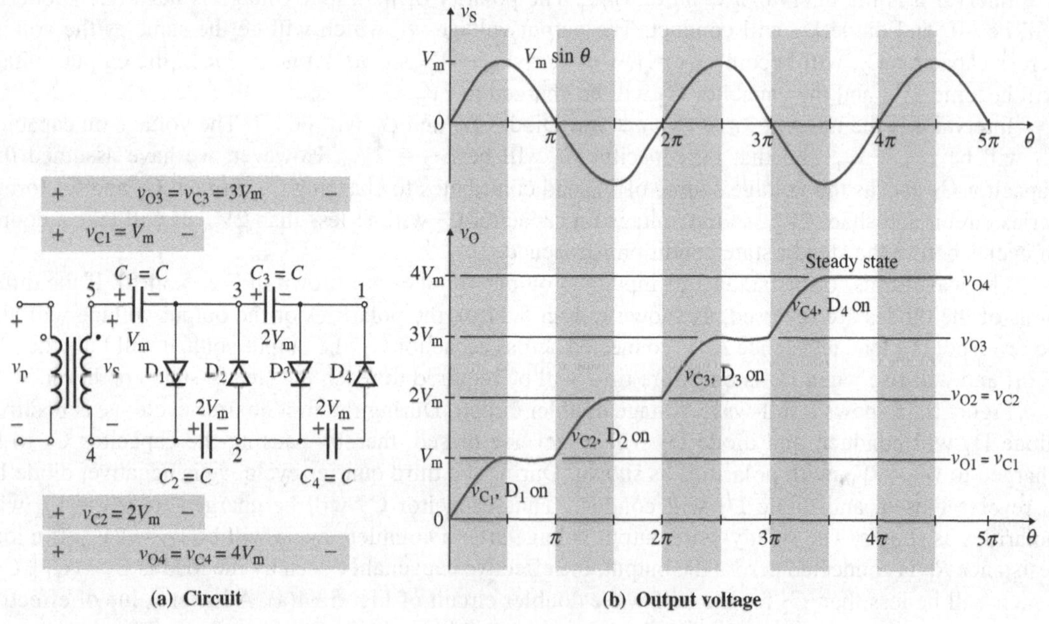

FIGURE 5.36 Voltage tripler and quadrupler

EXAMPLE 5.15

Voltage quadrupler circuit Use PSpice/SPICE to plot the output voltages v_{O2} and v_{O4} $(=v_{C4})$ for the voltage quadrupler in Fig. 5.36(a). Assume $v_S = 20 \sin 2000\pi t$ and $C_1 = C_2 = C_3 = C_4 = 0.1$ μF. Assume parameters of diode D1N4148:

```
IS=2.682N CJO=4P M=.3333 VJ=.5 BV=100 IBV=100U TT=11.54N
```

SOLUTION

The voltage quadrupler circuit for PSpice simulation is shown in Fig. 5.37. The PSpice plots of v_{O4}, shown in Fig. 5.38, give the peak output voltage as $V_{o4(peak)} = 76.59$ V, compared to the calculated value of $4V_m = 4 \times 20 = 80$ V. It takes a couple of cycles before steady-state conditions are reached.

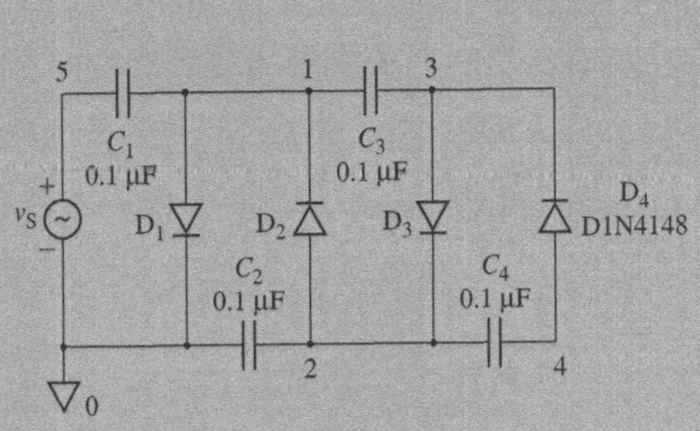

FIGURE 5.37 Voltage quadrupler circuit for PSpice simulation

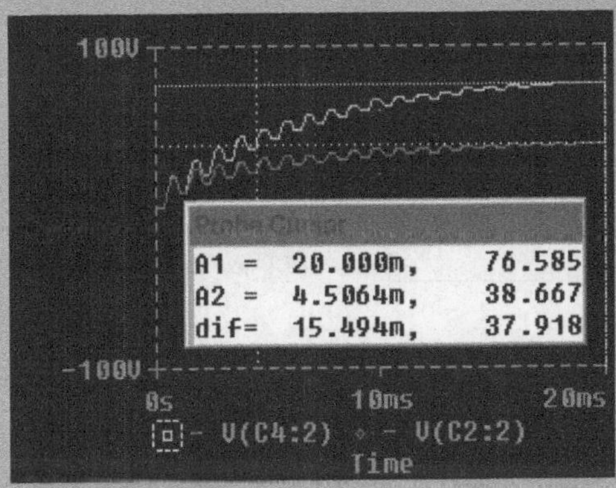

FIGURE 5.38 PSpice plots for Example 5.15

KEY POINTS OF SECTION 5.7

- A diode clamping circuit followed by a peak voltage detector can be used to multiply the peak input voltage V_m by a factor of 2, 3, or more.
- Each peak detector adds $2V_m$.

5.8 Diode Function Generators

Diodes can be employed to generate and synthesize driving-point functions, which refer to the *v-i* relations of two-port circuits. Some diode circuits for generating functions are shown in Fig. 5.39 [2, 5]. In deriving the transfer functions, it is important to keep in mind that a diode will conduct only when it is forward biased; it is off under reverse-biased conditions. The following guidelines will be helpful in analyzing the characteristics of diode function generators:

Step 1. To determine whether the diode is forward biased or reverse biased, assume that the diode is reverse biased and determine the anode-to-cathode voltage V_{AK} of the open diode.

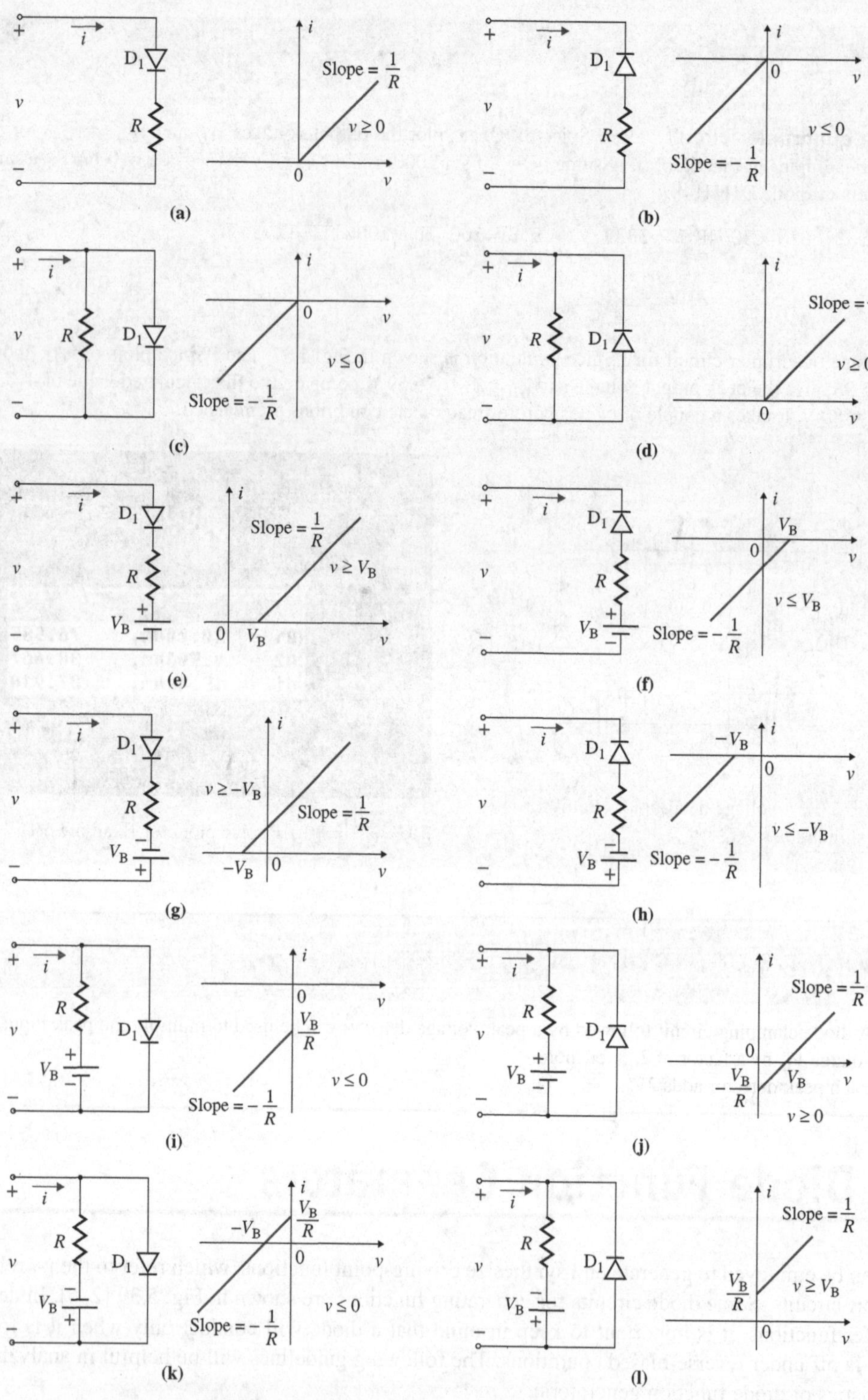

FIGURE 5.39 Diode circuits for function generation

Step 2. If V_{AK} is negative, the assumption that the diode is reverse biased is correct; proceed with the analysis.

Step 3. If V_{AK} is positive, the assumption that the diode is reverse biased is wrong. Replace the diode with a short circuit and reanalyze.

EXAMPLE 5.16

Finding the transfer function of a diode circuit A diode circuit is shown in Fig. 5.40(a). The circuit parameters are $R_1 = 5\ \text{k}\Omega$, $R_2 = 1.25\ \text{k}\Omega$, $R_3 = 1\ \text{k}\Omega$, $V_1 = 5$ V, and $V_2 = 8$ V.

(a) Plot the *v-i* relationship of the circuit.

(b) Use PSpice/SPICE to plot the transfer characteristic for $v_S = 0$ to 10 V. Assume parameters of diode D1N4148:

```
IS=2.682N CJO=4P M=.3333 VJ=.5 BV=100 IBV=100U TT=11.54N
```

SOLUTION

(a) If $v_S < V_1 = 5$ V, diodes D_1 and D_2 will be reverse biased. The input current i_S is described by

$$i_S = \frac{v_S}{R_1} = \frac{v_S}{5}\ \text{mA}$$

If $5 < v_S < 8$ V, diode D_1 conducts and diode D_2 is reverse biased. The input current i_S can be found from

$$i_S = i_1 + i_2 = \frac{v_S}{R_1} + \frac{v_S - V_1}{R_2} = v_S\left(\frac{1}{R_1} + \frac{1}{R_2}\right) - \frac{V_1}{R_2} = (v_S - 4)\ \text{mA}$$

If $v_S > 8$, both diodes D_1 and D_2 will conduct. The input current can be found from

$$i_S = \frac{v_S}{R_1} + \frac{v_S - V_1}{R_2} + \frac{v_S - V_2}{R_3} = v_S\left(\frac{1}{R_1} + \frac{1}{R_2} + \frac{1}{R_3}\right) - \frac{V_1}{R_2} - \frac{V_2}{R_3}$$

$$= 2v_S - 4 - 8 = (2v_S - 12)\ \text{mA}$$

The *v-i* plot of the relationship is shown in Fig. 5.40(b).

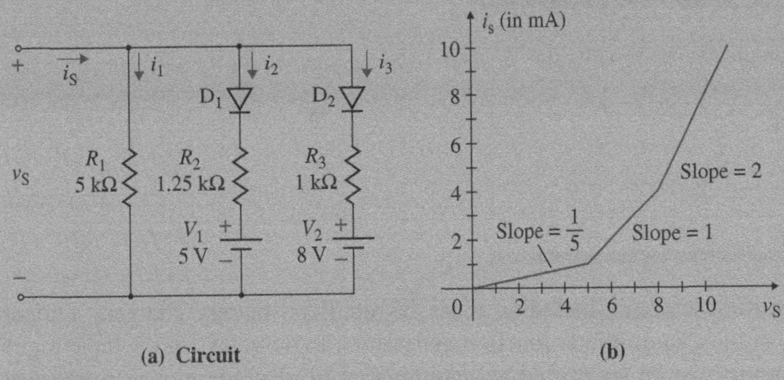

(a) Circuit (b)

FIGURE 5.40 Diode circuit for function generation

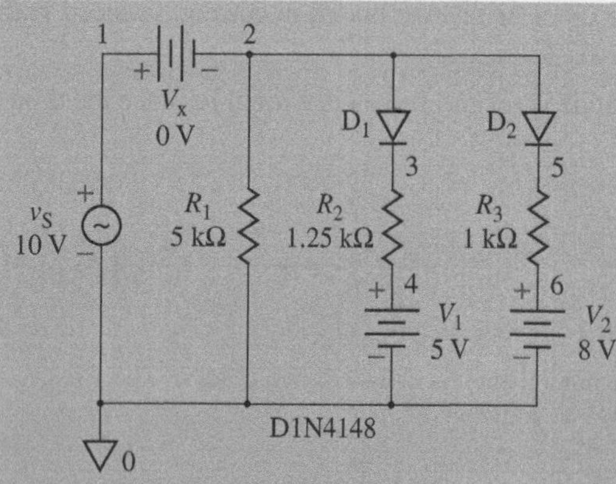

FIGURE 5.41 Function generator for PSpice simulation

(b) The function generator for PSpice simulation is shown in Fig. 5.41. The PSpice plot of i_S against v_S is shown in Fig. 5.42. The break voltages (8.23 V and 5.44 V) at which the diodes are switched into the circuits are higher than the estimated values because the diode drops were neglected in hand calculations, whereas PSpice uses real diodes.

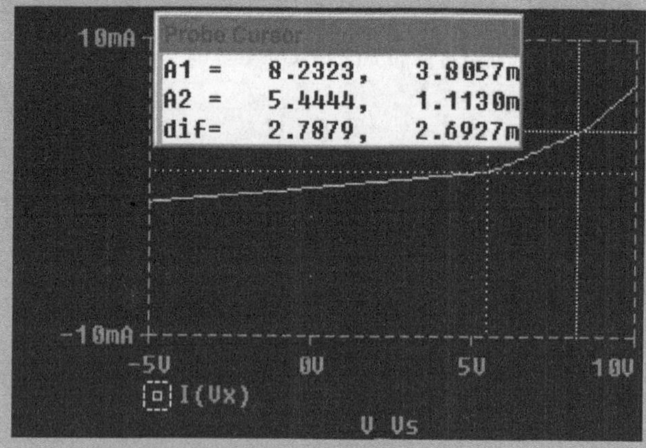

FIGURE 5.42 PSpice plot of transfer characteristic for Example 5.16

Summary

Diodes are used in many electronic circuits, including those of rectifiers, battery chargers, clippers, clampers, peak demodulators, voltage multipliers, function generators, logic gates, and voltage regulators. The analysis of diode circuits can be simplified by assuming an ideal diode model in which the resistance in the forward-biased condition is zero and the resistance in the reverse direction is very large, tending to infinity.

References

1. M. H. Rashid, *Power Electronics—Circuits, Devices and Applications.* Upper Saddle River, NJ: Prentice Hall, 2003.
2. R. R. Spencer and M. S. Ghausi, *Introduction to Electronic Circuit Design.* Upper Saddle River, NJ: Prentice Hall, 2006.
3. M. H. Rashid, *Introduction to SPICE Using OrCAD for Circuits and Electronics.* Englewood Cliffs, NJ: Prentice Hall, 2004.
4. B. S. Guru, *First Course in Electronics.* Deer Park, NY: Linus Publications, 2006.
5. M. S. Ghausi, *Electronic Devices and Circuits: Discrete and Integrated.* New York: Holt, Rinehart and Winston, 1985, p. 23.

Review Questions

1. What is a rectifier?
2. What is an AC–DC converter?
3. What is the efficiency of rectification?
4. What are the differences between half-wave and full-wave rectifiers?
5. What is the lowest frequency of harmonics in a half-wave rectifier?
6. What is the lowest frequency of harmonics in a full-wave rectifier?
7. What are the advantages of full-wave rectifiers?
8. What are the purposes of filters in rectifiers?
9. What is a DC filter?
10. What is an AC filter?
11. What is a clamper?
12. What is a clipper circuit?
13. What is a demodulator?
14. What is a voltage multiplier?
15. How is voltage multiplication accomplished?
16. What is the transfer characteristic of a diode circuit?

Problems

The symbol **D** indicates that a problem is a design problem.

5.2 Diode Rectifiers

5.1 The single-phase half-wave rectifier of Fig. 5.2(a) is supplied directly from a 120-V (rms), 60-Hz source through a step-down transformer with turns ratio $n = 10{:}1$. The load resistance R_L is 10 Ω. Determine **(a)** the average output voltage $V_{o(av)}$, **(b)** the average load current $I_{o(av)}$, **(c)** the rms load voltage $V_{o(rms)}$, **(d)** the rms load current $I_{o(rms)}$, **(e)** the ripple factor RF of the output voltage, **(f)** the rms ripple voltage $V_{r(rms)}$, **(g)** the average diode current $I_{D(av)}$, **(h)** the rms diode current $I_{D(rms)}$, **(i)** the peak inverse voltage PIV of the diode, **(j)** the average output power $P_{o(ac)}$, **(k)** the DC output power $P_{o(dc)}$, and **(l)** the frequency f_r of the output ripple voltage.

5.2 The single-phase half-wave rectifier of Fig. 5.1(a) is connected to a sinusoidal source of $V_s = 220$ V (rms), 50 Hz. Express the instantaneous output voltage $v_O(t)$ by a Fourier series.

5.3 The single-phase rectifier shown in Fig. 5.3(a) is employed as a battery charger. The battery capacity is 100 Wh, and the battery voltage is $E = 24$ V. The average charging current should be $I_{o(av)} = 5$ A. The primary AC input voltage is $V_p = 120$ V (rms), 60 Hz, and the transformer has a turns ratio of $n = 2:1$.

a. Calculate the conduction angle δ of the diode, the current-limiting resistance R, the power rating P_R of R, the charging time h in hours, the rectification efficiency η_R, and the peak inverse voltage PIV of the diode.

b. Use PSpice/SPICE to plot $P_{o(ac)}$ and $P_{o(dc)}$ as a function of time. Use default model parameters.

5.4 The input voltage to the single-phase bridge rectifier of Fig. 5.8(a) is shown in Fig. P5.4. Determine **(a)** the average voltage $V_{o(av)}$, **(b)** the rms output voltage $V_{o(rms)}$, and **(c)** the ripple factor RF of the output voltage. Assume a transformer with turns ratio $n = 1:1$.

FIGURE P5.4

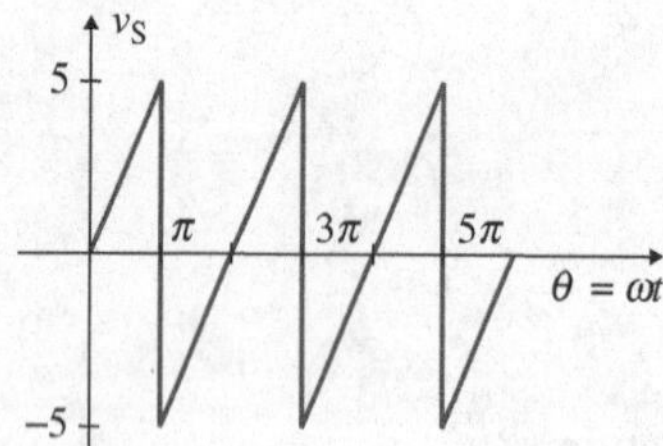

5.5 The single-phase full-wave center-tapped rectifier shown in Fig. 5.6(a) is supplied from a 220-V (rms), 50-Hz source through a step-down center-tapped transformer with turns ratio $n = 10:2$. The load resistance R_L is 10 Ω. Determine **(a)** the average output voltage $V_{o(av)}$, **(b)** the average load current $I_{o(av)}$, **(c)** the rms load voltage $V_{o(rms)}$, **(d)** the rms load current $I_{o(rms)}$, **(e)** the ripple factor RF of the output voltage, **(f)** the rms ripple voltage $V_{r(rms)}$, **(g)** the average diode current $I_{D(av)}$, **(h)** the rms diode current $I_{D(rms)}$, **(i)** the peak inverse voltage PIV of the diodes, **(j)** the average output power $P_{o(ac)}$, **(k)** the DC output power $P_{o(dc)}$, and **(l)** the frequency f_r of the output ripple voltage.

5.6 The single-phase full-wave rectifier of Fig. 5.6(a) is supplied from a 220-V (rms), 50-Hz source through a step-down center-tapped transformer with turns ratio $n = 10:2$.

a. Express the instantaneous output voltage $v_O(t)$ by a Fourier series.

b. Use PSpice/SPICE to calculate the harmonic components of the output voltage, up to and including the ninth harmonic. Use default model parameters of 1N4148 diodes.

5.7 The single-phase full-wave bridge rectifier of Fig. 5.8(a) is supplied directly from a 220-V (rms), 50-Hz source through a transformer with turns ratio $n = 10:1$. The load resistance R_L is 100 Ω. Determine **(a)** the average output voltage $V_{o(av)}$, **(b)** the average load current $I_{o(av)}$, **(c)** the rms load voltage $V_{o(rms)}$, **(d)** the rms load current $I_{o(rms)}$, **(e)** the ripple factor RF of the output voltage, **(f)** the rms ripple voltage $V_{r(rms)}$, **(g)** the average diode current $I_{D(av)}$, **(h)** the rms diode current $I_{D(rms)}$, **(i)** the peak inverse voltage PIV of the diode, **(j)** the average output power $P_{o(ac)}$, **(k)** the DC output power $P_{o(dc)}$, and **(l)** the frequency f_r of the output ripple voltage.

5.8 An AC voltmeter is constructed by using a DC meter and a bridge rectifier, as shown in Fig. 5.11(a). The meter has an internal resistance of $R_m = 50$ Ω, and its average current is $I_m = 200$ mA for a full-scale deflection. The current-limiting resistance is $R_s = 2.5$ kΩ.

a. Determine the rms value of the AC input voltage V_s that will give a full-scale deflection if the input voltage v_S is sinusoidal.

b. If this meter is used to measure the rms value of an input voltage with a triangular waveform, as shown in Fig. 5.11(b), calculate the necessary correction factor K to be applied to the meter reading.

5.9 A DC meter has an internal resistance of $R_m = 50\ \Omega$, and its full-scale deflection current is $I_m = 200$ mA.
D The meter should read an rms input voltage of $V_s = 250$ at the full-scale deflection.

a. Design an AC voltmeter that uses the DC meter and a bridge rectifier, as shown in Fig. 5.11(a).

b. Use PSpice/SPICE to check your results by plotting the average meter current. Use default model parameters of 1N4148 diodes.

5.10 An AC voltmeter is constructed by using a DC meter and a bridge rectifier as shown in Fig. P5.10. The DC meter has an internal resistance of $R_m = 250\ \Omega$, and the average meter current is $I_m = 1$ mA for full-scale deflection.

a. Determine the rms input V_s for full-scale deflection.

b. Use PSpice/SPICE to check your results by plotting the average meter current. Use default model parameters of 1N4148 diodes.

FIGURE P5.10

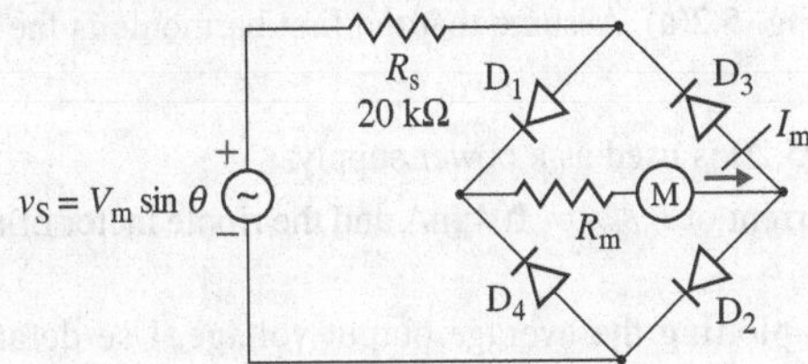

5.11 A single-phase bridge rectifier is shown in Fig. 5.9. The load resistance R_L is 2.5 kΩ, and the source resistance R_s is 1 kΩ.

a. Determine the transfer characteristic (v_O versus v_S) of the rectifier.

b. Use PSpice/SPICE to plot the transfer characteristic for $v_S = -10$ V to 10 V. Use default model parameters of 1N4148 diodes.

5.12 Repeat Prob. 5.11 for the half-wave rectifier of Fig. 5.1(a).

5.3 *Output Filters for Rectifiers*

5.13 The single-phase bridge rectifier shown in Fig. 5.12(a) is supplied directly from a 220-V (rms), 50-Hz source
D without any input transformer. The load resistance is $R_L = 1$ kΩ.

a. Design an L filter so that the rms ripple current $I_{r(rms)}$ is limited to less than 5% of $I_{o(av)}$. Assume that the second harmonic $I_{o2(rms)}$ is the dominant one and that the effects of higher-order harmonics are negligible.

b. Use PSpice/SPICE to check your design by plotting the output current. Use default model parameters of 1N4148 diodes.

5.14 Repeat Prob. 5.13 for the half-wave rectifier of Fig. 5.2(a). Assume that the first harmonic is the dominant
D one. Also assume a turns ratio of $n = 1{:}1$.

5.15 The single-phase full-wave bridge rectifier of Fig. 5.15(a) is supplied directly from a 120-V (rms), 60-Hz
D source without any input transformer. The load resistance is $R_L = 1$ kΩ. Assume that the second harmonic is the dominant one.

a. Design a C filter so that the rms ripple voltage $V_{r(rms)}$ is limited to less than 5% of $V_{o(av)}$.

b. With the value of C found in part (a), calculate the average output voltage $V_{o(av)}$, and the capacitor voltage if the load resistance R_L is disconnected.

c. Use PSpice/SPICE to check your design by plotting the instantaneous output voltage v_O. Use default model parameters of 1N4148 diodes.

5.16 D Repeat Prob. 5.15 for the half-wave rectifier of Fig. 5.2(a). Assume that the first harmonic is the dominant one.

5.17 Measurements of the output of a full-wave bridge rectifier give $V_{o(av)} = 150$ V, $I_{o(av)} = 120$ mA, and $V_{o(rms)} = 155$ V. The rectifier uses a C filter across the load resistance. The supply frequency f is 60 Hz.

a. Determine the ripple factor RF of the output voltage and the value of the filter capacitance C.

b. Use PSpice/SPICE to check your results by plotting the average output voltage. Use default model parameters of 1N4148 diodes.

5.18 D The single-phase bridge rectifier of Fig. 5.18(a) is supplied from a 120-V (rms), 60-Hz source without any input transformer. The load resistance is $R_L = 2$ kΩ. Assume that the second harmonic is the dominant one.

a. Design an LC filter so that the rms ripple voltage $V_{r(rms)}$ is limited to less than 5% of $V_{o(av)}$.

b. Use PSpice/SPICE to check your design by plotting the instantaneous output voltage v_O. Use default model parameters of 1N4148 diodes.

5.19 D Repeat Prob. 5.18 for the half-wave rectifier of Fig. 5.2(a). Assume that the first harmonic is the dominant one. Also assume a turns ratio of $n = 1{:}1$.

5.20 The single-phase bridge rectifier shown in Fig. P5.20 is used as a power supply.

a. Determine the DC output voltage for a load current of $I_{o(av)} = 104$ mA and the ripple factor of the output voltage for $R_L = 1$ kΩ.

b. Use PSpice/SPICE to check your results by plotting the average output voltage. Use default model parameters of 1N4148 diodes.

FIGURE P5.20

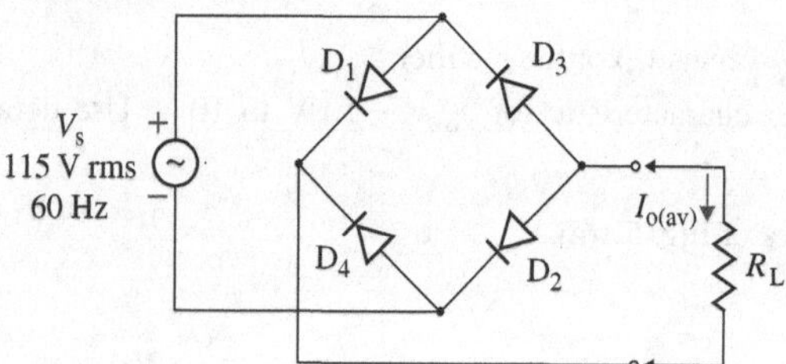

5.21 Repeat Prob. 5.20 for the load of Fig. P5.21 with $C = 100$ μF and $R_L = 1$ kΩ.

FIGURE P5.21

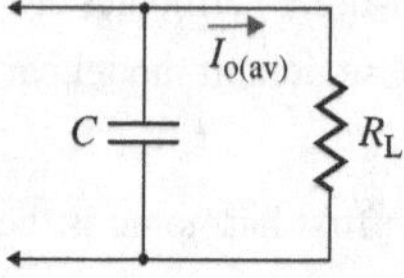

5.22 Repeat Prob. 5.20 for the load of Fig. P5.22 with $L = 5$ mH and $R_L = 1.5$ kΩ.

FIGURE P5.22

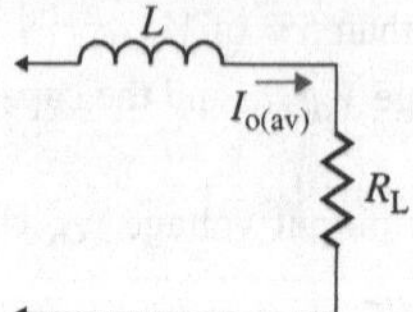

5.4 Diode Peak Detectors and Demodulators

5.23 The carrier frequency f_c of a radio signal is 250 kHz, and the modulating frequency f_m is 10 kHz. The load resistance R of the detector is 10 kΩ.

a. Design a demodulator for the waveform of Fig. 5.22(a) by determining the value of capacitance C in Fig. 5.21(a).

b. Use PSpice/SPICE to plot the output voltage v_O for a modulation index of $M = 0.5$ and a peak modulating voltage of $V_m = 20$ V. Use default model parameters.

5.24 Repeat Prob. 5.23(b) for a modulation index of $M = 1$.

D

5.5 Diode Clippers

5.25 The clipper circuit shown in Fig. 5.27(a) is supplied from the input voltage shown in Fig. 5.27(b). The battery voltage is $E_1 = 20$ V. The peak diode current $I_{D(peak)}$ is to be limited to 50 mA. Determine **(a)** the value of resistance R, **(b)** the average diode current $I_{D(av)}$ and the rms diode current $I_{D(rms)}$, and **(c)** the power rating P_R of the resistance R.

D

5.26 The clipper circuit shown in Fig. 5.25(a) is supplied from a sinusoidal input voltage of $v_s = 20 \sin(2000\pi t)$. The battery voltage is $E_1 = 5$ V. The peak diode current $I_{D(peak)}$ is to be limited to 10 mA.

D

a. Determine the value of resistance R, the average diode current $I_{D(av)}$ and the rms diode current $I_{D(rms)}$, and the power rating P_R of the resistance R.

b. Use PSpice/SPICE to plot the diode current. Use default model parameters of 1N4148 diodes.

5.27 The input voltage to the clipper circuit shown in Fig. P5.27(a) is $v_S = 5 \sin(2000\pi t)$. If $E_1 = 2$ V and $R = 10$ kΩ, plot **(a)** the output voltage $v_O(t)$ as a function of time, **(b)** the transfer characteristic of v_O versus v_S, and **(c)** the peak diode current $I_{D(peak)}$. Assume a diode voltage drop of $V_D = 0.7$ V.

FIGURE P5.27

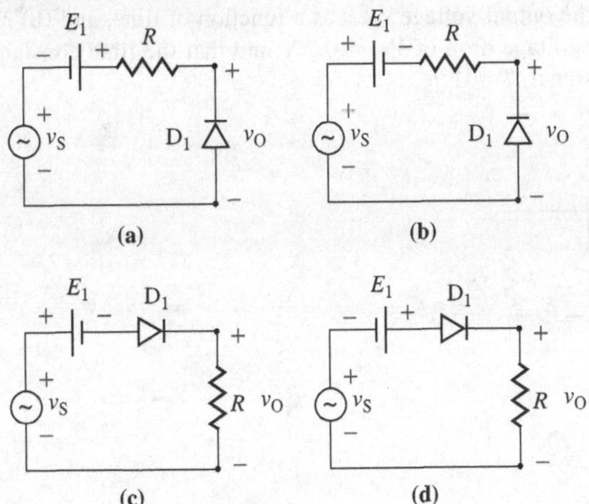

5.28 Repeat Prob. 5.27 for the circuit shown in Fig. P5.27(b).

5.29 Repeat Prob. 5.27 for the circuit shown in Fig. P5.27(c).

5.30 Repeat Prob. 5.27 for the circuit shown in Fig. P5.27(d).

5.6 Diode Clamping Circuits

5.31 D The input voltage v_S to the clamping circuit of Fig. 5.31(a) is a sinusoidal voltage $v_S = 30 \sin(2000\pi t)$. The peak diode current $I_{D(peak)}$ is to be limited to 0.5 A. Assume that $E_1 = 5$ V and that a limiting resistance R_s is connected in series with C.

a. Design the clamping circuit by determining the peak inverse voltage PIV of the diode and the values of R_s, R, and C.

b. Use PSpice/SPICE to plot the output voltage v_O. Use default model parameters of 1N4148 diodes.

5.32 D The input voltage v_S to the clamping circuit of Fig. 5.31(a) is $v_S = 20 \sin(2000\pi t)$. The peak diode current $I_{D(peak)}$ is to be limited to 0.5 A. Assume that $E_1 = 5$ V and that a series resistance R_s is connected in series with C to limit the diode current.

a. Design the clamping circuit by determining the peak inverse voltage PIV of the diode and the values of R_s, R, and C.

b. Use PSpice/SPICE to plot the output voltage v_O. Use default model parameters of 1N4148 diodes.

5.33 The input voltage to the clamping circuit shown in Fig. P5.33(a) is $v_S = 5 \sin(2000\pi t)$. If $E_1 = 2$ V, $C = 0.1\ \mu$F, and $R = 1$ MΩ, **(a)** plot the output voltage $v_O(t)$ as a function of time, and **(b)** repeat part (a) for Fig. P5.33(b). Assume a diode voltage drop of $V_D = 0.7$ V and that the time constant $\tau = RC$ is much larger than the period of the signal, $T = 1/f_S$.

FIGURE P5.33

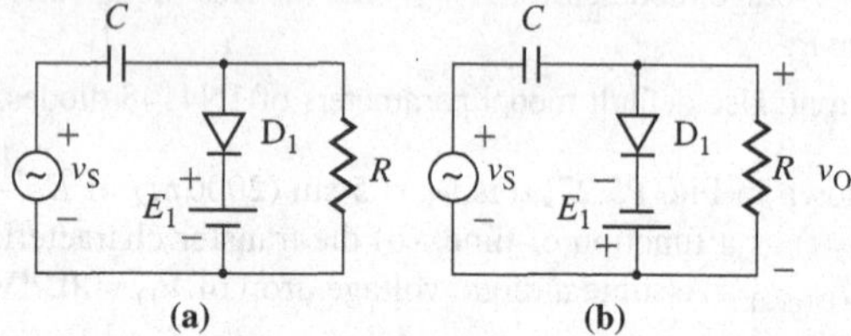

5.34 The input voltage to the clamping circuit shown in Fig. P5.34(a) is $v_S = 10 \sin(2000\pi t)$. If $E_1 = 5$ V, $C = 0.1\ \mu$F, and $R = 1$ MΩ, **(a)** plot the output voltage $v_O(t)$ as a function of time, and **(b)** repeat part (a) for Fig. P5.34(b). Assume a diode voltage drop of $V_D = 0.7$ V and that the time constant $\tau = RC$ is much larger than the period of the signal, $T = 1/f_S$.

FIGURE P5.34

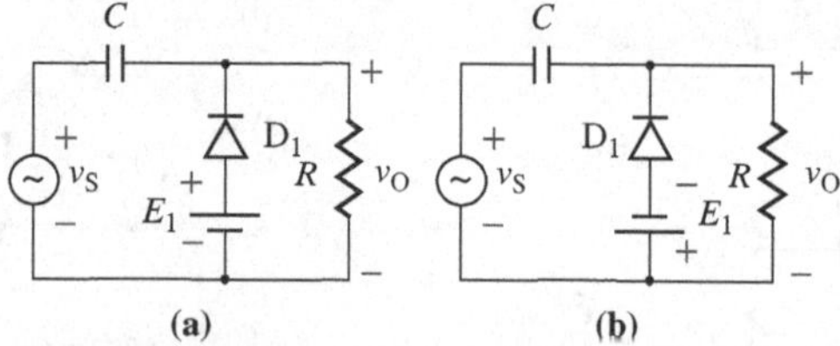

5.7 Diode Voltage Multipliers

5.35 Use PSpice/SPICE to plot the output voltage v_O for the voltage doubler in Fig. 5.34(c). Assume $v_S = 10 \sin 120\pi t$ and $C_1 = C_2 = C_3 = C_4 = 0.1\ \mu$F. Use default model parameters of 1N4148 diodes.

5.36 Use PSpice/SPICE to plot the output voltage v_{O4} $(= v_{C4})$ for the voltage quadrupler in Fig. 5.36(a). Assume $v_S = 10 \sin 120\pi t$ and $C_1 = C_2 = C_3 = C_4 = 0.01\ \mu$F. The resistances that are connected across diodes D_1, D_2, and D_3 are $R_1 = R_2 = R_3 = 5$ MΩ (not shown). Use default model parameters of 1N4148 diodes.

5.8 *Diode Function Generators*

5.37 The circuit parameters of the diode circuit in Fig. 5.40(a) are $R_1 = 10\ \text{k}\Omega$, $R_2 = 5\ \text{k}\Omega$, $R_3 = 2.5\ \text{k}\Omega$, $E_1 = 4$ V, and $E_2 = 10$ V.

a. Plot the *v-i* relationship of the circuit.

b. Use PSpice/SPICE to plot the transfer characteristic for $v_S = 0$ to 12 V. Use default model parameters of 1N4148 diodes.

5.38 A diode circuit is shown in Fig. P5.38. The circuit parameters are $R = 1\ \text{k}\Omega$ and $E = 4$ V.

a. Derive an expression for the *v-i* characteristic of the circuit. Plot the *v-i* characteristic.

b. Use PSpice/SPICE to check your results by plotting the *v-i* characteristic for $v_S = -5$ V to 10 V. Use default model parameters of 1N4148 diodes.

FIGURE P5.38

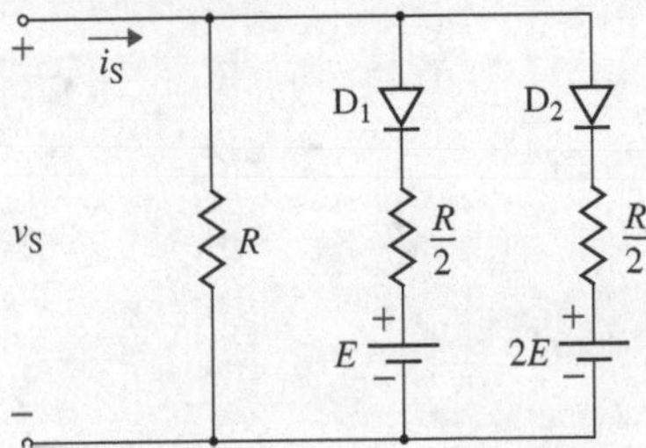

5.39 A *v-i* characteristic representing a square law is shown in Fig. P5.39.

D **a.** Design a diode circuit to generate this characteristic.

b. Use PSpice/SPICE to check your design by plotting the *v-i* characteristic. Use default model parameters of 1N4148 diodes.

FIGURE P5.39

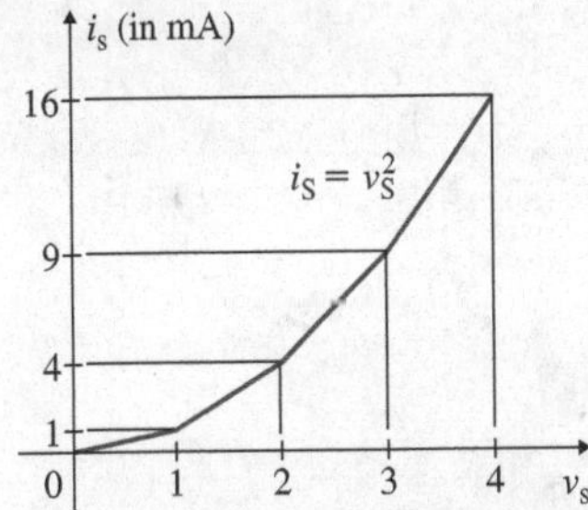

5.40 The input voltage to the diode clipper circuit shown in Fig. P5.40(a) is $v_S = 10 \sin (2000\pi t)$. If $R_1 = 10\ \text{k}\Omega$, $R_2 = 20\ \text{k}\Omega$, and $R_3 = 20\ \text{k}\Omega$, plot **(a)** the output voltage $v_O(t)$ as a function of time and **(b)** the transfer characteristic of v_O versus v_S. Assume an ideal diode drop of $V_D = 0$.

FIGURE P5.40

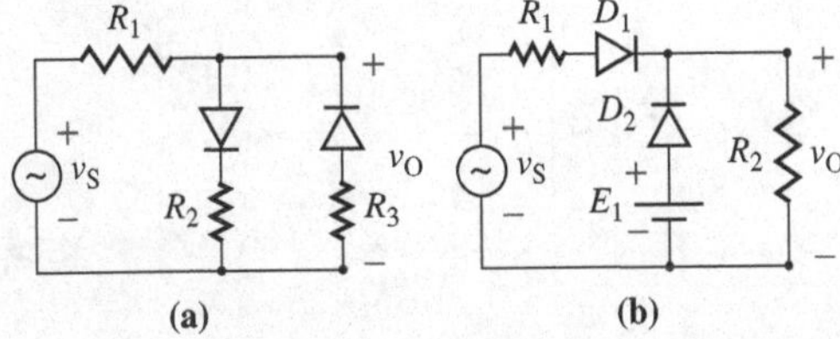

5.41 Repeat Prob. 5.40 for the circuit shown in Fig. P5.40(b) if $R_1 = 10\ \text{k}\Omega$, $R_2 = 40\ \text{k}\Omega$, and, $E_1 = 5$ V.

CHAPTER 6

SEMICONDUCTORS AND *pn* JUNCTION CHARACTERISTICS

Learning Outcomes

After completing this chapter, students should be able to do the following:

- Describe the physical structure and depletion region of the *pn* junction.
- Determine characteristics of the zero-biased, reverse-biased, and forward-biased *pn* junction.
- Determine the space charge widths, electric fields, and capacitances of a reverse-biased *pn* junction.
- Describe characteristics of the Schottky barrier junction.
- Determine the small-signal frequency model of a *pn* junction diode.

Symbols and Their Meanings

Symbol	Meaning
n_o, p_o	Electron and hole concentrations of a semiconductor material
n_a, n_d, n_i	Acceptor (hole), donor (electron), and intrinsic carrier concentrations
N_a, N_d	Net acceptor (hole) and donor (electron) carrier concentrations
V_{bi}, V_j	Built-in potential and junction potential
n_{po}, p_{no}	Minority carrier electrons at the edge of the *p*-region and of the *n*-region

Symbol	Meaning
C_j, C_d	Junction and diffusion capacitances
x_n, x_p, W	Space charge extension in the *n*-region, *p*-region, and total width
$\varepsilon_n(x)$, $\varepsilon_p(x)$	Electric field in the *n*-region and *p*-region
E_c, E_v, E_F, E_{Fi}	Conduction, valence, Fermi, and intrinsic energy bands
E_a, E_d, E_g	Acceptor, donor, and gap energy levels
$\phi_n(x)$, $\phi_p(x)$	Junction potentials in the *n*-region and *p*-region
$e\phi_{Fp}$, $e\phi_{Fn}$	Potential energy barriers in the *p*-region and *n*-region
v_D, v_F, v_R	Applied voltage of a *pn* junction, forward-biased voltage, and reverse-biased voltage

6.1 Introduction

We have seen in Chapters 4 and 5 that *pn* junction diodes can be used for signal conversion and processing. The characteristic of a practical diode that distinguishes it from an ideal one is that the practical diode experiences a finite voltage drop when it conducts and exhibits nonlinear characteristics. This drop is typically in the range of 0.5 V to 0.7 V. If the input voltage to a diode circuit is high enough, this small drop can be ignored. The voltage drop may, however, cause a significant error in electronic circuits, and the diode characteristic should be taken into account in evaluating the performance of diode circuits. To understand the characteristic of a practical diode, we need a clear understanding of its physical operation.

The *pn* junction is a basic building block in semiconductor devices, and the theory of the *pn* junction is still the fundamental concept in the physics of semiconductor devices. Most semiconductor devices contain at least one *pn* junction. A semiconductor diode, which has only one junction, is an example of *pn* devices. Other semiconductor devices are formed by combining two or more *pn* junctions in various configurations such as bipolar junction transistors, field-effect transistors, and silicon-controlled rectifiers. The characteristics of these devices depend on the *pn* characteristics under different biasing conditions: zero-biased, reverse-biased, and forward-biased. Semiconductor materials are the essential ingredients for *pn* junctions and semiconductor devices. The properties of high-purity, single-crystal materials are fundamental to the design of semiconductor devices.

6.2 Semiconductor Materials

Junction diodes are made of semiconductor materials [1]. A pure semiconductor is called an *intrinsic material* in which the concentrations of electrons and holes are equal. The currents induced in pure semiconductors are very small. The most commonly used semiconductors are silicon and germanium (Group IV in the periodic table as shown in Table 6.1), and gallium arsenide (Group V). Silicon materials cost less than germanium materials and allow diodes to operate at higher temperatures. For this reason, germanium diodes are rarely used. Gallium arsenide (GaAs) diodes can operate at higher switching speeds and higher frequencies than silicon diodes and hence are preferable. However, gallium arsenide materials are more expensive than silicon materials, and gallium arsenide diodes are more difficult to manufacture, so they are generally used only for high-frequency applications. GaAs devices are expected to become increasingly important in electronic circuits.

Semiconductors are a group of materials having conductivities between those of metals and insulators. One fundamental characteristic of semiconductor materials is that their conductivity can be varied over several orders of magnitude if we add controlled amounts of impurity atoms. To increase conductivity,

TABLE 6.1 A portion of the periodic table showing elements used in semiconductor materials

Period	Group II	Group III	Group IV	Group V	Group VI
2		B Boron	C Carbon	N Nitrogen	O Oxygen
3		Al Aluminum	Si Silicon	P Phosphorus	S Sulfur
4	Zn Zinc	Ga Gallium	Ge Germanium	As Arsenic	Se Selenium
5	Cd Cadmium	In Indium	Sn Tin	Sn Antimony	Te Tellurium
6	Hg Mercury				
Elementary semiconductors			Si Silicon Ge Germanium		
Compound semiconductors			SiC Silicon carbide SiGe Silicon germanium	GaAs Gallium arsenide	

controlled quantities of materials known as *impurities* are introduced into pure semiconductors, creating free electrons or holes. The current through a diode is the result of the flow of electrons and holes in a semiconductor when forces are applied. These electrons and holes are referred to as *carriers*. *Electrons* are negatively charged particles. A *hole* is the absence of an electron in a covalent bond and is like an independent positive charge. The electrons and holes flow in opposite directions, and the direction of the holes is the direction of the conventional current flow.

The process of adding carefully controlled amounts of impurities to pure semiconductors is known as *doping*. A semiconductor to which impurities have been added is referred to as *extrinsic*. Two types of impurities are normally used: *n*-type from Group V, such as antimony, phosphorus, and arsenic, and *p*-type from Group III, such as boron, gallium, and indium.

6.2.1 *n*-type Materials

The *n*-type impurities are pentavalent materials, with five electrons in the outermost shell of each atom. The addition of a controlled amount of an *n*-type impurity (having five valence electrons) to silicon or germanium (having four valence electrons) causes one electron to be loosely attached to the parent atom because only four electrons are needed to form a covalent bond within a silicon or germanium atom. If a small amount of energy, such as thermal energy, is added to the donor electron, the electron can become free, leaving behind a positively charged ion of the donor atom. At room temperature, there is sufficient energy to cause the redundant electron to break away from its parent atom; thus, a free electron is generated. This electron is free to move randomly within the semiconductor crystal. Thus, an *n*-type impurity donates free electrons to the semiconductor; for this reason, it is often referred to as a *donor impurity*. The resulting material is referred to as an *n*-type semiconductor (*n* for the negatively charged electron). An *n*-type semiconductor is shown in Fig. 6.1(a).

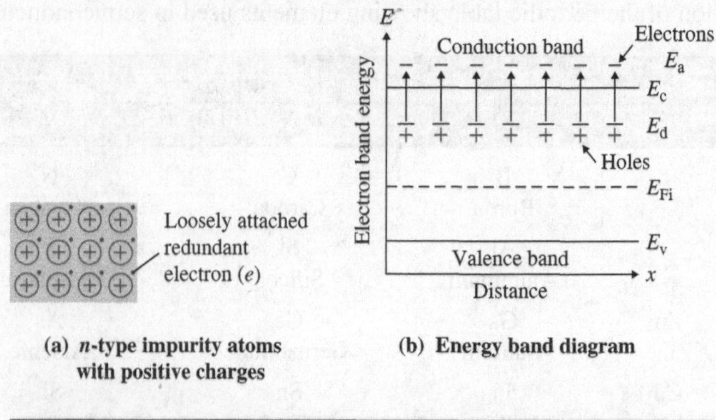

(a) *n*-type impurity atoms with positive charges **(b) Energy band diagram**

FIGURE 6.1 Positively charged atoms and energy band diagram of *n*-type semiconductors

With a sufficient amount of energy, the donor electrons can be elevated to the conduction band, making them free to move within the crystal. The impurity atom was originally neutral, and the removal of the redundant electron will cause the impurity atom to exhibit a positive charge equal to $+e$ and to remain fixed in the crystal lattice of the structure as shown in Fig. 6.1(a). Figure 6.1(b) shows the energy band levels of *n*-type semiconductors in complete ionized states. E_c is the conduction energy level, E_d is the donor energy level, E_v is the valence energy level, and E_{Fi} is the intrinsic Fermi energy level. E_{Fi} determines the statistical distribution of electrons, and its level is in the middle of E_c and E_v.

The relative dielectric constants and the effective masses of the semiconductor materials and their impurities are different. As a result, they have different ionization energies. Table 6.2 lists the impurity ionization energies in silicon, germanium, and gallium arsenide semiconductors.

6.2.2 *p*-type Materials

The *p*-type impurities are trivalent materials (Group III) with three valence electrons in the outer shell of each atom. The addition of a *p*-type impurity to silicon or germanium (Group IV) causes a vacancy

TABLE 6.2 Ionization energies in silicon, germanium, and gallium arsenide

		Ionization Energy (eV) of Materials		
Acceptors	Donors	Si	Ge	Gallium arsenide
Boron		0.045	0.0104	
Aluminum		0.06	0.0102	
	Phosphorus	0.045	0.012	
	Arsenic	0.05	0.0127	
Beryllium				0.028
Zinc				0.0307
Cadmium				0.0347
Silicon				0.0345
Germanium				0.0404
	Selenium			0.0059
	Tellurium			0.0058
	Silicon			0.0058
	Germanium			0.0061

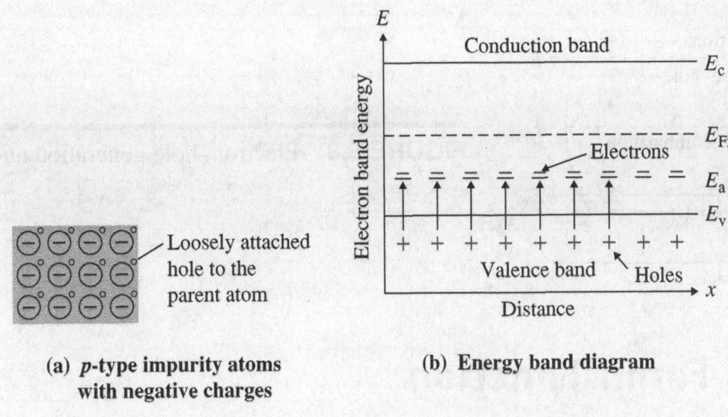

FIGURE 6.2 Negatively charged atoms and energy band diagram of *p*-type semiconductors

for one electron in the vicinity of the impurity atom because four electrons are necessary to complete covalent bonds. A vacancy for an electron is like a hole, which is equivalent to a positive charge $+e$. If an electron were to occupy this "empty" position, its energy would have to be greater than that of the valence electrons. If the valence electrons gain a small amount of thermal energy and move about in the crystal, the "empty" position becomes occupied and other valence electron positions become vacated, thereby creating holes in the semiconductor material. This type of semiconductor material is referred to as a *p*-type material (*p* for the positively charged hole). A *p*-type semiconductor is shown in Fig. 6.2(a).

At room temperature, there is sufficient energy to cause a nearby electron to move into the existing vacancy, in turn causing a vacancy elsewhere. In this way, the hole moves randomly within the semiconductor crystal. Thus, a *p*-type impurity accepts free electrons and is referred to as an *acceptor impurity*. With the electron it gains, the impurity atom exhibits a charge of $-e$ and remains fixed in the crystal lattice of the structure. With a sufficient amount of energy, the acceptor atom can generate holes in the valence band without generating electrons in the conduction band. Figure 6.2(b) shows the energy band levels of *p*-type semiconductors in complete ionized states where E_a is the acceptor energy level.

6.2.3 Majority and Minority Carriers

So far, we have assumed that materials are perfect; but practical materials are imperfect. The holes are also present in imperfect *n*-type semiconductor materials because of thermal agitation of electrons and holes within the materials. Therefore, in an *n*-type semiconductor, the electrons are the majority carriers and the holes are the minority carriers. Similarly, in a *p*-type semiconductor, the holes are the majority carriers and the electrons are the minority carriers. Doping and the application of energy can create electrons and holes (carriers). But within the semiconductor, there is also a recombination process by which electrons and holes (carriers) are annihilated. Any deviation from thermal equilibrium tends to change the electron and hole concentrations in a semiconductor. Any increase in energy (such as temperature or light) increases the rate at which electrons and holes are thermally generated; their concentrations change with time until new equilibrium values are reached. The simplified process of electron–hole generation and recombination is shown in Fig. 6.3.

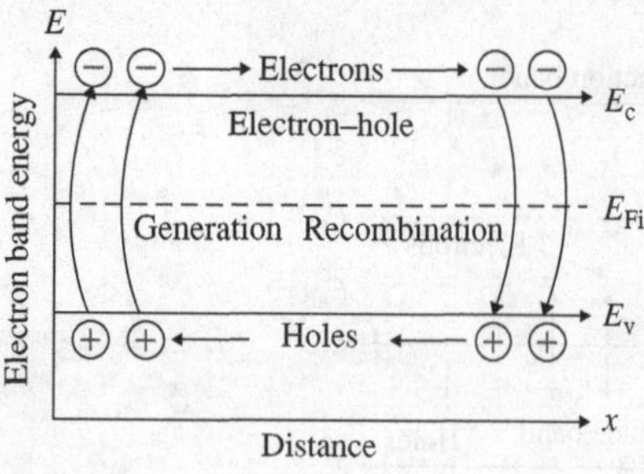

FIGURE 6.3 Electron–hole generation and recombination

6.2.4 The Fermi Function

The Fermi function $f(E)$ specifies how many of the existing states at the energy E will be filled with an electron, or equivalently under equilibrium conditions. That is, it specifies, under equilibrium conditions, the probability that an available state at an energy E will be occupied by an electron, and it is expressed mathematically [2, 3] as

$$f(E) = \frac{1}{1 + e^{(E-E_F)/kT}} \tag{6.1}$$

where E_F = Fermi energy or Fermi level
T = temperature in kelvin (K)
k = Boltzmann constant ($k = 8.617 \times 10^{-5}$ eV/K)

As the temperature approaches absolute zero, $T \approx 0$ K, the exponent term of Eq. (6.1) tends to infinity: $f(E) \rightarrow 0$ for $E > E_F$ and $f(E) \rightarrow 1$ for $E < E_F$. There is a sharp cutoff at the Fermi energy E_F. Therefore, all states at energies below E_F will be filled, and all states at energies above E_F will be empty. This is shown in Fig. 6.4(a).

As the system temperature increases above zero, $T > 0$ K, the exponent term of Eq. (6.1) has a finite value, and the function goes through a transition from a filled state to an empty state. For $E > E_F$, the function $f(E)$ decays exponentially to zero with increasing energy, and most states will be empty at the

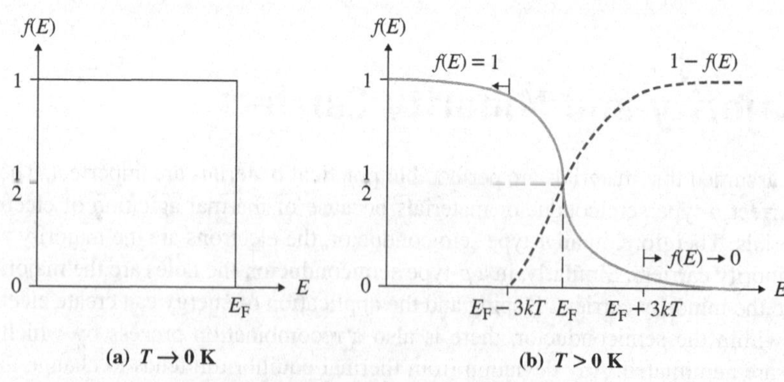

FIGURE 6.4 Energy dependence of the Fermi function

valence energy $E_v \simeq E = E_F + 3kT$ because the exponential term in Eq. (6.1) will be large and $f(E) \rightarrow 0$. For $E < E_F$, the function $f(E)$ increases exponentially with decreasing energy, and most states will be filled at the conduction energy $E_c \approx E = E_F - 3kT$ because the exponential term in Eq. (6.1) will be small and $f(E) \rightarrow 1$. Figure 6.4(b) shows the plots of $f(E)$ and $1 - f(E)$. Note that the Fermi function applies only under equilibrium conditions and is valid for all materials—insulators, semiconductors, and metals. It is simply a statistical function associated with electrons in general and does not depend on the characteristics and parameters of the semiconductors.

If $f(E)$ is the probability of electrons occupying states at a given energy E, then the probability that a state is empty (not filled) at a given energy E is equal to $1 - f(E)$. Thus, the probability that a state is filled at the conduction band edge (E_c) must be equal to the probability that a state is empty at the valence band edge (E_v). That is,

$$f(E_c) = 1 - f(E_v) \tag{6.2}$$

From Eq. (6.1) we get the probability function $f(E_c)$ at $E = E_c$ and $1 - f(E_v)$ at $E = E_v$:

$$f(E_c) = \frac{1}{1 + e^{(E_c - E_F)/kT}} \tag{6.3}$$

$$\simeq e^{-(E_c - E_F)/kT} \quad \text{for } (E_c - E_F) \geq 3kT \text{ and } e^{(E_c - E_F) \geq 3kT} >> 1 \tag{6.4}$$

$$1 - f(E_v) = 1 - \frac{1}{1 + e^{(E_v - E_F)/kT}} = \frac{1}{1 + e^{(E_F - E_v)/kT}} \tag{6.5}$$

$$\simeq e^{-(E_F - E_v)/kT} \quad \text{for } (E_F - E_v) \geq 3kT \text{ and } e^{(E_F - E_v) \geq 3kT} >> 1 \tag{6.6}$$

Equating Eq. (6.3) to Eq. (6.5), we get

$$\frac{1}{1 + e^{(E_c - E_F)/kT}} = \frac{1}{1 + e^{(E_F - E_v)/kT}} \tag{6.7}$$

which can be solved for the Fermi energy E_F:

$$E_F = \frac{E_c + E_v}{2} \tag{6.8}$$

Therefore, the Fermi energy level is positioned at the middle of the energy band.

6.2.5 Carrier Concentrations

The concentrations of electrons (n) and holes (p) depend on the amount of impurity doping and on the temperature. We can apply Eq. (6.4) to determine the thermal equilibrium electron concentration in the conduction band, which depends on the conduction band energy level and the temperature as given by

$$n_o = N_c \, e^{-(E_c - E_F)/kT} \tag{6.9}$$

where the parameter N_c is called the *effective density states function in the conduction band.* Its value depends on the effective mass values of the *n*-type semiconductor materials and the temperature as given by [2]

$$N_c = 2\left(\frac{2\pi m_n^* kT}{h^2}\right)^{3/2} \tag{6.10}$$

where m_n^* = the effective mass of a free electron and h = Planck's constant. Assuming $m_n^* \simeq m_o$ (mass of a free electron), $N_c = 2.5 \times 10^{19}\ \text{cm}^{-3}$ at $T = 300$ K for most semiconductors. Similarly, we can apply Eq. (6.6) to determine the thermal equilibrium hole concentration in the valence band, which depends on the valence band energy level and the temperature as given by

$$p_o = N_v\, e^{-(E_F - E_v)/kT} \tag{6.11}$$

where the parameter N_v is called the *effective density states function in the valence band.* Its value depends on the effective mass values of the *p*-type semiconductor materials and the temperature as given by

$$N_v = 2\left(\frac{2\pi m_p^* kT}{h^2}\right)^{3/2} \tag{6.12}$$

where $m_p^* \simeq m_o$ = the effective mass of a hole and h = Planck's constant. Assuming $m_p^* = m_o$ (mass of a free hole), $N_v = 1 \times 10^{19}\ \text{cm}^{-3}$ at $T = 300$ K for most semiconductors. The calculated values of N_c and N_v at $T = 300$ K are listed in Table 6.3. An intrinsic semiconductor will have a Fermi energy level called the *intrinsic Fermi energy,* $E_{Fi} = E_F$. From Eqs. (6.9) and (6.11), we can find the intrinsic concentrations of electrons and holes as

$$n_i = n_o = N_c\, e^{-(E_c - E_{Fi})/kT} \tag{6.13}$$

$$p_i = n_i = p_o = N_v\, e^{-(E_{Fi} - E_v)/kT} \tag{6.14}$$

If we take the product of n_i in Eq. (6.13) and p_i in Eq. (6.14), we can find

$$n_i p_i = n_i^2 = N_c\, e^{-(E_c - E_{Fi})/kT} \times N_v\, e^{-(E_{Fi} - E_v)/kT}$$

which can be simplified as follows:

$$n_i^2 = N_c N_v\, e^{-(E_c - E_v)/kT} = N_c N_v\, e^{-Eg/kT} \tag{6.15}$$

Here E_g is the conduction band energy, and n_i refers to either the intrinsic electron or hole concentration in the semiconductor material. The value of n_i is constant for a given semiconductor material at a constant temperature, and it is independent of the Fermi energy. The calculated values of n_i from Eq. (6.15) for $E_g = 1.12$ eV and $T = 300$ K are also listed in Table 6.3.

Under thermal equilibrium conditions at a given temperature, the product of the amount of electron concentration n_o and the amount of hole concentration p_o is always constant for a given semiconductor material. That is,

$$n_o\, p_o = n_i^2 \tag{6.16}$$

TABLE 6.3 Effective density of states function for conduction and valence bands

Materials at $T = 300$ K	N_c (cm^{-3})	N_v (cm^{-3})	m_m^*/m_o	m_p^*/m_o	n_i (cm^{-3}) at $E_g = 1.12$ eV
Silicon	2.8×10^{19}	1.04×10^{19}	1.08	0.56	1.5×10^{10}
Germanium	1.04×10^{19}	6.0×10^{18}	0.55	0.37	2.4×10^{13}
Gallium arsenide	4.7×10^{17}	7.0×10^{18}	0.067	0.48	1.8×10^{6}

For example, if $n_o = 1 \times 10^{16}$ cm^{-3} for silicon at $T = 300$ K, then

$$p_o = \frac{n_i^2}{n_o} = \frac{(1.5 \times 10^{10})^2}{1 \times 10^{16}} = 2.25 \times 10^4 \text{ cm}^{-3}$$

KEY POINTS OF SECTION 6.2

- Free electrons (in *n*-type material) and holes (in *p*-type material) are made available by adding a controlled amount of *n*-type impurities and *p*-type impurities to pure semiconductors, respectively.
- The Fermi function specifies the probability that an available state at a given energy will be occupied by an electron.
- The intrinsic concentration of a semiconductor material remains constant at a steady temperature, and it is independent of the Fermi energy.

6.3 Zero-Biased *pn* Junction

To consider the operation principle of a *pn* junction, we will assume that a *p*-type material is laid into one side of a single crystal of a semiconductor material and an *n*-type material is laid into the other side, as shown in Fig. 6.5(a). (This is not, however, the way to make a diode.) The doping profile of the impurity doping concentrations in the *p*-region (N_a) and *n*-region (N_d) is shown in Fig. 6.5(b) with the assumption that the doping concentration is uniform in each region.

At room temperature, the electrons, which are majority carriers in the *n*-region, diffuse from the *n*-type side to the *p*-type side; the holes, which are majority carrriers in the *p*-region, diffuse from the *p*-type side to the *n*-type side. The electrons and holes will recombine near the junction and thus cancel each other out. There will be opposite charges on each side of the junction, creating a *depletion region,* or *space charge region,* as shown in Fig. 6.5(c). Under thermal equilibrium conditions at a given temperature, no more electrons or holes will cross the junction.

Because opposite charges are present on each side of the junction, an electric field is established across the junction. The resultant junction potential barrier V_j, which arises because the *n*-type side is at a higher potential than the *p*-type side, prevents any flow of majority carriers to the other side. The variation of the potential across the junction is shown in Fig. 6.5(d). V_j is also called the *built-in potential* V_{bi}.

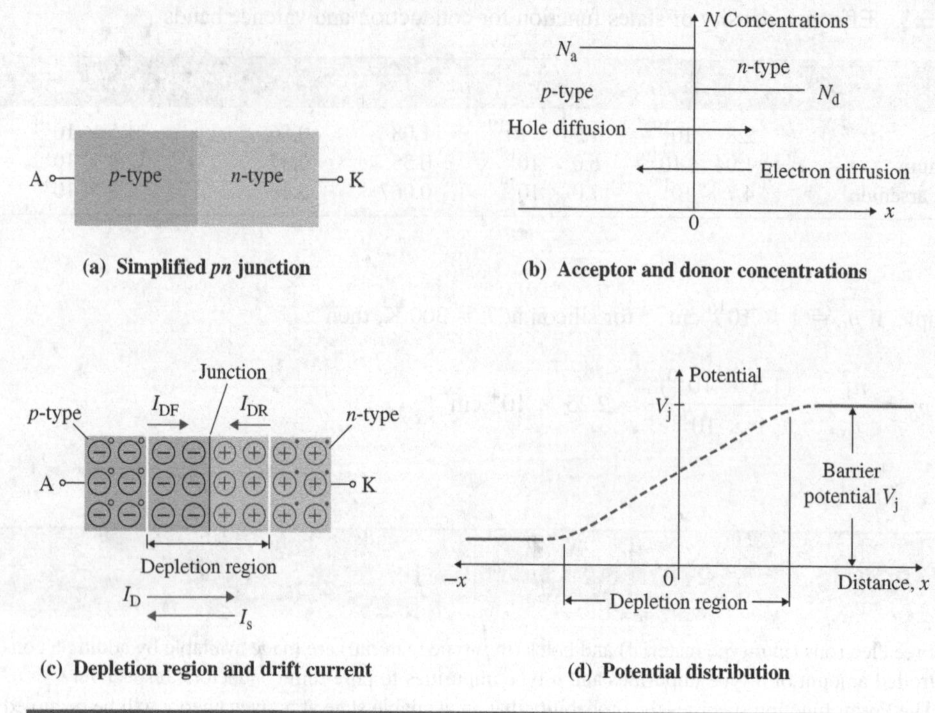

FIGURE 6.5 *pn* junction and depletion region

Because of the potential barrier V_j, the electrons, which are minority carriers in the *p*-side, will be swept across the junction to the *n*-side; the holes, which are minority carriers in the *n*-side, will be swept across the junction to the *p*-side. Therefore, a current caused by the minority carriers (holes) will flow from the *n*-side to the *p*-side; it is known as the *reverse drift current* I_{DR}. Similarly, a current known as the *forward diffusion current* I_{DF} will flow from the *p*-side to the *n*-side, caused by minority electrons. Under equilibrium conditions, the resultant current will be zero. Therefore, these two currents (I_{DF} and I_{DR}) are equal and flow in opposite directions. That is,

$$I_{DF} = -I_{DR} \tag{6.17}$$

6.3.1 Built-In Junction Potential

The energy bands in the neutral *p*- and *n*-regions on either side of the space charge region must bend due to the potential barriers $e\phi_{Fp}$ in the *p*-type and $e\phi_{Fn}$ in the *n*-type. However, the Fermi energy level (E_F) is constant throughout the entire system at thermal equilibrium, as shown in Fig. 6.6.

The intrinsic energy level E_{Fi} in the *p*- and *n*-regions is always equidistant from E_c and E_v. E_{Fi} determines the total junction potential V_{bi}, which is the difference between the intrinsic Fermi levels in the *p*- and *n*-regions as given by

$$V_{bi} = |\phi_{Fp}| + |\phi_{Fn}| \tag{6.18}$$

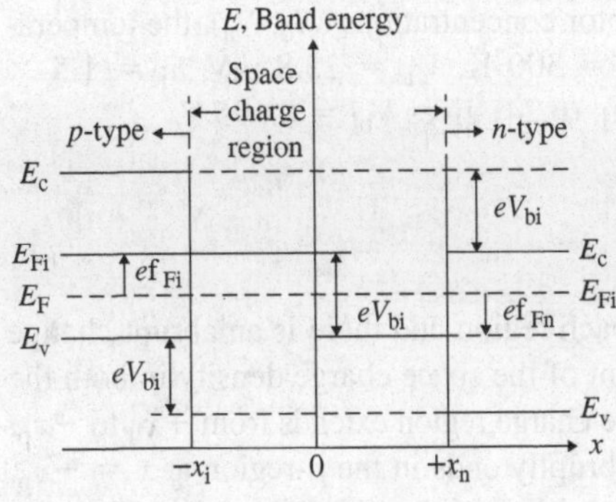

FIGURE 6.6 Energy band diagram of a *pn* junction at thermal equilibrium

The hole concentration in the conduction band in the *p*-region decays exponentially and can be determined from

$$n_p = n_i\, e^{-(E_F - E_{Fi})/kT} = n_i\, e^{-e\phi_{Fp}/kT} \tag{6.19}$$

Here n_i and E_{Fi} are the intrinsic carrier concentration and the intrinsic Fermi energy, respectively, in the *n*-region. ϕ_{Fp} is the potential barrier in the *p*-region. If we take a natural log of both sides, Eq. (6.19) gives

$$\phi_{Fp} = \frac{-kT}{e} \ln\left(\frac{n_p}{n_i}\right) \tag{6.20}$$

The electron concentration in the conduction band in the *n*-region can be determined from

$$n_n = n_i\, e^{-(E_F - E_{Fi})/kT} = n_i\, e^{+e\phi_{Fn}/kT} \tag{6.21}$$

where n_i and E_{Fi} are the intrinsic carrier concentration and the intrinsic Fermi energy, respectively, in the *n*-region. $+\phi_{Fn}$ is the potential barrier in the *n*-region. If we take a natural log of both sides, Eq. (6.21) gives

$$\phi_{Fn} = \frac{kT}{e} \ln\left(\frac{n_o}{n_i}\right) \tag{6.22}$$

Substituting ϕ_{Fp} from Eq. (6.20) and ϕ_{Fn} from Eq. (6.22) into Eq. (6.18) gives

$$V_{bi} = \frac{kT}{e} \ln\left(\frac{n_p}{n_i}\right) + \frac{kT}{e} \ln\left(\frac{n_n}{n_i}\right) = V_T \ln\left(\frac{n_n n_p}{n_i^2}\right) \tag{6.23}$$

where $V_T = kT/e$ is defined as the thermal voltage. Assuming that n_n equals the net donor concentration N_d of the *n*-region and n_p equals the net acceptor concentration N_a in the *p*-region, we can write Eq. (6.23) as

$$V_{bi} = \frac{kT}{e} \ln\left(\frac{N_d N_a}{n_i^2}\right) = V_T \ln\left(\frac{N_d N_a}{n_i^2}\right) \tag{6.24}$$

Therefore, the junction potential depends on the donor and acceptor concentrations (N_d, N_a), the temperature (K), and the intrinsic concentration n_i. For example, if T = 300 K, V_T = 25.8 mV, $n_i = 1.5 \times 10^{10}\text{cm}^{-3}$, $N_a = 2 \times 10^{16}$ cm^{-3}, and $N_d = 5 \times 10^{15}$ cm^{-3}, Eq. (6.24) gives V_{bi} = 0.695 V.

6.3.2 Electric Field Distribution

Assume that the doping concentration (N_a or N_d) is uniform in each region and there is an abrupt change in doping at the junction. Initially, there is a step function gradient of the space charge density in both the electron and hole concentrations, as shown in Fig. 6.7(a). The space charge region extends from $+x_n$ to $-x_p$. The distribution abruptly ends in the n-region at $x = +x_n$ and abruptly ends in the p-region at $x = -x_p$. We can determine the electric field by applying the one-dimensional Poisson's equation as given by

$$\frac{d^2\phi(x)}{dx} = \frac{\rho(x)}{\varepsilon_s} = \frac{d\varepsilon(x)}{dx} \tag{6.25}$$

where $\phi(x)$ = electric potential

$\varepsilon(x)$ = electric field

$\rho(x)$ = volume charge density, and

ε_s = permittivity of the semiconductor

ε_s equals to the product of the relative permeability ε_r = 11.7 and the permeability of the free air $\varepsilon_o = 8.85 \times 10^{-14}$. That is, $\varepsilon_s = \varepsilon_r\varepsilon_o$. Note that $\varepsilon(x)$ is the variable electric field as a function of x while ε denotes a fixed value of electric field.

From Fig. 6.7(a), the charge densities are

$$\begin{aligned} \rho(x) &= +eN_d \quad 0 < x < x_n \\ &= -eN_a \quad -x_p < x < 0 \end{aligned} \tag{6.26}$$

We can find the electric field in the p-region by integrating Eq. (6.25) as

$$\varepsilon_p(x) = \int \frac{\rho(x)}{\varepsilon_s} dx = \int \frac{-eN_a}{\varepsilon_s} dx = \frac{-eN_a}{\varepsilon_s} x + C_1 \tag{6.27}$$

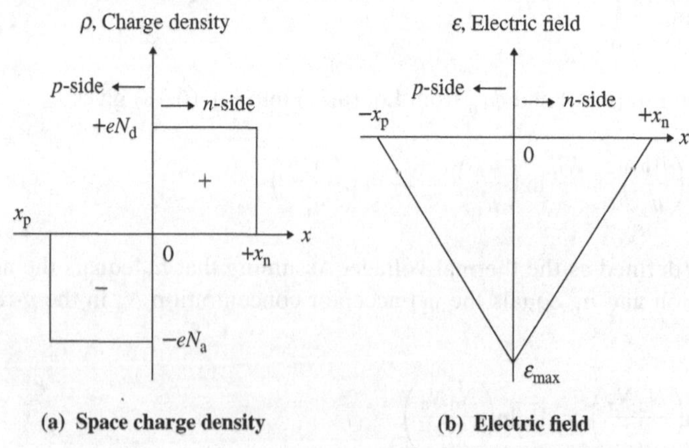

(a) Space charge density (b) Electric field

FIGURE 6.7 Space charge density and electric field

Here C_1 is a constant of integration that can be found from the final condition $\varepsilon = 0$ at $x = -x_p$; that is, if $\varepsilon_p(x = -x_p) = 0$, we find $C_1 = -eN_a x_p/\varepsilon_s$. Substituting C_1 in Eq. (6.27) gives

$$\varepsilon_p(x) = \frac{-eN_a}{\varepsilon_s}(x + x_p) \qquad -x_p < x < 0 \tag{6.28}$$

which is a linear function of distance x in the p-region. Similarly, we can find the electric field in the n-region from

$$\varepsilon_n(x) = \int \frac{\rho(x)}{\varepsilon_s} dx = \int \frac{eN_d}{\varepsilon_s} dx = \frac{eN_d}{\varepsilon_s} x + C_2 \tag{6.29}$$

Here C_2 is a constant of integration that can be found from the final condition $\varepsilon = 0$ at $x = x_n$; that is, if $\varepsilon_n(x = x_n) = 0$, we find $C_2 = -eN_d x_n/\varepsilon_s$. Substituting C_2 in Eq. (6.29) gives

$$\varepsilon_n(x) = \frac{-eN_d}{\varepsilon_s}(x_n - x) \qquad 0 < x < x_n \tag{6.30}$$

The electric field, which becomes maximum at $x = 0$, is given by

$$\varepsilon_{max} = \frac{-eN_d x_n}{\varepsilon_s} = \frac{-eN_a x_p}{\varepsilon_s} \tag{6.31}$$

since the electric field is continuous at the junction at $x = 0$. By setting the field in Eq. (6.28) equal to the field in Eq. (6.30) at $x = 0$, we get

$$N_a\, x_p = N_d\, x_n \tag{6.32}$$

Thus the number of negative charges per unit area in the p-region is equal to the number of positive charges per unit area in the n-region. An electric field exists in the depletion region, and the plot of the electric field in the depletion region is shown in Fig. 6.7(b). The electric field is a linear function of the distance x, and it becomes maximum at $x = 0$.

6.3.3 Junction Potential Distribution

The junction potential in the p-region can be found by integrating the electric field in Eq. (6.28)

$$\phi_p(x) = -\int \varepsilon(x)\, dx = \int \frac{eN_a}{\varepsilon_s}(x_p + x)\, dx = \frac{eN_a}{\varepsilon_s}\left(x_p x + \frac{x^2}{2}\right) + C_3 \tag{6.33}$$

Here C_3 is a constant of integration that can be found from the final condition $\varepsilon = 0$ at $x = -x_p$; that is, if $\varepsilon(x = -x_p) = 0$, we find $C_3 = eN_a x_p^2/2\varepsilon_s$. Substituting C_3 in Eq. (6.33) gives

$$\phi_p(x) = \frac{eN_a}{2\varepsilon_s}(x + x_p)^2 \qquad -x_p < x < 0 \tag{6.34}$$

Similarly, we can find the junction potential in the n-region from

$$\phi_n(x) = -\int \varepsilon(x) dx = \int \frac{eN_d}{\varepsilon_s}(x_n - x) dx = \frac{eN_d}{\varepsilon_s}\left(x_n x - \frac{x^2}{2}\right) + C_4 \tag{6.35}$$

Here C_4 is a constant of integration that can be found by equating $\phi_p(x)$ in Eq. (6.34) at $x = 0$ to $\phi_n(x)$ at $x = 0$ because the potential is a continuous function. Thus for $\phi_n(x = 0) = \phi_p(x = 0)$, we find $C_4 = \phi_p(x = 0) = eN_a x_p^2/2\varepsilon_s$. Substituting C_4 into Eq. (6.35) gives

$$\phi_n(x) = \frac{eN_d}{\varepsilon_s}\left(x_n x - \frac{x^2}{2}\right) + \frac{eN_a}{2\varepsilon_s} x_p^2 \qquad 0 < x < x_n \tag{6.36}$$

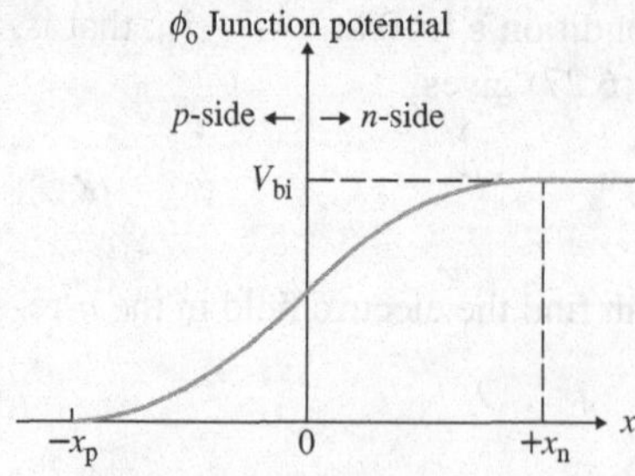

FIGURE 6.8 Junction potential distribution in the space charge region

The plot of the potential ϕ_n against the distance x is shown in Fig. 6.8, which shows a quadratic dependence on distance x. The magnitude of the junction potential at $x = x_n$ is equal to the built-in potential barrier V_{bi} or the junction potential V_j. Thus, for $x = x_n$ we can find the built-in potential from Eq. (6.36) as

$$V_{bi} = |\phi_n(x = x_n)| = \frac{e}{2\varepsilon_s}(N_d x_n^2 + N_a x_p^2) \tag{6.37}$$

Since the potential energy of an electron is related to the potential $\phi_n(x)$ by $E = -e\phi(x)$, the electron energy $e\phi_{Fn}$ (or the hole energy $e\phi_{Fp}$) also varies as a quadratic function of distance through the space charge region. The plot of the energy band diagram, which is a quadratic dependence on the distance x, is shown in Fig. 6.9. Thus, the conduction, valence, and intrinsic Fermi energy levels vary with distance in a semiconductor.

6.3.4 Space Charge Depletion Width

The distance of the space charge region that extends into the p-region can be found from Eq. (6.32) as

$$x_p = \frac{N_d\, x_n}{N_a} \tag{6.38}$$

Substituting x_p from Eq. (6.38) into Eq. (6.37) and solving for x_n, we get the space charge extension x_n in the n-region as

$$x_n = \sqrt{\frac{2\varepsilon_s V_{bi}}{e}\left(\frac{N_a}{N_d}\right)\left(\frac{1}{N_a + N_d}\right)} \tag{6.39}$$

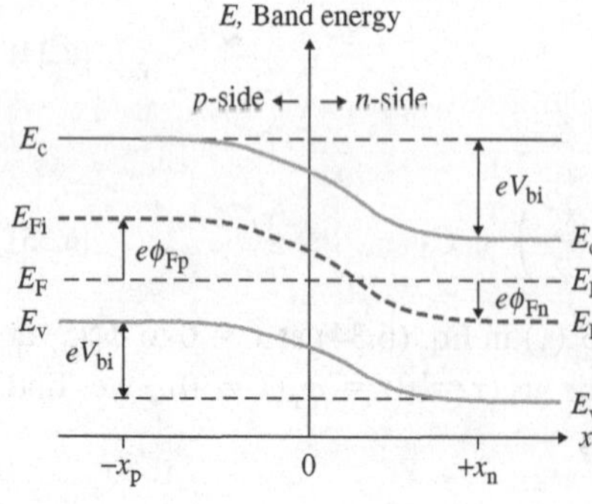

FIGURE 6.9 Energy band diagram of a pn junction in thermal equilibrium

Similarly, substituting $x_n = N_a x_p/N_d$ from Eq. (6.32) into Eq. (6.37), and then solving for x_p, we get the space charge extension x_p in the *p*-region as

$$x_p = \sqrt{\frac{2\varepsilon_s V_{bi}}{e}\left(\frac{N_d}{N_a}\right)\left(\frac{1}{N_a + N_d}\right)} \tag{6.40}$$

Therefore, the total width W of the depletion or space charge region is the sum of x_n and x_p:

$$W = x_n + x_p = \sqrt{\frac{2\varepsilon_s V_{bi}}{e}\left(\frac{N_a + N_d}{N_a N_d}\right)} \tag{6.41}$$

The space charge width W depends on the doping concentrations N_a and N_d. Once the built-in potential V_{bi} is determined from Eq. (6.24), the total space charge region width W can be determined from Eq. (6.41). Substituting for x_n or x_p, the maximum field in Eq. (6.31) can be related to the impurity concentrations N_d and N_a by

$$\mathcal{E}_{max} = \frac{-eN_d x_n}{\varepsilon_s} = \frac{-eN_a x_p}{\varepsilon_s} = -\sqrt{\frac{2eV_{bi}}{\varepsilon_s}\left(\frac{N_a N_d}{N_a + N_d}\right)} \tag{6.42}$$

which can also be written as a function of W:

$$\mathcal{E}_{max} = \frac{-2V_{bi}}{W} \tag{6.43}$$

EXAMPLE 6.1

Finding the space charge widths and the peak electric field in a *pn* junction The parameters of a uniformly doped *pn* junction for silicon semiconductors are $V_T = 26$ mV, $T = 25°$C, $N_a = 1 \times 10^{16}$ cm^{-3}, $N_d = 2 \times 10^{15}$ cm^{-3}, and $n_i = 1.5 \times 10^{16}$ cm^{-3}.

Find (a) the depletion width W and (b) the maximum field $\mathcal{E}_{max}$.

SOLUTION

$T = 25°$C, $N_a = 1 \times 10^{16}$ cm^{-3}, and $N_d = 2 \times 10^{15}$ cm^{-3}.

$\varepsilon_r = 11.7$ $\quad \varepsilon_o = 8.85 \times 10^{-14}$ $\quad e = 1.6 \times 10^{-19}$ $\quad k = 1.3806 \times 10^{-23}$

$T_k = 273 + T = 273 + 25 = 298$ K $\quad \varepsilon_s = \varepsilon_r \times \varepsilon_o = 11.7 \times 8.85 \times 10^{-14} = 1.035 \times 10^{-12}$

(a) From Eq. (6.24),

$$V_{bi} = 26 \times 10^{-3} \times \ln\left[\frac{1 \times 10^{16}\text{ cm}^{-3} \times 2 \times 10^{15}\text{ cm}^{-3}}{(1.5 \times 10^{16})^2\text{ cm}^3}\right] = 0.648 \text{ V}$$

From Eq. (6.39),

$$x_n = \sqrt{\frac{2 \times 1.035 \times 10^{-12} \times 0.648}{1.6 \times 10^{-19}}\left(\frac{1 \times 10^{16}\text{ cm}^{-3}}{2 \times 10^{15}\text{ cm}^{-3}}\right)\left(\frac{1}{1 \times 10^{16}\text{ cm}^{-3} + 2 \times 10^{15}\text{ cm}^{-3}}\right)} = 0.5913\ \mu\text{m}$$

From Eq. (6.40),

$$x_n = \sqrt{\frac{2 \times 1.035 \times 10^{-12} \times 0.648}{1.6 \times 10^{-19}}\left(\frac{2 \times 10^{15}\,\text{cm}^{-3}}{1 \times 10^{16}\,\text{cm}^{-3}}\right)\left(\frac{1}{1 \times 10^{16}\,\text{cm}^{-3} + 2 \times 10^{15}\,\text{cm}^{-3}}\right)} = 0.1183\ \mu\text{m}$$

Therefore, $W = x_n + x_p = 0.5913\ \mu\text{m} + 0.1183\ \mu\text{m} = 0.7096\ \mu\text{m}$.

(b) From Eq. (6.42),

$$\mathcal{E}_{max} = \frac{-eN_d x_n}{\varepsilon_s} = \frac{-1.6 \times 10^{-19} \times 2 \times 10^{15}\,\text{cm}^{-3} \times 0.5913\ \mu\text{m}}{1.035 \times 10^{-12}\,\text{cm}^{-1}} = -1.827 \times 10^4\ \text{V/cm}$$

KEY POINTS OF SECTION 6.3

- A semiconductor diode is formed by sandwiching a *p*-type material into one side and an *n*-type material in the other side of a single crystal.
- The built-in potential depends on the donor and acceptor concentrations, which are a strong function of temperature.
- An electric field exists in the depletion region. The width of the space charge region depends on the doping concentrations.

6.4 Reverse-Biased *pn* Junction

A *pn* junction is said to be reverse biased if the *n*-side is made positive with respect to the *p*-side, as depicted in Fig. 6.10(a). If the reverse voltage $v_R = v_D$ is increased, the potential barrier is increased from V_{bi} to $V_{bi} + V_R$ as shown in Fig. 6.10(b). The holes from the *p*-side and the electrons from the *n*-side cannot cross the junction, and the diffusion current I_{DF} due to the majority carriers will be negligible. Because of a higher potential barrier, however, the minority holes in the *n*-side will be swept easily across the junction to the *p*-side; the minority electrons in the *p*-side will be swept across the junction to the *n*-side. Thus, the current will flow solely due to the minority carriers. The reverse current flow will be due to the drift current I_{DR}, which is known as the reverse saturation (or leakage) current, denoted by I_S as in Eq. (4.1).

The number of minority carriers available is very small, and consequently the resulting current is also very small, on the order of pico-amperes. The production of minority carriers is dependent on the temperature. Thus, if the reverse voltage V_R is increased further, the diode current remains almost constant until a breakdown condition is reached. If the temperature increases, however, the reverse diode current also increases. The width of the depletion region grows with an increase in the reverse voltage.

Since there will not be an equilibrium condition in the *p*- and *n*-regions, the Fermi energy level will no longer be constant through the system. Figure 6.10(c) shows the energy band diagram of the *pn* junction. E_c and E_v are shifted by the total voltage $V_{PB} = V_{bi} + V_R$. As V_R pushes the energy levels, the Fermi level on the *n*-side E_{Fn} is now below the Fermi level on the *p*-side E_{Fp}. The difference between E_{Fp} and E_{Fn} is equal to eV_R—that is, $E_{Fp} - E_{Fn} = eV_R$.

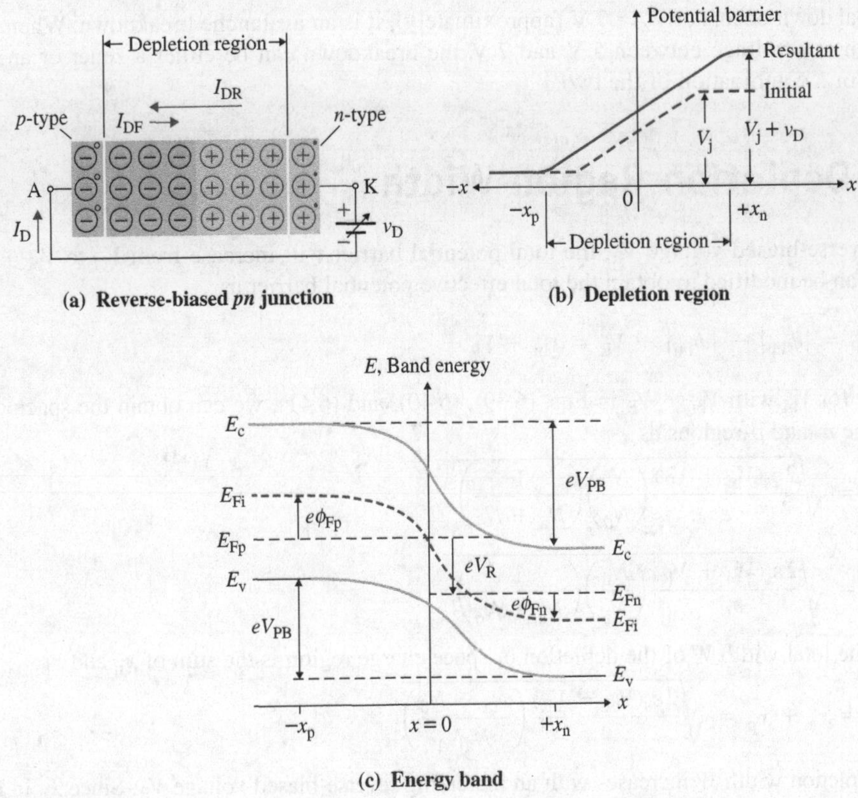

FIGURE 6.10 Reverse-biased *pn* junction

6.4.1 Breakdown Condition

If the reverse voltage is kept sufficiently high, the electric field in the depletion layer will be strong enough to break the covalent bonds of silicon (or germanium) atoms, producing a large number of electron–hole pairs throughout the semiconductor crystal. These electrons and holes give rise to a large reverse current flow. The *depletion region* (often called the *space charge region*) becomes so wide that collisions are less likely, but the even more intense electric field has the force to break the bonds directly. This phenomenon is called the *tunneling effect* or the *zener effect*. The mechanism is known as *zener breakdown,* in which case electrons and holes in turn cancel the negative and positive charges of the depletion region, and the junction potential barrier is virtually removed. The reverse current is then limited by the external circuit only, while the reverse terminal voltage remains almost constant at the zener voltage V_Z (see Sec. 4.7).

When the high electric field becomes strong enough, the electrons in the *p*-side will be accelerated through the crystal and will collide with the unbroken covalent bonds with a force sufficient to break them. The electrons generated by the collisions may gain enough kinetic energy to strike other unbroken bonds with sufficient force to break them as well. This cumulative effect, which will result in a large amount of uncontrolled current flow, is known as an *avalanche breakdown.*

In practice, the zener and avalanche effects are indistinguishable because both lead to a large reverse current. When a breakdown occurs at $V_Z < 5$ V (as in heavily doped junctions), it is a zener breakdown.

When a breakdown occurs at $V_Z > 7$ V (approximately), it is an avalanche breakdown. When a junction breaks down at a voltage between 5 V and 7 V, the breakdown can be either a zener or an avalanche breakdown or a combination of the two.

6.4.2 Depletion Region Width

With the reverse-biased voltage V_R, the total potential barrier will increase from V_{bi} to $V_{bi} + V_R$. Thus Eq. (6.18) can be modified to obtain the total effective potential barrier as

$$V_{PB} = |\phi_{Fp}| + |\phi_{Fn}| + V_R = V_{bi} + V_R \tag{6.44}$$

Substituting for V_{bi} with $V_{bi} + V_R$ in Eqs. (6.39), (6.40), and (6.41), we can obtain the space charge extension in the *n*- and *p*-regions as

$$x_n = \sqrt{\frac{2\varepsilon_s(V_{bi} + V_R)}{e}\left(\frac{N_a}{N_d}\right)\left(\frac{1}{N_a + N_d}\right)} \tag{6.45}$$

$$x_p = \sqrt{\frac{2\varepsilon_s(V_{bi} + V_R)}{e}\left(\frac{N_d}{N_a}\right)\left(\frac{1}{N_a + N_d}\right)} \tag{6.46}$$

Therefore, the total width W of the depletion or space charge region is the sum of x_n and x_p:

$$W = x_n + x_p = \sqrt{\frac{2\varepsilon_s(V_{bi} + V_R)}{e}\left(\frac{N_a + N_d}{N_a N_d}\right)} \tag{6.47}$$

Thus, the depletion width W increases with an increasing reverse-biased voltage V_R. Since x_n in Eq. (6.45) and x_p in Eq. (6.46) increase with reverse-biased voltage V_R, the magnitudes of the electric fields in Eq. (6.28) and Eq. (6.30) also increase. We can find the maximum field from Eq. (6.43) as a function of V_R and W:

$$\varepsilon_{max} = \frac{-2\,(V_{bi} + V_R)}{W} \tag{6.48}$$

Thus, the maximum field increases with V_R and decreases with W.

EXAMPLE 6.2

Finding the depletion width in a reverse-biased *pn* junction The parameters of a uniformly doped *pn* junction for silicon semiconductors are $V_R = 10$ V, $V_T = 26$ mV, $T = 25°C$, $N_a = 1 \times 10^{16}$ cm^{-3}, $N_d = 2 \times 10^{15}$ cm^{-3}, $e = 1.6 \times 10^{-19}$, and $k = 1.3806 \times 10^{-23}$.

Find (a) the depletion width W and (b) the maximum field ε_{max}.

SOLUTION

$T = 25°C$, $N_a = 1 \times 10^{16}$ cm^{-3}, and $N_d = 2 \times 10^{15}$ cm^{-3}.

$\varepsilon_r = 11.7$ $\varepsilon_o = 8.85 \times 10^{-14}$ $e = 1.6 \times 10^{-19}$ $k = 1.3806 \times 10^{-23}$

$T_k = 273 + T = 273 + 25 = 298$ K $\varepsilon_s = \varepsilon_r \times \varepsilon_o = 11.7 \times 8.85 \times 10^{-14} = 1.035 \times 10^{-12}$

(a) From Eq. (6.24),

$$V_{bi} = 26 \times 10^{-3} \times \ln\left[\frac{1 \times 10^{16}\,\text{cm}^{-3} \times 2 \times 10^{15}\,\text{cm}^{-3}}{(1.5 \times 10^{16}\,\text{cm}^{-3})^2}\right] = 0.648\ \text{V}$$

From Eq. (6.45),

$$x_n = \sqrt{\frac{2 \times 1.035 \times 10^{-12} \times (0.648 + 10)}{1.6 \times 10^{-19}}\left(\frac{1 \times 10^{16}\,\text{cm}^{-3}}{2 \times 10^{15}\,\text{cm}^{-3}}\right)\left(\frac{1}{1 \times 10^{16}\,\text{cm}^{-3} + 2 \times 10^{15}\,\text{cm}^{-3}}\right)} = 2.396\ \mu\text{m}$$

From Eq. (6.36),

$$x_n = \sqrt{\frac{2 \times 1.035 \times 10^{-12} \times (0.648 + 10)}{1.6 \times 10^{-19}}\left(\frac{2 \times 10^{15}\,\text{cm}^{-3}}{1 \times 10^{16}\,\text{cm}^{-3}}\right)\left(\frac{1}{1 \times 10^{16}\,\text{cm}^{-3} + 2 \times 10^{15}\,\text{cm}^{-3}}\right)} = 0.4793\ \mu\text{m}$$

Therefore, $W = x_n + x_p = 2.396\ \mu\text{m} + 0.4793\ \mu\text{m} = 2.876\ \mu\text{m}$.

(b) From Eq. (6.48),

$$\varepsilon_{max} = \frac{-1.6 \times 10^{-19} \times (0.648 + 10)\ \text{V}}{2.876 \times 10^{-6}\ \text{m}} = -7.406 \times 10^4\ \text{V/cm}$$

6.4.3 Junction Capacitance

Since a depletion region has positive charges in one side and negative charges in another side, there will be a capacitance associated with the *pn* junction. To find the junction capacitance, let us consider a small increase in the reverse voltage by dV_R, which will add an incremental positive charge $dQ = eN_d dx_n$ in the *n*-region and an incremental negative charge $-dQ = -eN_a dx_p$ in the *p*-region. These incremental changes are shown in Fig. 6.11.

Thus, the junction capacitance per square area with the variation of the reverse voltage becomes

$$C_j = \frac{dQ}{dV_R} = \frac{eN_d dx_n}{dV_R} = \frac{eN_a dx_p}{dV_R} \tag{6.49}$$

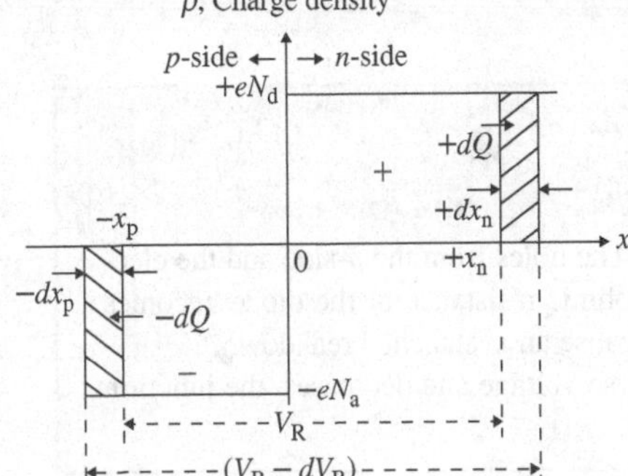

FIGURE 6.11 Incremental changes in the space charge width with an incremental change in reverse-biased voltage

Substituting x_n from Eq. (6.45), we get

$$C_j = eN_d \frac{dx_n}{dV_R} = eN_d \frac{d}{dV_R}\left[\frac{2\varepsilon_s(V_{bi}+V_R)}{e}\left(\frac{N_a}{N_d}\right)\left(\frac{1}{N_a+N_d}\right)\right]^{1/2} \tag{6.50}$$

After we complete the differentiation, this gives

$$C_j = \left[\frac{e\varepsilon_s N_a N_d}{(V_{bi}+V_R)(N_a+N_d)}\right]^{1/2} = \frac{K_j}{\sqrt{V_{bi}+V_R}} \tag{6.51}$$

where K_j is a constant for a specific *pn* junction. Equation (6.51) can be expressed as function of W in Eq. (6.47):

$$C_j = \frac{\varepsilon_s}{W} \tag{6.52}$$

Therefore, the junction capacitance C_j deceases with the reverse voltage V_R and the depletion width W. C_j is also referred to as the *depletion layer capacitance*. We can obtain the same expression for C_j if we use x_p from Eq. (6.46) in $dQ = eN_a dx_p$.

EXAMPLE 6.3

Finding the junction capacitance of a reverse-biased *pn* junction The parameters of a reverse-biased *pn* junction with uniform doping of silicon semiconductors are $V_T = 26$ mV, $V_R = 10$ V, $T = 25°C$, $N_a = 1 \times 10^{16}$ cm^{-3}, and $N_d = 2 \times 10^{15}$ cm^{-3}. Use the parameters of Example 6.2.

Calculate the junction capacitance if the cross-sectional area of the *pn* junction is $A_{pn} = 10^{-3}$ cm^2.

SOLUTION

$V_T = 26$ mV, $V_R = 10$ V, $T = 25°C$, $N_a = 1 \times 10^{16}$ cm^{-3}, $N_d = 2 \times 10^{15}$ cm^{-3}, and $A_{pn} = 10^{-3}$ cm^2.

From Eq. (6.51),

$$C_x = \sqrt{\frac{1.6\times10^{-19}\times1.035\times10^{-12}\times1\times10^{16}\,\text{cm}^{-3}\times2\times10^{15}\,\text{cm}^{-3}}{(0.648+10)\times(1\times10^{16}\,\text{cm}^{-3}+2\times10^{15}\,\text{cm}^{-3})}} = 3.601\ \text{nF/cm}^2$$

Therefore, $C_j = C_x A_{pn} = 3.601\ \text{nF} \times 10^{-3}\ \text{cm}^2 = 3.601$ pF.

KEY POINTS OF SECTION 6.4

- If a diode is reverse biased, the potential barrier is increased. The holes from the *p*-side and the electrons from the *n*-side cannot cross the junction. That is, the ohmic resistance of the diode becomes very high. A sufficiently high reverse voltage, however, may cause an avalanche breakdown.
- The width of the space charge region increases with the reverse voltage and decreases the junction capacitance.

6.5 Forward-Biased *pn* Junction

A *pn* junction is said to be forward biased if the *p*-side is made positive with respect to the *n*-side, as depicted in Fig. 6.12(a). If the forward voltage $v_D = V_F$ is increased, the potential barrier is reduced to $V_{bi} - V_F$, as shown in Fig. 6.12(b), and a large number of holes flow from the *p*-side to the *n*-side. Similarly, a large number of electrons flow from the *n*-side to the *p*-side. The resultant diode current becomes $I_D = I_{DF} - I_{DR}$. As the diode current I_D increases, the ohmic resistances of the *p*-side and the *n*-side cause a significant series voltage drop. If V_D is increased further, most of the increase in I_D will be lost as a series voltage drop. Thus, the width of the depletion region is reduced with the increase in the forward voltage. The potential barrier will not be reduced proportionally, but it can become zero.

In this case, the barrier height between the two regions is reduced. Figure 6.12(c) shows the energy band diagram of the *pn* junction. E_c and E_v are shifted by the total voltage $V_{PB} = V_{bi} - V_F$. As V_F pushes the energy levels, the Fermi level on the *n*-side (E_{Fn}) is now above the Fermi level on the *p*-side (E_{Fp}). The difference between E_{Fp} and E_{Fn} is equal to eV_F. That is, $E_{Fn} - E_{Fp} = eV_F$.

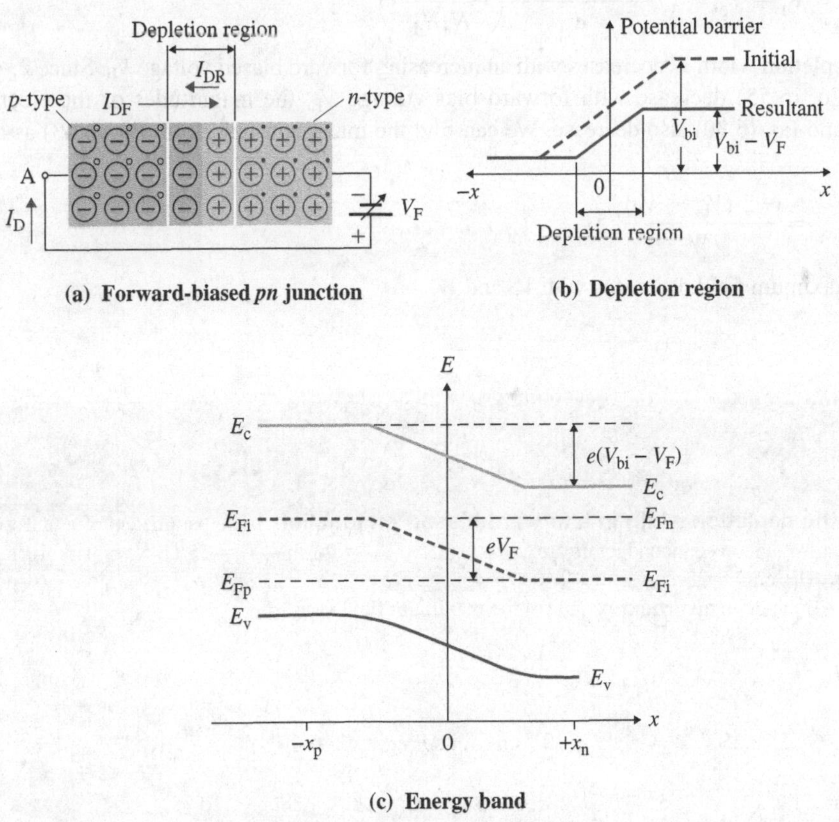

FIGURE 6.12 Forward-biased *pn* junction

6.5.1 Depletion Region Width

With the forward-biased voltage $v_D = V_F$, the total potential barrier will decrease from V_{bi} to $V_{bi} - V_F$. Thus, Eq. (6.18) can be modified to obtain the total effective potential as

$$V_{PB} = |\phi_{Fp}| + |\phi_{Fn}| - V_F = V_{bi} - V_F \quad (6.53)$$

Substituting for V_{bi} with $V_{bi} - V_F$ in Eqs. (6.39) through (6.41), we can obtain the space charge extension in the *n*- and *p*-regions as

$$x_n = \sqrt{\frac{2\varepsilon_s(V_{bi} - V_F)}{e}\left(\frac{N_a}{N_d}\right)\left(\frac{1}{N_a + N_d}\right)} \quad (6.54)$$

$$x_p = \sqrt{\frac{2\varepsilon_s(V_{bi} - V_F)}{e}\left(\frac{N_d}{N_a}\right)\left(\frac{1}{N_a + N_d}\right)} \quad (6.55)$$

Therefore, the total width W of the depletion or space charge region is the sum of x_n and x_p:

$$W = x_n + x_p = \sqrt{\frac{2\varepsilon_s(V_{bi} - V_F)}{e}\left(\frac{N_a + N_d}{N_a N_d}\right)} \quad (6.56)$$

Thus, the depletion width W decreases with an increasing forward biased voltage V_F. Since x_n in Eq. (6.54) and x_p in Eq. (6.55) decrease with forward bias voltage V_F, the magnitudes of the electric fields in Eq. (6.28) and Eq. (6.30) also decrease. We can find the maximum field from Eq. (6.43) as a function of V_F and W:

$$\varepsilon_{max} = \frac{-2\,(V_{bi} - V_F)}{W} \quad (6.57)$$

Thus, the maximum field decreases with V_F and W.

EXAMPLE 6.4

Finding the depletion width in a forward-biased *pn* junction The parameters of a uniformly doped *pn* junction for silicon semiconductors are $V_F = 0.60$ V, $V_T = 26$ mV, $T = 25°C$, $N_a = 1 \times 10^{16}\ cm^{-3}$, and $N_d = 2 \times 10^{15}\ cm^{-3}$.

Find (a) the depletion width W and (b) the maximum field ε_{max}.

SOLUTION

$V_F = 0.65$ V, $T = 25°C$, $N_a = 1 \times 10^{16}\ cm^{-3}$, and $N_d = 2 \times 10^{15}\ cm^{-3}$.

$\varepsilon_r = 11.7$ $\quad \varepsilon_o = 8.85 \times 10^{-14}$ $\quad e = 1.6 \times 10^{-19}$ $\quad k = 1.3806 \times 10^{-23}$

$T_k = 273 + T = 273 + 25 = 298$ K $\quad \varepsilon_s = \varepsilon_r \times \varepsilon_o = 11.7 \times 8.85 \times 10^{-14} = 1.035 \times 10^{-12}$

(a) From Eq. (6.24),

$$V_{bi} = 26 \times 10^{-3} \times \ln\left[\frac{1 \times 10^{16}\ \text{cm}^{-3} \times 2 \times 10^{15}\ \text{cm}^{-3}}{(1.5 \times 10^{16}\ \text{cm}^{-3})^2}\right] = 0.648\ \text{V}$$

From Eq. (6.54),

$$x_n = \sqrt{\frac{2 \times 1.035 \times 10^{-12} \times (0.648 - 0.6)}{1.6 \times 10^{-19}}\left(\frac{1 \times 10^{16}\ \text{cm}^{-3}}{2 \times 10^{15}\ \text{cm}^{-3}}\right)\left(\frac{1}{1 \times 10^{16}\ \text{cm}^{-3} + 2 \times 10^{15}\ \text{cm}^{-3}}\right)} = 0.1613\ \mu\text{m}$$

From Eq. (6.55),

$$x_p = \sqrt{\frac{2 \times 1.035 \times 10^{-12} \times (0.648 - 0.6)}{1.6 \times 10^{-19}}\left(\frac{2 \times 10^{15}\ \text{cm}^{-3}}{1 \times 10^{16}\ \text{cm}^{-3}}\right)\left(\frac{1}{1 \times 10^{16}\ \text{cm}^{-3} + 2 \times 10^{15}\ \text{cm}^{-3}}\right)} = 0.03226\ \mu\text{m}$$

Therefore, $W = x_n + x_p = 0.1613\ \mu\text{m} + 0.03226\ \mu\text{m} = 0.1936\ \mu\text{m}$.

(b) From Eq. (6.57),

$$\varepsilon_{max} = \frac{-1.6 \times 10^{-19} \times (0.648 - 0.6)}{0.1936 \times 10^{-6}\ \text{m}} = -4.983 \times 10^{3}\ \text{V/cm}.$$

6.5.2 Minority Carrier Charge Distribution

Since a large number of electrons diffuse from the *n*-side to the *p*-side and become minority carriers in the *p*-region, let us define that $n_{no} = N_d$ is the concentration of majority carrier electrons in the *n*-region once thermal equilibrium is reached. Since the product of n_{no} and n_{po} is constant according to Eq. (6.16), we can find the concentration of minority carrier electrons in the *p*-region as

$$n_{po} = \frac{n_i^2}{n_{no}} = \frac{n_i^2}{N_a} \tag{6.58}$$

Equation (6.24) gives the built-in potential as

$$V_{bi} = V_T \ln\left(\frac{N_d N_a}{n_i^2}\right)$$

which, after we substitute N_a from Eq. (6.58) and $n_{no} = N_d$, gives

$$V_{bi} = V_T \ln\left(\frac{N_d N_a}{n_i^2}\right) = V_T \ln\left(\frac{n_{no}}{n_{po}}\right) \tag{6.59}$$

Taking a natural log on both sides of Eq. (6.59), we can find the minority carrier electron concentration on the *p*-side, n_{po}, as a function of the majority carrier electron concentration on the *n*-side, n_{no}:

$$n_{po} = n_{no} e^{-V_{bi}/V_T} \tag{6.60}$$

For a forward-biased junction, we can substitute for V_{bi} in Eq. (6.60) with the effective voltage $V_{bi} - V_F$ to find the minority carrier electron concentration on the *p*-side, n_p:

$$n_p = n_{no} e^{-(V_{bi}-V_F)/V_T} = n_{no} e^{-V_{bi}/V_T} \times e^{V_F/V_T} \tag{6.61}$$

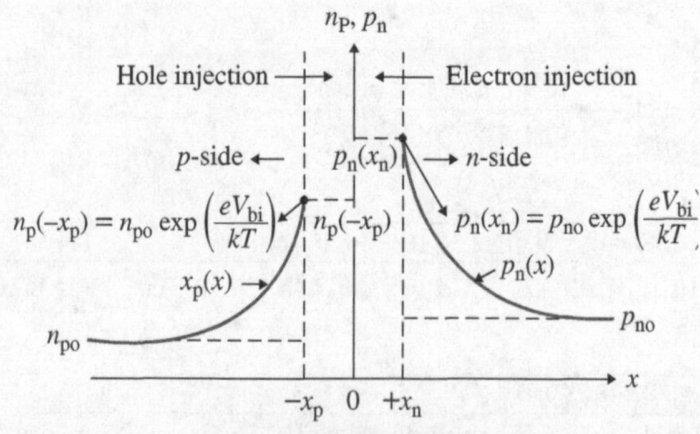

FIGURE 6.13 Excess minority carrier concentrations and charge distribution

This can be expressed as a function of n_{po} as

$$n_p = n_{po} \times e^{V_F/K_T} \tag{6.62}$$

Therefore, n_p is greater than n_{po} on the *p*-side, and a forward-biased *pn* junction is no longer in thermal equilibrium. Similarly, we can find the minority carrier hole concentration on the *n*-side, p_n:

$$p_n = p_{no} \times e^{V_F/V_T} \tag{6.63}$$

This also shows that p_n is greater than p_{no} on the *n*-side. Therefore, a forward-biased *pn* junction will create excess minority carriers at each edge of the space charge region of the *pn* junction, as shown in Fig. 6.13. Due to the exponential relationship, a relatively small forward-biased voltage can cause a significant increase in the minority carrier concentration. It is important to note that n_p and p_n decay exponentially with distance away from the junction to their thermal equilibrium values n_{po} and p_{no}, as also shown in Fig. 6.13.

EXAMPLE 6.5

Finding the minority carrier concentration of a forward-biased junction The parameters of a uniformly doped *pn* junction for silicon semiconductors are $V_F = 0.60$ V, $V_T = 26$ mV, $T = 25°C$, $N_a = 2 \times 10^{15}\ \text{cm}^{-3}$, and $N_d = 1 \times 10^{16}\ \text{cm}^{-3}$.

Find the minority carrier concentrations at the edge of the depletion region: (a) electrons in the *p*-side, n_p, and (b) holes in the *n*-side, p_n.

SOLUTION

$V_F = 0.60$ V, $V_T = 26$ mV, $T = 25°C$, $N_a = 2 \times 10^{15}\ \text{cm}^{-3}$, and $N_d = 1 \times 10^{16}\ \text{cm}^{-3}$.

$\varepsilon_r = 11.7 \quad \varepsilon_o = 8.85 \times 10^{-14} \quad e = 1.6 \times 10^{-19} \quad k = 1.3806 \times 10^{-23}$

$T_k = 273 + T = 273 + 25 = 298\ \text{K} \quad \varepsilon_s = \varepsilon_r \times \varepsilon_o = 11.7 \times 8.85 \times 10^{-14} = 1.035 \times 10^{-12}$

(a) From Eq. (6.58),

$$n_{po} = \frac{n_i^2}{N_a} = \frac{(1.5 \times 10^{16}\ \text{cm}^{-3})^2}{2 \times 10^{15}\ \text{cm}^{-3}} = 1.125 \times 10^5\ \text{cm}^{-3}$$

From Eq. (6.62),

$$n_p = n_{po} \times \exp\left(\frac{V_F}{V_T}\right) = 1.125 \times 10^5\ \text{cm}^{-3} \times \exp\left(\frac{0.6}{0.026}\right) = 1.531 \times 10^{15}\ \text{cm}^{-3}$$

(b) From Eq. (6.58),

$$p_{no} = \frac{n_i^2}{N_d} = \frac{(1.5 \times 10^{16}\ \text{cm}^{-3})^2}{1 \times 10^{16}\ \text{cm}^{-3}} = 2.25 \times 10^4\ \text{cm}^{-3}$$

From Eq. (6.63),

$$p_n = p_{no} \times \exp\left(\frac{V_F}{V_T}\right) = 2.25 \times 10^4\ \text{cm}^{-3} \times \exp\left(\frac{0.6}{0.026}\right) = 3.062 \times 10^{14}\ \text{cm}^{-3}$$

KEY POINTS OF SECTION 6.5

- If a diode is forward biased, the potential barrier is reduced and a large number of holes will flow from the *p*-side to the *n*-side. Similarly, a large number of electrons will flow from the *n*-side to the *p*-side. That is, the ohmic resistance of the diode becomes very small under forward-biased conditions.
- The width of the space charge region decreases with the forward voltage.
- A large number of electrons or holes diffuse from one side to the other, and they become minority carriers on the other side. This creates excess minority carriers at each edge of the space charge region of the *pn* junction.

6.6 Junction Current Density

It can be shown that the electron current density due to the charge flow from the *n*-region to the *p*-region at the edge of the *p*-region at $x = -x_p$ is given by [2–4]

$$J_n(x = -x_p) = \frac{eD_n n_{po}}{L_n}(e^{V_F/V_T} - 1) \tag{6.64}$$

where D_n is the minority electron diffusion density. L_n is the minority electron diffusion length and is related to D_n and the minority carrier life τ_{no} by $L_n^2 = D_n\tau_{no}$. The *minority carrier life* is defined as the average time for a minority electron in the *p*-region to recombine with a majority hole in the *p*-region. Similarly, the hole current density due to the charge flow from the *p*-region to the *n*-region at the edge of the *n*-region at $x = x_n$ is given by

$$J_p(x = x_n) = \frac{eD_p p_{no}}{L_p}(e^{V_F/V_T} - 1) \tag{6.65}$$

where D_p is the minority hole diffusion density. L_p is the minority hole diffusion length and is related to D_p and the minority hole carrier life τ_{po} by $L_p^2 = D_p\tau_{po}$. Therefore, the total current density in the *pn* junction is given by

$$J = J_n(x = -x_p) + J_p(x = x_n) = \left(\frac{eD_n n_{po}}{L_n} + \frac{eD_p p_{no}}{L_p}\right)(e^{V_F/V_T} - 1) \tag{6.66}$$

This can be written in a more general form as

$$J = J_s(e^{V_F/V_T} - 1) \tag{6.67}$$

where the parameter J_s is known as the *reverse saturation current density* and depends on the physical parameters of the *pn* junction as given by

$$J_s = \frac{eD_n n_{po}}{L_n} + \frac{eD_p p_{no}}{L_p} \tag{6.68}$$

Equation (6.67) for the junction current is applicable for both positive values (forward-biased condition) and negative values (reverse-biased condition) such that $v_D = -V_F$. The diode current can be obtained by multiplying the current density in Eq. (6.67) by the cross-sectional area of the *pn* junction A_{pn} and be expressed in the general form describing the Schottkey equation in Eq. (4.1) as

$$i_D = J \times A_{pn} = I_s(e^{\eta\, v_D/V_T} - 1) \tag{6.69}$$

Here the parameter I_s is known as the *reverse saturation current*, v_D is the applied voltage, and η is a constant whole value that varies from 1 to 2 depending on the manufacturing process of practical diodes. Equation (6.69) describes the characteristics of the Schottkey barrier junction.

▶ **NOTE** $v_D > 0$ for forward-biased conditions and $v_D < 0$ for reverse-biased conditions.

EXAMPLE 6.6

Finding the reverse saturation current The parameters of a reverse-biased *pn* junction with uniform doping of silicon semiconductors are $N_a = 1 \times 10^{16}\ \text{cm}^{-3}$, $N_d = 1 \times 10^{16}\ \text{cm}^{-3}$, $n_i = 1.5 \times 10^{10}\ \text{cm}^{-3}$, $D_n = 20\ \text{cm}^2/\text{s}$, $D_p = 10\ \text{cm}^2/\text{s}$, $\tau_{po} = \tau_{no} = 8 \times 10^{-6}$ s, and $A_{pn} = 10^{-3}\ \text{cm}^2$.

Find the reverse saturation current I_s.

SOLUTION

$N_a = 1 \times 10^{16}\ \text{cm}^{-3}$, $N_d = 1 \times 10^{16}\ \text{cm}^{-3}$, $n_i = 1.5 \times 10^{10}\ \text{cm}^{-3}$, $D_n = 20\ \text{cm}^2/\text{s}$, $D_p = 10\ \text{cm}^2/\text{s}$, $\tau_{po} = \tau_{no} = 8 \times 10^{-6}$ s, and $A_{pn} = 10^{-3}\ \text{cm}^2$.

$\varepsilon_r = 11.7$ $\quad \varepsilon_o = 8.85 \times 10^{-14}$ $\quad e = 1.6 \times 10^{-19}$ $\quad k = 1.3806 \times 10^{-23}$

From Eq. (6.58),

$$p_{no} = n_{po} = \frac{n_i^2}{N_a} = \frac{(1.5 \times 10^{16}\ \text{cm}^{-3})^2}{1 \times 10^{16}\ \text{cm}^{-3}} = 2.25 \times 10^4\ \text{cm}^{-3}$$

The minority electron diffusion length $L_n = \sqrt{D_n \tau_{no}} = \sqrt{20 \times 8 \times 10^{-6}\ \text{cm}^2} = 0.013$ cm. The minority hole diffusion length $L_p = \sqrt{D_p \tau_{po}} = \sqrt{10 \times 8 \times 10^{-6}\ \text{cm}^2} = 8.944 \times 10^{-3}$ cm.

From Eq. (6.68), we get

$$J_s = \frac{eD_n n_{po}}{L_n} + \frac{eD_p p_{no}}{L_p}$$

$$= \frac{1.6 \times 10^{-19} \times 20 \times 2.25 \times 10^4\ \text{cm}^{-3}}{0.013\ \text{m}} + \frac{1.6 \times 10^{-19} \times 10 \times 2.25 \times 10^4\ \text{cm}^{-3}}{8.944 \times 10^{-3}\ \text{cm}}$$

$$= 9.717 \times 10^{-12}\ \text{A/cm}^2$$

Therefore, the reverse saturation current is $I_s = J_s A_{pn} = 9.717 \times 10^{-12}\ \text{A/cm}^2 \times 10^{-3}\ \text{cm}^2 = 9.717 \times 10^{-15}$ A.

6.7 Temperature Dependence

According to Eq. (6.67), the current density J is a direct function of the reverse saturation current density J_s, which depends on the minority carrier concentrations n_{po} and p_{no}—which in turn are also proportional to n_i^2, which is a function of temperature. Therefore, we have

$$J_s \propto n_i^2 \propto T^3 e^{-E_g/kT} \tag{6.70}$$

where E_g is the electron energy, 1.12 eV. We can relate J_{s2} and J_{s1} corresponding to temperatures T_2 and T_1 by

$$\frac{J_{s2}}{J_{s1}} = \frac{T_2^3 e^{-E_g/kT_2}}{T_1^3 e^{-E_g/kT_1}} = \left(\frac{T_2}{T_1}\right)^3 \left(\frac{e^{-E_g/kT_2}}{e^{-E_g/kT_1}}\right) \tag{6.71}$$

Therefore, the reverse saturation current density J_s is sensitive to the temperature and increases rapidly with the temperature, as shown in Fig. 6.14(a) for a reverse-biased condition. The forward current J, which is a function of J_s and (eV_F/kT), is a function of temperature. As the temperature increases, the voltage drop decreases for the same amount of forward current, as shown in Fig. 6.14(b). For a constant forward voltage, the forward current increases with the temperature.

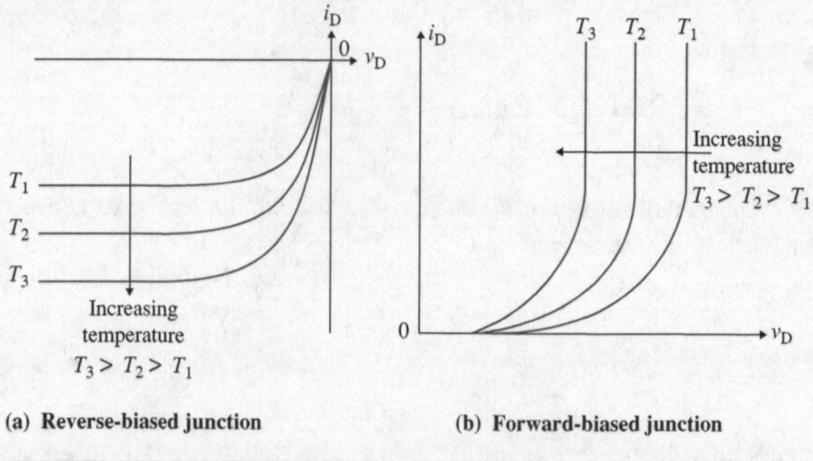

FIGURE 6.14 Temperature effects in a *pn* junction

6.8 High-Frequency AC Model

In Sec. 4.7 we considered the static behavior of a *pn* junction diode. A practical diode, however, exhibits some capacitive effects that need to be incorporated into any high-frequency model in order to get the time-dependent response of a diode circuit. We have seen that a depletion layer exists in the reverse-biased *pn* junction of diodes. That is, there is a region depleted of carriers, separating two regions of relatively good conductivity. Thus we have in essence a parallel-plate capacitor, with silicon as the dielectric. Also, there is an injection of a large number of minority carriers under forward-biased conditions. Therefore, there are two types of capacitances: depletion and diffusion.

6.8.1 Depletion Capacitance

A positively charged layer is separated from a negatively charged layer by a very small but finite distance. As the voltage across the *pn* junction changes, the charge stored in the depletion layer changes accordingly. This is shown in Fig. 6.15 for a nonlinear *q-v* relationship. The depletion capacitance relates the change in the charge (Δq) in the depleted region to the change in the bias voltage Δv_D, and it is given by

$$C_j = \left.\frac{dq_j}{dv_D}\right|_{\text{at estimated } Q\text{-point } v_D = -V_D} \tag{6.72}$$

This is derived in Eq. (6.51), and it can be expressed in general form as

$$C_j = \frac{C_{jo}}{(1 - v_D/V_j)^m} \tag{6.73}$$

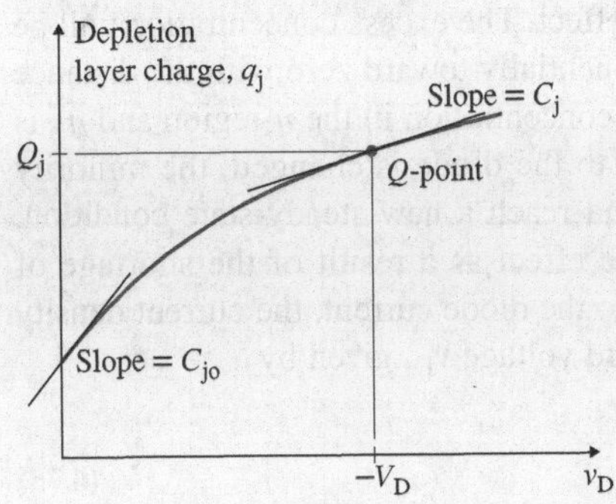

FIGURE 6.15 Charge–voltage relation of the depletion region

where m = junction gradient coefficient, whose value is in the range of 0.33 to 0.5

V_D = anode-to-cathode bias voltage, which will be positive in the forward direction and negative in the reverse direction

V_j = potential barrier with zero external voltage applied to the diode and is known as the *built-in potential* (It is a function of the type of semiconductor material, the degree of doping, and the junction temperature. For a silicon diode $V_j \approx$ 0.5 V to 0.9 V, and for a germanium diode $V_j \approx$ 0.2 V to 0.6 V.)

C_{jo} = depletion capacitance when the external voltage across the diode is zero

The depletion capacitance is also known as the *transition capacitance*. The value of C_j is directly proportional to the cross-section of the diode junction and is in the range of 0.1 pF to 100 pF. Notice from Eq. (6.72) that the depletion capacitance C_j can be varied by changing the reverse voltage $v_R = -v_D$ across the diode. The capability to change a capacitance by varying a voltage can be exploited in some applications. Diodes designed for such applications are called *varactors* or *varicaps*, depending on the applications. This depletion capacitance may be used for tuning FM radios, television circuits, microwave oscillators, and any other circuits in which a small variation in capacitance can effect a significant change in frequency. In these applications, a reverse-biased diode can be connected in parallel with an external capacitor of a parallel circuit consisting of resistor (R), indicator (L), and capacitor (C) circuit so that the resonant frequency f_p is given by

$$f_p = \frac{1}{2\pi\sqrt{L(C + C_j)}} \tag{6.74}$$

where C_j = depletion capacitance varied by the reverse-biased voltage $(-v_D)$ of the diode (Typical values of C_j are 10 pF to 100 pF at reverse voltages of 3 V to 25 V)

L = inductance of the parallel *RLC* circuit

C = capacitance of the parallel *RLC* circuit

6.8.2 Diffusion Capacitance

When the junction is forward biased, the depletion region becomes narrower and the depletion capacitance increases because the bias voltage v_D is positive. However, a large number of minority carriers are injected into the junction under the forward-biased condition. There will be an excess of minority charge carriers near the depletion layer, and this will cause a great charge storage effect. The excess concentration will be highest near the edge of the depletion layer and will decrease exponentially toward zero with the distance from the junction. This is shown in Fig. 6.13, where p_n is the hole concentration in the *n*-region and n_p is the electron concentration in the *p*-region. If the voltage applied to the diode is changed, the minority carrier charges stored in the *p*- and *n*-regions will also change and reach a new steady-state condition. Therefore, a forward-biased *pn* junction will exhibit a capacitive effect as a result of the shortage of minority carrier charges. Since these charges will be proportional to the diode current, the current density J_p in Eq. (6.65) can be applied to relate the charge q_m to the forward voltage v_D, given by

$$q_m = q_o(e^{v_D/V_T} - 1) \tag{6.75}$$

where q_o is the constant charge proportional to the leakage (or reverse saturation) current density J_s. Therefore, the *q-v* characteristic of a forward-biased diode will be nonlinear, and it can be modeled by a small-signal capacitance C_d known as the *diffusion capacitance*. That is,

$$C_d = \left.\frac{dq_m}{dv_D}\right|_{\text{at estimated } Q\text{-point } v_D = V_D} \tag{6.76}$$

which indicates that C_d is proportional to the value of $q_m + q_o$. In the reverse-biased condition, $C_d = 0$. In the forward direction, however, the value of C_d is approximately proportional to the DC bias current I_D (at the *Q*-point). That is, C_d is given by

$$C_d = K_d I_D \tag{6.77}$$

where K_d is a constant and C_d is directly proportional to the cross-section of the diode junction and is typically in the range of 10 pF to 100 pF.

6.8.3 Forward-Biased Model

A forward-biased diode will exhibit two capacitances: diffusion capacitance C_d and depletion-layer capacitance C_j, expressed by Eq. (6.52) for $v_D \geq 0$. These capacitances will affect the high-frequency applications of diodes. For the small-signal high-frequency model of a forward-biased diode, as shown in Fig. 6.16(a), the model parameters are given by

$$\frac{1}{r_d} = \left.\frac{di_D}{dv_D}\right|_{\text{at estimated } Q\text{-point } i_D = I_D} = \frac{d}{dv_D}[I_s(e^{\eta v_D/V_T} - 1)] \simeq \frac{I_D}{\eta V_T} \tag{6.78}$$

$$C_j = \frac{C_{jo}}{(1 - v_D/V_j)^m} \quad \text{for } v_D \geq 0 \tag{6.79}$$

$$C_d = K_d I_D$$

For example, if $C_{jo} = 4$ pF, $V_j = 0.75$ V, $m = 0.333$, and $V_D = 0.7158$ V, then $C_j = 11.18$ pF.

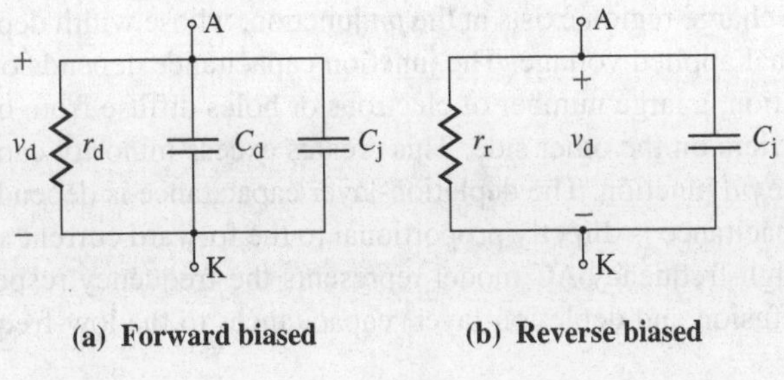

FIGURE 6.16 High-frequency AC diode models

6.8.4 Reverse-Biased Model

The small-signal AC resistance r_d in the reverse direction is very high, on the order of several megohms, and may be assumed to be very large, tending to infinity. The diffusion capacitance C_d, which depends on the diode current, is negligible in the reverse direction because the reverse current is very small. For the high-frequency AC model of a reverse-biased diode having r_r as the resistance in the reverse direction, as shown in Fig. 6.16(b), the model parameters are given by

$$r_r = \infty$$

$$C_j = \frac{C_{jo}}{(1 - v_D/V_j)^m} \quad \text{for } v_D \leq 0 \tag{6.80}$$

$$C_d = 0$$

For example, if $C_{jo} = 4$ pF, $V_j = 0.75$ V, $m = 0.333$, and $V_D = -20$ V, then $C_j = 1.32$ pF.

KEY POINT OF SECTION 6.8

- The high-frequency AC model represents the frequency response of the diode by including two junction (diffusion and depletion-layer) capacitances to the low-frequency AC model. The depletion-layer capacitance is dependent on the diode voltage. But the diffusion capacitance is directly proportional to the diode current and is present only in the forward direction.

Summary

A *pn* junction is formed by sandwiching a *p*-type material into one side and an *n*-type material in the other side of a single crystal. Free electrons (in *n*-type material) and holes (in *p*-type material) are made available by adding a controlled amount of *n*-type impurities and *p*-type impurities to pure semiconductors, respectively. The Fermi function specifies the probability that an available state will be occupied by an electron. The intrinsic concentration of a semiconductor material remains constant at a constant temperature.

A depletion or space charge region exists at the *pn* junction, whose width depends on the doping concentrations and the external applied voltage. The junction capacitance depends on the reverse voltage. In the forward-biased condition, a large number of electrons or holes diffuse from one side to the other side and become minority carriers on the other side. This creates excess minority carriers at each edge of the space charge region of the *pn* junction. The depletion-layer capacitance is dependent on the forward voltage. But the diffusion capacitance is directly proportional to the forward current and is present only in the forward direction. The high-frequency AC model represents the frequency response of the diode by including two junction (diffusion and depletion-layer) capacitances to the low-frequency AC model.

References

1. G. W. Neudeck, *The PN Junction Diode: Vol. 2 of the Modular Series on Solid State Devices*, 2nd ed. Reading, MA: Addison-Wesley, 1989.
2. D. A. Neamen, *Semiconductor Physics and Devices: Basic Principles*, 3rd ed. New York, NY: McGraw-Hill, 2003.
3. R. F. Pierret, *Semiconductor Device Fundamentals*. Reading, MA: Addison-Wesley, 1996.
4. B. G. Streetman and S. Banerjee, *Solid State Electronic Devices*, 5th ed. Upper Saddle River, NJ: Prentice Hall, 2006.

Review Questions

1. What is a donor impurity? What is an acceptor impurity?
2. What is doping?
3. What is the depletion region of a *pn* junction?
4. What are the minority carriers in *p*-type materials?
5. What are the majority carriers in *p*-type materials?
6. What are the minority carriers in *n*-type materials?
7. What are the majority carriers in *n*-type materials?
8. What is the effect of junction temperature on the diode characteristic?
9. What is the Fermi function?
10. What is the effect of temperature on the Fermi function?
11. What is the intrinsic electron or hole concentration in a semiconductor material?
12. What are the effects of reverse voltage on a *pn* junction?
13. What is the depletion region?
14. What is a built-in potential?
15. What is the effect of reverse voltage on the depletion region?
16. What is a breakdown condition of a *pn* junction?
17. What causes the junction capacitance of a *pn* junction?
18. What are the effects of forward voltage on a *pn* junction?
19. What is the minority carrier life?
20. Why is the reverse saturation current density sensitive to temperature?
21. What is the high-frequency AC model of a diode?
22. What is the depletion capacitance of a *pn* junction?
23. What is the diffusion capacitance of a *pn* junction?

Problems

6.2 Semiconductor Materials

6.1 Calculate the intrinsic carrier concentration n_i at $T = 200$ K, 400 K, and 600 K for **(a)** silicon, **(b)** germanium, and **(c)** gallium arsenide.

6.2 Two silicon semiconductor materials have the same properties but different gap band energies: $E_{gA} = 1.12$ eV and $E_{gB} = 1.25$ eV. Determine their intrinsic concentrations n_{iA} and n_{iB} and the ratio n_{iB}/n_{iA}.

6.3 If the maximum intrinsic concentration of silicon is to be limited to $n_i = 1.5 \times 10^{18}$ cm^{-3}, what will be the maximum permissible temperature if the gap band energy is $E_g = 1.15$ eV?

6.4 If the maximum intrinsic concentration of gallium arsenide is to be limited to $n_i = 1.5 \times 10^{18}$ cm^{-3}, what will be the maximum permissible temperature if the gap band energy is $E_g = 1.15$ eV?

6.5 Calculate the equilibrium electron concentration n_o of a silicon material if the gap band energy at $T = 350$ K is **(a)** $E_g = 0.75$ eV, **(b)** $E_g = 1.12$ eV, and **(c)** $E_g = 1.25$ eV.

6.6 Calculate the equilibrium electron concentration n_o for a gallium arsenide material if the gap band energy at $T = 350$ K is **(a)** $E_g = 0.75$ eV, **(b)** $E_g = 1.12$ eV, and **(c)** $E_g = 1.25$ eV.

6.7 Calculate the equilibrium hole concentration p_o of a silicon material if the gap band energy at $T = 350$ K is **(a)** $E_g = 0.75$ eV, **(b)** $E_g = 1.12$ eV, and **(c)** $E_g = 1.25$ eV.

6.8 Calculate the equilibrium hole concentration p_o for a gallium arsenide material if the gap band energy at $T = 350$ K is **(a)** $E_g = 0.75$ eV, **(b)** $E_g = 1.12$ eV, and **(c)** $E_g = 1.25$ eV.

6.9 The value of equilibrium electron concentration for a silicon material is $n_o = 1.5 \times 10^{17}$ cm^{-3} at $T = 30$°C. Determine the gap band energy E_g.

6.10 The value of equilibrium hole concentration for a silicon material is $n_o = 1.5 \times 10^{16}$ cm^{-3} at $T = 30$°C. Determine the gap band energy E_g.

6.11 Determine the intrinsic Fermi energy E_{Fi} for silicon if $E_g = 1.1$ eV and $T = 25$°C.

6.12 If the Fermi energy is 0.25 eV below the conduction band energy E_c and $N_c = 1.5 \times 10^{19}$ cm^{-3} at $T = 25$°C, **(a)** calculate the probability that an energy state in the conduction band at $(E_c + kT)$ is filled by an electron, and **(b)** calculate the thermal equilibrium electron concentration in silicon.

6.13 If the Fermi energy is 0.25 eV below the valence band energy E_v and $N_v = 1.5 \times 10^{19}$ cm^{-3} at $T = 25$°C, **(a)** calculate the probability that an energy state in the valence band at $(E_v - kT)$ is empty of an electron, and **(b)** calculate the thermal equilibrium hole concentration in silicon.

6.3 Zero-Biased pn Junction

6.14 The parameters of a uniformly doped *pn* junction for silicon semiconductors are $V_T = 26$ mV, $T = 25$°C, $N_a = 1 \times 10^{16}$ cm^{-3}, and $N_d = 2 \times 10^{15}$ cm^{-3}. Find **(a)** the depletion width W and **(b)** the maximum field $\mathcal{E}_{max}$.

6.15 The parameters of a uniformly doped *pn* junction for silicon semiconductors are $V_T = 26$ mV, $T = 25$°C, $N_a = 1 \times 10^{12}$ cm^{-3}, and $N_d = 10^{16}$ cm^{-3}. Find **(a)** V_{bi}, **(b)** x_n and x_p, **(c)** the depletion width W, and **(d)** the maximum field $\mathcal{E}_{max}$. Plot the electric field against the distance x through the junction.

6.16 Calculate the built-in potential V_{bi} of a uniformly doped *pn* junction for silicon semiconductors if $V_T = 26$ mV, $T = 25$°C, $N_a = 1 \times 10^{18}$ cm^{-3}, and $N_d = 2 \times 10^{15}$ cm^{-3}.

6.17 Plot the built-in potential V_{bi} against N_d for a uniformly doped *pn* junction for silicon semiconductors if V_T = 26 mV, T = 25°C, N_a = 1 × 10^{18} cm^{-3}, and 1 × 10^{14} ≤ N_d ≤ 1 × 10^{19} cm^{-3}.

6.18 Plot the built-in potential V_{bi} against N_d for a symmetrical silicon *pn* junction if V_T = 26 mV, T = 25°C, and 1 × 10^{14} ≤ N_a = N_d ≤ 1 × 10^{19} cm^{-3}.

6.19 Plot the built-in potential V_{bi} against temperature for a uniformly doped *pn* junction for silicon semiconductors if V_T = 26 mV, 25°C ≤ T ≤ 250°C, N_a = 1 × 10^{18} cm^{-3}, and N_d = 1 × 10^{19} cm^{-3}.

6.20 The parameters of a uniformly doped silicon *pn* junction are V_T = 26 mV, N_a = 1 × 10^{17} cm^{-3}, and N_d = 5 × 10^{15} cm^{-3}. Determine the temperature if the built-in potential barrier is V_{bi} = 0.56 V.

6.21 The parameters of a uniformly doped silicon *pn* junction are T = 25°C, N_a = 1 × 10^{16} cm^{-3}, and N_d = 5 × 10^{15} cm^{-3}. If the temperature changes by 15%, what will be the change in the built-in potential barrier V_{bi}?

6.22 The parameters of a uniformly doped silicon *pn* junction are T = 25°C, N_a = 1 × 10^{16} cm^{-3}, and N_d = 5 × 10^{15} cm^{-3}. If T, N_a, and N_d change by ±15%, what will be the minimum and maximum values of the built-in potential barrier V_{bi}?

6.4 Reverse-Biased pn Junction

6.23 The parameters of a uniformly doped *pn* junction for silicon semiconductors are V_R = 10 V, V_T = 26 mV, T = 25°C, N_a = 1 × 10^{16} cm^{-3}, and N_d = 2 × 10^{15} cm^{-3}. Find **(a)** the depletion width W and **(b)** the maximum field ε_{max}.

6.24 The parameters of a reverse-biased *pn* junction with uniform doping of silicon semiconductors are V_T = 26 mV, V_R = 15 V, T = 25°C, N_a = 1 × 10^{16} cm^{-3}, and N_d = 2 × 10^{15} cm^{-3}. Calculate the junction capacitance if the cross-sectional area of the *pn* junction is A_{pn} = 10^{-3} cm^2.

6.25 The parameters of a reverse-biased *pn* junction with uniform doping of silicon semiconductors are V_T = 26 mV, V_R = 10 V, T = 25°C, N_a = 1 × 10^{16} cm^{-3}, and N_d = 2 × 10^{15} cm^{-3}. Calculate the junction capacitance if the cross-sectional area of the *pn* junction is A_{pn} = 2 × 10^{-3} cm^2.

6.26 The parameters of a reverse-biased abrupt silicon *pn* junction are V_R = 12 V, T = 25°C, N_a = 10^{16} cm^{-3}, and N_d = 10^{15} cm^{-3}. Calculate **(a)** V_{bi}, **(b)** W, **(c)** the maximum field ε_{max}, and **(d)** the junction capacitance C_j if A_{pn} = 10^{-3} cm^2.

6.27 The parameters of a reverse-biased abrupt silicon *pn* junction are V_R = 12 V, T = 25°C, N_a = 10^{16} cm^{-3}, and N_d = 100 N_a. Calculate **(a)** V_{bi}, **(b)** W, **(c)** the maximum field ε_{max}, and **(d)** the junction capacitance C_j if A_{pn} = 2 × 10^{-3} cm^2.

6.28 The parameters of a reverse-biased abrupt silicon *pn* junction are V_R = 12 V, T = 25°C, N_a = 10^{16} cm^{-3}, and N_d = 100 N_a. Calculate **(a)** V_{bi}, **(b)** W, **(c)** the maximum field ε_{max}, and **(d)** the junction capacitance C_j. If V_R changes by ±20%, what will be the minimum and maximum values of the junction capacitance C_j if A_{pn} = 10^{-3} cm^2?

6.29 The parameters of a reverse-biased abrupt silicon *pn* junction are V_R = 12 V, T = 25°C, N_a = 10^{16} cm^{-3}, and N_d = 100 N_a. Calculate **(a)** V_{bi}, **(b)** W, **(c)** the maximum field ε_{max}, and **(d)** the junction capacitance C_j if A_{pn} = 10^{-3} cm^2. If N_a changes by ±15%, what will be the minimum and maximum values of the junction capacitance C_j?

6.30 The parameters of a reverse-biased abrupt silicon *pn* junction are V_R = 12 V, T = 25°C, N_a = 10^{16} cm^{-3}, and N_d = 2 × 10^{18} cm^{-3}. Calculate **(a)** V_{bi}, **(b)** W, **(c)** the maximum field ε_{max}, and **(d)** the junction capacitance C_j if A_{pn} = 10^{-3} cm^2. If N_a and N_d change by ±15%, what will be the minimum and maximum values of the maximum field ε_{max}?

6.31 The parameters of a reverse-biased *pn* junction with uniform doping of silicon semiconductors are $V_T = 26$ mV, $V_R = 10$ V, $T = 25°C$, $N_a = 1 \times 10^{16}\ cm^{-3}$, and $N_d = 2 \times 10^{15}\ cm^{-3}$. Calculate the reverse voltage V_R that will give a junction capacitance of $C_j = 5$ pF if $A_{pn} = 10^{-3}\ cm^2$.

6.32 A uniformly doped silicon *pn* junction operating at $T = 25°C$ is to be designed such that at a reverse-biased voltage of $V_R = 12$ V, the maximum field is limited to $\varepsilon_{max} = 5 \times 10^5$ V/cm. Determine the maximum doping concentration in the *n*-region.

6.33 A uniformly doped silicon *pn* junction operating at $T = 25°C$ has a reverse-biased voltage of $V_R = 12$ V. The charge in the *n*-region is to be limited to 15% of the total space charge, and the total junction capacitance is $C_j = 4$ pF if $A_{pn} = 6 \times 10^{-4}\ cm^2$. Determine **(a)** N_a, **(b)** N_d, and **(c)** V_{bi}.

6.5 Forward-Biased pn Junction

6.34 The parameters of a uniformly doped *pn* junction for silicon semiconductors are $V_F = 0.65$ V, $V_T = 26$ mV, $T = 25°C$, $N_a = 1 \times 10^{17}\ cm^{-3}$, and $N_d = 2 \times 10^{15}\ cm^{-3}$. Find **(a)** the depletion width W and **(b)** the maximum field ε_{max}.

6.35 The parameters of a uniformly doped *pn* junction for silicon semiconductors are $V_F = 0.5$ V, $V_T = 26$ mV, $T = 25°C$, $N_a = 2 \times 10^{16}\ cm^{-3}$, and $N_d = 2 \times 10^{15}\ cm^{-3}$. Find the minority carrier concentrations at the edge of the depletion region: **(a)** electrons in the *p*-side, n_p, and **(b)** holes in the *n*-side, p_n.

6.36 The parameters of a uniformly doped *pn* junction for silicon semiconductors are $V_F = 0.5$ V, $V_T = 26$ mV, $T = 25°C$, $N_a = 2 \times 10^{16}\ cm^{-3}$, and $N_d = 2 \times 10^{15}\ cm^{-3}$. If V_F changes by ±15%, calculate the minimum and maximum values of the minority carrier concentrations at the edge of the depletion region: **(a)** electrons in the *p*-side, n_p, and **(b)** holes in the *n*-side, p_n.

6.6 Junction Current Density

6.37 The parameters of a reverse-biased *pn* junction with uniform doping of silicon semiconductors are $N_a = 1 \times 10^{16}\ cm^{-3}$, $N_d = 2 \times 10^{15}\ cm^{-3}$, $n_i = 1.5 \times 10^{10}\ cm^{-3}$, $D_n = 20\ cm^2/s$, $D_p = 10\ cm^2/s$, $\tau_{po} = \tau_{no} = 8 \times 10^{-6}$ s, and $A_{pn} = 10^{-3}\ cm^2$. Find the reverse saturation current I_S.

6.38 The parameters of a reverse-biased *pn* junction with uniform doping of silicon semiconductors are $N_a = 1 \times 10^{16}\ cm^{-3}$, $N_d = 2 \times 10^{15}\ cm^{-3}$, $n_i = 1.5 \times 10^{10}\ cm^{-3}$, $D_n = 20\ cm^2/s$, $D_p = 10\ cm^2/s$, $\tau_{po} = \tau_{no} = 8 \times 10^{-6}$ s, and $A_{pn} = 10^{-3}\ cm^2$. If N_a and N_d change by ±15%, calculate the minimum and maximum values of the reverse saturation current I_S.

6.39 The parameters of a reverse-biased *pn* junction with uniform doping of silicon semiconductors are $N_a = 1 \times 10^{16}\ cm^{-3}$, $N_d = 50\ N_a$, $n_i = 1.5 \times 10^{10}\ cm^{-3}$, $D_n = 20\ cm^2/s$, $D_p = 10\ cm^2/s$, $\tau_{po} = \tau_{no} = 8 \times 10^{-6}$ s, and $A_{pn} = 10^{-3} cm^2$. Find the reverse saturation current I_S.

6.40 The parameters of a reverse-biased abrupt silicon *pn* junction are $V_F = 0.5$ V, $T = 25°C$, $N_a = 10^{16}\ cm^{-3}$, and $N_d = 2 \times 10^{18}\ cm^{-3}$. Calculate **(a)** W, **(b)** the maximum field ε_{max}, and **(c)** the junction capacitance C_j if $A_{pn} = 10^{-3}\ cm^2$. If N_a and N_d change by ±15%, what will be the minimum and maximum values of the maximum field ε_{max}?

6.41 The parameters of a reverse-biased *pn* junction with uniform doping of silicon semiconductors are $V_T = 26$ mV, $V_R = 10$ V, $T = 25°C$, $N_a = 1 \times 10^{16}\ cm^{-3}$, and $N_d = 2 \times 10^{15}\ cm^{-3}$. Calculate the reverse voltage V_R that will give a junction capacitance of $C_j = 5$ pF if $A_{pn} = 10^{-3}\ cm^2$.

6.42 A uniformly doped silicon *pn* junction operating at $T = 25°C$ is to be designed such that at a reverse-biased voltage of $V_R = 15$ V, the maximum field is limited to $\varepsilon_{max} = 5 \times 10^5$ V/cm. Determine the maximum doping concentration in the *n*-region.

6.43 A uniformly doped silicon *pn* junction operating at $T = 25°C$ has a reverse-biased voltage of $V_R = 12$ V. The charge in the *n*-region is to be limited to 15% of the total space charge, and the total junction capacitance is $C_j = 4$ pF if $A_{pn} = 6 \times 10^{-4}$ cm^2. Determine **(a)** N_a, **(b)** N_d, and **(c)** V_{bi}.

6.44 The parameters of a reverse-biased *pn* junction with uniform doping of silicon semiconductors are $N_a = 2 \times 10^{16}$ cm^{-3}, $N_d = 5 \times 10^{15}$ cm^{-3}, $n_i = 1.8 \times 10^{10}$ cm^{-3}, $D_n = 20$ cm^2/s, $D_p = 10$ cm^2/s, $\tau_{po} = \tau_{no} = 8 \times 10^{-6}$ s, and $A_{pn} = 10^{-3}$ cm^2. Find the reverse saturation current I_S.

6.45 The reverse saturation current of a forward-biased silicon *pn* junction diode is $I_S = 5 \times 10^{-14}$ A at $T = 25°C$. Determine the required diode voltage to induce a diode current of **(a)** $I_D = 1$ mA and **(b)** $I_D = 10$ mA.

6.46 The reverse saturation current of a forward-biased silicon *pn* junction diode is $I_S = 5 \times 10^{-14}$ A at $T = 25°C$. Determine the forward-biased diode current for **(a)** $V_F = 0.75$ V, **(b)** $V_F = 1.0$ V, and **(c)** $V_F = 1.2$ V.

6.47 The forward-biased current of a *pn* diode is $I_D = 10$ mA at $T = 25°C$. The GaAs *pn* junction at $T = 300$ K is $I_D = 15$ mA. The forward diode voltage is $V_F = 1.1$ V. Determine the reverse saturation current I_S.

6.7 Temperature Dependence

6.48 The saturation current is $I_S = 9.972 \times 10^{-15}$ A at $T = 25°C$. Find the value of I_S at $T = 50°C$. Assume $E_g = 1.15$ eV.

6.49 If the junction temperature changes by 5 times, what will be changes in the junction current density J_S, and the saturation current I_S? Assume $E_g = 1.15$ eV.

6.50 If the junction temperature changes by 10 times, what will be changes in the junction current density J_S, and the saturation current I_S? Assume $E_g = 1.15$ eV.

6.8 High-Frequency AC Model

6.51 If the parameters of a *pn* junction are $C_{jo} = 4$ pF, $V_j = 0.75$ V, $m = 0.333$, and $V_D = -50$ V, calculate the value of C.

6.52 If the parameters of a *pn* junction are $C_{jo} = 4$ pF, $V_j = 0.75$ V, $m = 0.333$, and $V_D = 0.7158$ V, calculate the value of C.

CHAPTER 7

METAL OXIDE SEMICONDUCTOR FIELD-EFFECT TRANSISTORS

Learning Outcomes

After completing this chapter, students should be able to do the following:

- Describe the operation of metal oxide semiconductor field-effect transistors (MOSFETs).
- List the types of MOSFETs and their characteristics.
- Analyze and design MOSFET biasing circuits.
- Determine the small-signal model parameters of MOSFETs.
- Analyze and design MOSFET amplifiers.
- List the circuit configurations of MOSFET amplifiers and their relative advantages and disadvantages.
- Determine the frequency model of MOSFETs.
- Determine the frequency responses of MOSFET amplifiers.

Symbols and Their Meanings

Symbol	Meaning
A_{vo}, G_{mo}	No-load voltage gain and transconductance of an amplifier
g_m, G_m	Transconductance of a MOSFET and an amplifier

Symbol	Meaning
g_{ds}, r_{ds}	Small-signal drain–source conductance and resistance of a MOSFET
i_d, I_D, i_D	AC, quiescent DC, and instantaneous DC drain currents
K_m, K_{pp}	MOS and process technology constants
K_p, K_n	MOS constants for PMOS and NMOS
L, W	Length and width of a MOSFET
R_i, R_o	Input and output resistances of an amplifier
r_o	Small-signal output resistance of a transistor
λ, V_M	Channel modulation length and voltage of a MOSFET
$v_O(t)$, $v_o(t)$	Instantaneous DC and AC output voltages
V_t, V_{tN}, V_{tP}	Threshold voltages of any MOSFET, NMOS, or PMOS
v_{ds}, V_{DS}, v_{DS}	Small-signal AC, quiescent DC, and instantaneous DC drain-to-source voltages
v_{gs}, V_{GS}, v_{GS}	Small-signal, quiescent DC, and instantaneous DC gate-to-source voltages

7.1 Introduction

In Chapter 2 we looked at an amplifier's characteristics from an input–output perspective and found the specifications of amplifiers that satisfied certain input and output requirements. Internally, amplifiers use one or more transistors as amplifying devices, and these transistors are biased from a single DC supply to operate properly at a desired (quiescent) Q-point. Using transistors, we can build amplifiers that give a voltage (or current) gain, a high input impedance, or a high (or low) output impedance. The terminal behavior of an amplifier depends on the types of devices used within the amplifier.

Transistors are active devices with highly nonlinear characteristics. Thus, to analyze and design a transistor circuit, we need models of transistors. Creating accurate models requires detailed knowledge of the physical operation of transistors and their parameters as well as a powerful analytical technique. A circuit can be analyzed easily using simple models, but there is generally a trade-off between accuracy and complexity. A simple model, however, is always useful to obtain the approximate values of circuit elements for use in a design exercise and the approximate performance of the elements for circuit evaluation. The details of transistor operation, characteristics, biasing, and modeling are outside the scope of this text [1–3]. In this chapter, we will consider the operation and external characteristics of field-effect transistors using simple linear models.

7.2 Metal Oxide Field-Effect Transistors

The basic concept of field-effect transistors (FETs) has been known since the 1930s; however, FETs did not find practical applications until the early 1960s. Since the late 1970s, MOSFETs have become very popular; they are being used increasingly in integrated circuits (ICs). The manufacturing of MOSFETs is relatively simple. A MOSFET device can be made small, and it occupies a small silicon area in an IC chip. MOSFETs are currently used for very-large-scale integrated (VLSI) circuits such as microprocessors and memory chips.

A metal oxide semiconductor field-effect transistor (MOSFET) is a unipolar device. The current flow in a MOSFET depends on one type of majority carrier (electrons or holes). The output current of MOSFETs is controlled by an electric field that depends on a gate control voltage. There are two types of MOSFETs: enhancement MOSFETs and depletion MOSFETs.

7.3 Enhancement MOSFETs

There are two types of enhancement MOSFETs: *n*-channel and *p*-channel. An *n*-channel enhancement MOSFET is often referred to as an NMOS. The physical structure of an NMOS showing its terminal is illustrated in Fig. 7.1(a); a schematic appears in Fig. 7.1(b). Since the *p*-type substrate and the two n^+-type junctions are reverse biased, there will be a depletion region as shown in Fig. 7.1(b) by shaded lines. Two n^+-type regions act as low-resistance connections to the source and the drain. An insulating layer of silicon dioxide is formed on top of the *p*-type substrate by oxidizing the silicon. Ohmic contacts are provided to the n^+-regions for connection to the external circuit by leaving two windows on the silicon dioxide and depositing a layer of aluminum. The substrate B is normally connected to the source terminal. An *n*-channel is induced under the influence of an electric field; there is no physical *n*-channel between the drain and the source of an NMOS, as shown by the darker shade in Fig. 7.1(b). The symbol for an NMOS is shown in Fig. 7.1(c), where the arrow points from the *p*-type region to the *n*-type region. An NMOS is often represented by the abbreviated symbol shown in Fig. 7.1(d) in which the arrowhead indicates the direction of the current.

A *p*-channel enhancement-type MOSFET, often referred to as a PMOS, is formed by two p^+-type regions on top of the *n*-type substrate, as shown in Fig. 7.2[(a) and (b)]. The *p*-regions offer low resistances. The symbol for a PMOS is the same as that for an NMOS, except that the direction of the arrow is reversed, as shown in Fig. 7.2(c). The abbreviated symbol is shown in Fig. 7.2(d).

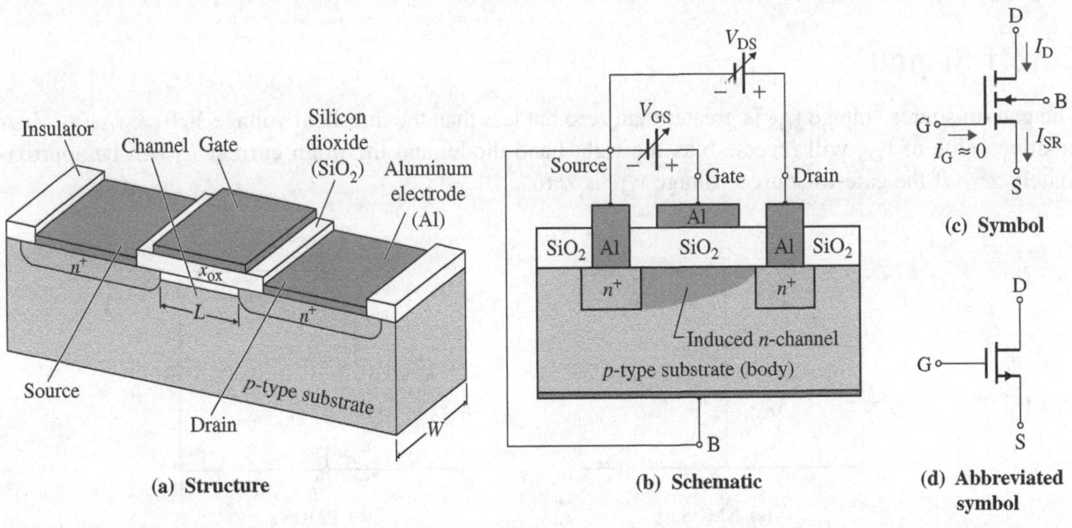

FIGURE 7.1 Structure and symbols of an *n*-channel enhancement MOSFET

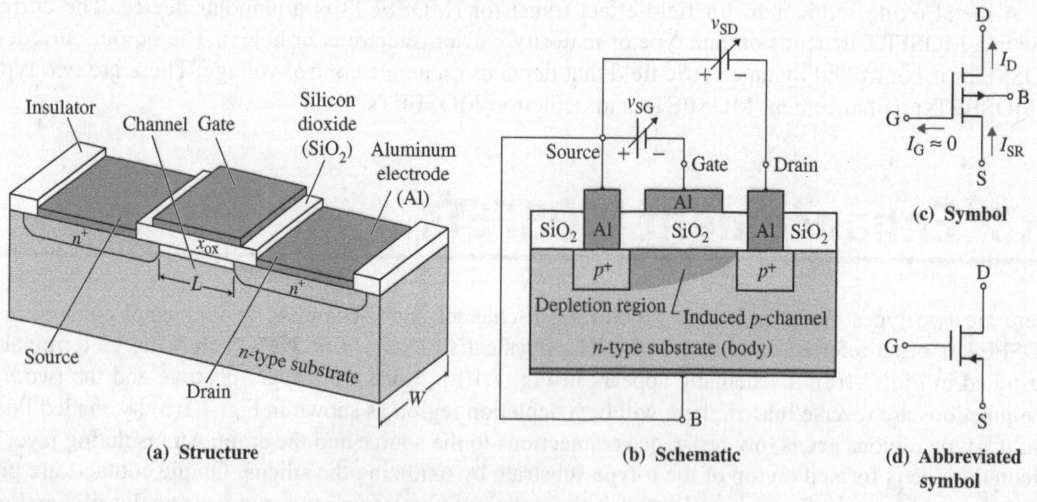

FIGURE 7.2 Structure and symbols of a *p*-channel enhancement MOSFET

7.3.1 Operation

An NMOS is operated with positive gate and drain voltages relative to the source, as shown in Fig. 7.3(a), whereas a PMOS is operated with negative gate and drain voltages relative to the source, as shown in Fig. 7.3(b). Their substrates are connected to the source terminal.

An NMOS may be viewed as consisting of two diode junctions that are formed between the substrate and the source and between the substrate and the drain, as shown in Fig. 7.4(a). The hypothetical diodes are in series and back to back, as shown in Fig. 7.4(b). The NMOS can operate in any of the four operating regions: cutoff region, linear ohmic, nonlinear ohmic, and saturation.

Cutoff Region

The gate-to-source voltage v_{GS} is greater than zero but less than the threshold voltage V_t:$0 \leq v_{GS} \leq V_t$. A positive value of v_{DS} will reverse-bias the right-hand diode, and the drain current i_D will be approximately zero if the gate-to-source voltage v_{GS} is zero.

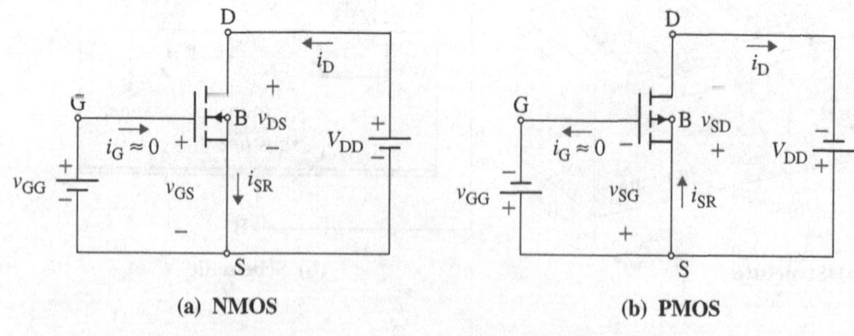

FIGURE 7.3 Biasing of an NMOS and a PMOS

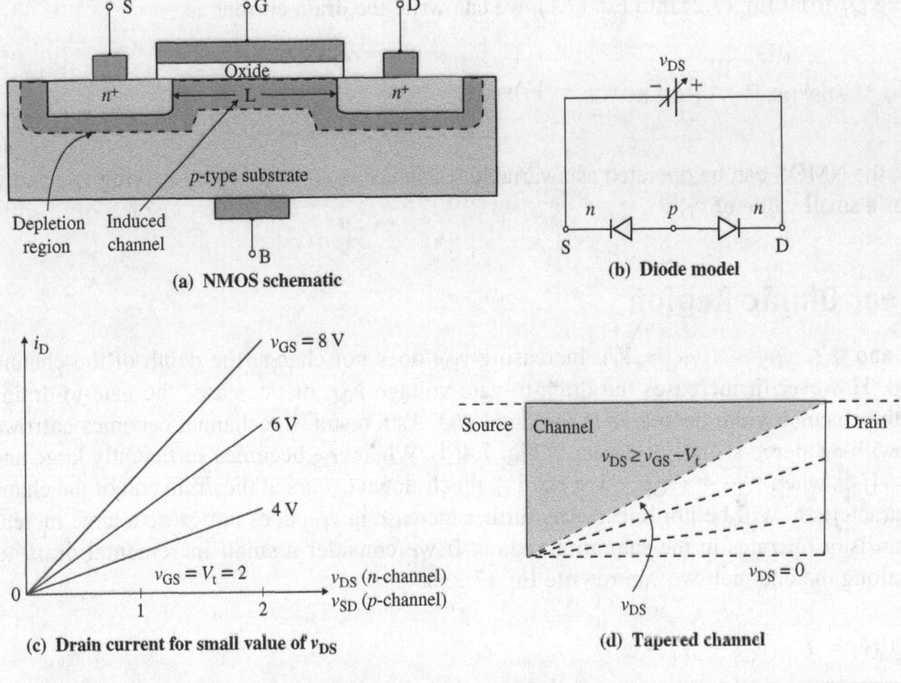

FIGURE 7.4 Effects of varying v_{GS} and v_{DS}

Linear Ohmic Region

$v_{GS} \geq V_t$ and $0 < v_{DS} << (v_{GS} - V_t)$. A positive value of v_{GS} will establish an electric field, which will attract negative carriers from the substrate and repel positive carriers. As a result, a layer of substrate near the oxide insulator becomes less *p*-type, and its conductivity is reduced. As v_{GS} increases, the surface near the insulator will attract more electrons than holes and will behave like an *n*-type channel. The minimum value of v_{GS} that is required to establish a channel is called the *threshold voltage* V_t. The drain current at $v_{GS} = V_t$ is very small. For $v_{GS} > V_t$, the drain current i_D increases almost linearly with v_{DS} for small values of v_{DS}, as shown in Fig. 7.4(c). If the drain-to-source voltage is low (usually less than 1 V), the drain current i_D can be calculated from Ohm's law ($i_D = v_{DS}/r_{DS}$). The conductance of the channel between the drain and the source can be found from

$$g_{DS} = \frac{1}{r_{DS}} = \frac{W}{L}\mu_n Q_n \tag{7.1}$$

where μ_n = mobility of the electrons in the reverse-biased (also called the inversion) layer under the oxide layer

Q_n = magnitude of the reverse-biased layer charge per unit area

W = channel width

L = channel length

We can find Q_n from the gate oxide capacitance C_{ox} and the voltage difference $(v_{GS} - V_t)$ as given by

$$Q_n = C_{ox}(v_{GS} - V_t) \tag{7.2}$$

Substituting Q_n from Eq. (7.2) into Eq. (7.1), we can write the drain current as

$$i_D = g_{DS} v_{DS} = \frac{W}{L} \mu_n C_{ox}(v_{GS} - V_t) v_{DS} \tag{7.3}$$

Therefore, the NMOS can be operated as a variable resistance $r_{DS}(= v_{DS}/i_D)$ by varying v_{GS} and will vary linearly for a small value of v_{DS}.

Nonlinear Ohmic Region

$v_{GS} \geq V_t$ and $0 < v_{DS} < (v_{GS} - V_t)$. Increasing v_{DS} does not change the depth of the channel at the source end. However, it increases the drain-to-gate voltage v_{DG} or decreases the gate-to-drain voltage v_{GD}, and the channel width decreases at the drain end. As a result, the channel becomes narrower at the drain end with a tapered shape, as shown in Fig. 7.4(d). When v_{DS} becomes sufficiently large and v_{GD} is less than V_t [i.e., when $v_{GD} = (v_{GS} - v_{DS}) \leq V_t$], pinch-down occurs at the drain end of the channel. The i_D-v_{DS} characteristic will be nonlinear. Any further increase in v_{DS} does not cause a large increase in i_D, and the transistor operates in the saturation region. If we consider a small incremental drain-to-source voltage v along the channel, we can rewrite Eq. (7.2) as

$$Q_n(v) = C_{ox}(v_{GS} - V_t - v)$$

which can be applied to obtain the drain current as given by

$$i_D = \frac{W}{L} \mu_n C_{ox} \int_0^{v_{DS}} (v_{GS} - V_t - v)\,dv = \frac{W}{L} \mu_n C_{ox} \left[(v_{GS} - V_t) v_{DS} - \frac{v_{DS}^2}{2} \right] \tag{7.4}$$

Equation (7.4) can also be written [4] as

$$i_D = \frac{K_m}{2}[2(v_{GS} - V_t) v_{DS} - v_{DS}^2] \tag{7.5}$$

where $K_m = (W/L)\mu_n C_{ox}$ is called the MOS constant whose value depends on the physical parameters. Equation (7.4) can be expressed in a more general form in terms of external voltages v_{GS} and v_{DS}:

$$i_D = K_n[2(v_{GS} - V_t) v_{DS} - v_{DS}^2] \tag{7.6}$$

Here K_n is a MOS constant given by

$$K_n = \frac{K_m}{2} = \frac{W}{L} \frac{\mu_n C_{ox}}{2} \tag{7.7}$$

where L = channel length (typically 10 μm) in m

W = channel width (typically 100 μm) in m

μ_n = surface mobility of electrons = 600 cm^2/(V-s)

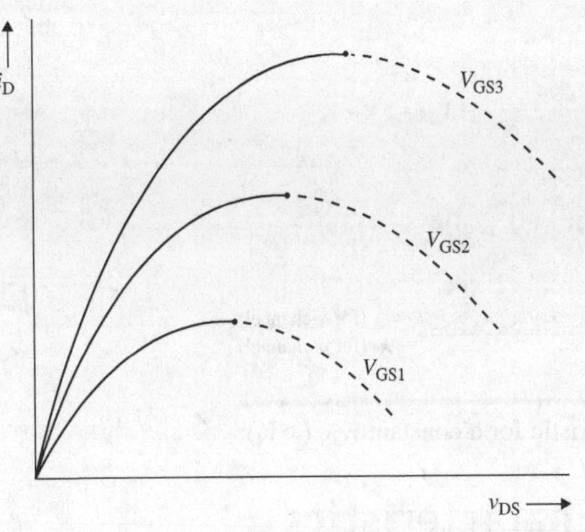

FIGURE 7.5 Plot of i_D versus v_{DS}

ε_o = permittivity of free space = 8.85×10^{-14} F/cm

ε_{ox} = dielectric constant of $SiO_2 = 4$

t_{ox} = thickness of the oxide

C_{ox} = MOSFET capacitance per unit area

For $t_{ox} = 0.10$ μm, C_{ox} is 3.54×10^{-8} F/cm^2.

K_{pp} is the product of μ_n and C_{ox} (i.e., $K_{pp} = \mu_n C_{ox} = \mu_n \varepsilon_{ox}/t_{ox}$), which depends on the process parameters and will be constant for a given technology. By choosing $W = 2L$, we can make the two constants equal: $K_m = K_n$.

Saturation Region

$v_{GS} \geq V_t$ and $v_{DS} \geq (v_{GS} - V_t)$. Figure 7.5 shows the plot of Eq. (7.6) for three values of gate-to-source voltages. We can find the v_{DS} for the peak drain current from the condition, $di_D/dv_D = 0$. That is,

$$\frac{di_D}{dv_{DS}} = \frac{K_n d}{dv_{DS}}[2(v_{GS} - V_t)v_{DS} - v_{DS}^2] = 0$$

which gives $v_{DS} = v_{GS} - V_t$ at which the saturation occurs—that is, $v_{DS(sat)} = v_{GS} - V_t$. Substituting $v_{DS} = v_{GS} - V_t$ in Eq. (7.6) gives the drain current in the saturation region:

$$i_D = K_n(v_{GS} - V_t)^2 \tag{7.8}$$

V_t should be substituted for by V_{tN}, the threshold voltage of an NMOS, or V_{tP}, the threshold voltage of a PMOS. Substituting $v_{GS} = (v_{DS} + V_t)$ in Eq. (7.8) gives the peak (saturation) drain current as

$$i_{D(sat)} = K_n(v_{DS} + V_t - V_t)^2 = K_n v_{DS}^2 \tag{7.9}$$

The complete i_D-v_{DS} characteristic for a constant v_{GS} is shown in Fig. 7.6. In practice, there is a very slight increase in drain current i_D as v_{DS} increases, and the slope of the i_D-v_{DS} characteristic has a finite value.

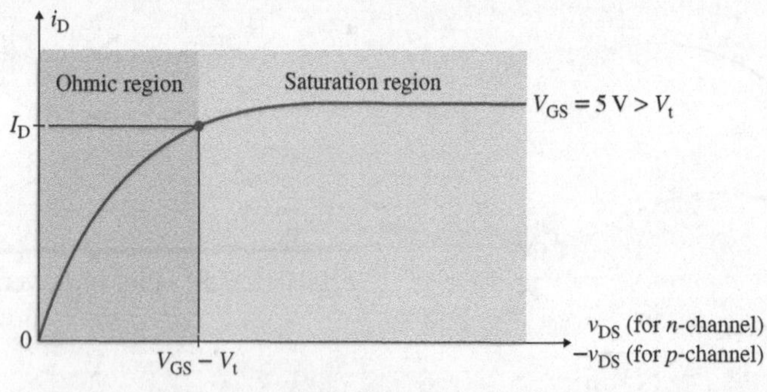

FIGURE 7.6 i_D-v_{DS} characteristic for a constant v_{GS} ($>V_t$)

7.3.2 Output and Transfer Characteristics

The drain characteristics of an NMOS are shown in Fig. 7.7(a), and the complete transfer characteristics are shown in Fig. 7.7(b) for an NMOS and a PMOS.

Increasing v_{DS} beyond the breakdown voltage, denoted by V_{BD}, causes an avalanche breakdown in the channel, and the drain current rises rapidly. This mode of operation must be avoided because a MOSFET can be destroyed by excessive power dissipation. Since the reverse voltage is highest at the drain end, the breakdown occurs at this end. The breakdown voltage specified by the manufacturer is typically in the range of 20 V to 100 V. Also, a large value of v_{GS} will cause a dielectric breakdown in the oxide layer of the device.

Since the gate is insulated from the effective channel in an NMOS, no gate current can flow and consequently the resistance between the gate and the source terminals is theoretically infinite. In practice, the resistance is finite but very large, on the order of 10^8 MΩ.

The output characteristics of an NMOS shown in Fig. 7.7(a) can be described by

$$\begin{aligned} i_D &= K_n[2(v_{GS} - V_t)v_{DS} - v_{DS}^2] \quad \text{for } v_{GS} > V_t \text{ and } v_{DS} < (v_{GS} - V_t) \\ &= K_n(v_{GS} - V_t)^2 \quad \text{for } v_{GS} > V_t \text{ and } v_{DS} \geq (v_{GS} - V_t) \end{aligned} \tag{7.10}$$

The equations for the NMOS can be applied to a PMOS if we substitute $V_t = -V_{tP}$ and $v_{DS} = -v_{DS}$. $V_t = V_{tN}$ for NMOS.

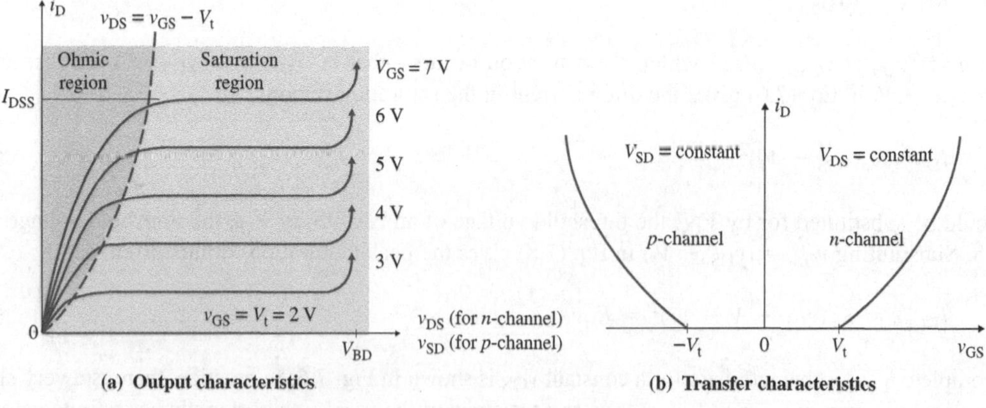

FIGURE 7.7 Drain and transfer characteristics of enhancement MOSFETs

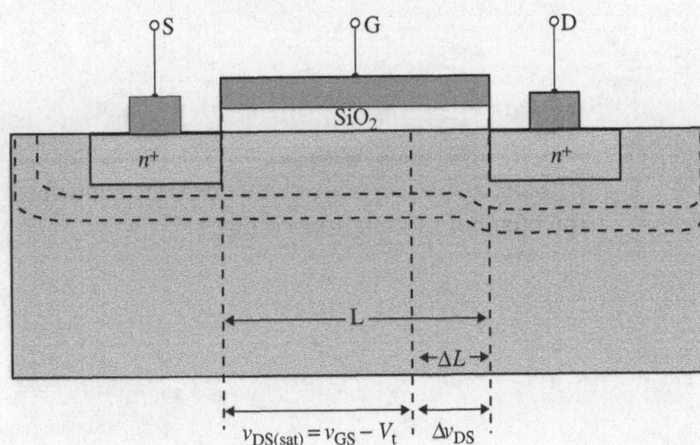

FIGURE 7.8 Channel length modulation of an n-channel MOSFET

7.3.3 Channel Length Modulation

If we increase the drain voltage v_{DS}, the voltage across the oxide layer decreases; therefore the inversion charge density decreases at the drain terminal as shown in Fig. 7.4(d). At $v_{DS} = v_{DS(sat)} = v_{GS} - V_t$, the inversion charge density at the drain terminal becomes zero. As we increase $v_{DS} > v_{DS(sat)}$, the zero density point moves toward the source terminal as shown in Fig. 7.8. Increasing v_{DS} increases the biasing voltage of the pn junction and causes the depletion region at the drain terminal to extend laterally into the channel, thereby reducing the effective channel length. As a result, the effective channel length is modulated by the drain-to-source voltage v_{DS}. The depletion width extending into the p-region of a pn junction with v_{DS} biasing can be found from Eq. (6.40) as

$$x_p = \sqrt{\frac{2\varepsilon_s}{qN_a}(|\phi_{Fp}| + v_{DS})} \tag{7.11}$$

where $|\phi_{Fp}|$ is the field potential due to the p-region given by

$$\phi_{Fp} = -\frac{kT}{q}\ln\left(\frac{N_a}{n_i}\right) = -V_T\ln\left(\frac{N_a}{n_i}\right) \tag{7.12}$$

We can find the extension of the space charge region

$$\Delta L = x_p(v_{DS(sat)} + \Delta v_{DS}) - x_p(v_{DS(sat)})$$

which after substituting $\Delta v_{DS} = v_{DS} - v_{DS(sat)}$ in Eq. (7.11) gives

$$\Delta L = \sqrt{\frac{2\varepsilon_s}{qN_a}}\left[\sqrt{|\phi_{Fp}| + v_{DS(sat)} + \Delta v_{DS}} - \sqrt{|\phi_{Fp}| + v_{DS(sat)}}\right] \tag{7.13}$$

Since the drain i_D is inversely proportional to the effective channel length, we get

$$i_D \propto \frac{1}{L - \Delta L} = \frac{1}{L(1 - \Delta L/L)} = \frac{1}{L}\left(1 + \frac{\Delta L}{L}\right) \tag{7.14}$$

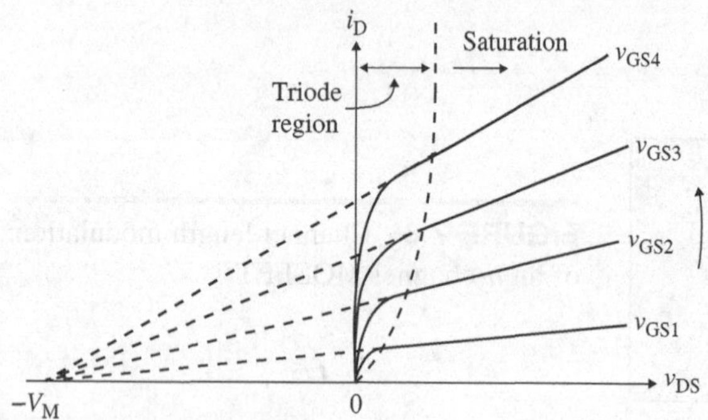

FIGURE 7.9 i_D-v_{DS} characteristics showing the channel modulation voltage

Since ΔL is a function of v_{DS}, the fractional change in the channel length is proportional to the drain-source biasing voltage. That is,

$$\frac{\Delta L}{L} = \lambda v_{DS} \tag{7.15}$$

Here λ is called the channel length modulation. We can include the channel length modulation effect in Eq. (7.15) to Eq. (7.8) as follows:

$$i_D = K_n(v_{GS} - V_t)^2(1 + \lambda v_{DS}) \tag{7.16}$$

This gives the slope of the output characteristics shown in Fig. 7.7(a). The plot of the i_D-v_{DS} characteristics for the saturation region is shown in Fig. 7.9. If we extrapolate the characteristics to the v_{DS}-axis, they intercept at a point V_M, which is known as the channel modulation voltage such that $V_M = 1/\lambda$.

EXAMPLE 7.1

Finding the channel modulation voltage Determine the channel modulation voltage V_M. The NMOS parameters are these: substrate impurity doping concentration $N_a = 2 \times 10^{16}$ cm^3, threshold voltage $V_{tN} = 0.5$ V, channel length $L = 10$ μm, $V_{GS} = 1.5$ V, and $V_{DS} = 5$ V.

SOLUTION

$V_T = 25.8$ mV, $N_a = 2 \times 10^{16}$ cm^3. $n_i = 1.5 \times 10^{10}$ cm^3, $V_{tN} = 0.5$ V, $L = 10$ μm, $V_{GS} = 1.5$ V, and $V_{DS} = 5$ V.

From Eq. (7.12),

$$\phi_{Fp} = -V_T \ln\left(\frac{N_a}{n_i}\right) = -25.8 \times 10^{-3} \times \ln\left(\frac{2 \times 10^{16}}{1.5 \times 10^{10}}\right) = -0.364$$

$$V_{DS(sat)} = V_{GS} - V_{tN} = 1.5 - 0.5 = 1 \text{ V}$$

$$\Delta v_{DS} = V_{DS} - V_{DS(sat)} = 5 - 1 = 4 \text{ V}$$

From Eq. (7.13),

$$\Delta L = \sqrt{\frac{2 \times 11.7 \times 8.86 \times 10^{-14}}{1.6 \times 10^{-19} \times 2 \times 10^{16}}}\left[\sqrt{|-0.364| + 1 + 4} - \sqrt{|-0.364| + 1}\right] = 0.2921\ \mu\text{m}$$

Let $x = \Delta L/L = 0.2921 \times 10^{-6}/(10 \times 10^{-6}) = 0.029$. The channel lambda is $\lambda = x/V_{DS} = 0.029/5 = 5.842 \times 10^{-3}$. Therefore, the modulation voltage is $V_M = 1/\lambda = 1/(5.842 \times 10^{-3}) = 171.19$ V.

7.3.4 Substrate Biasing Effects

The source-to-substrate *pn* junction must always be zero or reverse biased, so v_{SB} must always be greater than or equal to zero; otherwise electrons or holes will flow from the drain to the substrate rather than the source terminals. The body or the substrate of a MOSFET is often connected to the ground. In MOSFET circuits, the source and body may not be at the same potential as shown in Fig. 7.10, and applications of v_{SB} will increase the depletion region. In integrated circuits, however, the substrate is usually common to many MOS transistors. To maintain the cutoff condition for all the substrate-to-channel junctions, the substrate is usually connected to the most negative power supply in an NMOS circuit (the most positive in a PMOS circuit).

The reverse-biased voltage will widen the depletion region, thereby reducing the effective channel depth. Therefore, we need to apply more gate voltage to compensate for the channel reduction, and v_{SB} will affect the effective threshold voltage V_t of the MOSFET. It can be shown that increasing v_{SB} results in an increase in V_t as given by [5]

$$V_t = V_{to} + \frac{\sqrt{2q\varepsilon_s N_a}}{C_{ox}}\left[\sqrt{2|\phi_{Fp}| + V_{SB}} - \sqrt{2|\phi_{Fp}|}\right] \tag{7.17}$$

Here V_{SB} is the source-to-substrate voltage, and V_{to} is the initial threshold voltage with $v_{SB} = 0$. Note that v_{SB} must always be positive for an NMOS and negative for a PMOS. Also, V_t must always be positive for an NMOS and negative for a PMOS.

7.3.5 Complementary MOS (CMOS)

The complementary metal oxide semiconductor (CMOS) consists of an *n*-channel enhancement-mode device ($V_{tN} > 0$) in series with a *p*-channel enhancement-mode device ($V_{tP} < 0$). The cross-section of a

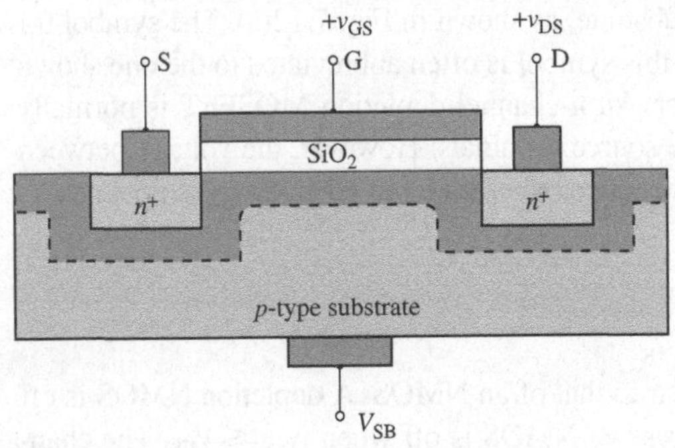

FIGURE 7.10 Applying source-to-substrate voltage to an NMOS

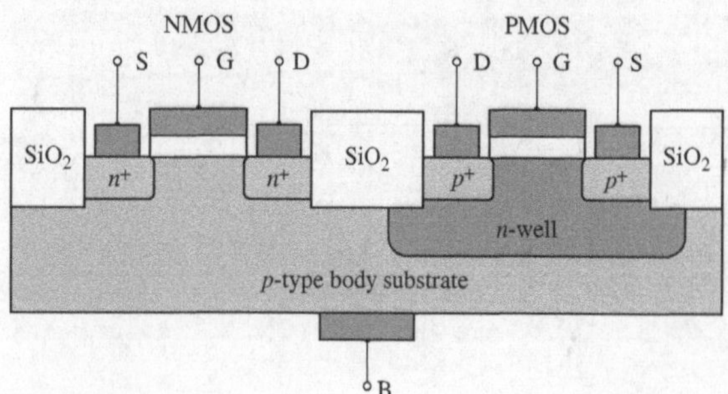

FIGURE 7.11 Cross section of a CMOS

CMOS is shown in Fig. 7.11. The NMOS transistor is implemented directly in the *p*-type substrate, while the PMOS transistor is fabricated in a specially created *n*-region, known as an *n*-well. The two devices are isolated from each other by a thick region of oxide that functions as an insulator. An external body terminal is also made from the *p*-type body and the *n*-well. Due to their unique advantages, such as very low power consumption, CMOS circuits are commonly used in integrated circuits. The CMOS inverter, which is the basis of CMOS digital electronics, is covered in detail in Sec. 15.7. CMOS technology has taken over many IC applications and continues to grow.

KEY POINTS OF SECTION 7.3

- An MOSFET is a voltage-controlled nonlinear device. A voltage between the gate and the source develops an electric field, which then controls the flow of drain current. Therefore, the drain current depends on the gate-to-source voltage, and an FET gives a transconductance gain.
- MOSFETs can be classified into two types: enhancement MOSFET and depletion MOSFETs. Each type can be either *n*-channel or *p*-channel.
- The output characteristic of a MOSFET can be divided into three regions: the cutoff region, in which the MOSFET is in the off state; the saturation region, in which the transistor exhibits a high output resistance and has a transconductance; and the ohmic region, in which the transistor offers a low resistance. A MOS is operated as an amplifier in the saturation region and as a switch in the ohmic region.

7.4 Depletion MOSFETs

The construction of an *n*-channel depletion MOSFET is very similar to that of an NMOS. An actual channel is formed by adding *n*-type impurity atoms to the *p*-type substrate, as shown in Fig. 7.12(a). The symbol for an *n*-channel depletion MOSFET is shown in Fig. 7.12(b); this symbol is often abbreviated to the one shown in Fig. 7.12(c). Note that the vertical line is bold or darker. An *n*-channel depletion MOSFET is normally operated with a positive voltage between the drain and the source terminals. However, the voltage between the gate and the source terminals can be positive, zero, or negative, whereas in an NMOS v_{GS} is positive.

7.4.1 Operation

The operation of an *n*-channel depletion MOSFET is similar to that of an NMOS. A depletion NMOS is off when its gate-to-source voltage v_{GS} is less than $-V_p$, whereas an NMOS is off when $v_{GS} \le V_{tN}$. The channel is fully established at $v_{GS} = 0$ for a depletion NMOS and at $v_{GS} = V_{tN}$ for an NMOS. Let us assume that

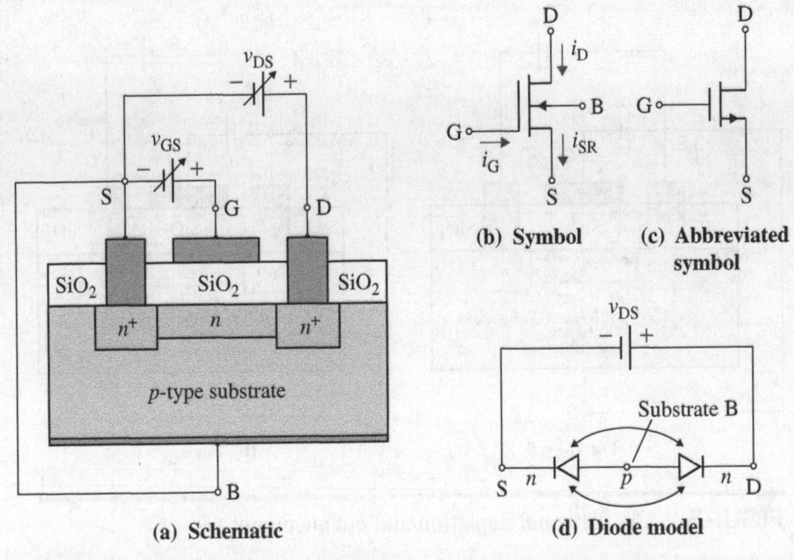

FIGURE 7.12 Schematic and symbols of an n-channel depletion MOSFET

the gate-to-source voltage is zero: $v_{GS} = 0$. If v_{DS} is increased from zero to some small value ($\approx$1 V), the drain current follows Ohm's law ($i_D = v_{DS}/r_{DS}$) and is directly proportional to v_{DS}. Any increase in the value of v_{DS} beyond $|V_p|$, known as the *pinch-down voltage,* does not increase the drain current significantly. The region beyond pinch-down is called the *saturation region.* The value of the drain current that occurs at $v_{DS} = |V_p|$ (with $v_{GS} = 0$) is termed the drain-to-source saturation current I_{DSS}. The complete i_D-v_{DS} characteristic for $v_{GS} = 0$ is shown in Fig. 7.13. In practice, there is a very slight increase in drain current i_D as v_{DS} increases beyond $|V_p|$, and the slope of the i_D-v_{DS} characteristic has a finite value. Saturation occurs at the value of v_{DS} at which the gate-to-channel voltage at the drain end equals V_p. That is,

$$v_{GD} = v_{GS} - v_{DS} = V_p \quad \text{or} \quad v_{DS} = v_{GS} - V_p \tag{7.18}$$

If v_{GS} is negative, some of the electrons in the n-channel area will be repelled from the channel and a depletion region will be created below the oxide layer, as shown in Fig. 7.14(a). This depletion region will result in a narrower channel. For $v_{GS} > 0$, a layer of substrate near the n-type channel becomes less p-type and its conductivity is enhanced as shown in Fig. 7.14(b). A positive value of v_{GS} increases the

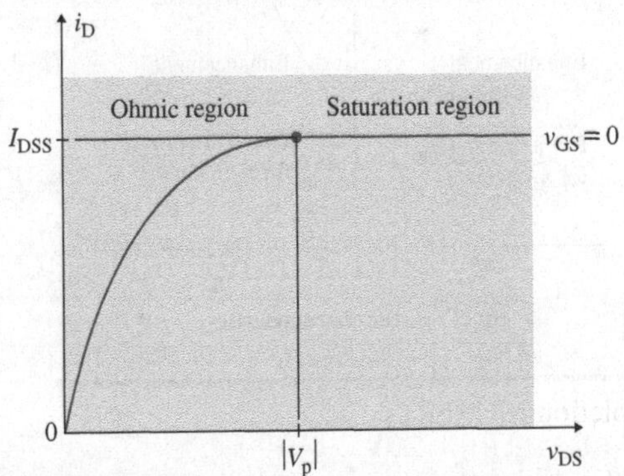

FIGURE 7.13 i_D-v_{DS} characteristic for a constant v_{GS} ($>V_p$)

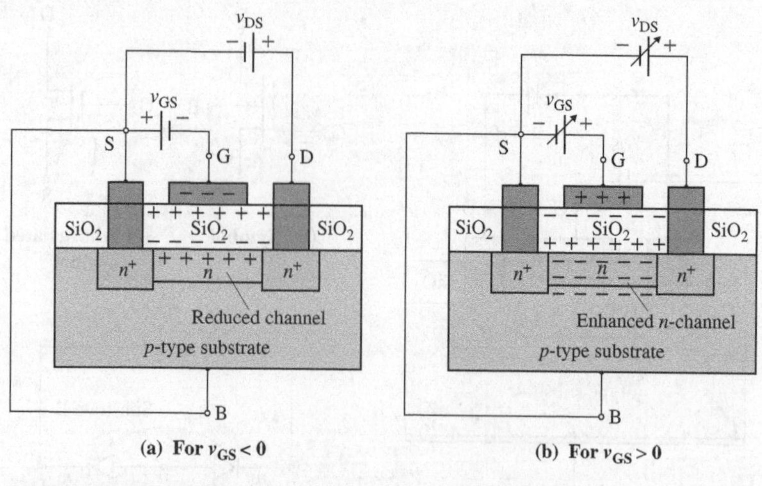

(a) For $v_{GS} < 0$ (b) For $v_{GS} > 0$

FIGURE 7.14 Channel depletion and enhancement

effective channel width in much the same way as in an NMOS. When the effective channel is increased, the transistor is said to be operating in the enhancement mode. The i_D-v_{DS} characteristics for various values of v_{GS} are shown in Fig. 7.15(a).

7.4.2 Output and Transfer Characteristics

The transfer characteristics are shown in Fig. 7.15(b) for an n-channel and a p-channel MOSFET. The output characteristics can be divided into three regions: ohmic, saturation, and cutoff.

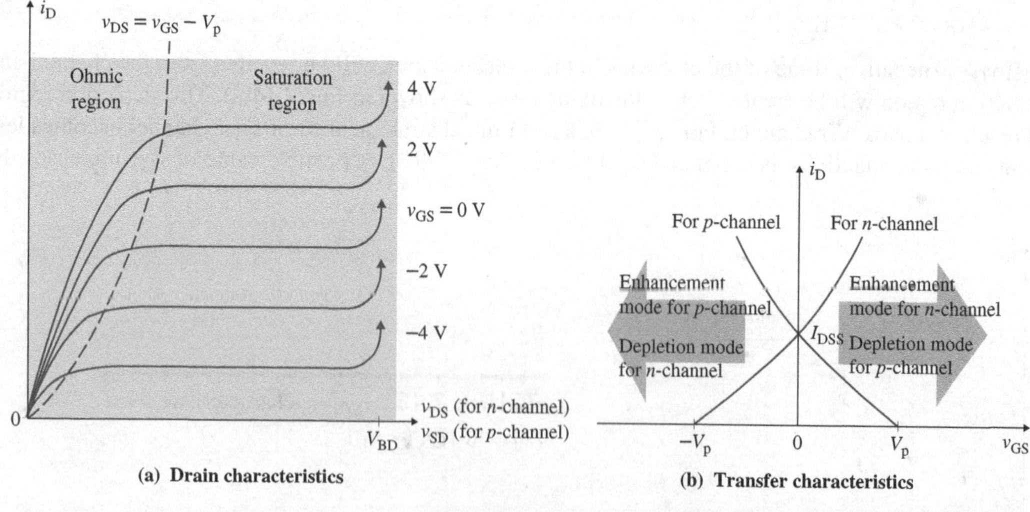

(a) Drain characteristics (b) Transfer characteristics

FIGURE 7.15 Drain and transfer characteristics of depletion MOSFETs

Ohmic Region

In the ohmic region, the drain-to-source voltage v_{DS} is low and the channel is not pinched down. The drain current i_D can be expressed as

$$i_D = K_n[2(v_{GS} - V_p)v_{DS} - v_{DS}^2] \quad \text{for } 0 < v_{DS} \le (v_{GS} - V_p) \tag{7.19}$$

which, for a small value of v_{DS} ($<<|V_p|$), can be reduced to

$$i_D = K_n[2(v_{GS} - V_p)v_{DS}] \tag{7.20}$$

where $K_n = I_{DSS}/V_p^2$.

Saturation Region

In the saturation region, $v_{DS} \geq (v_{GS} - V_p)$. The drain-to-source voltage v_{DS} is greater than the pinch-down voltage, and the drain current i_D is almost independent of v_{DS}. For operation in this region, $v_{DS} \geq (v_{GS} - V_p)$. Substituting the limiting condition $v_{DS} = (v_{GS} - V_p)$ in Eq. (7.19) gives the drain current i_D as

$$\begin{aligned} i_D &= K_n[2(v_{GS} - V_p)(v_{GS} - V_p) - (v_{GS} - V_p)^2] \\ &= K_n(v_{GS} - V_p)^2 \end{aligned} \tag{7.21}$$

Equation (7.21) represents the transfer characteristic, which is shown in Fig. 7.15(b) for both *n*- and *p*-channels. For a given value of i_D, Eq. (7.21) gives two values of v_{GS}, and only one value is the acceptable solution so that $v_{GS} > V_p$ for the *n*-channel and $v_{GS} < V_p$ for the *p*-channel. The pinch-down locus, which describes the boundary between the ohmic and saturation regions, can be obtained by substituting $v_{GS} = v_{DS} + V_p$ into Eq. (7.21):

$$i_D = K_n(v_{DS} + V_p - V_p)^2 = K_n v_{DS}^2 \tag{7.22}$$

This defines the pinch-down locus and forms a parabola similar to Eq. (7.9) and Fig. 7.5.

Cutoff Region

In the cutoff region, the gate-to-source voltage is less than the pinch-down voltage. That is, $v_{GS} < V_p$ for the *n*-channel and $v_{GS} > V_p$ for the *p*-channel, and the MOSFET is off. The drain current is zero: $i_D = 0$.

7.5 MOSFET Models and Amplifier

Since the drain currents of the enhancement and depletion MOSFETs depend on the gate–source voltage, they are known as voltage-dependent devices and exhibit similar output characteristics and, the same model can be applied to both of them with reasonable accuracy. An NMOS circuit with the transistor biased to operate in the saturation region is shown in Fig. 7.16(a). Using KVL around the drain-to-source loop gives

$$\begin{aligned} V_{DD} &= v_{DS} + R_D i_D \\ i_D &= \frac{V_{DD}}{R_D} - \frac{v_{DS}}{R_D} \end{aligned} \tag{7.23}$$

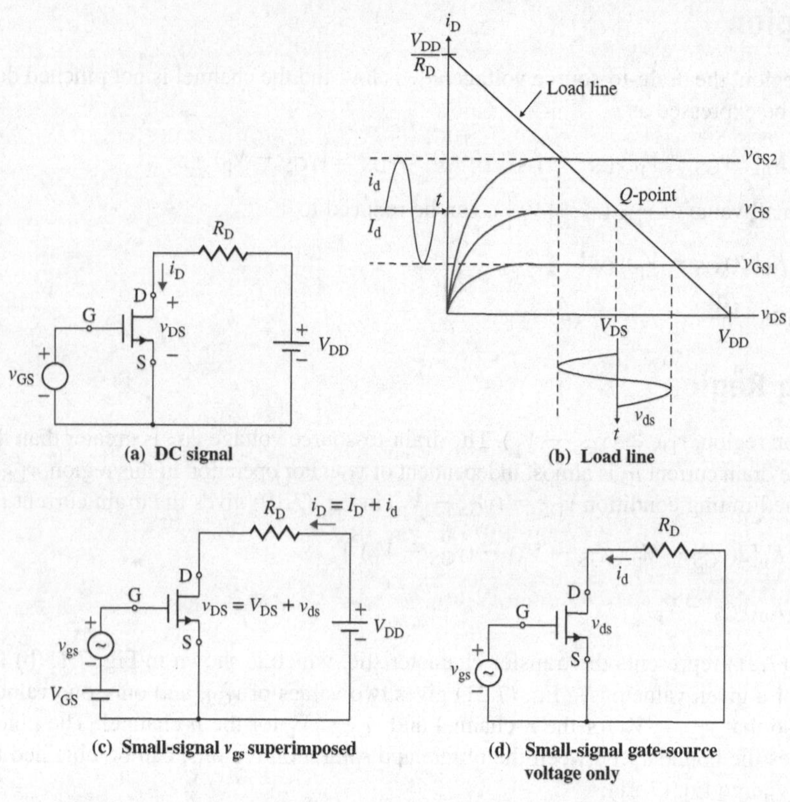

FIGURE 7.16 NMOS with a small-signal input voltage v_{gs}

which describes the load line, and intersects the i_D-axis at V_{DD}/R_D and the v_{DS}-axis at V_{DD} as shown in Fig. 7.16(b). The intersection of this load line with the i_D-v_{DS} characteristic gives the operating (or quiescent) point for a given value of V_{GS}. Let us assume that the drain current, drain-to-source voltage, and gate-to-source voltage have initial quiescent values of I_D, V_{DS}, and V_{GS}, respectively. In a MOSFET amplifier, an AC input signal is normally superimposed on the gate voltage. If a small AC signal v_{gs} is connected in series with V_{GS}, it will produce a small variation in the drain-to-source voltage v_{DS} and the drain current i_D. That is, if the gate-to-source voltage varies by a small amount, such that $v_{GS} = V_{GS} + v_{gs}$, there will be corresponding changes in the drain current and drain-to-source voltage such that $v_{DS} = V_{DS} + v_{ds}$ and $i_D = I_D + i_d$. This situation is shown in Fig. 7.16(b). The small variations of the drain current i_D, as i_d, and the drain-to-source voltage v_{DS}, as v_{ds}, around the operating point are shown in Fig. 7.16(b). The drain-to-source variation v_{ds} will equal the voltage gain times v_{gs}. If the values of i_d, v_{gs}, and v_{ds} are small, Fig. 7.16(b) can be represented by the small-signal circuit shown in Fig. 7.16(c). Therefore, we need two types of models for MOSFETs: a DC model and a small-signal model.

7.5.1 DC Models

The large-signal (DC) models of MOSFETs are nonlinear. The drain characteristics as shown in Fig. 7.7(a) and 7.15(b) of i_D as a function of v_{DS} for different values of v_{GS} describe the large-signal model of a MOSFET.

Since the gate-channel has an oxide layer, the gate current will be negligibly small. Thus, MOSFETs can be represented by the simple DC model of Fig. 7.17.

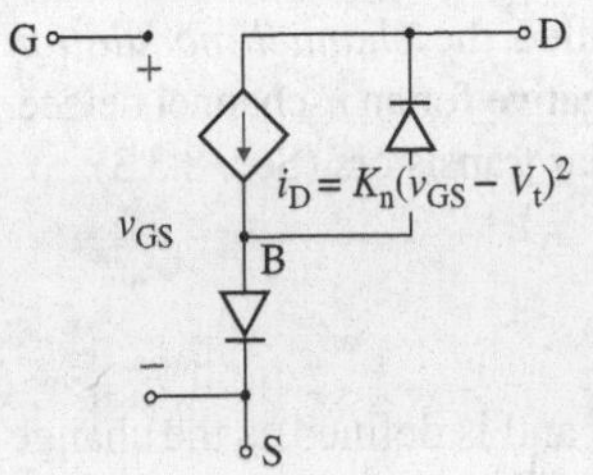

FIGURE 7.17 Large-signal model of n-channel MOSFETs

7.5.2 Small-Signal AC Models

The small-signal behavior of the MOSFET in Fig. 7.17 can be represented by a small-signal AC equivalent circuit consisting of a voltage-dependent current source $g_m v_{gs}$ in parallel with an output resistance r_o representing a finite slope of the i_D-v_{DS} characteristic. This circuit is shown in Fig. 7.18(a). Since the gate current i_g of MOSFETs is very small, tending to zero, the gate-to-source terminals are open circuits.

Applying the relations between Norton's and Thevenin's theorems, we can represent the current source in Fig. 7.18(a) by a voltage source, as shown in Fig. 7.18(b). We find v_{ds} from

$$v_{ds} = i_d r_o - r_o g_m v_{gs} = i_d r_o - \mu_g v_{gs} \tag{7.24}$$

where μ_g is the *open-circuit voltage gain* of the MOSFETs and is given by

$$\mu_g = r_o g_m \tag{7.25}$$

The circuits of Fig. 7.18[(a) and (b)] are referred to as the Norton and Thevenin circuits, respectively, and they are equivalent. r_o is the small-signal output resistance, and g_m is the transconductance gain of the MOSFET. Their values are dependent on the operating point and are quoted at a specified operating point (V_{DS}, I_D).

Small-Signal Output Resistance r_o

The small-signal output resistance is the inverse slope of the i_D-v_{DS} characteristic in the pinch-down or saturation region. We can use Eq. (7.16) to find the value of the output resistance r_o as given by

$$\frac{1}{r_o} = \frac{di_D}{dv_{DS}} = \frac{I_D}{|V_M|} = \lambda I_D \quad \text{for all MOSFETs} \tag{7.26}$$

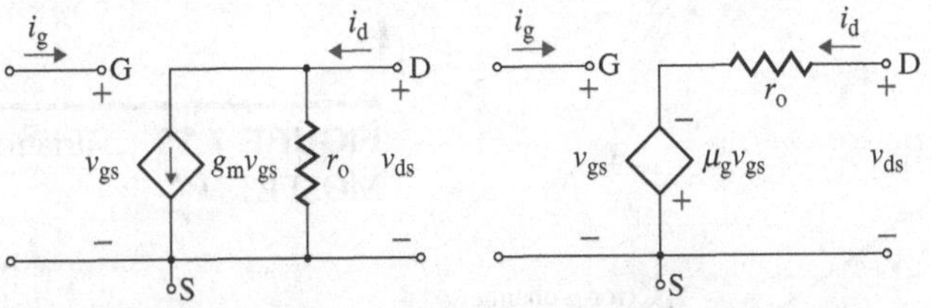

(a) Norton's equivalent **(b) Thevenin's equivalent**

FIGURE 7.18 Small-signal model of MOSFETs

Here V_M is called the *channel modulation voltage* and λ $(=1/|V_M|)$ is called the *channel modulation length* (see Fig. 7.9). The parameter V_M is positive for a *p*-channel device and negative for an *n*-channel device. Its typical magnitude is 100 V. V_M is analogous to the Early voltage V_A of bipolar transistors (Sec. 8.3.3).

Transconductance g_m

The transconductance is the slope of the transfer characteristic (i_D versus v_{GS}) and is defined as the change in the drain current corresponding to a change in the gate-to-source voltage. It is expressed by

$$g_m = \left.\frac{\delta i_D}{\delta v_{GS}}\right|_{v_{DS}=\text{constant}}$$

Assuming $i_D \approx I_D$, $v_{GS} \approx V_{GS}$, and $v_{DS} \approx V_{DS}$, the small-signal transconductance of an NMOS can be derived from Eq. (7.10):

$$g_m = \frac{\delta i_D}{\delta v_{GS}} = 2K_n(V_{GS} - V_t) \quad \text{for enhancement MOSFETs} \tag{7.27}$$

$$= g_{mo}\left(1 - \frac{V_{GS}}{V_t}\right) \quad \text{for enhancement MOSFETs} \tag{7.28}$$

where $g_{mo} = -2K_nV_t$. (7.29)

The small-signal transconductance of a depletion MOSFET can be derived from Eq. (7.21):

$$g_m = \frac{\delta i_D}{\delta v_{GS}} = 2K_n(v_{GS} - V_p) \quad \text{for depletion MOSFETs} \tag{7.30}$$

$$= g_{mo}\left(1 - \frac{v_{GS}}{V_p}\right) \quad \text{for depletion MOSFETs} \tag{7.31}$$

where $g_{mo} = -2K_nV_p = -2I_{DSS}/V_p$ (7.32)

g_{mo} is the transconductance corresponding to $v_{GS} = 0$, and it varies linearly with v_{GS}, as shown in Fig. 7.19. For $v_{GS} = 0$, the device is cut off; thus it is never operated with a value of g_{mo}. The pinch-down voltage V_p can be determined experimentally by plotting g_m versus v_{GS} and then extrapolating to the v_{GS}-axis. This is a very useful method for determining V_p and V_t for a MOSFET.

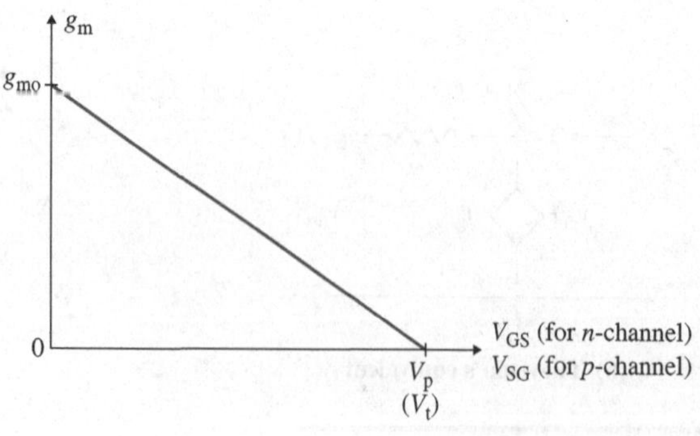

FIGURE 7.19 Variation of g_m with v_{GS} for MOSFETs

7.5.3 PSpice Models

The symbol for a MOSFET is M. The statements have the following general forms:

```
M<name> ND NG NS NB MMOD   for MOSFETs
```

where ND, NG, NS, and NB are the drain, gate, source, and bulk (or substrate) nodes, respectively. MMOD is the model name. The model statement has the following general forms:

```
.MODEL MMOD NMOS (P1=A1 P2=A2 P3=A3 ...PN=AN)   for n-channel MOSFETs
.MODEL MMOD PMOS (P1=A1 P2=A2 P3=A3 ...PN=AN)   for p-channel MOSFETs
```

Here NMOS and PMOS are the type symbols for *n*-channel and *p*-channel MOSFETs, respectively; and P1, P2, . . . , PN and A1, A2, . . . , AN are the parameters and their values, respectively.

Consider the NMOS of type 2N4351, whose parameters are $V_t = 1$ V to 5 V, and $g_m = 1$ mA/V at $i_D = 2$ mA and at $v_{DS} = 10$ V. Taking the geometric mean value, we get $V_t = \sqrt{1 \times 5} = 2.24$ V, which is specified in PSpice/SPICE by VTO=2.24 V. The constant K_n can be found from Eqs. (7.10) and (7.27):

$$i_D = K_n(v_{GS} - V_t)^2$$

$$g_m = 2K_n(v_{GS} - V_t)$$

These equations can be written in the form of a ratio as

$$\frac{g_m^2}{i_D} = \frac{4K_n^2(v_{GS} - V_t)^2}{K_n(v_{GS} - V_t)^2} = \frac{4K_n}{1}$$

which, for $g_m = 1$ mA/V and $i_D = 2$ mA, gives $K_n = 125\ \mu\text{A/V}^2$. The ratio W/L can be found from Eq. (7.7):

$$\frac{W}{L} = \frac{2K_n}{\mu_a C_{ox}} = \frac{2 \times 125 \times 10^{-6}}{600 \times 3.54 \times 10^{-8}} = 11.8$$

Assume $L = 10\ \mu$m; then $W = 118\ \mu$m. Also assume $|V_M| = 1/\lambda = 200$ V and $\lambda = 5\ \text{mV}^{-1}$. Then NMOS 2N4351 can be specified in PSpice/SPICE by the following statements [6, 7]:

```
M1 ND NG NS NB M2N4351
.MODEL M2N4351 NMOS (KP=125U VTO=2.24 L=10U W=118U LAMBDA=5M)
```

The MOS transistor has a length of 0.6 μm at minimum and can be expanded by integer increments of 0.3 μm. The minimum width is 0.9 μm and can be expanded by integer increments of 0.3 μm. Generally, attributes of an NMOS are L = 6 to 10 μm, W = 118 μm, AD = 720 to 283.2 μm, AS = 720 μm, PD = 302.4 to 120.4 μm, and PS = 302.4 to 120.4 μm. Generally, AS = AD = (2.4 μm × W), and PS = PD = (2.4 μm + W).

▶ **NOTE** The full data sheets for MOSFETs (e.g., NMOS of type 2N4351) can be found at the http://www.alldatasheet.com/ or by searching MOSFET data sheets at http://www.google.com.

7.5.4 Small-Signal Analysis

Once the *Q*-point is established and the small-signal parameters are determined, we can find the small-signal parameters of the amplifier in Fig. 7.16(a) in response to a small-signal voltage v_{gs}. For a small AC

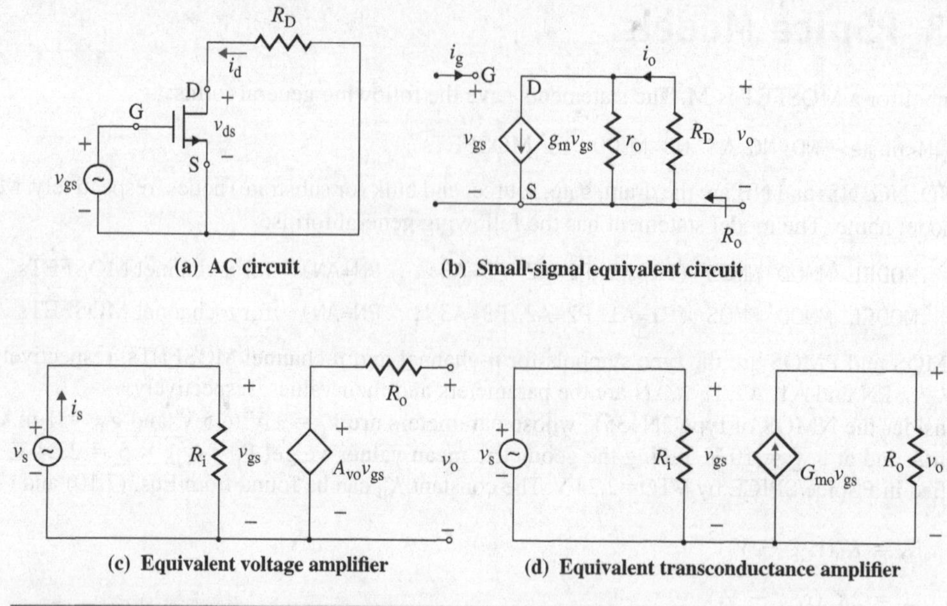

FIGURE 7.20 Small-signal AC equivalent circuits of the amplifier in Fig. 7.16(a)

signal, the DC supply offers zero impedance; V_{DD} and V_{GS} can be short-circuited. That is, one side of R_D is connected to the ground. The small-signal AC equivalent circuit of the amplifier is shown in Fig. 7.20(a). Replacing the transistor M_1 by its transconductance model of Fig. 7.18(a), the small-signal AC equivalent circuit is shown in Fig. 7.20(b). The following steps are involved in analyzing an amplifier circuit:

1. DC biasing analysis of the transistor circuit
2. Determination of the small-signal parameters g_m and r_o of the transistor
3. Determination of the AC equivalent circuit of the amplifier
4. Performing the small-signal analysis for finding R_i, A_{vo}, and R_o

From Fig. 7.20(b), the small-signal input resistance can be found from

$$R_i = \frac{v_{gs}}{i_g} = \infty$$

Thevenin's equivalent output resistance, looking from the output side for the condition $v_{gs} = 0$, can be found from

$$R_o = r_o \| R_D \tag{7.33}$$

The transconductance of the amplifier G_{mo}, which is the same as the transconductance of the transistor, is

$$G_{mo} = \frac{i_o}{v_{gs}} = -g_m \tag{7.34}$$

We can write the small-signal output voltage v_o as

$$v_o = -g_m(r_o \| R_D)v_{gs} \tag{7.35}$$

which gives the small-signal voltage gain A_{vo} as

$$A_{vo} = \frac{v_o}{v_{gs}} = -g_m(r_o \| R_D) \tag{7.36}$$

If we substitue $r_o = V_M/I_D$, Eq. (7.36) becomes

$$A_{vo} = -g_m\left(\frac{V_M R_D}{V_M + I_D R_D}\right) \tag{7.37}$$

Therefore, for obtaining a large voltage gain, the $g_m R_D$ product must be made large and the DC biasing drain current I_D should be small. This will require both a large DC supply voltage V_{DD} and a large value of resistance R_D. Figure 7.20[(c) and (d)] shows the equivalent voltage and transconductance amplifiers of the circuit in Fig. 7.20(a).

EXAMPLE 7.2

Finding the small-signal parameters of an NMOS amplifier The amplifier in Fig. 7.16(a) has $V_{GS} = 2$ V, $V_{DD} = 15$ V, and $R_D = 3.5$ kΩ. The NMOS parameters are $V_t = 1$ V, $K_n = 3.25$ mA/V^2, and $V_M = 1/\lambda = 100$ V.

(a) Find the DC biasing point V_{GS}, I_D, and V_{DS}.

(b) Find the small-signal transistor model parameters r_o and g_m.

(c) Find the small-signal amplifier parameters R_i, R_o, and A_{vo}.

(d) Use PSpice to plot the small-signal AC output voltage for 1-mV sinusoidal input signal at 1 kHz. The NMOS parameters are KP = 6.5 M, VTO = 1 V, L = 1 U, W = 1 U, and LAMBDA = 0.01. *Note*: PSpice uses $K_p = K_m = 2K_n$.

SOLUTION

(a) $K_m = K_p = 6.5 \times 10^{-3}$ for $W = L$. From Eq. (7.8),

$$I_D = K_n(V_{GS} - V_t)^2 = 3.25 \times 10^{-3}\ (2 - 1) = 3.25 \text{ mA}$$

$$V_{DS} = V_{DD} - R_D I_D = 15 - 3.5 \times 10^3 \times 3.25 \times 10^{-3} = 3.625 \text{ V}$$

(b) $r_o = V_M/I_D = 100/(3.25 \times 10^{-3}) = 30.77$ KΩ

$$g_m = 2 \times K_n(V_{GS} - V_t) = 2 \times 3.25 \times 10^{-3} \times (2 - 1) = 6.5 \text{ mA/V}$$

(c) $R_o = r_o \| R_D = 30.77\text{ k} \| 3.5\text{ k} = 3.143$ kΩ

$$G_{mo} = g_m = 6.5 \text{ mA/V}$$

$$A_{vo} = -g_m \times R_o = -6.5 \text{ mA/V} \times 3.143 \text{ k}\Omega = -20.426 \text{ V/V}$$

(d) Figure 7.21 shows the PSpice schematic and the PSpice plot for small-signal output voltage is shown in Fig. 7.22. The capacitor C_2 blocks the DC and passes the small-signal output, which shows a voltage of −19.84; this is close to the calculated value of −20.42.

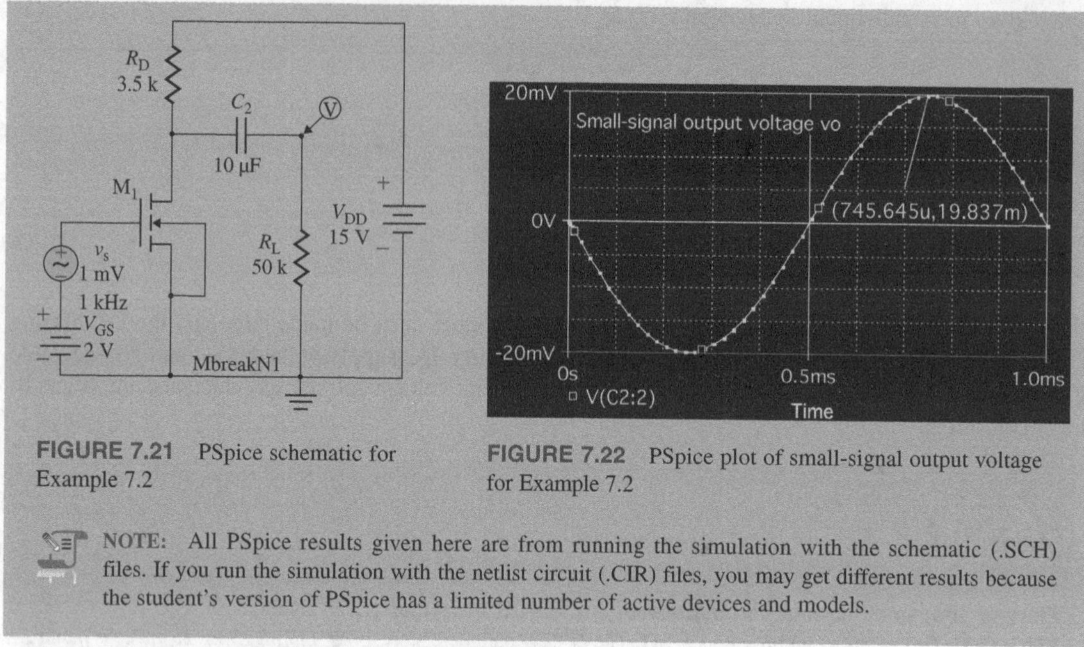

FIGURE 7.21 PSpice schematic for Example 7.2

FIGURE 7.22 PSpice plot of small-signal output voltage for Example 7.2

NOTE: All PSpice results given here are from running the simulation with the schematic (.SCH) files. If you run the simulation with the netlist circuit (.CIR) files, you may get different results because the student's version of PSpice has a limited number of active devices and models.

7.6 A MOSFET Switch

A MOSFET can be operated as a voltage-controlled switch. Figure 7.23(a) shows the circuit arrangement. A switch should have the characteristic of a low on-state voltage at the maximum current so that the switch is subjected to the minimum power loss. These conditions require that the transistor is operated in the ohmic (or triode) region, as shown in Fig. 7.23(b).

To operate the MOSFET in the ohmic region, the gate-to-source voltage must be sufficient to maintain the drain current. Assuming $v_I = v_{GS} > V_t$ and $v_{DS} = v_{GS} - V_t$ at the boundary condition between the triode and saturation regions, we can find the output voltage as given by

$$v_o = V_{DD} - R_D i_D = V_{DD} - R_D K_n (v_I - V_t)^2 \tag{7.38}$$

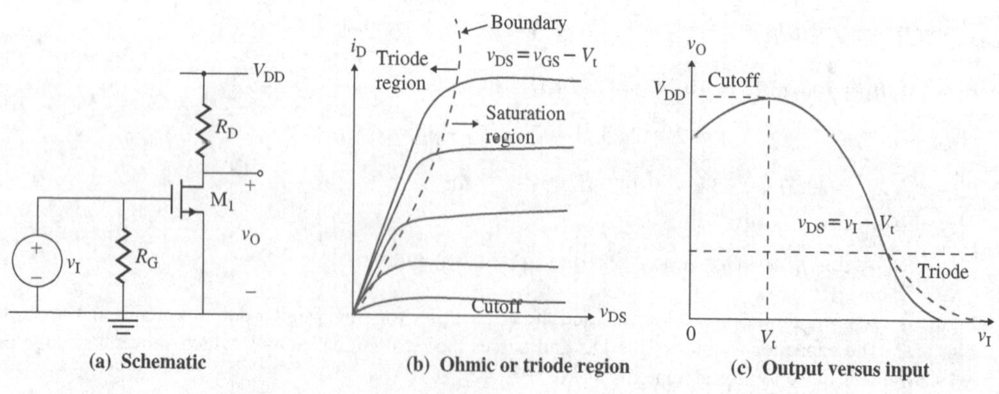

FIGURE 7.23 MOSFET switch

which describes the relationship between the input and output voltages as shown in Fig. 7.23(c). Substituting $v_O = v_{DS} = v_I - V_t$ in Eq. (7.38), we can determine the input voltage at the boundary condition:

$$V_{DD} - R_D K_n (v_I - V_t)^2 = v_I - V_t \tag{7.39}$$

This can be solved for the input voltage at the boundary condition as given by

$$v_{I(B)} = \left(\frac{-1 + \sqrt{4K_n R_D V_{DD}}}{2K_n R_D} + V_t \right) \tag{7.40}$$

Therefore, the MOSFET will operate in the ohmic region if $v_I > v_{I(B)}$ and in the saturation region if $v_I < v_{I(B)}$. For example, if $V_{DD} = 15$ V, $R_D = 3.5$ kΩ, $V_t = 1$ V, and $K_n = 3.25$ mA/V^2, Eq. (7.40) gives $v_{I(B)} = 2.104$ V. The slope of the $v_O - v_I$ described by Eq. (7.38) gives the voltage gain in the saturation region as

$$A_{vo} = \frac{dv_O}{dv_I} = \frac{d}{dv_I}[V_{DD} - R_D K_n (v_I - V_t)^2] = -2R_D K_n (v_I - V_t) \tag{7.41}$$

This gives the same result as Eq. (7.36) if we neglect the MOSFET output resistor $r_o \simeq \infty$. The maximum value of the drain current $I_{D(max)}$ is specified by the manufacturer data sheet, which limits the minimum value of drain resistance R_D. Assuming $V_{DS(sat)}$ is the drain-to-source saturation voltage, we can find the corresponding drain current $I_{D(sat)}$ as given by

$$I_{D(sat)} = \frac{V_{DD} - V_{DS(sat)}}{R_D} \tag{7.42}$$

which sets the limits of the drain current $I_{D(sat)} \le i_D \le I_{D(max)}$ and the corresponding drain resistance

$$R_{D(max)} \le R_D \le R_{D(min)} \tag{7.43}$$

7.7 DC Biasing of MOSFETs

It is necessary to bias a MOSFET at a stable operating point so that the biasing point does not change significantly with changes in the transistor parameters. Once the gate-to-source voltage v_{GS} has been set at a specified value, the MOS drain current i_D is then fixed. The drain-to-source voltage v_{DS} is dependent on i_D. Table 7.1 shows the parameters if their values are positive (+) or negative (−) quantities, and transfer characteristics for various types of MOSFETs.

Since the input gate is isolated electrically from the drain or source terminals, the gate voltage v_G can be set to any specified desired value independently of i_D, v_{DS}, and v_{SR}. The drain current depends on the gate-to-source voltage v_{GS}, which is the difference between the gate voltage v_G and the source (or substrate) voltage v_{SR}. That is, $v_{GS} = v_G - v_{SR}$. We can also write $v_{GS} = v_G$ for $v_{SR} = 0$, and $v_{GS} = -v_{SR}$ for $v_G = 0$. Therefore, we can bias a MOSFET at a specific v_{GS} by different biasing arrangements as shown in Fig. 7.24. Although there are many types of biasing circuits, we will consider the following types, which are most commonly used:

- Zero source resistance biasing
- Grounded gate terminal biasing
- Source resistance only biasing
- Source and drain resistance biasing

TABLE 7.1 Biasing conditions of MOSFETs

	n-channel		*p*-channel	
	Enhancement MOSFET	Depletion MOSFET	Enhancement MOSFET	Depletion MOSFET
K_n	$\frac{\mu_n C_{ox} W}{2L}$	$\frac{\mu_n C_{ox} W}{2L}$	$\frac{\mu_n C_{ox} W}{2L}$	$\frac{\mu_n C_{ox} W}{2L}$
V_t or V_p	+	−	−	+
v_{GS}	$>V_t$	$>V_p$	$<V_t$	$<V_p$
v_{DS}	+	+	−	−
i_D	+	+	−	−
V_{DD}	+	+	−	−
$\lambda = 1/V_M$	+	+	−	−

In the ohmic (or triode) region, $i_D = K_n[2(v_{GS} - V_t)v_{DS} - v_{DS}^2]$, where $v_{DS} < (v_{GS} - V_t)$ for *n*-channel and $v_{DS} > (v_{GS} - V_t)$ for *p*-channel.

In the saturation region, $i_D = K_n(v_{GS} - V_t)^2$, where $v_{DS} \geq (v_{GS} - V_t)$ for *n*-channel and $v_{DS} \leq (v_{GS} - V_t)$ for *p*-channel. *Note:* $V_p \approx V_t$.

▶ **NOTE** In the derivations of the drain currents for these biasing circuits, we will assume that MOSFETs operate in the saturation region and follow the relationship between i_D and v_{GS}: $i_D = K_n(v_{GS} - V_t)^2$.

7.7.1 MOSFET Biasing Circuit

The most common biasing circuit, which can implement the four arrangements in Fig. 7.24 if we select appropriate values of R_1, R_2, R_D, R_{SR}, V_{DD} and V_{SS}, is shown in Fig. 7.25(a). The value of v_{GS} can be adjusted by using a potential divider consisting of R_1 and R_2 as given by

$$v_G = \frac{R_2 V_{DD}}{R_1 + R_2} \tag{7.44}$$

Using KVL in the gate-to-source loop, and the same drain current i_D flows through the source terminal, we get

$$v_{GS} = v_G - R_{SR} i_D \tag{7.45}$$

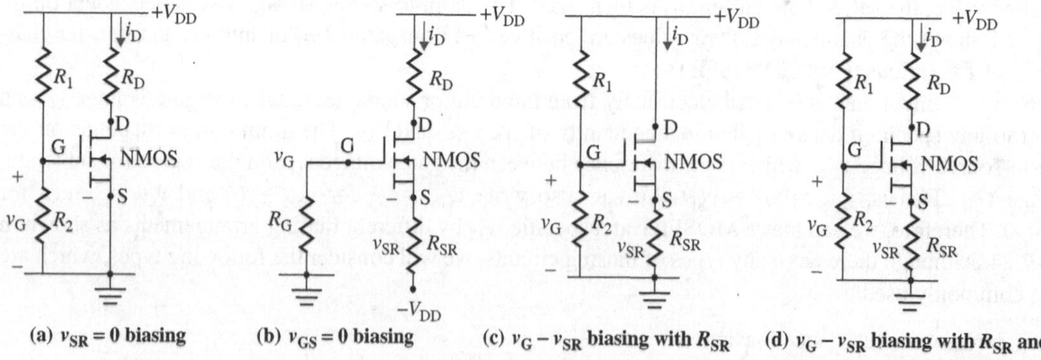

FIGURE 7.24 Arrangements for v_{GS} bias

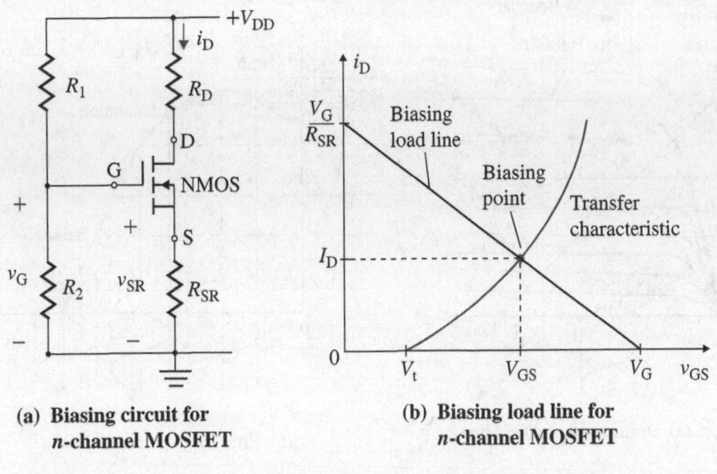

FIGURE 7.25 Biasing circuits for MOSFETs

which gives the biasing load line as given by

$$i_D = \frac{v_G}{R_{SR}} - \frac{v_{GS}}{R_{SR}} \tag{7.46}$$

The intersection of the biasing load line described by Eq. (7.46) with the transfer characteristic in Eq. (7.6) gives the operating point as shown in Fig. 7.25(b). Figure 7.25(b) also describes the input-output relationship. v_{GS} relates to i_D which in turn relates to v_{DS} (output voltage) of the MOSFET.

7.7.2 Design of MOSFET Biasing Circuit

Using KVL in the drain and the source loop in Fig. 7.25(a), we can write

$$v_{DS} = V_{DD} - R_D i_D - R_{SR} i_D = V_{DD} - (R_D + R_{SR}) i_D \tag{7.47}$$

This gives the drain-to-source load line as

$$i_D = \frac{V_{DD}}{R_D + R_{SR}} - \frac{v_{DS}}{R_D + R_{SR}} \tag{7.48}$$

This is the equation of a straight line and represents the load line, as shown in Fig.7.26(a). The intersection of the drain-to-source load line described by Eq. (7.48) with the MOS characteristic gives the operating point, defined by (V_{GS}, V_{DS}, I_D).

The given design parameters are K_n, V_t, V_M, and $I_{D(max)}$ of a MOSFET, along with V_{DD} and V_{SS}. Select suitable values of the DC biasing drain current I_{DQ} and the drain-to-source voltage V_{DS} so that i_D and v_{DS} can have the maximum swings in both positive and negative directions: $i_D = I_D \pm i_{d(peak)}$ and $v_{DS} = V_{DS} \pm v_{ds(peak)}$. Otherwise, the small-signal output will be distorted as shown in Fig. 7.26(b). To minimize distortion, I_D must therefore be less than $I_{D(max)}/2$, and the DC supply V_{DD} should be shared equally by all elements in the drain and the source loop. The guidelines for determining the biasing resistances in Fig. 7.25(a) for the different configurations shown in Fig. 7.24 are given in Table 7.2.

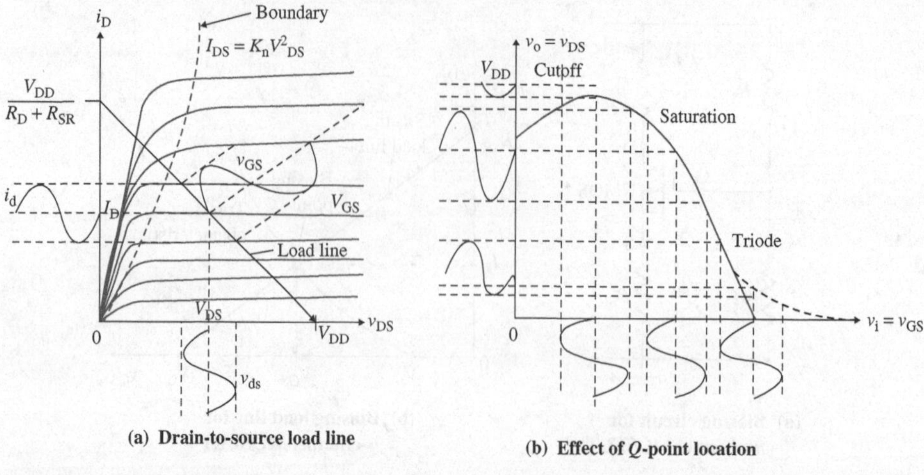

FIGURE 7.26 Drain-to-source load line and effects of operating point

TABLE 7.2 Guidelines for determining the biasing resistances

Biasing Parameters	Figure 7.24(a)	Figure 7.24(b)	Figure 7.24(c)	Figure 7.24(d)
V_{DS}	$\frac{V_{DD}}{2}$	$\frac{V_{DD}+V_{SS}}{3}$	$\frac{V_{DD}}{2}$	$\frac{V_{DD}}{3}$
V_{SR}	0	$\frac{V_{DD}+V_{SS}}{3}$	$\frac{V_{DD}}{2}$	$\frac{V_{DD}}{3}$
I_D	$\frac{I_{D(max)}}{3}$	$\frac{I_{D(max)}}{3}$	$\frac{I_{D(max)}}{3}$	$\frac{I_{D(max)}}{3}$
V_{GS}	$\pm\sqrt{I_D/K_n} + V_t$	$\pm\sqrt{I_D/K_n} + V_t$	$\pm\sqrt{I_D/K_n} + V_t$	$\pm\sqrt{I_D/K_n} + V_t$
V_G	V_{GS}	$V_{SR} + V_{GS}$	$V_{SR} + V_{GS}$	$V_{SR} + V_{GS}$
R_D	$\frac{V_{DD}}{2I_D}$	$\frac{V_{DD}+V_{SS}}{3I_D}$	0	$\frac{V_{DD}}{3I_C}$
R_{SR}	0	$\frac{V_{SS}-V_{GS}}{I_D}$	$\frac{V_{DD}}{2I_D}$	$\frac{V_{DD}}{3I_D}$
$\frac{R_1}{R_2}$				$\frac{V_{DD}}{V_G} - 1$

EXAMPLE 7.3

D

Designing a biasing circuit for an NMOS amplifier

(a) Design the biasing circuit shown in Fig. 7.25(a) for an NMOS. The DC supply voltage is $V_{DD} = 15$ V. The NMOS parameters are $V_t = 1$ V, $K_n = 3.25$ mA/V^2, $I_{D(max)} = 10$ mA, and $|V_M| = 1/\lambda = 100$ V.

(b) Calculate the small-signal parameters of the NMOS: g_m and r_o.
(c) Use SPICE to verify your design. Assume $K_p = 2K_n$ for $W = L = 1\ \mu$m.

SOLUTION

(a) Let us assume that $v_{DS} = V_{DD}/3 = 15/5 = 5$ V, and $i_D = I_{D(max)}/3 = 10$ mA$/3 = 3.33$ mA. $R_D = R_{SR} = v_{DS}/i_D = 5/(3.33 \times 10^{-3}) = 1.5$ kΩ.

Substituting $i_D = 3.33$ mA and $K_n = 3.25$ mA/V^2 into Eq. (7.10), $i_D = K_n\,(v_{GS} - V_t)^2$, gives $v_{GS} = 2.013$ V or -0.013 V. For the NMOS, v_{GS} must be greater than 1 V. Thus, the acceptable value is $v_{GS} = 2.013$ V. Using KVL around the gate-to-source loop,

$$v_G = v_{SR} + v_{GS} = 5 + 2.013 = 7.013 \text{ V}$$

$$\frac{R_1}{R_2} = \left(\frac{V_{DD}}{v_G}\right) - 1 = \frac{15}{7.013} - 1 = 1.139$$

Letting $R_2 = 100$ kΩ, we get $R_1 = 1.139R_2 = 1.139 \times 100$ k $= 113.9$ kΩ.

(b) From Eq. (7.27), $g_m = 2\,K_n\,(v_{GS} - V_t) = 2 \times 3.25$ m $\times (2.013 - 1) = 6.583$ mA/V. From Eq. (7.26), $r_o = 1/(\lambda i_D) = |V_M|/i_D = 100/3.33$ mA $= 30$ kΩ.

(c) The details of the DC bias calculations by the PSpice simulation are given here:

ID = 3.35E−03 (3.33 mA)	VGS = 1.99E+00 (2 V)	VDS = 4.96E+00 (5 V)
GM = 6.76E−03 (6.583 mA/V)	GDS = 3.19E−05 ($1/r_o = 1/30$ k $= 33.3\ \mu$A/V).	

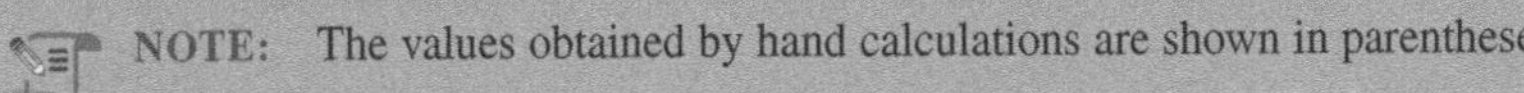

NOTE: The values obtained by hand calculations are shown in parentheses.

EXAMPLE 7.4

D

Designing a biasing circuit for a depletion NMOS amplifier

(a) Design a biasing circuit as shown in Fig. 7.25(a) for an n-channel depletion MOSFET. The DC supply voltage is $V_{DD} = 15$ V. The parameters are $I_{DSS} = 12.65$ mA and $V_p = -3.5$ V. Assume operation in the saturation region.
(b) Calculate the small-signal parameters g_m and r_o of the transistor.
(c) Use PSpice/SPICE to verify your design.

SOLUTION

(a) To accommodate the maximum AC swing and the variations in depletion MOS parameters, the following conditions as listed in Table 7.2 are recommended for biasing for the Q-point (I_D, V_{DS}):

$$i_D = \frac{I_{DSS}}{2} \tag{7.49}$$

$$v_{DS} = \frac{V_{DD}}{3} \tag{7.50}$$

That is,

$$i_D = \frac{I_{DSS}}{2} = \frac{12.65\text{ mA}}{2} = 6.3\text{ mA} \quad \text{and} \quad v_{DS} = \frac{V_{DD}}{3} = \frac{15}{3} = 5\text{ V}$$

Substituting $K_n = I_{DSS}/V_p^2$ in Eq. (7.21),

$$6.3\text{ mA} = 12.65\text{ mA} \times \left(1 + \frac{v_{GS}}{3.5}\right)^2 \quad \text{or} \quad \left(1 + \frac{v_{GS}}{3.5}\right)^2 = \pm 0.706$$

which gives $v_{GS} = -1.03$ V or -5.97 V. Since $v_{GS} > V_p\,(= -3.5\text{ V})$, the operational value of v_{GS} is -1.03 V.

Since v_{GS} of a depletion MOSFET is negative (-1.03 V), we do not need the biasing resistance $R_1(=\infty)$ because R_{SR} will cause a voltage drop of $v_{GS} = -R_{SR}i_D$. This arrangement is known as self-biasing of the depletion MOSFET as shown in Fig. 7.24(b).

We find that

$$R_{SR} = \frac{-v_{GS}}{i_D} = \frac{1.03\text{ V}}{6.3\text{ mA}} = 163.5\ \Omega$$

and its power rating is

$$P_{RSR} = (6.3 \times 10^{-3}\text{ A})^2 \times 163.5\ \Omega = 6.49\text{ mW}$$

Since $R_D i_D = V_{DD} - R_{SR}i_D - v_{DS} = 15 - 1.03 - 5 = 8.97$ V,

$$R_D = \frac{8.97\text{ V}}{6.3\text{ mA}} = 1424\ \Omega$$

and its power rating is

$$P_{RD} = (6.3 \times 10^{-3}\text{ A})^2 \times 1424\ \Omega = 56.52\text{ mW}$$

Since one side of R_2 is connected to the ground and the gate–source junction is like a reverse-biased diode, the DC current flowing through R_2 is very small, tending to zero. R_2 provides continuity of the circuit for the gate–source biasing voltage. In selecting the value of R_2, it is important to keep two things in mind: (1) R_2 should match the reverse-bias resistance of the gate–source junction, and (2) R_2 will carry current when an AC signal is applied to the gate terminal. A value of R_2 between 50 kΩ and 500 kΩ is generally suitable. Let $R_2 = 500$ kΩ.

(b) $K_n = I_{DSS}/V_p^2 = 12.65\text{ mA}/(3.5\text{ V})^2 = 1.033\text{ mA/V}^2$. From Eq. (7.30),

$$g_m = 2K_n(v_{GS} - V_p) = 2 \times 1.033\text{ m} \times (-1.03 + 3.5) = 5.10\text{ mA/V}$$

From Eq. (7.26),

$$r_o = \frac{1}{\lambda i_D} = \frac{1}{5.929\text{ m} \times 6.3\text{ mA}} = 26.77\text{ k}\Omega$$

(c) The biasing circuit for PSpice simulation is shown in Fig. 7.27. For PSpice simulation, we use $K_p = 2 \times K_n = 2 \times 1.033 = 2.066\text{ mA/V}^2$ for $W = L = 1\ \mu$m and $V_{to} = V_p = -3.5$ V. The details of the DC bias calculations by the PSpice simulation are given here:

ID = 6.46E−03 (6.3 mA) VGS = −1.06E+00 (−1.03 V) VDS = 4.74E+00 (5 V)
GM = 5.29E−03 (5.10 mA/V) GDS = 6.17E−05 ($1/r_o = 1/15.81\text{ k} = 63.25\ \mu\text{A/V}$)

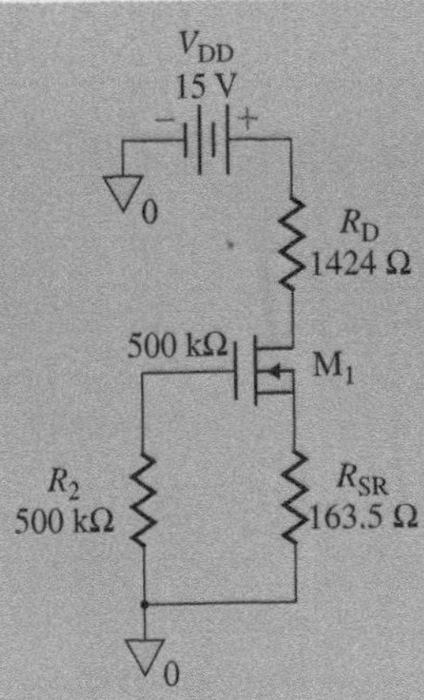

FIGURE 7.27 Biasing circuit for PSpice simulation for Example 7.4

NOTE: Notice from the output file of EX7-4.SCH that PSpice uses VTO=−3 V (instead of −3.5 V), LAMBDA=2.250000E-03 (instead of 5.929E-3 V^{-1}), and BETA=1.304000E-03 (instead of 1.033 mA/V^2). For this reason, the results from PSpice and hand calculations differ significantly. If we recalculated the values of R_D and R_{SR} with the PSpice parameters or changed the MOS parameters in the model statement, the results would be very close. If you run the simulation with EX7-4.CIR, the results will be closer to the hand calculations: I_D=6.14 mA and V_{GS}=1.11 V.

EXAMPLE 7.5

D **Design for limiting the drain current variation of an NMOS amplifier** Design a biasing circuit as shown in Fig. 7.25(a) for an NMOS for which V_t varies from 1 V to 1.5 V and K_n varies from 150 $\mu A/V^2$ to 100 $\mu A/V^2$. Limiting the variation in the drain current to 5 mA ± 20%, calculate the values of R_{SR}, R_1, R_2, and R_D. Assume V_{DD} = 15 V.

SOLUTION

V_{t1} = 1 V, V_{t2} = 1.5 V, K_{n1} = 150 $\mu A/V^2$, and K_{n2} = 100 $\mu A/V^2$. The two possible transfer characteristics that can result from the variations in the parameters are shown in Fig. 7.28. Using Eq. (7.10), we can describe these characteristics as follows:

$$I_{D1} = K_{n1}(V_{GS1} - V_{t1})^2 = 150 \times 10^{-6} \times (V_{GS1} - 1)^2$$

$$I_{D2} = K_{n2}(V_{GS2} - V_{t2})^2 = 100 \times 10^{-6} \times (V_{GS2} - 1.5)^2$$

For a drain current variation of I_{D1} = 5 mA + 20% = 5 mA × (1 + 0.2) = 6 mA, we have

$$6 \text{ mA} = 150 \ \mu\text{A} \times (V_{GS1} - 1)^2$$

which gives an operating value of V_{GS1} = 7.32 V.

For a drain current variation of I_{D2} = 5 mA − 20% = 5 mA × (1 − 0.2) = 4 mA, we have

$$4 \text{ mA} = 100 \ \mu\text{A} \times (V_{GS2} - 1.5)^2$$

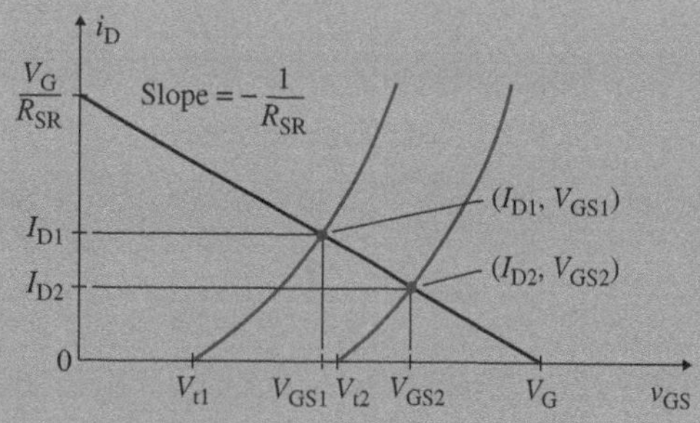

FIGURE 7.28 Two transfer characteristics

which gives an operating value of $V_{GS2} = 7.825$ V.

The slope of the biasing load line gives the value of R_{SR}:

$$R_{SR} = \frac{V_{GS2} - V_{GS1}}{I_{D1} - I_{D2}} = \frac{7.825 - 7.32}{6 - 4} \times 10^3 = 252.5\ \Omega$$

Applying Eq. (7.45) at the Q-point characteristic with v_{GS2} and i_{D2} gives

$$V_G = V_{GS2} + I_{D2}R_{SR} = 7.825 + 4 \times 10^{-3} \times 252.5 = 8.835\text{ V}$$

The values of R_1 and R_2 can be found from

$$V_G = \frac{R_2 V_{DD}}{R_1 + R_2} = \frac{R_2 \times 15}{R_1 + R_2} = 8.835\text{ V}$$

which gives $(1 + R_1/R_2) = 1.7$. Choose a suitable value of R_2, usually larger than 500 kΩ. Assuming $R_2 = 500$ kΩ, $R_1 = 350$ kΩ.

KEY POINT OF SECTION 7.7

- A MOSFET should be biased properly in order to activate the device and also to establish a DC operating point such that a small variation in the gate-to-source voltage causes a variation in the drain current. Like any amplifier, a MOSFET amplifier can be used as a buffer stage to offer a low output resistance and a high input resistance.

7.8 Common-Source (CS) Amplifiers

Figure 7.16(a) is an example of common-source (CS) amplifiers where the source terminal is common to both input and output terminals. Equation (7.36) gives the voltage gain as $A_{vo} = -g_m(r_o \| R_D) \simeq -g_m R_D$ for $r_o >> R_D$. The resistive biasing design limits the value of R_D and the voltage gain. Replacing R_D with an active current source that has an inherent high output resistance can increase the voltage gain significantly. Any resistance in the source terminal reduces the effective small-signal voltage gain, and we will evaluate the effect in the voltage gain. We will consider CS amplifiers with four types of load: (a) active current source load, (b) enhancement MOSFET load, (c) depletion MOSFET load, and (d) resistive load.

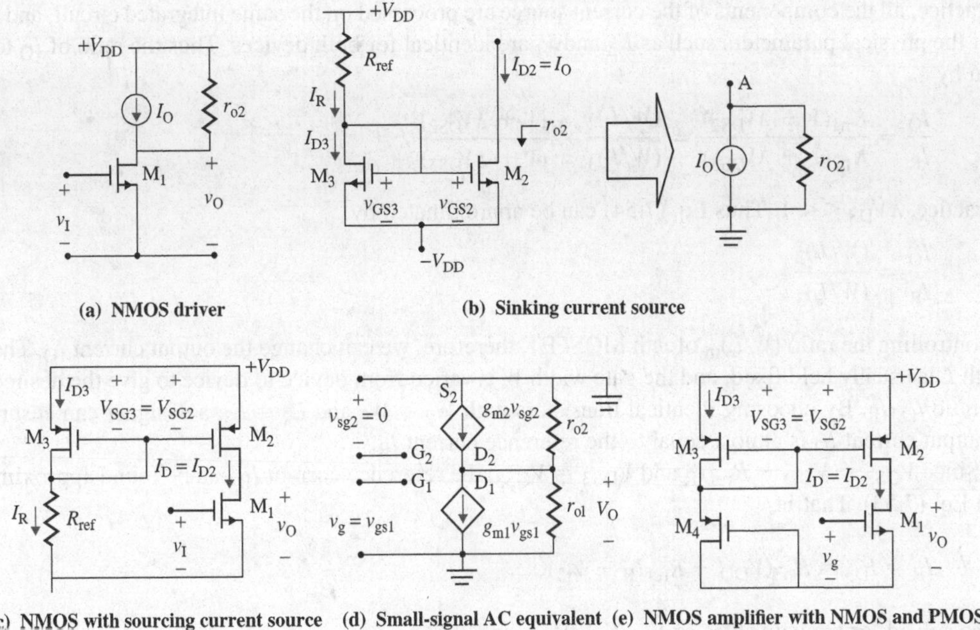

FIGURE 7.29 CS NMOS amplifier with a current source load

7.8.1 CS Amplifier with Current Source Load

A simple MOSFET amplifier with a current source is shown in Fig. 7.29(a). If r_{o1} and r_{o2} are the output resistances of the transistor M_1 and the current source I_O, respectively, we can find the small-signal voltage gain from Eq. (7.36) as given by

$$A_{vo} = \frac{v_O}{v_{gs}} = -g_m(r_{o1} \| r_{o2}) \tag{7.51}$$

A basic MOSFET current source is shown in Fig. 7.29(b), which can be represented with a sinking current source I_O with an output resistance as shown in Fig. 7.29(b): $r_{o2} = V_{M2}/I_{D2}$. Let us assume that the two transistors M_2 and M_3 are identical. Since their gate-to-source voltages are equal, their drain currents will be the same. That is, $I_{D2} = I_{D3}$. Thus the output current I_O ($=I_{D2}$) will be the mirror of I_{D3}. Since $V_{DS3} = V_{GS3}$, which is greater than or equal to ($V_{GS3} - V_{t3}$), M_3 will be in saturation. Let V_{t2} and V_{t3} be the threshold voltages of M_2 and M_3, respectively. For M_2 also to be in saturation, V_{DS2} must be greater than ($V_{GS2} - V_{t2}$). This condition reduces the voltage compliance range of the MOSFET current source and prevents it from operating from a low power supply (say, 1 V for a battery source).

▶ **NOTE** V_t is the threshold voltage of a MOSFET, whereas V_T is the thermal voltage.

The output current I_O, which is equal to the drain current of M_2, is given by

$$I_{D2} = I_O = K_{n2}(V_{GS2} - V_{t2})^2(1 + \lambda V_{DS2}) \tag{7.52}$$

Drain current I_{D2}, which is equal to the reference current I_R, is given by

$$I_{D3} = I_R = K_{n3}(V_{GS3} - V_{t3})^2(1 + \lambda V_{DS3}) \tag{7.53}$$

In practice, all the components of the current source are processed on the same integrated circuit, and hence all of the physical parameters such as K_n and V_t are identical for both devices. Thus the ratio of I_O to I_R is given by

$$\frac{I_O}{I_R} = \frac{K_{n1}(1 + \lambda V_{DS2})}{K_{n2}(1 + \lambda V_{DS3})} = \frac{(W/L)_2}{(W/L)_3} \times \frac{(1 + \lambda V_{DS2})}{(1 + \lambda V_{DS3})} \tag{7.54}$$

In practice, $\lambda V_{DS} << 1$. Thus Eq. (7.54) can be approximated by

$$\frac{I_O}{I_R} = \frac{(W/L)_2}{(W/L)_3} \tag{7.55}$$

By controlling the ratio $(W/L)_m$ of mth MOSFET, therefore, we can change the output current I_O. The gate length L is usually held fixed, and the gate width W is varied from device to device to give the desired current ratio I_O/I_R. By choosing identical transistors with $W_2 = W_3$ and $L_2 = L_3$, a designer can ensure that the output current I_O is almost equal to the reference current I_R.

Since $V_{GS3} = V_{DD} - R_{ref}I_R$ and $V_{DS3} = V_{GS3}$, the reference current I_R can be found approximately from Eq. (7.53). That is,

$$I_R = I_{D3} = K_{n3}(V_{DD} - R_{ref}I_R - V_{t3})^2 \tag{7.56}$$

can be solved for known values of V_{t3}, K_{n3}, V_{DD}, and R_{ref}.

Replacing I_O in Fig. 7.29(a) with the sourcing type of current source consisting of PMOS is shown in Fig. 7.29(c). This is accomplished by replacing M_2 and M_3 in Fig. 7.29(b) with PMOS. The small-signal equivalent circuit of Fig. 7.29(c) is shown in Fig.7.29(d), from which we can find the small-signal voltage gain as given by Eq. (7.51). For identical transistors $r_o = r_{o1} = r_{o2}$, Eq. (7.51) becomes

$$A_{vo} = -g_m(r_{o1} \| r_{o2}) = -g_m\frac{r_o}{2} = 2K_n(V_{GS} - V_t)\frac{1}{2\lambda I_D} \tag{7.57}$$

$$= \frac{2K_n(V_{GS} - V_t)}{2\lambda K_n(V_{GS} - V_t)^2} = \frac{1}{\lambda(V_{GS} - V_t)}$$

Therefore, we can conclude that the voltage gain A_{vo} is inversely proportional to λ, the biasing current I_D, and the gate-to-source voltage V_{GS}.

The reference resistance R_1 in Fig. 7.29(c) can be replaced by one or more PMOS, as shown in Fig. 7.29(e). Transistors $M_3, \ldots, M_n$ are used as voltage dividers to control the gate-to-source voltage of transistor M_2. If there are n cascode-connected PMOS, and all have identical characteristics, the gate-to-source voltage of the PMOS M_2 is given by

$$v_{GS2} = v_{GS3} = \frac{-V_{DD}}{n} \tag{7.58}$$

This gives the reference drain current i_D as

$$i_D = K_{n2}(v_{GS2} - V_{tP})^2 = K_{n2}\left(\frac{-V_{DD}}{n} - V_{tP}\right)^2 \tag{7.59}$$

Therefore, we can find the integer number of MOSFETs to obtain a specific reference current I_D or gate-to-source voltage V_{GS}. For example, if $V_{DD} = 12$ V, we need six PMOS MOSFETs to get $V_{SG} = 2$ V. We should note that V_{tP} in Eq. (7.59), which is the threshold voltage of a PMOS, has a negative value.

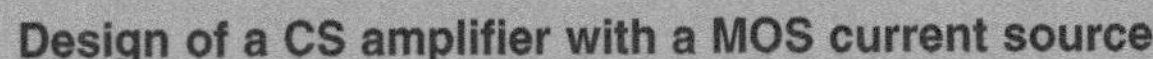

EXAMPLE 7.6

D

Design of a CS amplifier with a MOS current source

(a) Find the value of R_1 to obtain a biasing current of $I_D = 3.25$ mA and the small-signal voltage gain of the CS amplifier in Fig. 7.29(c). The DC supply voltage is $V_{DD} = 15$ V. The MOS parameters are $V_{tN} = -V_{tP} = 1$ V, $K_n = 3.25$ mA/V^2, $K_p = 6.5$ mA/V^2, for $W = L = 1$ μm and $|V_M| = 1/\lambda = 100$ V.

(b) Use SPICE to plot the small-signal output voltage v_o for a sinusoidal input signal v_s of 1 mV at 1 kHz.

SOLUTION

(a) For $I_D = 3.25$ mV, Eq. (7.8) gives 3.25 mA = 3.25 m × $(V_{GS3} - 1)^2$, which gives $V_{GS3} = -2$ V. Applying KVL through the loop via V_{DD}, R_1, and V_{GS3}, we can find the values of

$$R_{ref} = \frac{V_{DD} + V_{GS3}}{I_D} = \frac{15 - 2}{3.25\text{ m}} = 4\text{ k}\Omega$$

$$r_o = r_{o1} = r_{o2} = \frac{V_M}{I_D} = \frac{100}{3.25\text{ m}} = 30.77\text{ k}\Omega$$

$$g_m = 2K_n(V_{GS3} - V_{tN}) = 2 \times 3.25\text{ m} \times (2 - 1) = 6.5\text{ mA/V}$$

From Eq. (7.57), the voltage gain $A_{vo} = -g_m r_o/2 = -6.50\text{ m} \times 30.7\text{ k}/2 = -100$ V/V.

The DC gate voltage of M_1 is $V_{G1} = 2$ V, for which we can use a voltage divider as shown in Fig. 7.25(a). Therefore, $R_1/R_2 = V_{DD}/V_{G1} - 1 = 15/2 - 1 = 6.5$.

Let $R_2 = 100$ kΩ; then $R_1 - 6.5 \times R_2 = 6.5 \times 100\text{ k} = 650$ kΩ.

(b) The PSpice schematic is shown in Fig.7.30. The plot of the output voltage is shown in Fig. 7.31, which gives a voltage gain of −101 V/V; this is close to the calculated value of −100. Note that there is a phase shift of 180°.

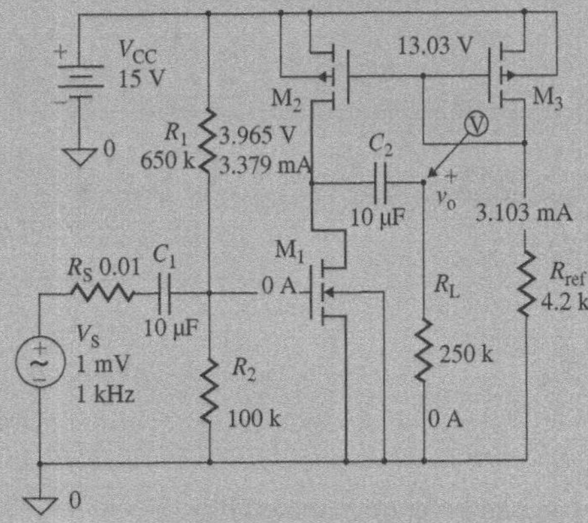

FIGURE 7.30 PSpice schematic for Example 7.6

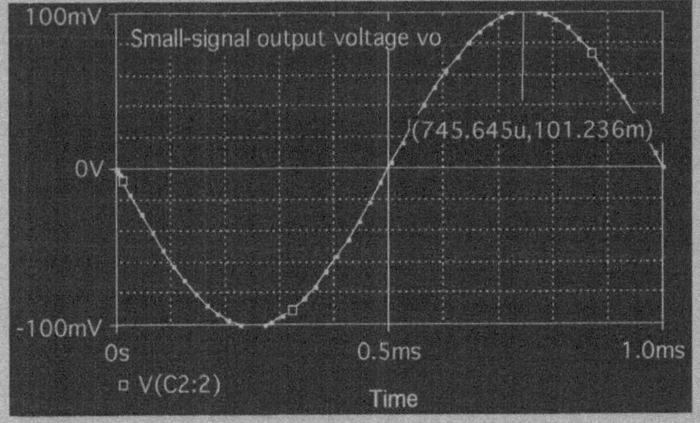

FIGURE 7.31 PSpice plot of small-signal output voltage for Example 7.6

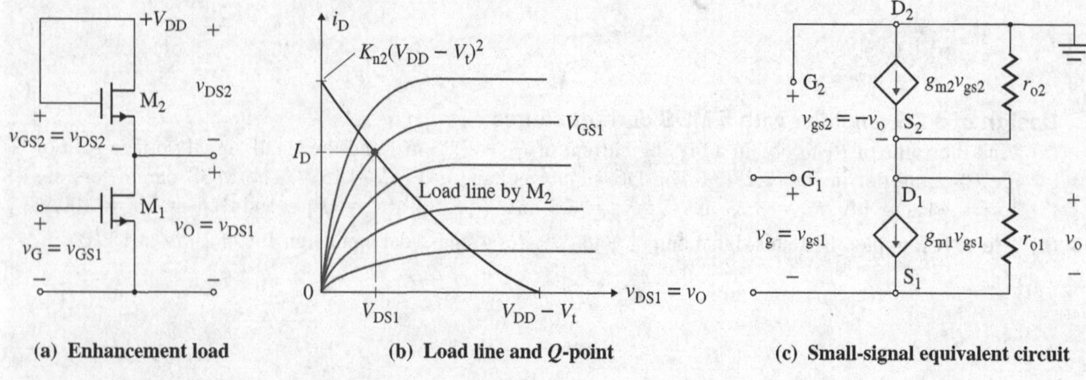

(a) Enhancement load **(b) Load line and Q-point** **(c) Small-signal equivalent circuit**

FIGURE 7.32 CS amplifier with enhancement load

7.8.2 CS Amplifier with Enhancement MOSFET Load

A CS amplifier with an NMOS driver and an NMOS active load, as shown in Fig. 7.32(a), is the simplest way of implementing an amplifier with NMOS technology. M_2 is diode-connected, and it behaves as a nonlinear resistive load. If the input voltage v_G is less than the threshold voltage V_t, then M_1 is off and no current flows in the circuit. If the input voltage v_G exceeds the threshold voltage V_t, then M_1 is turned on. Both M_1 and M_2 operate in the saturation region, and the circuit provides amplification. Since $v_{GS2} = v_{DS2} = V_{DD} - v_O$, the drain current i_D can be related to the output voltage v_O by

$$i_D = K_{n2}(v_{GS2} - V_t)^2 = K_{n2}(V_{DD} - v_O - V_t)^2$$

which gives $i_D = 0$ at $v_O = V_{DD} - V_t$ and $i_D = (V_{DD} - V_t)^2$ at $v_O = 0$. The i_D-versus-v_O ($=v_{DS1}$) characteristic is superimposed on the output characteristics of M_1 in Fig. 7.32(b), and the intersection of the two characteristics gives the operating point defined by I_D and V_{DS1}.

Replacing the transistors in Fig. 7.32(a) with their small-signal models gives the AC equivalent circuit shown in Fig. 7.32(c). Summing currents at the output node, we get

$$-g_{m2}v_o - \frac{v_o}{r_{o2}} - g_{m1}v_{gs1} - \frac{v_o}{r_{o1}} = 0$$

which gives the open-circuit voltage gain as

$$A_{vo} = \frac{v_o}{v_g} = \frac{v_o}{v_{gs1}} = \frac{-g_{m1}}{g_{m2} + 1/r_{o1} + 1/r_{o2}} \tag{7.60}$$

The equivalent output resistance can easily be shown to be

$$R_o = r_{o1} \parallel r_{o2} \parallel \left(\frac{1}{g_{m2}}\right) \tag{7.61}$$

For $g_{m2} >> 1/r_{o1}$ and $1/r_{o2}$, which is generally true, Eq. (7.61) can be approximated by

$$A_{vo} = -\frac{g_{m1}}{g_{m2}} = -\left[\frac{K_{n1}}{K_{n2}}\right]^{1/2} = -\left[\frac{W_1/L_1}{W_2/L_2}\right]^{1/2} \tag{7.62}$$

Because of the practical limitations of device geometries, the maximum voltage gain is in the range of 10 to 20. However, the small-signal voltage gain is independent of the DC operating point, and this amplifier gives a linear amplification over a broad band. For example, if $W_1 = 100\ \mu m$, $L_1 = 5\ \mu m$, $W_2 = 5\ \mu m$, and $L_2 = 25\ \mu m$, Eq. (7.62) gives $|A_{vo}| = 10$. It is worth noting that the load device M_2 will remain in the saturated mode of operation as long as the output voltage $v_O < (V_{DD} - V_t)$. Otherwise, the transistor will be in the cutoff region and will carry no current.

7.8.3 CS Amplifier with Depletion MOSFET Load

A depletion MOSFET can behave as a current source when the gate and source are shorted together, and it can be fabricated on the same IC chip as an enhancement MOSFET. This load device exhibits a very high output resistance as long as the device is operated in the saturation region. Therefore, to provide the large resistance required of a load for high voltage gain, a depletion MOSFET must be operated in the saturation region. A CS amplifier with an NMOS driver and a depletion active load is shown in Fig. 7.33(a). M_2 is diode-connected, and it behaves as a nonlinear resistive load. If the input voltage v_G is less than the threshold voltage V_t, then M_1 is off and no current flows in the circuit. If the input voltage v_G exceeds the threshold voltage V_t, then M_1 is turned on. Both M_1 and M_2 operate in the saturation region, and the circuit provides amplification. Since $v_{GS2} = 0$ and $v_O = V_{DD} - v_{DS2}$, the drain current i_D can be determined from

$$i_D = K_{n2}(v_{GS2} - V_t)^2 = K_{n2}(-V_t)^2 = K_{n2}V_t^2$$

The i_D-versus-v_O $(=V_{DD} - v_{DS2})$ characteristic is superimposed on the output characteristics of M_1 in Fig. 7.33(b), and the intersection of the two characteristics gives the operating point defined by I_D and V_{DS1}.

The AC equivalent circuit of the amplifier in Fig. 7.33(a) is shown in Fig. 7.33(c), from which we can find the open-circuit voltage gain

$$A_{vo} = -g_{m1}(r_{o1} \parallel r_{o2}) = -g_{m1}R_o \tag{7.63}$$

where $R_o = (r_{o1} \parallel r_{o2})$.

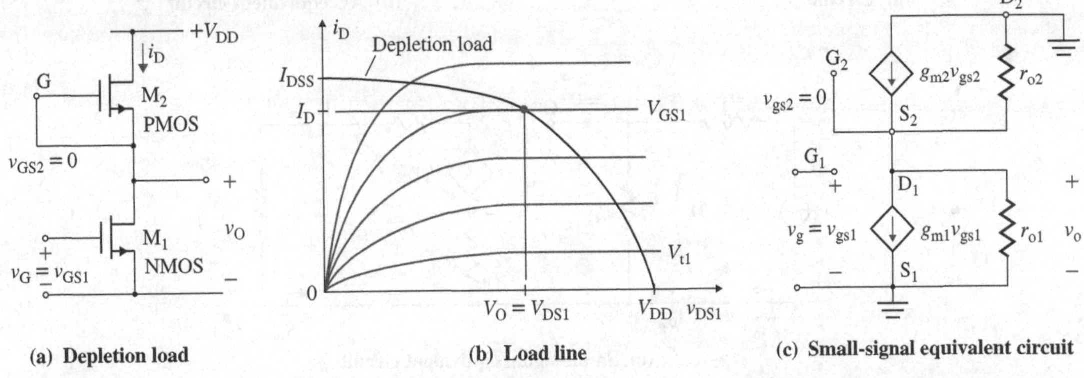

FIGURE 7.33 CS amplifier with depletion load

7.8.4 CS Amplifier with Resistive Load

A CS amplifier with resistive biasing is shown in Fig. 7.34(a). The amplifying device can be any type of MOSFET. The source resistance R_{SR} is split into two source resistors, R_{SR1} and R_{SR2}. R_{SR2} is shunted by a large capacitance C_S. Both R_{SR1} and R_{SR2} set the DC bias point, while R_{SR2} is effectively shorted for a small signal and gives the desired small-signal voltage gain. Let us assume that the coupling capacitors C_1, C_2, and C_S have high values so that they behave as short-circuited at the frequency of interest. Load resistance R_L is considered external to the amplifier and is not included. The DC biasing circuit is the same as the biasing circuit in Fig. 7.25(a).

The small-signal AC equivalent circuit of the amplifier is shown in Fig. 7.34(b). We could use either the small-signal Norton's equivalent model in Fig. 7.18(a) or Thevenin's equivalent model in Fig. 7.18(b). Due to the presence of R_{SR1} in the source branch, the analysis becomes simpler with the Thevenin's equivalent model. If we ignore the output resistance r_o, which we can generally do in most cases with reasonable accuracy, then the use of Norton's model is recommended. To obtain accurate results, we will replace the MOSFET by its small-signal model of Fig. 7.18(b); the amplifier circuit is shown in Fig. 7.34(c), which can be represented by an equivalent voltage amplifier as shown in Fig. 7.35(a) or by an equivalent transconductance amplifier as shown in Fig. 7.35(b).

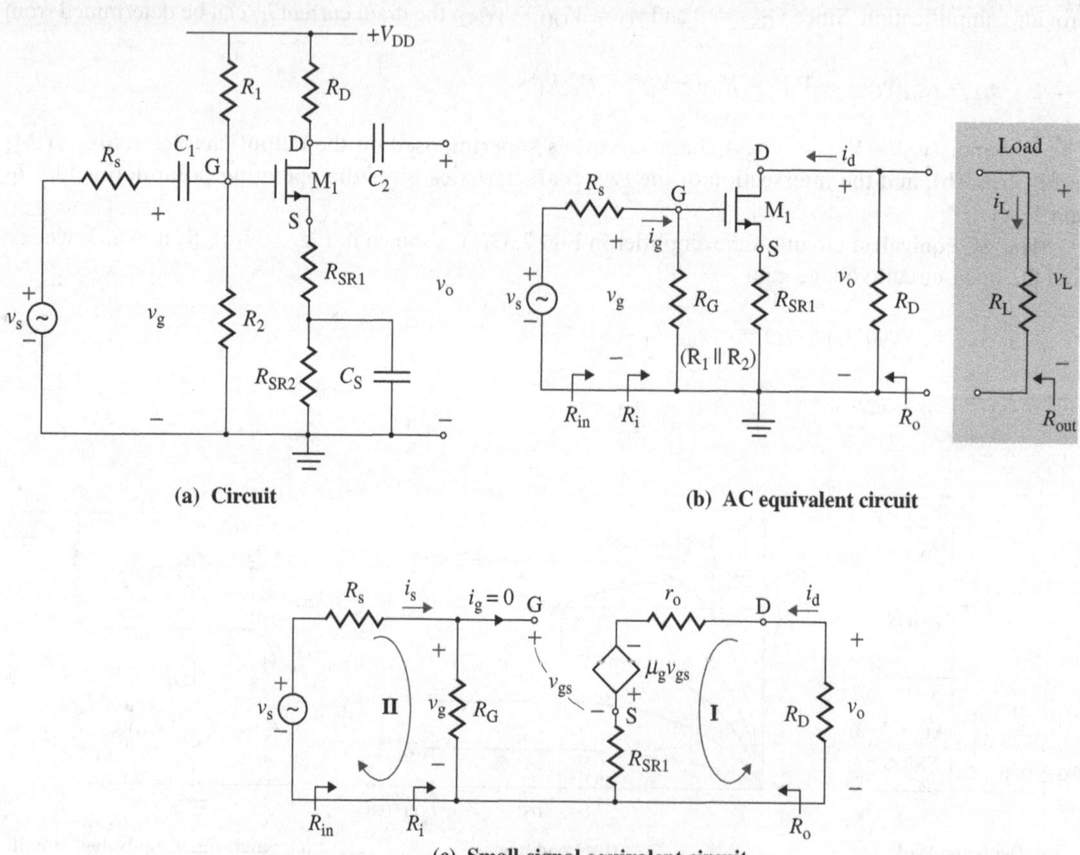

FIGURE 7.34 MOSFET amplifier with R_{SR} shunted by a capacitor

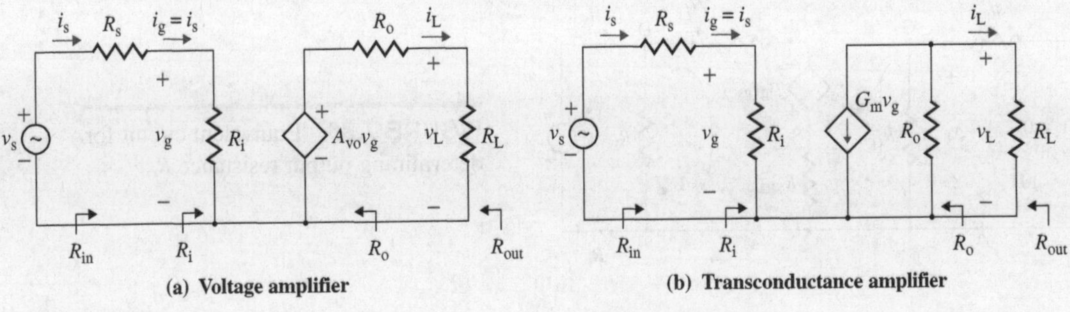

FIGURE 7.35 Equivalent voltage or transconductance representation

The DC analysis of a MOSFET amplifier must be performed prior to the small-signal analysis because the small-signal parameters depend on the DC operating point. The steps that are normally required to analyze a MOSFET amplifier are as follows:

Step 1. Draw the circuit diagram of the amplifier to be analyzed.

Step 2. Mark terminals G, D, and S for each MOSFET on the diagram. Locating these points is the beginning of drawing the equivalent circuit.

Step 3. Replace each MOSFET by its Thevenin (or Norton) model.

Step 4. Draw the other elements of the amplifier, keeping the original relative position of each element.

Step 5. Replace each DC voltage by its internal resistance. An ideal DC source should be replaced by a short circuit.

Input Resistance R_i $(= v_g/i_g)$

The input resistance R_i of the amplifier in Fig. 7.34(c) can be found from

$$R_i = \frac{v_g}{i_g} = R_G \tag{7.64}$$

The total input resistance R_{in} seen by the input signal v_s is

$$R_{in} = \frac{v_s}{i_s} = R_i + R_s$$

where R_s is the input resistance of the input signal source.

Output Resistance R_o

The output resistance R_o can be obtained by setting v_s equal to zero and then applying a test voltage v_x at the output side. This arrangement is shown in Fig. 7.36. Applying KVL around the gate, input, and source terminals (loop II) gives

$$v_{gs} = v_g - i_d R_{SR1} = -i_d R_{SR1}$$

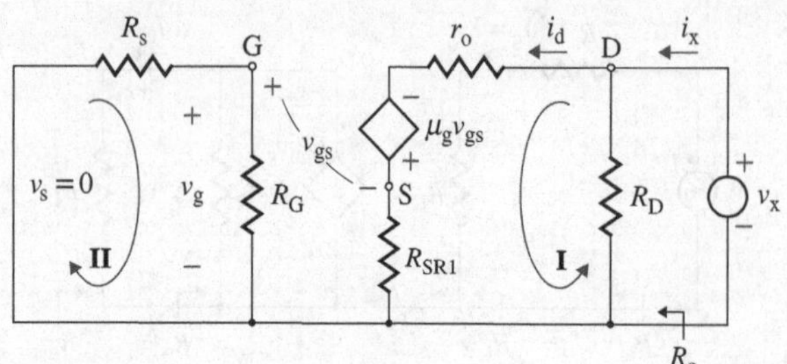

FIGURE 7.36 Equivalent circuit for determining output resistance R_o

Applying KVL around the drain, source, and test voltage source (loop I) gives

$$v_x = i_d r_o - \mu_g v_{gs} + i_d R_{SR1} = i_d r_o + \mu_g i_d R_{SR1} + i_d R_{SR1} = i_d r_o + (1 + \mu_g) R_{SR1} i_d$$

which yields

$$i_d = \frac{v_x}{r_o + (1 + \mu_g) R_{SR1}}$$

The test current i_x is given by

$$i_x = i_d + \frac{v_x}{R_D} = \frac{v_x}{r_o + (1 + \mu_g) R_{SR1}} + \frac{v_x}{R_D}$$

which gives the output resistance R_o as

$$R_o = \frac{v_x}{i_x} = [r_o + (1 + \mu_g) R_{SR1}] \| R_D \quad (7.65)$$

$$R_{out} = R_o \| R_L \quad (7.66)$$

Open-Circuit (or No-Load) Voltage Gain A_{vo} $(= v_o/v_g)$

By applying KVL around the loop formed by r_o, R_D, and the voltage-controlled voltage source in Fig. 7.34(c), we get

$$\mu_g v_{gs} = R_{SR} i_d + R_D i_d + r_o i_d \quad (7.67)$$

Substituting $v_{gs} = v_g - R_{SR} i_d$ into Eq. (7.67) gives the drain current i_d as

$$i_d = \frac{\mu_g v_g}{R_D + r_o + (1 + \mu_g) R_{SR1}} \quad (7.68)$$

The output voltage v_o can be found from

$$v_o = -R_D i_d \quad (7.69)$$

Substituting i_d from Eq. (7.68) into Eq. (7.69) gives the open-circuit voltage gain A_{vo} as

$$A_{vo} = \frac{v_o}{v_g} = \frac{-\mu_g R_D}{R_D + r_o + (1 + \mu_g) R_{SR1}} = -\frac{g_m r_o R_D}{R_D + r_o + (1 + g_m r_o) R_{SR1}} \quad (7.70)$$

which indicates that the resistance R_{SR} of the source terminal has an effect $(1 + g_m r_o)R_{SR1}$ and reduces the open-circuit voltage gain A_{vo} significantly. The voltage gain A_{vo} can be made large (a) by making $R_{SR1} = 0$, (b) by using a MOSFET with a large value of g_m, and (c) by choosing a high value of R_D. For $R_{SR1} = 0$, Eq. (7.70) gives the maximum open-circuit voltage gain as

$$A_{vo(max)} = -\frac{-\mu_g R_D}{R_D + r_o} = \frac{-g_m r_o R_D}{R_D + r_o} = \frac{-g_m R_D}{1 + R_D/r_o} \quad (7.71)$$

For $r_o >> R_D$, which is generally the case with a resistive biasing circuit, Eq. (7.71) gives $A_{vo(max)} \simeq -g_m R_D$.

EXAMPLE 7.7

D

Designing an NMOS amplifier to give a specified voltage gain

(a) Design an NMOS amplifier as shown in Fig. 7.34(a) to give a no-load voltage gain of $|A_{vo}| = v_o/v_g \geq 5$. The DC supply voltage is $V_{DD} = 15$ V. The NMOS parameters are $V_t = 1$ V, $K_n = 3.25$ mA/V^2, $K_p = 2K_n = 6.5$ mA/V^2 for $W = L = 1$ μm, $I_{D(max)} = 10$ mA, and $|V_M| = 1/\lambda = 100$ V.

(b) Use PSpice/SPICE to verify your results in part (a).

SOLUTION

(a) Step 1. Design the biasing circuit. The results of Example 7.3 give $R_D = 1.5$ kΩ, $R_{SR} = 1.5$ kΩ, $R_1 = 650$ kΩ, and $R_2 = 100$ kΩ.

Step 2. Find the small-signal parameters of the transistor. The results of Example 7.3 give $g_m = 6.583$ mA/V and $r_o = 30$ kΩ.

Step 3. Find the values of C_1, C_2, C_S, R_{SR1}, and R_{SR2}. Let us choose $C_1 = C_2 = C_S = 10$ μF. The worst-case maximum possible gain that we can obtain from the transistor operating at $i_D = 6$ mA can be found from Eq. (7.71):

$$|A_{vo(max)}| = \frac{g_m R_D}{1 + R_D/r_o} = \frac{1.5\text{ k} \times 6.583\text{ m}}{1 + 1.5\text{ k}/30\text{ k}} = 9.404\text{ V/V}$$

The desired gain is less than the maximum possible value, and we can proceed with the design. Otherwise we would need to choose another transistor with a higher value of g_m. The value of unbypassed emitter resistance R_{SR1} in Fig. 7.34(a) can be found from Eq. (7.70). That is,

$$R_D + r_o + (1 + \mu_g)R_{SR1} = \frac{\mu_g R_D}{|A_{vo}|} \quad (7.72)$$

which, for $|A_{vo}| = 5$, $R_D = 1.5$ kΩ, $r_o = 30$ kΩ, and $\mu_g = r_o g_m = 197.48$ V/V, gives $R_{SR1} = 139.79$ Ω and $R_{SR1} = R_{SR} - R_{SR1} = 1.5\text{ k} - 130.79 = 1.3$ kΩ.

(b) The PSpice schematic is shown in Fig. 7.37. The plot of the output voltage is shown in Fig. 7.38, which gives a voltage gain of −5.11 V/V. This is close to the calculated value of −5, but it is much too low in comparison to −100 with an NMOS amplifier with current source biasing (in Example 7.6). Note that there is a phase shift of 180°.

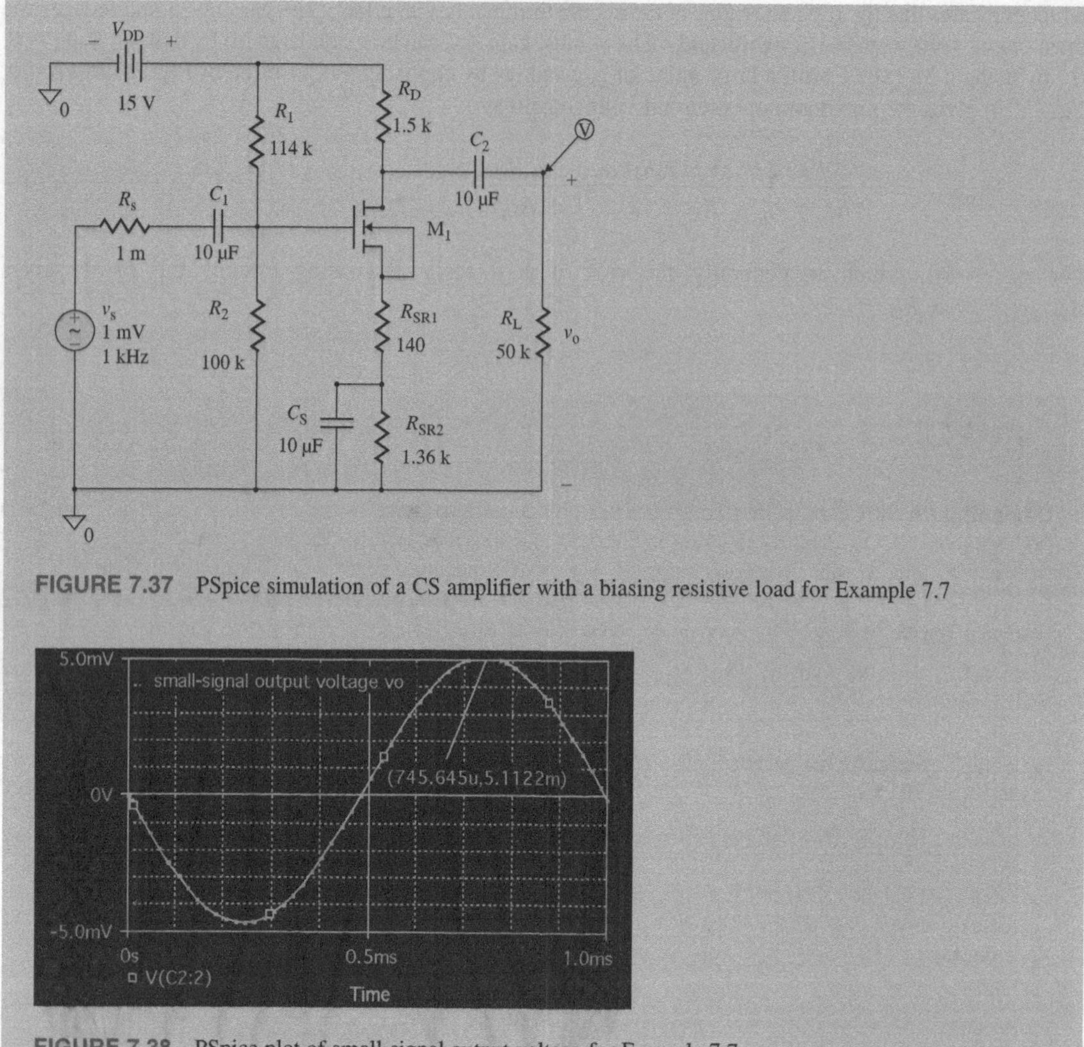

FIGURE 7.37 PSpice simulation of a CS amplifier with a biasing resistive load for Example 7.7

FIGURE 7.38 PSpice plot of small-signal output voltage for Example 7.7

TABLE 7.3 Summary of expressions for MOSFETs amplifiers

	CS Amplifier [Fig. 7.34(a)]	CD Amplifier [Fig. 7.42(a)]	CG Amplifier [Fig. 7.43(a)]	CS Amplifier with Active Load [Fig. 7.29(a)]
R_i (Ω)	R_G	R_G	$R_{SR} \parallel \left(\dfrac{r_o + R_D \parallel R_L}{1 + g_m r_o} \right)$	∞
R_o (Ω)	R_D	$\dfrac{r_o}{1 + g_m r_o} \parallel R_{SR}$	R_D	$r_{o2} \parallel r_{o1}$
A_{vo} (V/V)	$\dfrac{-g_m r_o R_D}{R_D + r_o + (1 + g_m r_o) R_{SR1}}$	$\dfrac{g_m (r_o \parallel R_{SR})}{1 + g_m (r_o \parallel R_{SR})}$	$\dfrac{R_D (1 + g_m r_o)}{r_o + R_D}$	$-g_{m1}(r_{o2} \parallel r_{o1})$

KEY POINTS OF SECTION 7.8

- The expressions for the input resistance R_i, the output resistance R_o, and the no-load voltage gain A_{vo} are summarized in Table 7.3.
- MOSFETs are commonly used in IC technology, operated with a MOS current source, a PMOS active load, or an NMOS active load.

7.9 Common-Drain Amplifiers

A general common-drain configuration is shown in Fig. 7.39(a). A common-drain amplifier has a very high input resistance and draws a very small gate current. It also offers a low output resistance and can be used as a buffer stage between a low resistance load (requiring a high current) and a signal source that can supply only a very small current. This configuration has a voltage gain approaching unity and is known as a *source follower*. We can derive an input and output relationship if we assume v_G is the input gate voltage and v_O is the output voltage at the source terminal. The gate-to-source v_{GS}, which controls the drain current, is given by

$$v_{GS} = v_G - v_O$$

The corresponding drain current, which must also flow through the source resistance R_{SR}, is as shown in biasing circut in Fig. 7.24(b).

$$i_D = K_n(v_{GS} - V_t)^2 = K_n(v_G - v_O - V_t)^2 = \frac{v_O}{R_{SR}} \tag{7.73}$$

which we can solve to find the output voltage for a specific value of v_G:

$$v_O(v_G) = \frac{2K(v_G - V_t) + 1 - \sqrt{[2K_n(v_G - V_t) + 1]^2 - 4K_n^2(v_G - V_t)^2}}{2K} \tag{7.74}$$

for $v_{GS} = (v_G - v_O) > V_t$

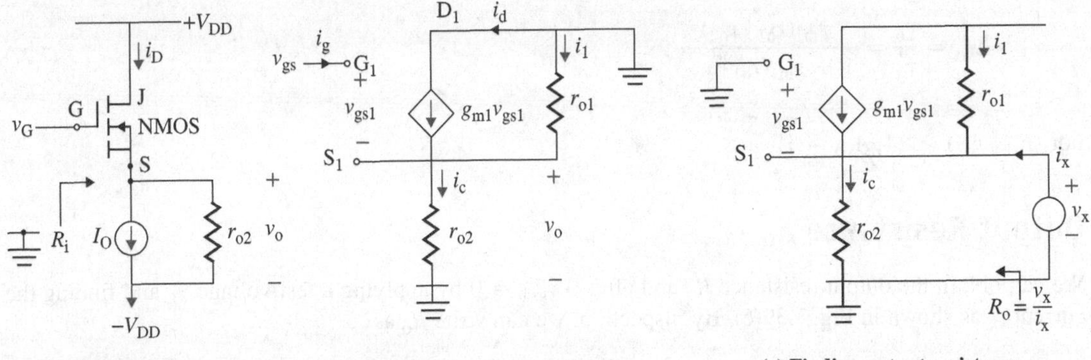

FIGURE 7.39 Source follower with current source load

Here $K = K_n R_{SR}$. For example, if $K_n = 3.25\ \text{mA/V}^2$, $R_{SR} = 23.1\ \text{k}\Omega$, and $V_t = 1$ V, we get $v_O(2) = 0.695$ V and $v_O(3) = 1.546$ V. This gives the voltage gain as

$$A_{VO} = \frac{\Delta v_O}{\Delta v_{GS}} = \left[\frac{v_O(3) - v_O(2)}{(3-2)}\right] = \frac{(1.546 - 0.695)}{1} = 0.851\ \text{V/V}$$

The value of R_{SR} should be large because the voltage gain becomes closer to unity as R_{SR} increases to very large, tending to infinity.

7.9.1 Active-Biased Source Follower

The source resistance in Fig. 7.24(c) can be replaced by a sinking current source as shown in Fig. 7.29(b). The simplified circuit is shown in Fig. 7.39(a). Replacing the transistor by its small-signal model, Fig. 7.39(b) shows the small-signal equivalent.

Input Resistance R_i

Since the gate current of a MOSFET is almost zero, $R_i = v_g/i_g = \infty$.

Voltage Gain A_{vo}

Since the drain current $i_d = g_{m1}v_{gs1}$ flows through the parallel combination of r_{o1} and r_{o2}, the small-signal output voltage v_o is given by

$$v_o = g_{m1}v_{gs}(r_{o1} \| r_{o2}) \tag{7.75}$$

Substituting $v_{gs} = v_g - v_o$ in Eq. (7.75), we get

$$v_o = g_{m1}(r_{o1} \| r_{o2})(v_g - v_o) \tag{7.76}$$

This, after simplification, gives the small-signal voltage gain A_{vo} as

$$A_{vo} = \frac{v_o}{v_g} = \frac{g_{m1}(r_{o1} \| r_{o2})}{1 + g_{m1}(r_{o1} \| r_{o2})} \tag{7.77}$$

For $(r_{o1} \| r_{o2}) >> 1$, $A_{vo} \simeq 1$.

Output Resistance R_o

We can obtain the output resistance R_o and after 0 $v_{gs1} = 0$ by applying a test voltage v_x and finding the current i_x as shown in Fig. 7.39(c). By inspection, we can write R_o as

$$R_o = \frac{v_x}{i_x} = \frac{1}{g_{m1}} \| r_{o1} \| r_{o2} \tag{7.78}$$

which can be approximated to $R_o \simeq 1/g_{m1}$ for $r_{o1}, r_{o2} >> 1/g_{m1}$.

EXAMPLE 7.8

D

Design of a CD amplifier with a MOS current source

(a) Use the sinking current in Fig. 7.29(b) to bias the source follower in Fig. 7.39(a) at a drain current of $I_D = 3.25$ mA. The DC supply voltage is $V_{DD} = 15$ V, and $R_{ref} = 4$ kΩ. The MOS parameters are $V_t = 1$ V, $K_n = 3.25$ mA/V^2, $K_p = 6.5$ mA/V^2, to or W = L, and $|V_M| = 1/\lambda = 100$V.

(b) Find the small-signal voltage A_{vo} and the output resistance R_o.

(c) Use SPICE to plot the small-signal output voltage v_o for a sinusoidal input signal v_s of 1 mV at 1 kHz.

SOLUTION

(a) From Example 7.6, we get $V_{GS} = 2$ V, $R_{ref} = 4$ kΩ, $r_o = r_{o1} = r_{o2} = 30.77$ kΩ, $g_{m1} = 6.5$ mA/V, $R_2 = 100$ kΩ, and $R_1 = 650$ kΩ.

(b) From Eq. (7.77),

$$A_{vo} = \frac{g_{m1}(r_{o1} \| r_{o2})}{1 + g_{m1}(r_{o1} \| r_{o2})} = \frac{6.5 \text{ m} \times (30.77 \text{ k} \| 30.77 \text{ k})}{1 + 6.5 \text{ m} \times (30.77 \text{ k} \| 30.77 \text{ k})} = 0.99$$

From Eq. (7.78),

$$R_o = \frac{1}{g_{m1}} \| r_{o1} \| r_{o2} = \frac{1}{6.5 \text{ m}} \| 30.77 \text{ k} \| 30.77 \text{ k} = 152.3 \ \Omega$$

(c) The PSpice schematic is shown in Fig. 7.40. The plot of the output voltage is shown in Fig. 7.41, which gives a voltage gain 0.988 V/V, which is close to the calculated value of 0.99 V/V. Note there is no phase shift.

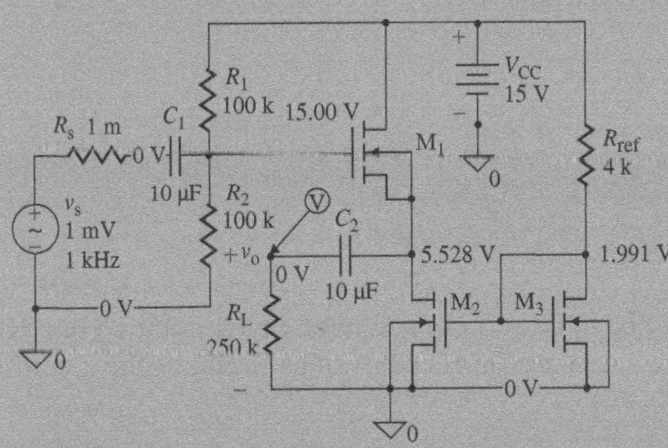

FIGURE 7.40 PSpice schematic for a source follower with an NMOS biasing for Example 7.8

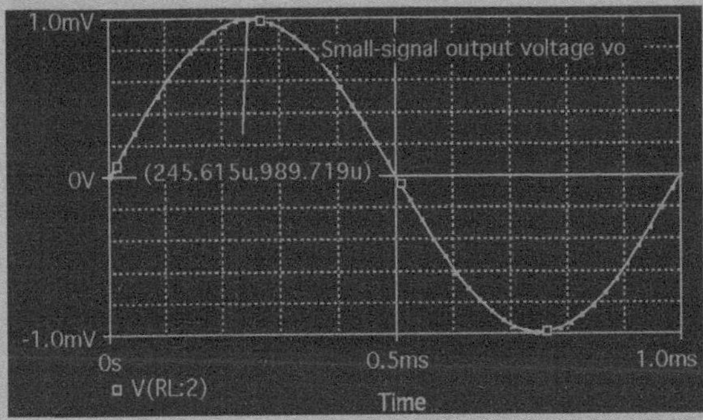

FIGURE 7.41 PSpice plot of small-signal output voltage for Example 7.8

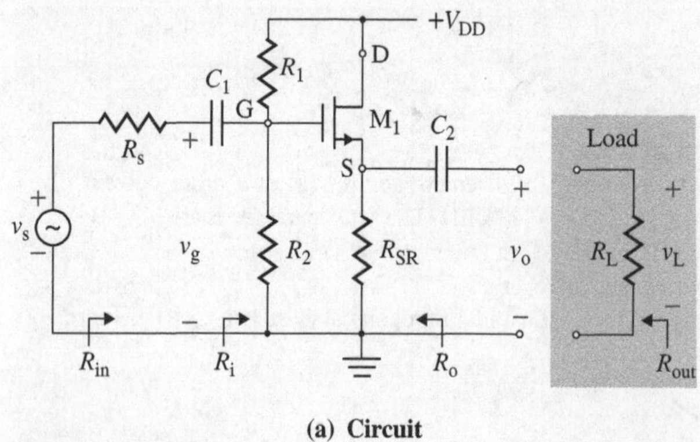

(a) Circuit

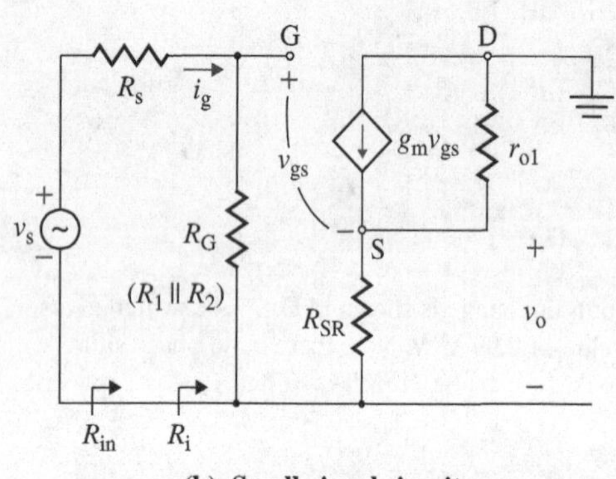

(b) Small-signal circuit

FIGURE 7.42 Common-drain amplifier

7.9.2 Resistive-Biased Source Follower

A source follower with resistive biasing is shown in Fig. 7.42(a). Let us assume that C_1 and C_2 are very large, tending to infinity. That is, $C_1 = C_2 \approx \infty$. The small-signal AC equivalent circuit of the amplifier is shown in Fig. 7.42(b).

Input Resistance R_i $(= v_g/i_g)$

The input resistance R_i is given by

$$R_i = \frac{v_g}{i_g} = R_1 \| R_2 = R_G$$

We can apply Eqs. (7.75) and (7.76) to obtain the output voltage, which, after simplification, gives the open-circuit voltage gain A_{vo} as

$$A_{vo} = \frac{v_o}{v_g} = \frac{g_m(r_{o1} \| R_{SR})}{1 + g_m(r_{o1} \| R_{SR})} \tag{7.79}$$

Output Resistance R_o

We can obtain the output resistance R_o by setting v_s equal to zero and then applying a test voltage v_x at the output side.

The output resistance R_o is given by

$$R_o = \frac{v_x}{i_x} = \frac{1}{g_m} \parallel r_{o1} \parallel R_{SR} \tag{7.80}$$

▶ **NOTE** The no-load voltage gain A_{vo} of a common-drain amplifier approaches unity. The input resistance R_i is very high. The output resistance R_o is low.

EXAMPLE 7.9

D **Designing a depletion MOSFET source follower** Design a source follower as shown in Fig. 7.42(a) to yield $R_i \geq 500$ kΩ and $i_D = 10$ mA. The MOS parameters are $V_p = -4$ V, $I_{DSS} = 20$ mA, and $V_M = -200$ V. Assume $V_{DD} = 20$ V.

SOLUTION

The design of a common-drain (CD) amplifier is very simple; it requires determining the values of R_{SR}. We know that

$$K_n = \frac{I_{DSS}}{V_p^2} = \frac{20 \text{ mA}}{(-4)^2} = 1.25 \text{ mA/V}^2$$

Step 1. For the depletion MOSFET, we can use the self-biasing circuit arrangement as shown in Fig. 7.24(b) where $R_1 = \infty$. Calculate the gate resistance R_2:

$$R_2 = R_i = 500 \text{ k}\Omega$$

Step 2. For known values of i_D, I_{DSS}, and V_p, calculate v_{GS} from Eq. (7.21), $i_D = K_n(v_{GS} - V_p)^2$:

$$10 \text{ mA} = 1.25 \text{ mA/V}^2 \times (v_{GS} + 4)^2$$

which gives $v_{GS} = -1.172$ V or -6.828 V. The acceptable value is $v_{GS} = -1.172$ V.

Step 3. For the known value of v_{GS}, calculate R_{SR}:

Using KVL through the loop formed by the gate, R_G and R_{SR} in Fig. 7.24(b), we get

$$0 = v_{GS} + R_{SR} i_D$$

Which gives

$$R_{SR} = -\frac{v_{GS}}{i_D} = -\left(\frac{-1.172 \text{ V}}{10 \text{ mA}}\right) = 117.2 \ \Omega$$

Step 4. Find the small-signal parameters of the transistor. From Eq. (7.30),

$$g_m = 2K_n(v_{GS} - V_p) = 2 \times 1.25 \text{ m} \times (-1.172 + 4) = 7.07 \text{ mA/V}$$

From Eq. (7.26),

$$r_{o1} = \frac{|V_M|}{i_D} = \frac{200\text{ V}}{10\text{ mA}} = 20\text{ k}\Omega$$

Thus,

$$\mu_g = g_m r_{o1} = 7.07 \times 20 = 141.1\text{ V/V}$$

Step 5. Find the values of C_1 and C_2. Let us choose $C_1 = C_2 = 10\ \mu\text{F}$.
Step 6. Calculate the output resistance R_o and the open-circuit voltage gain A_{vo}.

$$R_o = \frac{1}{g_m} \| r_{o1} \| R_{SR} = \frac{1}{7.07\text{ m}} \| 20\text{ k} \| 117.2 = 63.95\ \Omega$$

$$A_{vo} = \frac{g_m(R_{SR} \| r_{o1})}{1 + g_m(R_{SR} \| r_{o1})} = \frac{7.07\text{ m} \times (117.2 \| 20\text{ k})}{1 + 7.07\text{ m} \times (117.2 \| 20\text{ k})} = 0.451$$

KEY POINTS OF SECTION 7.9

- A common-drain amplifier also known as a source follower has a very high input resistance and draws a very small gate current. It also offers a low output resistance and can be used as a buffer stage between a low resistance load (requiring a high current) and a signal source that can supply only a very small current.
- A source follower with a sinking current source offers almost unity gain, a very input resistance and a low resistance.

7.10 Common-Gate Amplifiers

A common-gate (CG) amplifier is shown in Fig. 7.43(a). The circuit can be redrawn as shown in Fig. 7.43(b). The biasing of this circuit is identical to that of the common-source amplifier, and the DC bias circuit can be designed using the same technique. Let us assume that the values of C_1 and C_2 are very large, tending to infinity. That is, $C_1 = C_2 \approx \infty$. The small-signal AC equivalent circuit of the amplifier is shown in Fig. 7.44(a), which can be simplified to Fig. 7.44(b).

Input Resistance R_i $(= -v_{gs}/i_s)$

The input resistance R_i depends on R_D, which becomes parallel to the load resistance R_L. Thus, R_L must be included with R_D in the determination of R_i when the amplifier is operated with a load resistance R_L. Using KVL around the source–gate–drain loop of Fig. 7.44(b) gives an expression for the gate-to-source voltage:

$$-v_{gs} = \mu_g v_{gs} - (r_{o1} + R_D \| R_L) i_d$$

which yields

$$i_d = \frac{(1 + \mu_g) v_{gs}}{r_{o1} + R_D \| R_L}$$

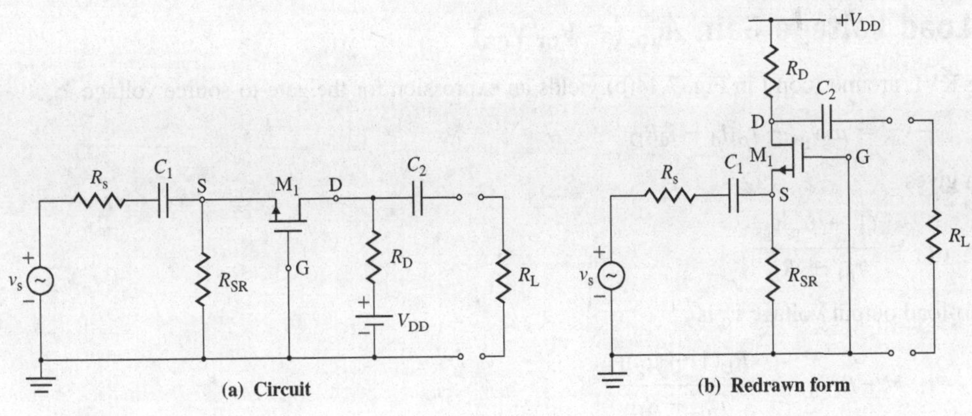

FIGURE 7.43 Common-gate amplifier

Using KCL at source node S in Fig. 7.44(b) yields an expression for the input current i_s:

$$i_s = \frac{-v_{gs}}{R_{SR}} - i_d = -\frac{-v_{gs}}{R_{SR}} - \frac{(1 + \mu_g)v_{gs}}{r_{o1} + R_D \| R_L}$$

which gives the input resistance R_i of the amplifier as

$$R_i = \frac{-v_{gs}}{i_s} = R_{SR} \left\| \left(\frac{r_{o1} + R_D \| R_L}{1 + \mu_g} \right) \right. \tag{7.81}$$

Since $\mu_g > 1$, the input resistance R_i becomes low. This is a limitation of the common-gate configuration, unless a low R_i (or Z_i) is desirable for impedance matching.

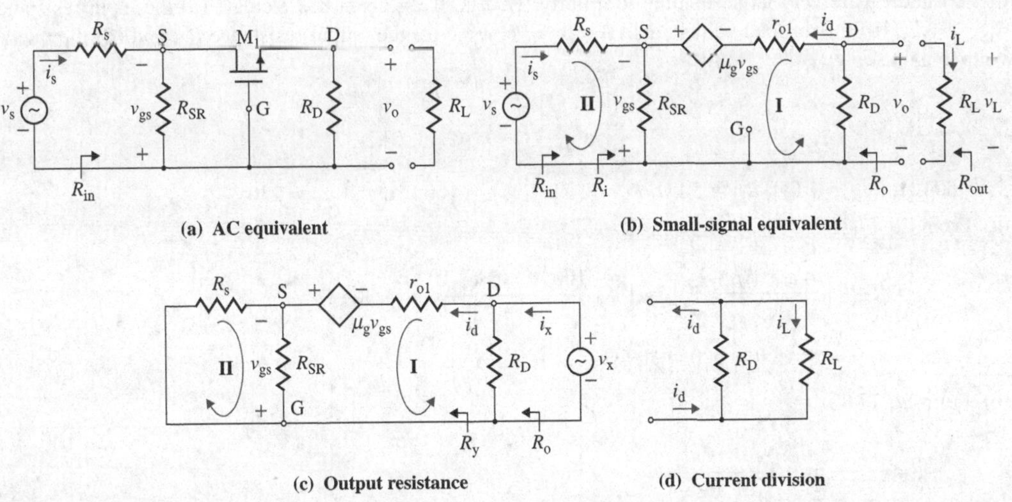

FIGURE 7.44 Small-signal AC equivalent circuits for a CG amplifier

No-Load Voltage Gain A_{vo} ($= v_o/v_{gs}$)

Using KVL around loop I in Fig. 7.44(b) yields an expression for the gate-to-source voltage v_{gs}:

$$-v_{gs} = \mu_g v_{gs} - r_{o1} i_d - i_d R_D$$

which gives

$$i_d = \frac{(1 + \mu_g) v_{gs}}{r_{o1} + R_D}$$

The no-load output voltage v_o is

$$v_o = -R_D i_d = -\frac{R_D(1 + \mu_g) v_s}{r_{o1} + R_D}$$

which gives the no-load voltage gain A_{vo} as

$$A_{vo} = \frac{v_o}{-v_{gs}} = \frac{R_D(1 + \mu_g)}{r_{o1} + R_D} \tag{7.82}$$

Output Resistance R_o

Assuming that the output resistance of the transistor is very large, tending to infinity (i.e., $r_{o1} \approx \infty$), the output resistance R_o can be found by inspection to be $R_o \approx R_C$.

EXAMPLE 7.10

Finding the parameters of a common-gate amplifier The CG amplifier of Fig. 7.44(a) has $R_s = 500\ \Omega$, $R_{SR} = 1\ \text{k}\Omega$, $R_D = 5\ \text{k}\Omega$, and $R_L = 10\ \text{k}\Omega$. The transistor parameters are $r_{o1} = 100\ \text{k}\Omega$ and $\mu_g = 230$. Assume that C_1 and C_2 are very large, tending to infinity. That is, $C_1 = C_2 \approx \infty$. Calculate **(a)** the input resistance $R_{in} = v_s/i_s$, **(b)** the no-load voltage gain $A_{vo} = v_o/(-v_{gs})$, **(c)** the output resistance R_o, and **(d)** the overall voltage gain $A_v = v_L/v_s$.

SOLUTION

$R_s = 500\ \Omega$, $R_{SR} = 1\ \text{k}\Omega$, $R_D = 5\ \text{k}\Omega$, $R_L = 10\ \text{k}\Omega$, $r_{o1} = 100\ \text{k}\Omega$, and $\mu_g = 230$.

(a) From Eq. (7.81),

$$R_i = R_{SR} \| \left(\frac{r_{o1} + R_D \| R_L}{1 + \mu_g} \right) = 1\ \text{k} \| \frac{100\ \text{k} + 5\ \text{k} \| 10\ \text{k}}{1 + 230} = 309\ \Omega$$

$$R_{in} = R_i + R_s = 309 + 500 = 809\ \Omega$$

(b) From Eq. (7.82),

$$A_{vo} = \frac{v_o}{-v_{gs}} = \frac{R_D(1 + \mu_g)}{r_{o1} + R_D}$$

$$= 5\ \text{k} \times \frac{1 + 230}{100\ \text{k} + 5\ \text{k}} = 11$$

(c) $R_o = R_D = 5\ k\Omega$.

(d) For the overall voltage gain A_v,

$$A_v = \frac{v_L}{v_s} = \frac{A_{vo}R_iR_L}{(R_i + R_s)(R_L + R_o)} = \frac{11 \times 309 \times 10\ k}{(309 + 500) \times (10\ k + 5\ k)} = 2.83$$

KEY POINT OF SECTION 7.10

- A common gate amplifier which has no Miller's effect is used in high-frequency applications. Both the input resistance and the voltage are low. There is no phase of the output voltage and it can be used for impedance matching.

7.11 Multistage Amplifiers

The design requirements of amplifiers normally specify an overall high voltage gain, a high input resistance, and a low output resistance. A single-transistor amplifier rarely satisfies the design requirements, and multi-stages are often used to satisfy the design specifications. To achieve the design specifications, multiple transistor stages are connected in such a way that the output of one stage is the input to the next stage and so on. The most common types of arrangements are (a) capacitor-coupled cascaded, (b) direct-coupled, and (c) cascoded.

7.11.1 Capacitor-Coupled Cascaded Amplifiers

In a capacitively coupled amplifier, the output of one stage is connected to the input of the next stage via a capacitor as shown in Fig. 7.45(a). The first stage is generally a common-source amplifier that is designed to offer the maximum voltage and a high input resistance R_i, which is inherent in MOS amplifiers. The source follower in the third stage satisfies the requirement of a low output resistance R_o. The second stage is a common-source amplifier which is needed to yield additional gain in meeting the overall voltage gain requirement A_{vo}. If A_{vo1} and A_{vo3} ($\simeq 1$) are the voltage gains of the first and third stages, respectively, then the required gain for the second stage is $A_{vo2} = A_{vo}/(A_{vo1} \times A_{vo3})$. The DC biasing point of each stage can be determined independently for each stage because the coupling capacitor provides DC isolation between the stages. The biasing drain current should be low to reduce the power drain from the DC voltage source.

Each stage can be represented by its parameters R_i, R_o, and A_{vo} as shown in Fig. 7.45(b). The output resistance of a stage acts as the source resistance of the following stage, and the input resistance of a stage is the load resistance of the preceding stage. There will be a loading effect due to the interaction between stages, and the effective voltage gain will be reduced (see Sec. 2.4). While designing an amplifier, we should keep in mind the gain reduction due to the loading effect and should start the design with a voltage gain higher than A_{vo} to satisfy the overall design requirement.

7.11.2 Direct-Coupled Amplifiers

In direct-coupled amplifiers, the output of one stage is directly connected to the input of the next stage. We can make the amplifier in Fig. 7.45(a) a direct-coupled amplifier if we remove the coupling capacitors C_2

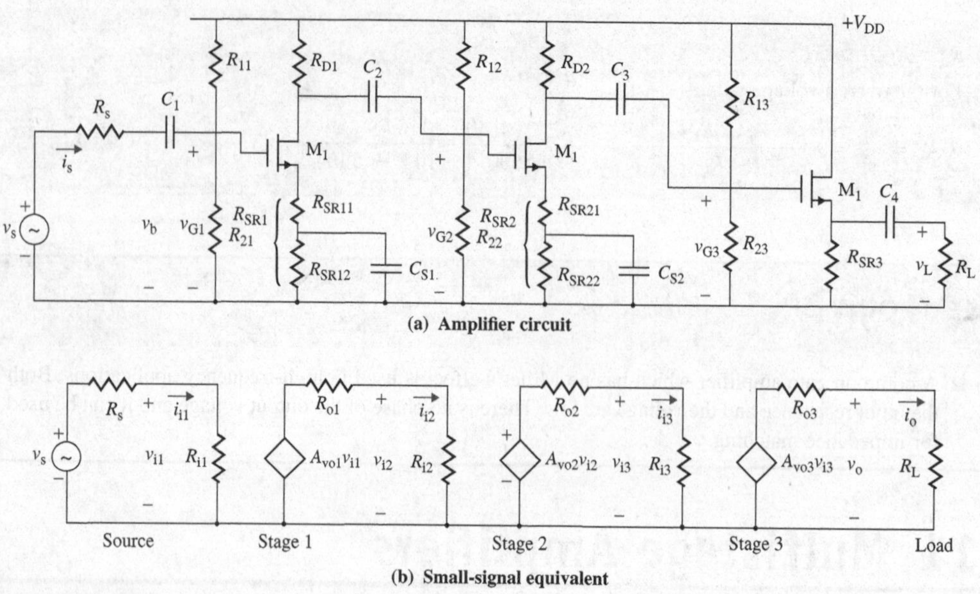

FIGURE 7.45 A three-stage capacitor-coupled cascaded amplifier

and C_3 and connect the following stage directly to the preceding stage. In this case, we can also remove the biasing resistances.

7.11.3 Cascoded Amplifiers

We can increase the effective output resistance of a transistor by connecting two transistors in a configuration commonly referred to as a *cascoded amplifier*. The input signal is applied to one transistor M_1 operating as a common-source amplifier, whose output is the input to the other transistor M_2 operating as a common-gate amplifier. This is shown in Fig.7.46(a). The input signal is applied to the common-source amplifier, and the output is obtained at the drain of the common-gate amplifier. Figure 7.46(a) is an example of a resistive biased cascoded amplifier. The cascading can be done with more than two transistors as shown in Fig. 7.46(e). The transistors $M_2, \ldots, M_4$ are biased by level-shifted transistors $M_{2B}, \ldots, M_{4B}$. This type of cascoding is commonly done in differential amplifiers (Chapter 9) to obtain large voltage gains.

DC Biasing

The DC equivalent circuit for determining the DC operating point of the transistors is shown in Fig. 7.46(b). We can simplify the analysis by assuming identical transistors $V_{t1} = V_{t2} = V_t$. Since the same drain current will flow through all the transistors, $V_{GS1} = V_{GS2} = V_{GS}$. Therefore, we can find the DC biasing gate voltages as given by

$$V_{G1} = \frac{R_1}{R_1 + R_2 + R_3} V_{DD} \tag{7.83}$$

$$V_{G2} = \frac{R_1 + R_2}{R_1 + R_2 + R_3} V_{DD} \tag{7.84}$$

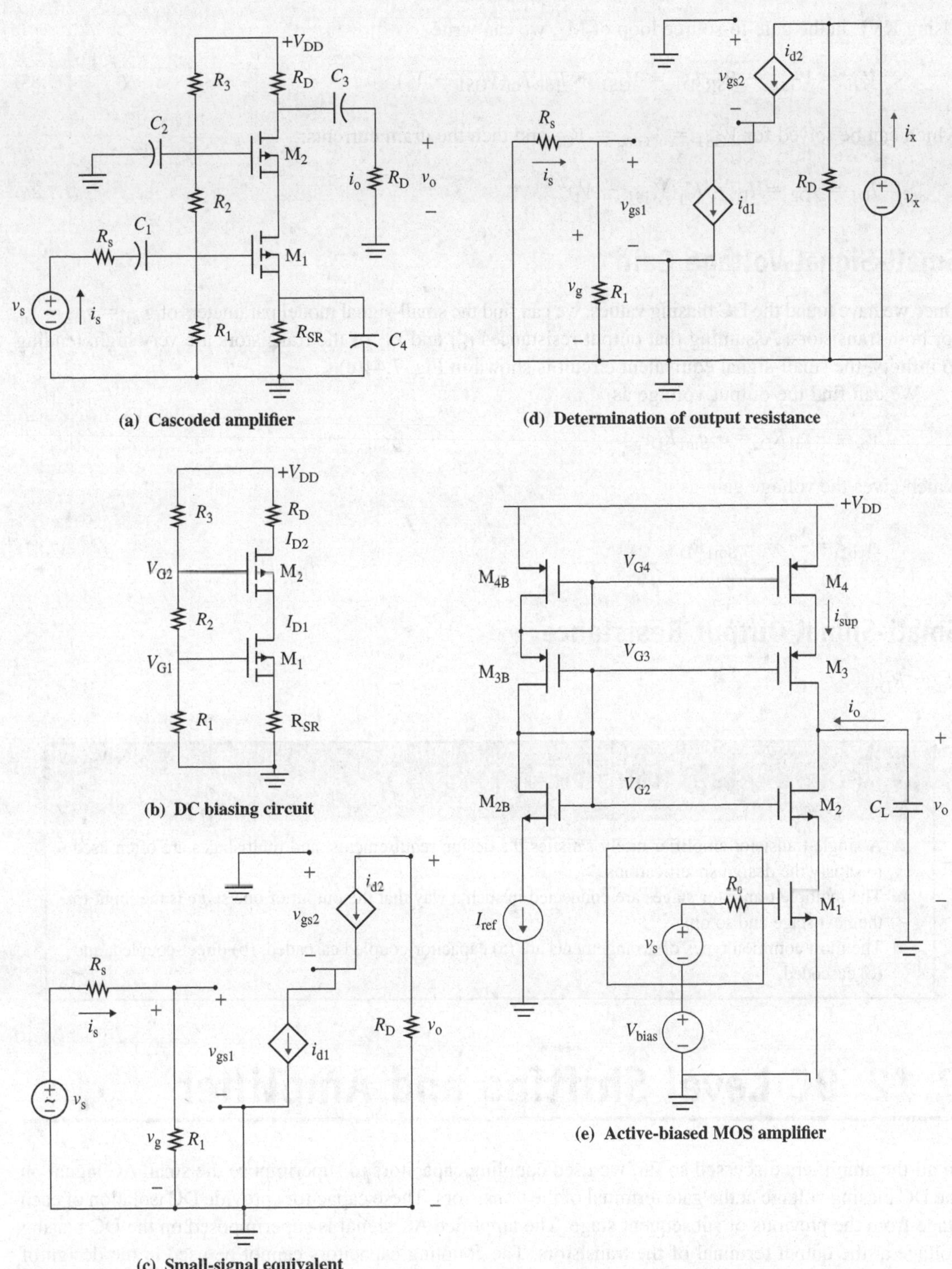

FIGURE 7.46 Cascoded amplifier

Using KVL in the gate-to-source loop of M_1, we can write

$$V_{G1} = V_{GS1} + R_{SR}I_{D1} = V_{GS1} + R_{SR}K_n(V_{GS1} - V_t)^2 \tag{7.85}$$

which can be solved for $V_{GS1} = V_{GS2} = V_{GS}$ and then the drain currents:

$$I_{D1} = I_{D2} = I_D = K_n(V_{GS1} - V_t)^2 \tag{7.86}$$

Small-Signal Voltage Gain

Once we have found the DC biasing values, we can find the small-signal model parameters of $g_{m1} = g_{m2} = g_m$ for both transistors. Assuming that output resistances r_{o1} and r_{o2} of the transistors are very high, tending to infinity, the small-signal equivalent circuit is shown in Fig. 7.44(c).

We can find the output voltage as

$$v_o = -i_{d1}R_D = -g_{m1}R_Dv_g$$

which gives the voltage gain as

$$A_{vo1} = \frac{v_o}{v_g} = -g_{m1}R_D \tag{7.87}$$

Small-Signal Output Resistance

$R_o = R_D$.

KEY POINTS OF SECTION 7.11

- A single-transistor amplifier rarely satisfies the design requirements, and multistages are often used to satisfy the design specifications.
- The multiple transistor stages are connected in such a way that the output of one stage is the input to the next stage and so on.
- The most common types of arrangements are (a) capacitor-coupled cascaded, (b) direct-coupled, and (c) cascoded.

7.12 DC Level Shifting and Amplifier

In all the amplifiers discussed so far, we used coupling capacitors to superimpose the small AC signal on the DC biasing voltage at the gate terminal of the transistors. These capacitors provide DC isolation of each stage from the previous or subsequent stage. The amplified AC signal is superimposed on the DC biasing voltage at the output terminal of the transistors. The coupling capacitors cannot be used in the design of amplifiers that amplify only DC signals. In some cases, it may be necessary to shift the quiescent voltage of one stage before applying its output to the following stage. Level-shifting circuits can adjust the DC bias levels between amplification stages. Level shifting is also required in order for the output to be close to zero in the quiescent state (at no input signal). The input resistance of the level-shifting stage should be high to prevent loading of the previous stage (usually the gain stage). Also, the output resistance should be low to effectively drive the subsequent stage.

7.12.1 Level-Shifting Methods

The source follower configuration is normally used to shift the level. The source follower has an inherent characteristic of level shifting by v_{GS} such that the output voltage is $v_O = v_G - v_{GS}$. Thus, the main idea is to create a voltage in the source terminal, and it can be accomplished by (a) a potential divider network, (b) a current source, and (c) a zener diode.

Potential Divider Level Shifting

This arrangement is shown in Fig. 7.47(a). The voltage shift is

$$v_O - v_G = -v_{GS} - R_1 i_D = -v_{GS} - R_1 K_n (v_{GS} - V_t)^2$$

which gives the output voltage as

$$v_O = v_G - v_{GS} - R_1 K_n (v_{GS} - V_t)^2 \tag{7.88}$$

Since V_{GS} is fixed for a specific drain current, a small change Δv_G will cause the same change to the output voltage; that is, $\Delta v_O = \Delta v_G$.

Current Source Level Shifting

Resistance R_2 in Fig. 7.47(a) can be replaced with a current source at a constant current I_O as shown in Fig. 7.47(b). The voltage shift is

$$v_O - v_G = -(v_{GS} + R_1 I_O) = -v_{GS} - R_1 K_n (v_{GS} - V_t)^2$$

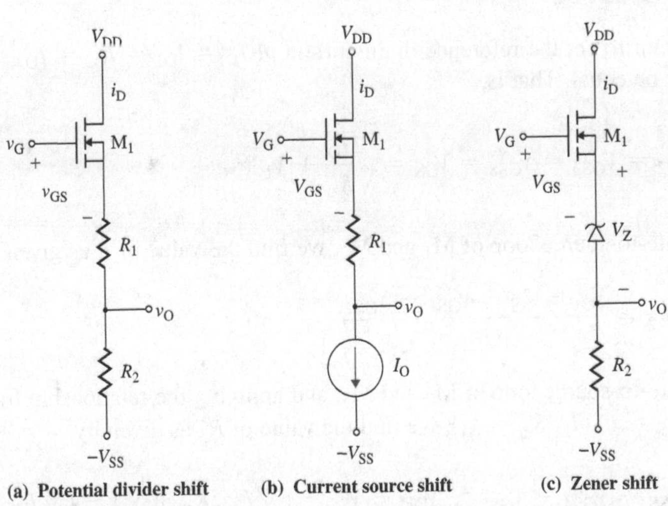

FIGURE 7.47 Level shifting

which gives the output voltage as

$$v_O = v_G - v_{GS} - R_1K_n(v_{GS} - V_t)^2 \quad (7.89)$$

Since the current through R_1 is fixed, the voltage drop across it is also fixed. It is important to note that v_O is independent of the negative DC supply voltage $-V_{SS}$.

Zener Level Shifting

Resistance R_1 in Fig. 7.47(a) can be replaced by a zener diode with zener voltage V_Z; this arrangement is shown in Fig. 7.47(c). The voltage shift is

$$v_O - v_S = -(v_{GS} + V_Z)$$

which gives the output voltage as

$$v_O = v_S - v_{GS} - V_Z \quad (7.90)$$

Since the zener voltage V_Z is fixed, the voltage drop across it is also fixed.

7.12.2 Level-Shifted MOS Amplifier

A MOS amplifier using level shifting is shown in Fig. 7.48(a) with four stages. The first stage generates the reference current for the second stage, which also acts as the reference current for the third stage. The fourth stage is the source follower. We will assume that all transistors are matched devices and have equal parameters: the current gains $K_{n1} = \cdots = K_{n8} = K_n$, $V_{t1} = \cdots = V_{t8} = V_t$, the modulation voltages, and $V_{M1} = \cdots = V_{M8} = V_M$.

Current Mirror Source

Assuming that we want to set the reference drain current at $I_D = I_{D1} = I_{D2} = I_{D3} = I_{D8}$, their gate-to-source voltages must be equal. That is,

$$V_{GS1} = V_{GS2} = V_{GS3} = V_{GS8} = V_{GS} = \sqrt{\frac{I_D}{K_n}} + V_t \quad (7.91)$$

Using KVL in the gate-to-source loop of M_1 and M_2, we find the value of R as given by

$$R = R_1 + R_2 = \frac{V_G + V_{SS} - V_{GS1} - V_{GS2}}{I_{D1}} \quad (7.92)$$

Using KVL in the gate-to-source loop of M_4 and M_5, and applying the relationship for the gate-to-source voltage of a MOS $V_{GS} = (\sqrt{I_D/K_n} + V_t)$, we find the value of R_3 as given by

$$R_3 = \frac{V_{DD} + V_{SS} - V_{GS4} - V_{GS5}}{I_{D4}} = \frac{V_{DD} + V_{SS} - (\sqrt{I_{D4}/K_n} + V_t) - (\sqrt{I_{D5}/K_n} + V_t)}{I_{D4}} \quad (7.93)$$

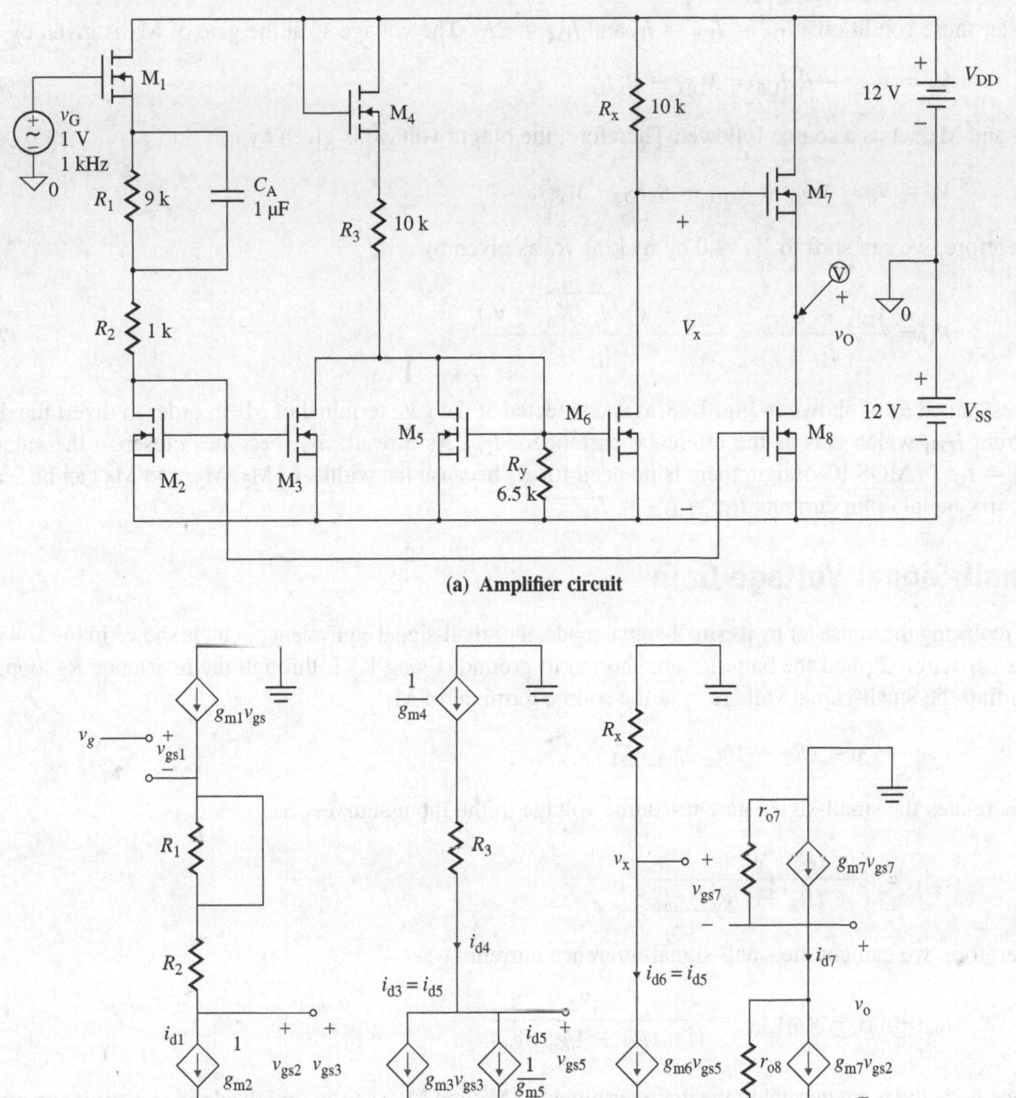

FIGURE 7.48 A MOS level-shifting amplifier

We can simplify this by assuming $I_{D5} \simeq I_{D3} = I_D$ and using the relation $I_{D4} = I_{D3} + I_{D5}$, as given by

$$R_3 = \frac{V_{DD} + V_{SS} - (\sqrt{2I_D/K_n} + V_t) - (\sqrt{I_D/K_n} + V_t)}{2I_D}$$

$$= \frac{V_{DD} + V_{SS} - 2.14\sqrt{I_D/K_n} - 2V_t}{2I_D} \tag{7.94}$$

Under these conditions, $I_{D5} = I_{D6} \simeq I_D$ and $I_{D4} \simeq 2I_D$. The voltage V_x at the gate of M_7 is given by

$$V_x = V_{DD} - R_x I_{D6} = V_{DD} - R_x I_D \tag{7.95}$$

M_7 and M_8 act as a source follower. Therefore, the output voltage is given by

$$V_o = V_x - V_{GS7} = V_{DD} - R_x I_D - V_{GS7} \tag{7.96}$$

Therefore, we can shift to $V_O = 0$ by making R_x as given by

$$R_x = \frac{V_{DD} - V_{GS7}}{I_D} = \frac{V_{DD} - (\sqrt{I_D/K_n} + V_t)}{I_D} \tag{7.97}$$

A resistance R_y as shown in Fig. 7.48(a) is connected at the gate terminal of M_5 in order to divert the drain current I_{D5}, which acts as the reference current for I_{D6}. As a result, I_{D5} becomes closer to the value of $I_{D3} = I_D$. In MOS IC design, there is no need for R_y because the widths of M_3, M_5, and M_6 can be scaled to carry equal drain currents $I_{D3} = I_{D5} = I_{D6}$.

Small-Signal Voltage Gain

By replacing the transistor by its small-signal mode, the small-signal equivalent circuit is shown in Fig. 7.48(b). The capacitor C_A and the batteries are shorted to ground. Using KVL through the resistance R_2 loop, we can find the small-signal voltage v_g at the source terminal of M_1:

$$v_g - v_{gs1} = (R_2 + 1/g_{m2})g_{m2}v_{gs1}$$

This relates the small-signal gate-to-source voltage to the input signal v_g as

$$v_{gs1} = \frac{v_g}{1 + (R_2 + 1/g_{m2})g_{m1}} \tag{7.98}$$

Therefore, we can get the small-signal reference current i_{d2}:

$$i_{d2} = i_{d3} = g_{m1}v_{gs1} = \frac{g_{m1}v_g}{1 + (R_2 + 1/g_{m2})g_{m1}} \tag{7.99}$$

Since i_{d3} is the input current to the drain terminals of M_3 and M_5, we can find the drain current i_{d5} by applying the current divider rule as given by

$$i_{d5} = i_{d6} = \frac{R_3 + 1/g_{m4}}{R_3 + 1/g_{m4} + 1/g_{m5}} i_{d2} \tag{7.100}$$

which, after we substitute i_{d2} from Eq. (7.99), becomes

$$i_{d5} = i_{d6} = \frac{R_3 + 1/g_{m4}}{R_3 + 1/g_{m4} + 1/g_{m5}} \times \frac{g_{m1}v_g}{1 + (R_2 + 1/g_{m2})g_{m1}} \tag{7.101}$$

Thus the small-signal voltage v_x becomes

$$v_x = R_x\, i_{d6} = \frac{R_3 + 1/g_{m4}}{R_3 + 1/g_{m4} + 1/g_{m5}} \times \frac{R_x\, g_{m1}v_g}{1 + (R_2 + 1/g_{m2})g_{m1}} \tag{7.102}$$

which gives the voltage gain between v_g and v_x as

$$A_{vx} = \frac{v_x}{v_g} = \frac{(R_3 + 1/g_{m4})R_x g_{m1}}{(R_3 + 1/g_{m4} + 1/g_{m5})[1 + (R_2 + 1/g_{m2})g_{m1}]} \tag{7.103}$$

Due to the source follower, the voltage gain A_{vx} is attenuated, and the no-load voltage gain A_{vo} becomes

$$A_{vo} = \frac{g_{m7}(r_{o7} \| r_{o8})}{1 + g_{m7}(r_{o7} \| r_{o8})} A_{vx} = \frac{(R_3 + 1/g_{m4})R_x g_{m1}}{(R_3 + 1/g_{m4} + 1/g_{m5})[1 + (R_2 + 1/g_{m2})g_{m1}]} \times \frac{g_{m7}(r_{o7} \| r_{o8})}{1 + g_{m7}(r_{o7} \| r_{o8})} \tag{7.104}$$

For $R_2 >> 1/g_{m2}$, $R_3 >> (1/g_{m4} + 1/g_{m5})$, and $(r_{o7} \| r_{o8}) >> 1$, Eq. (7.104) can be approximated to

$$A_{vo} \simeq \frac{R_x}{R_2} \tag{7.105}$$

It is important to note that any increase in v_g causes the current i_{d1} to increase, which is mirrored to i_{d3}; this in turn decreases i_{d5} by the same amount. i_{d6}, which is a mirror of i_{d5}, causes the voltage v_x to increase by $R_x i_{d6}$. The transistors, which act as current mirrors and shift the voltage levels, do not produce any voltage amplification. The voltage gain described by Eq. (7.104) is accomplished by shunting R_1 by the capacitor C for AC signals. The maximum voltage gain can be obtained by shunting both R_1 and R_2 by capacitor C_A; that is, for $R_2 = 0$, Eq. (7.104) gives the maximum voltage gain as

$$A_{vo(max)} = \frac{(R_3 + 1/g_{m4})R_x\, g_{m1}}{(R_3 + 1/g_{m4} + 1/g_{m5})(1 + g_{m1}/g_{m2})} \times \frac{g_{m7}(r_{o7} \| r_{o8})}{1 + g_{m7}(r_{o7} \| r_{o8})} \tag{7.106}$$

The design of this amplifier is very simple. It requires only finding the values of R_1, R_2, R_y, and $R_x(= R = R_1 + R_2)$ to give a specific voltage gain A_{vo}.

EXAMPLE 7.11

Finding the small-signal voltage gain of a level-shifted amplifier The parameters of the amplifier in Fig. 7.48(a) are $V_{DD} = 15$ V, $-V_{SS} = -15$ V, $R_1 = 9$ kΩ, $R_2 = 1$ kΩ, $R_3 = 4$ kΩ, $R_x = 10$ kΩ, and $R_y = 6.5$ kΩ. The circuit is biased at a DC gate-source voltage of $V_G = 1$ V. The MOS parameters are $V_t = 1$ V, $K_n = 3.25$ mA/V^2, $K_p = 6.5$ mA/V^2 for W = L, and $|V_M| = 1/\lambda = 100$ V.

(a) Find the small-signal voltage A_{vo} and the maximum possible gain.

(b) Use SPICE to plot the small-signal output voltage for a sinusoidal input signal of 1 mV at 1 kHz.

SOLUTION

From Eq. (8.92),

$$V_G + V_{SS} - V_{GS1} - V_{GS1} - (R_1 + R_2)K_n(V_{GS1} - V_t)^2 = 0 \text{ or}$$

$$1 + 12 - V_{GS1} - V_{GS1} - (1\text{ k} + 9\text{ k}) \times 3.25 \times 10^{-3} \times (V_{GS1} - 1)^2 = 0$$

which gives

$$V_{GS} = 1.552 \text{ V}$$

$$I_{D1} = K_n(V_{GS1} - V_t)^2 = 3.25 \text{ m} \times (1.552 - 1)^2 = 990 \text{ }\mu\text{A}$$

$$V_{GS4} = \sqrt{2I_{D1}/K_n} + V_t = \sqrt{2 \times 990 \text{ }\mu/3.25 \text{ m}} + 1 = 1.784 \text{ V}$$

$$g_{m2} = g_{m3} = g_{m5} = g_{m6} = g_{m7} = 2K_n(V_{GS1} - V_t) = 2 \times 3.25 \text{ m} \times (1.552 - 1) = 3.606 \text{ mA/V}$$

$$g_{m4} = 2K_n(V_{GS4} - V_t) = 2 \times 3.25 \text{ m} \times (1.784 - 1) = 5.099 \text{ mA/V}$$

$$r_{o7} = \frac{V_M}{I_{D7}} = \frac{100}{990 \text{ }\mu} = 100 \text{ k}\Omega$$

Substituting the values in Eq. (7.103), we get

$$A_{vx} = \frac{(R_3 + 1/g_{m4})R_x \, g_{m1}}{(R_3 + 1/g_{m4} + 1/g_{m5})[1 + (R_2 + 1/g_{m2})g_{m1}]}$$

$$= \frac{(10 \text{ k} + 1/5.099 \text{ m}) \times 10 \text{ k} \times 3.606 \text{ m}}{(10 \text{ k} + 1/5.099 \text{ m} + 1/3.606 \text{ m})[1 + (1 \text{ k} + 1/3.606 \text{ m}) \times 3.606 \text{ m}]} = 6.262 \text{ V/V}$$

Substituting the values in Eq. (7.104), we get

$$A_{vo} = \frac{g_{m7}(r_{o7} \| r_{o8})}{1 + g_{m7}(r_{o7} \| r_{o8})} A_{vx} = \frac{3.606 \text{ m} \times (100 \text{ k} \| 100 \text{ k})}{1 + 3.606 \text{ m} \times (100 \text{ k} \| 100 \text{ k})} \times 6.262 = 6.227 \text{ V/V}$$

Substituting the values for $R_2 = 0$ in Eq. (7.104) gives $A_{vo(max)} = 17.454$ V/V.

(b) The PSpice plot of the DC transfer function is shown in Fig. 7.49(a), which has an offset voltage of 36.93 mV at $v_G = 0$. The plot of the output voltage is shown in Fig. 7.49(b), which gives a voltage gain of 5.78, which is close to the calculated value of 6.227. Note that there is no phase shift and $r_{ds} = 1/r_o$. The PSpice biasing drain currents and the small-signal parameters are listed here:

Name	M_M1	M_M2	M_M3	M_M4	M_M5	M_M6	M_M7	M_M8
ID	8.98E−04	8.98E−04	8.99E−04	2.07E−03	9.34E−04	1.04E−03	9.92E−04	9.92E−04
VGS	1.49E+00	1.52E+00	1.52E+00	1.79E+00	1.53E+00	1.53E+00	1.52E+00	1.52E+00
GM	3.64E−03	3.44E−03	3.44E−03	5.23E−03	3.51E−03	3.93E−03	3.80E−03	3.80E−03
GDS	7.92E−06	8.85E−06	8.85E−06	2.03E−05	9.19E−06	9.19E−06	8.86E−06	8.85E−06

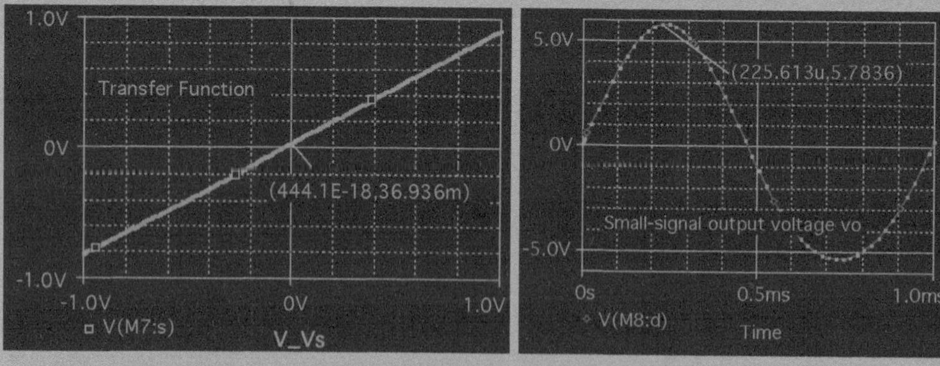

(a) DC transfer function **(b) Small-signal output voltage**

FIGURE 7.49 PSpice simulation of a level-shifted NMOS amplifier for Example 7.11

7.13 Frequency Response of MOSFET Amplifiers

The frequency response of MOS amplifiers will depend on the internal MOS junction capacitances and any external capacitances such as coupling and bypass capacitances. To determine the frequency characteristics, we need to add capacitances to the small-signal AC models of MOSFETs. In Secs. 2.7.4 and 2.7.5, we introduced short-circuit and zero-value methods. As examples, we will use these methods for determining the frequency response of MOSFETs, single-stage MOS amplifiers, and multistage amplifiers.

7.13.1 High-Frequency MOSFET Models

The small-signal high-frequency model of the MOSFETs of Fig. 7.50(a) in the saturation region is shown in Fig. 7.50(b). The gate-to-source capacitance C_{gs} and the gate-to-drain capacitance C_{gd} can be found approximately from [9, 10]

$$C_{gs} = \frac{C_{gs0}}{[1 + |V_{GS}|/V_{bi}]^{1/3}} \tag{7.107}$$

and

$$C_{gd} = \frac{C_{gd0}}{[1 + |V_{GD}|/V_{bi}]^{1/3}} \tag{7.108}$$

where V_{bi} = built-in potential with a zero applied voltage

C_{gs0} = value of C_{gs} at $V_{GS} = 0$ and is typically in the range of 1 pF to 4 pF

C_{gd0} = value of C_{gd} at $V_{GD} = 0$ and is typically in the range of 0.3 pF to 1 pF

C_{sb} and C_{bd} = depletion-layer capacitances from the source to the substrate and from the substrate to the drain, respectively

(Note that in order to avoid confusion between substrate and source terminals of a MOSFET, substrate is being abbreviated with a subscript b.) These capacitances can be found approximately from

$$C_{sb} = \frac{C_{sb0}}{[1 + |V_{SB}|/V_{bi}]^{1/2}} \tag{7.109}$$

and

$$C_{bd} = \frac{C_{bd0}}{[1 + V_{DB}|/V_{bi}]^{1/2}} \tag{7.110}$$

where V_{bi} is the built-in (or barrier) potential and is typically 0.6 V and C_{sb0} and C_{bd0} are the zero-biased capacitances and are typically 0.1 pF. The values of C_{sb} and C_{bd} range from 0.01 pF to 0.05 pF. To reduce the values of C_{sb} and C_{bd}, the substrate of a MOSFET is often connected to the negative DC supply voltage so that $|V_{SB}|$ and $|V_{DB}|$ have higher values.

C_{gb} is the parasitic oxide capacitance between the gate contact material and the substrate, and its value depends on the oxide thickness. It ranges from 0.004 fF to 0.15 fF per square micron but is typically 0.1 pF.

C_{gd} is the parasitic oxide capacitance between the gate and the drain. It is also called the *overlap capacitance* because the drain extends slightly under the gate electrode. Its typical value is in the range of 1 pF to 10 pF.

C_{gs} consists of two capacitances: C_{gsq} and C_{gs0}. C_{gs0} is the constant parasitic capacitance due to the overlap of the source region because the source extends slightly under the gate electrode. Its typical

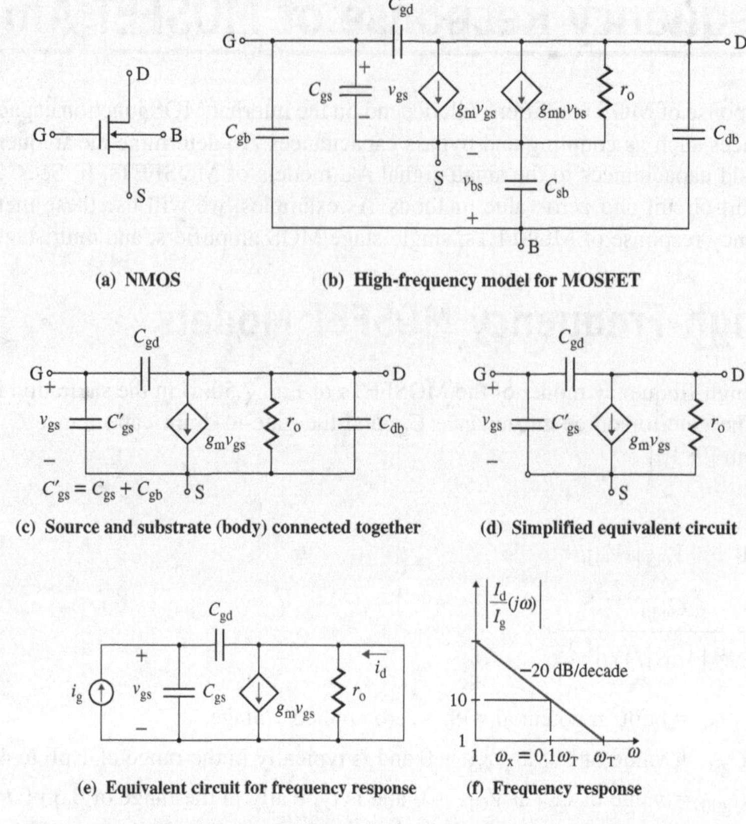

FIGURE 7.50 High-frequency model and response of a MOSFET

value is 10 fF. C_{gsq} is the gate-to-channel capacitance. The channel has a tapered shape and is pinched off at the drain, so C_{gsq} can be expressed as [1, 9]

$$C_{gsq} = \frac{2}{3}WLC_{ox} \tag{7.111}$$

where W = channel width

L = channel length

C_{ox} = capacitance per unit area, which is 3.54×10^{-8} F/cm^2 for an oxide thickness of $t_{ox} = 0.1$ μm

For example, if $W = 30$ μm, $L = 10$ μm, and $t_{ox} = 0.1$ μm, we get $C_{gsq} = 0.07$ pF $= 71$ fF. Table 7.4 shows the capacitances and output resistances for MOSFETs.

In some applications, the substrate is connected to the source, and the frequency model is reduced to Fig. 7.50(c). Capacitance C_{bd} can often be ignored, especially for hand calculations, and the model simplifies to Fig. 7.50(d). Let us apply a test current i_g to the gate of a MOSFET and short-circuit the drain terminal for AC signals. The high-frequency AC equivalent circuit in the saturation region is shown in Fig. 7.50(e). The voltage at the gate terminal in Laplace's domain of s is given by

$$V_{gs}(s) = \frac{1}{(C_{gs} + C_{gd})s} I_g(s) \tag{7.112}$$

TABLE 7.4 Parasitic capacitances and output resistances

	MOSFETs
C_{ds}	0.1–1 pF
C_{gs}	1–10 pF
C_{gd}	1–10 pF
r_o	1–50 kΩ
g_m	0.1–20 mA/V

Since r_o is very large and C_{gd} is very small, the currents through them will be very small. Thus,

$$I_d(s) = [g_m - sC_{gd}]V_{gs}(s)$$

Substituting $V_{gs}(s)$ from Eq. (7.112), we get

$$I_d(s) = \frac{g_m - sC_{gd}}{(C_{gs} + C_{gd})s} I_g(s) \tag{7.113}$$

For the frequencies at which the model in Fig. 7.50(e) is valid, $g_m >> \omega C_{gd}$, and we get the current gain $\beta_f(j\omega)$ in the frequency domain as

$$\beta_f(j\omega) = \frac{I_d(j\omega)}{I_g(j\omega)} = \frac{g_m}{(C'_{gs} + C_{gd})j\omega} \tag{7.114}$$

which indicates that the current gain will fall as the frequency increases, at a slope of −20 dB/decade. This relationship is shown in Fig. 7.50(f). The current gain will be unity, $|\beta_f(j\omega)| = 1$, and the unity-gain bandwidth ω_T is

$$\omega = \omega_T = \frac{g_m}{C'_{gs} + C_{gd}} = \frac{g_m}{C_{gs} + C_{gd} + C_{gb}} \quad \text{(in rad/s)} \tag{7.115}$$

or

$$f_T = \frac{g_m}{2\pi(C_{gs} + C_{gd} + C_{gb})} \quad \text{(in Hz)} \tag{7.116}$$

For MOSFETs, the value of frequency f_T ranges from 100 MHz to 2 GHz.

7.13.2 Small-Signal PSpice Model

The small-signal parameters of MOSFETs can be determined from the manufacturer's data sheet or from practical measurements [6–8]. Alternatively, PSpice/SPICE can calculate the DC biasing point and then generate the small-signal parameters. The small-signal AC equivalent circuits generated by PSpice for MOSFETs are shown in Fig. 7.51, where r_d and r_s are the parasitic resistances of the drain and source terminals, respectively.

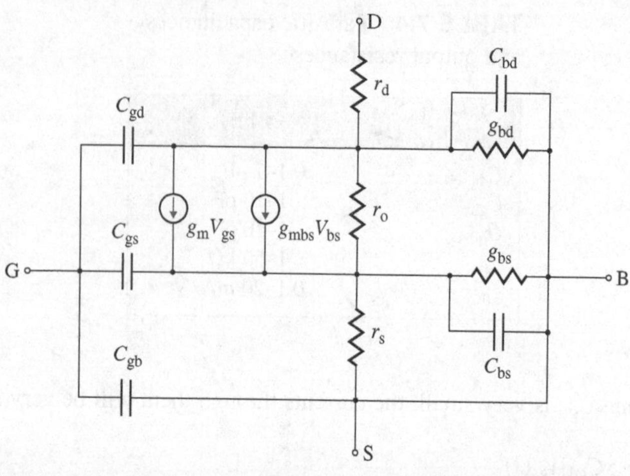

FIGURE 7.51 Small-signal PSpice model of MOSFETs

EXAMPLE 7.12

Finding the high-frequency model parameters of a depletion MOSFET The DC biasing values of the MOSFET are $I_D = 6.3$ mA, $V_{DS} = 5$ V, and $V_{GS} = -1.03$ V. The parameters of the MOSFET are $C_{gs0} = 2.4$ pF, $V_{bi} = 0.8$ V for C_{gs0}, $C_{gb0} = 1$ pF, $C_{gd0} = 1.6$ pF, $V_{bi} = 0.8$ V for C_{gd0}, $g_m = 4.98$ mA/V, and $r_o = 26.77$ kΩ.

(a) Calculate the capacitances of the MOS model in Fig. 7.50(d).

(b) Find the unity-gain bandwidth f_T.

SOLUTION

(a) From Eq. (7.107), $C_{gs} = 2.4\ \text{pF}/[1 + 1.03/0.8]^{1/3} = 1.8$ pF. The gate-drain voltage V_{GD} is

$$V_{GD} = V_{GS} + V_{SD} = V_{GS} - V_{DS} = -1.03 - 5 = -6.03\ \text{V}$$

From Eq. (7.108), $C_{gd} = 1.6\ \text{pF}/[1 + 6.03/0.8]^{1/3} = 0.78$ pF.

(b) From Eq. (7.116), the unity-gain bandwidth f_T is

$$f_T = \frac{g_m}{2\pi(C_{gs} + C_{gd})} = \frac{4.98\ \text{mA/V}}{2\pi \times (1.8\ \text{pF} + 0.78\ \text{pF})} = 307.2\ \text{MHz}$$

7.13.3 Common-Source Amplifiers

Once the DC biasing point of a MOSFET amplifier has been determined, the small-signal parameters can be determined, as discussed in Sec. 7.13.1. The high-frequency model of a MOSFET, shown in Fig. 7.52(a), can be simplified to Fig. 7.52(b) for low frequencies. We will apply the short-circuit and zero-value methods to determine the cutoff frequencies of common-source (CS) amplifiers, common-drain (CD) amplifiers, and common-gate (CG) amplifiers.

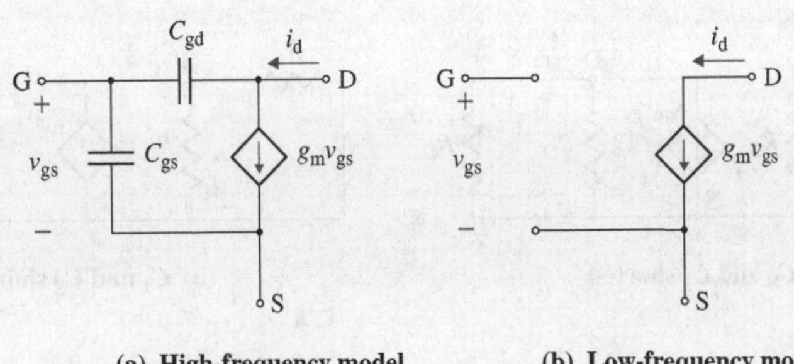

FIGURE 7.52 Small-signal high- and low-frequency models of a MOSFET

A common-source MOSFET amplifier is shown in Fig. 7.53(a).

Low Cutoff Frequencies

If the MOSFET in Fig. 7.53(a) is replaced by its small-signal model in Fig. 7.52(b), we get the low-frequency equivalent circuit shown in Fig. 7.53(b). There are three capacitors—two coupling capacitors, C_1 and C_2, and one bypass capacitor, C_S. If we assume C_2 and C_s are short-circuited, as shown in Fig. 7.54(a), Thevenin's equivalent resistance presented to C_1 is

$$R_{C1} = R_s + R_G \quad \textbf{(7.117)}$$

where $R_G = R_1 \| R_2$.

The equivalent circuit, with C_1 and C_s short-circuited, is shown in Fig. 7.54(b). Thevenin's equivalent resistance presented to C_2 is given by

Since $g_m v_{gs}$ behaves as open circuit at $v_{gs} = 0$.

$$R_{C2} = R_D + R_L \quad \textbf{(7.118)}$$

The equivalent circuit, with C_1 and C_2 short-circuited, is shown in Fig. 7.54(c). There is no voltage across R_s or R_G, so $v_{sr} = -v_{gs}$. Therefore, the resistance representing the current source is

$$R_t = \frac{v_{sr}}{-g_m v_{gs}} = \frac{-v_{gs}}{g_m v_{gs}} = \frac{1}{g_m}$$

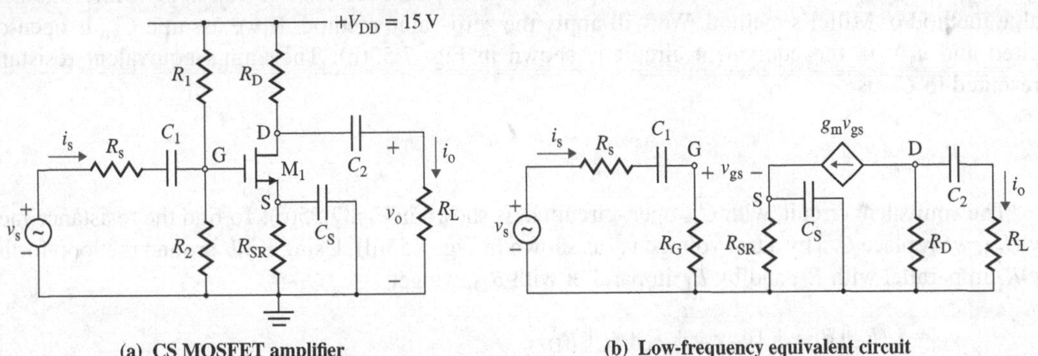

FIGURE 7.53 Common-source MOSFET amplifier

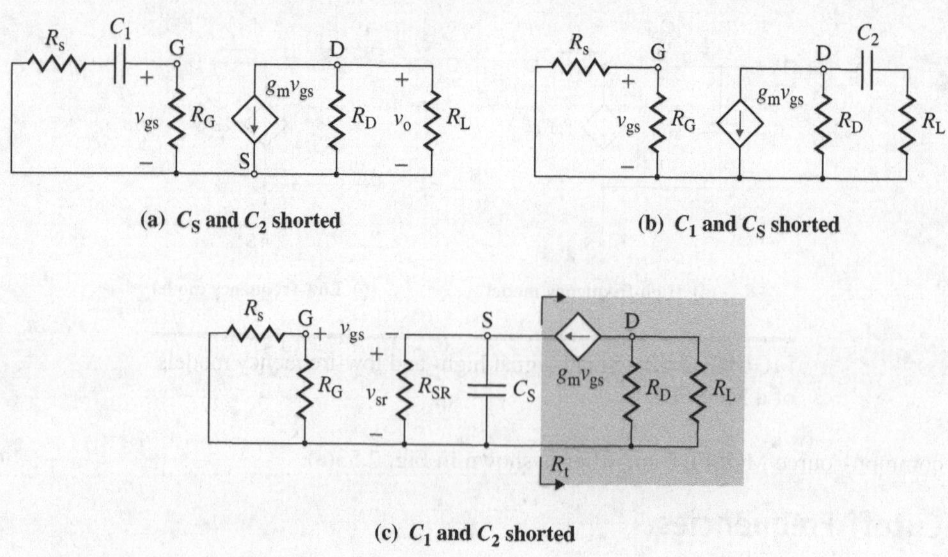

FIGURE 7.54 Equivalent circuits of a common-source FET for the short-circuit method

The Thevenin's equivalent C_s at the source terminal is

$$R_{CS} = R_{SR} \| R_t = R_{SR} \| \frac{1}{g_m} \tag{7.119}$$

In general, $R_{CS} < R_{C2} < R_{C1}$, and R_{CS} controls the low 3-dB frequency. Therefore, $f_L = f_{CS}$.

High Cutoff Frequencies

If the MOSFET of the CS amplifier in Fig. 7.53(a) is replaced by its high-frequency π model in Fig. 7.52(a), we get the high-frequency equivalent circuit shown in Fig. 7.55(a). Since C_{gd} is connected between the input and the output terminals and the output voltage is phase shifted, we can apply either the zero-value method or Miller's method. We will apply the zero-value method. If we assume C_{gd} is open-circuited and $v_s = 0$, the equivalent circuit is shown in Fig. 7.55(b). Thevenin's equivalent resistance presented to C_{gs} is

$$R_{Cgs} = R_s \| R_G \tag{7.120}$$

The equivalent circuit, with C_{gs} open-circuited, is shown in Fig. 7.55(c). To find the resistance faced by C_{gd}, we replace C_{gd} by a test voltage v_x, as shown in Fig. 7.55(d). Using KVL around the loop formed by R_s in parallel with R_G and by R_L in parallel with R_D, we get

$$v_x = i_x(R_s \| R_G) + (g_m v_{gs} + i_x)(R_L \| R_D)$$

$$= i_x(R_s \| R_G) + [g_m i_x(R_s \| R_G) + i_x](R_L \| R_D)$$

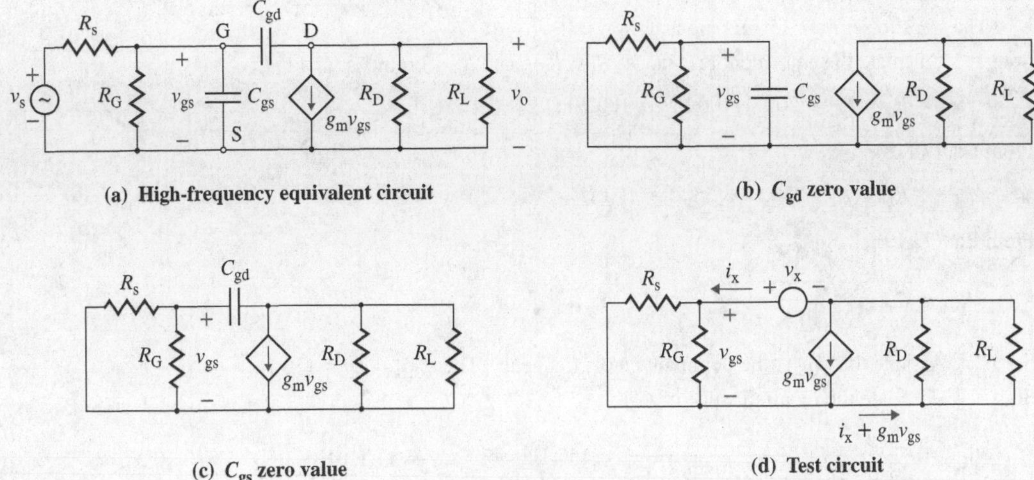

(a) High-frequency equivalent circuit

(b) C_{gd} zero value

(c) C_{gs} zero value

(d) Test circuit

FIGURE 7.55 High-frequency equivalent circuits of a common-source MOSFET amplifier

which gives the resistance faced by C_{gd} as

$$R_{Cgd} = \frac{v_x}{i_x} = R_s \| R_G + [1 + g_m(R_s \| R_G)](R_L \| R_D) \tag{7.121}$$

$$= (R_L \| R_D) + (R_s \| R_G)[1 + g_m(R_L \| R_D)]$$

Thus, the high 3-dB frequency is given by

$$f_H = \frac{1}{2\pi(R_{Cgs}C_{gs} + R_{Cgd}C_{gd})} \tag{7.122}$$

EXAMPLE 7.13

D

Designing a common-source amplifier to give a specified frequency response

(a) Design a common-source MOSFET amplifier as shown in Fig. 7.53(a) to give a low 3-dB frequency of $f_L = 150$ Hz and a high 3-dB frequency of $f_H = 2$ MHz. The circuit parameters are $C_{gd} = 2$ pF, $C_{gs} = 5$ pF, $R_s = 200\ \Omega$, $g_m = 10 \times 10^{-3}$ A/V, $R_{SR} = 2$ kΩ, $R_D = R_L = 5$ kΩ, $R_1 = 200$ kΩ, and $R_2 = 200$ kΩ.

(b) Use Miller's method to check the high-frequency design.

SOLUTION

$R_G = R_1 \| R_2 = 200\ \text{k}\Omega \| 200\ \text{k}\Omega = 100\ \text{k}\Omega.$

(a) The design will have two parts: one in which we set the low 3-dB frequency at $f_L = 150$ Hz and one in which we set the high 3-dB frequency at $f_H = 2$ MHz.

The steps to set $f_L = 150$ Hz are as follows:

Step 1. Calculate the equivalent resistances R_{C1}, R_{C2}, and R_{CS}. From Eq. (7.117),

$$R_{C1} = R_s + R_G = 200 + 100\ \text{k}\Omega = 100.2\ \text{k}\Omega$$

From Eq. (7.118),

$$R_{C2} = R_D + R_L = 5\ \text{k}\Omega + 5\ \text{k}\Omega = 10\ \text{k}\Omega$$

From Eq. (7.119),

$$R_{CS} = 2\ \text{k}\Omega \,\|\, \left(\frac{1}{10 \times 10^{-3}}\right) = 95.2\ \Omega$$

Step 2. Assume that f_{CS} is the dominant cutoff frequency. Then $f_{CS} = f_L = 150$ Hz.

Step 3. Calculate the required value of C_S:

$$f_{CS} = \frac{1}{2\pi R_{CS} C_S} = \frac{1}{2\pi \times 95.2 \times C_S} = 150\ \text{Hz} \quad \text{or} \quad C_S = 11.1\ \mu\text{F}$$

Step 4. Assume $f_{C2} = f_L/10 = 150/10 = 15$ Hz.

Step 5. Calculate the required value of C_2:

$$f_{C2} = \frac{1}{2\pi R_{C2} C_2} = \frac{1}{2\pi \times 10\ \text{k}\Omega \times C_2} = 15\ \text{Hz} \quad \text{or} \quad C_2 = 1.06\ \mu\text{F}$$

Step 6. Assume $f_{C1} = f_L/20 = 150/20 = 7.5$ Hz.

Step 7. Calculate the required value of C_1:

$$f_{C1} = \frac{1}{2\pi R_{C1} C_1} = \frac{1}{2\pi \times 100.2\ \text{k}\Omega \times C_1} = 7.5\ \text{Hz} \quad \text{or} \quad C_1 = 0.21\ \mu\text{F}$$

The steps to set $f_H = 500$ Hz are as follows:

Step 1. From Eq. (7.120),

$$R_{Cgs} = R_s \,\|\, R_G = 200\ \Omega \,\|\, 100\ \text{k}\Omega = 199.6\ \Omega$$

From Eq. (7.121),

$$R_{Cgd} = (5\ \text{k}\Omega \,\|\, 5\ \text{k}\Omega) + (200\ \Omega \,\|\, 100\ \text{k}\Omega) \times [1 + 10\ \text{mA/V} \times (5\ \text{k}\Omega \,\|\, 5\ \text{k}\Omega)] = 7.69\ \text{k}\Omega$$

Step 2. From Eq. (7.122),

$$f_H = \frac{1}{2\pi[199.6\ \Omega \times 5\ \text{pF} + 7.69\ \text{k}\Omega \times (C_{gd} + C_x)]} = 2\ \text{MHz} \quad \text{or} \quad C_{gd} + C_x = 10.2\ \text{pF}$$

which gives $C_x = 10.2 - 2 = 8.2$ pF. This is the value of the additional capacitor C_x that is to be connected between the gate and drain terminals.

(b) Applying Eq. (2.98), we have for the effective Miller's capacitance between the gate and source terminals

$$C_{eq} = (C_{gd} + C_x)[1 + g_m(R_L \,\|\, R_D)] + C_{gs} \tag{7.123}$$

$$= 10.2\ \text{pF} \times [1 + 10\ \text{mA/V} \times (5\ \text{k}\Omega \,\|\, 5\ \text{k}\Omega)] + 5\ \text{pF} = 270.2\ \text{pF}$$

The equivalent resistance faced by C_{eq} is $R_{eq} = R_{Cgs} = R_s \,\|\, R_G = 199.6\ \Omega$. Thus the high 3-dB frequency is

$$f_H = \frac{1}{2\pi C_{eq} R_{eq}} = \frac{1}{2\pi \times 270.2\ \text{pF} \times 199.6\ \Omega} = 2.95\ \text{MHz}$$

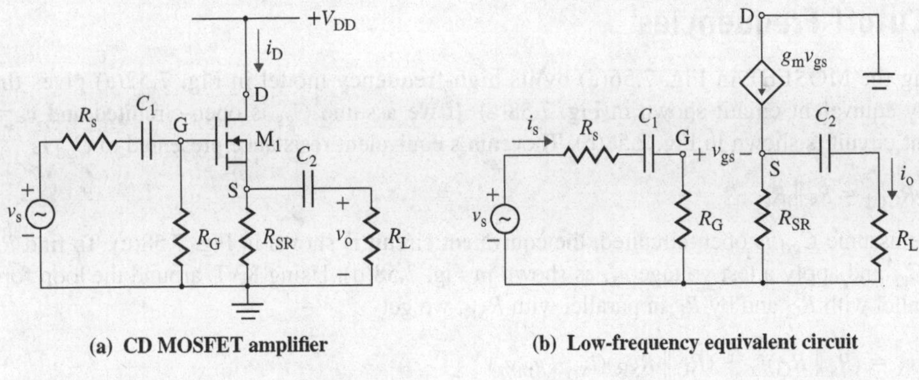

FIGURE 7.56 Common-drain MOSFET amplifier

7.13.4 Common-Drain Amplifiers

A common-drain MOSFET amplifier is shown in Fig. 7.56(a).

Low Cutoff Frequencies

Replacing the MOSFET in Fig. 7.56(a) by its small-signal model in Fig. 7.52(b) gives the low-frequency equivalent circuit shown in Fig. 7.56(b), which has two coupling capacitors C_1 and C_2. If we assume C_2 is short-circuited, as shown in Fig. 7.57(a), Thevenin's equivalent resistance presented to C_1 is

$$R_{C1} = R_s + R_G \tag{7.124}$$

If C_1 is short-circuited, the equivalent circuit is shown in Fig. 7.57(b). From Eq. (7.119), the output resistance is given by

$$R_o = R_{SR} \| \frac{1}{g_m} \tag{7.125}$$

Thevenin's equivalent resistance presented to C_2 is

$$R_{C2} = R_L + R_o \tag{7.126}$$

R_{C2}, which is normally less than R_{C1}, controls the low cutoff frequency.

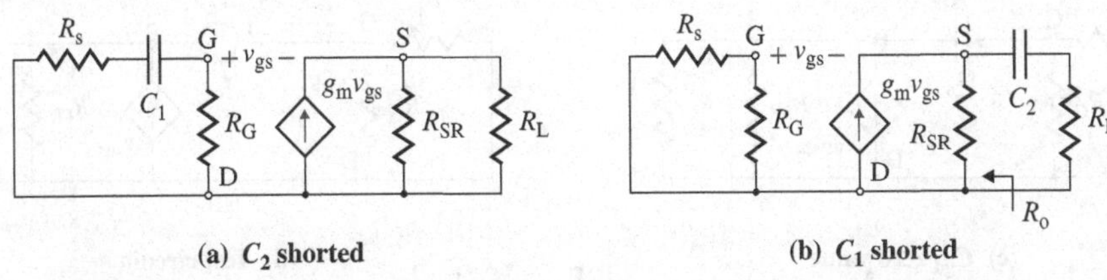

FIGURE 7.57 Equivalent circuits of a common-drain MOSFET amplifier for the short-circuit method

High Cutoff Frequencies

Replacing the MOSFET in Fig. 7.56(a) by its high-frequency model in Fig. 7.52(a) gives the high-frequency equivalent circuit shown in Fig. 7.58(a). If we assume C_{gs} is open-circuited and $v_s = 0$, the equivalent circuit is shown in Fig. 7.58(b). Thevenin's equivalent resistance presented to C_{gd} is

$$R_{Cgd} = R_s \parallel R_G \tag{7.127}$$

If we assume C_{gd} is open-circuited, the equivalent circuit is shown in Fig. 7.58(c). To find R_{Cgs}, we remove C_{gs} and apply a test voltage v_x, as shown in Fig. 7.58(d). Using KVL around the loop formed by R_s in parallel with R_G and by R_L in parallel with R_{SR}, we get

$$v_x = (R_s \parallel R_G)i_x + (R_L \parallel R_{SR})(i_x - g_m v_x)$$

which can be simplified to

$$i_x(R_s \parallel R_G + R_L \parallel R_{SR}) = v_x[1 + g_m(R_L \parallel R_{SR})]$$

or

$$\frac{i_x}{v_x} = \frac{1 + g_m(R_L \parallel R_{SR})}{R_s \parallel R_G + R_L \parallel R_{SR}}$$

Thus, Thevenin's equivalent resistance presented to C_{gs} is

$$R_{Cgs} = \frac{v_x}{i_x} = \frac{R_s \parallel R_G + R_L \parallel R_{SR}}{1 + g_m(R_L \parallel R_{SR})} \tag{7.128}$$

and the high 3-dB frequency is

$$f_H = \frac{1}{2\pi(R_{Cgd}C_{gd} + R_{Cgs}C_{gs})} \tag{7.129}$$

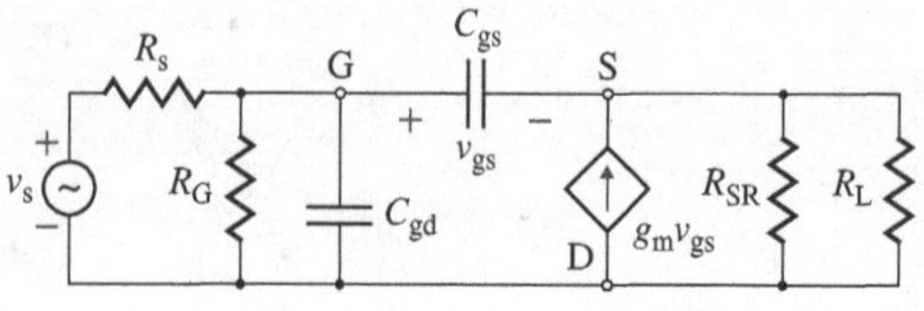

(a) High-frequency equivalent circuit

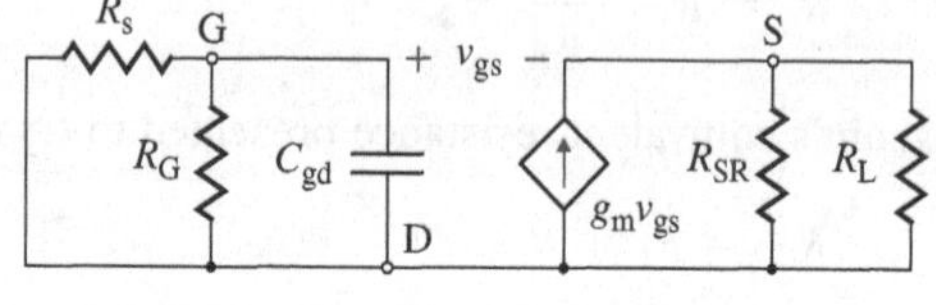

(b) C_{gs} zero value

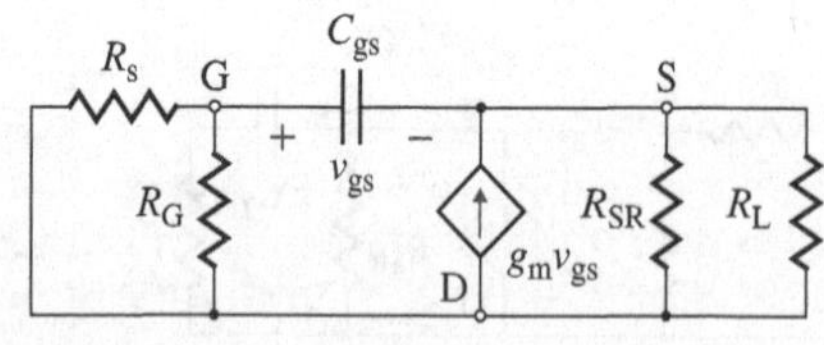

(c) C_{gd} zero value

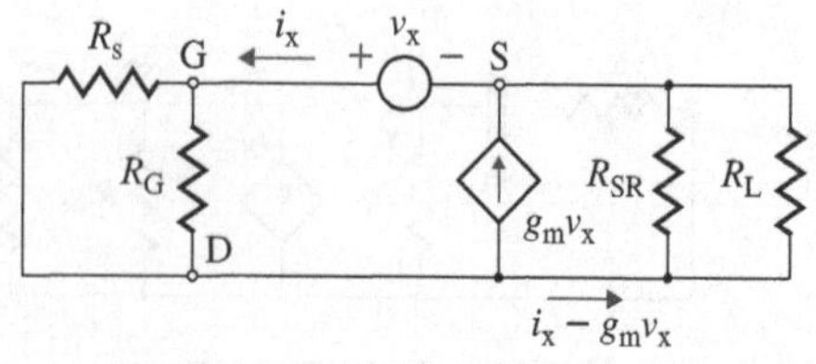

(d) Test circuit

FIGURE 7.58 High-frequency equivalent circuits of a common-drain MOSFET amplifier

EXAMPLE 7.14

Finding the high cutoff frequency of a common-drain MOSFET amplifier The circuit parameters of the MOSFET amplifier in Fig. 7.56(a) are $C_{gd} = 2$ pF, $C_{gs} = 5$ pF, $R_s = 200\ \Omega$, $g_m = 10 \times 10^{-3}$ A/V, $R_{SR} = 2\ \text{k}\Omega$, $R_L = 5\ \text{k}\Omega$, $R_1 = 200\ \text{k}\Omega$, and $R_2 = 200\ \text{k}\Omega$. Calculate the high 3-dB frequency f_H.

SOLUTION

$R_G = R_1 \| R_2 = 200\ \text{k}\Omega \| 200\ \text{k}\Omega = 100\ \text{k}\Omega$. From Eq. (7.127),

$$R_{Cgd} = 200\ \Omega \| 100\ \text{k}\Omega = 199.6\ \Omega$$

From Eq. (7.128),

$$R_{Cgs} = \frac{200\ \Omega \| 100\ \text{k}\Omega + 5\ \text{k}\Omega \| 2\ \text{k}\Omega}{1 + 10\ \text{m}\mho \times (5\ \text{k}\Omega \| 2\ \text{k}\Omega)} = 106.5\ \Omega$$

From Eq. (7.129), the high 3-dB frequency is

$$f_H = \frac{1}{2\pi(R_{Cgd}C_{gd} + R_{Cgs}C_{gs})} = \frac{1}{2\pi \times (199.6\ \Omega \times 2\ \text{pF} + 106.5\ \Omega \times 5\ \text{pF})} = 170.8\ \text{MHz}$$

7.13.5 Common-Gate Amplifiers

A common-gate MOSFET amplifier is shown in Fig. 7.59(a).

Low Cutoff Frequencies

Replacing the MOSFET in Fig. 7.59(a) by its low-frequency model gives the low-frequency equivalent circuit shown in Fig. 7.59(b), which contains two coupling capacitors C_1 and C_2. If we assume C_2 and C_G are short-circuited, the equivalent circuit is shown in Fig. 7.60(a). The resistance representing the current source is $1/g_m$, and Thevenin's equivalent resistance presented to C_1 is

$$R_{C1} = R_s + \left(R_{SR} \| \frac{1}{g_m}\right) \tag{7.130}$$

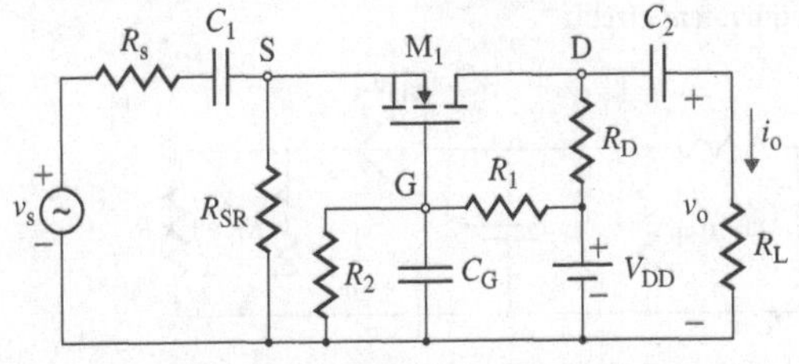

(a) CG MOSFET amplifier

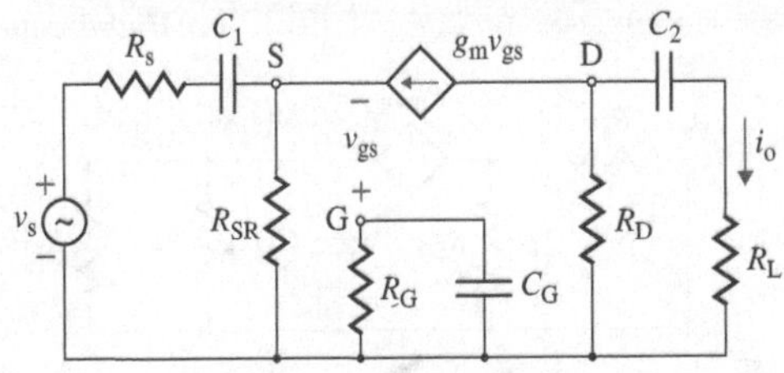

(b) Low-frequency equivalent circuit

FIGURE 7.59 Common-gate MOSFET amplifier

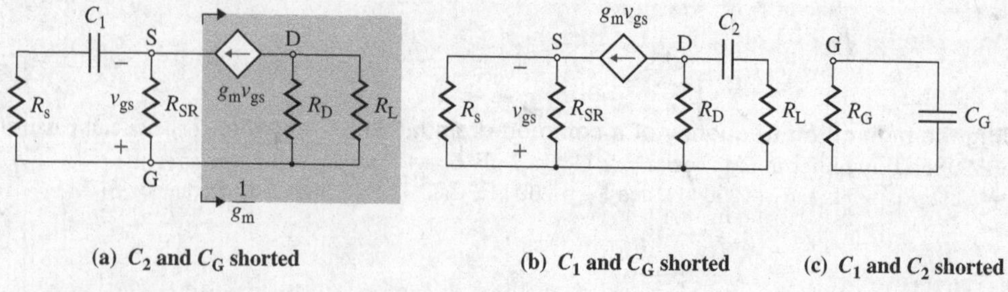

(a) C_2 and C_G shorted (b) C_1 and C_G shorted (c) C_1 and C_2 shorted

FIGURE 7.60 Equivalent circuits of a common-gate amplifier for the short-circuit method

If C_1 and C_G are short-circuited, as shown in Fig. 7.60(b), Thevenin's equivalent resistance presented to C_2 is

$$R_{C2} = R_D + R_L \tag{7.131}$$

If C_1 and C_2 are short-circuited, as shown in Fig. 7.60(c), Thevenin's equivalent resistance becomes

$$R_{CS} = R_G = R_1 \| R_2 \tag{7.132}$$

In general, $R_{CG} > R_{C2} > R_{C1}$, and R_{C1} controls the low cutoff frequency.

High Cutoff Frequencies

The high-frequency equivalent circuit of the common-gate amplifier in Fig. 7.59(a) is shown in Fig. 7.61(a). The equivalent circuit, with C_{gs} open-circuited, is shown in Fig. 7.61(b). $g_m v_{gs}$ behaves as open circuit at $v_{gs} = 0$. Thevenin's equivalent resistance presented to C_{gd} is

$$R_{Cgd} = R_L \| R_D \tag{7.133}$$

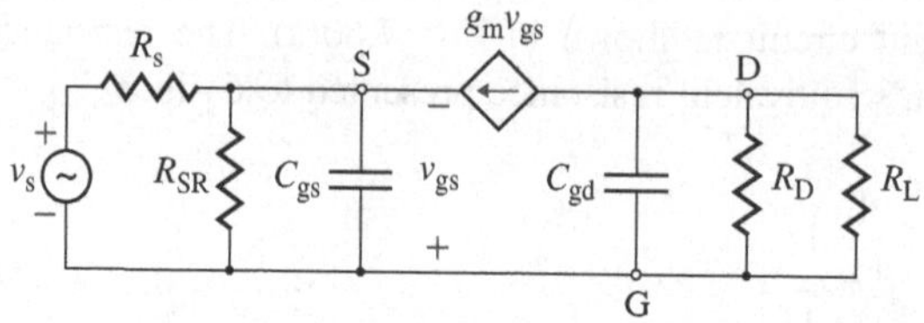

(a) High-frequency equivalent circuit

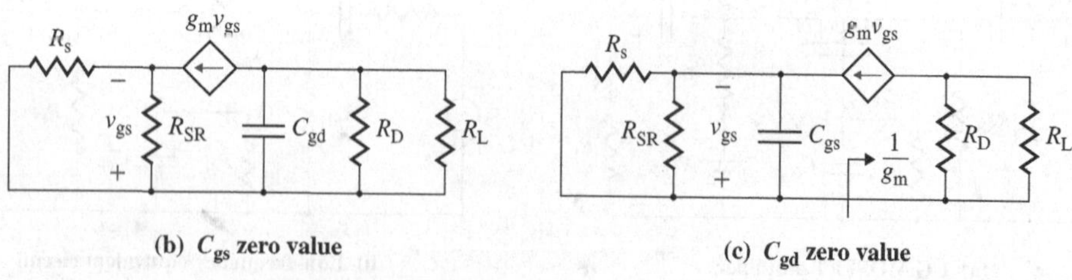

(b) C_{gs} zero value (c) C_{gd} zero value

FIGURE 7.61 High-frequency equivalent circuits of a common-gate amplifier

If we assume C_{gd} is open-circuited, the equivalent circuit is shown in Fig. 7.61(c). The resistance representing the current source is $1/g_m$, which forms a parallel circuit with R_s and R_{SR}. Thevenin's equivalent resistance presented to C_{gs} is given by

$$R_{Cgs} = R_s \| R_{SR} \| \frac{1}{g_m} \tag{7.134}$$

Thus, the high 3-dB frequency is given by

$$f_H = \frac{1}{2\pi(R_{Cgd}C_{gd} + R_{Cgs}C_{gs})} \tag{7.135}$$

▶ **NOTE** f_H is almost independent of the transistor gain g_m since $R_{Cgs} << R_{Cgd}$ and there is no Miller's capacitance multiplication effect. Common-gate amplifiers are used for high-frequency applications.

EXAMPLE 7.15

Finding the high cutoff frequency of a common-gate MOSFET amplifier The circuit parameters of the MOSFET amplifier in Fig. 7.59(a) are $C_{gd} = 2$ pF, $C_{gs} = 5$ pF, $R_s = 200\ \Omega$, $g_m = 10 \times 10^{-3}$ A/V, $R_{SR} = 2$ kΩ, $R_D = R_L = 5$ kΩ, $R_1 = 200$ kΩ, and $R_2 = 200$ kΩ. Calculate the high 3-dB frequency f_H.

SOLUTION

$R_G = R_1 \| R_2 = 200\text{ k}\Omega \| 200\text{ k}\Omega = 100\text{ k}\Omega$. From Eq. (7.133),

$$R_{Cgd} = 5\text{ k}\Omega \| 5\text{ k}\Omega = 2.5\text{ k}\Omega$$

From Eq. (7.134),

$$R_{Cgs} = R_s \| R_{SR} \| \frac{1}{g_m} = 200\ \Omega \| 2\text{ k}\Omega \| \frac{1}{10\text{ mA/V}} = 64.5\ \Omega$$

From Eq. (7.135), the high 3-dB frequency is

$$f_H = \frac{1}{2\pi(R_{Cgd}C_{gd} + R_{gs}C_{gs})} = \frac{1}{2\pi(2.5\text{ k}\Omega \times 2\text{ pF} + 64.5\ \Omega \times 5\text{ pF})} = 29.9\text{ MHz}$$

EXAMPLE 7.16

Finding the frequency response of a two-stage CD–CS MOSFET amplifier A two-stage MOSFET amplifier is shown in Fig. 7.62. The circuit parameters are $C_{gd1} = C_{gd2} = 2$ pF, $C_{gs1} = C_{gs2} = 5$ pF, $R_s = 200\ \Omega$, $g_{m1} = g_{m2} = 10 \times 10^{-3}$ A/V, $R_s = 200\ \Omega$, $R_{G1} = 50$ kΩ, $R_{SR1} = 250\ \Omega$, $R_{D2} = 5$ kΩ, $R_{SR2} = 150\ \Omega$, $R_L = 10$ kΩ, $C_1 = 1\ \mu$F, $C_2 = 10\ \mu$F, and $C_{S2} = 5.3\ \mu$F.

(a) Calculate the low 3-dB frequency f_L.

(b) Calculate the high 3-dB frequency f_H.

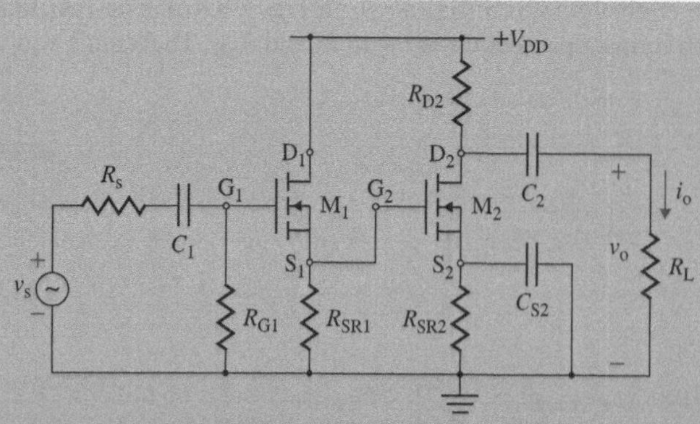

FIGURE 7.62 Two-stage MOSFET amplifier

SOLUTION

(a) The low-frequency equivalent circuit is shown in Fig. 7.63(a). There are two coupling capacitors, C_1 and C_2, and one source bypass capacitor, C_{S2}.

The time constant τ_1 due to C_1 is

$$\tau_1 = (R_s + R_{G1})C_1 \tag{7.136}$$

$$= (200\ \Omega + 50\ \text{k}\Omega) \times 1\ \mu\text{F} = 50.2\ \text{ms}$$

The time constant τ_2 due to C_2 is

$$\tau_2 = (R_{D2} + R_L)C_2 \tag{7.137}$$

$$= (5\ \text{k}\Omega + 10\ \text{k}\Omega) \times 10\ \mu\text{F} = 150\ \text{ms}$$

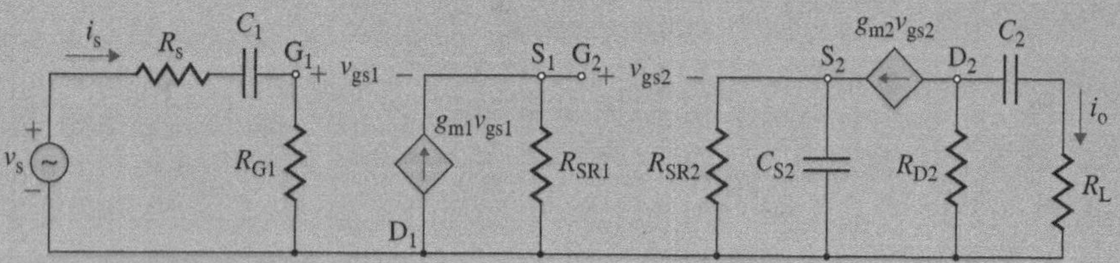

(a) Low-frequency equivalent circuit

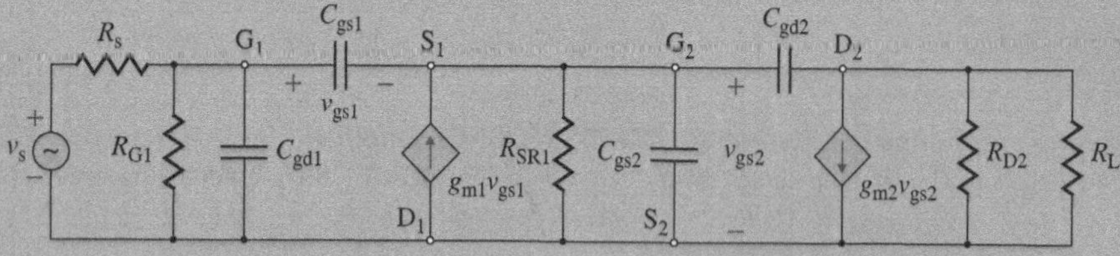

(b) High-frequency equivalent circuit

FIGURE 7.63 Equivalent circuits for Fig. 7.62

The time constant τ_3 due to C_{S2} is

$$\tau_3 = \left(R_{SR2} \| \frac{1}{g_{m2}}\right) C_{S2} \tag{7.138}$$

$$= [150\ \Omega \| (1000/10\ \text{A/V})] \times 5.3\ \mu\text{F} = 0.32\ \text{ms}$$

From Eq. (2.106), the low 3-dB frequency is

$$f_L = \frac{1}{2\pi}\left[\frac{1}{50.2\ \text{ms}} + \frac{1}{150\ \text{ms}} + \frac{1}{0.32\ \text{ms}}\right] = 502\ \text{Hz}$$

(b) The high-frequency equivalent circuit is shown in Fig. 7.63(b). Applying Eq. (7.128) gives Thevenin's equivalent resistance presented to C_{gs1} as

$$R_{gs1} = \frac{R_s \| R_{G1} + R_{SR1}}{1 + g_{m1}R_{SR1}} \tag{7.139}$$

$$= \frac{200\ \Omega \| 50\ \text{k}\Omega + 250\ \Omega}{1 + 10 \times 10^{-3}\ \text{A/V} \times 250\ \Omega} = 128.3\ \Omega$$

and the time constant τ_{gs1} is

$$\tau_{gs1} = R_{gs1}C_{gs1} \tag{7.140}$$

$$= 128.3\ \Omega \times 5\ \text{pF} = 0.642\ \text{ns}$$

If R_{gs2} is Thevenin's equivalent resistance faced by C_{gs2} with C_{gs1}, C_{gd1}, and C_{gd2} open-circuited, R_{gs2} will be the parallel combination of R_{SR1} and the output resistance of transistor M_1. That is,

$$R_{gs2} = R_{SR1} \| \frac{1}{g_{m1}} \tag{7.141}$$

$$= 250\ \Omega \| \frac{1000}{10\ \text{A/V}} = 71.4\ \Omega$$

and the time constant τ_{gs2} is

$$\tau_{gs2} = R_{gs2}C_{gs2} \tag{7.142}$$

$$= 71.4\ \Omega \times 5\ \text{pF} = 0.36\ \text{ns}$$

If R_{gd1} is Thevenin's equivalent resistance faced by C_{gd1} with C_{gs1}, C_{gs2}, and C_{gd2} open-circuited, the time constant τ_{gd1} is

$$\tau_{gd1} = (R_s \| R_{G1})C_{gd1} \tag{7.143}$$

$$= (200\ \Omega \| 50\ \text{k}\Omega) \times 2\ \text{pF} = 0.398\ \text{ns}$$

With C_{gs1}, C_{gs2}, and C_{gd1} open-circuited, Thevenin's equivalent resistance faced by C_{gd2} can be found by applying Eq. (2.116):

$$R_{gd2} = (R_{D2} \| R_L) + R_{gs2}[1 + g_{m2}(R_{D2} \| R_L)] \tag{7.144}$$

$$= (5\ \text{k}\Omega \| 10\ \text{k}\Omega) + 71.4\ \Omega \times [1 + 10 \times 10^{-3}\ \text{A/V} \times (5\ \text{k}\Omega \| 10\ \text{k}\Omega)] = 5.78\ \text{k}\Omega$$

and the time constant τ_{gd2} is

$$\tau_{gd2} = R_{gd2}C_{gd2} \tag{7.145}$$

$$= 5.78\ \text{k}\Omega \times 2\ \text{pF} = 11.57\ \text{ns}$$

Thus, the high 3-dB frequency is

$$f_H = \frac{1}{2\pi \times (0.642\ \text{n} + 0.36\ \text{n} + 0.398\ \text{n} + 11.57\ \text{n})} = 12.27\ \text{MHz}$$

KEY POINTS OF SECTION 7.13

- A MOSFET has depletion-layer capacitances from gate to source, gate to drain, and gate to substrate.
- A MOSFET has parasitic oxide capacitances from gate to source, gate to drain, gate to substrate, and drain to substrate. The gate-to-channel capacitance depends on the oxide thickness, channel length, and channel width.
- The transition frequency is limited by the internal capacitances.

7.14 Design of MOSFET Amplifiers

When an amplifier is being analyzed, the components are specified; however, when an amplifier is being designed, the designer must select the values of the circuit components. The design task can be simplified if a simple transistor model is used to find approximate values of the components. After the initial design stage, the next step is to analyze the amplifier with these approximate values and to compare the performance parameters with the desired values. Often the specifications are not met, and it is necessary to modify the component values. An amplifier is normally specified by the input resistance R_i, the output resistance R_o, and the voltage gain A_{vo}. These specifications are normally defined by the following values:

Source resistance R_s
DC supply voltages V_{DD} and V_{SS} for MOSFETs
Load resistance R_L
Overall voltage gain A_v $(=v_L/v_s)$ (at a specified R_L)
Output resistance R_o
Input resistance R_i

After the specifications of an amplifier have been established, we will develop the necessary design conditions and the steps in meeting design specifications.

We have noted that the technique of DC analysis differs from that of AC analysis. For DC analysis, the load line is set by the DC resistance R_{dc}. That is,

$$R_{dc} = \begin{cases} R_D + R_{SR} & \text{for the CS amplifier of Fig. 7.34(a)} \\ R_{SR} & \text{for the CD amplifier of Fig. 7.42(a)} \end{cases} \tag{7.146}$$

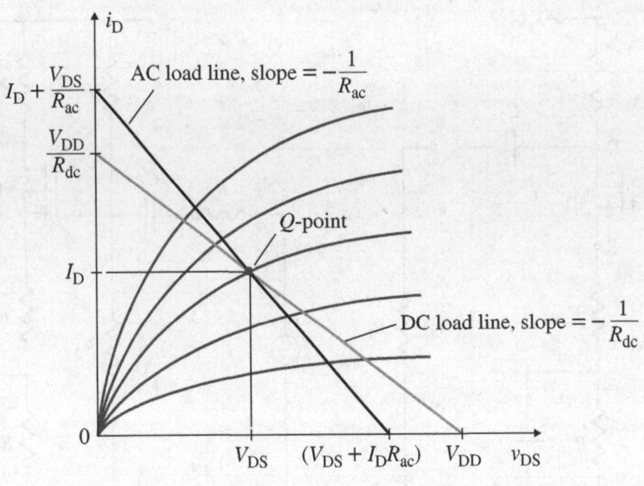

FIGURE 7.64 AC and DC load lines for CE amplifiers

For AC analysis, the load line is set by the AC resistance. That is,

$$R_{ac} = \begin{cases} R_D \| R_L & \text{for the CS amplifier of Fig. 7.34(a)} \\ R_{SR} \| R_L & \text{for the CD amplifier of Fig. 7.42(a)} \end{cases} \tag{7.147}$$

Under the no-load condition, the load resistance R_L is disconnected; the AC resistance R_{ac} equals R_D. Thus, there are two load lines that must be considered in designing an amplifier circuit. So far, we have considered the DC load line only while designing a biasing circuit. The AC and DC load lines for CS amplifiers are shown in Fig. 7.64. The Q-point, which is specified for a zero AC input signal, lies on both the AC and the DC load lines. The AC load line passes through the Q-point and has a slope of $-1/R_{ac}$. The slope of the AC line is greater in magnitude than that of the DC line. The AC load line may be described by

$$i_D - I_D = \frac{-(v_{DS} - V_{DS})}{R_{ac}}$$

which gives

$$i_D = -\frac{v_{DS}}{R_{ac}} + \left(\frac{V_{DS}}{R_{ac}} + I_D\right) \tag{7.148}$$

The maximum AC drain current $I_{D(max)}$, which occurs at $v_{DS} = 0$, can be found from Eq. (7.148):

$$I_{D(max)} = \frac{V_{DS}}{R_{ac}} + I_D \tag{7.149}$$

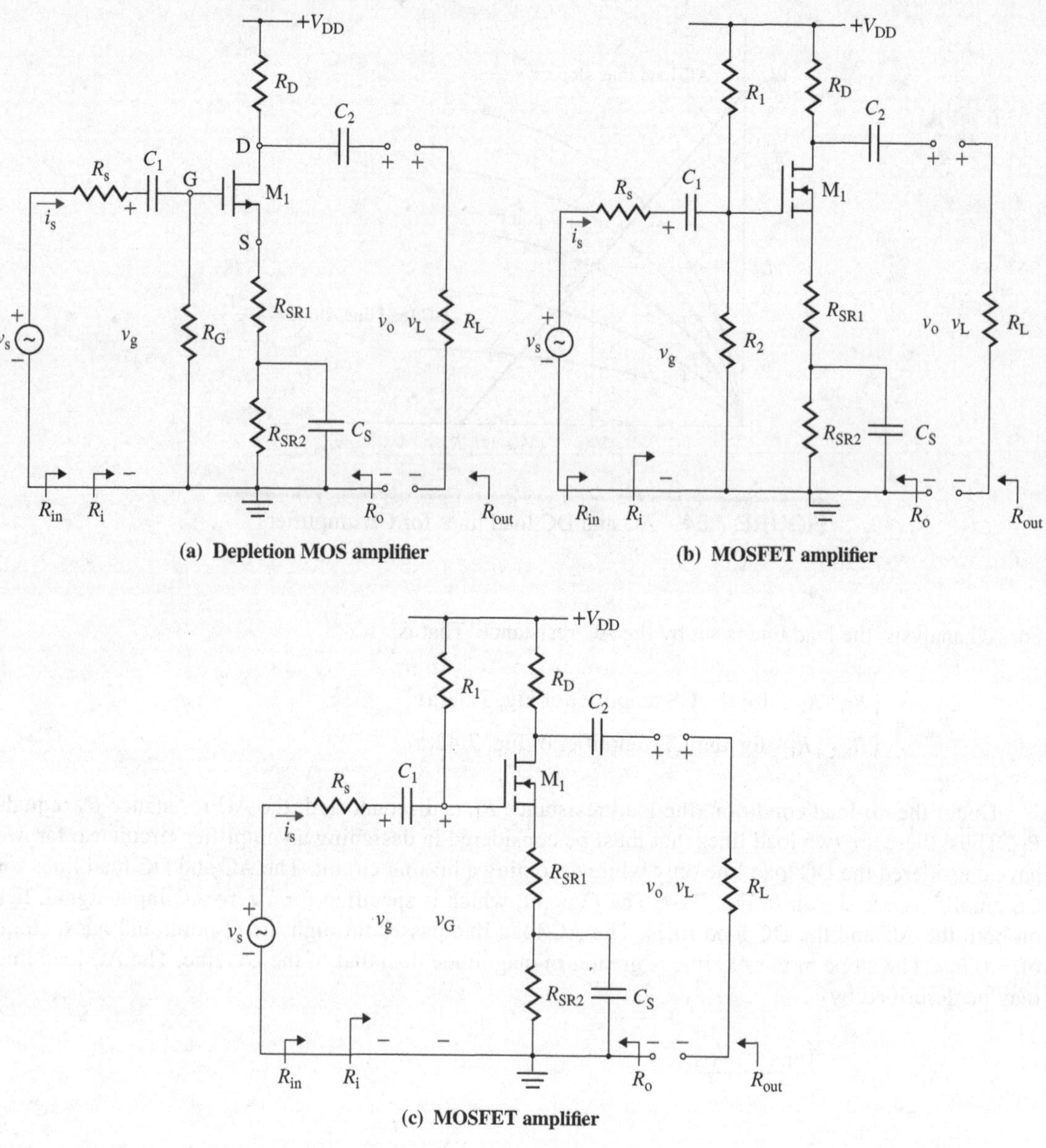

FIGURE 7.65 Circuit configurations for MOSFET amplifiers

From Eq. (7.149), the quiescent drain current I_D can be related to the AC and DC load lines by

$$\frac{V_{DD}}{I_D} = R_{dc} + R_{ac} \tag{7.150}$$

The input resistance of MOSFETs is high and can be selected independently of the voltage gain. MOSFET amplifiers are normally designed to provide a specified voltage gain A_v. Three possible circuit configurations are shown in Fig. 7.65. R_{SR} $(=R_{SR1} + R_{SR2})$ provides the required biasing voltage, and R_{SR1} gives the necessary voltage gain A_{vo}. After establishing the specifications of the amplifier, choose a suitable MOSFET and note its particular pinch-down voltage V_p (or threshold voltage V_t), drain current I_{DSS} (for $v_{GS} = 0$) (or MOSFET constant K_n), and channel modulation voltage V_M (or assume a typical value of 200 V). Then choose the drain current I_D at the Q-point. When choosing I_D, find the maximum value of $I_{D(max)}$ from the data sheet for the transistor you have in mind. Then choose $I_D \leq I_{D(max)}/2$ and the circuit topology of Fig. 7.65[(a), (b), or (c)].

The design steps required to accomplish the specifications are as follows:

Step 1. Using either Eq. (7.8) or Eq. (7.21), find the gate-to-source voltage V_{GS} for known values of I_D, I_{DSS}, V_t, and V_P.

$$I_D = \begin{cases} K_n(V_{GS} - V_t)^2 & \text{for enhancement MOSFETs} \\ K_n(V_{GS} - V_p)^2 & \text{for depletion MOSFETs} \end{cases}$$

Step 2. For the known value of V_{GS}, calculate R_{SR}. One method is to use

$$V_{GS} = R_{SR}\,|I_D| \quad \text{for depletion MOSFETs as in Fig. 7.65(a)}$$

For other configurations, use

$$V_{SR} = \frac{V_{DD}}{3} = R_{SR}I_D$$

which gives $R_{SR} = V_{DD}/(3I_D)$, and

$$V_{GS} = \begin{cases} \dfrac{R_2\,V_{DD}}{R_1 + R_2} - R_{SR}I_D & \text{for MOSFETs as in Fig. 7.65(b)} \\ V_{DD} - V_{SR} = V_{DD} - R_{SR}I_D & \text{for MOSFETs as in Fig. 7.65(c)} \end{cases}$$

Step 3. From Eq. (7.26), calculate the output resistance r_o of the MOSFET:

$$r_o = \frac{|V_M|}{i_D}$$

Step 4. From either Eq. (7.28) or Eq. (7.31), calculate the transconductance g_m of the MOSFET:

$$g_m = \begin{cases} g_{mo}\left(1 - \dfrac{V_{GS}}{|V_p|}\right) & \text{for depletion MOSFETs} \\ g_{mo}\left(1 - \dfrac{V_{GS}}{|V_t|}\right) & \text{for enhancement MOSFETs} \end{cases}$$

where

$$g_{mo} = \begin{cases} -2K_nV_p = \dfrac{2I_{DSS}}{|V_p|} & \text{for depletion MOSFETs} \\ -2K_n\,|V_t| & \text{for enhancement MOSFETs} \end{cases}$$

Step 5. Calculate the gate resistance R_G or resistances R_1 and R_2. Calculate R_G from

$$R_G = R_i \quad \text{for MOSFETs as in Fig. 7.65(a)}$$

We set

$$v_G = \frac{V_{DD}R_2}{R_1 + R_2} = \frac{V_{DD}}{1 + R_1/R_2}$$

$$R_G = \frac{R_1R_2}{R_1 + R_2}$$

which can be solved to calculate R_1 and R_2

$$R_1 = \frac{R_iV_{DD}}{V_G} \quad \text{for MOSFETs as in Fig. 7.65(b)}$$

$$R_2 = \frac{R_iV_{DD}}{V_{DD} - V_G} \quad \text{for MOSFETs as in Fig. 7.65(b)}$$

where $V_G = V_{SR} + V_{GS}$ and V_{SR} is the DC voltage at the source terminal.

Step 6. For known values of R_L, I_D, V_{DD}, and R_{SR}, calculate the drain resistance R_D. With $R_{dc} = R_D + R_{SR}$ and $R_{ac} = R_D \parallel R_L$, Eq. (7.150) gives

$$\frac{V_{DD}}{I_D} = R_{dc} + R_{ac} = R_D + R_{SR} + \frac{R_DR_L}{R_D + R_L}$$

Step 7. Assuming a voltage gain A_{vo}, let the no-load voltage gain A_{vo} be equal to A_v. That is, let $A_{vo} = A_v$ as the first approximation. From Eq. (7.70), the no-load voltage gain A_{vo} is given by

$$|A_{vo}| = \frac{v_o}{v_g} = \frac{\mu_gR_D}{R_D + r_o + (1 + \mu_g)R_{SR1}}$$

from which the source resistance R_{SR1} can be found:

$$R_{SR1} = \frac{\mu_gR_D - |A_{vo}|(R_D + r_o)}{|A_{vo}|(1 + \mu_g)}$$

where $\mu_g = g_mr_o$.

Step 8. Calculate the value of bypassed source resistance R_{SR2}:

$$R_{SR2} = R_{SR} - R_{SR1}$$

If $R_{SR2} < 0$, A_{vo} is too high; choose a transistor with a higher value of g_m.

Step 9. Using Eq. (7.65), calculate the output resistance R_o:

$$R_o = [r_o + (1 + \mu_g)R_{SR1}] \parallel R_D$$

Step 10. Calculate the voltage gain A_v:

$$A_v = \frac{v_L}{v_s} = \frac{A_{vo}R_iR_L}{(R_i + R_s)(R_L + R_o)}$$

Step 11. If the value of A_v in step 10 is not greater than or equal to the desired value of A_v, repeat steps 7 through 10 with progressively higher values of A_{vo} until you obtain the desired value for A_v in step 10. If the gain requirement cannot be obtained, choose a transistor with a higher value of g_m.

KEY POINTS OF SECTION 7.14

- In general, designing involves decision making and an iterative process. The design steps developed in this section will be helpful in finding component values to satisfy specifications.
- Designing an amplifier requires prior knowledge of desired specifications, choice of a MOSFET, and choice of a *Q*-point.
- Once the type of transistor and the *Q*-point have been chosen, the next step is to choose the biasing circuit and find its component values.
- The small-signal parameters, which are calculated from the values of the *Q*-point, are then used to find the emitter (or source) resistance needed to obtain the desired voltage gain or input resistance.

Summary

MOSFETs, which are voltage-dependent devices, are of two types: junction MOSFETs and MOSFETs. MOSFETs are of two types: enhancement and depletion. Each type can be either *p*-channel or *n*-channel. Depending on the value of the drain-to-source voltage, a MOSFET can operate in one of three regions: ohmic, saturation, or cutoff. In the ohmic region, a MOSFET is operated as a voltage-controlled device. In the saturation region, a MOSFET is operated as an amplifier. An enhancement MOSFET conducts only when the gate-to-source voltage exceeds the threshold voltage. The gate current of a MOSFET is very small (on the order of nA). A MOSFET can be modeled by a voltage-controlled current source. MOSFETs should be biased properly to set the gate-to-source voltage in appropriate polarity and magnitude. The *Q*-point should be stable, and a biasing circuit should be designed to minimize the effect of parameter variations. MOSFETs are widely used in very-large-scale integrated (VLSI) circuits.

References

1. R. T. Howe and C. G. Sodini, *Microelectronics—An Integrated Approach*. Englewood Cliffs, NJ: Prentice Hall, 1997.
2. D. A. Johns and K. Martin, *Analog Integrated Circuit Design*. New York: Wiley, 1997.
3. R. C. Jaeger and T. Blalock, *Microelectronic Circuit Design*. New York: McGraw-Hill, 2008.
4. D. A. Neamen, *Microelectronics: Circuit Analysis and Design*. New York: McGraw-Hill, 2007.

5. A. S. Sedra and K. C. Smith, *Microelectronic Circuits.* New York: Oxford University Press, 2004.
6. D. Foty, *MOSFET Modeling with SPICE.* Upper Saddle River, NJ: Prentice Hall, 1997.
7. M. H. Rashid, *Introduction to PSpice Using OrCAD for Circuits and Electronics.* Upper Saddle River, NJ: Prentice Hall, 2004.
8. Y. Tsividis, *Operation and Modeling of the MOS Transistor.* New York: McGraw-Hill, 1999.
9. A. R. Hambley, *Electronics—A Top-Down Approach to Computer-Aided Circuit Design.* New York: Macmillan Publishing, 1994.
10. B. Razavi, *RF Microelectronics.* Upper Saddle River, NJ: Prentice Hall, 1998.

Review Questions

1. What is a DC load line?
2. What is an AC load line?
3. What are the advantages of MOSFETs?
4. What are the types of MOSFETs?
5. What is an NMOS?
6. What is a PMOS?
7. What is the ohmic region of a MOSFET?
8. What are the effects of MOSFET characteristics on the biasing point?
9. What is the transconductance gain g_m of a MOSFET?
10. What is the small-signal output resistance r_o of a MOSFET?
11. What is the channel modulation voltage of a MOSFET?
12. What is the purpose of a source-bypassed capacitor?
13. What are the performance parameters of an amplifier?
14. What are the characteristics of CS-configuration amplifiers?
15. What are the characteristics of source followers?

Problems

The symbol **D** indicates that a problem is a design problem. The symbol **P** indicates that you can check the solution to a problem using PSpice/SPICE or Electronics Workbench.

7.2 and 7.4 Enhancement and Depletion MOSFETs

7.1 An NMOS has a channel width of $W = 40$ μm and a channel length of $L = 2$ μm; the thickness of the silicon dioxides is $t_{ox} = 10$ nm, the dielectric constant of the silicon dioxide layer is $\varepsilon_{ox} = 4$, and the mobility of the electrons in the inversion layer is $\mu_n = 500$ cm^2/(volt-sec). Determine the MOS constants K_n and K_p.

7.2 An NMOS has a substrate impurity doping concentration of $N_a = 2 \times 10^{16}$ cm^3, a threshold voltage of $V_{tN} = 1$ V, and a channel length of $L = 10$ μm; $V_{GS} = 2.5$ V and $V_{DS} = 5$ V. Determine the channel modulation voltage V_M.

7.3 An NMOS has a substrate impurity doping concentration of $N_a = 2 \times 10^{16}$ cm^3, a threshold voltage of $V_{tN} = 1$ V, and a channel length of $L = 10$ μm; $V_{GS} = 2.5$ V. Plot the channel modulation voltage V_M for $V_{DS} = 5$ V to 20 V.

7.4 An NMOS has a substrate impurity doping concentration of $N_a = 2 \times 10^{16}$ cm^3 and an intrinsic concentration of $n_i = 1.5 \times 10^{10}$ cm^{-3}; $T = 25$°C. **(a)** Determine the depletion width x_p extending to the *p*-region substrate for $V_{DS} = 10$ V, and (b) plot x_p.

7.5 An NMOS has a drain current of $I_{D1} = 1$ mA at $V_{GS1} = 1.5$ V and $I_{D2} = 2.5$ mA at $V_{GS2} = 2.5$ V. If the NMOS operates in the saturation region, determine **(a)** its threshold voltage V_t, **(b)** the MOS constant K_n, **(c)** the drain current I_D at $V_{GS} = 2$ V, and **(d)** the minimum drain-to-source voltage $V_{DS(sat)}$ $V_{GS} = 2$ V to operate in the saturation region.

7.6 A PMOS has a drain current of $I_{D1} = 1$ mA at $V_{GS1} = -1.5$ V; $I_{D2} = 2.5$ mA at $V_{GS2} = -2.5$ V. If the PMOS operates in the saturation region, determine **(a)** its threshold voltage V_t, **(b)** the MOS constant K_n, **(c)** the drain current I_D at $V_{GS} = -2$ V, and **(d)** the minimum drain-to-source voltage $V_{SD(sat)}$ at $V_{GS} = -2$ V to operate in the saturation region.

7.7 A depletion NMOS has a drain current of $I_{D1} = 1$ mA at $V_{GS1} = -2.5$ V, and $I_{D2} = 2.5$ mA at $V_{GS2} = -1$ V. If the NMOS operates in the saturation region, determine **(a)** its pinch-off voltage V_p, **(b)** the MOS constant K_n, **(c)** the drain current I_D at $V_{GS} = -1.5$ V, and **(d)** the minimum drain-to-source voltage $V_{DS(sat)}$ at $V_{GS} = -1.5$ V to operate in the saturation region.

7.8 A depletion PMOS has a drain current of $I_{D1} = 1$ mA at $V_{GS1} = 2.5$ V, and $I_{D2} = 2.5$ mA at $V_{GS2} = 1.5$ V. If the PMOS operates in the saturation region, determine **(a)** its pinch-off voltage V_p, **(b)** the MOS constant K_n, **(c)** the drain current I_D at $V_{GS} = 2$ V, and **(d)** the minimum drain-to-source voltage $V_{SD(sat)}$ at $V_{GS} = 2$ V to operate in the saturation region.

7.9 An NMOS has a substrate impurity doping concentration of $N_a = 2 \times 10^{16}$ cm^3 and an intrinsic concentration of $N_i = 1.5 \times 10^{10}$ cm^{-3}; $T = 25$°C, and the oxide thickness $t_{ox} = 0.10$ μm. The threshold voltage is $V_{to} = 1$ V at $V_{SB} = 0$ V. Plot V_t for $V_{SB} = 5$ V to 30 V.

7.10 An NMOS has a threshold voltage $V_t = 1$ V and a MOS constant $K_n = 0.5$ mA/V^2. It operates in the ohmic region and offers a drain-to-source resistance $R_{ds} = 50$ Ω at $V_{DS} = 2$ V. Determine the gate-to-source voltage V_{GS}.

7.11 An NMOS has a threshold voltage $V_t = 1$ V and MOS constant $K_n = 0.5$ mA/V^2. It operates in the ohmic region and offers a drain-to-source resistance $R_{ds} = 100$ Ω at $V_{GS} = 1.5$ V. Determine the drain-to-source voltage V_{DS}.

7.7 DC Biasing of MOSFETs

7.12 The NMOS biasing circuit in Fig. P7.12 has $R_D = 1.5$ kΩ, $R_G = 500$ kΩ, and $V_{DD} = 12$ V. The MOS parameters are $K_n = 0.5$ mA/V^2, $V_t = 1$ V, and $\lambda = 0.01$. Determine **(a)** the drain current I_D, **(b)** the gate-to-source voltage V_{GS}, **(c)** the drain-to-source voltage V_{DS}, **(d)** the small-signal transconductance g_m, and **(e)** the output resistance r_o.

FIGURE P7.12

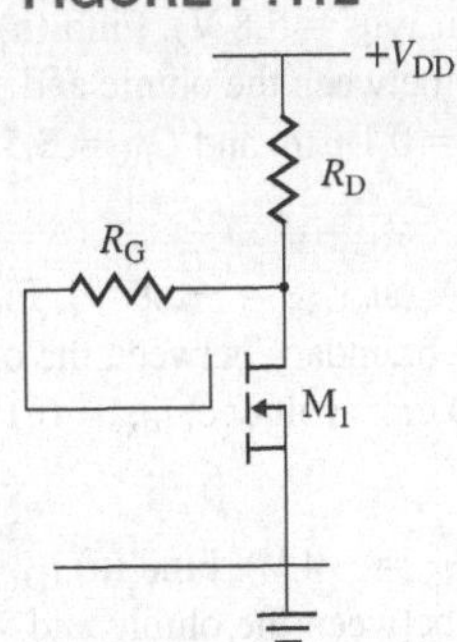

7.13 The NMOS biasing circuit in Fig. P7.13 has $R_D = 1.5$ kΩ, $R_{SR} = 500$ Ω, $R_G = 500$ kΩ, and $V_{DD} = 12$ V. The MOS parameters are $K_n = 0.5$ mA/V^2, $V_t = 1$ V, and $\lambda = 0.01$. Determine **(a)** the drain current I_D, **(b)** the gate-to-source voltage V_{GS}, **(c)** the drain-to-source voltage V_{DS}, **(d)** the small-signal transconductance g_m, and **(e)** the output resistance r_o.

FIGURE P7.13

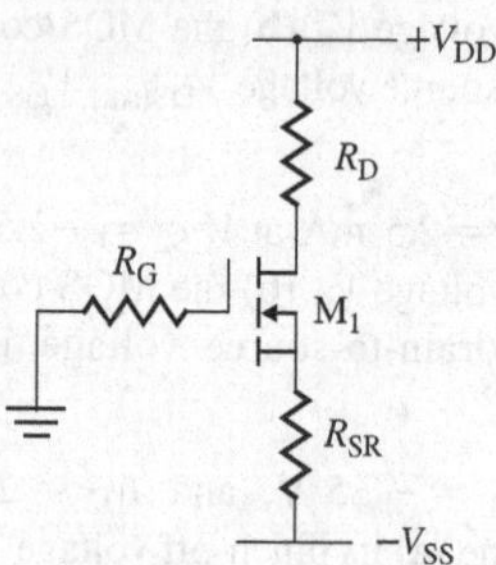

7.14 The MOS biasing circuit in Fig. P7.14 has $R_D = 1.5$ kΩ, $R_{SR} = 500$ Ω, $R_1 = 400$ kΩ, $R_2 = 600$ kΩ, and $V_{DD} = 12$ V. The MOS parameters are $K_n = 0.5$ mA/V², $V_t = -1.5$ V, and $\lambda = 0.01$. Determine **(a)** the drain current I_D, **(b)** the gate-to-source voltage V_{GS}, **(c)** the drain-to-source voltage V_{DS}, **(d)** the small-signal transconductance g_m, and **(e)** the output resistance r_o.

FIGURE P7.14

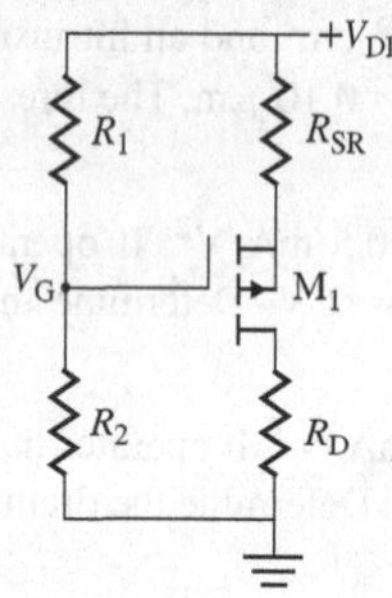

7.15 The pinch-down voltage of an *n*-channel depletion NMOS is $V_p = -5$ V, and the saturation current is $I_{DSS} =$ 40 mA. The value of v_{DS} is such that the transistor is operating in the saturation region. The drain current is $i_D = 15$ mA. Calculate the gate-to-source voltage v_{GS}.

7.16 The pinch-down voltage of a *p*-channel depletion NMOS is $V_p = 5$ V, and the saturation current is $I_{DSS} =$ -40 mA. The value of v_{DS} is such that the transistor is operating in the saturation region. The drain current is $i_D = -15$ mA. Calculate the gate-to-source voltage v_{GS}.

7.17 An *n*-channel enhancement MOSFET has $V_t = 3.5$ V and $i_D = 8$ mA (at $v_{GS} = 5.8$ V). Find **(a)** i_D when $v_{GS} = 5$ V, **(b)** v_{GS} when $i_D = 6$ mA, **(c)** the value of v_{DS} at the boundary between the ohmic and saturation regions if $i_D = 6$ mA, and **(d)** the ratio *W/L* if $\mu_n = 600$ cm²/volt-sec, $t_{ox} = 0.1$ µm, and $C_{ox} = 3.5 \times 10^{-11}$ F/cm². Assume operation in the saturation region.

7.18 A *p*-channel enhancement MOSFET has $V_t = -3.5$ V and $i_D = -8$ mA (at $v_{GS} = -5.8$ V). Find **(a)** i_D when $v_{GS} = -5$ V, **(b)** v_{GS} when $i_D = -6$ mA, **(c)** the value of v_{DS} at the boundary between the ohmic and saturation regions when $i_D = -6$ mA, and **(d)** the ratio *W/L* if $\mu_n = 600$ cm²/volt-sec, $t_{ox} = 0.1$ µm, and $C_{ox} = 3.5 \times 10^{-11}$ F/cm².

7.19 An *n*-channel depletion MOSFET has $V_p = -5$ V and $i_D = 0.5$ mA (at $v_{GS} = -4$ V). Find **(a)** i_D when v_{GS} $= -2$ V, **(b)** v_{GS} when $i_D = 6$ mA, **(c)** the value of v_{DS} at the boundary between the ohmic and saturation regions when $i_D = 6$ mA, and **(d)** the ratio *W/L* if $\mu_n = 600$ cm²/volt-sec and $C_{ox} = 3.5 \times 10^{-11}$ F/cm². Assume operation in the saturation region.

7.20 P The *n*-channel depletion NMOS circuit of Fig. 7.25(a) has $R_D = 1.5$ kΩ, $R_{SR} = 1$ kΩ, $R_1 = \infty$, $R_2 = 500$ kΩ, and $V_{DD} = 15$ V. Calculate I_D, V_{GS}, and V_{DS} if **(a)** $I_{DSS} = 25$ mA and $i_D = 0.5$ mA (at $v_{GS} = -6.5$ V) and **(b)** $I_{DSS} = 5$ mA and $i_D = 0.5$ mA (at $v_{GS} = -1.5$ V).

7.21 The biasing circuit for the n-channel depletion NMOS of Fig. 7.25(a) has $R_1 = 350$ kΩ, $R_2 = 100$ kΩ, $V_{DD} =$ 15 V, $R_D = 1.5$ kΩ, and $R_{SR} = 2.3$ kΩ. The transistor parameters are $I_{DSS} = 15$ mA and $V_p = -4.5$ V. Assume operation in the saturation region.

P

a. Calculate the values of I_D, V_{DS}, and V_{GS} at the Q-point.
b. Calculate the minimum value of R_{SR} so that $V_{GS} \leq 0$.
c. Use PSpice/SPICE to verify your design in part (a).

7.22 An n-channel depletion NMOS amplifier is shown in Fig. P7.22(a). The drain characteristic is shown in Fig. P7.22(b). The quiescent values are $I_D = 5$ mA, $V_{DS} = 10$ V, and $V_{GS} = -2$ V. Calculate the values of R_D and R_{SR}.

FIGURE P7.22

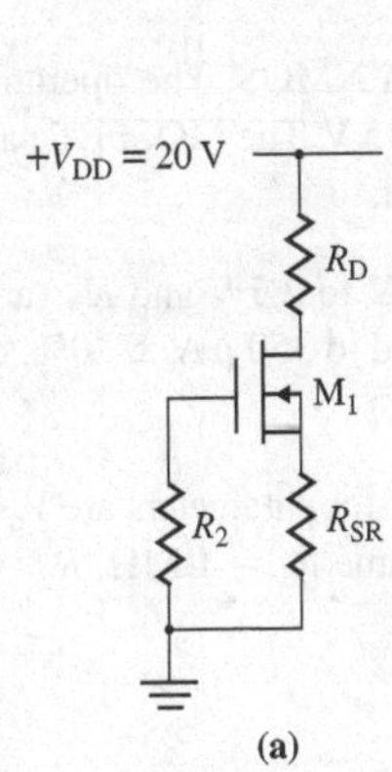

(a)

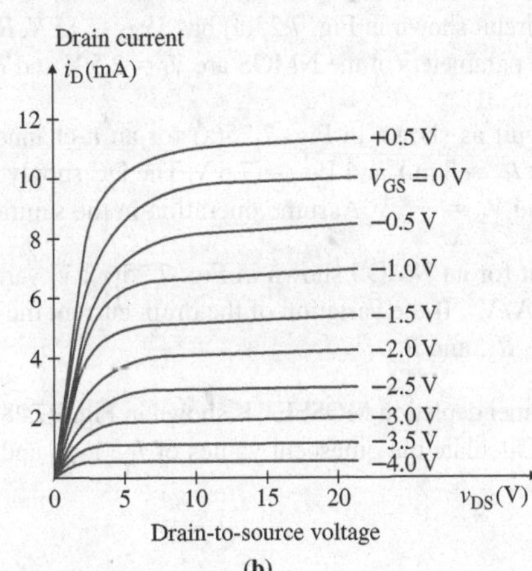

(b)

7.23 For the n-channel depletion NMOS circuit shown in Fig. P7.23, $R_D = 2.5$ kΩ and $V_{DD} = 18$ V. The parameters of the depletion NMOS are $V_p = -1.5$ V and $I_{DSS} = 5$ mA. Calculate the quiescent values of I_D, V_{DS}, and V_{GS}.

P

FIGURE P7.23

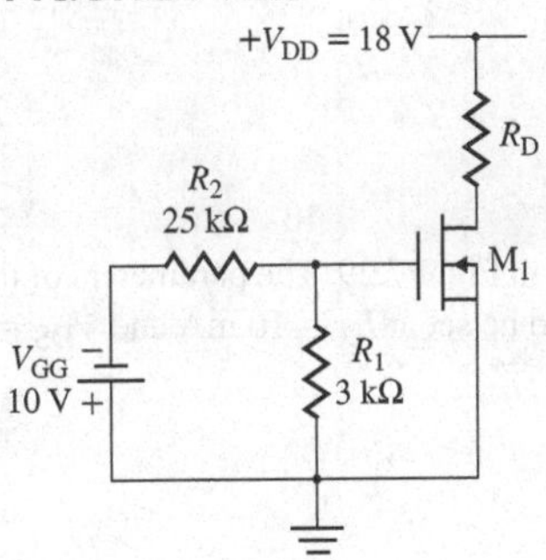

7.24 For the n-channel depletion NMOS circuit shown in Fig. P7.24, the quiescent values are $I_D = 7.5$ mA and $V_{DS} = 10$ V. The parameters of the MOSFET are $I_{DSS} = 10$ mA and $V_p = -5$ V. If the drain characteristic is described by

P

$$i_D = I_{DSS}\left[1 - \frac{V_{GS}}{V_p}\right]^2$$

calculate **(a)** the quiescent value of V_{GS} and **(b)** the values of R_{SR} and R_D. Assume $V_{DD} = 20$ V.

FIGURE P7.24

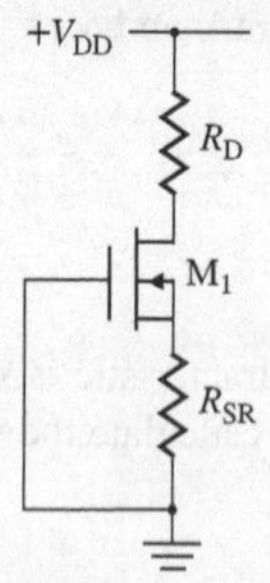

7.25 P The NMOS biasing circuit shown in Fig. 7.25(a) has $V_{DD} = 15$ V, $R_1 = 400$ kΩ, $R_2 = 150$ kΩ, $R_D = 2.5$ kΩ, and $R_{SR} = 4$ kΩ. The parameters of the NMOS are $V_t = 2.5$ V and $K_p = 1\ \text{mA/V}^2$. Calculate V_{DS} and V_{GS}.

7.26 D P Design a biasing circuit as shown in Fig. 7.25(a) for an n-channel depletion NMOS. The operating point must be maintained at $I_D = 8$ mA and $V_{DS} = 7.5$ V. The DC supply voltage is 15 V. The MOSFET parameters are $I_{DSS} = 15$ mA and $V_p = -5$ V. Assume operation in the saturation region.

7.27 D For the biasing circuit for an NMOS shown in Fig. 7.25(a), V_t varies from 1 V to 2.5 V and K_p varies from 200 $\mu\text{A/V}^2$ to 150 $\mu\text{A/V}^2$. If the variation of the drain current must be limited to 350 μA ± 20%, calculate the values of R_{SR}, R_1, R_2, and R_D.

7.28 P A circuit for an n-channel depletion MOSFET is shown in Fig. P7.28. The transistor parameters are $V_p = -5$ V and $I_{DSS} = 10$ mA. Calculate the quiescent values of I_D, V_{DS}, and V_{GS}. Assume $R_1 = 1$ MΩ, $R_2 = 60$ kΩ, and $R_D = 1$ kΩ.

FIGURE P7.28

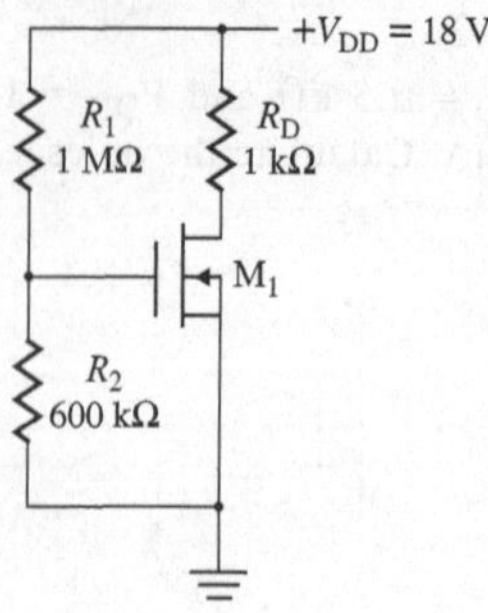

7.29 P A circuit for an n-channel enhancement-type MOSFET is shown in Fig. P7.29. The parameters of the NMOS are $V_t = 4$ V and $K_n = 1.2\ \text{mA/V}^2$. If the quiescent values are to be set at $I_D = 10$ mA and $V_{DS} = 8$ V, calculate the values of R_1, R_2, and R_D. Assume $V_{DD} = 20$ V.

FIGURE P7.29

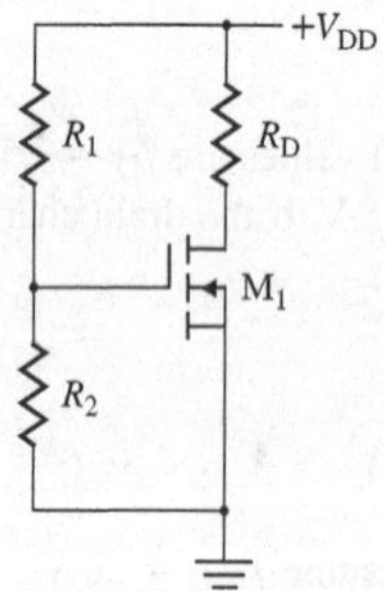

7.30 Plot the approximate transfer characteristic of the NMOS circuit of Fig. P7.30 for $V_i = 0$ to 5 V. The circuit
P parameters are $R_D = 25\ \text{k}\Omega$, $K_n = 20\ \mu\text{A/V}^2$ (v_O versus v_I), and $V_t = 2$ V.

FIGURE P7.30

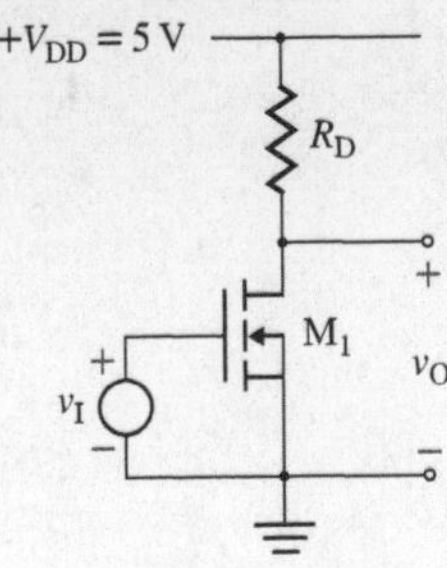

7.31 The parameters of the NMOS circuit shown in Fig. P7.31 are $K_n = 1\ \text{mA/V}^2$, $V_t = 2$ V, and $V_{DD} = 12$ V.
P Determine the values of V_o, I_D, and V_{DS}.

FIGURE P7.31

$+V_{DD} = 12$ V, M_1, R_{SR} 10 kΩ, I_D, V_O

7.32 The parameters of the NMOS circuit in Fig. P7.31 are $K_n = 1\ \text{mA/V}^2$, $V_t = 2$ V, and $V_{DD} = 12$ V. Deter-
P mine the value of R_{SR} so that $V_O = 5$ V.

7.33 The parameters of the MOSFET circuit shown in Fig. P7.33 are $K_n = 1.5\ \text{mA/V}^2$, $V_t = -2$ V, $R_{SR} = 1.5\ \text{k}\Omega$,
P and $V_{DD} = 12$ V. Determine the values of V_O, I_D, and V_{DS}.

FIGURE P7.33

$+V_{DD} = 12$ V, M_1, R_{SR}, V_O

7.8 – 7.10 MOSFET Amplifiers

7.34 The depletion NMOS amplifier of Fig. P7.34 has $R_s = 500\ \Omega$, $R_L = 10\ \text{k}\Omega$, $R_{SR} = R_D = 5\ \text{k}\Omega$, $R_G = 100\ \text{k}\Omega$,
P $I_{DSS} = 10$ mA, $V_p = -4$ V, $|V_M| = 200$ V, and $V_{DD} = 12$ V. Calculate **(a)** the input resistance $R_{in} = v_s/i_s$, **(b)** the no-load voltage gain $A_{vo} = v_o/v_g$, **(c)** the output resistance R_o, and **(d)** the overall voltage gain $A_v = v_L/v_s$.

FIGURE P7.34

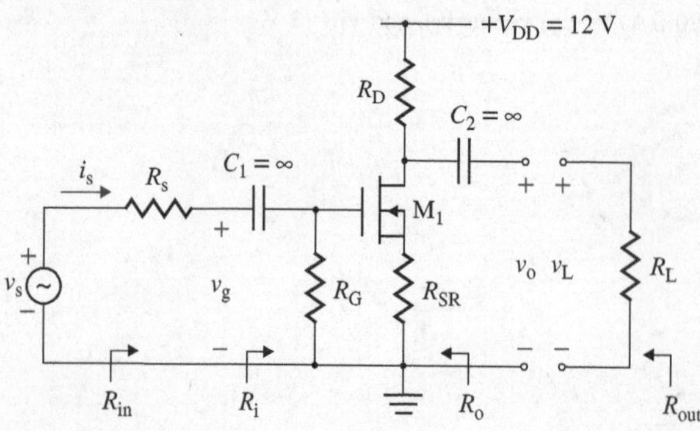

7.35 The MOSFET amplifier of Fig. P7.35 has $R_s = 500\ \Omega$, $R_D = R_L = 5\ \text{k}\Omega$, $R_{G1} = 7\ \text{M}\Omega$, $R_{G2} = 5\ \text{M}\Omega$, $K_p = 20\ \text{mA/V}^2$, $V_t = 3.5$ V, $|V_M| = 200$ V, and $V_{DD} = 12$ V. Calculate **(a)** the input resistance $R_{in} = v_s/i_s$, **(b)** the no-load voltage gain $A_{vo} = v_o/v_g$, **(c)** the output resistance R_o, and **(d)** the overall voltage gain $A_v = v_L/v_s$.

FIGURE P7.35

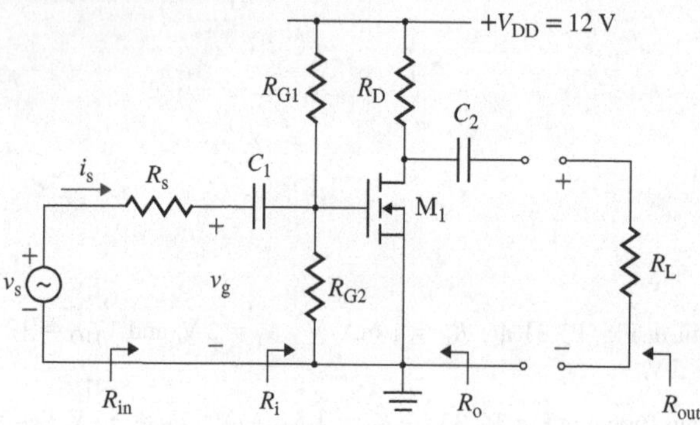

7.36 The NMOS amplifier of Fig. P7.36 has $V_{DD} = 15$ V, $R_s = 500\ \Omega$, $R_L = 10\ \text{k}\Omega$, $R_{SR} = 3\ \text{k}\Omega$, $R_D = 5\ \text{k}\Omega$, $R_{G1} = 700\ \text{k}\Omega$, $R_{G2} = 300\ \text{k}\Omega$, $V_M = -150$ V, $V_t = 2.4$ V, and $K_n = 2.042\ \text{mA/V}^2$. Calculate **(a)** the input resistance $R_{in} = v_s/i_s$, **(b)** the no-load voltage gain $A_{vo} = v_o/v_g$, **(c)** the output resistance R_o, and **(d)** the overall voltage gain $A_v = v_L/v_s$.

FIGURE P7.36

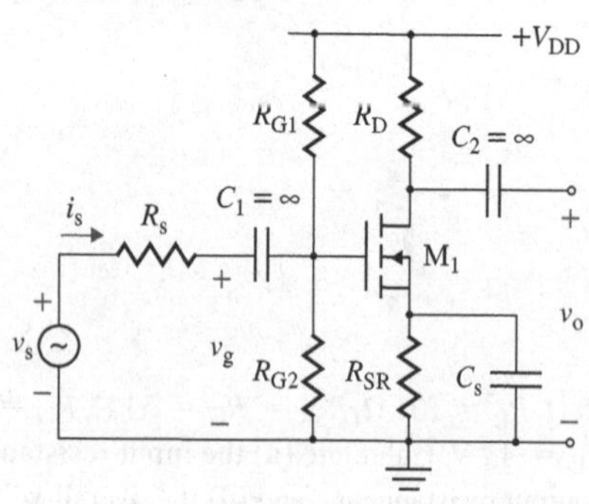

7.37 P The NMOS amplifier of Fig. P7.36 has $V_{DD} = 15$ V, $R_s = 1$ kΩ, $R_L = 5$ kΩ, $R_{SR} = 1$ kΩ, $R_D = 5$ kΩ, $R_{G1} = 400$ MΩ, $R_{G2} = 600$ MΩ, $V_M = -100$ V, $V_t = 2$ V, and $K_p = 10$ mA/V^2. Calculate **(a)** the input resistance $R_{in} = v_s/i_s$, **(b)** the no-load voltage gain $A_{vo} = v_o/v_g$, **(c)** the output resistance R_o, and **(d)** the overall voltage gain $A_v = v_L/v_s$.

7.38 The MOSFET amplifier of Fig. P7.38 has $R_s = 500$ Ω, $R_1 = 30$ kΩ, $R_2 = 50$ kΩ, $R_D = 10$ kΩ, and $R_L = 15$ kΩ. Assume $V_M = -200$ V, $V_t = 2$ V, and $K_n = 30$ mA/V^2. Calculate **(a)** the input resistance $R_{in} = v_s/i_s$, **(b)** the no-load voltage gain $A_{vo} = v_o/v_g$, **(c)** the output resistance R_o, and **(d)** the overall voltage gain $A_v = v_L/v_s$.

FIGURE P7.38

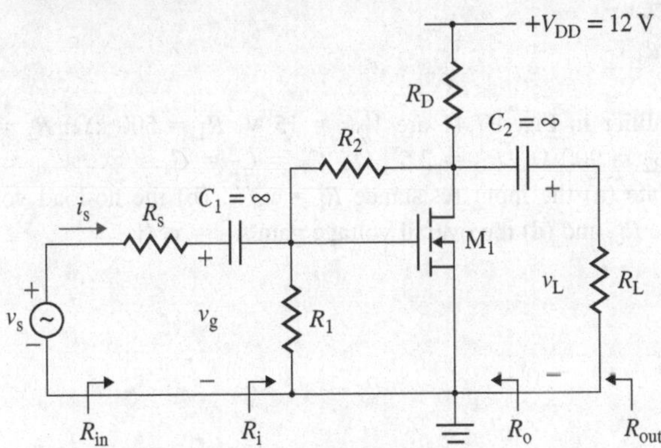

7.39 The parameters of the NMOS amplifier in Fig. P7.39 are $V_{DD} = 15$ V, $R_1 = 600$ kΩ, $R_2 = 400$ kΩ, $R_L = 20$ kΩ, $R_{SR1} = 100$ Ω, $R_{SR2} = 900$ Ω, $R_D = 2.5$ kΩ, $C_1 = C_2 = C_s \approx \infty$, $|V_M| \approx \infty$, $K_n = 1$ mA/V^2, and $V_t = 1.5$ V. Calculate **(a)** the input resistance $R_i = v_s/i_s$, **(b)** the no-load voltage gain $A_o = v_o/v_g$, **(c)** the output resistance R_o, and **(d)** the overall voltage gain $A_v = v_L/v_s$.

FIGURE P7.39

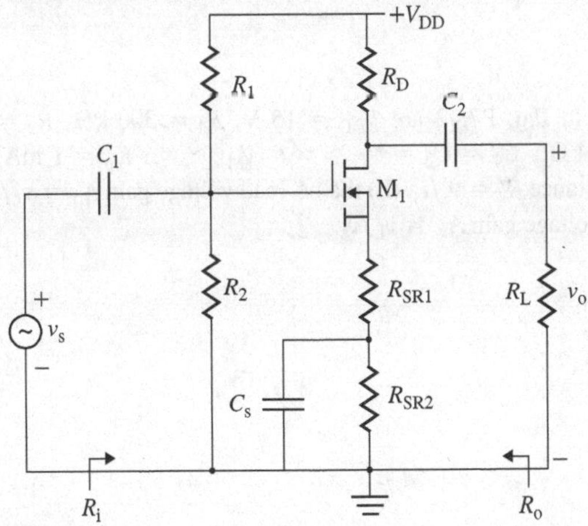

7.40 The parameters of the PMOS amplifier in Fig. P7.40 are $V_{DD} = 15$ V, $R_L = 50$ MΩ, $R_G = 500$ kΩ, $R_{SR1} = 500$ Ω, $R_D = 2.5$ kΩ, $C_1 = C_2 = C_s \approx \infty$, $|V_M| \approx \infty$, $K_n = 1$ mA/V^2, and $V_t = -1.5$ V. Calculate **(a)** the input resistance $R_i = v_s/i_s$, **(b)** the no-load voltage gain $A_o = v_o/v_g$, **(c)** the output resistance R_o, and **(d)** the overall voltage gain $A_v = v_L/v_s$.

FIGURE P7.40

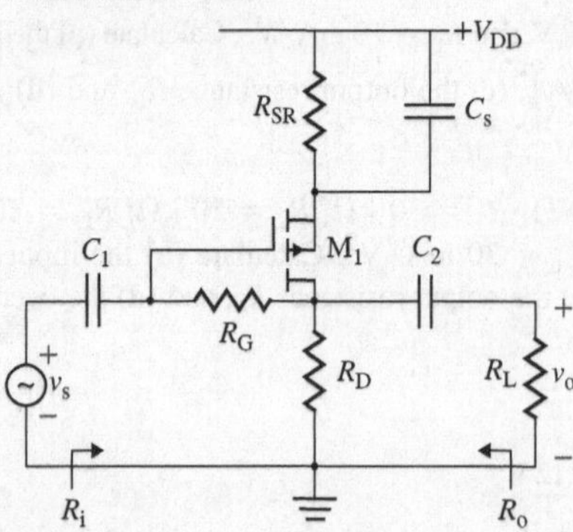

7.41 The parameters of the PMOS amplifier in Fig. P7.41 are $V_{DD} = 15$ V, $R_1 = 500$ kΩ, $R_2 = 800$ kΩ, $R_L = 20$ kΩ, $R_{SR1} = 100$ Ω, $R_{SR2} = 900$ Ω, $R_D = 2.5$ kΩ, $C_1 = C_2 = C_s \approx \infty$, $|V_M| \approx \infty$, $K_n = 1$ mA/V^2, and $V_t = -2$ V. Calculate **(a)** the input resistance $R_i = v_s/i_s$, **(b)** the no-load voltage gain $A_o = v_o/v_g$, **(c)** the output resistance R_o, and **(d)** the overall voltage gain $A_v = v_L/v_s$.

FIGURE P7.41

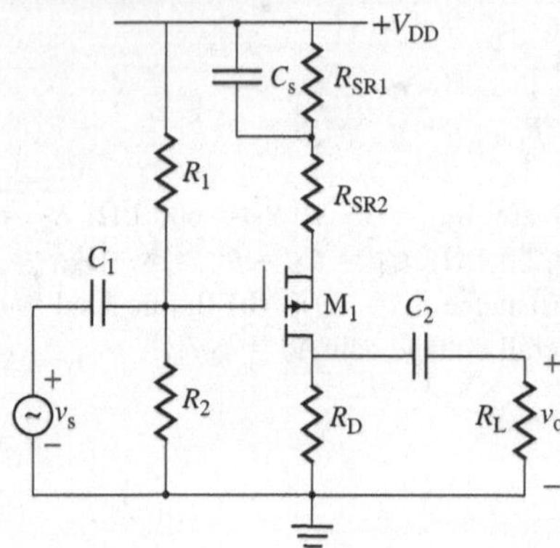

7.42 The parameters of the PMOS amplifier in Fig. P7.42 are $V_{DD} = 15$ V, $R_1 = 300$ kΩ, $R_2 = 700$ kΩ, $R_L = 20$ kΩ, $R_{SR} = 1$ kΩ, $R_D = 2.5$ kΩ, $C_1 = C_2 = C_G \approx \infty$, $|V_M| \approx \infty$, $K_p = 1$ mA/V^2, and $V_t = -1.5$ V. Calculate **(a)** the input resistance $R_i = v_s/i_s$, **(b)** the no-load voltage gain $A_o = v_o/v_g$, **(c)** the output resistance R_o, and **(d)** the overall voltage gain $A_v = v_L/v_s$.

FIGURE P7.42

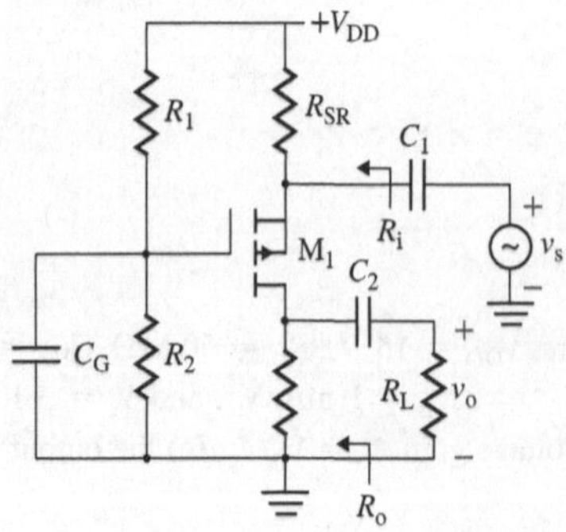

7.43 The parameters of the NMOS amplifier in Fig. P7.43 are $V_{DD} = 15$ V, $R_1 = 700$ kΩ, $R_2 = 300$ kΩ, $R_L = 20$ kΩ, $R_{SR} = 1$ kΩ, $R_D = 2.5$ kΩ, $C_1 = C_2 \approx \infty$, $|V_M| \approx \infty$, $K_n = 1$ mA/V^2, and $V_t = 1.7$ V. Calculate **(a)** the input resistance $R_i = v_s/i_s$, **(b)** the no-load voltage gain $A_o = v_o/v_g$, **(c)** the output resistance R_o, and **(d)** the overall voltage gain $A_v = v_L/v_s$.

FIGURE P7.43

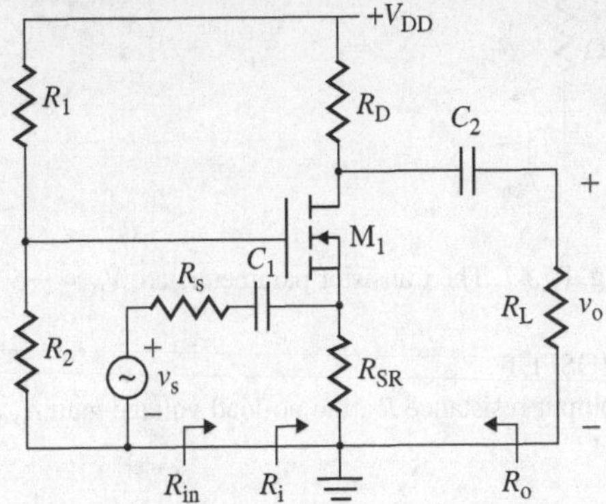

7.44 The source follower of Fig. 7.42(a) has $R_s = 1$ kΩ, $R_L = 1$ kΩ, $R_{SR} = 1$ kΩ, $R_1 = 700$ kΩ, $R_2 = 300$ kΩ, $I_{DSS} = 20$ mA, $V_p = -4$ V, $|V_M| = 200$ V, and $V_{DD} = 12$ V. Calculate **(a)** the input resistance $R_{in} = v_s/i_s$, **(b)** the no-load voltage gain $A_{vo} = v_o/v_g$, **(c)** the output resistance R_o, and **(d)** the overall voltage gain $A_v = v_L/v_s$.

7.45 The depletion MOS source follower of Fig. P7.45 has $R_s = 500$ Ω, $R_L = 10$ kΩ, $R_{SR} = 5$ kΩ, and $R_G = 10$ MΩ. Assume $V_p = -4$ V, $V_M = -100$ V, and $g_{mo} = 20$ mA/V. Calculate **(a)** the input resistance $R_{in} = v_s/i_s$, **(b)** the no-load voltage gain $A_{vo} = v_o/v_g$, **(c)** the output resistance R_o, and **(d)** the overall voltage gain $A_v = v_L/v_s$.

FIGURE P7.45

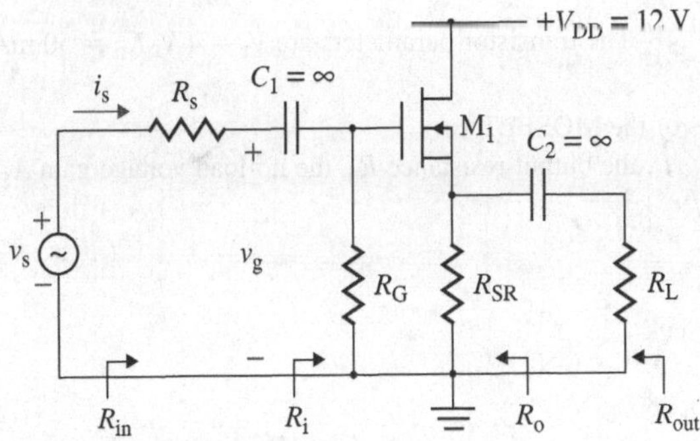

7.46 A depletion NMOS CS amplifier is shown in Fig. P7.46. The transistor parameters are $V_p = -5$ V, $I_{DSS} = 50$ mA, and $V_M = -150$ V.
P

a. Calculate the small-signal parameters of the MOSFET.

b. Calculate the input resistance $R_{in} = v_s/i_s$, the output resistance R_o, the no-load voltage gain $A_{vo} = v_o/v_g$, and the overall voltage gain $A_v = v_L/v_s$.

FIGURE P7.46

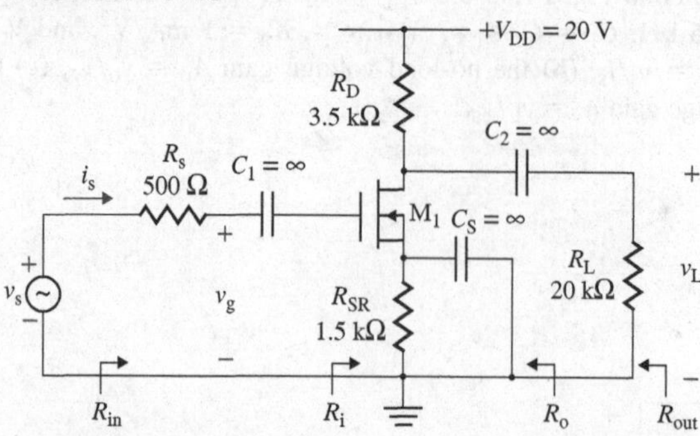

7.47 **P** A depletion NMOS source follower is shown in Fig. P7.47. The transistor parameters are $V_p = -5$ V, $I_{DSS} =$ 50 mA, and $V_M = -150$ V.

a. Calculate the small-signal parameters of the MOSFET.

b. Calculate the input resistance $R_{in} = v_s/i_s$, the output resistance R_o, the no-load voltage gain $A_{vo} = v_o/v_g$, and the overall voltage gain $A_v = v_L/v_s$.

FIGURE P7.47

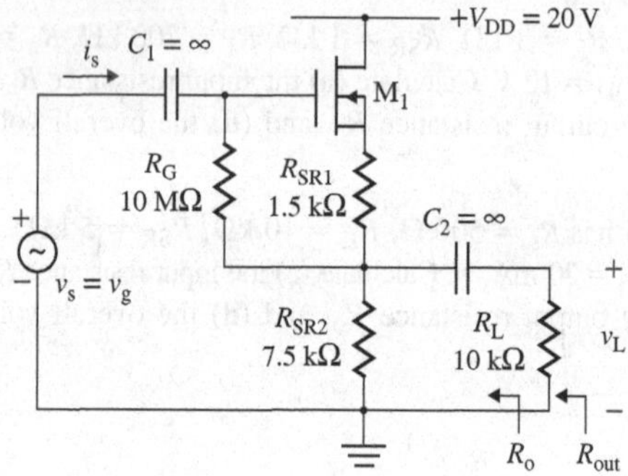

7.48 **P** An NMOS amplifier is shown in Fig. P7.48. The transistor parameters are $V_t = 4$ V, $K_n = 50$ mA/V_2, and $V_M = -150$ V.

a. Calculate the small-signal parameters of the MOSFET.

b. Calculate the input resistance $R_{in} = v_s/i_s$, the output resistance R_o, the no-load voltage gain $A_{vo} = v_o/v_g$, and the overall voltage gain $A_v = v_L/v_s$.

FIGURE P7.48

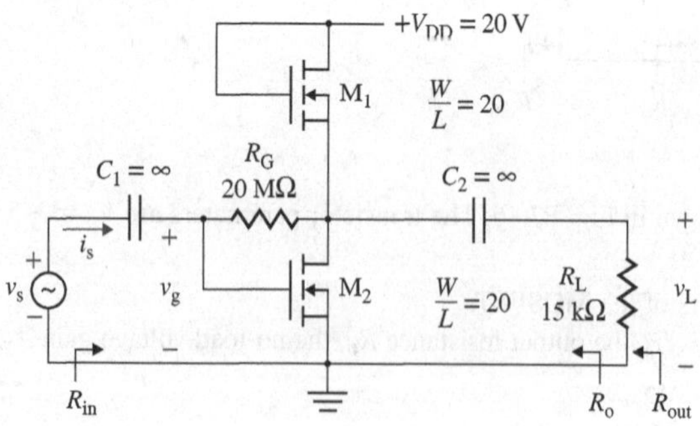

7.49 The parameters of the NMOS amplifier in Fig. P7.49 are $V_{DD} = 15$ V, $R_1 = 700$ kΩ, $R_2 = 300$ kΩ, $R_L = 20$ kΩ, $R_{SR} = 10$ kΩ, $C_1 = C_2 \approx \infty$, $|V_M| \approx \infty$, $K_n = 1$ mA/V^2, and $V_t = 1.7$ V. Calculate **(a)** the input resistance $R_i = v_s/i_s$, **(b)** the no-load voltage gain $A_o = v_o/v_g$, **(c)** the output resistance R_o, and **(d)** the overall voltage gain $A_v = v_L/v_s$.

FIGURE P7.49

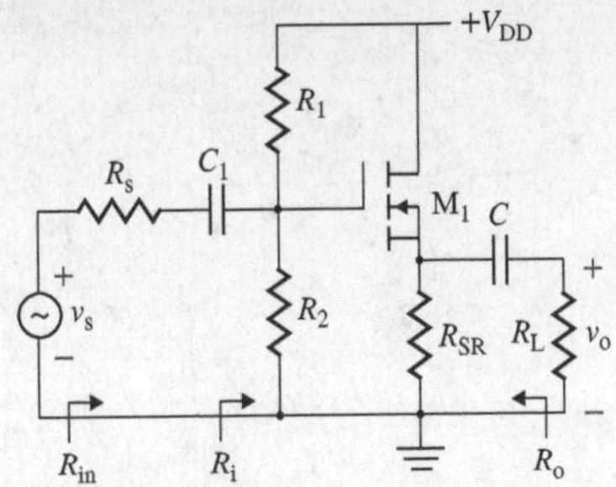

7.50 The parameters of the PMOS amplifier in Fig. P7.50 are $V_{DD} = 15$ V, $R_1 = 400$ kΩ, $R_2 = 600$ kΩ, $R_L = 20$ kΩ, $R_{SR} = 10$ kΩ, $C_1 = C_2 \approx \infty$, $|V_M| \approx \infty$, $K_n = 1$ mA/V^2, and $V_t = -2$ V. Calculate **(a)** the input resistance $R_i = v_s/i_s$, **(b)** the no-load voltage gain $A_o = v_o/v_g$, **(c)** the output resistance R_o, and **(d)** the overall voltage gain $A_v = v_L/v_s$.

FIGURE P7.50

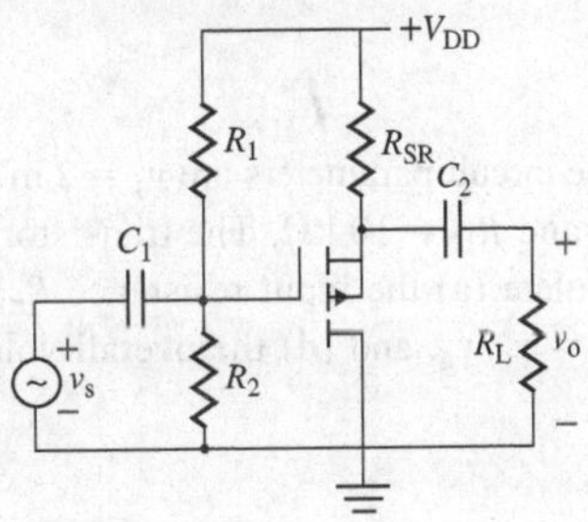

7.11 Multistage Amplifiers

7.51 The parameters of the MOS amplifier in Fig. P7.51 are $V_{DD} = 15$ V, $R_1 = 700$ kΩ, $R_2 = 300$ kΩ, $R_L = 20$ kΩ, $R_{SR1} = R_{SR2} = 1$ kΩ, $R_{D1} = R_{D2} = 2.5$ kΩ, $C_1 = C_2 \approx \infty$, $|V_M| \approx \infty$, $K_n = 1$ mA/V^2, $V_{tN} = 1.7$ V, and $V_{tP} = -2$ V. Calculate **(a)** the input resistance $R_i = v_s/i_s$, **(b)** the no-load voltage gain $A_o = v_o/v_g$, **(c)** the output resistance R_o, and **(d)** the overall voltage gain $A_v = v_L/v_s$.

FIGURE P7.51

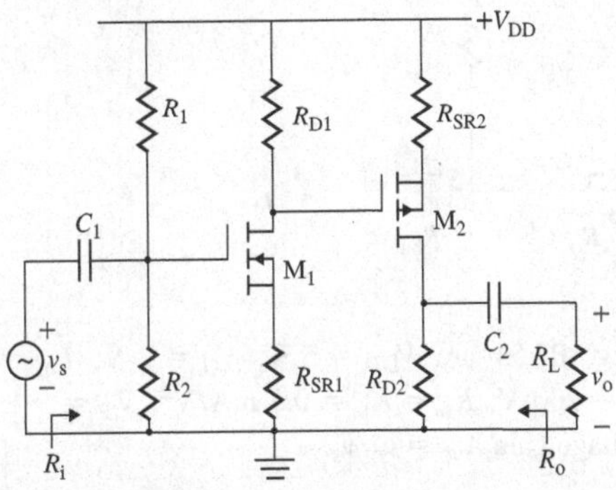

7.52 The parameters of the MOS amplifier in Fig. P7.52 are $V_{DD} = 15$ V, $R_1 = R_2 = R_3 = 500$ kΩ, $R_L = 20$ kΩ, $R_{SR} = 500$ Ω, $R_D = 2.5$ kΩ, $C_1 = C_2 = C_3 = C_4 \approx \infty$, $|V_M| \approx \infty$, $K_n = 1$ mA/V^2, and $V_t = 1$ V. Calculate **(a)** the input resistance $R_i = v_s/i_s$, **(b)** the no-load voltage gain $A_o = v_o/v_g$, **(c)** the output resistance R_o, and **(d)** the overall voltage gain $A_v = v_L/v_s$.

FIGURE P7.52

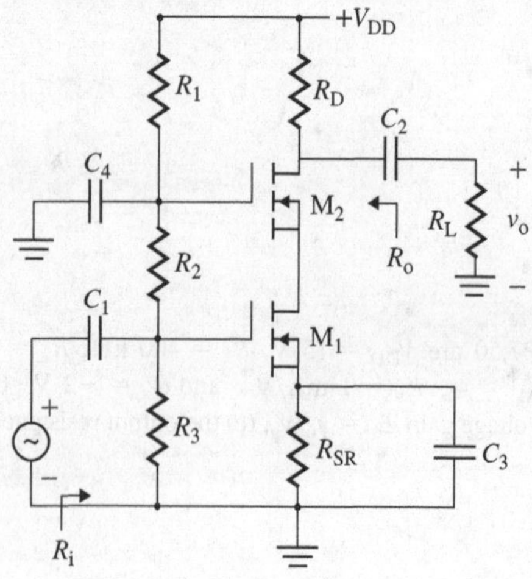

7.53 **P** A cascoded depletion MOS amplifier is shown in Fig. P7.53. The circuit parameters are $v_s = 2$ mV, $V_{DD} = 10$ V, $R_G = 20$ MΩ, $R_s = 500$ Ω, $R_{SR} = 500$ Ω, $R_D = 1$ kΩ, and $R_L = 10$ kΩ. The transistor parameters are $V_p = -4$ V, $I_{DSS} = 20$ mA, and $V_M = -150$ V. Calculate **(a)** the input resistance $R_{in} = v_s/i_s$, **(b)** the output resistance R_o, **(c)** the no-load voltage gain $A_{vo} = v_o/v_g$, and **(d)** the overall voltage gain $A_v = v_L/v_s$.

FIGURE P7.53

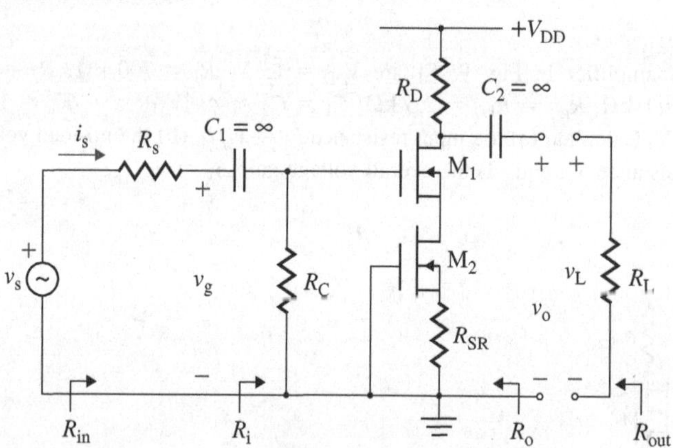

7.54 The parameters of the CMOS amplifier in Fig. P7.54 are $V_{DD} = 5$ V, $V_G = 2$ V, $I_{ref} = 0.5$ mA, $R_G = 500$ kΩ, $C_L \approx \infty$, $V_{MN} = -100$ V, $V_{MP} = 200$ V, $K_n = K_p = 0.5$ mA/V^2, $V_{tP} = -1.5$ V, and $V_{tN} = 1$ V. Calculate the small-signal no-load voltage gain $A_{vo} = v_o/v_s$.

FIGURE P7.54

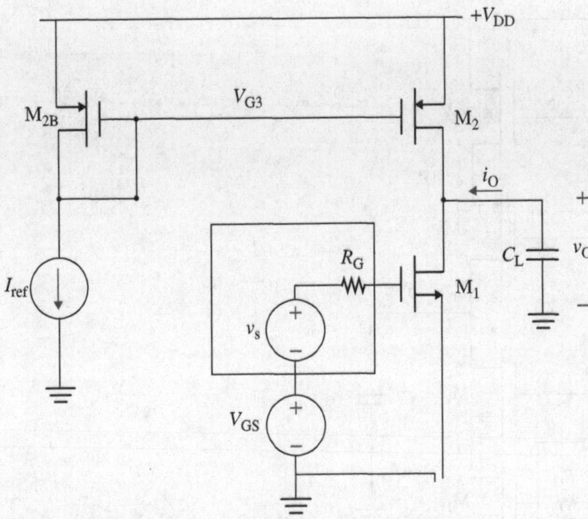

7.55 The parameters of the CMOS amplifier in Fig. P7.55 are $V_{DD} = 5$ V, $V_G = 2$ V, $I_{ref} = 0.5$ mA, $R_G - 500$ kΩ, $C_L \approx \infty$, $V_{MN} = -100$ V, $V_{MP} = 200$ V, $K_n = K_p = 0.5$ mA/V^2, $V_{tP} = -1.5$ V, and $V_{tN} = 1$ V. Calculate the small-signal no-load voltage gain $A_{vo} = v_o/v_s$.

FIGURE P7.55

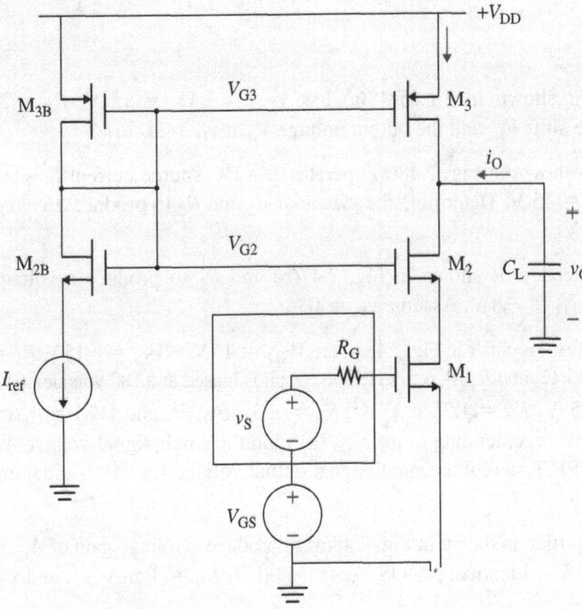

7.56 The parameters of the CMOS amplifier in Fig. P7.56 are $V_{DD} = 5$ V, $V_G = 2$ V, $I_{ref} = 0.5$ mA, $R_G = 500$ kΩ, $C_L \approx \infty$, $V_{MN} = -100$ V, $V_{MP} = 200$ V, $K_n = K_p = 0.5$ mA/V^2, $V_{tP} = -1.5$ V, and $V_{tN} = 1$ V. Calculate the small-signal no-load voltage gain $A_{vo} = v_o/v_s$.

FIGURE P7.56

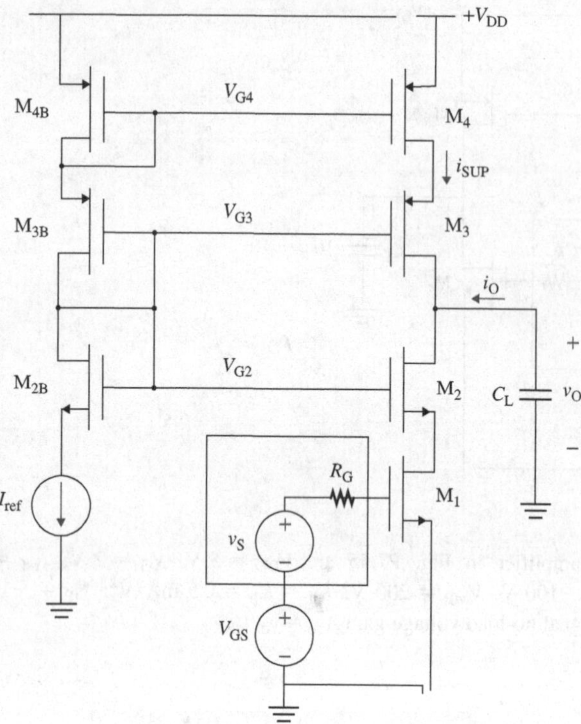

7.12 DC Level Shifting and MOS Amplifier

7.57 The potential level-shifting circuit shown in Fig. 7.48(a) has $V_{DD} = -V_{SS} = 15$ V, $R_1 = 2$ kΩ, and $R_2 = 4$ kΩ. Determine the voltage shift V_{sh} and the output voltage V_o at $v_G = 0$.

7.58 The potential level-shifting circuit shown in Fig. 7.48(a) operates at a DC source current $I_o = 1$ mA, and the DC voltages are $V_{DD} = -V_{SS} = 15$ V. Determine the values of R_1 and R_2 to produce a voltage shift of 3 V at an output voltage $v_o = -7$ V.

7.59 Determine the current source I_o needed as shown in Fig. 7.47(b) and R_1 to produce a voltage shift of $V_{sh} = 4$ V at an output voltage of $v_O = -8$ V. Assume $v_G = 0$.

7.60 The parameters of the MOS level-shifted amplifier in Fig. 7.48(a) are $V_{DD} = 15$ V, $-V_{SS} = -15$ V, $R_1 = 18$ kΩ, $R_2 = 2$ kΩ, $R_3 = 5$ kΩ, $R_x = 20$ kΩ, and $R_y = 6.5$ kΩ. The circuit is biased at a DC voltage of $V_G = 1$ V. The MOS parameters are $V_t = 1.5$ V, $K_n = 1.25$ mA/V^2, $K_p = 2.5$ mA/V^2, and $|V_M| = 1/\lambda = 200$ V. Assume the bypass capacitance C is large, tending to infinity. **(a)** Find the small-signal voltage A_{vo} and the maximum possible gain. **(b)** Use SPICE to plot the small-signal output voltage for a sinusoidal input signal of 1 mV at 1 kHz.

7.61 Design an NMOS level-shifting amplifier as shown in Fig. 7.48(a) to produce a voltage gain of $A_{vo} = 50$ V/V at a DC input signal of $v_s = 1$ mV. Use identical NMOS $V_M = -200$ V, $K_n = 1$ mA/V^2, and $V_{tN} = 1$ V. Assume $V_{DD} = 15$ V.

7.62 Design a PMOS level-shifting amplifier as shown in Fig. 7.48(a) to produce a voltage gain of $A_{vo} = 50$ V/V at a DC input signal of $v_s = 1$ mV. Use identical PMOS $V_{MP} = 200$ V, $K_n = 1$ mA/V^2, and $V_t = -1.5$ V. Assume $V_{DD} = 15$ V.

7.13 Frequency Response of MOSFET Amplifiers

7.63 A depletion NMOS is biased at $I_D = 4$ mA, $V_{DS} = 4$ V, and $V_{GS} = -2$ V. The parameters of the JFET are $C_{gs0} = 3.49$ pF, $C_{gd0} = 5.85$ pF, $g_m = 4.98$ mA/V, $r_o = 47$ kΩ, and $V_{bi} = 0.8$ V.
P

a. Calculate the capacitances of the MOSFET model in Fig. 8.48(d).

b. Find the unity-gain bandwidth ω_T.

c. Use PSpice/SPICE to generate the model parameters and plot the frequency characteristic (β_f versus frequency).

7.64 Repeat Prob. 7.63 for $I_D = 2$ mA, $V_{DS} = 4$ V, and $V_{GS} = -2.5$ V.
P

7.65 An NMOS transistor of type 2N4351 is biased at $I_D = 6$ mA, $V_{DS} = 5$ V, $V_{GS} = 8.6$ V, $V_{SB} = 1$ V, and $V_{DB} = 4$ V. The NMOS parameters are $K_p = 125\ \mu A/V^2$, $g_m = 4.98$ mA/V, $C_{gd} = 1.5$ pF, $C_{sb0} = 0.5$ pF, $C_{gs0} = 3.7$ pF at $V_{DB} = 10$ V, and $V_{bi} = 0.6$ V.
P

a. Calculate the capacitances of the MOSFET model in Fig. 8.48(d).

b. Find the unity-gain bandwidth ω_T.

7.66 Design a common-source depletion MOSFET amplifier as shown in Fig. P7.66 to give a midband gain of $20 \leq |A_{mid}| \leq 25$, $Z_{in(mid)} \geq 50$ kΩ, a low 3-dB frequency of $f_L \leq 10$ kHz, and a high 3-dB frequency of $f_H = 100$ kHz.
D P

FIGURE P7.66

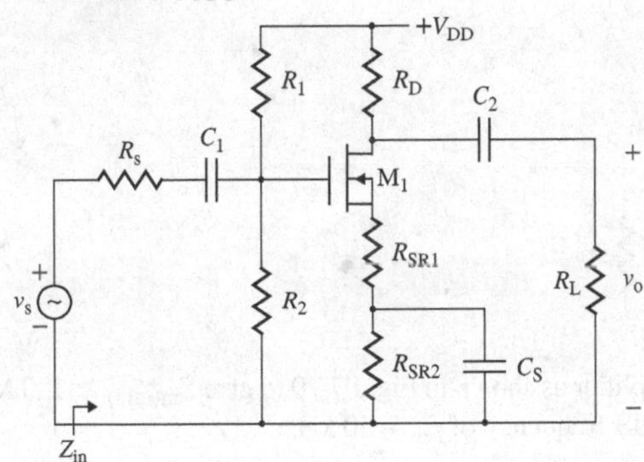

7.67 Design a common-source NMOS amplifier as shown in Fig. P7.67 to give a passband gain of $20 \leq |A_{PB}| \leq 30$, $Z_{in(mid)} \geq 100$ kΩ, a low 3-dB frequency of $f_L \leq 10$ kHz, and a high 3-dB frequency of $f_H = 200$ kHz.
D P

FIGURE P7.67

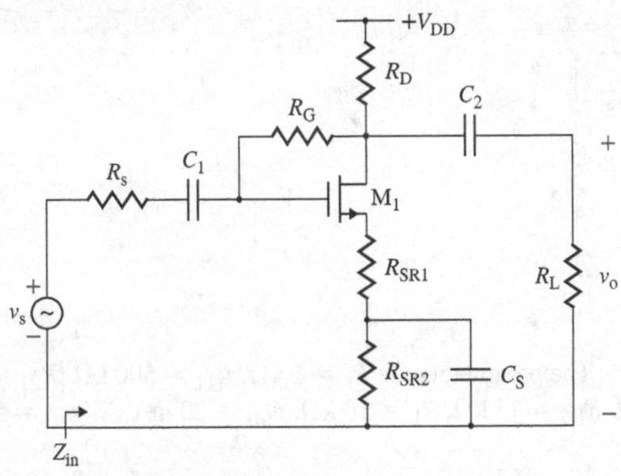

7.68 D P Design a common-source NMOS amplifier as shown in Fig. P7.68 to give a midband gain of $30 \le |A_{PB}| \le 35$, $Z_{in(mid)} \ge 100\ k\Omega$, a low 3-dB frequency of $f_L \le 20$ kHz, and a high 3-dB frequency of $f_H = 100$ kHz.

FIGURE P7.68

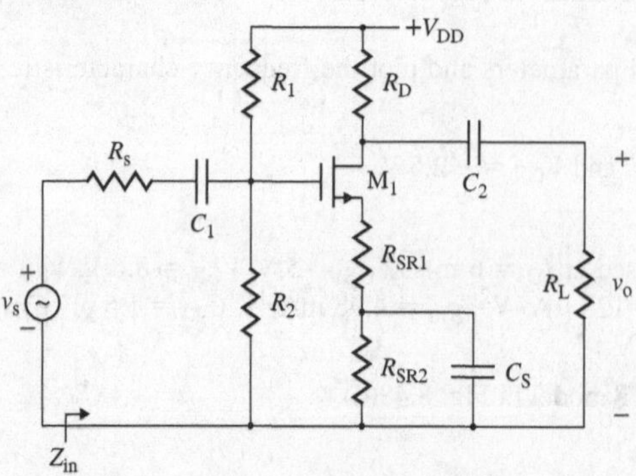

7.69 D P Design a common-drain depletion NMOS amplifier as shown in Fig. P7.69 to give $Z_{in(mid)} \ge 1\ M\Omega$, a low 3-dB frequency of $f_L \le 1$ kHz, and a high 3-dB frequency of $f_H = 50$ kHz.

FIGURE P7.69

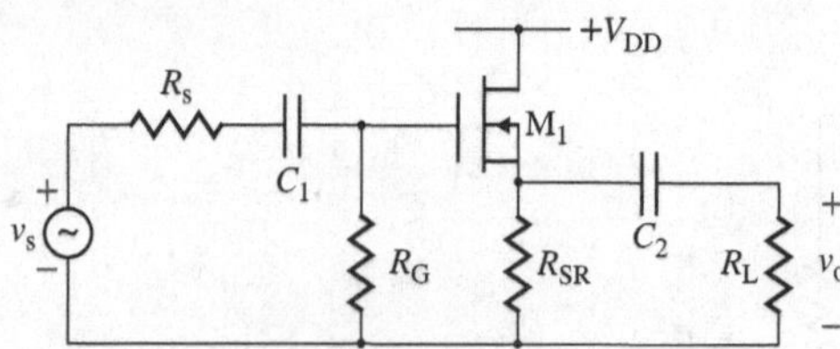

7.70 D P Design a common-drain depletion NMOS amplifier as shown in Fig. P7.70 to give $Z_{in(mid)} \ge 100\ M\Omega$, a low 3-dB frequency of $f_L \le 1$ kHz, and a high 3-dB frequency of $f_H = 50$ kHz.

FIGURE P7.70

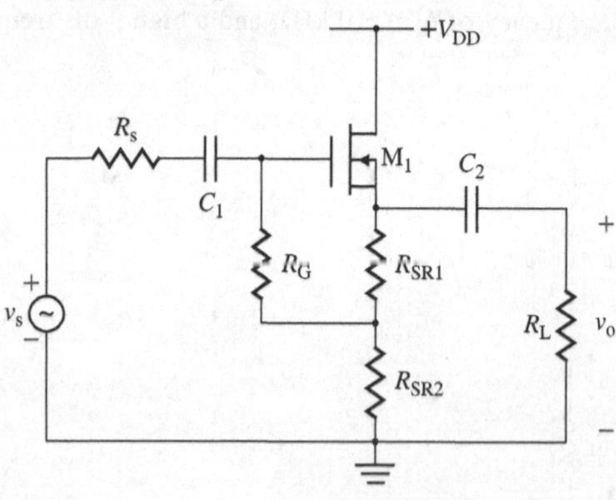

7.71 P A two-stage amplifier is shown in Fig. P7.71. The parameters are $R_s = 1\ k\Omega$, $R_{11} = 500\ k\Omega$, $R_{21} = 500\ k\Omega$, $R_{D1} = 10\ k\Omega$, $R_{12} = 500\ k\Omega$, $R_{22} = 500\ k\Omega$, $R_{D2} = 15\ k\Omega$, $R_L = 10\ k\Omega$, $g_{m1} = 20$ mA/V, $g_{m2} = 50$ mA/V,

$C_1 = 1\ \mu F$, $C_2 = 1\ \mu F$, $C_3 = 10\ \mu F$, $C_{gd1} = C_{gd2} = 2$ pF, and $C_{gs1} = C_{gs2} = 5$ pF. Calculate the low 3-dB frequency f_L and the high cutoff frequency f_H.

FIGURE P7.71

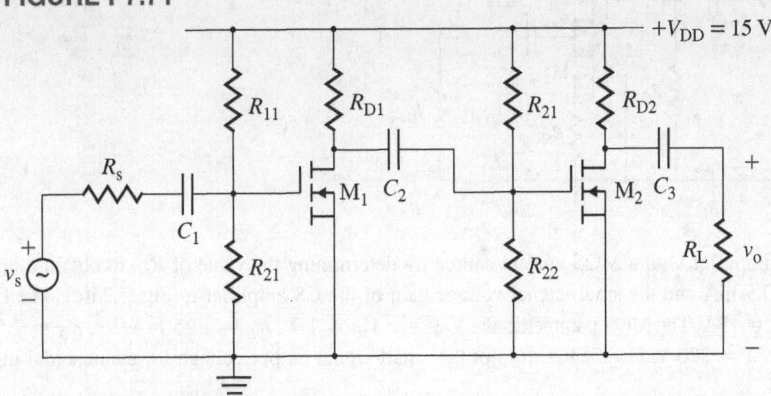

7.14 Design of MOSFET Amplifiers

7.72 Design a common-source depletion NMOS amplifier as shown in Fig. 7.65(a). The requirements are I_D = 10 mA, $A_v = -5$, and $R_i = 50$ kΩ. The FET parameters are $V_p = -4$ V, $I_{DSS} = 20$ mA, and $V_M = -200$ V. Assume $R_s = 500\ \Omega$, $V_{DD} = 20$ V, and $R_L = 50$ kΩ.

7.73 Design a common-source NMOS amplifier as shown in Fig. 7.65(b). The requirements are $A_v = -5$, R_i = 50 kΩ, and $I_D = 10$ mA. The MOSFET parameters are $V_t = 2$ V, $K_n = 40\ \text{mA/V}^2$, and $V_M = -200$ V. Assume $R_s = 0$, $V_{DD} = 20$ V, and $R_L = 50$ kΩ.

7.74 Design a common-source depletion NMOS amplifier as shown in 7.65(a). The requirements are $I_D = 20$ mA, $A_v = -4$, and $R_i = 50$ kΩ. The MOSFET parameters are $V_p = -5$ V, $I_{DSS} = 40$ mA, and $V_M = -100$ V. Assume $R_s = 500\ \Omega$, $V_{DD} = 20$ V, and $R_L = 5$ kΩ.

7.75 Design a common-source NMOS amplifier as shown in Fig. 7.65(b). The requirements are $A_v = -15$, $R_i = 10$ MΩ, and $I_D = 10$ mA. The MOSFET parameters are $V_t = 4$ V, $K_n = 50\ \text{mA/V}^2$, and $V_M = -100$ V. Assume $R_s = 1$ kΩ, $V_{DD} = 20$ V, and $R_L = 5$ kΩ.

7.76 Repeat Prob. 7.75 for the configuration shown in Fig. 7.65(c).

7.77 Design a source follower as shown in Fig. 7.42(a). The requirements are $R_i = 50$ kΩ and $I_D = 10$ mA. The MOSFET parameters are $V_p = -3$ V, $I_{DSS} = 40$ mA, and $V_M = -200$ V. Assume $R_s = 500\ \Omega$, $V_{DD} = 20$ V, and $R_L = 10$ kΩ.

7.78 Design a source follower as shown in Fig. 7.42(a) to yield $R_i = 50$ kΩ and $I_D = 10$ mA. The MOSFET parameters are $V_p = -4$ V, $I_{DSS} = 20$ mA, and $V_M = -200$ V. Assume $R_s = 0$, $V_{DD} = 20$ V, and $R_L = 10$ kΩ.

7.79 Design a cascoded amplifier as shown in Fig. P7.79 to give a voltage gain of $A_v = v_L/v_s = -5$. The MOSFET parameters are $V_p = -4$ V, $I_{DSS} = 10$ mA, and $V_M = -200$ V. Assume $V_{DD} = 15$ V and $R_s = 250\ \Omega$.

FIGURE P7.79

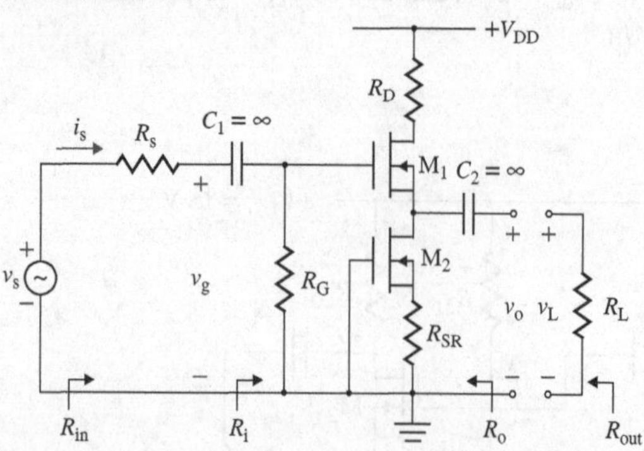

7.80 Design a CS amplifier with a MOS current source by determining the value of R_{ref} to obtain a biasing current of $I_D = 0.5$ mA and the small-signal voltage gain of the CS amplifier in Fig. 7.25(c). The DC supply voltage is $V_{DD} = 15$ V. The MOS parameters are $V_{tN} = -V_{tP} = 1$ V, $K_n = 1.25$ mA/V^2, $K_p = 2.5$ mA/V^2, and $|V_M| = 1/\lambda = 200$ V. Use SPICE to plot the small-signal output voltage for a sinusoidal input signal of 1 mV at 1 kHz.

7.81 Design a common-drain amplifier with a MOS current source as shown in Fig. 7.29(b) to bias the source follower in Fig. 7.39(a) at a drain current of $I_D = 3.25$ mA. The DC supply voltage is $V_{DD} = 15$ V, and $R_1 = 10$ kΩ. The MOS parameters are $V_t = 1.5$ V, $K_n = 1.25$ mA/V^2, $K_p = 2.5$ mA/V^2, and $|V_M| = 1/\lambda = 200$ V. Find the small-signal voltage A_{vo} and the output resistance R_o. Use SPICE to plot the small-signal output voltage for a sinusoidal input signal of 1 mV at 1 kHz.

7.82 Design a multistage NMOS amplifier to meet the following specifications: voltage gain $|A_v| = v_L/v_s = 600 \pm 5\%$ (with load), input resistance $R_i = v_s/i_s \geq 25$ kΩ, output resistance $R_o \leq 300$ Ω, load resistance $R_L = 25$ kΩ, source resistance $R_s = 1$ kΩ, DC supply $V_{DD} = 15$ V, input signal $v_s = 1$ mV to 5 mV (peak sinusoidal), 1 kHz. Use identical NMOS $V_M = -200$ V, $K_n = 1$ mA/V^2, and $V_{tN} = 1$ V. Assume $V_{DD} = 15$ V. (*Hints:* The first CS stage should meet the input resistance requirement; the third CD stage should meet the output resistance requirement; and the middle CS stage should attain the remaining gain requirement. Set the biasing drain current at $I_D = I_{D(max)}/3$ of the NMOS.)

7.83 Design a multistage PMOS amplifier to meet the following specifications: voltage gain $|A_v| = v_L/v_s = 600 \pm 5\%$ (with load), input resistance $R_i = v_s/i_s \geq 25$ kΩ, output resistance $R_o \leq 300$ Ω, load resistance $R_L = 25$ kΩ, source resistance $R_s = 1$ kΩ, DC supply $V_{DD} = 15$ V, input signal $v_s = 1$ mV to 5 mV (peak sinusoidal), 1 kHz, type 2N2222. (*Hints:* The first CS stage should meet the input resistance requirement; the third CD stage should meet the output resistance requirement; and the middle CS stage should attain the remaining gain requirement. Set the drain biasing current at $I_D = I_{D(max)}/3$ of the PMOS.)

7.84 Design a multistage depletion NMOS amplifier to meet the following specifications: voltage gain $|A_v| = v_L/v_s = 600 \pm 5\%$ (with load), input resistance, $R_i = v_s/i_s \geq 25$ kΩ, output resistance $R_o \leq 300$ Ω, load resistance $R_L = 25$ kΩ, source resistance $R_s = 1$ kΩ, DC supply $V_{DD} = 15$ V, input signal $v_g = 1$ mV to 5 mV (peak sinusoidal), 1 kHz. Use identical depletion NMOS $V_M = -200$ V, $K_n = 1.5$ mA/V^2, and $V_p = -3.5$ V. Assume $V_{DD} = 15$ V. (*Hints:* The first CS stage should meet the input resistance requirement; the third CD stage should meet the output resistance requirement; and the middle CS stage should attain the remaining gain requirement. Set the drain biasing current at $I_D = I_{D(max)}/3$ of the NMOS.)

CHAPTER 8

BIPOLAR JUNCTION TRANSISTORS AND AMPLIFIERS

Learning Outcomes

After completing this chapter, students should be able to do the following:

- Describe the operation of bipolar junction transistors (BJTs).
- List the types of bipolar transistors and their characteristics.
- List the circuit configurations of transistor amplifiers and their relative advantages and disadvantages.
- Analyze and design bipolar transistor biasing circuits.
- Determine the small-signal model parameters of bipolar transistors.
- Analyze and design bipolar transistor amplifiers.
- Design a BJT amplifier to meet certain specifications.
- Determine the low and high cutoff frequencies of bipolar transistor amplifiers.

Symbols and Their Meanings

Symbol	Meaning
$v_o(t)$, $v_O(t)$	Small-signal AC and instantaneous DC output voltages
v_b, v_B, V_B	Small-signal, instantaneous DC, and quiescent DC base voltages

Symbol	Meaning
v_{ce}, v_{CE}, V_{CE}	Small-signal AC, instantaneous DC, and quiescent DC collector–emitter voltages
v_{be}, v_{BE}, V_{BE}	Small-signal AC, instantaneous DC, and quiescent DC base–emitter voltages
i_c, i_C, I_C	Small-signal AC, instantaneous DC, and quiescent DC-collector currents
g_{ce}, r_{ce}, r_o	Small-signal collector–emitter conductance and resistance, and output resistance of a BJT
g_m, G_m	Transconductance of a BJT and an amplifier
A_{vo}, G_{mo}	No-load voltage gain and transconductance of an amplifier
λ, V_A	Modulation length and Early voltage of a BJT
β_f, β_F, α_F	Small-signal and DC forward-current gain and current ratio of a BJT
R_i, R_o, r_o	Input and output resistances of an amplifier and output resistance of a transistor

8.1 Introduction

In Chapter 2, we looked at an amplifier's characteristics from an input-output perspective and found the specifications of amplifiers that satisfied certain input and output requirements. Internally, amplifiers use one or more bipolar transistors as amplifying devices, and these transistors are biased from a single DC supply to operate properly at a desired *Q*-point. Using bipolar transistors, we can build amplifiers that give a voltage (or current) gain, a high input impedance, or a high (or low) output impedance. The terminal behavior of an amplifier depends on the types of devices used within the amplifier.

Bipolar transistors are active devices with highly nonlinear characteristics. Thus, to analyze and design a bipolar transistor circuit, we need models of transistors. Creating accurate models requires detailed knowledge of the physical operation of transistors and their parameters as well as a powerful analytical technique. A circuit can be analyzed easily using simple models, but there is generally a trade-off between accuracy and complexity. A simple model, however, is always useful to obtain the approximate values of circuit elements for use in a design exercise and the approximate performance of the elements for circuit evaluation. The details of bipolar transistor operation, characteristics, biasing, and modeling are outside the scope of this text [1–3]. In this chapter, we consider the operation and external characteristics of bipolar junction transistors using simple linear models.

8.2 Bipolar Junction Transistors

The bipolar junction transistor (BJT), developed in the 1960s, was the first device for amplification of signals. BJTs continue to play a key role in microelectronics, especially in analog electronics. Integrated circuit–fabrication techniques have led to small, high-speed devices. A BJT consists of a silicon (or germanium) crystal to which impurities have been added such that a layer of *p*-type (or *n*-type) silicon is sandwiched between two layers of *n*-type (or *p*-type) silicon. Therefore, there are two types of transistors: *npn* and *pnp*.

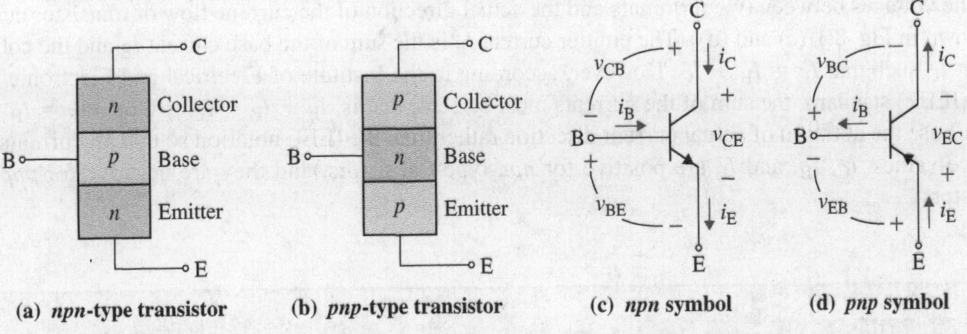

FIGURE 8.1 Basic structures and symbols of BJTs

The basic structures of *npn* and *pnp* transistors are shown in Fig. 8.1[(a) and (b)]. A BJT may be viewed as two *pn* junctions connected back to back. It is called *bipolar* because two polarity carriers (holes and electrons) carry charge in the device. A BJT is often referred to simply as a *transistor.* It has three terminals, known as the *emitter* (E), the *base* (B), and the *collector* (C). The symbols are shown in Fig. 8.1[(c) and (d)]. The direction of the arrowhead by the emitter determines whether the transistor is an *npn* or a *pnp* transistor, as illustrated in Fig. 8.1[(c) and (d)].

The block diagrams of Fig. 8.1 are highly simplified but useful to understand the concepts of basic transistor theory. The internal structure of actual bipolar transistors is more complex due to the fact that terminal connections are made at the surface, heavily doped n^+-buried layers must be included to minimize semiconductor resistances, and collector terminals of individual transistors must be isolated from each other to fabricate more than one bipolar transistor on a single piece of semiconductor material. Figure 8.2 shows a cross section of a conventional *npn* bipolar transistor fabricated in an integrated circuit configuration. In the epitaxial growth, a thin, single-crystal layer of material is grown on the surface of a single-crystal substrate, which acts as the seed, and the process takes place far below the melting temperature. The emitter and the collector regions are not symmetrical. The impurity-doping concentrations in the emitter and collector are different, and the geometry of these regions can also differ significantly.

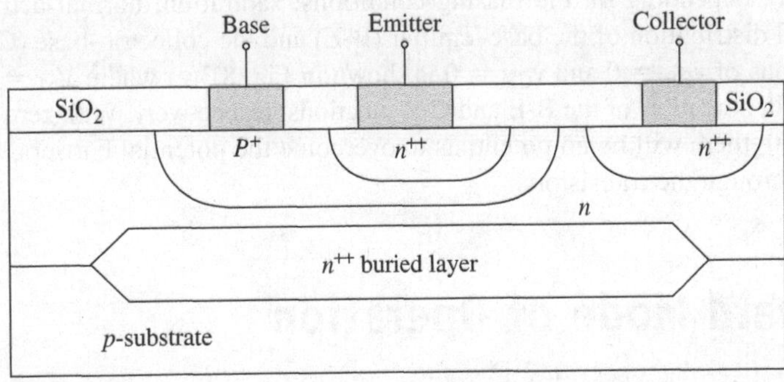

FIGURE 8.2 Cross section of a conventional integrated circuit *npn* bipolar transistor

The voltages between two terminals and the actual direction of the current-flow of transistor currents are shown in Fig. 8.1[(c) and (d)]. The emitter current I_E is the sum of the base current I_B and the collector current I_C such that $I_E = I_B + I_C$. However, according to the Institute of Electrical and Electronic Engineers (IEEE) standard, the sum of the currents must be zero; that is, $I_E + I_B + I_C = 0$ or $-I_E = I_B + I_C$. We will use the notation of actual current direction rather than the IEEE notation so that all currents have positive values. I_C, I_B, and I_E are positive for *npn*-type transistors, and they are negative for *pnp*-type transistors.

KEY POINTS OF SECTION 8.2

- The emitter and the collector regions are not symmetrical because the impurity-doping concentrations in the emitter and collector are different and the geometry of these regions can also differ significantly.
- We use the notation of actual current direction rather than the IEEE notation so that all currents have positive values. That is, I_C, I_B, and I_E are positive for *npn*-type transistors, and they are negative for *pnp*-type transistors.

8.3 Principles of BJT Operation

There are two *pn* junctions, which must be biased with external voltages to cause any current flow through any of the junctions, as discussed in Secs. 4.4 and 4.5. Recall from our discussion on semiconductor diodes that the current flows through a forward-biased *pn* junction due to the majority carriers and the current flows through a reverse-biased *pn* junction due to the minority carriers. The *npn* and *pnp* transistors are complementary devices. The principles of operation using the *npn* transistor are explained next, but the same basic principles and equations also apply to the *pnp* device. An *npn* transistor as shown in Fig. 8.3(a) is connected to two DC-voltage supplies v_{BE} and v_{CB} in order to cause a current flow. These are known as the *biasing voltages*. The transistor can operate in any of the four modes as shown in Fig. 8.3(b), depending on the biasing conditions: saturation, normal active, cutoff, and inverted. The potential distribution of the base–emitter (B-E) and the collector–base (C-B) junctions with zero-biasing conditions of $v_{BE} = 0$ and $v_{CB} = 0$ is shown in Fig. 8.3(c) where $V_{b1} = V_{bi(BE)}$ and $V_{b2} = V_{bi(CB)}$ are the built-in potentials of the B-E and C-B junctions, respectively. With zero-biasing conditions $v_{BE} = 0$ and $v_{CB} = 0$, there will be no potentials to overcome the potential barriers, and there will thus be no current flow through the transistors.

8.3.1 Forward Mode of Operation

The B-E *pn* junction is forward biased, and the base–collector (B-C) *pn* junction is reverse biased in the normal, active bias configuration as shown in Fig. 8.4(a). This configuration is called the *forward-active* operating mode. Using the *pn* junction theory developed in Sec. 6.5, the description of the device operation is as follows:

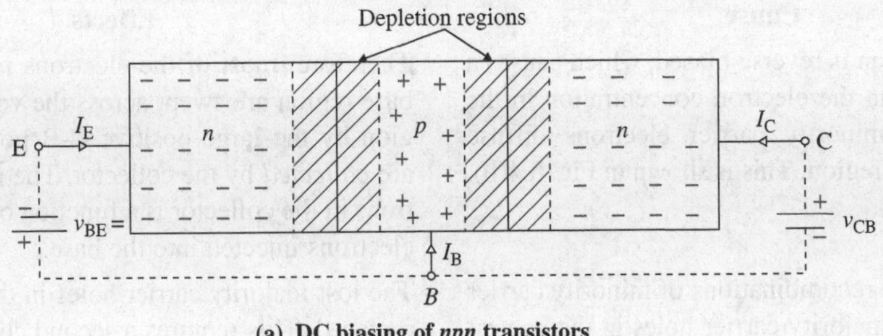

(a) DC biasing of *npn* transistors

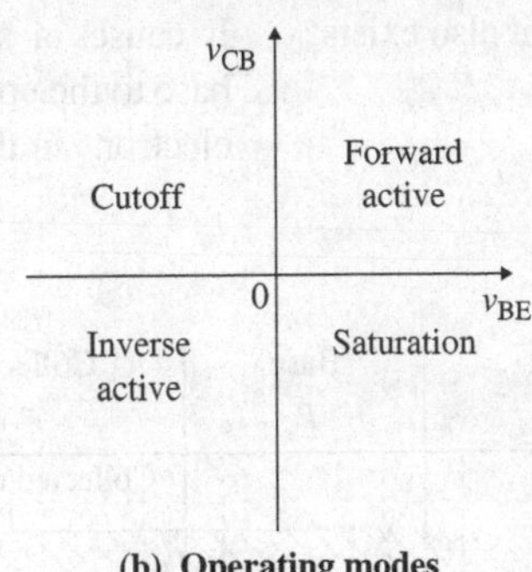

(b) Operating modes

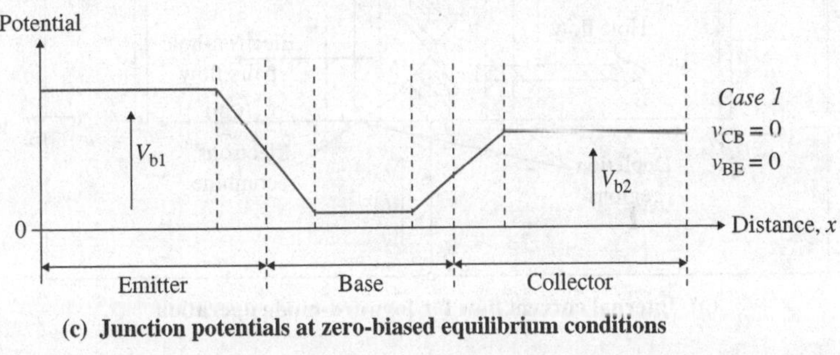

(c) Junction potentials at zero-biased equilibrium conditions

FIGURE 8.3 Biasing conditions for active-mode operation

Cause	Effects
The B-E junction is forward biased so electrons from the emitter will diffuse into the base region as shown in Fig. 8.4(a). The flow of electrons in the emitter is one major component of the emitter current.	Since the number of injected electrons involved is very much higher, an excess of electrons will be in the base region. The concentration of these minority carrier electrons is a function of the B-E voltage. Since by design, the impurity concentration in the base is very low, the number of holes in the base is very much smaller than that of electrons in the emitter, and the width of the base region is also made very small.

(continued)

Cause	Effects
The B-C junction is reverse biased, which causes a large gradient in the electron concentration in the base, so the minority carrier electrons diffuse across the base region. This is shown in Fig. 8.4(b).	Therefore, most of the electrons injected into the base region are swept across the very thin base region by the large positive C-B potential v_{CB} and are *collected* by the collector. The number of electrons in the collector is a function of the number of electrons injected into the base.
There are some recombinations of minority carrier electrons with majority carrier holes in the neutral base region.	The lost majority carrier holes in the base must be replaced. This requires a second component of the base current as shown in Fig. 8.4(a).
The reverse-biased B-C junction current also exists.	It causes a small reverse-biased current from the base to the collector I_{CBO} due to the minority carrier electrons in the base and holes in the collector.

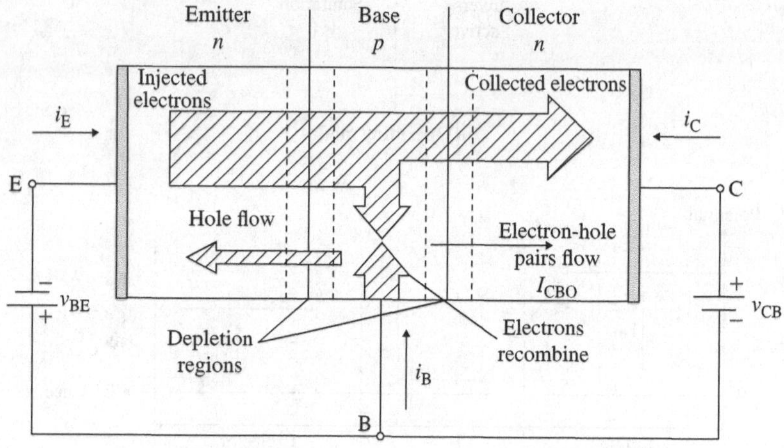

(a) Internal current flow for forward-mode operation

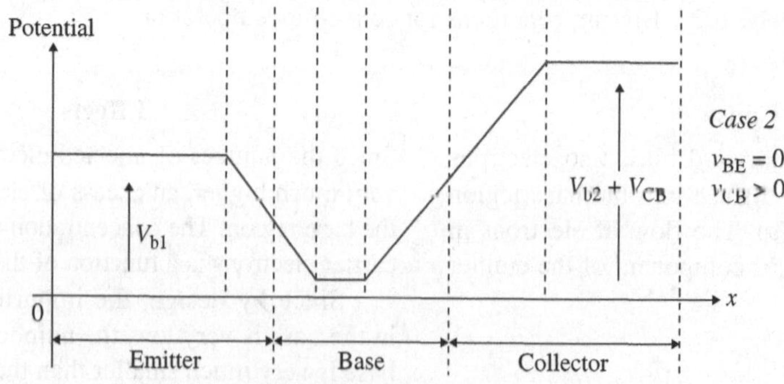

(b) Potential distribution for forward-mode conditions

FIGURE 8.4 Biasing and current flow for forward-mode conditions

Collector Current

The directions for the different components of the electron and hole currents are shown in Fig. 8.5. Due to the transistor action, the current at the collector terminal i_C is a function of the voltages v_{BE} and v_{CB} across the other two terminals. The total collector current, which is controlled by the B-E voltage, is the electron diffusion current $I_{E,n}$ minus the base electron recombination current $I_{B,n}$ and can be described by an exponential function from Eq. (6.69) as given by

$$i_C = I_{C,n} = I_{E,n} - I_{B,n} = I_S e^{v_{BE}/V_T} \tag{8.1}$$

where I_S is the saturation current, whose value ranges from 10^{-12} A to 10^{-16} A, depending on the collector saturation current density and the doping profiles and levels. V_T is the thermal voltage and equals kT/q, which is 25.8 mV at room temperature.

Emitter Current

The emitter current, as shown in Fig. 8.5, is due to the flow of electrons injected from the emitter into the base. This current, then, is ideally equal to the collector current given by Eq. (8.1). Since the B-E junction is forward biased, majority carrier holes in the base are injected across the B-E junction into the emitter. These injected holes produce a *pn* junction current $I_{E,p}$, also as indicated in Fig. 8.5. This current is only a B-E junction current, so this component of emitter current is not part of the collector current. The total emitter current is the sum of the electron diffusion current $I_{E,n}$ and the hole diffusion current. Note that there will be a B-E depletion layer recombination current $I_{B,n}$ which is negligible. The total emitter current, which is also controlled by the base–emitter voltage, can be described by

$$i_E = I_{E,n} + I_{E,p} + I_{B,n} = I_{E,n} + I_{E,p} = I_{SE}\, e^{v_{BE}/V_T} \tag{8.2}$$

where I_{SE} is the saturation current that depends on the emitter saturation current density and is related to the doping profiles and levels.

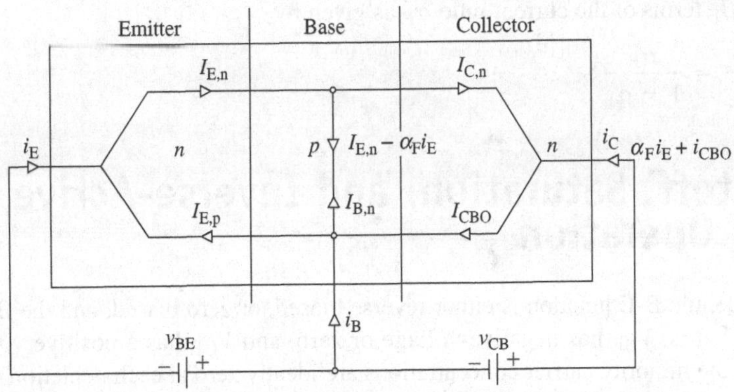

$I_{E,n}$ = Emitter current flow due to electrons
$I_{C,n}$ = Collector current flow due to electron recombination in the base
$I_{E,p}$ = Emitter current flow due to holes
$I_{B,n}$ = Base current flow due to electrons
I_{CBO} = Reverse saturation current from collector to base

FIGURE 8.5 Directions of electron and hole currents

Base Current

The base current is the sum of the hole diffusion current $I_{E,p}$ and the base recombination current $I_{B,n}$:

$$i_B = i_{B1} + i_{B2} = I_{E,p} + I_{B,n} = I_{SB}\, e^{v_{BE}/V_T} \tag{8.3}$$

where I_{SB} is the base saturation current.

Forward-Current Ratio

The forward-current ratio (or the transport factor), α_F, is defined as the ratio of the collector to the emitter current. Since all current components are functions of exp (v_{BE}/V_T), the ratio of collector current to emitter current is a constant. We can write

$$\frac{i_C}{i_E} \equiv \alpha_F \tag{8.4}$$

where α_F is the common-base forward-current ratio, $\alpha_F < 1$, but it should be as close to unity as possible. The collector current consists of two terms: (1) the dominant term being a fraction of the emitter current i_E, which is written as $\alpha_F i_E$, and (2) the second term being the reverse-biased saturation current I_{CBO} of the C-B junction diode. That is,

$$i_C = \alpha_F i_E + I_{CBO} \tag{8.5}$$

where I_{CBO} is the reverse saturation current from the collector to the base.

Forward-Current Gain

The forward-current gain β_F is defined as the ratio of the collector to the base current. Since all current components are functions of exp (v_{BE}/V_T), the ratio of collector current to the base current is also a constant. Since the base current equals the difference between the emitter and collector current, we can write the current gain β_F terms of the current ratio α_F as given by

$$\beta_F = \frac{i_C}{i_B} = \frac{\alpha_F}{1 - \alpha_F} \tag{8.6}$$

8.3.2 Cutoff, Saturation, and Inverse-Active Modes of Operation

In the cutoff mode, the B-E junction is either reverse biased, or zero biased, and the B-C junction is also reverse biased. That is, V_{BE} has negative voltage or zero, and V_{CB} has a positive voltage. For reverse-biased junctions, the minority carrier concentrations are ideally zero at each depletion edge. The potential barrier heights of both the B-E and B-C junctions are increased, so there is essentially no charge flow.

In the saturation mode, both junctions are forward biased. The B-E potential barrier is smaller than the potential barrier of the B-C junction. There is a gradient in the minority carrier concentration in the base to induce the collector current. Since both junctions are forward biased, the minority carrier concentrations are greater than the thermal equilibrium values at the depletion region edges. There will be a net flow of electrons from the emitter to the collector.

In the inverse-active mode, the B-E junction is reverse biased, and the B-C junction is forward biased. It is a mirror image of the forward-active mode. The potential barrier height of the B-E junction will increase while the potential barrier height of the B-C junction will decrease. Electrons from the collector will diffuse across the B-C junction into the base and then diffuse into the emitter. The bipolar transistor is not a symmetrical device and the characteristics will therefore be different from those of the active-mode operation. The B-C area is normally much larger than the B-E area, and as a result, not all of the injected electrons will be collected by the emitter. The relative doping concentrations in the base and collector are also different compared with those of the base and emitter. Therefore, we expect a significantly different characteristic between the forward-active and inverse-active modes of operation. The transistor is not normally operated in this mode.

8.3.3 Base Narrowing

We have assumed so far that the effective base width is essentially independent of the biasing voltages V_{BE} and V_{CB} of the emitter and collector junctions. The collector voltage affects the width of the space charge or depletion regions as shown in Fig. 8.6(a) for an *npn* transistor. Since the base region is usually lightly doped, the depletion region at the reverse-biased collector junction can extend significantly into the base region. As the collector voltage is increased, the space charge layer can take up more of the metallurgical base width w_B, and the effective base width w'_B is decreased. This effect is called *base narrowing*, or *base-width modulation*, and is known as the *Early effect* after J. M. Early, who first interpreted it.

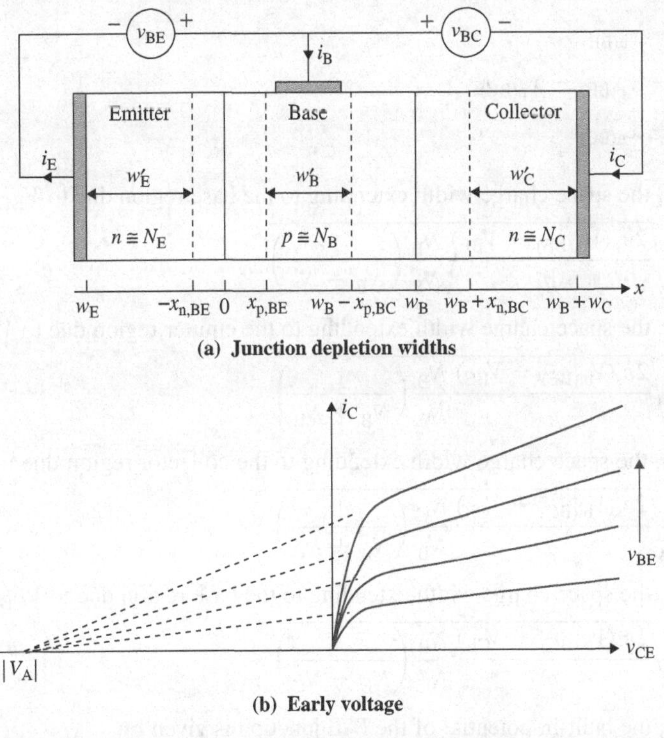

FIGURE 8.6 Effects of base junction narrowing

The decrease in the effective base width w'_B causes the collector current i_C to increase as well as the current gain β_F. As a result, the collector current i_C increases with the collector bias voltage V_{CB}. Figure 8.6(b) shows the variations of the collector current i_C against the collector–emitter voltage $V_{CE}(= V_{CB} + V_{BE})$ for various values of the B-E voltage V_{BE}. The collector current increases with an increased collector–emitter (C-E) voltage and an increased B-E voltage (or increased base current). The Early effect produces a nonzero slope of the i_C versus v_{CE} characteristics and gives a finite output conductance. For an ideal characteristic with collector current independent of the collector voltage v_{CE}, the slope of the line will be zero; thus, the output conductance will be zero. The slope introduced by the Early effect is almost linear with i_C and v_{CE} characteristics. If the collector current characteristics are extrapolated to zero collector current, the curves intercept the voltage axis at a point known as the *Early voltage*. The Early voltage V_A is positive for an *npn* transistor and negative for a *pnp* transistor. The typical values of Early voltage are in the range of 100 V to 300 V. From Fig 8.6(b), we can write the output conductance as

$$g_o = \frac{1}{r_o} = \frac{di_C}{dv_{CE}} = \frac{I_C}{V_{CE} + V_A} \tag{8.7}$$

where r_o is the output resistance of the transistor. If we include the finite slope of the i_C versus v_{CE} characteristics due to the Early effect, the collector current in Eq. (8.1) can be modified to

$$i_C = I_{SC}\, e^{v_{BE}/V_T}\left(1 + \frac{V_{CE}}{V_A}\right) \tag{8.8}$$

If w_E, w_B, and w_C are the metallurgical widths of the emitter, base, and collector regions, respectively, we can calculate their corresponding effective widths w'_E, w'_B, and w'_C as follows:

$$w'_E = w_E - x_{n(BE)} \tag{8.9}$$

$$w'_B = w_B - x_{p(BE)} - x_{p(BC)} \tag{8.10}$$

$$w'_C = w_C - x_{n(BC)} \tag{8.11}$$

Applying Eq. (6.54), the space charge width extending to the base region due to V_{BE} is given by

$$x_{n(BE)} = \sqrt{\frac{2\varepsilon_s\,(V_{bi(BE)} - V_{BE})}{q}\frac{N_E}{N_B}\left(\frac{1}{N_B + N_E}\right)} \tag{8.12}$$

Applying Eq. (6.55), the space charge width extending to the emitter region due to V_{BE} is given by

$$x_{p(EB)} = \sqrt{\frac{2\varepsilon_s(V_{bi(BE)} - V_{BE})}{q}\frac{N_B}{N_E}\left(\frac{1}{N_B + N_E}\right)} \tag{8.13}$$

Applying Eq. (6.45), the space charge width extending to the collector region due to V_{CB} is given by

$$x_{n(BC)} = \sqrt{\frac{2\varepsilon_s(V_{bi(BC)} + V_{CB})}{q}\frac{N_C}{N_B}\left(\frac{1}{N_B + N_C}\right)} \tag{8.14}$$

Applying Eq. (6.46), the space charge width extending to the base region due to V_{CB} is given by

$$x_{p(CB)} = \sqrt{\frac{2\varepsilon_s(V_{bi(BC)} + V_{CB})}{q}\frac{N_B}{N_C}\left(\frac{1}{N_B + N_C}\right)} \tag{8.15}$$

Applying Eq. (6.23), the built-in potential of the B-E junction is given by

$$V_{bi(BE)} = V_T\ \ln\left(\frac{N_B N_E}{n_i^2}\right) \tag{8.16}$$

Applying Eq. (6.23), the built-in potential of the B-C junction is given by

$$V_{bi(BC)} = V_T \ln\left(\frac{N_B N_C}{n_i^2}\right) \tag{8.17}$$

N_E, N_B, and N_C are the emitter, base, and collector doping, respectively, in negative cubic centimeters.

The total depletion, or space charge width, $w_{d(BC)}$ of the B-C junction is the sum of two depletion components given by

$$w_{d(BC)} = x_{p(CB)} + x_{n(BC)} \tag{8.18}$$

Using Eqs. (8.14) and (8.15), the width of the collector junction depletion region can be found from

$$w_{d(BC)} = \sqrt{\frac{2\varepsilon_s(V_{bi(BC)} + V_{CB})}{q}\frac{N_B}{N_C}\left(\frac{1}{N_B + N_C}\right)} + \sqrt{\frac{2\varepsilon_s(V_{bi(BC)} + V_{CB})}{q}\frac{N_C}{N_B}\left(\frac{1}{N_B + N_C}\right)} \tag{8.19}$$

$$= \sqrt{\frac{2\varepsilon_s(V_{bi(BC)} + V_{CB})}{q}\left(\frac{N_B + N_C}{N_B N_C}\right)}$$

Since $V_{CB} >> V_{bi(BC)}$, Eq. (8.19) can be simplified to

$$w_{d(BC)} = \sqrt{\frac{2\varepsilon_s}{q}\left(\frac{N_B + N_C}{N_B N_C}\right)}\sqrt{V_{CB}} \tag{8.20}$$

which shows that the depletion width is proportional to $\sqrt{V_{CB}}$. If the reverse-bias voltage on the collector junction is increased far enough, it is possible to decrease the base width w_B to the extent such that the effective base width w_B' becomes almost nonexistent. This is known as the *punch-through condition* in which the holes are swept directly from the emitter region to the collector and transistor action is lost. Punch-through is a breakdown effect that is generally avoided in circuit design.

EXAMPLE 8.1

Finding the depletion region width Calculate the width of the B-C depletion region if the C-B voltages are $V_{CB} = 2$ V, 6 V, 12 V. The physical parameters are $N_C = 2 \times 10^{16}$ cm^{-3}, $N_B = 5 \times 10^{15}$ cm^{-3}, $V_T = 25.8$ mV, $T = 25°$C, $w_B = 0.7$ μm, $\xi_s = 11.7 \times 8.85 \times 10^{-14}$, $q = 1.6 \times 10^{-19}$, and $n_i = 1.5 \times 10^{10}$ cm^{-3}.

SOLUTION

Substituting the values in Eq. (8.17), we get the built-in potential as

$$V_{bi(BC)} = V_T \ln\left(\frac{N_B N_C}{n_i^2}\right) = 25.8 \times 10^{-3}\text{ V} \ln\left[\frac{2 \times 10^{16}\text{ cm}^{-3} \times 5 \times 10^{15}\text{ cm}^{-3}}{(1.5 \times 10^{10}\text{ cm}^{-3})^2}\right] = 0.695\text{ V}$$

Substituting the values in Eq. (8.19), we get the width of the depletion region with $V_{CB} = 2$ V as

$$w_{d(BC)} = \sqrt{\frac{2 \times 11.7 \times 8.85 \times 10^{-14}(0.695 + 2)\text{ V}}{1.6 \times 10^{-19}}\left(\frac{2 \times 10^{16}\text{ cm}^{-3} + 5 \times 10^{15}\text{ cm}^{-3}}{2 \times 10^{16} \times 5 \times 10^{15}\text{ cm}^{-3}}\right)}$$

$$= 9.338 \times 10^{-5}\text{ cm} = 0.9338\ \mu\text{m}$$

For $V_{CB} = 6$ V, we get $w_{d(BC)} = 1.472 \times 10^{-4}$ cm $= 1.472$ μm. For $V_{CB} = 12$ V, we get $w_{d(BC)} = 2.027 \times 10^{-4}$ cm $= 2.027$ μm.

8.3.4 Physical Parameters of Saturation Current I_S and Current Gain β_F

Figure 8.7(a) shows the minority carrier concentrations through the *npn* transistor. The potential barrier between the emitter and the base is reduced due to the forward-biased condition discussed in Sec. 6.5, so electrons from the emitter diffuse across the B-E space charge region. The electrons diffuse across the base and are swept into the collector by the electric field in the B-C space charge region. The majority of these electrons reach the collector and create the major component of the collector current. Figure 8.7(b) shows

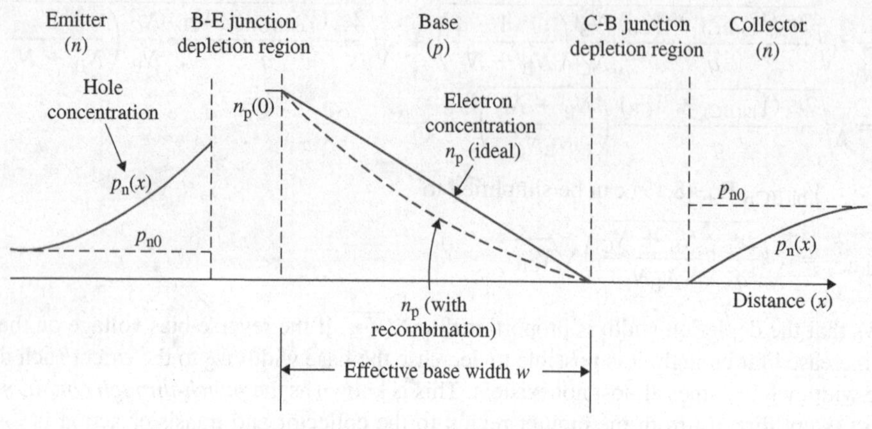

(a) **Minority carrier distribution**

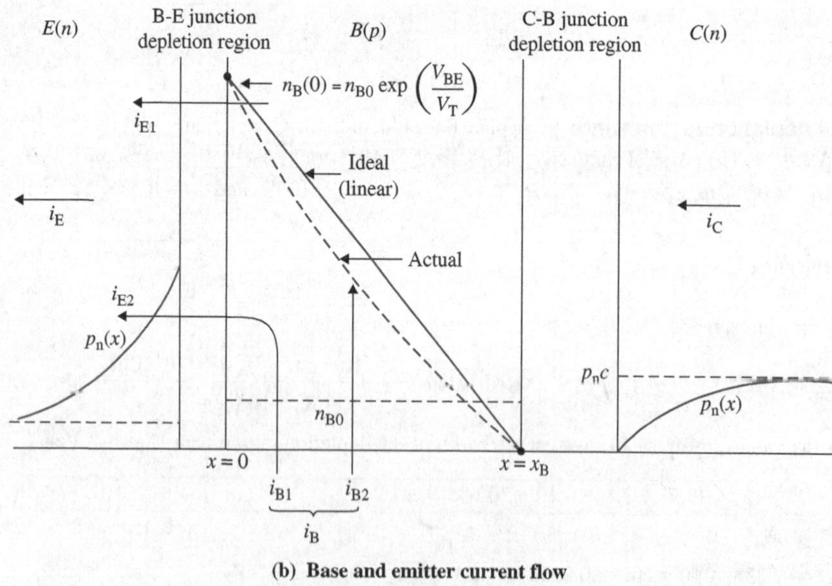

(b) **Base and emitter current flow**

FIGURE 8.7 Minority carrier distribution and current flows

the flow of the B-E currents. From Eq. (6.62), the concentration of the minority carrier electrons at the edge of the space charge in the *p*-region is given by

$$n_p(x = 0) = n_{po}e^{v_{BE}/V_T} \tag{8.21}$$

From Eq. (6.63), the concentration of the minority carrier holes at the edge of the space charge in the *n*-region is given by

$$p_n(x = 0) = p_{no}e^{v_{BE}/V_T} \tag{8.22}$$

where $n_{po} = n_i^2/N_B$ is the thermal-equilibrium concentration of the minority carrier electrons in the *p*-region base and $p_{no} = n_i^2/N_E$ is the thermal-equilibrium concentration of the minority carrier holes in the *n*-region emitter.

Collector Saturation Current I_{SC}

If we assume linear electron distribution of the electrons in the base, we find the electron diffusion current $I_{E,n}$ (in the direction of *x*), which is the major component of the emitter current, as

$$I_{E,n} = qD_BA_{BE}\frac{dn_B(x)}{dx} = qD_BA_{BE}\frac{n_B(0) - 0}{0 - x_B} = -\frac{qD_BA_{BE}}{x_B}n_{BO}e^{v_{BE}/V_T} \tag{8.23}$$

where A_{BE} is the cross-sectional area of the B-E junction, D_B is the electron diffusivity in the base, n_{BO} is the thermal-equilibrium electron concentration in the base, V_T is the thermal voltage, and q is the magnitude of the electron charge. The negative slope of the minority carrier concentration causes a negative current; that is, the actual current flows from right to left in the negative direction of *x*. By neglecting the base-electron recombination current $I_{B,n}$ in Eq. (8.1) and equating Eq. (8.23) with Eq. (8.1), we can write the *collector saturation current,* known simply as the *saturation current,* as

$$I_S = I_{SC} = \frac{qD_BA_{BE}}{x_B}n_{po} = \frac{qD_BA_{BE}}{x_B}\left(\frac{n_i^2}{N_B}\right) \tag{8.24}$$

where n_i is the intrinsic carrier density and N_B is the doping concentration in the base. It is important to note that the saturation current I_S is inversely proportional to the base width x_B and directly proportional to the area A_{BE} of the emitter–base (E-B) junction. Because I_S is proportional to n_i^2, it approximately doubles for every 5°C rise in temperature. Since I_S is a direct function of the emitter area, transistors having different emitter areas will carry different emitter currents in relation to the emitter sizes for the same amount of applied v_{BE}. For example, let us consider two transistors that are identical but one of them having the E-B junction area, say, twice that of the other. The transistor with the larger junction area will have the saturation current twice that of the smaller one. Therefore, for the same value of v_{BE}, the larger device will have a collector current twice that of the smaller device. This concept is known as *emitter scaling,* which is frequently employed in integrated circuit design.

Base Saturation Current I_{SB}

The base current will have two components: (1) the hole diffusion current $I_{E,p} = i_{B1}$ from the base to the emitter and (2) the base recombination current $I_{B,n} = i_{B2}$ in order to replace the holes lost from the base through the recombination. If we neglect the base recombination current, the base recombination current $I_{B,n}$ in Eq. (8.3), the base current approximates to the hole diffusion current $I_{E,p}$ as given by

$$i_{B1} = I_{E,p} = \frac{qD_E A_{BE}}{L_E}\left(\frac{n_i^2}{N_E}\right)e^{v_{BE}/V_T} \tag{8.25}$$

where D_E is the hole diffusivity constant in the emitter, L_E is the hole diffusion length in the emitter, and N_E is the doping concentration of the emitter.

The emitter to the base recombination current $I_{B,n}$ is due to holes that have to be supplied by the external base circuit, to replace the holes lost from the base through the recombination. If we define τ_b as the average time for a minority electron in the base to recombine with a majority hole in the base and Q_n is the minority carrier charge stored in the base that recombines with holes, we find the base current $I_{B,n}$ to replenish the holes from the external circuit as given by

$$i_{B2} = I_{B,n} = \frac{Q_n}{\tau_b} \tag{8.26}$$

where τ_b is also known as the *minority carrier lifetime.* From Fig. 8.7(b), we can find Q_n, which is approximately the area of the triangle under the straight-line charge distribution, as given by

$$Q_n = qA_{BE} \times \frac{1}{2} n_B(0) \times x_B \tag{8.27}$$

Substituting for $n_B(0)$ from Eq. (8.21) and $n_{po} = n_i^2/N_B$ into Eq. (8.27), we get

$$Q_n = qA_{BE}x_B \times \frac{1}{2}\left(\frac{n_i^2}{N_B}\right)e^{v_{BE}/V_T} \tag{8.28}$$

which, after substituting in Eq. (8.26), gives

$$i_{B2} = \frac{1}{2}\frac{qA_{BE}x_B}{\tau_b} \times \left(\frac{n_i^2}{N_B}\right)e^{v_{BE}/V_T} \tag{8.29}$$

From Eqs. (8.25) and (8.29), we can find the total base current as given by

$$i_B = i_{B1} + i_{B2} = \left[\frac{qD_E A_{BE}}{L_E}\left(\frac{n_i^2}{N_E}\right) + \frac{1}{2}\frac{qA_{BE}x_B}{\tau_b} \times \left(\frac{n_i^2}{N_B}\right)\right]e^{v_{BE}/V_T} \tag{8.30}$$

which gives the base saturation current I_{SB} as given by

$$I_{SB} = \frac{qD_E A_{BE}}{L_E}\left(\frac{n_i^2}{N_E}\right) + \frac{1}{2}\frac{qA_{BE}x_B}{\tau_b} \times \left(\frac{n_i^2}{N_B}\right)$$

Using I_S from Eq. (8.24), we can find the relation between I_{SB} and $I_S(=I_{SC})$ as

$$I_{SB} = I_S\left(\frac{D_E}{D_B}\frac{N_B}{N_E}\frac{x_B}{L_E} + \frac{1}{2}\frac{x_B^2}{D_B \tau_b}\right) \tag{8.31}$$

Current Gain β_F

Since $i_C = \beta_F i_B$ and $I_S = \beta_F I_{SB}$, Eq. (8.31) gives the current gain β_F

$$\beta_F = \left(\frac{D_E}{D_B}\frac{N_B}{N_E}\frac{x_B}{L_E} + \frac{1}{2}\frac{x_B^2}{D_B \tau_b}\right)^{-1} \tag{8.32}$$

Therefore, the value of β_F should be a constant for a particular transistor and depends on the physical parameters of a particular transistor. Its value is highly influenced by two factors: (1) the width of the base region, x_B, and (2) the relative dopings of the base region and the emitter region, N_B/N_E. To obtain a high value of β_F, the transistors should be designed by making the base thin (x_B small) and lightly doped and by making the emitter heavily doped (N_B/N_E small).

KEY POINTS OF SECTION 8.3

- A BJT can operate in any of the four operating modes depending on the biasing conditions: saturation, normal active, cutoff, and inverted. For an amplification, the B-E junction is forward biased and the C-B junction is reverse biased, while for operation in the saturation region, both B-E and C-B junctions are forward biased.
- The major physical parameters of a BJT are the forward current gain, the forward current ratio, the saturation current, and the Early voltage.
- The collector voltage affects the width of the space charge or depletion regions and the width of the depletion region depends on the C-B voltage.

8.4 Input and Output Characteristics

To properly initiate current flow, a transistor must be biased. Figure 8.8 illustrates an example of biasing using two DC supplies, V_{CC} and V_{BB}. This arrangement is not used in practice; it is shown only to illustrate the transistor characteristics. A practical biasing circuit uses only one DC supply for transistor biasing; this arrangement is discussed later in this section. R_C serves as a load resistance.

However, the arrangement shown in Fig. 8.8[(a) or (b)] is useful in the development of the concept of transistor models and signal amplification.

Each of the three terminals of a transistor may be classified as an input terminal, an output terminal, or a common terminal. There are three possible configurations: (1) common emitter (CE), in which the emitter is the common terminal; (2) common collector (CC) or emitter follower, in which the collector is the common terminal; and (3) common base (CB), in which the base is the common terminal.

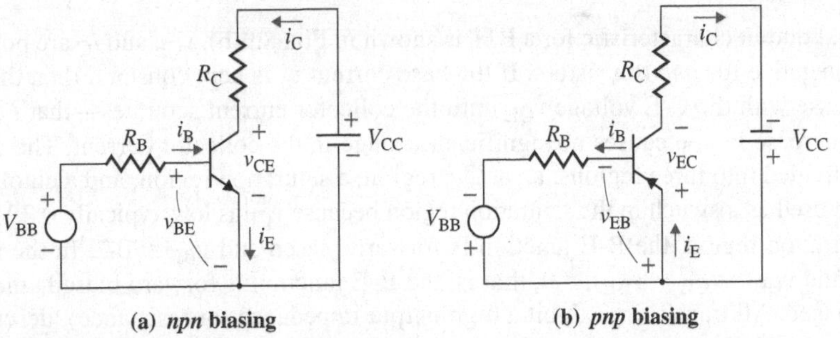

(a) *npn* biasing

(b) *pnp* biasing

FIGURE 8.8 Biasing of transistors

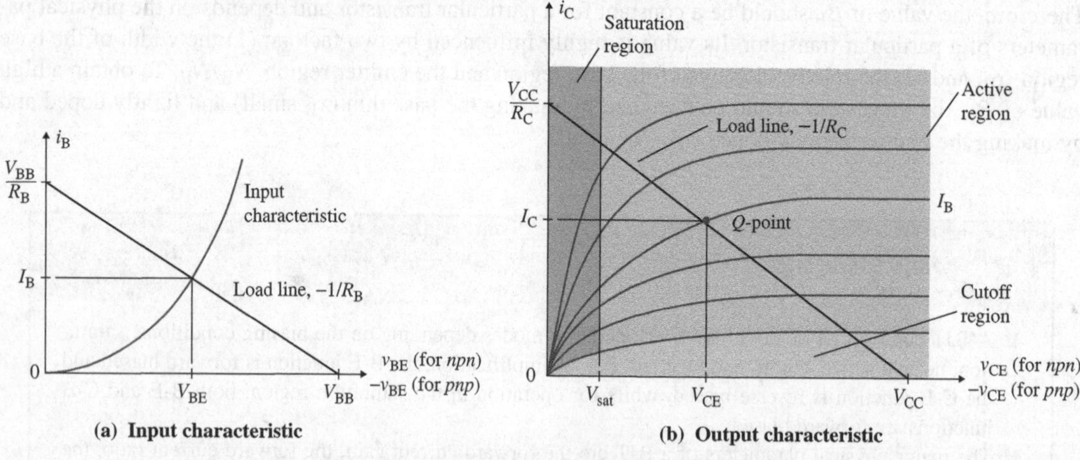

FIGURE 8.9 Input and output characteristics

The CB configuration is not as commonly used as the other two. A transistor can be described by two characteristics: an input characteristic and an output characteristic. The input characteristic is similar to that of a forward-biased diode if the emitter is the common terminal; the input characteristic for *npn* and *pnp* transistors is shown in Fig. 8.9(a), which can be described mathematically by Eq. (8.33) as follows:

$$i_B = \left(\frac{I_S}{\beta_F}\right) e^{v_{BE}/V_T} \tag{8.33}$$

Applying Kirchhoff's voltage law (KVL) as the base loop, we write

$$V_{BB} = R_B i_B + v_{BE} \tag{8.34}$$

which can be solved for the base current i_B as given by

$$i_B = \frac{V_{BB} - v_{BE}}{R_B} \tag{8.35}$$

Equation (8.35), which describes the base load line for the input characteristic as shown in Fig. 8.9(a), gives $v_{BE} = 0$ at $i_B = V_{BB}/R_B$ and $v_{BE} = V_{BB}$ at $i_B = 0$. The intersection of the base load line with the input characteristic gives the base operating point defined by I_B and V_{BE}. Equations (8.33) and (8.35) can be solved to find the DC biasing B-E voltage V_{BE} and also the DC base current I_B for known values of V_{BB} and R_B.

A typical output characteristic for a BJT is shown in Fig. 8.9(b). v_{CE} and i_C are positive for *npn* transistors and negative for *pnp* transistors. If the base current i_B is kept constant, then the collector current i_C will increase with the C-E voltage v_{CE} until the collector current saturates—that is, reaches a level at which any increase in v_{CE} causes no significant change in the collector current. The output characteristic may be divided into three regions: an active region, a saturation region, and a cutoff region. The transistor can be used as a switch in the saturation region because v_{CE} is low, typically 0.3 V. In both the active and the saturation region, the B-E junction is forward biased and $v_{BE} \approx 0.7$. In the active region, $0 < v_{BE} < v_{CE}$ and $v_{CB}(=v_{CE} - v_{BE}) > 0$; that is, the B-E junction is forward biased, and the C-B junction is reverse biased. All transistors exhibit a high output impedance (or resistance), described by Eq. (8.7). Operation in the active region can give an amplification of signals with a minimum amount of distortion, because the output characteristic is approximately linear.

A transistor is a current-controlled device. The collector current i_C is related to the base current i_B by a *forward-current amplification factor* β_F, which is defined as

$$\beta_F = \left.\frac{i_C}{i_B}\right|_{v_{CE}=\text{constant}} \tag{8.36}$$

Once base current i_B is determined, the collector current i_C and the emitter current i_E can be found as follows:

$$i_C = \beta_F i_B \tag{8.37}$$

$$i_E = i_B + i_C = i_B + \beta_F i_B = (1 + \beta_F) i_B \tag{8.38}$$

Using KVL around the loop formed by V_{CC}, R_C, and the collector-emitter, we can relate the collector current i_C to v_{CE} by

$$V_{CC} = v_{CE} + i_C R_C$$

which gives the dependence of the collector current on the load resistance R_C and which can be rearranged to yield the following relation, known as the *load-line equation*:

$$i_C = \frac{V_{CC}}{R_C} - \frac{V_{CE}}{R_C} \tag{8.39}$$

Equation (8.39) gives $v_{CE} = 0$ at $i_C = V_{CC}/R_C$ and $v_{CE} = V_{CC}$ at $i_C = 0$. The intersection of the load line with the output characteristic gives the operating point (or *Q*-point), which is defined by three parameters: I_B, I_C, and V_{CE}. Thus, for a given value of i_B, the value of i_C can be found, and then the load line gives the value of v_{CE}, as shown in Fig. 8.9(b).

KEY POINTS OF SECTION 8.4

- Each of the three terminals of a transistor may be classified as an input terminal, an output terminal, or a common terminal. There are three possible configurations: (1) common emitter (CE), in which the emitter is the common terminal; (2) common collector (CC) or emitter follower, in which the collector is the common terminal; and (3) common base (CB), in which the base is the common terminal.
- The output characteristic of a BJT can be divided into three regions: (1) a cutoff region in which the transistor is off, (2) an active region in which the transistor exhibits a high output resistance and has a current amplification, and (3) a saturation region in which the transistor offers a low resistance.

8.5 BJT Circuit Models

The purpose of an amplifier is to convert an input signal of small amplitude into an output signal of different amplitude while minimizing any distortion introduced by the amplifier. If the input is a sine wave, the output should also be a sine wave. If an AC small-signal v_{be} is superimposed on the DC biasing voltage V_{BE} at the base of the transistor, the base current I_B will change by a small amount i_b, thereby causing an amplified change i_c ($\approx i_b$ times the current gain) in the collector current I_C. This change will cause the operating point to move up and down along the load line around the Q-point. Too large an AC signal, however, will drive the transistor both into the saturation region (to the left of the v_{CE}-axis) and into the cutoff region (to the right of the v_{CE}-axis). Therefore, the design and analysis of an amplifier involves two signals: a DC signal and an AC signal. The DC analysis finds the Q-point defined by I_C, I_B, and V_{CE}. For an AC analysis, a *small-signal AC model* of a BJT around the Q-point is required.

8.5.1 Linear DC Model

Linear DC models are used for determining the operating point (or Q-point) of a BJT. The B-E junction, which is forward biased in the active region, can be represented by a forward-biased diode, as shown in Fig. 8.10(a). The C-B junction, which is reverse biased, can be represented by an open circuit. The base current varies with the base-to-emitter voltage, as shown in the input characteristic in Fig. 8.9(a). The input characteristic is replaced by a piecewise linear model with resistance R_{BE} in series with a voltage source V_{BE} whose value ranges from 0.5 V to 0.8 V, as shown in Fig. 8.10(b). The finite slope of the output characteristic can be represented by adding an output resistor r_o between the collector and emitter terminals. For most applications, this model can be approximated by Fig. 8.10(c) by assuming $R_{BE} = 0$ and $r_o = \infty$. It is commonly used for obtaining quick results.

8.5.2 Small-Signal AC Model

Linear DC models are used for determining the Q-point; however, an AC model is used for determining the voltage or power gain when the transistor is operated as an amplifier in the active region. If we apply a small sinusoidal input voltage $v_{be} = V_m \sin \omega t$ while operating in the active region, the base potential will be $v_{BE} = V_{BE} + v_{be}$, and the corresponding base current will be $i_B = I_B + i_b$. The corresponding collector current will be $i_C = I_C + i_c$, as shown in Fig. 8.11(a). The small-signal AC resistance r_π seen by v_{be} will be the inverse slope of the $i_B - v_{BE}$ characteristic at the Q-point (I_B, V_{BE}), as shown in Fig. 8.11(b). That is, we can obtain r_π by differentiating i_B:

$$\frac{1}{r_\pi} = \frac{i_b}{v_{be}} = \left.\frac{di_B}{dv_{BE}}\right|_{\text{at } Q\text{-point}} = \frac{I_B}{V_T} = \frac{I_B}{25.8 \text{ mV}} \tag{8.40}$$

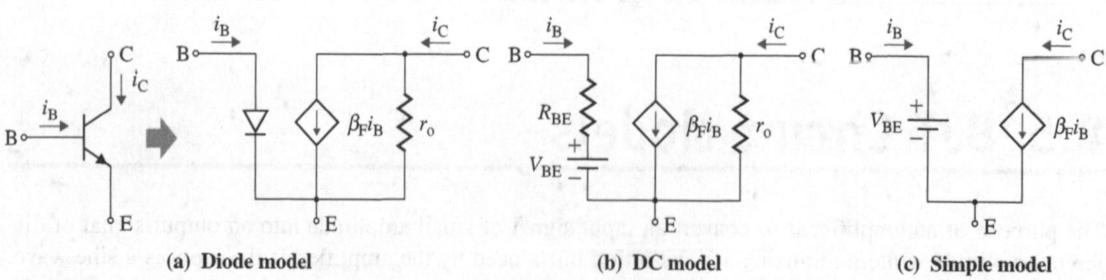

FIGURE 8.10 Linear DC models of bipolar transistors

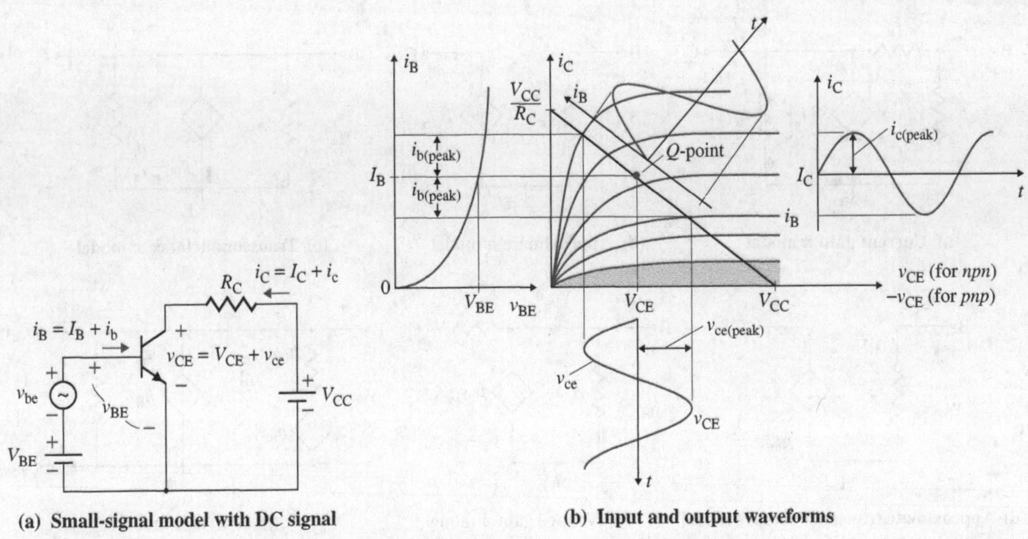

FIGURE 8.11 BJT with a small-signal input voltage

If the base current i_B swings between $I_B + i_{b(peak)}$ and $I_B - i_{b(peak)}$, the collector current i_C will swing between $I_C + i_{c(peak)}$ and $I_C - i_{c(peak)}$. The C-E voltage v_{CE} will vary accordingly from $V_{CE} - v_{ce(peak)}$ to $V_{CE} + v_{ce(peak)}$, as illustrated also in Fig. 8.11(b). The small-signal collector current i_c will depend on the small-signal AC current gain β_f, defined by

$$\beta_f = \frac{i_c}{i_b} = \frac{\Delta i_C}{\Delta i_B}\bigg|_{\text{at } Q\text{-point}} \tag{8.41}$$

which may be considered approximately equal to the DC current gain β_F for most applications. That is, $\beta_F = \beta_f$. We will make this assumption throughout.

The collector current can be related to the B-E voltage by transconductance g_m, defined by

$$g_m = \frac{i_c}{v_{be}} = \frac{\beta_F di_B}{dv_{BE}}\bigg|_{\text{at } Q\text{-point}} = \frac{\beta_f I_B}{V_T} = \frac{I_C}{V_T} = \frac{\beta_f}{r_\pi} \tag{8.42}$$

where the derivative is evaluated at the Q-point. The output characteristic in the active region exhibits a finite slope representing an output resistance defined by Eq. (8.7)

$$\frac{1}{r_o} = \frac{i_c}{v_{ce}} = \frac{di_C}{dv_{CE}}\bigg|_{\text{at } Q\text{-point}} = \frac{I_C}{V_A + V_{CE}} = \frac{I_C}{V_A} \quad \text{for } V_A >> V_{CE} \tag{8.43}$$

where V_A is a constant called the *Early voltage* whose value ranges from 100 V to 200 V, depending on the transistor [4]. The value of r_o is large (on the order of 50 kΩ) and can be neglected for most analyses.

Any increase in V_{CE} will increase the width of the collector depletion layer; consequently, the effective base width will be reduced, causing a reduction in I_B. The decrease in I_B due to an increase in V_{CE} can be modeled by a *C-B resistance* r_μ. The value of r_μ can be approximated by $r_\mu = 10 r_o \beta_f$, which is very large compared to r_π and r_o and is not normally included in the transistor model, especially for hand calculations.

Thus, the small-signal behavior of a transistor can be modeled by an input resistance r_π, a base current–dependent collector current $i_c = \beta_f i_b$ along with an output resistance r_o, and a C-B resistance r_μ. Since the C-B junction is reverse biased, r_μ can be neglected by assuming $r_\mu = \infty$. This model, shown in Fig. 8.12(a), can be approximated by Fig. 8.12(b). The transconductance representations are

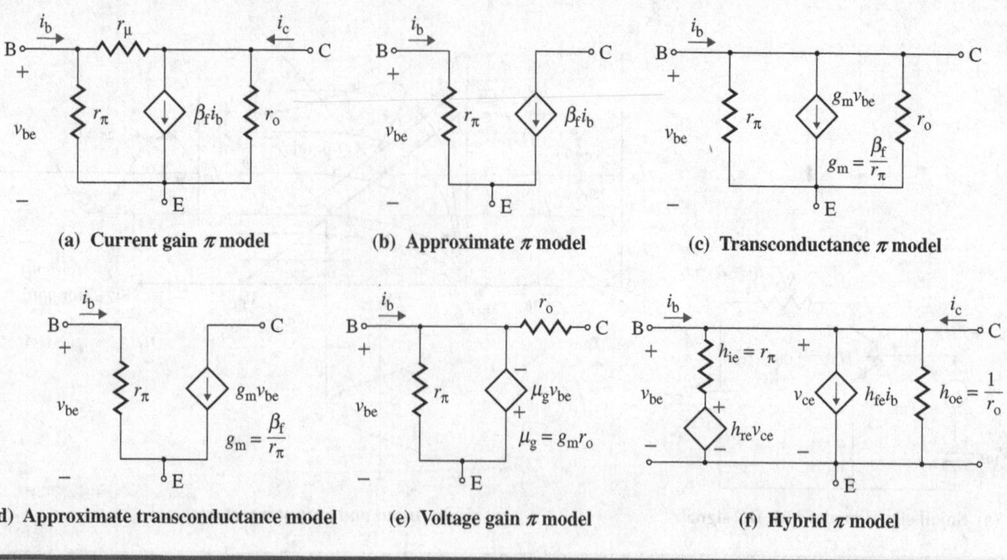

FIGURE 8.12 Small-signal AC model of a BJT

shown in Fig. 8.12[(c) and (d)]. If Norton's current source is converted to Thevenin's voltage source, Fig. 8.12(c) can be represented by Fig. 8.12(e), where $\mu_g = g_m r_o$. Note that the units of the model parameters in Fig. 8.12(a) are different.

▶ **NOTE** r_π is the small-signal base–emitter resistance r_{be}. It uses the subscript π because it is the input resistance of the model, which looks like the symbol π and is also known as the π model.

8.5.3 Small-Signal Hybrid Model

The manufacturers of BJTs usually specify the common-emitter hybrid parameters corresponding to the hybrid model shown in Fig. 8.12(f). The parameters are as follows: (See also Appendix C.)

h_{ie} ($\equiv r_\pi$) is the *short-circuit input resistance* (or simply the input resistance).

h_{fe} ($\equiv \beta_f$) is the *short-circuit forward-transfer current ratio* (or small-signal current gain).

h_{re} is the *open-circuit reverse-voltage ratio* (or voltage-feedback ratio), which takes into account the effect of v_{CE} on i_B. This ratio is very small; its value is typically 0.5×10^{-4}. r_μ represents the effect of h_{re}.

h_{oe} ($\equiv 1/r_o$) is the *open-circuit output admittance* (or simply the output admittance) of the C-E junction. It is also very small; its value is typically 10^{-6} ℧.

Often h_{re} and h_{oe} can be omitted from a circuit model without significant loss of accuracy, especially in hand calculations. The subscript e on the h parameters indicates that these hybrid parameters are derived for a common-emitter configuration.

8.5.4 PSpice/SPICE Model

PSpice/SPICE generates a complex BJT model, provided a number of physical parameters are given. The symbol for a BJT is Q, and it is described by the statement [5]

```
Q<name>  NC  NB  NE  QMOD
```

where NC, NB, and NE are the collector, base, and emitter nodes, respectively. QMOD is the model name, which can be up to eight characters long. The model statement for an *npn* transistor has the general form

```
.MODEL  QMOD  NPN (P1=A1 P2=A2 P3=A3 .......PN=AN)
```

The model statement for a *pnp* transistor has the general form

```
.MODEL  QMOD  PNP (P1=A1 P2=A2 P3=A3 .......PN=AN)
```

In these model statements, NPN and PNP are the type symbols for *npn* and *pnp* transistors, respectively. P1, P2, . . . , PN and A1, A2, . . . , AN are the parameters and their values, respectively.

As an example, let us derive two parameters, I_S and β_F, for transistor Q2N2222. Reading from the plot of v_{BE} versus i_C on the data sheet for Q2N2222, we get $v_{BE} = 0.7$ V at $i_C = 20$ mA. Inserting these values into Eq. (8.1) yields

$$20 \text{ mA} = I_S \exp\left(\frac{0.7 \text{ V}}{25.8 \text{ mV}}\right)$$

which gives $I_S = 3.295 \times 10^{-14}$ A. The DC gain β_F for $i_C = 150$ mA can vary between 100 and 300. This variation is not defined, however, and can change randomly from one transistor to another of the same type. As a working approximation, the *geometric mean* value is usually used; that is, $\beta_F = \sqrt{100 \times 300} = 173$. Since the value for Early voltage is not given, let us assume that $V_A = 200$ V. With these values of I_S, β_F, and V_A, the transistor Q2N2222 can be specified in PSpice/SPICE by the following statements:

```
Q1  NC  NB  NE  QMOD
.MODEL  Q2N2222  NPN (IS=3.295E-14  BF=173  VA=200)
```

▶ **NOTE** The full data sheets for BJTs (e.g., *npn*-type 2N2222 and *pnp*-type 2N2907A) can be found at http://www.alldatasheet.com/ or by searching BJT datasheet at http://www.google.com.

8.5.5 Small-Signal Analysis

Once the *Q*-point is established and the small-signal parameters are determined, we can find the small-signal parameters of the amplifier in Fig. 8.11(a) in response to a small-signal voltage v_{be}. For a small AC signal, the DC supply offers zero impedance; V_{CC} and V_{BB} can be short-circuited; that is, one side of R_C is connected to the ground. The small-signal AC equivalent circuit of the amplifier is shown in Fig. 8.13(a). Replacing the transistor Q_1 by its transconductance model of Fig. 8.12(c), the small-signal AC equivalent circuit is shown in Fig. 8.13(b). The following steps are involved in analyzing an amplifier circuit:

1. Analyzing the DC biasing of the transistor circuit
2. Determining the small-signal parameters g_m, r_π, and r_o of the transistor model in Fig. 8.12(c)
3. Determining the AC equivalent circuit of the amplifier
4. Performing the small-signal analysis for finding R_i, A_{vo}, and R_o

From Fig. 8.13(b), the small-signal input resistance can be found from

$$R_i = \frac{v_{be}}{i_b} = r_\pi$$

Thevenin's equivalent output resistance, looking from the output side for the condition $v_{be} = 0$, and $g_m v_{be}$ behaving as open circuited, can be found by inspection as

$$R_o = r_o \parallel R_C \tag{8.44}$$

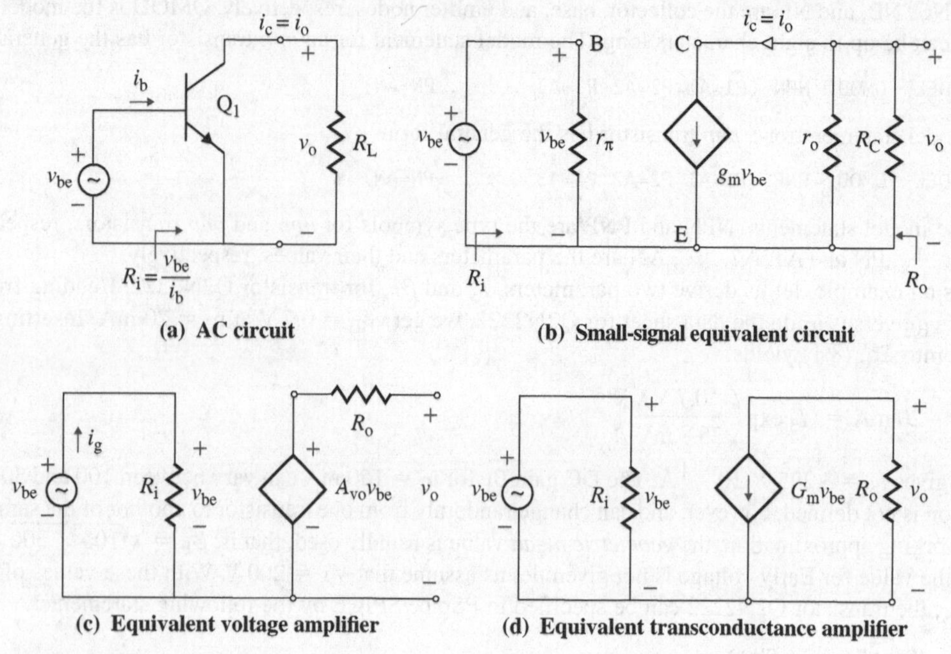

FIGURE 8.13 Small-signal AC equivalent circuits of the amplifier in Fig. 8.11(a)

▶ **NOTE** For the sake of simplicity, we often use the small-signal parameters r_π, r_o, and g_m instead of $r_{\pi 1}$, r_{o1}, and g_{m1} for transistor Q_1

The transconductance of the amplifier, which is the same as the transconductance of the transistor, is

$$G_{mo} = \frac{i_o}{v_{be}} = g_m \tag{8.45}$$

We can write the small-signal output voltage v_o as

$$v_o = -i_o(r_o \parallel R_C) = -g_m(r_o \parallel R_C)v_{be} \tag{8.46}$$

which gives the small-signal voltage gain as

$$A_{vo} = \frac{v_o}{v_{be}} = -g_m(r_o \parallel R_C) = -\left(\frac{I_C}{V_T}\right)\left(\frac{r_o R_C}{r_o + R_C}\right) \tag{8.47}$$

Substituting $r_o = V_A/I_C$, Eq. (8.47) becomes

$$A_{vo} = -\left(\frac{I_C}{V_T}\right)\left(\frac{V_A R_C}{V_A + I_C R_C}\right) \approx \frac{I_C R_C}{V_T} \quad \text{(for } V_A >> R_C I_C\text{)} \tag{8.48}$$

Therefore, for a large voltage gain, the $I_C R_C$ product must also be made large. This requires both a large DC supply voltage V_{CC} and a large value of resistance R_C.

EXAMPLE 8.2

Finding the small-signal parameters of an amplifier The amplifiers in Fig. 8.11(a) are $V_{BE} = 0.68$ V, $V_{CC} = 15$ V, and $R_C = 1$ kΩ. The transistor parameters are $\beta_F = 100$, $V_A = 200$ V, $V_T = 25.8$ mV, and $I_S = 3.3 \times 10^{-14}$ A.

(a) Find the DC-biasing point I_B, I_C, and V_{CE}.

(b) Find the small-signal transistor model parameters r_π, r_o, and g_m.

(c) Find the small-signal amplifier parameters R_i, R_o, and A_{vo}.

SOLUTION

(a) From Eq. (8.1), $I_C = I_S \exp(v_{BE}/V_T) = 3.3 \times 10^{-14} \times \exp[0.682/(25.8 \times 10^{-3})] = 9.97$ mA, $I_B = I_C/\beta_F = 9.97 \text{ mA}/100 = 99.7\ \mu\text{A}$, and $V_{CE} = V_{CC} - R_C I_C = 15 \text{ V} - 9.97 \text{ mA} \times 1 \text{ k}\Omega = 5.03$ V.

(b) From Eq. (8.40), $r_\pi = V_T/I_B = 25.8 \text{ mV}/99.7\ \mu\text{A} = 258.8\ \Omega$.

From Eq. (8.42), $g_m = I_C/V_T = 9.97 \text{ mA}/25.8 \text{ mV} = 386$ mA/V.

From Eq. (8.43), $r_o = V_A/I_C = 200 \text{ V}/9.97 \text{ mA} = 20.06$ kΩ.

From Eq. (8.7), $r_o = (V_{CE} + V_A)/I_C = (5.03 + 200) \text{ V}/9.97 \text{ mA} = 20.56$ kΩ.

(c) $R_i = r_\pi = 258.8\ \Omega$, $R_o = r_o \| R_C = 952.5\ \Omega$, $G_{mo} = g_m = 386$ mA/V, and $A_{vo} = -g_m(r_o \| R_C) = -386 \text{ mA/V} \times (20.56 \text{ k}\Omega \| 1 \text{ k}) = -368.5$ V/V.

KEY POINTS OF SECTION 8.5

- For analysis of a BJT amplifier, the transistor must be represented by its DC and small-signal AC models. Therefore, two types of analysis are performed: AC analysis and DC analysis.
- Linear DC models are used for determining the Q point; however, an AC model is used for determining the voltage or power gain when the transistor is operated as an ampllifier in the active region. The parameters of the small-signal models depend on the DC-biasing point.
- The manufacturers of BJTs usually specify the common-emitter hybrid parameters corresponding to the hybrid model whose parameters can be determined from the other small-signal model parameters.

8.6 The BJT Switch

The BJT can be operated as a switch that will have the characteristic of a low on-state voltage at the maximum current so that the switch is subjected to the minimum power loss. This condition requires that the transistor is operated in the saturation region and the B-C junction is reverse biased such that V_{CE} is a low $V_{CE} = V_{CB} + V_{BE} = -V_{BC} + V_{BE} = V_{CE(sat)}$. Figure 8.8[(a) or (b)] shows the circuit arrangement for operating the BJT as a switch. Since $v_{CE} = v_{CB} + v_{BE}$, and also using $v_{CE} = V_{CC} - R_C i_C$ from the collector loop, we get

$$v_{CB} = v_{CE} - v_{BE} = V_{CC} - R_C i_C - v_{BE} \tag{8.49}$$

which means that v_{CE} must be less than v_{BE}, $v_{CE} < v_{BE}$, for operating the BJT as a switch. This condition can be satisfied by varying the product $R_C i_C$. The value of v_{BE} that will make $v_{CB} \simeq 0$ can be determined from the following condition:

$$V_{CC} - v_{BE} = R_C i_C = R_C I_S e^{v_{BE}/V_T} \tag{8.50}$$

The maximum value of the collector current $I_{C(max)}$ is specified by the manufacturer data sheet, which limits the minimum value of collector resistance R_C. Assuming that $V_{CE(sat)}$ is the C-E saturation voltage, we can find the corresponding collector saturation current as given by

$$I_{C(sat)} = \frac{V_{CC} - V_{CE(sat)}}{R_C} \tag{8.51}$$

which sets the limits of the collect or current $I_{C(sat)} \leq i_C \leq I_{C(max)}$ and the corresponding collector resistance $R_{C(max)} \leq R_C \leq R_{C(min)}$.

Using KVL in the base loop, we get the base current as

$$i_B = \frac{V_{BB} - v_{BE}}{R_B} \tag{8.52}$$

which must be larger than the minimum base current $I_{B(min)}$ to drive the transistor into saturation; that is,

$$I_{B(min)} > I_B = \frac{V_{BB} - v_{BE}}{R_B} \tag{8.53}$$

To operate the transistor in the saturation region, the base current must be sufficient enough to maintain the collector saturation current. That is, i_B must be greater than the value $I_{C(sat)}/\beta_F$ corresponding collector current $I_{C(sat)}$; that is,

$$\frac{V_{BB} - v_{BE}}{R_B} > \frac{V_{CC} - V_{CE(sat)}}{\beta_F R_C} \tag{8.54}$$

Too much base current will drive the transistor hard into saturation, giving a low value of $V_{CE(sat)}$, but it will take a longer time to switch from the on-state to the off-state due to a larger amount of charge storage in the depletion regions. On the other hand, too little of base current may not be enough to keep the transistor into saturation to obtain a low C-E voltage; it needs to make a right judgment. It is recommended to use a 125% overdrive factor (ODF); that is,

$$I_{B(max)} = \frac{\text{ODF}\ I_{C(sat)}}{\beta_F} \tag{8.55}$$

or

$$\frac{V_{BB} - v_{BE}}{R_B}\beta_F = \frac{V_{CC} - V_{CE(sat)}}{R_C} \times \text{ODF} \tag{8.56}$$

Therefore, the condition for the maximum value of the base current in Eq. (8.53) is

$$I_{B(min)} = I_{B(sat)} = \frac{V_{BB} - v_{BE}}{R_{B(max)}} \tag{8.57}$$

which gives limits to the base current $I_{B(min)} (= I_{C(max)}/\beta_F) \leq i_B \leq I_{B(max)}$ and the corresponding base resistance $R_{B(max)} \geq R_B \leq R_{B(min)}$.

KEY POINTS OF SECTION 8.6

- A BJT can be operated as a switch that will have the characteristic of a low onstate voltage at the maximum current so that the switch is subjected to the minimum power loss.
- The transistor is operated in the saturation region and the B-C junction is reverse biased. To operate the transistor in the saturation region, the base current must be sufficent to maintain the collector saturation current.

8.7 DC Biasing of Bipolar Junction Transistors

If a transistor is used for the amplification of voltage (or current), it is necessary to bias the device. The main reasons for biasing are to turn the device on and, in particular, to place the operating point in the region of its characteristic where the device operates most linearly so that any change in the input signal causes a proportional change in the output signal. In practice, a fixed DC supply is normally used, and the circuit elements are selected so as to bias the C-B and E-B junctions in appropriate magnitude and polarity.

The determination of the DC-biasing point described by (I_B, I_C, V_{CE}) is the first step in the analysis of the transistor circuit. Once the values of I_B and I_C are found, we can find g_m, r_π, and r_o; that is, $g_m = I_C/V_T$, $r_\pi = 25.8\ \text{mV}/I_B$, and $r_o \simeq V_A/I_C$. Since the B-E junction behaves like a diode, the transistor needs a B-E voltage of $V_{BE} \simeq 0.7$ V to conduct. If we apply more than 0.7 V, the transistor will be damaged due to excessive current. Resistors are used to limit the transistor currents as shown in Fig. 8.14(a) with base resistor R_B and collector resistor R_C, Fig. 8.14(b) with emitter resistor R_E, and Fig. 8.14(c) with collector resistor R_C and emitter resistor R_B. Although there are many types of biasing circuits, we will consider the following types, which are most commonly used:

Active current–source biasing
Single–base resistor biasing
Emitter resistance–feedback biasing
Emitter-follower biasing
Two–base resistor biasing

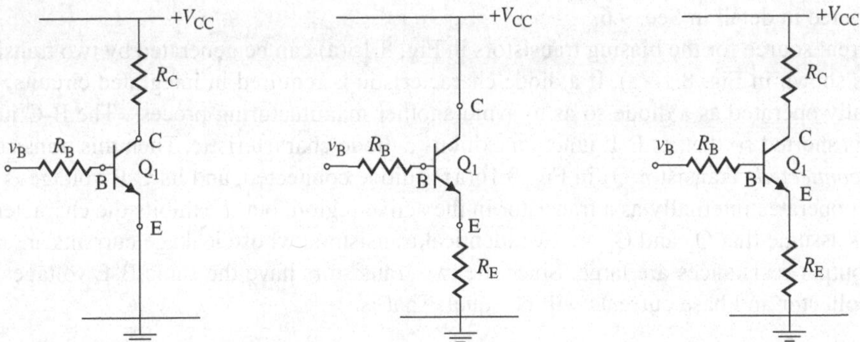

(a) **Base and collector resistors R_B and R_C** (b) **Emitter resistor R_E** (c) **Collector and emitter resistors R_C and R_B**

FIGURE 8.14 Resistors for limiting transistor currents

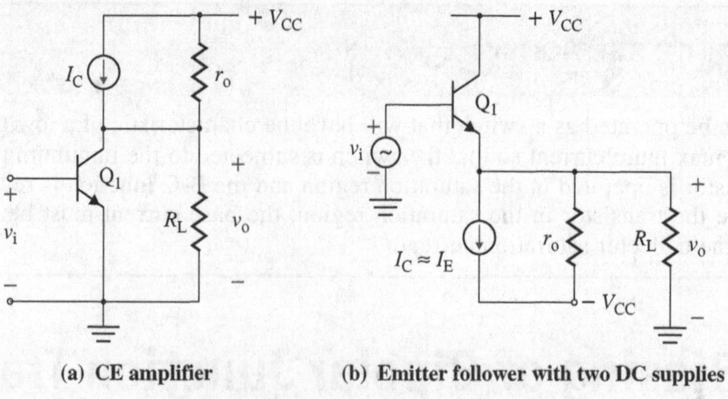

FIGURE 8.15 Amplifier with a biasing current source

▶ **NOTE** In the derivations of the currents for these biasing circuits, we will assume the following relations between the transistor currents: $i_c = \beta_F i_B$, $i_E = (1 + \beta_F)i_B$, and $i_C = \alpha_F i_E$.

8.7.1 Active Current–Source Biasing

According to Eq. (8.48), the voltage gain can be increased by having a large value of the collector resistance R_C in the amplifier of Fig. 8.11(a). R_C can be replaced by a current source, which normally has a high output resistance, thereby producing a high voltage gain, as shown in Fig. 8.15(a). To allow for a wide output voltage swing, an amplifier is often connected to two DC supplies as shown in Fig. 5.15(b). The current I_C of the active load flows out of the current source circuit into Q_1, and this type of source is referred to as a *sourcing current source*. The source current in Fig. 8.15(b) flows from Q_1 *into* the current source, and this type of constant current source is often referred to as a *current sinking source*. An ideal current source should have a constant current and a very large output resistance r_o, tending to infinity. Since a current source has a large output resistance r_o, the voltage gain of the amplifier will also be large. Current sources are used for biasing transistors in integrated circuits. There are several types of BJT current sources, and these are covered in detail in Sec. 9.6.

The current source for the biasing transistors in Fig. 8.15(a) can be generated by two transistors and a resistor, as shown in Fig. 8.16(a). If a diode characteristic is required in integrated circuits, a transistor is generally operated as a diode so as to avoid another manufacturing process. The B-C junction of a transistor is shorted so that its B-E junction exhibits a diode characteristic. Then this transistor is said to be *diode connected*. Transistor Q_3 in Fig. 8.16(a) is diode connected, and its C-B voltage is forced to zero. Q_3 still operates internally as a transistor in the active region, but it exhibits the characteristic of a diode. Let us assume that Q_2 and Q_3 are two identical transistors, whose leakage currents are negligible and whose output resistances are large. Since the two transistors have the same B-E voltages ($V_{BE2} = V_{BE3}$), the collector and base currents will be equal. That is,

$$I_{C2} = I_{C3} \quad \text{and} \quad I_{B2} = I_{B3}$$

Applying Kirchhoff's current law (KCL) at the collector of Q_3, we get the reference current:

$$I_{ref} = I_{C3} + I_{B2} + I_{B3} = I_{C3} + 2I_{B3}$$

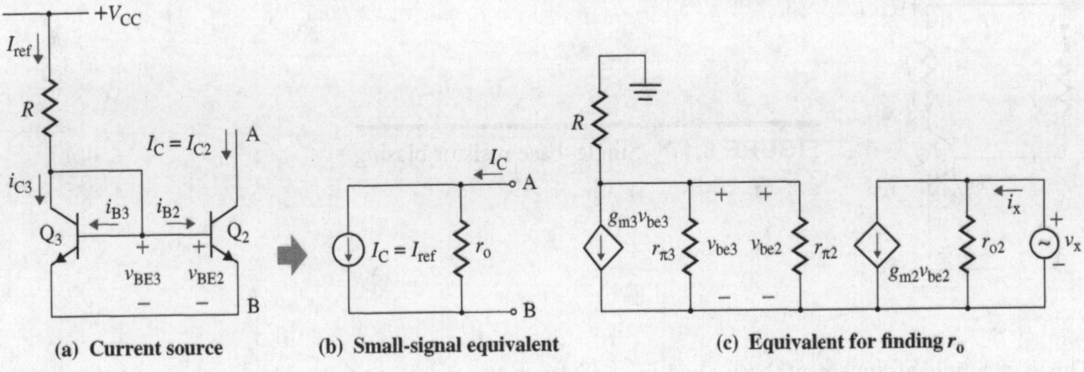

FIGURE 8.16 Transistor current source

Since $I_{C3} = \beta_F I_{B3}$,

$$I_{ref} = I_{C3} + 2I_{B3} = I_{C3} + \frac{2I_{C3}}{\beta_F}$$

which gives the collector current I_{C3} as

$$I_{C3} = I_{C2}\frac{I_{ref}}{1 + 2/\beta_F} \tag{8.58}$$

If $\beta_F >> 2$, which is usually the case, Eq. (8.58) can be approximated by

$$I_{C3} \approx I_{ref} = \frac{V_{CC} - V_{BE3}}{R} = I_{C2} \tag{8.59}$$

Thus, for two identical transistors, the reference and output currents are equal. In practice, however, the transistors may not be identical, and the two collector currents will have a constant ratio. The small-signal AC equivalent circuit is shown in Fig. 8.16(b). The equivalent circuit for finding r_o is shown in Fig. 8.16(c), where the output resistance r_o is the same as r_{o2}:

$$r_o = \frac{v_x}{i_x} = r_{o2} = \frac{V_A}{I_{C2}} \quad \text{(for } V_A = V_{A2} = V_{A3}\text{)} \tag{8.60}$$

8.7.2 Single–Base Resistor Biasing

This type of biasing circuit, shown in Fig. 8.17, is the same as that in Fig. 8.14(a). Using KVL around the B-E loop, we can find the base current i_B, the collector current i_C, and C-E voltage v_{CE}, for known value of V_{BE} that is typically 0.7 V, as given by

$$i_B = \frac{V_{CC} - v_{BE}}{R_B}$$

$$i_C = \beta_F i_B = \frac{\beta_F(V_{CC} - v_{BE})}{R_B} \tag{8.61}$$

$$v_{CE} = V_{CC} - R_C\, i_C = V_{CC} - \beta_F R_C i_B \tag{8.62}$$

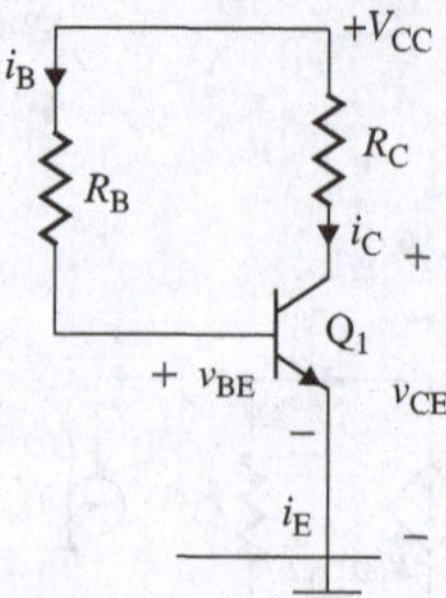

FIGURE 8.17 Single–base resistor biasing

This is a simple arrangement, but i_C and v_{CE} are dependent directly on β_F, which can vary over a wide range for transistors of the same type.

8.7.3 Emitter Resistance–Feedback Biasing

This type of biasing circuit is shown in Fig. 8.18. Using KVL around the B-E loop, we can write

$$V_{CC} = R_B i_B + v_{BE} + R_E i_E = R_B i_B + v_{BE} + R_E(1 + \beta_F) i_B \quad [\text{Use } i_E = (1 + \beta_F) i_B.]$$

which gives the base current, for known value of V_{BE}, as

$$i_B = \frac{V_{CC} - v_{BE}}{R_B + R_E(1 + \beta_F)} \tag{8.63}$$

$$i_C = \beta_F i_B = \frac{\beta_F (V_{CC} - v_{BE})}{R_B + R_E(1 + \beta_F)} \tag{8.64}$$

$$v_{CE} = V_{CC} - R_C i_C - R_E i_E = V_{CC} - R_C \beta_F i_B - (1 + \beta_F) R_E i_B \tag{8.65}$$

For $R_E(1 + \beta_F) >> R_B$ and $\beta_F >> 1$, Eq. (8.64) can be approximated to

$$i_C \approx \frac{V_{CC} - v_{BE}}{R_E}$$

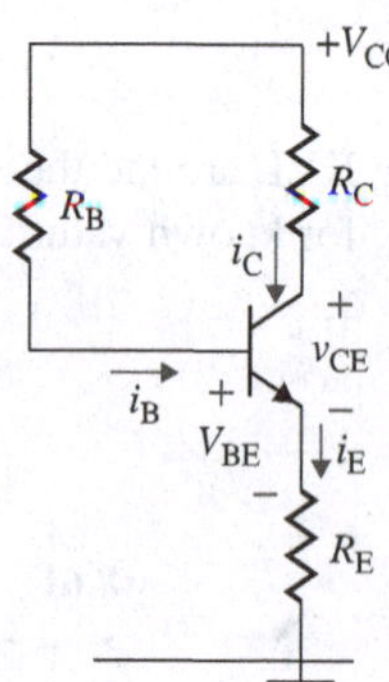

FIGURE 8.18 Emitter resistance–feedback biasing

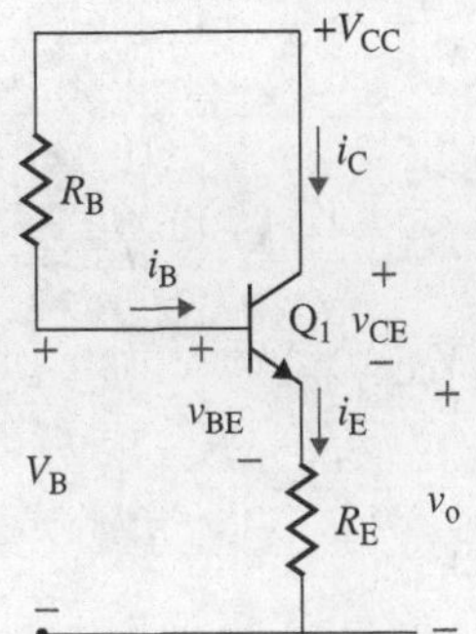

FIGURE 8.19 Emitter-follower biasing

This is also a simple circuit. By proper biasing design, i_C and v_{CE} can be made almost independent of β_F, which can vary over a wide range for transistors of the same type. It is important to note the emitter resistance R_E appears in series with R_B in the base circuit as $(1 + \beta_F)R_E$ with a multiplying factor of $(1 + \beta_F)$. Similarly, the base resistance R_B will appear in series with R_E in the emitter circuit as $R_E/(1 + \beta_F)$ with a multiplying factor of $1/(1 + \beta_F)$.

8.7.4 Emitter-Follower Biasing

This type of biasing circuit is shown in Fig. 8.19. This circuit is known as the emitter follower because the output voltage at the emitter terminal is $v_O = v_B - v_{BE}(\simeq 0.7\text{ V})$. If the base voltage changes by Δv_B, then the output voltage will also change by the same amount $\Delta v_O = \Delta v_B$. Using KVL around the B-E loop, we can find the base current i_B, the collector current i_C, and the C-E voltage v_{CE} as given by

$$i_B = \frac{V_{CC} - v_{BE}}{R_B + R_E(1 + \beta_F)} \tag{8.66}$$

$$i_C = \beta_F i_B = \frac{\beta_F(V_{CC} - v_{BE})}{R_B + R_E(1 + \beta_F)} \tag{8.67}$$

$$v_{CE} = V_{CC} - R_E i_E = V_{CC} - (1 + \beta_F)R_F i_B \tag{8.68}$$

8.7.5 Two–Base Resistor Biasing

This type of biasing circuit is shown in Fig. 8.20(a). The analysis can be simplified by replacing R_1 and R_2 with Thevenin's equivalent voltage V_{Th} and resistance R_{Th} as shown in Fig. 8.20(b). The replacement gives

$$V_{Th} = \frac{R_2}{R_1 + R_2} V_{CC} \tag{8.69}$$

$$R_{Th} = \frac{R_1 R_2}{R_1 + R_2} \tag{8.70}$$

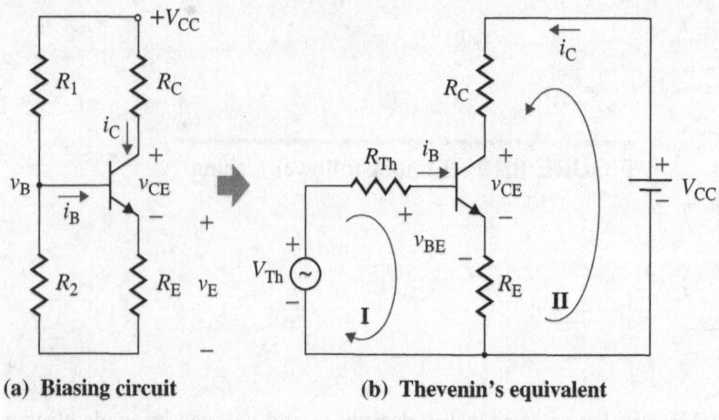

FIGURE 8.20 DC-biasing circuit

Equations (8.69) and (8.70) can be solved for R_1 as

$$R_1 = \frac{R_{Th}V_{CC}}{V_{Th}} \tag{8.71}$$

Substituting R_1 from Eq. (8.71) into Eq. (8.70), we get

$$R_2 = \frac{R_{Th}V_{CC}}{V_{CC} - V_{Th}} \tag{8.72}$$

The biasing analysis is similar to Fig. 8.18 if we substitute R_{Th} for R_B and the base is connected to V_{Th} instead of V_{CC}. Using KVL around the B-E loop, we can find the base current i_B, the collector current i_C, and the C-E voltage v_{CE} as given by

$$i_B = \frac{V_{Th} - V_{BE}}{R_{Th} + R_E(1 + \beta_F)} \tag{8.73}$$

$$i_C = \beta_F i_B = \frac{\beta_F(V_{Th} - V_{BE})}{R_{Th} + R_E(1 + \beta_F)} \tag{8.74}$$

$$v_{CE} = V_{CC} - R_C i_C - R_E i_E = V_{CC} - R_C\beta_F i_B - (1 + \beta_F)R_F i_B \tag{8.75}$$

$R_E(1 + \beta_F) >> R_{Th}$ and $\beta_F >> 1$, Eq. (8.74) can be approximated to

$$i_C \approx \frac{V_{Th} - V_{BE}}{R_E} \tag{8.76}$$

The voltage at the base v_B is fixed by potential divider consisting of R_1 and R_2. Since the B-E voltage is the difference between the base and emitter voltages, $v_{BE} = v_B - v_E$, any changes in the base voltage v_B and the emitter voltage v_E due to variations of transistor parameters cause minimum changes in $v_{BE} = v_B - v_E$ and thereby the transistor currents.

8.7.6 Biasing Circuit Design

Using the relationship $i_E = i_C/\alpha_F$, we can rewrite Eqs. (8.65) or (8.75) in terms of i_C as given by

$$v_{CE} = V_{CC} - R_C i_C - R_E i_E = V_{CC} - \left(R_C + + \frac{R_E}{\alpha_F}\right) i_C \tag{8.77}$$

The value of α_F ranges from 0.9 to 0.99 and is related to β_F by

$$\alpha_F = \frac{i_E}{i_C} = \frac{\beta_F}{1 + \beta_F} \tag{8.78}$$

In practice, $\beta_F >> 1$ and $\alpha_F \simeq 1$. Thus, Eq. (8.77) can be approximated to

$$v_{CE} = V_{CC} - (R_C + R_E) i_C \tag{8.79}$$

which is the equation of a straight line and represents the load line, as shown in Fig. 8.21.

Since an AC voltage is normally superimposed on the operating base-to-emitter voltage V_{BE} in order to operate the transistor as an amplifier, the *Q*-point is subjected to a swing in either direction. Therefore, the *Q*-point should be positioned so that it can provide enough range to accommodate the maximum voltage swing and it is least sensitive to variations in the DC gain β_F.

The given parameters are β_F and $I_{C(max)}$ of a transistor, $V_{BE} = 0.7$ V (typically), and V_{CC}. Select suitable values of the DC biasing collector current I_{CQ} and the C-E voltage V_{CEQ} so that i_C and v_{CE} can have the maximum swings in both positive and negative directions, $i_C = I_{CQ} \pm i_{c(peak)}$ and $v_{CE} = V_{CEQ} \pm v_{ce(peak)}$. There, I_{CQ} must be less than $I_{C(max)}/2$, and the DC supply V_{CC} should equally be shared equally by all elements in the C-E loop. The guidelines for determining the biasing resistances of different circuits are summarized in Table 8.1.

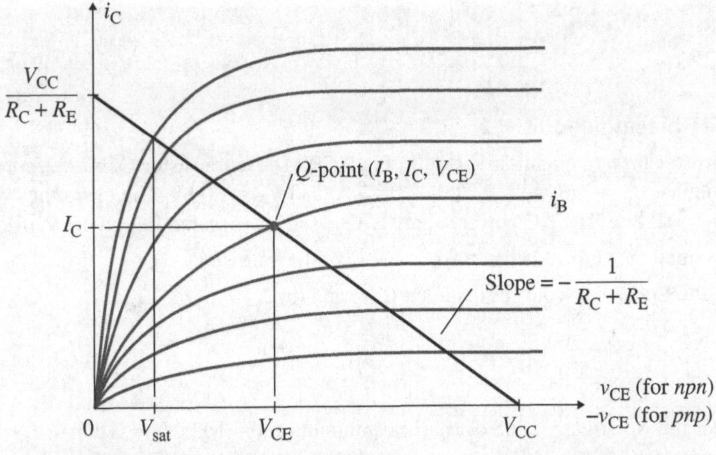

FIGURE 8.21 Load line and *Q*-point

TABLE 8.1 Summary of biasing design guidelines

Biasing Parameters	Figure 8.17	Figure 8.18	Figure 8.19	Figure 8.20
V_{CEQ}	$\frac{V_{CC}}{2}$	$\frac{V_{CC}}{3}$	$\frac{V_{CC}}{2}$	$\frac{V_{CC}}{3}$
V_{EQ}	$\frac{V_{CC}}{2}$	$\frac{V_{CC}}{3}$	$\frac{V_{CC}}{2}$	$\frac{V_{CC}}{3}$
I_{CQ}	$\frac{I_{C(max)}}{3}$	$\frac{I_{C(max)}}{3}$	$\frac{I_{C(max)}}{3}$	$\frac{I_{C(max)}}{3}$
R_C	$\frac{V_{CC}}{2I_{CQ}}$	$\frac{V_{CC}}{3I_{CQ}}$	0	$\frac{V_{CC}}{3I_{CQ}}$
R_E	0	$\frac{V_{CC}}{3I_{CQ}\alpha_F}$	$\frac{V_{CC}}{2I_{CQ}\alpha_F}$	$\frac{V_{CC}}{3I_{CQ}\alpha_F}$
R_B	$\frac{V_{CC} - V_{BE}}{I_{BQ}}$	$\frac{V_{CC} - V_{BE} - V_{EQ}}{I_{BQ}}$	$\frac{V_{CC} - V_{BE} - V_{EQ}}{I_{BQ}}$	
R_{Th}				$0.1 \times (1 + \beta_F)R_E$
R_1				$R_1 = \frac{R_{Th}V_{CC}}{V_{Th}}$
R_2				$R_2 = \frac{R_{Th}V_{CC}}{V_{CC} - V_{Th}}$

▶ **NOTE** Manufacturers usually specify three values for a parameter: minimum, nominal, and maximum. For example, the beta (β_F) of Q2N2222 has three values: minimum $\beta_F = 100$, nominal $\beta_F = 173$, and maximum $\beta_F = 300$. It is the designer's task to choose the appropriate value of the transistor parameter(s) to find the component values. The minimum value of β_F is normally used to yield the worst-case design of the biasing circuit—that is, to obtain the desired Q-point at the worst value of β_F.

EXAMPLE 8.3

D

Designing a BJT biasing circuit

(a) Design a transistor biasing circuit as shown in Fig. 8.20(a). Use transistor Q2N2222, for which minimum $\beta_F = 100$, nominal $\beta_F = 173$, $I_S = 3.295 \times 10^{-14}$ A, and $V_A = 200$ V. The operating collector current is to be set at $I_C = 10$ mA. The DC power supply is $V_{CC} = 15$ V. Assume $V_{BE} = 0.7$ V.

(b) Calculate the small-signal parameters r_π, g_m, and r_o of the transistor.

(c) Use PSpice/SPICE to verify your results in parts (a) and (b).

SOLUTION

We will design for the worst-case value of β_F (i.e., minimum $\beta_F = 100$). $I_C = 10$ mA, $I_B = 10 \text{ mA}/100 = 0.1$ mA, and $V_{CC} = 15$ V.

(a) Step 1. Calculate the values of α_F and I_E. From Eq. (8.78),

$$\alpha_F = \frac{\beta_F}{(1+\beta_F)} = \frac{100}{(1+100)} = 0.99$$

From Eq. (8.4),

$$I_E = \frac{I_C}{\alpha_F} = \frac{I_C}{0.99} = \frac{10 \text{ mA}}{0.99} = 10.1 \text{ mA}$$

Step 2. Calculate the value of V_E. From Table 8.1 (Fig. 8.20),

$$V_E = \frac{V_{CC}}{3} = \frac{15}{3} = 5 \text{ V}$$

Step 3. Calculate the value of R_E and its power rating:

$$R_E = \frac{V_E}{I_E} = \frac{5 \text{ V}}{10.1 \text{ mA}} = 495 \ \Omega$$

The power rating of R_E is

$$P_{RE} = I_E^2 R_E = (10.1 \times 10^{-3})^2 \times 495 = 50.49 \text{ mW}$$

Step 4. Calculate the value of V_{CE}. From Table 8.1 (Fig. 8.20),

$$V_{CE} = \frac{V_{CC}}{3} = \frac{15}{3} = 5 \text{ V}$$

Step 5. Calculate the value of R_C and its power rating.

$$I_C R_C = V_{CC} - V_E - V_{CE} = 15 - 5 - 5 = 5 \text{ V}$$

$$R_C = \frac{5}{I_C} = \frac{5}{10 \text{ mA}} = 500 \ \Omega$$

The power rating of R_C is

$$P_{RC} = I_C^2 R_C = (10 \times 10^{-3})^2 \times 500 = 50 \text{ mW}$$

Step 6. Calculate the values of R_{Th} and V_{Th}. From Table 8.1 (Fig. 8.20),

$$R_{Th} = \frac{(1+\beta_F)R_E}{10} = \frac{(1+100) \times 495}{10} = 5 \text{ k}\Omega$$

From Fig. 8.20(b),

$$V_{Th} = V_E + V_{BE} + R_{Th} I_B = V_E + 0.7 \text{ V} + 5 \text{ k}\Omega \times 0.1 \text{ mA} = 5 + 0.7 + 0.5 = 6.2 \text{ V}$$

Step 7. Calculate the value of R_1 and its power rating. From Eq. (8.71),

$$R_1 = \frac{R_{Th} V_{CC}}{V_{Th}} = \frac{5 \text{ k} \times 15}{6.2} = 12.1 \text{ k}\Omega$$

The power rating of R_1 is

$$P_{R1} = \frac{(V_{CC} - V_{Th})^2}{R_1} = \frac{(15 - 6.2)^2}{12.1 \text{ k}} = 6.4 \text{ mW}$$

Step 8. Calculate the value of R_2 and its power rating. From Eq. (8.72),

$$R_2 = \frac{R_{Th} V_{CC}}{V_{CC} - V_{Th}} = \frac{5 \text{ k} \times 15}{15 - 6.2} = 8.52 \text{ k}\Omega$$

The power rating of R_2 is

$$P_{R2} = \frac{V_{Th}^2}{R_2} = \frac{6.2^2}{8.52\text{ k}} = 4.51\text{ mW}$$

(b) From Eq. (8.40),

$$r_\pi = \frac{25.8\text{ mV}}{I_B} = \frac{\beta_F \times 25.8\text{ mV}}{I_C} = \frac{100 \times 25.8\text{ mV}}{10\text{ mA}} = 258\ \Omega$$

From Eq. (8.42),

$$g_m = \frac{I_C}{V_T} = \frac{10\text{ mA}}{25.8\text{ mV}} = 387.6\text{ mA/V}$$

From Eq. (8.43),

$$r_o = \frac{V_{CE} + V_A}{I_C} = \frac{(5 + 200)\text{ V}}{10\text{ mA}} = 20.5\text{ k}\Omega$$

(c) The DC-biasing circuit for PSpice simulation is shown in Fig. 8.22.

The results of the .OP command (for EX8-3.SCH) are automatically printed on the output file. (The values obtained from hand calculations are shown in parentheses.)

```
**** SMALL SIGNAL BIAS SOLUTION   TEMPERATURE = 27.000 DEG C
NODE     VOLTAGE    NODE        VOLTAGE    NODE    VOLTAGE    NODE    VOLTAGE
(1)       5.2519    (2)         15.0000    (3)     10.4270    (4)      4.5710
IB=8.91E-05
IC=9.15E-03           (10 mA)
VBE=6.81E-01          (0.7 V)
VBC=-5.18E+00
VCE=5.86E+00          (5 V)
BETADC=1.03E+02       (100)
GM=3.54E-01           (0.3876)
RPI=2.90E+02          (258 Ω)
RO=2.24E+04           (20 kΩ)
```

NOTE: All PSpice results given here are from running the simulation with the schematic (.SCH) files. If you run the simulation with the netlist circuit (.CIR) files, you may get different results, as the student's version of PSpice has a limited number of active devices and models.

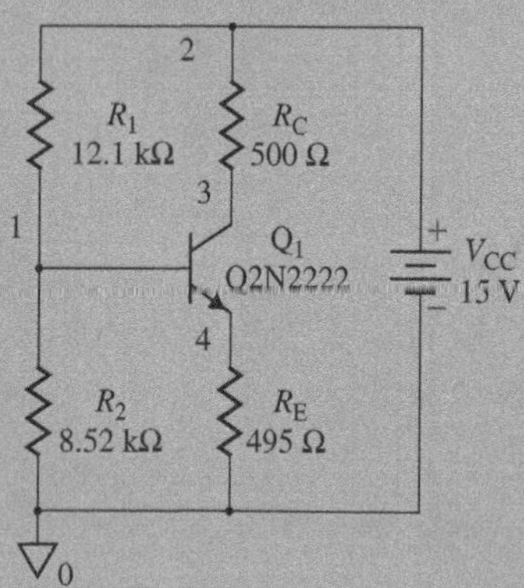

FIGURE 8.22 DC-biasing circuit for PSpice simulation

KEY POINTS OF SECTION 8.7

- The main reasons for biasing are to turn the device on and, in particular, to place the operating point in the region of its characteristic where the device operates most linearly, so that any change in the input signal causes proportional changes in the output signal.
- In practice, a fixed DC supply is normally used, and the circuit elements are selected so as to bias the C-B and E-B junctions in appropriate magnitude and polarity.
- There are many types of biasing circuits; the most commonly used ones are active current–source biasing and two–base resistor biasing in order to ensure stable Q-point of the transistor parameters and the biasing resistors.

8.8 Common-Emitter Amplifiers

Once the Q-point has been established by a biasing circuit, an input voltage can be applied through *coupling capacitors,* as shown in Fig. 8.23. C_1 and C_2 isolate the DC signals of the biasing circuit from the input signal v_s and the load resistance R_L, respectively. If the input signal v_s were connected directly to the base without C_1, the source resistance R_s would form a parallel circuit with R_2, and the base potential V_B would be disturbed. Similarly, the collector potential V_C would depend on R_L if C_2 were removed.

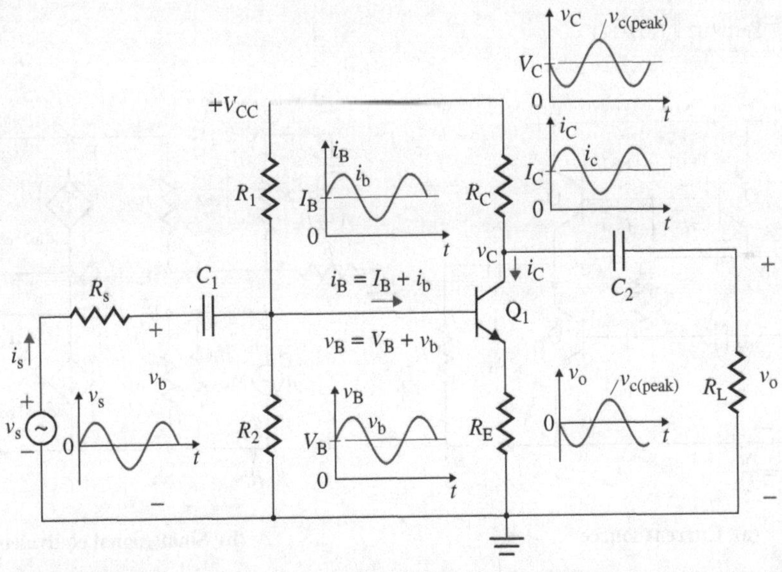

FIGURE 8.23 Common-emitter amplifier circuit

Let us assume that the capacitors have large values so that they are virtually shorted at the frequency of the input signal v_s. With a sinusoidal input voltage $v_s = V_m \sin \omega t$, the base potential will be $v_B = V_B + v_b$. If the base current i_B swings between $I_B + i_{b(peak)}$ and $I_B - i_{b(peak)}$, the collector current i_C will swing between $I_C + i_{c(peak)}$ and $I_C - i_{c(peak)}$. The C-E voltage v_{CE} will vary accordingly from $V_{CE} - v_{ce(peak)}$ to $V_{CE} + v_{ce(peak)}$, and the collector voltage v_C will vary from $V_C - R_C(I_C + i_{c(peak)})$ to $V_C - R_C(I_C - i_{c(peak)})$. These waveforms are depicted in Fig. 8.23. Since C_2 will block any DC signal, the output voltage will vary from $-(R_C \| R_L)i_{c(max)}$ to $(R_C \| R_L)i_{c(min)}$.

8.8.1 Active-Biased Common-Emitter Amplifier

A CE amplifier with a current source is shown in Fig. 8.24(a). The current source consists of *pnp* transistors to generate a sourcing current source, and its output resistance acts as the load of transistor Q_1. Since the collector load element is a *pnp* transistor instead of a resistor, it is said to be *active.* Since a DC supply offers zero impedance to an AC signal, V_{CC} behaves as short-circuited; that is, one side of both Q_2, Q_3, and R_B is connected to the ground. Replacing the transistors by their small-signal models gives the AC equivalent circuit shown in Fig. 8.24(b). The small-signal analysis is similar to the basic amplifier circuit in Fig. 8.13(b) if we replace R_C by r_{o2} and r_π by $r_{\pi 1}$ in parallel with R_B. For identical transistors of matched characteristics, we can assume $V_{A1} = V_{A2} = V_A$, $r_{o1} = r_{o2} = r_o$, and $g_{m1} = g_{m2} = g_m$. The small-signal input resistance can be found from

$$R_i = \frac{v_b}{i_s} = r_{\pi 1} \| R_B \tag{8.80}$$

The output resistance R_o seen looking into the output is the parallel combination of the two transistor output resistances. That is,

$$R_o = r_{o2} \| r_{o1} \tag{8.81}$$

For $v_{eb2} = 0$, $g_m v_{eb2}$ is open circuited. The output voltage v_o is

$$v_o = -g_{m1}(r_{o2} \| r_{o1})v_{be1}$$

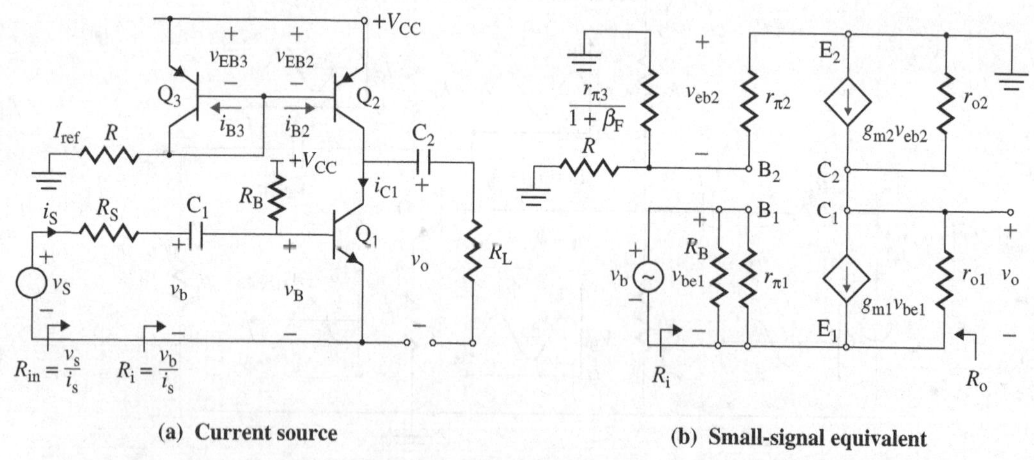

(a) Current source (b) Small-signal equivalent

FIGURE 8.24 Common-emitter amplifier with a current source

which gives the no-load voltage gain A_{vo} as

$$A_{vo} = \frac{v_o}{v_b} = -g_{m1}(r_{o2} \parallel r_{o1}) \tag{8.82}$$

which, after substituting $r_{o1} = r_{o2} = V_A/I_C$ and $g_{m1} = g_{m2} = I_C/V_T$, can be simplified to

$$A_{vo} = -\frac{I_C}{V_T}\frac{V_A}{2I_C} = \frac{V_A}{2V_T} \tag{8.83}$$

It is interesting to note that the no-load voltage gain is independent of the DC biasing collector current, but it depends directly on the Early voltage and decreases inversely with the thermal voltage V_T. For larger voltage gain, we should use transistors with a higher value of Early voltage V_A.

EXAMPLE 8.4

D

Designing a common-emitter amplifier with an active current source

(a) Design a CE amplifier with an active current source, as shown in Fig. 8.24(a). Use transistors Q2N2222 and Q2N2907, for which nominal $\beta_F = \beta_f = 173$, $I_S = 3.295 \times 10^{-14}$ A, and $V_A = 100$ V. The operating collector current is set at $I_C = 10$ mA. The DC power supply is $V_{CC} = 15$ V. Assume $V_{BE} = 0.7$ V.

(b) Use PSpice/SPICE to verify your results in part (a).

SOLUTION

(a) Step 1. Design the biasing current source. $I_C = 10$ mA, and $V_{CC} = 15$ V. From Eq. (8.59), we can find the value of R in order to set the biasing current to $I_{ref} = 10$ mA.

$$R = \frac{V_{CC} - V_{BE3}}{I_{ref}} = \frac{15 - 0.7}{10\text{ mA}} = 1.43\text{ k}\Omega$$

Step 2. Find the small-signal parameters of the transistors. From Eq. (8.40),

$$r_{\pi 1} = r_{\pi 2} = \frac{25.8\text{ mV}}{I_B} = \frac{\beta_F \times 25.8\text{ mV}}{I_C} = \frac{173 \times 25.8\text{ mV}}{10\text{ mA}} = 446\ \Omega$$

From Eq. (8.42),

$$g_{m1} = g_{m2} = \frac{I_C}{V_T} = \frac{10\text{ mA}}{25.8\text{ mV}} = 387.6\text{ mA/V}$$

From Eq. (8.43),

$$r_{o1} = r_{o2} = \frac{V_A}{I_C} = \frac{100\text{ V}}{10\text{ mA}} = 10\text{ k}\Omega$$

Step 3. Evaluate the values of the input resistance R_i $(=v_b/i_b)$, the open-circuit voltage gain A_{vo}, and the output resistance R_o. We know that

$$R_i = r_{\pi 1} = 446\ \Omega$$

From Eq. (8.81),

$$R_o = r_{o2} \parallel r_{o1} = 10\text{ k} \parallel 10\text{ k} = 5\text{ k}\Omega$$

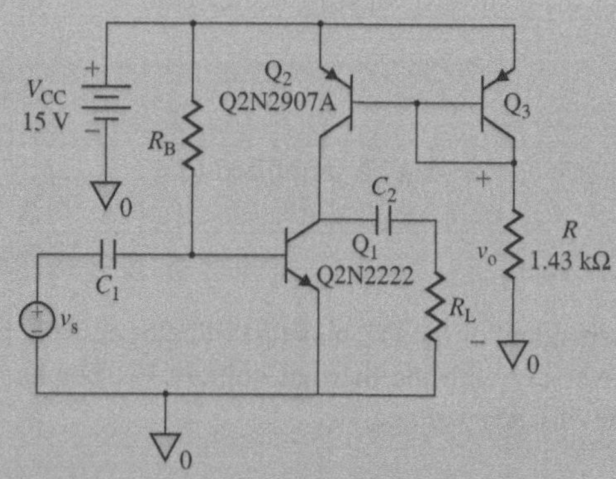

FIGURE 8.25 A common-emitter amplifier with an active load for PSpice simulation

From Eq. (8.82),

$$A_{vo} = -g_{m1}(r_{o2} \| r_{o1}) = -387.6 \text{ mA/V} \times 5 \text{ k}\Omega = -1938$$

(b) From Eq. (8.1), we can find the value of v_{BE1} needed to give $I_C = 10$ mA for $I_S = 3.295 \times 10^{-14}$ A:

$$10 \text{ mA} = 3.295 \times 10^{-14} \times \exp\left(\frac{v_{BE1}}{25.8 \text{ mV}}\right)$$

which gives $v_{BE1} = 0.682$ V.

Using Eq. (8.61), we can find the biasing resistor R_B:

$$R_B = \frac{\beta_F(V_{CC} - V_{BE})}{I_C} = \frac{173 \times (15 - 0.682)}{10 \times 10^{-3}} = 247.7 \text{ k}\Omega \quad \text{(Use 250 k}\Omega\text{.)}$$

The CE amplifier with an active load for PSpice simulation is shown in Fig. 8.25.

The PSpice plot of the transfer function v_o [≡V(Q1:C)] against v_S is shown in Fig. 8.26. Notice that the operating range of the input voltage is very small—that is, 8.563 mV (709.863 mV − 701.30 mV). Thus, the small-signal gain becomes −15 V/8.563 mV = −1752. Details of the .TF analysis are given below. The values obtained from hand calculations are shown in parentheses. The value of v_S (=v_{BE1}) was adjusted to 0.705 V instead of 0.682 V in order to operate in the linear range of the amplifier and to illustrate the benefit of using an active load. Since the voltage gain is very large, any small change in v_S could drive the amplifier into saturation. Thus, if we build the amplifier and test it in the laboratory with a value of $v_S = 0.682$ V, it might not work; we need to adjust v_S. Using transistor circuit model, we make the range of input voltage from 681 mV to 685 mV.

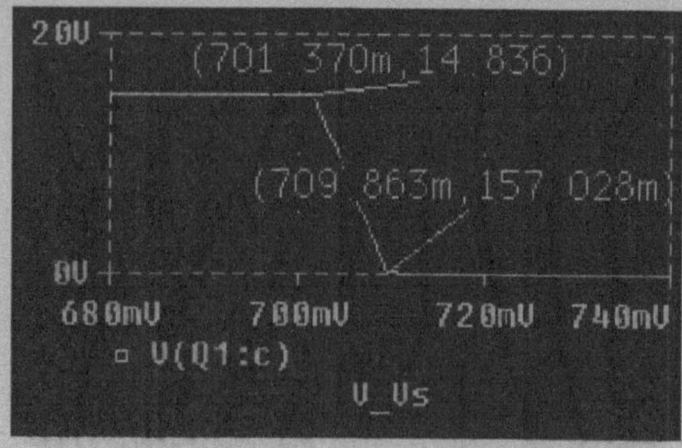

FIGURE 8.26 PSpice plot for Example 8.4

```
****  SMALL-SIGNAL CHARACTERISTICS
V(Q1:C)/VS=-1.796E+03                        Avo = −1796  (−1938)
INPUT RESISTANCE AT VS=5.262E+02             Ri = 526 Ω     (446 Ω)
OUTPUT RESISTANCE AT V(Q1:C)=4.726E+03       Ro = 4.726 kΩ (5 kΩ)
The results of the .OP command are as follows.
NAME        Q1            Q2             Q3              For Q1
MODEL       Q2N2222       Q2N2907A       Q2N2907A
IB          5.39E-05      -4.38E-05      -4.38E-05       (IB = 57.8 μA)
IC          1.03E-02      -1.03E-02      -9.85E-03       (IC = 10 mA)
VBE         7.05E-01      -7.86E-01      -7.86E-01       (VBE = 0.705 V)
GM          3.87E-01       3.96E-01       3.77E-01       (gm = 0.3876 A/V)
RPI         5.16E+02       5.96E+02       5.96E+02       (rπ = 446 Ω)
RO          7.91E+03       1.17E+04       1.17E+04       (ro = 10 kΩ)
```

8.8.2 Resistive-Biased Common-Emitter Amplifier

A CE amplifier with an emitter resistance is shown in Fig. 8.27 where C_1 and C_2 are the coupling capacitors to isolate the small-signal input and output voltages from the DC biasing. C_E will act as a short circuit and bypasses the small AC signal, which results in a higher voltage gain but lowers the small-signal input resistance. Since a DC supply offers zero impedance to an AC signal, V_{CC} can be short-circuited. That is, one side of both R_C and R_B is connected to the ground. The AC equivalent circuit of the amplifier is shown in Fig. 8.28(a), which is similar to Fig. 8.18 except the DC supply V_{CC} and the capacitors C_E, C_1, and C_2 are shorted. Replacing the transistor Q_1 by its model of Fig. 8.28(c), the small-signal AC equivalent circuit is shown in Fig. 8.28[(b) or (c)], which can be represented by an equivalent voltage amplifier as shown in Fig. 8.28(d). We will consider R_L as an external element to the amplifier so that the effect of loading can be determined. Thus, R_L is not included in Fig. 8.28(b).

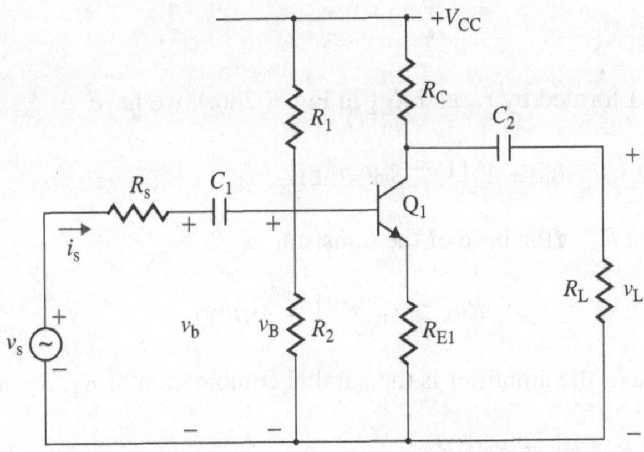

FIGURE 8.27 Common-emitter BJT amplifier with one emitter resistor

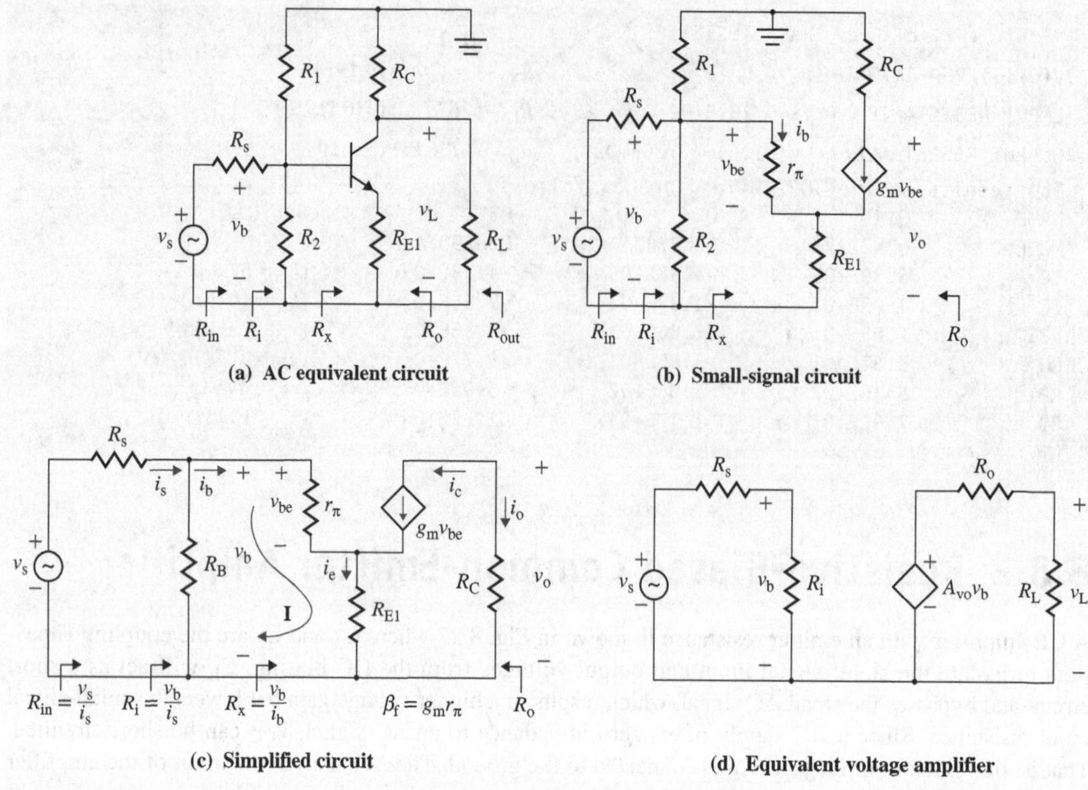

FIGURE 8.28 Equivalent circuits of a common-emitter amplifier

We will derive the input resistance R_i, the output resistance R_o, and the no-load voltage gain A_{vo} of common-emitter BJT amplifier. We will include the biasing resistance R_B, which will reduce the value of R_i.

Input Resistance R_i

Using KVL around loop I formed by r_π and R_{E1} in Fig. 8.28(c), we have

$$v_b = i_b r_\pi + R_{E1} i_e = i_b[r_\pi + (1 + g_m r_\pi)R_{E1}] \tag{8.84}$$

which gives the resistance R_x at the base of the transistor as

$$R_x = \frac{v_b}{i_b} = r_\pi + (1 + g_m r_\pi)R_{E1} = r_\pi + (1 + \beta_f)R_{E1} \tag{8.85}$$

Thus, the input resistance of the amplifier is the parallel combination of R_1, R_2, and R_x. That is,

$$R_i = \frac{v_b}{i_s} = R_1 \parallel R_2 \parallel R_x = R_B \parallel R_x \tag{8.86}$$

where $\quad R_B = R_1 \parallel R_2 \quad$ (8.87)

Thus, R_i depends on R_E, R_1, and R_2. Their values can be chosen to give the input resistance required of the amplifier.

Output Resistance R_o

Output resistance R_o, which is Thevenin's resistance, can be calculated from Fig. 8.28(c) if v_s is shorted and a test voltage v_x is applied across R_C. Since $v_s = 0$, the dependent source current will be zero—that is, the circuit will be open. The output resistance will simply be R_C. That is,

$$R_o = R_C \tag{8.88}$$

Open-Circuit (or No-Load) Voltage Gain A_{vo}

The open-circuit output voltage from Fig. 8.28(c) is

$$v_o = -R_C i_c = -R_C g_m v_{be} \tag{8.89}$$

The B-E voltage v_{be}, which controls the collector current, can be related to r_π by

$$v_{be} = r_\pi i_b \tag{8.90}$$

Substituting i_b from Eq. (8.84) into Eq. (8.90) yields

$$v_{be} = \frac{r_\pi}{r_\pi + (1 + g_m r_\pi)R_{E1}} v_b \tag{8.91}$$

Substituting v_{be} from Eq. (8.91) into Eq. (8.89) gives the output voltage

$$v_o = -R_C g_m \frac{r_\pi}{r_\pi + (1 + g_m r_\pi)R_{E1}} v_b$$

which gives the *open-circuit voltage gain* A_{vo} as

$$A_{vo} = \frac{v_o}{v_b} = \frac{-g_m r_\pi R_C}{r_\pi + (1 + g_m r_\pi)R_{E1}} = \frac{-\beta_f R_C}{r_\pi + (1 + \beta_f)R_{E1}} \tag{8.92}$$

This equation indicates that the voltage gain A_{vo} can be made large (1) by making $R_{E1} = 0$, (2) by using a transistor with a large value of g_m (or β_f), and (3) by choosing a high value of R_C. For $R_{E1} = R_E - R_{E2} = 0$, Eq. (8.92) gives the maximum open-circuit voltage gain as

$$A_{vo(max)} = -\frac{g_m r_\pi R_C}{r_\pi} = -g_m R_C = -\frac{\beta_f R_C}{r_\pi} \tag{8.93}$$

Making $R_{E1} = 0$ will decrease the input resistance R_x in Eq. (8.85), and the amplifier will draw more current from the input source, but the DC-biasing point also depends on R_C and R_E. These conflicting constraints—a higher value of R_{E1} for a larger input resistance and a lower value for a larger voltage gain—can be satisfied by using two emitter resistors R_{E1} and R_{E2}, as shown in Fig. 8.29. R_{E1} and R_{E2} set the DC-biasing point, and R_{E1} gives the desired AC input resistance or voltage gain. For DC-biasing calculations, however, R_E $(=R_{E1} + R_{E2})$ should be used. It is often necessary to compromise among the design specifications for the biasing point, the input resistance, and the open-circuit voltage gain. It is not always possible to satisfy all the design specifications with one amplifier stage.

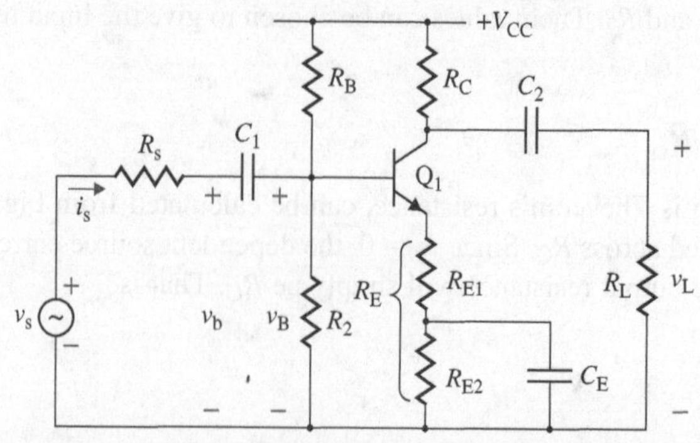

FIGURE 8.29 Common-emitter BJT amplifier with two emitter resistors

EXAMPLE 8.5

D

Designing a common-emitter BJT amplifier

(a) Design a CE amplifier as shown in Fig. 8.29 to give a voltage gain of $|A_{vo}| = v_o/v_b \geq 20$. Use transistor Q2N2222, for which minimum $\beta_F = 100$, nominal $\beta_F = 173$, $I_S = 3.295 \times 10^{-14}$ A, and $V_A = 100$ V. The operating DC collector current is to be set at $I_C = 10$ mA. The DC power supply is $V_{CC} = 15$ V. Assume $V_{BE} = 0.7$ V.

(b) Use PSpice/SPICE to verify your results in part (a).

SOLUTION

(a) Step 1. Design the biasing circuit. The results of Example 8.3 give $R_C = 500\ \Omega$, $R_E = 495\ \Omega$, $R_1 = 12.1\ \text{k}\Omega$, and $R_2 = 8.52\ \text{k}\Omega$.

Step 2. Find the small-signal parameters of the transistor. The results of Example 8.3 give $r_\pi = 258\ \Omega$, $g_m = 387.6$ mA/V, $\beta_F = 100$, and $r_o = 20\ \text{k}\Omega$ (which can be ignored for hand calculations).

Step 3. Find the values of C_1, C_2, C_E, and R_{E1}. Let us choose $C_1 = C_2 = C_E = 10\ \mu$F. The worst-case maximum possible gain that we can obtain from transistor Q2N2222 operating at $I_C = 10$ mA can be found from Eq. (8.93):

$$|A_{vo(max)}| = \frac{\beta_f R_C}{r_\pi} = \frac{100 \times 500}{258} = 193.8\ \text{V/V}$$

The desired gain is less than the maximum possible value, so we can proceed with the design. Otherwise, we would need to choose another transistor with a higher value of β_f. The value of unbypassed emitter resistance R_{E1} in Fig. 8.29 can be found from Eq. (8.92). That is,

$$r_\pi + (1 + \beta_f)R_{E1} = \frac{\beta_f R_C}{|A_{vo}|} \tag{8.94}$$

which, for $|A_{vo}| = 20$, $\beta_f = 100$, $R_C = 500\ \Omega$, and $r_\pi = 258\ \Omega$, gives $R_{E1} = 22.2\ \Omega$ and $R_{E2} = R_E - R_{E1} = 495 - 22.2 = 472.8\ \Omega$.

(b) The transistor circuit for PSpice simulation is shown in Fig. 8.30. Note that a signal source resistance $R_s = 0.1\ \Omega$ is connected at the input side to make the PSpice circuit more general.

The results of .OP analysis obtained from the output file follow. (The values obtained from hand calculations are shown in parentheses.)

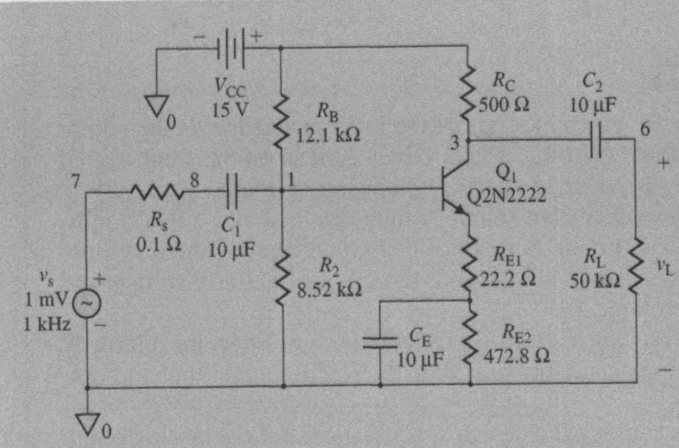

FIGURE 8.30 Transistor circuit for PSpice simulation

IB	8.91E-05	$I_B = 89.1\ \mu A$	(100 μA)
IC	9.15E-03	$I_C = 9.1$ mA	(10 mA)
VBE	6.81E-01	$V_{BE} = 0.681$ V	(0.7 V)
VBC	-5.18E+00	$V_{CB} = 5.15$ V	
VCE	5.86E+00	$V_{CE} = 5.86$ V	(5 V)
GM	3.54E-01	$g_m = 0.354$ A/V	(0.3876 A/V)
RPI	2.90E+02	$r_\pi = 280\ \Omega$	(258 Ω)
RO	2.24E+04	$r_o = 22.4$ kΩ	(20 kΩ)

The PSpice plots of the base voltage v_B = V(C1:2), the collector voltage v_C = V(RC:1), the input voltage v_s = V(vs:+), and the load voltage v_o = V(RL:2) are shown in Fig. 8.31. Notice that v_B and v_C have a DC value with an AC signal superimposed on them such that $v_B = V_B + v_s$ and $v_C = V_C + v_o$. Capacitor C_1 superimposes v_s on V_B, whereas capacitor C_2 separates the amplified AC voltage (i.e., the output voltage v_o) from v_C. The voltage gain v_o/v_s is 16.77, which is less than the desired value of 20. Thus, the design calculation should be repeated until the desired gain is obtained. We might try reducing the value of R_{E1} and increasing the value of R_{E2} by the same amount so that the biasing point remains fixed. Even if we modify the design to satisfy the specifications, the results can be expected to differ from those that would be obtained in the laboratory, although not significantly due to the variable value of β_F.

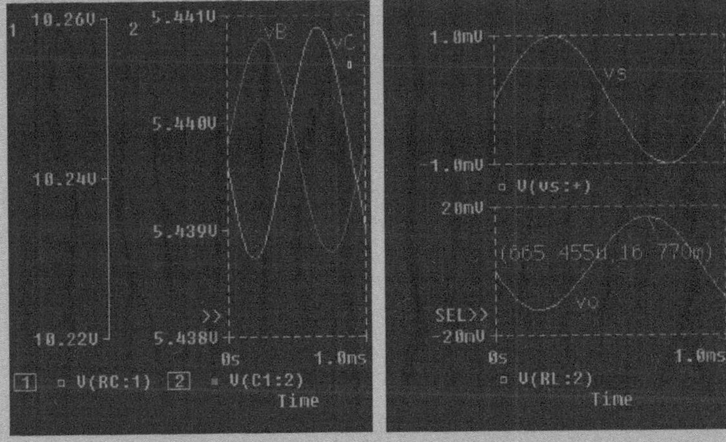

FIGURE 8.31 PSpice plots for Example 8.5

KEY POINTS OF SECTION 8.8

- The specifications of an amplifier are given in terms of the input resistance R_i, the output resistance R_o, and the no-load voltage gain A_{vo}. Their values depend on the small-signal models of the transistor, which in turn depend on the DC-biasing Q-point.
- BJT can be used to generate a sourcing current source, which can then bias a BJT amplifier and act as a high-resistance load, giving a high no-load voltage gain, which is independent of the DC biasing collector current, but depends directly on the Early voltage and decreases inversely with thermal voltage V_T.
- The voltage gain and the input resistance of a resistive-biased amplifier are relatively low. The amplifier requires coupling capacitors to isolate the small-signal input and output voltages from the DC biasing.
- It is often necessary to compromise among the design specifications for the biasing point, the input resistance, and the no-load voltage gain. It is not always possible to satisfy all the design specifications with one amplifier stage.

8.9 Emitter Followers

A common-collector amplifier is generally known as an *emitter follower* because the emitter voltage follows the voltage at the base terminal. Such an amplifier has a low output resistance and a high input resistance. It is commonly used as a buffer stage between a load and the source.

8.9.1 Active-Biased Emitter Follower

This arrangement is similar to the circuit in Fig. 8.19, except that the emitter resistance R_E is replaced by the sinking current source using *npn* transistors shown in Fig. 8.16(a). This is shown in Fig. 8.32(a). Replacing the transistors by their signal model, the AC equivalent circuit is shown in Fig. 8.32(b), which can be simplified to Fig. 8.32(c).

Input Resistance R_i

When the load resistance R_L is connected to the amplifier, R_L becomes parallel to R_E and will affect the input resistance. Unlike the case of a CE amplifier, in an emitter follower, R_L should be included in finding the input resistance R_i. Using KVL around the B-E loop I in Fig. 8.32(c), we get

$$v_b = i_b r_\pi + i_e(r_{o1} \| r_{o2} \| R_L) = i_b[r_\pi + (1 + \beta_f)(r_{o1} \| r_{o2} \| R_L)] \tag{8.95}$$

which gives the resistance R_x at the base of the transistor as

$$R_x = \frac{v_b}{i_b} = r_\pi + (1 + \beta_f)(r_{o1} \| r_{o2} \| R_L) \tag{8.96}$$

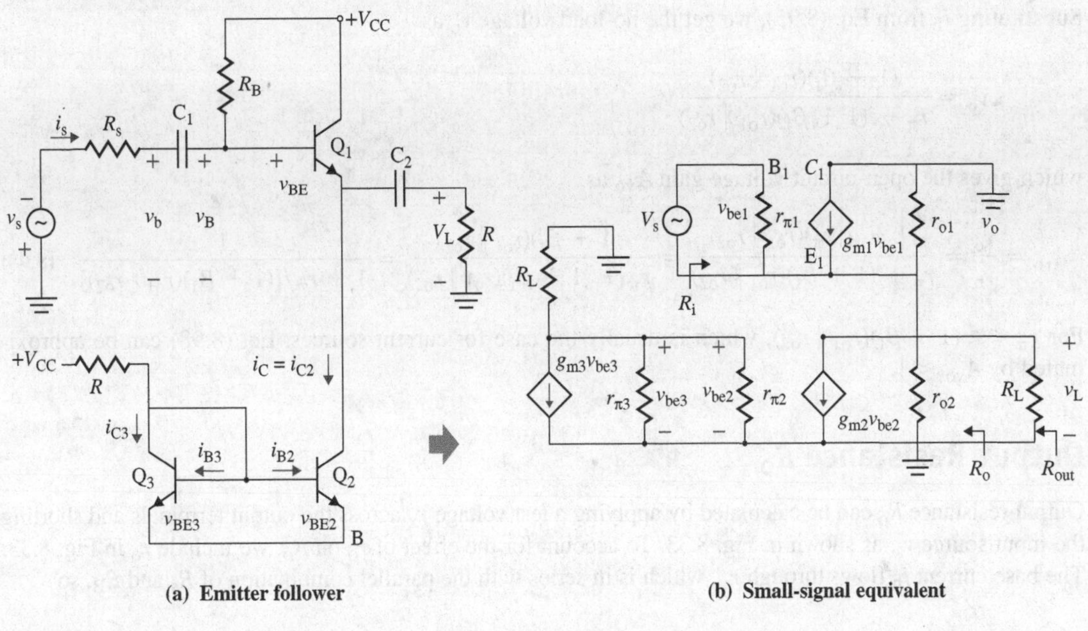

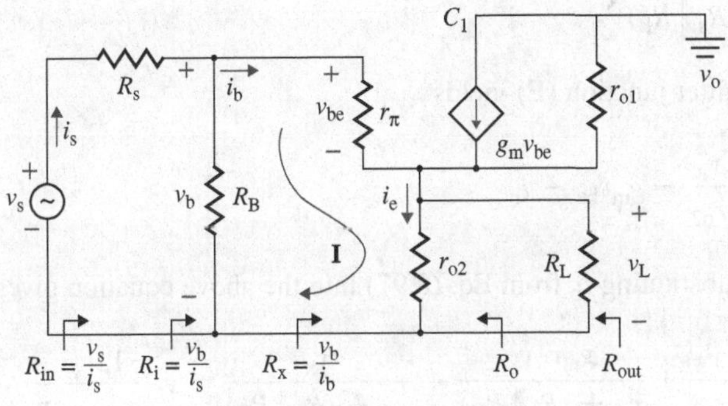

FIGURE 8.32 Active-biased emitter followers

Input resistance R_i, which is the parallel combination of R_B and R_x, is

$$R_i = \frac{v_b}{i_s} = R_B \parallel R_x \tag{8.97}$$

Open-Circuit (or No-Load) Voltage Gain A_{vo}

The open-circuit output voltage from Fig. 8.32(c) v_o is

$$v_o = i_e R_E = (i_b + g_m v_{be})(r_{o1} \parallel r_{o2}) \quad \text{where } v_{be} = v_{be1}$$

Since $v_{be} = r_\pi i_b$, we get

$$v_o = (i_b + g_m v_{be})R_E = (1 + g_m r_\pi) i_b (r_{o1} \parallel r_{o2}) = (1 + \beta_f) i_b (r_{o1} \parallel r_{o2})$$

Substituting i_b from Eq. (8.95), we get the no-load voltage v_o as

$$v_o = \frac{(1 + \beta_f)(r_{o1} \parallel r_{o2})}{r_\pi + (1 + \beta_f)(r_{o1} \parallel r_{o2})} v_b$$

which gives the open-circuit voltage gain A_{vo} as

$$A_{vo} = \frac{v_o}{v_b} = \frac{(1 + g_m r_\pi)(r_{o1} \parallel r_{o2})}{r_\pi + (1 + \beta_f)(r_{o1} \parallel r_{o2})} = \frac{(1 + \beta_f)(r_{o1} \parallel r_{o2})}{r_\pi + (1 + \beta_f)(r_{o1} \parallel r_{o2})} = \frac{1}{1 + r_\pi/[(1 + \beta_f)(r_{o1} \parallel r_{o2})]} \tag{8.98}$$

For $r_\pi << (1 + \beta_f)(r_{o1} \parallel r_{o2})$, which is usually the case for current sources, Eq. (8.98) can be approximated by $A_{vo} \approx 1$.

Output Resistance R_o

Output resistance R_o can be calculated by applying a test voltage v_x across the output terminals and shorting the input source v_s, as shown in Fig. 8.33. To account for the effect of r_o on R_o, we include r_o in Fig. 8.33. The base current i_b flows through r_π, which is in series with the parallel combination of R_s and R_B, so

$$i_b = \frac{-v_x}{r_\pi + (R_s \parallel R_B)} \tag{8.99}$$

Using KCL at the emitter junction (E) yields

$$i_x = \frac{v_x}{r_{o1}} + \frac{v_x}{r_{o2}} - g_m v_{be} - i_b$$

Since $v_{be} = -i_b r_\pi$, substituting i_b from Eq. (8.97) into the above equation gives

$$i_x = \frac{v_x}{r_{o1}} + \frac{v_x}{r_{o2}} + \frac{g_m r_\pi v_x}{r_\pi + (R_s \parallel R_B)} + \frac{v_x}{r_\pi + (R_s \parallel R_B)} = v_x\left[\frac{1}{r_{o1}} + \frac{1}{r_{o2}} + \frac{1 + g_m r_\pi}{r_\pi + (R_s \parallel R_B)}\right]$$

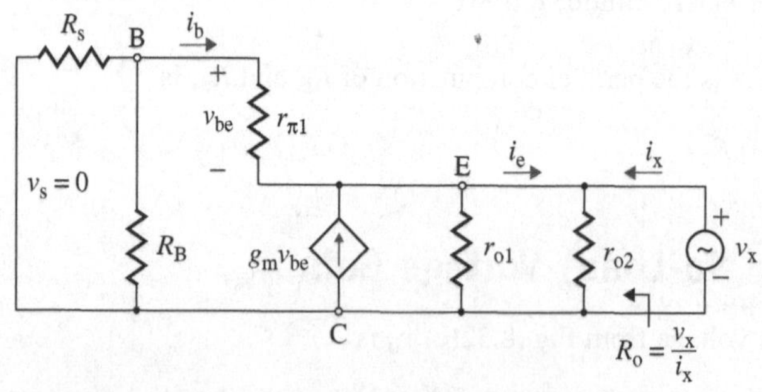

FIGURE 8.33 Equivalent circuit for determining output resistance R_o

which gives the output resistance R_o as

$$R_o = \frac{v_x}{i_x} = r_{o1} \parallel r_{o2} \parallel \frac{r_\pi + (R_s \parallel R_B)}{1 + g_m r_\pi} \tag{8.100}$$

Thus, R_o is the parallel combination of r_{o1}, r_{o2}, and $[r_\pi + (R_s \parallel R_B)]$ reflected from the i_b branch into the i_e branch. Since $\beta_f >> 1$ and $R_s << R_B$, the output resistance r_{o2} can be approximated by $R_o \approx (r_\pi + R_s)/\beta_f$.

Note from Eqs. (8.96) and (8.98) that the output resistance r_{o2} of the current source should be as high as possible to give a higher input resistance R_i and a higher voltage gain A_{vo}. Since r_{o2} is generally high, the active-biased emitter follower gives a higher input resistance and almost close-to-unity voltage gain.

An ideal current source should maintain a constant current at an infinite output resistance under all operating conditions. To achieve this goal, a number of current sources are developed in Chapter 9. The commonly used current sources are the basic current source (which was applied in Secs. 8.8.1 and 8.9.1), the modified basic source, the Widlar current source, the cascaded current source, and the Wilson current source.

▶ **NOTE** For a current source of $I_{reg} = 5\ \mu A$, a resistor of $R = 5.72\ M\Omega$ would be required. Resistors of such high values are costly in terms of the die area. Resistors over 50 kΩ are generally avoided for integrated circuit applications. Thus, this current source is not suitable for generating a current of less than about 0.6 mA at $V_{CC} = 30$ V and 0.3 mA at $V_{CC} = 15$ V.

8.9.2 Resistive-Biased Emitter Follower

The coupling capacitors C_1 and C_2 are connected to the circuit in Fig. 8.19 to feed the input signal v_s. This arrangement is shown in Fig. 8.34(a) in which R_B sets the biasing base current. The equivalent circuit for DC analysis is shown in Fig. 8.34(b). Replacing the transistor by its model of Fig. 8.10(a), the small-signal AC equivalent circuit of the emitter follower is shown in Fig. 8.34(c). Figure 8.34(d) shows the emitter follower using the two–base resistor biasing as shown in Fig. 8.20.

The analysis of this circuit is similar to that of the circuit in Fig. 8.32(a), and we can use the same equations by substituting r_{o2} by R_E. From Eqs. (8.96) and (8.97), we can find the input resistance as given by

$$R_i = R_B \parallel [r_\pi + (1 + \beta_f)(r_{o1} \parallel R_E \parallel R_L)] \tag{8.101}$$

From Eq. (8.98), we can find the no-load voltage gain as given by

$$A_{vo} = \frac{(1 + g_m r_\pi)(r_{o1} \parallel R_E)}{1 + (1 + g_m r_\pi)(r_{o1} \parallel R_E)} = \frac{(1 + \beta_f)(r_{o1} \parallel R_E)}{1 + (1 + \beta_f)(r_{o1} \parallel R_E)} \tag{8.102}$$

From Eq. (8.100), we can find the output resistance as given by

$$R_o = (r_{o1} \parallel R_E) \parallel \frac{r_\pi + (R_s \parallel R_B)}{1 + \beta_f} \tag{8.103}$$

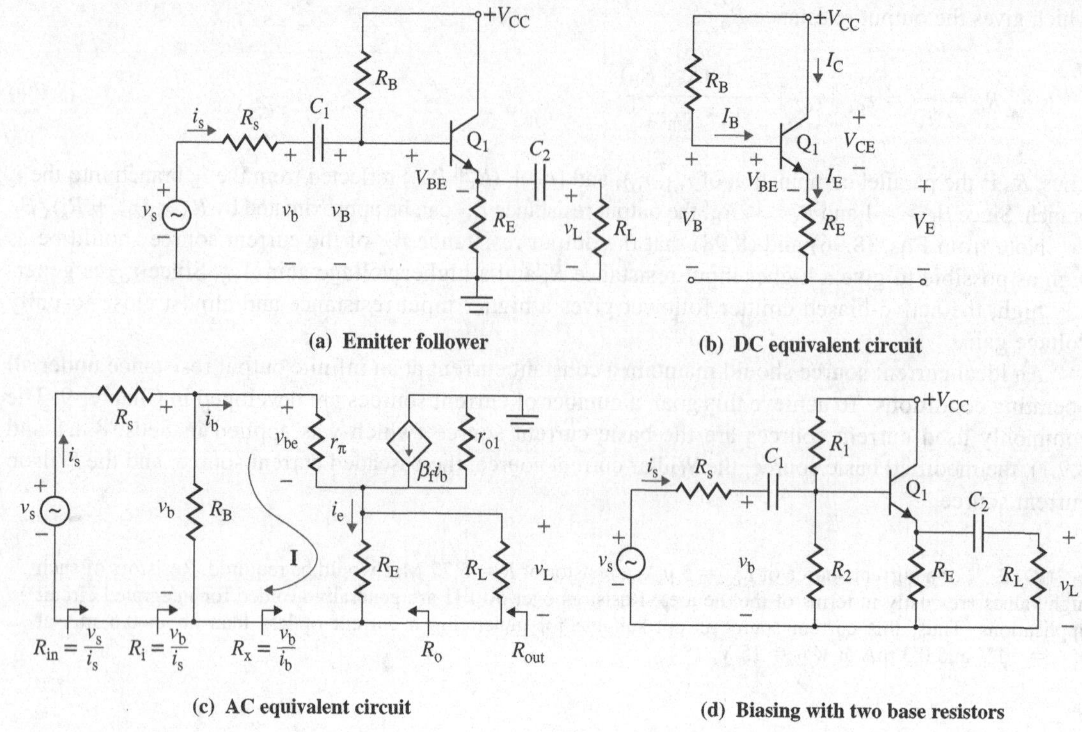

FIGURE 8.34 Emitter follower

EXAMPLE 8.6

D

Designing an emitter follower

(a) Design an emitter follower with the topology shown in Fig. 8.34(a). Use transistor Q2N2222, for which minimum $\beta_F = 100$, nominal $\beta_F = 173$, $I_S = 3.295 \times 10^{-14}$ A, and $V_A = 200$ V. The operating collector current is set at $I_C = 10$ mA. The DC power supply is $V_{CC} = 15$ V. Assume $V_{BE} = 0.7$ V, $R_L = 5$ kΩ, and $R_s = 250$ Ω.

(b) Use PSpice/SPICE to verify your results in part (a).

SOLUTION

(a) **Step 1.** Design the biasing circuit. $I_C = 10$ mA, and $V_{CC} = 15$ V. We will design for the worst-case value of $\beta_F = 100$.

$$I_E = \frac{(1 + \beta_F)I_C}{\beta_F} = \frac{101 \times 10 \text{ mA}}{100} = 10.1 \text{ mA}$$

$$I_B = \frac{I_C}{\beta_F} = \frac{10 \text{ mA}}{100} = 0.1 \text{ mA}$$

From Table 8.1 for Fig. 8.19, $V_E = V_{CC}/2 = 15/2 = 7.5$ V, which gives the value of

$$R_E = \frac{V_E}{I_E} = \frac{7.5}{10.1 \text{ mA}} = 743\ \Omega$$

The power rating of R_E is

$$P_{RE} = I_E^2 R_E = (10.1 \text{ mA})^2 \times 743 = 758 \text{ mW}$$

The base voltage V_B becomes

$$V_B = V_E + V_{BE} = 7.5 + 0.7 = 8.2 \text{ V}$$

The value of R_B can be found from

$$R_B = \frac{V_{CC} - V_B}{I_B} = \frac{(V_{CC} - V_B)\beta_F}{I_C} \tag{8.104}$$

$$= \frac{(15 - 8.2) \text{ V} \times 100}{10 \text{ mA}} = 68 \text{ k}\Omega$$

The power rating of R_B is

$$P_{RB} = I_B^2 R_B = (0.1 \text{ mA})^2 \times 68 \text{ k}\Omega = 0.68 \text{ mW}$$

Step 2. Find the small-signal parameters of the transistor. The results of Example 8.3 give $r_\pi = 258\ \Omega$, $g_m = 387.6$ mA/V, $\beta_F = 100$, and $r_{o1} = 20$ kΩ.

Step 3. Find the values of C_1 and C_2. Let us choose $C_1 = C_2 = 10\ \mu$F.

Step 4. Evaluate the values of the input resistance R_i, the open-circuit voltage gain A_{vo}, and the output resistance R_o. From Eq. (8.96),

$$R_x = r_\pi + (1 + \beta_f)(r_{o1} \| R_E \| R_L) = 258 + (1 + 100) \times (20 \text{ k} \| 743 \| 5 \text{ k}) = 65.59 \text{ k}\Omega$$

From Eq. (8.97),

$$R_i = R_B \| R_x = 68 \text{ k} \| 65.59 \text{ k} = 33.4 \text{ k}\Omega$$

$$R_{in} = \frac{v_s}{i_s} = R_i + R_s = 33.4 \text{ k} + 250 = 33.6 \text{ k}\Omega$$

From Eq. (8.102),

$$A_{vo} = \frac{(1 + \beta_f)(R_E \| r_{o1})}{[r_\pi + (1 + \beta_f)(R_E \| r_{o1})]}$$

$$= \frac{(1 + 100) \times (743 \| 20 \text{ k})}{[258 + (1 + 100) \times (743 \| 20 \text{ k})]} = 0.9966$$

From Eq. (8.103),

$$R_o = R_E \| r_{o1} \| \frac{r_\pi + (R_s \| R_B)}{1 + \beta_f} = 743 \| 20 \text{ k} \| \frac{258 + 250 \| 68 \text{ k}}{1 + 100} = 5\ \Omega$$

The output resistance including R_L is

$$R_{out} = R_L \| R_o = 5 \text{ k} \| 5 \approx 4.99\ \Omega$$

(b) The emitter-follower circuit for PSpice simulation is shown in Fig. 8.35.

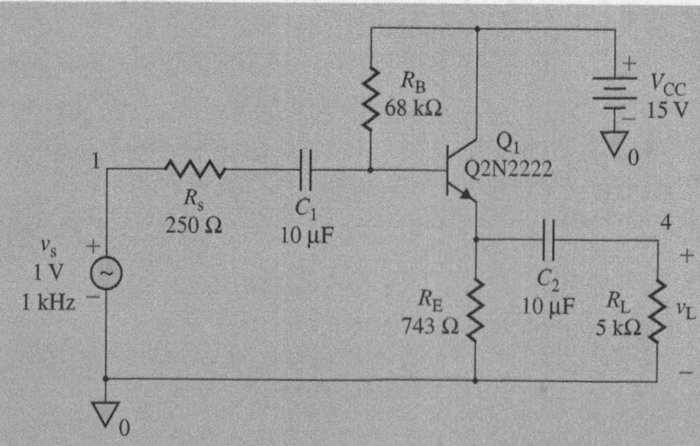

FIGURE 8.35 Emitter-follower circuit for PSpice simulation

The PSpice plots, which are shown in Fig. 8.36 for $v_s = 1$ V, give $v_o = 987$ mV and $R_{in} = V_{s(rms)}/I_{s(rms)} = 41.98$ kΩ (expected value is 33.6 kΩ), and the voltage gain is $A_v = v_o/v_s = 0.987$ (expected value is 0.9966). If we run the simulation with a very large value of R_L, tending to infinity (say, $R_L = 10$ GΩ), the output voltage will be the maximum $v_{o(max)}$. Then if we connect the normal load (say, $R_L = 5$ kΩ) and run the simulation, the output voltage should drop because of the current flow through the output resistance R_o of the amplifier. PSpice simulation gives $v_o = 987$ mV and $v_{o(max)} = 990$ mV for $R_L = 10$ GΩ. Thus, R_o can be found from

$$\Delta v_o = v_{o(max)} - v_o = R_o i_L = \frac{R_o v_o}{R_L}$$

which gives

$$R_o = \frac{R_L(v_{o(max)} - v_o)}{v_o} = \frac{5\text{ k}\Omega \times (990 - 987)\text{ mV}}{987\text{ mV}} = 15\ \Omega$$

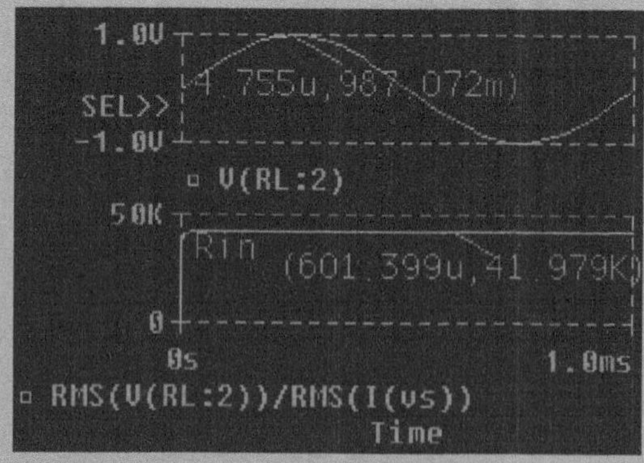

FIGURE 8.36 PSpice plots for Example 8.6

NOTE: This example simulates the steps that would normally be used to measure the output resistance of an amplifier in the laboratory. As expected, the simulated results differ from the design values, and the design calculations should be modified. Even if we modify the design to satisfy the design specifications, the results can be expected to differ from those that would be obtained in the laboratory, although not significantly.

If we ran a PSpice simulation of the linear circuit shown in Fig. 8.34(c), the results would be closer to the expected values, but they would not take into account the nonlinear behavior of the transistor. The results of .TF analysis are as follows: (The values obtained from hand calculations are shown in parentheses to the right.)

```
**** SMALL-SIGNAL CHARACTERISTICS
V(4)/VS=9.884E-01
INPUT RESISTANCE AT VS=3.310E+04
OUTPUT RESISTANCE AT V(4)=4.981E+00
```

$A_v = 0.9884$ (0.9966)
$R_{in} = 33.1\ k\Omega$ ($33.6\ k\Omega$)
$R_{out} = 4.981\ \Omega$ ($4.99\ \Omega$)

KEY POINTS OF SECTION 8.9

- A common-collector amplifier is generally known as an emitter follower because the emitter voltage follows the voltage at the base terminal with ideally a unity voltage gain. This type of amplifier has low output resistance and a high input resistance. It is commonly used as a buffer stage between a load and the source.
- The voltage gain of a resistive-biased emitter follower is close to unity, and an active-biased amplifier with a sinking current source can yield almost unity voltage gain with a very low output resistance.

8.10 Common-Base Amplifiers

In a common-base (CB) amplifier, the input signal is applied to the emitter terminal. That is, the base is common to both the input and the output terminal. Such an amplifier has a low input resistance. There is no change in the phase shift, however, between the input and output signals; that is, the output signal is in phase with the input signal. Similar to other amplifier types, we could use either active or resistive biasing. A CB amplifier with resistive biasing is shown in Fig. 8.37(a). The configuration of this circuit may appear different from that of the common emitter, but it is not. The circuit can be redrawn as shown in Fig. 8.37(b), where the input signal v_s is connected to the emitter terminal via a coupling capacitor C_1. Thus, the biasing of this circuit is identical to that of the common emitter, and the technique discussed earlier can be applied to design the DC-biasing circuit.

Let us assume that C_1, C_2, and output resistance r_{o1} of Q_1 are very large, tending to infinity. That is, $C_1 = C_2 = \infty$, and $r_o = \infty$. The small-signal AC equivalent circuit of the amplifier in Fig. 8.37(a) is shown in Fig. 8.38(a), which can be simplified to Fig. 8.38(b). R_L is considered an external element and is not included in Fig. 8.38(a). This amplifier can be represented by the equivalent voltage and transconductance amplifiers shown in Fig. 8.38[(c) and (d)], respectively.

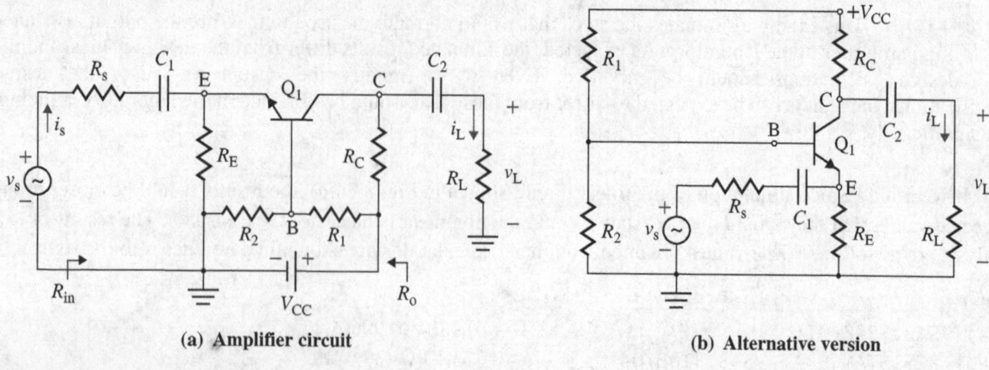

FIGURE 8.37 Common-base amplifier

8.10.1 Input Resistance R_i

Since $v_e = -i_b(r_\pi + R_B)$ in Fig. 8.38(b), the voltage v_e can be related to the base–emitter voltage v_{be} by

$$v_{be} = -\frac{r_\pi v_e}{r_\pi + R_B} \tag{8.105}$$

where $R_B = R_1 \| R_2$. Using KCL at the emitter junction E of Fig. 8.38(b) and substituting for v_{be}, we get

$$i_e \approx -i_b - g_m v_{be} = \frac{v_e}{r_\pi + R_B} - g_m v_{be} = \frac{v_e}{r_\pi + R_B} + \frac{g_m r_\pi v_e}{r_\pi + R_B} \approx \frac{1 + \beta_f}{r_\pi + R_B} v_e$$

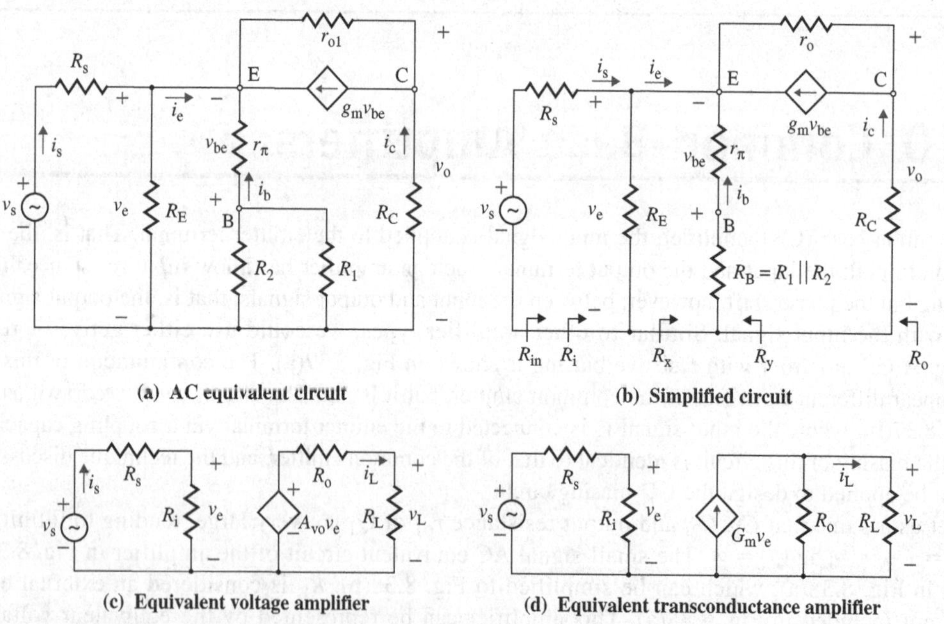

FIGURE 8.38 Small-signal AC equivalent circuits of a CB amplifier

which gives the resistance R_x at the emitter terminal as

$$R_x = \frac{v_e}{i_e} = \frac{r_\pi + R_B}{1 + g_m r_\pi} = \frac{r_\pi + R_B}{1 + \beta_f}$$

Input resistance R_i, which is the parallel combination of R_E and R_x, is

$$R_i = \frac{v_e}{i_s} = R_E \parallel R_x = R_E \parallel \frac{r_\pi + R_B}{1 + \beta_f} \quad (8.106)$$

Since $(r_\pi + R_B)/(1 + \beta_f)$ will have a small value, the input resistance R_i is usually low. This is the major disadvantage of a CB amplifier. Thus, $R_{in} = R_i + R_s$.

8.10.2 No-Load Voltage Gain A_{vo}

The no-load output voltage from Fig. 8.38(b) v_o is

$$v_o = -i_c R_C = -R_C g_m v_{be}$$

Substituting v_{be} from Eq. (8.105), we get

$$v_o = R_C g_m \frac{r_\pi}{r_\pi + R_B} v_e$$

which gives the *no-load voltage gain* A_{vo} as

$$A_{vo} = \frac{v_o}{v_e} = \frac{g_m r_\pi R_C}{r_\pi + R_B} = \frac{\beta_f R_C}{r_\pi + R_B} \quad (8.107)$$

The no-load voltage gain A_{vo} can be increased by making $R_B = 0$ if a bypass capacitor C_B is connected between the base and the ground, as shown in Fig. 8.39. Equation (8.107) gives the maximum no-load voltage gain as

$$A_{vo(max)} = \frac{\beta_f R_C}{r_\pi} = g_m R_C \quad (8.108)$$

8.10.3 Output Resistance R_o

Assuming that the output resistance of the transistor is very large, tending to infinity (i.e., $r_{o1} \approx \infty$), the output resistance R_o can be found by inspection of Fig. 8.38(b) to be $R_o \approx R_C$.

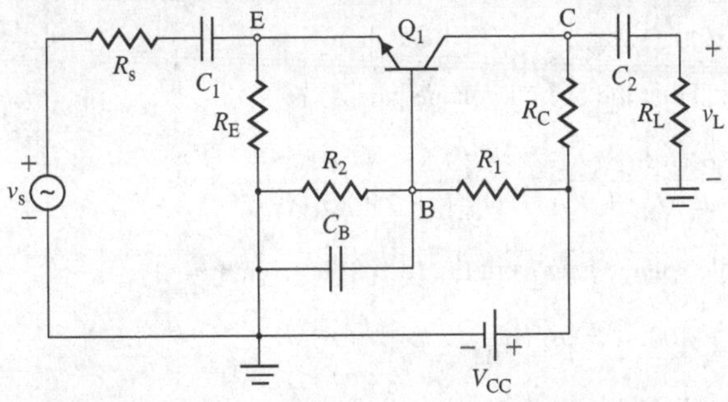

FIGURE 8.39 Common-base configuration with bypass capacitor C_B

▶ **NOTE** Because of the low input resistance, the overall voltage gain A_v is reduced considerably. The collector-base capacitor C_μ, which has a low capacitance, appears between the output and base terminals, not between the output and input terminals. If a capacitor is connected between the input and output terminals of an amplifier, the capacitance is subject to Miller's multiplication effect, as discussed in Sec. 2.6. In a CB amplifier, C_μ is not subject to Miller's multiplication effect, and thus CB amplifiers are used for high-frequency applications.

EXAMPLE 8.7

Finding the parameters of a common-base amplifier The CB amplifier of Fig. 8.37(a) has $R_1 = 13.16\ \text{k}\Omega$, $R_2 = 8.06\ \text{k}\Omega$, $R_E = 495\ \Omega$, $R_C = 500\ \Omega$, $R_L = 5\ \text{k}\Omega$, and $R_s = 250\ \Omega$. The parameters of the transistor are $r_\pi = 258\ \Omega$, $\beta_f = 100$, and $r_{o1} = \infty$. Assume that $C_1 = C_2 = \infty$.

(a) Calculate the input resistance $R_{in}\ (=v_s/i_s)$, the no-load voltage gain $A_{vo}\ (=v_o/v_e)$, the output resistance R_o, the overall voltage gain $A_v\ (=v_L/v_s)$, and the maximum permissible voltage gain $A_{vo(max)}$.

(b) Use PSpice/SPICE to verify your results in part (a).

SOLUTION

(a) We first find R_B and R_x.

$$R_B = R_1 \| R_2 = 13.16\ \text{k} \| 8.06\ \text{k} = 5\ \text{k}\Omega$$

$$R_x = \frac{r_\pi + R_B}{1 + \beta_f} = \frac{258 + 5\ \text{k}}{101} = 52.1\ \Omega$$

From Eq. (8.106),

$$R_i = R_E \| R_x = 495 \| 52.1 = 47.1\ \Omega$$

Thus

$$R_{in} = R_i + R_s = 47.1 + 250 = 297.1\ \Omega$$

From Eq. (8.107),

$$A_{vo} = \frac{\beta_f R_C}{r_\pi + R_B} = \frac{100 \times 500}{258 + 5\ \text{k}} = 9.5$$

and $\quad R_o = R_C = 500\ \Omega$

Using Fig. 8.38(c), we find that the overall voltage gain A_v is

$$A_v = \frac{v_L}{v_s} = \frac{A_{vo} R_i R_L}{(R_i + R_s)(R_L + R_o)} = \frac{9.5 \times 47.1 \times 5\ \text{k}}{(47.1 + 250)(5\ \text{k} + 500)} = 1.37$$

The maximum permissible voltage gain from Eq. (8.108) is given by

$$A_{vo(max)} = \frac{\beta_f R_C}{r_\pi} = \frac{100 \times 500}{258} = 194$$

(b) The emitter-follower circuit for PSpice simulation is shown in Fig. 8.40.

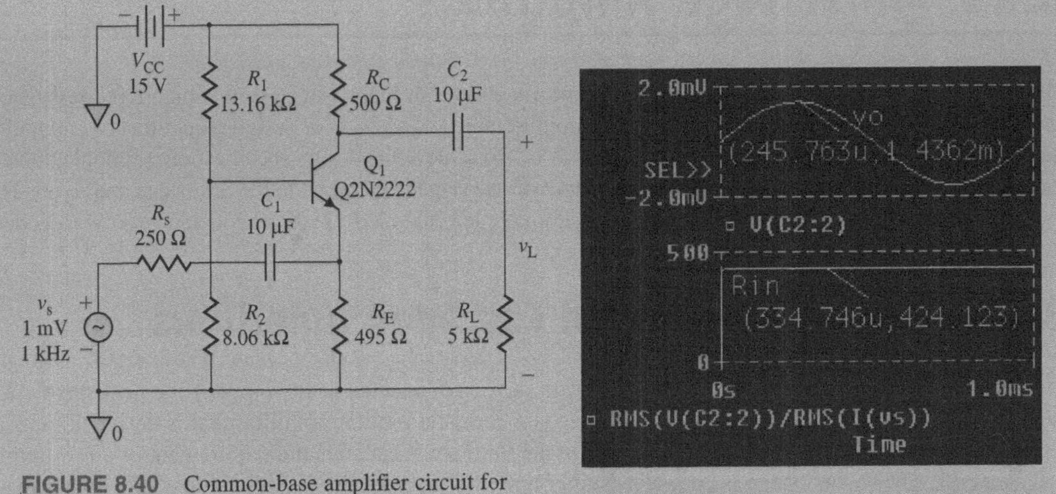

FIGURE 8.40 Common-base amplifier circuit for PSpice simulation

FIGURE 8.41 PSpice plots for Example 8.7

The PSpice plots, which are shown in Fig. 8.41 for $v_s = 1$ mV, give $v_o = 1.44$ mV and $R_{in} = V_{s(rms)}/I_{s(rms)} = 424\ \Omega$ (expected value is 297 Ω), and the voltage gain is $A_v = v_o/v_s = 1.43$ (expected value is 1.37). Thus, the results are close to the expected values.

KEY POINTS OF SECTION 8.10

- In a CB amplifier, the input signal is applied to the emitter terminal. That is, the base is common to both the input and the output terminal. Similar to other amplifier types, we could use either active or resistive biasing.
- This type of amplifier has a low input resistance. There is no phase shift, however, between the input and output signals; that is, the output signal is in phase with the input signal.
- The expressions for BJT amplifiers in terms of their parameters, the input resistance R_i, the output resistance R_o, and the no-load voltage gain A_{vo}, are summarized in Table 8.2.

TABLE 8.2 Summary of expressions for BJT amplifiers

	CE Amplifier (Fig. 8.29)	Emitter Follower (Fig. 8.34[a])	CB Amplifier (Fig. 8.37[a])	CE Amplifier with Active Load (Fig. 8.24[a])
R_i (Ω)	$R_B \parallel [r_\pi + (1+\beta_f)R_{E1}]$	$R_B \parallel [r_\pi + (1+\beta_f)(r_{o1} \parallel R_E \parallel R_L)]$	$R_E \parallel \dfrac{r_\pi + R_B}{\beta_f}$	$r_\pi \parallel R_B$
R_o (Ω)	R_C	$R_E \parallel r_{o1} \parallel \dfrac{r_\pi + (R_s \parallel R_B)}{1+\beta_f}$	R_C	$r_{o2} \parallel r_{o1}$
A_{vo} (V/V)	$\dfrac{-\beta_f R_C}{r_\pi + (1+\beta_f)R_{E1}}$	$\dfrac{(1+\beta_f)(R_E \parallel r_{o1})}{r_\pi + (1+\beta_f)(R_E \parallel r_{o1})}$	$\dfrac{\beta_f R_C}{r_\pi + R_B}$	$-g_{m1}(r_{o2} \parallel r_{o1})$

8.11 Multistage Amplifiers

The design requirements of amplifiers normally specify an overall high voltage gain, a high input resistance, and a low output resistance. A single-transistor amplifier rarely satisfies these design requirements, and multistages are often used to satisfy the design specifications. To achieve the design specifications, multiple transistor stages are connected in such a way that the output of one stage is the input to the next stage and so on. The most common types of arrangements are (1) capacitor-coupled cascaded, (2) direct coupled, and (3) cascoded.

8.11.1 Capacitor-Coupled Cascaded Amplifiers

In a *capacitor-coupled amplifier*, the output of one stage is connected to the input of the next stage via a capacitor as shown in Fig. 8.42(a). The first stage is a common-emitter amplifier that is designed to offer a high input resistance R_i. The emitter follower in the third stage satisfies the requirement of a low output resistance R_o. The second stage is common-emitter amplifier to meet the overall voltage gain requirement A_{vo}. If A_{vo1} and $A_{vo3}(\simeq 1)$ are the voltage gains of first and third stages, respectively, then the required gain for the second stage is $A_{vo2} = A_{vo}/(A_{vo1} \times A_{vo3})$. The DC-biasing point of each stage can be determined independently for each stage because the coupling capacitor provides DC isolation between the two stages. The biasing collector current should be low to reduce the power drain from the DC voltage source.

Each stage can be represented by its parameters R_1, R_o, and A_{vo} as shown in Fig. 8.42(b). The output resistance of a stage acts as the source resistance of the following stage, and the input resistance of

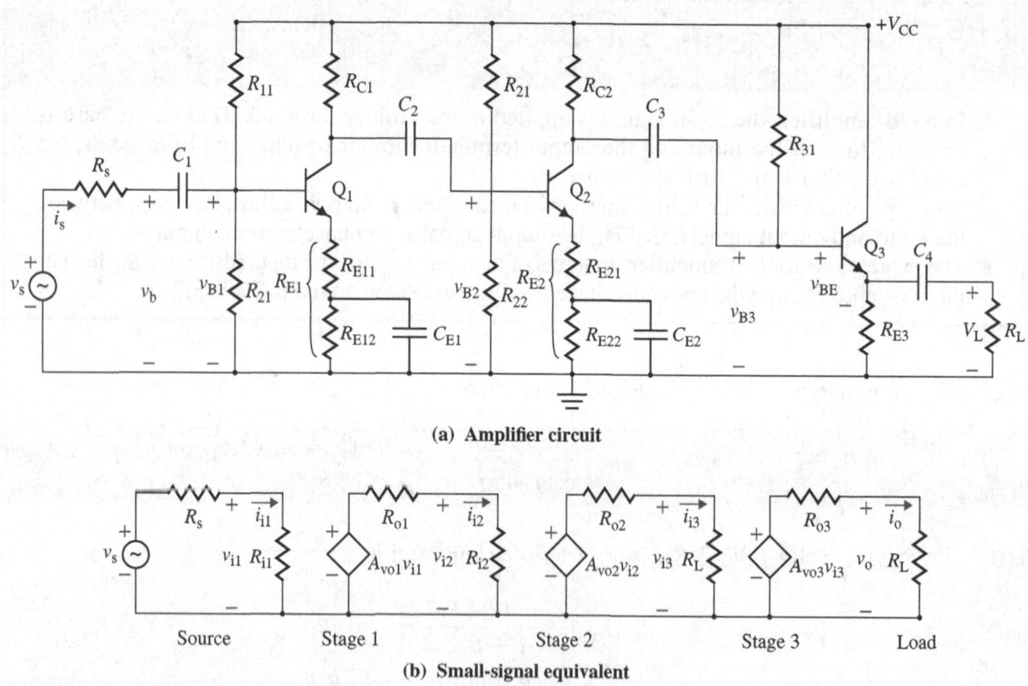

FIGURE 8.42 A three-stage capacitor-coupled cascaded amplifier

a stage is the load resistance of the preceding stage. Thus, there will be a loading effect due to the interaction between stages, and the effective voltage gain will be reduced (see Sec. 2.4). When designing an amplifier, keep in mind the gain reduction due to loading effect and start the design with a voltage gain higher than A_{vo} to satisfy the design requirement.

8.11.2 Direct-Coupled Amplifiers

In *direct-coupled amplifiers*, the output of one stage is directly connected to the input of the next stage. We can make the amplifier in Fig. 8.42(a) a direct-coupled amplifier if we remove the coupling capacitors C_1 and C_2 and connect the following stage directly to the preceding stage. In this case, we can also remove the biasing resistances R_{21}, R_{22}, and R_{31}.

8.11.3 Cascoded Amplifiers

The effective output resistance of a transistor can be increased by connecting two transistors in a configuration commonly referred to as a *cascoded amplifier*. The input signal is applied to one transistor whose output is the input to the other transistor, as shown in Fig. 8.43(a). The input signal is applied to the common-emitter amplifier, and the output is obtained at the collector of the common-base amplifier of transistor Q_2.

DC Biasing

The DC equivalent circuit for determining the DC-operating point of the transistors is shown in Fig. 8.43(b). The analysis can be simplified by assuming $\beta_{F1} = \beta_{F2} = \beta_F >> 1$. That is, the base current of a transistor is negligible, and the base currents can be ignored in comparison to the currents through the biasing resistances $I_{B1} << I_3$ and $I_{B2} << I_2$. Therefore, we can find the DC biasing base voltages as given by

$$V_{B1} = \frac{R_{B1}}{R_{B1} + R_{B2} + R_{B3}} V_{CC} \tag{8.109}$$

$$V_{B2} = \frac{R_{B1} + R_{B2}}{R_{B1} + R_{B2} + R_{B3}} V_{CC} \tag{8.110}$$

which can be used to find the collector and emitter currents:

$$I_{E1} \simeq I_{C1} = I_{E2} \simeq I_{C2} = \frac{V_{B1} - V_{BE}}{R_E} \tag{8.111}$$

The C-E voltages are given by

$$V_{CE1} = V_{B2} - V_{BE2} - (V_{B1} - V_{BE1}) = V_{B2} - V_{B1} \quad (\text{for } V_{BE1} = V_{BE2}) \tag{8.112}$$

$$V_{CE2} = V_{CC} - R_C I_{C2} - (V_{B2} - V_{BE2}) \tag{8.113}$$

Small-Signal Voltage Gain

Once we have found the DC-biasing values, we can find the small-signal mode parameters r_π, r_o, and g_m for both transistors. The small-signal equivalent circuit is shown in Fig. 8.43(c). The small-signal input resistance is given by

$$R_i = \frac{v_s}{i_s} = R_B \parallel r_{\pi 1} \tag{8.114}$$

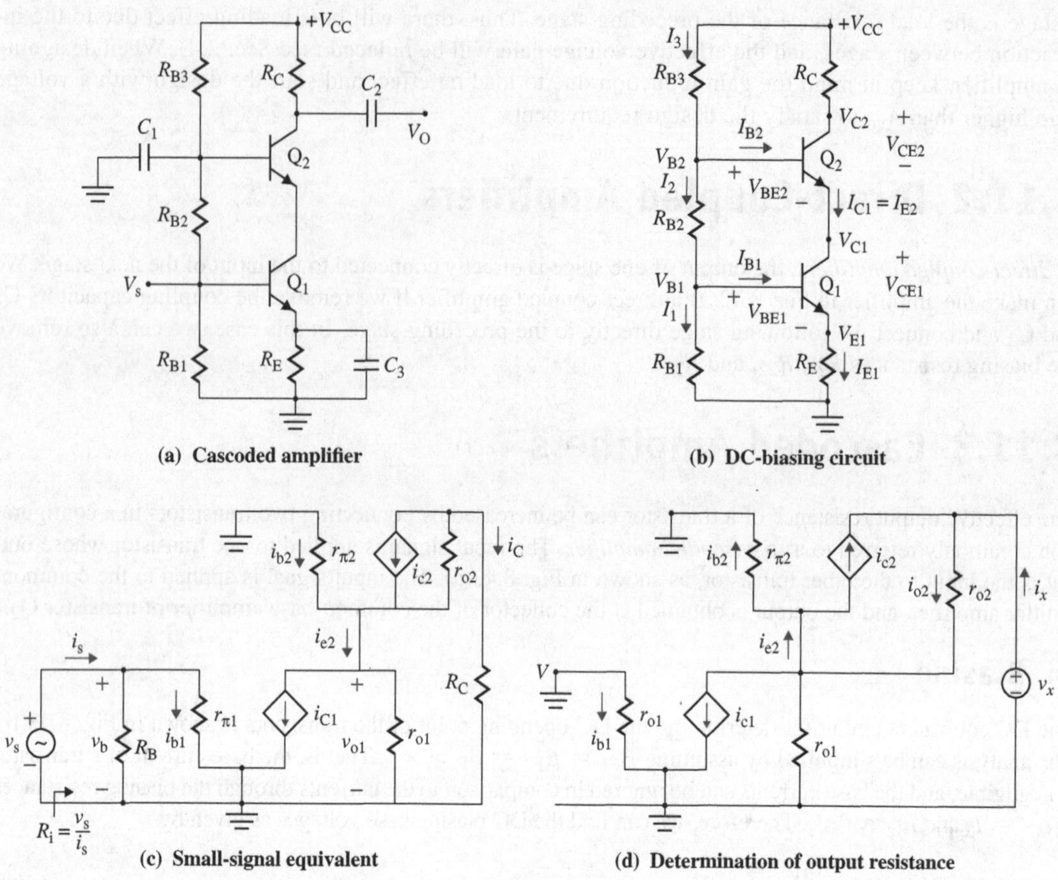

FIGURE 8.43 Cascoded amplifiers

Since $i_{c2} = g_{m2}v_{be2}$, $i_{e2} = i_{c2}/\alpha_F$, and $v_{o1} = -v_{be2}$, equating the currents $i_{c1} = i_{e2}$ gives

$$g_{m1}v_{be1} = i_{e2} = \frac{g_{m2}v_{be2}}{\alpha_F} = \frac{-g_{m2}v_{vo1}}{\alpha_F}$$

which gives the voltage gain of the first stage as

$$A_{vo1} = \frac{v_{o1}}{v_{be1}} = -\alpha_F \quad (\text{for } g_{m1} = g_{m2}) \tag{8.115}$$

Since the collector current $i_{c2} = \alpha_F i_{e2}$, we can find the output voltage as

$$v_o = -i_{c2}R_C = -R_C\alpha_F i_{e2} = -R_C\alpha_F i_{c1} = -R_C\alpha_F g_{m1}v_{be1}$$

which gives the small-signal overall voltage gain A_{vo} as

$$A_{vo} = \frac{v_o}{v_{be1}} = -\alpha_F g_{m1}R_C \tag{8.116}$$

Therefore, the voltage gain of the second stage is

$$A_{vo2} = \frac{A_{vo}}{A_{vo2}} = -\frac{\alpha_F g_{m1} R_C}{\alpha_F} = -g_{m1} R_C \tag{8.117}$$

Small-Signal Output Resistance

The equivalent circuit for determining the output resistance is shown in Fig. 8.43(d). Since $v_{be1} = 0$, the current source $g_{m1} v_{be1} = 0$ is open-circuited. Using KVL, we can write

$$v_x = r_{o2}(i_x - g_{m2} v_{be2}) + v_{o1}$$

After substituting $v_{o1} = i_x(r_{o1} \parallel r_{\pi 2})$ and $v_{be2} = -v_{vo1}$, we can get the output resistance, after simplification, as given by

$$R_o = \frac{v_x}{i_x} = (r_{o1} \parallel r_{\pi 2}) + r_{o2}[1 + g_{m2}(r_{o1} \parallel r_{\pi 2})] \tag{8.118}$$

$$\simeq r_{o2}(1 + g_{m2} r_{\pi 2}) = r_{o2}(1 + \beta_F) \quad (\text{for } r_{o1} >> r_{\pi 2})$$

which is the same as the output resistance R_o of the Widlar current source described by Eq. (9.30) in Sec. 9.4 if we substitute R_2 with r_{o1}.

KEY POINTS OF SECTION 8.11

- The design requirements of amplifiers normally specify an overall high voltage gain, a high input resistance, and a low output resistance. A single-transistor amplifier rarely satisfies these design requirements and multistages are often used to satisfy the design specifications.
- The output of one stage in a capacitively coupled amplifier is connected to the input of the next stage via a capacitor. In direct-coupled amplifiers, the output of one stage is directly connected to the input of the next stage.
- The effective output resistance of a transistor can be increased by connecting two transistors in a configuration commonly referred to as a cascoded amplifier. The input signal is applied to one transistor whose output is the input to the other transistor.

8.12 The Darlington Pair Transistor

In all types of biasing circuits, the input resistance, the output resistance, and the voltage gain are dependent on the transistor current gain β_F. A compound transistor configuration, known as a *Darlington pair*, is often used to give a much higher input resistance and a much lower input bias current than a single transistor would provide. It consists of two cascaded transistors as shown in Fig. 8.44(a); the internal structure is shown in Fig. 8.44(b). A Darlington pair can be represented as an equivalent single transistor as shown in Fig. 8.45.

The effective B-E voltage is

$$V_{BE} = V_{BE1} + V_{BE2} = V_T \ln\left(\frac{I_{C1}}{I_{S1}}\right) + V_T \ln\left(\frac{I_{C2}}{I_{S2}}\right) = V_T \ln\left(\frac{I_{C1} I_{C2}}{I_{S1} I_{S2}}\right) \tag{8.119}$$

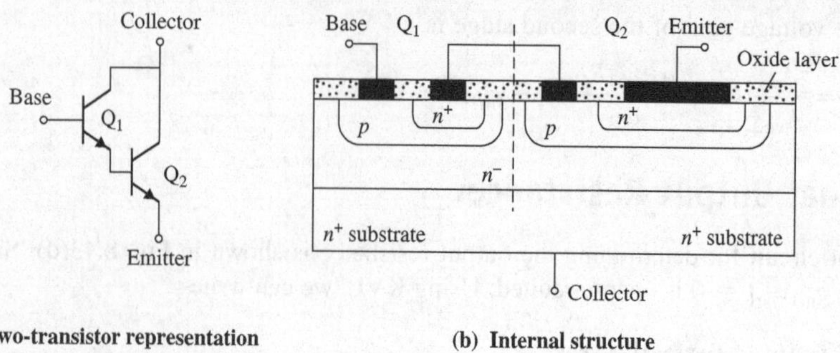

(a) Two-transistor representation **(b) Internal structure**

FIGURE 8.44 Darlington pair transistor

Since $I_C = I_{C2} = \beta_F I_{C1}$, Eq. (8.119) becomes

$$V_{BE} = V_T \ln\left(\frac{I_C^2}{\beta_F I_{S1} I_{S2}}\right) \quad \textbf{(8.120)}$$

Solving for I_C, we get

$$I_C = \sqrt{\beta_F I_{S1} I_{S2}} \exp\left(\frac{V_{BE}}{2V_T}\right) = I_S \exp\left(\frac{V_{BE}}{V_T'}\right) \quad \textbf{(8.121)}$$

where $I_S = \sqrt{\beta_F I_{S1} I_{S2}}$ = effective saturation current and $V_T' = 2V_T$ = effective thermal voltage. The collector current I_C can be related to I_{B1} by

$$I_C = I_{C2} = \beta_F I_{B2} = (1 + \beta_F) I_{C1} = \beta_F(1 + \beta_F) I_{B1} \approx \beta_F^2 I_{B1} \quad \textbf{(8.122)}$$

Thus, the effective input resistance of the compound pair is given by

$$r_\pi' = \frac{V_T'}{I_{B1}} = \beta_F^2 \frac{2V_T}{I_C} \quad \textbf{(8.123)}$$

which will be $2\beta_F$ times greater than that for a single device. For a single equivalent transistor Q_T, $r_\pi' = 2\beta_F r_\pi$. Thus, if $I_C = 200\ \mu A$, $\beta_F = 100$, and $V_T = 26$ mV,

$$r_\pi = \frac{\beta_F V_T}{I_C} = \frac{100 \times 26 \text{ mV}}{200\ \mu A} = 13\ k\Omega$$

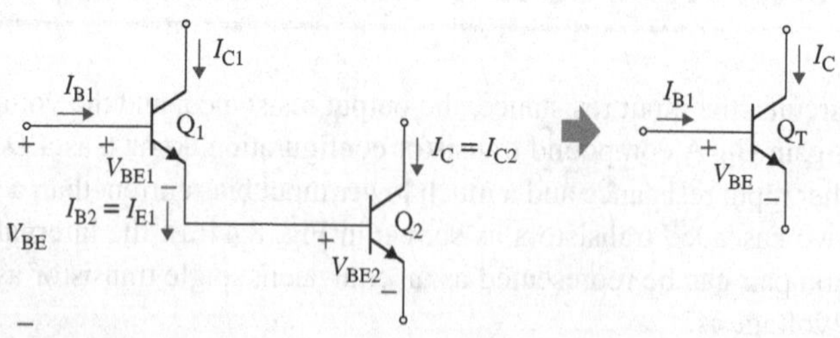

FIGURE 8.45 Darlington pair single-transistor equivalent

for a single transistor and

$$r'_\pi = \frac{100^2 \times 2 \times 26 \text{ mV}}{200\ \mu\text{A}} = 2.6 \text{ M}\Omega$$

for a Darlington pair. The input offset voltage V_{OS}, however, will increase (generally $\sqrt{2}$ times) as a result of the increase in the effective thermal voltage.

EXAMPLE 8.8

Finding the effective parameters of a Darlington pair The Darlington pair shown in Fig. 8.46 is biased in such a way that the collector biasing current I_{C2} of Q_2 is 1 mA. The current gains of the two transistors are the same, $\beta_{F1} = \beta_{F2} = \beta_F = 100$, and the Early voltage is $V_A = 75$ V. Calculate **(a)** the effective input resistance r_π, **(b)** the effective transconductance g_m, **(c)** the effective current gain $\beta_{F(\text{eff})}$, and **(d)** the effective output resistance r_o.

SOLUTION

The two transistors in Fig. 8.46 can be replaced by an equivalent transistor, shown in Fig. 8.47(a), which may be regarded as the subcircuit of the two transistors and can be modeled by the circuit in Fig. 8.47(b). Replacing each transistor by its model gives the small-signal equivalent circuit of a Darlington pair shown in Fig. 8.47(c).

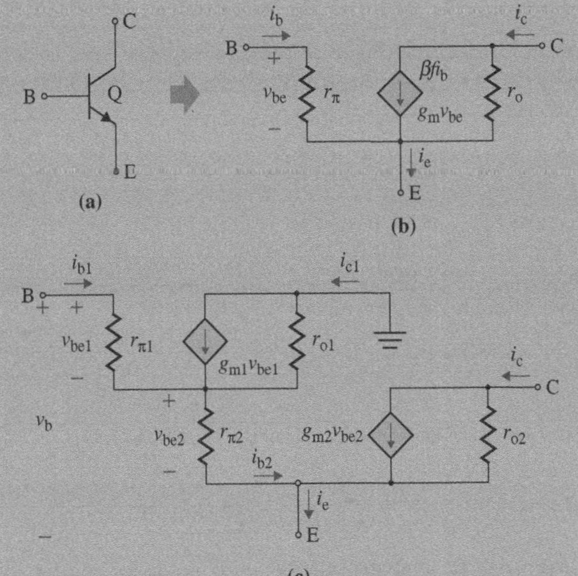

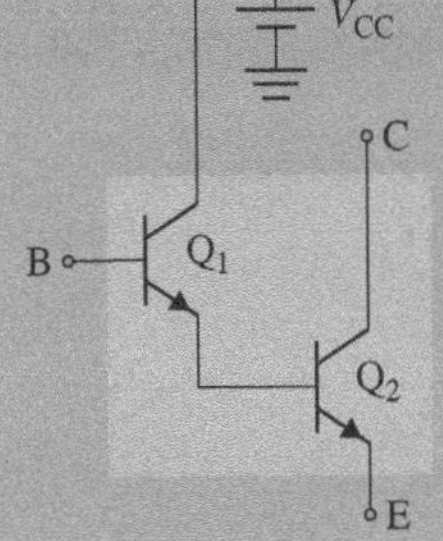

FIGURE 8.46 Two transistors connected as a Darlington pair

FIGURE 8.47 Equivalent model of Darlington pair

Let us assume that $r_o \approx \infty$. The two collector and base currents of the transistors will be different. According to Eq. (8.14), r_π depends inversely on the collector current. Thus, the small-signal input resistances $r_{\pi1}$ and $r_{\pi2}$ of the transistors will be different. Assuming we have $V_T = 25.8$ mV,

$$g_{m2} = \frac{I_{C2}}{V_T} = \frac{1\text{ mA}}{25.8\text{ mV}} = 38.76\text{ mA/V}$$

$$r_{\pi2} = \frac{\beta_F}{g_{m2}} = \frac{\beta_F V_T}{I_{C2}} = \frac{100 \times 25.8\text{ mV}}{1\text{ mA}} = 2.58\text{ k}\Omega$$

$$I_{B2} = I_{E1} = \frac{I_{C2}}{\beta_F} = \frac{1\text{ mA}}{100} = 10\ \mu\text{A}$$

$$r_{o2} = \frac{V_A}{I_{C2}} = \frac{75\text{ V}}{1\text{ mA}} = 75\text{ k}\Omega$$

$$I_{C1} = \frac{I_{B2}\beta_F}{1 + \beta_F} = \frac{10\ \mu\text{A} \times 100}{1 + 100} = 9.9\ \mu\text{A}$$

$$g_{m1} = \frac{I_{C1}}{V_T} = \frac{9.9\ \mu\text{A}}{25.8\text{ mV}} = 383.7\ \mu\text{A/V}$$

$$r_{\pi1} = \frac{\beta_F}{g_{m1}} = \frac{100}{383.7\ \mu\text{A}} = 260.6\text{ k}\Omega$$

$$r_{o1} = \frac{V_A}{I_{C1}} = \frac{75\text{ V}}{9.9\ \mu\text{A}} = 7.6\text{ M}\Omega$$

(a) Figure 8.47(c) is similar to Fig. 8.28(b). Thus, Eq. (8.85) can be applied to Fig. 8.47(c) if r_π and R_E are replaced by $r_{\pi1}$ and $r_{\pi2}$, respectively. Replacing R_{E1} in Eq. (8.85) with $r_{\pi2}$ (in parallel with r_{o1}) gives the input resistance

$$r_\pi = r_{\pi1} + (1 + \beta_F)(r_{\pi2} \parallel r_{o1}) \tag{8.124}$$

$$= 260.6\text{ k}\Omega + (1 + 100) \times (2.58\text{ k}\Omega \parallel 7.6\text{ M}\Omega) = 521.1\text{ k}\Omega$$

(b) The voltage v_{be2} in Fig. 8.47(c) is identical to v_o in Fig. 8.34(c) provided $r_{\pi2}$ is substituted for R_E. Thus, v_{be2} can be related to the input voltage v_b by Eq. (8.102):

$$v_{be2} = \frac{(1 + \beta_{F1})r_{\pi2}}{r_{\pi1} + (1 + \beta_{F1})r_{\pi2}} v_b \tag{8.125}$$

The collector current i_c of the second transistor can be found from

$$i_c = g_{m2}v_{be2} = g_{m2}\frac{(1 + \beta_{F1})r_{\pi2}}{r_{\pi1} + (1 + \beta_{F1})r_{\pi2}} v_b$$

which gives the equivalent transconductance g_m as

$$g_m = \frac{i_c}{v_b} = g_{m2}\frac{(1 + \beta_{F1})r_{\pi2}}{r_{\pi1} + (1 + \beta_{F1})r_{\pi2}} = g_{m2}\frac{1}{1 + r_{\pi1}/[(1 + \beta_{F1})r_{\pi2}]} \tag{8.126}$$

Since the emitter current I_{E1} of Q_1 is equal to the base current I_{B2} of Q_2, the biasing base current I_{B2} of Q_2 will be related to the biasing base current I_{B1} of Q_1 by

$$I_{B2} = I_{E1} = (1 + \beta_{F1})I_{B1}$$

According to Eq. (8.21), r_π is inversely proportional to the biasing base current I_B. Thus, $r_{\pi1}$ and $r_{\pi2}$ will be related by $r_{\pi1} = (1 + \beta_{F1})r_{\pi2}$. Therefore, Eq. (8.126) can be simplified to

$$g_m = \frac{g_{m2}}{2} \tag{8.127}$$

$$= \frac{38.76}{2} = 19.38 \text{ mA/V}$$

(c) The collector current i_{c2} of Q_2 can also be written as

$$i_c = \beta_{F2} i_{b2}$$

Since $i_{b2} = (1 + \beta_{F1})i_{b1}$,

$$i_c = \beta_{F2} i_{b2} = \beta_{F2}(1 + \beta_{F1})i_{b1}$$

Thus, the effective current gain $\beta_{F(eff)}$, which is the ratio of i_c to i_b, is

$$\beta_{F(eff)} = \frac{i_c}{i_{b1}} = \beta_{F2}(1 + \beta_{F1}) \tag{8.128}$$

$$= \beta_{F2}(1 + \beta_{F1}) = 100 \times (1 + 100) = 10{,}100$$

(d) The effective output resistance r_o is

$$r_o = r_{o2} = 75 \text{ k}\Omega \tag{8.129}$$

Thus, a model that can represent the two CC or CE transistors is shown in Fig. 8.47(b).

KEY POINT OF SECTION 8.12

- A compound transistor configuration known as a *Darlington pair*, which increases the effective current gain β_F, is often used to give a much higher input resistance and a much lower input bias current than a single transistor would provide.

8.13 DC Level Shifting and Amplifier

In all the amplifiers discussed so far, we used coupling capacitors to superimpose the small AC signal on the DC biasing voltage at the base terminal of the transistors. These capacitors provide DC isolations of each stage from the previous or subsequent stage. The amplified AC signal is superimposed on the DC biasing voltage at the output terminal of the transistors. The coupling capacitors cannot be used in the design of amplifiers that only amplify DC signals. In some cases, it may be necessary to shift the quiescent voltage of one stage before applying its output to the following stage. Level-shifting circuits can adjust the DC-bias levels between amplification stages. Level shifting is also required to make the output close to zero in the quiescent state with no input signal. The input resistance of the level-shifting stage should be high to prevent loading of the previous gain stage. Also, the output resistance should be low to effectively drive the subsequent stage.

8.13.1 Level-Shifting Methods

The emitter-follower configuration is normally used to level shift. The emitter follower has an inherent characteristic of level shifting by $V_{BE} \simeq 0.7$ V such that the output voltage is $v_O = v_S - V_{BE}$. Thus, the idea is to create a voltage in the emitter terminal, and it can be accomplished by (1) a potential divider network, (2) a current source, and (3) a zener diode.

Potential Divider Level Shifting

This arrangement is shown in Fig. 8.48(a). The voltage shift is

$$v_O - v_S = -(V_{BE} + R_1 i_E)$$

which, after substituting for $i_E = (v_S - V_{BE} + V_{EE})/(R_1 + R_2)$ and simplifying, gives the output voltage as

$$v_O = \frac{R_2}{R_1 + R_2}(v_S - V_{BE}) - \frac{R_1 V_{EE}}{R_1 + R_2} \tag{8.130}$$

Although this circuit provides a voltage shift, it also attenuates the input signal by a factor $R_2/(R_1 + R_2)$.

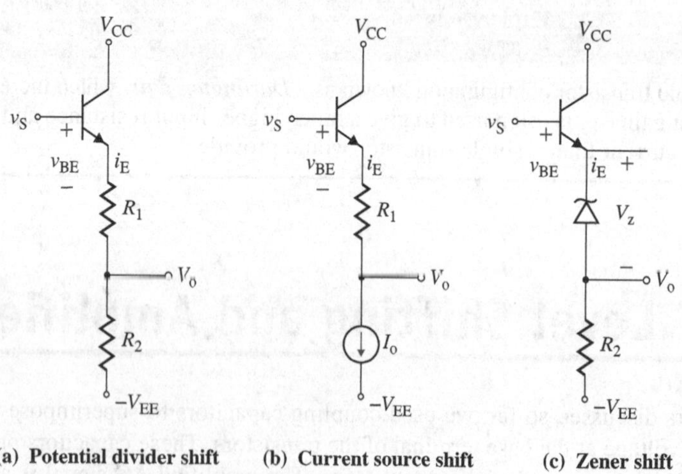

(a) Potential divider shift (b) Current source shift (c) Zener shift

FIGURE 8.48 Level shifting

Current Source Level Shifting

The attenuation of Fig. 8.48(a) can be prevented by replacing R_2 with a current source at a constant current I_O. This arrangement is shown in Fig. 8.48(b). The voltage shift is

$$v_O - v_S = -(V_{BE} + R_1 I_O)$$

which gives the output voltage as

$$v_O = v_S - V_{BE} - R_1 I_O \tag{8.131}$$

Since the current through R_1 is fixed, the voltage drop across it is also fixed. It is important to note that v_O is independent of the negative DC supply voltage $-V_{EE}$.

Zener Level Shifting

Resistance R_1 in Fig. 8.48(a) is replaced by a zener diode with zener voltage V_Z. This arrangement is shown in Fig. 8.48(c). The voltage shift is

$$v_O - v_S = -(V_{BE} + V_Z)$$

which gives the output voltage as

$$v_O = v_S - V_{BE} - V_Z \tag{8.132}$$

Since the zener voltage V_Z is fixed, the voltage drop across it is also fixed.

8.13.2 Level-Shifted DC Amplifier

A BJT amplifier using level shifting is shown in Fig. 8.49(a) with four stages [6]. The first stage generates the reference current for the second stage, which acts as the reference current for the third stage. The fourth stage is the emitter follower. Assume that all transistors are matched devices and have equal parameters: the current gains $\beta_{F1} = \cdots = \beta_{F7} = \beta_F$, the Early voltages $V_{A1} = \cdots = V_{A7} = V_A$, and the B-E voltage $V_{BE1} = \cdots = V_{BE7} = V_{BE}$. To simplify the derivations, assume that $\beta_F >> 1$ and the base currents of all transistors are small as compared to the collector and emitter currents, and these can be neglected.

Current Mirror Source

Using KVL in the B-E loops of Q_1, Q_2, and Q_3, Q_4, we can find the reference currents I_1, I_2, and I_R as given by

$$I_1 = \frac{V_1 + V_{EE} - V_{BE1} - V_{BE2}}{R_A + R_B} \tag{8.133}$$

$$I_R = \frac{V_2 + V_{EE} - V_{BE4} - V_{BE5}}{R} \tag{8.134}$$

$$I_2 = I_R - I_{C3} \tag{8.135}$$

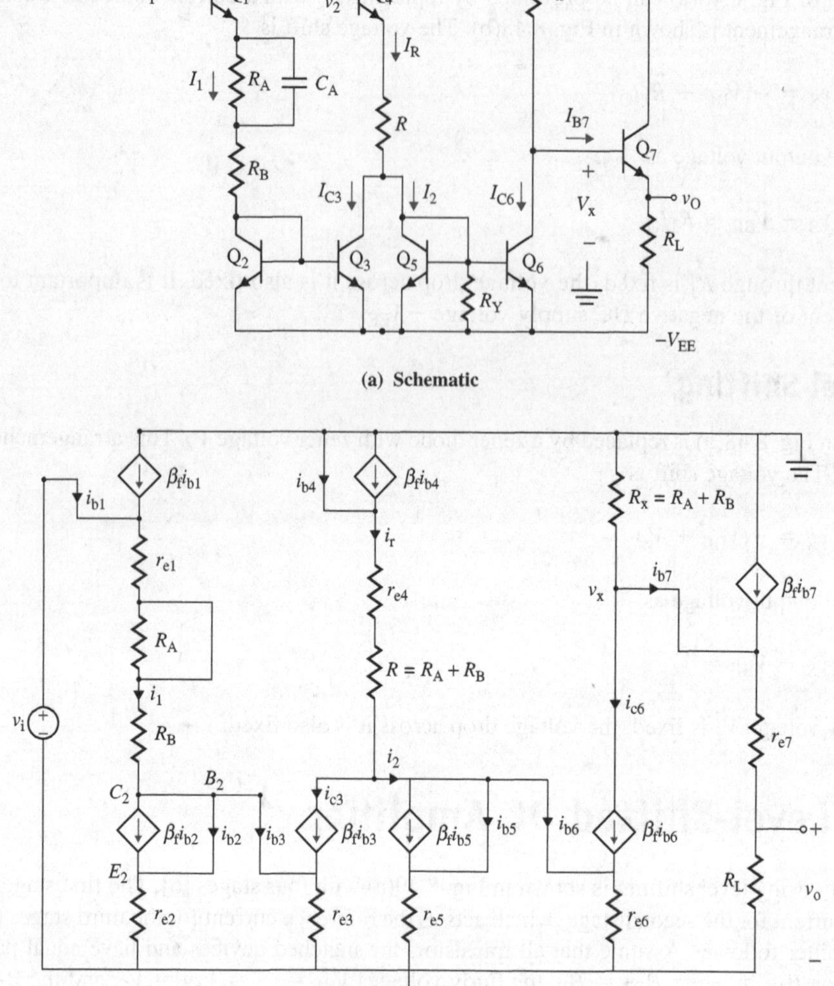

FIGURE 8.49 A BJT level-shifting amplifier

Using Eq. (8.58), we can find the biasing current sources:

$$I_{C3} = \frac{I_1}{1 + 2/\beta_F} \simeq I_1 \tag{8.136}$$

$$I_{C6} = \frac{I_2}{1 + 2/\beta_F} \simeq I_2 \tag{8.137}$$

$$I_{E1} = I_{E2} \simeq I_1 \quad I_{E3} \simeq I_{C3} \quad I_{E4} = I_R \quad I_{E5} \simeq I_2$$

DC Output Voltage

For $I_{C6} >> I_{B7}$ and $I_{C6} \simeq I_2$, we can find the voltage at the base of the output transistor Q_7 as given by

$$V_x = V_{CC} - R_x(I_{C6} + I_{B7}) = V_{CC} - R_x I_{C6} = V_{CC} - R_x I_2$$

which, after substituting I_2 from Eq. (8.135), can be written as

$$V_x = V_{CC} - R_x(I_R - I_{C3}) = V_{CC} - R_x(I_R - I_1) \quad (8.138)$$

which, after substituting I_1 from Eq. (8.133) and substituting I_R from Eq. (8.134), gives

$$V_x = V_{CC} - R_x\left(\frac{V_2 + V_{EE} - V_{BE4} - V_{BE5}}{R}\right) + R_x\left(\frac{V_1 + V_{EE} - V_{BE1} - V_{BE2}}{R_A + R_B}\right) \quad (8.139)$$

We can simplify Eq. (8.139) if we choose resistors such that $R_A + R_B = R = R_x$ and the B-E voltage drops are matched. That is, Eq. (8.139) is simplified to

$$V_x = V_{CC} - V_2 + V_1 \quad (8.140)$$

Therefore, the voltage V_x at the emitter of transistor Q_7 is shifted by V_{BE7}, and the output voltage becomes

$$V_O = V_x - V_{BE7} = V_{CC} + (V_1 - V_2) - V_{BE7} \quad (8.141)$$

If the base of transistor Q_4 is connected to V_{CC} as shown in Fig. 8.49, then $V_2 = V_{CC}$. We get the output voltage as given by

$$V_O = V_{CC} - V_{CC} + V_1 - V_{BE7} = V_1 - V_{BE7} \quad (8.142)$$

Therefore, there is a DC offset voltage due to V_{BE7}, which can be canceled by decreasing the current I_{C6} through R_x by an amount

$$\Delta I_{C6} = \frac{V_{BE7}}{R_x} \quad (8.143)$$

This can be accomplished by placing a compensating resistor R_y from the base of Q_4 to the negative supply ($-V_{EE}$). This will decrease the reference current I_2 by an amount

$$\Delta I_2 = \frac{V_{BE5}}{R_y} \quad (8.144)$$

Since the decrease in I_{C6} should be equal to the decrease in I_2, we can find the relation between R_x and R_y as

$$\frac{V_{BE7}}{R_x} = \frac{V_{BE5}}{R_y} \quad (8.145)$$

Therefore, I_2 in Eq. (8.135) should be corrected as given by

$$I_2 = I_R - I_{C3} - \frac{V_{BE5}}{R_y} \quad (8.146)$$

Under these conditions, the output voltage in Eq. (8.142) becomes

$$V_O = V_1 \quad (8.147)$$

In integrated circuit design, the emitter of Q_7 can be scaled to produce a B-E junction drop identical to that of Q_5, and R_y can be made equal to R_x in order to obtain the condition in Eq. (8.145).

Small-Signal Voltage Gain

By replacing the transistors by its small-signal model, the small-signal equivalent circuit is shown in Fig. 8.49(b). The capacitor C_A and the batteries (V_{CC} and V_{EE}) are shorted to ground. Using KVL through the resistance R_B loop, we can find the small-signal reference current i_1 and its mirrored current i_{c3} as

$$i_1 = i_{c3} = \frac{v_1}{r_{e1} + r_{e2} + R_B} \tag{8.148}$$

Using KVL through the resistance R loop and the relation $i_2 = i_R - i_{c3}$, we can find the relation for the small-signal reference current i_R as

$$0 = (r_{e4} + R + r_{e5})i_R - r_{e5}i_{c3} = (r_{e4} + R + r_{e5})(i_{c3} + i_2) - r_{e5}i_{c3}$$

which gives i_2 as

$$i_2 = -\frac{r_{e4} + R}{r_{e4} + r_{e5} + R} i_{c3} \approx -i_{c3} \quad \text{for} \quad R >> r_{e4}, r_{e5} \tag{8.149}$$

The small-signal current i_2 is mirrored to i_{c6}, which produces the voltage v_x at the base of transistor Q_7 as given by

$$v_x = -i_{c6}R_x = i_{c3}R_x$$

Assuming a unity-gain emitter follower, we get the small-signal output voltage v_o as

$$v_o = i_{c3}R_x$$

which, after substituting i_{c3} from Eq. (8.148), gives the output voltage v_o as

$$v_o = i_{c3}R_x = \frac{R_x}{r_{e1} + r_{e2} + R_B} v_1$$

which gives the no-load voltage gain A_{vo} as

$$A_{vo} = \frac{v_o}{v_1} = \frac{R_x}{r_{e1} + r_{e2} + R_B} \tag{8.150}$$

For $R_x = R_A + R_B$ and $R_B >> (r_{e1} + r_{e2})$, Eq. (8.150) can be simplified to

$$A_{vo} = \frac{R_A + R_B}{R_B} = 1 + \frac{R_A}{R_B} \tag{8.151}$$

It is important to note that any increase in v_1 causes the current i_1 to increase, which is mirrored to i_{c3}, which in turn decreases i_2 by the same amount. Current i_{c6}, which is a mirror of i_2, causes the voltage v_x to increase by $R_x i_{c6}$. The transistors, which act as current mirrors and shift the voltage levels, do not produce any voltage amplification. The voltage gain described by Eq. (8.151) is accomplished by shorting R_A by the capacitor C_A for AC signals. The maximum voltage gain can be obtained by shunting both R_A and R_B by capacitor C_A. For making $r_{e1} = r_{e2}$ and $R_B = 0$, Eq. (8.150) gives the maximum voltage gain as

$$A_{vo(max)} = \frac{R_x}{2r_{e1}} = \frac{R_x I_1}{2V_T} \tag{8.152}$$

where the product $R_x I_1$ is the DC voltage drop across R_x and should not exceed half of the DC supply voltage $V_{CC}/2$ in order to minimize signal distortion. Therefore, Eq. (8.152) gives the maximum permissible voltage gain as

$$A_{vo(perm)} = \frac{R_x}{2r_{e1}} = \frac{R_x V_{CC}}{4V_T} \tag{8.153}$$

The design of this amplifier is very simple. It requires finding only the values of R_A, R_B, R_y, and $R_x(= R = R_A + R_B)$ to obtain a specific voltage gain A_{vo}.

- Level-shifting circuits, which can adjust the DC bias levels between amplification stages, are often used to make the output close to zero in the quiescent state with no input signal.
- A level-shifted amplifier can also provide a voltage gain. Level shifting can be accomplished by a potential divider, current source, and zener diode.

8.14 Frequency Model and Response of Bipolar Junction Transistors

In deriving the small-signal BJT model in Fig. 8.12(a), we assumed that the B-E and C-E junctions had no capacitances. A model that includes these capacitances will represent the frequency characteristic of BJTs. In this section, we develop the frequency and PSpice/SPICE models and then determine the frequency characteristic of BJTs.

8.14.1 High-Frequency Model

An accurate small-signal high-frequency π model that includes capacitances and parasitic resistances is shown in Fig. 8.50(a). The resistances r_b, r_c, and r_e are the series parasitic resistances in the base, collector, and emitter contacts, respectively. The typical values of these resistances are $r_b = 50\ \Omega$ to $500\ \Omega$, $r_c = 20\ \Omega$ to $500\ \Omega$, and $r_e = 1\ \Omega$ to $3\ \Omega$. The value of $r_\mu \approx 10\beta_f r_o$ is very large, typically 10 MΩ.

A change in the input voltage v_{be} will cause a change in the total minority carrier charge q_e in the base. Because of charge-neutrality requirements, there will be an equal amount of change in the total majority carrier charge q_h in the base. Because of the change in the charge, there will be a capacitance involved. This capacitance, known as the *base-charging capacitance*, is defined by

$$C_b = \frac{q_h}{v_{be}} \tag{8.154}$$

(Refer to the diffusion capacitance C_d for the diode because the B-E junction is similar to the diode junction.) A certain amount of time, called the *base transit time*, is required for a minority carrier to

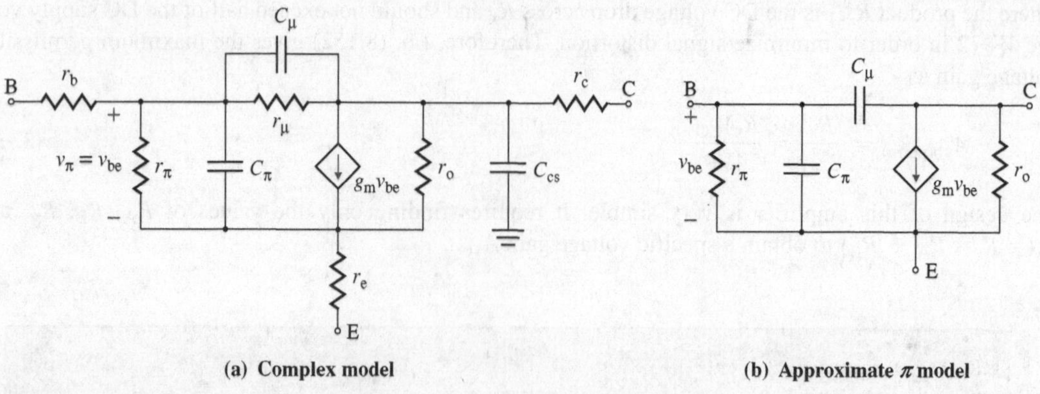

FIGURE 8.50 Small-signal high-frequency π model of a BJT

cross the base. If q_e is the charge in transit and I_C is the collector current, the *forward base transit time* τ_F is defined by

$$\tau_F = \frac{q_e}{I_C}$$

Thus, τ_F is the average time each carrier spends crossing the base. The change in minority charge for a change in collector current is

$$\Delta q_e = \tau_F \Delta I_C \tag{8.155}$$

The change in minority charge must be equal to the change in majority charge q_h. That is, $\Delta q_e = \Delta q_h$. Therefore,

$$\Delta q_h = \Delta q_e = \tau_F \Delta I_C$$

In terms of small-signal quantities, we can write

$$q_h = q_e = \tau_F i_c \tag{8.156}$$

Substituting q_h from Eq. (8.156) into Eq. (8.154) yields

$$C_b = \frac{q_h}{v_{be}} = \frac{\tau_F i_c}{v_{be}} \tag{8.157}$$

Substituting $i_c = g_m v_{be}$ from Eq. (8.42) into Eq. (8.157) yields

$$C_b = \tau_F g_m = \tau_F \frac{I_C}{V_T} \tag{8.158}$$

Thus, C_b is proportional to the collector biasing current I_C.

In addition to the base-charging capacitance, there will be a *B-E depletion capacitance* C_{je}, which is defined by [7]

$$C_{je} = \frac{C_{je0}}{[1 - V_{BE}/V_{je}]^{1/3}} \tag{8.159}$$

where C_{je0} is the value of C_{je} for $V_{BE} = 0$ and typically is in the range of 0.2 pF to 1 pF.

V_{je}, which is the *built-in potential* across the junction with zero applied voltage, can be shown in Eq. 6.7 to be [1]

$$V_{je} = V_T \ln\left(\frac{N_a N_d}{n_i^2}\right) \tag{8.160}$$

where N_a is the doping density of p-type material in atoms per cubic centimeter, N_d is the doping density of n-type material in atoms per cubic centimeter, and n_i is the intrinsic carrier concentration in a pure sample of semiconductor. At 25°C (or 300 K), $n_i \approx 1.5 \times 10^{10}$ for silicon. For $N_a = 10^{15}$, $N_d = 10^{16}$, and $V_T = 25.8$ mV, the built-in potential is

$$V_{je} = 25.8 \times 10^{-3} \times \ln\left[\frac{10^{15} \times 10^{16}}{(1.5^2 \times 10^{20})}\right] = 632.6 \text{ mV at } 25°\text{C}$$

C_{je} depends on the B-E voltage V_{BE} ($\leq V_{je}$) and the temperature because of V_T. Any increase in V_{BE} will cause C_{je} to increase. Thus, a reverse-biased B-E junction will exhibit a lower value of C_{je}.

The *B-E input capacitance* C_π is the sum of C_b and C_{je}. That is,

$$C_\pi = C_b + C_{je}$$

The *C-B junction capacitance* can be found approximately from

$$C_\mu = \frac{C_{\mu 0}}{[1 + V_{CB}/V_{jc}]^{1/3}} \tag{8.161}$$

where $V_{jc} = V_{je}$ and $C_{\mu 0}$ is the value of C_μ for $V_{CB} = 0$ and typically is in the range of 0.2 pF to 1 pF. A higher value of V_{CB} will cause C_μ to decrease. Thus, a BJT operating as a switch will have a low value of V_{CB}, usually less than 0.7 V, and will exhibit a higher value of C_μ than a BJT operating as an amplifying device with $V_{CB} > 0.7$ V.

There is also a capacitance from the collector to the substrate (body) of the transistor. The substrate is usually connected to the ground. The collector-substrate capacitance can be found approximately from

$$C_{cs} = \frac{C_{cs0}}{[1 + V_{CS}/V_{js}]^{1/3}} \tag{8.162}$$

where $V_{js} = V_{je}$ and C_{cs0} is the value of C_{cs} for $V_{CS} = 0$ and typically is in the range of 1 pF to 3 pF.

Since r_μ is very large and the collector-substrate capacitance C_{cs} is very small, their effects can be neglected. Although the C-B capacitance C_μ is small, it has a magnified influence on the frequency response as a result of the *Miller effect*. The simplified equivalent circuit, which neglects r_μ, C_{cs}, r_b, r_c, and r_e, is shown in Fig. 8.50(b).

▶ **NOTE** Manufacturers specify the common-emitter hybrid (h) parameters of a BJT rather than the π-model parameters. However, the h parameters can be converted to π-model parameters (see Appendix D).

8.14.2 Small-Signal PSpice/SPICE Model

When simulating electronic circuits, PSpice/SPICE first calculates the DC-biasing point and generates the small-signal parameters for AC and transient analysis. The AC equivalent circuit generated by PSpice is shown in Fig. 8.51 [5].

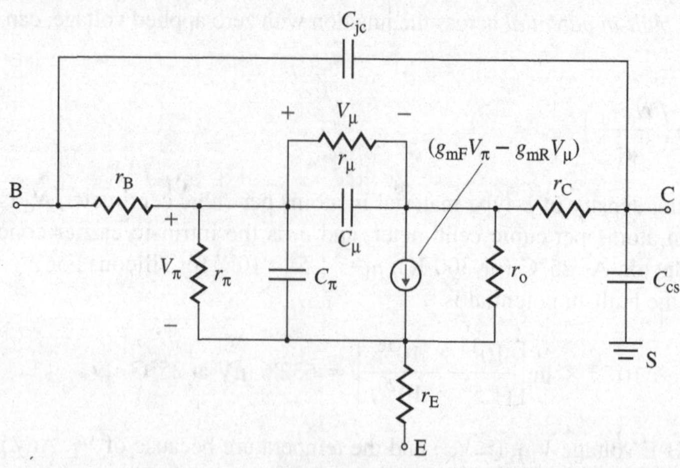

FIGURE 8.51 Small-signal PSpice model of a BJT

8.14.3 Frequency Response of BJTs

Since the BJT model contains capacitances, the current gain β_f will depend on the frequency. The dependency of β_f on the frequency is normally given in the data sheet; the value of C_π is not usually specified, but it can be determined from the expression for β_f as a function of frequency. Let us apply a test current i_b to the base of a BJT and short-circuit the collector terminal for AC signals. This arrangement is shown in Fig. 8.52(a), and the high-frequency AC equivalent circuit is shown in Fig. 8.52(b). The voltage at the base terminal is given by

$$V_{be}(s) = \left[r_\pi \,\|\, \frac{1}{(C_\pi + C_\mu)s} \right] I_b(s) = \frac{r_\pi}{1 + r_\pi(C_\pi + C_\mu)s} I_b(s) \tag{8.163}$$

Since r_o is very large and C_μ is very small, the current through them will be very small. That is,

$$I_c(s) = (g_m - sC_\mu)V_{be}(s) \tag{8.164}$$

Substituting $V_{be}(s)$ from Eq. (8.163), we get

$$I_c(s) = \frac{(g_m - sC_\mu)r_\pi}{1 + r_\pi(C_\pi + C_\mu)s} I_b(s)$$

At the frequencies at which the model in Fig. 8.52(b) is valid, $g_m >> \omega C_\mu$, and we get the current gain $\beta_f(j\omega)$ in the frequency domain (for $s = j\omega$) as

$$\beta_f(j\omega) = \frac{I_c(j\omega)}{I_b(j\omega)} = \frac{r_\pi g_m}{1 + r_\pi(C_\pi + C_\mu)j\omega} = \frac{\beta_f}{1 + \beta_f(C_\pi + C_\mu)j\omega/g_m} \tag{8.165}$$

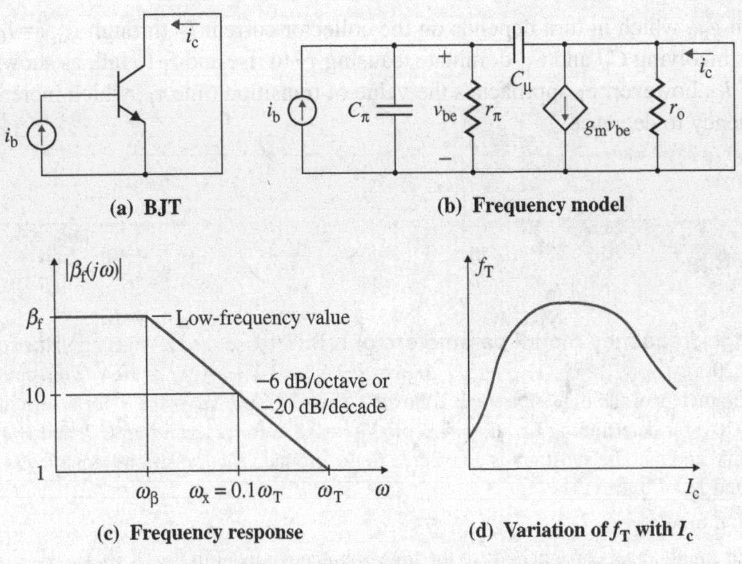

FIGURE 8.52 Frequency response of a BJT

which indicates that the current gain will fall as the frequency increases, at a slope of −20 dB/decade. This relationship is shown in Fig. 8.52(c) with a 3-dB frequency given by

$$\omega_\beta = = \frac{g_m}{\beta_f(C_\pi + C_\mu)} \tag{8.166}$$

The current gain will be unity, $|\beta_f(j\omega)| = 1$, when

$$\omega = \omega_T = \frac{g_m}{C_\pi + C_\mu} \quad \text{(in rad/s)} \tag{8.167}$$

or

$$f_T = \frac{g_m}{2\pi(C_\pi + C_\mu)} \quad \text{(in Hz)} \tag{8.168}$$

where ω_T or f_T is called the *transition frequency* and is a measure of the useful frequency of the transistor when used as an amplifier. ω_T is also the *unity-gain bandwidth* of the transistor because from Eqs. (8.166) and (8.167) we get

$$\omega_T = \omega_\beta \beta_f \tag{8.169}$$

Unity-gain bandwidth is usually specified by the manufacturer, with typical values ranging from 100 MHz to a few gigahertz ω_T is normally determined by measuring the frequency ω_x when $|\beta_f(j\omega_x)| =$ 10 or 5. That is, $\omega_T = \omega_x|\beta_f(j\omega_x)|$. The transition period τ_T corresponding to ω_T is given by

$$\tau_T = \frac{1}{\omega_T} = \frac{C_\pi + C_\mu}{g_m} = \frac{C_b + C_{je} + C_\mu}{g_m} = \frac{C_b}{g_m} + \frac{C_{je}}{g_m} + \frac{C_\mu}{g_m}$$

$$= \tau_F + \frac{C_{je}}{g_m} + \frac{C_\mu}{g_m} \tag{8.170}$$

which depends on g_m, which in turn depends on the collector current I_C through g_m $(=I_C/V_T)$. As I_C decreases, the terms involving C_{je} and C_μ dominate, causing τ_T to rise and f_T to fall, as shown in Fig. 8.52(d). At high values of I_C, however, τ_T approaches the value of transition time τ_F, which increases with current, causing the frequency to decrease.

EXAMPLE 8.9

Finding the high-frequency model parameters of a BJT Use the DC-biasing values of the transistor circuit in Fig. 8.20(a): $I_C = 10$ mA, $V_{CE} = 5$ V, $V_{BE} = 0.7$ V, and $V_{CS} = V_C = 10$ V. This circuit is shown in Fig. 8.53. The parameters of the transistor are as follows: $C_{je0} = 29.6$ pF, $V_{je} = 0.8$ V for determining C_{je}, $C_{\mu 0} =$ 19.4 pF, $V_{jc} = 0.8$ V for determining C_μ, $C_{cs0} = 3$ pF, $V_{js} = 0.8$ V for determining C_{cs}, and $\beta_f = 100$. Assume that $V_T = 25.8$ mV and that the substrate is connected to the ground. The transition frequency is $f_T = 300$ MHz at $V_{CE} = 20$ V and $I_C = 20$ mA.

(a) Find transition time τ_F.

(b) Calculate the small-signal capacitances of the high-frequency model in Fig. 8.50(a).

(c) Use PSpice/SPICE to generate the model parameters.

SOLUTION

(a) The transition period is $\tau_T = 1/2\pi f_T = 1/(2\pi \times 300 \text{ MHz}) = 530.5$ ps. The transition frequency $f_T =$ 300 MHz is specified at $I_C = 20$ mA. The transconductance g_m (at $I_C = 20$ mA) becomes

$$g_m = \frac{I_C}{V_T} = \frac{20 \text{ mA}}{25.8 \text{ mV}} = 775.2 \text{ mA/V}$$

We know that

$$V_{CB} = V_{CE} - V_{BE} = 20 - 0.7 = 19.3 \text{ V}$$

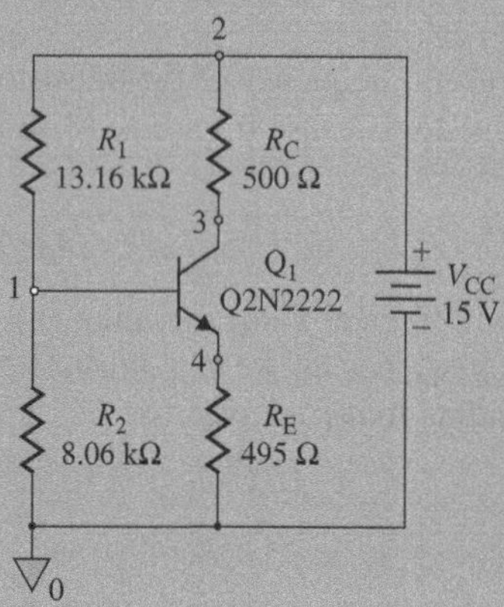

FIGURE 8.53 DC-biasing circuit for a BJT

Equation (8.161) gives

$$C_\mu = \frac{19.4\text{ pF}}{(1 + 19.3/0.8)^{1/3}} = 6.62\text{ pF}$$

From Eq. (8.170) we can find the transit time τ_F:

$$530.5\text{ ps} = \tau_F + \frac{25\text{ pF}}{0.7752} + \frac{6.62\text{ pF}}{0.7752}$$

which gives $\tau_F = \text{TF} = 489.7$ ps.

(b) We know that

$$V_{CB} = V_{CE} + V_{EB} = V_{CE} - V_{BE} = 5 - 0.7 = 4.3\text{ V}$$

From Eq. (8.159),

$$C_{je} = \frac{C_{je0}}{(1 - V_{BE}/V_{je})^{1/3}} = \frac{29.6\text{ pF}}{(1 - 0.7/0.8)^{1/3}} = 59.2\text{ pF}$$

From Eq. (8.161),

$$C_\pi = \frac{C_{\pi 0}}{(1 + V_{CB}/V_{jc})^{1/3}} = \frac{19.4\text{ pF}}{(1 + 4.3/0.8)^{1/3}} = 10.47\text{ pF}$$

From Eq. (8.162),

$$C_{cs} = \frac{C_{cs0}}{(1 + V_{CS}/V_{js})^{1/3}} = \frac{3\text{ pF}}{(1 + 10/0.8)^{1/3}} = 1.26\text{ pF}$$

From Eq. (8.158),

$$C_b = \frac{\tau_F I_C}{V_T} = \frac{489.7 \times 10^{-12} \times 10\text{ mA}}{25.8\text{ mV}} = 189.8\text{ pF}$$

Then

$$C_\pi = C_b + C_{je} = 189.8\text{ pF} + 59.2\text{ pF} = 249\text{ pF}$$

(c) Since the capacitances of a BJT depend on the junction voltages, PSpice/SPICE parameters for BJTs are specified at the zero-biased conditions. PSpice/SPICE first calculates the biasing voltages by finding the *Q*-point and then adjusts the values of junction capacitances accordingly. We will add the PSpice zero-biased parameters affecting the capacitances. That is, CJE = C_{je0} = 29.6 pF, CJC = C_{JC} = 19.4 pF, CJS = C_{cs} = 1.26 pF, VJE = V_{je} = 0.8 V, VJC = V_{jc} = 0.8 V, VJS = V_{js} = 0.8 V, and TF = τ_F = 489.7 ps. The model statement for the transistor Q2N2222 in Example 8.5 is as follows for $V_{cs} = 0$:

```
.MODEL Q2N2222 NPN (BF=100 IS=3.295E-14 VA=200 CJE=29.6pF
+  CJC=19.4pF CJS=1.3pF VJE=0.8 VJC=0.8 VJS=0.8 TF=489.7ps)
```

The results of .OP analysis, which are obtained from the output file, are as follows: (The values obtained from hand calculations are shown in parentheses; the results obtained from the PSpice model are shown in the left-hand column.)

GM	3.54E-01	g_m = 0.354 A/V (0.3876 A/V)
RPI	2.90E+02	r_π = 290 Ω (258 Ω)
RO	2.24E+04	r_o = 22.4 kΩ (20 kΩ)
CBE	2.19E-10	C_π = 219 pF (243.6 pF)
CBC	1.00E-11	C_μ = 10 pF (10.47 pF)
CJS	1.26E-12	C_{cs} = 1.26 pF (1.26pF)
FT	2.46E+08	f_T = 246 MHz (308.49 MHz)

NOTE: There are many factors that affect the parameters of BJTs. The hand calculations are expected to give approximate values only. Even the PSpice results will differ from results of measurements on practical BJTs.

KEY POINTS OF SECTION 8.14

- The maximum useful frequency of a BJT is called the transition frequency, and it is limited by the internal capacitances C_π and C_μ of the BJT. The collector–base capacitance C_μ is small; however, it has a magnified influence on the frequency response as a result of the Miller effect.
- The base transit time is the average time each majority carrier spends crossing the base.
- The transition frequency, also known as the unity-gain bandwidth of the transistor, is a measure of the useful frequency and is fixed for a particular transistor.

8.15 Frequency Response of BJT Amplifiers

In Secs. 2.5.4 and 2.5.5, we introduced the short-circuit and zero-value methods. As examples, we will use these methods for determining the frequency response of BJT amplifiers.

8.15.1 Common-Emitter BJT Amplifiers

A common-emitter BJT amplifier is shown in Fig. 8.54(a). The transisor can be replaced by its simple high-frequency π model, shown in Fig. 8.54(b). The values of C_π and C_μ are low (on the order of 10 pF), and these capacitors can be considered to be open-circuited at a low frequency. Thus, Fig. 8.54(b) is reduced to Fig. 8.54(c) at a low frequency. If the transistor can be replaced by its small-signal AC model, the frequency response will depend on the time constants of the model's capacitors. The values of C_1, C_2, and C_E are generally much larger than those of C_π and C_μ. The amplifier will exhibit a midband characteristic. A typical frequency response profile is shown in Fig. 8.54(d), where f_L is the low 3-dB frequency, f_H is the high 3-dB frequency, and A_{PB} is the passband gain. The low break frequencies will depend mostly on C_1, C_2, and C_E; the high break frequencies will depend on C_π and C_μ. An extra capacitance C_x is connected between the collector and the base to give the desired high break frequency.

Low Cutoff Frequencies

The common-emitter (CE) amplifier in Fig. 8.54(a) is expected to operate at low frequencies such that C_π and C_μ will have small values and will behave as if they were open-circuited. The low-frequency equivalent circuit shown in Fig. 8.55 has three capacitors—two coupling capacitors C_1 and C_2 and a bypass capacitor C_E.

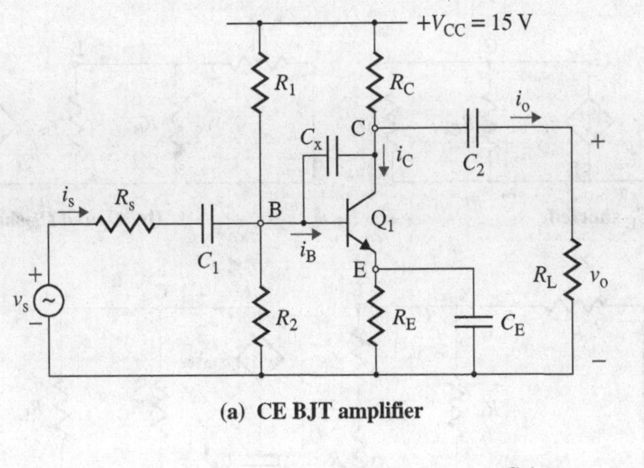

(a) CE BJT amplifier

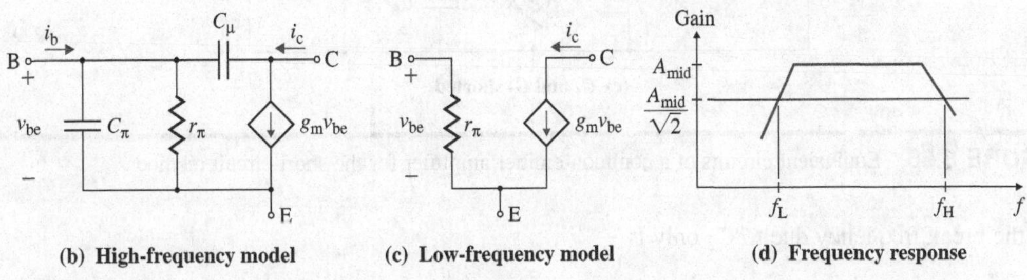

(b) High-frequency model (c) Low-frequency model (d) Frequency response

FIGURE 8.54 Common-emitter BJT amplifier

Let us consider the effects of C_1 only; C_2 and C_E are short-circuited. This situation is shown in Fig. 8.56(a). Thevenin's equivalent resistance presented to C_1 is

$$R_{C1} = R_s + R_B \,\|\, r_\pi \tag{8.171}$$

where $R_B = (R_1 \,\|\, R_2)$. Thus, the break frequency due to C_1 only is

$$f_{C1} = \frac{1}{2\pi R_{C1} C_1} \tag{8.172}$$

The equivalent circuit, with C_1 and C_E short-circuited, is shown in Fig. 8.56(b). Thevenin's equivalent resistance is given by

$$R_{C2} = R_C + R_L \tag{8.173}$$

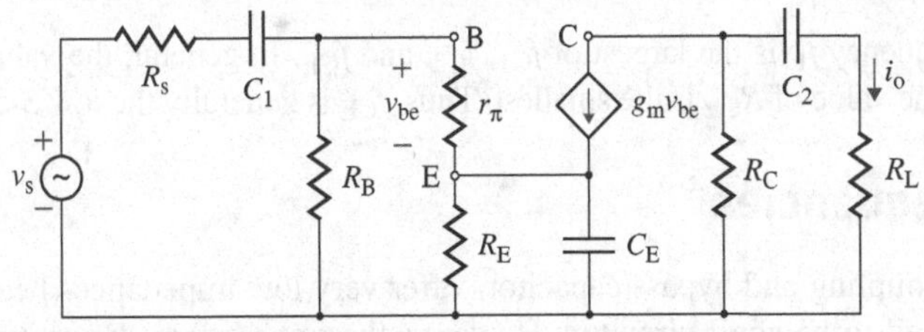

FIGURE 8.55 Low-frequency AC equivalent circuit of a common-emitter amplifier

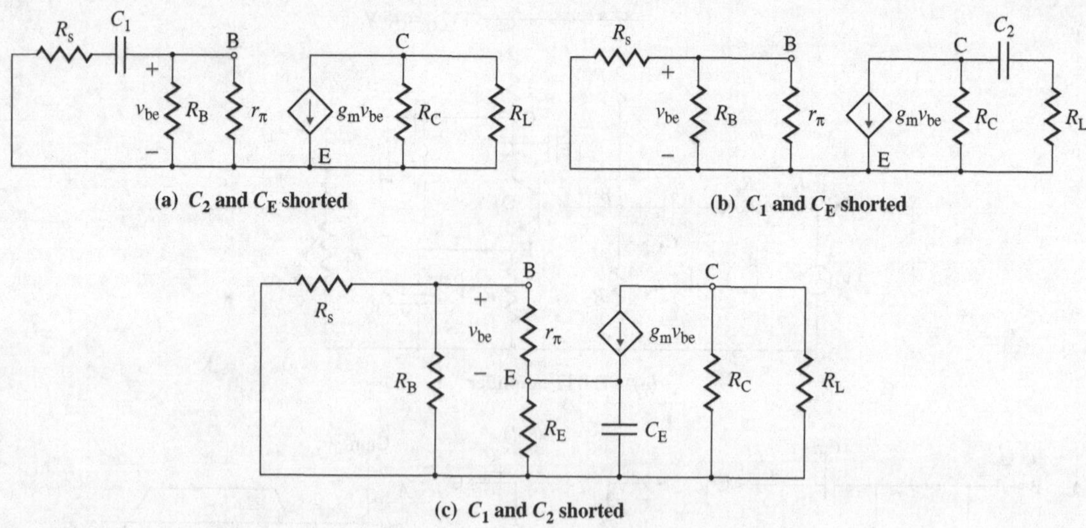

FIGURE 8.56 Equivalent circuits of a common-emitter amplifier for the short-circuit method

and the break frequency due to C_2 only is

$$f_{C2} = \frac{1}{2\pi R_{C2} C_2} \tag{8.174}$$

The equivalent circuit, with C_1 and C_2 short-circuited, is shown in Fig. 8.56(c). Dividing a resistance in the base circuit by $(1 + \beta_f)$ $(=1 + g_m r_\pi)$ will give the equivalent emitter resistance. Thus, Thevenin's equivalent resistance can be found by paralleling R_E with the equivalent B-E resistance. That is,

$$R_{CE} = R_E \parallel \frac{r_\pi + (R_s \parallel R_B)}{1 + \beta_f} \tag{8.175}$$

If $R_s < R_B$ and r_π, $R_E > 1\ \text{k}\Omega$, and $\beta_f >> 1$, Eq. (8.175) can be approximated by

$$R_{CE} \approx \frac{r_\pi}{\beta_f} = \frac{1}{g_m} \tag{8.176}$$

The break frequency due to C_E only is

$$f_{CE} = \frac{1}{2\pi R_{CE} C_E} \tag{8.177}$$

The low 3-dB frequency f_L is the largest of f_{C1}, f_{C2}, and f_{CE}. In general, the value of C_E is much larger than that of C_1 or C_2; the value of R_{CE} is the smallest. Thus, f_{CE} is generally the low 3-dB frequency: $f_L = f_{CE}$.

High Cutoff Frequencies

At high frequencies, coupling and bypass capacitors offer very low impedances because of their high values and can be assumed to be short-circuited. However, the impedances due to transistor capacitors are comparable to those of other circuit elements and hence affect the voltage gain.

If the transistor of the CE amplifier in Fig. 8.54(a) is replaced by the high-frequency model shown in Fig. 8.54(b), the result is the high-frequency equivalent circuit shown in Fig. 8.57. C_x is the extra

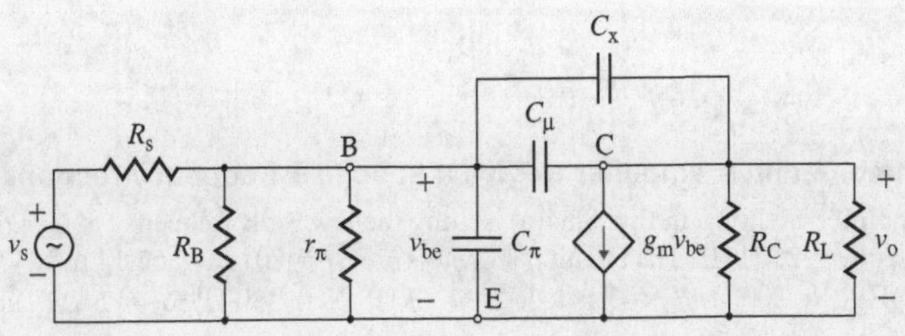

FIGURE 8.57 High-frequency circuit of a common-emitter amplifier

capacitance connected between the collector and the base of the transistor. The equivalent circuit, with C_μ open-circuited, is shown in Fig. 8.58(a). The resistance faced by C_π and C_x is given by

$$R_{C\pi} = r_\pi \parallel (R_s \parallel R_B) \tag{8.178}$$

The equivalent circuit, with C_π open-circuited, is shown in Fig. 8.58(b). Let us replace C_μ by voltage source v_x, as shown in Fig. 8.58(c). Using KVL, we get

$$v_x = v_{be} + (R_L \parallel R_C)(i_x + g_m v_{be}) = R_i' i_x + (R_L \parallel R_C)(i_x + g_m i_x R_i']$$
$$= [(R_L \parallel R_C) + R_i'(1 + g_m R_L \parallel R_C)]i_x$$

which gives Thevenin's equivalent resistance faced by C_μ as

$$R_{C\mu} = \frac{v_x}{i_x} = R_L \parallel R_C + R_i'[1 + g_m(R_L \parallel R_C)] \tag{8.179}$$

where $R_i' = (r_\pi \parallel R_B \parallel R_s)$. Thus, the high 3-dB frequency f_H is given by

$$f_H = \frac{1}{2\pi[R_{C\pi}C_\pi + R_{C\mu}(C_\mu + C_x)]} \tag{8.180}$$

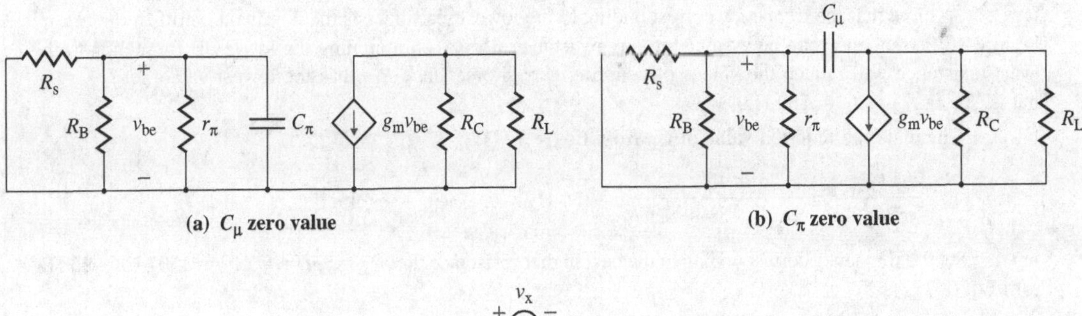

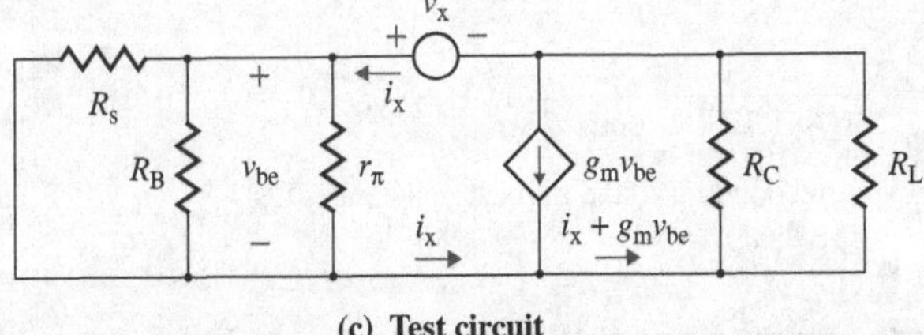

(c) Test circuit

FIGURE 8.58 High-frequency equivalent circuits for the zero-value method

Designing a common-emitter amplifier to give a specified frequency response

(a) Design a CE amplifier as shown in Fig. 8.54(a), setting the low 3-dB frequency at $f_L = 150$ Hz and the high 3-dB frequency at $f_H = 250$ kHz. The circuit parameters are $\beta_f = 80$, $g_m = 57.14\ \mu A/V$, $r_\pi = 1.4\ k\Omega$, $C_\pi = 15$ pF, $C_\mu = 1$ pF, $R_s = 200\ \Omega$, $R_1 = 7\ k\Omega$, $R_2 = 4.3\ k\Omega$, $R_E = 330\ \Omega$, $R_C = 5\ k\Omega$, and $R_L = 5\ k\Omega$.

(b) Determine the passband gain A_{PB}.

(c) Use Miller's capacitor method to check the high-frequency design.

(d) Use PSpice/SPICE to plot the frequency response from 100 Hz to 1 MHz with decade increments and 100 points per decade.

SOLUTION

Let $R_B = R_1 \| R_2 = 7\ k\Omega \| 4.3\ k\Omega = 2.66\ k\Omega$. Then

$$R_i' = r_\pi \| R_B \| R_s = R_{C\pi} = 164\ \Omega$$

(a) The design will have two parts: one in which we set the low 3-dB frequency at $f_L = 150$ Hz and one in which we set the high 3-dB frequency at $f_H = 250$ kHz.

The steps to set $f_L = 150$ Hz are as follows:

Step 1. Calculate the equivalent resistances R_{C1}, R_{C2}, and R_{CE}. From Eq. (8.171),

$$R_{C1} = R_s + R_B \| r_\pi = 200\ \Omega + 2.66\ k\Omega \| 1.4\ k\Omega = 1.12\ k\Omega$$

From Eq. (8.173),

$$R_{C2} = R_C + R_L = 5\ k\Omega + 5\ k\Omega = 10\ k\Omega$$

From Eq. (8.175),

$$R_{CE} = 330 \| \left[\frac{1.4\ k\Omega + 200\ \Omega \| 2.66\ k\Omega}{1 + 80}\right] = 18.48\ \Omega$$

Step 2. Assume that the frequency corresponding to the lowest resistance is the dominant cutoff frequency f_L. The low 3-dB frequency can be assigned to any resistance. However, assigning the low 3-dB frequency to the lowest resistance will reduce the values of coupling capacitors. Since R_{CE} has the lowest value, let $f_{CE} = f_L$. That is, $f_{CE} = f_{L1} = f_L = 150$ Hz.

Step 3. Calculate the required value of C_E from Eq. (8.177):

$$f_{CE} = \frac{1}{2\pi R_{CE} C_E} = \frac{1}{2\pi \times 18.48 \times C_E} = 150\ \text{H}_\text{Z} \quad \text{or} \quad C_E = 57.4\ \mu F$$

Step 4. Set the frequency corresponding to the next higher resistance. Let $f_{L2} = f_{C1} = f_L/10 = 150/10 = 15$ Hz. From Eq. (8.172),

$$f_{C1} = \frac{1}{2\pi R_{C1} C_1} = \frac{1}{2\pi \times 1.12\ k\Omega \times C_1} = 15\ \text{H}_\text{z} \quad \text{or} \quad C_1 = 9.5\ \mu F$$

Step 5. Set the frequency corresponding to the highest resistance. Let $f_{L3} = f_{C2} = f_L/20 = 150/20 = 7.5$ Hz. From Eq. (8.174),

$$f_{C2} = \frac{1}{2\pi R_{C2} C_2} = \frac{1}{2\pi \times 10\ k\Omega \times C_2} = 7.5\ \text{Hz} \quad \text{or} \quad C_2 = 2.1\ \mu F$$

The steps to set $f_H = 250$ kHz are as follows:

Step 1. From Eq. (8.178),

$$R_{C\pi} = r_\pi \| (R_s \| R_B) = 1.4\ \text{k}\Omega \| (200 \| 2.66\ \text{k}\Omega) = 164\ \Omega$$

From Eq. (8.179),

$$R_{C\mu} = (5\ \text{k}\Omega \| 5\ \text{k}\Omega) + 164\ \Omega \times [1 + 57.14\ \text{m℧} \times (5\ \text{k}\Omega \| 5\ \text{k}\Omega)] = 26.1\ \text{k}\Omega$$

Step 2. From Eq. (8.180), the high 3-dB frequency f_H is

$$f_H = \frac{1}{2\pi[164\ \Omega \times 15\ \text{pF} + (26.1\ \text{k}\Omega)(C_\mu + C_x)]}\ 250\ \text{kHz} \quad \text{or} \quad C_\mu + C_x = C_{\text{eff}} = 24.3\ \text{pF}$$

which gives $C_x = 24.3 - 1 = 23.3$ pF. This is the value of the additional capacitor C_x that is to be connected between the collector and base terminals of the transistor.

(b) From Eq. (2.114), the passband voltage gain is

$$A_{PB} = (-57.14 \times 10^{-3})(5\ \text{k}\Omega \| 5\ \text{k}\Omega) \times \frac{2.66\ \text{k}\Omega \| 1.4\ \text{k}\Omega}{200 + (2.66\ \text{k}\Omega \| 1.4\ \text{k}\Omega)} = -117.3\ \text{V/V}$$

(c) Applying Eq. (2.98), we have for the effective capacitance between the base and the emitter terminals

$$\begin{aligned} C_{eq} &= (C_\mu + C_x)[1 + g_m(R_L \| R_C)] + C_\pi \\ &= 24.3\ \text{pF} \times [1 + 57.14\ \text{m℧} \times (5\ \text{k}\Omega \| 5\ \text{k}\Omega)] + 15\ \text{pF} = 3.51\ \text{nF} \end{aligned}$$

The equivalent resistance faced by C_{eq} is $R_{eq} = R_i' = r_\pi \| R_B \| R_s = 164\ \Omega$, so

$$f_H = \frac{1}{2\pi C_{eq} R_{eq}} = \frac{1}{2\pi \times 3.51\ \text{nF} \times 164\ \Omega} = 276.5\ \text{kHz}$$

Miller's capacitor method gives a higher frequency than the zero-value method does. The actual frequency is higher than the one obtained by the zero-value method but lower than the one obtained by Miller's capacitor method. Thus, design by the zero-value method provides a more conservative estimate, although the SPICE results match the zero-value prediction.

(d) The AC equivalent circuit for PSpice simulation is shown in Fig. 8.59, where r_o has been added to include the output resistance of the transistor and also to make the circuit more general.

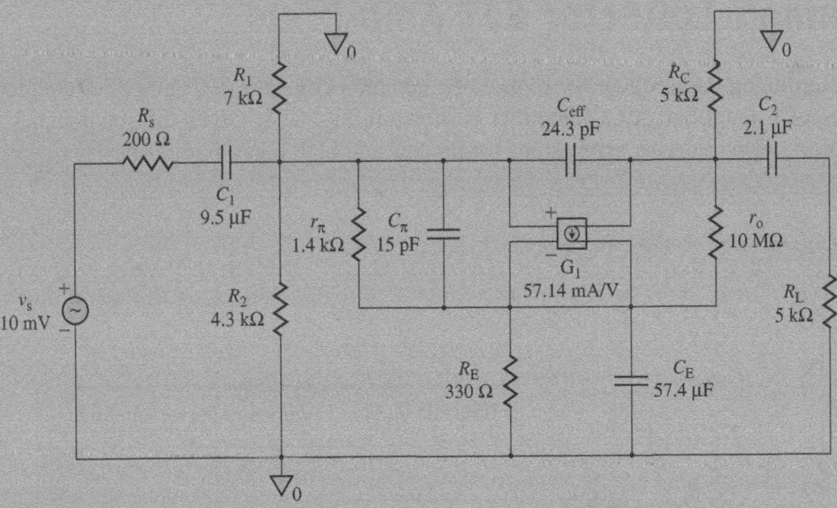

FIGURE 8.59 Small-signal equivalent circuit for PSpice simulation

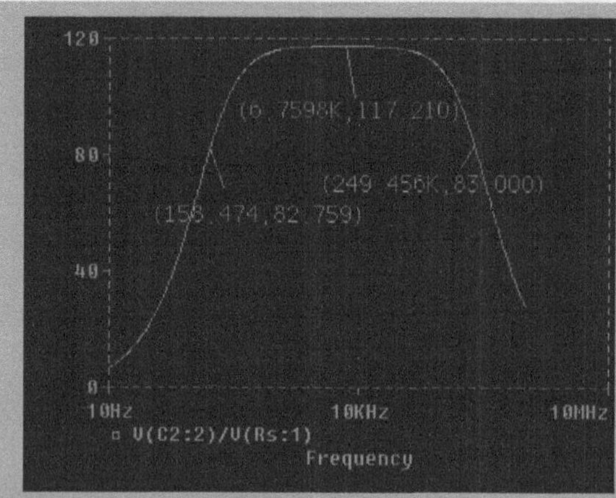

FIGURE 8.60 Frequency response for Example 8.10

The PSpice plot of the frequency response is shown in Fig. 8.60, which gives the passband gain as $|A_{PB}| = 117.2$ (expected value is 117.3). The low 3-dB frequency is approximately $f_L = 158.5$ Hz (expected value is 150 Hz), and the high 3-dB frequency is approximately $f_H = 249.5$ kHz (expected value is 250 kHz). The design value of f_H is 250 kHz, and that of f_L is 150 Hz. Thus, the results are very close.

NOTE: The main objective of this PSpice simulation was to verify the design methods used to set the low and high cutoff frequencies. Thus, the small-signal model rather than the actual PSpice transistor model was used in the simulation. If we designed an amplifier with the small-signal model and then ran the simulation with an actual PSpice transistor, we would expect to get an error. But, the results should be close.

8.15.2 Common-Collector BJT Amplifiers

The techniques for determining the frequency response of a common-collector (CC) amplifier are identical to those used with a CE amplifier. A CC amplifier is shown in Fig. 8.61. Given the small-signal AC model of the transistor(s), we will derive expressions for the low and high break frequencies.

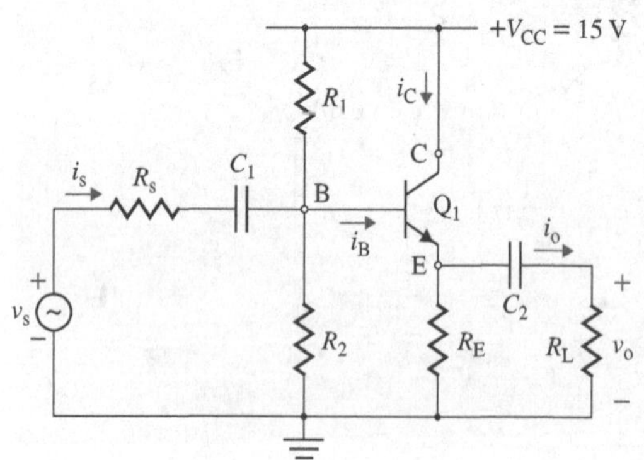

FIGURE 8.61 Common-collector amplifier

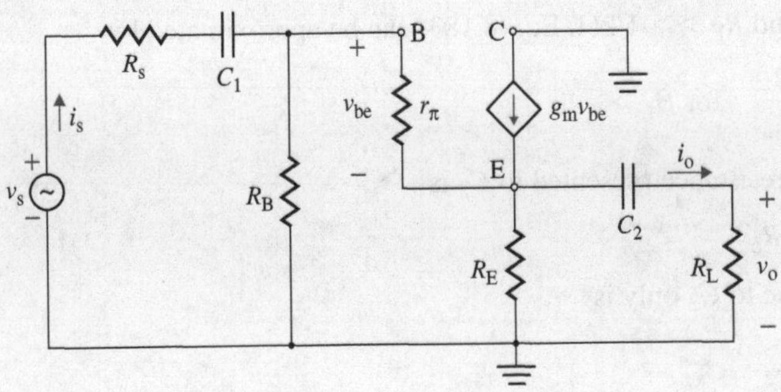

(a) Low-frequency equivalent circuit

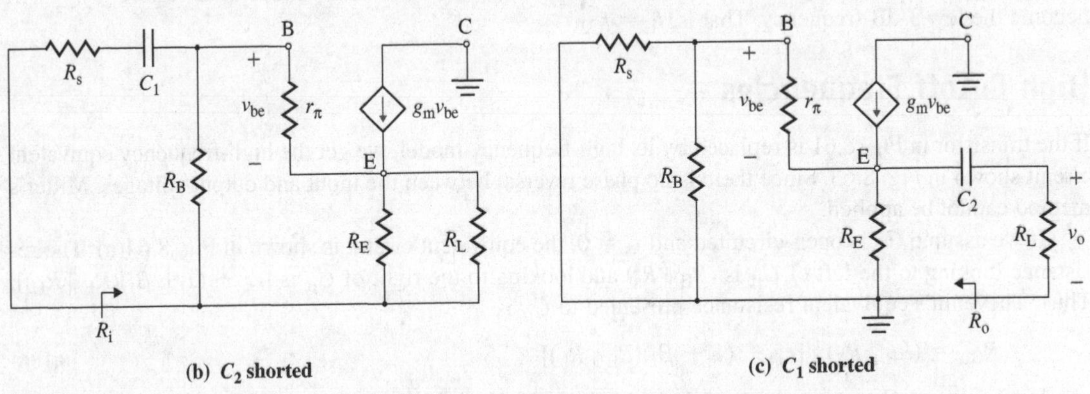

(b) C_2 shorted

(c) C_1 shorted

FIGURE 8.62 Equivalent circuits of a common-collector amplifier for the short-circuit method

Low Cutoff Frequencies

The low-frequency equivalent circuit obtained by replacing the transistor in Fig. 8.61 by its small-signal low-frequency model is shown in Fig. 8.62(a). There will be two low break frequencies corresponding to the coupling capacitances C_1 and C_2. Assuming that C_2 is short-circuited, as shown in Fig. 8.62(b), R_E becomes parallel to R_L. Converting the effective emitter resistance ($R_E \| R_L$) into the base terminal and using Eq. (8.101), we get the equivalent input resistance

$$R_i = R_B \| [r_\pi + (1 + \beta_f)(R_E \| R_L)] \tag{8.181}$$

where $R_B = R_1 \| R_2$. Thevenin's equivalent resistance presented to C_1 is

$$R_{C1} = R_s + R_i$$

so the 3-dB frequency due to C_1 only is

$$f_{C1} = \frac{1}{2\pi R_{C1} C_1} \tag{8.182}$$

The equivalent circuit, with C_1 short-circuited, is shown in Fig. 8.62(c). Converting the effective base resistance $[r_\pi + (R_s \| R_B)]$ into the emitter terminal and using Eq. (8.103), we get the equivalent output resistance

$$R_o = R_E \| \frac{r_\pi + (R_s \| R_B)}{1 + \beta_f} \tag{8.183}$$

If $R_s << R_B$ and r_π, and $R_E >> 1\text{ k}\Omega$, Eq. (8.183) can be approximated by

$$R_o \approx \frac{r_\pi}{\beta_f} = \frac{1}{g_m} \quad \text{for } \beta_f >> 1 \tag{8.184}$$

Thevenin's equivalent resistance presented to C_2 is

$$R_{C2} = R_L + R_o$$

The 3-dB frequency due to C_2 only is

$$f_{C2} = \frac{1}{2\pi R_{C2} C_2} \tag{8.185}$$

In a CC amplifier, R_{C2} is generally much lower than R_{C1}. For the same values of C_1 and C_2, f_{C2} will become the low 3-dB frequency. That is, $f_L = f_{C2}$.

High Cutoff Frequencies

If the transistor in Fig. 8.61 is replaced by its high-frequency model, we get the high-frequency equivalent circuit shown in Fig. 8.63. Since there is no phase reversal between the input and output voltages, Miller's method cannot be applied.

If we assume C_π is open-circuited and $v_s = 0$, the equivalent circuit is shown in Fig. 8.64(a). The resistance looking to the left of C_μ is $(R_B \| R_s)$ and looking to the right of C_μ is $[r_\pi + (1 + \beta_f)(R_E \| R_L)]$. Thus, Thevenin's equivalent resistance presented to C_μ is

$$R_{C\mu} = (R_B \| R_S) \| [r_\pi + (1 + \beta_f)(R_E \| R_L)] \tag{8.186}$$

If we assume C_μ is open-circuited, the equivalent circuit is shown in Fig. 8.64(b). To find $R_{C\pi}$, let us remove C_π and apply a test voltage v_x, as shown in Fig. 8.64(c). Using KVL around the loop formed by R_B in parallel with R_s and by R_L in parallel with R_E, we get

$$v_x = (R_B \| R_s)\left(i_x - \frac{v_x}{r_\pi}\right) + (R_L \| R_E)\left(i_x - \frac{v_x}{r_\pi} - g_m v_x\right)$$

which can be simplified to

$$i_x(R_B \| R_s + R_L \| R_E) = v_x\left[1 + \frac{R_B \| R_s}{r_\pi} + \frac{R_L \| R_E}{r_\pi} + g_m(R_L \| R_E)\right]$$

or

$$\frac{i_x}{v_x} = \frac{1}{r_\pi} + \frac{1 + g_m(R_L \| R_E)}{R_B \| R_s + R_L \| R_E}$$

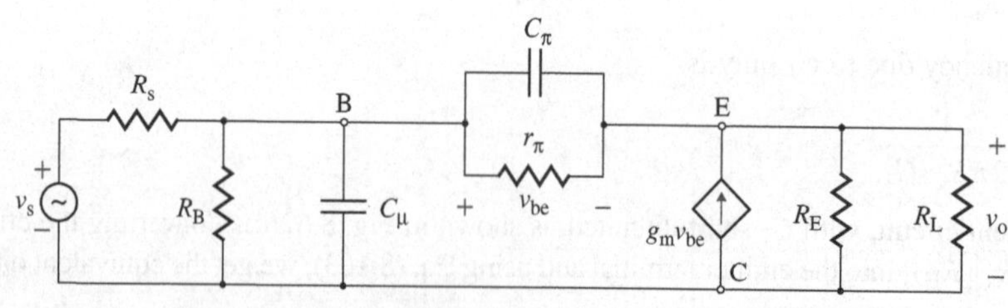

FIGURE 8.63 High-frequency circuit of a common-collector amplifier

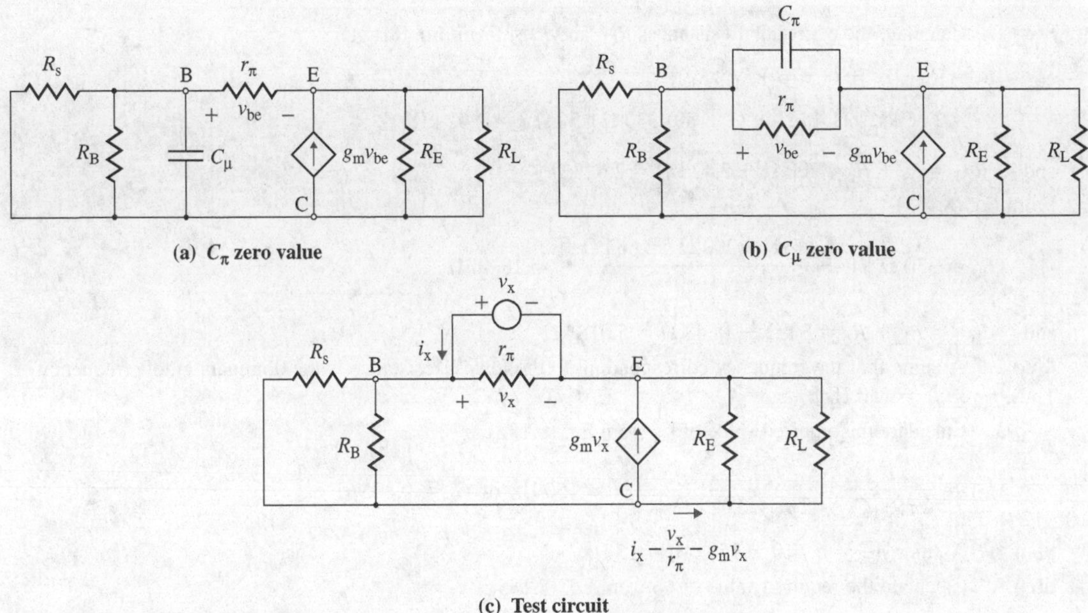

FIGURE 8.64 High-frequency equivalent circuits

Thus, Thevenin's equivalent resistance presented to C_π is

$$R_{C\pi} = \frac{v_x}{i_x} = r_\pi \parallel \frac{R_B \parallel R_s + R_L \parallel R_E}{1 + g_m(R_L \parallel R_E)} \tag{8.187}$$

and the high 3-dB frequency is

$$f_H = \frac{1}{2\pi(R_{C\pi}C_\pi + R_{C\mu}C_\mu)} \tag{8.188}$$

EXAMPLE 8.11

D

Designing a common-collector amplifier to give a specified frequency response

(a) Design a CC BJT amplifier as shown in Fig. 8.61 to give a low 3-dB frequency of $f_L = 150$ Hz. Calculate the values of C_1 and C_2. The circuit parameters of the CC BJT amplifier are $C_\pi = 15$ pF, $C_\mu = 1$ pF, $g_m = 57.14$ m℧, $\beta_f = 80$, $R_s = 200\ \Omega$, $r_\pi = 1.4$ kΩ, $R_E = 330\ \Omega$, $R_1 = 7$ kΩ, $R_2 = 4.3$ kΩ, and $R_L = 5$ kΩ.

(b) Calculate the high 3-dB cutoff frequency f_H.

SOLUTION

$R_B = R_1 \parallel R_2 = 7\text{ k}\Omega \parallel 4.3\text{ k}\Omega = 2.66\text{ k}\Omega$.

(a) The design steps for the low 3-dB frequency are as follows:

Step 1. Calculate the equivalent resistances R_{C1} and R_{C2}. From Eq. (8.181),

$$R_i = R_B \| [r_\pi + (1+\beta_f)(R_E \| R_L)]$$
$$= 2.66\text{ k}\Omega \| [1.4\text{ k}\Omega + (1+80)(330\ \Omega \| 5\text{ k}\Omega)] = 2.42\text{ k}\Omega$$

and $R_{C1} = R_s + R_i = 200\ \Omega + 2.42\text{ k}\Omega = 2.62\text{ k}\Omega$

From Eq. (8.183),

$$R_o = 330\ \Omega \| \left[\frac{(1.4\text{ k}\Omega + 200\ \Omega \| 2.66\text{ k}\Omega)}{(1+80)}\right] = 18.48\ \Omega$$

and $R_{C2} = R_L + R_o = 5\text{ k}\Omega + 18.48\ \Omega = 5.018\text{ k}\Omega$

Step 2. Assume that the frequency corresponding to the lowest resistance is the dominant cutoff frequency. Thus, $f_{C1} = f_L = 150$ Hz.

Step 3. Calculate the required value of C_1 from Eq. (8.182):

$$f_{C1} = \frac{1}{2\pi R_{C1} C_1} = \frac{1}{2\pi \times 2.62\text{ k}\Omega \times C_1} = 150\text{ Hz} \quad \text{or} \quad C_1 = 0.405\ \mu\text{F}$$

Step 4. Assume $f_{C2} = f_L/10 = 150/10 = 15$ Hz.

Step 5. Calculate the required value of C_2 from Eq. (8.185):

$$f_{C2} = \frac{1}{2\pi R_{C2} C_2} = \frac{1}{2\pi \times 5.018\text{ k}\Omega \times C_2} = 15\text{ Hz} \quad \text{or} \quad C_2 = 2.11\ \mu\text{F}$$

(b) From Eqs. (8.186) and (8.187),

$$R_{C\mu} = (2.66\text{ k}\Omega \| 200\ \Omega) \| [1.4\text{ k}\Omega + (1+80)(330\ \Omega \| 5\text{ k}\Omega)] = 184.7\ \Omega$$

$$R_{C\pi} = 1.4\text{ k}\Omega \| \frac{(2.66\text{ k}\Omega \| 200\ \Omega) + (5\text{ k}\Omega \| 330\ \Omega)}{1 + 57.14\text{ m℧} \times (5\text{ k}\Omega \| 330\ \Omega)} = 26.03\ \Omega$$

From Eq. (8.188), the high 3-dB frequency is

$$f_H = \frac{1}{2\pi(R_{C\pi}C_\pi + R_{C\mu}C_\mu)} = \frac{1}{2\pi(26.03\ \Omega \times 15\text{ pF} + 184.7\ \Omega \times 1\text{ pF})} = 276.7\text{ MHz}$$

8.15.3 Common-Base BJT Amplifiers

A common-base (CB) amplifier has a higher 3-dB frequency than either a CE or a CC amplifier. A CB amplifier is shown in Fig. 8.65. We will derive expressions for the low and high cutoff frequencies.

Low Cutoff Frequencies

The low-frequency equivalent circuit of the CB amplifier in Fig. 8.65 is shown in Fig. 8.66(a). There are two coupling capacitors, C_1 and C_2, and one bypass capacitor, C_B. If we assume C_2 and C_B are short-circuited, the equivalent circuit is shown in Fig. 8.66(b). The resistance R_t representing the current source is

$$R_t = \frac{v_{be}}{g_m v_{be}} = \frac{1}{g_m} = \frac{r_\pi}{\beta_f}$$

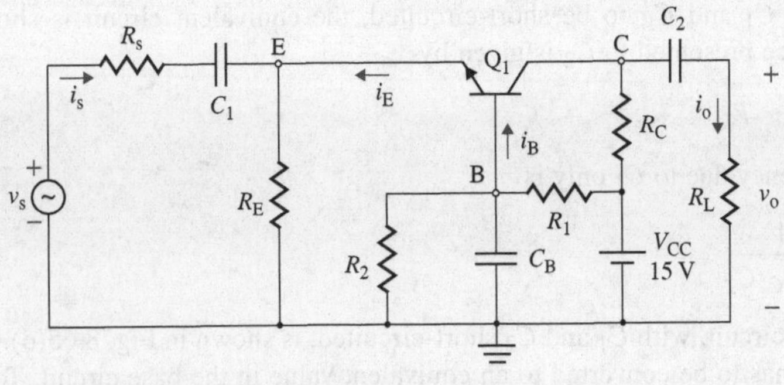

FIGURE 8.65 Common-base amplifier

Thus, the input resistance R_i at the emitter terminal is

$$R_i = R_E \parallel r_\pi \parallel R_t = R_E \parallel r_\pi \parallel \frac{r_\pi}{\beta_f} = R_E \parallel \frac{r_\pi}{1 + \beta_f} \tag{8.189}$$

and Thevenin's resistance presented to C_1 is

$$R_{C1} = R_s + R_i$$

Thus, the break frequency due to C_1 only is

$$f_{C1} = \frac{1}{2\pi R_{C1} C_1} \tag{8.190}$$

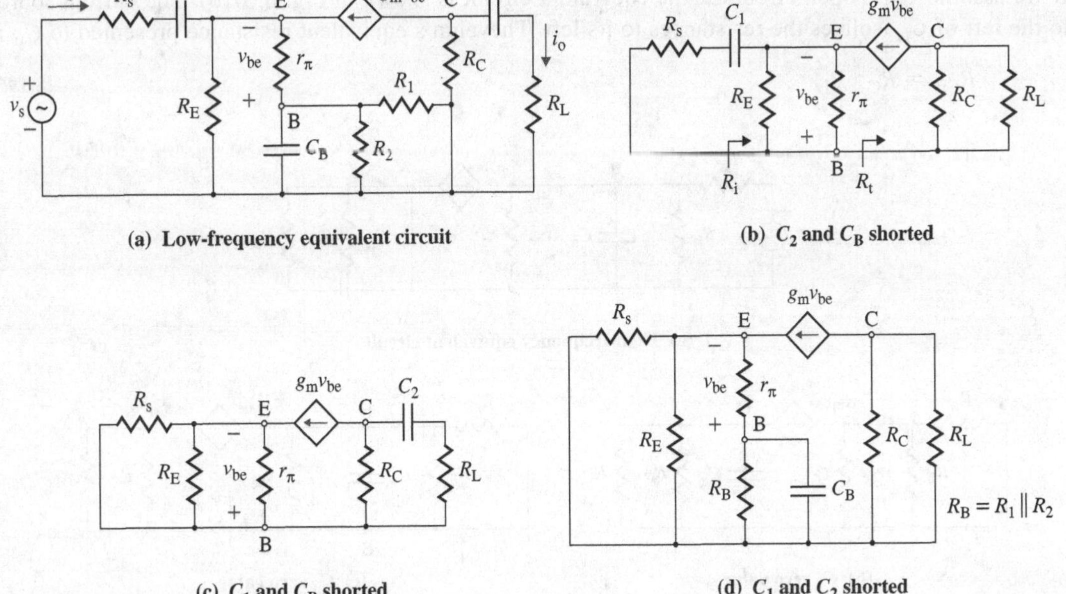

FIGURE 8.66 Equivalent circuits of a common-base amplifier for the short-circuit method

If we consider C_1 and C_B to be short-circuited, the equivalent circuit is shown in Fig. 8.65(c). Thevenin's resistance presented to C_2 is given by

$$R_{C2} = R_C + R_L \tag{8.191}$$

and the break frequency due to C_2 only is

$$f_{C2} = \frac{1}{2\pi R_{C2} C_2} \tag{8.192}$$

The equivalent circuit, with C_1 and C_2 short-circuited, is shown in Fig. 8.65(d). $R_s \parallel R_E$, which is in the emitter circuit, has to be converted to an equivalent value in the base circuit. To find the resistance presented to C_B, we multiply $(R_s \parallel R_E)$ by $(1 + \beta_f)$ and then combine the result in series with r_π; the combination forms a parallel circuit with R_B $(=R_1 \parallel R_2)$. That is, Thevenin's resistance becomes

$$R_{CB} = R_B \parallel [r_\pi + (1 + \beta_f)(R_s \parallel R_E)] \tag{8.193}$$

The break frequency due to C_B only is

$$f_{CB} = \frac{1}{2\pi R_{CB} C_B} \tag{8.194}$$

The low 3-dB frequency f_L is the largest of f_{C1}, f_{C2}, and f_{CE}. In general, the value of R_{C1} is on the order of a few hundred ohms, whereas R_{C2} and R_{CB} are on the order of kiloohms. To limit the values of the coupling capacitors, f_{C1} is normally chosen as the low 3-dB frequency; that is, $f_{C1} = f_L$.

High Cutoff Frequencies

The high-frequency equivalent circuit for the CB BJT amplifier in Fig. 8.65 is shown in Fig. 8.67(a). If we assume C_π is open-circuited, the equivalent circuit is shown in Fig. 8.67(b). The current source to the left of C_μ isolates the resistances to its left. Thevenin's equivalent resistance presented to C_μ is

$$R_{C\mu} = R_C \parallel R_L \tag{8.195}$$

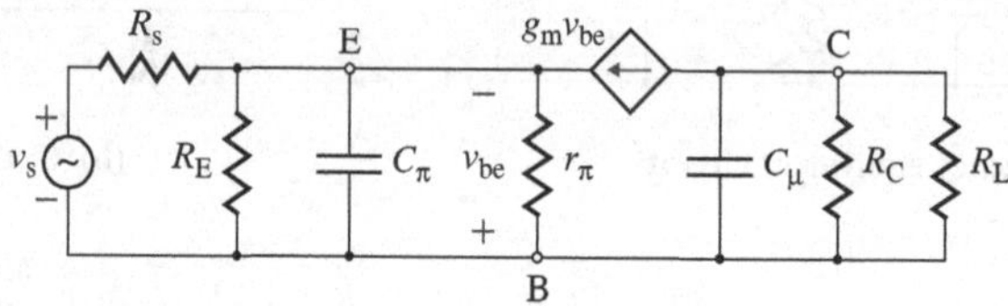

(a) High-frequency equivalent circuit

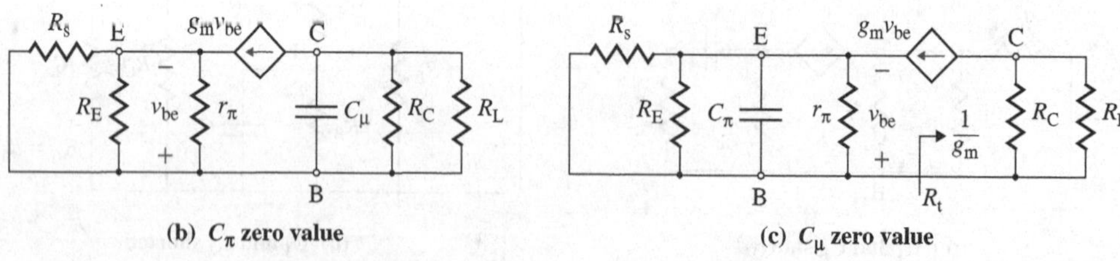

(b) C_π zero value

(c) C_μ zero value

FIGURE 8.67 High-frequency equivalent circuits of a common-base amplifier

If we assume C_μ is open-circuited, the equivalent circuit is shown in Fig. 8.67(c). The resistance offered by the current source is $1/g_m$, which forms a parallel circuit with R_E, R_s, and r_π. Thevenin's equivalent resistance presented to C_π is given by

$$R_{C\pi} = (R_s \parallel R_E) \parallel r_\pi \parallel \frac{1}{g_m} \tag{8.196}$$

Thus, the high 3-dB frequency is

$$f_H = \frac{1}{2\pi(R_{C\pi}C_\pi + R_{C\mu}C_\mu)} \tag{8.197}$$

In general, $1/g_m$ has a small value and is much less than R_s, R_E, or r_π; that is, $R_{C\pi} \approx 1/g_m$. For $R_{C\mu} >> R_{C\pi}$, which is normally the case, Eq. (8.197) can be approximated by

$$f_H \approx \frac{1}{2\pi R_{C\mu}C_\mu} \tag{8.198}$$

▶ **NOTE** f_H is independent of the transistor transconductance gain g_m, and there is no Miller's capacitance multiplication effect. CB amplifiers are used for high-frequency applications.

EXAMPLE 8.12

D **Designing a common-base amplifier to give a specified frequency response**

(a) Design a CB BJT amplifier as shown in Fig. 8.65 to give a low 3-dB frequency of $f_L = 150$ Hz. Calculate the values of C_1, C_2, and C_B. The circuit parameters are $C_\pi = 15$ pF, $C_\mu = 1$ pF, $g_m = 57.14$ m℧, $\beta_f = 80$, $R_s = 200\ \Omega$, $r_\pi = 1.4$ kΩ, $R_E = 330\ \Omega$, $R_C = 5$ kΩ, $R_1 = 7$ kΩ, $R_2 = 4.3$ kΩ, and $R_L = 5$ kΩ.

(b) Calculate the high 3-dB frequency f_H.

SOLUTION

$R_B = R_1 \parallel R_2 = 7\text{ k}\Omega \parallel 4.3\text{ k}\Omega = 2.66\text{ k}\Omega$

(a) The design steps are as follows:

Step 1. Calculate the equivalent resistances R_{C1}, R_{C2}, and R_{CB}.
From Eq. (8.189),

$$R_i = 330\ \Omega \parallel \left[\frac{1.4\text{ k}\Omega}{1+80}\right] = 16.4\ \Omega$$

and $R_{C1} = R_s + R_i = 200 + 16.4 = 216.4\ \Omega$
From Eq. (8.191),

$$R_{C2} = R_C + R_L = 5\text{ k}\Omega + 5\text{ k}\Omega = 10\text{ k}\Omega$$

From Eq. (8.193),

$$R_{CB} = 2.66\text{ k}\Omega \parallel [1.4\text{ k}\Omega + (1+80)(200\ \Omega \parallel 330\ \Omega)] = 2.16\text{ k}\Omega$$

Step 2. At the lowest resistance, the dominant cutoff frequency is $f_{C1} = f_L = 150$ Hz.

Step 3. Calculate the required value of C_1 from Eq. (8.190):

$$f_{C1} = \frac{1}{2\pi R_{C1}C_1} = \frac{1}{2\pi \times 216.4 \times C_1} = 150 \text{ Hz} \quad \text{or} \quad C_1 = 4.9\ \mu\text{F}$$

Step 4. At the next higher resistance $R_{CB} = 2.16\ \text{k}\Omega$, $f_{CB} = f_L/10 = 150/10 = 15$ Hz.

Step 5. Calculate the required value of C_B from Eq. (8.194):

$$f_{CB} = \frac{1}{2\pi R_{CB}C_B} = \frac{1}{2\pi \times 2.16\ \text{k}\Omega \times C_B} = 15 \text{ Hz} \quad \text{or} \quad C_B = 4.91\ \mu\text{F}$$

Step 6. At the highest resistance $R_{C2} = 10\ \text{k}\Omega$, $f_{C3} = f_L/20 = 150/20 = 7.5$ Hz.

Step 7. Calculate the required value of C_2 from Eq. (8.192):

$$f_{C2} = \frac{1}{2\pi R_{C2}C_2} = \frac{1}{2\pi \times 10\ \text{k}\Omega \times C_2} = 7.5 \text{ Hz} \quad \text{or} \quad C_2 = 2.12\ \mu\text{F}$$

(b) From Eq. (8.195),

$$R_{C\mu} = R_C \| R_L = 5\ \text{k}\Omega \| 5\ \text{k}\Omega = 2.5\ \text{k}\Omega$$

From Eq. (8.196),

$$R_{C\pi} = (200\ \Omega \| 330\ \Omega) \| 1.4\ \text{k}\Omega \| \frac{1}{57.14\ \text{m}\mho} = 15.2\ \Omega$$

From Eq. (8.197), the high 3-dB frequency is

$$f_H = \frac{1}{2\pi(R_{C\pi}C_\pi + R_{C\mu}C_\mu)} = \frac{1}{2\pi(15.2\ \Omega \times 15\text{ pF} + 2.5\ \text{k}\Omega \times 1\text{ pF})} = 58.34 \text{ MHz}$$

8.15.4 Multistage Amplifiers

Multistage amplifiers are often used to meet voltage gain, frequency range, input impedance, and/or output impedance requirements. In this section, we apply the short-circuit and zero-value methods to determine the cutoff frequencies of multistage amplifiers. Some equations will be similar to those used in the preceding sections because the equivalent circuits for the amplifiers are similar to those encountered previously.

When a capacitor C_μ is connected between the base (or gate) and the collector (or drain) of a transistor, it greatly influences the high 3-dB frequency.

EXAMPLE 8.13

Finding the frequency response of a two-stage CE-CE BJT amplifier A two-stage CE-CE BJT amplifier is shown in Fig. 8.68. The circuit parameters are $C_{\pi1} = C_{\pi2} = 15$ pF, $C_{\mu1} = C_{\mu2} = 1$ pF, $g_{m1} = g_{m2} = 57.14$ m℧, $R_s = 200\ \Omega$, $R_{11} = 22\ \text{k}\Omega$, $R_{21} = 47\ \text{k}\Omega$, $R_{C1} = 8\ \text{k}\Omega$, $R_{E1} = 5\ \text{k}\Omega$, $R_{12} = 22\ \text{k}\Omega$, $R_{22} = 47\ \text{k}\Omega$, $R_{C2} = 8\ \text{k}\Omega$, $R_{E2} = 5\ \text{k}\Omega$, $R_L = 5\ \text{k}\Omega$, $r_{\pi1} = r_{\pi2} = 1.4\ \text{k}\Omega$, $\beta_{f1} = 100$, $\beta_{f2} = 150$, $C_1 = 10\ \mu$F, $C_2 = 5\ \mu$F, $C_3 = 10\ \mu$F, $C_{E1} = 50\ \mu$F, and $C_{E2} = 50\ \mu$F.

(a) Calculate the low 3-dB frequency f_L.

(b) Calculate the high 3-dB frequency f_H.

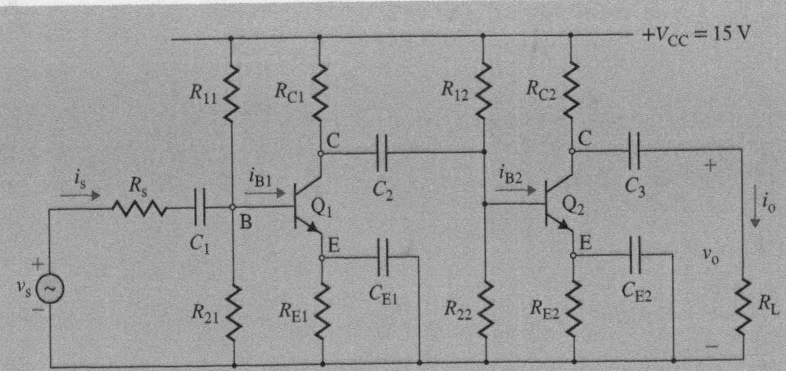

FIGURE 8.68 Two-stage CE-CE BJT amplifier

SOLUTION

We have

$$R_{B1} - R_{11} \parallel R_{21} = 22\ \text{k}\Omega \parallel 47\ \text{k}\Omega = 15\ \text{k}\Omega$$

and $R_{B2} = R_{12} \parallel R_{22} = 22\ \text{k}\Omega \parallel 47\ \text{k}\Omega = 15\ \text{k}\Omega$

(a) The circuit has five external capacitors—three coupling capacitors and two emitter bypass capacitors. The low-frequency AC equivalent circuit is shown in Fig. 8.69(a).

The time constant τ_1 due to C_1 only is

$$\tau_1 = [R_s + (R_{B1} \parallel r_{\pi 1})]C_1$$

$$= [200\ \Omega + (15\ \text{k}\Omega \parallel 1.4\ \text{k}\Omega)] \times 10\ \mu\text{F} = 14.8\ \text{ms}$$

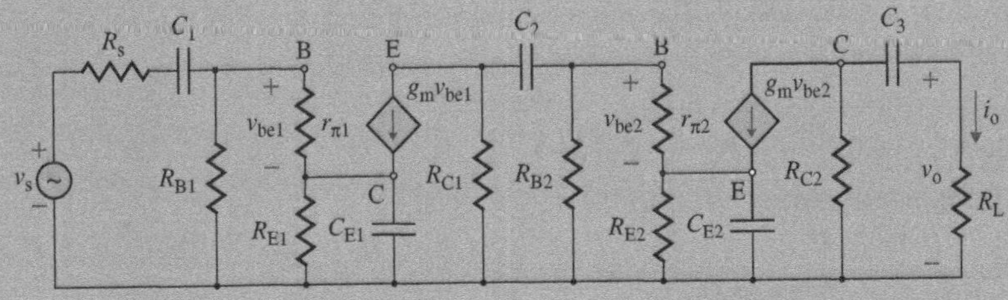

(a) Low-frequency equivalent circuit

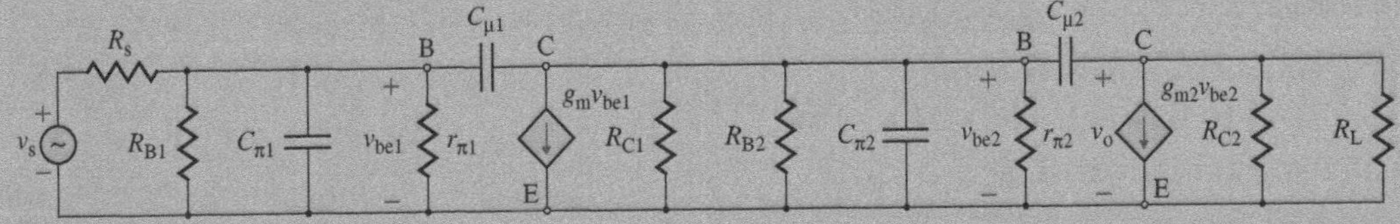

(b) High-frequency equivalent circuit

FIGURE 8.69 Equivalent circuits for Fig. 8.68

The time constant τ_2 due to C_2 only is

$$\tau_2 = [R_{C1} + (R_{B2} \| r_{\pi 2})]C_2$$
$$= [8\text{ k}\Omega + (15\text{ k}\Omega \| 1.4\text{ k}\Omega)] \times 5\ \mu\text{F} = 46.4\text{ ms}$$

The time constant τ_3 due to C_3 only is

$$\tau_3 = [R_{C2} + R_L]C_3$$
$$= [8\text{ k}\Omega + 5\text{ k}\Omega] \times 10\ \mu\text{F} = 130\text{ ms}$$

The time constant τ_4 due to C_{E1} only is

$$\tau_4 = \left[R_{E1} \left\| \frac{r_{\pi 1} + (R_s \| R_{B1})}{1 + \beta_{f1}}\right.\right]C_{E1}$$
$$= \left[5\text{ k}\Omega \left\| \frac{1.4\text{ k}\Omega + (200\ \Omega \| 15\text{ k}\Omega)}{1 + 100}\right.\right] \times 50\ \mu\text{F} = 0.79\text{ms}$$

The time constant τ_5 due to C_{E2} only is

$$\tau_5 = \left[R_{E2} \left\| \frac{r_{\pi 2} + (R_{C1} \| R_{B2})}{1 + \beta_{f2}}\right.\right]C_{E2}$$
$$= \left[5\text{ k}\Omega \left\| \frac{1.4\text{ k}\Omega + (8\text{ k}\Omega \| 15\text{ k}\Omega)}{1 + 150}\right.\right] \times 50\ \mu\text{F} = 2.17\text{ms}$$

From Eq. (2.106), the low 3-dB frequency f_L is

$$f_L = \frac{1}{2\pi}\left[\frac{1}{14.8\text{ ms}} + \frac{1}{46.4\text{ ms}} + \frac{1}{130\text{ ms}} + \frac{1}{0.79\text{ ms}} + \frac{1}{2.17\text{ ms}}\right] = 290.2\text{ Hz}$$

If we consider only the smallest time constant $\tau_4 = 0.79$ ms, we get

$$f_L = \frac{1}{2\pi\tau_4} = \frac{1}{2\pi \times 0.79\text{ ms}} = 201.5\text{ Hz}$$

(b) Replacing the transistors by their high-frequency model gives the high-frequency equivalent circuit shown in Fig. 8.69(b). If we assume $R_{\pi 1}$ is Thevenin's equivalent resistance faced by $C_{\pi 1}$ with $C_{\mu 1}$, $C_{\mu 2}$, and $C_{\pi 2}$ open-circuited, the time constant $\tau_{\pi 1}$ can be found from

$$\tau_{\pi 1} = R_{\pi 1}C_{\pi 1} = [r_{\pi 1} \| (R_s \| R_{B1})]C_{\pi 1} \tag{8.199}$$
$$= [1.4\text{ k}\Omega \| (200\ \Omega \| 15\text{ k}\Omega)] \times 15\text{ pF} = 2.6\text{ ns}$$

If $R_{\pi 2}$ is Thevenin's equivalent resistance faced by $C_{\pi 2}$ with $C_{\mu 1}$, $C_{\mu 2}$, and $C_{\pi 1}$ open-circuited, the time constant $\tau_{\pi 2}$ can be found from

$$\tau_{\pi 2} = R_{\pi 2}C_{\pi 2} = (r_{\pi 2} \| R_{B2} \| R_{C1})C_{\pi 2} \tag{8.200}$$
$$= [1.4\text{ k}\Omega \| 15\text{ k}\Omega \| 8\text{ k}\Omega] \times 15\text{ pF} = 16.6\text{ ns}$$

With $C_{\mu 2}$, $C_{\pi 1}$, and $C_{\pi 2}$ open-circuited, the effective load resistance of $C_{\mu 1}$ is

$$R_{L1(\text{eff})} = r_{\pi 2} \| R_{B2} \| R_{C1} \tag{8.201}$$
$$= 1.4\text{ k}\Omega \| 15\text{ k}\Omega \| 8\text{ k}\Omega = 1.1\text{ k}\Omega$$

and its effective input side resistance of $C_{\mu 1}$ is

$$R_{\pi 1(\text{eff})} = r_{\pi 1} \| R_s \| R_{B1} \tag{8.202}$$
$$= 1.4\text{ k}\Omega \| 200\ \Omega \| 15\text{ k}\Omega = 173\ \Omega$$

From Eq. (2.116), the time constant presented to $C_{\mu 1}$ is

$$\tau_{\mu 1} = [R_{L1(eff)} + R_{\pi 1(eff)}(1 + g_{m1}R_{L1(eff)})]C_{\mu 1} \quad (8.203)$$
$$= [1.1\text{ k}\Omega + 173 \times (1 + 57.14\text{ m℧} \times 1.1\text{ k}\Omega)] \times 1\text{ pF} = 12.2\text{ ns}$$

With $C_{\mu 1}$, $C_{\pi 1}$, and $C_{\pi 2}$ open-circuited, the effective load resistance of $C_{\mu 2}$ is

$$R_{L2(eff)} = R_L \| R_{C2} \quad (8.204)$$
$$= 5\text{ k}\Omega \| 8\text{ k}\Omega = 3.08\text{ k}\Omega$$

and its effective input side resistance of $C_{\mu 2}$ is

$$R_{\pi 2(eff)} = r_{\pi 2} \| R_{B2} \| R_{C1} = R_{L1(eff)} \quad (8.205)$$
$$= 1.4\text{ k}\Omega \| 15\text{ k}\Omega \| 8\text{ k}\Omega = 1.1\text{ k}\Omega$$

From Eq. (2.116), the time constant presented to $C_{\mu 2}$ is

$$\tau_{\mu 2} = [R_{L2(eff)} + R_{\pi 2(eff)}(1 + g_{m2}R_{L2(eff)})]C_{\mu 2} \quad (8.206)$$
$$= [3.08\text{ k}\Omega + 1.1\text{ k}\Omega \times (1 + 57.14\text{ m℧} \times 3.08\text{ k}\Omega)] \times 1\text{ pF} = 197.8\text{ ns}$$

From Eq. (2.112), the high 3-dB frequency f_H is

$$f_H = \frac{1}{2\pi}\left[\frac{1}{\tau_{\pi 1} + \tau_{\pi 2} + \tau_{\mu 1} + \tau_{\mu 2}}\right] = \frac{1}{2\pi}\left[\frac{10^9}{2.6 + 16.6 + 12.2 + 197.8}\right] = 694.4\text{ kHz}$$

EXAMPLE 8.14

Finding the frequency response of a two-stage CE-CB BJT amplifier A two-stage CE-CB amplifier is shown in Fig. 8.70. The circuit parameters are $C_{\pi 1} = C_{\pi 2} = 15$ pF, $C_{\mu 1} = C_{\mu 2} = 1$ pF, $R_s = 200\ \Omega$, $R_{11} = 22$ kΩ, $R_{21} = 47$ kΩ, $R_{C1} = 15$ kΩ, $R_{E1} = 9$ kΩ, $R_{12} = 22$ kΩ, $R_{22} = 47$ kΩ, $R_{C2} = 15$ kΩ, $R_{E2} = 9$ kΩ, $R_L = 10$ kΩ, $r_{\pi 1} = r_{\pi 2} = 1.4$ kΩ, $\beta_{f1} = 100$, $\beta_{f2} = 150$, $g_{m1} = 71.4$ m℧, $g_{m2} = 107.1$ m℧, $C_1 = 1\ \mu$F, $C_2 = 10\ \mu$F, $C_3 = 1\ \mu$F, $C_E = 10\ \mu$F, and $C_B = 10\ \mu$F.

(a) Calculate the low 3-dB frequency f_L.

(b) Calculate the high 3-dB frequency f_H.

SOLUTION

We have

$$R_{B1} = R_{11} \| R_{21} = 22\text{ k}\Omega \| 47\text{ k}\Omega = 15\text{ k}\Omega$$
$$R_{B2} = R_{12} \| R_{22} = 22\text{ k}\Omega \| 47\text{ k}\Omega = 15\text{ k}\Omega$$

(a) The low-frequency equivalent circuit is shown in Fig. 8.71(a).
The time constant τ_1 due to C_1 only is

$$\tau_1 = [R_s + (R_{B1} \| r_{\pi 1})]C_1$$
$$= [200\ \Omega + (15\text{ k}\Omega \| 1.4\text{ k}\Omega)] \times 1\ \mu\text{F} = 1.48\text{ ms}$$

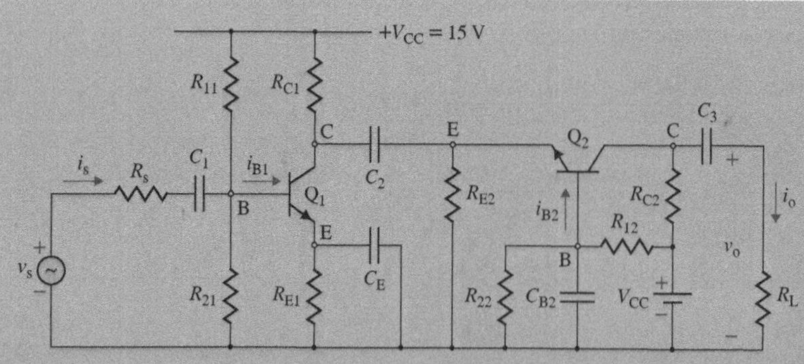

FIGURE 8.70 Two-stage CE-CB BJT amplifier

The time constant τ_2 due to C_2 only is

$$\tau_2 = \left[R_{C1} + (R_{E2} \parallel r_{\pi 2}) \parallel \frac{1}{g_{m2}}\right] C_2$$

$$= \left[15\text{ k}\Omega + (9\text{ k}\Omega \parallel 1.4\text{ k}\Omega) \parallel \frac{1000}{107.1\ \mho}\right] \times 10\ \mu\text{F} = 150.1\text{ ms}$$

The time constant τ_3 due to C_3 only is

$$\tau_3 = [R_{C2} + R_L]C_3$$

$$= [15\text{ k}\Omega + 10\text{ k}\Omega] \times 1\ \mu\text{F} = 25\text{ ms}$$

The time constant τ_4 due to C_{E1} only is

$$\tau_4 = \left[R_{E1} \parallel \frac{r_{\pi 1} + (R_s \parallel R_{B1})}{1 + \beta_{f1}}\right] C_E$$

$$= \left[9\text{ k}\Omega \parallel \frac{1.4\text{ k}\Omega + (200\ \Omega \parallel 15\text{ k}\Omega)}{1 + 100}\right] \times 10\ \mu\text{F} = 0.16\text{ ms}$$

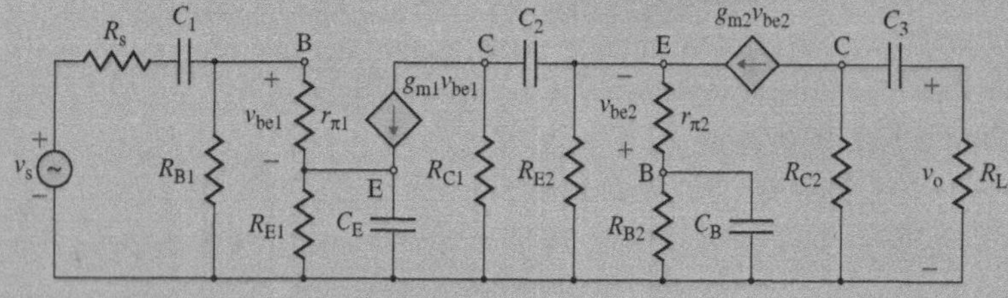

(a) Low-frequency equivalent circuit

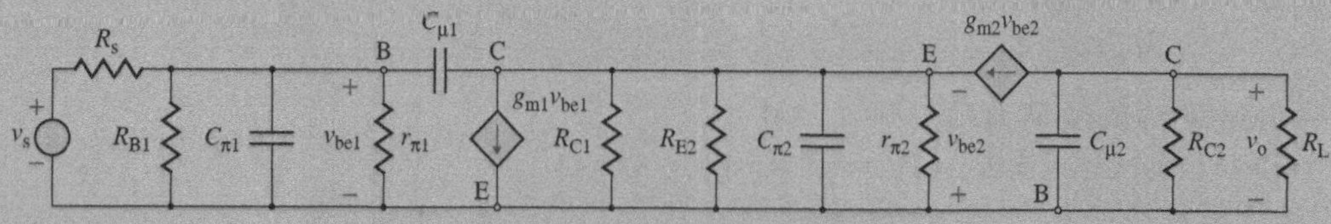

(b) High-frequency equivalent circuit

FIGURE 8.71 Equivalent circuits for Fig. 8.70

The time constant τ_5 due to C_B only is

$$\tau_5 = [R_{B2} \| r_{\pi 2} + (1 + \beta_{f2})(R_{C1} \| R_{E2})]C_{B2}$$
$$= [15\text{ k}\Omega \| 1.4\text{ k}\Omega + (1 + 150)(15\text{ k}\Omega \| 9\text{ k}\Omega)] \times 10\ \mu\text{F} = 147.2\text{ ms}$$

From Eq. (2.106), the low 3-dB frequency f_L is

$$f_L = \frac{1}{2\pi}\left[\frac{1}{1.48\text{ ms}} + \frac{1}{150.1\text{ ms}} + \frac{1}{25\text{ ms}} + \frac{1}{0.16\text{ ms}} + \frac{1}{147.2\text{ ms}}\right] = 1111\text{ Hz}$$

If we consider only the smallest time constant $\tau_4 = 0.16$ ms, we get

$$f_L = \frac{1}{2\pi\tau_4} = \frac{1}{2\pi \times 0.16\text{ ms}} = 995\text{ Hz}$$

(b) Replacing the transistors by their high-frequency models gives the high-frequency equivalent circuit shown in Fig. 8.71(b). If $R_{\pi 1}$ is Thevenin's equivalent resistance faced by $C_{\pi 1}$ with $C_{\mu 1}$, $C_{\mu 2}$, and $C_{\pi 2}$ open-circuited, the time constant $\tau_{\pi 1}$ is

$$\tau_{\pi 1} = R_{\pi 1}C_{\pi 1} = (r_{\pi 1} \| R_s \| R_{B1})C_{\pi 1} \tag{8.207}$$
$$= (1.4\text{ k}\Omega \| 200\ \Omega \| 15\text{ k}\Omega) \times 15\text{ pF} = 2.6\text{ ns}$$

If $R_{\pi 2}$ is Thevenin's equivalent resistance faced by $C_{\pi 2}$ with $C_{\mu 1}$, $C_{\mu 2}$, and $C_{\pi 1}$ open-circuited, the time constant $\tau_{\pi 2}$ can be found from

$$\tau_{\pi 2} = R_{\pi 2}C_{\pi 2} = \left[r_{\pi 2} \| R_{E2} \| R_{C1} \| \frac{1}{g_{m2}}\right]C_{\pi 2} \tag{8.208}$$

$$= \left[1.4\text{ k}\Omega \| 9\text{ k}\Omega \| 15\text{ k}\Omega \| \frac{1000}{107.1\ \mho}\right] \times 15\text{ pF} = 0.14\text{ ns}$$

With $C_{\mu 2}$, $C_{\pi 1}$, and $C_{\pi 2}$ open-circuited, the effective load resistance of $C_{\mu 1}$ is

$$R_{L1(\text{eff})} = r_{\pi 2} \| R_{E2} \| R_{C1} \| \frac{1}{g_{m2}} \tag{8.209}$$

$$= 1.4\text{ k}\Omega \| 9\text{ k}\Omega \| 15\text{ k}\Omega \| \left(\frac{1000}{107.1\ \mho}\right) = 9.3\ \Omega$$

and its effective input side resistance of $C_{\mu 1}$ is

$$R_{\pi 1(\text{eff})} = r_{\pi 1} \| R_s \| R_{B1} \tag{8.210}$$
$$= 1.4\text{ k}\Omega \| 200\ \Omega \| 15\text{ k}\Omega = 173\ \Omega$$

From Eq. (2.116), the time constant presented to $C_{\mu 1}$ is

$$\tau_{\mu 1} = [R_{L1(\text{eff})} + R_{\pi 1(\text{eff})}(1 + g_{m1}R_{L1(\text{eff})})]C_{\mu 1} \tag{8.211}$$
$$= [9.3\ \Omega + 173\ \Omega \times (1 + 71.4\text{ m}\mho \times 9.3\ \Omega)] \times 1\text{ pF} = 0.3\text{ ns}$$

With $C_{\mu 1}$, $C_{\pi 1}$, and $C_{\pi 2}$ open-circuited, the time constant presented to $C_{\mu 2}$ is

$$\tau_{\mu 2} = (R_L \| R_{C2})C_{\mu 2} \tag{8.212}$$
$$= (10\text{ k}\Omega \| 15\text{ k}\Omega) \times 1\text{ pF} = 6.0\text{ ns}$$

From Eq. (2.112), the high cutoff frequency is

$$f_H = \frac{1}{2\pi}\left[\frac{1}{\tau_{\pi 1} + \tau_{\pi 2} + \tau_{\mu 1} + \tau_{\mu 2}}\right] = \frac{1}{2\pi}\left[\frac{10^9}{2.6 + .14 + 0.3 + 6.0}\right] = 17.6 \text{ MHz}$$

which is much higher than $f_H = 694.4$ kHz for the CE-CE BJT amplifier in Example 8.13.

KEY POINTS OF SECTION 8.15

- The bypass capacitor C_E usually sets the low 3-dB frequency. Because of the Miller effect, the capacitor C_μ influences the high 3-dB frequency.
- For a small-signal input, capacitor C_μ appears between the base and the ground and there is no Miller effect. As a result, a CC amplifier can operate at a much higher frequency than a CE amplifier. However, the voltage gain of a CC amplifier is approximately unity.
- There is no Miller effect in a CB amplifier, and the high 3-dB frequency limit is higher that that of a CE amplifier. But the voltage gain is lower than that of a CE amplifier.

8.16 MOSFETs versus BJTs

An MOSFETs has the following advantages over a BJT:

1. It has an extremely high input resistance, on the order of megaohms.
2. It has no offset voltage when it is used as a switch, whereas a BJT requires a minimum base–emitter voltage V_{BE}.
3. It is relatively immune to ionizing radiation, whereas a BJT is very sensitive because its beta value is particularly affected.
4. It is less "noisy" than a BJT and thus more suitable for input stages of low-level amplifiers. It is used extensively in FM receivers.
5. It provides better thermal stability than a BJT—that is, the parameters of MOSFETs are less sensitive to temperature changes.

MOSFETs have a smaller gain bandwidth than BJTs and are more susceptible to damage in handling. The gain bandwidth is the frequency at which the gain becomes unity.

8.17 Design of Amplifiers

When an amplifier is being analyzed, the components are specified; however, when an amplifier is being designed, the designer must select the values of the circuit components. The design task can be simplified if a simple transistor model is used to find approximate values of the components. After the initial design stage, the next step is to analyze the amplifier with these approximate values and to compare the performance parameters with the desired values. Often the specifications are not met, and it is necessary to modify the component values. An amplifier is normally specified by the input resistance R_i, the output resistance R_o, and the voltage gain A_{vo}. These specifications are normally defined by the following values:

Source resistance R_s
DC supply voltage V_{CC} for BJTs
Load resistance R_L
Overall voltage gain A_v ($= v_L/v_s$) (at a specified R_L)
Input resistance at the base of the transistor R_i

After the specifications of an amplifier have been established, the next step is to decide on the type of transistor to be used. In the following analysis, we develop the necessary design conditions and the steps in meeting design specifications.

8.17.1 BJT Amplifier Design

Once a decision has been made to design a BJT amplifier, choose a suitable BJT and note its particular current gain β_f ($=\beta_F$) and Early voltage V_A (or assume a typical value of 200 V). Then choose a collector current I_C at the Q-point. The manufacturer normally provides curves showing the variations in current gain h_{fe} (β_f) against the collector current. I_C may be chosen from the manufacturer's data sheet so that the current gain β_f is maximum. Depending on the type of transistor (*npn* or *pnp*), V_{CC} and I_C will have positive or negative values; the value of V_A may be specified as a negative number. However, we use only the *magnitudes* of V_{CC}, I_C, and V_A.

We have noted that the technique of DC analysis differs from that of AC analysis. For DC analysis, the load line is set by the DC resistance R_{dc}. That is,

$$R_{dc} = \begin{cases} R_C + R_E & \text{for the CE amplifier of Fig. 8.28(a)} \\ R_E & \text{for the CC amplifier of Fig. 8.34(a)} \end{cases} \tag{8.213}$$

For AC analysis, the load line is set by the AC resistance. That is,

$$R_{ac} = \begin{cases} R_C \parallel R_L & \text{for the CE amplifier of Fig. 8.28(a)} \\ R_E \parallel R_L & \text{for the CC amplifier of Fig. 8.34(a)} \end{cases} \tag{8.214}$$

Under the no-load condition, the load resistance R_L is disconnected; the AC resistance R_{ac} equals R_C. Thus, there are two load lines that must be considered in designing an amplifier circuit. So far, we have considered the DC load line only while designing a biasing circuit. The AC and DC load lines for CE amplifiers are shown in Fig. 8.72. The Q-point, which is specified for a zero AC input signal, lies on both the AC and the DC load lines. The AC load line passes through the Q-point and has a slope of $-1/R_{ac}$. The slope of the AC line is greater in magnitude than that of the DC line. The AC load line may be described by

$$i_C - I_C = \frac{-(v_{CE} - V_{CE})}{R_{ac}}$$

which gives

$$i_C = -\frac{v_{CE}}{R_{ac}} + \left(\frac{V_{CE}}{R_{ac}} + I_C\right) \tag{8.215}$$

The maximum AC collector current $I_{C(max)}$, which occurs at $v_{CE} = 0$, can be found from Eq. (8.215):

$$I_{C(max)} = \frac{V_{CE}}{R_{ac}} + I_C \tag{8.216}$$

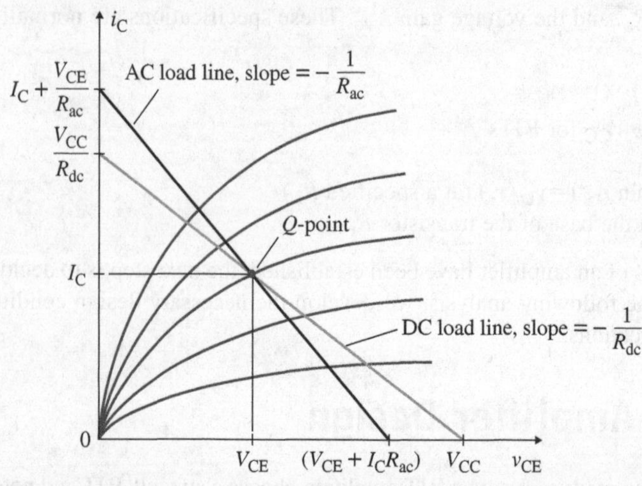

FIGURE 8.72 AC and DC load lines for CE amplifiers

An amplifier should be designed to accommodate the maximum AC swing along the AC load line, for which $i_{C(max)}$ must be twice the value of I_C, that is,

$$i_{C(max)} = 2I_C = \frac{V_{CE}}{R_{ac}} + I_C$$

which gives

$$I_C = \frac{V_{CE}}{R_{ac}} \tag{8.217}$$

Assuming $I_C \approx I_E$, Eq. (8.75) gives the DC load line

$$V_{CC} = V_{CE} + (R_C + R_E)I_C = V_{CE} + R_{dc}I_C$$

Substituting V_{CE} from Eq. (8.217) into the above equation yields

$$V_{CC} = V_{CE} + R_{dc}I_C = R_{ac}I_C + R_{dc}I_C = (R_{ac} + R_{dc})I_C$$

which gives the collector biasing current I_C as

$$I_C = \frac{V_{CC}}{R_{ac} + R_{dc}} \tag{8.218}$$

Thus, the Q-point is determined by both the AC and the DC resistances, which are dependent on R_C, R_E, and R_L. A higher voltage gain can be obtained at the expense of a low input resistance and a high output resistance. Therefore, BJT amplifiers are normally designed for a specified voltage gain or for a specified input resistance.

Designing for Specified Voltage Gain

When the voltage gain A_v $(=v_L/v_s)$ of the amplifier is specified, generally the input resistance R_i is not of major concern. Such an amplifier, shown in Fig. 8.29, acts as the middle stage of a multistage amplifier, providing as much gain as possible. The steps required to complete the design objectives are as follows:

Step 1. Calculate $V_E = |V_{CC}|/3$.

Step 2. Calculate $R_E = V_E/I_E = V_E\beta_F/[(1 + \beta_F)|I_C|]$.

Step 3. Calculate the voltage V_B at the transistor base: $V_B = V_E + V_{BE} = V_E + 0.7$.

Step 4. Calculate $R_B = 0.1(1 + \beta_f)R_E$.

Step 5. Calculate the values of R_1 and R_2:

$$R_1 = \frac{R_B|V_{CC}|}{|V_B|}$$

$$R_2 = \frac{R_B}{1 - |V_B/V_{CC}|}$$

Step 6. Calculate the value of R_C for known values of R_L, R_E, V_{CC}, and I_C. R_{ac} is the parallel combination of R_C and R_L. From Eqs. (8.214) and (8.218), we get

$$\frac{|V_{CC}|}{|I_C|} = R_{dc} + R_{ac} = R_E + R_C + \frac{R_C R_L}{R_C + R_L}$$

Step 7. Calculate the values of r_π and r_o.

$$r_\pi = \frac{25.8 \text{ mV}}{|I_B|} = \frac{25.8 \text{ mV}}{|I_C|}\beta_f$$

$$r_o \approx \frac{|V_A|}{|I_C|}$$

Step 8. As the first approximation, let the no-load voltage gain $|A_{vo}|$ (for $R_L = \infty$) equal $|A_v|$. From Eq. (8.92), the no-load voltage gain $|A_{vo}|$ is given by

$$|A_{vo}| = \frac{g_m r_\pi R_C}{r_\pi + (1 + \beta_f)R_{E1}} = \frac{\beta_F R_C}{R_x}$$

from which the resistance R_x in Eq. (8.85) at the transistor base can be found:

$$R_x = \frac{\beta_F R_C}{|A_{vo}|}$$

Step 9. Calculate the required value of emitter resistance R_{E1} from

$$R_{E1} = \frac{R_x - r_\pi}{1 + \beta_f}$$

If $R_{E1} < 0$, the desired $|A_v|$ is too large.

Step 10. Calculate the value of bypassed emitter resistance R_{E2}:

$$R_{E2} = R_E - R_{E1}$$

If $R_{E2} < 0$, the desired $|A_v|$ is too small; choose a transistor of lower current gain β_F.

Step 11. Calculate the output resistance $R_o = R_C$.

Step 12. Calculate the voltage gain $A_v = v_L/v_s$:

$$|A_v| = \frac{A_{vo}R_iR_L}{(R_i + R_s)(R_L + R_o)}$$

Step 13. If the value of $|A_v|$ in step 12 is not greater than or equal to the desired absolute value of A_v, repeat steps 8 through 12 with progressively higher values of $|A_{vo}|$ until you obtain the desired value for overall voltage gain A_v in step 12.

Designing for Specified Input Resistance

When the input resistance R_i of the amplifier is specified, generally the voltage gain A_v $(=v_L/v_s)$ is not of major concern. Such an amplifier normally acts as the input stage of a multistage amplifier. The first seven steps in completing the design objectives are the same as described above. The next two steps are as follows:

Step 8. Knowing, from Eq. (8.86), that the input resistance R_i is given by

$$R_i = R_B \parallel R_x = \frac{R_BR_x}{R_B + R_x}$$

find the required value of resistance R_x at the base of the transistor from

$$R_x = \frac{R_i}{1 - R_i/R_B} \quad \text{for } R_B \geq R_i$$

If $R_x < 0$, the desired R_i is too high; choose a lower value of R_i or a higher value of R_B (by repeating steps 1 to 4 with a transistor of higher current gain β_f and a lower collector current I_C).

Step 9. Calculate the required value of unbypassed emitter resistance R_{E1} from

$$R_x = r_\pi + R_{E1}(1 + \beta_f) \quad \text{or} \quad R_{E1} = \frac{R_x - r_\pi}{1 + \beta_f}$$

If $R_{E1} < 0$, choose a transistor of higher current gain β_F and lower biasing current I_C.

Steps 10 to 12 are the same as those used in designing for a specified voltage gain.

KEY POINTS OF SECTION 8.17

- In general, designing involves decision making and an iterative process. The design steps developed in this section will be helpful in finding component values to satisfy specifications.
- Designing an amplifier requires prior knowledge of desired specifications, choice of a BJT, and choice of a Q-point.
- Once the type of transistor and the Q-point have been chosen, the next step is to choose the biasing circuit and find its component values.
- The small-signal parameters, which are calculated from the values of the Q-point, are then used to find the emitter (or source) resistance needed to obtain the desired voltage gain or input resistance.

Summary

Bipolar junction transistors (BJTs) are active devices, and they are of two types: *npn* and *pnp*. BJTs are current-controlled devices; the output depends on the input current. A BJT can operate in any one of three regions: the cutoff, active, or saturation region. The forward current gain β_F, which is a very important parameter, is the ratio of the collector current to the base current. The biasing circuit sets the operating point such that the effects of parameter variations are minimized and allows for the superposition of AC signals with minimum distortion. BJTs may be represented by linear or nonlinear models. The linear models, which give approximate results, are commonly used for initial design and analysis. The nonlinear models are normally used for computer-aided design and analysis, especially with PSpice/SPICE.

A common-emitter amplifier is used for voltage amplification. Emitter resistance increases input resistance, but it reduces voltage gain. A compromise is normally required between high input resistance and high voltage gain requirements. A common-collector amplifier, which is known as an emitter follower, offers a high input resistance and a low output resistance, with a gain approaching unity. An amplifier can have two load lines: an AC load line and a DC load line. The AC load line is affected by external load resistance. Designing an amplifier normally requires specifying the input resistance, the output resistance, and the voltage gain.

Bypass and coupling capacitors control the low 3-dB frequency; capacitors of the small-signal transistor models control the high 3-dB frequency. Analysis of low break frequencies can be simplified by the short-circuit method, in which the time constant due to one capacitor is determined by assuming that the other capacitors are effectively short-circuited. This method can be extended to the analysis of multistage amplifiers.

References

1. R. T. Howe and C. G. Sodini, *Microelectronics—An Integrated Approach.* Englewood Cliffs, NJ: Prentice Hall, 1997.
2. R. C. Jaeger and T. Blalock, *Microelectronic Circuit Design* New York: McGraw-Hill, 2008.
3. B. Razavi, *RF Microelectronics.* Upper Saddle River, NJ: Prentice Hall, 1998.
4. D. A. Johns and K. Martin, *Analog Integrated Circuit Design.* New York: Wiley, 1997.
5. M. H. Rashid, *Introduction to PSpice Using OrCAD for Circuits and Electronics.* Upper Saddle River, NJ: Prentice-Hall Inc., 3rd edition, 2003.
6. D. T. Comer and D. J. Comer, "A New Amplifier Circuit with Both Practical and Tutorial Value", *IEEE Transactions on Education,* Vol. 43, No. 1, February 2000, pp 25–29.
7. D. A. Neamen, *Microelectronics: Circuit Analysis and Design.* New York: McGraw-Hill, 2007.

Review Questions

1. What are the types of BJTs?
2. What are the differences between *npn*- and *pnp*-type BJTs?
3. What are the possible regions of BJT operation?
4. What is a short-circuit amplification factor?

5. What is a forward amplification factor?
6. What are the characteristics of an active region?
7. What are the characteristics of a saturation region?
8. What is the purpose of biasing a BJT?
9. What is a load line?
10. What is the relationship between power dissipation and junction temperature?
11. What are the linear models of BJTs?
12. What is a transistor saturation current?
13. What is the small-signal current gain of a BJT?
14. What is the small-signal input resistance of a BJT?
15. What is the small-signal output resistance of a BJT?
16. What is the Early voltage?
17. What is the purpose of an emitter-bypassed capacitor?
18. What are the performance parameters of an amplifier?
19. What are the characteristics of CE amplifiers?
20. What are the characteristics of CC amplifiers?
21. What is a DC load line?
22. What is an AC load line?
23. What are the advantages of FETs over BJTs?
24. What is the transition frequency of a transistor?
25. What is the transit time of a BJT?

Problems

The symbol **D** indicates that a problem is a design problem. The symbol **P** indicates that you can check the solution to a problem using PSpice/SPICE or Electronics Workbench.

8.3 Principles of BJT Operation

8.1 Calculate the width of the B-C depletion region of an *npn* transistor if the C-B voltage is $V_{CB} = 20$ V. The physical parameters are $N_C = 10^{16}$ cm^{-3}, $N_B = 1 \times 10^{15}$ cm^{-3}, $T = 25°C$, $w_B = 0.7$ μm, $\varepsilon_s = 11.7 \times 8.85 \times 10^{-14}$, $q = 1.6 \times 10^{-19}$, and $n_i = 1.5 \times 10^{10}$.

8.2 Calculate the neutral base width of an *npn* transistor if the C-B voltage is $V_{CB} = 12$ V. The physical parameters are $N_C = 10^{16}$ cm^{-3}, $N_B = 1 \times 10^{15}$ cm^{-3}, $T = 25°C$, $w_B = 0.7$ μm, $\varepsilon_s = 11.7 \times 8.85 \times 10^{-14}$, $q = 1.6 \times 10^{-19}$, and $n_i = 1.5 \times 10^{10}$.

8.3 Plot the width of the B-C depletion region of an *npn* transistor if the C-B voltage is varied from $V_{CB} = 1$ V to 20 V. The physical parameters are $N_C = 10^{16}$ cm^{-3}, $N_B = 1 \times 10^{15}$ cm^{-3}, $T = 25°C$, $w_B = 0.7$ μm, $\varepsilon_s = 11.7 \times 8.85 \times 10^{-14}$, $q = 1.6 \times 10^{-19}$, and $n_i = 1.5 \times 10^{10}$.

8.4 Calculate the width of the B-C depletion region of an *npn* transistor if the C-B voltage is $V_{CB} = 15$ V. The physical parameters are $N_C = N_B = 10^{16}$ cm^{-3}, $T = 25°C$, $w_B = 0.7$ μm, $\varepsilon_s = 11.7 \times 8.85 \times 10^{-14}$, $q = 1.6 \times 10^{-19}$, and $n_i = 1.5 \times 10^{10}$.

8.5 The physical parameters of an *npn* transistor are $N_C = 10^{16}$ cm^{-3}, $N_B = 1 \times 10^{15}$ cm^{-3}, $T = 25°C$, $w_B = 0.7$ μm, $\varepsilon_s = 11.7 \times 8.85 \times 10^{-14}$, $q = 1.6 \times 10^{-19}$, and $n_i = 1.5 \times 10^{10}$. Determine the maximum C-B voltage to limit the width of the B-C region to $\pm 10\%$ of the base width w_B.

8.6 The physical parameters of an *npn* transistor are $N_C = N_B = 10^{16}$ cm^{-3}, $T = 25°C$, $w_B = 0.7$ μm, $\varepsilon_s = 11.7 \times 8.85 \times 10^{-14}$, $q = 1.6 \times 10^{-19}$, and $n_i = 1.5 \times 10^{10}$. Determine the maximum C-B voltage to limit the width of the B-C region to $\pm 15\%$ of the base width w_B.

8.7 The physical parameters of an *npn* transistor are $N_E = 10^{16}\ \text{cm}^{-3}$, $N_B = 1 \times 10^{15}\ \text{cm}^{-3}$, $n_i = 1.5 \times 10^{10}\ \text{cm}^{-3}$, $D_E = 20\ \text{cm}^2/\text{s}$, $D_B = 10\ \text{cm}^2/\text{s}$, $x_B = 0.7\ \mu\text{m}$, $L_E = 0.013$ cm, $\tau_B = 8 \times 10^{-6}$ s, and $A_{BE} = 10^{-3}\ \text{cm}^2$. Calculate **(a)** the collector reverse saturation current I_{SC}, **(b)** the current gain β_F, and **(c)** the base saturation current I_{SB}.

8.8 The physical parameters of an *npn* transistor are $N_E = 10^{16}\ \text{cm}^{-3}$, $N_B = 1 \times 10^{15}\ \text{cm}^{-3}$, $n_i = 1.5 \times 10^{10}\ \text{cm}^{-3}$, $D_E = 20\ \text{cm}^2/\text{s}$, $D_B = 10\ \text{cm}^2/\text{s}$, $x_B = 0.7\ \mu\text{m}$, $L_E = 0.013$ cm, $\tau_B = 8 \times 10^{-6}$ s, and $A_{BE} = 10^{-3}\ \text{cm}^2$.

a. Calculate the emitter doping density to obtain a current gain $\beta_F = 150$.

b. Plot the current gain β_F against the doping ratio $x = N_E/N_B = 1$ to 10.

8.9 The physical parameters of an *npn* transistor are $N_E = 10^{16}\ \text{cm}^{-3}$, $N_B = 1 \times 10^{15}\ \text{cm}^{-3}$, $n_i = 1.5 \times 10^{10}\ \text{cm}^{-3}$, $D_E = 20\ \text{cm}^2/\text{s}$, $D_B = 10\ \text{cm}^2/\text{s}$, $L_E = 0.013$ cm, $\tau_B = 8 \times 10^{-6}$ s, and $A_{BE} = 10^{-3}\ \text{cm}^2$.

a. Calculate the base width x_B to obtain a current gain of $\beta_F = 100$.

b. Plot the current gain β_F against the base width $x_B = 0.1\ \mu\text{m}$ to $1\ \mu\text{m}$.

8.10 The physical parameters of an *npn* transistor are $N_E = 10^{16}\ \text{cm}^{-3}$, $N_B = 1 \times 10^{15}\ \text{cm}^{-3}$, $n_i = 1.5 \times 10^{10}\ \text{cm}^{-3}$, $D_E = 20\ \text{cm}^2/\text{s}$, $D_B = 10\ \text{cm}^2/\text{s}$, $x_B = 0.7\ \mu\text{m}$, $L_E = 0.013$ cm, $\tau_B = 8 \times 10^{-6}$ s, and $A_{BE} = 10^{-3}\ \text{cm}^2$. Calculate the B-E voltage to give a collector current of $I_C = 1.2$ mA if $V_T = 25.8$ mV, $V_{CE} = 15$ V, and $|V_A| = 200$ V.

8.11 The physical parameters of a *pnp* transistor are $N_E = 10^{15}\ \text{cm}^{-3}$, $N_B = 5 \times 10^{16}\ \text{cm}^{-3}$, $n_i = 1.5 \times 10^{10}\ \text{cm}^{-3}$, $D_E = 20\ \text{cm}^2/\text{s}$, $D_B = 10\ \text{cm}^2/\text{s}$, $x_B = 0.7\ \mu\text{m}$, $L_E = 0.013$ cm, $\tau_B = 8 \times 10^{-6}$ s, and $A_{BE} = 10^{-3}\ \text{cm}^2$. Calculate the E-B voltage to give a collector current of $I_C = 1.5$ mA if $V_T = 25.8$ mV, $V_{CE} = -12$ V, and $|V_A| = 200$ V.

8.4 Input and Output Characteristics

8.12 The parameters of an *npn* transistor are $\alpha_F = 0.9934$ and $I_B = 25\ \mu\text{A}$. Determine **(a)** the forward current gain β_F, **(b)** the collector current i_C, and **(c)** the emitter current i_E.

8.13 The parameters of the transistor in Fig. P8.13 are $\beta_F = 150$, $V_{BE} = 0.7$ V, and $V_{CE(sat)} = 0.3$ V. If $R_C = 1.5\ \text{k}\Omega$, determine the critical value of R_B so that the transistor operates **(a)** in the active (amplifier) region and **(b)** in the saturation region.

FIGURE P8.13

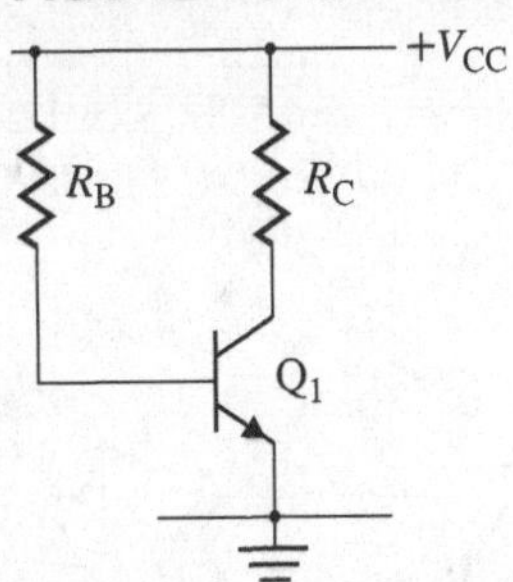

8.14 The parameters of the transistor in Fig. P8.14 are $\beta_F = 150$, $V_{BE} = 0.7$ V, and $V_{CE(sat)} = 0.3$ V. If $R_C = 1.5\ \text{k}\Omega$ and $R_E = 500\ \Omega$, determine the critical value of R_B so that the transistor operates **(a)** in the active (amplifier) region and **(b)** in the saturation region.

FIGURE P8.14

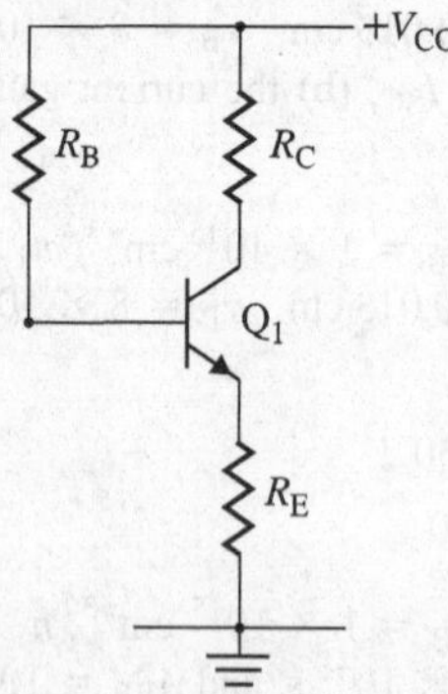

8.15 The parameters of the transistor in Fig. P8.15 are $\beta_F = 150$, $V_{BE} = 0.7$ V, and $V_{CE(sat)} = 0.3$ V. If $R_C = 1.5$ kΩ, determine the critical value of R_B so that the transistor operates **(a)** in the active (amplifier) region and **(b)** in the saturation region.

FIGURE P8.15

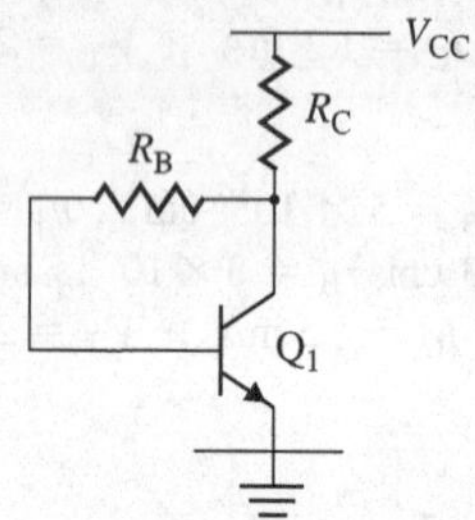

8.7 DC Biasing of Bipolar Junction Transistors

8.16 The parameters of the transistor in Fig. P8.16 are $\beta_F = 150$, $V_{BE} = 0.7$ V, and $V_{CE(sat)} = 0.3$ V. If $V_{CC} = 12$ V, $V_{EE} = 12$ V, and $R_C = 1.5$ kΩ, determine **(a)** the base current I_B, **(b)** the collector current I_C, and **(c)** the C-E voltage V_{CE}.

FIGURE P8.16

+V_{CC}
R_C
Q_1
R_E
$-V_{EE}$

8.17 The parameters of the transistor in Fig. P8.17 are $\beta_F = 150$, $V_{BE} = 0.7$ V, and $V_{CE(sat)} = 0.3$ V. If $V_{CC} = 12$ V, $R_B = 200$ kΩ, $R_C = 1.5$ kΩ, and $R_E = 500$ Ω, determine **(a)** the base current I_B, **(b)** the collector current I_C, and **(c)** the C-E voltage V_{CE}.

FIGURE P8.17

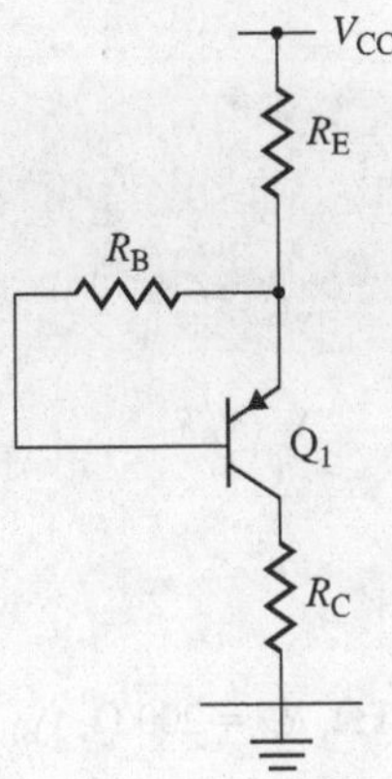

8.18 The parameters of the transistor in Fig. P8.18 are $\beta_F = 150$, $V_{BE} = 0.7$ V, and $V_{CE(sat)} = 0.3$ V. If $V_{CC} = 15$ V, $V_B = 5$ V, $R_B = 200$ kΩ, $R_C = 1.5$ kΩ, and $R_E = 500$ Ω, determine **(a)** the base current I_B, **(b)** the collector current I_C, and **(c)** the C-E voltage V_{CE}.

FIGURE P8.18

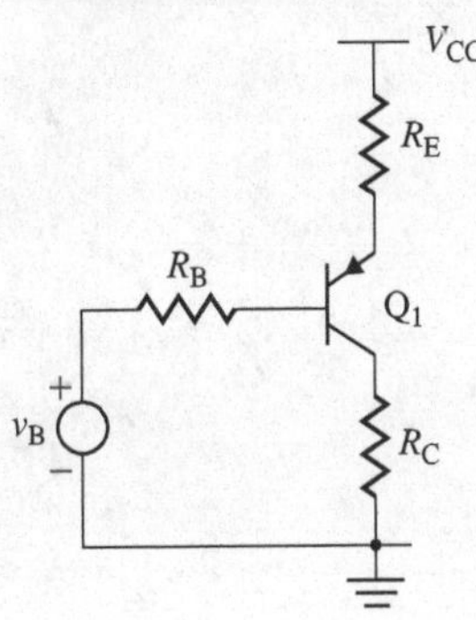

8.19 P The parameters of the *pnp* BJT circuit in Fig. P8.19 are $R_C = 10$ kΩ, $R_E = 1$ kΩ, $V_{CC} = 15$ V, $V_{EE} = 5$ V, $V_{EB} = 0.6$ V, and $\alpha_F = 0.992$. Calculate I_B, I_C, I_E, V_{CE}, and V_{CB} at the Q-point.

FIGURE P8.19

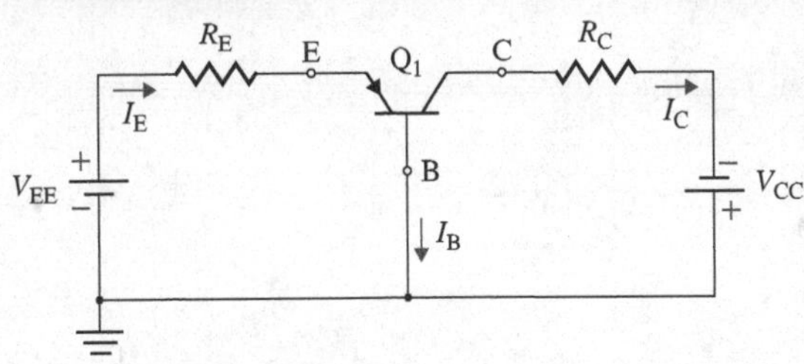

8.20 P The parameters of the *npn* transistor circuit in Fig. P8.20 are $R_1 = 100$ kΩ, $R_C = 1$ kΩ, $R_E = 200$ Ω, $V_{BE} = 0.7$ V, and $V_{CC} = 12$ V.

a. Calculate I_B, I_C, I_E, and V_{CE} at the operating point if $\beta_F = 50$ and if $\beta_F = 250$.

b. Repeat part (a) if $R_E = 0$.

FIGURE P8.20

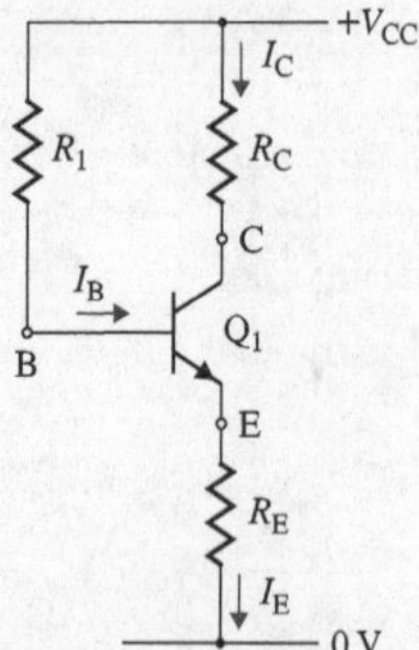

8.21 P The parameters of the transistor circuit in Fig. P8.21 are $R_1 = 10\ \text{k}\Omega$, $R_C = 1\ \text{k}\Omega$, $R_E = 200\ \Omega$, $V_{BE} = 0.7$ V, and $V_{CC} = 12$ V.

a. Calculate I_B, I_C, I_E, and V_{CE} at the *Q*-point if $\beta_F = 50$ and if $\beta_F = 250$.

b. Repeat part (a) if $R_E = 0$.

FIGURE P8.21

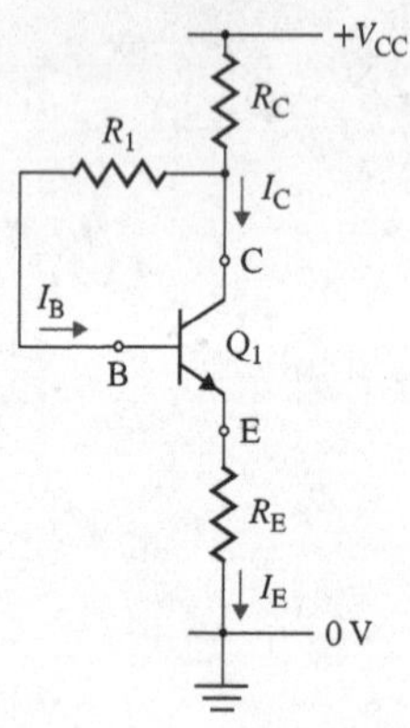

8.22 D P Design a biasing circuit as shown in Fig. 8.20(a). Calculate the values and power ratings of R_E, R_C, R_1, and R_2 and the total power dissipation P_T of the circuit. The power supply is $V_{CC} = 30$ V. The quiescent values are $I_C = 2$ mA and $V_{CE} = 12.6$ V. The nominal value of β_F is 50. Assume $V_{BE} = 0.5$ V and $r_\mu = \infty$.

8.23 P The *npn* transistor circuit of Fig. P8.23 has $V_{BE} = 0.5$ V and $\beta_F = 80$. Determine the value of R_1 that gives $I_C = 4$ mA and the corresponding value of V_{CE}.

FIGURE P8.23

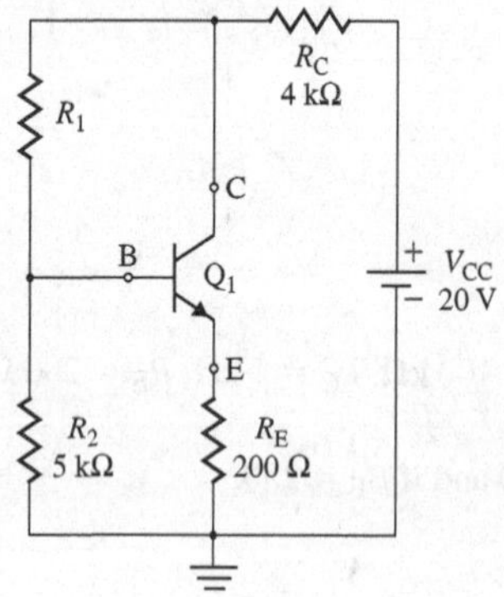

8.24 The *pnp* transistor circuit of Fig. P8.24 has $\beta_F = 100$ and $V_{EB} = 0.7$ V. Calculate I_C and V_{CE}.

FIGURE P8.24

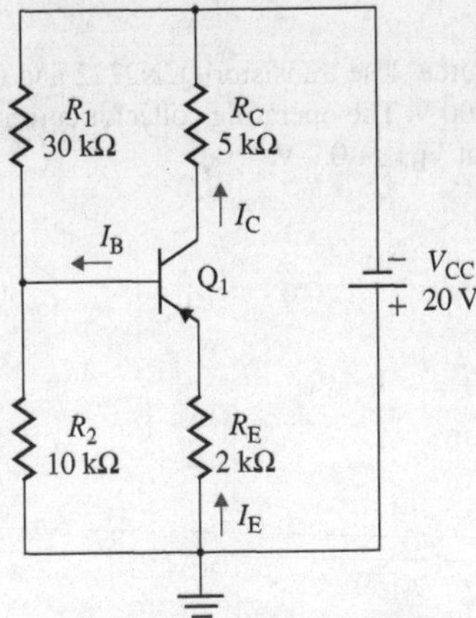

8.8–8.10 Common-Emitter Amplifiers, Emitter Followers, and Common-Base Amplifiers

8.25 The parameters of the amplifier circuit in Fig. P8.25 are $V_{CC} = 5$ V, $R_C = 500\ \Omega$, $R_1 = 6.5\ \text{k}\Omega$, $R_2 = 2.5\ \text{k}\Omega$, $R_E = 450\ \Omega$, $R_s = 500\ \Omega$, $R_L = 5\ \text{k}\Omega$, and $C_1 = C_2 = \infty$. Assume $\beta_F = 100$ and $V_A = 200$ V.

FIGURE P8.25

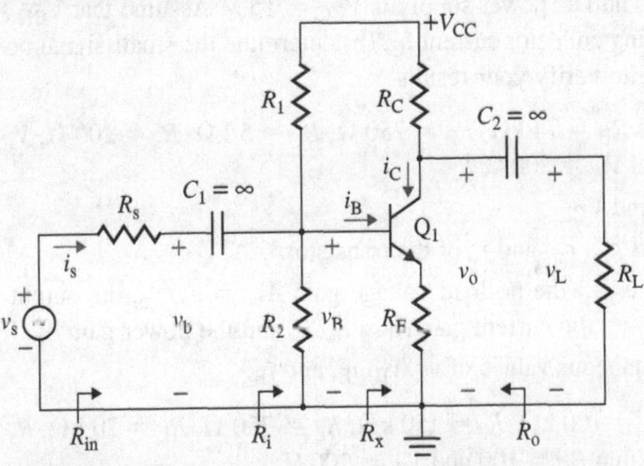

a. Find the Q-point defined by I_B, I_C, and V_{CE}.

b. Calculate the small-signal parameters g_m, r_π, and r_o of the transistor.

c. Calculate the input resistance $R_{in} = v_s/i_s$, the no-load voltage gain $A_{vo} = v_o/v_b$, the output resistance R_o, the overall voltage gain $A_v = v_L/v_s$, the current gain $A_i = i_o/i_s$, and the power gain A_p.

d. Use PSpice/SPICE to plot the instantaneous values of v_L, v_C, v_B, i_C, and i_B.

8.26 The BJT amplifier of Fig. 8.23 has $R_1 = 5.5\ \text{k}\Omega$, $R_2 = 1.5\ \text{k}\Omega$, $R_C = 1.5\ \text{k}\Omega$, $R_E = 150\ \Omega$, $R_L = 5\ \text{k}\Omega$, $R_s = 200\ \Omega$, and $V_{CC} = 18$ V. Assume $\beta_F = 100$ and $V_A = 200$ V.

a. Find the Q-point defined by I_B, I_C, and V_{CE}.

b. Calculate the small-signal parameters g_m, r_π, and r_o of the transistor.

c. Calculate the input resistance $R_{in} = v_s/i_s$, the no-load voltage gain $A_{vo} = v_o/v_b$, the output resistance R_o, the overall voltage gain $A_v = v_L/v_s$, the current gain $A_i = i_o/i_s$, and the power gain A_p.

d. Use PSpice/SPICE to plot the instantaneous values of v_L, v_C, v_B, i_C, and i_B.

8.27 The parameters of the CE amplifier with an active current source as shown in Fig. 8.24(a) are $\beta_F = 173$, $I_S = 3.295 \times 10^{-14}$ A, and $V_A = 200$ V. The DC power supply is $V_{CC} = 15$ V. Assume that $V_{BE} = 0.7$ V. If $R = 14.3$ kΩ, **(a)** determine the operating collector current I_C, **(b)** determine the small-signal no-load voltage gain A_{vo}, and **(c)** use PSpice/SPICE to verify your results.

8.28 Design a CE amplifier of Fig. P8.28 with an active current source. Use transistor Q2N2222 and Q2N2907 whose nominal $\beta_F = 173$, $I_S = 3.295 \times 10^{-14}$ A, and $V_A = 100$ V. The operating collector current is set at $I_C = 1$ mA. The DC power supply is $V_{CC} = 15$ V. Assume that $V_{BE} = 0.7$ V.

FIGURE P8.28

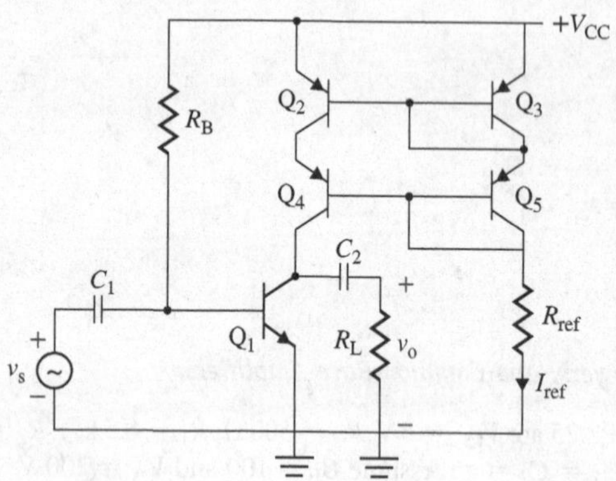

8.29 The parameters of the CE amplifier with an active current source as shown in Fig. P8.28 are $\beta_F = 173$, $I_S = 3.295 \times 10^{-14}$ A, and $V_A = 200$ V. The DC power supply is $V_{CC} = 15$ V. Assume that $V_{BE} = 0.7$ V. If $R = 14.3$ kΩ, **(a)** determine the operating collector current I_C, **(b)** determine the small-signal no-load voltage gain A_{vo}, and **(c)** use PSpice/SPICE to verify your results.

8.30 P The emitter follower of Fig. 8.34(a) has $R_B = 74$ kΩ, $R_E = 750$ Ω, $R_L = 5$ kΩ, $R_s = 200$ Ω, $V_{CC} = 18$ V, and $V_{BE} = 0.7$ V. Assume $\beta_F = 100$ and $V_A = 200$ V.

a. Find the Q-point defined by I_B, I_C, and V_{CE}.

b. Calculate the small-signal parameters g_m, r_π, and r_o of the transistor.

c. Calculate the input resistance $R_{in} = v_s/i_s$, the no-load voltage gain $A_{vo} = v_o/v_b$, the output resistance R_o, the overall voltage gain $A_v = v_L/v_s$, the current gain $A_i = i_o/i_s$, and the power gain A_p.

d. Use PSpice/SPICE to plot the instantaneous values of v_L, v_B, i_E, and i_B.

8.31 The emitter follower of Fig. P8.31 has $R_1 = 100$ kΩ, $R_2 = 150$ kΩ, $R_E = 750$ Ω, $R_L = 20$ kΩ, $R_s = 200$ Ω, $V_{CC} = 15$ V, and $V_{BE} = 0.7$ V. Assume that $\beta_F = 100$ and $V_A = 200$ V.

FIGURE P8.31

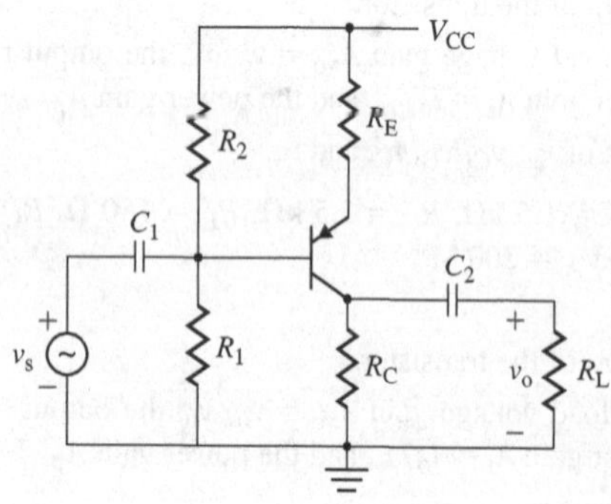

a. Find the Q-point defined by I_B, I_C, and V_{CE}.
b. Calculate the small-signal parameters of the transistor g_m, r_π, and r_o.
c. Calculate (i) the input resistance R_{in}, (ii) the no-load voltage gain A_o $(=v_o/v_b)$, (iii) the output resistance R_o, (iv) the overall voltage gain A_v $(=v_L/v_s)$, (v) the current gain A_i, and (vi) the power gain A_p.

8.32 An emitter follower is biased by a transistor current source, as shown in Fig. P8.32(a). Assume all transistors are identical, with $\beta_F = 100$, $V_{BE} = 0.7$ V, and $V_A = 200$ V, and assume $R = 5$ kΩ and $R_L = 1$ kΩ.

a. Find the equivalent biasing current I_O and resistance r_o, as shown in Fig. P8.32(b).
b. Calculate the small-signal input resistance R_i and the output resistance R_o of the emitter follower.

FIGURE P8.32

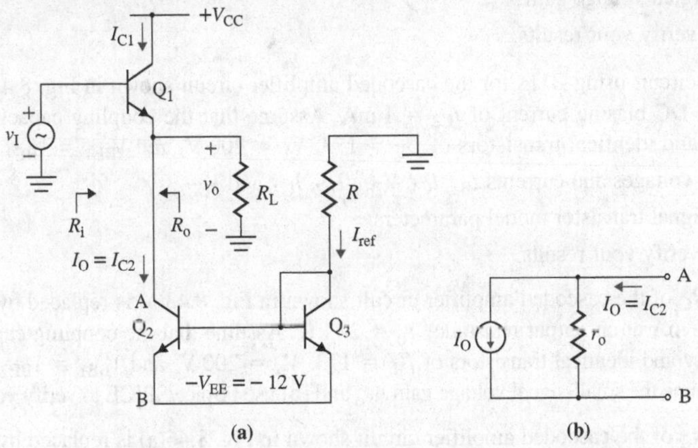

8.33 The biasing current for the emitter follower in Fig. P8.32(a) can be generated by the circuits in Fig. P8.33. Find the equivalent biasing current I_O and resistance r_o. Assume a transistor with $\beta_F = 100$, $V_{BE} = 0.7$ V, and $V_A = 200$ V and a diode drop of $V_D = 0.7$ V.

FIGURE P8.33

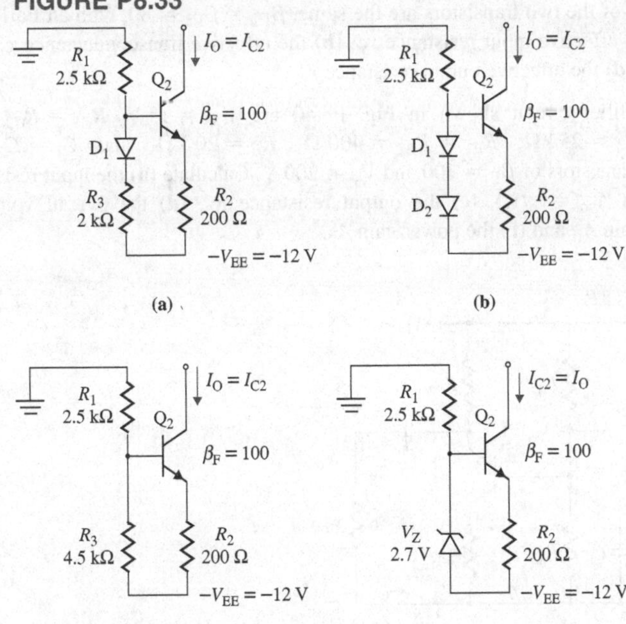

8.11–8.12 Multistage Amplifiers and the Darlington Pair Transistor

8.34 The cascoded amplifier circuit shown in Fig. 8.43(a) has $R_{B1} = R_{B2} = R_{B3} = 50\ \text{k}\Omega$ and $R_C = R_E = 1\ \text{k}\Omega$. Assume that the coupling capacitances are large tending to infinity and identical transistors of $\beta_F = 173$, $V_A = 200$ V, and $V_{BE1} = V_{BE2} = 0.7$ V.

a. Determine the biasing voltages and currents $I_{B1}, I_{C1}, V_{CE1}, I_{B2}, I_{C2}$, and V_{CE2}.

b. Use PSpice/SPICE to verify your results.

8.35 The cascoded amplifier circuit shown in Fig. 8.43(a) has $R_{B1} = R_{B2} = R_{B3} = 50\ \text{k}\Omega$ and $R_C = R_E = 1\ \text{k}\Omega$. Assume that the coupling capacitances are large tending to infinity and identical transistors of $\beta_F = 173$, $V_A = 200$ V, and $V_{BE1} = V_{BE2} = 0.7$ V.

a. Determine the small-signal voltage gain A_{vo}.

b. Use PSpice/SPICE to verify your results.

8.36 Design an active biasing circuit using BJTs for the cascoded amplifier circuit shown in Fig. 8.43(a) to replace R_C by producing a DC biasing current of $I_{C2} = 1$ mA. Assume that the coupling capacitances are large tending to infinity and identical transistors of $\beta_F = 173$, $V_A = 200$ V, and $V_{BE1} = V_{BE2} = 0.7$ V.

a. Determine the biasing voltages and currents $I_{B1}, I_{C1}, V_{CE1}, I_{B2}, I_{C2}$, and V_{CE2}.

b. Determine the small-signal transistor model parameters.

c. Use PSpice/SPICE to verify your results.

8.37 The collector resistance R_C of the cascoded amplifier circuit shown in Fig. 8.43(a) is replaced by an active current source of $I_{C2} = 1$ mA at an output resistance $r_o = 20\ \text{k}\Omega$. Assume that the coupling capacitances are large tending to infinity and identical transistors of $\beta_F = 173$, $V_A = 200$ V, and $V_{BE1} = V_{BE2} = 0.7$ V. If $R_E = 500\ \Omega$, **(a)** determine the small-signal voltage gain A_{vo} and **(b)** use PSpice/SPICE to verify your results.

8.38 The collector resistance R_C of the cascoded amplifier circuit shown in Fig. 8.46(a) is replaced by an active current source. Assume that the coupling capacitances are large tending to infinity and identical transistors of $\beta_F = 173$, $V_A = 200$ V, and $V_{BE1} = V_{BE2} = 0.7$ V. If $R_E = 0$, **(a)** determine the DC-biasing circuit needed to give a small-signal voltage gain $A_{vo} = 250$ and **(b)** use PSpice/SPICE to verify your results.

8.39 The Darlington pair shown in Fig. 8.44(a) is biased in such a way that the collector bias current I_{C2} of Q_2 is 400 μA. The current gains of the two transistors are the same, $\beta_{F1} = \beta_{F2} = 80$, with an Early voltage $V_A = 50$ V. Calculate **(a)** the effective input resistance r_π, **(b)** the effective transconductance g_m, **(c)** the effective current gain β, and **(d)** the effective output resistance r_o.

8.40 The parameters of the amplifier circuit shown in Fig. P8.40 are $V_{CC} = 15$ V, $R_{C1} = R_{C2} = 1\ \text{k}\Omega$, $R_1 = R_3 = 65\ \text{k}\Omega$, $R_2 = R_4 = 25\ \text{k}\Omega$, $R_{E1} = R_{E2} = 400\ \Omega$, $R_L = 20\ \text{k}\Omega$, and $C_1 = C_2 = C_3 = C_E \approx \infty$. Assume, identical transistors of $\beta_F = 100$ and $V_A = 200$ V. Calculate **(a)** the input resistance R_i, **(b)** the no-load voltage gain $A_{vo}(= v_o/v_s)$, **(c)** the output resistance R_o, **(d)** the overall voltage gain $A_v(= v_L/v_s)$, **(e)** the current gain A_i, and **(f)** the power gain A_p.

FIGURE P8.40

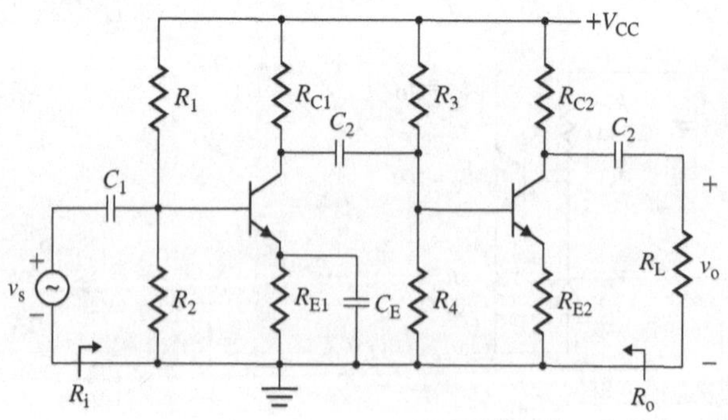

8.41 The parameters of the amplifier circuit shown in Fig. P8.41 are $V_{CC} = 15$ V, $R_{C1} = R_{C2} = 1$ kΩ, $R_1 = R_3 = 130$ kΩ, $R_2 = R_4 = 50$ kΩ, $R_{E1} = R_{E2} = 500$ Ω, $R_L = 20$ kΩ, and $C_1 = C_2 = C_3 = C_{E1} = C_{E2} \approx \infty$. Assume identical transistors of $\beta_F = 100$ and $V_A = 200$ V. Calculate **(a)** the input resistance R_i, **(b)** the no-load voltage gain $A_{vo}(=v_o/v_s)$, **(c)** the output resistance R_o, **(d)** the overall voltage gain $A_v(=v_L/v_s)$, **(e)** the current gain A_i, and **(f)** the power gain A_p.

FIGURE P8.41

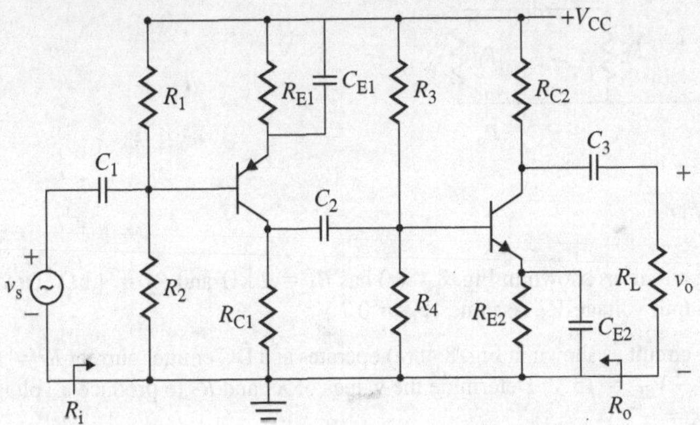

8.42 The parameters of the amplifier circuit shown in Fig. P8.42 are $V_{CC} = 15$ V, $R_{C1} = R_{C2} = 1$ kΩ, $R_1 = 65$ kΩ, $R_2 = 25$ kΩ, $R_E = 500$ Ω, $R_L = 20$ kΩ, and $C_1 = C_2 = C_E \approx \infty$. Assume identical transistors of $\beta_F = 100$ and $V_A = 200$ V. Calculate **(a)** the input resistance R_i, **(b)** the no-load voltage gain $A_{vo}(= v_o/v_s)$, **(c)** the output resistance R_o, **(d)** the overall voltage gain $A_v(= v_L/v_s)$, **(e)** the current gain A_i, and **(f)** the power gain A_p.

FIGURE P8.42

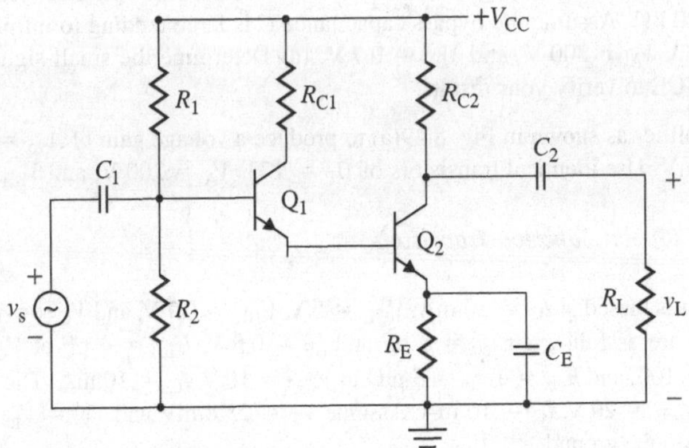

8.43 The parameters of the amplifier circuit shown in Fig. P8.43 are $V_{CC} = 15$ V, $R_1 = 65$ kΩ, $R_2 = 25$ kΩ, $R_E = 500$ Ω, $R_L = 20$ kΩ, and $C_1 = C_2 \approx \infty$. Assume identical transistors of $\beta_F = 100$ and $V_A = 200$ V. Calculate **(a)** the input resistance R_i, **(b)** the no-load voltage gain $A_{vo}(=v_o/v_s)$, **(c)** the output resistance R_o, **(d)** the overall voltage gain $A_v(=v_L/v_s)$, **(e)** the current gain A_i, and **(f)** the power gain A_p.

FIGURE P8.43

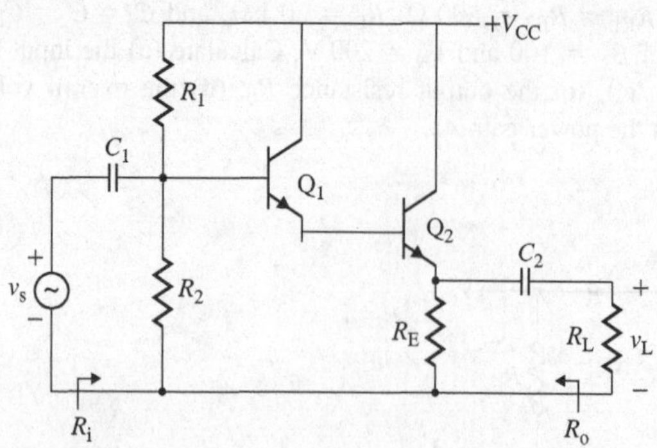

8.13 DC Level Shifting and Amplifier

8.44 The potential level-shifting circuit as shown in Fig. 8.48(a) has $R_1 = 2\ k\Omega$ and $R_2 = 4\ k\Omega$. Determine the voltage shift V_{sh} and the output voltage V_o. Assume $V_{BE} = 0.7$ V.

8.45 The potential level-shifting circuit as shown in Fig. 8.48(a) operates at a DC emitter current $I_E = 1$ mA, and the DC voltages are $V_{CC} = -V_{EE} = 15$ V. Determine the values of R_1 and R_2 to produce a voltage shift of 3 V at an output voltage $V_O = -7$ V.

8.46 Determine the current source I_o needed and R_1 in Fig. 8.48(b) to produce a voltage shift of $V_{sh} = 4$ V at an output voltage of $V_O = -8$ V. Assume $V_{BE} = 0.7$ V.

8.47 The BJT level-shifting amplifier as shown in Fig. 8.49(a) has $R = 20\ k\Omega$, $R_A = 15\ k\Omega$, $R_B = 5\ k\Omega$, $R_x = R_y = 20\ k\Omega$, and $R_L = 20\ k\Omega$. Assume the bypass capacitance C_A is large tending to infinity and all identical transistors of $\beta_F = 173$, $V_A = 200$ V, and $V_{BE} = 0.7$ V.

a. Determine the biasing collector currents $I_{C1}, I_{C2}, I_{C3}, I_{C4}, I_{C5}, I_{C6}$, and I_{C7}.

b. Determine the DC voltage V_{B7} at the collector of Q_6.

c. Use PSpice/SPICE to verify your results.

8.48 The BJT level-shifting amplifier as shown in Fig. 8.49(a) has $R = 20\ k\Omega$, $R_A = 15\ k\Omega$, $R_B = 5\ k\Omega$, $R_x = R_y = 20\ k\Omega$, and $R_L = 20\ k\Omega$. Assume the bypass capacitance C is large tending to infinity and all identical transistors of $\beta_F = 173$, $V_A = 200$ V, and $V_{BE} = 0.7$ V. **(a)** Determine the small-signal voltage gain A_{vo} and **(b)** use PSpice/SPICE to verify your results.

8.49 Design a BJT level-shifting amplifier as shown in Fig. 8.49(a) to produce a voltage gain of $A_{vo} = 100$ V/V at a DC input signal of $v_s = 1$ mV. Use identical transistors of $\beta_F = 173$, $V_A = 200$ V, and $V_{BE} = 0.7$ V.

8.14 Frequency Model and Response of Bipolar Junction Transistors

8.50 An *npn* transistor of type 2N3904 is biased at $I_C = 20$ mA, $V_{CE} = 5$ V, $V_{BE} = 0.7$ V, and $V_{CS} = V_C = 7$ V.
P The parameters of the transistor are as follows: $C_{je0} = 8$ pF at $V_{BE} = 0.5$ V, $C_{\mu 0} = 4$ pF at $V_{CB} = 5$ V, $C_{cs0} = 4$ pF at $V_{CS} = 8$ V, $\beta_f = 100$, and $h_{oe} = 1/r_o = 5\ \mu\mho$ at $V_{CE} = 10$ V, $I_C = 10$ mA. The transition frequency is $f_T = 300$ MHz at $V_{CE} = 20$ V, $I_C = 10$ mA. Assume $V_T = 25.8$ mV and $V_{je} = V_{jc} = V_{js} = 0.8$ V. The substrate is connected to the ground.

a. Calculate the small-signal capacitances of the high-frequency model in Fig. 8.52(b).

b. Find transition time τ_F.

c. Calculate f_T of the transistor at the operating point.

8.51 Repeat Prob. 8.50 for $I_C = 2$ mA, $V_{CE} = 4$ V, $V_{BE} = 0.7$ V, and $V_{CS} = V_C = 3$ V.
P

8.52 A *pnp* transistor of type 2N3905 is biased at $I_C = 50$ mA, $V_{CE} = -6$ V, $V_{BE} = -0.7$ V, and $V_{CS} = V_C = -10$ V.
P The parameters of the transistor are as follows: $C_{je0} = 10$ pF at $V_{BE} = -0.5$ V, $C_{\mu 0} = 4.5$ pF at $V_{CB} = -5$ V, $C_{cs0} = 4$ pF at $V_{CS} = 8$V, $\beta_f = 50$, and $h_{oe} = 1/r_o = 5\ \mu℧$ at $V_{CE} = 10$ V, $I_C = -1$ mA. The transition frequency is $f_T = 200$ MHz at $V_{CE} = -20$ V, $I_C = -10$ mA. Assume $V_T = 25.8$ mV and $V_{je} = V_{jc} = V_{js} = 0.8$ V. The substrate is connected to the ground.

a. Calculate the small-signal capacitances of the high-frequency model in Fig. 8.49(a).

b. Find transition time τ_F.

c. Calculate f_T of the transistor at the operating point.

8.53 Repeat Prob. 8.52 for $I_C = 5$ mA, $V_{CE} = 5$ V, $V_{BE} = 0.7$ V, and $V_{CS} = V_C = 6$ V.
P

8.15 Frequency Response of BJT Amplifiers

For Probs. 8.54–8.59 involving BJT amplifiers, use transistors whose parameters are $\beta_f = 100$, $C_{je} = 8$ pF at $V_{BE} = 0.5$ V, $C_\mu = 4$ pF at $V_{CB} = 5$ V, $C_{cs} = 4$ pF at $V_{CS} = 8$ V, $\beta_f = 100$, $V_{je} = V_{jc} = V_{js} = 0.8$ V, and $h_{oe} = 1/r_o = 5\ \mu℧$ at $V_{CE} = 10$ V. The transition frequency is $f_T = 300$ MHz at $V_{CE} = 20$ V, $I_C = 10$ mA. The substrate is connected to the ground. Assume $I_C = 5$ mA (unless specified), $V_{CC} = 15$ V, $V_{BE} = 0.7$ V, $R_s = 1$ kΩ, and $R_L = 10$ kΩ. Use PSpice/SPICE to check your design by plotting the frequency response and give an approximate cost estimate.

8.54 Design a CE amplifier as shown in Fig. P8.54 to give a passband gain of $40 \le |A_{PB}| \le 50$, a low 3-dB
D P frequency of $f_L \le 1$ kHz, and a high 3-dB frequency of $f_H = 50$ kHz.

FIGURE P8.54

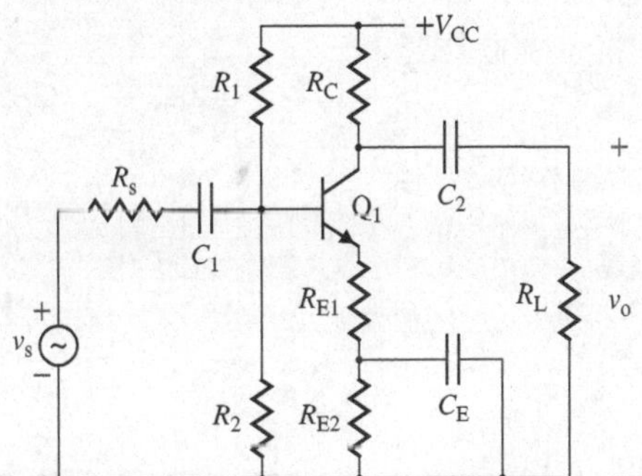

8.55 Design a CB amplifier as shown in Fig. P8.55 to give a passband gain of $20 \le |A_{PB}| \le 30$, a low 3-dB fre-
D P quency of $f_L \le 1$ kHz, and a high 3-dB frequency of $f_H = 100$ kHz. Assume $R_s = 15$ kΩ and $R_L = 10$ kΩ.

FIGURE P8.55

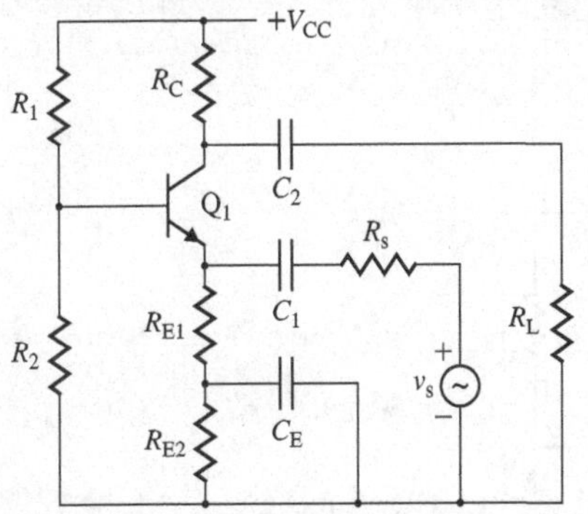

8.56 Design a CE amplifier as shown in Fig. P8.56 to give a passband gain of $50 \leq |A_{PB}| \leq 60$, a low 3-dB frequency of $f_L \leq 1$ kHz, and a high 3-dB frequency of $f_H = 50$ kHz.

D P

FIGURE P8.56

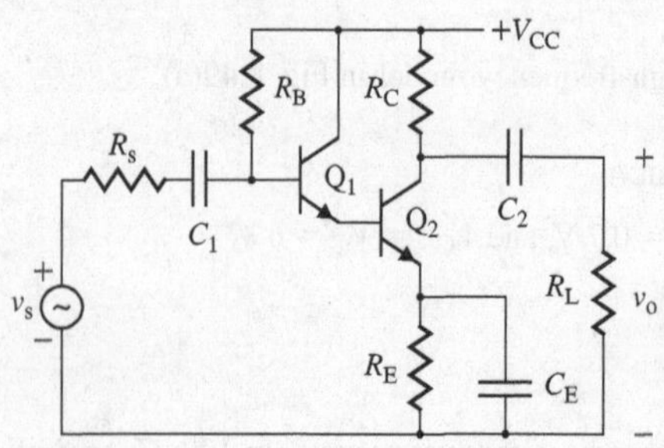

8.57 Design a CC-CE amplifier as shown in Fig. P8.57 to give a passband gain of $25 \leq |A_{PB}| \leq 35$, $Z_{in(mid)} \geq$ 50 kΩ, a low 3-dB frequency of $f_L \leq 5$ kHz, and a high 3-dB frequency of $f_H = 50$ kHz.

D P

FIGURE P8.57

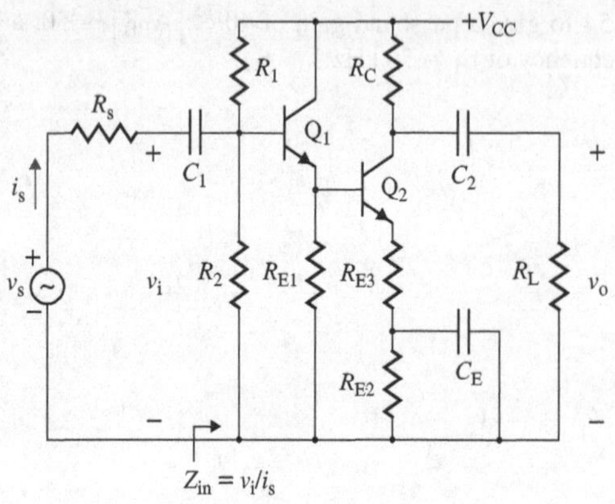

8.58 Design a CE-CC amplifier as shown in Fig. P8.58 to give a passband gain of $20 \leq |A_{PB}| \leq 30$, $Z_{i1(mid)} \leq$ 100 Ω, a low 3-dB frequency of $f_L \leq 1$ kHz, and a high 3-dB frequency of $f_H = 100$ kHz.

D P

FIGURE P8.58

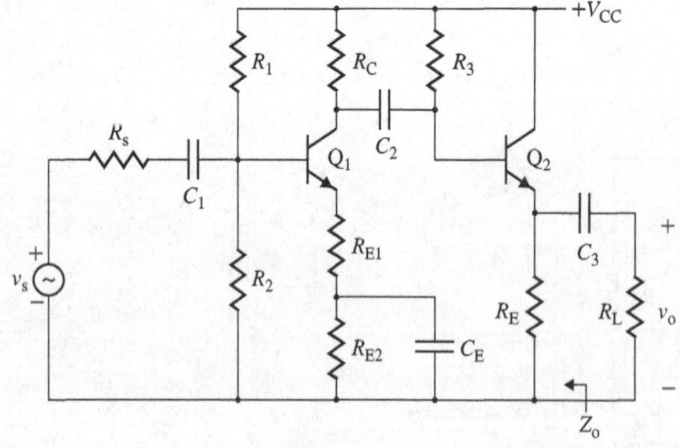

8.59 For the CC-CE amplifier in Fig. P8.59, find the passband gain $|A_{PB}|$, $Z_{in(mid)}$, the low 3-dB frequency f_L, and the high 3-dB frequency f_H. Assume $C_o = 1\ \mu F$.

FIGURE P8.59

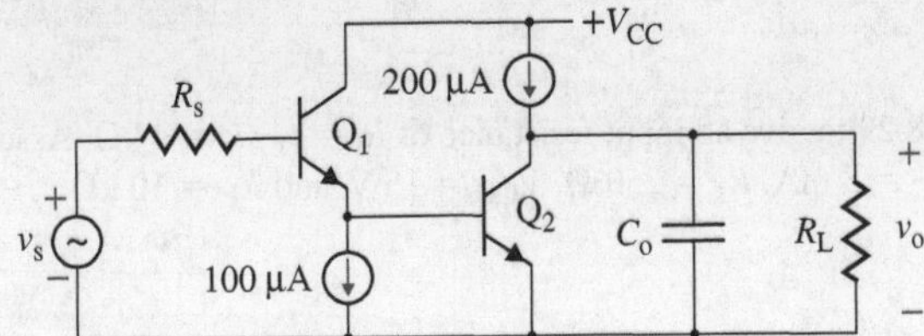

8.60 A two-stage amplifier is shown in Fig. P8.60. The parameters are $R_s = 5\ k\Omega$, $R_{11} = 70\ k\Omega$, $R_{21} = 45\ k\Omega$, $R_C = 5\ k\Omega$, $R_E = 1\ k\Omega$, $R_{sr} = 2\ k\Omega$, $R_{12} = 1\ M\Omega$, $R_{22} = 2\ M\Omega$, $R_D = 10\ k\Omega$, $R_L = 10\ k\Omega$, $r_{\pi 1} = 1.4\ k\Omega$, $\beta_{f1} = 50$, $g_{m1} = 35.7\ m\mho$, $g_{m2} = 107.1\ m\mho$, $C_1 = 2\ \mu F$, $C_2 = 5\ \mu F$, $C_3 = 1\ \mu F$, $C_G = 1\ \mu F$, $C_E = 10\ \mu F$, $C_\pi = 15\ pF$, $C_\mu = 1\ pF$, $C_{gd} = 2\ pF$, and $C_{gs} = 5\ pF$. Calculate the low 3-dB frequency f_L and the high cut-off frequency f_H.

FIGURE P8.60

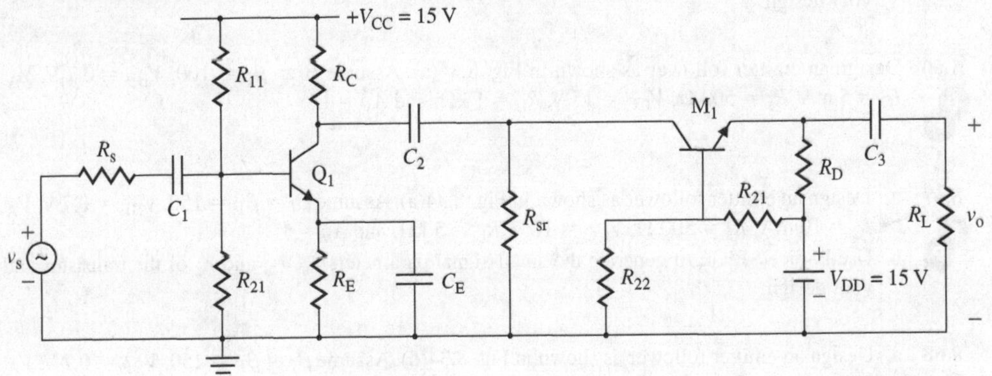

8.61 A two-stage amplifier is shown in Fig. P8.61. The parameters are $R_s = 500\ \Omega$, $R_B = 47\ k\Omega$, $R_C = 10\ k\Omega$, $R_L = 10\ k\Omega$, $r_{\pi 1} = r_{\pi 2} = 1.4\ k\Omega$, $\beta_{f1} = \beta_{f2} = 150$, $C_1 = 10\ \mu F$, $C_2 = 10\ \mu F$, $C_{\pi 1} = C_{\pi 2} = 15\ pF$, and $C_{\mu 1} = C_{\mu 2} = 15\ pF$. Calculate the low 3-dB frequency f_L and the high cutoff frequency f_H.

FIGURE P8.61

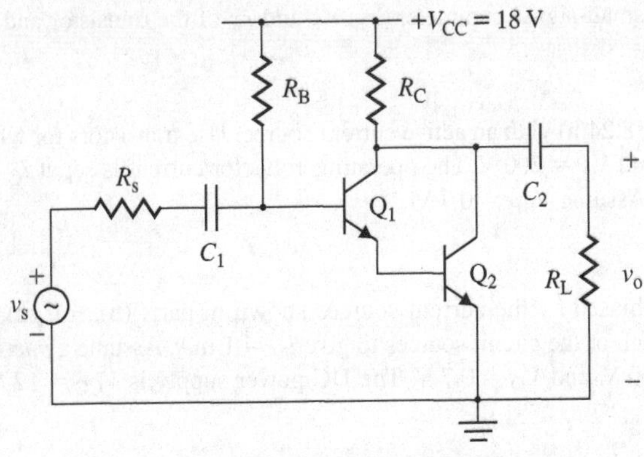

8.17 Design of Amplifiers

8.62 D P Design a CE amplifier as shown in Fig. 8.29 to give a voltage gain of $A_v = v_L/v_s = -20$. Assume $\beta_f = \beta_F = 100$, $V_{BE} = 0.7$ V, $V_A = 200$ V, $I_C = 10$ mA, $V_{CC} = 15$ V, $R_s = 500\ \Omega$, and $R_L = 20$ kΩ.

8.63 D P Design a CE amplifier as shown in Fig. 8.29 to give an input resistance of $R_i = v_b/i_s = 4$ kΩ. Assume $\beta_f = \beta_F = 100$, $V_{BE} = 0.7$ V, $V_A = 200$ V, $I_C = 5$ mA, $R_s = 250\ \Omega$, $V_{CC} = 15$ V, and $R_L = 10$ kΩ.

8.64 D P
a. Design a CE amplifier as shown in Fig. 8.29 to give a voltage gain of $A_v = v_L/v_s = -25$. Assume $\beta_f = \beta_F = 150$, $V_{BE} = 0.7$ V, $V_A = 200$ V, $I_C = 15$ mA, $V_{CC} = 18$ V, $R_s = 250\ \Omega$, and $R_L = 5$ kΩ.
b. Use PSpice/SPICE to generate the small-signal parameters r_π, r_o, and r_μ of the transistor and to verify your design.

8.65 D P
a. Design a CE amplifier as shown in Fig. 8.29 to give an input resistance of $R_i = v_b/i_s \geq 3.5$ kΩ. Assume $\beta_f = \beta_F = 150$, $V_{BE} = 0.7$ V, $V_A = 200$ V, $I_C = 15$ mA, $V_{CC} = 18$ V, and $R_L = 5$ kΩ.
b. Use PSpice/SPICE to generate the small-signal parameters r_π, r_o, and r_μ of the transistor and to verify your design.

8.66 D P Design an emitter follower as shown in Fig. 8.34(a). Assume $\beta_f = \beta_F = 100$, $V_{BE} = 0.7$ V, $V_A = 200$ V, $I_C = 5$ mA, $R_s = 500\ \Omega$, $V_{CC} = 15$ V, $R_L = 1$ kΩ, and $A_v \approx 1$.

8.67 D P
a. Design an emitter follower as shown in Fig. 8.34(a). Assume $\beta_f = \beta_F = 150$, $V_{BE} = 0.7$ V, $V_A = 150$ V, $I_C = 10$ mA, $R_s = 500\ \Omega$, $V_{CC} = 18$ V, $R_L = 5$ kΩ, and $A_v \approx 1$.
b. Use PSpice/SPICE to generate the small-signal parameters r_π, r_o, and r_μ of the transistor and to verify your design.

8.68 D P
a. Design an emitter follower as shown in Fig. 8.34(d). Assume $\beta_f = \beta_F = 150$, $V_{BE} = 0.7$ V, $V_A = 150$ V, $I_C = 10$ mA, $R_s = 500\ \Omega$, $V_{CC} = 18$ V, $R_L = 5$ kΩ, and $A_v \approx 1$.
b. Use PSpice/SPICE to generate the small-signal parameters r_π, r_o, and r_μ of the transistor and to verify your design.

8.69 D P
a. Design an emitter follower as shown in Fig. 8.34(d) to give an input resistance of $R_i = v_b/i_s \geq 15$ kΩ. Assume $\beta_f = \beta_F = 150$, $V_{BE} = 0.7$ V, $V_A = 150$ V, $I_C = 15$ mA, $R_s = 500\ \Omega$, $V_{CC} = 18$ V, and $R_L = 5$ kΩ.
b. Use PSpice/SPICE to generate the small-signal parameters r_π, r_o, and r_μ of the transistor and to verify your design.

8.70 D P Design a CE amplifier as shown in Fig. 8.24(a) with an active current source. Use transistors for which minimum $\beta_F = 200$, nominal $\beta_F = 250$, and $V_A = 200$ V. The operating collector current is set at $I_C = 1$ mA. The DC power supply is $V_{CC} = 10$ V. Assume $V_{BE} = 0.7$ V.

8.71 D P The CE amplifier of Fig. P8.71(a) is biased by the current sources shown in parts (b), (c), (d), and (e). Determine the circuit parameters for each of the circuit sources to give $I_O = 1$ mA. Assume a *pnp* transistor of $\beta_F = 100$, $V_{BE} = -0.7$ V, $V_A = 200$ V, and $V_D = 0.7$ V. The DC power supply is $V_{CC} = 12$ V. (Note: There is no unique solution.)

FIGURE P8.71

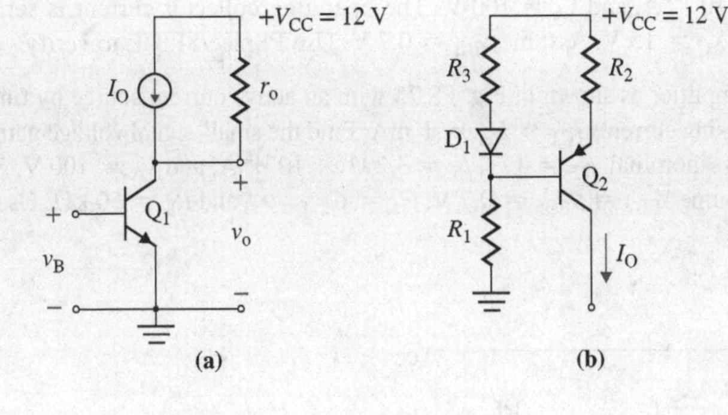

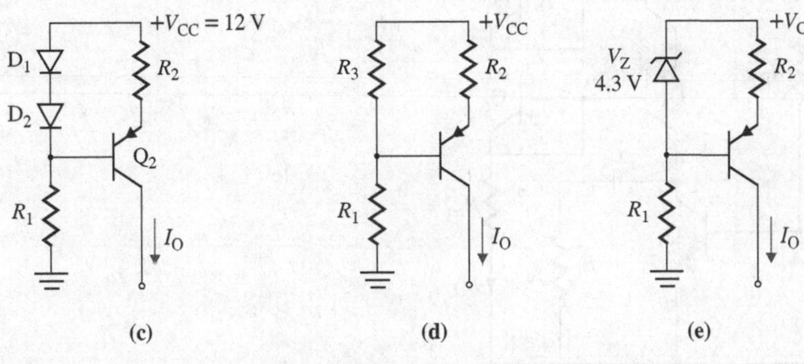

8.72 Design a common emitter–common base (CE-CB) amplifier as shown in Fig. P8.72 to give a voltage gain of $A_v = v_L/v_s = -12$. Assume $V_{CC} = 15$ V and $R_s = 250\ \Omega$. Use bipolar transistors of type 2N2222 or 2N3904.

FIGURE P8.72

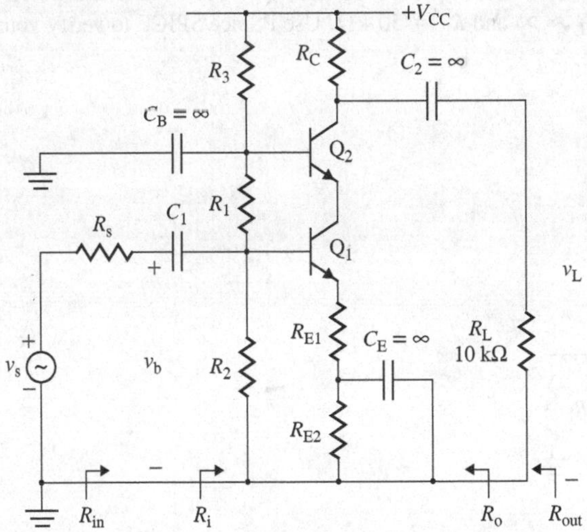

8.73 Design a CE amplifier of Fig. 8.24(a) with an active current source. Use BJT transistors whose nominal $\beta_F = 173$, $I_s = 3.295 \times 10^{-14}$ A, and $V_A = 100$ V. The operating collector current is set at $I_C = 10$ mA. The DC power supply is $V_{CC} = 15$ V. Assume $V_{BE} = 0.7$ V. Use PSpice/SPICE to verify your results.

8.74 Design an emitter follower of Fig. 8.32(a) with an active current source. Use BJT transistors whose nominal $\beta_F = 173$, $I_s = 3.295 \times 10^{-14}$ A, and $V_A = 100$ V. The operating collector current is set at $I_C = 1$ mA. The DC power supply is $V_{CC} = 15$ V. Assume $V_{BE} = 0.7$ V. Use PSpice/SPICE to verify your results.

8.75 Design an active-biased amplifier as shown in Fig. P8.75 with an active current source by finding the values of R_B and R_{ref} to obtain biasing currents $I_{C1} \approx I_{ref} = 1$ mA. Find the small-signal voltage gain. Assume identical BJT transistors whose nominal $\beta_F = 173$, $I_S = 3.295 \times 10^{-14}$ A, and $V_A = 100$ V. The DC power supply is $V_{CC} = 15$ V. Assume $V_{EB3} = V_{EB5} = 0.7$ V, $C_1 = C_2 \approx \infty$, and $R_L = 50$ kΩ. Use PSpice/SPICE to verify your results.

FIGURE P8.75

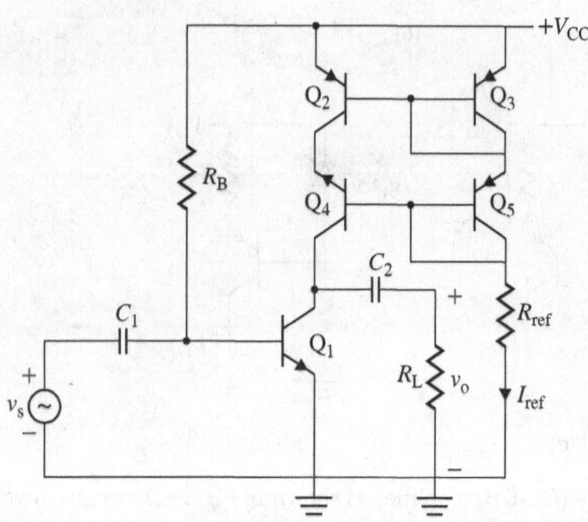

8.76 Design an active-biased amplifier as shown in Fig. P8.76 with an active current source by finding the value of R_B to obtain biasing currents $I_{C1} \approx I_{ref} = 1$ mA. Find the small-signal voltage gain. Assume identical BJT transistors whose nominal $\beta_F = 173$, $I_s = 3.295 \times 10^{-14}$ A, and $V_A = 100$ V. The DC power supply is $V_{CC} = 15$ V. Assume $V_{BB} = 15$ V $C_L \approx \infty$ and $R_L = 50$ kΩ. Use PSpice/SPICE to verify your results.

FIGURE P8.76

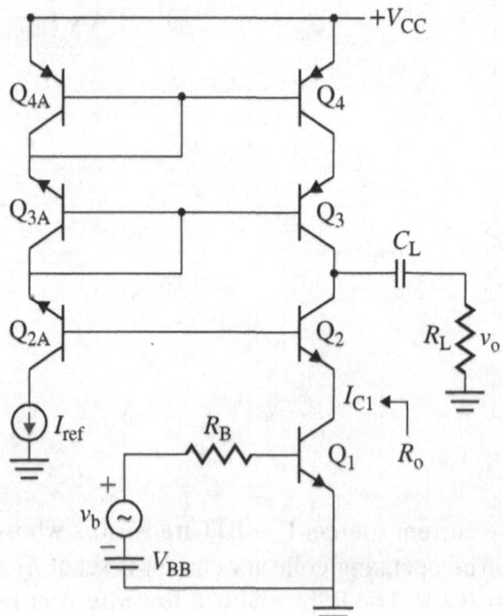

8.77 Design a multistage BJT amplifier to meet the following specifications: voltage gain $|A_v| = v_L/v_s =$ $600 \pm 5\%$ (with load); input resistance $R_i = v_s/i_s \geq 25$ kΩ; output resistance $R_o \leq 300$ Ω; load resistance $R_L = 25$ kΩ; source resistance $R_s = 1$ kΩ; DC supply $V_{CC} = 15$ V; input signal $v_s = 1$ mV to 5 mV (peak sinusoidal), 1 kHz. Use identical *npn* Q2N2222 transistors of $I_{C(max)} = 10$ mA, $\beta_F = 256$, $I_S = 14.34 \times 10^{-15}$ A, and $V_A = 74$ V. (*Hints:* Use the first CE stage to meet the input resistance requirement, the third CC stage to meet the output resistance, and the middle CE stage to attain the remaining gain requirement. Set the drain-biasing current at $I_C \leq I_{C(max)}/3$.)

8.78 Design a multistage BJT amplifier to meet the following specifications: voltage gain $|A_v| = v_L/v_s =$ $600 \pm 5\%$ (with load); input resistance $R_i = v_s/i_s \geq 25$ kΩ; output resistance $R_o \leq 300$ Ω; load resistance $R_L = 25$ kΩ; source resistance $R_s = 1$ kΩ; DC supply $V_{CC} = 15$ V; input signal $v_s = 1$ mV to 5 mV (peak sinusoidal), 1 kHz. Use identical *pnp* Q2N2907A transistors of $I_{C(max)} = 10$ mA, $\beta_F = 231$, $I_S = 651 \times 10^{-18}$ A, and $V_A = 115$ V. (*Hints:* Use the first CE stage to meet the input resistance requirement, the third CC stage to meet the output resistance, and the middle CE stage to attain the remaining gain requirement. Set the drain-biasing current at $I_C \leq I_{C(max)}/3$.)

CHAPTER 9

DIFFERENTIAL AMPLIFIERS

Learning Outcomes

After completing this chapter, students should be able to do the following:

- Design, analyze, and evaluate different types of MOSFET constant-current sources for biasing MOSFET amplifiers.
- Design, analyze, and evaluate different types of BJT constant-current sources for biasing BJT amplifiers.
- Describe and analyze the characteristics of differential amplifiers and their DC and small-signal characteristics.
- Identify the parameters influencing the differential and the common-mode gains of differential amplifiers.
- Analyze and evaluate cascode-connected transistors in order to obtain higher differential voltage gains.

Symbols and Their Meanings

Symbol	Meaning
A_d, A_c	Small-signal differential and common-mode voltage gains
CMRR	Common-mode rejection ratio of an amplifier
v_{id}, v_{ic}	Small-signal differential and common-mode signal
v_{o1}, v_{o2}	Small-signal output voltages due to input voltages at inverting and noninverting terminals

Symbol	Meaning
r_{o1}, r_{o2}	Small-signal output resistances of transistors
V_M, V_A	MOSFET channel modulation and BJT Early voltages
V_t, V_p	Threshold voltage of enhancement MOSFET and pinch-off voltage of depletion MOSFET

9.1 Introduction

Differential amplifiers are commonly used as an input stage in various types of analog ICs, such as operational amplifiers, voltage comparators, voltage regulators, video amplifiers, power amplifiers, and balanced modulators and demodulators [1–3]. A differential amplifier is a very important transistor stage and determines many of the performance characteristics of an IC. In ICs, including differential amplifiers, it is unnecessary to bias transistors by setting the values of biasing resistors. Because of variations in resistor values, power supply, and temperature, the quiescent point of transistors changes. Transistors can be used to generate the characteristics of DC constant-current sources. Transistors can also be used to produce an output voltage source that is independent of its load or, equivalently, of the output current. This chapter covers the operation, analysis, and characteristics of differential amplifiers using BJTs and MOSFETs. It also covers active current sources and voltage sources.

9.2 Internal Structure of Differential Amplifiers

A differential amplifier acts as an input stage; its output voltage is proportional to the difference between its two input voltages v_1 and v_2. It has a high voltage gain and is directly DC coupled to the input voltages and the load. As we will see later in this chapter, the voltage gain of a differential amplifier depends directly on the output resistance of the current source acting as an active load.

In amplifiers with discrete components, passive components such as resistors and capacitors are less expensive than active devices such as transistors (i.e., MOSFETs and BJTs); thus, in multistage amplifiers, interstage coupling is accomplished with capacitors. However, in monolithic circuits, the die area is the principal determining cost factor. Capacitors of the values and sizes used in amplifiers made with discrete components cannot be included in ICs and must be external to the chip. But using external capacitors increases the pin count of the package and the cost of the IC. A DC-coupled circuit is used to eliminate capacitors. The cheapest component in an IC is the one that can be fabricated within the least area, usually the transistor. The optimal IC has as little resistance as possible and more transistors.

9.2.1 Characteristics of Differential Amplifiers

The differential stage can be represented by an equivalent amplifier, as shown in Fig. 9.1(a). If the two input voltages are equal, a differential amplifier gives an output voltage of almost zero. Its voltage gain is very large, so the input voltage is low, typically less than 50 mV. Thus, we can consider the input voltages as small signals with zero DC components. That is, $v_{G1} = v_{g1}$ and $v_{G2} = v_{g2}$.

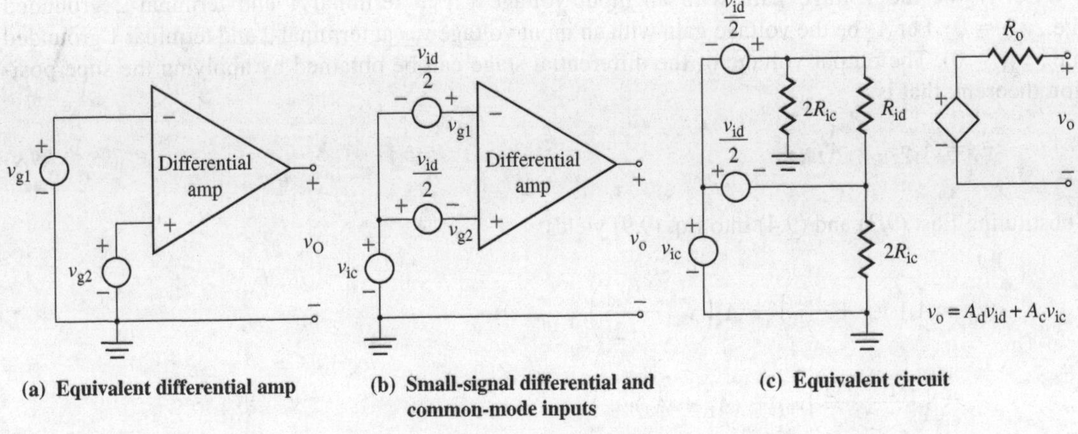

FIGURE 9.1 Small-signal equivalent circuit with differential and common-mode inputs

Let us define a differential voltage v_{id} as

$$v_{id} = v_{g1} - v_{g2} \tag{9.1}$$

and a common-mode voltage v_{ic} as

$$v_{ic} = \frac{v_{g1} + v_{g2}}{2} \tag{9.2}$$

From Eqs. (9.1) and (9.2), the two input voltages can be expressed as

$$v_{g1} = v_{ic} + \frac{v_{id}}{2} \tag{9.3}$$

and $$v_{g2} = v_{ic} - \frac{v_{id}}{2} \tag{9.4}$$

By replacing the input signals with the equivalent differential and common-mode signals, we can represent the differential stage by an equivalent amplifier, as shown in Fig. 9.1(b). Let v_{o1} be the output voltage due to v_{g1} only, and let v_{o2} be the output voltage due to v_{g2} only. Then we can define a differential output voltage v_{od} as

$$v_{od} = v_{o1} - v_{o2} \tag{9.5}$$

and a common-mode output voltage as

$$v_{oc} = \frac{v_{o1} + v_{o2}}{2} \tag{9.6}$$

From Eqs. (9.5) and (9.6), the two output voltages can be expressed as

$$v_{o1} = v_{oc} + \frac{v_{od}}{2} \tag{9.7}$$

and $$v_{o2} = v_{oc} - \frac{v_{od}}{2} \tag{9.8}$$

Let A_1 be the voltage gain with an input voltage v_{g1} at terminal 1 and terminal 2 grounded (i.e., $v_{g2} = 0$). Let A_2 be the voltage gain with an input voltage v_{g2} at terminal 2 and terminal 1 grounded (i.e., $v_{g1} = 0$). The output voltage of the differential stage can be obtained by applying the superposition theorem; that is,

$$v_o = A_1 v_{g1} + A_2 v_{g2} \tag{9.9}$$

Substituting Eqs. (9.3) and (9.4) into Eq. (9.9) yields

$$\begin{aligned} v_o &= A_1\left(v_{ic} + \frac{v_{id}}{2}\right) + A_2\left(v_{ic} - \frac{v_{id}}{2}\right) \\ &= \left(\frac{A_1 - A_2}{2}\right)v_{id} + (A_1 + A_2)v_{ic} \\ &= A_d v_{id} + A_c v_{ic} \qquad (9.10) \\ &= A_d\left(v_{id} + \frac{A_c}{A_d}v_{ic}\right) \qquad (9.11) \end{aligned}$$

where $A_d = (A_1 - A_2)/2$ = differential voltage gain and $A_c = A_1 + A_2$ = common-mode voltage gain.

The output voltage v_o in Eq. (9.11) is due to a common-mode input voltage v_{ic} and a differential input voltage v_{id}. If A_d is much greater than A_c, the output voltage will be almost independent of the common-mode signal v_{ic}. A differential amplifier is expected to amplify the differential voltage as much as possible while rejecting (not amplifying) common-mode signals such as noise or other unwanted signals, which will be present in both terminals.

The ability of an amplifier to reject common-mode signals is defined by a performance criterion called the *common-mode rejection ratio* (CMRR), which is defined by

$$\text{CMRR} = \left|\frac{A_d}{A_c}\right| \tag{9.12}$$

$$= 20\log\left|\frac{A_d}{A_c}\right| \quad \text{(in dB)} \tag{9.13}$$

Substituting Eq. (9.12) into Eq. (9.11) gives the output voltage

$$v_o = A_d\left(v_{id} + \frac{1}{\text{CMRR}}v_{ic}\right) \tag{9.14}$$

which shows that, to reduce the effect of v_{ic} on the output voltage v_o—that is, to get v_{oc} to approach zero—the value of CMRR must be very large, tending to infinity for an ideal amplifier. Thus, a differential amplifier should behave differently for common-mode and differential signals. The small-signal equivalent circuit is shown in Fig. 9.1(c). R_{id} and R_{ic} are the input resistances due to the differential and common-mode signals, respectively. The parameters of a differential amplifier are A_d (ideally ∞), A_c (0), CMRR (∞), R_{id} (∞), and R_{ic} (∞). In the following sections, we will determine the circuit elements affecting these parameters.

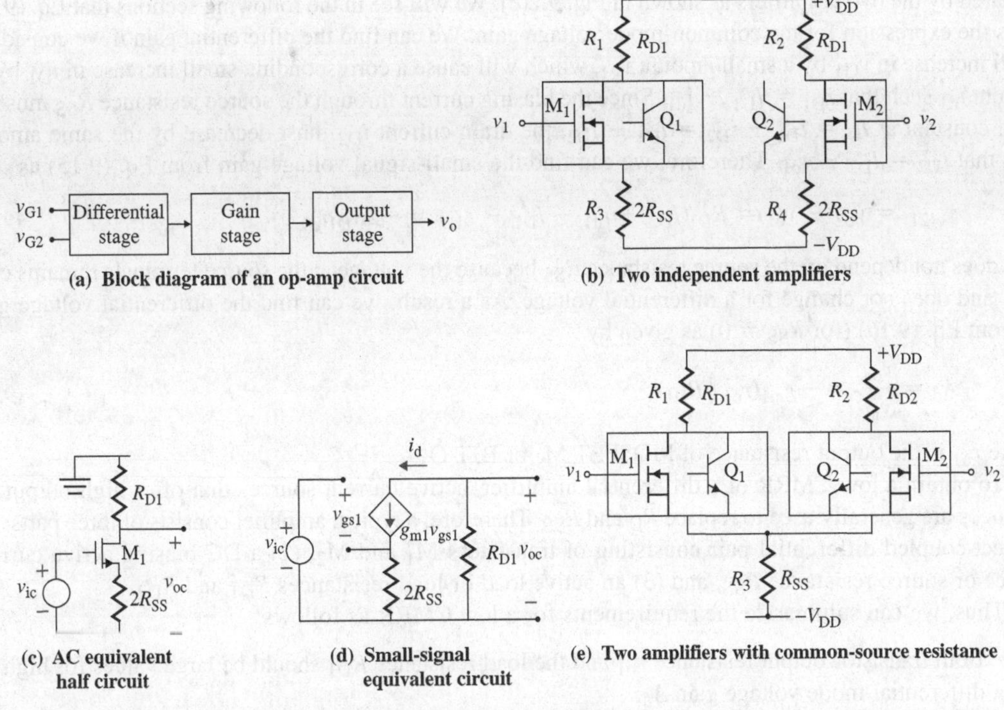

FIGURE 9.2 Basic structure of a differential amplifier

9.2.2 Internal Structure of Differential Amplifiers

A differential amplifier [4] acts as an input stage of an op-amp circuit as shown in Fig. 9.2(a). It serves as a direct DC-coupled differential stage. It may be viewed as consisting of two common-source (CS) amplifiers as shown in Fig. 9.2(b). If the two amplifiers are identical—that is, $R_{D1} = R_{D2}$—their DC biasing drain currents will be the same, and $i_{D1} = i_{D2}$. The DC output voltages are given by

$$v_{o1} = V_{DD} - R_{D1}i_{D1} \quad \text{and} \quad v_{o2} = V_{DD} - R_{D2}i_{D2}$$

The output voltage between the two output terminals of M_2 and M_1 is

$$v_{o(21)} = v_{o2} - v_{o1} = V_{DD} - R_{D2}i_{D2} - V_{DD} + R_{D1}i_{D1} = R_{D1}(i_{D2} - i_{D1}) \tag{9.15}$$

Therefore, we get $v_{o(21)} = v_{o(12)} = 0$ for identical amplifiers of the condition $i_{D1} = i_{D2}$. The small-signal AC equivalent half-circuit is shown in Fig. 9.2(c), which, after replacing the transistor by it transconductance model, becomes as shown in Fig. 9.2(d). Assuming that the output of the transistor M_1 is $r_{o1} >> R_{D1}$ and using Eq. (7.70), we can find the small-signal common-mode voltage gain as given by

$$A_c = \frac{v_{oc}}{v_{ic}} = -\frac{g_{m1}R_{D1}}{1 + 2g_{m1}R_{SS}} \tag{9.16}$$

The derivation of Eq. (9.16) assumes that no interaction exists between the two amplifiers, but there is an interaction between them. The source (or emitter) resistance, whose value is $R_3 \| R_4 (= 2R_{SS} \| 2R_{SS} = R_{SS})$, is shared by the two amplifiers as shown in Fig. 9.2(e). We will see in the following sections that Eq. (9.16) gives the expression for the common-mode voltage gain. We can find the differential gain if we consider a small increase in v_{G1} by a small amount v_{g1}, which will cause a corresponding small increase in i_{D1} by an amount i_{d1} such that $i_{D1} = I_{D1} + i_{d1}$. Since the biasing current through the source resistance R_{SS} must remain constant at $I_O = i_{D1} + i_{D2} = I_{D1} + I_{D2}$, the drain current i_{D2} must decrease by the same amount such that $i_{D2} = I_{D2} - i_{d1}$. Therefore, we can find the small-signal voltage gain from Eq. (9.15) as

$$v_{o(21)} = v_{o2} - v_{o1} = R_{D1}(I_{D2} - i_{d1} - I_{D1} - i_{d1}) = -2R_{D1}i_{d1} \tag{9.17}$$

This does not depend on the source resistance R_{SS} because the voltage at the source terminals remains constant and does not change for a differential voltage. As a result, we can find the differential voltage gain A_d from Eq. (9.16) (for $R_{SS} = 0$) as given by

$$A_d = \frac{v_{od}}{v_{id}} = -g_{m1}(r_{o1} \| R_{D1}) \tag{9.18}$$

where r_{o1} is the output resistance of MOSFET M_1 or BJT Q_1.

To obtain a low CMRR of a differential amplifier, active current sources that offer high output resistances are generally used to replace R_D and R_{SS}. Therefore, a typical amplifier consists of three parts: (1) a direct-coupled differential pair consisting of transistors M_1 and M_2, (2) a DC biasing active current source or source resistance R_{SS}, and (3) an active load or load resistances R_{D1} and R_{D2}.

Thus, we can summarize the requirements for a low CMRR as follows:

- Both transistor output resistance r_{o1} and the load resistance R_{D1} should be large values for high differential-mode voltage gain A_d.
- The source resistance R should be large values for a low common-mode voltage gain A_c.

KEY POINTS OF SECTION 9.2

- A differential amplifier consists of an active biasing circuit, an active load, and a differential transistor pair.
- The performance of a differential amplifier is measured by a differential gain A_d that occurs in response to a differential voltage between two input terminals, a common-mode gain A_c that occurs in response to a voltage common to both input terminals, and a common-mode rejection ratio CMRR.
- The CMRR is the ratio of the differential gain to the common-mode gain, and it is a measure of the ability of an amplifier to amplify the differential signal and reject common-mode signals.

9.3 MOSFET Current Sources

In Secs. 7.8.1 and 8.7.1, we saw the effects of active current sources in increasing the voltage gain of an amplifier. Transistor current sources are widely used in analog ICs both as biasing elements and as loads for amplifying stages. Current sources are less sensitive to variations in DC power supply and temperature. Especially, for a small value of bias current, the current sources are more economical than resistors in terms of die area required for resistors. A current source can be designed by using either MOSFETs or BJTs [5].

MOSFET current sources are analogous to BJT current sources. We can convert a BJT current source to an equivalent MOSFET current source by assuming that the β_F of the BJTs is infinite. Since MOSFETs

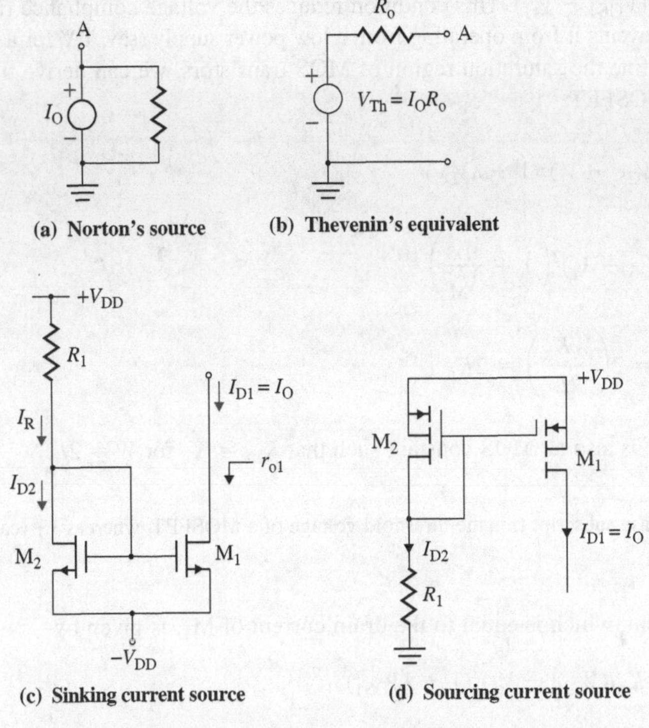

FIGURE 9.3 Basic MOSFET current source

do not draw any gate current, there is no need for gate current compensation as there is with BJTs. The choice of a BJT source or a MOSFET source generally depends on the type of integrated circuit involved (e.g., bipolar or MOS). BJT sources have some advantages over MOSFET sources, such as a wider compliance range and a higher output resistance. However, a higher output resistance can be obtained by cascode-like connections of MOSFETs.

Norton's equivalent of a current source I_O having an output resistance R_o is shown in Fig. 9.3(a), and its Thevenin's equivalent is in Fig. 9.3(b). An ideal current source should maintain a constant current at an infinite output resistance under all operating conditions. To achieve this goal, a number of current sources are developed. The commonly used current sources are the basic current source, which was applied in Secs. 7.8.1 and 8.7.1, the modified basic source, the cascoded current source, and the Wilson current source.

9.3.1 Basic Current Source

A basic MOSFET current source is shown in Fig. 9.3(c). The direction of the current flow in Fig. 9.3(c) is *into* the current source circuit; this type of constant-current source is often referred to as a *current sink*. Let us assume that the two transistors M_1 and M_2 are identical. Since their gate–source voltages are equal, their drain currents will be the same; that is, $I_{D1} = I_{D2}$. Thus, the output current I_O ($=I_{D1}$) will be the mirror of I_{D2}. Since $V_{DS2} = V_{GS2}$, M_2 will be in saturation. Let V_{t1} and V_{t2} be the threshold voltages of M_1 and M_2, respectively. For M_1 also to be in saturation, V_{DS1}, which is greater than or equal to ($V_{GS2} - V_{t2}$), must be greater than $(V_{GS1} - V_{t1})$. This condition reduces the voltage compliance range of the MOSFET current source and prevents it from operating from a low power supply (say, 1 V for a battery source). Using the equations that define the saturation region of MOS transistors, we can derive the drain current of an enhancement-type MOSFET:

$$I_D = \frac{WK_x}{2L}(V_{GS} - V_t)^2(1 + \lambda V_{DS}) \tag{9.19}$$

$$= \frac{K_m}{2}(V_{GS} - V_t)^2\left(1 + \frac{V_{DS}}{V_M}\right) \tag{9.20}$$

$$K_m = \frac{WK_x}{L} = 2\left(\frac{WK_x}{2L}\right) = 2K_n \tag{9.21}$$

where $K_n = WK_x/2L$ is also an MOS constant such that $K_m = K_n$ for $W = 2L$.

▶ **NOTE** V_t (lowercase subscript t) is the threshold voltage of a MOSFET, whereas V_T (capital subscript T) is the thermal voltage.

The output current, which is equal to the drain current of M_1, is given by

$$I_{D1} = I_O = K_{n1}(V_{GS1} - V_{t1})^2(1 + \lambda V_{DS1}) \tag{9.22}$$

Drain current I_{D2}, which is equal to the reference current I_R, is given by

$$I_{D2} = I_R = K_{n2}(V_{GS2} - V_{t2})^2(1 + \lambda V_{DS2}) \tag{9.23}$$

In practice, all the components of the current source are processed on the same IC, and hence all physical parameters such as K_x and V_t are identical for both devices. Thus, the ratio of I_O to I_R is given by

$$\frac{I_O}{I_R} = \frac{K_{n1}(1 + \lambda V_{DS1})}{K_{n2}(1 + \lambda V_{DS2})} = \frac{(W/L)_1}{(W/L)_2} \times \frac{(1 + \lambda V_{DS1})}{(1 + \lambda V_{DS2})} \tag{9.24}$$

In practice, $\lambda V_{DS} << 1$. Thus, Eq. (9.24) can be approximated by

$$\frac{I_O}{I_R} = \frac{(W/L)_1}{(W/L)_2} \tag{9.25}$$

By controlling the ratio W/L, therefore, we can change the output current. The gate length L is usually held fixed, and the gate width W is varied from device to device to give the desired current ratio I_O/I_R. By choosing identical transistors with $W_1 = W_2$ and $L_1 = L_2$, a designer can ensure that the output current I_O is almost equal to the reference current I_R.

Since $V_{GS2} = V_{DD} - R_1I_R$ and $V_{DS2} = V_{GS2}$, the reference current I_R can be found approximately from Eq. (9.23); that is,

$$I_R = I_{D2} = K_{n2}(V_{DD} - R_1I_R - V_{t2})^2 \tag{9.26}$$

can be solved for known values of V_{t2}, K_{n2}, V_{DD}, and R_1. The current source in Fig. 9.3(c) behaves as a current sink rather than a source. A current source equivalent to this current sink can be obtained by using p-channel MOS (PMOS) transistors as shown in Fig. 9.3(d).

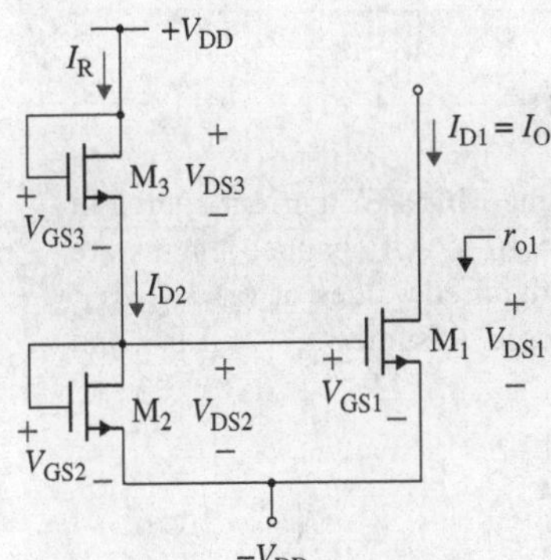

FIGURE 9.4 Basic MOSFET current source without resistance

9.3.2 Modified Basic Current Source

The reference resistance R_1 can be replaced by another MOSFET M_3, as shown in Fig. 9.4. Transistors M_2 and M_3 are used as voltage dividers to control the gate–source voltage of transistor M_1. If M_1 and M_2 are identical, the output current I_O exactly mirrors the drain current through M_2 and M_3. The value of V_{GS1} should be made as low as possible without taking M_1 out of the saturation region.

Since $V_{GS1} = V_{GS2}$, the drain current I_{D1} $(= I_O)$ is equal to the drain current I_{D2} and is given by

$$I_{D1} = I_{D2} = I_R = K_{n1}(V_{GS1} - V_{t1})^2(1 + \lambda V_{GS1}) \tag{9.27}$$

Since $V_{DS2} = V_{GS2}$, the drain current I_{D2} is equal to the reference current I_R and is given by

$$I_{D2} = I_R = K_{n2}(V_{GS2} - V_{t2})^2(1 + \lambda V_{GS2}) \tag{9.28}$$

Since $V_{GS3} = V_{DD} - V_{GS2}$, the drain current of M_3 is given by

$$\begin{aligned} I_{D3} = I_R &= K_{n3}(V_{GS3} - V_{t3})^2(1 + \lambda V_{DS3}) \\ &= K_{n3}(V_{DD} - V_{GS2} - V_{t3})^2[1 + \lambda(V_{DD} - V_{GS2})] \end{aligned} \tag{9.29}$$

Since $I_{D2} = I_{D3} = I_R$, from Eqs. (9.28) and (9.29) we get

$$\frac{K_{n2}(V_{GS2} - V_{t2})^2(1 + \lambda V_{GS2})}{K_{n3}(V_{DD} - V_{GS2} - V_{t3})^2[1 + \lambda(V_{DD} - V_{GS2})]} = 1 \tag{9.30}$$

Thus, by controlling the constants K_{n2} and K_{n3}, we can obtain the desired value of $V_{GS2} = V_{GS1}$, which will give the desired output current.

Output Resistance R_o

The small-signal drain-source resistance r_{ds1} can be derived from Eq. (9.20):

$$\frac{1}{r_{ds1}} = \frac{\delta i_{D1}}{\delta v_{DS1}} = \frac{K_{n1}}{V_M}(V_{GS} - V_t)^2 \approx \frac{I_{D1}}{V_M} \tag{9.31}$$

Thus, the small-signal output resistance of the current source becomes

$$R_o = r_{ds1} = \frac{V_M}{I_{D1}} = \frac{1}{\lambda I_{D1}} = r_{o1} \tag{9.32}$$

which is relatively small. This small output resistance is a disadvantage of having only one MOSFET M_1 at the output side of a current source.

EXAMPLE 9.1

D

Designing a simple MOSFET current source The parameters of the MOSFET current source in Fig. 9.4 are $V_t = 1$ V, $I_O = 50$ μA, $I_R = 40$ μA, $V_{DD} = 10$ V, and $V_M = 10$ V. All channel lengths are equal, $L_1 = L_2 = L_3 = L = 10$ μm, and $K_x = 20$ μA/V^2. Calculate the required values of **(a)** K_{n1}, W_1, **(b)** K_{n2}, W_2, **(c)** K_{n3}, W_3, and **(d)** the output resistance R_o of the current source. Assume $V_{GS1} = 1.5$ V and $V_{DS1} = 5$ V.

SOLUTION

(a) From Eq. (9.22),

$$50 \times 10^{-6} = K_{n1}(1.5 - 1)^2\left(1 + \frac{5}{10}\right)$$

which gives $K_{n1} = 133.3$ μA/V^2. From Eq. (9.21),

$$133.3 \times 10^{-6} = \frac{W_1 \times 20 \times 10^{-6}}{10 \times 10^{-6}}$$

which gives $W_1 = 66.65$ μm.

(b) $V_{DS2} = V_{GS2} = V_{GS1} = 1.5$ V. From Eq. (9.28),

$$40 \times 10^{-6} = K_{n2}(1.5 - 1)^2\left(1 + \frac{1.5}{10}\right)$$

which gives $K_{n2} = 139.1$ μA/V^2. From Eq. (9.21),

$$139.1 \times 10^{-6} = \frac{W_2 \times 20 \times 10^{-6}}{10 \times 10^{-6}}$$

which gives $W_2 = 69.55$ μm.

(c) $V_{GS3} = V_{DS3} = V_{DD} - V_{DS2} = V_{DD} - V_{GS1} = 10 - 1.5 = 8.5$ V. From Eq. (9.29),

$$40 \times 10^{-6} = K_{n3}(8.5 - 1)^2\left(1 + \frac{8.5}{10}\right)$$

which gives $K_{n3} = 0.384$ μA/V^2. From Eq. (9.21),

$$0.384 \times 10^{-6} = \frac{W_3 \times 20 \times 10^{-6}}{10 \times 10^{-6}}$$

which gives $W_3 = 0.192$ μm. In practice, because of manufacturing limitations, the minimum value of L or W is 10 μm. Since the value of W_3 is smaller than 10 μm, V_M must be increased to make W_3 at least 10 μm.

(d) From Eq. (9.32), the output resistance R_o is

$$R_o = r_{ds1} = \frac{V_M}{K_{n1}(V_{GS1} - V_t)^2} = \frac{10}{133.3 \times 10^{-6} \times (1.5 - 1)^2} = 300 \text{ k}\Omega$$

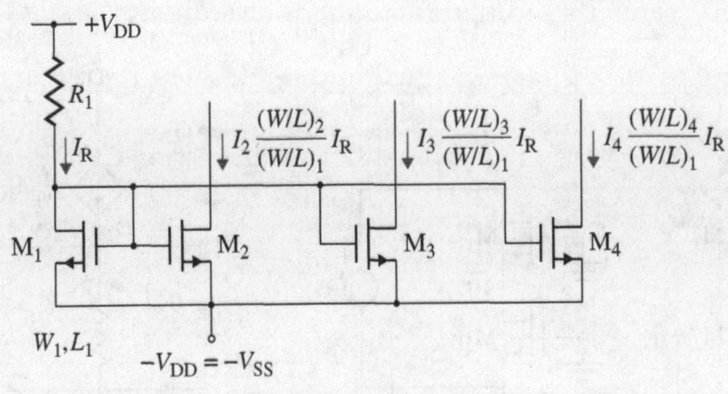

FIGURE 9.5 Multiple MOSFET current sources

9.3.3 Multiple Current Sources

Since there is no gate current in a MOSFET, a number of MOSFETs can be connected to a single reference MOSFET M_1, as shown in Fig. 9.5. Different output currents can be obtained by suitably adjusting the width-to-length ratios of MOSFETs (i.e., M_2, M_3, and M_4). In practice, the gate length L is normally kept constant, and the gate widths (W) of M_2, M_3, and M_4 are varied, to give the desired output currents. Thus, for equal L, Eq. (9.25) gives the relationship of the output currents I_2, I_3, and I_4 to I_R as

$$I_2 = \left(\frac{W_2}{W_1}\right) I_R$$

$$I_3 = \left(\frac{W_3}{W_1}\right) I_R$$

$$I_4 = \left(\frac{W_4}{W_1}\right) I_R$$

Output Resistance R_o

The small-signal output resistance of the current source is

$$R_o = r_{ds1} = \frac{V_M}{I_{D1}} = \frac{1}{\lambda I_{D1}} = r_{o1}$$

9.3.4 Cascode Current Source

The output resistance of the basic current source in Fig. 9.3(c) can be increased by adding two more MOSFETs in a cascodelike connection, as shown in Fig. 9.6(a). The analysis of the circuit is straightforward. The small-signal circuit for finding the output resistance is shown in Fig. 9.6(b), and its small-signal equivalent is shown in Fig. 9.6(c); r_{o2} is the output resistance of transistor M_2. Using KVL and the relation $v_{gs1} = -r_{o2} i_x$, we get

$$\begin{aligned} v_x &= r_{o1} i_1 + r_{o2} i_x \\ &= r_{o1}(i_x - g_{m1} v_{gs1}) + r_{o2} i_x \\ &= r_{o1}(i_x + g_{m1} r_{o2} i_x) + r_{o2} i_x \end{aligned}$$

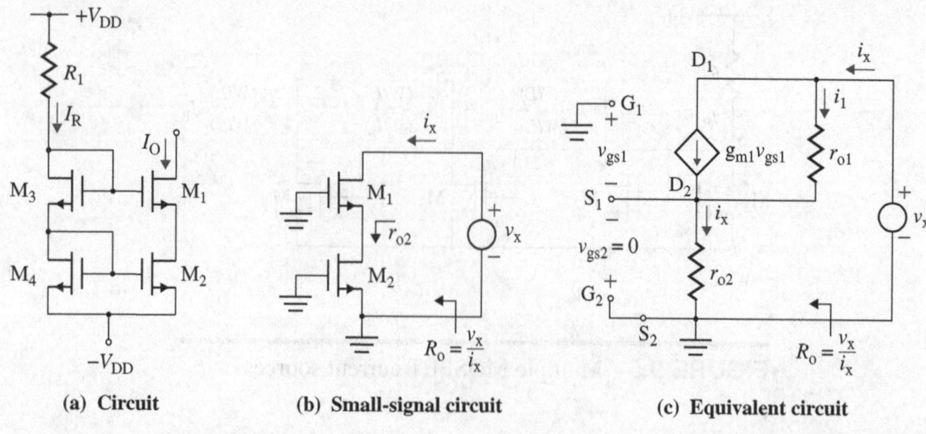

FIGURE 9.6 Cascode current source

which gives the output resistance R_o of the current source as

$$R_o = r_{o1}(1 + g_{m1}r_{o2}) + r_{o2} \tag{9.33}$$

For identical transistors, $r_{o1} = r_{o2} = r_o$, and R_o becomes

$$R_o = r_o(2 + g_{m1}r_o) \tag{9.34}$$

$$\approx g_{m1}r_o^2 \tag{9.35}$$

Thus, the output resistance can be significantly increased, to a level comparable to that of a BJT source. However, the voltage compliance range will be reduced because of the two drain-source voltages in series (i.e., $V_{DS1} + V_{DS2}$).

9.3.5 Wilson Current Source

The MOSFET version of the Wilson current source is shown in Fig. 9.7(a). The equivalent circuit for finding the output resistance R_o is shown in Fig. 9.7(b). We have

$$v_{gs3} = \frac{i_x}{g_{m2}}$$

and $\quad v_{gs1} + v_{gs3} = -g_{m3}v_{gs3}r_{o3}$

which can be simplified to relate v_{gs1} to v_{gs3} and v_{gs2} by

$$v_{gs1} = -(1 + g_{m3}r_{o3})v_{gs3} = -\frac{(1 + g_{m3}r_{o3})i_x}{g_{m2}}$$

Applying KVL to Fig. 9.7(b), we get

$$v_x = (i_x - g_{m1}v_{gs1})r_{o1} + \frac{i_x}{g_{m2}}$$

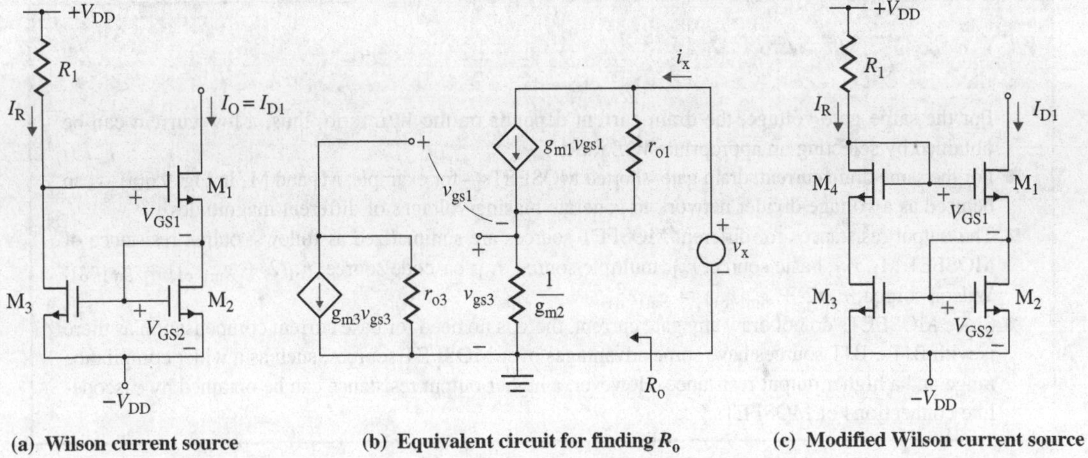

(a) Wilson current source (b) Equivalent circuit for finding R_o (c) Modified Wilson current source

FIGURE 9.7 Wilson current source

which, after substituting for v_{gs1} and simplifying, gives the output resistance R_o as

$$R_o = \frac{v_x}{i_x} = r_{o1} + \frac{1}{g_{m2}} + \frac{g_{m1}}{g_{m2}} r_{o1}(1 + g_{m3} r_{o3})$$

$$\approx r_{o1} + r_{o1}(1 + g_{m3} r_{o3}) \quad (\text{for } g_{m1} = g_{m2} = g_{m3}) \tag{9.36}$$

where r_{o2} and r_{o1} are the output resistances of transistors M_2 and M_1, respectively.

The problem with this circuit is that the drain voltages V_{D1} and V_{D3} of M_1 and M_3 are unequal. As a result, their drain currents I_{D1} and I_{D3} are also unequal. This problem can be solved by adding one diode-connected MOSFET, as shown in Fig. 9.7(c). This modification ensures that M_1 and M_3 have equal drain voltages and thus equal drain currents.

9.3.6 Design of Active Current Sources

The specifications for designing a current source will include the output current I_Q, the output resistance R_o, and the DC supply voltage V_{DD}. The design sequence is as follows:

Step 1. Determine the design specifications: output current and output resistance.

Step 2. Decide on the type of device to use—either BJTs or MOSFETs.

Step 3. Choose the circuit topology best suited to the specifications. Use simple transistor models for hand analysis to find the circuit-level solution, including component values and specifications of BJTs or MOSFETs.

Step 4. Use the standard values of components—for example, $R_1 = 5.6\ \text{M}\Omega \pm 5\%$ instead of $5.72\ \text{M}\Omega$, $R_2 = 30\ \text{k}\Omega \pm 5\%$ instead of $29.3\ \text{k}\Omega$, and $R_3 = 27\ \text{k}\Omega \pm 5\%$ instead of $27.5\ \text{k}\Omega$. Evaluate your design and modify the values, if necessary.

Step 5. Use PSpice/SPICE verification, employing complex circuit models to calculate the worst-case results due to component and parameter variations. Modify your design, if necessary.

KEY POINTS OF SECTION 9.3

- For the same gate voltage, the drain current depends on the W/L ratio; thus, a low current can be obtained by selecting an appropriate W/L ratio.
- For the same drain current, drain gate–shorted MOSFETs—for example, M_3 and M_4 in Fig. 9.6(a)—can be used as a voltage divider network to generate biasing voltages of different magnitudes.
- The output resistances for different MOSFET sources are summarized as follows: output resistance of MOSFET M_1, r_{o1}; basic source, r_{o1}; multiple source, r_{o1}; cascode source, $r_{o1}(2 + g_{m1}r_{o1}) \simeq g_{m1}r_{o1}^2$; Wilson source, $r_{o1}(2 + g_{m1}r_{o1}) \simeq g_{m1}r_{o1}^2$.
- Since MOSFETs do not draw any gate current, there is no need for base current compensation as there is with BJTs. BJT sources have some advantages over MOSFET sources, such as a wider compliance range and a higher output resistance. However, a higher output resistance can be obtained by cascode-like connections of MOSFETs.

9.4 MOS Differential Amplifiers

During the past few years, MOS technology has developed considerably. MOS transistors are being used increasingly in analog integrated circuits. It is relatively easy to connect MOS transistors in cascode form to control the drain current and give high output resistance. MOS differential pairs are the building blocks in MOS ICs.

9.4.1 NMOS Differential Pair

An n-channel NMOS pair is shown in Fig. 9.8. The DC biasing is normally done by a MOS current source. Although a resistor R_D is shown as the load in Fig. 9.8, a MOS active current mirror is normally used as the load.

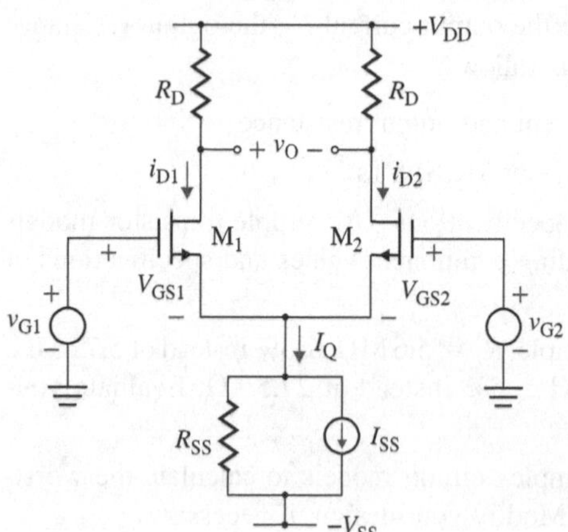

FIGURE 9.8 MOS differential pair

DC Transfer Characteristics

The following analysis is performed for an n-channel MOSFET pair, but it is equally applicable to a p-channel pair with appropriate sign changes. The analysis can be simplified by making the following assumptions:

1. The output resistances of the MOSFETs are infinite: $r_o = \infty$.
2. The MOSFETs are identical and operate in the saturation region. The threshold voltages are the same, $V_{t1} = V_{t2} = V_t$, and the constants are equal, $K_{n1} = K_{n2} = K_n$.
3. The output resistance of the transistor current source is infinite: $R_{SS} = \infty$.

Assuming that the drain current is related to v_{GS} by the approximate square law relationship in Eq. (7.10), we can write

$$i_D = K_n(v_{GS} - V_t)^2 \tag{9.37}$$

Taking the square root of both sides of Eq. (9.37), we can write the square root of the drain currents as

$$\sqrt{i_{D1}} = \sqrt{K_n}(v_{GS1} - V_t) \tag{9.38}$$

$$\sqrt{i_{D2}} = \sqrt{K_n}(v_{GS2} - V_t) \tag{9.39}$$

We can subtract $\sqrt{i_{D2}}$ from $\sqrt{i_{D1}}$ to find a relation for the differential voltage $v_{id} = v_{GS1} - v_{GS2}$:

$$\sqrt{i_{D1}} - \sqrt{i_{D2}} = \sqrt{K_n}(v_{GS1} - V_t) - \sqrt{K_n}(v_{GS2} - V_t) = \sqrt{K_n}(v_{GS1} - v_{GS2}) \tag{9.40}$$
$$= \sqrt{K_n}v_{id}$$

The sum of i_{D1} and i_{D2} must equal I_Q; that is,

$$I_Q = i_{D1} + i_{D2} \tag{9.41}$$

Substituting Eq. (9.41) into Eq. (9.40) and solving the resultant quadratic, we find the drain currents:

$$i_{D1} = \frac{I_Q}{2} + \sqrt{2K_nI_Q}\left(\frac{v_{id}}{2}\right)\left[1 - \frac{(v_{id}/2)^2}{(I_Q/2K_n)}\right]^{1/2} \tag{9.42}$$

$$i_{D2} = \frac{I_Q}{2} - \sqrt{2K_nI_Q}\left(\frac{v_{id}}{2}\right)\left[1 - \frac{(v_{id}/2)^2}{(I_Q/2K_n)}\right]^{1/2} \tag{9.43}$$

At the quiescent point $v_{id} = 0$, we get

$$i_{D1} = i_{D2} = \frac{I_Q}{2}$$

$$v_{GS1} = v_{GS2} = V_{GS}$$

$$I_Q = 2I_D = 2K_n(V_{GS} - V_t)^2$$

$$\sqrt{2K_n I_Q} = 2K_n \frac{(V_{GS} - V_t)^2}{(V_{GS} - V_t)} = \frac{I_Q}{(V_{GS} - V_t)}$$

$$\left(\frac{I_Q}{2K_n}\right) = (V_{GS} - V_t)^2$$

Substituting these relations into Eqs. (9.42) and (9.43), we can rewrite $i_{D1} = i_{D2}$ as

$$i_{D1} = \frac{I_Q}{2} + \left(\frac{I_Q}{V_{GS} - V_t}\right)\left(\frac{v_{id}}{2}\right)\left[1 - \left(\frac{v_{id}/2}{V_{GS} - V_t}\right)^2\right]^{1/2} \tag{9.44}$$

$$i_{D2} = \frac{I_Q}{2} - \left(\frac{I_Q}{V_{GS} - V_t}\right)\left(\frac{v_{id}}{2}\right)\left[1 - \left(\frac{v_{id}/2}{V_{GS} - V_t}\right)^2\right]^{1/2} \tag{9.45}$$

For $v_{id}/2 << (V_{GS} - V_t)$, i_{D1} and i_{D2} can be approximated by

$$i_{D1} = \frac{I_Q}{2} + \left(\frac{I_Q}{V_{GS} - V_t}\right)\left(\frac{v_{id}}{2}\right) \tag{9.46}$$

$$i_{D2} = \frac{I_Q}{2} - \left(\frac{I_Q}{V_{GS} - V_t}\right)\left(\frac{v_{id}}{2}\right) \tag{9.47}$$

Thus, the change in drain current from the quiescent value of $I_Q/2$ is given by

$$\Delta I_D = \left(\frac{I_Q}{V_{GS} - V_t}\right)\left(\frac{v_{id}}{2}\right) \tag{9.48}$$

which can be normalized with respect to the maximum value $I_Q/2$ as

$$\frac{\Delta I_D}{I_Q/2} = \frac{v_{id}}{V_{GS} - V_t} = \frac{v_{id}}{\sqrt{I_Q/2K_n}} \tag{9.49}$$

If v_{id} is sufficiently large, all of the biasing current I_D must flow through only one of the MOSFETs. The range of v_{id} for which both transistors conduct can be found from Eq. (9.49) under the condition $\Delta I_D = I_Q/2$. That is,

$$v_{id} \le \left(\frac{I_Q}{2K_n}\right)^{1/2} \tag{9.50}$$

This equation gives the value of v_{id} for which the current I_Q is carried by one of the two transistors. Thus, outside the range defined by Eq. (9.50), currents i_{D1} and i_{D2} will be either zero or I_Q. The plots of the normalized currents i_{D1} and i_{D2} against the differential voltage v_{id}/v_n are shown in Fig. 9.9, where $v_n = \sqrt{I_Q/2K_n}$.

The output voltages of a MOSFET pair are as follows:

$$v_{O1} = V_{DD} - i_{D1}R_D \tag{9.51}$$

$$v_{O2} = V_{DD} - i_{D2}R_D \tag{9.52}$$

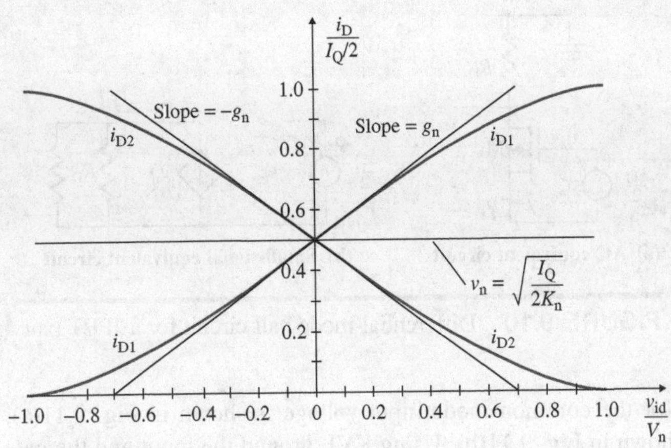

FIGURE 9.9 Normalized DC transfer characteristic of MOSFET pair

The differential DC output voltage is

$$v_{od} = v_{O1} - v_{O2} = R_D(i_{D2} - i_{D1}) = -R_D \Delta I_D$$

Substituting for ΔI_D from Eq. (9.47), we get

$$v_{od} = -R_D\left(\frac{I_Q}{V_{GS} - V_t}\right)\left(\frac{v_{id}}{2}\right) \tag{9.53}$$

$$= -R_D\left(\frac{K_n I_Q}{2}\right)^{1/2} v_{id}$$

which shows the relation between the output voltage v_{od} and the differential voltage v_{id}. If v_{id} is zero, v_{od} is also zero.

Small-Signal Analysis

From Eq. (7.27), the transconductance of the small-signal model is given by

$$g_m = 2K_n(V_{GS} - V_t) \tag{9.54}$$

$$= 2\sqrt{2K_n I_Q}$$

which shows that for a MOSFET a higher value of g_m requires a higher value of the biasing current I_Q.

The half circuit for differential input voltage is shown in Fig. 9.10(a), and its small-signal equivalent circuit is shown in Fig. 9.10(b). The ($r_{o1} \| R_D$) differential voltage gain A_d for a single-ended output can easily be derived as

$$A_d = \frac{v_{od}/2}{v_{id}/2} = -g_m(r_{o1} \| R_D) \tag{9.55}$$

which, for $r_{o1} >> R_D$, can be approximated to $A_d = -g_m R_D$.

▶ **NOTE** If we define $v_{od} = v_{o2} - v_{o1}$ and $v_{gd} = v_{G1} - v_{G2}$, then $A_d = g_m(R_D \| r_{o1})$.

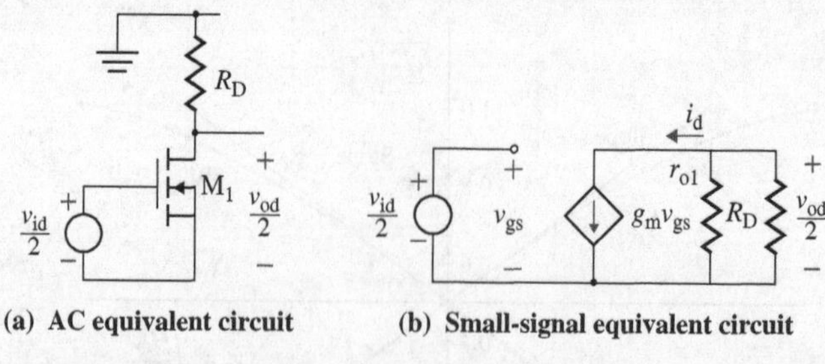

FIGURE 9.10 Differential-mode half circuit for a JFET pair

The half circuit for the common-mode input voltage is shown in Fig. 9.11(a), and its small-signal equivalent circuit is shown in Fig. 9.11(b). Using KVL around the input and the gate–source loop, we get

$$v_{gs} = v_{ic} - v_{gs} g_m 2R_{SS}$$

which leads to the following relationship between v_{gs} and v_{ic}:

$$v_{gs}(1 + g_m 2R_{SS}) = v_{ic} \tag{9.56}$$

The common-mode output voltage is given by

$$v_{oc} = -R_D i_d = -R_D g_m v_{gs}$$

Substituting v_{gs} from Eq. (9.56) into the above equation, we get

$$v_{oc} = v_{ic}\left[\frac{-g_m R_D}{1 + g_m 2R_{SS}}\right]$$

which gives the common-mode voltage gain A_c (for a single-ended output) as

$$A_c = \frac{v_{oc}}{v_{ic}} = \frac{-g_m R_D}{1 + g_m 2R_{SS}} \tag{9.57}$$

▶ **NOTE** To simplify the analysis, the effect of transistor output resistance r_{o1} is not included in Eq. (9.57).

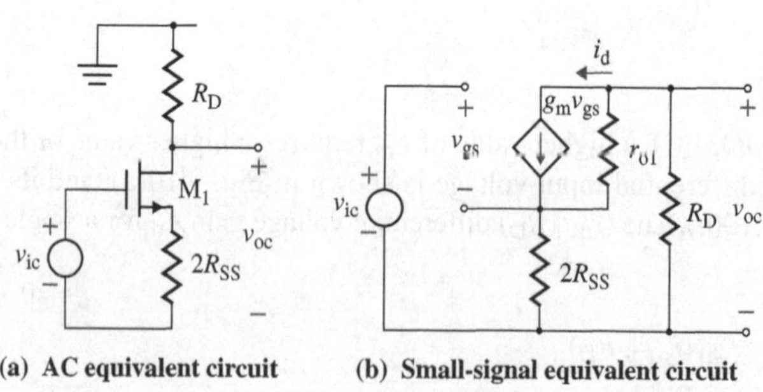

FIGURE 9.11 Common-mode half circuit for depletion MOSFET pair

From Eqs. (9.55) and (9.57) for $r_{o1} >> R_D$, we can find CMRR as

$$\text{CMRR} = \frac{A_d}{A_c} = 1 + 2g_m R_{SS} \tag{9.58}$$

which is valid only for single-ended output; that is, the output is taken out from one of the output terminals.

The common-mode and differential input resistances are given by

$$R_{ic} = R_{id} = \infty \tag{9.59}$$

▶ **NOTE** To simplify the analysis, the effect of transistor output resistance r_o is not included in Eqs. (9.57) and (9.58).

EXAMPLE 9.2

Analyzing a MOS differential pair with an active current source The parameters of the MOS differential pair in Fig. 9.8 are $R_{SS} = 50\text{ k}\Omega$, $I_Q = 10\text{ mA}$, $V_{DD} = 30\text{ V}$, and $R_D = 5\text{ k}\Omega$. The NMOSs are identical and have $K_n = 1.25\text{ mA/V}^2$ and $V_t = 1.0\text{ V}$. Assume $V_M = 100\text{ V}$.

(a) Calculate the DC drain currents through the MOSFETs if $v_{id} = 10\text{ mV}$.

(b) Assuming $I_{D1} = I_{D2}$, calculate A_d, A_c, and CMRR; R_{id} and R_{ic}; and the small-signal output voltage if $v_{g1} = 10\text{ mV}$ and $v_{g2} = 20\text{ mV}$.

(c) Find the drain voltage V_D.

SOLUTION

(a) For $v_{id} = 10\text{ mV}$, Eq. (9.42) gives the DC drain current i_{D1} for transistor M_1 as

$$i_{D1} = \frac{10\text{ m}}{2} + \sqrt{2 \times 1.25\text{ m} \times 10\text{ m}} \times \left(\frac{10\text{ m}}{2}\right)\left[1 - \frac{(10\text{ m}/2)^2}{10\text{ m}/(2 \times 1.25\text{ m})}\right]^{1/2} = 5.25\text{ mA}$$

$$i_{D2} = I_D - i_{D1} = 10\text{ mA} - 5.25\text{ mA} = 4.75\text{ mA}$$

(b) We know that $I_{D1} = I_{D2} = I_Q/2 = 10\text{ mA}/2 = 5\text{ mA}$. From Eq. (9.54),

$$g_m = \sqrt{2 \times K_n I_Q} = \sqrt{2 \times 1.25\text{ m} \times 10\text{ m}} = 5\text{ mA/V}$$

$$r_{o1} = \frac{V_M}{I_D} = \frac{100}{5\text{ m}} = 20\text{ k}\Omega$$

From Eq. (9.55), the single-ended differential voltage gain A_d is

$$A_d = -g_m(R_D \| r_{o1}) = -5\text{ m} \times (5\text{ k} \| 20\text{ k}) = -20\text{ V/V}$$

From Eq. (9.57), the single-ended common-mode voltage gain A_c is

$$A_c = \frac{-g_m R_D}{1 + g_m 2R_{SS}} = \frac{-5\text{ m} \times 5\text{ k}}{1 + 5\text{ m} \times 2 \times 50\text{ k}} = -0.0399\text{ V/V}$$

Thus, CMRR = $|A_d/A_c| = 20/0.0399 = 501$ (or 54 dB).

From Eq. (9.59),

$$R_{id} = R_{ic} = \infty$$

We know that

$$v_{id} = v_{g2} - v_{g1} = 20\text{ mV} - 10\text{ mV} = 10\text{ mV}$$

and $$v_{ic} = \frac{v_{g1} + v_{g2}}{2} = \frac{10\text{ mV} + 20\text{ mV}}{2} = 15\text{ mV}$$

Using Eq. (9.10), we have

$$v_o = A_d v_{id} + A_c v_{ic} = -20 \times 10\text{ mV} - 0.0399 \times 15\text{ mV} = -201\text{ mV}$$

(c) The DC drain voltage at the drain terminal of a transistor is

$$V_D = V_{DD} - I_D R_D = 30\text{ V} - 5\text{ mA} \times 5\text{ k}\Omega = 5\text{ V}$$

Thus, for $A_d = -20$, the maximum differential voltage will be $v_{id} = 5/20 = 250$ mV. Therefore, V_{DD} must be greater than $I_D R_D$ in order to allow output voltage swing due to the input voltages.

EXAMPLE 9.3

Analyzing an NMOS differential pair with an active current source Repeat Example 9.2 if the transistor current source is replaced by the resistance $R_{SS} = 50$ kΩ; that is, $I_{SS} = 0$. Assume $V_M = 100$ V.

SOLUTION

(a) The DC drain current and the gate–source voltage of the MOSFETs can be determined from the DC common-mode half circuit for $v_{g1} = v_{g2} = 0$, shown in Fig. 9.12:

$$V_{GS} + 2I_D R_{SS} = V_{SS}$$

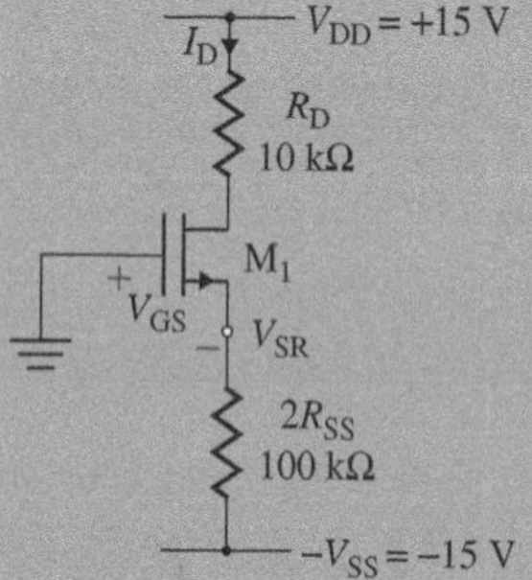

FIGURE 9.12 Common-mode half circuit for DC biasing of MOS pair

Substituting $i_D = I_D$ and $v_{GS} = V_{GS}$ from Eq. (9.37) into the preceding equation, we get

$$V_{GS} + 2R_{SS}K_n(V_{GS} - V_t)^2 = V_{SS}$$

Substituting the values, we get

$$V_{GS} + 2 \times 50\text{ k} \times (V_{GS} - 1)^2 = 15$$

This quadratic equation yields the solution for the gate–source voltage $V_{GS} = 1.478$ V, or 0.514 V. Since $V_{GS} > V_t$, the acceptable value is $V_{GS} = 1.478$ V, which gives the drain current as

$$I_{D1} = K_n(V_{GS} - V_t)^2 = 1.25\text{ m} \times (1.478 - 1)^2 = 285.6\ \mu\text{A}$$

Therefore, the voltage at the source with respect to the ground is

$$V_{SR} = -V_{GS} = -1.478\text{ V} \quad \text{and} \quad I_Q = 2I_D = 2 \times 285.6\ \mu = 571.2\ \mu\text{A}$$

(b) From Eq. (9.54),

$$g_m = \sqrt{2K_nI_Q} = \sqrt{2 \times 1.25\text{ m} \times 571.2\ \mu} = 1.194\text{ mA/V}$$

$$r_{o1} = \frac{V_M}{I_D} = \frac{100}{285.6\ \mu} = 350\text{ k}\Omega$$

From Eq. (9.55),

$$A_d = -g_m(R_D \| r_{o1}) = -1.194\text{ m} \times (5\text{ k} \| 350\text{ k}) = -5.89\text{ V/V}$$

From Eq. (9.57),

$$A_c = \frac{-g_mR_D}{1 + g_m \times 2R_{SS}} = \frac{-1.194\text{ m} \times 5\text{ k}}{1 + 1.194\text{ m} \times 2 \times 50\text{ k}} = -0.0489$$

Thus, CMRR $= |A_d/A_c| = 5.89/0.0489 = 120.4$, or 41.61 dB. From Eq. (9.59),

$$R_{id} = R_{ic} = \infty$$

We know that

$$v_{id} = v_{g2} - v_{g1} = 20\text{ mV} - 10\text{ mV} = 10\text{ mV}$$

$$v_{ic} = \frac{v_{g1} + v_{g2}}{2} = \frac{10\text{ mV} + 20\text{ mV}}{2} = 15\text{ mV}$$

Using Eq. (9.10), we have

$$v_o = A_dv_{id} + A_cv_{ic} = -5.887 \times 10\text{ mV} - 0.489 \times 15\text{ mV} = -59.6\text{ mV}$$

Thus, the output voltage and the voltage gain are much lower than with current-source biasing because the biasing current is low.

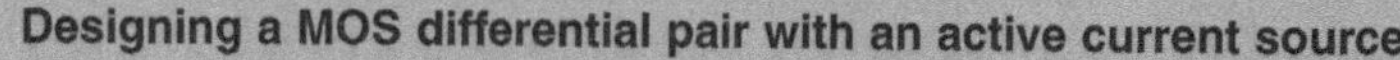

EXAMPLE 9.4

D

Designing a MOS differential pair with an active current source

(a) Design a MOS differential pair as shown in Fig. 9.13, in which one input terminal is grounded. The output is taken from the drain of transistor M_1. The DC biasing current is $I_Q = 10$ mA, and $V_{DD} = V_{SS} = 15$ V. The MOSFETs are identical and have $K_n = 1.25\ \text{mA/V}^2$ and $V_t = 1$ V. A small-signal voltage gain of $A_1 - 10$ V/V is required. Assume $V_M = \infty$.

(b) Calculate the design values of A_d, A_c, and CMRR.

(c) Find the maximum permissible value of R_D and the corresponding voltage gain A_d.

SOLUTION

(a) $I_Q = 10$ mA, and $I_{D1} = I_{D2} = I_Q/2 = 10\ \text{mA}/2 = 5$ mA. From Eq. (9.54),

$$g_m = \sqrt{2 \times K_n I_Q} = \sqrt{2 \times 1.25\ \text{m} \times 10\ \text{m}} = 5\ \text{mA/V}$$

From Eq. (9.37), the DC gate–source voltage is

$$V_{GS1} = -\left(\sqrt{\frac{I_{D1}}{K_n}} + V_t\right) = -\left(\sqrt{\frac{5\ \text{mA}}{1.25\ \text{m}}} + 1\right) = -2.236\ \text{V}$$

Therefore, the voltage at the source terminal with respect to the ground is $V_{SR} = -V_{GS} = -2.236$ V, and

$$R_{SS} = \frac{V_{SR} - V_{SS}}{I_{SS}} = \frac{(-2.236 + 15)\ \text{V}}{10\ \text{mA}} = 1.3\ \text{k}\Omega$$

(b) $v_{G1} = V_{G1} + v_{g1}$, $V_{G1} = 0$, and $v_{G2} = v_{g2} = 0$. Then

$$v_{id} = v_{g1} - v_{g2} = v_{g1}$$

$$v_{ic} = \frac{v_{g1} + v_{g2}}{2} = \frac{v_{g1}}{2}$$

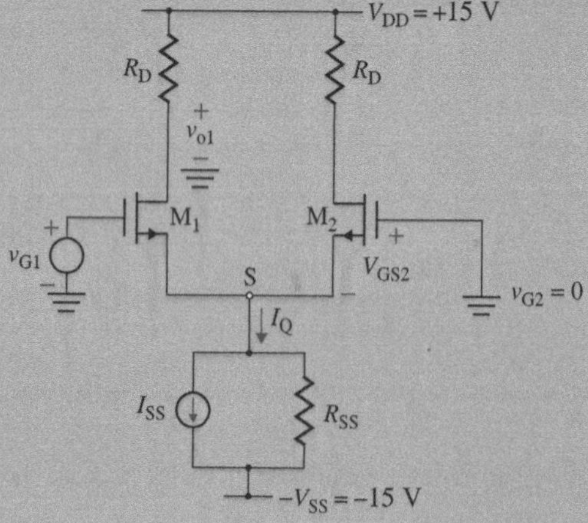

FIGURE 9.13 MOS differential pair with a single input

From Eq. (9.7),

$$v_{o1} = v_{oc} + \frac{v_{od}}{2} = A_c v_{ic} + A_d v_{id} = A_c \frac{v_{g1}}{2} + A_d v_{g1} = \frac{v_{g1}}{2}(A_c + 2A_d)$$

Substituting for A_d and A_c from Eqs. (9.55) and (9.57) gives the voltage gain A_1 as

$$A_1 = \frac{v_{o1}}{v_{g1}} = \frac{1}{2}(A_c + 2A_d) = -\frac{1}{2}\left[\frac{g_m R_D}{1 + 2g_m R_{SS}} + 2g_m R_D\right]$$

Substituting $A_1 = -10$, $g_m = 5$ mA/V, and $R_{SS} = 1.3$ kΩ into the above equation gives $R_D = 1.93$ kΩ.

$$r_{o1} = \frac{V_M}{I_{D1}} = \infty$$

From Eq. (9.55),

$$A_d = -g_m(R_D \parallel r_{o1}) = -5\text{ m} \times (1.93\text{ k} \parallel 20\text{ k}) = -8.8\text{ V/V}$$

From Eq. (9.57),

$$A_c = \frac{-g_m R_D}{1 + g_m \times 2R_{SS}} = \frac{-5\text{ m} \times 1.93\text{ k}}{1 + 5\text{ m} \times 2 \times 1.7\text{ k}} = -0.64\text{ V/V}$$

Thus, CMRR = $|A_d/A_c| = 8.8/0.64 = 13.76$, or 22.78 dB.

(c) For $I_D = I_Q/2 = 5$ mA and $V_{DD} = 15$ V, the maximum value of R_D is

$$R_{D(max)} = \frac{V_{DD}}{I_D} = \frac{15\text{ V}}{5\text{ mA}} = 3\text{ k}\Omega$$

which gives the maximum value of

$$A_{d(max)} = -g_m R_{D(max)} = -5\text{ mA} \times 3\text{ k}\Omega = -15\text{ V/V}$$

9.4.2 MOS Differential Pair with Active Load

MOS differential amplifiers are normally used with current mirror active loads. A commonly used configuration is shown in Fig. 9.14. The current mirror consists of transistors M_3 and M_4. Increasing the input voltage to M_1 by $v_{id}/2$ will cause the drain current of M_1 to increase by an amount $g_m v_{id}/2$. This increase will cause a similar increase in the drain current of M_4 due to the current mirror effect and also a decrease in the drain current of M_2. Since M_1 and M_2 are NMOS transistors and their complements (M_3 and M_4) are PMOS types, this configuration is known as a *CMOS amplifier*; the manufacturing process by which the amplifiers are produced is known as *CMOS technology*.

The small-signal equivalent of the output side of the CMOS amplifier is shown in Fig. 9.15. The output resistance R_o is the parallel combination of r_{o2} and r_{o4}. That is,

$$R_o = r_{o2} \parallel r_{o4} \tag{9.60}$$

where

$$r_{o2} = \frac{2V_M}{I_Q} = \text{output resistance of transistor } M_2$$

$$r_{o4} = \frac{2V_M}{I_Q} = \text{output resistance of transistor } M_4$$

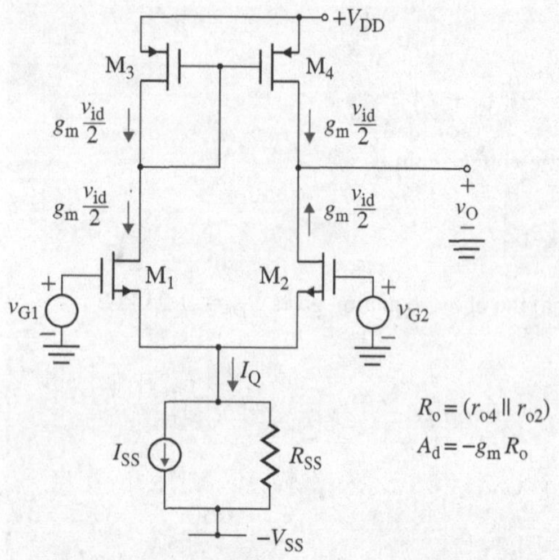

FIGURE 9.14 CMOS differential amplifier

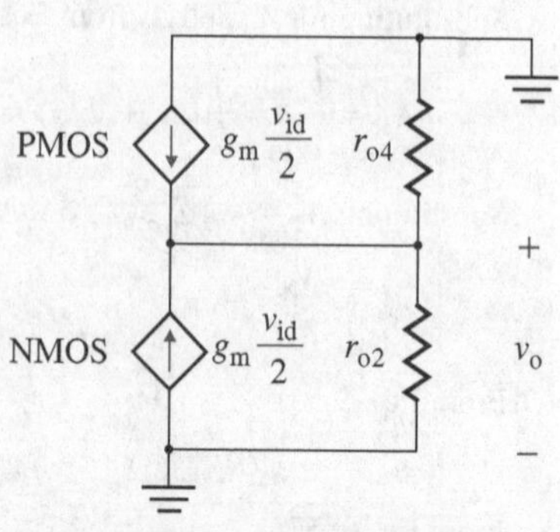

FIGURE 9.15 Small-signal equivalent of the CMOS amplifier

Using Eq. (9.55), we find the differential voltage gain A_d:

$$A_d = \frac{v_o}{v_{id}} = -g_m(r_{o2} \parallel r_{o4}) \tag{9.61}$$

where g_m, which is the transconductance of MOSFET M_2, is given by Eq. (9.54). Substituting $r_{o4} = r_{o2} = 2V_M/I_Q$ and $g_m = \sqrt{2K_n I_Q}$ and simplifying, we get

$$A_d = -\sqrt{\frac{2K_n}{I_Q}}\,V_M \tag{9.62}$$

which gives a higher voltage gain for a lower value of I_Q.

EXAMPLE 9.5

D

Designing a CMOS amplifier

(a) Design the CMOS amplifier shown in Fig. 9.16 by determining the *W/L* ratios of the MOSFETs and the threshold voltage V_t. The differential voltage gain should be $A_d = 60$ V/V at biasing current $I_Q = 10$ μA. Assume identical transistors whose channel modulation voltage is $V_M = 20$ V, channel constant is $K_x = 20$ μA/V², and channel length is $L = 10$ μm. The *W/L* ratio of the current source is 2, and $K_x = 10$ μA/V². Assume $V_{DD} = V_{SS} = 5$ V.

(b) Use PSpice/SPICE to find the small-signal differential voltage gain A_d for $v_{G1} = 1$ mV and $v_{G2} = 0$.

SOLUTION

(a) $A_d = 60$, $I_Q = 10$ μA, $V_M = 20$ V, $K_x = 20$ μA/V², and $L = 10$ μm. From Eq. (9.62), we find the MOS constant K_n as

$$K_{n2} = \left(\frac{I_Q}{2}\right)\left(\frac{A_d}{V_M}\right)^2 = \left(\frac{10\ \mu\text{A}}{2}\right)\left(\frac{60}{20}\right)^2 = 45\ \mu\text{A/V}^2$$

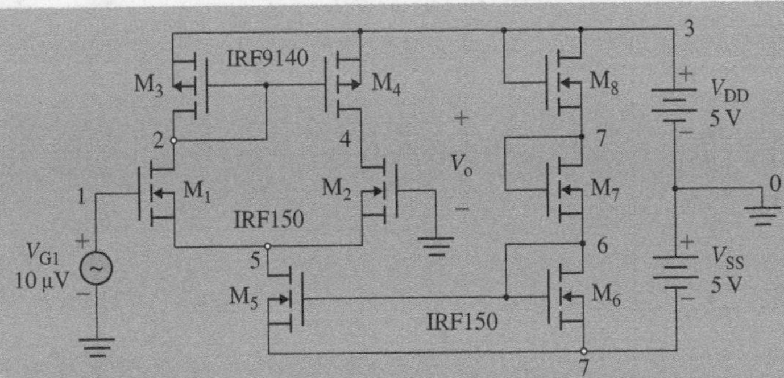

FIGURE 9.16 CMOS amplifier

From Eq. (9.21), we find the W/L ratio as

$$\frac{W}{L} = \frac{K_n}{K_x} = \frac{45\ \mu}{20\ \mu} = 2.25$$

Since $L = 10\ \mu$m, the desired value is $W = 2.25 \times 10\ \mu\text{m} = 22.5\ \mu$m. For identical transistors for the current source,

$$V_{GS6} = \frac{V_{DD} + V_{SS}}{3} = \frac{(5 + 5)\text{ V}}{3} = 3.33\text{ V}$$

Since $W = 2L$ for the MOSFETs of the current source, the MOS constant is

$$K_{n6} = \frac{W_6 K_x}{2L_6} = \frac{2 \times 10}{2}\ \mu\text{A/V}^2 = 10\ \mu\text{A/V}^2$$

The biasing current is given by

$$I_Q = K_{n6}(V_{GS6} - V_t)^2$$

which, for $V_{GS6} = 3.33$ V, $I_Q = 10\ \mu$A, and $K_{n6} = 10\ \mu\text{A/V}^2$, gives $V_t = 2.33$ V. Then

$$r_{o2} = r_{o4} = \frac{2V_M}{I_Q} = \frac{2 \times 20}{10\ \mu} = 4\text{ M}\Omega$$

$$R_o = r_{o2} \| r_{o4} = 2\text{ M}\Omega$$

$$r_{o5} = \frac{V_M}{I_Q} = \frac{20}{10\ \mu} = 2\text{ M}\Omega$$

(b) The CMOS amplifier for PSpice simulation is shown in Fig. 9.17. The plot of the output voltage is shown in Fig. 9.18, which gives a voltage gain 81.72 V/V (expected value 60). Note there is no phase shift of 180° because the output is taken from the drain terminal of the pair.

The results of simulation (.TF analysis) are as follows. (The expected values are listed on the right.)

**** SMALL-SIGNAL CHARACTERISTICS	
V(4)/VID1=8.189E+01=81.89	($A_{id} = 60$ V/V)
INPUT RESISTANCE AT VG1=1.000E+20	($R_{id} = \infty$)
OUTPUT RESISTANCE AT V(4)=2.221E+06=2.221 MΩ	($R_{od} = 2$ MΩ)
ID1=5.44E-06=5.44 μA	($I_{D1} = 5.44\ \mu$A)
ID=I_{D5}=1.09E-05=10.9 μA	($I_Q = 10\ \mu$A)

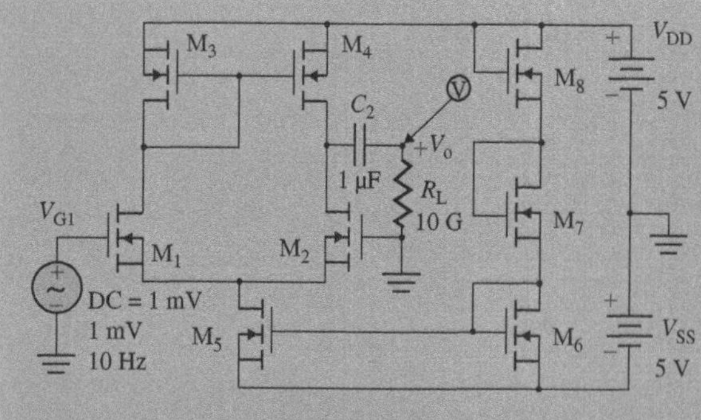

FIGURE 9.17 CMOS amplifier for a PSpice simulation for Example 9.5

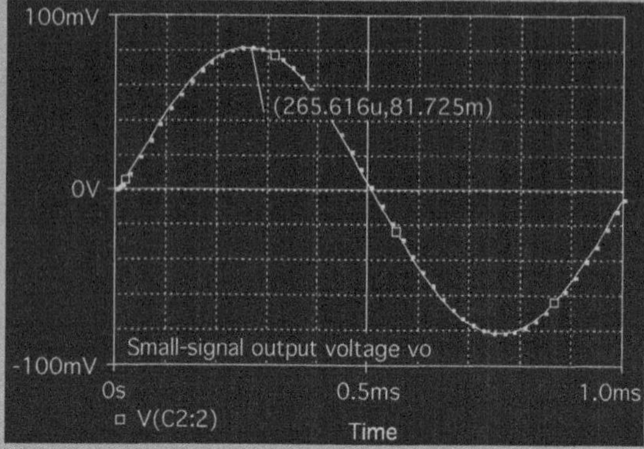

FIGURE 9.18 PSpice plot of small-signal output voltage for Example 9.5

NOTE: As expected, the PSpice results depend on the values of the *W/L* ratio for the MOSFETs. If you run PSpice from the schematic, you will need to change the model parameters of the MOSFETs; otherwise, the results will be different from those shown above.

9.4.3 Cascoded MOS Differential Amplifier

We can notice from Eq. (9.61) that the differential gain increases with the output resistance R_o of the active amplifier load. MOSFETs can be connected in cascode configuration as shown in Fig. 9.6(a) to increase the output resistance and to improve the frequency response. A common modification to Fig. 9.14(a) is shown in Fig. 9.19(a). Transistors M_1 and M_2 are connected in a common-source configuration and form a common-source differential pair, whereas M_3 and M_4 are connected in a common-gate configuration and form a common-gate differential stage. M_6 and M_8 act as the load. For identical MOSFETs, the DC biasing currents are equal, $I_{D1} = I_{D2} = I_{D4} = I_{D6} = I_{D8} = I_Q/2$. All MOSFETs have equal output resistances, $r_{o1} = r_{o2} = r_{o4} = r_{o6} = r_{o8} = V_M/I_{D1}$, and equal transconductances, $g_{m1} = g_{m2} = g_{m4} = g_{m6} = g_{m8} = 2K_n(V_{GS1} - V_t)$. The small-signal half-circuit is shown in Fig. 9.19(b). Using Eq. (9.34), we can find the effective load resistance offered by M_6 and M_8 as given by

$$r'_{o6} = r_{o6}(2 + g_{m1}r_{o6}) \tag{9.63}$$

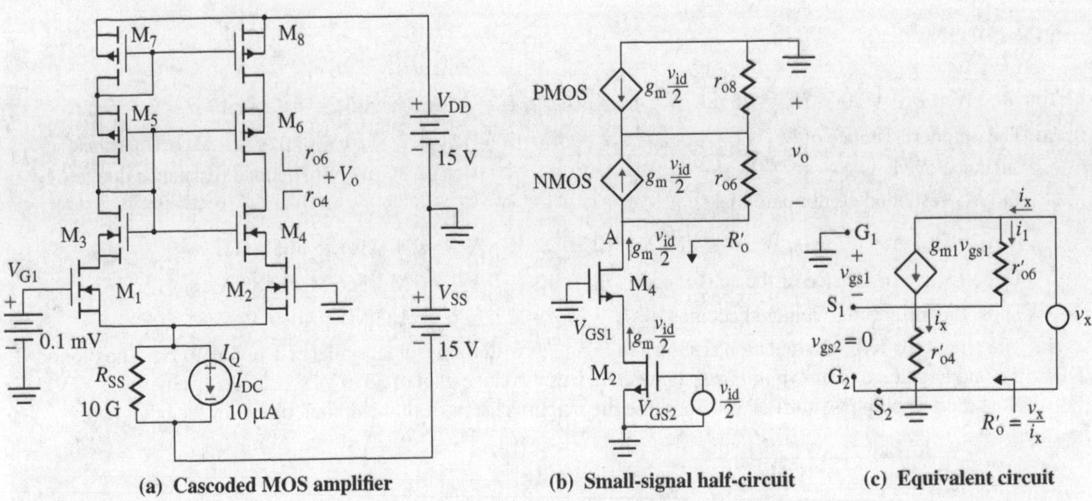

FIGURE 9.19 Cascoded MOS differential amplifier

Similarly, we can find the effective resistance offered by M_2 and M_4 as given by

$$r'_{o4} = r_{o4}(2 + g_{m1}r_{o4}) \tag{9.64}$$

Therefore, the effective output resistance at the output terminal as shown in Fig. 9.19(c) is given by

$$R_o = r'_{o4} \parallel r'_{o6} \tag{9.65}$$

Applying Eq. (9.61) we can find the small-signal differential output voltage as given by

$$A_d = \frac{v_{od}}{v_{id}} = -g_{m1}R_o = -g_{m1}(r'_{o4} \parallel r'_{o6}) \tag{9.66}$$

This should give a much larger differential gain compared to that of Fig. 9.14(a). For example, if $|V_M| = 20$ V, $I_{D1} = 5$ μA, $g_{m1} = 31.5$ μA/V, $r_{o4} = r_{o6} = 5$ MΩ, $r'_{o6} = r'_{o8} = 514$ MΩ, and $R_o = 257$ MΩ, we get $A_d = 8126$ V/V.

EXAMPLE 9.6

Analyzing cascoded MOS amplifiers The DC biasing current of the MOS amplifier shown in Fig. 9.19(a) is kept constant at $I_Q = 10$ μA. All MOS transistors are identical: $V_M = 20$ V, $K_n = 25$ μA/V^2, $W = 30$ μm, and $L = 10$ μm.

(a) Determine the differential voltage gain A_d for single-ended output at the drain terminal of M_4.

(b) Use PSpice to verify the result.

SOLUTION

You are given $V_M = 20$ V, $K_n = 25\ \mu\text{A/V}^2$, $W = 30\ \mu\text{m}$, $L = 10\ \mu\text{m}$, and $I_D = I_Q/2 = 10\ \mu\text{A}/2 = 5\ \mu\text{A}$.

(a) The output resistance of M_2 is $r_{o2} = r_{o4} = r_{o6} = r_{o8} = 2V_M/I_Q = 2 \times 20\ \text{V}/10\ \mu\text{A} = 4\ \text{M}\Omega$. The transconductance of M_2 is $g_{m2} = \sqrt{2K_nI_Q} = \sqrt{2 \times 25\ \mu \times 10\ \mu} = 22.36\ \mu\text{A/V}$. The load resistance due to M_2 and M_4 cascoded combination is

$$r'_{o4} = r'_{o6} = r_{o4}(2 + g_{m2}r_{o4}) = 4\ \text{M}\Omega \times (2 + 22.36\ \mu\text{A/V} \times 4\ \text{M}\Omega) = 366\ \text{M}\Omega.$$

The effective resistance of the active load is $R_o = r'_{o4} \| r'_{o6} = 366\ \text{M} \| 366\ \text{M} = 183\ \text{M}\Omega$.

Thus, the differential voltage becomes $A_d = g_{m2}R_o = 22.36\ \mu \times 183\ \text{M} = 4089$ V/V.

(b) The circuit for PSpice simulation is shown in Fig. 9.20 with a sinusoidal signal of 1 μV at 10 Hz. The plot of the output voltage is shown in Fig. 9.21, which gives a voltage gain of 5370 V/V (expected value 4089 V/V). Note there is no phase shift of 180° because the output is taken from the drain terminal of the other pair.

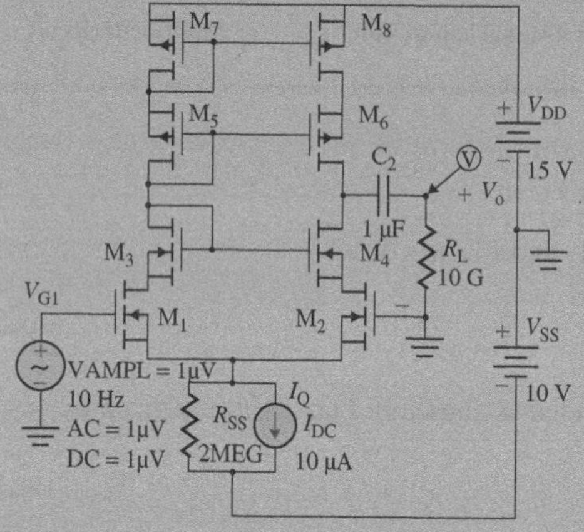

FIGURE 9.20 PSpice schematic for Example 9.6

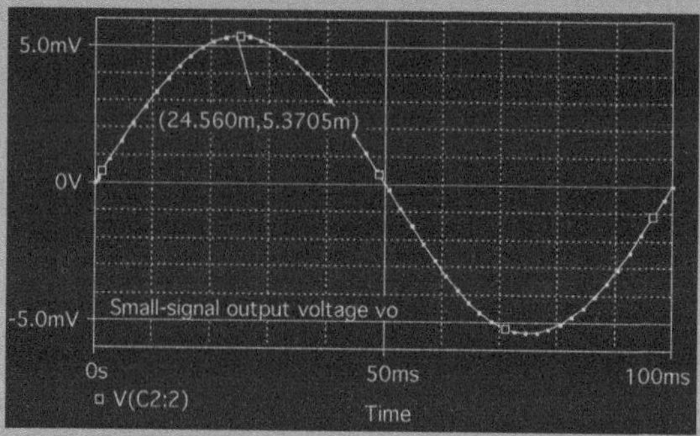

FIGURE 9.21 PSpice plot of small-signal output voltage for Example 9.6

KEY POINTS OF SECTION 9.4

- A MOS amplifier exhibits a linear DC characteristic and has a very high input resistance, tending to infinity.
- It is relatively easy to connect MOS transistors in cascode form in order to control the drain current and give high output resistance. A high-voltage gain can be obtained with a cascode connection.

9.5 Depletion MOS Differential Amplifiers

The depletion MOS amplifiers are similar to the enhancement MOSs, except that the gate–source voltage of a depletion MOS can range from a negative value to a positive value. The drain current becomes the maximum $i_{D(max)} = I_{DSS}$ when $v_{GS} = 0$.

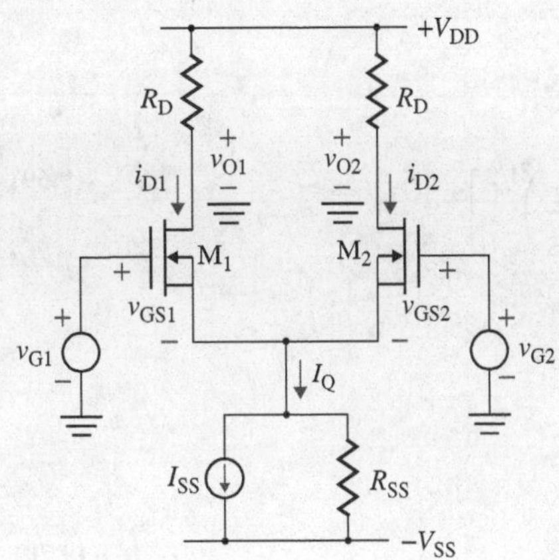

FIGURE 9.22 Depletion MOSFET differential pair

9.5.1 Depletion MOS Differential Pair with Resistive Load

A depletion NMOS source-coupled pair is shown in Fig. 9.22. Although the DC biasing circuit can be either a simple resistor (in which case the equivalent current generator will be zero) or a transistor current source, a current source is generally used.

DC Transfer Characteristics

The following analysis is performed for an *n*-channel MOS pair, but it is equally applicable to a *p*-channel pair with appropriate sign changes. The analysis can be simplified by making the following assumptions:

1. The output resistances of the MOSFETs are infinite: $r_d = \infty$.
2. The MOSFETs are identical and operate in the saturation region. The pinch-down voltages are the same, $V_{p1} = V_{p2} = V_p$, and the drain currents (with drain $V_{GS} = 0$) are equal, $I_{DSS1} = I_{DSS2} = I_{DSS}$.
3. The output resistance of the transistor current source is infinite: $R_{SS} = \infty$.

Using KVL around the loop formed by the two input voltages and two gate–source junctions, we get

$$v_{G1} - v_{GS1} + v_{GS2} - v_{G2} = 0 \tag{9.67}$$

Assuming that the drain current is related to v_{GS} by the approximate square law relationship in Eq. (7.21), we get

$$i_D = I_{DSS}\left(1 - \frac{v_{GS}}{V_p}\right)^2$$

from which we can find

$$\frac{v_{GS}}{V_p} = 1 - \left(\frac{i_D}{I_{DSS}}\right)^{1/2} \tag{9.68}$$

Substituting Eq. (9.68) into Eq. (9.67) yields

$$\begin{aligned} -v_{G1} + v_{G2} &= -v_{GS1} + v_{GS2} \\ &= -V_p\left[1 - \left(\frac{i_{D1}}{I_{DSS}}\right)^{1/2}\right] + V_p\left[1 - \left(\frac{i_{D2}}{I_{DSS}}\right)^{1/2}\right] \\ &= V_p\left(\frac{i_{D1}}{I_{DSS}}\right)^{1/2} - V_p\left(\frac{i_{D2}}{I_{DSS}}\right)^{1/2} \end{aligned} \tag{9.69}$$

which gives

$$-\frac{v_{G1} + v_{G2}}{V_p} = \frac{v_{id}}{V_p} = \left(\frac{i_{D1}}{I_{DSS}}\right)^{1/2} - \left(\frac{i_{D2}}{I_{DSS}}\right)^{1/2} \tag{9.70}$$

where $v_{id} = v_{G2} - v_{G1}$ is the differential DC voltage.

Using KCL at the source nodes of the transistors gives

$$I_Q = i_{D1} + i_{D2} \tag{9.71}$$

Substituting Eq. (9.71) into Eq. (9.70) and solving the resultant quadratic, we get

$$i_{D1} = \frac{I_Q}{2} + \frac{I_Q}{2}\left(\frac{v_{id}}{V_p}\right)\left[2\left(\frac{I_{DSS}}{I_Q}\right) - \left(\frac{v_{id}}{V_p}\right)^2\left(\frac{I_{DSS}}{I_Q}\right)^2\right]^{1/2} \tag{9.72}$$

$$i_{D2} = \frac{I_Q}{2} - \frac{I_Q}{2}\left(\frac{v_{id}}{V_p}\right)\left[2\left(\frac{I_{DSS}}{I_Q}\right) - \left(\frac{v_{id}}{V_p}\right)^2\left(\frac{I_{DSS}}{I_Q}\right)^2\right]^{1/2} \tag{9.73}$$

If v_{id} is sufficiently large, all of the biasing current I_Q must flow through only one of the MOSFETs. The range of v_{id} for which both transistors conduct can be found from Eq. (9.72) with the condition $i_{D2} = 0$; that is,

$$\frac{I_Q}{2} - \frac{I_Q}{2}\left(\frac{v_{id}}{V_p}\right)\left[2\left(\frac{I_{DSS}}{I_Q}\right) - \left(\frac{v_{id}}{V_p}\right)^2\left(\frac{I_{DSS}}{I_Q}\right)^2\right]^{1/2} = 0$$

which gives

$$\left|\frac{v_{id}}{V_p}\right| \le \sqrt{\frac{I_Q}{I_{DSS}}} \tag{9.74}$$

This equation gives the value of v_{id} for which the current I_Q is carried by one of the two transistors. Typical drain currents for various values of I_Q are shown in Fig. 9.23. The range of v_{id} for which the circuit exhibits a linear characteristic is approximately equal to the pinch-down voltage $|V_p|$ (typically 2 V to 5 V), compared to V_T (26 mV) for BJTs.

The output voltages for a MOS differential pair are

$$v_{O1} = V_{DD} - i_{D1}R_D \tag{9.75}$$

$$v_{O2} = V_{DD} - i_{D2}R_D \tag{9.76}$$

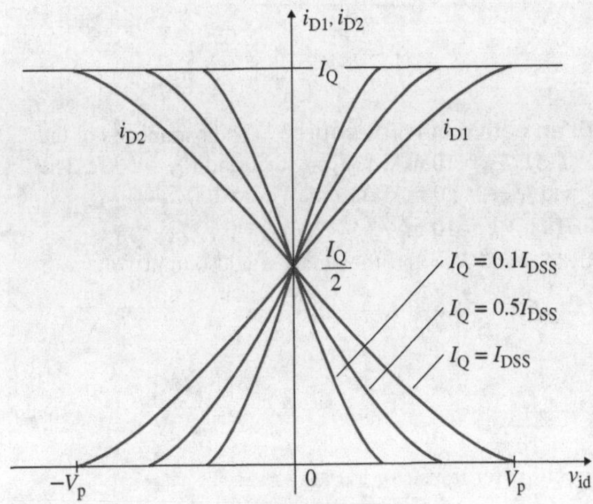

FIGURE 9.23 DC transfer characteristic of depletion NMOS pairs

The differential DC output voltage is

$$v_{od} = v_{O1} - v_{O2} = R_D(i_{D2} - i_{D1})$$

Substituting i_{D1} and i_{D2} from Eqs. (9.72) and (9.73) into the above equation and simplifying, we get

$$v_{od} = -\frac{I_Q R_D}{V_p} v_{id}\left[2\left(\frac{I_{DSS}}{I_Q}\right) - \left(\frac{v_{id}}{V_p}\right)^2\left(\frac{I_{DSS}}{I_Q}\right)^2\right]^{1/2} \tag{9.77}$$

As with the any differential circuit, if v_{id} is zero, v_{od} is also zero. A MOS-coupled pair allows direct coupling of cascoded stages without introducting DC offsets.

Small-Signal Analysis

From Eq. (7.78), the transconductance is given by

$$g_m = \left|-\frac{2I_{DSS}}{V_p}\left(1 - \frac{V_{GS}}{V_p}\right)\right|$$

$$= \frac{2}{|V_p|}[|I_D I_{DSS}|]^{1/2} \tag{9.78}$$

The small-signal analyses for differential and common-mode gains are the same as the enhancement MOS pair. From Eqs. (9.55) and (9.57), we get

$$A_d = -g_m(R_D \parallel r_{o1}) \tag{9.79}$$

$$A_c = \frac{-g_m R_D}{1 + g_m 2R_{SS}} \tag{9.80}$$

▶ **NOTE** To simplify the analysis, the effect of transistor output resistance r_o is not included in Eq. (9.80).

EXAMPLE 9.7

Analyzing a depletion MOS differential pair with an active current source The parameters of the depletion MOS differential pair in Fig. 9.22 are $R_{SS} = 50\text{ k}\Omega$, $I_Q = 10\text{ mA}$, $V_{DD} = 30\text{ V}$, and $R_D = 5\text{ k}\Omega$. The depletion MOSFETs are identical and have $V_p = -4\text{ V}$ and $I_{DSS} = 20\text{ mA}$. Assume $V_M = 100\text{ V}$.

(a) Calculate the DC drain currents through the MOSFETs if $v_{id} = 10\text{ mV}$.

(b) Assuming $I_{D1} = I_{D2}$, calculate A_d, A_c, and CMRR; R_{id} and R_{ic}; and the small-signal output voltage if $v_{g1} = 10\text{ mV}$ and $v_{g2} = 20\text{ mV}$.

(c) Find the drain voltage V_D.

SOLUTION

(a) For $v_{id} = 10\text{ mV}$, Eq. (9.72) gives the DC drain current i_{D1} for transistor M_1 as

$$i_{D1} = \frac{10\text{ m}}{2}\left\{1 + \frac{100\text{ m}}{-4}\left[2\left(\frac{20\text{ m}}{10\text{ m}}\right) - \left(\frac{100\text{ m}}{-4}\right)^2\left(\frac{20\text{ m}}{10\text{ m}}\right)^2\right]^{1/2}\right\} = 5.25\text{ mA}$$

$$i_{D2} = I_Q - i_{D1} = 10\text{ mA} - 5.25\text{ mA} = 4.75\text{ mA}$$

(b) We know that $I_{D1} = I_{D2} = I_Q/2 = 10\text{ mA}/2 = 5\text{ mA}$. From Eq. (9.78),

$$g_m = \frac{2[|I_{D1}I_{DSS}|]^{1/2}}{|V_p|} = \left(\frac{2}{4}\right) \times 2\ \overline{5\text{ mA} \times 20\text{ mA}} = 5\text{ mA/V}$$

$$r_{o1} = \frac{V_M}{I_D} = \frac{100}{5\text{ m}} = 20\text{ k}\Omega$$

From Eq. (9.79), the single-ended differential voltage gain A_d is

$$A_d = -g_m(R_D \| r_{o1}) = -5\text{ m} \times (5\text{ k} \| 20\text{ k}) = -20\text{ V/V}$$

From Eq. (9.80), the single-ended common-mode voltage gain A_c is

$$A_c = \frac{-g_m R_D}{1 + g_m 2R_{SS}} = \frac{-5\text{ m} \times 5\text{ k}}{1 + (5\text{ m} \times 2 \times 50\text{ k})} = -0.0399\text{ V/V}$$

Thus, CMRR $= |A_d/A_c| = 20/0.0399 = 501$ (or 54 dB). From Eq. (9.59),

$$R_{id} = R_{ic} = \infty$$

We know that

$$v_{id} = v_{g2} - v_{g1} = 20\text{ mV} - 10\text{ mV} = 10\text{ mV}$$

and $$v_{ic} = \frac{v_{g1} + v_{g2}}{2} = \frac{10\text{ mV} + 20\text{ mV}}{2} = 15\text{ mV}$$

Using Eq. (9.10), we have

$$v_o = A_d v_{id} + A_c v_{ic} = -20 \times 10\text{ mV} - 0.0399 \times 15\text{ mV} = -250.6\text{ mV}$$

(c) The DC drain voltage at the drain terminal of a transistor is

$$V_D = V_{DD} - I_D R_D = 30\text{ V} - 5\text{ mA} \times 5\text{ k}\Omega = 5\text{ V}$$

Thus, for $A_d = -20$, the maximum differential voltage will be $v_{id} = 5/20 = 250\text{ mV}$. Therefore, V_{DD} must be greater than $I_D R_D$ in order to allow output voltage swing due to the input voltages.

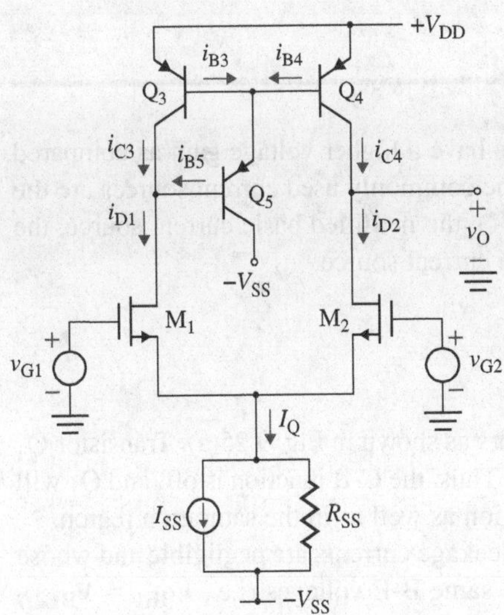

FIGURE 9.24 Depletion MOS differential amplifier with current mirror active load

9.5.2 Depletion MOS Differential Pair with Active Load

Like MOS differential amplifiers, depletion MOS differential amplifiers use a current mirror active load to achieve a large voltage gain. The active load can be made either with MOSFETs similar to that in Fig. 9.24 or with BJTs. A depletion MOS differential amplifier with a basic current source as the BJT active load is shown in Fig. 9.24. From Eq. (9.60), we know that the output resistance R_o is the parallel combination of r_{o2} and r_{o4}; that is,

$$R_o = r_{o2} \| r_{o4} \tag{9.81}$$

where $r_{o2} = 2V_M/I_Q$ = output resistance of transistor M_2 and $r_{o4} = 2V_A/I_Q$ = output resistance of transistor Q_4.

Using Eq. (9.61), we find the differential voltage gain A_d:

$$A_d = \frac{v_o}{v_{id}} = -g_m(r_{o2} \| r_{o4}) \tag{9.82}$$

where g_m, which is the transconductance of MOSFET M_2, is given by Eq. (9.78).

KEY POINTS OF SECTION 9.5

- A MOSFET amplifier has a very high input resistance, in the range of $10^9\ \Omega$ to $10^{12}\ \Omega$.
- An active load considerably increases the differential gain of a MOSFET amplifier.
- The range of v_{id} for which the circuit exhibits a linear DC characteristic is much higher for MOSFETs than for BJTs. For MOSFETs, this range is approximately equal to the pinch-down voltage $|V_p|$ (typically 2 V to 5 V), compared to V_T (26 mV) for BJTs.

9.6 BJT Current Sources

The BJT current sources are similar to MOS sources. Since BJTs have a higher voltage gain as compared to MOSFETs, BJT sources can offer higher output resistances. The commonly used current sources are the basic current sources (which were applied in Secs. 7.8.1 and 8.7.1), the modified basic current source, the Widlar current source, the cascode current source, and the Wilson current source.

9.6.1 Basic Current Source

The simplest current source consists of a resistor and two transistors as shown in Fig. 9.25(a). Transistor Q_1 is diode connected, and its C-B voltage is forced to zero: $v_{CB} = 0$. Thus, the C-B junction is off, and Q_1 will operate in the active region. Transistor Q_2 can be in the active region as well as in the saturation region.

Let us assume that Q_1 and Q_2 are identical transistors whose leakage currents are negligible and whose output resistances are infinite. Since the two transistors have the same B-E voltages (i.e., $V_{BE1} = V_{BE2}$), the collector and base currents are equal: $I_{C1} = I_{C2}$ and $I_{B1} = I_{B2}$. Applying KCL at the collector of Q_1 gives the reference current:

$$I_R = I_{C1} + I_{B1} + I_{B2} = I_{C1} + 2I_{B1}$$

Since $I_{C1} = \beta_F I_{B1}$,

$$I_R = I_{C1} + 2I_{B1} = I_{C1} + \frac{2I_{C1}}{\beta_F}$$

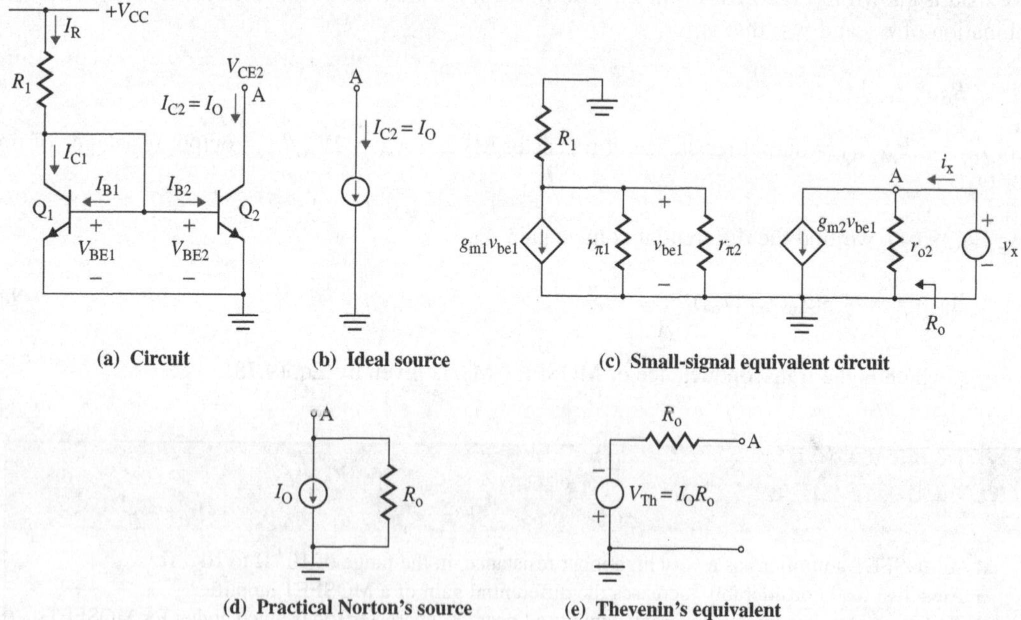

FIGURE 9.25 Basic current source

which gives the collector current I_{C1} as

$$I_{C1} = I_{C2} = \frac{I_R}{1 + 2/\beta_F} \tag{9.83}$$

$$= \frac{V_{CC} - V_{BE1}}{R_1} \times \frac{1}{1 + 2/\beta_F} \tag{9.84}$$

If the DC current gain $\beta_F >> 2$, which is usually the case, Eq. (9.84) is reduced to

$$I_{C1} = I_{C2} \approx I_R$$

Thus, for two identical transistors, the reference and output currents are almost equal. I_{C2}, which is the mirror image of I_{C1}, is known as the *mirror current* of I_{C1}. For transistors with small values of β_F, the current ratio will not be unity. In practice, however, the transistors may not be identical, and the two collector currents will have a constant ratio. The equivalent circuit for the ideal current source is shown in Fig. 9.25(b).

For a transistor with finite output resistance, the effect of the Early voltage V_A should be taken into account, and the collector current in Eq. (8.8) can be modified to

$$I_C = I_S\left[\exp\left(\frac{V_{BE}}{V_T}\right) - 1\right]\left(1 + \frac{V_{CE}}{V_A}\right) \tag{9.85}$$

If we take into account the variation in the collector current due to the C-E voltage, the ratio of the two collector currents can be found from

$$\frac{I_{C2}}{I_{C1}} = \frac{1 + V_{CE2}/V_A}{1 + V_{CE1}/V_A} \tag{9.86}$$

Output Resistance R_o

The small-signal AC equivalent circuit for determining the output resistance is shown in Fig. 9.25(c). Output resistance R_o is the same as r_{o2}; that is,

$$R_o = \frac{v_x}{i_x} = r_{o2} = \frac{V_A}{I_{C2}} \tag{9.87}$$

Thevenin's equivalent voltage is given by

$$V_{Th} = I_O R_o = I_{C2} R_o = I_{C2}\frac{V_A}{I_{C2}} = V_A \tag{9.88}$$

Norton's and Thevenin's equivalents of the current source are shown in Fig. 9.25[(d) and (e)]. If the output of the current source is open-circuited, a voltage of $-V_{Th}$ is expected to appear across transistor Q_1. However, this will not actually happen because transistor Q_1 will saturate when the voltage across the current source (i.e., the C-E voltage of Q_1) reaches zero.

The current source in Fig. 9.25(a) behaves as a current sink rather than a source. A current source equivalent to this current sink can be obtained by using *pnp* transistors and a negative power supply. This arrangement is shown in Fig. 9.26.

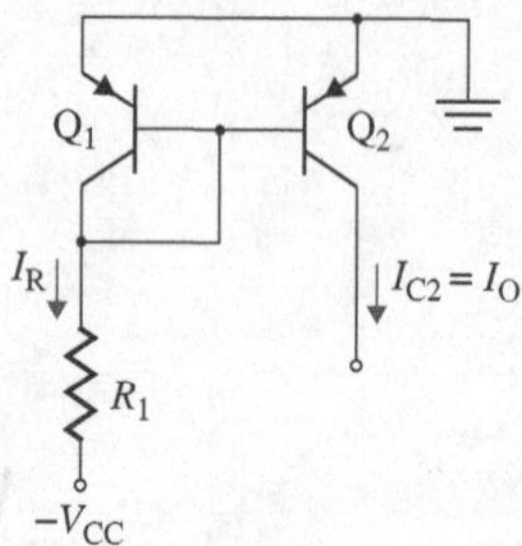

FIGURE 9.26 Current source using *pnp* transistors

EXAMPLE 9.8

D

Designing a simple basic current source

(a) Design the basic current source in Fig. 9.25(a) to give an output current of $I_O = 5\ \mu A$. The transistor parameters are $\beta_F = 100$, $V_{CC} = 30$ V, $V_{BE1} = V_{BE2} = V_{CE1} = 0.7$ V, and $V_A = 150$ V.

(b) Calculate the output resistance R_o, Thevenin's equivalent voltage V_{Th}, and the collector current ratio if $V_{CE2} = 20$ V.

SOLUTION

$V_{BE1} = V_{BE2} = V_{CE1} = 0.7$ V, and $V_A = 150$ V.

(a) From Eq. (9.83),

$$I_O = I_{C2} = I_{C1} = \frac{I_R}{1 + 2/\beta_F}$$

which, for $I_O = 5\ \mu A$, gives $I_R = (5\ \mu A)(1 + 2/\beta_F) = 5.1\ \mu A$. From Eq. (9.84),

$$R_1 = \frac{V_{CC} - V_{BE1}}{I_R} = \frac{30\text{ V} - 0.7\text{ V}}{5.1\ \mu A} = 5.75\text{ M}\Omega$$

(b) From Eq. (9.87),

$$R_o = \frac{V_A}{I_{C2}} = \frac{150\text{ V}}{5\ \mu A} = 30\text{ M}\Omega$$

From Eq. (9.88),

$$V_{Th} = V_A = 150\text{ V}$$

From Eq. (9.86),

$$\frac{I_{C2}}{I_{C1}} = \frac{1 + V_{CE2}/V_A}{1 + V_{CE1}/V_A} = \frac{1 + 20/150}{1 + 0.7/150} = 1.128$$

Thus, $I_{C2} = 1.128 \times I_{C1} = 1.128 \times 5\ \mu A = 5.64\ \mu A$, which agrees, with a degree of error, with the desired value of $I_O = 5\ \mu A$.

▶ **NOTES**

1. For a current source of 5 μA, a resistor of 5.75 MΩ would be required. Resistors of such high value are very costly in terms of the die area. Resistors over 50 kΩ are generally avoided for IC applications. Thus, this current source is not suitable for generating a current of less than about 0.6 mA at V_{CC} = 30 V and 0.3 mA at V_{CC} = 15 V.
2. If the output of the current source is open-circuited, a voltage of $-V_{Th} = -150$ V will not appear across transistor Q_1. Rather, the voltage will be $V_{CE1(sat)} \approx 0.2$ V.

9.6.2 Modified Basic Current Source

Notice from Eq. (9.83) that the collector current I_{C2} ($=I_{C1}$) differs from the reference current I_R by a factor of $(1 + 2/\beta_F)$. For low-gain transistors (especially *pnp* types), I_{C2} can differ significantly from I_R. The error can be reduced by adding another transistor so that I_{C2} becomes less dependent on the transistor parameter β_F. This type of circuit is shown in Fig. 9.27(a). Applying KCL at the emitter of transistor Q_3 gives

$$I_{E3} = I_{B1} + I_{B2} = \frac{I_{C1}}{\beta_F} + \frac{I_{C2}}{\beta_F} \tag{9.89}$$

Since $V_{BE1} = V_{BE2}$, it follows that $I_{C1} = I_{C2}$. Thus, Eq. (9.89) becomes

$$I_{E3} = \frac{2}{\beta_F} I_{C2}$$

The base current of Q_3 is related to I_{E3} by

$$I_{B3} = \frac{I_{E3}}{1 + \beta_F} = \frac{2}{\beta_F(1 + \beta_F)} I_{C2} \tag{9.90}$$

Using KCL at the collector of Q_1 gives

$$I_R = I_{C1} + I_{B3} = I_{C1} + \frac{2}{\beta_F(1 + \beta_F)} I_{C2} \tag{9.91}$$

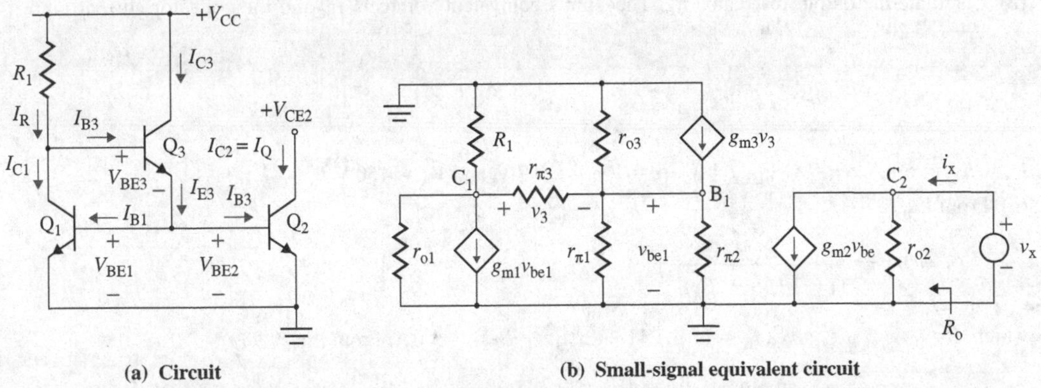

(a) Circuit

(b) Small-signal equivalent circuit

FIGURE 9.27 Modified basic current source

Since $I_{C1} = I_{C2}$, Eq. (9.91) gives the output current I_O as

$$I_O = I_{C2} = \frac{I_R}{1 + 2/(\beta_F^2 + \beta_F)} \tag{9.92}$$

which indicates that the reference current I_R is related to the output current I_O by a factor of only $[1 + 2/(\beta_F^2 + \beta_F)]$. The reference current can be found from

$$I_R = \frac{V_{CC} - V_{BE1} - V_{BE3}}{R_1} \tag{9.93}$$

In the derivation of Eq. (9.92), the output resistances of the transistors are ignored. However, for a finite transistor output resistance, the collector current ratio in Eq. (9.86) is applicable.

Output Resistance R_o

The small-signal equivalent circuit for determining the output resistance is shown in Fig. 9.27(b). The output resistance R_o is the same as r_{o2}; that is,

$$R_o = \frac{v_x}{i_x} = r_{o2} = \frac{V_A}{I_{C2}} \tag{9.94}$$

Thevenin's equivalent voltage is given by

$$V_{Th} = I_O R_o = I_{C2} R_o = I_{C2} \frac{V_A}{I_{C2}} = V_A \tag{9.95}$$

EXAMPLE 9.9

D

Designing a simple modified basic current source

(a) Design the modified basic current source in Fig. 9.27(a) to give an output current of $I_O = 5\ \mu A$. The transistor parameters are $\beta_F = 100$, $V_{CC} = 30$ V, $V_{BE1} = V_{BE2} = V_{BE3} = 0.7$ V, and $V_A = 150$ V.

(b) Calculate the output resistance R_o, Thevenin's equivalent voltage V_{Th}, and the collector current ratio if $V_{CE2} = 20$ V.

SOLUTION

$I_O = I_{C2} = 5\ \mu A$, $V_{BE1} = V_{BE2} = V_{BE3} = V_{CE1} = 0.7$ V, and $V_A = 150$ V.

(a) From Eq. (9.92),

$$I_O = I_{C2} = \frac{I_R}{1 + 2/(\beta_F^2 + \beta_F)}$$

which, for $I_O = 5\ \mu A$, gives $I_R = (5\ \mu A)[1 + 2/(\beta_F^2 + \beta_F)] = 5\ \mu A$. From Eq. (9.93),

$$R_1 = \frac{V_{CC} - V_{BE1} - V_{BE3}}{I_R} = \frac{30 - 0.7 - 0.7}{5\ \mu A}$$

which gives the required value of the resistor as $R_1 = 5.72\ \text{M}\Omega$.

(b) From Eq. (9.94),

$$R_o = \frac{V_A}{I_{C2}} = \frac{150}{5\ \mu\text{A}} = 30\ \text{M}\Omega$$

From Eq. (9.95),

$$V_{Th} = V_A = 150\ \text{V}$$

From Eq. (9.86),

$$\frac{I_{C2}}{I_{C1}} = \frac{1 + V_{CE2}/V_A}{1 + (V_{BE1} + V_{BE3})/V_A} = \frac{1 + 20/150}{1 + (0.7 + 0.7)/150} = 1.123$$

Neglecting I_{B3}, we have $I_{C1} \approx I_O = 5\ \mu\text{A}$. Thus, $I_{C2} = 1.123 \times I_{C1} = 1.123 \times 5\ \mu\text{A} = 5.62\ \mu\text{A}$, which agrees, with a degree of error, with the desired value of $I_O = 5\ \mu\text{A}$.

9.6.3 Widlar Current Source

Biasing currents of low magnitudes, typically on the order of 5 μA, are required in a variety of applications. Currents of low magnitude can be obtained by inserting a resistance of moderate value in series with the emitter Q_2 in Fig. 9.25(a). A circuit with this modification, as shown in Fig. 9.28(a), is known as a *Widlar current source*. As a result of the addition of R_2 into the circuit, I_{C2} is no longer equal to I_R, and the value of I_{C2} can be made much smaller than that of I_{C1}. This circuit can give currents in the microampere range, with acceptable circuit resistance values of less than 50 kΩ.

Using KVL around the B-E loop in Fig. 9.28(a) gives

$$V_{BE1} - V_{BE2} - (I_{C2} + I_{B2})R_2 = 0$$

Since $I_{C2} >> I_{B2}$,

$$V_{BE1} - V_{BE2} - I_{C2}R_2\left(1 + \frac{1}{\beta_F}\right) = 0 \tag{9.96}$$

Assuming that the transistors have infinite output resistances, $V_A = \infty$, Eq. (9.85) becomes

$$I_C = I_S \exp\left(\frac{V_{BE}}{V_T}\right)$$

which gives

$$V_{BE} = V_T \ln\left(\frac{I_C}{I_S}\right) \tag{9.97}$$

Applying V_{BE} from Eq. (9.97) in Eq. (9.96), we get

$$V_T \ln\left(\frac{I_{C1}}{I_{S1}}\right) - V_T \ln\left(\frac{I_{C2}}{I_{S2}}\right) - I_{C2}R_2 = 0 \quad (\text{for } \beta_F >> 1)$$

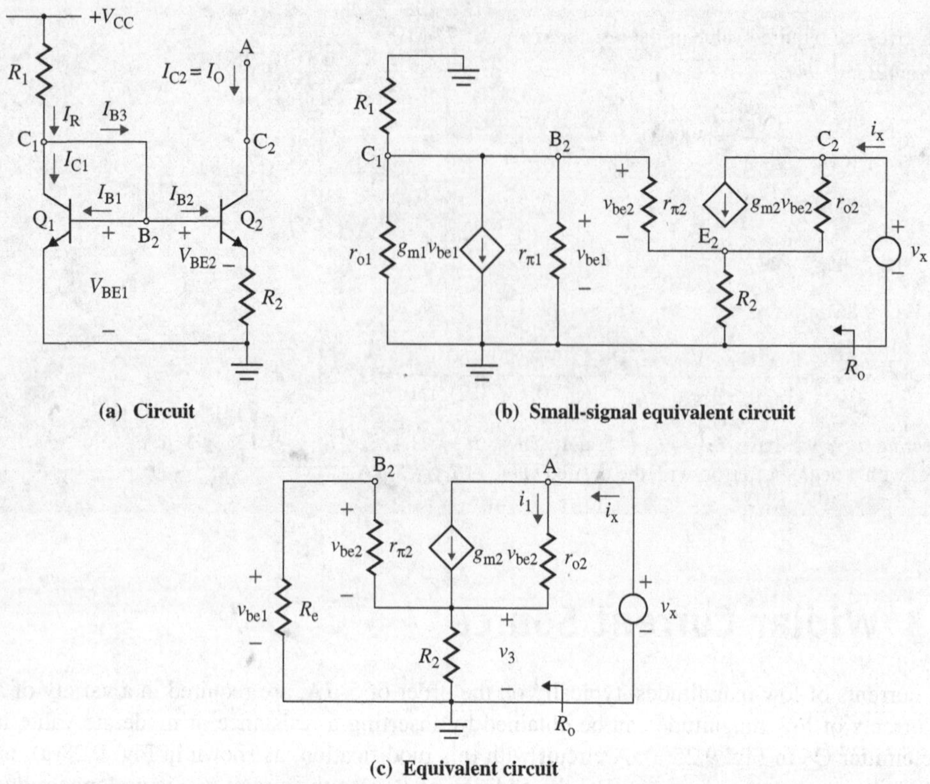

(a) Circuit

(b) Small-signal equivalent circuit

(c) Equivalent circuit

FIGURE 9.28 Widlar current source

which, for identical transistors with $I_{S1} = I_{S2}$, can be simplified to

$$V_T \ln\left(\frac{I_{C1}}{I_{S1}} \times \frac{I_{S2}}{I_{C2}}\right) - I_{C2}R_2 = 0$$

This equation gives the relation of I_{C2} to I_{C1} and R_2 as

$$V_T \ln\left(\frac{I_{C1}}{I_{C2}}\right) = I_{C2}R_2 \tag{9.98}$$

The reference current I_R can be found from

$$I_R = \frac{V_{CC} - V_{BE1}}{R_1} \tag{9.99}$$

which is also related to the base and collector currents; that is,

$$I_R = I_{C1} + I_{B1} + I_{B2}$$

$$= I_{C1}\left(1 + \frac{1}{\beta_F}\right) + \frac{I_{C2}}{\beta_F} \tag{9.100}$$

Substituting I_{C1} from Eq. (9.98) into the preceding equation, we get I_R in terms of I_{C2} and R_2 as

$$I_R = \frac{1+\beta_F}{\beta_F} I_{C2} \exp\left(\frac{I_{C2}R_2}{V_T}\right) + \frac{I_{C2}}{\beta_F} \tag{9.101}$$

Thus, I_{C2} is a nonlinear function of I_R and R_2. To find I_{C2}, we must solve this transcendental equation by trial and error, using known values of I_R and R_2. However, for design purposes, I_{C2} and I_R are known, and it is necessary only to find the value of R_2.

Output Resistance R_o

The small-signal equivalent circuit for determining the output resistance is shown in Fig. 9.28(b). This circuit can be reduced to Fig. 9.28(c), where R_e is the parallel equivalent of $r_{\pi 1}$, $1/g_{m1}$, r_{o1}, and R_1; that is,

$$R_e = r_{\pi 1} \left\| \frac{1}{g_{m1}} \right\| R_1 \| r_{o1} \tag{9.102}$$

Using Eq. (8.42), we can relate the input resistance $r_{\pi 1}$ to the transconductance g_{m1} and the small-signal current gain $\beta_f (\approx \beta_F)$:

$$r_{\pi 1} = \frac{\beta_f(\approx \beta_F)}{g_{m1}}$$

Since $\beta_f >> 1$, $r_{\pi 1} >> 1/g_{m1}$, and in general $R_1 >> 1/g_{m1}$, Eq. (9.102) can be approximated by

$$R_e \approx \frac{1}{g_{m1}} \tag{9.103}$$

The collector current flows through the parallel combination of R_2 and $(r_{\pi 2} + R_e)$, so

$$v_3 = i_x[R_2 \| (r_{\pi 2} + R_e)] \tag{9.104}$$

The voltage v_3 is related to v_2 by

$$v_2 = -\frac{v_3 r_{\pi 2}}{r_{\pi 2} + R_e} = -\frac{i_x r_{\pi 2}[R_2 \| (r_{\pi 2} + R_e)]}{r_{\pi 2} + R_e} \tag{9.105}$$

The current i_1 thus becomes

$$i_1 = i_x - g_{m2} v_2$$

The test voltage v_x is

$$\begin{aligned} v_x &= v_3 + r_{o2} i_1 = v_3 + r_{o2} i_x - r_{o2} g_{m2} v_2 \\ &= v_3 + i_x r_{o2} + \frac{r_{o2} g_{m2} v_3 r_{\pi 2}}{r_{\pi 2} + R_e} \end{aligned} \tag{9.106}$$

Substituting v_3 from Eq. (9.104) into Eq. (9.106) and simplifying, we get the resistance at the collector of the transistor:

$$R_o = \frac{v_x}{i_x} = R_2 \| (r_{\pi 2} + R_e) + r_{o2}\left[1 + \frac{g_{m2} r_{\pi 2}}{r_{\pi 2} + R_e} R_2 \| (r_{\pi 2} + R_e)\right] \tag{9.107}$$

Since r_{o2} is large, the first term is much smaller than the second one. If the first term is neglected, Eq. (9.107) can be reduced to

$$R_o \approx r_{o2}\left[1 + \frac{g_{m2}r_{\pi 2}}{r_{\pi 2} + R_e} R_2 \parallel (r_{\pi 2} + R_e)\right] \tag{9.108}$$

From Eq. (8.42), we get

$$\frac{1}{g_{m1}} = \frac{1}{I_{C1}/V_T} = \frac{V_T}{I_{C1}} \tag{9.109}$$

$$r_{\pi 2} = \frac{\beta_F}{g_{m2}} = \frac{\beta_F}{I_{C2}/V_T} = V_T\frac{\beta_F}{I_{C2}} = \beta_F\frac{V_T}{I_{C1}} \times \frac{I_{C1}}{I_{C2}} = \frac{\beta_F}{g_{m1}} \times \frac{I_{C1}}{I_{C2}} \tag{9.110}$$

Since $I_{C1} >> I_{C2}$, $r_{\pi 2} >> 1/g_{m1}$ and $r_{\pi 2} >> R_e$; that is,

$$R_e + r_{\pi 2} \approx r_{\pi 2} \tag{9.111}$$

Substituting R_e from Eq. (9.103) into Eq. (9.107) gives

$$R_o \approx r_{o2}[1 + g_{m2}(R_2 \parallel r_{\pi 2})] \tag{9.112}$$

Since $\beta_F >> 1$ and $r_{\pi 2} = \beta_F/g_{m2}$, Eq. (9.112) can be rewritten as

$$R_o = r_{o2}\frac{g_{m2}R_2(1 + \beta_F) + \beta_F}{g_{m2}R_2 + \beta_F} = r_{o2}\left[\frac{1 + g_{m2}R_2}{1 + g_{m2}R_2/\beta_F}\right] \tag{9.113}$$

Generally, $\beta_F >> g_{m2}R_2$, so Eq. (9.113) can be approximated by

$$R_o \approx r_{o2}(1 + g_{m2}R_2) \tag{9.114}$$

$$\approx r_{o2}\left(1 + \frac{I_{C2}R_2}{V_T}\right) \tag{9.115}$$

Thus, R_o depends on $I_{C2}R_2$, which is the DC voltage drop across R_2. The larger this drop is made, the higher the output resistance becomes. For the Widlar source, $I_{C2}R_2$ is limited to several hundred millivolts for practical current ratios, and the corresponding value of V_{Th} is limited to about $10V_A$.

EXAMPLE 9.10

D

Designing a Widlar current source

(a) Design the Widlar current source in Fig. 9.28(a) to give $I_O = 5\ \mu A$ and $I_R = 1$ mA. The parameters are $V_{CC} = 30$ V, $V_{BE1} = 0.7$ V, $V_T = 26$ mV, $V_A = 150$ V, and $\beta_F = 100$.

(b) Calculate the output resistance R_o and Thevenin's voltage V_{Th}. Assume $V_T = 26$ mV.

SOLUTION

$I_{C2} = 5\ \mu\text{A}$, and $V_T = 26$ mV.

(a) From Eq. (9.99),

$$1\text{ mA} = \frac{30 - 0.7}{R_1}$$

so $R_1 = 29.3\ \text{k}\Omega$. From Eq. (9.100),

$$1\text{ mA} = I_{C1}\left(1 + \frac{1}{100}\right) + \frac{5\ \mu\text{A}}{100}$$

which gives $I_{C1} \approx 990\ \mu\text{A}$. From Eq. (9.98),

$$26\text{ mV} \times \ln\left(\frac{990\ \mu\text{A}}{5\ \mu\text{A}}\right) = 5\ \mu\text{A} \times R_2$$

so $R_2 = 27.5\ \text{k}\Omega$.

(b) $r_{o2} = V_A/I_{C2} = 150/(5\ \mu\text{A}) = 30\ \text{M}\Omega$. From Eq. (9.110),

$$r_{\pi 2} = \frac{V_T \beta_F}{I_{C2}} = 26\text{ m} \times \frac{100}{5\ \mu} = 520\ \text{k}\Omega$$

Also from Eq. (9.109),

$$g_{m2} = \frac{I_{C2}}{V_T} = \frac{5\ \mu\text{A}}{26\text{ mV}} = 192.3\ \mu\text{A/V}$$

From Eq. (9.112),

$$R_o \approx 30\ \text{M}\Omega \times [1 + 192.3\ \mu\text{A/V} \times (27.5\ \text{k}\Omega \parallel 520\ \text{k}\Omega)] = 180.68\ \text{M}\Omega$$

Using the approximation in Eq. (9.115), we have

$$R_o \approx 30\ \text{M}\Omega \times \left(1 + \frac{5\ \mu\text{A} \times 27.5\ \text{k}\Omega}{26\text{ mV}}\right) = 188.66\ \text{M}\Omega$$

and $V_{Th} = R_o I_{C2} = 188.65\ \text{M}\Omega \times 5\ \mu\text{A} = 943.3$ V

NOTE: The Widlar current source gives a low output current at high output resistance, and Thevenin's equivalent voltage is very high.

9.6.4 Cascode Current Source

The emitter resistance R_2 in the Widlar current source in Fig. 9.28(a) can be replaced by a basic current source consisting of two transistors Q_3 and Q_4. This arrangement, shown in Fig. 9.29, will give a larger output resistance. In a cascodelike connection, two or more transistors are connected in series so that their collector biasing currents are almost identical (e.g., transistors Q_1 and Q_3), whereas in a cascodelike connection, the transistors operate in parallel fashion so that one transistor drives the other (e.g., Q_3 and Q_4). According to Eq. (9.55), the larger the output resistance, the greater the voltage gain of an amplifier

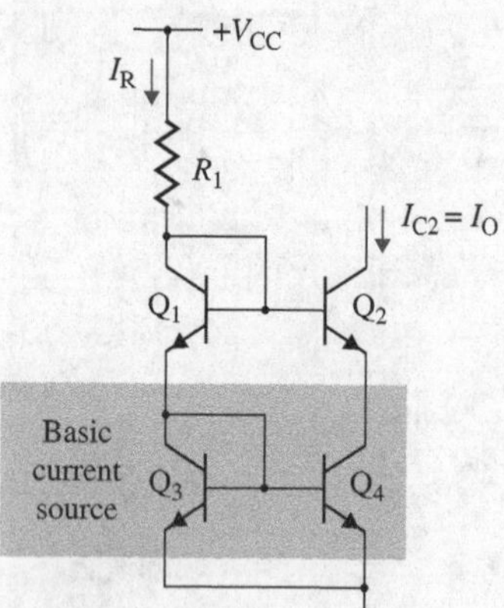

FIGURE 9.29 Cascode current source

becomes. Substituting the output resistance r_{o4} of transistor Q_4 for R_2 in Eq. (9.112) gives the output resistance of this cascode source:

$$R_o \approx r_{o2}[1 + g_{m2}(r_{o4} \| r_{\pi 2})] = r_{o2}(1 + g_{m2}r_{\pi 2}) = r_{o2}(1 + \beta_F) \tag{9.116}$$

9.6.5 Wilson Current Source

A Wilson current source, shown in Fig. 9.30(a), also gives a high output resistance. However, the output current is approximately equal to the reference current. Base current I_{B2}, which is the difference between the reference current I_R and the collector current I_{C1}, is multiplied by $(1 + \beta_F)$ to give I_{E2}, which flows in the diode-connected transistor Q_3 and causes a collector current of the same magnitude to flow in Q_1. There is a feedback path that regulates I_{C1}, which is approximately equal to I_{E2} and I_{C2}. Thus, I_{C2} remains very nearly equal to I_{C1} and nearly constant, giving a high output resistance:

$$I_{E2} = (1 + \beta_F)I_{B2} = (I_R - I_{C1})(1 + \beta_F)$$

Assuming that the transistors have infinite output resistances, $V_A = \infty$, and that the transistors are identical so that $I_{C1} = I_{C3}$, we can write

$$I_{E2} = I_{C3} + I_{B3} + I_{B1} = I_{C3}\left(1 + \frac{1}{\beta_F}\right) + \frac{I_{C1}}{\beta_F}$$
$$= I_{C3}\left(1 + \frac{2}{\beta_F}\right) \tag{9.117}$$

Using Eq. (9.117), we can relate I_{C2} to I_{E2} and I_{C3}:

$$I_{C2} = I_{E2}\frac{\beta_F}{1 + \beta_F} = I_{C3}\left(1 + \frac{2}{\beta_F}\right)\frac{\beta_F}{1 + \beta_F} = I_{C3}\frac{2 + \beta_F}{1 + \beta_F}$$

which gives

$$I_{C3} = I_{C2}\frac{1 + \beta_F}{2 + \beta_F} \tag{9.118}$$

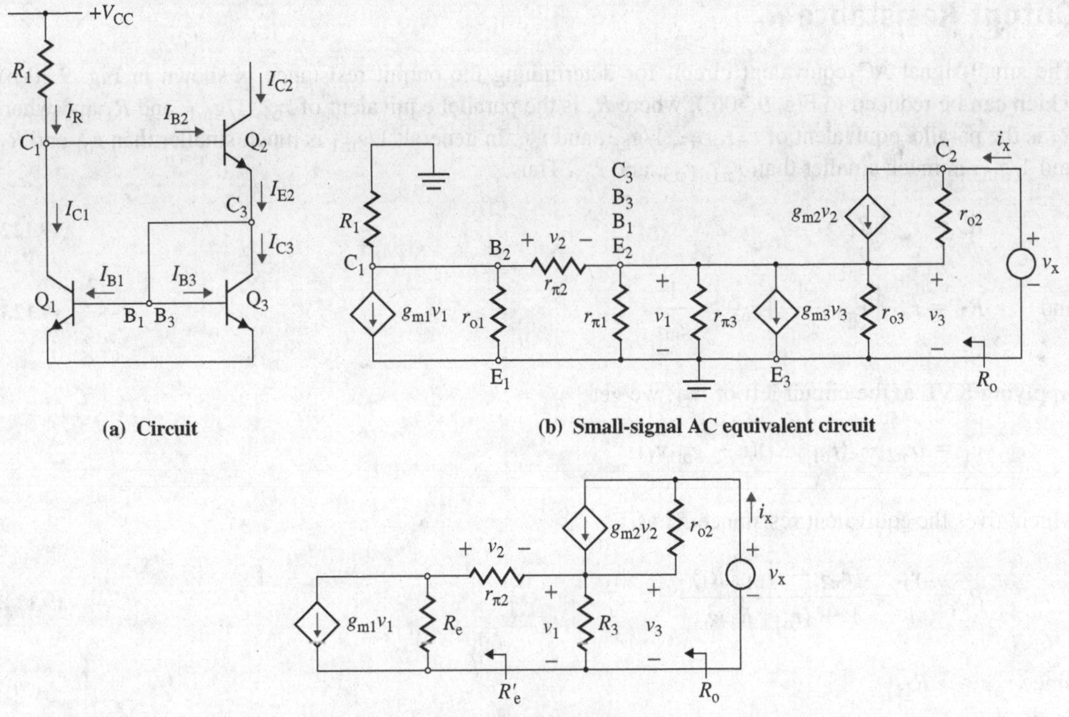

FIGURE 9.30 Wilson current source

I_{C1} is related to the reference current I_R by

$$I_{C1} = I_R - I_{B2} = I_R - \frac{I_{C2}}{\beta_F} \tag{9.119}$$

Since $I_{C3} = I_{C1}$, equating Eq. (9.118) to Eq. (9.119) yields

$$I_{C1} = I_{C3} = I_{C2}\frac{1 + \beta_F}{2 + \beta_F} = I_R - \frac{I_{C2}}{\beta_F}$$

which gives the output current $I_O = I_{C2}$ as

$$I_O = I_{C2} = I_R\left[\frac{(2 + \beta_F)\beta_F}{\beta_F^2 + 2\beta_F + 2}\right] = I_R\left[1 - \frac{2}{\beta_F^2 + 2\beta_F + 2}\right] \tag{9.120}$$

Since $\beta_F >> 1$, Eq. (9.120) can be approximated by

$$I_O \approx I_R \tag{9.121}$$

Thus, the output current almost equals the reference current and is less sensitive to the current gain β_F, which varies in response to temperature changes.

Output Resistance R_o

The small-signal AC equivalent circuit for determining the output resistance is shown in Fig. 9.30(b), which can be reduced to Fig. 9.30(c), where R_e is the parallel equivalent of r_{o1}, $1/g_{m1}$, and R_1 and where R_3 is the parallel equivalent of $r_{\pi 1}$, $r_{\pi 3}$, $1/g_{m3}$, and r_{o3}. In general, $1/g_{m1}$ is much smaller than r_{o1} and R_1, and $1/g_{m3}$ is much smaller than $r_{\pi 1}$, $r_{\pi 3}$, and r_{o3}. Thus,

$$R_e = r_{o1} \| R_1 \tag{9.122}$$

and $$R_3 = r_{\pi 1} \| r_{\pi 3} \| \frac{1}{g_{m3}} \| r_{o3} \approx \frac{1}{g_{m3}} \tag{9.123}$$

Applying KVL to the circuit left of $r_{\pi 1}$, we get

$$v_1 = i r_{\pi 2} + (r_{o1} \| R_1)(i - g_{m1} v_1)$$

which gives the equivalent resistance R_e' as

$$R_e' = \frac{v_1}{i} = \frac{r_{\pi 2} + (r_{o1} \| R_1)}{1 + (r_{o1} \| R_1) g_{m1}} \tag{9.124}$$

Since $r_{o1} >> R_1$,

$$R_e' = \frac{r_{\pi 2} + R_1}{1 + g_{m1} R_1}$$

which, for $R_1 >> r_{\pi 2}$ and $g_{m1} R_1 >> 1$, can be simplified to

$$R_e' \approx \frac{1}{g_{m1}}$$

The output resistance can be found from Eq. (9.107) by substituting R_3 for R_2 and R_e' for $(r_{\pi 2} + R_e)$. That is,

$$R_o = \frac{v_x}{i_x} = R_3 \| R_e' + r_{o2}\left[1 + \frac{g_{m2} r_{\pi 2}}{R_e'}(R_2 \| R_e')\right] \tag{9.125}$$

Since r_{o2} is large, the first term is much smaller than the second one and can be ignored. Equation (9.124) can be approximated by

$$R_o \approx r_{o2}\left[1 + \frac{g_{m2} r_{\pi 2}}{R_e'}(R_3 \| R_e')\right] \tag{9.126}$$

Assuming $(R_3 \| R_e')/R_e' = 1/2$, $R_e' = 1/g_{m1}$, and $g_{m1} = g_{m3}$, Eq. (9.126) becomes

$$R_o \approx r_{o2}\left(1 + \frac{g_{m2} r_{\pi 2}}{2}\right)$$

$$\approx r_{o2}\left(1 + \frac{\beta_F}{2}\right) \tag{9.127}$$

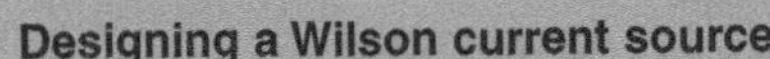

EXAMPLE 9.11

D

Designing a Wilson current source

(a) Design the Wilson current source in Fig. 9.30(a) to give $I_O = 5\ \mu A$. The parameters are $V_{CC} = 30$ V, $V_{BE1} = V_{BE2} = V_{BE3} = 0.7$ V, $V_T = 26$ mV, $V_A = 150$ V, and $\beta_F = 100$. Assume that all transistors are identical.

(b) Calculate the output resistance R_o and Thevenin's equivalent voltage V_{Th}.

(c) Use PSpice/SPICE to calculate the output current, the output resistance, and the reference current for $\beta_F = 100$ and 400. Assume that all transistors are identical, and $V_{CE} = 10$ V.

SOLUTION

$I_{C2} = 5\ \mu A$, and $V_T = 26$ mV.

(a) From Eq. (9.120),

$$5\ \mu A = I_R\left[1 - \frac{2}{(100^2 + 2 \times 100 + 2)}\right]$$

so $I_R \approx 5\ \mu A$. The reference current is

$$I_R = \frac{V_{CC} - V_{BE1} - V_{BE2}}{R_1}$$

That is,

$$5\ \mu A = \frac{30 - 0.7 - 0.7}{R_1}$$

which gives $R_1 = 5.72\ M\Omega$.

(b) From Eq. (9.118),

$$I_{C3} = I_{C1} = \frac{5\ \mu A \times (1 + 100)}{2 + 100} = 4.95\ \mu A$$

Then $r_{o1} = r_{o3} = \dfrac{V_A}{I_{C1}} = \dfrac{150}{4.95\ \mu} = 30.3\ M\Omega$

$$r_{o2} = \frac{V_A}{I_{C2}} = \frac{150}{5\ \mu} = 30\ M\Omega$$

Since $1/g_{m1} = 1/g_{m3} = V_T/I_{C1} = 26\ mV/4.95\ \mu A = 5.25\ k\Omega$,

$$g_{m1} = g_{m3} = 190.4\ \mu A/V$$

Since $1/g_{m2} = 26\ mV/5\ \mu A = 5.2\ k\Omega$,

$$g_{m2} = 192.3\ \mu A/V$$

Then $r_{\pi 1} = r_{\pi 3} = \dfrac{\beta_F}{g_{m1}} = 100 \times 5.25\text{ k} = 525\text{ k}\Omega$

$$r_{\pi 2} = 100 \times 5.2\text{ k} = 520\text{ k}\Omega$$

From Eq. (9.124),

$$R_e' = \frac{520\text{ k}\Omega + (30.3\text{ M}\Omega \parallel 5.72\text{ M}\Omega)}{[1 + (30.3\text{ M}\Omega \parallel 5.72\text{ M}\Omega) \times 190.4\ \mu\text{A/V}]}$$

$$= 5.81\text{ k}\Omega$$

From Eq. (9.123),

$$R_3 = 525\text{ k} \parallel 525\text{ k} \parallel 5.25\text{ k} \parallel 30\text{ M} \approx 5.15\text{ k}\Omega$$

From Eq. (9.126),

$$R_o = (30\text{ M})\left[1 + \frac{192.3\ \mu \times 520\text{ k}}{5.81\text{ k}} \times 5.15\text{ k} \parallel 5.81\text{ k}\right] = 1.44\text{ G}\Omega$$

and $V_{Th} = 1.44\text{ G} \times 5\ \mu = 7.2\text{ kV}$

NOTE: If we use Eq. (9.127),

$$R_o \approx (30\text{ M})\left(1 + \frac{100}{2}\right) = 1.53\text{ G}\Omega$$

(c) The Wilson current source for PSpice simulation is shown in Fig. 9.31. We will use parametric sweep for the model parameter β_F of the transistors. The voltage source V_y acts as an ammeter for the output current.

The results of simulation (.TF analysis) are as follows. (The hand calculations are shown in parentheses.) For $\beta_F = 100$, the simulation gives

```
VOLTAGE SOURCE CURRENTS
NAME    CURRENT
VCC     -5.022E-06
```

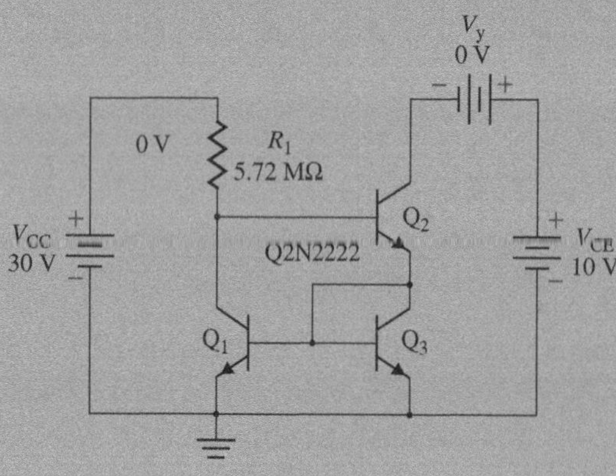

FIGURE 9.31 Wilson current source for PSpice simulation

```
VX       5.022E-06        (I_R = 5 µA)
VY       5.006E-06        (I_R = 5 µA)
VCE     -5.006E-06
**** BIPOLAR JUNCTION TRANSISTORS
NAME     Q1          Q2          Q3
IB       4.95E-08    4.73E-08    4.95E-08
IC       4.98E-06    5.01E-06    4.95E-06
VBE      6.37E-01    6.36E-01    6.37E-01
VBC     -6.36E-01   -8.73E+00    0.00E+00
VCE      1.27E+00    9.36E+00    6.37E-01
BETADC   1.00E+02    1.06E+02    1.00E+02
GM       1.92E-04    1.94E-04    1.92E-04
RPI      5.22E+05    5.47E+05    5.22E+05
RO       3.03E+07    3.17E+07    3.03E+07
**** SMALL-SIGNAL CHARACTERISTICS
I(VY)/VCC=1.739E-07
INPUT RESISTANCE AT VCC=5.730E+06
OUTPUT RESISTANCE AT I(VY)=1.602E+09
```

($r_{\pi 1}$ = 525 kΩ, $r_{\pi 2}$ = 520 kΩ, $r_{\pi 3}$ = 525 kΩ)
(r_{o1} = 30.3 MΩ, r_{o2} = 30 MΩ, r_{o3} = 30.3 MΩ)
(R_1 = 5.72 MΩ)
(R_o = 1.44 GΩ)

For $\beta_F = 400$, the simulation gives

```
I(VY)/VCC=1.738E-07
INPUT RESISTANCE AT VCC=5.730E+06
OUTPUT RESISTANCE AT I(VY)=5.441E+09
```

(R_o = 6.03 GΩ)

NOTE: R_o changes from 1.602E+09 Ω (for β_F = 100) to 5.441E+09 Ω (for β_F = 400 Ω) and depends on β_F, as expected.

9.6.6 Multiple Current Sources

A DC reference current can be generated in one location and reproduced in another location for biasing amplifier circuits in ICs. A group of current sources with only one reference current is shown in Fig. 9.32. This is an extension of the modified basic current source in Fig. 9.25. Transistor Q_1 and resistor R_1 serve as the reference for current-sink transistors Q_3 through Q_6. Transistor Q_2 supplies the total base currents for the transistors and makes the collector current of Q_1 almost equal to the reference current I_R; that is,

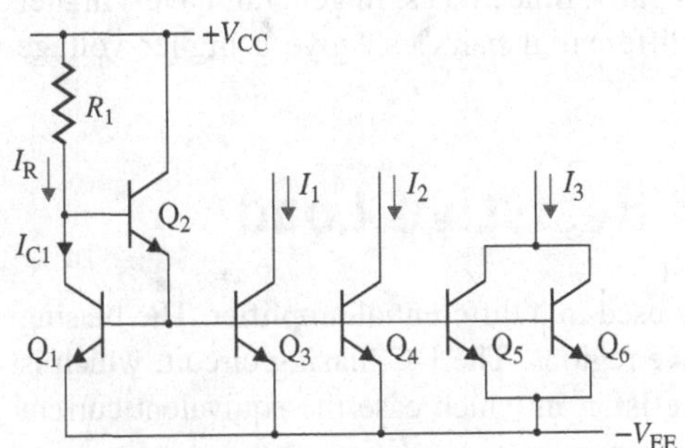

FIGURE 9.32 Multiple current sources

TABLE 9.1 Comparison of BJT current sources

Current Source Type	Output Resistance R_o	Beta (β_F) Dependency of I_o	Comments
Basic source	r_o	$-2/\beta_F$	Not suitable for low currents (i.e., typically less than 0.3 mA)
Modified source	r_o	$-2/(\beta_F^2 + \beta_F)$	Not suitable for low currents (i.e., typically less than 0.3 mA)
Widlar source	$r_{o2}(1 + g_{m2}R_2)$	Nonlinear	Current as low as 5 μA; higher output resistance
Cascode source	$r_{o2}(1 + \beta_F)$	$-2/\beta_F$	Not suitable for low currents; higher output resistance
Wilson source	$r_{o2}(1 + \beta_F/2)$	$-2/(\beta_F^2 + 2\beta_F + 2)$	Not suitable for low currents; higher output resistance

$I_R \approx I_{C1}$. The collector currents I_1 and I_2 will be mirrors of current I_R. Since two transistors Q_5 and Q_6 are connected in parallel, I_3 will be two times I_R (i.e., $I_3 = 2I_R$). The parallel combination of Q_5 and Q_6 should be equivalent to a single transistor whose E-B junction has double the area of Q_1. Therefore, the emitter areas of transistors can be scaled in ICs so as to provide multiples of the reference current simply by designing the transistors so that they have an area ratio equal to the desired multiple.

- Transistors can generate the characteristics of a constant-current source. An ideal current source has a very high output resistance, and its output current is not sensitive to the transistor parameter β_F.
- Table 9.1 compares various BJT current sources. The output resistances for different BJT sources are also summarized in the table.

9.7 BJT Differential Amplifiers

The BJT differential pairs are analogous to MOS differential pairs. Since BJTs, in general, have a higher transconductance compared to identical MOSFETs, BJT differential pairs can give a higher voltage gain, but they have limitations of a lower input resistance.

9.7.1 BJT Differential Pair with Resistive Load

An emitter-coupled pair, as shown in Fig. 9.33, is commonly used in a differential amplifier. The biasing current should be such that the transistors operate in the active regions. The DC biasing circuit, which is shown as a constant-current source, can be either a simple resistor, in which case the equivalent current generator will be zero, or a transistor current source, which is generally used in ICs.

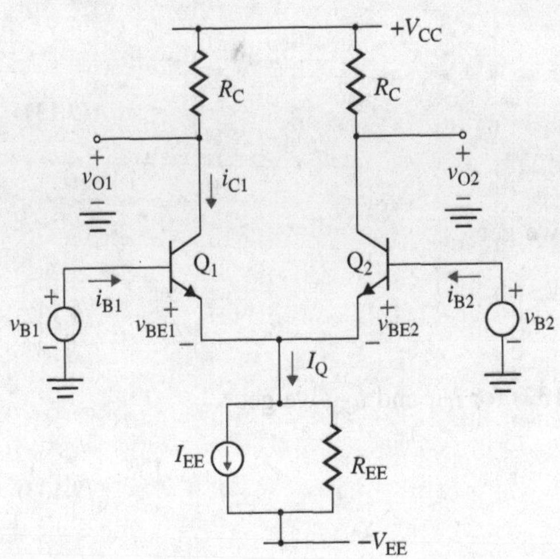

FIGURE 9.33 Emitter-coupled differential pair

DC Transfer Characteristics

The DC transfer characteristic, which gives the relation between the input and output voltages, can be determined from the large-signal analysis, and it should be linear over a wide range. The analysis can be simplified by making the following assumptions:

1. The output resistances of the transistors are infinite: $r_o = \infty$.
2. The output resistance of the transistor current source is infinite: $R_{EE} = \infty$.

Using KVL around the loop formed by the two input voltages and the two B-E junctions, we get

$$v_{B1} - v_{BE1} + v_{BE2} - v_{B2} = 0 \tag{9.128}$$

Assuming v_{BE1}, $v_{BE2} >> V_T$ and equal leakage currents $I_{S1} = I_{S2} = I_S$ and using the transistor current equations, we get

$$v_{BE1} = V_T \ln\left(\frac{i_{C1}}{I_{S1}}\right) = V_T \ln\left(\frac{i_{C1}}{I_S}\right) \tag{9.129}$$

$$v_{BE2} = V_T \ln\left(\frac{i_{C2}}{I_{S2}}\right) = V_T \ln\left(\frac{i_{C2}}{I_S}\right) \tag{9.130}$$

Substituting Eqs. (9.129) and (9.130) into Eq. (9.128) gives

$$v_{B1} - V_T \ln\left(\frac{i_{C1}}{I_S}\right) + V_T \ln\left(\frac{i_{C2}}{I_S}\right) - v_{B2} = 0$$

That is,

$$v_{B1} - v_{B2} = V_T\left[\ln\left(\frac{i_{C1}}{I_S}\right) - \ln\left(\frac{i_{C2}}{I_S}\right)\right]$$

which can be simplified to

$$\frac{i_{C1}}{i_{C2}} = \exp\left(\frac{v_{B1} - v_{B2}}{V_T}\right) = \exp\left(\frac{v_{id}}{V_T}\right) \tag{9.131}$$

where $v_{id} = v_{B1} - v_{B2}$ is the differential DC input voltage.

Using KCL at the emitter terminal of the transistors, we get

$$I_Q = \frac{1}{\alpha}(i_{C1} + i_{C2}) \tag{9.132}$$

where $\alpha = \beta_F/(1 + \beta_F) \approx 1$. Solving Eqs. (9.131) and (9.132) for i_{C1} and i_{C2}, we get

$$i_{C1} = \frac{\alpha I_Q}{1 + \exp(-v_{id}/V_T)} \tag{9.133}$$

$$\approx \alpha I_Q \quad \text{for } v_{id} >> V_T \tag{9.134}$$

$$i_{C2} = \frac{\alpha I_Q}{1 + \exp(v_{id}/V_T)} \tag{9.135}$$

$$\approx 0 \quad \text{for } v_{id} >> V_T \tag{9.136}$$

Thus, if i_{C1} increases, i_{C2} decreases such that $i_{C1} + i_{C2} = \alpha I_Q = \alpha I_{EE}$ remains constant.

The plots of the two collector currents are shown as a function of v_{id} in Fig. 9.34(a). Notice that, for $v_{id} >> V_T$, i_{C1} and i_{C2} become independent of v_{id} and all the currents flow through one of the transistors. For $v_{id} \leq V_T$, i_{C1} and i_{C2} have an approximately linear relation. The differential voltage change Δv_{id} required to shift the current distribution from $i_{C1} = 0.9I_Q$ and $i_{C2} = 0.1I_Q$ to the opposite case, $i_{C1} = 0.1I_Q$ and $i_{C2} = 0.9I_Q$, is called the *transition voltage*; it will have a value of approximately $2V_T = 52.6$ mV.

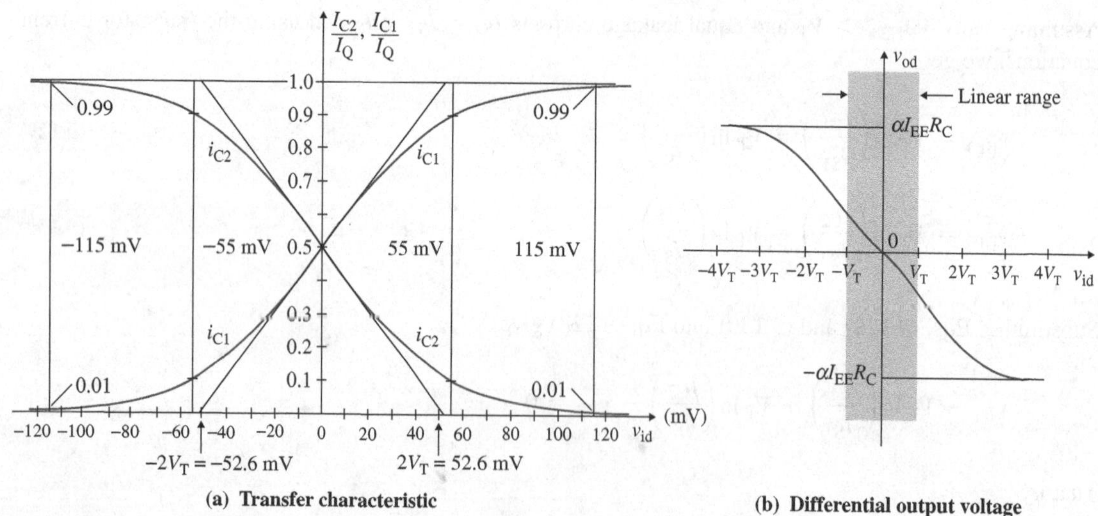

FIGURE 9.34 Transfer characteristics of emitter-coupled pair

The DC output voltages are

$$v_{O1} = V_{CC} - i_{C1}R_C \tag{9.137}$$

$$v_{O2} = V_{CC} - i_{C2}R_C \tag{9.138}$$

The differential DC output voltage is

$$v_{od} = v_{O1} - v_{O2} = R_C(i_{C2} - i_{C1})$$

Substituting Eqs. (9.133) and (9.135) into the above equation and simplifying yields

$$v_{od} = \alpha I_{EE}R_C \tanh\left(-\frac{v_{id}}{2V_T}\right) \tag{9.139}$$

Since tanh $x \approx x$ for a small value of x, Eq. (9.139) can be approximated by

$$v_{od} \approx -\alpha I_{EE}R_C\left(\frac{v_{id}}{2V_T}\right) \tag{9.140}$$

The plot of v_{od} as a function of v_{id} is shown in Fig. 9.34(b). If v_{id} is zero, v_{od} is also zero, and this feature allows direct coupling of cascoded stages without introducing DC offsets. Thus, the amplifier is a true differential, or difference, amplifier, responding only to the difference in the voltages applied to the two input terminals. If $v_{B1} = -v_{B2}$, then $v_{ic} = (v_{B1} + v_{B2})/2$ is zero and there will be only a differential voltage. If, on the other hand, $v_{B1} = v_{B2}$, then v_{id} is zero and there will be a pure common-mode voltage (i.e., no output voltage). However, the range of the differential input voltage v_{id} over which the emitter-coupled pair exhibits a linear characteristic is very small, typically two or three times V_T. This range can be extended by inserting emitter degeneration resistors, as shown in Fig. 9.35(a); in this case, the range over which the characteristic is linear is approximately equal to I_QR_E. The factor by which the voltage gain is reduced is approximately the same as the factor by which the input range is increased. The variations of v_{od} with several values of R_E are shown in Fig. 9.35(b).

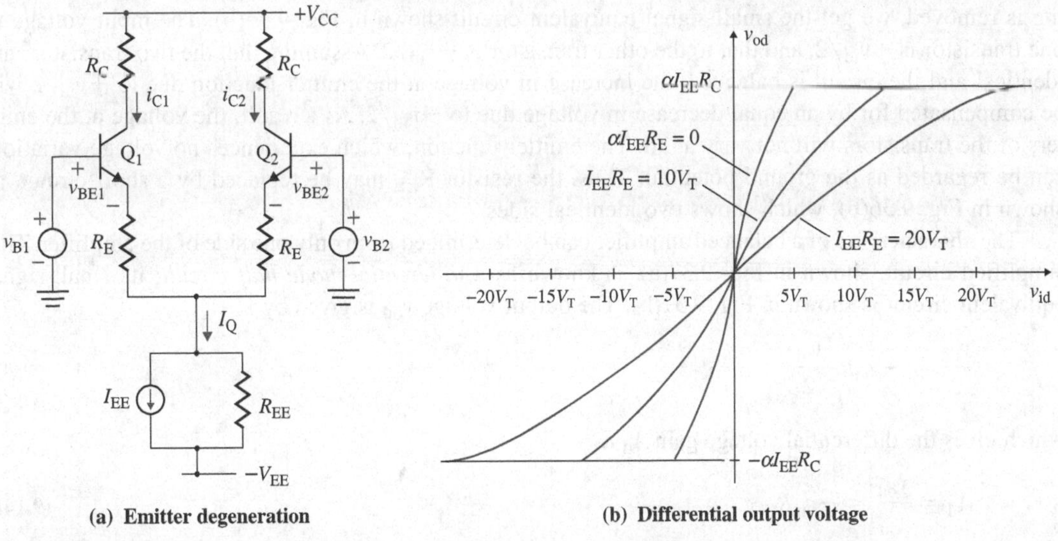

FIGURE 9.35 Emitter-coupled pair with degeneration resistors

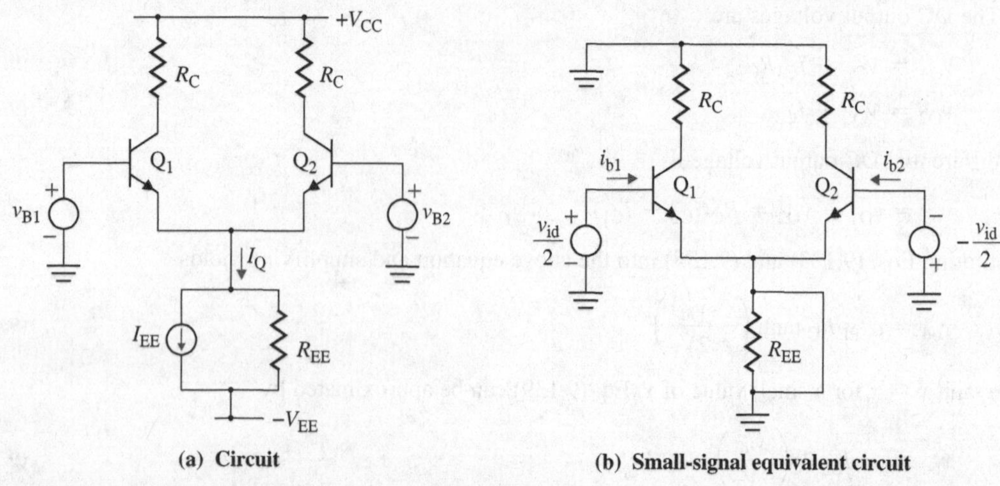

FIGURE 9.36 Emitter-coupled pair with differential input

Small-Signal Analysis

In studying a BJT differential amplifier, it is often of interest to examine the small-signal behavior for small DC differential voltages near zero, when the amplifier operates in the linear portion of the transfer characteristic. Circuit properties such as differential voltage gain A_d, common-mode gain A_c, common-mode rejection ratio (CMMR), common-mode input resistance R_{ic}, and differential mode input resistance R_{id} can be determined from the small-signal analysis.

Small Differential Signal Let us assume that the common-mode signal is zero, $v_{ic} = 0$, and only the differential input voltage v_{id} is applied. This situation is shown in Fig. 9.36(a). If the DC differential biasing is removed, we get the small-signal equivalent circuit shown in Fig. 9.36(b). The input voltage to one transistor is $+v_{id}/2$, and that to the other transistor is $-v_{id}/2$. Assuming that the two transistors are identical and the circuit is balanced, the increase in voltage at the emitter junction due to $+v_{id}/2$ will be compensated for by an equal decrease in voltage due to $-v_{id}/2$. As a result, the voltage at the emitters of the transistors will not vary at all. The emitter junction, which experiences no voltage variation, can be regarded as the ground potential. Thus, the resistor R_{EE} may be replaced by a short circuit, as shown in Fig. 9.36(b), which shows two identical sides.

The characteristic of a balanced amplifier can be determined from only one side of the amplifier. This simplified circuit, shown in Fig. 9.37(a), is known as a *differential-mode half circuit*; its small-signal equivalent circuit is shown in Fig. 9.37(b). The output voltage v_{od} is given by

$$\frac{v_{od}}{2} = -g_m R_C \frac{v_{id}}{2}$$

which gives the differential voltage gain A_d as

$$A_d = \frac{v_{od}}{v_{id}} = -g_m R_C \tag{9.141}$$

▶ **NOTE** If we include the output resistance r_{o1} of Q_1, Eq. (9.141) becomes $A_d = -g_m(R_C \parallel r_{o1})$.

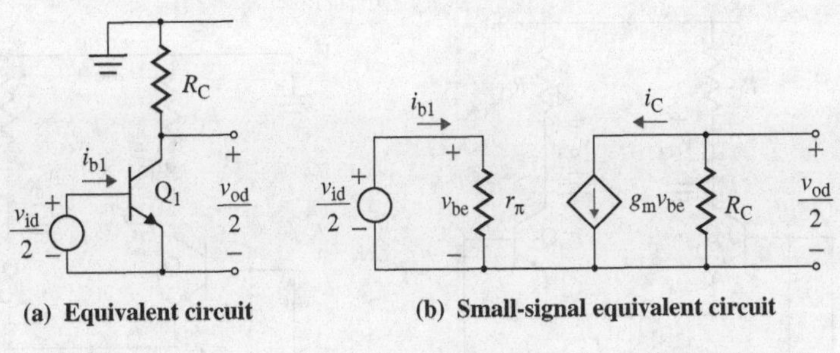

(a) Equivalent circuit (b) Small-signal equivalent circuit

FIGURE 9.37 Differential-mode half circuit

High values of R_C and g_m are required to give a high value of A_d. An active load (discussed in Sec. 9.6) rather than a discrete resistance ensures a high value of R_C and hence of A_d. From Fig. 9.37(b), we can write

$$\frac{v_{id}}{2} = i_{b1} r_\pi$$

which gives the differential input resistance R_{id} as

$$R_{id} = \frac{v_{id}}{i_{b1}} = 2r_\pi \tag{9.142}$$

A high value of r_π ($=V_T/I_C$)—that is, a low collector biasing current ($I_{C1} = I_{C2} = I_Q/2$)—is required to achieve a higher value of R_{id}.

Small Common-Mode Signal An emitter-coupled pair with only common-mode input v_{ic} is shown in Fig. 9.38(a). If the DC common-mode biasing is removed, the result is the small-signal equivalent circuit shown in Fig. 9.38(b). Assuming that the two transistors are identical, the collector currents must be identical, and the voltage at the emitter junction will increase by the same amount in response to inputs at both transistors. Since the voltage across R_{EE} will be the same for both inputs, the resistor R_{EE} can be split into two parallel resistors, each of value $2R_{EE}$, as shown in Fig. 9.38(c). As a result of symmetry, no current will flow through the lead that connects the two sides, and $i_x = 0$. Thus, this lead can be disconnected without affecting the circuit behavior; as shown in Fig. 9.38(d), the two half circuits may be considered to be completely independent. The common-mode behavior can be determined from only one side, as shown in Fig. 9.39(a). The small-signal equivalent circuit is shown in Fig. 9.39(b), from which we get

$$v_{ic} = i_{b1} r_\pi + i_b(1 + \beta_F)2R_{EE} \tag{9.143}$$

which gives the common-mode input resistance R_{ic} as

$$R_{ic} = \frac{v_{ic}}{i_{b1}} = r_\pi + (1 + \beta_F)2R_{EE} \tag{9.144}$$

Equation (9.144) will give a high value for R_{ic}, so R_{ic} should be calculated using r_μ in the BJT model. The current through r_μ can be found from

$$i_\mu = \frac{v_{ic} - v_o}{r_\mu} = \frac{v_{ic} + \beta_F R_C i_{b1}}{r_\mu}$$

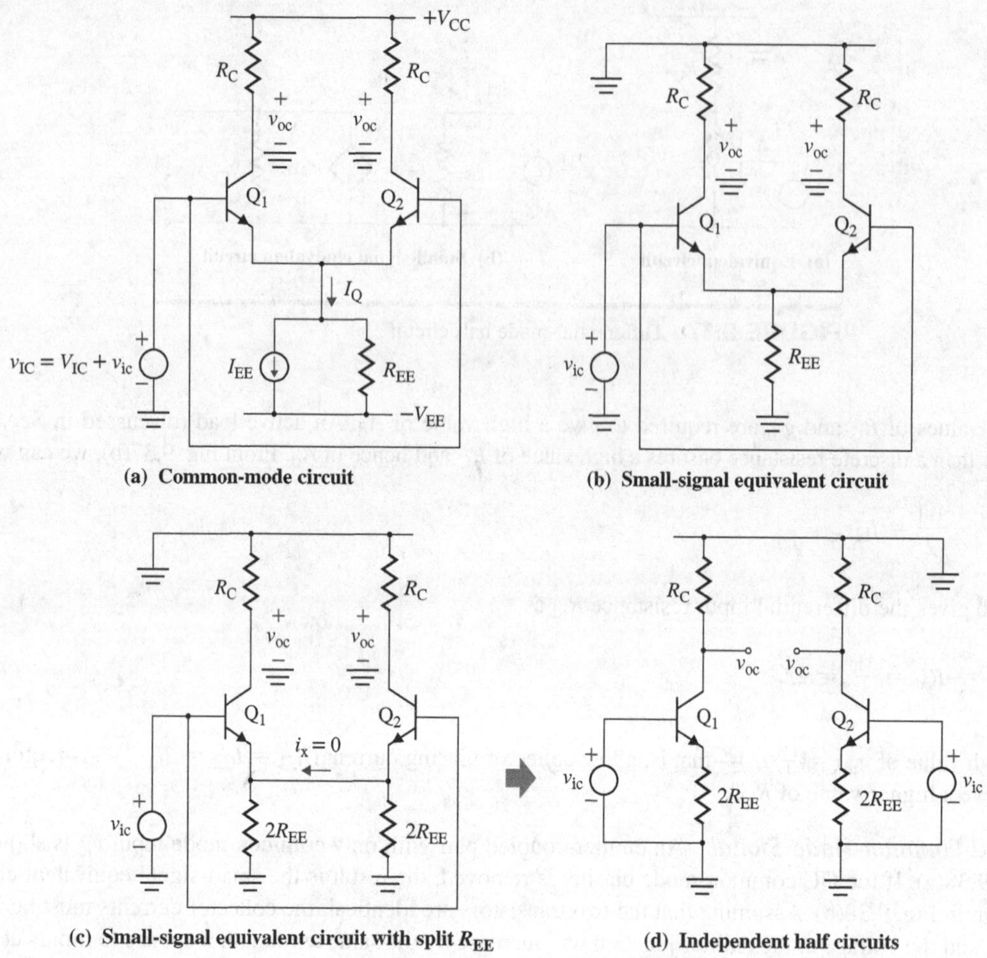

(a) Common-mode circuit

(b) Small-signal equivalent circuit

(c) Small-signal equivalent circuit with split R_{EE}

(d) Independent half circuits

FIGURE 9.38 Emitter-coupled pair with common-mode input

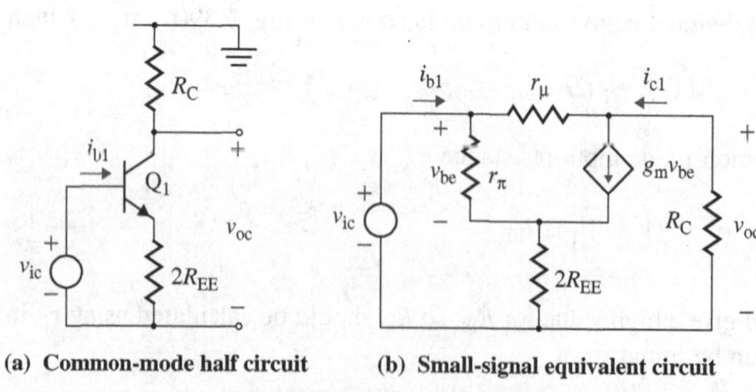

(a) Common-mode half circuit

(b) Small-signal equivalent circuit

FIGURE 9.39 Common-mode half circuit

which, after substitution of $i_{b1} = v_{ic}/[r_\pi + (1 + \beta_F)2R_{EE}]$, gives the effective resistance as

$$R_{i\mu} = \frac{r_\mu}{1 + \beta_F R_C/[r_\pi + (1 + \beta_F)2R_{EE}]}$$

Thus, the common-mode resistance R_{ic} becomes

$$R_{ic} = [r_\pi + 2R_{EE}(1 + \beta_F)] \parallel \frac{r_\mu}{r_\pi + (1 + \beta_F)2R_{EE}}$$

The common-mode output voltage v_{oc} is

$$v_{oc} = -R_C i_{c1} = -R_C \beta_F i_{b1} = v_{ic}\left[\frac{-\beta_F R_C}{r_\pi + (1 + \beta_F)2R_{EE}}\right]$$

which gives the common-mode voltage gain A_c as

$$A_c = \frac{v_{oc}}{v_{ic}} = \frac{-\beta_F R_C}{r_\pi + (1 + \beta_F)2R_{EE}} \tag{9.145}$$

$$= \frac{-g_m R_C}{1 + 2g_m R_{EE}(1 + 1/\beta_F)} \tag{9.146}$$

Small-Signal CMRR From Eqs. (9.141) and (9.146), we can find the CMRR as

$$\text{CMRR} = \left|\frac{A_d}{A_c}\right| = 1 + 2g_m R_{EE}\left(1 + \frac{1}{\beta_F}\right) \tag{9.147}$$

$$\approx 1 + 2g_m R_{EE} \quad \text{for } \beta_F >> 1 \tag{9.148}$$

which indicates that a high output resistance R_{EE} on the biasing current source will improve the CMRR; that is, the value of R_{EE} should be as large as possible. To obtain a high value of g_m $(=I_C/V_T)$, the collector biasing current $(I_{C1} = I_{C2} = I_Q/2 = I_{EE}/2)$ should be made large.

The small-signal input currents, which will flow when both v_{id} and v_{ic} are applied, can be found by superposition. Since R_{ic} is common to both i_{b1} and i_{b2}, we get

$$i_{b1} = \frac{v_{ic}}{2R_{ic}} + \frac{v_{id}}{R_{id}} \tag{9.149}$$

$$i_{b2} = \frac{v_{ic}}{2R_{ic}} + \frac{v_{id}}{R_{id}} \tag{9.150}$$

The input resistance can thus be represented by the π-equivalent circuit of Fig. 9.40(a), where R_{ic} is assumed to be much larger than R_{id}. The T-equivalent circuit is shown in Fig. 9.40(b), and its values are shown in Fig. 9.40(c).

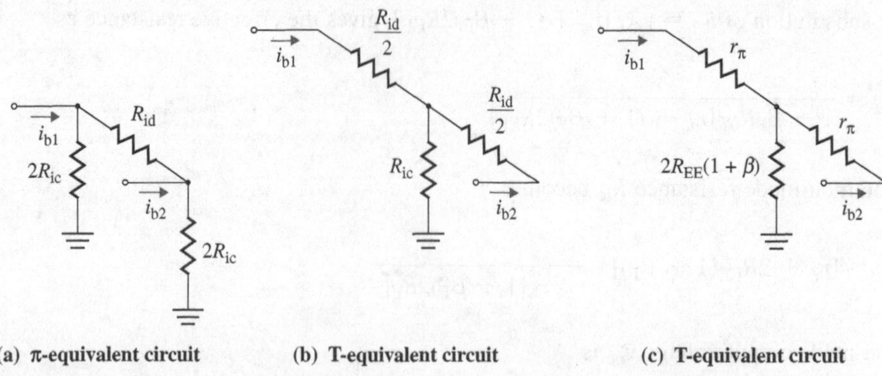

(a) π-equivalent circuit (b) T-equivalent circuit (c) T-equivalent circuit

FIGURE 9.40 π- and T-equivalent circuits

EXAMPLE 9.12

Finding the performance parameters of an emitter-coupled pair The parameters of the emitter-coupled pair in Fig. 9.36(a) are $\beta_F = 100$, $R_{EE} = 50\ \text{k}\Omega$, $I_Q = 1$ mA, $V_{CC} = 15$ V, and $R_C = 10\ \text{k}\Omega$.

(a) Calculate the DC collector currents through the transistors if $v_{id} = 5$ mV.

(b) Assuming $i_{C1} = i_{C2}$, calculate A_d, A_c, and CMRR; R_{id} and R_{ic}; and the small-signal output voltage if $v_{B1} = 20$ mV and $v_{B2} = 10$ mV. Assume $V_T \approx 26$ mV.

SOLUTION

$\alpha = \beta_F/(1 + \beta_F) = 100/(1 + 100) = 0.99$.

(a) From Eq. (9.133),

$$i_{C1} = \frac{\alpha I_Q}{1 + \exp(-v_{id}/V_T)} = \frac{0.99 \times 1\,\text{mA}}{1 + \exp(-5\text{ m}/26\text{ m})} = 0.543\text{ mA}$$

and $\quad i_{C2} = 1\text{ m} - 0.543\text{ m} = 0.457\text{ mA}$

(b) We know that $i_{C1} = i_{C2} = 1\text{ mA}/2 = 0.5$ mA. Thus,

$$g_m = \frac{i_{C1}}{V_T} = \frac{0.5\text{ mA}}{26\text{ mV}} = 19.23\text{ mA/V}$$

From Eq. (9.141),

$$A_d = -g_m R_C = -19.23\text{ m} \times 10\text{ k} = -192.3\text{ V/V}$$

From Eq. (9.146),

$$A_c = \frac{-g_m R_C}{1 + 2g_m R_{EE}(1 + 1/\beta_F)}$$

$$= \frac{-19.23\text{ mA/V} \times 10\text{ k}\Omega}{1 + 2 \times 19.23\text{ mA/V} \times 50\text{ k}\Omega \times (1 + 1/100)} = -0.099\text{ V/V}$$

Thus, CMRR $= |A_d/A_c| = 192.3/0.099 = 1942.4$ (or 65.77 dB). From Eq. (9.148),

$$\text{CMRR} \approx 1 + 2g_m R_{EE} = 1 + 2 \times 19.23\text{ m} \times 50\text{ k} = 1924$$

We know that $r_\pi = \beta_F/g_m = 100/19.23\text{ m} = 5.2\text{ k}\Omega$. From Eq. (9.142),

$$R_{id} = 2r_\pi = 2 \times 5.2\text{ k} = 10.4\text{ k}\Omega$$

From Eq. (9.144),

$$R_{ic} = r_\pi + (1 + \beta_F)2R_{EE} = 5.2\text{ k} + (1 + 100) \times 2 \times 50\text{ k} = 10.1\text{ M}\Omega$$

We know that $v_{id} = 20 - 10 = 10$ mV, and $v_{ic} = (20 + 10)/2 = 15$ mV. From Eq. (9.10),

$$v_o = A_d v_{id} + A_c v_{ic} = -192.3 \times 10\text{ mV} - 0.099 \times 15\text{ mV} = -1924.5\text{ mV}$$

▶ **NOTE** To apply Eq. (9.10) and other equations in Sec. 9.2, we must have $|v_{id}| \leq V_T$. The DC collector voltage of a transistor is

$$V_C = V_{CC} - I_C R_C = 15\text{ V} - 0.5\text{ mA} \times 10\text{ k}\Omega = 10\text{ V}$$

Thus, for $|A_d| = 192.3$, the maximum differential voltage will be $v_{id} = 10/192.3 = 52$ mV. Therefore, to allow output voltage swing due to the input voltages, V_{CC} must be greater than $I_C R_C$.

EXAMPLE 9.13

D

Designing an emitter-coupled pair

(a) Design an emitter-coupled pair as shown in Fig. 9.41 in which one input terminal is grounded. The output is taken from the collector of transistor Q_1. The biasing current is $I_{EE} = 1$ mA, and $V_{CC} = V_{EE} = 15$ V. The transistors are identical. Assume $V_{BE} = 0.7$ V, $V_T = 26$ mV, $V_A = \infty$, and $\beta_F = 100$. A small-signal voltage gain of $A_1 = -250$ V/V is required.

(b) Calculate the design values of A_d, A_c, and CMRR.

SOLUTION

(a) $I_{EE} = 1$ mA. Since $v_{B2} = 0$, we can write

$$R_{EE} I_{EE} + V_{BE} = V_{EE} = 15\text{ V}$$

Thus,

$$R_{EE} = \frac{15 - 0.7}{1\text{ m}} = 14.3\text{ k}\Omega$$

We have

$$i_{C1} = i_{C2} = \frac{I_{EE}}{2} = \frac{1\text{ mA}}{2} = 0.5\text{ mA}$$

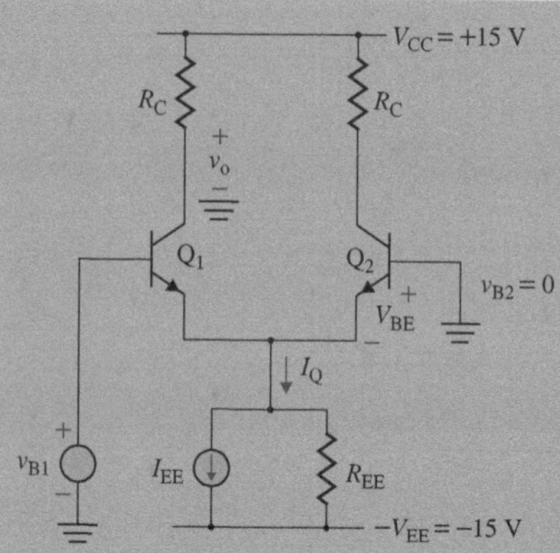

FIGURE 9.41 Emitter-coupled pair with single input

and $g_m = \frac{i_{C1}}{V_T} = \frac{0.5\text{ mA}}{26\text{ mV}} = 19.23\text{ mA/V}$

Since $v_{B2} = V_{B2} + v_{b2} = 0$, $v_{b2} = 0$. Then

$$v_{id} = v_{b1} - v_{b2} = v_{b1}$$

and $v_{ic} = \frac{v_{b1} + v_{b2}}{2} = \frac{v_{b1}}{2}$

From Eq. (9.10), we get

$$v_{o1} = A_c v_{ic} + \frac{A_d v_{id}}{2} = A_c \frac{v_{b1}}{2} + \frac{A_d v_{b1}}{2} = \frac{v_{b1}}{2}(A_c + A_d) \tag{9.151}$$

Substituting for A_d and A_c from Eqs. (9.141) and (9.146) gives the voltage gain A_1 as

$$A_1 = \frac{v_{o1}}{v_{b1}} = \frac{1}{2}(A_c + A_d) = -\frac{1}{2}\left[\frac{g_m R_C}{1 + 2g_m R_{EE}(1 + 1/\beta_F)} + g_m R_C\right] \tag{9.152}$$

Substituting $A_1 = -250$, $g_m = 19.23$ mA/V, $R_{EE} = 14.3$ kΩ, and $\beta_F = 100$ into Eq. (9.152) gives $R_C = 25.95$ kΩ.

(b) From Eq. (9.141),

$$A_d = -g_m R_C = -19.23\text{ m} \times 25.95\text{ k} = -499.01$$

From Eq. (9.146),

$$A_c = \frac{-19.23\text{ mA/V} \times 25.95\text{ k}\Omega}{1 + 2 \times 19.23\text{ mA/V} \times 14.3\text{ k}\Omega \times (1 + 1/100)} = -0.897$$

Thus, CMRR $= |A_d/A_c| = 499.01/0.897 = 556.3$ (or 54.91 dB).

9.7.2 BJT Differential Amplifiers with Basic Current Mirror Active Load

We saw in Sec. 9.7.1 that the differential gain of a differential pair with a resistive load R_C is $-g_m R_C = -R_C I_C/V_T = -R_C I_Q/2V_T$. In differential amplifiers, a very small value of biasing current I_Q, in the microampere range, is often used. As a result, a very large value of R_C, on the order of megaohms, will be required to give a substantial voltage gain. However, a large value of R_C will cause a large DC voltage drop, reducing the collector voltage to $V_{CC} - R_C I_Q/2$, which will be substantially less than V_{CC}. This low collector voltage will reduce the allowable input voltage range of the amplifier.

Active devices such as transistors occupy much less silicon area than medium-size or large resistors. In practical amplifiers, the load resistor R_C is normally replaced by a constant-current source, which offers a very high load resistance to the amplifier and hence can give a high voltage gain. This type of load, known as an *active load*, has a small voltage drop, typically 0.7 V, and hence allows a wider input voltage range.

A differential amplifier with a basic current source as the active load is shown in Fig. 9.42. The active load consists of transistors Q_3 and Q_4. Since their B-E voltages are the same, their collector currents will be equal. That is, $i_{C3} \approx i_{C4}$. Thus, the current through Q_4 will be the mirror of the current through Q_3. Under quiescent conditions, the differential amplifiers will be balanced such that $I_{C1} = I_{C2}$. Since $I_{C1} = I_{C2}$ and $I_{C3} = I_{C4}$, we can find the quiescent load current:

$$I_O = I_{B3} + I_{B4} = \frac{I_{C3} + I_{C4}}{\beta_F} \approx \frac{I_{C1} + I_{C2}}{\beta_F} = \frac{\alpha I_Q}{\beta_F} = \frac{I_Q}{1 + \beta_F}$$

Since $\beta_F >> 1$, I_O will be very small.

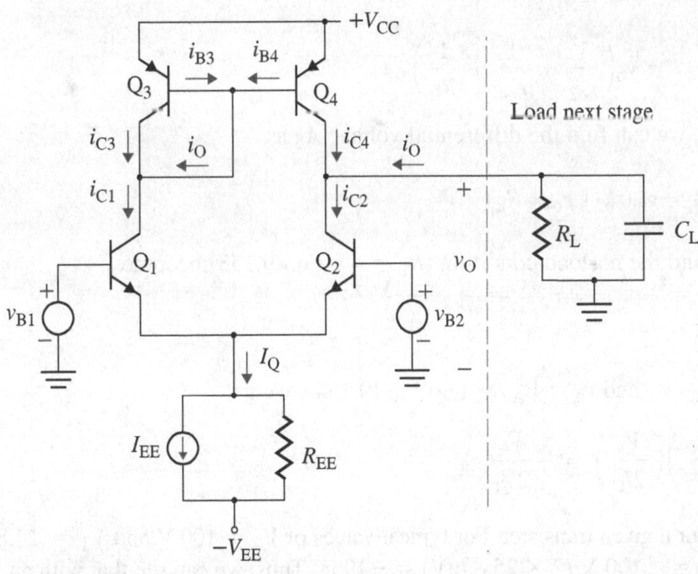

FIGURE 9.42 Differential amplifier with a basic current mirror active load

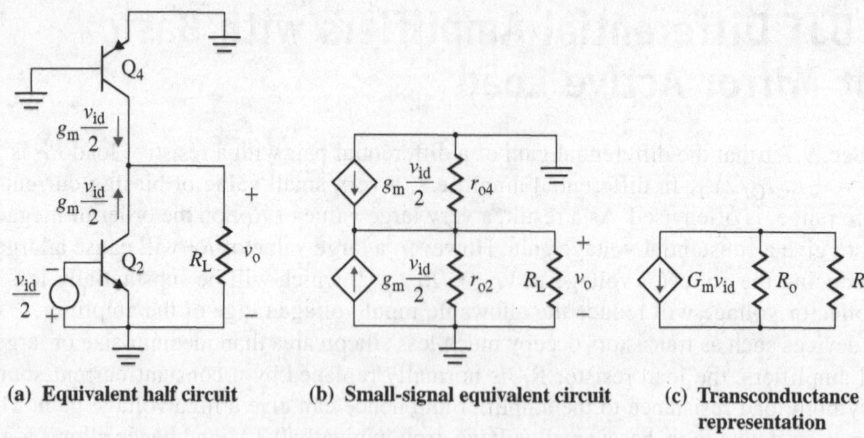

FIGURE 9.43 Equivalent basic current mirror circuit

Small-Signal Analysis

Small Differential Signal If the input voltages change by a small differential amount v_{id}, the collector current of Q_1 will change by a small amount $g_m v_{id}/2$, and the collector current of Q_3 will change by the same amount. Since i_{C4} is a mirror of i_{C3}, the collector current of Q_4 will change by $g_m v_{id}/2$. The B-E voltage of Q_2 will decrease by $v_{id}/2$, causing its collector current to change by an amount $-g_m v_{id}/2$. The equivalent half circuit is shown in Fig. 9.43(a), and its small-signal equivalent appears in Fig. 9.43(b).

Using KCL at the collector of transistors Q_2 and Q_4, we can find the output voltage v_o in terms of v_{id}:

$$\frac{-2 \times g_m v_{id}}{2} = \frac{v_o}{r_{o2}} + \frac{v_o}{r_{o4}} + \frac{v_o}{R_L}$$

$$= v_o\left(\frac{1}{r_{o2}} + \frac{1}{r_{o4}} + \frac{1}{R_L}\right)$$

After simplification, we can find the differential voltage A_d as

$$A_d = \frac{v_o}{v_{id}} = -g_m(r_{o2} \parallel r_{o4} \parallel R_L) \tag{9.153}$$

For $r_{o2} = r_{o4} = r_o$ and the no-load condition ($R_L = \infty$), Eq. (9.153) becomes

$$A_d = -\frac{g_m r_o}{2} \tag{9.154}$$

Substituting $g_m = I_C/V_T$ and $r_o = V_A/I_C$ into Eq. (9.154), we get

$$A_d = -\left(\frac{I_C}{V_T}\right)\left(\frac{V_A}{2I_C}\right) = -\frac{V_A}{2V_T} \tag{9.155}$$

which is a constant for a given transistor. For typical values of $V_A = 100$ V and $V_T = 25.8$ mV, the differential gain becomes $A_d = -100$ V/(2 × 25.8 mV) = −1938. Thus, we can see that with an active load, a very large voltage gain can be obtained with only a single amplifier stage. Also, the gain A_d depends only on the physical parameters V_A and V_T. Since V_T is temperature dependent, A_d will be temperature dependent too.

The differential input resistance R_{id} is given by

$$R_{id} = \frac{v_{id}}{i_d} = 2r_\pi \quad (\text{for } I_C > 0) \tag{9.156}$$

Equation (9.156) is not valid for the no-load condition—that is, when $I_C = 0$. Substituting $r_\pi = \beta_F V_T / I_C$ into Eq. (9.156) gives

$$R_{id} = 2\frac{\beta_F V_T}{I_C} = 2 \times \frac{\beta_F V_T}{I_Q/2} = \frac{4\beta_F V_T}{I_Q} \tag{9.157}$$

which indicates that a lower biasing current I_Q will give a higher value of R_{id} while still maintaining a high voltage gain A_d. A very low value of I_Q, however, will affect the frequency and transient responses of the amplifier, which is undesirable. If a low biasing current is desired to achieve a high input resistance, then a MOSFET differential amplifier is preferable because the amplifier can be operated at a relatively higher value of biasing current without affecting the frequency and transient responses.

The output resistance R_o is the parallel combination of r_{o2} and r_{o4}; that is,

$$R_o = r_{o2} \parallel r_{o4} \tag{9.158}$$

$$= \frac{r_o}{2}$$

(since $r_{o2} = r_{o4}$).

A differential amplifier is normally followed by other stages. The input resistance R_L of the next stage acts as a load of the amplifier and hence influences the overall voltage gain. The amplifier is often represented as a transconductance amplifier so that the short-circuit current and the effect of load resistance on the output voltage can be determined easily. This arrangement is shown in Fig. 9.43(c). The total current is $2g_m v_{id}/2 = g_m v_{id}$, which gives the effective G_m as

$$G_m = g_m = \frac{I_C}{V_T} = \frac{I_Q}{2V_T} \tag{9.159}$$

Small Common-Mode Signal The common-mode input resistance R_{ic} can be found from Eq. (9.144). The approximate value of the common-mode gain A_c can be found from Eq. (9.145) by replacing R_C by R_{o4}. For practical amplifiers, A_c is generally very small and can be ignored in finding the output voltage.

9.7.3 Differential Amplifier with Modified Current Mirror

Let us use the modified basic current shown in Fig. 9.27(a) as the active load. This arrangement is shown in Fig. 9.44. The active load consists of transistors Q_3, Q_4, and Q_5. The addition of Q_5 makes the ratio i_{C4}/i_{C3} independent of current gain β_F, and i_{C4} approximates a true mirror of i_{C3}. Transistors Q_6 and Q_7 belong to the second stage and act as the load of the differential amplifier. Q_6 and Q_7 form a compound transistor (Darlington pair), in which the emitter current of Q_6 becomes the base current of Q_7. As a result, the effective current gain becomes β_F^2 such that $i_{C7} = \beta_F^2 i_{B6}$.

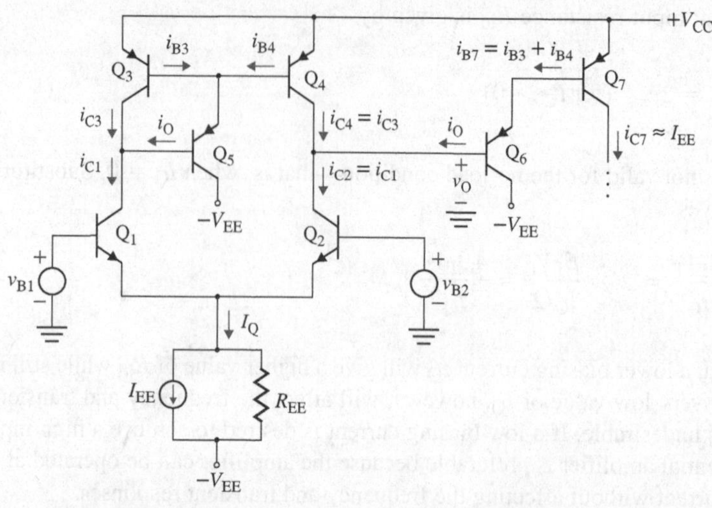

FIGURE 9.44 Differential amplifier with a modified current mirror active load

Under quiescent conditions with identical transistors, $I_{C1} + I_{C2} = I_Q$, $I_{C1} = I_{C2}$, and $I_{C3} = I_{C4}$. Thus, $I_{B5} = I_{B6}$, so $I_{E5} = I_{E6}$; that is,

$$I_{B7} = I_{E6} = I_{E5} = I_{B3} + I_{B4} = \frac{I_{C3} + I_{C4}}{\beta_F} = \frac{I_Q}{\beta_F}$$

Thus, the collector current of Q_7 becomes $I_{C7} = \beta_F I_{B7} = I_Q$. This current mirror causes the quiescent current of the next stage to have the same value as the differential amplifier while having the same performance as the amplifier of Fig. 9.42.

EXAMPLE 9.14

Analyzing a BJT differential amplifier with a current mirror active load The parameters of the differential amplifier in Fig. 9.44 are $\beta_F = 100$, $I_Q = 20\ \mu A$, and $V_{CC} = 15$ V. Calculate A_d, R_{id}, R_o, and the overall voltage gain with load $A_{d(load)}$. Assume $V_T = 26$ mV and $V_A = 100$ V.

SOLUTION

We have

$$I_{C1} = I_{C2} = I_{C4} = \frac{20\ \mu A}{2} = 10\ \mu A$$

$$I_{B6} = \frac{I_Q}{\beta_{F7}\beta_{F8}} = \frac{20\ \mu A}{100 \times 100} = 2\ nA$$

$$g_m = \frac{I_{C2}}{V_T} = \frac{10\ \mu A}{26\ mV} = 387.6\ \mu A/V$$

$$r_{o2} = r_{o3} = r_o = \frac{V_A}{I_C} = \frac{100\ V}{10\ \mu A} = 10\ M\Omega$$

$$r_\pi = \frac{\beta_F V_T}{I_C} = \frac{100 \times 26\ mV}{10\ \mu A} = 260\ k\Omega$$

From Eq. (9.155),

$$A_d = -\frac{V_A}{2V_T} = -\frac{100\ V}{2 \times 26\ mV} = -1923\ V/V$$

From Eq. (9.156),

$$R_{id} = 2r_\pi = 2 \times 260\ k\Omega = 520\ k\Omega$$

and $$R_o = r_{o2} \| r_{o4} = \frac{r_o}{2} = \frac{10\ M\Omega}{2} = 5\ M\Omega$$

Since Q_6 and Q_7 form a compound transistor, its effective B-E voltage is that of two B-E junctions in series. Thus,

$$R_L = 2r_{\pi 6} = \frac{2V_T}{I_{B6}} = \frac{2 \times 26\ mV}{2\ nA} = 26\ M\Omega$$

The effective transconductance is $G_m = g_m = 387.6\ \mu A/V$. Thus, the overall voltage gain with load $A_{d(load)}$ is

$$A_{d(load)} = -G_m(R_o \| R_L) = -387.6\ \mu A/V \times (5\ M\Omega \| 26\ M\Omega) = -1625\ V/V$$

9.7.4 Cascode Differential Amplifier

Notice from Eq. (9.154) that the differential gain increases with the output resistance R_o of the differential amplifier. Transistors are often connected in cascode configurations to increase the output resistance and also to improve the frequency response. A common modification to Fig. 9.44 is shown in Fig. 9.45. Transistors Q_5 and Q_6 are connected in a common-base configuration and form a common-base differential stage. Note that Q_5 and Q_6 must be biased by connecting the base terminals of Q_5 and Q_6 to the DC supply V_{CC} through a potential divider.

The equivalent half circuit is shown in Fig. 9.46(a), which can be simplified to Fig. 9.46(b). Q_4 is replaced by its equivalent circuit. R'_o, which is the equivalent output resistance of the Q_2 and Q_6 combination, can be determined from the test circuit shown in Fig. 9.46(c). The emitter resistance of Q_6 is the parallel combination of $r_{\pi 6}$ and r_{o2}; that is,

$$R'_E = r_{\pi 6} \| r_{o2} \approx r_{\pi 6} \tag{9.160}$$

(since $r_{o2} >> r_{\pi 6}$). Using Eq. (9.114), we get R'_o:

$$\begin{aligned} R'_o &= r_{o6}(1 + g_{m6}R'_E) = r_{o6}(1 + g_{m6}r_{\pi 6}) \\ &= r_{o6}(1 + \beta_{F6}) \approx \beta_{F6} r_{o6} \end{aligned} \tag{9.161}$$

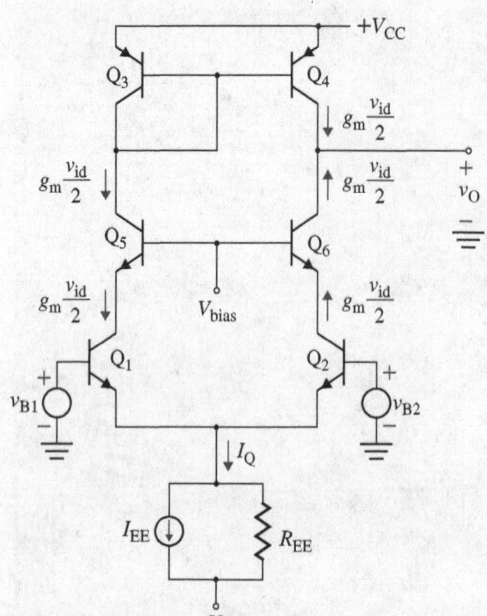

FIGURE 9.45 Cascode differential amplifier

Since all devices are biased at the same current $r_{o6} = r_{o3} = r_{o2} = r_o$, all β_F are the same. Thus,

$$R'_o = \beta_F r_o \tag{9.162}$$

The output resistance of the amplifier becomes

$$R_o = r_o \| R'_o = r_o \| \beta_F r_o \approx r_o \tag{9.163}$$

The differential gain becomes

$$A_d = -g_m R_o = -g_m r_o \tag{9.164}$$

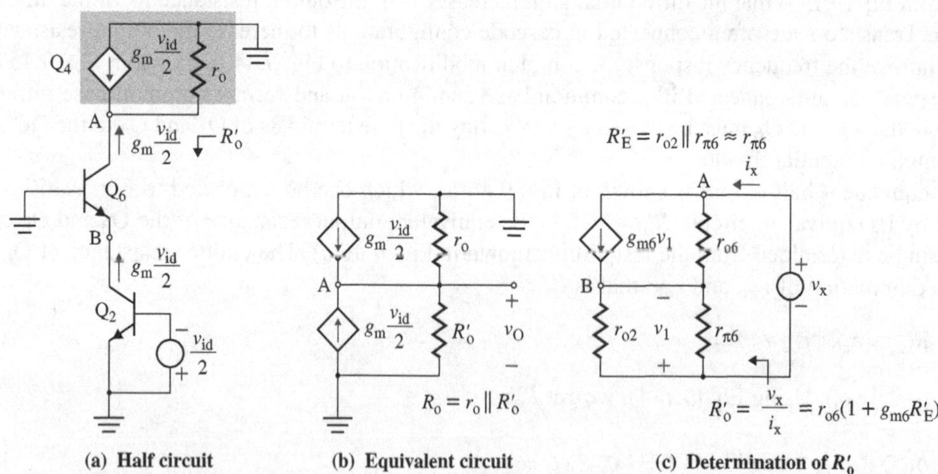

FIGURE 9.46 Circuit for determination of output resistance R'_o

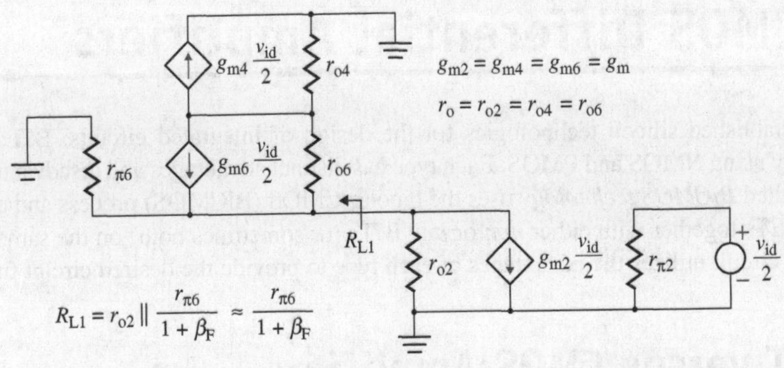

FIGURE 9.47 Circuit for determination of load resistance R_{L1} at the collector of Q_2

Substituting $g_m = I_C/V_T$ and $r_o = V_A/I_C$ into Eq. (9.164), we get

$$A_d = -\frac{V_A}{V_T} \tag{9.165}$$

Thus, the differential gain is twice that of the modified current mirror amplifier in Fig. 9.44. Transistor Q_6 offers a very high resistance seen from the output side.

Transistor Q_6 offers a load resistance R_{L1} seen from the collector of Q_2. R_{L1} can be determined from the test circuit shown in Fig. 9.47; it is the parallel combination of r_{o2}, r_{o6}, and $r_{\pi6}$ (referred to as the emitter of transistor Q_6). That is,

$$R_{L1} = r_{o2} \,\Big\|\, \frac{r_{\pi6}}{1+\beta_F} \,\Big\|\, r_{o6} \equiv \frac{r_{\pi6}}{1+\beta_F} \tag{9.166}$$

which will have a low value. Thus, transistor Q_5 (or Q_6) acts as the *current buffer*, accepting the signal current $(g_m v_{id}/2)$ from the collector of Q_1 at a low resistance R_{L1} and delivering an almost equal current $(g_m v_{id}/2)$ to the load at a very high resistance R'_o. The low resistance at the collector of Q_2 improves the frequency response. Additionally, the internal capacitances of Q_6 (from collector to base and emitter to base) are connected to the ground, and there is no Miller multiplication effect. As a result, the high cutoff frequency is increased.

KEY POINTS OF SECTION 9.7

- The DC transfer characteristic of a BJT differential pair is nonlinear. However, a BJT differential pair is normally operated in the linear region where $v_{id} < V_T$.
- The values of the load resistance R_C and the current source resistance R_{EE} should be large for large values of differential gain and CMRR, respectively. A discrete resistor limits the maximum differential input voltage range.
- The voltage gain is $g_m R_C$ for the difference output and $g_m R_C/2$ for the single-sided output. A small biasing current I_Q increases the transconductance and the voltage gain; however, the differential input resistance is reduced.
- An active load increases the differential gain considerably. The gain is directly proportional to the ratio V_A/V_T. Also, a cascodelike connection increases the gain.

9.8 BiCMOS Differential Amplifiers

There are two established silicon technologies for the design of integrated circuits: BJT technology and CMOS technology using NMOS and PMOS. Each type has distinct advantages and disadvantages. An emerging technology called *BiCMOS technology* uses the bipolar-CMOS (BiCMOS) process and combines *n*- and *p*-channel MOSFETs together with either *npn* or *pnp* BJTs (or sometimes both) on the same semiconductor chip. A BiCMOS circuit utilizes the advantages of each type to provide the desired circuit functions.

9.8.1 BJT versus CMOS Amplifiers

We will begin by considering the basic BJT and CMOS amplifiers. Figure 9.48(a) shows a BJT amplifier with active load that is similar to the half circuit of a BJT differential pair. The output resistance is given by

$$R_o = r_o = \frac{V_A}{I_C} \tag{9.167}$$

where V_A is the Early voltage. Typically, $V_A = 50$ V and $I_C = 5\ \mu$A, so $r_o = 50\text{ V}/(5\ \mu\text{A}) = 10\text{ M}\Omega$. Assuming that the current source load has infinite resistance, the voltage gain is given by

$$A_d = -g_m R_o = -g_m r_o$$

$$= -\frac{I_C}{V_T} \times \frac{V_A}{I_C} = -\frac{V_A}{V_T} \tag{9.168}$$

which is independent of the biasing current I_C. Since typically $V_A = 50$ V and $V_T = 25.8$ mV at room temperature, the intrinsic gain of a BJT amplifier is $A_d = -1938$ V/V. The input resistance is

$$R_i = r_\pi = \beta_F \frac{V_T}{I_C} \tag{9.169}$$

which is generally low. Typically, $\beta_F = 60$ and $I_C = 5\ \mu$A, so $R_i = 60 \times 25\text{ mV}/5\ \mu\text{A} = 300\text{ k}\Omega$. Although lowering the value of I_C will increase R_i, it will lower the g_m of the transistor and hence the upper frequency limit of the amplifier—that is, f_T in Eq. (8.168).

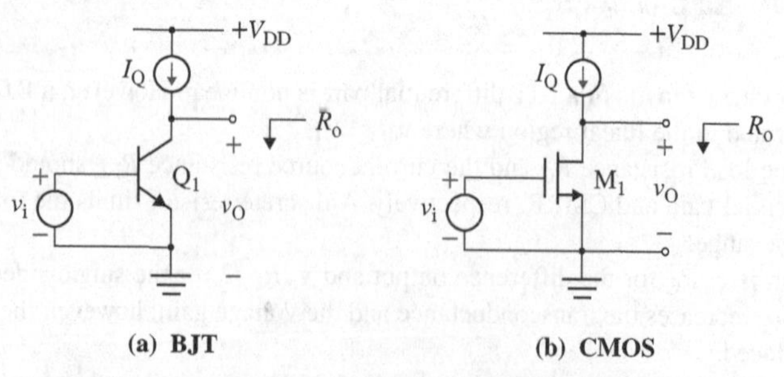

FIGURE 9.48 Basic BJT and CMOS amplifiers

Now consider the MOSFET amplifier shown in Fig. 9.48(b). The output resistance is given by

$$R_o = r_o = \frac{V_M}{I_D} \tag{9.170}$$

where V_M is the channel modulation voltage. Typically, $V_A = 20$ V and $I_D = 5$ μA, so $r_o = 20$ V/(5 μA) = 4 MΩ. Assuming that the current source load has infinite resistance, the voltage gain is given by

$$A_d = -g_m R_o = -g_m r_o$$

$$= -2\sqrt{I_D K_n}\left(\frac{V_M}{I_D}\right) = -2\left(\frac{K_n}{I_D}\right)^{1/2} V_M \tag{9.171}$$

Thus, the gain is inversely proportional to $\sqrt{I_D}$ and will increase as the biasing current is lowered. Decreasing the DC biasing current, however, reduces the amplifier bandwidth. For example, if $I_D = 5$ μA, $V_M = 20$ V, and $K_n = 25$ μA/V^2, $A_d = -89$ V/V.

In summary, for the same value of biasing current I_D, the values of g_m and r_o are much larger for a BJT amplifier than for a MOSFET amplifier—typically 2.5 times as large. The V_A of a BJT amplifier (typically 50 V) is greater than the V_M of a MOSFET (typically 20 V). The voltage gain of a BJT amplifier is greater than that of a MOSFET amplifier by a factor of about 10. However, a MOSFET amplifier has a practically infinite input resistance.

9.8.2 BiCMOS Amplifiers

A BiCMOS amplifier combines the best features of BJT and MOSFET amplifiers. It consists of cascode-like connections of BJTs and MOSFETs. The basic half circuit for a BiCMOS configuration is shown in Fig. 9.49(a). The MOSFET M_1 acts as the driving device, and the BJT Q_1 acts as the load. The cascode configuration is shown in Fig. 9.49(b). The BJT Q_1 acts as the driving device, and the MOSFET M_1 and

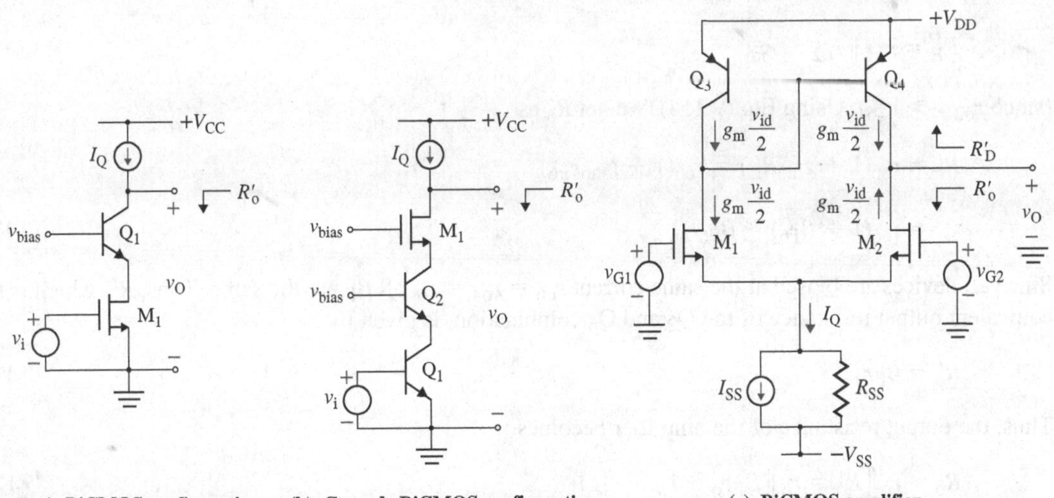

FIGURE 9.49 BiCMOS amplifier

the BJT Q_2 act as the load. A BiCMOS amplifier is shown in Fig. 9.49(c); this amplifier is identical to the CMOS amplifier in Fig. 9.14. As its abbreviation indicates, the current mirror in a BiCMOS is bipolar rather than unipolar, as in MOS devices. Transistors M_1 and M_2 are the amplifying devices. Transistor Q_4 acts as the load of transistor M_2. The output resistance R_o is the parallel combination of the output resistance $r_{o2} = 2V_M/I_Q$ for transistor M_2 and the output resistance $r_{o4} = 2V_A/I_Q$ for transistor Q_4; that is,

$$R_o = R'_o \| R'_D = r_{o2} \| r_{o4} \tag{9.172}$$

$$= \frac{2V_A V_M}{I_Q(V_A + V_M)} \tag{9.173}$$

The differential voltage gain A_d is given by

$$A_d = \frac{v_o}{v_{id}} = -g_m(r_{o2} \| r_{o4}) \tag{9.174}$$

where g_m, which is the transconductance of the driving MOSFET M_2, is given by Eq. (9.54). Substituting $(r_{o4} \| r_{o2}) = 2V_A V_M/I_Q(V_A + V_M)$ and $g_m = \sqrt{2K_n I_Q}$ into Eq. (9.174), we get

$$A_d = -\sqrt{2K_n I_Q}\frac{2V_M V_A}{I_Q(V_A + V_M)} \tag{9.175}$$

Since $r_{o4} > r_{o2}$, R_o and A_d will be greater for a BiCMOS amplifier than for a CMOS amplifier.

9.8.3 Cascode BiCMOS Amplifiers

Like the cascode BJT amplifier in Fig. 9.45, BiCMOS amplifiers can use cascodelike transistors to increase the voltage gain. This arrangement is shown in Fig. 9.50. Transistors Q_5 and Q_6 are connected in a CB configuration and form a CB differential stage. As shown in Fig. 9.46(c), the emitter resistance R'_E of Q_6 is the parallel combination of $r_{\pi 6}$ and r_{o2}; that is,

$$R'_E = r_{\pi 6} \| r_{o2} \approx r_{\pi 6} \tag{9.176}$$

(since $r_{o2} >> r_{\pi 6}$). Using Eq. (9.114), we get R'_o as

$$R'_o = r_{o6}(1 + g_{m6}R'_E) = r_{o6}(1 + g_{m6}r_{\pi 6})$$

$$= r_{o6}(1 + \beta_{F6}) \approx \beta_{F6}r_{o6} \tag{9.177}$$

Since all devices are biased at the same current $r_{o6} = r_{o4} = r_o$, all β_F are the same. Thus, R'_o, which is the equivalent output resistance of the Q_2 and Q_6 combination, is given by

$$R'_o = \beta_F r_o \tag{9.178}$$

Thus, the output resistance of the amplifier becomes

$$R_o = r_o \| R'_o = r_o \| \beta_F r_o \approx r_o \tag{9.179}$$

The differential voltage gain becomes

$$A_d = -g_m R_o = -g_{m1} r_o \tag{9.180}$$

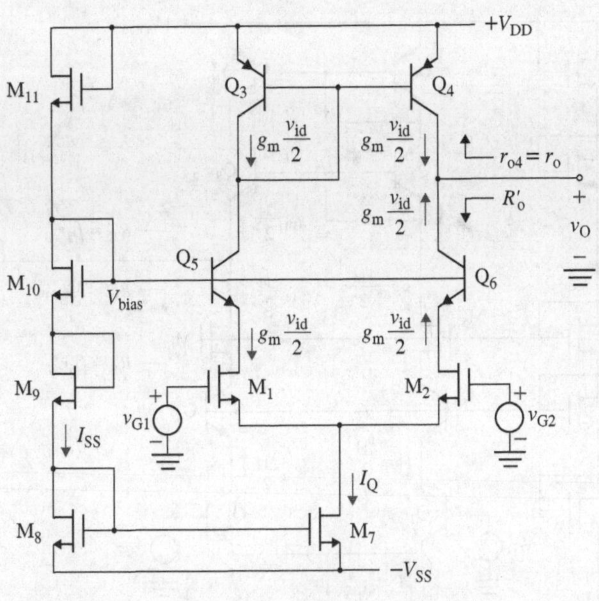

FIGURE 9.50 Cascode BiCMOS amplifier

Substituting $g_m = \sqrt{2K_nI_Q}$ and $r_o = 2V_A/I_Q$ into the preceding equation, we get

$$A_d = -\sqrt{2K_nI_Q}\frac{2V_A}{I_Q} = -2V_A\sqrt{\frac{2K_n}{I_Q}} \tag{9.181}$$

which gives a greater voltage gain than Eq. (9.175).

9.8.4 Double-Cascode BiCMOS Amplifiers

A very high gain can be obtained by double cascoding, as shown in Fig. 9.51. There are two cascode connections: the first connects BJTs Q_1, Q_2, Q_3, and Q_4, and the second connects MOSFETs M_5 and M_6. The resistance R'_S looking from the collector of Q_4 will be

$$R'_S = r_{o4}(1 + \beta_{F4}) \simeq \beta_{F4}r_{o4} \tag{9.182}$$

which will be the source resistance of M_6. Using Eq. (9.33), we can find the output resistance R'_o of M_6 as

$$R'_o \simeq r_{o6}(1 + g_{m6}R'_S) = r_{o6}(1 + g_{m6}\beta_{F4}r_{o4}) \tag{9.183}$$

$$\simeq r_{o6}g_{m6}\beta_{F4}r_{o4} \tag{9.184}$$

Like Q_2 and Q_4, transistors Q_8 and Q_{10} are also connected in cascode, and their equations will be similar to those for Q_2 and Q_4. From Eq. (9.177), the resistance looking from the collector of Q_8 is

$$R'_D = r_{o8}(1 + \beta_{F8}) \simeq \beta_{F8}r_{o8} \tag{9.185}$$

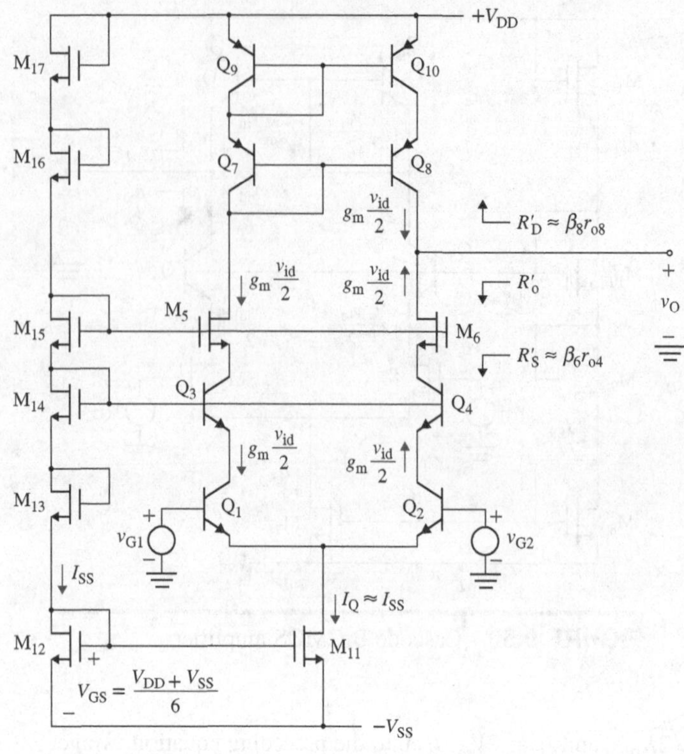

FIGURE 9.51 Double-cascode BiCMOS amplifier

Thus, the output resistance of the amplifier becomes

$$R_o = R'_D \| R'_S = (r_{o6}g_{m6}\beta_{F4}r_{o4}) \| (\beta_{F8}r_{o8}) \tag{9.186}$$

$$= \beta_F r_o \quad (\text{for } r_{o4} = r_{o8} = r_o,\ \beta_{F8} = \beta_F) \tag{9.187}$$

The differential voltage gain becomes

$$A_d = -g_m R_o = -g_m(r_{o6}g_{m6}\beta_{F4}r_{o4}) \| (\beta_{F8}r_{o8}) \tag{9.188}$$

$$= -g_m \beta_F r_o \tag{9.189}$$

where $g_m = I_Q/2V_T$. Equation (9.189) shows a considerable increase in the voltage gain.

EXAMPLE 9.15

Analyzing BiCMOS amplifiers The DC biasing current of a BiCMOS amplifier is kept constant at $I_Q = 10\ \mu A$. All bipolar transistors are identical, with $V_A = 50$ V and $\beta_F = 40$. Also, the MOS transistors are identical, with $V_M = 20$ V, $K_n = 25\ \mu A/V^2$, $W = 30\ \mu m$, and $L = 10\ \mu m$. Assume $V_T = 25.8$ mV.

Determine the differential voltage gain A_d for single-ended output **(a)** for the BiCMOS amplifier in Fig. 9.49(c), **(b)** for the cascode BiCMOS amplifier in Fig. 9.50, and **(c)** for the double-cascode BiCMOS amplifier in Fig. 9.51.

SOLUTION

$V_A = 50$ V, $\beta_F = 40$, $V_M = 20$ V, $K_n = 25\ \mu A/V^2$, $W = 30\ \mu m$, $L = 10\ \mu m$, and $I_D = I_C = I_Q/2 = 10\ \mu A/2 = 5\ \mu A$.

(a) We have

$$r_{o2} = \frac{2V_M}{I_Q} = \frac{2 \times 20\ \text{V}}{10\ \mu\text{A}} = 4\ \text{M}\Omega$$

$$r_{o4} = \frac{2V_A}{I_Q} = \frac{2 \times 50\ \text{V}}{10\ \mu\text{A}} = 10\ \text{M}\Omega$$

$$R_o = r_{o2} \| r_{o4} = 4\ \text{M}\Omega \| 10\ \text{M}\Omega = 2.86\ \text{M}\Omega$$

$$g_{m2} = \sqrt{2K_n I_Q} = \sqrt{2 \times 25\ \mu \times 10\ \mu} = 22.36\ \mu\text{A/V}$$

Thus, the differential voltage becomes

$$A_d = -g_{m2}R_o = -22.36\ \mu \times 2.86\ \text{M} = -63.9$$

(b) We have

$$r_{o4} = r_{o6} = \frac{2V_A}{I_Q} = \frac{2 \times 50\ \text{V}}{10\ \mu\text{A}} = 10\ \text{M}\Omega$$

$$R_o' = \beta_F r_{o6} = 40 \times 10\ \text{M}\Omega = 400\ \text{M}\Omega$$

$$R_o = r_{o4} \| R_o' = 10\ \text{M} \| 400\ \text{M} = 10\ \text{M}\Omega$$

$$g_{m2} = \sqrt{2K_n I_Q} = \sqrt{2 \times 25\ \mu \times 10\ \mu} = 22.36\ \mu\text{A/V}$$

Thus, the differential voltage becomes

$$A_d = -g_{m2}R_o = -22.36\ \mu \times 10\ \text{M} = -223.6\ \text{V/V}$$

(c) From Eq. (9.187),

$$R_o = \beta_F r_{o4} = 40 \times 10\ \text{M} = 400\ \text{M}\Omega$$

$$g_{m2} = \frac{I_Q}{2V_T} = \frac{10\ \mu\text{A}}{2 \times 25.8\ \text{mV}} = 193.8\ \mu\text{A/V}$$

Thus, the differential voltage gain becomes

$$A_d = -g_{m2}R_o = -193.8\ \mu\text{A/V} \times 400\ \text{M}\Omega = -77{,}520\ \text{V/V}$$

which is considerably larger than for the other two configurations.

KEY POINTS OF SECTION 9.8

- A BiCMOS amplifier combines the advantages of BJT and MOS technologies to achieve the desirable circuit functions of infinite input resistance and a very large voltage gain and CMRR.
- BJT and MOS transistors can be connected in cascode to give an extremely large output resistance and voltage gain.

9.9 Frequency Response of Differential Amplifiers

The transistors (either MOSFETs or BJTs) in differential amplifiers have capacitance—hence, the gain of such amplifiers will be frequency dependent [5–7]. The techniques in Chapter 2 for analyzing frequency response can be applied to find the frequency response of differential amplifiers. In this section, we consider only the differential gain with passive and active loads. R_C represents the output resistance of the active source.

9.9.1 Frequency Response with Resistive Load

Let us first consider the differential half circuit shown in Fig. 9.52(a). Replacing the transistor by its frequency model gives the circuit in Fig. 9.52(b). The transistor can be either a BJT or a MOSFET. We can apply the same model and analysis for MOSFETs by substituting $r_\pi = \infty$, $C_\pi \approx C_{gs}$ and $C_\mu \approx C_{gd}$ in the derivations. Since the input signal is $v_{id}/2$, the r_π, C_π, and C_μ parameters of the transistor model are sealed by a factor of 2. Since $g_m(R_C \| r_{o1}) >> 1$, C_μ will dominate the high cutoff frequency ω_H. If we replace $C_\mu/2$ by its effective Miller capacitance C_M, we get the equivalent circuit shown in Fig. 9.52(c) and Eq. (2.98); that is,

$$C_M = \left(\frac{C_\mu}{2}\right)[1 + g_m(R_C \| r_{o1})] \tag{9.190}$$

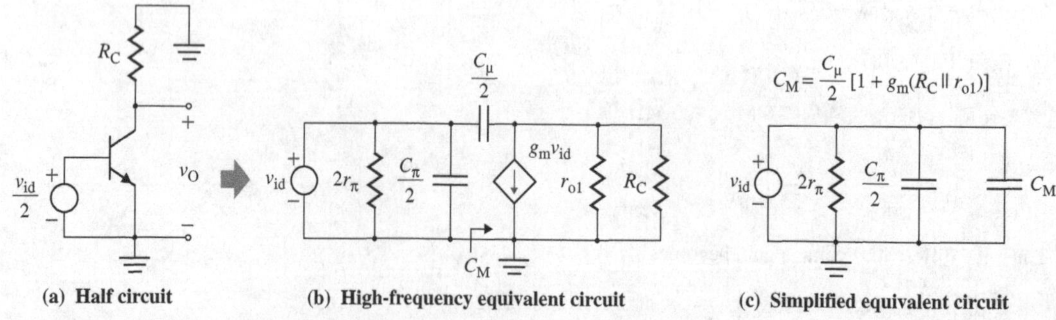

FIGURE 9.52 Differential half circuit and high-frequency equivalent

Thus, the high cutoff frequency ω_H is given by

$$\omega_H = \frac{1}{2r_\pi(C_\pi/2 + C_M)} = \frac{1}{2r_\pi\{C_\pi/2 + (C_\mu/2)[1 + g_m(R_C \| r_{o1})]\}} \tag{9.191}$$

which, for $R_C = r_{o1} = r_o$ and $g_m(R_C \| r_{o1}) >> 1$, can be approximated by

$$\omega_H = \frac{1}{2r_\pi(C_\mu/2)g_m r_o/2} = \frac{2}{r_\pi C_\mu g_m r_o} = \frac{2}{r_\pi C_\mu g_m R_C} \tag{9.192}$$

The signal source resistance $R_s = 0$. The effective resistance for $C_\pi/2$ will be $R_{C\pi} = (2r_\pi \| R_s) = 0$. Thus, the break frequency due to $C_\pi/2$ will be at infinity and will not influence ω_H in Eq. (9.192). Therefore, the frequency-dependent differential gain is given by

$$A_d(j\omega) = \frac{A_{do}}{1 + j\omega/\omega_H} \tag{9.193}$$

where $A_{do} = g_m(r_{o2} \| r_{o4})$ is the low-frequency differential gain with differential output voltage.

9.9.2 Frequency Response with Active Load

Now let us consider the output circuit for the amplifier with an active load, as shown in Fig. 9.42. The AC equivalent of that circuit is shown in Fig. 9.53(a). Replacing the transistors by their frequency model, we get Fig. 9.53(b). Since the effective transconductance is $2g_m$, the Miller capacitance C_M from Eq. (2.98) becomes

$$C_M = \left(\frac{C_\mu}{2}\right)[1 + 2g_m(r_{o2} \| r_{o4})] \tag{9.194}$$

Thus, the high cutoff frequency ω_H is given by

$$\omega_H = \frac{1}{2r_\pi\{C_\pi/2 + (C_\mu/2)[1 + 2g_m(r_{o2} \| r_{o4})]\}} \tag{9.195}$$

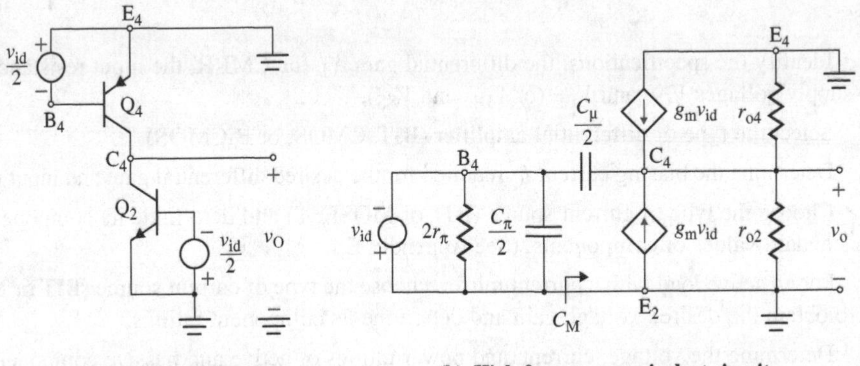

FIGURE 9.53 Differential amplifier with active load and high-frequency equivalent

which, for $r_{o2} = r_{o4} = r_o$ and $2g_m(r_{o2} \| r_{o4}) >> 1$, can be approximated by

$$\omega_H = \frac{1}{2r_\pi/(C_\mu/2)g_m r_o} = \frac{1}{r_\pi C_\mu g_m r_o} \tag{9.196}$$

Therefore, the frequency-dependent differential gain is given by

$$A_d(j\omega) = \frac{A_{do}}{1 + j\omega/\omega_H} \tag{9.197}$$

where $A_{do} = -2g_m(r_{o2} \| r_{o4})$ is the low-frequency differential gain with an active load and a single-sided output.

Notice from Eq. (9.192) that the high cutoff frequency is double that for an amplifier with an active load and single-sided output. However, this high cutoff frequency for an amplifier with a passive load is obtained at the expense of the output voltage gain. For this reason, the first stage of a wide-band amplifier often uses a balanced circuit with differential voltage when the CMRR is not the prime consideration.

KEY POINTS OF SECTION 9.9

- The transistors in differential amplifiers have capacitance; hence, the gain of such amplifiers will be frequency dependent and exhibit a high-pass characteristic.
- The high cutoff frequency for an amplifier with a passive load is obtained at the expense of the output voltage gain. For this reason, the first stage of a wide-band amplifier often uses a balanced circuit with differential voltage.

9.10 Design of Differential Amplifiers

Differential amplifiers are used as the input stage and are designed for a high differential gain, a high CMRR, and high differential and common-mode input resistances. The design of a differential stage involves the following steps:

Step 1. Identify the specifications: the differential gain A_d, the CMRR, the input resistance R_{id}, and the DC supply voltages V_{CC} and V_{EE} (or V_{DD} and V_{SS}).

Step 2. Select the type of differential amplifier (BJT, CMOS, or BiCMOS).

Step 3. Determine the biasing current I_Q required for the desired differential gain and input resistance.

Step 4. Choose the type of current source (BJT or MOSFET) and determine its component ratings. Use the standard values of components. (See Appendix E.)

Step 5. For an active load with a current mirror, choose the type of current source (BJT or MOSFET) needed to obtain the desired voltage gain and determine its component ratings.

Step 6. Determine the voltage, current, and power ratings of active and passive components.

Step 7. Analyze and evaluate the complete differential amplifier to check for the desired specifications.

Step 8. Use PSpice/SPICE to simulate and verify your design, using the standard values of components with their tolerances.

Summary

Active current sources are commonly employed in integrated circuits to bias transistor circuits at appropriate operating points. There are various types of current sources. The output current of a good source is independent of the transistor parameters and offers a high output resistance. The design of BJT current sources can be simplified by ignoring the base currents of transistors and by assuming a constant voltage drop between the base and the emitter. Diode-connected BJTs and MOSFETs rather than diodes are normally used in current sources to give matching mirror characteristics.

A current source can be either a source or a sink, depending on how it is connected. Current mirrors are often used as an active load to increase the output resistance and the voltage gain of a differential amplifier. Also, cascodelike connections increase the voltage gain.

Emitter-coupled (or source-coupled) amplifiers offer the advantage of direct coupling of cascoded stages. They are usually used as the input stage of differential amplifiers to give a high input resistance and a high CMRR. However, the range of the differential input voltage is very small (typically two to three times V_T for the BJT amplifier). The CMRR depends on the output resistance of the biasing current source. Thus, a current source with a high value of output resistance is highly desirable. Source-coupled differential pairs offer a higher input resistance than emitter-coupled pairs, but they have a low CMRR.

References

1. P. E. Allen and D. R. Holberg, *CMOS Analog Circuit Design*. New York: Oxford University Press, 1996.
2. P. M. Chirlian, *Analysis and Design of Integrated Electronic Circuits*. New York: Wiley, 1986.
3. L. J. Giacoletto, *Differential Amplifiers*. New York: Wiley, 1970.
4. V. H. Grinich and H. G. Jackson, *Introduction to Integrated Circuits*. New York: McGraw-Hill, 1975.
5. P. R. Gray, P. J. Hurst, S. H. Lewis, and R. G. Meyer, *Analysis and Design of Analog Integrated Circuits*. New York: Wiley, 2001.
6. R. C. Jaeger and T. Blalock, *Microelectronic Circuit Design*. New York: McGraw-Hill, 2008.
7. B. Razavi, *Design of Analog CMOS Integrated Circuits*. New York: McGraw-Hill, 2001.

Review Questions

1. What are the advantages and disadvantages of a basic current source?
2. What are the advantages and disadvantages of a modified current source?
3. What are the advantages and disadvantages of a Widlar current source?
4. What are the advantages and disadvantages of a cascode current source?
5. What are the advantages and disadvantages of a Wilson current source?
6. What are the differences between Widlar and Wilson current sources?
7. What is a current mirror load?
8. What is the common-mode rejection ratio (CMRR)?
9. What are the advantages of an emitter-coupled pair?
10. What is the DC characteristic of an emitter-coupled pair?
11. What design criteria will yield a large value of CMRR in an emitter-coupled pair?
12. What are the advantages of a source-coupled pair?

13. What is the DC characteristic of a source-coupled pair?
14. What design criteria will yield a large value of CMRR in a source-coupled pair?
15. What is the purpose of cascodelike connections of transistors?

Problems

The symbol **D** indicates that a problem is a design problem. The symbol **P** indicates that you can check the solution to a problem using PSpice/SPICE.

Assume that the device parameters are as follows: diodes, $I_S = 10^{-13}$ A and $I_{D(min)} = 1$ mA to ensure conduction; transistors, $\beta_F = h_{fe} = 50$, $V_{BE} = 0.7$ V, $I_S = 10^{-14}$ A, and $V_{CE(sat)} = 0.2$ V.

9.3 MOSFET Current Sources

9.1 The parameters of the MOSFET current source in Fig. 9.4 are $V_t = 1$ V, $I_O = 20$ μA, $I_R = 20$ μA, $V_{DD} = 15$ V, and $V_M = 40$ V. The channel lengths are $L_1 = L_2 = 10$ μm and $L_3 = 100$ μm, and $K_x = 20$ μA/V^2. Calculate the required values of **(a)** K_{n1}, W_1, **(b)** K_{n2}, W_2, **(c)** K_{n3}, W_3, and **(d)** the output resistance R_o of the current source. Assume $V_{GS1} = 1.5$ V and $V_{DS1} = 5$ V.

9.2 D **a.** Design the cascode current source in Fig. 9.6(a) to give $I_O = 10$ μA. Assume $V_{DD} = 10$ V. All MOSFETs are identical and have $L = 20$ μm, $W = 60$ μm, $V_t = 1$ V, $K_x = 20$ μA/V^2, and $V_M = 40$ V.
b. Calculate the output resistance R_o and Thevenin's equivalent voltage V_{Th}.

9.3 D **a.** Design the modified Wilson current source in Fig. 9.7(c) to give $I_O = 10$ μA. Assume $V_{DD} = 10$ V and $V_M = 40$ V. All MOSFETs are identical and have $L = 10$ μm, $W = 40$ μm, $K_x = 20$ μA/V^2, and $V_M = 40$ V.
b. Calculate the output resistance R_o and Thevenin's equivalent voltage V_{Th}.

9.4 D Design a MOSFET current source so that $R_o \geq 50$ kΩ at an output current of $I_O = 1$ mA.

9.5 D Design a MOSFET current source so that $R_o \geq 500$ kΩ at an output current of $I_O = 0.1$ mA.

9.6 D The Widlar current source shown in Fig. P9.6 has $I_{ref} = 50$ μA, $R = 2$ kΩ, and $V_{DD} = 12$ V. The MOS parameters are $K_n = 100$ μA/V^2, $V_t = 1$ V, $|V_M| = 100$ V, and $(W/L)_1 = (W/L)_2 = 20$. Determine **(a)** the output current I_o, **(b)** the output resistance r_{o2}, and **(c)** the value of R_{ref}.

FIGURE P9.6

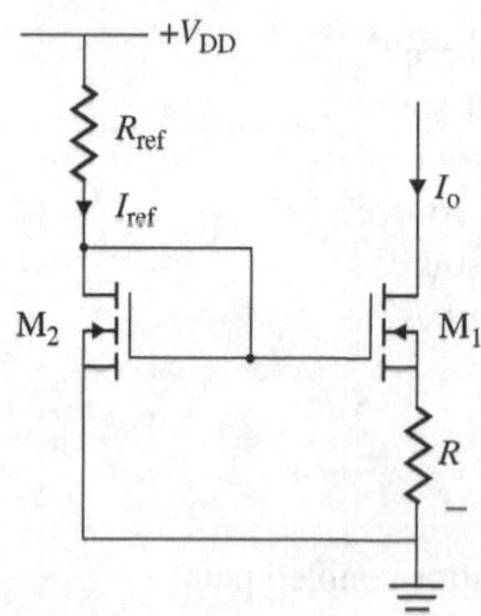

9.7 Design the modified Wilson current source in Fig. P9.7 to give $I_O = 10\ \mu A$ by determining the resistance R_{ref}. Assume $V_{DD} = 12$ V. All MOSFETs are identical and have $L = 10\ \mu m$, $W = 40\ \mu m$, $K_x = 50\ \mu A/V^2$, and $V_M = 100$ V. Calculate the output resistance R_o and Thevenin's equivalent voltage V_{Th}.

FIGURE P9.7

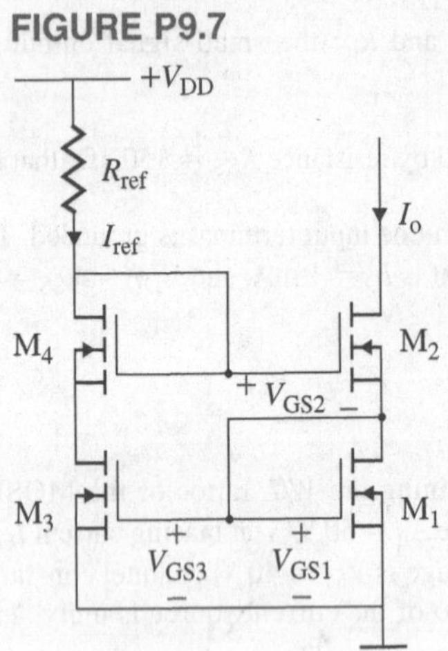

9.8 The current source shown in Fig. P9.8 has $I_{ref} = 150\ \mu A$ and $V_{DD} = 12$ V. The NMOS parameters are $K_n = 200\ \mu A/V^2$, $V_t = 1$ V, $V_M = -100$ V, and $(W/L)_1 = (W/L)_2 = (W/L)_3 = 20$. Determine **(a)** the gate–source voltages (V_{GS1}, V_{GS2}, V_{GS3}), **(b)** the drain–source voltages (V_{DS1}, V_{DS2}, V_{DS3}) if the devices are to operate in the active region, and **(c)** the value of R_{ref}.

FIGURE P9.8

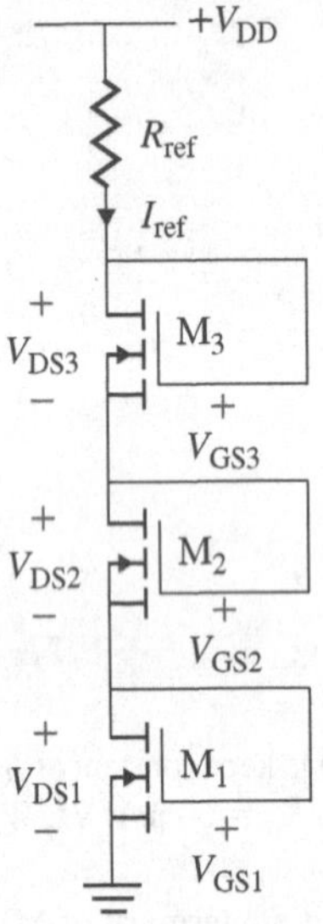

9.9 Repeat Prob. 9.8 if the current source in Fig. P9.8 has only two MOSFETs.

9.4 MOS Differential Amplifiers

9.10 The parameters of the MOS differential pair in Fig. 9.8 are $R_{SS} = 150\ \text{k}\Omega$, $I_Q = 1$ mA, $V_{DD} = 15$ V, and $R_D = 5\ \text{k}\Omega$. The MOSFETs are identical and have $K_n = 1.5\ \text{mA/V}^2$, $W/L = 1$, and $V_t = 1.2$ V.

a. Calculate the DC drain currents through the MOSFETs if $v_{id} = 10$ mV.

b. Assuming $I_{D1} = I_{D2}$, calculate A_d, A_c, and CMRR; R_{id} and R_{ic}; the small-signal output voltage if $v_{g1} = 5$ mV and $v_{g2} = 10$ mV; and the drain voltage V_D.

9.11 Repeat Prob. 9.10 if the active biasing current source is replaced by resistance $R_{SS} = 150\ \text{k}\Omega$; that is, $I_Q = 0$.

9.12 D Design a MOS differential pair as shown in Fig. 9.13 in which one input terminal is grounded. The output is taken from the drain of transistor M_1. The DC biasing current is $I_Q = 1$ mA, and $V_{DD} = V_{SS} = 15$ V. The MOSFETs are identical and have $K_n = 1.5\ \text{mA/V}^2$ and $V_t = 1.2$ V.

a. For a small-signal voltage gain of $A_1 = -20$ V/V.

b. Calculate the design values of A_d, A_c, and CMRR.

9.13 D Design the CMOS amplifier shown in Fig. 9.16 by determining the W/L ratios of the MOSFETs and the threshold voltage V_t. The differential voltage gain should be $A_d = 50$ V/V at biasing current $I_Q = 1$ mA. Assume identical transistors whose channel modulation voltage is $V_M = 40$ V, channel constant is $K_x = 10\ \mu\text{A/V}^2$, and channel length is $L = 10\ \mu$m. The W/L ratio of the current source is unity, and $V_{DD} = V_{SS} = 10$ V.

9.14 A CMOS amplifier is shown in Fig. P9.14. The parameters for the NMOS are $V_t = +2$ V, $V_M = -40$ V, and $V_{GS} = +4$ V at $I_D = 1$ mA; the parameters for the PMOS are $V_t = -3$ V, $V_M = 40$ V, and $V_{GS} = -6$ V at $I_D = 1$ mA. Calculate **(a)** A_d, A_c, and CMRR and **(b)** R_{id} and R_{ic}.

FIGURE P9.14

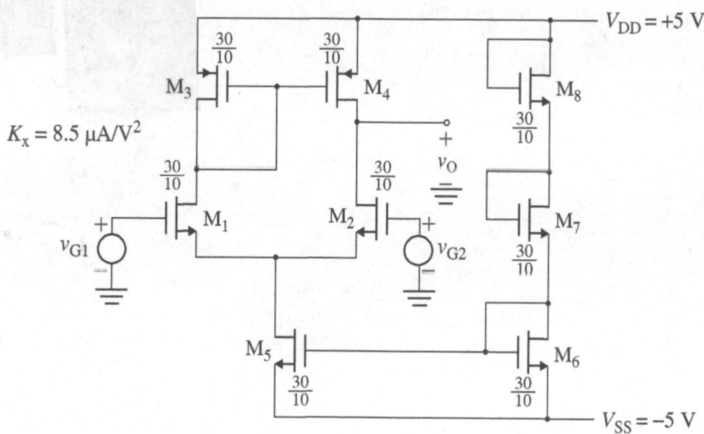

9.15 D The DC biasing current of the cascoded MOS amplifier shown in Fig. 9.19(a) is kept constant at $I_Q = 50\ \mu$A and $V_{DD} = V_{SS} = 5$ V. All MOS transistors are identical and have $V_M = 50$ V, $K_n = 50\ \mu\text{A/V}^2$, $W = 50\ \mu$m, and $L = 10\ \mu$m.

a. Determine the differential voltage gain A_d for single-ended output at the drain terminal of M_4.

b. Use PSpice to verify the result.

9.16 The MOS parameters of the CMOS amplifier shown in Figure P9.16 are $K_n = 100\ \mu A/V^2$, $V_{tN} = 1$ V, $V_{tP} = 1.5$ V, $|V_M| = 100$ V, $L = 10\ \mu m$, $(W/L)_1 = 60$, $(W/L)_7 = (W/L)_8 = 120$, and $(W/L)_2 = (W/L)_3 = (W/L)_4 = (W/L)_5 = (W/L)_6 = 30$. The DC supply voltages are $V_{DD} = V_{SS} = 5$ V, the reference current is to be set at $I_{ref} = 100\ \mu A$, and $C_L \approx \infty$.

a. Determine the value of R_{ref} and V_{GS6}.

b. Calculate A_d, A_c, and CMRR.

c. Determine the value of capacitor C_x to limit the upper frequency at $f_H = 500$ kHz.

FIGURE P9.16

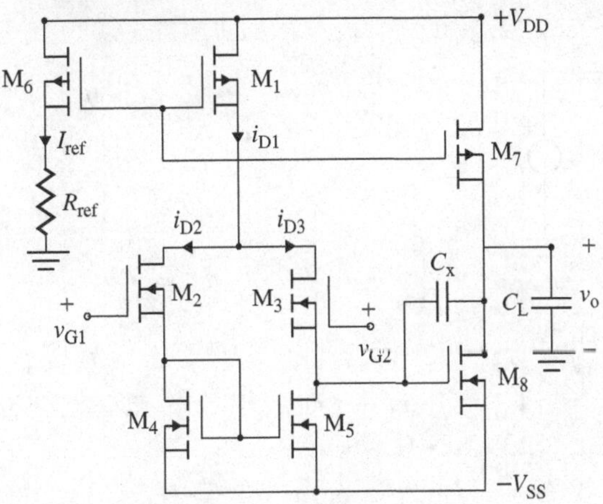

9.17 D Design a CMOS amplifier as shown in Fig. P9.16 by determining the W/L ratios of the MOSFETs and the threshold voltage V_t. The small-signal voltage gain should be $A_d = 250$ V/V at biasing current $I_{ref} = 100\ \mu A$. Assume identical transistors whose channel modulation voltage is $V_M = 70$ V, channel constant is $K_x = 10\ \mu A/V^2$, and channel length $L = 10\ \mu m$. Also, the width $(W/L)_6 = 1$ and $V_{DD} = V_{SS} = 5$ V.

9.5 *Depletion MOS Differential Amplifiers*

9.18 The parameters of the depletion MOS differential pair in Fig. 9.22 are $R_{SS} = 50\ k\Omega$, $I_Q = 1$ mA, $V_{DD} = V_{SS} = 30$ V, and $R_D = 2\ k\Omega$. The MOSFETs are identical and have $V_p = -4$ V and $I_{DSS} = 20$ mA.

a. Calculate the DC drain currents through the MOSFETs if $v_{id} = 30$ mV.

b. Assuming $I_{D1} = I_{D2}$, calculate A_d, A_c, and CMRR; R_{id} and R_{ic}; and the small-signal output voltage if $v_{g1} = 50$ mV and $v_{g2} = 20$ mV.

9.19 Repeat Prob. 9.18 if the transistor current source is replaced by resistance $R_{SS} = 100\ k\Omega$; that is, $I_{SS} = 0$.

9.20 D P **a.** Design a depletion MOS differential pair as shown in Fig. 9.13 in which one input terminal is grounded. The output is taken from the drain of transistor M_2. The DC biasing current is $I_Q = 5$ mA, and $V_{DD} = V_{SS} = 15$ V. The MOSFETs are identical and have $V_p = -4$ V and $I_{DSS} = 20$ mA. A small-signal voltage gain of $|A_2| = 20$ V/V is required.

b. Calculate the design values of A_d, A_c, and CMRR.

9.21 A depletion MOS amplifier is shown in Fig. P9.21. The MOS parameters are $V_p = 4$ V, $I_{DSS} = -400$ μA, $V_M = 40$ V, $I_Q = 200$ μA, and $V_{DD} = V_{SS} = 15$ V.

a. Calculate the DC drain currents through the MOSFETs if $v_{id} = 10$ mV.

b. Assuming $I_{D1} = I_{D2}$, calculate A_d, A_c, and CMRR; R_{id} and R_{ic}; and the small-signal output voltage if $v_{g1} = 20$ mV and $v_{g2} = 10$ mV.

FIGURE P9.21

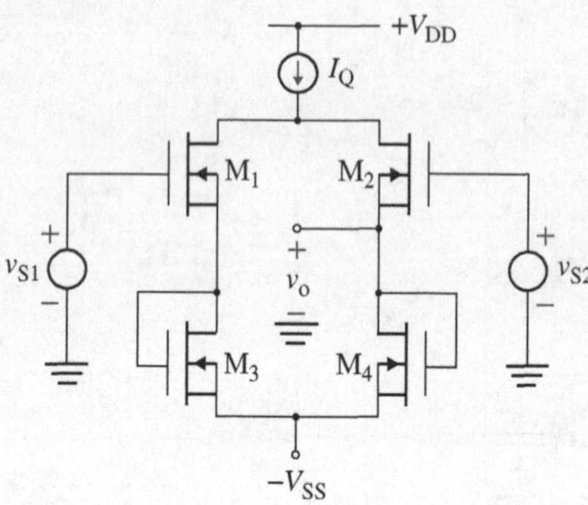

9.6 BJT Current Sources

9.22 The parameters of the basic current source in Fig. 9.25(a) are $\beta_F = 150$, $R_1 = 20$ kΩ, $V_{CC} = 15$ V, $V_{BE1} = V_{BE2} = 0.7$ V, and $V_A = 100$ V. Calculate **(a)** the output current $I_O = I_{C2}$, **(b)** the output resistance R_o, **(c)** Thevenin's equivalent voltage V_{Th}, and **(d)** the collector current ratio I_{C2}/I_{C1} if $V_{CE2} = 30$ V.

9.23 D P a. Design the basic current source in Fig. 9.25(a) to give an output current of $I_O = 200$ μA. The transistor parameters are $\beta_F = 100$, $V_{CC} = 30$ V, $V_{BE1} = V_{BE2} = V_{CE1} = 0.7$ V, and $V_A = 150$ V.

b. Calculate the output resistance R_o, Thevenin's equivalent voltage V_{Th}, and the collector current ratio if $V_{CE2} = 30$ V.

9.24 The parameters of the modified current source in Fig. 9.27(a) are $\beta_F = 150$, $R_1 = 10$ kΩ, $V_{CC} = 15$ V, $V_{BE1} = V_{BE2} = V_{BE3} = 0.7$ V, and $V_A = 100$ V. Calculate **(a)** the output current $I_O = I_{C2}$, **(b)** the output resistance R_o, **(c)** Thevenin's equivalent voltage V_{Th}, and **(d)** the collector current ratio I_{C2}/I_{C1} if $V_{CE2} = 30$ V.

9.25 D P a. Design the modified basic current source in Fig. 9.27(a) to give an output current of $I_O = 50$ μA. The transistor parameters are $\beta_F = 150$, $V_{CC} = 30$ V, $V_{BE1} = V_{BE2} = V_{BE3} = 0.7$ V, and $V_A = 100$ V.

b. Calculate the output resistance R_o, Thevenin's equivalent voltage V_{Th}, and the collector current ratio if $V_{CE2} = 20$ V.

9.26 The multiple transistors of the current source in Fig. P9.26 have $\beta_F = 150$, $R_1 = 10$ kΩ, $V_{CC} = 15$ V, and $V_A = 100$ V. The B-E voltages are equal, $V_{BE} = 0.7$ V. Calculate **(a)** the output current I_O, **(b)** the output resistance R_o, **(c)** Thevenin's equivalent voltage V_{Th}, and **(d)** the collector current ratio if $V_{CE2} = 15$ V.

FIGURE P9.26

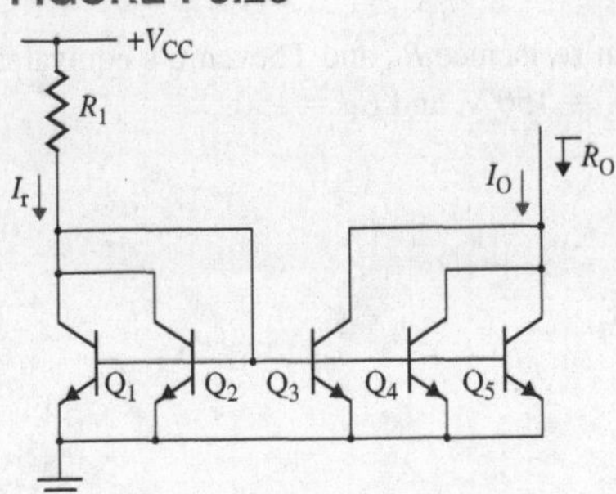

9.27 D P **a.** Design the Widlar current source in Fig. 9.28(a) to give $I_O = 10\ \mu A$ and $I_R = 2$ mA. The parameters are $V_{CC} = 30$ V, $V_{BE1} = 0.7$ V, $V_T = 26$ mV, $V_A = 150$ V, and $\beta_F = 100$.

b. Calculate the output resistance R_o and Thevenin's equivalent voltage V_{Th}.

9.28 D P Design a Widlar current source as shown in Fig. 9.28(a) to produce a 10-μA output current. Assume $\beta_F = 100$, $V_{CC} = 30$ V, and $R_1 = 30$ kΩ. Calculate the output resistance R_o.

9.29 Determine the output current I_O and the output resistance R_o of the current source circuit in Fig. P9.29. Assume $V_{CC} = 30$ V, $R_1 = 20$ kΩ, $R_2 = 10$ kΩ, $V_{BE} = 0.7$ V, $V_A = 150$ V, and $\beta_F = 100$.

FIGURE P9.29

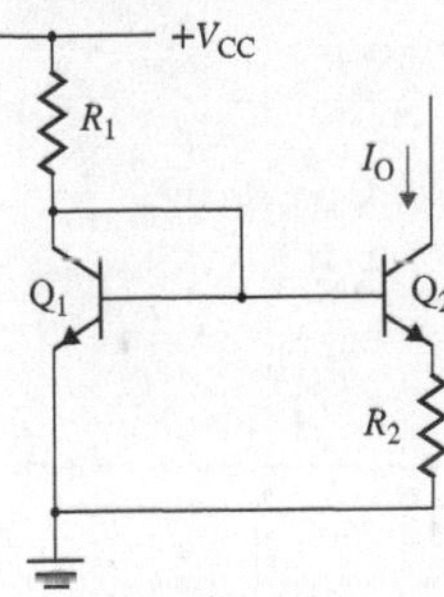

9.30 D P **a.** Design the Wilson current source in Fig. 9.30(a) to give $I_O = 10\ \mu A$. The parameters are $V_{CC} = 30$ V, $V_{BE} = 0.7$ V, $V_T = 26$ mV, $V_A = 100$ V, and $\beta_F = 150$.

b. Calculate the output resistance R_o and Thevenin's equivalent voltage V_{Th}.

9.31 For the Wilson current source in Fig. P9.31, determine the output current I_O and the output resistance R_o. Assume $V_{CC} = 20$ V, $V_{BE} = 0.7$ V, $V_T = 26$ mV, $V_A = 150$ V, and $\beta_F = 150$.

FIGURE P9.31

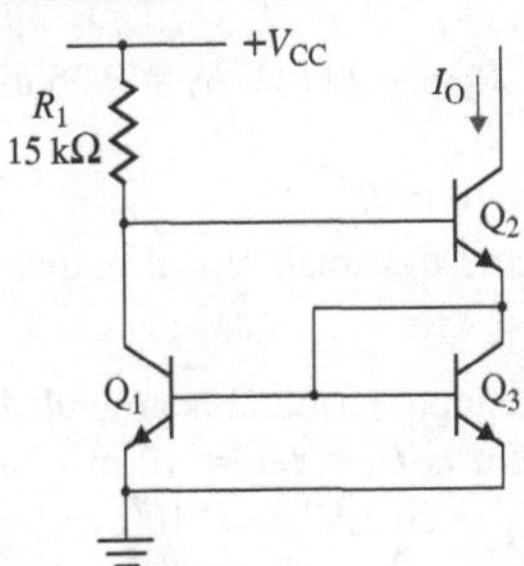

9.32 Repeat Prob. 9.31 for $V_{CC} = 30$ V.

9.33 For the current source in Fig. P9.33, determine the output resistance R_o and Thevenin's equivalent voltage V_{Th}. Assume $V_{CC} = 30$ V, $V_{BE} = 0.7$ V, $V_T = 26$ mV, $V_A = 150$ V, and $\beta_F = 150$.

FIGURE P9.33

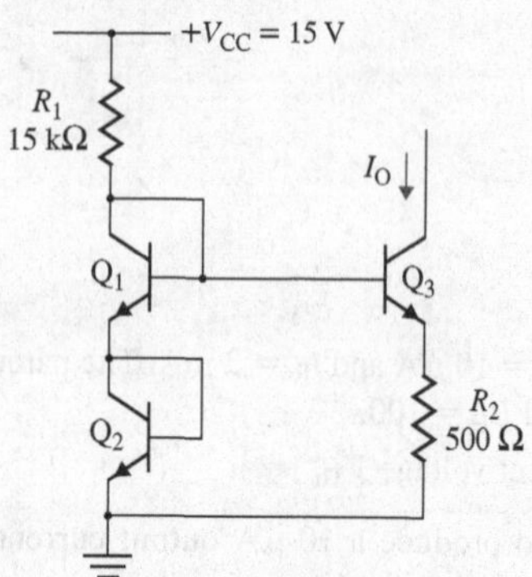

9.34 Determine the sensitivity S of output current I_O to supply voltage V_{CC} for the circuit in Fig. P9.34. S is defined as

$$S = \frac{V_{CC}/I_O}{\delta I_O/\delta V_{CC}}$$

FIGURE P9.34

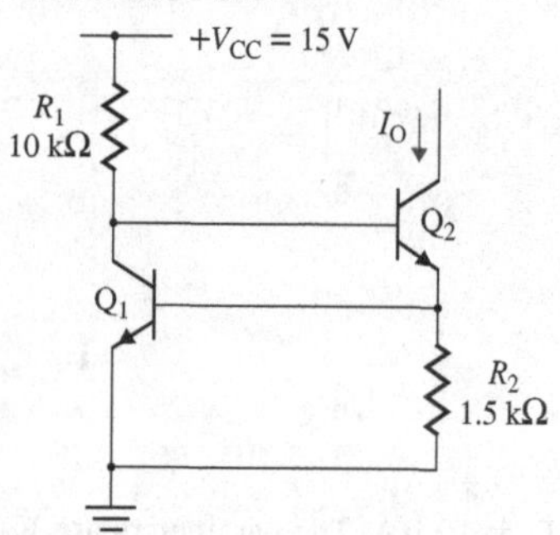

9.35 D Design a BJT current source so that $R_o \geq 50$ kΩ at an output current of $I_O = 1$ mA.

9.36 D Design a BJT current source so that $R_o \geq 500$ kΩ at an output current of $I_O = 1$ mA.

9.7 BJT Differential Amplifiers

9.37 P The parameters of the emitter-coupled pair in Fig. 9.33 are $\beta_F = 150$, $R_{EE} = 20$ kΩ, $I_Q = 0.25$ mA, $V_{CC} = 12$ V, and $R_C = 10$ kΩ.

a. Calculate the DC collector currents through the transistors if $v_{id} = 10$ mV.

b. Assuming $I_{C1} = I_{C2}$, calculate A_d, A_c, and CMRR; R_{id} and R_{ic}; and the small-signal output voltage if $v_{B1} = 30$ mV and $v_{B2} = 20$ mV. Assume $V_T = 26$ mV.

9.38 D **a.** Design an emitter-coupled pair as shown in Fig. 9.41, in which one input terminal is grounded. The output is taken from the collector of transistor Q_2. The biasing current is $I_Q = I_{EE} = 10$ mA, and $V_{CC} = -V_{EE} = 12$ V. The transistors are identical. Assume $V_{BE} = 0.7$ V, $V_T = 26$ mV, $\beta_F = 100$, and $V_A = 40$ V. A small-signal voltage gain of $A_1 = -150$ V/V is required.

b. Calculate the design values of A_d, A_c, and CMRR.

9.39 A differential amplifier is shown in Fig. P9.39. The transistors are identical. Assume $V_{BE} = 0.7$ V, $V_T = 26$ mV, $\beta_F = 50$, and $V_A = 40$ V. Calculate the values of A_d, R_{id}, A_c, R_{ic}, and CMRR.

FIGURE P9.39

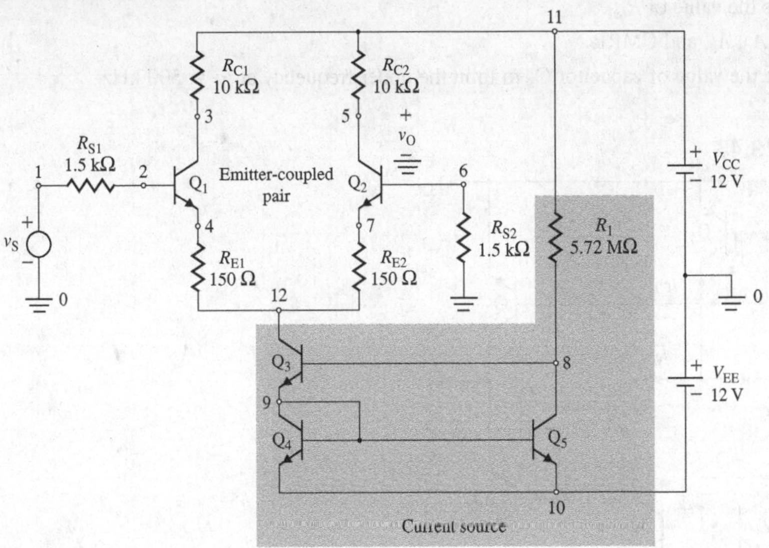

9.40 The parameters of the differential amplifier in Fig. 9.44 are $\beta_{F(npn)} = 100$, $\beta_{F(pnp)} = 50$, $V_A = 40$ V, $I_Q = 10$ μA, and $V_{CC} = 10$ V. Calculate A_d, R_{id}, R_o, and the overall voltage gain with load $A_{d(load)}$. Assume $V_T = 26$ mV.

9.41 The parameters of the differential amplifier in Fig. 9.45 are $\beta_{F(npn)} = 100$, $\beta_{F(pnp)} = 50$, $V_A = 40$ V, $I_Q = 5$ μA, and $V_{CC} = 10$ V. Calculate A_d, R_{id}, R_o, and the overall voltage gain with load $A_{d(load)}$. Assume $V_T = 26$ mV.

9.42 A differential amplifier is shown in Fig. P9.42. The transistors are identical. Assume $V_{BE} = 0.7$ V, $V_T = 26$ mV, $\beta_{F(npn)} = 100$, $\beta_{F(pnp)} = 50$, $V_A = 40$ V, and $V_{CC} = 10$ V. Calculate the values of R_1, R_2, A_d, R_{id}, A_c, R_{ic}, and CMRR.

FIGURE P9.42

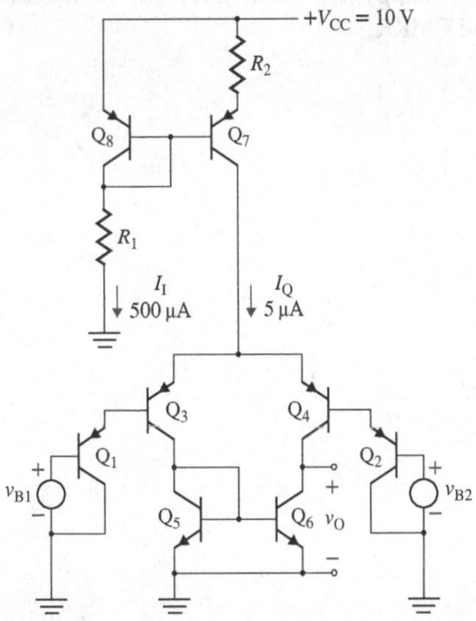

9.43 The parameters of the BJT amplifier shown in Fig. P9.43 are $\beta_{FN} = 100$, $\beta_{FP} = 120$, $|V_A| = 100$ V, and $V_{BEN} = -V_{BEP} = 0.65$ V. The DC supply voltages are $V_{CC} = V_{EE} = 5$ V, the reference current is to be set at $I_{ref} = 100$ μA, and $R_L = 50$ kΩ.

a. Determine the value of R_{ref}.

b. Calculate A_d, A_c, and CMRR.

c. Determine the value of capacitor C_x to limit the upper frequency at $f_H = 500$ kHz.

FIGURE P9.43

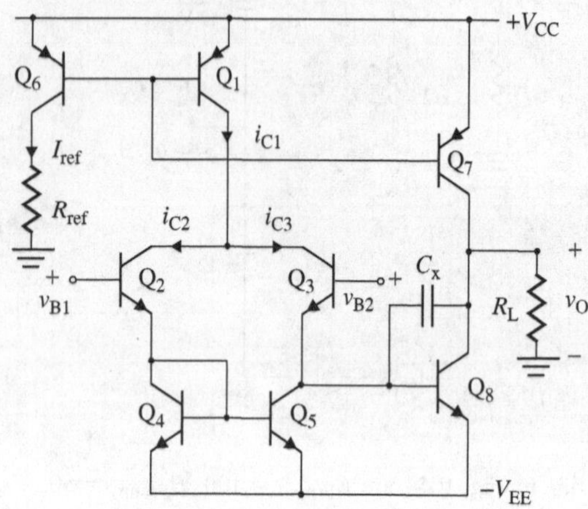

9.44 (D) Design a differential amplifier as shown in Fig. 9.44 to satisfy the following conditions: a differential voltage gain of $A_o \geq 1500$ V/V, CMRR of ≥ 3500, and load of $R_L = 50$ kΩ, with a coupling capacitor of $C_L = \infty$. Use all BJTs and the cascoded current source, similar to the one in Fig. 9.29. Use PSpice to verify your design with a $v_{id} = 5$ μV and $v_{ic} = 10$ mV. Supply voltages are $V_{CC} = -V_{EE} = 15$ V.

9.8 BiCMOS Differential Amplifiers

9.45 A BiCMOS amplifier is shown in Fig. P9.45. The depletion PMOS parameters are $V_p = 4$ V, $I_{DSS} = 400$ μA, and $V_M = 100$ V. The BJT parameters are $\beta_{F(npn)} = 100$, $\beta_{F(pnp)} = 50$, and $V_A = 40$ V. Assume $V_{DD} = V_{EE} = 15$ V and $I_Q = 200$ μA. Calculate A_d, A_c, and CMRR.

FIGURE P9.45

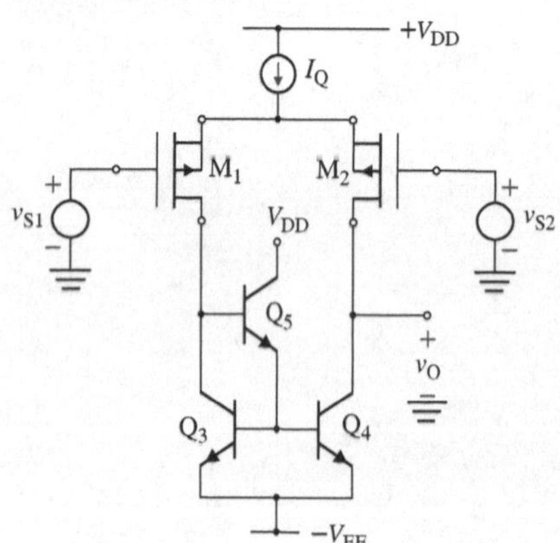

9.46 The DC biasing current of a BiCMOS amplifier is kept constant at $I_Q = 5$ μA. All bipolar transistors are identical, with $\beta_{F(npn)} = 100$, $\beta_{F(pnp)} = 50$, and $V_A = 40$ V. Also, the MOS transistors are identical, with $|V_M| = 20$ V. For the NMOS, $V_t = +2$ V and $V_{GS} = +4$ V at $I_D = 1$ mA; for the PMOS, $V_t = -3$ V and $V_{GS} = -6$ V at $I_D = 1$ mA. Determine the differential voltage gain A_d for single-ended output **(a)** for the basic BiCMOS amplifier in Fig. 9.49(c), **(b)** for the cascode BiCMOS amplifier in Fig. 9.50, and **(c)** for the double-cascode BiCMOS amplifier in Fig. 9.51.

9.47 A BiCMOS amplifier is shown in Fig. P9.47. The PMOS parameters are $V_t = -3$ V and $V_{GS} = -6$ V at $I_D = 1$ mA. The BJT parameters are $\beta_{F(npn)} = 100$, $\beta_{F(pnp)} = 50$, and $V_A = 40$ V. Assume $V_{DD} = -V_{EE} = 15$ V and $I_Q = 200$ μA. Calculate A_d, A_c, and CMRR.

FIGURE P9.47

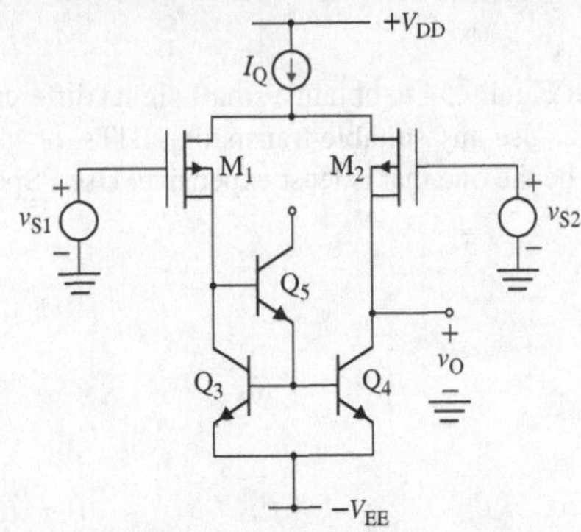

9.48 (D) Design a differential amplifier as shown in Fig. 9.51 to satisfy the following conditions: a differential voltage gain of $A_o \geq 1500$ V/V, CMRR of ≥ 3500, and load of $R_L = 50$ kΩ, with a coupling capacitor of $C_L = \infty$. Use all BJTs and NMOSs. Use PSpice to verify your design with $v_d = 5$ μV and $v_c = 10$ mV. Supply voltages are $V_{CC} = V_{EE} = 12$ V.

9.9 *Frequency Response of Differential Amplifiers*

9.49 The emitter-coupled pair in Fig. 9.33 has $R_{EE} = 20$ kΩ, $I_Q = 5$ mA, $V_{CC} = 12$ V, and $R_C = 10$ kΩ. The small-signal transistor parameters are $C_\pi = 5$ pF, $C_\mu = 2$ pF, $\beta_{F(npn)} = 100$, and $\beta_{F(pnp)} = 50$. Find the frequency-dependent gains $A_d(j\omega)$ and $A_c(j\omega)$.

9.50 The emitter-coupled pair in Fig. P9.50 has $I_Q = 5$ mA, $V_{CC} = -V_{EE} = 15$ V, and $R_C = 10$ kΩ. The small-signal transistor parameters are $C_\pi = 5$ pF, $C_\mu = 2$ pF, $\beta_{F(npn)} = 100$, and $\beta_{F(pnp)} = 50$. Find the frequency-dependent gains $A_d(j\omega)$ and $A_c(j\omega)$.

FIGURE P9.50

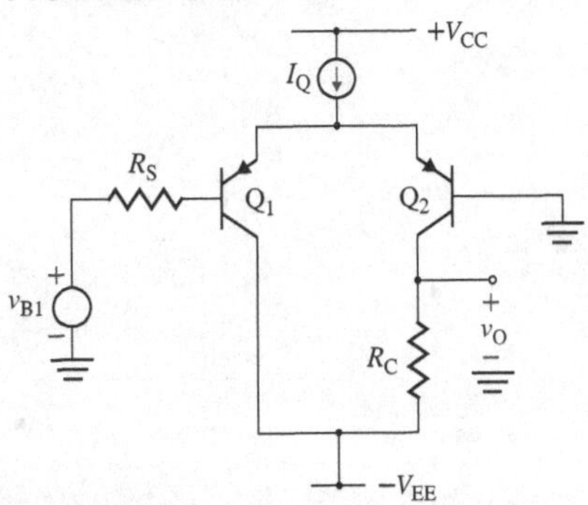

9.51 The differential amplifier in Fig. 9.42 has $V_A = 40$ V, $I_Q = 10$ μA at $R_{EE} = 50$ kΩ, and $V_{CC} = -V_{EE} = 15$ V. The small-signal transistor parameters are $C_\pi = 5$ pF, $C_\mu = 2$ pF, $\beta_{F(npn)} = 100$, and $\beta_{F(pnp)} = 50$. Find the frequency-dependent gains $A_d(j\omega)$ and $A_c(j\omega)$.

9.52 The source-coupled pair in Fig. 9.8 has $R_{SS} = 50$ kΩ, $I_Q = 10$ mA, $V_{DD} = 30$ V, and $R_D = 5$ kΩ. The MOSFETs are identical and have $K_n = 1.5$ mA/V^2, $W/L = 1$, $|V_M| = 100$ V, and $V_t = 1.2$ V. The small-signal MOS parameters are $C_{gs} = 5$ pF and $C_{gd} = 2$ pF. Find the frequency-dependent gains $A_d(j\omega)$ and $A_c(j\omega)$.

9.53 The MOS amplifier shown in Fig. 9.14 has $R_{SS} = 50$ kΩ, $I_Q = 1$ mA, and $V_{DD} = 15$ V. The MOSFETs are identical and have $K_n = 1.5$ mA/V^2, $W/L = 1$, $|V_M| = 100$ V, and $V_t = 1.2$ V. The small-signal MOS parameters are $C_{gs} = 5$ pF and $C_{gd} = 2$ pF. Find the frequency-dependent gains $A_d(j\omega)$ and $A_c(j\omega)$.

9.54 D P Design a differential amplifier configuration as shown in Fig. P9.54 to obtain a small-signal differential gain, with $A_d = 250$ V/V, CMRR $\geq 10^3$, and $f_b \geq 250$ kHz. Use any suitable transistors, BJTs, or MOSFETs. Assume $V_{CC} = V_{EE} = 12$ V. The configuration should be the one that is least expensive. Use PSpice to verify your design and give the component estimates.

FIGURE P9.54

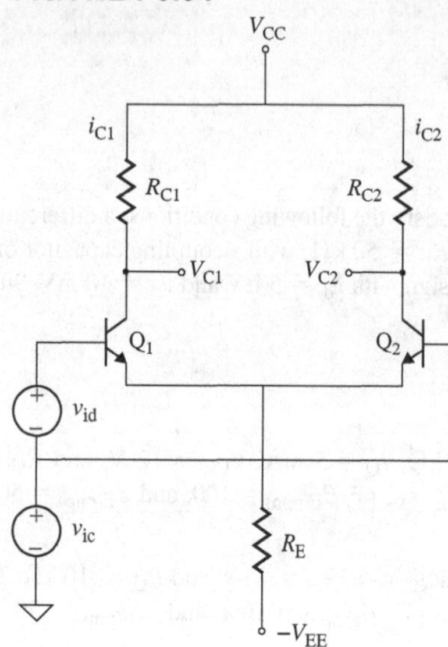

CHAPTER 10

FEEDBACK AMPLIFIERS

Learning Outcomes

After completing this chapter, students should be able to do the following:

- Describe the types and the properties of feedback amplifiers.
- Describe the different feedback configurations, their properties, and circuit implementations.
- Select the feedback configuration and analyze feedback amplifiers to meet specific requirements.
- Design a feedback network to meet desired closed-loop gain, the input impedance, and the output impedance.
- Determine the stability conditions of feedback amplifiers.
- Apply the compensation techniques to stabilize an unstable amplifier.

Symbols and Their Meanings

Symbol	**Meaning**
A, A_f	Open-loop and closed-loop gain of an amplifier
A_o, A_{of}	Low-frequency open-loop and closed-loop gain of an amplifier
i_i, i_e, i_f	Input, error, and feedback current signals
R_i, R_{if}	Input resistances of an amplifier without feedback and with feedback

Symbol	Meaning
R_o, R_{of}	Output resistances of an amplifier with feedback and without feedback
R_{ie}, R_{oe}	Equivalent input resistance and output resistance of an amplifier
v_i, v_e, v_f	Input, error, and feedback voltage signals
ω_o, ω_p, ω_g	Oscillation, pole, and crossover frequencies of an amplifier
f, f_D	Pole and dominant pole frequency of an amplifier
ϕ, T_L	Phase angle and transmission loop
β, β_F	Feedback factor and current gain of a BJT

10.1 Introduction

Feedback is commonly used in amplifier circuits. A signal that is proportional to the output is compared with an input or a reference signal so that a desired output is obtained from the amplifier. The difference between the input and the feedback signals, called the *error signal*, is amplified by the amplifier. There are two types of feedback:

- In *negative feedback*, the output signal (or a fraction of it) is continuously fed back to the input side and is subtracted from the input signal to create an error signal, which is then corrected by the amplifier to produce the desired output signal.
- In *positive feedback*, the output signal (or a fraction of it) is continuously fed back to the input side and added to the input signal to create a larger error signal, which is then amplified to produce a larger output until the output reaches the saturation voltage limit of the amplifier.

In negative feedback, the signal that is fed back to the input side is known as the *feedback signal*, and its polarity is opposite that of the input signal (i.e., it is out of phase by 180° with respect to the input signal). Negative feedback in an amplifier has four major benefits: (1) It stabilizes the overall gain of the amplifier with respect to parameter variations due to temperature, supply voltage, and so on; (2) it increases or decreases the input and output impedances; (3) it reduces the distortion and the effect of nonlinearity; and (4) it increases the bandwidth. There are two disadvantages of negative feedback: (1) The overall gain is reduced almost in direct proportion to the benefits, and it is often necessary to compensate for the decrease in gain by adding an extra amplifier stage; (2) the circuit may tend to oscillate, in which case careful design is required to overcome this problem. Negative feedback is also known as *degenerative feedback* because it degenerates (or reduces) the output signal.

The op-amp circuits in Chapter 2 use negative feedback. The amplifier gain A_f is almost independent of the op-amp gain A; it depends on the external circuit elements only. For example, the gain of the inverting amplifier in Fig. 3.11 is $-R_F/R_1$, which is independent of the op-amp gain A, and the input impedance is approximately R_1. The gain of the noninverting op-amp amplifier in Fig. 3.9 is $(1 + R_F/R_1)$, and its input impedance is very large. The output impedance of both amplifiers is very small.

In positive feedback, the feedback signal is in phase with the input signal. Thus, the error signal is the algebraic sum of the input and feedback signals, and it is amplified by the amplifier. Thus, the output may continue to increase, resulting in an unstable situation, and the circuit may oscillate between the limits of the power supplies at the resonant frequency of the amplifier. Positive feedback is often referred to as *regenerative feedback* because it increases the output signal. Positive feedback is generally applied in oscillator circuits, which we will study in Chapter 13. Note that positive feedback does not necessarily imply oscillations. In fact, positive feedback is quite useful in some applications, such as active filters.

10.2 Feedback

Consider the noninverting op-amp amplifier shown in Fig. 10.1(a). The voltages v_s, v_f, and v_e are related as follows:

$$v_e = v_s - v_f$$

$$v_o = Av_e$$

$$v_f = \frac{R_1}{R_1 + R_F} v_o$$

The input and output relationships described by these equations can be represented by a block diagram as shown in Fig. 10.1(b). The voltage v_e, which is the difference between v_s and v_f, is amplified by the voltage gain A. The *feedback signal* v_f is proportional to the output voltage and is fed back to the input side. Thus, the amplifier feeds the output voltage back to the input side and compares the voltages. There are two circuits: the amplifier circuit and the feedback circuit (or network) consisting of R_1 and R_F. The voltages v_s, v_f, and v_e form a series circuit at the input side as shown in Fig. 10.1(a), whereas v_o is applied directly to the feedback network. That is, the noninverting amplifier uses series-shunt, or *voltage-sensing/voltage-comparing*, feedback.

Now consider the inverting op-amp amplifier shown in Fig. 10.2(a). There is a common node at the input side where the input current i_i and the feedback current i_f meet. The voltages and currents are related as follows:

$$i_e = i_i - i_f$$

$$v_e = -R_i i_e$$

$$v_o = Av_e$$

$$i_f = \frac{-v_e - v_o}{R_F} \approx \frac{-v_o}{R_F}$$

since $v_e \approx 0$ and $v_o >> v_e$. The input and output relationships described by these equations are shown by the block diagram in Fig. 10.2(b). The current i_e, which is the difference between i_i and i_f, is amplified by the transimpedance gain $-R_iA$. The *feedback current signal* i_f is proportional to the output voltage.

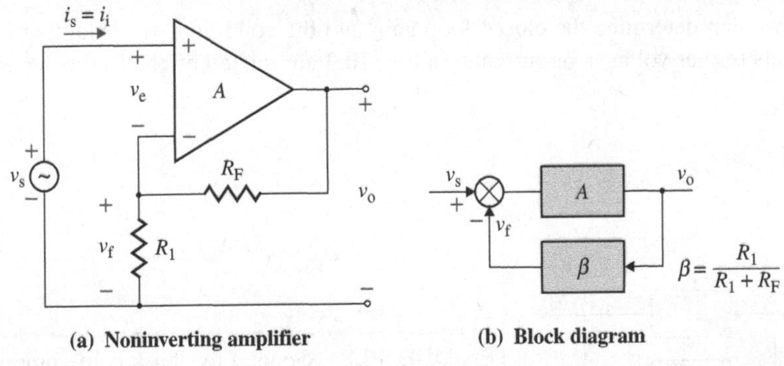

FIGURE 10.1 Feedback representation of noninverting op-amp amplifier

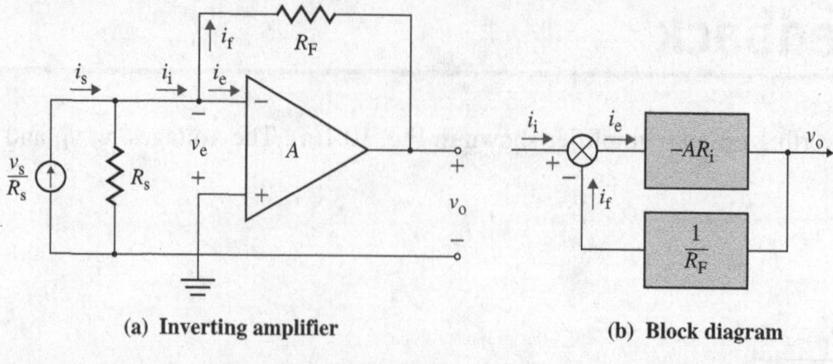

FIGURE 10.2 Feedback representation of inverting op-amp amplifier

Thus, the amplifier feeds the output voltage back to the input side and compares the currents. Here, the feedback network consists of R_F. The currents i_i, i_f, and i_e form a shunt connection at the input side, whereas v_o is applied directly to the feedback network. Thus, the inverting amplifier uses shunt-shunt, or *voltage-sensing/current-comparing*, feedback.

10.3 Characteristics of Feedback

In the inverting and noninverting amplifiers in Figs. 10.1 and 10.2, the output voltage is fed back directly to the input side and is compared to either the voltage or the current at the input side. Feedback can be represented by the general configuration shown in Fig. 10.3, where β is called the *feedback ratio* (or *factor*) and A is the amplifier gain. The units of A could be V/V, A/V, A/A, or V/A, and the units of β will be the reciprocal of those of A. For the noninverting amplifier, A is in V/V and β is in V/V; for the inverting amplifier, A is in V/A and β is in A/V.

▶ **NOTE** β is the feedback ratio, whereas β_f $(=\beta_F)$ is the forward current gain of a bipolar transistor. To distinguish the feedback factor β from the BJT transistor current gain β_F, in this chapter we use the hybrid parameter h_{fe} for the BJT current gain.

10.3.1 Closed-Loop Gain

From Fig. 10.3, we can determine the closed-loop gain and the conditions for obtaining negligible error. The various signals (either voltages or currents) in Fig. 10.3 are related by the following equations:

$$S_o = AS_e \tag{10.1}$$

$$S_e = S_i - S_f \tag{10.2}$$

$$S_f = \beta S_o \tag{10.3}$$

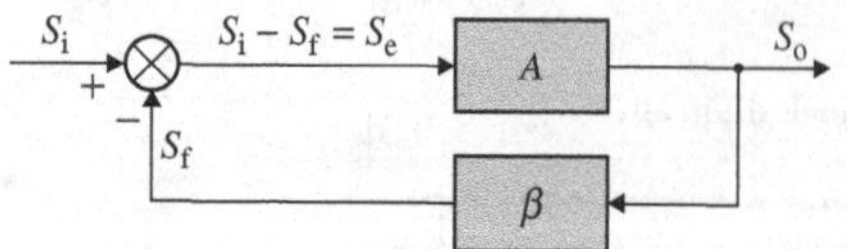

FIGURE 10.3 General feedback configuration

where A = open-loop gain of the amplifier
S_o = output signal
S_e = error signal
S_f = feedback signal
S_i = input signal
β = feedback factor

Substituting S_e from Eq. (10.2) into Eq. (10.1) gives

$$S_o = AS_e = AS_i - AS_f \tag{10.4}$$

Substituting S_f from Eq. (10.3) into Eq. (10.4) yields

$$S_o = AS_i - \beta AS_o$$

which gives the overall gain A_f with negative feedback as

$$A_f = \frac{S_o}{S_i} = \frac{A}{1 + \beta A} \tag{10.5}$$

A_f is often known as the *closed-loop gain*. Equation (10.5) is derived for negative feedback. βA is the gain around the feedback loop, known as the *loop gain* or the *loop transmission*. Let us define $T_L = \beta A$. If $T_L >> 1$, Eq. (10.5) becomes

$$A_f \approx \frac{1}{\beta} \tag{10.6}$$

That is, for large values of loop gain T_L, the closed-loop gain A_f is independent of the open-loop gain A and depends on the feedback factor β only.

Substituting S_f from Eq. (10.3) and S_o from Eq. (10.5) into Eq. (10.2) gives the error signal S_e:

$$S_e = S_i - S_f = S_i - \beta S_o = S_i - \frac{\beta AS_i}{1 + \beta A} = \frac{S_i}{1 + \beta A} = \frac{S_i}{1 + T_L} \tag{10.7}$$

As T_L becomes much greater than 1, S_e becomes much smaller than S_i, and $S_i \approx S_f$. Substituting S_o from Eq. (10.5) into Eq. (10.3) gives

$$S_f = \frac{\beta A}{1 + \beta A} S_i = \frac{T_L}{1 + T_L} S_i \tag{10.8}$$

If $T_L >> 1$, $S_f \approx S_i = \beta S_o$. That is, the output signal S_o is the amplified version of the input signal S_i, provided $\beta < 1$.

With positive feedback, the sign of βA changes and the closed-loop gain becomes

$$A_f = \frac{S_o}{S_i} = \frac{A}{1 - \beta A} \tag{10.9}$$

This confirms that the output will continue to increase until the saturation limit. Therefore, the amplifier can be made to oscillate by making $\beta A = 1$.

10.3.2 Gain Sensitivity

In most practical amplifiers, the open-loop gain A is dependent on temperature and the operating conditions of active devices. The effect of variations in the open-loop gain A can be determined from the sensitivity of the closed-loop gain A_f. Differentiating A_f in Eq. (10.5) with respect to A gives

$$\frac{dA_f}{dA} = \frac{(1+\beta A) - \beta A}{(1+\beta A)^2} = \frac{1}{(1+\beta A)^2} \tag{10.10}$$

If A changes by δA, then A_f will change by δA_f. Thus, Eq. (10.10) yields

$$\delta A_f = \frac{\delta A}{(1+\beta A)^2} \quad \text{(assuming } \delta A_f \approx dA_f) \tag{10.11}$$

which gives the approximate value of δA_f for finite increments in δA. The fractional change in A_f is given by

$$\frac{\delta A_f}{A_f} = \frac{1+\beta A}{A}\,\frac{\delta A}{(1+\beta A)^2} = \frac{\delta A/A}{1+\beta A} \quad \text{(assuming } \delta A_f \approx dA_f) \tag{10.12}$$

which shows that a fractional change in A of $(\delta A/A)$ causes a fractional change in A_f of $(\delta A_f/A_f)$ such that

$$\frac{\delta A_f}{A_f} = \frac{1}{1+\beta A}\left(\frac{\delta A}{A}\right)$$

If $\delta A/A$ is, say, 10%, the change in A_f is

$$\frac{\delta A_f}{A_f} = \frac{10}{1+\beta A}\%$$

only. Thus, the sensitivity of the closed-loop gain A_f to the open-loop gain A is defined as

$$S_A^{A_f} = \frac{\delta A_f/A_f}{\delta A/A} = \frac{1}{1+\beta A} \tag{10.13}$$

For $\beta A >> 1$, which is generally the case, the sensitivity of A_f to A becomes very small. Thus, a significant change in A will cause only a small change in A_f.

10.3.3 Feedback Factor Sensitivity

We can see from Eq. (10.6) that the closed-loop gain A_f depends on the feedback factor β only. The effect of variations of the feedback factor β on the gain A_f can be determined. Differentiating A_f in Eq. (10.5) with respect to β gives

$$\frac{dA_f}{d\beta} = -\frac{A^2}{(1+\beta A)^2} = -A_f^2 \tag{10.14}$$

If β changes by $\delta\beta$, then A_f will change by δA_f. From Eq. (10.14), we get

$$\delta A_f = -A_f^2\delta\beta \quad \text{(assuming } \delta A_f \approx dA_f) \tag{10.15}$$

Thus, the fractional change in A_f is given by

$$\frac{\delta A_f}{A_f} = -A_f\delta\beta \tag{10.16}$$

The sensitivity of closed-loop gain A_f to the feedback factor β is defined as

$$S_{\beta}^{A_f} = \frac{\delta A_f/A_f}{\delta\beta/\beta} = -A_f\beta = -\frac{\beta A}{1+\beta A} \tag{10.17}$$

For $\beta A >> 1$, Eq. (10.17) can be reduced to

$$S_{\beta}^{A_f} = -1 \tag{10.18}$$

Therefore, the gain A_f is directly sensitive to any change in the feedback factor β. The negative sign in Eq. (10.18) signifies that an increase in β will cause a decrease in A_f.

▶ **NOTE** Although the value of A can be positive or negative depending on the circuit configuration, we will use only the absolute value of A in equations for negative feedback such as Eq. (10.5).

EXAMPLE 10.1

Finding the effect of changes in the open-loop gain on the closed-loop gain The open-loop gain of an amplifier is $A = 250$, and the feedback factor is $\beta = 0.8$.

(a) Determine the closed-loop gain $A_f = S_o/S_i$.

(b) If the open-loop gain A changes by +20%, determine the percentage change in the closed-loop gain A_f and its value.

(c) If the feedback factor β changes by −20%, determine the percentage change in the closed-loop gain A_f and its value.

SOLUTION

$A = 250$, $\beta = 0.8$, and $T_L = \beta A = 250 \times 0.8 = 200$.

(a) From Eq. (10.5),

$$A_f = \frac{250}{1+200} = 1.2438$$

(b) $\delta A/A = 20\%$. From Eq. (10.13),

$$\frac{\delta A_f}{A_f} = \frac{20\%}{1+200} = 0.1\%$$

$$\delta A_f = 0.1\% \times 1.2438 = 0.00124$$

$$A_f = 1.2438 + 0.00124 = 1.245$$

(c) $\delta\beta/\beta = 20\%$. From Eq. (10.18),

$$\frac{\delta A_f}{A_f} = -20\%$$

$$\delta A_f = -20\% \times 1.2438 = -0.249$$

$$A_f = 1.2439 - 0.249 = 0.995$$

NOTE: The overall gain A_f does not change much with a wide variation in the open-loop gain A. But the gain A_f changes directly with the feedback factor β. In designing a feedback amplifier, special care should be taken to ensure that the variation in the feedback factor is kept to a minimum.

10.3.4 Frequency Response

Negative feedback increases the bandwidth of an amplifier. To prove this, let us consider a simple amplifier whose open-loop gain A is dependent on the frequency, which can be expressed in Laplace's domain as

$$A(s) = \frac{A_o}{1 + s/(2\pi f_H)} \tag{10.19}$$

where A_o is open-loop low-frequency gain and f_H is open-loop 3-dB break frequency, in Hertz. From Eq. (10.5), the overall gain is given by

$$A_f(s) = \frac{A(s)}{1 + \beta A(s)} \tag{10.20}$$

where the feedback factor β is independent of frequency. Substituting $A(s)$ from Eq. (10.19) into Eq. (10.20) yields

$$A_f(s) \frac{A_o/[1 + s/(2\pi f_H)]}{1 + \beta A_o/[1 + s/(2\pi f_H)]} = \frac{A_o}{1 + \beta A_o} \frac{1}{1 + s/[2\pi f_H(1 + \beta A_o)]} \tag{10.21}$$

which gives the low-frequency closed-loop gain A_{of} as

$$A_{of} = \frac{A_o}{1 + \beta A_o} = \frac{A_o}{1 + T_{Lo}} \tag{10.22}$$

where $T_{Lo} = \beta A_o$ is called the *low-frequency loop gain*. From Eq. (10.21), the 3-dB break frequency f_{Hf} with feedback becomes

$$f_{Hf} = f_H(1 + \beta A_o) \tag{10.23}$$

Thus, *without feedback*, the following equations apply:

Low-frequency gain $= A_o$

Bandwidth BW $= f_H$

Gain–bandwidth product GBW $= A_o f_H$ (10.24)

With feedback, these equations apply:

Low-frequency gain $A_{of} = \dfrac{A_o}{1 + \beta A_o}$

Bandwidth $\mathrm{BW_f} = f_H(1 + \beta A_o)$

Gain–bandwidth product $\mathrm{GBW_f} = A_{of}\mathrm{BW_f} = A_o f_H$ (10.25)

We can conclude from Eqs. (10.22) and (10.23) that feedback reduces the low-frequency gain by a factor of $(1 + \beta A_o)$ but increases the 3-dB frequency by the same amount $(1 + \beta A_o)$. However, the gain–bandwidth product remains constant at $A_o f_H$. Negative feedback allows the designer to trade gain for bandwidth, and it is widely used as a method for designing broadband amplifiers. The gain reduction is generally compensated for by adding more amplifying stages, which may also be feedback amplifiers. The plots of magnitude against frequency for A_f and A are shown in Fig. 10.4.

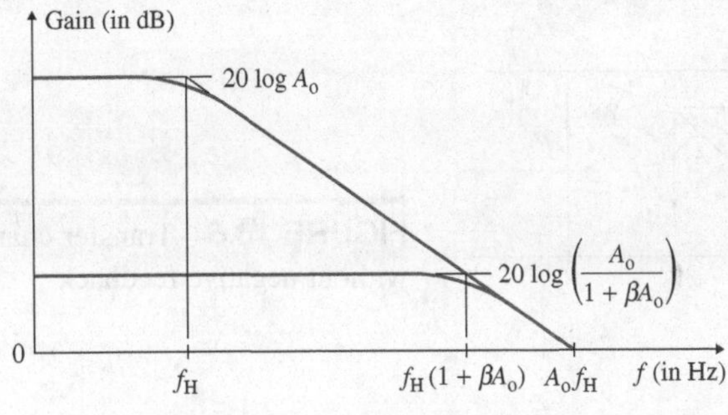

FIGURE 10.4 Plots of magnitude against frequency

EXAMPLE 10.2

Finding the effect of feedback on the frequency of an amplifier The feedback factor of a closed-loop amplifier is $\beta = 0.8$. The open-loop gain is expressed as

$$A(s) = \frac{250}{1 + s/(2\pi \times 100)}$$

Determine **(a)** the closed-loop low-frequency gain A_{of}, **(b)** the closed-loop bandwidth BW_f, and **(c)** the gain–bandwidth product GBW.

SOLUTION

$A_o = 250$, $\beta = 0.8$, and $f_H = 100$.

(a) From Eq. (10.22),

$$A_{of} = \frac{250}{(1 + 250 \times 0.8)} = 1.24378$$

(b) From Eq. (10.23),

$$B_{wf} = 100 \times (1 + 250 \times 0.8) = 20.1 \text{ kHz}$$

(c) From Eq. (10.24),

$$\text{Gain–bandwidth product GBW} = A_o f_H = 250 \times 100 = 25 \times 10^3$$

10.3.5 Distortion

An amplifier contains nonlinear devices such as transistors. As a result, the plot of the output signal S_o against the input signal S_i will not be linear. Thus, if the input signal is a sinusoidal waveform, the output signal will not be sinusoidal; that is, the output signal will be distorted. The effect of distortion in an amplifier is to reduce the gain of the open-loop transfer function. The closed-loop gain A_f, however, remains almost independent of the open-loop gain, as shown by Eq. (10.6). Therefore, negative feedback can reduce the effect of slope changes on the open-loop transfer function.

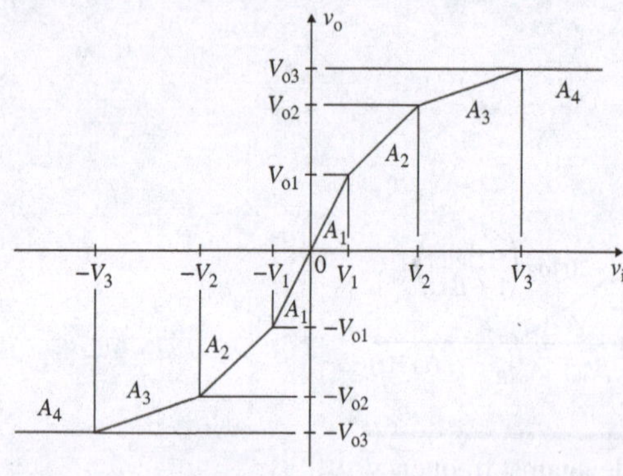

FIGURE 10.5 Transfer characteristic without negative feedback

Let us consider an amplifier whose transfer characteristic (v_o versus v_i) is nonlinear, as shown in Fig. 10.5. There are four regions of constant gain: A_1, A_2, A_3, and A_4. If negative feedback is applied with a feedback factor of β, Eq. (10.6) can be used as follows to calculate the closed-loop gains corresponding to the four regions:

$$A_{f1} = \frac{A_1}{1 + \beta A_1} \approx \frac{1}{\beta} \quad \text{for } \beta A_1 >> 1$$

$$A_{f2} = \frac{A_2}{1 + \beta A_2} \approx \frac{1}{\beta} \quad \text{for } \beta A_2 >> 1$$

$$A_{f3} = \frac{A_3}{1 + \beta A_3} \approx \frac{1}{\beta} \quad \text{for } \beta A_3 >> 1$$

$$A_{f4} = \frac{A_4}{1 + \beta A_4} \approx \frac{1}{\beta} \quad \text{for } \beta A_4 >> 1$$

Therefore, the slopes of the transfer characteristic in the four regions will be almost equal, and the transfer characteristic will be as shown in Fig. 10.6. The transfer characteristic with negative feedback is much less nonlinear than that of the original amplifier without negative feedback.

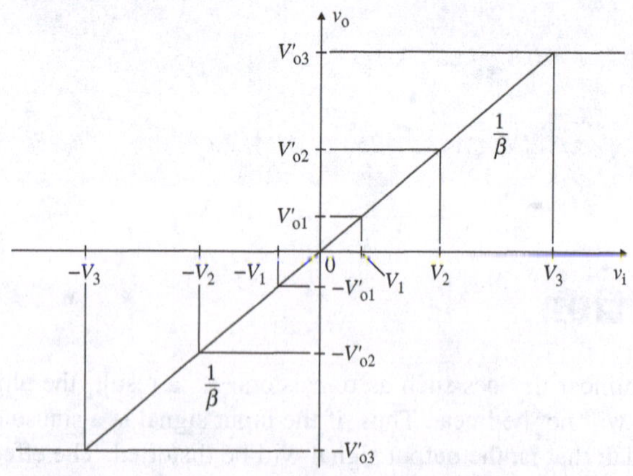

FIGURE 10.6 Transfer characteristic with negative feedback

EXAMPLE 10.3

Finding the effect of amplifier nonlinearity on the closed-loop gain The transfer characteristic of an amplifier without feedback is approximated by the following values of open-loop gain for given ranges of input voltage v_i:

$$A = \begin{cases} 1000 & \text{for } 0 < v_i \le 0.5 \text{ mV} \\ 500 & \text{for } 0.5 \text{ mV} < v_i \le 1 \text{ mV} \\ 250 & \text{for } 1 \text{ mV} < v_i \le 2 \text{ mV} \\ 0 & \text{for } v_i > 2 \text{ mV} \end{cases}$$

If the feedback factor is $\beta = 0.5$, determine the closed-loop gains of the transfer characteristic.

SOLUTION

From Eq. (10.5), the closed-loop gains become

$$A_f = \begin{cases} \dfrac{1000}{1 + 0.5 \times 1000} = 1.996 & \text{for } 0 < v_i \le 0.5 \text{ mV} \\ \dfrac{500}{1 + 0.5 \times 500} = 1.992 & \text{for } 0.5 \text{ mV} < v_i \le 1 \text{ mV} \\ \dfrac{250}{1 + 0.5 \times 250} = 1.984 & \text{for } 1 \text{ mV} < v_i \le 2 \text{ mV} \\ 0 & \text{for } v_i > 2 \text{ mV} \end{cases}$$

NOTE: As the input voltage v_i varies from 0 to 2 mV, the slope of output voltage versus input voltage (i.e., gain) varies from 1000 to 250 in a nonlinear fashion. But the closed-loop gain A_f remains almost constant.

KEY POINTS OF SECTION 10.3

- Feedback reduces the gain of a feedback amplifier. However, the bandwidth is widened proportionally. Also, feedback reduces the effect of amplifier distortion, nonlinearity, and variations in amplifier parameters.
- For a large value of loop-gain T_L, the closed-loop gain A_f is inversely proportional to the feedback factor β. That is, A_f is sensitive to changes in the feedback network parameters.

10.4 Feedback Topologies

The feedback configuration in Fig. 10.3 represents a general form; it does not indicate whether the input and output signals are voltages or currents. In practical amplifiers, the input and output signals can be either voltages or currents. If the output voltage is the feedback signal, it can be either compared with the input voltage to generate the error voltage signal (as shown in Fig. 10.1) or compared with the input current to generate the error current signal (as shown in Fig. 10.2). Similarly, the output current can be fed back and either compared with the input voltage to generate the error voltage signal or compared with the input current to generate the error current signal. Therefore, there are four feedback configurations, depending on whether the input and output signals are voltages or currents [1].

10.4.1 Feedback Configurations

There are four feedback configurations: series-shunt, series-series, shunt-shunt, and shunt-series. A potential divider, that is, a resistor (or capacitor) in series with another resistor (or capacitor) can sense a voltage, and placing a small resistor in the wire and sensing the voltage across it can sense a current. For a voltage subtraction, apply the input and the feedback signals to two distinct nodes, whereas for current subtraction, apply them to a single node. If the feedback signal is taken from the output terminal—for example, as in Fig. 10.7(c)—then the output side is a shunt feedback; if the feedback signal is taken from the voltage across a resistor—for example, as in Fig. 10.8(c)—then it is a series connection on the output sides.

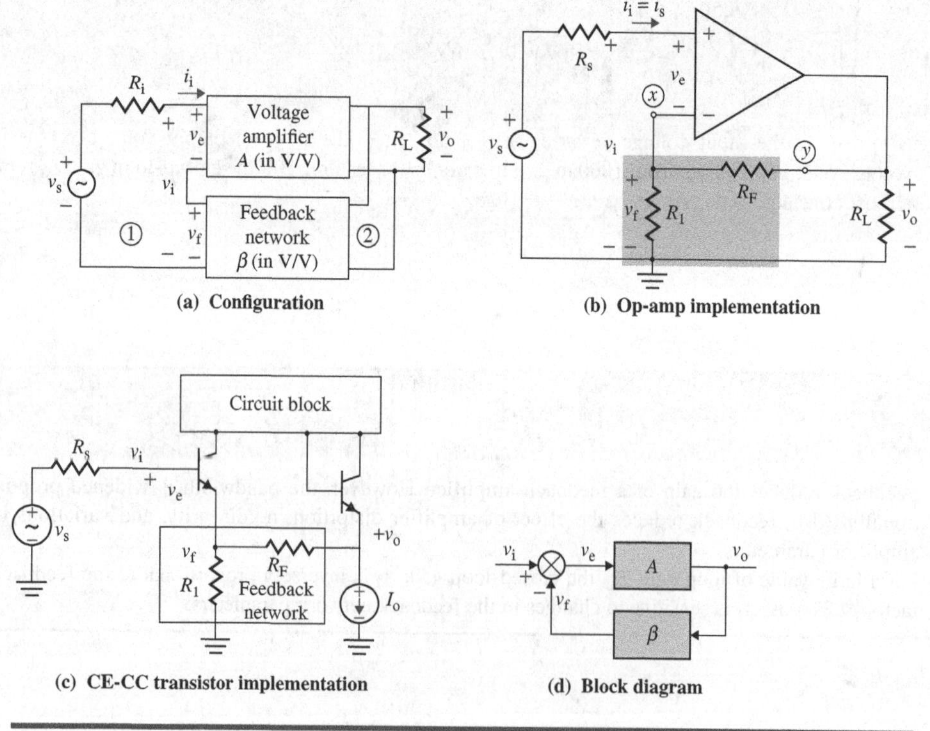

FIGURE 10.7 Series-shunt feedback

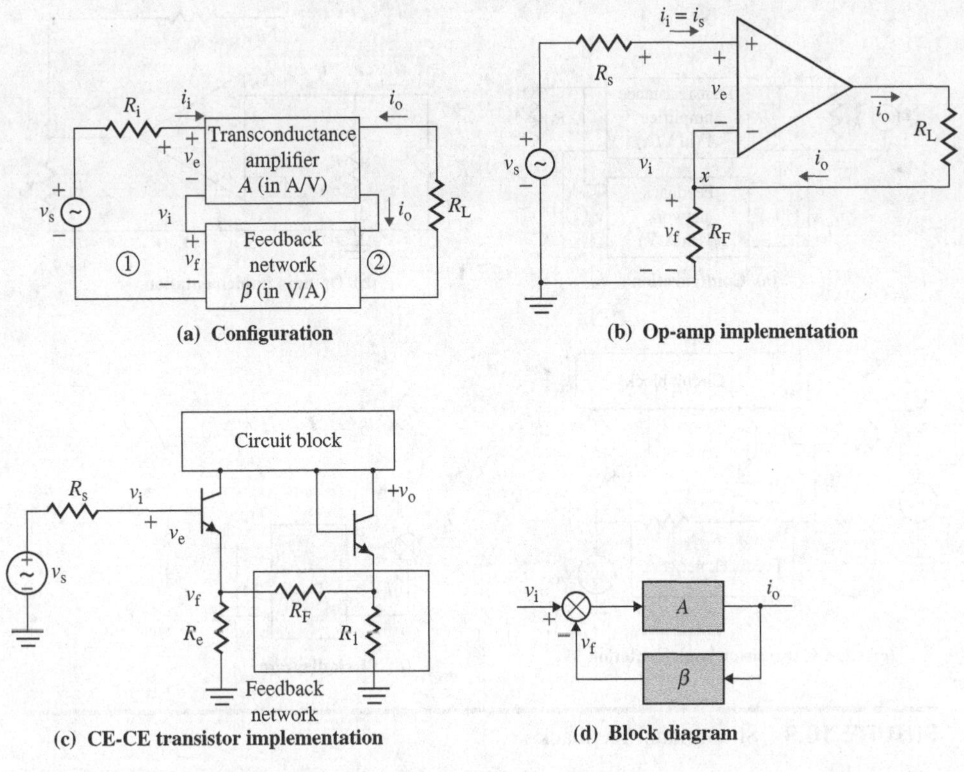

FIGURE 10.8 Series-series feedback

In series-shunt (*voltage-sensing/voltage-comparing*) feedback, shown in Fig. 10.7(a), the output voltage v_o is the input to the feedback network, and the feedback voltage v_f is proportional to the output voltage v_o. The feedback network forms a series circuit with the input voltage v_i but a parallel circuit with the output voltage v_o. The input current i_i flows through the loop formed by the input voltage, the amplifier, and the feedback network; that is, $v_i - v_f = v_e$. A series-shunt implementation using an op-amp is shown in Fig. 10.7(a). The discrete transistor implementation is shown in Fig. 10.7(c) and the feedback block as shown in Fig. 10.7(d).

In series-series (*current-sensing/voltage-comparing*) feedback, shown in Fig. 10.8(a), the output current i_o is the input to the feedback network, and the feedback voltage v_f is proportional to the output current i_o. The feedback network forms a series circuit with the input voltage and the output current. The input current flows through the loop formed by the input voltage, the amplifier, and the feedback network; that is, $v_i - v_f = v_e$. A series-series implementation using an op-amp is shown in Fig. 10.8(b). The discrete transistor implementation is shown in Fig. 10.8(c) and the feedback block as shown in Fig. 10.8(d).

In shunt-shunt (*voltage-sensing/current-comparing*) feedback, shown in Fig. 10.9(a), the output voltage v_o is also the input to the feedback network, and the feedback current i_f is proportional to the output voltage v_o. The feedback network is in parallel with both the input and the output voltages. The input current i_i is shared by the amplifier and the feedback network; that is, $i_i - i_f = i_e$. A shunt-shunt implementation using an op-amp is shown in Fig. 10.9(b). The discrete transistor implementation is shown in Fig. 10.9(c) and the feedback block in Fig. 10.9(d).

In shunt-series (*current-sensing/current-comparing*) feedback, shown in Fig. 10.10(a), the output current i_o is the input to the feedback network, and the feedback current i_f is proportional to the output

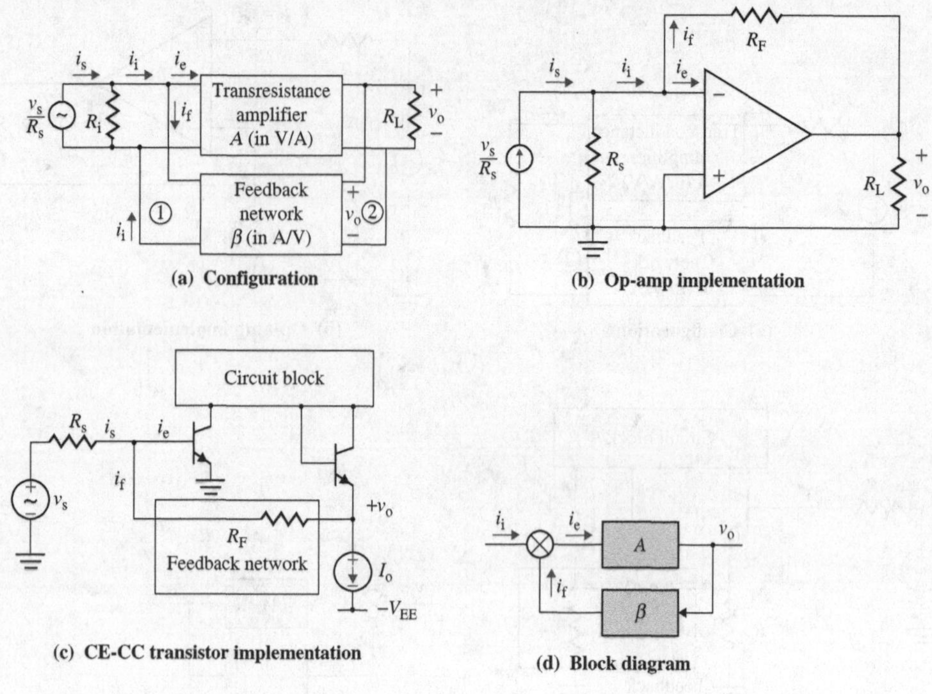

(a) Configuration (b) Op-amp implementation

(c) CE-CC transistor implementation (d) Block diagram

FIGURE 10.9 Shunt-shunt feedback

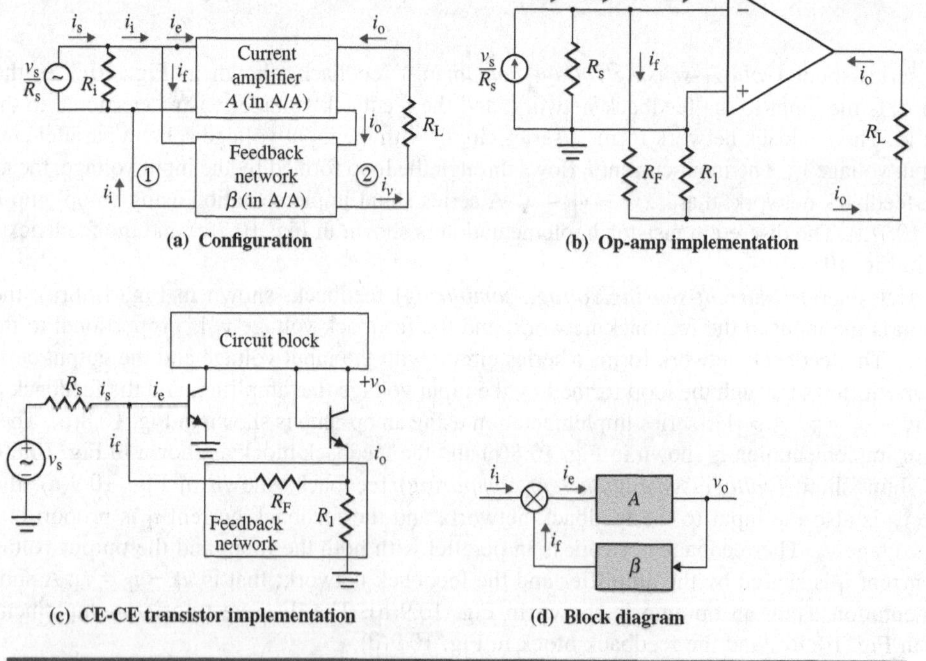

(a) Configuration (b) Op-amp implementation

(c) CE-CE transistor implementation (d) Block diagram

FIGURE 10.10 Shunt-series feedback

TABLE 10.1 Feedback relationships

	Gain	Input Resistance	Output Resistance
Without feedback	A	R_i	R_o
Series-shunt A (V/V) β (V/V)	$A_f = \dfrac{A}{1+\beta A}$	$R_{if} = R_i(1+\beta A)$	$R_{of} = \dfrac{R_o}{1+\beta A}$
Series-series A (A/V or ℧) β (V/A or Ω)	$A_f = \dfrac{A}{1+\beta A}$	$R_{if} = R_i(1+\beta A)$	$R_{of} = R_o(1+\beta A)$
Shunt-shunt A (V/A or Ω) β (A/V or ℧)	$A_f = \dfrac{A}{1+\beta A}$	$R_{if} = \dfrac{R_i}{1+\beta A}$	$R_{of} = \dfrac{R_o}{1+\beta A}$
Shunt-series A (A/A) β (A/A)	$A_f = \dfrac{A}{1+\beta A}$	$R_{if} = \dfrac{R_i}{1+\beta A}$	$R_{of} = R_o(1+\beta A)$

current i_o. The feedback network is in parallel with the input voltage but in series with the output current. The input current is shared by the amplifier and the feedback network; that is, $i_i - i_f = i_e$. A shunt-series implementation using an op-amp is shown in Fig. 10.10(b). The discrete transistor implementation is shown in Fig. 10.10(c) and the feedback block in Fig. 10.10(d).

10.4.2 Feedback Relationships

There are two circuits in a feedback amplifier: the amplifier circuit (or A circuit) and the feedback circuit (or β circuit). The effective gain is always decreased by a factor of $(1 + \beta A)$. In series-type arrangements, both A and β circuits are connected in series, and the effective resistance is increased by a factor of $(1 + \beta A)$. In shunt-type arrangements, A and β circuits are connected in parallel, and the effective resistance is decreased by a factor of $(1 + \beta A)$. The effects of different types of feedback are summarized in Table 10.1. Depending on the type of feedback, an amplifier is normally represented by one of four amplifier topologies: voltage, current, transconductance, or transresistance. A in Eq. (10.5) simply represents gain, which could be a voltage gain, a current gain, transconductance, or transresistance of the amplifier under the open-loop condition. Thus, A could be in units of V/V, A/A, A/V, or V/A.

KEY POINTS OF SECTION 10.4

- A feedback amplifier can be connected in one of four possible configurations: series-shunt, series-series, shunt-shunt, or shunt-series.
- With series feedback, the effective resistance is increased by a factor of $(1 + \beta A)$, whereas with shunt feedback the effective resistance is reduced by a factor of $(1 + \beta A)$.

10.5 Analysis of Feedback Amplifiers

In op-amp circuits, the open-loop gain A is independent of the feedback factor β; that is, the feedback network does not influence A. The first step in analyzing a feedback amplifier is to identify the main amplifier and its feedback network. However, in BJT and MOSFET amplifiers, the feedback is realized internally, and the feedback network cannot be separated from the main amplifier without affecting the open-loop gain A. Thus, the feedback network does influence A. The analysis of a feedback amplifier can be simplified by following these steps:

Step 1. Identify the feedback network.

Step 2. Identify the type of feedback on the input and output sides.

Step 3. Take into account the effects of the feedback network on the open-loop gain A by modifying the amplifier as follows:

a. Short-circuit the shunt feedback side to the ground so that there is no voltage signal to the feedback network. For example, terminal y of R_F in Fig. 10.7(b) would be connected to the ground so that R_1 became parallel to R_F.

b. Sever the series feedback side so that there is no current signal to the feedback network. For example, the inverting terminal x of the op-amp in Fig. 10.8(b) would be disconnected so that there was no current flowing into the feedback circuit and R_1 became in series with R_F.

Step 4. Represent the modified amplifier (from step 3) using one of the following equivalent amplifier topologies:

a. Voltage amplifier for series-shunt feedback
b. Transconductance amplifier for series-series feedback
c. Transresistance amplifier for shunt-shunt feedback
d. Current amplifier for shunt-series feedback

Calculate the values of the input resistance R_i, the output resistance R_o, and the open-loop gain A (representing transconductance, voltage, transresistance, or current) of the amplifier.

Step 5. The output of the amplifier is the input to the feedback network. Find the feedback factor β from one of the following two-port representations of the feedback network:

a. Voltage gain (V/V) representation for series-shunt feedback
b. Transresistance (V/A) representation for series-series feedback
c. Transconductance (A/V) representation for shunt-shunt feedback
d. Current gain (A/A) representation for shunt-series feedback

Step 6. Calculate the input resistance with feedback from one of the following equations:

$$R_{if} = R_i(1 + \beta A) \quad \text{for series-series and series-shunt feedback} \tag{10.26}$$

$$R_{if} = \frac{R_i}{1 + \beta A} \quad \text{for shunt-series and shunt-shunt feedback} \tag{10.27}$$

Step 7. Calculate the output resistance with feedback from one of the following equations:

$$R_{of} = R_o(1 + \beta A) \quad \text{for shunt-series and series-series feedback} \tag{10.28}$$

$$R_{of} = \frac{R_o}{1 + \beta A} \quad \text{for series-shunt and shunt-shunt feedback} \tag{10.29}$$

TABLE 10.2 V-I representations of amplifiers and feedback networks

Feedback Type	Amplifier	Units of A	Feedback Network	Units of β
Series-shunt	V-V	V/V	V-V	V/V
Shunt-series	I-I	A/A	I-I	A/A
Series-series	V-I	℧	I-V	Ω
Shunt-shunt	I-V	Ω	V-I	℧

Step 8. Use the following feedback equation to find the closed-loop gain A_f:

$$A_f = \frac{A}{1 + \beta A} \tag{10.30}$$

where β is the feedback factor representing the voltage gain, current gain, transconductance, or transresistance of the feedback network.

The V-I representations of the amplifier and its feedback network for different types of feedback are shown in Table 10.2. It is important to note that the units of A and β are different in each type of representation. To use the generalized equations for R_{if}, R_{of}, and A_f, the specific representation is essential.

▶ **NOTE** Since the units of A can be different depending on the type of feedback configuration, we will use the symbol μ_g for the open-loop voltage gain of the amplifier.

10.6 Series-Shunt Feedback

Series-shunt feedback is normally applied to a voltage amplifier. The amplifier in Fig. 10.7(a) is replaced by a voltage amplifier with an input resistance of R_i, an output resistance of R_o, and an open-loop voltage gain of A (in V/V). This arrangement is shown in Fig. 10.11. The feedback voltage v_f is in series with the input voltage v_i and is proportional to the output voltage v_o. The feedback network can be considered as a two-port network, and it can be modeled as a voltage gain circuit, as in Fig. 10.11, with an input resistance of R_y, an output resistance of R_x, and an open-loop voltage gain of β.

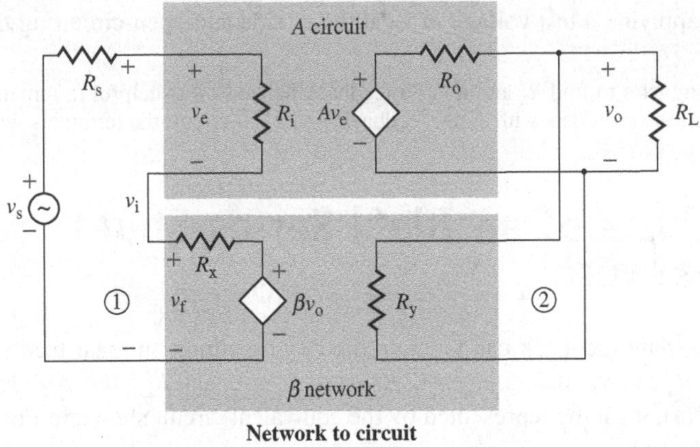

FIGURE 10.11 Series-shunt configuration

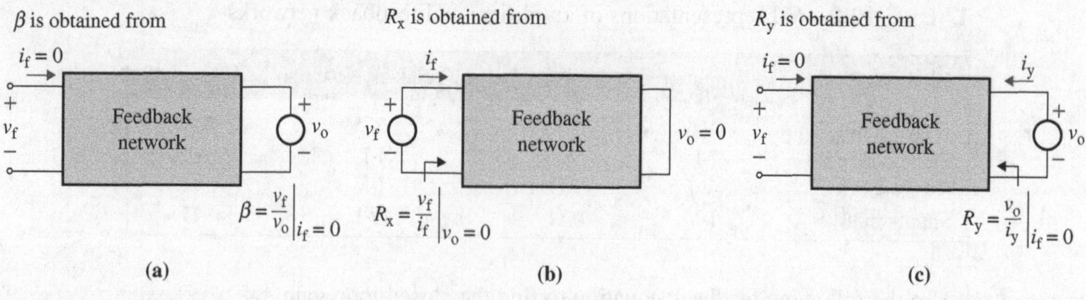

FIGURE 10.12 Test conditions for determining the parameters of a series-shunt feedback network

Determination of the model parameters requires separating the feedback network and representing it as a two-port network of four terminals. The test conditions for determining the parameters of the feedback network are shown in Fig. 10.12. The model parameters are obtained from three equations. The first equation is

$$R_x = \frac{v_f}{i_f}\Big|_{v_o=0} \quad \text{(short-circuited output side)} \tag{10.31}$$

which is obtained by applying a test voltage of v_f at the v_f side and shorting the v_o side. Note that the voltage across a short circuit is zero, and no current flows through an open circuit. The second equation is

$$R_y = \frac{v_o}{i_y}\Big|_{i_f=0} \quad \text{(open-circuited input side)} \tag{10.32}$$

which is obtained by applying a test voltage of v_o at the v_o side and open-circuiting the v_f side. The third equation is

$$\beta = \frac{v_f}{v_o}\Big|_{i_f=0} \quad \text{(open-circuited input side)} \tag{10.33}$$

which is obtained by applying a test voltage of v_o at the v_o side and open-circuiting the v_f side.

▶ **NOTE** In performing tests to find R_x and R_y of a feedback network, it is helpful to remember the following general rule: Short-circuit the terminals with shunt feedback, and open-circuit the terminals with series feedback.

10.6.1 Analysis of an Ideal Series-Shunt Feedback Network

The analysis of series-shunt feedback can be simplified by assuming an ideal feedback network and neglecting the effects of R_s and R_L; that is, $R_x = 0$, $R_y = \infty$, $R_s = 0$, and $R_L = \infty$. An ideal feedback network is shown in Fig. 10.13(a); it can be represented by the equivalent circuit shown in Fig. 10.13(b). The feedback factor β is in V/V, and

$$v_o = Av_e \tag{10.34}$$

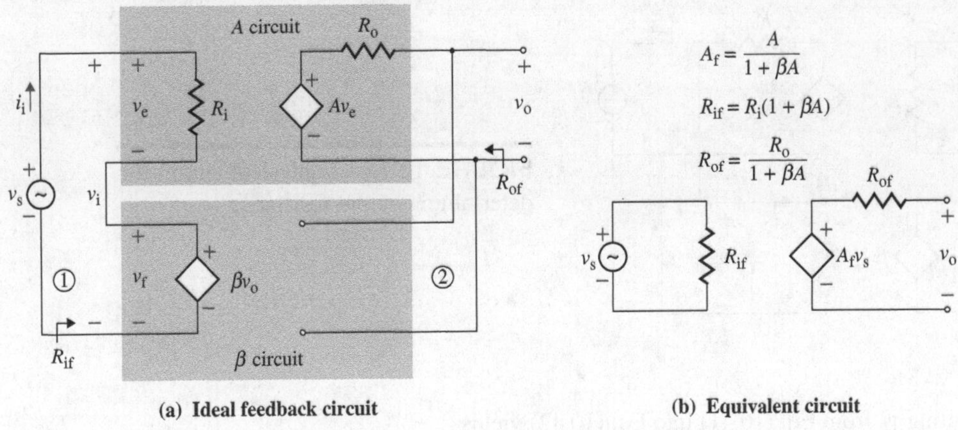

(a) Ideal feedback circuit (b) Equivalent circuit

FIGURE 10.13 Series-shunt configuration with an ideal feedback network

The feedback signal v_f, which is proportional to the output voltage v_o, is

$$v_f = \beta v_o \tag{10.35}$$

and
$$v_e = v_s - v_f \tag{10.36}$$

From Eqs. (10.34), (10.35), and (10.36), the equation for the closed-loop voltage gain A_f becomes

$$A_f = \frac{A}{1 + \beta A} \tag{10.37}$$

which is similar to Eq. (10.5). From Eq. (10.36),

$$v_s = v_e + v_f$$

Substituting v_f from Eq. (10.35) and v_o from Eq. (10.34) into the above equation gives

$$v_s = v_e + \beta v_o = v_e + \beta A v_e = v_e(1 + \beta A) \tag{10.38}$$

The input current i_i is

$$i_i = \frac{v_e}{R_i} \tag{10.39}$$

Substituting v_s from Eq. (10.38) gives the input resistance R_{if} with feedback:

$$R_{if} = \frac{v_s}{i_i} = \frac{v_e(1 + \beta A)}{v_e/R_i} = (1 + \beta A)R_i \tag{10.40}$$

The input resistance R_{if} is always increased by a factor of $(1 + \beta A)$ with series feedback at the input side.

The output resistance with feedback, which is Thevenin's equivalent resistance, can be obtained by applying a test voltage v_x to the output side and shorting the input source. The equivalent circuit for determining Thevenin's equivalent resistance is shown in Fig. 10.14. We have

$$v_e + v_f = v_e + \beta v_x = 0 \quad \text{or} \quad v_e = -\beta v_x \tag{10.41}$$

and
$$i_x = \frac{v_x - Av_e}{R_o} \tag{10.42}$$

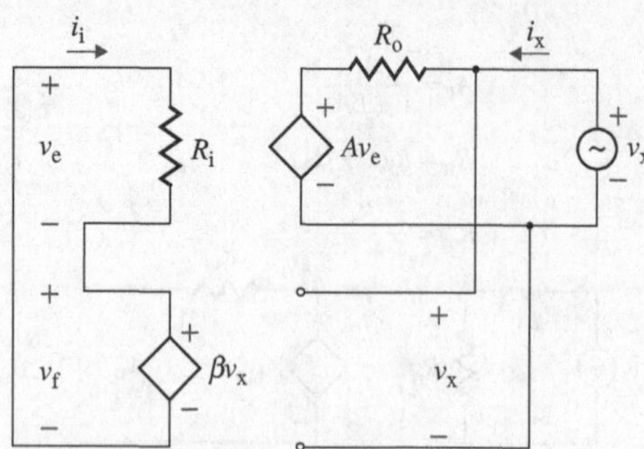

FIGURE 10.14 Equivalent circuit for determining output resistance

Substituting v_e from Eq. (10.41) into Eq. (10.42) yields

$$i_x = \frac{v_x - A(-\beta v_x)}{R_o} = \frac{(1 + \beta A)v_x}{R_o} \tag{10.43}$$

which gives the output resistance R_{of} with feedback as

$$R_{of} = \frac{R_o}{1 + \beta A} \tag{10.44}$$

Thus, the output resistance R_{of} at the output side is reduced by a factor of $(1 + \beta A)$. Shunt feedback at the output side always lowers the output resistance by a factor of $(1 + \beta A)$.

Series-shunt feedback increases the input resistance by $(1 + \beta A)$ and reduces the output resistance by $(1 + \beta A)$. This type of feedback is normally applied to a voltage amplifier. The input impedance, output impedance, and overall voltage gain can be written in generalized form in Laplace's domain of s as follows:

$$Z_{if}(s) = [1 + \beta A(s)]Z_i(s) \tag{10.45}$$

$$Z_{of}(s) = \frac{Z_o(s)}{1 + \beta A(s)} \tag{10.46}$$

$$A_f(s) = \frac{A(s)}{1 + \beta A(s)} \tag{10.47}$$

▶ **NOTES**

1. A is the open-loop voltage gain of the amplifier, in V/V.
2. β is the voltage gain of the feedback network and is less than or equal to 1 V/V.
3. $T_L = \beta A$ is the loop gain, which is dimensionless.
4. If the source has an impedance, then the overall voltage gain is reduced. The effective input voltage v_i to the amplifier can be found from

$$v_i = \frac{Z_i}{Z_s + Z_i} v_s \tag{10.48}$$

where Z_s is source impedance, Z_i is input impedance of the amplifier, and v_s is signal source voltage.

10.6.2 Analysis of a Practical Series-Shunt Feedback Network

The feedback network in Fig. 10.11 has a finite input resistance R_y and an output resistance R_x, which load the original amplifier, thereby affecting the performance of the feedback amplifier. The analysis in the previous section did not take into account the loading effect of the feedback network. The loading effect can be taken into account by including R_s, R_x, R_y, and R_L in the A circuit, as shown in Fig. 10.15(a). The open-loop parameters are modified accordingly, as shown in Fig. 10.15(b). These modifications allow us to apply the equations for ideal feedback.

The equivalent input resistance R_{ie} is given by

$$R_{ie} = R_i + R_x + R_s \tag{10.49}$$

where R_s is the source resistance. The equivalent output resistance R_{oe} is given by

$$R_{oe} = R_o \,\|\, R_y \,\|\, R_L \tag{10.50}$$

Using the voltage divider rule in Fig. 10.15(a), we see that the output voltage v_o is given by

$$v_o = \frac{R_y \,\|\, R_L}{(R_y \,\|\, R_L) + R_o} A v_e \tag{10.51}$$

where v_e is the voltage across R_i but not across R_{ie}. We need to find the voltage across R_i. Using the voltage divider rule gives v_e as

$$v_e = \frac{R_i}{R_i + R_x + R_s} v_{e1}$$

Substituting v_e into Eq. (10.51) gives the modified open-loop gain A_e:

$$A_e = \frac{v_o}{v_{e1}} = \frac{R_y \,\|\, R_L}{(R_y \,\|\, R_L) + R_o} \times \frac{R_i}{R_i + R_x + R_s} A \tag{10.52}$$

If R_{ie}, R_{oe}, and A_e are substituted for R_i, R_o, and A, respectively, Eqs. (10.34) through (10.48) can be applied to calculate the closed-loop parameters R_{if}, R_{of}, and A_f.

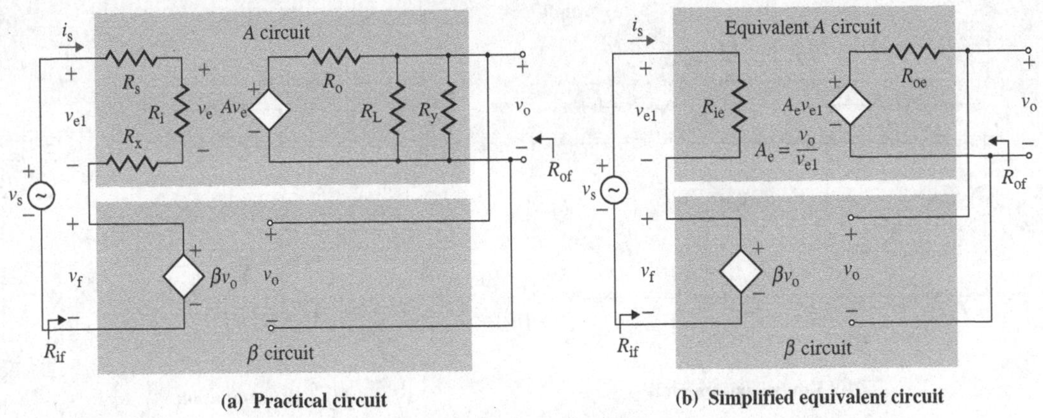

FIGURE 10.15 Practical series-shunt feedback

EXAMPLE 10.4

Finding the performance of a noninverting amplifier with series-shunt feedback The noninverting amplifier shown in Fig. 10.7(b) has $R_L = 10\ \text{k}\Omega$ and $R_s = 5\ \text{k}\Omega$. The feedback resistors are $R_1 = 10\ \text{k}\Omega$ and $R_F = 90\ \text{k}\Omega$. The op-amp parameters are $R_i = 2\ \text{M}\Omega$ and $R_o = 75\ \Omega$, and the open-loop voltage gain is $\mu_g = 2 \times 10^5$.

(a) Determine the input resistance seen by the source $R_{if} = v_s/i_s$, the output resistance R_{of}, and the closed-loop voltage gain $A_f = v_o/v_s$.

(b) Use PSpice/SPICE to verify your results.

SOLUTION

$R_L = 10\ \text{k}\Omega$, $R_s = 5\ \text{k}\Omega$, $R_1 = 10\ \text{k}\Omega$, $R_F = 90\ \text{k}\Omega$, $R_i = 2\ \text{M}\Omega$, $R_o = 75\ \Omega$, and $\mu_g = 2 \times 10^5$. Replacing the op-amp by its equivalent circuit gives the amplifier shown in Fig. 10.16(a).

(a) The steps in analyzing the feedback network are as follows:

Step 1. R_1 and R_F, which constitute the feedback network as shown in Fig. 10.16(a), produce a feedback voltage v_f proportional to the output voltage v_o.

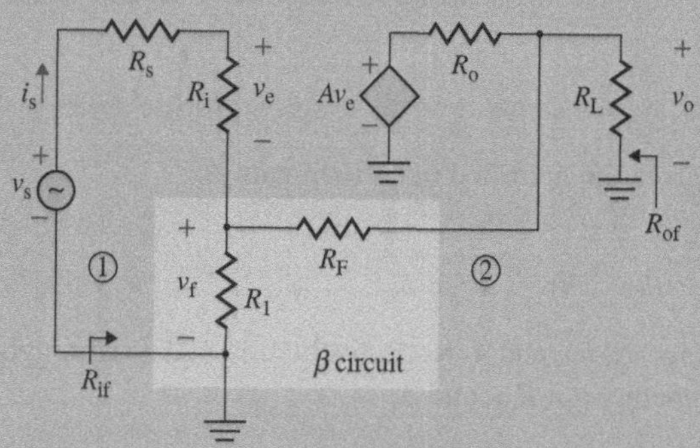

(a) Amplifier

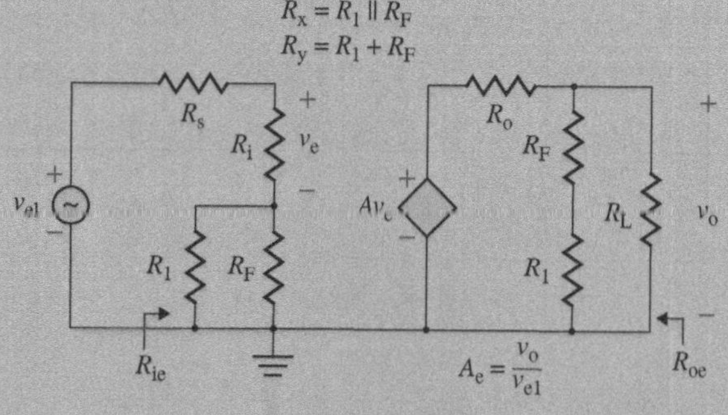

(b) Equivalent A circuit

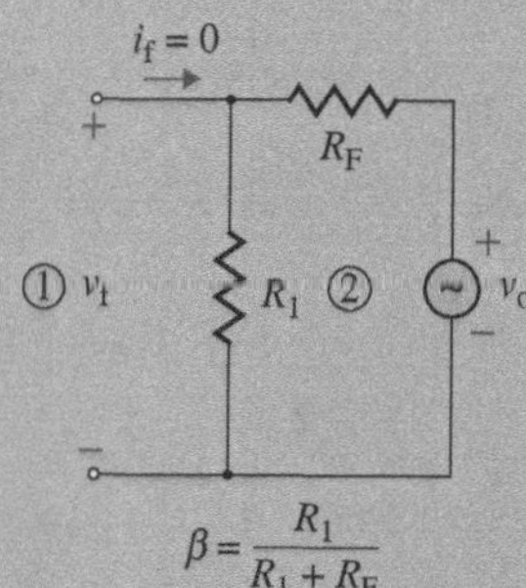

(c) β determination

FIGURE 10.16 Noninverting amplifier with series-shunt feedback

Step 2. The amplifier uses series-shunt feedback. Thus, A must be in V/V; $A = \mu_g = 2 \times 10^5$.

Step 3. The effect of the feedback network at the input side is taken into account by short-circuiting the shunt feedback at the output side. Similarly, the effect at the output side is taken into account by severing the series feedback at the input side. We ground R_F on side 2 and sever the feedback circuit on side 1 from R_i. This modification is shown in Fig. 10.16(b). Then

$$R_x = R_1 \| R_F = 10\text{ k} \| 90\text{ k} = 9\text{ k}\Omega$$

$$R_y = R_1 + R_F = 10\text{ k} + 90\text{ k} = 100\text{ k}\Omega$$

Step 4. If we represent the amplifier of Fig. 10.16(b) by an equivalent voltage amplifier, the input resistance is

$$R_{ie} = R_s + R_i + (R_1 \| R_F) = R_s + R_i + R_x = 5\text{ k} + 2\text{ M} + 9\text{ k} = 2014\text{ k}\Omega$$

and the output resistance is

$$R_{oe} = R_o \| (R_1 + R_F) \| R_L = 75 \| (10\text{ k} + 90\text{ k}) \| 10\text{ k} = 74.4\ \Omega$$

From Eq. (10.52), the modified open-loop gain A_e is

$$A_e = \frac{(R_1 + R_F) \| R_L}{(R_1 + R_F) \| R_L + R_o} \times \frac{R_i}{R_s + R_i + (R_1 \| R_F)} A$$

$$= \frac{[(10\text{ k} + 90\text{ k}) \| 10\text{ k}] \times 2000\text{ k} \times 2 \times 10^5}{[(10\text{ k} + 90\text{ k}) \| 10\text{ k} + 75](5\text{ k} + 2000\text{ k} + 9\text{ k})} = 1.9698 \times 10^5$$

Step 5. From Fig. 10.16(c), the feedback factor β is given by

$$\beta = \left.\frac{v_f}{v_o}\right|_{i_f=0} = \frac{R_1}{R_1 + R_F} = \frac{10\text{ k}}{10\text{ k} + 90\text{ k}} = 0.1\text{ V/V}$$

Step 6. The input resistance (seen by the source) with feedback is

$$R_{if} = \frac{v_s}{i_s} = R_{ie}(1 + \beta A_e) = 2014\text{ k} \times (1 + 0.1 \times 1.9698 \times 10^5) = 39.67\text{ G}\Omega$$

Step 7. The output resistance with feedback is

$$R_{of} = \frac{R_{oe}}{1 + \beta A_e} = \frac{74.4}{1 + 0.1 \times 1.9698 \times 10^5} = 3.778\text{ m}\Omega$$

Step 8. The closed-loop voltage gain A_f is

$$A_f = \frac{v_o}{v_s} = \frac{A_e}{1 + \beta A_e} = \frac{1.9698 \times 10^5}{1 + 0.1 \times 1.9698 \times 10^5} = 9.999\text{ V/V}$$

which is very close to the closed-loop gain we would get if we were to use Eq. (3.18):

$$A_f = 1 + \frac{R_F}{R_1} = 1 + \frac{90\text{ k}}{10\text{ k}} = 10$$

(b) The series-shunt feedback circuit for PSpice simulation is shown in Fig. 10.17. The results of PSpice simulation are shown below, with hand calculations to the right:

PSpice	Hand calculation
`V(5)/VS=9.999E+00=9.999`	$A_f = 9.99$
`INPUT RESISTANCE AT VS=3.967E+10=39.67 GΩ`	$R_{if} = 39.67\text{ G}\Omega$
`OUTPUT RESISTANCE AT V(5)=3.776E-02=37.7 mΩ`	$R_{of} = 37.8\text{ m}\Omega$

The PSpice results are very close to the hand-calculated values.

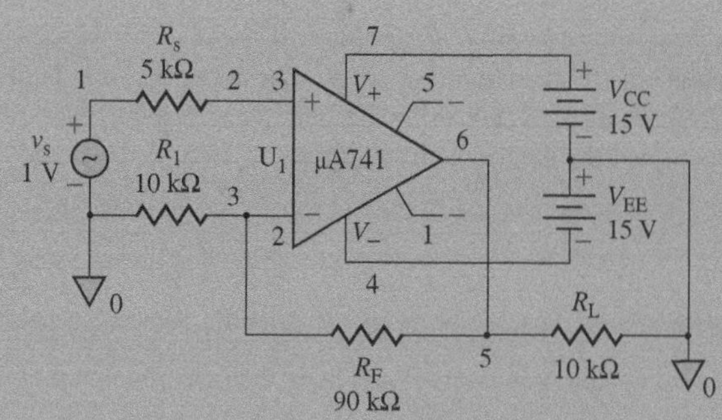

FIGURE 10.17 Series-shunt feedback network for PSpice simulation

NOTE: The main objective of PSpice/SPICE simulation in this chapter is to verify the techniques for analyzing feedback amplifiers. To verify hand-calculated results, we will use the simple DC op-amp model shown in Fig. 3.7 and the simple π model for BJTs. If we were to run simulations using the model provided in PSpice/SPICE, the results would differ slightly.

EXAMPLE 10.5

Finding the performance of a BJT amplifier with series-shunt feedback The AC equivalent circuit of a BJT amplifier is shown in Fig. 10.18(a). The transistor can be modeled as shown in Fig. 10.18(b). The DC bias currents of the transistors are $I_{C1} = 0.5$ mA, $I_{C2} = 1$ mA, and $I_{C3} = 5$ mA. The transistor parameters are $h_{fe} = h_{fe1} = h_{fe2} = h_{fe3} = 100$ and $r_\mu = r_o = \infty$.

(a) Use the techniques of feedback analysis to calculate the input resistance R_{if}, the output resistance R_{of}, and the closed-loop voltage gain A_f.

(b) Use PSpice/SPICE to check your results.

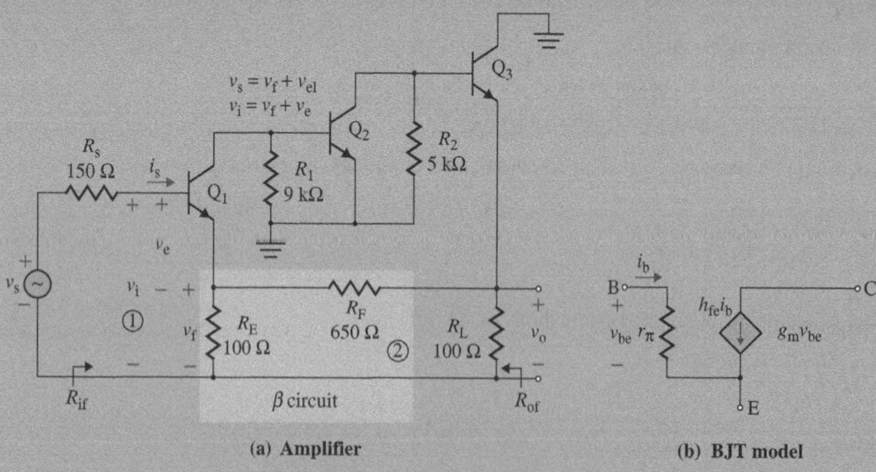

FIGURE 10.18 Three-stage amplifier with series-shunt feedback

NOTE: The following analysis can also be applied to MOSFET amplifiers by substituting a very large value of the B-E resistance r_π tending to infinity, $1/r_\pi \simeq 0$.

SOLUTION

For $I_{C1} = 0.5$ mA, Eqs. (8.40) and (8.42) give

$$r_{\pi 1} = \frac{h_{fe} \times 25.8 \text{ mV}}{I_{C1}} = \frac{100 \times 25.8 \text{ mV}}{0.5 \text{ mA}} = 5.16 \text{ k}\Omega$$

$$g_{m1} = \frac{I_{C1}}{V_T} = \frac{0.5 \text{ mA}}{25.8 \text{ mV}} = 19.38 \text{ mA/V}$$

For $I_{C2} = 1$ mA,

$$r_{\pi 2} = \frac{h_{fe} \times 25.8 \text{ mV}}{I_{C2}} = \frac{100 \times 25.8 \text{ mV}}{1 \text{ mA}} = 2.58 \text{ k}\Omega$$

$$g_{m2} = \frac{I_{C2}}{V_T} = \frac{1 \text{ mA}}{25.8 \text{ mV}} = 38.76 \text{ mA/V}$$

For $I_{C3} = 5$ mA,

$$r_{\pi 3} = \frac{h_{fe} \times 25.8 \text{ mV}}{I_{C3}} = \frac{100 \times 25.8 \text{ mV}}{5 \text{ mA}} = 516 \ \Omega$$

$$g_{m3} = \frac{I_{C3}}{V_T} = \frac{5 \text{ mA}}{25.8 \text{ mV}} = 193.8 \text{ mA/V}$$

(a) The steps in analyzing the feedback network are as follows:

Step 1. R_E and R_F, which constitute the feedback network, produce a feedback voltage proportional to the output voltage. The input voltage v_i is compared with the feedback signal v_f. The error voltage $v_e = v_i - v_f$ is the B-E voltage of transistor Q_1. The block diagram representing the feedback mechanism is shown in Fig. 10.19(a). The feedback network is shown in Fig. 10.19(b).

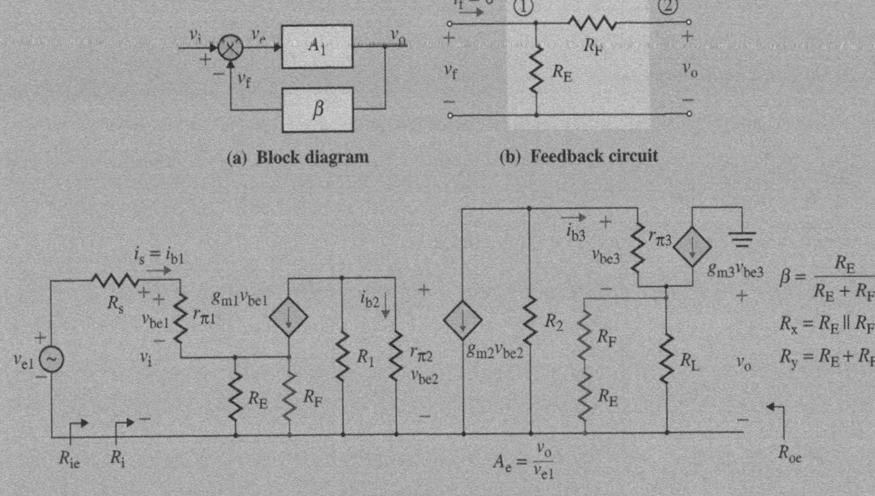

FIGURE 10.19 Equivalent circuits for Example 10.5

Step 2. The amplifier uses series-shunt feedback. Thus, the voltage gain A must be expressed in V/V.

Step 3. The effect of the feedback network is taken into account at the input side by shorting R_F at side 2 to the ground, as shown in Fig. 10.19(b), and at the output side by removing the feedback network from the emitter of Q_1. These modifications are shown in the small-signal AC equivalent circuit in Fig. 10.19(c).

Step 4. If we represent the amplifier of Fig. 10.19(c) by an equivalent voltage amplifier, the input resistance at the base of Q_1 is

$$R_i = \frac{v_i}{i_{b1}} = r_{\pi 1} + (1 + g_{m1} r_{\pi 1})(R_E \| R_F) = 5.16 \text{ k} + 101 \times (100 \| 650) = 13.91 \text{ k}\Omega$$

The input resistance is

$$R_{ie} = \frac{v_{el}}{i_{b1}} = R_s + R_i = R_s + r_{\pi 1} + (1 + g_{m1} r_{\pi 1})(R_E \| R_F) = 150 + 13.91 \text{ k} = 14.06 \text{ k}\Omega$$

The output resistance is

$$R_{oe} = R_L \| (R_F + R_E) \| \left(\frac{R_2 + r_{\pi 3}}{1 + g_{m1} r_{\pi 1}}\right) = 100 \| 750 \| \left(\frac{5 \text{ k} + 516}{101}\right) = 33.74 \ \Omega$$

Thus,

$$V_{be1} = \left(\frac{r_{\pi 1}}{R_i}\right) v_i = \left(\frac{5.16 \text{ k}}{13.91 \text{ k}}\right) v_i = 0.371 v_i$$

$$v_{b2} = v_{be2} = -(R_1 \| r_{\pi 2}) g_{m1} v_{be1} = -(9 \text{ k} \| 2.58 \text{ k}) \times 19.38 \text{ m} \times 0.371 v_i = -14.412 v_i$$

$$v_{b3} = -\{R_2 \| [r_{\pi 3} + (1 + g_{m3} r_{\pi 3})(R_L \| (R_F + R_E))]\} g_{m2} v_{be2}$$

$$= -\{5 \text{ k} \| [516 + (1 + 193.8 \text{ m} \times 516)(100 \| 750)]\} \times 38.76 \text{ m} \times (-14.412 v_i) = 1.825 \times 10^3 v_i$$

$$v_o = \frac{v_{b3} \times (1 + g_{m3} r_{\pi 3})[R_L \| (R_F + R_E)]}{\{r_{\pi 3} + (1 + g_{m3} r_{\pi 3})[R_L \| (R_F + R_E)]\}}$$

$$= \frac{(1.825 \times 10^3 \times v_i) \times (1 + 193.8 \text{ m} \times 516)(100 \| 750)}{[516 + (1 + 193.8 \text{ m} \times 516)(100 \| 750)]} = 1725.7 v_i$$

Therefore, the open-loop voltage gain A is given by $A = v_o / v_i = 1725.7$ V/V

and $$A_e = \frac{v_o}{v_{el}} = \frac{A R_i}{R_s + R_i} = \frac{1725.7 \times 13.91 \text{ k}}{150 + 13.91 \text{ k}} = 1707.3 \text{ V/V}$$

Step 5. From Fig. 10.19(b) and Eq. (10.33), the feedback factor β is given by

$$\beta = \left.\frac{v_f}{v_o}\right|_{i_i = 0} = \frac{R_E}{R_E + R_F} = \frac{100}{100 + 650} = 0.1333 \text{ V/V}$$

Step 6. The input resistance (seen by the source) with feedback is

$$R_{if} = \frac{v_s}{i_s} = R_{ie}(1 + \beta A_e) = 14.06 \text{ k}\Omega \times (1 + 0.1333 \times 1707.3) = 3.21 \text{ M}\Omega$$

Step 7. The output resistance with feedback is

$$R_{of} = \frac{R_{oe}}{1 + \beta A_e} = \frac{33.74}{1 + 0.1333 \times 1707.3} = 0.147 \ \Omega$$

Step 8. The closed-loop voltage gain A_f is

$$A_f = \frac{v_o}{v_s} = \frac{A_e}{1 + \beta A_e} = \frac{1707.3}{1 + 0.1333 \times 1707.3} = 7.47 \text{ V/V}$$

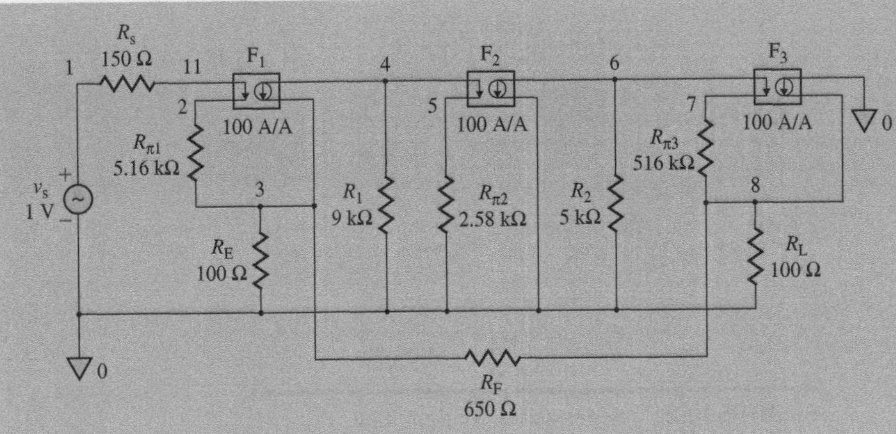

FIGURE 10.20 Series-shunt feedback network for PSpice simulation

(b) The series-shunt feedback network for PSpice simulation is shown in Fig. 10.20. The results of PSpice simulation (.TF analysis) are shown below, with hand calculations to the right:

`V(8)/VS=7.468E+00=7.468 V/V`	$A_f = 7.47$ V/V
`INPUT RESISTANCE AT VS=3.214E+06=3.214 MΩ`	$R_{if} = 3.21$ MΩ
`OUTPUT RESISTANCE AT V(8)=1.460E-01=0.146 Ω`	$R_{of} = 0.147$ Ω

Note the close agreement between the results obtained by PSpice and the hand calculations.

KEY POINTS OF SECTION 10.6

- Series-shunt feedback is applied to voltage amplifiers. This type of feedback increases the input resistance and reduces the output resistance by a factor of $(1 + \beta A)$.
- The amplifier is represented by a voltage amplifier and the feedback network as a voltage gain. Both A and β are in units of V/V.
- The loading effect of a feedback network can be taken into account by remembering the following general rule: Short-circuit the terminals with shunt feedback, and open-circuit the terminals with series feedback.

10.7 Series-Series Feedback

Series-series feedback is normally applied to a transconductance amplifier. For the series-series feedback amplifier in Fig. 10.8(a), we obtain the equivalent circuit shown in Fig. 10.21 if we represent the op-amp by its transconductance model. A is the open-loop transconductance gain of the op-amp in A/V. That is, $Av_e = \mu_g v_e/R_o$ and $A = \mu_g/R_o$. The feedback voltage v_f is proportional to the load current i_o. The feedback network can also be modeled in transimpedance form with an input resistance of R_y, an output resistance of R_x, and a transimpedance of β in V/A.

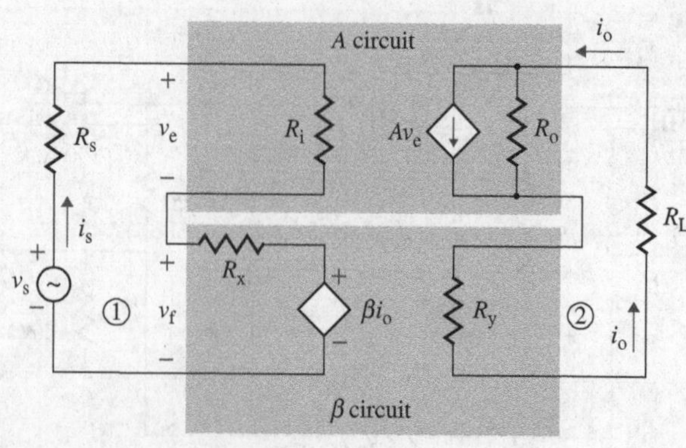

FIGURE 10.21 Series-series configuration

The test conditions for determining the parameters of the feedback network are shown in Fig. 10.22. The parameters of the model can be obtained from three equations. The first equation is

$$R_x = \left.\frac{v_f}{i_f}\right|_{i_o=0} \quad \text{(open-circuited output side)} \tag{10.53}$$

which is obtained by applying a test voltage of v_f at side 1 and open-circuiting side 2. The second equation is

$$R_y = \left.\frac{v_y}{i_o}\right|_{i_f=0} \quad \text{(open-circuited input side)} \tag{10.54}$$

which is obtained by applying a test voltage of v_y at side 2 and open-circuiting side 1. The third equation is

$$\beta = \left.\frac{v_f}{i_o}\right|_{i_f=0} \quad \text{(open-circuited input side)} \tag{10.55}$$

which is obtained by applying a test voltage of v_y at side 2 and open-circuiting side 1.

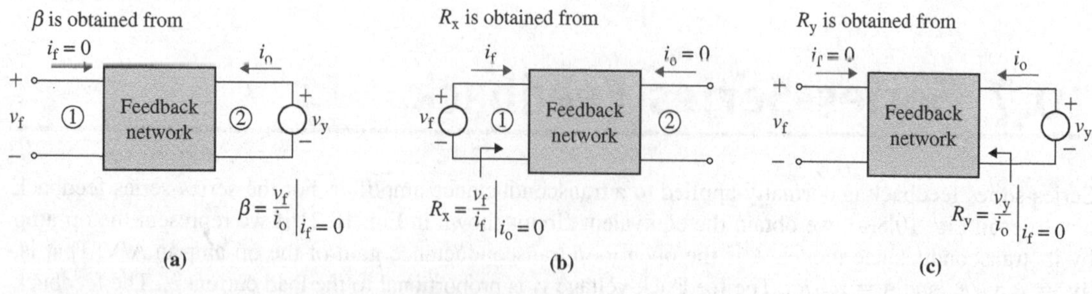

FIGURE 10.22 Test conditions for determining the parameters of a series-series feedback network

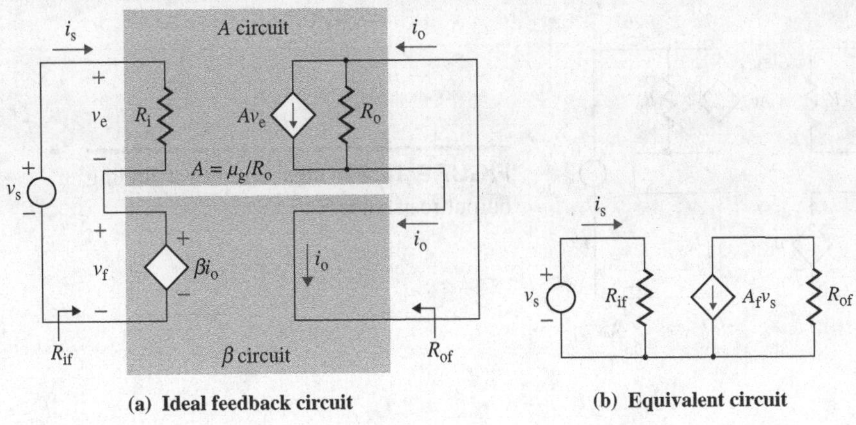

(a) Ideal feedback circuit (b) Equivalent circuit

FIGURE 10.23 Ideal series-series feedback network

10.7.1 Analysis of an Ideal Series-Series Feedback Network

Let us assume an ideal series-series feedback network—that is, $R_x = 0$, $R_y = 0$, $R_s = 0$, and $R_L = 0$. The feedback amplifier in Fig. 10.21 can be simplified to the one in Fig. 10.23(a), which can be represented by the equivalent circuit shown in Fig. 10.23(b). For this circuit,

$$i_o = Av_e$$

$$v_f = \beta i_o$$

$$v_s = v_e + v_f = \frac{i_o}{A} + \beta i_o = \frac{1 + \beta A}{A} i_o$$

which gives the closed-loop transconductance gain A_f as

$$A_f = \frac{i_o}{v_s} = \frac{A}{1 + \beta A} \tag{10.56}$$

Using KVL around the input side, we get

$$v_s = R_i i_s + v_f = R_i i_s + \beta i_o = R_i i_s + \beta A v_e = R_i i_s + \beta A R_i i_s$$

which gives the closed-loop input resistance R_{if} as

$$R_{if} = \frac{v_s}{i_s} = R_i(1 + \beta A) \tag{10.57}$$

To determine the output resistance with feedback, let us apply a test voltage v_x, as shown in Fig. 10.24:

$$v_x = R_o(i_o - Av_e) = R_o(i_o + \beta A i_o) = R_o(1 + \beta A) i_o$$

which gives the closed-loop output resistance R_{of} as

$$R_{of} = \frac{v_x}{i_x} = R_o(1 + \beta A) \tag{10.58}$$

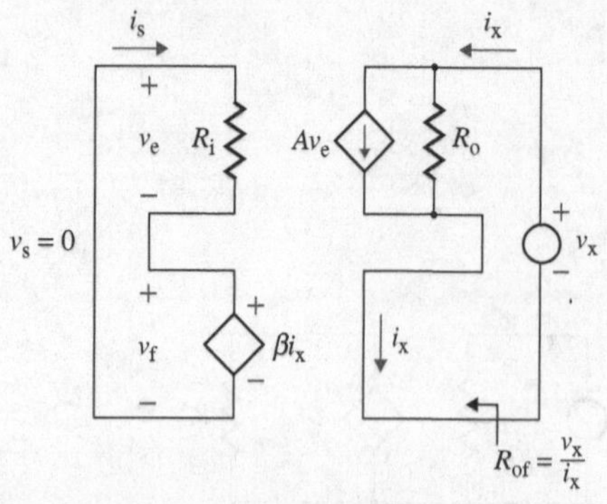

FIGURE 10.24 Test circuit for finding output resistance R_{of}

▶ **NOTES**

1. A is the open-loop transconductance of the amplifier, in A/V.
2. β is the transresistance of the feedback network, in V/A.
3. $T_L = \beta A$ is the loop gain, which is dimensionless.

10.7.2 Analysis of a Practical Series-Series Feedback Network

The analysis in the previous section did not take into account the loading effect of the feedback network. The open-loop parameters of the amplifier in Fig. 10.25(a) can be modified to include the loading effect due to R_s, R_x, R_y, and R_L, providing the equivalent ideal feedback network in Fig. 10.25(b). The modified parameters are given by

$$R_{ie} = R_s + R_i + R_x \tag{10.59}$$

$$R_{oe} = R_o + R_y + R_L \tag{10.60}$$

▶ **NOTE** It may appear from Fig. 10.25(b) that R_o would have to be in parallel with $(R_y + R_L)$ for us to find R_{oe} in Eq. (10.60), but if we convert the current source Av_e to a voltage source, we see that the effective resistance used to find the current i_o becomes $(R_o + R_y + R_L)$.

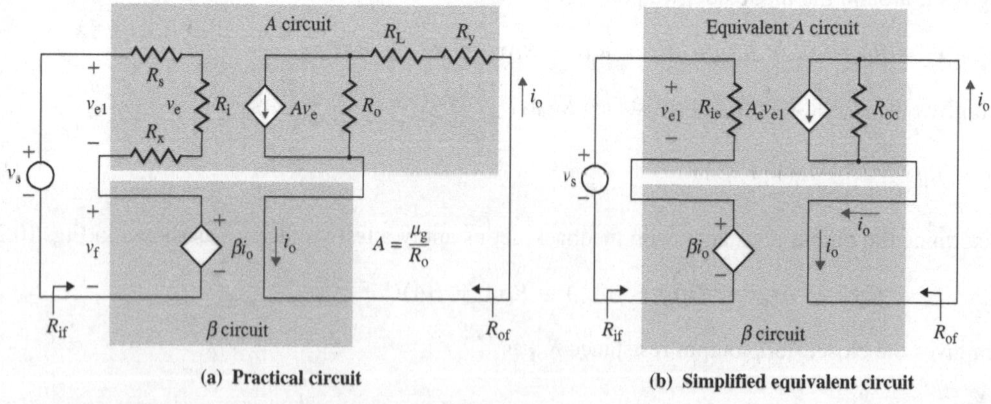

FIGURE 10.25 Practical series-series feedback amplifier

$$i_o = \frac{R_o}{R_o + R_L + R_y} A v_e \tag{10.61}$$

$$v_e = \frac{R_i}{R_i + R_x + R_s} v_{e1} \tag{10.62}$$

Substituting v_e from Eq. (10.62) into Eq. (10.61) gives the modified open-loop transconductance A_e:

$$A_e = \frac{i_o}{v_{e1}} = \frac{R_o R_i}{(R_o + R_L + R_y)(R_s + R_i + R_x)} A \tag{10.63}$$

If the values of R_i, R_o, and A are replaced by R_{ie}, R_{oe}, and A_e, respectively, then Eqs. (10.56) through (10.58) can be applied to calculate the closed-loop parameters R_{if}, R_{of}, and A_f.

EXAMPLE 10.6

Finding the performance of a noninverting amplifier with series-series feedback The noninverting amplifier shown in Fig. 10.8(b) has $R_L = 4\ \Omega$ and $R_s = 5\ \text{k}\Omega$. The feedback resistance is $R_F = 5\ \Omega$. The op-amp parameters are $R_i = 2\ \text{M}\Omega$ and $R_o = 75\ \Omega$, and the open-loop voltage gain is $\mu_g = 2 \times 10^5$.

(a) Determine the input resistance seen by the source $R_{if} = v_s/i_s$, the output resistance R_{of}, and the closed-loop transconductance gain $A_f = i_o/v_s$.

(b) Use PSpice/SPICE to verify your results.

SOLUTION

$R_L = 4\ \Omega$, $R_s = 5\ \text{k}\Omega$, $R_F = 5\ \Omega$, $R_i = 2\ \text{M}\Omega$, $R_o = 75\ \Omega$, and $\mu_g = 2 \times 10^5$. Replacing the op-amp in Fig. 10.8(b) by its equivalent circuit gives the amplifier shown in Fig. 10.26(a).

(a) The steps in analyzing the feedback network are as follows:

Step 1. R_F constitutes the feedback network, and it produces a feedback voltage v_f proportional to the output current i_o.

Step 2. The amplifier uses series-series feedback. Thus, A must be in A/V:

$$A = \frac{\mu_g}{R_o} = \frac{2 \times 10^5}{75} = 2.67\ \text{kA/V}$$

Step 3. The effect of the feedback network is taken into account by severing R_F from the op-amp at side 1 and from R_L at side 2. This modification is shown in Fig. 10.26(b). Then

$$R_x = R_F = 5\ \Omega$$
$$R_y = R_F = 5\ \Omega$$

Step 4. If we represent the amplifier of Fig. 10.26(b) by an equivalent voltage amplifier, the input resistance is

$$R_{ie} = R_s + R_i + R_x = R_s + R_i + R_F = 5\ \text{k} + 2\ \text{M} + 5 \approx 2005\ \text{k}\Omega$$

and the output resistance is

$$R_{oe} = R_o + R_F + R_L = 75 + 5 + 4 = 84\ \Omega$$

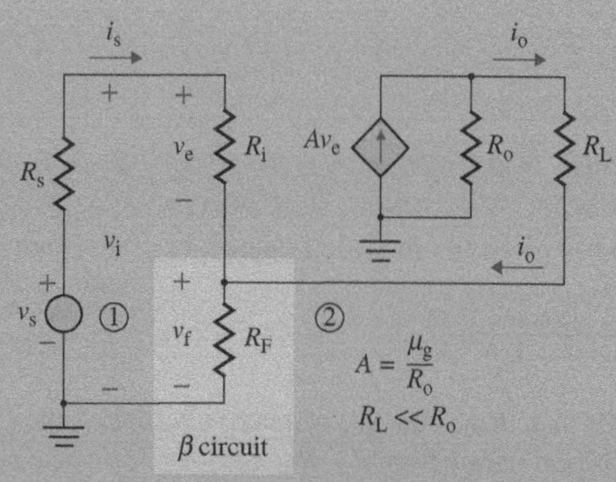

(a) Amplifier

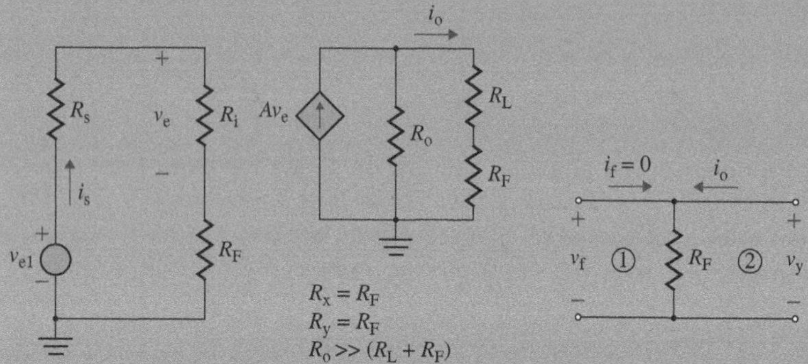

(b) Equivalent circuit

(c) β determination

FIGURE 10.26 Noninverting amplifier with series-series feedback

We find the modified open-loop transconductance gain A_e from Eq. (10.63):

$$A_e = \frac{R_o R_i}{(R_o + R_L + R_F)(R_s + R_i + R_F)} A$$

$$= \frac{75 \times 2000\text{ k} \times 2.67\text{ kA/V}}{(75 + 4 + 5)(5\text{ k} + 2000\text{ k} + 5)} = 2.375 \times 10^3\text{ A/V}$$

Step 5. From Fig. 10.26(c), the feedback factor β is given by

$$\beta = \left.\frac{v_f}{i_y}\right|_{i_o=0} = R_F = 5\ \Omega$$

Step 6. The input resistance (seen by the source) with feedback is

$$R_{if} = \frac{v_s}{i_s} = R_{ie}(1 + \beta A_e) = 2005\text{ k} \times (1 + 5 \times 2.375 \times 10^3) = 23.81\text{ G}\Omega$$

Step 7. The output resistance with feedback is

$$R_{of} = R_{oe}(1 + \beta A_e) = 84 \times (1 + 5 \times 2.375 \times 10^3) = 997.6\text{ k}\Omega$$

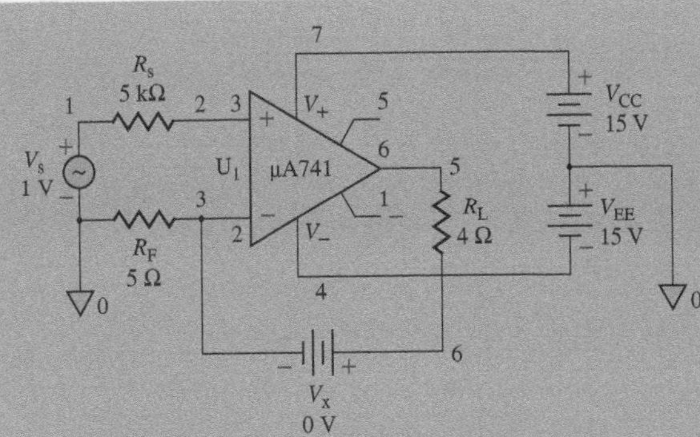

FIGURE 10.27 Series-series feedback network for PSpice simulation

Step 8. The closed-loop transconductance gain A_f is

$$A_f = \frac{i_o}{v_s} = \frac{A_e}{1 + \beta A_e} = \frac{2.375 \times 10^3}{1 + 5 \times 2.375 \times 10^3} = 200 \text{ mA/V}$$

(b) The series-series feedback circuit for PSpice simulation is shown in Fig. 10.27. The results of PSpice/SPICE simulation (.TF analysis) are shown below, with hand calculations to the right:

`I(VX)/VS=2.000E-01=0.2`	$A_f = 200$ mA/V
`INPUT RESISTANCE AT VS=2.381E+10=23.81 GΩ`	$R_{if} = 23.81$ GΩ
`OUTPUT RESISTANCE AT I(VX)=9.976E+05=997.6 kΩ`	$R_{of} = 997.6$ kΩ

The PSpice results are very close to the hand-calculated values.

EXAMPLE 10.7

Finding the performance of a BJT amplifier with series-series feedback The AC equivalent circuit of a feedback amplifier is shown in Fig. 10.28. The DC bias currents of the transistors are $I_{C1} = 0.5$ mA, $I_{C2} = 1$ mA, and $I_{C3} = 5$ mA. The transistor parameters are $h_{fe} = h_{fe1} = h_{fe2} = h_{fe3} = 100$ and $r_\mu = r_o = \infty$.

(a) Use the techniques of feedback analysis to calculate the input resistance R_{if}, the output resistance R_{of}, and the closed-loop transconductance gain $A_f = i_o/v_s$.

(b) Use PSpice/SPICE to check your results.

NOTE: The following analysis can also be applied to MOSFET amplifiers by substituting a very large value of the B-E resistance r_π tending to infinity, $1/r_\pi \simeq 0$.

SOLUTION

For $I_{C1} = 0.5$ mA, Eqs. (8.40) and (8.42) give

$$r_{\pi 1} = \frac{h_{fe} \times 25.8 \text{ mV}}{I_{C1}} = \frac{100 \times 25.8 \text{ mV}}{0.5 \text{ mA}} = 5.16 \text{ k}\Omega$$

$$g_{m1} = \frac{I_{C1}}{V_T} = \frac{0.5 \text{ mA}}{25.8 \text{ mV}} = 19.38 \text{ mA/V}$$

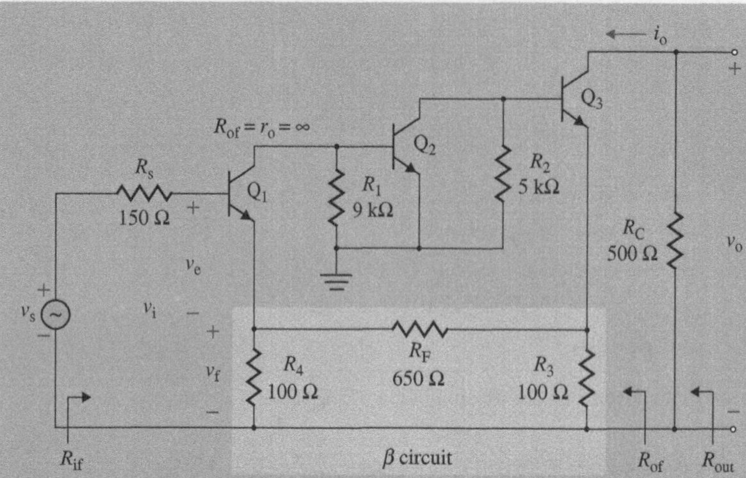

FIGURE 10.28 Three-stage amplifier with series-series feedback

For $I_{C2} = 1$ mA,

$$r_{\pi 2} = \frac{h_{fe} \times 25.8 \text{ mV}}{I_{C2}} = \frac{100 \times 25.8 \text{ mV}}{1 \text{ mA}} = 2.58 \text{ k}\Omega$$

$$g_{m2} = \frac{I_{C2}}{V_T} = \frac{1 \text{ mA}}{25.8 \text{ mV}} = 38.76 \text{ mA/V}$$

For $I_{C3} = 5$ mA,

$$r_{\pi 3} = \frac{h_{fe} \times 25.8 \text{ mV}}{I_{C3}} = \frac{100 \times 25.8 \text{ mV}}{5 \text{ mA}} = 516 \ \Omega$$

$$g_{m3} = \frac{I_{C3}}{V_T} = \frac{5 \text{ mA}}{25.8 \text{ mV}} = 193.8 \text{ mA/V}$$

(a) The steps in analyzing the feedback network are as follows:

Step 1. R_3, R_F, and R_4 constitute the feedback network. The input voltage v_i is compared with the feedback signal v_f. An error voltage $v_e = v_i - v_f$ is the input to the base of the transistor Q_3. The block diagram representing the feedback mechanism is shown in Fig. 10.29(a). The feedback network is shown in Fig. 10.29(b).

Step 2. The amplifier uses series-series feedback. Thus, the transconductance gain A must be expressed in A/V.

Step 3. The effect of the feedback network is taken into account at the input side by removing the network from the emitter of Q_1 at side 2, as shown in Fig. 10.29(b), and at the output side by removing the network from the emitter of Q_1 at side 1. These modifications are shown in the small-signal equivalent circuit in Fig. 10.29(c).

Step 4. If we represent the amplifier in Fig. 10.29(c) by an equivalent transconductance amplifier, the input resistance is

$$R_i = r_{\pi 1} + (1 + g_{m1} r_{\pi 1})[R_4 \| (R_F + R_3)] = 5.16 \text{ k} + 101 \times (100 \| 750) = 14.07 \text{ k}\Omega$$

$$R_{ie} = R_i + R_s = 14.07 \text{ k} + 150 = 14.22 \text{ k}\Omega$$

The output resistance seen at the collector of Q_3 is

$$R_o = r_o = \infty$$

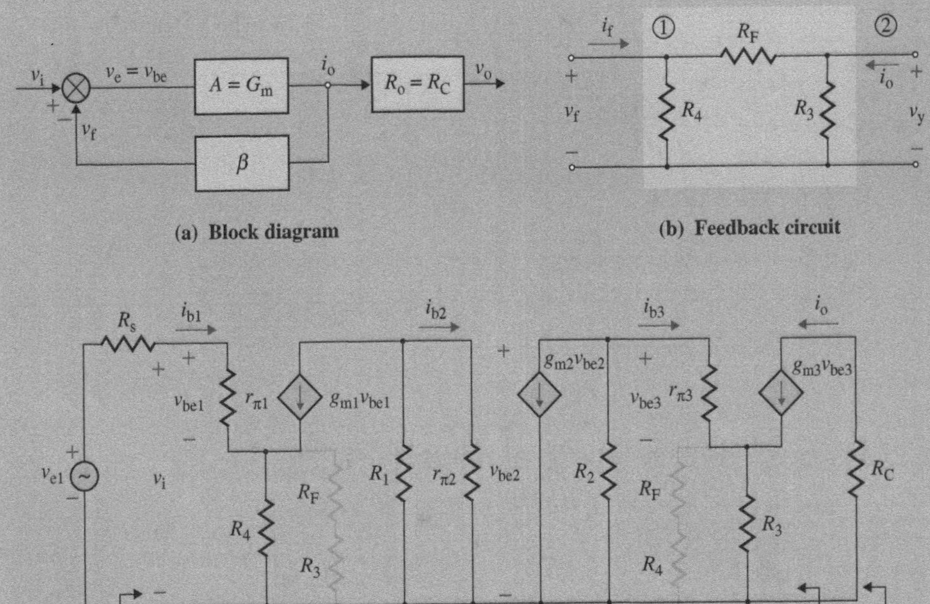

(c) Small-signal equivalent circuit

FIGURE 10.29 Equivalent circuits for Example 10.7

From Fig. 10.29(c), we get

$$V_{be1} = \left(\frac{r_{\pi1}}{R_i}\right)v_i = \left(\frac{5.16\text{ k}}{14.07\text{ k}}\right)v_i = 0.367v_i$$

$$v_{b2} = v_{be2} = -(R_1 \| r_{\pi2})g_{m1}v_{be1} = -(9\text{ k}\|2.58\text{ k}) \times 19.38\text{ m} \times 0.367v_i = -14.25v_i$$

$$v_{b3} = -\{R_2 \| [r_{\pi3} + (1 + g_{m3}r_{\pi3})(R_3 \| (R_F + R_4))]\}g_{m2}v_{be2}$$

$$= -\{5\text{ k}\|[516 + (1 + 193.8\text{ m} \times 516)(100\|750)]\} \times 38.76\text{ m} \times (-14.25v_i)$$

$$= 1.805 \times 10^3 v_i$$

$$v_o = \frac{v_{b3} \times r_{\pi3}}{\{r_{\pi3} + (1 + g_{m3}r_{\pi3})[R_3 \| (R_F + R_4)]\}}$$

$$= \frac{(1.825 \times 10^3 \times v_i) \times 516}{[516 + (1 + 193.8\text{ m} \times 516)(100\|750)]} = 98.766v_i$$

$$i_o = g_{m3}v_{be3} = 193.8\text{ m} \times 98.766v_i = 19.14v_i$$

Therefore, the open-loop transconductance A is given by $A = i_o/v_i = 19.14\text{ A/V}$.

Step 5. From Eq. (10.55), the feedback factor β is given by

$$\beta = \left.\frac{v_f}{i_o}\right|_{i_f=0} = \frac{R_3R_4}{R_3 + R_4 + R_F} = \frac{100 \times 100}{100 + 100 + 650} = 11.76\text{ V/A}$$

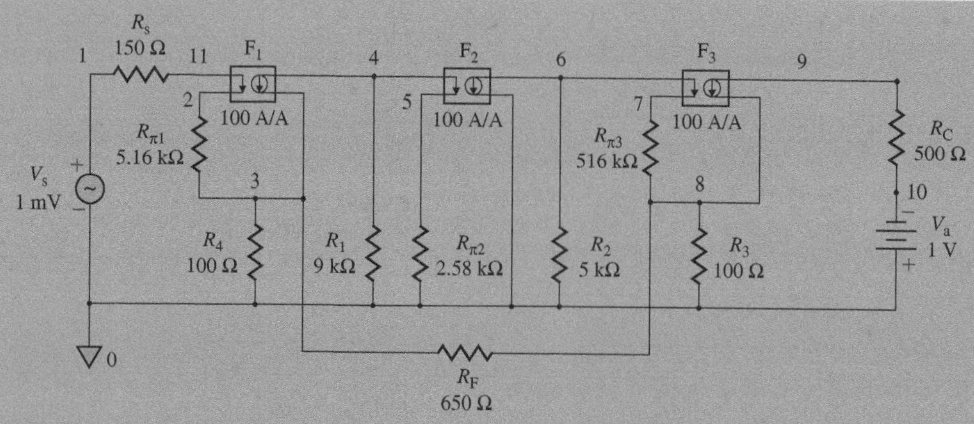

FIGURE 10.30 Series-series feedback network for PSpice simulation

Step 6. The input resistance with feedback is

$$R_{if} = \frac{v_s}{i_s} = R_{ie}(1 + \beta A_e) = 14.22\text{ k}\Omega \times (1 + 11.76 \times 18.94) = 3.18\text{ M}\Omega$$

Step 7. The output resistance with feedback is

$$R_{of} = R_o(1 + \beta A) = \infty$$

Since the output is taken across R_C, the output resistance of the amplifier is

$$R_{out} = R_{of} \| R_C = R_C = 500\ \Omega$$

Step 8. The closed-loop transconductance gain A_f is

$$A_f = \frac{i_o}{v_s} = \frac{18.94}{1 + 11.76 \times 18.94} = 84.66\text{ mA/V}$$

(b) The series-series feedback amplifier for PSpice simulation is shown in Fig. 10.30. The results of PSpice simulation (.TF analysis) are shown below, with hand calculations to the right.

`I(VU)/VS=8.379E-02=83.79 mA/V`	$A_f = 84.66$ mA/V
`INPUT RESISTANCE AT VS=3.215E+06=3.215 MΩ`	$R_{if} \approx 3.18$ MΩ
`OUTPUT RESISTANCE AT I(VU)=8.639E+13 Ω`	$R_{of} = \infty$

The PSpice results are very close to the hand-calculated values.

KEY POINTS OF SECTION 10.7

- Series-series feedback is applied to transconductance amplifiers. This type of feedback increases both the input resistance and the output resistance by a factor of $(1 + \beta A)$.
- The amplifier is represented by a transconductance amplifier and the feedback network as a transresistance gain. A is in units of A/V, and β is in units of V/A.

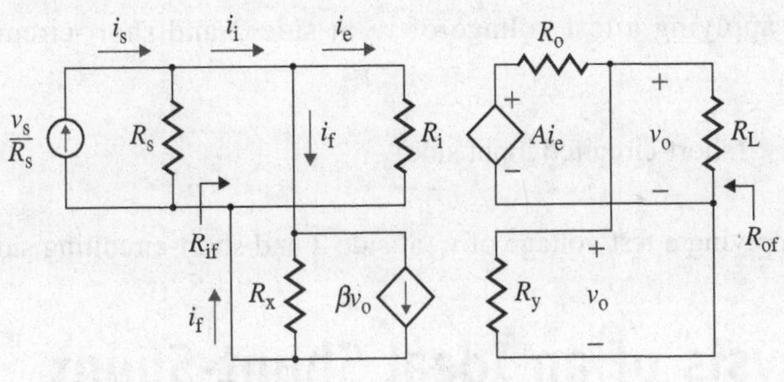

FIGURE 10.31 Shunt-shunt feedback

10.8 Shunt-Shunt Feedback

In shunt-shunt feedback, as shown in Fig. 10.9(a), the feedback circuit is in parallel with the amplifier. Although any amplifier can be represented by any of the four types, the analysis of shunt-shunt feedback can be simplified by representing the amplifier in transresistance form. This is shown in Fig. 10.31, in which the amplifier has an input resistance of R_i, an output resistance of R_o, and an open-loop transresistance gain of A (in V/A). That is, $A = \mu_g R_i$. The feedback current i_f is proportional to the output voltage v_o. The feedback network is modeled in transconductance form with an input resistance of R_y, an output resistance of R_x, and an open-loop transconductance gain of β.

The test conditions for determining the parameters of the feedback network are shown in Fig. 10.32. Given that the voltage across a short circuit is zero and no current flows through an open circuit, the model parameters can be defined by three equations. The first equation is

$$R_x = \left.\frac{v_f}{i_f}\right|_{v_o=0} \quad \text{(short-circuited output side)} \tag{10.64}$$

which is obtained by applying a test voltage of v_f at side 1 and short-circuiting side 2. The second equation is

$$R_y = \left.\frac{v_o}{i_y}\right|_{v_f=0} \quad \text{(short-circuited input side)} \tag{10.65}$$

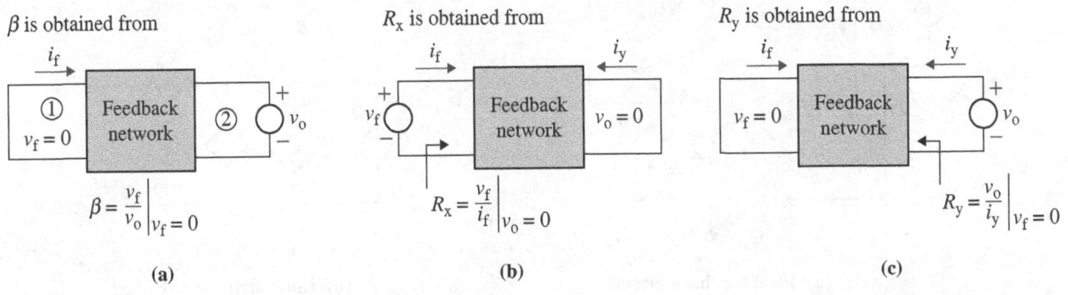

FIGURE 10.32 Test conditions for determining the parameters of a shunt-shunt feedback network

which is obtained by applying a test voltage of v_o at side 2 and short-circuiting side 1. The third equation is

$$\beta = \left.\frac{i_f}{v_o}\right|_{v_f=0} \qquad \text{(short-circuited input side)} \tag{10.66}$$

which is obtained by applying a test voltage of v_o at side 2 and short-circuiting side 1.

10.8.1 Analysis of an Ideal Shunt-Shunt Feedback Network

Let us assume an ideal shunt-shunt feedback network—that is, $R_s = \infty$, $R_x = \infty$, $R_y = \infty$, and $R_L = 0$. The feedback amplifier in Fig. 10.31 is simplified to that shown in Fig. 10.33(a), which can be represented by the equivalent circuit shown in Fig. 10.33(b). The output voltage v_o becomes

$$v_o = Ai_e \tag{10.67}$$

The feedback current i_f is proportional to the output voltage v_o. That is,

$$i_f = \beta v_o \tag{10.68}$$

and

$$i_e = i_s - i_f \tag{10.69}$$

Substituting i_e from Eq. (10.69) and i_f from Eq. (10.68) into Eq. (10.67), we get the closed-loop transresistance A_f as

$$A_f = \frac{v_o}{i_s} = \frac{A}{1 + \beta A} \tag{10.70}$$

From Eq. (10.69),

$$i_s = i_e + i_f$$

Substituting i_f from Eq. (10.68) and v_o from Eq. (10.67) into the above equation gives

$$i_s = i_e + \beta v_o = i_e + \beta A i_e = i_e(1 + \beta A) \tag{10.71}$$

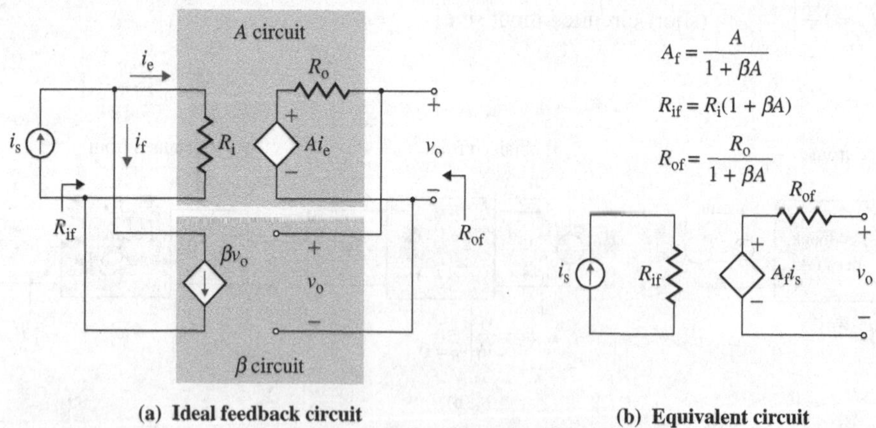

FIGURE 10.33 Ideal shunt-shunt feedback circuit

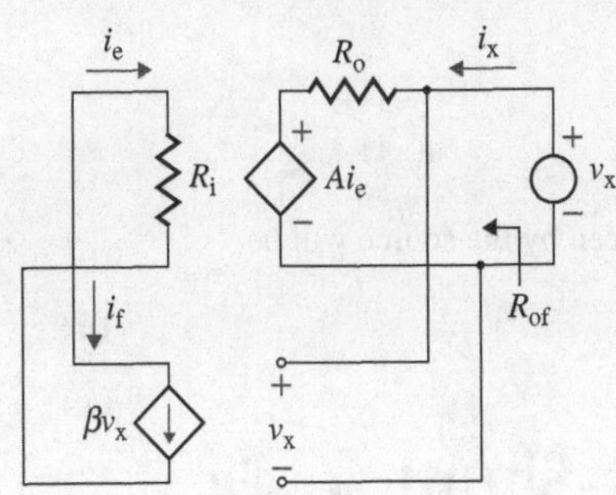

FIGURE 10.34 Equivalent circuit for determining output resistance

The error current i_e is related to v_i by

$$v_i = R_i i_e \tag{10.72}$$

Using i_s from Eq. (10.71) and v_i from Eq. (10.72), we can find the input resistance with feedback R_{if}:

$$R_{if} = \frac{v_i}{i_s} = \frac{i_e R_i}{i_e(1 + \beta A)} = \frac{R_i}{1 + \beta A} \tag{10.73}$$

The input resistance of an amplifier with shunt feedback at the input side is always decreased by a factor of $(1 + \beta A)$.

The output resistance with feedback R_{of}, which is Thevenin's equivalent resistance, can be obtained by applying a test voltage v_x to the output side and open-circuiting the input current source. The equivalent circuit for determining Thevenin's equivalent output resistance is shown in Fig. 10.34. We have

$$i_e = -i_f = -\beta v_x \tag{10.74}$$

and
$$i_x = \frac{v_x - 4i_e}{R_o} \tag{10.75}$$

Substituting i_e from Eq. (10.74) into Eq. (10.75) yields

$$i_x = \frac{v_x + \beta A v_x}{R_o} = \frac{(1 + \beta A)v_x}{R_o} \tag{10.76}$$

which gives the output resistance with feedback R_{of} as

$$R_{of} = \frac{R_o}{1 + \beta A} \tag{10.77}$$

Thus, shunt feedback at the output always lowers the output resistance by a factor of $(1 + \beta A)$.

The input impedance, the output impedance, and the overall gain can be written in generalized form in Laplace's domain as follows:

$$Z_{if}(s) = \frac{Z_i(s)}{1 + \beta A(s)} \tag{10.78}$$

$$Z_{of}(s) = \frac{Z_o(s)}{1 + \beta A(s)} \tag{10.79}$$

$$A_f(s) = \frac{A(s)}{1 + \beta A(s)} \tag{10.80}$$

▶ **NOTES**

1. A is the open-loop transresistance of the amplifier, in V/A.
2. β is the transconductance of the feedback network, in A/V.
3. $T_L = \beta A$ is the loop gain, which is dimensionless.
4. If the source has an input impedance of Z_s, then the total impedance Z_{in} seen by the source will be

$$Z_{in}(s) = \frac{v_s}{i_s} = Z_{if}(s) + Z_s(s) \tag{10.81}$$

10.8.2 Analysis of a Practical Shunt-Shunt Feedback Network

The open-loop parameters of the amplifier in Fig. 10.35(a) can be modified to include the loading effect due to R_s, R_x, R_y, and R_L, producing the equivalent ideal feedback circuit in Fig. 10.35(b). The modified parameters are given by

$$R_{ie} = R_s \| R_i \| R_x \tag{10.82}$$

$$R_{oe} = R_o \| R_y \| R_L \tag{10.83}$$

and
$$v_o = \frac{R_y \| R_L}{(R_y \| R_L) + R_o} A i_e \tag{10.84}$$

where i_e is the current through R_i only, not through R_i and $R_s \| R_x$. Thus, by the current divider rule, i_e is given by

$$i_e = \frac{R_s \| R_x}{R_s \| R_x + R_i} i_{e1} \tag{10.85}$$

Substituting i_e from Eq. (10.85) into Eq. (10.84) gives the modified open-loop transresistance gain A_e:

$$A_e = \frac{v_o}{i_{e1}} = \frac{R_y \| R_L}{R_y \| R_L + R_o} \times \frac{R_s \| R_x}{R_s \| R_x + R_i} A \tag{10.86}$$

With these values of R_{ie}, R_{oe}, and A_e, we can use the equations in the preceding section to calculate the closed-loop parameters R_{if}, R_{of}, and A_f.

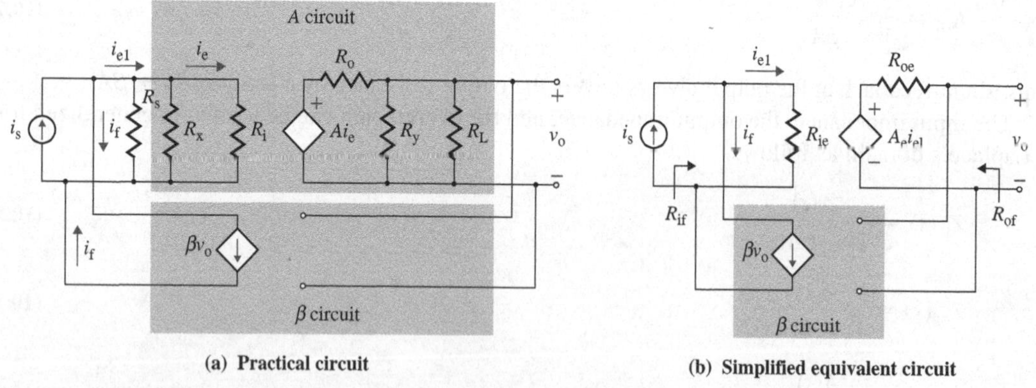

FIGURE 10.35 Practical shunt-shunt feedback amplifier

EXAMPLE 10.8

Finding the performance of an inverting amplifier with shunt-shunt feedback The inverting amplifier shown in Fig. 10.9(b) has $R_s = 2\ \text{k}\Omega$, $R_L = 5\ \text{k}\Omega$, and $R_F = 8\ \text{k}\Omega$. The op-amp parameters are $R_i = 2\ \text{M}\Omega$ and $R_o = 75\ \Omega$, and the open-loop voltage gain is $\mu_g = 2 \times 10^5$.

(a) Determine the input resistance seen by the source $R_{if} = v_i/i_s$, the output resistance R_{of}, the closed-loop transresistance $A_f = v_o/i_s$, and the voltage gain $A_{vf} = v_o/v_s$.

(b) Use PSpice/SPICE to verify your results.

SOLUTION

$R_L = 5\ \text{k}\Omega$, $R_s = 2\ \text{k}\Omega$, $R_F = 8\ \text{k}\Omega$, $R_i = 2\ \text{M}\Omega$, $R_o = 75\ \Omega$, and $\mu_g = 2 \times 10^5$. Replacing the op-amp in Fig. 10.9(b) by its equivalent circuit gives the amplifier shown in Fig. 10.36(a).

(a) The steps in analyzing the feedback network are as follows:

Step 1. R_F constitutes the feedback network, and it produces a feedback current i_f proportional to the output voltage v_o.

Step 2. The amplifier uses shunt-shunt feedback. Thus, A must be in V/A. Converting the voltage-controlled voltage source to a current-controlled voltage source, we get

$$v_o = -\mu_g v_e = -\mu_g R_i i_e = A i_e$$

which gives the open-loop transresistance A:

$$A = -\mu_g R_i = -2 \times 10^5 \times 2 \times 10^6 = -4 \times 10^{11}\ \text{V/A}$$

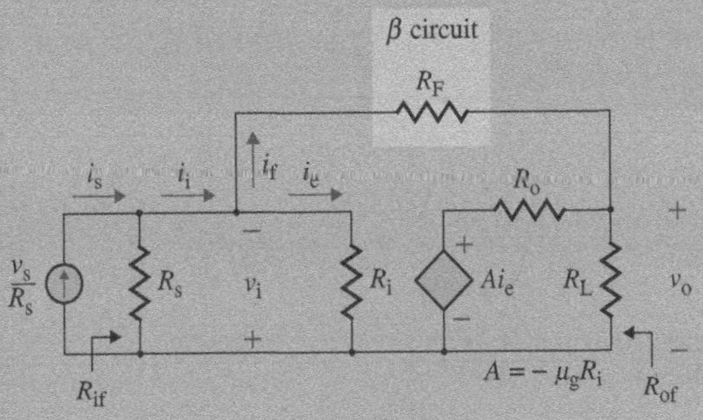

(a) Amplifier

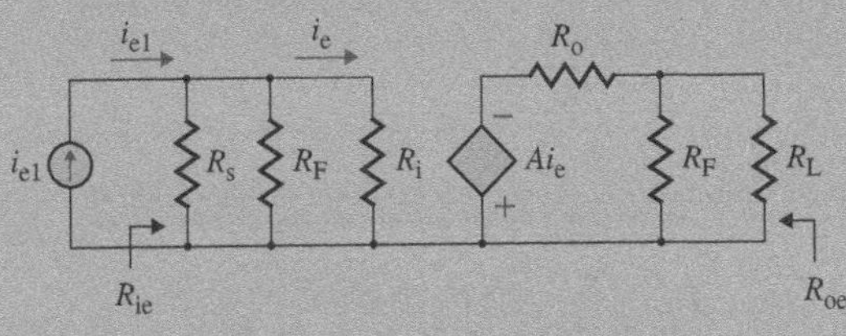

(b) Equivalent circuit

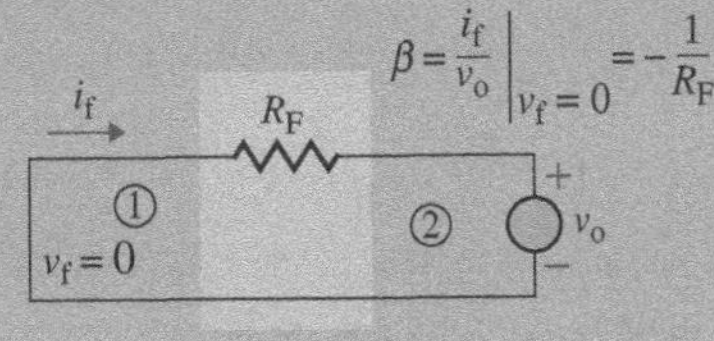

(c) β determination

FIGURE 10.36 Inverting op-amp amplifier

Step 3. The effect of the feedback network is taken into account by shorting R_F to the ground at sides 1 and 2. This arrangement is shown in Fig. 10.36(b). Then

$$R_x = R_F = 8\text{ k}\Omega$$

and $$R_y = R_F = 8\text{ k}\Omega$$

Step 4. If we represent the amplifier of Fig. 10.36(a) by an equivalent transresistance amplifier, as shown in Fig. 10.36(b), we have

$$R_{ie} = R_s \| R_i \| R_F = 2\text{ k} \| 2\text{ M} \| 8\text{ k} = 1.6\text{ k}\Omega$$

and $$R_{oe} = R_o \| R_F \| R_L = 75 \| 8\text{ k} \| 5\text{ k} = 73.2\ \Omega$$

From Eq. (10.86), we find that the modified open-loop transresistance gain A_e is

$$A_e = \frac{v_o}{i_{e1}} = \frac{-R_F \| R_L}{(R_F \| R_L) + R_o} \times \frac{R_s \| R_F}{R_s \| R_F + R_i} A$$

$$= \frac{-(8\text{ k} \| 5\text{ k}) \times (2\text{ k} \| 8\text{ k}) \times 4 \times 10^{11}}{(8\text{ k} \| 5\text{ k} + 75)(2\text{ k} \| 8\text{ k} + 2\text{ M})} = -312.1\text{ M}\Omega$$

Step 5. From Fig. 10.36(c), the feedback factor β is given by

$$\beta = \left.\frac{i_f}{v_o}\right|_{v_f=0} = -\frac{1}{R_F} = -125\ \mu\mho$$

The loop gain is

$$T_L = \beta A_e = 125\ \mu\mho \times 312.1 \times 10^6 = 39.02\text{ k}\Omega$$

Step 6. The input resistance at the input side of the op-amp is

$$R_{if} = \frac{R_{ie}}{1 + \beta A_e} = \frac{R_{ie}}{1 + T_L}$$

$$= \frac{1.6\text{ k}\Omega}{1 + 39.02 \times 10^3} = 41.0\text{ m}\Omega$$

Step 7. The output resistance with feedback is

$$R_{of} = \frac{R_{oe}}{1 + \beta A_e} = \frac{R_{oe}}{1 + T_L} = \frac{73.2\ \Omega}{1 + 39.02 \times 10^3} = 1.876\text{ m}\Omega$$

Step 8. The closed-loop transresistance gain A_f is

$$A_f = \frac{v_o}{i_s} = \frac{A_e}{1 + \beta A_e} = \frac{A_e}{1 + T_L}$$

$$= \frac{-312.1 \times 10^6}{1 + 39.02 \times 10^3} = -8\text{ k}\Omega$$

$$v_o = \left(\frac{v_o}{v_s}\right)v_s = \left(\frac{v_o}{i_s}\right)\left(\frac{i_s}{v_s}\right)v_s = \left(\frac{-8\text{ k}}{2\text{ k}}\right)v_s = -4v_s$$

Therefore, the overall voltage gain is $v_o/v_s = -4$.

NOTE: Substituting $R_1 = R_s$ in Eq. (3.40) gives the voltage gain of the inverting amplifier as $R_F/R_s = -8\text{ k}/2\text{ k} = -4$.

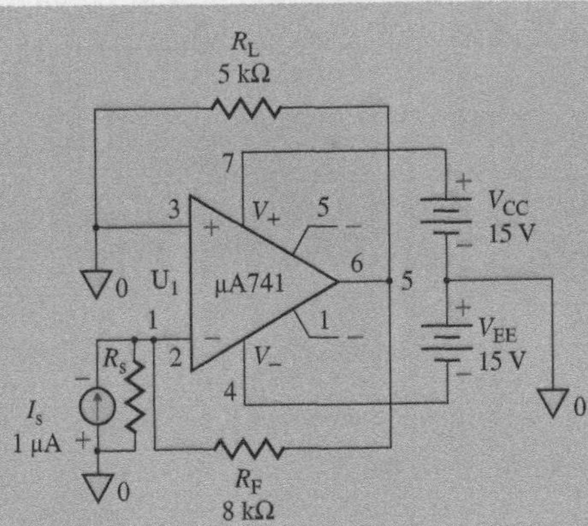

FIGURE 10.37 Shunt-shunt feedback circuit for PSpice simulation

(b) The shunt-shunt feedback circuit for PSpice simulation is shown in Fig. 10.37. The results of the simulation (.TF analysis) are shown below, with hand calculations to the right:

```
V(5)/Is=-8.000E+03=-8 kΩ                              
INPUT RESISTANCE AT IS=4.097E-02=40.97 mΩ             
OUTPUT RESISTANCE AT V(5)=1.876E-03=1.876 mΩ          
```

$A_f = -8\ k\Omega$

$R_{if} = 41.0\ m\Omega$

$R_{of} = 1.876\ m\Omega$

The PSpice results are very close to the hand-calculated values.

EXAMPLE 10.9

Finding the performance of a BJT amplifier with shunt-shunt feedback The parameters of the amplifier in Fig. 10.38 are $R_{C1} = 5\ k\Omega$, $R_E = 2.5\ k\Omega$, $R_{C2} = 5\ k\Omega$, $R_F = 4\ k\Omega$, and $R_s = 200\ \Omega$. The DC bias currents of the transistors are $I_{C1} = 0.5$ mA and $I_{C2} = 1$ mA. The transistor parameters are $h_{fe} = h_{fe1} = h_{fe2} = 150$ and $r_\mu = r_o = \infty$.

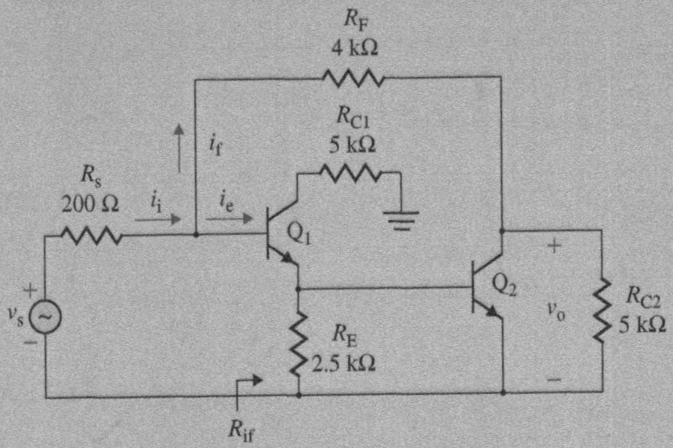

FIGURE 10.38 AC equivalent of two-stage amplifier with shunt-shunt feedback

(a) Use the techniques of feedback analysis to calculate the input resistance R_{if}, the output resistance R_{of}, and the closed-loop transresistance gain A_f.

(b) Use PSpice/SPICE to check your results.

NOTE: The following analysis can also be applied to MOSFET amplifiers by substituting a very large value of the B-E resistance r_π tending to infinity, $1/r_\pi \simeq 0$.

SOLUTION

For $I_{C1} = 0.5$ mA, Eqs. (8.40) and (8.42) give

$$r_{\pi 1} = \frac{h_{fe} \times 25.8 \text{ mV}}{I_{C1}} = \frac{150 \times 25.8 \text{ mV}}{0.5 \text{ mA}} = 7.74 \text{ k}\Omega$$

$$g_{m1} = \frac{I_{C1}}{V_T} = \frac{0.5 \text{ mA}}{25.8 \text{ mV}} = 19.38 \text{ mA/V}$$

For $I_{C2} = 1$ mA,

$$r_{\pi 2} = \frac{h_{fe} \times 25.8 \text{ mV}}{I_{C2}} = \frac{150 \times 25.8 \text{ mV}}{1 \text{ mA}} = 3.87 \text{ k}\Omega$$

$$g_{m2} = \frac{I_{C2}}{V_T} = \frac{1 \text{ mA}}{25.8 \text{ mV}} = 38.76 \text{ mA/V}$$

The feedback current i_f is proportional to the output voltage v_o. The effective input current to the amplifier is $i_e = i_i - i_f$.

(a) The steps in analyzing the feedback network are as follows:

Step 1. R_F acts as the feedback network. The feedback current i_f is proportional to the output voltage v_o. The effective input current to the amplifier is $i_e = i_i - i_f$. The functional block diagram is shown in Fig. 10.39(a).

Step 2. The amplifier uses shunt-shunt feedback, as shown in Fig. 10.39(b). Thus, the units of the gain A must be V/A.

Step 3. The effect of the feedback network can be taken into account at the input side by short-circuiting R_F to the ground at side 2 in Fig. 10.39(b), and at the output side by short-circuiting R_F to the ground at side 1. These modifications are shown in the small-signal AC equivalent circuit in Fig. 10.39(c).

Step 4. If we represent the amplifier in Fig. 10.39(c) by an equivalent transresistance amplifier, the resistance at the base of the amplifier is

$$R_b = \frac{v_i}{i_{b1}} = r_{\pi 1} + (1 + g_{m1} r_{\pi 1})(R_E \| r_{\pi 2}) = 7.74 \text{ k} + 151(2.5 \text{ k} \| 3.87 \text{ k}) = 237 \text{ k}\Omega$$

The input resistance of the amplifier is

$$R_{ie} = \frac{v_i}{i_s} = R_s \| R_F \| [r_{\pi 1} + (1 + g_{m1} r_{\pi 1})(R_E \| r_{\pi 2})]$$
$$= 200 \| 4 \text{ k} \| [7.74 \text{ k} + 151(2.5 \text{ k} \| 3.87 \text{ k})] = 190.3 \ \Omega$$

The output resistance of the amplifier is

$$R_{oe} = R_F \| R_{C2} = 4 \text{ k} \| 5 \text{ k} = 2.22 \text{ k}\Omega$$

Thus,

$$v_{b1} = R_{ie} i_s = 190.3 i_s$$

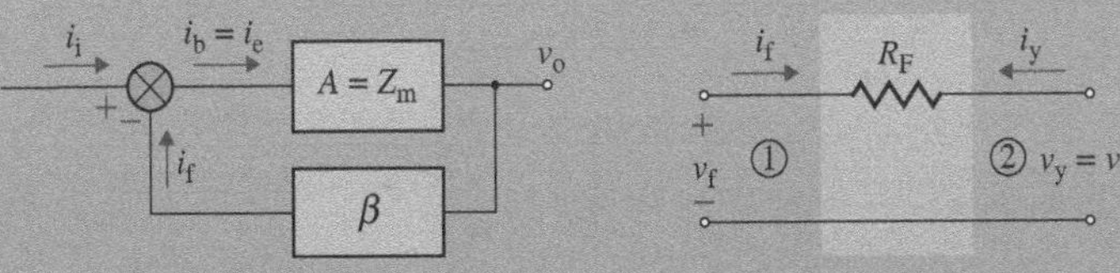

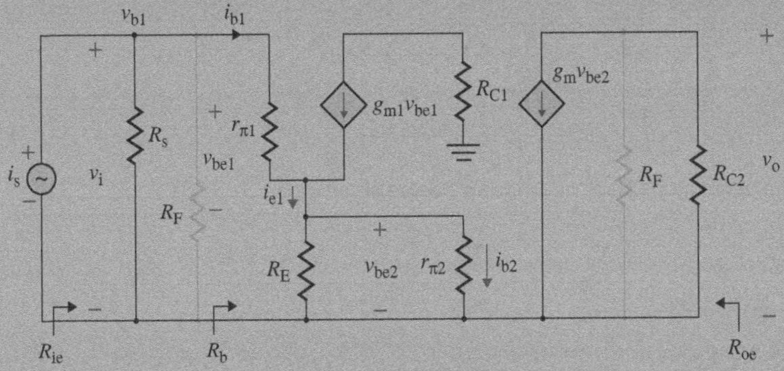

FIGURE 10.39 Equivalent circuits for Example 10.9

$$v_{be2} = \frac{v_{b1} \times (1 + g_{m1}r_{\pi 1})(R_E \| r_{\pi 2})}{[r_{\pi 1} + (1 + g_{m1}r_{\pi 1})(R_E \| r_{\pi 2})]}$$

$$= \frac{(190.3 \times i_s) \times (1 + 193.8 \text{ m} \times 7.74 \text{ k})(2.5 \text{ k} \| 3.87 \text{ k})}{[7.74 \text{ k} + (1 + 193.8 \text{ m} \times 7.74 \text{ k})(2.5 \text{ k} \| 3.87 \text{ k})]} = 184.11 i_s$$

$$i_o = g_{m2}v_{be2} = 38.76 \text{ m} \times 184.11 i_s = 7.136 i_s$$

$$v_o = -(R_F \| R_{C2}) i_o = -(4 \text{ k} \| 5 \text{ k}) \times 7.136 i_s = -15.865 \times 10^3 \times i_s$$

Therefore, the open-loop transresistance A is given by $A = v_o/i_o = -15.865$ kA/V.

Step 5. Using Eq. (10.66), we find that the feedback factor β is

$$\beta = \left.\frac{i_f}{v_o}\right|_{v_f=0} = -\frac{1}{R_F} = -\frac{1}{4 \text{ k}} = -0.25 \text{ mA/V}$$

and

$$\beta A_e = -15.865 \text{ kV/A} \times \left(-\frac{1}{4 \text{ kV/A}}\right) = 3.966$$

Step 6. The input resistance seen by the current source is

$$R_{if} = \frac{v_i}{i_s} = \frac{R_{ie}}{1 + \beta A_e} = \frac{190.3\ \Omega}{1 + 0.25 \text{ mA/V} \times 15.865 \text{ kV/A}} = 38.3\ \Omega$$

Step 7. The output resistance with feedback is

$$R_{of} = \frac{R_{oe}}{1 + \beta A_e} = \frac{2.22 \text{ k}\Omega}{1 + 0.25 \text{ mA/V} \times 15.865 \text{ kV/A}} = 447\ \Omega$$

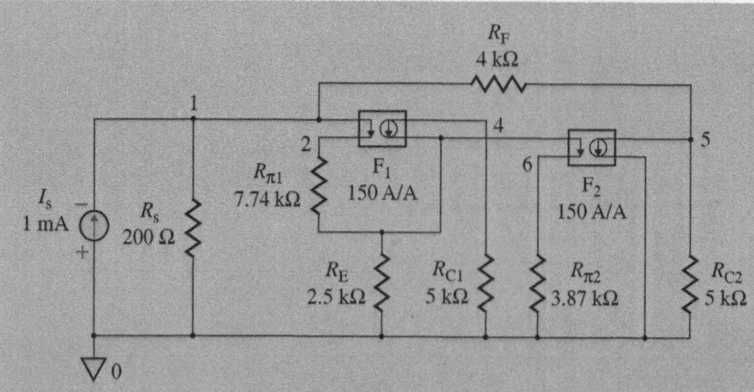

FIGURE 10.40 Two-stage shunt-shunt feedback network for PSpice simulation

Step 8. The closed-loop transresistance gain A_f is

$$A_f = \frac{v_o}{i_s} = \frac{-15.865 \text{ kV/A}}{1 + 0.25 \text{ mA/V} \times 15.865 \text{ kV/A}} = -3.195 \text{ kV/A}$$

(b) The shunt-shunt feedback amplifier for PSpice simulation is shown in Fig. 10.40. The results of PSpice/SPICE simulation (.TF analysis) are shown below, with hand calculations to the right:

`V(5)/Is=-3.19E+03=-3.19 kV/A`	$A_f = -3.195$ kV/A
`INPUT RESISTANCE AT Is=3.854E+01=38.54 Ω`	$R_{in} = 38.3\ \Omega$
`OUTPUT RESISTANCE AT V(5)=4.5E+02=450 Ω`	$R_{of} = 447\ \Omega$

The PSpice results are very close to the hand-calculated values.

KEY POINTS OF SECTION 10.8

- Shunt-shunt feedback is applied to transresistance amplifiers. This type of feedback reduces both the input and the output resistances by a factor of $(1 + \beta A)$.
- The amplifier is represented by a transresistance amplifier and the feedback network as a transconductance gain. A is in units of V/A, and β is in units of A/V.

10.9 Shunt-Series Feedback

In shunt-series feedback, as shown in Fig. 10.10(a), the feedback network is in parallel with the amplifier at the input side and in series with the amplifier at the output side. The amplifier is represented as a current amplifier. This is shown in Fig. 10.41, in which the amplifier has an input resistance of R_i, an output resistance of R_o, and an open-loop current gain of A (in A/A); that is, $A = \mu_g R_i / R_o$.

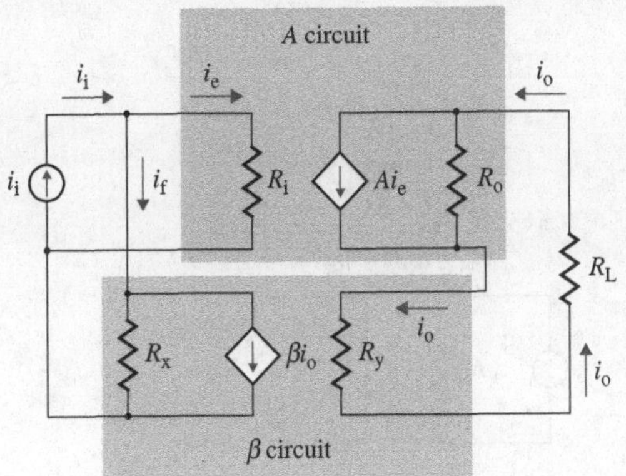

FIGURE 10.41 Shunt-series feedback amplifier

The feedback current i_f is proportional to the output current i_o. The feedback network is modeled in gain form, with an input resistance of R_y, an output resistance of R_x, and a current gain of β. The test conditions for determining the parameters of the feedback network are shown in Fig. 10.42. The parameters of the model are defined by three equations. The first equation is

$$R_x = \left.\frac{v_f}{i_f}\right|_{i_y=0} \quad \text{(open-circuited output side)} \tag{10.87}$$

which is obtained by applying a test voltage of v_f at side 1 and open-circuiting side 2. The second equation is

$$R_y = \left.\frac{v_o}{i_y}\right|_{v_f=0} \quad \text{(short-circuited input side)} \tag{10.88}$$

which is obtained by applying a test voltage of v_y at side 2 and short-circuiting side 1. The third equation is

$$\beta = \left.\frac{i_f}{i_y}\right|_{v_f=0} \quad \text{(short-circuited input side)} \tag{10.89}$$

which is obtained by applying a test voltage of v_y at side 2 and short-circuiting side 1.

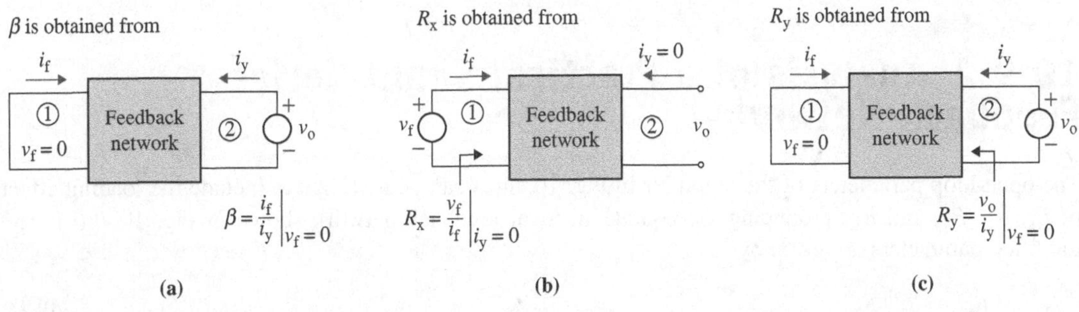

FIGURE 10.42 Test conditions for determining the parameters of a shunt-series feedback circuit

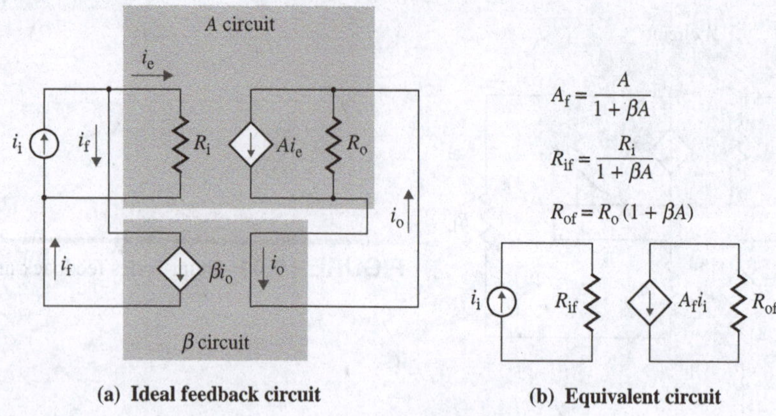

(a) Ideal feedback circuit

(b) Equivalent circuit

FIGURE 10.43 Ideal shunt-series feedback amplifier

10.9.1 Analysis of an Ideal Shunt-Series Feedback Network

Let us assume an ideal feedback network—that is, $R_X = \infty$, $R_y = 0$, and $R_L = 0$. The feedback amplifier in Fig. 10.41 can be simplified to the one in Fig. 10.43. The feedback factor β is in A/A.

If $R_L << R_o$, it can be shown that

$$R_{if} = \frac{v_i}{i_i} = \frac{R_i}{1 + \beta A} \tag{10.90}$$

$$R_{of} = R_o(1 + \beta A) \tag{10.91}$$

$$A_f = \frac{i_o}{i_i} = \frac{A}{1 + \beta A} \tag{10.92}$$

▶ **NOTES**

1. A is the open-loop current gain of the amplifier, in A/A.
2. β is the current gain of the feedback network, in A/A.
3. $T_L = \beta A$ is the loop gain, which is dimensionless.

10.9.2 Analysis of a Practical Shunt-Series Feedback Network

The open-loop parameters of the amplifier in Fig. 10.44(a) can be modified to include the loading effect of R_s, R_x, R_y, and R_L, producing the equivalent ideal feedback network shown in Fig. 10.44(b). The modified parameters are given by

$$R_{ie} = R_i \,\|\, R_x \tag{10.93}$$

and

$$R_{oe} = R_o + R_y + R_L \tag{10.94}$$

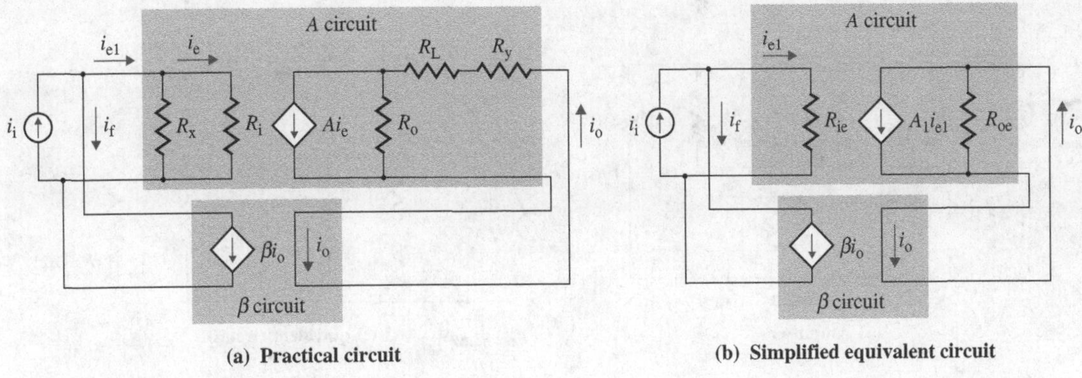

(a) Practical circuit **(b) Simplified equivalent circuit**

FIGURE 10.44 Practical shunt-series feedback amplifier

It may appear from Fig. 10.44(b) that R_o would have to be in parallel with $(R_y + R_L)$ for us to find R_{oe} in Eq. (10.94), but if we convert the current source Ai_e to a voltage source, we see that the effective resistance used to find the current i_o becomes $(R_o + R_y + R_L)$:

$$i_o = \frac{R_o}{R_y + R_o + R_L} Ai_e \tag{10.95}$$

$$i_e = \frac{R_x}{R_i + R_x} i_{e1} \tag{10.96}$$

From Eqs. (10.95) and (10.96), we can derive the modified open-loop current gain A_e:

$$A_e = \frac{i_o}{i_{e1}} = \frac{R_o R_x}{(R_y + R_L + R_o)(R_i + R_x)} A \tag{10.97}$$

If the values of R_i, R_o, and A are replaced by R_{ie}, R_{oe}, and A_e, respectively, Eqs. (10.90) through (10.92) can be applied to calculate the closed-loop parameters R_{if}, R_{of}, and A_f.

EXAMPLE 10.10

Finding the performance of an inverting amplifier with shunt-series feedback The parameters of the shunt-series feedback amplifier shown in Fig. 10.10(b) are $R_L = 5\ \Omega$, $R_s = 2.5\ \text{k}\Omega$, $R_F = 200\ \Omega$, and $R_1 = 5\ \Omega$. The op-amp parameters are $R_i = 2\ \text{M}\Omega$, $R_o = 75\ \Omega$, and $\mu_g = 2 \times 10^5$.

(a) Determine the input resistance at the op-amp input $R_{if} = v_i/i_i$, the output resistance R_{of}, and the closed-loop current gain $A_f = i_o/i_i$.

(b) Use PSpice/SPICE to check your results.

SOLUTION

$R_s = 2.5\ \text{k}\Omega$, $R_F = 200\ \Omega$, $R_1 = 5\ \Omega$, $R_L = 5\ \Omega$, $R_i = 2\ \text{M}\Omega$, $R_o = 75\ \Omega$, and $\mu_g = 2 \times 10^5$. Replacing the op-amp by its equivalent circuit gives the amplifier shown in Fig. 10.45(a).

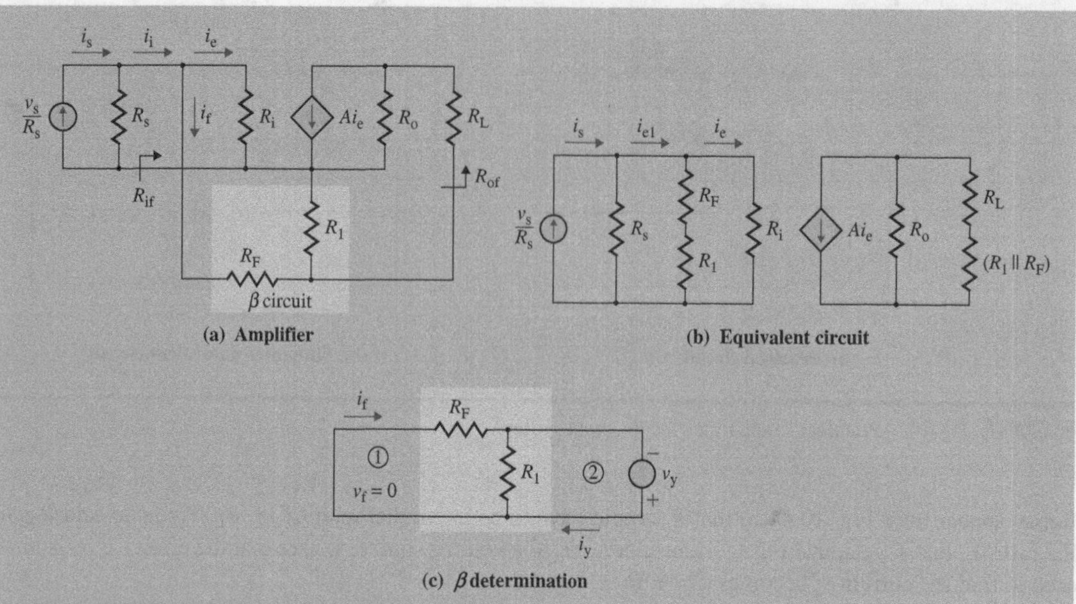

FIGURE 10.45 Op-amp with shunt-series feedback

(a) Converting the voltage-controlled voltage source to a current-controlled current source, we get

$$i_o = \frac{\mu_g v_i}{R_o} = \frac{\mu_g R_i i_e}{R_o} = A i_e$$

which gives

$$A = \frac{\mu_g R_i}{R_o} = \frac{2 \times 10^5 \times 2 \times 10^6}{75} = 53.33 \times 10^8 \text{ A/A}$$

The effect of the feedback network is taken into account by severing R_1 from the load at side 2 and shorting R_F to the ground at side 1, as shown in Fig. 10.45(b). Then

$$R_x = R_F + R_1 = 200 + 5 = 205\ \Omega$$

$$R_y = R_F \| R_1 = 200 \| 5 = 4.88\ \Omega$$

$$R_{ie} = R_i \| (R_F + R_1) = R_i \| R_x = 2\text{ M} \| 205 = 204.98\ \Omega$$

$$R_{oe} = R_o + R_y + R_L = \text{R}_o + (R_F \| R_1) + R_L = 75 + 4.88 + 5 = 84.88\ \Omega$$

Equation (10.89) gives the feedback factor β:

$$\beta = \left.\frac{i_f}{i_y}\right|_{v_f=0} = \frac{R_1}{R_F + R_1} = \frac{5}{200 + 5} = 24.39 \times 10^{-3} \text{ A/A}$$

Equation (10.97) gives the modified open-loop current gain A_e:

$$A_e = \frac{75 \times 205 \times 53.33 \times 10^8}{(4.88 + 5 + 75)(2\text{ M} + 205)} = 483 \times 10^3 \text{ A/A}$$

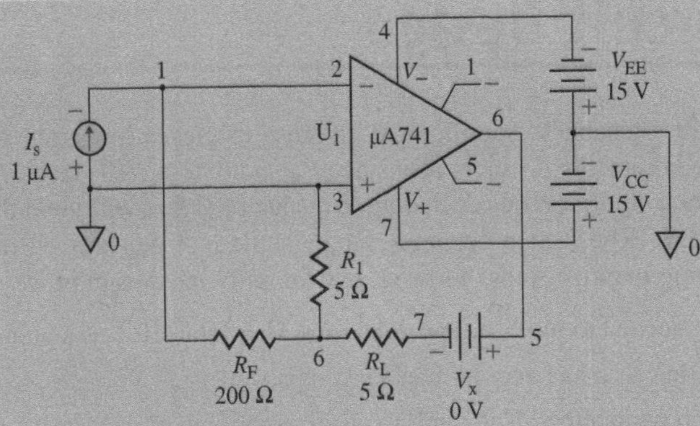

FIGURE 10.46 Shunt-series feedback circuit for PSpice simulation

The loop gain is

$$T_L = \beta A_e = 24.39 \times 10^{-3} \times 483 \times 10^3 = 11.76 \times 10^3$$

The resistances are

$$R_{if} = \frac{R_{ie}}{1 + T_L} = \frac{204.98}{1 + 11.76 \times 10^3} = 17.4 \text{ m}\Omega$$

$$R_{in} = \frac{v_s}{i_s} = R_{if} + R_s = 17.4 \text{ m} + 2.5 \text{ k} \approx 2.5 \text{ k}\Omega$$

$$R_{of} = R_{oe} \times (1 + T_L) = 84.88 \times (1 + 11.76 \times 10^3) = 999.6 \text{ k}\Omega$$

Then

$$A_f = \frac{i_o}{i_i} = \frac{483 \times 10^3}{1 + 11.763 \times 10^3} \approx 41$$

(b) The shunt-series feedback circuit for PSpice simulation is shown in Fig. 10.46. The results of PSpice simulation (.TF analysis) are shown below, with hand calculations to the right:

`I(VX)/Is=4.100E+01=41`	$A_f = 41$
`INPUT RESISTANCE AT Is=1.740E-02=17.4 mΩ`	$R_{if} = 17.4$ mΩ
`OUTPUT RESISTANCE AT I(VX)=1.000E+06=1 MΩ`	$R_{of} = 999.6$ kΩ

The PSpice results are very close to the hand-calculated values.

KEY POINTS OF SECTION 10.9

- Shunt-series feedback is applied to current amplifiers. This type of feedback reduces the input resistance and increases the output resistance by a factor of $(1 + \beta A)$.
- The amplifier is represented by a current amplifier and the feedback network as a current gain. Both A and β are in units of A/A.

10.10 Feedback Circuit Design

The feedback factor β is the key parameter of a feedback amplifier and modifies its closed-loop gain A_f. The type of feedback used depends on the requirements for input resistance R_{if} and output resistance R_{of}. If A is independent of β, designing a feedback amplifier requires finding the value of β that will yield the desired value of A_f, R_{if}, R_{of}, or bandwidth BW. The design becomes cumbersome if A depends on the feedback network—that is, on β. The following iterative steps, however, will simplify the design process:

Step 1. Decide on the type of feedback needed to meet the specifications. Use Table 10.1 as a guide.

Step 2. Disconnect the feedback link; that is, make sure no feedback is present.

Step 3. Find the approximate open-loop parameters A, R_i, and R_o of the amplifier.

Step 4. Find the values of feedback factor β and feedback resistance(s) that will satisfy the closed-loop requirement. Use the relations in Table 10.1.

Step 5. Using the feedback resistance(s), recalculate the open-loop parameters A, R_i, and R_o.

Step 6. Find the closed-loop parameters A_f, R_{if}, and R_{of}.

Step 7. Repeat steps 3 through 5 until the desired closed-loop condition is satisfied. Normally, a number of iterations will be required.

EXAMPLE 10.11

D **Designing a series-shunt feedback circuit** Feedback is applied to a voltage amplifier whose open-loop parameters are $R_i = 4.5\ \text{k}\Omega$, $R_o = 500\ \Omega$, low-frequency voltage gain $A = -450$ V/V, and bandwidth BW $= f_H = 10$ kHz. The load resistance is $R_L = 10\ \text{k}\Omega$. Determine the values of the feedback network so that the following specifications are satisfied:

(a) Bandwidth with feedback $f_{Hf} = 1$ MHz, $R_{if} > R_i$, and $R_{of} < R_o$

(b) $R_{if} = 50R_i$ and $R_{of} < R_o$

(c) $R_{if} > R_i$ and $R_{of} = R_o/250$

SOLUTION

$R_i = 4.5\ \text{k}\Omega$, $R_o = 500\ \Omega$, $A = -450$ V/V, and $f_H = 10$ kHz.

(a) Since the input resistance R_{if} should increase and the output resistance R_{of} should decrease, we can see from Table 10.1 that the feedback must be of the series-shunt type. Thus, we can use the equations derived in Sec. 10.6. The feedback network consists of R_1 and R_F, as shown in Fig. 10.47. Since A is negative, the value of v_f will be negative. Thus, v_f is added to v_s rather than subtracted from it (i.e., $v_s + v_f = v_e$), and this is accomplished by connecting the feedback signal to the negative terminal of v_s. To minimize the loading effects, $(R_1 + R_F)$ must be much larger than R_L. The condition

$$(R_1 + R_F) >> R_L \tag{10.98}$$

is generally satisfied by choosing $R_1 + R_F = 10R_L$. Since the gain–bandwidth product remains constant, Eq. (10.24) gives $Af_H = A_f f_{Hf}$. That is,

$$450 \times 10\ \text{kHz} = A_f \times 1\ \text{MHz}$$

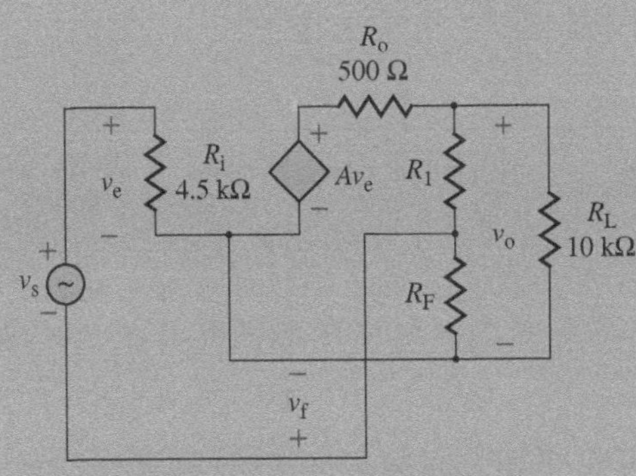

FIGURE 10.47 Amplifier with series-shunt feedback

which gives $A_f = 450/100 = 4.5$ V/V. From Eq. (10.37),

$$4.5 = \frac{450}{1 + 450\beta}$$

which gives $\beta = 0.22$. The feedback factor β is related to R_1 and R_F by

$$\beta = \frac{R_F}{R_1 + R_F} = \frac{R_F}{10R_L} \tag{10.99}$$

which, for $\beta = 0.22$ and $R_L = 10\ \text{k}\Omega$, gives

$$R_F = 10\beta R_L = 10 \times 0.22 \times 10\ \text{k} = 22\ \text{k}\Omega$$

and $\quad R_1 = 10R_L - R_F = 100\ \text{k} - 22\ \text{k} = 78\ \text{k}\Omega$

(b) Since $R_{if} = 50R_i$ and $R_{if} = R_i(1 + \beta A)$ from Eq. (10.40), we get

$$50 = 1 + \beta A \qquad \text{(for negative feedback)}$$

which, for $|A| = 450$, gives $\beta = 0.109$. For $\beta = 0.109$ and $R_L = 10\ \text{k}\Omega$, Eq. (10.99) gives

$$R_F = 10\beta R_L = 10 \times 0.109 \times 10\ \text{k} \approx 11\ \text{k}\Omega$$

and $\quad R_1 = 10R_L - R_F = 100\ \text{k} - 11\ \text{k} = 89\ \text{k}\Omega$

(c) Since $R_{of} = R_o/250$ and $R_{of} = R_o/(1 + \beta A)$ from Eq. (10.44), we get

$$250 = 1 + \beta A$$

which, for $|A| = 450$, gives $\beta = 0.5533$. For $\beta = 0.5533$ and $R_L = 10\ \text{k}\Omega$, Eq. (10.99) gives

$$R_F = 10\beta R_L = 10 \times 0.5533 \times 10\ \text{k} \approx 55\ \text{k}\Omega$$

and $\quad R_1 = 10R_L - R_F = 100\ \text{k} - 55\ \text{k} = 45\ \text{k}\Omega$

NOTE: The above solution gives approximate values for the designer to start with. More accurate values of A_f, R_{if}, and R_{of} could be found by considering the loading effects of the feedback network and the load resistance (as in Example 10.4). The design steps should be repeated until the desired specifications are met.

EXAMPLE 10.12

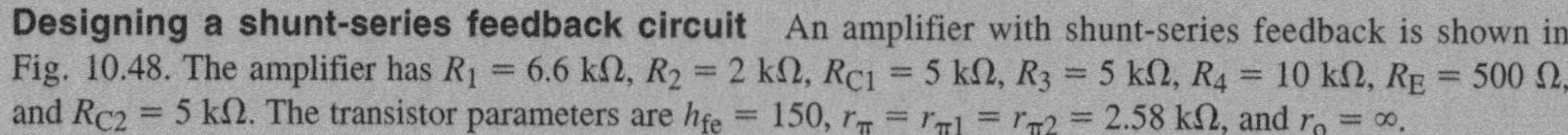

D **Designing a shunt-series feedback circuit** An amplifier with shunt-series feedback is shown in Fig. 10.48. The amplifier has $R_1 = 6.6\ \text{k}\Omega$, $R_2 = 2\ \text{k}\Omega$, $R_{C1} = 5\ \text{k}\Omega$, $R_3 = 5\ \text{k}\Omega$, $R_4 = 10\ \text{k}\Omega$, $R_E = 500\ \Omega$, and $R_{C2} = 5\ \text{k}\Omega$. The transistor parameters are $h_{fe} = 150$, $r_\pi = r_{\pi 1} = r_{\pi 2} = 2.58\ \text{k}\Omega$, and $r_o = \infty$.

(a) Determine the value of feedback resistor R_F so that the closed-loop current gain A_f is 10% of the open-loop current gain A.

(b) Use PSpice/SPICE to check your results.

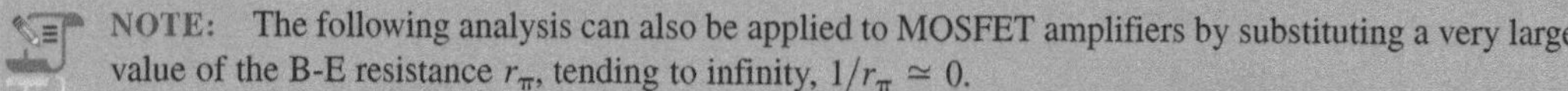

NOTE: The following analysis can also be applied to MOSFET amplifiers by substituting a very large value of the B-E resistance r_π, tending to infinity, $1/r_\pi \simeq 0$.

SOLUTION

The feedback current i_f is proportional to the emitter voltage $v_e = v_f$, which in turn is proportional to the output current $i_o \approx -i_c$. The feedback mechanism is shown in Fig. 10.49(a) and the feedback network in Fig. 10.49(b). Replacing the transistors by their small-signal model gives the small-signal AC equivalent circuit of the amplifier shown in Fig. 10.49(c),

(a) We have

$$g_m = \frac{h_{fe}}{r_\pi} = \frac{150}{2.58\ \text{k}} = 5.814\ \text{mA/V}$$

$$R_{B1} = R_1 \| R_2 = 6.6\ \text{k} \| 2\ \text{k} = 1.53\ \text{k}\Omega$$

$$R_{B2} = R_3 \| R_4 \| R_{C1} = 5\ \text{k} \| 10\ \text{k} \| 5\ \text{k} = 2\ \text{k}\Omega$$

Step 1. Assume that there is no feedback—that is, $R_F = \infty$.

Step 2. Find the open-loop parameters A, R_i, and R_o. The input resistance of the amplifier is

$$R_i = \frac{v_s}{i_s} = R_{B1} \| r_{\pi 1} = 1.53\ \text{k} \| 2.58\ \text{k} = 962.4\ \Omega$$

The output resistance of the amplifier is

$$R_o = r_o = \infty \quad \text{and} \quad R_{oe} = R_o$$

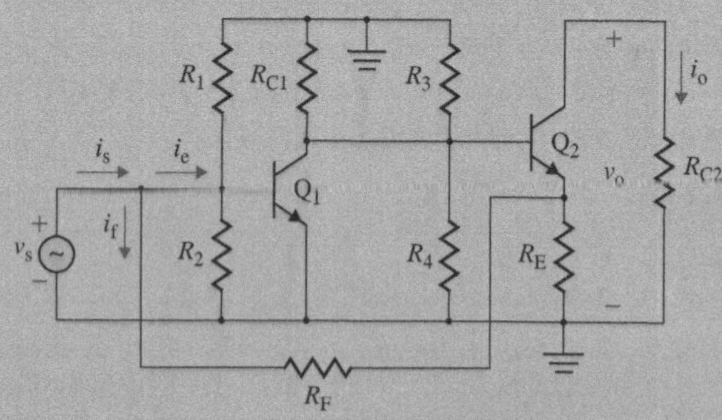

FIGURE 10.48 Two-stage amplifier with shunt-series feedback

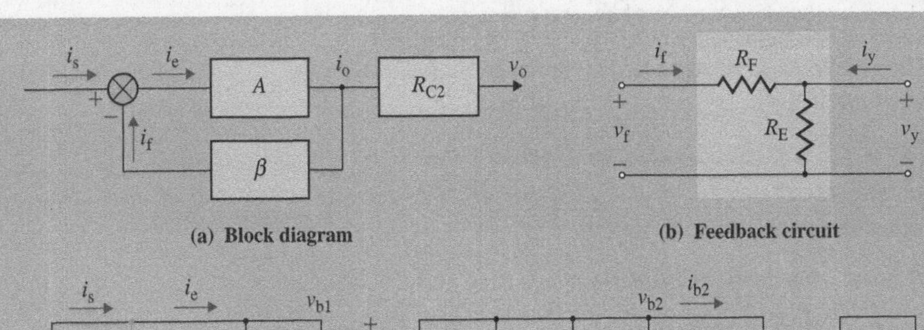

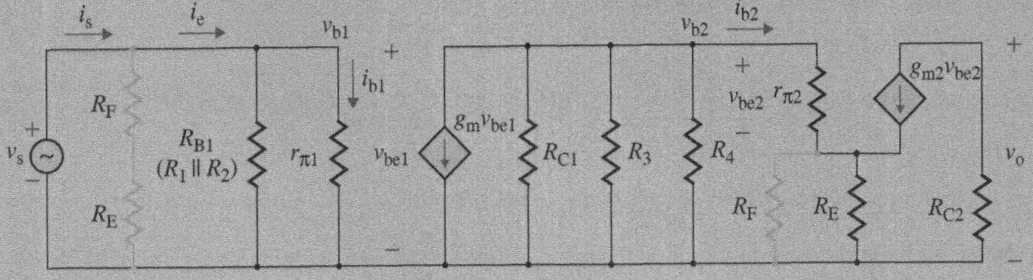

FIGURE 10.49 Equivalent circuits for Example 10.12

From Fig. 10.49(c), we get

$$v_{b1} = v_{be1} = R_i i_s = 962.4 i_s$$

$$v_{b2} = -\{R_{B2} \| [r_{\pi 2} + (1 + g_{m2} r_{\pi 2}) R_E]\} g_{m1} v_{be1}$$

$$= -\{2\text{ k} \| [2.58\text{ k} + (1 + 58.14\text{ m} \times 2.58\text{ k}) \times 500]\} \times 58.14\text{ m} \times (962.4 i_s) = -109.1 \times 10^3 i_s$$

$$v_{be2} = \frac{v_{b2} \times r_{\pi 2}}{[r_{\pi 2} + (1 + g_{m2} r_{\pi 2}) R_E]}$$

$$= \frac{(-109.1 \times 10^3 i_s) \times 2.58\text{ k}}{[2.58\text{ k} + (1 + 58.14\text{ m} \times 2.58\text{ k}) \times 500]} = -3.605 i_s$$

$$i_o = g_{m2} v_{be2} = 58.14\text{ m} \times (-3.605 i_s) = -209.2 i_s$$

Therefore, the open-loop current gain A is given by $A = i_o / i_s = -209.2$ A/A.

Step 3. Find the values of feedback factor β and resistance R_F. Since the closed-loop current gain A_f is 10% of the open-loop current gain A, we can write

$$A_f = \frac{|A|}{1 + \beta |A|} = 0.1\text{ A}$$

That is,

$$1 + \beta |A| = 10 \quad \text{or} \quad \beta\, |A| = 9 \quad \text{and} \quad \beta = \frac{9}{|A|} = \frac{9}{209.2} = 0.043$$

Since the feedback network in Fig. 10.49(b) must be represented by a current amplifier, the feedback factor β is given by

$$\beta = \frac{i_f}{i_y}\bigg|_{v_f=0} = \frac{R_E}{R_E + R_F} = \frac{500}{500 + R_F}$$

which, for $\beta = 0.043$, gives $R_F = 11.13$ kΩ.

Step 4. Find the new open-loop parameters A, R_i, and R_o. R_F is included by short-circuiting the shunt feedback side and severing the series feedback side. This arrangement is shown in Fig. 10.49(c) by the gray lines. The input resistance of the amplifier is

$$R_i = \frac{v_s}{i_s} = (R_F + R_E) \| R_{B1} \| r_{\pi 1} = (11.13\text{ k} + 500) \| 1.53\text{ k} \| 2.58\text{ k} = 887.2\ \Omega$$

The output resistance of the amplifier is

$$R_o = r_o = \infty \quad \text{and} \quad R_{oe} = R_o = \infty$$

From Fig. 10.49(c), we get

$$v_{b1} = v_{be1} = R_i i_s = 887.2 i_s$$

$$v_{b2} = -\{R_{B2} \| [r_{\pi 2} + (1 + g_{m2} r_{\pi 2})(R_E \| R_F)]\} g_{m1} v_{be1}$$

$$= -\{2\text{ k} \| [2.58\text{ k} + (1 + 58.14\text{ m} \times 2.58\text{ k}) \times (11.13\text{ k} \| 500)]\} \times 58.14\text{ m} \times (887.2 i_s)$$

$$= -100.7 \times 10^3 i_s$$

$$v_{be2} = \frac{v_{b2} \times r_{\pi 2}}{[r_{\pi 2} + (1 + g_{m2} r_{\pi 2})(R_E \| R_F)]}$$

$$= \frac{(-100.7 \times 10^3 i_s) \times 2.58\text{ k}}{[2.58\text{ k} + (1 + 58.14\text{ m} \times 2.58\text{ k}) \times (11.13\text{ k} \| 500)]} = -3.471 \times 10^3 i_s$$

$$i_o = g_{m2} v_{be2} = 58.14\text{ m} \times (-3.471 \times 10^3 i_s) = -201.4 i_s$$

Therefore, the open-loop current gain A is given by $A = i_o/i_s = -201.4$ A/A.
The closed-loop parameters are as follows:

$$R_{if} = \frac{R_i}{1 + \beta A} = \frac{887.2}{1 + 0.043\text{ A/A} \times 201.4\text{ A/A}} = 91.8\ \Omega$$

The feedback will not affect the output resistance; that is, $R_{of} = \infty$ and

$$A_f = \frac{201.4}{1 + 0.043\text{ A/A} \times 201.4\text{ A/A}} = 20.85\text{ A/A}$$

Step 5. Repeating steps 3 to 5 for the second iteration with $A = 201.4$ A/A, we get the following values:

$$\beta = \frac{9}{|A|} = \frac{9}{201.4} = 0.0447$$

$$R_F = 10.69\text{ k}\Omega$$

$$R_i = \frac{v_s}{i_s} = (R_F + R_E) \| R_{B1} \| r_{\pi 1} = (10.69\text{ k} + 500) \| 1.53\text{ k} \| 2.58\text{ k} = 886.2\ \Omega$$

$$R_o = r_o = \infty$$

$$R_{oe} = R_o = \infty$$

$$v_{b1} = v_{be1} = R_i i_s = 886.2 i_s$$

$$v_{b2} = -\{R_{B2} \| [r_{\pi 2} + (1 + g_{m2} r_{\pi 2})(R_E \| R_F)]\} g_{m1} v_{be1}$$

$$= -\{2\text{ k} \| [2.58\text{ k} + (1 + 58.14\text{ m} \times 2.58\text{ k}) \times (11.13\text{ k} \| 500)]\} \times 58.14\text{ m} \times (886.2 i_s)$$

$$= -100.4 \times 10^3 i_s$$

$$v_{be2} = \frac{v_{b2} \times r_{\pi 2}}{[r_{\pi 2} + (1 + g_{m2} r_{\pi 2})(R_E \| R_F)]}$$

$$= \frac{(-100.4 \times 10^3 i_s) \times 2.58\text{ k}}{[2.58\text{ k} + (1 + 58.14\text{ m} \times 2.58\text{ k}) \times (11.13\text{ k} \| 500)]} = -3.466 \times 10^3 i_s$$

$$i_o = g_{m2} v_{be2} = 58.14\text{ m} \times (-3.466 \times 10^3 i_s) = -201.1 i_s$$

Therefore, the open-loop current gain A is given by $A = i_o/i_s = -201.1$ A/A.

$$R_{if} = \frac{R_i}{1 + \beta A} = \frac{886.2}{1 + 0.0447\text{ A/A} \times 201.1\text{ A/A}} = 88.5\ \Omega$$

The output resistance with feedback is

$$R_{of} = R_o(1 + \beta A) = \infty$$

Since the output is taken across R_C, the output resistance of the amplifier is

$$R_{out} = R_{of} \| R_C = R_C = 5\text{ k}\Omega$$

and $$A_f = \frac{201.1}{1 + 0.0447\text{ A/A} \times 201.1\text{ A/A}} = 20.13\text{ A/A}$$

which is 10% of $A = 201.1$. Thus, there is no need for further iterations.

(b) The shunt-series feedback amplifier for PSpice simulation is shown in Fig. 10.50. The results of PSpice simulation (.TF analysis) are shown below, with hand calculations to the right:

`I(VZ)/IS=2.002E+01=20.02`	$A_f = 20.13$ A/A
`INPUT RESISTANCE AT IS=8.803E+01=88.03 Ω`	$R_{if} = 88.5\ \Omega$
`OUTPUT RESISTANCE AT I(VZ)=7.991E+13=79.91 TΩ`	$R_{of} = \infty$

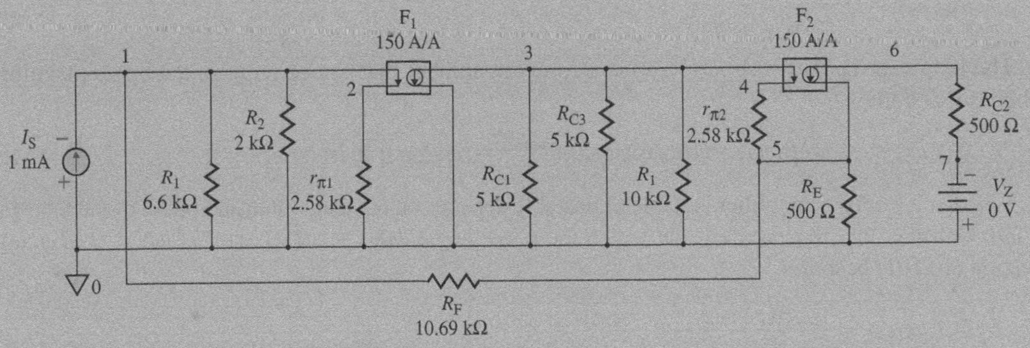

FIGURE 10.50 Two-stage shunt-series feedback amplifier for PSpice simulation

NOTE: The output current is measured through a dummy voltage source V_Z connected across R_{C2}. If V_Z were connected in series with R_{C2}, PSpice would give a very large output resistance as a result of the ideal current source of transistor Q_2.

KEY POINTS OF SECTION 10.10

- Designing feedback amplifiers requires determining the type of feedback and the feedback network.
- It is also necessary to find the component values of the network. A designer normally begins by finding the value of feedback factor β that gives the desired closed-loop gain, with the assumption of an ideal feedback network. Once initial estimates of the component values have been made, the normal analysis is performed to verify the closed-loop gain. Normally, several iterations are required to come to the final solution.

10.11 Stability Analysis

Negative feedback modifies the gain, the input resistance, and the output resistance of an amplifier. It also improves the performance parameters; for example, it reduces both the sensitivity of the gain to amplifier parameter changes and the distortion due to nonlinearities. However, negative feedback may become positive, thereby causing oscillation and instability. So far in this chapter we have assumed that the feedback network is resistive and the feedback factor β remains constant. But β can depend on frequency. In such cases, the closed-loop transfer function of a negative feedback circuit in Laplace's domain of s is given by

$$A_f(s) = \frac{S_o(s)}{S_i(s)} = \frac{A(s)}{1 + A(s)\beta(s)} \tag{10.100}$$

10.11.1 Closed-Loop Frequency and Stability

The open-loop gain $A(s)$ and the feedback factor $\beta(s)$ are dependent on the frequency. For physical systems, $s = j\omega$, and Eq. (10.100) can be written in the frequency domain as follows:

$$A_f(j\omega) = \frac{A(j\omega)}{1 + A(j\omega)\beta(j\omega)} \tag{10.101}$$

The loop gain $T_L(j\omega) = A(j\omega)\beta(j\omega)$ is a complex number that can be represented by its magnitude and phase, as follows:

$$T_L(j\omega) = A(j\omega)\beta(j\omega) = |A(j\omega)\beta(j\omega)|e^{j\phi(\omega)} = |T_L(j\omega)| \angle\phi \tag{10.102}$$

Whether a feedback amplifier is stable or unstable depends on the magnitude and phase of the loop gain $T_L(j\omega)$. Consider the frequency ω_{180} at which the phase angle $\phi(\omega) = \pm 180°$, so $T_L(j\omega) = -|T_L(j\omega)|$. Equation (10.101) becomes

$$A_f(j\omega_{180}) = \frac{A(j\omega_{180})}{1 - |T_L(j\omega_{180})|} \tag{10.103}$$

and the feedback becomes positive. However, the value of $A_f(j\omega_{180})$ will depend on the following conditions:

1. If $|T_L(j\omega_{180})| < 1$, the denominator $(1 - |T_L(j\omega_{180})|)$ will be less than unity; that is, $|A_f(j\omega_{180})| > |A(j\omega_{180})|$. In this case, the feedback system will be *stable*.
2. If $|T_L(j\omega_{180})| = 1$, the denominator $(1 - |T_L(j\omega_{180})|)$ will be zero, and $A_f(j\omega_{180})$ will be infinite; that is, the amplifier will have an output with zero input voltage, and the loop will oscillate without

any external input signal. In this case, the feedback system will be an *oscillator*. Let us assume that there is no input signal in Fig. 10.3; that is, $S_i = 0$. We get

$$S_f = A(j\omega_{180})\beta(j\omega_{180})S_e = |T_L(j\omega_{180})|S_e = -S_e \quad \textbf{(10.104)}$$

Since S_f is multiplied by -1 in the summer block at the input side, the feedback causes the signal S_e at the amplifier input to be sustained. Thus, there will be sinusoidal signals of frequency ω_{180} at the input and output of the amplifier. Under these conditions, the amplifier is said to oscillate at the frequency ω_{180}.

3. If $|T_L(j\omega_{180})| > 1$, the denominator $(1 - |T_L(j\omega_{180})|)$ will be greater than unity; that is, $|A_f(j\omega_{180})| < |A(j\omega_{180})|$. In this case, the feedback system will be *unstable*. We have

$$S_f = A(j\omega_{180})\beta(j\omega_{180})S_e = |T_L(j\omega_{180})|S_e \quad \textbf{(10.105)}$$

Since S_f is multiplied by -1 in the summer block at the input side, the amplifier oscillates, and the oscillations grow in amplitude until some nonlinearity reduces the magnitude of the loop gain $|T_L(j\omega_{180})|$ to exactly unity. In practical amplifiers, nonlinearity is always present in some form, and sustained oscillations will be obtained. This type of feedback condition is employed in oscillator circuits. Oscillators use positive feedback with a loop gain greater than unity; nonlinearity reduces the loop gain to unity.

We have seen that oscillations can occur in a negative feedback amplifier, depending on the frequency. The stability of an amplifier is determined by its response to an input or a disturbance. An amplifier is said to be *absolutely stable* if the output eventually settles down after a small disturbance. An amplifier is *absolutely unstable* if a small disturbance causes the output to build up (i.e., to increase continuously) until it reaches the saturation limits of the amplifier. The stability of an amplifier depends on its poles.

10.11.2 Poles and Instability

In the above analysis, the denominator of Eq. (10.100) was the key parameter in characterizing the response of an amplifier. The closed-loop gain $A_f(s)$ of an amplifier will be infinite if

$$1 + A(s)\beta(s) = 0 \quad \textbf{(10.106)}$$

Equation (10.106) is called the *characteristic equation*, and its roots determine the response of the amplifier. For example, if

$$1 + A(s)\beta(s) = s^2 + 2s + 5 = 0$$

the roots are $s = -1 \pm j2$. The roots of the characteristic equation are the poles of a system.

10.11.3 Transient Response and Stability

Consider an amplifier with a pole pair at $s = \sigma_o \pm j\omega_n$. If there is any disturbance (such as the closing of the DC power-supply switch), the equation of the transient response (after conversion from Laplace's s domain to the time domain) will contain terms of the following form:

$$v_o(t) = e^{\sigma_o t}[e^{+j\omega_n t} + e^{-j\omega_n t}] = 2e^{\sigma_o t}\cos(\omega_n t)$$

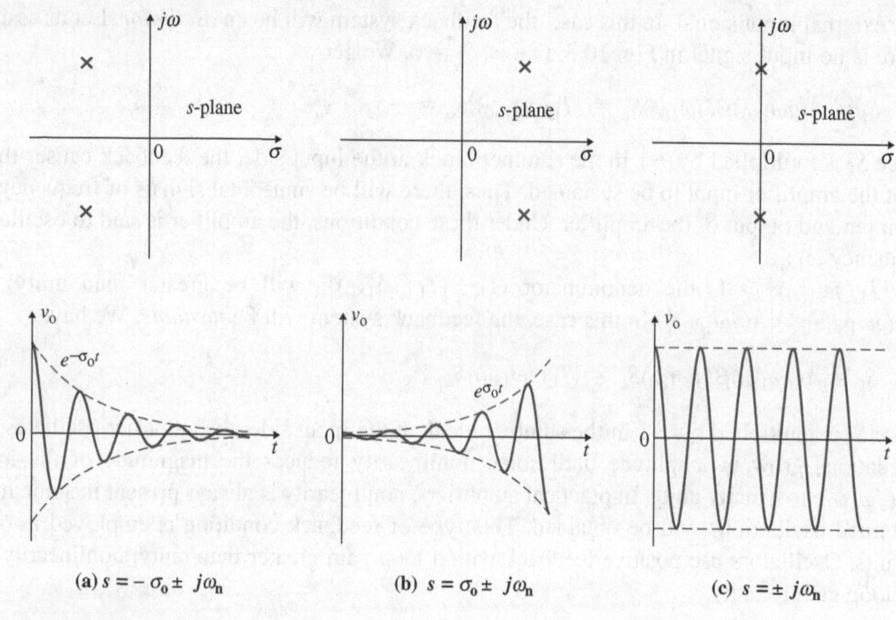

FIGURE 10.51 Relationship between pole location and transient response

This is a sinusoidal output with an envelope of exp $(\sigma_o t)$. Thus, the relationship between stability and the roots of this characteristic equation may be stated as follows:

1. If the poles are in the left half of the *s*-plane, as shown in Fig. 10.51(a), such that the denominator is of the form $(s - \sigma_o + j\omega_n)(s - \sigma_o - j\omega_n)$, all roots of the characteristic equation will have negative real parts, and σ_o will be negative. Thus, the response due to initial conditions or disturbances will decay exponentially to zero as time approaches infinity. The plot of such a response is also shown in Fig. 10.51(a). Such a system will be *stable*.
2. If the poles are in the right half of the *s*-plane, as shown in Fig. 10.51(b), such that the denominator is of the form $(s + \sigma_o + j\omega_n)(s + \sigma_o - j\omega_n)$, all roots will have positive real parts, and σ_o will be positive. The response will increase exponentially in magnitude as time increases until some nonlinearity limits its growth. The plot of such a response is also shown in Fig. 10.51(b). Such an amplifier will be *unstable*.
3. If the poles are on the $j\omega$-axis, as shown in Fig. 10.51(c), such that the denominator is of the form $(s + j\omega_n)(s - j\omega_n)$, the characteristic equation will not have roots with real parts, and σ_o will be zero. The response will be sustained oscillations, as shown in the plot in Fig. 10.51(c). The output will be sinusoidal in response to an initial condition or a disturbance. For a feedback amplifier, this will mean instability. However, for an oscillator circuit, this will mean the normal output.

10.11.4 Closed-Loop Poles and Stability

The location of the poles of a closed-loop system on a complex plane determines the transient response of a system. If the poles lie in the left half plane, all time-domain exponential terms decay to zero. If the poles lie in the right half plane, the system is likely to oscillate because its time-domain response will be growing exponentially. The plot of the poles as the loop gain varies on a complex plane, is called the *root locus* and it indicates how close the systems are to oscillate the system. Let us take an amplifier with one pole of the open-loop gain relation as given by

$$A(s) = \frac{A_o}{1 + s/\omega_o} \tag{10.107}$$

Therefore, the closed-loop gain becomes

$$A_f(s) = \frac{A(s)}{1 + \beta A(s)} = \frac{A_o/1 + \beta A_o}{1 + s/\omega_o(1 + \beta A_o)} \tag{10.108}$$

This gives the closed-loop pole $\omega_p = -\omega_o(1 + \beta A_o)$, which has a real-valued pole in the left half plane. The pole as shown in Fig. 10.52(a) moves away from the origin as the loop gain increases.

Let us consider an open-loop gain with two poles as given by

$$A(s) = \frac{A_o}{\left(1 + s/\omega_{p1}\right)\left(1 + s/\omega_{p1}\right)} \tag{10.109}$$

This gives the closed-loop gain as

$$\begin{aligned} A_f(s) &= \frac{A(s)}{1 + \beta A(s)} = \frac{A_o}{\left(1 + s/\omega_{p1}\right)\left(1 + s/\omega_{p1}\right) + \beta A_o} \\ &= \frac{A_o\omega_{p1}\omega_{p2}}{s^2 + (\omega_{p1} + \omega_{p2})s + (1 + \beta A_o)\omega_{p1}\omega_{p2}} \end{aligned} \tag{10.110}$$

This can be solved for the closed-loop two poles as given by

$$s_1, s_2 = \frac{-(\omega_{p1} + \omega_{p2}) \pm \sqrt{(\omega_{p1} + \omega_{p2})^2 - 4(1 + \beta A_o)\omega_{p1}\omega_{p2}}}{2} \tag{10.111}$$

for $\beta = 0$, $s_1 = -\omega_{p1}$, and $s_2 = -\omega_{p2}$. As β increases, the poles move toward each other and become equal at $\beta = \beta_1$. For $\beta > \beta_1$, they become complex. The pole plot is shown in Fig. 10.52(b).

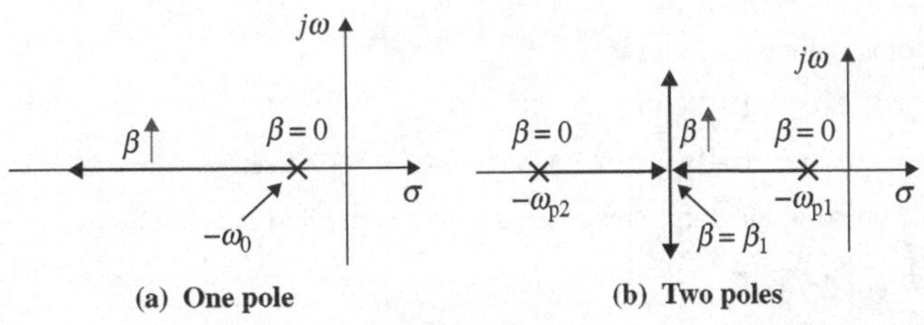

FIGURE 10.52 Closed-loop pole plots

EXAMPLE 10.13

Finding the stability due to a step input The open-loop gain of an amplifier is given by

$$A(s) = \frac{s}{s^2 - 2}$$

Determine the stability of the closed-loop response due to a step input signal. Assume feedback factor $\beta(s) = 1$.

SOLUTION

For a step input in Laplace's domain of s, $S_i(s) = 1/s$. From Eq. (10.5), the closed-loop gain is

$$A_f(s) = \frac{S_o(s)}{S_i(s)} = \frac{s/(s^2 - 2)}{1 + s/(s^2 - 2)} = \frac{s}{s^2 + s - 2} = \frac{s}{(s - 1)(s + 2)}$$

For a step input, $S_i(s) = 1/s$ and the output response is given by

$$S_o(s) = A_f(s)S_i(s) = \frac{A_f(s)}{s} = \frac{1}{(s - 1)(s + 2)} = \frac{1}{3}\left(\frac{1}{s - 1} - \frac{1}{s + 2}\right)$$

Thus, in the time domain, the output response due to a step input becomes

$$s_o(t) = \frac{1}{3}(e^t - e^{-2t})$$

Therefore, at $t = \infty$, $s_o(t) = \infty$. The amplifier will be unstable because it has a positive pole, $s = 1$.

EXAMPLE 10.14

Finding the step response of a feedback amplifier The open-loop gain of an amplifier is given by

$$A(s) = \frac{1}{s^2 + 3}$$

Determine the closed-loop step response of the amplifier. Assume feedback factor $\beta(s) = 1$.

SOLUTION

For a step input in Laplace's domain of s, $S_i(s) = 1/s$. From Eq. (10.5), the closed-loop gain is

$$A_f(s) = \frac{S_o(s)}{S_i(s)} = \frac{1/(s^2 + 3)}{1 + 1/(s^2 + 3)} = \frac{1}{s^2 + 4}$$

For a step input, the closed-loop response is

$$S_o(s) = A_f(s)S_f(s) = \frac{1}{(s^2 + 4)s} = \frac{1}{4}\left(\frac{1}{s} - \frac{s}{s^2 + 4}\right)$$

Therefore, in the time domain, the output response due to a step input is given by

$$s_o(t) = \frac{1}{4}(1 - \cos 2t)$$

Therefore, the response due to a step input is *oscillatory*. The amplifier will be unstable because it has poles on the $j\omega$-axis, $s = \pm j2$.

10.11.5 Nyquist Stability Criterion

If the loop gain $T_L(j\omega) = A(j\omega)\beta(j\omega)$ in Eq. (10.102) becomes -1, the closed-loop gain will tend to be infinite, and the amplifier will be unstable. Therefore, a feedback amplifier will be unstable if the loop gain $T_L(j\omega)$ becomes

$$T_L(j\omega) = A(j\omega)\beta(j\omega) = -1 \tag{10.112}$$

To satisfy the condition of Eq. (10.112), the magnitude and phase angle of the loop gain must be unity and $\pm 180°$, respectively; that is,

$$|T_L(j\omega)| = 1 \tag{10.113}$$

$$\phi = \angle T_L(j\omega) = \pm 180° \tag{10.114}$$

Therefore, investigating the stability of an amplifier requires determining the frequency response and characterizing the magnitude and phase angle against the frequency. The frequency response can be represented by a series of phasors, each at a different frequency. Joining the extremities (i.e., the magnitude points) of these phasors gives the frequency locus, as shown in Fig. 10.53, which is a plot of magnitude versus phase in polar coordinates as the frequency ω is varied from 0 to $\pm\infty$. This plot is known as a *Nyquist plot*. The $T_L(j\omega)$ plot for negative frequencies is the mirror image through the Re-axis (real axis) of the plot for positive frequencies. The Nyquist plot intersects the negative Re-axis at the frequency ω_{180}. If the intersection occurs to the left of the point $(-1, 0)$, the magnitude of the loop gain at this frequency is greater than unity and the amplifier will be unstable. On the other hand, if the intersection occurs to the right of the point $(-1, 0)$, the amplifier will be stable. The point on the locus where the phase is $-180°$ is called the *phase crossover frequency* ω_p $(=\omega_{180})$; the point on the locus where the gain is unity is called the *gain crossover frequency* ω_g.

The stability of a system can be determined from the Nyquist criterion for stability, which can be stated as follows: If the Nyquist plot encircles the point $(-1, 0)$, the amplifier is unstable. This criterion tests for the poles of loop gain $T_L(s)$ in the right half plane [2]. If the Nyquist plot encircles the point $(-1, 0)$, the amplifier has poles in the right half plane, and the circuit will oscillate. The number of times the plot encircles the point $(-1, 0)$ gives the number of poles in the right half plane. There are two in Fig. 10.53.

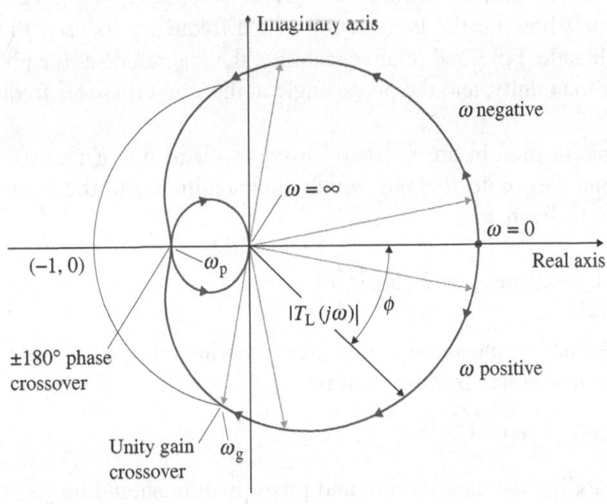

FIGURE 10.53 Typical Nyquist plot, or polar plot

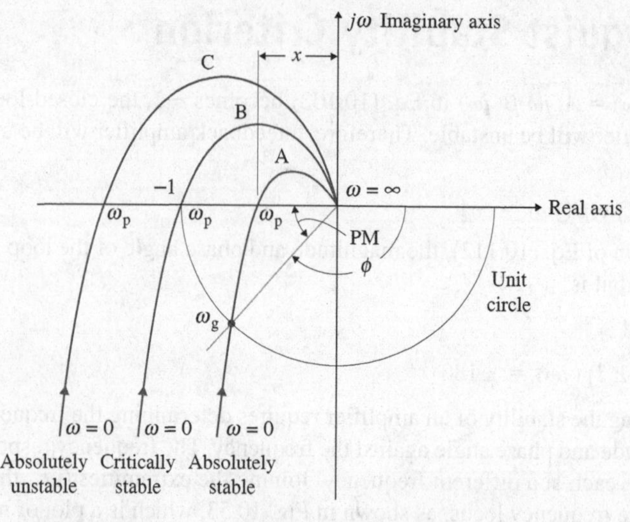

FIGURE 10.54 Nyquist plots

The Nyquist plots of amplifiers with three different characteristics are shown in Fig. 10.54. For plot A, the loop gain is less than unity. Therefore, oscillations will not build up on the closed loop, and the amplifier will be absolutely stable. For plot B through $(-1, 0)$, the loop gain is unity, and the poles are on the imaginary axis. The amplifier will be critically stable on the closed loop. For plot C, the loop gain is greater than unity, and the amplifier will be absolutely unstable on the closed loop.

10.11.6 Relative Stability

The term "absolutely stable" or "absolutely unstable" tells us what will eventually happen to a system. However, it does not say *how* stable or unstable the system is. *Relative stability* is a measure of the degree of stability, and it indicates how far the intersection of the frequency locus with the Re-axis is from the point $(-1, 0)$ on the right side. For good relative stability, the magnitude at the phase crossover frequency should have a value less than unity, and the phase angle at the gain crossover frequency should not have a value near $\pm 180°$.

Gain margin and phase margin are normally used as measures of relative stability. *Gain margin* (GM) is defined as the number of decibels by which the magnitude x of the loop gain falls short of unity when the phase angle is 180°; that is,

$$\text{GM} = 20 \log 1 - 20 \log x = 20 \log\left(\frac{1}{x}\right) \tag{10.115}$$

Phase margin (PM) is defined as the amount of degrees by which the phase angle ϕ of the loop gain falls short of $\pm 180°$ when the magnitude is unity; that is,

$$\text{PM} = \phi_{\text{m}} = 180 - |\phi| \tag{10.116}$$

For adequate relative stability, the gain margin and phase margin should be greater than 10 dB and 45°, respectively.

EXAMPLE 10.15

Finding the phase margin and gain margin of a feedback amplifier Determine the phase margin and gain margin of a feedback amplifier whose loop gain is given by

$$T_L(j\omega) = \frac{2}{(1 + j\omega)^3}$$

SOLUTION

The locus will cross the negative Re-axis when the phase is $\phi = -180°$ at $\omega = \omega_{180} = \omega_p$. Thus,

$$3 \tan^{-1}(\omega_p) = 180° \quad \text{or} \quad \tan^{-1}\omega_p = 60° \quad \text{or} \quad \omega_p = \sqrt{3} \text{ rad/s}$$

The magnitude at $\omega = \omega_p$ becomes

$$x = |T_L(j\omega)| = \frac{2}{(\sqrt{1+3})^3} = 0.25$$

That is, $x < 1$. Thus, the frequency locus will not encircle the point $(-1, 0)$, and the amplifier will be absolutely stable on the closed loop. Using Eq. (10.115), we have

$$\text{GM} = 20 \log\left(\frac{1}{0.25}\right) = 12.04 \text{ dB}$$

The gain crossover frequency ω_g is found when the gain is unity:

$$x = |T_L(j\omega_g)| = \frac{2}{(\sqrt{1+\omega_g^2})^3} = 1$$

so $\quad (1 + \omega_g^2)^3 = 2^2 \quad$ or $\quad 1 + \omega_g^2 = 1.5874 \quad$ or $\quad \omega_g = 0.7664$ rad/s

The phase angle ϕ (at $\omega = \omega_g$) is

$$\phi_g = 3 \tan^{-1}\omega_g = 3 \tan^{-1}(0.7664) = 3 \times 37.47° = 112.4°$$

Using Eq. (10.116), we have

$$\text{PM} = \phi_m = 180° - 112.4° = 67.6°$$

10.11.7 Effects of Phase Margin

At the gain crossover frequency ω_g, the magnitude of the loop gain is unity. That is,

$$|T_L(j\omega_g)| = |A(j\omega_g)|\beta = 1$$

or $$|A(j\omega_g)| = \frac{1}{\beta} \tag{10.117}$$

where β is assumed to remain constant and is independent of frequency. The phase margin influences the transient and frequency responses of a feedback amplifier, and its effects can be determined from Eq. (10.101). The effects of the phase margin will be illustrated through the analysis of four cases.

Case 1. The phase margin is PM = $\phi_m = 30°$, and $|\phi| = 180° - 30° = 150°$. Substituting Eq. (10.117) into Eq. (10.101) yields

$$A_f(j\omega_g) = \frac{A(j\omega_g)}{1 + 1\angle\phi} = \frac{A(j\omega_g)}{1 + e^{j\phi}} = \frac{A(j\omega_g)}{1 + e^{-j150°}} = \frac{A(j\omega_g)}{1 - 0.866 - j0.5} = \frac{A(j\omega_g)}{0.134 - j0.5}$$

which gives the magnitude of the closed-loop gain as

$$|A_f(j\omega_g)| = \frac{|A(j\omega_g)|}{0.517} = 1.93|A(j\omega_g)| = \frac{1.93}{\beta} \tag{10.118}$$

Therefore, there will be a peak of 1.93 times the low-frequency gain of $1/\beta$.

Case 2. The phase margin is PM = $\phi_m = 45°$, and $\phi = 180° - 45° = 135°$.

$$A_f(j\omega_g) = \frac{A(j\omega_g)}{1 + e^{-j135°}} = \frac{A(j\omega_g)}{1 - 0.71 - j0.71} = \frac{A(j\omega_g)}{0.29 - j0.707}$$

$$\text{and} \quad |A_f(j\omega_g)| = \frac{|A(j\omega_g)|}{0.765} = 1.306|A(j\omega_g)| = \frac{1.306}{\beta} \tag{10.119}$$

In this case, there will be a peak of 1.306 times the low-frequency gain of $1/\beta$.

Case 3. The phase margin is PM = $\phi_m = 60°$, and $\phi = 180° - 60° = 120°$.

$$A_f(j\omega_g) = \frac{A(j\omega_g)}{1 + e^{-j120°}} = \frac{A(j\omega_g)}{1 - 0.5 - j0.866} = \frac{A(j\omega_g)}{0.5 - j0.866}$$

$$\text{and} \quad |A_f(j\omega_g)| = |A(j\omega_g)| = \frac{1}{\beta} \tag{10.120}$$

In this case, there will be no peak above the low-frequency gain of $1/\beta$.

Case 4. The phase margin is PM = $\phi_m = 90°$, and $\phi = 180° - 90° = 90°$.

$$A_f(j\omega_g) = \frac{A(j\omega_g)}{1 + e^{-j90°}} = \frac{A(j\omega_g)}{1 - j1.0}$$

$$\text{and} \quad |A_f(j\omega_g)| = \frac{|A(j\omega_g)|}{\sqrt{2}} = 0.707|A(j\omega_g)| = \frac{0.707}{\beta} \tag{10.121}$$

In this case, there will be a gain reduction below the low-frequency gain of $1/\beta$.

The frequency responses of a transfer function with one pole and a transfer function with two poles are shown in Fig. 10.55 for various values of the phase margin. For a 90° phase margin, the transfer function has only one pole. In a two-pole system, as the phase margin is reduced, the gain peak increases until the gain approaches infinity, and oscillation occurs at a phase margin of $\phi_m = 0$. After the gain peak (at a normalized frequency of $f/f_g = 1$), the gain decays with a slope of -40 dB/decade because there are two poles in the transfer function.

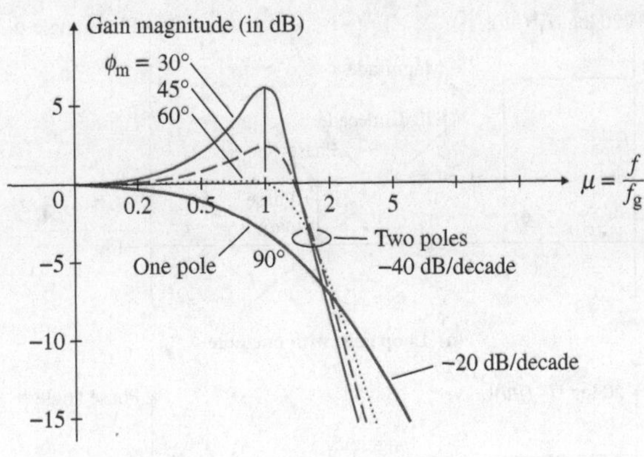

FIGURE 10.55 Effect of phase margin on frequency response

10.11.8 Stability Using Bode Plots

A *Bode plot*, which is a plot of magnitude and phase against frequency, is a very convenient method of determining the stability of an amplifier. The loop gain is plotted in decibels, and the frequency is plotted on a logarithmic scale. Consider the loop gain with a single pole frequency given by

$$T_L(j\omega) = A(j\omega)\beta(j\omega) = \frac{A_o}{1 + j\omega/\omega_{p1}} = \frac{A_o}{1 + jf/f_{p1}} \tag{10.122}$$

whose phase angle is $\phi = \angle T_L(j\omega) = -\tan^{-1}(f/f_{p1})$. The typical Bode plot is shown in Fig. 10.56(a). The magnitude of the loop gain is 20 log $|T_L(j\omega)|$ until $f = f_{p1}$, at which point the loop gain falls at a rate of 20 dB/decade and the phase angle is $\phi = -45°$. *Phase margin* is the difference between 180° and the phase angle when the gain is 0 dB. *Gain margin*, which is the value of the magnitude in dB when the phase angle is −180°, can be read from the plot. Thus, for a single-pole amplifier, $\phi = -90°$ at $T_L(j\omega_g) = 1$, and the phase margin is $\phi_m = 180 - |\phi| = 90°$. There is no phase crossover, and the gain margin is infinite. An amplifier with negative feedback will always be stable.

Consider the loop gain with three-pole frequencies given by

$$T_L(j\omega) = A(j\omega)\beta(j\omega) = \frac{A_o}{(1 + jf/f_{p1})(1 + jf/f_{p2})(1 + jf/f_{p3})} \tag{10.123}$$

whose phase angle can be found from

$$\phi = \angle T_L(j\omega) = -\tan^{-1}\left(\frac{f}{f_{p1}}\right) - \tan^{-1}\left(\frac{f}{f_{p2}}\right) - \tan^{-1}\left(\frac{f}{f_{p3}}\right) \tag{10.124}$$

The typical Bode plot is shown in Fig. 10.56(b). Assuming that the poles are widely separated (i.e., by a decade: $f_{p3} = 10f_{p2} = 100f_{p1}$), the phase angle ϕ is approximately −45° at the first pole frequency f_{p1}, (−90° − 45° =) −135° at the second pole frequency f_{p2}, and (−180° − 45° =) −225° at the third pole frequency f_{p3}. The phase angle at $T_L(j\omega_g) = 1$ is $\phi = -270°$, which gives a phase margin of $\phi_m = 180° - 270° = -90°$; that is, there is no phase margin, and the amplifier will be absolutely unstable. However, at $\omega = \omega_p = \omega_{180}$, the loop gain is positive; that is, there is a gain margin.

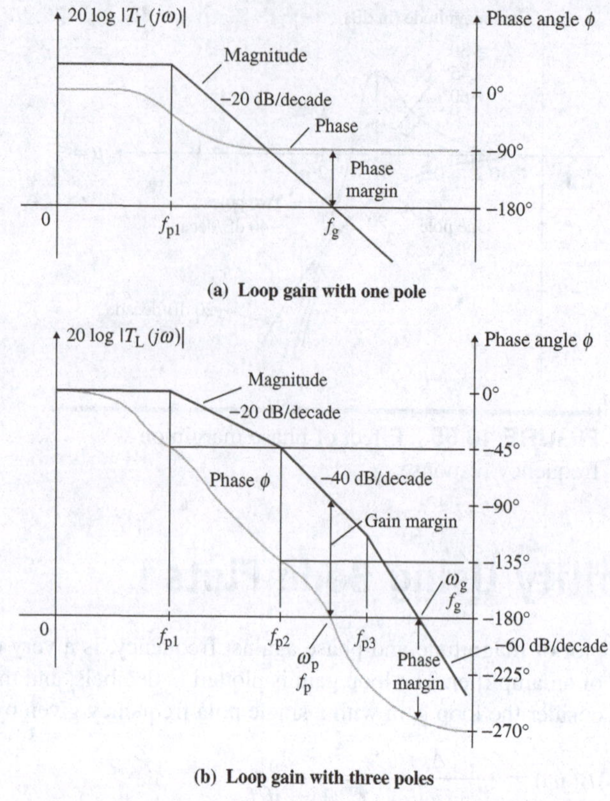

FIGURE 10.56 Typical Bode plots

Rather than plotting the loop gain $20 \log |T_L(j\omega)|$ directly, we can choose to plot $20 \log |A(j\omega)|$ and $20 \log |1/\beta(j\omega)|$ separately. This approach is shown in Fig. 10.57 for two resistive values of β. The difference between the two curves gives

$$20 \log |A(j\omega)| - 20 \log \left|\frac{1}{\beta}(j\omega)\right| = 20 \log |A(j\omega)| + 20 \log |\beta(j\omega)| = 20 \log |T_L(j\omega)|$$

The sum of the two phase angles gives

$$\angle A(j\omega) - \angle \frac{1}{\beta}(j\omega) = \phi$$

This approach has the advantage of allowing us to investigate the stability of an amplifier for a variety of feedback networks simply by drawing the lines for $20 \log |1/\beta(j\omega)|$. For zero gain margin,

$$20 \log |A(j\omega)| - 20 \log \left|\frac{1}{\beta}(j\omega)\right| = 0$$

That is, the intersection of the $20 \log |1/\beta(j\omega)|$ plot with the $20 \log |A(j\omega)|$ plot gives the critical value of β.

We can see from Fig. 10.57 that the −180° phase occurs on the −40 dB/decade segment of the Bode plot. Therefore, for stability, the $20 \log |1/\beta(j\omega)|$ plot should intersect the $20 \log |A(j\omega)|$ plot at a point on the −20 dB/decade segment. As a general rule, the difference of slopes at the intersection—which is called the *rate of closure*—should not exceed 20 dB/decade.

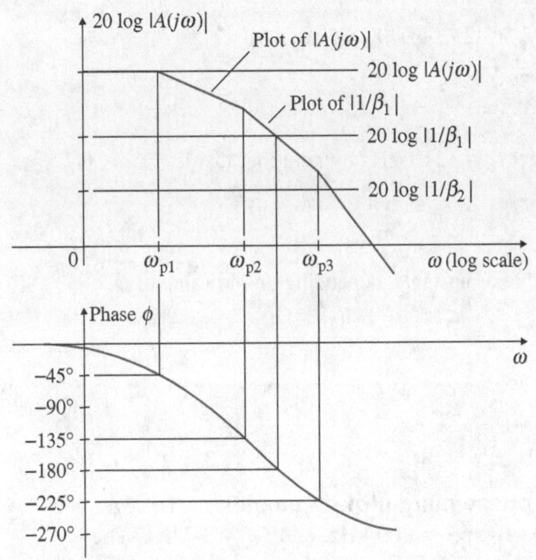

FIGURE 10.57 Separate Bode plots for $A(j\omega)$ and $1/\beta(j\omega)$

EXAMPLE 10.16

Finding the phase crossover frequency and the phase margin of an amplifier The open-loop gain of an amplifier has break frequencies at $f_{p1} = 10$ kHz, $f_{p2} = 100$ kHz, and $f_{p3} = 1$ MHz. The low-frequency (or DC) gain is $A_o = 2 \times 10^5$. Calculate **(a)** the phase crossover frequency f_p and **(b)** the phase margin ϕ_m for $\beta = 0.01$, 0.001, and 0.0001.

SOLUTION

The magnitude of the open-loop gain is given by

$$A(j\omega) = \frac{2 \times 10^5}{(1 + jf/10^4)(1 + jf/10^5)(1 + jf/10^6)}$$

$$|A(j\omega)| = \frac{2 \times 10^5}{[1 + (f/10^4)^2]^{1/2}[1 + (f/10^5)^2]^{1/2}[1 + (f/10^6)^2]^{1/2}}$$

and the phase angle is

$$\phi = -\tan^{-1}\left(\frac{f}{10^4}\right) - \tan^{-1}\left(\frac{f}{10^5}\right) - \tan^{-1}\left(\frac{f}{10^6}\right)$$

(a) The frequency at which $\phi = -180°$ is found by iteration as $f_p = f_{180} = 333$ kHz.

(b) For $\beta = 0.01$ and $|A(j\omega)| = 1/\beta = 100$, we get $f_g = 1144.8$ kHz by iteration; that is,

$$\phi = -223.4° \quad \text{and} \quad \phi_m = 180 - |\phi| = -43.4°$$

Thus, the amplifier will be unstable.

For $\beta = 0.001$ and $|A(j\omega)| = 1/\beta = 1000$, we get $f_g = 423.3$ kHz by iteration; that is,

$$\phi = -188.3° \quad \text{and} \quad \phi_m = 180 - |\phi| = -8.3°$$

Thus, the amplifier will be unstable.

For $\beta = 0.0001$ and $|A(j\omega)| = 1/\beta = 10{,}000$, we get $f_g = 124.1$ kHz by iteration; that is,

$$\phi = -143.6° \quad \text{and} \quad \phi_m = 180 - |\phi| = 36.4°$$

Thus, the amplifier will be stable.

We can conclude that a feedback amplifier can be stable or unstable, depending on the value of β.

EXAMPLE 10.17

Finding the phase crossover frequency and the phase margin of an amplifier The open-loop gain of an amplifier has break frequencies at $f_{p1} = 100$ kHz, $f_{p2} = 200$ kHz, and $f_{p3} = 1$ MHz. The low-frequency (or DC) gain is $A_o = 800$, and the feedback factor is $\beta = 0.5$. Calculate the gain crossover frequency f_g and the phase margin ϕ_m.

SOLUTION

The low-frequency loop gain is $A_o\beta = 800 \times 0.5 = 400$. The magnitude of the loop gain is given by

$$|T_L(j\omega)| = x = \frac{400}{[1 + (f/f_{P1})^2]^{1/2}[1 + (f/f_{p2})^2]^{1/2}[1 + (f/f_{p3})^2]^{1/2}}$$

and the phase angle is

$$\phi = -\tan^{-1}\left(\frac{f}{f_{p1}}\right) - \tan^{-1}\left(\frac{f}{f_{p2}}\right) - \tan^{-1}\left(\frac{f}{f_{p3}}\right)$$

At the gain crossover $|T_L(j\omega_g)| = 1$, the gain crossover frequency f_g can be determined from the Bode plot or by iteration. By iteration, the value of frequency that gives a loop gain of unity is found to be $f_g = 1917.03$ kHz, and the corresponding value of phase angle is $\phi = -233.5°$. Thus, the phase margin is

$$\phi_m = 180° - |\phi| = 180° - 233.5° = -53.5°$$

which is negative, so the amplifier will be unstable.

KEY POINTS OF SECTION 10.11

- An amplifier with negative feedback can be stable or unstable, depending on the frequency and the feedback factor β. If the loop gain is $|T_L(j\omega)| \geq 1$ and its phase angle is $\phi = \pm 180°$, the amplifier will be unstable.
- Nyquist and Bode plots can be used to determine the stability of a feedback amplifier. The degree of stability is normally measured by the gain margin and the phase margin. A phase margin of 45° is usually adequate to limit the peak to 30% of the low-frequency gain.

10.12 Compensation Techniques

We know that if an amplifier has more than two poles, the phase angle of the loop gain could exceed $-180°$ beyond a certain frequency. An amplifier with negative feedback can be unstable, depending on the frequency ω and the amount of feedback β. The process of stabilizing an unstable feedback amplifier is called *compensation*. The basic amplifier of a feedback circuit should be designed with as few stages as possible because each stage of gain adds more poles to the transfer function, making the compensation problem more difficult. An amplifier can be stabilized by adding a dominant pole, by changing the dominant pole, by Miller compensation, or by modifying the feedback path [3–5].

10.12.1 Addition of a Dominant Pole

A dominant pole can be introduced into the amplifier so that the phase shift is less than $-180°$ when the loop gain is unity. Consider a feedback amplifier with loop gain of the form

$$T_L(j\omega) = A(j\omega)\beta = \frac{A_o}{(1 + j\omega/\omega_{p1})(1 + j\omega/\omega_{p2})(1 + j\omega/\omega_{p3})} \tag{10.125}$$

where $\omega_{p1} = 2\pi f_{p1}$ rad/s, $\omega_{p2} = 2\pi f_{p2}$ rad/s, and $\omega_{p3} = 2\pi f_{p3}$ rad/s. The Bode plot is shown in Fig. 10.58(a). To compensate the amplifier, a new dominant pole $\omega_D = 2\pi f_D$ is introduced so that $\omega_D < \omega_{p1} < \omega_{p2} < \omega_{p3}$, and the resultant loop gain becomes

$$T_L(j\omega) = A(j\omega)\beta = \frac{A_o}{(1 + j\omega/\omega_D)(1 + j\omega/\omega_{p1})(1 + j\omega/\omega_{p2})(1 + j\omega/\omega_{p3})} \tag{10.126}$$

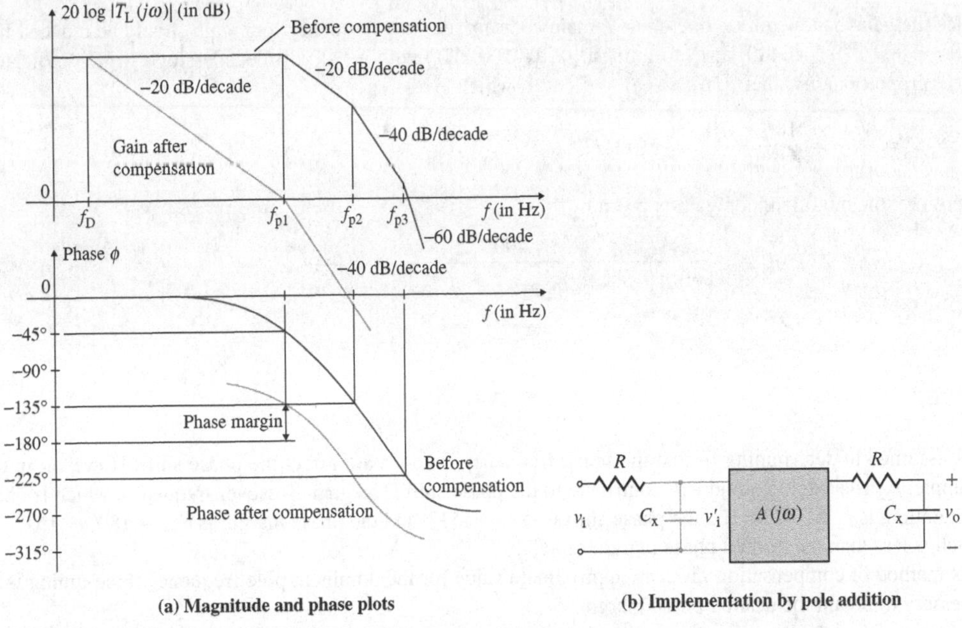

FIGURE 10.58 Compensation by addition of a dominant pole

The Bode plot of this modified loop gain is indicated in Fig. 10.58(a) by lighter lines. The introduction of the dominant pole causes the loop gain to fall at a rate of 20 dB/decade until frequency f_{p1} is reached. If the frequency f_D of the dominant pole is chosen so that the loop gain is unity at frequency f_{p1}, then the phase shift at frequency f_{p1} due to the dominant pole f_D is $-90°$, and the phase shift due to the first pole f_{p1} is $-45°$. At $f = f_{p1}$, the total phase shift is $(-90° - 45° =) -135°$, and the phase margin is $\phi_m = 180° - |\phi| = 180° - 135° = 45°$, which means that the amplifier will be stable. The original amplifier would have been unstable in the feedback connection. This compensation is achieved at the expense of bandwidth reduction. The uncompensated unity-gain bandwidth f_{p1} is much higher than the compensated unity-gain bandwidth f_D. This method of compensation involves a sacrifice of the frequency capability of the amplifier and is often known as *narrowbanding*.

A dominant pole can be implemented by adding a capacitor C_x in such a manner that it adds a break frequency to the basic amplifier. With this approach, illustrated in Fig. 10.58(b) showing two possible locations, f_D is given by

$$f_D = \frac{1}{2\pi R C_x} \tag{10.127}$$

where R is the resistance seen by capacitor C_x.

EXAMPLE 10.18

Stabilizing an amplifier by adding a dominant pole Stabilize the amplifier in Example 10.17 by adding a dominant pole so that the phase margin is 45°.

SOLUTION

Since the gain–bandwidth product must remain constant, the low-frequency loop gain should be reduced from $A_o\beta$ (or $800 \times 0.5 = 400$) at f_D to unity at f_{p1} (=100 kHz) with a -20 dB/decade slope, which indicates a direct proportionality; that is, $f_D \times A_o\beta = f_{p1} \times 1$, which gives

$$f_D = \frac{f_{p1}}{A_o\beta} = \frac{100 \times 10^3}{400} = 250 \text{ Hz}$$

Therefore, the modified loop gain is given by

$$T_L(j\omega) = A(j\omega)\beta = \frac{400}{(1 + jf/f_D)(1 + jf/f_{p1})(1 + jf/f_{p2})(1 + jf/f_{p3})}$$

NOTES

1. It is assumed in determining f_D that the break frequency f_{p2} does not affect the phase shift. However, in this example, f_{p2} is close to f_{p1} and *will* contribute to the phase shift. The gain crossover frequency, which is obtained by iteration, is $f_g = 74{,}795$ Hz, the phase shift is $\phi = -151°$, and the phase margin is $\phi_m = 180° - 151° = 29°$, which is less than the desired phase margin of 45°.
2. This method of compensation gives an approximate value for the dominant pole frequency. Fine-tuning is necessary to obtain the desired phase margin.
3. The value of f_D, which can be adjusted to obtain a phase margin of 45°, is 152 Hz at a gain crossover frequency of $f_g = 52.106$ kHz.

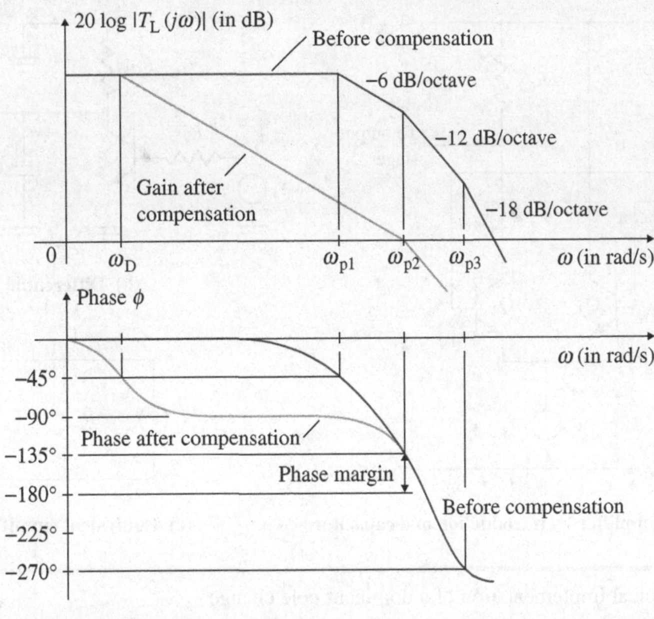

FIGURE 10.59 Compensation by changing the dominant pole

10.12.2 Changing the Dominant Pole

In the compensation method just discussed in Sec. 10.12.1, a dominant pole was added to the amplifier, and the original amplifier poles were assumed to be unaffected. This approach reduces the bandwidth considerably. A second method of compensation is to change the dominant pole, adding a capacitor to the amplifier in such a way that the original dominant frequency f_{p1} is reduced so that it performs the compensation function. That is, the original pole f_{p1} is moved to the left so that $f_D = f_{p1}$. This modification is shown in Fig. 10.59. For a 45° phase margin in a unity feedback amplifier, f_D must cause the gain to fall to unity at f_{p2}. Thus, f_{p2} will become the unity-gain bandwidth. Since f_{p2} is five or ten times the frequency f_{p1}, this method gives a substantial improvement in bandwidth.

The dominant pole is changed by adding a capacitor internally to the amplifier. One way of doing so is shown in Fig. 10.60(a) for the differential stage of an op-amp. The differential half circuit is shown in Fig. 10.60(b), and the small-signal equivalent circuit is shown in Fig. 10.60(c). The current i_s is the collector current of Q_2, C_i is the internal capacitance, and R_i is the effective resistance; that is, $C_i = C_{\pi 4}$, and $R_i = R_{C1} \| r_{o2} \| r_{\pi 4}$. Thus, the dominant pole frequency can be found from

$$f_D = f_{p1} = \frac{1}{2\pi R_i(C_i + C_x)} \tag{10.128}$$

where C_x is the additional capacitance that is added to give the desired dominant frequency. The disadvantage of this method is that the value of C_x is usually quite large (typically >1000 pF). Thus, it will be difficult—if not impossible—to realize a compensating capacitor in an IC chip. The maximum practical size of a monolithic capacitor is about 100 pF.

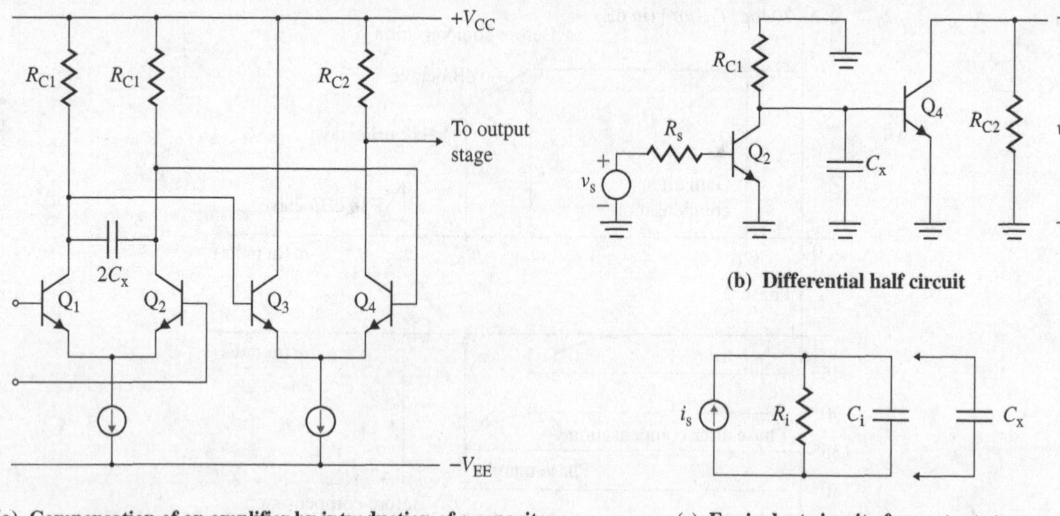

FIGURE 10.60 Typical implementation of a dominant pole change

We have assumed that the addition of C_x will reduce the original dominant pole frequency f_{p1} so that it performs the compensation function. However, the higher-frequency poles of the amplifier will also be changed by the addition of C_x. In practice, the effect of C_x on the pole positions is usually evaluated by computer simulation. Another estimate of C_x is made on the basis of the new data, and this process of trial and error continues until the desired stability condition is reached, usually after several iterations.

EXAMPLE 10.19

Finding the compensating capacitance to modify the dominant pole The equivalent circuit of an output stage is shown in Fig. 10.60(c), where $R_i = 24\ \text{k}\Omega$ and $C_i = 20$ pF.

(a) Find the break frequency f_p.

(b) Find the additional capacitance C_x that will move the break frequency to $f_D = 40$ kHz.

SOLUTION

(a) The break frequency f_p is given by

$$f_p = \frac{1}{2\pi C_i R_i} = \frac{1}{2\pi \times 20 \times 10^{-12} \times 24 \times 10^3} = 331.6\ \text{kHz}$$

(b) An additional capacitance C_x will move the break frequency to f_D; that is,

$$f_D = \frac{1}{2\pi(C_i + C_x)R_i}$$

which, for $f_D = 40$ kHz, gives $C_i + C_x = 165.8$ pF; that is, $C_x = 165.8 - 20 = 145.8$ pF.

10.12.3 Miller Compensation and Pole Splitting

In a third compensation method, a small capacitance (say, C_x) is connected between the input and output of a gain stage in a multistage amplifier. Compensation is achieved using Miller multiplication of the capacitance. Figure 10.61(a) illustrates this form of compensation for the μA741 op-amp. C_x constitutes a shunt-shunt feedback, and the feedback factor β depends on the frequency. The Darlington pair consisting of Q_{16} and Q_{17} can be replaced by a single equivalent transistor Q, as shown in Fig. 10.61(b); the simplified equivalent circuit is shown in Fig. 10.61(c). R_i and C_i represent the total resistance and total capacitance, respectively, between node B and the ground. R_o and C_o represent the total resistance and total capacitance, respectively, between node C and the ground. C_i includes the Miller capacitance due to C_μ of the transistor and the output capacitance of the preceding stage. Similarly, C_o includes the Miller capacitance due to C_μ and the input capacitance of the succeeding stage. In the absence of the compensating capacitance (i.e., when $C_x = 0$), the two poles will be

$$f_{p1} = \frac{1}{2\pi C_i R_i} \tag{10.129}$$

and

$$f_{p2} = \frac{1}{2\pi C_o R_o} \tag{10.130}$$

The analysis of Fig. 10.61(c) will be similar to that of Fig. 2.23(b). Using Eqs. (2.93) and (2.94), we get as the new poles

$$\omega'_{p1} \approx \frac{1}{g_m C_x R_i R_o} \tag{10.131}$$

$$\omega'_{p2} \approx \frac{g_m C_x}{C_i C_o + C_x(C_i + C_o)} \tag{10.132}$$

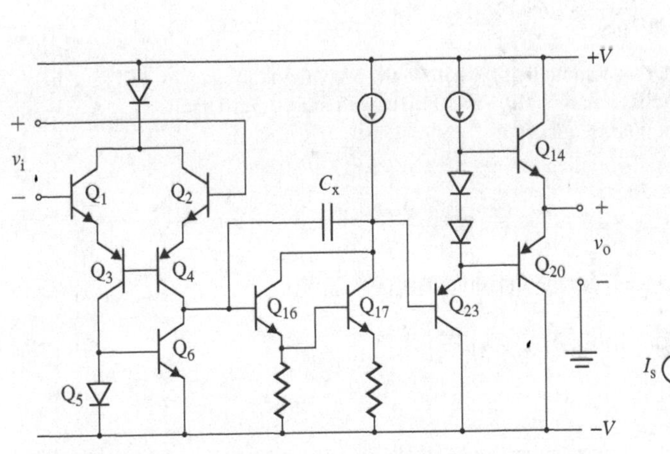

(a) μA741 op-amp showing Miller compensation capacitance C_x

(b) Equivalent transistor

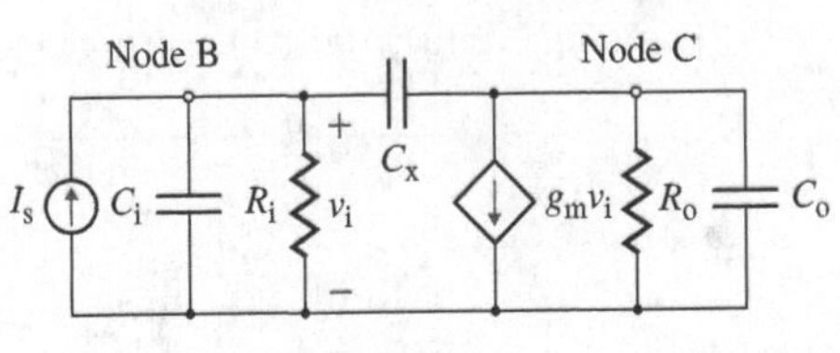

(c) Equivalent circuit

FIGURE 10.61 Miller compensation and pole spliting

If $C_x >> C_o$, Eq. (10.132) can be approximated by

$$\omega'_{p2} = \frac{g_m}{C_i(C_o/C_x) + (C_i + C_o)} \approx \frac{g_m}{C_i + C_o} \quad (10.133)$$

From the preceding equations, we can see that as C_x is increased, ω'_{p1} is reduced, and ω'_{p2} is increased. This separation of poles is called *pole splitting*. The increase in ω'_{p2} will move the pole frequency for a 45° phase margin to the right and will hence widen the bandwidth. In Eq. (10.131), C_x is multiplied by a factor $g_m R_o$, and the effective capacitance is much larger: $g_m R_o C_x$. Thus, the value of C_x will be much smaller than it is when compensation is achieved by adding or changing a pole.

EXAMPLE 10.20

D **Finding the compensating capacitance by pole splitting** The open-loop gain of an amplifier is 2×10^5 and has break frequencies at $f_{p1} = 100$ kHz, $f_{p2} = 1$ MHz, and $f_{p3} = 10$ MHz. The gain stage of the amplifier has the equivalent circuit shown in Fig. 10.61(c), whose parameters are $g_m = 100$ mA/V, $C_i = 50$ pF, and $C_o = 10$ pF. Determine the value of compensating capacitance C_x that will give a closed-loop phase margin of 45° with a resistive feedback of up to $\beta = 1$.

SOLUTION

$g_m = 100 \times 10^{-3}$, $C_i = 50$ pF, $C_o = 10$ pF, $f_{p1} = 100$ kHz, $f_{p2} = 1$ MHz, and $f_{p3} = 10$ MHz. We can find the values of R_i and R_o as follows:

$$R_i = \frac{1}{2\pi f_{p1} C_i} = \frac{1}{2\pi \times 100 \text{ kHz} \times 50 \text{ pF}} = 31.83 \text{ k}\Omega$$

$$R_o = \frac{1}{2\pi f_{p2} C_o} = \frac{1}{2\pi \times 1\text{MHz} \times 10 \text{ pF}} = 15.92 \text{ k}\Omega$$

From Eq. (10.133), we can find the modified value f'_{p2}; that is,

$$f'_{p2} \approx \frac{g_m}{2\pi(C_i + C_o)} = \frac{100 \times 10^{-3} \text{ A/V}}{2\pi(50 \text{ pF} + 10 \text{ pF})} = 265.3 \text{ MHz}$$

which is more than f_{p3} (=10 MHz). Thus, let us assume that f_{p3} will be the second pole frequency and find the compensation capacitance C_x to set the 45° phase margin at $f_{p3} = 10$ MHz with unity gain. That is, $f_D \times A_o\beta = f_{p3} \times 1$, which gives the modified dominant pole frequency as

$$f'_{p1} \approx f_D = \frac{f_{p3}}{A_o\beta} = \frac{10 \times 10^6}{2 \times 10^5 \times 1} = 50 \text{ Hz}$$

Thus, $f'_{p1} \approx f_D = 50$ Hz.

From Eq. (10.131), we get the capacitance C for the first dominant pole f'_{p1}:

$$C_x \approx \frac{1}{2\pi f'_{p1} g_m R_i R_o} = \frac{1}{2\pi \times 50 \text{ Hz} \times 100 \text{ mA/V} \times 31.83 \text{ k}\Omega \times 15.92 \text{ k}\Omega} = 62.8 \text{ pF}$$

From Eq. (2.91), we get

$$f'_{p1} = \frac{1}{2\pi[R_i(C_i + C_x) + R_o(C_x + C_o) + g_m C_x R_i R_o]} = 49.94 \text{ Hz}$$

which is close to 50 Hz.

10.12.4 Modification of the Feedback Path

Op-amps are normally compensated by adding a capacitor internally. This type of compensation, however, wastes bandwidth because the bandwidth is reduced considerably. For example, the dominant pole frequency of the μA741 op-amp is only 10 Hz.

Compensation can also be accomplished by modifying the feedback network so that the feedback factor β becomes frequency dependent and has a zero that cancels the pole of the original amplifier. This method is generally used in the compensation of fixed-gain amplifiers, where achieving a wide bandwidth is of prime concern.

Let us consider a feedback network with one pole and one zero. Then the loop gain of an amplifier with three poles will be of the form

$$A(j\omega)\beta(j\omega) = \frac{A_o(1 + j\omega/\omega_z)}{(1 + j\omega/\omega_p)(1 + j\omega/\omega_{p1})(1 + j\omega/\omega_{p2})(1 + j\omega/\omega_{p3})} \quad (10.134)$$

where ω_p and ω_z are the pole and the zero of the feedback network, respectively.

A typical implementation is shown in Fig. 10.62(a) for a shunt-series feedback amplifier. The feedback network includes a capacitor C_F. The loading effects of C_F at the input and output sides of the amplifier are shown in Fig. 10.62(b). The capacitor C_F will have only a minor effect on the amplifier transfer

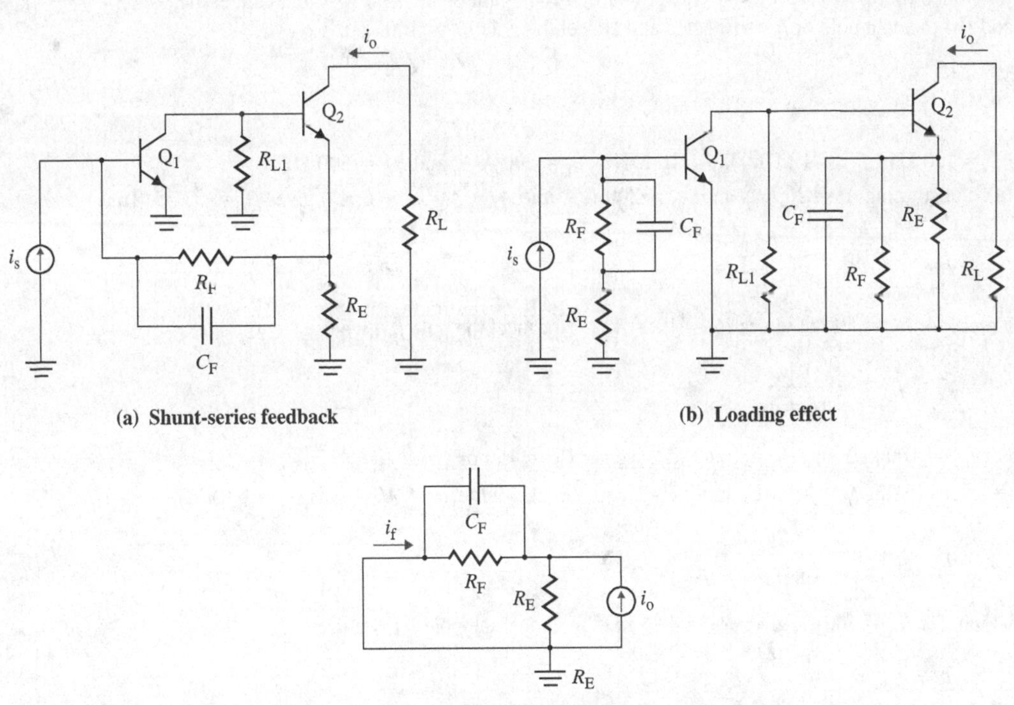

FIGURE 10.62 Compensation by feedback path modification

function $A(s)$. The feedback network is shown in Fig. 10.62(c), from which we can find the feedback transfer function as

$$\beta(s) = \frac{i_f}{i_o} = \frac{R_E}{R_E + R_F} \times \frac{1 + sR_FC_F}{1 + sC_FR_FR_E/(R_E + R_F)} \quad \textbf{(10.135)}$$

$$= \frac{\beta_o(1 + s/\omega_z)}{1 + s/\omega_p} \quad \textbf{(10.136)}$$

where $\beta_o = R_E/(R_E + R_F)$ = low-frequency feedback factor
$\omega_z = 1/(R_FC_F)$ = zero of the feedback network
$\omega_p = (R_E + R_F)/(R_ER_FC_F)$ = pole of the feedback network

EXAMPLE 10.21

D

Compensating by feedback path modification The open-loop gain of the amplifier in Example 10.12 has break frequencies at f_{p1} = 100 kHz, f_{p2} = 1 MHz, and f_{p3} = 10 MHz. The low-frequency gain is A_o = 200 A/A, and the emitter resistance is R_E = 500 Ω. Determine the values of compensating capacitance C_F and resistance R_F **(a)** to give a low-frequency closed-loop gain of A_f = 20 A/A and cancel the pole f_{p1} = 100 kHz and **(b)** to add a pole of f_p = 10 MHz and cancel the pole f_{p1} = 100 kHz.

SOLUTION

f_{p1} = 100 kHz, f_{p2} = 1 MHz, f_{p3} = 10 MHz, A_o = 200 A/A, and R_E = 500 Ω.

(a) Substituting A_f = 20 A/A and A_o = 200 A/A into $A_f = A_o/(1 + \beta_oA_o)$ gives β_o = 0.045. Thus,

$$\beta_o = 0.045 = \frac{R_E}{R_E + R_F}$$

which, for R_E = 500 Ω, gives R_F = 10.61 kΩ. To cancel the pole f_{p1}, we use

$$f_z = f_{p1} = \frac{1}{2\pi R_FC_F}$$

For f_{p1} = 100 kHz and R_F = 10.61 kΩ, we get C_F = 150 pF.

(b) Substituting f_p = 10 MHz and $f_z = f_{p1}$ = 100 kHz into $f_p = f_z/\beta_o$ gives β_o = 0.01. Thus,

$$\beta_o = 0.01 = \frac{R_E}{R_E + R_F}$$

which, for R_E = 500 Ω, gives R_F = 49.5 kΩ. To cancel the pole f_{p1}, we use

$$f_z = f_{p1} = \frac{1}{2\pi R_FC_F}$$

For f_{p1} = 100 kHz and R_F = 49.5 kΩ, we get C_F = 32.15 pF.

EXAMPLE 10.22

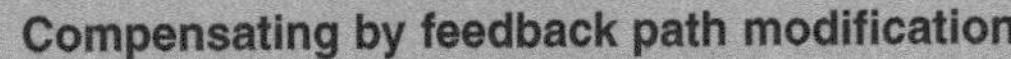

Compensating by feedback path modification

(a) A common-emitter amplifier with shunt-shunt feedback is shown in Fig. 10.63. The capacitance C_F and resistance R_F form the feedback network. The voltage gain without feedback is $A = |A_v| = v_o/v_s \approx 100$. Design the feedback network to meet the following specifications:

The bandwidth with feedback must be increased by a factor of 10; that is, $\text{BW}_f = 10\text{BW}$.
The voltage gain with feedback must be $|A_f| = v_o/v_s = 10$.

(b) Use PSpice/SPICE to check your results.

SOLUTION

(a) Shunt-shunt feedback will reduce both the input resistance and the output resistance, but it should widen the bandwidth. The value of C_F should be such that it is virtually short-circuited over the frequency range of the amplifier. The feedback is taken from the output side because the output is out of phase with the input voltage v_S. Since the gain of the amplifier is $A = 10^2$, the closed-loop gain will be almost independent of A. We get

$$\frac{v_o}{i_s} = -\frac{1}{\beta} = -R_F \tag{10.137}$$

and $\quad i_s = \dfrac{v_s}{R_s}$

which gives the closed-loop gain as

$$A_f = \frac{v_o}{v_s} = -\frac{R_F}{R_s} \approx -\frac{1}{\beta R_s} \quad \text{(V/V)} \tag{10.138}$$

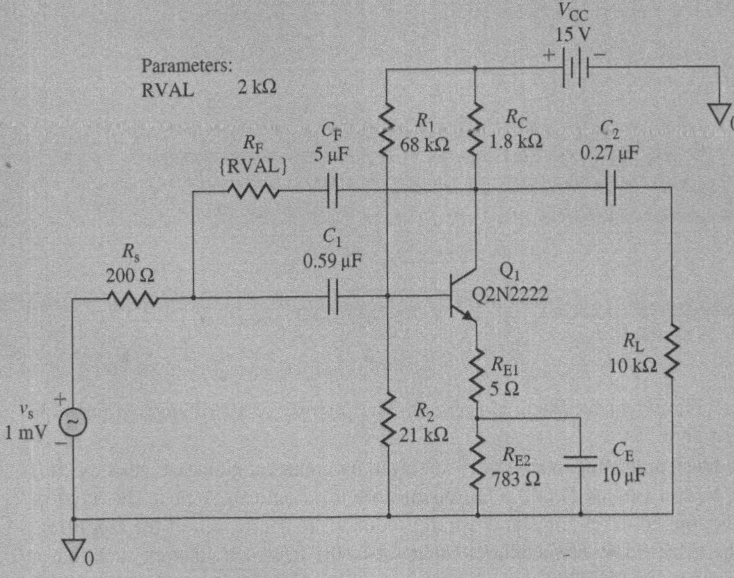

FIGURE 10.63 Common-emitter amplifier with shunt-shunt feedback

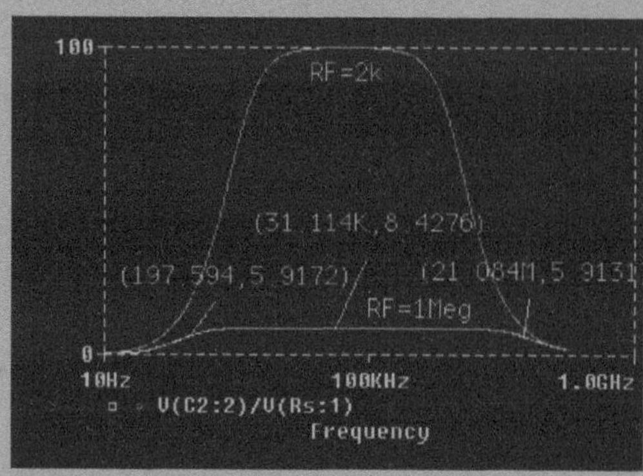

FIGURE 10.64 Frequency response for Example 10.22

For shunt-shunt feedback, $BW_f = BW(1 + \beta A)$. The gain must decrease by a factor of 10. Thus, $\beta R_s = 1/10 = 0.1$.

$$\frac{1}{\beta R_s} = \frac{R_F}{R_s} = 10$$

which, for $R_s = 200\ \Omega$, gives $R_F = 2\ k\Omega$. We want to choose a capacitor that will ensure that C_F is virtually short-circuited at low frequency, $f_L = 1$ kHz: let's choose $C_F = 5\ \mu F$. Thus,

$$X_{CF} \leq \frac{1}{2\pi \times 1\ \text{kHz} \times 5\ \mu\text{F}} = 31.8\ \Omega$$

(b) The PSpice plots of the frequency response are shown in Fig. 10.64 for $R_F = 2\ k\Omega$ and 1 MΩ. For $R_F = 1\ M\Omega$ (for almost no feedback, $\beta \approx 0$), we get $|A| = 100$, $f_L = 1089$ Hz, $f_H = 1838$ kHz, and

$$BW = f_H - f_L = 1838\ \text{k} - 1.09\ \text{k} = 1836.91\ \text{kHz}$$

With feedback of $R_F = 2\ k\Omega$ (for $\beta = 0.1$), we get $|A_f| = 8.43$ (expected value is 10), $f_{Lf} = 198$ Hz, $f_{Hf} = 21{,}084$ kHz, and

$$BW_f = f_{Hf} - f_{Lf} = 21{,}084\ \text{k} - 0.2\ \text{k} = 21{,}083.8\ \text{kHz}$$

(expected value is 10BW = 10 × 1836.91 kHz = 18,369 kHz). The difference between the PSpice values and the expected values is caused by the fact that we did not include the effects of resistances such as R_F and R_C in hand calculations.

KEY POINTS OF SECTION 10.12

- An amplifier may be unstable when feedback is applied. The process of stabilizing an unstable amplifier is called compensation.
- The compensation of feedback amplifiers is normally achieved by connecting an external capacitor to the basic amplifier in such a way as to add a dominant pole or to split the poles. This type of compensation is normally applied to op-amps and usually reduces the bandwidth of the amplifier.
- Compensation can also be achieved by connecting a capacitor to the feedback network such that it adds a pole and a zero to the loop gain. This type of compensation is normally applied to fixed-gain amplifiers so as to yield a wide bandwidth.

TABLE 10.3 Effects of feedback topologies

	Series-shunt	Shunt-shunt	Shunt-series	Series-series
Z_{if}	$R_i(1 + \beta A)$	$R_i/(1 + \beta A)$	$R_i/(1 + \beta A)$	$R_i(1 + \beta A)$
Z_{of}	$R_o/(1 + \beta A)$	$R_o/(1 + \beta A)$	$R_o(1 + \beta A)$	$R_o(1 + \beta A)$
A_f	$A/(1 + \beta A)$	$A/(1 + \beta A)$	$A/(1 + \beta A)$	$A/(1 + \beta A)$

Summary

There are two types of feedback: negative feedback and positive feedback. Negative feedback is normally used in amplifier circuits, and positive feedback is applied exclusively in oscillators. Negative feedback has certain advantages, such as stabilization of overall gain with respect to parameter variations, reduction of distortion, reduction of the effects of nonlinearity, and increase in bandwidth. However, these advantages are obtained at the expense of gain reduction, and additional amplifier stages may be required to make up the gain reduction. If the loop gain $\beta A >> 1$, the overall (or closed-loop) gain depends inversely on the feedback factor β and is directly sensitive to changes in the feedback factor.

The gain–bandwidth product of feedback amplifiers remains constant. If the gain is reduced by negative feedback, then the bandwidth is increased by the same amount. Depending on its implementation in electronic circuits, feedback can be classified as one of four types: series-shunt, shunt-shunt, series-series, or shunt-series. A shunt connection reduces the input (or output) impedance by a factor of $(1 + \beta A)$, and a series connection increases the input (or output) impedance by a factor of $(1 + \beta A)$. The closed-loop gain is always decreased by a factor of $(1 + \beta A)$. Table 10.3 summarizes the effects of various feedback topologies.

The stability of an amplifier depends on the poles of the transfer function. For a stable amplifier, the characteristic equation should not have any roots with positive real parts. The Nyquist criterion is one of the methods for determining the stability of feedback systems. The stability of an amplifier is normally measured in terms of phase margin and gain margin. A Bode plot can also be used to determine the stability of an amplifier. An inherently unstable amplifier can be made stable by introducing a dominant pole in the transfer function; this is generally achieved in electronic circuits by connecting a feedback capacitor.

References

1. B. Razavi, *Design of Analog CMOS Integrated Circuits*. New York: McGraw-Hill, 2001.
2. B. C. Kuo, *Automatic Control Systems*. Englewood Cliffs, NJ: Prentice Hall, 1982.
3. P. R. Gray and R. G. Meyer, *Analysis and Design of Integrated Circuits*. New York: Wiley, 1992, Chapters 8 and 9.
4. P. M. Chirlian, *Analysis and Design of Analog Integrated Electronic Circuits*. New York: Harper & Row, 1981, Chapter 16.
5. P. R. Gray, P. J. Hurst, S. H. Lewis, and R. G. Meyer, *Analysis and Design of Analog Integrated Circuits*. New York: Wiley, 2001.

Review Questions

1. What are the two types of feedback?
2. What are the advantages of feedback?
3. What are the disadvantages of feedback?
4. What is loop transmission?
5. What is gain sensitivity?
6. What is feedback factor sensitivity?
7. What is the difference between open-loop gain and closed-loop gain?
8. What are the four types of feedback topologies?
9. What are the features of series-shunt feedback?
10. What are the features of shunt-shunt feedback?
11. What are the features of shunt-series feedback?
12. What are the features of series-series feedback?
13. What are the effects of the feedback network on amplifier performance?
14. What are the effects of the source impedance on amplifier performance?
15. What are the effects of the load impedance on amplifier performance?
16. How do you take into account the β dependency of A?
17. How do you find the modified gain A of a feedback amplifier?
18. What are the steps in designing a feedback network?
19. What is a characteristic equation?
20. What are the effects of the poles on the stability of an amplifier?
21. What is the Nyquist criterion?
22. What are the conditions for instability?
23. What is gain crossover frequency?
24. What is phase crossover frequency?
25. What is relative stability?
26. What is a phase margin?
27. What is a gain margin?
28. What is the effect of phase margin on system response?
29. What is a Bode plot?
30. What is compensation?
31. What is the common method of compensation in amplifiers?
32. How is a dominant pole introduced in electronic circuits?

Problems

The symbol **D** indicates that a problem is a design problem. The symbol **P** indicates that you can check the solution to a problem using PSpice/SPICE or Electronics Workbench.

10.3 *Characteristics of Feedback*

10.1 Three voltage amplifiers are cascaded, as shown in Fig. P10.1.

a. Determine the value of β to give a closed-loop gain of $A_f = -100$.

b. If the gain of each stage increases by 10%, determine A_f. Use the β of part (a).

FIGURE P10.1

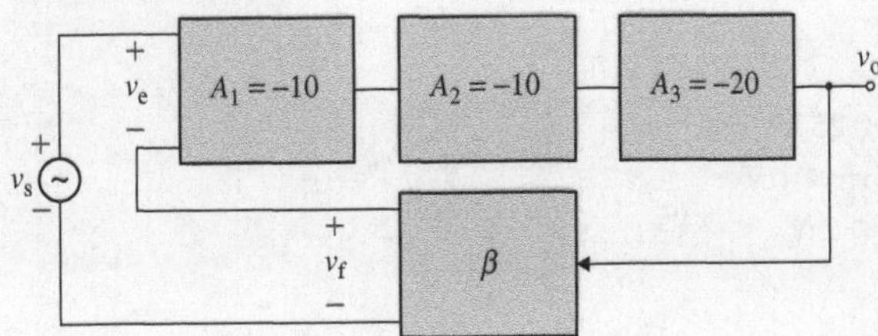

10.2 The open-loop gain of an amplifier with negative feedback is $A = 50$, and the feedback factor is $\beta = 0.5$.

a. Determine the closed-loop gain A_f.

b. If the open-loop gain A changes by $+15\%$, determine the percentage change in the closed-loop gain A_f and its value.

c. If the feedback factor β changes by $+15\%$, determine the percentage change in the closed-loop gain A_f and its value.

10.3 The open-loop gain of an amplifier is $A = 50$, and the feedback factor is $\beta = 0.8$. If the open-loop gain A changes by -20% and the feedback factor β changes by $+15\%$, determine the closed-loop gain A_f.

10.4 Two feedback amplifiers are connected in series. Each amplifier has a distortion of 20%. Determine the overall gain and the overall distortion if **(a)** each amplifier has its own feedback, as shown in Fig. P10.4(a), and **(b)** only one feedback is applied to the cascaded amplifiers, as shown in Fig. P10.4(b).

FIGURE P10.4

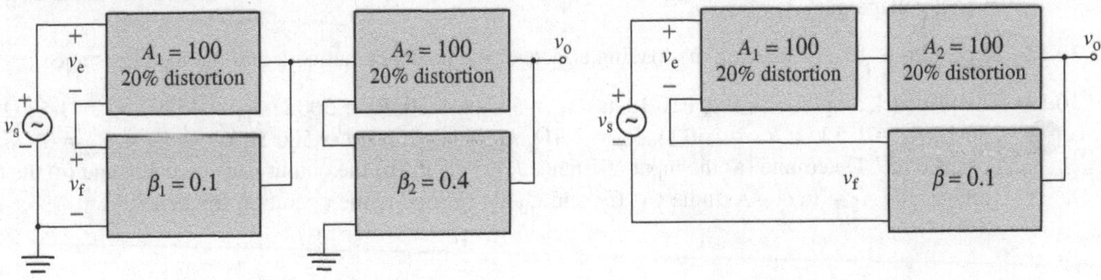

(a) Cascaded feedback **(b) Feedback on cascaded amplifiers**

10.5 A feedback amplifier is to have a closed-loop gain of $A_f = 60$ dB and a sensitivity of 10% to the open-loop gain A. Determine the open-loop gain with a unity feedback $\beta = 1$.

10.6 The feedback factor of an amplifier is $\beta = 0.5$. The open-loop gain A, which is frequency dependent, can be expressed as

$$A(j\omega) = \frac{2 \times 10^5}{1 + jf/10}$$

Determine **(a)** the closed-loop low-frequency gain A_{of}, **(b)** the closed-loop bandwidth BW, and **(c)** the gain–bandwidth product GBW.

10.7 The feedback factor of an amplifier is $\beta = 0.8$. The open-loop gain A can be expressed in Laplace's domain of s as

$$A(s) = \frac{250s}{(1 + 0.1s)(1 + 0.001s)}$$

Determine **(a)** the closed-loop low-frequency gain A_{of}, **(b)** the closed-loop bandwidth BW, and **(c)** the gain–bandwidth product GBW.

10.8 The transfer characteristic of an amplifier is described by the following values of open-loop gain A for given ranges of input voltage v_2:

$$A = \begin{cases} 50 & \text{for } 0 < v_i \leq 0.25\text{ V} \\ 40 & \text{for } 0.25\text{ V} < v_i \leq 1\text{ V} \\ 10 & \text{for } 1\text{ V} < v_i \leq 2\text{ V} \\ 0 & \text{for } v_i > 2\text{ V} \end{cases}$$

If the feedback factor is $\beta = 0.8$, determine the closed-loop gains of the transfer characteristic.

10.6 Series-Shunt Feedback

10.9 P The noninverting amplifier shown in Fig. 10.7(a) has $R_1 = 40$ kΩ, $R_f = 10$ kΩ, $R_L = 15$ kΩ, and $R_s = 0$. The op-amp parameters are $R_i = 2$ MΩ and $R_o = 50$ Ω, and the open-loop voltage gain is $\mu_g = 2 \times 10^5$. Determine **(a)** the input resistance seen by the source $R_{if} = v_s/i_s$, **(b)** the output resistance R_{of}, and **(c)** the closed-loop voltage gain $A_f = v_o/v_s$.

10.10 Repeat Prob. 10.9, ignoring the loading effects of the feedback network and the load resistance.

10.11 P A voltage amplifier with negative feedback is shown in Fig. 10.47. The amplifier has an open-loop voltage gain of $A = 250$, an input resistance of $R_i = 4.5$ kΩ, and an output resistance of $R_o = 500$ Ω. The load resistance is $R_L = 10$ kΩ. The feedback circuit has $R_1 = 24$ kΩ and $R_F = 8$ kΩ. The source has a resistance of $R_s = 2$ kΩ. Determine **(a)** the input resistance $R_{if} = v_s/i_s$, **(b)** the output resistance R_{of}, and **(c)** the overall voltage gain $A_f = v_o/v_s$.

10.12 Repeat Prob. 10.11, ignoring the loading effects of the feedback network and the load resistance.

10.13 P The feedback amplifier in Fig. P10.13 has $A_1 = 50$, $A_2 = 60$, $R_s = 500$ Ω, $R_1 = 15$ kΩ, $R_2 = 1.5$ kΩ, $R_3 = 250$ Ω, $R_4 = 1.5$ kΩ, $R_5 = 250$ Ω, $R_6 = 2$ kΩ, $R_L = 4.7$ kΩ, $R_F = 500$ Ω, $C_1 = C_2 = C_3 = 0.1$ µF, and $v_s = 100$ mV. Determine **(a)** the input resistance $R_{if} = v_s/i_s$, **(b)** the output resistance R_{of}, and **(c)** the overall voltage gain $A_f = v_o/v_s$. Assume C_1, C_2, and C_3 are shorted at the operating frequency.

FIGURE P10.13

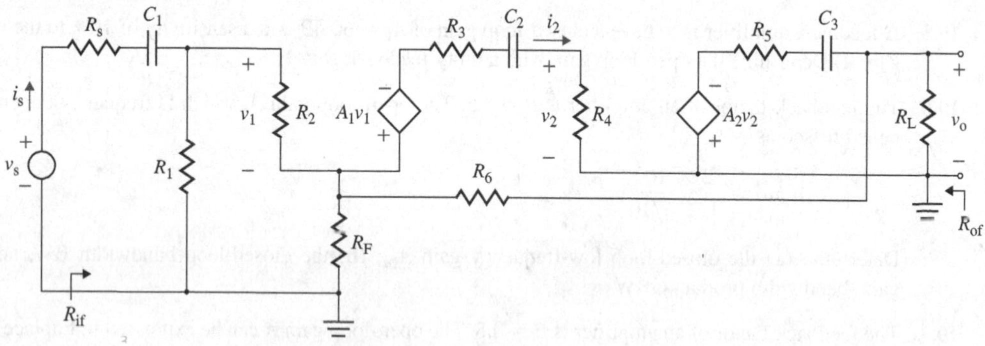

10.14 Repeat Prob. 10.13, ignoring the loading effects of the feedback network and the load resistance.

10.15 Use PSpice/SPICE to plot output impedance against frequency for the feedback amplifier in Prob. 10.13. The frequency should be varied from 10 Hz to 1 MHz in increments of 1 decade and 10 points per decade.

10.16 P The AC equivalent circuit of the CE amplifier in Fig. P10.16 has $R_1 = 10$ kΩ, $R_2 = 1.5$ kΩ, $R_C = 1.5$ kΩ, $R_E = 250$ Ω, $R_s = 200$ Ω, $R_3 = 24$ kΩ, $R_4 = 8$ kΩ, and $R_L = 1$ kΩ. The transistor parameters are $h_{fe} = 150$, $r_\pi = 2.5$ kΩ, and $r_o = 25$ kΩ. Use the techniques of feedback analysis to calculate the **(a)** input resistance R_{if}, **(b)** the output resistance R_{of}, and **(c)** the closed-loop voltage gain A_f.

FIGURE P10.16

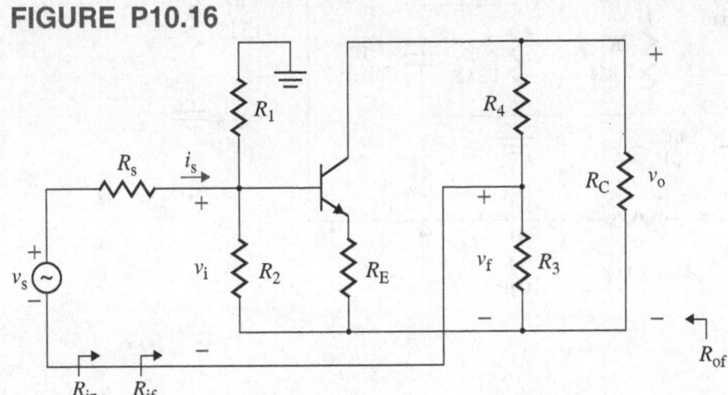

10.17 P Use the techniques of feedback analysis to calculate the input resistance R_{if}, the output resistance R_{of}, and the closed-loop voltage gain A_f of the amplifier in Fig. 10.18(a) with series-shunt feedback. The DC bias currents of the transistors are $I_{C1} = 0.1$ mA, $I_{C2} = 0.5$ mA, and $I_{C3} = 2$ mA. The transistor parameters are $h_{fe} = h_{fe1} = h_{fe2} = h_{fe3} = 150$, $r_o = 25$ kΩ, and $r_\mu = \infty$.

10.18 D The emitter follower in Fig. P10.18 has $R_B = 75$ kΩ, $R_E = 750$ Ω, $R_L = 10$ kΩ, and $R_s = 250$ Ω. The transistor parameters are $h_{fe} = 150$, $r_\pi = 250$ Ω, and $r_o = \infty$. Draw a block diagram of the feedback mechanism. Use the techniques of feedback analysis to calculate **(a)** the input resistance R_{if}, **(b)** the output resistance R_{of}, and **(c)** the closed-loop voltage gain A_f.

FIGURE P10.18

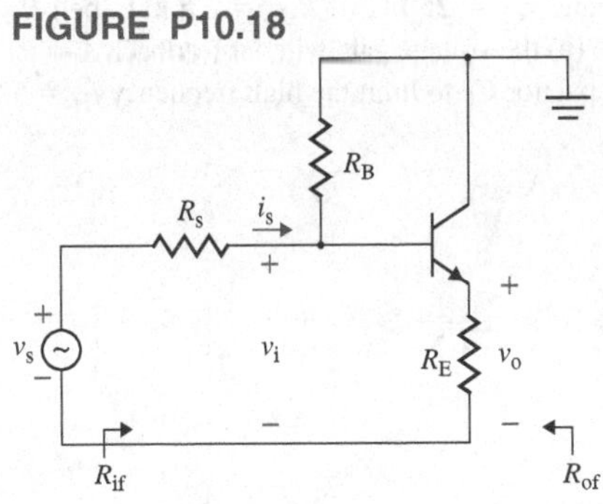

10.19 P Use the techniques of feedback analysis to calculate the input resistance R_{if}, the output resistance R_{of}, and the closed-loop voltage gain A_f of the amplifier in Fig. P10.19. The transistor parameters are $h_{fe} = h_{fe1} = h_{fe2} = 10$, $r_{\pi 1} = r_{\pi 2} = 250$ Ω, $r_o = 1.5$ kΩ, and $r_\mu = \infty$.

FIGURE P10.19

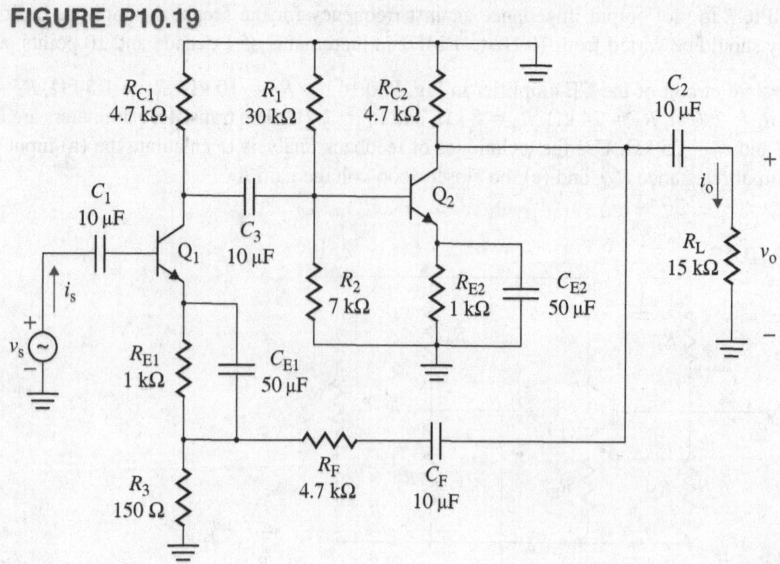

10.20 The AC equivalent circuit of a feedback amplifier is shown in Fig. 10.18(a). The DC bias currents of the transistors are $I_{C1} = 0.5$ mA, $I_{C2} = 1$ mA, and $I_{C3} = 5$ mA. The transistor parameters are $h_{fe} = 100$, $r_o =$ 25 kΩ, and $r_\mu = \infty$. Use the techniques of feedback analysis to calculate **(a)** the input resistance R_{if}, **(b)** the output resistance R_{of}, and **(c)** the closed-loop voltage gain A_f.

10.21 For the amplifier in Fig. 10.18(a), determine the value of the feedback resistor R_F so that the closed-loop voltage gain A_f is 25% of the open-loop voltage gain A. The DC bias currents of the transistors are $I_{C1} =$ 0.5 mA, $I_{C2} = 1$ mA, and $I_{C3} = 5$ mA. The transistor parameters are $h_{fe} = h_{fe1} = h_{fe2} = h_{fe3} = 100$, $r_o = 25$ kΩ, and $r_\mu = \infty$.

10.22 The MOS amplifier shown in Fig. P10.22 is biased to have the following small-signal MOS parameters: $g_{m1} = 1.2$ mA/V, $r_{o1} = 25$ kΩ, $g_{m2} = 1.6$ mA/V, and $r_{o1} = 25$ kΩ. If $R_{D1} = 1.5$ kΩ, then $R_{D2} = 1$ kΩ, $R_{SR} = 500$ Ω, $R_F = 5$ kΩ, and $C_F = 20$ pF. Determine **(a)** the voltage gain without feedback $A = v_o/v_s$, **(b)** the voltage gain with feedback A_f, and **(c)** the feedback capacitor C_F to limit the high frequency $f_H = 50$ kHz.

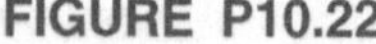

FIGURE P10.22

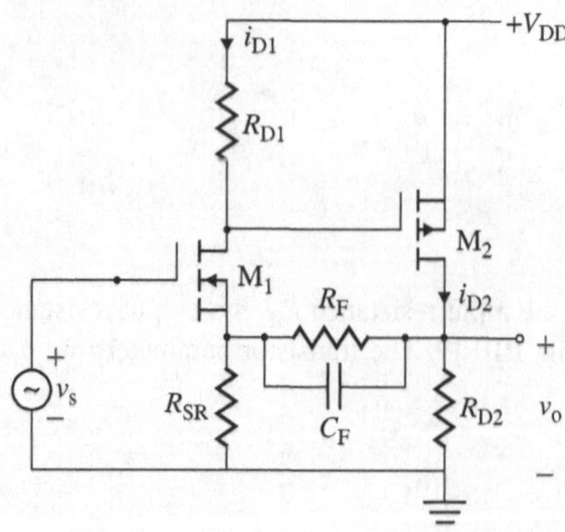

10.23 The MOS amplifier shown in Fig. P10.23 is biased to have the following small-signal MOS parameters: $g_{m1} = g_{m2} = 1$ mA/V, $r_{o1} = r_{o2} = 20$ kΩ, $g_{m3} = g_{m4} = 1.2$ mA/V, and $r_{o3} = r_{o4} = 22$ kΩ. If $C_F = 6$ nF and $C_1 = 4$ nF, determine **(a)** the voltage gain without feedback $A = v_o/v_s$, **(b)** the voltage gain with feedback A_f, and **(c)** the high cutoff frequency f_H.

FIGURE P10.23

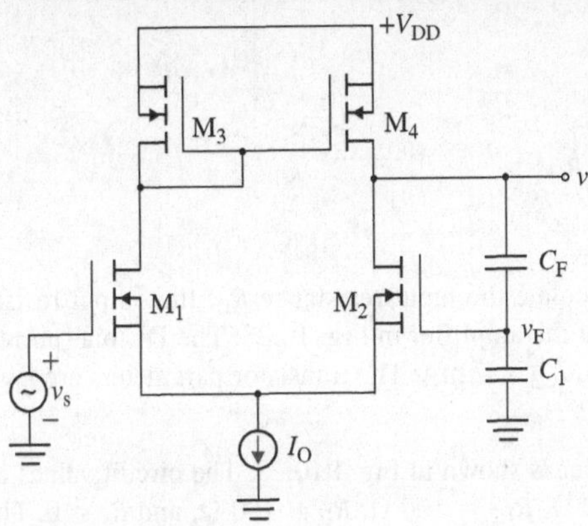

10.7 Series-Series Feedback

10.24 P A transconductance amplifier with negative feedback is shown in Fig. P10.24. The amplifier has an open-loop transconductance of $A = 50 \times 10^{-3}$ A/V, an input resistance of $R_i = 25$ kΩ, and an output resistance of $R_o = 50$ kΩ. The feedback circuit has $R_F = 2.5$ kΩ. The source resistance is $R_s = 1$ kΩ, and the load resistance is $R_L = 500$ Ω. Determine **(a)** the input resistance $R_{if} = v_s/i_s$, **(b)** the output resistance R_{of}, and **(c)** the overall transconductance gain $A_f = i_o/v_s$.

FIGURE P10.24

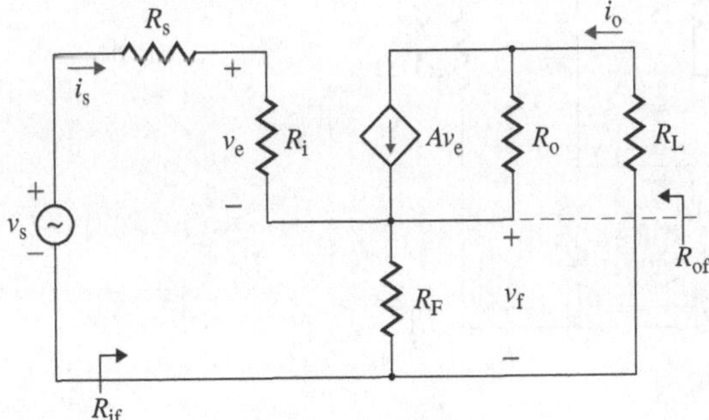

10.25 Repeat Prob. 10.24, ignoring the loading effects of the feedback network and the load resistance; that is, assume $R_s = 0$ and $R_L = 0$.

10.26 P Use the techniques of feedback analysis to determine the input and output resistance of the CE transistor amplifier in Fig. P10.26. The circuit parameters are $R_s = 500$ Ω, $R_E = 250$ Ω, $R_2 = 15$ kΩ, $R_1 = 5$ kΩ, $R_C = 5$ kΩ, and $R_L = 10$ kΩ. The π-model parameters are $r_o = 25$ kΩ, $h_{fe} = 150$, $r_\pi = 250$ Ω, $g_m = 0.3876$ A/V, and $r_\mu = \infty$.

FIGURE P10.26

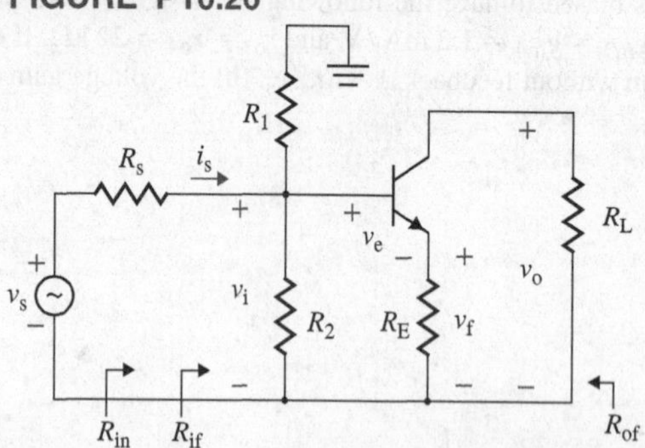

10.27 Use the techniques of feedback analysis to calculate the input resistance R_{if}, the output resistance R_{of}, and the closed-loop transconductance gain A_f of the amplifier in Fig. 10.28. The DC bias currents of the transistors are $I_{C1} = 0.1$ mA, $I_{C2} = 0.5$ mA, and $I_{C3} = 2$ mA. The transistor parameters are $h_{fe} = h_{fe1} = h_{fe2} = h_{fe3} = 150$, $r_o = 25$ kΩ, and $r_\mu = \infty$.

10.28 The AC equivalent circuit of a feedback amplifier is shown in Fig. P10.28. The circuit values are $R_{C1} = 2.5$ kΩ, $R_{C2} = 5$ kΩ, $R_{C3} = 1.5$ kΩ, $R_{E1} = 100$ Ω, $R_{E2} = 100$ Ω, $R_F = 750$ Ω, and $R_s = 0$. The transistor parameters are $h_{fe} = 100$, $r_\pi = 2.5$ kΩ, $r_o = 25$ kΩ, and $r_\mu = \infty$. Use the techniques of feedback analysis to calculate **(a)** the input resistance R_{if}, **(b)** the output resistance R_{of}, and **(c)** the closed-loop voltage gain A_f.

FIGURE P10.28

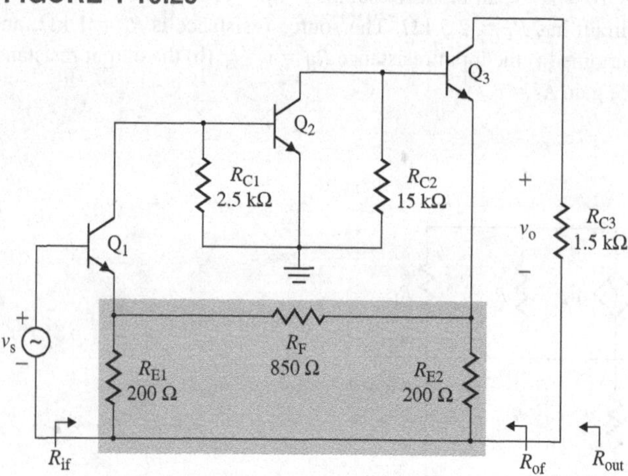

10.29 For the amplifier in Fig. 10.28 determine the value of the feedback resistor R_F so that the closed-loop transconductance gain A_f is 25% of the open-loop transconductance gain A. The DC bias currents of the transistors are $I_{C1} = 0.5$ mA, $I_{C2} = 1$ mA, and $I_{C3} = 5$ mA. The transistor parameters are $h_{fe} = h_{fe1} = h_{fe2} = h_{fe3} = 100$, $r_o = 25$ kΩ, and $r_\mu = \infty$.

10.30 For the amplifier in Fig. P10.30, determine the value of the feedback resistor R_F so that the closed-loop transconductance gain A_f is 5 mA/V. The transistor parameters are $h_{fe} = 100$, $r_\pi = 250\ \Omega$, $r_o = 50\ \text{k}\Omega$, and $r_\mu = \infty$. Assume $R_i = 2\ \text{M}\Omega$, $R_o = 50\ \Omega$, and $A = 10^5$ V/V.

FIGURE P10.30

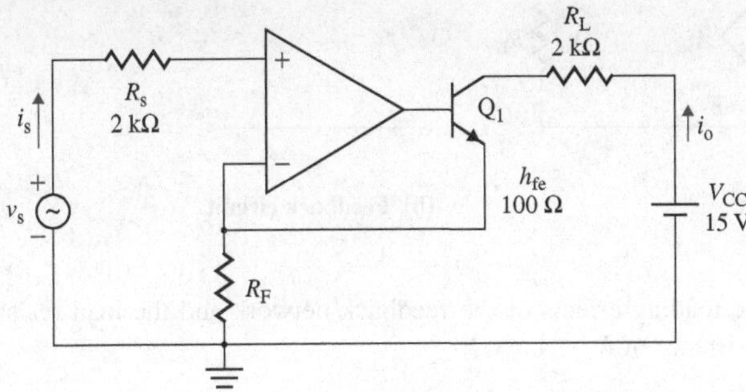

10.31 The MOS amplifier shown in Fig. P10.31 is biased to have the following small-signal MOS parameters: $g_{m1} = 1.2\ \text{mA/V}$, $r_{o1} = 25\ \text{k}\Omega$, $g_{m2} = 1.6\ \text{mA/V}$, and $r_{o2} = 25\ \text{k}\Omega$. If $R_D = 1.5\ \text{k}\Omega$, then $R_{SR1} = 500\ \Omega$, $R_{SR2} = 2\ \text{k}\Omega$, and $R_F = 8\ \text{k}\Omega$. Determine **(a)** the voltage gain without feedback $A = i_o/v_s$, **(b)** the voltage gain with feedback A_f, and **(c)** the feedback capacitor C_F to limit the high frequency $f_H = 50$ kHz.

FIGURE P10.31

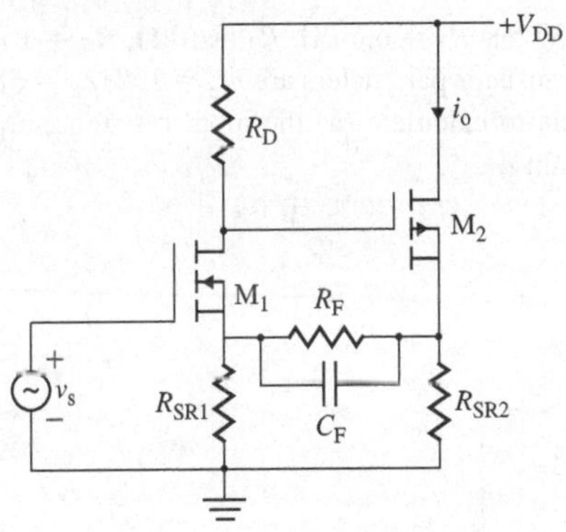

10.8 *Shunt-Shunt Feedback*

10.32 The inverting op-amp amplifier shown in Fig. 10.9(c) has $R_F = 40\ \text{k}\Omega$ and $R_1 = 10\ \text{k}\Omega$. The op-amp has an input resistance of $R_i = 5\ \text{M}\Omega$, an output resistance of $R_o = 50\ \Omega$, and an open-loop voltage gain of $A = 2 \times 10^5$ V/V. The load resistance is $R_L = 15\ \text{k}\Omega$. Determine **(a)** the input resistance seen by the source $R_{in} = v_s/i_s$, **(b)** the output resistance R_{of}, **(c)** the closed-loop transresistance $A_f = v_o/i_i$, and **(d)** the overall voltage gain $A_{vf} = v_o/v_s$. Assume source resistance $R_s = 1\ \text{k}\Omega$.

10.33 A transresistance amplifier with negative feedback is shown in Fig. P10.33. The open-loop transresistance is $A = 750\ \text{k}\Omega$, the input resistance is $R_i = 5.5\ \text{k}\Omega$, and the output resistance is $R_o = 500\ \Omega$. The feedback circuit has $R_F = 47\ \text{k}\Omega$. The source has a resistance of $R_s = 0$. Determine **(a)** the input resistance seen by the source $R_{in} = v_s/i_s$, **(b)** the output resistance R_{of}, and **(c)** the overall voltage gain $A_{vf} = v_o/v_s$. Assume load resistance $R_L = 1\ \text{k}\Omega$.

FIGURE P10.33

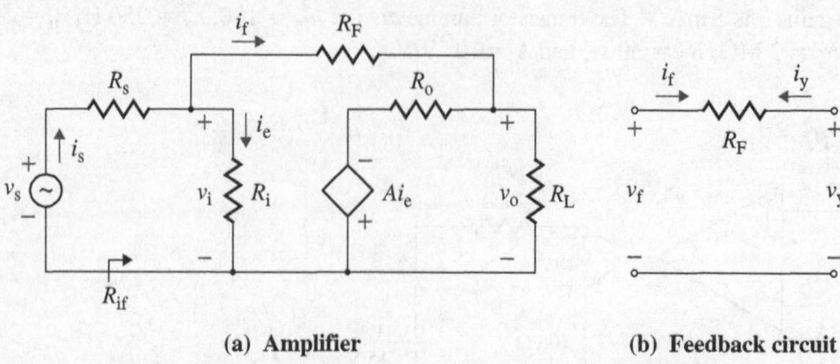

(a) Amplifier **(b) Feedback circuit**

10.34 Repeat Prob. 10.33, ignoring the loading effects of the feedback network and the load resistance and assuming that the source has a resistance of R_s = 1.5 kΩ.

10.35 P Use the techniques of feedback analysis to calculate the input resistance R_{if}, the output resistance R_{of}, and the closed-loop transresistance gain A_f of the amplifier in Fig. 10.38. The circuit values are R_{C1} = 5 kΩ, R_E = 2.5 kΩ, R_{C2} = 5 kΩ, R_F = 4 kΩ, and R_s = 200 Ω. The transistor parameters are h_{fe} = 150, r_π = 2 kΩ, r_o = 25 kΩ, and $r_\mu = \infty$.

10.36 P The AC equivalent circuit of a feedback amplifier is shown in Fig. 10.38. The circuit values are R_{C1} = 5 kΩ, R_E = 2.5 kΩ, R_{C2} = 5 kΩ, R_F = 4 kΩ, and R_s = 200 Ω. The transistor parameters are h_{fe} = 100, r_π = 2 kΩ, r_o = 25 kΩ, and $r_\mu = \infty$. Use the techniques of feedback analysis to calculate **(a)** the input resistance R_{if}, **(b)** the output resistance R_{of}, and **(c)** the closed-loop voltage gain A_f.

10.37 P The AC equivalent circuit of the amplifier in Fig. P10.37 has R_1 = 6.6 kΩ, R_2 = 1 kΩ, R_C = 1 kΩ, R_E = 100 Ω, R_s = 500 Ω, R_F = 8 kΩ, and R_L = 5 kΩ. The transistor parameters are h_{fe} = 100, r_π = 581 Ω, and r_o = 22.5 kΩ. Use the techniques of feedback analysis to calculate **(a)** the input resistance R_{if}, **(b)** the output resistance R_{of}, and **(c)** the closed-loop voltage gain A_f.

FIGURE P10.37

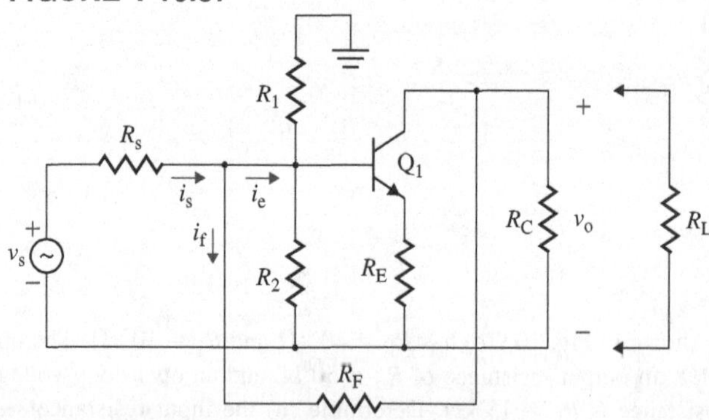

10.38 P The AC equivalent circuit of a feedback amplifier is shown in Fig. P10.38. The circuit values are R_s = 1 kΩ, R_C = 10 kΩ, and R_F = 24 kΩ. The transistor parameters are h_{fe} = 150, r_π = 500 Ω, r_o = 25 kΩ, and $r_\mu = \infty$. Use the techniques of feedback analysis to calculate **(a)** the input resistance R_{if}, **(b)** the output resistance R_{of}, and **(c)** the closed-loop voltage gain A_f.

FIGURE P10.38

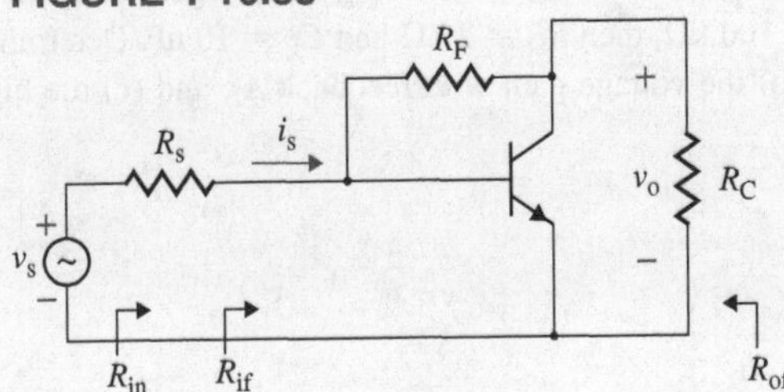

10.39 (P) The AC equivalent circuit of a feedback amplifier is shown in Fig. P10.39. The circuit values are $R_D = 10\ \text{k}\Omega$, $R_F = 50\ \text{k}\Omega$, and $R_s = 1\ \text{k}\Omega$. The transistor parameters are $g_m = 1\ \text{mA/V}$, $r_d = 50\ \text{k}\Omega$, and $r_\mu = 25\ \text{k}\Omega$. Use the techniques of feedback analysis to calculate **(a)** the input resistance R_{if}, **(b)** the output resistance R_{of}, and **(c)** the closed-loop voltage gain A_f.

FIGURE P10.39

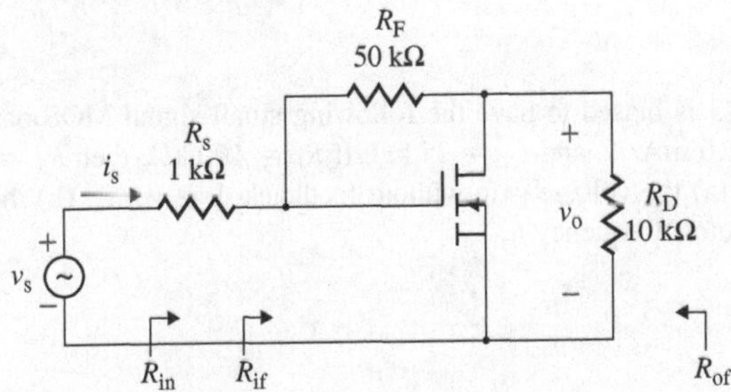

10.40 (D) (P) Determine the value of the feedback resistor R_F so that the closed-loop voltage gain A_f of the amplifier in Fig. 10.38 is 25% of the open-loop voltage gain A. The circuit values are $R_{C1} = 5\ \text{k}\Omega$, $R_E = 2.5\ \text{k}\Omega$, $R_{C2} = 5\ \text{k}\Omega$, and $R_s = 200\ \Omega$. The transistor parameters are $h_{fe} = 150$, $r_\pi = 2\ \text{k}\Omega$, $r_o = 25\ \text{k}\Omega$, and $r_\mu = \infty$.

10.41 (D) (P) Determine the value of the feedback resistor R_F so that the closed-loop current gain A_f of the amplifier in Fig. P10.41 is 10% of the open-loop voltage gain A. The transistor parameters are $h_{fe} = 150$, $r_\pi = 2\ \text{k}\Omega$, $r_o - 25\ \text{k}\Omega$, and $r_\mu = \infty$. Assume that the coupling capacitors can be considered as short-circuited at the operating frequency range.

FIGURE P10.41

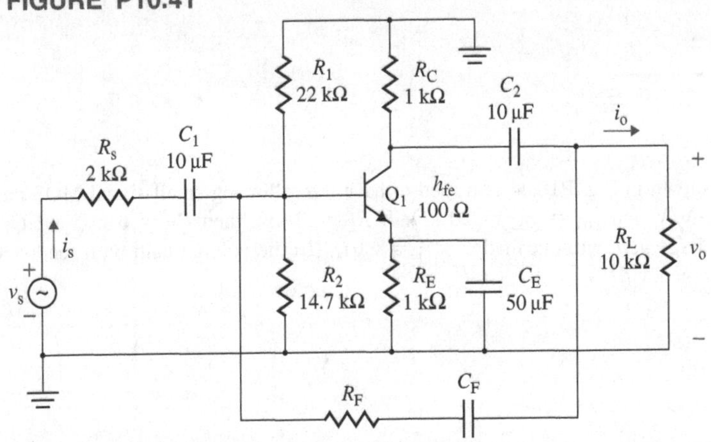

10.42 The MOS amplifier shown in Fig. P10.42 is biased to have the following small-signal MOS parameters: $g_{m1} = 1.2\ \text{mA/V}$ and $r_{o1} = 25\ \text{k}\Omega$. If $R_F = 100\ \text{k}\Omega$, then $R_D = 2\ \text{k}\Omega$ and $C_F = 10\ \text{nF}$. Determine **(a)** the voltage gain without feedback $A = v_o/i_s$, **(b)** the voltage gain with feedback A_f, and **(c)** the high cutoff frequency f_H.

FIGURE P10.42

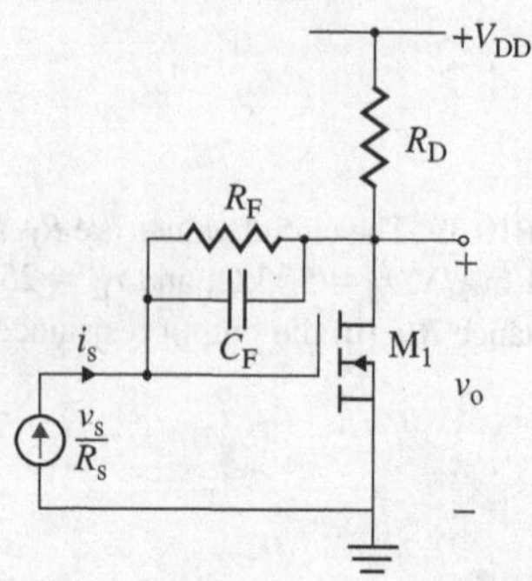

10.43 The MOS amplifier shown in Fig. P10.43 is biased to have the following small-signal MOS parameters: $g_{m1} = 1.2\ \text{mA/V}$, $r_{o1} = 25\ \text{k}\Omega$, $g_{m2} = 1.6\ \text{mA/V}$, and $r_{o2} = 25\ \text{k}\Omega$. If $R_F = 100\ \text{k}\Omega$, then $R_1 = 400\ \text{k}\Omega$, $R_s = 1\ \text{k}\Omega$, and $C_F = 10\ \text{nF}$. Determine **(a)** the voltage gain without feedback $A = v_o/v_s$, **(b)** the voltage gain with feedback A_f, and **(c)** the high cutoff frequency f_H.

FIGURE P10.43

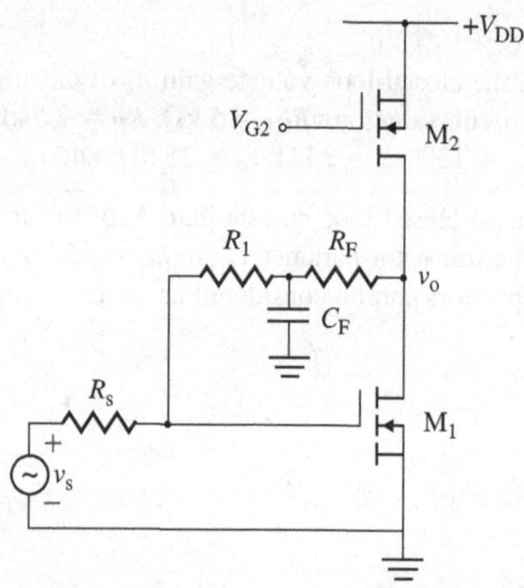

10.44 The MOS amplifier shown in Fig. P10.44 is biased to have the following small-signal MOS parameters: $g_{m1} = g_{m2} = 1.2\ \text{mA/V}$ and $r_{o1} = r_{o2} = 20\ \text{k}\Omega$. If $R_D = 2\ \text{k}\Omega$, then $C_F = 6\ \text{nF}$, and $C_1 = 2\ \text{nF}$. Determine **(a)** the voltage gain without feedback $A = v_o/i_s$, **(b)** the voltage gain with feedback A_f, and **(c)** the high cutoff frequency f_H.

FIGURE P10.44

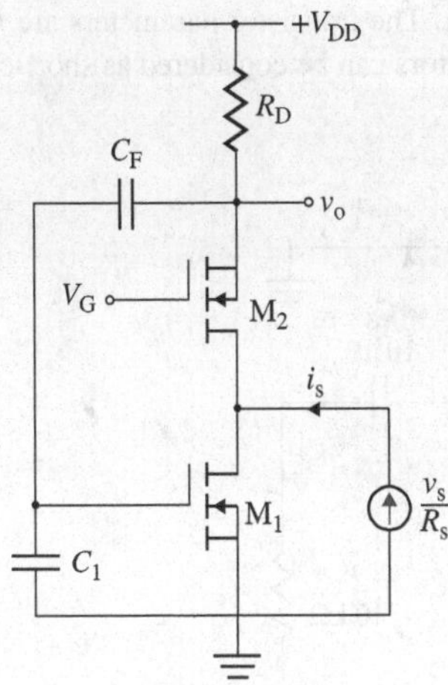

10.9 *Shunt-Series Feedback*

10.45 P The current amplifier in Fig. P10.45 has negative feedback. The open-loop current gain is $A = 60$, the input resistance is $R_i = 500$ kΩ, and the output resistance is $R_o = 27$ kΩ. The feedback circuit has $R_F = 20$ kΩ and $R_1 = 2.5$ kΩ. The source resistance is $R_s = 500$ Ω, and the load resistance is $R_L = 100$ Ω. Determine **(a)** the input resistance $R_{if} = v_s/i_s$, **(b)** the output resistance R_{of}, and **(c)** the overall current gain $A_f = i_o/i_s$.

FIGURE P10.45

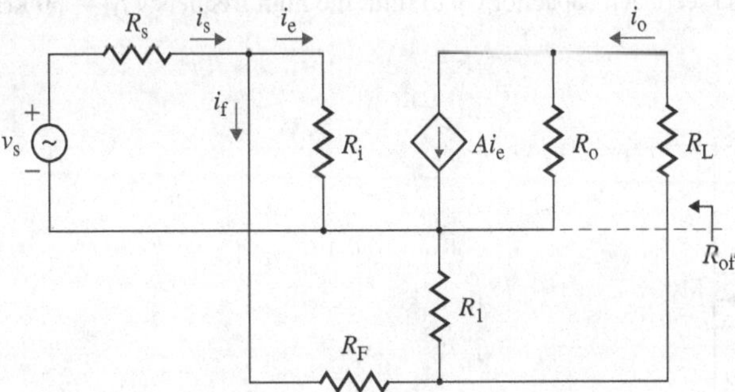

10.46 Repeat Prob. 10.45, ignoring the loading effects of the feedback network and the load resistance and assuming that the source has a resistance of $R_s = 5$ kΩ.

10.47 P The CE amplifier in Fig. 10.48 has $R_F = 10$ kΩ, $R_1 = 6.6$ kΩ, $R_2 = 1$ kΩ, $R_{C1} = 5$ kΩ, $R_E = 500$ Ω, $R_3 = 5$ kΩ, $R_4 = 10$ kΩ, $R_{C2} = 5$ kΩ, and $R_s = 200$ Ω. The transistor parameters are $h_{fe} = 150$, $r_\pi = 258$ Ω, and $r_o = 25$ kΩ. Use the techniques of feedback analysis to calculate **(a)** the input resistance R_{if}, **(b)** the output resistance R_{of}, and **(c)** the closed-loop voltage gain A_f.

10.48 D P Determine the value of the feedback resistor R_F so that the closed-loop current gain A_f of the amplifier in Fig. 10.48 with shunt-series feedback is 25% of the open-loop current gain A. The amplifier has $R_1 = 6.6$ kΩ, $R_2 = 2$ kΩ, $R_{C1} = 5$ kΩ, $R_E = 500$ Ω, $R_3 = 5$ kΩ, $R_4 = 10$ kΩ, $R_{C2} = 5$ kΩ, and $R_s = 0$. The transistor parameters are $h_{fe} = 100$, $r_\pi = 2.58$ kΩ, and $r_o = 25$ kΩ.

10.49 Determine the value of the feedback resistor R_F so that the midfrequency closed-loop current gain A_f of the amplifier in Fig. P10.49 is 10% of the open-loop current gain A. The transistor parameters are $h_{fe} = 100$, $r_\pi = 2.58\ \text{k}\Omega$, and $r_o = 25\ \text{k}\Omega$. Assume that the coupling capacitors can be considered as short-circuited at the operating frequency range.

FIGURE P10.49

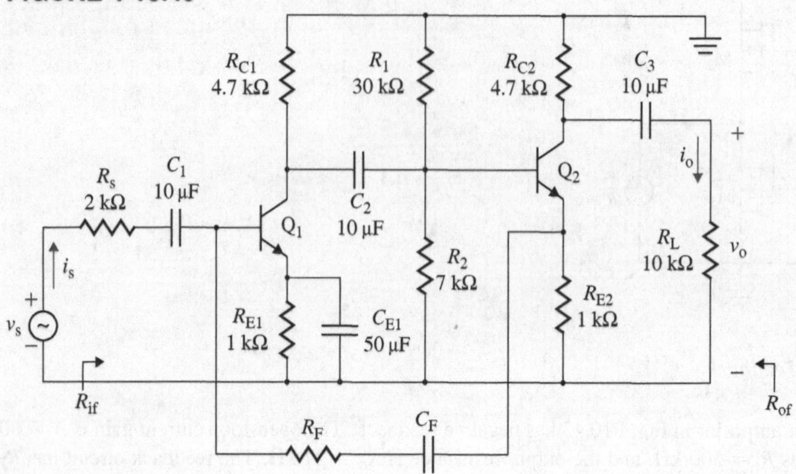

10.50 The MOS amplifier shown in Fig. P10.50 is biased to have the following small-signal MOS parameters: $g_{m1} = 1.2\ \text{mA/V}$, $r_{o1} = 25\ \text{k}\Omega$, $g_{m2} = 1.6\ \text{mA/V}$, and $r_{o2} = 25\ \text{k}\Omega$. If $R_D = 1.5\ \text{k}\Omega$, then $R_{SR1} = 500\ \text{k}\Omega$, $R_{SR2} = 2\ \text{k}\Omega$, and $R_F = 8\ \text{k}\Omega$. Determine **(a)** the voltage gain without feedback $A = i_o/v_s$, **(b)** the voltage gain with feedback A_f, and **(c)** the feedback capacitor C_F to limit the high frequency $f_H = 50$ kHz.

FIGURE P10.50

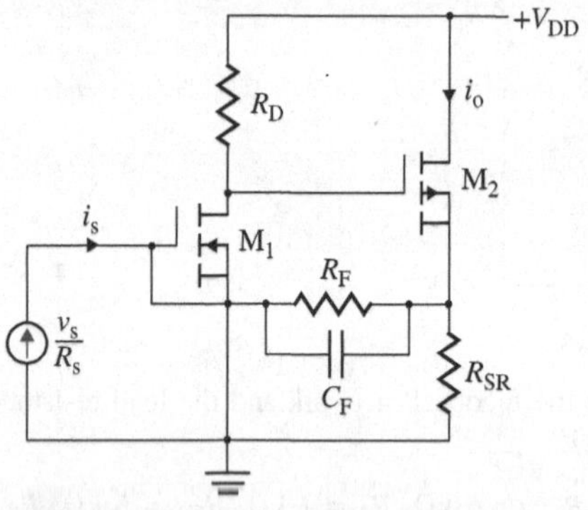

10.11 *Stability Analysis*

10.51 The open-loop gain of an amplifier is given by

$$A(s) = \frac{6}{s^2 + 2s - 30}$$

Determine the closed-loop response due to a step input. Assume feedback factor $\beta(s) = 1$.

10.52 The open-loop gain of an amplifier is given by

$$A(s) = \frac{s}{s^2 + 100}$$

Determine the closed-loop response due to a step input. Assume feedback factor $\beta(s) = 1$.

10.53 The loop gain of a feedback amplifier is given by

$$T_L(j\omega) = \frac{A_o}{j\omega(1 + j\omega T_1)(1 + j\omega T_2)}$$

where A_o is a gain constant and T_1 and T_2 are time constants. Determine the phase margin and gain margin of the amplifier.

10.54 The loop gain of a feedback amplifier is given by

$$T_L(j\omega) = \frac{A_o}{j\omega[(j\omega)^2 K_2 + j\omega K_1 + 1]}$$

where A_o is a gain constant and K_1 and K_2 are constants. Determine the phase margin and gain margin of the amplifier.

10.55 The loop gain of a feedback amplifier is given by

$$T_L(s) = \frac{10}{s(s + 1)(s + 2)}$$

Determine the phase margin and gain margin of the amplifier.

10.56 If the phase margin of an amplifier is PM = 40° and the magnitude of the open-loop gain is $|A(j\omega)| = 50$, find the magnitude of the closed-loop gain $|A_f(j\omega)|$.

10.57 The open-loop gain of an amplifier has break frequencies at $f_{p1} = 10$ kHz, $f_{p2} = 100$ kHz, and $f_{p3} = 1$ MHz. The low-frequency gain is $A_o = 250$, and the feedback factor is $\beta = 0.9$. Calculate the gain margin GM and the phase margin PM.

10.12 *Compensation Techniques*

10.58 For the amplifier in Prob. 10.57, determine the frequency at the dominant pole so that the phase margin is PM = 45°.

10.59 P The feedback amplifier in Fig. 10.60(a) has $g_m = 40 \times 10^{-3}$, $R_i = 3.5$ kΩ, $C_i = 10$ pF, $R_o = 24$ kΩ, and $C_o = 5$ pF.

a. Calculate the two pole frequencies for $C_x = 0$ and the value of feedback capacitance C_x so that the frequency of the dominant pole is $f_D = 1.5$ kHz.

b. Use PSpice/SPICE to plot the closed-loop transimpedance A_f and the input impedance Z_i against frequency.

10.60 D P The equivalent circuit of an output stage is shown in Fig. 10.60(c), where $R_i = 22$ kΩ and $C_i = 18$ pF.

a. Find the break frequency f_p.

b. Find the additional capacitance C_x that will move the break frequency to $f_D = 50$ kHz.

10.61 D P The open-loop gain of the multistage CE amplifier in Fig. 8.68 has break frequencies at $f_{p1} = 805$ kHz, $f_{p2} = 9.6$ MHz, $f_{p3} = 13$ MHz, and $f_{p4} = 61$ MHz. The midfrequency gain is $A_{mid} = 9594$. Determine the value of compensating capacitance C_x that will give a closed-loop phase margin of 45° with a resistive feedback of up to $\beta = 1$.

10.62 The open-loop gain of the multistage MOSFET amplifier in Fig. 7.61(a) has break frequencies at f_{p1} = 13.75 kHz, f_{p2} = 248 MHz, and f_{p3} = 400 MHz. The midfrequency gain is A_{mid} = 311. Determine the value of compensating capacitance C_x that will give a closed-loop phase margin of 45° with a resistive feedback of $\beta = 1$.

10.63 The open-loop gain of the CE amplifier in Fig. 8.53 has a midfrequency gain of $A_{mid} = -300$ V/V. Determine the values of compensating capacitance C_F and resistance R_F in a shunt-shunt feedback network to give a midfrequency closed-loop gain of $A_f = -30$ V/V. The low break frequency should be less than 1 kHz. Assume source resistance R_s = 250 Ω.

10.64 The open-loop gain of the amplifier in Example 10.12 has break frequencies at f_{p1} = 10 kHz, f_{p2} = 100 kHz, and f_{p3} = 1 MHz. The low-frequency gain is A_o = 100 A/A, and the emitter resistance is R_E = 200 Ω. Determine the values of compensating capacitance C_F and resistance R_F **(a)** to give a low-frequency closed-loop gain of A_f = 20 A/A and cancel the pole f_{p1} = 10 kHz and **(b)** to add a pole of f_p = 10 MHz and cancel the pole f_{p1} = 10 kHz.

10.65 The transistor Q_4 of the amplifier output stage as shown in Fig. 10.60(a) has an input resistance r_{be4} = 20 kΩ and an input capacitance C_{i4} = 40 pF.

a. Find the break frequency f_p.

b. Find the value of additional compensating capacitance C_x which will modify the dominant pole and will move the break frequency to f_D = 25 kHz.

10.66 The differential amplifier shown in Fig. 10.60(a) has an open-loop gain with break frequencies at f_{p1} = 50 kHz, f_{p2} = 1 MHz, and f_{p3} = 10 MHz. The gain stage of the amplifier has the equivalent circuit shown in Fig. 10.60(c) whose parameters are g_m = 10 mA/V, C_i = 20 pF, and C_o = 5 pF. Determine the value of compensating capacitance C_x by pole splitting that will give a closed-loop phase margin of 45° with a resistive feedback of up to $\beta = 1$.

10.67 The amplifier shown in Fig. P10.67 has an open-loop gain with break frequencies at f_{p1} = 150 kHz and f_{p2} = 1 MHz. The parameters of Q_4 are g_{m3} = 1 mA/V, and the output capacitance of C_{o3} = 10 pF. The input capacitance of Q_2 is C_{i2} = 10 pF. Determine the value of compensating capacitance C_x by pole splitting that will give a closed-loop phase margin of 50° with a resistive feedback of up to $\beta = 1$.

FIGURE P10.67

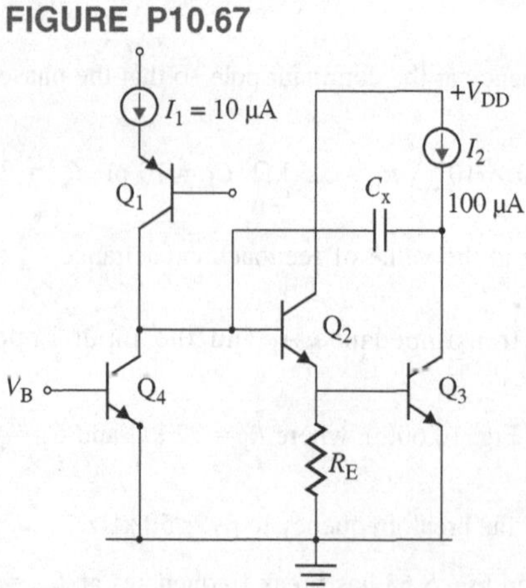

10.68 The open-loop gain of the amplifier shown in Fig. 10.61(a) has break frequencies at $f_{p1} = 50$ kHz, $f_{p2} = 1$ MHz, and $f_{p3} = 10$ MHz. The low-frequency gain is $A_o = 100$ A/A, and the emitter resistance $R_E = 400\ \Omega$. Determine the values of compensating capacitance C_F and resistance R_F by feedback path modification **(a)** to give a low-frequency closed-loop gain of $A_f = 20$ A/A and to cancel the pole $f_{p1} = 100$ kHz and **(b)** to add a pole of $f_p = 50$ MHz and to cancel the pole $f_{p1} = 50$ kHz.

10.69 A common-source amplifier with shunt-shunt feedback is shown in Fig. P10.69. The capacitor C_F and resistor R_F form the feedback network. The voltage gain without feedback is $A = |A_v| = v_o/v_s \approx 20$.

a. Design the feedback network to meet the following specifications: The bandwidth with feedback must be increased by 10 times; that is, $BW_f = 10$ BW. The voltage gain with feedback is $|A_f| = v_o/v_s = 5$.

b. Use PSpice to verify your results.

FIGURE P10.69

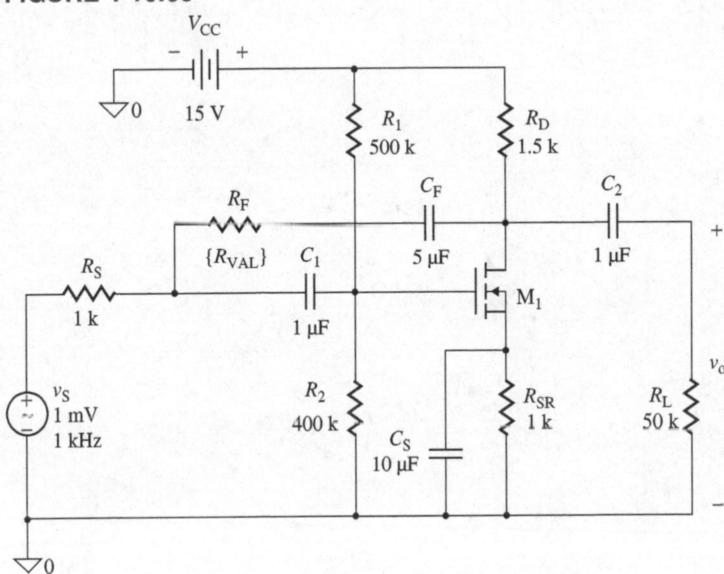

CHAPTER 11

POWER AMPLIFIERS

Learning Outcomes

After completing this chapter, students should be able to do the following:

- Describe the types of power amplifiers and their transfer characteristics.
- Select a power amplifier to meet certain requirements.
- Analyze and design power amplifiers.
- Design power amplifiers to meet certain specifications.
- Apply the methods for eliminating crossover distortion and reducing offsets and nonlinearities on the output voltage.
- Apply an active current source to bias the output stage.
- Identify and describe the internal structure of IC power op-amps.

Symbols and Their Meanings

Symbol	Meaning
v_i, v_o	Input and output voltages of an amplifier
I_p, V_p, V_{pp}	Peak current, peak voltage, and peak-to-peak voltage of a transistor
I_C, I_D	Collector and drain currents of a transistor
I_Q, I_R	Quiescent and reference currents of an amplifier
α, k	Conduction angle and duty cycle of a transistor

Symbol	Meaning
P_C, P_S, P_L	Collector power, DC supply power, and load power of an amplifier
i_S, i_O	Input signal and output currents of an amplifier
V_{BEN}, V_{BEP}	Base–emitter (B-E) voltages of an *npn* and a *pnp* transistor
h_{fe}	Hybrid current gain parameter of a transistor
η	Efficiency of an amplifier
ω_n	Natural frequency of an *LC* filter, radians per second

11.1 Introduction

The amplifiers in Chapters 7 and 8 were operated as input and/or intermediate stages to obtain a large voltage gain or current gain. The transistors within the amplifiers were operated in the active region so that their small-signal models were valid. These stages were not required to provide appreciable amounts of power, and the distortion of the output signal was negligible because the transistors operated in the active region.

The requirements for the output stages of audio-frequency power amplifiers are significantly different from those of small-signal low-power amplifiers. An output stage must deliver an appreciable amount of power and be capable of driving low-impedance loads such as loudspeakers. The distortion of the output signal must also be low. Distortion is measured by a quality factor known as *total harmonic distortion* (THD), which is the rms value of the harmonic components of the output signal, excluding the fundamental, expressed as a percentage of the rms of the fundamental component. The THD of high-fidelity audio amplifiers is usually less than 0.1%.

The DC power requirement of an audio amplifier must be as small as possible so that the efficiency of the amplifier is as high as possible. Increasing the efficiency of the amplifier reduces the amount of power dissipated by the transistors and the amount of power drawn from DC supplies, thereby reducing the cost of the power supply and prolonging the life of batteries in battery-powered amplifiers. Also, a low DC power requirement helps to keep the internal junction temperature of the transistors well below the maximum allowable temperature (in the range of 150°C to 200°C for silicon devices). As a result, a low DC power requirement minimizes the size of heat sinks and can eliminate the need for cooling fans. Therefore, an output stage should deliver the required amount of power to the load *efficiently*.

11.2 Classification of Power Amplifiers

Power amplifiers [1, 2] are generally classified into six types: A, B, AB, and C for analog designs and classes D and E for switching designs. This classification is based on the shape of the drain current i_D for a MOSFET or collector current i_C waveform for a BJT in response to a sinusoidal input signal. In an analog amplifier, the input signal to the amplifying devices causes a proportional output current to flow out of the output terminal, whereas in switching amplifiers the output current of the amplifying device is pulsating type. The output current and the load power come from the DC power supply. The load of power amplifiers is generally a speaker. The waveforms of the collector (or the drain) currents for various types of amplifiers are shown in Fig. 11.1. Table 11.1 shows the conduction angle and duty cycle of the input signal for analog circuits. The duty cycle for the class D and E amplifiers is generally variable up to 50% of the switching period.

In a class A amplifier, the DC biasing collector current I_C of a transistor is higher than the peak amplitude of the AC output current I_p. Thus, the transistor in a class A amplifier conducts during the entire cycle of the input signal, and the conduction angle is $\theta = \omega t = 360°$. That is, the collector current of a

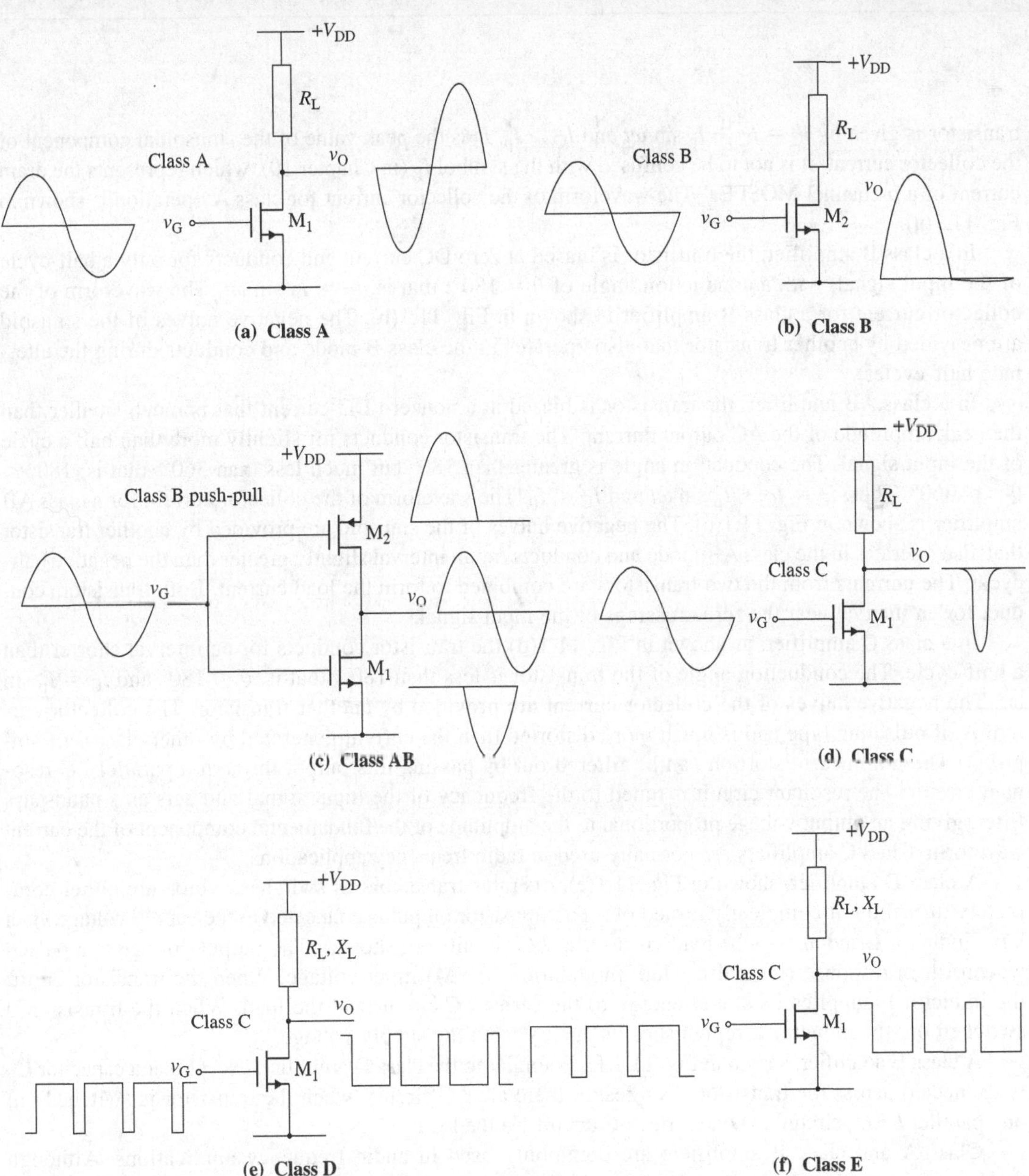

FIGURE 11.1 Output voltages for various classes of amplifiers

TABLE 11.1 Conduction angle and duty cycle of power transistors

	Class A	Class AB	Class B	Class C	Class D	Class E
Conduction angle α	$\alpha = 2\pi$ or 360°	$\pi < \alpha < 2\pi$ or 181° to 359°	$\alpha = \pi$ or 180°	$\alpha < 180°$ or 0° to 179°	$\alpha < 180°$ or 0° to 179°	$\alpha < 180°$ or 0° to 179°
Duty cycle k	$k = 100\%$	$50\% < k < 100\%$	$k = 50\%$	$k < 50\%$	$k < 50\%$	$k < 50\%$

transistor is given by $i_C = I_C + I_p \sin \omega t$ and $I_C > I_p$. I_p is the peak value of the sinusoidal component of the collector current; it is not to be confused with the symbol I_P (in Chapter 10), which represents the drain current of a *p*-channel MOSFET. The waveform of the collector current for class A operation is shown in Fig. 11.1(a).

In a class B amplifier, the transistor is biased at zero DC current and conducts for only a half-cycle of the input signal, with a conduction angle of $\theta = 180°$; that is, $i_C = I_p \sin \omega t$. The waveform of the collector current for a class B amplifier is shown in Fig. 11.1(b). The negative halves of the sinusoid are provided by another transistor that also operates in the class B mode and conducts during the alternate half-cycles.

In a class AB amplifier, the transistor is biased at a nonzero DC current that is much smaller than the peak amplitude of the AC output current. The transistor conducts for slightly more than half a cycle of the input signal. The conduction angle is greater than 180° but much less than 360°; that is, $180° < \theta << 360°$. Thus, $i_C = I_C + I_p \sin \omega t$ and $I_C < I_p$. The waveform of the collector current for a class AB amplifier is shown in Fig. 11.1(c). The negative halves of the sinusoid are provided by another transistor that also operates in the class AB mode and conducts for an interval slightly greater than the negative half-cycle. The currents from the two transistors are combined to form the load current. Both transistors conduct for an interval near the zero crossings of the input signal.

In a class C amplifier, as shown in Fig. 11.1(d) the transistor conducts for an interval shorter than a half-cycle. The conduction angle of the transistor is less than 180°; that is, $\theta < 180°$ and $i_C = I_p \sin \omega t$. The negative halves of the collector current are provided by another transistor. The collector current is of pulsating type and is much more distorted than the current generated by other classes of amplifier. The nonlinear distortion can be filtered out by passing this output through a parallel *LC*-resonant circuit. The resonant circuit is tuned to the frequency of the input signal and acts as a band-pass filter, giving an output voltage proportional to the amplitude of the fundamental component of the current waveform. Class C amplifiers are normally used in radio frequency applications.

A class D amplifier, shown in Fig. 11.1(e), operates transistors as switches, which are either completely turned on or completely turned off. The transistor output is connected to the supply voltage via a large inductor L and also to the load via a serial *LC* circuit (not shown). The output voltage is a pulsed waveform in response to a pulse-width modulation (PWM) input voltage. When the transistor is off, the inductor L supplies its stored energy to the series *LC* circuit and the load. When the transistor is switched on, the inductor L replenishes its energy from the supply voltage.

A class E amplifier, shown in Fig. 11.1(f), is similar to the class D amplifier, except that a capacitor C_p is connected across the transistor. As a result, there are *LC* circuits when the transistor is switched on: one parallel $L-C_p$ circuit and one series *LC* circuit via the load.

Class A and class B amplifiers are commonly used in audio-frequency applications. Although there are many other types of amplifiers, we will consider the following kinds: emitter followers, class A amplifiers, class B push-pull amplifiers, complementary class AB push-pull amplifiers, quasi-complementary class AB push-pull amplifiers, transformer-coupled class AB push-pull amplifiers, and power op-amps.

KEY POINTS OF SECTION 11.2

- A power amplifier can be classified into one of six groups—A, B, AB, C, D, or E—depending on the conduction interval of the transistors used in the amplifier.
- To avoid distortion due to clipping, the maximum peak collector current of a transistor is limited to a specified value.

11.3 Power Transistors

The amplifying devices can be either MOSFETs or BJTs. The MOSFETs offer superior performances over the BJTs [3–5]. They require virtually zero input current, and they have faster switching times, no secondary breakdown, and stable gain and response time over a wide temperature range. The BJT current gain β_F can vary widely with temperature, and the variation of the MOSFET transconductance g_m with temperature is less than the variation of the BJT current gain β_F. As a result, power MOSFETs are replacing power BJTs in most applications.

There are no significant differences in the internal construction of small-signal and power BJTs. Although the mechanism of operation of power MOSFETs is same as that of small-signal MOSFETs, the power MOSFETs differ in their internal construction from the small-signal transistors in order to have more channel width for obtaining more current-carrying capability. To achieve a large channel width with good characteristics, power MOSFETs are fabricated with a repetitive pattern of small cells operating in parallel. The voltage ratings of power MOSFETs range from 50 V to 100 V, with current ratings from 10 A to 30 A.

There are two basic power MOSFET structures. The first is called a *DMOS* device, which uses a double-diffusion process; its cross section is shown in Fig. 11.2(a). It has two parallel current paths from the drain to the source. The *p*-substrate region is diffused deeper than the *n*+-source. The *n*-drift region must be moderately doped so that the drain breakdown voltage is sufficiently large and the thickness of the *n*-drift region is made as thin as possible to minimize drain resistance. The second structure, as shown in Fig. 11.2(b), uses a vertical channel known as a VMOS structure. In this case, the *p*-substrate diffusion is performed over the entire surface, and a V-shaped groove is then formed, extending through the *n*-drift region. For a high efficiency, on-resistance R_{on}, which should be low, can be found from

$$R_{DS(on)} = R_{SC} + R_{CH} + R_{DC} \tag{11.1}$$

where R_{SC} = source terminal contact resistance
R_{CH} = channel resistance
R_{DC} = drain terminal contact resistance

The values of the contact resistances R_{SC} and R_{DC} are proportional to the semiconductor resistivity. From Eq. (7.3), we can find the channel resistance R_{CH} in the linear region of operation as given by

$$R_{CH} = \frac{v_{DS}}{i_D} = \frac{L}{W\mu_n C_{ox}(v_{GS} - V_t)} \tag{11.2}$$

This explains why the channel length should be low and the width should be high to reduce on-resistance. An increase in the drain current increases the power loss and the junction temperature, which in turn increases the threshold voltage V_t and R_{CH}, thereby limiting the drain current.

For the sake of illustration of power amplifiers, we will show and analyze circuits using BJTs. The BJT analysis can be applied to MOSFET circuits by substituting $\beta_F = 0$ for the BJT current gain. Table 11.2 lists the circuit parameters for replacing a BJT by a MOSFET in a circuit.

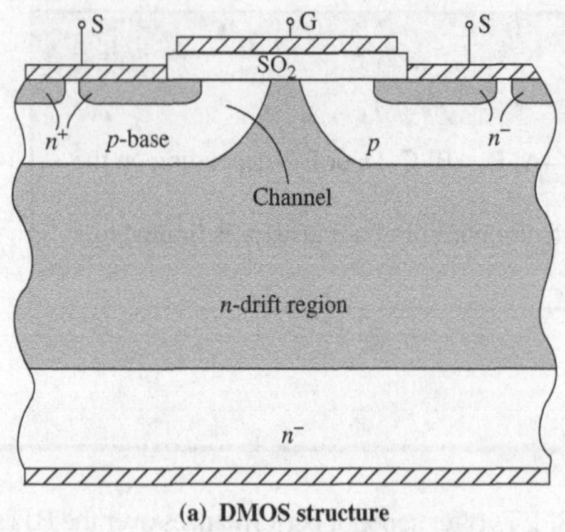

(a) DMOS structure

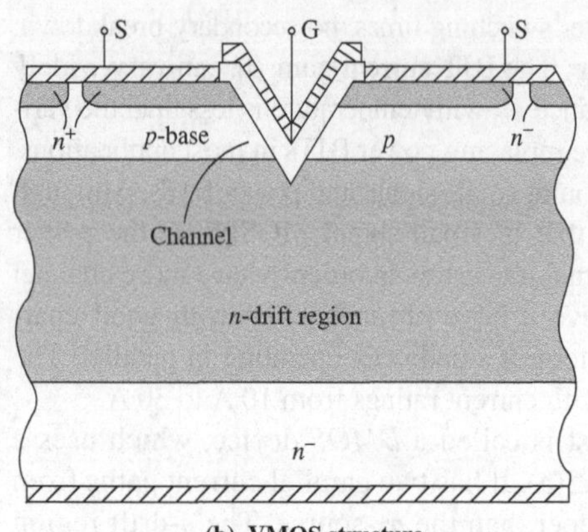

(b) VMOS structure

FIGURE 11.2 Cross section of a power MOSFET

TABLE 11.2 Circuit parameters of MOSFETs and BJTs

Circuit Parameters	MOSFETs	BJTs
Supply voltage	Power supply $\pm V_{DD}$	$\pm V_{CC}$
Output current	Drain current i_D	Collector current i_C
Driving voltage	Gate–source voltage v_{GS}	Base–emitter voltage v_{BE}
Input voltage	Gate voltage v_G	Base voltage v_B
Transistor input current	$i_G \cong 0$	Base current i_B
Current ratio	$\dfrac{i_D}{i_G} \simeq \infty$	$\dfrac{i_C}{i_B} = \beta_F$
Diode-connected transistor		

KEY POINTS OF SECTION 11.3

- The power MOSFETs offer superior performances over the power BJTs. MOSFETs require virtually zero input current and have faster switching times, no secondary breakdown, and stable gain and response time over a wide temperature range.
- There are two basic power MOSFET structures. The channel length of MOSFETs is made low, and the width is made high to reduce on-resistance.

11.4 Class A Amplifiers

The common-source (common-emitter) amplifiers in Secs. 7.8 and 7.9 and the source and emitter followers in Secs. 8.8 and 8.9 fall into the class A category. The transistors of class A amplifiers are always turned on. Therefore, the transistor can become hot, with most of the power provided by the power supply being dissipated as heat. Although efficiency is poor (around 20%), the accuracy of the signal amplifications is quite high. We have analyzed the amplifiers to derive the small-signal parameters without any consideration to the circuit efficiency. For power amplifiers, efficiency is a major performance parameter.

11.4.1 Emitter Followers

An emitter follower is a class A amplifier; its circuit diagram is shown in Fig. 11.3(a). Section 8.9 discussed the characteristics of an emitter follower: a very low output impedance, a very high input impedance, and a voltage gain of almost unity at a large value of load resistance. The voltage gain and the DC current of transistor Q_1 are affected by the values of load resistance R_L. The peak-to-peak voltage swing is less than V_{CC}. If the emitter resistance R_E in Fig. 11.3(a) can be replaced by a current source, as shown in Fig. 11.3(b), the peak-to-peak voltage swing can be increased to a value larger than V_{CC}. The voltage gain can be maintained at almost unity even with a small load resistance, on the order of 100 Ω.

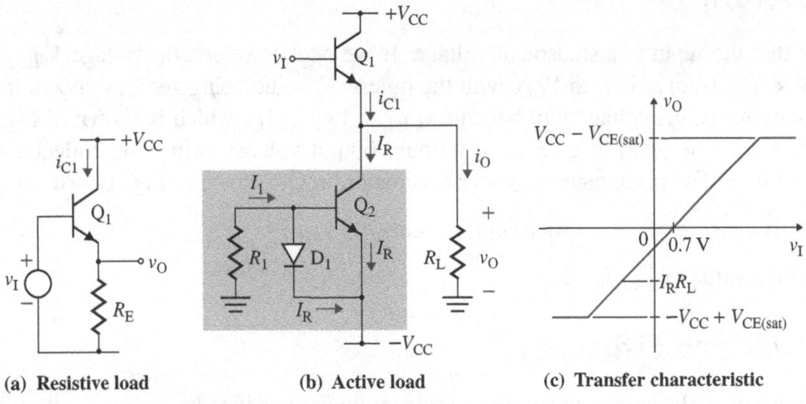

FIGURE 11.3 Emitter follower

Transfer Characteristic

Assuming that the diode drop is $V_{D1} = 0.7$ V, the B-E drop of a transistor is $V_{BE} = 0.7$ V, and the transistor current gain $\beta_F >> 1$, the reference current I_R can be approximated by I_1. Using KVL for the current source, we get

$$0 = I_R R_1 + V_{BE} - V_{CC}$$

which gives the reference current I_R as

$$I_R = \frac{V_{CC} - V_{BE}}{R_1} \tag{11.3}$$

The output voltage is given by

$$v_O = v_1 - V_{BE} \tag{11.4}$$

which yields the transfer characteristic shown in Fig. 11.3(c). At $v_I = 0$ (i.e., $v_O = -V_{BE}$), the characteristic has an offset voltage. Thus, for $i_{C1} > 0$, Q_1 will be on, and the positive peak value of output voltage is

$$+V_{O(max)} = V_{CC} - V_{CE(sat)} \tag{11.5}$$

Q_1 will be turned off when $i_{C1} = 0$ and $i_O = -I_R$. The output voltage can also be written as

$$v_O = i_O R_L = -I_R R_L$$

$$= -\frac{V_{CC} - V_{BE}}{R_1} R_L \tag{11.6}$$

If the value of R_L is less than R_1, the peak negative value of the output voltage will be less than the peak negative value of $V_{CC} - V_{CE(sat)}$ and the output will not be symmetrical. This will cause distortion (or clipping). Thus, the condition that avoids distortion and obtains the maximum output voltage swing is given by

$$R_L \geq R_1 \tag{11.7}$$

and the maximum voltage swing (peak to peak) without clipping is

$$V_{pp} = 2(V_{CC} - V_{CE(sat)}) \tag{11.8}$$

Signal Waveforms

Let us assume that the input is a sinusoidal voltage. If we neglect saturation voltage $V_{CE(sat)}$, the output voltage v_O can swing from $-V_{CC}$ to V_{CC}, with the quiescent value being zero, as shown in Fig. 11.4(a). The collector–emitter (C-E) voltage will become $v_{CE1} = V_{CC} - v_O$, which is shown in Fig. 11.4(b). Assuming that $I_R = I_Q$ is selected to give the maximum output voltage swing, the collector current i_{C1} is shown in Fig. 11.4(c). The instantaneous power dissipation in Q_1, shown in Fig. 11.4(d), is given by

$$P_{D1} \approx v_{CE1} i_{C1} = V_{CC}(1 - \sin \omega t) I_R (1 + \sin \omega t) \tag{11.9}$$

and has an average value of $V_{CC} I_R / 2$.

Output Power and Efficiency

The power efficiency of the output stage of an emitter follower is defined by

$$\eta = \frac{\text{Load power } P_L}{\text{Supply power } P_S} \tag{11.10}$$

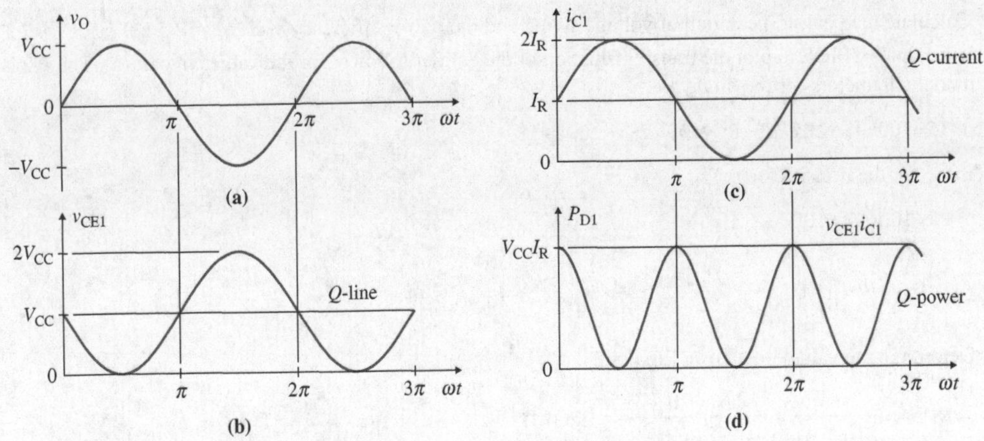

FIGURE 11.4 Signal waveforms of an emitter follower

Assuming that the output voltage is sinusoidal with a peak value of V_p, the average load power will be

$$P_L = \frac{V_p^2}{2R_L} \tag{11.11}$$

The average current drawn by transistor Q_1 will be I_R, and thus the average power drawn from the positive supply will be $V_{CC}I_R$. Since the current in transistor Q_2 remains constant at I_R, the power drawn from the negative supply will also be $V_{CC}I_R$, if we neglect the power drawn by the current source consisting of the diode D_1 and resistor R_1. Thus, the total average supply power will be

$$P_S = 2V_{CC}I_R \tag{11.12}$$

From Eqs. (11.11) and (11.12), we get the power efficiency as

$$\eta = \frac{V_p^2}{4R_L V_{CC} I_R} = \frac{1}{4}\left(\frac{V_p}{R_L I_R}\right)\left(\frac{V_p}{V_{CC}}\right) \tag{11.13}$$

which will give the maximum efficiency when

$$V_p = V_{CC} \leq R_L I_R$$

That is, $\eta_{max} = 25\%$, which is rather low. In practice, the peak output voltage is limited to less than V_{CC} to avoid transistor saturation and associated nonlinear distortion. Thus, the efficiency actually ranges from 10% to 20%. Emitter followers are generally used as output stages for high-frequency (≈10 MHz), low-power (≤1 W) amplifiers.

EXAMPLE 11.1

D

Designing an emitter follower

(a) Design the emitter follower of the circuit in Fig. 11.3(b). Assume $V_{CC} = 12$ V, $V_{BE} = 0.7$ V, $V_{CE(sat)} = 0.5$ V, $I_R = 5$ mA, and $R_L = 650\ \Omega$. Assume identical transistors of current gain $h_{fe} = \beta_F = 100$. Note that h_{fe} is the hybrid parameter and both h_{fe} and β_F represent the current gain of a BJT.

(b) Determine the critical value of load resistance to avoid clipping (or distortion).

(c) Calculate the peak-to-peak output voltage swing if $R_L = 650\ \Omega$.

(d) Calculate the peak-to-peak output voltage swing and the power efficiency η if $R_L = 2.5\ \text{k}\Omega$.

(e) Use PSpice/SPICE to plot the transfer function of the emitter follower for the values in part (a). The PSpice model parameters for the diode are

```
IS=100E-15 RS=16 BV=100 IBV=100E-15
```

and those for the transistors are

```
BF=100 VA=100
```

SOLUTION

(a) Determine the value of R_1 from Eq. (11.3):

$$R_1 = \frac{V_{CC} - V_{BE}}{I_R} = \frac{12\ \text{V} - 0.7\ \text{V}}{5\ \text{mA}} = 2260\ \Omega$$

(b) From Eq. (11.7), we find the critical value of load resistance to avoid clipping:

$$R_{L(crit)} = R_1 = 2260\ \Omega$$

(c) For $R_L = 650\ \Omega$, the negative peak output voltage is

$$-V_{O(max)} = -I_R R_L = -5\ \text{mA} \times 650\ \Omega = -3.25\ \text{V}$$

The positive peak output voltage is

$$+V_{O(max)} = V_{CC} - V_{CE(sat)} = 12\ \text{V} - 0.5\ \text{V} = 11.5\ \text{V}$$

Therefore, the peak-to-peak output voltage swing will be from 11.5 V to −3.25 V.

(d) For $R_L = 2.5\ \text{k}\Omega$ (which is greater than $R_{L(crit)} = 2260\ \Omega$), the negative peak output voltage will be limited to

$$-V_{O(max)} = -V_{CC} + V_{CE(sat)} = -12\ \text{V} + 0.5\ \text{V} = -11.5\ \text{V}$$

Therefore, the peak-to-peak output voltage swing will be from −11.5 V to +11.5 V. Thus, Eq. (11.13) gives

$$\eta = \frac{(11.5\ \text{V})^2}{4 \times 2.5\ \text{k}\Omega \times 12\ \text{V} \times 5\ \text{mA}} = 22\%$$

(e) The emitter-follower circuit for PSpice simulation is shown in Fig. 11.5.

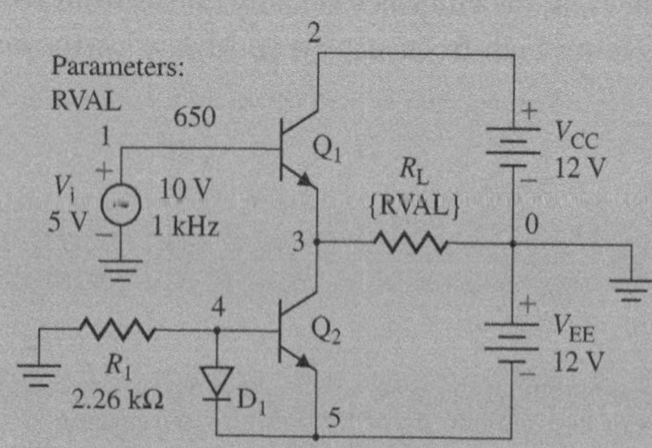

FIGURE 11.5 Emitter-follower circuit for PSpice simulation

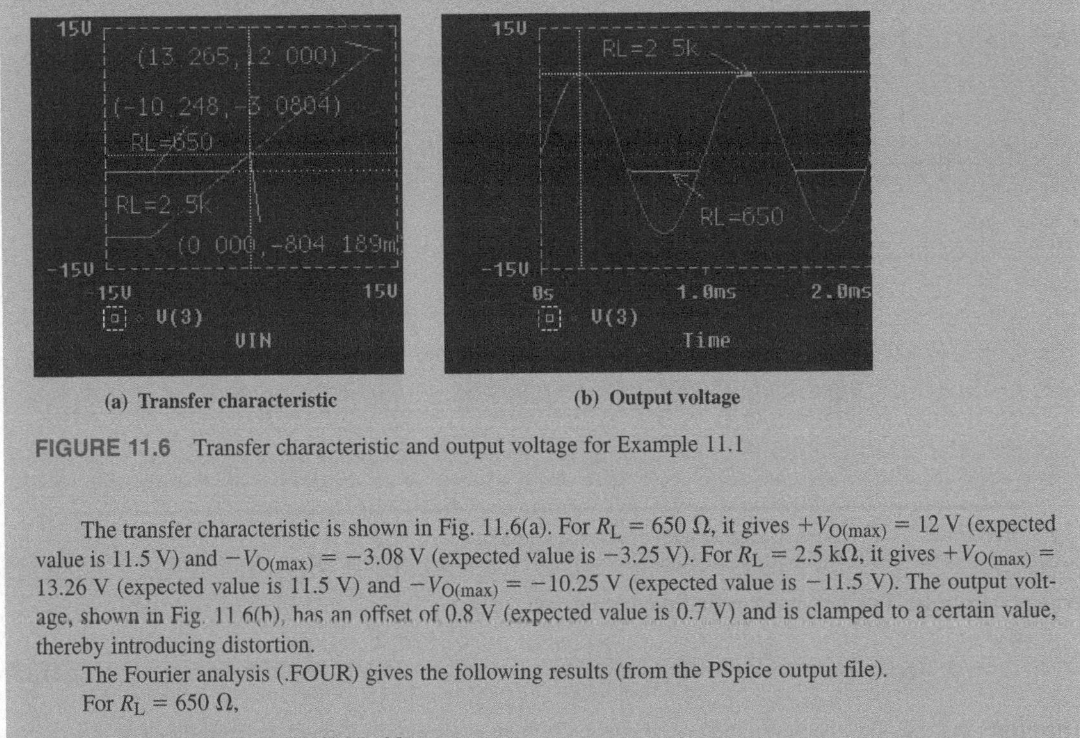

(a) Transfer characteristic (b) Output voltage

FIGURE 11.6 Transfer characteristic and output voltage for Example 11.1

The transfer characteristic is shown in Fig. 11.6(a). For $R_L = 650\ \Omega$, it gives $+V_{O(max)} = 12$ V (expected value is 11.5 V) and $-V_{O(max)} = -3.08$ V (expected value is -3.25 V). For $R_L = 2.5\ k\Omega$, it gives $+V_{O(max)} =$ 13.26 V (expected value is 11.5 V) and $-V_{O(max)} = -10.25$ V (expected value is -11.5 V). The output voltage, shown in Fig. 11.6(b), has an offset of 0.8 V (expected value is 0.7 V) and is clamped to a certain value, thereby introducing distortion.

The Fourier analysis (.FOUR) gives the following results (from the PSpice output file).

For $R_L = 650\ \Omega$,

```
DC COMPONENT=1.300972E+00
TOTAL HARMONIC DISTORTION=3.131888E+01 PERCENT
```

For $R_L = 2.5\ k\Omega$,

```
DC COMPONENT=-7.972313E-01
TOTAL HARMONIC DISTORTION=1.895552E-01 PERCENT
```

11.4.2 Basic Common-Emitter Amplifier

The simplest kind of class A amplifier is shown in Fig. 11.7(a). This configuration lacks biasing stability (i.e., emitter resistance R_E) and is not suitable for power amplifiers. However, we will use this circuit to derive the power efficiency of class A amplifiers. Let us assume that the nonlinearity introduced by the transistor is negligible so that the output signals will be sinusoidal for sinusoidal input signals. These assumptions will simplify the various power calculations. The waveforms of the collector current and voltages are shown in Fig. 11.7(b).

Transfer Characteristic

The input voltage v_I is related to the collector current I_C by

$$v_I = V_{BE} = V_T \ln\left(\frac{i_C}{I_S}\right) \tag{11.14}$$

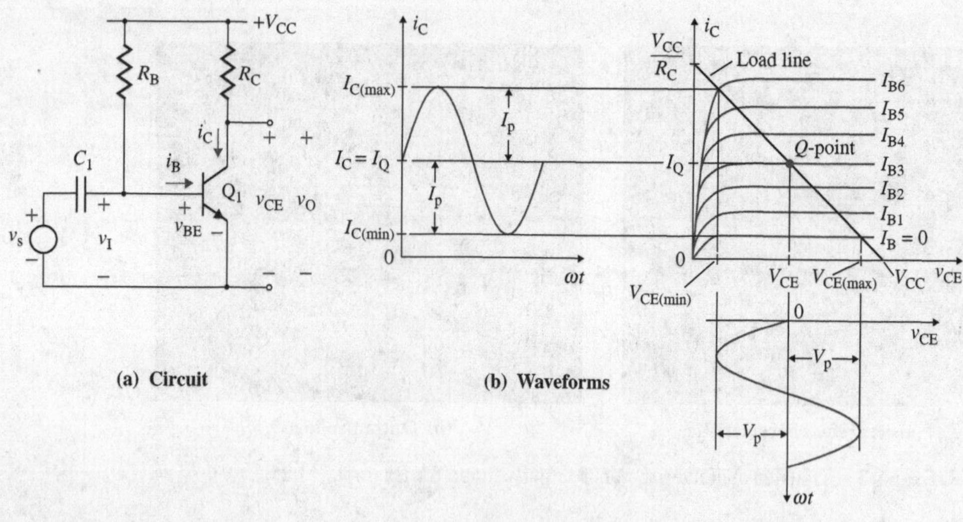

FIGURE 11.7 Basic common-emitter class A amplifier

Using i_C from Eq. (11.14), we can find the output voltage as

$$v_O = V_{CC} - R_C i_C = V_{CC} - R_C I_S \ln\left(\frac{v_I}{V_T}\right) \tag{11.15}$$

Therefore, the transfer characteristic (v_O versus v_I), which is shown in Fig. 11.8, is nonlinear.

Output Power and Efficiency

The average DC power required from the power supply is given by

$$P_S = V_{CC} I_C \tag{11.16}$$

The average load or output power is given by

$$P_L = \left(\frac{V_p}{\sqrt{2}}\right)\left(\frac{I_p}{\sqrt{2}}\right) = \frac{V_p I_p}{2} \tag{11.17}$$

$$= \frac{I_p R_L I_P}{2} = \frac{I_p^2 R_L}{2} \tag{11.18}$$

▶ **NOTE** $\sqrt{2}$ converts the peak value to an rms value.

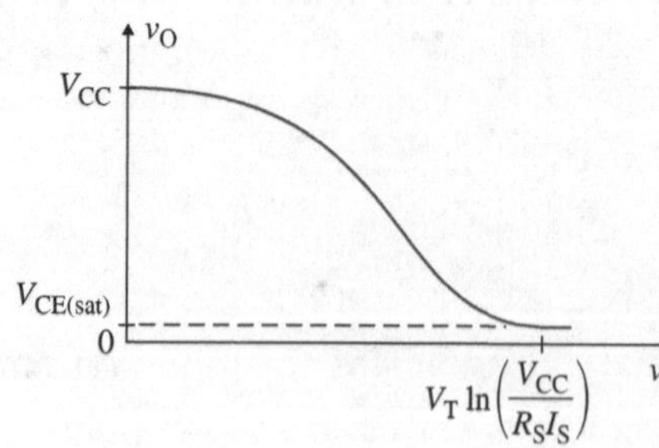

FIGURE 11.8 Transfer characteristic of a common-emitter class A amplifier

where V_p and I_p are the peak values of the AC output voltage and current, respectively. Using the minimum and maximum values of the output signals, we can express V_p and I_p as

$$V_p = \frac{V_{CE(max)} - V_{CE(min)}}{2} \tag{11.19}$$

$$I_p = \frac{I_{C(max)} - I_{C(min)}}{2} \tag{11.20}$$

where V_{CE} ideally extends over its full range. Using Eqs. (11.19) and (11.20), we can express Eq. (11.17) as

$$P_L = \frac{V_p I_p}{2} = \frac{(V_{CE(max)} - V_{CE(min)})(I_{C(max)} - I_{C(min)})}{8} \tag{11.21}$$

which will give the maximum load power $P_{L(max)}$ when $V_{CE(min)} = 0$, $I_{C(min)} = 0$, $V_{CE(max)} = V_{CC}$, and $I_{C(max)} = 2I_C$. Thus, Eq. (11.21) gives $P_{L(max)}$ as

$$P_{L(max)} = \frac{V_{CC}(2I_C)}{8} = \frac{V_{CC}I_C}{4} \tag{11.22}$$

The conversion efficiency, which is defined as the ratio of load power to the DC source power, is expressed as

$$\eta = \frac{P_L}{P_S} \times 100\% \tag{11.23}$$

Substituting Eqs. (11.16) and (11.21) into Eq. (11.23) gives the maximum efficiency as

$$\eta_{max} = \frac{(V_{CE(max)} - V_{CE(min)})(I_{C(max)} - I_{C(min)})}{8V_{CC}I_C} \tag{11.24}$$

which, for $V_{CE(min)} = 0$, $I_{C(min)} = 0$, $V_{CE(max)} = V_{CC}$, and $I_{C(max)} = 2I_C$, becomes

$$\eta_{max} = \frac{V_{CE(max)}I_{CE(max)}}{8V_{CC}I_C} = \frac{V_{CC}(2I_C)}{8V_{CC}I_C} = \frac{1}{4} = 25\% \tag{11.25}$$

Hence, the maximum efficiency for a class A amplifier under ideal conditions is 25%. Although in practice the actual efficiency will be less than 25%, this percentage is often used as a guideline for determining the biasing requirement I_C. For example, if $V_{CC} = 30$ V and $P_{L(max)} = 50$ W, then

$$P_S = \frac{P_{L(max)}}{\eta_{(max)}} = \frac{50}{0.25} = 200 \text{ W}$$

and

$$I_C = \frac{P_S}{V_{CC}} = \frac{200}{30} = 6.67 \text{ A}$$

The quality of an amplifier is often measured by the *figure of merit* F_m, which is defined by

$$F_m = \frac{\text{Maximum collector dissipation}}{\text{Maximum output power}} = \frac{P_{C(max)}}{P_{L(max)}} \tag{11.26}$$

The maximum collector dissipation is given by

$$P_{C(max)} = \frac{V_{CC}I_C}{2} \tag{11.27}$$

Substituting Eqs. (11.22) and (11.27) into Eq. (11.26) yields

$$F_m = \frac{V_{CC}I_C/2}{V_{CC}I_C/4} = 2 \tag{11.28}$$

Thus, the collector power dissipation is twice the maximum output power. That is, for a maximum output of 50 W, the collector must be able to dissipate at least 100 W. This requirement is the major disadvantage of class A amplifiers because it necessitates the use of a large and expensive heat sink to cool the transistors.

11.4.3 Common-Emitter Amplifiers with Active Load

Because of their high voltage gain, common-emitter stages are often used as output-stage drivers in integrated circuit design. A common-emitter stage is shown in Fig. 11.9(a). A current source consisting of Q_2 and Q_3 establishes the reference current I_R, which is given by

$$I_R = \frac{V_{CC} - V_{EB2}(= V_{EB3})}{R_1} \tag{11.29}$$

With no load, $R_L = \infty$, $i_O = 0$, and $i_{C1} = I_Q = I_R$. Thus, the load current i_O is given by

$$i_O = I_R - i_{C1} \tag{11.30}$$

where the collector current i_{C1} is related to the input voltage v_I by

$$i_{C1} = I_S \exp\left(\frac{v_I}{V_T}\right) \tag{11.31}$$

Transfer Characteristic

The output voltage v_O is given by

$$\begin{aligned} v_O &= R_L i_O = R_L(I_R - i_{C1}) \\ &= R_L\left[I_R - I_S \exp\left(\frac{v_I}{V_T}\right)\right] \end{aligned} \tag{11.32}$$

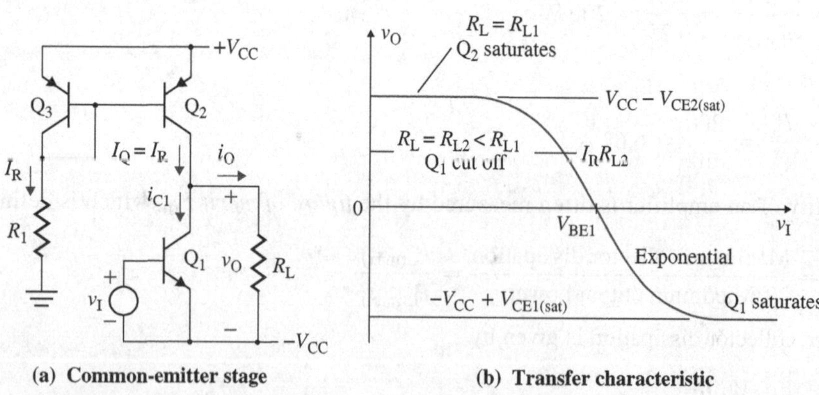

FIGURE 11.9 Common-emitter class A amplifier and its transfer characteristic

which is the transfer characteristic (v_O versus v_I), shown in Fig. 11.9(b). When $i_{C1} = 0$, transistor Q_1 is cut off and transistor Q_2 supplies I_Q to the load. The output voltage becomes

$$v_O = R_L I_R = R_L \frac{V_{CC} - V_{BE2}}{R_1} \tag{11.33}$$

If this value of v_O is less than the maximum possible positive output voltage of $V_{CC} - V_{CE2(sat)}$, distortion will occur, as shown in Fig. 11.9(b). Thus, the condition for the maximum positive output voltage swing is $R_L \geq R_1$. As the input voltage v_I is increased, the current in transistor Q_1 increases and v_O becomes negative according to Eq. (11.32), until Q_1 saturates and the output voltage becomes

$$v_O = -V_{CC} + V_{CE1(sat)}$$

The transfer characteristic is basically a simple exponential and shows distortion on the output voltage. The input voltage required to produce the maximum output voltage swing is typically a few tens of millivolts or less.

Output Power and Efficiency

All the equations derived for the emitter follower and the basic common-emitter circuits apply to this common-emitter stage. Thus, the maximum efficiency is $\eta_{max} = 25\%$, and the figure of merit is $F_m = 2$. The breakdown voltage rating of Q_1 and Q_2 must be $2V_{CC}$.

11.4.4 Transformer-Coupled Load Common-Emitter Amplifier

The efficiency of an amplifier can be improved with a transformer-coupled load. A class A amplifier with a transformer-coupled load is shown in Fig. 11.10(a). The elimination of the collector resistance R_C, used for DC biasing in Fig. 11.7(a), accounts for the increase in efficiency. The transformer at the output stage provides an impedance match in order to transfer maximum power to the load. A load such as the impedance of a loudspeaker is usually very small, typically 4 Ω to 16 Ω.

The voltage and current relations of the output transformer are

$$V_{pL} = \left(\frac{n_p}{n_s}\right) V_{sL} \quad \text{and} \quad I_{pL} = \left(\frac{n_s}{n_p}\right) I_{sL}$$

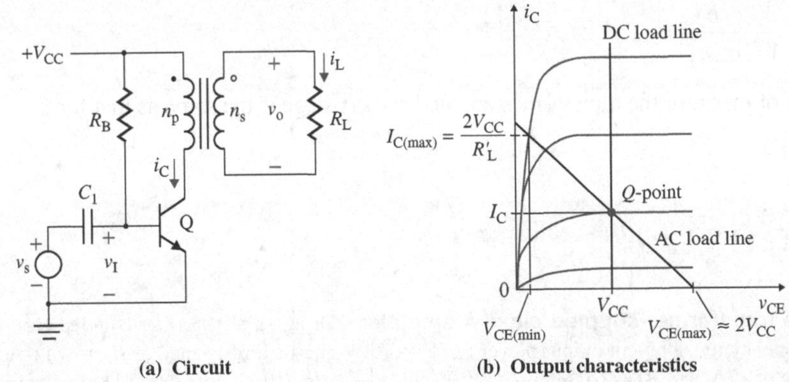

(a) Circuit

(b) Output characteristics

FIGURE 11.10 Class A amplifier with transformer-coupled load

where n_s and n_p = secondary and primary windings, respectively
V_{sL} and V_{pL} = secondary and primary voltages, respectively
I_{sL} and I_{pL} = the secondary and primary currents, respectively

The effective load resistance referred to the primary side can be found from

$$R'_L = \left(\frac{V_{pL}}{I_{pL}}\right) = \left(\frac{n_p}{n_s}\right)^2\left(\frac{V_{sL}}{I_{sL}}\right) = \left(\frac{n_p}{n_s}\right)^2 R_L \tag{11.34}$$

The AC (dynamic) load line is determined by R'_L. The DC (static) load line is almost vertical because of the very small primary resistance of the transformer. Assuming $V_{CE(min)} = 0$ in Eq. (11.19) and $I_{C(min)} = 0$ in Eq. (11.20), the peak values of the output voltage and current at the primary side of the transformer are

$$V_p = \frac{V_{CE(max)}}{2} = V_{CC} \tag{11.35}$$

$$I_p = \frac{I_{C(max)}}{2} = I_C \tag{11.36}$$

Equation (11.21) gives the maximum efficiency as

$$\eta_{max} = \frac{V_{CE(max)}I_{C(max)}}{8V_{CC}I_C} = \frac{2V_{CC}(2I_C)}{8V_{CC}I_C} = \frac{1}{2} = 50\% \tag{11.37}$$

Hence, the maximum efficiency of a class A stage is doubled by using a transformer coupled to the load. The value of V_p for a transformer-coupled stage is V_{CC}; for the basic common-emitter amplifier, it is only $V_{CC}/2$.

Equation (11.18) gives the maximum load power as

$$P_{L(max)} = \frac{I_p^2 R'_L}{2} = \frac{V_{CC}^2}{2R'_L} \tag{11.38}$$

The maximum collector dissipation is given by

$$P_{C(max)} = V_{CC}I_C = V_{CC}I_p = \frac{V_{CC}^2}{R'_L} \tag{11.39}$$

Substituting Eqs. (11.38) and (11.39) into Eq. (11.26) yields

$$F_m = \frac{V_{CC}^2/R'_L}{V_{CC}^2/2R'_L} = 2 \tag{11.40}$$

Thus, the figure of merit for the transformer-coupled class A stage is the same as that for the basic common-emitter stage.

EXAMPLE 11.2

D **Designing a transformer-coupled class A amplifier** Design a transformer-coupled class A amplifier with high efficiency to supply an output power of $P_L = 10$ W at a load resistance of $R_L = 4\ \Omega$. Assume a DC supply voltage of 12 V and BJTs of $\beta_F = h_{fe} = 100$ and $V_{CE(sat)} = 0.7$ V. Note that h_{fe} is the hybrid parameter and both β_f ($\approx\beta_F$) and h_{fe} represent the current gain of a BJT.

SOLUTION

The design steps are as follows:

Step 1. Determine the maximum collector-to-emitter voltage of the transistors:

$$V_{CE(max)} \geq 2V_{CC} = 2 \times 12 = 24 \text{ V}$$

Step 2. From Eq. (11.40), calculate the collector power dissipation, which must be at least twice the AC power:

$$P_C \geq F_m P_L = 2P_L = 2 \times 10 = 20 \text{ W}$$

Step 3. Calculate the value of the quiescent collector current I_C:

$$I_C = \frac{P_C}{V_{CC}} = \frac{20}{12} = 1.67 \text{ A}$$

Step 4. Calculate the slope of the load line to find the AC load resistance R'_L:

$$R'_L = \frac{V_{CC}}{I_C} = \frac{12}{1.67} = 7.19\ \Omega$$

Step 5. From Eq. (11.34), calculate the required turns ratio of the transformer:

$$\frac{n_p}{n_s} = \left(\frac{R'_L}{R_L}\right)^{1/2} = \left(\frac{7.19}{4}\right)^{1/2} = 1.34$$

Step 6. Calculate the peak collector current:

$$I_{C(max)} = 2I_C = 2 \times 1.67 = 3.34 \text{ A}$$

Step 7. Calculate the quiescent base current I_B:

$$I_B = \frac{I_C}{h_{fe}} = \frac{1.67}{100} = 16.7 \text{ mA}$$

Step 8. Calculate the base resistance R_B:

$$R_B = \frac{V_{CC} - V_{CE(sat)}}{I_B} = \frac{12 \text{ V} - 0.7 \text{ V}}{16.7 \text{ mA}} = 677\ \Omega$$

KEY POINTS OF SECTION 11.4

- An emitter follower biased with an active current source is the most commonly used output stage.
- An emitter follower has a high input impedance and a low gain of almost unity. However, it exhibits an offset voltage of approximately $-V_{BE} \approx -0.7$ V at $V_i = 0$.
- In a class A amplifier, the transistor conducts a continuous DC biasing current. As a result, the maximum power efficiency is only 25%. The figure of merit, which is the ratio of maximum collector power dissipation to maximum output power, is 2.
- The power efficiency of a class A stage can be increased to 50% with a transformer-coupled load.

11.5 Class B Push-Pull Amplifiers

In a class B push-pull amplifier, two complementary transistors (one *npn* transistor and one *pnp* transistor) are employed to perform the push-pull operation [1, 6]. The efficiency is much higher than that of a class A amplifier, and the accuracy of the output is improved. It is the most commonly used audio-power amplifier. This section discusses two types of class B amplifiers.

11.5.1 Complementary Push-Pull Amplifiers

A complementary push-pull amplifier is shown in Fig. 11.11(a). For $v_I > 0$, transistor Q_P remains off and transistor Q_N operates as an emitter follower. For a sufficiently large value of v_I, Q_N saturates and the maximum positive output voltage becomes

$$V_{CE(max)} = V_{CC} - V_{CE1(sat)}$$

For $v_I < 0$, transistor Q_N remains off and transistor Q_P operates as an emitter follower. For a sufficiently large negative value of v_I, Q_P saturates and the maximum negative output voltage becomes

$$-V_{CE(max)} = -(V_{CC} - V_{CE2(sat)}) = -V_{CC} + V_{CE2(sat)}$$

Assuming identical transistors of $V_{BE1} = V_{BE2} = V_{BE}$, the output voltage is given by

$$v_O = v_I - V_{BE} \quad \text{for} \quad -0.7\text{ V} \geq v_I \geq 0.7\text{ V} \tag{11.41}$$

which gives the transfer characteristic of v_O versus v_I shown in Fig. 11.11(b). However, during the interval $-0.7\text{ V} \leq v_I \leq 0.7\text{ V}$, both Q_P and Q_N remain off, and $v_O = 0$. This causes a dead zone and crossover distortion on the output voltage, as illustrated in Fig. 11.12.

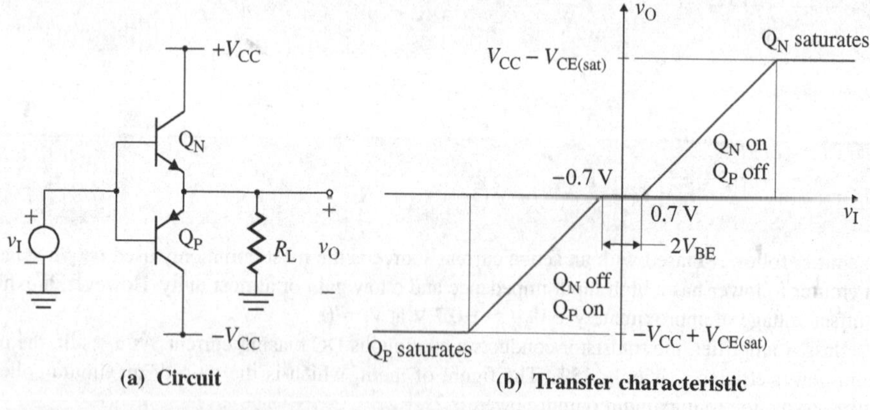

FIGURE 11.11 Complementary class B push-pull amplifier

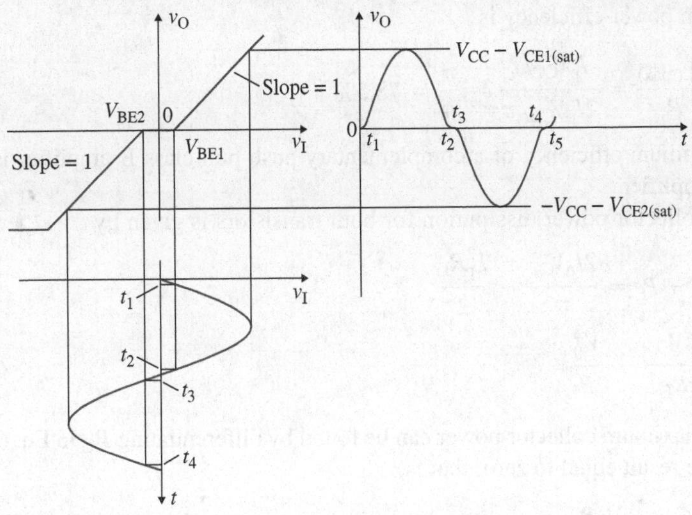

FIGURE 11.12 Crossover distortion on input and output waveforms

Output Power and Efficiency

Let us assume that $V_{CE1(sat)} = V_{CE2(sat)} = V_{CE(sat)} = 0$ and $I_{C(min)} = 0$. Assuming a sinusoidal variation of the collector current $i_{C1} = I_p \sin \omega t$, the average collector current of a transistor can be found from

$$I_{C1} = \frac{1}{2\pi}\int_0^{\pi} i_{c1}\,dt = \frac{1}{2\pi}\int_0^{\pi} I_p \sin(\omega t)\,d(\omega t) = \frac{I_p}{\pi} \tag{11.42}$$

The average current drawn from the DC supply source by transistors Q_N and Q_P is

$$I_{dc} = 2I_{C1} = \frac{2I_p}{\pi} \tag{11.43}$$

Thus, the average input power supplied from the DC source is

$$P_S = I_{dc}V_{CC} = \frac{2I_pV_{CC}}{\pi} \tag{11.44}$$

From Eq. (11.17), the output power is given by

$$P_L = \frac{I_p^2 R_L}{2} = \frac{I_pV_p}{2}$$

Thus, the power efficiency becomes

$$\eta = \frac{P_L}{P_S} = \frac{I_pV_p/2}{2I_pV_{CC}/\pi} = \frac{\pi}{4}\left(\frac{V_p}{V_{CC}}\right) \tag{11.45}$$

which gives $\eta = 50\%$ at $V_p = 2V_{CC}/\pi$ and $\eta = 78.5\%$ at $V_p = V_{CC}$. At $V_p = V_{CC}$, the maximum output power is given by

$$P_{L(max)} = \frac{I_p^2R_L}{2} = \frac{I_pV_p}{2} = \frac{I_pV_{CC}}{2} = \frac{V_{CC}^2}{2R_L} \tag{11.46}$$

Thus, the maximum power efficiency is

$$\eta_{\max} = \frac{P_{L(\max)}}{P_S} = \frac{I_p V_{CC}/2}{2I_p V_{CC}/\pi} = \frac{\pi}{4} = 78.5\%$$

Therefore, the maximum efficiency of a complementary push-pull class B amplifier is much higher than that of a class A amplifier.

The average collector power dissipation for both transistors is given by

$$2P_C = P_S - P_L = \frac{2I_p V_{CC}}{\pi} - \frac{I_p^2 R_L}{2} \tag{11.47}$$

$$= \frac{2V_p V_{CC}}{\pi R_L} - \frac{V_p^2}{2R_L} \tag{11.48}$$

The condition for maximum collector power can be found by differentiating P_C in Eq. (11.47) with respect to I_p and setting the result equal to zero; that is,

$$\frac{dP_C}{dI_p} = \frac{2V_{CC}}{\pi} - \frac{2I_p R_L}{2} = 0$$

which gives the peak current for maximum collector power dissipation as

$$I_{p(\max)} = \frac{2V_{CC}}{\pi R_L} \tag{11.49}$$

and the corresponding maximum peak voltage as

$$V_{p(\max)} = I_p R_L = \frac{2V_{CC}}{\pi} \tag{11.50}$$

Substituting $I_{p(\max)}$ from Eq. (11.49) and $V_{p(\max)}$ from Eq. (11.50) into Eq. (11.47) gives the maximum collector dissipation as

$$2P_{C(\max)} = \frac{4V_{CC}^2}{\pi^2 R_L} - \frac{2V_{CC}^2}{\pi^2 R_L} = \frac{2V_{CC}^2}{\pi^2 R_L} \tag{11.51}$$

Normalizing P_C in Eq. (11.48) with respect to $P_{C(\max)}$ in Eq. (11.51), we get

$$\frac{P_C}{P_{C(\max)}} = \pi\left(\frac{V_p}{V_{CC}}\right) - \frac{\pi^2}{4}\left(\frac{V_p}{V_{CC}}\right)^2 \tag{11.52}$$

which becomes 100% at $V_p = 2V_{CC}/\pi$ and 67.4% at $V_p = V_{CC}$. The normalized plot of P_C (with respect to $P_{C(\max)}$) versus peak voltage V_p is shown in Fig. 11.13. The power dissipation is highest at 50% efficiency.

We can obtain the figure of merit from Eqs. (11.46) and (11.51) as follows:

$$F_m = \frac{P_{C(\max)}}{P_{L(\max)}} = \frac{V_{CC}^2/\pi^2 R_L}{V_{CC}^2/2R_L} = \frac{2}{\pi^2} \approx \frac{1}{5} = \frac{2}{10} = 20\% \tag{11.53}$$

Thus, the figure of merit for class B amplifiers exceeds that of class A amplifiers by a factor of 10. The power dissipation rating of the individual transistors is only approximately one-fifth of the output power, and this results in much smaller heat sinks, which are normally needed to keep the junction temperature of power transistors within the maximum permissible limit.

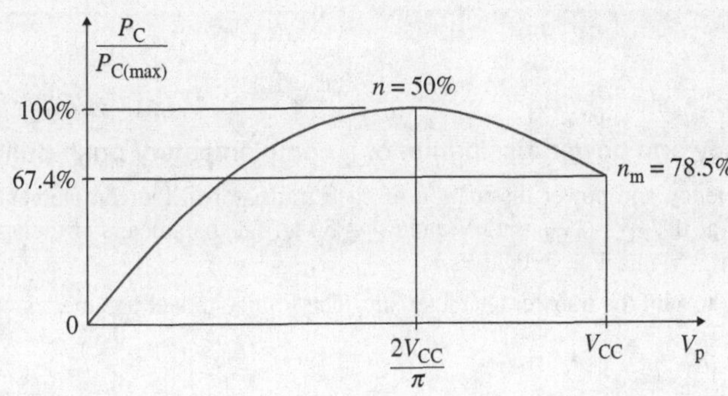

FIGURE 11.13 Power dissipation versus peak output voltage

Dead-Zone Minimization

The dead zone can be reduced to practically zero by using feedback with an op-amp, as shown in Fig. 11.14(a). The op-amp is connected in unity-gain mode with series-shunt feedback. With this arrangement, either Q_P or Q_N will be on if v_I and v_O differ by $\pm V_{BE}/A$, where A is the open-loop gain of the op-amp. Thus, for $A = 10^5$ and $V_{BE} = 0.7$, the dead zone will be reduced to less than $\pm(0.7/10^5) = \pm 7\ \mu V$. The transfer characteristic is shown in Fig. 11.14(b). Resistance R_1 limits the current drawn by the transistors from the op-amp output. It also provides the base current necessary for the load current I_L. Since the emitter current of a transistor is related to the base current by a factor of $(1 + h_{fe})$, the maximum value of R_1 can be found approximately from

$$I_B(1 + h_{fe}) = \frac{V_{BE}(1 + h_{fe})}{R_1} \geq \frac{V_{O(max)}}{R_L} = \frac{V_{CC} - V_{CE(sat)}}{R_L}$$

which gives the maximum value of R_1 as

$$R_1 \leq \frac{V_{BE}(1 + h_{fe})R_L}{V_{CC} - V_{CE(sat)}} \tag{11.54}$$

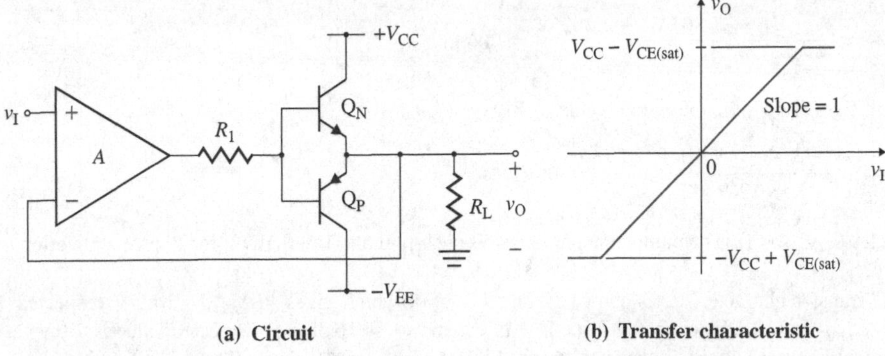

FIGURE 11.14 Minimization or elimination of the dead zone with feedback

EXAMPLE 11.3

Finding the efficiency and power dissipation of a complementary push-pull amplifier

(a) Calculate the efficiency and power dissipation of each transistor in the complementary push-pull output stage in Fig. 11.14(a) if $V_{CC} = V_{EE} = 12$ V and $R_L = 50\ \Omega$. The parameters of the transistors are $\beta_F = h_{fe} = 100$, $V_{CE(sat)} = 0.2$ V, and $V_{BE} = 0.72$ V.

(b) Use PSpice/SPICE to plot the transfer characteristic. The PSpice model parameters of the transistors are

```
BF=100 VJE=0.7V
```

SOLUTION

(a) The peak load voltage is

$$V_p = V_{CC} - V_{CE(sat)} = 12 - 0.2 = 11.8 \text{ V}$$

The peak load current is

$$I_p = \frac{V_p}{R_L} = \frac{11.8 \text{ V}}{50\ \Omega} = 0.236 \text{ A}$$

From Eq. (11.44), the DC power from the supply source is

$$P_S = \frac{2I_pV_{CC}}{\pi} = \frac{2 \times 0.236 \text{ A} \times 12 \text{ V}}{\pi} = 1.803 \text{ W}$$

From Eq. (11.17), the output power is

$$P_L = \frac{I_pV_p}{2} = \frac{0.236 \text{ A} \times 11.8 \text{ V}}{2} = 1.392 \text{ W}$$

Thus, the power efficiency is

$$\eta = \frac{P_L}{P_S} = \frac{1.392 \text{ W}}{1.803 \text{ W}} = 77.2\%$$

The power dissipation of each transistor can be found from

$$P_C = \frac{P_S - P_L}{2} = \frac{1.803 \text{ W} - 1.392 \text{ W}}{2} = 206 \text{ mW}$$

(b) From Eq. (11.54), the maximum value of R_1 is

$$R_1 \le \frac{0.72 \times (1 + 100) \times 50}{12 - 0.2} = 308\ \Omega$$

We will let $R_1 = 300\ \Omega$. The complementary class B push-pull amplifier circuit for PSpice simulation is shown in Fig. 11.15.

The transfer characteristic is shown in Fig. 11.16(a), which gives $V_{O(max)} = 10.1$ V (expected value is 11.8), $V_{O(min)} = -10.1$ V (expected value is -11.8), and $v_O = 16.38\ \mu$V (expected value is 7 μV) at $v_I = 0$. The input and output voltages are shown in Fig. 11.16(b) for $v_I = 10 \sin (2000\pi t)$.

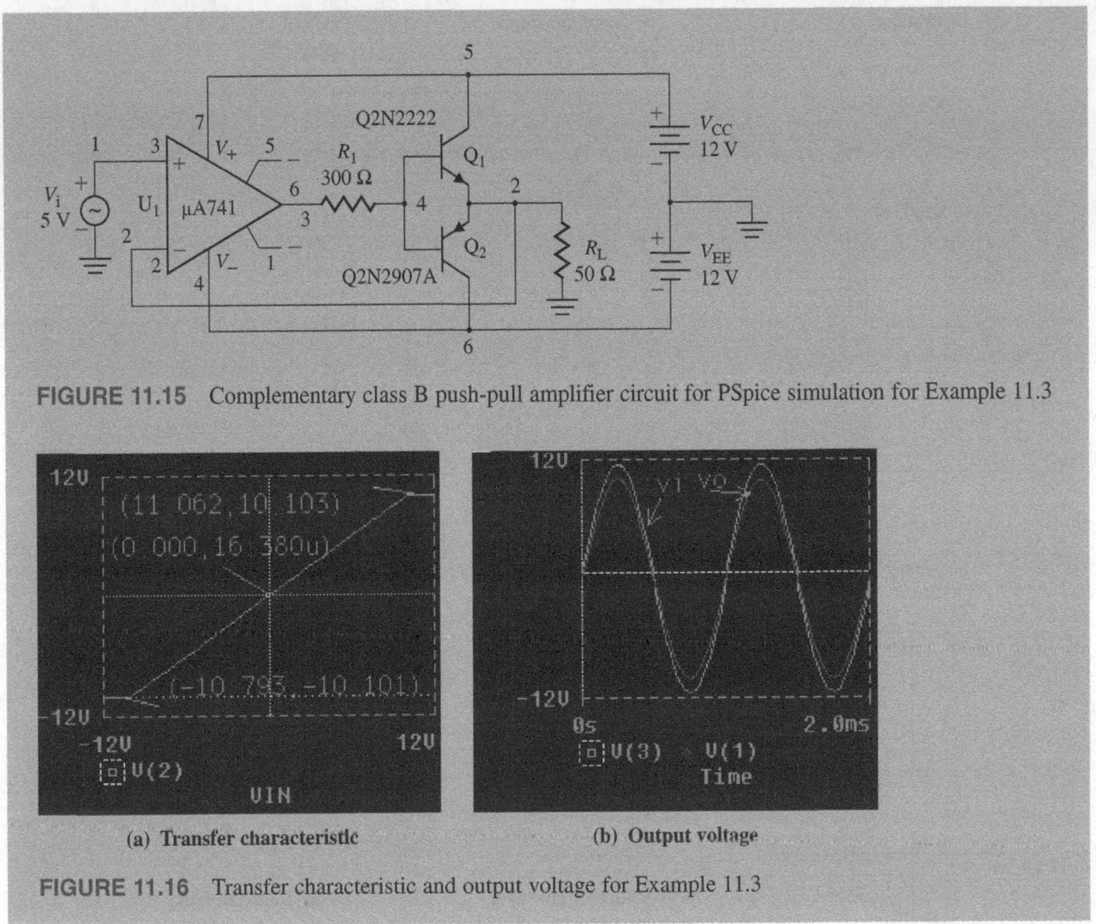

FIGURE 11.15 Complementary class B push-pull amplifier circuit for PSpice simulation for Example 11.3

(a) Transfer characteristic (b) Output voltage

FIGURE 11.16 Transfer characteristic and output voltage for Example 11.3

11.5.2 Transformer-Coupled Load Push-Pull Amplifier

The power efficiency of an amplifier can be improved considerably by using a transformer-coupled class B push-pull configuration, as shown in Fig. 11.17(a). The amplifier has three stages: input transformer TX_1, the gain stage of transistors Q_1 and Q_2, and output transformer TX_2. Resistance R_B and a battery V_{BB} are used to provide DC-biasing voltage V_{BE} for the transistors. Two transistors are employed to perform the push-pull operation. For $v_I > 0$, the base of transistor Q_1 is positive and that of Q_2 becomes negative because of the transformer action. The *npn* transistor Q_2 remains off, and the *npn* transistor Q_1 operates as an amplifier. For $v_I < 0$, the base of transistor Q_2 is positive and that of Q_1 becomes negative because of the transformer action. Transistor Q_1 remains off, and transistor Q_2 operates as an amplifier. The input transformer TX_1 of the input stage supplies a virtually distortion-free input signal and matches the output impedance of the driver stage to the input impedance of the output stage.

Signal Waveforms

Assuming the input current is sinusoidal, the collector currents of the transistors are as shown in Fig. 11.17(b). The load current shown in Fig. 11.17(c) is composed of the two collector currents i_{C1} and i_{C2}. The load

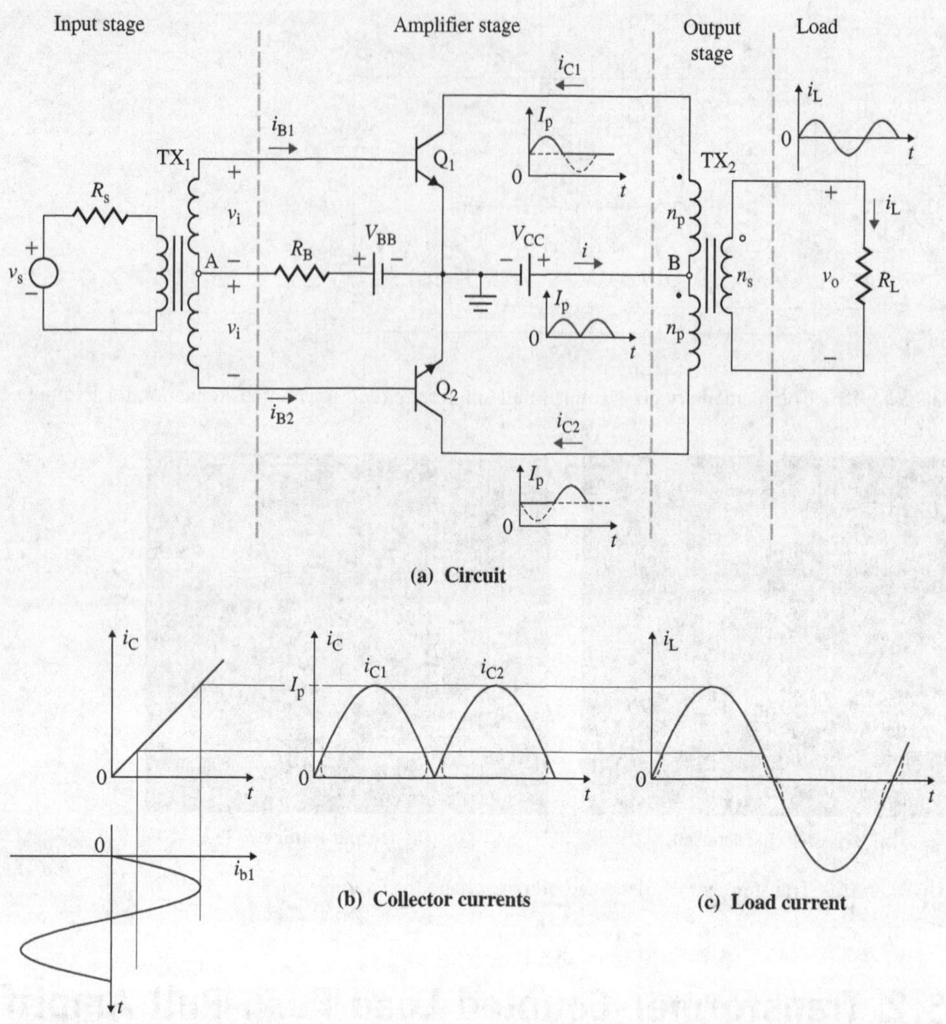

FIGURE 11.17 Transformer-coupled class B push-pull amplifier

current is distorted near zero crossings because the transistors are nonlinear devices and because the base current $i_B = 0$ (and hence $i_C = 0$) for $V_{BE} \leq 0.7$ V. This distortion, shown in Fig. 11.17(c) by dashed lines, is usually referred to as *crossover distortion.*

Output Power and Efficiency

The AC load line for a single transistor (Q_N) is shown in Fig. 11.18. The maximum peak current of a transistor is V_{CC}/R_L'. The average current drawn from the DC supply source by transistors Q_N and Q_P is

$$I_{dc} = 2I_{C1} = \frac{2I_p}{\pi} \tag{11.55}$$

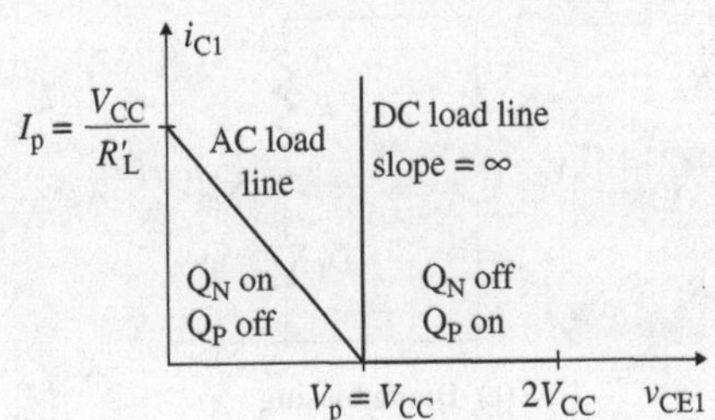

FIGURE 11.18 Load line of a single transistor

Thus, the average input power supplied from the DC source is

$$P_{dc} = I_{dc}V_{CC} = \frac{2I_pV_{CC}}{\pi} \tag{11.56}$$

From Eq. (11.18), the maximum output power is

$$P_{L(max)} = \frac{I_p^2R'_L}{2} = \frac{I_pV_p}{2} = \frac{I_pV_{CC}}{2} = \frac{V_{CC}^2}{2R'_L} \tag{11.57}$$

where the effective load resistance R'_L (referred to as the primary side of TX_2) is given by

$$R'_L = \left(\frac{n_p}{n_s}\right)^2 R_L \tag{11.58}$$

Thus, the maximum power efficiency is

$$\eta_{max} = \frac{P_{L(max)}}{P_{dc}} = \frac{I_pV_{CC}/2}{2I_pV_{CC}/\pi} = \frac{\pi}{4} = 78.5\% \tag{11.59}$$

Therefore, the maximum efficiency of a transformer-coupled push-pull class B amplifier is much higher than that of a class A amplifier.

The average collector power dissipation for both transistors is given by

$$2P_C = P_{dc} - P_L = \frac{2I_pV_{CC}}{\pi} - \frac{I_p^2R'_L}{2} \tag{11.60}$$

$$= \frac{2V_pV_{CC}}{\pi R'_L} - \frac{V_p^2}{2R'_L} \tag{11.61}$$

From Eq. (11.49), the peak current for maximum collector power dissipation is

$$I_{p(max)} = \frac{2V_{CC}}{\pi R'_L} \tag{11.62}$$

From Eq. (11.50), the peak voltage for maximum collector power dissipation is

$$V_{p(max)} = I_pR'_L = \frac{2V_{CC}}{\pi} \tag{11.63}$$

Substituting Eqs. (11.62) and (11.63) into Eq. (11.51) gives the maximum collector dissipation as

$$2P_{C(max)} = \frac{4V_{CC}^2}{\pi^2R'_L} - \frac{2V_{CC}^2}{\pi^2R'_L} = \frac{2V_{CC}^2}{\pi^2R'_L} \tag{11.64}$$

The power efficiency becomes 50% at $V_p = V_{p(max)}$. At $V_p = V_{CC}$, the maximum efficiency of 78.5% occurs.

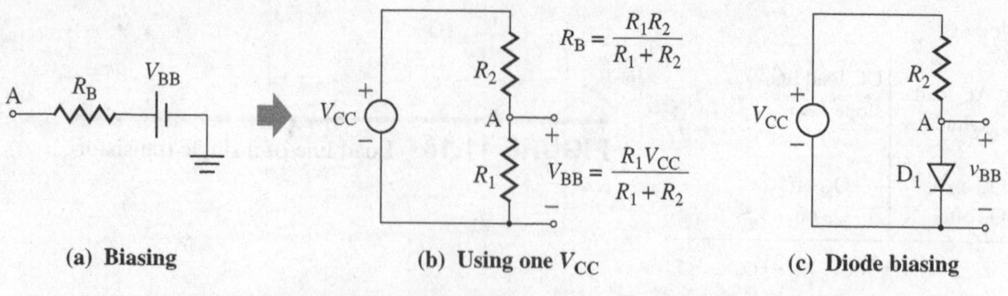

FIGURE 11.19 DC biasing of transformer-coupled class B push-pull amplifier

We can obtain the figure of merit from Eqs. (11.57) and (11.64) as follows:

$$F_m = \frac{P_{C(max)}}{P_{L(max)}} = \frac{V_{CC}^2/\pi^2 R_L'}{V_{CC}^2/2R_L'} = \frac{2}{\pi^2} \approx \frac{1}{5} = 20\% \tag{11.65}$$

which is the same as that for a complementary push-pull amplifier. The figure of merit for transformer-coupled amplifiers exceeds that of class A amplifiers by a factor of 10.

DC Biasing

Resistor R_B and battery V_{BB}, shown in Fig. 11.19(a), provide the B-E DC voltage V_{BE}, which is approximately 0.7 V for a silicon transistor. In practice, a V_{CC} supply with a suitable voltage divider is used rather than a separate supply, as shown in Fig. 11.19(b). R_1 and R_2 are chosen so that $V_{BE} \approx 0.7$ V (for silicon transistors). The parallel combination of R_1 and R_2 is kept as small as possible so that voltage drop $2R_BI_{B1} << V_{BE}$, which is generally satisfied by choosing $2R_BI_{B1} = 0.1V_{BE}$. Since the diode drop is similar to the B-E voltage of a transistor, a silicon diode is often used instead of resistance R_1, as shown in Fig. 11.19(c).

EXAMPLE 11.4

D

Designing a transformer-coupled class B amplifier Design a transformer-coupled class B push-pull amplifier, as shown in Fig. 11.17(a), to supply a maximum output power of $P_{L(max)} = 10$ W at a load resistance of $R_L = 4\ \Omega$. Assume a DC supply voltage of 15 V and transistors of $\beta_F = h_{fe} = 100$ and $V_{BE} = 0.7$ V.

SOLUTION

The design steps are as follows:

Step 1. Determine the maximum collector-to-emitter voltage of the transistors:

$$V_{CE(max)} \geq 2V_{CC} = 2 \times 15 \text{ V} = 30 \text{ V}$$

Step 2. From Eq. (11.57), calculate the effective load resistance:

$$R_L' = \frac{V_{CC}^2}{2P_{L(max)}} = \frac{(15 \text{ V})^2}{2 \times 10 \text{ W}} = 11.25\ \Omega$$

Step 3. Calculate the peak current of each transistor:

$$I_p = \frac{V_{CC}}{R'_L} = \frac{15\text{ V}}{11.25\ \Omega} = 1.33\text{ A}$$

Step 4. From Eq. (11.55), calculate the average current of each transistor:

$$I_{C1} = \frac{I_p}{\pi} = \frac{1.33\text{ A}}{\pi} = 0.424\text{ A}$$

Step 5. From Eq. (11.64), calculate the maximum collector power dissipation:

$$P_C = \left(\frac{2}{\pi^2}\right)P_{L(max)} = \left(\frac{2}{\pi^2}\right) \times 10\text{ W} = 2\text{ W}$$

Step 6. Calculate the DC power from the source:

$$P_S = 2I_{C1}V_{CC} = 2 \times 0.424\text{ A} \times 15\text{ V} = 12.72\text{ W}$$

Step 7. From Eq. (11.58), calculate the required turns ratio of the transformer:

$$\frac{n_p}{n_s} = \left(\frac{R'_L}{R_L}\right)^{1/2} = \left(\frac{11.25\ \Omega}{4\ \Omega}\right)^{1/2} = 1.68$$

Step 8. Calculate the required quiescent base current I_{B1}:

$$I_{B1} = \frac{I_{C1}}{h_{fe}} = \frac{0.424\text{ A}}{100} = 4.24\text{ mA}$$

Step 9. Calculate the biasing resistances R_1 and R_2. We have

$$V_{BB} - V_{BE} = R_B(I_{B1} + I_{B2}) = R_B(2I_{B1})$$

If we let $R_B(2I_{B1}) = 10\%$ of $V_{BE} = 0.1V_{BE} = 0.1 \times 0.7\text{ V} = 0.07\text{ V}$, then

$$R_B = \frac{0.07\text{ V}}{2I_{B1}} = 8.25\ \Omega$$

and $\quad V_{BB} = V_{BE} + R_B(2I_{B1}) = 0.7\text{ V} + 0.07\text{ V} = 0.77\text{ V}$

Since $\quad V_{BB} = \dfrac{R_1 V_{CC}}{R_1 + R_2} = 0.77\text{ V}$

and $\quad R_B = \dfrac{R_1 R_2}{R_1 + R_2} = 8.25\ \Omega$

Solving for R_1 and R_2, we get $\quad R_1 = \dfrac{R_B V_{CC}}{V_{CC} - V_{BB}} = \dfrac{8.25\ \Omega \times 15\text{ V}}{15\text{ V} - 0.77\text{ V}} = 8.7\ \Omega$

$$R_2 = \frac{R_B V_{CC}}{V_{BB}} = \frac{8.25\ \Omega \times 15\text{ V}}{0.77\text{ V}} = 161\ \Omega$$

KEY POINTS OF SECTION 11.5

- In a class B push-pull amplifier, a *pnp* and an *npn* transistor form a pair, and each transistor conducts for only 180°. The quiescent DC-biasing current is zero. As a result, the maximum power efficiency is 78.5%, and the maximum figure of merit is only 20%.
- Because of B-E voltage drops, a push-pull amplifier exhibits a dead zone in the transfer characteristic, which increases distortion of the output voltage. The crossover distortion and nonlinearities can be reduced practically to zero by applying feedback.

11.6 Complementary Class AB Push-Pull Amplifiers

The crossover distortion of a complementary class B push-pull amplifier is minimized or eliminated in a class AB amplifier, in which the transistors operate in the active region when the input voltage v_I is small ($v_I \approx 0$ V). The transistors are biased in such a way that each transistor conducts for a small quiescent current I_Q at $v_I = 0$ V. A biasing circuit is shown in Fig. 11.20(a). A biasing voltage V_{BB} is applied between the bases of Q_N and Q_P. For $v_I = 0$, a voltage $V_{BB}/2$ appears across the B-E junction of each Q_N and Q_P. Choosing $V_{BB}/2 = V_{BEN} = V_{EBP}$ will ensure that both transistors will be on the verge of conducting. That is, $v_O = 0$ for $v_I = 0$. A small positive input voltage v_I will then cause Q_N to conduct; similarly, a small negative input voltage will cause Q_P to conduct.

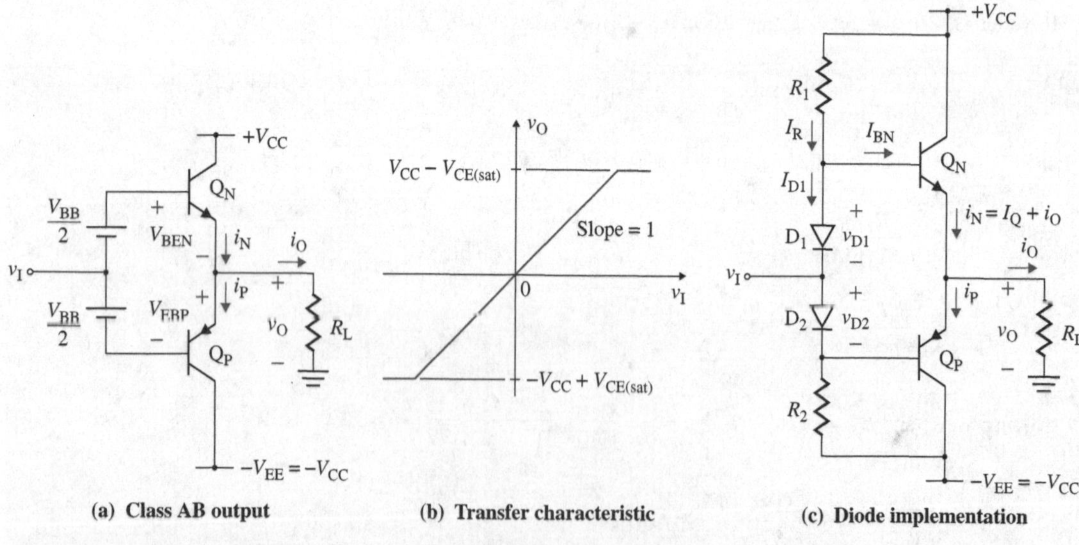

FIGURE 11.20 Elimination of the dead zone in a class AB amplifier

11.6.1 Transfer Characteristic

The output voltage v_O is given by

$$v_O = v_I + \frac{V_{BB}}{2} - V_{BEN} (= V_{EBP}) \tag{11.66}$$

which, for identical transistors of $V_{BEN} = V_{EBP}$ and $V_{BB}/2 = V_{BEN}$, gives $v_O = v_I$. Therefore, most of the crossover distortion is eliminated. The transfer characteristic is shown in Fig. 11.20(b). For positive v_O, a current i_O flows through R_L; that is,

$$i_N = i_P + i_O \tag{11.67}$$

Any increase in i_N will cause a corresponding increase in V_{BEN} above the quiescent value of $V_{BB}/2$. Since V_{BB} must remain constant, the increase in V_{BEN} will cause an equal decrease in V_{EBP} and hence in i_P. Thus,

$$V_{BB} = V_{BEN} + V_{EBP} \tag{11.68}$$

which, expressed in terms of saturation current I_S, becomes

$$2V_T \ln\left(\frac{I_Q}{I_S}\right) = V_T \ln\left(\frac{i_N}{I_S}\right) + V_T \ln\left(\frac{i_P}{I_S}\right)$$

After simplification, we get

$$I_Q^2 = i_N i_P \tag{11.69}$$

$$= i_N(i_N - i_O) = i_N^2 - i_N i_O \tag{11.70}$$

which can be solved for the current i_N for a given quiescent current I_Q. Thus, as i_N increases, i_P decreases by the same ratio. However, their product remains constant. As v_I becomes positive, Q_N acts as an emitter follower delivering output power, and Q_P conducts only a very small current. When v_I becomes negative, the opposite occurs: Q_P acts as an emitter follower, and v_O follows the input signal v_I. The circuit operates in class AB mode because both transistors remain on and operate in the active region.

11.6.2 Output Power and Efficiency

The power relationships in class AB amplifiers are identical to those in class B amplifiers, except that the class AB circuit dissipates a quiescent power of $I_Q V_{CC}$ per transistor. Thus, from Eq. (11.44), we can find the average power supplied from the DC source as

$$P_S = \frac{2I_p V_{CC}}{\pi} + I_Q V_{CC} = V_{CC}\left(I_Q + \frac{2I_p}{\pi}\right) \tag{11.71}$$

11.6.3 Biasing with Diodes

The biasing circuit in Fig. 11.20(a) has a serious problem when the temperatures of Q_N and Q_P increase as a result of their power dissipation. Recall that the value of V_{BE} for a given current falls with temperature at approximately 2.5 mV/°C. Thus, if the biasing voltage $V_{BB}/2$ remains constant with temperature,

V_{BE} ($=V_{BB}/2$) is also held constant, and the collector current will increase as temperature increases. The increase in the collector current increases the power dissipation, in turn increasing the collector current and causing the temperature to rise further. This phenomenon, in which a positive feedback mechanism leads to excessive temperature rise, is called *thermal runaway*. Thermal runaway can ultimately lead to the destruction of the transistors unless they are protected.

To avoid thermal runaway, the biasing voltages must decrease as the temperature increases. One solution is to use diodes that have a compensating effect, as shown in Fig. 11.20(c). The diodes must be in close contact with the output transistors so that their temperature will increase by the same amount as that of Q_N and Q_P. Therefore, in discrete circuits, the diodes should be mounted on the metal of Q_N or Q_P. Since resistances R_1 and R_2 provide the quiescent current I_Q for the transistors and also ensure that the diodes conduct, to guarantee the base biasing current for Q_N when the load current becomes maximum we must have

$$I_R = I_{D1} + \frac{i_N}{1 + h_{fe}} \approx I_{D1} + \frac{I_Q + i_O}{1 + h_{fe}}$$

Thus, the values of R_1 and R_2 can be found from

$$R_1 = R_2 = \frac{V_{CC} - V_{D1}(=V_{D2} = V_{BB}/2)}{I_{D1(\min)} + (I_Q + i_{O(\max)})/(1 + h_{fe})} \tag{11.72}$$

where $I_Q = I_S \exp(V_{BB}/2V_T)$ and $I_{D1(\min)}$ is the minimum current needed to ensure diode conduction. Because I_Q is usually smaller than $i_{O(\max)}$, I_Q can often be neglected in finding the values of R_1 and R_2.

EXAMPLE 11.5

D

Designing a biasing circuit for a class AB amplifier

(a) Design a biasing circuit for the class AB amplifier of Fig. 11.20(c) to supply the maximum output voltage at a load resistance of $R_L = 50\ \Omega$. The quiescent biasing current I_Q is 2 mA. Assume a DC supply voltage of 12 V. The diode parameters are $I_S = 10^{-13}$ A, $V_{D1} = V_{D2} = 0.7$ V, and $I_{D(\min)} = 1$ mA to ensure conduction. The transistor parameters are $\beta_F = h_{fe} = 50$, $V_{BE} = 0.7$ V, $I_S = 10^{-14}$ A, and $V_{CE(sat)} = 0.2$ V.

(b) Find the biasing voltage V_{BB} for $v_O = 0$ and 11.8 V.

SOLUTION

(a) The maximum peak load voltage is

$$V_{p(\max)} = V_{CC} - V_{CE(sat)} = 12 - 0.2 = 11.8 \text{ V}$$

The maximum peak load current is

$$I_{p(\max)} = \frac{V_{p(\max)}}{R_L} = \frac{11.8}{50} = 236 \text{ mA}$$

From Eq. (11.71), the maximum DC power from the supply source is

$$P_S \approx \frac{2I_{p(\max)}V_{CC}}{\pi} + I_Q V_{CC} = \frac{2 \times 236 \text{ mA} \times 12 \text{ V}}{\pi} + 2 \text{ mA} \times 12 \text{ V} = 1.83 \text{ W}$$

From Eq. (11.17), the output power is

$$P_{L(max)} = \frac{I_{p(max)}V_{p(max)}}{2} = 236\text{ mA} \times \frac{11.8\text{ V}}{2} = 1.39\text{ W}$$

Thus, the maximum power efficiency is

$$\eta_{max} = \frac{P_{L(max)}}{P_S} = \frac{1.39}{1.83} = 76.1\%$$

The power dissipation of each transistor can be found from

$$P_C = \frac{P_S - P_{L(max)}}{2} = \frac{1.83\text{ W} - 1.39\text{ W}}{2} = 220\text{ mW}$$

From Eq. (11.72), we get

$$R_1 = R_2 = \frac{12\text{ V} - 0.7\text{ V}}{1\text{ mA} + (2\text{ mA} + 236\text{ mA})/(1 + 50)} = 1.99\text{ k}\Omega$$

(b) We have

$$I_R = I_{D1(min)} + \frac{I_Q + I_{p(max)}}{1 + h_{fe}} = 1\text{ mA} + \frac{2\text{ mA} + 236\text{ mA}}{1 + 50} = 5.667\text{ mA}$$

For $v_O = 0$, $i_N = I_Q = 2$ mA. Thus, the base current of the *npn* transistor is

$$I_{BN} = \frac{i_N}{1 + h_{fe}} = \frac{2\text{ mA}}{1 + 50} = 0.039\text{ mA}$$

and $\quad I_{D1} = I_R - I_{BN} = 5.667\text{ mA} - 0.039\text{ mA} = 5.628\text{ mA}$

Therefore, the biasing voltage V_{BB} becomes

$$V_{BB} = 2V_T \ln\left(\frac{I_{D1}}{I_S}\right) = 2 \times 25.8\text{ mV} \times \ln\left(\frac{5.628\text{ mA}}{10^{-13}\text{ A}}\right) = 1.277\text{ V}$$

For $v_O = 11.8$, $i_N = I_Q + I_{p(max)} = 2\text{ mA} + 236\text{ mA} = 238\text{ mA}$. Thus,

$$I_{BN} = \frac{i_N}{1 + h_{fe}} = \frac{238\text{ mA}}{1 + 50} = 4.67\text{ mA}$$

and $\quad I_{D1} = I_R - I_{BN} = 5.667\text{ mA} - 4.67\text{ mA} = 1\text{ mA}$

Therefore, for $I_{D1} = 1$ mA, the biasing voltage V_{BB} becomes $V_{BB} = 1.19$ V.

11.6.4 Biasing with Diodes and an Active Current Source

The biasing technique in Fig. 11.20(a) is generally used in integrated circuits; however, an active current source is normally used rather than discrete resistance. This arrangement is shown in Fig. 11.21(a). In integrated circuits, collector-shorted transistors are usually used instead of diodes. If Q_N and Q_P are to handle large amounts of power, their geometry must also be large. However, the diodes can be smaller devices such that $I_R = I_Q/n$, where n is the ratio of the emitter junction area of Q_N and Q_P to the junction area of D_1 and D_2. That is, the saturation current I_S of Q_N and Q_P can be n times that of the biasing diodes.

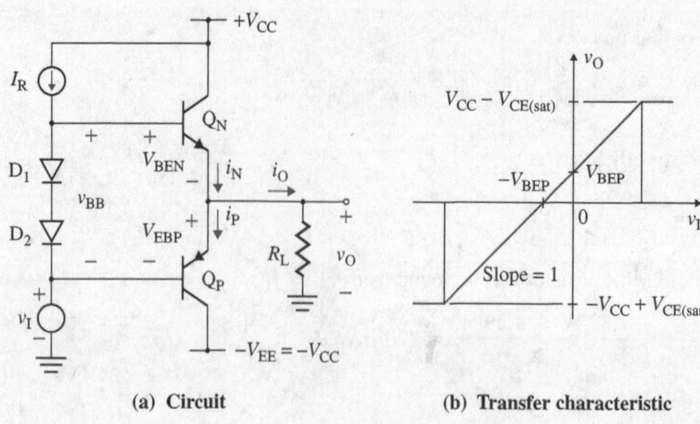

FIGURE 11.21 Biasing of a class AB amplifier with diodes and an active current source

Transfer Characteristic

The voltage between the bases of Q_P and Q_N is the same as the voltage drop across the two diodes. That is,

$$V_{BB} = V_{D1} + V_{D2} \approx 0.7 + 0.7 = 1.4 \text{ V}$$

The base-to-emitter voltage of Q_N is given by

$$V_{BEN} = V_{BB} - V_{EBP} = 1.4 - V_{EBP} \tag{11.73}$$

Thus, the B-E junctions of both Q_N and Q_P are always forward biased. Because of diodes D_1 and D_2, Q_N and Q_P remain in the active region when $v_I = 0$ V. The output voltage v_O is given by

$$v_O = v_I + V_{BB} - V_{BEN} \tag{11.74}$$

$$= v_I + V_{EBP} = v_I - V_{BEP}$$

which yields the transfer characteristic shown in Fig. 11.21(b). The dead zone is eliminated. However, there is an offset voltage of V_{EBP}, which can be reduced practically to zero by applying feedback similar to that applied to the class B amplifier in Fig. 11.14.

EXAMPLE 11.6

D **Designing an active current–source biasing circuit for a class AB amplifier**

(a) Design an active current source for the class AB amplifier in Fig. 11.21(a) in order to provide the biasing current of $I_R = 5.67$ mA needed for Example 11.5. Assume $V_{CC} = V_{EE} = 12$ V, $V_{BE} = 0.7$ V, and $R_L = 50\ \Omega$.

(b) Use PSpice/SPICE to plot the transfer characteristic and the instantaneous i_N, i_P, and i_O for $v_I = 5 \sin (2000\pi t)$. The PSpice model parameters for the transistors are

```
IS=1E-14 BF=50 VJE=0.7
```

and for the diodes are

```
IS=1E-13 BV=100
```

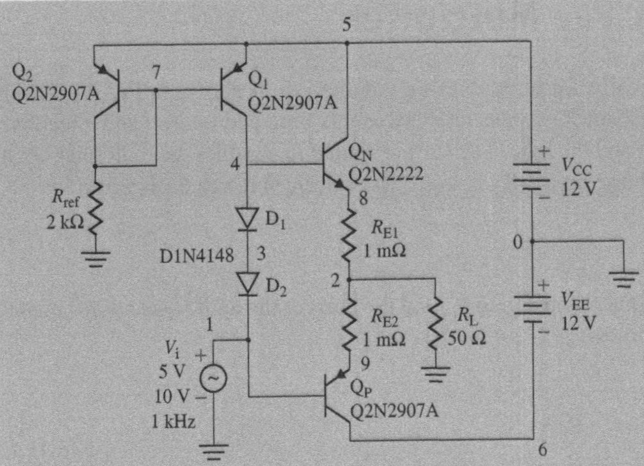

FIGURE 11.22 Complementary class AB push-pull amplifier circuit for PSpice simulation

SOLUTION

(a) The biasing current source I_R, shown in Fig. 11.22, can be produced by two *pnp* transistors Q_1 and Q_2 and a resistance R_1. Thus,

$$I_R \approx I_{ref} = \frac{V_{CC} - V_{EBP}}{R_{ref}} \tag{11.75}$$

which, for $I_R = 5.67$ mA, gives

$$R_{ref} = \frac{V_{CC} - V_{EBP}}{I_R} = \frac{12\text{ V} - 0.7\text{ V}}{5.67\text{ mA}} \approx 2\text{ k}\Omega$$

(b) In practice, emitter resistances R_{E1} and R_{E2}, shown in Fig. 11.22, are connected to ensure the stability of the biasing point.

The transfer characteristic for the circuit in Fig. 11.22 is shown in Fig. 11.23(a). It gives $V_{O(max)} \equiv$ V(2) = 11.139 V and offset voltages of $v_O = 560.4$ mV at $v_I = 0$ and $v_O = 0$ at $v_I = -640.9$ mV. The plots of i_N, i_P, and i_O are shown in Fig. 11.23(b), which give $i_{N(peak)} \equiv$ I(RE1) = 109.66 mA, $i_{P(peak)} \equiv$ I(RE2) = −83.92 mA, and $i_{O(peak)} \equiv$ I(RL) = 109.59 mA to −83.83 mA. Thus, because of the offset voltage, the load current is not symmetrical.

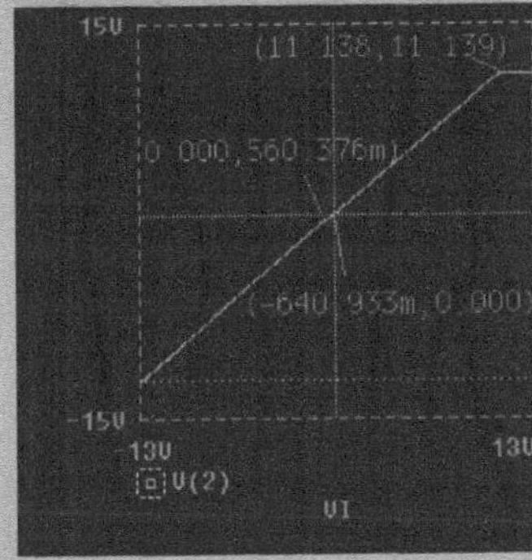

(a) Transfer characteristic

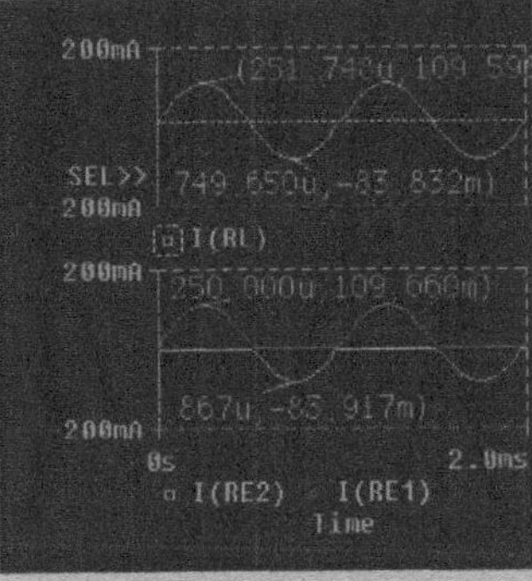

(b) Output voltage

FIGURE 11.23 Transfer characteristic and output voltage for Example 11.6

11.6.5 Biasing with a V_{BE} Multiplier

A V_{BE} *multiplier circuit* that can automatically adjust the biasing voltage V_{BB} is shown in Fig. 11.24. The circuit consists of a transistor Q_1 with a resistor R_1 connected between its base and emitter and a feedback resistor R_F connected between the collector and the base. The current source I_R supplies the multiplier circuit and the base current for Q_N. Since the voltage across R_1 is V_{BE1}, the current through R_1 is given by

$$I_1 = \frac{V_{BE1}}{R_1} \tag{11.76}$$

The base current of Q_1 is generally negligible compared to I_1, and the current through R_F is approximately equal to I_1. Thus, the biasing voltage becomes

$$\begin{aligned} V_{BB} &= I_1(R_1 + R_F) \\ &= \frac{V_{BE1}}{R_1}(R_1 + R_F) = V_{BE1}\left(1 + \frac{R_F}{R_1}\right) \end{aligned} \tag{11.77}$$

Therefore, the circuit multiplies V_{BE1} by the factor $(1 + R_F/R_1)$—hence the name V_{BE} multiplier. By selecting the ratio R_F/R_1, one can set the value of V_{BB} required to give a desired quiescent current I_Q. For $R_F/R_1 = 1$, $V_{BB} = 2V_{BE1}$. The value of V_{BE1} is related to i_{C1} by

$$V_{BE1} = V_T \ln\left(\frac{i_{C1}}{I_{S1}}\right) \tag{11.78}$$

where I_{S1} is the saturation current of Q_1 and

$$i_{C1} = I_R - I_1 - \frac{I_Q + i_O}{1 + h_{fe}} \tag{11.79}$$

Under quiescent conditions, $i_O = 0$ and the base current of Q_N is normally small enough that it can be neglected. That is, $I_Q/(1 + h_{fe}) \approx 0$, and Q_1 carries the maximum current: $I_{C1(max)} \approx I_R - I_1$. However, at the peak value of v_O, the base current of Q_N will be maximum and the current available to the multiplier will be minimum; that is,

$$I_{C1(min)} = I_R - I_1 - \frac{I_Q + i_{O(max)}}{1 + h_{fe}} \tag{11.80}$$

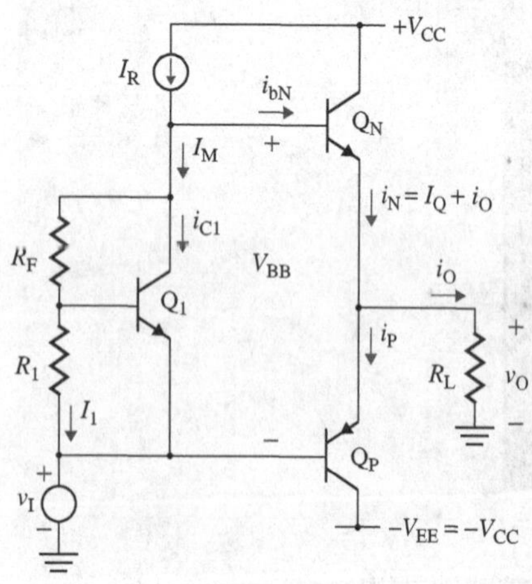

FIGURE 11.24 Biasing of a class AB amplifier with a V_{BE} multiplier

Thus, i_{C1} can vary widely from $I_{C1(max)}$ to $I_{C1(min)}$. However, according to Eq. (11.78), a large change in i_{C1} will cause only a small change in V_{BE1}. Thus, I_1 and V_{BB} will remain almost constant. As with diode biasing, the transistor Q_1 must be in close contact with Q_N and Q_P in order to provide a thermal compensating effect.

EXAMPLE 11.7

D

Designing a V_{BE} multiplier for a class AB amplifier

(a) Design a V_{BE} multiplier for the class AB amplifier in Fig. 11.24 in order to provide the biasing current of $I_R = 5.67$ mA needed for Example 11.5. Assume $h_{fe} = 50$, $I_S = 10^{-14}$ A, $V_{CC} = V_{EE} = 12$ V, $V_{BE} = 0.7$ V, and $R_L = 50\ \Omega$. Assume a minimum current of $I_{M(min)} = 1$ mA to the multiplier, $I_Q = 2$ mA, and peak load current $I_{p(max)} = 236$ mA.

(b) Use PSpice/SPICE to plot the transfer characteristic and the instantaneous i_N, i_P, and i_O for $v_I = 5$ sin $(2000\pi t)$. The PSpice model parameters for the transistors are

```
IS=1E-14 BF=50 VJE=0.7
```

and for the diodes are

```
IS=1E-13 BV=100
```

SOLUTION

(a) Since the current source must provide the base current when the load current is maximum,

$$I_R = I_{M(min)} + \frac{I_Q + I_{p(max)}}{1 + h_{fe}} = 1\text{ mA} + \frac{2\text{ mA} + 236\text{ mA}}{1 + 50} = 5.67\text{ mA}$$

The biasing voltage V_{BB} required to yield a quiescent current of $I_Q = 2$ mA is

$$V_{BB} = 2V_T \ln\left(\frac{I_Q}{I_S}\right) = 2 \times 25.8\text{ mV} \times \ln\left(\frac{2\text{ mA}}{10^{-14}\text{ mA}}\right) = 1.343\text{ V}$$

The minimum current through the multiplier must be $I_{M(min)} = 1$ mA. Let $I_{1(min)} = I_{M(min)}/2 = 0.5$ mA and $I_{C(min)} = I_{M(min)}/2 = 0.5$ mA. If $I_{C(min)}$ is too small, transistor Q_1 will be off, which is not desirable. Equation (11.77) gives

$$R_1 + R_F = \frac{V_{BB}}{I_{1(min)}} = \frac{1.343\text{ V}}{0.5\text{ mA}} \approx 2.7\text{ k}\Omega$$

The current source must be designed to supply $I_R = 5.67$ mA. However, when the output voltage is zero, then $i_O = 0$, and $I_R = 5.67$ mA must flow through the multiplier. That is, transistor Q_1 must carry $i_{C1} = I_R - I_{1(min)} = 5.67\text{ mA} - 0.5\text{ mA} = 5.17$ mA, and the corresponding B-E voltage will be

$$V_{BE1} = V_T \ln\left(\frac{i_{C1}}{I_S}\right) = 25.8\text{ mV} \times \ln\left(\frac{5.17\text{ mA}}{10^{-14}\text{ mA}}\right) = 0.696\text{ V}$$

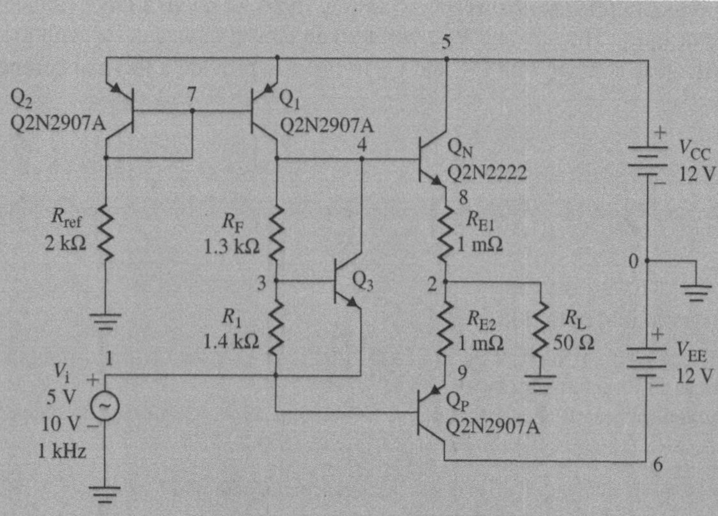

FIGURE 11.25 Complementary class AB push-pull amplifier circuit, with V_{BE} multiplier, for PSpice simulation

Thus, the value of R_1 can be found from Eq. (11.76) as

$$R_1 = \frac{V_{BE1}}{I_1} = \frac{0.696\text{ V}}{0.5\text{ mA}} = 1.39\text{ k}\Omega \approx 1.4\text{ k}\Omega$$

Therefore, the value of R_F becomes

$$R_F = 2.7\text{ k}\Omega - R_1 = 2.7\text{ k}\Omega - 1.4\text{ k}\Omega = 1.3\text{ k}\Omega$$

(b) The circuit file for the PSpice simulation is similar to that for Example 11.6, except that the diodes are replaced by the V_{BE} multiplier. This configuration is shown in Fig. 11.25.

The transfer characteristic is shown in Fig. 11.26(a). It gives $v_{O(max)} = 11.13$ V and offset voltages of $v_O = 794.3$ mV at $v_I = 0$ and $v_O = 0$ at $v_I = -796$ mV. The plots of the input and output voltages are shown in Fig. 11.26(b). Their waveforms are almost identical, except that the magnitudes are shifted by 763 mV.

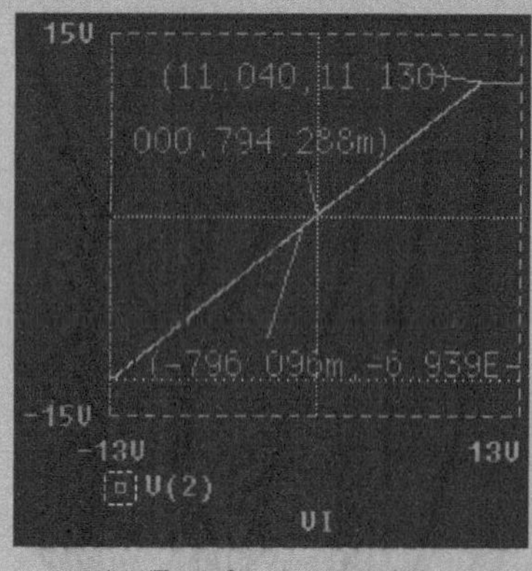

(a) Transfer characteristic

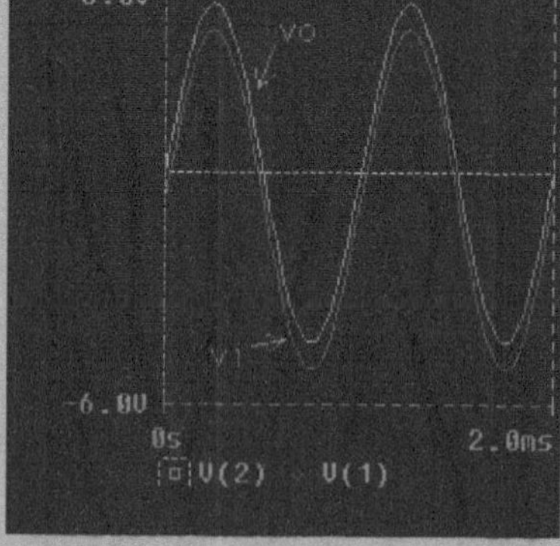

(b) Output voltage

FIGURE 11.26 Transfer characteristic and output voltage for Example 11.7

11.6.6 Quasi-Complementary Class AB Amplifiers

Because *pnp* transistors have limited current-carrying capability, the complementary output stage is suitable only for delivering load power on the order of a few hundred milliwatts or less. If output power of several watts or more is required, *npn* transistors should be used. A composite *pnp* transistor can be made from a *pnp* transistor Q_P and a high-power *npn* transistor Q_{N1}. This arrangement, called a quasi-complementary output stage, is shown in Fig. 11.27(a).

The pair Q_P-Q_{N1} is equivalent to a *pnp* transistor, as shown in Fig. 11.27(b). The collector current of Q_P is given by

$$I_{CP} = I_S \exp\left(\frac{V_{EBP}}{V_T}\right) \tag{11.81}$$

The composite collector current I_C is the emitter current of Q_{N1}; that is,

$$I_C = (1 + h_{fe})I_{CP} = (1 + h_{fe})I_S \exp\left(\frac{V_{EBP}}{V_T}\right) \tag{11.82}$$

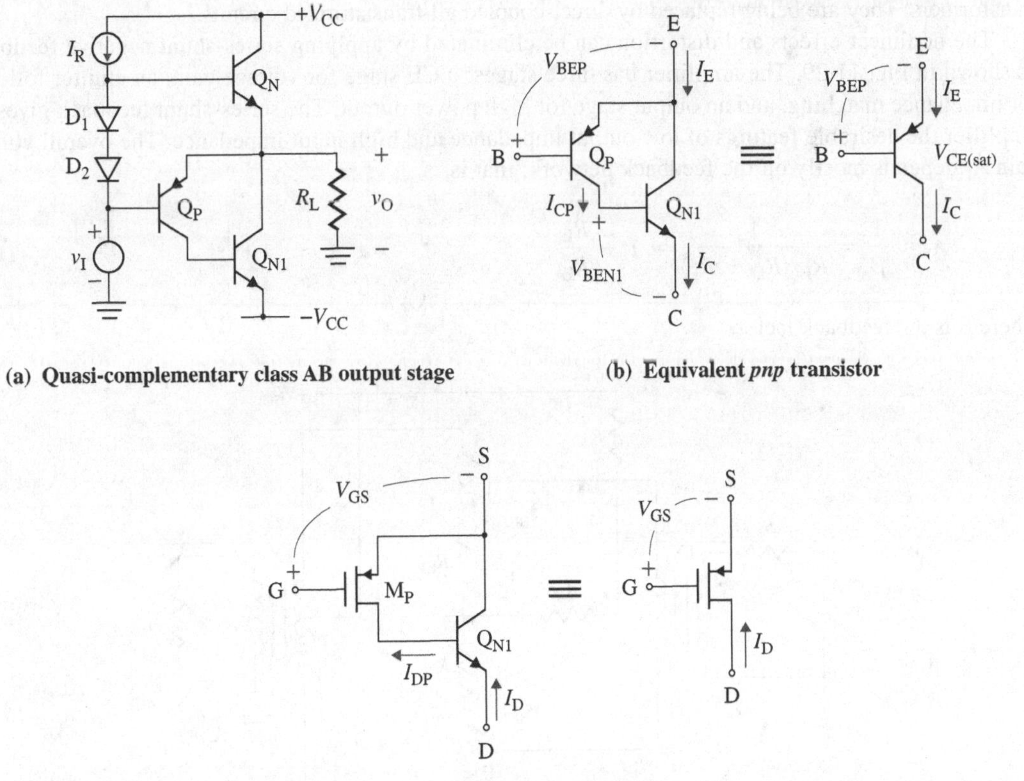

(a) Quasi-complementary class AB output stage

(b) Equivalent *pnp* transistor

(c) Equivalent PMOS

FIGURE 11.27 Quasi-complementary class AB output stage

which is the same as the relationship for a normal *pnp* transistor. However, the *npn* transistor carries most of the current and the *pnp* transistor carries only a small amount of the current. The saturation voltage of the composite *pnp* will be $V_{CEP(sat)} + V_{BEN1}$, which is higher than that of a normal *pnp* transistor.

A composite *pnp* transistor can be replaced by a MOS-bipolar combination [4], known as a composite PMOS, as shown in Fig. 11.27(c). From Eq. (7.7), the overall transfer characteristic of the composite PMOS is given by

$$I_D = -(1 + h_{fe})I_{DP} = -(1 + h_{fe})\frac{\mu_n C_o}{2}\left(\frac{W}{L}\right)(V_{GS} - V_t)^2 \quad (11.83)$$

Thus, the composite PMOS has a W/L ratio that is $(1 + h_{fe})$ times larger than that of a normal PMOS device.

11.6.7 Transformer-Coupled Class AB Amplifiers

The class AB circuit shown in Fig. 11.28 is identical to the class B circuit in Fig. 11.17(a), except that it is biased slightly into conduction so that a quiescent current I_Q flows through Q_1 and Q_2. This current is achieved by making V_{BB} slightly greater than $V_{BE} = V_{BE1}$ $(= V_{BE2} \approx 0.7 \text{ V})$. Resistors R_1 and R_2 can be selected to give the desired value of V_{BB}. Although transformer-coupled amplifiers offer high power efficiency, they suffer from nonlinearities and distortion introduced by the nonlinear characteristics of the transformers. They are being replaced by direct-coupled all-transistorized circuits.

The nonlinear effects and distortion can be eliminated by applying series-shunt negative feedback, as shown in Fig. 11.29. The amplifier has three stages: a CE stage for voltage gain, an emitter follower for impedance matching, and an output stage for high power output. The series-shunt feedback gives the amplifier the desirable features of low output impedance and high input impedance. The overall voltage gain A_f depends mostly on the feedback network; that is,

$$A_f \approx \frac{1}{\beta} = \frac{1}{R_E/(R_E + R_F)} = 1 + \frac{R_F}{R_E} \quad (11.84)$$

where β is the feedback factor.

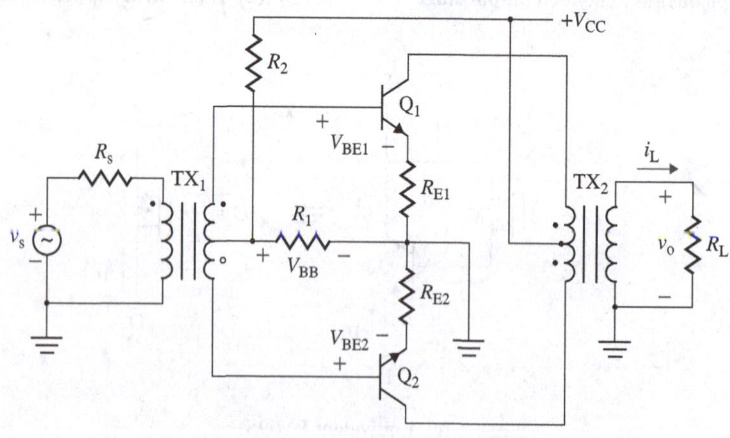

FIGURE 11.28 Transformer-coupled class AB push-pull amplifier

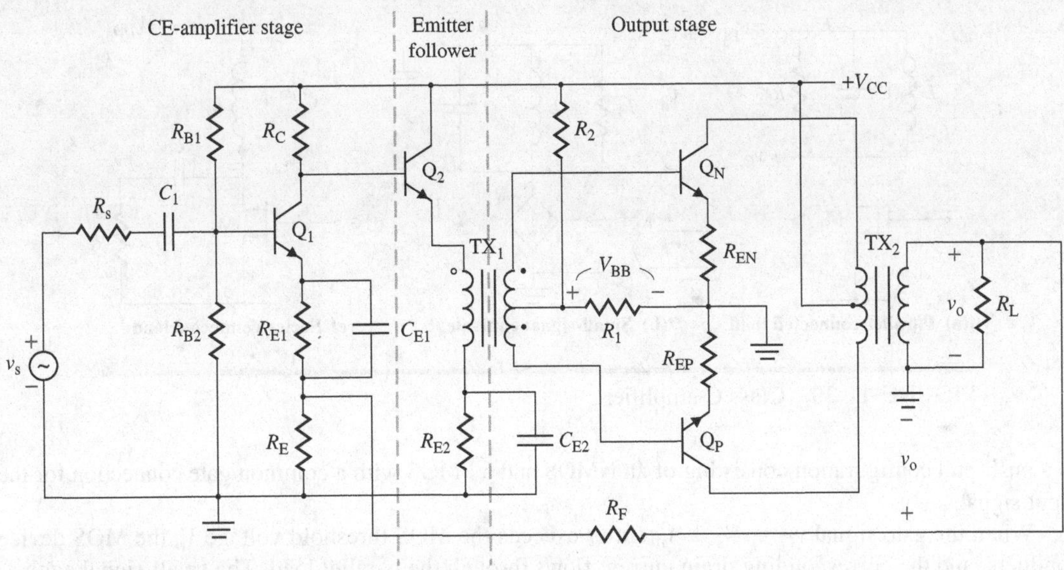

FIGURE 11.29 Transformer-coupled class A amplifier with series-shunt feedback

KEY POINTS OF SECTION 11.6

- The complementary class AB amplifier is the most commonly used output stage. Its circuit operation is similar to that of a class B amplifier, except that the transistors have a slightly positive bias so that a quiescent DC current flows even when the input voltage is zero.
- The B-E DC voltage of each transistor is usually set to approximately V_{BE}, which is the voltage required to yield the desired quiescent current. The amplifier is commonly biased with diodes and an active current source or with a V_{BE} multiplier.
- Class AB amplifiers exhibit an offset output voltage at zero input voltage. However, feedback can be applied to reduce the offset voltage.
- A quasi-complementary amplifier uses a composite *pnp* transistor, which can deliver higher output power than a normal *pnp* device.

11.7 Class C Amplifiers

Class C amplifiers conduct less than 50% of the input signal, and the efficiencies can be as high as 90%, but the output has a high amount of distortion. These amplifiers can find applications in radio frequency transmitters where the distortion can be considerably reduced by using tuned loads on the amplifying devices. Figure 11.30(a) shows the simplified arrangement of a class C amplifier with a parallel-connected load resistance R_L. To isolate the load from the DC power supply and for impedance matching with the external load, R_L is generally connected through a transformer. The class C amplifier can also be operated

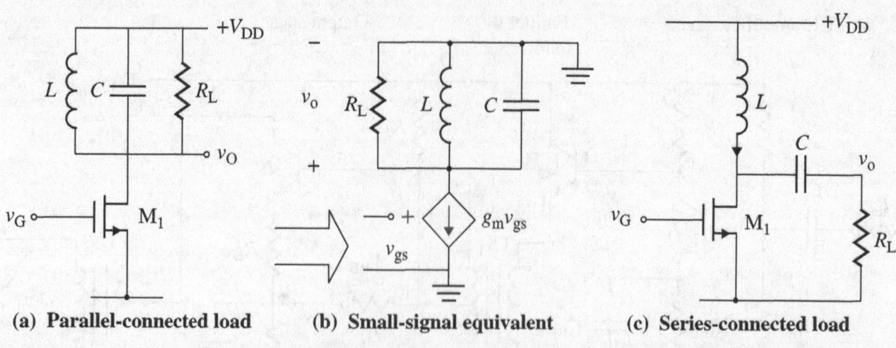

FIGURE 11.30 Class C amplifier

as a push-pull configuration consisting of an NMOS and a PMOS with a common gate connection for the input signal.

When the gate signal $v_G = V_G + V_m \sin \omega t$ exceeds the MOS threshold voltage V_t, the MOS device conducts, and the corresponding drain current flows through the parallel load. The small-signal equivalent circuit is shown in Fig. 11.30(b). In the linear region, the transistor will behave as a switch with a drain–source resistance, which can be found from Eq. (11.2) as given by

$$R_{DS} = \frac{v_{DS}}{i_D} \simeq \frac{L}{W\mu_n C_{ox}(V_G + V_m \sin \omega t - V_t)} \tag{11.85}$$

The AC gate–source voltage $v_{gs} = V_m \sin \omega t$ will vary the drain current ($i_d = g_m v_{gs}$), which is the source for the *RLC* load. From Fig. 11.30(b), we can find the small-signal AC output voltage as given by

$$v_o = g_m v_{gs}(R_L \parallel X_L \parallel X_C) \tag{11.86}$$

where $X_L(= j\omega L)$ and $X_C(= 1/j\omega C)$ are the impedances of *L* and *C*, respectively. If the gate–source voltage is a pulsed DC voltage of PWM, the drain current will also be of the same form.

The values of *L* and *C* can be selected either to oscillate at a specified damping factor or to form a tuned circuit by choosing the values of *L* and *C* such that $\omega L = 1/\omega C$ at the tuned frequency ω_n. The tuned circuit will resonate only at particular frequencies, so the unwanted frequencies are dramatically suppressed, and the desirable full signal (sine wave) will be abstracted by the tuned load. Provided that the transmitter is not required to operate over a very wide band of frequencies, this arrangement works extremely well. Other residual harmonics can be removed using a filter.

The load R_L also can be connected to form a series resonant circuit as shown in Fig. 11.30(c). When the transistor is switched on, the drain and the inductor currents will rise through the path formed by V_{DD}, *L*, and M_1. When the transistor is switched off, the drain current will be zero, and the inductor current will fall through the path formed by V_{DD}, *L*, *C*, and R_L. The energy stored in the inductor during the transistor on-time will be transferred to the load R_L via the capacitor C. The damping factor of series and parallel *RLC* circuits (see Appendix B) can be found from

$$\delta = \frac{1}{2R_L}\sqrt{\frac{L}{C}} \quad \text{(for series resonant)} \tag{11.87}$$

$$= \frac{1}{2R_L}\sqrt{\frac{L}{C}} \quad \text{(for parallel resonant)}$$

The values of L and C are related to the damping ratio δ and the resonant frequency f_n as given by

$$L = \frac{R_L}{2\delta\omega_n} \quad \text{(for series resonant)} \tag{11.88}$$

$$= \frac{2\delta R_L}{\omega_n} \quad \text{(for parallel resonant)}$$

$$C = \frac{2\delta}{R_L\omega_n} \quad \text{(for series resonant)} \tag{11.89}$$

$$= \frac{1}{2\delta R_L\omega_n} \quad \text{(for parallel resonant)}$$

EXAMPLE 11.8

Finding the values of *L* and *C* for a class C amplifier The gate voltage of the class C amplifier in Fig. 11.30(a) is $v_G = 2 + 1.5 \sin(2\pi f_s t)$ with $f_s = 1$ MHz. The DC supply voltages are $V_{DD} = \pm 15$ V. The load resistance $R_L = 50\ \Omega$, which is connected through a transformer of turns ratio of $n = 1{:}1$.

(a) Find the values of L and C to form a tuned resonant circuit.

(b) Use PSpice to plot the voltage $v_o(t)$, the transistor drain current $i_D(t)$, and the transistor drain voltage $v_D(t)$ for varying the gate voltage component $V_m = 3$ V to 4 V with an increment of 0.5 V. Use the PSpice parameters of the NMOS IRF150 whose $V_t = 2.84$ V.

SOLUTION

(a) For $n = 1{:}1$, the effective load resistance is $R = R_L/n^2 = 50\ \Omega$. From Eq. (11.87), we can find the damping factor for a tuned circuit of $L = C$, $\delta = (1/2R)\sqrt{L/C} = 1/2 \times 50 = 0.01$. For $f_s = 1 \times 10^6$, the resonant frequency $f_o = f_s = 1 \times 10^6$, which is related to L and C by $f_o = 1/(2\pi\sqrt{LC})$. From Eqs. (11.88) and (11.89), we get

$$L = \frac{2\delta R}{2\pi f_s} = \frac{2 \times 0.01 \times 50}{2 \times \pi \times 10^6} = 159.2\text{ nH} \simeq 160\text{ nH}$$

$$C = \frac{1}{2\delta R 2\pi f_s} = \frac{1}{2 \times 0.01 \times 50 \times 2 \times \pi \times 10^6} = 159.2\text{ nF} \simeq 160\text{ nF}$$

(b) The PSpice schematic is shown in Fig. 11.31. A small resistance $R_x = 1\ \mu\Omega$ is connected to avoid a convergence problem of the zero-resistance DC loop. The PSpice plot of the transistor drain current $i_D(t)$ is shown in Fig. 11.32(a), the transistor drain voltage $v_D(t)$ in Fig. 11.32(b), and the output load voltage $v_o(t)$ in Fig. 11.32(c) for $V_m = 3$ V, 3.5 V, and 4 V. For $v_G(t) < V_t = 2.84$ V, the transistor is off. The output voltage increases with the gate voltage as expected as shown in Fig. 11.32(c). The drain current becomes negative at a higher value because the inductive load L induces a negative voltage (Ldi/dt) effect due to the abrupt turning off of the transistor; this negative voltage forward biases the built-in source–channel *pn* junction. The peak drain voltage reaches $2V_{DD}$ due to the Ldi/dt effect.

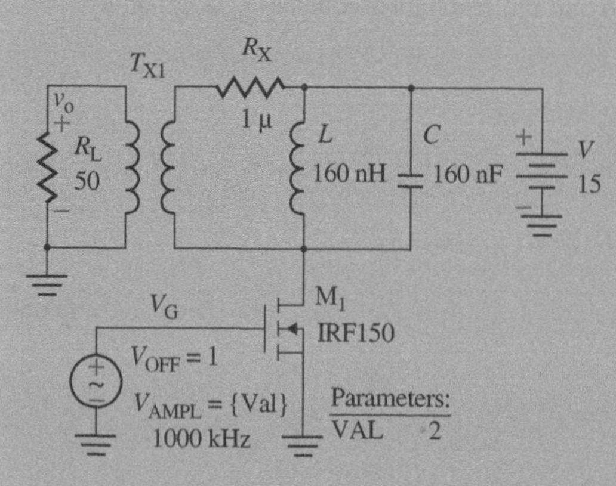

FIGURE 11.31 PSpice schematic for Example 11.8

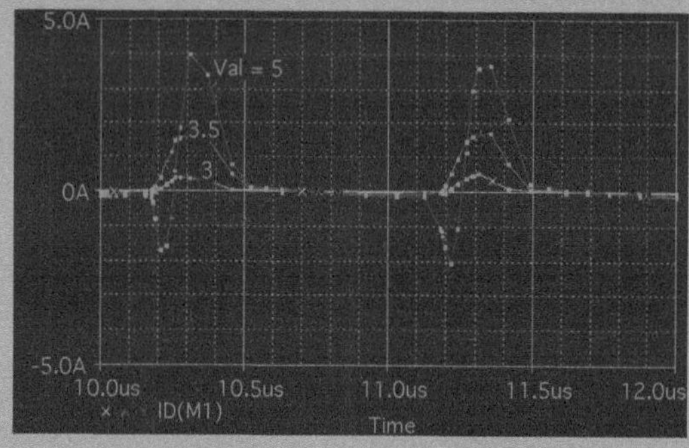

(a) Transistor drain current

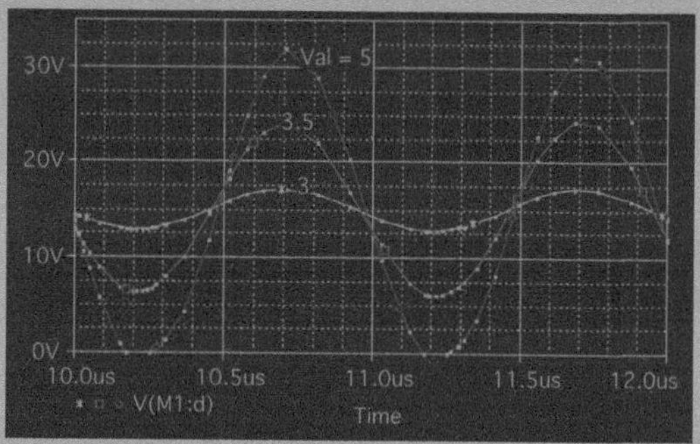

(b) Transistor drain voltage

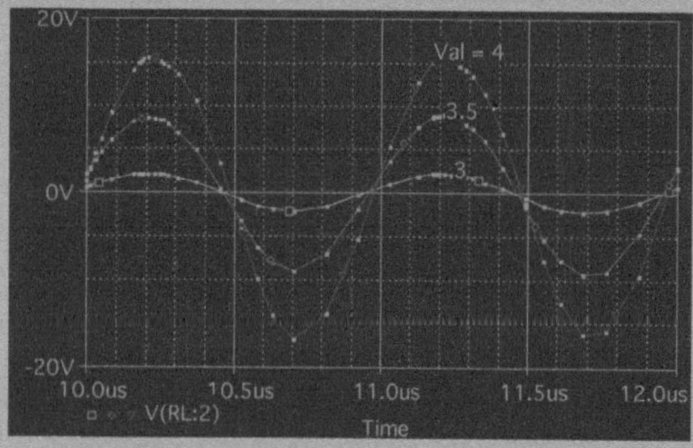

(c) Output load voltage

FIGURE 11.32 PSpice voltages and currents of class C amplifier for Example 11.8

11.8 Class D Amplifiers

A class D amplifier as shown in Fig. 11.33(a) operates transistors M_1 and M_2 as complementary switches—which are either completely turned on or completely turned off with a PWM gate signal—which are generated by comparing the sinusoidal reference (or input) signal voltage v_r with a triangular wave carrier signal v_{cr}. The technique for the generation of the PWM waveform is shown in Fig. 11.33(b), which can be implemented with an op-amp comparator. When the gate–source voltage is positive, the NMOS is switched on while the PMOS is turned off, and the voltage at the transistor drain is $+V_{DD}$. When the gate–source voltage is negative, the PMOS is switched on while the NMOS is turned off, and the transistor output is $-V_{SS}$.

The output voltage at the drain terminal is a pulsed waveform switching between $+V_{DD}$ and V_{SS} at the carrier frequency f_{cr}, which is a multiple (in the range of 10 to 30) of the signal frequency f_s. Since the PWM waveform was generated from a sinusoidal reference signal, a low-pass *LC* filter as shown in Fig. 11.33(c) is used to restore the original sinusoidal component from the high-frequency modulated voltage. One additional *LC* filter may be used to reduce the amount of distortion. The class D amplifier is highly efficient (around 95%), which can be even closer to 100%, and finds applications in such portable devices as laptop computers and MP3 players. Class D amplifiers also find applications as high-quality audio amplifiers. The pulse width of the modulated output depends on the ratio of the peak reference signal to the peak carrier signal, commonly known as the *modulation index* [7], as given by

$$m = \frac{V_r}{V_{cr}} \tag{11.90}$$

The characteristic of the *LC* filter can be determined from the input–output relationship in the Laplace's domain of *s* as given by

$$G_f(s) = \frac{V_o(s)}{V_i(s)} = \frac{\omega_n^2}{s^2 + 2\delta\omega_n s + \omega_n^2} \tag{11.91}$$

where the natural resonant frequency is $\omega_n = 1/\sqrt{LC}$, and the damping factor $\delta = (1/2R)\sqrt{L/C}$. By substituting $s = j\omega$ in Eq. (11.91), we can find the filter transfer function in the frequency domain as given by

$$G_f(j\omega) = \frac{\omega_n^2}{(j\omega)^2 + j\omega 2\delta\omega_n + \omega_n^2} = \frac{\omega_n^2}{\omega_n^2 - \omega^2 + j\omega 2\delta\omega_n} \tag{11.92}$$

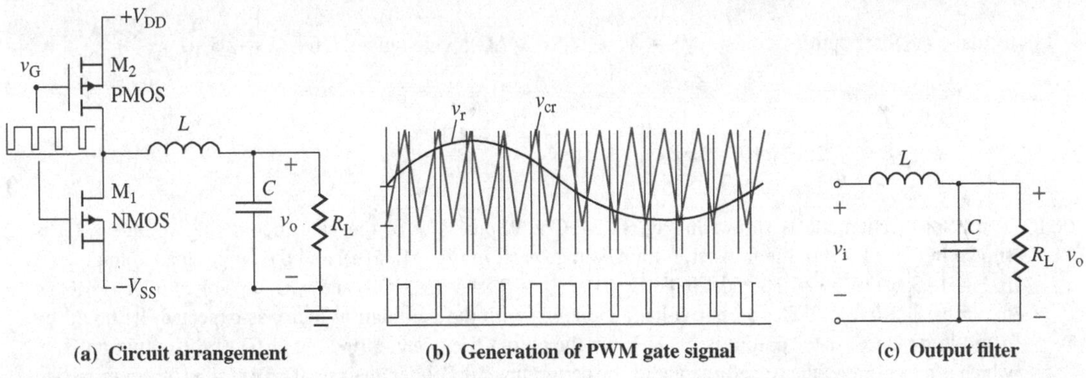

(a) Circuit arrangement (b) Generation of PWM gate signal (c) Output filter

FIGURE 11.33 Class D amplifier

which can be expressed in polar form with normalized frequency ratio $u = \omega/\omega_n$ as given by

$$G_f(j\omega) = \frac{1}{1 - u^2 + j2\delta u} = \frac{1}{\sqrt{(1 - u^2)^2 + (2\delta u)^2}} \angle - \tan^{-1}\left(\frac{1 - u^2}{2\delta u}\right) \quad (11.93)$$

which shows that $u = \omega/\omega_n$ affects the amount of phase shift present at the output. A high ω_n will decrease the phase shift, whereas a low ω_n will increase it. To minimize the amount of ripple present at the output, ω_n has to be smaller than the carrier frequency $\omega_{cr}(=1/2\pi f_{cr})$, and it should ideally be equal to the reference frequency $\omega_r(=2\pi f_r)$.

EXAMPLE 11.9

Finding the values of *L* and *C* for a class D amplifier The gate voltage of the class D amplifier in Fig. 11.33(a) is generated by modulating a sinusoidal reference $V_r = 1$ V at $f_r = 20$ kHz with a carrier signal $V_{cr} = 1$ V at $f_{cr} = 30f_s$. The DC supply voltages are $V_{DD} = \pm 5$ V. The load resistance $R_L = 8\ \Omega$ speaker.

(a) Find the values of L and C to form a tuned resonant circuit.

(b) Use PSpice to plot the gate-modulating voltage $v_G(t)$, the transistor output voltage $v_D(t)$, and the transistor output voltage $v_D(t)$ by a modulation index of $m = 0.5$ and 0.9. Use the PSpice parameters of the NMOS IRF150 whose $V_t = 2.84$ V and the PMOS IRF9140 whose $V_t = 3.67$ V.

SOLUTION

(a) The lower-order harmonic component is $f_L = 2f_s = 2 \times 20 \times 10^3 = 40 \times 10^3$. For the impedance of the filter capacitor, C must be much smaller than R_L so that the higher-order harmonic components pass through C. Using a 10:1 ratio, we can write $R_L >> 1/(2\pi f_L C)$ or

$$C = \frac{10}{2\pi f_L R_L} = \frac{10}{2\pi \times 40 \times 10^3 \times 8} \simeq 2.5\ \mu\text{F}$$

Let us make the filter cutoff frequency $f_n = f_L = 1/(2\pi\sqrt{LC})$, which gives L for $C = 2.5\ \mu$F as

$$L = \frac{1}{C(2\pi f_n)^2} = \frac{1}{2.5 \times 10^{-6}(2\pi \times 40 \times 10^3)^2} \simeq 6.4\ \mu\text{H}$$

(b) The PSpice schematic is shown in Fig. 11.34. The PSpice plot of the gate modulating voltage $v_G(t)$ is shown in Fig. 11.35(a), the transistor drain voltage $v_D(t)$ in Fig. 11.35(b), and the output load voltage $v_o(t)$ in Fig. 11.35(c) for $m = 0.8$ and 1.0. For $v_G(t) < V_t = 2.84$ V (MMOS) and $v_G(t) > V_t = 3.67$ V (PMOS), the transistors are off. The output voltage increases with the modulation index as expected. It should be noted that the transistor output is $2V_{DD}$. From the output file, we can find the THD of the output voltage, which is a measure of the waveform quality, by performing the Fourier analysis: THD = 4.847% for $m = 0.8$, and THD = 3.0% for $m = 1$.

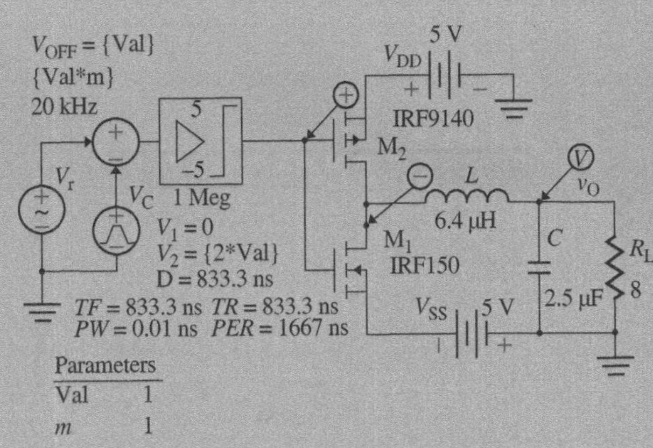

FIGURE 11.34 PSpice schematic for Example 11.9

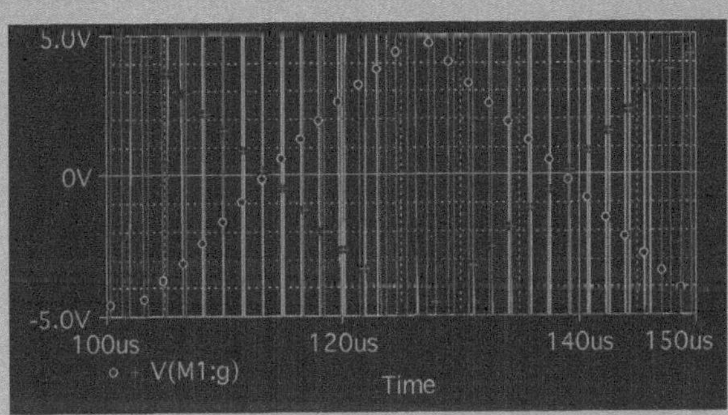

(a) Gate modulated voltage

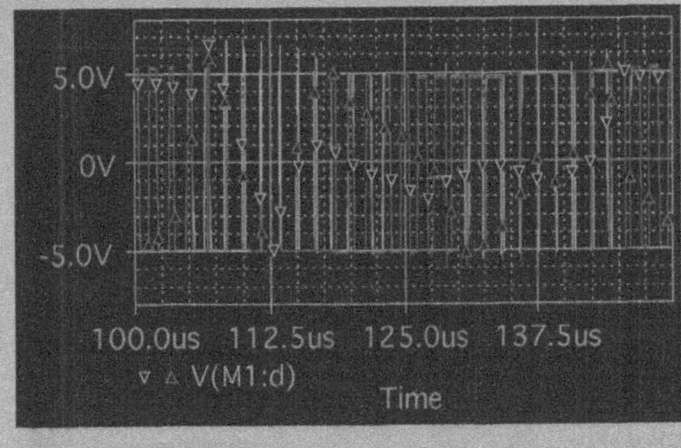

(b) Transistor drain voltage

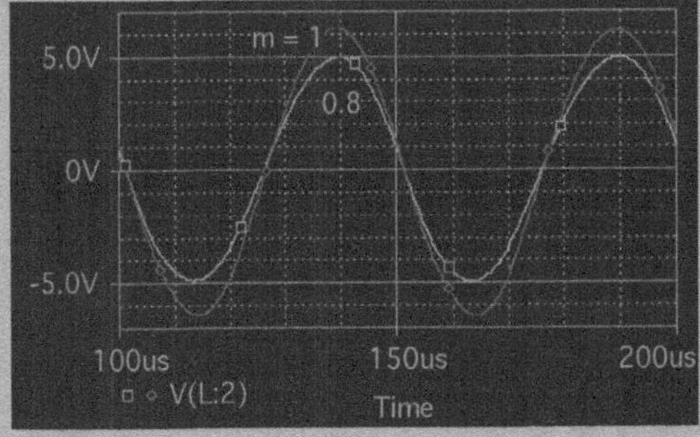

(c) Output load voltage

FIGURE 11.35 PSpice voltages and currents of a class D amplifier for Example 11.9

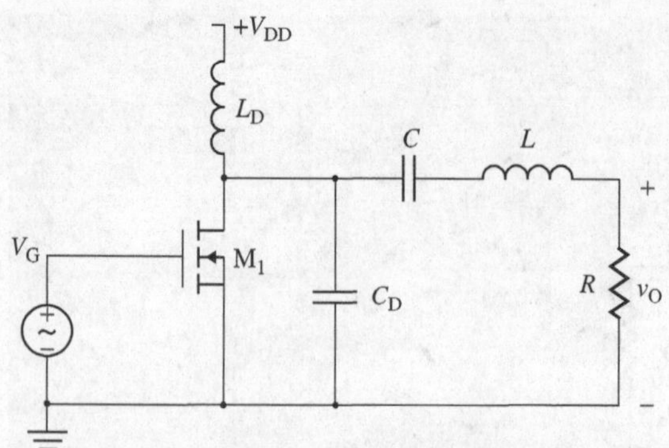

FIGURE 11.36 Class E amplifier

11.9 Class E Amplifiers

As shown in Fig. 11.36, a class E amplifier is similar to the class D amplifier, except that a capacitor C_D is connected across the transistor. It uses only one transistor as a switch and has low switching losses, yielding a high efficiency of more than 95%. It is used for applications requiring less than 100 W. When the transistor is switched on, the current through the inductor L_D rises; when the transistor is turned off, L_D forms a resonant circuit with C_D and oscillates on. As a result, the transistor output is almost resonant voltage. To minimize switching losses and increase efficiency, the transistor should be gated when the resonance current through the inductor L_D reaches its minimum low (ideally zero). That means the transistor will be turned on at zero current and will be subjected to minimum switching losses, increasing the amplifier efficiency.

To obtain an almost sinusoidal current through the load, the values of L and C are chosen to have a high quality factor, $Q \geq 7$, and a low damping ratio, usually $\delta \leq 0.072$. The optimum parameters to give the maximum efficiency can be found from [7]

$$L_D = \frac{0.4001R}{\omega_s} \tag{11.94}$$

$$C_D = \frac{2.165}{R\omega_s} \tag{11.95}$$

$$\omega_s L_D - \frac{1}{\omega_s C_D} = 0.3533R \tag{11.96}$$

EXAMPLE 11.10

Finding the values of L_D, C_D, L, and C for a class E amplifier The gate voltage of the class E amplifier in Fig. 11.36 is a sinusoidal voltage of 4 V (peak) at 100 kHz. The DC supply voltages are $V_{DD} = 12$ V. The load resistance $R = 10\ \Omega$.

(a) Find the optimum values of L_D, C_D, L, and C.

(b) Use PSpice to plot the gate voltage $v_G(t)$, the transistor drain current $i_D(t)$, the transistor output voltage $v_D(t)$, and the output voltage $v_o(t)$ for peak gate voltages of 4 V and 4.5 V. Use the PSpice parameters of the NMOS IRF150 whose $V_t = 2.84$ V.

SOLUTION

(a) For $f_s = 100$ kHz,

$$\omega_s = 2\pi f_s = 2\pi \times 100\text{ k} = 6.283 \times 10^5 \text{ rad/s}$$

From Eq. (11.94),

$$L_D = \frac{0.4001R}{\omega_s} = \frac{0.4001 \times 10}{6.283 \times 10^5} \simeq 6.4\ \mu\text{H}$$

From Eq. (11.95),

$$C_D = \frac{2.165}{R\omega_s} = \frac{2.165}{10 \times 6.283 \times 10^5} \simeq 345\text{ nF}$$

For choosing $Q = 7$,

$$L_D = \frac{QR}{\omega_s} = \frac{7 \times 10}{6.283 \times 10^5} = 111\ \mu\text{H}$$

From Eq. (11.96),

$$C_D = \frac{1}{(\omega_s L - 0.3533\ R)\omega_s} = \frac{1}{(6.283 \times 10^5 \times 111 \times 10^{-6} - 0.3533 \times 10) \times 6.283 \times 10^5} \simeq 24\text{ nF}$$

(b) The PSpice schematic is shown in Fig. 11.37. The PSpice plots of the gate voltage $v_G(t)$ and the drain current i_D are shown in Fig. 11.38(a), the transistor output voltage $v_D(t)$ and the inductor current in Fig. 11.38(b), and the output load voltage $v_o(t)$ in Fig. 11.38(c) for peak gate voltages of 4 V and 4.5 V. For $v_G(t) < V_t = 2.84$ V (MMOS), the transistors are off. The output voltage increases with the gate voltage as expected, and the transistor output is $2V_{DD}$. From the output file, we can find the THD of the output voltage, which is a measure of the waveform quality, by performing the Fourier analysis: THD = 4.279% for Val = 4 V and THD = 3.078% for Val = 4.5 V.

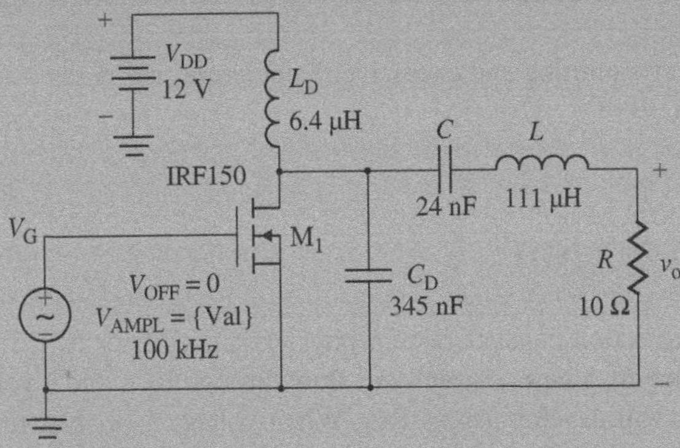

FIGURE 11.37 PSpice schematic for Example 11.10

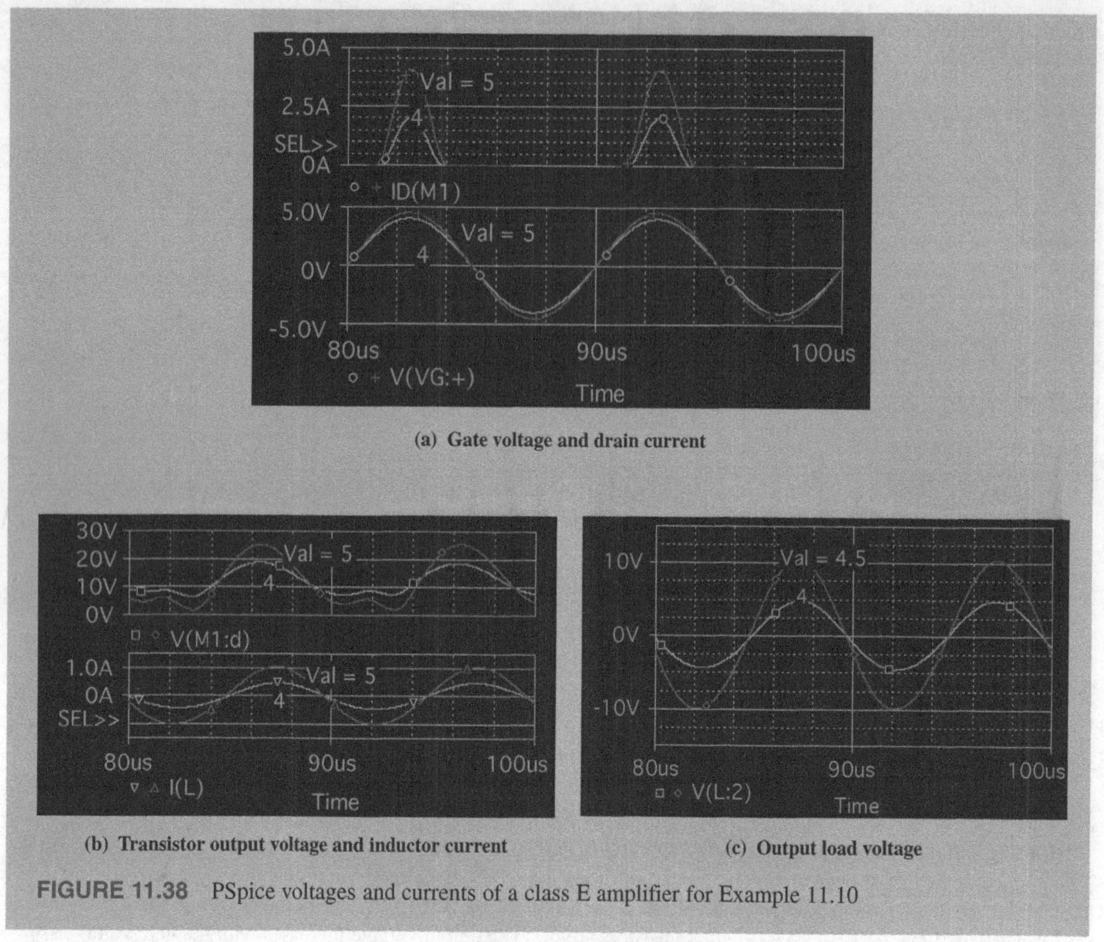

(a) Gate voltage and drain current

(b) Transistor output voltage and inductor current

(c) Output load voltage

FIGURE 11.38 PSpice voltages and currents of a class E amplifier for Example 11.10

11.10 Short-Circuit and Thermal Protection

An output stage is normally protected against short-circuiting and excessive temperature rise. A class AB amplifier with such protection is shown in Fig. 11.39.

11.10.1 Short-Circuit Protection

The circuit used in Fig. 11.39 for short-circuit protection consists of transistor Q_1 and resistor R_{E1}. If a short circuit occurs at the load while Q_N is conducting, a large current will flow through R_{E1}, and a voltage V_{RE1} proportional to the short-circuit current will develop across R_{E1}. When voltage V_{RE1} becomes large enough, transistor Q_1 will turn on and carry most of the biasing current I_{R1}. Thus, the base current of Q_N will be reduced to a safe level. The voltage drops across the emitter resistors will reduce the output

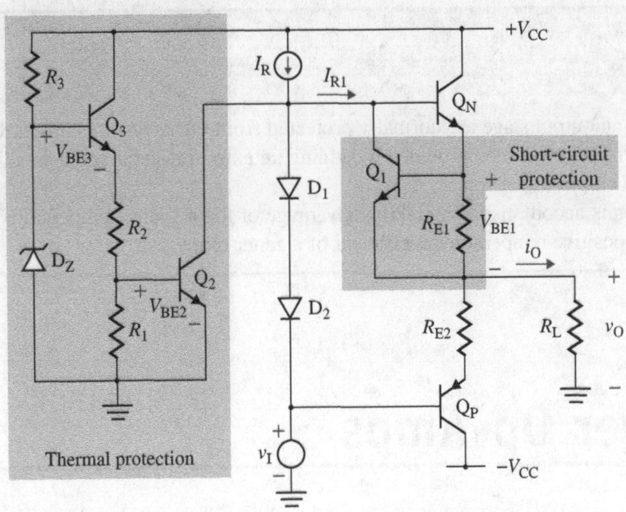

FIGURE 11.39 Short-circuit and thermal protection in a class AB amplifier

voltage by the same amount. Therefore, the values of R_{E1} and R_{E2} should be as low as possible (on the order of milliohms). Their values are determined from

$$R_{RE1} = R_{RE2} = \frac{V_{BE1}}{i_{O(short)}} \tag{11.97}$$

where $i_{O(short)}$ is the permissible short-circuiting current to turn on transistor Q_1. For $V_{BE1} = 0.7$ V and $i_{O(short)} = 200$ mA, $R_{RE1} = R_{RE2} = 0.7/200$ mA $= 3.5\ \Omega$. Although R_{E1} and R_{E2} reduce the output voltage swing, they will give biasing stability to the quiescent current I_Q and protect Q_N and Q_P against thermal runaway (i.e., excessive junction temperature rise).

11.10.2 Thermal Protection

The circuit used in Fig. 11.39 for thermal protection consists of two transistors (Q_2 and Q_3), three resistors (R_1, R_2, and R_3), and a zener diode. Normally, transistor Q_2 is off. If the temperature rises, V_{BE3} will fall because of the negative temperature coefficient of Q_3, and the zener voltage V_Z will rise because of the positive temperature coefficient of zener diode D_Z. As a result, the voltage at the emitter of Q_3 will rise, and the voltage at the base of Q_2 will also rise. If the temperature rise is adequate, Q_2 will turn on, diverting the reference current I_R from the amplifier and hence shutting down the amplifier. The maximum permissible temperature rise ΔT can be determined from

$$[\Delta T(K_{DZ} + K_{Q3}) + V_{BE3}]\frac{R_1}{R_1 + R_2} = V_{BE2} = V_T \ln\left(\frac{I_R}{I_S}\right) \tag{11.98}$$

where K_{DZ} = temperature coefficient of the zener diode, in V/°C
K_{Q3} = temperature coefficient of transistor Q_3, in V/°C
V_{BE3} = quiescent B-E voltage of Q_3

KEY POINTS OF SECTION 11.10

- The transistors of an output stage are normally protected from the excessive current that results from short-circuiting. Protection can be provided by adding an extra transistor and a small collector resistor for each of the output transistors.
- Thermal protection is accomplished by taking advantage of the negative temperature coefficient of a transistor and the positive temperature coefficient of a zener diode.

11.11 Power Op-Amps

Op-amps have some desirable characteristics, such as a very high open-loop gain ($>10^5$), a very high input impedance (up to $10^9\ \Omega$), and a very low input biasing current. However, the AC output power of op-amps is generally low. High power can be obtained from a power amplifier consisting of an op-amp followed by a class AB buffer. The general structure of a power op-amp is shown in Fig. 11.40. The buffer stage consists of transistors Q_1, Q_2, Q_3, and Q_4. R_1 and R_2 bias transistors Q_1 and Q_2 such that $-V_{BE1} + V_{BE3} \approx 0$ and $V_{BE2} - V_{BE4} = 0$. Transistor Q_3 supplies the positive load current until the voltage across R_3 is sufficiently large to turn on Q_5. Then Q_5 supplies additional load current. Similarly, transistors Q_4 and Q_6 supply the negative load current. The stage formed by Q_5 and Q_6 supplies the additional load current and acts as a *current booster*. Emitter resistors R_{E1} and R_{E2} are used for biasing stability.

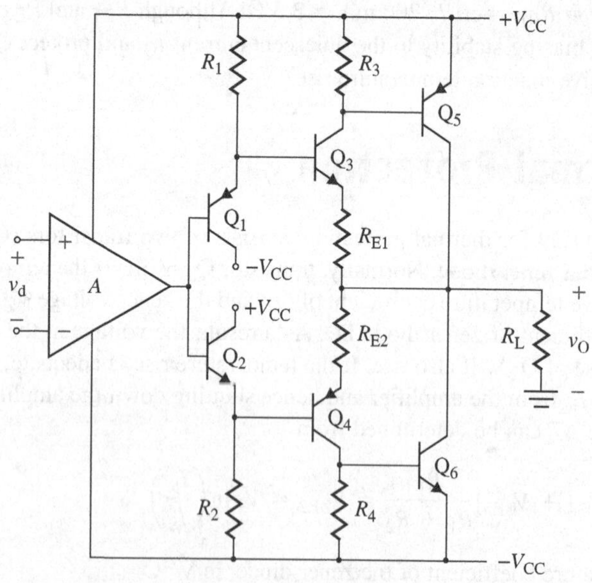

FIGURE 11.40 General structure of a power op-amp

11.11.1 IC Power Amplifiers

A variety of power amplifiers that combine a conventional op-amp chip with a current booster are available commercially. Some have internal negative feedback already applied to give a fixed closed-loop voltage gain; others do not have on-chip feedback. We will consider two such representative op-amps: the LH0021 op-amp and the LM380 op-amp, both manufactured by National Semiconductor.

Power Op-Amp LH0021

The schematic of power op-amp LH0021 is shown in Fig. 11.41. The LH0021 is designed to operate from a power supply of ±25 V. It is capable of a peak output voltage swing of about 12 V into a 10-Ω load over the entire frequency range of 15 kHz. The distortion of the output voltage is less than 1.6%. The circuit can be divided into three stages: a differential stage, a gain stage, and an output stage.

The differential input stage consists of transistors Q_1 through Q_4 biased by Q_7, which sinks the base currents of Q_3 and Q_4. Transistors Q_5 and Q_6 serve as the current mirror active load.

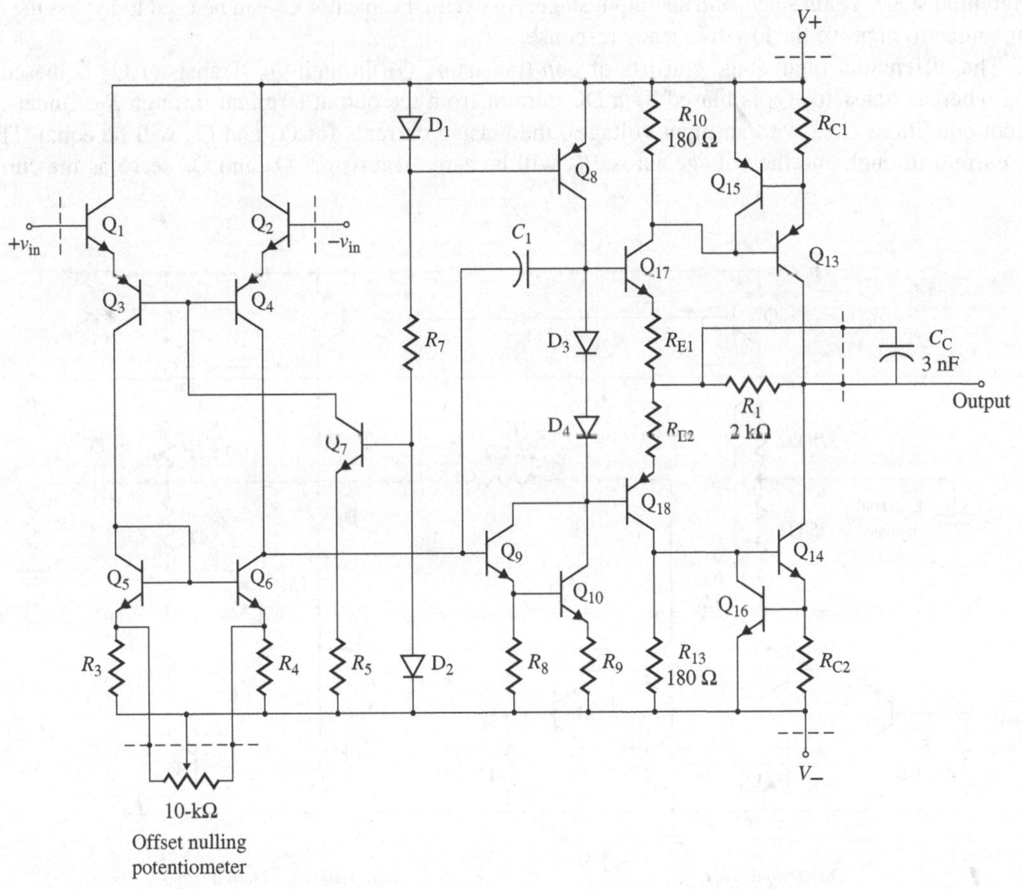

FIGURE 11.41 Power op-amp LH0021 (Courtesy of National Semiconductor, Inc.)

The gain stage is a common-emitter configuration and consists of Q_9 and Q_{10} connected as a Darlington pair. Transistor Q_8 serves as an active current source for this stage. Capacitor C_1 is the pole-splitting compensating capacitor connected in shunt-shunt feedback.

The output stage is a complementary class AB push-pull circuit. It consists of transistors Q_{13}, Q_{14}, Q_{15}, Q_{16}, Q_{17}, and Q_{18}. Diodes D_3 and D_4 provide the biasing voltage for class AB operation in order to minimize crossover distortion. Transistors Q_{13} and Q_{14} act as the current booster. Resistors R_{C1} and R_{C2} limit the currents through Q_{13} and Q_{14}, respectively, by turning on Q_{15} and Q_{16}. Resistor R_1 protects Q_{17} and Q_{18} by limiting the current flow through them.

A small external capacitor C_C is connected to offer a low impedance to a capacitive load. The combination of R_1 and the load capacitor does not form a low-pass *RC* network, and any phase delay of the output voltage can be avoided.

Power Op-Amp LM380

The schematic of power op-amp LM380 is shown in Fig. 11.42. The LM380 is designed to operate from a single power supply in the range of 12 V to 22 V. The output power can be as high as 5 W onto a 10-Ω load. The distortion of the output voltage is less than 3%. The circuit can be divided into three stages: a differential stage, a gain stage, and an output stage. An external capacitor C_x can be used to bypass the current source to improve the low-frequency response.

The differential input stage consists of *pnp* transistors Q_3 through Q_6. Transistor Q_3 is biased by Q_{10}, whereas transistor Q_4 is biased by a DC current from the output terminal through R_2. Under quiescent conditions (i.e., with an input voltage), the biasing currents for Q_3 and Q_4 will be equal. Thus, the current through and the voltage across R_3 will be zero. Transistors Q_5 and Q_6 serve as the current

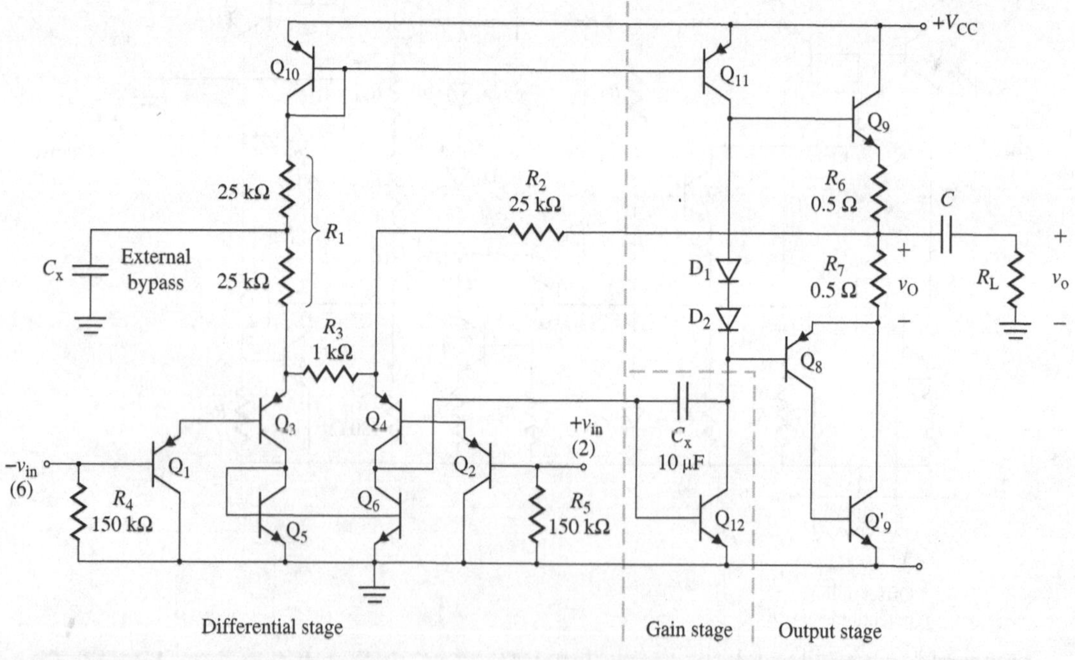

FIGURE 11.42 Power op-amp LM380 (Courtesy of National Semiconductor, Inc.)

mirror active load for this stage. The *pnp* transistors Q_1 and Q_2 act as emitter followers for input buffering. Resistors R_1 and R_2 provide DC paths to the ground for the base currents of Q_3 and Q_4.

The gain stage consisting of Q_{12} has the common-emitter configuration. Transistor Q_{11} serves as a current source active load for the gain stage. Capacitor C is the pole-splitting compensating capacitor intended to yield a wide bandwidth.

The output stage is a quasi-complementary class AB push-pull circuit. It consists of transistors Q_8, Q_9, and Q_9'. Diodes D_1 and D_2 provide the biasing voltage for class AB operation. Emitter resistors R_6 and R_7 give biasing stability.

Resistor R_2 provides DC feedback from the DC output voltage v_O to the emitter of Q_4. If for some reason v_O increases, then there will be a corresponding increase in the current through R_2 and the emitter current I_{E4} of Q_4, causing an increase in the collector current of Q_4. As a result, the voltage at the base of Q_{12} and its base current will increase. This, in turn, will increase the collector current of Q_{12} and reduce the base current of Q_9. Thus, v_O will be reduced.

To find v_O, let us assume that all transistors are identical and the base currents are negligible compared to the emitter currents. The emitter biasing current of Q_3 can be found approximately from

$$I_{E3} = \frac{V_{CC} - V_{EB10} - V_{EB3} - V_{EB1}}{R_1} = \frac{V_{CC} - 3V_{EB}}{R_1} \tag{11.99}$$

Also, the emitter current of Q_4 can be found from

$$I_{E4} = \frac{v_O - V_{EB4} - V_{EB2}}{R_2} = \frac{v_O - 2V_{EB}}{R_2} \tag{11.100}$$

where v_O is the DC output voltage. For $I_{E3} = I_{E4}$ (so that no current flows through R_3), we get

$$\frac{V_{CC} - 3V_{EB}}{R_1} = \frac{v_O - 2V_{EB}}{R_2}$$

which, for $R_1 = 2R_2$, as shown in Fig. 11.42, gives the DC output voltage as

$$v_O = \frac{V_{CC} + V_{EB}}{2} = \frac{V_{CC} - V_{BE}}{2} \tag{11.101}$$

Thus, for $V_{BE} << V_{CC}$, which is usually the case, the DC output voltage is approximately half the supply voltage V_{CC}; that is, $v_O \approx V_{CC}/2$.

11.11.2 Bridge Amplifier

The output power can be doubled by using two power op-amps, as shown in Fig. 11.43. This arrangement, called a *bridge amplifier*, is commonly used in high-power applications. The input voltage v_I is applied to the noninverting input of one amplifier and also to the inverting input of the other, so that the output voltages are 180° out of phase. Thus, the output of the noninverting amplifier is

$$v_{O1} = \left(1 + \frac{R_F}{R_1}\right) v_I \tag{11.102}$$

The output voltage of the inverting amplifier is

$$v_{O2} = -\frac{R_3}{R_2} v_I \tag{11.103}$$

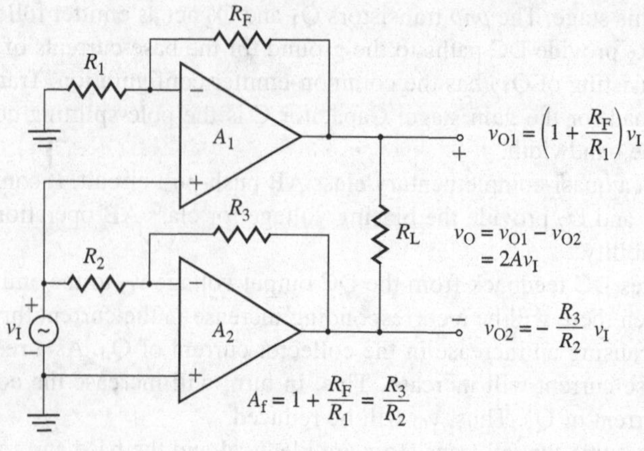

FIGURE 11.43 Bridge amplifier

The voltage across the load becomes

$$v_O = v_{O1} - v_{O2} = \left(1 + \frac{R_F}{R_1}\right) v_I + \frac{R_3}{R_2} v_I$$

which, for $R_3/R_2 = 1 + R_F/R_1 = A_f$, becomes

$$v_O = 2A_f v_I \tag{11.104}$$

where A_f is the closed-loop voltage gain of each amplifier.

KEY POINTS OF SECTION 11.11

- IC power amplifiers are known as power op-amps. They consist of a differential stage, a gain stage, and an output stage with a current booster.
- Power op-amps are normally used with feedback and are compensated for frequency response by internal capacitance in the gain stage.

11.12 Thermal Considerations

Power transistors dissipate a large amount of power. The power dissipation is converted to heat, which causes the temperature of the collection junction to rise. The physical structure, packaging, and specifications of transistors differ depending on their current-handling capability and power dissipation. The current rating of power transistors can go as high as 500 A, with power dissipation of up to 200 W, especially when the transistors are used as switching elements for power converters [6]. Power transistors must be protected from excessive temperature rise. The junction temperature T_J must be kept within a specified maximum $T_{J(max)}$ to avoid damage to the transistor. For silicon transistors, $T_{J(max)}$ is in the range of 150°C to 200°C.

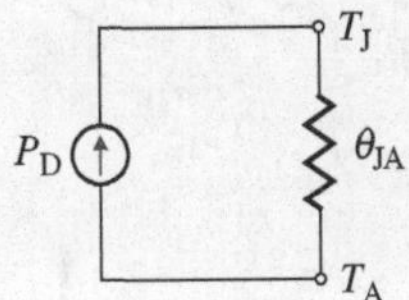

FIGURE 11.44 Electrical equivalent of thermal process

11.12.1 Thermal Resistance

If a transistor operates in the open air without any cooling arrangement, the heat will be transferred from the transistor junction to the ambient. Thermal resistance is a measure of the heat transfer. It is the temperature drop divided by power dissipation under steady-state conditions. Thus, the units are °C/W. The thermal resistance θ_{JA} for heat flow from the junction to the ambient is given by

$$\theta_{JA} = \frac{T_J - T_A}{P_D} \text{ (in °C/W)} \tag{11.105}$$

where T_J = junction temperature, in °C
T_A = ambient temperature, in °C

Equation (11.105) represents the heat-transfer process and is analogous to Ohm's law. Power dissipation corresponds to current, temperature difference to voltage difference, and thermal resistance to electrical resistance. Thus, the thermal process can be represented by an analogous electrical circuit, as shown in Fig. 11.44.

11.12.2 Heat Sink and Heat Flow

To keep the junction temperature below $T_{J(max)}$, the transistor is normally mounted on a heat sink, which facilitates the removal of heat from the device to the surrounding air. Typical heat sinks with devices attached are shown in Fig. 11.45. Heat is transferred from the device to the air by one of three methods:

1. Conduction from junction to case with thermal resistance θ_{JC} and from case to heat sink with thermal resistance θ_{CS}. The values of θ_{JC} and θ_{CS} depend on the cross section, length, and temperature difference across the conducting medium. The imperfect matching of adjacent surfaces will increase θ_{CS}.

FIGURE 11.45 Transistors mounted on heat sinks

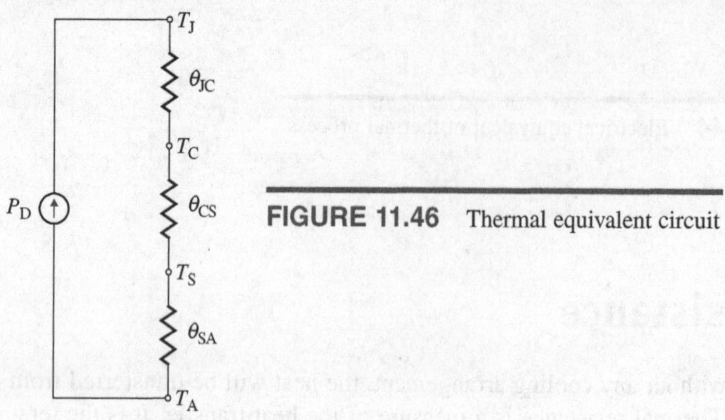

FIGURE 11.46 Thermal equivalent circuit

The value of θ_{CS} can be made small by coating the mated surfaces of the transistor and sink with a thermally conducting compound.

2. Convection from case to ambient with thermal resistance θ_{CA} and from heat sink to ambient with thermal resistance θ_{SA}. The values of θ_{CA} and θ_{SA} depend on surface condition, type of convecting fluid, velocity and characteristic of the fluid, and temperature difference between surface and fluid.
3. Radiation from cooling fins to the air. The heat transfer will depend on surface emissivity and area, as well as temperature difference between the radiating fins and the air.

The thermal equivalent circuit of a transistor on a heat sink is shown in Fig. 11.46. The power dissipation is related to the junction temperature T_J and the ambient temperature T_A by

$$T_J - T_A = P_D(\theta_{JC} + \theta_{CS} + \theta_{SA}) \tag{11.106}$$

where θ_{JC} = thermal resistance from junction to case, in °C/W

θ_{CS} = thermal resistance from case to sink, in °C/W

θ_{SA} = thermal resistance from sink to ambient, in °C/W

The values of θ_{JC} and θ_{CS} are specified in the manufacturer's data sheet. The circuit designer has to specify the required value of θ_{SA} for the heat sink. The value of θ_{SA} found from Eq. (11.106) may be too small to match any standard heat sink with natural air cooling. In this case, it might be necessary to cool the heat sink using liquid or forced air from a fan.

11.12.3 Power Dissipation versus Temperature

The ambient temperature T_A and the case temperature T_C are related to the power dissipation P_D by

$$T_C - T_A = P_D(\theta_{CS} + \theta_{SA}) \tag{11.107}$$

The transistor manufacturer normally specifies the maximum junction temperature $T_{J(max)}$, the maximum power dissipation $P_{D(max)}$ at a specified case temperature T_{C0} (usually 25°C), and the thermal resistance θ_{JC}. The manufacturer often provides a power-derating curve, as shown in Fig. 11.47, which gives the maximum allowable power dissipation if the case temperature is higher than 25°C. T_{C0} is the case temperature

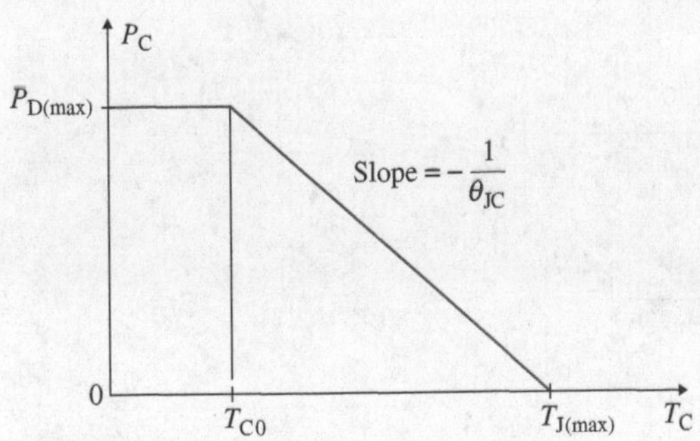

FIGURE 11.47 Power derating curve

at which the derating begins. $T_{C(max)}$ is the maximum value of the case temperature in degrees Celsius and is equal to $T_{J(max)}$. The power dissipation at a case temperature of T_C can be found from

$$P_D(T = T_C) = P_{D(max)} - \frac{P_{D(max)}}{T_{J(max)} - T_{CO}}(T_C - T_{CO}) \quad \text{for } T_C \geq T_{CO} \tag{11.108}$$

EXAMPLE 11.11

Finding the power dissipation of a transistor The power dissipation of a transistor is specified as $P_{D(max)} = 150$ W at $T_{CO} = 25°C$. The transistor is mounted on a heat sink. The thermal resistances are $\theta_{JC} = 0.5°C/W$, $\theta_{CS} = 0.2°C/W$, and $\theta_{SA} = 1.5°C/W$. If $T_{J(max)} = 200°C$ and $T_A = 45°C$, calculate the maximum permissible power dissipation of the transistor.

SOLUTION

We can find the maximum power dissipation P_D from Eq. (11.106):

$$200 - 45 = P_D(0.5 + 0.2 + 1.5) = 2.2P_{D(max)}$$

which gives

$$P_D = \frac{200 - 45}{2.2} = 70.5 \text{ W}$$

From $P_D = 70.5$ W, Eq. (11.107) gives the case temperature as

$$T_C = 45 + 70.5 \times (0.2 + 1.5) = 164.9°C$$

The corresponding power dissipation becomes

$$P_D(T_C = 164.9°C) = 150 - \frac{150}{200 - 25} \times (164.9 - 25) = 30.1 \text{ W}$$

In other words, for $P_D = 70.5$ W, $T_C = 164.9°C$, which in turn limits the power dissipation to 30.1 W. Therefore, P_D cannot be 70.5 W. We need to find the actual power. Substituting T_C from Eq. (11.107) into Eq. (11.108) yields

$$P_D = P_{D(max)} - \frac{P_{D(max)}}{T_{J(max)} - T_{CO}}[T_A + P_D(\theta_{CS} + \theta_{SA})]$$

which gives

$$P_D\left[1+\frac{P_{D(max)}}{T_{J(max)}-T_{CO}}(\theta_{CS}+\theta_{SA})\right]=P_{D(max)}\left[1-\frac{T_A}{T_{J(max)}-T_{CO}}\right]$$

Since

$$P_D\left[1+\frac{150}{200-25}(0.2+1.5)\right]=150\times\left[1-\frac{45}{200-25}\right]$$

the permissible power dissipation is $P_D = 45.35$ W.

▶ **NOTE** $\theta_{JC} = P_{D(max)}/(T_{J(max)} - T_{CO}) = 150/(200 - 25) = 0.857°C/W$. But in Example 11.11, $\theta_{JC} = 0.5°C/W$ was specified in order to illustrate the method for finding the actual power dissipation.

KEY POINTS OF SECTION 11.12

- Power dissipates from a transistor in the form of heat, which causes the temperature of the collection junction to rise. Power transistors or power op-amps are normally mounted on a heat sink, which provides a low thermal impedance for the heat flow.
- The maximum power dissipation is specified by the manufacturer at a specified case temperature T_{CO} (usually 25°C), which is not normally attainable. It is often necessary to determine the allowable power dissipation from a derating curve.

11.13 Design of Power Amplifiers

Since class B and class AB amplifiers eliminate the dead zone, they are generally used as the output stages of practical amplifiers. Thus, the design of a power amplifier is mainly the design of the output stage, and it involves the following steps:

Step 1. Identify the specifications of the output stage (e.g., output power P_L, load resistance R_L, and DC supply voltages V_{CC} and V_{EE} or V_{DD} and V_{SS}).

Step 2. Select the type of output operation, usually class B or class AB operation.

Step 3. Determine the voltage and current ratings of all transistors.

Step 4. Determine the values and power ratings of all resistors. Also, determine the turns ratio(s) of transformers in case of a transformer-coupled load.

Step 5. Select the type of DC-biasing circuit. Determine the specifications of active and passive components.

Step 6. Select the power transistors that will meet the voltage, current, and power requirements. Find their maximum junction temperature $T_{J(max)}$ and thermal resistances θ_{JC} and θ_{CS}.

Step 7. Determine the power dissipation of the transistors, and find the desired thermal resistance of the heat sink.

Step 8. Use PSpice/SPICE to simulate and verify your design, using the standard values of components with their tolerances.

Summary

A power amplifier generally forms the output stage of an audio-frequency amplifier. The requirements of power amplifiers are different from those of small-signal low-power amplifiers. Power amplifiers must deliver an appreciable amount of power to a low-impedance load, while at the same time creating very little distortion in the output signal. Power amplifiers are generally classified into six types: class A, class B, class AB, class C, class D, and class E. The efficiency of a class A amplifier is only 25%, that of a class B type is 50%, and that of a class AB push-pull type is 78.5%. Class D and E amplifiers can give 90% or higher efficiency. Complementary class AB push-pull amplifiers eliminate or reduce the distortion and dead zone in the output signal. The output stage is normally biased by an active current source in order to supply quiescent biasing current at zero input signal. A quasi-complementary amplifier, which uses a composite *pnp* transistor consisting of a *pnp* and an *npn* transistor, can increase the output power. Power IC op-amps generally consist of an op-amp followed by a class AB buffer. A power transistor must be mounted on a heat sink to limit the junction temperature to an acceptable value.

References

1. P. R. Gray, P. J. Hurst, S. H. Lewis, and R. G. Meyer, *Analysis and Design of Analog Integrated Circuits*. New York: Wiley, 2001.
2. H. L. Krauss, C. W. Bostian, and F. H. Raab, *Solid State Radio Engineering*. New York: Wiley, 1981.
3. E. S. Oxner, *Power FETs and Their Applications*. Englewood Cliffs, NJ: Prentice Hall, 1982.
4. D. Foty, *MOSFET Modeling with SPICE*. Upper Saddle River, NJ: Prentice Hall, 1997.
5. M. H. Rashid, *Introduction to PSpice Using OrCAD for Circuits and Electronics*. Upper Saddle River, NJ: Prentice Hall, 2003.
6. S. Soclof, *Design and Applications of Analog Integrated Circuits*. Englewood Cliffs, NJ: Prentice Hall, 1991, Chapter 12.
7. M. H. Rashid, *Power Electronics—Circuits, Devices, and Applications*. Englewood Cliffs, NJ: Prentice Hall, 2003.

Review Questions

1. What are the six types of power amplifiers?
2. What are the major differences among the six types of power amplifiers?
3. What are the advantages of an emitter follower with active current–source biasing?
4. What is the limiting design condition for avoiding clipping on the output voltage?
5. What is a figure of merit for an amplifier?
6. What is a common-emitter class A amplifier?
7. What are the advantages and disadvantages of a common-emitter class A amplifier?
8. What is the maximum efficiency of a common-emitter class A amplifier?
9. What is a figure of merit for a common-emitter class A amplifier?
10. What is a transformer-coupled load class A amplifier?
11. What are the advantages and disadvantages of a transformer-coupled load class A amplifier?
12. What is the maximum efficiency of a transformer-coupled load class A amplifier?
13. What is a figure of merit for a transformer-coupled load class A amplifier?
14. What is a complementary class B push-pull amplifier?
15. What are the advantages and disadvantages of a complementary class B push-pull amplifier?

16. What is the maximum efficiency of a complementary class B push-pull amplifier?
17. What is a figure of merit for a complementary class B push-pull amplifier?
18. What is the cause of crossover distortion in a complementary class B push-pull amplifier?
19. What is a transformer-coupled load class B push-pull amplifier?
20. What are the advantages and disadvantages of a transformer-coupled load class B push-pull amplifier?
21. What is the maximum efficiency of a transformer-coupled load class B push-pull amplifier?
22. What is a figure of merit for a transformer-coupled load class B push-pull amplifier?
23. What are the methods of eliminating or minimizing crossover distortion in a complementary class B push-pull amplifier?
24. What is a complementary class AB push-pull amplifier?
25. What are the advantages and disadvantages of a complementary class AB push-pull amplifier?
26. What is the maximum efficiency of a complementary class AB push-pull amplifier?
27. What is a figure of merit for a complementary class AB push-pull amplifier?
28. What are the methods for DC biasing an output stage?
29. What are the methods for providing short-circuit and thermal protection for an output stage?
30. What is an IC power amplifier?
31. What is the thermal equivalent circuit of a transistor?
32. What is the power derating curve of a power transistor?
33. What is the purpose of a heat sink for a power transistor?
34. What is class C amplifier?
35. What is class D amplifier?
36. What is class E amplifier?

Problems

The symbol **D** indicates that a problem is a design problem. The symbol **P** indicates that you can check the solution to a problem using PSpice/SPICE or Electronics Workbench.

Assume that the PSpice model parameters for the diodes are

```
IS=100E-15 BV=100 IBV=100E-13
```

and for the transistors are

```
IS=100E-15 IBV=100E-14 BF=100 VJE=0.8 VA=100
```

(unless specified).

11.4 Class A Amplifiers

11.1 a. Design an emitter follower as shown in Fig. 11.3(b). Assume $V_{CC} = 15$ V, $V_{BE} = 0.7$ V, $V_{CE(sat)} = 0.5$ V, $I_R = 10$ mA, and $R_L = 1$ kΩ. Assume identical transistors of $\beta_F = h_{fe} = 100$. Find the voltage, current, and power rating of the transistors.

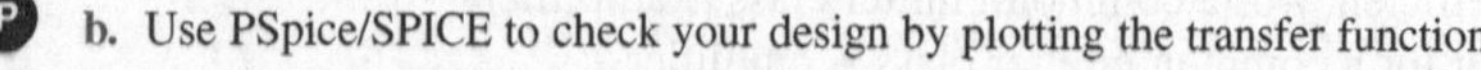

b. Use PSpice/SPICE to check your design by plotting the transfer function.

11.2 For the emitter follower in Prob. 11.1, determine the critical value of load resistance to avoid clipping (or distortion), and calculate the peak-to-peak output voltage swing V_{pp} if $R_L = 2$ kΩ.

11.3 The parameters of the emitter follower in Fig. 11.3(b) are $V_{CC} = 12$ V, $R_1 = 2.5$ kΩ, and $R_L = 750$ Ω. The transistors are identical, and their parameters are $V_{BE} = 0.7$ V, $V_{CE(sat)} = 0.5$ V, and $\beta_F = h_{fe} = 100$. Determine (**a**) the peak voltage and current of transistor Q_1 and (**b**) the average output power P_L.

11.4 For the emitter follower in Prob. 11.3, calculate **(a)** the power efficiency, **(b)** the value of R_L for maximum efficiency, and **(c)** the corresponding value of efficiency η_{max}.

11.5 D P **a.** Design a class A amplifier, as shown in Fig. 11.7(a), for high efficiency. The audio output power is 48 W. The available supply voltage is 12 V. Assume transistors of $\beta_F = h_{fe} = 100$. Find the voltage, current, and power rating of the transistors.

b. Use PSpice/SPICE to check your design by plotting the collector current and voltage.

11.6 The parameters of the class A amplifier in Fig. 11.9(a) are $V_{CC} = 15$ V, $R_1 = 15$ kΩ, and $R_L = 1$ kΩ. The transistors are identical, and their parameters are $V_{BE} = 0.7$ V, $V_{CE(sat)} = 0.5$ V, and $\beta_F = h_{fe} = 100$. Determine **(a)** the peak voltage and current of transistor Q_1, **(b)** the average output power P_L, and **(c)** the efficiency η.

11.7 D P **a.** Design a transformer-coupled load class A amplifier, as shown in Fig. 11.10(a), to have high efficiency and to supply an output power of $P_L = 32$ W at a load resistance of $R_L = 8$ Ω. Assume a DC supply voltage of $V_{CC} = 15$ V. Find the voltage, current, and power rating of the transistors and the transformer turns ratio. Assume $\beta_F = 100$ and $V_{CE(sat)} = 0.5$ V.

b. Use PSpice/SPICE to check your design by plotting the collector current and voltage.

11.8 The parameters of the transformer-coupled load class A amplifier in Fig. 11.10(a) are $V_{CC} = 15$ V and $R_L = 16$ Ω. The transistor parameters are $V_{BE} = 0.7$ V, $V_{CE(sat)} = 0.5$ V, and $h_{fe} = 100$. The transformer turns ratio is 2:1. Determine **(a)** the peak voltage and current of transistor Q_1, **(b)** the average output power P_L, and **(c)** the efficiency η.

11.5 *Class B Push-Pull Amplifiers*

11.9 D P **a.** Design a complementary class B push-pull amplifier, as shown in Fig. 11.11(a), to supply an output power of $P_L = 16$ W at a load resistance of $R_L = 4$ Ω. Assume a DC supply voltage of $V_{CC} = 15$ V and transistors of $\beta_F = h_{fe} = 100$ and $V_{BE} = 0.7$ V. Find the voltage, current, and power rating of the transistors.

b. Use PSpice/SPICE to check your design by plotting the transfer function, the output voltage, and the load current.

11.10 For the amplifier in Prob. 11.9, calculate the efficiency and power dissipation of each transistor.

11.11 Calculate the power efficiency η and power dissipation P_D of each transistor in the complementary class B push-pull output stage in Fig. 11.11(a) if $V_{CC} = 15$ V and $R_L = 10$ Ω. The parameters of the transistors are $\beta_F = h_{fe} = 150$, $V_{CE(sat)} = 0.2$ V, and $V_{BE} = 0.7$.

11.12 Calculate the power efficiency η and power dissipation P_D of each transistor in the complementary class B push-pull output stage in Fig. 11.11(a) if $V_{CC} = 12$ V and $R_L = 50$ Ω. The parameters of the transistors are $\beta_F = h_{fe} = 100$, $V_{CE(sat)} = 0.2$ V, and $V_{BE} = 0.7$.

11.13 D P Design a transformer-coupled class B push-pull amplifier, as shown in Fig. 11.17(a), to supply an output power of $P_L = 32$ W at a load resistance of $R_L = 8$ Ω. Assume a DC supply voltage of $V_{CC} = 15$ V and transistors of $\beta_F = h_{fe} = 100$ and $V_{BE} = 0.7$ V.

11.14 The parameters of the transformer-coupled class B push-pull amplifier in Fig. 11.17(a) are $V_{CC} = 15$ V and $R_L = 8$ Ω. The transistor parameters are $V_{BE} = 0.7$ V, $V_{CE(sat)} = 0.5$ V, and $\beta_F = h_{fe} = 100$. The transformer turns ratio is 2:1. Determine **(a)** the peak voltage and current of transistor Q_1, **(b)** the average output power P_L, and **(c)** the efficiency η.

11.6 *Complementary Class AB Push-Pull Amplifiers*

11.15 D P **a.** Design a biasing circuit for the class AB amplifier of Fig. 11.20(c) to supply a maximum output power of $P_{L(max)} = 20$ W. The quiescent biasing current I_Q is 2 mA. Assume a DC supply voltage of $V_{CC} = 15$ V. The diode parameters are $I_S = 10^{-13}$ A and $I_{D(min)} = 1$ mA to ensure conduction. The transistor parameters are $h_{fe} = 50$, $V_{BE} = 0.7$ V, and $V_{CE(sat)} = 0.2$ V.

b. Find the biasing voltage V_{BB} for $v_O = 0$ and 11.8 V.

11.16 D P **a.** Design an active current source for the class AB amplifier in Fig. 11.21(a) in order to provide a biasing current of $I_R = 10$ mA. Assume $V_{CC} = 15$ V, $V_{BE} = 0.7$ V, and $R_L = 50\ \Omega$. The diode parameters are $I_S = 10^{-13}$ A and $I_{D(min)} = 1$ mA to ensure conduction. The transistor parameters are $h_{fe} = 50$, $V_{BE} = 0.7$ V, and $V_{CE(sat)} = 0.2$ V.

b. Use PSpice/SPICE to plot the transfer characteristic and the instantaneous i_N, i_P, and i_O for $v_I = 5 \sin(2000\pi t)$.

11.17 D P **a.** Design a V_{BE} multiplier for the class AB amplifier in Fig. 11.24 in order to provide a biasing current of $I_R = 10$ mA. Assume $V_{CC} = 15$ V, $V_{BE} = 0.7$ V, and $R_L = 50\ \Omega$. Assume a minimum current of $I_{M(min)} = 1$ mA to the multiplier and $I_Q = 2$ mA. The transistor parameters are $\beta_F = h_{fe} = 50$, $V_{BE} = 0.7$ V, and $V_{CE(sat)} = 0.2$ V.

b. Use PSpice/SPICE to plot the transfer characteristic and the instantaneous i_N, i_P, and i_O for $v_I = 5 \sin(2000\pi t)$.

11.7–11.9 *Class C, D, and E Amplifiers*

11.18 The gate voltage of the class C amplifier in Fig. 11.30(a) is $v_G = 2.5 + 1.0 \sin(2\pi f_s t)$ with $f_s = 1.5$ MHz. The DC supply voltages are $V_{DD} = \pm 15$ V. The load resistance $R_L = 50\ \Omega$, which is connected through a transformer of a turns ratio of $n = 1{:}1$.

a. Find the values of L and C to form a tuned resonant circuit.

b. Use PSpice to plot the voltage $v_o(t)$, the transistor drain current $i_D(t)$, and the transistor drain voltage $v_D(t)$for varying the gain voltage component $V_m = 3$ V to 4 V with an increment of 0.5 V.

c. Use the PSpice parameters of the NMOS IRF150 whose $V_t = 2.84$ V.

11.19 The gate voltage of the class C amplifier in Fig. 11.30(a) is $v_G = 2.5 + 1.0 \sin(2\pi f_s t)$ with $f_s = 1.5$ MHz. The DC supply voltages are $V_{DD} = \pm 15$ V. The load resistance $R_L = 150\ \Omega$, which is connected through a transformer of a turns ratio of $n = 2{:}1$.

a. Find the values of L and C to form a tuned resonant circuit.

b. Use PSpice to plot the voltage $v_o(t)$, the transistor drain current $i_D(t)$, and the transistor drain voltage $v_D(t)$ for varying the gain voltage component $V_m = 3$ V to 4 V with an increment of 0.5 V.

c. Use the PSpice parameters of the NMOS IRF150 whose $V_t = 2.84$ V.

11.20 The gate voltage of the class C amplifier in Fig. 11.30(a) is $v_G = 2.5 + 1.5 \sin(2\pi f_s t)$ with $f_s = 1.5$ MHz. The DC supply voltages are $V_{DD} = \pm 15$ V. The load resistance $R_L = 50\ \Omega$, which is connected through a transformer of a turns ratio of $n = 2{:}1$.

a. Find the values of L and C to form a tuned resonant circuit.

b. Use PSpice to plot the voltage $v_o(t)$, the transistor drain current $i_D(t)$, and the transistor drain voltage $v_D(t)$ for varying the gain voltage component $V_m = 3$ V to 4 V with an increment of 0.5 V.

c. Use the PSpice parameters of the NMOS IRF150 whose $V_t = 2.84$ V.

11.21 The gate voltage of the class D amplifier in Fig. 11.33(a) is generated by modulating a sinusoidal reference $V_r = 1$ V at 25 kHz with a carrier signal $V_{cr} = 1$ V at $f_{cr} = 25f_s$. The DC supply voltages are $V_{DD} = \pm 5$ V. The load resistance $R_L = 8\ \Omega$ speaker.

a. Find the values of L and C to form a tuned resonant circuit.

b. Use PSpice to plot the gate-modulating voltage $v_G(t)$, the transistor output voltage $v_D(t)$, and the transistor output voltage $v_D(t)$ by a modulation index of $m = 0.5$ and 0.9.

c. Use the PSpice parameters of the NMOS IRF150 whose $V_t = 2.84$ V and the PMOS IRF9140 whose $V_t = 3.67$ V.

11.22 The gate voltage of the class D amplifier in Fig. 11.33(a) is generated by modulating a sinusoidal reference $V_r = 1$ V at 40 kHz with a carrier signal $V_{cr} = 1$ V at $f_{cr} = 30f_s$. The DC supply voltages are $V_{DD} = \pm 5$ V. The load resistance $R_L = 16\ \Omega$ speaker.

a. Find the values of L and C to form a tuned resonant circuit.

b. Use PSpice to plot the gate-modulating voltage $v_G(t)$, the transistor output voltage $v_D(t)$, and the transistor output voltage $v_D(t)$ by a modulation index of $m = 0.5$ and 0.9.

c. Use the PSpice parameters of the NMOS IRF150 whose $V_t = 2.84$ V and the PMOS IRF9140 whose $V_t = 3.67$ V.

11.23 The gate voltage of the class D amplifier in Fig. 11.33(a) is generated by modulating a sinusoidal reference $V_r = 1$ V at 20 kHz with a carrier signal $V_{cr} = 1$ V at $f_{cr} = 20f_s$. The DC supply voltages are $V_{DD} = \pm 5$ V. The load resistance $R_L = 16\ \Omega$ speaker.

a. Find the values of L and C to form a tuned resonant circuit.

b. Use PSpice to plot the gate-modulating voltage $v_G(t)$, the transistor output voltage $v_D(t)$, and the transistor output voltage $v_D(t)$ by a modulation index of $m = 0.5$ and 0.9.

c. Use the PSpice parameters of the NMOS IRF150 whose $V_t = 2.84$ V and the PMOS IRF9140 whose $V_t = 3.67$ V.

11.24 The gate voltage of the class E amplifier in Fig. 11.36 is a sinusoidal voltage of 5 V (peak) at 100 kHz. The DC supply voltages are $V_{DD} = 15$ V. The load resistance $R = 16\ \Omega$.

a. Find the optimum values of L_D, C_D, L, and C.

b. Use PSpice to plot the gate voltage $v_G(t)$, the transistor drain current $i_D(t)$, the transistor output voltage $v_D(t)$, and the output voltage $v_o(t)$ for peak gate voltages of 4 V and 4.5 V.

c. Use the PSpice parameters of the NMOS IRF150 whose $V_t = 2.84$ V.

11.25 The gate voltage of the class E amplifier in Fig. 11.36 is a sinusoidal voltage of 4 V (peak) at 50 kHz. The DC supply voltages are $V_{DD} = 12$ V. The load resistance $R = 8\ \Omega$.

a. Find the optimum values of L_D, C_D, L, and C.

b. Use PSpice to plot the gate voltage $v_G(t)$, the transistor drain current $i_D(t)$, the transistor output voltage $v_D(t)$, and the output voltage $v_o(t)$ for peak gate voltages of 4 V and 4.5 V.

c. Use the PSpice parameters of the NMOS IRF150 whose $V_t = 2.84$ V.

11.26 The gate voltage of the class E amplifier in Fig. 11.36 is a sinusoidal voltage of 4 V (peak) at 150 kHz. The DC supply voltages are $V_{DD} = 12$ V. The load resistance $R = 8\ \Omega$.

a. Find the optimum values of L_D, C_D, L, and C.

b. Use PSpice to plot the gate voltage $v_G(t)$, the transistor drain current $i_D(t)$, the transistor output voltage $v_D(t)$, and the output voltage $v_o(t)$ for peak gate voltages of 4 V and 4.5 V.

c. Use the PSpice parameters of the NMOS IRF150 whose $V_t = 2.84$ V.

11.10 *Short-Circuit and Thermal Protection*

11.27 D Design a thermal protection circuit as shown in Fig. 11.39 to limit the maximum temperature rise to $\Delta T = 110°C$. The temperature coefficients of the diode and the transistors are $K_{DZ} = K_{Q3} = 2.5$ mV/°C. The normal biasing current is $I_R = 5$ mA, and the collector current of Q_3 is 1 mA. Assume $V_{BEQ2} = 0.7$ V and $V_{CC} = 15$ V.

11.11 *Power Op-Amps*

11.28 D Use LM380 power op-amps to design a bridge amplifier as shown in Fig. 11.43. The output power is $P_L = 20$ W at $R_L = 4\ \Omega$. Assume $V_{CC} = 15$ V and $v_I = 100$ mV (peak).

11.29 The gain and output stages of an LM380 power op-amp are shown in Fig. P11.29. Determine the small-signal differential voltage gain $A_{id} = v_O/v_I$, the differential input resistance R_{id}, and the output resistance R_o. The diode parameters are $I_S = 10^{-13}$ A and $I_{D(min)} = 1$ mA to ensure conduction. The transistor parameters are $\beta_F = h_{fe} = 50$, $V_{BE} = 0.7$ V, $V_A = 40$ V, and $V_{CE(sat)} = 0.2$ V. Assume a DC supply voltage of $V_{CC} = 15$ V and $R_L = 10\ \mathrm{k}\Omega$.

FIGURE P11.29

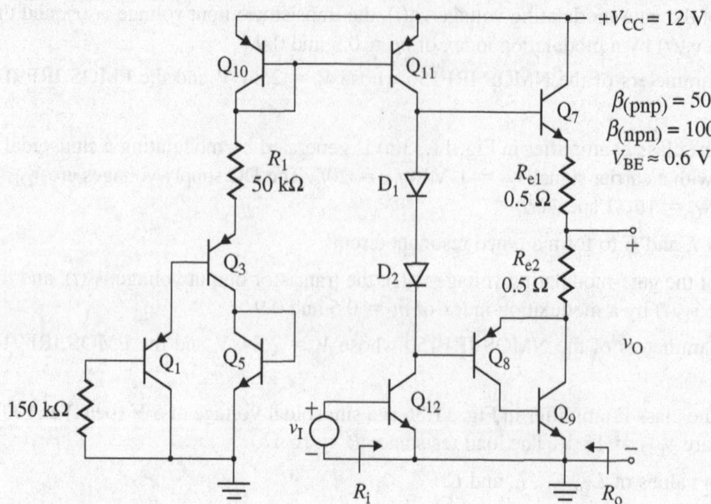

11.30 The power op-amp shown in Fig. 11.40 uses DC supply voltage $V_{CC} = -V_{EE} = 15$ V and transistors of type Q2N2222 having $I_s = 14.34 \times 10^{-15}$ and $\beta_F = 255.9$.

a. Determine the output voltage v_O and the input resistance R_i for an input voltage of $v_{in} = 5$ V.

b. Use PSpice to verify the results in part (a) and determine the high cutoff frequency f_H by plotting the frequency response.

11.31 D Design the bridge amplifier shown in Fig. 11.43 to obtain a voltage gain $A_f = 250$ V/V. Assume $V_{CC} = -V_{EE} = -12$ V. Use PSpice to verify the results.

11.12 *Thermal Considerations*

11.32 The power dissipation of a transistor is specified as $P_{D(max)} = 250$ W at $T_{CO} = 25$°C. The transistor is mounted on a heat sink. The thermal resistances are $\theta_{JC} = 0.7$°C/W, $\theta_{CS} = 0.2$°C/W, and $\theta_{SA} = 0.8$°C/W. If $T_{J(max)} = 200$°C and $T_A = 45$°C, calculate the maximum permissible power dissipation of the transistor.

11.33 The power dissipation of a transistor is specified as $P_{D(max)} = 290$ W at $T_{CO} = 25$°C. The transistor is mounted on a heat sink. The thermal resistances of the transistor are $\theta_{JC} = 0.6$°C/W and $\theta_{CS} = 0.2$°C/W. If $T_{J(max)} =$ 200°C, $T_C = 150$°C, and $T_A = 45$°C, calculate the required thermal resistance of the heat sink θ_{SA}.

CHAPTER 12

ACTIVE FILTERS

Learning Outcomes

After completing this chapter, students should be able to do the following:

- List the advantages of active filters over passive filters.
- Describe the characteristics and types of active filters.
- Determine the pole locations for the Butterworth function.
- Analyze active filters.
- Select an active filter to meet desired frequency requirements.
- Design an active filter to meet desired specifications.

Symbols and Their Meanings

Symbol	Meaning
V_i, V_o	Input and output voltages of a filter
BW	Bandwidth of a filter
f_{clk}	Clock frequency, in hertz
$p_1, \ldots, p_n$, $z_1, \ldots, z_n$	Poles and zeros of a transfer function
$H(s)$, $D(s)$, $N(s)$	Transfer function, and denominator and numerator of a transfer function
H_n	Transfer function of an nth-order filter
Q, K	Quality factor and constant parameter of a filter

Symbol	Meaning
f_o, f_c, f_L, f_H	Cutoff, center, low cutoff, and high cutoff frequencies of a filter
R_n, C_n	Normalized resistance and capacitance of a filter
θ_p, θ_z	Pole and zero angles

12.1 Introduction

In electrical engineering, a filter is a frequency-selective circuit that passes a specified band of frequencies and blocks or attenuates signals of frequencies outside this band. These signals are usually voltages. Filters that employ only passive elements such as capacitors, inductors, and resistors are called *passive filters*. Filters that make use of the properties of op-amps in addition to resistors and capacitors are called *active filters* or more often *analog filters*, in contrast to *digital filters*. Both analog and digital filters can be implemented in the same IC chip. This chapter introduces active filters and deals with the analysis and design of simple circuit topologies. Because of their practical importance, analog filters are often covered in a single course [1, 2].

12.2 Active versus Passive Filters

Both active and passive filters are used in electronic circuits. However, active filters offer the following advantages over passive filters:

- *Flexibility of gain and frequency adjustment*: Since op-amps can provide a voltage gain, the input signal in active filters is not attenuated as it is in passive filters. It is easy to adjust or tune active filters.
- *No loading effect*: Because of the high input resistance and low output resistance of op-amps, active filters do not cause loading of the input source or the load.
- *Cost and size*: Active filters are less expensive than passive filters because of the availability of low-cost op-amps and the absence of inductors.
- *Parasitics*: Parasitics are reduced in active filters because of their smaller size.
- *Digital integration*: Analog filters and digital circuitry can be implemented on the same IC chip.
- *Filtering functions*: Active filters can realize a wider range of filtering functions than passive filters.
- *Gain*: An active filter can provide gain, whereas a passive filter often exhibits a significant loss.

Active filters also have some disadvantages:

- *Bandwidth*: Active components have a finite bandwidth, which limits the applications of active filters to the audio-frequency range. Passive filters do not have such an upper-frequency limitation and can be used up to approximately 500 MHz.
- *Drifts*: Active filters are sensitive to component drifts due to manufacturing tolerances or environmental changes; in contrast, passive filters are less affected by such factors.
- *Power supplies*: Active filters require power supplies, whereas passive filters do not.
- *Distortion*: Active filters can handle only a limited range of signal magnitudes; beyond this range, they introduce unacceptable distortion.
- *Noise*: Active filters use resistors and active elements, which produce electrical noise.

In general, the advantages of active filters outweigh their disadvantages in voice and data communication applications. Active filters are used in almost all sophisticated electronic systems for communication and signal processing, such as television, telephone, radar, space satellite, and biomedical equipment. However, passive filters are still widely used.

12.3 Types of Active Filters

Let $V_i \angle 0$ be the input voltage to the filtering circuit shown in Fig. 12.1. The output voltage V_o and its phase shift θ will depend on the frequency ω. If two voltages are converted to Laplace's domain of s, the ratio of the output voltage $V_o(s)$ to the input voltage $V_i(s)$ is referred to as the *voltage transfer function* $H(s)$:

$$H(s) = \frac{V_o(s)}{V_i(s)}$$

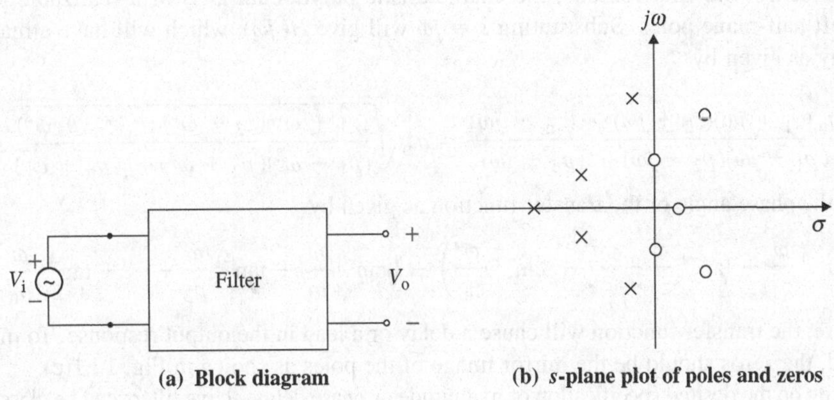

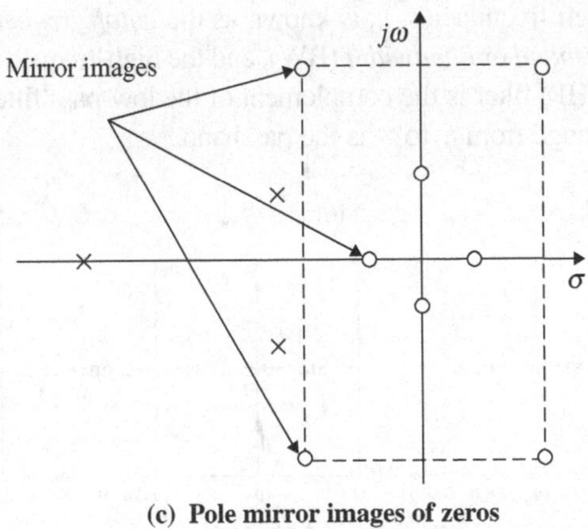

FIGURE 12.1 Filtering circuit

The transfer function $H(s)$ can be expressed in the general form

$$H(s) = \frac{a_m s^m + \cdots + a_2 s^2 + a_1 s + a_0}{s^n + \cdots + b_2 s^2 + b_1 s + b_0} \quad (\text{for } n \geq m) \tag{12.1}$$

whose coefficients are determined so as to meet the desired filter specifications. Substituting $s = j\omega$ will give $H(j\omega)$, which will have a magnitude and a phase delay. The denominator and the numerator of Eq. (12.1) are polynomials in s with real coefficients. If these polynomials are factored, $H(s)$ can be written as [3, 4]

$$H(s) = \frac{N(s)}{D(s)} = \frac{a_m(s + z_1)(s + z_2)\cdots(s + z_m)}{(s + p_1)(s + p_2)\cdots(s + p_n)} \tag{12.2}$$

The expression $z_1, z_2, \ldots, z_m$ is referred to as the *zeros* of $H(s)$ because $H(s) = 0$ when $s = -z_m$ and the expression $p_1, p_2, \ldots, p_n$ is referred to as the *poles* of $H(s)$ because $H(s) = \infty$ when $s = -p_n$. The poles and zeros can be plotted in the complex s-plane as shown in Fig. 12.1(b), where $s = \sigma + j\omega$ is considered as one set. Since the coefficients of $H(s)$ are all real, the poles and zeros occur in conjugate pairs.

For stability all poles must lie in the left half-plane. When the poles lie on the $j\omega$-axis and are simple, the network oscillates; when the poles lie in the right half-plane, the responses grow exponentially with time. Thus, for stability reasons, the characteristic polynomial $D(s)$ of a realizable system must have only left half-plane poles. Substituting $s = j\omega$ will give $H(j\omega)$, which will have a magnitude and a phase delay as given by

$$H(j\omega) = \frac{a_m(z_1 + j\omega)(z_2 + j\omega)\cdots(z_m + j\omega)}{(p_1 + j\omega)(p_2 + j\omega)\cdots(p_n + j\omega)} = a_m\sqrt{\frac{(z_1^2 + \omega^2)(z_2 + \omega^2)\cdots(z_m + \omega^2)}{(p_1 + \omega^2)(p_2 + \omega^2)\cdots(p_n + \omega^2)}}\angle\theta_H \tag{12.3}$$

where θ_H is the phase angle of the transfer function as given by

$$\theta_H = \left(\tan^{-1}\frac{\omega}{z_1} + \tan^{-1}\frac{\omega}{z_2} + \cdots + \tan^{-1}\frac{\omega}{z_m}\right) - \left(\tan^{-1}\frac{\omega}{p_1} + \tan^{-1}\frac{\omega}{p_2} + \cdots + \tan^{-1}\frac{\omega}{p_n}\right) \tag{12.4}$$

Therefore, the transfer function will cause a delay or a lead in the output response. To minimize this delay or lead, the zeros should be the mirror image of the poles as shown in Fig. 12.1(c).

Depending on the desired specification of magnitude or phase delay, active filters can be classified as low-pass filters, high-pass filters, band-pass filters, band-reject filters, or all-pass filters. The ideal characteristics of these filters are shown in Fig. 12.2. A *low-pass* (LP) filter passes frequencies from DC to a desired frequency f_o ($=\omega_o/2\pi$) and attenuates high frequencies. f_o is known as the *cutoff frequency*. The low-frequency range from 0 to f_o is known as the *passband* or *bandwidth* (BW), and the high-frequency range from f_o to ∞ is known as the *stop band*. A *high-pass* (HP) filter is the complement of the low-pass filter; the frequency range from 0 to f_o is the stop band, and the range from f_o to ∞ is the passband.

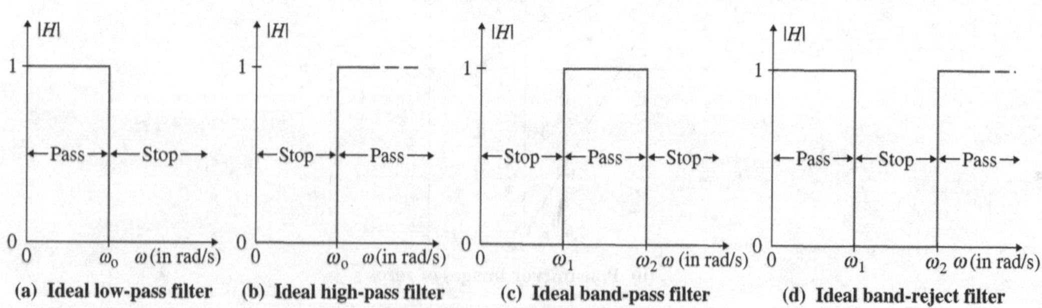

FIGURE 12.2 Ideal filter characteristics

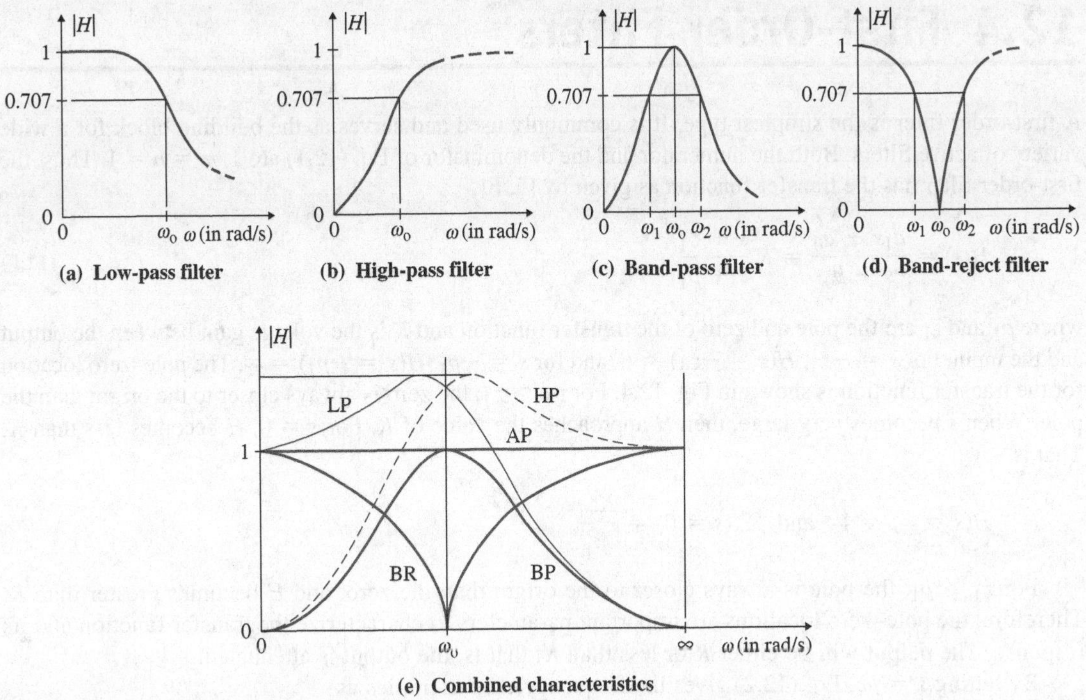

FIGURE 12.3 Realistic filter characteristics

A *band-pass* (BP) filter passes frequencies from f_1 to f_2 and stops all other frequencies. A band-reject (BR) filter is the complement of a BP filter; the frequencies from f_1 to f_2 are stopped, and all other frequencies are passed. BR filters are sometimes known as *band-stop filters*. An *all-pass* (AP) filter passes all frequencies from 0 to ∞, but it provides a phase delay.

It is impossible to create filters with the ideal characteristics shown in Fig. 12.2. Instead of abrupt changes from pass to stop behavior and from stop to pass behavior, practical filters exhibit a gradual transition from stop band to passband. Realistic filter characteristics are shown in Fig. 12.3 [(a)–(d)]. All characteristics are combined in Fig. 12.3(e). The cutoff frequency corresponds to the frequency at which the gain is 70.7% of its maximum value. The sharpness of the transition or the rate at which the characteristic changes is known as the *roll-off* or the *falloff rate*. If the frequency is plotted on a logarithmic scale, the plot is known as a *Bode plot*, and the roll-off, or asymptotic slope, is measured in multiples of ±6 dB per octave or ±20 dB per decade.

KEY POINTS OF SECTION 12.3

- Depending on the frequency characteristic, filters can be classified as low-pass, high-pass, band-pass, band-reject, or all-pass.
- It is not possible to create filters with the ideal characteristics of abrupt changes from pass to stop behavior and from stop to pass behavior. Practical filters exhibit a gradual transition from stop band to passband.

12.4 First-Order Filters

A first-order filter is the simplest type. It is commonly used and serves as the building block for a wide variety of active filters. Both the numerator and the denominator of Eq. (12.1) are 1, $m = n = 1$. Thus, the first-order filter has the transfer function as given by [5, 6]

$$H(s) = \frac{a_1 s + a_0}{s + b_0} = K\frac{s + z_1}{s + p_1} \tag{12.5}$$

where p_1 and z_1 are the pole and zero of the transfer function and K is the voltage gain between the output and the input. For $s = -z_1$, $H(s = -z_1) = 0$, and for $s = -p_1$, $H(s = -p_1) = \infty$. The pole–zero location for the transfer function is shown in Fig. 12.4. For $p_1 > z_1$, the zero is always closer to the origin than the pole. When s becomes very large, then H approaches the value of K. For $s = 0$, H becomes less than K. That is,

$$H(s = \infty) = 1 \quad \text{and} \quad H(s = 0) = \frac{Kz_1}{p_1}$$

For $z_1 > p_1$, the pole is always closer to the origin than the zero, and H becomes greater than K. Therefore, the pole–zero locations are important parameters to characterize the transfer function and its response. The output will be either K or less than K; that is, the output is attenuated.

By letting $s = j\omega$, Eq. (12.2) gives the frequency characteristics as

$$H(j\omega) = K\frac{z_1 + j\omega}{p_1 + j\omega} = \frac{Kz_1}{p_1} \times \frac{1 + (j\omega/z_1)}{1 + (j\omega/p_1)} \tag{12.6}$$

which gives the magnitude as

$$|H(j\omega)| = \frac{Kz_1}{p_1}\sqrt{\frac{1 - (\omega/z_1)^2}{1 - (\omega/p_1)^2}} \tag{12.7}$$

The phase angle θ is given by

$$\theta = \theta_z - \theta_p = \tan^{-1}\left(\frac{\omega}{z_1}\right) - \tan^{-1}\left(\frac{\omega}{p_1}\right) \tag{12.8}$$

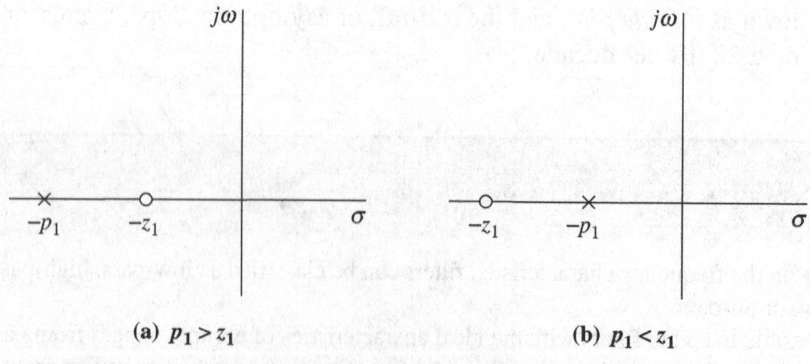

FIGURE 12.4 Pole and zero locations of first-order filters

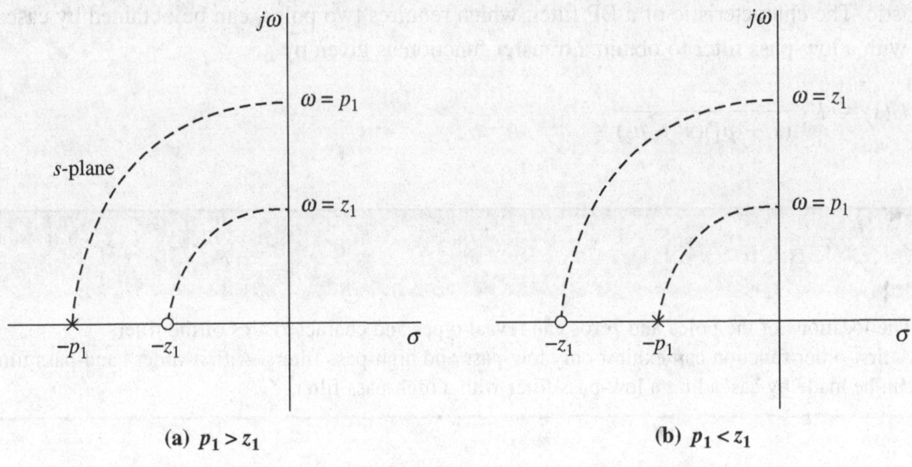

FIGURE 12.5 Poles and zeros for $s = z_1$ and $s = z_p$

The phase angle θ can be leading, having a positive value, or lagging, having a negative value. The phase angle for zero becomes $\theta_z = \tan^{-1}(\omega/z_1) = 45°$ at $\omega = z_1$, as shown in Fig. 12.5(a), and the phase angle for pole is $\theta_p = \tan^{-1}(\omega/z_p) = 45°$ at $\omega = z_p$, as shown in Fig. 12.5(b). Depending on the values of z_1 and p_1, Eq. (12.3) can exhibit the characteristics of a lead low-pass filter, a lag low-pass filter, a lead HP filter, and a lag HP filter. The conditions for z_1 and p_1 to obtain these filters are shown in Table 12.1. A pole 1 in the denominator gives the low-pass characteristic, whereas the addition of a zero at the origin gives the HP characteristic. If $p_1 > z_1$, it makes an HP characteristic, whereas $p_1 < z_1$ makes a low-pass

TABLE 12.1 Poles and zeros for high- and low-pass filters

Pole and Zero	T(s)	Classification
jω σ	$\dfrac{s + z_1}{s + p_1}$ $(p_1 > z_1)$	Lag high-pass
jω σ	$\dfrac{s}{s + p_1}$ $\boxed{z = 0}$	Lead high-pass
jω σ	$\dfrac{s + z_1}{s + p_1}$ $(z_1 > p_1)$	Lead low-pass
jω σ	$\dfrac{p_1}{s + p_1}$ $\boxed{z = 0}$	Lag low-pass

characteristic. The characteristic of a BP filter, which requires two poles, can be obtained by cascading an HP filter with a low-pass filter to obtain a transfer function as given by

$$H(s) = K_1 \frac{s + z_1}{(s + p_1)(s + p_2)} \tag{12.9}$$

KEY POINTS OF SECTION 12.4

- The locations of the poles and zeros can reveal types and characteristics of the filters.
- A first-order function can exhibit only low-pass and high-pass filters. A first-order band-pass filter can be made by cascading a low-pass filter with a high-pass filter.

12.5 The Biquadratic Function

For an active filter with $n > 2$, Eq. (12.1) will become complex. Thus, a second-order transfer function (i.e., a function with $n = 2$) is most commonly used. For $n = 2$, Eq. (12.1) becomes

$$H(s) = \frac{a_2 s^2 + a_1 s + a_0}{s^2 + b_1 s + b_0} = K \frac{(s + z_1)(s + z_2)}{(s + p_1)(s + p_2)} \tag{12.10}$$

The poles (or zeros) can be determined from the roots of a quadratic equation from

$$p_1, p_2 = \frac{-b_1 \pm \sqrt{b_1^2 - 4b_0}}{2} = -\frac{b_1}{2} \pm \sqrt{\left(\frac{b_1}{2}\right)^2 - b_0} \tag{12.11}$$

where b_0 is generally greater than $b_1^2/4$ and the poles are complex such that $p_1, p_2 = \alpha_p \pm j\beta_p$, which can be substituted in Eq. (12.7) to give the transfer function of the form as given by

$$H(s) = K \frac{(s + \alpha_z + j\beta_z)(s + \alpha_z - j\beta_z)}{(s + \alpha_p + j\beta_p)(s + \alpha_p - j\beta_p)} = K \frac{s^2 + 2\alpha_z s + \omega_z^2}{s^2 + 2\alpha_p s + \omega_p^2} \tag{12.12}$$

$$= K \frac{s^2 + (\omega_z/Q_z)s + \omega_z^2}{s^2 + (\omega_p/Q_p)s + \omega_p^2} \tag{12.13}$$

where $Q_p = \omega_p/2\alpha_p = \omega_p/2\text{Re}(p_1)$ is called the *pole quality factor* and $Q_z = \omega_z/2\alpha_z = \omega_z/2\text{Re}(z_1)$ is called the *zero quality factor*. $\omega_p(=\sqrt{b_0})$ and $\omega_z(=\sqrt{a_0})$ are the pole and zero natural resonant frequencies, respectively. The imaginary part β_p is related to α_p and ω_p by $\omega_p^2 = \alpha_p^2 + \beta_p^2$. This can be solved for β_p as given by

$$\beta_p = \sqrt{\omega_p^2 - \alpha_p^2} = \omega_p \sqrt{1 - \frac{1}{4Q_p^2}} \tag{12.14}$$

The pole angle ϕ_p as shown in Fig. 12.6(a) can be found from

$$\phi_p = \cos^{-1}\left(\frac{\alpha_p}{\omega_p}\right) = \cos^{-1}\left(\frac{1}{2Q_p}\right) \tag{12.15}$$

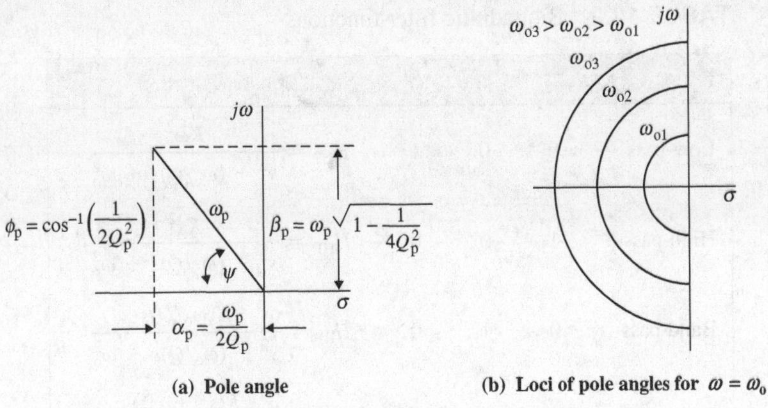

FIGURE 12.6 Pole angles and their loci

which shows that the phase delay depends on the quality factor Q_p, whose value is chosen to be greater than 1. If $Q_p > 5$, Eq. (12.14) is approximated to $\beta_p \simeq \omega_p$ with an error less than 1%. As a rule of thumb, $Q_p = 7$ is generally used for second filter designs.

Equation (12.10) is known as the *biquadratic function*, which serves as the building block for a wide variety of active filters and has the general form given by

$$H(s) = K\frac{k_2s^2 + k_1(\omega_o/Q)s + k_0\omega_o^2}{s^2 + (\omega_o/Q)s + \omega_o^2} \tag{12.16}$$

where ω_o = *undamped natural* (or *resonant*) *frequency*

Q = *quality factor* or *figure of merit*

K = *DC gain*

Constants k_2, k_1, and k_0 are ± 1 or 0. Table 12.2 shows their possible values for each type of filter. Substituting $s = j\omega$ into Eq. (12.16) will give the frequency domain $H(j\omega)$, which will have a magnitude and a phase delay:

$$H(j\omega) = \frac{-k_2\omega^2 + jk_1(\omega_o/Q)\omega + k_0\omega_o^2}{-\omega^2 + j(\omega_o/Q)\omega + \omega_o^2} = \frac{(k_0\omega_o^2 - k_2\omega^2) + jk_1(\omega_o/Q)\omega}{(\omega_o^2 - \omega^2) + j(\omega_o/Q)\omega} \tag{12.17}$$

where $\omega = 2\pi f$, in radians per second, and f = supply frequency, in hertz.

Normalizing the frequency ω with respect to the natural frequency ω_o so that $u = \omega/\omega_o$, Eq. (12.15) becomes

$$H(ju) = \frac{(k_0 - k_2u^2) + (jk_1u/Q)}{(1 - u^2) + (ju/Q)} \tag{12.18}$$

where ω_o = natural frequency, in radians per second.

It can be shown (Appendix B) that Q is related to the bandwidth BW and to ω_o by

$$Q = \frac{\omega_o}{\text{BW}} = \frac{\omega_o}{\omega_H - \omega_L} \tag{12.19}$$

where ω_H = high cutoff frequency, in radians per second, and ω_L = low cutoff frequency, in radians per second.

TABLE 12.2 Biquadratic filter functions

Filter	k_2	k_1	k_0	Transfer Function
Low-pass	0	0	1	$H_{LP} = \dfrac{K\omega_o^2}{s^2 + (\omega_o/Q)s + \omega_o^2}$
High-pass	1	0	0	$H_{HP} = \dfrac{Ks^2}{s^2 + (\omega_o/Q)s + \omega_o^2}$
Band-pass	0	1	0	$H_{BP} = \dfrac{K(\omega_o/Q)s}{s^2 + (\omega_o/Q)s + \omega_o^2}$
Band-reject	1	0	1	$H_{BR} = \dfrac{K(s^2 + \omega_o^2)}{s^2 + (\omega_o/Q)s + \omega_o^2}$
All-pass	1	−1	1	$H_{AP} = K\dfrac{s^2 - (\omega_o/Q)s + \omega_o^2}{s^2 + (\omega_o/Q)s + \omega_o^2}$

Depending on the values of the constants k_0, k_1, and k_2, there are four possible cases whose transfer function and phase angles in frequency domain can be described as follows:

Case 1. For $k_0 = 1$ and $k_1 = k_2 = 0$, there is no zero. Equation (12.18) gives the low-pass characteristic as shown in Table 12.3(a), and its transfer function in frequency domain is given by

$$H_{LP}(s) = K\frac{1}{(1 - u^2) + (ju/Q)} = \left|\frac{K}{\sqrt{(1 - u^2)^2 + (u/Q)^2}}\right| \angle - \tan^{-1}\left(\frac{u/Q}{1 - u^2}\right) \tag{12.20}$$

Case 2. For $k_1 = 1$ and $k_0 = k_2 = 0$, there is one zero on the $j\omega$-axis. Equation (12.18) gives the BP characteristic as shown in Table 12.3(b), and its transfer function in frequency domain is given by

$$H_{BP}(s) = K\frac{ju/Q}{(1 - u^2) + (ju/Q)} = \left|\frac{(u/Q)K}{\sqrt{(1 - u^2)^2 + (u/Q)^2}}\right| \angle \left[\frac{\pi}{2} - \tan^{-1}\left(\frac{u/Q}{1 - u^2}\right)\right] \tag{12.21}$$

Case 3. For $k_0 = k_2 = 1$ and $k_1 = 0$, there are two zeros at the origin. Equation (12.18) gives the BR characteristic as shown in Table 12.3(c), and its transfer function in frequency domain is given by

$$H_{BR}(s) = K\frac{(1 - u^2)}{(1 - u^2) + (ju/Q)} = \left|\frac{(1 - u^2)K}{\sqrt{(1 - u^2)^2 + (u/Q)^2}}\right| \angle - \tan^{-1}\left(\frac{u/Q}{1 - u^2}\right) \tag{12.22}$$

Case 4. For $k_2 = 1$ and $k_0 = k_1 = 0$, there are two zeros at the origin. Equation (12.18) gives the HP characteristic as shown in Table 12.3(d), and its transfer function in frequency domain is given by

$$H_{HP}(s) = K\frac{-u^2}{(1 - u^2) + (ju/Q)} = \left|\frac{u^2K}{\sqrt{(1 - u^2)^2 + (u/Q)^2}}\right| \angle \left[\pi - \tan^{-1}\left(\frac{u/Q}{1 - u^2}\right)\right] \tag{12.23}$$

TABLE 12.3 Characteristics of second-order filters

Filter Type	Transfer Function	Frequency Response	Poles and Zeros
(a) Low-pass	$H_{LP}(s) = \dfrac{\omega_o^2}{s^2 + (\omega_o/Q)s + \omega_o^2}$	ω	$j\omega$, σ
(b) Low-band-pass	$H_{BP}(s) = \dfrac{(\omega_o/Q)s}{s^2 + (\omega_o/Q)s + \omega_o^2}$	ω	$j\omega$, σ
(c) Band-reject notch	$H_{BS}(s) = \dfrac{s^2 + \omega_o^2}{s^2 + (\omega_o/Q)s + \omega_o^2}$	ω	$j\omega$, σ
(d) High-pass	$H_{HP}(s) = \dfrac{s^2}{s^2 + (\omega_o/Q)s + \omega_o^2}$	ω	$j\omega$, 2, σ
(e) All-pass	$H_{AP}(s) = \dfrac{s^2 - (\omega_o/Q)s + \omega_o^2}{s^2 + (\omega_o/Q)s + \omega_o^2}$	ω	$j\omega$, σ

Case 5. For $k_0 = k_2 = 1$ and $k_1 = -1$, two zeros are identical to the two poles. Equation (12.18) gives the AP characteristic as shown in Table 12.3(e), and its transfer function in frequency domain is given by

$$H_{AP}(s) = K\frac{(1 - u^2) + (ju/Q)}{(1 - u^2) + (ju/Q)} = |K| \angle 0° \quad \textbf{(12.24)}$$

KEY POINTS OF SECTION 12.5

- The quality factor Q is a measure of the bandwidth of a filter. The lower the value of Q, the more selective the filter will be.
- The denominator of all quadratic filter functions is the same, but the numerator depends on the type of filter.

12.6 Butterworth Filters

The denominator of a filter transfer function determines the poles and the falloff rate of the frequency response. Notice from Table 12.2 that the denominator of the biquadratic function has the same form for all types of filters. Butterworth filters [7] are derived from the magnitude-squared function

$$|H_n(j\omega)|^2 = \frac{1}{(1 + \omega/\omega_o)^{2n}} \tag{12.25}$$

which gives the magnitude of the transfer function as

$$|H_n(j\omega)| = \frac{1}{[1 + (\omega/\omega_o)^{2n}]^{1/2}} \tag{12.26}$$

Plots of this response, known as the *Butterworth response*, are shown in Fig. 12.7 for n = 1, 2, 4, 6, 8, and 10. This type of response has the following properties:

1. $|H_n(j0)| = 1$ for all n (voltage gain at zero frequency—that is, DC voltage gain at $\omega = 0$).
2. $|H_n(j\omega_o)| = 1/\sqrt{2} \approx 0.707$ for all n (voltage gain at $\omega = \omega_o$).
3. $|H_n(j\omega_o)|$ exhibits n-pole roll-off for $\omega > \omega_o$.
4. It can be shown that all but one of the derivatives of $|H_n(j\omega)|$ equal zero near $\omega = 0$. That is, the *maximally flat response* is at $\omega = 0$.
5. For $n > 10$, the response becomes close to the ideal characteristic of abrupt change from passband to stop band.

By substituting $\omega = s/j$ into Eq. (12.25), we can express the transfer function for Butterworth filters in the s domain by

$$|H_n(s)|^2 = \left|\frac{1}{1 + (-1)^n(s/\omega_o)^{2n}}\right| \tag{12.27}$$

$$= \left|\frac{1}{D_n(s)D_n(-s)}\right| \tag{12.28}$$

where $D_n(s)$ is a polynomial in s, all of whose roots have negative real parts, and $|D_n(s)| = |D_n(-s)|$.

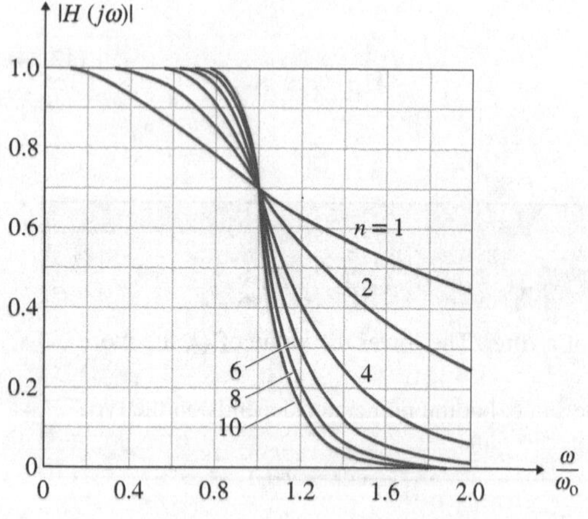

FIGURE 12.7 Butterworth response

12.6.1 Butterworth Function for $n = 1$

For $n = 1$ and letting $\omega_o = 1$, Eq. (12.27) becomes

$$|H_1(s)|^2 = \left|\frac{1}{1 - s^2}\right| = \left|\frac{1}{D_1(s)D_1(-s)}\right| \tag{12.29}$$

Factorizing $1 - s^2$, we get $D_1(s)$ and $D_1(-s)$ as

$$D_1(s)D_1(-s) = 1 - s^2 = (1 + s)(1 - s)$$

which gives

$$D_1(s) = (1 + s) \quad \text{and} \quad D_1(-s) = (1 - s)$$

Since $|D_1(s)| = |D_1(-s)|$, Eq. (12.24) gives the Butterworth function with negative real parts. That is, for $D_1(s)$ only, we get the general form

$$H_1(s) = \frac{1}{(s/\omega_o) + 1} = \frac{\omega_o}{s + \omega_o}$$

The pole angles are $\phi_p = 0°$ and $180°$. For $n = 1$, it becomes a first-order function that is not generally used. For the Butterworth filter, the minimum is a second-order function, $n \geq 2$.

12.6.2 Butterworth Function for $n = 2$

If we let $\omega_o = 1$, for $n = 2$ Eq. (12.27) becomes

$$|H_2(s)|^2 = \left|\frac{1}{1 + s^4}\right| = \left|\frac{1}{D_2(s)D_2(-s)}\right| \tag{12.30}$$

Factoring $1 + s^4$ gives $D_2(s)D_2(-s)$ as

$$D_2(s)D_2(-s) = s^4 + 1 = \left(s - \frac{-1 - j}{\sqrt{2}}\right)\left(s - \frac{1 + j}{\sqrt{2}}\right)\left(s - \frac{1 - j}{\sqrt{2}}\right)\left(s - \frac{1 + j}{\sqrt{2}}\right)$$

which gives

$$D_2(s) = \left(s - \frac{-1 - j}{\sqrt{2}}\right)\left(s - \frac{-1 + j}{\sqrt{2}}\right) = s^2 + \sqrt{2}s + 1$$

and $$D_2(-s) = \left(s - \frac{1 - j}{\sqrt{2}}\right)\left(s - \frac{1 + j}{\sqrt{2}}\right) = s^2 - \sqrt{2}s + 1$$

Since $|D_2(s)| = |D_2(-s)|$, Eq. (12.30) gives the Butterworth function with negative real parts. That is, for $D_2(s)$ only, we get the general form

$$H_2(s) = \frac{1}{(s/\omega_o)^2 + \sqrt{2}(s/\omega_o) + 1} = \frac{\omega_o^2}{s^2 + \sqrt{2}\omega_o s + \omega_o^2} \tag{12.31}$$

which has $Q = 1/\sqrt{2} = 0.707$. Thus, for $n = 2$, a Butterworth filter will exhibit the frequency characteristic of a second-order system (Appendix B), and the frequency response will fall at a rate of -40 dB/decade or -12 dB/octave.

The angles for all k (=2n = 4) poles are

$$\phi_k = \frac{180° + 360° \times k}{2 \times n} = 45°, 135°, 225°, 315°$$

12.6.3 Butterworth Function for $n = 3$

If we let $\omega_o = 1$, for $n = 3$ Eq. (12.27) becomes

$$|H_3(s)|^2 = \left|\frac{1}{1 - s^6}\right| = \left|\frac{1}{D_3(s)D_3(-s)}\right| \tag{12.32}$$

Factoring $1 - s^6$, we get

$$D_3(s)D_3(-s) = 1 - s^6 = (s^2 + s + 1)(s^2 - s + 1)(s + 1)(-s + 1)$$

which gives $D_3(s)$, whose roots have negative real parts, as

$$D_3(s) = (s^2 + s + 1)(s + 1) = s^3 + 2s^2 + 2s + 1$$

The transfer function for $n = 3$ is given by

$$H_3(s) = \frac{1}{(s/\omega_o)^3 + 2(s/\omega_o)^2 + 2(s/\omega_o) + 1} \tag{12.33}$$

$$= \frac{\omega_o^3}{s^3 + 2\omega_o s^2 + 2\omega_o^2 s + \omega_o^3} \tag{12.34}$$

Thus, for $n = 3$, a Butterworth filter will exhibit the frequency characteristic of a third-order system, and the frequency response will fall at a rate of −60 dB/decade or −18 dB/octave.

The angles for all k (= 2n = 6) poles are

$$\phi_k = \frac{360° \times k}{2 \times n} = 0°, 60°, 120°, 180°, 240°, 300°$$

12.6.4 Butterworth Function for Higher-Order Filters

A Butterworth filter will exhibit the frequency characteristic of an nth-order system and will fall at a rate of $20n$ dB/decade or $6n$ dB/octave. The denominator $D_n(s)$ depends on the LHS poles that will be phase shifted by 180° with respect to the poles of $D_n(s)$ on the RHS. Each pole p_k has it complex conjugate $\bar{p}_k$ as shown in Fig. 12.8 in the s-plane for $n = 1$ to 5. Taking the LHS real-axis as the reference, we can find the LHS pole angles for an nth-order filter from the following expression:

$$\phi_k = 180° - 90°\left(\frac{2k + n - 1}{n}\right) \quad \text{(for } k = 1, 2, \ldots, n\text{)} \tag{12.35}$$

Assuming $\omega_o = 1$, the corresponding real and imaginary parts of the poles are given by

$$\alpha_k = \cos\phi_k \quad \text{and} \quad \beta_k = \sin\phi_k$$

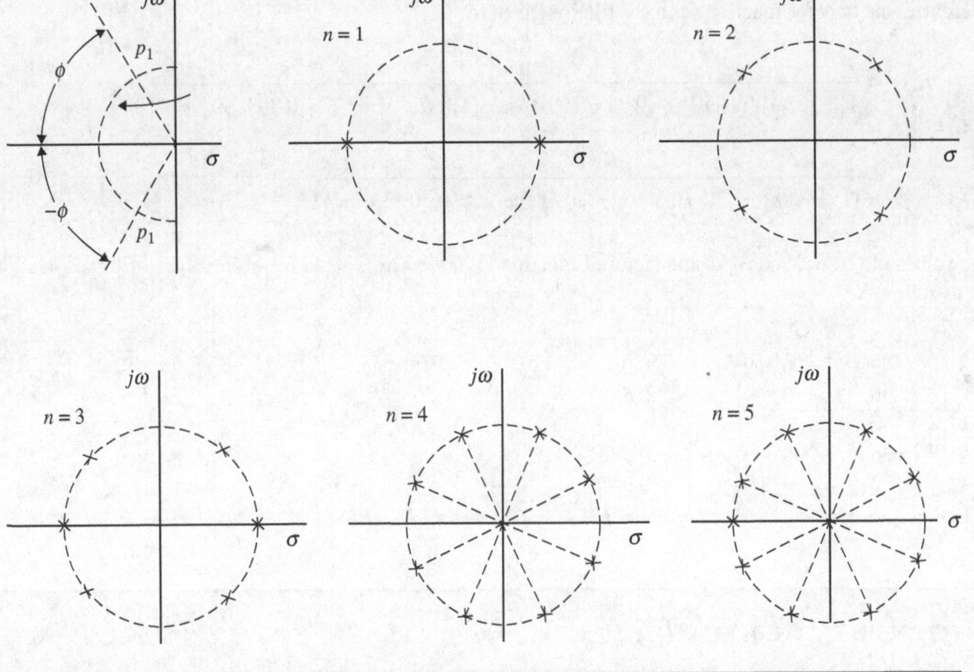

FIGURE 12.8 Pole locations for the Butterworth function

Substituting α_k in Eq. (12.12), we can find the function $D(s)$, which is the reciprocal of $H(s)$, from

$$D_n(s) = (1 + s)\prod_k (s^2 + 2\alpha_k s + 1) \quad \text{(for odd values of } k = 1, 3, \ldots, n) \tag{12.36}$$

$$= \prod_k (s^2 + 2\alpha_k s + 1) \quad \text{(for even values of } k = 2, 4, \ldots, n)$$

EXAMPLE 12.1

Finding the transfer function for the Butterworth response Find the transfer function for the fifth-order Butterworth response of a BP filter.

SOLUTION

$n = 5$. From Eq. (12.35) we get the pole angles $\phi_k = 0°, \pm 36°$, and $\pm 72°$. The real values are $\alpha_k = \cos\phi_k = 1, 0.809, 0.309, 0.809$, and 0.309. Therefore,

$$D_5(s) = (1 + s)(s^2 + 2\alpha_1 s + 1)(s^2 + 2\alpha_2 s + 1)$$

$$= (1 + s)(s^2 + 2 \times 0.309s + 1)(s^2 + 2 \times 0.809s + 1)$$

Therefore, the transfer function for a BP filter is given by

$$H(s) = \frac{Ks^5}{1 + (s/\omega_o)[(s/\omega_o)^2 + 2 \times 0.309s/\omega_o + 1][(s/\omega_o)^2 + 2 \times 0.809s/\omega_o + 1]}$$

$$= \frac{Ks^5\omega_o^5}{(1 + s)(s^2 + 2 \times 0.309s + \omega_o^2)(s^2 + 2 \times 0.809s + \omega_o^2)}$$

The values of Q_k, α_k, and β_k of the factorial function $\Pi_k(s^2 + 2\alpha_k s + 1)$ for even values of $k = 2, 4, \ldots, n$ are as follows:

k	ϕ_k	α_k	β_k
1	72	0.309	0.951
2	36	0.809	0.588
3	0	1	0
4	−36	0.809	−0.588
5	−72	0.309	−0.951

KEY POINTS OF SECTION 12.6

- Butterworth filters can give a maximally flat response.
- For $n > 10$, the response becomes close to the ideal characteristic of abrupt change from passband to stop band. However, a filter with $n = 2$ is quite satisfactory for most applications.

12.7 Transfer Function Realization

The poles and zeros of a transfer function for active filters are generally obtained from series and parallel connection impedances as shown in Fig. 12.9. The transfer function describing the relationship between the output and input voltages can be described by [8]

$$H(s) = \frac{V_o(s)}{V_i(s)} = \frac{Z_2(s)}{Z_1(s) + Z_2(s)} \tag{12.37}$$

where the impedances $Z_1(s)$ and $Z_2(s)$ consist of series and parallel connections of Rs and Cs. The basic elements for generating poles and zeros are listed in Table 12.4.

The transfer function is a dimensionless quantity. It is convenient to normalize all circuit elements so that they become unitless quantities. The relationships among the circuit elements remain the same, irrespective

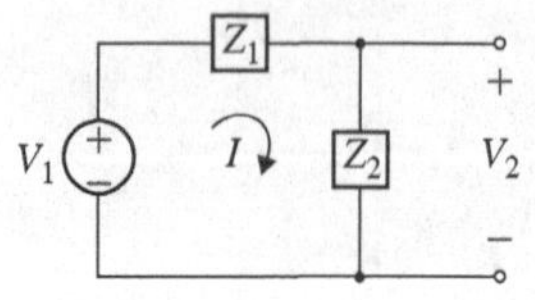

FIGURE 12.9 Impedances for obtaining transfer function

TABLE 12.4 Circuit elements for generating poles and zeros

Z_1 or Z_2 Elements	Impedance	Admittance
R (resistor)	R	$\frac{1}{R}$
C (capacitor)	$\frac{1}{sC}$	sC
R and C in parallel	$\frac{1}{sC + 1/R}$	$sC + \frac{1}{R}$
R and C in series	$R + \frac{1}{sC}$	$\frac{1}{R + (1/sC)}$

of their values. Normalization allows scaling of all values and has the advantages of (a) removing the dimensions from the circuit variables, (b) the relative relationship independent of impedance level, and (c) allowing the designer to select convenient and practical values of circuit elements. Frequency normalization is done by dividing the frequency variable by a normalizing frequency, usually ω_o, and the impedances are normalized by dividing all impedances in the circuit by a normalizing resistance R_o. The normalized s becomes $s_n = s/\omega_o$, and the normalized frequency becomes $\omega_n = \omega/\omega_o$. Thus, the normalized impedances for resistance R and capacitance C become

$$R_n = \frac{R}{R_o} \quad \text{and} \quad \frac{1}{s_n C_n} = \left(\frac{1}{s/\omega_o}\right)\left(\frac{1}{\omega_o C R_o}\right)$$

Therefore, the normalized element values are given by

$$R_n = \frac{R}{R_o} \quad \text{and} \quad C_n = C\omega_o R_o \tag{12.38}$$

12.8 Low-Pass Filters

Depending on the order of the biquadratic polynomial in Eq. 12.16, low-pass filters can be classified into two types: first-order and second-order.

12.8.1 First-Order Low-Pass Filters

The transfer function of a first-order low-pass filter has the general form

$$H(s) = \frac{K\omega_o}{s + \omega_o} \tag{12.39}$$

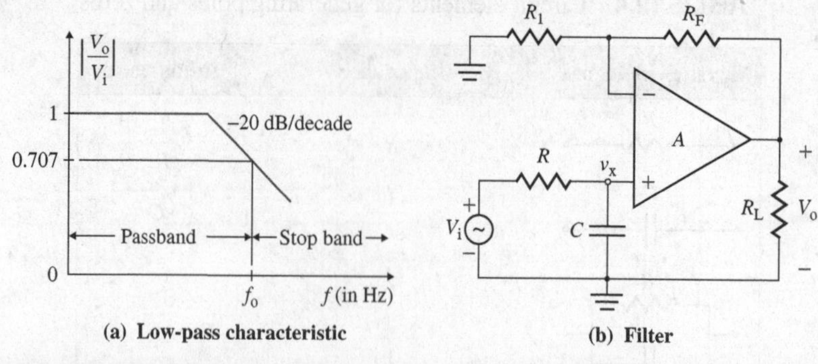

FIGURE 12.10 First-order low-pass filter with $K = 1$

A typical frequency characteristic is shown in Fig. 12.10(a). A first-order filter that uses an RC network for filtering is shown in Fig. 12.10(b). The op-amp operates as a noninverting amplifier, which has the characteristics of a very high input impedance and a very low output impedance.

The voltage (V_x in Laplace's domain of s) at the noninverting terminal of the op-amp can be found by the voltage divider rule:

$$V_x(s) = \frac{1/sC}{R + 1/sC} V_i(s) = \frac{1}{1 + sRC} V_i(s)$$

The output voltage of the noninverting amplifier is

$$V_o(s) = \left(1 + \frac{R_F}{R_1}\right) V_x(s) = \left(1 + \frac{R_F}{R_1}\right) \frac{1}{1 + sRC} V_i(s)$$

which gives the voltage transfer function $H(s)$ as

$$H(s) = \frac{V_o(s)}{V_i(s)} = \frac{K}{1 + sRC} \tag{12.40}$$

where the DC gain is

$$K = 1 + \frac{R_F}{R_1} \tag{12.41}$$

Substituting $s = j\omega$ into Eq. (12.40), we get

$$H(j\omega) = \frac{V_o(j\omega)}{V_i(j\omega)} = \frac{K}{1 + j\omega RC} \tag{12.42}$$

which gives the cutoff frequency f_o at 3-dB gain as

$$f_o = \frac{1}{2\pi RC} \tag{12.43}$$

The magnitude and phase angle of the filter gain can be found from

$$|H(j\omega)| = \frac{K}{[1 + (\omega/\omega_o)^2]^{1/2}} = \frac{K}{[1 + (f/f_o)^2]^{1/2}} \tag{12.44}$$

and $\quad \phi = -\tan^{-1}(f/f_o)$ (12.45)

where f = frequency of the input signal, in hertz.

EXAMPLE 12.2

D

Designing a first-order low-pass filter

(a) Design a first-order low-pass filter to give a high cutoff frequency of $f_o = 1$ kHz with a pass-band gain of 4. If the desired frequency is changed to $f_n = 1.5$ kHz, calculate the new value of R_n.

(b) Use PSpice/SPICE to plot the frequency response of the filter designed in part (a) from 10 Hz to 10 kHz.

SOLUTION

(a) The high cutoff frequency is $f_o = 1$ kHz. Choose a value of C less than or equal to 1 μF: let $C = 0.01$ μF. Using Eq. (12.43), calculate the value of R:

$$R = \frac{1}{2\pi f_o C} = \frac{1}{2\pi \times 1\text{ kHz} \times 0.01\ \mu\text{F}} = 15{,}916\ \Omega \quad \text{(use a 20-k}\Omega\text{ potentiometer)}$$

Choose values of R_1 and R_F to meet the pass-band gain K. From Eq. (12.41), $K = 1 + R_F/R_1$. Since $K = 4$,

$$\frac{R_F}{R_1} = 4 - 1 = 3$$

If we let $R_1 = 10$ kΩ, $R_F = 30$ kΩ.

Calculate the frequency scaling factor, $\text{FSF} = f_o/f_n$:

$$\text{FSF} = \frac{f_o}{f_n} = \frac{1\text{ kHz}}{1.5\text{ kHz}} = 0.67$$

Calculate the new value of $R_n = \text{FSF} \times R$:

$$R_n = \text{FSF} \times R = 0.67 \times 15{,}916 = 10{,}664\ \Omega \quad \text{(use a 15-k}\Omega\text{ potentiometer)}$$

NOTE: The exact numbers of the resistances are used to demonstrate the validity of the calculated values with the PSpice simulation results. In a practical design, the commercially available standard values will be used (i.e., 16 kΩ instead of 15,916 Ω; see Appendix E).

(b) A low-pass filter with the calculated values of the circuit parameters and the LF411 op-amp is shown in Fig. 12.11.

The plot of the voltage gain is shown in Fig. 12.12, which gives $K = 4.0$ (expected value is 4) and $f_o \approx 998$ Hz (expected value is 1 kHz) at $|H(j\omega)| = 0.707 \times 4 = 2.828$. Thus, the results are close to the expected values. If we use a linear op-amp model in Fig. 3.8, the results will differ slightly.

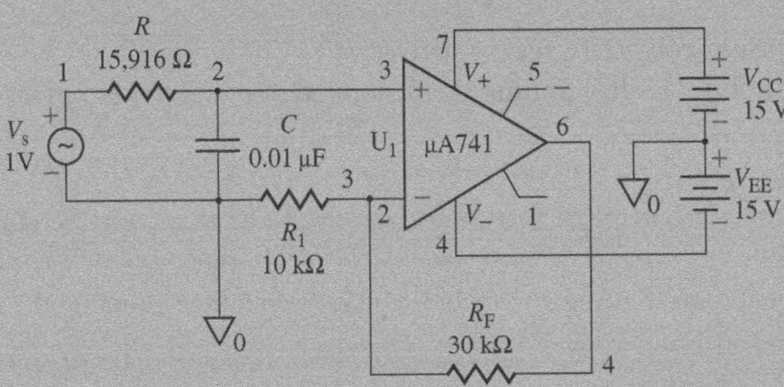

FIGURE 12.11 Low-pass filter for PSpice simulation

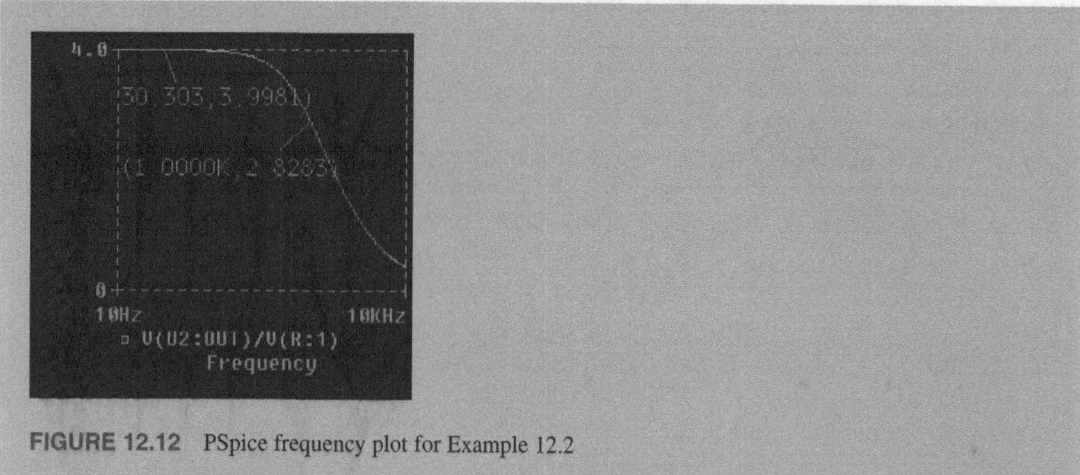

FIGURE 12.12 PSpice frequency plot for Example 12.2

12.8.2 Second-Order Low-Pass Filters

The roll-off of a first-order filter is only -20 dB/decade in the stop band. A second-order filter exhibits a stop-band roll-off of -40 dB/decade and thus is preferable to a first-order filter. In addition, a second-order filter can be the building block for higher-order filters ($n = 4, 6, \ldots$). Substituting $k_2 = k_1 = 0$ and $k_0 = 1$ into Eq. (12.16), we get the general form

$$H(s) = \frac{K\omega_o^2}{s^2 + (\omega_o/Q)s + \omega_o^2} \tag{12.46}$$

where K is the DC gain. A typical frequency characteristic is shown in Fig. 12.13(a); for high values of Q, overshoots will be exhibited at the resonant frequency f_o. For frequencies above f_o, the gain rolls off at the rate of -40 dB/decade. A first-order filter can be converted to a second-order filter by adding an additional RC network, known as the *Sallen–Key* circuit, as shown in Fig. 12.13(b). The input RC network is shown in Fig. 12.13(c); the equivalent circuit appears in Fig. 12.13(d). The transfer function of the filter network is

$$H(s) = \frac{V_o(s)}{V_i(s)} = \frac{K/R_2R_3C_2C_3}{s^2 + s(R_3C_3 + R_2C_3 + R_2C_2 - KR_2C_2)/R_2R_3C_2C_3 + 1/R_2R_3C_2C_3} \tag{12.47}$$

where $K = (1 + R_F/R_1)$ is the DC gain. (See Prob. 12.16 for the derivation.)

Equation (12.47) is similar in form to Eq. (12.46). Setting the denominator equal to zero gives the characteristic equation

$$s^2 + s\frac{R_3C_3 + R_2C_3 + R_2C_2 - KR_2C_2}{R_2R_3C_2C_3} + \frac{1}{R_2R_3C_2C_3} = 0 \tag{12.48}$$

which will have two real parts and two equal roots. Setting $s = j\omega$ in Eq. (12.48) and then equating the real parts to zero, we get

$$-\omega^2 + \frac{1}{R_2R_3C_2C_3} = 0$$

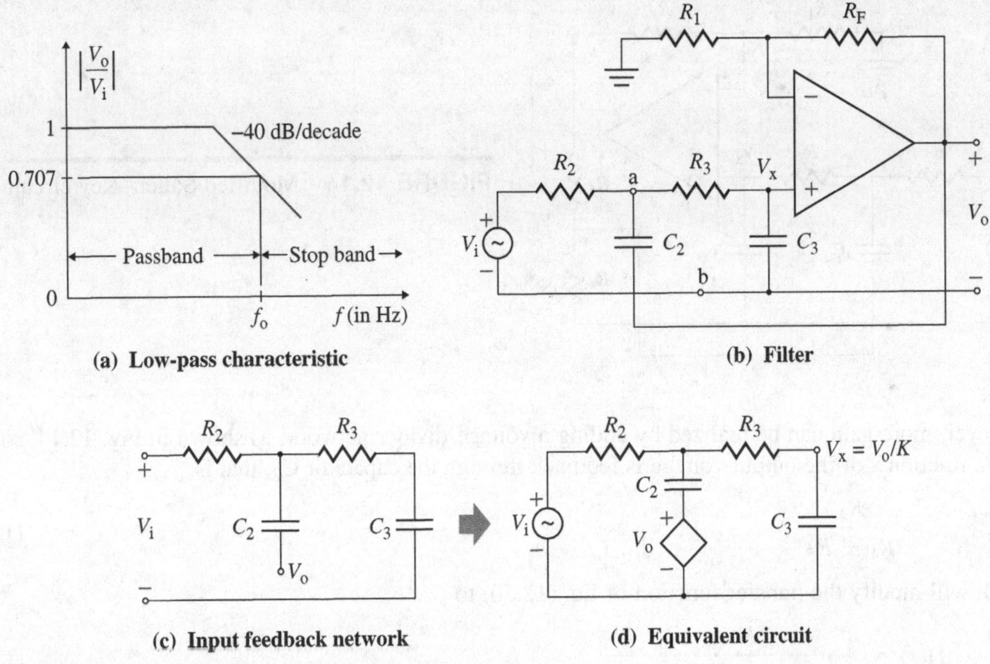

FIGURE 12.13 Second-order low-pass filter with $K = 1$

which gives the cutoff frequency as

$$f_o = \frac{\omega_o}{2\pi} = \frac{1}{2\pi\sqrt{R_2R_3C_2C_3}} \tag{12.49}$$

To simplify the design of second-order filters, equal resistances and capacitances are normally used—that is, $R_1 = R_2 = R_3 = R$, $C_2 = C_3 = C$. Then Eq. (12.47) can be simplified to

$$H(s) = \frac{K\omega_o^2}{s^2 + (3 - K)\omega_o s + \omega_o^2} \tag{12.50}$$

Comparing the denominator of Eq. (12.50) with that of Eq. (12.46) shows that Q can be related to K by

$$Q = \frac{1}{3 - K} \tag{12.51}$$

or

$$K = 3 - \frac{1}{Q} \tag{12.52}$$

The frequency response of a second-order system at the 3-dB point will depend on the damping factor ζ such that $Q = 1/2\zeta$. A Q-value of $1/\sqrt{2}$ $(= 0.707)$, which represents a compromise between the peak magnitude and the bandwidth, causes the filter to exhibit the characteristics of a flat passband as well as a stop band, and gives a fixed DC gain of $K = 1.586$:

$$K = 1 + \frac{R_F}{R_1} = 3 - \sqrt{2} = 1.586 \tag{12.53}$$

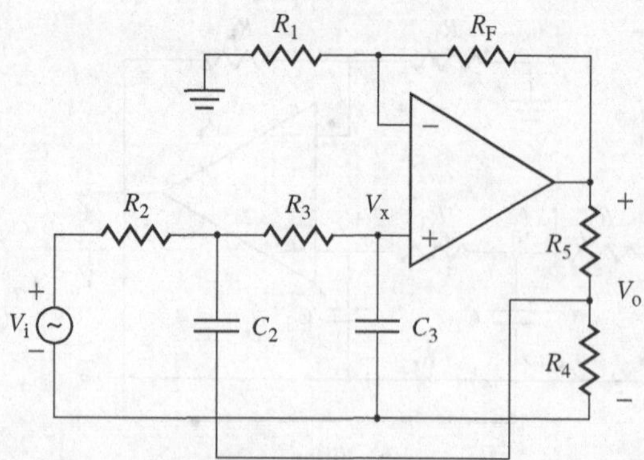

FIGURE 12.14 Modified Sallen–Key circuit

However, more gain can be realized by adding a voltage-divider network, as shown in Fig. 12.14, so that only a fraction x of the output voltage is fed back through the capacitor C_2; that is,

$$x = \frac{R_4}{R_4 + R_5} \tag{12.54}$$

which will modify the transfer function of Eq. (12.50) to

$$H(s) = \frac{K\omega_o^2}{s^2 + (3 - xK)\omega_o s + \omega_o^2} \tag{12.55}$$

and the quality factor Q of Eq. (12.51) to

$$Q = \frac{1}{3 - xK} \tag{12.56}$$

Thus, for $Q = 0.707$, $xK = 1.586$, allowing a designer to realize more DC gain K by choosing a lower value of x, where $x < 1$.

EXAMPLE 12.3

D **Designing a second-order low-pass filter**

(a) Design a second-order low-pass filter as in Fig. 12.14, to give a high cutoff frequency of $f_H = f_o = 1$ kHz, a pass-band gain of $K = 4$, and $Q = 0.707$, 1, 2, and ∞.

(b) Use PSpice/SPICE to plot the frequency response of the output voltage of the filter designed in part (a) from 10 Hz to 10 kHz.

SOLUTION

(a) To simplify the design calculations, let $R_1 = R_2 = R_3 = R_4 = R$ and let $C_2 = C_3 = C$. Choose a value of C less than or equal to 1 μF: Let $C = 0.01$ μF. For $R_2 = R_3 = R$ and $C_2 = C_3 = C$, Eq. (12.49) is reduced to

$$f_o = \frac{1}{2\pi RC}$$

which gives the value of R as

$$R = \frac{1}{2\pi f_o C} = \frac{1}{2\pi \times 1 \text{ kHz} \times 0.01 \text{ μF}} = 15{,}916\ \Omega \quad \text{(use a 20-kΩ potentiometer)}$$

Then $\quad R_F = (K - 1)R_1 = (4 - 1) \times 15{,}916 = 47{,}748\ \Omega$

For $Q = 0.707$ and $K = 4$, Eq. (12.56) gives $x = 1.586/K = 1.586/4 = 0.396$. From Eq. (12.54), we get

$$\frac{R_5}{R_4} = \frac{1}{x} - 1 = \frac{1 - x}{x} \tag{12.57}$$

which, for $x = 0.396$ and $R_4 = R = 15{,}916\ \Omega$, gives

$$R_5 = 1.525 \times 15{,}916 = 24{,}275\ \Omega \quad \text{(use a 30-kΩ potentiometer)}$$

For $Q = 1$ and $K = 4$, Eq. (12.56) gives $3 - xK = 1$ or $x = 2/K = 0.5$, and

$$R_5 = R = 15{,}916\ \Omega$$

For $Q = 2$ and $K = 4$, Eq. (12.56) gives $3 - xK = 1/2$ or $x = 2.5/K = 0.625$, and

$$R_5 = 0.6R = 9550\ \Omega$$

For $Q = \infty$ and $K = 4$, Eq. (12.56) gives $3 - xK = 1/Q = 0$ or $x = 3/K = 0.75$, and

$$R_5 = 0.333R = 5305\ \Omega$$

NOTE: The exact numbers of the resistances are used to demonstrate the validity of the calculated values with the PSpice simulation results. In a practical design, the commercially available standard values will be used (i.e., 48 kΩ instead of 47,748 Ω and 25 kΩ instead of 24,275 Ω; see Appendix E).

(b) The low-pass filter, with the designed values of the circuit parameters and a simple DC model of the op-amp, is shown in Fig. 12.15.

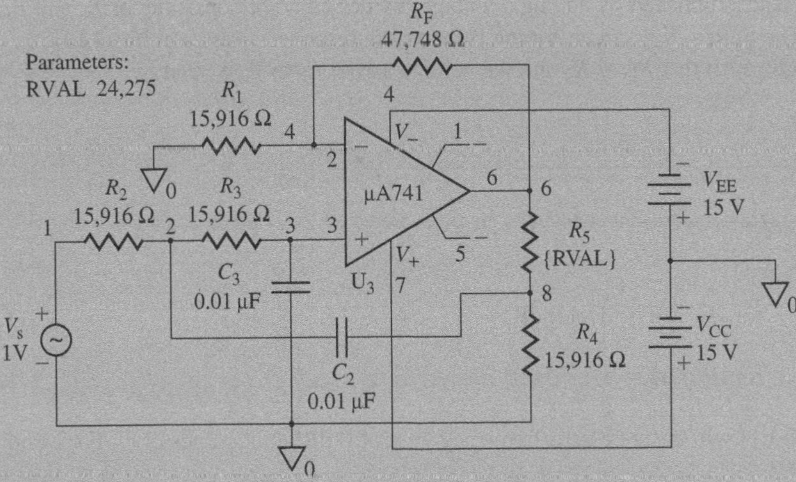

FIGURE 12.15 Second-order low-pass filter for PSpice simulation

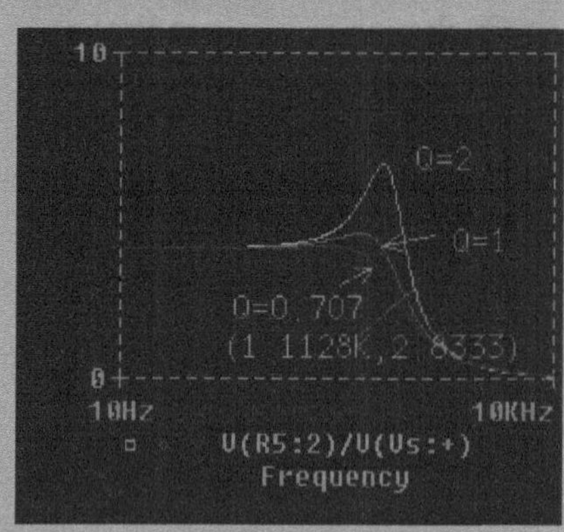

FIGURE 12.16 PSpice frequency response for Example 12.3

The PSpice plot of the voltage gain A_v [=V(R5:2)/V(Vs:+)] (using linear op-amp model in Fig. 3.7) is shown in Fig. 12.16. For $Q = 0.707$, we get $f_o = 758$ Hz (expected value is 1 kHz) at a gain of 2.833 (estimated value is $4 \times 0.707 = 2.828$). The error in the frequency is caused by the finite frequency-dependent gain of the op-amp. If we use an ideal op-amp in Fig. 3.7, the simulation will be very close to the expected value. The peaking of the gain increases with higher values of Q; however, the bandwidth also increases slightly ($f_o = 1113$ Hz for $Q = 2$). The plot for $Q = \infty$ is not shown.

12.8.3 Butterworth Low-Pass Filters

The Butterworth response requires that $|H(j0)| = 1$ (or 0 dB); the transfer function in Eq. (12.50) for the Sallen–Key circuit gives $|H(j0)| = K$ to achieve a Butterworth response with Sallen–Key topology. Therefore, we must reduce the gain by $1/K$. Consider the portion of the circuit to the left of the terminals a and b in Fig. 12.13(b). The resistance R_2 is in series with the input voltage V_i, as shown in Fig. 12.17(a). The gain reduction can be accomplished by adding a voltage-divider network consisting of R_a and R_b, as shown in Fig. 12.17(b). The Sallen–Key circuit for the Butterworth response appears in Fig. 12.17(c). The values of R_a and R_b must be such that $R_{in} = R_2$ and the voltage across R_b is V_i/K; that is,

$$\frac{R_a R_b}{R_a + R_b} = R_2 \tag{12.58}$$

$$\frac{R_b}{R_a + R_b} = \frac{1}{K} \tag{12.59}$$

Solving for R_a and R_b, we get

$$R_a = KR_2 \qquad \text{for } |H(j0)| = 1 \text{(or 0 dB)} \tag{12.60}$$

$$R_b = \frac{K}{K-1} R_2 \quad \text{for } |H(j0)| = 1 \text{ (or 0 dB)} \tag{12.61}$$

Equations (12.60) and (12.61) will ensure a zero-frequency gain of 0 dB at all values of Q. For example, if $K = 4$ and $R_2 = 15{,}916\ \Omega$,

$$R_a = 4 \times 15{,}916 = 63{,}664\ \Omega \quad \text{and} \quad R_b = \frac{4 \times 15{,}916}{(4-1)} = 21{,}221\ \Omega$$

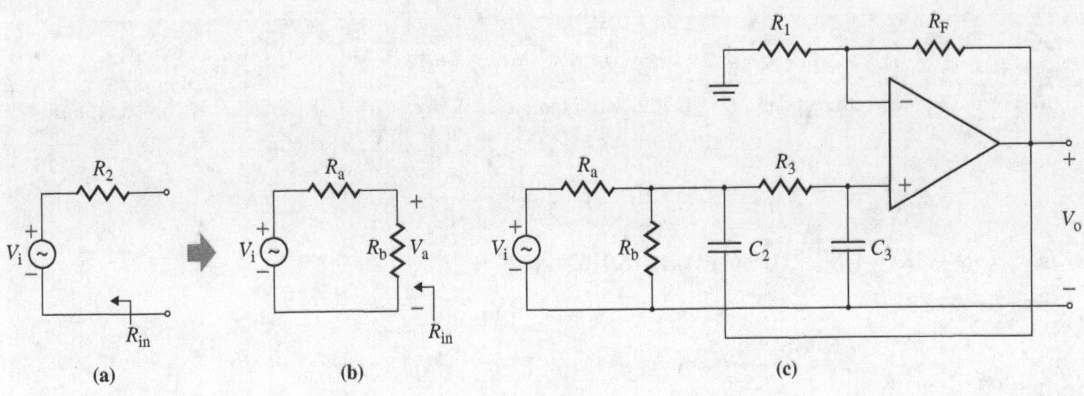

FIGURE 12.17 Sallen–Key circuit for the Butterworth response

It is more desirable, however, to have a gain of 0 dB at the resonant frequency ω_o—that is, $|H(j\omega_o)| = 1$ (or 0 dB). Substituting $s = j\omega_o$ into Eq. (12.50) gives a gain magnitude of $K/(3 - K)$, which gives the required gain reduction of $(3 - K)/K$; that is,

$$\frac{R_b}{R_a + R_b} = \frac{3 - K}{K} \tag{12.62}$$

Solving Eqs. (12.58) and (12.62) for R_a and R_b, we get

$$R_a = R_2 \frac{K}{3 - K} \quad \text{for } |H(j\omega_o)| = 1 \text{ (or 0 dB)} \tag{12.63}$$

$$R_b = R_2 \frac{K}{2K - 3} \quad \text{for } |H(j\omega_o)| = 1 \text{ (or 0 dB)} \tag{12.64}$$

Therefore, we can design an active filter to yield a gain of 0 dB at either $\omega = 0$ or $\omega = \omega_o$. In the case where $|H(j\omega_o)| = 1$ (or 0 dB) is specified, the zero-frequency gain will be reduced by a factor $(3 - K)/K$; that is,

$$|H(j0)| = 3 - K \quad \text{for } |H(j\omega_o)| = 1 \text{ (or 0 dB)} \tag{12.65}$$

For $Q = \sqrt{2}$ and $K = 3 - 1/Q = 1.586$, Eq. (12.65) gives $|H(j0)| = 3 - K = 1.414$, provided that we design the filter for $|H(j\omega_o)| = 1$ (or 0 dB).

EXAMPLE 12.4

D

Designing a second-order low-pass Butterworth filter for $|H(j\omega_o)| = 1$

(a) Design a second-order Butterworth low-pass filter as in Fig. 12.17(c) to yield $|H(j\omega_o)| = 1$ (or 0 dB), a cutoff frequency of $f_o = 1$ kHz, and $Q = 0.707$.

(b) Use PSpice/SPICE to plot the frequency response of the output voltage of the filter designed in part (a) from 10 Hz to 10 kHz.

SOLUTION

(a) For the Butterworth response, $Q = 0.707$, and Example 12.3 gives $C = 0.01\ \mu F$ and $R = 15{,}916\ \Omega$. From Eq. (12.52),

$$K = 3 - \frac{1}{Q} = 3 - \frac{1}{0.707} = 1.586$$

and $R_F = (K - 1)R_1 = (1.586 - 1) \times 15{,}916 = 9327\ \Omega$

From Eq. (12.63),

$$R_a = \frac{RK}{3 - K} = \frac{15{,}916 \times 1.586}{3 - 1.586} = 17{,}852\ \Omega$$

From Eq. (12.64),

$$R_b = \frac{RK}{2K - 3} = \frac{15{,}916 \times 1.586}{2 \times 1.586 - 3} = 146{,}760\ \Omega$$

From Eq. (12.65),

$$|H(j0)| = 3 - K = 3 - 1.586 = 1.414$$

NOTE: The exact numbers of the resistances are used to demonstrate the validity of the calculated values with the PSpice simulation results. In a practical design, the commercially available standard values will be used (i.e., 147 kΩ instead of 146,760 Ω and 18 kΩ instead of 17,852 Ω).

(b) For PSpice simulation, the circuit in Fig. 12.15 can be modified by removing R_4 and R_5, replacing R_2 by R_a, and adding R_b. This modified circuit is shown in Fig. 12.18.

The PSpice plot of the voltage gain is shown in Fig. 12.19, which gives $|H(j\omega_o)| = 1.0$ at $f_o = 1$ kHz and $|H(j0)| = 1.414$, both of which correspond to the expected values.

NOTE: The simulation was run with the DC linear model described in Sec. 6.3.

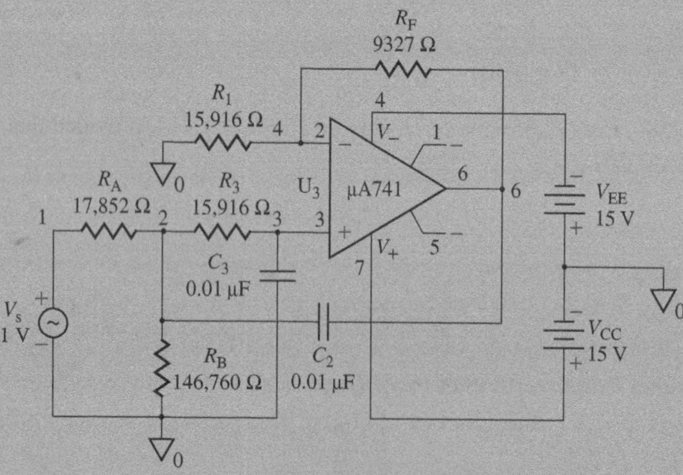

FIGURE 12.18 Second-order low-pass Butterworth filter for PSpice simulation

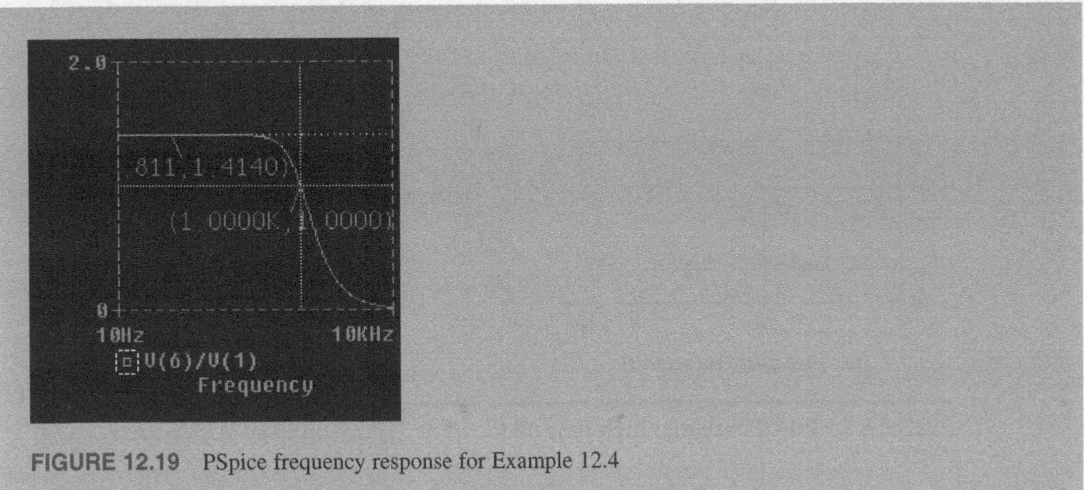

FIGURE 12.19 PSpice frequency response for Example 12.4

KEY POINTS OF SECTION 12.8

- A second-order filter with a fall rate of 40 dB/decade is preferred over a first-order filter with a fall rate of 20 dB/decade. First- and second-order filters can be used as building blocks for higher-order filters.
- The Sallen–Key circuit is a commonly used second-order filter. This circuit can be designed to exhibit the characteristics of a flat passband as well as a stop band and can be modified to give pass-band gain as well as the Butterworth response.

12.9 High-Pass Filters

High-pass filters can be classified broadly into two types: first-order and second-order. Higher-order filters can be synthesized from these two basic types. Since the frequency scale of a low-pass filter is 0 to f_o and that of a high-pass filter is f_o to ∞, their frequency scales have a reciprocal relationship. Therefore, if we can design a low-pass filter, we can convert it to a high-pass filter by applying an *RC-CR* transformation. This transformation can be accomplished by replacing R_n by C_n and C_n by R_n. The op-amp, which is modeled as a voltage-controlled voltage source, is not affected by this transformation. The resistors that are used to set the DC gain of the op-amp circuit are not affected either.

12.9.1 First-Order High-Pass Filters

The transfer function of a first-order high-pass filter has the general form

$$H(s) = \frac{sK}{s + \omega_o} \tag{12.66}$$

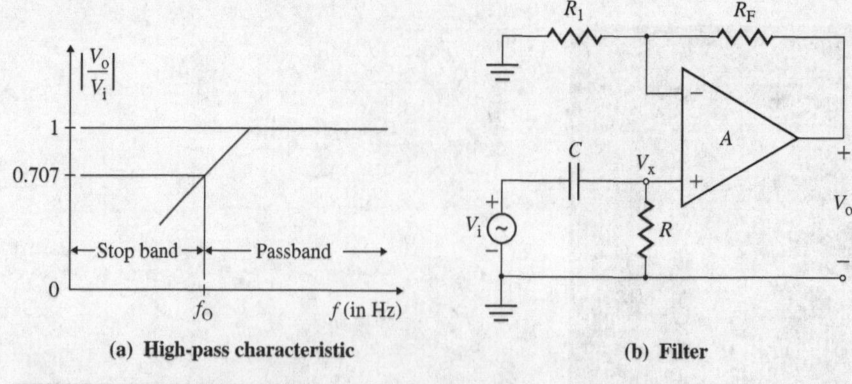

FIGURE 12.20 First-order high-pass filter

A typical high-pass frequency characteristic is shown in Fig. 12.20(a). A first-order high-pass filter can be formed by interchanging the frequency-dependent resistor and capacitor of the low-pass filter of Fig. 12.10(b). This arrangement is shown in Fig. 12.20(b). The voltage at the noninverting terminal of the op-amp can be found by the voltage divider rule. That is,

$$V_x(s) = \frac{R}{R + 1/sC} V_i(s) = \frac{S}{s + 1/RC} V_i(s)$$

The output voltage of the noninverting amplifier is

$$V_o(s) = \left(1 + \frac{R_F}{R_1}\right) V_x(s) = \left(1 + \frac{R_F}{R_1}\right) \frac{s}{s + 1/RC} v_i(s)$$

which gives the voltage gain as

$$H(s) = \frac{V_o(s)}{V_i(s)} = \frac{sK}{s + 1/RC} \tag{12.67}$$

where $K = 1 + R_F/R_1$ is the DC *voltage gain.*

Substituting $s = j\omega$ into Eq. (12.67), we get

$$H(j\omega) = \frac{V_o(j\omega)}{V_i(j\omega)} = \frac{j\omega K}{j\omega + 1/RC} = \frac{j\omega K}{j\omega + \omega_o} \tag{12.68}$$

which gives the cutoff frequency f_o at 3-dB gain as

$$f_o = \frac{\omega_o}{2\pi} = \frac{1}{2\pi RC} \tag{12.69}$$

as in Eq. (12.43). The magnitude and phase angle of the filter gain can be found from

$$|H(j\omega)| = \frac{(\omega/\omega_o)K}{[1 + (\omega/\omega_o)^2]^{1/2}} = \frac{(f/f_o)K}{[1 + (f/f_o)^2]^{1/2}} \tag{12.70}$$

and $\phi = 90° - \tan^{-1}(f/f_o)$ (12.71)

This filter passes all signals with frequencies higher than f_o. However, the high-frequency limit is determined by the bandwidth of the op-amp itself. The gain–bandwidth product of a practical μ741-type op-amp is 1 MHz.

EXAMPLE 12.5

D

Designing a first-order high-pass filter Design a first-order high-pass filter with a cutoff frequency of $f_o = 1$ kHz and a pass-band gain of 4.

SOLUTION

High-pass filters are formed simply by interchanging R and C of the input RC network, so the design and frequency scaling procedures for low-pass filters are also applicable. Since $f_o = 1$ kHz, we can use the values of R and C that were determined for the low-pass filter of Example 12.2—that is,

$C = 0.01\ \mu\text{F}$

$R = 15{,}916\ \Omega$ (use a 20-kΩ potentiometer)

Similarly, we use $R_1 = 10$ kΩ and $R_F = 30$ kΩ to yield $K = 4$.

A PSpice simulation that confirms the design values can be run by interchanging the locations of R and C in Fig. 12.11 so that the statements for R and C read as follows:

```
C  1  2  0.01UF   ; For C connected between nodes 1 and 2
R  2  0  15916    ; For R connected between nodes 2 and 0
```

12.9.2 Second-Order High-Pass Filters

A second-order high-pass filter has a stop-band characteristic of 40 dB/decade rise. The general form of a second-order high-pass filter is

$$H(s) = \frac{s^2 K}{s^2 + (\omega_o/Q)s + \omega_o^2} \tag{12.72}$$

where K is the high-frequency gain. Figure 12.21(a) shows a typical frequency response. As in the case of the first-order filter, a second-order high-pass filter can be formed from a second-order low-pass filter by interchanging the frequency-dominant resistors and capacitors. Figure 12.21(b) shows a second-order high-pass filter derived from the Sallen–Key circuit of Fig. 12.13(b). The transfer function can be derived by applying the RC-to-CR transformation and substituting $1/s$ for s in Eq. (12.47). For $R_1 = R_2 = R_3 = R$

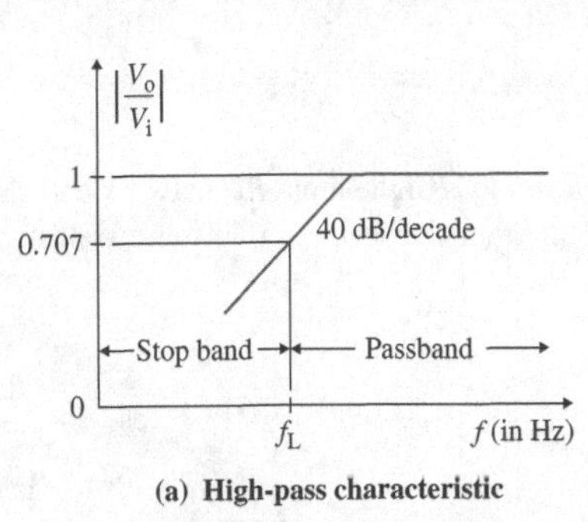

(a) High-pass characteristic

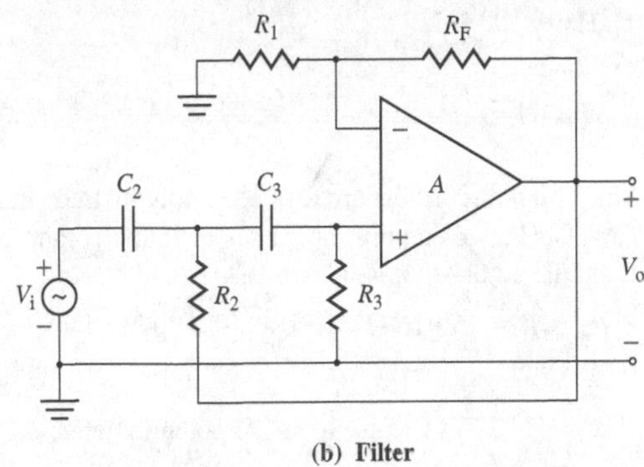

(b) Filter

FIGURE 12.21 Second-order high-pass filter

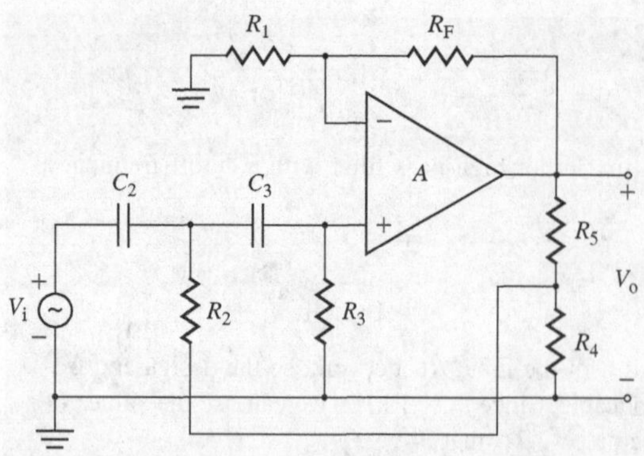

FIGURE 12.22 Modified second-order high-pass filter

and $C_2 = C_3 = C$, the transfer function becomes

$$H(s) = \frac{s^2 K}{s^2 + (3 - K)\omega_o s + \omega_o^2} \tag{12.73}$$

and Eq. (12.49) gives the cutoff frequency as

$$f_o = \frac{\omega_o}{2\pi} = \frac{1}{2\pi\sqrt{R_2 R_3 C_2 C_3}} = \frac{1}{2\pi RC} \tag{12.74}$$

Q and K of the circuit remain the same. A voltage-divider network can be added, as shown in Fig. 12.22, so that only a fraction x of the output voltage is fed back through resistor R_2. The transfer function of Eq. (12.73) then becomes

$$H(s) = \frac{s^2 K}{s^2 + (3 - xK)\omega_o s + \omega_o^2} \tag{12.75}$$

EXAMPLE 12.6

D

Designing a second-order high-pass filter

(a) Design a second-order high-pass filter as in Fig. 12.22, with a cutoff frequency of $f_o = 1$ kHz, a pass-band gain of $K = 4$, and $Q = 0.707$, 1, 2, and ∞.

(b) Use PSpice/SPICE to plot the frequency response of the output voltage of the filter designed in part (a) from 10 Hz to 100 kHz.

SOLUTION

(a) Since high-pass filters are formed simply by interchanging the Rs and Cs of the input RC network and since $f_o = 1$ kHz, we can use the values of R and C that were determined for the second-order low-pass filter of Example 12.3—that is, $C = 0.01$ μF, and

$R_4 = R = 15{,}916\ \Omega$ (use a 20-kΩ potentiometer)

For $Q = 0.707$,

$R_5 = 24{,}275\ \Omega$ (use a 30-kΩ potentiometer)

For $Q = 1$,

$R_5 = R = 15{,}916\ \Omega$

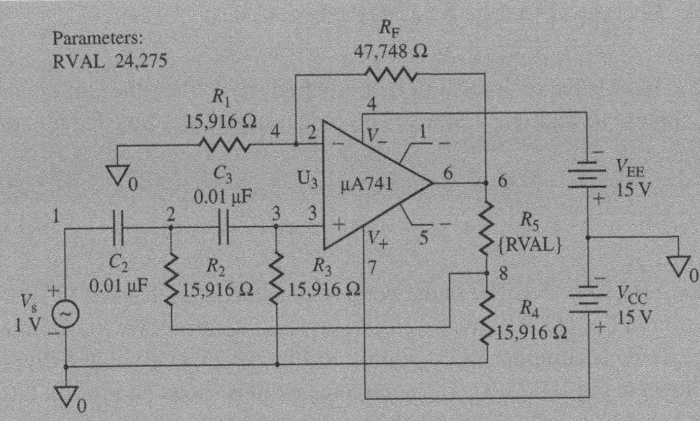

FIGURE 12.23 Second-order high-pass filter for PSpice simulation

For $Q = 2$,

$$R_5 = 0.6R = 9550\ \Omega$$

For $Q = \infty$,

$$R_5 = 0.3333R = 5305\ \Omega$$

NOTE: The exact numbers of the resistances are used to demonstrate the validity of the calculated values with the PSpice simulation results. In a practical design, the commercially available standard values will be used (i.e., 10 kΩ instead of 9550 Ω; see Appendix E).

(b) Figure 12.23 shows the circuit obtained by interchanging the locations of R and C in Fig. 12.15. The PSpice plots are shown in Fig. 12.24. As expected, the voltage gain shows increased peaking for a higher value of Q. The PSpice statements for R and C read as follows:

```
C2  1  2  0.01UF   ; For C2 connected between nodes 1 and 2
R2  2  8  15916    ; For R2 connected between nodes 2 and 8
C3  2  3  0.01UF   ; For C3 connected between nodes 2 and 3
R3  3  0  15916    ; For R3 connected between nodes 3 and 0
```

NOTE: We will notice a high-end roll-off due to the internal capacitances of the µA741 op-amp.

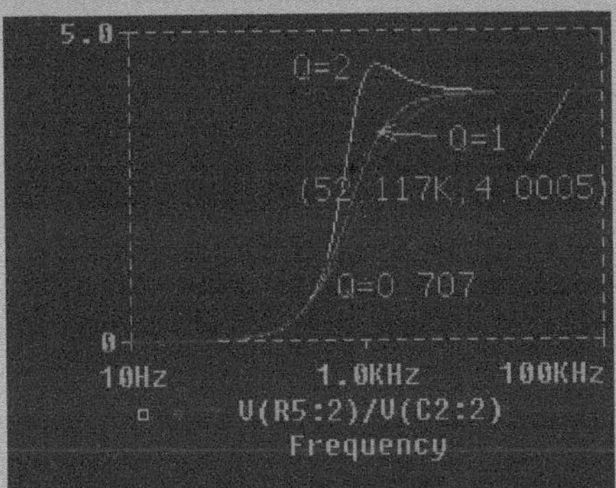

FIGURE 12.24 PSpice plots of frequency response for Example 12.6

12.9.3 Butterworth High-Pass Filters

Since the frequency scale of a low-pass filter is the reciprocal of that of a high-pass filter, the Butterworth response of Eq. (12.26) can also be applied to high-pass filters. The magnitude of the transfer function becomes

$$|H_n(j\omega)| = \frac{1}{[1 + (\omega_o/\omega)^{2n}]^{1/2}} \tag{12.76}$$

where $|H_n(j\infty| = 1$ for all n, rather than $|H_n(j0)| = 1$. The Butterworth response requires that $|H(j\infty)| = 1$ (or 0 dB); however, the transfer function in Eq. (12.73) gives $|H(j\infty)| = K$. Therefore, we must reduce the gain by $1/K$. The gain reduction can be accomplished by adding to Fig. 12.25(a) a voltage-divider network consisting of C_a and C_b, as shown in Fig. 12.25(b). The complete circuit is shown in Fig. 12.25(c). The values of C_a and C_b must be such that $C_{in} = C_2$ and the voltage across C_b is V_i/K; that is,

$$C_a + C_b = C_2 \tag{12.77}$$

$$\frac{C_a}{C_a + C_b} = \frac{1}{K} \tag{12.78}$$

Solving for C_a and C_b, we get

$$C_a = \frac{C_2}{K} \qquad [\text{for } |H(j\infty)| = 1 \text{ (or 0 dB)}] \tag{12.79}$$

$$C_b = C_2\frac{K - 1}{K} \qquad [\text{for } |H(j\infty)| = 1 \text{ (or 0 dB)}] \tag{12.80}$$

Equations (12.79) and (12.80) will ensure a high-frequency gain of 0 dB at all values of Q. For $C_2 = 0.01\ \mu\text{F}$ and $K = 4$, we get

$$C_a = \frac{0.01\ \mu\text{F}}{4} = 2.5\ \text{nF} \quad \text{and} \quad C_b = \frac{0.01\ \mu\text{F} \times (4 - 1)}{4} = 7.5\ \text{nF}$$

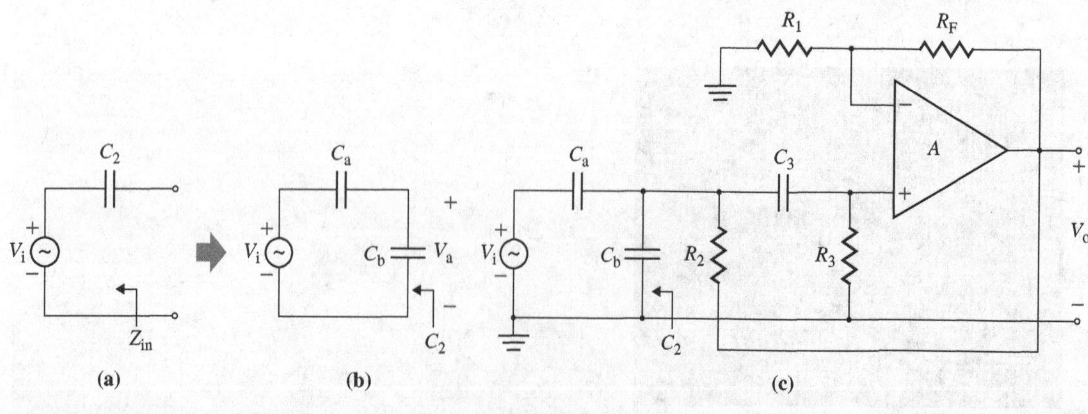

FIGURE 12.25 Butterworth second-order high-pass filter

As with low-pass filters, however, it is more desirable to have a gain of 0 dB at the resonant frequency ω_o—that is, for $|H(j\omega_o)| = 1$ (or 0 dB). Substituting $s = j\omega_o$ into Eq. (12.73) yields a gain magnitude of $K/(3 - K)$, which gives the required gain reduction of $(3 - K)/K$. Thus, Eq. (12.78) becomes

$$\frac{C_a}{C_a + C_b} = \frac{3 - K}{K} \tag{12.81}$$

Solving Eqs. (12.77) and (12.81) for C_a and C_b, we get

$$C_a = C_2\frac{3 - K}{K} \qquad [\text{for } |H(j\omega_o)| = 1 \text{ (or 0 dB)}] \tag{12.82}$$

$$C_b = C_2\frac{2K - 3}{K} \qquad [\text{for } |H(j\omega_o)| = 1 \text{ (or 0 dB)}] \tag{12.83}$$

Therefore, we can design a Butterworth high-pass filter to yield a gain of 0 dB at either $\omega = \infty$ or $\omega = \omega_o$. However, in the case where $|H(j\omega_o)| = 1$ (or 0 dB) is specified, the high-frequency gain will be reduced by a factor $(3 - K)/K$; that is,

$$|H(j\infty)| = 3 - K \quad [\text{for } |H(j\omega_o)| = 1 \text{ (or 0 dB)}] \tag{12.84}$$

For $Q = \sqrt{2}$ and $K = 3 - 1/Q = 1.586$, Eq. (12.84) gives $|H(j\omega_o)| = 3 - K = 1.414$, provided that we design the filter for $|H(j\omega_o)| = 1$ (or 0 dB).

EXAMPLE 12.7

D

Designing a second-order high-pass Butterworth filter for $|H(j\infty)| = 1$

(a) Design a second-order Butterworth high-pass filter as in Fig. 12.25(c) to yield $|H(j\infty)| = 1$ (or 0 dB), a cutoff frequency of $f_o = 1$ kHz, and $Q = 0.707$.

(b) Use PSpice/SPICE to plot the frequency response of the output voltage of the filter designed in part (a) from 10 Hz to 100 kHz.

SOLUTION

(a) For $Q = 0.707$, Example 12.6 gives $C = 0.01$ µF and $R = 15{,}916\ \Omega$. From Eq. (12.52),

$$K = \frac{3 - 1}{Q} = \frac{3 - 1}{0.707} = 1.586$$

and $\quad R_F = (K - 1)R_1 = (1.586 - 1) \times 15{,}916 = 9327\ \Omega$

From Eq. (12.79),

$$C_a = \frac{C}{K} = \frac{0.01\ \mu\text{F}}{1.586} = 6.305 \text{ nF}$$

From Eq. (12.80),

$$C_b = \frac{C(K - 1)}{K} = \frac{0.01\ \mu\text{F} \times (1.586 - 1)}{1.586} = 3.695 \text{ nF}$$

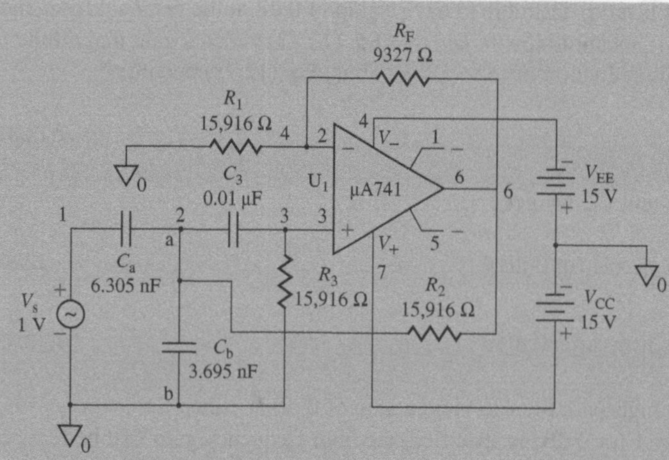

FIGURE 12.26 Second-order high-pass Butterworth filter for PSpice simulation

NOTE: The exact numbers of the resistances and capacitances are used to demonstrate the validity of the calculated values with the PSpice simulation results. In a practical design, the commercially available standard values will be used (i.e., 9.3 kΩ instead of 9327 Ω and C_b = 4 nF instead of 3.695 nF; see Appendix E).

(b) For PSpice simulation, the circuit of Fig. 12.18 can be transformed into the circuit of Fig. 12.26 by removing R_4 and R_5, interchanging the locations of R and C, replacing C_2 by C_a, and adding C_b between nodes a and b. The PSpice statements for R and C read as follows:

```
CA  1  2  6.305NF   ; For CA connected between nodes 1 and 2
CB  2  0  3.695NF   ; For CB connected between nodes 2 and 0
R2  2  6  15916     ; For R2 connected between nodes 2 and 6
C3  2  3  0.01UF    ; For C3 connected between nodes 2 and 3
R3  3  0  15916     ; For R3 connected between nodes 3 and 0
```

The PSpice plot of the gain is shown in Fig. 12.27, which gives $|H(j\infty)| = 1.0$ at $\omega = \infty$.

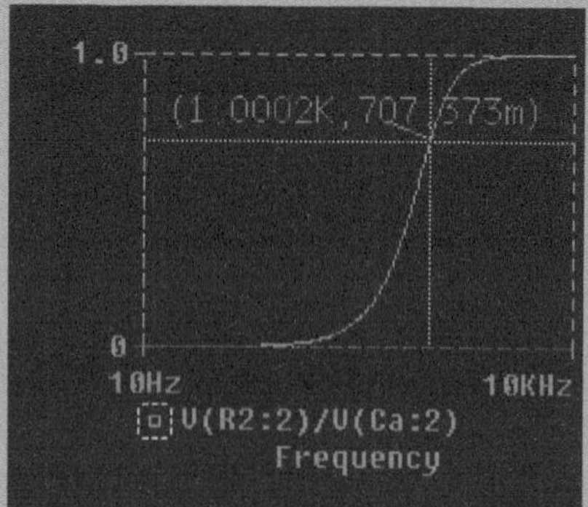

FIGURE 12.27 PSpice plot of frequency response for Example 12.7

KEY POINTS OF SECTION 12.9

- A low-pass filter can be converted to a high-pass filter by applying the *RC*-to-*CR* transformation: R_n is replaced by C_n, and C_n is replaced by R_n.
- The Sallen–Key low-pass circuit can be modified to exhibit second-order high-pass characteristics with a pass-band gain as well as a Butterworth response.

12.10 Band-Pass Filters

A band-pass filter has a passband between two cutoff frequencies f_L and f_H such that $f_H > f_L$. Any frequency outside this range is attenuated. The transfer function of a BP filter has the general form

$$H_{BP}(s) = \frac{K_{PB}(\omega_C/Q)s}{s^2 + (\omega_C/Q)s + \omega_C^2} \tag{12.85}$$

where K_{PB} is the pass-band gain and ω_C is the center frequency in radians per second. There are two types of band-pass filters: wide band pass and narrow band pass. Although there is no dividing line between the two, it is possible to identify them from the value of the *quality factor Q*. A filter may be classified as wide band pass if $Q \le 10$ and narrow band pass if $Q > 10$. The higher the value of Q, the more selective the filter or the narrower its bandwidth (BW) will be. Thus, Q is a measure of the selectivity of a filter. The relationship of Q to 3-dB bandwidth and center frequency f_C is given by

$$Q = \frac{\omega_C}{\text{BW}} = \frac{f_C}{f_H - f_L} \tag{12.86}$$

For a wide-band-pass filter, the center frequency f_C can be defined as

$$f_C = \sqrt{f_L f_H} \tag{12.87}$$

where f_L is the low cutoff frequency, in hertz, and f_H is the high cutoff frequency, in hertz. In a narrow-band-pass filter, the output peaks at the center frequency f_C.

12.10.1 Wide-Band-Pass Filters

The frequency characteristic of a wide-band-pass filter is shown in Fig. 12.28(a), where $f_H > f_L$. This characteristic can be obtained by implementing Eq. (12.85), which may not give a flat midband gain over a wide bandwidth. An alternative arrangement is to use two filters: one low-pass filter and one high-pass filter. The output is obtained by multiplying the low-frequency response by the high-frequency response, as shown in Fig. 12.28(b); this solution can be implemented simply by cascading the first-order (or second-order)

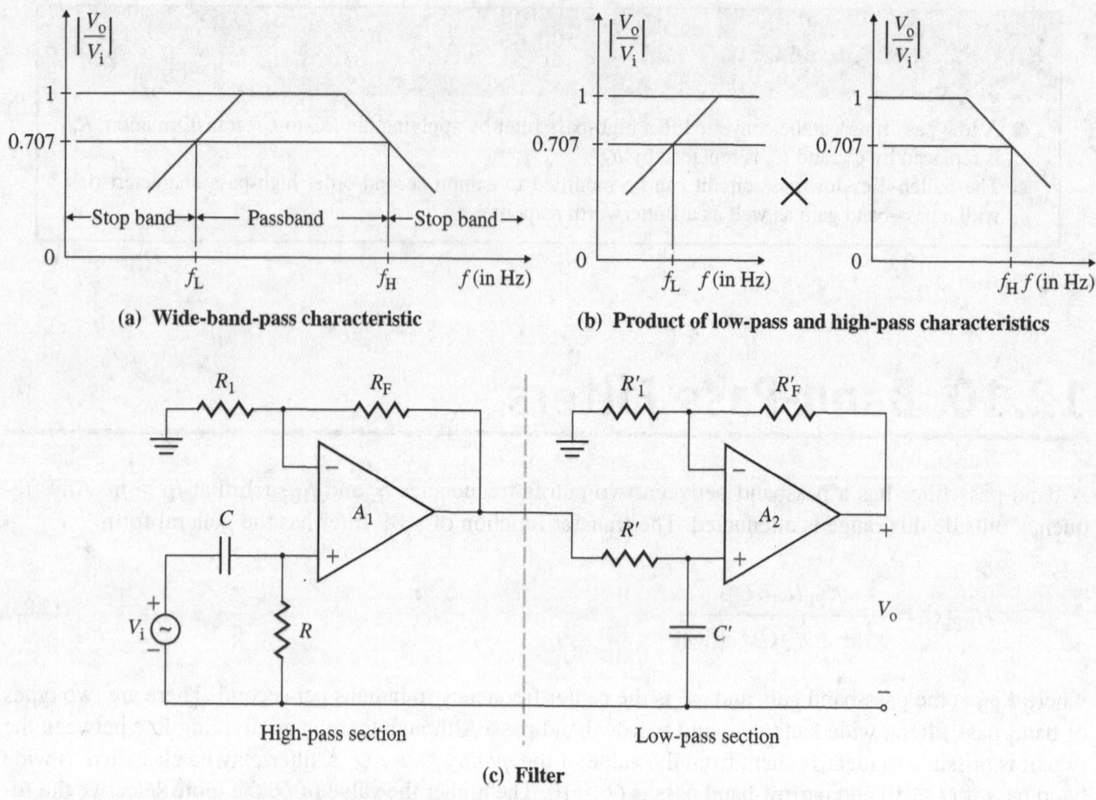

FIGURE 12.28 Wide-band-pass filter

high-pass and low-pass sections. The order of the band-pass filter depends on the order of the high-pass and low-pass sections. This arrangement has the advantage that the falloff, rise, and midband gain can be set independently. However, it requires more op-amps and components.

Figure 12.28(c) shows a ±20 dB/decade wide-band-pass filter implemented with first-order high-pass and first-order low-pass filters. In this case, the magnitude of the voltage gain is equal to the product of the voltage gain magnitudes of the high-pass and low-pass filters. From Eqs. (12.40) and (12.67), the transfer function of the wide-midband filter for first-order implementation becomes

$$H(s) = \frac{K_{PB}\omega_H s}{(s + \omega_L)(s + \omega_H)} \tag{12.88}$$

Using Eqs. (12.46) and (12.72) gives the transfer function for second-order implementation:

$$H(s) = \frac{K_{PB}\omega_H^2 s^2}{[s^2 + (\omega_L/Q)s + \omega_L^2][s^2 + (\omega_H/Q)s + \omega_H^2]} \tag{12.89}$$

where K_{PB} = overall pass-band gain = high-pass gain K_H × low-pass gain K_L.

EXAMPLE 12.8

D

Designing a wide-band-pass filter

(a) Design a wide-band-pass filter with $f_L = 10$ kHz, $f_H = 1$ MHz, and a pass-band gain of $K_{PB} = 16$.

(b) Calculate the value of Q for the filter.

(c) Use PSpice/SPICE to plot the frequency response of the filter designed in part (a) from 100 Hz to 10 MHz.

SOLUTION

(a) Let the gain of the high-pass section be $K_H = 4$. For the first-order high-pass section, $f_L = 10$ kHz. Following the steps in Example 12.5, we let $C = 1$ nF. Then

$$R = \frac{1}{2\pi \times 10\,\text{kHz} \times 1\,\text{nF}} = 15.915\,\text{k}\Omega$$

$$\text{and} \qquad K_H = 1 + \frac{R_F}{R_1} = 4 \qquad \text{or} \qquad \frac{R_F}{R_1} = 4 - 1 = 3$$

If we let $R_1 = 10\,\text{k}\Omega$, $R_F = 3R_1 = 30\,\text{k}\Omega$.

For the first-order low-pass section, $f_H = 1$ MHz and the desired gain is $K_L = K_{PB}/K_H = 16/4 = 4$. Following the steps in Example 12.2, we let $C' = 10$ pF. Then

$$R' = \frac{1}{2\pi \times 1\,\text{MHz} \times 10\,\text{pF}} = 15.915\,\text{k}\Omega$$

$$\text{and} \qquad K_L = 1 + \frac{R'_F}{R'_1} = 4 \qquad \text{or} \qquad \frac{R'_F}{R'_1} = 4 - 1 = 3$$

If we let $R'_1 = 10\,\text{k}\Omega$, $R'_F = 3R'_1 = 30\,\text{k}\Omega$.

(b) From Eq. (12.87),

$$f_C = \sqrt{10\,\text{kHz} \times 1\,\text{MHz}} = 100\,\text{kHz}$$

$$\text{and} \qquad \text{BW} = 1\,\text{MHz} - 10\,\text{kHz} = 990\,\text{kHz}$$

From Eq. (12.86), we can find

$$Q = \frac{100\,\text{kHz}}{1\,\text{MHz} - 10\,\text{kHz}} = 0.101$$

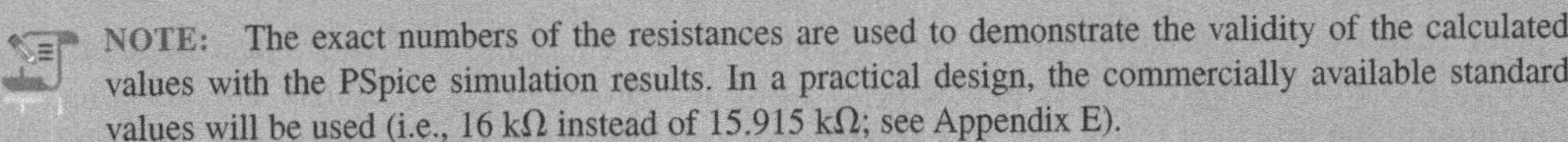

NOTE: The exact numbers of the resistances are used to demonstrate the validity of the calculated values with the PSpice simulation results. In a practical design, the commercially available standard values will be used (i.e., 16 kΩ instead of 15.915 kΩ; see Appendix E).

(c) The wide-band-pass filter with the calculated values is shown in Fig. 12.29.

The frequency response (with a linear model in Fig. 3.8) is shown in Fig. 12.30, which gives $K_{PB} = 15.842$ (expected value is 16), $f_L = 10.04$ kHz (expected value is 10 kHz), and $f_H = 997$ kHz (expected value is 1 MHz). For a low value of bandwidth, the response due to the high-pass filter may not reach the expected gain before the low-pass filter becomes effective. As a result, the pass-band gain may be much lower than 16.

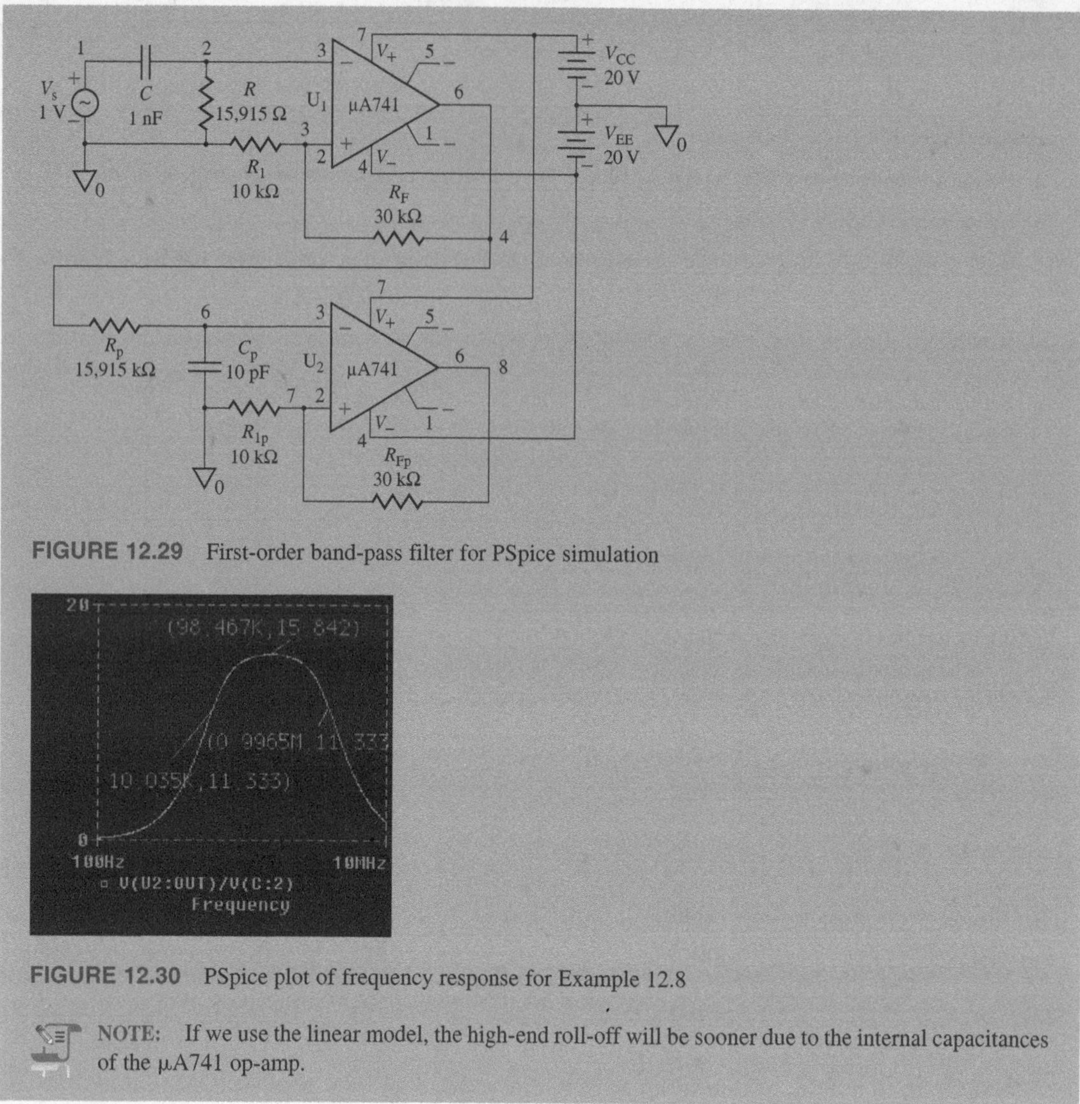

FIGURE 12.29 First-order band-pass filter for PSpice simulation

FIGURE 12.30 PSpice plot of frequency response for Example 12.8

NOTE: If we use the linear model, the high-end roll-off will be sooner due to the internal capacitances of the μA741 op-amp.

12.10.2 Narrow-Band-Pass Filters

A typical frequency response of a narrow-band-pass filter is shown in Fig. 12.31(a). This characteristic can be derived by setting a high Q-value for the band-pass filter shown in Fig. 12.31(b). This filter uses only one op-amp in the inverting mode. Because it has two feedback paths, it is also known as a *multiple feedback filter*. For a low Q-value, it can also exhibit the characteristic of a wide-band-pass filter.

A narrow-band-pass filter is generally designed for specific values of f_C and Q or f_C and BW. The op-amp, along with C_2 and R_2, can be regarded as an inverting differentiator such that $V_o(s) = (-sC_2R_2)V_x(s)$; the equivalent filter circuit is shown in Fig. 12.31(c). The transfer function of the filter network is

$$H_{BP}(s) = \frac{V_o(s)}{V_i(s)} = \frac{(-1/R_1C_1)s}{s^2 + (1/R_2)(1/C_1 + 1/C_2)s + 1/R_1R_2C_1C_2} \tag{12.90}$$

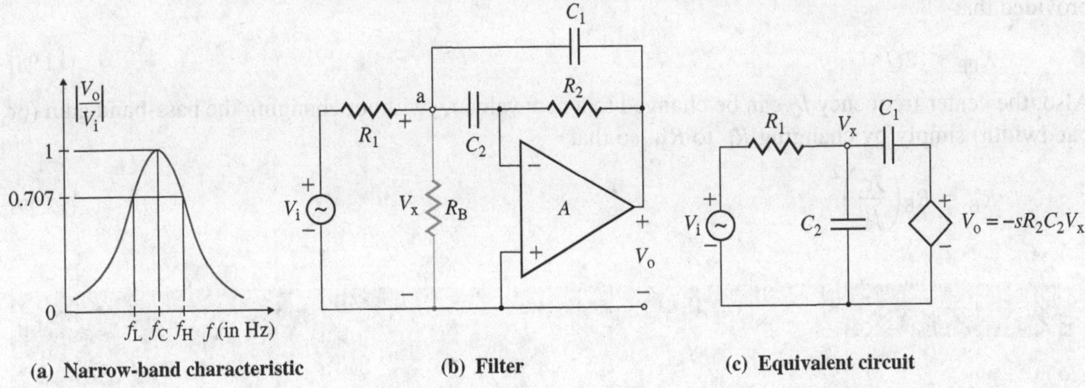

FIGURE 12.31 Narrow-band-pass filter

which is similar in form to Eq. (12.85). (See Prob. 12.29 for the derivation.) For $C_1 = C_2 = C$, Eq. (12.90) gives

$$\omega_C = \frac{1}{\sqrt{R_1R_2C_1C_2}} = \frac{1}{C\sqrt{R_1R_2}} \tag{12.91}$$

$$Q = \frac{1}{2}\sqrt{\frac{R_2}{R_1}} \tag{12.92}$$

$$K_{PB}\left(\frac{\omega_C}{Q}\right) = \frac{1}{R_1C_1} \tag{12.93}$$

Solving these equations, we can find the component values:

$$R_1 = \frac{Q}{2\pi f_C C K_{PB}} \tag{12.94}$$

$$R_2 = \frac{Q}{\pi f_C C} \tag{12.95}$$

$$K_{PB} = \frac{R_2}{2R_1} = 2Q^2 \tag{12.96}$$

Resistance R_1 can be replaced by R_A, and resistance R_B can be connected between nodes a and 0 so that the design specification $|H_{BP}(j\omega_C)| = 1$ (or 0 dB) is met for the Butterworth response. The method of calculating the values of R_A and R_B for a gain reduction of $1/K_{PB}$ $(=1/2Q^2)$ is explained in Sec. 12.8.

Notice from Eq. (12.96) that, for a known value of Q, the value of K_{PB} is fixed. It is, however, possible to have different values of K_{PB} and Q by choosing only the value of R_B without changing the value of R_1. The new value of the gain K_{PB} is related to $2Q^2$ by

$$\frac{K_{PB}}{2Q^2} = \frac{R_B}{R_1 + R_B}$$

which gives the value of R_B as

$$R_B = \frac{R_1K_{PB}}{2Q^2 - K_{PB}} = \frac{Q}{2\pi f_C C(2Q^2 - K_{PB})} \tag{12.97}$$

provided that

$$K_{PB} < 2Q^2 \tag{12.98}$$

Also, the center frequency f_C can be changed to a new value f'_C without changing the pass-band gain (or bandwidth) simply by changing R_B to R'_B, so that

$$R'_B = R_B\left(\frac{f_C}{f'_C}\right)^2 \tag{12.99}$$

EXAMPLE 12.9

D

Designing a narrow-band-pass filter

(a) Design a narrow-band-pass filter as in Fig. 12.31(b) such that $f_C = 1$ kHz, $Q = 4$, and $K_{PB} = 8$.

(b) Calculate the value of R_B required to change the center frequency from 1 kHz to 1.5 kHz.

(c) Use PSpice/SPICE to plot the frequency response of the narrow-band-pass filter designed in part (a) from 100 Hz to 1 MHz.

SOLUTION

(a) $f_C = 1$ kHz and $Q = 4$. Let $C_1 = C_2 = C = 0.0047\ \mu$F. Check that the condition in Eq. (12.98) is satisfied; that is, $2Q^2 = 2 \times 4^2 = 32$, which is greater than $K_{PB} = 8$. Thus, we must use R_B in Fig. 12.32. Using Eqs. (12.94), (12.95), (12.96), and (12.97), we get

$$R_1 = \frac{Q}{2\pi f_C C K_{PB}} = \frac{4}{2\pi \times 1\ \text{kHz} \times 0.0047\ \mu\text{F} \times 8} = 16.93\ \text{k}\Omega$$

$$R_2 = \frac{Q}{\pi f_C C} = \frac{4}{\pi \times 1\ \text{kHz} \times 0.0047\ \mu\text{F}} = 270.9\ \text{k}\Omega$$

$$K_{PB} = \frac{R_2}{2R_1} = \frac{270.9\ \text{k}\Omega}{2 \times 16.93\ \text{k}\Omega} = 8$$

$$R_B = \frac{Q}{2\pi f_C C(2Q^2 - K_{PB})} = \frac{4}{2\pi \times 1\text{kHz} \times 0.0047\ \mu\text{F} \times (2 \times 4^2 - 8)} = 5.64\ \text{k}\Omega$$

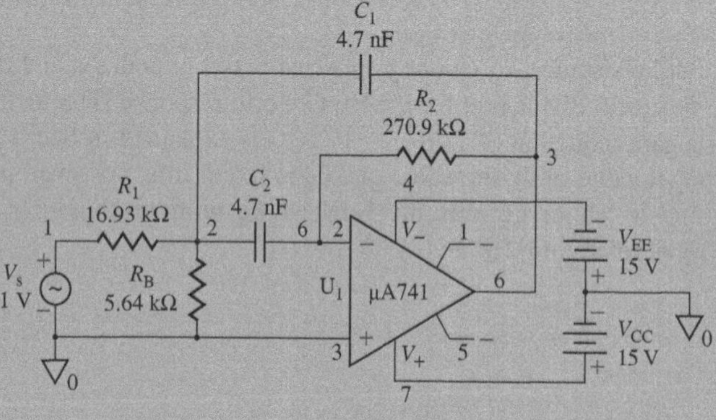

FIGURE 12.32 Narrow-band-pass filter for PSpice simulation

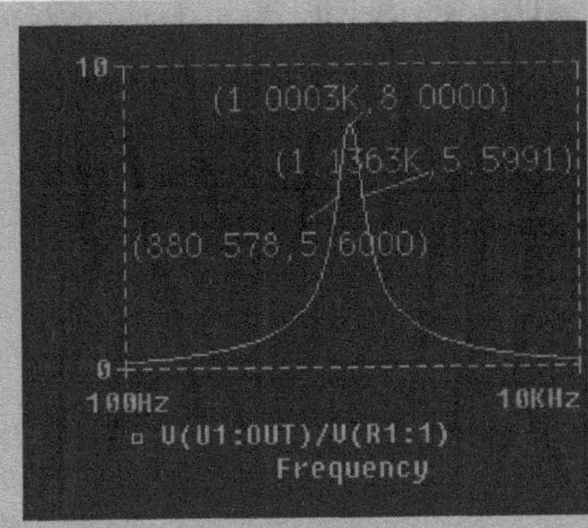

FIGURE 12.33 PSpice plot of frequency response for Example 12.9

(b) From Eq. (12.99), we find that the new value of R'_B is

$$R'_B = R_B\left(\frac{f_C}{f'_C}\right)^2 = 5.64\ \text{k}\Omega\left(\frac{1\ \text{kHz}}{1.5\ \text{kHz}}\right)^2 = 2.51\ \text{k}\Omega$$

NOTE: The exact numbers of the resistances are used to demonstrate the validity of the calculated values with the PSpice simulation results. In a practical design, the commercially available standard values will be used (i.e., 17 kΩ instead of 16.93 kΩ and 270 kΩ instead of 270.9 kΩ; see Appendix E).

(c) The narrow-band-pass filter with the designed values is shown in Fig. 12.32.

The frequency response is shown in Fig. 12.33, which gives $f_C \approx 1$ kHz (expected value is 1 kHz) and $K_{PB} = 8$.

KEY POINTS OF SECTION 12.10

- The wide-band-pass characteristic can be obtained by cascading a high-pass filter with a low-pass filter.
- A narrow-band-pass filter has a sharply tuned center frequency and can be implemented with only one op-amp in inverting mode of operation.

12.11 Band-Reject Filters

A band-reject filter attenuates signals in the stop band and passes those outside this band. It is also called a *band-stop* or *band-elimination filter*. The transfer function of a second-order band-reject filter has the general form

$$H_{BR}(s) = \frac{K_{PB}(s^2 + \omega_C^2)}{s^2 + (\omega_C/Q)s + \omega_C^2} \tag{12.100}$$

where K_{PB} is the pass-band gain. Band-reject filters can be classified as wide band reject or narrow band reject. A narrow-band-reject filter is commonly called a *notch filter*. Because of its higher Q (>10), the bandwidth of a narrow-band-reject filter is much smaller than that of a wide-band-reject filter.

12.11.1 Wide-Band-Reject Filters

The frequency characteristic of a wide-band-reject filter is shown in Fig. 12.34(a). This characteristic can be obtained by adding a low-pass response to a high-pass response, as shown in Fig. 12.34(b); the solution can be implemented by summing the responses of a first-order (or second-order) high-pass section and low-pass section through a summing amplifier. This arrangement is shown in Fig. 12.34(c). The order of the band-reject filter depends on the order of the high-pass and low-pass sections. For a band-reject response to be realized, the cutoff frequency f_L of the high-pass filter must be larger than the cutoff frequency f_H of the low-pass filter. In addition, the pass-band gains of the high-pass and low-pass sections must be equal. With an inverting summer (A_3), the output will be inverted. Note that R_{OM} in Fig. 12.34(c) has no function but to minimize the op-amp offsets (see Chapter 14).

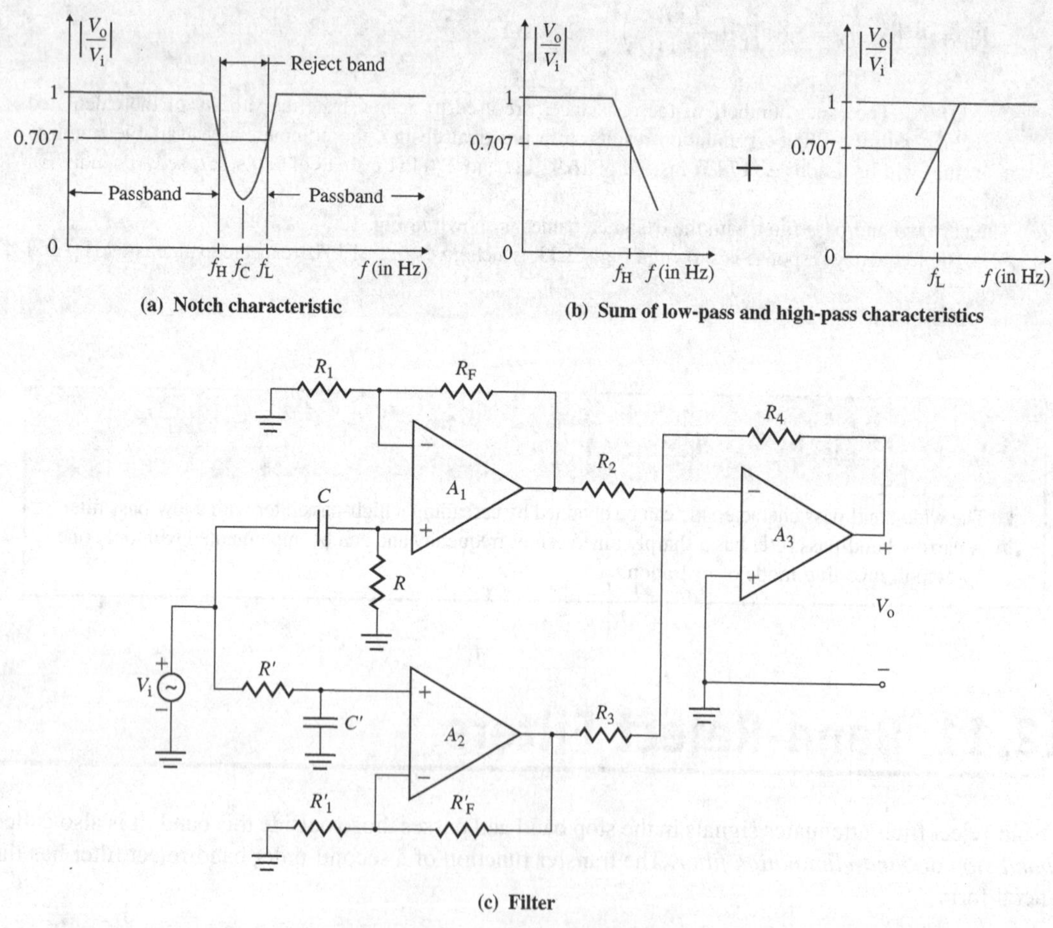

FIGURE 12.34 Wide-band-reject filter

EXAMPLE 12.10

D

Designing a wide-band-reject filter

(a) Design a wide-band-reject filter as shown in Fig. 12.34(c) with f_L = 100 kHz, f_H = 10 kHz, and a pass-band gain of K_{PB} = 4.

(b) Calculate the value of Q for the filter.

(c) Use PSpice/SPICE to plot the frequency response of the filter designed in part (a) from 10 Hz to 10 MHz.

SOLUTION

(a) In Example 12.8, we designed a wide-band-pass filter with f_L = 10 kHz and f_H = 1 MHz. In this example, we have f_L = 100 kHz and f_H = 10 kHz; that is, $f_L > f_H$. However, we can follow the design steps in Example 12.8 to find the component values, provided that we interchange the high-pass and low-pass sections. Thus, for the high-pass section of f_L = 100 kHz, C = 100 pF and R = 15.915 kΩ, and for the low-pass section of f_H = 10 kHz, C' = 1 nF and R' = 15.915 kΩ. For a pass-band gain of K_{PB} = 4, use $R_1 = R_1'$ = 10 kΩ and $R_F = R_F'$ = 30 kΩ. For the summing amplifier, set a gain of 1. Choose $R_2 = R_3 = R_4$ = 10 kΩ.

(b) From Eq. (12.87),

$$f_C = \sqrt{10\text{ kHz} \times 100\text{ kHz}} = 31.623\text{ kHz}$$

and BW = 100 kHz − 10 kHz = 90 kHz

From Eq. (12.86), we can find

$$Q = \frac{31.623\text{ kHz}}{(100\text{ kHz} - 10\text{ kHz})} = 0.351$$

(c) The circuit for PSpice simulation of the wide-band-reject filter is shown in Fig. 12.35.

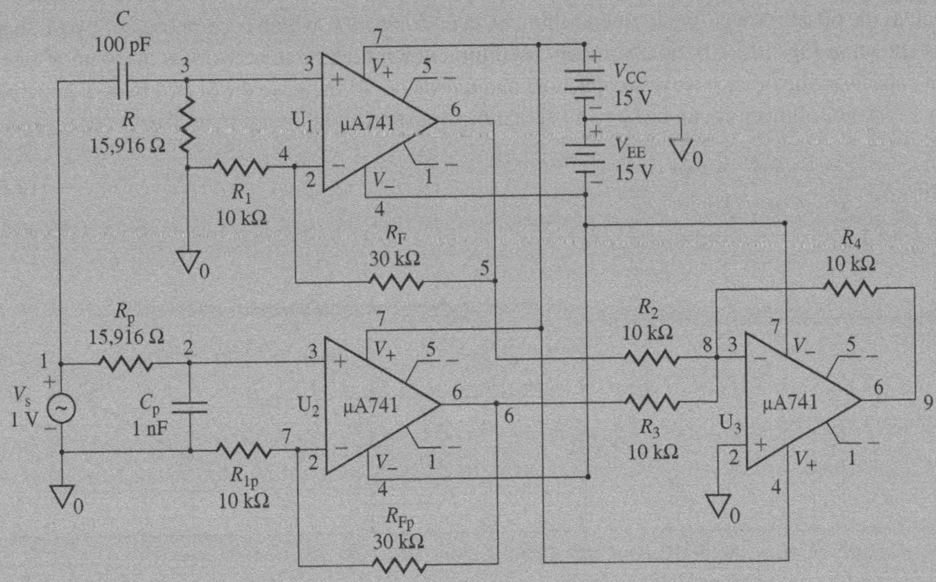

FIGURE 12.35 Wide-band-reject filter for PSpice simulation

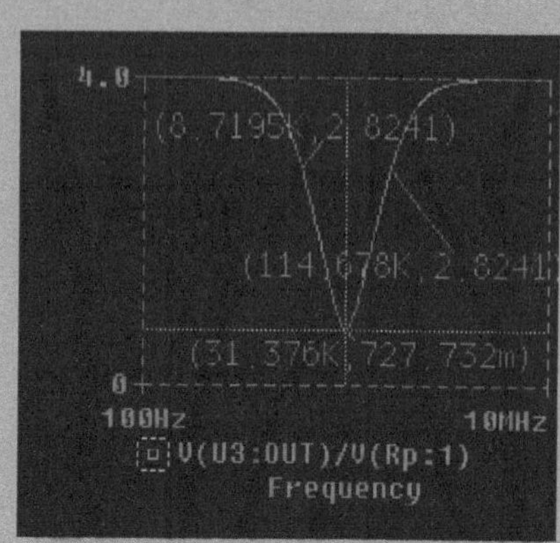

FIGURE 12.36 PSpice plot of frequency response for Example 12.10 (using linear op-amp model)

The frequency response is shown in Fig. 12.36, which gives $f_C \approx 31.376$ kHz (expected value is 31.623 kHz) and $K_{PB} = 4$ (expected value is 4).

NOTE: If we use the linear op-amp model, there will be a serious high-end roll-off with μA741 op-amp.

12.11.2 Narrow-Band-Reject Filters

A typical frequency response of a narrow-band-reject filter is shown in Fig. 12.37(a). This filter, often called a *notch filter*, is commonly used in communication and biomedical instruments to eliminate undesired frequencies such as the 60-Hz power line frequency hum. A *twin-T network*, which is composed of two T-shaped networks, as shown in Fig. 12.37(b), is commonly used for a notch filter. One network is made up of two resistors and a capacitor; the other uses two capacitors and a resistor. To increase the Q of a twin-T network, it is used with a voltage follower. It can be shown [9] that the transfer function of a twin-T network is given by

$$H_{NF}(s) = \frac{K_{PB}(s^2 + \omega_n^2)}{s^2 + (\omega_o/Q)s + \omega_o^2} \tag{12.101}$$

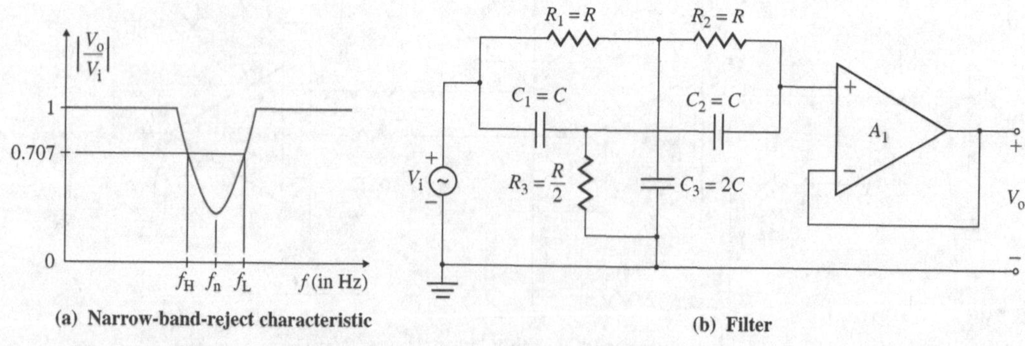

FIGURE 12.37 Narrow-band-reject filter

where $\omega_n = 1/RC$

$\omega_o = 1/\sqrt{3}RC$

$Q = \sqrt{3/4}$

$K_{PB} = 1$

Therefore, the *notch-out frequency*, which is the frequency at which maximum attenuation occurs, is given by

$$f_n = \frac{1}{2\pi RC} \tag{12.102}$$

EXAMPLE 12.11

D

Designing a narrow-band-reject (notch) filter

(a) Design a notch filter as in Fig. 12.37(b) with $f_n = 60$ Hz.

(b) Use PSpice/SPICE to plot the frequency response of the filter designed in part (a) from 1 Hz to 1 kHz.

SOLUTION

(a) $f_n = 60$ Hz. Choose a value of C less than or equal to 1 μF: Let $C = 0.047$ μF. Then, from Eq. (12.102),

$$R = \frac{1}{2\pi f_n C} = \frac{1}{2\pi \times 60\text{ Hz} \times 0.047\text{ μF}} = 56.44\text{ k}\Omega$$ (use a 59-kΩ standard resistor of 10% tolerance)

$$R_3 = \frac{R}{2} = 28.22\text{ k}\Omega$$ (use two 59-kΩ resistors in parallel)

$$C_3 = 2C = 0.094\text{ μF}$$ (use two 0.047-μF capacitors in parallel)

(b) The circuit of the notch filter for PSpice simulation is shown in Fig. 12.38.

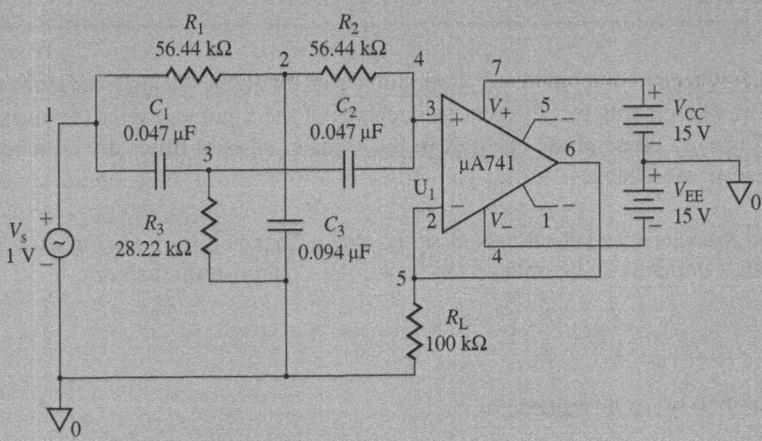

FIGURE 12.38 Narrow-band-reject filter for PSpice simulation

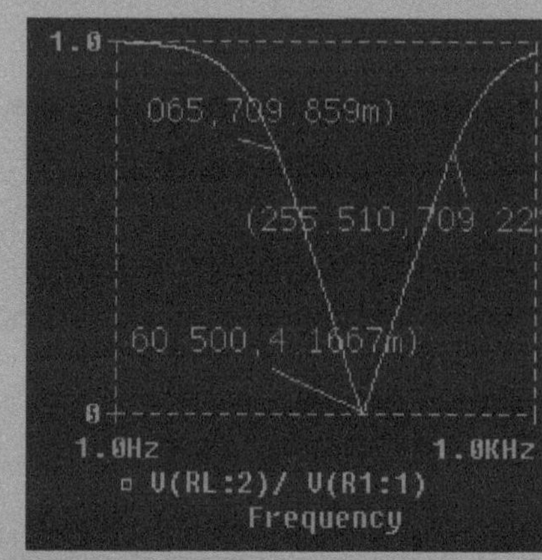

FIGURE 12.39 PSpice plot of frequency response for Example 12.11

The frequency response of the filter is shown in Fig. 12.39, which gives $f_n = 60.9$ Hz (expected value is 60 Hz) and $K_{PB} = 1$ (expected value is 1).

KEY POINTS OF SECTION 12.11

- The wide-band-reject characteristic can be obtained by adding the output of a low-pass filter to that of a high-pass filter through a summing amplifier.
- A narrow-band-reject filter has a sharply tuned reject frequency and can be implemented with only one op-amp in the noninverting mode of operation.

12.12 All-Pass Filters

An all-pass filter passes all frequency components of the input signals without attenuation. However, this filter provides predictable phase shifts for different frequencies of the input signals. Transmission lines (e.g., telephone wires) usually cause phase changes in the signals; all-pass filters are commonly used to compensate for these phase changes. An all-pass filter is also called a *delay equalizer* or a *phase corrector*.

Figure 12.40(a) shows the characteristic of an all-pass filter; the circuit diagram is shown in Fig. 12.40(b). The output voltage in Laplace's domain can be obtained by using the superposition theorem:

$$V_o(s) = -\frac{R_F}{R_1}V_i(s) + \frac{1/sC}{R + 1/sC}\left(1 + \frac{R_F}{R_1}\right)V_i(s) \tag{12.103}$$

If we assume $R_F = R_1$, Eq. (12.103) can be reduced to

$$V_o(s) = -V_i(s) + \frac{2}{1 + sRC}V_i(s)$$

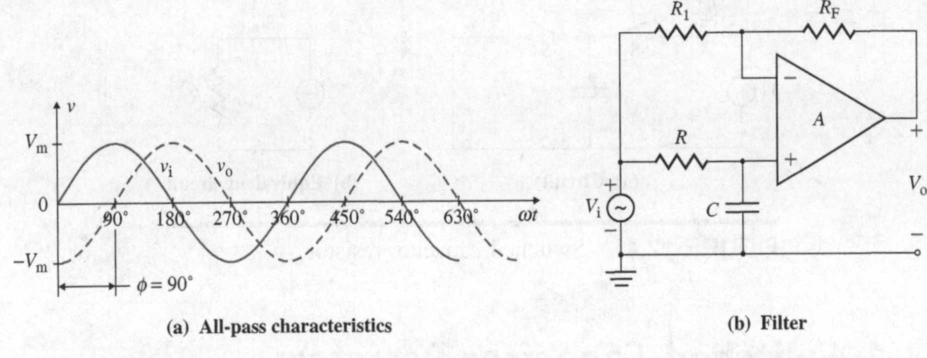

(a) All-pass characteristics (b) Filter

FIGURE 12.40 All-pass filter

which gives the voltage gain as

$$H(s) = \frac{V_o(s)}{V_i(s)} = \frac{1 - sRC}{1 + sRC} \tag{12.104}$$

Substituting $s = j\omega$ into Eq. (12.104) gives the magnitude of the voltage gain as

$$|H(j\omega)| = 1$$

and the phase angle ϕ as

$$\phi = -2\tan^{-1}(\omega RC) = -2\tan^{-1}(2\pi fRC) \tag{12.105}$$

Equation (12.105) indicates that, for fixed values of R and C, the phase angle ϕ can change from 0° to −180° as the frequency f of the input signal is varied from 0 to ∞. For example, if $R = 21$ kΩ and $C = 0.1$ μF, we will get $\phi = -64.4°$ at 60 Hz. If the positions of R and C are interchanged, the phase shift ϕ will be positive. That is, the output signal leads the input signal.

KEY POINT OF SECTION 12.12

- An all-pass filter does not give any gain attenuation, but it provides predictable phase shifts for different frequencies of the input signals.

12.13 Switched-Capacitor Filters

Switched-capacitor filters use on-chip capacitors and MOS switches to simulate resistors. The cutoff frequencies are proportional to and determined by the external clock frequency. In addition, the cutoff or center frequency can be programmed to fall anywhere within an extremely wide range of frequencies—typically more than a 200,000:1 range. Switched-capacitor filters are becoming increasingly popular since they require no external reactive components, capacitors, or inductors. They offer the advantages of low cost, fewer external components, high accuracy, and excellent temperature stability. However, they generate more noise than standard active filters.

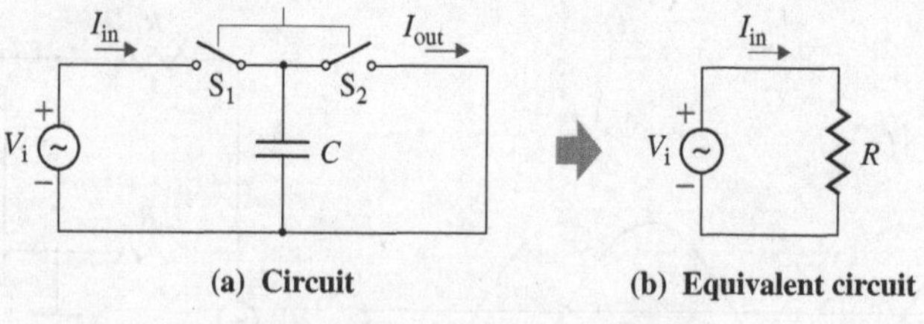

FIGURE 12.41 Switched-capacitor resistor

12.13.1 Switched-Capacitor Resistors

In all the filters discussed so far, discrete resistors and capacitors were connected to one or more op-amps to obtain the desired cutoff frequencies and voltage gain. Use of discrete resistors is avoided in integrating circuits to reduce chip size; instead, resistor behavior is simulated by using active switches. A resistor is usually simulated by a capacitor and switches. The value of this *simulated resistor* is inversely proportional to the rate at which the switches are opened or closed.

Consider a capacitor with two switches, as shown in Fig. 12.41. The switches are actually MOS transistors that are alternately opened and closed. When S_1 is closed and S_2 is open, the input voltage is applied to the capacitor. Therefore, the total charge on the capacitor is

$$q = V_i C \tag{12.106}$$

When S_1 is open and S_2 is closed, the charge q flows to the ground. If the switches are ideal (i.e., they open and close instantaneously and have zero resistance when they close), the capacitor C will charge and discharge instantly. The charging current I_{in} and the discharging current I_{out} of the capacitor are shown in Fig. 12.42. If the switches are opened and closed at a faster rate, the current pulses will have the same magnitude but will occur more often. That is, the average current will be higher at a higher switching rate. The average current flowing through the capacitor of Fig. 12.41 is given by

$$I_{av} = \frac{q}{T} = \frac{V_i C}{T} = V_i C f_{clk} \tag{12.107}$$

where q = capacitor's charge

T = time between closings of S_1 or closings of S_2, in seconds

$f_{clk} = 1/T$ = clock frequency, in hertz

The equivalent resistance seen by the input voltage is

$$R = \frac{V_i}{I_{av}} = \frac{V_i}{V_i C f_{clk}} = \frac{1}{C f_{clk}} \tag{12.108}$$

which indicates that the value of R is a function of C and f_{clk}. For a fixed value of C, the value of R can be adjusted by adjusting f_{clk}. Therefore, a switched-capacitor resistor, also known as a *clock-tunable resistor*, can be built in IC form with a capacitor and two MOS switches. Note that any change in V_i must occur at a rate much slower than f_{clk}, especially when V_i is an AC signal.

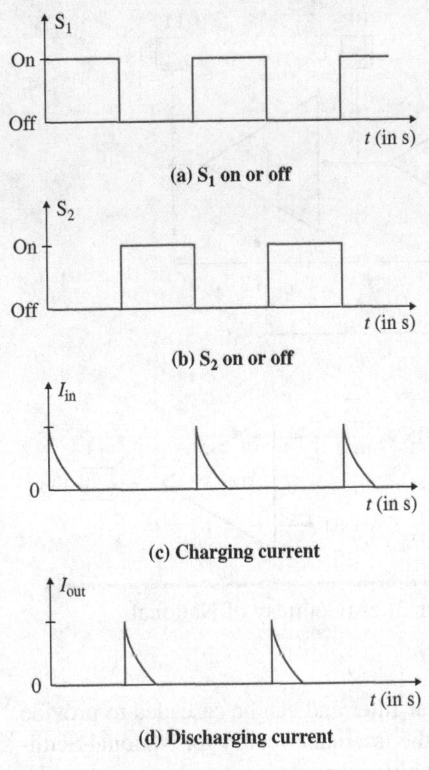

FIGURE 12.42 Current into and out of switched-capacitor resistor

12.13.2 Switched-Capacitor Integrators

A simulated resistor can be used as a part of an IC to build a switched-capacitor integrator, as shown in Fig. 12.43. The switches S_1 and S_2 must never be closed at the same time. That means the clock waveform driving the MOS switches must not overlap if the filter is to operate properly.

12.13.3 Universal Switched-Capacitor Filters

A *universal filter* combines many features in an op amp and can be used to synthesize any of the normal filter types: band-pass, low-pass, high-pass, notch, and all-pass. Universal filters are available commercially (e.g., the type FLT-U2 manufactured by Datel-Intersil). A switched-capacitor filter is a type

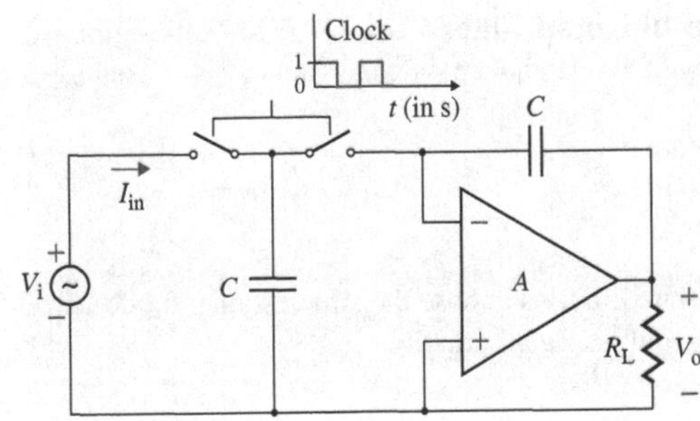

FIGURE 12.43 Switched-capacitor integrator

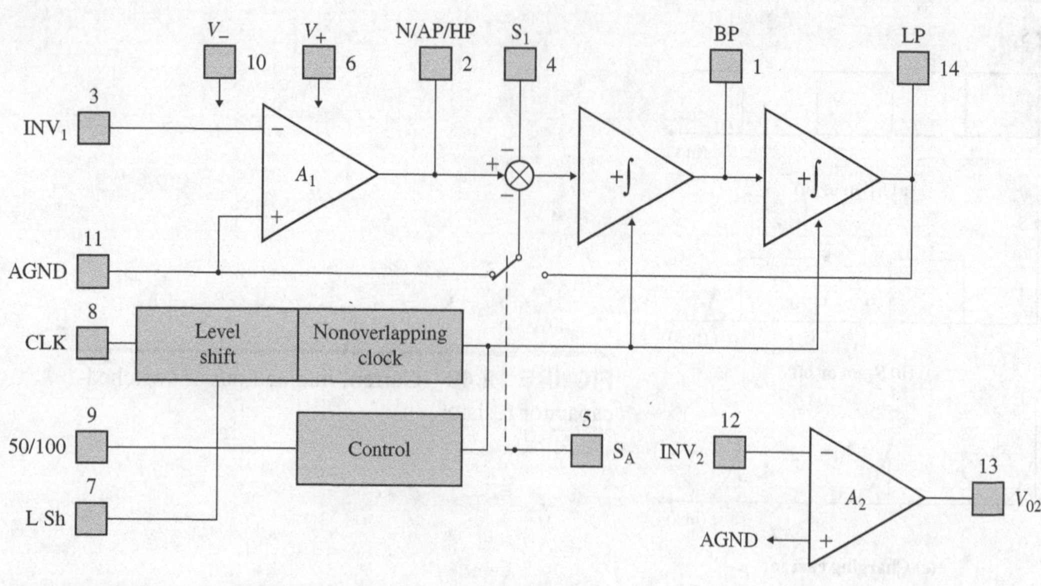

FIGURE 12.44 MF5 universal monolithic switched-capacitor filter (Courtesy of National Semiconductor, Inc.)

of universal active filter. It has the characteristics of a second-order filter and can be cascaded to provide very steep attenuation slopes. Figure 12.44 is a block diagram of the internal circuitry for National Semiconductor's MF5. The basic filter consists of an op-amp, two positive integrators, and a summing node. An MOS switch, controlled by a logic voltage on pin 5 (S_A), connects one of the inputs of the first integrator either to the ground or to the output of the second integrator, thus allowing more application flexibility. The MF5 includes a pin (9) that sets the ratio of the clock frequency (f_{clk}) to the center frequency (f_C) at either 50:1 or 100:1. The maximum recommended clock frequency is 1 MHz, which results in a maximum center frequency f_C of 20 kHz at 50:1 or 10 kHz at 100:1, provided the product Qf_o is less than 200 kHz. An extra uncommitted op-amp is available for additional signal processing. A very convenient feature of the MF5 is that f_o can be controlled independently of Q and the pass-band gain. Without affecting the other characteristics, one can tune f_o simply by varying f_{clk}. The selection of the external resistor values is very simple, so the design procedure is much easier than for typical RC active filters.

EXAMPLE 12.12

D **Designing a second-order Butterworth filter using a universal filter** Using the MF5, design a second-order Butterworth low-pass filter with a cutoff frequency of 1 kHz and a pass-band gain of −4. Assume a power supply of ±5 V and a CMOS clock.

SOLUTION

Step 1. Choose the mode in which the MF5 filter will be operated. Let us choose the simplest one: mode 1, which has low-pass, band-pass, and notch output and inverts the output signal polarity.

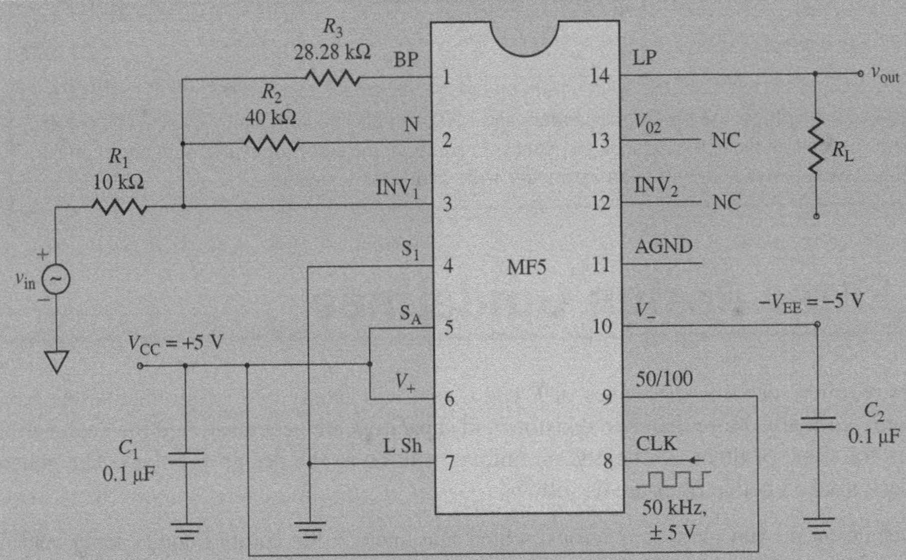

FIGURE 12.45 MF5 configured as a second-order low-pass filter

Step 2. Determine the values of the external resistors. The MF5 requires three external resistors that determine Q and the gain of the filter. The external resistors are connected as shown in Fig. 12.45. For mode 1, the relationship among Q, K_{LP}, and the external resistors is given by the data sheet (on which only three out of six possible modes are shown) as

$$Q = \frac{f_C}{BW} = \frac{R_3}{R_2} \tag{12.109}$$

$$K_{LP} = -\frac{R_2}{R_1} \tag{12.110}$$

In this mode, the input impedance of the filter is equal to R_1 since the input signal is applied to INV (pin 3) through R_1. To provide a fairly high input impedance, let $R_1 = 10\ \text{k}\Omega$. From Eq. (12.110), we get

$$R_2 = -K_{LP}R_1 = -(-4) \times 10\ \text{k}\Omega = 40\ \text{k}\Omega$$

For a second-order Butterworth low-pass filter, $Q = 0.707$. Therefore, Eq. (12.109) gives

$$R_3 = QR_2 = 0.707 \times 40\ \text{k}\Omega = 28.28\ \text{k}\Omega$$

Step 3. Choose the power supplies and complete their connections. Since a power supply of ±5 V is required, V_+ (pin 6) is connected to +5 V, V_- (pin 10) is connected to −5 V, and AGND (pin 11) is connected to the ground. To eliminate any ripples, two 0.1-μF capacitors are connected across the power supplies.

Step 4. Choose the clock frequency f_{clk}. The 50/100 (pin 9) must be connected to V_+ (pin 6) for a ratio of 50:1 or to V_- (pin 10) for a ratio of 100:1. Let us choose an f_{clk}-to-f_o ratio of 50:1. That means the 50/100 (pin 9) must be connected to V_+ (pin 6). Since the cutoff frequency is 1 kHz, the external clock frequency is $f_{clk} = 50 \times 1\ \text{kHz} = 50\ \text{kHz}$.

Step 5. For a CMOS clock, the L Sh (pin 7) should be connected to the ground (pin 11). The low-pass filter S_A (pin 5) is connected to V_+ (pin 6), and S_1 (pin 4) is connected to the ground (pin 11).

The complete circuit for the second-order low-pass filter is shown in Fig. 12.45.

KEY POINT OF SECTION 12.13

- Switched-capacitor filters use on-chip capacitors and MOS switches to simulate resistors. The cutoff frequencies depend on the external clock frequency. In addition, the cutoff or center frequency can be programmed to fall anywhere within an extremely wide range of frequencies.

12.14 Filter Design Guidelines

Designing filters requires selecting the values of R and C that will satisfy two requirements: the bandwidth and the gain. Normally, more than two resistors and capacitors are necessary, and the designer has to assume values for some of them, so there is no unique solution to the design problem. The general guidelines for designing an active filter are as follows:

Step 1. Decide on the design specifications, which may include the cutoff frequencies f_L and f_H, pass-band gain K_{PB}, bandwidth BW, damping factor of $\zeta = 0.707$ for a flat response, $|H(j\omega_o)| = 0.707$, and $|H(j0)| = 1$.

Step 2. Assume a suitable value for the capacitor. The recommended values of C are 1 μF to 5 pF. (Mylar or tantalum capacitors are recommended because they give better performance than other types of capacitors.)

Step 3. Having assumed a value for the capacitor, find the value for the resistor that will satisfy the bandwidth or frequency requirement.

Step 4. If the value of R does not fall within the practical range of 1 kΩ to 500 kΩ, choose a different value of C.

Step 5. Find values for the other resistances that will satisfy the gain requirements and fall in the range of 1 kΩ to 500 kΩ.

Step 6. If necessary, change the filter's cutoff frequency. The procedure for converting the original cutoff frequency f_o to the new cutoff frequency f_n is called *frequency scaling*. It is accomplished by multiplying the value of R or C (but not both) by the ratio of the original frequency f_o to the new cutoff frequency f_n. The new value of R or C can be found from

$$R_n\,(\text{or } C_n) = \frac{\text{Original cutoff frequency } f_o}{\text{New cutoff frequency } f_n}\,R\ (\text{or } C) \tag{12.111}$$

KEY POINTS OF SECTION 12.14

- Designing filters requires selecting values of R and C to meet the specifications for bandwidth and DC gain. Normally more than two resistors and capacitors are used, and values must be assumed for some of them (normally C). There is no unique solution to a design problem.
- Once a filter has been designed for one cutoff frequency, new values of R and C can be determined by multiplying the value of R or C (but not both) by the ratio of the original frequency f_o to the new cutoff frequency f_n.

Summary

Active filters offer many advantages over passive filters. The many types of active filters—low-pass, high-pass, band-pass, band-reject, and all-pass—are based on the frequency characteristics. A second-order filter has a sharper stop band and is preferable to a first-order filter. An all-pass filter gives a phase shift that is proportional to the input signal frequency.

Universal filters are very popular because of their flexibility in synthesizing frequency characteristics with a very high accuracy. A switched-capacitor filter is a type of universal filter that uses on-chip capacitors and MOS switches to simulate resistors. Its cutoff frequency is proportional to and determined by the external clock frequency.

References

1. W. K. Chen, *Passive and Active Filters—Theory and Implementation*. New York: Wiley, 1986.
2. M. H. Rashid, *Introduction to SPICE using* OrCAD *for Circuits and Electronics*. Englewood Cliffs, NJ: Prentice Hall, 2003, Chapter 10.
3. M. E. Van Valkenburg, *Analog Filter Design*. New York: CBS College Publishing, 1982.
4. R. Schaumann and M. E. Van Valkenburg, *Design of Analog Filters*. New York: Oxford University Press, 2001.
5. L. P. Huelsman and P. E. Allen, *Introduction to the Theory and Design of Active Filters*. New York: McGraw-Hill, 1980.
6. A. V. Oppenheim and A. S. Willsky, with S. H. Nawab, *Signals and Systems*. Upper Saddle River, NJ: Prentice Hall, 1997.
7. R. Schaumann, M. S. Ghausi, and K. R. Laker, *Design of Analog Filters—Passive, Active RC, and Switched Capacitor*. Englewood Cliffs, NJ: Prentice Hall, 1990.
8. R. A. Gayakwad, *Op-Amps and Linear Integrated Circuits*. Englewood Cliffs, NJ: Prentice Hall, 1993.
9. G. C. Temes and L. Lapatra, *Introduction to Circuit Synthesis and Design*. New York: McGraw-Hill, 1977.

Review Questions

1. What is an active filter?
2. What are the advantages of active filters over passive ones?
3. What are the types of active filters?
4. What are the passband and the stop band of a filter?
5. What is a cutoff frequency?
6. What is the Butterworth response of a filter?
7. What are the differences between first-order and second-order filters?
8. What is frequency scaling of filters?
9. What is a notch filter?
10. What is a notch-out frequency?
11. What is an all-pass filter?
12. What is a universal filter?
13. What is a switched-capacitor resistor?
14. What is a switched-capacitor filter?
15. What is a clock-tunable resistor?

Problems

The symbol **D** indicates that a problem is a design problem. The symbol **P** indicates that you can check the solution to a problem using PSpice/SPICE or Electronics Workbench. For PSpice/SPICE simulation, assume op-amps with parameters $R_i = 2\ \text{M}\Omega$, $R_o = 75\ \Omega$, and $A_o = 2 \times 10^5$.

12.4 First-Order Filters

12.1 Determine **(a)** the transfer function of the network shown in Fig. P12.1 and **(b)** its poles and zeros.

FIGURE P12.1

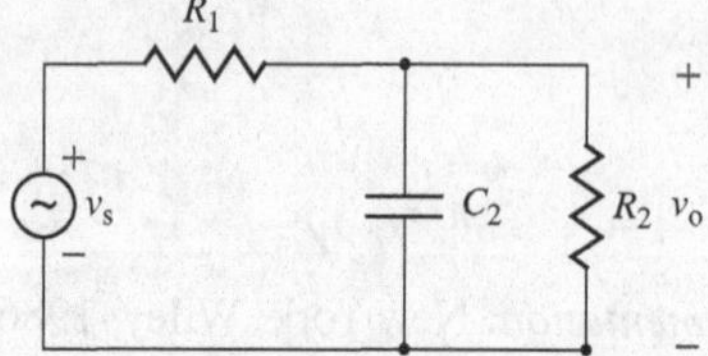

12.2 Determine **(a)** the transfer function of the network shown in Fig. P12.2 and **(b)** its poles and zeros.

FIGURE P12.2

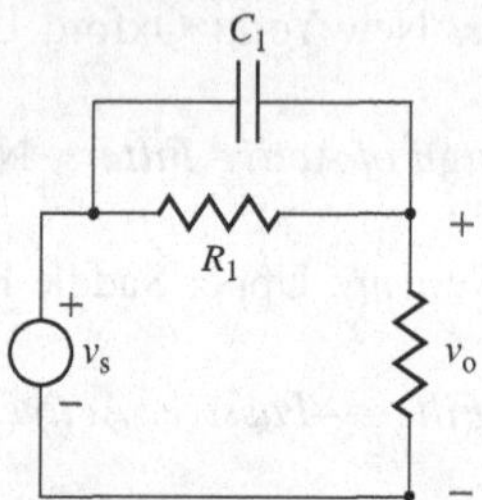

12.3 Determine **(a)** the transfer function of the network shown in Fig. P12.3 and **(b)** its poles and zeros.

FIGURE P12.3

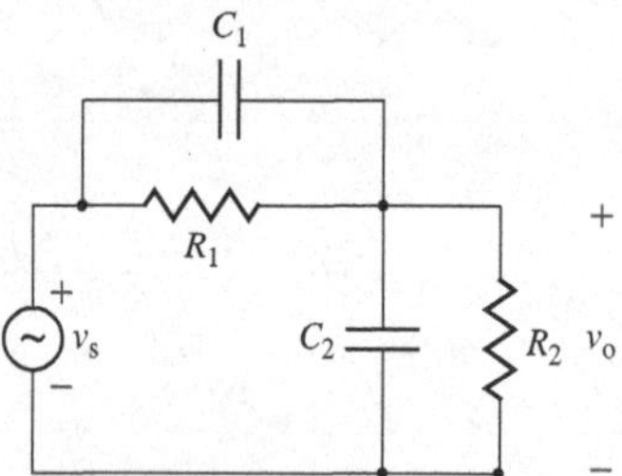

12.4 Determine **(a)** the transfer function of the network shown in Fig. P12.4 and **(b)** its poles and zeros.

FIGURE P12.4

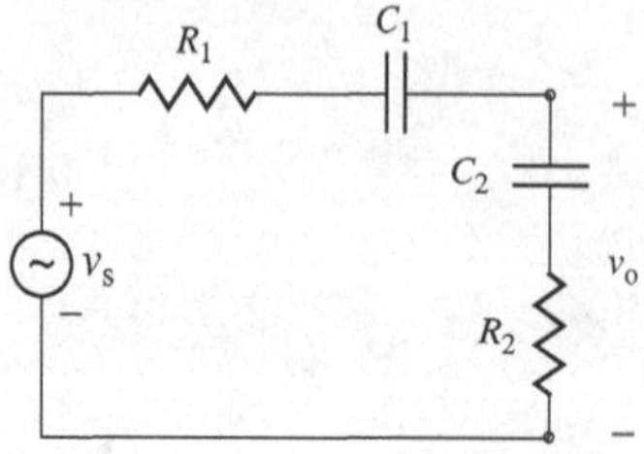

12.5 *The Biquadratic Function*

12.5 Determine **(a)** the pole and zero quality factors Q_p and Q_z, **(b)** the pole and zero resonant frequencies ω_p and ω_z, **(c)** the pole factor β_p, and **(d)** the pole angle ϕ_p. The transfer function has the general form as given by

$$H(s) = \frac{5s^2 + 15s + 100}{s^2 + 20s + 200}$$

12.6 Plot the frequency and phase responses of the low-pass transfer function $H_{LP}(j\omega)$ in Eq. (12.20) versus $u = 0.5$ to 5 for $Q = 0.5, 0.707$, and 1 and $K = 1$.

12.7 Plot the frequency and phase responses of the high-pass transfer function $H_{HP}(j\omega)$ in Eq. (12.21) versus $u = 0.5$ to 5 for $Q = 0.5, 0.707$, and 1 and $K = 1$.

12.8 Plot the frequency and phase responses of the band-pass transfer function $H_{BP}(j\omega)$ in Eq. (12.22) versus $u = 0.5$ to 5 for $Q = 0.5, 0.707$, and 1 and $K = 1$.

12.9 Plot the frequency and phase responses of the band-reject transfer function $H_{BP}(j\omega)$ in Eq. (12.23) versus $u = 0.5$ to 5 for $Q = 0.5, 0.707$, and 1 and $K = 1$.

12.6 *Butterworth Filters*

12.10 Determine the transfer function for fourth-order Butterworth response of a band-pass filter.

12.11 Determine the transfer function for sixth-order Butterworth response of a band-pass filter.

12.12 Determine the transfer function for seventh-order Butterworth response of a band-pass filter.

12.13 Determine the transfer function for eighth-order Butterworth response of a band-pass filter.

12.14 Determine the transfer function for ninth-order Butterworth response of a band-pass filter.

12.8 *Low-Pass Filters*

12.15 Design a first-order low-pass filter as in Fig. 12.10(b) to give a low cutoff frequency of $f_o = 2$ kHz with a pass-band gain of 1. If the desired frequency is changed to $f_n = 1.5$ kHz, calculate the new value of R_n.

12.16 Derive the transfer function $H(s)$ of the network in Fig. 12.13(d).

12.17 Design a second-order low-pass filter as in Fig. 12.14 to give a low cutoff frequency of $f_o = 10$ kHz, a pass-band gain of $K = 5$, and $Q = 0.707$, 1, and ∞.

12.18 Design a second-order Butterworth low-pass filter as in Fig. 12.17(c) to yield $|H(j\omega_o)| = 1$ (or 0 dB), a cut-off frequency of $f_o = 10$ kHz, and $Q = 0.707$.

12.19 Design a second-order Butterworth filter as in Fig. 12.14 to yield $|H(j0)| = 1$ (or 0 dB), a cutoff frequency of $f_o = 10$ kHz, and $Q = 0.707$.

12.20 Design a third-order Butterworth low-pass filter as in Fig. P12.20 to give a high cutoff frequency of $f_o =$ 10 kHz and a pass-band gain of 10. The transfer function has the general form

$$H_3(s) = \frac{10\omega_o^3}{s^3 + 2\omega_o s^2 + 2\omega_o^2 s + \omega_o^3}$$

FIGURE P12.20

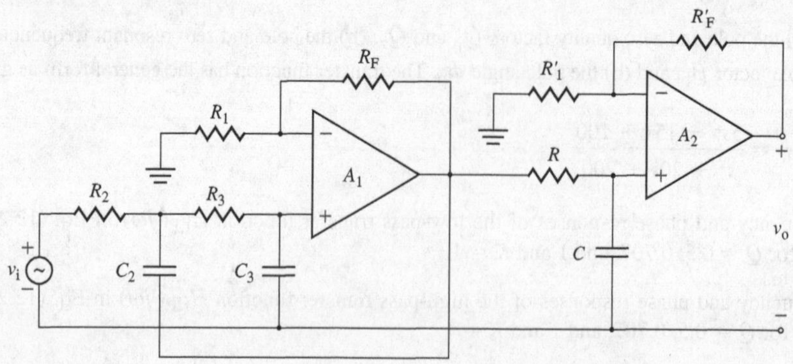

12.21 Design a fourth-order Butterworth low-pass filter as in Fig. P12.21 to give a high cutoff frequency of $f_o = 10$ kHz and a pass-band gain of 25. The transfer function has the general form

$$H_4(s) = \frac{25\omega_o^4}{(s^2 + \sqrt{2}\omega_o s + \omega_o^2)^2}$$

FIGURE P12.21

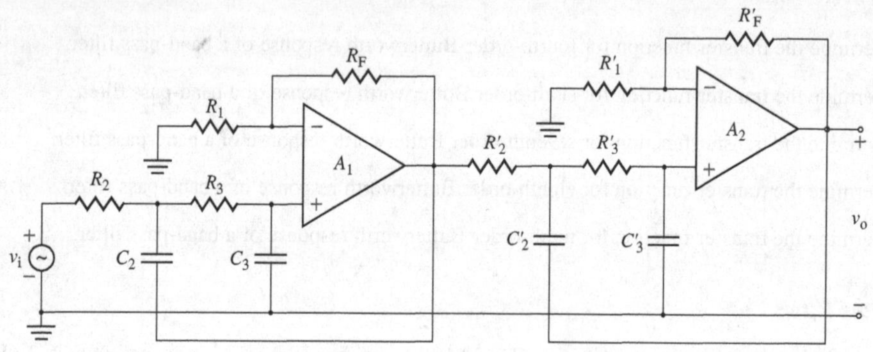

12.9 *High-Pass Filters*

12.22 Design a first-order high-pass filter as in Fig. 12.20(b) to give a low cutoff frequency of $f_o = 400$ Hz and a pass-band gain of $K = 2$. If the desired frequency is changed to $f_n = 1$ kHz, calculate the new value of R_n.

12.23 Design a second-order high-pass filter as in Fig. 12.22 to give a low cutoff frequency of $f_o = 2$ kHz and a pass-band gain of 2. If the desired frequency is changed to $f_n = 3.5$ kHz, calculate the new value of R_n.

12.24 Design a second-order Butterworth high-pass filter as in Fig. 12.25(c) to yield $|H(j\infty)| = 1$ (or 0 dB), a cut-off frequency of $f_o = 10$ kHz, and $Q = 0.707$.

12.25 Design a second-order Butterworth high-pass filter as in Fig. 12.25(c) to yield $|H(j\omega_o)| = 1$ (or 0 dB), a cutoff frequency of $f_o = 10$ kHz, and $Q = 0.707$.

12.26 Design a third-order Butterworth high-pass filter as in Fig. P12.26 to give a low cutoff frequency of $f_o =$ 10 kHz and a pass-band gain of 10. The transfer function has the general form

$$H_3(s) = \frac{10s^3}{s^3 + 2\omega_o s^2 + 2\omega_o^2 s + \omega_o^3}$$

FIGURE P12.26

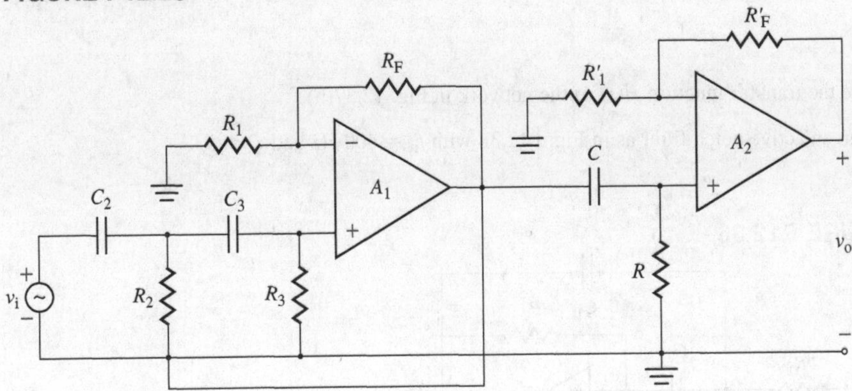

12.10 *Band-Pass Filters*

12.27 Design a wide-band-pass filter with f_L = 400 Hz, f_H = 2 kHz, and a pass-band gain of K_{PB} = 4. Calculate the value of Q for the filter.

12.28 Design a wide-band-pass filter with f_L = 1 kHz, f_H = 10 kHz, and a pass-band gain of K_{PB} = 20. Calculate the value of Q for the filter.

12.29 Derive the transfer function $H(s)$ of the network in Fig. 12.31(c).

12.30 Design a band-pass filter as in Fig. P12.30 to give f_C = 5 kHz, Q = 20, and K_{PB} = 40.

FIGURE P12.30

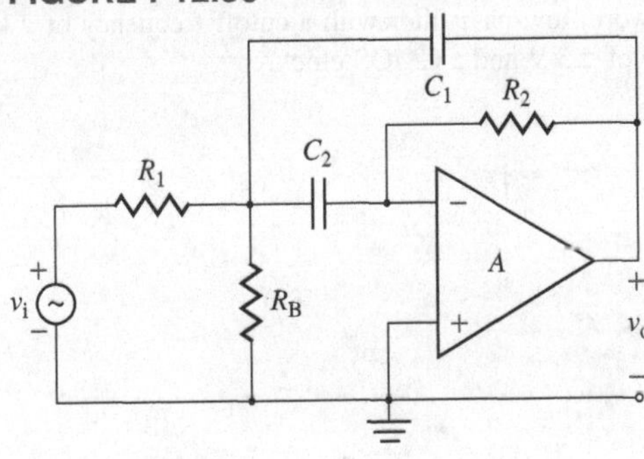

12.31 a. Design a narrow-band-pass filter as in Fig. 12.31(b) such that f_C = 2 kHz, Q = 20, and K_{PB} = 10.

b. Calculate the value of R_B required to change the center frequency from 2 kHz to 5.5 kHz.

12.11 *Band-Reject Filters*

12.32 Design a wide-band-reject filter as in Fig. 12.34(c) to give f_H = 400 kHz, f_L = 2 kHz, and K_{PB} = 10. Calculate the value of Q for the filter.

12.33 Design a wide-band-reject filter with a falloff rate of 40 dB/decade to give f_H = 400 kHz, f_L = 2 kHz, and K_{PB} = 40.

12.34 Design an active notch filter as in Fig. 12.37(b) with $f_n = 400$ Hz.

12.35 Derive the transfer function $H(s)$ of the network in Fig. 12.37(b).

12.36 Design an active notch filter as in Fig. P12.36 with $f_n = 400$ Hz and $Q = 5$.

FIGURE P12.36

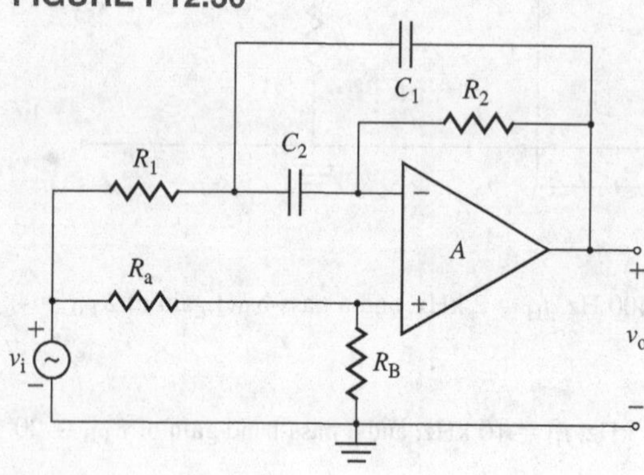

12.12 *All-Pass Filters*

12.37 Design an all-pass filter as in Fig. 12.40(b) so that the phase shift is $\phi = \pm150°$ at 60 Hz.

12.38 D Using the MF5, design a second-order Butterworth low-pass filter with a cutoff frequency of 2 kHz and a pass-band gain of −2. Assume a power supply of ±5 V and a CMOS clock.

CHAPTER 13

OSCILLATORS

Learning Outcomes

After completing this chapter, students should be able to do the following:

- Describe the operating principle of oscillators for generating sinusoidal voltage and the conditions for sustained oscillations.
- Describe the characteristics and types of oscillators.
- Select an oscillator circuit to meet specific frequency requirements.
- Design an oscillator circuit to meet desired frequency specifications.

Symbols and Their Meanings

Symbol	Meaning
A, A_{f}	Open-loop and closed-loop voltage gains of an amplifier
f_{o}, ω_{o}	Oscillation frequencies in hertz and radians per second
g_{m}, r_{π}, r_{o}	Transconductance, input resistance, and output resistance of a transistor
h_{fe}	Hybrid current gain parameter of a BJT transistor
K_{p}, V_{t}	MOSFET constant and threshold voltage

Symbol	Meaning
L_p, R_p	Equivalent primary side load inductance and resistance of a transformer-coupled load
$\beta(s)$, $H(s)$	Feedback factor and transfer function of an amplifier
$v_o(t)$	Instantaneous output voltage of an oscillator

13.1 Introduction

We know from Sec. 10.11 that an amplifier with negative feedback will be unstable if the magnitude of the loop gain is greater than or equal to 1 and its phase shift is $\pm 180°$. Under these conditions, the feedback becomes positive and the output of the amplifier oscillates. An oscillator is a circuit that generates a repetitive waveform of fixed amplitude at a fixed frequency without any external input signal. A waveform of this characteristic can be obtained by applying positive feedback in amplifiers. Positive feedback provides enough feedback signal to maintain oscillations. Although oscillations are very undesirable in linear amplifier circuits, oscillators are designed specifically to produce controlled and predictable oscillation. Thus, the strategy for designing oscillators is quite different from that for designing linear amplifiers. Occasionally, oscillators have inputs that are used to control the frequency or to synchronize the oscillations with an external reference. Oscillators are used in many electronic circuits, such as in radios, televisions, computers, and communications equipment.

13.2 Principles of Oscillators

An oscillator is an amplifier with positive feedback. The block diagram of an amplifier with positive feedback, shown in Fig. 13.1(a), suggests the following relationships:

$$v_e = v_i + v_f$$

$$v_o = Av_e$$

$$v_f = \beta v_o$$

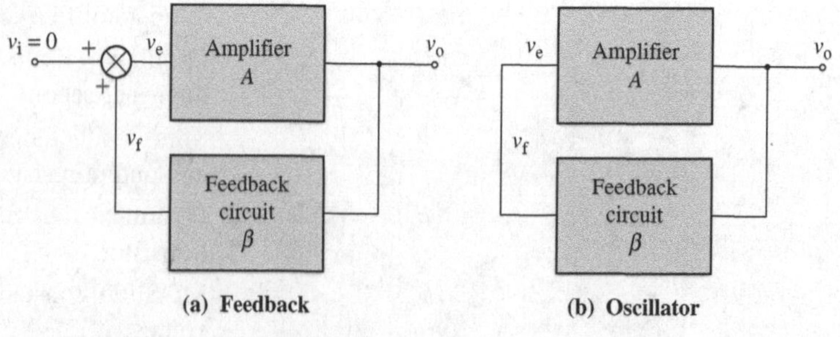

FIGURE 13.1 Oscillator block diagram

Using these relationships, we get the closed-loop voltage gain A_f:

$$A_f = \frac{v_o}{v_i} = \frac{A}{1 - A\beta} \tag{13.1}$$

13.2.1 Conditions for Oscillations

A_f in Eq. (13.1) can be made very large by making $1 - A\beta = 0$. That is, an output of reasonable magnitude can be obtained with a very small-value input signal, tending to zero, as shown in Fig. 13.1(b). Thus, the amplifier will be unstable when $1 - A\beta = 0$, which gives the loop gain as

$$A\beta = 1 \tag{13.2}$$

Expressing Eq. (13.2) in polar form, we get

$$A\beta = 1\angle 0^\circ \quad \text{or} \quad 1\angle 360^\circ \tag{13.3}$$

The above discussion leads to the following design criteria for oscillators:

1. The magnitude of the loop gain $|A\beta|$ must be unity or slightly larger at the desired oscillation frequency.
2. The total phase shift ϕ of the loop gain must be equal to 0° or 360° at the same frequency.
3. The first two conditions must not be satisfied at other frequencies. This condition is normally met by carefully selecting the component values.
4. The first two conditions must continue to be satisfied as parameter values change in response to component tolerance, temperature change, aging, and device replacement. Meeting this criterion often requires special design considerations.

If an amplifier provides a phase shift of 180°, the feedback circuit must provide an additional phase shift of 180° so that the total phase shift around the loop is 360°. The type of waveform generated by an oscillator depends on the types of components used in the circuit; hence the waveform may be sinusoidal, square, or triangular. The frequency of oscillation is determined by the feedback components:

- *RC* components generate a sinusoidal waveform at audio frequencies—that is, in the range from several hertz (Hz) to several kilohertz (kHz).
- *LC* components generate a square wave at radio frequencies—that is, in the range from 100 kHz to 100 MHz.
- Crystals generate a triangular or sawtooth wave over a wide range—that is, in the range from 10 kHz to 10 MHz.

Oscillators can be classified into many types depending on the feedback components, amplifiers, and circuit topologies used [1]. This chapter covers the following types of oscillators: phase-shift oscillators, quadrature oscillators, three-phase oscillators, Wien-bridge oscillators, Colpitts oscillators, Hartley oscillators, crystal oscillators, and active-filter tuned oscillators.

EXAMPLE 13.1

Finding the gain and phase for oscillation A block diagram of an oscillator is shown in Fig. 13.2. Determine the values of the gain A and the phase angle θ that will produce a steady-state oscillation.

SOLUTION

Applying the condition of Eq. (13.3), we know that the loop gain must be

$$A \angle 180^\circ \times 0.01 \angle 0^\circ \times 0.5 \angle \theta \times 10 \angle 0^\circ = 1 \angle 360^\circ$$

from which we get

$$A \times 0.01 \times 0.5 \times 10 = 1 \quad \text{or} \quad A = \frac{1}{0.01 \times 0.5 \times 10} = 20$$

and $\quad 180 + 0 + \theta + 0 = 360 \quad \text{or} \quad \theta = 360 - 180 = 180^\circ$

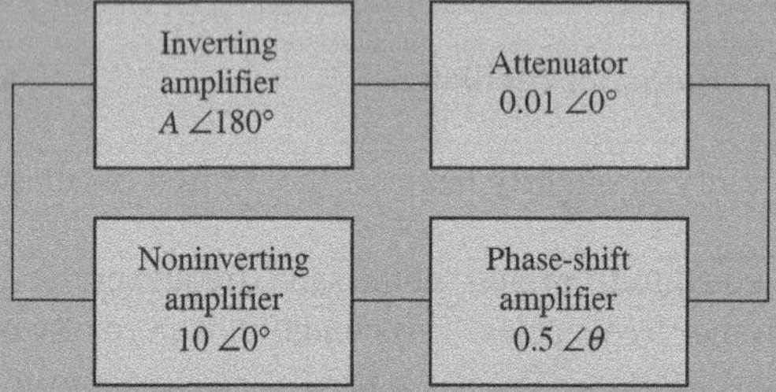

FIGURE 13.2 Oscillator circuit

EXAMPLE 13.2

Finding the frequency and conditions to sustain oscillation The amplifier shown in Fig. 13.3(a) has a voltage gain of $A = 50$, input resistance $R_i = 10\ \text{k}\Omega$, and output resistance $R_o = 200\ \Omega$. Find the resonant frequency ω_o and the values of R and R_3 that will sustain the oscillation.

SOLUTION

The voltage-controlled current source representation is shown in Fig. 13.3(b), where transconductance $G_m = A/R_o$. R_i is in parallel with R, and we let $R_1 = R_i \| R$. Also, R_o is in series with R_3, and we let $R_F = R_o + R_3$. The equivalent circuit is shown in Fig. 13.4, which represents the amplifier as an ideal voltage amplifier. The feedback transfer function β of the feedback circuit is given by

$$\beta(j\omega) = \frac{V_f}{V_o}(j\omega) = \frac{j\omega L \| (-j/\omega C) \| R_1}{j\omega L \| (-j/\omega C) \| R_1 + R_F}$$

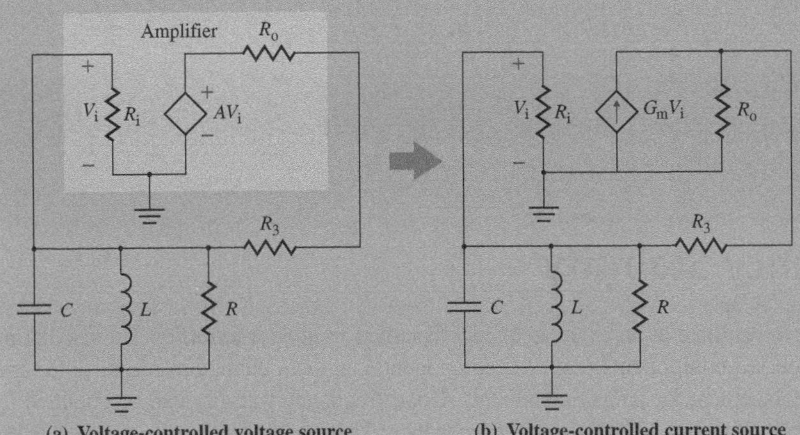

(a) Voltage-controlled voltage source (b) Voltage-controlled current source

FIGURE 13.3 *LC*-feedback amplifier

which can be simplified to give the loop gain as

$$A\beta(j\omega) = \frac{j\omega LA}{R_F(1 - \omega^2 LC) + j\omega L(1 + R_F/R_1)} \tag{13.4}$$

This will provide a 0° phase shift at the resonant frequency ω_o given by

$$\omega_o = \frac{1}{\sqrt{LC}} \quad \text{(in rad/s)} \tag{13.5}$$

At this frequency, the magnitude of $A\beta(j\omega)$ becomes

$$|A\beta(j\omega)| = \frac{A}{1 + R_F/R_1} \tag{13.6}$$

which must be equal to 1 and gives the condition for oscillation as

$$\frac{R_F}{R_1} = A - 1 \tag{13.7}$$

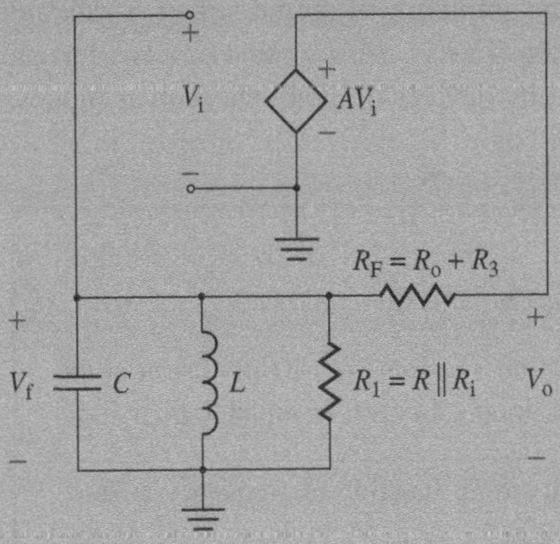

FIGURE 13.4 Equivalent circuit for Example 13.2

That is, for $A = 50$, $R_F/R_1 = 49$. Let $R_1 = R_i \| R = 5\ \text{k}\Omega$; then, for $R_i = 10\ \text{k}\Omega$, $R = 10\ \text{k}\Omega$. Therefore,

$$R_F = R_o + R_3 = 49R_1 = 245\ \text{k}\Omega$$

Thus, for $R_o = 200\ \Omega$, $R_3 = 244.8\ \text{k}\Omega$.

13.2.2 Frequency Stability

The ability of an oscillator to oscillate at an exact frequency is called *frequency stability*. The oscillating frequency is a function of circuit components (e.g., *LC* components) and can change in response to temperature changes, device replacement, or parasitic elements. Good frequency stability can be obtained by making the phase shift a strong function of frequency at resonance. That is, $|d\phi/d\omega|$ (at $\omega = \omega_o$) is made large so that only a slight change in ω is required to correct any phase shift and restore the loop gain to zero phase shift.

The *quality factor* (or *figure of merit*) Q of a circuit also determines the frequency stability. The higher the Q factor, the better the stability will be because the variation in phase shift with frequency near resonance is greater. Crystal oscillators are far more stable than *RC* or *LC* oscillators, especially at higher frequencies. The equivalent electrical circuit of a crystal has a very high Q-value, leading to a high value of $d\phi/d\omega$. *LC* and crystal oscillators are generally used for the generation of high-frequency signals; *RC* oscillators are used mostly for audio-frequency applications.

13.2.3 Amplitude Stability

Like the frequency, the gain of practical amplifiers can change in response to changes in parameters such as temperature, age, and operating point. Therefore, $|A\beta|$ might drop below unity. If the magnitude of $A\beta$ falls below unity, an oscillating circuit ceases oscillating. In practice, an oscillator is designed with a value of $|A\beta|$ that is slightly higher than unity—say, by 5%—at the oscillating frequency. The greater the value of $|A\beta|$, the greater will be the amplitude of the output signal and the amount of its distortion. This distortion will usually lower the gain A to the value required to sustain oscillation.

For good stability, the change in the gain A with a change in the amplitude of output voltage v_o should be made large; an increase in amplitude must result in a decrease in gain. That is, dA/dv_o must be a large negative number. An oscillator is often stabilized by adding nonlinear limiting devices or elements such as diodes.

KEY POINTS OF SECTION 13.2

- To sustain oscillations, the magnitude of the loop gain $|A\beta|$ must be unity or slightly larger at the desired oscillation frequency, and the total phase shift ϕ of the loop gain must be equal to 0° or 360° at the same frequency.
- For good frequency stability, the phase shift must be made a strong function of frequency at resonance; that is, the Q factor should be high.
- For good stability, the change in the gain with a change in the amplitude of output voltage v_o should be made large; an increase in amplitude must result in a decrease in gain.

13.3 Audio-Frequency Oscillators

An audio-frequency (AF) oscillator uses *RC* components for generating a sinusoidal waveform in the frequency range from several hertz (Hz) to several kilohertz (kHz). There are different ways to make an amplifier oscillate. We will consider the following types of AF oscillators: phase-shift oscillator, quadrature oscillator, three-phase oscillator, Wien-bridge oscillator, and ring oscillator. The op-amp Wien bridge is the most common and easy to implement. The MOS oscillators allow CMOS implementation.

13.3.1 Phase-Shift Oscillators

A phase-shift oscillator consists of an inverting amplifier with a positive feedback circuit [1]. The amplifier gives a phase shift of 180°, and the feedback circuit gives another phase shift of 180°, so the total phase shift around the loop is 360°. A phase-shift oscillator consisting of an inverting op-amp amplifier with positive feedback is shown in Fig. 13.5(a). The feedback circuit provides voltage feedback from the output back to the input of the amplifier. Any signal that appears at the inverting terminal is shifted by 180° at the output. Therefore, an additional 180° shift is required for oscillation at a specific frequency f_o in order to give a total phase shift around the loop of 360°. Since the feedback network consists of resistors and capacitors, as shown in Fig. 13.5(b), this type of oscillator is also known as an *RC* oscillator. The transfer function of the feedback network in Laplace's domain of *s* is given by

$$\beta(s) = \frac{V_f(s)}{V_o(s)} = \frac{R^3C^3s^3}{R^3C^3s^3 + 6R^2C^2s^2 + 5RCs + 1} \tag{13.8}$$

(see Prob. 13.3). The closed-loop voltage gain of the op-amp circuit is

$$A(s) = \frac{V_o(s)}{V_f(s)} = -\frac{R_F}{R_1} \tag{13.9}$$

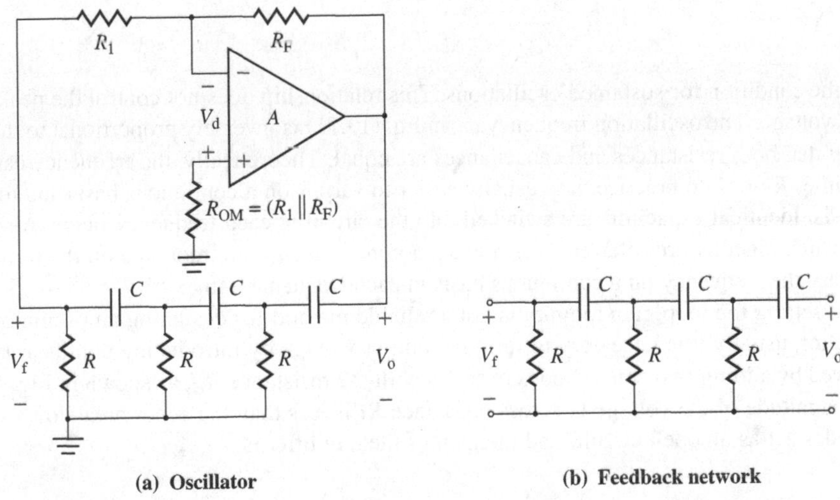

FIGURE 13.5 Phase-shift oscillator

Since $A\beta = 1$ for an oscillator, from Eqs. (13.8) and (13.9), we get

$$-\frac{R_F}{R_1}\left[\frac{R^3C^3s^3}{R^3C^3s^3 + 6R^2C^2s^2 + 5RCs + 1}\right] = 1 \tag{13.10}$$

Substituting $s = j\omega$ into Eq. (13.10) and canceling the elements in the denominator, we get

$$-R_F(-jR^3C^3\omega^3) = R_1(-jR^3C^3\omega^3 - 6R^2C^2\omega^2 + j5RC\omega + 1)$$

Equating the real parts to zero, we get

$$R_1(-6R^2C^2\omega^2 + 1) = 0$$

which gives the oscillation frequency ω_o as

$$\omega_o = \omega = 2\pi f_o = \frac{1}{\sqrt{6}RC} \quad \text{(in rad/s)} \tag{13.11}$$

where f_o is the frequency in hertz. Equating the imaginary parts on both sides yields

$$-R_F(-jR^3C^3\omega^3) = R_1(-jR^3C^3\omega^3 + j5RC\omega)$$

which gives

$$R_F = R_1\left[\frac{5}{R^2C^2\omega^2} - 1\right] \tag{13.12}$$

Substituting the value of $\omega = \omega_o$ from Eq. (13.11) into Eq. (13.12) yields

$$\frac{R_F}{R_1} = 29 \tag{13.13}$$

which gives the condition for sustained oscillations. This relationship does not control the peak amplitude of the output voltage. The oscillation frequency ω_o in Eq. (13.11) is inversely proportional to the RC product, assuming that both resistances and capacitances are equal. Theoretically, the frequency can be varied by varying either R or C. In practice, it is usually easier to vary R on a continuous basis and to vary C on a discrete basis. Identical capacitors are switched into the circuit at each frequency range. Also, identical resistances, which together are referred to as a *gauged potentiometer*, are mounted on the same shaft and are used to vary the frequency on a continuous basis in each frequency range.

Note that setting the loop gain to unity is not a reliable method for designing an oscillator. To stabilize an oscillator, usually it is necessary to limit the output voltage by introducing nonlinearity. Stability can be achieved by adding two zener diodes in series with the resistance R_B, as shown in Fig. 13.6(a). As long as the magnitude of the voltage v_f across resistance R_1 is less than the zener breakdown voltage V_Z, the zener diodes act as an open circuit, and the gain of the amplifier is

$$|A_1| = \frac{R_F}{R_1} \tag{13.14}$$

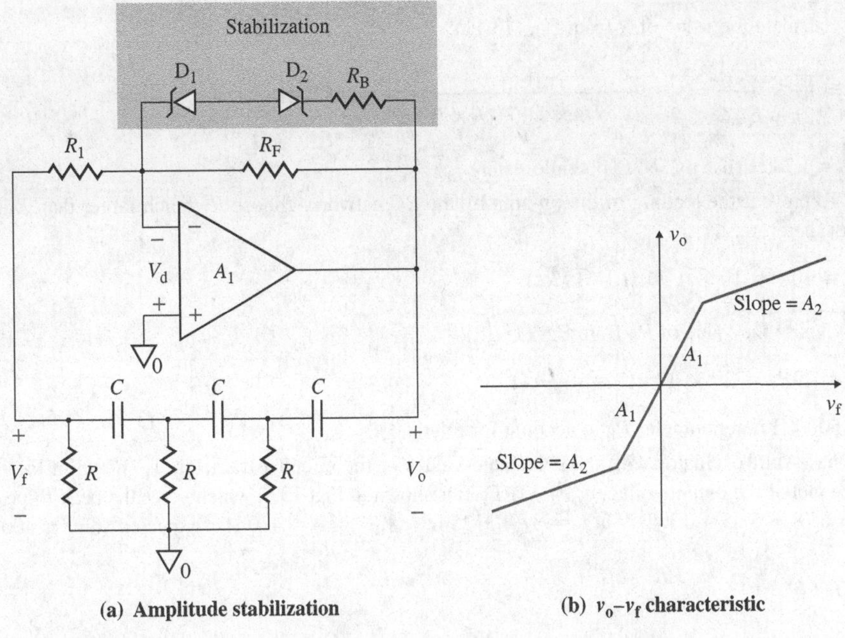

(a) Amplitude stabilization **(b) v_o–v_f characteristic**

FIGURE 13.6 Stabilization of a phase-shift oscillator

As soon as the magnitude of v_f starts to increase above V_Z, the zener diodes conduct, and the resistor R_B suddenly becomes in parallel with R_F so that the gain is reduced. The new gain becomes

$$|A_2| = \frac{R_F \| R_B}{R_1} \tag{13.15}$$

which is less than $|A_1|$. Furthermore, if the output amplitude starts to decrease, the gain $|A|$ is increased again. The v_o–v_f characteristic of the amplifier is shown in Fig. 13.6(b).

EXAMPLE 13.3

D **Designing a phase-shift oscillator**

(a) Design the phase-shift oscillator shown in Fig. 13.5(a) so that the oscillating frequency is $f_o = 400$ Hz.

(b) Use PSpice/SPICE to plot the transient response of the output voltage $v_o(t)$ in part (a) from 0 to 4 ms. Assume $V_{CC} = V_{EE} = 12$ V.

SOLUTION

(a) The following steps can be used to complete the design.

Step 1. Choose a suitable value of C: Let $C = 0.1$ μF.

Step 2. Calculate the value of R from Eq. (13.11):

$$R = \frac{1}{2\pi\sqrt{6}f_oC} = \frac{1}{2\pi \times \sqrt{6} \times 400\ \text{Hz} \times 0.1\ \mu\text{F}} = 1624\ \Omega$$

Choose $R = 1.7\ \text{k}\Omega$ (use a 2.7-kΩ potentiometer).

Step 3. To prevent the loading of the op-amp by the RC network, choose R_1 much larger than R by making $R_1 \geq 10R$. Therefore, let

$$R_1 = 10R = 10 \times 1.7\ \text{k}\Omega = 17\ \text{k}\Omega$$

Step 4. Choose the value of R_F from Eq. (13.13):

$$R_F = 29R_1 = 29 \times 17\ \text{k}\Omega = 493\ \text{k}\Omega$$

Choose a 500-kΩ potentiometer R_F to account for tolerance.

(b) The phase-shift oscillator with the calculated values of the circuit parameters is shown in Fig. 13.7. The PSpice plot of the output voltage $v_o \equiv$ V(RF:2) is shown in Fig. 13.8, which gives the peak-to-peak voltage $V_{pp} = 6.78 + 6.65 = 13.43$ V at $f_o = 1/(4.3125\ \text{m} - 1.875\ \text{m}) = 410$ Hz (expected value is 400 Hz).

NOTES:

1. If you observe carefully, you will notice that the amplitude of the output voltage is falling slowly and the oscillation will not be sustained for a long time. A nonlinear device is often necessary to stabilize the oscillator.
2. An initial voltage of 1 V has been assigned to the capacitor C in order to start the oscillator, and the UIC (use initial condition) is used in transient analysis. Otherwise, PSpice will first calculate the biasing values and then use those values to find the solutions, and the circuit will not oscillate. In practice, random noise or transients can cause the oscillations to begin, and they are sustained by the feedback of the appropriate signal.

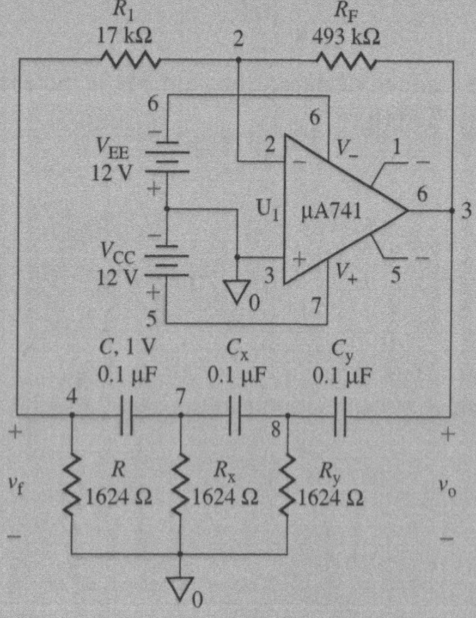

FIGURE 13.7 Phase-shift oscillator for PSpice simulation

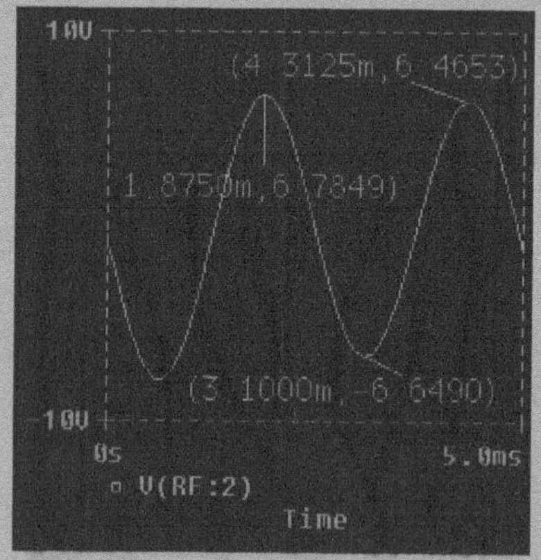

FIGURE 13.8 PSpice plot of output voltage for Example 13.3

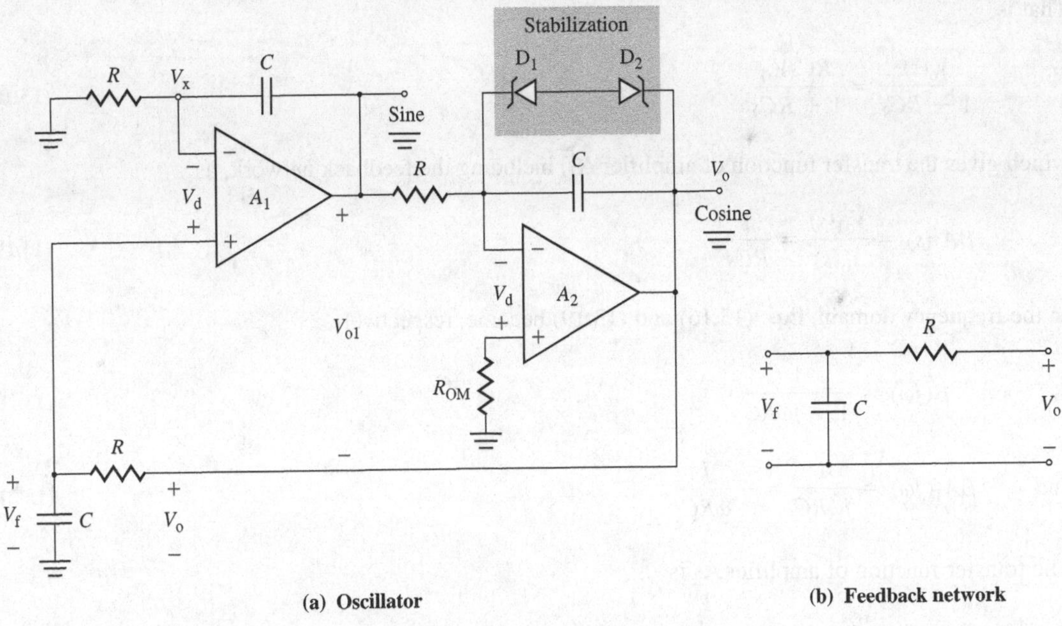

FIGURE 13.9 Quadrature oscillator

13.3.2 Quadrature Oscillators

A quadrature oscillator, as shown in Fig. 13.9(a), generates two signals (sine and cosine) that are in quadrature—that is, out of phase by 90°. The actual location of sine and cosine signals is arbitrary. In Fig. 13.9(a), the output of amplifier A_1 is labeled as sine and that of amplifier A_2 as cosine. This oscillator requires a dual op-amp. Amplifier A_2 operates as an inverting integrator and provides a phase shift of $-270°$ (or 90°); amplifier A_1, in combination with the feedback network, operates as a noninverting integrator and provides the remaining $-90°$ (or 270°) to give the total phase shift of 360° that is required to satisfy the condition of oscillation. The transfer function of the feedback network shown in Fig. 13.9(b) in Laplace's domain of s is given by

$$\beta(s) = \frac{V_f(s)}{V_o(s)} = \frac{1/Cs}{R + 1/Cs} = \frac{1}{1 + RCs} \tag{13.16}$$

If V_{o1} is the voltage at the output of amplifier A_1, the voltage V_x at its inverting terminal is given by

$$V_x = \frac{RV_{o1}}{R + 1/Cs} = \frac{RCsV_{o1}}{1 + RCs} \tag{13.17}$$

Since the differential voltage between the op-amp terminals is very small ($V_d \approx 0$), we can write

$$V_f = V_x - V_d \approx V_x$$

That is,

$$\frac{V_o(s)}{1 + RCs} = \frac{RCsV_{o1}}{1 + RCs} \tag{13.18}$$

which gives the transfer function of amplifier A_1, including the feedback network, as

$$\beta A_1(s) = \frac{V_{o1}(s)}{V_o(s)} = \frac{1}{RCs} \tag{13.19}$$

In the frequency domain, Eqs. (13.16) and (13.19) become, respectively,

$$\beta(j\omega) = \frac{1}{1 + j\omega RC} \tag{13.20}$$

and $$\beta A_1(j\omega) = \frac{1}{j\omega RC} = -\frac{j}{\omega RC} \tag{13.21}$$

The transfer function of amplifier A_2 is

$$A_2(j\omega) = \frac{V_o}{V_{o1}}(j\omega) = -\frac{1}{j\omega RC} = \frac{j}{\omega RC}$$

which gives a phase shift of 90°. Thus, $\beta G_1(j\omega)$ must give a phase shift of $-90°$.

To provide a phase shift of $-90°$, $|\beta G_1(j\omega)|$ in Eq. (13.21) must equal unity. That is, $\omega RC = 1$, and the frequency of oscillation is given by

$$f_o = \frac{1}{2\pi RC} \quad \text{(in Hz)} \tag{13.22}$$

At this frequency, the magnitude of $\beta(j\omega)$ in Eq. (13.20) becomes

$$\beta = |\beta(j\omega)| = \left|\frac{1}{1 + j1}\right| = \frac{1}{\sqrt{2}} \tag{13.23}$$

The loop gain becomes

$$\beta A(j\omega) = A_2(j\omega)\beta A_1(j\omega) = \frac{j}{\omega RC} \times \frac{1}{j\omega RC} = \frac{1}{\omega^2 RC} = 1$$

Therefore, the overall closed-loop gain A_v of amplifiers A_1 and A_2 is given by

$$A_f = \frac{1}{\beta} = \sqrt{2} = 1.4142 \tag{13.24}$$

which implies a constant gain of 1.4142. The design of a quadrature oscillator is very simple. For $f_o = 200$ Hz and assuming $C = 0.1\ \mu$F, Eq. (13.22) gives $R = 7958\ \Omega$ (use a 10-kΩ potentiometer). This oscillator can be stabilized by connecting two zener diodes back to back across one of the integrating capacitors, as shown in Fig. 13.9(a) by the shaded area.

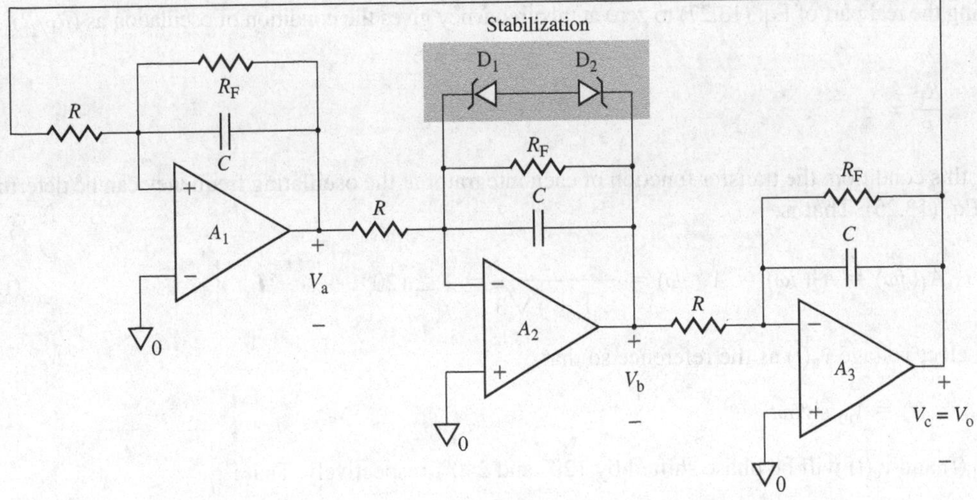

FIGURE 13.10 Three-phase oscillator

13.3.3 Three-Phase Oscillators

A three-phase oscillator generates three sinusoidal voltages of equal magnitude, but displaced by 120° from each other. They have the same form as the voltages in a three-phase power system and are normally used for generating control signals synchronized to the power system. A three-phase oscillator consisting of three "lossy" integrator circuits connected in cascade with unity feedback is shown in Fig. 13.10. The transfer function of each of the integrators is given by

$$A_1(s) = A_2(s) = A_3(s) = -\frac{R_F \| (1/Cs)}{R} = \frac{-R_F/R}{1 + R_F Cs} \tag{13.25}$$

Since $\beta = 1$ for unity feedback, the loop gain is given by

$$\beta A(s) = A_1(s)A_2(s)A_3(s) = \frac{-(R_F/R)^3}{(R_F Cs)^3 + 3(R_F Cs)^2 + 3(R_F Cs) + 1} \tag{13.26}$$

Then the characteristic equation, which is the numerator of $1 - \beta A(s) = 0$, is

$$(R_F Cs)^3 + 3(R_F Cs)^2 + 3(R_F Cs) + 1 + \left(\frac{R_F}{R}\right)^3 = 0 \tag{13.27}$$

Substituting $s = j\omega$ into Eq. (13.27) and then equating the imaginary part to zero, we get the frequency of oscillation ω_o as

$$\omega_o = \frac{\sqrt{3}}{R_F C} \quad \text{(in rad/s)} \tag{13.28}$$

Equating the real part of Eq. (13.27) to zero at this frequency gives the condition of oscillation as $(R_F/R)^3 = 8$; that is,

$$\frac{R_F}{R} = 2 \tag{13.29}$$

Under this condition, the transfer function of each integrator at the oscillating frequency can be determined from Eq. (13.25). That is,

$$A_1(j\omega) = A_2(j\omega) = A_3(j\omega) = -\frac{2}{1 + j\sqrt{3}} = 1 \angle 120° \tag{13.30}$$

If we select voltage $v_a(t)$ as the reference so that

$$v_a(t) = V_m \sin \omega t \tag{13.31}$$

then $v_b(t)$ and $v_c(t)$ will be phase shifted by 120° and 240°, respectively. That is,

$$v_b(t) = V_m \sin(\omega t + 120°) \tag{13.32}$$

and

$$v_c(t) = V_m \sin(\omega t + 240°) = V_m \sin(\omega t - 120°) \tag{13.33}$$

For $f_o = 60$ Hz and assuming $C = 0.1\ \mu\text{F}$, Eq. (13.28) gives

$$R_F = \frac{\sqrt{3}}{2\pi \times 60 \times 0.1\ \mu\text{F}} = 45.94\ \text{k}\Omega \quad \text{(use a 50-k}\Omega\text{ potentiometer)}$$

and Eq. (13.29) gives $R = R_F/2 = 22.97\ \text{k}\Omega$. This oscillator can be stabilized by connecting two zener diodes back to back across one of the integrating capacitors, as shown in Fig. 13.10 by the shaded area.

13.3.4 Wien-Bridge Oscillators

A Wien bridge, which is used for making measurements of unknown resistors or capacitors, is shown in Fig. 13.11(a). The bridge has a series *RC* network in one arm and a parallel *RC* network in the adjoining arm. R_1 and R_F are connected in two other arms. While the bridge is making measurements, either R_1 or R_F acts as a calibrated resistor; the resistance is varied until the null voltage $V_d = 0$ is found. If all component values are known except one, the value of that one can be determined from the following relation:

$$\frac{R_2}{R_3} + \frac{C_2}{C_1} = \frac{R_F}{R_1} \tag{13.34}$$

If an op-amp is inserted into the basic bridge, as shown in Fig. 13.11(b), the bridge is known as a *Wien-bridge oscillator*, provided the elements are adjusted so that $R_2 = R_3 = R$ and $C_1 = C_2 = C$. The op-amp, along with R_1 and R_F, operates as a noninverting amplifier, as shown in Fig. 13.12(a). The Wien-bridge oscillator is one of the most commonly used AF oscillators.

The transfer function of the feedback network shown in Fig. 13.12(b) is given by

$$\beta(s) = \frac{V_f(s)}{V_o(s)} = \frac{RCs}{R^2C^2s^2 + 3RCs + 1} \tag{13.35}$$

(see Prob. 13.9).

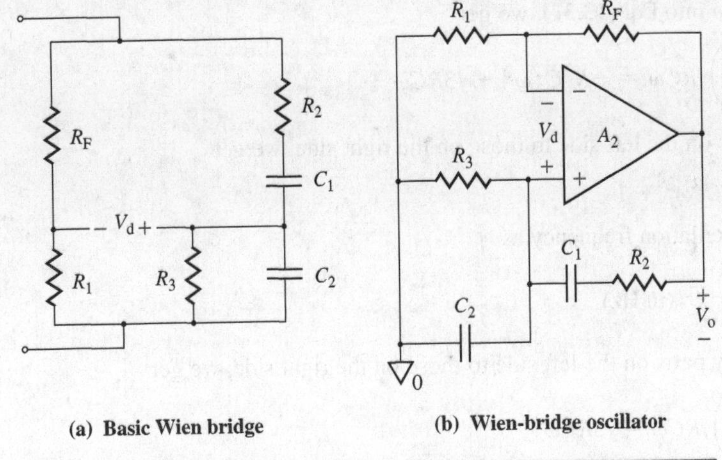

(a) Basic Wien bridge (b) Wien-bridge oscillator

FIGURE 13.11 Wien bridge

The closed-loop voltage gain of the noninverting amplifier is given by

$$A(s) = \frac{V_o(s)}{V_f(s)} = 1 + \frac{R_F}{R_1} \tag{13.36}$$

For an oscillator, $A\beta = 1$. Using Eqs. (13.35) and (13.36), we get

$$\left(1 + \frac{R_F}{R_1}\right)\frac{RCs}{R^2C^2s^2 + 3RCs + 1} = 1 \tag{13.37}$$

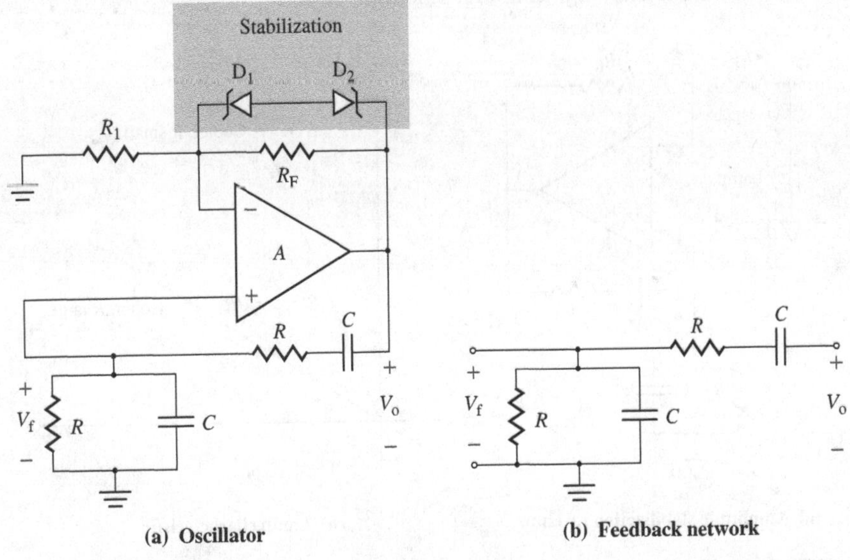

(a) Oscillator (b) Feedback network

FIGURE 13.12 Wien-bridge oscillator

Substituting $s = j\omega$ into Eq. (13.37), we get

$$\left(1 + \frac{R_F}{R_1}\right) jRC\omega = -R^2C^2\omega^2 + j3RC\omega + 1$$

Equating real parts on the left side to those on the right side, we get

$$0 = -R^2C^2\omega^2 + 1$$

which gives the oscillation frequency as

$$f_o = \frac{1}{2\pi RC} \quad \text{(in Hz)} \tag{13.38}$$

Equating imaginary parts on the left side to those on the right side, we get

$$\left(1 + \frac{R_F}{R_1}\right) jRC\omega = j3RC\omega$$

which gives the condition for oscillation as

$$1 + \frac{R_F}{R_1} = 3$$

or

$$\frac{R_F}{R_1} = 2 \tag{13.39}$$

For stabilization, a power-sensitive resistor such as a lamp or a thermistor is usually used to adjust dynamically the loop gain of the oscillator. Figure 13.13(a) shows the use of a small incandescent lamp, whose resistance characteristic is shown in Fig. 13.13(b). When the filament of the lamp is cold, the resistance is small and the gain A is large. But when the lamp filament becomes hot, the resistance becomes larger, and the gain A becomes small. This automatic adjustment of the gain causes distortion of the amplifier to be low and stabilizes the oscillator. The nonlinear characteristic can be provided by two back-to-back zener diodes in series with a resistor R_B, as in Fig. 13.6(a).

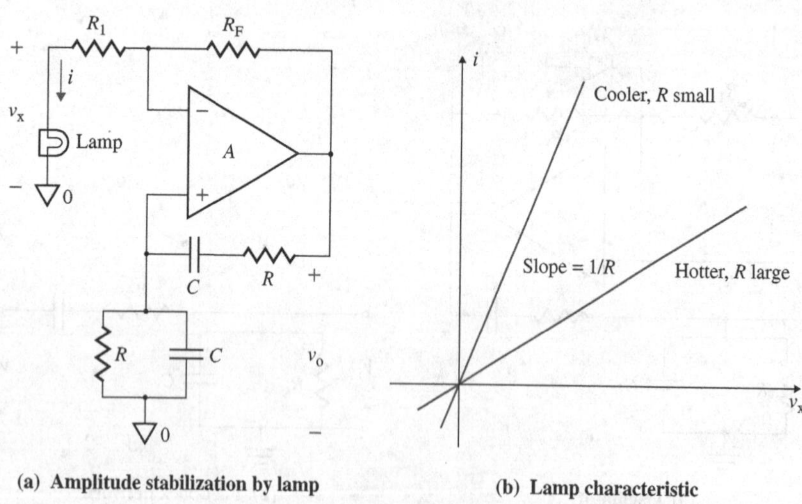

(a) Amplitude stabilization by lamp **(b) Lamp characteristic**

FIGURE 13.13 Stabilization of a Wien-bridge oscillator

EXAMPLE 13.4

D

Designing a Wien-bridge oscillator

(a) Design the Wien-bridge oscillator in Fig. 13.12(a) so that $f_o = 1$ kHz.

(b) Use PSpice/SPICE to plot the transient response of the output voltage $v_o(t)$ in part (a) from 0 to 2 ms. Assume $V_{CC} = V_{EE} = 12$ V.

SOLUTION

(a) The following steps can be used to complete the design.

Step 1. Choose a suitable value of C: Let $C = 0.01$ μF.

Step 2. Calculate the value of R from Eq. (13.38):

$$R = \frac{1}{2\pi f_o C} = \frac{1}{2\pi \times 1\ \text{kHz} \times 0.01\ \mu\text{F}} = 15{,}915\ \Omega$$

Choose $R = 16$ kΩ.

Step 3. Choose the value of R_F from Eq. (13.39). Letting $R_1 = 10$ kΩ, we have

$$R_F = 2R_1 = 2 \times 10\ \text{k}\Omega = 20\ \text{k}\Omega$$

NOTE: The exact numbers of the resistances are used to demonstrate the validity of the calculated values with the PSpice simulation results. In a practical design, the commercially available standard values will be used (i.e., 16 kΩ instead of 15,916 Ω; see Appendix E).

(b) The Wien-bridge oscillator with the desired values is shown in Fig. 13.14. The PSpice plot of the output voltage $v_o \equiv$ V(U1:OUT) is shown in Fig. 13.15, which gives the peak-to-peak voltage $V_{pp} = 5.94 + 5.97 = 11.91$ V at $f_o = 1/(1.269\text{ m} - 0.247\text{ m}) = 978$ Hz (expected value is 1 kHz).

NOTE: An initial voltage of 1 V has been assigned to the capacitor C_p in order to start the oscillator, and the UIC (use initial condition) is used in transient analysis. In practice, random noise or transients can cause the oscillations to begin, and they are sustained by the feedback of the appropriate signal.

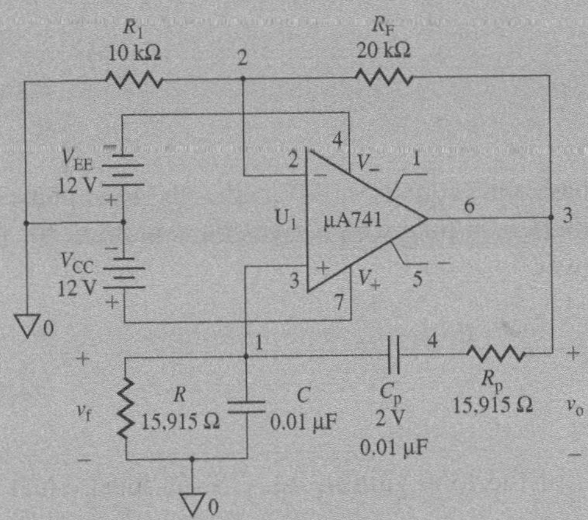

FIGURE 13.14 Wien-bridge oscillator for PSpice simulation

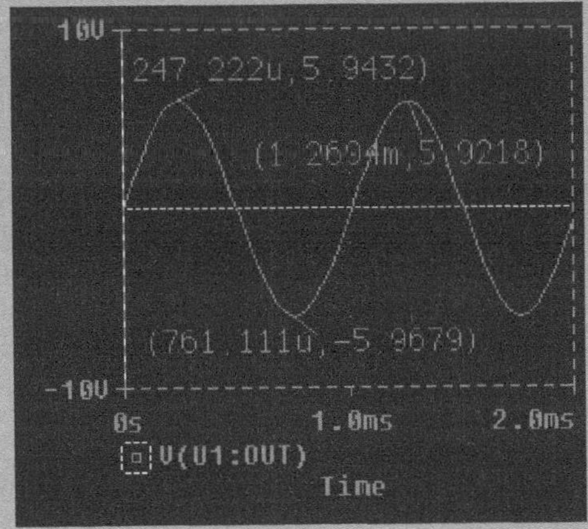

FIGURE 13.15 PSpice plot of output voltage for Example 13.4

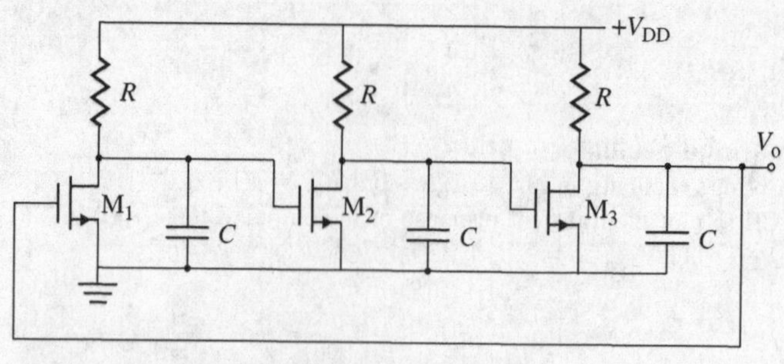

FIGURE 13.16 Three-stage ring oscillator

13.3.5 Ring Oscillators

A ring oscillator consists of a number of gain stages in a loop in order to make the total phase shift of 180°. A three-stage ring oscillator is shown in Fig. 13.16. By neglecting the MOS capacitance, we can find the transfer function of each stage as given by

$$A(s) = -g_m\left(R \parallel \frac{1}{sC}\right) = -\frac{g_m R}{1 + RCs} = -\frac{A_o}{1 + RCs} \tag{13.40}$$

which in frequency domain becomes

$$A(j\omega) = -\frac{A_o}{1 + (j\omega/\omega_n)} \tag{13.41}$$

where $A_o = g_m R$ is the small-signal low-frequency gain and $\omega_n = 1/RC$ is the natural frequency. Since three stages are cascaded, the open-loop transfer function becomes

$$H(j\omega) = -\frac{A_o^3}{1 + (j\omega/\omega_n)^3} \tag{13.42}$$

The circuit oscillates only if the frequency-dependent phase shift equals $-180°$ so that the total phase of $H(j\omega) = 360°$ or 0°. Therefore, each stage must contribute $\theta = 60°$, which gives the condition for the oscillation frequency ω_o as $\tan^{-1}(\omega_o/\omega_n) = 60°$. This gives

$$\omega_o = \sqrt{3}\,\omega_n = \frac{\sqrt{3}}{RC} \tag{13.43}$$

At the oscillation frequency of $\omega = \omega_o$, the magnitude of the loop gain in Eq. (13.42) must equal 1; that is,

$$|H(j\omega)| = \frac{A_o^3}{[\sqrt{1 + (\omega_o/\omega_n)^2}]^3} = 1 \tag{13.44}$$

which, for $\omega_o = \sqrt{3}\,\omega_n$, gives the minimum value of A_o to sustain the oscillation as

$$A_o = g_m R \geq 2 \tag{13.45}$$

EXAMPLE 13.5

Designing a MOS ring oscillator

(a) Design the ring oscillator in Fig. 13.16 so that $f_o = 1$ kHz. The MOSFET parameters are MOS constant $K_p = 25\ \mu A/V^2$, the threshold voltage $V_t = 1$ V, length $L = 10\ \mu m$, width $W = 20\ \mu m$, and modulation length $\lambda = 0.01$.

(b) Use PSpice [2] to plot the transient response of the output voltage $v_o(t)$ in part (a) from 0 to 2 ms. Assume $V_{DD} = 5$ V.

SOLUTION

(a) Assuming that the MOS is operating in the saturation region, the transconductance of each MOS is $g_m = 2K_p(v_{GS} - V_t)$, which can give the desired value of R to obtain the required gain A_o as given by

$$R = \frac{A_o}{g_m} = \frac{A_o}{2K_p(v_{GS} - V_t)} \tag{13.46}$$

Using KVL in the loop formed by V_{DD}, R, and the gate–source voltage of M_2, we can write

$$V_{DD} = Ri_{D1} + v_{GS1} = \frac{A_o}{2K_p(v_{GS1} - V_t)} \times K_p(v_{GS1} - V_t)^2 + v_{GS1}$$

which, for identical MOSFETs of $v_{GS} = v_{GS1} = v_{GS2}$, gives the gate–source voltage as

$$v_{GS} = \frac{V_{DD} + (A_o/2)V_t}{1 + (A_o/2)} \tag{13.47}$$

Letting $A_o = 4$ (higher than 2), Eq. (13.47) gives the DC biasing gate–source voltage

$$v_{GS} = \frac{5 + (4/2) \times 1}{1 + (4/2)} = 2.333 \text{ V}$$

$$i_D = K_p(v_{GS1} - V_t)^2 = 25\ \mu \times (2.333 - 1)^2 = 44.44\ \mu A$$

$$g_m = 2K_p(v_{GS} - V_t) = 2 \times 25\ \mu \times (2.333 - 1) = 66.67\ \mu A/V$$

$$R = \frac{A_o}{g_m} = \frac{4}{66.67\ \mu A/V} = 60\ k\Omega$$

For $f_o = 1$ kHz, Eq. (13.43) gives

$$C = \frac{\sqrt{3}}{2\pi f_o R} = \frac{\sqrt{3}}{2\pi \times 1\text{ k} \times 60\text{ k}} = 4.59\text{ nF}$$

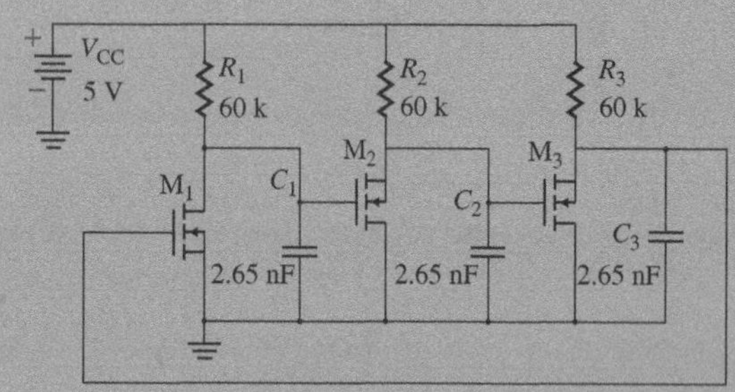

FIGURE 13.17 PSpice schematic for Example 13.5

(b) The PSpice schematic with desired values is shown in Fig. 13.17. The PSpice plot of the output voltage $v_o \equiv$ V(C3:2) is shown in Fig. 13.18, which gives the peak-to-peak voltage of $V_{pp} = (3.7974 - 1.4154) = 2.38$ V at $f_o = 1/(2.6499 \text{ ms} - 1.7135 \text{ ms}) = 1046$ Hz (expected 1 kHz). Note that an initial voltage of 0 has been assigned to the capacitor C_1 in order to start the oscillator, and the UIC (use initial condition) is used in transient analysis. In practice, random noise or transients can cause the oscillations to begin, and they are sustained by the feedback of the appropriate signal.

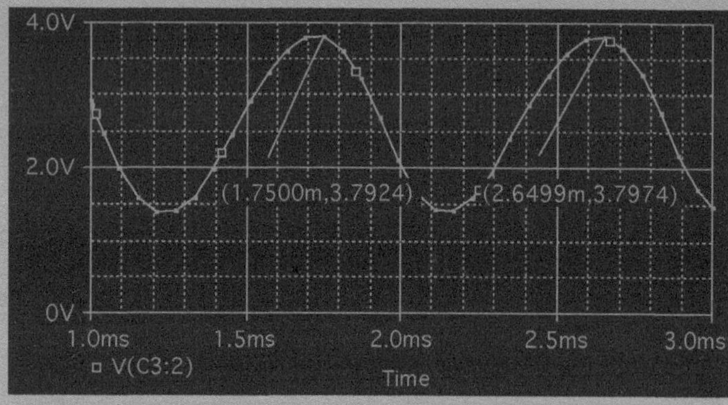

FIGURE 13.18 Output voltage waveform for Example 13.5

KEY POINTS OF SECTION 13.3

- A phase-shift oscillator uses an inverting amplifier and a phase-shifting network to satisfy the requirements of unity loop gain with 0° or 360° phase shift. The oscillation frequency ω_o is inversely proportional to the *RC* product of the feedback network. A nonlinear device is often introduced, however, to stabilize the oscillator.
- A quadrature oscillator uses two op-amp inverting integrators and an *RC* phase shifter. The output could be sine or cosine. The oscillation frequency ω_o is inversely proportional to the *RC* product of the feedback network.
- A three-phase oscillator uses three inverting integrators with a unity feedback loop. The oscillation frequency ω_o is inversely proportional to the *RC* product of the feedback network.
- A Wien-bridge oscillator uses a noninverting amplifier and an *RC* phase-shifting network. The oscillation frequency ω_o is inversely proportional to the *RC* product of the feedback network.

13.4 Radio Frequency Oscillators

A radio frequency (RF) oscillator uses *LC* components for generating a waveform in the frequency range from 100 kHz to 100 MHz. There are different ways to make amplifiers by using op-amps, BJTs, and MOSFETs; choose the *LC* components to create conditions for oscillations. We consider the following types of RF oscillators: Colpitts oscillator, Hartley oscillator, and two-stage MOS oscillator. The op-amp Colpitts oscillator is the most common and easy to implement. The MOS oscillators allow CMOS implementation.

13.4.1 Colpitts Oscillators

A Colpitts oscillator is a tuned *LC*-type oscillator, as shown in Fig. 13.19(a). *LC* oscillators have the advantage of having relatively small reactive elements. They exhibit higher Q than *RC* oscillators, but they are difficult to tune over a wide range. For a positive feedback circuit to operate as an oscillator, the loop gain must be zero; that is,

$$1 - A\beta = 0$$

which is really the characteristic equation of the circuit. Therefore, the condition for oscillation can be found from the characteristic equation without deriving the transfer function. Nodal analysis can be applied to find the determinant, which is then set to zero.

The op-amp operates as an inverting amplifier of gain $A = R_F/R_1$. If the amplifier is replaced by its equivalent circuit, Fig. 13.19(a) can be simplified to Fig. 13.19(b). If the voltage source Av_f is replaced by its equivalent current source $g_m v_f$, Fig. 13.19(b) can be reduced to Fig. 13.19(c). Using nodal analysis in Fig. 13.19(c) in Laplace's domain of s, we can write

$$\left[sC_2 + \frac{1}{R_L} + \frac{1}{sL}\right]V_o(s) + \left[g_m - \frac{1}{sL}\right]V_f(s) = 0 \quad \text{(at node B)} \tag{13.48}$$

$$-\frac{1}{sL}V_o(s) + \left[sC_1 + \frac{1}{sL} + \frac{1}{R_1}\right]V_f(s) = 0 \quad \text{(at node A)} \tag{13.49}$$

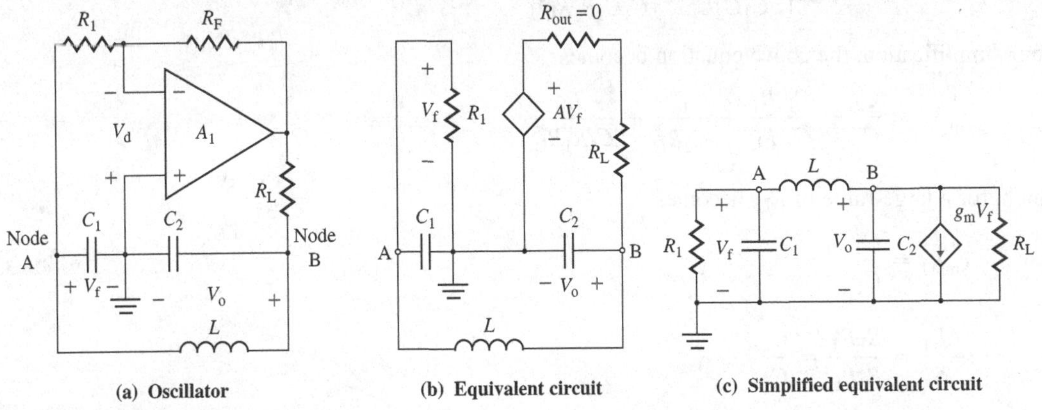

FIGURE 13.19 Colpitts oscillator

To find the condition for oscillation, we set the determinant to zero; that is,

$$\left(sC_2 + \frac{1}{R_L} + \frac{1}{sL}\right)\left(sC_1 + \frac{1}{sL} + \frac{1}{R_1}\right) + \left(g_m - \frac{1}{sL}\right)\frac{1}{sL} = 0$$

which, after simplification, yields

$$s^3C_1C_2LR_1R_L + s^2L(C_1R_1 + C_2R_L) + s(C_1R_1R_L + C_2R_1R_L + L) + (R_1 + R_L + g_mR_1R_L) = 0 \tag{13.50}$$

where $g_m = A/R_L = R_F/(R_1R_L)$. Substituting $s = j\omega$ and equating the imaginary parts to zero, we get

$$-j\omega^3(C_1C_2LR_1R_L) + j\omega(C_1R_1R_L + C_2R_1R_L + L) = 0$$

which gives the frequency of oscillation ω_o as

$$\omega_o = \left[\frac{C_1 + C_2}{C_1C_2L} + \frac{1}{C_1C_2R_1R_L}\right]^{1/2} \quad \text{(in rad/s)} \tag{13.51}$$

Assuming R_L is large, such that $R_1R_L > 1/(C_1C_2)$, Eq. (13.51) can be approximated by

$$\omega_o = \left[\frac{C_1 + C_2}{C_1C_2L}\right]^{1/2} \quad \text{(in rad/s)} \tag{13.52}$$

Similarly, equating the real parts of Eq. (13.50) to zero yields

$$-\omega^2L(C_1R_1 + C_2R_L) + (R_1 + R_L + g_mR_1R_L) = 0$$

which gives

$$\omega^2L(C_1R_1 + C_2R_L) = R_1 + R_L + g_mR_1R_L$$

Substituting the value of $\omega = \omega_o$ from Eq. (13.51) gives

$$L(C_1R_1 + C_2R_L)\left[\frac{C_1 + C_2}{C_1C_2L} + \frac{1}{C_1C_2R_1R_L}\right] = R_1 + R_L + g_mR_1R_L$$

After simplification, the above equation becomes

$$g_mR_1 \approx \frac{C_2}{C_1} + \frac{C_1}{C_2}\frac{R_1}{R_L} + \frac{L}{C_2R_L^2} + \frac{L}{C_1R_1R_L}$$

which, for a large value of R_L, becomes

$$g_mR_1 = \frac{C_2}{C_1} \tag{13.53}$$

or

$$\frac{AR_1}{R_L} = \frac{R_FR_1}{R_1R_L} = \frac{C_2}{C_1}$$

That is, $$\frac{R_F}{R_L} = \frac{C_2}{C_1} \tag{13.54}$$

which is independent of R_1 and gives the relationship among R_L, R_F, C_1, and C_2. Equation (13.53) gives the minimum value of g_m (or R_F/R_1R_L) required to sustain the oscillation with a constant amplitude. If g_m is smaller than this value, the oscillation will die exponentially to zero. On the other hand, if g_m is larger than this value, the amplitude will grow exponentially until the nonlinearity of the op-amp limits the amplitude. Therefore, to ensure oscillation, the value of g_m must exceed the minimum value.

In the above analysis, we used a simple op-amp model and neglected the loss in the resistance of the inductor. As a result, we obtained relatively simple expressions for the frequency and the condition to sustain oscillation. If a complex op-amp model including the inductor loss were used, the oscillation frequency would depend (generally only slightly) on other circuit parameters. Usually, the inductor or one of the capacitors is made adjustable so that the frequency can be initially tuned to the desired value.

EXAMPLE 13.6

D **Designing a Colpitts oscillator** Design the Colpitts oscillator of Fig. 13.19(a) so that the oscillating frequency is $f_o = 150$ kHz.

SOLUTION

Step 1. Choose suitable values of C_1 and C_2: Let $C_1 = 0.01$ μF and $C_2 = 0.1$ μF; that is, $C_2/C_1 = 0.1/0.01 = 10$.

Step 2. Calculate the value of L from Eq. (13.52):

$$L = \frac{C_1 + C_2}{4\pi^2 C_1 C_2 f_o^2} = \frac{0.01\ \mu\text{F} + 0.1\ \mu\text{F}}{4\pi^2 \times 0.01\ \mu\text{F} \times 0.1\ \mu\text{F} \times (150\ \text{kHz})^2} = 124\ \mu\text{H}$$

Step 3. Choose the values of R_F and R_L from Eq. (13.54):

$$\frac{R_F}{R_L} = \frac{C_2}{C_1} = \frac{0.1}{0.01} = 10$$

Let $R_L = 100$ kΩ. Therefore, $R_F = 10R_L = 1$ MΩ.

Step 4. Choose a value of A: Let $A = 10 = R_F/R_1$. Then

$$R_1 = \frac{R_F}{A} = \frac{1\ \text{M}\Omega}{10} = 100\ \text{k}\Omega$$

Step 5. Check the values of g_m and f_o:

$$g_m = \frac{A}{R_1} = \frac{10}{100\ \text{k}\Omega} = 0.1\ \text{mA/V}$$

Equation (13.51) gives $\omega_o = 941.86$ krad/s, and $f_o = \omega_o/2\pi = 149.9$ kHz.

EXAMPLE 13.7

Finding the oscillation frequency of a Colpitts BJT-tuned oscillator A Colpitts BJT oscillator is shown in Fig. 13.20. The circuit parameters are $r_\pi = 1.1\ \text{k}\Omega$, $h_{fe} = 100$, $L = 1.5$ mH, $C_1 = 1$ nF, $C_2 = 99$ nF, and $R_L = 10\ \text{k}\Omega$. R_2, R_3, and R_E set the DC-biasing circuit. L, C_1, and C_2 form a parallel resonant circuit. The *RF* choke offers very high impedance at the frequency of oscillation and acts as a constant-current source feeding the *LC* resonant circuit.

(a) Calculate the frequency of oscillation f_o.

(b) Check to make sure the condition for oscillation is satisfied.

(c) Calculate the value of R_2.

SOLUTION

Since the *RF* choke offers a very high impedance, it can be considered as a short circuit for small-signal analysis. Thus, the AC equivalent circuit is as shown in Fig. 13.21(a). Replacing the transistor by its transconductance model (voltage-controlled current source) gives the small-signal AC equivalent circuit shown in Fig. 13.21(b), which is similar to the circuit shown in Fig. 13.21(c). Thus, the analysis of the Colpitts op-amp oscillator in Sec. 13.4.1 is applicable in this case:

$$g_m = \frac{h_{fe}}{r_\pi} = \frac{100}{1.1\ \text{k}\Omega} = 90.91\ \text{mA/V}$$

(a) From Eq. (13.52), the frequency of oscillation is

$$f_o = \frac{1}{2\pi}\left[\frac{C_1 + C_2}{C_1 C_2 L}\right]^{1/2} = \frac{1}{2\pi}\left[\frac{1\ \text{nF} + 99\ \text{nF}}{1\ \text{nF} \times 99\ \text{nF} \times 1.5\ \text{mH}}\right]^{1/2} = 130.6\ \text{kHz}$$

(b) For $R_2 >> r_\pi$, $R_1 = r_\pi \| R_2 \approx r_\pi = h_{fe}/g_m$. Therefore, Eq. (13.53) becomes

$$g_m R_1 \approx g_m r_\pi = \frac{g_m h_{fe}}{g_m} = \frac{C_2}{C_1}$$

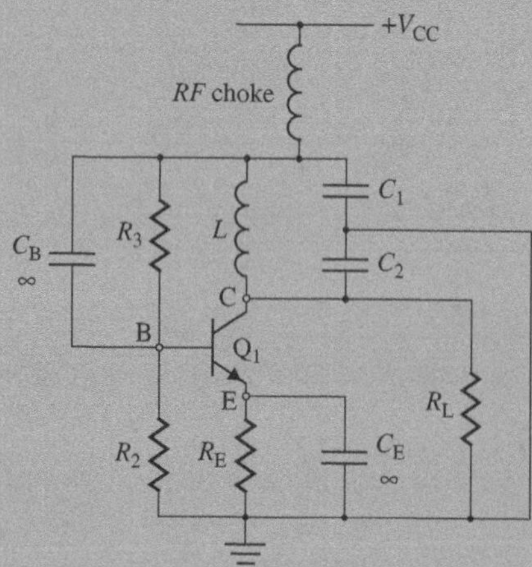

FIGURE 13.20 Colpitts BJT oscillator

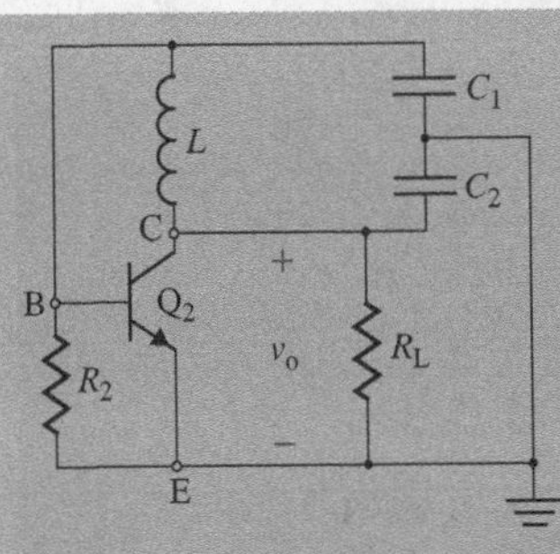

(a) AC equivalent circuit

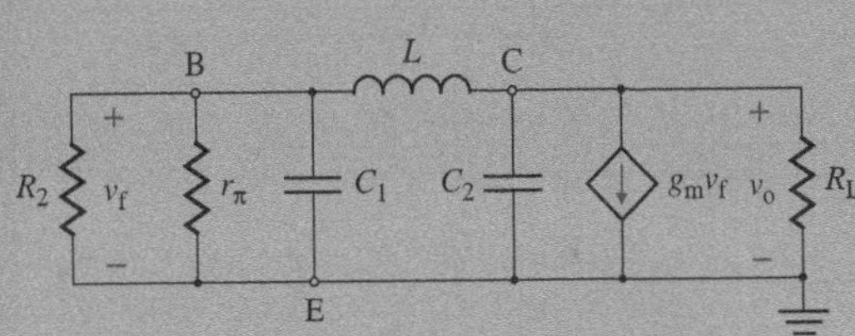

(b) Small-signal AC equivalent circuit

FIGURE 13.21 Equivalent circuits for Example 13.7

which gives

$$h_{fe} = \frac{C_2}{C_1} \tag{13.55}$$

$$= \frac{99}{1} = 99$$

This value is approximately equal to the value of transistor $h_{fe} = 100$. Thus, the condition for oscillation is satisfied.

(c) From Eq. (13.53),

$$R_1 = \frac{C_2}{C_1 g_m} = \frac{99}{g_m} = \frac{99 r_\pi}{h_{fe}} = 99 \times \frac{1.1\,\text{k}\Omega}{100} = 1089\ \Omega$$

Since $R_1 = r_\pi \| R_2 = 1089\ \Omega$,

$$\frac{r_\pi R_2}{r_\pi + R_2} = 1089$$

which gives $R_2 = 108.9\ \text{k}\Omega$.

EXAMPLE 13.8

Finding the oscillation frequency of a Colpitts BJT oscillator A Colpitts BJT oscillator is shown in Fig. 13.22. The circuit parameters are $r_\pi = 2.7\ \text{k}\Omega$, $h_{ie} = 150$, $L = 10\ \mu\text{H}$, $C_1 = 1$ nF, $C_2 = 100$ nF, and $R_L = 10\ \text{k}\Omega$. The BJT acts like a normal amplifier, and the LC resonant signal is fed back to the input base in order to sustain the oscillation. Therefore, the configuration is different from the oscillator circuit in Fig. 13.20.

(a) Find the approximate value of the frequency of oscillation f_o.

(b) Use PSpice/SPICE to plot the transient response of the output voltage $v_o(t)$ from 20 ms to 2 ms.

SOLUTION

(a) From Eq. (13.52), the frequency of oscillation is

$$f_o = \frac{1}{2\pi}\left[\frac{C_1 + C_2}{C_1 C_2 L}\right]^{1/2} = \frac{1}{2\pi}\left[\frac{1\ \text{nF} + 100\ \text{nF}}{1\ \text{nF} \times 100\ \text{nF} \times 10\ \mu\text{H}}\right]^{1/2} = 1.6\ \text{MHz}$$

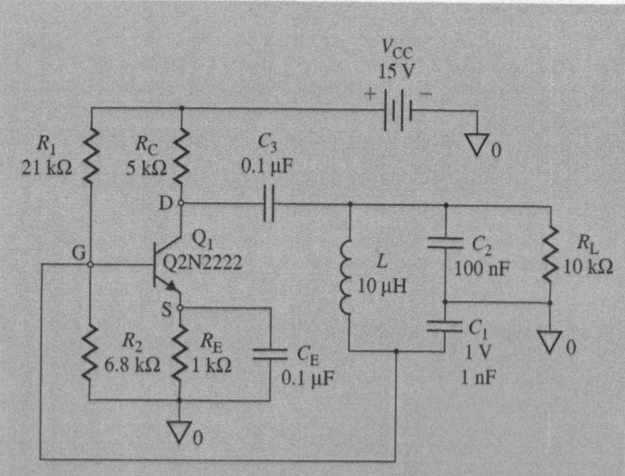

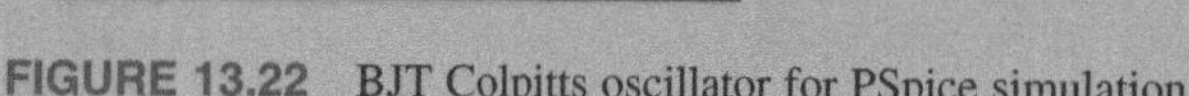

FIGURE 13.22 BJT Colpitts oscillator for PSpice simulation

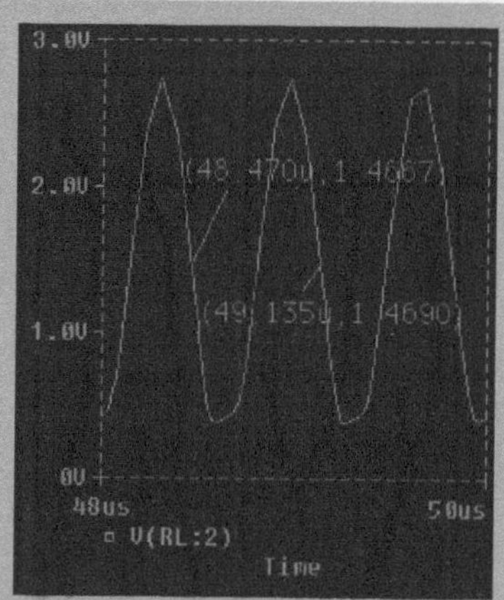

FIGURE 13.23 PSpice plot of output voltage for Example 13.8

(b) The PSpice plot of the output voltage $v_o \equiv$ V(RL:2) is shown in Fig. 13.23, which gives $f_o = 1/(49.149\ \mu s - 48.505\ \mu s) = 1.5$ MHz (expected value is 1.6 MHz).

NOTE: An initial voltage of 1 V has been assigned to the capacitor C_2 in order to start the oscillator, and the UIC (use initial condition) is used in transient analysis.

EXAMPLE 13.9

Finding the oscillation frequency of an *LC*-tuned MOSFET oscillator An *LC*-tuned MOSFET oscillator is shown in Fig. 13.24. Find the values of *L*, *C*, and *n* for an oscillation frequency of $f_o = 150$ kHz. The parameters of the MOSFET are $g_m = 5$ mA/V, $r_d = 25$ kΩ, and $R_G = 10$ kΩ.

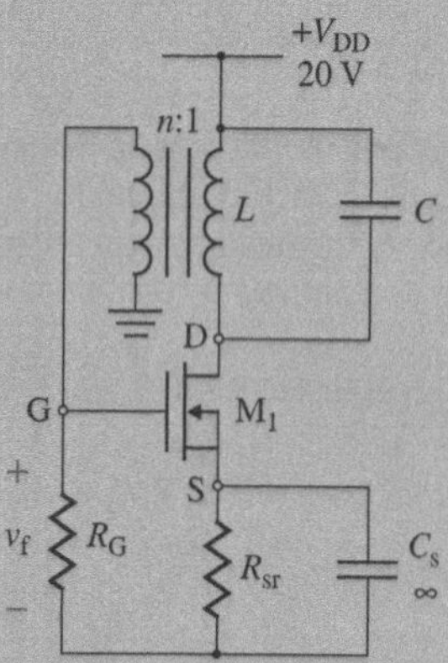

FIGURE 13.24 Colpitts MOSFET oscillator

SOLUTION

This oscillator uses the transformer to feed back the oscillating signal to the gate input, to sustain the oscillation. The small-signal AC equivalent circuit is shown in Fig. 13.25(a). Replacing the MOSFET by its transconductance model gives the small-signal equivalent circuit shown in Fig. 13.25(b), which can be simplified to Fig. 13.25(c). The transfer characteristic of v_o versus v_f in Laplace's domain can be written as

$$\frac{V_o(s)}{g_m V_f(s)} = Z(s) = \frac{1}{1/R_1 + sC + 1/sL}$$

which gives the voltage gain A as

$$A(s) = \frac{V_o(s)}{V_f(s)} = \frac{g_m}{(1/R_1) + sC + (1/sL)} \tag{13.56}$$

$$\text{where} \quad R_1 = r_d \| \frac{R_G}{n^2} \tag{13.57}$$

Substituting $s = j\omega$ into Eq. (13.56), we get

$$A(j\omega) = \frac{g_m}{(1/R_1) + j(\omega C - 1/\omega L)} \tag{13.58}$$

For oscillation, $A = |A(j\omega)| = 1\angle 0°$. Therefore, the imaginary part of the denominator must equal zero; that is,

$$j\left(\omega C - \frac{1}{\omega L}\right) = 0$$

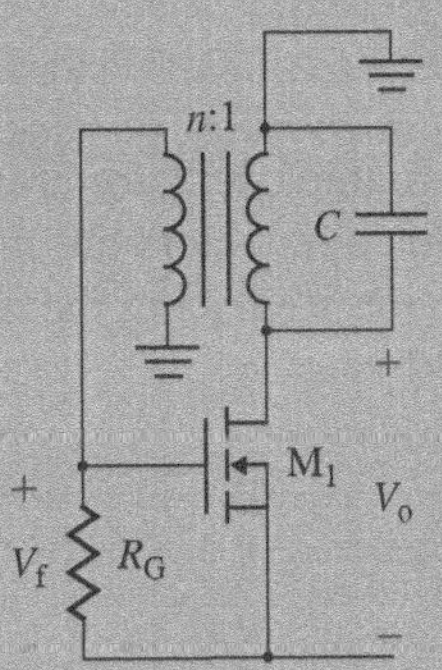

(a) AC equivalent circuit

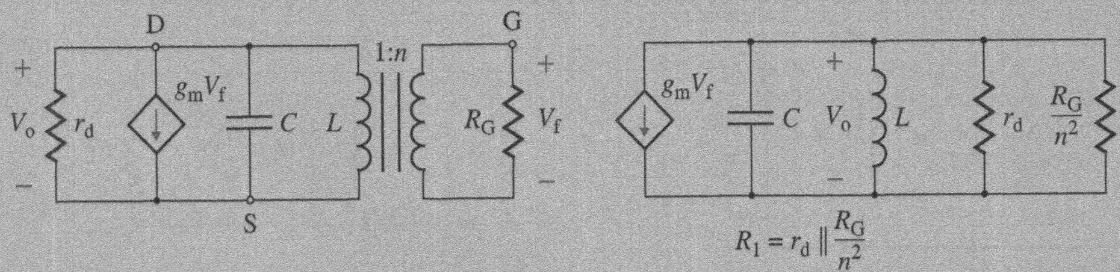

(b) Small-signal AC equivalent circuit

(c) Simplified equivalent circuit

FIGURE 13.25 Equivalent circuits for Example 13.9

which gives the oscillation frequency f_o as

$$f_o = \frac{\omega_o}{2\pi} = \frac{1}{2\pi\sqrt{LC}} \tag{13.59}$$

At this frequency, the gain must be unity; that is,

$$|A(j\omega)| = g_m R_1 = 1 \tag{13.60}$$

Step 1. Choose a suitable value of C: Let $C = 0.01$ μF.

Step 2. Calculate the value of L from Eq. (13.59):

$$L = \frac{1}{4\pi^2 C f_o^2} = \frac{1}{4\pi^2 \times 0.01\ \mu\text{F} \times (150\ \text{kHz})^2} = 112.6\ \mu\text{H}$$

Step 3. Find the value of R_1. From Eq. (13.60),

$$R_1 = \frac{1}{g_m} = \frac{1}{5\ \text{mA/V}} = \frac{1000}{5\ \text{A/V}} = 200\ \Omega$$

Step 4. Using Eq. (13.57), calculate the value of the turns ratio n: Since

$$200 = 25\ \text{k}\Omega \| \left(\frac{10\ \text{k}\Omega}{n^2}\right)$$

$n^2 = 49.6$ and $n = 7.04$. Thus, $n = 7$.

13.4.2 Hartley Oscillators

If the inductor and the capacitors of a Colpitts op-amp oscillator are interchanged, it becomes a Hartley op-amp oscillator, as shown in Fig. 13.26(a). Since inductors are more expensive than capacitors, this oscillator is less desirable than a Colpitts oscillator. Replacing the amplifier with its equivalent current source $g_m V_f$ reduces Fig. 13.26(a) to Fig. 13.26(b).

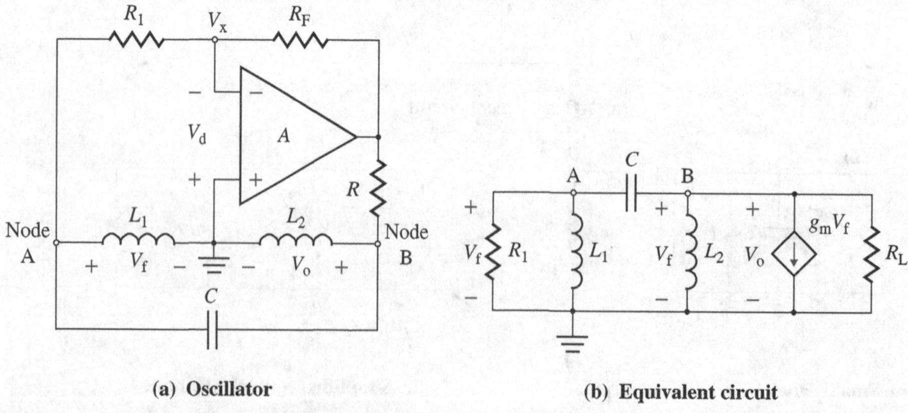

FIGURE 13.26 Hartley oscillator

Using nodal analysis in Fig. 13.26(b), we can write

$$\left[sC + \frac{1}{R_L} + \frac{1}{sL_2}\right]V_o(s) + (g_m - sC)V_f(s) = 0 \quad \text{(at node B)} \tag{13.61}$$

$$-sCV_o(s) + \left[sC + \frac{1}{sL_1} + \frac{1}{R_1}\right]V_f(s) = 0 \quad \text{(at node A)} \tag{13.62}$$

To find the condition for oscillation, we set the determinant to zero; that is,

$$\left(sC + \frac{1}{R_L} + \frac{1}{sL_2}\right)\left(sC + \frac{1}{sL_1} + \frac{1}{R_1}\right) + (g_m - sC)sC = 0$$

which, after simplification, yields

$$s^3CL_1L_2(R_1 + R_L + g_mR_1R_L) + s^2[CR_1R_L(L_1 + L_2) + L_1L_2] + s(L_1R_L + L_2R_1) + R_1R_L = 0 \tag{13.63}$$

where $g_m = A/R_L = R_F/(R_1R_L)$. Substituting $s = j\omega$ and equating the real parts of Eq. (13.63) to zero, we get

$$-\omega^2[CR_1R_L(L_1 + L_2) + L_1L_2] + R_1R_L = 0$$

which gives the frequency of oscillation ω_o as

$$\omega_o = \frac{1}{[C(L_1 + L_2) + L_1L_2/R_1R_L]^{1/2}} \quad \text{(in rad/s)} \tag{13.64}$$

For $C(L_1 + L_2) >> L_1L_2/R_1R_L$, Eq. (13.64) can be approximated by

$$f_o = \frac{1}{2\pi}\left[\frac{1}{C(L_1 + L_2)}\right]^{1/2} \quad \text{(in Hz)} \tag{13.65}$$

Equating the imaginary parts of Eq. (13.63) to zero, we get

$$-j\omega^3CL_1L_2(R_1 + R_L + g_mR_1R_L) + j\omega(L_1R_L + L_2R_1) = 0$$

Substituting $\omega = 2\pi f_o$ into Eq. (13.65), we get

$$\frac{1}{C(L_1 + L_2)} \times CL_1L_2(R_1 + R_L + g_mR_1R_L) = L_1R_L + L_2R_1$$

which, solved for g_mR_1, gives

$$g_mR_1 = \frac{L_1}{L_2} + \frac{R_1L_2}{R_LL_1} \tag{13.66}$$

For a large value of R_L, Eq. (13.66) gives the approximate value of g_m:

$$g_m \approx \frac{L_2}{R_1L_1} \tag{13.67}$$

Equation (13.67) gives the minimum value of g_m required to sustain the oscillation with a constant amplitude. To ensure oscillation, the value of g_m must exceed the minimum value. The capacitor or one of the inductors is usually made adjustable so that the frequency can be initially trimmed to the desired value.

EXAMPLE 13.10

D

Designing a Hartley oscillator

(a) Design the Hartley oscillator shown in Fig. 13.27 so that $f_o = 5$ MHz. Use a depletion NMOS whose parameters are $K_p = 250\ \mu A/V^2$ and $V_p = -2.83$ V. The load resistance is $R = 150\ \Omega$, and the power supply voltage is $V_{DD} = 15$ V. This oscillator uses the capacitor C to feed back the oscillating signal to the gate input, to sustain the oscillation.

(b) Use PSpice to plot the transient response of the output voltage $v_o(t)$ in part (a) from 10 μs to 10.5 μs. Assume $V_{DD} = 15$ V.

SOLUTION

The small-signal AC equivalent circuit is shown in Fig. 13.28, which is similar to Fig. 13.26(b):

$$g_m = 2K_p\,|V_p| = 2 \times 250\ \mu A/V^2 \times |-2.83| = 1.415\ mA/V$$

Step 1. Choose suitable values of L_1 and L_2: Let $L_1 = L_2 = 10\ \mu H$.

Step 2. Calculate the value of C from Eq. (13.65):

$$C = \frac{1}{4\pi^2 f_o^2 (L_1 + L_2)} = \frac{1}{4\pi^2 \times (5\ \text{MHz})^2 (10\ \mu H + 10\ \mu H)} = 50.66\ \text{pF}$$

Step 3. Find the value of effective load resistance R_L. The Q-point of the circuit is $V_{DSQ} = V_{DD} = 15$ V, and $V_{GSQ} = 0$. The value of g_m varies from 3 mA/V to 6.5 mA/V. To ensure oscillation, we choose $g_m = 3$ mA/V. Since R_1 is infinity in Fig. 13.28, Eq. (13.66) can be reduced to

$$g_m = \frac{L_1}{R_1 L_2} + \frac{L_2}{R_L L_1} = \frac{L_2}{R_L L_1}$$

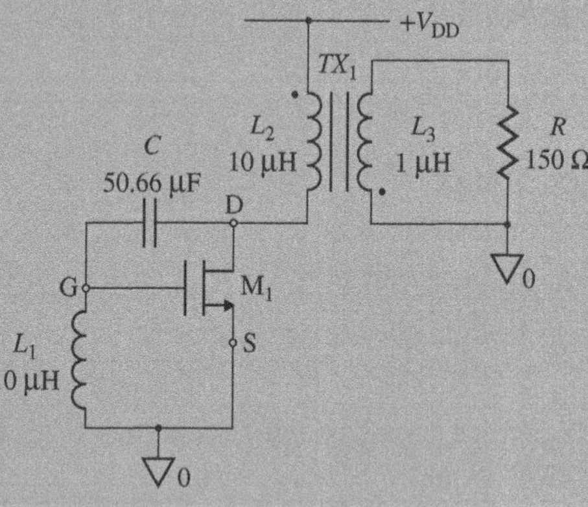

FIGURE 13.27 Hartley JFET oscillator

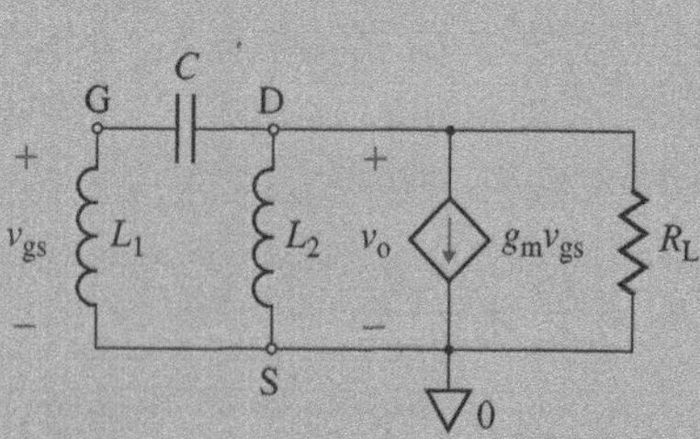

FIGURE 13.28 Simplified equivalent oscillator for Example 13.10

which gives the value of effective load resistance R_L as

$$R_L = \frac{L_2}{g_m L_1} = \frac{10\ \mu\text{H}}{1.415\ \text{mA/V} \times 10\ \mu\text{H}} = 707\ \Omega$$

This is the lowest value of R_L at which sustained oscillation can occur for $g_m = 1.415$ mA/V. Therefore, we select a higher value—say, $R_L = 1.5$ kΩ.

Step 4. Calculate the value of turns ratio n, which is related to the load resistance R and the effective load resistance R_L, by

$$n = \sqrt{\frac{R}{R_L}} = \sqrt{\frac{100\ \Omega}{1.5\ \text{k}\Omega}} = 0.316$$

Step 5. Calculate the inductance of the transformer secondary L_3, which is related to L_2, by

$$L_3 = n^2 L_2 = 0.316^2 \times 10\ \mu\text{H} = 1\ \mu\text{H}$$

(b) The PSpice schematic with the desired values is shown in Fig. 13.29. The PSpice plot of the output voltage $v_o \equiv$ V(R:2) is shown in Fig. 13.30, which gives the peak-to-peak voltage of $V_{pp} = +3.75$ V to -4.0 V at $f_o = 1/(10.326\ \mu\text{s} - 10.124\ \mu\text{s}) = 4.95$ MHz (expected 5 MHz). Note that an initial voltage of 2 V has been assigned to the capacitor C in order to start the oscillator, and UIC (use initial condition) is used in transient analysis. In practice, random noise or transients can cause the oscillations to begin, and they are sustained by the feedback of the appropriate signal.

NOTE: Increasing the value of R will increase the output voltage and the readers are encouraged to simulate the circuit with different values of R. * symbols in Figs. 13.27 and 13.29 are the dot symbols of the transformer and refer to the direction of the positive induced voltages in the windings.

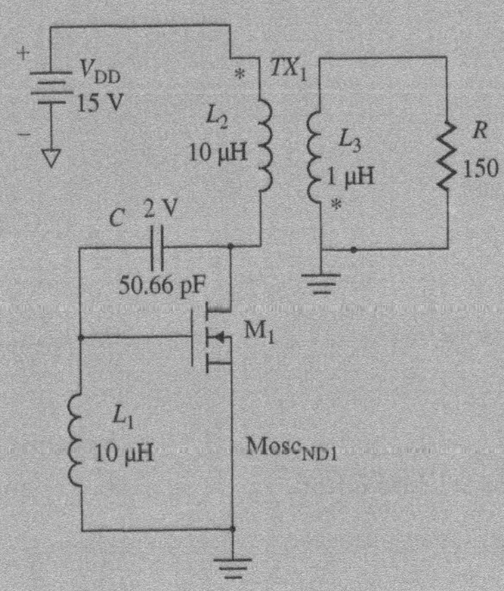

FIGURE 13.29 PSpice schematic for Example 13.10

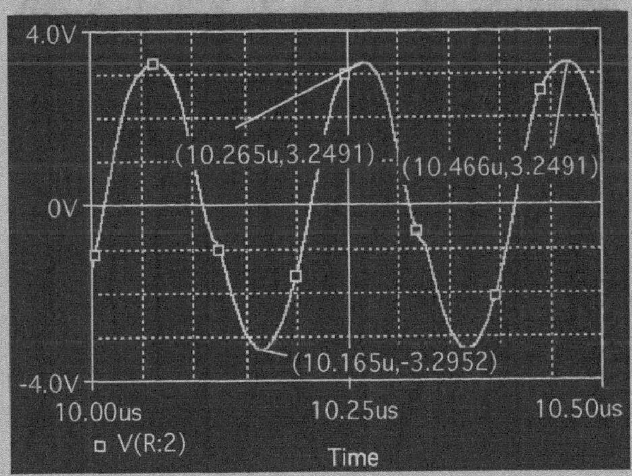

FIGURE 13.30 Output voltage waveform for Example 13.10

13.4.3 Two-Stage MOS Oscillators

In the analyses of the Colpitts and Hartley oscillators, we assumed that inductors are lossless, with no resistances. In practice, inductors suffer from a resistive component that can be modeled as a parallel resistance with an ideal inductor. The parallel equivalent makes the analysis simpler. The series impedance can

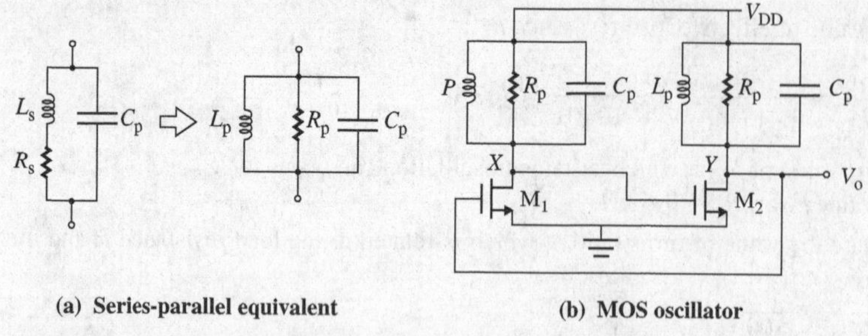

FIGURE 13.31 Two-stage MOS oscillator

be transformed to an equivalent parallel topology as shown in Fig. 13.31(a). A two-stage MOS oscillator [3] is shown in Fig. 13.31(b). The drain of one transistor is connected to the gate of the other transistor, and it is also called the cross-coupled oscillator. The frequency-dependent phase around the loop is zero (0°) because each stage contributes to a zero frequency-dependent phase shift at resonance. The equivalent R_p and L_p can be determined by equating their impedances.

$$j\omega L_s + R_s = \frac{R_p(j\omega L_p)}{R_p + j\omega L_p} \tag{13.68}$$

or
$$(L_sR_p + L_pR_s)j\omega L_s + R_sR_p - \omega^2 L_sL_p = j\omega L_sR_p \tag{13.69}$$

Equating the real and imaginary parts on both sides, we get

$$R_sR_p - \omega^2 L_sL_p = 0 \quad \text{and} \quad L_sR_p + L_pR_s = L_sR_p \tag{13.70}$$

Solving for R_p and L_p, we get

$$L_p = L_s\left(1 + \frac{R_s^2}{L_s^2\omega^2}\right) = L_s\left(1 + \frac{1}{Q^2}\right) \tag{13.71}$$

$$R_p = \frac{\omega^2 L_sL_p}{R_s} \tag{13.72}$$

where $Q = \omega L_s/R_s$ is the quality factor of an inductor. For $Q > 3$, $L_p \simeq L_s$ and $R_p \simeq \omega^2 L_s^2/R_s$. Assuming identical transistors, their transconductances and the output resistances will be equal, $g_m = g_{m1} = g_{m2}$ and $r_o = r_{o1} = r_{o2}$. The open-loop voltage gain is

$$A_o = -g_{m1}(R_p \| r_{o1}) \times -g_{m1}(R_p \| r_{o1}) = g_m^2(R_p \| r_o)^2 \tag{13.73}$$

To satisfy the oscillation condition, it requires a gain of 1 only—that is, $A_o \geq 1$—which is an advantage of this oscillator. The oscillation frequency f_o is given by

$$f_o = \frac{1}{2\pi\sqrt{C_pL_p}} \tag{13.74}$$

The MOS oscillator in Fig. 13.31(b) can be made to operate as a Colpitts oscillator by applying a voltage feedback from the drain to the source terminal of the MOSFET, as shown in Fig.13.32(a). Capacitors C_1

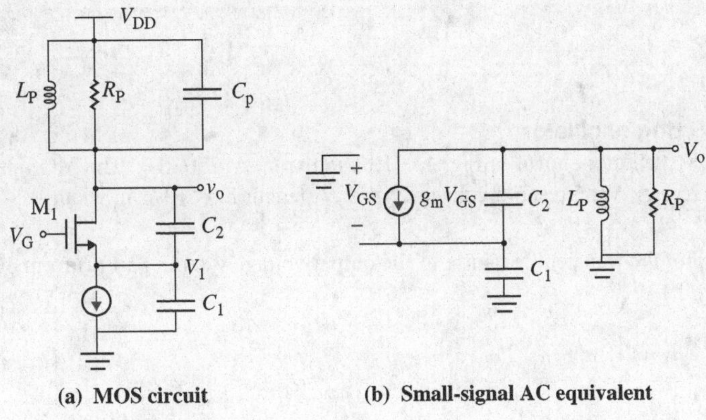

FIGURE 13.32 MOS Colpitts oscillator

and C_2 form a potential divide. The equivalent capacitance C_F offered by C_1 and C_2 is in parallel with C_p. To simplify the analysis, we will omit C_p to derive the oscillation conditions, and the small-signal AC equivalent circuit is shown in Fig. 13.32(b). Using the equivalent circuit in Fig. 13.19(c), we can express the equivalent impedance of M_1, C_1, and C_2 as

$$Z_M = \frac{1}{sC_1} + \frac{1}{sC_2} + \frac{1}{sC_1}\frac{1}{sC_2} g_m = \frac{s(C_1 + C_2) + g_m}{s^2 C_1 C_2} \tag{13.75}$$

Therefore, the equivalent impedance of Z_M, L_p, and R_p is given by

$$\frac{1}{Z_T} = \frac{1}{Z_M} + \frac{1}{sL_p} + \frac{1}{R_p} = \frac{s^2 C_1 C_2}{s(C_1 + C_2) + g_m} + \frac{1}{sL_p} + \frac{1}{R_p}$$

$$= \frac{s^3 C_1 C_2 L_p R_p + s^2 (C_1 + C_2) L_p + s(C_1 + C_2) R_p + s g_m L_p + g_m R_p}{[s(C_1 + C_2) + g_m] s L_p R_p} \tag{13.76}$$

For resonant oscillation, $Z_T(j\omega) = 0$, which gives the following expression:

$$(j\omega)^3 C_1 C_2 L_p R_p + (j\omega)^2 (C_1 + C_2) L_p + j\omega (C_1 + C_2) R_p + j\omega g_m L_p + g_m R_p = 0$$

Equating the imaginary part gives the resonant frequency as

$$\omega = \sqrt{\frac{g_m L_p + (C_1 + C_2) R_p}{C_1 C_2 L_p R_p}} \simeq \sqrt{\frac{(C_1 + C_2)}{C_1 C_2 L_p}}$$

Equating the real part gives the gain as

$$g_m R_p = \omega^2 (C_1 + C_2) L_p$$

which, for $\omega^2 = C_1 + C_2 / C_1 C_2 L_p$, becomes

$$g_m R_p = \frac{(C_1 + C_2)^2}{C_1 C_2} \tag{13.77}$$

EXAMPLE 13.11

D

Designing a MOS ring oscillator

(a) Design the MOS oscillator shown in Fig. 13.31(b) so that $f_o = 250$ kHz. The MOS parameters are MOS constant $K_p = 25\ \mu A/V^2$, threshold voltage $V_t = 1$ V, length $L = 10\ \mu m$, width $W = 10\ \mu m$, and modulation $\lambda = 0.01$.

(b) Use PSpice to plot the transient response of the output voltage $v_o(t)$ in part (a) from 4000 μs to 4020 μs. Assume $V_{CC} = 12$ V.

SOLUTION

(a) The DC-biasing analysis is similar to the MOS ring oscillator in Fig. 13.16. Choosing a voltage gain of $A_o = 2$, Eq. (13.47) gives the DC biasing gate–source voltage

$$v_{GS} = \frac{5 + (2/2) \times 1}{1 + (2/2)} = 6.5 \text{ V}$$

$$i_D = K_p(v_{GS1} - V_t)^2 = 25\ \mu A/V^2 \times (6.5 - 1)^2 = 765.2\ \mu A$$

$$g_m = 2K_p(v_{GS} - V_t) = 2 \times 25\ \mu A/V^2 \times (6.5 - 1) = 275\ \mu A/V$$

$$R_p = \frac{\sqrt{A_o}}{g_m} = \frac{\sqrt{2}}{275\ \mu A/V} = 5.1 \text{ k}\Omega$$

Let $C_p = 0.01\ \mu F$. For $f_o = 250$ kHz, Eq. (13.74) gives

$$L_p = \frac{1}{(2\pi f_o)^2 C_p} = \frac{1}{(2\pi \times 250 \text{ kHz})^2 \times 0.01\ \mu F} = 40\ \mu H$$

(b) The PSpice schematic with the desired values is shown in Fig. 13.33. The PSpice plot of the output voltage $v_o \equiv$ V(M2:d) is shown in Fig. 13.34, which gives the peak-to-peak voltage of $V_{pp} = 23{,}635$ V at $f_o = 1/(4.0049 \text{ ms} - 4.0009 \text{ ms}) = 250$ kHz (expected 250 kHz). Note that an initial voltage of 1.5 V has been assigned to the capacitor C_{p1} in order to start the oscillator, and UIC (use initial condition) is used in transient analysis. In practice, random noise or transients can cause the oscillations to begin, and they are sustained by the feedback of the appropriate signal.

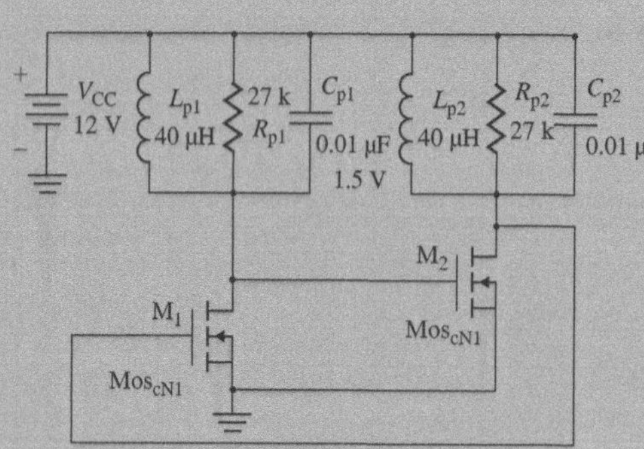

FIGURE 13.33 PSpice schematic for Example 13.11

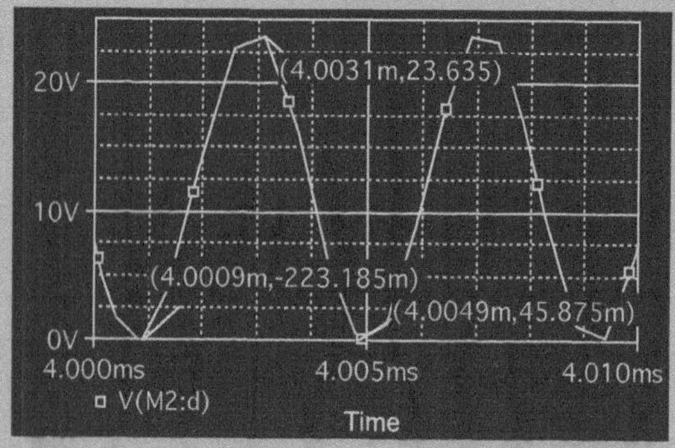

FIGURE 13.34 Output voltage waveform for Example 13.11

KEY POINTS OF SECTION 13.4

- A Colpitts oscillator uses an inverting amplifier and a phase-shifting network consisting of two capacitors and one inductor. The oscillation frequency ω_o is inversely proportional to the LC product.
- A Hartley oscillator is a derivation of a Colpitts oscillator. The phase-shifting network consists of two inductors and a capacitor. The oscillation frequency ω_o is inversely proportional to the LC product.
- In a cross-coupled oscillator, the drain of a MOS transistor is connected to the gate of another transistor. The frequency-dependent phase around the loop is zero (0°) because each stage contributes to zero frequency-dependent phase shift at resonance. A MOS oscillator can be operated as a Colpitts oscillator by applying a voltage feedback from the drain to the source terminal.

13.5 Crystal Oscillators

Because of their excellent frequency stability, quartz crystals are commonly used to control the frequency of oscillation. If the inductor L of the Colpitts oscillator in Fig. 13.19(a) is changed to a crystal, the oscillator is called a *crystal oscillator*. Crystal oscillators are commonly used in digital signal processing. The symbol for a vibrating piezoelectric crystal is shown in Fig. 13.35(a); its circuit model is shown in Fig. 13.35(b), which can be simplified to Fig. 13.35(c). The *quality factor Q* of a crystal can be as high as several hundred thousands. C_p represents the electrostatic capacitance between the two parallel plates of the crystal. L has a large value (as high as hundreds of henries) and is determined from $L \approx 1/C_s\omega_o^2$, where ω_o is the resonant frequency of the crystal. R_s can be as high as a few hundred thousand ohms and is determined from $R_s \approx \omega_o L/Q$, where the quality factor Q is in the range of 10^4 to 10^6. Typical values for a 2-MHz quartz crystal are $Q = 80 \times 10^3$, $C_p/C_s = 350$, $L = 520$ mH, $C_s = 0.0122$ pF, and $R_s = 82\ \Omega$. Table 13.1 shows typical component values for common cuts of quartz oscillator crystals.

Since Q is very high in the typical quartz crystal, we may neglect R_s. The crystal impedance is given by

$$Z(s) = \frac{1}{sC_p + sL + 1/sC_s} = \frac{1}{sC_p}\left[\frac{s^2 + 1/LC_s}{s^2 + (C_p + C_s)/(LC_sC_p)}\right]$$

$$= \frac{1}{sC_p}\left[\frac{s^2 + \omega_s^2}{s^2 + \omega_p^2}\right] \tag{13.78}$$

TABLE 13.1 Common cuts of quartz oscillator crystals (RCA Corp.)

Frequency	32 kHz	280 kHz	525 kHz	2 MHz	10 MHz
Cut	XY bar	DT	DT	AT	AT
R_s	40 kΩ	1820 Ω	1400 Ω	82 Ω	5 Ω
L	4800 H	25.9 H	12.7 H	0.52 H	12 mH
C_s, in pF	0.00491	0.0126	0.00724	0.0122	0.0145
C_p, in pF	2.85	5.62	3.44	4.27	4.35
C_p/C_s	580	450	475	350	300
Q	25,000	25,000	30,000	80,000	150,000

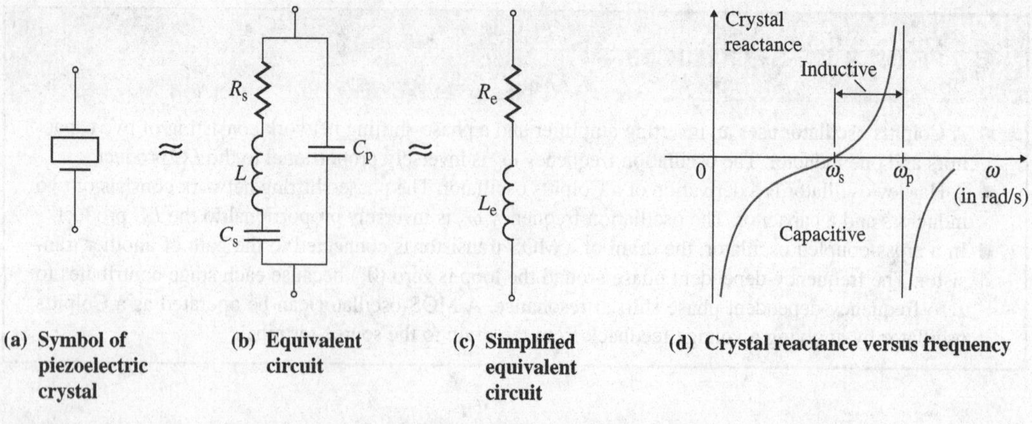

FIGURE 13.35 Symbol and circuit model of piezoelectric crystal

If we substitute $s = j\omega$, the impedance in Eq. (13.78) becomes

$$Z(j\omega) = -\left(\frac{j}{\omega C_p}\right)\left(\frac{\omega^2 - \omega_s^2}{\omega^2 - \omega_p^2}\right) \tag{13.79}$$

Therefore, the crystal exhibits two resonant frequencies: series resonance at

$$\omega_s = \frac{1}{\sqrt{LC_s}} \tag{13.80}$$

and parallel resonance at

$$\omega_p = \left[\frac{C_s + C_p}{C_s C_p L}\right]^{1/2} \tag{13.81}$$

Note that $\omega_p > \omega_s$. However, since $C_p >> C_s$, the two resonance frequencies are very close. The plot of crystal reactance against frequency in Fig. 13.35(d) illustrates that the crystal exhibits the characteristic of an inductor over the narrow frequency range between ω_s and ω_p.

It is possible to have a variety of crystal oscillators. A Colpitts-derived op-amp crystal oscillator is shown in Fig. 13.36(a); its equivalent circuit is shown in Fig. 13.36(b). This circuit should oscillate at the resonance frequency of the crystal inductance L with the series equivalent of C_s and $C_p + C_1C_2/(C_1 + C_2)$. Since C_s is much smaller than C_p, C_1, or C_2, it will be dominant and the oscillating frequency can be approximately found from

$$\omega_o \approx \frac{1}{\sqrt{LC_s}} \tag{13.82}$$

Using nodal analysis in Fig. 13.36(b), we can write

$$\left[sC_2 + \frac{1}{R_e + sL_e} + \frac{1}{R_L}\right]V_o(s) + \left[g_m - \frac{1}{R_e + sL_e}\right]V_f(s) = 0 \tag{13.83}$$

$$-\frac{1}{R_e + sL_e}V_o(s) + \left[sC_1 + \frac{1}{R_e + sL_e} + \frac{1}{R_1}\right]V_f(s) = 0 \tag{13.84}$$

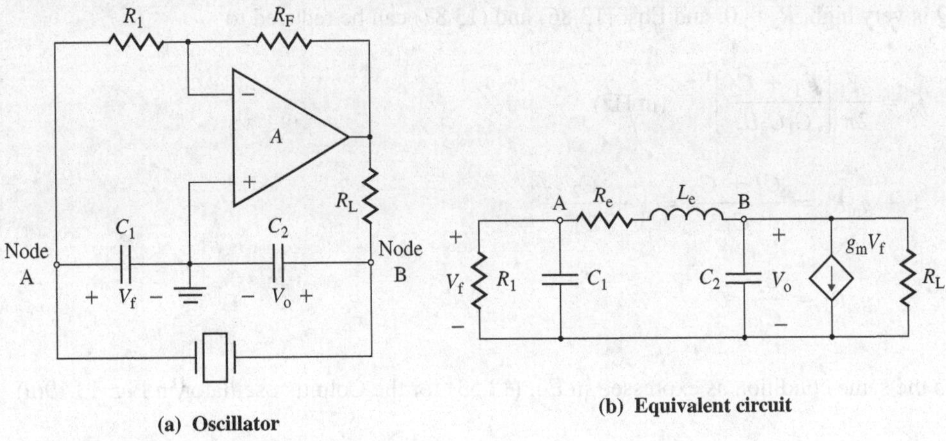

FIGURE 13.36 Crystal oscillator

Assuming R_L is large, tending to infinity, we set the determinant to zero to find the condition for oscillation; that is,

$$\left(sC_2 + \frac{1}{R_e + sL_e}\right)\left(sC_1 + \frac{1}{R_e + sL_e} + \frac{1}{R_1}\right) + \left(g_m - \frac{1}{R_e + sL_e}\right)\frac{1}{R_e + sL_e} = 0$$

which, after simplification, yields

$$s^3C_1C_2L_eR_1 + s^2(C_1C_2R_eR_1 + C_2L_e) + s(C_1R_1 + C_2R_1 + C_2R_e) + 1 + g_mR_1 = 0 \tag{13.85}$$

where $g_m = A/R_L = R_F/(R_1R_L)$. Substituting $s = j\omega$ and equating the imaginary parts of Eq. (13.85) to zero, we get

$$-j\omega^3(C_1C_2L_eR_1) + j\omega(C_1R_1 + C_2R_1 + C_2R_e) = 0$$

which gives the frequency of oscillation f_o as

$$f_o = \frac{1}{2\pi}\left[\frac{C_1R_1 + C_2R_1 + C_2R_e}{C_1C_2L_eR_1}\right]^{1/2} \quad \text{(in Hz)} \tag{13.86}$$

Similarly, equating the real parts of Eq. (13.85) to zero, we get

$$-\omega^2(C_1C_2R_eR_1 + C_2L_e) + 1 + g_mR_1 = 0$$

which gives

$$1 + g_mR_1 = \omega^2(C_1C_2R_eR_1 + C_2L_e)$$

After substitution of the value of $\omega = \omega_o = 2\pi f_o$ from Eq. (13.86), the above equation becomes

$$1 + g_mR_1 = \frac{(C_1R_1 + C_2R_1 + C_2R_e)(C_1C_2R_eR_1 + C_2L_e)}{C_1C_2L_eR_1} \tag{13.87}$$

Since Q is very high, $R_e \approx 0$, and Eqs. (13.86) and (13.87) can be reduced to

$$f_o = \frac{1}{2\pi}\left[\frac{C_1 + C_2}{C_1 C_2 L_e}\right]^{1/2} \quad \text{(in Hz)} \tag{13.88}$$

and
$$1 + g_m R_1 = \frac{C_1 + C_2}{C_1} = 1 + \frac{C_2}{C_1}$$

or
$$g_m R_1 = \frac{C_2}{C_1} \tag{13.89}$$

which is the same condition as expressed in Eq. (13.53) for the Colpitts oscillator in Fig. 13.19(a).

EXAMPLE 13.12

Finding the oscillation frequency of a crystal oscillator The op-amp oscillator in Fig. 13.36(a) uses a 2-MHz crystal and has $C_1 = 0.01\ \mu\text{F}$, $C_2 = 0.1\ \mu\text{F}$, $R_L = 100\ \text{k}\Omega$, $R_1 = 100\ \text{k}\Omega$, and $R_F = 1\ \text{M}\Omega$.
(a) Find the frequency of oscillation f_o.
(b) Use PSpice/SPICE to verify the frequency in part (a). Assume $V_{CC} = V_{EE} = 15$ V.

SOLUTION

(a) For a 2-MHz crystal, $C_s = 0.0122$ pF, $C_p = 4.27$ pF, $R_s = 82\ \Omega$, and $L = 0.52$ H. Let

$$C_{eqp} = \frac{C_p + C_1 C_2}{C_1 + C_2}$$

$$= \frac{4.27\ \text{pF} + 0.01\ \mu\text{F} \times 0.1\ \mu\text{F}}{0.01\ \mu\text{F} + 0.1\ \mu\text{F}} = 9095\ \text{pF}$$

The effective capacitance C_{eq} is given by C_s in series with $C_{eqp} = C_p + C_1C_2/(C_1 + C_2)$:

$$C_{eq} = \frac{C_s \times C_{eqp}}{C_s + C_{eqp}}$$

$$= \frac{0.0122\ \text{pF} \times 9095\ \text{pF}}{0.0122\ \text{pF} + 9095\ \text{pF}} \approx 0.0122\ \text{pF}$$

Thus, the frequency of oscillation f_o becomes

$$f_o = \frac{1}{2\pi}\left[\frac{1}{C_{eq}L}\right]^{1/2} = \frac{10^6}{2\pi\sqrt{0.0122 \times 0.52}} = 1.998\ \text{MHz}$$

(b) The circuit for PSpice simulation is shown in Fig. 13.37. The PSpice plot of the output voltage across C_s [i.e., $v_o \equiv$ V(Cs:2)] is shown in Fig. 13.38, which gives $f_o = 1/(763.146\ \text{n} - 243.137\ \text{n}) = 1.923$ MHz, close to the calculated frequency of 1.998 MHz. The amplitude of the output voltage is stable, not falling.

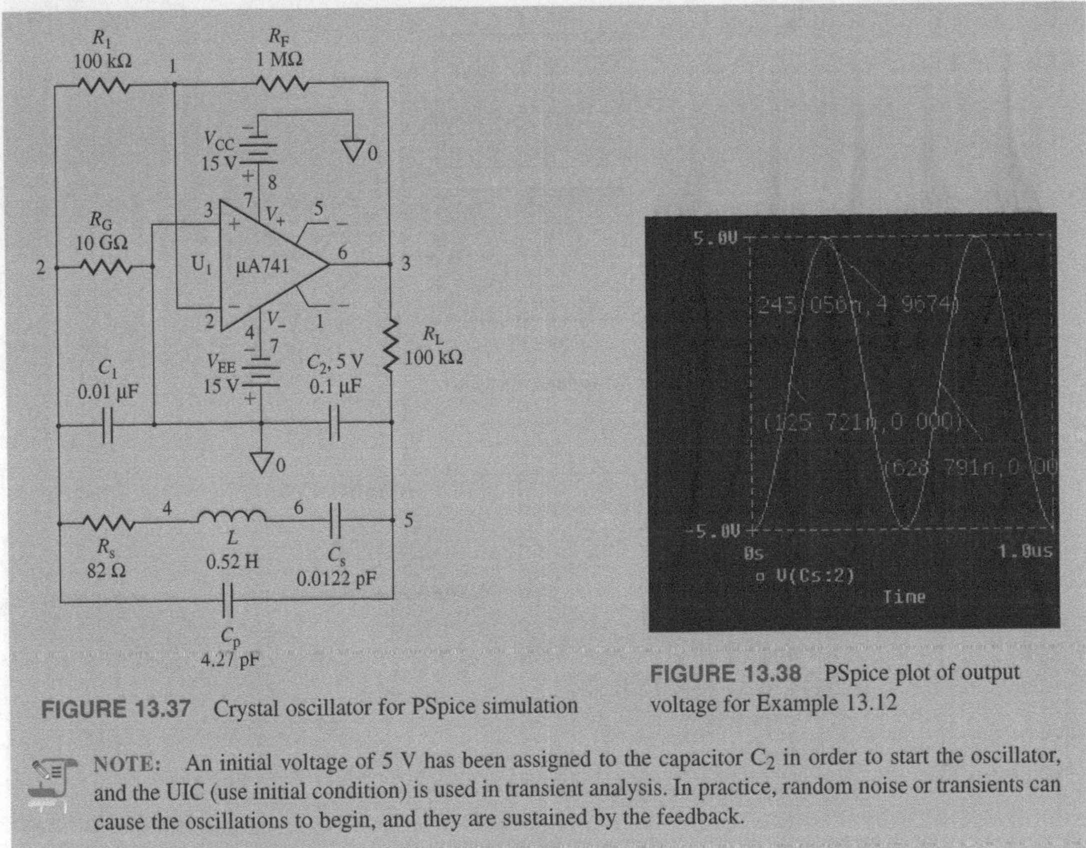

FIGURE 13.37 Crystal oscillator for PSpice simulation

FIGURE 13.38 PSpice plot of output voltage for Example 13.12

NOTE: An initial voltage of 5 V has been assigned to the capacitor C_2 in order to start the oscillator, and the UIC (use initial condition) is used in transient analysis. In practice, random noise or transients can cause the oscillations to begin, and they are sustained by the feedback.

KEY POINT OF SECTION 13.5

- A crystal oscillator has the same circuit topology as a Colpitts oscillator except that it uses a crystal instead of an inductor. A crystal has strong frequency stability.

13.6 Active-Filter Tuned Oscillators

An active band-pass filter with a high Q-value can be operated as an oscillator provided that a positive feedback is applied. This type of oscillator, which consists of a narrow-band filter and a limiter, is illustrated in Fig. 13.39(a). To understand the operation of the circuit, let us assume that the oscillation has already started. The output of the filter v_o is a sine wave whose frequency is the center frequency of the filter f_o. This sine wave is fed to a limiter, which produces a square-wave output v_f of frequency f_o. The peak amplitude of the square wave is determined by the type of limiting devices. The square wave is in turn fed back to the band-pass filter, which filters the harmonics and produces a sinusoidal output v_o at the fundamental frequency f_o. The quality of the sine wave is a direct function of the selectivity (Q-factor) of the band-pass filter. The design of this type of oscillator is very simple, and the oscillator has independent frequency control.

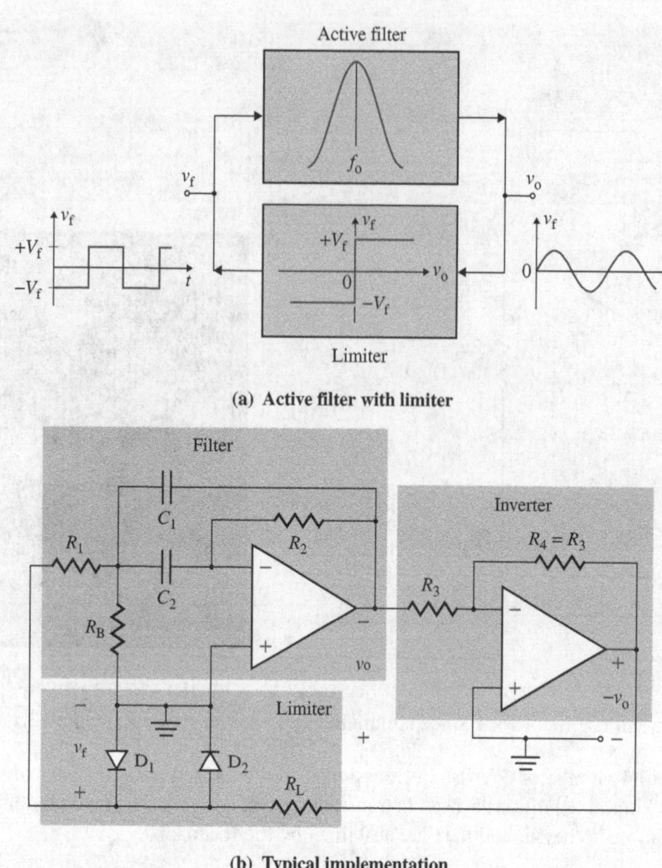

(a) Active filter with limiter

(b) Typical implementation

FIGURE 13.39 Active-filter tuned oscillator

A typical practical implementation of an active-filter tuned oscillator using a narrow-band filter is shown in Fig. 13.39(b). An inverter is added to the output of the filter in order to provide positive feedback. A simple diode limiter along with resistance R_L is used to generate a square wave, which is fed back to the input of the filter.

EXAMPLE 13.13

PSpice simulation of an active-filter tuned oscillator Modify the narrow-band filter that was designed in Example 12.10 so that it operates as an oscillator, and use PSpice/SPICE to plot the output voltage.

SOLUTION

The circuit for PSpice simulation is shown in Fig. 13.40. Instead of an inverter, a voltage-controlled voltage source with -1 gain is used to provide the positive feedback required for oscillation.

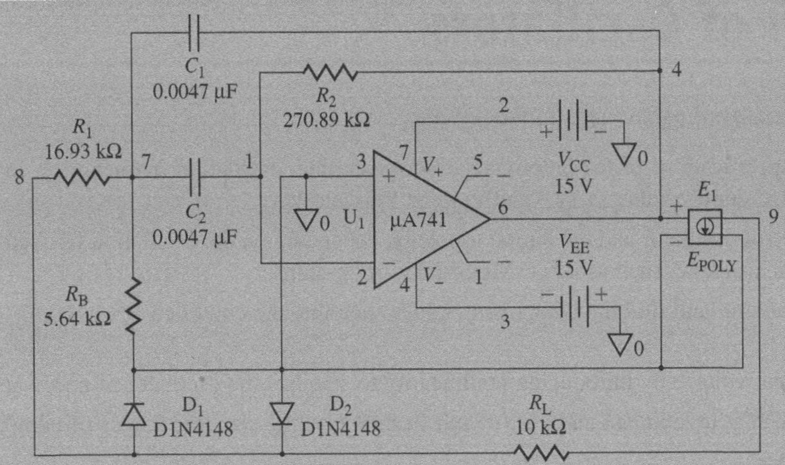

FIGURE 13.40 Active-filter tuned oscillator for PSpice simulation

The PSpice plots of the output voltage $v_O \equiv$ V(E1:3) at the output of the op-amp and the voltage across the diodes $v_{D1} \equiv$ V(RL:1) are shown in Fig. 13.41. We get the oscillation frequency $f_o = 1/(5.913 \text{ m} - 4.8548 \text{ m}) = 962$ Hz (expected value is 1 kHz), the peak amplitude of the output voltage is ± 5.17 V, and the peak amplitude of the diode limiter is ± 557 mV.

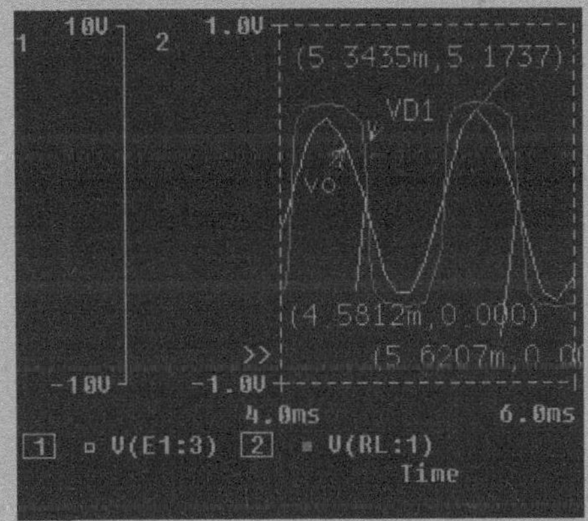

FIGURE 13.41 PSpice plot of output voltage for Example 13.13

KEY POINT OF SECTION 13.6

- An active-filter tuned oscillator uses a narrow-band filter with positive feedback. A limiter is used in the feedback path to provide square-wave input signals to the filter. This type of oscillator has the advantages of independent frequency and amplitude control.

13.7 Design of Oscillators

The design of a sinusoidal oscillator involves the following steps:

Step 1. Identify the specifications of the output stage—for example, oscillation frequency f_o, load resistance R_L, and the DC supply voltages V_{CC} and V_{EE} (or V_{DD} and V_{SS}).

Step 2. Select the type of oscillator and the circuit topology, depending on the oscillation frequency and the types of devices available, such as BJTs, MOSFETs, or op-amps.

Step 3. Analyze the circuit, and find the component values such that the condition $A\beta = 1 \angle 0°$ or $\angle 360°$ is satisfied.

Step 4. Limit the output voltage by introducing nonlinearity to stabilize the oscillator, if necessary.

Step 5. Use PSpice/SPICE to simulate and verify your design. Use the standard values of components with their tolerances.

EXAMPLE 13.14

Worst-case analysis of the phase-shift oscillator in Example 13.3 Use PSpice/SPICE to find worst-case output voltage and frequency ranges of the phase-shift oscillator in Example 13.3. Use standard component values: $C = 0.1\ \mu F \pm 10\%$, $R = 1.6\ k\Omega \pm 5\%$, $R_1 = 17\ k\Omega \pm 5\%$, and $R_F = 490\ k\Omega \pm 5\%$.

SOLUTION

To assign tolerances to resistors and capacitors, we will use model RMOD for resistors and CMOD for capacitors. Also, we will add a statement for the worst-case analysis (.WCASE) [2].

The PSpice plots of the worst-case maximum and nominal output voltages $v_o \equiv$ V(3) are shown in Fig. 13.42(a), which give the worst-case peak values of 8.7 V and −9.1 V and an oscillation frequency of 1/2.322 ms = 430 Hz (nominal value is 422 Hz). The plots of the worst-case minimum and nominal output voltages $v_o \equiv$ V(3) are shown in Fig. 13.42(b), which give the worst-case peak values of 8.05 V and −9.45 V and an oscillation frequency of 1/2.246 ms = 445 Hz (nominal value is 422 Hz). Thus, the output frequency can vary from 430 Hz to 445 Hz based on component tolerances.

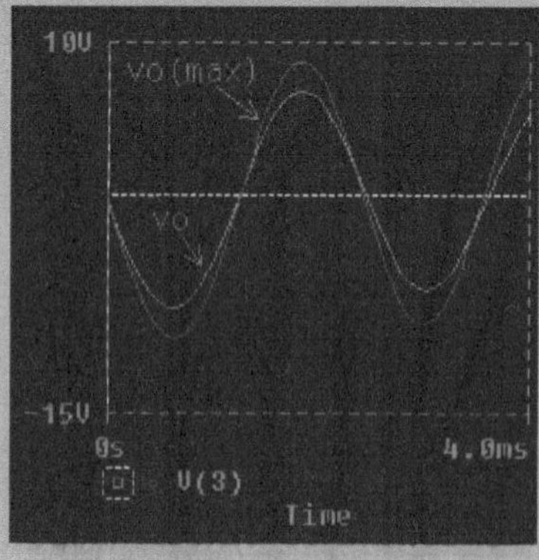

(a) Maximum and nominal values

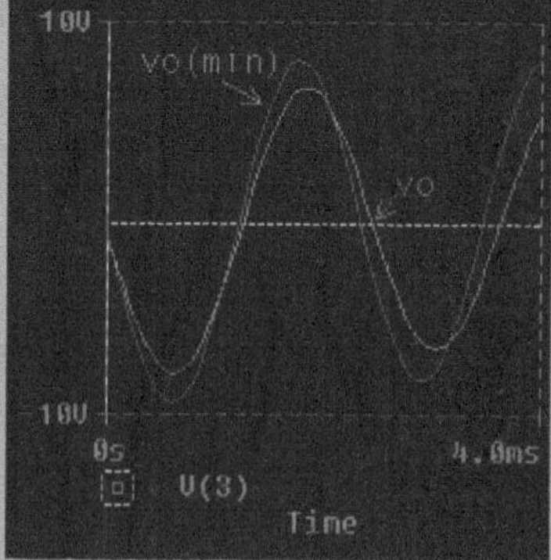

(b) Minimum and nominal values

FIGURE 13.42 Worst-case output of the phase-shift oscillator for Example 13.14

Summary

Oscillators use positive feedback and are commonly employed in electronics circuits. There are many types of oscillators. However, *RC*, *LC*, and crystal oscillators are the most commonly used. For a feedback circuit to operate as an oscillator, the magnitude and the phase shift of the loop gain must be unity and 0° (or 360°), respectively. Frequency stability is an important criterion in defining the quality of an oscillator. Crystal oscillators have the highest frequency stability. The frequency of oscillation and the conditions for oscillation can be determined from the transfer function or the determinant of a circuit.

References

1. R. A. Gayakwad, *Op-Amps and Linear Integrated Circuits*. Englewood Cliffs, NJ: Prentice Hall, 1993.
2. M. H. Rashid, *Introduction to PSpice Using OrCAD for Circuits and Electronics*. Upper Saddle River, NJ: Prentice Hall, 2004.
3. B. Razavi, *Design of Analog CMOS Integrated Circuits*. New York: McGraw-Hill, 2001.

Review Questions

1. What is an oscillator?
2. What are the two conditions for oscillation?
3. What are the major types of oscillators?
4. What is an *RC* oscillator?
5. What is an *LC* oscillator?
6. What is a crystal oscillator?
7. What is the frequency stability of oscillators?
8. What is the figure of merit of an oscillator?
9. What is a phase-shift oscillator?
10. What is a Wien-bridge oscillator?
11. What is a quadrature oscillator?
12. What is a Colpitts oscillator?
13. What is a Hartley oscillator?
14. What is a crystal oscillator?
15. What is an active-filter tuned oscillator?

Problems

The symbol **D** indicates that a problem is a design problem. The symbol **P** indicates that you can check the solution to a problem using PSpice/SPICE or Electronics Workbench.

13.2 *Principles of Oscillators*

13.1 For the circuit in Fig. P13.1, determine the values of A and θ that will produce a steady-state sinusoidal oscillation.

FIGURE P13.1

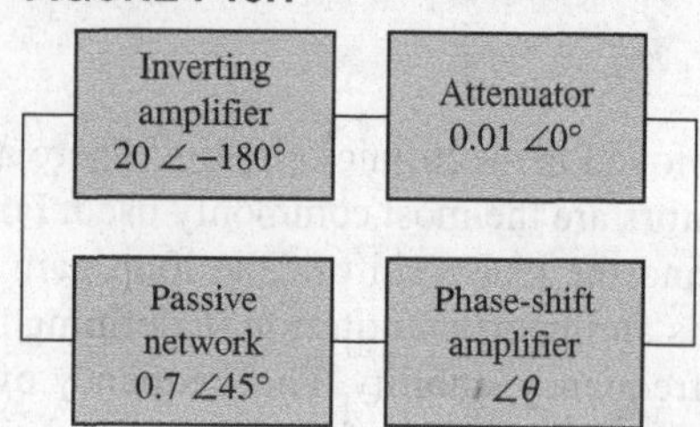

13.2 The amplifier in Fig. 13.3 has a voltage gain of $A = 200$, input resistance $R_i = 50\ \text{k}\Omega$, and output resistance $R_o = 500\ \Omega$. Find the values of R, R_3, C, and L so that the oscillation frequency is $f_o = 5$ kHz.

D P

13.3 Audio-Frequency Oscillators

13.3 Derive the transfer function $\beta(s)$ of the feedback network shown in Fig. 13.5(b).

13.4 Design a phase-shift oscillator as shown in Fig. 13.5(a) so that $f_o = 1$ kHz.

13.5 Find the values of R and C for the phase-shift oscillator in Fig. P13.5 so that the oscillation frequency is $f_o = 5$ kHz.

D

FIGURE P13.5

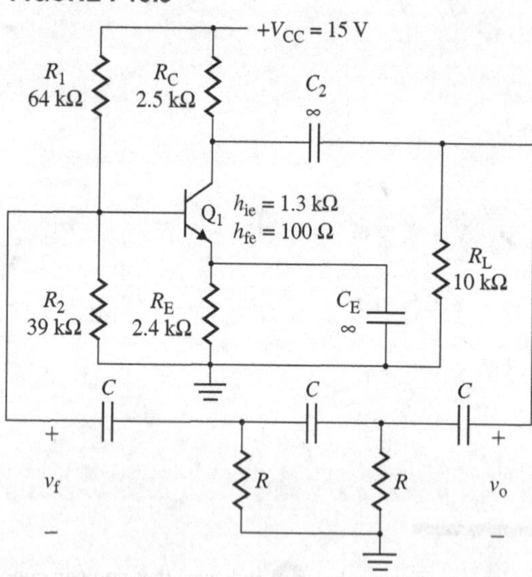

13.6 Find the values of R, R_3, C, and L for the phase-shift oscillator in Fig. P13.6 so that the oscillation frequency is $f_o = 5$ kHz.

D

FIGURE P13.6

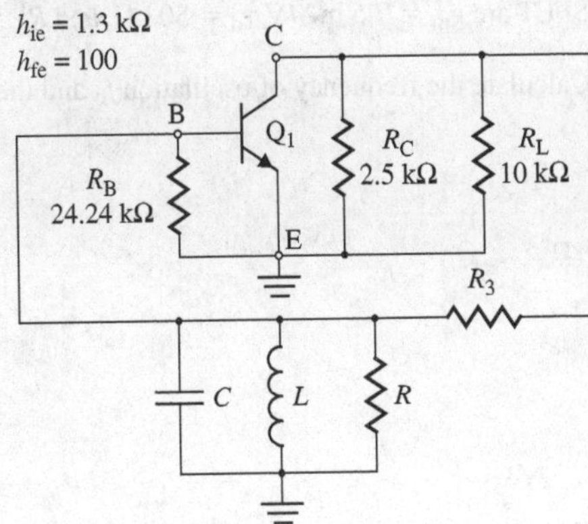

13.7 Design a quadrature oscillator as shown in Fig. 13.9(a) so that $f_o = 500$ Hz.

13.8 Design a three-phase oscillator as shown in Fig. 13.10 so that $f_o = 50$ Hz.

13.9 Derive the transfer function $\beta(s)$ of the feedback network in Fig. 13.12(b).

13.10 Design a Wien-bridge oscillator as shown in Fig. 13.12(a) so that $f_o = 5$ kHz.

13.11 Design the MOS ring oscillator in Fig. 13.16 so that $f_o = 2$ kHz. The MOSFET parameters are MOS constant $K_p = 25\ \mu\text{A/V}^2$, threshold voltage $V_t = 1$ V, length $L = 10\ \mu$m, width $W = 20\ \mu$m, and modulation length $\lambda = 0.01$. Use PSpice to plot the transient response of the output voltage $v_o(t)$ in part (a) from 0 to 1 ms. Assume $V_{DD} = 5$ V.

13.12 Determine the oscillation frequency f_o for the ring oscillator in Fig. 13.16 if $R = 40$ kΩ and $C = 6$ nF. The MOSFET parameters are MOS constant $K_p = 25\ \mu\text{A/V}^2$, threshold voltage $V_t = 1$ V, length $L = 10\ \mu$m, width $W = 20\ \mu$m, and modulation length $\lambda = 0.01$. Use PSpice to plot the transient response of the output voltage $v_o(t)$ to verify the result. Assume $V_{DD} = 5$ V.

13.4 *Radio Frequency Oscillators*

13.13 Design a Colpitts oscillator as shown in Fig. 13.19(a) so that the oscillation frequency is $f_o = 500$ kHz.

13.14 A Colpitts BJT oscillator is shown in Fig. 13.20. The circuit parameters are $r_\pi = h_{ie} = 500\ \Omega$, $h_{fe} = 200$, $L = 1.5$ mH, $C_1 = 10$ nF, $C_2 = 10$ nF, and $R_L = 5$ kΩ. Calculate the frequency of oscillation f_o and the value of R_1 required to sustain the oscillation.

13.15 Design a Colpitts BJT oscillator as shown in Fig. 13.20 so that $f_o = 250$ kHz. The circuit parameters are $r_\pi = h_{ie} = 500\ \Omega$, $h_{fe} = 200$, and $R_L = 5$ kΩ.

D P

13.16 An *LC*-tuned MOSFET oscillator is shown in Fig. 13.24. Find the values of *L*, *C*, and *n* for an oscillation frequency of $f_o = 100$ kHz. The parameters of the MOSFET are $g_m = 7.5$ mA/V, $r_d = 50$ kΩ, and $R_G = 20$ kΩ.

13.17 A Colpitts BJT oscillator is shown in Fig. P13.17. Calculate the frequency of oscillation f_o and the value of R_{E1} required to sustain the oscillation.

FIGURE P13.17

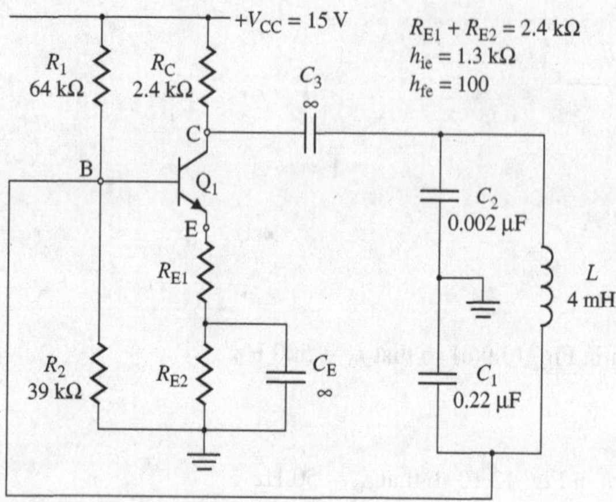

13.18 Design a Colpitts oscillator as shown in Fig. 13.19(a) so that the oscillation frequency is $f_o = 5$ kHz. Assume $V_{CC} = V_{EE} = 12$ V.

13.19 The Colpitts oscillator of Fig. 13.19(a) has $C_1 = 400$ pF, $C_2 = 200$ pF, and $L = 1$ mH. Determine the frequency of oscillation f_o and the minimum value of gain $A = R_F/R_1$ needed to sustain the oscillation.

13.20 Determine the frequency of oscillation for the Colpitts MOSFET oscillator in Fig. P13.20(a). The MOSFET can be replaced by its transconductance model, shown in Fig. P13.20(b). The parameters are $r_d = 25$ kΩ, $g_m = 5$ mA/V, $R_G = 1$ MΩ, $L = 1.5$ mH, $C_1 = 10$ nF, and $C_2 = 10$ nF. Calculate the frequency of oscillation and check to make sure the condition for oscillation is satisfied.

FIGURE P13.20

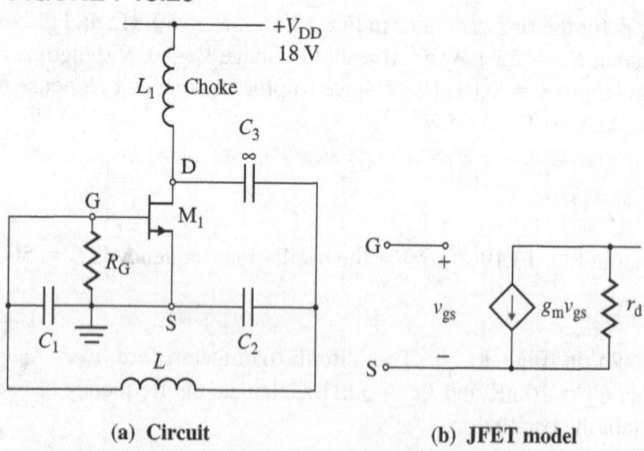

(a) Circuit

(b) JFET model

13.21 Design a Hartley oscillator as shown in Fig. 13.27 so that $f_o = 500$ kHz. Use a 2N3821 *n*-channel JFET whose parameters are $I_{DSS} = 0.5$ mA to 2.5 mA, $V_p = -4$ V, and $g_m = 1.5$ mA/V to 4.5 mA/V. The load resistance is $R = 50$ Ω, and the power supply voltage is $V_{DD} = 15$ V.

13.22 Design the MOS oscillator in Fig. 13.32(b) so that $f_o = 200$ kHz. The MOS parameters are MOS constant $K_p = 25\ \mu A/V^2$, threshold voltage $V_t = 1$ V, length $L = 10\ \mu m$, width $W = 10\ \mu m$, and modulation $\lambda = 0.01$. Use PSpice to plot the transient response of the output voltage $v_o(t)$ in part (a) from 4000 μs to 4020 μs. Assume $V_{CC} = 12$ V.

13.23 Determine the oscillation frequency f_o for the ring oscillator in Fig. 13.31(a) if $R_p = 3.5\ k\Omega$ and $L_p = 50\ \mu H$. The MOSFET parameters are MOS constant $K_p = 25\ \mu A/V^2$, threshold voltage $V_t = 1$ V, length $L = 10\ \mu m$, width $W = 20\ \mu m$, and modulation length $\lambda = 0.01$. Use PSpice to plot the transient response of the output voltage $v_o(t)$ to verify the result. Assume $V_{DD} = 12$ V.

13.5 Crystal Oscillators

13.24 Design a crystal oscillator as shown in Fig. 13.36(a) so that $f_o = 14$ MHz. The crystal parameters are $f_o = 10$ MHz, $Q = 150 \times 10^3$, $C_s = 0.0145$ pF, $C_p = 4.35$ pF, and $L = 12$ mH. Assume a transconductance gain of $|g_m| = 1$ mA/V for the amplifier.

13.6 Active-Filter Tuned Oscillators

13.25 The equivalent circuit of a tuned oscillator is shown in Fig. P13.25. Derive the expressions for the oscillation condition and the frequency of oscillation.

FIGURE P13.25

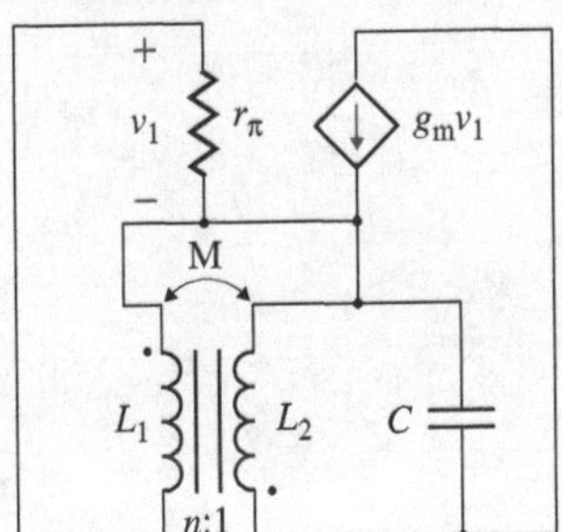

13.26 Design an active-filter tuned oscillator as shown in Fig. 13.39(b) so that $f_o = 10$ kHz. Assume $R_L = 10\ k\Omega$ and $V_{CC} = V_{EE} = 12$ V.

CHAPTER 14

OPERATIONAL AMPLIFIERS

Learning Outcomes

After completing this chapter, students should be able to do the following:

- Describe the internal structure and types of op-amps (CMOS, bipolar, and BiCMOS) and the effects of amplifier configurations (difference and single-ended output) on the op-amp performance.
- Identify the circuit parameters of the internal design that affect the op-amp performance (offset voltage, offset current, and unity-gain frequency).
- List the typical values of performance parameters for different types of op-amps.
- Analyze op-amp circuits to determine the DC-biasing conditions and the performance parameters.

Symbols and Their Meanings

Symbol	Meaning
i_{B1}, i_{B2}	Instantaneous base currents of transistors 1 and 2
i_{Bn}, i_{Cn}, i_{En}	Base, collector, and emitter currents of an *n*th BJT Q_n
G_m	Transconductance of an amplifier
I_{Sn}	Saturation current of an *n*th BJT Q_n
K_{pn}, V_{tn}	MOS constant and threshold voltage of an *n*th MOSFET M_n
r_{on}	Output resistance of an *n*th BJT Q_n or MOSFET M_n

Symbol	Meaning
R_i, R_o	Input and output resistances of an amplifier stage
v_{B1}, v_{B2}	Instantaneous voltages at the base of transistors 1 and 2
V_{io}, V_{oo}	Input and output offset voltages
$V_{CC}(V_{DD})$, $V_{EE}(V_{SS})$	Positive and negative DC supply voltages
v_{GSn}, i_{Dn}	Gate–source voltage and drain current of an *n*th MOSFET M_n
$(W/L)_n$, $(W/L)_p$	Width-to-length ratios of an NMOS and a PMOS
β_{Fn}, g_{mn}	Current gain of an *n*th BJT Q_n and transconductance of an *n*th MOSFET M_n

14.1 Introduction

So far we have discussed separately the analysis and design of transistor amplifiers, differential amplifiers, and output stages. An operational amplifier normally consists of these stages. In this chapter, we examine the internal circuitry of 10 commercially available op-amps. Much of the circuitry will be closely related to that of other ICs. We analyze in detail one of the oldest but most popular amplifiers, the LM741.

14.2 Internal Structure of Op-Amps

The general configuration of an op-amp is shown in Fig. 14.1 [1]. All stages are direct coupled—that is, there are no coupling or bypass capacitors. Since capacitors and resistors of over 50 kΩ occupy large areas on IC chips and exhibit parasitic effects, they are usually avoided in ICs. Therefore, op-amp circuits are designed using transistors with matching characteristics. Mismatches do exist, however, and cause offset voltages.

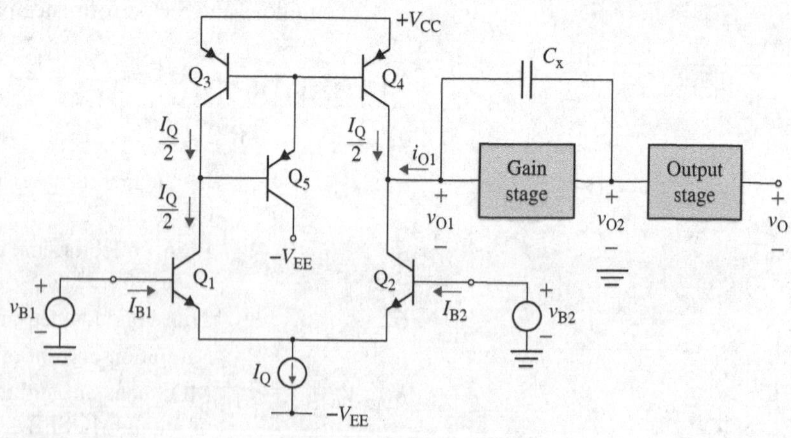

FIGURE 14.1 General configuration of an op-amp

The voltage gain of each of the differential and gain stages is normally in the range from 300 to 1000. The output stage is an emitter follower, and its gain is unity. Thus, the overall open-loop voltage gain of an op-amp is on the order of 10^5 to 10^6. The output stage provides a low-output impedance so that it can drive a load with relatively low values of load resistance. It is normally operated in class AB mode to reduce crossover distortion. In general, BJT op-amps have a larger voltage gain, whereas MOSFET op-amps have higher input resistances.

For stability of the op-amp, the phase shift of each stage is kept to a minimum. Since each stage contributes to the phase shift, the total number of stages is generally limited to three. A compensation capacitor C_x is normally connected across the second stage.

▶ **NOTE** Unless specified, we will assume that the base–emitter voltage is $V_{BE} = 0.6$ V.

KEY POINT OF SECTION 14.2

- In general, an op-amp circuit consists of a differential input stage, a gain stage, an output stage, and protection circuitry. Each stage uses active biasing and an active load.

14.3 Parameters and Characteristics of Practical Op-Amps

Most op-amps contain a differential-coupled pair as an input stage. Practical op-amps exhibit characteristics that deviate significantly from the ideal characteristics [2, 3]. These deviations are discussed in detail in Chapter 6. Here we consider some parameters that depend on the internal design of the op-amp. The effects of deviations from the ideal can be incorporated in the equivalent circuit of the op amp, as shown in Fig. 14.2 where $R_{cm} = R_{ic}$ is the common-mode input resistance.

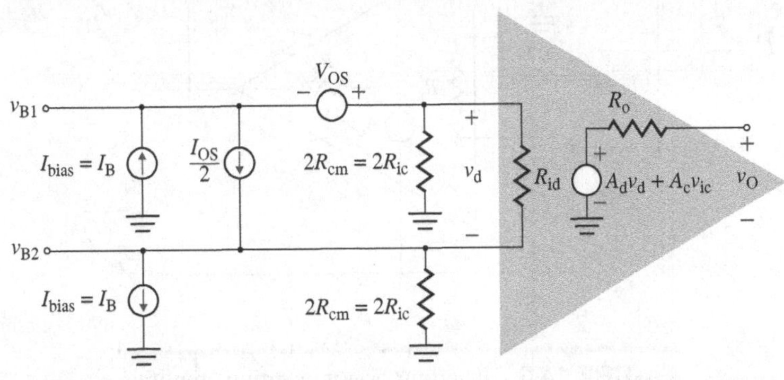

FIGURE 14.2 Equivalent circuit of an op-amp

14.3.1 Input Biasing Current

The input biasing current I_B is defined as the average of the two DC input currents to the bases of Q_1 and Q_2; that is,

$$I_B = \frac{I_{B1} + I_{B2}}{2} \tag{14.1}$$

$$\approx \frac{I_C}{\beta_F} = \frac{I_Q}{2\beta_F} \quad \text{(for BJT input stage only)} \tag{14.2}$$

The polarity of I_B is shown for an *npn* transistor input stage in Fig. 14.2. For a *pnp* input stage, I_B would flow out of the amplifier terminals. Transistors in an ideal op-amp are assumed to have a large value of β_F, tending to infinity, and to draw zero DC input current. In a practical op-amp, however, β_F has a finite value. Thus, I_B has a small but finite value. The typical magnitudes of the biasing currents are 10 pA to 100 nA for BJT input devices and 1 nA to 10 pA for MOSFET input devices. This input biasing current will cause a small DC-output voltage when the external input voltage is zero.

The effect of biasing current I_B can be determined for the inverting or noninverting amplifier, as shown Fig. 14.3. Assuming $I_{B1} = I_{B2} = I_B$, then $v_d = R_i\,(I_{B1} - I_{B2}) = 0$. Since $v_d = 0$ due to the feedback of V_{ob}, there will be no current flowing through R_1; that is, $i_S = 0$, and the biasing current I_B will flow through R_F. Thus,

$$v_d = 0 = -I_B R_F + V_{ob}$$

which gives the output voltage due to input biasing current I_B as

$$V_{ob} = R_F I_B \tag{14.3}$$

Thus, the output offset voltage due to the input biasing current I_B depends directly on the feedback resistance R_F. To minimize the effect of I_B, the value of R_F should be small. However, the ratio R_F/R_1 determines the voltage gain v_O/v_S. The output voltage due to I_B does not depend on whether the input voltage source v_S is connected. Thus, Eq. (14.3) is applicable for both inverting and noninverting amplifiers.

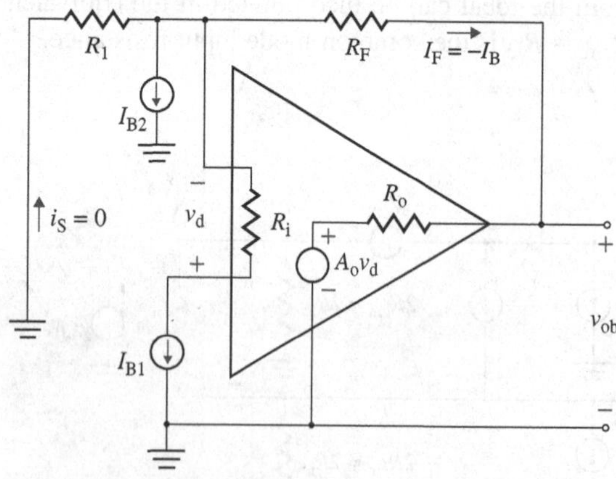

FIGURE 14.3 Inverting or noninverting amplifier with input biasing current

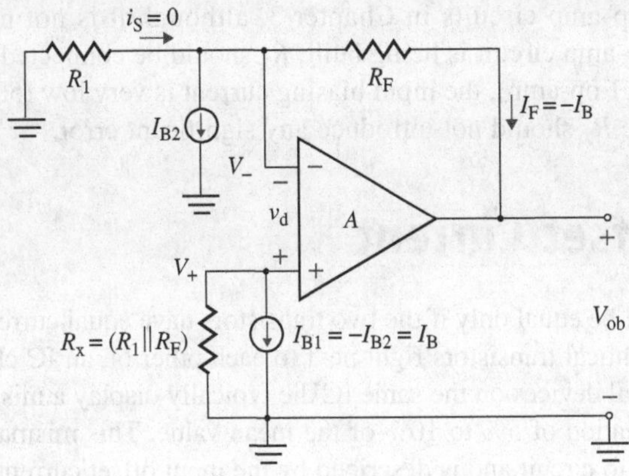

FIGURE 14.4 Effect of offset-minimizing resistance R_x

The effect of I_B on the output voltage can be eliminated or minimized by making Thevenin's equivalent resistance at the $(-)$ terminal equal to that at the $(+)$ terminal. This arrangement can be implemented by connecting a resistance R_x to the noninverting terminal, as shown in Fig. 14.4, such that

$$V_+ = V_- \quad \text{or} \quad R_{Th+} = R_{Th-}$$

which gives the *offset-minimizing resistance* R_x as

$$R_x = \frac{R_1 R_F}{R_1 + R_F} = R_1 \| R_F \tag{14.4}$$

For example, if $R_F = 100\ \text{k}\Omega$ and $I_B = 500$ nA, then $V_{ob} = 100\ \text{k}\Omega \times 500 \times 10^{-9} = 50$ mV without R_x, and $V_{ob} = 0$ if $R_x = (10\ \text{k}\Omega \| 100\ \text{k}\Omega) = 9.091\ \text{k}\Omega$. Therefore, by connecting resistance R_x, which is equal to the parallel combination of R_1 and R_F, we can minimize the output voltage due to the input biasing current. Since I_{B1} and I_{B2} are not exactly equal, V_{ob} will be minimized but not completely eliminated. The connections of the input voltage v_S and the input/output relations are shown in Fig. 14.5 for inverting and noninverting amplifiers.

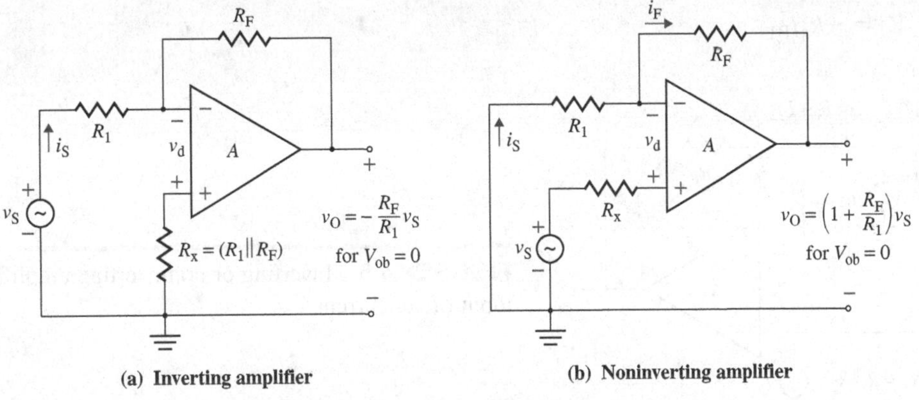

FIGURE 14.5 Amplifiers with offset-minimizing resistance R_x

R_X was included in the op-amp circuits in Chapter 3, although it is not necessary for an ideal op-amp. However, while an op-amp circuit is being built, R_X should be connected. For op-amps with an FET input stage (e.g., the LF411 op-amp), the input biasing current is very low (50 pA), and the absence of offset-minimizing resistance R_X should not introduce any significant error.

14.3.2 Input Offset Current

The DC input base currents will be equal only if the two transistors have equal current gains (betas). However, even two theoretically identical transistors right next to each other on an IC chip will not be exactly identical. Geometrically identical devices on the same IC die typically display a mismatch that is normally distributed with a standard deviation of 5% to 10% of the mean value. This mismatch in the two biasing currents is random from circuit to circuit and is described by the input offset current I_{OS}, which is defined as the difference between the two base currents of the transistors:

$$I_{OS} = |I_{B1} - I_{B2}| \tag{14.5}$$

I_{OS} arises from mismatches in the area and β_F of the transistors. Beta and collector current values typically deviate by 10% and 1%, respectively. Thus, I_{OS} can be found approximately from

$$I_{OS} = \frac{(0.1 + 0.01)I_C}{\beta_F} = \frac{0.11 I_Q}{2\beta_F} = \frac{0.055 I_Q}{\beta_F} \tag{14.6}$$

The value of the bias-minimizing resistance R_X in Eq. (14.4) was derived by assuming equal input-biasing currents: $I_{B1} = I_{B2} = I_B$. In practice, these currents are not equal because of internal imbalances within the op-amp circuit. The input offset current I_{io} is a measure of the degree of mismatching, and it is defined by

$$I_{io} = I_{B1} - I_{B2} \tag{14.7}$$

For the μA741C op-amp, I_{io} is quoted as having a maximum absolute value of 200 nA and a typical value of 30 nA; it can be either positive or negative. An amplifier with offset-minimizing resistance R_X is shown in Fig. 14.6. The effective input voltage at the noninverting terminal is given by

$$V_+ = -R_X I_{B1} \tag{14.8}$$

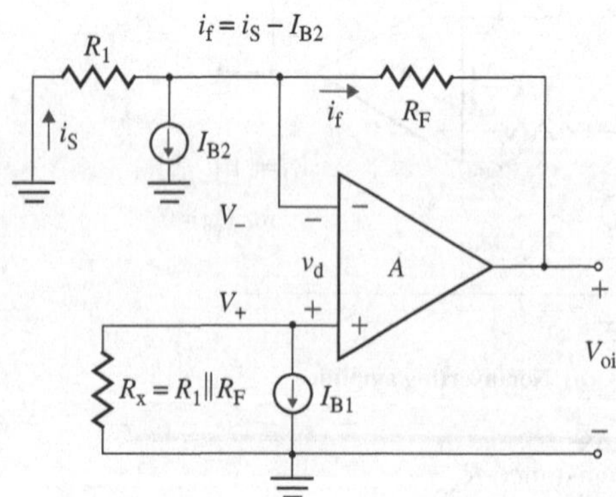

FIGURE 14.6 Inverting or noninverting amplifier with input offset current

Since the differential voltage between the input terminals is $v_d \approx 0$, $V_- = V_+$. Assuming the op-amp draws negligible current and applying KCL at the inverting terminal, we get

$$\frac{V_+}{R_1} + I_{B2} + \frac{V_+ - V_{oi}}{R_F} = 0$$

which gives

$$\frac{V_{oi}}{R_F} = \frac{V_+}{R_1 \parallel R_F} + I_{B2} \tag{14.9}$$

Substituting $V_+ = -R_x I_{B1}$, we get the output offset voltage

$$V_{oi} = R_F\left(\frac{-R_x}{R_1 \parallel R_F} I_{B1} + I_{B2}\right) \tag{14.10}$$

Substituting R_x $(=R_1 \parallel R_F)$ from Eq. (14.4) into Eq. (14.10) gives the output voltage as

$$V_{oi} = R_F(I_{B2} - I_{B1}) = -R_F I_{io} \tag{14.11}$$

For example, if $R_F = 100$ kΩ and $I_{io} = \pm 200$ nA, then

$$V_{oi} = \pm 100 \times 10^3 \times 200 \times 10^{-9} = \pm 20 \text{ mV} \quad \text{(DC)}$$

To minimize the effect of I_{io}, the value of R_F should be small.

▶ **NOTE** If R_x is zero or is absent, Eq. (14.8) is to be used. If $R_x = R_1 \parallel R_F$, Eq. (14.11) is to be used.

EXAMPLE 14.1

Finding the effects of offsets on the output of an op-amp integrator The inverting integrator in Fig. 14.7 has $R_1 = 1$ kΩ, $R_x = 1$ kΩ, $C_F = 0.1$ μF, $V_{CC} = 15$ V, $-V_{EE} = -15$ V, and maximum saturation voltage = ±14 V. The op-amp parameters are $V_{io} = 6$ mV, $I_B = 500$ nA, and $I_{io} = 200$ nA at 25°C.

(a) Determine the total output offset voltage v_{of}.

(b) Repeat part (a) if $R_x = 0$

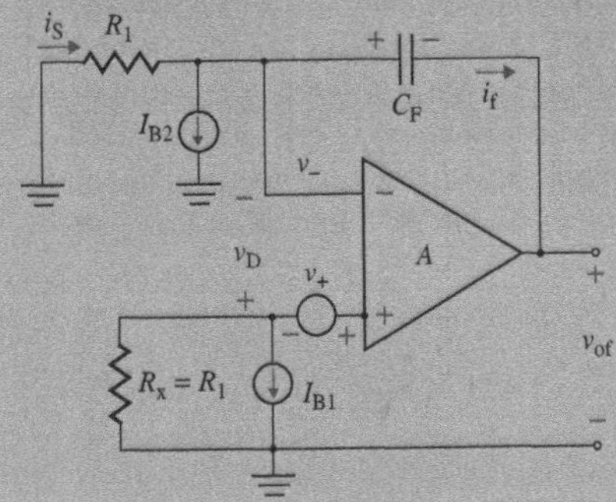

FIGURE 14.7 Circuit for Example 14.1

SOLUTION

$R_1 = 1\ \text{k}\Omega$, $R_x = 1\ \text{k}\Omega$, $V_{io} = 6$ mV, $I_B = 500$ nA, and $I_{io} = 200$ nA. The equivalent circuit of the integrator with input offset voltage and input biasing currents is also shown in Fig. 14.7.

(a) $R_x = R_1$. The output voltage will be due to V_{io} and I_{io}. With $V_{io} = 0$, we get

$$v_+ = -R_x I_{B1} = -R_1 I_{B1}$$

$$v_- = v_+ = -R_1 I_{B1}$$

$$i_S = -\frac{v_-}{R_1}$$

$$i_f = i_S - I_{B2} = I_{B1} - I_{B2}$$

With $I_{B1} = I_{B2} = 0$, we get

$$i_f = \frac{-V_{io}}{R_1}$$

Applying the superposition theorem, we get for the current flowing through the capacitor C_F

$$i_f = i_S + I_{B1} - I_{B2} = \frac{-V_{io}}{R_1} + I_{io} \tag{14.12}$$

where $I_{io} = (I_{B1} - I_{B2})$ is the input offset current. The total output offset voltage due to the capacitor current i_f can be found from

$$-v_{of} = \frac{1}{C_F}\int i_f\, dt + (-V_{io} + R_x I_{B1}) + v_C(t = 0) \tag{14.13}$$

where $v_C(t = 0)$ is the initial capacitor voltage. Substituting i_f from Eq. (14.12) into Eq. (14.13) gives the total output offset voltage as

$$-v_{of} = \frac{1}{C_F}\int \left(\frac{-V_{io}}{R_1} + I_{io}\right) dt + (-V_{io} + R_x I_{B1}) + v_C(t = 0)$$

$$v_{of} = \frac{V_{io}}{C_F R_1} t - \frac{I_{io}}{C_F} t + V_{io} - R_x I_{B1} - v_C(t = 0) \tag{14.14}$$

This equation indicates that the output offset voltage will rise linearly until the output reaches the saturation voltage of the amplifier, which is ±14 V in this example. If the power supplies are turned on and enough time is allowed, the output will build up to the saturation voltage, even without any external input signal to the integrator. For this reason, this is not a practical circuit (as discussed in Sec. 3.5); it needs a DC feedback resistor R_F, as shown in Fig. 3.16. After the power supply is switched on, the time required for the total output offset voltage to reach the saturation level $v_{of} = 14$ V can be found from Eq. (14.14) with $v_C(t = 0) = 0$:

$$14 = \left(\frac{6 \times 10^{-3}}{0.1 \times 10^{-6} \times 1 \times 10^{3}} - \frac{200 \times 10^{-9}}{0.1 \times 10^{-6}}\right) t + 6 \times 10^{-3} - 1\ \text{k}\Omega \times 600 \times 10^{-9} - 0$$

which gives $t = 241.3$ ms.

(b) If $R_x = 0$, the total output offset voltage will be due to the input offset voltage, the input biasing current, $-I_{B2} = -I_B + (I_{io}/2)$, and the input offset current, and it can be found from Eq. (14.14) by replacing I_{io} by I_{B2}. That is,

$$v_{of} = \left(\frac{V_{io}}{C_F R_1} + \frac{I_B - I_{io}/2}{C_F}\right)t + V_{io} - v_C(t = 0) \tag{14.15}$$

The time required for the output voltage to reach the saturation level $v_{of} = 14$ V can be found from Eq. (14.15) with $v_C(t = 0) = 0$:

$$14 = \left(\frac{6 \times 10^{-3}}{0.1 \times 10^{-6} \times 1 \times 10^3} + \frac{400 \times 10^{-9}}{0.1 \times 10^{-6}}\right)t + 6 \times 10^{-3} + 0$$

which gives $t = 218.7$ ms.

14.3.3 Input Offset Voltage

If the input terminals of an op-amp are tied together and connected to the ground, as shown in Fig. 14.8(a), a certain DC voltage exits at the output. This voltage is called the *output offset voltage* V_{oo}. The input offset voltage is the differential input voltage that exists between two terminals without any external inputs applied. In other words, it may be regarded as the input voltage that should be applied between the input terminals to force the output voltage to zero, as shown in Fig. 14.8(b). If V_{oo} is divided by the voltage gain A_o of the op-amp, the result is the *input offset voltage* V_{io}. Assuming that the output voltage is not saturated, V_{io} can be determined from

$$V_{io} = \frac{V_{oo}}{A_o} \quad \text{for } |V_{oo}| \leq |V_{sat}| \tag{14.16}$$

Input offset voltage is quoted as an absolute value, and it may be positive or negative. The maximum value of V_{io} for the μA741C is ±6 mV. If the output voltage is saturated to V_{sat}, as is usually the case, Eq. (14.16) is valid only if $|V_{oo}| \leq |V_{sat}|$. The polarity of V_{io} is unpredictable, and so is the output offset voltage V_{oo}.

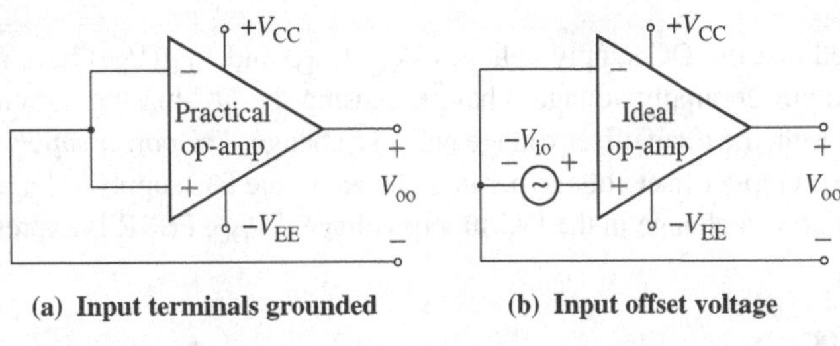

FIGURE 14.8 Input offset voltage

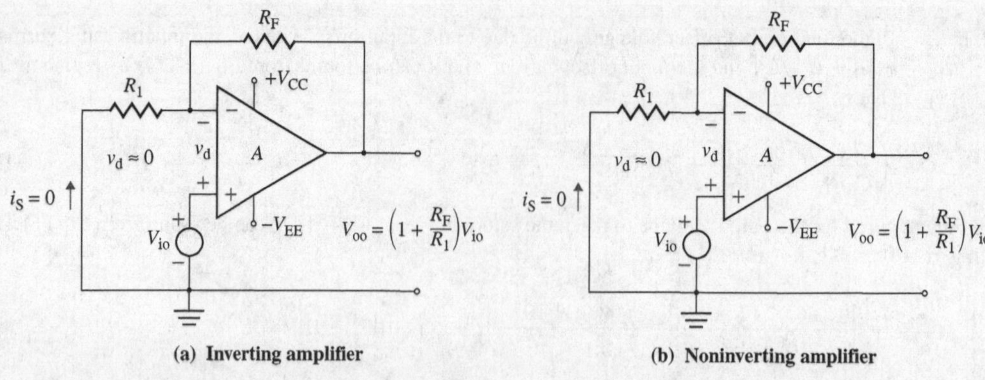

FIGURE 14.9 Inverting and noninverting amplifiers with input offset voltage

The output offset voltage is caused by internal mismatching in the input stage. A simple differential pair, as shown in Fig. 9.2(b), consists of two transistors Q_1 and Q_2. Any differential signal between the input terminals is amplified and gives the output voltage V_{oo}. In practice, the characteristics of the two transistors will not be exactly the same; therefore, the collector-biasing currents I_{C1} and I_{C2} will differ. As a result, even without any input voltages, there could be a differential output voltage, which is amplified in subsequent stages and possibly aggravated by more mismatching.

The effect of the input offset voltage can be determined for the inverting and noninverting amplifiers in Fig. 14.9. For both configurations, V_{io} may be considered as the input to the noninverting terminal, since there is no other input signal. Applying Eq. (3.18) gives the output offset voltage:

$$V_{oo} = \left(1 + \frac{R_F}{R_1}\right)V_{io} \tag{14.17}$$

For example, if $R_1 = 10\ \text{k}\Omega$, $R_F = 100\ \text{k}\Omega$, and $V_{io} = \pm 6$ mV, then

$$V_{oo} = \pm\left(1 + \frac{100\ \text{k}\Omega}{10\ \text{k}\Omega}\right) \times 6\ \text{mV} = \pm 66\ \text{mV}$$

That is, the output voltage V_{oo} can be ± 66 mV (DC) without any external input signal v_S applied.

14.3.4 Power Supply Rejection Ratio

So far we have assumed that the DC supply voltages $V_{CC}(V_{DD})$ and $V_{EE}(V_{SS})$ have no effect on the output voltage. In practice, the power supply voltages change, causing the DC-biasing currents of the internal transistors to change. As a result, the input offset voltage will also change. The *power supply rejection ratio* (PSRR) is defined as the change in input offset voltage per unit change in the DC supply voltage. If ΔV_{io} is the change in input offset voltage due to a change in the DC supply voltage ΔV_{DC}, PSRR is expressed as

$$\text{PSRR} = \frac{\Delta V_{io}}{\Delta V_{DC}} \tag{14.18}$$

$$= 20 \log\left|\frac{\Delta V_{io}}{\Delta V_{DC}}\right| \tag{14.19}$$

This ratio is also known as the *supply voltage rejection ratio* (SVRR) or the *power supply sensitivity* (PSS). The maximum value of PSRR for the μA741C op-amp is 150 μV/V. For example, if the DC supply voltages change from $V_{DC} = \pm 15$ V to ± 12 V and PSRR = 150 μV/V, then

$$\Delta V_{DC} = 2 \times 15 - 2 \times 12 = 6 \text{ V}$$

and $$\Delta V_{io} = \text{PSRR}\ \Delta V_{DC} = 150\ \mu\text{V} \times 6 = 900\ \mu\text{V}$$

14.3.5 Thermal Voltage Drift

In previous sections, we assumed that the input offset voltage V_{io}, input biasing current I_B, and input offset current I_{io} remain constant. A practical op-amp consists of devices such as diodes and transistors whose parameters change with temperature. *Thermal drift* is a measure of the change in an offset parameter due to a unit change in temperature. *Thermal voltage drift* is defined as the rate of change of input offset voltage V_{io} per unit change in temperature, and it is expressed as

$$D_v = \frac{\Delta V_{io}}{\Delta T} \quad (\text{V/°C}) \tag{14.20}$$

Thermal-biasing current drift is defined as the rate of change of input biasing current I_B per unit change in temperature, and it is expressed as

$$D_b = \frac{\Delta I_B}{\Delta T} \quad (\text{A/°C}) \tag{14.21}$$

Thermal input offset current drift is defined as the rate of change of input offset current I_{io} per unit change in temperature, and it is expressed as

$$D_i = \frac{\Delta I_{io}}{\Delta T} \quad (\text{A/°C}) \tag{14.22}$$

Thus, the output voltage due to drifts can be found from

$$V_{od} = \left(1 + \frac{R_F}{R_1}\right)\Delta V_{io} + R_F \Delta I_{io} = \left(1 + \frac{R_F}{R_1}\right) D_v \Delta T + R_F D_i \Delta T \tag{14.23}$$

$$= \left(1 + \frac{R_F}{R_1}\right) D_v \Delta T + R_F D_b \Delta T \quad \text{for } R_x = 0 \tag{14.24}$$

EXAMPLE 14.2

Finding the effects of thermal drift on the output of an inverting op-amp circuit The inverting amplifier in Fig. 14.5(a) has $R_1 = 10$ kΩ, $R_F = 100$ kΩ, and $R_x = R_F \| R_1 = 9.091$ kΩ. The op-amp parameters are $V_{io} = \pm 6$ mV, $I_B = 500$ nA, $I_{io} = \pm 200$ nA, and PSRR = 150 μV/V. The thermal drifts are D_v = 15 μV/°C, $D_i = 0.5$ nA/°C, and $D_b = 0.5$ nA/°C at 25°C. The temperature is 55°C. The DC supply voltages change from $V_{CC} = 15$ V to 12 V and $-V_{EE} = -15$ V to -12 V. The input voltage is $v_S = 100$ mV (DC). Determine the output voltage v_O if **(a)** $R_x = R_F \| R_1 = 9.091$ kΩ and **(b)** $R_x = 0$.

SOLUTION

$R_1 = 10\ \text{k}\Omega$, $R_F = 100\ \text{k}\Omega$, $V_{io} = \pm 6$ mV, $I_B = 500$ nA, $I_{io} = \pm 200$ nA, $D_v = 15\ \mu\text{V/°C}$, $D_i = 0.5$ nA/°C, $D_b = 0.5$ nA/°C, and $v_S = 100$ mV. Then

$$\Delta T = 55 - 25 = 30°\text{C}$$

$$\Delta V_{DC} = (15 + 15) - (12 + 12) = 6\ \text{V}$$

$$\Delta V_{io} = D_v\,\Delta T + \text{PSRR}\,\Delta V_{DC} = 15 \times 10^{-6} \times 30 + 150 \times 10^{-6} \times 6 = 1.35\ \text{mV}$$

$$\Delta I_{io} = D_i\,\Delta T = 0.5 \times 10^{-9} \times 30 = 15\ \text{nA}$$

$$\Delta I_B = D_b\,\Delta T = 0.5 \times 10^{-9} \times 30 = 15\ \text{nA}$$

(a) With offset-minimizing resistance R_x, the total output voltage of the inverting amplifier is given by

$$v_O = -\frac{R_F}{R_1}v_S \pm \left(1 + \frac{R_F}{R_1}\right)(V_{io} + \Delta V_{io}) \pm R_F(I_{io} + \Delta I_{io}) \tag{14.25}$$

$$= -\left(\frac{100\ \text{k}\Omega}{10\ \text{k}\Omega}\right) \times 100 \times 10^{-3} \pm \left(1 + \frac{100\ \text{k}\Omega}{10\ \text{k}\Omega}\right) \times (6 + 1.35) \times 10^{-3}$$

$$\pm\ 100 \times 10^3 \times (200 + 15) \times 10^{-9}$$

$$= -1000\ \text{mV} \pm 80.85\ \text{mV} \pm 21.5\ \text{mV}$$

$$= -1102.35\ \text{mV (min)} \quad \text{or} \quad -897.65\ \text{mV (max)}$$

(b) With $R_x = 0$, the total output voltage of the inverting amplifier is given by

$$v_O = -\frac{R_F}{R_1}v_S \pm \left(1 + \frac{R_F}{R_1}\right)(V_{io} + \Delta V_{io}) + R_F(I_B + \Delta I_B) \tag{14.26}$$

$$= -\left(\frac{100\ \text{k}\Omega}{10\ \text{k}\Omega}\right) \times 100 \times 10^{-3} \pm \left(1 + \frac{100\ \text{k}\Omega}{10\ \text{k}\Omega}\right) \times (6 + 1.35) \times 10^{-3}$$

$$+\ 100 \times 10^3 \times (500 + 15) \times 10^{-9}$$

$$= -1000\ \text{mV} \pm 80.85\ \text{mV} + 51.5\ \text{mV}$$

$$= -1029.4\ \text{mV (min)} \quad \text{or} \quad -867.65\ \text{mV (max)}$$

14.3.6 Determining the Thermal Voltage Drift

Because of mismatches, an output voltage will exist even when the external input is zero. The differential voltage that must be applied to the input terminals of an amplifier to drive the output to zero is called the *input offset voltage* V_{io}. This input offset voltage is a function of temperature. The rate of change of the input offset voltage V_{io} per unit change in temperature is known as the *thermal voltage drift*, and it is expressed as

$$D_v = \frac{\Delta V_{io}}{\Delta T} \quad \text{(in V/°C)} \tag{14.27}$$

BJT Amplifiers

The difference in the B-E voltages will cause an offset voltage. Using Eqs. (9.129) and (9.130), we can express this offset voltage as

$$v_{BE2} - v_{BE1} = V_T\left[\ln\frac{i_{C2}}{I_{S2}} - \ln\frac{i_{C1}}{I_{S1}}\right] = V_T \ln\left(\frac{i_{C2}}{I_{S2}} \times \frac{I_{S1}}{i_{C1}}\right) \quad (14.28)$$

Since the mismatch in the collector current is generally small (i.e., $i_{C1} \approx i_{C2}$), we can find the input offset voltage V_{io} from

$$V_{io} = v_{BE2} - v_{BE1} = V_T \ln\left(\frac{I_{S1}}{I_{S2}}\right) \quad (14.29)$$

Thus, the saturation current I_S is the prime contributing factor to V_{io} in a BJT amplifier and is proportional to the transistor base width W_B. Its value can vary from one transistor to another. Assuming that $W_{B1} = W_B$ and $W_{B2} = W_B + \Delta W_B$ are the base widths of transistors Q_1 and Q_2, respectively, Eq. (14.29) becomes

$$V_{io} = V_T \ln\left(\frac{W_B}{W_B + \Delta W_B}\right) = V_T \ln\left(\frac{1}{1 + \Delta W_B/W_B}\right) \quad (14.30)$$

For $W_B >> \Delta W_B$, which is usually the case, Eq. (14.30) can be approximated by

$$V_{io} = V_T\left(\frac{\Delta W_B}{W_B}\right) \quad (14.31)$$

The beta of a transistor is inversely proportional to the base width W_B. Therefore, any change in W_B will cause an almost equal change in beta; that is, $\Delta\beta_F/\beta_F = \Delta W_B/W_B$. Since the change in the base width is generally within 10%, the offset voltage can be found approximately from

$$V_{io} = 0.1V_T \quad (14.32)$$

For BJT input devices, this value is typically 2 mV to 5 mV (for compound devices) and can often be nullified with an external potentiometer.

The ratio $\Delta W_B/W_B$ is relatively independent of temperature. Since $V_T = kT/q$, $dV_T/dT = k/q = V_T/T$. Thus, we can find the thermal drift from Eq. (14.31) as follows:

$$D_v = \frac{dV_{io}}{dT} = \frac{V_T}{T}\left(\frac{\Delta W_B}{W_B}\right) = \frac{V_{io}}{T} \quad (14.33)$$

For example, if $V_{io} = 2.6$ mV and $T = 25°\text{C} = 273 + 25 = 298$ K,

$$D_v = \frac{2.6 \text{ mV}}{298} = 8.72\ \mu\text{V/K}$$

CMOS Amplifiers

From Eq. (9.54), the transconductance of a MOSFET is given by

$$g_m = 2K_p(V_{GS} - V_t) = \frac{2I_D}{V_{GS} - V_t} \quad (14.34)$$

If ΔI_D is the difference between the drain currents of MOSFETs, an input offset voltage $\Delta V_{GS} = V_{OS}$ must be applied between the two gates to cancel the difference. Any increase in ΔV_{GS} will cause an increase in ΔI_D. However, any differential increase ΔV_t in the threshold voltages will cause a decrease in drain current. That is, Eq. (14.19) gives

$$\Delta I_D = \frac{g_m}{2}(\Delta V_{GS} - \Delta V_t) = \frac{g_m}{2}(V_{io} - \Delta V_t)$$

which gives the input offset voltage V_{io} as

$$V_{io} = \Delta V_t + \frac{\Delta I_D}{I_D}(V_{GS} - V_t) \tag{14.35}$$

The ratio $\Delta I_D / I_D$ will depend on any change in the W/L ratio of the MOSFET. Thus, Eq. (14.35) can be written in terms of the W/L ratio:

$$V_{io} = \Delta V_t + \frac{\Delta(W/L)}{W/L}(V_{GS} - V_t) \tag{14.36}$$

In CMOS amplifiers, the value of ΔV_t alone, which is absent in BJTs, can be as high as 2 mV—a magnitude as large as V_{io} for BJTs. The value of $(V_{GS} - V_t)$ is usually much greater than V_T. Thus, the offset voltage of CMOS amplifiers will generally be considerably larger than that of BJT amplifiers. To have a low value of V_{io}, MOSFETs should be operated with a low value of $(V_{GS} - V_t)$.

For MOSFET amplifiers, thermal drift cannot be correlated with the offset voltage. The thermal drift of $\Delta V_t / dT$ can be quite significant.

14.3.7 Offset Voltage Adjustment

We saw in the previous section that an op-amp can have an output voltage V_{oo} without any external input signal. Op-amps are generally compensated internally; they have built-in offset adjustment terminals, as shown in Fig. 14.10. The output voltage can be adjusted to zero by an offset null potentiometer.

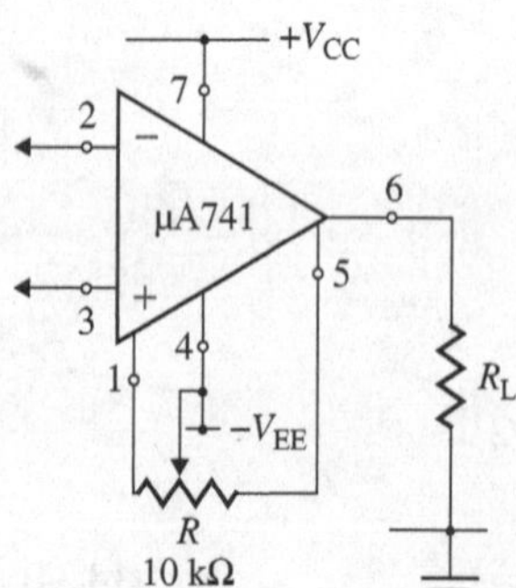

FIGURE 14.10 Op-amp with compensating terminals

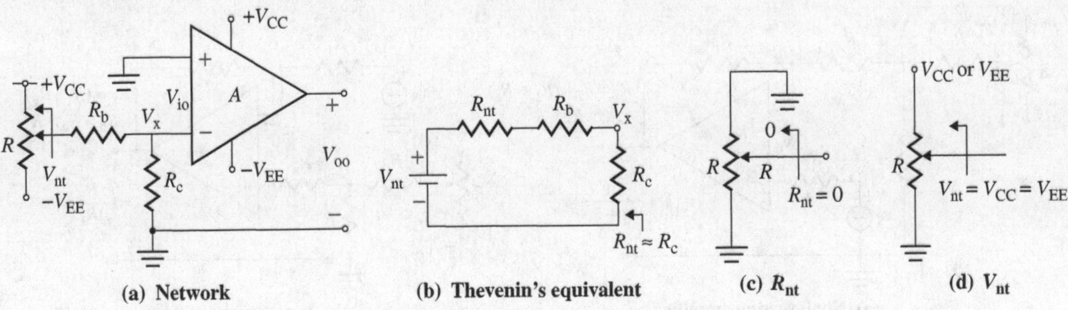

FIGURE 14.11 Offset-compensating network (external connection)

The recommended value of the potentiometer is normally quoted in the data sheet; it is 10 kΩ for the μA741 series. By varying the potentiometer, the output offset voltage can be adjusted to zero within a certain input offset voltage adjustment range (±15 mV for the μA741 op-amp).

It is possible to compensate for the offset voltage by injecting a small voltage into the (+) terminal or the (−) terminal of an op-amp. An offset-compensating network is shown in Fig. 14.11(a). The potentiometer R is varied to produce an input offset voltage V_{io}, which should be just adequate to negate the output offset voltage. Thevenin's equivalent circuit of the network is shown in Fig. 14.11(b); the equivalent resistance of the compensating network is $R_{nt} = 0$, shown in Fig. 14.11(c), and the equivalent voltage is $V_{nt} \simeq V_{CC} = V_{EE}$, shown in Fig. 14.11(d). From the voltage division rule, the voltage V_x, which should be equal to V_{io}, is given by assuming $R_o >> R_{nt}$:

$$V_x \approx V_{io} = \frac{R_c V_{nt}}{R_{nt} + R_b + R_c} \tag{14.37}$$

The values for the network should be such that they do not alter the normal operation of the amplifier. That is, $R_b > R_{nt}$ so that R_b does not affect V_{nt} significantly; $R_b >> R_c$ so that the op-amp biasing current flows mostly through R_c (usually ≤100 Ω). The following relations are recommended:

$$R_b \geq 10R_{nt} \qquad [\text{where } R_{nt(max)} = (R/2) \parallel (R/2) = R/4]$$

$$R_b \geq 1000R_c$$

For $R_b > R_{nt} > R_c$, Eq. (14.37) can be approximated by

$$V_{io} \approx \frac{R_c V_{nt}}{R_b} = \frac{R_c V_{CC}(= V_{EE})}{R_b} \tag{14.38}$$

The compensating network can be used in inverting, noninverting, and differential amplifiers, as shown in Fig. 14.12. For a voltage follower, Fig. 14.12(a) can be modified by making $R_1 = 0$ and $R_F = \infty$ and then connecting one side of R_c to the output side instead of connecting to the ground.

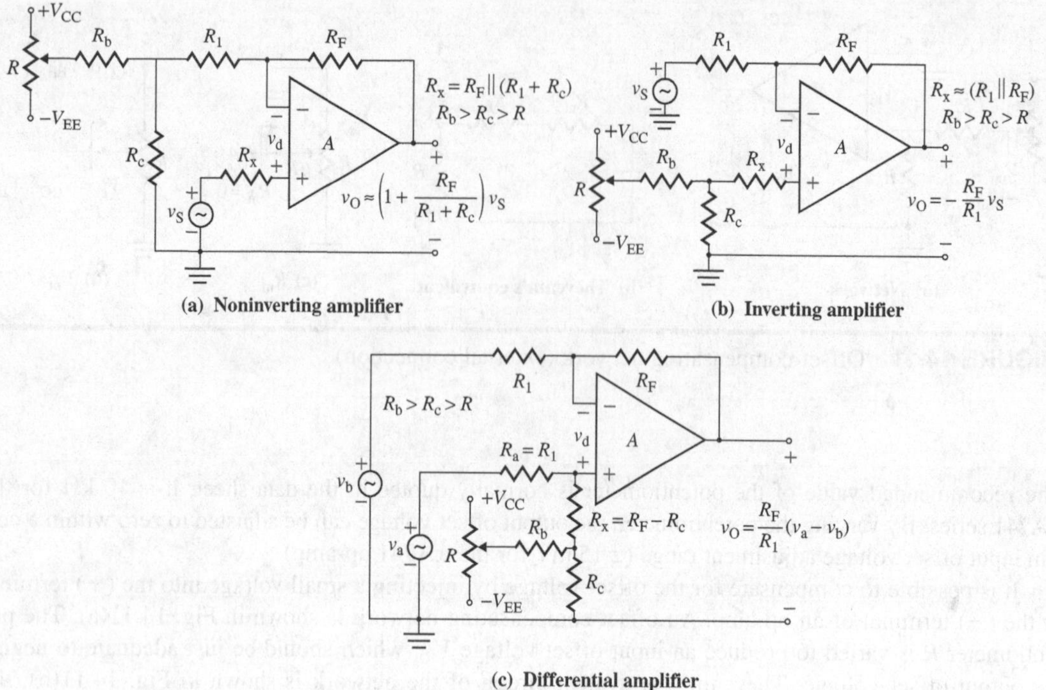

FIGURE 14.12 Op-amp amplifiers with offset-compensating network

EXAMPLE 14.3

D

Designing an offset-compensating network The noninverting amplifier in Fig. 14.12(a) has $R_1 = 10\ \text{k}\Omega$, $R_F = 100\ \text{k}\Omega$, and $R_x = R_F \| R_1 = 9.091\ \text{k}\Omega$. Design the offset-compensating network. The op-amp parameters are $V_{io} = 6$ mV, $I_B = 500$ nA, $I_{io} = 200$ nA, and PSRR = 150 μV/V. The DC supply voltages are $V_{CC} = 15$ V and $-V_{EE} = -15$ V.

SOLUTION

Since the offset due to the biasing current I_B is minimized by the resistance R_x, the output offset voltage will be contributed mostly by V_{io}. For $V_{io} = 6$ mV and $V_{nt} = V_{CC} = 15$ V, Eq. (14.38) gives 6 mV = $15R_c/R_b$.

Letting $R_c = 10\ \Omega$, we get

$$R_b = \frac{15R_c}{6\text{ mV}} = 25\ \text{k}\Omega$$

Letting $R_b = 10R_{nt(max)} = 10(R/4)$, we get

$$R = \frac{4R_b}{10} = 4 \times \frac{25\ \text{k}\Omega}{10} = 10\ \text{k}\Omega \qquad \text{(potentiometer)}$$

The network will change the voltage gain from

$$1 + \frac{R_F}{R_1} = 1 + \frac{100\ \text{k}\Omega}{10\ \text{k}\Omega} = 11$$

to $$1 + \frac{R_F}{R_1 + R_c} = 1 + \frac{100\ \text{k}\Omega}{10{,}010\ \Omega} = 10.99$$

causing a 0.09% error.

14.3.8 Common-Mode Rejection Ratio

The CMRR, which is defined as the ratio of the differential voltage gain to the common-mode voltage gain, may also be defined as the change in input offset voltage per unit change in common-mode voltage. Let $v_{ic} = 0$ and apply v_{id} to drive the output voltage to zero. v_{id} should thus be equal to the input offset voltage V_{OS}. If we keep v_{id} constant and increase v_{ic} by an amount Δv_{ic}, the output voltage will change by an amount

$$\Delta v_O = A_c \Delta v_{ic} \tag{14.39}$$

To drive the output voltage to zero, we have to change v_{id} by an amount Δv_{id}, where

$$\Delta v_{id} = \frac{\Delta v_O}{A_d} = \frac{A_c \Delta v_{ic}}{A_d} = \Delta V_{OS} \tag{14.40}$$

Equation (14.40) indicates that any change in v_{ic} causes a corresponding change in V_{OS}. From Eq. (14.40), we get

$$\text{CMRR} = \frac{A_d}{A_c} = \left.\frac{\Delta v_{ic}}{\Delta v_{id}}\right|_{v_o=0} = \frac{\Delta v_{ic}}{\Delta V_{OS}} = \left.\frac{\delta v_{ic}}{\delta V_{OS}}\right|_{v_o=0} \tag{14.41}$$

which shows that the input offset voltage is dependent on the CMRR and the common-mode signal. For CMRR $= 10^5$ (or 100 dB) and $\Delta v_{ic} = 15$ V, the change in the input offset voltage will be $\Delta V_{OS} = 150\ \mu$V.

14.3.9 Input Resistance

The input resistance for a MOSFET input stage is very high, in the range of $10^9\ \Omega$ to $10^{12}\ \Omega$. For a BJT input stage, however, the input resistance is typically in the range of 100 kΩ to 1 MΩ. Usually, the voltage gain is large enough that this input resistance has little effect on the circuit performance. In some differential amplifiers, a compound transistor configuration known as a *Darlington pair* is used to give a much higher input resistance and a much lower input biasing current than a single transistor would provide. A Darlington pair is shown in Fig. 14.13. The effective B-E voltage of the equivalent transistor Q_T in Fig. 14.13(b) is

$$\begin{aligned} V_{BE} &= V_{BE1} + V_{BE2} = V_T \ln\left(\frac{I_{C1}}{I_{S1}}\right) + V_T \ln\left(\frac{I_{C2}}{I_{S2}}\right) \\ &= V_T \ln\left(\frac{I_{C1} I_{C2}}{I_{S1} I_{S2}}\right) \end{aligned} \tag{14.42}$$

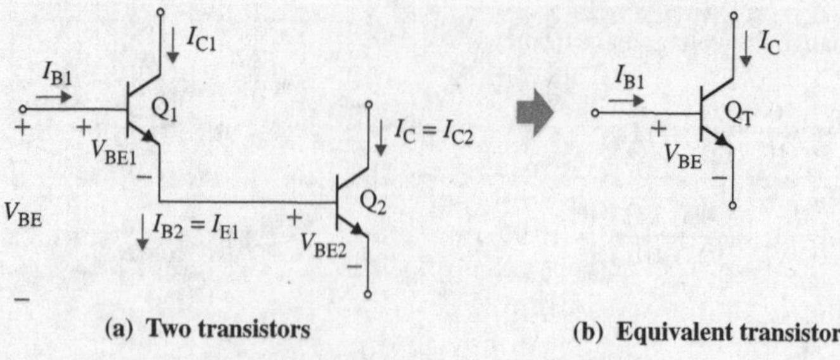

FIGURE 14.13 Darlington pair

Since $I_C = I_{C2} = I_{C1}\beta_F\beta_F/(1 + \beta_F) \approx \beta_F I_{C1}$, Eq. (14.42) becomes

$$V_{BE} = V_T \ln\left(\frac{I_C^2}{\beta_F I_{S1} I_{S2}}\right) = 2V_T \ln\left(\frac{I_C}{\beta_F I_{S1} I_{S2}}\right) \tag{14.43}$$

Solving for I_C, we get

$$I_C = \sqrt{\beta_F I_{S1} I_{S2}}\left(\frac{V_{BE}}{2V_T}\right)$$

$$= I_S \exp\left(\frac{V_{BE}}{V'_T}\right) \tag{14.44}$$

where $I_S = \sqrt{\beta_F I_{S1} I_{S2}}$ = effective saturation current and $V'_T = 2V_T$ = effective thermal voltage.

The collector current I_C of equivalent transistor Q_T can be related to I_{B1} by

$$I_C = I_{C2} = \beta_F I_{B2} = (1 + \beta_F)I_{C1} = \beta_F(1 + \beta_F)I_{B1} \simeq \beta_F^2 I_{B1} \tag{14.45}$$

Thus, the effective input resistance of the compound pair is given by

$$r'_\pi = \frac{V'_T}{I_{B1}} \simeq \beta_F^2 \frac{2V_T}{I_C} \tag{14.46}$$

which will be $2\beta_F$ times greater than that for a single device. For a single equivalent transistor, $r'_\pi = 2\beta_F r_\pi$. Thus, if $I_C = 200\ \mu A$, $\beta_F = 100$, and $V_T = 26$ mV,

$$r_\pi = \frac{\beta_F V_T}{I_C} = 100 \times \frac{26\text{ mV}}{200\ \mu\text{A}} = 13\text{ k}\Omega$$

for a single transistor and

$$r'_\pi = 100^2 \times 2 \times \frac{26\text{ mV}}{200\ \mu\text{A}} = 2.6\text{ M}\Omega$$

for a Darlington pair. The input offset voltage V_{OS}, however, will increase (generally $\sqrt{2}$ times) as a result of the increase in the effective thermal voltage.

14.3.10 Output Resistance

The output stage is usually an emitter follower in class AB operation, and hence it gives a low output resistance, on the order of 40 Ω to 100 Ω. This resistance is low enough that it does not strongly affect performance. If $I_{C(out)}$ is the collector-biasing current of the output stage, then r_π of the output transistor becomes $r_\pi = \beta_F V_T / I_{C(out)}$, which, if converted to the emitter terminal by a dividing factor of $(1 + \beta_f)$, will give approximately the output resistance of the op-amp. That is,

$$R_{out} = \frac{\beta_F V_T}{I_{C(out)}(1 + \beta_F)} \simeq \frac{V_T}{I_{C(out)}} \tag{14.47}$$

The collector-biasing current of the output stage (which is in milliamperes) is generally much greater than that of the differential stage (which is in microamperes). For example, if $\beta_F = 100$, $V_T = 26$ mV, and $I_{C(out)} = 1$ mA, $R_{out} = 26$ mV/1 mA = 26 Ω.

14.3.11 Frequency Response

Because of the parasitic capacitances and the minority carrier charge storage in the devices within the op-amps, the voltage gain decreases at high frequency. The frequency at which the open-loop voltage gain falls to unity is defined as the *unity-gain bandwidth*, and it is in the range of 1 MHz to 20 MHz. The differential stage in Fig. 14.1 can be represented by a voltage-controlled current source, as shown in Fig. 14.14(a). C_1 is the effective capacitance due to the output capacitance of the differential stage and the input capacitance of the second stage. R_1 is the effective resistance due to the output resistance of the differential stage and the input resistance of the second stage. G_{m1} is the transconductance of the differential stage.

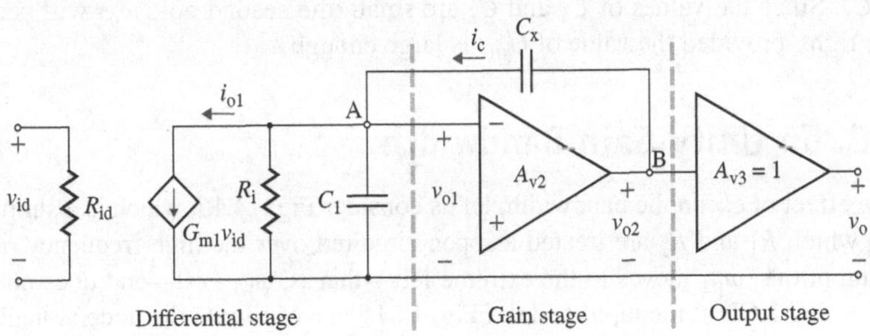

(a) Small-signal equivalent circuit of op-amp

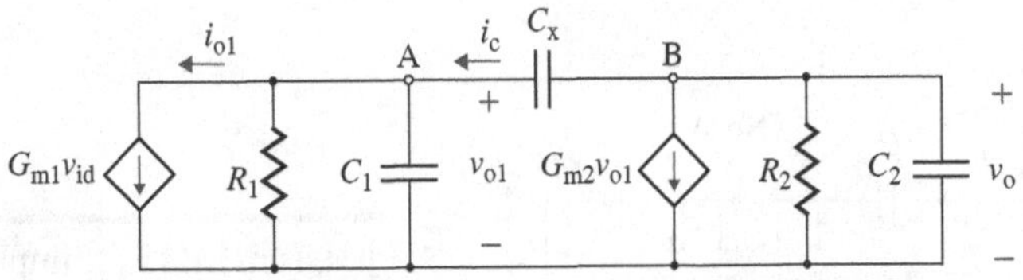

(b) Small-signal high-frequency equivalent circuit

FIGURE 14.14 High-frequency equivalent circuit of an op-amp

The second stage generally has a common-emitter or common-source configuration with an active load for a large voltage gain. It can also be represented by another voltage-controlled current source, as shown in Fig. 14.14(b). G_{m2} is the transconductance of the second stage. C_2 and R_2 are the effective capacitance and the output resistance of the second stage, respectively. A_{v2} is the voltage gain of the second stage and is negative—that is, $A_{v2} = -G_{m2}R_2$.

The gain of the output stage is generally unity: $A_{v3} = 1$. Since the gain of the second stage is negative, the capacitance C_x will exhibit Miller's effect and split the poles. Also, C_x will influence both the frequency response and the slew rate considerably. From Eqs. (10.131) and (10.132), we can find the new poles:

$$\omega_{p1} \approx \frac{1}{G_{m2}C_xR_1R_2} \tag{14.48}$$

$$\omega_{p2} \approx \frac{G_{m2}C_x}{C_1C_2 + C_x(C_1 + C_2)} \tag{14.49}$$

If $C_x >> C_1$ and C_2, Eq. (14.49) can be approximated by

$$\omega_{p2} \approx \frac{G_{m2}C_x}{C_x(C_1 + C_2)} = \frac{G_{m2}}{C_1 + C_2} \tag{14.50}$$

The first pole is due to the Miller capacitance

$$C_M = C_x(1 + G_{m2}R_2) \simeq C_xG_{m2}R_2 \quad (\text{for } G_{m2}R_2 >> 1)$$

which is much larger than C_1. To make ω_{p1} act as the unity-gain frequency ω_u, we can select the appropriate value of C_x. Since the values of C_1 and C_2 are small, the second pole ω_{p2} will become very large and move to the right, provided the value of G_{m2} is large enough.

Effects of C_x on Unity-Gain Bandwidth

To determine the effect of C_x on the bandwidth, let us consider Fig. 14.15, which is a simplified version of Fig. 14.14(b) in which R_1 and R_2 are treated as open circuited over the high-frequency range of interest. Under these assumptions, ω_{p1} moves to the extreme left—that is, $\omega_{p1} \approx 0$—and does not give the unity-gain bandwidth. Using KCL at the input node of Fig. 14.15, we can write the node voltages as

$$(v_o - v_{o1})(\omega C_x) = G_{m1}v_{id} + v_{o1}(\omega C_1) \tag{14.51}$$

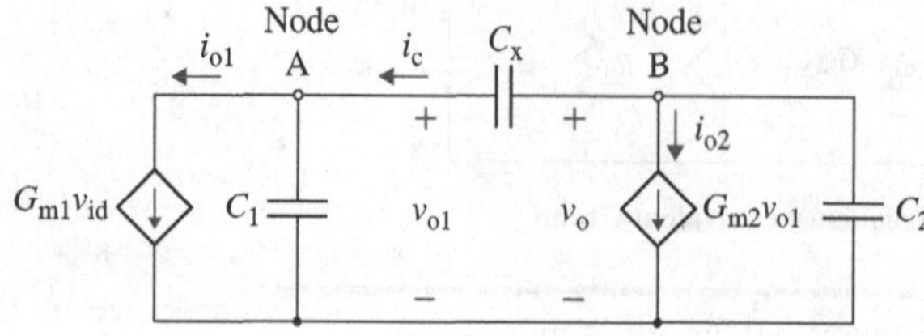

FIGURE 14.15 Simplified high-frequency equivalent circuit of an op-amp

where $G_{m1} = I_Q/2V_T$ is the transconductance of the differential stage. I_Q is the DC biasing current of the differential amplifier shown in Fig. 14.1.

Substituting $v_o = -A_{v2}v_{o1}$ (i.e., $v_{o1} = -v_o/A_{v2}$) into Eq. (14.51) and simplifying, we get the overall voltage gain A_o:

$$A_o(\omega) = \frac{v_o}{v_{id}} = \frac{G_{m1}}{(1 + 1/A_{v2})(\omega C_x) + (\omega C_1)/A_{v2}} \quad (14.52)$$

$$= \frac{G_{m1}/(\omega C_x)}{(1 + 1/A_{v2}) + (C_1/C_x)/A_{v2}} \quad (14.53)$$

where A_{v2} is the voltage gain of the second stage. For $A_{v2} >> 1$, which is usually the case, Eq. (14.53) can be simplified to

$$A_o(\omega) = \frac{G_{m1}}{\omega C_x} \quad (14.54)$$

At the unity-gain frequency ω_u, $|A(\omega)| = 1$. Thus, ω_u is given by

$$\omega_u = \frac{G_{m1}}{C_x} \quad (14.55)$$

which gives the corresponding frequency (i.e., the unity-gain bandwidth) f_u as

$$f_u = \frac{G_{m1}}{2\pi C_x} \quad (14.56)$$

$$= \frac{I_Q}{4\pi V_T C_x} \quad (14.57)$$

Thus, f_u is directly proportional to the biasing current I_Q of the differential stage and inversely proportional to the compensating capacitance C_x. A specific value of C_x is purposely added to the circuit, either on the chip for compensated op-amps or as an external capacitor, to set the desired value of f_u. For example, if $I_Q = 20$ μA, $V_T = 26$ mV, and $C_x = 50$ pF, we get $f_u = 1.22$ MHz.

Effects of C_x on Zeros

The capacitance C_x also has a Miller's effect on the output side of the second stage, given by

$$C_N = C_x\left(1 + \frac{1}{G_{m2}R_2}\right) \approx C_x \quad (\text{for } G_{m2}R_2 >> 1)$$

and introduces a right-half-plane zero on the transfer function. The zero ω_z can be found from Fig. 14.16 by making the output voltage $v_o \approx 0$, as shown in Fig. 14.16(a). We get

$$G_{m2}v_{o1} = (v_{o1} - v_o)(\omega C_x) = v_{o1}(\omega C_x)$$

which gives the zero frequency ω_z as

$$\omega_z = \frac{G_{m2}}{C_x} \quad (14.58)$$

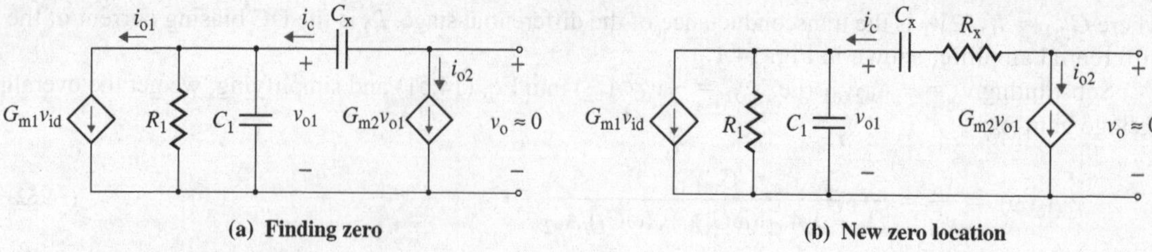

FIGURE 14.16 Equivalent circuit for determining zeros

If G_{m2} is large, which is usually the case for BJT amplifiers, then the zero will be at a very high frequency. However, if G_{m2} of the second stage is of the same magnitude as G_{m1} of the first stage, which is generally the case for CMOS amplifiers, the zero frequency will be close to the unity-gain frequency ω_u.

Since a zero introduces a phase shift, the phase margin of the amplifier will be decreased, affecting the amplifier stability. However, it is possible to remedy the stability problem by adding a resistance R_x in series with C_x, as shown in Fig. 14.16(b). The combination of R_x and C_x can move the zero and make ω_z a very large frequency. The zero frequency ω_z can be found by making the output voltage $v_o \cong 0$; that is,

$$G_{m2}v_{o1} = \frac{(v_{o1} - v_o)}{R_x + 1/(j\omega C_x)}$$

which gives the new zero frequency ω_z as

$$\omega_z = \frac{1}{C_x(1/G_{m2} - R_x)} \tag{14.59}$$

Thus, as R_x approaches $1/G_{m2}$, the zero frequency ω_z tends to approach infinity. It is important to note that, by making $R_x > 1/G_{m2}$, we can locate the zero frequency at the negative real axis, which will increase the phase margin.

According to Eq. (14.50), the second pole ω_{p2} may be close to the unity-gain frequency ω_u for a low value of G_{m2}. As a result, ω_{p2} may introduce an appreciable phase shift and thus decrease the phase margin. This problem can be resolved by increasing the value of C_x in order to split the poles further.

14.3.12 Slew Rate

Op-amps are limited by the *slew rate* (SR), which specifies the maximum rate at which the output voltage can change without introducing any significant amount of distortion. That is, SR = $(dv_o/dt)_{max}$. Like the unity-gain frequency, the SR depends on the capacitance C_x. Figure 14.15 illustrates that the current i_c through the capacitor is related to the output voltage v_o by

$$i_c = C_x\frac{d}{dt}(v_o - v_{o1}) \tag{14.60}$$

$$= C_x\frac{d}{dt}\left(v_o + \frac{v_o}{A_{v2}}\right)$$

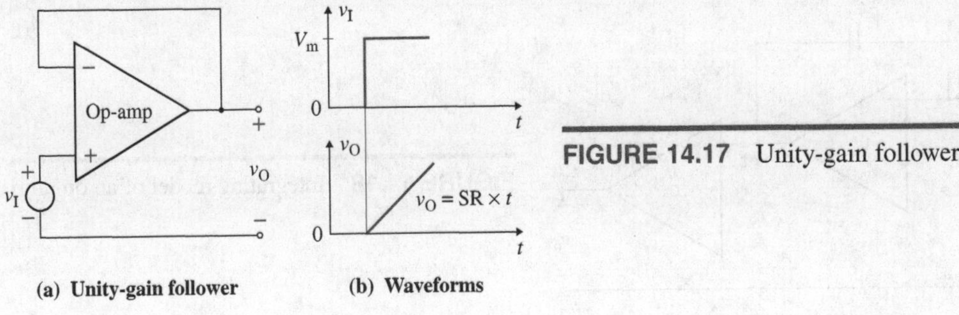

FIGURE 14.17 Unity-gain follower

(since $v_{o1} = -v_o/A_{v2}$), which gives dv_o/dt as

$$\frac{dv_o}{dt} = \frac{i_c}{C_x(1 + 1/A_{v2})} \tag{14.61}$$

Let us assume that $A_{v2} >> 1$—that is, v_{o1} tending to zero, $v_{o1} \approx 0$—and that i_c can be approximated by the output current of the differential amplifier (i.e., $i_c = i_{o1} = G_{m1}v_{id}$). Then Eq. (14.61) can be approximated by

$$\frac{dv_o}{dt} \simeq \frac{i_c}{C_x} = \frac{i_{o1}}{C_x} \tag{14.62}$$

Consider the unity-gain follower shown in Fig. 14.17(a). If a step input $v_I = V_m$ (say, 10 V) is applied to it, in zero time the output will not change, as shown in Fig. 14.17(b). Thus, V_m will appear as the differential voltage between the two input terminals, and the differential stage in Fig. 14.1 will be overdriven. Transistors Q_1 and Q_3 in Fig. 14.1 will carry the whole biasing current I_Q, and transistor Q_2 will be cut off. Q_4, however, is the mirror of Q_3, and it will produce a current of I_Q, which will flow into the second stage.

The biasing current limits the maximum value of the output current i_{o1} of the differential stage to the value I_Q in one direction and the value $-I_Q$ in the other direction; that is, $i_{o1(max)} = \pm I_Q$. Thus, the SR corresponding to $i_{o1(max)}$ is given by

$$SR = \left.\frac{dv_o}{dt}\right|_{max} = \frac{i_{o1(max)}}{C_x} = \pm\frac{I_Q}{C_x} \tag{14.63}$$

Like the unity-gain frequency f_u, the slew rate is directly proportional to I_Q and inversely proportional to C_x. If the gain of the second stage is large, then the op-amp behaves as an integrator, as shown in Fig. 14.18.

Relation between SR and f_u

Substituting $I_Q = 4\pi V_T C_x f_u$ from Eq. (14.57) into Eq. (14.63), we can relate the positive SR (known simply as SR) to f_u; that is,

$$SR = \frac{4\pi V_T C_x f_u}{C_x} = 4\pi V_T f_u \tag{14.64}$$

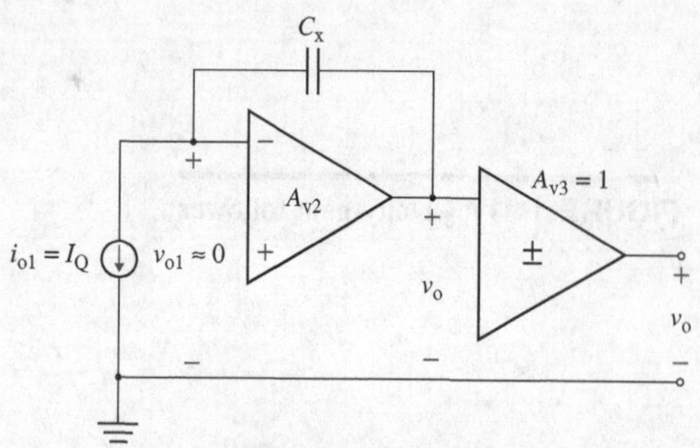

FIGURE 14.18 Integrating model of an op-amp

or

$$f_u = \frac{SR}{4\pi V_T} \tag{14.65}$$

Thus, there is a direct relation between the SR and the unity-gain frequency of an op-amp. For example, if $I_Q = 20\ \mu A$, $V_T = 26$ mV, and $C_x = 50$ pF, then Eq. (14.54) gives $f_u = 1.22$ MHz and Eq. (14.63) gives

$$SR = \frac{I_Q}{C_x} = \frac{20\ \mu A}{50\ pF} = 0.4\ V/\mu s$$

KEY POINTS OF SECTION 14.3

- The CMRR of an op-amp is very large (typically 10^5), and the output voltage due to a common-mode signal is negligible.
- The voltage gain of an op-amp decreases with frequency; however, the gain–bandwidth product remains constant. That is, if the gain decreases, the bandwidth increases.
- The slew rate SR of an op-amp limits the maximum input frequency at which the op-amp can amplify a signal without significant distortion. For minimum distortion, the SR of the input signal should be less than that of the op-amp.
- An imperfect op-amp produces an output offset voltage caused by parameters such as V_{io}, I_{io}, I_B, PSRR, and thermal drift. A resistor is usually connected at the (+) terminal to minimize offset due to DC-biasing currents.
- A compensating network (either internal or external) may be connected to negate the offset voltage.
- The input offset voltage of MOSFET amplifiers is considerably higher than that of BJT amplifiers. This difference is due to mismatches in base widths for BJTs, in I_{DSS} and V_p for MOSFETs, and in V_t and W/L for MOSFETs.
- A Darlington BJT pair is commonly used to provide high input resistance, low input base current, and high current gain.
- The unity-gain frequency f_u depends directly on the biasing current I_Q of the differential stage and inversely on the compensation capacitance C_x. The slew rate SR is directly proportional to f_u.
- If the transconductance of the gain stage is on the same order as that of the first stage, the zero frequency will be close to the unity-gain frequency and the phase margin will be reduced. A resistor R_x is usually connected in series with C_x to move the zero frequency to infinity.

14.4 CMOS Op-Amps

The internal structure of operational amplifiers has, in general, three stages: a differential stage, a gain stage, and an output stage. The gain and output stages are often combined into a single stage. This type of amplifier has a very high input resistance, and the voltage gain is generally less than that of other types of comparable amplifiers. In this section, we illustrate three CMOS amplifiers: one basic type and two commercial types.

14.4.1 Basic CMOS Op-Amp

A basic CMOS amplifier with a differential stage and a gain stage is shown in Fig. 14.19 [4, 5]. PMOS transistors M_6, M_7, and M_8 are used to generate biasing current sources. NMOS transistors M_1, M_2, M_3, and M_4 form the differential stage. NMOS M_1 operates as a common-source amplifier. C_x is used for frequency compensation by introducing a dominant pole (see Sec. 10.12). The main task in the design of CMOS amplifiers is to determine the width-to-length (W/L) ratios of the NMOS and PMOS. Usually, W/L ratios of M_3, M_4, and M_8 are half of those for other MOSFETs. The differential pair is the "heart" of op-amps and can be formed by cascoding MOSFETs to yield high gain and low CMRR (see Sec. 9.2.1).

Using Eq. (9.61), the voltage gain of the differential stage is given by

$$A_1 = -(g_{m2} + g_{m4})(r_{o2} \parallel r_{o4})$$

and the voltage gain of the gain (second) stage is given by

$$A_2 = -g_{m5}(r_{o5} \parallel r_{o6})$$

Therefore, the overall low-frequency voltage gain of the amplifier is

$$A_o = A_1 A_2 = (g_{m2} + g_{m4})(r_{o2} \parallel r_{o4}) g_{m5}(r_{o5} \parallel r_{o6}) \tag{14.66}$$

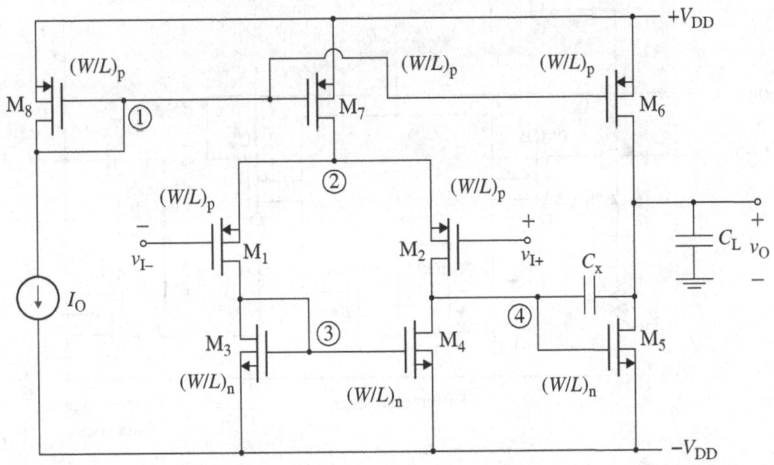

FIGURE 14.19 Basic CMOS amplifier

TABLE 14.1 Parameters of op-amp MC14573

Parameter	Minimum	Typical	Maximum	Units
DC supply voltages V_{DD}			± 18	V
Input offset voltage V_{OS}		160	600	μV
Thermal drift D_v		1		μV/°C
Input biasing current I_B			1.0	nA
Input offset current I_{OS}			200	pA
Differential input resistance R_{id}		10^{12}		Ω
Common-mode input resistance R_{ic}		10^{12}		Ω
Output resistance R_o		50		Ω
Input capacitance C_i		1.0		pF
Open-loop voltage gain A_o		90		dB
Common-mode rejection ratio (CMRR)	75	95		dB
Unity-gain bandwidth f_u		70		MHz
Slew rate (SR)		2.5		V/μs
Power supply rejection ratio (PSRR)	75	97		dB

14.4.2 CMOS Op-Amp MC14573

The simplified schematic of op-amp MC14573 is shown in Fig. 14.20. Its internal structure can be divided into a differential CMOS stage, a gain stage, and a biasing circuit. This op-amp has a very high input resistance, but the gain is lower than that of other op-amps, as expected. Some parameters of the MC14573 op-amp are listed in Table 14.1.

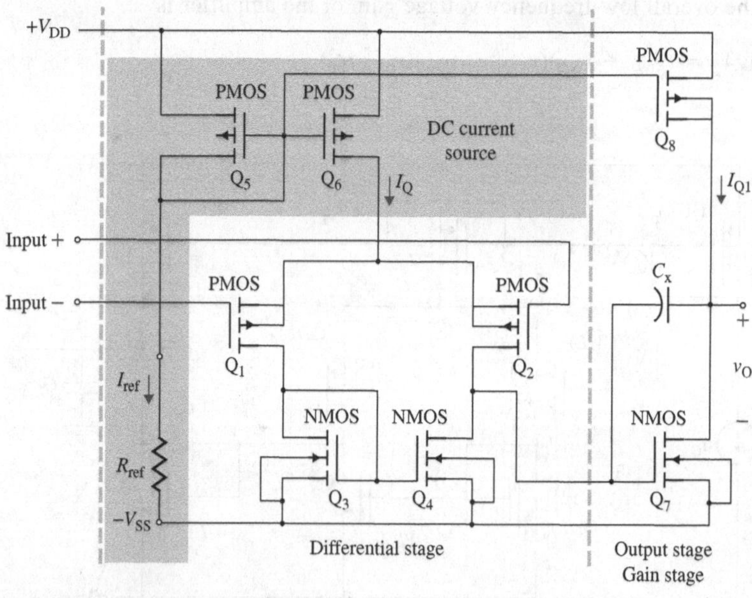

FIGURE 14.20 Schematic for op-amp MC14573 (Copyright of Motorola. Used by permission.)

Differential Stage

The input stage is a PMOS differential amplifier consisting of transistors Q_1 and Q_2. Its load is a current mirror consisting of NMOS transistors Q_3 and Q_4.

Gain Stage

The gain stage is also the output stage. Q_7 is operated as a common-source amplifier with Q_8 as the active load. C_x is the compensation capacitor connected to Q_7.

DC Biasing

The differential stage is biased by transistors Q_5 and Q_6 as a current source. An external resistance R_{ref} sets the DC biasing current I_Q, which can be found approximately from

$$I_Q = I_{ref} = \frac{V_{DD} + V_{SS} - V_{SG}}{R_{ref}} \tag{14.67}$$

$$= K_p(V_{GS} - V_t)^2$$

The value of V_{GS5} can be solved for known values of V_{DD}, V_{SS}, V_t, and K_p. Transistor Q_8 serves as the load to the output stage. Since the gate voltages of Q_5, Q_6, and Q_8 are the same, the biasing drain current Q_8 equals $I_{Q1} = I_Q$.

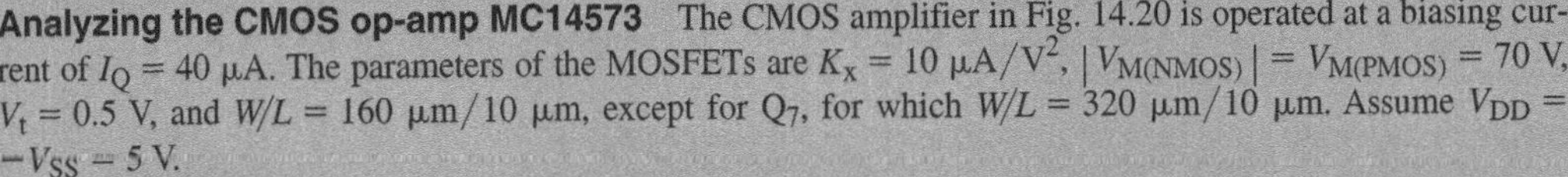

Analyzing the CMOS op-amp MC14573 The CMOS amplifier in Fig. 14.20 is operated at a biasing current of $I_Q = 40\ \mu A$. The parameters of the MOSFETs are $K_x = 10\ \mu A/V^2$, $|V_{M(NMOS)}| = V_{M(PMOS)} = 70$ V, $V_t = 0.5$ V, and $W/L = 160\ \mu m/10\ \mu m$, except for Q_7, for which $W/L = 320\ \mu m/10\ \mu m$. Assume $V_{DD} = -V_{SS} = 5$ V.

(a) Find V_{GS}, g_m, and r_o for all MOSFETs.

(b) Find the low-frequency voltage gain of the amplifier A_{vo}.

(c) Find the value of the external resistance R_{ref}.

(d) Find the value of compensation capacitance C_x that gives a unity-gain bandwidth of 1 MHz and the corresponding slew rate.

(e) Find the value of resistance R_x to be connected in series with C_x in order to move the zero frequency to infinity.

(f) Find the common-mode input voltage range.

(g) Find the output voltage range.

SOLUTION

(a) K_p, V_{GS}, g_m, and r_o are calculated using the following equations:

$$K_p = \frac{K_x W}{L} \tag{14.68}$$

TABLE 14.2 Calculated values for Example 14.4

	Q_1	Q_2	Q_3	Q_4	Q_5	Q_6	Q_7	Q_8
W/L (in μm/μm)	160/10	160/10	160/10	160/10	160/10	160/10	320/10	160/10
K_p (in μA/V^2)	160	160	160	160	160	160	320	160
I_D (in μA)	20	20	20	20	40	40	40	40
V_{GS} (in V)	0.854	0.854	0.854	0.854	1	1	0.854	1
g_m (in μA/V)	113.1	113.1	113.1	113.1	160	160	226.3	160
r_o (in MΩ)	3.5	3.5	3.5	3.5	1.75	1.75	1.75	1.75

$$V_{GS} = V_t + \sqrt{\frac{I_D}{K_p}} \tag{14.69}$$

$$g_m = 2K_p(V_{GS} - V_t) = 2\sqrt{K_p I_D} \tag{14.70}$$

$$r_o = \frac{|V_M|}{I_D} \tag{14.71}$$

Their values are shown in Table 14.2.

(b) Using Eq. (7.34) the voltage gain of the first differential stage with the active load is given by

$$A_{v1} = -g_{m4}(r_{o2} \parallel r_{o4}) \tag{14.72}$$

$$= -113.1\ \mu \times (3.5\ \text{M} \parallel 3.5\ \text{M}) = -198$$

The voltage gain of the second stage with the active load is given by

$$A_{v2} = -g_{m7}(r_{o7} \parallel r_{o8}) \tag{14.73}$$

$$= -226.3\ \mu \times (1.75\ \text{M} \parallel 1.75\ \text{M}) = -198$$

Thus, the overall voltage gain is $A_{vo} = A_{v1}A_{v2} = 198 \times 198 = 39{,}204$ (or 91.87 dB).

(c) From Eq. (14.67), we get the value of the external resistance R_{ref} as

$$R_{ref} = \frac{V_{DD} - V_{SS} - V_{GS5}}{I_Q} = \frac{(5 + 5 - 1)\ \text{V}}{40\ \mu\text{A}} = 225\ \text{k}\Omega$$

(d) For $f_u = 1$ MHz, Eq. (14.56) gives the value of compensation capacitance C_x as

$$C_x = \frac{G_{m1}}{2\pi f_u} = \frac{G_{m2}}{2\pi f_u} = \frac{113.1\ \mu\text{A/V}}{2\pi \times 1\ \text{MHz}} = 18\ \text{pF}$$

From Eq. (14.63), we get the slew rate as

$$\text{SR} = \frac{I_Q}{C_x} = \frac{40\ \mu\text{A}}{18\ \text{pF}} = 2.22\ \text{V}/\mu\text{s}$$

(e) From Eq. (14.59), the value of resistance R_x for $\omega_z = 0$ is

$$R_x = \frac{1}{G_{m2}} = \frac{1}{g_{m7}} = \frac{1}{226.3\ \mu\text{A/V}} = 4.42\ \text{k}\Omega$$

(f) Transistor Q_6 will leave saturation when the common-mode input voltage becomes

$$v_{ic(max)} = V_{DD} - |V_{GS6}| + |V_t| - |V_{GS1}| = 5 - 1 + 0.5 - 0.854 = 3.646 \text{ V}$$

Transistors Q_1 and Q_2 will leave saturation when the common-mode input voltage falls below the voltage at the drain of Q_1 by $|V_t|$—that is, when

$$v_{ic(min)} = -V_{SS} + |V_{GS3}| - |V_t| = -5 + 1 - 0.5 = -4.5 \text{ V}$$

Thus, the common-mode input voltage range is −4.5 V to 3.646 V.

(g) Transistor Q_8 will leave saturation when the output voltage becomes

$$v_{o(max)} = V_{DD} - |V_{GS8}| + |V_t| = 5 - 1 + 0.5 = 4.5 \text{ V}$$

Transistor Q_7 will leave saturation when the output voltage becomes

$$v_{o(min)} = -V_{SS} + |V_{GS7}| - |V_t| = -5 + 0.854 - 0.5 = -4.646 \text{ V}$$

Thus, the output voltage range is −4.646 V to 4.5 V.

14.4.3 CMOS Op-Amp TLC1078

The simplified schematic of the TLC1078 op-amp is shown in Fig. 14.21. Its structure is similar to that of op-amp MC14573 in Fig. 14.20. The internal structure can be divided into a differential CMOS stage, a gain stage, an output stage, and a biasing circuit. Some parameters of the TLC1078 op-amp are listed in Table 14.3.

TABLE 14.3 Parameters of op amp TLC1078

Parameter	Minimum	Typical	Maximum	Units
DC supply voltages V_{DD}			18	V
Input offset voltage V_{OS}		180	600	μV
Thermal drift D_v		1		μV/°C
Input biasing current I_B		0.7		pA
Input offset current I_{OS}		0.1		pA
Differential input resistance R_{id}		1.5		TΩ
Common-mode input resistance R_{ic}		1.5		TΩ
Output resistance R_o		50		Ω
Input capacitance C_i		1.0		pF
Open-loop voltage gain A_o	110	120		dB
Common-mode rejection ratio (CMRR)	75	97		dB
Unity-gain bandwidth f_u		110		MHz
Slew rate (SR)		47		V/ms
Power supply rejection ratio (PSRR)	75	97		dB

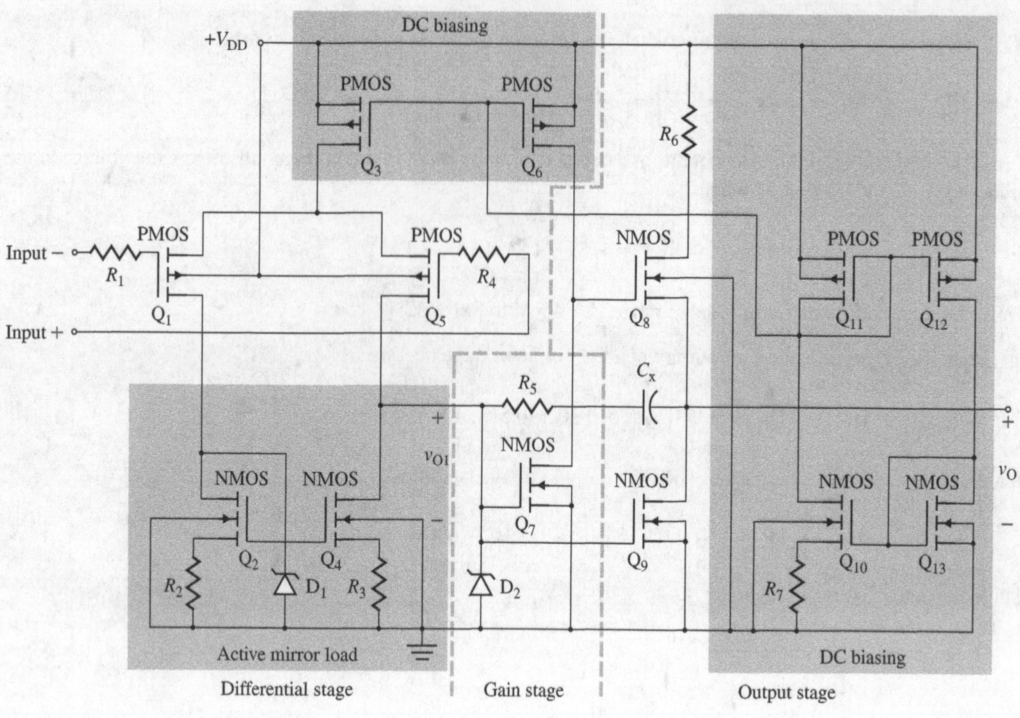

FIGURE 14.21 Schematic for op-amp TLC1078 (Reprinted by permission of Texas Instruments)

Differential Stage

The input stage is a PMOS differential amplifier consisting of transistors Q_1 and Q_5. Its load is a current mirror consisting of NMOS transistors Q_2 and Q_4, which have sources R_2 and R_3 to give high output resistances.

Gain Stage

The gain stage uses NMOS Q_7 in a common-source configuration with PMOS Q_6 as the current source active load. The combination of C_x and R_5 is a pole-zero circuit, which can reduce the lowest pole or break frequency to a low value to ensure stability. Also, it can produce a zero to cancel out one of the second poles of the amplifier's open-loop frequency response.

Output Stage

The output stage consists of a pair of NMOS transistors that operate in the class AB push-pull mode. Transistor Q_8 is the source follower, sourcing current to the load during the interval when the signal at the output of Q_7 goes up above the quiescent value. Transistor Q_9 acts as the common-source amplifier, sinking current from the load during the interval when the signal at the output of the differential stage goes down below the quiescent value.

A decreasing signal at the output of the differential stage will be amplified by Q_7 with a phase shift of 180° and then will appear through the source follower of Q_8 as an increasing signal to the load. However, an increasing signal will be amplified by the common-source amplifier of Q_9 with a phase shift of 180° and then will appear as a decreasing signal to the load. Thus, the voltages applied to the gates of Q_8 and Q_9 will be phase shifted by 180°. The voltage gain through the two paths will be the same, however.

DC Biasing

The differential amplifier is biased by transistors Q_3 and Q_6 as a current source whose gate voltage is set by the voltage divider circuit consisting of transistors Q_{10} through Q_{13}. All transistors are identical, except for Q_{10} and Q_{13}, whose channel widths differ from that of the others.

EXAMPLE 14.5

D

Analyzing the CMOS op-amp TLC1078 The CMOS amplifier in Fig. 14.21 is operated at a biasing current of $I_Q = 40\ \mu A$. The parameters of the MOSFETs are $K_x = 10\ \mu A/V^2$, $|V_{M(NMOS)}| = V_{M(PMOS)} = 70$ V, $V_t = 0.5$ V, and $W/L = 160\ \mu m/10\ \mu m$, except for Q_{10}, for which $W/L = 40\ \mu m/10\ \mu m$. Find the value of the resistance R_7. Assume $V_{DD} = V_{SS} = 5$ V.

SOLUTION

We have

$$K_p = \frac{K_x W}{L} = 10\ \mu \times \frac{160}{10} = 160\ \mu A/V^2 \qquad \text{(for all MOSFETs expect } Q_{10}\text{)}$$

$$K_{p10} = \frac{K_x W}{L} = 10\ \mu \times \frac{40}{10} = 40\ \mu A/V^2 \qquad \text{(for } Q_{10}\text{)}$$

For $I_Q = 40\ \mu A$ and $K_p = 160\ \mu A/V^2$, Eq. (14.69) gives

$$V_{GS} = V_t + \sqrt{I_D/K_p} = 0.5 + \sqrt{40/160} = 1 \text{ V}$$

Thus,

$$V_{GS3} = V_{GS6} = V_{GS11} = V_{GS12} = V_{GS13} = 1 \text{ V}$$

Also,

$$I_{D3} = I_{D6} = I_{D11} = I_{D12} = I_{D10} = I_{D13} = I_Q$$

The drain current I_{D13} of Q_{13} is given by

$$I_{D13} = K_{p13}(V_{GS13} - V_t)^2 \tag{14.74}$$

Since $V_{GS10} = V_{GS13} - I_Q R_7$, the drain current I_{D10} of Q_{10} is given by

$$I_{D10} = K_{p10}(V_{GS10} - V_t)^2 = K_{p10}(V_{GS13} - I_Q R_7 - V_t)^2 \tag{14.75}$$

The V_{GS} values of Q_{10} and Q_{13} are different, but their drain currents are equal; that is,

$$I_{D13} = I_{D10} \quad \text{and} \quad K_{p13}(V_{GS13} - V_t)^2 = K_{p10}(V_{GS13} - I_Q R_7 - V_t)^2$$

which relates R_7 to I_Q as follows:

$$R_7 = \frac{V_{GS13} - V_t}{I_Q}\left[\left(\frac{K_{p13}}{K_{p10}}\right)^{1/2} - 1\right] \tag{14.76}$$

Since K_p is proportional to the ratio W/L, Eq. (14.76) can be expressed as

$$R_7 = \frac{V_{GS13} - V_t}{I_Q}\left[\left(\frac{W_{13}}{W_{10}}\right)^{1/2} - 1\right] \tag{14.77}$$

which, for $I_Q = 40\ \mu A$, $W_{10} = 40\ \mu m$, $W_{13} = 160\ \mu m$, $V_{GS13} = 1$ V, and $V_t = 0.5$ V, gives $R_7 = 12.5\ k\Omega$. Thus, the ratio $\sqrt{W_{13}/W_{10}}$ sets the value of resistance R_7 or I_Q.

KEY POINTS OF SECTION 14.4

- CMOS op-amps can have either two or three stages. The gain and output stages are often combined into a single stage. This type of amplifier has a very high input resistance, and the voltage gain is generally less than that of other comparable types of amplifiers.
- Since MOSFETs have lower transconductance than other amplifiers, the gain stage often requires zero-pole compensation (see Fig. 14.21).

14.5 BJT Op-Amps

Like MOSFET op-amps, BJT amplifiers have three stages: a differential stage, a gain stage, and an output stage. They have the disadvantages of a lower input resistance and a higher input biasing current. However, BJTs provide higher voltage gain [6–8]. In this section, we consider two popular op-amps.

14.5.1 BJT Op-Amp LM124

The LM124 op-amp is designed to operate from a power supply of ±16 V, but it can operate from a single power supply as low as 5 V. The simplified schematic is shown in Fig. 14.22. Some parameters of the LM124 op-amp are listed in Table 14.4.

Differential Stage

The differential input stage consists of bipolar transistors Q_1 through Q_4. They are connected in Darlington pairs to give high input resistances and low input biasing currents. These BJTs drive the current mirror load consisting of transistors Q_8 and Q_9 for a high differential gain.

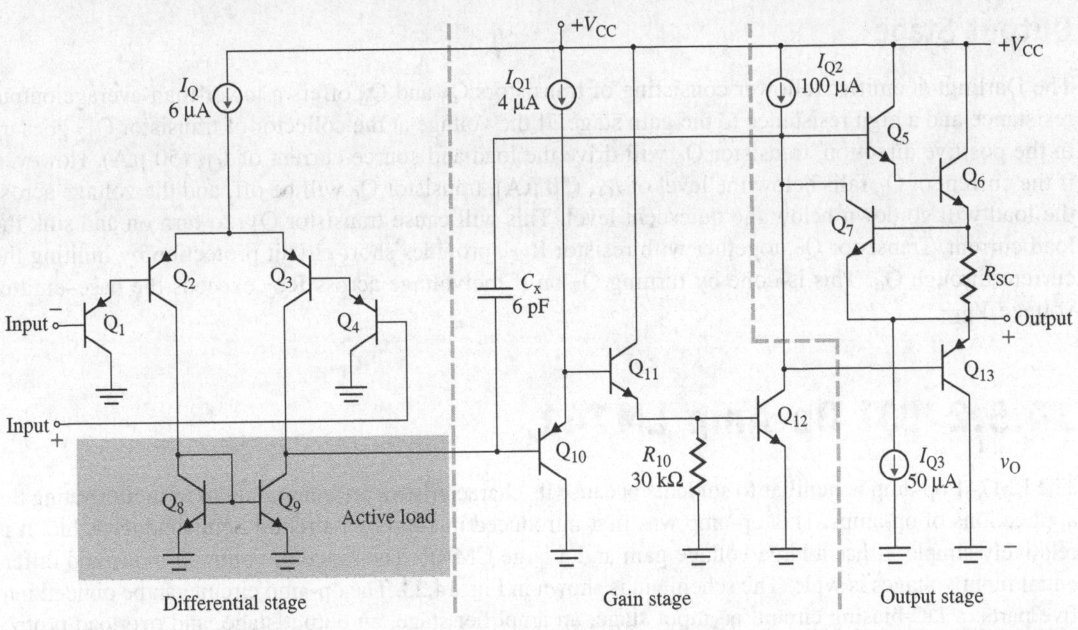

FIGURE 14.22 Simplified schematic for op-amp LM124 (Courtesy of National Semiconductor, Inc.)

Gain Stage

The gain stage is a common-emitter amplifier consisting of *pnp* transistor Q_{10} (with an active load), which is followed by a Darlington pair consisting of transistors Q_{11} and Q_{12}. The beta of a *pnp* transistor is generally low. The Darlington pair offers a high load to transistor Q_{10}. This arrangement ensures a high voltage gain (100 dB). The number of poles in the op-amp increases with the number of transistor stages. C_x is used for feedback compensation.

TABLE 14.4 Parameters of op-amp LM124

Parameter	Minimum	Typical	Maximum	Units
DC supply voltages V_S			+16	V
Input offset voltage V_{OS}		1.0	2.0	mV
Thermal drift D_v		100	200	μV/°C
Input biasing current I_B		20	50	nA
Input offset current I_{OS}		2.0	10	nA
Differential input resistance R_{id}		2		MΩ
Common-mode input resistance R_{ic}		2		MΩ
Output resistance R_o		75		Ω
Input capacitance C_i		4.0		pF
Open-loop voltage gain A_o	50	100		V/mV
Common-mode rejection ratio (CMRR)	70	85		dB
Unity-gain bandwidth f_u		1.0		MHz
Slew rate (SR)		3.0		V/μs
Power supply rejection ratio (PSRR)	650	100		dB

Output Stage

The Darlington emitter follower consisting of transistors Q_5 and Q_6 offers a lower-than-average output resistance and a high resistance to the gain stage. If the voltage at the collector of transistor Q_{12} goes up in the positive direction, transistor Q_6 will drive the load and source current of I_{Q3} (50 μA). However, if the current of Q_6 falls below the level of I_{Q3} (50 μA), transistor Q_6 will be off, and the voltage across the load will go down below the quiescent level. This will cause transistor Q_{13} to turn on and sink the load current. Transistor Q_7, together with resistor R_{SC}, provides short-circuit protection by limiting the current through Q_6. This is done by turning Q_7 on if the voltage across R_{SC} exceeds the base–emitter voltage V_{BE7}.

14.5.2 BJT Op-Amp LM741

The LM741 op-amp is familiar to students because its characteristics are commonly used in illustrating the applications of op-amps. This op-amp was first introduced in 1966 by Fairchild Semiconductor, Inc. It is relatively simple. It has a large voltage gain and a large CMRR. The range of common-mode and differential input voltages is wide. The schematic is shown in Fig. 14.23. The op-amp circuit can be divided into five parts: a DC-biasing circuit, an input stage, an amplifier stage, an output stage, and overload protection. Some parameters of the LM741 op-amp are listed in Table 14.5.

Differential Stage

The differential input stage consists of bipolar transistors Q_1 through Q_4. They form common-emitter and common-base configurations (see Fig. 9.42) for higher bandwidth and gain. This stage is biased by the current source consisting of Q_8, and it drives the current mirror active load consisting of Q_5, Q_6, and Q_7. The biasing feedback loop formed by Q_8 and Q_9 stabilizes the biasing currents in each of the input transistors at approximately one-half of the collector current of Q_{10}.

TABLE 14.5 Parameters of op-amp LM741

Parameter	Minimum	Typical	Maximum	Units
DC supply voltages V_{CC}			±22	V
Input offset voltage V_{OS}		1.0	5.0	mV
Thermal drift D_v		15		μV/°C
Input biasing current I_B		80	550	nA
Input offset current I_{OS}		20	200	nA
Differential input resistance R_{id}	0.3	2		MΩ
Common-mode input resistance R_{ic}		2		MΩ
Output resistance R_o		75		Ω
Input capacitance C_i		4.0		pF
Open-loop voltage gain A_o	50	200		V/mV
Common-mode rejection ratio (CMRR)	70	90		dB
Unity-gain bandwidth f_u		1.0		MHz
Slew rate (SR)		0.5		V/μs
Power supply rejection ratio (PSRR)	70	90		dB

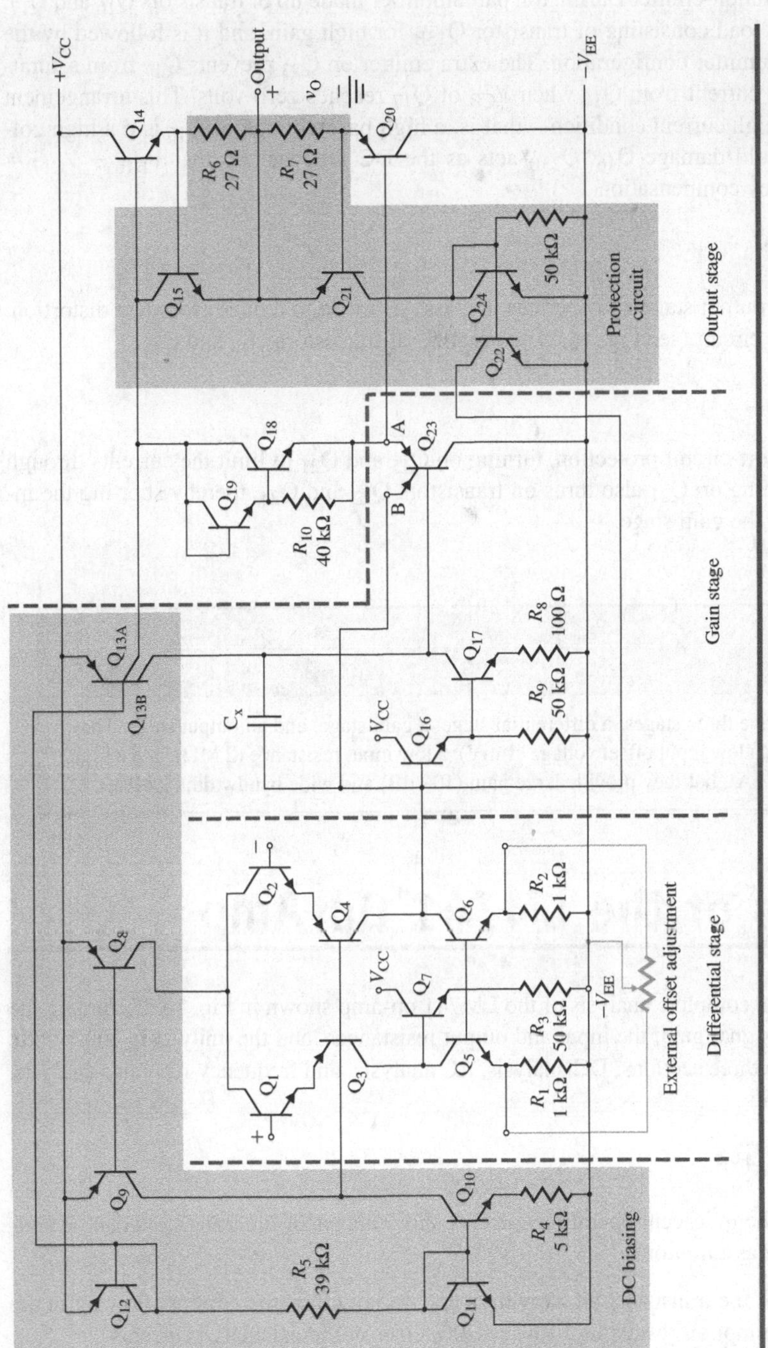

FIGURE 14.23 Schematic for op-amp LM741 (Courtesy of National Semiconductor, Inc.)

Gain Stage

The gain stage consists of a common-emitter Darlington pair amplifier made up of transistors Q_{16} and Q_{17}. Transistor Q_{17} drives an active load consisting of transistor Q_{13B} for high gain, and it is followed by the *pnp* transistor Q_{23} in common-emitter configuration. The extra emitter on Q_{23} prevents Q_{17} from saturating by diverting the base drive current from Q_{16} when V_{CB} of Q_{17} reaches zero volts. This arrangement eliminates the possibility of a high current condition—that is, a high base voltage of Q_{16} and a high collector voltage of Q_7—that could damage Q_{16}. Q_{23B} acts as the DC feedback so that $V_{B16} - V_{C17} = V_{EB23B}$. C_x is used for frequency compensation.

Output Stage

The output stage is a push-pull output stage and operates in class AB mode to reduce crossover distortion. It is biased by a V_{BE} multiplier circuit (see Fig. 14.24) consisting of transistors Q_{18} and Q_{19}.

Protection Circuitry

Resistors R_6 and R_7 provide short-circuit protection, turning on Q_{15} and Q_{21} to limit the currents through Q_{14} and Q_{20}, respectively. Turning on Q_{21} also turns on transistors Q_{22} and Q_{24}, thereby shorting the input signal to the base of Q_{16} in the gain stage.

KEY POINT OF SECTION 14.5

- BJT amplifiers usually have three stages: a differential stage, a gain stage, and an output stage. They have the disadvantages of a low input offset voltage (1 mV), a low input resistance (2 MΩ), and a high input biasing current (60 nA), but they provide large gain (100 dB) and wide bandwidth (1 MHz).

14.6 Analysis of the LM741 Op-Amp

As an example, let us carry out a complete analysis of the LM741 op-amp shown in Fig. 14.23, finding the DC biasing currents, the small-signal gain, the input and output resistances, and the unity-gain bandwidth. The analysis can be divided into three sections: DC analysis, AC analysis, and frequency-response analysis.

14.6.1 DC Analysis

The DC analysis to determine the quiescent operating currents and voltages of the transistors can be simplified by making the following assumptions:

1. The output resistances of the transistors are very high and do not affect the currents flowing in the circuit. Usually, this assumption results in a 10% to 20% error in the calculated currents.
2. The output voltage is maintained at a constant specified value by the internal feedback loop. Let us assume that the output voltage is zero. This assumption is necessary because of the very high gain,

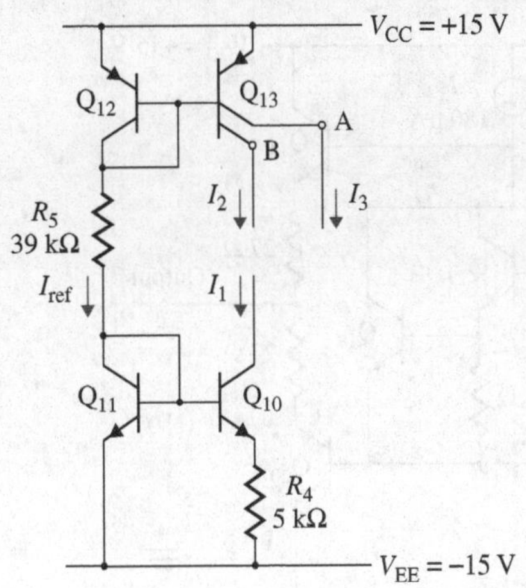

FIGURE 14.24 Biasing circuit for the LM741 op-amp

typically 10^5, which results in a very low input voltage. If the output voltage is calculated with two input terminals grounded, any change in the beta or output resistances could cause a large change in the output voltage, and the transistors could be operating in the saturation region rather than in the active region as expected.

3. The *npn* transistors have large betas; let us assume $\beta_{F(npn)} = 250$.
4. The betas of *pnp* transistors are much lower than those of *npn* transistors; let us assume $\beta_{F(pnp)} = 50$. Assume $V_T = 26$ mV for all transistors.

Biasing Circuit

The currents in the biasing current sources of Q_{10} and Q_{13AB} can be calculated from Fig. 14.24. Let us assume that all transistors are operating in the forward-active region, the base currents are negligible, $V_T = 26$ mV, and $V_{BE} = 0.7$ V. The reference current can be calculated as follows:

$$I_{ref} = \frac{V_{CC} - V_{EE} - V_{BE11} - V_{EB12}}{R_5} \tag{14.78}$$

$$= \frac{(15 + 15 - 2 \times 0.7)\text{ V}}{39\text{ k}\Omega} = 0.73\text{ mA}$$

The transistors Q_{10} and Q_{11} form a Widlar current source. Using Eq. (9.98), we can find the output current I_1 from

$$V_T \ln\left(\frac{I_{ref}}{I_1}\right) = R_4 I_1 \tag{14.79}$$

$$V_T \ln\left(\frac{0.73\text{ mA}}{I_1}\right) = 5\text{ k}\Omega \times I_1$$

which, by trial-and-error solution, gives $I_1 = 19\ \mu$A.

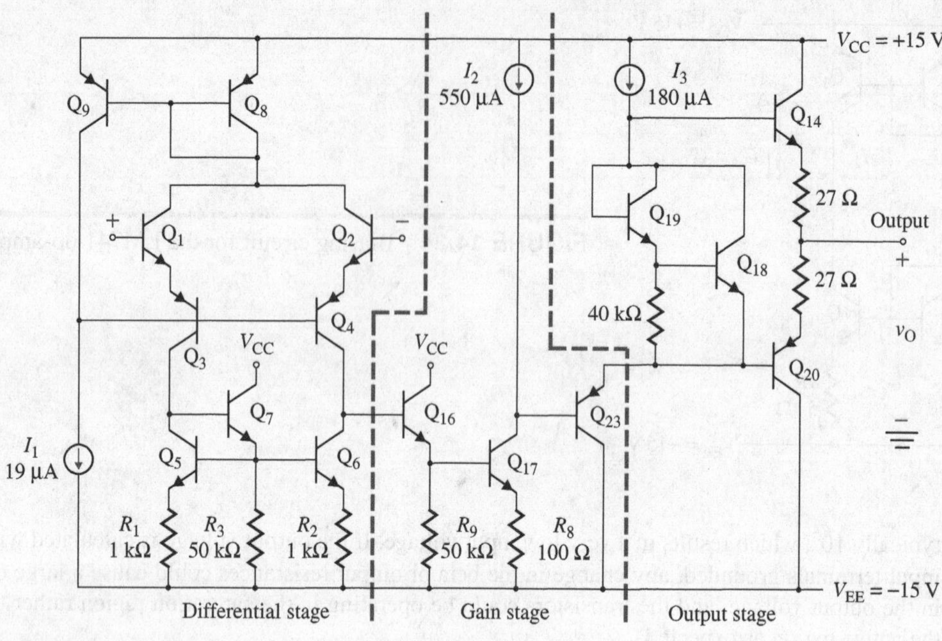

FIGURE 14.25 The LM741 op-amp with biasing current sources

Transistor Q_{13} is a multicollector lateral *pnp* device. Thus, currents I_2 and I_3 are three-fourths and one-fourth of the reference current I_{ref}, respectively:

$$I_2 = \left(\frac{3}{4}\right) \times 0.73 \text{ mA} = 0.55 \text{ mA}$$

$$I_3 = \left(\frac{1}{4}\right) \times 0.73 \text{ mA} = 0.18 \text{ mA}$$

If we replace the biasing circuit in Fig. 14.24 by the equivalent current sources of I_1, I_2, and I_3, the circuit in Fig. 14.23 can be simplified to that in Fig. 14.25.

Input Stage

The input stage provides a high input resistance for differential and common-mode input signals, the differential-to-single output, and some voltage gain. The input stage of the LM741 with biasing current sources is shown in Fig. 14.26. Since the *npn* transistors have large values of beta, the base currents are negligible compared to the collector currents. For identical transistors, $I_{C9} = I_{C8}$.

The current I_T, which is the sum of I_{C8} plus the base currents of Q_8 and Q_9, can be found from

$$I_T = I_{C8} + I_{B8} + I_{B9} = I_{C9}\left(1 + \frac{2}{\beta_{F(pnp)}}\right) \tag{14.80}$$

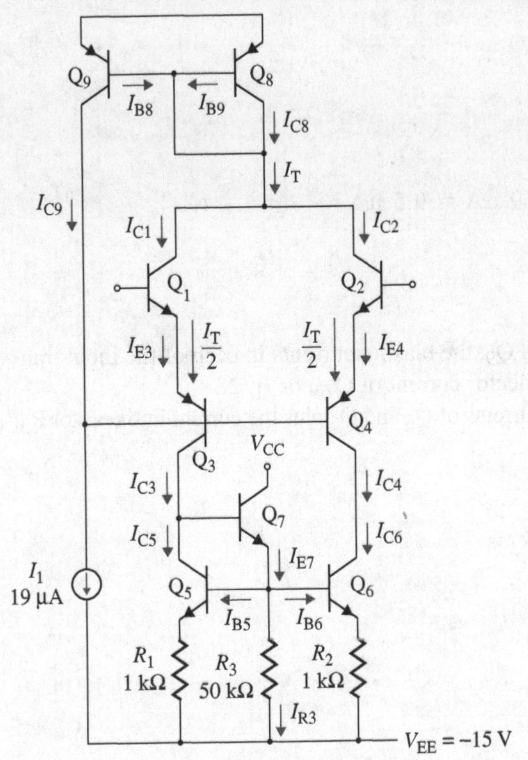

FIGURE 14.26 Input stage of the LM741 op-amp

where $\beta_{F(pnp)}$ is the current gain for *pnp* transistors. Neglecting the base currents of Q_1 and Q_2, we find that the emitter currents of Q_3 and Q_4 are

$$I_{E3} = I_{E4} = \frac{I_T}{2} = \frac{I_{C9}}{2}\left(1 + \frac{2}{\beta_{F(pnp)}}\right) \tag{14.81}$$

The sum of the base currents of Q_3 and Q_4 and the collector current of Q_9 must be equal to the biasing current of $I_1 = 19$ μA. Thus,

$$I_1 = I_{C9} + \frac{I_{E3}}{1 + \beta_{F(pnp)}} + \frac{I_{E4}}{1 + \beta_{F(pnp)}} = I_{C9}\left[1 + \frac{1 + 2/\beta_{F(pnp)}}{1 + \beta_{F(pnp)}}\right] \tag{14.82}$$

which can be simplified to $I_1 = I_{C9}[(2 + \beta_{F(pnp)})/(1 + \beta_{F(pnp)})] \simeq I_{C9}$. Substituting I_{C9} from Eq. (14.82) into Eq. (14.80) yields

$$I_T \simeq I_1\left(1 + \frac{2}{\beta_{F(pnp)}}\right) \tag{14.83}$$

$$\simeq (19\ \mu\text{A})\left(1 + \frac{2}{50}\right) = 19.38\ \mu\text{A}$$

Then

$$I_{C1} = I_{C2} = \frac{19.38\ \mu A}{2} = 9.68\ \mu A$$

$$I_{C3} = I_{C4} = \frac{\beta_{F(pnp)}}{1 + \beta_{F(pnp)}} I_{E3} = \frac{50}{1 + 50} \times 9.69\ \mu A = 9.5\ \mu A$$

$$I_{C5} = I_{C6} \approx I_{C3} = 9.5\ \mu A$$

Because of the biasing feedback loop formed by Q_8 and Q_9, the biasing currents in each of the input transistors Q_1 and Q_2 are approximately one-half of the collector current of Q_{10}, or $I_1/2$.

The emitter current I_{E7} of Q_7 is the sum of the base currents of Q_5 and Q_6 plus the current into resistor R_3:

$$I_{E7} = I_{B5} + I_{B6} + I_{R3} \approx I_{R3}$$

Assuming $I_{E5} \approx I_{C5}$ and $I_{E6} \approx I_{C6}$, the voltage across R_3 is

$$V_{R3} = V_{BE5} + R_1 I_{E5} = V_{BE6} + R_2 I_{E6}$$

$$= V_T \ln\left(\frac{I_{C5}}{I_S}\right) + R_1 I_{E5} \tag{14.84}$$

$$= 26\ \text{mV} \times \ln\left(\frac{9.5\ \mu A}{10^{-14}\ A}\right) + 1\ k\Omega \times 9.5\ \mu A = 537.5\ \text{mV} + 9.5\ \text{mV} = 547\ \text{mV}$$

The collector current of Q_7 is

$$I_{C7} \approx I_{E7} = \frac{V_{R3}}{R_3} = \frac{547\ \text{mV}}{50\ k\Omega} = 10.9\ \mu A$$

Gain Stage

The gain stage provides a high voltage gain with a very high input and output resistance. The amplifier stage is shown in Fig. 14.27. Since it is assumed that the output voltage of the amplifier is zero, the base current of Q_{23} is zero and the collector current of Q_{17} is $I_{C17} = I_2 = 550\ \mu A$. The voltage at the base of Q_{17}

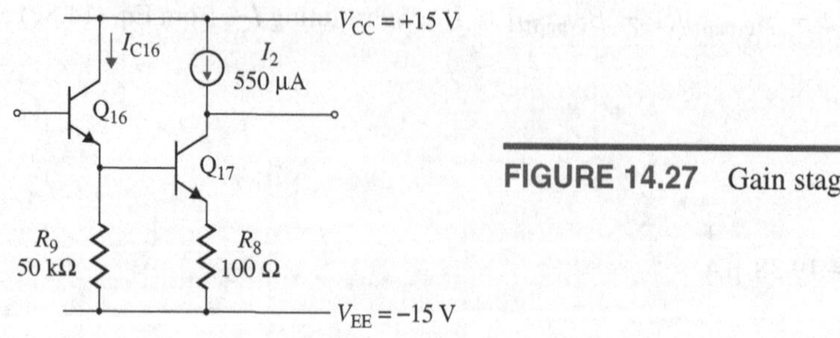

FIGURE 14.27 Gain stage

is equal to the base–emitter voltage of Q_{17} plus the voltage drop across resistor R_8. Assuming $I_{E17} = I_{C17}$ and $I_S = 10^{-14}$ A, the voltage at the base of Q_{17} with respect to $-V_{EE}$ is

$$\begin{aligned} V_{B17} &= V_{BE17} + R_8 I_{E17} \\ &= V_T \ln\left(\frac{I_{C17}}{I_S}\right) + R_8 I_{E17} \qquad \textbf{(14.85)} \\ &= V_T \ln\left(\frac{550\ \mu\text{A}}{10^{-14}\ \text{A}}\right) + 100\ \Omega \times 550\ \mu\text{A} = 643\ \text{mV} + 55\ \text{mV} = 698\ \text{mV} \end{aligned}$$

The current through R_9 is

$$I_{R9} = \frac{V_{B17}}{R_9} = \frac{698\ \text{mV}}{50\ \text{k}\Omega} = 13.96\ \mu\text{A}$$

The base current of Q_{B17} is

$$I_{B17} = \frac{I_{C17}}{\beta_{F(npn)}} = \frac{550\ \mu\text{A}}{250} = 2.2\ \mu\text{A}$$

The collector current of Q_{16} is

$$I_{C16} \approx I_{E16} = I_{B17} + I_{R9} = 2.2\ \mu\text{A} + 13.94\ \mu\text{A} = 16.14\ \mu\text{A}$$

Output Stage

The output stage supplies a high load current and offers a low output resistance. This stage is shown in Fig. 14.28. Assuming that the base currents are negligible, the collector current of Q_{23} is $I_{C23} = I_3 = 180\ \mu\text{A}$. Further assuming that the circuit is connected with feedback in such a way that the output

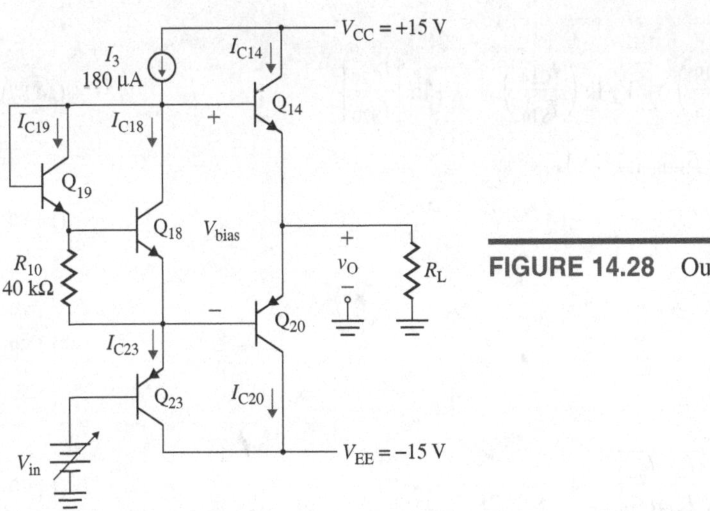

FIGURE 14.28 Output stage

voltage is driven to zero and the output current is also zero, I_{C14} and I_{C20} are approximately equal in magnitude. Thus,

$$I_{C14} = I_{C20} \tag{14.86}$$

Let us estimate that $V_{BE18} = 0.6$ V and the collector current of Q_{19} is given approximately by

$$I_{C19} \approx \frac{V_{BE18}}{R_{10}} = \frac{0.6}{40\ \text{k}\Omega} = 15\ \mu\text{A}$$

Then

$$I_{C18} = I_{C23} - I_{C19} = 180\ \mu\text{A} - 15\ \mu\text{A} = 165\ \mu\text{A}$$

Now we can calculate a more accurate value for V_{BE18}:

$$V_{BE18} = V_T \ln\left(\frac{I_{C18}}{I_S}\right) = (26\ \text{mV}) \ln\left(\frac{165\ \mu\text{A}}{10^{-14}\ \text{A}}\right) = 611.7\ \text{mV}$$

The collector current of Q_{19} is

$$I_{C19} = \frac{I_{C18}}{\beta_{F(npn)}} + \frac{V_{BE18}}{R_{10}} = \frac{165\ \mu\text{A}}{250} + \frac{611.7\ \text{mV}}{40\ \text{k}\Omega} = 16\ \mu\text{A}$$

and

$$I_{C18} = I_3 - I_{C19} = 180\ \mu\text{V} - 16\ \mu\text{V} = 164\ \mu\text{A}$$

which is very close to the original estimate of 165 μA. If there had been a significant difference between this value and the original estimate, we would have used this value to find new values of V_{BE18}, I_{C19}, and I_{C18}, continuing the iterations until the desired value was found.

Using KVL around the loop formed by Q_{14}, Q_{20}, Q_{18}, and Q_{19} in Fig. 14.28, we get

$$V_{BE18} + V_{BE19} = V_{BE14} + |V_{BE20}|$$

which can be written as

$$V_T \ln\left(\frac{I_{C18}}{I_{S18}}\right) + V_T \ln\left(\frac{I_{C19}}{I_{S19}}\right) = V_T \ln\left(\frac{I_{C14}}{I_{S14}}\right) + V_T \ln\left|\frac{I_{C20}}{I_{S20}}\right| \tag{14.87}$$

For an output voltage of $v_O = 0$ and $\beta_{F(npn)} >> 1$,

$$|I_{C14}| = |I_{C20}| \tag{14.88}$$

and Eq. (14.87) can be simplified to

$$\frac{I_{C18}I_{C19}}{I_{S18}I_{S19}} = \frac{I_{C14}^2}{I_{S14}I_{S20}} \tag{14.89}$$

from which we get

$$I_{C14} = I_{C20} = \sqrt{I_{C18}I_{C19}}\sqrt{\frac{I_{S14}I_{S20}}{I_{S18}I_{S19}}} \tag{14.90}$$

Thus, the collector currents depend on the I_S values, which depend on the physical geometry of the transistors. Q_{14} and Q_{20} are normally designed to carry much larger current than other transistors. The specific geometries used by different manufacturers may be different, but the I_S value of Q_{14} and Q_{20} is typically three times that of Q_{18} and Q_{19}. Thus,

$$I_{S14} = I_{S20} \approx 3I_{S18} = 3I_{S19} \tag{14.91}$$

Substituting I_{S14} from Eq. (14.91) into Eq. (14.90) yields

$$\begin{aligned} I_{C14} = I_{C20} &= 3\sqrt{I_{C18}I_{C19}} \\ &= 3 \times \sqrt{(164\ \mu\text{A})(16\ \mu\text{A})} = 153.7\ \mu\text{A} \end{aligned} \tag{14.92}$$

Overload Protection

Transistor Q_{15} (in Fig. 14.23) turns on only when the voltage across R_6 exceeds 550 mV at an output-sourcing current of $550\ \text{mV}/R_6 = 550\ \text{mV}/27 = 20$ mA. When Q_{15} turns on, it limits the current to the base of Q_{14} and the output current cannot increase further. Thus, Q_{15} provides short-circuit protection, preventing damage to the op-amp due to excess current flow and power dissipation. These problems can occur if the output is shorted to a negative power supply. Similarly, transistors Q_{21}, Q_{22}, and Q_{24} protect transistor Q_{20} in the case of sinking current. When Q_{21} turns on, it protects Q_{20} by limiting the current to the base of Q_{20}, thereby turning on Q_{22} and Q_{24}.

If the inverting terminal were overdriven such that its voltage became more positive than that of the noninverting terminal, Q_1 would turn off. As a result, Q_6 would also be off, and the current into the base of Q_{16} would be $I_T = 19.4\ \mu$A. This current would be amplified by the beta of Q_{16}, which could be as high as 1000, giving a collector current of $I_{C16} = 1000 \times 19.4\ \mu\text{A} \approx 19.4$ mA that would flow into the base of Q_{17}. Q_{17} would thus become saturated. Saturation would result in a power dissipation in Q_{16} of

$$I_{C16}(V_{CC} + V_{EE}) = 19.4\ \text{mA} \times (15 + 15)\ \text{V} \approx 580\ \text{mW}$$

The extra emitter on Q_{23} prevents Q_{17} from developing a high current condition that could damage Q_{16}.

14.6.2 Small-Signal AC Analysis

Small-signal analysis is performed to determine the input resistance, the output resistance, the transconductance, and the voltage gain. The op-amp circuit may be broken up into three stages: the input stage, the gain stage, and the output stage. Since the op-amp may be considered as three cascaded stages, we will represent the input and gain stages by their transconductance equivalents to simplify the analysis to determine the overall gain.

Notice from Eq. (14.41) that the CMRR is the change in the common-mode voltage per unit change in the input offset voltage ΔV_{OS}, which can be determined from Eq. (14.40) if A_d is known. Thus, we need to determine only A_d. The small-signal parameters of the transistors are

$$r_{\pi16} = \frac{\beta_{F16}V_T}{I_{C16}} = \frac{250 \times 26\ \text{mV}}{16.1\ \mu\text{A}} = 403.7\ \text{k}\Omega$$

$$r_{\pi 17} = \frac{\beta_{F17} V_T}{I_{C17}} = \frac{250 \times 26 \text{ mV}}{550 \text{ µA}} = 11.82 \text{ k}\Omega$$

$$r_{\pi 23} = \frac{\beta_{F23} V_T}{I_{C23}} = \frac{50 \times 26 \text{ mV}}{180 \text{ µA}} = 7.2 \text{ k}\Omega$$

$$g_{m3} = g_{m4} = g_{m5} = g_{m6} = \frac{I_{C3}}{V_T} = \frac{I_{C4}}{V_T} = \frac{I_{C6}}{V_T} = \frac{9.5 \text{ µA}}{26 \text{ mV}} = 365.4 \text{ µA/V}$$

$$g_{m13A} = g_{m23} = \frac{I_{C13A}}{V_T} = \frac{180 \text{ µA}}{26 \text{ mV}} = 6.92 \text{ mA/V}$$

$$g_{m13B} = g_{m17} = \frac{I_{C17}}{V_T} = \frac{550 \text{ µA}}{26 \text{ mV}} = 21.15 \text{ mA/V}$$

The output resistance r_o is given by

$$r_o = \frac{V_A}{I_C} = \frac{V_A}{V_T g_m} = \frac{1}{\eta_n g_m} \tag{14.93}$$

where $\eta_n = V_T/V_A$ is a transistor constant. Assuming $\eta_n = 5 \times 10^{-4}$ for *pnp* transistors, Eq. (14.93) gives

$$r_{o4} = \frac{1}{\eta_n g_{m4}} = \frac{1}{5 \times 10^{-4} \times 365.4 \text{ µA/V}} = 5.47 \text{ M}\Omega$$

$$r_{o13A} = r_{o23} = \frac{1}{\eta_n g_{m13A}} = \frac{1}{5 \times 10^{-4} \times 6.92 \text{ mA/V}} = 289 \text{ k}\Omega$$

$$r_{o13B} = \frac{1}{\eta_n g_{m13B}} = \frac{1}{5 \times 10^{-4} \times 21.15 \text{ mA/V}} = 94.56 \text{ k}\Omega$$

Assuming $\eta_n = 2 \times 10^{-4}$ for *npn* transistors, Eq. (14.93) gives

$$r_{o6} = \frac{1}{\eta_n g_{m6}} = \frac{1}{2 \times 10^{-4} \times 365.4 \text{ µA/V}} = 13.68 \text{ M}\Omega$$

$$r_{o17} = \frac{1}{\eta_n g_{m17}} = \frac{1}{2 \times 10^{-4} \times 21.15 \text{ mA/V}} = 236.4 \text{ k}\Omega$$

Input Stage

Let us assume that $v_{ic} = 0$ and only $v_{id}/2$ is applied. The AC equivalent circuit for the differential-mode signal is shown in Fig. 14.29(a). The input voltage to transistor Q_1 is $+v_{id}/2$ and that to transistor Q_2 is $-v_{id}/2$. Assuming that the transistors are identical and the circuit is balanced, an increase in voltage at the base junction of Q_3 and Q_4 due to $+v_{id}/2$ will be compensated by an equal decrease in voltage due to $-v_{id}/2$. The voltage at the base terminals of *pnp* transistors Q_3 and Q_4 will not vary at all. As a result, the bases of Q_3 and Q_4 are effectively at ground potential.

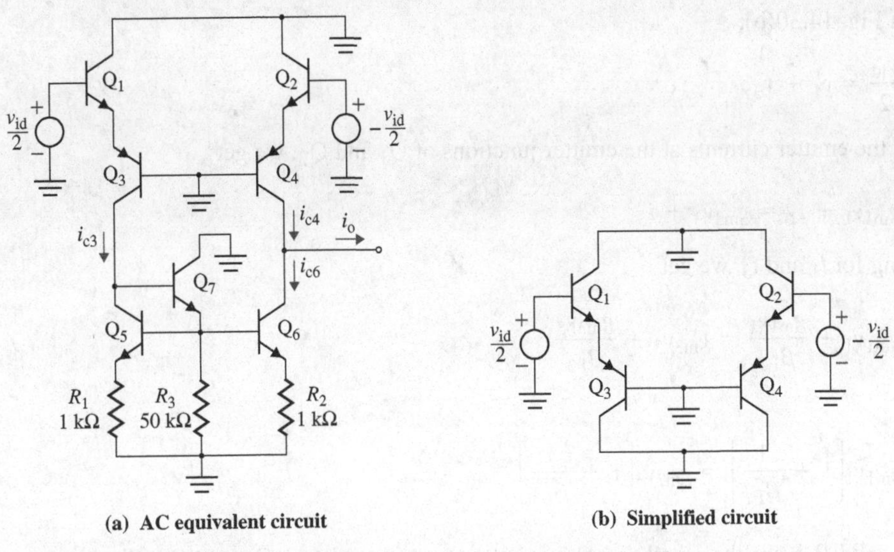

FIGURE 14.29 AC equivalent of input stage with differential input

The transconductance of the input stage can be found by shorting the output to the ground and calculating the resulting current. Since $i_{c3} = i_{c5}$ and i_{c6} is the current mirror of i_{c5}, the current in the active load circuit (i_{c6}) will be equal in magnitude to the collector current of Q$_3$ (i_{c3}); that is,

$$i_{c6} = i_{c3} \tag{14.94}$$

Thus, the output current under short-circuit conditions is

$$i_o = i_{c4} - i_{c6} = i_{c4} - i_{c3} \tag{14.95}$$

Under these conditions, Fig. 14.29(a) can be reduced to Fig. 14.29(b), which has two identical sides. The half-circuit AC equivalent is shown in Fig. 14.30(a), and its small-signal equivalent circuit is shown in Fig. 14.30(b), which can be simplified to Fig. 14.30(c).

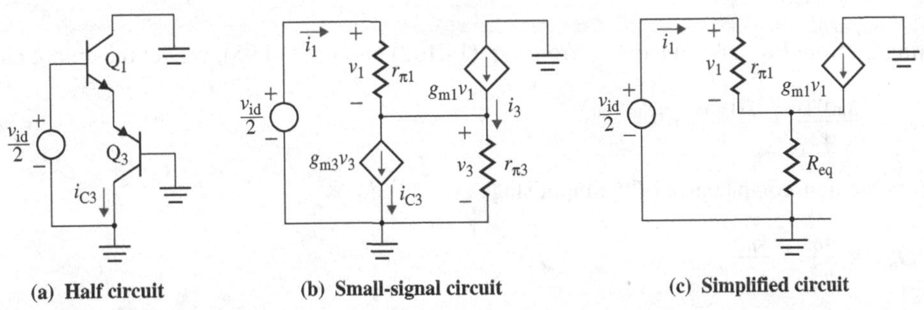

FIGURE 14.30 Half-circuit AC equivalent for differential input

From Fig. 14.30(b),

$$\frac{v_{id}}{2} = v_1 + v_3 \tag{14.96}$$

Summing the emitter currents at the emitter junctions of Q_1 and Q_3, we get

$$g_{m1}v_1 + i_1 = g_{m3}v_3 + i_3 \tag{14.97}$$

Substituting for i_1 and i_3, we get

$$g_{m1}v_1 + \frac{g_{m1}v_1}{\beta_{F1}} = g_{m3}v_3 + \frac{g_{m3}v_3}{\beta_{F3}}$$

or

$$g_{m1}v_1\left[1 + \frac{1}{\beta_{F1}}\right] = g_{m3}v_3\left[1 + \frac{1}{\beta_{F3}}\right] \tag{14.98}$$

where β_{F1} and β_{F3} are the small-signal current gains of transistors Q_1 and Q_3. Since $|I_{C1}| = |I_{C3}|$, $g_{m1} \approx g_{m3}$. Assuming that $\beta_{F1} >> 1$ and $\beta_{F3} >> 1$, Eq. (14.98) is reduced to

$$v_1 = v_3 \tag{14.99}$$

and Eq. (14.96) becomes

$$\frac{v_{id}}{2} = v_1 + v_3 = v_3 + v_3 = 2v_3$$

or

$$v_3 = \frac{v_{id}}{4} \tag{14.100}$$

Using Eq. (14.100), we can find the collector current of Q_3:

$$i_{c3} = g_{m3}v_3 = \frac{g_{m3}v_{id}}{4} \tag{14.101}$$

From the symmetry of the circuit in Fig. 14.29(a), we get

$$i_{c4} = -i_{c3} = \frac{g_{m3}v_{id}}{4} \tag{14.102}$$

Substituting i_{c3} from Eq. (14.101) and i_{c4} from Eq. (14.102) into Eq. (14.95), we get the output current as

$$i_o = \frac{g_{m3}v_{id}}{4} + \frac{g_{m3}v_{id}}{4} = \frac{g_{m3}v_{id}}{2} \tag{14.103}$$

which gives the transconductance of the input stage as

$$G_{m1} = \frac{i_o}{v_{id}} = \frac{g_{m3}}{2} \tag{14.104}$$

$$= \frac{I_{C3}}{2V_T} = \frac{9.5\ \mu\text{A}}{2 \times 26\ \text{mV}} = \frac{1}{5.47\ \text{k}\Omega} = 182.7\ \mu\text{A/V}$$

Q_3 can be replaced by the equivalent resistance R_{eq} seen looking from the emitter of Q_3, as shown in Fig. 14.30(c). R_{eq} can be found from

$$R_{eq} = \frac{v_3}{g_{m3}v_3 + v_3/r_{\pi 3}} = \frac{1}{g_{m3} + 1/r_{\pi 3}} = \frac{1}{g_{m3}(1 + 1/\beta_{F3})} \tag{14.105}$$

$$\approx \frac{1}{g_{m3}} = \frac{r_{\pi 3}}{\beta_{F3}} = \frac{r_{\pi 1}}{\beta_{F1}} \tag{14.106}$$

Using Eq. (8.85), we can relate v_{id} to i_1 as follows:

$$\frac{v_{id}/2}{i_1} = r_{\pi 1} + R_{eq}(1 + \beta_{F1}) \tag{14.107}$$

Substituting $R_{eq} = r_{\pi 1}/\beta_{F1}$ from Eq. (14.106) into Eq. (14.107) gives the input resistance as

$$R_{id} = \frac{v_{id}}{i_1} = 2\left[r_{\pi 1} + \frac{r_{\pi 1}}{\beta_{F1}}(1 + \beta_{F1})\right] \tag{14.108}$$

$$\approx 4r_{\pi 1}$$

$$= 4\frac{\beta_{F1}V_T}{I_{C1}} = 4 \times \frac{250 \times 26 \text{ mV}}{9.69\ \mu\text{A}} = 2.68 \text{ M}\Omega$$

Notice from Eq. (14.108) that the input resistance is four times the input resistance of the input transistors. Also note that when v_{id} changes, the output voltage changes and produces feedback to the input through the output resistance of Q_4. As a result, the input resistances seen from the two input terminals will not be exactly the same. This effect is neglected in the derivation of R_{id}.

The output resistance R_{o1} of the differential stage can be determined by setting the input voltage to zero and applying a test voltage v_x, as shown in Fig. 14.31(a). The analysis can be simplified by making the following assumptions:

1. Thevenin's equivalent resistance R_{th6} at the base of Q_6 is very small compared to r_π of Q_6, so the base of Q_6 is grounded. In reality, the voltage at this point is very small, and this point may be considered a virtual ground without greatly affecting the results. That is, $v_{B6} \approx 0$. Figure 14.31(a) can be viewed as shown in Fig. 14.31(b).
2. The output resistance of Q_2 is large and does not affect the results.

Figure 14.31(b) can be further simplified to the half circuit shown in Fig. 14.31(c). If Q_2 is replaced by $1/g_{m2}$, Fig. 14.31(c) can be represented by the equivalent circuit in Fig. 14.32(a). If R_{Q4} and R_{Q6} are the effective resistances seen from the collectors of Q_4 and Q_6, respectively, Fig. 14.32(a) can be represented by Fig. 14.32(b). Using Eq. (9.113), we get

$$R_{Q4} = r_{o4}\left[\frac{1 + g_{m4}(1/g_{m2})}{1 + g_{m4}/\beta_{F4}g_{m2}}\right] \tag{14.109}$$

$$\approx 2r_{o4} \quad (\text{for } g_{m4} = g_{m2} \text{ and } \beta_{F4} >> 1)$$

$$= 2 \times 5.47 \text{ M}\Omega = 10.94 \text{ M}\Omega$$

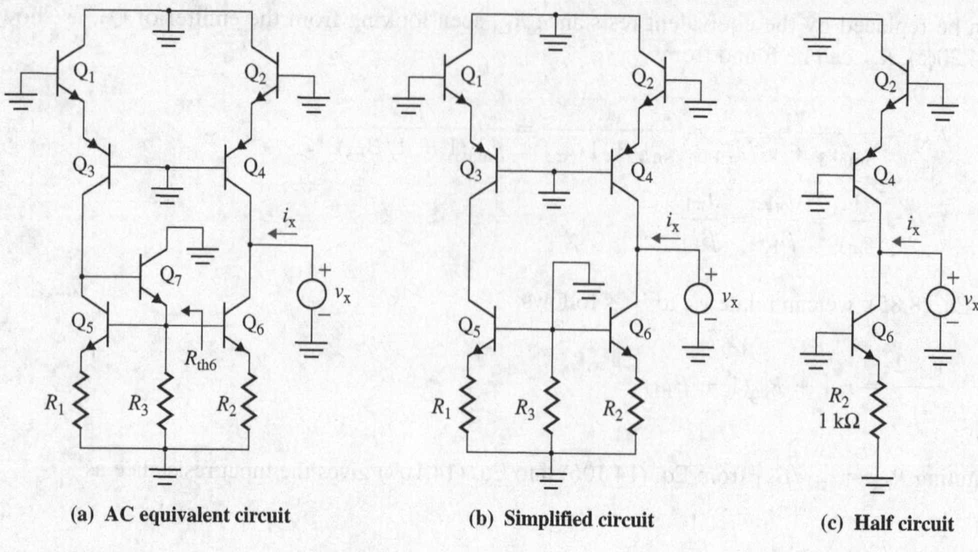

(a) AC equivalent circuit (b) Simplified circuit (c) Half circuit

FIGURE 14.31 Test circuit for calculation of R_{o1}

$$R_{Q6} = r_{o6}\left[\frac{1 + g_{m6}R_2}{1 + g_{m6}R_2/\beta_{F6}}\right] \tag{14.110}$$

$$= r_{o6}\left[\frac{1 + 365.4\ \mu\text{A/V} \times 1\ \text{k}\Omega}{1 + 365.4\ \mu\text{A/V} \times 1\ \text{k}\Omega/250}\right] = 1.36r_{o6}$$

$$= 1.36 \times 13.68\ \text{M}\Omega = 18.65\ \text{M}\Omega$$

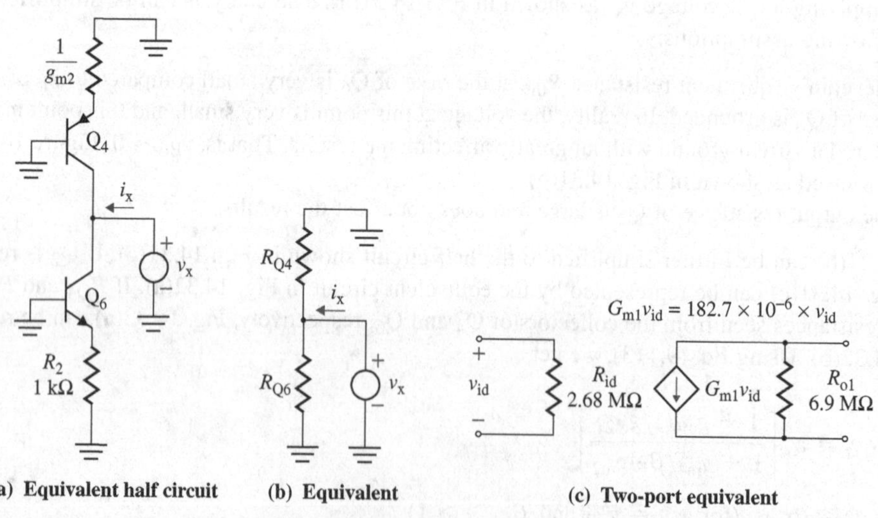

(a) Equivalent half circuit (b) Equivalent (c) Two-port equivalent

FIGURE 14.32 Two-port equivalent of the input stage

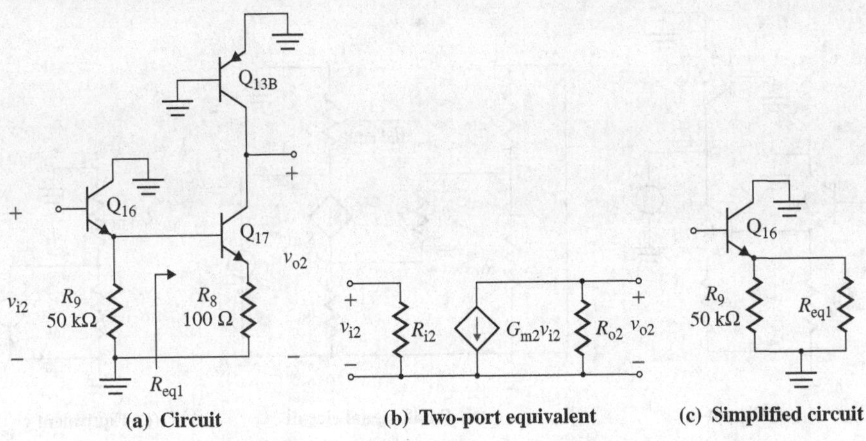

FIGURE 14.33 Small-signal AC equivalent circuit for the gain stage

The output resistance of the input stage is

$$R_{o1} = R_{Q4} \| R_{Q6} = 10.94\ \text{M}\Omega \| 18.65\ \text{M}\Omega = 6.9\ \text{M}\Omega$$

The two-port equivalent circuit of the input stage is shown in Fig. 14.32(c).

Gain Stage

The AC equivalent for the gain stage, shown in Fig. 14.33(a), can be represented by the equivalent circuit shown in Fig. 14.33(b). If R_{eq1} is Thevenin's equivalent resistance seen looking into the base of transistor Q_{17}, Fig. 14.33(a) can be reduced to the form shown in Fig. 14.33(c). From Eq. (8.85), we can find the input resistance of a common-emitter amplifier with a resistance in the emitter as follows:

$$R_{eq1} = r_{\pi 17} + (1 + \beta_{F17})R_8 \tag{14.111}$$

$$= 11.82\ \text{k}\Omega + (1 + 250) \times 100\ \Omega = 36.9\ \text{k}\Omega$$

Using Eq. (8.85), the input resistance of the gain stage is

$$R_{i2} = r_{\pi 16} + (1 + \beta_{F16})(R_{eq1} \| R_9) \tag{14.112}$$

$$= 403.7\ \text{k}\Omega + (1 + 250)(36.9\ \text{k}\Omega \| 50\ \text{k}\Omega) = 5.73\ \text{M}\Omega$$

Assuming that the voltage gain of the emitter follower Q_{16} is unity, the transconductance G_{m2} of this stage is that of the common-emitter amplifier of Q_{17} with resistance in the emitter; that is, the transconductance of the gain stage is

$$G_{m2} = \frac{g_{m17}}{1 + g_{m17}R_8} \tag{14.113}$$

$$= \frac{21.15\ \text{mA/V}}{1 + 21.15\ \text{mA/V} \times 100\ \Omega} = \frac{1}{14\ \Omega} = 6.79\ \text{mA/V}$$

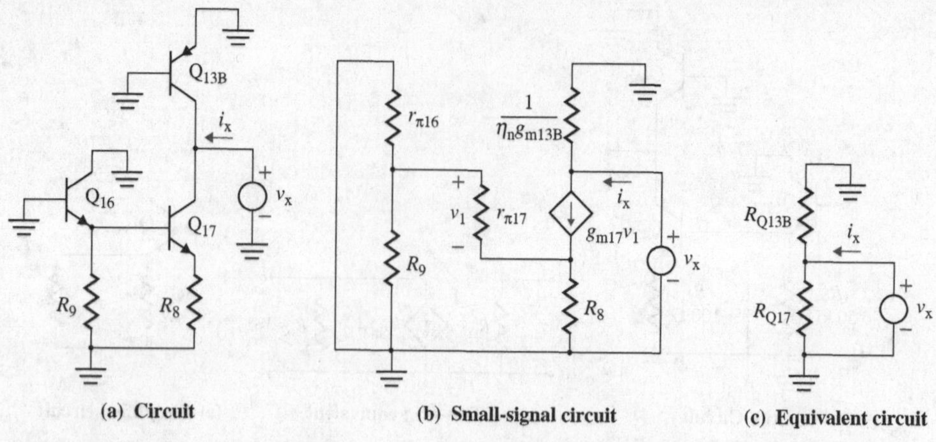

FIGURE 14.34 AC equivalent circuit for calculation of R_{o2}

The circuit for determining the output resistance is shown in Fig. 14.34(a). The small-signal AC equivalent is shown in Fig. 14.34(b). If R_{Q13B} and R_{Q17} are the effective resistances seen from the collectors of Q_{13B} and Q_{17}, respectively, Fig. 14.34(b) can be represented by the equivalent circuit in Fig. 14.34(c) and

$$R_{Q13B} = r_{o13B} = \frac{1}{\eta_n g_{m4}} = 94.56\ \text{k}\Omega$$

Using Eq. (9.113), we get

$$R_{Q17} \approx r_{o17}\left[\frac{1 + g_{m17}R_8}{1 + g_{m17}R_8/\beta_{F17}}\right] \tag{14.114}$$

$$\approx r_{o17}\left[\frac{1 + 21.15\ \text{mA/V} \times 100\ \Omega}{1 + 21.15\ \text{mA/V} \times 100\ \Omega/250}\right] = 3.09 r_{o17}$$

$$= 3.09 \times 236.4\ \text{k}\Omega = 730.5\ \text{k}\Omega$$

The output resistance of the gain stage is

$$R_{o2} = R_{Q13B} \,\|\, R_{Q17} = 94.56\ \text{k}\Omega \,\|\, 730.5\ \text{k}\Omega = 83.72\ \text{k}\Omega$$

Output Stage

Since V_{CE} of transistor Q_{18} is the sum of the V_{BE} voltages of Q_{18} and Q_{19}, transistors Q_{18} and Q_{19} can be replaced by two diodes, as shown in Fig. 14.35(a). The output could be either sourcing or sinking, depending on the output voltage and the load. As a result, the input and resistances of this stage greatly depend on the particular values of output voltage and current. We will make the following assumptions:

1. The output is sourcing—that is, the output current is flowing out of the output stage.
2. The load current is $i_L = 2$ mA at $R_L = 5$ kΩ.
3. Transistor Q_{14} is in the active region.
4. Transistor Q_{20} is conducting a very small amount of current and is considered to be off.

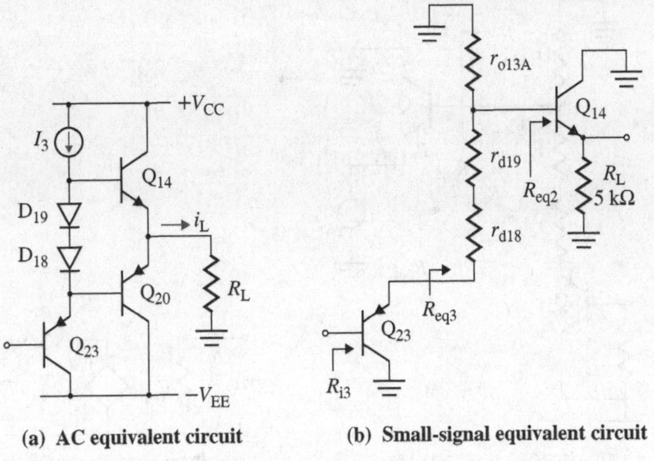

FIGURE 14.35 AC equivalent circuit for the output stage

The small-signal AC equivalent of Fig. 14.35(a) is shown in Fig. 14.35(b). Notice that the circuit consists of two emitter followers in series and that the voltage gain is approximately unity. Thus, the voltage gain of the output stage is

$$A_3 \approx 1 \tag{14.115}$$

Since Q_{14} carries the load current i_L and a no-load DC-biasing collector current of Q_{14} of 153.7 μA,

$$I_{C14} \simeq I_{E4} = 2 \text{ mA} + 153.7 \text{ μA} = 2.15 \text{ mA}$$

Thus,

$$r_{\pi 14} = \frac{\beta_{F14} V_T}{I_{C14}} = \frac{250 \times 26 \text{ mV}}{2.15 \text{ mA}} = 3.02 \text{ k}\Omega$$

Since Q_{23} and diodes D_{18} and D_{19} operate approximately at a current of $I_3 \simeq 180$ μA,

$$r_{d18} = r_{d19} = \frac{V_T}{I_{d18}} = \frac{26 \text{ mV}}{180 \text{ μA}} = 144\ \Omega$$

If R_{eq2} is the resistance seen looking at the base of Q_{14}, we can use Eq. (8.85) to find R_{eq2}:

$$R_{eq2} = r_{\pi 14} + (1 + \beta_{F14}) R_L \tag{14.116}$$

$$= 3.02 \text{ k}\Omega + (1 + 250) \times 5 \text{ k}\Omega = 1258 \text{ k}\Omega$$

Thevenin's equivalent resistance seen looking from the emitter of Q_{23} can be found from

$$R_{eq3} = r_{d18} + r_{d19} + (r_{o13A} \| R_{eq2}) \tag{14.117}$$

$$= 144\ \Omega + 144\ \Omega + (289 \text{ k}\Omega \| 1258 \text{ k}\Omega) = 235.3 \text{ k}\Omega$$

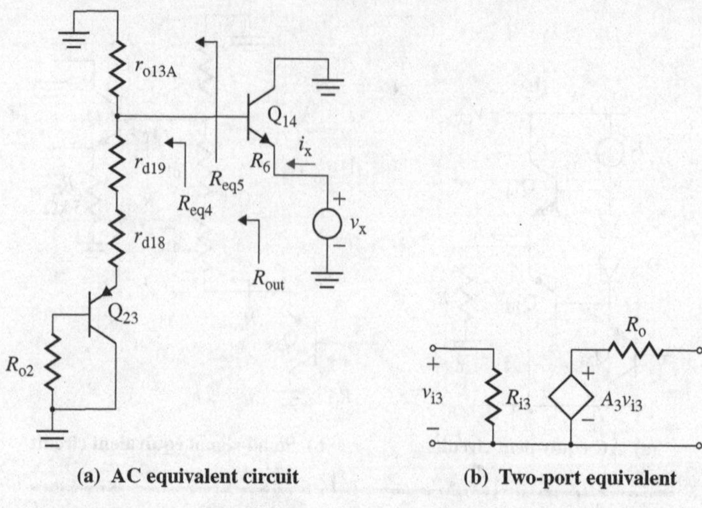

FIGURE 14.36 AC equivalent circuit for calculation of R_o

Using Eq. (8.85), we can find the input resistance of the stage:

$$R_{i3} = r_{\pi 23} + (1 + \beta_{F23})R_{eq3} \tag{14.118}$$

$$= 7.2\ \text{k}\Omega + (1 + 50) \times 235.3\ \text{k}\Omega = 12\ \text{M}\Omega$$

Notice that the input resistance of the output stage (12 MΩ) is much larger than the output resistance of the preceding stage (83.72 kΩ). As a result, the overall gain of the amplifier is contributed by the differential and gain stages, the overall gain is not affected by the variations in external load resistance.

The output resistance of the output stage can be determined from the AC equivalent circuit shown in Fig. 14.36(a), which includes the output resistance R_{o2} of the preceding stage. Converting R_{o2} at the base of Q_{23} to its emitter gives the equivalent resistance seen looking from the base of Q_{14} toward Q_{23}:

$$R_{eq4} = r_{d18} + r_{d19} + \frac{R_{o2} + r_{\pi 23}}{1 + \beta_{F23}} \tag{14.119}$$

$$= 144 + 144 + \frac{83.72\ \text{k}\Omega + 7.2\ \text{k}\Omega}{1 + 50} = 2.07\ \text{k}\Omega$$

The resistance seen looking from the base of Q_{14} to the left is

$$R_{eq5} = r_{o13A} \,\|\, R_{eq4} \tag{14.120}$$

$$= 289\ \text{k}\Omega \,\|\, 2.07\ \text{k}\Omega = 2.06\ \text{k}\Omega$$

The resistance seen looking into the output terminal is

$$R_{out} = \frac{R_{eq5} + r_{\pi 14}}{1 + \beta_{F14}} \tag{14.121}$$

$$= \frac{2.06\ \text{k}\Omega + 3.02\ \text{k}\Omega}{1 + 250} = 20.2\ \Omega$$

The current-limiting resistance R_6 must be added to R_{out} to find the actual output resistance of the op-amp:

$$R_o = R_{out} + R_6 \tag{14.122}$$
$$= 20.2 + 27 = 47.2\ \Omega$$

The two-port equivalent circuit of the output stage is shown in Fig. 14.36(b).

14.6.3 Frequency-Response Analysis

For $G_{m1} = 182.7\ \mu A/V$ and $C_x = 30$ pF, Eq. (14.56) gives the unity-gain frequency (or bandwidth) as

$$f_u = \frac{G_{m1}}{2\pi C_x} = \frac{182.7\ \mu A/V}{2\pi \times 30\ pF} = 969.3\ kHz$$

From Eq. (14.64), the slew rate is

$$SR = 4\pi V_T f_u = 4\pi \times 26\ mV \times 969.3\ kHz = 0.317\ V/\mu s$$

The manufacturer-specified values are $f_u = 1$ MHz and SR = 0.5 V/μs. The discrepancy between the calculated values and the manufacturer-specified values is caused by the fact that Eq. (14.64) gives an approximate value of SR and does not take into account the input and output resistances of preceding and subsequent amplifier stages.

14.6.4 Small-Signal Equivalent Circuit

The small signal equivalent of the complete circuit is shown in Fig. 14.37. The voltage gain is

$$A_v = G_{m1}(R_{o1} \parallel R_{i2})\, G_{m2}(R_{o2} \parallel R_{i3}) \tag{14.123}$$
$$= 182.7\ \mu A/V \times (6.9\ M\Omega \parallel 5.73\ M\Omega) \times 6.79\ mA/V \times (83.72\ k\Omega \parallel 12.0\ M\Omega)$$
$$= 571.9 \times 564.5 = 323{,}000$$

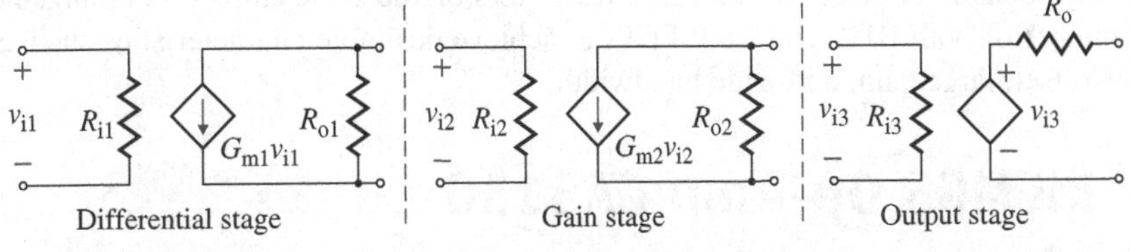

FIGURE 14.37 Small-signal equivalent circuit for the LM741 op-amp

where G_{m1} = the transconductance of the first (differential) stage
G_{m2} = the transconductance of the second (gain) stage
R_{o1} = the output resistance of the first (differential) stage
R_{o2} = the output resistance of the second (gain) stage
R_{i3} = the input resistance of the third (output) stage

Input and output resistances are

$$R_{id} = 2.68 \text{ M}\Omega \quad \text{and} \quad R_o = 47.2 \ \Omega$$

▶ **NOTES**

1. The input stage and the gain stage contribute about the same gain: 571.9 and 564.5, respectively. The gain stage loads the input stage and reduces its gain by about half. This loading makes the op-amp beta dependent, causing the gain to vary with temperature and fabrication-process tolerances.
2. The output stage does not significantly load the gain stage, and the voltage gain is almost independent of the external load resistance.
3. Although the results of the analysis are approximate, they give insight into the operation and performance of the op-amp.
4. The output resistances of transistors were neglected in the DC analysis. If these were taken into consideration, the biasing currents would change, thereby changing the small-signal output resistance and the results.
5. The variation in the transistor beta was neglected. But the transistor current gain falls at low collector current levels because of recombination in the emitter–base space charge layer.

KEY POINTS OF SECTION 14.6

- This section illustrated a three-part analysis of a complete op-amp circuit, the LM741: DC analysis to find the DC biasing currents, small-signal analysis to find the voltage gain, and analysis of the frequency response to find the unity-gain bandwidth for all stages of the op-amp.
- The values obtained from hand calculations correspond well to the values specified in the manufacturer's data sheet. That is, A_v = 323,000 (compared to 200,000), R_{id} = 2.68 MΩ (compared to 2 MΩ), R_o = 47.2 Ω (compared to 75 Ω), f_u = 969.3 kHz (compared to 1 MHz), and SR = 0.317 V/μs (compared to 0.5 V/μs).

14.7 BiCMOS Op-Amps

BiCMOS op-amps contain both CMOS and BJT transistors on the same chip. This arrangement incorporates the advantages of both BJTs and MOSFETs to achieve desirable characteristics such as high input resistance, low offset, large gain, and wide bandwidth.

14.7.1 BiCMOS Op-Amp CA3130

The simplified schematic of the BiCMOS op-amp CA3130 is shown in Fig. 14.38. The internal structure can be divided into three stages: a differential MOS stage, a gain stage, and an output stage. Some parameters of the CA3130 op-amp are listed in Table 14.6.

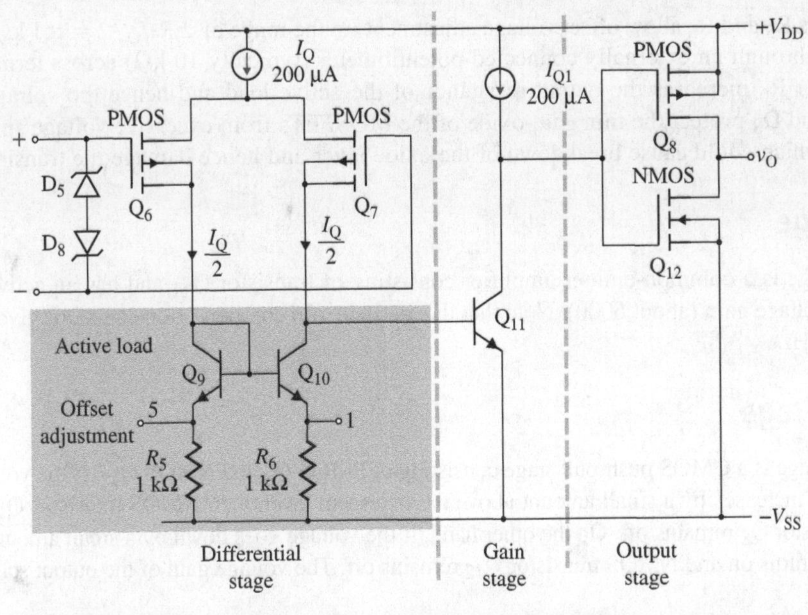

FIGURE 14.38 Simplified schematic for op-amp RCA-CA3130 (Courtesy of Harris Corporation, Semiconductor Sector)

Differential Stage

The input stage is a PMOS differential amplifier consisting of transistors Q_6 and Q_7. This stage is biased by a current source of I_Q, and it drives a current mirror load consisting of BJT transistors Q_9 and Q_{10}. Although a BJT active load offers high resistance, the voltage gain is only about 5 because of the relatively low transconductance of MOSFETs.

TABLE 14.6 Parameters of op-amp CA3130

Parameter	Minimum	Typical	Maximum	Units
DC supply voltages V_{DD}			±18	V
Input offset voltage V_{OS}		8.0	15.0	mV
Thermal drift D_v		15	30	μV/°C
Input biasing current I_B		5	50	pA
Input offset current I_{OS}		0.5	30	pA
Differential input resistance R_{id}		1.5		TΩ
Common-mode input resistance R_{ic}		1.5		TΩ
Output resistance R_o		75		Ω
Input capacitance C_i		4.3		pF
Open-loop voltage gain A_o	50	320		V/mV
Common-mode rejection ratio (CMRR)	70	90		dB
Unity-gain bandwidth f_u		15		MHz
Slew rate (SR)		10		V/μs
Power supply rejection ratio (PSRR)	70	90		dB

Resistors R_5 and R_6 allow offset voltage adjustment (in the range of $\pm R_5 I_Q/2 = \pm 1\ \text{k}\Omega \times 100\ \mu\text{A} = \pm 100$ mV) through an externally connected potentiometer (typically 10 kΩ) across terminals 1 and 5. Also, resistor R_6 increases the output resistance of the active load and hence the voltage gain. Zener diodes D_5 and D_8 protect the thin gate-oxide of the MOSFETs from excessive voltage spikes and static discharge, which could cause breakdown of the oxide layer and hence damage the transistors.

Gain Stage

The gain stage is a common-emitter amplifier consisting of transistor Q_{11} and has an active current load for a large voltage gain (about 6000). Note that the absence of a compensation capacitor gives a high bandwidth (15 MHz).

Output Stage

The output stage is a CMOS push-pull stage consisting of PMOS Q_8 and NMOS Q_{12}. If the voltage at the collector of Q_{11} increases by a small amount above the quiescent level, then NMOS transistor Q_{12} turns on and PMOS transistor Q_8 remains off. On the other hand, if the voltage goes down by a small amount, then PMOS transistor Q_8 turns on and NMOS transistor Q_{12} remains off. The voltage gain of the output stage is about 30.

14.7.2 BiCMOS Op-Amp CA3140

The simplified schematic of the BiCMOS op-amp CA3140 is shown in Fig. 14.39. Its internal structure is similar to that of op-amp CA3130 in Fig. 14.38, except for the output stage and the addition of a compensation capacitor C_x in the second stage. The voltage gain of the differential stage is about 10. The offset

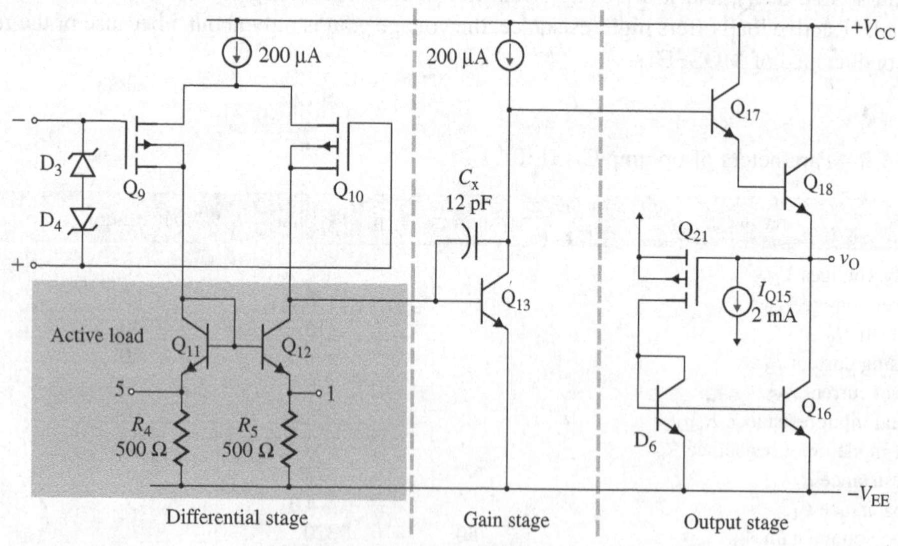

FIGURE 14.39 Simplified schematic for op-amp RCA-CA3140 (Courtesy of Harris Corporation, Semiconductor Sector)

TABLE 14.7 Parameters of op-amp CA3140

Parameter	Minimum	Typical	Maximum	Units
DC supply voltages V_{DD}			± 18	V
Input offset voltage V_{OS}		8.0	15.0	mV
Thermal drift D_v		10	30	μV/°C
Input biasing current I_B		10	50	pA
Input offset current I_{OS}		0.5	30	pA
Differential input resistance R_{id}		1.5		TΩ
Common-mode input resistance R_{ic}		1.5		TΩ
Output resistance R_o		50		Ω
Input capacitance C_i		1		pF
Open-loop voltage gain A_o	20	100		V/mV
Common-mode rejection ratio (CMRR)	70	90		dB
Unity-gain bandwidth f_u		4.5		MHz
Slew rate (SR)		9		V/μs
Power supply rejection ratio (PSRR)	76	80		dB

voltage adjustment is in the range of $\pm R_5 I_Q/2 = \pm 500\ \Omega \times 100\ \mu A = \pm 50$ mV. C_x provides feedback stability, but it reduces the bandwidth. The Darlington *npn* emitter follower in the output stage increases the effective load resistance seen by transistor Q_{13} in the gain stage and hence increases the voltage gain to about 10. The voltage gain of the gain stage is about 10,000. Some parameters of the CA3140 op-amp are listed in Table 14.7.

Output Stage

The Darlington emitter follower offers a lower-than-average output resistance and a high resistance to the gain stage. If the voltage at the collector of transistor Q_{13} goes up in the positive direction above the quiescent value, transistor Q_{18} will drive the load and source current of I_{Q15} (2 mA). However, if the current of Q_{18} falls below the level of I_{Q15} (2 mA), the voltage across the load will go down below the quiescent level. Transistor Q_{18} will always remain on. This will cause MOSFET Q_{21} to turn on. Since the collector current of Q_{16} will be the mirror of the drain current of MOSFET Q_{21}, transistor Q_{16} will sink the load current.

14.7.3 BiCMOS Op-Amp LH0022

The simplified schematic of the LH0022 op-amp is shown in Fig. 14.40, which is a modification of the original circuit by replacing junction field-effect transistors Q_1 and Q_2 with depletion-type NMOS. The LH0022 is designed to operate from a $\pm$15-V power supply. It is capable of a peak output voltage swing of about 12 V into a 1-kΩ load over the entire frequency range of 1 MHz. The circuit can be divided into a differential stage, a gain stage, an output stage, overload protection, and a DC-biasing circuit. Some parameters of the LH0022 are listed in Table 14.8.

Differential Stage

The differential input stage consists of Q_1 through Q_4. Transistors Q_1 and Q_2 are depletion NMOSs that operate in a common-drain (or source-follower) configuration. These NMOSs drive transistors Q_3 and Q_4,

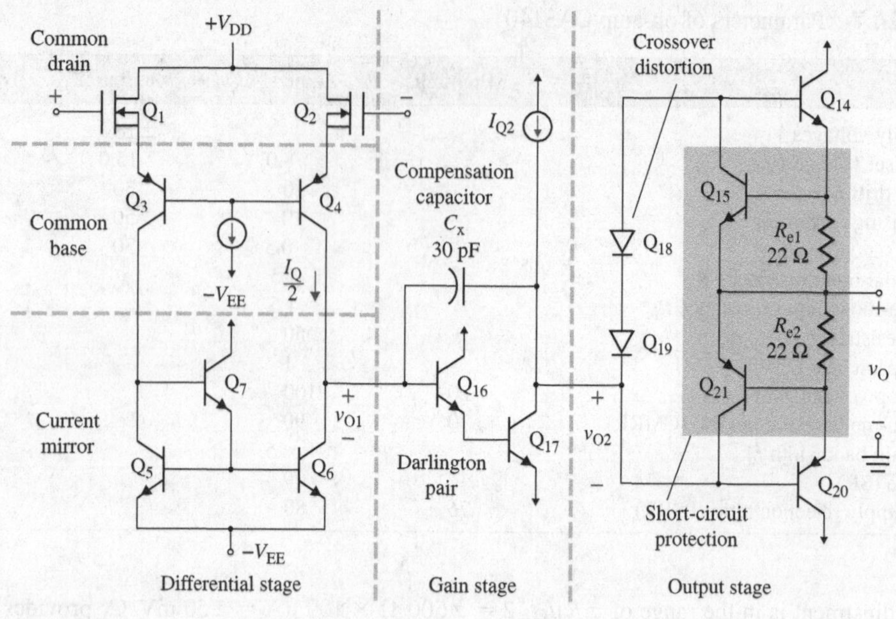

FIGURE 14.40 Simplified schematic for op-amp LH0022 (Courtesy of National Semiconductor, Inc.)

which operate in a common-base configuration. The current mirror, consisting of transistors Q_5, Q_6, and Q_7, acts as the active load.

The combination of common-drain and common-base configurations provides a very high input resistance and a very low input biasing current from the MOSFETs while giving a large voltage gain from the BJTs. Also, the common-drain configuration performs the function of shifting voltage levels

TABLE 14.8 Parameters of op-amp LH0022

Parameter	Minimum	Typical	Maximum	Units
DC supply voltages V_{CC}, V_{EE}			±15	V
Input offset voltage V_{OS}		6.0	20	mV
Thermal drift D_v		10		μV/°C
Input biasing current I_B		15	50	pA
Input offset current I_{OS}		2	10	pA
Differential input resistance R_{id}		10^{12}		Ω
Common-mode input resistance R_{ic}		10^{12}		Ω
Output resistance R_o		75		Ω
Input capacitance C_i		4.0		pF
Open-loop voltage gain A_o	25	100		V/mV
Common-mode rejection ratio (CMRR)	70	80		dB
Unity-gain bandwidth f_u		1.0		MHz
Slew rate (SR)	1.0	3.0		V/μs
Power supply rejection ratio (PSRR)	70	80		dB

toward the negative supply voltage $-V_{EE}$ so that the output signal can be shifted upward in the positive direction in the following stages. Note that the DC level of the output voltage of this stage will be approximately

$$-V_{EE} + V_{BE16} + V_{BE17} \equiv -V_{EE} + 1.2 \text{ V}$$

Gain Stage

The gain stage has the common-collector and common-emitter configurations forming a Darlington pair made up of Q_{16} and Q_{17}. This arrangement offers a high load resistance to the differential stage and hence provides a large voltage gain. Capacitor C_x is the pole-splitting compensating capacitor connected in shunt-shunt feedback, and it controls the unity-gain bandwidth and the slew rate.

Output Stage

The output stage is a complementary class AB push-pull circuit consisting of transistors Q_{14} and Q_{20}. Diode-connected transistors Q_{18} and Q_{19} provide the biasing voltage for class AB operation in order to reduce crossover distortion.

Protection Circuitry

Transistors Q_{15} and Q_{21} act as the current booster. Resistors R_{e1} and R_{e2} provide short-circuit protection by turning on Q_{15} and Q_{21} so as to limit the currents through Q_{14} and Q_{20}, respectively.

Biasing Circuitry

The complete schematic, including the biasing circuitry, is shown in Fig. 14.41, which has an extra protection circuit (consisting of transistors Q_{23} and Q_{24}) to short the input terminal of Q_{16} (in the gain stage) to ground through Q_{23}. Transistors Q_{10} and Q_{11} form a Widlar current mirror, which biases transistors Q_3 and Q_4 of the common-base configuration. The current source consisting of transistors Q_{12} and Q_{13} biases the gain and output stages. The reference current I_{ref} can be found from

$$I_{ref} = \frac{V_{CC} + V_{EE} - V_{BE11} - V_{EB12}}{R_5} \tag{14.124}$$

Transistor Q_{13} is a multicollector lateral *pnp* device. Its geometry is shown in Fig. 14.42(a), and its symbol is shown in Fig. 14.42(b). The collector ring has been split into two parts—one part faces on three-fourths of the emitter periphery and collects the holes injected from that periphery, and the second one faces on one-fourth of the emitter periphery and collects the holes from that periphery. The structure is analogous to two *pnp* transistors whose base–emitter junctions are connected in parallel, one with an I_S that is one-fourth that of a standard *pnp* transistor and the other with an I_S that is three-fourths that of a standard *pnp* transistor. This electrical equivalence is shown in Fig. 14.42(c). Thus, the currents I_{Q2} and I_{Q3} are three-fourths and one-fourth of the reference current I_{ref}, respectively. That is,

$$I_{Q2} = \frac{3I_{ref}}{4} \quad \text{and} \quad I_{Q3} = \frac{I_{ref}}{4}$$

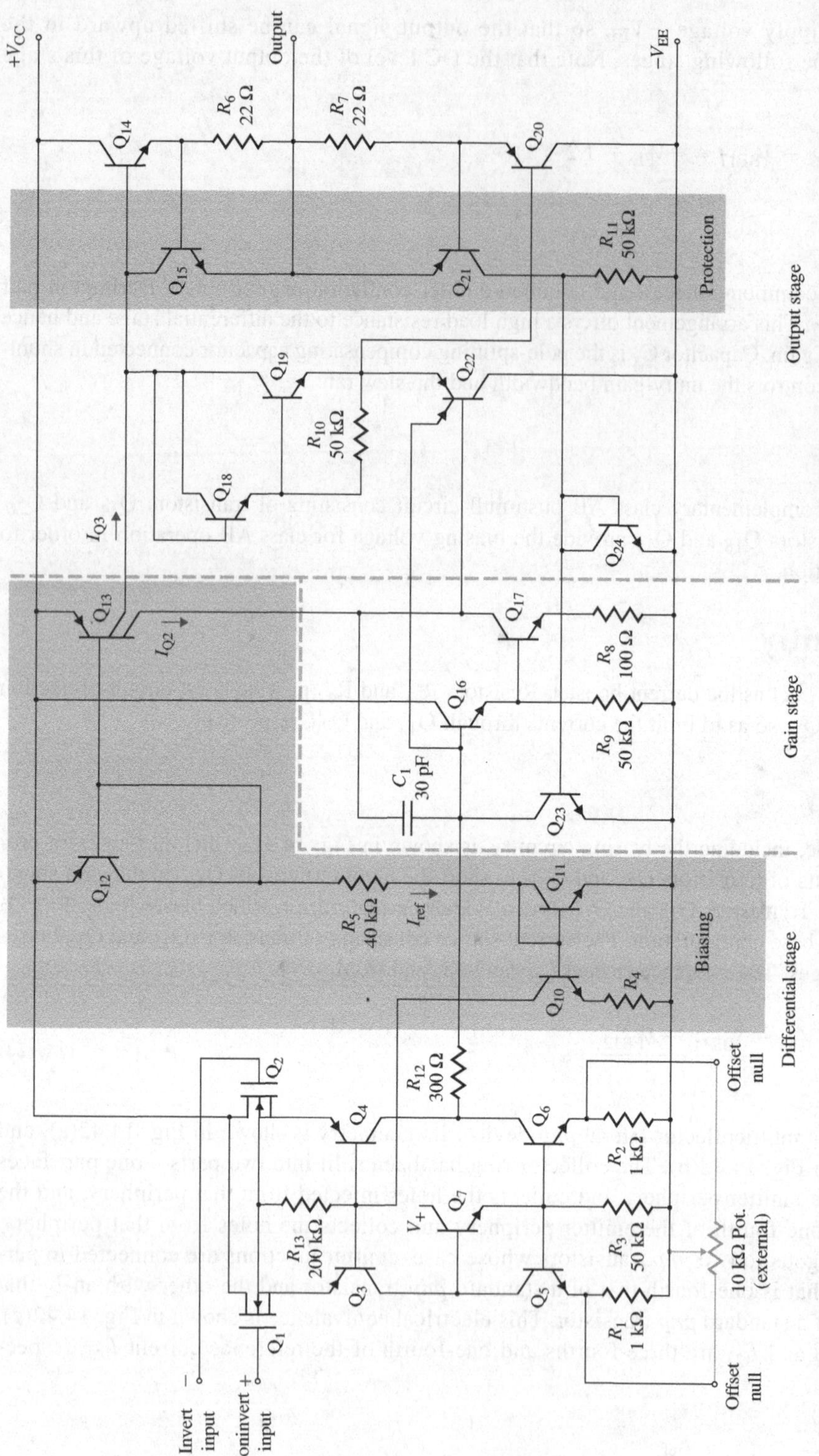

FIGURE 14.41 Schematic for op-amp LH0022 (Courtesy of National Semiconductor, Inc.)

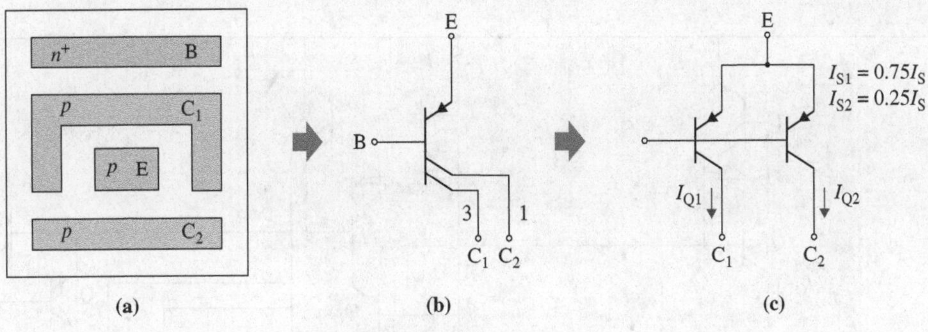

FIGURE 14.42 Electrical equivalent for multicollector lateral *pnp* transistor

14.7.4 BiCMOS Op-Amp LF411

The simplified schematic of the LF411 op-amp is shown in Fig. 14.43, which is a modification of the original circuit by replacing junction field-effect transistors J_1 and J_2 with depletion-type PMOS. Its internal structure is similar to that of the LH0022 op-amp. The differential input stage consists of *p*-channel depletion transistors M_1 and M_2 with a current mirror load. This stage has low offset, low drift, and a high unity-gain bandwidth. The complete schematic appears in Fig. 14.44, which shows in detail the biasing circuitry. There is a transistor Q_3 in the drain of M_2, but none in that of M_1. Depletion NMOS M_2 operates in a common-source configuration, whereas MOSFET M_1 operates in a common-drain configuration. This type of arrangement gives a lower voltage gain but a wider bandwidth.

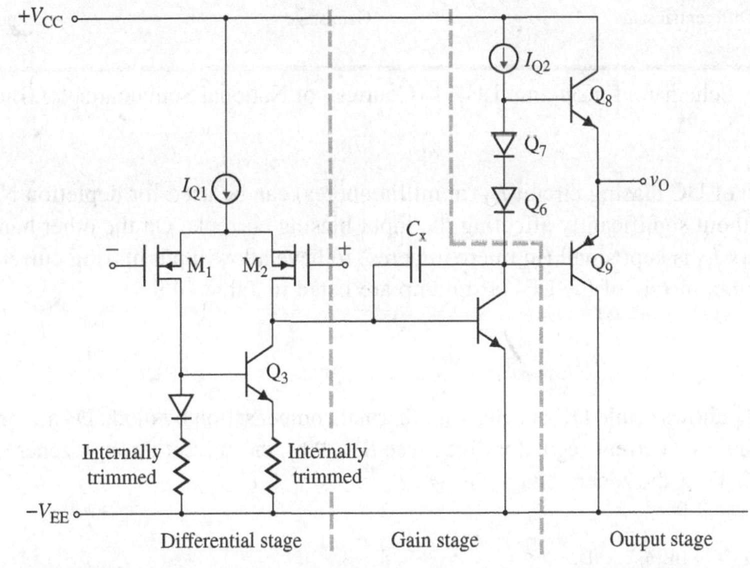

FIGURE 14.43 Simplified schematic for op-amp LF411 (Courtesy of National Semiconductor, Inc.)

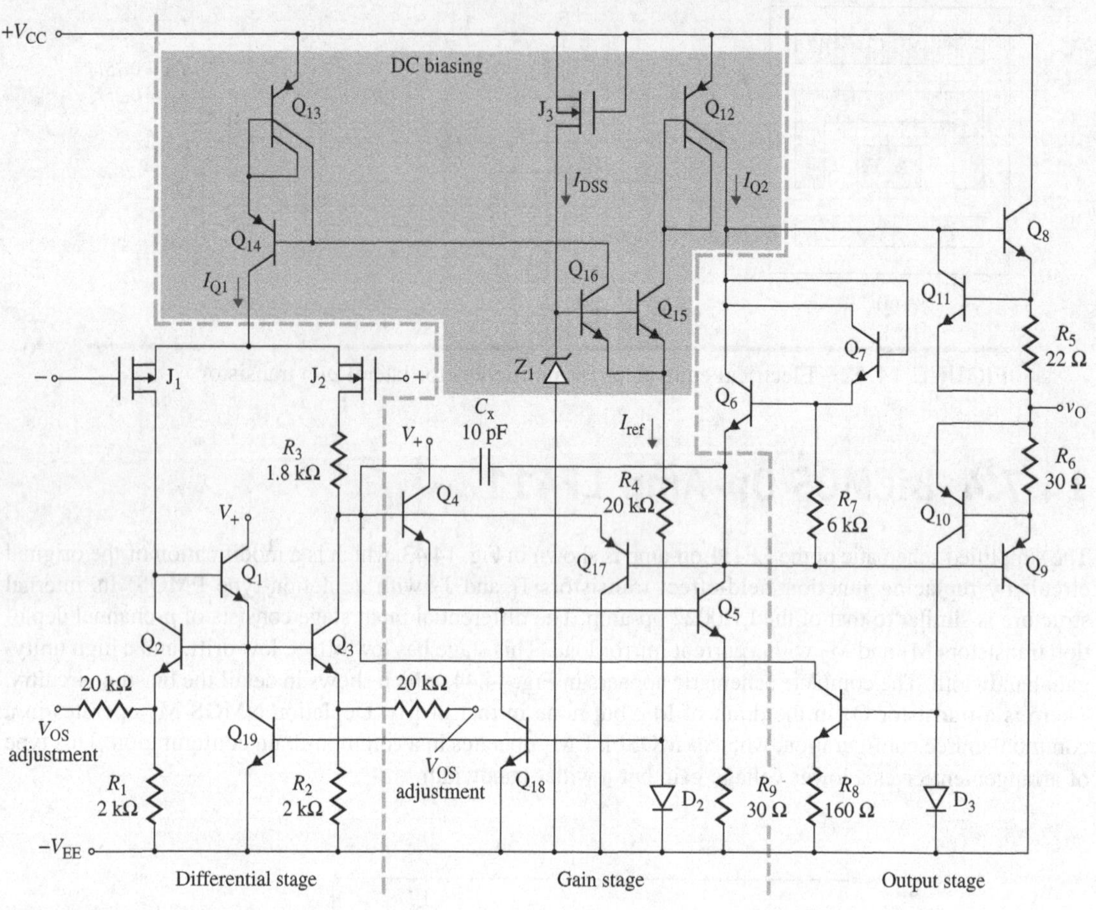

FIGURE 14.44 Schematic for op-amp LF411 (Courtesy of National Semiconductor, Inc.)

A large value of DC biasing circuit I_Q (in milliamperes) can be used for depletion NMOS differential amplifiers without significantly affecting the input biasing currents. On the other hand, for BJT differential amplifiers I_Q is kept small (in microamperes) to have a low input biasing current or high input resistance. Some parameters of the LF411 op-amp are listed in Table 14.9.

DC Biasing

The zener diode Z_1 allows stable DC biasing with thermal compensation by diode D_2 and resistor R_4. With MOSFET M_3 acting as a current-regulator diode (see Sec. 9.6), the current through zener diode Z_1 is kept constant at I_{DSS}. If V_Z is the zener voltage, then

$$I_{ref} = \frac{V_Z - V_{BE16} - V_{D2}}{R_4} \tag{14.125}$$

which in turn sets the DC biasing currents I_{Q1} and I_{Q2}.

TABLE 14.9 Parameters of op-amp LF411

Parameter	Minimum	Typical	Maximum	Units
DC supply voltages V_{CC}, V_{EE}			±18	V
Input offset voltage V_{OS}		0.8	2.0	mV
Thermal drift D_v		7	20	μV/°C
Input biasing current I_B		50	200	pA
Input offset current I_{OS}		25	100	pA
Differential input resistance R_{id}		10^{12}		Ω
Common-mode input resistance R_{ic}		10^{12}		Ω
Output resistance R_o		75		Ω
Input capacitance C_i		4.0		pF
Open-loop voltage gain A_o	25	200		V/mV
Common-mode rejection ratio (CMRR)	70	100		dB
Unity-gain bandwidth f_u	2.7	4.0		MHz
Slew rate (SR)	8.0	15		V/μs
Power supply rejection ratio (PSRR)	70	100		dB

Thermal Protection

Transistor Q_{19} is normally off. If the temperature rises, V_{BE16} will fall because of the negative temperature coefficient of Q_{16}, and the zener voltage V_Z will rise because of the positive temperature coefficient of zener diode Z_Z. As a result, the voltage at the emitter of Q_{16} will rise; thus, the voltage at the anode of D_2 will also rise. If the temperature rise is adequate, Q_{18} and Q_{19} will turn on and reduce the gain and the output voltage of the amplifier.

14.7.5 BiCMOS Op-Amp LH0062

The simplified schematic of the LH0062 op-amp is shown in Fig. 14.45, which is a modification of the original circuit by replacing junction field-effect transistors Q_1 and Q_2 with depletion-type NMOS. The internal structure is similar to that of op-amp LH0022.

Differential Stage

The differential input stage consists of n-channel transistors Q_1 and Q_2 with an active load. This is a normal source-coupled differential pair with a difference output, which gives a lower gain but a wider bandwidth. The input resistance is very high for the MOSFET input stage. The DC biasing current I_Q is set by the current source consisting of transistor Q_3 and zener diode D_1. Thus,

$$I_Q \equiv \frac{V_Z - V_{BE3}}{R_5} \tag{14.126}$$

$$= \frac{1.2\text{ V} - 0.6\text{ V}}{600\ \Omega} = 1\text{ mA}$$

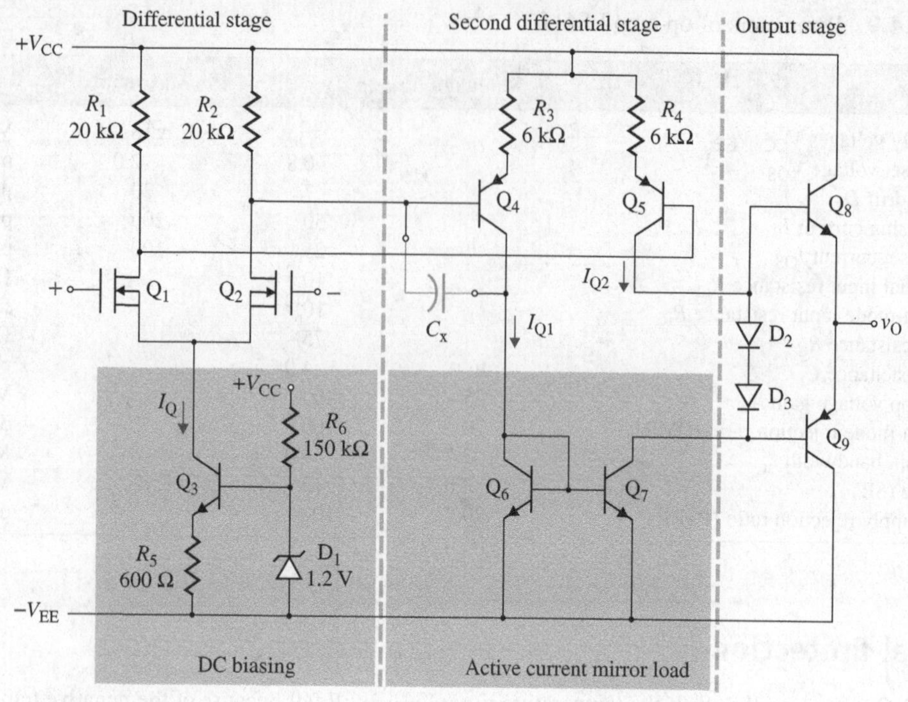

FIGURE 14.45 Simplified schematic for op-amp LH0062 (Courtesy of National Semiconductor, Inc.)

Gain Stage

The gain stage is a common-emitter coupled pair, which drives a current mirror load consisting of transistors Q_6 and Q_7. The single-ended output gives a larger voltage gain. The biasing current I_{Q1} can be found from

$$I_{Q1}R_3 + V_{EB4} = \frac{R_2 I_Q}{2}$$

That is,

$$I_{Q1} = \frac{R_2 I_Q/2 - V_{EB4}}{R_3} \tag{14.127}$$

$$= \frac{20\text{ k}\Omega \times 0.5\text{ mA} - 0.6\text{ V}}{6\text{ k}\Omega} = 1.6\text{ mA}$$

The collector current of Q_7 is the mirror of that of Q_6; that is, I_{Q2} is the mirror of I_{Q1}, and $I_{Q2} = I_{Q1} = 1.6$ mA, which is also the biasing current for the output stage. The high gain and wider bandwidth are obtained by cascading two differential stages. The slew rate is also high because of the higher DC biasing current. Some parameters of the LH0062 op-amp are listed in Table 14.10.

TABLE 14.10 Parameters of op-amp LH0062

Parameter	Minimum	Typical	Maximum	Units
DC supply voltages V_{CC}, V_{EE}			±18	V
Input offset voltage V_{OS}		10	15	mV
Thermal drift D_v		10	35	μV/°C
Input biasing current I_B		10	65	pA
Input offset current I_{OS}		1	5	pA
Differential input resistance R_{id}		10^{12}		Ω
Common-mode input resistance R_{ic}		10^{12}		Ω
Output resistance R_o		75		Ω
Input capacitance C_i		4.0		pF
Open-loop voltage gain A_o	25	160		V/mV
Common-mode rejection ratio (CMRR)	70	90		dB
Unity-gain bandwidth f_u		15		MHz
Slew rate (SR)	50	75		V/μs
Power supply rejection ratio (PSRR)	70	90		dB

14.7.6 BiCMOS Op-Amp LH0032

The LH0032 is an ultrafast op-amp. Its schematic is shown in Fig. 14.46, which is a modification of the original circuit by replacing junction field-effect transistors Q_1 and Q_2 with depletion-type NMOS. The internal structure is similar to that of op-amp LH0062, except that the second differential stage uses common-base

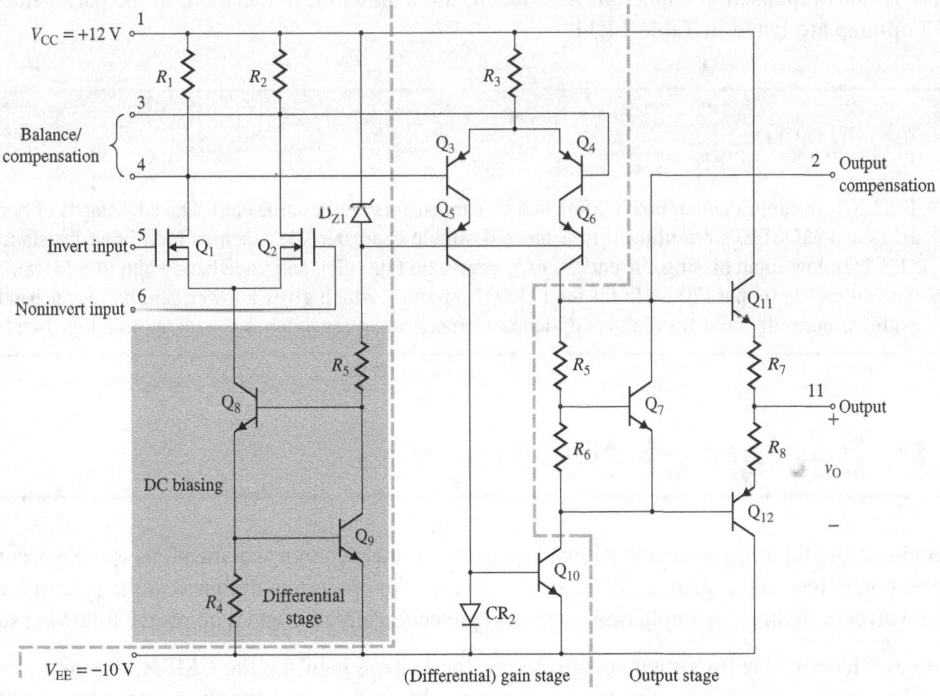

FIGURE 14.46 Schematic for op-amp LH0032 (Courtesy of National Semiconductor, Inc.)

TABLE 14.11 Parameters of op-amp LH0032

Parameter	Minimum	Typical	Maximum	Units
DC supply voltages V_S			± 18	V
Input offset voltage V_{OS}		1.0	2.0	mV
Thermal drift D_v		15	30	μV/°C
Input biasing current I_B		50	150	pA
Input offset current I_{OS}		10	30	pA
Differential input resistance R_{id}		10^{12}		Ω
Common-mode input resistance R_{ic}		10^{12}		Ω
Output resistance R_o		75		Ω
Input capacitance C_i		4.0		pF
Open-loop voltage gain A_o	60	70		dB
Common-mode rejection ratio (CMRR)	50	60		dB
Unity-gain bandwidth f_u		70		MHz
Slew rate (SR)	350	500		V/μs
Power supply rejection ratio (PSRR)	50	60		dB

cascode BJTs for a higher voltage gain. The zener diode D_{Z1} gives a stable reference voltage for the biasing circuit consisting of transistors Q_8 and Q_9. In general, a differential stage with a difference output (i.e., the first differential stage) gives a lower voltage gain but a wider bandwidth. A differential pair with a single-ended output (i.e., the second differential stage) gives a higher voltage gain but a lower bandwidth. However, the common-base cascode connection gives a wider bandwidth.

The LH0032 op-amp has very high gain, a very large slew rate, and a very large bandwidth. Note that there is no compensation capacitor. As a result, the bandwidth is widened. Some parameters of the LH0032 op-amp are listed in Table 14.11.

KEY POINTS OF SECTION 14.7

- BiCMOS op-amps contain both CMOS and BJT transistors on the same chip. The advantages of both BJTs and MOSFETs are utilized to achieve desirable characteristics such as high input resistance (1.5 TΩ), low input biasing current (10 pA), large gain (100 dB), and wide bandwidth (4.5 MHz).
- The difference output (70 MHz for the LH0032 op-amp), which gives a lower gain but wider bandwidth, is normally used for ultrafast op-amps. Compensation capacitance is avoided (see Fig. 14.11).

14.8 Design of Op-Amps

An operational amplifier is a complete IC that is expected to meet certain specifications with respect to input resistance, output resistance, gain, CMRR, and bandwidth. An op-amp is designed at the system level. The process involves designing the amplifying stages and protection circuitry and requires the following steps [9]:

Step 1. Identify the important specifications: the voltage gain A_d, the CMRR, the input resistance R_{id}, the output resistance R_o, the unity-gain bandwidth, and the DC supply voltages V_{CC} and V_{EE} (or V_{DD} and V_{SS}).

Step 2. Select the type of op-amp (BJT, CMOS, or BiCMOS).

Step 3. Choose the circuit configurations and stages.

Step 4. Determine the biasing current requirement I_Q for the desired specifications.

Step 5. Choose the type of current source (BJT or MOSFET) and determine its component ratings.

Step 6. Choose the active load with a current mirror (BJT or MOSFET) in order to obtain the desired voltage gain and determine its component ratings.

Step 7. Choose the type of gain stage (BJT or MOSFET) and determine its component ratings.

Step 8. Choose the type of output stage (BJT or MOSFET) and determine its component ratings.

Step 9. Choose the type of frequency compensation (pole or pole-zero circuit) and determine its component ratings.

Step 10. Determine the voltage, current, and power ratings of active and passive components.

Step 11. Analyze and evaluate the complete differential amplifier to see that it meets the desired specifications.

Step 12. Use PSpice/SPICE to simulate and verify your design. If it would be implemented with discrete devices, use standard values of components, with tolerances of, say, 5%.

Summary

The open-loop voltage gain of an op-amp does not remain constant, but rather decreases with frequency. The frequency response of an internally compensated op-amp has the characteristic of a single time-constant network. The slew rate of the input signal should be less than that of the op-amp in order to avoid distortion of the output voltage. The common-mode rejection ratio is a measure of the ability of an op-amp to reject common-mode signals; the ratio should be as high as possible.

The characteristics of practical op-amps differ from those of ideal ones. The output of an op-amp is affected by parameters such as input offset voltage, input offset current, input biasing current, thermal drift, the power supply rejection ratio, and the input frequency. The effect of input biasing currents can be minimized by adding an offset-minimizing resistor.

The internal structure of an op-amp usually consists of differential, gain, and output stages. There are many possible combinations of op-amp stages, depending on whether the amplifier is a bipolar, MOSFET, or CMOS op-amp. In general, a MOSFET op-amp gives a very high input resistance but less gain, whereas a bipolar op-amp gives a higher gain but lower input resistance. A Darlington BJT pair is commonly used for high current gain and input resistance. A difference output of a differential stage gives less gain but yields wider bandwidth.

Op-amps exhibit offset voltages and currents, which can be minimized by appropriate internal design. The input offset voltage is dependent on the CMRR and the common-mode signal. A pole or pole-zero compensation circuit is commonly used for frequency compensation, but the compensation comes at the expense of unity-gain bandwidth.

The small-signal voltage gain and the input and output resistances of the LM741 op-amp can be determined for various load conditions. Since the voltage gain is very high, the amplifier offers offset voltages and currents because of variations in the transistor parameters and circuit resistances. Although the analysis required to derive equations describing the characteristics of current sources and differential amplifiers can be simplified with some assumptions, computer-aided analysis is generally required to evaluate the actual performance of an op-amp at the final design stage. The analysis of other op-amps would be similar to the analysis of the LM741 op-amp in this chapter.

References

1. A. B. Grebene, *Bipolar and MOS Analog Integrated Circuit Design*. New York: Wiley, 2003.
2. R. A. Gayakwad, *Op-Amps and Linear Integrated Circuits*. Englewood Cliffs, NJ: Prentice Hall, 1993.
3. W. D. Stanley, *Operational Amplifiers with Linear Integrated Circuits*. Upper Saddle River, NJ: Prentice Hall, 2002.
4. D. G. Ong, *Modern MOS Technology*. New York: McGraw-Hill, 1984.
5. P. R. Gray, P. J. Hurst, S. H. Lewis, and R. G. Meyer, *Analysis and Design of Analog Integrated Circuits*. New York: Wiley, 2001.
6. S. Soclof, *Design and Applications of Analog Integrated Circuits*. Englewood Cliffs, NJ: Prentice Hall, 1991.
7. R. F. Coughlin and F. F. Driscoll, *Operational Amplifiers and Linear Integrated Circuits*. Upper Saddle River, NJ: Prentice Hall, 2001.
8. R. C. Jaeger and T. Blalock, *Microelectronic Circuit Design*. New York: McGraw-Hill, 2008.
9. F. C. Franco, *Design with Operational Amplifiers and Analog Integrated Circuits*. New York: McGraw-Hill, 2002.

Review Questions

1. What are the main stages of an op-amp?
2. What is the input biasing current of an op-amp?
3. What is the input offset current of an op-amp, and what factors influence its value?
4. What is the input offset voltage of an op-amp, and what factors influence its value?
5. What is the CMRR of an op-amp?
6. What factors influence the unity-gain bandwidth of an op-amp?
7. What factors influence the slew rate of an op-amp?
8. What is the relation between slew rate and unity-gain bandwidth?
9. What is the typical value of the input resistance of an op-amp?
10. What is the typical value of the output resistance of an op-amp?
11. Ideally, what should be the differential voltage gain of an op-amp?
12. Ideally, what should be the common-mode voltage gain of an op-amp?
13. What is the unity-gain bandwidth of an op-amp?
14. What is the effect of rise time on the frequency response of an op-amp?
15. What is a slew rate?
16. What is the slew rate of a step input voltage?
17. What is the slew rate of a sinusoidal input voltage?
18. What is the feedback factor?
19. What is the typical break frequency of an op-amp?
20. What is the effect of input offset voltage on the output of inverting and noninverting amplifiers?
21. What is the effect of input biasing currents on the output of inverting and noninverting amplifiers?

22. What is the effect of input offset current on the output of inverting and noninverting amplifiers?
23. What is the common method for minimizing the effect of input biasing currents?
24. What is the cause of thermal drift?
25. What is the effect of thermal drift?
26. What is the effect of output offset voltage on an integrator?
27. What is the PSRR?
28. What is pole-zero compensation of an op-amp?
29. What circuit configurations are used for ultrafast op-amps?
30. What are the advantages and disadvantages of MOSFET op-amps?
31. What are the advantages and disadvantages of bipolar op-amps?
32. What are the advantages and disadvantages of CMOS op-amps?
33. What are the advantages and disadvantages of BiCMOS op-amps?
34. What is the function of short-circuit protection in op-amps?
35. What are the typical values of input resistance, output resistance, and voltage gain for the LM741 op-amp?

Problems

The symbol **D** indicates that a problem is a design problem. The symbol **P** indicates that you can check the solution to a problem using PSpice/SPICE or Electronics Workbench.

14.3 Parameters and Characteristics of Practical Op-Amps

14.1 The input voltages of an op-amp are $v_1 = 120\ \mu\text{V}$ and $v_2 = 80\ \mu\text{V}$. The op-amp parameters are CMRR = 90 dB and $A_d = A_o = 2 \times 10^5$. Determine **(a)** the differential voltage v_d, **(b)** the common-mode voltage v_c, **(c)** the magnitude of the common-mode gain A_c, and **(d)** the output voltage v_O.

14.2 The rise time of an op-amp is $t_r = 0.3\ \mu\text{s}$. What is the maximum frequency limit of the op-amp?

14.3 The inverting amplifier in Fig. 14.9(a) has $R_1 = 15\ \text{k}\Omega$ and $R_F = 50\ \text{k}\Omega$. The input offset voltage is $V_{io} = \pm 6$ mV at 25°C. Determine the output offset voltage V_{oo}.

14.4 The input biasing current I_B for the amplifier in Fig. 14.3 is $I_B = 500$ nA (DC) at 25°C. If $R_1 = 15\ \text{k}\Omega$ and $R_F = 50\ \text{k}\Omega$, determine **(a)** the output offset voltage due to input biasing current I_B and **(b)** the offset-minimizing resistance R_x.

14.5 The maximum input offset current of the amplifier in Fig. 14.4 is $I_{io} = \pm 200$ nA at 25°C. If $R_1 = 15\ \text{k}\Omega$ and $R_F = 50\ \text{k}\Omega$, determine the output offset voltage due to the input offset current.

14.6 The inverting amplifier in Fig. 14.5(a) has $R_1 = 15\ \text{k}\Omega$ and $R_F = 50\ \text{k}\Omega$. The op-amp has $V_{io} = \pm 6$ mV, $I_B = 500$ nA, and $I_{io} = \pm 300$ nA at 25°C. Determine the total output offset voltage v_{of} if **(a)** $R_x\ (= R_F \| R_1) = 11.54\ \text{k}\Omega$ and **(b)** $R_x = 0$. Assume $v_S = 0$.

14.7 The noninverting amplifier in Fig. 14.5(b) has $R_1 = 15\ \text{k}\Omega$ and $R_F = 150\ \text{k}\Omega$. The op-amp has $V_{io} = \pm 6$ mV, $I_B = 500$ nA, and $I_{io} = \pm 300$ nA at 25°C. Determine **(a)** the total output offset voltage if $R_x = R_F \| R_1 = 11.54\ \text{k}\Omega$ and **(b)** the output offset voltage v_{of} if $R_x = 0$. Assume $v_S = 0$.

14.8 The integrator in Fig. 14.7 has $R_1 = 10\ \text{k}\Omega$, $R_x = 10\ \text{k}\Omega$, $C_1 = 0.1\ \mu\text{F}$, $V_{CC} = 15$ V, $-V_{EE} = -15$ V, and maximum saturation voltage $= \pm 11$ V. The op-amp has input offset voltage $V_{io} = 6$ mV, input biasing current $I_B = 500$ nA, and input offset current $I_{io} = 300$ nA at 25°C.

a. Determine the time required for the op-amp output offset voltage to reach the saturation limit of ±14 V.
b. Repeat part (a) if $R_x = 0$.

14.9 The inverting amplifier in Fig. 14.5(a) has $R_1 = 15\ \text{k}\Omega$, $R_F = 50\ \text{k}\Omega$, and $R_x = R_F \| R_1 = 11.54\ \text{k}\Omega$. The op-amp has $V_{io} = \pm 6$ mV, $I_B = 500$ nA, and $I_{io} = \pm 300$ nA. The thermal drifts are $D_v = 15\ \mu\text{V}/°\text{C}$, $D_i = 0.5\ \text{nA}/°\text{C}$, and $D_b = 0.5\ \text{nA}/°\text{C}$ at 25°C. The temperature is 55°C. Determine **(a)** the output offset voltage due to drifts V_{od} and **(b)** the total output voltage v_{of} if the input voltage is $v_S = 150$ mV (DC).

14.10 The noninverting amplifier in Fig. 14.5(b) has $R_1 = 15\ \text{k}\Omega$, $R_F = 50\ \text{k}\Omega$, and $R_x = R_F \| R_1 = 11.54\ \text{k}\Omega$. The op-amp has $V_{io} = \pm 6$ mV, $I_B = 500$ nA, and $I_{io} = \pm 300$ nA. The thermal drifts are $D_v = 15\ \mu\text{V}/°\text{C}$, $D_i = 0.5\ \text{nA}/°\text{C}$, and $D_b = 0.5\ \text{nA}/°\text{C}$ at 25°C. The temperature is 55°C. Determine **(a)** the output offset voltage due to drifts V_{od} and **(b)** the total output voltage v_o if the input voltage is $v_S = 150$ mV (DC).

14.11 The inverting amplifier in Fig. 14.5(a) has $R_1 = 15\ \text{k}\Omega$ and $R_F = 150\ \text{k}\Omega$. The supply voltages change from ± 12 V to ± 10 V, and PSRR = 150 μV/V at 25°C. Determine **(a)** the input offset voltage V_{io} due to changes in the supply voltages and **(b)** the corresponding output offset voltage V_{oo}.

14.12 The noninverting amplifier in Fig. 14.5(b) has $R_1 = 15\ \text{k}\Omega$ and $R_F = 50\ \text{k}\Omega$. The supply voltages change from ± 12 V to ± 10 V, and PSRR = 150 μV/V at 25°C. Determine **(a)** the input offset voltage V_{io} due to changes in the supply voltages and **(b)** the corresponding output offset voltage V_{oo}.

14.13 The inverting amplifier in Fig. 14.5(a) has $R_1 = 10\ \text{k}\Omega$, $R_F = 100\ \text{k}\Omega$, and $R_x = R_F \| R_1 = 9.091\ \text{k}\Omega$. The op-amp parameters are $V_{io} = 0.8$ mV, $I_B = 200$ nA, $I_{io} = 100$ nA, and PSRR = 150 μV/V. The drifts are $D_v = 15\ \mu\text{V}/°\text{C}$, $D_i = 0.5\ \text{nA}/°\text{C}$, and $D_b = 0.5\ \text{nA}/°\text{C}$ at 25°C. The temperature is 55°C. The DC supply voltages change from $V_{CC} = 12$ V to 10 V and $-V_{EE} = -12$ V to -10 V. The input voltage is $v_S = 100$ mV (DC). Determine the output voltage v_O if **(a)** $R_x = R_F \| R_1 = 9.091\ \text{k}\Omega$ and **(b)** $R_x = 0$.

14.14 (D) The noninverting amplifier in Fig. 14.12(a) has $R_1 = 10\ \text{k}\Omega$, $R_F = 150\ \text{k}\Omega$, and $R_x = R_F \| R_1 = 9.091\ \text{k}\Omega$. Design the offset compensating network. The op-amp parameters are $V_{io} = 0.8$ mV, $I_B = 200$ nA, $I_{io} = 100$ nA, and PSRR = 150 μV/V. The DC supply voltages are $V_{CC} = 12$ V and $-V_{EE} = -12$ V.

14.15 (D) The inverting amplifier in Fig. 14.12(b) has $R_1 = 10\ \text{k}\Omega$, $R_F = 50\ \text{k}\Omega$, and $R_x = R_F \| R_1 = 8.33\ \text{k}\Omega$. Design the offset compensating network. The op-amp parameters are $V_{io} = 0.8$ mV, $I_B = 200$ nA, $I_{io} = 100$ nA, and PSRR = 150 μV/V. The DC supply voltages are $V_{CC} = 12$ V and $-V_{EE} = -12$ V.

14.16 (D) The differential amplifier in Fig. 14.12(c) has $R_a = R_1 = 12\ \text{k}\Omega$ and $R_F = R_x = 24\ \text{k}\Omega$. Design the offset compensating network. The op-amp parameters are $V_{io} = 6$ mV, $I_B = 500$ nA, $I_{io} = 200$ nA, and PSRR = 150 μV/V. The DC supply voltages are $V_{CC} = 12$ V and $-V_{EE} = -12$ V.

14.17 The Darlington pair shown in Fig. 14.13 is biased in such a way that the collector-biasing current I_{C2} of Q_2 is 400 μA. The current gains of the two transistors are the same, $\beta_{F1} = \beta_{F2} = 80$, and the Early voltage is $V_A = 50$ V. Calculate **(a)** the effective input resistance r_π, **(b)** the effective transconductance g_m, **(c)** the effective current gain $\beta_{F(\text{eff})}$, and **(d)** the effective output resistance r_o.

14.4 CMOS Op-Amps

14.18 The CMOS of the op-amp in Fig. 14.20 has $K_x = 10\ \mu\text{A}/\text{V}^2$, $W/L = 160\ \mu\text{m}/10\ \mu\text{m}$, and $V_t = 0.5$ V. The biasing current is $I_Q = 40\ \mu$A. If V_t changes by 2% and W/L by 1%, find the input offset voltage V_{OS}.

14.19 (D) The CMOS amplifier in Fig. 14.20 is operated at a biasing current of $I_Q = 50\ \mu$A. The parameters of the MOSFETs are $K_x = 10\ \mu\text{A}/\text{V}^2$, $|V_{M(\text{NMOS})}| = V_{M(\text{PMOS})} = 60$ V, $V_t = 1$ V, and $W/L = 80\ \mu\text{m}/10\ \mu\text{m}$, except for Q_7, for which $W/L = 160\ \mu\text{m}/10\ \mu\text{m}$. Assume $V_{DD} = V_{SS} = 5$ V.

a. Find V_{GS}, g_m, and r_o for all MOSFETs.

b. Find the low-frequency voltage gain of the amplifier A_{vo}.

c. Find the value of the external resistance R_{ref}.

d. Find the value of compensation capacitance C_x that gives a unity-gain bandwidth of 1 MHz and the corresponding slew rate.

e. Find the value of resistance R_x to be connected in series with C_x in order to move the zero frequency to infinity.

f. Find the common-mode input voltage range.

g. Find the output voltage range.

14.5 BJT Op-Amps

14.20 The BJTs for the op-amp shown in Fig. P14.20 have $\beta_{F(npn)} = 100$, $\beta_{F(pnp)} = 50$, $V_{A(pnp)} = V_{A(npn)} = 80$ V, $V_{BE} = 0.6$ V, and $I_S = 10^{-14}$ A. Find **(a)** the input biasing current I_B, **(b)** the input resistance R_i, **(c)** the unity-gain bandwidth f_u and the slew rate, and **(d)** the overall low-frequency voltage gain A_{vo}. Assume $V_{CC} = -V_{EE} = 15$ V.

FIGURE P14.20

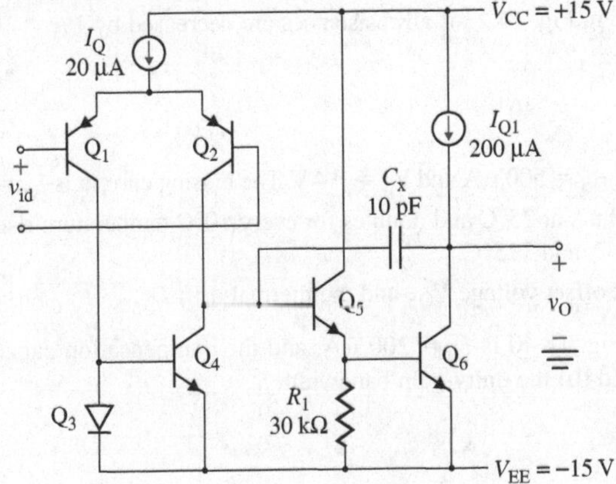

14.21 The BJTs for the op-amp shown in Fig. P14.21 have $\beta_{F(npn)} = 100$, $\beta_{F(pnp)} = 50$, $V_{A(pnp)} = V_{A(npn)} = 80$ V, $V_{BE} = 0.6$ V, and $I_S = 10^{-14}$ A. Find **(a)** the input biasing current I_B, **(b)** the input resistance R_i, and **(c)** the overall low-frequency voltage gain A_{vo}. Assume $V_{CC} = V_{EE} = 15$ V.

FIGURE P14.21

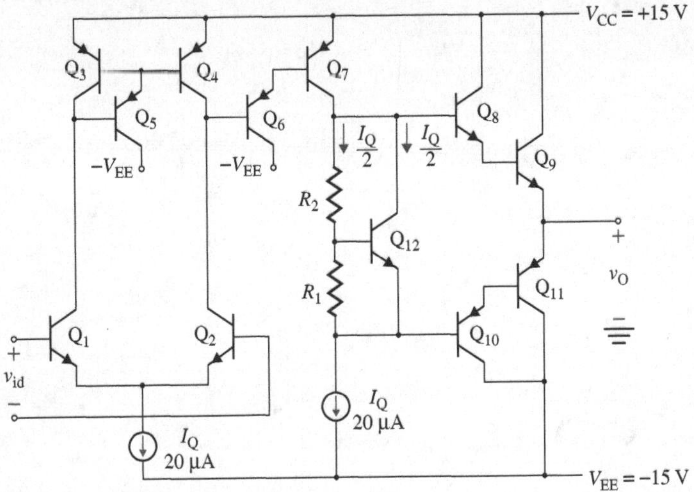

14.22 The BJTs for the LM124 op-amp shown in Fig. 14.22 have $\beta_{F(npn)} = 100$, $\beta_{F(pnp)} = 50$, $V_{A(pnp)} = V_{A(npn)} = 80$ V, $V_{BE} = 0.6$ V, and $I_S = 10^{-14}$ A. Assume $V_{CC} = 12$ V. Find **(a)** the input biasing current I_B, **(b)** the input resistance R_i, **(c)** the unity-gain bandwidth f_u and the slew rate, **(d)** the overall low-frequency voltage gain A_{vo}, and **(e)** the value of R_{SC} for a short-circuit current limit of 25 mA.

14.6 *Analysis of the LM741 Op-Amp*

14.23 Calculate the gain of the LM741 op-amp in Fig. 14.23 if R_8 is reduced from 100 Ω to 0.

14.24 Calculate the gain of the LM741 op-amp in Fig. 14.23 if the current gain for *npn* transistors is changed from $\beta_F = 250$ to $1.1 \times 250 = 275.0$ and the saturation current is changed from $I_S = 10^{-14}$ A to 1.05×10^{-14} A.

14.25 Calculate the gain of the LM741 op-amp in Fig. 14.23 if all resistances are increased by 1%.

14.26 Calculate the gain of the LM741 op-amp in Fig. 14.23 if all resistances are decreased by 1%.

14.7 *BiCMOS Op-Amps*

14.27 The BiCMOS op-amp in Fig. 14.40 has $I_{DDS} = 500$ μA and $V_p = -4$ V. The biasing current is $I_Q = 200$ μA.

a. If the input biasing current is $I_B = 1.0$ nA at 25°C and doubles for every 10°C temperature rise, find the input biasing current I_B at 75°C, 100°C, and 125°C.

b. If I_{DDS} changes by 2%, find the input offset voltage V_{OS} and the thermal drift D_v.

14.28 The biasing current for the op-amp in Fig. 14.40 is $I_Q = 200$ μA, and the compensation capacitance is $C_x = 30$ pF. Find **(a)** the slew rate SR and **(b)** the unity-gain bandwidth f_u.

CHAPTER 15

INTRODUCTION TO DIGITAL ELECTRONICS

Learning Outcomes

After completing this chapter, students should be able to do the following:

- Describe the definition of logic states.
- List the performance parameters of logic gates.
- List the logic families and describe their internal circuitry.
- List the relative advantages and disadvantages of logic families.
- Design and analyze simple logic gates.

Symbols and Their Meanings

Symbol	Meaning
I_{IL}, I_{IH}	Input low and input high currents of a logic gate
I_{OL}, I_{OH}	Output low and output high currents of a logic gate
I_{Bn}, I_{Cn}, I_{En}	Base, collector, and emitter currents of an *n*th BJT
i_{Dn}, v_{GSn}	Drain current and gate–source voltage of an *n*th MOSFET
i_{DL}, v_{DSL}	Drain current and drain–source voltage of a load MOSFET
K_L, K_n	Constants of a load MOSFET and a driver NMOS of a logic gate

Symbol	Meaning
N	Fan-out of a logic gate
NM_L, NM_H	Low logic–noise margin and high logic–noise margin of a logic gate
P_{on}, P_{off}, P_{static}	On-state, off-state, and static low of a logic gate
R_{dN}, R_{dP}	Drain–source resistances of an NMOS and a PMOS
t_{pdo} or t_{pLH}	Propagation delay time from low to high
t_{pd1} or t_{pHL}	Propagation delay time from high to low
v_I, v_O	Input and output voltages of a logic gate
V_{tL}, V_{tD}	Threshold voltages of load and driver MOSFETs of a logic gate
V_{IL}, V_{IH}	Input low and input high voltages of a logic gate
V_{OL}, V_{OH}	Output low and output high voltages of a logic gate
V_{TW}, V_{LS}	Transition width and logic swing voltages of a logic gate

15.1 Introduction

If we observe the output (v-i) characteristic of a transistor closely, we notice that there are two distinct regions: a low-resistance region and a high-resistance region. In analog electronics, transistors are operated in the active region as amplifying devices, so they exhibit the characteristic of a high output resistance. However, with proper biasing conditions, transistors can also exhibit the characteristic of a low output resistance—that is, a low output voltage. In digital electronics, transistors are operated as on (low output) and off switches.

15.2 Logic States

Electronic circuits used to perform logic functions are known as *digital logic circuits* or *logic gates*. They use two binary variables, or states: 0 (low) and 1 (high). The binary states are normally represented by two distinct voltages: voltage V_H for logic 1 and voltage V_L for logic 0. If the logic 1 voltage is higher than the logic 0 voltage (e.g., $V_H = 5$ and $V_L = 0$), the circuit is said to use *positive logic*. If the logic 1 voltage is lower than the logic 0 voltage (e.g., $V_H = -5$ and $V_L = 0$), the circuit is said to use *negative logic*. In this chapter, we consider positive logic.

To accommodate variations in component tolerances, temperature, and noise in a logic circuit, two voltage ranges are usually used to define the two logic states. The ranges are illustrated in Fig. 15.1, in which V_{H1} is the lowest voltage that will always be recognized as logic 1 and V_{L2} is the highest voltage that will always be recognized as logic 0. If the voltage lies in the range V_{H2} to V_{H1}, the state of the digital circuit is interpreted as logic 1. If the voltage lies in the range V_{L2} to V_{L1}, the state of the digital circuit is interpreted as logic 0. The two voltage regions are separated by an *undefined*, or *excluded, region*. This is a forbidden band, and the signal voltage is not permitted to lie in this region. The difference $V_{H1} - V_{L2}$ is called the *transition region*.

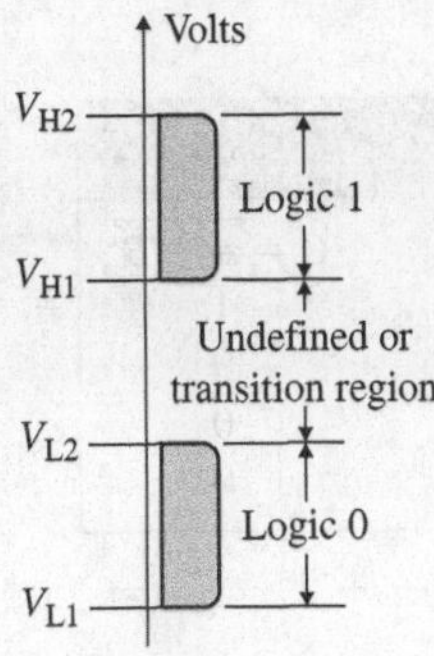

FIGURE 15.1 Voltage ranges and binary variables

KEY POINT OF SECTION 15.2

- **The binary states of a logic circuit are normally represented by two distinct voltages: V_H for logic 1 and V_L for logic 0. V_H is positive for positive logic and negative for negative logic.**

15.3 Logic Gates

The commonly used logic gates are NOT (inverter), AND, NAND, OR, and NOR. Consider a voltage-controlled switch with a resistance, as shown in Fig. 15.2(a), which is controlled by the input signal v_I applied between terminals 1 and 2. The output is taken across the switch between terminal 3 and the ground terminal 2. When v_I is low (around 0), the switch is open, and the output voltage v_O is high (equal to the DC supply voltage V_{CC}). When v_I is high enough to close the switch (generally above a specified threshold voltage), the switch closes, and the output voltage v_O is low (around 0). The input and output voltages are shown in Fig. 15.2(b). The output is the logical inversion of the input signal, and this circuit is known as an *inverter* or a *NOT gate*. Inverters are the building blocks of logic gates.

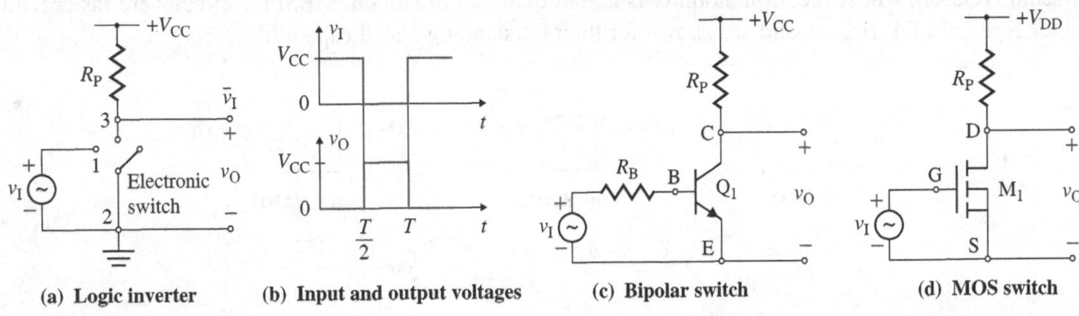

FIGURE 15.2 Inverter

TABLE 15.1 Truth table for NOT, AND, NAND, OR, and NOR

Input	NOT	Input	Input	AND	NAND	OR	NOR
A	$C = \overline{A}$	A	B	$C = AB$	$C = \overline{AB}$	$C = A + B$	$C = \overline{A + B}$
0	1	0	0	0	1	0	1
1	0	0	1	0	1	1	0
		1	0	0	1	1	0
		1	1	1	0	1	0

The switch can be realized by either a bipolar transistor, as shown in Fig. 15.2(c), or a MOS transistor, as shown in Fig. 15.2(d). Transistors M_1 and Q_1 are switched between two states: nonconducting (or off) and conducting (or on). R_P is called the *pull-up resistance* because the output is pulled up toward the positive supply voltage V_{CC} or V_{DD} when the switching transistor is off.

An inverter has only one input voltage. Logic gates usually combine one or more logic variable inputs to produce an output. All of the possible combinations of the input variables and the corresponding outputs are normally listed in a table called a *truth table*. The truth table for NOT, AND, NAND, OR, and NOR is shown in Table 15.1. If both inputs are high, the output becomes high in an AND gate and low in a NAND gate. If at least one input is high, the output becomes high in an OR gate and low in a NOR gate. The symbols for these logic gates are shown in Fig. 15.3.

Digital circuits are available exclusively in integrated circuits (ICs) and can be classified into families. Each member of a family is made with the same technology, has similar structure, and exhibits the same basic features. There are two MOS-logic families (NMOS, using only *n*-channel MOSFETs, and CMOS, using both *n*- and *p*-channel MOSFETs in a complementary configuration), and two BJT families, (a transistor-transistor logic [TTL] family and an emitter-coupled logic [ECL] family). Each of these families has unique advantages and disadvantages. The choice of a logic family is based on considerations such as logic functions, logic flexibility, speed, noise immunity, operating temperature range, power dissipation, and cost.

MOS transistor logic circuits have advantages over bipolar logic gates because MOSFETs are simpler to fabricate and occupy less space in integrated form than BJTs. A MOSFET can be connected to act as a resistive load to replace a diffused resistor (a resistor made by diffusion during the IC-manufacturing process) in bipolar integrated circuits. The packing density of MOSFETs is extremely high, and MOSFET circuits can be fabricated for LSI, VLSI, and ULSI applications. MOSFETs and metal–Schottky barrier FETs (MESFETs) can also be used in digital integrated circuits. MESFETs are usually fabricated of gallium arsenide (GaAs), whose electron mobility is higher than that of silicon. MESFET circuits are faster than other types of FET circuits and are known for their outstanding speed capabilities.

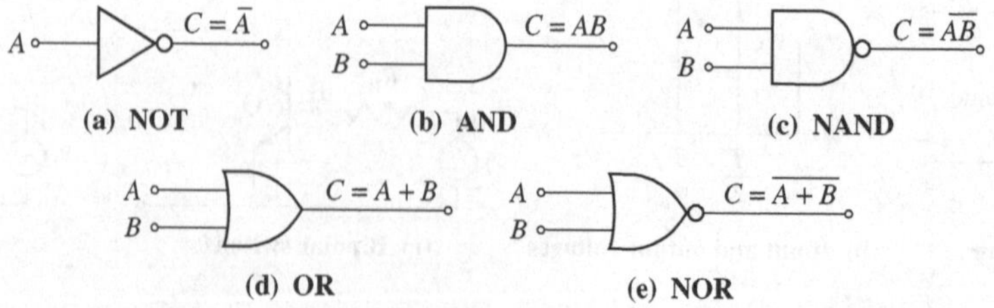

A — $C = \overline{A}$

(a) NOT

A, B — $C = AB$

(b) AND

A, B — $C = \overline{AB}$

(c) NAND

A, B — $C = A + B$

(d) OR

A, B — $C = \overline{A + B}$

(e) NOR

FIGURE 15.3 Symbols for logic gates

TABLE 15.2 Classification of digital integrated circuits

Degree of Integration	Number of Gates
Small-scale integration (SSI)	Fewer than 10
Medium-scale integration (MSI)	From 10 to 100
Large-scale integration (LSI)	From 100 to 1000
Very-large-scale integration (VLSI)	From 1000 to 10^5
Ultra-large-scale integration (ULSI)	More than 10^5

Depending on the complexity of the internal circuitry of the IC chip, digital IC packages can be classified by degree of integration, as shown in Table 15.2.

- The commonly used logic gates are NOT (inverter), AND, NAND, OR, and NOR. Inverters are the building blocks of logic gates. There are many families of logic gates, among them NMOS, CMOS, TTL, and ECL gates.

15.4 Performance Parameters of Logic Gates

Figure 15.2(b) shows the characteristics of an ideal inverter [1–3]. However, the performance of actual inverters differs significantly from the ideal in the following ways:

1. The switch is not ideal; that is, when the switch is closed, it has a finite voltage drop rather than a short circuit.
2. The switch may not open and close instantaneously because of a time delay between the application of the input signal and the propagation of the desired output signal.
3. The input terminal of the inverter usually draws some current from the driving source.
4. The inverter usually drives a load or acts as a source to the next stage(s), and it also should be capable of supplying the driving current.

Manufacturers' data sheets on logic gates specify many performance parameters. As examples, typical data for TTL and CMOS devices are shown in Table 15.3, where t_{pd0} (also abbreviated as t_{pLH}) is the propagation delay time from low to high and t_{pd1} (also abbreviated as t_{pHL}) is the propagation delay time from high to low.

TABLE 15.3 Parameters of 54L/74L TTL and 54C/74C CMOS logic gates

Family	V_{CC}	V_{IL} max	I_{IL} max (mA)	V_{IH} min	I_{IH} 2.4 V (μA)	V_{OL} max	I_{OL} (μA)	V_{OH} min	I_{OH} (μA)	t_{pd0} TYP (ns)	t_{pd1} TYP (ns)	P_D/gate (μW)	P_D/gate 1 MHz, 50 pF (mW)
54L/74L	5	0.7	0.18	2.0	10	0.3	2000	2.4	100	31	35	1000	2.25
54C/74C	5	0.8	—	3.5	—	0.4	360	2.4	100	60	45	0.01	1.25
54C/74C	10	2.0	—	8.0	—	1.0	10	8.0	10	25	30	0.03	5

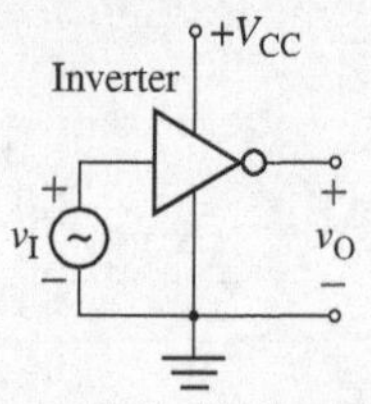

(a) Inverter connection

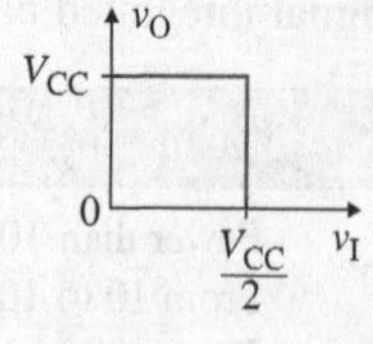

(b) Transfer characteristic

FIGURE 15.4 Transfer characteristic of an ideal inverter

15.4.1 Voltage Transfer Characteristic

The *voltage transfer characteristic* (VTC) gives the relationship between the input voltage v_I and the output voltage v_O. An inverter with a single power supply V_{CC} and its VTC are shown in Figs. 15.4(a) and 15.4(b), respectively. The inverter has a threshold voltage of $V_{tI} = V_{CC}/2$; that is, the output is high (at $v_O = V_{CC}$) for $v_I < V_{CC}/2$ and low (at $v_O = 0$) for $v_I > V_{CC}/2$. The transition from low to high and vice versa is very sharp at $v_I = V_{CC}/2$. Thus, the incremental voltage gain is $dv_O/dv_I = 0$ for $v_I < V_{CC}/2$ and it is $dv_O/dv_I = \infty$ for $v_I = V_{CC}/2$. Therefore, the inverter exhibits a nonlinear characteristic. If it is connected with feedback, it will oscillate between the low and high states.

A practical inverter does not have a finite threshold voltage; rather, it goes through a *transition region* from the high state to the low state. The VTC of a practical inverter is shown in Fig. 15.5; the VTC has three distinct regions: the low-input region, $v_I < V_{IL}$; the transition region, $V_{IL} \le v_I \le V_{IH}$; and the high-input region, $v_I > V_{IH}$.

The VTC has two transitions, one at $v_I = V_{IL}$ and one at $v_I = V_{IH}$, with two corresponding output voltages, V_{OH} and V_{OL}. The transition voltages are defined as the points at which the slope of the VTC is -1 (that is, $dv_O/dv_I = -1$). The following variables are applicable to all logic circuits:

V_{OH} (high-level output voltage) is the *minimum* output voltage that will establish a high level (logic 1). Data sheets guarantee that the output voltage will exceed this level at all specified operating conditions.

V_{OL} (low-level output voltage) is the *maximum* output voltage that will establish a low level (logic 0). Data sheets guarantee that the output voltage will not exceed this level at all specified operating conditions.

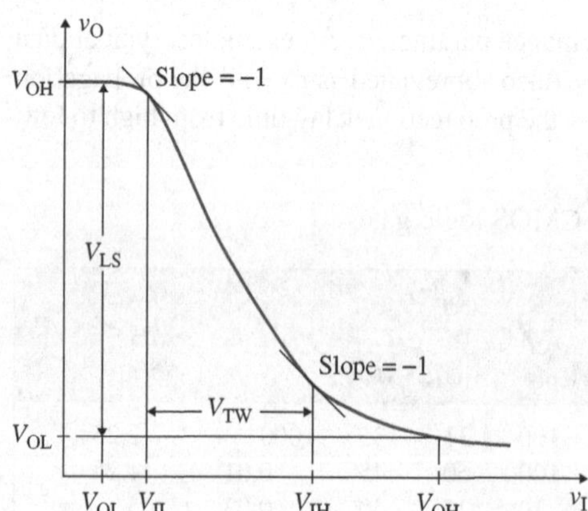

FIGURE 15.5 Transfer characteristic of a practical inverter

V_{IL} (low-level input voltage) is the *maximum* positive voltage that can be applied to an input terminal of a gate and still be recognized as logic low (0).

V_{IH} (high-level input voltage) is the *minimum* positive voltage that can be applied to an input terminal of a gate before the transition starts from the logic high (1) to logic low (0).

In the transition region, the output is undefined. The width of the transition region is a measure of ambiguity, and it is defined by

$$V_{TW} = V_{IH} - V_{IL} \tag{15.1}$$

A low value of V_{TW} is desirable to reduce ambiguity in the input logic state. *Logic swing* is also a measure of the ambiguity in the logic state, and it is defined by

$$V_{LS} = V_{OH} - V_{OL} \tag{15.2}$$

A high value of V_{LS} is desirable to reduce ambiguity and increase noise immunity. A gate for which $V_{OH} = 2.4$ V, $V_{OL} = 0.3$ V, $V_{IH} = 2$ V, and $V_{IL} = 0.7$ V has

$$V_{TW} = V_{IH} - V_{IL} = 2 - 0.7 = 1.3 \text{ V}$$

$$V_{LS} = V_{OH} - V_{OL} = 2.4 - 0.3 = 2.1 \text{ V}$$

15.4.2 Noise Margins

Noise generally is present in logic circuits, superimposed on input signals. Noise simply refers to extraneous signals, which may arise from inadequate regulation or decoupling of the power supply, electromagnetic radiation, inductive or capacitive coupling from other parts of the system, or line drops. One of the advantages of logic circuits is their tolerance for variations in the input signal, which results from the fact that the input signal is interpreted simply as high or low. *Noise immunity* is a measure of the tolerance for variations in the signal level, and it is that voltage that, applied to the input, will cause the output to change its state. Noise immunity is an important device characteristic. However, *noise margin* is of more use to the designer; it defines the amount of noise a system can tolerate under any circumstances and still maintain the integrity of the logic levels. That is, noise margin measures the ability of a gate to maintain its logic state under varying voltage levels.

In digital circuits, one gate usually drives another; that is, the output of the first gate is the input to the following gate. Thus, the gate whose output is high at V_{OH} will drive an identical gate whose high-level input voltage is V_{IH}. This concept is illustrated in Fig. 15.6(a). The difference $V_{OH} - V_{IH}$ represents a margin of safety at the logic high output. It is called the *logic 1*, or *high, noise margin* and is given by

$$NM_H = V_{OH} - V_{IH} \tag{15.3}$$

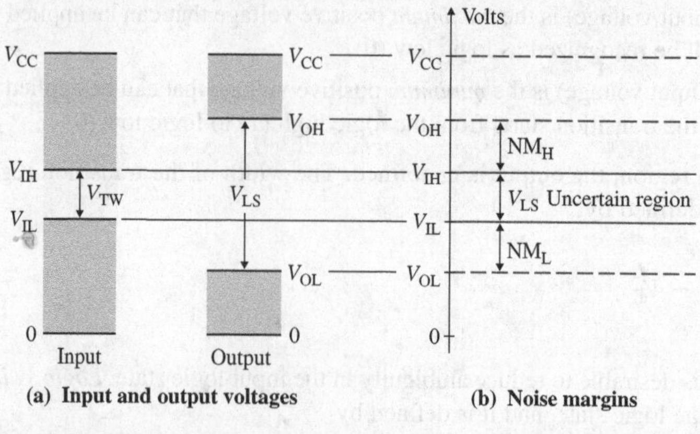

FIGURE 15.6 Noise margins

Similarly, the noise margin at the logic low output is called the *logic 0*, or *low*, *noise margin* and is denoted by

$$NM_L = V_{IL} - V_{OL} \tag{15.4}$$

Therefore, the noise margin is the difference between the guaranteed logic 1 (or 0) level output voltage and the guaranteed logic 1 (or 0) level input voltage. The *absolute noise margin* is the smaller of the two noise margins; that is,

$$NM = \min(NM_L, NM_H) \tag{15.5}$$

The noise margins are shown in Fig. 15.6(b). A gate with a high logic swing V_{LS} and a small transition width V_{TW} will have good noise immunity. There will be no uncertain or undefined region if $V_{IL} = V_{IH} = (V_{OH} + V_{OL})/2$, and the noise margins will be equal: $NM_L = NM_H$. However, this will cause the transition region to be eliminated with an abrupt switch in the ideal VTC, as shown in Fig. 15.4(b). For example, a gate for which $V_{OH} = 2.5$ V, $V_{OL} = 0.3$ V, $V_{IH} = 2$ V, and $V_{IL} = 0.7$ V has

$$NM_H = V_{OH} - V_{IH} = 2.5 - 2 = 0.5 \text{ V}$$

$$NM_L = V_{IL} - V_{OL} = 0.7 - 0.3 = 0.4 \text{ V}$$

$$NM = \min(NM_L, NM_H) = \min(0.4, 0.5) = 0.4 \text{ V}$$

15.4.3 Fan-Out and Fan-In

A gate draws an input current from the input signal and also delivers current to the load gate(s). Note that the current flowing out of a terminal will have a negative value. The gate acts as the load for the input signal and as the driver for the load gates. Just as there are four voltages, there are four currents associated with a gate and its load, and they are defined as follows:

I_{OH} (high-level output current) is the current flowing into the output terminal when it is in the high state (logic 1).

I_{OL} (low-level output current) is the current flowing into the output terminal when it is in the low state (logic 0).

I_{IL} (low-level input current) is the current flowing into the input terminal when a specified low-level voltage (logic 0) is applied to the input.

I_{IH} (high-level input current) is the current flowing into the input terminal when a specified high-level voltage (logic 1) is applied to the input.

A gate must be capable of accepting more than one input. The number of independent input nodes is known as the *fan-in*. The output of one gate must be capable of driving more than one input of subsequent gates. The number N of inputs that can be driven by a gate is known as the *fan-out*. Fan-out is illustrated in Fig. 15.7. More precisely, *fan-out* is defined as the maximum number of load gates of similar design that can be connected to the output of a logic gate (i.e., the driver gate) without changing its logic state. Since logic gates draw a different amount of current in the logic low and logic high states, the fan-out is the smaller of the numbers for the logic low and logic high states. That is, fan-out N is equal to either I_{OL}/I_{IL} or I_{OH}/I_{IH}, whichever gives the lower natural number. N is given by

$$N \equiv \min\left(\frac{I_{OL}}{I_{IL}}, \frac{I_{OH}}{I_{IH}}\right) \tag{15.6}$$

For example, if $I_{OL} = 2$ mA, $I_{IL} = 0.1$ mA, $I_{OH} = 100$ μA, and $I_{IH} = 10$ μA,

$$N \equiv \min\left[\left(\frac{2\text{ mA}}{0.1\text{ mA}}\right), \left(\frac{100\ \mu\text{A}}{10\ \mu\text{A}}\right)\right] = \min(20,10) = 10$$

The fan-out of a logic gate is usually quoted in terms of the number of inverters the gate can drive. This helps in comparing gates and determining the loading effects of the gate in a digital circuit with multiple gates of the same family.

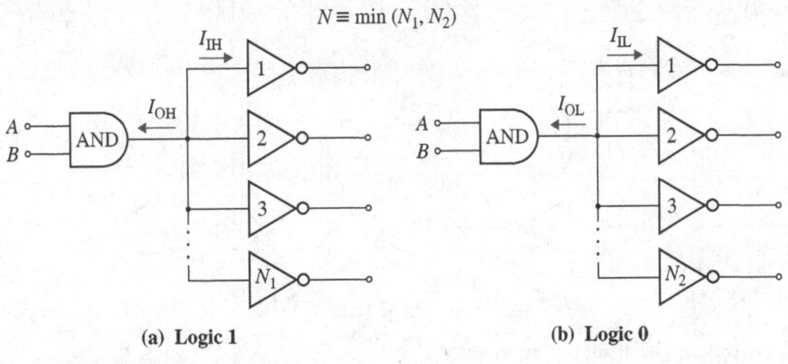

FIGURE 15.7 Fan-out of a gate

EXAMPLE 15.1

D

Designing a simple inverter An inverter, as shown in Fig. 15.8, drives identical inverters and has $V_{OL} = 0.3$ V, $I_{OL} = 2$ mA, $V_{OH} = 2.4$ V, $I_{OH} = 100$ μA, and $V_{CC} = 5$ V. The input currents drawn by each load inverter are $I_{IH} = 0.18$ mA (at logic high) and $I_{IL} = 10$ μA (at logic low).

(a) If there are five load inverters, determine the value of pull-up resistance R_P that will ensure a logic 1 output of $V_{OH} = 2.4$ V.

(b) If $R_P = 4$ kΩ, find the fan-out N.

SOLUTION

(a) At high output, all load inverters are connected to the driving inverter, and each draws 0.18 mA. Because of the resistance R_P, the output voltage will be lower than V_{CC}. Thus,

$$v_O = V_{CC} - R_P(NI_{IH} + I_{OH})$$

To ensure that $v_O \geq V_{OH} = 2.4$ V, we calculate the value of R_P (for $N = 5$) as

$$R_P \leq \frac{V_{CC} - V_{OH}}{NI_{IH} + I_{OH}} = \frac{(5 - 2.4)\text{ V}}{5 \times 0.18\text{ mA} + 100\text{ μA}} = 2.6\text{ kΩ}$$

At low output, each load inverter draws $I_{IL} = 10$ μA. Thus,

$$V_{CC} = V_{OL} + R_P(I_{OL} + NI_{IL})$$

To ensure that $R_P \geq I_{OL} = 2$ mA, we calculate the value of R_P as

$$R_P \geq \frac{V_{CC} - V_{OL}}{I_{OL} + NI_{IL}} = \frac{(5 - 0.3)\text{ V}}{2\text{ mA} + 5 \times 10\text{ μA}} = 2.29\text{ kΩ}$$

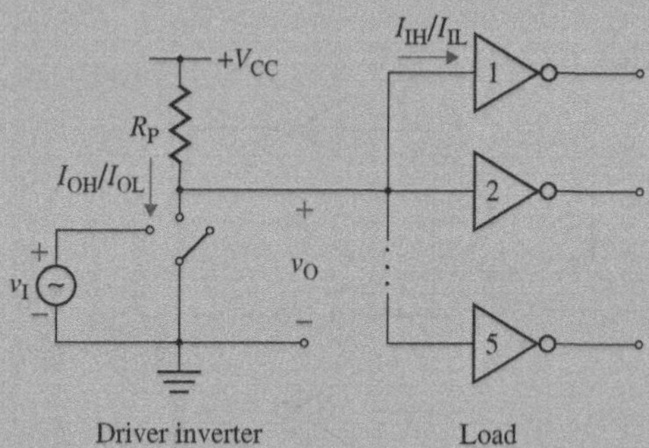

FIGURE 15.8 Logic high fan-out of an inverter

which depends mostly on the value of I_{OL} because $I_{IL} << I_{OL}$. Therefore, the value of R_P should be in the range $2.29\ k\Omega \leq R_P \leq 2.6\ k\Omega$. Let us choose $R_P = 2.5\ k\Omega$.

(b) For $R_P = 4\ k\Omega$, we get the value of N (for logic high) as

$$N = \frac{V_{CC} - V_{OH}}{R_P I_{IH}} - \frac{I_{OH}}{I_{IH}} = \frac{(5 - 2.4)\ \text{V}}{4\ \text{k}\Omega \times 0.18\ \text{mA}} - \frac{0.1\ \text{mA}}{0.18\ \text{mA}} = 3.06$$

Since fractional loads are not possible, the fan-out is $N = 3$. If we choose $N = 4$, v_O will be less than $V_{OH} = 2.4$ V, and the inverter will be in the transition (or logic 0) region.

15.4.4 Propagation Delay

A switching device such as a bipolar transistor exhibits junction capacitances. As a result, the output of an inverter may not respond instantaneously to the input signal. In addition, the load gate(s) offers a certain amount of capacitance C_L to the driving inverter, as shown Fig. 15.9(a). C_L is the equivalent input capacitance of the load gate(s), including any capacitance due to the wiring connection. Therefore, the input and output responses of the inverter will exhibit finite rise time t_r and fall time t_f, as shown in Fig. 15.9(b). The *rise time* t_r is the time required for the waveform to rise from 10% to 90% of its final (high) value. Similarly, the *fall time* t_f is the time required for the waveform to decrease from 90% to 10% of its final (low) value.

The speed of operation of a gate depends on how fast a change in the input propagates through it and causes a change at the output. There will be a delay time between the input and output waveform; this time is commonly known as the *propagation delay time* t_{pd}. The propagation delay time is defined as the time between when the input pulse waveform is at 50% of its logic high value and when the corresponding output pulse waveform is at 50% of its logic high value. For two edges (i.e., falling and rising), there are two delay times, denoted as t_{pd1} (or t_{pHL}) for the high-to-low logic and t_{pd0} (or t_{pLH}) for the low-to-high logic. The average of t_{pd1} and t_{pd0} is the average propagation time t_{pd}:

$$t_{pd} = \frac{t_{pd1} + t_{pd0}}{2} \tag{15.7}$$

This value is commonly used as a figure of merit to compare the performance of different logic families. For hand calculations of the propagation delays, the input voltage can be assumed to be ideal, as shown in Fig. 15.9(c). Typical values of t_{pd} range from 0.5 ns to 10 ns. Cycle time t_{cyc}, another parameter used to compare the performance of logic families, is the time between identical points of successive cycles in a signal waveform. The clock frequency f_{clk}, which is the reciprocal of the cycle time, is more often used. Practical digital systems are usually designed to operate with a cycle time 20 to 50 times the propagation time of a single gate.

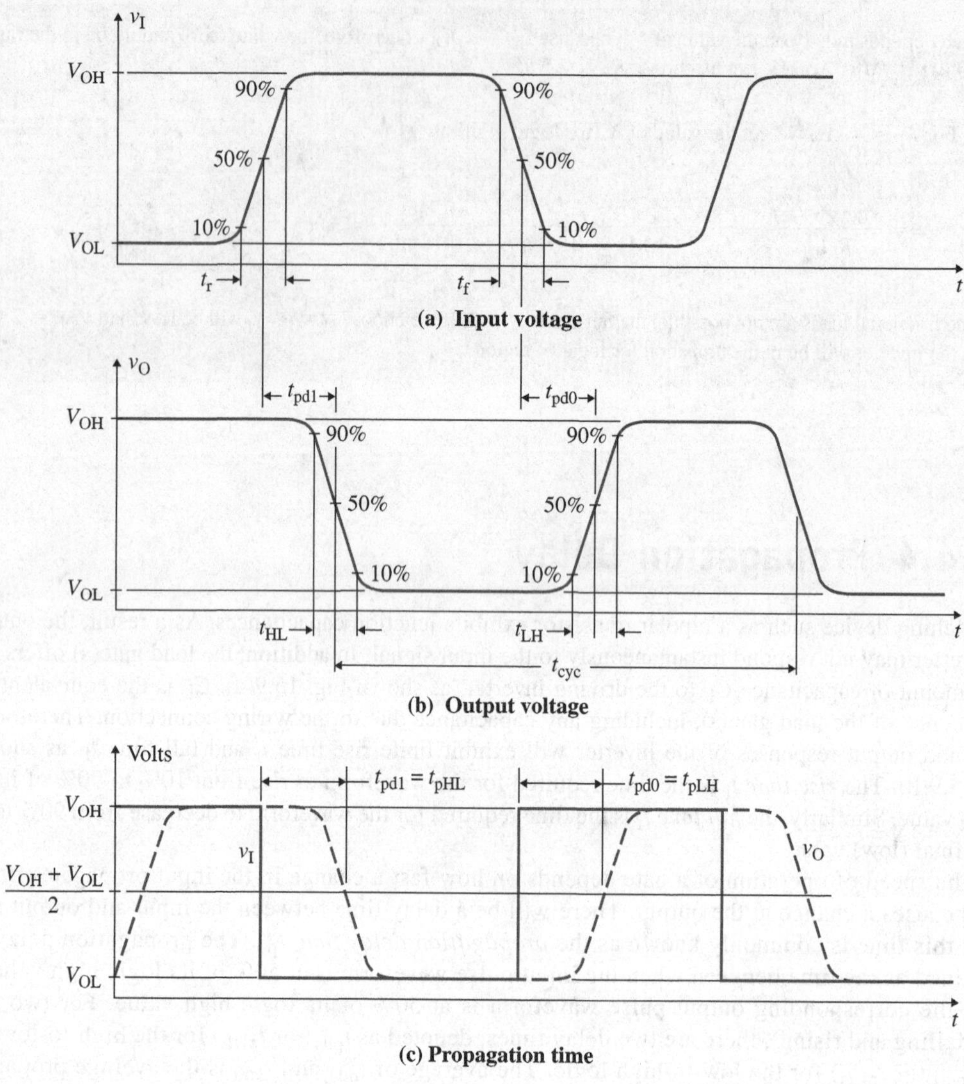

FIGURE 15.9 Propagation times (Note: V_{OH} may not be equal to V_{CC}; V_{OL} may not be equal to 0.)

15.4.5 Power Dissipation

The amount of power (P_D) consumed by a digital circuit is also an important parameter. Knowing this parameter enables the designer to determine the amount of current that will be drawn from the power supply. The power dissipation has static and dynamic components. As an example, consider the inverter in Fig. 15.10(a), where C_L is the load capacitance (normally the input capacitance of another gate or the wiring capacitance or internal capacitance of the switching device itself).

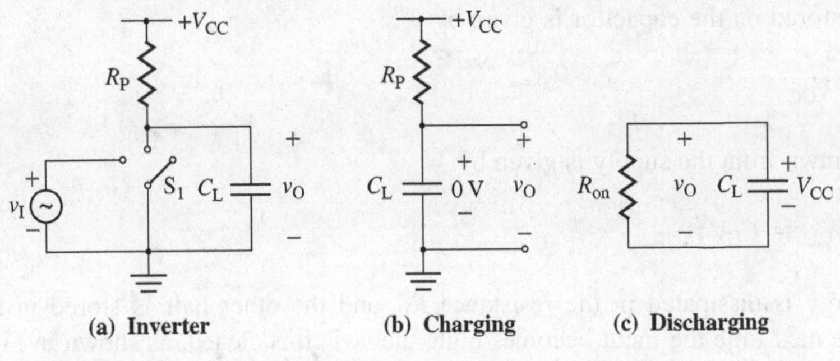

FIGURE 15.10 Charging and discharging of capacitor C_L

Static Power

The power consumed by an inverter depends on its logic state. When the switch S_1 is closed at logic 0, the current is drawn from the supply. The power delivered by the supply is called the *static*, or *quiescent*, *power*. Thus, the static power at logic 0 is given by

$$P_{on} = \frac{V_{CC}^2}{R_P + R_{on}} \tag{15.8}$$

where R_{on} is the on-state switch resistance, whose value is usually low, and $R_{on} << R_P$. When the switch is open at logic 1, a small leakage current I_{leak} flows through the switch. The static power at logic 1 is given by

$$P_{off} = \frac{V_{CC}^2}{R_P + R_{off}} \tag{15.9}$$

where R_{off} is the off-state switch resistance, whose value is usually very large, and $R_{off} >> R_P$. Thus, for the idealized inverter, $P_{on} \approx V_{CC}^2/R_P$ and $P_{off} \approx 0$. The static power is usually expressed as an average value. Assuming that a gate, on average, spends half the time in each state, the *average static power dissipation* becomes

$$P_{static} = \frac{1}{2}(P_{on} + P_{off}) \tag{15.10}$$

$$\approx \frac{V_{CC}^2}{2R_P} \quad \left(\text{for } P_{on} \approx \frac{V_{CC}^2}{R_P} \text{ and } P_{off} \approx 0\right) \tag{15.11}$$

Dynamic Power

An inverter also consumes power each time it changes state. Let us assume that at $t = 0^-$, the input is high and the switch is closed. Thus, the capacitor C_L is discharged, and it has no charge, as shown in Fig. 15.10(a). When the switch is opened at $t = 0^+$, the capacitor will charge exponentially to the supply voltage V_{CC} (approximately) through R_P, as shown in Fig. 15.10(b). The charging current will also flow through R_P, and thus power will be dissipated in R_P.

The charge stored on the capacitor is given by

$$Q = C_L V_{CC} \tag{15.12}$$

and the energy drawn from the supply is given by

$$E = QV_{CC} = C_L V_{CC}^2 \tag{15.13}$$

Half of this energy is dissipated in the resistance R_P, and the other half is stored in the capacitor as $(1/2)C_L V_{CC}^2$. The next time the input becomes high, the switch is closed, as shown in Fig. 15.10(c), and the capacitor discharges through the switch resistance R_{on}. That is, the energy stored in the capacitor is dissipated as heat in R_{on}. Therefore, every time the capacitor C_L is charged or discharged, an amount of energy must be provided by the power supply. The energy per cycle is $C_L V_{CC}^2$. Since energy per unit time is the power, the *dynamic power dissipation* is given by

$$P_{dynamic} = f_{clk} C_L V_{CC}^2 \tag{15.14}$$

where f_{clk} is the clock frequency of the inverter in hertz. Therefore, the total power that must be supplied by the power supply is given by

$$P_D = P_{static} + P_{dynamic} \tag{15.15}$$

Since P_D is dependent on the clock frequency, power dissipation can be a severe problem in digital circuits with frequencies over 100 MHz.

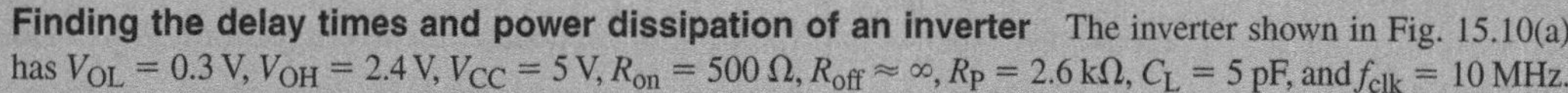

EXAMPLE 15.2

D **Finding the delay times and power dissipation of an inverter** The inverter shown in Fig. 15.10(a) has $V_{OL} = 0.3$ V, $V_{OH} = 2.4$ V, $V_{CC} = 5$ V, $R_{on} = 500\ \Omega$, $R_{off} \approx \infty$, $R_P = 2.6$ kΩ, $C_L = 5$ pF, and $f_{clk} = 10$ MHz.

(a) Find the delay times for the output voltage to rise from 0.3 V to 2.4 V (t_{pd0}) and to fall from 5 V to 0.3 V (t_{pd1}).

(b) Find the power dissipation P_D.

SOLUTION

(a) When the switch is open, the capacitor C_L charges exponentially from 0.3 V to 5 V. The output voltage $v_O(t)$ can be expressed in the general form

$$v_O(t) = v_O(t = \infty) + [v_O(t = 0) - v_O(t = \infty)]e^{-t/\tau} \tag{15.16}$$

For $v_O(t = 0) = 0.3$ V and $v_O(t = \infty) = 5$ V, Eq. (15.16) becomes

$$v_O(t) = 5 - 4.7e^{-t/\tau}$$

which, for $v_O(t) = 2.4$ V and $\tau = R_P C_L = 2.6\ \text{k}\Omega \times 5$ pF = 13 ns, gives the delay time $t_{pd0} = -\tau \ln [(5 - 2.4)/(5 - 0.3)] = 7.7$ ns. When the switch is closed, the capacitor C_L discharges exponentially from 5 V to 0.3 V. The output voltage $v_O(t)$ can be expressed in the general form

$$v_O(t) = v_O(t = 0)e^{-t/\tau}$$

For $v_O(t = 0) = 2.4$ V, the above equation becomes

$$v_O(t) = 2.4e^{-t/\tau}$$

which, for $v_O(t)$ to fall from 2.4 V to 0.3 V and $\tau = R_{on} C_L = 0.5\ \text{k}\Omega \times 5$ pF = 2.5 ns, gives the delay time $t_{pd1} = -\tau \ln (0.3/2.4) = 5.2$ ns.

(b) Substituting the numerical values in Eq. (15.10) gives

$$P_{static} = \frac{V_{CC}^2}{2\,(R_P + R_{on})} = 4.03 \text{ mW}$$

From Eq. (15.14),

$$P_{dynamic} = f_{clk} C_L V_{CC}^2 = 10 \text{ MHz} \times 5 \text{ pF} \times 5^2 = 1.25 \text{ mW}$$

Thus, $P_D = 4.03 + 1.25 = 5.28$ mW.

15.4.6 Delay-Power Product

It is desirable to have both a low propagation delay time (high speed) and low power dissipation. However, these two requirements conflict with each other. For example, if the power dissipation is reduced by decreasing the supply current, the delay will increase. The *delay-power product* (DP) is the product of average propagation delay time (t_{pd}) and power dissipation (P_D):

$$\text{DP} = t_{pd} P_D \qquad (15.17)$$

DP is a figure of merit for comparing logic gates. A small value of DP indicates that a circuit has a fast switching speed and dissipates very little power. In practice, a trade-off between power dissipation and switching speed is usually required. Typical values of DP range from 5 pJ to 50 pJ.

KEY POINT OF SECTION 15.4

- The performance specifications of a logic gate normally include a description of the voltage transfer characteristic, noise margins, fan-out, fan-in, propagation delay, power dissipation, and delay-power product.

15.5 NMOS Inverters

NMOS inverters are the building blocks of NMOS digital circuits [4–6]. They use only *n*-channel MOSFETs, which have low channel resistance because of the greater mobility of electrons in an *n*-channel. If all the MOSFETs are enhancement-type devices, the fabrication procedure becomes simpler. But combining an enhancement-type device (as the driver) and a depletion-type device (as the load) increases the switching speed. An enhancement-type transistor is used as the driver because (1) it will be off when the input voltage v_I is low and (2) its drain and gate voltages have the same polarity. This feature allows direct coupling between stages (i.e., one stage can be connected to the next one without any coupling capacitor). Although depletion-load inverters are generally used in integrated circuits for high-speed switching, we will analyze inverters with both enhancement and depletion MOSFETs.

15.5.1 NMOS Inverter with Enhancement Load

In ICs, an inverter uses a transistor as the load because it requires much less chip area (typically 50 times less than for a diffused resistor with the same value). The resistor in Fig. 15.2(a) can be replaced by a MOSFET.

Enhancement Load

Consider an enhancement-type *n*-channel MOSFET (NMOS) whose gate is connected to the drain terminal, as shown in Fig. 15.11(a). For $v_{GSL} = v_{DSL} \le V_{tL}$ (threshold voltage), the drain current will be zero; that is,

$$i_{DL} = 0$$

$$v_{DSL} = v_{GSL}$$

where L in the subscript after D, DS, and GS refers to the load MOSFET.

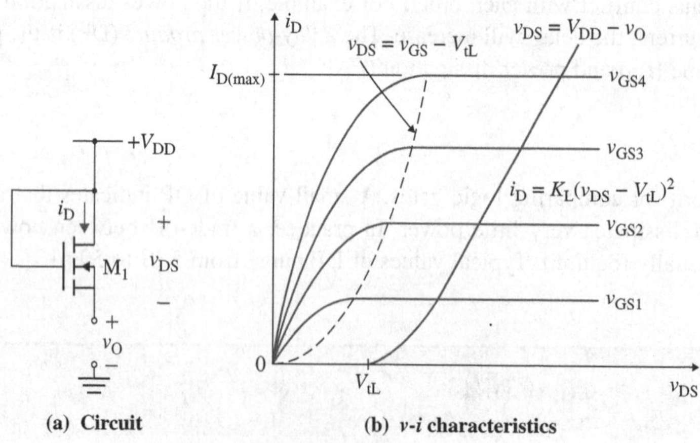

FIGURE 15.11 *n*-channel enhancement MOSFET as a load with $v_{GS} = v_{DS}$

For $v_{GSL} = v_{DSL} > V_{tL}$, a drain current will flow and the transistor will always operate in the saturation (pinch-down) mode under the following condition:

$$v_{DS} > (v_{GSL} - V_{tL}) = v_{DSL} - V_{tL} = V_{DS(sat)}$$

The corresponding drain current is given by

$$i_{DL} = K_L(v_{GSL} - V_{tL})^2 \tag{15.18}$$

where K_L is the MOS constant and V_{tL} is the threshold voltage of the load MOSFET. The output characteristics, which are shown in Fig. 15.11(b), indicate that the transistor will act as a nonlinear resistor if operated in the saturation region.

Static Characteristics

An NMOS inverter with an enhancement load is shown in Fig. 15.12(a). The substrates of the MOSFETs (M_D and M_L) are connected to the ground. The substrate in an IC logic gate is common to all devices, and therefore it is connected to the ground (or to the most negative potential). Consequently, there is a nonzero source-to-body voltage V_{SB} (which varies with logic levels) for the device. There is also a body-to-source capacitance, which influences both the frequency and the transient response of the device. However, this effect, called the *body effect*, is not present if the substrate (i.e., the body) is connected directly to the source. To simplify the analysis, we will assume that the substrates are connected to the source terminals so that $V_{SB} = 0$ for both M_D and M_L.

The output characteristics of the load and driver transistors are shown in Fig. 15.12(b). Before the application of high input, $v_I = V_{OL}$ and $v_O = V_{OH}$. The inverter is operating at point *A*. When v_I goes high (to V_{OH}), transistor M_D turns on, and the operating point jumps from *A* to *B* and then moves along the *i-v* curve of M_D until it finally reaches the quiescent point *Q* so that $v_O = V_{OL}$. When v_I goes from high to low (to V_{OL}), the operating point jumps from *Q* to *A* and then moves along the *i-v* curve of M_L until it finally reaches $v_O = V_{OH}$. Depending on the input voltage v_I, the circuit operation can be divided into three regions, as shown in Fig. 15.12(c): region I, region II, and region III.

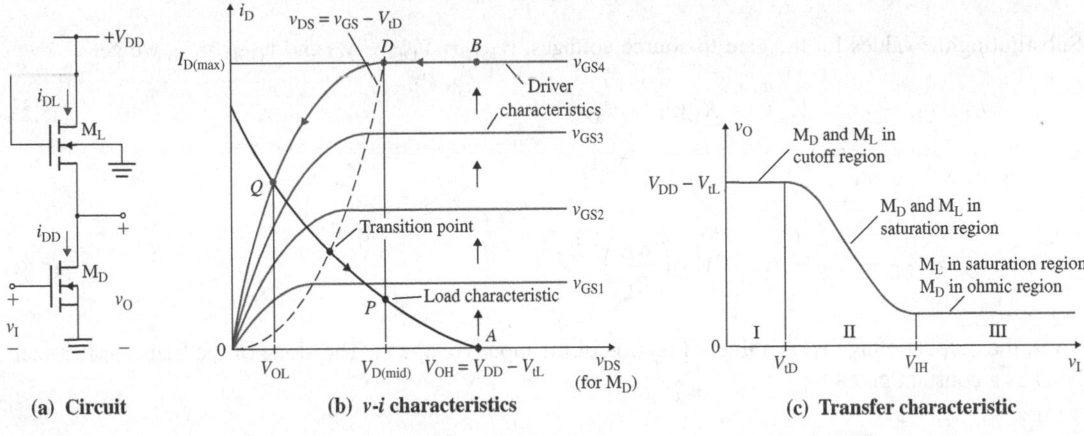

FIGURE 15.12 NMOS inverter with enhancement load

In region I, the input voltage $v_I \le V_{tD}$, where V_{tD} is the threshold voltage of the drive transistor M_D. Thus, M_D is cut off, so the drain currents of the two transistors must be zero. For the load transistor M_L,

$$i_{DL} = 0 = K_L(v_{GSL} - V_{tL})^2 \quad (15.19)$$

where i_{DL} = drain current of the load transistor M_L
v_{GSL} = gate-to-source voltage of the load transistor M_L
V_{tL} = threshold voltage of the load transistor M_L
K_L = MOS constant parameter of the load transistor M_L

Using KVL, we can write

$$v_{GSL} = V_{DD} - v_O \quad (15.20)$$

Substituting v_{GSL} from Eq. (15.20) into Eq. (15.19) yields

$$0 = K_L(V_{DD} - v_O - V_{tL})^2$$

which gives the output voltage as

$$v_O = V_{OH} = V_{DD} - V_{tL} \quad (15.21)$$

Therefore, the high output voltage V_{OH} is less than V_{DD} by the amount of the threshold voltage of the load transistor M_L. The input voltage corresponding to $v_O = V_{OH}$ is

$$V_{IL} = V_{tD} \quad (15.22)$$

In region II, $v_I > V_{tD}$. As v_I becomes slightly greater than V_{tD}, M_D begins to conduct and operates in the saturation region. The drain currents of the two transistors will be equal; that is, $i_{DL} = i_{DD}$, which, for both transistors in saturation, gives

$$K_L(v_{GSL} - V_{tL})^2 = K_D(v_{GSD} - V_{tD})^2$$

▶ **NOTE** D in the subscripts after D, DS, and GS refers to the driver MOSFET.

Substituting the values for the gate-to-source voltages, $v_{GSL} = V_{DD} - v_O$ and $v_{GSD} = v_I$, we get

$$K_L(V_{DD} - v_O - V_{tL})^2 = K_D(v_I - V_{tD})^2 \quad (15.23)$$

Solving for v_O, we get

$$v_O = V_{DD} - V_{tL} - (v_I - V_{tD})\left(\frac{K_D}{K_L}\right)^{1/2} \quad (15.24)$$

Thus, the output voltage v_O is a linear function of the input voltage v_I. The slope of the transfer characteristics is a constant given by

$$\frac{dv_O}{dv_I} = -\sqrt{\frac{K_D}{K_L}} = -\sqrt{K_R} \quad (15.25)$$

where K_R is known as the *geometry ratio*. The slope changes abruptly from 0 to $-\sqrt{K_R}$ at $V_{IL} = V_{tD}$. The slope, which is inversely proportional to the transition width V_{TW}, can be increased and the transition region can be narrowed with a large value of K_D/K_L. Since the MOS constant parameters (K_D and K_L) are proportional to the ratio *W/L*, the slope can be rewritten as

$$\frac{dv_O}{dv_I} = -\left[\frac{(W/L)_D}{(W/L)_L}\right]^{1/2} \tag{15.26}$$

The voltage $V_{D(mid)} = v_I = v_O$ can be found from Eq. (15.23):

$$K_L(V_{DD} - V_{MD} - V_{tL})^2 = K_D(V_{MD} - V_{tD})^2$$

which can be solved for V_{MD} as follows:

$$V_{MD} = \frac{V_{DD} - V_{tL} + V_{tD}\sqrt{K_R}}{1 + \sqrt{K_R}} \tag{15.27}$$

If v_I is increased sufficiently, M_D will operate at the edge of the saturation region. If $v_{DSD(sat)}$ is the drain–source voltage at the transition point,

$$v_{DSD(sat)} = v_{GSD} - V_{tD} \tag{15.28}$$

Since $v_{DSD(sat)} = v_O$ and $v_{GSD} = v_I$, Eq. (15.28) can be written as

$$v_O = v_I - V_{tD} \tag{15.29}$$

Equating v_O in Eq. (15.24) to the output voltage in Eq. (15.29) gives the input voltage $V_{I(tran)}$ at the transition point between the saturation and nonsaturation (ohmic) regions; that is,

$$v_O = V_{I(tran)} - V_{tD} = V_{DD} - V_{tL} - (V_{I(tran)} - V_{tD})\sqrt{K_R}$$

which, solved for $V_{I(tran)}$, yields

$$V_{I(tran)} = \frac{V_{DD} - V_{tL} + V_{tD}(1 + \sqrt{K_R})}{1 + \sqrt{K_R}} \tag{15.30}$$

In region III, $v_I > V_{I(tran)}$. M_L and M_D operate in the saturation and nonsaturation (ohmic) regions, respectively. From Eq. (7.6), the drain current i_{DD} is given by

$$i_{DD} = K_D[2(v_{GSD} - V_{tD})v_{DSD} - v_{DSD}^2] \tag{15.31}$$

Since the two drain currents must be the same, we can find the relation between the input and the output voltages. Thus, from Eqs. (15.18) and (15.31), we get

$$K_L(v_{GSL} - V_{tL})^2 = K_D[2(v_{GSD} - V_{tD})v_{DSD} - v_{DSD}^2] \tag{15.32}$$

Substituting the values for the gate-to-source voltages, $v_{GSL} = V_{DD} - v_O$, $v_{GSD} = v_I$, and $v_{DSD} = v_O$, we get the relationship between the input and output voltages:

$$K_L(V_{DD} - v_O - V_{tL})^2 = K_D[2(v_I - V_{tD})v_O - v_O^2] \tag{15.33}$$

If we use $dv_O/dv_I = -1$, Eq. (15.33) gives

$$v_I = V_{tD} + 2v_O + \frac{v_O + V_{tL} - V_{DD}}{K_R} \tag{15.34}$$

Solving Eqs. (15.33) and (15.34) for the logic high input voltage $V_I = V_{IH}$ and the corresponding V_O, we get

$$V_{IH} = \frac{(V_{DD} - V_{tL})(2 + 1/K_R)}{\sqrt{1 + 3K_R}} + V_{tD} - \frac{V_{DD} - V_{tL}}{K_R} \tag{15.35}$$

$$V_O = \frac{V_{DD} - V_{tL}}{\sqrt{1 + 3K_R}} \tag{15.36}$$

▶ **NOTE** V_O in Eq. (15.36) is not V_{OL}, which is found by substituting $v_I = V_{OH} = V_{DD} - V_{tL}$ and $v_O = V_{OL}$ in Eq. (15.33) and then solving the quadratic equation for V_{OL}.

The output voltage and the drain currents depend on the input voltage v_I. The transfer characteristics are shown in Fig. 15.13 for various values of the ratio K_D/K_L. As the ratio K_D/K_L becomes larger, a steeper characteristic is obtained, which is highly desirable in digital circuits. At a larger value of K_D/K_L, the low output voltage V_{OL} becomes smaller. To guarantee turnoff of the succeeding stages, the low output voltage V_{OL} should be less than the threshold voltage of the driver transistors.

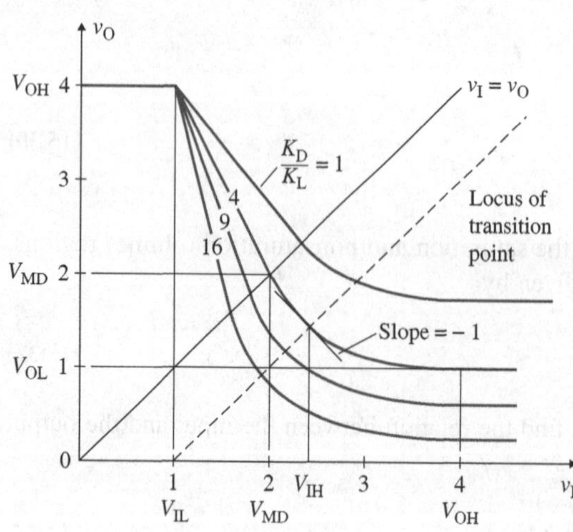

FIGURE 15.13 Transfer characteristics for various values of the ratio K_D/K_L

Body Effect

The previous equations were derived by neglecting the body effect in the load transistor M_L. In IC inverters, the body effect increases the threshold voltage. A higher value of V_{IL} causes V_{OH} to drop significantly, and the noise margin is decreased. In addition, a large geometry ratio K_R (large K_D and low K_L) is required to achieve a steep VTC. The threshold voltage V_t is related to V_{SB} (see Sec. 7.3.4) by

$$V_t = V_{t0} + \gamma\left|\sqrt{V_{SB} + 2\phi_f} - \sqrt{2\phi_f}\right| \tag{15.37}$$

where V_{t0} = *threshold voltage* at $V_{SB} = 0$, typically 1 V to 1.5 V
γ = *fabrication process constant*, typically $0.3\sqrt{\text{V}}$ to $1\sqrt{\text{V}}$
$2\phi_f$ = *equilibrium electrostatic potential* of the *p*-type body material, typically 0.6 V

Using Eq. (15.37), $V_{tD0} = V_{t0}$ for M_D, and $V_{SB} = v_O$ for M_L, we find that V_{tL} is given by

$$V_{tL} = V_{tD0} + \gamma\left|\sqrt{v_O + 2\phi_f} - \sqrt{2\phi_f}\right| \tag{15.38}$$

This relationship for $V_{tL}(v_O)$, which is a nonlinear function of v_O, can be used to find the transfer characteristic of the inverter. However, several iterations are required, and the process can be tedious. It is rarely necessary to obtain a detailed analysis for circuit design, and such a task is usually left for computer-aided simulation.

EXAMPLE 15.3

D **Designing an enhancement-load NMOS inverter** Design an enhancement-load NMOS inverter, as shown in Fig. 15.12(a), to obtain a noise margin of $NM_L \geq 0.8$ V. The threshold voltages are $V_{tL} = V_{tD} = 1$ V, and $K_L = 20\ \mu\text{A/V}^2$. The supply voltage is $V_{DD} = 5$ V. Assume load capacitance $C_L = 0.5$ pF and clock frequency $f_{clk} = 5$ MHz.

(a) Find the design parameter $K_R = K_D/K_L$, neglecting the body effect.

(b) Calculate NM_L if the body effect is included. Assume $\gamma = 0.5\sqrt{\text{V}}$ and $2\phi_f = 0.6$ V.

(c) Calculate the low-to-high propagation time t_{pLH}.

(d) Calculate the high-to-low propagation time t_{pHL}.

(e) Calculate the delay-power product (DP).

SOLUTION

$NM_L = V_{IL} - V_{OL} \geq 0.8$ V. Thus, for $V_{IL} = V_{tD} = 1$ V,

$$V_{OL} = V_{IL} - NM_L \leq 0.2 \text{ V}$$

Technically, $V_{OH} = V_{DD} - V_{tD} = 5 - 1 = 4$ V. However, to simplify the analysis for finding an expression for K_R, we will assume $V_{OH} = V_{IH} = V_{DD} = 5$ V. For $V_{IH} = 5$ V, M_L and M_D operate in the saturation and nonsaturation (ohmic) regions, respectively. Thus, for $v_I = V_{IH} = V_{OH}$ and $v_O = V_{OL}$, Eq. (15.33) yields

$$K_D[2(V_{OH} - V_{tD})V_{OL} - V_{OL}^2] = K_L(V_{DD} - V_{OL} - V_{tL})^2 \tag{15.39}$$

which gives the desired value of the geometry ratio K_R as

$$K_R = \frac{K_D}{K_L} = \frac{(V_{DD} - V_{OL} - V_{tL})^2}{2(V_{OH} - V_{tD})V_{OL} - V_{OL}^2} \tag{15.40}$$

(a) Without the body effect, $V_{tD} = 1$ V. For $V_{OL} = 0.2$ V and $V_{OH} = 5$ V, Eq. (15.40) gives $K_R = 9.26$. Since a higher value gives a lower value of V_{OL}, we choose $K_R = 10$. For this value of K_R, we can find V_{OL} by solving the quadratic equation in Eq. (15.39):

$$10[2(5 - 1)V_{OL} - V_{OL}^2] = (5 - V_{OL} - 1)^2$$

which gives $V_{OL} = 0.19$ V or 7.81 V (not physically meaningful). Thus, $V_{OL} = 0.19$ V, which is less than 0.2 V.

(b) With the body effect of $V_O = 0.2$ V, Eq. (15.38) gives

$$V_{tL} = V_{tD} + \gamma|\sqrt{v_O + 2\phi_f} - \sqrt{2\phi_f}| \tag{15.41}$$

$$= 1 + 0.5|\sqrt{0.2 + 0.6} - \sqrt{0.6}| = 1.06 \text{ V}$$

With this value of $V_{tL} = 1.06$ V, Eq. (15.39) gives

$$10[2(5 - 1)V_{OL} - V_{OL}^2] = (5 - 1.06 - V_{OL})^2$$

which gives $V_{OL} = 0.18$ V. Thus,

$$NM_L = V_{IL} - V_{OL} = 1 - 0.18 = 0.82 \text{ V}$$

which is better than the specified 0.8 V.

(c) The NMOS inverter with a load capacitor C_L is shown in Fig. 15.14(a). As v_I goes low to V_{OL}, M_D turns off immediately, and the capacitor C_L is charged up by the drain current i_{DL}. The operating point, shown in Fig. 15.14(b), moves along the load line of M_L from point Q of V_{OL} to midpoint P of $V_{O(mid)} = (V_{OH} + V_{OL})/2$. The charging of the capacitor is shown in Fig. 15.14(c). The output voltage is given by

$$C_L\frac{dv_O}{dt} = i_{DL} = K_L(V_{DD} - v_O - V_{tL})^2 \tag{15.42}$$

which can be integrated from V_{OL} to $V_{O(mid)}$ to give the low-to-high propagation time t_{pLH}, as shown in Fig. 15.14(d). To simplify the analysis, we will find the average value of the charging current I_{CLH} and then find an approximate value of t_{pLH}; that is,

$$t_{pLH} = \frac{C_L(V_{O(mid)} - V_{OL})}{I_{CLH}} \tag{15.43}$$

where I_{CLH} is given by

$$I_{CLH} = \frac{i_C(V_{OL}) + i_C(V_{O(mid)})}{2} \tag{15.44}$$

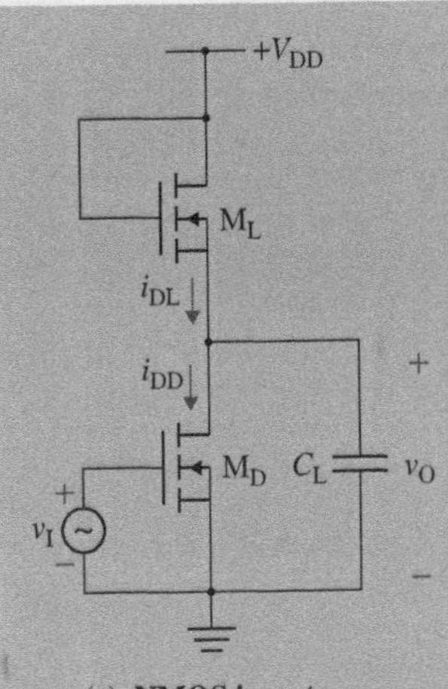

(a) NMOS inverter

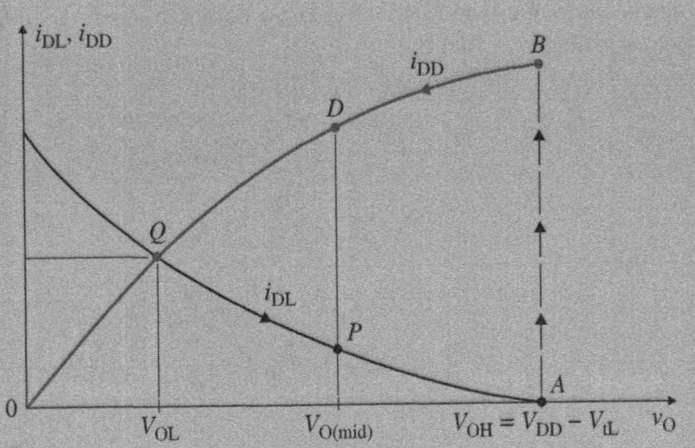

(b) Transition points

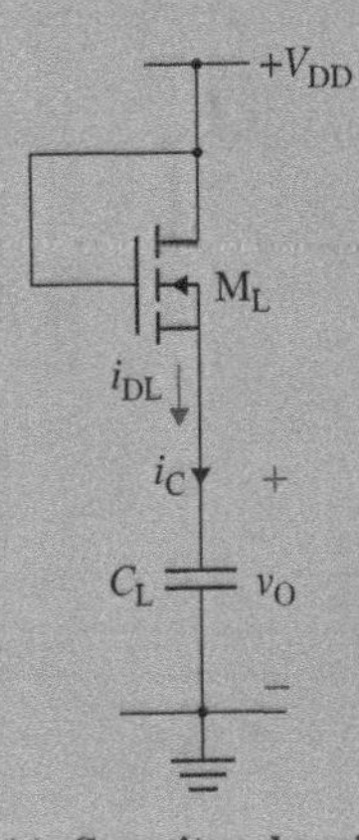

(c) Capacitor charging

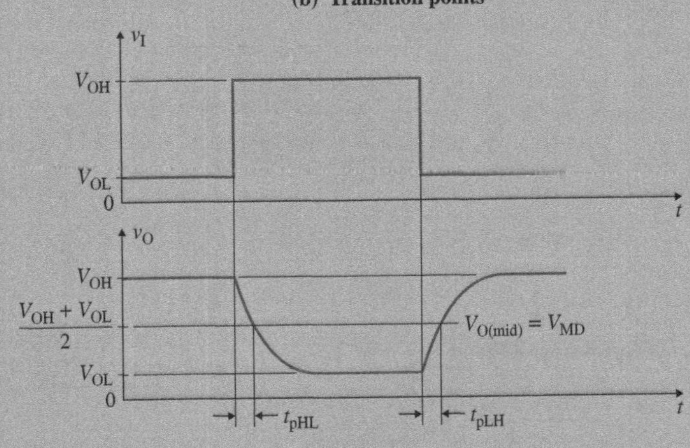

(d) Propagation delay times

FIGURE 15.14 Switching speed of enhancement-load NMOS inverter

For $V_{OL} = 0.2$ V,

$$V_{OH} = V_{DD} - V_{tL} = 5 - 1 = 4 \text{ V}$$

$$V_{O(mid)} = \frac{V_{OH} + V_{OL}}{2} = 2.1 \text{ V}$$

Then $i_C(V_{OL}) = i_{DL}(V_{OL}) = K_L(V_{DD} - v_O - V_{tL})^2 = 20\ \mu \times (5 - 0.2 - 1)^2 = 288.8\ \mu\text{A}$

$$i_C(V_{O(mid)}) = K_L(V_{DD} - V_{O(mid)} - V_{tL})^2 = 20\ \mu \times (5 - 2.1 - 1)^2 = 72.2\ \mu\text{A}$$

$$I_{CLH} = \frac{288.8\ \mu + 72.2\ \mu}{2} = 180.5\ \mu\text{A}$$

Thus, $t_{pLH} = 0.5 \text{ pF} \times \dfrac{(2.1 - 0.2) \text{ V}}{180.5\ \mu\text{A}} = 5.26 \text{ ns}$

(d) As v_I goes high to V_{OH}, M_D turns on and the capacitor C_L discharges through M_D. The operating point moves along the load line of M_D from point B of V_{OH} to midpoint D of $V_{O(mid)} = (V_{OH} + V_{OL})/2$. The output voltage is given by

$$C_L \frac{dv_O}{dt} = i_{DD} - i_{DL} \tag{15.45}$$

where $i_{DD} = \begin{cases} K_D (V_{OH} - V_{tD})^2 & \text{for } (V_{OH} - V_{tD}) \geq v_O > V_{O(mid)} \\ K_D[2(V_{OH} - V_{tD})v_O - v_O^2] & \text{for } v_O \leq V_{O(mid)} \end{cases}$

Using the average value of the capacitor discharging current i_{CHL}, we can find the approximate value of t_{pHL} from

$$t_{pHL} = \frac{C_L(V_{OH} - V_{O(mid)})}{I_{CHL}} \tag{15.46}$$

where i_{CHL} is given by

$$i_{CHL} = \frac{i_{DD}(V_{OH}) + i_{DD}(V_{O(mid)}) - i_{DL}(V_{O(mid)})}{2} \tag{15.47}$$

For $K_D = K_R K_L = 200\ \mu A/V^2$,

$$V_{OL} = 0.2\text{ V}$$

$$V_{OH} = V_{DD} - V_{tL} = 5 - 1 = 4\text{ V}$$

$$V_{O(mid)} = \frac{V_{OH} + V_{OL}}{2} = 2.1\text{ V}$$

Thus, $i_{DD}(V_{OH}) = K_D(V_{OH} - V_{tD})^2 = 200\ \mu \times (4 - 1)^2 = 1800\ \mu A$

$$i_{DD}(V_{O(mid)}) = K_D[2(V_{OH} - V_{tD})V_O - V_O^2] = 200\ \mu \times [2 \times (4 - 1) \times 2.1 - 2.1^2] = 1638\ \mu A$$

$$i_{DL}(V_{O(mid)}) = K_L(V_{DD} - V_O - V_{tL})^2 = 20\ \mu \times (5 - 2.1 - 1)^2 = 72.2\ \mu A$$

$$i_{CHL} = \frac{1800 + 1638 - 72.2}{2} = 1683\ \mu A$$

$$t_{pHL} = 0.5\text{ pF} \times \frac{(4 - 2.1)\text{ V}}{1683\ \mu A} = 0.56\text{ ns}$$

which is much shorter than $t_{pLH} = 5$ ns. The value of $i_{DL}(V_{O(mid)})$, which is much smaller than that of $i_{DD}(V_{OH})$, can often be neglected.

(e) The propagation time t_{pd}, which is the average of t_{pLH} and t_{pHL}, is

$$t_{pd} = \frac{t_{pLH} + t_{pHL}}{2} = \frac{5.26\text{ ns} + 0.56\text{ ns}}{2} = 2.91\text{ ns}$$

Then $P_{static} = \frac{V_{DD}\, i_D(V_{DL})}{2} = \frac{V_{DD}\, K_L(V_{DD} - V_{OL} - V_{tL})^2}{2}$ (15.48)

$$= \frac{5 \times 20\ \mu \times (5 - 0.2 - 1)^2}{2} = 722\ \mu W$$

and $P_{dynamic} = f_{clk} C_L V_{DD}^2 = 5\ \text{MHz} \times 0.5\ \text{pF} \times 5^2 = 62.5\ \mu W$

so $P_D = P_{static} + P_{dynamic} = 722\ \mu + 62.5\ \mu = 784.5\ \mu W$

Therefore, the delay-power product is

$$DP = P_D \times t_{pd} = 784.5\ \mu W \times 2.91\ \text{ns} = 2.28\ \text{pJ}$$

15.5.2 NMOS Inverter with Depletion Load

The load in an inverter can be a depletion-type MOSFET. Consider an n-channel depletion MOSFET whose gate is connected to the source terminal, as shown in Fig. 15.15(a). With this connection, $v_{GS} = 0$; the output characteristic is shown in Fig. 15.15(b), which indicates that v_{DS} must be zero to obtain a zero drain current.

Static Characteristics

An NMOS inverter with a depletion-load transistor M_L is shown in Fig. 15.16(a), in which the substrates of the MOSFETs are connected to the ground. To simplify the analysis, we will assume that the substrates are connected to the source terminals so that $V_{SB} = 0$ for both M_D and M_L. The output characteristics of the load and driver transistors are shown in Fig. 15.16(b). If $v_I = V_{OL}$, then $v_O = V_{OH}$. When v_I goes high (to V_{OH}), transistor M_D turns on, and the operating point jumps from A to B and then moves along the i-v curve of M_D until it finally reaches the quiescent point Q so that $v_O = V_{OL}$. When v_I goes from high to low (to V_{OL}), the operating point jumps from Q to A and then moves along the i-v curve of M_L until it finally reaches $v_O = V_{OH}$. Depending on the input voltage v_I, the VTC can be divided into four regions, as shown in Fig. 15.16(c).

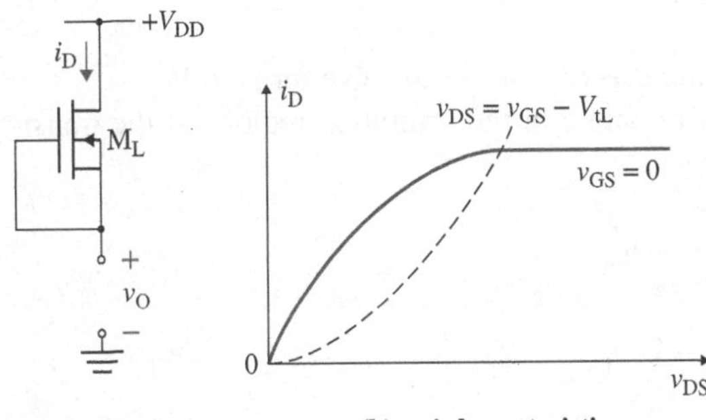

FIGURE 15.15 n-channel depletion-type MOSFET with $v_{GS} = 0$

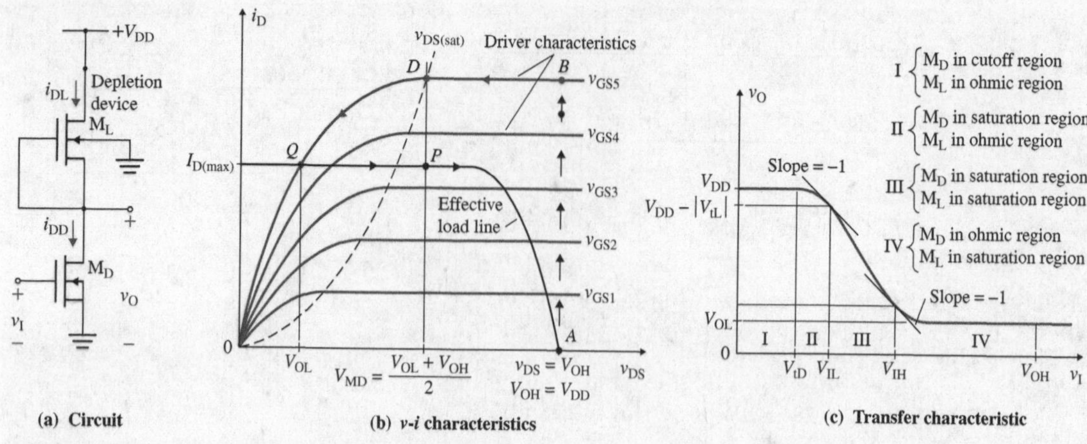

FIGURE 15.16 NMOS inverter with depletion load

In region I, the input voltage $v_I \le V_{tD}$, where V_{tD} is the threshold voltage of the drive transistor M_D. M_D is cut off, so the drain currents of the two transistors must be zero. M_L operates in the nonsaturation (ohmic) region. The drain current of load transistor M_L is $i_{DL} = 0$, which, for $v_{DSL} = 0$, gives the output voltage as

$$v_O = V_{OH} = V_{DD} - v_{DSL} = V_{DD}$$

Thus, V_{OH} is not reduced by V_{tL}, as it is in the case of an enhancement-load inverter.

In region II, $v_I > V_{tD}$. As v_I becomes slightly greater than V_{tD}, M_D begins to conduct and operates in the saturation region. M_L is still in the nonsaturation region. The two drain currents must be equal; that is, $i_{DL} = i_{DD}$, which, for M_L in the nonsaturation region and M_D in the saturation region, gives

$$K_L[2(v_{GSL} - V_{tL})v_{DSL} - v_{DSL}^2] = K_D(v_{GSD} - V_{tD})^2$$

Substituting the values for $v_{GSL} = 0$, $v_{DSL} = V_{DD} - v_O$, and $v_{GSD} = v_I$, we get the relationship between the input and output voltages as

$$K_L[-2V_{tL}(V_{DD} - v_O) - (V_{DD} - v_O)^2] = K_D(v_I - V_{tD})^2 \tag{15.49}$$

V_{IL} can be found by differentiating Eq. (15.49) and setting $dv_O/dv_I = -1$ to solve for $v_I = V_{IL}$.

If v_I is increased sufficiently, both M_D and M_L will operate in the saturation region. At the transition point from nonsaturation to saturation for M_L, we get

$$v_{DSL} = V_{DD} - v_O = v_{GSL} - V_{tL} = 0 - V_{tL} = -V_{tL}$$

which gives v_O at the edge of the transition as

$$V_{O(tran1)} = V_O = V_{DD} + V_{tL} \tag{15.50}$$

Substituting v_O from Eq. (15.50) into Eq. (15.49) gives the corresponding input voltage $V_I = V_{I(tran1)}$; that is,

$$K_L[-2V_{tL}(V_{DD} - V_{DD} - V_{tL}) - (V_{DD} - V_{DD} - V_{tL})^2] = K_D(v_I - V_{tD})^2$$

which, for $v_I > V_{tD}$, gives

$$V_{I(tran1)} = v_I = V_{tD} - V_{tL}\left(\frac{K_L}{K_D}\right)^{1/2} = V_{tD} - V_{tL}\sqrt{\frac{1}{K_R}} \tag{15.51}$$

▶ **NOTE** $V_{I(tran1)}$ can also be found from

$$K_L(v_{GSL} - V_{tL})^2 = K_D(v_{GSD} - V_{tD})^2$$

for $v_{GSL} = 0$ and $v_{GSD} = v_I = V_{I(tran1)}$.

In region III, $v_I = V_{I(tran1)}$. Both M_L and M_D operate in the saturation region. Since the two currents must be equal, $i_{DL} = i_{DD}$, or

$$K_L(v_{GSL} - V_{tL})^2 = K_D(v_{GSD} - V_{tD})^2$$

Substituting $v_{GSL} = 0$ and $v_{GSD} = v_I$, we get the input voltage when both transistors are in saturation:

$$v_I = V_{tD} - V_{tL}\left(\frac{K_L}{K_D}\right)^{1/2} = V_{tD} - V_{tL}\sqrt{\frac{1}{K_R}}$$

The corresponding value of drain current is

$$i_{DL} = i_{DD} = K_L(v_{GSL} - V_{tL})^2 = K_L(-V_{tL})^2 = K_L V_{tL}^2 \tag{15.52}$$

which indicates that the drain current is independent of v_I and remains constant. At this current level, M_L will continue to operate in the saturation region and M_D will be forced into nonsaturation. There will be a quick transition as M_D switches from the saturation to the nonsaturation region. The output voltage will change to

$$V_{O(tran2)} = v_O = v_{DSD} = v_{GSD} - V_{tD} = v_I - V_{tD} \tag{15.53}$$

The input voltage at this transition is the same as $V_{I(tran1)}$; that is,

$$V_{I(tran2)} = V_{I(tran1)} = V_{tD} - V_{tL}\left(\frac{K_L}{K_D}\right)^{1/2} \tag{15.54}$$

Note that at $v_I = V_{I(tran1)} = V_{I(tran2)}$, there will be two transitions: the first as M_D switches from saturation to nonsaturation at an output voltage of $v_O = V_{DD} + V_{tL}$ and then the second as M_L switches from nonsaturation to saturation at an output voltage of $v_O = v_I - V_{tD}$.

In region IV, $v_I > V_{I(tran1)} = V_{I(tran2)}$. The drive transistor M_D goes into nonsaturation, and M_L operates in the saturation region. Since the two currents must be equal, $i_{DL} = i_{DD}$, which, for M_L in saturation and M_D in nonsaturation, gives

$$K_L(v_{GSL} - V_{tL})^2 = K_D[2(v_{GSD} - V_{tD})v_{DSD} - v_{DSD}^2]$$

Substituting $v_{GSL} = 0$, $v_{DSD} = v_O$, and $v_{GSL} = v_I$, we get the relationship between the input and output voltages:

$$K_L(-V_{tL})^2 = K_D[2(v_I - V_{tD})v_O - v_O^2] \tag{15.55}$$

If we set $dv_O/dv_I = -1$, Eq. (15.55) yields

$$v_I = 2v_O + V_{tD}$$

which, after substitution into Eq. (15.55), gives

$$V_{IH} = \frac{2|V_{tL}|}{\sqrt{3K_R}} + V_{tD} \tag{15.56}$$

$$v_O = \frac{|V_{tL}|}{\sqrt{3K_R}} \tag{15.57}$$

The typical loci for the transition points for both the load and the drive transistors are shown in Fig. 15.17. The typical transfer characteristics are shown in Fig. 15.18 for several values of the ratio K_D/K_L. The depletion load provides a constant current once M_L goes into saturation.

Body Effect

With the body effect, V_{tL} varies with $v_{SB} = v_O$ by

$$V_{tL} = V_{tL0} + \gamma|\sqrt{v_O + 2\phi_f} - \sqrt{2\phi_f}| \tag{15.58}$$

where V_{tL0} is the threshold voltage at $v_{SB} = 0$. Since $|V_{tL}|$ decreases with falling v_O, the load current will decrease slightly and v_O will still drop rapidly with v_I.

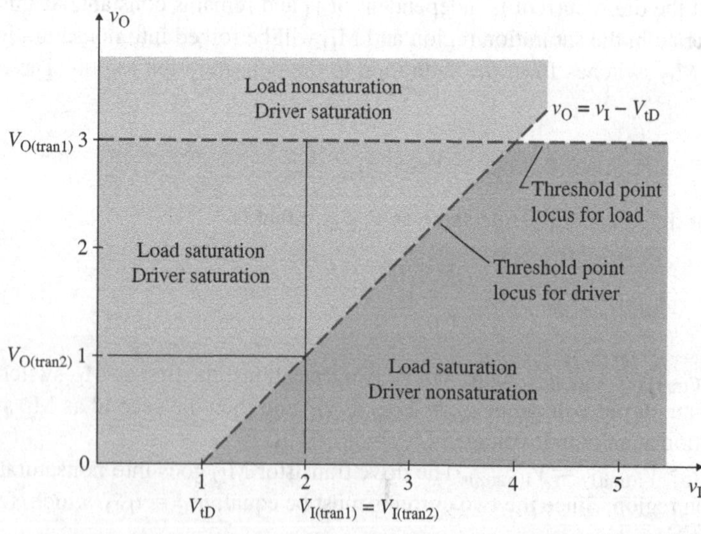

FIGURE 15.17 Transition loci for driver and load transistors

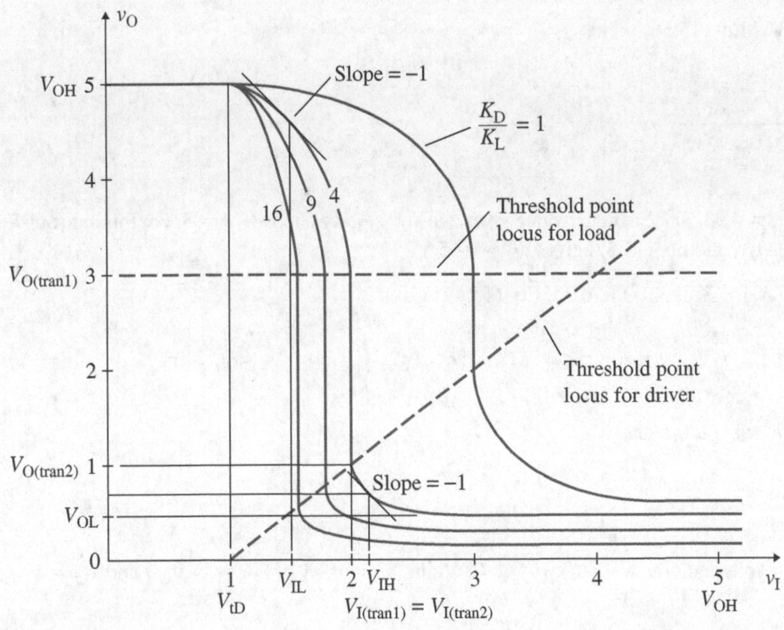

FIGURE 15.18 Transfer characteristics for various values of the ratio K_D/K_L

EXAMPLE 15.4

D

Designing a depletion-load NMOS inverter Design a depletion-load NMOS inverter, as shown in Fig. 15.16(a), to obtain a noise margin of $NM_H > 2.7$ V. The threshold voltages are $V_{tL} = -2.5$ V, $V_{tD} = 1$ V, and $K_L = 20\ \mu A/V^2$. The supply voltage is $V_{DD} = 5$ V. Assume $V_{OH} = V_{DD} = 5$ V, load capacitance $C_L = 0.5$ pF, and frequency $f_{clk} = 5$ MHz.

(a) Find the design parameter $K_R = K_D/K_L$, neglecting the body effect.

(b) Calculate NM_L if the body effect is neglected.

(c) Calculate NM_H if the body effect is included. Assume $\gamma = 0.5\ \sqrt{V}$ and $2\phi_f = 0.6$ V.

(d) Calculate the low-to-high propagation time t_{pLH}.

(e) Calculate the high-to-low propagation time t_{pHL}.

(f) Calculate the delay-power product (DP).

(g) Use PSpice/SPICE to verify your results.

SOLUTION

(a) $NM_H = V_{OH} - V_{IH} \geq 2.7$ V. Thus, for $V_{OH} = V_{DD} = 5$ V,

$$V_{IH} = V_{OH} - NM_H \leq 2.3 \text{ V}$$

For $V_{IH} \leq 2.3$ V, Eq. (15.56) yields

$$K_R = \frac{(2V_{tL})^2}{3(V_{IH} - V_{tD})^2} \tag{15.59}$$

which gives $K_R = 4.93$. Since a higher value gives a lower V_{IH}, we choose $K_R = 5$. For this value of K_R, Eq. (15.56) gives $V_{IH} = 2.29$ V, and Eq. (15.57) gives $v_O = 0.65$ V.

(b) For $K_D = K_L \times K_R = 100\ \mu\text{A/V}^2$, Eq. (15.49) gives

$$20\ \mu \times [(-2) \times (-2.5) \times (5 - v_O) - (5 - v_O)^2] = 20\ \mu \times 5(v_I - 1)^2 \tag{15.60}$$

Setting $dv_O/dv_I = -1$, we get

$$v_I = 0.2v_O + 0.5$$

Solving these two equations, we get $V_{IL} = 1.46$ V and $v_O = 4.78$ V. For $v_I = V_{OH}$ and $v_O = V_{OL}$, Eq. (15.55) yields

$$K_D[2(V_{OH} - V_{tD})V_{OL} - V_{OL}^2] = K_L V_{tL}^2 \tag{15.61}$$

which, for $V_{OH} = 5$ V and $K_R = 5$, gives $V_{OL} = 0.16$ V. Thus,

$$NM_L = V_{IL} - V_{OL} = 1.46 - 0.16 = 1.3 \text{ V}$$

(c) With the body effect, at $v_O = 0.65$ V, Eq. (15.58) gives

$$V_{tL} = V_{tL0} + \gamma|\sqrt{v_O + 2\phi_f} - \sqrt{2\phi_f}| \tag{15.62}$$

$$= -2.5 + 0.5|\sqrt{0.65 + 0.6} - \sqrt{0.6}| = -2.33 \text{ V}$$

With this value, Eq. (15.56) gives $V_{IH} = 2.20$ V. Thus,

$$NM_H = V_{OH} - V_{IH} = 5 - 2.20 = 2.80 \text{ V}$$

which is better than the specified 2.7 V.

(d) The inverter with a load capacitor C_L is shown in Fig. 15.19(a). As v_I goes low to V_{OL}, M_D turns off immediately and the capacitor C_L is charged up by the drain current i_{DL}. The operating point, shown in Fig. 15.19(b), moves along the load line of M_L from point Q of V_{OL} to midpoint P of $V_{O(mid)} = (V_{OH} + V_{OL})/2$. The charging of the capacitor is shown in Fig. 15.19(c). The output voltage is given by

$$C_L \frac{dv_O}{dt} = i_{DL} = K_L V_{tL}^2 \tag{15.63}$$

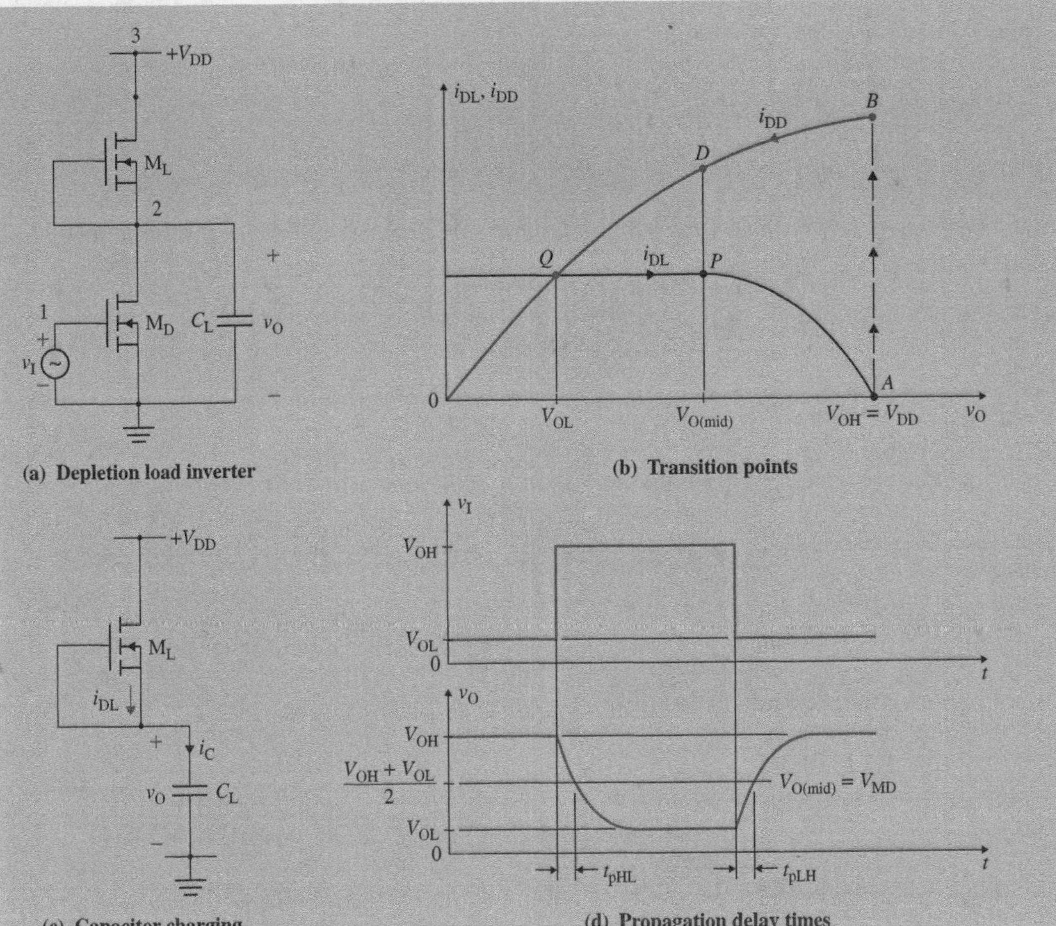

(a) Depletion load inverter

(b) Transition points

(c) Capacitor charging

(d) Propagation delay times

FIGURE 15.19 Switching speed of depletion-load NMOS inverter

which, after integration from V_{OL} to $V_{O(mid)}$, gives t_{pLH} as

$$t_{pLH} = \frac{C_L(V_{OH} - V_{OL})}{2K_L V_{tL}^2} \tag{15.64}$$

$$= \frac{0.5\text{ pF} \times (5 - 0.16)}{2 \times 20\ \mu \times 2.5^2} = 9.7\text{ ns}$$

(e) As v_I goes high to V_{OH}, M_D turns on and the capacitor C_L discharges through M_D. The operating point moves along the load line of M_D from point B of V_{OH} to midpoint D of $V_{O(mid)}$. The output voltage is given by

$$C_L\frac{dv_O}{dt} = i_{DD} - i_{DL} = i_{DD} - K_L V_{tL}^2 \tag{15.65}$$

$V_{OL} = 0.16$ V, $V_{OH} = V_{DD} = 5$ V, and

$$V_{O(mid)} = \frac{V_{OH} + V_{OL}}{2} = 2.6 \text{ V}$$

Thus, $$i_{DD}(V_{OH}) = K_D(V_{OH} - V_{tD})^2 = 100\ \mu \times (5 - 1)^2 = 1600\ \mu\text{A}$$

$$i_{DD}(V_{O(mid)}) = K_D[2(V_{OH} - V_{tD})v_O - v_O^2] = 100\ \mu \times [2 \times (5 - 1) \times 2.6 - 2.6^2] = 1404\ \mu\text{A}$$

$$i_{DL}(V_{O(mid)}) = K_L V_{tL}^2 = 20\ \mu \times 2.5^2 = 125\ \mu\text{A}$$

$$i_{CHL} = \frac{1600\ \mu + 1404\ \mu - 125\ \mu}{2} = 1440\ \mu\text{A}$$

Using Eq. (15.46), we have

$$t_{pHL} = 0.5 \text{ pF} \times \frac{(5 - 2.6) \text{ V}}{1440\ \mu\text{A}} = 0.83 \text{ ns}$$

which is much shorter than $t_{pLH} = 9.7$ ns. Assuming that i_{DD} is much greater than i_{DL}, Eq. (15.65) can be simplified to

$$C_L \frac{dv_O}{dt} = i_{DD} = K_D(V_{DD} - V_{tD})^2$$

which, after integration between V_{OH} and $(V_{OH} + V_{OL})/2$, gives t_{pHL} as

$$t_{pHL} = \frac{C_L(V_{OH} - V_{OL})}{2K_D(V_{DD} - V_{tD})^2} \tag{15.66}$$

$$= \frac{0.5 \text{ pF} \times (5 - 0.16)}{2 \times 100\ \mu \times (5 - 1)^2} = 0.76 \text{ ns}$$

(close to 0.83 ns).

(f) The propagation time t_{pd}, which is the average of t_{pLH} and t_{pHL}, is

$$t_{pd} = \frac{t_{pLH} + t_{pHL}}{2} = \frac{9.7 \text{ ns} + 0.76 \text{ ns}}{2} = 5.2 \text{ ns}$$

The current at $v_O = V_{OL}$ becomes

$$i_{DL} = i_{DD} = K_L V_{tL}^2 = 125\ \mu\text{A}$$

$$P_{static} = \frac{5 \times 125\ \mu\text{A}}{2} = 312.5\ \mu\text{W}$$

Since $P_{dynamic} = f_{clk} C_L V_{DD}^2 = 5 \text{ MHz} \times 0.5 \text{ pF} \times 5^2 = 62.5\ \mu\text{W}$,

$$P_D = P_{static} + P_{dynamic} = 312.5\ \mu + 62.5\ \mu = 375\ \mu\text{W}$$

Therefore, the delay-power product is

$$\text{DP} = P_D \times t_{pd} = 375\ \mu\text{W} \times 5.2 \text{ ns} = 1.95 \text{ pJ}$$

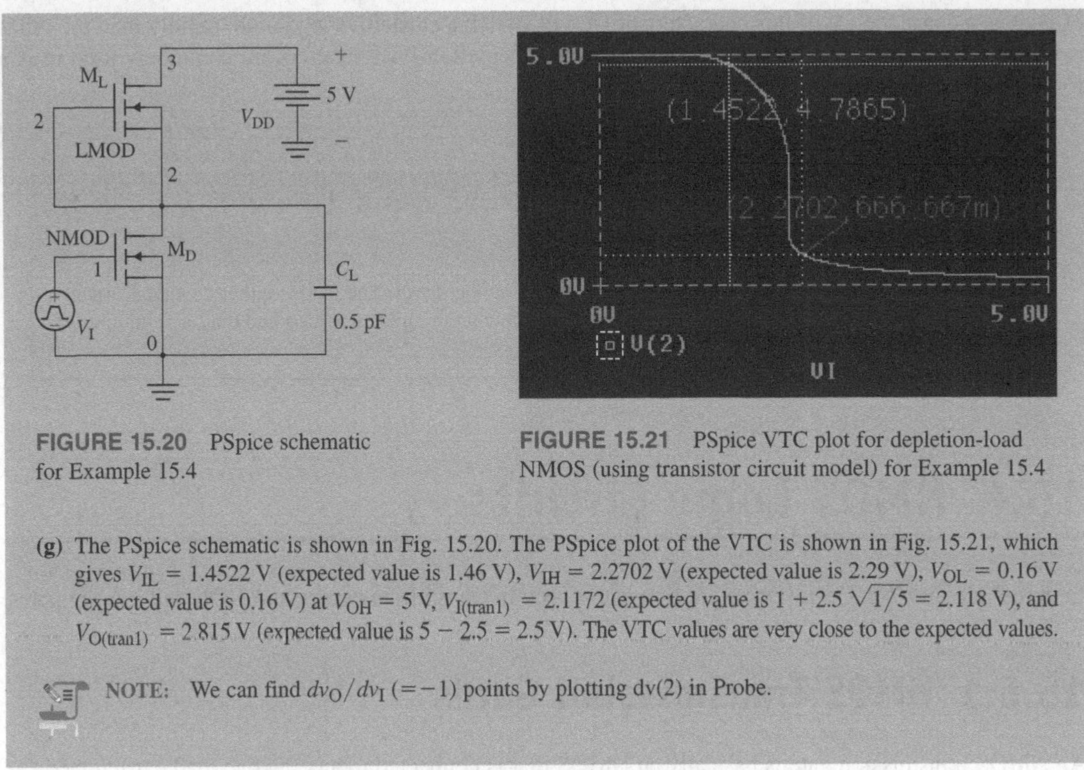

FIGURE 15.20 PSpice schematic for Example 15.4

FIGURE 15.21 PSpice VTC plot for depletion-load NMOS (using transistor circuit model) for Example 15.4

(g) The PSpice schematic is shown in Fig. 15.20. The PSpice plot of the VTC is shown in Fig. 15.21, which gives $V_{IL} = 1.4522$ V (expected value is 1.46 V), $V_{IH} = 2.2702$ V (expected value is 2.29 V), $V_{OL} = 0.16$ V (expected value is 0.16 V) at $V_{OH} = 5$ V, $V_{I(\text{tran1})} = 2.1172$ (expected value is $1 + 2.5\sqrt{1/5} = 2.118$ V), and $V_{O(\text{tran1})} = 2.815$ V (expected value is $5 - 2.5 = 2.5$ V). The VTC values are very close to the expected values.

NOTE: We can find $dv_O/dv_I\,(=-1)$ points by plotting dv(2) in Probe.

15.5.3 Comparison of NMOS Inverters

The load lines for three basic types of inverters are shown in Fig. 15.22, superimposed on the drain characteristics of the driver transistor. The resistive load line exhibits linear characteristics, and its high output voltage is dependent on drain resistance R_D. An inverter with an enhancement load has a lower drain current for the same value of v_{DS}. An inverter with a depletion load shows a constant current over a wide range

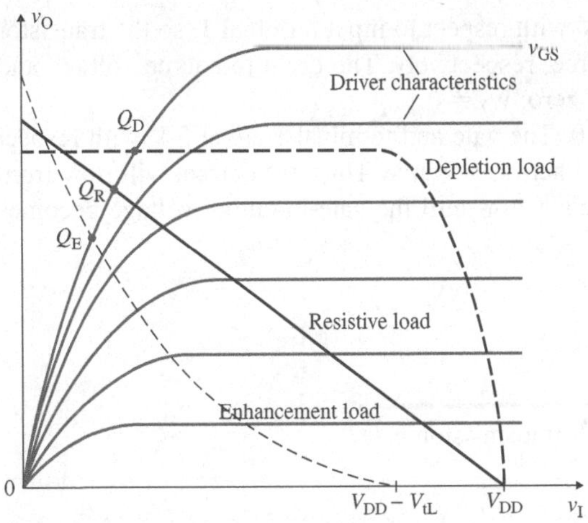

FIGURE 15.22 Load lines for three NMOS inverters

of v_{DS}; as a consequence of this characteristic, it will switch a capacitive load more rapidly than the other two inverters. Also, it exhibits a higher noise margin. Depletion-load inverters are commonly used in ICs for high-speed switching.

- An NMOS inverter uses an enhancement MOSFET as the driver. The load is either an enhancement or a depletion MOSFET. Depletion-load inverters allow more rapid switching and thus are commonly used in integrated circuits.

15.6 NMOS Logic Circuits

In digital circuits, NMOS inverters are frequently used in transmission gates and NOR and NAND gates.

15.6.1 NMOS Transmission Gates

An NMOS transmission gate is basically an NMOS that is connected to an effective load capacitance C_L, as shown in Fig. 15.23. The substrate is connected to the most negative potential in the circuit rather than to the source terminal. The MOSFET may be assumed to be completely bilateral; that is, the drain and source terminals are identical.

To examine the operation of NMOS transmission gates, we will assume that the substrate is connected to zero (that is, $v_{SUB} = 0$) and consider the following cases based on the level of input voltage v_I:

In case 1, $v_G = 0$ and $v_I = 0$ or 5 V. The gate will not be positive with respect to either input terminal 1 or output terminal 2, and the transistor will always be cut off. Thus, the input and output terminals will be isolated from each other. The input voltage v_I may have any value without affecting the output voltage v_O.

In case 2, $v_G = 5$ V and $v_I = 0$. The gate is at 5 V with respect to input terminal 1, so the transistor is turned on. Terminals 1 and 2 act as drain and source, respectively. The drain-to-source voltage and drain current will be zero. The output voltage will be zero: $v_O = 0$.

In case 3, $v_G = 5$ V, $v_I = 5$ V, and initially $v_O = 0$. The gate and terminal 1 are at 5 V with respect to terminal 2. Terminal 1 acts as a drain, and terminal 2 acts as a source. Thus, the current will flow from terminal 1 to terminal 2, and it will charge the load capacitor until the gate-to-output voltage becomes

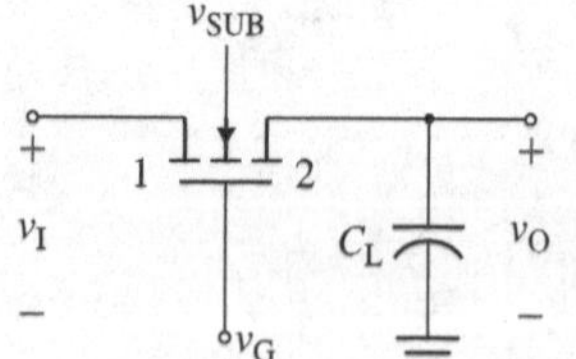

FIGURE 15.23 NMOS transmission gate

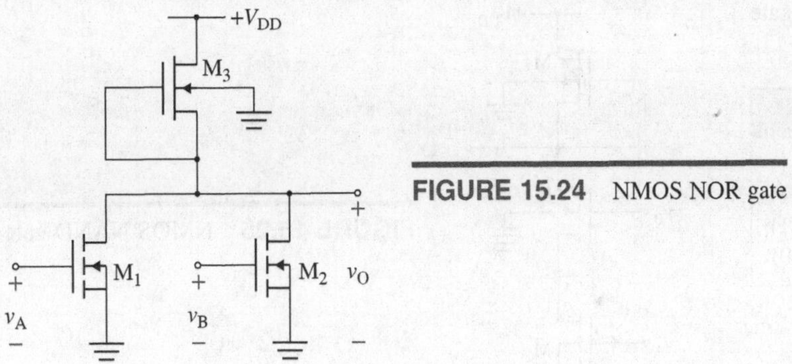

FIGURE 15.24 NMOS NOR gate

equal to the threshold voltage V_t of the transistor; that is, $v_G - v_O = V_t$. For a transistor with $V_t = 1$ V, $v_O = v_G - V_t = 5 - 1 = 4$ V.

In case 4, $v_G = 5$ V, $v_I = 0$, and initially $v_O = v_G - V_t = 4$ V. The gate is at 5 V, and terminal 2 is at 4 V with respect to terminal 1. Terminal 2 acts as a drain, and terminal 1 acts as a source. The transistor will be on. Thus, the current will flow from terminal 2 to terminal 1, and it will discharge the load capacitor; that is, $v_O = 0$.

In case 5, $v_G = 0$, $v_I = 0$ or 5 V, and initially $v_O = v_G - V_t = 4$ V. When v_G goes to zero, the situation is similar to that in case 1, and the transistor will be cut off. The input and output terminals are isolated.

Therefore, as long as v_G is high, the transistor transmits the input voltage to the output. If v_G is low, the transistor is turned off, and the input and output terminals are isolated. Depending on the levels of the gate and input voltages, the transmission gate can be operated as (1) an analog switch, (2) a sample-and-hold circuit that converts analog signals to their digital equivalents, or (3) a pass transistor for steering logic signals.

15.6.2 NMOS NOR Gates

An NMOS NOR logic gate is shown in Fig. 15.24. Two parallel NMOS transistors are connected in series with a depletion-type load. The transfer characteristic is similar to that of a depletion-load inverter. If both input voltages v_A and v_B are less than the threshold voltage of the driver transistors, $v_O = V_{OH}$ (for logic 1). If $v_A = V_{OH}$ (for logic 1), then M_1 will turn on, and the output voltage will drop to low output, $v_O = V_{OL}$. If both v_A and v_B are high (at V_{OH}), then both M_1 and M_2 will turn on, and the output voltage will be low (at V_{OL}). The exact value of the output voltage will depend on the transistor parameters, as discussed in Sec. 15.5, and it will vary slightly depending on whether one or both driver transistors are turned on. The logic function for an NMOS NOR gate is shown in Table 15.4.

TABLE 15.4 NMOS NOR gate logic function

v_A	v_B	v_O
V_{OL} (0)	V_{OL} (0)	V_{OH} (1)
V_{OL} (0)	V_{OH} (1)	V_{OL} (0)
V_{OH} (1)	V_{OL} (0)	V_{OL} (0)
V_{OH} (1)	V_{OH} (1)	V_{OL} (0)

TABLE 15.5 NMOS NAND gate logic function

v_A	v_B	v_O
V_{OL} (0)	V_{OL} (0)	V_{OH} (1)
V_{OL} (0)	V_{OH} (1)	V_{OH} (1)
V_{OH} (1)	V_{OL} (0)	V_{OH} (1)
V_{OH} (1)	V_{OH} (1)	V_{OL} (0)

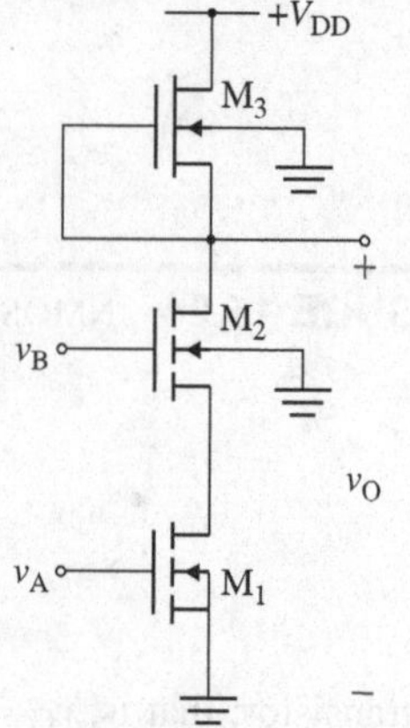

FIGURE 15.25 NMOS NAND gate

15.6.3 NMOS NAND Gates

An NMOS NAND gate is shown in Fig. 15.25. Two transistors are connected in series with a depletion-type load. If both input voltages v_A and v_B are less than the threshold voltage of the driver transistors, $v_O = V_{OH}$ (for logic 1). If $v_A = V_{OH}$ and v_B is still less than the threshold voltage V_{t2} of M_2, then M_2 is still cut off, and $v_O = V_{OH}$. If both v_A and v_B are high (at V_{OH}), then both M_1 and M_2 will turn on, and the output voltage will be low (at V_{OL}). The exact value of the output voltage will depend on the transistor parameters. The logic function for an NMOS NAND gate is shown in Table 15.5.

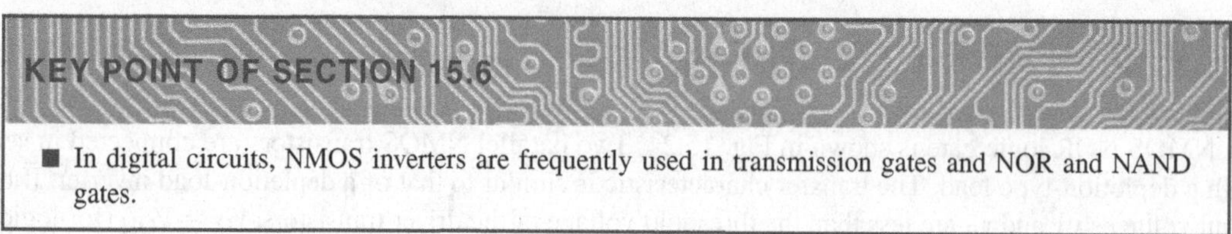

KEY POINT OF SECTION 15.6

- In digital circuits, NMOS inverters are frequently used in transmission gates and NOR and NAND gates.

15.7 CMOS Inverters

Complementary, or CMOS, circuits use both *n*-channel and *p*-channel enhancement-type MOSFETs in the same circuit. Because of their very low power consumption, CMOS circuits are commonly used in ICs. A CMOS inverter is shown in Fig. 15.26(a). Transistor M_P is a *p*-channel device, and transistor M_N is an *n*-channel device. Each substrate is tied to its source. The input signal is connected to both transistor gates, and the output terminal is common to both drain terminals. The load characteristics of two CMOS devices are shown in Fig. 15.26(b) for two extreme inputs: $v_I = 0$ and $v_I = V_{DD}$. For $v_I = 0$, M_N is off and its drain current is zero. M_P has the characteristic corresponding to $v_{GSP} = V_{DD}$. For $v_I = V_{DD}$, M_P is off and its drain current is zero. M_N has the characteristic corresponding to $v_{GSN} = V_{DD}$. Thus, their drain currents are zero at these inputs, and the current drawn from the supply is zero.

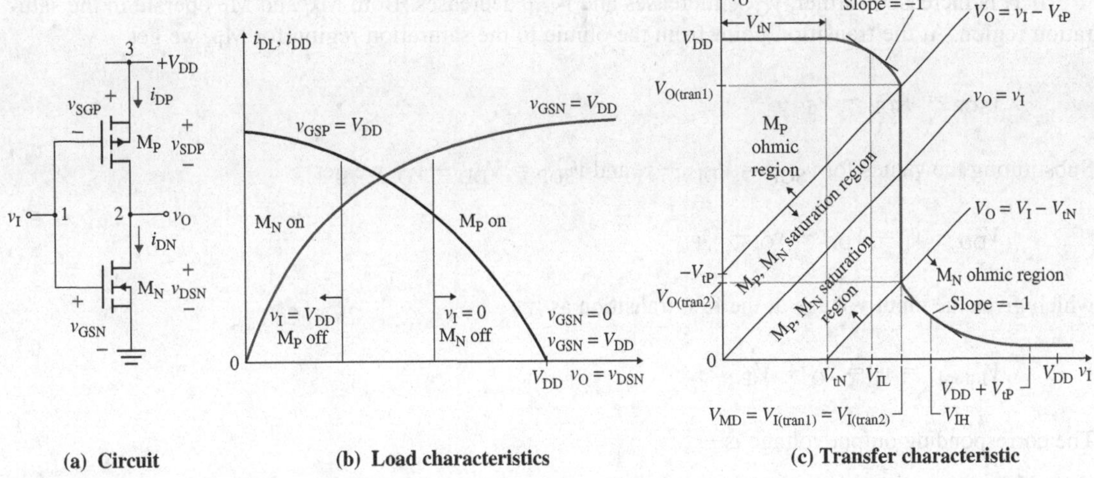

FIGURE 15.26 CMOS inverter

As the input voltage v_I is varied from zero to the maximum value V_{DD}, the output voltage v_O falls from V_{DD} to zero. The transfer characteristic is shown in Fig. 15.26(c). Depending on the input voltage v_I, the VTC can be divided into five regions.

In region I, $0 \leq v_I < V_{tN}$, where V_{tN} is the threshold voltage of transistor M_N. Since $v_{GSN} = v_I < V_{tN}$, M_N remains off. At a low value of v_I, $V_{SGP} = V_{DD} - v_I$ is high and positive. M_P is turned on, and it is driven into the nonsaturation (ohmic) region. Since the channel resistance of the off-transistor M_N is very much greater than that of the on-transistor M_P and since M_N and M_P form a voltage divider, the output voltage is $v_O \approx V_{DD}$, as shown in Fig. 15.26(c). The various voltages are $v_{GSN} = v_I$, $v_{GSP} = -V_{DD} + v_I$, and $v_O = V_{DD}$.

In region II, $V_{tN} \leq v_I \leq V_{MD} = V_{I(tran1)}$ and $v_{GSP} = (-V_{DD} + v_I) < V_{tP}$, where V_{tP} is the threshold voltage of transistor M_P and $V_{I(tran1)}$ is defined by Eq. (15.68). As v_I becomes greater than or equal to V_{tN}, M_N conducts and operates in the saturation region, for which v_{GSN} is described by $V_{tN} < v_{GSN}\ (= v_I) < (v_{DSN} + V_{tN})$.

For $v_{GSP} < V_{tP}$, M_P remains in the ohmic region. With M_N in the saturation region and M_P in the ohmic region, the two drain currents must be equal; that is, $i_{DN} = i_{DP}$. Applying the expressions for the saturation and ohmic regions gives

$$K_n(v_{GSN} - V_{tN})^2 = K_p[2(v_{SGP} + V_{tP})v_{SDP} - v_{SDP}^2]$$

where K_n and K_p are the constants for n-type and p-type transistors, respectively. Substituting $v_{GSN} = v_I$, $v_{SGP} = V_{DD} - v_I$, and $v_{SDP} = V_{DD} - v_O$ into the above equation, we get the relationship between v_I and v_O as

$$K_n(v_I - V_{tN})^2 = K_p[2(V_{DD} - v_I + V_{tP})(V_{DD} - v_O) - (V_{DD} - v_O)^2] \tag{15.67}$$

V_{IL} can be found by differentiating Eq. (15.67) and setting $dv_O/dv_I = -1$ to solve for $v_I = V_{IL}$.

If v_I is increased further, v_{GSN} increases and v_{SGP} decreases. Both M_N and M_P operate in the saturation region. At the transition point from the ohmic to the saturation region for M_P, we get

$$v_{SGP} = v_{SDP} - V_{tP}$$

Substituting the values for $v_{SGP} = V_{DD} - v_I$ and $v_{SDP} = V_{DD} - v_O$, we get

$$V_{DD} - v_I = V_{DD} - v_O - V_{tP}$$

which gives the input voltage at the first transition as

$$V_{I(tran1)} = v_I = v_O + V_{tP}$$

The corresponding output voltage is

$$V_{O(tran1)} = v_O = v_I - V_{tP}$$

whose plot is a straight line that intercepts the output axis at $-V_{tP}$ (a positive quantity), as shown in Fig. 15.26(c). The intersection with the transfer characteristic gives $V_{I(tran1)}$ and $V_{O(tran1)}$. To solve for the value of $V_{I(tran1)}$ or $V_{O(tran1)}$, one of these quantities must be known.

In region III, $v_I = V_{I(tran1)}$. Both M_N and M_P operate in the saturation region. Since the two drain currents must be equal, $i_{DN} = i_{DP}$, or

$$K_n(v_{GSN} - V_{tN})^2 = K_p(v_{SGP} + V_{tP})^2$$

Substituting $v_{GSN} = v_I$ and $v_{SGP} = V_{DD} - v_I$, we get the input voltage at the transition of M_P from the ohmic to the saturation region:

$$V_{I(tran1)} = V_{MD} = \frac{V_{DD} + V_{tP} + V_{tN}\sqrt{K_n/K_p}}{1 + \sqrt{K_n/K_p}} = \frac{V_{DD} + V_{tP} + V_{tN}\sqrt{K_R}}{1 + \sqrt{K_R}} \tag{15.68}$$

which is independent of output voltage v_O. For identical transistors, $K_n = K_p$ and $V_{tN} = |V_{tP}|$, and Eq. (15.68) is reduced to

$$V_{I(tran1)} = V_{MD} = \frac{V_{DD}}{2} \tag{15.69}$$

which is desirable to maximize the noise immunity of the circuit. Once we determine the value of $V_{I(tran1)}$ from Eq. (15.69), we can find the output voltage at the edge of the transition of M_P from

$$V_{O(tran1)} = v_O = V_{I(tran1)} - V_{tP}$$

This segment ends when M_N enters the ohmic region, which is defined by

$$v_{GSN} = v_{DSN} + V_{tN}$$

Substituting $v_{GSN} = v_I$ and $v_{DSN} = v_O$ into the preceding equation yields

$$v_I = v_O + V_{tN}$$

which gives the input voltage at the transition of M_N from the saturation region to the ohmic region as

$$V_{I(tran2)} = v_I = v_O + V_{tN}$$

The corresponding output voltage is given by

$$v_O = v_I - V_{tN}$$

which intercepts the input axis at V_{tN}, as shown in Fig. 15.26(c). The intersection with the transfer characteristic gives $V_{I(tran2)}$ and $V_{O(tran2)}$. Since $V_{I(tran1)}$ is independent of v_O and $v_I = V_{I(tran1)} = V_{I(tran2)}$, the quantities $V_{O(tran1)}$ and $V_{O(tran2)}$ must be different. There will be two transitions—the first for M_P from the ohmic region to the saturation region at an output voltage of $v_O = v_I + V_{tP}$ and then the second for M_N from the saturation region to the ohmic region at an output voltage of $V_{O(tran2)} = V_{I(tran1)} - V_{tN}$.

In region IV, $V_{I(tran1)} = V_{I(tran2)} \le v_I \le (V_{DD} + V_{tP})$. M_N goes into the ohmic region, and M_P still operates in the saturation region. Since the two drain currents must be equal, $i_{DN} = i_{DP}$, which, for M_P in the saturation region and M_N in the ohmic region, gives

$$K_n[2(v_{GSN} - V_{tN})v_{DSN} - v_{DSN}^2] = K_p(v_{SGP} + V_{tP})^2$$

Substituting $v_{GSN} = v_I$, $v_{DSN} = v_O$, and $v_{SGP} = V_{DD} - v_I$, we can express the relationship between v_O and v_I as

$$K_n[2(v_I - V_{tN})v_O - v_O^2] = K_p(V_{DD} - v_I + V_{tP})^2 \tag{15.70}$$

V_{IH} can be found by differentiating Eq. (15.70) and setting $dv_O/dv_I = -1$ to solve for $v_I = V_{IH}$.

In region V, $v_I > (V_{DD} + V_{tP})$. M_P is in the cutoff region, and M_N is in the ohmic region. There will be virtually no current through the transistors, and the output voltage will be zero; that is,

$$i_{DN} = i_{DP} = 0$$

$$v_O = 0$$

EXAMPLE 15.5

D **Designing a CMOS inverter** Design a CMOS inverter, as shown in Fig. 15.26(a), to operate at a transition voltage $V_{MD} = 2.5$ V. The threshold voltages are $V_{tP} = -1$ V, $V_{tN} = 1$ V, and $K_p = 20\ \mu A/V^2$. The supply voltage is $V_{DD} = 5$ V. Assume $V_{IH} = V_{OH} = 5$ V, load capacitance $C_L = 0.5$ pF, and frequency $f_{clk} = 5$ MHz.

(a) Find the design parameter $K_R = K_n/K_p$, neglecting the body effect.

(b) Calculate NM_L and NM_H.

(c) Calculate the propagation delay t_{pd}.

(d) Calculate the delay-power product (DP).

(e) Use PSpice/SPICE to verify your results.

SOLUTION

(a) Solving Eq. (15.68) for K_R, we have

$$K_R = \frac{(W/L)_N}{(W/L)_P} = \left[\frac{V_{DD} + |V_{tP}| - V_M}{V_M - V_{tN}}\right]^2 \tag{15.71}$$

which, for $V_{MD} = 2.5$ V, gives $K_R = 1$. Thus, $K_n = K_p = 20\ \mu A/V^2$.

(b) $V_{OH} = V_{DD} = 5$ V. Then

$$V_{O(tran1)} = V_{MD} - V_{tP} = 2.5 + 1 = 3.5 \text{ V}$$

$$V_{O(tran2)} = V_{MD} - V_{tN} = 2.5 - 1 = 1.5 \text{ V}$$

Substituting numerical values of V_{DD}, V_{tN}, V_{tP}, K_n and K_p in Eq. (15.67), we get

$$v_I^2 = 14 - 8v_I + 2v_O + 2v_I v_O - v_O^2$$

which, at $dv_O/dv_I = -1$, gives $v_O = v_I + 2.5$. Solving these two equations, we get $V_{IL} = v_I = 2.13$ V at $v_O = 4.63$ V.

Substituting numerical values of V_{DD}, V_{tN}, V_{tP}, K_n and K_p in Eq. (15.70), we get

$$16 - 8v_I + v_I^2 = 2v_I v_O - 2v_O - v_O^2$$

which, at $dv_O/dv_I = -1$, gives $v_I = v_O + 2.5$. Solving these two equations, we get $V_{IH} = v_I = 2.88$ V at $v_O = 0.38$ V.

At $v_I = V_{DD} + V_{tP} = 5 - 1 = 4$, M_P turns off and $V_{OL} = v_O = 0$. Thus,

$$NM_L = V_{IL} - V_{OL} = 2.13 - 0 = 2.13 \text{ V}$$

$$NM_H = V_{OH} - V_{IH} = 5 - 2.88 = 2.12 \text{ V}$$

(c) The inverter with a load capacitor C_L has an arrangement similar to that shown in Fig. 15.19(a). As v_I goes low to V_{OL}, M_N turns off immediately, and capacitor C_L is charged up by the drain current i_{DP}. Since M_P is in saturation and $v_I = V_{OL} = 0$, the output voltage is related to the charging current by

$$C_L \frac{dv_O}{dt} = i_{DP} = K_p(V_{DD} - v_I + V_{tP})^2$$

which, after integration between V_{DD} and $V_{DD}/2$, gives t_{pLH} as

$$t_{pLH} = \frac{C_L V_{DD}}{2K_p(V_{DD} - |V_{tP}|)^2} \tag{15.72}$$

$$= \frac{0.5 \text{ pF} \times 5 \text{ V}}{2 \times 20\ \mu A/V^2 \times (5 \text{ V} - 1 \text{ V})^2} = 3.91 \text{ ns}$$

Because of the topology of the CMOS inverter, t_{pHL} will have the same value as t_{pLH}. Thus, the propagation time $t_{pd} \approx t_{pLH} = 3.91$ ns. In practice, a symmetric CMOS does not occupy the minimum chip area, and thus not all CMOS designs are symmetric.

(d) A CMOS inverter draws a negligible current (on the order of nanoamperes) from the power supply in both high and low states. Hence, the static power dissipation is almost zero: $P_{static} = 0$. This is a distinct advantage

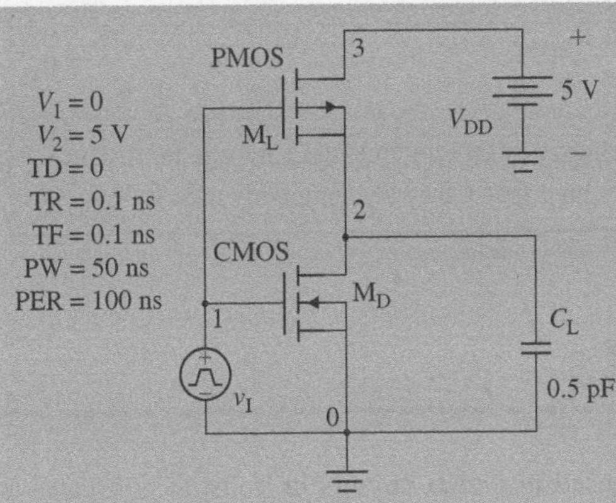

FIGURE 15.27 PSpice schematic for Example 15.5

for portable CMOS equipment because standby operation of the equipment will not discharge the battery. We have

$$P_{dynamic} = f_{clk} C_L V_{DD}^2 = 5 \text{ MHz} \times 0.5 \text{ pF} \times 5^2 = 62.5 \ \mu\text{W}$$

so

$$P_D = P_{static} + P_{dynamic} = 0 + 62.5 \ \mu\text{W} = 62.5 \ \mu\text{W}$$

Therefore, the delay-power product is

$$DP = P_D \times t_{pd} = 62.5 \ \mu\text{W} \times 3.91 \text{ ns} = 0.244 \text{ pJ}$$

(e) The PSpice schematic is shown in Fig.15.27. The PSpice plot of the VTC is shown in Fig. 15.28(a), which gives $V_{IL} = 2.1245$ V (expected value is 2.13 V), $V_{IH} = 2.8755$ V (expected value is 2.88 V), $V_{OL} = 0$ (expected value is 0) at $V_{OH} = 5$ V, and $V_{I(tran1)} = 2.5$ (expected value is 2.5 V). The VTC values are very close to the hand calculations, as expected. The transient response is shown in Fig. 15.28(b), which gives $t_{pHL} = 4.18$ ns and $t_{pLH} = 4.255$ ns (for $L = 50$ μm and $W = 100$ μm). The transient performance, however, will depend on the values of length (L) and width (W). A typical value of the transconductance parameter for the NMOS process is $\mu_n \epsilon / t_{ox} = 40 \ \mu\text{A/V}^2$. [Note: We can find $dv_O/dv_I = -1$ points by plotting dv(2) in Probe.]

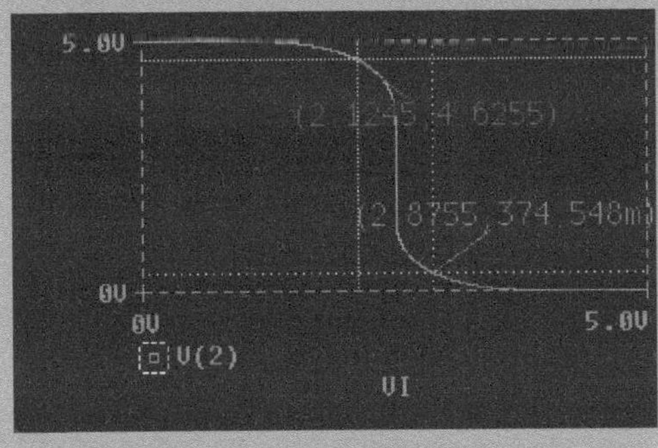

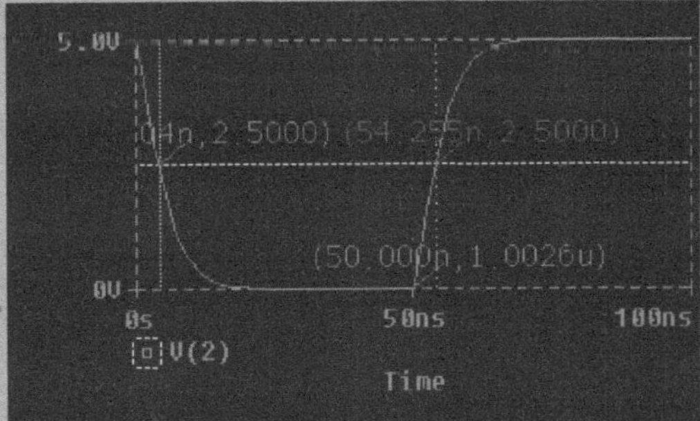

(a) VTC

(b) Transient response

FIGURE 15.28 PSpice plots for CMOS inverter (using transistor circuit model) for Example 15.5

- CMOS circuits use both *n*-channel and *p*-channel enhancement MOSFETs in the same circuit. Because they have very low power consumption and offer very high speed, they are commonly used in ICs.

15.8 CMOS Logic Circuits

Like NMOS inverters, CMOS inverters are frequently used in digital circuits in transmission gates and NOR and NAND gates.

15.8.1 CMOS Transmission Gates

A CMOS transmission gate consists of an NMOS and a PMOS that are connected in parallel and feed an effective load capacitance C_L. This arrangement is shown in Fig. 15.29. The substrate of the NMOS (M_N) is usually connected to the most negative potential (assumed to be the ground in Fig. 15.29), and the substrate of the PMOS (M_P) is connected to the most positive potential (usually the positive supply voltage V_{DD}) in the circuit rather than to the source terminal. MOSFETs may be assumed to be completely bilateral; that is, the drain and source terminals of each transistor are identical. The gate control voltage $\overline{v}_G = v_{cnt}$ of the PMOS is the complement of the gate voltage of the NMOS. Assuming that v_I operates between 0 and 5 V, $V_{DD} = 5$ V, $V_{tN} = 1$ V, and $V_{tP} = -1$ V, we will consider the following cases based on the level of input voltage v_I.

In case 1, $v_G = 0$, $v_{cnt} = 5$ V, $v_I = 0$ or 5 V, and initially $v_O = 0$ or 5 V. The gate of M_N is never positive with respect to either terminal, so M_N is always cut off. The gate of M_P is never negative with respect to either terminal, so M_N is always cut off. Thus, the input and output terminals are isolated from each other. The input voltage v_I may take on any value without affecting the output voltage v_O.

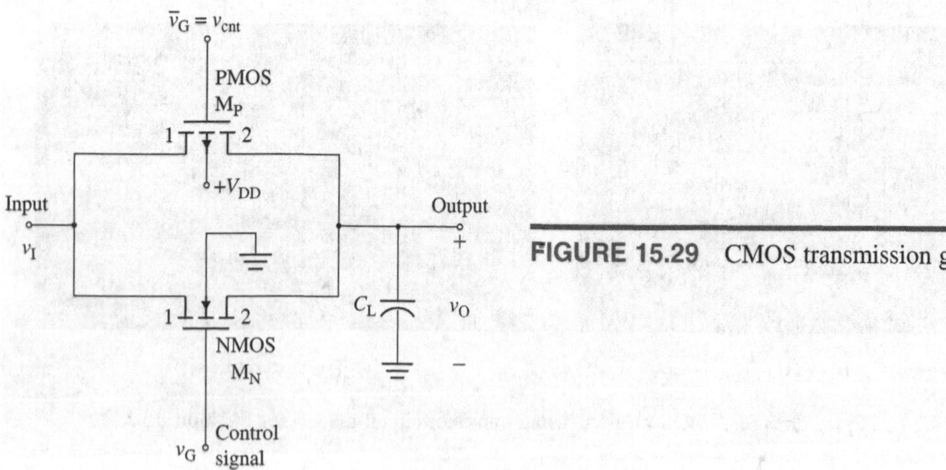

FIGURE 15.29 CMOS transmission gate

In case 2, $v_G = 5$ V, $v_{cnt} = 0$, and $v_I = 0$. The gate-to-terminal-1 voltage of M_N is 5 V, so M_N will be turned on. Terminals 1 and 2 of M_N act as the drain and the source, respectively. The drain-to-source voltage and drain current are zero. The output voltage is zero: $v_O = 0$. With $v_{cnt} = 0$ and $v_I = v_O = 0$, M_P will be cut off.

In case 3, $v_G = 5$ V, $v_{cnt} = 0$, $v_I = 5$ V, and initially $v_O = 0$. The gate and terminal 1 of M_N are at 5 V with respect to terminal 2 of M_N. Terminal 1 and terminal 2 of M_N act as the drain and the source, respectively. The drain current of M_N flows from terminal 1 to terminal 2 and charges the load capacitor until the gate-to-output voltage becomes equal to the threshold voltage V_{tN}; that is,

$$v_{GSN} = v_G - v_O = V_{tN}$$

For a transistor with $V_{tN} = 1$ V, $v_O = v_G - V_{tN} = 5 - 1 = 4$ V. When v_O rises to $v_G - V_{tN} = 5 - 1 = 4$ V, the gate-to-source voltage of M_N is equal to the threshold voltage V_{tN}, and M_N turns off. However, with $v_{cnt} = 0$ on the gate of M_P, M_P is turned on, with terminal 1 and terminal 2 acting as the drain and the source, respectively. After M_N is turned off, M_P is still turned on; the capacitor is charged all the way up to v_I, so the output voltage becomes $v_O = v_I$, at which point the drain-to-source voltage on M_P is zero and M_P is turned off. It is important to note that the output of a CMOS transmission gate is the full value of v_I.

In case 4, $v_G = 5$ V, $v_{cnt} = 0$, $v_I = 5$ V, and initially $v_O = v_I$. The gate and terminal 2 of M_N are at 5 V with respect to terminal 1 of M_N. Terminal 1 and terminal 2 of M_N act as the source and the drain, respectively. M_N is always turned on, and the drain current flows from terminal 2 to terminal 1, causing the load capacitor to be discharged completely to zero, so $v_O = 0$. M_P, whose terminal 2 acts as the source, always remains turned off.

In case 5, $v_G = 0$, $v_{cnt} = 5$ V, $v_I = 0$ or 5 V, and initially $v_O = v_I$. When v_G goes to zero, the situation is similar to case 1, and both transistors will be cut off. The input and output terminals are isolated.

The advantage of a CMOS transmission gate is that the output voltage v_O is always equal to v_I when the transmission gate is turned on, so $v_O = v_I$. The logic function can be described by

$$v_O = \begin{cases} v_I & \text{if } v_G \text{ is high (at logic 1)} \\ v_O & \text{if } v_G \text{ is low (at logic 0)} \end{cases}$$

With an NMOS transmission gate, the output voltage is reduced by the threshold voltage, so $v_O = v_I - V_{tN}$. The major disadvantage of the CMOS gate is that it requires both a gate voltage v_G and its complement for successful operation.

Propagation Delay

For $v_I = V_{DD}$ and $v_O = 0$ (initially), both M_N and M_P are in saturation. Capacitor C_L, which arises from the interconnecting area of the two MOSFETs and the load, charges to V_{DD}. Thus, the *charging time constant* τ_{LH} (for output low to high) can be found from the channel resistance of the complementary pair; that is,

$$\tau_{LH} = C_L(R_L \parallel R_{dN} \parallel R_{dP}) \tag{15.73}$$

where R_L is the load resistance. R_{dN}, which is the static resistance of the NMOS, is given by

$$R_{dN} = \frac{v_{DSN}}{i_{DN}} = \frac{V_{DD} - v_O}{K_n(V_{DD} - v_O - V_{tN})^2} \tag{15.74}$$

R_{dP}, which is the static resistance of the PMOS, is given by

$$R_{dP} = \frac{v_{SDP}}{i_{DP}} = \frac{V_{DD} - v_O}{K_p(V_{DD} - |V_{tP}|)^2} \tag{15.75}$$

The charging time constant τ_{LH} is usually higher than the discharging time. Assuming $t_{pLH} \approx t_{pHL}$ and $t_{pd} \approx t_{pLH}$, its value can be estimated from τ_{LH}. Thus, the propagation time t_{pd} to charge the capacitor C_L from 0 to $V_{DD}/2$ is given by

$$t_{pd} = \tau_{LH} \ln 2 = 0.69315\tau_{LH} \tag{15.76}$$

15.8.2 CMOS NOR and NAND Gates

A two-input NOR logic gate is shown in Fig. 15.30. Two parallel NMOSs are connected in series with two PMOSs. The substrates of the PMOSs are connected to the most positive potential V_{DD}; the substrates of the NMOSs are connected to the most negative potential—that is, the ground. The transfer characteristic is similar to that of a CMOS inverter. If both input voltages v_A and v_B are less than the threshold voltage (assuming $v_A = v_B = V_{OL}$), M_{N1} and M_{N2} are cut off. At the same time, the *p*-channel transistors M_{P1} and M_{P2} are turned on. Thus, the output voltage becomes high: $v_O = V_{OH}$ (at logic 1).

If $v_A = V_{OH}$, then M_{N1} is turned on, and M_{P1} is turned off. In this case, the output voltage drops to low: $v_O = V_{OL}$. If both v_A and v_B are equal to V_{OH}, then both M_{N1} and M_{N2} turn on, and the output voltage becomes low. For $v_A = v_B = V_{OL}$, both M_{N1} and M_{N2} are cut off, and the drain currents will be zero. The logic function is the same as that for an NMOS, shown in Table 15.4. If one or both of the logic inputs are at logic high, then at least one PMOS is cut off, and the drain current(s) is again zero. Therefore, the steady-state current is zero, and the power dissipation is essentially zero. Current flow and power dissipation occur only during the transition from one state to another.

A two-input NAND logic gate is shown in Fig. 15.31. Two NMOSs are connected in series, and two PMOSs are connected in parallel. The transfer characteristic is similar to that of a CMOS inverter. If both

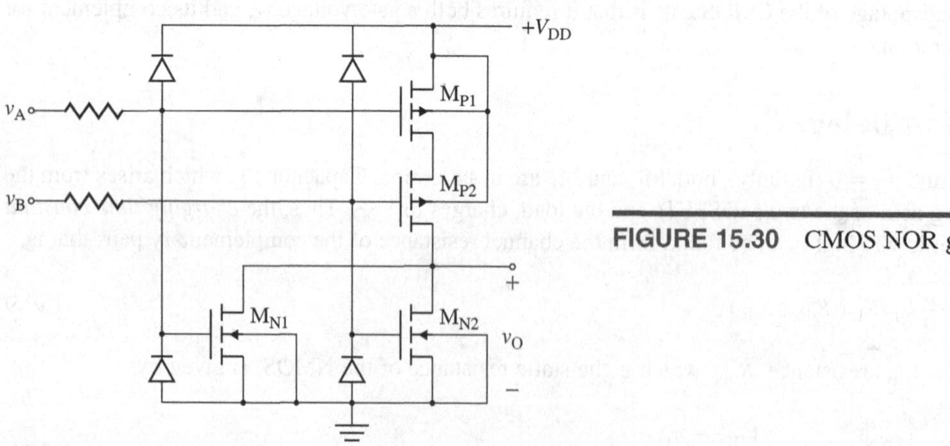

FIGURE 15.30 CMOS NOR gate

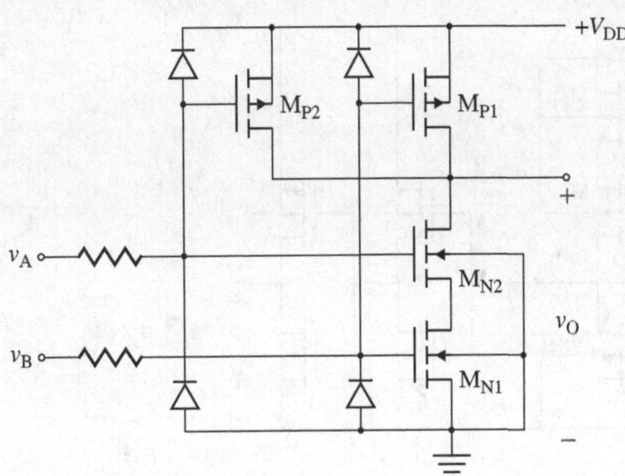

FIGURE 15.31 CMOS NAND gate

input voltages v_A and v_B are less than the threshold voltage (assuming $v_A = v_B = V_{OL}$ at logic 0), M_{N1} and M_{N2} are turned off, and M_{P1} and M_{P2} are turned on. In this case, the output voltage becomes high: $v_O = V_{OH}$ (at logic 1). If $v_A = V_{OH}$ and $v_B = V_{OL}$, then M_{N1} is turned on, M_{N2} is cut off, M_{P1} is off, and M_{P2} is turned on. Because of the high impedance from the drain to the source of M_{N2}, the output voltage becomes high: $v_O = V_{OH}$. If $v_A = v_B = V_{OH}$, then both M_{N1} and M_{N2} are turned on, and both M_{P1} and M_{P2} are turned off. In this case, the output voltage is low: $v_O = V_{OL}$. The logic function is the same as that for an NMOS, shown in Table 15.5.

CMOS gates have extremely high input resistance, on the order of hundreds of megohms, which is desirable for the driving circuitry. Unfortunately, they are susceptible to damage to the thin gate-oxide layer (typically 0.1 μm for a metal gate and less for a polysilicon gate), which has a breakdown voltage of 50 V to 100 V. Since the capacitance of a typical human body is 100 pF to 300 pF, a person walking across a waxed laboratory floor or brushing against a garment may generate static voltage in excess of 10 kV! Stray electrostatic discharge from a person handling the CMOS can easily release enough energy to cause permanent damage. A charged body can release tens of kilowatts of power over hundreds of nanoseconds. To protect a CMOS, clamping diodes, shown in Fig. 15.30, are built in to CMOS gates to limit any input voltages that fall outside the range from V_{SS} to V_{DD}. Also, any distributed input resistance (R_s), which is typically 1.5 kΩ for a metal gate and 250 Ω for a polysilicon gate, limits the transient gate current.

15.8.3 CMOS Families

The CMOS family of gates dates from the late 1960s. The 74Cxx logic families are second-generation CMOS circuits. These are polysilicon rather than metal gates, and the devices are smaller and faster than earlier designs. In addition, outputs are double buffered, as shown in Fig. 15.32. A 74Cxx gate has two cascaded inverters to isolate the logic from the output. As a result, the voltage gain is increased with a very steep VTC.

The 74HCxx series is the third generation of the CMOS family, which continues the trend toward smaller size and considerably less power, typically $P_D = 500$ μW and $t_{pd} = 10$ ns. The 74ACxx series is the fourth generation of the CMOS family, which is faster at the same power, typically $P_D = 500$ μW and $t_{pd} = 4$ ns.

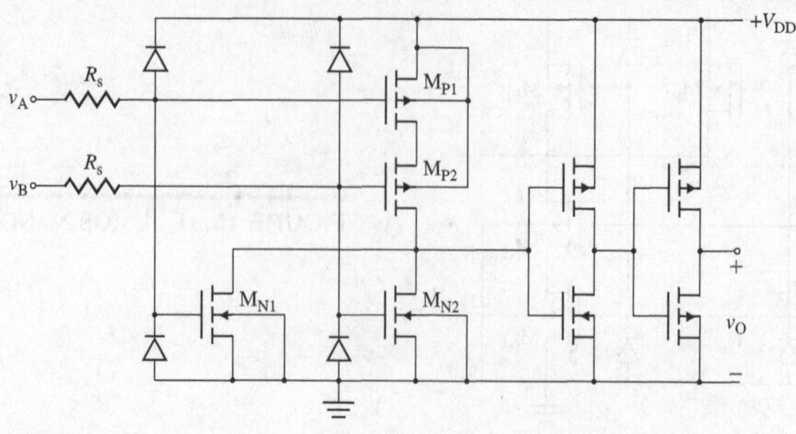

FIGURE 15.32 Buffered CMOS NOR gate

- Like NMOS inverters, CMOS inverters are frequently used in digital circuits in transmission gates and NOR and NAND gates.

15.9 Comparison of CMOS and NMOS Gates

The major advantages and disadvantages of CMOS and NMOS gates are listed below:

1. The output voltage of a CMOS gate is the full input voltage v_I without the drop due to the threshold voltage, as in the case of NMOS gates.
2. A CMOS gate requires more transistors than an NMOS gate to perform similar logic functions.
3. CMOS gates consume very little power, thereby enabling very-large-scale integration (VLSI). NMOS gates consume more power than CMOS gates and have thermal limitations that make them less attractive for VLSI.
4. CMOS gates draw spikes of current during the transition from one state to another; the current peaks occur when both NMOS and PMOS transistors are in saturation.
5. CMOS gates occupy a larger area and have larger capacitances than NMOS gates.

15.10 BJT Inverters

BJT switches are the building blocks of bipolar logic circuits [7, 2, 5]. The earliest logic device using BJTs was the basic inverter, developed in the 1960s. This device was followed by the *resistor-transistor logic* (RTL), *diode-transistor logic* (DTL), and *transistor-transistor logic* (TTL) families of logic circuits. In RTL, DTL, and TTL families, the BJTs operate by switching between the saturation (on) and cutoff (off) regions of operation; hence, they are generally called *saturation-logic* families. Because their BJT saturation-logic circuits experience delay due to the storage time in saturated devices, RTL and DTL circuits are no longer used. In some BJTs (e.g., Schottky BJTs), the time delay required to bring a transistor out of saturation is avoided by preventing the BJT from becoming saturated. Although TTL logic circuits are being challenged by CMOS circuits, TTL technology has improved over the years and continues to be popular. The gate delay of a modern TTL circuit may be as low as 1.5 ns.

There are two other families of logic circuits: the emitter-coupled logic (ECL) circuit and the integrated-injection logic (I^2L) circuit. They are operated by switching a constant current between two parts of a circuit and avoiding transistor saturation. The gate delay of an ECL circuit can be less than 1 ns, and ECLs find applications in digital communication circuits and high-speed circuits in supercomputers. I^2L circuits, however, have lost ground to CMOS logic circuits in many applications.

15.10.1 Voltage Transfer Characteristics

A BJT with a collector (pull-up) resistance R_C is shown in Fig. 15.33(a). When the input v_I is low, such that the base-to-emitter voltage is less than the forward-bias cut-in voltage, $v_{BE} < V_{BE(cut\text{-}in)}$, transistor Q_1 is off, and the output v_O is high; that is, $v_O = V_{OH} = V_{CC}$. The transistor operates at point A, as shown in Fig. 15.33(b). When v_I goes high to V_{OH}, the operating point jumps to point B and moves along the output characteristic (in the active region) toward the quiescent point Q (in the saturation region). The output voltage becomes low—that is, the collector–emitter saturation voltage, $V_{CE(sat)}$. The VTC is shown in Fig. 15.33(c).

The collector current at saturation is given by

$$I_{C(sat)} = \frac{V_{CC} - V_{CE(sat)}}{R_C} \tag{15.77}$$

and the corresponding base current at saturation is given by

$$I_{B(sat)} = \frac{I_{C(sat)}}{\beta_F} \quad \text{(at } V_{CE(sat)}\text{)}$$

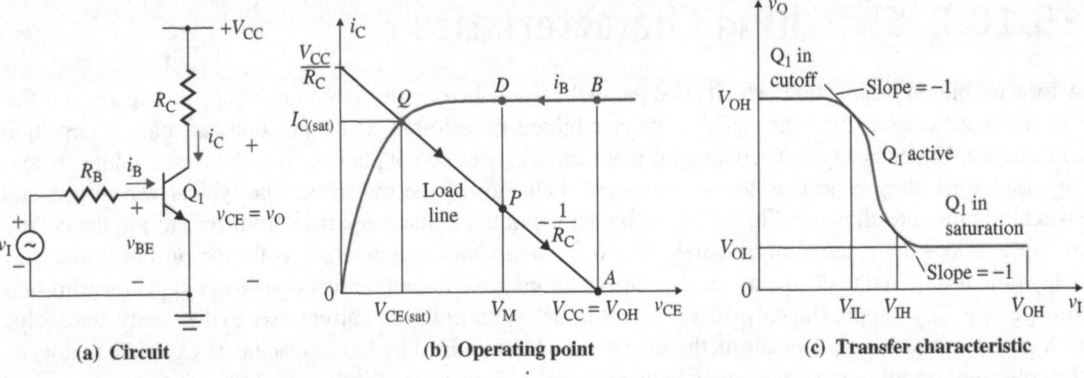

FIGURE 15.33 BJT inverter

where β_F is the forward current gain of the transistor. Normally, the circuit is designed with I_B higher than $I_{B(sat)}$ to ensure that Q_1 is driven into saturation. I_B can be found from

$$I_B = \frac{v_I - V_{BE(sat)}}{R_B}$$

The ratio of I_B to $I_{B(sat)}$ is called the *overdrive factor* k_{ODF}:

$$k_{ODF} = \frac{I_B}{I_{B(sat)}}$$

The ratio of $I_{C(sat)}$ to I_B is called the *forced* β_F, given by

$$\beta_{F(forced)} = \frac{I_{C(sat)}}{I_B} = \frac{I_{B(sat)}\beta_F}{I_B} = \frac{\beta_F}{k_{ODF}}$$

$V_{CE(sat)}$, which changes slightly with collector current $I_{C(sat)}$, can be found from

$$V_{CE(sat)} = V_T \ln \frac{I_B\beta_F + I_C\beta_F(1 - \alpha_R)}{I_B\beta_F\alpha_R - I_C\alpha_R}$$

which, for a typical value of reverse current gain $\alpha_R = 0.1$, $I_C = I_{C(sat)}$, and $I_B = I_{B(sat)}$, becomes

$$V_{CE(sat)} = V_T \ln\left(\frac{10 + 9I_{C(sat)}/I_{B(sat)}}{1 - I_{C(sat)}/\beta_F I_{B(sat)}}\right) = V_T \ln\left(\frac{10 + 9\beta_{F(forced)}}{1 - \beta_{F(forced)}/\beta_F}\right) \tag{15.78}$$

(See Appendix D.) The value of $V_{CE(sat)}$ usually falls in the range of 0.1 V to 0.3 V. The base-to-emitter junction characteristic is similar to that of a diode, and $V_{BE(sat)}$ usually falls in the range of 0.65 V to 0.8 V.

15.10.2 Switching Characteristics

A forward-biased *pn* junction exhibits two parallel capacitances: a depletion-layer capacitance and a diffusion capacitance. On the other hand, a reverse-biased or zero-biased *pn* junction has only a depletion capacitance. Under steady-state conditions, these capacitances do not play any role. However, under switching conditions, they contribute to the on and off behavior of the transistor. The typical waveforms and switching times are shown in Fig. 15.34. As the base–emitter voltage v_{BE} rises from zero to V_B, the collector current does not respond immediately. There is a *delay time* t_d before any collector current flows. This delay time is required to charge up the capacitance of the base-to-emitter junction (BEJ) to the forward-bias voltage $V_{BE(cut\text{-}in)}$ (approximately 0.7 V). After this delay, the collector current rises to the steady-state value of $I_{C(sat)}$. The *rise time* t_r depends on the time constant determined by base capacitance C_B. Note that when the collector current i_C rises, the output voltage v_O falls. Thus, t_r for i_C in Fig. 15.34(b) corresponds to t_f for v_O in Fig. 15.34(c).

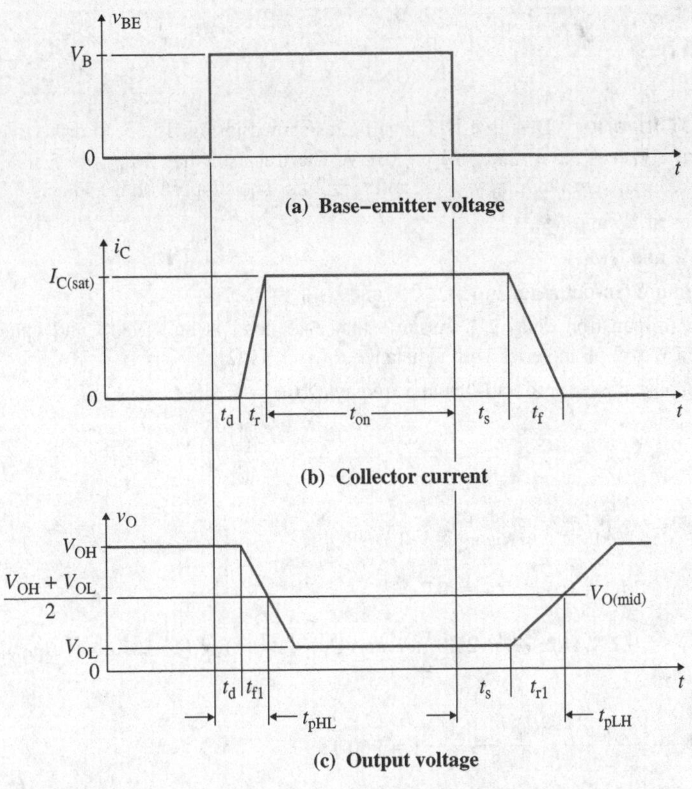

FIGURE 15.34 Switching times of bipolar transistors

The base current is normally more than that required to saturate the transistor. As a result, the excess minority carrier charge is stored in the base region. The higher the overdrive factor k_{ODF}, the greater will be the amount of extra charge stored in the base. This extra charge, which is called the *saturating charge*, is proportional to the excess base drive Δi_B given by

$$\Delta i_B = I_B - I_{B(sat)} = k_{ODF} I_{B(sat)} - I_{B(sat)} = I_{B(sat)}(k_{ODF} - 1)$$

The saturating charge is given by

$$Q_s = \tau_s \, \Delta i_B = \tau_s I_{B(sat)}(k_{ODF} - 1)$$

where τ_s is known as the *storage time constant* of the transistor. When v_{BE} falls to zero, the collector current does not change for a time t_s, called the *storage time*, which is the time required to remove the saturating charge from the base. Once the extra charge is removed, the BEJ capacitance discharges to zero. *Fall time* t_f depends on the time constant, which is determined by the capacitance of the reverse-biased BEJ.

The propagation time t_{pHL} consists of the delay time t_d and the time t_{f1} for the output to fall from V_{OH} to $(V_{OH} + V_{OL})/2$; that is, $t_{pHL} = t_d + t_{f1}$. The propagation time t_{pLH} consists of the storage time t_s and the time t_{r1} for the output to rise from V_{OL} to $(V_{OH} + V_{OL})/2$. That is, $t_{pLH} = t_s + t_{r1}$.

EXAMPLE 15.6

D

Designing a BJT inverter Design a BJT inverter, as shown in Fig. 15.35, to drive five identical inverters ($N = 5$) and to give $V_{OH} = 3.5$ V and $NM_L = 0.4$ V. The transistor has $I_{C(max)} = 5$ mA, $\beta_F = 100$ to 150, $V_{BE(cut\text{-}in)} = 0.6$ V, $V_{BE(sat)} = 0.8$ V, $t_d = 1$ ns, and $t_s = 2$ ns. The supply voltage V_{CC} is 5 V.

(a) Find the values of R_C and R_B.

(b) Calculate k_{ODF} and NM_H.

(c) Find the maximum fan-out N for $V_{IH} = 2.0$ V and $k_{ODF} = 1$.

(d) Calculate the propagation delay t_{pd}. Assume that each load is an NMOS and can be represented by a capacitance $C_B = 2$ pF in series with resistance $R_B = 1.8$ kΩ.

(e) Calculate the delay-power product (DP) at a frequency $f_{clk} = 5$ MHz.

SOLUTION

(a) $N = 5$, $V_{OH} = 3.5$ V, $V_{IL} = V_{BE(cut\text{-}in)} = 0.6$ V, and

$$V_{OL} = V_{IL} - NM_L = 0.6 - 0.4 = 0.2 \text{ V}$$

For $V_{CE(sat)} = V_{OL} = 0.2$ V (for $V_T = 25.8$ mV and $\beta_F = 100$), Eq. (15.78) gives $\beta_{F(forced)} = 71.8$. Using Eq. (15.77), we have

$$R_C \geq \frac{(V_{CC} - V_{CE(max)})}{I_{C(sat)}} = \frac{(5 - 0.2) \text{ V}}{5 \text{ mA}} = 960 \ \Omega$$

We choose $R_C = 1$ kΩ. When the input is low, Q_1 is off and $v_O = V_{OH}$. The base current I_B of each load stage is

$$I_B = \frac{V_{OH} - V_{BE(sat)}}{R_B}$$

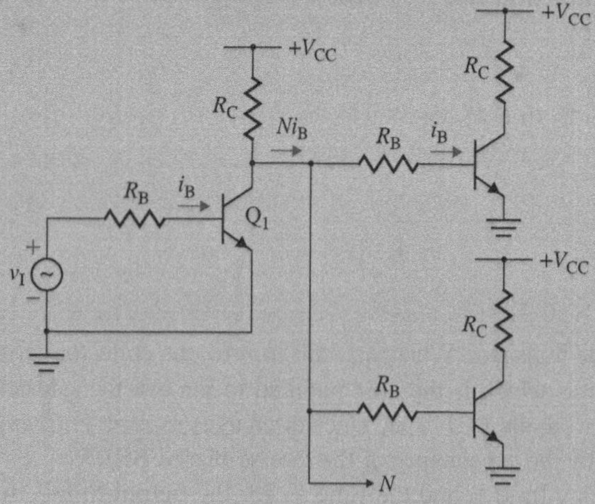

FIGURE 15.35 BJT inverter driving identical inverters

and $V_{OH} = V_{CC} - NI_B R_C$. Eliminating I_B and solving for V_{OH}, we get

$$V_{OH} = \frac{V_{CC} + NR_C V_{BE(sat)}/R_B}{1 + NR_C/R_B} = \frac{R_B V_{CC} + NR_C V_{BE(sat)}}{R_B + NR_C} \tag{15.79}$$

which, for $V_{OH} = 3.5$ V, $N = 5$, and $V_{BE(sat)} = 0.8$ V, gives $R_B = 9R_C = 9$ kΩ.

(b) We have

$$I_B = \frac{I_{C(sat)}}{\beta_{F(forced)}} = \frac{5\text{ mA}}{71.8} = 70\ \mu\text{A}$$

$$I_{B(sat)} = \frac{I_{C(sat)}}{\beta_F} = \frac{5\text{ mA}}{100} = 50\ \mu\text{A}$$

$$k_{ODF} = \frac{I_B}{I_{B(sat)}} = \frac{70}{50} = 1.40$$

$$V_{IH} = R_B I_B + V_{BE(sat)} \tag{15.80}$$

$$= 9\text{ k}\Omega \times 70\ \mu\text{A} + 0.8\text{ V} = 1.43\text{ V}$$

Thus, $NM_H = V_{OH} - V_{IH} = 3.5 - 1.43 = 2.07$ V. This noise margin is a measure of the safety factor that allows the load transistors to remain saturated despite changes in supply voltage, temperature, and manufacturing tolerances.

(c) When Q_1 is off, the output voltage v_O becomes

$$v_O = V_{CC} - NI_B R_C = V_{CC} - Nk_{ODF} I_{B(sat)} R_C = V_{CC} - N\frac{k_{ODF}(V_{CC} - V_{CE(sat)})}{\beta_F}$$

For regeneration of logic levels at the load gates, $v_O \geq V_{IH}$, and we find the fan-out N as

$$N \leq \frac{\beta_F(V_{CC} - V_{IH})}{k_{ODF}(V_{CC} - V_{CE(sat)})} \tag{15.81}$$

$$\leq 100 \times \frac{5 - 1.4}{1.39 \times (5 - 0.2)} = 53.4$$

Hence, the maximum fan-out is $N = 53$.

(d) When Q_1 turns off, each load is represented by a capacitance C_B and a resistance R_B. The equivalent circuit is shown in Fig. 15.36(a). For identical loads, all branches are effectively in parallel as far as the output node is concerned, and Fig. 15.36(a) can be simplified to Fig. 15.36(b). That is, the time constant for low to high can be found from

$$\tau_{r1} \approx NC_B\left(R_C + \frac{R_B}{N}\right) \tag{15.82}$$

$$= 5 \times 2\text{ pF} \times \left(1\text{ k}\Omega + \frac{9\text{ k}\Omega}{5}\right) = 28\text{ ns}$$

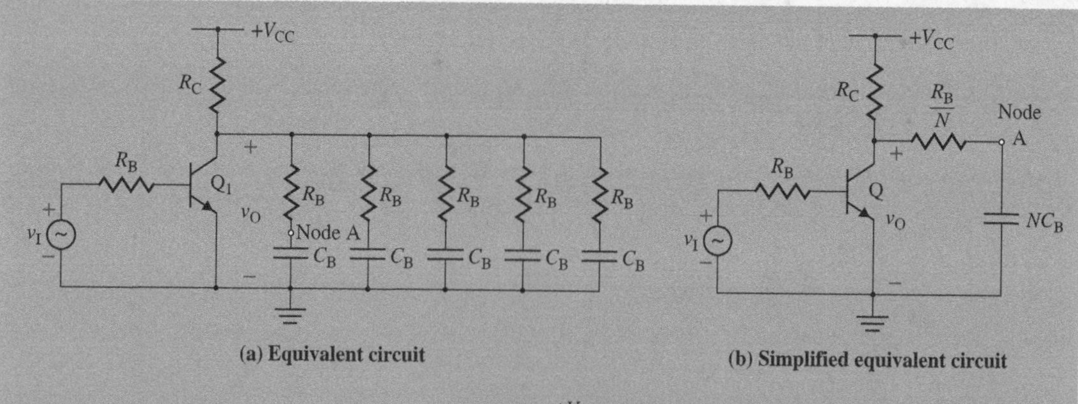

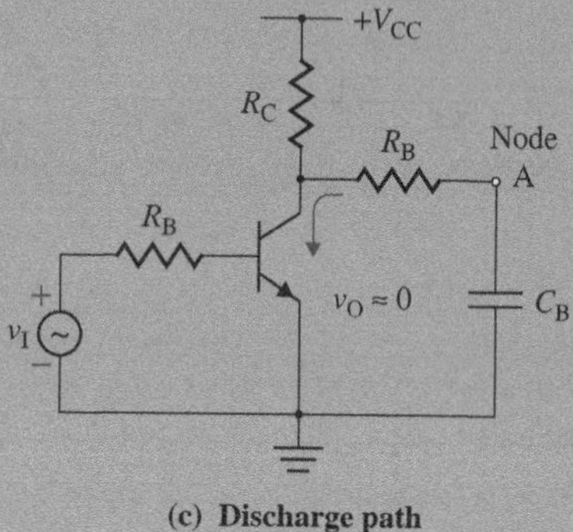

(c) Discharge path

FIGURE 15.36 Equivalent circuits for Example 15.6

The output voltage v_O during charging can be found from

$$v_O = V_{CC} - \frac{R_C V_{CC}}{R_C + R_B/N} e^{-t/\tau_{r1}}$$

from which we can find the time required for v_O to rise from $V_{OL} = 0.2$ V to $V_{O(mid)}$, which is equal to $(V_{OH} + V_{OL})/2 = 1.85$ V; that is,

$$t_{r1} = t_2(\text{for } v_O = V_{O(mid)}) - t_1(\text{for } v_O = V_{OL}) = \tau_{r1} \ln \frac{V_{CC} - V_{OL}}{V_{CC} - V_{O(mid)}} \tag{15.83}$$

$$= 28 \text{ ns} \times \ln\left[\frac{(5 - 0.2)}{(5 - 1.85)}\right] = 11.79 \text{ ns}$$

Thus, $t_{pLH} = t_s + t_{r1} = 2 + 11.79 = 13.79$ ns.

When Q_1 is turned on to saturation, each capacitor discharges through the transistor. The equivalent circuit is shown in Fig. 15.36(c). The time constant for high to low can be found from

$$\tau_{f1} \approx C_B R_B \tag{15.84}$$

$$= 2 \text{ pF} \times 9 \text{ k}\Omega = 18 \text{ ns}$$

The output voltage v_O during discharging can be found from

$$v_O = V_{OH} e^{-t/\tau_{f1}}$$

from which we can find the time for v_O to fall from $V_{OH} = 3.5$ V to $(V_{OH} + V_{OL})/2 = 1.85$ V; that is,

$$t_{f1} = \tau_{f1} \ln\left(\frac{2V_{OH}}{V_{OH} + V_{OL}}\right) \tag{15.85}$$

$$= 18 \text{ ns} \times \ln\left(2 \times \frac{3.5}{3.5 + 0.2}\right) = 11.48 \text{ ns}$$

Thus, $t_{pHL} = t_d + t_{f1} = 1 + 11.48 = 12.48$ ns. The propagation time is

$$t_{pd} = \frac{t_{pHL} + t_{pLH}}{2} = 13.135 \text{ ns}$$

(e) The total power loss in the two junctions is found as follows:

$$P_{static} = V_{CE(sat)} I_{C(sat)} + V_{BE(sat)} I_B$$

$$= 0.2 \times 5 \text{ mA} + 0.8 \times 70 \ \mu\text{A} = 1.056 \text{ mW}$$

$$P_{dynamic} = N f_{clk} C_B V_{OH}^2 = 5 \times 5 \text{ MHz} \times 2 \text{ pF} \times 3.5^2 = 0.61 \text{ mW}$$

$$P_D = P_{static} + P_{dynamic} = 1.67 \text{ mW}$$

Therefore, the delay-power product is

$$DP = P_D \times t_{pd} = 1.67 \text{ mW} \times 13.135 \text{ ns} = 21.9 \text{ pJ}$$

▶ **NOTE** The value of DP is much higher than that of a CMOS or an NMOS inverter.

KEY POINT OF SECTION 15.10

- A BJT is operated as a switching device. It is usually overdriven to ensure operation in the saturation region. This causes a reduction in the switching speed due to charge recovery. The higher the amount of overdrive, the greater will be the amount of extra charge stored in the base.

15.11 Transistor-Transistor Logic Gates

Equation (15.81) shows that, to achieve a high fan-out, β_F should be as high as possible. In 1965 the first TTL family was introduced specifically to increase switching speed without sacrificing fan-out or noise margin and without increasing power dissipation. TTL gates were leading DTL gates in sales by 1970 and dominated the digital IC market for more than 10 years. Many developments in the process and manufacturing technology as well as in circuit design techniques have led to several new generations (families) of TTL gates. Each generation has its advantages and disadvantages relative to the previous generations. As the technology has advanced, it has become economically feasible to obtain high performance by using

TABLE 15.6 Parameters of 54/74 series TTL logic gates

Parameter	74	74H	74L	74S	74LS	74AS	74ALS	74F
V_{IL}, in V	0.8	0.8	0.8	0.8	0.8	0.8	0.8	0.8
V_{IH}, in V	2.0	2.0	2.0	2.0	2.0	2.0	2.0	2.0
V_{OL}, in V	0.4	0.4	0.4	0.5	0.5	0.5	0.5	0.5
V_{OH}, in V	2.4	2.4	2.4	2.7	2.7	2.7	2.7	2.7
t_{pd}, in ns at $C_L = 50$ pF	10	6	30	3	10	1.5	4	2.5
P_D, in mW	10	25	1	20	2	8	1	5
PD, in pJ	100	150	30	60	20	12	4	12.5

more transistors per gate, and several families have emerged, among which are standard TTL gates, high-speed (H) TTL gates, low-power (L) TTL gates, Schottky (S) TTL gates, low-power Schottky (LS) gates, advanced Schottky (AS) gates, advanced low-power Schottky (ALS) gates, and Fairchild advanced Schottky (F or Fast) gates.

The performance parameters of TTL 54/74 families are summarized in Table 15.6. In early TTL families, the improvement in switching speed resulted in more power dissipation. The Schottky series, however, has both lower power dissipation and improved speed. Currently, TTL applications focus on low-power Schottky (LS) and advanced low-power Schottky (ALS) families of gates. The ALS family offers the highest speed (although it is surpassed by the ECL family and rivaled by the CMOS family). As examples, we will analyze the TTL, and ALS types.

15.11.1 Standard TTL Gates

The input stage of a TTL gate uses a multiemitter transistor Q_1, as shown in Fig. 15.37(a). In isoplanar ICs, multiemitters are normally fabricated within the same base region. Figure 15.37(b) shows a cross section of a three-emitter transistor. Q_1 can be viewed as three diodes from the base to the emitter and one from the base to the collector, as shown in Fig. 15.37(c).

A TTL NAND gate with a multiemitter transistor input is shown in Fig. 15.38. The combination of Q_4, D_1, and Q_3 is called the *totem-pole output stage*. The transistor Q_2 forms a phase splitter since the collector and emitter voltages are 180° out of phase. The circuit operation can be divided into two modes: mode 1 and mode 2.

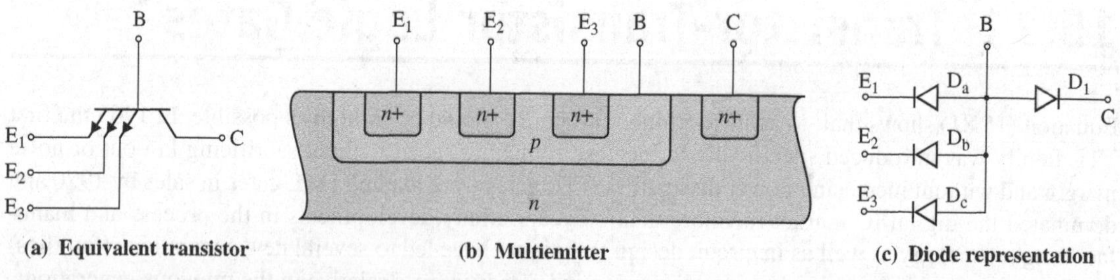

FIGURE 15.37 Multiemitter bipolar transistor

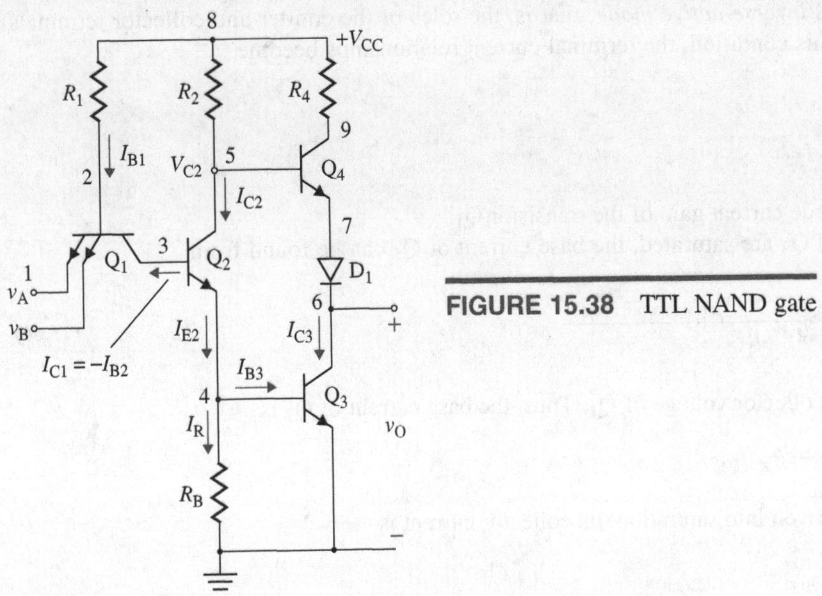

FIGURE 15.38 TTL NAND gate

During mode 1, the input voltages v_A and v_B are high, at V_{OH}; that is, $v_A = v_B = V_{OH}$. The TTL NAND gate is at low-state output voltage, as shown in Fig. 15.39, with N similar load gates. I_{C3} becomes the collector saturation current I_{CS} of Q_3. The two emitter junctions of Q_1 are reverse biased. The voltage v_{B1} at the base of Q_1 is large enough to forward-bias the base–collector junction of Q_1 and drive Q_2 into saturation. Since the base–collector junction is forward biased and the BEJ is reverse biased,

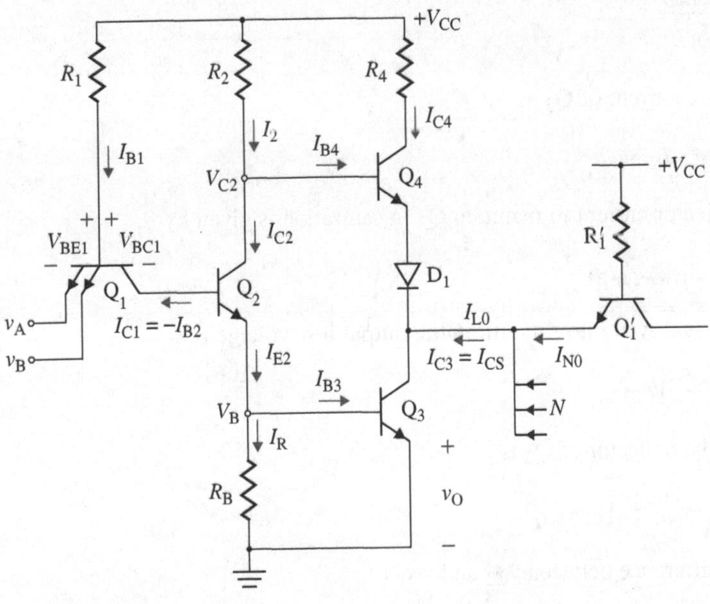

FIGURE 15.39 TTL gate in the output low state

Q_1 will be operating in its *inverse-active mode*; that is, the roles of the emitter and collector terminals are interchanged. Under this condition, the terminal current relationships become

$$I_{E1} = -\beta_R I_{B1}$$

$$I_{C1} = -(1 + \beta_R) I_{B1}$$

where β_R is the inverse-mode current gain of the transistor Q_1.

Assuming that Q_2 and Q_3 are saturated, the base current of Q_1 can be found from

$$I_{B1} = \frac{V_{CC} - V_{BE2(sat)} - V_{BE3(sat)} - V_{BC1}}{R_1} \tag{15.86}$$

where V_{BC1} is the base-to-collector voltage of Q_1. Thus, the base current of Q_2 is

$$I_{B2} = -I_{C1} = (1 + \beta_R) I_{B1}$$

Since the transistor Q_2 is driven into saturation, its collector current is

$$I_{C2} = \frac{V_{CC} - V_{BE3(sat)} - V_{CE2(sat)}}{R_2} \tag{15.87}$$

Thus, the emitter current of Q_2 is

$$I_{E2} = I_{B2} + I_{C2}$$

and the current through the recovery resistance R_B is

$$I_R = \frac{V_{BE3(sat)}}{R_B}$$

which gives the base current of Q_3 as

$$I_{B3} = I_{E2} - I_R$$

The maximum collector current to maintain Q_3 in saturation is given by

$$I_{C3(sat)} = \beta_{F(forced)} I_{B3}$$

Since transistor Q_3 is driven into saturation, the output low voltage is

$$v_O = V_{OL} = V_{CE3(sat)}$$

and the voltage at the collector of Q_2 is

$$V_{C2} = V_{BE3(sat)} + V_{CE2(sat)}$$

Thus, the voltage difference between v_{C2} and v_O is

$$v_{C2} - v_O = V_{BE3(sat)} + V_{CE2(sat)} - V_{CE3(sat)} \approx V_{BE3(sat)}$$

which is the voltage across the BEJ of Q_4 and D_1 and will not be sufficient to turn on both Q_4 and D_1. With Q_3 in saturation and the output voltage in its low state, Q_4 will be off. Thus, Q_3 will sink to the (low-state) load current I_{LO} given by

$$I_{LO} = N_L I_{NO}$$

where I_{NO} is the individual load current for each fan-out at low output (i.e., $I_{NO} = I_{IL}$) and is obtained by multiplying the base current of Q'_1 by β_R; that is,

$$I_{NO} = \beta_R \left[\frac{V_{CC} - V_{CE3(sat)} - V_{BE1(sat)}}{R_1} \right] \quad (15.88)$$

Thus, the low-output fan-out is given by

$$N_L = \frac{I_{C3(sat)}}{I_{NO}}$$

During mode 2, at least one of the inputs v_A or v_B is low, at V_{OL}; that is, v_A (or v_B) = V_{OL}. The TTL NAND gate in the output high state is shown in Fig. 15.40. The BEJ is forward biased through R_1 and V_{CC}. The base current I_{B1} causes an emitter current through the particular emitter that is connected to the low input. Transistor action forces collector current into Q_1. But the collector current of Q_1, which equals the reverse-biased saturation current out of the base of Q_2, is usually much smaller than its base current. As a result, Q_1 will be in saturation. With v_A (or v_B) = V_{OL}, the base current of Q_1 becomes

$$I_{B1} = \frac{V_{CC} - V_{BE1(sat)}}{R_1}$$

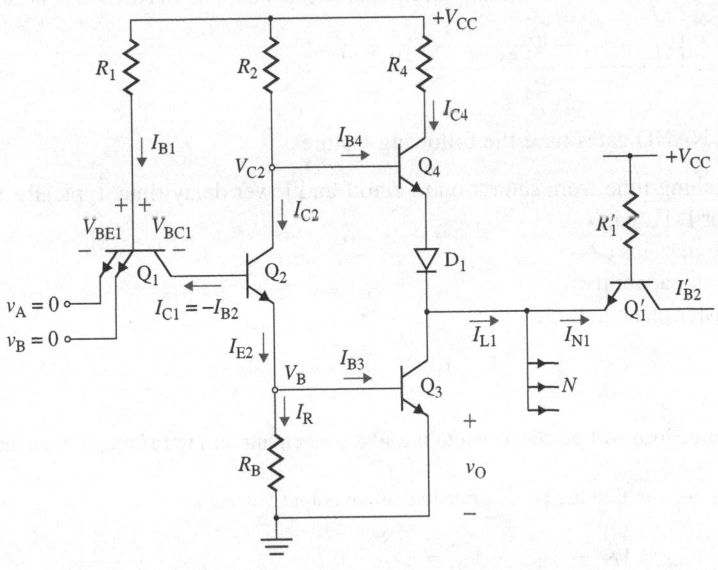

FIGURE 15.40 TTL gate in the output high state

Transistors Q_2 and Q_3 are cut off. Transistor Q_4 and diode D_1 supply the (high-state) load current I_{L1} given by

$$I_{L1} = N_H I_{N1}$$

where I_{N1} is the individual load current for each fan-out at high output (i.e., $I_{N1} = I_{IH}$). Since Q_1 is operating in the inverse-action mode, its base current must be multiplied by β_R to give $I_{N1} = \beta_R I_{B1}$; that is,

$$I_{N1} = \beta_R \left[\frac{V_{CC} - V_{BE2(sat)} - V_{BE3(sat)} - V_{BC1}}{R_1} \right] \tag{15.89}$$

With Q_4 in the active region, the base current of Q_4 is

$$I_{B4} = \frac{I_{L1}}{1 + \beta_F} = \frac{N_H I_{N1}}{1 + \beta_F}$$

Using KVL around the loop formed by V_{CC}, R_2, Q_4, and D_1, we get

$$\begin{aligned} V_{CC} &= R_2 I_{B4} + V_{BE4} + V_{D1} + v_O \\ &= \frac{R_2 N_H I_{N1}}{1 + \beta_F} + V_{BE4} + V_{D1} + v_O \end{aligned} \tag{15.90}$$

which gives the value of the high-output fan-out as

$$N_H = \frac{(1 + \beta_F)(V_{CC} - V_{BE4} - V_{D1} - v_O)}{R_2 I_{N1}} \tag{15.91}$$

If R_4 were not present, the collector current of Q_4 would be $\beta_F I_{B4}$, which would be highly undesirable. To limit the collector current of Q_4 to an acceptable value, R_4 is introduced. The maximum collector current of Q_4 is

$$I_{C4(max)} = \frac{V_{CC} - v_O - V_{CE4(sat)} - V_{D1}}{R_4} \tag{15.92}$$

In summary, TTL NAND gates have the following features:

1. Short switching time from saturation to cutoff and lower delay time, typically 10 ns as compared to 40 ns for DTL gates
2. High noise margins
3. High fan-out capability
4. Sharp transfer characteristic

▶ **NOTES**

1. In practice, the same load will be connected to the TTL gate circuit during the output high and output low states; that is, $N_L = N_H = N$.
2. From Eq. (15.90), we can find the permissible maximum output voltage as

$$v_{O(max)} = V_{OH} \approx V_{CC} - V_{BE4} - V_{D1} = V_{CC} - 0.7 - 0.7 = V_{CC} \left[\frac{1 - 1.4}{V_{CC}} \right]$$

For $V_{CC} = 5$ V, $V_{OH} = 5 - 1.4 = 3.6$ V.

EXAMPLE 15.7

D

Designing a TTL NAND gate

(a) Design the TTL NAND gate of the circuit in Fig. 15.38. It has two inputs and feeds four identical NAND gates. The output voltage at output high is $V_{OH} = 3.5$ V, and $I_{N1} = 72.5$ μA. Assume $V_{CC} = 5$ V, $V_D = 0.7$ V, $V_{CE(sat)} = 0.2$ V, $V_{BE} = 0.7$ V, $V_{BC} = 0.7$ V, $\beta_{F(forced)} = 10$, $\beta_R = 0.1$, $I_{C3(sat)} = 1$ mA, and $I_{C4(max)} = 1$ mA. Capacitance of each load is, $C_i = 5$ pF.

(b) Use PSpice/SPICE to check your design by plotting the transfer function. Determine NM_L and NM_H. Model parameters for diodes are

```
RS=4 TT=0.1NS
```

and for transistors are

```
BF=10 BR=0.1 TF=0.1NS TR=10NS VJC=0.85 VAF=50
```

SOLUTION

(a) The design can be carried out using the following steps:

Step 1. Using Eq. (15.89), calculate the value of R_1:

$$R_1 = \beta_R \frac{V_{CC} - 2V_{BE(sat)} - V_{BC}}{I_{N1}} = \frac{0.1 \times [5 - (2 \times 0.7) - 0.7]\text{ V}}{72.5\ \mu\text{A}} = 4\text{ k}\Omega$$

Step 2. Using Eq. (15.91) for $v_O = V_{OH}$, find the value of R_2 to meet the output requirement:

$$R_2 = \frac{(1 + \beta_{F(forced)})(V_{CC} - V_{BE4} - V_{D1} - V_{OH})}{NI_{N1}}$$

$$= \frac{(1 + 10) \times (5 - 0.7 - 0.7 - 3.5)\text{ V}}{4 \times 72.5\ \mu\text{A}} = 3.79\text{ k}\Omega$$

Step 3. Using Eq. (15.86), find the value of I_{B1} at output high:

$$I_{B1} = \frac{(5 - 0.7 - 0.7 - 0.7)\text{ V}}{4\text{ k}\Omega} = 725\ \mu\text{A}$$

Step 4. Using Eq. (15.88), calculate the low-state load current:

$$I_{N0} = \beta_R\left[\frac{V_{CC} - V_{CE3(sat)} - V_{BE1(sat)}}{R_1}\right] = 0.1\left[\frac{(5 - 0.2 - 0.7)\text{ V}}{4\text{ k}\Omega}\right] = 102.5\ \mu\text{A}$$

The total load current at low state for $N = 4$ is

$$I_{L0} = 4 \times 102.5\ \mu\text{A} = 410\ \mu\text{A}$$

Step 5. Calculate the high-state I_{B2}, I_{C2}, and I_{E2}. The high-state I_{B2} is

$$I_{B2} = -I_{C1} = (1 + \beta_R)I_{B1} = (1 + 0.1) \times 725\ \mu A = 798\ \mu A$$

The high-state I_{C2} is

$$I_{C2} = \frac{V_{CC} - V_{BE3(sat)} - V_{CE2(sat)}}{R_2} = \frac{(5 - 0.7 - 0.2)\ V}{3.79\ k\Omega} = 1082\ \mu A$$

The high-state I_{E2} is

$$I_{E2} = I_{B2} + I_{C2} = 798\ \mu A + 1082\ \mu A = 1.88\ mA$$

Step 6. Assume that most of I_{E2} flows through the base of Q_3 rather than through R_B. Let $I_R = 0.4I_{E2} = 0.4 \times 1.88\ m = 0.75\ mA$. Then calculate the value of R_B:

$$R_B = \frac{V_{BE3(sat)}}{I_R} = \frac{0.7\ V}{0.75\ mA} = 933\ \Omega$$

Choose $R_B = 1\ k\Omega$.

Step 7. Calculate the maximum permissible collector low-state current to drive Q_3 in saturation:

$$I_{B3} = I_{E2} - I_R = 1.88\ m - 0.75\ m = 1.13\ mA$$

$$I_{C3(max)} = \beta_F I_{B3} = 10 \times 1.13\ mA = 11.3\ mA$$

which is higher than $I_{LO} = 410\ \mu A$, from step 4. Thus, the design should be satisfactory. (Otherwise steps 1 through 7 should be repeated with a lower value of N, a higher value of $\beta_{F(sat)}$, or a lower value of I_{N1}.) The overdrive factor is

$$k_{ODF} = \frac{I_{C3(max)}}{I_{LO}} = \frac{11.3\ m}{410\ \mu} = 27.6$$

Step 8. Calculate the base current of Q_4 at output high:

$$I_{B4} = \frac{NI_{N1}}{1 + \beta_F} = \frac{4 \times 72.5\ \mu A}{1 + 10} = 26.4\ \mu A$$

Step 9. Using Eq. (15.92), calculate the value of R_4 for $v_O = V_{OH}$:

$$R_4 = \frac{V_{CC} - V_{OH} - V_{CE4(sat)} - V_{D1}}{I_{C4(max)}} = \frac{(5 - 3.5 - 0.2 - 0.7)\ V}{1\ mA} = 600\ \Omega$$

The power rating of R_4 is $P_{R4} = (1\ mA)^2 R_4 = (1\ mA)^2 \times 600 = 0.6\ mW$.

Step 10. Calculate the power rating of R_2 due to I_{C2} or I_{B2}, whichever has the higher value. Since $I_{C2} = 1082\ \mu A$ and $I_{B2} = 798\ \mu A$, use I_{C2}.

$$P_{R2} = (1.082\ mA)^2 R_2 = (1.082\ mA)^2 \times 3.79\ k\Omega = 4.44\ mW$$

Step 11. Calculate the power rating of R_1 due to I_{B1} at output high or I_{B1} at output low, whichever has the higher value. Since the base current of Q_1 at output high is

$$I_{B1} = \frac{V_{CC} - V_{BE2(sat)}}{R_1} = \frac{(5 - 0.7)\text{ V}}{4\text{ k}\Omega} = 1.075\text{ mA}$$

and I_{B1} at output low is 725 μA (from step 3), use I_{B1} at output high.

$$P_{R1} = (1.075\text{ mA})^2 R_2 = (1.075\text{ mA})^2 \times 4\text{ k}\Omega = 4.6\text{ mW}$$

(b) Each transistor has junction capacitances that will affect the switching speed. To examine the propagation time, we will assume that a load resistance of

$$R_L = \frac{V_{OH}}{NI_{N1}} = \frac{3.5\text{ V}}{4 \times 72.5\text{ }\mu\text{A}} = 12\text{ k}\Omega$$

and an equivalent capacitance of

$$C_L = 4 \times 5\text{ pF} = 20\text{ pF}$$

is connected to the output.

The PSpice schematic is shown in Fig. 15.41. The PSpice plots of the characteristics, which are shown in Fig. 15.42, give $V_{OH} = 3.53$ V (expected value is 3.5 V), $V_{IL} = 0.6$ V at $V_O = 3.49$ V, $V_{OL} = 42$ mV at $V_O = 5$ V, and $V_{IH} = 1.5$ V at $V_O = 62$ mV.

$$NM_L = V_{IL} - V_{OL} = 0.6 - 0.042\text{V} \approx 0.56\text{ V}$$

$$NM_H = V_{OH} - V_{IH} = 3.53 - 1.5 = 2.03\text{ V}$$

$t_{pHL} = 4.53$ ns, $t_{pLH} = 6.68$ ns, and $t_{pd} = 5.61$ ns.

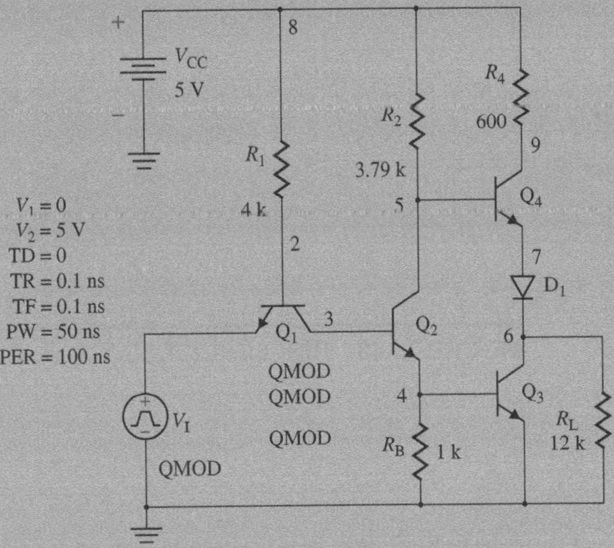

FIGURE 15.41 PSpice schematic for Example 15.7

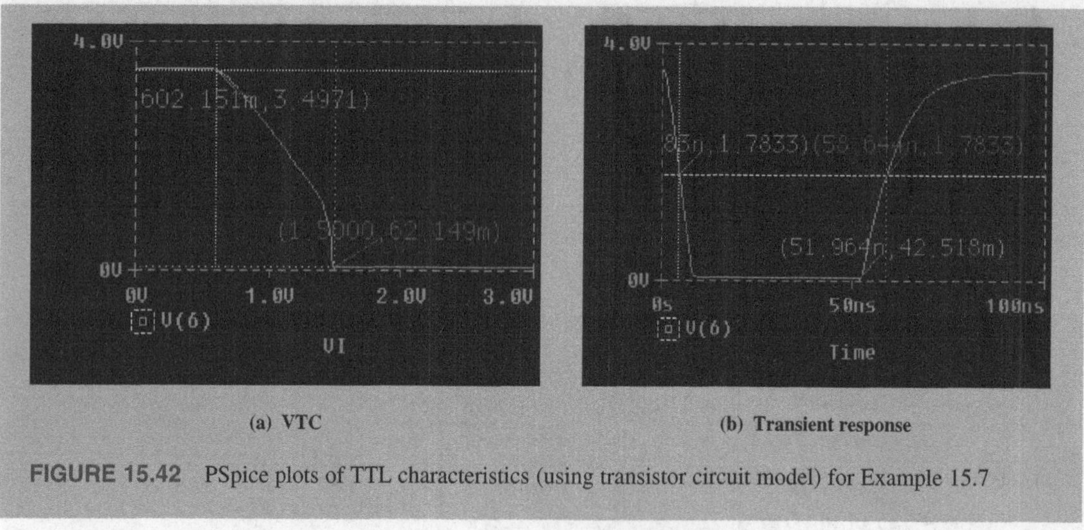

(a) VTC

(b) Transient response

FIGURE 15.42 PSpice plots of TTL characteristics (using transistor circuit model) for Example 15.7

15.11.2 High-Speed TTL NAND Gates

Notice from Eq. (15.90) that the voltage drop across R_2 caused by the load current flow reduces the output high voltage v_O. The output voltage can be increased by replacing Q_4 in Fig. 15.38 by a Darlington pair, as shown in Fig. 15.43. Transistors Q_4 and Q_5 form the Darlington pair. The BEJ of Q_5 offers the voltage offset and serves the purpose of diode D_1 in Fig. 15.38. Resistor R_5 is included to aid the reverse-biased recovery current of Q_4. The operation and analysis of this gate are similar to those of the gate in Fig. 15.38.

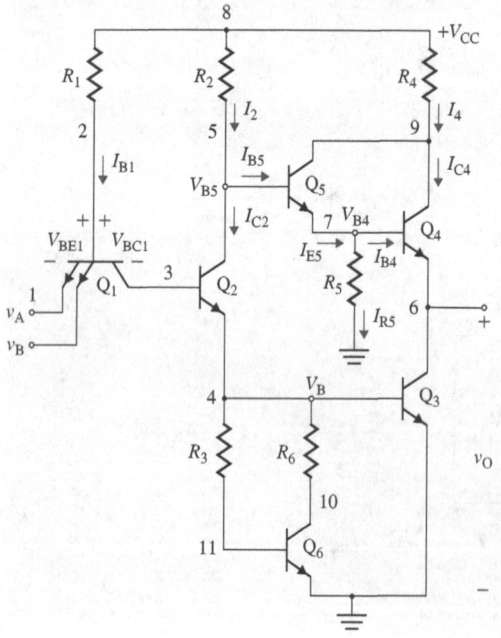

FIGURE 15.43 High-speed TTL NAND gate

The current through R_5 is

$$I_{R5} = \frac{V_{B4}}{R_5} = \frac{v_O + V_{BE4}}{R_5}$$

The emitter current of Q_5 is

$$I_{E5} = I_{B4} + I_{R5}$$

which gives its base current as

$$I_{B5} = \frac{I_{E5}}{1 + \beta_{F5}} = \frac{I_{B4} + I_{R5}}{1 + \beta_{F5}} = \frac{I_{E4}}{(1 + \beta_{F4})(1 + \beta_{F5})} + \frac{I_{R5}}{1 + \beta_{F5}}$$

Equation (15.90) can be modified to give the output voltage as

$$V_{CC} = R_2 I_{B5} + V_{BE4} + V_{BE5} + v_O$$

or

$$v_O = V_{CC} - (R_2 I_{B5} + V_{BE4} + V_{BE5}) \tag{15.93}$$

Also, resistance R_B is replaced by an active base recovery circuit consisting of R_3, R_6, and Q_6. This recovery circuit reduces the amount of diverting current at the base of Q_3. As a result, the VTC becomes sharper, the noise margin is enhanced, and the delay time is reduced. The recovery circuit and its equivalent are shown in Fig. 15.44[(a) and (b)]. I_{BR} is the base recovery current of Q_3.

Assuming that Q_6 is close to saturation but not saturated, the collector current of Q_6 is given by

$$I_{C6} = \frac{I_{BR}}{1 + 1/\beta_{F6}}$$

which, near the edge of saturation, must be equal to

$$I_{C6} = \frac{V_{BE3} - V_{CE6(sat)}}{R_6}$$

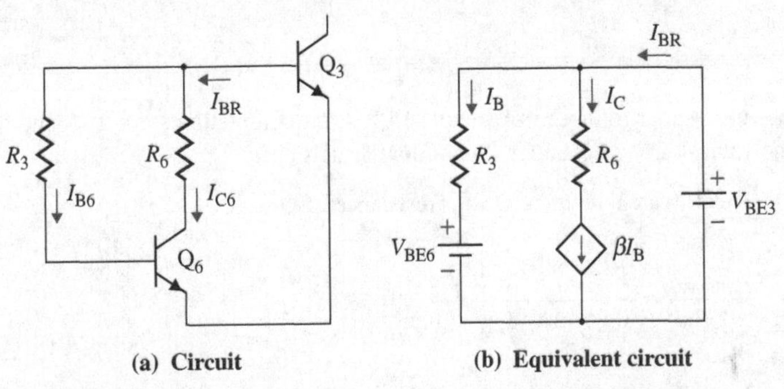

FIGURE 15.44 Base recovery circuit

Equating these two and assuming $\beta_F = \beta_{F(forced)}$, we can find the value of R_6 as

$$R_6 = \frac{V_{BE3} - V_{CE6(sat)}}{I_{BR}/(1 + 1/\beta_{F(forced)})} \tag{15.94}$$

Suppose that R_6 is chosen to yield the same value of I_{BR} as was obtained when a passive resistance R_B was used; that is, $I_{BR} = V_{BE3}/R_B$. Substituting into Eq. (15.94) yields

$$R_6 = \frac{R_B(V_{BE3} - V_{CE6(sat)})}{V_{BE3}/(1 + 1/\beta_{F(forced)})} \tag{15.95}$$

In summary, high-speed TTL gates have the following features:

1. Short propagation delay time
2. Low noise immunity
3. Sharp voltage transfer characteristic
4. High fan-out
5. Maximum output voltage limited to, typically, 3.6 V

EXAMPLE 15.8

D

Designing a TTL NAND gate

(a) Design the TTL NAND gate of the circuit in Fig. 15.43. It has two inputs and feeds four identical NAND gates. The output voltage at output high is $V_{OH} = 3.5$ V, and $I_{N1} = 72.5$ μA. Assume $V_{CC} = 5$ V, $V_D = 0.7$ V, $V_{CE(sat)} = 0.2$ V, $V_{BE} = 0.7$ V, $\beta_{F(forced)} = 10$, $\beta_R = 0.1$, $I_{C3(sat)} = 1$ mA, $I_{C4(max)} = 1$ mA, and $I_{BR} = 750$ μA. Capacitance of each load is, $C_i = 5$ pF.

(b) Use PSpice/SPICE to check your design by plotting the transfer characteristic. Determine NM_L and NM_H. Model parameters for transistors are

```
BF=10 BR=0.1 TF=0.1NS TR=10NS VJC=0.85 VAF=50
```

SOLUTION

(a) The design steps are similar to those in Example 15.7. After completing steps 1, 6, and 9 in Example 15.7 to determine the values of R_1, R_B, and R_4, we follow these steps.

Step 1. Using Eq. (15.94), calculate the collector resistance of Q_6:

$$R_6 = \frac{V_{BE3} - V_{CE6(sat)}}{I_{BR}/(1 + 1/\beta_F)} = \frac{(0.7 - 0.2)\text{ V}}{0.75\text{ mA}/(1 + 1/10)} = 733\ \Omega$$

Choose $R_3 = 2R_6 = 2 \times 733 = 1466\ \Omega$.

Step 2. Choose a value of I_{R5}: Let $I_{R5} = 1130\ \mu A$ (the same as the value for the TTL) ≈ 1.2 mA. For $v_O = V_{OH}$, find the value of R_5:

$$R_5 = \frac{V_{OH} + V_{BE4}}{I_{R5}} = \frac{(3.5 + 0.7)\text{ V}}{1.2\text{ mA}} = 3.5\text{ k}\Omega$$

Step 3. Calculate the base current of Q_4 at output high and the base current of Q_5:

$$I_{B4} = \frac{N_H I_{N1}}{1 + \beta_{F4}} = \frac{4 \times 72.5\ \mu\text{A}}{1 + 10} = 26.4\ \mu\text{A}$$

$$I_{B5} = \frac{I_{B4} + I_{R5}}{1 + \beta_{F5}} = \frac{26.4\ \mu\text{A} + 1.2\text{ mA}}{1 + 10} = 111.5\ \mu\text{A}$$

Step 4. Using Eq. (15.93) for $v_O = V_{OH}$, find the value of R_2 that meets the output requirement:

$$R_2 = \frac{V_{CC} - V_{BE4} - V_{BE5} - V_{OH}}{I_{B5}} = \frac{(5 - 0.7 - 0.7 - 3.5)\text{ V}}{111.5\ \mu\text{A}} = 897\ \Omega$$

Step 5. Using Eq. (15.89), calculate the value of R_1:

$$R_1 = \beta_R \left[\frac{V_{CC} - 2V_{BE(sat)} - V_{BC1}}{I_{N1}}\right] = 0.1\left[\frac{(5 - 2 \times 0.7 - 0.7)\text{ V}}{72.5\ \mu\text{A}}\right] = 4\text{ k}\Omega$$

Step 6. Using Eq. (15.86), find the value of I_{B1} at output low, and calculate the high-state I_{B2}:

$$I_{B1} = \frac{V_{CC} - V_{BE2(sat)} - V_{BE3(sat)} - V_{BC1}}{R_1} = \frac{(5 - 0.7 - 0.7 - 0.7)\text{ V}}{4\text{ k}\Omega} = 725\ \mu\text{A}$$

$$I_{B2} = -I_{C1} = (1 + \beta_R)I_{B1} = (1 + 0.1) \times 725\ \mu\text{A} = 798\ \mu\text{A}$$

Step 7. Calculate the high-state I_{C2}, I_{E2}, and I_{B3}. The high-state I_{C2} for low output is

$$I_{C2} = \frac{V_{CC} - V_{BE3(sat)} - V_{CE2(sat)}}{R_2} = \frac{(5 - 0.7 - 0.2)\text{ V}}{897} = 4.57\text{ mA}$$

In the high state,

$$I_{E2} = I_{B2} + I_{C2} = 798\ \mu\text{A} + 4.57\text{ mA} = 5.37\text{ mA}$$

For the base current,

$$I_{B3} = I_{E2} - I_R = 5.37\text{ mA} - 0.75\text{ mA} = 4.62\text{ mA}$$

Step 8. Calculate the maximum permissible collector low-state current to drive Q_3 in saturation:

$$I_{C3(max)} = \beta_{F(forced)} I_{B3} = 10 \times 4.62\text{ mA} = 46.2\text{ mA}$$

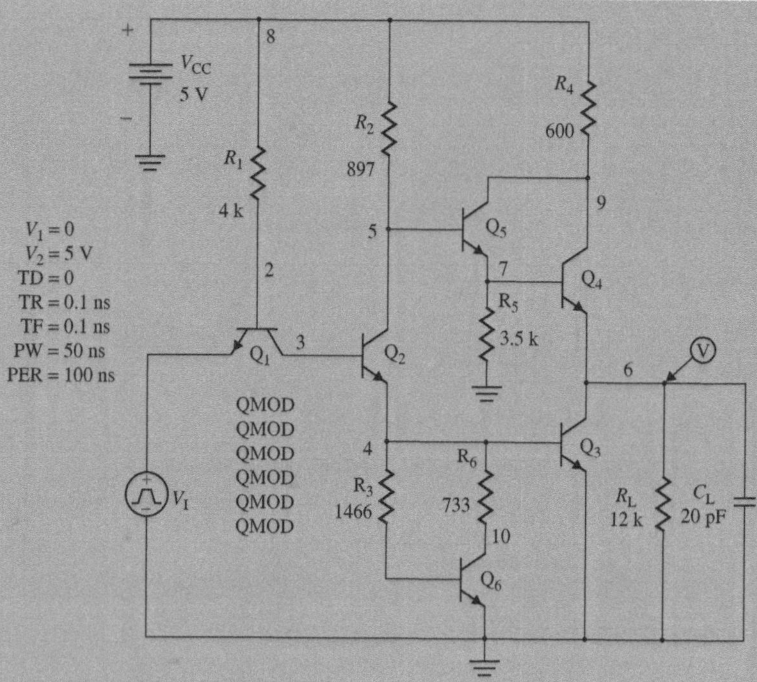

FIGURE 15.45 PSpice schematic for Example 15.8

which is higher than $I_{LO} = 410\ \mu A$, from step 4 of Example 15.7. Thus, the design should be satisfactory. (Otherwise, steps 1 through 8 should be repeated with a lower value of N, a higher value of $\beta_{F(sat)}$, or a lower value of I_{N1}.) The overdrive factor is

$$k_{ODF} = \frac{I_{C3(sat)}}{I_{LO}} = \frac{46.2\text{ mA}}{410\ \mu\text{A}} = 113$$

which is high; that is, the fan-out can be much more than $N = 4$.

(b) To examine the propagation time, we will assume that a load resistance of

$$R_L = \frac{V_{OH}}{NI_{LO}} = \frac{3.5\text{ V}}{4 \times 72.5\ \mu\text{A}} = 12\text{ k}\Omega$$

and an equivalent capacitance of

$$C_L = 4 \times 5\text{ pF} = 20\text{ pF}$$

is connected to the output.

The PSpice schematic is shown in Fig. 15.45. The PSpice plots, which are shown in Fig. 15.46, give $V_{OH} = 3.3$ V (expected value is 3.5 V), $V_{IL} = 1.35$ V at $V_O = 3.3$ V, $V_{OL} = 62$ mV at $v_O = 3.3$ V, and $V_{IH} = 1.6$ V at $V_O = 62$ mV.

$$NM_L = V_{IL} - V_{OL} = (1.35 - 0.062)\text{ V} \approx 1.29\text{ V}$$

$$NM_H = V_{OH} - V_{IH} = 3.3 - 1.6 = 1.7\text{V}$$

$t_{pHL} = 1.93$ ns, $t_{pLH} = (57.57\text{ ns} - 51.77\text{ ns}) = 5.8$ ns, and $t_{pd} = (t_{pHL} + t_{pLH})/2 = 3.87$ ns.

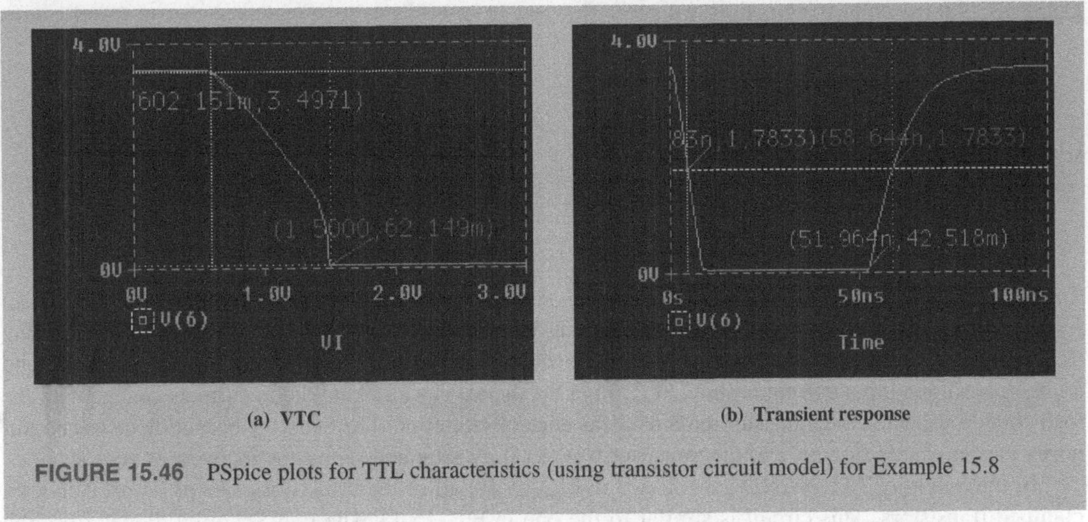

(a) VTC (b) Transient response

FIGURE 15.46 PSpice plots for TTL characteristics (using transistor circuit model) for Example 15.8

15.11.3 Schottky TTL NAND Gates

In TTL gates, the transistors are driven into saturation. Since the delay time of a TTL gate is a strong function of the storage time of the saturated transistors, a nonsaturating logic gate has an advantage. A Schottky clamped transistor, which is prevented from being driven into saturation, can switch faster than a saturated transistor. Schottky clamped transistors are incorporated in many transistor logic gates.

A Schottky clamped transistor is basically a bipolar transistor with a built-in clamped Schottky diode, as shown in Fig. 15.47(a). Its symbol is shown in Fig. 15.47(b). The forward voltage drop of a Schottky diode is low, typically 0.3 V. When transistor Q_1 is in its active region of operation, the base–collector junction is reverse biased. The clamped diode is also reverse biased and does not affect the circuit operation. Q_1 behaves as a normal *npn* transistor. As transistor Q_1 goes into saturation, the base–collector junction becomes forward biased, and it is clamped to the 0.3-V Schottky diode voltage. The excess base current is shunted through the diode, and the transistor is prevented from going heavily into saturation.

Assuming that transistor Q_1 is clamped at the edge of saturation, $I_C = \beta I_B$. The diode current can be related to the input and base currents by

$$I_D = I_I - I_B = I_I - \frac{I_C}{\beta_F}$$

(a) Schottky clamped transistor (b) Symbol

FIGURE 15.47 Schottky clamped transistor

Using KCL at the collector junction, we get

$$I_C = I_D + I_L = I_I - \frac{I_C}{\beta_F} + I_L$$

which yields

$$I_C = \frac{I_I + I_L}{1 + 1/\beta_F}$$

Thus, for an increased value of load current I_L, the value of I_C is increased and the value of diode current I_D is reduced. That is, the major part of the input current is diverted into the base of the transistor, keeping the transistor at the edge of saturation. For a small value of load current, the value of I_C becomes small, and a large part of the input current is shunted through the diode. The base and diode currents change with the load conditions, while the transistor remains at the edge of saturation. The Schottky barrier diode has no minority carrier charge storage, and the transistor is never fully saturated. Thus, the recovery is very quick.

In the Schottky TTL NAND gate shown in Fig. 15.48, all of the transistors except Q_4 are Schottky clamped transistors. This circuit is similar to the one in Fig. 15.43. The two Schottky diodes from the input terminals to the ground act as clamps, to suppress any ringing that might occur from voltage transients and clamp any negative undershoots at approximately −0.3 V.

The analysis of a Schottky TTL gate is similar to that of a standard TTL gate. When the output transistor Q_3 is on, $V_{BE} = 0.7$ V, the voltage drop across the Schottky diode is clamped to $V_{BC} = 0.3$ V, and

$$V_{CE} = V_{CB} + V_{BE} = -V_{BC} + V_{BE} = -0.3 + 0.7 = 0.4 \text{ V}$$

Thus, the output voltage of a Schottky gate in its output low state is slightly higher than the value of $V_{CE(sat)}$ for standard TTL gates. The output voltage in the output high state is essentially the same as that of the standard TTL gate.

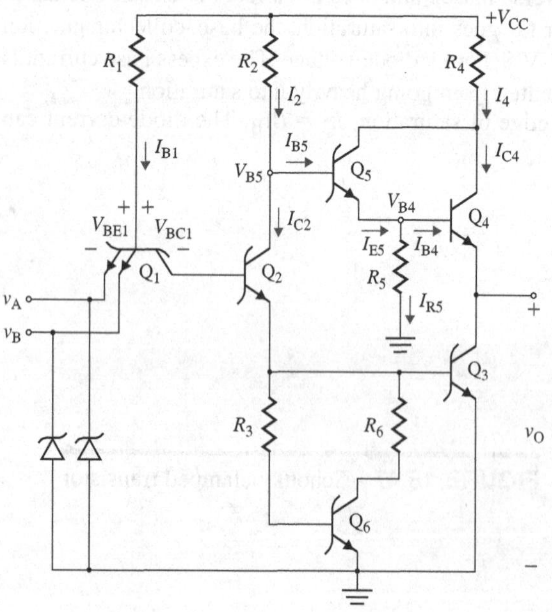

FIGURE 15.48 Schottky TTL NAND gate

In summary, Schottky TTL gates have the following features:

1. Minimal delay time. Because a Schottky clamped transistor operates at a low saturation, the delay time is approximately 2 ns to 5 ns, compared to 10 ns to 15 ns in standard and high-speed TTL gates; that is, the propagation delay time is reduced by a factor of 5 to 10.
2. Sharper voltage transfer characteristic than standard TTL gates
3. Low noise immunity
4. Slightly higher output voltage in the output low state than standard TTL gates, typically 0.4 V

The success of the Schottky gate led to the development of other Schottky gates: low-power Schottky (LS), advanced Schottky (AS), and advanced low-power Schottky (ALS) gates.

KEY POINT OF SECTION 15.11

- The TTL families have gone through many developmental stages, which have resulted in improved switching speed and lower power dissipation. Applications focus on the low-power Schottky (LS) and advanced low-power Schottky (ALS) families because they offer the highest speed.

15.12 Emitter-Coupled Logic OR/NOR Gates

Storage time is the dominant parameter affecting the delay time of transistors that are driven into saturation. Storage time can be minimized by preventing the transistors from operating at saturation. Schottky TTL gates minimize transistor saturation by clamping the base–collector junction at the edge of the saturation condition. This significantly reduces the storage time and delay time, but slight saturation of the Schottky transistors adds to the switching time. The transistors in ECL gates are never driven into saturation, and so the storage time is virtually zero. The first ECL nonsaturated logic gates were introduced in 1962 by Motorola under the family name MECL I. Since then, the MECL gates have progressed through several generations: MECL II, MECL III, MECL 10K, and MECL 10KH.

An ECL uses an *emitter-coupled pair* as a current switch circuit, as shown in Fig. 15.49(a). It consists of two identical transistors Q_1 and Q_2, two matched resistors $R_C = R_{C1} = R_{C2}$, and a current

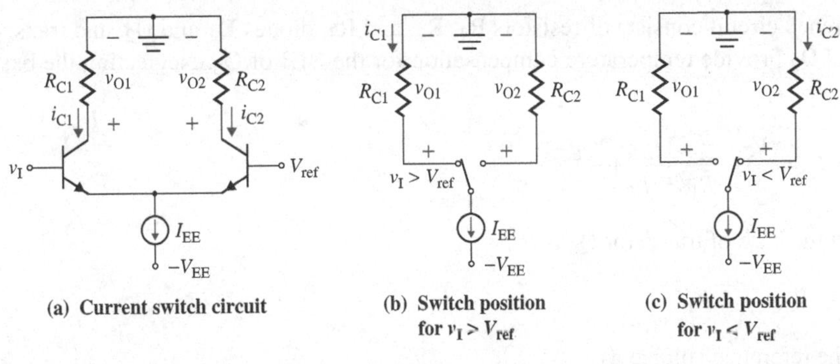

FIGURE 15.49 Current switch circuit for ECL

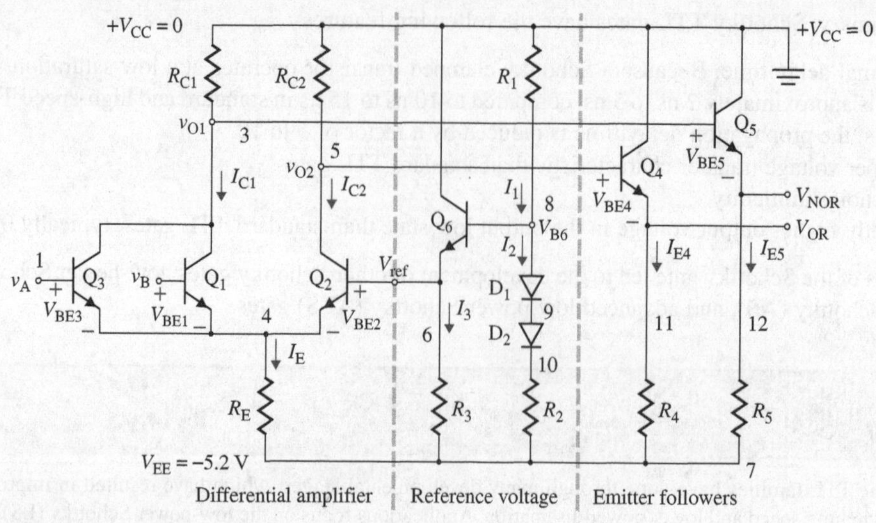

FIGURE 15.50 ECL OR/NOR gate

source I_{EE}. The input voltage v_I, which is applied to the base of Q_1, is compared to the reference voltage V_{ref}, which is applied to the base of Q_2. If v_I is greater than V_{ref} by a few hundred millivolts, the source current I_{EE} flows through Q_1. Thus, the output voltage becomes $v_{O1} = v_{C1} = -R_C i_{CI} = -R_C I_{EE}$, as shown in Fig. 15.49(b). On the other hand, if v_I is less than V_{ref} by a few hundred millivolts, the source current I_{EE} flows through Q_2. The output voltage becomes $v_{O2} = v_{C2} = -R_C i_{C2} = -R_C I_{EE}$, as shown in Fig. 15.49(c). Thus, the input voltage v_I causes the current I_{EE} to flow through either Q_1 or Q_2.

An ECL OR/NOR gate, which is based on a differential pair, is shown in Fig. 15.50. Input transistors Q_1 and Q_3 are connected in parallel. If the differential input voltage $v_d = v_A$ (or v_B) $- V_{ref}$ is more than approximately 100 mV, the output voltage v_{O1} is directly proportional to v_d. Similarly, the output voltage v_{O2} is directly proportional to the differential voltage $-v_d = V_{ref} - v_A$ (or v_B). However, for the transistors to operate as switches, the differential voltage must be greater than approximately 120 mV. Transistors Q_4 with R_4 and Q_5 with R_5 operate as emitter followers. The collector terminals are normally placed at zero voltage because it can be proved analytically that placing the ground near the collectors of the transistors results in less noise sensitivity. For this reason, the supply voltages are generally $V_{CC} = 0$ and $V_{EE} = -5.2$ V.

The reference circuit consists of resistors R_1, R_2, and R_3, diodes D_1 and D_2, and transistor Q_6. The diodes D_1 and D_2 provide temperature compensation for the BEJ of Q_6. Neglecting the base current of Q_6, we get

$$I_1 = I_2 = \frac{V_{CC} - 2V_D - V_{EE}}{R_1 + R_2}$$

The voltage at the base of transistor Q_6 is

$$V_{B6} = V_{CC} - I_1 R_1$$

which gives the reference voltage as

$$V_{ref} = V_{B6} - V_{BE6}$$

The emitter current of Q_6 is given by

$$I_{E6} \approx \frac{V_{ref} - V_{EE}}{R_3}$$

The circuit operation can be divided into two modes.

During mode 1, either input v_A or input v_B is at logic high (V_{OH}); that is, v_A (or v_B) = $V_{CC} - 0.7$ V. The OR logic is at the v_{O2} output, and the NOR logic is at the v_{O1} output. Transistor Q_2 is off. That is, $i_{C2} = 0$, and $v_{O2} = V_{CC}$. The output voltage V_{OR} is given by

$$V_{OR} = V_{O2} - 0.7 \text{ V} = V_{CC} - 0.7 \text{ V}$$

If $v_A = V_{OH} = V_{CC} - 0.7$ V, then Q_1 is on, and we get

$$V_E = v_A - V_{BE1}$$

which gives the emitter current I_E as

$$I_E = \frac{V_E - V_{EE}}{R_E}$$

Assuming that $i_{C1} \approx I_E$, the voltage v_{O1} can be found from

$$v_{O1} = V_{CC} - i_{C1}R_{C1}$$

and the output V_{NOR} becomes

$$V_{NOR} = v_{O1} - V_{BE5}$$

The current I_{E4}, which is the emitter current of Q_4, is

$$I_{E4} = \frac{V_{E4} - V_{EE}}{R_4} = \frac{V_{OR} - V_{EE}}{R_4}$$

and the emitter current of Q_5 is

$$I_{E5} = \frac{V_{E5} - V_{EE}}{R_5} = \frac{V_{NOR} - V_{EE}}{R_5}$$

During mode 2, inputs v_A and v_B are at logic low (V_{OL}). Transistors Q_1 and Q_3 are off; that is, $i_{C1} = 0$, and $v_{O1} = V_{CC}$. The output voltage V_{NOR} is given by

$$V_{NOR} = V_{OH} = v_{O1} - 0.7 \text{ V} = V_{CC} - 0.7 \text{ V}$$

Transistor Q_2 is on, and we get

$$V_E = V_{ref} - V_{BE2}$$

which gives the emitter current I_E as

$$I_E = \frac{V_E - V_{EE}}{R_E}$$

Assuming that $i_{C2} \approx I_E$, the voltage v_{O2} becomes

$$v_{O2} = V_{B4} = V_{CC} - i_{C2}R_{C2}$$

and the output V_{OR} is

$$V_{OR} = v_{O2} - V_{BE4}$$

The current I_{E4}, which is the emitter current of Q_4, is

$$I_{E4} = \frac{V_{E4} - V_{EE}}{R_4} = \frac{V_{OR} - V_{EE}}{R_4}$$

and the emitter current of Q_5 is

$$I_{E5} = \frac{V_{E5} - V_{EE}}{R_5} = \frac{V_{NOR} - V_{EE}}{R_5}$$

Defining logic high as

$$V_{OH} = V_{CC} - 0.7 = -0.7 \text{ V}$$

and substituting the numerical values, we get the logic low as

$$V_{OL} = v_{O1} - V_{BE5} = V_{CC} - i_{C1}R_{C1} - V_{BE5} = V_{NOR} = -1.63 \text{ V}$$

gives the input and output voltages shown in Table 15.7.

In summary, ECL OR/NOR gates have the following features:

1. No transistor saturation and negligible delay time
2. High power dissipation, typically 50 mW to 70 mW (compared to 2 mW to 10 mW for Schottky TTL circuits)
3. Availability of complementary outputs, which eliminates the need to include separate inverters to provide these complementary outputs
4. High fan-out, typically in the range of 50 to 100
5. Low noise margin, since the logic high and logic low output voltages are only approximately −0.7 V and −1.63 V, respectively
6. Sharp voltage transfer characteristic

TABLE 15.7 ECL/OR gate logic function

v_A	v_B	V_{OR}	V_{NOR}
V_{OL} (for 0)	V_{OL} (for 0)	V_{OL} (for 0)	V_{OH} (for 1)
V_{OH} (for 1)	V_{OL} (for 0)	V_{OH} (for 1)	V_{OL} (for 0)
V_{OL} (for 0)	V_{OH} (for 1)	V_{OH} (for 1)	V_{OL} (for 0)
V_{OH} (for 1)	V_{OH} (for 1)	V_{OH} (for 1)	V_{OL} (for 0)

EXAMPLE 15.9

D

Designing an ECL OR/NOR gate

(a) Design the ECL OR/NOR gate of the circuit in Fig. 15.50. It has two inputs. The desired collector currents are $I_{C1} = 3$ mA and $I_{C4} = I_{C5} = 3$ mA. At logic high, $V_{OR} = -0.7$ V, $V_{NOR} = -1.63$ V, and v_A (or v_B) $= V_{OR} = -0.7$ V. Assume $V_{BE} = 0.7$ V, $V_{EE} = -5.2$ V, $V_{BE} = 0.7$ V, and $\beta_F = 100$.

(b) Calculate the maximum fan-out with similar ECL gates if V_{OR} is allowed to fall from $V_{OR(max)} = -0.7$ V to $V_{OR(min)} = -0.75$ V.

(c) Use PSpice/SPICE to check your design by plotting the transfer function. Calculate NM_L and NM_H. Use PSpice/SPICE to find P_{static} if $v_A = v_B = -0.7$ V.

SOLUTION

(a) Assuming v_A (or v_B) $= V_{OR} = -0.7$ V, the steps to complete the design are as follows:

Step 1. Calculate V_{CC}, v_A, and V_E. The value of V_{CC} is

$$V_{CC} = V_{OR} + 0.7 = -0.7 + 0.7 = 0$$

The input voltage for logic high is

$$v_A = V_{OH} = V_{CC} - 0.7 = 0 - 0.7 = -0.7 \text{ V}$$

The emitter voltage is

$$V_E = v_A - V_{BE1} = -0.7 - 0.7 = -1.4 \text{ V}$$

Step 2. Calculate the emitter resistance R_E:

$$R_E = \frac{V_E - V_{EE}}{I_E \simeq I_C} = \frac{-1.4 \text{ V} - (-5.2 \text{ V})}{3 \text{ mA}} = \frac{(-1.4 + 5.2) \text{ V}}{3 \text{ mA}} = 1.27 \text{ k}\Omega$$

Step 3. Calculate the voltage v_{O1}, the collector resistance R_{C1}, R_4, and R_5:

$$v_{O1} = V_{NOR} + V_{BE5} = -1.63 + 0.7 = -0.93 \text{ V}$$

$$R_{C1} = \frac{V_{CC} - v_{O1}}{I_{C1}} = \frac{V_{CC} - v_{O1}}{I_E} = \frac{0 - (-0.93 \text{ V})}{3 \text{ mA}} = \frac{0.93 \text{ V}}{3 \text{ mA}} = 310 \ \Omega$$

$$R_4 = \frac{V_{OR} - V_{EE}}{I_{E4} \simeq I_{C3}} = \frac{-0.7 \text{ V} - (-5.2 \text{ V})}{3 \text{ mA}} = 1.5 \text{ k}\Omega$$

$$R_5 = \frac{V_{NOR} - V_{EE}}{I_{E5} \simeq I_{C3}} = \frac{-1.63 \text{ V} - (-5.2 \text{ V})}{3 \text{ mA}} = 1.19 \text{ k}\Omega$$

Step 4. Since the input voltages v_A and v_B are greater than V_{ref} when the circuit is in the logic high state and less than V_{ref} when it is in the logic low state, set V_{ref} at the midpoint between the logic low and logic high levels and calculate the voltage at the base of Q_6:

$$V_{ref} = \frac{V_{NOR} + V_{OR}}{2} = \frac{-1.63 - 0.7}{2} = -1.165 \text{ V}$$

$$V_{B6} = V_{ref} + V_{BE6} = -1.165 + 0.7 = -0.465 \text{ V}$$

Step 5. Calculate the values of R_1, R_2, and I_1. Using KVL, we can write

$$I_1(R_1 + R_2) = -V_{EE} - 2V_D$$

$$I_1 R_1 = -V_{B6}$$

which gives

$$\frac{I_1(R_1 + R_2)}{I_1 R_1} = \frac{-V_{EE} - 2V_D}{-V_{B6}}$$

or $$1 + \frac{R_2}{R_1} = \frac{-V_{EE} - 2V_D}{-V_{B6}} = \frac{-(-5.2) - 2 \times 0.7}{-(-0.465)} = 8.172$$

Choose $R_1 = 300\ \Omega$. Then

$$R_2 = (8.172 - 1)R_1 = (8.172 - 1) \times 300 = 2.15 \text{ k}\Omega$$

Calculate the value of I_1 as

$$I_1 = I_2 = -\frac{V_{B6}}{R_1} = -\frac{-0.465}{300} = 1.55 \text{ mA}$$

Step 6. Calculate the value of R_3. For good temperature compensation, the current through the emitter of Q_6 should be the same as the current through diodes D_1 and D_2; that is,

$$I_3 = I_2 = \frac{V_{ref} - V_{BE}}{R_3} = \frac{-1.165 - (-5.2)}{R_3} = 1.55 \text{ mA}$$

or $R_3 = 2.6\ \text{k}\Omega$.

Step 7. Choose R_{C2} to be slightly more than R_{C1} (approximately 3% more for resistors of 1% tolerance)—that is, 2% more than the tolerance value. Thus,

$$R_{C2} = 1.03 \times R_{C1} = 1.03 \times 310 = 319.3\ \Omega$$

Let $R_{C2} = 320\ \Omega$.

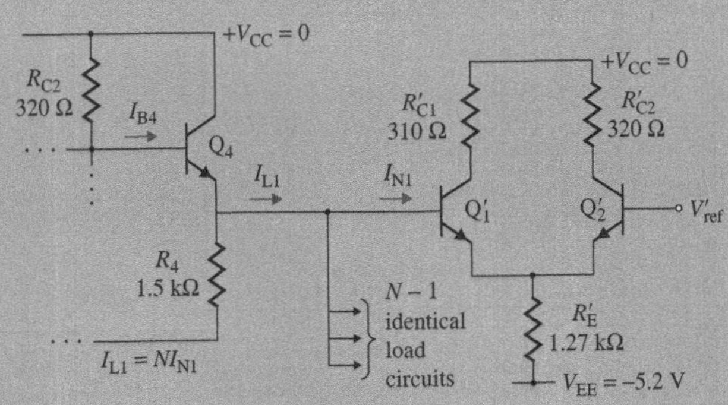

FIGURE 15.51 ECL gate driver and ECL load gate circuits

(b) The fan-out can be determined from Fig. 15.51, which shows the emitter-follower output stage of an ECL circuit driving a difference amplifier input stage of an ECL load. This circuit is shown for V_{OR} at the logic high level. The load transistor Q_1 is on, and the load emitter current is given by

$$I_E = \frac{V_{OR(max)} - V_{BE} - V_{EE}}{R_E}$$

$$= \frac{-0.7\text{ V} - 0.7\text{ V} - (-5.2\text{ V})}{1.27\text{ k}\Omega} = 2.99\text{ mA}$$

The input base current for each fan-out is given by

$$I_{N1} = \frac{I_E}{1+\beta_F} = \frac{2.99\text{ mA}}{1+100} = 29.6\ \mu\text{A}$$

Thus, the total load current is

$$I_{L1} = NI_{N1} = N \times 29.6\ \mu\text{A}$$

The emitter current of Q_4 is

$$I_{E4} = \frac{V_{OR(min)} - V_{EE}}{R_4} = \frac{-0.75\text{ V} - (-5.2\text{ V})}{1.5\text{ k}\Omega} = 2.97\text{ mA}$$

The base current I_{B4} required to supply the load current I_L and the current I_{E4} is

$$I_{B4} = \frac{I_{E4} + I_L}{1+\beta_F} = \frac{V_{CC} - V_{BE4} - V_{OR(min)}}{R_{C2}}$$

which gives

$$\frac{2.97\text{ mA} + (N \times 29.6\ \mu\text{A})}{1+100} = \frac{0 - 0.7 - (-0.75)}{320}$$

or $N = 432$.

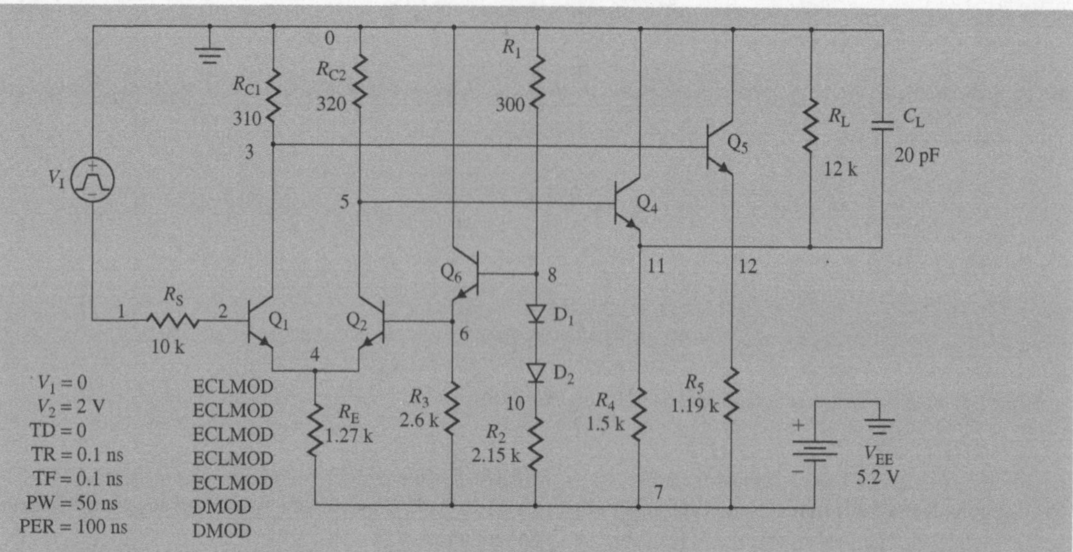

FIGURE 15.52 PSpice schematic for Example 15.9

(c) The PSpice schematic is shown in Fig. 15.52. The PSpice plots shown in Fig. 15.53 give $V_{OH} = -0.81$ V (expected value is -0.7 V), $V_{IL} = 0.95$ V at $v_O = -0.855$ V, $V_{OL} = -1.59$ V at $V_I = 2$ V, and $V_{IH} = 1.33$ V at $V_O = -1.55$ V. Note that V(11) and V(12) are the voltages at the OR and NOR terminals, respectively.

$$\mathrm{NM_L} = V_{IL} - |V_{OL}| = 0.95 - 1.59 = -0.64 \text{ V}$$

$$\mathrm{NM_H} = |V_{OH}| - V_{IH} = 0.81 - 1.33 = -0.52 \text{ V}$$

$t_{pHL} = 1.05$ ns, $t_{pLH} = 1.22$ ns, and $t_{pd} = 1.13$ ns. The PSpice output file gives $P_{static} = 62$ mW.

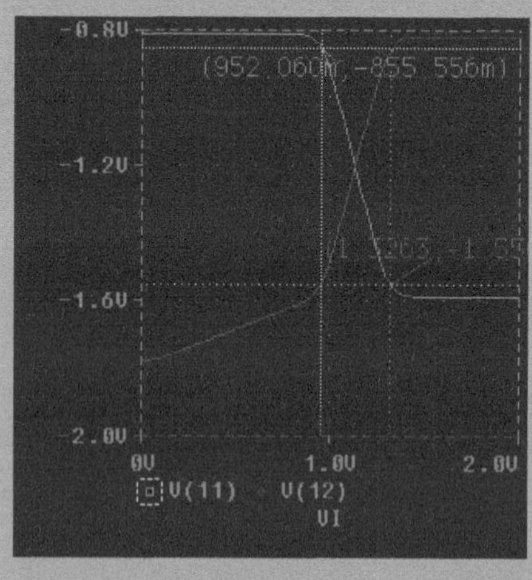

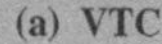
(a) VTC

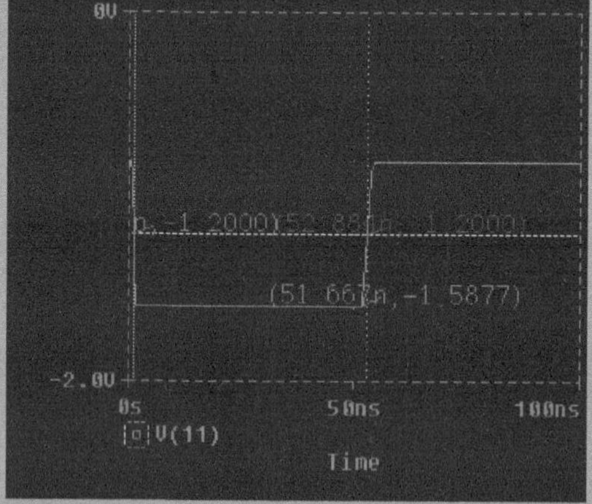

(b) Transient response

FIGURE 15.53 PSpice plots for ECL characteristics (using transistor circuit model) for Example 15.9

KEY POINT OF SECTION 15.12

- The transistors in ECL gates are never driven into saturation, and thus the storage time is virtually zero. Since their introduction in 1962 by Motorola, the MECL gates have progressed through several generations: MECL I, MECL II, MECL III, MECL 10K, and MECL 10KH.

15.13 BiCMOS Inverters

The CMOS inverter is a low-power, compact inverter that exhibits a high input resistance. Its bipolar circuits, however, have low propagation delays. A BiCMOS combines a CMOS with a BJT output buffer and thus incorporates the best features of both technologies. A BiCMOS inverter is shown in Fig. 15.54. The BJT totem-pole output stage provides the high current capability to charge the load capacitance C_L rapidly while maintaining the low-power advantage of the CMOS. In practice, R_1 and R_2 are polysilicon resistors or MOS resistors. To examine the charging and discharging of C_L, we will divide the operation into two modes.

During mode 1, the input voltage v_I is low (at V_{OL}). The NMOS M_N is off. Q_1 is also off because its base is effectively grounded through R_1. The PMOS M_P is on, with $v_{GSP} = -V_{DD}$, and it provides the base current to Q_2. Thus, the load capacitance C_L is charged up through Q_2 (in the active region) to approximately $V_{DD} - V_{BE2}$. At $v_O = V_{DD} - V_{BE2}$, Q_2 is cut off, and C_L then continues to charge toward V_{DD} through M_P and R_2. Thus, V_{BE2} decreases to approximately 0, and $v_O \approx V_{OH} = V_{DD}$. The base of Q_2 also discharges through R_2.

During mode 2, the input voltage v_I goes high (to V_{OH}). Both M_P and Q_2 are off. M_N is turned on. Capacitor C_L discharges through R_2 and M_N. As a result of the current flow through R_1, Q_1 is turned on, and the collector current of Q_1 causes C_L to discharge rapidly until V_{BE1} drops below cut-in. After that, Q_1 turns off, and C_L continues to discharge through R_2, M_N, and R_1 to approximately zero. That is, $V_{OL} \approx 0$. The base of Q_1 also discharges through R_1.

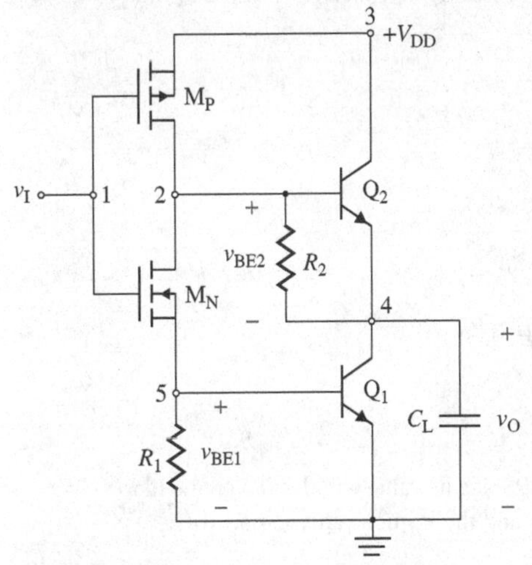

FIGURE 15.54 BiCMOS inverter

15.13.1 Propagation Delay

The propagation delay time of a BiCMOS is reduced because of the increased charging and discharging currents available from the BJTs. When the input switches from logic high to logic low, M_P supplies the charging current I_B for Q_2. Because of Q_1, the minimum value of v_I is V_{BE}. Thus, $V_{SGP} = V_{DD} - V_{BE}$, and I_B is given by

$$I_B = K_p(V_{DD} - V_{BE} - |V_{tP}|)^2$$

The charging current supplied by Q_2 in the active mode is

$$I_E = \beta_F I_B$$

Thus, the time t_{pLH} required to charge C_L from V_{OL} to $(V_{OL} + V_{OH})/2 \approx V_{DD}/2$ (neglecting the junction capacitances of the transistors) is given by

$$t_{pLH} \approx \frac{(V_{OL} + V_{OH})C_L}{2I_E} = \frac{V_{DD}C_L}{2\beta_F K_p(V_{DD} - V_{BE} - |V_{tP}|)^2} \tag{15.96}$$

If the input switches from logic low to logic high, M_N will provide the discharging path of C_L. Thus, the time t_{pHL} to discharge C_L from $V_{OH} = (V_{OL} + V_{OH})/2 \approx V_{DD}/2$ to V_{OL} is given by

$$t_{pHL} = \frac{(V_{OL} + V_{OH})C_L}{2I_E} = \frac{V_{DD}C_L}{2\beta_F K_n(V_{DD} - V_{BE} - V_{tN})^2} \tag{15.97}$$

When the BJTs are identical and the threshold voltages of the MOSFETs have the same magnitude, $t_{pLH} = t_{pHL}$ and $t_{pd} = t_{pLH}$.

EXAMPLE 15.10

Finding the propagation delays of a BiCMOS inverter The BiCMOS inverter shown in Fig. 15.54 has $V_{DD} = 5$ V, $V_{BE} = 0.7$ V, $R_1 = 1$ kΩ, $R_2 = 4$ kΩ, $C_L = 0.5$ pF, and $\beta_F = 100$. The threshold voltages are $V_{tP} = -1$ V, $V_{tN} = 1$ V, and $K_p = 20$ μA/V^2. Assume $V_{OL} = 0$ and $V_{OH} = 5$ V.

(a) Calculate t_{pLH}, t_{pHL}, and t_{pd}.

(b) Use PSpice/SPICE to plot the VTC and transient response.

SOLUTION

(a) Using Eq. (15.96), we have

$$t_{pLH} = \frac{5\text{ V} \times 0.5\text{ pF}}{2 \times 100 \times 20\ \mu\text{A/V}^2 \times (5\text{ V} - 0.7\text{ V} - 1\text{ V})^2} = 0.06\text{ ns}$$

$$t_{pd} = t_{pHL} = t_{pLH} = 0.06\text{ ns}$$

Although t_{pLH} is independent of the L/W ratio of the MOSFETs, in PSpice its value will depend on the ratio L/W. This causes the difference between the calculated value in part (a) and the PSpice value in part (b).

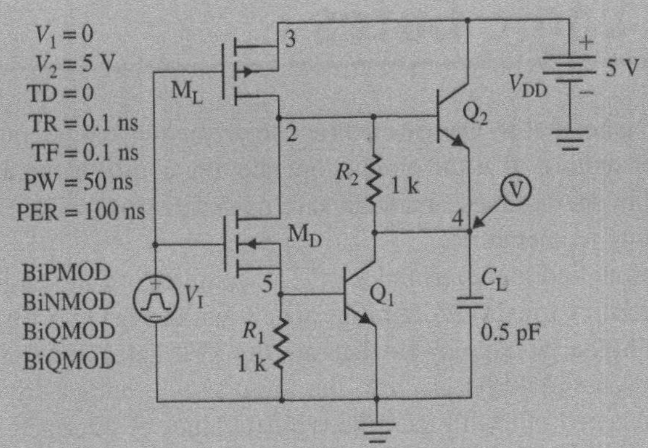

FIGURE 15.55 PSpice schematic for Example 15.10

(b) The PSpice schematic is shown in Fig. 15.55. The PSpice plot of the VTC, shown in Fig. 15.56(a), gives $V_{IL} = 2.125$ V, $V_{IH} = 2.875$ V, $V_{OL} = 0$ at $V_{OH} = 5$ V, and $V_{I(tran1)} = 2.5$ V (expected value is 2.5 V). The transient response is shown in Fig. 15.56(b), which gives $t_{pHL} = 2.93$ ns, $t_{pLH} = 0.65$ ns, and $t_{pd} = 1.79$ ns, which is lower than $t_{pd} = 3.91$ ns for the CMOS in Example 15.5.

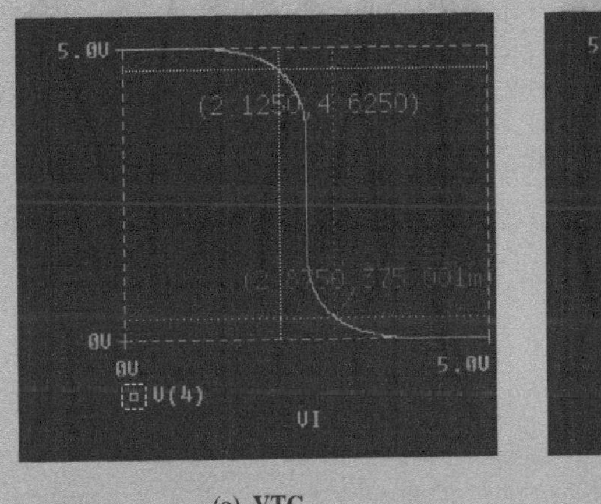

(a) VTC

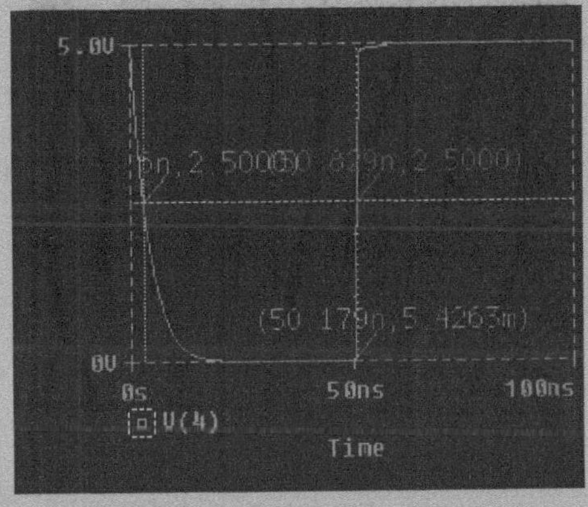

(b) Transient response

FIGURE 15.56 PSpice plots for BiCMOS inverter (using transistor circuit model) for Example 15.10

NOTE: Figure 15.56 illustrates the very sharp rise and fall times that occur while the BJTs are conducting. The output signal is nearly V_{DD}, and the noise margins for this circuit are high, both being more than 2 V.

KEY POINT OF SECTION 15.13

- A BiCMOS combines a CMOS with a BJT output buffer and thus incorporates the best features of both technologies.

15.14 Interfacing of Logic Gates

In practice, the load of a logic gate is another logic gate; that is, one gate acts as the driver and another one as the load. The interfacing of different logic gates requires that the circuits operate at a common supply voltage and have logic-level compatibility. Also, the devices must maintain safe power dissipation levels and good noise immunity over the required operating temperatures.

The voltage characteristics required at the output and input terminals of TTL families are shown in Fig. 15.57(a) for $V_{CC} = 5$ V; the voltage characteristics for CMOS families are shown in Fig. 15.57(b). The CMOS gates are designed to switch states at higher voltage levels than are the TTL gates. In interfacing one type of logic gate with another, attention should be given to the logic swing, output drive capability, DC input current, noise immunity, and speed of each type. The typical values of interfacing parameters such as V_{OL}, V_{OH}, I_{OL}, I_{OH}, V_{IL}, V_{IH}, I_{IL}, and I_{IH} are shown in Table 15.3 for CMOS and TTL devices.

15.14.1 TTL Driving CMOS

When a TTL device is used to drive a CMOS, the output drive capability of the driving device and the switching levels and input currents of the driven devices are important considerations. The input currents for a CMOS are very small in both the 1 and the 0 state, typically $I_{IL} = I_{IH} = 10$ pA, and the thresholds for a CMOS are typically $V_{IL(max)} = 1.5$ V and $V_{IH(min)} = 3.5$ V. Thus, to obtain some noise immunity, the output of the TTL driver must be no more than $V_{OL} = 1.5$ V at logic 0 and no less than $V_{OH} = 3.5$ V at logic 1. Depending on whether the device is in the 0 or the 1 state, the driver will be sinking or sourcing current.

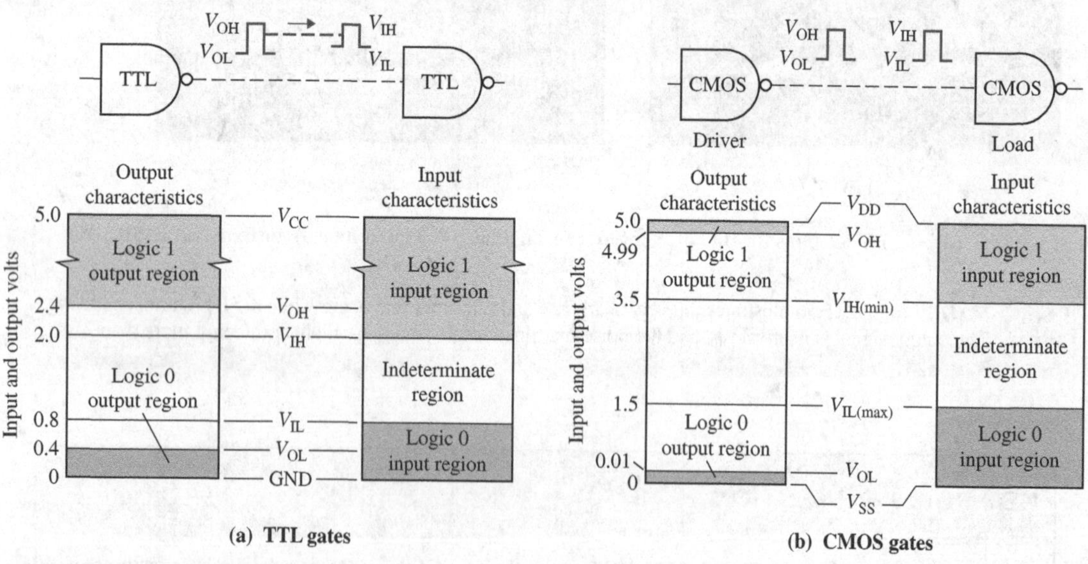

FIGURE 15.57 Input and output characteristics of logic gates

Current Sinking

When the output of the TTL device is in the low state (at V_{OL}), the collector of Q_1, shown in Fig. 15.58(a), is essentially at ground potential. Transistor Q_1 must go into saturation to ensure a stable 0 level, typically 0.3 V. To attain this voltage level, there should be a high impedance path from the output to the power supply V_{CC}. Since the CMOS has extremely high input impedances, typically 10^{11} Ω, it will not contribute any significant current through Q_1. Thus, the current-sinking capability is not a problem. The voltage level is not a problem either, as CMOS devices have high noise immunity, usually greater than 1 V.

Current Sourcing

When Q_1, shown in Fig. 15.58(b), is off (at logic 1 output V_{OH}), there is a current flow from the V_{CC} terminal of the driver, through a pull-up resistor R_P, into the input stages of the load; that is, the driving gate acts as the current source for the load. The total load current (NI_{IH}) should not reduce the output level below V_{IH} required by the CMOS load. A driver with a built-in pull-up resistor R_P, as shown in Fig. 15.58(b), presents no problem in the interface with the CMOS. However, a driver with an open collector requires an external pull-up resistor R_P, as shown in Fig. 15.58(c).

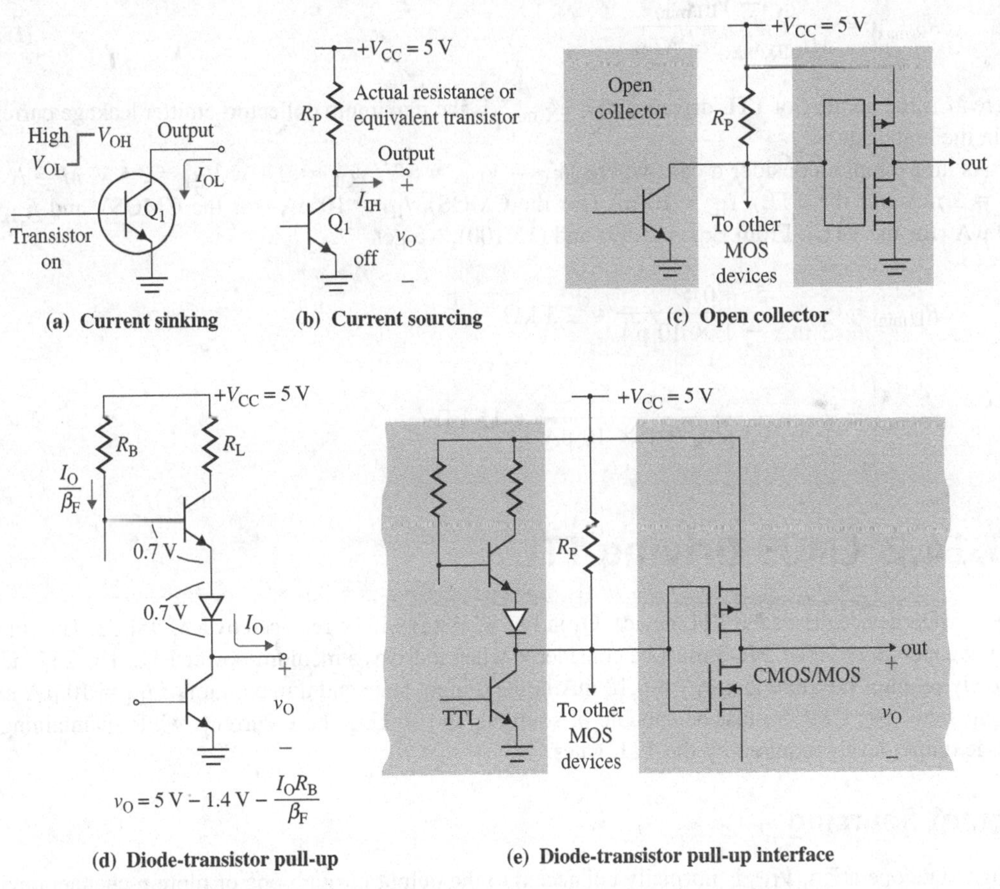

FIGURE 15.58 TTL devices driving CMOS devices

Consider the diode-transistor arrangement shown in Fig. 15.58(d), which will reduce the output voltage v_O given by

$$v_O = V_{CC} - V_{BE} - V_D - \frac{i_O R_B}{\beta_F} \tag{15.98}$$

Depending on the load current i_O, the minimum output level of $V_{OH} = 2.4$ V may not ensure an acceptable state of $V_{IH(min)} = 3.5$ V for the CMOS device, and this could cause a problem in the logic 1 state. However, the minimum level of $V_{OH} = 2.4$ V for a TTL gate is often specified at a given load current ($I_{OH} = 200$ μA). Since the CMOS draws very little current, an output level of $V_{OH} = 3.5$ V will be available, but there is no noise immunity. Therefore, a pull-up resistor R_P should be added from the output terminal of the driver, as shown in Fig. 15.58(e). The minimum value of R_P can be found from

$$R_{P(min)} = \frac{V_{DD} - V_{OL(max)}}{I_{OL} - NI_{IL}} \tag{15.99}$$

where N is the number of CMOS load gates. The maximum value of the external pull-up resistor can be found from

$$R_{P(max)} = \frac{V_{CC} - V_{IH(min)}}{MI_{CEX(max)} - NI_{IH}} \tag{15.100}$$

where M is the number of TTL drivers and $I_{CEX(max)}$ is the maximum collector–emitter leakage current of Q_1 in the high state.

As an example, consider a case where $V_{CC} = V_{DD} = 5$ V, $V_{OL} = 0.4$ V, $V_{IH} = 3.5$ V, $M = N = 1$, $I_{OL} = 2$ mA (for the TTL), $I_{IL} = 10$ pA (for the CMOS), $I_{IH} = 10$ pA (for the CMOS), and $I_{CEXI} = 100$ μA (for the TTL). From Eqs. (15.99) and (15.100), we get

$$R_{P(min)} = \frac{5 - 0.4}{2\text{ mA} - 1 \times 10\text{ pA}} \approx 2.3\text{ k}\Omega$$

$$R_{P(max)} = \frac{5 - 3.5}{1 \times 100\ \mu\text{A} - 1 \times 10\text{ pA}} \approx 15\text{ k}\Omega$$

15.14.2 CMOS Driving TTL

If the CMOS device drives the TTL device, Q_1 in Fig. 15.58(a) will be replaced by a MOSFET. The current-sinking capability of the CMOS must be considered when it drives a medium-power TTL. The TTL device typically requires no more than $I_{IL} = 0.18$ mA in the 0 input state and a maximum of $I_{IH} = 10$ μA in the 1 input state. The CMOS must be capable of sourcing and sinking these currents while maintaining the voltage output levels required by the TTL gates.

Current Sourcing

In high-state operation, V_{DD} is normally connected to the output through one or more p-channel devices, which must be able to source the total load leakage current of TTL load stages. The I_{OH} of the CMOS stage must match the leakage currents NI_{IH} (for the logic 1 state) with a fan-out of N.

Current Sinking

When the output of the CMOS is in the low state (at V_{OL}), an n-channel device is on and the output is approximately at ground potential. The CMOS device sinks the current flowing from the TTL input load stage. The I_{OL} of the CMOS stage must match the input currents NI_{IL}(for the logic 0 state) with a fan-out of N.

- The load of a logic gate is often another logic gate. It is essential that the logic gates be compatible in terms of supply voltage, logic levels, and noise immunity.

15.15 Comparison of Logic Gates

Broadly speaking there are three logic families: TTL, ECL, and CMOS. Each family has its advantages and disadvantages. The delay-power product (DP), which is the product of power dissipation P_D and propagation delay t_{pd}, is the key parameter for high-speed switching. Figure 15.59 compares delay time with power dissipation for various logic families. Note that the first generation of the CMOS family is labeled

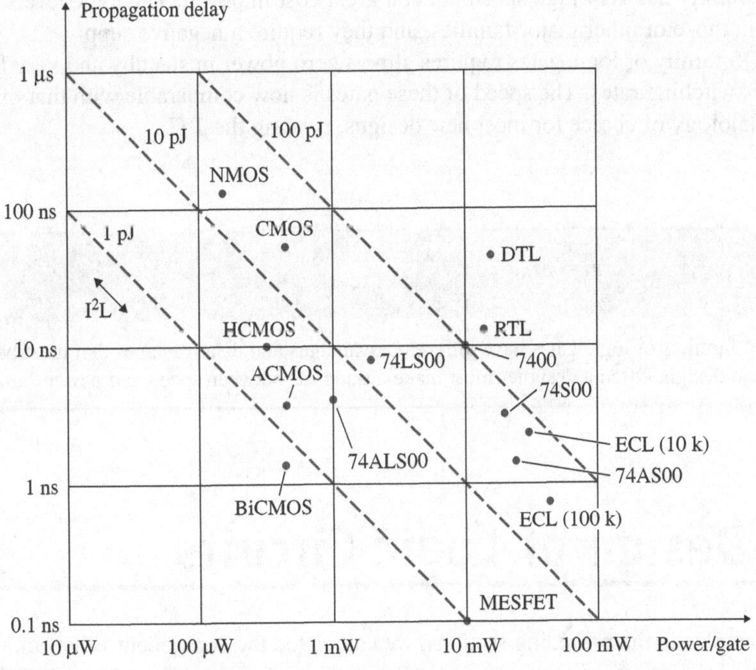

FIGURE 15.59 Comparison of delay time and power dissipation for logic families

TABLE 15.8 Comparison of logic families

Parameter	TTL 74LSxx	ECL 10K	CMOS 400B
V_{OH}, in V	3.4	−0.9	5
V_{OL}, in V	0.25	−1.7	0
NM_H, in V	1.4	0.36	2.25
NM_L, in V	0.6	0.33	2.25
P_D/gate	2 mW	24 mW	1.5 μW × f_{clk}
t_{pd}, in ns	9.5 at 15 pF	2	30 + 1.7/pF

CMOS; HCMOS refers to high-speed CMOS and ACMOS to advanced-type CMOS. The diode-transistor logic (DTL) and resistor-transistor logic (RTL) gates were developed earlier and have the highest delay-power product. Integrated-injection logic (I^2L) gates have low power but are relatively slower. The MESGET (gallium arsenide) gates are extremely fast because of the high mobility of electrons in GaAs—typically five to six times higher than that of electrons in *n*-type silicon. Gate delays as low as 20 ps to 100 ps have been reported, but this speed must be weighed against the power dissipation, which is typically 10 mW.

The typical values of various parameters for the TTL, ECL, and CMOS families are shown in Table 15.8. We can draw the following conclusions from these data:

- The TTL (LS) family of logic gates has high speed and low power dissipation and is compatible with many applications. In addition, the technology is mature and hence low in cost.
- The ECL family has very high speed but at a great cost in power. The logic levels of these gates differ from those of other major families, and they require a negative supply.
- The CMOS family of logic gates requires almost zero power in standby and very low power at moderate switching rates. The speed of these gates is now comparable with that of the TTL. This is the technology of choice for most new designs, rivaling the TTL.

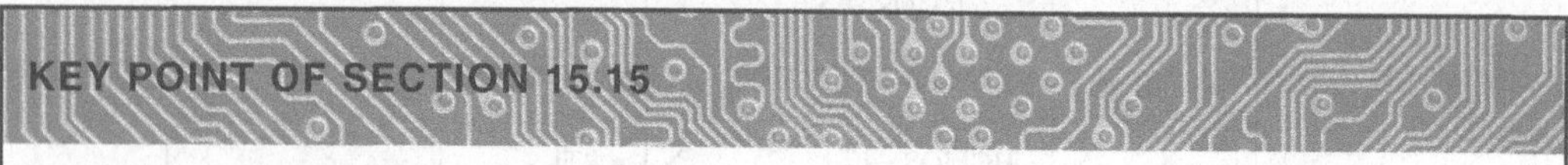

- Different families of logic gates have different advantages and disadvantages that must be taken into account in design. Often a designer must make a trade-off between speed and power dissipation.

15.16 Design of Logic Circuits

In the design examples in the preceding sections, we calculated the component values of MOS and bipolar inverters and found their performance parameters, such as V_{OL}, V_{OH}, t_{pd}, NM_L, and NM_H. Since bipolar and MOS gates are available in IC circuits, designs are seldom done using discrete devices. However, these examples provided an idea of the inside operation of the gates. Circuit design using gates usually involves applications of logic gates, and the following points should be kept in mind:

- Logic gates are highly nonlinear, and their output is considered as either high or low. The width of the undefined input voltage range should be kept to a minimum so that the noise margin becomes as large as possible.
- The output voltage levels of a gate must be compatible with the input voltage levels of the same or similar gates; that is, a design must address interfacing issues.
- The output of one gate must be capable of driving the inputs of more than one gate; that is, fan-out of a gate should be as large as possible for a circuit consisting of multiple gates.
- The logic gate should consume only as much power as is needed to operate at the required speed. Also, the DC supply must meet the design requirement.

EXAMPLE 15.11

D

Designing a clock circuit

(a) Design a clock circuit to produce a pulse output at a frequency of $f_{clk} = 100$ kHz. The DC supply voltage is 5 V.

(b) Use PSpice/SPICE to check your design by plotting the output voltage.

SOLUTION

(a) Given that $T = 1/f_{clk} = 1/100\text{ kHz} = 10\ \mu\text{s}$, the design steps are carried out as follows:

Step 1. Choose the circuit topology. Figure 15.60(a) shows a circuit that uses inverters; NOR gates could be used instead.

Step 2. Choose the gate types. Although use of CMOS inverters would simplify the analysis, we will use bipolar inverters of type 7404. We will derive a general expression for the clock frequency that can be applied to CMOS inverters also. From the data sheet, the inverter parameters are $V_{CC} = 5$ V, $V_{OH} = 3.4$ V, $V_{OL} = 0.5$ V, $V_{IH} = 2$ V, $V_{IL} = 0.8$ V, $I_{IL} = 0.1$ mA, and $I_{IH} = 20\ \mu$A.

Step 3. Analyze the circuit. Suppose that, at $t = 0$, v_{O1} (that is, the output of inverter U_1) is high at V_{OH}, v_O is low at V_{OL}, and v_I (the input of inverter U_1) is at $V_{IH} - V_{OH}$ (from the previous cycle). Capacitor C will charge exponentially, with a time constant of $\tau = RC$. The equivalent circuit during charging is shown in Fig. 15.60(b). The input voltage v_I will have a DC component and an exponentially decaying component. Thus,

$$v_I(t) - V_{OH} = RI_{IL} + [v_I(0^+) - v_O(0^+)]e^{-t/\tau} = RI_{IL} + (V_{IH} - 2V_{OH})e^{-t/\tau} \qquad (\text{for } 0 \le t \le T_2)$$

At $t = T_2$, v_I rises to V_{IL}, and we can find T_2 from the condition $v_I(t = T_2) = V_{IL}$:

$$T_2 = RC \ln\left(\frac{V_{IH} - 2V_{OH}}{V_{IL} - V_{OH} - RI_{IL}}\right) \tag{15.101}$$

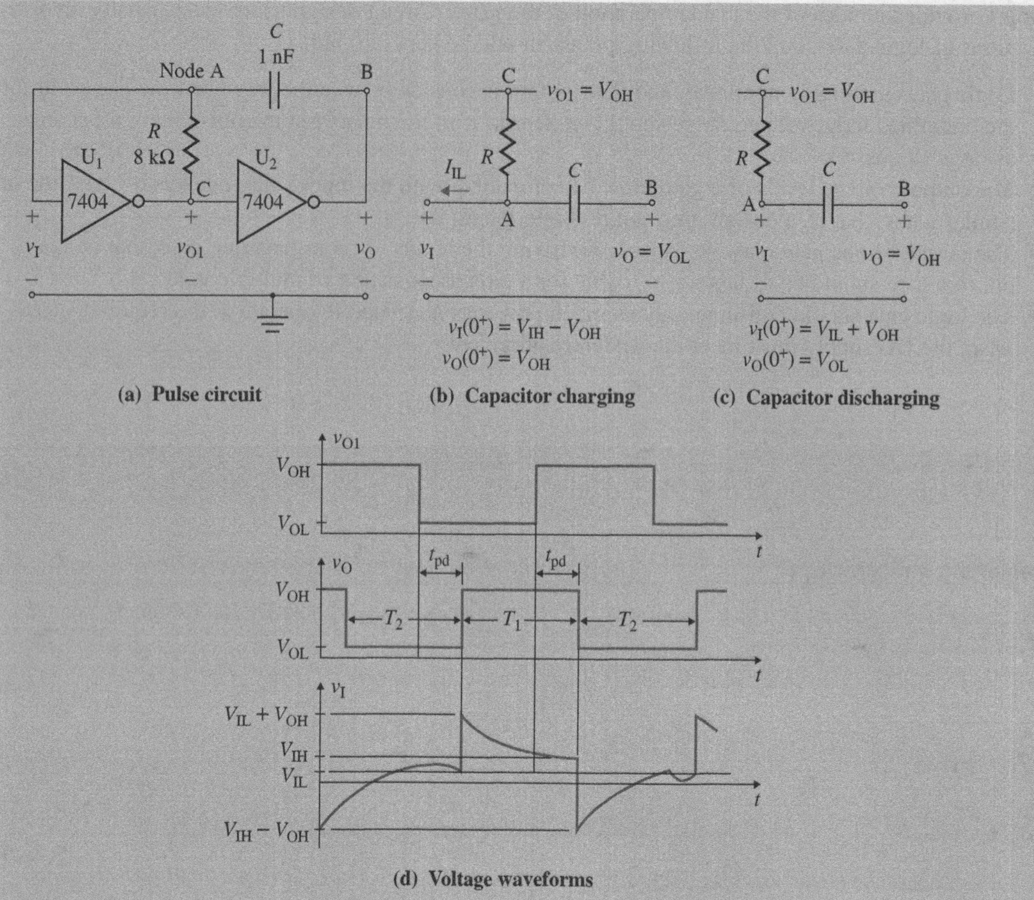

(a) Pulse circuit (b) Capacitor charging (c) Capacitor discharging

(d) Voltage waveforms

FIGURE 15.60 Pulse circuit using inverters

where I_{IL} is the input current in the logic low state. When v_I reaches V_{IL} at $t = T_2$, v_{O1} goes low to V_{OL} and v_O rises to V_{OH} after a delay time of t_{pd}. The capacitor feeds this change in v_O back to the input side. As a result, v_I jumps from V_{IL} to $V_{IL} + V_{OH}$.

Now, with $v_I = V_{IL} + V_{OH}$, $v_{O1} = V_{OL}$, and $v_O = V_{OH}$, the capacitor C will discharge exponentially, with a time constant of $\tau = RC$. The equivalent circuit during discharging is shown in Fig. 15.60(c). The input voltage v_I will have a DC component of V_{OL} and a decaying component. Redefining the time origin $t = 0$ (at $t = T_2$) gives the input voltage v_I:

$$v_I(t) = V_{OL} + [v_I(0^+) - v_O(0^+)]e^{-t/\tau} = V_{OL} + (V_{OH} + V_{IL} - V_{OL})e^{-t/\tau} \quad \text{(for } 0 \le t \le T_1\text{)}$$

At $t = T_1$, v_I falls to V_{IH}, and we can find T_1 from the condition $v_I(t = T_1) = V_{IH}$:

$$T_1 = RC \ln\left(\frac{V_{OH} + V_{IL} - V_{OL}}{V_{IH} - V_{OL}}\right) \tag{15.102}$$

When v_I falls to V_{IH} at $t = T_1$, v_{O1} becomes high at V_{OH} and v_O falls to V_{OL} after a delay time of t_{pd}. Because the capacitor feeds this change in v_O back to the input side, v_I falls from V_{IH} to $V_{IH} - V_{OH}$, and the cycle is repeated. The voltage waveforms are shown in Fig. 15.60(d). The period T of the output voltage is given by

$$T = T_1 + T_2 = RC\left(\ln\frac{V_{OH} + V_{IL} - V_{OL}}{V_{IH} - V_{OL}} + \ln\frac{V_{IH} - 2V_{OH}}{V_{IL} - V_{OH} - RI_{IL}}\right) \tag{15.103}$$

NOTE: For a CMOS gate, $V_{OL} \approx 0$, $V_{OH} = V_{DD}$, $I_{IL} \equiv 0$, and $V_{IL} = V_{IH} = V_t$ (threshold voltages of the NMOS). Thus, Eq. (15.103) becomes

$$T = T_1 + T_2 = RC\left(\ln\frac{V_{DD} + V_{IL}}{V_{IH}} + \ln\frac{V_{IH} - 2V_{OH}}{V_{IL} - V_{OH}}\right) \tag{15.104}$$

Step 4. Find the values of R and C. Using Eq. (15.103), we get

$$10\ \mu\text{A} = RC\left(\ln\frac{3.4 + 0.8 - 0.5}{2 - 0.5} + \ln\frac{2 - 2 \times 3.4}{0.8 - 3.4 - R \times 0.1\ \text{mA}}\right)$$

Choose a suitable value of C: Let $C = 1$ nF. Solving for R by iteration, we get $R \approx 8$ kΩ. Then $\tau = RC = 8$ μs, $T_1 = 7.22$ μs, and $T_2 = 2.76$ μs.

(b) The PSpice simulation is shown in Fig. 15.61. The PSpice plots shown in Fig. 15.62 give $T = 7.53$ μs, $T_1 = 6.41$ μs, and $T_2 = 1.12$ μs. Thus, a higher value of R (say, $R = 10$ kΩ) will be needed.

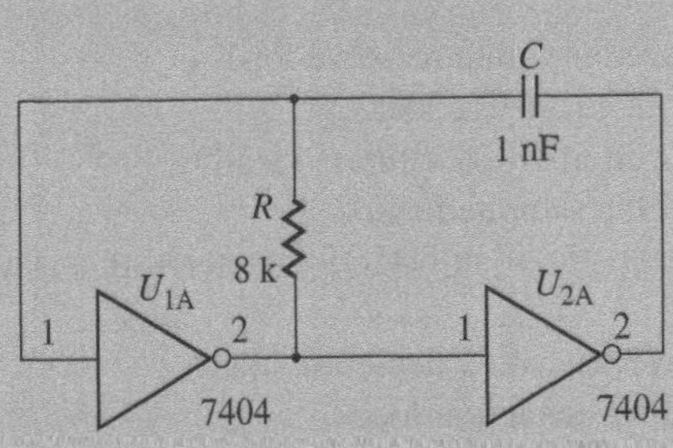

FIGURE 15.61 PSpice schematic for Example 15.11

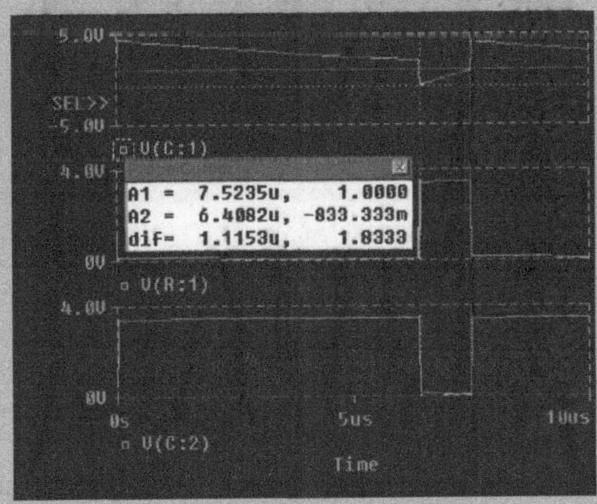

FIGURE 15.62 PSpice plots for clock circuit: $v_{O1} \equiv$ V(R:1) and $v_O \equiv$ V(C:2) for Example 15.11

Summary

Electronic circuits are used to perform various logic functions, such as NOT, AND, OR, NOR, and NAND. The transistors are operated as on and off switches. This chapter analyzed the characteristics of NMOS and CMOS inverters, which are the basis of digital circuits and have applications in NAND and NOR gates. The use of a depletion-load transistor allows the logic level to be the same as the supply voltage and also gives the shortest NMOS switching times. NMOS and CMOS transmission gates are commonly used in steering logic function. The power consumption of CMOS gates is very small, essentially zero. In general, CMOS gates have many advantages over NMOS gates.

DTL gates are simple, but their switching speed is low because they include saturated transistors. TTL gates are better than DTL gates in terms of switching speed, component density, and fan-out. I^2L gates have higher component density, higher fan-out, and lower power dissipation than TTL gates. Schottky TTL gates minimize transistor saturation and have higher switching speeds than standard TTL gates. In ECL gates, the transistors are operated in the active region, and saturation is completely avoided, thereby reducing propagation delay time. However, ECL gates have more power dissipation than TTL and I^2L gates. Each gate has its advantages and limitations, and the applications engineer has to decide which types to use in a specific application.

References

1. D. A. Hodges and H. G. Jackson, *Analysis and Design of Digital Integrated Circuits*. New York: McGraw-Hill, 1988.
2. G. M. Glasford, *Digital Electronic Circuits*. Englewood Cliffs, NJ: Prentice Hall, 1988.
3. J. E. Ayers, *Digital Integrated Circuits: Analysis and Design*. Boca Raton, FL: CRC Press, 2004.
4. F. Maloberti, *Analog Design for CMOS VLSI Systems*. Boston: Kluwer Academic, 2001.
5. J. M. Rabaey, A. Chandrakasan, and B. Nikolic, *Digital Integrated Circuits: A Design Perspective*. Upper Saddle River, NJ: Pearson Education, 2003.
6. B. Razavi, *Design of Analog CMOS Integrated Circuits*. New York: McGraw-Hill, 2001.
7. K. Gopalan, *Introduction to Digital Microelectronic Circuits*. Chicago: R. Irwin, 1996.

Review Questions

1. What are the advantages and disadvantages of an NMOS inverter with a resistive load?
2. What are the intervals of operation of an NMOS inverter with a resistive load?
3. What are the advantages and disadvantages of an NMOS inverter with a saturated load?
4. What are the intervals of operation of an NMOS inverter with a saturated load?
5. What is the effect of the ratio K_D/K_L on the transfer characteristic of an NMOS inverter with a saturated load?
6. What are the advantages and disadvantages of an NMOS inverter with a depletion load?
7. What are the intervals of operation of an NMOS inverter with a depletion load?
8. What is the effect of the ratio K_D/K_L on the transfer characteristic of an NMOS inverter with a depletion load?

9. What is the function of an NMOS transmission gate?
10. What are the advantages and disadvantages of a CMOS inverter?
11. What are the intervals of operation of a CMOS inverter?
12. What is the function of a CMOS transmission gate?
13. What are the differences between CMOS and NMOS transmission gates?
14. What are the advantages and disadvantages of CMOS and NMOS gates?
15. What is the saturation of a transistor?
16. What is the overdrive factor of a transistor?
17. What is the forced β_F of a transistor?
18. What is delay time?
19. What is the saturating charge of a transistor?
20. What is the storage time of a transistor?
21. What is the fan-out of a gate?
22. What is the noise margin of a gate?
23. What is the pull-down resistance of a gate?
24. What is a multiemitter transistor?
25. What are the advantages and limitations of a TTL gate?
26. What is a totem-pole output stage?
27. What is the active-biased recovery circuit?
28. What are the advantages and limitations of high-speed TTL gates?
29. What is a Schottky clamped transistor?
30. What are the advantages and limitations of Schottky TTL gates?
31. What are the advantages and limitations of ECL gates?
32. What is the purpose of connecting the collectors of an ECL gate to the ground?
33. What problems result from saturation in transistor gates?

Problems

The symbol **D** indicates that a problem is a design problem. The symbol **P** indicates that you can check the solution to a problem using PSpice/SPICE or Electronics Workbench.

15.4 Performance Parameters of Logic Gates

15.1 **D** An inverter, as shown in Fig. 15.8, drives identical inverters and has $V_{OL} = 0.4$ V, $I_{OL} = 1$ mA, $V_{OH} = 2.4$ V, $I_{OH} = 100$ μA, and $V_{CC} = 5$ V. The input currents drawn by each load inverter are $I_{IL} = 0.1$ mA (at logic low) and $I_{IH} = 10$ μA (at logic high).

a. If there are three load inverters, determine the value of pull-up resistance R_P that will ensure a logic 1 output of $V_{OH} = 2.4$ V.

b. If $R_P = 2.5$ kΩ, find the fan-out N.

15.2 The inverter shown in Fig. 15.10(a) has $V_{OL} = 0.4$ V, $V_{OH} = 2.4$ V, $V_{CC} = 5$ V, $R_{on} = 200$ Ω, $R_{off} \approx \infty$, $R_P = 2.5$ kΩ, $C_L = 2$ pF, and $f_{clk} = 100$ kHz.

a. Find the delay times for the output voltage to rise from 0.4 V to 2.4 V and to fall from 5 V to 0.4 V.

b. Find the power dissipation P_D.

15.5 NMOS Inverters

15.3 The parameters of the NMOS inverter in Fig. P15.3 are $V_{tL} = 1.5$ V, $V_{DD} = 5$ V, $R_D = 1.5$ kΩ, and $K_L = 0.5$ mA/V^2.

a. Determine the input voltage at the transition point $V_{I(tran1)}$.

b. Calculate the output voltage v_O, the drain current i_D, and the power dissipation P_D for $v_I = 2.5$ V and for $v_I = 5$ V.

FIGURE P15.3

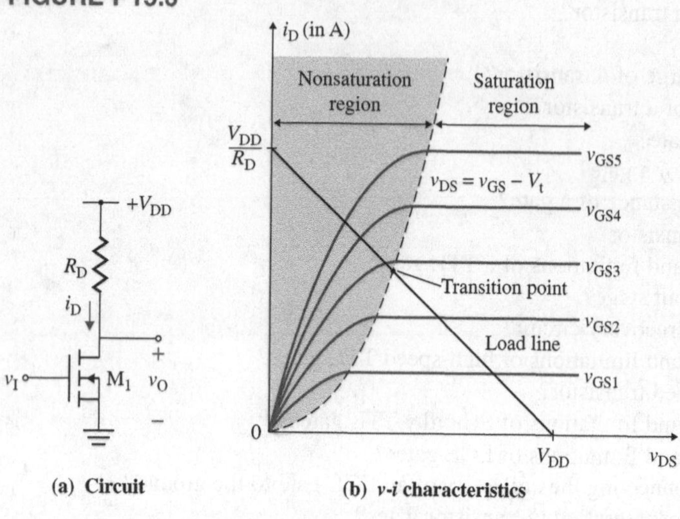

(a) Circuit (b) *v-i* characteristics

15.4 The parameters of the NMOS inverter in Fig. P15.3 are $V_{tL} = 1$ V, $V_{DD} = 5$ V, and $R_D = 1.5$ kΩ. Determine the conduction parameter K_L so that the output voltage v_O is 0.5 V at $v_I = 5$ V.

15.5 The parameters of the NMOS inverter in Fig. 15.12(a) with a saturated load are $V_{tL} = V_{tD} = 1.5$ V, $V_{DD} = 5$ V, $K_L = 0.1$ mA/V^2, and $K_D = 50$ μA/V^2.

a. Determine the input voltage at the transition point $V_{I(tran)}$.

b. Calculate the output voltage, the drain current, and the power dissipation P_D for $v_I = 2.5$ V and for $v_I = 5$ V.

15.6 The parameters of the NMOS inverter in Fig. 15.12(a) with a saturated load are $V_{tL} = V_{tD} = 1$ V and $V_{DD} = 5$ V. Determine the ratio K_D/K_L so that **(a)** $v_O = 0.25$ V at $v_I = 5$ V and **(b)** $v_O = 0.25$ V at $v_I = 4.5$ V.

15.7 The parameters of the NMOS inverter in Fig. 15.12(a) with an enhancement load are $V_{tL} = V_{tD} = 1$ V, $V_{DD} = 5$ V, $K_L = 0.25$ mA/V^2, and $K_D = 1$ mA/V^2.

a. Determine the input voltage at the transition point $V_{I(tran)}$.

b. Calculate the output voltage and the drain current for $v_I = 2$ V and for $v_I = 5$ V.

15.8 D P Design an enhancement-load NMOS inverter, as shown in Fig. 15.12(a), to obtain a noise margin of $NM_L \geq 0.9$ V. The threshold voltages are $V_{tL} = V_{tD} = 1$ V, and $K_L = 40$ μA/V^2. The supply voltage is $V_{DD} = 5$ V. Assume $V_{OH} = V_{IH} = 5$ V, load capacitance $C_L = 2$ pF, and clock frequency $f_{clk} = 1$ MHz.

a. Find the design parameter $K_R = K_D/K_L$, neglecting the body effect.

b. Calculate NM_L if the body effect is included. Assume $\gamma = 0.5\sqrt{V}$ and $2\phi_f = 0.6$ V.

c. Calculate the low-to-high propagation time t_{pLH}.

d. Calculate the high-to-low propagation time t_{pHL}.

e. Calculate the delay-power product (DP).

15.9 The parameters of the NMOS inverter in Fig. 15.16(a) with a depletion load are $V_{tL} = -1.5$ V, $V_{tD} = 1$ V, $V_{DD} = 5$ V, $K_D = 1.5$ mA/V^2, and $K_L = 0.1$ mA/V^2.

a. Determine the input and output voltages at the first and second transition points.

b. Calculate the output voltage, the drain current, and the power dissipation P_D for $v_I = 2.5$ V and for $v_I = 5$ V.

15.10 The parameters of the NMOS inverter in Fig. 15.16(a) with a depletion load are $V_{tL} = -1.5$ V, $V_{tD} = 1$ V, and $V_{DD} = 5$ V. Determine the ratio K_D/K_L so that **(a)** $v_O = 0.25$ V at $v_I = 5$ V and **(b)** $v_O = 0.25$ V at $v_I = 4.5$ V.

15.11 The parameters of the NMOS inverter in Fig. 15.16(a) with a depletion load are $V_{tD} = 1$ V, $V_{DD} = 5$ V, $K_D = 1.5$ mA/V^2, and $K_L = 0.1$ mA/V^2. Determine the value of V_{tL} so that $v_O = 0.15$ V at $v_I = 5$ V.

15.12 The parameters of the NMOS inverter in Fig. 15.16(a) with a depletion load are $V_{tL} = -2$ V, $V_{tD} = 1$ V, $V_{DD} = 5$ V, $K_D = 1$ mA/V^2, and $K_L = 0.25$ mA/V^2.

a. Determine the input and output voltages at the first and second transition points.

b. Calculate the output voltage and the drain currents for $v_I = 1.5$ V and for $v_I = 5$ V.

15.13 Design a depletion-load NMOS inverter, as shown in Fig. 15.16(a), to obtain a noise margin of $NM_H \geq 2.5$ V. The threshold voltages are $V_{tL} = -1.5$ V, $V_{tD} = 1$ V, and $K_L = 40$ μA/V^2. The supply voltage is $V_{DD} = 5$ V. Assume $V_{OH} = V_{DD} = 5$ V, load capacitance $C_L = 2$ pF, and frequency $f_{clk} = 1$ MHz.

a. Find the design parameter $K_R = K_D/K_L$, neglecting the body effect.

b. Calculate NM_L if the body effect is neglected.

c. Calculate NM_H if the body effect is included. Assume $\gamma = 0.5\sqrt{V}$ and $2\phi_f = 0.6$ V.

d. Calculate the low-to-high propagation time t_{pLH}.

e. Calculate the high-to-low propagation time t_{pHL}.

f. Calculate the delay-power product (DP).

15.6 *NMOS Logic Circuits*

15.14 The parameters of the NMOS transmission gate in Fig. 15.23 are $V_t = 1.5$ V, $K = 0.1$ mA/V^2, $C_L = 5$ pF, and $V_G = 5$ V.

a. What is the steady-state output voltage if the input voltage is changed in the following sequence: $v_I = 0$, $v_I = 5$ V, and $v_I = 3$ V?

b. What would the steady-state output voltage be if v_G were switched to 0?

15.15 An NMOS *RS* flip-flop circuit is shown in Fig. P15.15. The input conditions, which are sequential from time 1 to time 4, are also listed in Fig. P15.15. Determine the state of each transistor (on or off) at each time and the logic outputs at Q_A and Q_B for each input condition.

FIGURE P15.15

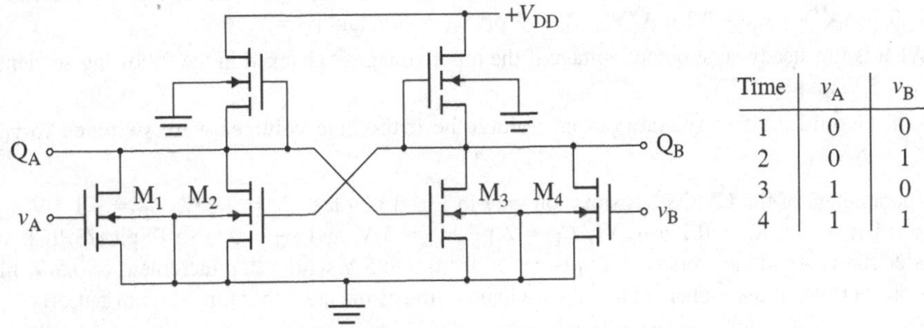

Time	v_A	v_B
1	0	0
2	0	1
3	1	0
4	1	1

15.7 CMOS Inverters

15.16 The parameters of the CMOS inverter in Fig. 15.26(a) are $V_{tN} = 1.5$ V, $V_{tP} = -1.5$ V, $V_{DD} = 5$ V, $K_n = 1$ mA/V^2, and $K_p = 100$ mA/V^2.

a. Determine the input and output voltages at the first and second transition points.

b. Calculate the output voltage and the drain currents for $v_I = 1.5$ V and for $v_I = 4.5$ V.

15.17 The parameters of the CMOS inverter in Fig. 15.26(a) are $V_{tN} = 1.5$ V, $V_{tP} = -1.5$ V, $V_{DD} = 5$ V, $K_n = 0.2$ mA/V^2, and $K_p = 100$ mA/V^2.

a. Determine the input and output voltages at the first and second transition points.

b. Calculate the output voltage and the drain currents for $v_I = 1.5$ V and for $v_I = 4.5$ V.

15.18 The parameters of the CMOS inverter in Fig. 15.26(a) are $V_{tN} = 1$ V, $V_{tP} = -1$ V, $V_{DD} = 5$ V, $K_n = 1$ mA/V^2, and $K_p = 1$ mA/V^2.

a. Determine the input and output voltages at the first and second transition points.

b. Calculate the output voltage and the drain currents for $v_I = 1.5$ V and for $v_I = 3.5$ V.

15.19 The parameters of the CMOS inverter in Fig. 15.26(a) are $V_{tN} = 1.5$ V, $V_{tP} = -1.5$ V, and $V_{DD} = 5$ V. Determine the ratio K_n/K_p so that the transition occurs at an input voltage of $v_I = V_{I(tran1)} = V_{DD}/2$.

15.20 The parameters of the CMOS inverter in Fig. 15.26(a) are $V_{tN} = 1.5$ V, $V_{tP} = -1$ V, and $V_{DD} = 5$ V. Calculate the output voltages if the input voltage is varied from 0 to 5 V with a step of 0.25 V. Assume $K_n/K_p = 1$.

15.21 The parameters of the CMOS inverter in Fig. 15.26(a) are $V_{tN} = 1.5$ V, $V_{tP} = -1.5$ V, and $V_{DD} = 5$ V. Calculate the output voltages if the input voltage is varied from 0 to 5 V with a step of 0.5 V. The ratio $K_n/K_p = 1$.

15.22 The parameters of the CMOS inverter in Fig. 15.26(a) are $V_{tN} = 1$ V, $V_{tP} = -1$ V, $V_{DD} = 5$ V, $K_n = 1$ mA/V^2, and $K_p = 1$ mA/V^2.

a. Determine the input and output voltages at the first and second transition points.

b. Calculate the output voltages and the drain currents for $v_I = 1.5$ V and for $v_I = 3.5$ V.

15.23 D P Design a CMOS inverter, as shown in Fig. 15.26(a), to operate at a transition voltage $V_M = 2.5$ V. The threshold voltages are $V_{tP} = -1.5$ V, $V_{tN} = 1$ V, and $K_p = 40$ μA/V^2. The supply voltage is $V_{DD} = 5$ V. Assume $V_{IH} = V_{OH} = 5$ V, load capacitance $C_L = 2$ pF, and frequency $f_{clk} = 1$ MHz.

a. Find the design parameter $K_R = K_n/K_p$, neglecting the body effect.

b. Calculate NM_L and NM_H.

c. Calculate the propagation delay t_{pd}.

d. Calculate the delay-power product (DP).

15.8 CMOS Logic Circuits

15.24 The parameters of the CMOS transmission gate in Fig. 15.29 are $V_{tN} = 1.5$ V, $V_{tP} = -1.5$ V, $V_{DD} = 5$ V, $K_n = 0.1$ mA/V^2, $K_p = 0.1$ mA/V^2, $C_L = 2$ pF, $\overline{v}_G = 5$ V, and $v_G = 0$.

a. What is the steady-state output voltage if the input voltage is changed in the following sequence: $v_I = 0$, $v_I = 5$ V, and $v_I = 3$ V?

b. What would the steady-state output voltage be if the gate voltages were switched to $\overline{v}_G = 0$ and $v_G = 5$ V?

15.25 The parameters of the CMOS transmission gate in Fig. 15.29 are $V_{tN} = 1.5$ V, $V_{tP} = -1.5$ V, $V_{DD} = 5$ V, $K_n = 0.1$ mA/V^2, $K_p = 0.1$ mA/V^2, $C_L = 2$ pF, $\overline{v}_G = 5$ V, and $v_G = 0$. Use PSpice/SPICE to plot the transfer characteristic v_O versus v_I if v_I is varied from 0 to 5 V with a step increment of 0.5 V. Indicate the regions over which the *n*-channel and the *p*-channel transistors are either turned on or cut off.

15.26 A CMOS *RS* flip-flop circuit is shown in Fig. P15.26. The input conditions, which are sequential from time 1 to time 4, are also listed. Determine the state of each transistor (on or off) at each time and the logic outputs at Q_A and Q_B for each input condition.

FIGURE P15.26

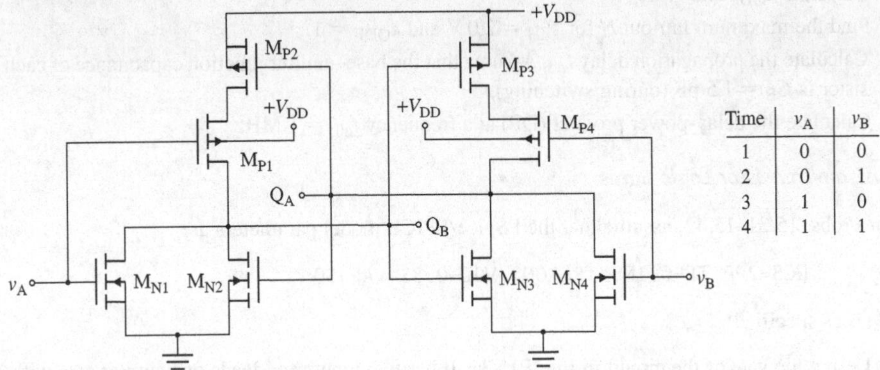

Time	v_A	v_B
1	0	0
2	0	1
3	1	0
4	1	1

15.27 A CMOS logic circuit is shown in Fig. P15.27. The input conditions, which are sequential from time 1 to time 4, are also listed. Determine the state of each transistor (on or off) at each time and the logic output v_O.

FIGURE P15.27

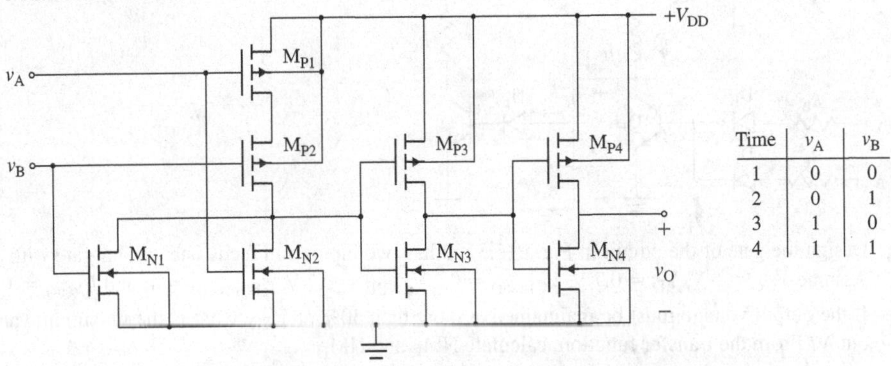

Time	v_A	v_B
1	0	0
2	0	1
3	1	0
4	1	1

15.10 *BJT Inverters*

For Probs. 15.28–15.30, assume that the PSpice/SPICE model parameters are

```
CCS=2PF TF=0.1NS TR=10NS VJC=0.85 VAF=50
```

(β_F is as specified).

15.28 The bipolar transistor in Fig. 15.33(a) is specified to have β_F in the range of 10 to 50. The load resistance is $R_C = 10\ \text{k}\Omega$. The DC supply voltage is $V_{CC} = 5$ V, and the input voltage to the base circuit is $v_I = 5$ V. If $V_{CE(sat)} = 0.1$ V and $V_{BE(sat)} = 0.7$ V, find **(a)** the value of R_B that results in saturation with an overdrive factor of 5, **(b)** the forced β_F, and **(c)** the power loss in the transistor P_D.

15.29 The bipolar transistor in Fig. 15.33(a) is specified to have β_F in the range of 8 to 40. The load resistance is $R_C = 1\ \text{k}\Omega$. The DC supply voltage is $V_{CC} = 5$ V, and the input voltage to the base circuit is $v_I = 5$ V. If $V_{CE(sat)} = 0.2$ V and $V_{BE(sat)} = 0.7$ V, find **(a)** the value of R_B that results in saturation with an overdrive factor of 5, **(b)** the forced β_F, and **(c)** the power loss in the transistor P_D.

15.30 D P Design a BJT inverter, as shown in Fig. 15.35, to drive four identical inverters ($N = 4$) and to give $V_{OH} =$ 2.5 V and $NM_L = 0.2$ V. The transistor has $I_{C(max)} = 4$ mA, $\beta_F = 80$ to 120, $V_{BE(cut\text{-}in)} = 0.5$ V, $V_{BE(sat)} =$ 0.7 V, $t_d = 2$ ns, and $t_s = 2$ ns. The supply voltage V_{CC} is 5 V.

a. Find the values of R_C and R_B.

b. Calculate k_{ODF} and NM_H.

c. Find the maximum fan-out N for $V_{IH} = 2.0$ V and $k_{ODF} = 1$.

d. Calculate the propagation delay t_{pd}. Assume that the base–emitter junction capacitance of each load transistor is $C_B = 1.5$ pF (during switching).

e. Calculate the delay-power product (DP) at a frequency $f_{clk} = 1$ MHz.

15.11 *Transistor-Transistor Logic Gates*

For Probs. 15.31–15.42, assume that the PSpice/SPICE model parameters are

```
CCS=2PF TF=0.1NS TR=10NS VJC=0.85 VAF=50
```

(β_F is as specified).

15.31 D P **a.** Design the gate of the circuit in Fig. P15.31. It has two inputs and feeds one similar gate with one input. Assume $V_{CC} = 5$ V, $V_D = 0.7$ V, $V_{CE(sat)} = 0.1$ V, $V_{BE} = 0.7$ V, $\beta_{F(forced)} = 10$, and $I_{C(sat)} = 1$ mA.

b. If the output voltage must be maintained at more than 30% of V_{CC}, what is the maximum value of fan-out N?

FIGURE P15.31

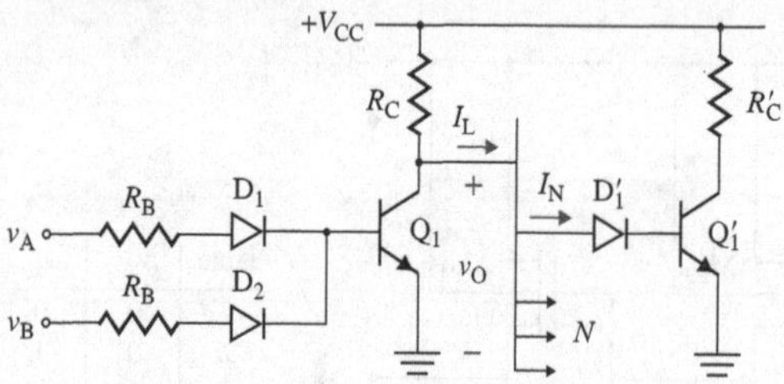

15.32 D P **a.** Design the gate of the circuit in Fig. P15.31. It has two inputs and feeds one similar gate with one input. Assume $V_{CC} = 5$ V, $V_D = 0.7$ V, $V_{CE(sat)} = 0.2$ V, $V_{BE} = 0.7$ V, $\beta_{F(forced)} = 4$, and $I_{C(sat)} = 1$ mA.

b. If the output voltage must be maintained at more than 30% of V_{CC}, what is the maximum value of fan-out N? From the transfer function, calculate NM_L and NM_H.

15.33 D P Design the TTL NAND gate of the circuit in Fig. P15.33. It has two inputs and feeds three similar NAND gates. Assume $V_{CC} = 5$ V, $V_D = 0.7$ V, $V_{CE(sat)} = 0.2$ V, $V_{BE} = 0.7$ V, $\beta_{F(forced)} = 10$, and $I_{C(sat)} = 1$ mA. From the transfer function, calculate NM_L and NM_H.

FIGURE P15.33

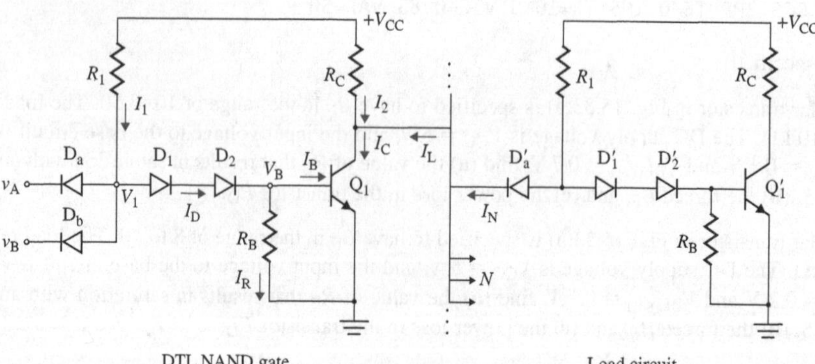

15.34 A TTL gate is shown in Fig. P15.34.

P **a.** Calculate the currents I_1, I_2, I_C, and I_B and the voltages V_I and V_O for input conditions $v_A = v_B = 0$ and $v_A = v_B = 5$ V.

b. Calculate the fan-out for the output low condition. Assume $I_{NO} = 50$ μA and $I_{LO} = 200$ μA.

FIGURE P15.34

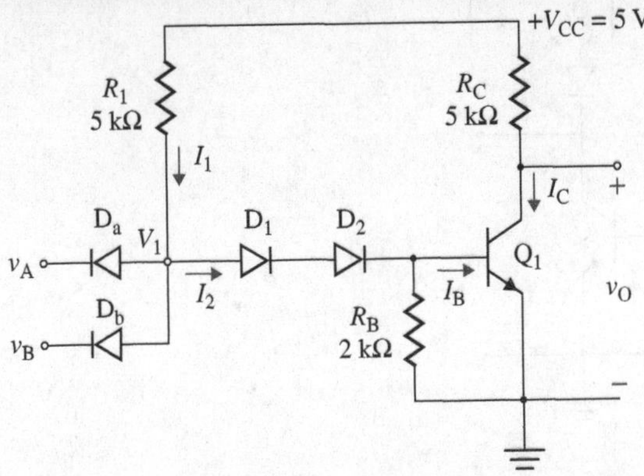

15.35 Design the TTL NAND gate of the circuit in Fig. 15.38. It has two inputs and feeds two similar NAND gates. The output voltage at output high is $V_{OH} = 3.5$ V, and $I_{N1} = 650$ μA. Assume $V_{CC} = 5$ V, $V_D = 0.7$ V, $V_{CE(sat)} = 0.2$ V, $V_{BE} = 0.7$ V, $\beta_{F(forced)} = 10$, $\beta_R = 0.2$, $I_{C3(sat)} = 1$ mA, and $I_{C4(max)} = 2$ mA. From the transfer function, calculate NM_L and NM_H.

D P

15.36 A TTL gate is shown in Fig. P15.36. Calculate the output voltage and all currents of the transistors and fan-out for the conditions $v_A = v_B = 5$ V and $v_A = v_B = 0$. Assume $V_{CC} = 5$ V, $V_{CE(sat)} = 0.2$ V, $V_{BE} = 0.7$ V, $V_{BC1} = 0.7$ V, $\beta_{F(forced)} = 10$, and $\beta_R = 0.4$.

P

FIGURE P15.36

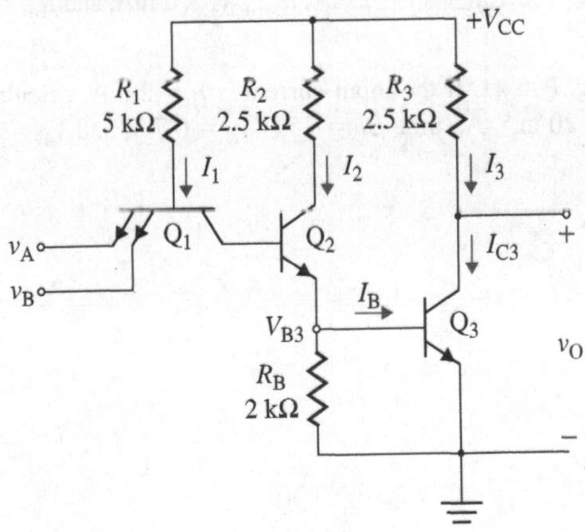

15.37 A TTL gate is shown in Fig. P15.37. Calculate the output voltage and all currents of the transistors and fan-out for the conditions $v_A = v_B = 5$ V and $v_A = v_B = 0$. Assume $V_{CC} = 5$ V, $V_{OH} = 3$ V, $V_{CE(sat)} = 0.2$ V, $V_{BE} = 0.7$ V, $V_{BC1} = 0.7$ V, $\beta_{F(forced)} = 10$, and $\beta_R = 0.4$.

P

FIGURE P15.37

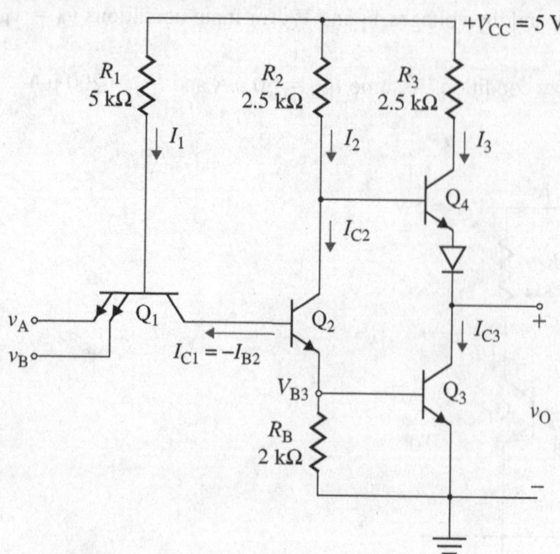

15.38 (D, P) Design the TTL NAND gate of the circuit in Fig. 15.36. It has two inputs and feeds four identical NAND gates. The output voltage at output high is $V_{OH} = 2.8$ V, and $I_{N1} = 72.5$ μA. Assume $V_{CC} = 5$ V, $V_D = 0.7$ V, $V_{CE(sat)} = 0.3$ V, $V_{BE} = 0.8$ V, $\beta_{F(forced)} = 20$, $\beta_R = 0.1$, $I_{C3(sat)} = 2$ mA, and $I_{C4(max)} = 2$ mA. From the VTC, find NM_L and NM_H.

15.39 (D, P) Design the high-speed TTL NAND gate of the circuit in Fig. 15.43. It has two inputs and feeds four similar NAND gates. The output voltage at output high is $V_{OH} = 3.5$ V, and $I_{N1} = 650$ μA. Assume $V_{CC} = 5$ V, $V_{CE(sat)} = 0.2$ V, $V_{BE} = 0.7$ V, $\beta_{F(forced)} = 10$, $\beta_R = 0.2$, $I_{C3(max)} = 1$ mA, $I_{C4(max)} \approx 2$ mA, and $I_{BR} = 0.7$ mA. From the VTC, calculate NM_L and NM_H.

15.40 (D, P) Design the TTL NAND gate of the circuit in Fig. 15.43. It has two inputs and feeds four identical NAND gates. The output voltage at output high is $V_{OH} = 2.8$ V, and $I_{N1} = 65$ μA. Assume $V_{CC} = 5$ V, $V_D = 0.7$ V, $V_{CE(sat)} = 0.3$ V, $V_{BE} = 0.7$ V, $\beta_{F(forced)} = 40$, $\beta_R = 0.1$, $I_{C3(sat)} = 2$ mA, $I_{C4(max)} = 2$ mA, and $I_{BR} = 700$ μA. From the VTC, calculate NM_L and NM_H.

15.41 A Schottky clamped transistor is shown in Fig. P15.41. If the input current is $I_I = 2$ mA, calculate I_D, I_B, and I_C for the conditions $I_L = 5$ mA and $I_L = 20$ mA. Assume $\beta_F = 25$, $V_{BE} = 0.7$ V, and $V_D = 0.3$ V.

FIGURE P15.41

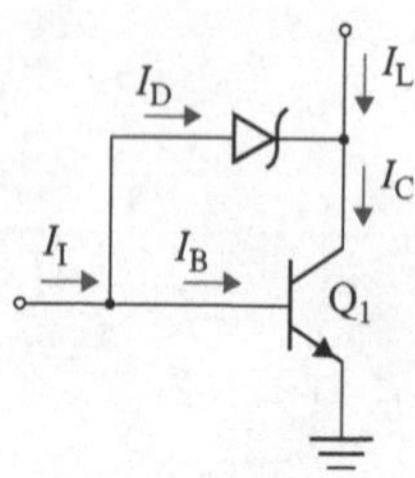

15.42 A Schottky clamped transistor is shown in Fig. P15.42.

a. If the input current is $I_I = 2$ mA, calculate I_D, I_B, and I_C for the condition $I_L = 0$.

b. Determine the maximum value of load current I_L that the transistor can sink and still remain at the edge of saturation. Assume $\beta_F = 25$, $V_{BE} = 0.7$ V, $V_{CE(sat)} = 0.2$ V, and $V_D = 0.3$ V.

FIGURE P15.42

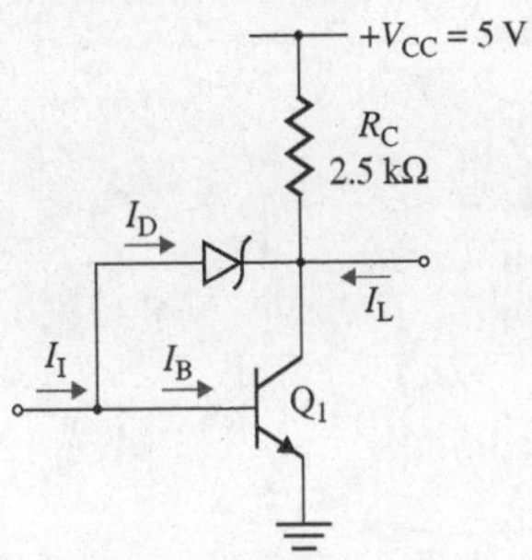

15.12 Emitter-Coupled Logic OR/NOR Gates

15.43 D P **a.** Design the ECL OR/NOR gate of the circuit in Fig. 15.50. It has two inputs. The desired collector currents are $I_{C1} = 2$ mA and $I_{C4} = I_{C5} = 1$ mA. At logic 1, $V_{OR} = -0.7$ V, $V_{NOR} = -1.5$ V, and v_A (or v_B) $= V_{OR} = -0.7$ V. Assume $V_{EE} = -5.2$ V, $V_{BE} = 0.7$ V, and $\beta_F = 50$.

b. Calculate the maximum fan-out with similar ECL gates if V_{OR} is allowed to fall from $V_{OR(max)} = -0.7$ V to $V_{OR(min)} = -0.75$ V.

15.44 D Repeat Prob. 15.43 if $V_{CC} = 3$ V, $V_{EE} = -3.2$ V, $V_{OR} = 0.5$ V, $V_{NOR} = 1.5$ V, and v_A (or v_B) $= V_{OR} = 0.5$ V.

15.45 D P An ECL logic gate with one input is shown in Fig. P15.45. Calculate the resistances R_{C1}, R_{C2}, R_E, and R_4. The desired emitter current of all transistors is 2 mA. At logic 1, the outputs are $V_{OR} = 2.5$ V and $V_{NOR} = 3.5$ V. Assume $V_{CC} = 5$ V, $V_{EE} = 0$, $V_{BE} = 0.7$ V, and $\beta_F = 200$. From the transfer function, calculate NM_L and NM_H.

FIGURE P15.45

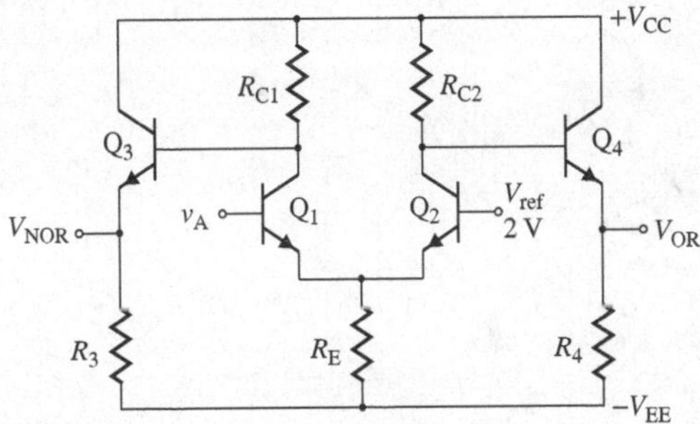

15.46 D P **a.** Design the ECL OR/NOR gate of the circuit in Fig. 15.50. It has two inputs. The desired collector currents are $I_{C1} = 2$ mA and $I_{C4} = I_{C5} = 1$ mA. At logic high, $V_{OR} = -0.7$ V, $V_{NOR} = -1.63$ V, and v_A (or v_B) $= V_{OR} = -0.7$ V. Assume $V_{EE} = -5.2$ V, $V_{BE} = 0.7$ V, and $\beta_F = 50$.

b. Calculate the maximum fan-out with similar ECL gates if V_{OR} is allowed to fall from $V_{OR(max)} = -0.7$ V to $V_{OR(min)} = -0.75$ V.

c. From the VTC, calculate NM_L and NM_H.

15.13 BiCMOS Inverters

15.47 The BiCMOS inverter shown in Fig. 15.54 has $V_{DD} = 5$ V, $V_{BE} = 0.7$ V, $R_1 = 2$ kΩ, $R_2 = 6$ kΩ, $C_1 = 2$ pF, and $\beta_F = 150$. The threshold voltages are $V_{tP} = -1.5$ V, $V_{tN} = 1.5$ V, and $K_p = K_n = 40$ μA/V^2. Assume $V_{OL} = 0$ and $V_{OH} = 5$ V.

a. Calculate t_{pLH}, t_{pHL}, and t_{pd}.

b. Use PSpice/SPICE to plot the VTC and transient response.

CHAPTER 16

INTEGRATED ANALOG CIRCUITS AND APPLICATIONS

Learning Outcomes

After completing this chapter, students should be able to do the following:

- Design and analyze precision rectifiers.
- Design and analyze nonlinear voltage limiters.
- List the differences between comparators and op-amps.
- Apply comparators and op-amps in generating waveforms such as the zero-crossing detector, Schmitt trigger, and square-wave, triangular-wave, and sawtooth wave generators.
- Describe the internal structure and the principle of operation of commonly used analog-integrated circuits such as the NE/SE-566 VCO, the 555 timer, NE/SE-565 PLL, Teledyne 9400 V/F and F/V converters, LF198 sample-and-hold amplifier, MC1408 and NE/SE-5018 D/A converters, and NE/SE-5034 A/D converter.
- Apply some ICs and design circuits for waveform generation, conversion, detection, or signal processing.

Symbols and Their Meanings

Symbol	**Meaning**
A, A_f	Open-loop and closed-loop voltage gains
β	Feedback factor

Symbol	Meaning
f_o, T, k	Output frequency, period, and duty cycle
f_L	Lock frequency
i_S, i_f	Input signal and feedback currents
v_S, v_O	Input and output signal voltages
v_+, v_-	Noninverting and inverting terminal voltages
v_{st}, v_{CN}	Switching and control signal voltages
$v_{S(max)}$, $v_{O(max)}$	Maximum input and output voltages
V_L, V_H	Reference low and high voltages
V_{ref}, V_{sat}	Reference and saturation voltages
V_{Lt}, V_{Ht}	Low and high threshold voltages
$+V_{th}$, $-V_{th}$	Positive and negative threshold voltages
V_{CC}, V_{EE}	Positive and negative DC supply voltages

16.1 Introduction

Electronic circuits are commonly employed in generating waveforms of various shapes for control and interfacing purposes. This chapter examines the basic principles of operation of circuits used to generate waveforms and the applications of some commonly used ICs [1–3].

16.2 Circuits with Op-Amps and Diodes

Many applications—such as peak signal detectors, precision rectifiers, comparator circuits, and limiters—require nonlinear functions. A diode, however, typically has a finite voltage drop of 0.7 V, which distorts the output voltage of a circuit with one or more diodes, especially for low-voltage signals. Op-amp circuits with diodes can reduce the effect of diode drop and are used for precision signal processing.

Let us take the op-amp circuit shown in Fig. 16.1(a) in which the diode provides a unidirectional current flow. If the input voltage v_S is negative, the output voltage v_{O1} of the op-amp becomes negative, causing the diode to be reverse biased; no current flows through the load R_L because the current i_i flowing into the op-amp is zero. Thus, the output voltage is zero: $v_O = 0$ for $v_S \leq 0$. But the voltage v_{O1} will reach the negative saturation limit of the op-amp. If the input voltage v_S is positive, the output voltage v_{O1} of the op-amp becomes positive, causing the diode to be forward biased and supply the load current i_L. Since $v_d \approx 0$ and $i_i = 0$, the output voltage is $v_O = v_S$ for $v_S \geq 0$. The transfer characteristic of v_O versus v_S is shown in Fig. 16.1(b).

For the diode to start conduction, only a very small input voltage is required: $v_{S(min)} = V_D/A_o$, where V_D is the diode drop (typically 0.7 V) and A_o is the open-circuit gain of the op-amp (typically 2×10^5). This circuit exhibits the characteristic of a diode. However, the effect of the diode drop is negligible. Thus, this circuit is called a *superdiode*. The drawback of this circuit is that, for negative values of v_S, the voltage v_{O1} will swing to the negative saturation limit, thereby slowing the speed of op-amp operation.

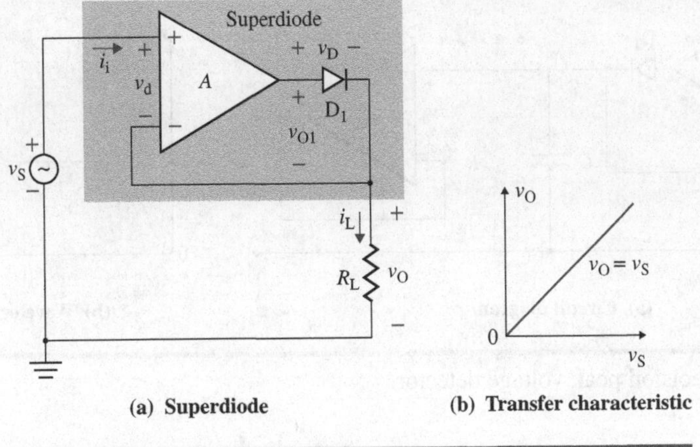

FIGURE 16.1 Superdiode

16.2.1 Most Positive Signal Detectors

The value of the most positive of a number of input signals can be detected by the circuit shown in Fig. 16.2. The inputs to the unity follower are the input signals to be detected. The diode that has the highest positive signal will conduct, and that signal will appear on the output of the circuit. The current source I_{DC} keeps the diode current constant irrespective of the value of the input signal, and it maintains a constant diode drop. As a result, the voltage drop across the conducting diode does not vary with variations in the input signal. Without the current source, the current and voltage of the conducting diode will change with the level of the most positive voltage, and the output voltage will vary. The unity follower acts as a buffer stage, providing a very high resistance at the diodes and a very low resistance at the output. If each diode is replaced by the superdiode shown in Fig. 16.1, there will be no voltage drop at the diodes; however, this arrangement will increase the complexity of the circuit.

16.2.2 Precision Peak Voltage Detectors

A precision peak voltage detector is used in many applications, such as monitoring the peak temperature of a day (or month) or initiating the action of a device if the signal exceeds a reference value. The circuit of Fig. 16.3 will detect the peak input voltage. The superdiode circuit allows the capacitor to charge to the

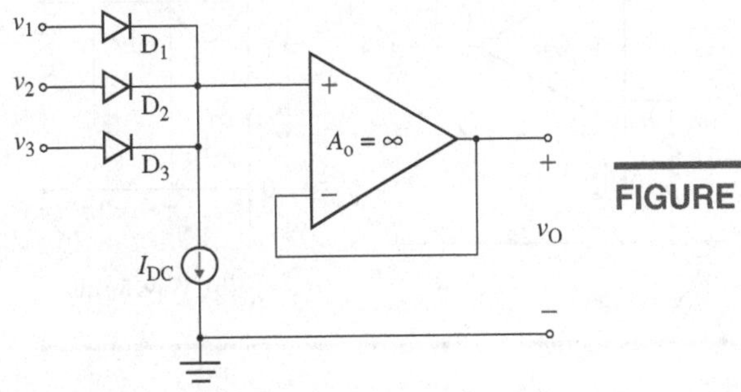

FIGURE 16.2 Positive signal detector

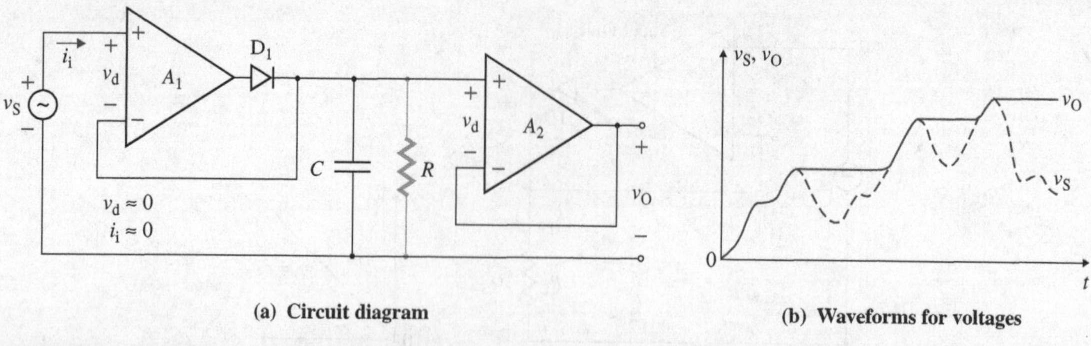

(a) Circuit diagram

(b) Waveforms for voltages

FIGURE 16.3 Precision peak voltage detector

peak input signal. The unity follower offers a high resistance to the capacitor and supplies the capacitor voltage to the output without discharging any charge of the capacitor. Resistance R, when connected, will allow the capacitor to discharge slowly so that the circuit can adjust to a lower input voltage. Otherwise, the capacitor will maintain its previous higher voltage and will not indicate the correct peak voltage.

16.2.3 Precision Half-Wave Rectifiers

A diode requires a minimum voltage, typically 0.7 V, to conduct. In a single-phase half-wave rectifier, one diode conducts. If the input voltage is less than 0.7 V, the output of the rectifier will be zero. Therefore, diode rectifiers are not suitable for rectification of low voltage. An op-amp circuit with two diodes, as shown in Fig. 16.4(a), can rectify a very small voltage in the range of microvolts. The circuit operation can be divided into two intervals: interval 1 and interval 2. We will consider the circuit operation with a sinusoidal input voltage $v_S = V_m \sin \omega t$.

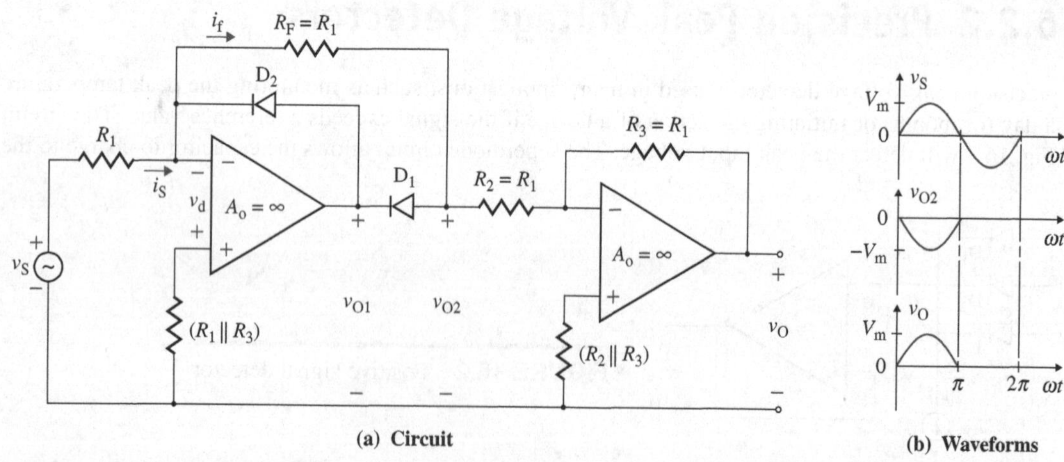

(a) Circuit

(b) Waveforms

FIGURE 16.4 Precision half-wave rectifier

During interval 1, $0 \leq \omega t \leq \pi$. The input voltage is positive. The voltage v_{O1} at the output of the first op-amp is negative, and diode D_2 is off. Diode D_1 conducts, and current i_f through R_F equals input current i_S. Since the current flowing into the op-amp is zero, $i_S = i_f$ and

$$i_S = i_f = \frac{v_S}{R_I}$$

$$v_d = R_F i_f + v_{O2}$$

For an ideal op-amp, $v_d = 0$, and voltage v_{O2} becomes

$$v_{O2} = -R_F i_f = -R_I i_F = -R_I i_S \quad (\text{for } R_F = R_I)$$
$$= -v_S$$

Thus, the output voltage at the output of the second inverting op-amp is

$$v_O = -v_{O2} = v_S \quad (\text{for } v_S \geq 0, R_3 = R_2 = R_1, \text{ and } v_S \geq 0)$$

During interval 2, $\pi \leq \omega t \leq 2\pi$. The input voltage is negative. The voltage v_{O1} at the output of the first op-amp is positive, and diode D_2 conducts. As a result, the voltage v_{O1} is clamped to approximately the voltage of one diode. Diode D_1 remains off, and no current flows through R_F. The voltage v_{O2} becomes zero. Thus, the output voltage at the output of the second op-amp is $v_O = -v_{O2} = 0$.

The output voltage v_O is almost independent of the diode characteristics. This is because diode D_1 is included in series with the op-amp. Since the op-amp gain is very high, tending to infinity, the effect of the diode and its voltage drop becomes insignificant. The output waveforms are shown in Fig. 16.4(b).

If the directions of the diodes are reversed, the output voltage will correspond to the negative part of the input voltage; there is no need for the second inverting op-amp. This arrangement is shown in Fig. 16.5(a). When the input voltage v_S is positive, the voltage v_{O1} becomes negative, making diode D_2 conduct and diode D_1 turn off. As a result, the output voltage becomes zero; that is, the output voltage $v_O = 0$ for $v_S \geq 0$. On the other hand, when the input voltage v_S is negative, the voltage v_{O1} becomes positive, making diode D_2 turn off and diode D_1 conduct. As a result, the output voltage will be equal and opposite to the input voltage; that is, the output voltage $v_O = -v_S$ for $v_S < 0$. The transfer characteristic is shown in Fig. 16.5(b), and the voltage waveforms are shown in Fig. 16.5(c).

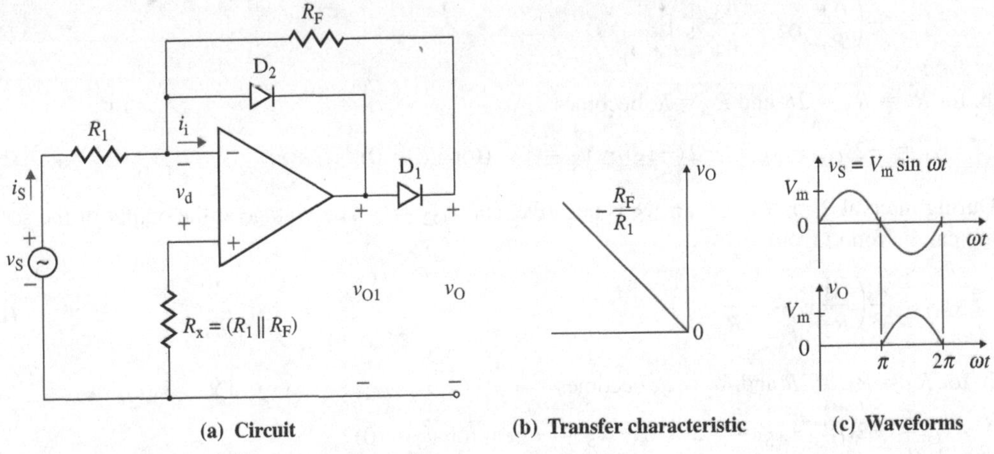

FIGURE 16.5 Alternate precision half-wave rectifier

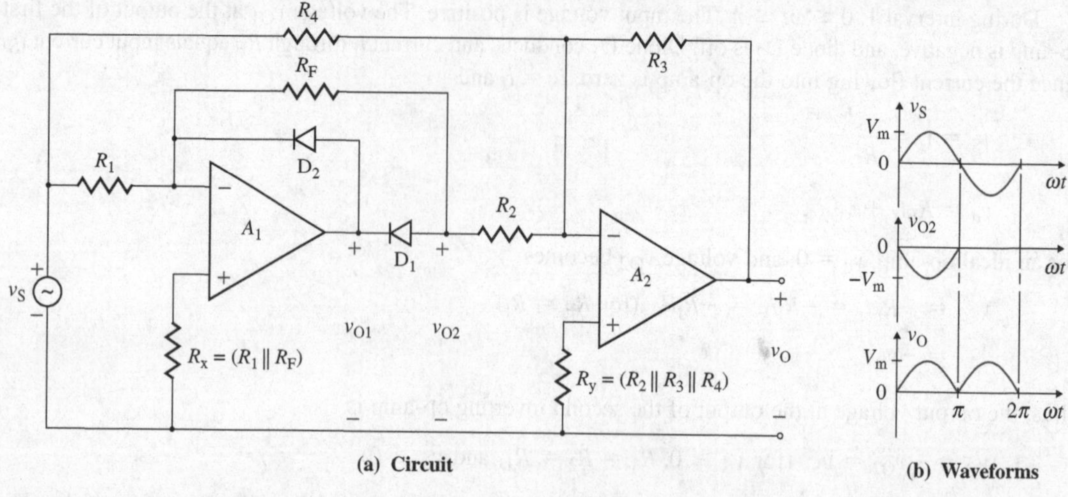

FIGURE 16.6 Precision full-wave rectifier

16.2.4 Precision Full-Wave Rectifiers

The half-wave rectifier of Fig. 16.5(a) can be modified to operate as a precision full-wave rectifier if we use the following algebraic relationship:

$$v_O = 2v_S - v_S = v_S \quad \text{(for the positive interval of the input voltage)}$$

This situation is shown in Fig. 16.6(a). Let us consider the case with $R_1 = R_2 = R_F = R$ and $R_3 = R_4 = 2R$. We will divide the circuit operation into two intervals, interval 1 and interval 2, and use a sinusoidal input voltage $v_S = V_m \sin \omega t$.

During interval 1, $0 \leq \omega t \leq \pi$. v_S is positive, and $v_{O2} = -v_S$. The voltage at the output of the second op-amp can be found from

$$v_O = -\left(\frac{R_3}{R_2}v_{O2} + \frac{R_3}{R_4}v_S\right) \tag{16.1}$$

which, for $R_3 = R_4 = 2R$ and $R_2 = R$, becomes

$$v_O = -2v_{O2} - v_S = -2(-v_S) - v_S = v_S \quad (\text{for } v_S \geq 0)$$

During interval 2, $\pi \leq \omega t \leq 2\pi$. v_S is negative, and $v_{O2} = 0$. The voltage at the output of the second op-amp can be found from

$$v_O = -\left(\frac{R_3}{R_2}v_{O2} + \frac{R_3}{R_4}v_S\right) \tag{16.2}$$

which, for $R_3 = R_4 = 2R$ and $R_2 = R$, becomes

$$v_O = -2v_{O2} - v_S = -2 \times 0 - v_S = -v_S \quad (\text{for } v_S < 0)$$

Thus, the output voltage is the inverted version of the input voltage, as shown in Fig. 16.6(b).

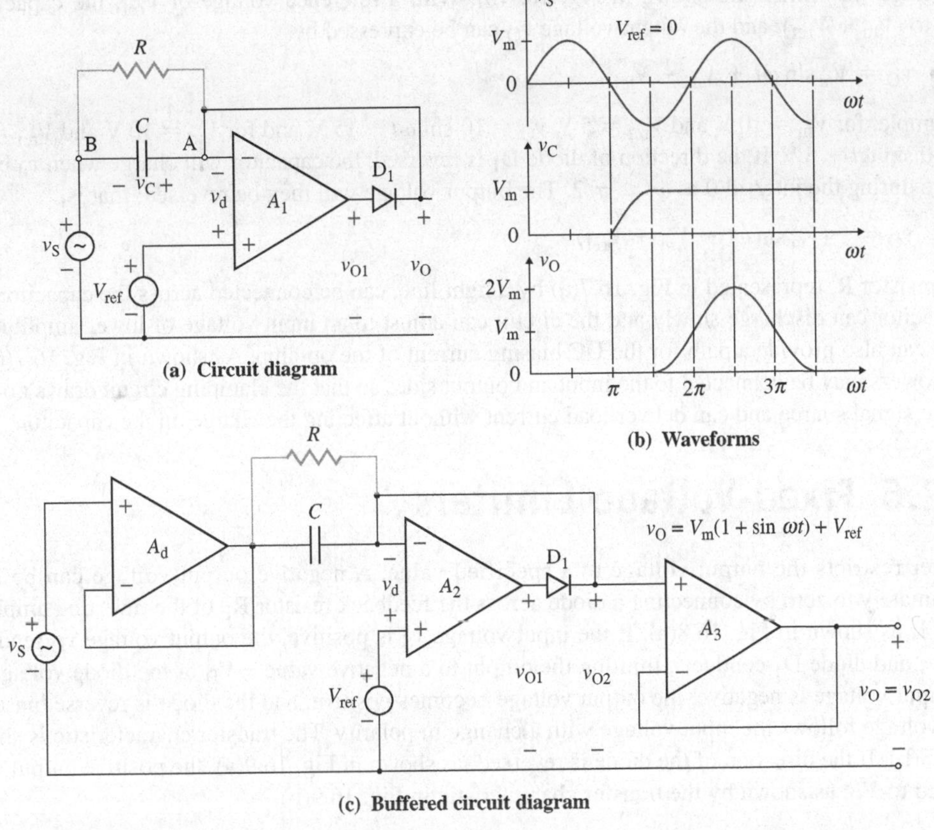

FIGURE 16.7 Precision clamping circuit

16.2.5 Precision Clamping Circuits

A precision clamping circuit is sometimes used in signal processing to add just enough DC voltage to the input voltage so that the sum never crosses the zero level. Figure 16.7(a) shows a superdiode circuit with an input voltage v_S and a DC reference voltage V_{ref}. As long as the op-amp voltage v_d is positive, v_{O1} will be positive, and diode D_1 will conduct, thereby making the (−) terminal behave as a virtual ground. To consider the circuit operation, let us take a sinusoidal input voltage $v_S = V_m \sin \omega t$ and $V_{ref} = 0$.

During the interval $0 \le \omega t \le \pi$, v_S is positive, v_d is negative, v_{O1} is negative, and diode D_1 is reverse biased. No current flows through the capacitor, and it cannot charge. During the interval $\pi < \omega t \le 3\pi/2$, v_S is negative, v_d is positive, v_{O1} is positive, and diode D_1 conducts. The (−) terminal behaves as a virtual ground. A charging current flows through the capacitor, with point A being at a higher potential than point B. The capacitor is charged to the peak negative voltage V_m. For the interval $\omega t > 3\pi/2$,

$$v_d = -(V_m + v_S) = -V_m(1 + \sin \omega t)$$

and diode D_1 remains off; that is, the output voltage v_O is

$$v_O = -v_d = V_m(1 + \sin \omega t)$$

The voltage waveforms are shown in Fig. 16.7(b). With a reference voltage of V_{ref}, the capacitor will charge to $(V_m + V_{ref})$, and the output voltage v_O can be expressed by

$$v_O = V_m \sin \omega t + V_m + V_{ref} \tag{16.3}$$

For example, for $V_m = 10$ V and $V_{ref} = 5$ V, $v_O = 10 \sin \omega t + 15$ V, and for $V_m = 10$ V and $V_{ref} = -5$ V, $v_O = 10 \sin \omega t + 5$ V. If the direction of diode D_1 is reversed, the capacitor will charge when v_d becomes negative during the interval $0 \leq \omega t \leq \pi/2$. The output voltage will then be reversed; that is,

$$v_O = -(V_m \sin \omega t + V_m + V_{ref})$$

A resistor R, represented in Fig. 16.7(a) by a light line, can be connected across the capacitor so that the capacitor can discharge slowly and the circuit can adjust to an input voltage of lower amplitude. The resistor can also provide a path for the DC biasing current of the op-amp. As shown in Fig. 16.7(c), voltage followers may be connected to the input and output sides so that the clamping circuit draws no current from the signal source and can deliver load current without affecting the charge on the capacitor.

16.2.6 Fixed-Voltage Limiters

A limiter restricts the output voltage to a specified value. A negative output voltage can be limited approximately to zero by connecting a diode across the feedback resistor R_F of the inverting amplifier in Fig. 3.11, as shown in Fig. 16.8(a). If the input voltage v_S is positive, the output voltage v_O tends to be negative, and diode D_1 conducts, limiting the output to a negative value $-V_D$ of the diode voltage drop. If the input voltage is negative, the output voltage becomes positive, and the diode is reverse biased. The output voltage follows the input voltage with a change in polarity. The transfer characteristic is shown in Fig. 16.8(b). If the direction of the diode is reversed, as shown in Fig. 16.9(a), the positive output voltage is limited to V_D, as shown by the transfer characteristic in Fig. 16.9(b).

16.2.7 Adjustable Voltage Limiters

Output voltage can be limited to an adjustable level by choosing resistors with appropriate values. A circuit that limits the negative output voltage is shown in Fig. 16.10(a). If the output is positive, diode D_1 is reverse biased, and the circuit operates as an inverting amplifier. If the output voltage is negative, its transfer

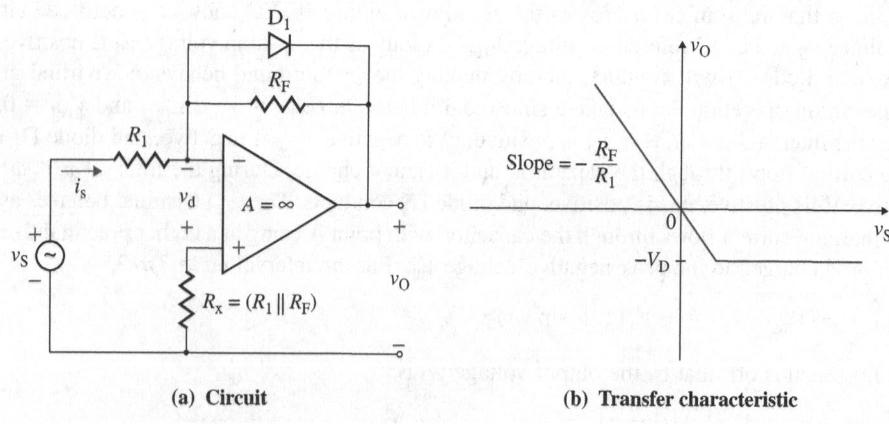

(a) Circuit

(b) Transfer characteristic

FIGURE 16.8 Negative voltage limiter

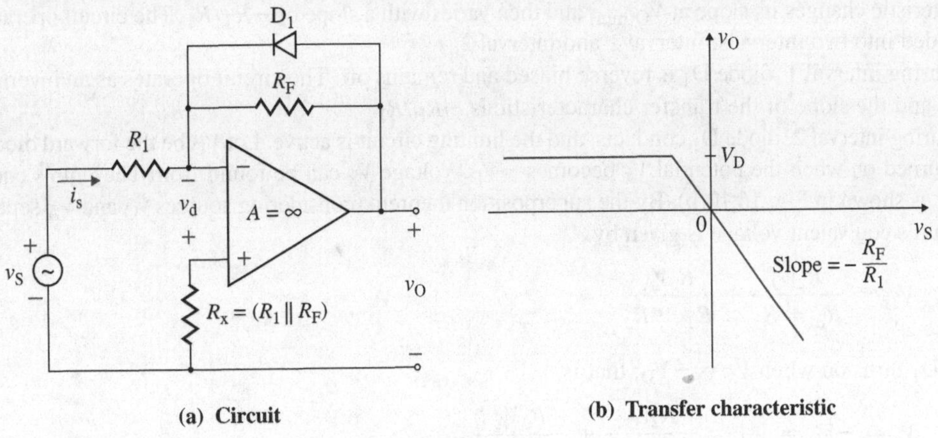

FIGURE 16.9 Positive voltage limiter

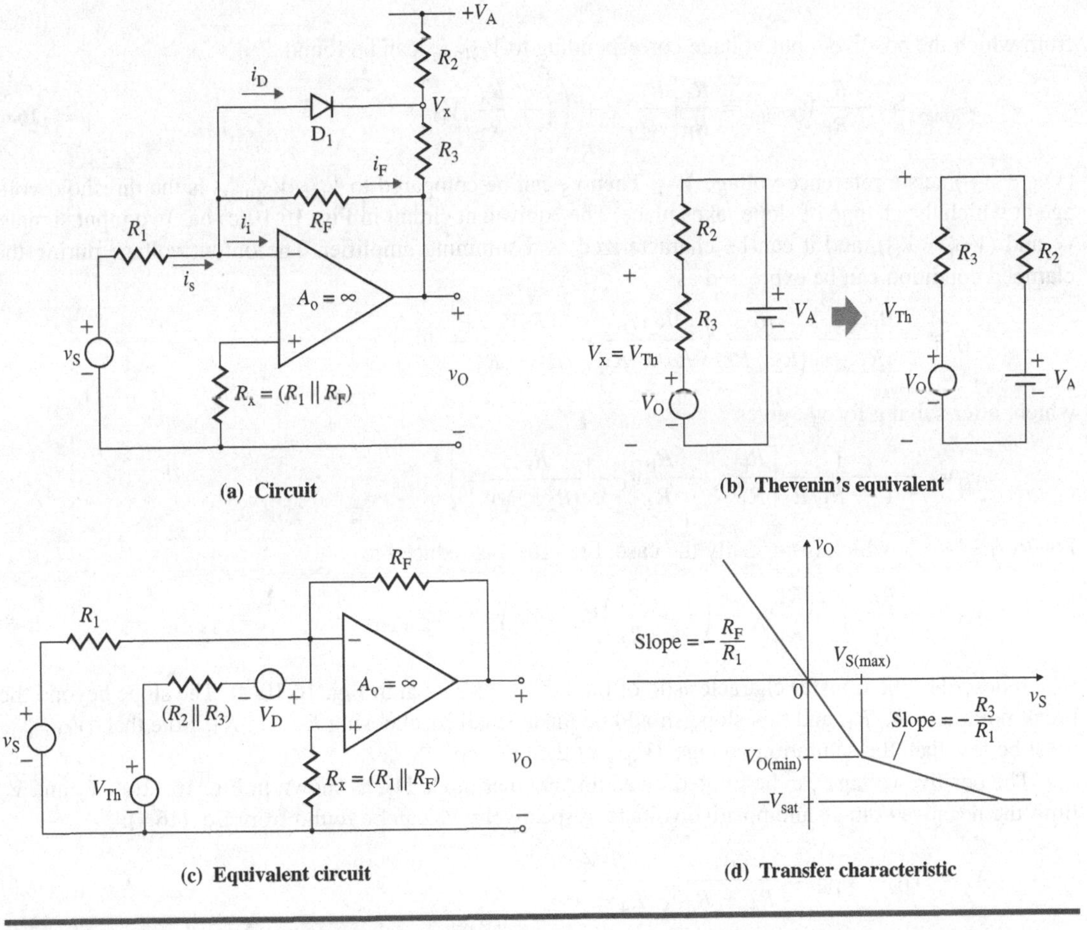

(a) Circuit

(b) Thevenin's equivalent

(c) Equivalent circuit

(d) Transfer characteristic

FIGURE 16.10 Adjustable negative voltage limiter

characteristic changes its slope at $V_{O(min)}$ and then varies with a slope of $-R_3/R_1$. The circuit operation can be divided into two intervals: interval 1 and interval 2.

During interval 1, diode D_1 is reverse biased and remains off. The circuit operates as an inverting amplifier, and the slope of the transfer characteristic is $-R_F/R_1$.

During interval 2, diode D_1 conducts, and the limiting circuit is active. Let V_D be the forward diode drop. D_1 is turned on when the potential V_x becomes $-V_D$. Voltage V_x can be found from Thevenin's equivalent circuit, as shown in Fig. 16.10(b). By the superposition theorem (considering sources v_O and V_A separately), Thevenin's equivalent voltage is given by

$$V_{Th} = \frac{R_2 v_O}{R_2 + R_3} + \frac{R_3 V_A}{R_2 + R_3}$$

Diode D_1 turns on when $V_x = -V_D$; that is,

$$V_x = -V_D = V_{Th} = \frac{R_2 v_O}{R_2 + R_3} + \frac{R_3 V_A}{R_2 + R_3} \tag{16.4}$$

which gives the negative clamping output voltage $V_{O(min)}$ as

$$V_{O(min)} = v_O = -V_D - (V_A + V_D)\frac{R_3}{R_2} = -V_D\left(1 + \frac{R_3}{R_2}\right) - V_A\frac{R_3}{R_2} \tag{16.5}$$

from which the positive input voltage corresponding to $V_{O(min)}$ can be found:

$$V_{S(max)} = -\frac{R_1}{R_F}V_{O(min)} = \frac{R_1}{R_F}\left[\frac{R_3}{R_2}V_A + \left(1 + \frac{R_3}{R_2}\right)V_D\right] \tag{16.6}$$

$(V_{Th} + V_D)$ acts a reference voltage V_{ref}. Then v_S can be compared to V_{ref}. $V_{S(max)}$ is the threshold voltage at which the change of slope takes place. The equivalent circuit in Fig. 16.10(c) has two input signals, v_S and $(V_{Th} + V_D)$, and it can be characterized as a summing amplifier. The output voltage during the clamped condition can be expressed as

$$v_O = -\frac{R_F v_S}{R_1} - \frac{R_F}{(R_2 \parallel R_3)}\left(\frac{R_2 v_O}{R_2 + R_3} + \frac{R_3 V_A}{R_2 + R_3} + V_D\right)$$

which, after solving for v_O, gives

$$v_O = -\frac{1}{1 + R_F/R_3}\left[\frac{R_F}{R_1}v_S + \frac{R_F}{R_2}V_A + \frac{R_F}{(R_2 \parallel R_3)}V_D\right] \tag{16.7}$$

For $R_F/R_3 >> 1$, which is normally the case, Eq. (16.7) is reduced to

$$v_O = -\frac{R_3}{R_1}v_S - \frac{R_3}{R_2}V_A - \left(1 + \frac{R_3}{R_2}\right)V_D \quad (\text{for } v_S > V_{S(max)}) \tag{16.8}$$

which describes the transfer characteristic of the limiter, as shown in Fig. 16.10(d). The slope beyond the break point is $-R_3/R_1$, and this slope should be made small by choosing $R_1 >> R_3$. Note that $|V_{O(min)}|$ must be less than the saturation voltage $|V_{sat}|$ of the op-amp.

The positive voltage can be limited by adding another diode D_2, as shown in Fig. 16.11(a). V_x and V_y limit the negative voltage and positive voltage, respectively. V_y can be found from Eq. (16.4):

$$V_y = V_D = V_{Th1} = \frac{R_4 v_O}{R_4 + R_5} - \frac{R_5 V_B}{R_4 + R_5} \tag{16.9}$$

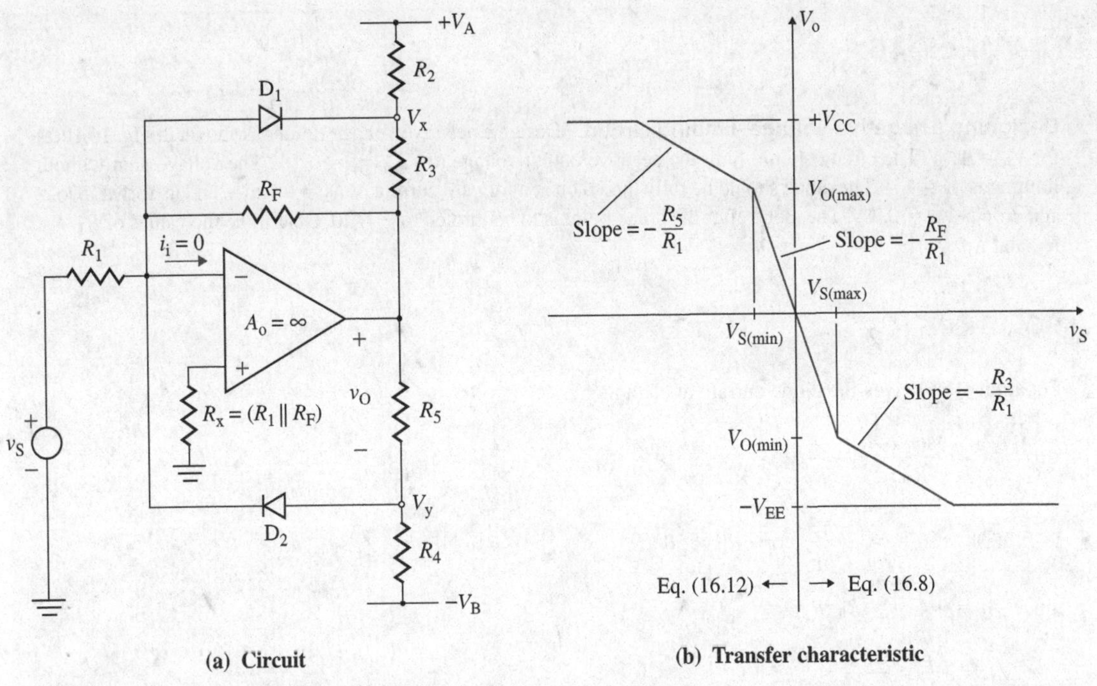

(a) Circuit (b) Transfer characteristic

FIGURE 16.11 Adjustable positive and negative voltage limiter

Similarly, the positive clamping voltage can be found from Eq. (16.5):

$$V_{O(max)} = v_O = V_D + (V_B + V_D)\frac{R_5}{R_4} = V_D\left(1 + \frac{R_5}{R_4}\right) + V_B\frac{R_5}{R_4} \qquad \textbf{(16.10)}$$

The negative input voltage corresponding to $V_{O(max)}$ can be found from

$$V_{S(min)} = -\frac{R_1}{R_F}V_{O(max)} = -\frac{R_1}{R_F}\left[\frac{R_5}{R_4}V_B + \left(1 + \frac{R_5}{R_4}\right)V_D\right] \qquad \textbf{(16.11)}$$

If V_{Th1} is Thevenin's equivalent voltage due to V_B, $(V_{Th1} - V_D)$ acts as reference voltage V_{ref1}. Then v_S can be compared to V_{ref1}. $V_{S(min)}$ is the threshold voltage at which the change of slope takes place. Using Eq. (16.7), we can find the output voltage during the positive voltage clamping to be

$$v_O = -\frac{R_5}{R_1}v_S + \frac{R_5}{R_4}V_B + \left(1 + \frac{R_5}{R_4}\right)V_D \quad (\text{for } v_S < V_{S(min)}) \qquad \textbf{(16.12)}$$

which describes the transfer characteristic of an adjustable positive and negative voltage limiter, as shown in Fig. 16.11(b). This is a practical limiter and is commonly used. It is also called a *soft limiter* since the output voltage will increase slightly if the input voltage is increased beyond the break points. All DC supply voltages of the limiter are generally made the same magnitude; that is, $V_A = V_B = V_{CC} = V_{EE}$.

EXAMPLE 16.1

D

Designing a negative voltage–limiting circuit Design a negative voltage limiter as shown in Fig. 16.10(a) for $V_A = 12$ V. The circuit should limit the negative output voltage to $V_{O(min)} = -5$ V. The voltage gain without limiting is $A_f = -4$. The diode should be fully turned on at a forward current of $i_D = 0.1$ mA, and its forward voltage drop is $V_D = 0.7$ V. The slope after the break point is to be limited to $-1/20$. Determine the values of R_1, R_2, R_3, and R_F.

SOLUTION

Equation (16.6) gives the diode current at clamping as

$$i_D = \frac{V_{S(max)}}{R_1} = -\frac{1}{R_F} V_{O(min)}$$

so $$0.1 \times 10^{-3} = -\frac{-5}{R_F} \quad \text{or} \quad R_F = 50 \text{ k}\Omega$$

Since $A_f = -4 = -R_F/R_1$,

$$R_1 = \frac{R_F}{4} = \frac{50 \text{ k}}{4} = 12.5 \text{ k}\Omega$$

Since slope $S = -R_3/R_1 = -1/20$,

$$R_3 = \frac{R_1}{20} = \frac{12.5 \text{ k}}{20} = 625 \ \Omega$$

From Eq. (16.5),

$$-5 = -0.7 - (12 + 0.7)\frac{R_3}{R_2}$$

$$\frac{R_3}{R_2} = 0.3386$$

$$R_2 = \frac{R_3}{0.3386} = \frac{625}{0.3386} = 1846 \ \Omega$$

EXAMPLE 16.2

Finding the limiting voltages of an op-amp limiting circuit The adjustable limiter in Fig. 16.11(a) has $R_1 = 15$ kΩ, $R_F = 60$ kΩ, $R_2 = 4$ kΩ, $R_3 = 1$ kΩ, $R_4 = 5$ kΩ, $R_5 = 1$ kΩ, $V_A = 15$ V, $-V_B = -15$ V, and $V_D = 0.7$ V. Determine **(a)** the positive clamping voltage $V_{O(max)}$ and the corresponding input voltage $V_{S(min)}$, **(b)** the negative clamping voltage $V_{O(min)}$ and the corresponding input voltage $V_{S(max)}$, and **(c)** the output voltage when the input voltage is $v_S = 5$ V.

SOLUTION

$R_1 = 15\ \text{k}\Omega$, $R_F = 60\ \text{k}\Omega$, $R_2 = 4\ \text{k}\Omega$, $R_3 = 1\ \text{k}\Omega$, $R_4 = 5\ \text{k}\Omega$, $R_5 = 1\ \text{k}\Omega$, $V_A = 15$ V, $-V_B = -15$ V, $V_D = 0.7$ V, and $v_S = 5$ V.

(a) From Eq. (16.10),

$$V_{O(\max)} = 0.7 + (15 + 0.7) \times \frac{1\ \text{k}}{5\ \text{k}} = 3.84\ \text{V}$$

From Eq. (16.11),

$$V_{S(\min)} = -V_{O(\max)}\frac{R_1}{R_F} = -3.84 \times \frac{15\text{k}}{60\ \text{k}} = -0.96\ \text{V}$$

(b) From Eq. (16.5),

$$V_{O(\min)} = -0.7 - (15 + 0.7) \times \frac{1\ \text{k}}{4\ \text{k}} = -4.625\ \text{V}$$

From Eq. (16.6),

$$V_{S(\max)} = -V_{O(\min)}\frac{R_1}{R_F} = 4.625 \times \frac{15\ \text{k}}{60\ \text{k}} = 1.15625\ \text{V}$$

(c) From Eq. (16.8),

$$v_O = -5 \times \frac{1\ \text{k}}{15\ \text{k}} - 15 \times \frac{1\ \text{k}}{4\ \text{k}} - \left(1 + \frac{1\ \text{k}}{4\ \text{k}}\right) \times 0.7 = -4.958\ \text{V}$$

EXAMPLE 16.3

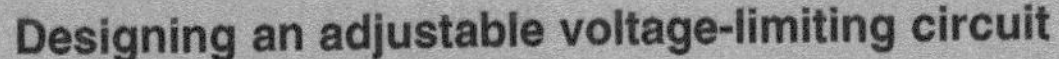

Designing an adjustable voltage-limiting circuit

(a) Design an adjustable limiter as shown in Fig. 16.11(a) to satisfy the following specifications: $V_A = 12$ V, $-V_B = -12$ V, $V_{O(\min)} = -5$ V, $V_{O(\max)} = 6$ V, voltage gain $A_f = -4$, slope after break with a positive input (for $v_S > 0$) $= S_1 = -1/20$, and slope after break with a negative input (for $v_S < 0$) $= S_2 = -1/25$. Assume that diodes are fully on at a diode current of $i_D = 0.1$ mA and the corresponding on-state diode voltage is $V_D = 0.7$ V.

(b) Use PSpice/SPICE [4] to plot the transfer characteristic v_O versus $v_S = \text{V}(6)$. Assume $V_{CC} = 12$ V and $-V_{EE} = -12$ V.

SOLUTION

(a) The steps in completing the design are as follows:

Step 1. The diode current can be related to $V_{O(\min)}$ and R_F by

$$i_D = -\frac{V_{O(\min)}}{R_F} \quad \text{or} \quad 0.1\ \text{mA} = -\frac{-5}{R_F}$$

which gives $R_F = 50\ \text{k}\Omega$.

Step 2. The voltage gain A_f is given by

$$A_f = -\frac{R_F}{R_1} \quad \text{or} \quad -4 = -\frac{50\ \text{k}\Omega}{R_1}$$

which gives $R_1 = 12.5\ \text{k}\Omega$.

Step 3. The slope after break with a positive input (for $v_S > 0$) is

$$S_1 = -\frac{R_3}{R_1} \quad \text{or} \quad -\frac{1}{20} = -\frac{R_3}{12.5\ \text{k}\Omega}$$

which gives $R_3 = 625\ \Omega$.

Step 4. The slope after break with a negative input (for $v_S < 0$) is

$$S_2 = -\frac{R_5}{R_1} \quad \text{or} \quad -\frac{1}{25} = -\frac{R_5}{12.5\ \text{k}\Omega}$$

which gives $R_5 = 500\ \Omega$.

Step 5. The value of R_2 can be found from Eq. (16.5):

$$V_{O(\min)} = -V_D - (V_A + V_D)\frac{R_3}{R_2} \quad \text{or} \quad -5 = -0.7 - (12 + 0.7)\frac{625}{R_2}$$

which gives $R_2 = 1846\ \Omega$.

Step 6. The value of R_4 can be found from Eq. (16.10):

$$V_{O(\max)} = V_D + (V_B + V_D)\frac{R_5}{R_4} \quad \text{or} \quad 6 = 0.7 + (12 + 0.7)\frac{500}{R_4}$$

which gives $R_4 = 1198\ \Omega$.

(b) The limiter circuit for PSpice simulation is shown in Fig. 16.12. The transfer characteristic (for voltage-controlled voltage source model) is shown in Fig. 16.13, which gives $V_{S(\max)} = 1.41$ V, $V_{S(\min)} = -1.72$ V, $V_{O(\max)} = 6.7$ V (expected value is 6 V), and $V_{O(\min)} = -5.5$ V (expected value is -5 V). The discrepancies between the design values and the PSpice results may be attributed to the use of an ideal op-amp versus a macromodel and the use of a specific diode versus a diode of type D1N4148. If we run the simulation using voltage-controlled voltage source model, the PSpice and calculated values will agree more closely.

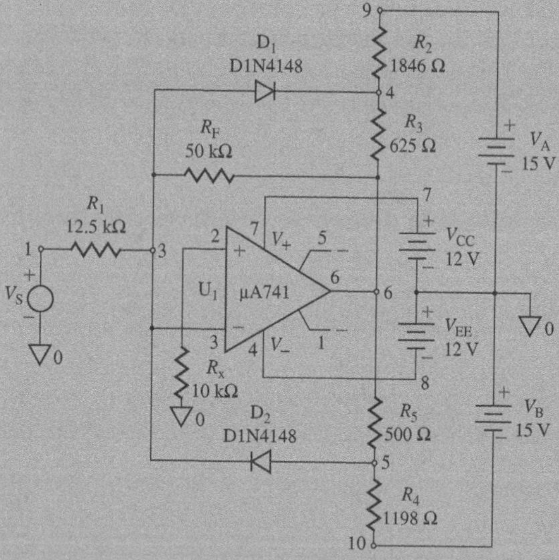

FIGURE 16.12 Limiter circuit for PSpice simulation

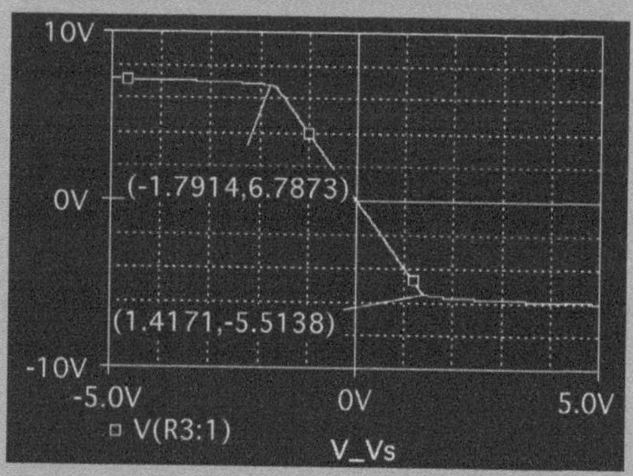

FIGURE 16.13 PSpice transfer characteristic for Example 16.3

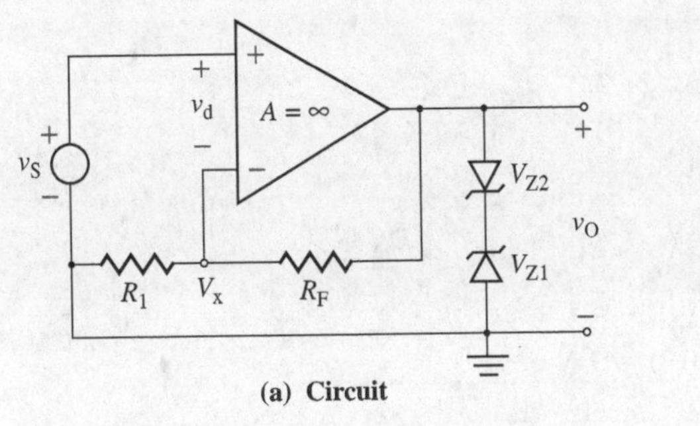

(a) Circuit

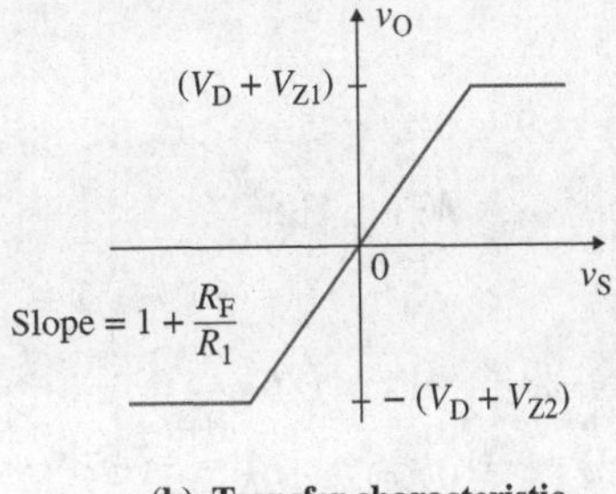

(b) Transfer characteristic

FIGURE 16.14 Output voltage–clamping circuit with zener diodes

16.2.8 Zener Voltage Limiters

The output can be limited by adding two zener diodes across the output terminals of a noninverting or an inverting amplifier. An arrangement for a noninverting amplifier circuit is shown in Fig. 16.14(a). Zener diodes limit the output voltage between $-(V_{Z2} + V_D)$ and $(V_{Z1} + V_D)$, where V_D is the voltage drop of a zener diode in the forward direction. Without zener action, the output voltage can be found by using Eq. (3.18) for the noninverting amplifier:

$$v_O = \left(1 + \frac{R_F}{R_1}\right) v_S$$

If the output voltage is to be limited to a maximum voltage $V_{O(max)} = v_O = (V_{Z1} + V_D)$ and a minimum voltage $V_{O(min)} = v_O = -(V_{Z2} + V_D)$, it can be expressed as

$$v_O = \begin{cases} V_{Z1} + V_D = V_{O(max)} & \text{(for } v_S > V_{S(max)}\text{)} \\ -(V_{Z2} + V_D) = V_{O(min)} & \text{(for } v_S < V_{S(min)}\text{)} \\ \left(1 + \frac{R_F}{R_1}\right) v_S & \text{(for } V_{S(min)} \le v_S \le V_{S(min)}\text{)} \end{cases}$$

The transfer characteristic is shown in Fig. 16.14(b).

EXAMPLE 16.4

Designing a zener voltage–clamping circuit Design an output voltage–clamping circuit as shown in Fig. 16.14(a) so that the normal slope of the transfer characteristic is $S = v_O/v_S = 10$, $V_{O(max)} = 5.7$ V, and $V_{O(min)} = -7.7$ V. Determine the zener voltages V_{Z1} and V_{Z2}. Assume $V_D = 0.7$ V.

SOLUTION

$V_{O(max)} = 5.7$ V, and $V_{O(min)} = -7.7$ V. Since $S = 1 + R_F/R_1 = 10$,

$$\frac{R_F}{R_1} = 10 - 1 = 9$$

If $R_1 = 5\ \text{k}\Omega$,

$$R_F = 9 \times 5\ \text{k} = 45\ \text{k}\Omega$$

Since $V_{O(max)} = (V_{Z1} + V_D)$,

$$V_{Z1} = V_{O(max)} - V_D = 5.7 - 0.7 = 5\ \text{V}$$

Since $V_{O(min)} = -(V_{Z2} + V_D)$,

$$V_{Z2} = -(V_{O(min)} + V_D) = -(-7.7 + 0.7) = 7\ \text{V}$$

16.2.9 Hard Limiters

The limiter in Fig. 16.11(a) can be made to switch between $V_{O(min)}$ and $V_{O(max)}$ if the input voltage v_S becomes greater than zero or less than zero, respectively. This can be accomplished by making the feedback resistance R_F in Fig. 16.11(a) very large, tending to infinity. That is, for $R_F = \infty$, the gain of the circuit becomes very large, tending to infinity, and the output voltage becomes $V_{O(min)}$ if $v_S > 0$ and $V_{O(max)}$ if $v_S < 0$. The circuit shown in Fig. 16.15(a) compares the input signal v_S with the reference signal $V_{ref} = 0$, and if the input signal is greater than or less than zero, the output changes its value. This circuit is known as a *hard limiter*, although it exhibits finite slopes beyond the breaks.

The equations given earlier for adjustable voltage limiters can be applied to the limiter in Fig. 16.15(a), except that $R_F = \infty$ and the gain $A_f = -R_F/R_1$ is very large, tending to infinity. The transfer characteristic is shown in Fig. 16.15(b). The voltage limiting can also be accomplished by connecting two zener diodes, as shown in Fig. 16.16(a), for which $V_{O(max)} = (V_{Z1} + V_D)$ and $V_{O(min)} = -(V_{Z2} + V_D)$. Practical circuits exhibit finite slopes beyond the breaks because of the finite resistances of the zener diodes. These finite slopes are represented in Fig. 16.16(b) by light lines.

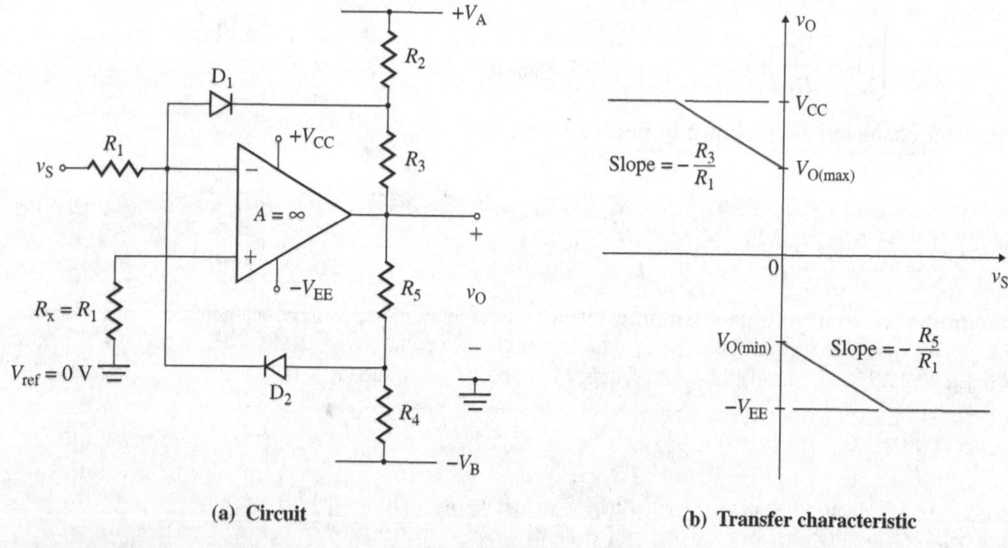

FIGURE 16.15 Hard limiter

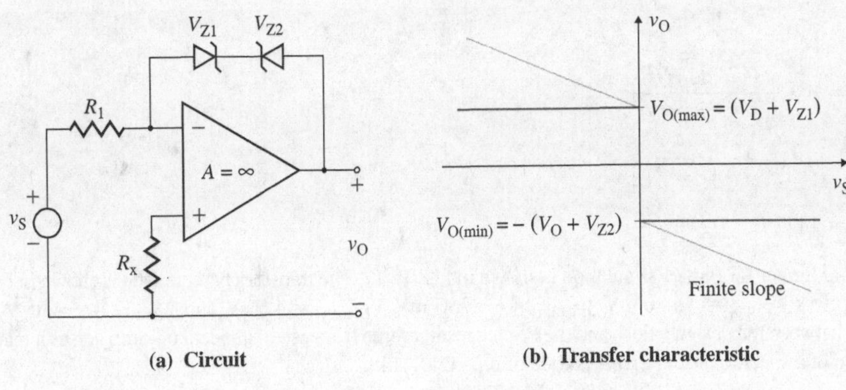

FIGURE 16.16 Zener hard limiter

EXAMPLE 16.5

D

Designing a hard voltage limiter

(a) Design a hard limiter like the one in Fig. 16.15(a). The circuit should limit the negative output voltage to $V_{O(min)} = -5$ V and the positive output voltage to $V_{O(max)} = 5$ V. The magnitude of the slopes after the break points should be less than or equal to 1/20. The diode drop is $V_D = 0.7$ V. The DC supplies are given by $V_A = V_B = 12$ V. Determine the values of R_1, R_2, R_3, R_4, and R_5.

(b) Use PSpice/SPICE to plot the transfer characteristic. Assume $V_{CC} = 12$ V, $-V_{EE} = -12$ V, and $v_S = -6$ V to 6 V.

SOLUTION

(a) Choose $R_1 = 10$ kΩ. Since the slope $|S_1| = R_3/R_1 = 1/20$,

$$R_3 = \frac{R_1}{20} = \frac{10\ \text{k}\Omega}{20} = 500\ \Omega$$

From Eq. (16.5),

$$-5 = -0.7 - \frac{(12 + 0.7)R_3}{R_2}$$

$$\frac{R_3}{R_2} = 0.3386$$

$$R_2 = 1477\ \Omega$$

Since the slope $|S_2| = R_5/R_1 = 1/20$,

$$R_5 = \frac{R_1}{20} = \frac{10\ \text{k}\Omega}{20} = 500\ \Omega$$

From Eq. (16.10),

$$5 = 0.7 + \frac{(12 + 0.7)R_5}{R_4}$$

$$\frac{R_5}{R_4} = 0.3386$$

$$R_4 = 1477\ \Omega$$

(b) The hard limiter for PSpice simulation is shown in Fig. 16.17. The transfer characteristic is shown in Fig. 16.18, which gives $V_{S(max)} = 55.86$ mV, $V_{S(min)} = -27.93$ mV, $V_{O(max)} = 4.50$ V, and $V_{O(min)} = -4.55$ V. Discrepancies between the design values and the PSpice results come from use of an ideal op-amp versus a macromodel and use of a specific diode versus a diode of type D1N4148.

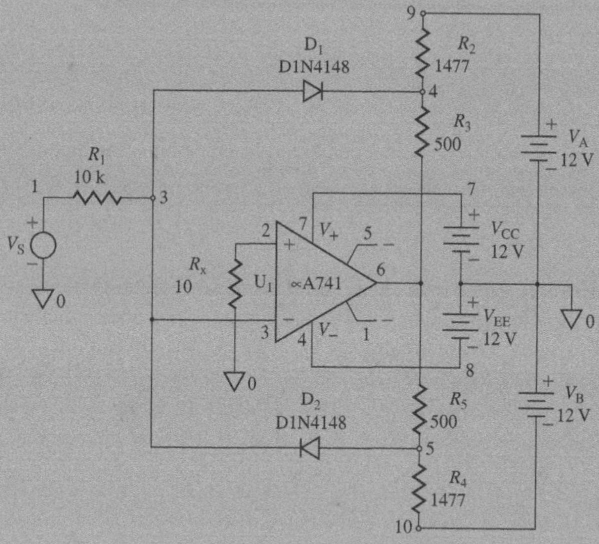

FIGURE 16.17 Comparator circuit for PSpice simulation

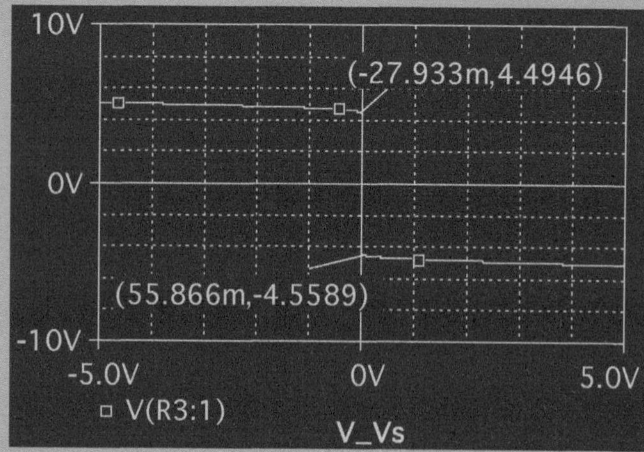

FIGURE 16.18 PSpice transfer characteristic for Example 16.5

KEY POINTS OF SECTION 16.2

- A diode has a voltage drop, typically 0.7 V, which affects the waveforms of voltage and currents in a circuit. However, when a diode is placed inside the feedback path of an op-amp circuit, the effective voltage drop becomes negligible, on the order of microvolts. The diode is then called a superdiode. Superdiodes are generally used for precision signal processing.
- The output voltage of an op-amp circuit can be clamped to a certain level by connecting a diode in parallel with the feedback resistance R_F. If we make $R_F = \infty$, the voltage gain $A_f = \infty$ and the output voltage swings from V_{sat} to $-V_{sat}$ as the input voltage v_S crosses zero.

16.3 Comparators

A comparator compares a signal voltage v_S on one input terminal with a known voltage, called the *reference voltage* V_{ref}, on the other input terminal. The symbol of a comparator, which is similar to that of an op-amp, is shown in Fig. 16.19(a). A comparator gives a digital output voltage v_O. Thus, it can be considered a simple one-bit analog-to-digital (A/D) converter, which produces a digital 1 output ($v_O = V_H$) if the input voltage v_S is above the reference level V_{ref} and a digital 0 output ($v_O = V_L$) if the input voltage v_S falls below the reference level V_{ref}. The output levels V_L and V_H may be of opposite polarity (i.e., V_H positive and V_L negative or vice versa), or both V_L and V_H may be either positive or negative. The transfer characteristic of an ideal comparator is shown in Fig. 16.19(b). The output may be symmetric or asymmetric.

A practical comparator has a finite voltage gain in the range from 3000 to 200,000 and takes a finite amount of time (in the range from 10 ns to 1 μs) to make a transition from one level to another (e.g., V_L to V_H). The transfer characteristic of a practical comparator is shown in Fig. 16.19(c). The input voltage swing required to produce the output voltage transition is in the range of about 0.1 mV to 4 mV. The output of a comparator must switch rapidly between the levels. The bandwidth must be wide because the wider the bandwidth, the faster the switching speed will be. Some typical parameters (listed here for the LM111 comparator) are as follows:

- Operates from a single 5-V power supply
- Input current: 150 nA (maximum)
- Offset current: 20 nA (maximum)
- Differential input voltage: ±30 V
- Voltage gain: 200 V/mV (typical)

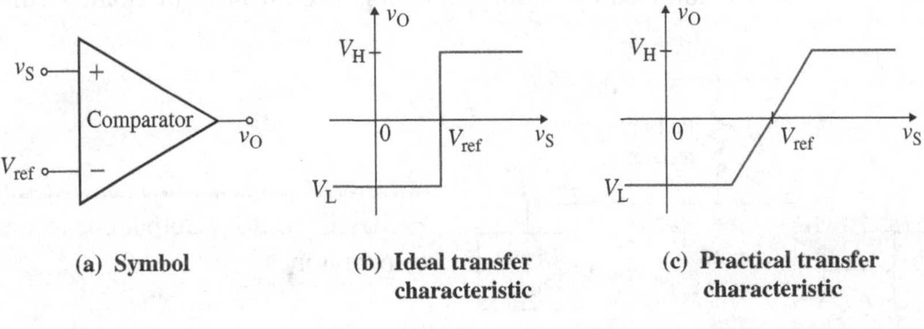

FIGURE 16.19 Symbol and transfer characteristics of a comparator

TABLE 16.1 Comparators versus op-amps

Op-Amps	Comparators
Operation is in closed-loop mode. To avoid an unstable oscillatory response, a sacrifice is usually made in bandwidth, rise time, and slew rate.	Operation is in open-loop mode. No sacrifice is needed in frequency characteristics, and a very fast response time can be obtained.
The output voltage is designed to be zero when the differential input voltage is zero.	The output voltage operates between two fixed output levels: V_L (low) and V_H (high).
The output voltage saturates about 1 V or 2 V away from the positive and negative supply voltages (V_{CC} and V_{EE}).	The low- and high-output levels can be changed for ease in interfacing with digital logic circuits.

16.3.1 Comparators versus Op-Amps

A comparator is designed to operate under open-loop conditions, usually as a switching device, whereas an op-amp is generally operated under closed-loop conditions as a linear amplifier. Otherwise, comparators are very similar to op-amps. Like an op-amp, a comparator has an offset voltage (typically 4 mV), a biasing current (typically 150 nA), and an offset current (typically 20 nA). The characteristics of comparators and op-amps are listed in Table 16.1.

16.3.2 Output-Side Connection

Comparators are often used as an interface between digital and analog signals. The power supply at the analog side (V_{CC} and V_{EE}, typically ±15 V) is different from that at the digital side (V'_{CC}, typically 0 to 5 V). Comparators generally have an open-collector output stage, which allows separate power supplies for the analog and digital parts. A block diagram of the LM111 comparator is shown in Fig. 16.20. The output terminals are the collector and the emitter of an *npn* transistor. If the input voltage to the transistor is low, the transistor is off, and the output logic level is 1; that is, the output is high (5 V), and the current flows through the pull-up resistor R_P to the digital load. On the other hand, if the input voltage to the transistor is high, the transistor is driven into saturation, and the output logic level is 0; that is, the output is low at the saturation voltage of the transistor (typically 0.2 V), and no current flows through the pull-up resistor R_P to the digital load.

16.3.3 Threshold Comparators

The voltage at which a comparator changes from one level to another is called the *crossover* (or *threshold*) voltage. Its value can be adjusted by adding resistors, as shown in the noninverting comparator

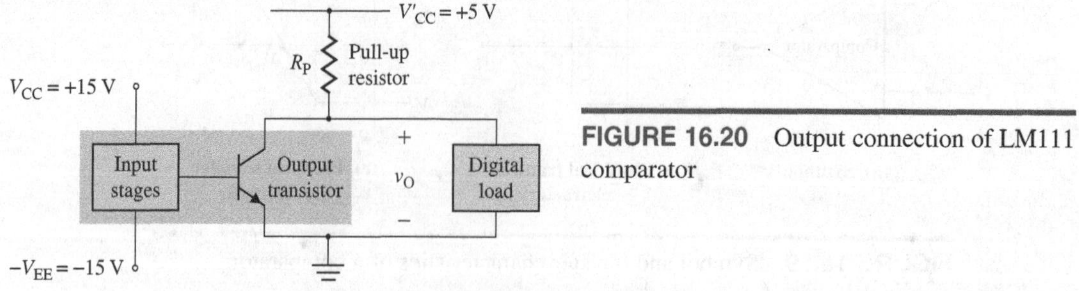

FIGURE 16.20 Output connection of LM111 comparator

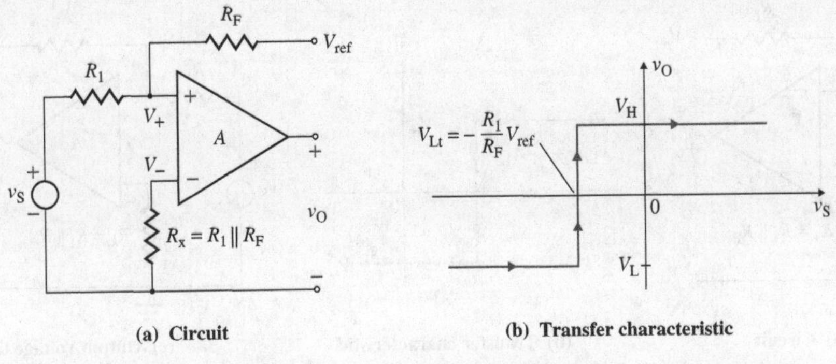

FIGURE 16.21 Noninverting threshold comparator

in Fig. 16.21(a). From the superposition theorem, the voltage V_+ at the noninverting terminal is given by

$$V_+ = \frac{R_1}{R_1 + R_F}V_{ref} + \frac{R_F}{R_1 + R_F}v_S \tag{16.13}$$

Ideally, the crossover will occur when $V_+ = 0$; that is,

$$R_1 V_{ref} + R_F v_S = 0$$

which gives the low threshold voltage of the comparator $V_{Lt} = v_S$ (for changing from low to high) as

$$V_{Lt} = -\frac{R_1}{R_F}V_{ref} \tag{16.14}$$

Thus, the output voltage becomes high (V_H) at the positive saturation voltage ($+V_{sat}$) when $V_+ > 0$ (i.e., $v_S > V_{Lt}$). The transfer characteristic is shown in Fig. 16.21(b).

If the input signal v_S is connected to the inverting terminal, as shown in Fig. 16.22(a), the output will change from high (V_H) to low (V_L). This situation is shown in Fig. 16.22(b). The high threshold voltage of the comparator $V_{Ht} = v_S$ (for changing from high to low) is given by

$$V_{Ht} = \frac{R_1}{R_1 + R_F}V_{ref} \tag{16.15}$$

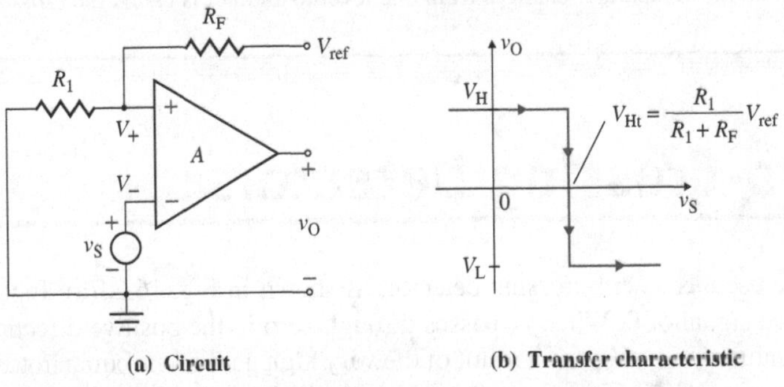

FIGURE 16.22 Inverting configuration for noninverting threshold comparator

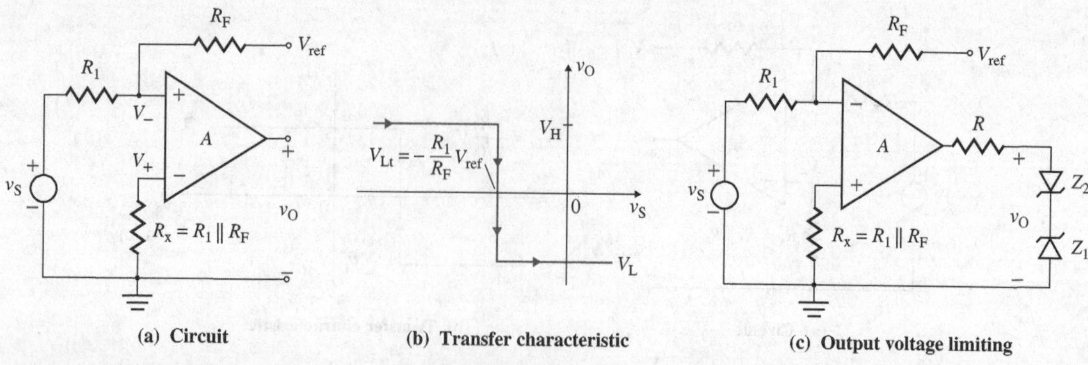

(a) Circuit (b) Transfer characteristic (c) Output voltage limiting

FIGURE 16.23 Inverting threshold comparator

Thus, the output voltage becomes low (V_L) at the negative saturation voltage ($-V_{sat}$) when $v_S > V_+$ (i.e., $v_S > V_{Ht}$). The transfer characteristic is shown in Fig. 16.22(b).

If both the input signal v_S and the reference signal V_{ref} are connected to the inverting terminal, as shown in Fig. 16.23(a), the output will be the inversion of the output in Fig. 16.21(b). That is, the output will change from high (V_H) to low (V_L) when the input is $v_S = V_{Lt}$. This situation is shown in Fig. 16.23(b).

In both inverting and noninverting configurations, the output voltage is limited to the saturation voltage of the comparator. The output voltage can, however, be set to specified limits by external limiters such as zener diodes connected across the output terminals of Figs. 16.21(a), 16.22(a), and 16.23(a). This approach is illustrated in Fig. 16.23(c), in which resistance R is connected to limit the current through the zener diodes.

KEY POINTS OF SECTION 16.3

- A comparator compares a signal voltage on one input terminal with a known voltage, called the *reference voltage*, on the other input terminal. It is designed to operate under open-loop conditions, usually as a switching device. Comparators are often used as an interface between digital and analog signals.
- A comparator can be used as a *threshold comparator* in either inverting or noninverting mode. The voltage at which the comparator changes from one level to another is called the *crossover* (or *threshold*) voltage.

16.4 Zero-Crossing Detectors

A comparator can be used as a zero-crossing detector, as shown in Fig. 16.24(a). Input signal v_S is compared with a reference signal of 0. When v_S passes through zero in the positive direction, the output v_O is driven into negative saturation ($-V_{sat}$) as a result of the very high gain of the comparator. Conversely, when v_S passes through zero in the negative direction, the output is driven into positive saturation ($+V_{sat}$). The input and output waveforms are shown in Fig. 16.24(b).

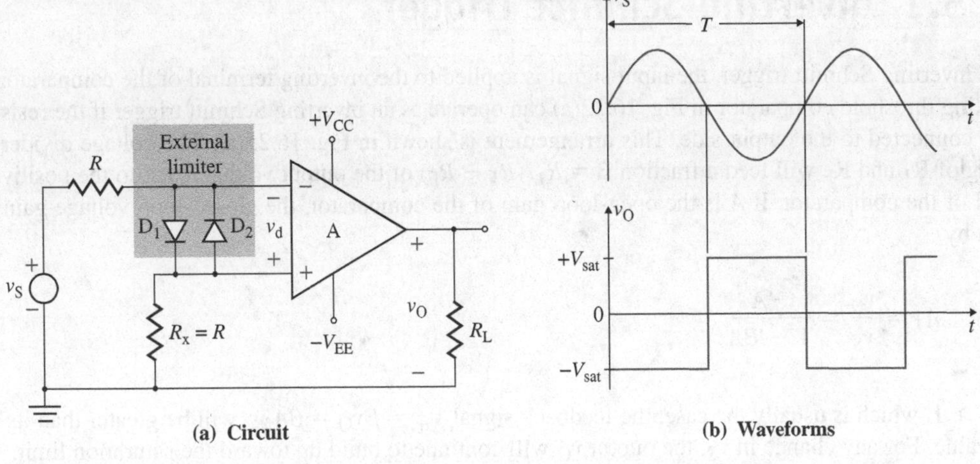

FIGURE 16.24 Zero-crossing detector

Diodes D_1 and D_2 in Fig. 16.24(a) protect the comparator from damage due to excessive input voltage v_S. Because of these diodes, the differential input voltage v_d of the comparator is clamped to approximately 0.7 V or −0.7 V. These diodes, called *clamp diodes*, are external to the comparator. It is up to the designer to determine if the diodes are needed to protect the circuit. Resistance R is connected in series with input signal v_S to limit the current through D_1 and D_2. Resistance R_x is used to reduce the effect of comparator offset problems.

If v_S is such that it crosses zero very slowly, then v_O may not switch quickly from one saturation voltage to the other. Instead, v_O may fluctuate between two saturation voltages $+V_{sat}$ and $-V_{sat}$ as a result of input offset voltage or noise signals at the comparator input terminals. Therefore, a zero-crossing detector will not be suitable for a low-frequency signal or a signal with noise superimposed on it.

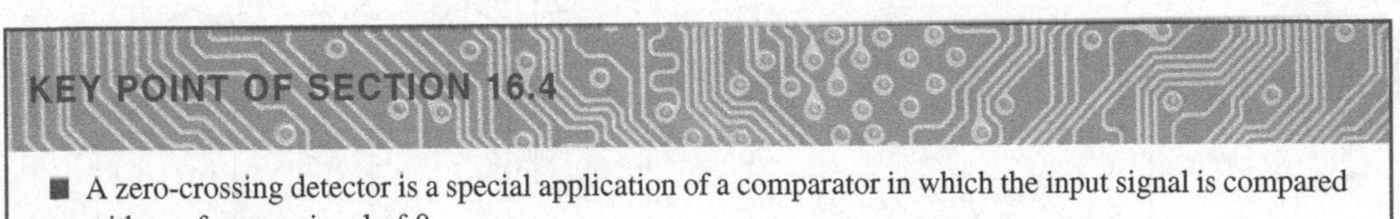

- A zero-crossing detector is a special application of a comparator in which the input signal is compared with a reference signal of 0.

16.5 Schmitt Triggers

A *Schmitt trigger* compares a regular or irregular waveform with a reference signal and converts the waveform to a square or pulse wave. A Schmitt trigger is often known as a *squaring circuit*. It is also known as a *bistable multivibrator* because it has two stable states, low and high. It can remain in one state indefinitely; it moves to the other stable state only when a triggering signal is applied. Schmitt triggers can be classified into two types depending on the type of op-amp configuration used: inverting or noninverting.

16.5.1 Inverting Schmitt Trigger

In an inverting Schmitt trigger, the input signal is applied to the inverting terminal of the comparator. The inverting threshold comparator in Fig. 16.22(a) can operate as an inverting Schmitt trigger if the resistance R_F is connected to the output side. This arrangement is shown in Fig. 16.25(a). The voltage divider consisting of R_1 and R_F will feed a fraction $\beta = R_1/(R_1 + R_F)$ of the output voltage back to the positive terminal of the comparator. If A is the open-loop gain of the comparator, the closed-loop voltage gain A_f is given by

$$A_f = \frac{v_O}{v_S} = \frac{-A}{1 - \beta A} \tag{16.16}$$

If $\beta A > 1$, which is usually the case, the feedback signal $V_+ = \beta v_O = \beta A_f v_S$ will be greater than its original value. For any change in v_S, the ouptut v_O will continue to build up toward the saturation limit.

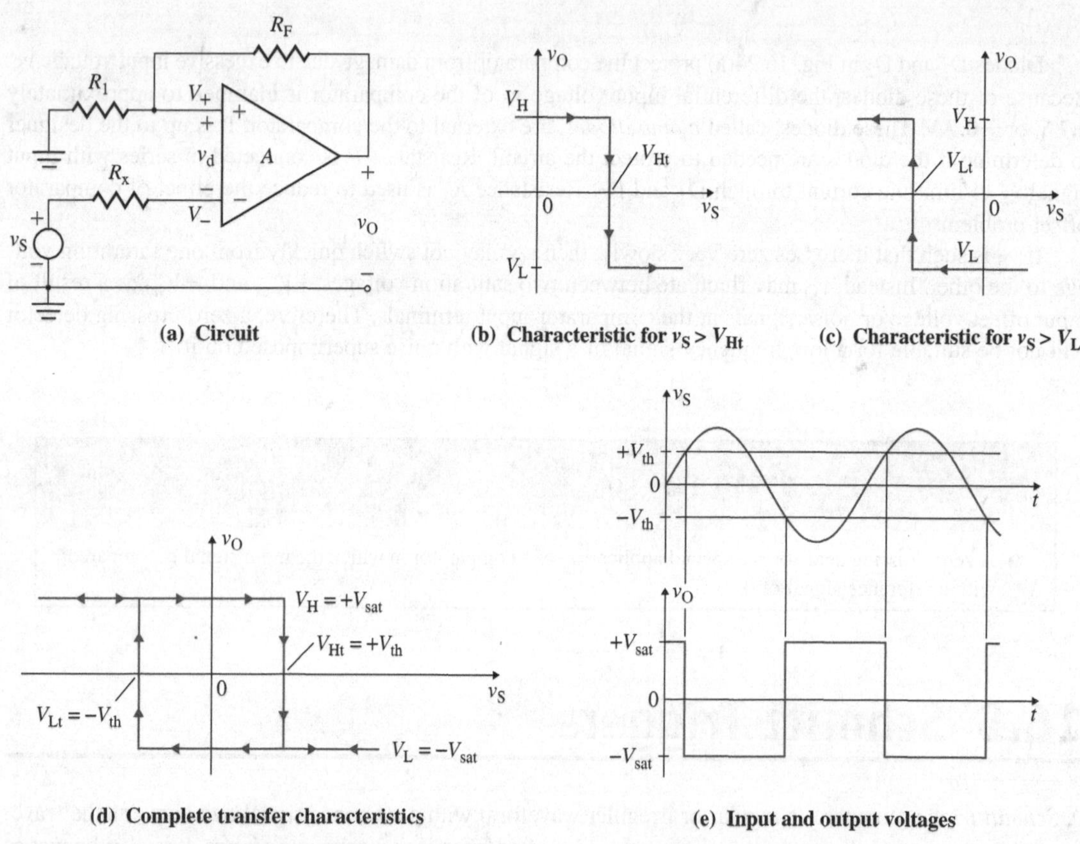

FIGURE 16.25 Schmitt trigger

Transfer Characteristics

To start with, let us assume that v_S is negative and the output is at the positive saturation voltage, $V_H = +V_{sat}$. If v_S is increased from 0, there will be no change in the output until v_S reaches a value of $v_S = V_+ = \beta V_{sat}$. If v_S begins to exceed $V_{Ht} = \beta V_{sat}$, the differential voltage v_d, which will be negative, will be amplified by the voltage gain A of the comparator; that is, v_O will be negative, thereby making V_+ negative also. The result will be an increase in the magnitude of the differential voltage v_d, and v_O will become more negative. This regenerative process will continue until eventually the comparator saturates, with its output voltage equal to the negative saturation voltage, $v_O = V_L = -V_{sat}$, and $V_+ = -\beta V_{sat}$. Increasing v_S beyond $v_S = \beta V_{sat}$ will have no effect on the state of the output voltage. The transfer characteristic for increasing v_S is shown in Fig. 16.25(b).

If v_S is decreased further while the output is low, there will be no change in the output until v_S goes negative, with a value of $v_S = V_+ = -\beta V_{sat}$. If v_S begins to exceed $V_{Lt} = -\beta V_{sat}$, the differential voltage v_d, which will be positive, will be amplified by the gain of the comparator; that is, V_+ will be positive. The result will be an increase in the differential voltage v_d, and v_O will become more positive. This regenerative process will continue until eventually the comparator saturates, with its output voltage equal to the positive saturation voltage, $v_O = V_H = +V_{sat}$, and $V_+ = \beta V_{sat}$. Decreasing v_S further ($v_S \leq -\beta V_{sat}$) will have no effect on the state of the output voltage. The transfer characteristic for decreasing v_S is shown in Fig. 16.25(c).

The complete transfer characteristics are shown in Fig. 16.25(d). A Schmitt trigger exhibits a *hysteresis*, or *deadband*, condition. That is, when the input of the Schmitt trigger exceeds $V_{Ht} = +V_{th}$, its output switches from $+V_{sat}$ to $-V_{sat}$, and when the input goes below $V_{Lt} = -V_{th}$, the output reverts to its original state, $+V_{sat}$.

Every time input voltage v_S exceeds certain levels, called the *positive (or upper) threshold voltage* $+V_{th}$ and the *negative (or lower) threshold voltage* $-V_{th}$, it changes the state of output voltage v_O. If the input signal is a sine wave, the output will be a square wave, as shown in Fig. 16.25(e). $+V_{th}$ and $-V_{th}$ are given by

$$+V_{th} = V_{Ht} = \frac{R_1}{R_1 + R_F}(+V_{sat}) \tag{16.17}$$

$$-V_{th} = V_{Lt} = \frac{R_1}{R_1 + R_F}(-V_{sat}) \tag{16.18}$$

where $V_{sat} = |+V_{sat}| = |-V_{sat}|$ and $V_{th} = |+V_{th}| = |-V_{th}|$.

Effect of Positive Feedback

R_F provides positive feedback. As soon as the output voltage begins to change, positive feedback increases the differential voltage v_d, which, in turn, further changes the output voltage. Once a transition is initiated by a change in the input signal v_S, the positive feedback forces the comparator to complete the transition from one state to another rapidly and to operate in saturation, either positive or negative. Positive feedback leads to rapid transition of the output. Oscillations, which normally occur in the active region and hence prevail for a short time, are avoided.

EXAMPLE 16.6

D **Designing a Schmitt trigger with a hysteresis band**

(a) Design a Schmitt trigger as in Fig. 16.25(a) so that $V_{th} = |+V_{th}| = |-V_{th}| = 5$ V. Assume $V_{sat} = |-V_{sat}| = 14$ V.

(b) Use PSpice/SPICE [4] to plot the hysteresis characteristic for $v_S = 10 \sin(800\pi t)$.

SOLUTION

(a) The steps required to design the Schmitt trigger are as follows:

Step 1. Find the values of R_1 and R_F. From Eq. (16.17),

$$1 + \frac{R_F}{R_1} = \frac{V_{sat}}{V_{th}} \tag{16.19}$$

$$= \frac{14}{5} = 2.8$$

so $R_F/R_1 = 2.8 - 1 = 1.8$. Let $R_1 = 10$ kΩ; then

$$R_F = 1.8 \times R_1 = 18 \text{ k}\Omega \quad \text{(use a 20-k}\Omega\text{ potentiometer)}$$

Step 2. Choose the value of offset minimizing resistance R_x:

$$R_x = R_1 \| R_F = 10 \text{ k}\Omega \| 18 \text{ k}\Omega = 6.43 \text{ k}\Omega$$

(b) The circuit for PSpice simulation is shown in Fig. 16.26. The comparator is simulated by the PSpice macromodel of the LM111. To obtain the negative output voltage swing, terminal 1 is connected to the negative power supply instead of to the ground.

The transfer characteristic and the output voltage v_O = V(U1:OUT) (using EX16-6.SCH) are shown in Fig. 16.27[(a) and (b)], respectively. The simulated values are $V_{Ht} = 5.02$ V (expected value is 5 V), $V_{Lt} = -5.28$ V (expected value is -5 V), $V_H = 14.05$ V (expected value is 14 V), and $V_L = -14.83$ V (expected value is -14 V).

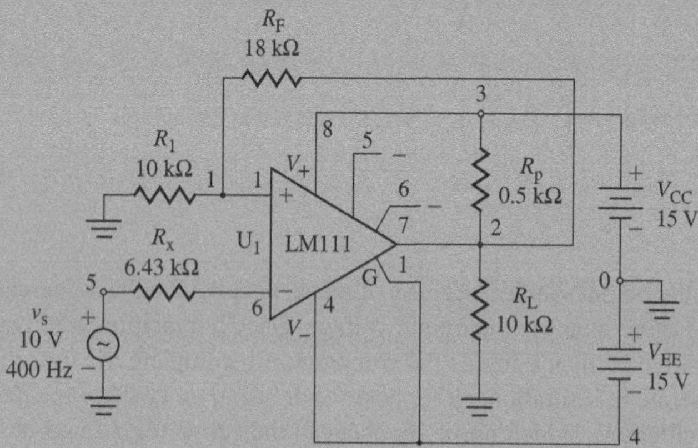

FIGURE 16.26 Schmitt trigger circuit for PSpice simulation

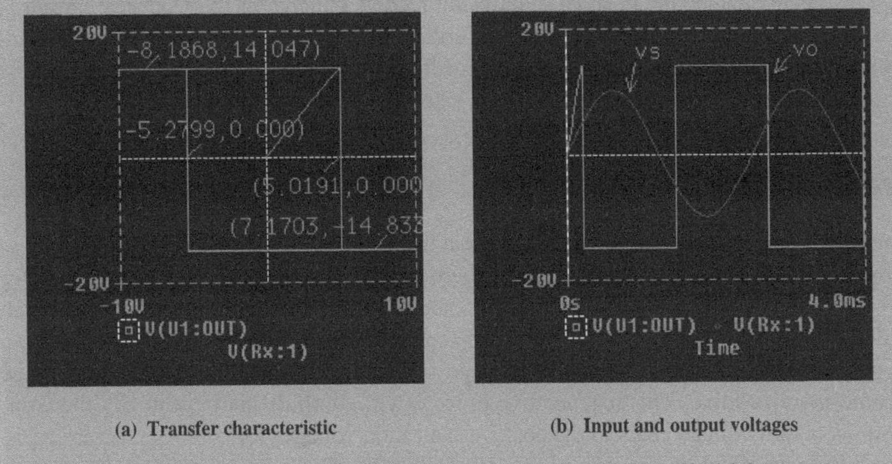

(a) Transfer characteristic (b) Input and output voltages

FIGURE 16.27 Transfer characteristic and output voltage for Example 16.6

At the beginning, the output voltage varies linearly with the input voltage until the input reaches the threshold level, after which the transfer characteristic follows the normal hysteresis band. As a result, the transfer characteristic starts from the origin ($v_O = 0$ and $v_S = 0$). When the output transistor is off, the pull-up resistor R_P forms a potential divider with the load resistor R_L. As a result, the positive output voltage will depend on R_P, whose value should be made small compared to that of R_L. Notice that the transition from low to high and vice versa is very sharp because of the high slew rate of the comparator.

16.5.2 Noninverting Schmitt Trigger

In a noninverting Schmitt trigger, the input signal is applied to the noninverting terminal of the comparator, and the transfer characteristics are inverted. The noninverting threshold comparator in Fig. 16.21(a) can be operated as a noninverting Schmitt trigger if the resistance R_F is connected to the output side. This arrangement is shown in Fig. 16.28(a). Resistance R_F will feed a current signal, a fraction $\beta = 1/R_F$ of the output voltage, back to the positive terminal of the comparator, providing positive shunt-shunt feedback. As soon

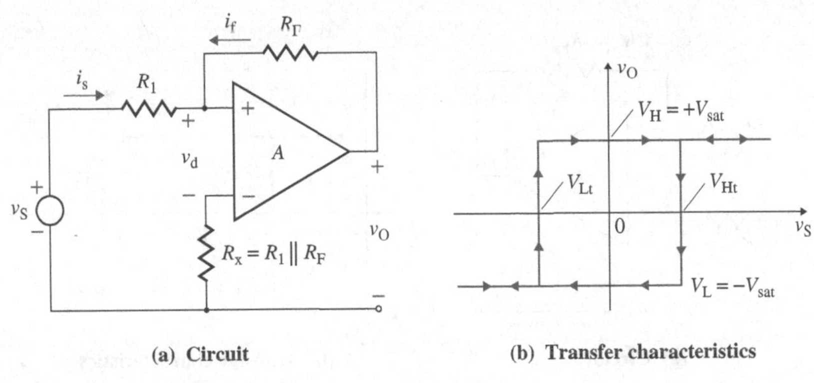

(a) Circuit (b) Transfer characteristics

FIGURE 16.28 Noninverting Schmitt trigger

as the output voltage begins to change, the positive shunt-shunt feedback increases the feedback current i_f, which in turn increases the differential voltage v_d and hence further changes the output voltage. Once a transition is initiated by a change in the input signal v_S, positive feedback forces the comparator to complete the transition from one state to another rapidly and to operate in saturation, either positive or negative.

Transfer Characteristics

To start with, let us assume that v_S is negative and the output is at the negative saturation voltage, $V_L = -V_{sat}$. If v_S is increased from a relatively large negative value, there will be no change in the output until v_S reaches a value of $v_S = V_{Lt} = -V_{sat}R_1/R_F$. If v_S begins to exceed V_{Lt}, the differential voltage v_d, which will be positive, will be amplified by the voltage gain A of the comparator. That is, v_O will be positive, thereby making v_+ positive also. The result will be an increase in the magnitude of the differential voltage v_d, and v_O will become more positive. This regenerative process will continue until eventually the comparator saturates, with its output voltage equal to the positive saturation voltage, $V_H = +V_{sat}$. Increasing v_S further ($v_S \geq V_{Lt}$) will have no effect on the state of the output voltage.

If v_S is decreased while the output is high, there will be no change in the output until v_S decreases to a value of $v_S = V_{Ht} = +V_{sat}R_1/R_F$. If v_S begins to decrease beyond V_{Ht}, the differential voltage v_d will be negative and will be amplified by the gain of the comparator. As a result, v_O will become more negative. This regenerative process will continue until eventually the comparator saturates, with its output voltage equal to the negative saturation voltage $-V_{sat}$. Decreasing v_S further ($v_S \leq V_{Ht}$) will have no effect on the state of the output voltage. The complete transfer characteristics are shown in Fig. 16.28(b).

16.5.3 Schmitt Trigger with Reference Voltage

The *switching voltage* of a Schmitt trigger circuit is defined as the average of V_{Lt} and V_{Ht}. For the circuits in Figs. 16.25(a) and 16.28(a), $V_{Lt} = -V_{Ht}$, and hence the switching voltage V_{st}, which is the width of the hysteresis band, is zero; that is, $V_{st} = (V_{Lt} + V_{Ht})/2 = 0$. However, some applications require shifting the crossover voltage in either the positive or the negative direction along the v_S-axis. This can be accomplished by adding a reference voltage V_{ref} to the circuit in Fig. 16.25(a), as shown in Fig. 16.29(a) for a

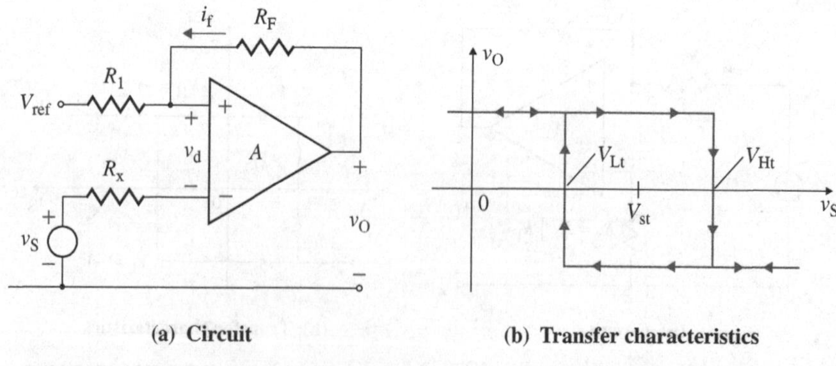

(a) Circuit

(b) Transfer characteristics

FIGURE 16.29 Noninverting Schmitt trigger with reference voltage

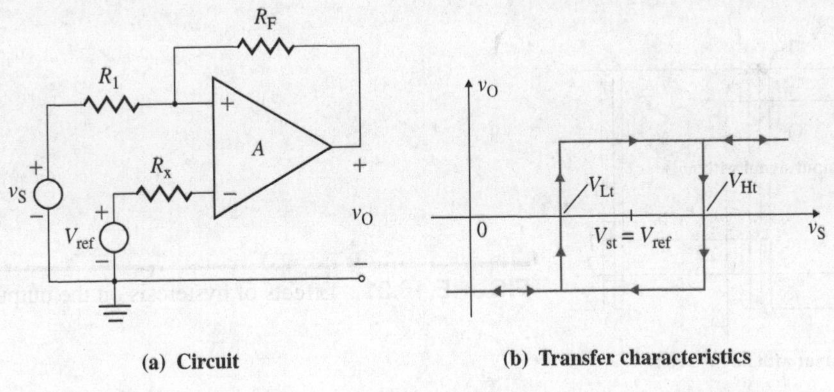

(a) Circuit

(b) Transfer characteristics

FIGURE 16.30 Noninverting Schmitt trigger with reference voltage

noninverting Schmitt trigger. The complete transfer characteristics are shown in Fig. 16.29(b). Assuming that V_{Lt} and V_{Ht} are symmetric about the zero-axis, the switching voltage is given by

$$V_{st} = \frac{R_F}{R_1 + R_F} V_{ref} \tag{16.20}$$

Thus, the upper and lower crossover voltages become

$$V_{Ht} = V_{st} + \frac{R_1}{R_1 + R_F}(+V_{sat}) \tag{16.21}$$

$$V_{Lt} = V_{st} + \frac{R_1}{R_1 + R_F}(-V_{sat}) \tag{16.22}$$

The direction of the hysteresis loop in Fig. 16.29(b) can be reversed by applying a reference voltage V_{ref} to the circuit in Fig. 16.28(a). This arrangement is shown in Fig. 16.30(a), and the corresponding transfer characteristics are shown in Fig. 16.30(b).

16.5.4 Effects of Hysteresis on the Output Voltage

To understand the effect of the *hysteresis*, or *deadband*, condition, consider a sinusoidal signal with a noise signal superimposed on it, as shown in Fig. 16.31(a). If there is no hysteresis, the output voltage will change to its saturation limit when the input signal v_S crosses zero, as shown in Fig. 16.31(b). However, if the output is made to change when the input signal exceeds specified voltage limits V_{Ht} and V_{Lt}, there will be less switching of the output voltage, as shown in Fig. 16.31(c). As a result, any unwanted signal (e.g., noise) will not make the output change. Also, a deadband can be used to reduce the number of contact-bounces in a system such as a temperature-control system, in which the heating element turns on or off when the temperature falls below or rises above the set value.

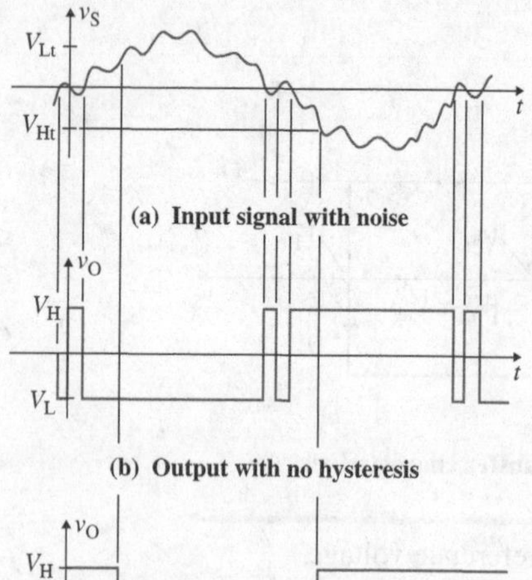

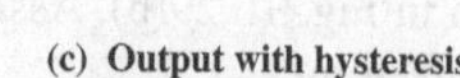

FIGURE 16.31 Effects of hysteresis on the output voltage

EXAMPLE 16.7

D

Designing a Schmitt trigger with a shifted hysteresis band

(a) Design a Schmitt trigger as in Fig. 16.29(a) so that $V_{Ht} = 7$ V and $V_{Lt} = 3$ V. Assume $V_{sat} = |-V_{sat}| = 14$ V and an input frequency of $f = 400$ Hz. Determine the values of R_1, R_F, and V_{ref}.

(b) Use PSpice/SPICE to plot the hysteresis characteristic for $v_S = 10 \sin(800\pi t)$. Use the macromodel of the μA741 op-amp.

SOLUTION

(a) The Schmitt trigger can be designed using the following steps:

Step 1. Find the values of R_1 and R_F. From Eqs. (16.21) and (16.22), we can find the input width of the hysteresis band (HB) as follows:

$$\text{HB} = V_{Ht} - V_{Lt} = \frac{2R_1}{R_1 + R_F} V_{sat}$$

which gives

$$1 + \frac{R_F}{R_1} = \frac{2V_{sat}}{V_{Ht} - V_{Lt}} = \frac{2 \times 14}{7 - 3} = 7$$

Let $R_1 = 10$ kΩ; then $R_F = (7 - 1) \times R_1 = 60$ kΩ.

Step 2. Determine the value of reference voltage V_{ref}.

$$V_{st} = \frac{R_F}{R_1 + R_F} V_{ref} = \frac{V_{Ht} + V_{Lt}}{2} = \frac{7 + 3}{2} = 5 \text{ V}$$

which gives $V_{ref} = 5.83$ V.

(b) We will use op-amp μA741 rather than comparator LM111 to illustrate the advantage of a comparator in a Schmitt trigger. The circuit for PSpice simulation is shown in Fig. 16.32.

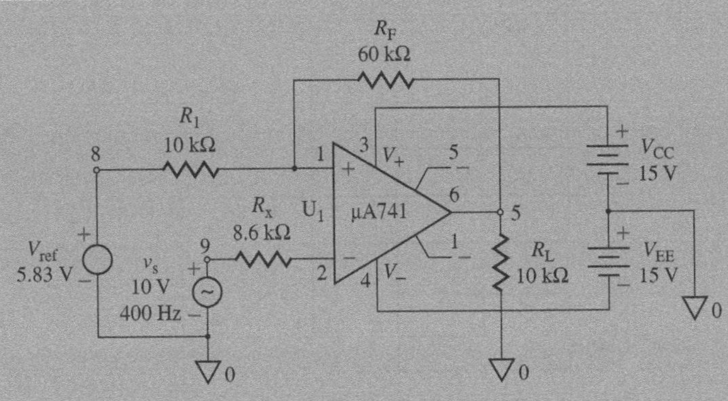

FIGURE 16.32 Schmitt trigger circuit for PSpice simulation

The transfer characteristic and the output voltage (using EX16-7.SCH) are shown in Fig. 16.33[(a) and (b)], which gives V_{Ht} = 7.03 V (expected value is 7 V), V_{Lt} = 3.03 V (expected value is 3 V), and V_{sat} = 14.6 V (expected value is 14 V). The transition from low to high and vice versa is not very sharp, as a result of the low slew rate of the op-amp (compared to that of a comparator, shown in Fig. 16.27).

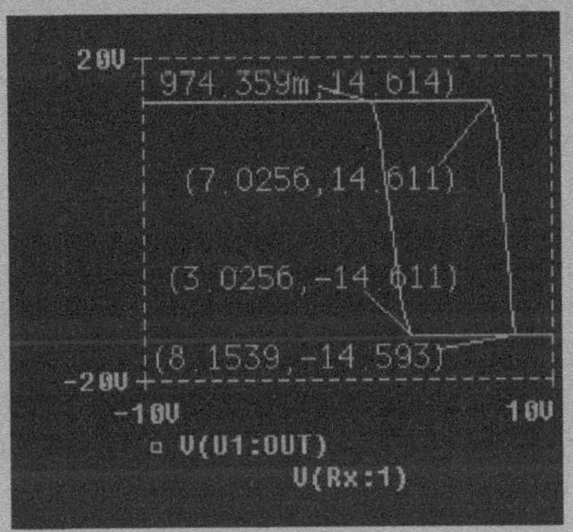

(a) Transfer characteristic

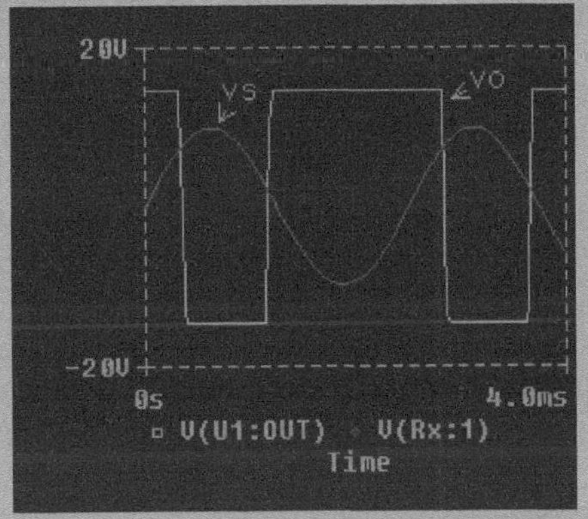

(b) Input and output voltages

FIGURE 16.33 Transfer characteristic and input and output voltages for Example 16.7

KEY POINTS OF SECTION 16.5

- A *Schmitt trigger* compares a regular or irregular waveform with a reference signal and converts the waveform to a square or pulse wave. A Schmitt trigger is often known as a *squaring circuit*. It is also known as a *bistable multivibrator* because it has two stable states, low and high. Schmitt triggers can be classified into two types depending on the type of op-amp configuration used: inverting or noninverting.
- A Schmitt trigger exhibits a *hysteresis*, or *deadband*, condition; that is, when the input of the Schmitt trigger exceeds $+V_{th}$, its output switches from $+V_{sat}$ to $-V_{sat}$, and when the input goes below $-V_{th}$, the output reverts to its original state, $+V_{sat}$.
- When a general-purpose op-amp (e.g., μA741) is used to generate the characteristic of a hysteresis loop, the transition from low to high and vice versa is not as sharp as it is with a comparator.

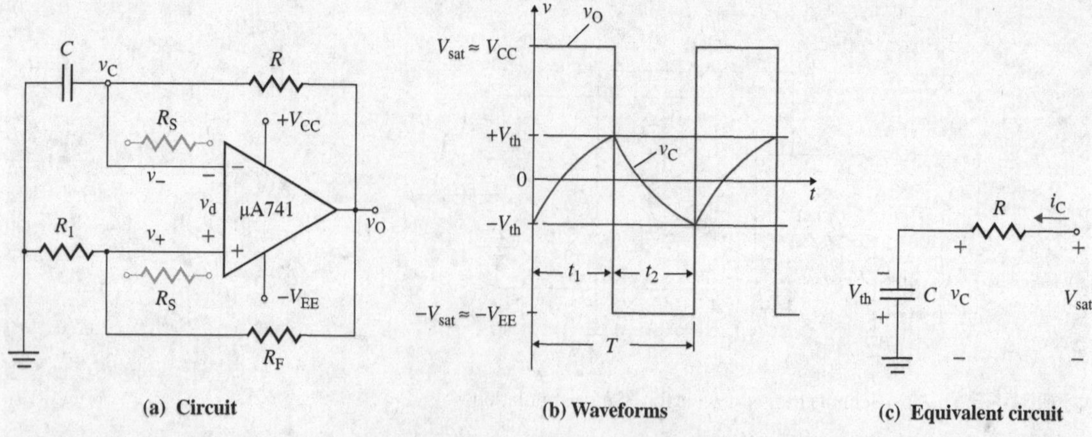

FIGURE 16.34 Square-wave generator

16.6 Square-Wave Generators

A square wave can be generated if the output of an op-amp is forced to swing repetitively between positive saturation $+V_{sat}$ and negative saturation $-V_{sat}$. This can be accomplished by connecting a bistable multivibrator (or Schmitt trigger) with an *RC* circuit in the feedback loop. The circuit implementation is shown in Fig. 16.34(a). This square-wave generator is also called a *free-running* or *astable multivibrator* because the output does not have any stable state. The output of the op-amp will be in either positive saturation or negative saturation, depending on whether the differential input voltage v_d is positive or negative.

Assuming that the voltage across capacitor C is zero, the voltage at the inverting terminal is zero initially; that is, $v_- = 0$ at the instant the DC supply voltages V_{CC} and V_{EE} are turned on. However, at the same instant, the voltage v_+ at the noninverting terminal will have a very small value that will depend on the output offset voltage V_{OO}; that is,

$$v_d = v_+ - v_- = v_+ = V_{OO}$$

which has a positive value. But v_d will be amplified as a result of the very large gain of the op-amp (typically 2×10^5) and will drive the output of the op-amp to positive saturation $+V_{sat}$. Capacitor C will start charging toward $+V_{sat}$ through R. However, as soon as the voltage across C, which is equal to v_-, is slightly more than v_+, then $v_d = v_+ - v_-$ will become negative, and the output of the op-amp will switch to negative saturation $-V_{sat}$. The operation of the circuit can be divided into two modes: mode 1 for $v_d > 0$ and mode 2 for $v_d < 0$.

During mode 1, $v_d > 0$ and the output voltage of the op-amp is at positive saturation $+V_{sat}$. The voltage v_+ becomes

$$v_+ = \frac{R_1}{R_1 + R_F}(+V_{sat}) \tag{16.23}$$

Capacitor C will again start charging toward $+V_{sat}$ through R. As soon as the voltage across C is slightly more than v_+, then $v_d = v_+ - v_-$ will become negative, and the output of the op-amp will be forced to switch to negative saturation $-V_{sat}$.

During mode 2, $v_d < 0$ and the op-amp's output voltage is at negative saturation $-V_{sat}$. The voltage v_+ follows the voltage divider rule and can be found from

$$v_+ = \frac{R_1}{R_1 + R_F}(-V_{sat}) \tag{16.24}$$

As long as v_d is negative, the output will remain in negative saturation. The capacitor will discharge and then recharge toward $-V_{sat}$ through R. When the voltage across C is slightly more negative than v_+, then $v_d = v_+ - v_-$ will become positive, and the output of the op-amp will be forced to switch to positive saturation $+V_{sat}$. Then mode 1 begins again, and the cycle is repeated.

Voltage v_+ acts as the reference. The capacitor voltage v_- tries to follow v_+. However, as soon as the magnitude of v_- becomes slightly greater than that of v_+, v_+ switches its polarity. As a result, the output swings from positive to negative and vice versa. The waveforms of the output voltage and capacitor voltage are shown in Fig. 16.34(b).

Assuming that $+V_{sat}$ is the output voltage and the capacitor has an initial voltage of $-V_{th}$ during mode 1, the equivalent circuit during the charging period is as shown in Fig. 16.34(c). Using KVL, we can write

$$V_{sat} = Ri_C + \frac{1}{C}\int i_C dt - V_{th}$$

which gives the charging current $i_C(t)$ as

$$i_C(t) = \frac{V_{sat} + V_{th}}{R} e^{-t/RC} \tag{16.25}$$

where
$$V_{th} = \frac{R_1}{R_1 + R_F} V_{sat} \tag{16.26}$$

The capacitor voltage $v_C(t)$ can be found from

$$v_C(t) = V_{sat} - (V_{sat} + V_{th})e^{-t/RC} \tag{16.27}$$

At $t = t_1$, the capacitor is recharged to V_{th}. That is, $v_C(t = t_1) = V_{th}$. From Eq. (16.27), we get

$$V_{th} = V_{sat} - (V_{sat} + V_{th})e^{-t_1/RC} \tag{16.28}$$

which gives

$$t_1 = -RC \ln \frac{V_{sat} - V_{th}}{V_{sat} + V_{th}} = -RC \ln \frac{V_{sat} - R_1 V_{sat}/(R_1 + R_F)}{V_{sat} + R_1 V_{sat}/(R_1 + R_F)}$$

$$= -RC \ln \frac{R_F}{2R_1 + R_F} = RC \ln \frac{2R_1 + R_F}{R_F} \tag{16.29}$$

The period T of the output voltage is given by

$$T = t_1 + t_2 = 2t_1 = 2RC \ln \frac{2R_1 + R_F}{R_F} = 2RC \ln \left(1 + \frac{2R_1}{R_F}\right) \tag{16.30}$$

Thus, the frequency of the output voltage f_o is given by

$$f_o = \frac{1}{T} = \frac{1}{2RC \ln (1 + 2R_1/R_F)} \tag{16.31}$$

Equation (16.31) shows that the output frequency depends not only on the time constant $\tau = RC$ but also on the relationship between R_1 and R_F. If $R_F \approx 1.164R_1$, Eq. (16.31) is reduced to

$$f_o = \frac{1}{2RC} \tag{16.32}$$

▶ **NOTES**

1. The inputs of the op-amp are subjected to large differential voltages. To prevent excessive differential current flow to the op-amp, a series resistance R_S, typically on the order of 100 kΩ, can be connected to each of the inverting and noninverting terminals of the op-amp.
2. The peak-to-peak output voltage can be reduced by connecting a pair of zener diodes back to back at the output terminal.

EXAMPLE 16.8

D

Designing a square-wave generator

(a) Design the square-wave generator shown in Fig. 16.34(a) so that $f_o = 5$ kHz. Assume $+V_{sat} = |-V_{sat}| = 14$ V.
(b) Use PSpice/SPICE to check your design. Use the macromodel of the LM111 for sharper transition.

SOLUTION

(a) The steps used to design the square-wave generator are as follows:

Step 1. Choose the value of R_1: Let $R_1 = 10$ kΩ.

Step 2. To simplify the design, choose $R_F = 1.164R_1$ and find R_F:

$$R_F = 1.164R_1 = 1.164 \times 10\text{ k}\Omega = 11.64\text{ k}\Omega \quad \text{(use a 20-k}\Omega\text{ potentiometer)}$$

Step 3. Choose a value of C: Let $C = 0.01$ μF.

Step 4. Find the value of R from Eq. (16.32):

$$R = \frac{1}{2Cf_o} = \frac{1}{2 \times 0.01\ \mu\text{F} \times 5\text{ kHz}} = 10\text{ k}\Omega$$

(b) The circuit for PSpice simulation is shown in Fig. 16.35.

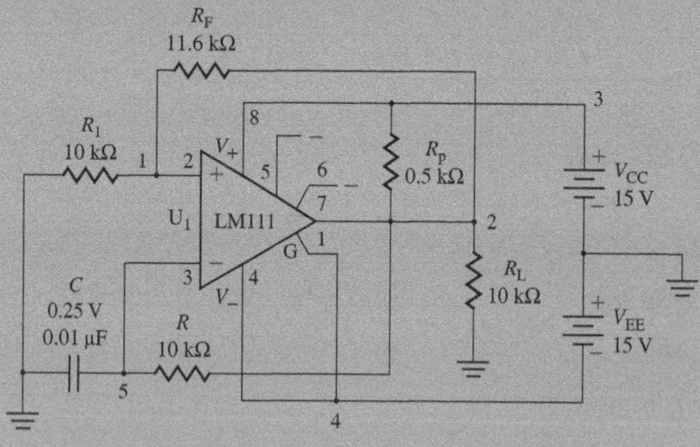

FIGURE 16.35 Square-wave generator for PSpice simulation

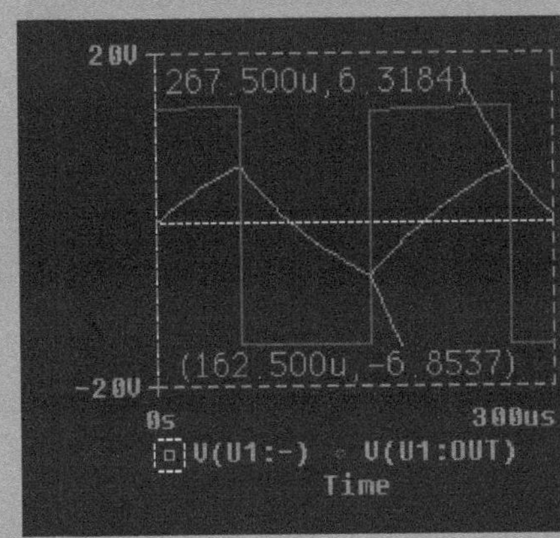

FIGURE 16.36 Output voltage waveforms for Example 16.8

The output waveforms (using EX16-8.SCH) are shown in Fig. 16.36. The results are $V_{sat} = 13.53$ V (expected value is 14 V), $-V_{sat} = -14.83$ V (expected value is -14 V), $V_{th} = 6.3$ V, $-V_{th} = -6.8$ V, and $T = 2 \times (267.5 - 162.5) = 210$ μs (expected value is 200 μs). The output voltage does not change sharply from one state to another. It is not a full square wave because of the slew rate and bandwidth limits of the op-amp.

KEY POINT OF SECTION 16.6

- A square wave can be generated by forcing the output of an op-amp to swing repetitively between positive saturation $+V_{sat}$ and negative saturation $-V_{sat}$. This is accomplished by connecting a bistable multivibrator (or Schmitt trigger) in the feedback loop. This square-wave generator is also called a *free-running* or *astable multivibrator.*

16.7 Triangular-Wave Generators

A triangular-wave generator can be produced by integrating the square-wave output of a Schmitt trigger circuit. This can be accomplished by cascading an integrator with a bistable multivibrator (or Schmitt trigger), as shown in Fig. 16.37(a), which consists of a comparator A_1 and an integrator A_2. Comparator A_1 continuously compares the noninverting input v_+ at point P with the inverting input v_- (i.e., 0 V). Thus, the differential voltage is $v_d = v_+ - v_- = v_+$. Because of the very large gain of the op-amp, the output of A_1 will be at negative saturation $-V_{sat}$ or positive saturation $+V_{sat}$ when v_+ goes slightly below or above 0, respectively.

To examine the principle of operation of the triangular-wave generator, let us assume that when the supply voltages V_{CC} and $-V_{EE}$ are switched on, the voltage at the noninverting terminal begins slightly above 0 as a result of the input offset voltage of the op-amp. Because of the high gain of the op-amp, the output of A_1 will be switched to positive saturation $+V_{sat}$. Therefore, the output of the op-amp is forced to swing repetitively between positive saturation $+V_{sat}$ and negative saturation $-V_{sat}$. The output voltages of A_1 and A_2 are shown in Fig. 16.37(b). The operation of the circuit can be divided into two modes.

During mode 1, $v_+ > 0$ and the output of A_1 is at positive saturation $+V_{sat}$, which is the input to the inverting integrator A_2. The output of A_2 will be a negative-going ramp. Thus, one side of R_F will be at

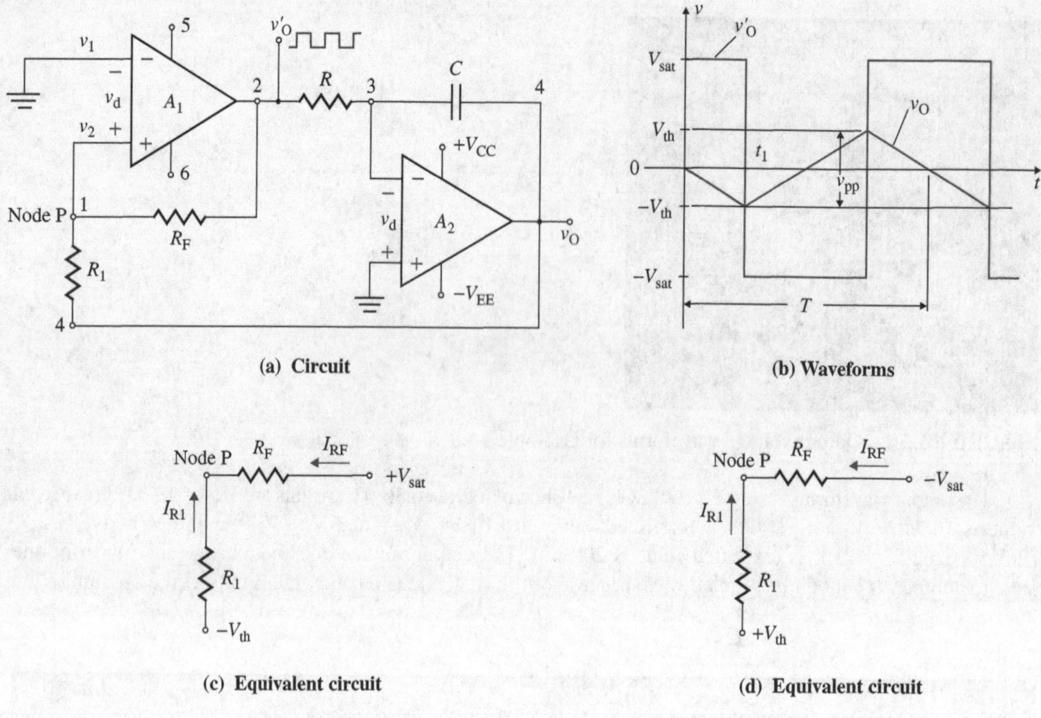

FIGURE 16.37 Triangular-wave generator

$+V_{sat}$, and one side of R_1 will be at the negative-going ramp of A_2. When the negative-going ramp exceeds a certain value $-V_{th}$, the voltage at point P will be slightly below 0, and the output of A_1 will switch from positive saturation $+V_{sat}$ to negative saturation $-V_{sat}$.

The equivalent circuit for determining the condition under which the circuit switches over to negative saturation is shown in Fig. 16.37(c). Since the current flowing into the op-amp is negligible, $I_{R1} = -I_{RF}$; that is,

$$\frac{-V_{th}}{R_1} = \frac{-V_{sat}}{R_F}$$

which gives the condition for $v_+ < 0$ as

$$-V_{th} = \frac{R_1}{R_F}(-V_{sat}) \tag{16.33}$$

During mode 2, $v_+ < 0$ and the output of A_1 is at negative saturation $-V_{sat}$, which is the input to the integrator A_2. The output of A_2 will be a positive-going ramp. When the positive-going ramp exceeds a certain value $+V_{th}$, the voltage at point P will be slightly above 0, and the output of A_1 will switch from negative saturation $-V_{sat}$ to positive saturation $+V_{sat}$. When this occurs, mode 1 starts again, and the cycle is repeated.

The equivalent circuit for determining the condition under which the circuit switches over to positive saturation is shown in Fig. 16.37(d). Neglecting the current flowing into the op-amp, we have $I_{R1} = -I_{RF}$; that is,

$$\frac{V_{th}}{R_1} = \frac{V_{sat}}{R_F}$$

which gives the condition for $v_+ > 0$ as

$$V_{th} = \frac{R_1}{R_F}(+V_{sat}) \tag{16.34}$$

where $V_{sat} = |+V_{sat}| = |-V_{sat}|$. The peak-to-peak output amplitude of the triangular wave is given by

$$v_{pp} = V_{th} - (-V_{th}) = \frac{2R_1}{R_F}V_{sat} \tag{16.35}$$

The period and the frequency of the output voltage can be determined from the time that is required to charge the capacitor from $-V_{th}$ to V_{th} or from V_{th} to $-V_{th}$. Let us consider the capacitor voltage during mode 2, when the input voltage to the integrator is $-V_{sat}$; that is,

$$v_C(t) = -\frac{1}{RC}\int v_O' \, dt - V_{th} = -\frac{1}{RC}\int (-V_{sat})\, dt - V_{th}$$

$$= \frac{V_{sat}}{RC}t - V_{th} \tag{16.36}$$

At half period, $t = t_1 = T/2$ and $v_C(t = T/2) = V_{th}$, and Eq. (16.36) yields

$$V_{th} = \frac{V_{sat}}{RC}\left(\frac{T}{2}\right) - V_{th}$$

which gives the period as

$$T = \frac{4RCV_{th}}{V_{sat}} \tag{16.37}$$

Substituting the value of V_{th} from Eq. (16.34), we have for the period of the triangular wave

$$T = \frac{4RC}{V_{sat}} \times \frac{R_1 V_{sat}}{R_F} = \frac{4RCR_1}{R_F} \tag{16.38}$$

which gives the frequency of oscillation as

$$f_o = \frac{R_F}{4RCR_1} \tag{16.39}$$

EXAMPLE 16.9

D

Designing a triangular-wave generator

(a) Design the triangular-wave generator shown in Fig. 16.37(a) so that $f_o = 4$ kHz and $V_{th} = |-V_{th}| = 5$ V. Assume $+V_{sat} = |-V_{sat}| = 14$ V.

(b) Use PSpice/SPICE to check your design.

SOLUTION

(a) The steps used to design the triangular-wave generator are as follows:

Step 1. Find the values of R_1 and R_F. Equation (16.34) gives

$$\frac{R_1}{R_F} = \frac{V_{th}}{V_{sat}} = \frac{5}{14} = 0.36$$

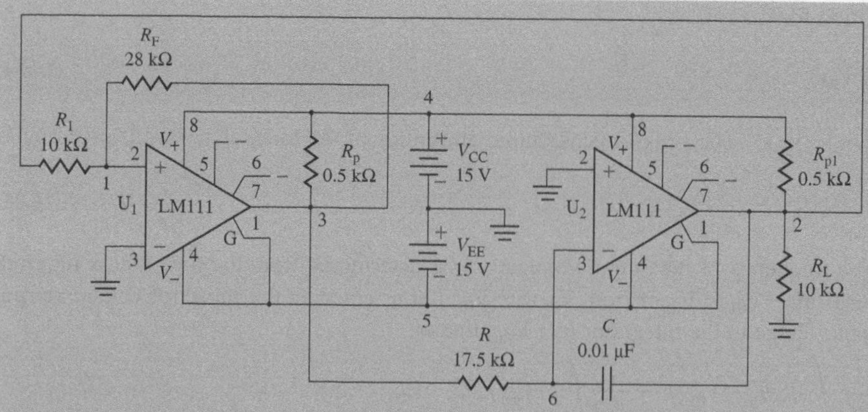

FIGURE 16.38 Triangular-wave generator for PSpice simulation

Let R_1 = 10 kΩ; then

$$R_F = \frac{R_1}{0.36} = \frac{10\text{ k}\Omega}{0.36} = 28\text{ k}\Omega \quad \text{(use a 30-k}\Omega\text{ potentiometer)}$$

Step 2. Choose a suitable value of C: Let $C = 0.01\ \mu$F.

Step 3. Find the value of R. Equation (16.39) gives

$$R = \frac{R_F}{4f_o C R_1} \tag{16.40}$$

$$= \frac{28\text{ k}\Omega}{4 \times 4\text{ kHz} \times 0.01\ \mu\text{F} \times 10\text{ k}\Omega} = 17.5\text{ k}\Omega \quad \text{(use a 20-k}\Omega\text{ potentiometer)}$$

(b) The circuit for PSpice simulation is shown in Fig. 16.38. The op-amp is simulated by PSpice macromodel LM111.

The PSpice plots of the voltages at the output of amplifiers A_1 and A_2 (using EX16-9.SCH) are shown in Fig. 16.39. The results are V_{sat} = 14.4 V (expected value is 14 V), $-V_{sat} = -14.8$ V (expected value is -14 V), V_{th} = 5.3 V (expected value is 5 V), $-V_{th} = -5.2$ V (expected value is -5 V), and $T = 312.1 - 62.6 = 249.5\ \mu$s (expected value is 250 μs). If the integrator time constant $\tau = RC$ is made much smaller than the period of the output waveform, the triangular wave can be made very close to a sine wave.

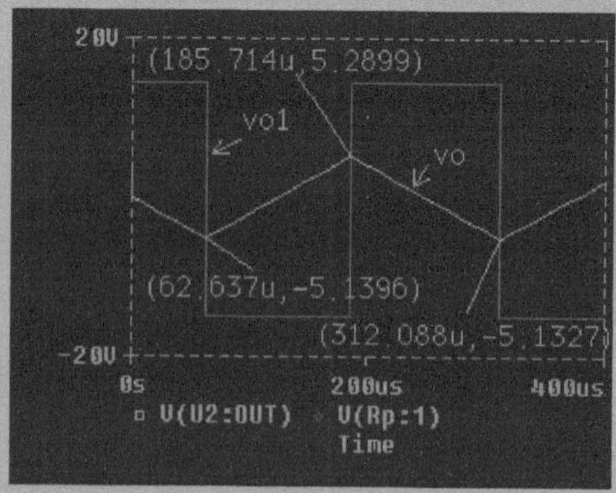

FIGURE 16.39 Output waveforms for Example 16.9

KEY POINTS OF SECTION 16.7

- A triangular-wave generator can be produced by integrating the square-wave output of a Schmitt trigger. This can be accomplished by cascading an integrator with a bistable multivibrator (or Schmitt trigger).
- If the integrator time constant $\tau = RC$ is made much smaller than the period of the output waveform, the triangular wave can be made very close to a sine wave.

16.8 Sawtooth-Wave Generators

In a triangular waveform, the rise time is always equal to the fall time. That is, the same amount of time is required for a triangular wave to swing from $-V_{th}$ to $+V_{th}$ as from $+V_{th}$ to $-V_{th}$. On the other hand, a sawtooth waveform has unequal rise and fall times. The rise time may be faster than the fall time or vice versa. The triangular-wave generator in Fig. 16.37(a) may be converted to a sawtooth generator by adding a variable DC voltage V_{ref} to the noninverting terminal of the op-amp. The addition of V_{ref}, which acts as a reference signal for the integrator A_2, can be accomplished by using a potentiometer, as shown in Fig. 16.40(a). The voltage waveforms are shown in Fig. 16.40(b). As with the triangular-wave generator, the operation of the circuit can be divided into two modes.

During mode 1, $v_+ > 0$ and the output of A_1 is at positive saturation $+V_{sat}$, which is the input signal to the integrator A_2. The equivalent circuit for the operation of the integrator is shown in Fig. 16.40(c). At the beginning of this mode, the output voltage is V_{th}, and the voltage at the inverting terminal is $v_1 \approx V_{ref}$. Thus, the initial voltage on the capacitor is

$$v_C(t = 0) = v_1 - v_O = V_{ref} - V_{th}$$

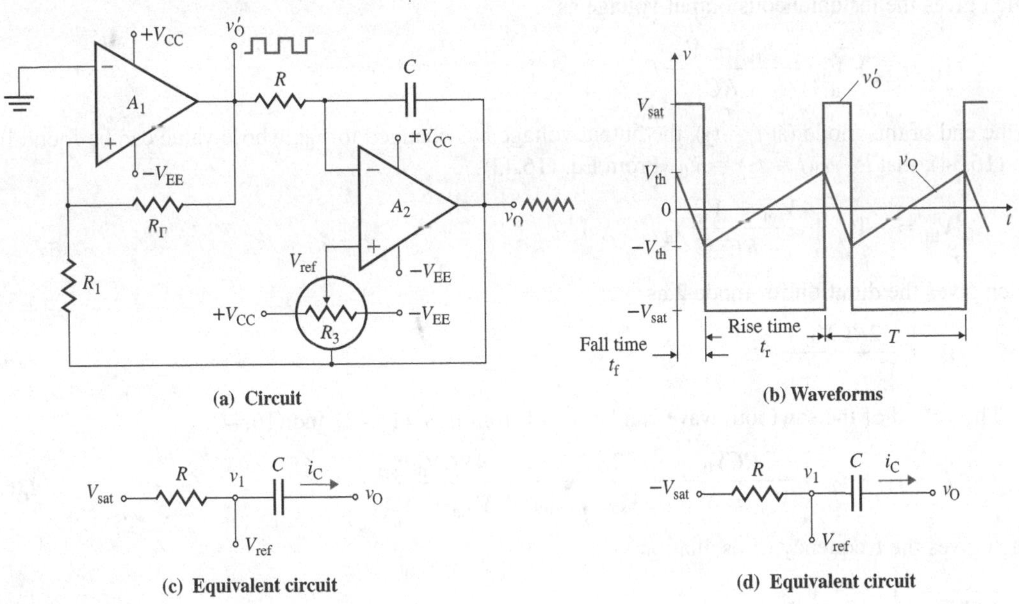

FIGURE 16.40 Sawtooth-wave generator

The instantaneous voltage across the capacitor $v_C(t)$ is given by

$$V_{ref} - v_O(t) = \frac{1}{C}\int i_C\,dt + v_C(t=0) = \frac{1}{C}\int \frac{V_{sat} - V_{ref}}{R}\,dt + V_{ref} - V_{th}$$

which gives the instantaneous output voltage as

$$v_O(t) = V_{th} - \frac{V_{sat} - V_{ref}}{RC}\,t \qquad (16.41)$$

At the end of this mode (at $t = t_1$), the output voltage has changed to $-V_{th}$, whose value can be found from Eq. (16.33). That is, $v_O(t = t_1) = -V_{th}$. From Eq. (16.41),

$$-V_{th} = V_{th} - \frac{V_{sat} - V_{ref}}{RC}\,t_1$$

which gives the duration of mode 1 as

$$t_1 = \frac{2RCV_{th}}{V_{sat} - V_{ref}} \qquad (16.42)$$

During mode 2, $v_+ < 0$ and the output of A_1 is at negative saturation $-V_{sat}$, which is the input to the integrator A_2. The equivalent circuit for the operation of the integrator is shown in Fig. 16.40(d). At the beginning of this mode, the output voltage is $-V_{th}$, and the initial voltage on the capacitor is

$$v_C(t=0) = v_1 - v_O = V_{ref} + V_{th}$$

If we redefine the time origin $t = 0$ as the beginning of this mode, the instantaneous voltage across the capacitor is given by

$$V_{ref} - v_O(t) = \frac{1}{C}\int i_C\,dt + v_C(t=0) = \frac{1}{C}\int \frac{-V_{sat} - V_{ref}}{R}\,dt + V_{ref} + V_{th}$$

which gives the instantaneous output voltage as

$$v_O(t) = -V_{th} + \frac{V_{sat} + V_{ref}}{RC}\,t \qquad (16.43)$$

At the end of this mode (at $t = t_2$), the output voltage has changed to V_{th}, whose value can be found from Eq. (16.34); that is, $v_O(t = t_2) = V_{th}$. From Eq. (16.43),

$$V_{th} = -V_{th} + \frac{V_{sat} + V_{ref}}{RC}\,t_2$$

which gives the duration for mode 2 as

$$t_2 = \frac{2RCV_{th}}{V_{sat} + V_{ref}} \qquad (16.44)$$

The period of the sawtooth wave can be found from Eqs. (16.42) and (16.44):

$$T = t_1 + t_2 = \frac{2RCV_{th}}{V_{sat} - V_{ref}} + \frac{2RCV_{th}}{V_{sat} + V_{ref}} = \frac{4RCV_{th}V_{sat}}{V_{sat}^2 - V_{ref}^2} \qquad (16.45)$$

which gives the frequency of oscillation as

$$f_o = \frac{V_{sat}^2 - V_{ref}^2}{4RCV_{th}V_{sat}} \qquad (16.46)$$

Duty cycle k, which is defined as the ratio of t_1 to T, can be determined from Eqs. (16.42) and (16.45) as follows:

$$k = \frac{t_1}{T} = \frac{2RCV_{th}}{V_{sat} - V_{ref}} \times \frac{V_{sat}^2 - V_{ref}^2}{4RCV_{th}V_{sat}} = \frac{V_{sat} + V_{ref}}{2V_{sat}}$$

$$= \frac{1}{2}\left[1 + \frac{V_{ref}}{V_{sat}}\right] \tag{16.47}$$

▶ **NOTE** This circuit allows a designer more control over the shape of the output waveform because k, T, and V_{th} can be varied.

EXAMPLE 16.10

D **Designing a sawtooth-wave generator** Design the sawtooth-wave generator shown in Fig. 16.40(a) so that $f_o = 4$ kHz, $V_{th} = 5$ V, and the circuit has a duty cycle of $k = 0.25$. Assume $V_{sat} = |-V_{sat}| = 14$ V.

SOLUTION

The steps used to design the sawtooth-wave generator are as follows:

Step 1. Find the value of V_{ref} required to obtain the desired duty cycle k. From Eq. (16.47),

$$V_{ref} = (2k - 1)V_{sat} \tag{16.48}$$

$$= (2 \times 0.25 - 1) \times 14 = -7 \text{ V}$$

Step 2. Find the values of R_1 and R_F. Equation (16.34) gives

$$\frac{R_1}{R_F} = \frac{V_{th}}{V_{sat}} = \frac{5}{14} = 0.36$$

Let $R_1 = 10$ kΩ; then

$$R_F = \frac{R_1}{0.36} = \frac{10 \text{ k}\Omega}{0.36} = 28 \text{ k}\Omega \quad \text{(use a 30-k}\Omega\text{ potentiometer)}$$

Step 3. Choose a suitable value of C: Let $C = 0.01$ µF.

Step 4. Find the value of R. Equation (16.46) gives

$$R = \frac{V_{sat}^2 - V_{ref}^2}{4f_oCV_{th}V_{sat}} \tag{16.49}$$

$$= \frac{14^2 \text{ V}^2 - (-7)^2 \text{ V}^2}{4 \times 4 \text{ kHz} \times 0.01 \text{ µF} \times 5 \text{ V} \times 14 \text{ V}} = 13.1 \text{ k}\Omega \quad \text{(use a 15-k}\Omega\text{ potentiometer)}$$

KEY POINT OF SECTION 16.8

- A sawtooth waveform has unequal rise and fall times. The rise time may be faster than the fall time or vice versa. A triangular-wave generator can be converted to a sawtooth-wave generator by adding a variable DC voltage V_{ref} to the noninverting terminal of the op-amp.

16.9 Voltage-Controlled Oscillators

A *voltage-controlled oscillator* (VCO) is an oscillator circuit in which the oscillation frequency is controlled by an externally applied voltage [5, 6]. A linear relationship between the oscillation frequency f_o and the control voltage v_C is often required in a VCO. A VCO is also called a *voltage-to-frequency* (V/F) *converter*. VCOs are employed in many applications such as frequency modulation (FM), tone generation, and frequency-shift keying (FSK).

To convert a voltage to a frequency, a capacitor is normally charged and then discharged at a constant current whose value depends on an externally applied voltage. The charging begins when the capacitor voltage falls to a lower threshold voltage V_L. Similarly, the discharging begins when the capacitor voltage rises to an upper threshold voltage V_H.

Figure 16.41(a) shows a simplified block diagram of the operation of a VCO. The current sources are used to charge and discharge capacitor C_1. The input to the Schmitt trigger is the capacitor voltage; the output of the Schmitt trigger has two threshold voltage switching levels V_L and V_H, which control the closing and opening of the current switch. Thus, depending on the capacitor voltage $v_C(t)$, the current switch connects the capacitor either to the top current source for charging or to the bottom current source for discharging. The waveform of the capacitor voltage is shown in Fig. 16.41(b); it has two modes, charging and discharging.

16.9.1 Charging Mode

During charging mode, the capacitor is charged by the top current source I_Q from the lower threshold voltage V_L to the upper trigger level V_H. The time required to charge the capacitor from V_L to V_H is given by

$$\Delta t_1 = \frac{C_1}{I_Q}\Delta v_C = \frac{C_1}{I_Q}(V_H - V_L) \tag{16.50}$$

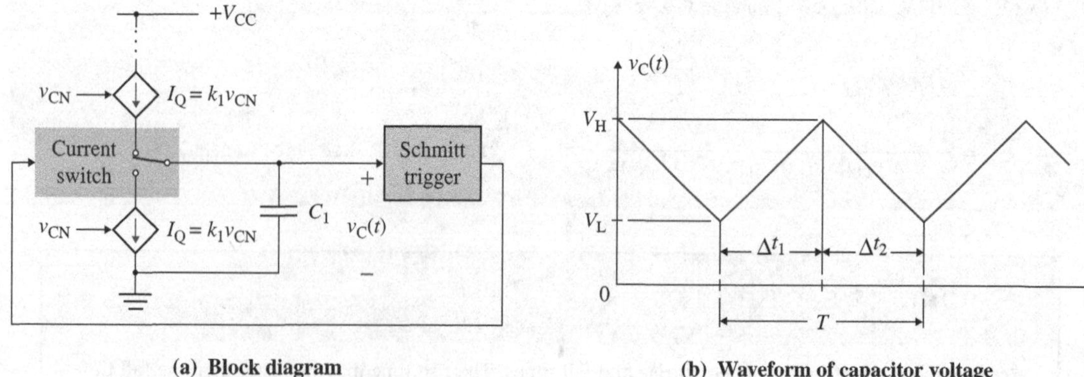

(a) Block diagram

(b) Waveform of capacitor voltage

FIGURE 16.41 Principle of a voltage-controlled oscillator

16.9.2 Discharging Mode

The discharging mode begins when the capacitor is charged to the upper trigger level V_H. At this point the current switch disconnects the top current source from C_1 and connects the bottom current source to C_1. The capacitor is then discharged by the bottom current source down to the lower trigger level V_L. The time required to discharge the capacitor from V_H to V_L is given by

$$\Delta t_2 = -\frac{C_1}{I_Q}\Delta v_C = -\frac{C_1}{I_Q}(V_L - V_H) = \frac{C_1}{I_Q}(V_H - V_L) \tag{16.51}$$

As long as the charging and discharging currents are maintained at the same magnitude of I_Q, $\Delta t_1 = \Delta t_2$. The period of oscillation T is given by

$$T = \Delta t_1 + \Delta t_2 = \frac{2C_1(V_H - V_L)}{I_Q} \tag{16.52}$$

which gives the frequency of oscillation f_o as

$$f_o = \frac{1}{T} = \frac{I_Q}{2C_1(V_H - V_L)} \tag{16.53}$$

Let us assume that the voltage-controlled current sources have a linear voltage-to-current transfer relationship; that is,

$$I_Q = G_m(v_{CN} + V_{CO}) \tag{16.54}$$

where G_m = transconductance of the current source, in A/V
v_{CN} = applied control voltage, in V
V_{CO} = constant voltage, in V

Therefore, the oscillation frequency will be a linear function of the control voltage v_{CN}; that is,

$$f_o = \frac{I_Q}{2C_1(V_H - V_L)} = \frac{G_m(v_{CN} + V_{CO})}{2C_1(V_H - V_L)} \tag{16.55}$$

which gives the voltage-to-frequency transfer coefficient K_{vf} as

$$K_{vf} = \frac{df_o}{dv_{CN}} = \frac{G_m}{2C_1(V_H - V_L)} \tag{16.56}$$

16.9.3 Circuit Implementation

A circuit implementation for the charging and discharging of the capacitor is shown in Fig. 16.42. It can be divided into two parts: a voltage-controlled current source and a current switch.

Voltage-Controlled Current Source

The voltage-controlled current source consists of an *npn* transistor Q_1 and a *pnp* transistor Q_2. The voltage at the emitter of Q_2 is given by

$$V_{E2} = v_{CN} - V_{BE1} + V_{BE2}$$

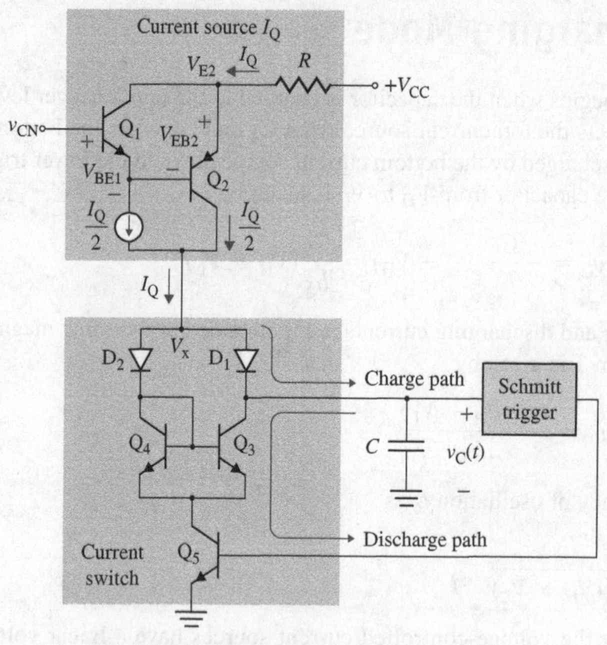

FIGURE 16.42 Circuit implementation

which gives the current source I_Q flowing through R as

$$I_Q = \frac{V_{CC} - V_{E2}}{R} = \frac{V_{CC} - v_{CN} + V_{BE1} - V_{EB2}}{R} \tag{16.57}$$

Since V_{BE1} (for an *npn* transistor) $\approx V_{EB2}$ (for a *pnp* transistor) within the range from 10 mV to 50 mV, I_Q in Eq. (16.57) can be approximated by

$$I_Q \approx \frac{V_{CC} - v_{CN}}{R} \tag{16.58}$$

which gives a linear relationship between the current source I_Q and the control voltage v_{CN}.

Current Switch

The current switch consists of diodes D_1 and D_2 and transistors Q_3, Q_4, and Q_5. Transistor Q_5 is controlled by the Schmitt trigger and is operated as an on or off switch. When Q_5 is off, capacitor C is charged by the current source I_Q via diode D_1. Diode D_2 and transistors Q_3 and Q_4 are off. When Q_5 is turned on (in saturation) by the Schmitt trigger, D_2, Q_3, and Q_4 are also turned on. As a result, the current source I_Q flows through D_2 and Q_4 instead of via diode D_1. The voltage at the anode of diode D_2 becomes

$$V_x = V_{D2(anode)} = V_{D2} + V_{BE4} + V_{CES(sat)} = 0.6 + 0.6 + 0.2 = 1.4 \text{ V}$$

However, the voltage at the cathode of diode D_1 is the capacitor voltage $v_C(t)$, which will be larger than 1.4 V. Thus, diode D_1 will be reverse biased (i.e., off). Capacitor C will discharge through Q_3 at a rate of I_Q. The current through Q_3 is equal to the current through Q_4. Thus, transistor Q_5 will carry a current of $2I_Q$.

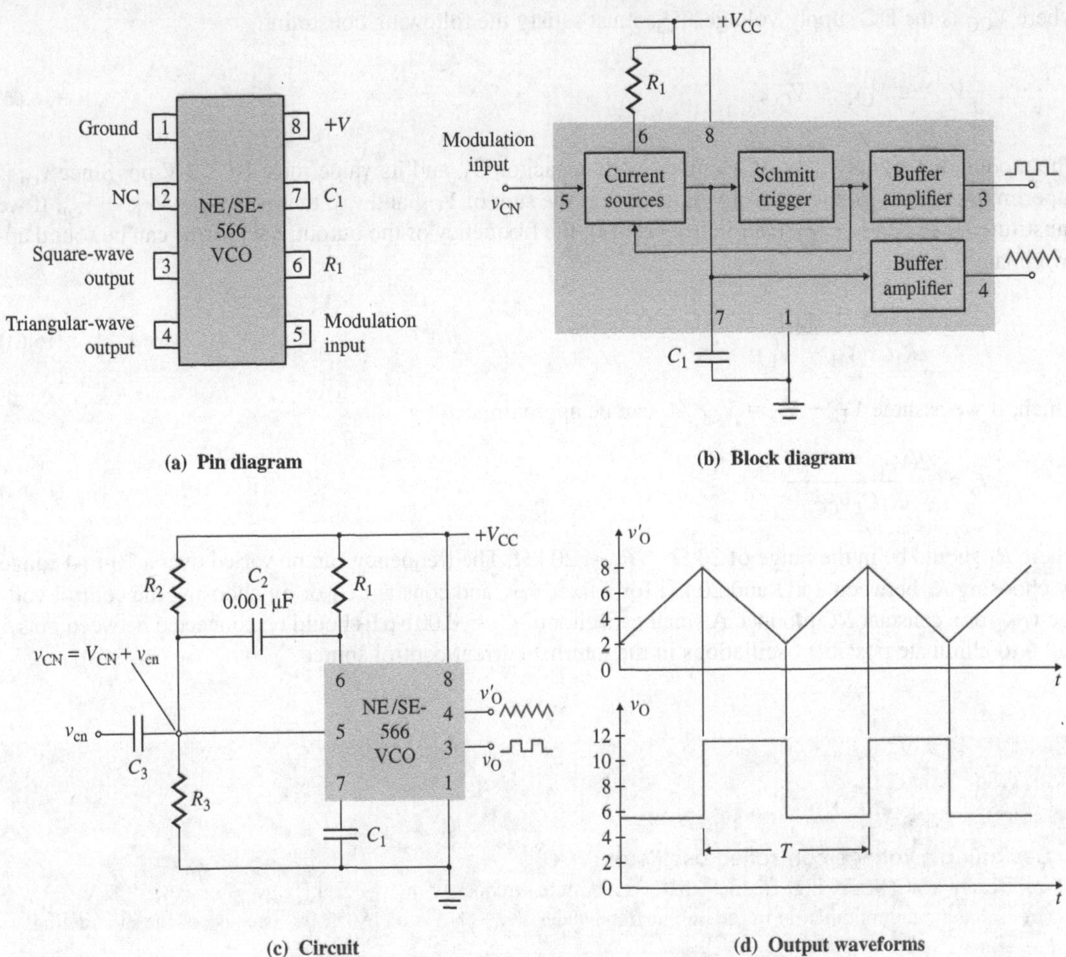

FIGURE 16.43 Voltage-controlled oscillator NE/SE-566 (Courtesy of Philips Semiconductors)

16.9.4 The NE/SE-566 VCO

A typical example of a VCO-integrated circuit is the NE/SE-566 VCO, whose pin diagram is shown in Fig. 16.43(a) and whose internal block diagram is shown in Fig. 16.43(b). This VCO produces simultaneous square-wave and triangular-wave outputs at frequencies of up to 1 MHz. Both outputs are buffered so that the output impedance of each is 50 Ω. The typical amplitude of the square wave is 5.4 V pp (peak to peak), and that of the triangular wave is 2.4 V pp. A typical connection diagram is shown in Fig. 16.43(c), and the typical outputs are shown in Fig. 16.43(d).

The output frequency is determined by an external resistor R_1, capacitor C_1, and the voltage v_{CN} applied to control terminal 5. The nominal DC value V_{CN} of v_{CN} is set by the voltage divider formed by R_2 and R_3 and must be in the range of

$$V_{CN} = \frac{R_3}{R_2 + R_3} V_{CC} \tag{16.59}$$

where V_{CC} is the DC supply voltage. V_{CN} must satisfy the following constraint:

$$\frac{3}{4}V_{CC} \leq V_{CN} \leq V_{CC} \tag{16.60}$$

The modulating signal v_{cn} is AC-coupled with capacitor C_3, and its value must be <3 V pp. Since v_{cn} is superimposed on V_{CN}, the control voltage v_{CN} is the sum of V_{CN} and v_{cn}; that is, $v_{CN} = V_{CN} + v_{cn}$. If we substitute $I_Q = (V_{CC} - v_{CN})/R_1$ in Eq. (16.53), the frequency of the output waveforms can be found approximately from

$$f_o = \frac{V_{CC} - v_{CN}}{2R_1C_1(V_H - V_L)} \tag{16.61}$$

which, if we assume $V_H - V_L = V_{CC}/4$, can be approximated by

$$f_o \approx \frac{2(V_{CC} - v_{CN})}{R_1C_1V_{CC}} \tag{16.62}$$

where R_1 should be in the range of $2\ \text{k}\Omega < R_1 < 20\ \text{k}\Omega$. The frequency can be varied over a 10-to-1 range by choosing R_1 between 2 kΩ and 20 kΩ for a fixed v_{CN} and constant C_1 or by choosing the control voltage v_{CN} for a constant RC product. A small capacitor of $C_2 = 0.001\ \mu\text{F}$ should be connected between pins 5 and 6 to eliminate possible oscillations in the internal current control source.

EXAMPLE 16.11

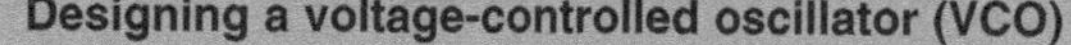

Designing a voltage-controlled oscillator (VCO)

(a) Design a VCO as in Fig. 16.43(c) that has a nominal frequency of $f_o = 20$ kHz. Assume $V_{CC} = 12$ V.

(b) Calculate the modulation in the output frequencies if v_{CN} is varied by ±10% because of the modulating signal v_{cn}.

SOLUTION

(a) The steps used to design the VCO are as follows:

Step 1. Find the limiting values of V_{CN}. From Eq. (16.60), $9\ \text{V} \leq V_{CN} \leq 12\ \text{V}$.

Step 2. Choose a suitable value of C_1: Let $C_1 = 0.001\ \mu\text{F}$.

Step 3. Choose a value of R_1 between 2 kΩ and 20 kΩ: Let $R_1 = 10\ \text{k}\Omega$.

Step 4. Find the value of v_{CN}. From Eq. (16.62), v_{CN} is given by

$$v_{CN} = V_{CC}\left(1 - \frac{R_1C_1f_o}{2}\right) \tag{16.63}$$

$$= 12\left(1 - \frac{10\ \text{k}\Omega \times 0.001\ \mu\text{F} \times 20\ \text{kHz}}{2}\right) = 12 \times 0.9 = 10.8\ \text{V}$$

which falls within the range specified in step 1. If the calculated value falls outside of the specified range, choose a different value for R_1 and/or C_1 and recalculate v_{CN}.

Step 5. Find the values of R_2 and R_3. Letting $V_{CN} = v_{CN} = 10.8$ V from step 4, Eq. (16.59) gives

$$1 + \frac{R_2}{R_3} = \frac{V_{CC}}{V_{CN}} \tag{16.64}$$

$$= \frac{12}{10.8} = 1.11$$

so $R_2/R_3 = 1.11 - 1 = 0.11$. Let $R_3 = 100\ \text{k}\Omega$; then

$$R_2 = 0.11 \times R_3 = 11\ \text{k}\Omega$$

(b) For a 10% increase in v_{CN},

$$v_{CN}^{+} = V_{CN} + v_{cn} = 1.1 \times v_{CN} = 1.1 \times 10.8 = 11.88\ \text{V}$$

The corresponding value of output frequency can be calculated from Eq. (16.62):

$$f_{o1} \approx \frac{2 \times (12 - 11.88)\ \text{V}}{10\ \text{k}\Omega \times 0.001\ \mu\text{F} \times 12\ \text{V}} = 2\ \text{kHz}$$

For a 10% decrease in v_{CN},

$$v_{CN}^{-} = V_{CN} - v_{cn} = 0.9 \times V_C = 0.9 \times 10.8 = 9.72\ \text{V}$$

The output frequency is

$$f_{o2} \approx \frac{2 \times (12 - 9.72)\ \text{V}}{10\ \text{k}\Omega \times 0.001\ \mu\text{F} \times 12\ \text{V}} = 38\ \text{kHz}$$

Thus, the change in the output frequency is

$$\Delta f_o = f_{o2} - f_{o1} = 38\ \text{kHz} - 2\ \text{kHz} = 36\ \text{kHz}$$

Using Eq. (16.62), we can find the V/F transfer coefficient K_{vf}:

$$K_{vf} = \frac{df_o}{dv_{CN}} = -\frac{2}{R_1 C_1 V_{CC}} \tag{16.65}$$

$$= -\frac{2}{10\ \text{k}\Omega \times 0.001\ \mu\text{F} \times 12\ \text{V}} = -16.67\ \text{kHz/V}$$

Thus, for $\Delta v_{CN} = 2v_{CN} = -2 \times 0.1 \times 10.8 = -2.16$ V,

$$\Delta f_o = f_{o2} - f_{o1} = K_{vf}\Delta v_{CN} = -16.67\ \text{kHz/V} \times (-2.16\ \text{V}) = 36.01\ \text{kHz}$$

which is the same as the value obtained by calculating individual frequencies.

NOTE: If the modulating signal is a sine wave so that $v_{cn} = V_m \sin \omega t$, then the control voltage becomes $v_{CN} = V_{CN} + v_m \sin \omega t$. During the positive half-cycle of the modulating signal, the control voltage will increase and the frequency f_o of the output voltage will decrease. However, during the negative half-cycle of the modulating signal, the control voltage will decrease and the frequency f_o of the output voltage will increase.

KEY POINTS OF SECTION 16.9

- A *voltage-controlled oscillator* (VCO) is an oscillator circuit in which the oscillation frequency is controlled by an externally applied voltage.
- To convert a voltage to frequency, a capacitor is normally charged and then discharged at a constant current whose value depends on an externally applied voltage. The output of a Schmitt trigger controls the charging and discharging time of the capacitor.
- The NE/SE-566 VCO can produce simultaneous square-wave and triangular-wave outputs at frequencies of up to 1 MHz. The output frequency is determined by an external resistor R_1, capacitor C_1, and the voltage v_{CN} applied to control terminal 5.

16.10 The 555 Timer

The 555 timer, introduced by Signetics Corporation in early 1970, is one of the most versatile integrated circuits. The 555 is a monolithic timing circuit that can produce highly accurate and highly stable delays or oscillation. It is used in many applications, such as monostable and astable multivibrators, digital logic probes, analog frequency meters, tachometers, infrared transmitters, and burglar and toxic gas alarms. Several versions of the 555 timer are available from various manufacturers. In addition to looking briefly at its internal structure, we consider two common applications of the 555 timer: monostable multivibrator and astable multivibrator.

16.10.1 Functional Block Diagram

The pin diagram of a 555 timer is shown in Fig. 16.44(a); its functional block diagram is shown in Fig. 16.44(b). The timer consists of two comparators CM_1 and CM_2, an *RS* flip-flop, a discharge transistor Q_1, and a resistive voltage divider string. The voltage divider sets the voltage at the inverting terminal of CM_1 to $2V_{CC}/3$ and the voltage at the noninverting terminal of CM_2 to $V_{CC}/3$.

The reset, threshold, and trigger inputs control the state of the flip-flop. If the reset input is low, the Q output of the flip-flop is low, and $\overline{Q}$ is high. With $\overline{Q}$ high, current flows through the base of transistor Q_1, and the transistor is switched on (in saturation). This generally provides a path for an external capacitor to discharge its voltage.

The reset input has the highest priority in setting the state of the flip-flop. Thus, Q is low if the reset input is low, regardless of the inputs to the comparators. If the reset is not in use, then it is connected to the positive DC supply V_{CC} so that it does not affect the state of the flip-flop.

If the trigger input becomes lower than the voltage at the noninverting input of CM_2 (i.e., $<V_{CC}/3$), the output of CM_2 (i.e., the S input to the flip-flop) will be high. As a result, the Q output of the flip-flop will be set to high. Thus, $\overline{Q}$ will be low, and the discharge transistor Q_1 will be off.

If the threshold input becomes higher than the voltage at the inverting input of CM_1 (i.e., $>2V_{CC}/3$), the output of CM_1 will be high. As a result, the Q output of the flip-flop will be reset to low. Thus, $\overline{Q}$ will be high, and the discharge transistor Q_1 will be on (in saturation), providing a discharge path.

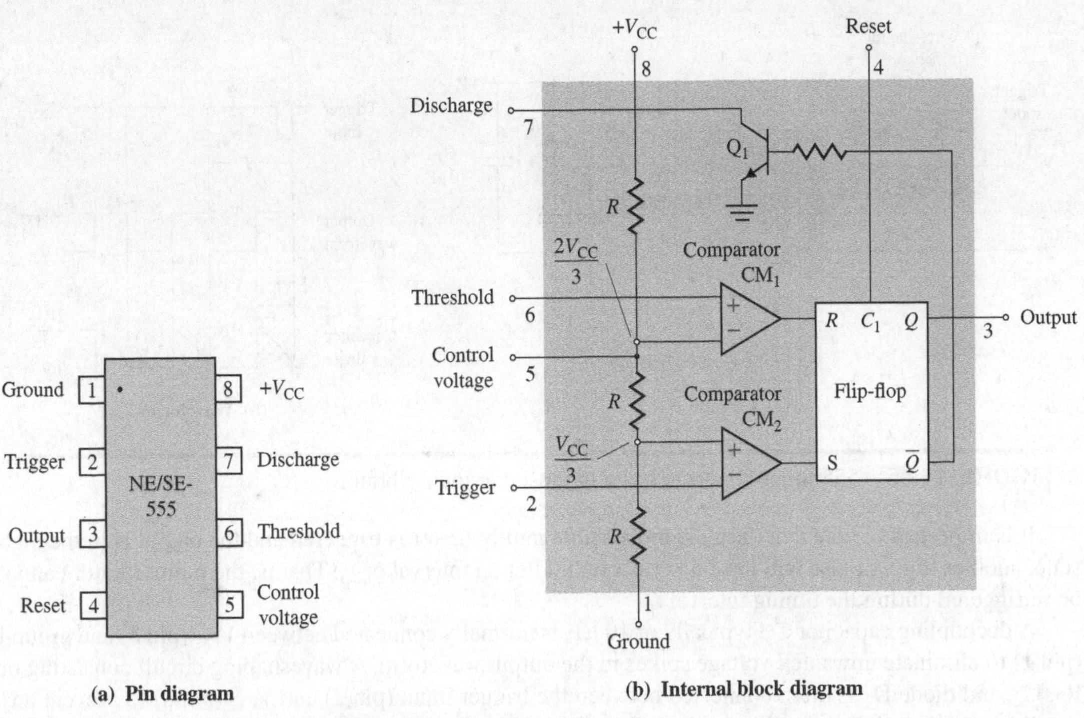

FIGURE 16.44 Functional block diagram of 555 timer (Courtesy of Philips Semiconductors)

16.10.2 Monostable Multivibrators

A monostable multivibrator is a *one-shot* pulse-generating circuit. Normally, its output is zero—that is, at logic low level in the stable state. This circuit has only one stable state at output low—hence the name *monostable*. The circuit configuration of the 555 timer for monostable operation is shown in Fig. 16.45(a). The discharge and threshold terminals are connected together. The external pulse v_I is applied to the trigger terminal through coupling capacitor C_2.

If the external pulse v_I is high, the Q output of the flip-flop is low; that is, the output of the timer is low. At the negative edge of the trigger signal, the flip-flop will be set to high. Thus, the output will be switched to high ($\approx V_{CC}$). It will remain high until the capacitor is charged to the threshold voltage of $2V_{CC}/3$, at which time the flip-flop will be reset and the output will return to zero.

The duration of the output pulse (t_p) is determined by the RC network connected externally to the 555 timer. At the end of the timing interval t_p, the output automatically reverts to its stable state of logic low. The output remains low until another negative-going trigger pulse is applied. Then the cycle repeats. The waveforms for the trigger input voltage $v_I(t)$, output voltage $v_O(t)$, and capacitor voltage $v_C(t)$ are shown in Fig. 16.45(b).

The width of the triggering pulse must be smaller than the expected pulse width of the output waveform. Also, the trigger pulse must be a negative-going signal and have an amplitude larger than $V_{CC}/3$. The time during which the output remains high is given by

$$t_p = 1.1RC \tag{16.66}$$

where R and C are the external resistance and capacitance, respectively.

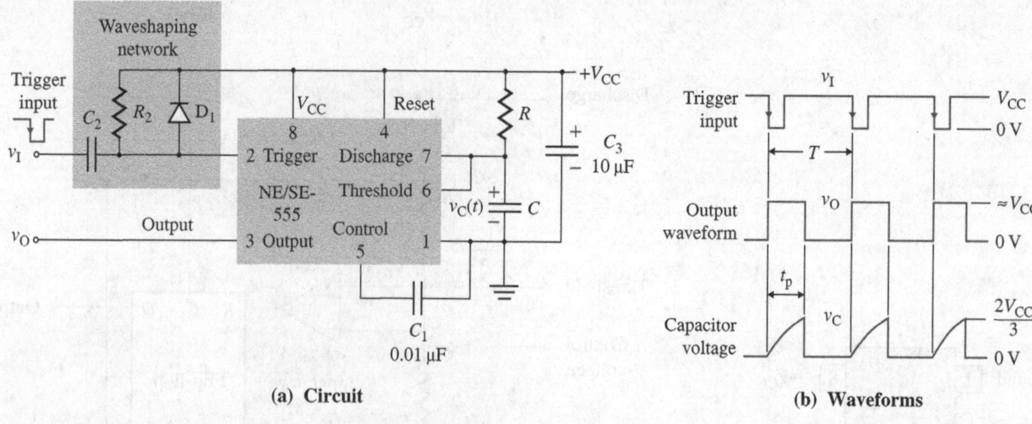

FIGURE 16.45 555 timer connected as a monostable multivibrator

It is important to note that once the monostable multivibrator is triggered and the output is in the high state, another trigger pulse will have no effect until after an interval of t_p. That is, the multivibrator cannot be retriggered during the timing interval t_p.

A decoupling capacitor C_3, typically of 10 μF, is normally connected between V_{CC} (pin 8) and ground (pin 1) to eliminate unwanted voltage spikes in the output waveform. A waveshaping circuit consisting of R_2, C_2, and diode D_1 is often connected between the trigger input (pin 2) and V_{CC} (pin 8) to prevent any possible mistriggering by positive pulse edges. This circuit is shown in Fig. 16.45(a) within the shaded area. The values of R_2 and C_2 should be selected so that the output pulse width t_p is much larger than the time constant R_2C_2. This condition is generally met by maintaining the following relationship:

$$t_p = 10R_2C_2 \tag{16.67}$$

EXAMPLE 16.12

D

Designing a monostable multivibrator Design a monostable multivibrator as in Fig. 16.45(a) so that $t_p = 5$ ms. Assume $V_{CC} = 12$ V.

SOLUTION

The steps used to design a monostable multivibrator are as follows:

Step 1. Choose a suitable value of C: Let $C = 0.1$ μF.

Step 2. Find the value of R. From Eq. (16.66),

$$R = \frac{t_p}{1.1C} = \frac{5 \text{ ms}}{1.1 \times 0.1 \text{ μF}} = 45.5 \text{ kΩ} \quad \text{(use a 50-kΩ potentiometer)}$$

Step 3. Choose a suitable value of C_2: Let $C_2 = 0.01$ μF.

Step 4. Find the value of R_2. From Eq. (16.67),

$$R_2 = \frac{t_p}{10C_2} = \frac{5 \text{ ms}}{10 \times 0.01 \text{ μF}} = 50 \text{ kΩ}$$

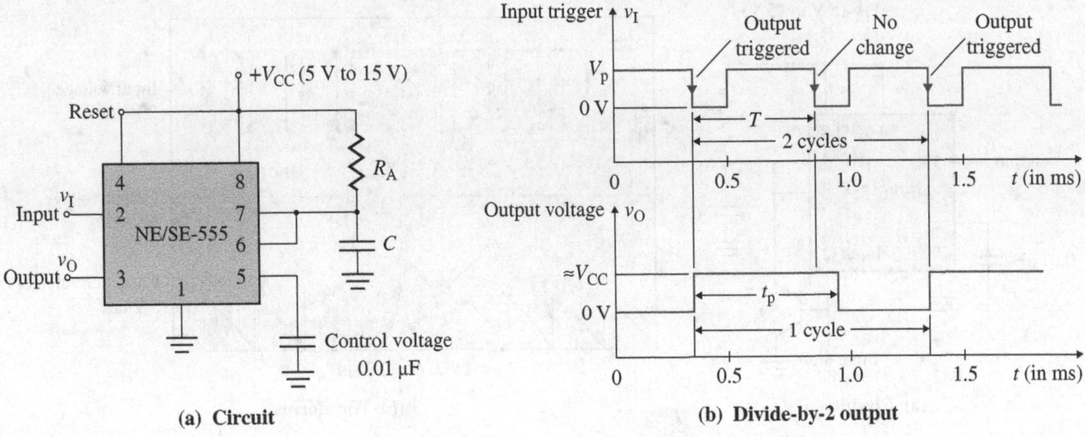

FIGURE 16.46 Monostable multivibrator as a frequency divider

16.10.3 Applications of Monostable Multivibrators

Monostable multivibrators can be used in many applications such as frequency dividers, missing-pulse detectors, and pulse stretchers.

Frequency Divider

If the frequency of the input signal is known, adjusting the length of the timing cycle t_p will permit a monostable multivibrator to be used as a frequency divider. The circuit configuration for a frequency divider is shown in Fig. 16.46(a). This application makes use of the fact that the monostable multivibrator cannot be retriggered during the timing interval.

For a monostable multivibrator to be used as a divide-by-2 circuit, the timing interval must be slightly larger (say, by 20%) than the period of the trigger signal T, as shown in Fig. 16.46(b); that is, $t_p = 1.2T$. At the first falling edge, the output is set to high. However, at the second falling edge, the capacitor is still charging and has not yet reached the threshold value of $2V_{CC}/3$. As a result, the second triggering signal has no effect on the output, and the output is set by the alternate trigger pulses.

For a monostable multivibrator to be used as a divide-by-3 circuit, t_p must be slightly larger than twice the period of the trigger signal. Thus, for a divide-by-n circuit, t_p must be slightly larger than $(n-1)T$; that is, $t_p = [0.2 + (n-1)]T$. For example, for a divide-by-2 circuit, if $f_o = 5$ kHz and $T = 1/f_o = 200$ μs, then $t_p = 1.2T = 240$ μs.

Missing-Pulse Detector

In some applications, a train of regular pulses is needed for the normal operation of a circuit or system. Any missing pulse may cause malfunction. The configuration of a monostable multivibrator that can detect any missing pulse is shown in Fig. 16.47(a).

The timing cycle is continuously reset by the input pulse train, but the time duration is not enough to complete the timing cycle. When the trigger pulse becomes low, transistor Q_1 is turned on, and it provides a discharge path for capacitor C. As a result, the capacitor voltage cannot reach the threshold voltage $2V_{CC}/3$. A change in frequency, or a missing pulse, allows completion of the timing cycle so that v_C

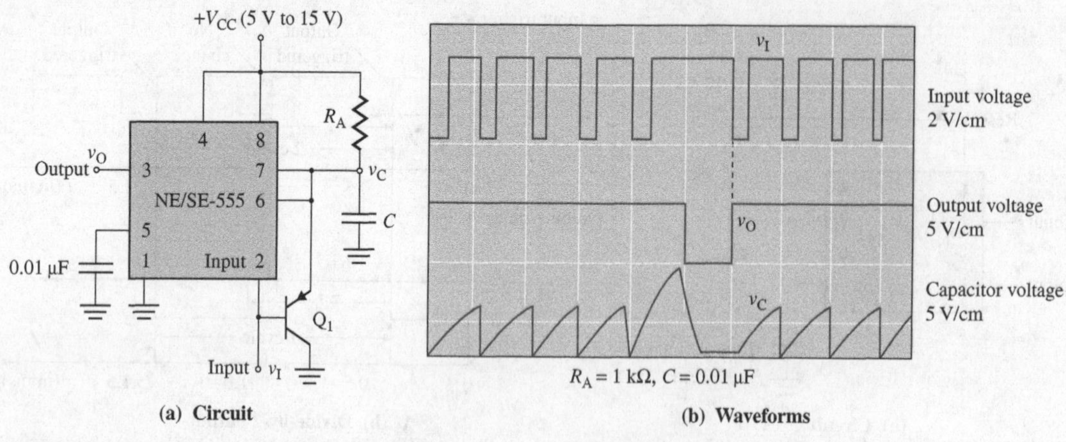

FIGURE 16.47 Monostable multivibrator as a missing-pulse detector

reaches $2V_{CC}/3$, causing a change in the output level due to the discharge of C through the internal transistor Q_1 in Fig. 16.44(b). The time delay t_p should be slightly longer than the normal time between input pulses. The waveforms are shown in Fig. 16.47(b).

Pulse Widener

A narrow pulse is not desirable for driving an LED display, as the flashing of the LED will not be visible to the eyes if the on time is infinitesimally small compared to the off time. A narrow pulse can be widened by a monostable multivibrator. This application is possible because of the fact that the timing interval t_p is longer than the negative pulse width of the trigger input.

The circuit configuration for a pulse widener is shown in Fig. 16.48(a). At the falling trigger edge, the output will be high, and it will remain high until the capacitor voltage reaches $2V_{CC}/3$ after time t_p. The waveforms of the input and output voltages are shown in Fig. 16.48(b). Since the output pulse can be viewed as the stretched version of the narrow input signal, this configuration is also known as a *pulse stretcher.*

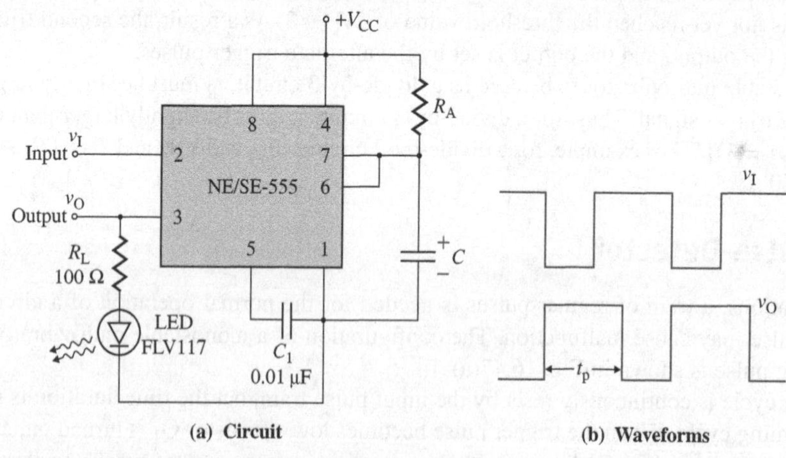

FIGURE 16.48 Monostable multivibrator as a pulse widener

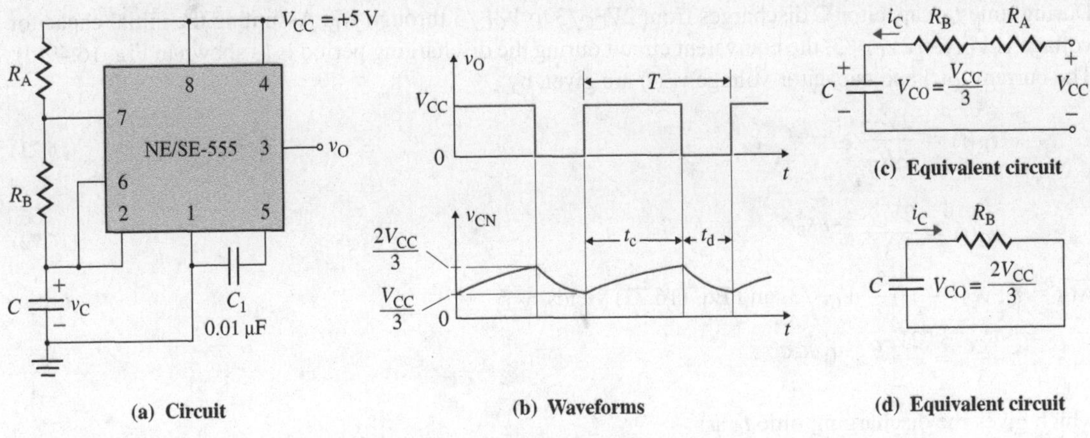

FIGURE 16.49 555 timer connected as an astable multivibrator

16.10.4 Astable Multivibrators

An astable multivibrator is a rectangular-wave-generating circuit. Because this circuit does not require an external trigger to change the state of the output, it is often called a *free-running multivibrator*. A 555 timer connected as an astable multivibrator is shown in Fig. 16.49(a). The duration of the high or low output is determined by resistors R_A and R_B and capacitor C. When the output is high, capacitor C starts charging toward V_{CC} through R_A and R_B. As soon as the capacitor voltage equals $2V_{CC}/3$, the output switches to low, and capacitor C discharges through R_B and the internal circuit of the timer. When the capacitor voltage equals $V_{CC}/3$, the output goes high, and the capacitor charges through R_A and R_B. Then the cycle repeats. The waveforms for the output voltage and the voltage across the capacitor are shown in Fig. 16.49(b).

The capacitor is periodically charged and discharged between $2V_{CC}/3$ and $V_{CC}/3$. Assuming the initial capacitor voltage is $V_{CO} = V_{CC}/3$, the equivalent circuit during the charging period is as shown in Fig. 16.49(c). The charging current $i_C(t)$ and the capacitor voltage $v_C(t)$ are given by

$$i_C(t) = \frac{2V_{CC}}{3(R_A + R_B)} e^{-t/(R_A+R_B)C} \tag{16.68}$$

$$v_C(t) = V_{CC} - \frac{2V_{CC}}{3} e^{-t/(R_A+R_B)C} \tag{16.69}$$

At $t = t_c$, $v_C(t = t_c) = 2V_{CC}/3$, and Eq. (16.69) yields

$$\frac{2V_{CC}}{3} = V_{CC} - \frac{2V_{CC}}{3} e^{-t_c/(R_A+R_B)C}$$

which gives the charging time t_c as

$$t_c = C(R_A + R_B)\ln(2) = 0.69C(R_A + R_B) \tag{16.70}$$

During time t_d, capacitor C discharges from $2V_{CC}/3$ to $V_{CC}/3$ through R_B. Assuming the initial capacitor voltage is $V_{CO} = 2V_{CC}/3$, the equivalent circuit during the discharging period is as shown in Fig. 16.49(d). The current $i_C(t)$ and capacitor voltage $v_C(t)$ are given by

$$i_C(t) = \frac{2V_{CC}}{3R_B} e^{-t/R_BC} \tag{16.71}$$

$$v_C(t) = \frac{2V_{CC}}{3} e^{-t/R_BC} \tag{16.72}$$

At $t = t_d$, $v_C(t = t_d) = V_{CC}/3$, and Eq. (16.72) yields

$$\frac{V_{CC}}{3} = \frac{2V_{CC}}{3} e^{-t_d/R_BC}$$

which gives the discharging time t_d as

$$t_d = CR_B \ln(2) = 0.69CR_B \tag{16.73}$$

Thus, the period of the output waveform is given by

$$T = t_c + t_d = 0.69C(R_A + R_B) + 0.69CR_B = 0.69C(R_A + 2R_B) \tag{16.74}$$

and the frequency of the output voltage is therefore given by

$$f_o = \frac{1}{T} = \frac{1}{0.69C(R_A + 2R_B)} = \frac{1.45}{C(R_A + 2R_B)} \tag{16.75}$$

Duty cycle k, which is the ratio of charging time t_c to period T, can be found from Eqs. (16.70) and (16.74):

$$k = \frac{t_c}{T} = \frac{R_A + R_B}{R_A + 2R_B} \tag{16.76}$$

Thus, the duty cycle k can be set by selecting R_A or R_B.

EXAMPLE 16.13

D **Designing an astable multivibrator** Design an astable multivibrator as in Fig. 16.49(a) so that $k = 75\%$ and $f_o = 2.5$ kHz. Assume $V_{CC} = 12$ V. Use PSpice to verify by simulating, EX16-13.SCH.

SOLUTION

$k = 75\% = 0.75$, and $T = 1/f_o = 1/2.5$ kHz $= 400$ μs. The steps used to design an astable multivibrator are as follows:

Step 1. Find the charging time t_c and the discharging time t_d:

$$t_c = kT \tag{16.77}$$

$$= 0.75 \times 400\ \mu s = 300\ \mu s$$

$$t_d = (1 - k)T \tag{16.78}$$

$$= (1 - 0.75) \times 400\ \mu s = 100\ \mu s$$

Step 2. Choose a suitable value of C: Let $C = 0.1\ \mu F$.

Step 3. Find the value of R_B. From Eq. (16.73),

$$R_B = \frac{t_d}{0.69C} \tag{16.79}$$

$$= \frac{100\ \mu s}{0.69 \times 0.1\ \mu F} = 1449\ \Omega$$

Step 4. Find the value of R_A. From Eq. (16.70),

$$R_A = \frac{t_c}{0.69C} - R_B \tag{16.80}$$

$$= \frac{300\ \mu s}{0.69 \times 0.1\ \mu F} - 1449 = 2899\ \Omega$$

EXAMPLE 16.14

D **PSpice/SPICE simulation of an astable multivibrator** An astable multivibrator can be used as a voltage-controlled oscillator (VCO) if an external control voltage is applied to terminal 5. This arrangement is shown in Fig. 16.50 for $v_C = 6 + 4 \sin (2000\pi t)$. Use PSpice/SPICE to plot the output voltage $v_O(t)$ from 0 to 2 ms with an increment of 10 ns.

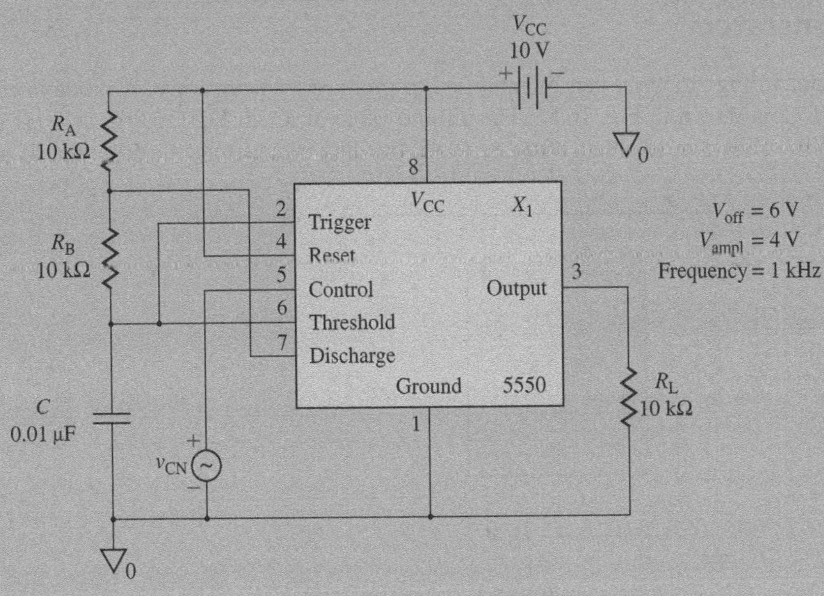

FIGURE 16.50 Astable multivibrator as a VCO for PSpice simulation

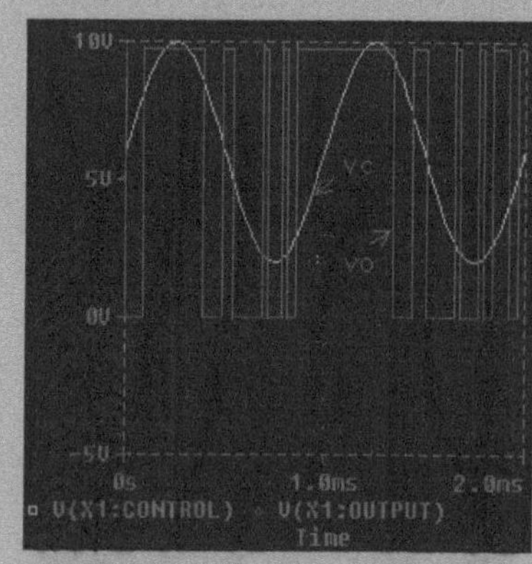

FIGURE 16.51 Control and output voltages for Example 16.14

SOLUTION

The PSpice plots of the control and output voltages are shown in Fig. 16.51. As expected, the frequency becomes lower—that is, the period becomes larger—as the control voltage increases in magnitude.

16.10.5 Applications of Astable Multivibrators

As examples of applications of an astable multivibrator, we consider a square-wave generator, a ramp generator, and an FSK modulator.

Square-Wave Generator

The astable multivibrator in Fig. 16.49(a) can be modified to produce a square wave. A diode is connected across resistor R_B, as shown in Fig. 16.52. For a finite value of R_A in Eq. (16.70), $t_c > t_d$, and the duty cycle is $k > 50\%$. However, to obtain a square wave, the duty cycle k must be 50%; that is, the

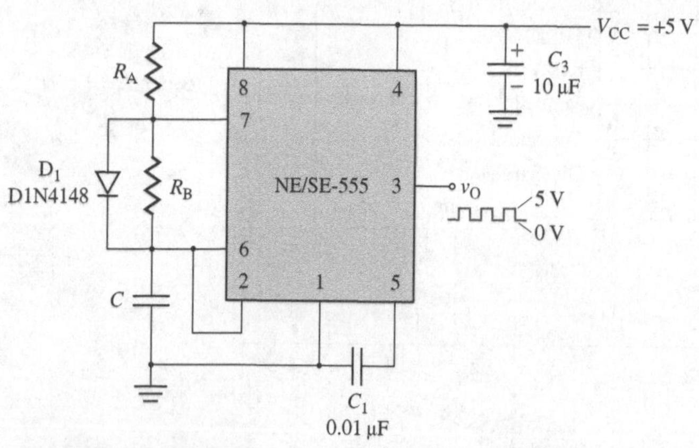

FIGURE 16.52 555 timer connected as a square-wave generator

value of R_A should be set to zero. With $R_A = 0$, terminal 7 is connected directly to V_{CC}. During the discharging time, capacitor C discharges through the internal circuit of the timer, and an external current is applied by V_{CC} to the same internal path. The current flowing through the internal path may be large enough to damage the timer. This situation can be avoided by connecting a diode across R_B so that R_B is bypassed during the charging time. In that case, Eqs. (16.74) and (16.75) are reduced, respectively, to

$$T = 0.69C(R_A + R_B) \quad (\text{for } k = 0.5) \tag{16.81}$$

$$f_o = \frac{1}{T} = \frac{1.45}{C(R_A + R_B)} \quad (\text{for } k = 0.5) \tag{16.82}$$

▶ **NOTE** We need both R_A and D_1. Making $R_A = 0$, the circuit will not work.

EXAMPLE 16.15

D **Designing a square-wave generator** Design a square-wave generator as in Fig. 16.52 so that $k = 50\%$ and $f_o = 2.5$ kHz. Assume $V_{CC} = 12$ V. Use PSpice to verify by simulating, EX16-15.SCH.

SOLUTION

$k = 50\% = 0.5$, and $T = 1/f_o = 1/2.5 \text{ kHz} = 400\ \mu\text{s}$. The steps used to design the square-wave generator are as follows:

Step 1. Find the charging time t_c and the discharging time t_d:

$$t_c = t_d = kT \tag{16.83}$$

$$= 0.5 \times 400\ \mu\text{s} = 200\ \mu\text{s}$$

Step 2. Choose a suitable value of C: Let $C = 0.1\ \mu\text{F}$.

Step 3. Find the value of R_B. From Eq. (16.73),

$$R_B = \frac{t_d}{0.69C} \tag{16.84}$$

$$= \frac{200\ \mu\text{s}}{0.69 \times 0.1\ \mu\text{F}} = 2899\ \Omega$$

Step 4. Find the value of R_A. From Eq. (16.82),

$$R_A = \frac{1.45}{Cf_o} - R_B \tag{16.85}$$

$$= \frac{1.45}{0.1\ \mu\text{F} \times 2.5\text{ kHz}} - 2899\ \Omega = 2901\ \Omega$$

Ramp Generator

The astable multivibrator in Fig. 16.49(a) can be used as a free-running ramp generator. This is accomplished by charging the capacitor through a constant current source and discharging through the internal circuit of the timer. That is, resistors R_A and R_B are replaced by a current source, as shown in Fig. 16.53(a). The waveforms for the output voltage and the capacitor voltage are shown in Fig. 16.53(b).

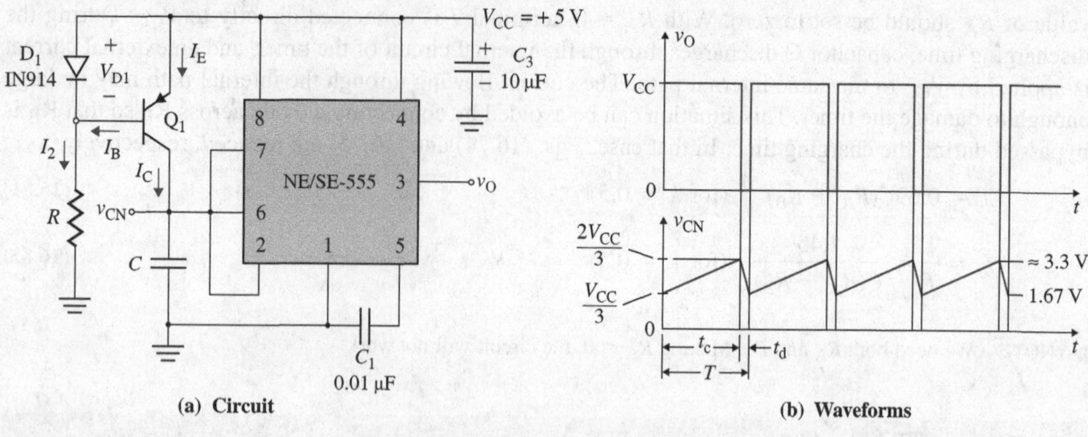

FIGURE 16.53 555 timer connected as a ramp generator

The collector current, which is the charging current, is given by

$$I_C = I_E - I_B$$

Assuming that the voltage drop of diode D_1 is approximately equal to the base–emitter voltage V_{BE} of the transistor, the diode current $I_D \approx I_E$. Thus,

$$I_C = I_D - I_B = I_2$$

$$= \frac{V_B}{R} = \frac{V_{CC} - V_{BE}}{R} \tag{16.86}$$

The capacitor charges from $V_{CC}/3$ to $2V_{CC}/3$ at a constant current of I_C. For a charging time of t_c, the change in the capacitor voltage Δv_C is given by

$$\Delta v_C = \frac{2V_{CC}}{3} - \frac{V_{CC}}{3} = \frac{1}{C}\int_0^{t_c} I_C\,dt = \frac{1}{C} I_C t_c$$

which gives the charging time t_c as

$$t_c = \frac{CV_{CC}}{3I_C} \tag{16.87}$$

The charging time is related to the duty cycle k and period T by

$$kT = t_c$$

$$= \frac{CV_{CC}}{3I_C} \tag{16.88}$$

Therefore, the free-running frequency of the ramp generator is given by

$$f_o = \frac{1}{T} = \frac{3kI_C}{CV_{CC}} \tag{16.89}$$

$$= \frac{3k(V_{CC} - V_{BE})}{CRV_{CC}} \tag{16.90}$$

If the discharging time t_d of the capacitor is negligible compared to its charging time t_c, then $k \approx 1$ and the free-running frequency becomes

$$f_o = \frac{3(V_{CC} - V_{BE})}{CRV_{CC}} \tag{16.91}$$

EXAMPLE 16.16

D **Designing a ramp generator** Design a ramp generator using the circuit in Fig. 16.53(a) so that $k = 50\%$ and $f_o = 2.5$ kHz. Assume $V_{CC} = 12$ V, $V_{BE} = 0.7$ V, and a transistor of $\beta_F = 150$.

SOLUTION

$k = 50\% = 0.5$, and $T = 1/f_o = 1/2.5\text{ kHz} = 400\ \mu\text{s}$. The steps used to design the ramp generator are as follows:

Step 1. Find the charging time t_c and the discharging time t_d:

$$t_c = t_d = kT \tag{16.92}$$

$$= 0.5 \times 400\ \mu\text{s} = 200\ \mu\text{s}$$

Step 2. Choose a suitable value of C: Let $C = 0.1\ \mu$F.

Step 3. Find the value of R. From Eq. (16.90),

$$R = \frac{3k(V_{CC} - V_{BE})}{V_{CC}Cf_o} \tag{16.93}$$

$$= \frac{3 \times 0.5 \times (12 - 0.7)\text{ V}}{12 \times 0.1\ \mu\text{F} \times 2.5\text{ kHz}} = 5.65\text{ k}\Omega$$

Step 4. Find the collector current I_C of the transistor. From Eq. (16.86),

$$I_C = \frac{V_{CC} - V_{BE}}{R} = \frac{(12 - 0.7)\text{ V}}{5.65\text{ k}\Omega} = 2\text{ mA}$$

Step 5. Find the current I_D through the diode:

$$I_D = I_E = I_C + I_B = I_C\frac{1 + \beta_F}{\beta_F} = 2\text{ mA} \times \frac{1 + 150}{150} \approx 2.01\text{ mA}$$

FSK Modulator

In computer peripheral and radio (wireless) communication, the binary data or code is transmitted by means of a carrier frequency that shifts between two preset frequencies. This technique for data transmission is called *frequency-shift keying* (FSK). The frequency shift is usually accomplished by driving a VCO with the binary data signal so that the 0 to 1 states (commonly called *space* and *mark*) of the binary data signal produce two frequencies, known as the *space* and *mark frequencies*. For example, when teletypewriter information is transmitted using a modulator/demodulator (*modem*, for short), a 1070-Hz (for mark) and 1270-Hz (for space) pair will represent the original signal, whereas a 2025-Hz (for mark) and 2250-Hz (for space) pair will represent the answer signal.

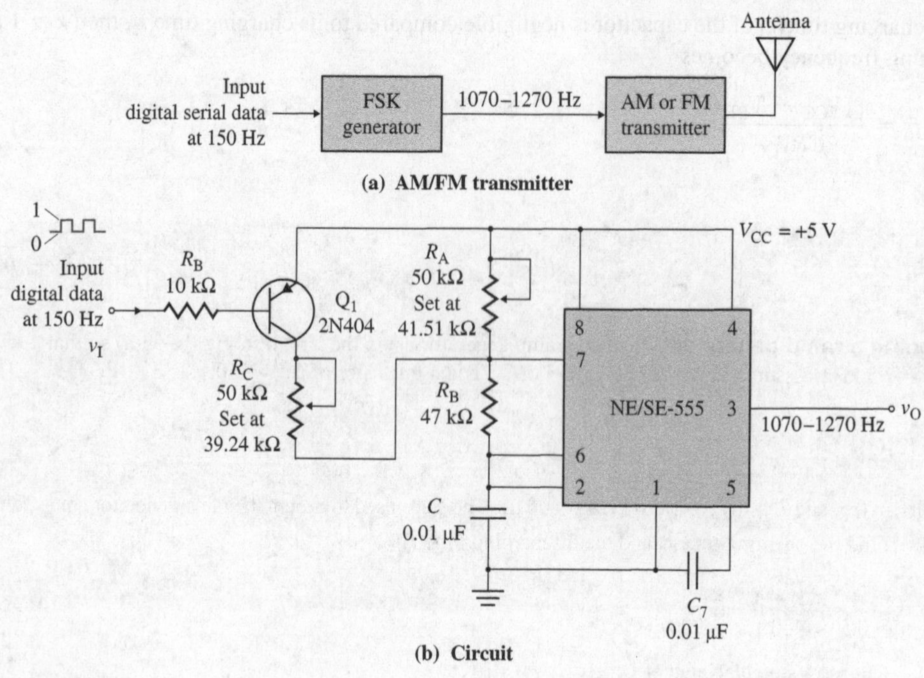

FIGURE 16.54 Astable multivibrator as an FSK modulator (Courtesy of Philips Semiconductors)

FSK modulators are often used in AM/FM transmitters, as shown in Fig. 16.54(a). The 555 astable multivibrator can be used as an FSK generator; the connection is shown in Fig. 16.54(b). The on or off condition of transistor Q_1 will depend on the input signal. Thus, the output frequency depends on the logic state of the digital input signal. A signal frequency of 150 Hz is commonly used for data transmission.

When the input signal is 1, transistor Q_1 is off, and the 555 operates in its normal mode as an astable multivibrator. Thus, the output frequency corresponding to logic 1 can be found from Eq. (16.75) as

$$f_{o(\text{mark})} = \frac{1.45}{C(R_A + 2R_B)}$$

The values for C, R_A, and R_B can be selected so as to give 1070 Hz.

When the input signal is 0, transistor Q_1 is turned on (in saturation). As a result, R_C is effectively connected in parallel with R_B. This reduces the charging time of capacitor C and increases the output frequency. Thus, the output frequency corresponding to logic 0 can be found from

$$f_{o(\text{space})} = \frac{1.45}{C(R_A \| R_C + 2R_B)}$$

The value for R_C can be selected so as to give 1270 Hz.

Thus, with properly selected values of C, R_A, R_B, and R_C, the 555 astable multivibrator can produce frequencies of 1070 Hz and 1270 Hz corresponding to 1 and 0, respectively. The difference between the FSK signals of 1270 Hz and 1070 Hz (i.e., 200 Hz) is called the *frequency shift*.

KEY POINTS OF SECTION 16.10

- The 555 timer is one of the most versatile ICs. It is used as a monostable or astable multivibrator in many applications.
- A monostable multivibrator is a *one-shot* pulse-generating circuit. This circuit has only one stable state at output low—hence the name *monostable*. The 555 can be configured as a monostable multivibrator, which can be used as a frequency divider, a missing-pulse detector, or a pulse widener.
- An astable multivibrator is a rectangular-wave-generating circuit. Because it does not require an external trigger to change the state of the output, it is often called a *free-running* multivibrator. The 555 can be configured as an astable multivibrator, which can be used as a square-wave generator, a ramp generator, or an FSK modulator.

16.11 Phase-Lock Loops

The phase-lock loop (PLL) is one of the fundamental building blocks of electronic circuits used in such applications as motor-speed controllers, FM stereo decoders, tracking filters, frequency-synthesized transmitters and receivers, and FSK decoders. A block diagram of a phase-lock loop is shown in Fig. 16.55(a). The loop consists of a phase detector, a low-pass filter, and a VCO.

The phase detector (or comparator) compares the phase of the input voltage with that of the VCO output voltage and produces a DC or low-frequency voltage proportional to their phase difference. The output of the phase detector, which is called the *error voltage*, is applied to a low-pass filter. The filter removes any high-frequency components and produces a smooth DC voltage. This DC voltage is then applied to the control input of the VCO, whose output frequency is proportional to the DC value. If the frequency of the input voltage shifts slightly, the phase difference between the input signal and the VCO output voltage will begin to increase with time. This will change the control voltage to the VCO in such a way as to bring the VCO frequency back to the same value as the input voltage. The VCO frequency is continuously adjusted until it is equal to the input frequency.

The operation of a PLL involves three modes: a *free-running mode*, a *capture mode*, and a *phase-lock mode*. During the free-running mode, there is no input frequency (or voltage), and the VCO runs at a fixed frequency corresponding to zero-applied input voltage. This frequency is called the *center*, or *free-running*, *frequency* f_O. Once an input frequency is applied, the VCO frequency starts to change, and the PLL is said

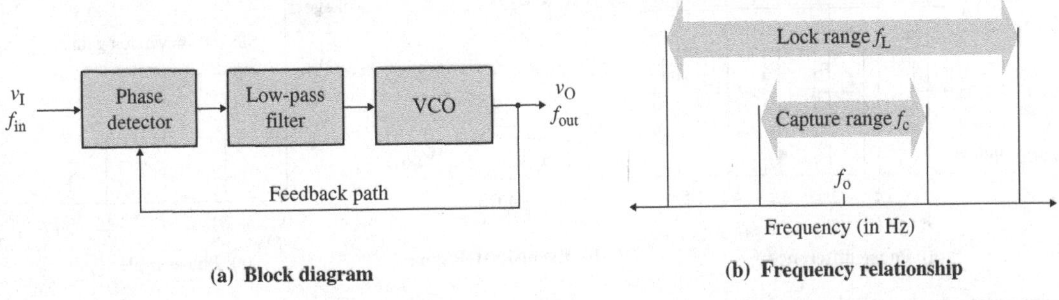

FIGURE 16.55 Block diagram of a phase-lock loop

to be in the capture mode. The VCO frequency changes continuously to match the input frequency. When the input frequency becomes equal to the output frequency, the PLL is said to be in the phase-lock mode. The feedback loop maintains the lock when the input signal frequency changes.

The center frequency f_o is the free-running frequency of the VCO. The *lock range* f_L is defined as the range of input frequencies around the center frequency for which the loop can maintain lock. The *capture range* f_C is defined as the range of input frequencies around the center frequency for which the loop will become locked from an unlocked condition. The relations among f_o, f_L, and f_C are shown in Fig. 16.55(b).

16.11.1 Phase Detector

A phase detector takes two input voltages and produces a DC voltage proportional to their phase difference. To understand the principle of operation, consider two voltages v_{I1} and v_{I2}, as shown in Fig. 16.56(a), with a phase difference of ϕ. An output voltage is obtained when they differ in phase—that is, when only one input is high. A phase detector can be implemented using either an exclusive-OR gate, as shown in Fig. 16.56(b), or an analog multiplier [7]. Integrating the output voltage will give an average output voltage, which will be a linear function of the phase difference ϕ, as shown in Fig. 16.56(c). The average output voltage $V_{O(DC)}$ can be expressed as

$$V_{O(DC)} = \begin{cases} \dfrac{V_{CC}}{\pi}\phi & (\text{for } 0 \le \phi \le \pi) \\ \dfrac{V_{CC}}{\pi}(2\pi - \phi) & (\text{for } \pi \le \phi \le 2\pi) \end{cases} \quad \textbf{(16.94)}$$

The phase difference can also be detected by using an edge-triggered *RS* flip-flop. Two input signals are shown in Fig. 16.57(a). If these signals are passed through an edge-triggered *RS* flip-flop as shown in Fig. 16.57(b), the output voltage will be as shown in Fig. 16.57(a). Integrating the output voltage will give an average output voltage, as shown in Fig. 16.57(c). The average output voltage $V_{O(DC)}$ is given by

$$V_{O(DC)} = \frac{V_{CC}}{2\pi}\phi \quad (\text{for } 0 \le \phi \le 2\pi) \quad \textbf{(16.95)}$$

Phase detectors can be broadly divided into two types: digital detectors and analog detectors. The digital detectors are simple to implement with digital devices. However, they are sensitive to the harmonic

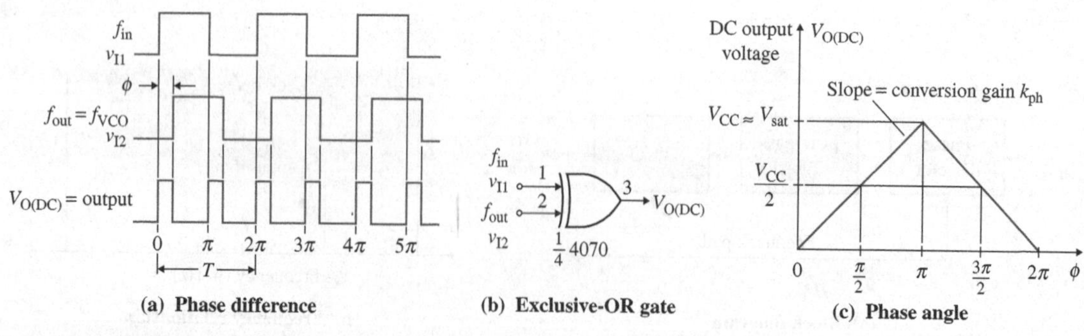

FIGURE 16.56 Phase detector

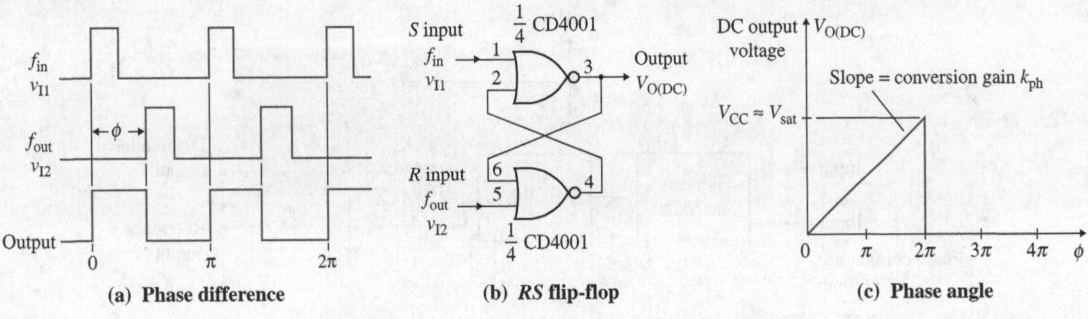

FIGURE 16.57 Edge-triggered phase detector

content of the input signal and changes in the duty cycles of the input signal and the VCO output voltage. Analog detectors are monolithic types such as CMOS MC4344/4044. They respond only to transitions in the input signals. Thus, sensitivity to harmonic content and duty cycle is not a problem. The output voltage is independent of variations in the amplitude and duty cycle of the input waveform. Analog detectors are generally preferred over digital detectors, especially in applications in which accuracy is a critical factor.

16.11.2 Integrated Circuit PLL

The NE/SE-565 PLL is one of the most commonly used IC devices. The elements of the PLL in Fig. 16.55(a) are built into the 565 IC. The internal block diagram of the 565 is shown in Fig. 16.58(a), and the pin configuration is shown in Fig. 16.58(b). A typical connection diagram for the NE/SE-565 PLL is shown in Fig. 16.58(c). A small capacitor C_3, typically of 0.001 μF, is connected between pins 7 and 8 to eliminate possible oscillations. The center frequency of the PLL is given approximately by

$$f_o \approx \frac{1.2}{4R_1C_1} \tag{16.96}$$

where R_1 and C_1 are an external resistance and a capacitance connected to pins 8 and 9, respectively. C_1 can have any value, but R_1 must have a value between 2 kΩ and 20 kΩ. A capacitor C_2 is connected between pins 7 and 10 to form a first-order low-pass filter with an internal resistance of 3.6 kΩ. The filter capacitor C_2 should be large enough to eliminate variations in the demodulated output voltage at pin 7 in order to stabilize the VCO frequency.

The 565 PLL can typically lock to and track an input signal over a bandwidth of ±60% of the center frequency f_o. The lock range f_L is given by

$$f_L = \frac{8f_o}{V_{CC} - V_{EE}} \tag{16.97}$$

where V_{CC} and $-V_{EE}$ are the positive and negative power supplies, in volts, respectively. The capture range f_c is given by

$$f_c = \left[\frac{f_L}{2\pi \times 3.6 \times 10^3 C_2}\right]^{1/2} \quad (C_2 \text{ in farads}) \tag{16.98}$$

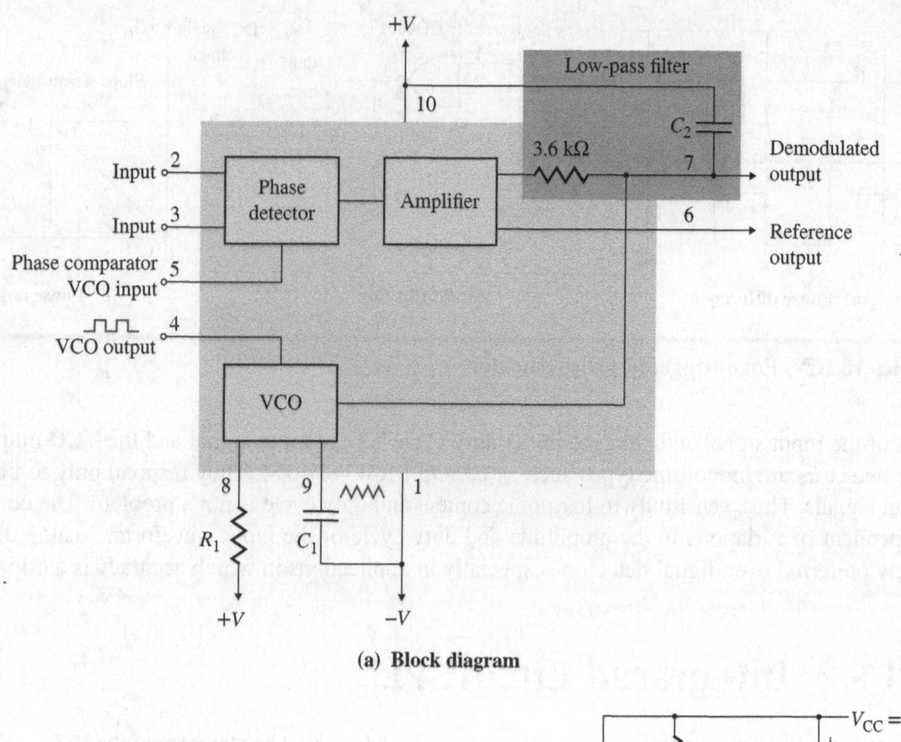

(a) Block diagram

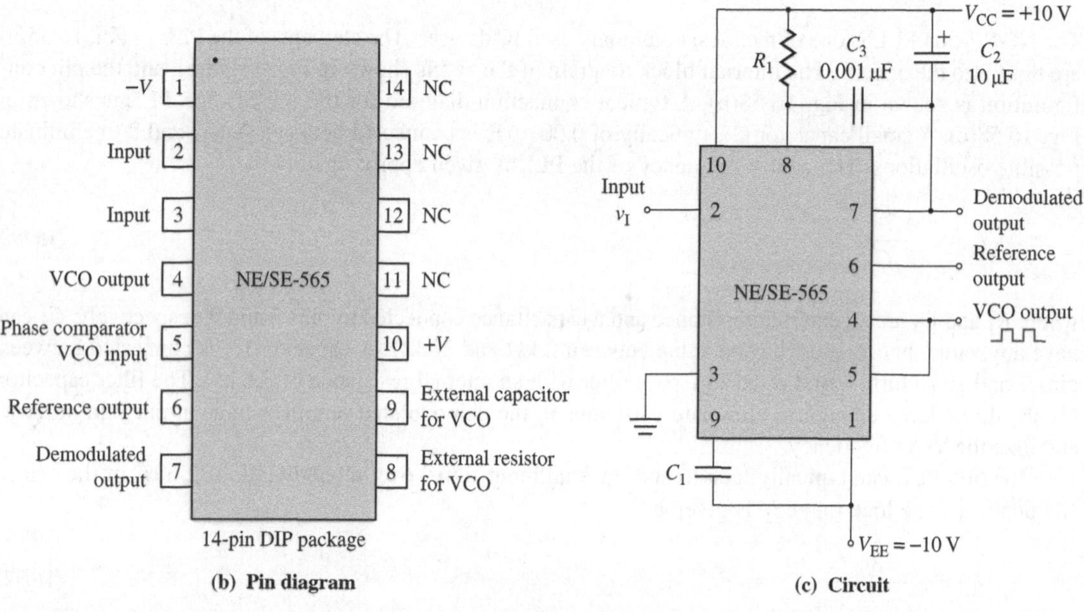

FIGURE 16.58 NE/SE-565 PLL connection diagram

EXAMPLE 16.17

D **Designing a PLL** Design a PLL as shown in Fig. 16.58(c) so that $f_o = 2.5$ kHz and $f_c = 50$ Hz. Assume $V_{CC} = -V_{EE} = 12$ V.

SOLUTION

The steps used to design the 565 PLL are as follows:

Step 1. Choose a suitable value of C_1: Let $C_1 = 0.01\ \mu$F.

Step 2. Find the value of R_1. From Eq. (16.96),

$$R_1 = \frac{1.2}{4C_1 f_o} = \frac{1.2}{4 \times 0.01\ \mu\text{F} \times 2.5\ \text{kHz}} = 12\ \text{k}\Omega$$

Step 3. Find the lock range f_L. From Eq. (16.97),

$$f_L = \frac{8 \times 2.5\ \text{kHz}}{12 - (-12)} = 833\ \text{Hz}$$

Step 4. Find the value of C_2. From Eq. (16.98),

$$C_2 = \frac{f_L}{2\pi \times 3.6 \times 10^3 f_c^2} = \frac{833}{2\pi \times 3.6 \times 10^3 \times 50^2} = 14.17\ \mu\text{F}$$

Choose $C_2 = 14\ \mu$F.

16.11.3 Applications of the 565 PLL

As examples of applications of the 565 PLL, we consider a frequency multiplier, an FSK demodulator, and an SCA (subsidiary carrier authorization) decoder [8].

Frequency Multiplier

A block diagram of a frequency multiplier using the 565 PLL is shown in Fig. 16.59(a). A frequency divider is inserted between the VCO and the phase detector. Since the output frequency of the divider is locked to the input frequency, the VCO will actually be running at a multiple of the input frequency; that is, $f_o = Nf_{in}$, where N is an integer. The amount of multiplication desired can be obtained by selecting the appropriate divide-by network. A typical connection of the 565 PLL to give an output frequency of $f_o = 5f_{in}$ is shown in Fig. 16.59(b).

To set up the circuit, you must know the frequency limits of the input signal. The free-running frequency of the VCO then can be adjusted by means of R_1 and C_1 so that the output frequency of the divider is midway between the input frequency limits. That is, for $f_{in} = 400$ Hz to 4 kHz, the output frequency would be $f_o = 2$ kHz to 20 kHz, with a midway frequency of $f_{o(mid)} = 11$ kHz. The filter capacitance C_2 should be large enough (typically 10 μF) to eliminate variations in the demodulated output voltage (at pin 7) in order to stabilize the VCO frequency. The output of the VCO will be a square wave whose frequency will be the multiple of the input frequency as long as the loop is in lock. The input and output waveforms are shown in Fig. 16.59(c).

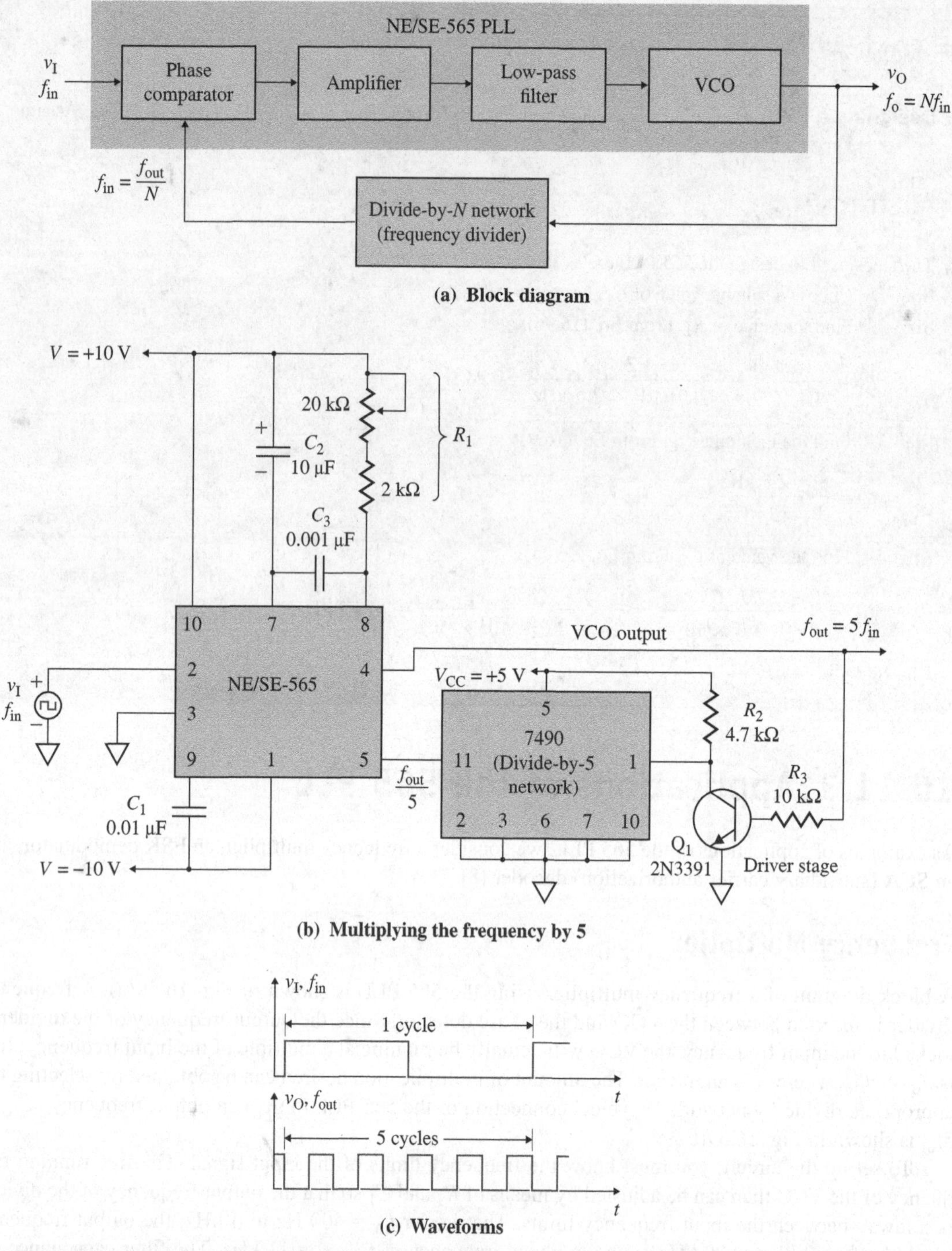

FIGURE 16.59 565 PLL as a frequency multiplier

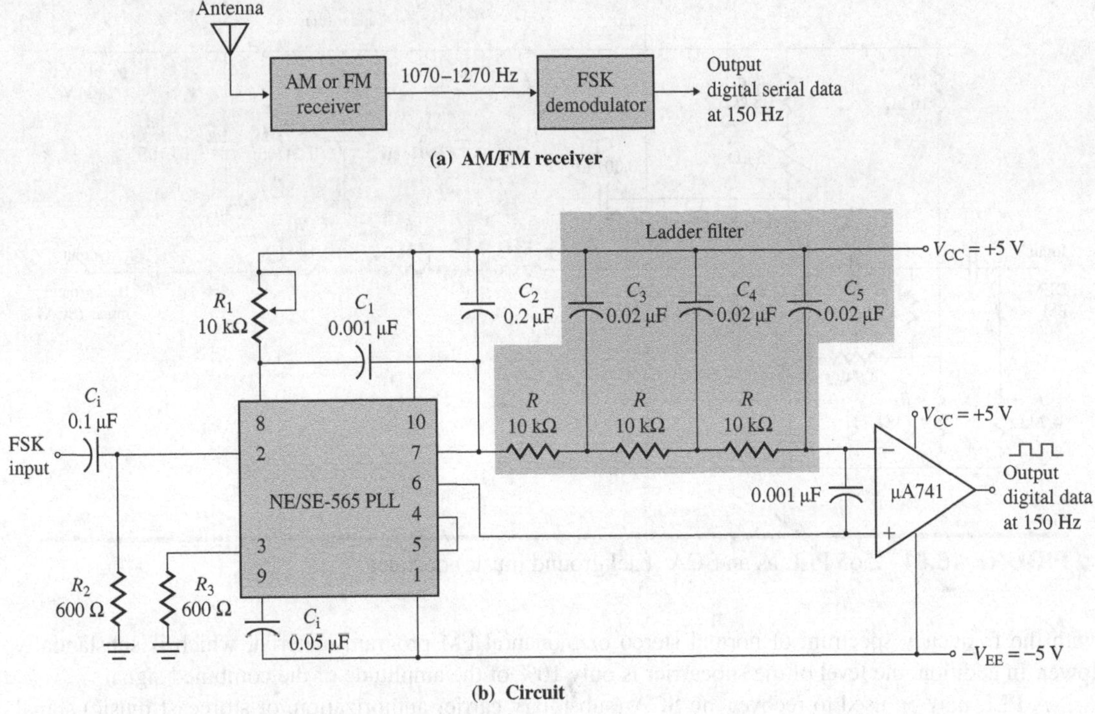

FIGURE 16.60 565 PLL as an FSK demodulator

FSK Demodulator

An FSK demodulator is often used in AM/FM receivers, as shown in Fig. 16.60(a). One very useful application of the 565 PLL is as an FSK demodulator to receive FSK signals of 1070 Hz and 1270 Hz; the configuration is shown in Fig. 16.60(b). As the signal appears at the input, the PLL locks to the input frequency and tracks it between the two frequencies, with a corresponding DC shift at the output. The input signal is connected through a coupling capacitor C_i to block the DC level from the FSK receiver. Both input terminals are connected to the ground through identical resistors R_2 and R_3.

The loop filter capacitor C_2 determines the dynamic characteristics of the demodulator, and its value should be smaller than usual to eliminate overshoot on the output pulse. A three-stage *RC*-ladder low-pass filter is used to remove the carrier component from the output. The high cutoff frequency of the ladder filter—that is, $f_H = 1/2\pi RC$—should be approximately halfway between the maximum keying rate (150 Hz) and twice the input frequency (2 × 1070 Hz is approximately 2200 Hz).

The output signal of 150 Hz can be made logic compatible by connecting a voltage comparator between the output and pin 6 of the 565 PLL. R_2 and C_1 determine the free-running frequency of the VCO. The free-running frequency is adjusted with R_1 so as to produce a slightly positive voltage with f_o = 1070 Hz.

SCA (Background Music) Decoder

Some FM stations are authorized by the FCC to broadcast uninterrupted background music for commercial use, using a frequency-modulated subcarrier of 67 kHz. This frequency was chosen so as not to interfere

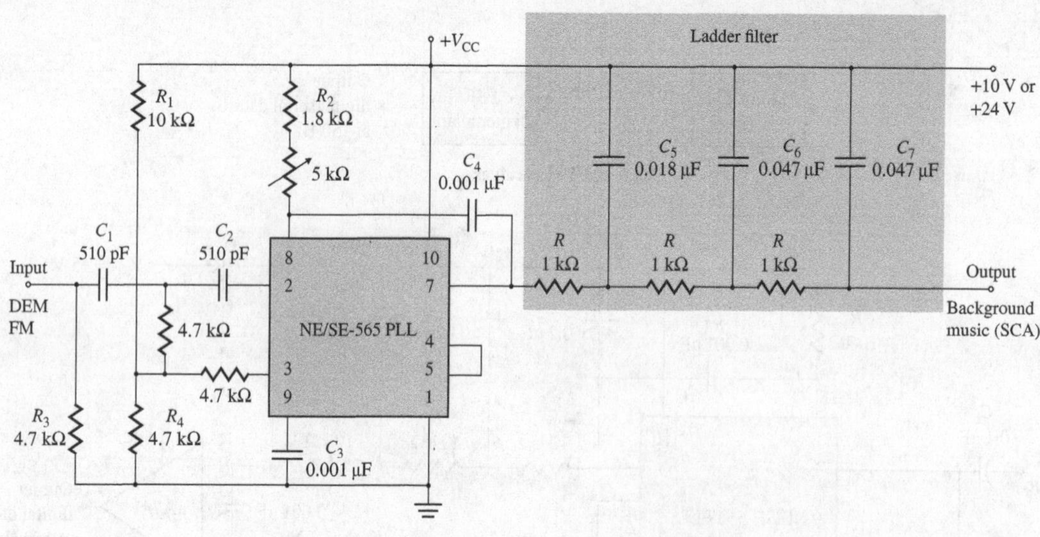

FIGURE 16.61 565 PLL as an SCA (background music) decoder

with the frequency spectrum of normal stereo or monaural FM program material, which is substantially lower. In addition, the level of the subcarrier is only 10% of the amplitude of the combined signal.

A PLL may be used to recover the SCA (subsidiary carrier authorization, or storecast music) signal from the combined signal of many commercial FM broadcast stations. This application involves demodulation of a frequency-modulated subcarrier of the main channel. The SCA signal can be filtered out and demodulated by the 565 PLL without the use of any resonant circuits. A connection diagram is shown in Fig. 16.61. The PLL is tuned to 67 kHz with a 5-kΩ potentiometer; only approximate tuning is required, since the loop will seek the signal.

The demodulated output (pin 7) is passed through a three-stage low-pass filter to provide deemphasis and attenuate the high-frequency noise that often accompanies SCA transmissions. Since no capacitor is provided directly at pin 7, the circuit operates as a first-order loop. The demodulated output signal is on the order of 50 mV, and the frequency response extends to 7 kHz. By connecting the circuit of Fig. 16.61 to a point between the FM discriminator and the deemphasis filter of a commercial-band (home) FM receiver and tuning the receiver to a station that broadcasts an SCA signal, one can obtain hours of commercial-free background music.

KEY POINTS OF SECTION 16.11

- A phase-lock loop (PLL) consists of a phase detector, a low-pass filter, and a voltage-controlled oscillator. PLLs find applications as frequency multipliers, FSK demodulators, and SCA (background music) decoders.
- The operation of a PLL involves three modes: a free-running mode, a capture mode, and a phase-lock mode. The VCO frequency is continuously adjusted until it is equal to the input frequency. When the input frequency becomes equal to the output frequency, the PLL is said to be in the phase-lock mode.
- The 565 PLL can typically lock to and track an input signal over a bandwidth of ±60% of the center frequency.

16.12 Voltage-to-Frequency and Frequency-to-Voltage Converters

The NE/SE-566 VCO discussed in Sec. 16.9 can be used as a voltage-to-frequency converter. In many applications, it is also necessary to convert frequency to voltage. The TelCom 9400 series converters can be used as either *voltage-to-frequency* (V/F) or *frequency-to-voltage* (F/V) *converters*, and they can produce pulse and square-wave outputs with a frequency range of 1 Hz to 100 kHz. For V/F conversion, the device accepts an analog input signal and generates an output pulse train whose frequency is linearly proportional to the input voltage. For F/V conversion, the device accepts any input frequency waveform and provides a linearly proportional voltage output. The complete V/F or F/V conversion requires only the addition of two capacitors, three resistors, and a reference voltage. The 9400 series consists of CMOS devices and bipolar devices that can operate on single or dual supply voltages.

16.12.1 V/F Converter

The 9400 V/F converter operates on the principle of charge balancing. The functional block diagram is shown in Fig. 16.62(a). The input voltage v_I is converted to a current I_{in} by the input resistor R_{in}. This current $I_{in} = v_I/R_{in}$ is then converted to a charge by the internal integrating capacitor C_{int} and gives a linearly decreasing voltage v_{O3} at the output of the op-amp integrator; that is,

$$v_{O3} = -\frac{I_{in}}{C_{int}}t = -\frac{v_I}{R_{in}C_{int}}t \tag{16.99}$$

As soon as voltage v_{O3} falls below the threshold level of the threshold detector, the switch closes and causes reference voltage V_{ref} to be applied to reference capacitor C_{ref} for a time long enough to charge the capacitor to reference voltage V_{ref}. This action also reduces the charge on the integrating capacitor by a fixed amount ($q = C_{ref}V_{ref}$), causing the integrator output to step up by a certain amount. At the end of the charging period, C_{ref} is shorted out, dissipating the charge stored on the reference capacitor so that the system is ready to repeat the cycle when the output passes again through the zero axis.

The continued charging of the integrating capacitor C_{int} by the input voltage is balanced out by fixed charges from the reference voltage. As the input voltage is increased, the number of reference pulses required to maintain balance increases, causing the output frequency to increase also. Since each charge increment is fixed (i.e., $q = C_{ref}V_{ref}$), the increase in frequency with voltage is linear. The output frequency f_o is related to the input voltage by

$$f_o = \frac{v_I}{|V_{ref}|\,R_{in}C_{ref}} \tag{16.100}$$

where v_I = input voltage
$|V_{ref}|$ = reference voltage
C_{ref} = reference capacitance

The pin diagram of the 9400 V/F converter is shown in Fig. 16.62(b), and its internal block diagram is shown in Fig. 16.62(c). The threshold detector senses the output of the integrator. The output of the detector triggers a 3-μs network when its input voltage passes through the threshold. The nominal threshold of

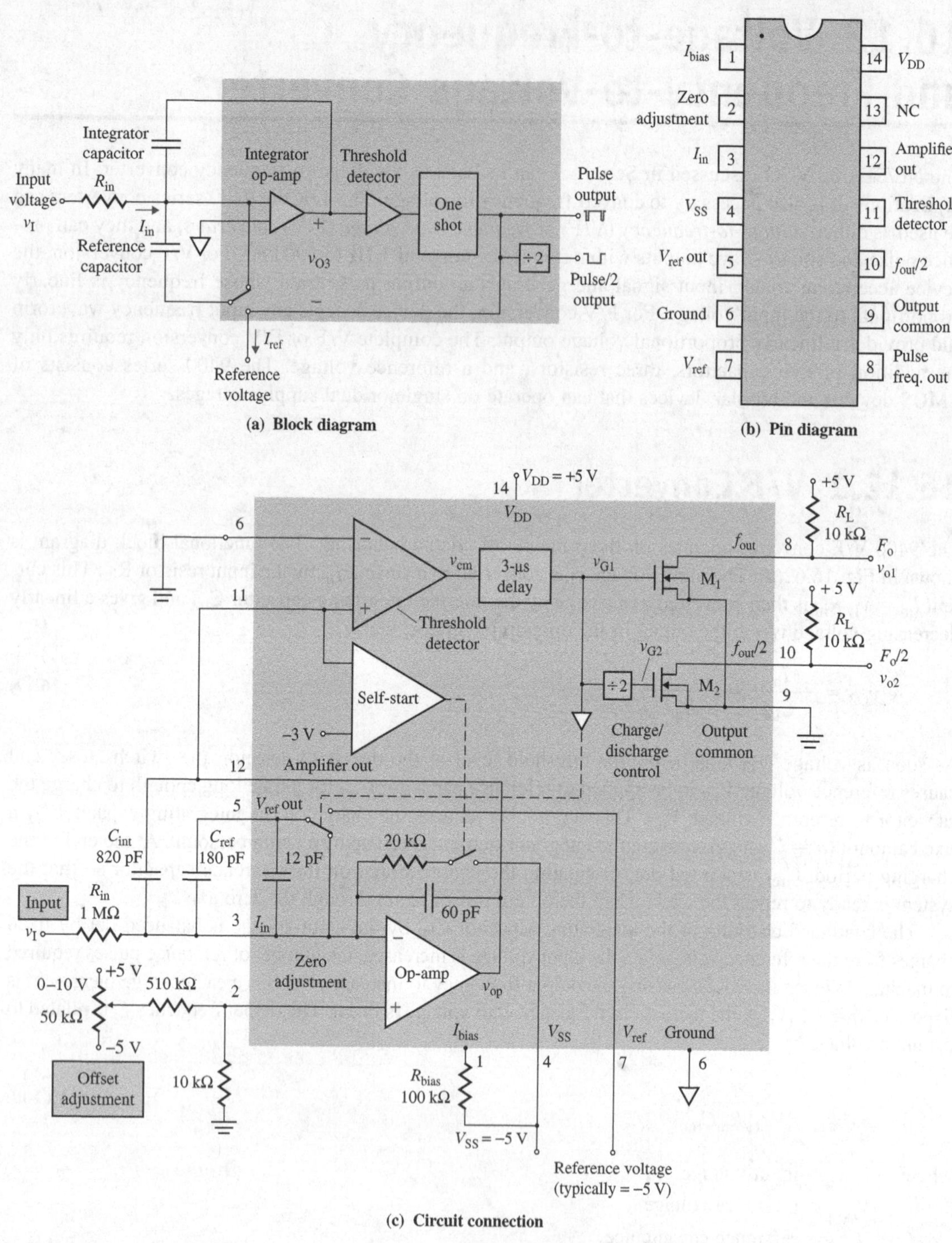

FIGURE 16.62 9400 V/F converter (Courtesy of TelCom Semiconductor, Inc.)

the detector is halfway between the power supplies, or $(V_{DD} + V_{SS})/2 \pm 400$ mV. The output of the 3-μs network is applied to the output transistor M_1, the divide-by-2 network, and the C_{ref} charge/discharge control circuit.

The self-start circuit ensures that the V/F converter operates properly when the power is first applied. If the integrator output is below the threshold voltage (i.e., 0 V) of the threshold detector and C_{ref} is already charged, then a positive voltage step will not occur when the power is turned on. The integrator output will continue to decrease until it crosses the −3.0-V threshold of the self-start comparator. When this happens, an internal resistor of 20 kΩ is connected to the op-amp integrator input, thereby forcing the output to become positive. As soon as the op-amp output becomes positive, the self-start circuit is disabled, and the 9400 operates in its normal mode.

Pulse f_{out} $(=f_o)$ output is an open-drain n-channel FET that provides a pulse waveform whose frequency is proportional to the input voltage v_I. Pulse $f_{out}/2$ $(=f_o/2)$ output is an open-drain n-channel FET that provides a square wave whose frequency is one-half that of the pulse waveform. This output will change state on the rising edge of f_o. Both f_o and $f_o/2$ outputs require a pull-up resistor and interface directly with MOS, CMOS, and TTL logic circuits.

The waveforms of the V/F converter are shown in Fig. 16.63. Three microseconds after the output V_d of the detector switches to low, the output V_{G1} of the delay circuit switches from low to high. When V_{G1}

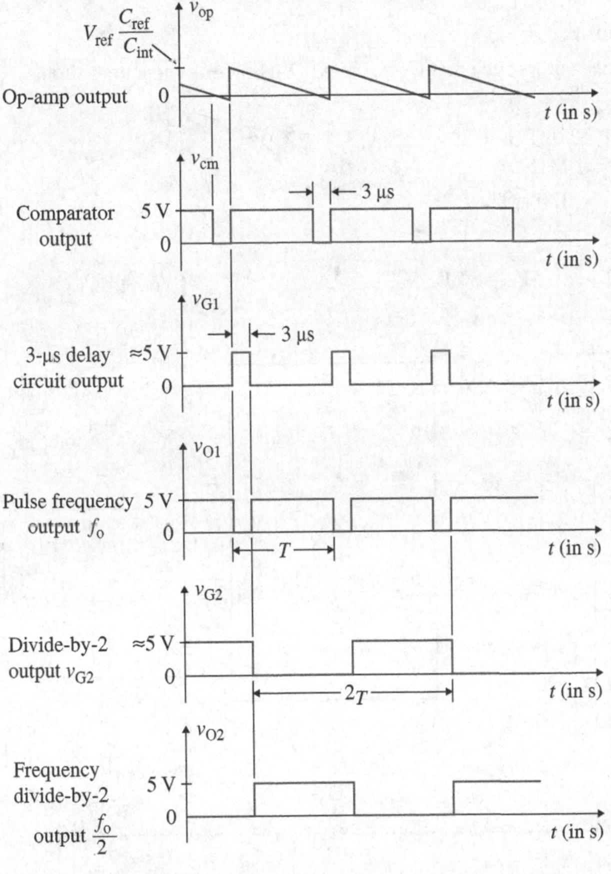

FIGURE 16.63 Waveforms of the 9400 V/F converter

is low, transistor M_1 is off and f_{out} is high (i.e., 5 V). The divide-by-2 network is a negative-edge-triggered flip-flop whose output V_{G2} is a complement (inversion) of V_{G1}. Thus, transistor M_2 will be on, and $f_{out}/2$ will be low. With V_{G1} low, the charge/discharge control is disabled and capacitor C_{ref} remains discharged (i.e., shorted out).

EXAMPLE 16.18

D **Designing a V/F converter** Using the 9400 as shown in Fig. 16.64, design a V/F converter so that $f_o = 5$ kHz at $v_I = 5$ V. The input voltage v_I can vary between 10 mV and 10 V. Assume $V_{DD} = -V_{SS} = 5$ V.

SOLUTION

The steps used to design the V/F converter are as follows:

Step 1. Choose V_{DD} and V_{SS} such that

$$4\text{ V} \le V_{DD} \le 7.5\text{ V} \quad \text{and} \quad -7.5\text{ V} \le V_{SS} \le -4\text{ V}$$

Choose $V_{DD} = 5$ V and $V_{SS} = -5$ V.

Step 2. Choose the capacitors such that $C_3 = C_4 = 0.1$ μF. These capacitors should be close to pins 4 and 14, respectively.

Step 3. Choose the reference voltage $V_{ref} = V_{SS} = -5$ V.

Step 4. Choose $R_{in} = 1$ MΩ.

Step 5. Choose $R_{bias} = 100$ kΩ.

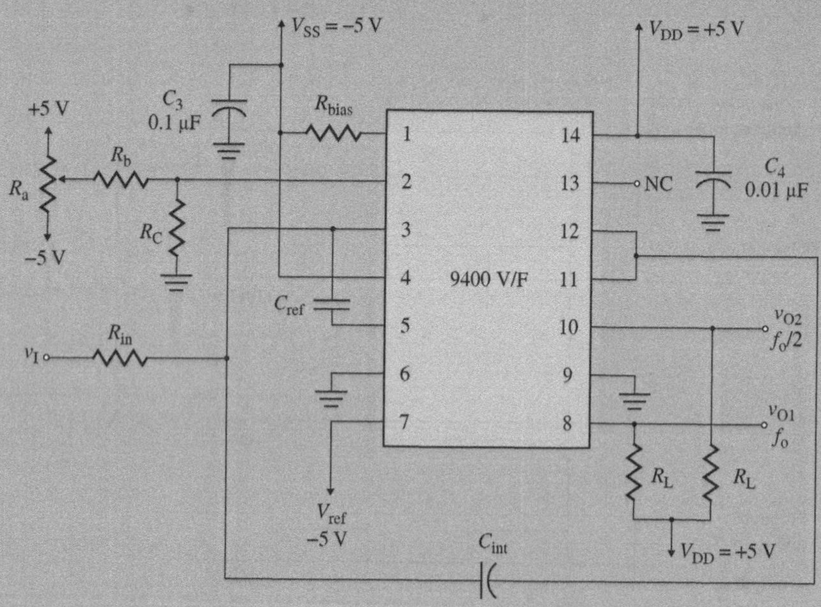

FIGURE 16.64 TelCom 9400 converter connected as a V/F converter

Step 6. Choose the pull-up resistance $R_L = 10\ \text{k}\Omega$.
Step 7. Choose C_{ref} such that $C_{ref} < 500$ pF. From Eq. (16.100),

$$C_{ref} = \frac{v_I}{|V_{ref}|R_{in}f_o} = \frac{5\ \text{V}}{5\ \text{V} \times 1\ \text{M}\Omega \times 5\ \text{kHz}} = 200\ \text{pF}$$

C_{ref} should be located as close as possible to pins 3 and 5. Glass-film capacitors are recommended for high accuracy.
Step 8. Choose C_{int} such that $4C_{ref} \leq C_{int} \leq 10C_{ref}$. Assume

$$C_{int} = 5 \times C_{ref} = 5 \times 200\ \text{pF} = 1\,\text{nF}$$

C_{int} should be located as close as possible to pins 3 and 12.
Step 9. Determine the values of the offset resistors. Since $R_c \leq R_a \leq R_b$, assume R_a is a 50-kΩ potentiometer, $R_b = 450\ \text{k}\Omega$, and $R_c = 10\ \text{k}\Omega$.
Step 10. Determine the minimum frequency corresponding to the minimum input voltage $v_I = 10$ mV. From Eq. (16.100),

$$f_{o(min)} = \frac{10\ \text{mV}}{5\ \text{V} \times 1\ \text{M}\Omega \times 200\ \text{pF}} = 10\ \text{Hz}$$

Step 11. Set the input voltage to the minimum value $v_I = 10$ mV, and adjust potentiometer R_a to obtain the corresponding minimum output frequency $f_{o(min)} = 10$ Hz.
Step 12. Determine the maximum frequency corresponding to the maximum input voltage $v_I = 10$ V. From Eq. (16.100),

$$f_{o(max)} = \frac{10\ \text{V}}{5\ \text{V} \times 1\ \text{M}\Omega \times 200\ \text{pF}} = 10\ \text{kHz}$$

Step 13. Set the input voltage to the maximum value $v_I = 10$ V, and adjust R_{in}, V_{ref}, or C_{ref} to obtain the corresponding maximum output frequency $f_{o(max)} = 10$ kHz.

16.12.2 F/V Converter

When used as an F/V converter, the 9400 generates an output voltage that is linearly proportional to the input frequency f_{in}. The internal block diagram of the 9400 F/V converter is shown in Fig. 16.65(a). The input signal is differentiated by an *RC* network whose output is then applied to the (+) input of the threshold detector (i.e., at pin 11). The threshold detector has about ± 200 mV of hysteresis. Each time the input to the detector at pin 11 crosses zero in the negative direction, its output goes to low. Three microseconds later, the charge/discharge circuit is enabled, instantaneously connecting the reference capacitor C_{ref}, which remains discharged, to the reference voltage V_{ref}. This causes a precise amount of charge ($q = C_{ref}V_{ref}$) to be dispensed into the op-amp's summing junction. This charge in turn flows through feedback resistor R_{int} and generates a voltage pulse at the output of the op-amp. Capacitor C_{int} across R_{int} averages these pulses into a DC signal that is linearly proportional to the input frequency. The waveforms of the F/V converter are shown in Fig. 16.65(b). For charge q dispensed to capacitor C_{int} in time period T, we get

$$q = iT = \frac{V_O}{R_{int}}T$$

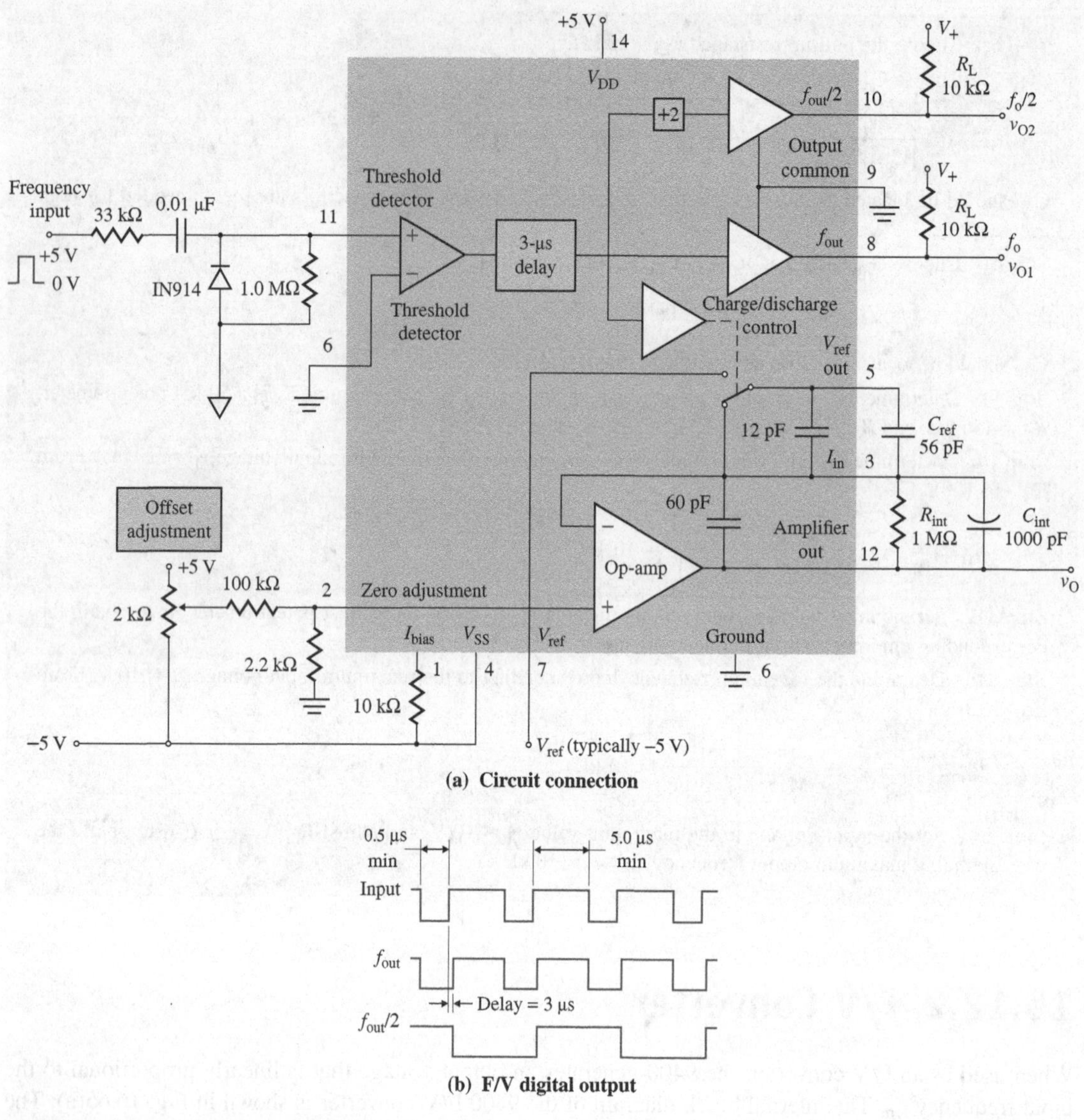

FIGURE 16.65 Internal block diagram of 9400 F/V converter (Courtesy of TelCom Semiconductor, Inc.)

which, for $q = C_{ref}V_{ref}$, relates average output voltage V_O to input frequency f_{in} as follows:

$$V_O = |V_{ref}|R_{int}C_{ref}f_{in} \tag{16.101}$$

where f_{in} = input frequency, in hertz

$|V_{ref}|$ = reference voltage, in volts

R_{int} = internal integrating resistance, in ohms

C_{ref} = reference capacitance, in farads

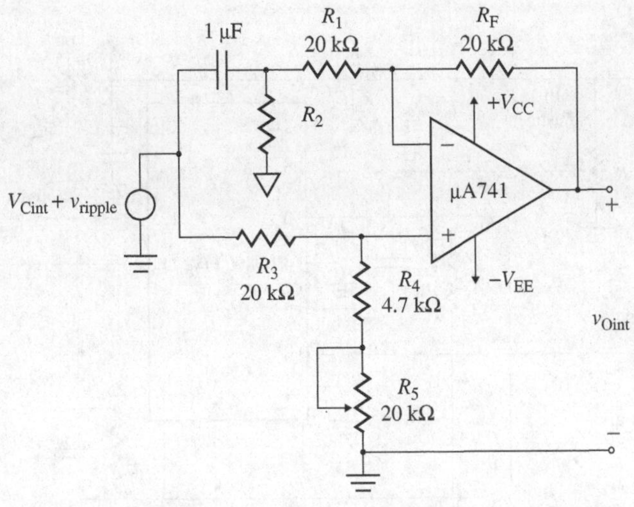

FIGURE 16.66 Ripple elimination in an F/V converter

The F/V converter will accept any input wave shape. However, the positive pulse width of the detector input (pin 11) must be at least 5 μs, and the negative pulse width must be greater than 0.5 μs. When the input frequency is less than 1 kHz, the duty cycle should be greater than 20% to ensure that C_{ref} is fully charged and discharged.

The output voltage V_O will have a certain amount of ripple, which is inversely proportional to C_{int} and the input frequency f_{in}. Therefore, for low frequencies, C_{int} can be increased in the range from 1 μF to 100 μF to reduce the ripple. To eliminate the ripple on V_O, an op-amp circuit in the common-mode configuration may be connected at the output of the F/V converter. This arrangement is shown in Fig. 16.66. Since the AC ripple content appears at both the (+) and the (−) terminals of the op-amp, the AC ripple will be canceled, and the output will have only DC voltage.

EXAMPLE 16.19

D **Designing an F/V converter** Using the 9400 as shown in Fig. 16.67, design an F/V converter so that $V_O = 2.5$ V at $f_{in} = 5$ kHz. The input frequency f_{in} can vary between 0 and 10 kHz. Assume $V_{DD} = 5$ V, $-V_{SS} = 0$.

SOLUTION

The steps used to design the F/V converter are as follows.

Step 1. Choose V_{DD} and V_{SS} such that

$$4\text{ V} \le V_{DD} \le 7.5\text{ V}$$

Choose $V_{DD} = 5$ V.

Step 2. Choose the capacitors such that $C_4 = 0.1$ μF. These capacitors should be close to pins 4 and 14, respectively.

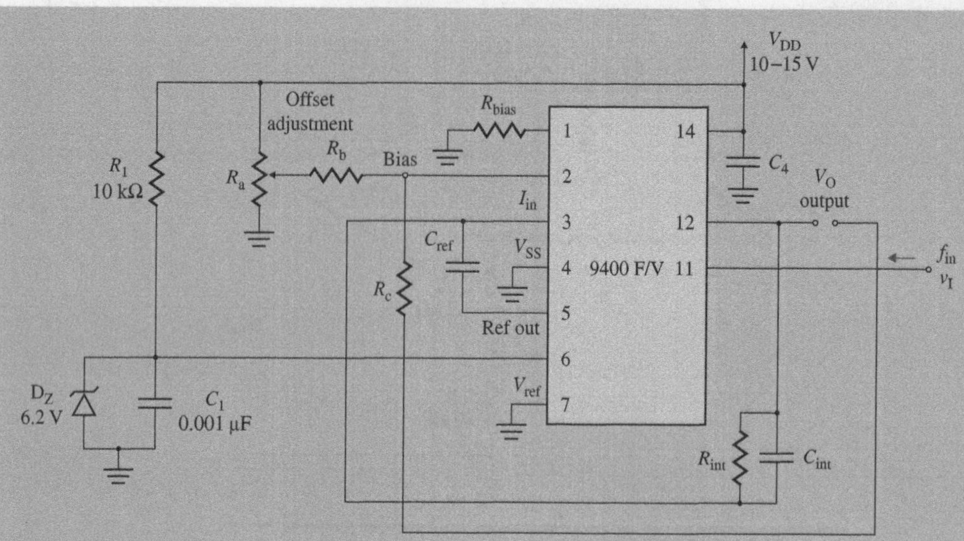

FIGURE 16.67 TelCom 9400 converter connected as an F/V converter

Step 3. Choose the reference voltage $V_{ref} = 0$ V.

Step 4. Choose $R_{int} = 1$ MΩ.

Step 5. Choose $R_{bias} = 100$ kΩ.

Step 6. Choose the pull-up resistance $R_L = 10$ kΩ.

Step 7. Choose C_{ref}. From Eq. (16.101),

$$C_{ref} = \frac{V_O}{|V_{ref}|R_{int}f_{in}} = \frac{2.5\text{ V}}{5 \times 1\text{ M}\Omega \times 5\text{ kHz}} = 100\text{ pF}$$

C_{ref} should be located as close as possible to pins 3 and 5. Glass-film capacitors are recommended for high accuracy.

Step 8. Choose C_{int}. Let

$$C_{int} = 10 \times C_{ref} = 10 \times 100\text{ pF} = 1\text{ nF}$$

C_{int} should be located as close as possible to pins 3 and 12. Since the amount of ripple on the output voltage is inversely proportional to C_{int} and the input frequency, C_{int} can be increased to lower the ripple. Acceptable values of C_{int} for low frequencies are 1 μF to 100 μF.

Step 9. Determine the values of the offset resistors. Since $R_c \le R_a \le R_b$, assume R_a is a 50-kΩ potentiometer, $R_b = 450$ kΩ, and $R_c = 10$ kΩ.

Step 10. With no input signal applied ($f_{in} = 0$), adjust the potentiometer R_a to obtain the minimum output voltage $V_{O(min)} = 0$.

Step 11. Determine the maximum output voltage corresponding to the maximum input frequency $f_{in} = 10$ kHz. From Eq. (16.101),

$$V_{O(max)} = 5 \times 1\text{ M}\Omega \times 100\text{ pF} \times 10\text{ kHz} = 5\text{ V}$$

Step 12. Set the input frequency to the maximum value $f_{in} = 10$ kHz, and adjust C_{ref} so that V_O is approximately 2.5 V.

KEY POINTS OF SECTION 16.12

- The TelCom 9400 converter can be used as either a voltage-to-frequency (V/F) converter or a frequency-to-voltage (F/V) converter, and it can produce pulse and square-wave outputs with a frequency range of 1 Hz to 100 kHz.
- In a V/F converter, the input voltage is converted to charge by an op-amp integrator and gives a linearly decreasing output voltage. As soon as this voltage falls below a threshold level, a threshold detector causes the output to step up by a certain amount so that the system is ready to repeat the cycle when the output passes again through the zero axis.
- In an F/V converter, a precise amount of charge dispensed into the op-amp's summing junction generates a voltage pulse at the output of the op-amp. A capacitor is then used to average these pulses into a DC signal that is linearly proportional to the input frequency.

16.13 Sample-and-Hold Circuits

Sample-and-hold (SAH) circuits are used as an interface between an analog signal and a digital circuit in a wide variety of applications such as data acquisition, analog-to-digital (A/D) conversion, and synchronous data demodulation. An SAH circuit is used to *sample* an analog signal at a particular instant of time and *hold* the value of the sample as long as required. The sampling instants and hold duration are determined by a logic control signal. The hold duration depends on the type of application. For example, in A/D conversion, samples must be held long enough for the conversion to be completed.

The principle of operation of a sample-and-hold circuit can be explained with Fig. 16.68(a), which shows a circuit consisting of capacitor C, switch S_1, and internal resistor R_s. The capacitor is used to hold the sample. The switch provides a means of rapidly charging the capacitor to the sample voltage and then removing the input so that the capacitor can retain the desired voltage. When the control signal v_{CN} is high, the switch is closed. If the time constant R_sC is very small, the output voltage v_O will be very close to the

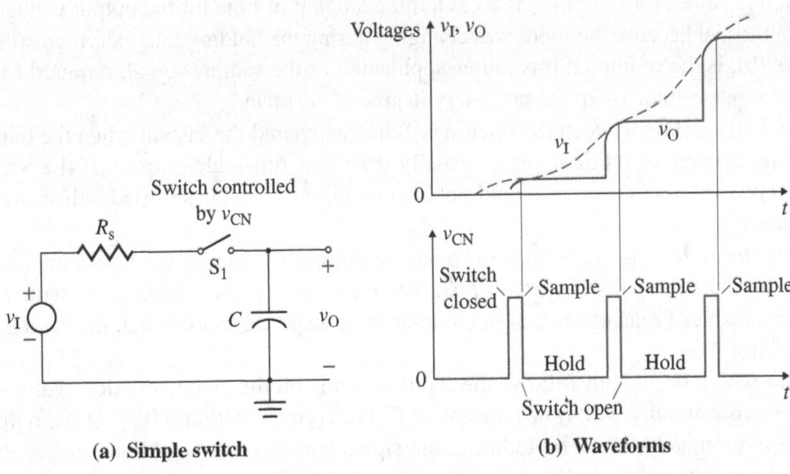

FIGURE 16.68 Principle of a sample-and-hold circuit

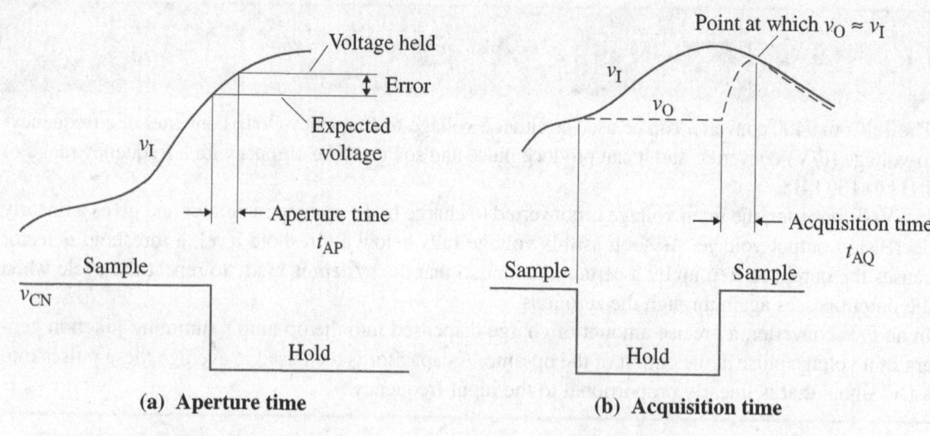

FIGURE 16.69 Aperture time and acquisition time

input voltage v_I and will be equal to it at the instant the control signal becomes low and the switch is opened. The idealized waveforms for output voltage v_O, input voltage v_I, and control voltage v_{CN} are shown in Fig. 16.68(b).

In practice, the capacitor can neither charge instantly nor hold a constant voltage. Also, the switch cannot open and close instantaneously. As a result, the practical output waveform will differ from the ideal one. Among the important specifications given by the manufacturers of SAH circuits are aperture time, acquisition time, settling time, and drop.

Aperture time t_{AP}, shown in Fig. 16.69(a), is the maximum time required for the SAH circuit to open. It is the delay between application of the control signal to open the switch and the instant when the switch actually does open. This time depends on the type of switch, but typically ranges from 4 μs to 20 μs. The t_{AP} of FET switches is in the range of 50 ns to 100 ns. The aperture time should be much less than the sampling period (i.e., the reciprocal of the sampling rate). Since the input signal is changing continuously, the hold voltage will change slightly during the aperture time, causing an error in the hold voltage.

Once the switch is closed for sampling, it takes a finite amount of time for the output voltage to become identical to the input signal because the input was changing during the holding interval. *Acquisition time* t_{AQ}, shown in Fig. 16.69(b), is the minimum time, after application of the sample signal, required for the output voltage to reach the input voltage (with the necessary degree of accuracy).

Settling time t_s is the delay between the opening of the switch and the instant when the output reaches within a specified percentage of its final value (usually 0.99% of full-scale output). If the SAH circuit is followed by an A/D converter, conversion should not begin until the signal has settled; otherwise, the wrong signal will be converted.

Drop, or *output decay rate*, is the voltage drop across capacitor C during the hold time. It is inversely proportional to the capacitance, since $dv_O/dt = I/C$, where I is the capacitor leakage current. This leakage current can arise as a result of biasing current in an op-amp, leakage current through the switch, or internal leakage in the capacitor.

The speed with which the output follows the input depends on the characteristics of the input signal v_I. v_O will follow v_I exponentially with time constant R_sC. For v_O to be within 0.01% of the output, its time period should be approximately $9R_sC$. In addition, the signal source must be able to supply the charging current required by capacitor C. Usually, the analog signal is buffered from the switch by a unity-gain op-amp follower in order to ensure a low value of R_s.

16.13.1 SAH Op-Amp Circuits

An SAH circuit can be implemented using an op-amp and a switch, as shown in Fig. 16.70(a). When switch S_1 is closed, the circuit operates as an *RC* filter. For a step input voltage of V_I, the output voltage $v_O(t)$ can be found from

$$v_O(t) = -\frac{R_F}{R_1} V_I(1 - e^{-t/R_F C}) \tag{16.102}$$

For v_O to reach V_I in the shortest time, the time constant $R_F C$ must be shorter than the sample interval so that the output can track the input. When switch S_1 is opened, the capacitor will hold its voltage of $-V_I$. To minimize the output voltage drop, the op-amp should have a low input biasing current (as does an op-amp with a FET input stage, for example). Also, a high-quality capacitor with a low leakage current should be used.

Switch S_1 can be replaced by a transistor M_1 (i.e., a *p*-type depletion MOSFET), as shown in Fig. 16.70(b). If the control voltage v_{CN} is low (say, 0 V), the FET M_1 will be on (i.e., the switch will be closed), and the capacitor will be in sample mode, charging to V_I. If the control voltage v_{CN} is high (say, +5 V), the FET will be off (i.e., the switch will be open), and the capacitor will be in hold mode.

Diode D_1 clamps the voltage at node A to 0.7 V. When M_1 is on, the diode effectively becomes connected across the FET (i.e., between its drain and source). Since the voltage drop across the FET is low, the voltage across the diode will also be small—much less than 0.7 V. Thus, the diode will have no effect during the sampling time.

16.13.2 SAH Integrated Circuits

SAH integrated circuits such as the LF198 use BiFET technology to obtain ultrahigh DC accuracy (within 0.01%) with fast signal acquisition (4 μs) and a low drop (3 mV/s). The functional block diagram of the LF198 is shown in Fig. 16.71(a), and its connection diagram is shown in Fig. 16.71(b). The manufacturers give curves showing the variation in acquisition time t_{AQ} with hold capacitance C_h. For example, $t_{AQ} = 4$ μs for hold capacitance $C_h = 1000$ pF, and $t_{AQ} = 20$ μs for $C_h = 0.01$ μF.

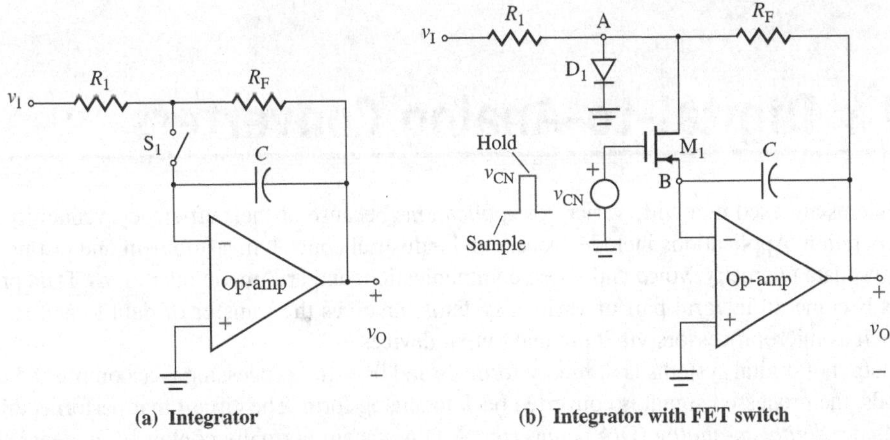

FIGURE 16.70 Inverting SAH op-amp circuit

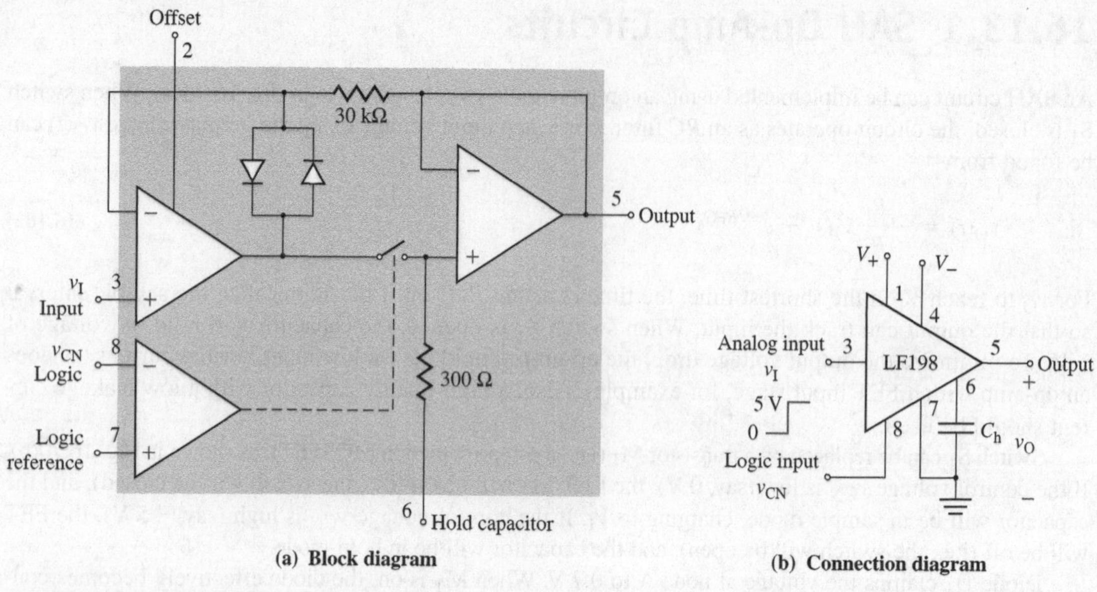

FIGURE 16.71 Sample-and-hold LF198 (Courtesy of National Semiconductor, Inc.)

KEY POINTS OF SECTION 16.13

- An SAH circuit uses a capacitor to sample an analog signal at a particular instant of time and hold the value of the sample as long as required. A switch is closed to charge the capacitor rapidly to the sample voltage and then is opened to remove the input so that the capacitor can retain the desired voltage.
- The specifications of an SAH circuit include acquisition time, aperture time, settling time, and drop, or output decay rate.

16.14 Digital-to-Analog Converters

Digital systems are used in a wide variety of applications because of their efficiency, reliability, and economical operation. Applications include process and industrial control, measurement and testing, graphics and displays, data telemetry, voice and video communication, and arithmetic operations. Data processing, which has become an integral part of various systems, involves the transfer of data to and from digital devices such as microprocessors via input and output devices.

The output of digital systems is in binary form: 1s and 0s. After processing is accomplished using digital methods, the processed signal is converted back to analog form. The circuit that performs this conversion is called a *digital-to-analog* (D/A) *converter*. A D/A system normally contains four separate parts: a reference quantity; a set of binary switches to simulate binary coefficients $B_0, \ldots, B_N$; a resistive network; and an output summing means.

16.14.1 Weighted-Resistor D/A Converter

A simple D/A converter is shown in Fig. 16.72(a). This converter can convert a 4-bit parallel digital word ($B_0B_1B_2B_3$) to an analog voltage that is proportional to the binary number corresponding to the digital word. Four switches are used to simulate the binary inputs. (In practice, a 4-bit binary counter may be used instead.) The logic voltages, which represent the individual bits B_0, B_1, B_2, and B_3, are used to operate switches S_0, S_1, S_2, and S_3, respectively. When a B is a 1, the corresponding switch is connected to reference voltage V_{ref}; when a B is a 0, the corresponding switch is grounded.

The inverting terminal of the op-amp is at virtual ground (i.e., $V_d \approx 0$), so the total current I_S is given by

$$I_S = V_{ref}\left(\frac{B_3}{R_3} + \frac{B_2}{R_2} + \frac{B_1}{R_1} + \frac{B_0}{R_0}\right)$$

Since the current flowing into the op-amp is negligible, $I_S \approx I_F$. Thus, the analog output voltage is given by

$$V_O = -R_F I_F = -R_F V_{ref}\left(\frac{B_3}{R_3} + \frac{B_2}{R_2} + \frac{B_1}{R_1} + \frac{B_0}{R_0}\right) \tag{16.103}$$

Resistors are weighted so that successive resistor values are related by a factor of 2 and the value of each individual resistor is inversely proportional to the numerical significance of the appropriate binary digit. That is,

$$\text{LSB (least significant bit)} \rightarrow \quad R_0 = \frac{R}{2^0} = R$$

$$R_1 = \frac{R}{2^1} = \frac{R}{2}$$

$$R_2 = \frac{R}{2^2} = \frac{R}{4}$$

$$\text{MSB (most significant bit)} \rightarrow \quad R_3 = \frac{R}{2^3} = \frac{R}{8}$$

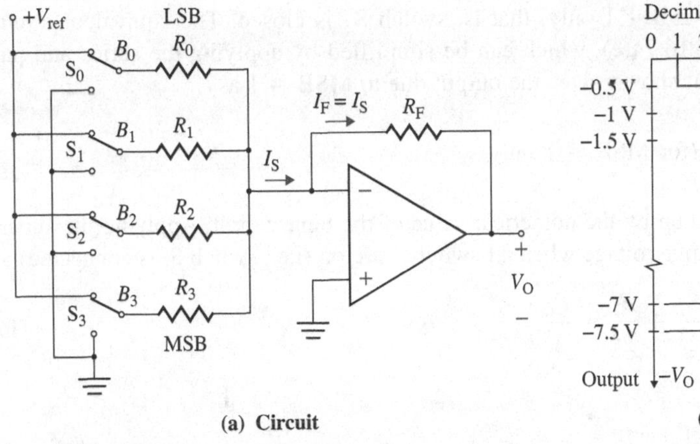

FIGURE 16.72 Weighted-resistor D/A converter

Substituting these weighted resistor values into Eq. (16.103) gives the analog output voltage V_O as

$$V_O = -\frac{V_{ref}R_F}{R}(2^3B_3 + 2^2B_2 + 2^1B_1 + 2^0B_0) \tag{16.104}$$

where $B_i = 1$ if switch S_i is connected to V_{ref} and $B_i = 0$ if switch S_i is grounded. For an input of $B_3B_2B_1B_0 = 1111$, $V_O = -15V_{ref}R_F/R$; for $B_3B_2B_1B_0 = 0110$, $V_O = -6V_{ref}R_F/R$; and for $B_3B_2B_1B_0 = 0001$, $V_O = -V_{ref}R_F/R$. Thus, the output V_O is directly proportional to the numerical value of the binary number $B_3B_2B_1B_0$. Since there are 16 (i.e., 2^4) combinations of the binary inputs B_3, B_2, B_1, and B_0, the analog output will have 16 possible corresponding values. For $V_{ref} = 5$ V and $R = 10R_F$, Eq. (16.104) gives V_O as

$$V_O = -0.5 \times (2^3B_3 + 2^2B_2 + 2^1B_1 + 2^0B_0)$$

The plot is shown in Fig. 16.72(b). The major disadvantage of this D/A converter is the wide variety of resistor values required to weight the network. If the resistor values change in response to temperature changes, it will be difficult to obtain identical tracking characteristics. As a result, the accuracy and the stability of the D/A will be degraded.

16.14.2 *R*-2*R* Ladder Network D/A Converter

The *R*-2*R* D/A ladder converter, shown in Fig. 16.73(a), has only two resistor values *R* and 2*R*, rather than a wide range of resistor values. The plot of the output (which is known as a "resistance ladder") is shown in Fig. 16.73(b). Figure 16.73(c) exhibits the property that the equivalent resistance, looking into any of the terminals *X*, *Y*, S_3, S_2, S_1, or S_0 with the remainder of the terminals grounded, is 3*R*.

Consider the circuit with LSB = 1 only; that is, switch S_0 is closed. The equivalent circuit for LSB = 1 only is shown in Fig. 16.74(a). Successive Thevenin's conversions lead, through the circuits shown in Fig. 16.74[(b) and (c)], to the final circuit shown in Fig. 16.74(d), which gives the output due to LSB = 1 as

$$V_O = -\frac{V_{ref}R_F}{3R}\left(\frac{B_0}{2^4}\right) \quad \text{(for LSB = 1 only)}$$

Now consider the circuit with MSB = 1 only; that is, switch S_3 is closed. The equivalent circuit for MSB = 1 only is shown in Fig. 16.74(e), which can be simplified by applying the series and parallel rule for *R*s. The simplified circuit shown gives the output due to MSB = 1 as

$$V_O = -\frac{V_{ref}R_F}{3R}\left(\frac{B_3}{2^1}\right) \quad \text{(for MSB = 1 only)}$$

Thus, the output voltage is scaled up by the numerical value of the binary digit. Applying the superposition theorem, we can find the output voltage when all switches are on (i.e., switch S_i is connected) as

$$V_O = -\frac{V_{ref}R_F}{3R}\left(\frac{B_3}{2^1} + \frac{B_2}{2^2} + \frac{B_1}{2^3} + \frac{B_0}{2^4}\right) \tag{16.105}$$

which can be simplified to

$$V_O = -\frac{V_{ref}R_F}{48R}(2^3B_3 + 2^2B_2 + 2^1B_1 + 2^0B_0) \tag{16.106}$$

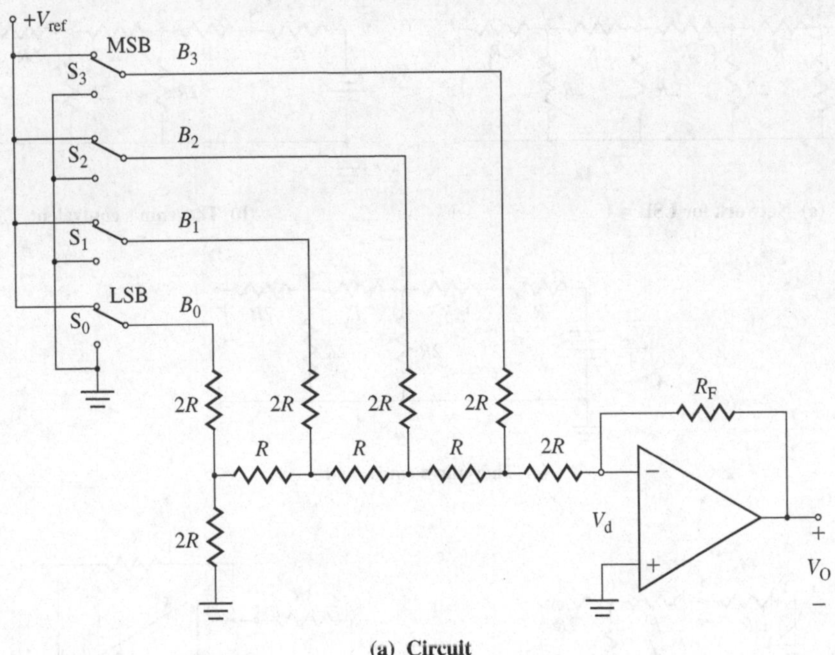

(a) Circuit

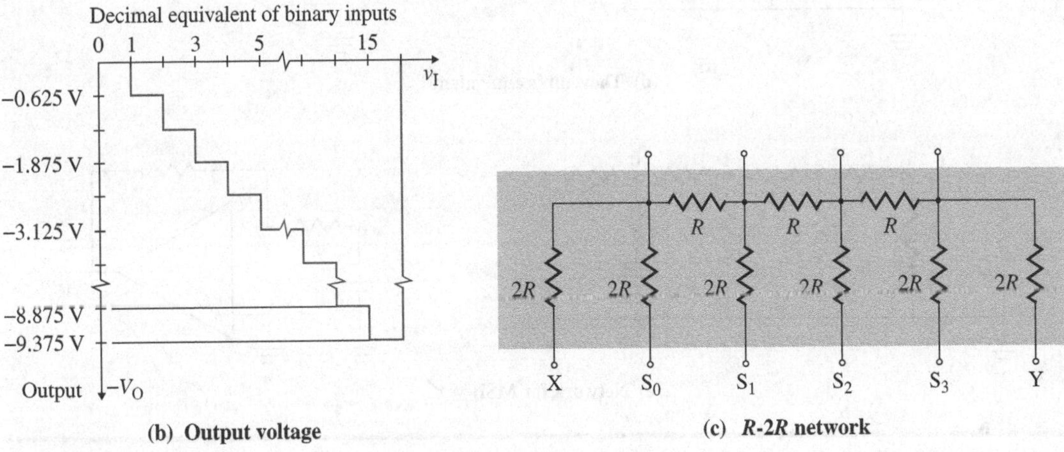

(b) Output voltage

(c) R-2R network

FIGURE 16.73 R-2R ladder D/A converter

where $B_i = 1$ if switch S_i is connected to V_{ref} and $B_i = 0$ if switch S_i is grounded. For an input of $B_3B_2B_1B_0 = 1111$, $V_O = -15V_{ref}R_F/48R$; for $B_3B_2B_1B_0 = 0110$, $V_O = -6V_{ref}R_F/48R$; and for $B_3B_2B_1B_0 = 0001$, $V_O = -V_{ref}R_F/48R$. For $V_{ref} = 5$ V and $6R = R_F$, Eq. (16.106) gives V_O as

$$V_O = -\left(\frac{5}{48}\right) \times (2^3B_3 + 2^2B_2 + 2^1B_1 + 2^0B_0)$$

whose plot is shown in Fig. 16.73(b) for $V_{O(max)} = 10$ V.

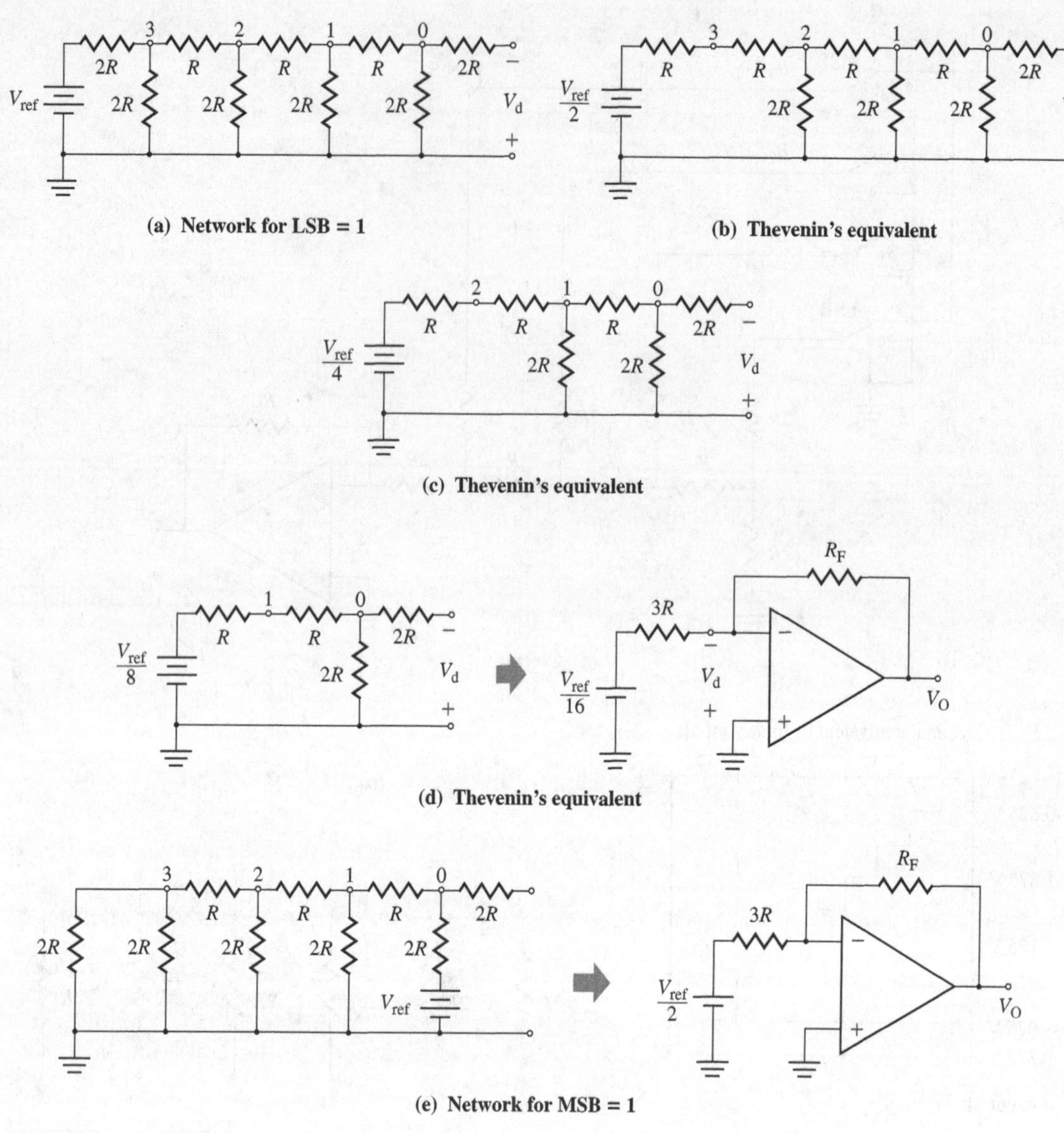

FIGURE 16.74 Equivalent *R*-2*R* ladder for LSB = 1 only and MSB = 1 only

16.14.3 Integrated Circuit D/A Converters

Switches in IC D/A converters are made either of BJTs or of MOSFETs. They are generally one of two types: voltage driven or current driven. Voltage-driven converters, which use BJTs or MOSFETs as on or off switches, are generally used for relatively low-speed low-resolution applications. In a current-driven converter, switching is accomplished using emitter-coupled logic (ECL) current switches, which do not saturate but are driven from the active region to cutoff. This type of converter is capable of much faster operation than the voltage-driven type. IC D/A converters of 8, 10, 12, 14, and 16 bits are commercially available with either a current output, a voltage output, or both a current and a voltage output.

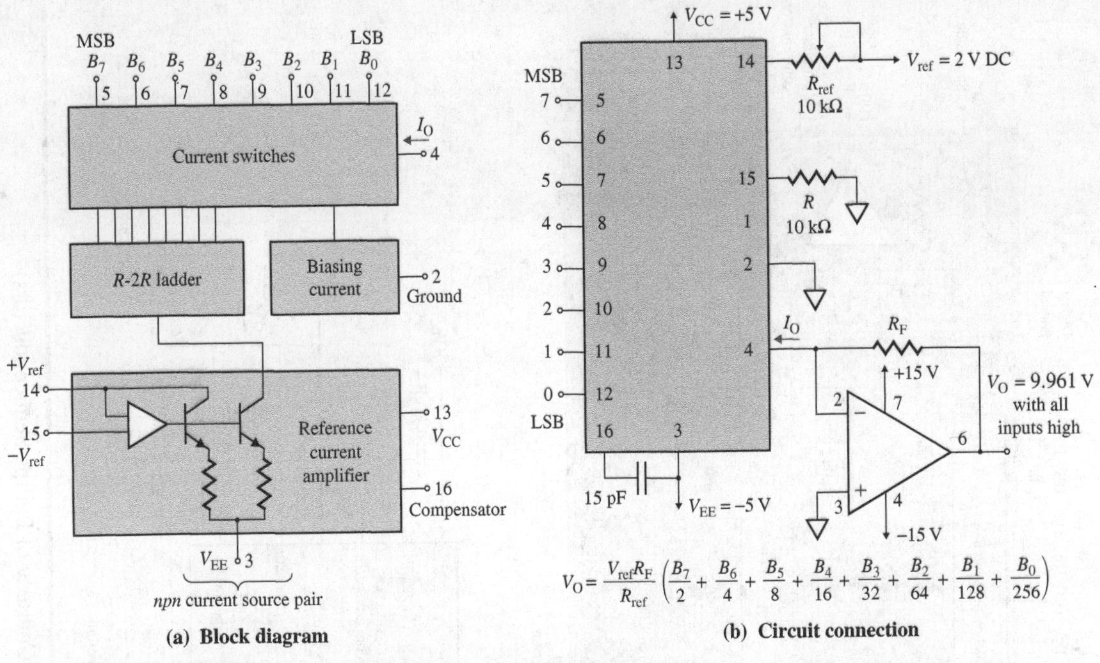

FIGURE 16.75 MC1408 D/A converter with current output

The MC1408 is an example of a D/A converter with current output. It is a low-cost, high-speed converter designed for use in applications where the output current is a linear product of an 8-bit digital word and an analog reference voltage. Its internal block diagram, shown in Fig. 16.75(a), consists of four parts: current switches, an *R*-2*R* ladder, a biasing current network, and a reference current amplifier. The connection diagram is shown in Fig. 16.75(b). The output current is converted to a voltage by an instantaneous current-to-voltage (I/V) op-amp converter.

The NE/SE-5018 is an example of a D/A converter with voltage output. It gives an output voltage that is a linear product of an 8-bit digital word and an analog reference voltage. Its internal block diagram is shown in Fig. 16.76(a). A typical configuration of the 5018 is shown in Fig. 16.76(b).

The manufacturer's specifications for a D/A converter normally include the following parameters. *Resolution* is determined by the number of input bits of the D/A converter. An 8-bit converter has 2^8 possible output levels, so its resolution is $1/2^8 = 1/256 = 0.39\%$. For a 4-bit converter, the resolution is $1/2^4 = 1/16 = 6.25\%$. Thus, resolution is the value of the LSB. *Accuracy* is defined in terms of the maximum deviation of the D/A output from an ideal straight line drawn from zero to full-scale output. *Nonlinearity*, or *linearity error*, is the difference between the actual output of the D/A converter and its ideal straight-line output. The error is normally expressed as a percentage of the full-scale range. *Gain error* is any error in gain, usually caused by deviations in the feedback resistor on the I/V converter. *Offset error* is any error caused by the fact that the output of the D/A converter is not zero when the binary inputs are all zero. This error stems from input offsets (in voltages and currents) of the op-amp as well as the D/A converter. *Settling time* is the time required for the output of the D/A converter to reach within $\pm 1/2$ LSB of the final value for a given digital input—that is, go from zero to full-scale output. *Stability* is a measure of the independence of the converter parameters from variations in external conditions such as temperature and supply voltage.

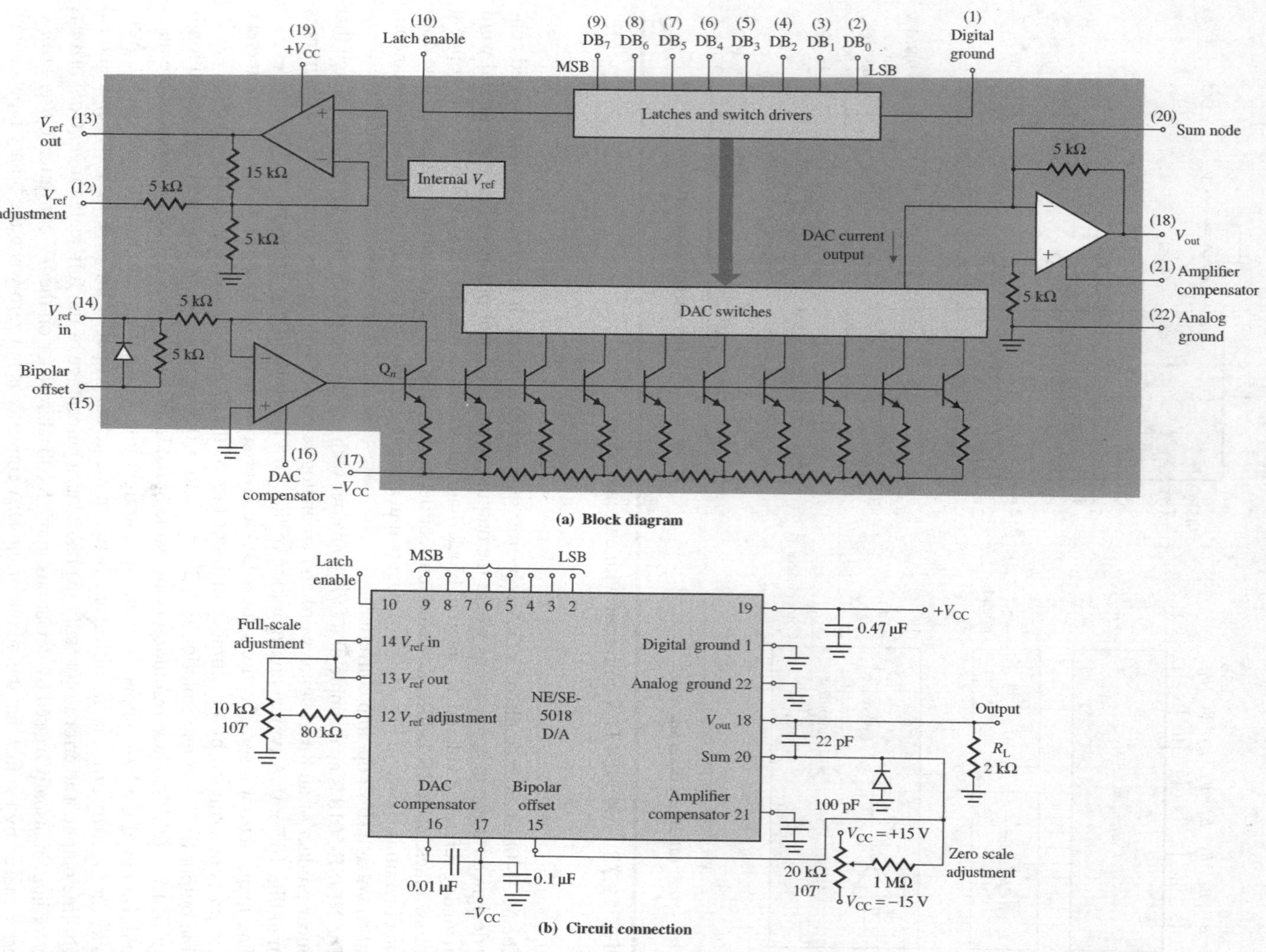

FIGURE 16.76 NE/SE-5018 D/A converter with voltage output (Courtesy of Philips Semiconductors)

KEY POINTS OF SECTION 16.14

- A D/A converter can convert a digital word to an analog voltage that is proportional to the binary number corresponding to the digital word.
- The specifications for a D/A converter include resolution, accuracy, nonlinearity (or linearity error), gain error, offset error, settling time, and stability. The resolution of an N-bit converter is $1/2^N$.

16.15 Analog-to-Digital Converters

A large number of physical devices generate output signals that are analog or continuous variables; examples include temperature and pressure gauges and flow transducers. For digital processing, the input signal must be converted into a binary form of 1s and 0s. The circuit that performs this conversion is called an analog-to-digital (A/D) converter. There are many types of A/D converters, depending on the type of conversion technique used, such as counting, tracking (up-down), successive approximation, single-ramp integrating, or dual-ramp integrating. The successive-approximation technique is the one most commonly used, mainly because it offers excellent trade-offs in resolution, speed, accuracy, and cost.

16.15.1 Successive-Approximation A/D Converter

A successive-approximation A/D converter operates by successively dividing in half the voltage range of the converter. The simplified block diagram of a 4-bit A/D converter is shown in Fig. 16.77(a). The converter consists of five parts: an analog comparator, a 4-bit register that has independent set and reset capability for each stage, a 4-bit D/A converter, a ring counter, and a logic control. The ring counter provides a timing (or clock) signal to control the operation of the converter. The logic control synchronizes the operation of the converter with the clock. The combination of the logic control, 4-bit register, and ring counter is often known as the *successive-approximation register* (SAR).

The comparator converts analog voltages to digital signals. It has two inputs, V_a and V_b, and gives a binary voltage. If $V_a > V_b$, the output is high (logic 1); if $V_a < V_b$, the output is low (logic 0). Thus, the comparator output V_{com} is

$$V_{com} = \text{sgn}\,(V_a - V_b) = \begin{cases} 1 & \text{for } V_a > V_b \\ 0 & \text{for } V_a < V_b \end{cases}$$

An SAH circuit is commonly used to hold the input voltage constant during the conversion process. There is no need for an SAH circuit if the input signal varies slowly enough and has a low enough noise level that the input will not change during the conversion.

The algorithm for the operation of a successive-approximation A/D converter can be best described by an example. The steps in converting an analog voltage of, say, 10 V are as follows:

Step 1. The first pulse from the ring counter sets the D/A converter, 4-bit register, and ring counter so that MSB = 1 and all others are 0; that is, $B_3 = 1$, and $B_2 = B_1 = B_0 = 0$. Thus, for $B_3B_2B_1B_0 = 1000$, the output V_b of the D/A is 8 V, which is compared by the comparator. If $V_a \geq 8$ V, the MSB in the register (B_3) is maintained at 1; otherwise, it is set to 0. At the end of step 1, $B_3 = 1$ for $V_a = 10$ V.

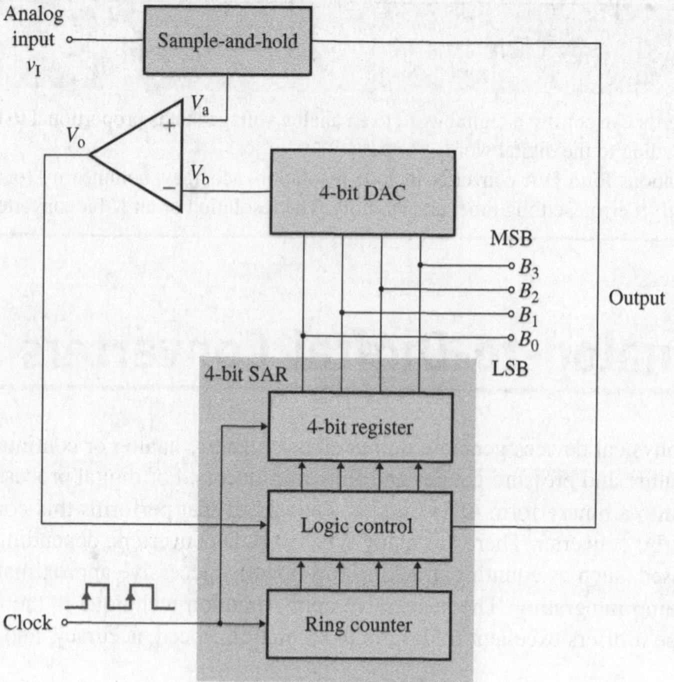

(a) 4-bit A/D converter

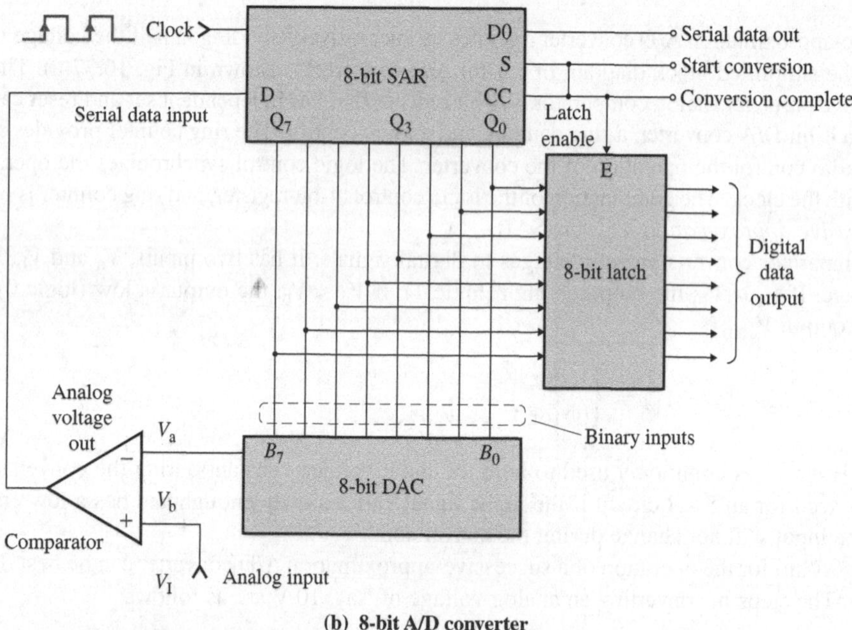

(b) 8-bit A/D converter

FIGURE 16.77 Successive-approximation A/D converter

Step 2. The second pulse from the ring counter sets $B_2 = 1$. B_1 and B_0 remain at 0, and B_3 remains at either 1 or 0, depending on the condition in step 1; that is, $B_3 = B_2 = 1$, and $B_1 = B_0 = 0$. Thus, for $B_3B_2B_1B_0 = 1100$, the output V_b of the D/A is 12 V, which is compared by the comparator. If $V_a \geq 12$ V, the B_2 in the register is maintained at 1; otherwise, it is set to 0. At the end of step 2, $B_2 = 0$ for $V_a = 10$ V.

Step 3. The third pulse from the ring counter sets $B_1 = 1$. B_0 remains at 0. B_2 and B_3 remain as they were at the end of step 2. That is, $B_3 = 1$, $B_2 = 0$, $B_1 = 1$, and $B_0 = 0$. Thus, for $B_3B_2B_1B_0 = 1010$, the output V_b of the D/A is 10 V, which is compared by the comparator. If $V_a \geq 10$ V, the B_1 in the register is maintained at 1; otherwise, it is set to 0. At the end of step 3, $B_1 = 1$ for $V_a = 10$ V.

Step 4. The fourth pulse from the ring counter sets $B_0 = 1$. B_3, B_2, and B_1 remain as they were at the end of step 3. That is, $B_3 = 1$, $B_2 = 0$, $B_1 = 1$, and $B_0 = 1$. Thus, for $B_3B_2B_1B_0 = 1011$, the output V_b of the D/A is 11 V, which is compared by the comparator. If $V_a \geq 11$ V, the B_0 in the register is maintained at 1; otherwise, it is set to 0. That is, $B_0 = 0$ for $V_a = 10$ V.

At the end of the fourth step, the desired number, which is in the counter, will give Read output. The results of the conversion steps are shown in Table 16.2. For an N-bit A/D converter, the conversion process will take N clock periods. That is, for an 8-bit A/D converter and a 10-MHz clock, the conversion will take $8/(10 \times 10^6) = 8 \times 10^{-7} = 800$ ns.

The successive-approximation technique can be extended to the higher-bit converter shown in Fig. 16.77(b), in which the SAR also performs the functions of logic control and ring counter. The conversion complete (CC) signal enables the latch. Digital data appear at the output of the latch and are also available serially as the SAR determines each bit. The cycle of the conversion process is normally repeated continuously, and the CC signal is connected to the Start-Conversion input.

16.15.2 Integrated Circuit A/D Converters

There are many types of IC A/D converters, such as the integrating A/D converter, the integrating A/D converter with three-stage outputs, and the tracking A/D converter with latched output. Also, the output can be in straight binary, binary-coded decimal (BCD), complementary binary (1s or 2s), or sign-magnitude binary form.

The NE5034 is an example of an IC A/D converter. Its internal block diagram is shown in Fig. 16.78(a). It is a high-speed microprocessor-compatible 8-bit A/D converter that uses the successive-approximation technique. It includes a comparator, a reference D/A converter, an SAR, an internal clock, and three-stage buffers all on the same chip. The connection diagram for the NE5034 is shown in Fig. 16.78(b). Upon receipt of the Start pulse, successive bits are applied to the input of the internal 8-bit current D/A converter by the

TABLE 16.2 Successive-approximation process for $V_a = 10$ V

Step	V_b	B_3	B_2	B_1	B_0	Comparisons	Answer
1	8 V	1	0	0	0	Is $V_a \geq 8$ V?	Yes
2	12 V	1	1	0	0	Is $V_a \geq 12$ V?	No
3	10 V	1	0	1	0	Is $V_a \geq 10$ V?	Yes
4	11 V	1	0	1	1	Is $V_a \geq 11$ V?	No
	10 V	1	0	1	0	Read output	

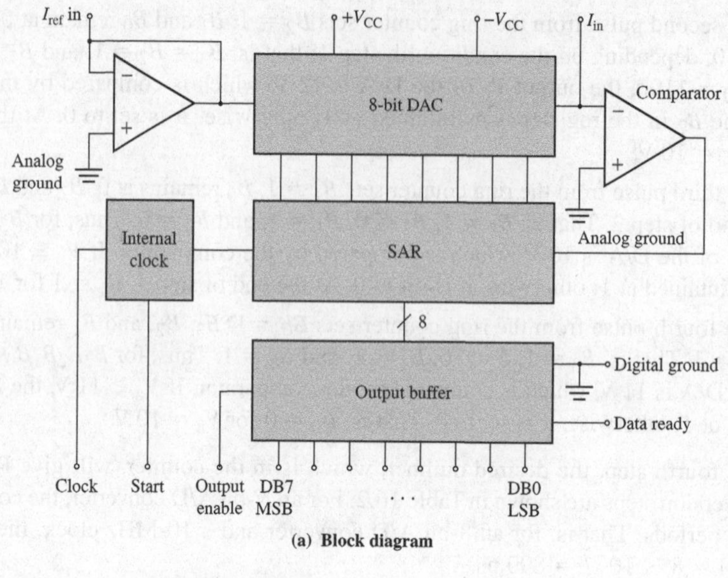

(a) Block diagram

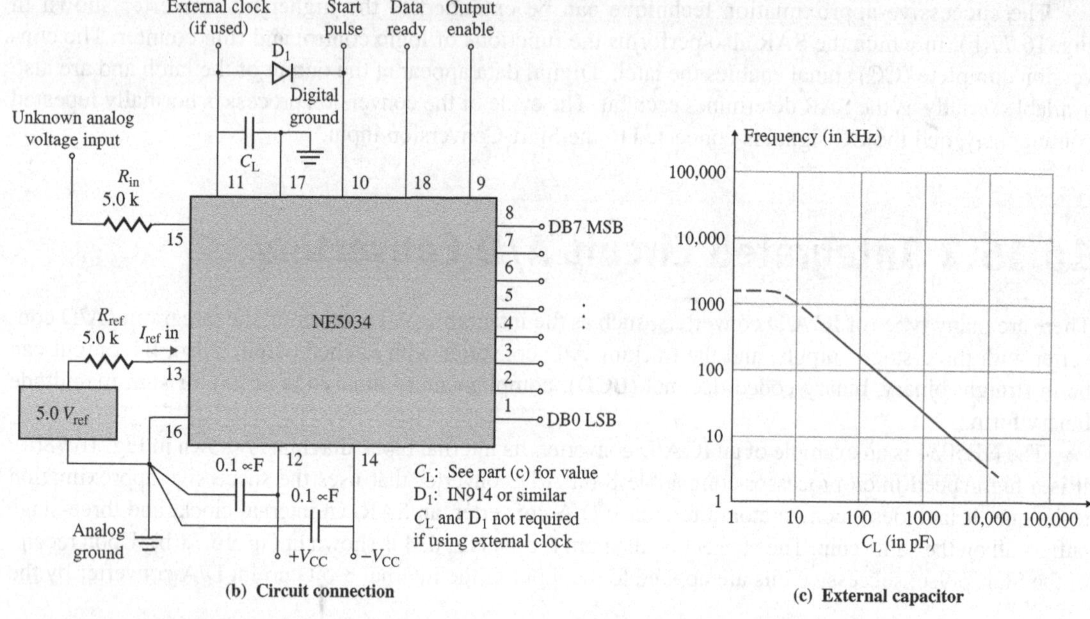

(b) Circuit connection

(c) External capacitor

FIGURE 16.78 NE5034 8-bit A/D converter

I^2L SAR, beginning with the MSB (DB7). During the successive approximations, the sequence Data-Ready (DR) remains at 1. When the Output-Enable (OE) input is at logic 1, the data outputs assume a high-impedance status. When OE is at logic 0, the data are placed on the outputs. External capacitor C_L sets the internal clock frequency, as shown in Fig. 16.78(c). For $C_L = 100$ pF, for example, $f_{clock} = 120$ kHz.

The manufacturer's specifications for an A/D converter normally include the following parameters: *Input signal* is the maximum allowable analog input voltage range and may be unipolar or bipolar.

Conversion speed is the speed at which the A/D converter can make repetitive data conversions. The conversion time for successive-approximation converters ranges from 1 μs to 100 μs, whereas for an ultrafast parallel converter the time is in the range of 10 ns to 60 ns. *Quantizing error* is the error inherent in the conversion process because of the finite resolution of the discrete output. It is usually $\pm 1/2$ LSB. For a 10-bit converter with an analog input range of 0 to 10 V, the quantizing error will be $1/2^{10} \times 10\text{ V} \approx 10\text{ mV}$. *Accuracy* is the deviation of the actual bit transition value from the ideal transition value at any level over the range of the A/D converter. Accuracy includes errors from both the analog and the digital parts. With a digital error of 10 mV and a quantizing error of 10 mV, the overall error becomes 20 mV. With this amount of error, the converter will operate as a 9-bit A/D converter because a 9-bit converter has a quantizing error of $1/2^9 \times 10\text{ V} \approx 20\text{ mV}$.

KEY POINTS OF SECTION 16.15

- An A/D converter can convert an analog signal to a digital word that is proportional to the analog signal. Although there are many conversion techniques, the successive-approximation technique is the one most commonly used, mainly because of its excellent trade-offs in resolution, speed, accuracy, and cost.
- The specifications for an A/D converter include input signal range, conversion speed, quantizing error, and accuracy.

16.16 Circuit Design Using Analog Integrated Circuits

There are many analog ICs for general- and special-purpose applications. They include operational amplifiers, voltage comparators, instrument amplifiers, timers, buffers, interfacing circuits, voltage/frequency converters, data conversion circuits, power conversion and control circuits, and voltage regulators. The circuit design for an application using an IC is very simple and requires the selection of external components only. The steps involved are as follows:

Step 1. Identify the function(s) to be performed and the specifications, including available power supplies, range of input and output signals, and operating frequency range.

Step 2. Find a suitable IC that can perform the desired function(s) and look for application examples and/or guidelines for that IC. Usually the manufacturer provides application examples and guidelines.

Step 3. Determine the values of external components (usually capacitors and resistors). Generally, the manufacturer provides selection charts or curves. Unless otherwise specified, use standard values of components, with tolerances of, say, 5%.

Step 4. Simulate the circuit with a simulator such as PSpice/SPICE or Electronics Workbench, if the IC is supported by the simulator.

Step 5. Build and test the circuit, if possible.

Summary

Op-amp circuits with diodes can generate nonlinear functions for applications such as peak signal detectors, precision rectifiers, and comparator circuits. Op-amp circuits with diodes can reduce the effects of diode drop for low-voltage signals and are used for precision signal processing. The diode is placed in the feedback path of the op-amp and behaves as a superdiode with a negligible voltage drop on the order of microvolts.

Various waveforms are often required in electronic and control circuits. There are many integrated circuits that can be used to generate these waveforms. The design of wave generators is very simple and requires the selection of the external circuit components only. ICs such as op-amps, comparators, the NE/SE-566 VCO, the 555 timer, the NE/SE-565 PLL, and the 9400 series converters can be used to generate various waveforms.

References

1. J. R. Hufault, *Op-Amp Network Design*. New York: Wiley, 1986.
2. F. W. Hughes, *Op-Amp Handbook*. Englewood Cliffs, NJ: Prentice Hall, 1986.
3. C. F. Wojslow, *Operational Amplifiers*. New York: Wiley, 1986.
4. M. H. Rashid, *Introduction to PSpice Using OrCAD for Circuits and Electronics*. Englewood Cliffs, NJ: Prentice Hall, 2004, Chapter 10.
5. P. R. Gray, P. J. Hurst, S. H. Lewis, and R. G. Meyer, *Analysis and Design of Analog Integrated Circuits*. New York: Wiley, 2001.
6. J. M. Jacob, *Applications and Design with Analog Integrated Circuits*. Englewood Cliffs, NJ: Regents/ Prentice Hall, 1993.
7. S. Soclof, *Design and Applications of Analog Integrated Circuits*. Englewood Cliffs, NJ: Prentice Hall, 1991, Chapters 15–17.
8. R. A. Gayakwad, *Op-Amps and Linear Integrated Circuits*. Englewood Cliffs, NJ: Prentice Hall, 1993.

Review Questions

1. What is a voltage limiter?
2. What is a soft limiter?
3. What is a clamper?
4. What is a comparator?
5. What is a limiting comparator?
6. What are the advantages of precision rectifiers?
7. What is a superdiode?
8. What are the differences between a comparator and an op-amp?
9. What is the principle of operation of a zero-crossing detector?
10. What is a Schmitt trigger?
11. What circuit parameters determine the upper and lower threshold voltages of a Schmitt trigger?
12. What are the effects of hysteresis on output voltage?

13. What is the principle of operation of a square-wave generator?
14. Why is a square-wave generator called an astable multivibrator?
15. What is saturation of an op-amp?
16. What is the purpose of series resistors in the input terminals of an op-amp in a square-wave generator?
17. What circuit parameters determine the frequency of a square wave?
18. What is the principle of operation of a triangular-wave generator?
19. What circuit parameters determine the frequency of a triangular wave?
20. What is the principle of operation of a sawtooth-wave generator?
21. What circuit parameters determine the frequency of a sawtooth wave?
22. What is the duty cycle of a sawtooth wave?
23. What is a voltage-controlled oscillator (VCO)?
24. What is the principle of operation of a VCO?
25. What circuit parameters determine the output frequency of a VCO?
26. What is the 555 timer?
27. What is a monostable multivibrator?
28. What advantages does the 555 timer connected as an astable multivibrator have over an op-amp astable multivibrator?
29. What circuit parameters determine the output frequency of the 555 timer connected as ramp generator?
30. What is a phase-lock loop (PLL)?
31. What are the main components of a PLL?
32. What is the principle of operation of a PLL?
33. What is the free-running mode of a PLL?
34. What is the capture mode of a PLL?
35. What is the phase-lock mode of a PLL?
36. What is the capture range of a PLL?
37. What is the lock range of a PLL?
38. What are the relationships among the free-running frequency, capture frequency, and lock frequency of a PLL?
39. What are some applications of the TelCom 9400 series converter?
40. What is a sample-and-hold circuit?
41. What are the main parts of a sample-and-hold circuit?
42. What is a digital-to-analog (D/A) converter?
43. What are the main parts of a D/A converter?
44. What is an analog-to-digital (A/D) converter?
45. What are the main parts of an A/D converter?

Problems

The symbol D indicates that a problem is a design problem. The symbol P indicates that you can check the solution to a problem using PSpice/SPICE or Electronics Workbench.

16.2 Circuits with Op-Amps and Diodes

16.1 P **a.** The precision full-wave rectifier in Fig. 16.6(a) has $R_1 = R_2 = R_F = R = 10\ \text{k}\Omega$, $R_3 = 40\ \text{k}\Omega$, and $R_4 = 40\ \text{k}\Omega$. The input voltage is $v_S = 2\sin(377\pi t)$. Plot the transfer characteristic and draw the waveform of the output voltage v_O.

b. Use PSpice/SPICE to plot the transfer characteristic and the output voltage.

16.2 D P **a.** Design a precision full-wave rectifier as shown in Fig. 16.6(a) to provide a voltage gain of $A_f = v_O/v_S = 50$. The input voltage is $v_S = 0.01\sin(200\pi t)$.

b. Use PSpice/SPICE to check your design by plotting the output voltage.

16.3 D P **a.** Design a negative voltage limiter using the circuit in Fig. 16.10(a) by determining the values of R_1, R_2, R_3, and R_F. The supply DC voltage is $V_A = 15$ V. The circuit should limit the negative output voltage to $V_{O(min)} = -8$ V. The voltage gain without limiting is $A_f = -5$. The diode is fully turned on at a forward current of $i_D = 0.1$ mA, and its corresponding forward voltage drop is $V_D = 0.7$ V. The slope after the break point is to be limited to $S_1 = -1/30$.

b. Use PSpice/SPICE to plot the transfer characteristic for part (a).

16.4 The adjustable limiter in Fig. 16.11(a) has $R_1 = 12\ \text{k}\Omega$, $R_F = 70\ \text{k}\Omega$, $R_2 = 8\ \text{k}\Omega$, $R_3 = 1\ \text{k}\Omega$, $R_4 = 6\ \text{k}\Omega$, $R_5 = 1\ \text{k}\Omega$, $V_A = 15$ V, $-V_B = -15$ V, and $V_D = 0.7$ V. Determine **(a)** the positive clamping voltage $V_{O(max)}$ and the corresponding input voltage $V_{S(min)}$, **(b)** the negative clamping voltage $V_{O(min)}$ and the corresponding input voltage $V_{S(max)}$, and **(c)** the output voltage when the input voltage is $v_S = 5$ V.

16.5 D P **a.** Design an adjustable voltage limiter as shown in Fig. 16.11(a) to satisfy the following specifications: $V_{O(min)} = -5$ V, $V_{O(max)} = 5$ V, voltage gain $A_f = -10$. The slope is $S_1 = -1/20$ after break for $v_S > 0$, and the slope is $S_2 = -1/20$ after break for $v_S < 0$. The DC supply voltages are $V_A = 12$ V and $-V_B = -12$ V. The on-state diode current is $i_D = 0.2$ mA, and the corresponding on-state diode voltage is $V_D = 0.7$ V.

b. Use PSpice/SPICE to plot the transfer characteristic. Assume $V_{CC} = 12$ V and $-V_{EE} = -12$ V. Use the PSpice/SPICE op-amp macromodel.

16.6 D P Design an output voltage–clamping circuit as shown in Fig. 16.14(a) so that the slope of the transfer characteristic is $S = v_O/v_S = 20$, $V_{O(max)} = 6.7$ V, and $V_{O(min)} = -8.7$ V. Determine the zener voltages V_{Z1} and V_{Z2}. Assume $V_D = 0.7$ V.

16.7 D P **a.** Design a hard limiter as shown in Fig. 16.15(a) by determining the values of R_1, R_2, R_3, R_4, and R_5. The circuit should limit the negative output voltage to $V_{O(min)} = -6$ V and the positive voltage to $V_{O(max)} = 6$ V. The magnitude of the slopes after the break points should be less than or equal to $1/25$. The diode drop is $V_D = 0.7$ V at $I_D = 0.1$ mA. The DC supplies are given by $V_A = -V_B = 15$ V.

b. Use PSpice/SPICE to plot the transfer characteristic. Assume $V_{CC} = 15$ V, $-V_{EE} = -15$ V, and $v_S = -5$ V to 5 V. Use the PSpice/SPICE op-amp macromodel.

16.8 D P **a.** Design a hard limiter as shown in Fig. 16.15(a) by determining the values of R_1, R_2, R_3, R_4, and R_5. The circuit should limit the negative output voltage to $V_{O(min)} = -5$ V and the positive voltage to $V_{O(max)} = 5$ V. The magnitude of the slopes after the break points should be less than or equal to $1/50$. The diode drop is $v_D = 0.7$ V at $i_D = 0.1$ mA. The DC supplies are given by $V_A = -V_B = 15$ V.

b. Use PSpice/SPICE to plot the transfer characteristic. Assume $V_{CC} = 15$ V, $-V_{EE} = -15$ V, and $v_S = -5$ V to 5 V. Use the PSpice/SPICE op-amp macromodel.

16.9 D P **a.** Design a hard limiter as shown in Fig. 16.15(a) by determining the values of R_1, R_2, R_3, R_4, and R_5. The circuit should limit the negative output voltage to $V_{O(min)} = -7$ V and the positive voltage to $V_{O(max)} = 9$ V. The magnitude of the slopes after the break points should be less than or equal to $1/50$. The diode drop is $v_D = 0.7$ V at $i_D = 0.1$ mA. The DC supplies are given by $V_A = -V_B = 15$ V.

b. Use PSpice/SPICE to plot the transfer characteristic. Assume $V_{CC} = 15$ V, $-V_{EE} = -15$ V, and $v_S = -5$ V to 5 V. Use the PSpice/SPICE op-amp macromodel.

16.3 & 16.5 *Comparators and Schmitt Triggers*

For Probs. 16.10 through 16.14, use comparator LM111 and $v_S = 10 \sin (2000\pi t)$ to plot the hysteresis characteristic using PSpice/SPICE.

16.10 Design a Schmitt trigger as in Fig. 16.25(a) so that $V_{th} = |+V_{th}| = |-V_{th}| = 5$ V. Assume $V_{sat} = |-V_{sat}| =$ 12 V.

16.11 The parameters of the Schmitt trigger in Fig. 16.25(a) are $R_1 = 100\ \Omega$ and $R_F = 47\ \text{k}\Omega$. Calculate the threshold voltages $+V_{th}$ and $-V_{th}$. Assume $V_{sat} = |-V_{sat}| = 12$ V.

16.12 Design an inverting Schmitt trigger with the reference voltage of Fig. 16.25(a) so that $V_{Ht} = -8$ V and $V_{Lt} =$ -4 V. Assume $V_{sat} = |-V_{sat}| = 12$ V.

16.13 Design a noninverting Schmitt trigger as in Fig. 16.28(a) so that $V_{th} = |+V_{th}| = |-V_{th}| = 5$ V. Assume $V_{sat} = |-V_{sat}| = 12$ V.

16.14 Design a noninverting Schmitt trigger with the reference voltage of Fig. 16.29(a) so that $V_{Ht} = 8$ V and $V_{Lt} = 4$ V. Assume $V_{sat} = |-V_{sat}| = 12$ V.

16.6–16.8 *Square-, Triangular-, and Sawtooth-Wave Generators*

For Probs. 16.15 through 16.19, use op-amp LF411 to plot the output using PSpice/SPICE.

16.15 Design the square-wave generator shown in Fig. 16.34(a) so that $f_o = 2$ kHz. Assume $V_{sat} = |-V_{sat}| = 10$ V.

16.16 The parameters of the square-wave generator of Fig. 16.34(a) are $R_1 = 10\ \text{k}\Omega$, $R_F = 15\ \text{k}\Omega$, $R = 10\ \text{k}\Omega$, and $C = 0.047\ \mu\text{F}$. Calculate the output frequency f_o.

16.17 Design the triangular-wave generator shown in Fig. 16.37(a) so that $f_o = 2$ kHz and $V_{th} = 5$ V. Assume $V_{sat} =$ $|-V_{sat}| = 12$ V.

16.18 The parameters of the triangular-wave generator of Fig. 16.37(a) are $R_1 = 10\ \text{k}\Omega$, $R_F = 40\ \text{k}\Omega$, $R = 10\ \text{k}\Omega$, and $C = 0.047\ \mu\text{F}$. Calculate the output frequency f_o.

16.19 Design the sawtooth-wave generator shown in Fig. 16.40(a) so that $f_o = 5$ kHz, $V_{th} = 5$ V, and the circuit has a duty cycle of $k = t_1/T = 0.4$. Assume $V_{sat} = |-V_{sat}| = 12$ V.

16.9 *Voltage-Controlled Oscillators*

16.20 a. Design a VCO as shown in Fig. 16.43(c) that has a nominal frequency of $f_o = 10$ kHz. Assume $V_{CC} = 15$ V.
b. Calculate the modulation in the output frequencies if v_{CN} is varied by $\pm 10\%$.

16.21 The parameters of the VCO in Fig. 16.43(c) are $R_A = 2.5\ \text{k}\Omega$, $R_1 = R_B = 10\ \text{k}\Omega$, and $C = 0.01\ \mu\text{F}$.

a. Calculate the nominal frequency of the output waveform f_o.
b. Calculate the modulation in the output frequencies if v_{CN} is varied by $\pm 10\%$. Assume $V_{CC} = 12$ V.

16.10 *The 555 Timer*

For Probs. 16.22 through 16.29, use the 555 timer to plot the output by PSpice/SPICE.

16.22 Design a monostable multivibrator as in Fig. 16.45(a) so that $t_p = 2$ ms. Assume $V_{CC} = 15$ V.
D P

16.23 Design an astable multivibrator as in Fig. 16.49(a) so that $k = 80\%$ and $f_o = 5$ kHz. Assume $V_{CC} = 15$ V.
D P

16.24 The parameters of the astable multivibrator in Fig. 16.49(a) are $R_A = 2.2$ kΩ, $R_B = 3.9$ kΩ, and $C = 0.1$ μF. Determine **(a)** the charging time t_c, **(b)** the discharging time t_d, and **(c)** the free-running frequency f_o.
P

16.25 Design a square-wave generator as in Fig. 16.52 so that $k = 50\%$ and $f_o = 5$ kHz. Assume $V_{CC} = 15$ V.

16.26 The parameters of the square-wave generator in Fig. 16.52 are $R_A = 2.7$ kΩ, $R_B = 4.7$ kΩ, and $C = 1$ μF. Determine **(a)** the charging time t_c, **(b)** the discharging time t_d, and **(c)** the free-running frequency f_o.
P

16.27 Design a ramp generator as in Fig. 16.53(a) so that $k = 50\%$ and $f_o = 5$ kHz. Assume $V_{CC} = 15$ V, $V_{BE} = 0.7$ V, and a transistor of $\beta_F = 100$.
D P

16.28 The parameters of the ramp generator in Fig. 16.53(a) are $R = 10$ kΩ, $V_{CC} = 15$ V, $V_{BE} = 0.7$ V, and a transistor of $\beta_F = 100$. Determine the free-running frequency f_o.
P

16.29 Design the FSK modulator shown in Fig. 16.54(a) to produce frequencies of 1270 Hz and 1570 Hz corresponding to 1 (mark) and 0 (space), respectively.
D P

16.11 *Phase-Lock Loops*

16.30 Design a PLL as shown in Fig. 16.58(c) so that $f_o = 5$ kHz and $f_c = \pm 50$ Hz. Assume $V_{CC} = -V_{EE} = 15$ V.
D

16.31 The parameters of the PLL in Fig. 16.58(c) are $R_1 = 12$ kΩ, $C_1 = 0.01$ μF, $C_2 = 10$ μF, and $V_{CC} = -V_{EE} = 15$ V. Determine **(a)** the free-running frequency f_o, **(b)** the lock frequency f_L, and **(c)** the capture range f_c.

16.12 *Voltage-to-Frequency and Frequency-to-Voltage Converters*

16.32 Design a V/F converter as shown in Fig. 16.64 so that $f_o = 2.5$ kHz at $v_I = 5$ V. The input voltage v_I can vary between 10 mV and 10 V. Assume $V_{DD} = -V_{SS} = 5$ V.
D

16.33 Design an F/V converter as shown in Fig. 16.67 so that $V_O = 2.5$ V at $f_{in} = 10$ kHz. The input frequency f_{in} can vary between 0 and 20 kHz. Assume $V_{DD} = -V_{SS} = 5$ V.
D

16.13 *Sample-and-Hold Circuits*

16.34 Design the SAH circuit shown in Fig. 16.71(a) so that the drop is within 0.5%. The leakage current in the hold mode is 1 nA, and the hold voltage is $V_h = 5$ V. The internal hold time is $t_h = 100$ μs. Find the holding capacitance C_h.
D

16.35 Design the SAH circuit shown in Fig. 16.71(a) so that the output tracks the input within 0.5%. The internal hold time is $t_h = 100$ ns. The input biasing current of the op-amp is $I_B = 10$ nA. Assume $R_F = R_1 = 20$ kΩ and $v_I = 5$ V.
D

16.36 Design the SAH circuit shown in Fig. 16.71(a) so that the drop is within 0.5%. The internal hold time is $t_h = 0.1$ ms, and the hold voltage is $V_h = 5$ V. The input biasing current of the op-amp is $I_B = 200$ nA. Assume $R_F = R_1 = 20$ kΩ.
D

16.14 *Digital-to-Analog Converters*

16.37 A 10-bit D/A converter of type 2R ladder, as shown in Fig. 16.73(a), has an input of 00 1001 1001, and the reference voltage is 5 V. Find the analog output voltage V_O.

16.38 A D/A converter is to have a full-scale output of 5 V and a resolution of less than 20 mV. What is the bit size?

16.39 The 8-bit D/A converter shown in Fig. 16.73(a) is to have a full-scale output of 10 V with a reference voltage of 5 V. Find the values of R_F and R.

16.15 *Analog-to-Digital Converters*

16.40 An 8-bit A/D converter has a reference voltage of 10 V. Find **(a)** the analog input corresponding to the binary outputs 1010 1010 and 0101 0101, **(b)** the binary output if $V_a = 3$ V, and **(c)** the resolution of the converter.

16.41 Construct a table similar to Table 16.2 for a 4-bit A/D converter if $V_a = 5$ V with a reference voltage of 16 V.

16.42 Construct a table similar to Table 16.2 for an 8-bit A/D converter if $V_a = 4$ V with a reference voltage of 10 V.

APPENDIX A

INTRODUCTION TO OrCAD

A.1 Introduction

This appendix provides an introduction to the electronic circuit simulation software OrCAD [1, 2]. The objectives of this appendix are the following:

- Becoming familiar with the electronic circuit simulation software OrCAD Capture
- Learning how to draw electrical circuits by OrCAD Capture

After installation instructions presented in Sec. A.2, Sec. A.3 provides an overview of the software package. Then Secs. A.4 to A.8 cover the basic steps in the circuit analysis process:

- Drawing the circuit
- Selecting the type of analysis
- Simulating the circuit
- Displaying the results of the simulation

The remaining sections, Secs. A.9 to A.13, deal with specific operations:

- Copying and capturing schematics
- Varying parameters
- Performing frequency response analysis
- Modeling devices and elements
- Creating netlists
- Adding library files

A.2 Installing the Software

The instructions for installing the OrCAD software are printed on the OrCAD installation disk or the CD-ROM. The software can be downloaded from the Cadence website at http://www.cadence.com/products/orcad/Pages/downloads.aspx

The following steps, which pertain to OrCAD Version 9.2, are applicable to other versions as well:

1. Place the schematics CD-ROM in the CD drive.
2. From Windows, enter the **File Manager** and click the left mouse button on the CD drive.
3. Click on **setup.exe**, **File**, **Run**, and **OK**.
4. Click on **OK** to select the products to install **Capture** and **PSpice**, as shown in Fig. A.1(a).
5. Click on **OK** to select the default **C:OrcadLite**.
6. Click on **Yes** to choose **Program Folder: Orcad Family Release 9.2 Lite Edition**, as shown in Fig. A.1(b).
7. Create the OrCAD Capture Lite and PSpice A/D icons from the Start menu; then go to **Programs**, **Orcad Family**, and **Capture Lite** menu. These icons are shown in Fig. A.2.
8. Click the left mouse button on the Capture Lite icon once, and the window of Capture Lite will open.

▶ **NOTE** Although there are later versions of OrCAD, the 9.2 version has proven to be bug free and works well.

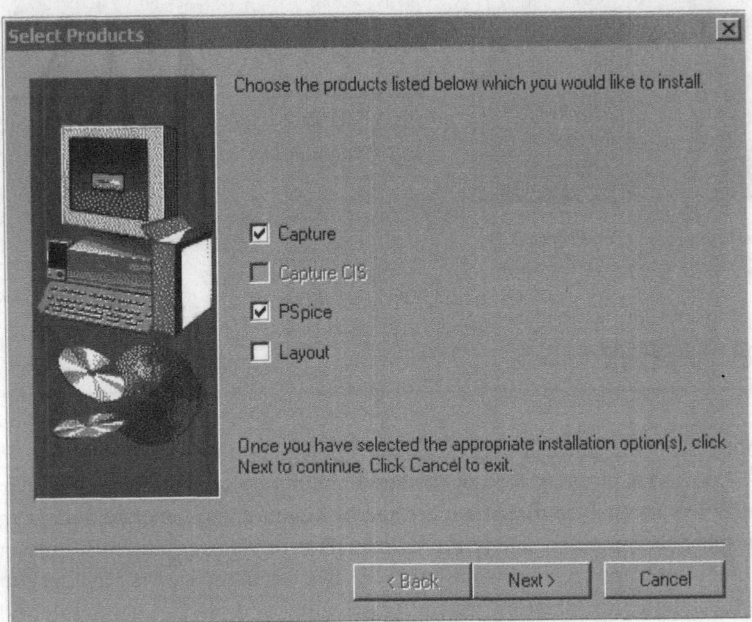

(a) Selecting products

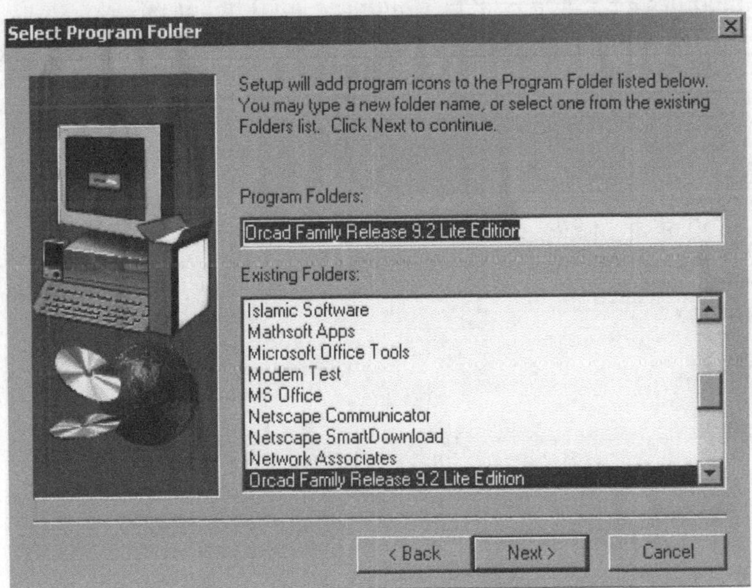

(b) Selecting the program folder

FIGURE A.1 Installation setup for OrCAD Capture

(a) OrCAD Capture Lite

(b) PSpice A/D

FIGURE A.2 Icons for OrCAD Capture Lite and PSpice A/D

A.3 Overview

The OrCAD Capture software package has three major interactive programs: Capture, PSpice A/D, and Probe. *Capture* is a powerful program that lets us build circuits by drawing them within a window on the monitor. *PSpice A/D* lets us analyze the circuit created by Capture and generate voltage and current solutions. *Probe* is a graphic postprocessor and lets us display plots of voltages, currents, impedance, and power.

The general layout of Capture is shown in Fig. A.3. The top menu shows 10 main choices. The right-side menu shows the schematic "drawing" menu for selecting and placing parts. File, Edit, View, Place, and PSpice are most frequently used. For any help, click on the Help menu.

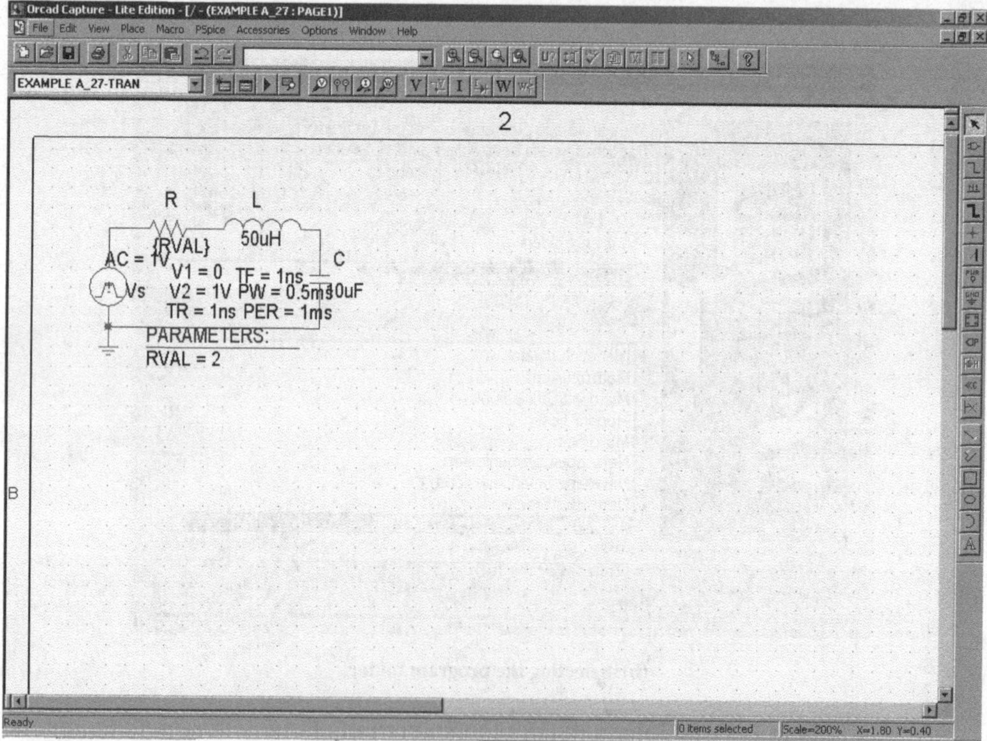

FIGURE A.3 General layout of OrCAD Capture

TABLE A.1 PSpice mouse operations

Button	Action	Function
Left	Single click	Select an item
	Double click	End a mode
	Double click on selected object	Edit a selection
	Single click on selected object and hold	Drag a selection
	Shift + single click	Extend a selection
Right	Single click	Abort the mode
	Double click	Repeat an action

The mouse follows an *object–action* sequence. First select an object and then perform an action, as in the following:

1. Click the mouse on a menu title so that it stays open. Then click on the command that you want.
2. A single click on the left mouse button selects an item.
3. A double click on the left mouse button performs an action such as to end a mode or edit a selection.
4. To drag a selected item, click the left button, hold it down, and move the mouse. Release the left button when placed.
5. To end a mode or edit a selection, click the right mouse button once.
6. To repeat an action, double click the right button.

These mouse operations are summarized in Table A.1.

A.4 The Circuit Analysis Process

As an example, let us draw and analyze the pulse response of an *RLC* circuit as shown in Fig. A.4. The steps to draw and analyze a circuit are as follows:

1. Draw the circuit under Capture.
2. Select the mode of analysis under PSpice.
3. Simulate the circuit under PSpice A/D.
4. Display the results under Probe.

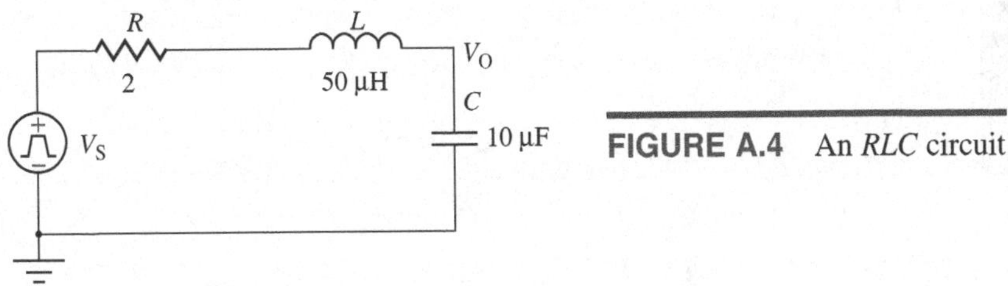

FIGURE A.4 An *RLC* circuit

A.5 Drawing the Circuit

The steps in drawing a circuit are as follows:

Step 1. Begin a new project.
Step 2. Get components from the Get New Part menu and place them on the drawing board.
Step 3. Rotate components as desired.
Step 4. Wire the components together.
Step 5. Label the components and add text as desired.
Step 6. Set attributes of the components.
Step 7. View the schematic.
Step 8. Save the circuit.

A.5.1 Beginning a New Project

Let us begin a new project to draw and analyze; call it Example B-3. Open the File menu, choose **New**, and then select **Project** as shown in Fig. A.5.

The New Project menu opens as shown in Fig. A.6. Give the file name of the new project (e.g., Example B-3), select **Analog or Mixed A/D**, and give the location of this new file (e.g., **C:\Rashid\PT3**). Next select **Create a blank project**, as shown in Fig. A.7.

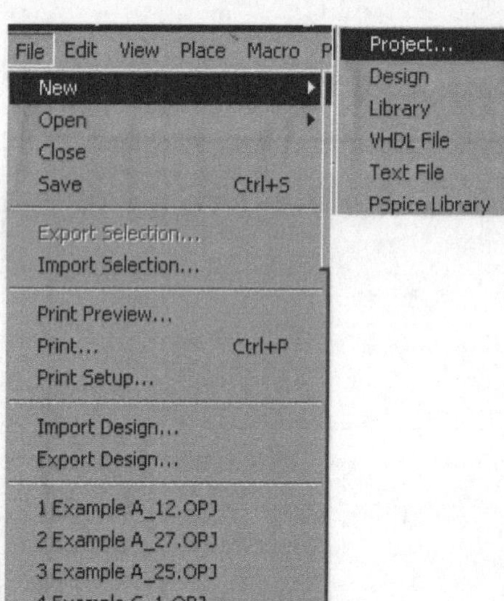

FIGURE A.5 File menu for a new project

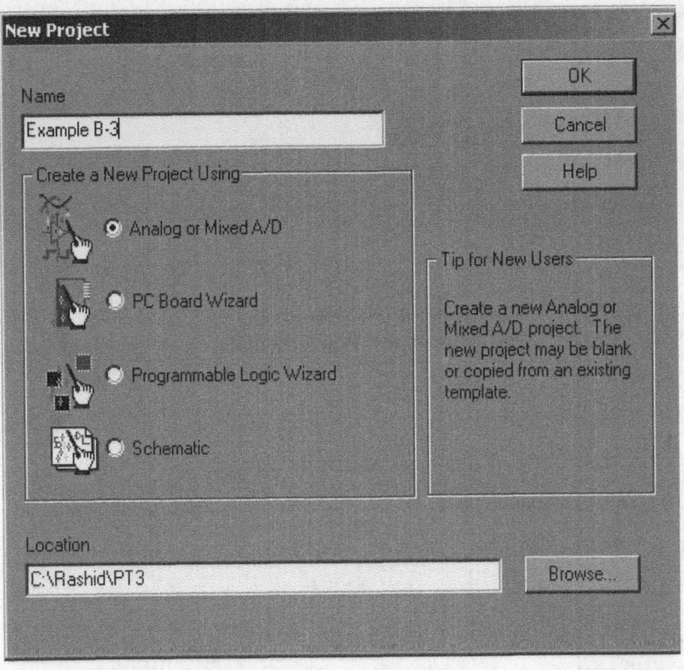

FIGURE A.6 Project menu

A.5.2 Getting and Placing Components

Let us start by placing a pulse source, a resistor, an inductor, a capacitor, and a ground on the drawing board. From the capture editor, you can place parts from the component libraries onto your schematic.

1. Use the Part command in the Place menu shown in Fig. A.8 or from the Place part menu on the right side. Alternately, choose the **Place part** command from the Draw menu on the right side. Choose **Browse** to browse the list of libraries or use the search command to enter the name of a known part, resistor R, as shown in Fig. A.9.
2. Choose **OK** or double click after selecting the part. The chosen part becomes the "current part" and is ready to be placed on your schematic.

FIGURE A.7 Create a blank project

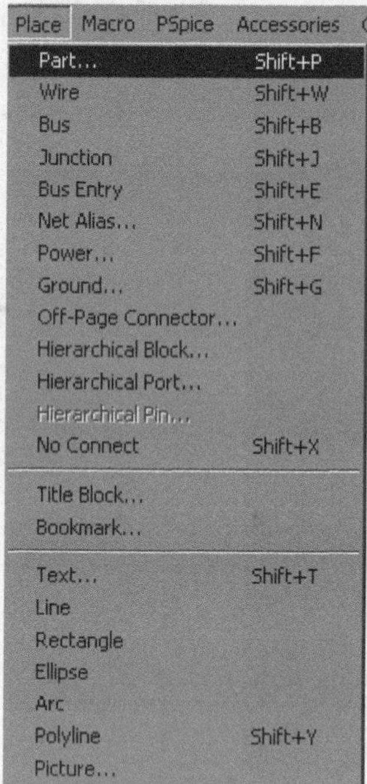

FIGURE A.8 Place menu

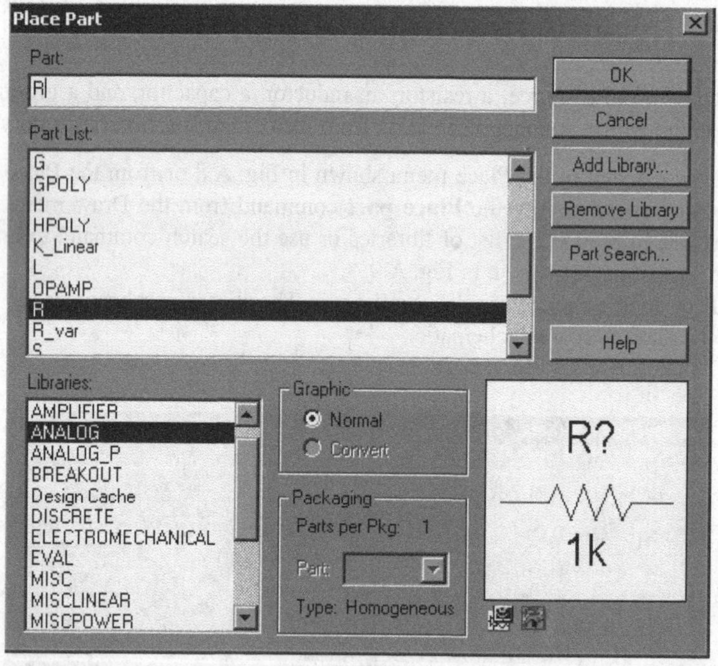

FIGURE A.9 Place part menu

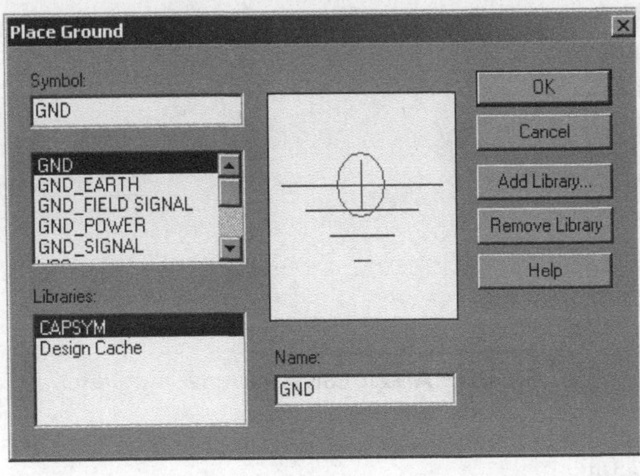

FIGURE A.10 Ground parts for an *RLC* circuit

3. The cursor is replaced by the shape of the part.
4. Click left to place an instance of the part; double click to place the part and end the mode or click right to end the mode without placing the part. To drag, point to an object, press and hold the left mouse button, and move the mouse. When the object is where you want it, release the mouse button.
5. To move a component, point to it, press and hold the left mouse button, and drag it to a new location.
6. To remove a component, select it, and choose **Delete** from the Edit menu.
7. Place a pulse source (VPULSE) from the **source.slb** library, a resistor (R), an inductor (L), and a capacitor (C) from the **analog.slb** library and a ground symbol **(GND)** from the Place menu or Place Ground (zero 0) menu on the right-side menu as shown in Fig. A.10. Arrange them as shown in Fig. A.11.

A.5.3 Rotating Components

Now rotate the capacitor so that it can be wired neatly into the circuit. Each time that you rotate a component, it turns clockwise 90°. The Flip command flips a selected object(s) to produce a mirror image of the object.

1. To rotate the capacitor (or other component), select it and choose **Rotate** from the Edit menu as shown in Fig. A.12. If you select an area of your schematic, the area is rotated around the center of the selection box. This is done as shown in Fig. A.13.
2. To flip a component, select it and choose **Flip** from the Edit menu.

FIGURE A.11 Parts for an *RLC* circuit

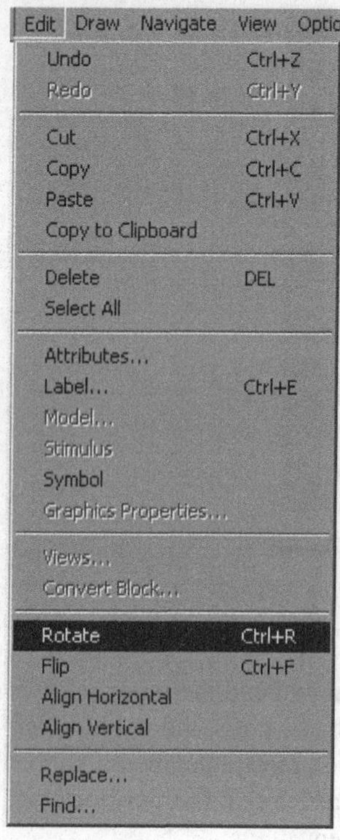

FIGURE A.12 Edit menu

▶ **HELP** If the Rotate command is dimmed, the capacitor is not selected. Try again by pointing to it so that the pointer becomes a hand and then click the left mouse button.

3. To deselect the selected capacitor (or other selected component), click it with the right mouse button or click an empty spot with the left mouse button.
4. To drag or rotate two or more components at once, first select them by drawing a rectangle around the components and then drag or rotate the rectangle.

▶ **HELP** To draw a rectangle around components, point above and beside one of the components that you want to select. Press and hold the left mouse button and drag diagonally until the rest of the components are in the rectangle that appears.

5. To deselect one of the selected components, click it with the right mouse button. To deselect everything on the marked rectangle, click an empty spot with the left mouse button.

R_1 1 k
L_1 10 μH
V_1
C_1 1 n

FIGURE A.13 Rotating components

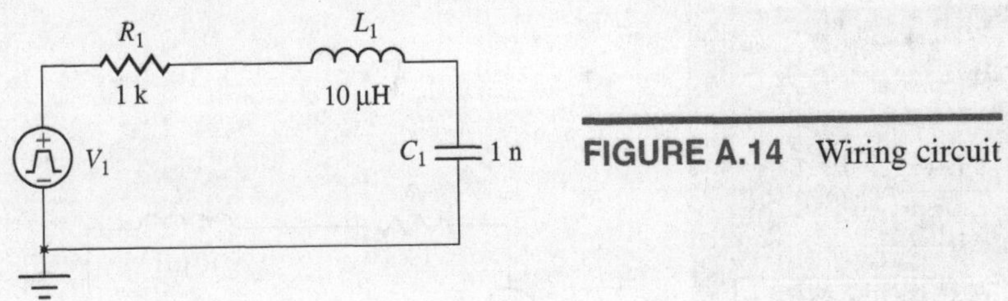

FIGURE A.14 Wiring circuit

A.5.4 Wiring Components

Once components are placed on the drawing board, you need to connect them. You can use the schematic editor to draw wires on your schematic and/or make vertices.

1. To draw a wire, choose **Wire** from the Place menu or Wire menu on the right side.
2. The cursor changes to a "wire" mode and is displayed as crosshairs. Click left to start drawing.
3. Move the mouse in any direction to extend the wire.
4. Click left again to end the wiring of a part. Click left once to start and click left again to stop. Continue until wiring of all parts is completed.
5. Press the **ESC** key or double click left or click right to end the wiring mode (the cursor will change from crosshairs to normal mode).

Now wire the circuit as shown in Fig. A.14.

▶ **TIP** To bring back the last command used (for wire), double click the right mouse button. In the lower right-hand corner, you will see the Wire command after the word *Cmd.*

A.5.5 Labeling Components and Adding Text

We can place a label on selected wires, bus segments, or ports. Wire and bus segments or ports may display multiple labels; however, all labels for a segment will contain the same text. Each component in a circuit can be labeled. We will assign levels R, L, C, V_S, and V_O.

1. To edit a label, select a wire, bus segment, or port for labeling.
2. Double click on the label to bring up the dialog box for **Display Properties**.
3. Enter the text for the label.
4. Click **OK**.
5. Change all levels R, L, C, V_S, and V_O as shown in Fig. A.15.

▶ **TIP** To move the level, left click on it and move to a desired location.

6. Next click on **Text** in the Place menu (see Fig. A.8).

You can place text anywhere on your schematic and size it to suit your needs.

1. To add text to your schematic, choose **Text** from the Place menu or **Place text** on the right-side menu.
2. In the dialog box, type in the desired text.
3. To change the font size, modify the **Font Size** shown in the dialog box to suit your needs.
4. Click **OK**.
5. Move the text to the desired location on the schematic and click left to place the text. Click right to end the mode.

Now place the output voltage, V_O as shown in Fig. A.16.

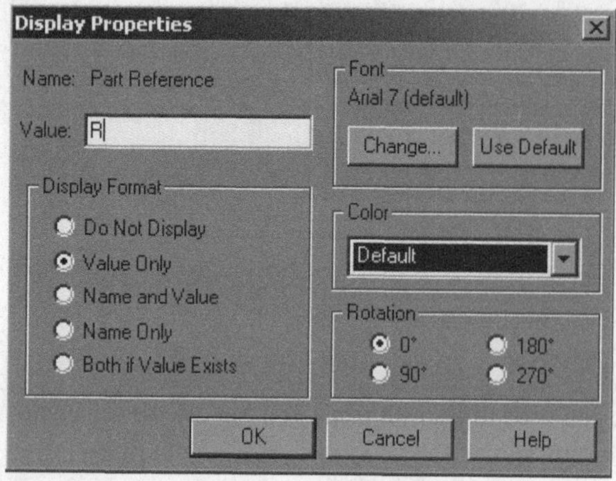

(a) Display Properties menu

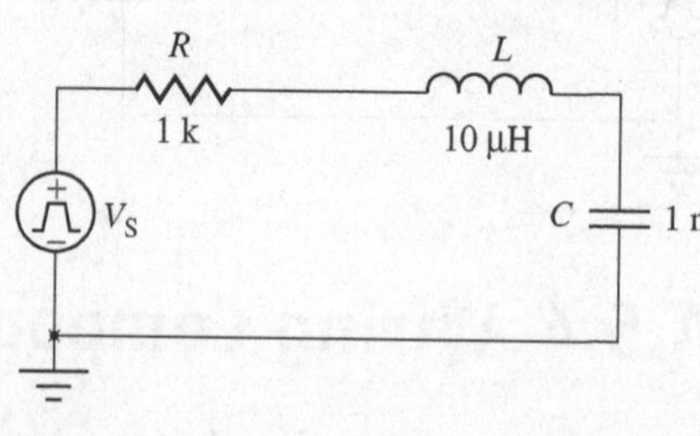

(b) *RLC* circuit with desired labels

FIGURE A.15 Labeling components

A.5.6 Setting Attributes

An attribute of a schematic item consists of a name and value pair.

1. To edit the properties of a selected object(s), select the object(s) to edit.
2. Choose **Properties** from the Edit menu or double click on the attribute text to bring up the Display Properties dialog box directly as shown in Fig. A.17.
3. Select an individual attribute: A dialog box appears in which you can enter a new value for the attribute. You can also select an entire part: A dialog box appears showing all attributes that can be edited for that part—for example, **Value Only**.

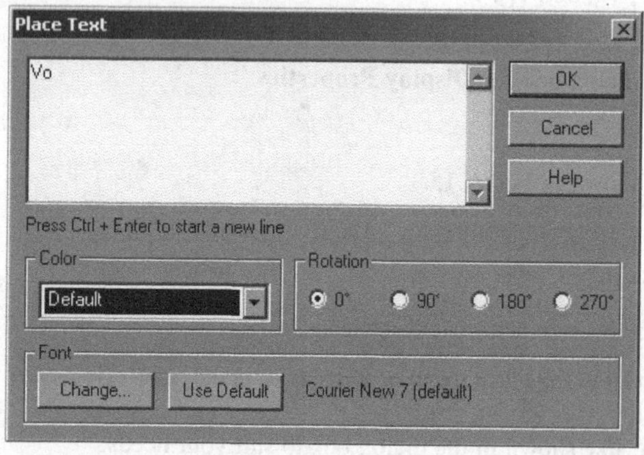

(a) Place Text dialog box

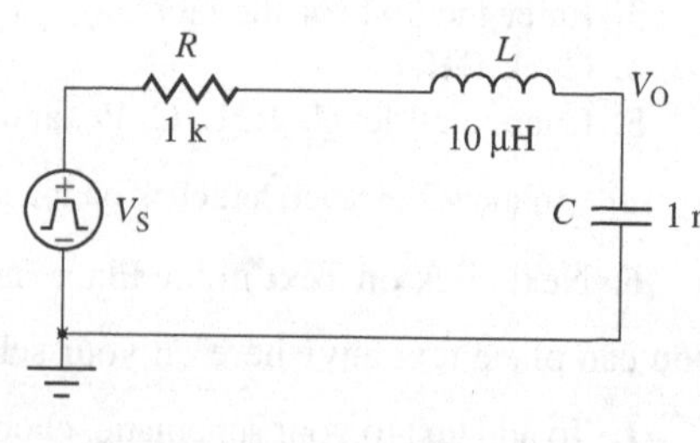

(b) *RLC* circuit with desired text

FIGURE A.16 Adding text

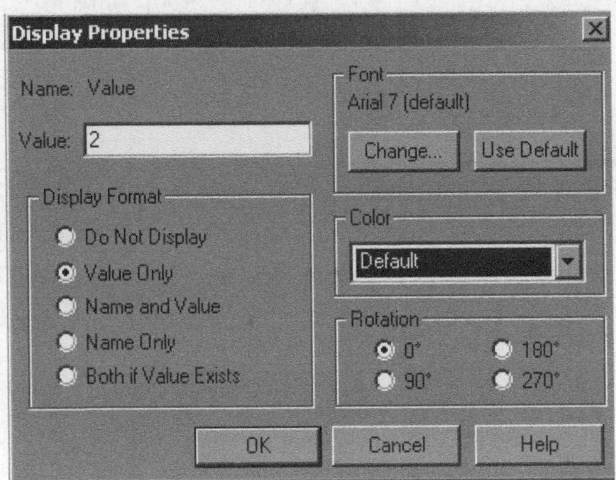

(a) Display Properties dialog box

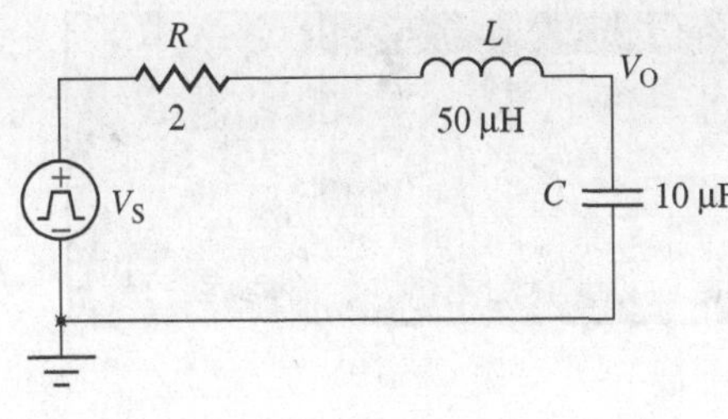

(b) *RLC* circuit showing attributes

FIGURE A.17 Setting attributes

4. Set $R = 2\ \Omega$, $L = 50\ \mu$H, and $C = 10\ \mu$F.

▶ **TIP** A quick way to set a component's value is to double click it.

▶ **HELP** Changes made to attributes in the schematic editor occur only on the particular part instance. Changes do not affect the underlying symbol in the library.

5. To change the value of an attribute, select it. The name and value should appear in the properties of the Edit menu.
   a. Change the value in the **Value** edit field and click **OK**.
   b. To delete an attribute, select it from the list and press **Delete**.
   c. To change whether the attribute name and/or value is shown on the schematic, select it and click **OK**.
   d. To add a new attribute, type the name and value in the **Edit** fields and press **OK**. Set the pulse source V_S as shown in Fig. A.18. Click on the attributes and then type V1 = 0, V2 = 1 V, TD (delay time) = 0, TR (rise time) = 1 ns, TF (fall time) = 1 ns, PW (pulse width) = 0.5 ms, PER (period) = 1 ms.
6. Click **OK** to incorporate the change.

▶ **TIP** Protected attributes, marked with an asterisk (*), cannot be changed using the Schematic Editor.

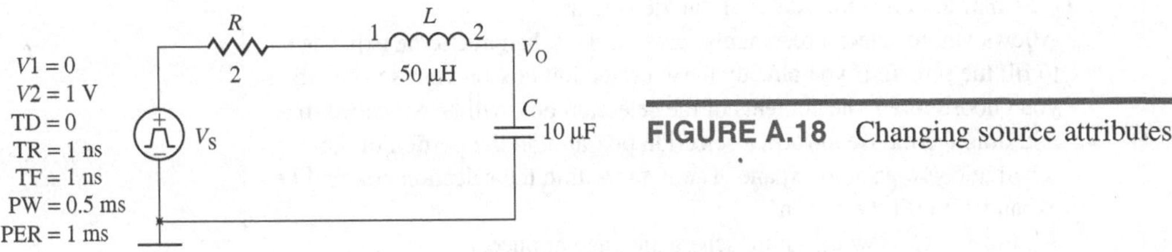

FIGURE A.18 Changing source attributes

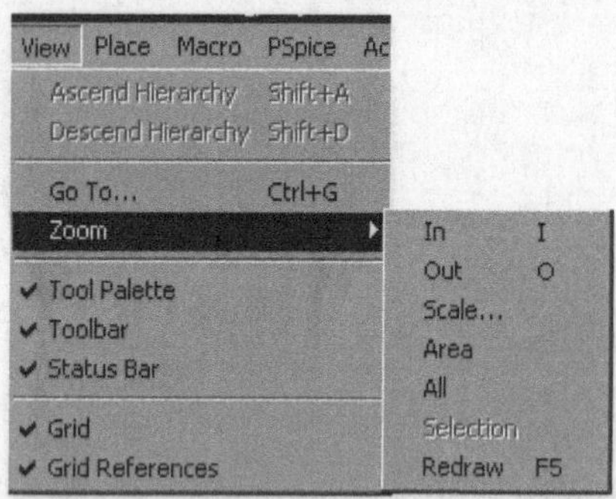

FIGURE A.19 View menu

A.5.7 Viewing the Schematic

You can change the viewing scale on your schematic via the View menu, shown in Fig. A.19. Table A.2 explains the functions of commands in the View menu.

A.5.8 Saving the Circuit File

Your schematic files are automatically saved into the current directory, unless otherwise specified.

- To save your schematic file, choose **Save** from the File menu, as shown in Fig. A.20, to save changes to the current file. You do not need to type a file name extension.

A circuit name must be a valid DOS file name. All schematic files are automatically given the file name extension OPJ. For example, the file name FIGB_3 will be saved as FIG B_3.OPJ.

TABLE A.2 View menu commands

Command	Function
Fit	Resets the viewing scale so that all parts, wires, and text can be seen on the screen.
In	Allows you to view an area on the schematic at closer range (i.e., magnify). After you select this command, a crosshair appears on the screen. Move the crosshair to the area you want to zoom in on.
Out	Changes the viewing scale so that you can view the schematic from a greater distance (i.e., view more of the schematic on the screen). After you select this command, a crosshair appears on the screen. Move the crosshair to define the center of the viewing area.
Area	Allows you to select a rectangular area on the schematic to be expanded to fill the screen. If you already have a selection box on your screen when you choose **Area**, the contents of the selection box will be expanded. If not, drag the mouse to form a selection box around the portion of the schematic you want to expand. The items within the selection box will be expanded to fill the screen.
Entire Page	Allows you to view the entire schematic page at once.

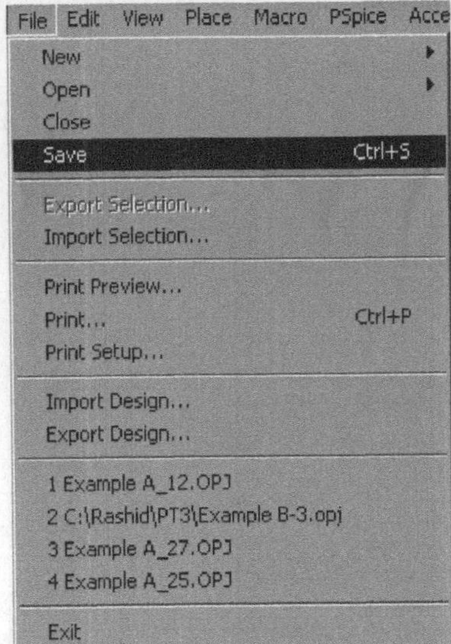

FIGURE A.20 File menu

A.6 Selecting the Type of Analysis

PSpice allows *DC sweep analysis, AC* (frequency response) *sweep analysis*, and *transient analysis*. The Setup command specifies which types of simulation analyses are enabled and allows the user to set up the parameters for selected analyses (described next).

When a signal is first applied to a circuit, there is a short-lived transient state before the circuit settles down to its usual responses. For the sample *RLC* circuit, we will conduct a transient analysis to examine the charging and discharging voltage of the capacitor.

1. Choose **New Simulation Profile** for a new simulation file or **Edit Simulation Profile** for an existing file from the PSpice menu, shown in Fig. A.21, and the Analysis Setup dialog box will open.
2. Enable transient analysis by clicking once in the **Enabled** space so that a check appears in the space, as shown in Fig. A.22.

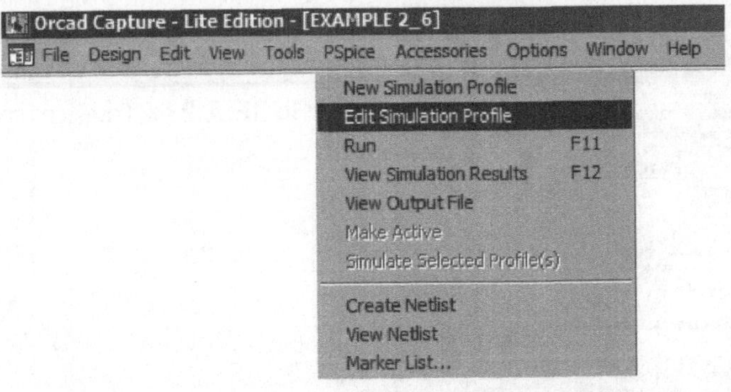

FIGURE A.21 PSpice setup for analysis

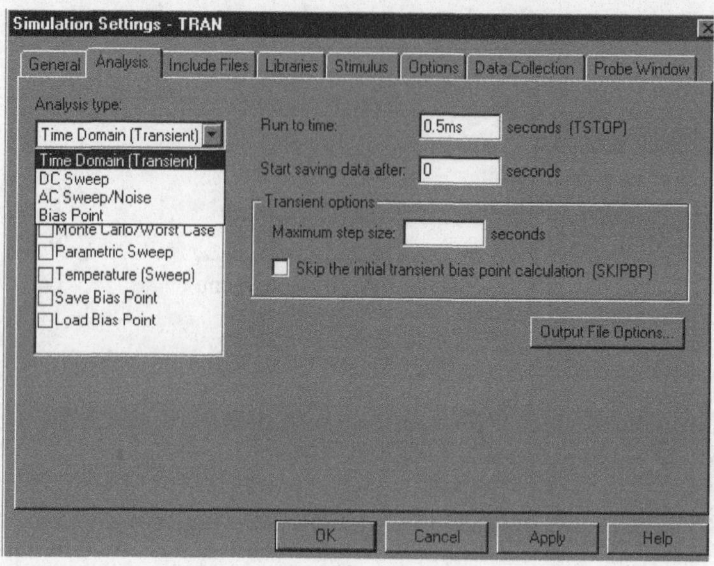

FIGURE A.22 Selecting Analysis type

3. Click on **Transient** to open the Transient specifications dialog box.
4. Type the print particulars, such as a **Print values** of 10 ns and a **Run to time** of 0.5 ms, as shown in Fig. A.23.
5. Click **OK**.

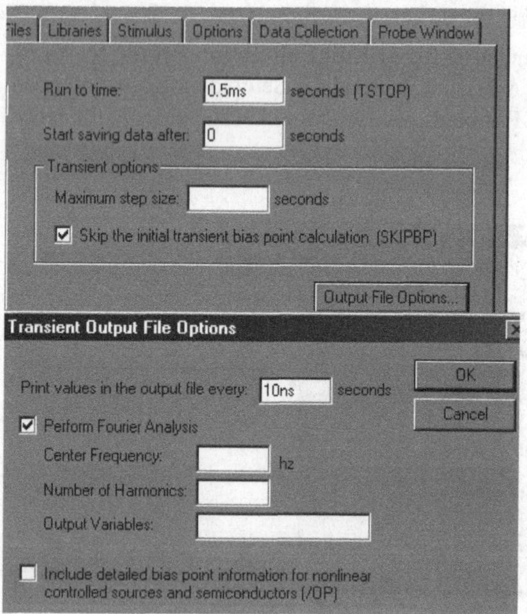

FIGURE A.23 Transient specifications

A.7 Simulation with PSpice

Now we are ready to simulate the circuit with PSpice. To begin the simulation, left click on **Simulate** in the Analysis menu.

During the circuit simulation process, PSpice creates and accesses a number of files. The first file is the *Schematics* file (*.SCH*), generated when a circuit drawn on the screen is saved. When the Schematics file is analyzed, three new files are generated: the *Circuit* file (*.CIR*), the *Netlist* file (*.NET*), and the *Alias* file (*.ALS*). The Circuit file (the master file) contains the *simulation directives* and references to the Netlist, Alias, and Model files. The Netlist file contains a Kirchhoff-like set of equations that lists parts and how they are connected; the equations relate the voltages and currents to the circuit elements through node numbers. The Alias file lists alternative names for circuit nodes. The Model file lists the characteristics and model statement of each component.

▶ **TIP** If there is any problem such as a missing attribute name or value, PSpice will indicate the error, and the simulation will be aborted. You can find the error message in the output file. For example, a message about an error in the file FIG1.1 will appear in FIG1_1.OUT.

When PSpice is run, each simulation directive in the Circuit (master) file specifies the information to be sent to the Output and Data files.

- The *Output* file (*.OUT*) is an ASCII file that holds the "audit trail" for the simulation. It contains a wide variety of information, including the original netlist, all output variables, and various tables.
- The *Data* file (*.DAT*) is sent to Probe, which uses the binary information to generate plots and graphs within the Probe window.

While a simulation is running, you can check the status of the simulation, as shown in Fig. A.24. When the simulation is completed, PSpice will display the message "Transient Analysis finished."

▶ **TIP** *Time step* is the internal simulation step for convergence to give a specified accuracy, whereas *Print step* (which was specified in the Transient specifications dialog box) is for printing or plotting the output variables.

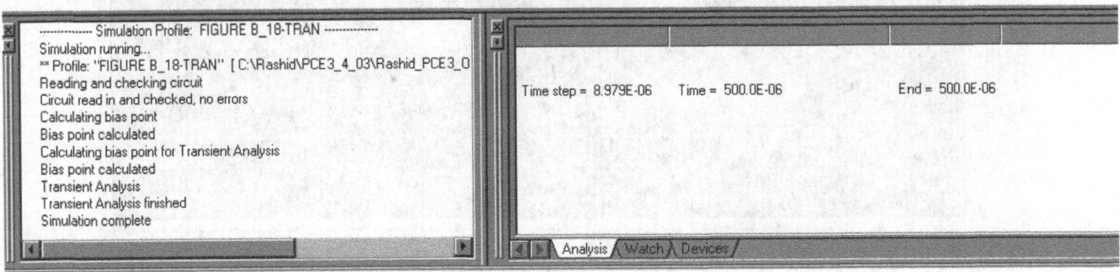

FIGURE A.24 PSpice analysis status box

A.8 Displaying the Results of a Simulation

Probe is a graphics postprocessor that allows the simulation results to be displayed in graphical form. After the calculations are completed, assuming no errors are found, PSpice sends the database it generated (FIG1_1.DAT) to Probe, which displays a graph.

1. To use Probe, choose **Run Probe** from the selecting Analysis type in Fig. A.22 menu (or use the F12 function key). As shown in Fig. A.25, Probe opens with an initial (default) graph in which the x-axis is automatically set to the transient variable, time.

▶ **TIP** To run Probe automatically after a simulation, choose **Automatically Run Probe After Simulation** from **Probe Setup** in the Analysis menu.

2. Choose **Trace** on the Probe menu. Within the large box at the top of the dialog box that appears, you will see the list of default *trace variables* that you can choose from, as shown in Fig. A.26. Before moving to the third step, specify the plot variables(s). For this example, we choose the following variables:
   a. V(R:2): V specifies the variable type (voltage, which is referenced to the ground), R specifies a component (resistor), and 2 specifies one end (node 2) of the component. A 1 indicates what was the left-hand side and a 2 what was the right-hand side when the component was initially placed horizontally. After one counterclockwise rotation, node 2 would be at the top.
   b. V(Vs:+): This is the voltage at the positive terminal of the voltage source V_S.
   c. I(R): This is the current flowing through resistance R from terminal 1 (left) to terminal 2 (right).

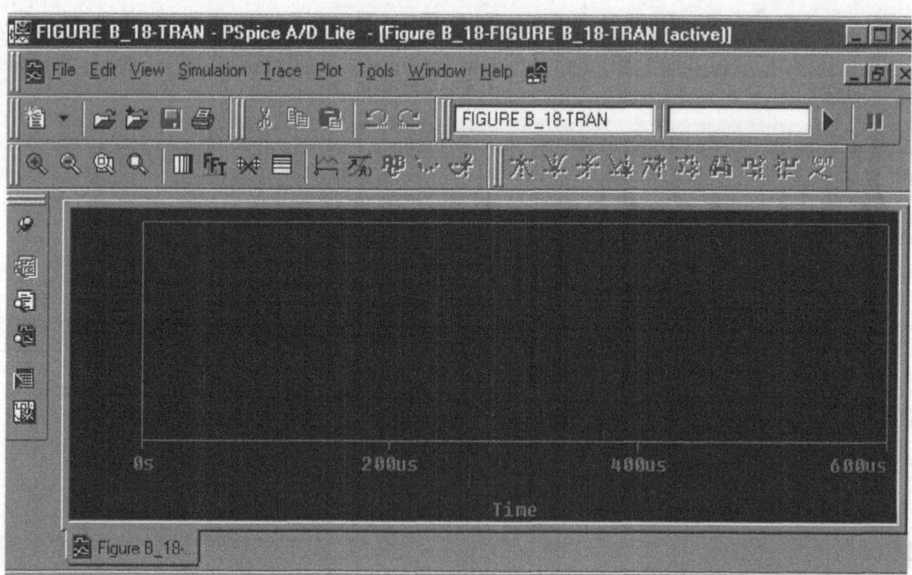

FIGURE A.25 Probe menu

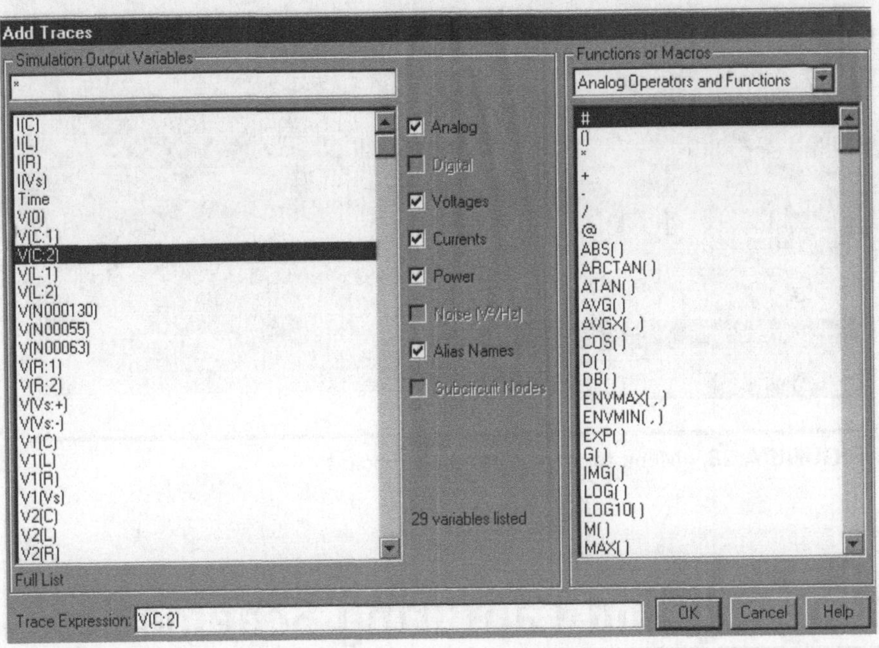

FIGURE A.26 Probe trace variables

3. After specifying the plot variables, type **V(C:2)** in the trace command box. This instructs Probe to display a graph of the voltage at terminal 2 of capacitor C, which is our desired output voltage. Click **OK** and you will see the plot of the output voltage, V(C:2), shown in Fig. A.27.
4. While copying to the clipboard and pasting to another document, the background and foreground colors can be changed as shown in the dialog box in Fig. A.28.

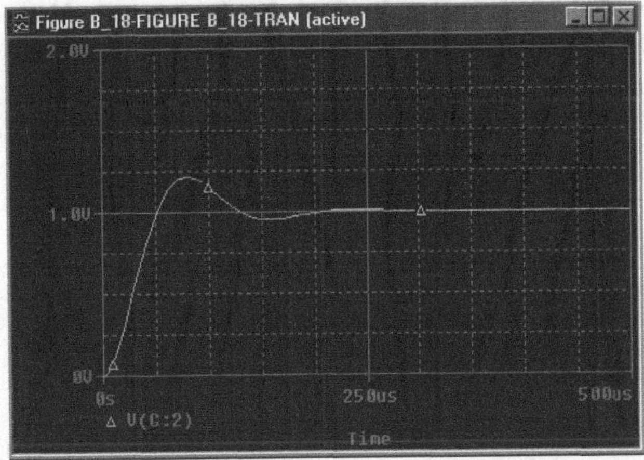

FIGURE A.27 Transient plot of the output voltage

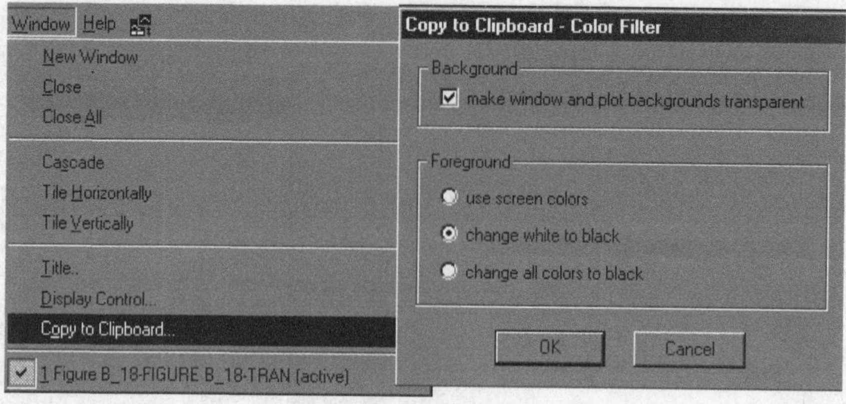

FIGURE A.28 Menu for copying to the clipboard

A.9 Copying and Capturing Schematics

You can use the Copy to Clipboard command from the Edit menu to copy one or more items from your schematic into another Windows program. With the mouse, select a rectangular area on the schematic to copy. Choose **Copy to Clipboard** from the Edit menu. Open the Windows program into which you want to paste the item(s). Use the Paste command from the newly opened Windows program to paste the item into the new file.

The Cut command deletes items from your schematic and places them in the Paste buffer. The Copy command copies items from your schematic to the Paste buffer. The Paste command places the contents of the Paste buffer on the schematic.

To cut or copy one or more items on your schematic, first select the item(s) on your schematic to be cut or copied. Then choose **Cut** or **Copy** from the Edit menu. To paste a cut or copied item onto your schematic, choose **Paste** from the Edit menu. Then place the cursor on the schematic where you'd like the cut or copied item to appear, and left click to place the item; right click to end the mode.

A.10 Varying Parameters

PSpice allows variation of component values or device parameters. We will plot the output voltage of our sample *RLC* circuit for three values: $R = 1\ \Omega$, $2\ \Omega$, and $10\ \Omega$.

1. To begin, get the part PARAM from the special.slb library file.
2. To change the value of *R* to a variable name such as RVAL, double click on the value of *R* and type **{RVAL}**.
3. To change the attribute of the part PARAM, double click on **PARAM** to open the Attributes dialog box, shown in Fig. A.29.
4. Choose **NAME1=RVAL** and **VALUE1=2** and click each time on **Save Attr**.

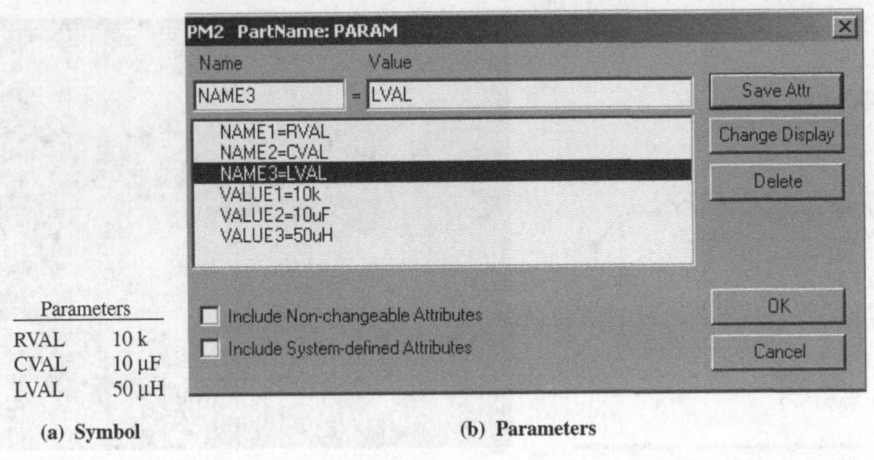

(a) Symbol (b) Parameters

FIGURE A.29 Dialog box for PARAM

5. Choose **Setup** from the Analysis menu; then click once in the enabled space to enable parametric analysis.
6. Click on **Parametric** to open the box shown in Fig. A.30.
7. Enter or enable the Parametric specifications as follows: For Sweep variable, choose **Global parameter**; for Parameter name, type in **RVAL**; for Sweep Type, choose **Value list**; and for the values, type in 1 2 10.

▶ **TIP** You can define up to three variables in the same PARAM.

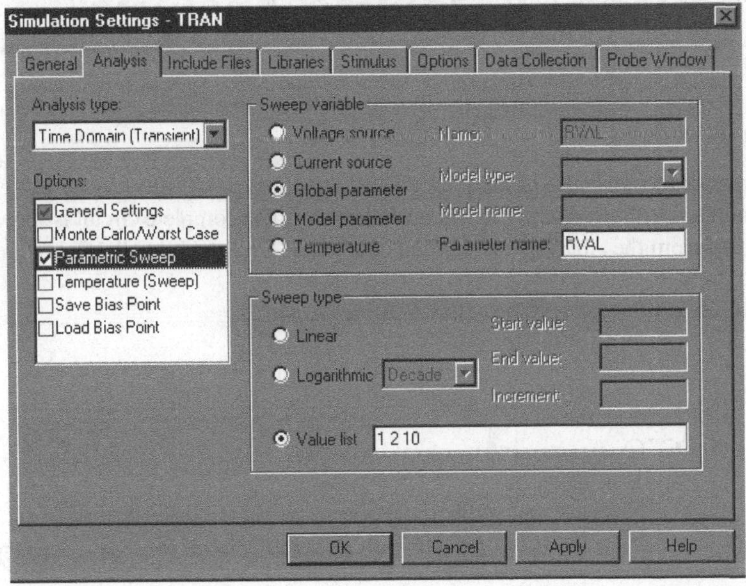

FIGURE A.30 Parametric specifications

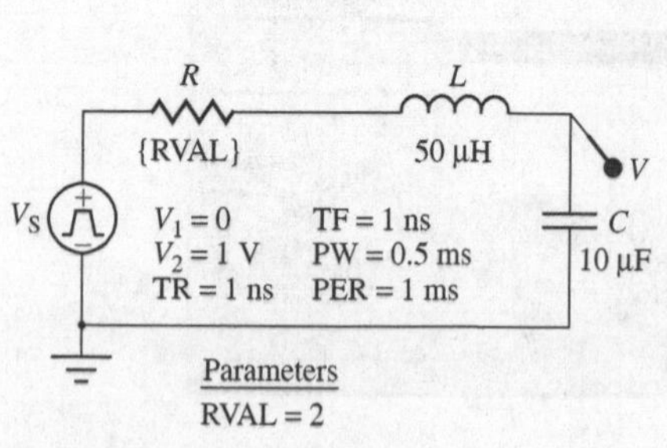

FIGURE A.31 *RLC* circuit with PARAM

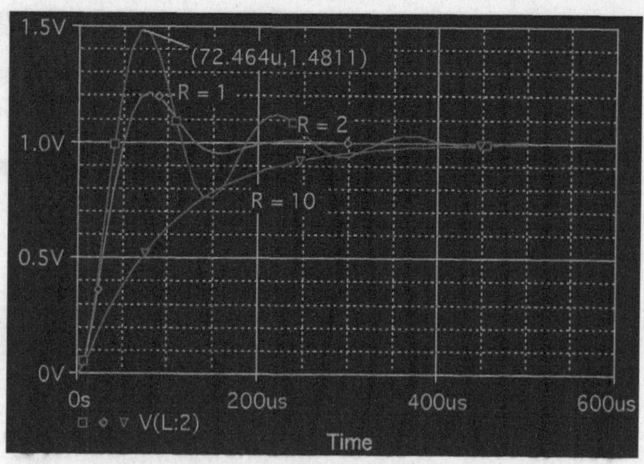

FIGURE A.32 Transient response for $R = 1\ \Omega, 2\ \Omega, 10\ \Omega$

8. Run the simulation for the complete circuit, shown in Fig. A.31. Use **Probe** to plot the output variable, V(C:2), as shown in Fig. A.32.

▶ **TIP** The labels R=1, R=2, and R=10 in Fig. A.32 were typed by choosing (in sequence) **Tools, Label**, and **Text** from the Probe menu. You can copy the Probe plot(s) to other Windows programs by choosing (in sequence) **Tools** and **Copy to Clipboard** from the Probe menu.

A.11 Frequency Response Analysis

As an example of frequency response analysis, we will plot the output voltage and phase angle of the example *RLC* circuit for three values: $R = 1\ \Omega$, $2\ \Omega$, and $10\ \Omega$.

1. To begin, change the attributes of the source V_S, shown in Fig. A.17, to 1 V AC by selecting the part and changing the attributes in the Part Name dialog box. The complete circuit is shown in Fig. A.33.
2. Choose **Setup** from the Analysis menu; then enable **AC Sweep** in the dialog box that appears (see Fig. A.22).
3. Click on **AC Sweep** to open the box shown in Fig. A.34.

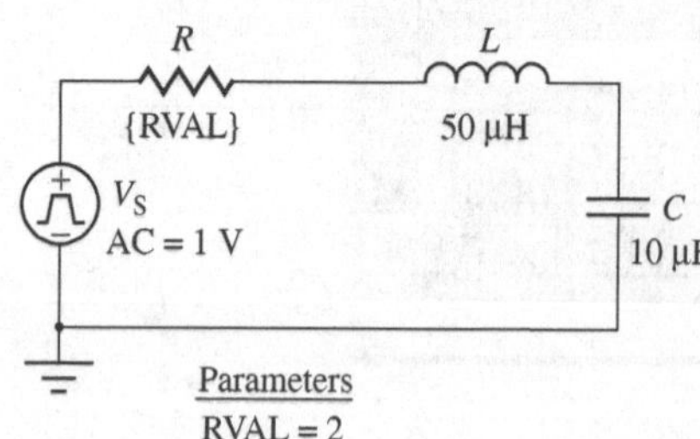

FIGURE A.33 *RLC* circuit for frequency response

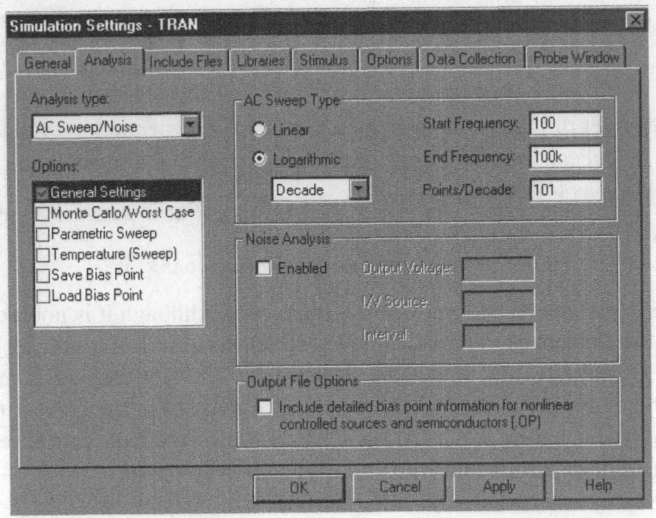

FIGURE A.34 Dialog box for frequency response analysis

4. Enter or enable the **AC Sweep** specifications as follows: For AC **Sweep Type**, choose **Decade**; for Start Frequency, type in 100; for End Frequency, type in 100k; for Points/Decade, type in 101.
5. Run the simulation for the complete circuit, shown in Fig. A.33. Use **Probe** to plot the output voltage, V(C:2), and the phase of the output voltage, VP(C:2), as shown in Fig. A.35.

▶ **TIP** If you add DB, as in VDB(C:2), you will get the output voltage in decibels.

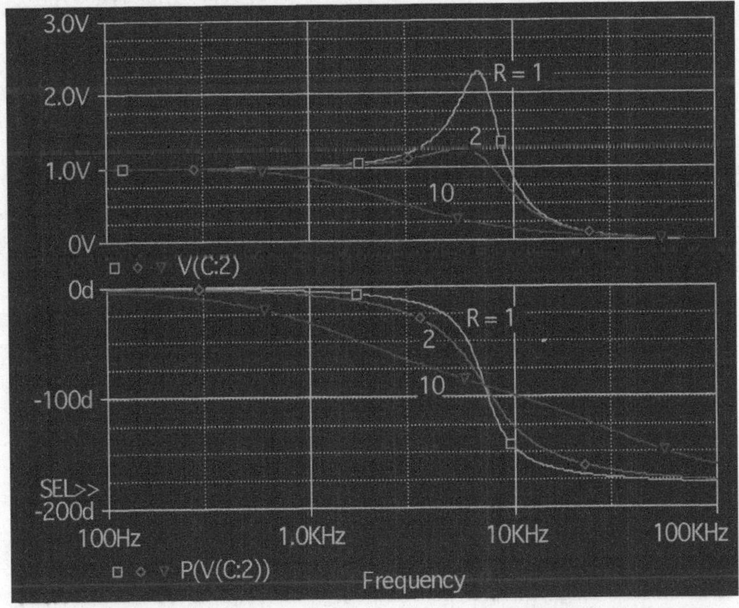

FIGURE A.35 Frequency response for $R = 1\ \Omega$, $2\ \Omega$, and $10\ \Omega$

A.12 Modeling Devices and Elements

A model that specifies a set of parameters for an element can be generated in PSpice using the .MODEL command. The same model can be used by one or more elements in the same circuit. The general form of the model statement [1] is as follows:

```
.MODEL   MNAME   TYPE (P1=A1 P2=A2
+        P3=A3 . . . PN=AN  [<tolerance specifications>])
```

MNAME is the name of the model and must start with a letter. Although it is not required, it is advisable to use the symbol of the element as the first letter (e.g., R for resistor, L for inductor). P1, P2, . . ., PN are the element parameters, and A1, A2, . . ., AN are their values. TYPE is the type name of the element and must be one of the types shown in Table A.3. An element must have the correct type name; that is, a resistor must have the type name RES, not IND or CAP. However, a circuit with several model names can include more than one model of the same type.

Tolerance specifications are used with .MC or .WORSE analysis only. They may be appended to each parameter using the following format:

```
[DEV/<distribution name> <value in % from 0 to 9>]
[LOT/<distribution name> <value in % from 0 to 9>]
```

where <distribution name> is either UNIFORM (in which case uniformly distributed deviations are generated over the range of ±<value>) or GAUSS (in which case deviations with Gaussian distribution are generated over the range ±4 and <value> specifies the ±1 deviation).

Following are some sample model statements:

```
.MODEL RLOAD  RES (R=1 TC1=0.02 TC2=0.005)
.MODEL RLOAD  RES (R=1 DEV/GAUSS 0.5% LOT/UNIFORM 10%)
.MODEL CPASS  CAP (C=1 VC1=0.01 VC2=0.002 TC1=0.02 TC2=0.005)
```

TABLE A.3 Type names of elements

Type Name	Element
RES	Resistor
CAP	Capacitor
D	Diode
IND	Inductor
NPN	*npn* bipolar junction transistor
PNP	*pnp* bipolar junction transistor
NJF	*n*-channel junction FET
PJF	*p*-channel junction FET
NMOS	*n*-channel MOSFET
PMOS	*p*-channel MOSFET
GASFET	*n*-channel GaAs MESFET
VSWITCH	Voltage-controlled switch
ISWITCH	Current-controlled switch
CORE	Nonlinear magnetic core (transformer)

```
.MODEL    LFILTER    IND (L=1 IL1=0.1 IL2=0.002 TC1=0.02 TC2=0.005)
.MODEL    DNOM       D (IS=1E-9)
.MODEL    DLOAD      D (IS=1E-9 DEV 0.5% LOT 10%)
.MODEL    QMOD       NPN (BF=50 IS=1E-9)
```

A.12.1 Resistors

The symbol for a resistor is R. The name of a resistor must start with R, and the model statement takes the following general form:

```
R<name>   N+   N-   RNAME   RVALUE
```

A resistor does not have a polarity, and so the order of the nodes does not matter. However, the current is assumed to flow from the node designated by N+ as the positive node through the resistor to the node designated by N− as the negative node. RNAME is the model name that defines the parameters of the resistor. RVALUE is the nominal value of the resistance.

▶ **NOTE** Some versions of PSpice and SPICE do not make this assumption and thus do not allow you to refer to currents through a resistor. For example, such versions will not allow you to use a notation such as I(RL) to indicate the current through RL.

The model parameters for resistors are shown in Table A.4. If RNAME is omitted, RVALUE is the resistance in ohms; RVALUE can be positive or negative but must *not* be zero. If RNAME is included and TCE is *not* specified, the resistance as a function of temperature is calculated from

$$\text{RES} = \text{RVALUE} * \text{R} * [1 + \text{TC1} * (\text{T} - \text{T0}) + \text{TC2} * (\text{T} - \text{T0})^2]$$

If RNAME is included and TCE is specified, the resistance as a function of temperature is calculated from

$$\text{RES} = \text{RVALUE} * \text{R} * 1.01^{\text{TCE} * (\text{T}-\text{T0})}$$

where T and T0 are the operating temperature and the room temperature, respectively, in degrees Celsius.

Following are some sample resistor statements:

```
RL        5      6     5K
RLOAD     10     13    ARES 1MEG
.MODEL    RMOD         RES (R=1   TC1=0.02   TC2=0.005)
RINPUT    13     17    RRES   2K
.MODEL    ARES         RES (R=1   TCE=1.5)
```

TABLE A.4 Model parameters for resistors

Name	Meaning	Units	Default
R	Resistance multiplier		1
TC1	Linear temperature coefficient	$°C^{-1}$	0
TC2	Quadratic temperature coefficient	$°C^{-2}$	0
TCE	Exponential temperature coefficient	%/°C	0

TABLE A.5 Model parameters for capacitors

Name	Meaning	Units	Default
C	Capacitance multiplier		1
VC1	Linear voltage coefficient	V^{-1}	0
VC2	Quadratic voltage coefficient	V^{-2}	0
TC1	Linear temperature coefficient	$°C^{-1}$	0
TC2	Quadratic temperature coefficient	$°C^{-2}$	0

A.12.2 Capacitors

The symbol for a capacitor is C. The name of a capacitor must start with C, and the model statement takes the following general form:

```
C<name>  N+  N-  CNAME  CVALUE  IC=VO
```

N+ is the positive node, and N− is the negative node. The voltage of node N+ is assumed to be positive with respect to node N−, and the current flows from node N+ through the capacitor to node N−. CNAME is the model name, and CVALUE is the nominal value of the capacitor. IC defines the initial (time zero) voltage of the capacitor, VO.

The model parameters for capacitors are shown in Table A.5. If CNAME is omitted, CVALUE is the capacitance in farads; CVALUE can be positive or negative but must *not* be zero. If CNAME is included, the capacitance as a function of voltage and temperature is calculated from

$$\text{CAP} = \text{CVALUE} * \text{C} * (1 + \text{VC1} * \text{V} + \text{VC2} * \text{V}^2)[1 + \text{TC1} * (\text{T} - \text{T0}) + \text{TC2} * (\text{T} - \text{T0})^2]$$

where T and T0 are the operating temperature and the room temperature, respectively, in degrees Celsius.

Following are some sample capacitor statements:

```
C1       2     6    0.01UF
CLOAD    10    13   10PF  IC=1.5V
CINPUT   14    16   DCAP  5PF
CX       10    25   DCAP  10NF  IC=3.5V
.MODEL   DCAP       CAP (C=1  VC1=0.01  VC2=0.002  TC1=0.02  TC2=0.005)
```

▶ **TIP** The initial conditions (if any) apply only if you specify the UIC (use initial condition) option under the .TRAN command.

A.12.3 Inductors

The symbol for an inductor is L. The name of an inductor must start with L, and the model statement takes the following general form:

```
L<name> N+  N-  LNAME  LVALUE  IC=IO
```

N+ is the positive node, and N− is the negative node. The voltage of N+ is assumed to be positive with respect to node N−, and the current flows from node N+ through the inductor to node N−. LNAME is the model name, and LVALUE is the nominal value of the inductor. IC defines the initial (time zero) current of the inductor IO.

TABLE A.6 Model parameters for inductors

Name	Meaning	Units	Default
L	Inductance multiplier		1
IL1	Linear current coefficient	A^{-1}	0
IL2	Quadratic current coefficient	A^{-2}	0
TC1	Linear temperature coefficient	$°C^{-1}$	0
TC2	Quadratic temperature coefficient	$°C^{-2}$	0

The model parameters for inductors are shown in Table A.6. If LNAME is omitted, LVALUE is the inductance in henries; LVALUE can be positive or negative but must *not* be zero. If LNAME is included, the inductance as a function of current and temperature is calculated from

$$\text{IND} = \text{LVALUE} * \text{L} * (1 + \text{IL1} * \text{I} + \text{IL2} * \text{I}^2)[1 + \text{TC1} * (\text{T} - \text{T0}) + \text{TC2} * (\text{T} - \text{T0})^2]$$

where T and T0 are the operating temperature and the room temperature, respectively, in degrees Celsius.

Following are some sample inductor statements:

```
LE          3         5        5MH
LLOAD       10        14       2UH    IC=0.1MA
LLINE       12        14       LMOD   2MH
LCHOKE      15        29       LMOD   5UH    IC=0.4A
.MODEL      LMOD               IND (L=1  IL1=0.1  IL2=0.002 TC1=0.02  TC2=0.005)
```

▶ **TIP** The initial conditions (if any) apply only if you specify the UIC (use initial condition) option using the .TRAN command.

A.12.4 Diodes

The model statement for a diode has the following general form:

```
.MODEL   DNAME   D (P1=A1 P2=A2 P3=A3 . . . PN=AN)
```

DNAME is the name of the model and can begin with any character, but its size is normally limited to eight characters. D is the type symbol for diodes. P1, P2, . . ., PN are the model parameters, and A1, A2, . . ., AN are their values. The model parameters are listed in Table A.7.

A.12.5 Bipolar Transistors

The model statement for *npn* transistors has the following general form:

```
.MODEL QNAME NPN (P1=A1 P2=A2 P3=A3 . . . PN=AN)
```

The general form of the model statement for *pnp* transistors is

```
.MODEL QNAME PNP (P1=A1 P2=A2 P3=A3 . . . PN=AN)
```

QNAME is the name of the BJT model, and NPN and PNP are the type symbols for *npn* and *pnp* transistors, respectively. QNAME can begin with any character, but its size is normally limited to eight characters.

TABLE A.7 Model parameters for diodes

Name	Model Parameter	Units	Default	Typical
IS	Saturation current (I_S)	A	1E-14	1E-14
RS	Parasitic resistance (R_S) (series lead and bulk resistance)	Ω	0	10
N	Emission coefficient (n)		1	1
TT	Transit time	s	0	0.1 ns
CJO	Zero-bias *pn* capacitance (C_{jo})	F	0	2 pF
VJ	Junction potential (V_j)	V	1	0.6
M	Junction grading coefficient		0.5	0.5
EG	Activation energy (0.67 for Shockley and 1.11 for silicon)	eV	1.11	1.11
XTI	IS temperature exponent		3	3
KF	Flicker noise coefficient		0	
AF	Flicker noise exponent		1	
FC	Forward-bias depletion capacitance coefficient		0.5	
BV	Reverse breakdown voltage	V	∞	50
IBV	Reverse breakdown current (reverse current at BV)	A	1E-10	

P1, P2, . . ., PN are the parameters, and A1, A2, . . ., AN are their values. Table A.8 shows the model parameters for BJTs.

A.12.6 JFETs

The model statement for an *n*-channel JFET has the following general form:

```
.MODEL JNAME NJF (P1=A1 P2=A2 P3=A3 . . . PN=AN)
```

The general form of the model statement for a *p*-channel JFET is

```
.MODEL JNAME PJF (P1=A1 P2=A2 P3=A3 . . . PN=AN)
```

JNAME is the name of the model, and NJF and PJF are the type symbols for *n*-channel and *p*-channel JFETs, respectively. P1, P2, . . ., PN are the parameters, and A1, A2, . . ., AN are their values. Table A.9 lists the model parameters for JFETs.

A.12.7 MOSFETs

The symbol for a metal oxide semiconductor field-effect transistor (MOSFET) is M. The name of a MOSFET must start with M, and the model statement takes the following general form:

```
M<name>  ND  NG  NS  NB  MNAME
+        [L=<value>] [W=<value>]
```

where ND, NG, NS, and NB are the drain, gate, source, and bulk (or substrate) nodes, respectively. MNAME is the model name. The positive current is the current flowing into a terminal; that is, the current flows from the drain node through the device to the source node in an *n*-channel MOSFET.

TABLE A.8 Model parameters for BJTs

Name	Model Parameter	Units	Default	Typical
IS	*pn* saturation current	A	1E-16	1E-16
BF	Ideal maximum forward beta		100	100
NF	Forward current emission coefficient		1	1
VAF(VA)	Forward Early voltage	V	∞	100
IKF(IK)	Corner for forward beta high-current roll-off	A	∞	10 MA
NE	Base–emitter leakage emission coefficient		1.5	2
BR	Ideal maximum reverse beta		1	0.1
NR	Reverse current emission coefficient		1	
IKR	Corner for reverse beta high-current roll-off	A	∞	100 MA
RB	Zero-bias (maximum) base resistance (base spreading resistance)	Ω	0	100
RE	Emitter ohmic resistance	Ω	0	1
RC	Collector ohmic resistance (collector lead and bulk resistance)	Ω	0	10
CJE	Base–emitter zero-bias *pn* capacitance (C_{jeo})	F	0	2 pF
VJE(PE)	Base–emitter built-in potential (V_{jbe})	V	0.75	0.7
MJE(ME)	Base–emitter *pn* grading factor		0.33	0.33
CJC	Base–collector zero-bias *pn* capacitance ($C_{\mu o}$)	F	0	1 pF
VJC(PC)	Base–collector built-in potential (V_{jc})	V	0.75	0.5
MJC(MC)	Base–collector *pn* grading factor		0.33	0.33
CJS(CCS)	Collector–substrate zero-bias *pn* capacitance	F	0	2 pF
MJS(MS)	Collector–substrate *pn* grading factor			0
FC	Forward-bias depletion capacitor coefficient		0.5	
TF	Ideal forward transit time	s	0	0.1 ns
TR	Ideal reverse transit time	s	0	10 ns
EG	Bandgap voltage (barrier height)	eV	1.11	1.11
XTI(PT)	IS temperature effect exponent (temperature coefficient for IS)		3	
KF	Flicker noise coefficient		0	6.6E-16
AF	Flicker noise exponent		1	1

TABLE A.9 Model parameters for JFETs

Name	Model Parameter	Units	Default	Typical
VTO	Threshold voltage	V	−2	−2
BETA	Transconductance coefficient	A/V^2	1E-4	1E-3
LAMBDA	Channel-length modulation	V^{-1}	0	1E-4
RD	Drain ohmic resistance	Ω	0	100
RS	Source ohmic resistance	Ω	0	100
IS	Gate *pn* saturation current	A	1E-14	1E-14
PB	Gate *pn* potential	V	1	0.6
CGD	Gate–drain zero-bias *pn* capacitance (C_{gdo})	F	0	5 pF
CGS	Gate–source zero-bias *pn* capacitance (C_{gso})	F	0	1 pF
FC	Forward-bias depletion capacitance coefficient		0.5	
KF	Flicker noise coefficient		0	
AF	Flicker noise exponent		1	

TABLE A.10 Model parameters for MOSFETs

Name	Model Parameter	Units	Default	Typical
LEVEL	Model type (Shichman-Hodges)		1	
L	Channel length	m	DEFL	
W	Channel width	m	DEFW	
VTO	Zero-bias threshold voltage	V	0	0.1
KP	Transconductance	A/V^2	2E-5	2.5E-5
GAMMA	Bulk threshold parameter	$V^{1/2}$	0	0.35
LAMBDA	Channel-length modulation (LEVEL=1)	V^{-1}	0	0.02
RD	Drain ohmic resistance	Ω	0	10
RS	Source ohmic resistance	Ω	0	10
IS	Bulk *pn* saturation current	A	1E-14	1E-15
JS	Bulk *pn* saturation current/area	A/m^2	0	1E-8
PB	Bulk *pn* potential	V	0.8	0.75
CBD	Bulk-drain zero-bias *pn* capacitance	F	0	5 pF
CBS	Bulk-source zero-bias *pn* capacitance	F	0	2 pF
CJ	Bulk *pn* zero-bias bottom capacitance/length	F/m^2	0	
MJ	Bulk *pn* bottom grading coefficient		0.5	
FC	Bulk *pn* forward-bias capacitance coefficient		0.5	
CGSO	Gate–source overlap capacitance/channel width	F/m	0	
CGDO	Gate–drain overlap capacitance/channel width	F/m	0	
CGBO	Gate-bulk overlap capacitance/channel length	F/m	0	
NSUB	Substrate doping density	$1/cm^3$	0	
TOX	Oxide thickness	m	∞	
UO	Surface mobility	$cm^2/V \cdot s$	600	
KF	Flicker noise coefficient		0	1E-26
AF	Flicker noise exponent		1	1.2

The model statement for an *n*-channel MOSFET has the following general form:

```
.MODEL  MNAME  NMOS (P1=A1 P2=A2 P3=A3 . . . PN=AN)
```

The general form of the model statement for a *p*-channel MOSFET is

```
.MODEL  MNAME  PMOS (P1=A1 P2=A2 P3=A3 . . . PN=AN)
```

NMOS and PMOS are the type symbols for *n*-channel and *p*-channel MOSFETs, respectively. MNAME can begin with any character, but its size is normally limited to eight characters. P1, P2, . . ., PN are the parameters, and A1, A2, . . ., AN are their values. Table A.10 lists the model parameters for MOSFETs.

A.13 Creating Netlists

Once you have drawn a schematic, you can create a netlist to use with other PSpice/SPICE software applications. After drawing your schematic, choose **Create Netlist** from the Analysis menu, shown in Fig. A.36. The netlist for Figs. A.31 and A.33 follows. It can be found in the file EXA-25.NET.

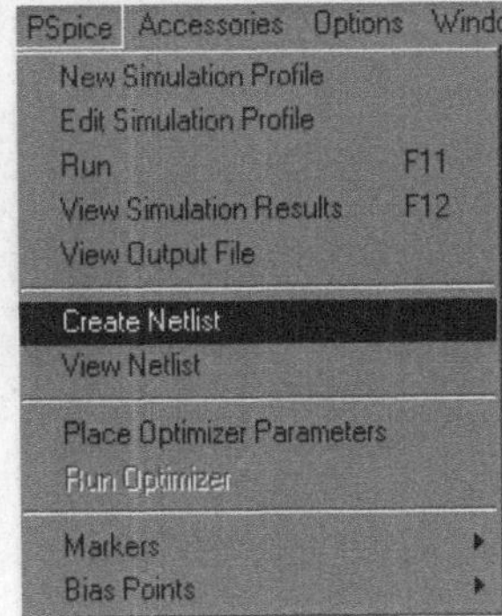

FIGURE A.36 Dialog box for creating netlist

```
* Schematics Netlist
C_C     0 $N_0001   10uF                ; C is connected between nodes 0 and
                                        ; $N_0001
L_L     $N_0002 $N_0001   50uH          ; L is connected between nodes $N_0002
                                        ; and $N_0001
R_R     $N_0003 $N_0002 {RVAL}          ; R is connected between nodes $N_0003
                                        ; and $N_0002
V_Vs    $N_0003 0   AC   1V             ; Vs is connected between nodes $N_0003
                                        ; and 0
+PULSE 0 1V 0 1ns 1ns 0.5ms 1ms         ; Pulse voltage specifications
```

▶ **TIP** When you open a schematic file (e.g., EX5-1.SCH) for the first time, PSpice may indicate an error with the message "Not Finding Netlist." If this happens, first create the netlist by choosing **Create Netlist** from the Analysis menu, shown in Fig. A.36.

If the commands for transient analysis and AC analysis are included, the netlist for the PSpice circuit file becomes as follows:

```
Frequency Response ; The first line is the title line; PSpice always
                   ; ignores this statement
* Circuit Description
C_C     0 $N_0001   10uF                ; C is connected between nodes 0 and
                                        ; $N_0001
L_L     $N_0002 $N_0001   50uH          ; L is connected between nodes $N_0002
                                        ; and $N_0001
R_R     $N_0003 $N_0002 {RVAL}          ; R is connected between nodes $N_0003
                                        ; and $N_0002
* Source Descriptions for both ac and pulse sources
V_Vs    $N_0003 0   AC   1V             ; Vs is connected between nodes $N_0003
                                        ; and 0
+ PULSE 0 1V 0 1ns 1ns 0.5ms 1ms        ; Pulse voltage specifications
* Analysis Descriptions for both ac and pulse sources
.AC   DEC 101 100HZ 100KHZ              ; ac analysis from 100 Hz to 100 kHz
                                        ; with decade increments of 101 points
                                        ; per decade
.TRAN  10ns 0.5ms                       ; Transient analysis from 0 to 0.5 ms
                                        ; with 10 ns printing/plotting interval
.PROBE                                  ; Graphics post-processor
.END                                    ; This is the last line and must always
                                        ; be included
```

TABLE A.11 Circuit elements and sources

Circuit Elements and Sources	First Letter	Model Type
Bipolar junction transistor	B	NPN/PNP
Capacitor	C	CAP
Current-controlled current source	F	
Current-controlled switch	W	VSWITCH
Current-controlled voltage source	H	
Diode	D	D
Exponential source	EXP	
GaAs MES field-effect transistor	B	GASFET
Ground	AGND	
Independent current source	I	
Independent DC voltage source	V	VDC
Inductor	L	IND/CORE
Junction field-effect transistor	J	NJF/PJF
MOS field-effect transistor	M	NMOS/PMOS
Mutual inductors (transformer)	K	
Piecewise linear source	PWL	
Polynomial source	POLY(n)	
Pulse voltage source	PULSE	VPULSE
Resistor	R	RES
Single-frequency frequency-modulation source	SFFM	
Sinusoidal voltage source	SIN	VSIN
Transmission line	T	
Voltage-controlled current source	G	
Voltage-controlled switch	S	VSWITCH
Voltage-controlled voltage source	E	EVALUE

The first letter and the model type for each element name are listed in Table A.11. The PSpice/SPICE commands are listed in Table A.12.

A.14 Adding Library Files

The directory of the PSpice circuit simulator (default name C:\MSIMEV61) has a subdirectory LIB, which contains library files of schematics and device models. This library has only a limited number of transistor schematics and models. If you want to use devices or elements with different model parameters, use elements or devices from the symbol library BREAKOUT.SLB, as shown in Fig. A.37. Use Notepad to modify the model statement in the BREAKOUT.LIB file—for example,

```
C:\OrcadLite\Capture\library\PSpice/Breakout.Lib
```

You can either use the library file RASHID.LIB that comes with the book or create your own library file containing the model statement of a device with the same model name as in the PSpice schematic—for example,

```
.MODEL    Q2N2222    Q (BF=100)                ; BJT Q2N2222
.MODEL    D1N4148    D (IS=10E-15 BV=100)      ; Diode D1N4148
```

This way you keep the same model name and use the PSpice schematic symbols, but change the model parameters.

TABLE A.12 PSpice/SPICE commands

Analysis or Function	Command
Absolute value (operator)	ABS
AC/frequency analysis	.AC
DC operating point	.OP
DC sweep	.DC
Difference (operator)	DIF
End of subcircuit	.ENDS
Fourier analysis	.FOUR
Frequency response transfer function	FREQ
Function definition	.FUNC
Gain limit (operator)	GLIMIT
Global nodes	.GLOBAL
Graphics postprocessor	.PROBE
Include file	.INC
Initial conditions	.IC
Library file	.LIB
Model definition	.MODEL
Multiplier (operator)	MULTI
Node setting	.NODESET
Noise analysis	.NOISE
Options	.OPTIONS
Parameter definition	PARAM
Parameter variation	.PARAM
Parametric analysis	.STEP
Plot output	.PLOT
Print output	.PRINT
Probe	.PROBE
Sensitivity analysis	.SENS
Subcircuit call	X_Call
Subcircuit definition	.SUBCKT
Summation (operator)	SUM
Table (operator)	TABLE
Temperature	.TEMP
Transfer function	.TF
Transient analysis	.TRAN
Value	VALUE
Value of voltage-controlled voltage source	EVALUE
Width	.WIDTH

▶ **TIP** If you wish to change the model parameters of any device, use a word processing program in ASCII (DOS) text format (or DOS Editor or Notepad in PSpice) to edit the file EVAL.LIB and change the model parameters.

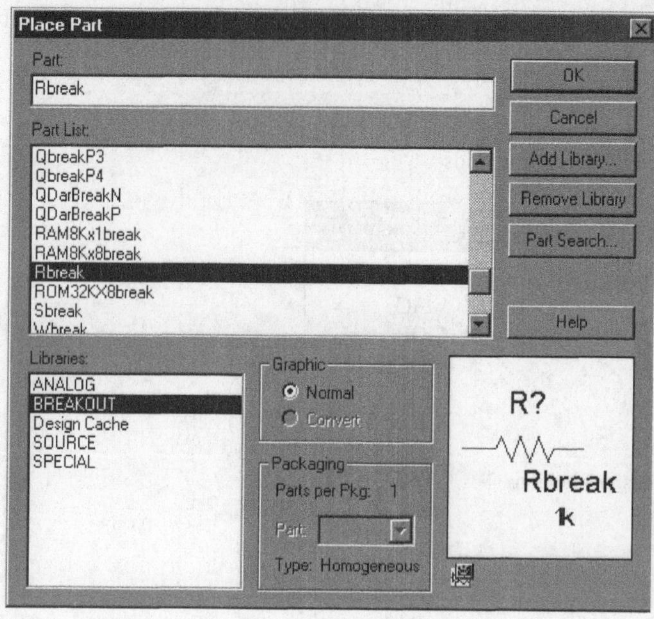

FIGURE A.37 Menu for adding parts

We can add the library file RASHID_MODEL.LIB while running the simulation. First, choose **Library and Include Files** from the Simulation Settings menu, shown in Fig. A.38. Then browse through the library files and select the file RASHID_MODEL.LIB. Click on **Add Library*** and then on **OK**. PSpice will look for device models in this library file while running the simulation. The Include Files dialog box is shown in Fig. A.39.

Simulation Settings - TRAN

General | Analysis | Include Files | Libraries | Stimulus | Options | Data Collection | Probe Window

Analysis type: Time Domain (Transient)

Run to time: 0.5ms seconds (TSTOP)

Start saving data after: 0 seconds

Options: General Settings, Monte Carlo/Worst Case, Parametric Sweep, Temperature (Sweep), Save Bias Point, Load Bias Point

Transient options — Maximum step size: seconds; Skip the initial transient bias point calculation (SKIPBP)

Output File Options...

OK | Cancel | Apply | Help

FIGURE A.38 Simulation Settings menu

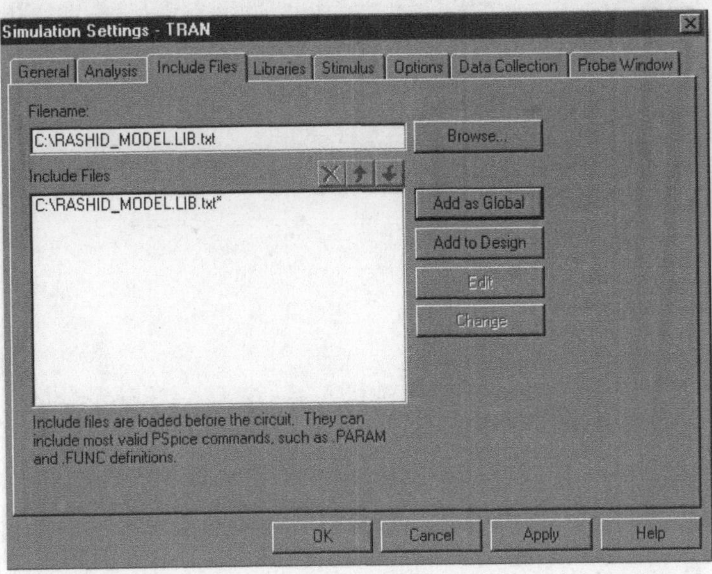

FIGURE A.39 Dialog box for adding library files

References

1. M. H. Rashid, *Introduction to PSpice Using OrCAD for Circuits and Electronics*. Upper Saddle River, NJ: Prentice Hall, 2003.
2. M. E. Herniter, *Schematic Capture with MicroSim PSpice*. Englewood Cliffs, NJ: Prentice Hall, 1996.

APPENDIX B

REVIEW OF BASIC CIRCUITS

B.1 Introduction

Electronic devices that are used as parts of electronic circuits are normally modeled by equivalent circuits. Doing a performance evaluation and design of any electronic circuit requires a knowledge of circuit analysis. This appendix reviews the basic circuit theorems and analysis techniques that are commonly used for electronic circuits.

B.2 Kirchhoff's Current Law

Kirchhoff's current law (KCL) states that the sum of all currents at a node must be zero; that is,

$$\sum I_n = 0$$

where I_n is the current flowing into nth node and $n = 1, 2, 3, \ldots, \infty$.

EXAMPLE B.1

Finding the currents in two parallel resistors For the circuit shown in Fig. B.1, find currents I_1 and I_2.

SOLUTION

Using KCL at node 1 gives

$$I_S - I_1 - I_2 = 0 \quad \text{or} \quad I_S = I_1 + I_2 \tag{B.1}$$

Since the voltage V_S across R_1 is the same as that across R_2, we can write $V_S = R_1I_1 = R_2I_2$, which gives $I_2 = R_1I_1/R_2$. Substituting for I_2, we get

$$I_S = I_1 + \frac{R_1}{R_2}I_1 = \frac{R_2 + R_1}{R_2}I_1$$

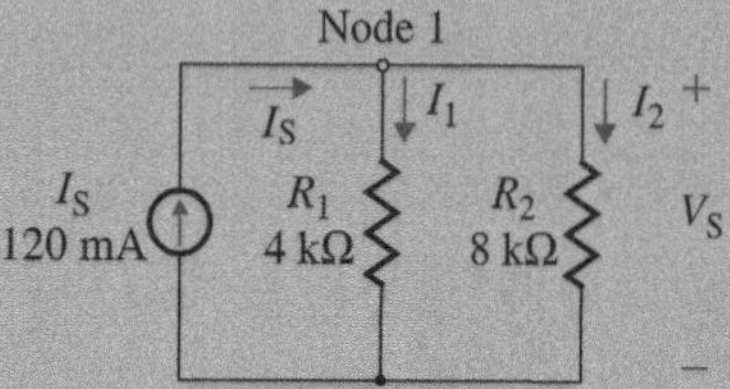

FIGURE B.1 Current distribution in two resistors

which gives the current I_1 through R_1 as

$$I_1 = \frac{R_2}{R_1 + R_2} I_S \tag{B.2}$$

$$= \frac{8\ \text{k}\Omega}{4\ \text{k}\Omega + 8\ \text{k}\Omega} \times 120\ \text{mA} = 80\ \text{mA}$$

Similarly, substituting $I_1 = R_2 I_2 / R_1$ in Eq. (B.1) and simplifying, we get the current I_2 through R_2 as

$$I_2 = \frac{R_1}{R_1 + R_2} I_S \tag{B.3}$$

$$= \frac{4\ \text{k}\Omega}{4\ \text{k}\Omega + 8\ \text{k}\Omega} \times 120\ \text{mA} = 40\ \text{mA}$$

The branch currents I_1 and I_2 can be found by applying Ohm's law as given by

$$V_S = I_S(R_1 \| R_2) = \frac{R_1 R_2}{R_1 + R_2} I_S$$

which can be used to find $I_1 = V_S/R_1$ as in Eq. (B.2) and $I_2 = V_S/R_2$ as in Eq. (B.3).

NOTE: Equations (B.2) and (B.3) give the current distribution in two resistors, and the two equations together are often known as the *current divider rule*.

EXAMPLE B.2

Finding the currents in three parallel resistors For the circuit shown in Fig. B.2, find the currents I_1, I_2, and I_3.

SOLUTION

The current source I_S is divided into I_1 through R_1, I_2 through R_2, and I_3 through R_3. Using KCL at node 1 gives

$$I_S - I_1 - I_2 - I_3 = 0 \quad \text{or} \quad I_S = I_1 + I_2 + I_3 \tag{B.4}$$

Since the voltage V_S across R_1 is the same as that across R_2 and R_3, we can write $V_S = R_1 I_1 = R_2 I_2 = R_3 I_3$, which gives $I_2 = R_1 I_1 / R_2$ and $I_3 = R_1 I_1 / R_3$. Substituting for I_2 and I_3 in Eq. (B.4), we get

$$I_S = I_1 + \frac{R_1}{R_2} I_1 + \frac{R_1}{R_3} I_1 = \frac{R_1 R_2 + R_2 R_3 + R_3 R_1}{R_2 R_3} I_1$$

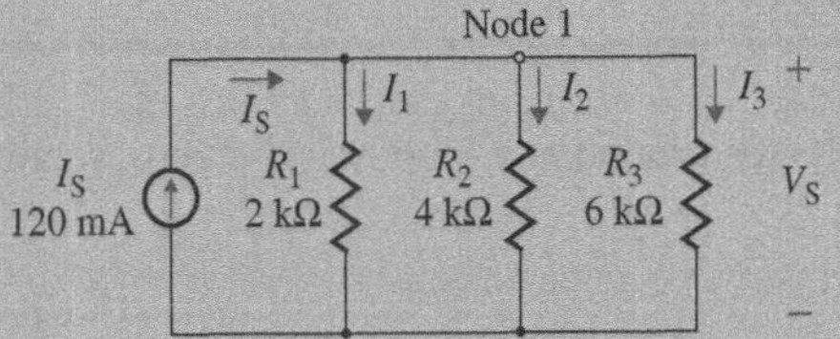

FIGURE B.2 Current distribution in three resistors

which gives the current I_1 through R_1 as

$$I_1 = \frac{1/R_1}{1/R_1 + 1/R_2 + 1/R_3} I_S = \frac{R_2R_3}{R_1R_2 + R_2R_3 + R_3R_1} I_S \tag{B.5}$$

$$= \frac{4\text{ k}\Omega \times 6\text{ k}\Omega}{2\text{ k}\Omega \times 4\text{ k}\Omega + 4\text{ k}\Omega \times 6\text{ k}\Omega + 6\text{ k}\Omega \times 2\text{ k}\Omega} \times 120\text{ mA} = 65.45\text{ mA}$$

Substituting $I_1 = R_2I_2/R_1$ and $I_3 = R_2I_2/R_3$ in Eq. (B.4) yields the current I_2 through R_2 as

$$I_2 = \frac{1/R_2}{1/R_1 + 1/R_2 + 1/R_3} I_S = \frac{R_1R_3}{R_1R_2 + R_2R_3 + R_3R_1} I_S \tag{B.6}$$

$$= \frac{2\text{ k}\Omega \times 6\text{ k}\Omega}{2\text{ k}\Omega \times 4\text{ k}\Omega + 4\text{ k}\Omega \times 6\text{ k}\Omega + 6\text{ k}\Omega \times 2\text{ k}\Omega} \times 120\text{ mA} = 32.73\text{ mA}$$

Similarly, substituting $I_1 = R_3I_3/R_1$ and $I_2 = R_3I_3/R_2$ in Eq. (B.4) yields the current I_3 through R_3 as

$$I_3 = \frac{1/R_3}{1/R_1 + 1/R_2 + 1/R_3} I_S = \frac{R_1R_2}{R_1R_2 + R_2R_3 + R_3R_1} I_S \tag{B.7}$$

$$= \frac{2\text{ k}\Omega \times 4\text{ k}\Omega}{2\text{ k}\Omega \times 4\text{ k}\Omega + 4\text{ k}\Omega \times 6\text{ k}\Omega + 6\text{ k}\Omega \times 2\text{ k}\Omega} \times 120\text{ mA} = 21.82\text{ mA}$$

The branch currents I_1, I_2, and I_3 can be found by applying Ohm's law as given by

$$V_S = I_S(R_1 \| R_2 \| R_3) = \frac{1}{(1/R_1) + (1/R_2) + (1/R_3)} I_S$$

which can be used to find $I_1 = V_S/R_1$ as in Eq. (B.5), $I_2 = V_S/R_2$ as in Eq. (B.6), and $I_3 = V_S/R_3$ as in Eq. (B.7).

NOTE: If there are n parallel resistances, the current I_n through the nth resistance R_n can be expressed in general form as

$$I_n = \frac{1/R_n}{(1/R_1) + (1/R_2) + \cdots + (1/R_m)} I_S$$

B.3 Kirchhoff's Voltage Law

Kirchhoff's voltage law (KVL) states that the sum of the voltages around any loop must be zero; that is,

$$\sum V_n = 0$$

where V_n is the voltage across the nth segment of a loop and $n = 1, 2, 3, \ldots, \infty$.

EXAMPLE B.3

Finding the voltage distribution in two resistors In Fig. B.3, the voltage V_S is divided into V_1 across R_1 and V_2 across R_2. Find the voltages V_1 and V_2.

SOLUTION

Using KVL around loop I gives

$$V_S - V_1 - V_2 = 0 \quad \text{or} \quad V_S = V_1 + V_2 \tag{B.8}$$

Since the current I_S through R_1 is the same as that through R_2, we can write $V_1 = R_1 I_S$ and $V_2 = R_2 I_S$. Substituting for V_1 and V_2 in Eq. (B.8), we get

$$V_S = V_1 + V_2 = R_1 I_S + R_2 I_S = (R_1 + R_2) I_S$$

which gives the current I_S as

$$I_S = \frac{V_S}{R_1 + R_2}$$

$$= \frac{24 \text{ V}}{4 \text{ k}\Omega + 8 \text{ k}\Omega} = 2 \text{ mA}$$

Therefore, the voltage V_1 across R_1 can be found from

$$V_1 = R_1 I_S = \frac{R_1}{R_1 + R_2} V_S \tag{B.9}$$

$$= \frac{4 \text{ k}\Omega}{4 \text{ k}\Omega + 8 \text{ k}\Omega} \times 24 \text{ V} = 8 \text{ V}$$

Similarly, the voltage V_2 across R_2 can be found from

$$V_2 = R_2 I_S = \frac{R_2}{R_1 + R_2} V_S \tag{B.10}$$

$$= \frac{8 \text{ k}\Omega}{4 \text{ k}\Omega + 8 \text{ k}\Omega} \times 24 \text{ V} = 16 \text{ V}$$

NOTE: The loop current I_S can be found directly by applying Ohm's law, $I_S = V_S/(R_1 + R_2)$. Equations (B.9) and (B.10) give the voltage distribution in two resistors only when the current through R_1 and R_2 is the same. This distribution of voltage is often known as the *voltage* (or *potential*) *divider rule*.

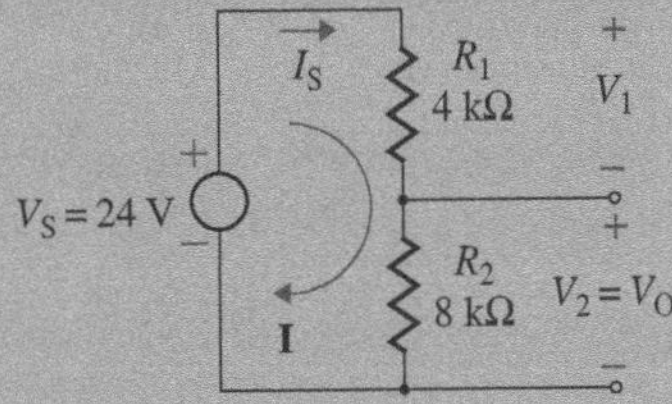

FIGURE B.3 Voltage distribution in two resistors

EXAMPLE B.4

Analyzing a circuit with a current-dependent current source For the circuit shown in Fig. B.4 with a current-dependent current source, find the currents I_B, I_C, and I_E and voltage V_C. Assume $R_{Th} = 15\ k\Omega$, $r_\pi = 1\ k\Omega$, $R_C = 2\ k\Omega$, $R_E = 500\ \Omega$, $\beta_F = 100$, $V_{CC} = 30$ V, and $V_{Th} = 5$ V.

SOLUTION

Using KCL at node 1, we get

$$I_E = I_B + I_C = I_B + \beta_F I_B = (1 + \beta_F)I_B \tag{B.11}$$

Using KVL around loop I, we get

$$V_{Th} = R_{Th}I_B + r_\pi I_B + R_E I_E = R_{Th}I_B + r_\pi I_B + R_E(1 + \beta_F)I_B$$

which gives I_B as

$$I_B = \frac{V_{Th}}{R_{Th} + r_\pi + R_E(1 + \beta_F)} \tag{B.12}$$

$$= \frac{5\ \text{V}}{15\ k\Omega + 1\ k\Omega + 500\ \Omega \times (1 + 100)} = 75.19\ \mu\text{A}$$

Current I_C, which is dependent only on I_B, can be found from

$$I_C = \beta_F I_B = \frac{\beta_F V_{Th}}{R_{Th} + R_\pi + R_E(1 + \beta_F)} \tag{B.13}$$

$$= \frac{100 \times 5\ \text{V}}{15\ k\Omega + 1\ k\Omega + 500\ \Omega \times (1 + 100)} = 7519\ \mu\text{A}$$

Then

$$I_E = I_B + I_C = 75.19\ \mu\text{A} + 7519\ \mu\text{A} = 7594\ \mu\text{A}$$

$$V_C = V_{CC} - I_C R_C = 30\ \text{V} - 7519\ \mu\text{A} \times 2\ k\Omega = 14.96\ \text{V}$$

Voltage V_E can be found from

$$V_E = R_E I_E = R_E(I_C + I_B) = R_E(\beta_F I_B + I_B) = R_E I_B(1 + \beta_F)$$

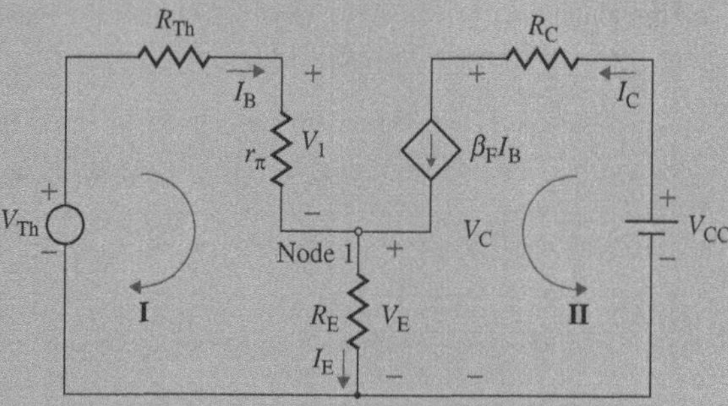

FIGURE B.4 Circuit with current-dependent current source

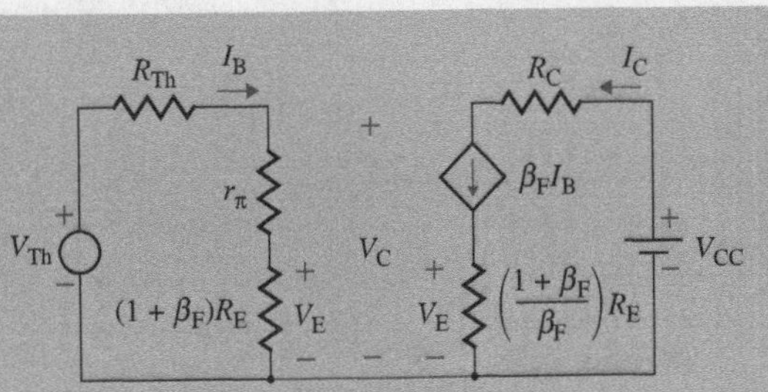

FIGURE B.5 Splitting of current R_E

Since $I_C = \beta_F I_B$, we get

$$V_E = R_E I_B(1 + \beta_F) = R_E I_C[(1 + \beta_F)/\beta_F)]$$

Thus, R_E offers a resistance of $R_E(1 + \beta_F)$ to current I_B in loop I and a resistance of $R_E(1 + \beta_F)/\beta_F$ to current I_C in loop II. Therefore, R_E can be split, or "reflected," into loop I and loop II by adjusting its value such that V_E is preserved on loop I and loop II. This arrangement is shown in Fig. B.5.

NOTE: The current-dependent current source $\beta_F I_B$ is often represented by a voltage-dependent voltage source $g_m V_1$ such that $\beta_F I_B = \beta_F V_1/r_\pi = (\beta_F/r_\pi)V_1$. g_m, which is known as the *transconductance*, is related to β_F by $g_m = \beta_F/r_\pi$. Therefore, by substituting $\beta_F = g_m r_\pi$, we can apply the above equations for a voltage-dependent voltage source that is often used as the small-signal model of bipolar and field-effect transistors.

B.4 Superposition Theorem

The superposition theorem states that the current through or voltage across any element in a linear network is equal to the algebraic sum of the currents or voltages produced independently by each source. To calculate the effect of one source, the other independent sources are removed by short-circuiting the voltage sources and open circuiting the current sources. Any internal resistance associated with the removed voltage sources must be considered, however.

EXAMPLE B.5

Finding the output voltage using the superposition theorem The circuit in Fig. B.6 has a DC source $V_{DC} = 10$ V, an AC source $v_{ac} = 15 \sin(377\pi t)$, $R_1 = 2$ kΩ, and $R_2 = 3$ kΩ. Use the superposition theorem to determine the instantaneous output voltage v_O.

SOLUTION

$V_{DC} = 10$ V, $v_{ac} = 15 \sin(377\pi t)$, $R_1 = 2$ kΩ, and $R_2 = 3$ kΩ. The DC equivalent circuit with source V_{DC} only is shown in Fig. B.7(a); the output voltage due to V_{DC} is

$$V_{O1} = \frac{R_2}{R_1 + R_2} V_{DC} = \frac{3\text{ k}}{2\text{ k} + 3\text{ k}} \times 10 = 6\text{ V}$$

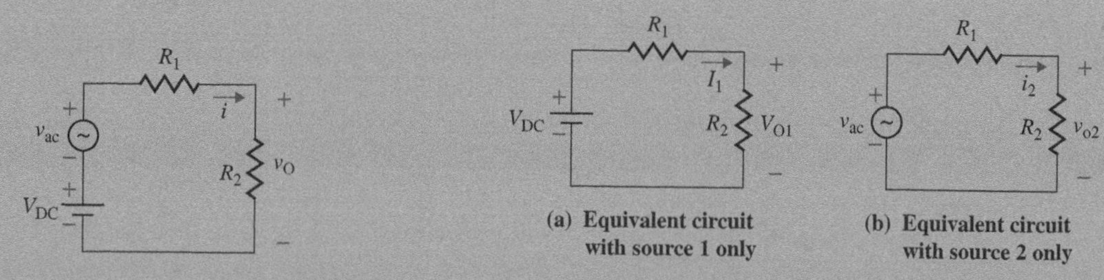

FIGURE B.6 Circuit for Example B.5

(a) Equivalent circuit with source 1 only

(b) Equivalent circuit with source 2 only

FIGURE B.7 Equivalent circuits for Example B.5

Figure B.7(b) shows the AC equivalent circuit with source v_{ac} only; the output voltage due to v_{ac} is

$$v_{o2} = \frac{R_2}{R_1 + R_2} v_{ac} = \frac{3\text{ k}}{2\text{ k} + 3\text{ k}} \times 15 \sin(377\pi t) = 9 \sin(377\pi t)$$

Therefore, the resultant output voltage v_O can be found by combining the output voltages due to individual sources; that is,

$$v_O = V_{O1} + v_{o2} = 6 + 9 \sin(377\pi t) = 3 \times [2 + 3 \sin(377\pi t)]$$

EXAMPLE B.6

Finding the output voltage using the superposition theorem A circuit with three input voltages V_{S1}, V_{S2}, and V_{S3} is shown in Fig. B.8. Use the superposition theorem to determine the output voltage V_O. Use $R_1 = 2\text{ k}\Omega$, $R_2 = 4\text{ k}\Omega$, $R_3 = 6\text{ k}\Omega$, $V_{S1} = 10\text{ V}$, $V_{S2} = 12\text{ V}$, and $V_{S3} = 15\text{ V}$.

SOLUTION

The equivalent circuit with source V_{S1} only is shown in Fig. B.9(a). Applying the voltage divider rule, we can find the output voltage due to V_{S1} as

$$V_{O1} = \frac{R_2 \| R_3}{R_1 + R_2 \| R_3} V_{S1} = \frac{4\text{ k} \| 6\text{ k}}{2\text{ k} + (4\text{ k} \| 6\text{ k})} \times 10 = 5.45\text{ V}$$

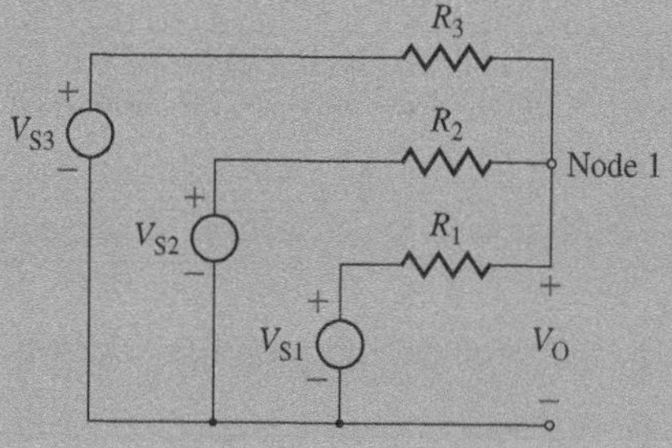

FIGURE B.8 Circuit for Example B.6

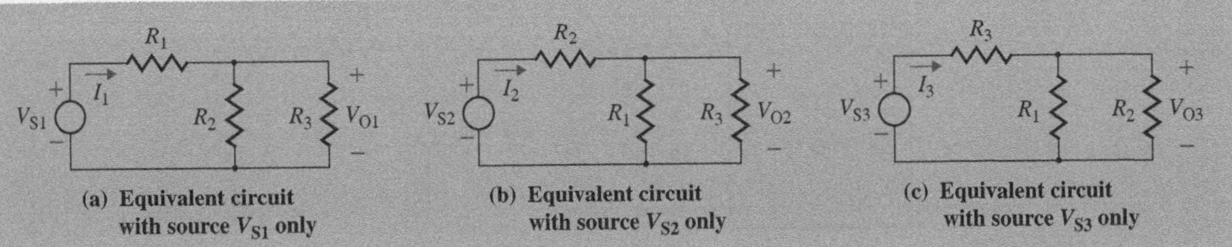

FIGURE B.9 Equivalent circuits for Example B.6

The equivalent circuit with source V_{S2} only is shown in Fig. B.9(b); the output voltage due to V_{S2} is

$$V_{O2} = \frac{R_1 \| R_3}{R_2 + R_1 \| R_3} V_{S2} = \frac{2\text{ k} \| 6\text{ k}}{4\text{ k} + (2\text{ k} \| 6\text{ k})} \times 12 = 3.27\text{ V}$$

The equivalent circuit with source V_{S3} only is shown in Fig. B.9(c); the output voltage due to V_{S3} is

$$V_{O3} = \frac{R_1 \| R_2}{R_3 + R_1 \| R_2} V_{S3} = \frac{2\text{ k} \| 4\text{ k}}{6\text{ k} + (2\text{ k} \| 4\text{ k})} \times 15 = 2.73\text{ V}$$

Therefore, the resultant output voltage V_O can be found by combining the output voltages due to individual sources; that is,

$$V_O = V_{O1} + V_{O2} + V_{O3} = 5.45 + 3.27 + 2.73 = 11.45\text{ V}$$

An alternative approach is to apply KVL at node 1 and look for V_O:

$$V_O = \frac{\text{Currents into node 1 if it were at ground potential}}{\text{Conductances radiating from node 1}}$$

$$= \frac{V_{S1}/R_1 + V_{S2}/R_2 + V_{S3}/R_3}{1/R_1 + 1/R_2 + 1/R_3} = \frac{10/2\text{ k} + 12/4\text{ k} + 15/6\text{ k}}{1/2\text{ k} + 1/4\text{ k} + 1/6\text{ k}} = 11.45\text{ V}$$

B.5 Thevenin's Theorem

Thevenin's theorem states that any two-terminal linear DC (or AC) network can be replaced by an equivalent circuit consisting of a voltage source and a series resistance (or impedance). This theorem is commonly used to find the voltage (or current) of a linear network with one or more sources. It allows us to concentrate on a specific portion of the network by replacing the remaining network by an equivalent circuit. In the case of sinusoidal AC circuits, the reactances are frequency dependent, and Thevenin's equivalent circuit is valid for only one frequency.

Figure B.10(a) shows a general DC network; Thevenin's equivalent circuit is shown in Fig. B.10(b). The steps in determining an equivalent voltage source V_{Th} and an equivalent resistance R_{Th} for Thevenin's equivalent circuit are as follows:

Step 1. Decide on the part of the network for which you desire a Thevenin's representation and mark the terminals, as shown in Fig. B.10(a).

Step 2. Remove that part of the network. In Fig. B.10(a), the load resistance R_L is to be removed.

Step 3. Mark the open-circuit terminals of the remaining network—namely, a and b.

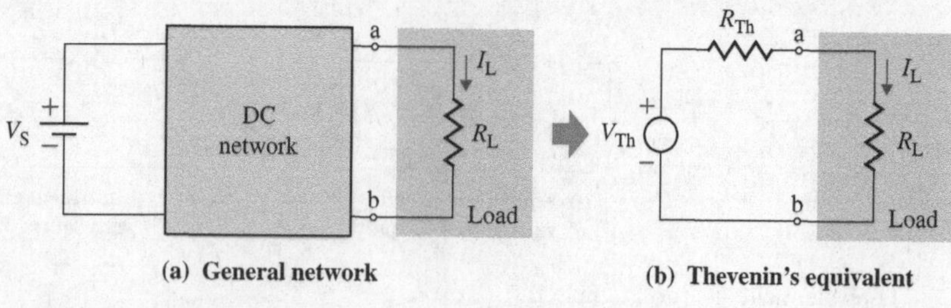

FIGURE B.10 Thevenin's equivalent circuit

Step 4. Determine the open-circuit voltage V_{Th} between terminals a and b.

Step 5. Set all independent sources to zero (voltage sources are replaced by short circuits and current sources by open circuits). Apply a test voltage V_X across terminals a and b. The ratio of V_X to its current I_X gives Thevenin's resistance R_{Th}.

EXAMPLE B.7

Finding Thevenin's equivalent circuit Represent the network shown in Fig. B.11(a) by Thevenin's equivalent, as shown in Fig. B.11(b). Assume $V_{CC} = 12$ V, $R_1 = 15$ kΩ, and $R_2 = 7.5$ kΩ.

SOLUTION

The open-circuit voltage, which is Thevenin's voltage between terminals a and b, can be found from the voltage divider rule in Eq. (B.10); that is,

$$V_{Th} = \frac{R_2}{R_1 + R_2} V_{CC} \tag{B.14}$$

$$= \frac{7.5\ \text{k}}{15\ \text{k} + 7.5\ \text{k}} \times 12 = 4\ \text{V}$$

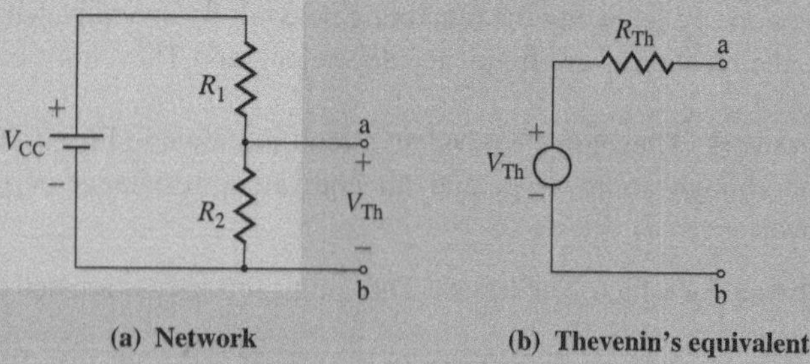

FIGURE B.11 Network for Example B.7

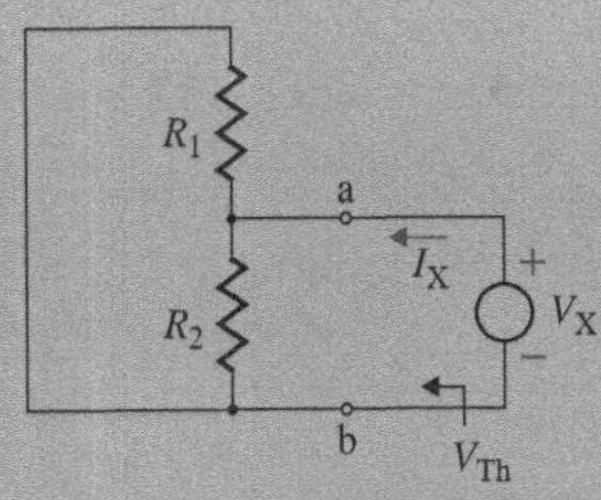

FIGURE B.12 Equivalent circuits

If source V_{CC} is set to zero and test voltage V_X is applied across terminals a and b, the circuit for determining R_{Th} is shown in Fig. B.12. R_{Th} becomes the parallel combination of R_1 and R_2. That is,

$$R_{Th} = \frac{V_X}{I_X} = R_1 \| R_2 \tag{B.15}$$

$$= 15 \text{ k} \| 7.5 \text{ k} = 5 \text{ k}\Omega$$

EXAMPLE B.8

Finding Thevenin's equivalent circuit Represent the network shown in Fig. B.13(a) by Thevenin's equivalent, as shown in Fig. B.13(b). Assume $V_{CC} = 12$ V, $V_A = 9$ V, $R_1 = 15$ kΩ, and $R_2 = 7.5$ kΩ.

SOLUTION

Since there are two voltage sources V_{CC} and V_A, we will apply the superposition theorem to find V_{Th}. The equivalent circuit with source V_{CC} only is shown in Fig. B.14(a). Applying the voltage divider rule, we can find the output voltage due to V_{CC} only as

$$V_{O1} = \frac{R_2}{R_1 + R_2} V_{CC} = \frac{7.5 \text{ k}}{15 \text{ k} + 7.5 \text{ k}} \times 12 = 4 \text{ V}$$

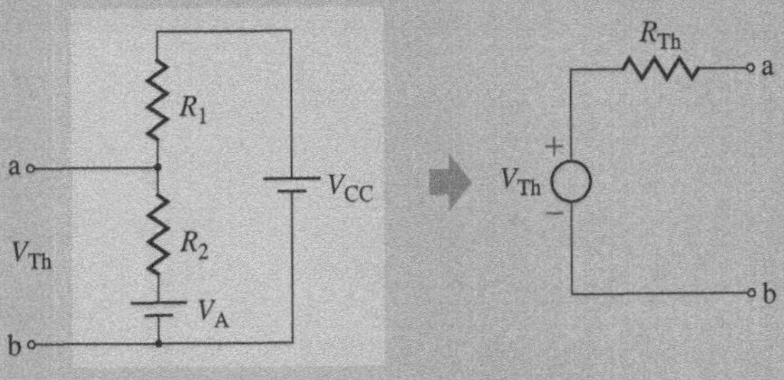

(a) Network (b) Thevenin's equivalent

FIGURE B.13 Network for Example B.8

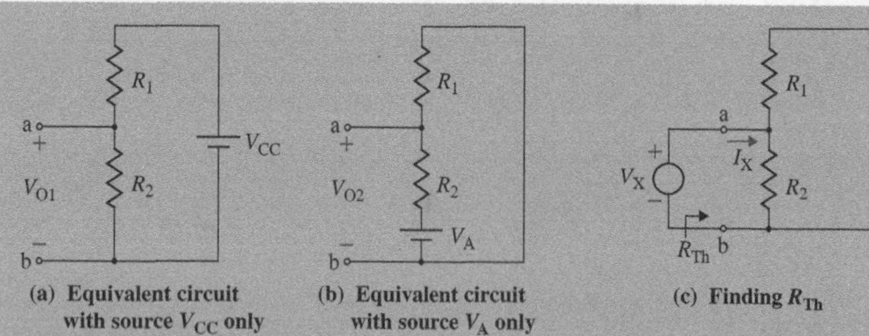

(a) Equivalent circuit with source V_{CC} only (b) Equivalent circuit with source V_A only (c) Finding R_{Th}

FIGURE B.14 Equivalent circuits

The circuit with source V_A only is shown in Fig. B.14(b); the output voltage due to V_A is

$$V_{O2} = \frac{R_1}{R_1 + R_2} V_A = \frac{15 \text{ k}}{15 \text{ k} + 7.5 \text{ k}} \times 9 = 6 \text{ V}$$

The resultant output voltage V_O, which equals V_{Th}, can be found by combining the output voltages due to individual sources; that is,

$$\begin{aligned} V_{Th} &= V_O = V_{O1} + V_{O2} \\ &= \frac{R_2}{R_1 + R_2} V_{CC} + \frac{R_1}{R_1 + R_2} V_A \\ &= 4 + 6 = 10 \text{ V} \end{aligned} \tag{B.16}$$

If sources V_A and V_{CC} are set to zero and test voltage V_X is applied across terminals a and b, the circuit for determining R_{Th} is shown in Fig. B.14(c). R_{Th} becomes the parallel combination of R_1 and R_2; that is,

$$\begin{aligned} R_{Th} &= \frac{V_X}{I_X} = R_1 \| R_2 \\ &= 15 \text{ k} \| 7.5 \text{ k} = 5 \text{ k}\Omega \end{aligned} \tag{B.17}$$

It is evident that Eqs. (B.15) and (B.17) are the same.

EXAMPLE B.9

Representing a network by Thevenin's equivalent circuit Represent the network shown in Fig. B.15 by Thevenin's equivalent circuit. The circuit values are $R_i = 1.5 \text{ k}\Omega$, $R_C = 25 \text{ k}\Omega$, $\beta_F = 50$, $h_r = 3 \times 10^{-4}$, and $V_s = 5 \text{ mV}$.

(a) Calculate the parameters of Thevenin's equivalent circuit.

(b) Use PSpice/SPICE to check your results.

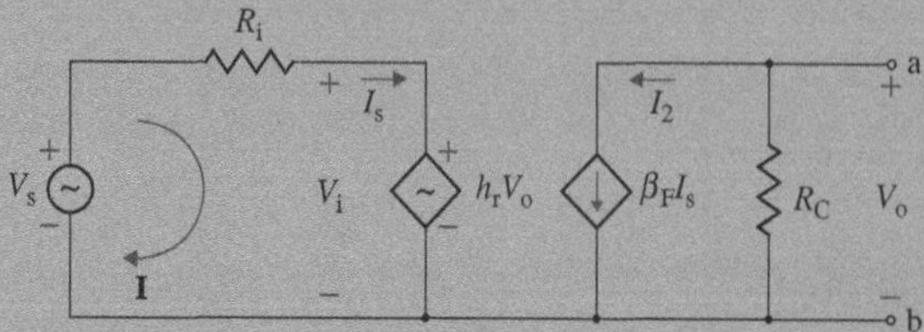

FIGURE B.15 Network for Example B.9

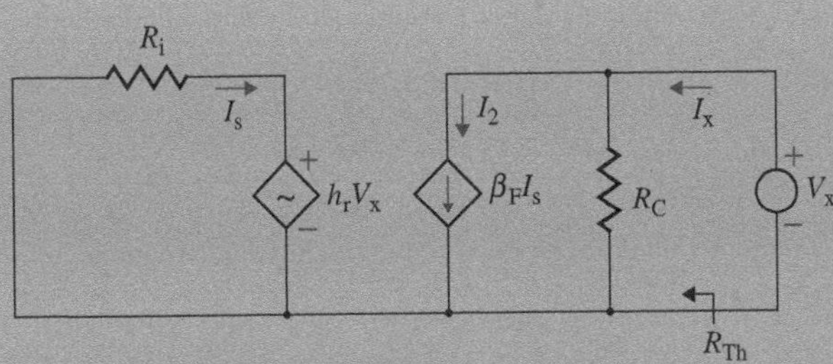

FIGURE B.16 Circuit for determining Thevenin's resistance

SOLUTION

(a) $R_i = 1.5\ \text{k}\Omega$, $R_C = 25\ \text{k}\Omega$, $\beta_F = 50$, $h_r = 3 \times 10^{-4}$, and $V_s = 5$ mV. The output voltage V_o between terminals a and b is

$$V_o = -I_2 R_C = -\beta_F I_s R_C \quad \text{(B.18)}$$

The input current I_s can be found from loop I as

$$I_s = \frac{V_s - h_r V_o}{R_i} \quad \text{(B.19)}$$

Substituting I_s from Eq. (B.19) into Eq. (B.18) gives the output voltage V_o as

$$V_{Th} = V_o = \frac{-\beta_F R_C}{R_i - \beta_F h_r R_C} V_s \quad \text{(B.20)}$$

$$= \frac{-50 \times 25\ \text{k} \times 5\ \text{m}}{1.5\ \text{k} - 50 \times 3 \times 10^{-4} \times 25\ \text{k}} = -5.5556\ \text{V}$$

Thevenin's resistance R_{Th} can be determined from the circuit in Fig. B.16, which is obtained by short-circuiting the independent source V_s. Considering V_x and I_x as the test voltage and current, respectively, we get

$$I_s = -\frac{h_r V_x}{R_i} \quad \text{(B.21)}$$

$$I_x = \beta_F I_s + \frac{V_x}{R_C} \quad \text{(B.22)}$$

Substituting I_s from Eq. (B.21) into Eq. (B.22) gives

$$I_x = -\frac{\beta_F h_r V_x}{R_i} + \frac{V_x}{R_C} = \frac{R_i - \beta_F h_r R_C}{R_i R_C} V_x$$

which gives Thevenin's resistance R_{Th} as

$$R_{Th} = \frac{V_x}{I_x} = \frac{R_i R_C}{R_i - \beta_F h_r R_C} \quad \text{(B.23)}$$

$$= \frac{1.5\ \text{k} \times 25\ \text{k}}{1.5\ \text{k} - 50 \times 3 \times 10^{-4} \times 25\ \text{k}} = 33.33\ \text{k}\Omega$$

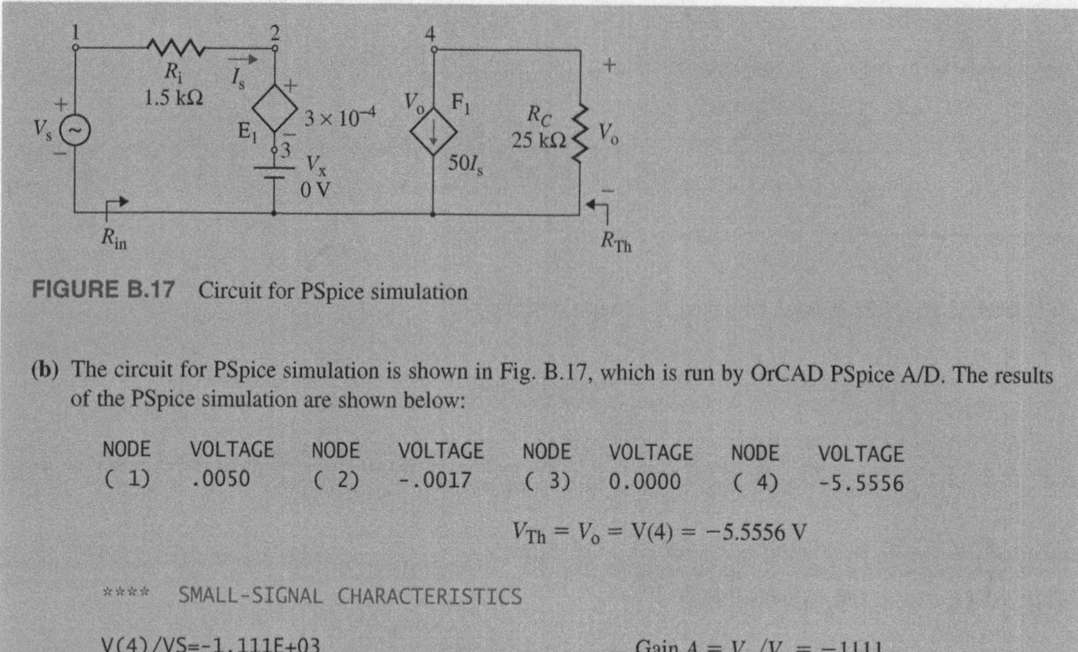

FIGURE B.17 Circuit for PSpice simulation

(b) The circuit for PSpice simulation is shown in Fig. B.17, which is run by OrCAD PSpice A/D. The results of the PSpice simulation are shown below:

```
NODE   VOLTAGE    NODE   VOLTAGE    NODE   VOLTAGE    NODE   VOLTAGE
( 1)   .0050      ( 2)   -.0017     ( 3)   0.0000     ( 4)   -5.5556
```

$$V_{Th} = V_o = V(4) = -5.5556 \text{ V}$$

```
****   SMALL-SIGNAL CHARACTERISTICS

V(4)/VS=-1.111E+03

INPUT RESISTANCE AT VS=1.125E+03

OUTPUT RESISTANCE AT V(4)=3.333E+04
```

Gain $A = V_o/V_s = -1111$

$R_{in} = V_s/I_s = 1.125 \text{ k}\Omega$

$R_{Th} = 33.33 \text{ k}\Omega$

B.6 Norton's Theorem

Norton's theorem states that any two-terminal linear DC (or AC) network can be replaced by an equivalent circuit consisting of a current source and a parallel resistance (or impedance). Norton's equivalent circuit can be determined from Thevenin's equivalent circuit; the relationship between them is shown in Fig. B.18. Norton's resistance R_N is identical to Thevenin's resistance R_{Th}, and Norton's current equals the short-circuit current at the terminals of interest.

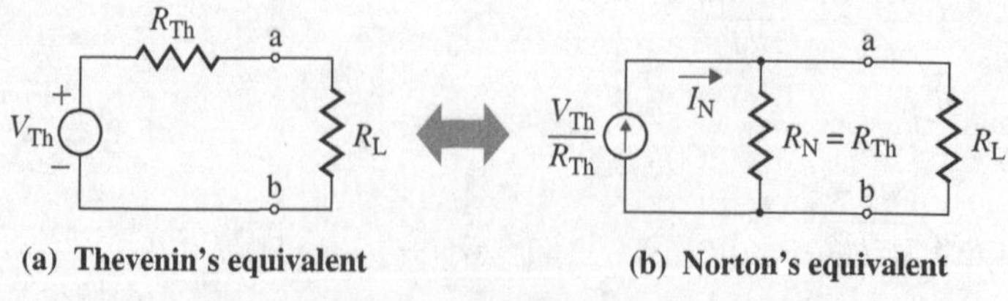

FIGURE B.18 Thevenin's and Norton's equivalent circuits

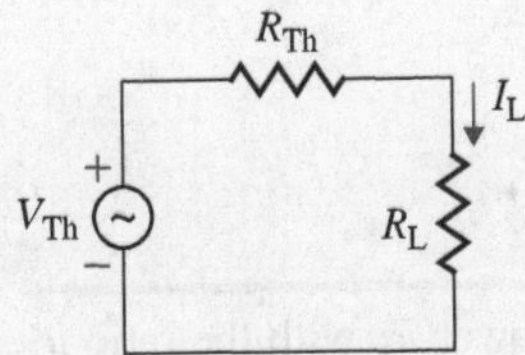

FIGURE B.19 Thevenin's equivalent circuit with a resistive load

B.7 Maximum Power Transfer Theorem

In electronic circuits, it is often necessary to transmit maximum power to the load. Let us consider the circuit in Fig. B.19, which could be Thevenin's equivalent circuit of a network. The power P_L delivered to load resistance R_L can be found from

$$P_L = I_L^2 R_L = \left[\frac{V_{Th}}{R_{Th} + R_L}\right]^2 R_L$$

$$= \frac{V_{Th}^2}{R_{Th}} \times \frac{1}{(1 + R_L/R_{Th})^2} \times \frac{R_L}{R_{Th}} \tag{B.24}$$

For a given circuit, V_{Th} and R_{Th} will be fixed. Therefore, the load power P_L depends on the load resistance R_L. If we set $R_L = uR_{Th}$, Eq. (B.24) becomes

$$P_L = \frac{V_{Th}^2}{R_{Th}} \frac{u}{(1 + u)^2}$$

$$= \frac{u}{(1 + u)^2} P$$

where $P = V_{Th}^2/R_{Th}$. Normalizing P_L with respect to P, we get the normalized power P_n as

$$P_n = \frac{P_L}{P} = \frac{u}{(1 + u)^2} \tag{B.25}$$

Figure B.20 shows the variation of normalized power P_n with u. The power P_n becomes maximum at $u = 1$; that is, $R_{Th} = uR_L = R_L$. The value of R_L for a maximum power transfer can also be determined from the condition $dP_L/dR_L = 0$. From Eq. (B.24),

$$\frac{dP_L}{dR_L} = V_{Th}^2 \left[\frac{(R_{Th} + R_L)^2 - 2R_L(R_{Th} + R_L)}{(R_{Th} + R_L)^4}\right] = 0$$

or

$$(R_{Th} + R_L)^2 - 2R_L(R_{Th} + R_L) = 0$$

$$R_L = \pm R_{Th}$$

Since R_{Th} cannot be negative,

$$R_L = R_{Th} \tag{B.26}$$

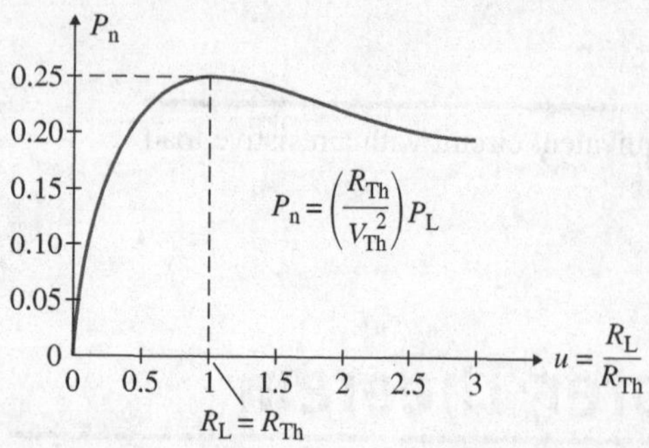

FIGURE B.20 Normalized power P_n with the ratio u

Thus, the maximum power transfer occurs when the load resistance R_L is equal to Thevenin's resistance R_{Th} of the network. For Norton's equivalent circuit of Fig. B.18(b), the maximum power will be delivered to the load when

$$R_N = R_L \tag{B.27}$$

Substituting R_L from Eq. (B.26) into Eq. (B.24) gives the maximum power P_{max} delivered to the load as

$$P_{max} = \frac{V_{Th}^2 R_L}{4R_L^2} = \frac{V_{Th}^2}{4R_L} \tag{B.28}$$

The input power P_{in} supplied by the source V_s is

$$P_{in} = \frac{V_{Th}^2}{R_{Th} + R_L} = \frac{V_{Th}^2}{2R_L} \tag{B.29}$$

Thus, the efficiency η at the maximum power transfer condition is

$$\eta = \frac{P_{max}}{P_{in}} \times 100\% = \frac{V_{Th}^2}{4R_L} \times \frac{2R_L}{V_{Th}^2} \times 100\% = 50\%$$

Therefore, the efficiency is always 50% at the maximum power transfer condition. In electronic circuits, the amount of power being transferred is usually small, and efficiency is often not of primary concern. However, efficiency is of major concern in circuits involving high power—for example, in power systems.

B.8 Transient Response of First-Order Circuits

The transient response gives the instantaneous value of an output voltage (or current) for a specified instantaneous input voltage (or current). The response due to a step-signal input is commonly used in evaluating electronic circuits because such a response allows us to predict the response due to other signals such as a pulse or square-wave input.

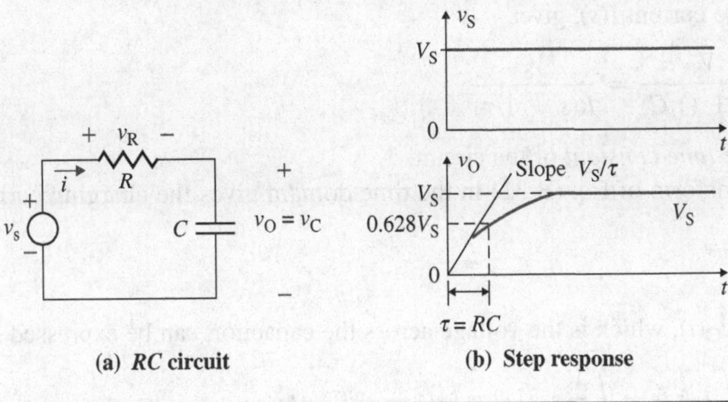

(a) *RC* circuit (b) Step response

FIGURE B.21 Series *RC* circuit

B.8.1 Step Response of Series *RC* Circuits

Consider the series *RC* circuit in Fig. B.21(a) with a step input voltage V_S. The output voltage v_O is taken across capacitor C. For $t \geq 0$, the charging current i of the capacitor can be found from

$$V_S = v_R + v_C = Ri + \frac{1}{C}\int i\,dt + v_C(t = 0) \tag{B.30}$$

with initial capacitor voltage $v_C(t = 0) = 0$.

Using the Laplace transformations in Table B.1, we can transform Eq. (B.30) into Laplace's domain of s as follows:

$$\frac{V_S}{s} = RI(s) + \frac{1}{Cs}I(s)$$

TABLE B.1 Some Laplace transformations

$f(t)$	$F(s)$
1	$\frac{1}{s}$
t	$\frac{1}{s^2}$
$e^{-\alpha t}$	$\frac{1}{s + \alpha}$
$\sin \alpha t$	$\frac{\alpha}{s^2 + \alpha^2}$
$\cos \alpha t$	$\frac{s}{s^2 + \alpha^2}$
$f'(t)$	$sF(s) - F(0)$
$f''(t)$	$s^2F(s) - sF(s) - F'(0)$

which, solved for the current $I(s)$, gives

$$I(s) = \frac{V_S}{sR + (1/C)} = \frac{V_S}{R(s + 1/\tau)} \tag{B.31}$$

where $\tau = RC$ is the *time constant* of the circuit.

The inverse transform of Eq. (B.31) in the time domain gives the charging current:

$$i(t) = \frac{V_S}{R} e^{-t/\tau} \tag{B.32}$$

The output voltage $v_O(t)$, which is the voltage across the capacitor, can be expressed as

$$v_O(t) = \frac{1}{C}\int_0^t i\,dt = \frac{1}{C}\int_0^t \frac{V_S}{R} e^{-t/\tau}\,dt = V_S(1 - e^{-t/\tau}) \tag{B.33}$$

In the steady state (at $t = \infty$), Eq. (B.32) gives

$$i(t = \infty) = 0$$

From Eq. (B.33),

$$v_O(t = \infty) = V_S \tag{B.34}$$

At $t = \tau$, Eq. (B.33) gives

$$v_O(t = \tau) = V_S(1 - e^{-1}) = 0.632 V_S \tag{B.35}$$

The initial slope of the tangent to $v_O(t)$ can be found from Eq. (B.33):

$$\left.\frac{dv_O}{dt}\right|_{t=0} = \left.\frac{V_S}{\tau} e^{-t/\tau}\right|_{t=0} = \frac{V_S}{\tau} = \frac{V_S}{RC} \tag{B.36}$$

The transient response of $v_O(t)$, which rises expontentially due to a step input, is shown in Fig. B.21(b).

B.8.2 Step Response of Series *CR* Circuits

In a *CR* circuit, the output voltage is taken across resistance *R* instead of capacitance *C*, as shown in Fig. B.22(a). The output voltage v_O, which is the voltage across resistance *R*, can be found from Eq. (B.32); that is,

$$v_O(t) = Ri(t) = V_S e^{-t/\tau} \tag{B.37}$$

which, in the steady state (at $t = \infty$), gives

$$i(t = \infty) = 0$$

$$v_O(t = \infty) = 0$$

At $t = \tau$, Eq. (B.37) gives

$$v_O(t = \tau) = V_S e^{-1} = 0.368 V_S \tag{B.38}$$

From Eq. (B.37), the initial slope of the tangent to $v_O(t)$ is

$$\left.\frac{dv_O}{dt}\right|_{t=0} = -\left.\frac{V_S}{\tau} e^{-t/\tau}\right|_{t=0} = -\frac{V_S}{\tau} = -\frac{V_S}{RC} \tag{B.39}$$

The response of $v_O(t)$, which falls exponentially due to a step voltage input, is shown in Fig. B.22(b).

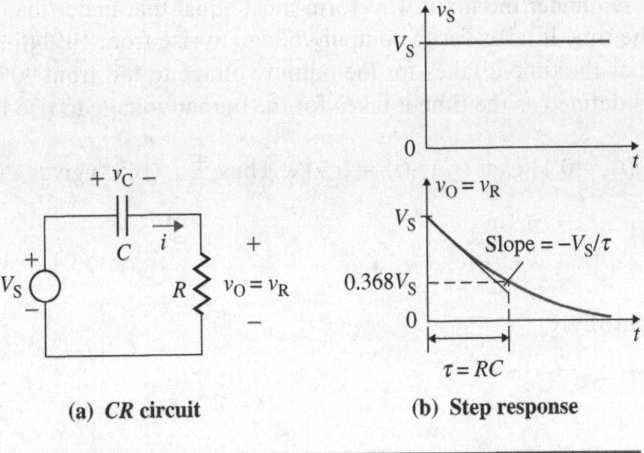

FIGURE B.22 Step response of a series *CR* circuit

B.8.3 Pulse Response of Series *RC* Circuits

An input pulse v_S of duration T, shown in Fig. B.23(a), is applied to the circuit of Fig. B.21(a). The response due to a pulse signal will depend on the ratio of time constant τ to duration T. We consider three cases: $\tau = T$, $\tau << T$, and $\tau >> T$.

In case 1, $\tau = T$, the output voltage $v_O(t)$ has just enough time to reach a near steady-state value of V_S. The capacitor C is charged exponentially to approximately voltage V_S. When the input voltage $v_S(t)$ falls to zero at $t = T$, the output (or capacitor) voltage $v_O(t)$ falls exponentially to zero, as shown

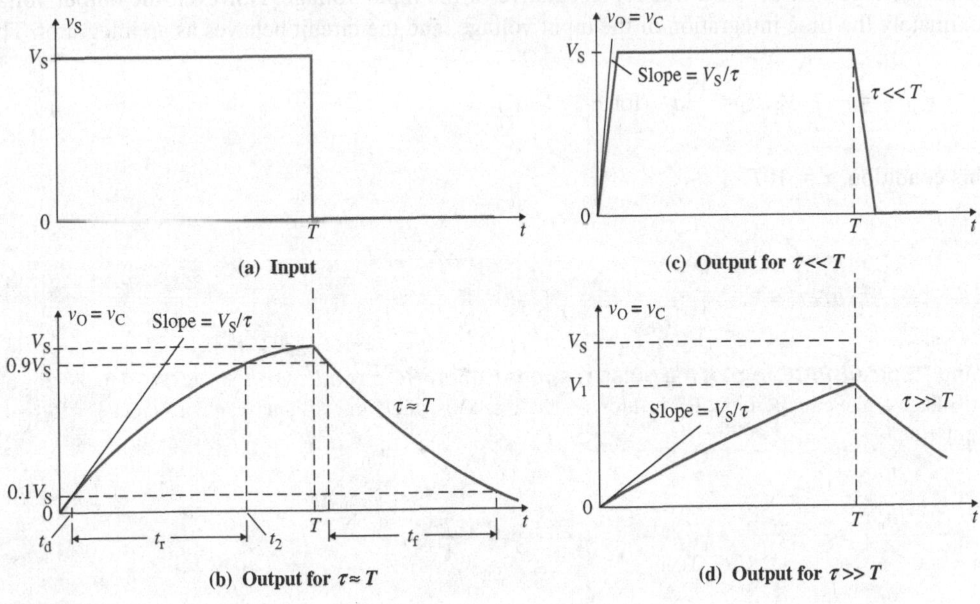

FIGURE B.23 Pulse response of an *RC* circuit

in Fig. B.23(b). The area under the input waveform must equal that under the output waveform. *Rise time* t_r is defined as the time it takes for the output voltage to rise from 10% to 90% of the final value. *Fall time* t_f is defined as the time it takes for the output voltage to fall from 90% to 10% of the initial value. *Delay time* t_d is defined as the time it takes for the output voltage to rise from 0% to 10% of the final value.

At $t = t_1 = t_d$, $v_O(t) = 0.1V_S$; at t_2, $v_O(t) = 0.9V_S$. Thus, Eq. (B.33) gives

$$0.1V_S = V_S(1 - e^{-t_1/\tau})$$
$$e^{-t_1/\tau} = 0.9$$
$$t_1 = -\tau \ln(0.9)$$

and

$$0.9V_S = V_S(1 - e^{-t_2/\tau})$$
$$e^{-t_2/\tau} = 0.1$$
$$t_2 = -\tau \ln(0.1)$$

The rise time t_r, which is equal to the fall time t_f, can be found as follows:

$$\begin{aligned} t_r = t_f &= t_2 - t_1 \\ &= -\tau \ln(0.1) + \tau \ln(0.9) = \tau \ln(9) \approx 2.2\tau \end{aligned} \tag{B.40}$$

In case 2, $\tau << T$, t_r and t_f are much smaller than T. The output voltage $v_O(t)$ represents the input signal more closely, as shown in Fig. B.23(c). This condition is generally satisfied by choosing circuit parameters such that $10\tau = T$.

In case 3, $\tau >> T$, there is not enough time for the output voltage $v_O(t)$ to reach the steady-state value of V_S. The output voltage at $t = T$ is V_1, which is much smaller than V_S, as shown in Fig. B.23(d). The output voltage starts to decay exponentially toward zero before reaching its maximum value. Thus, the output voltage will not be a true representative of the input voltage. However, the output voltage is approximately the time integration of the input voltage, and the circuit behaves as an integrator. That is,

$$v_O(t) = \frac{1}{\tau}\int_0^t V_S\,dt = \frac{V_S}{\tau}t \quad (\text{for } \tau >> T)$$

For this condition, $\tau = 10T$.

EXAMPLE B.10

Using PSpice/SPICE to plot the pulse response of an *RC* circuit Use PSpice/SPICE to plot the output voltage of the circuit in Fig. B.21(a) for $\tau = 0.1$ ms, 1 ms, and 5 ms. Assume $T = 2$ ms and $v_S = V_S = 1$ V of pulse input.

SOLUTION

For $\tau = 0.1$ ms, let $C = 0.1\ \mu$F. Then

$$R = \frac{\tau}{C} = \frac{0.1 \text{ ms}}{0.1\ \mu\text{F}} = 1\ \text{k}\Omega$$

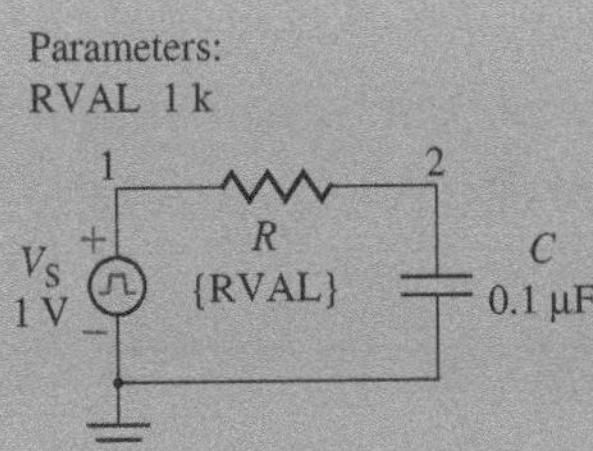

FIGURE B.24 *RC* circuit for PSpice simulation

For $\tau = 1$ ms, let $C = 0.1\ \mu$F. Then

$$R = \frac{\tau}{C} = \frac{1 \text{ ms}}{0.1\ \mu\text{F}} = 10 \text{ k}\Omega$$

For $\tau = 5$ ms, let $C = 0.1\ \mu$F. Then

$$R = \frac{\tau}{C} = \frac{5 \text{ ms}}{0.1\ \mu\text{F}} = 50 \text{ k}\Omega$$

The series *RC* circuit for PSpice simulation is shown in Fig. B.24 with a pulse input voltage. The list of the circuit file, which is run by OrCAD PSpice A/D, is as follows.

```
Example B.10   Pulse Response of Series RC Circuit
.PARAM RVAL=1K
.STEP PARAM RVAL LIST  1K  10K  50K
VS  1  0  PULSE (0V  1V  0  1NS  1NS  2MS  4MS)  ; Pulse input voltage
R  1  2  {RVAL}
C  2  0  0.1UF
.TRAN  0.1MS  4MS                               ; Transient analysis
.PROBE
.END
```

The PSpice plots of the output voltage $v_O(t)$ for three values of the time constant are shown in Fig. B.25. The lower the time constant τ, the faster the output voltage rises and falls.

 **NOTE:** You can use the PSpice Parametric command for variable *R* to vary the time constant.

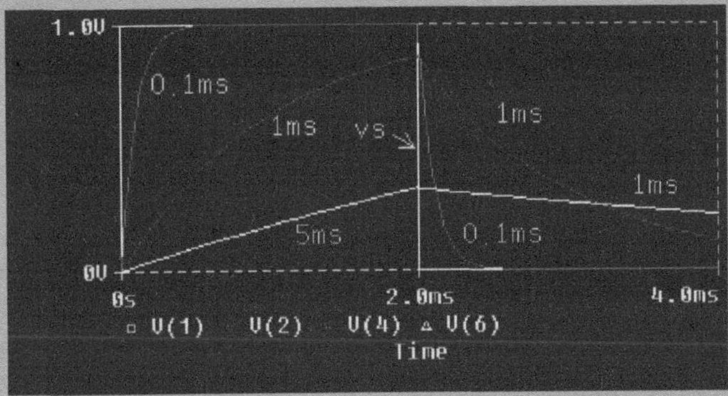

FIGURE B.25 Plots of $v_O(t)$ for Example B.10

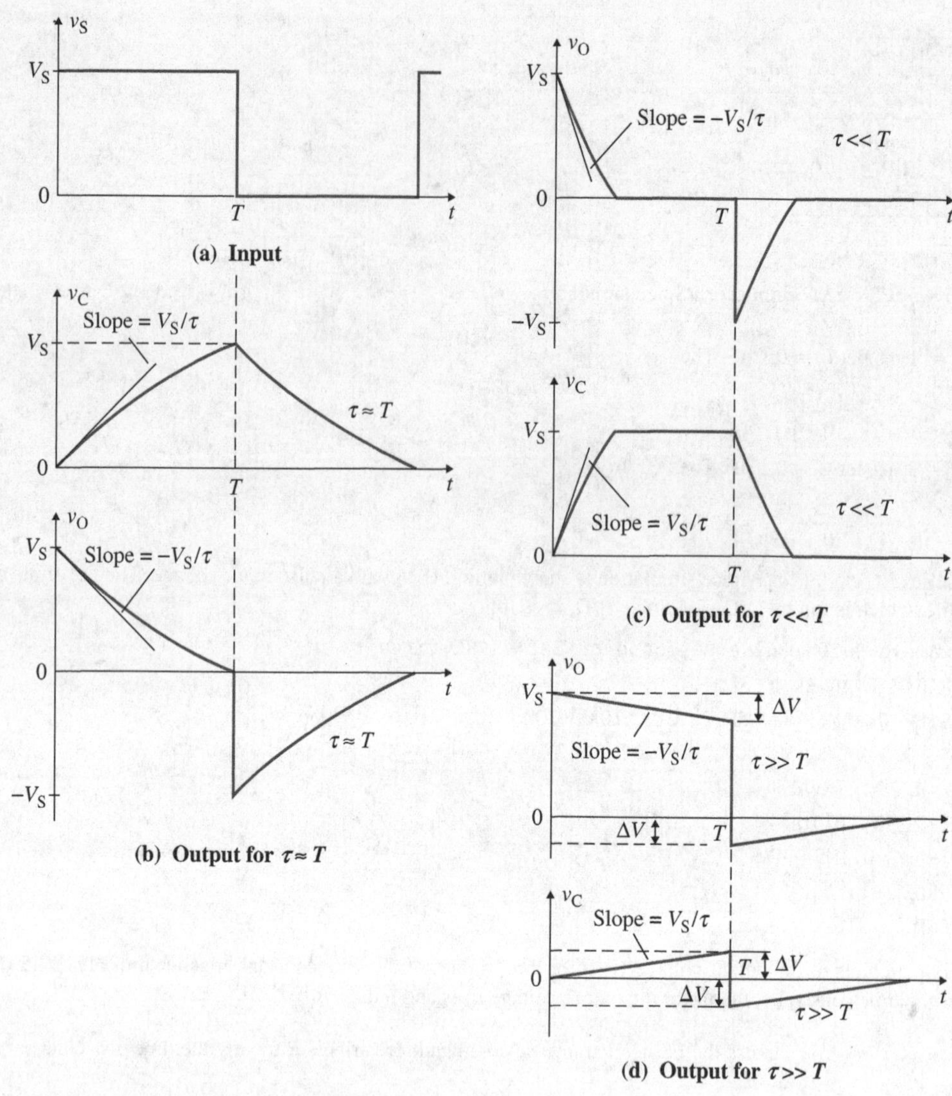

FIGURE B.26 Pulse response of a series *CR* circuit

B.8.4 Pulse Response of Series *CR* Circuits

An input pulse v_S of duration T, shown in Fig. B.26(a), is applied to the circuit of Fig. B.22(a). The response due to a pulse signal will depend on the ratio of time constant τ to duration T. Let us consider three cases: $\tau = T$, $\tau << T$, and $\tau >> T$.

In case 1, $\tau = T$, the capacitor voltage $v_C(t)$ starts to increase exponentially while the output voltage $v_O(t)$ starts to decay exponentially from V_S. This situation is shown in Fig. B.26(b). At $t = T$, the input signal v_S falls to zero and the capacitor discharges exponentially through the resistance R and the input source v_S. The output voltage v_O decays exponentially from a negative value toward zero.

In case 2, $\tau << T$, the output voltage $v_O(t)$ decays exponentially with a small time constant to zero. During the period $0 \leq t \leq T$, the capacitor charges exponentially to reach a steady-state value of V_S. For $t > T$, the capacitor discharges exponentially with a small time constant through the resistance R and the input source v_S. The output voltage $v_O(t)$ decays exponentially from a negative value toward zero. The waveforms for $v_O(t)$ and $v_C(t)$ are shown in Fig. B.26(c).

In case 3, $\tau >> T$, output voltage v_O falls only a small amount. The portion of the exponential curve v_O from $t = 0$ to $t = T$ will be almost linear, as shown in Fig. B.26(d). The fall in the output voltage v_O can be found approximately from Fig. B.26(d) as

$$\Delta V = \frac{V_S}{\tau} T \tag{B.41}$$

The *sag* S of the output voltage is defined as

$$S = \frac{\Delta V}{V_S} = \frac{V_S T/\tau}{V_S} = \frac{T}{\tau} = \frac{T}{RC} \tag{B.42}$$

EXAMPLE B.11

Using PSpice/SPICE to plot the pulse response of a *CR* circuit Use PSpice/SPICE to plot the output voltage $v_O(t)$ of the circuit in Fig. B.22(a) for $\tau = 0.1$ ms, 1 ms, and 5 ms. Assume $T = 2$ ms and $v_S = V_S = 1$ V of pulse input.

SOLUTION

For $\tau = 0.1$ ms, let $C = 0.1$ μF. Then

$$R = \frac{\tau}{C} = \frac{0.1 \text{ ms}}{0.1\ \mu\text{F}} = 1\ \text{k}\Omega$$

For $\tau = 1$ ms, let $C = 0.1$ μF. Then

$$R = \frac{\tau}{C} = \frac{1 \text{ ms}}{0.1\ \mu\text{F}} = 10\ \text{k}\Omega$$

For $\tau = 5$ ms, let $C = 0.1$ μF. Then

$$R = \frac{\tau}{C} = \frac{5 \text{ ms}}{0.1\ \mu\text{F}} = 50\ \text{k}\Omega$$

The series *CR* circuit for PSpice simulation is shown in Fig. B.27.

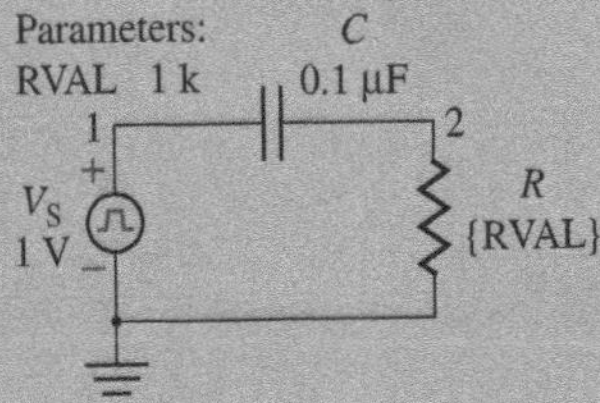

FIGURE B.27 *CR* circuit for PSpice simulation

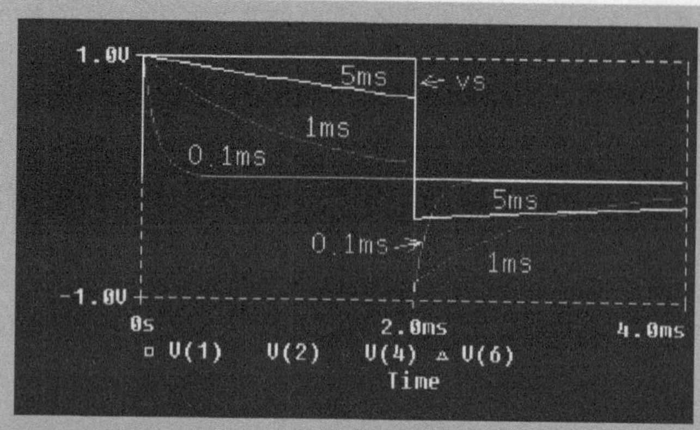

FIGURE B.28 Plots of $v_O(t)$ for Example B.11

The list of the circuit file is as follows.

```
Example B.11  Pulse Response of Series CR Circuit
.PARAM RVAL=1K
.STEP PARAM RVAL LIST  1K  10K  50K
VS  1  0  PULSE (0V  1V  0  1NS  1NS  2MS  4MS)    ; Pulse input voltage
R  2  0  {RVAL}
C  1  2  0.1UF
.TRAN  0.1MS  4MS                                   ; Transient analysis
.PROBE
.END
```

The plots of the output voltage $v_O(t)$ for three values of the time constant are shown in Fig. B.28.

EXAMPLE B.12

Pulse response of a parallel *RC* circuit A constant-current source $i_S = I_S$, shown in Fig. B.29(a), feeds a parallel *RC* circuit with $C = 0.1\ \mu F$ and $R = 100\ k\Omega$, as shown in Fig. B.29(b). The input is a pulse current of duration $T = 0.5$ ms. Determine **(a)** the instantaneous current $i_C(t)$ through capacitance *C*, **(b)** the instantaneous current $i_R(t)$ through resistance *R*, and **(c)** the sag *S* of the capacitor current.

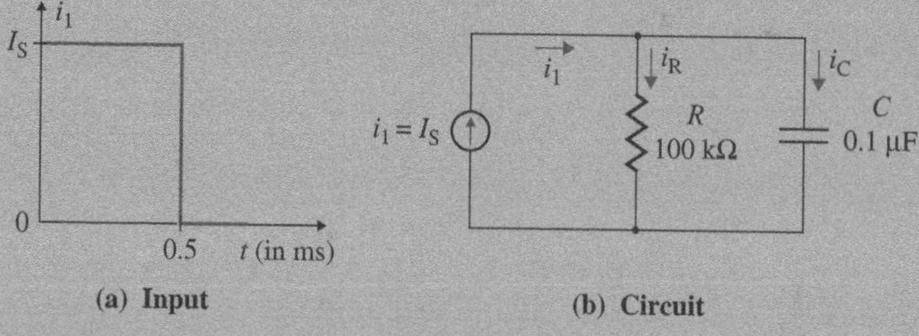

FIGURE B.29 Parallel *RC* circuit with constant current source I_S

SOLUTION

For a pulse signal, the input current in Laplace's domain is $I_S(s) = I_S/s$.

(a) Using the current divider rule, we can find the capacitor current I_C in Laplace's domain:

$$I_C(s) = \frac{R}{R + 1/Cs} I_S(s) = \frac{s}{s + 1/RC} I_S(s) = \frac{s}{s + 1/RC} \times \frac{I_S}{s} = \frac{I_S}{s + 1/RC}$$

$$= I_S \frac{I}{s + 1/RC} \tag{B.43}$$

The inverse transform of $I_C(s)$ in Eq. (B.43) gives

$$i_C(t) = I_S e^{-t/\tau} \tag{B.44}$$

where $\tau = RC$.

(b) The instantaneous current $i_R(t)$ through resistance R is

$$i_R(t) = I_S - i_C(t) = I_S(1 - e^{-t/\tau}) \tag{B.45}$$

(c) $\tau = RC = 100 \times 10^3 \times 0.1 \times 10^{-6} = 10$ ms, and $T = 0.5$ ms. Therefore, $\tau >> T$, and Eq. (B.42) gives

$$S = T/\tau = 0.5/10 = 5\%$$

B.8.5 Step Response of Series *RL* Circuits

A series *RL* circuit with a step-function input is shown in Fig. B.30(a). The output voltage v_O is taken across inductance *L*. The current *i* through the inductor can be deduced from

$$V_S = v_L + v_R = L\frac{di}{dt} + Ri \tag{B.46}$$

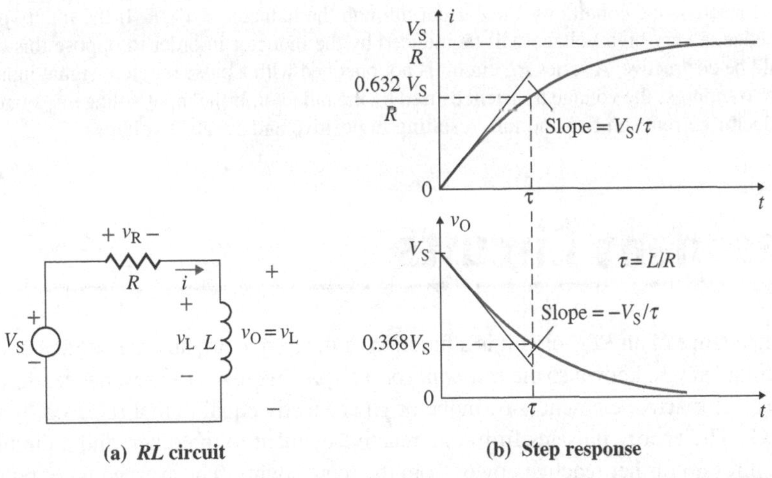

FIGURE B.30 Step response of a series *RL* circuit

with an initial inductor current of zero, $i(t = 0) = 0$. In Laplace's domain of s, Eq. (B.46) becomes

$$\frac{V_S}{s} = LsI(s) + RI(s)$$

which, solved for $I(s)$, gives

$$I(s) = \frac{V_S}{s(sL + R)} = \frac{V_S}{Ls(s + 1/\tau)} = \frac{V_S}{R}\left[\frac{1}{s} - \frac{1}{(s + 1/\tau)}\right] \tag{B.47}$$

where $\tau = L/R$ is the *time constant* of an RL circuit. Taking the inverse transform of $I(s)$ in Eq. (B.47) gives the instantaneous current as

$$i(t) = \frac{V_S}{R}(1 - e^{-t/\tau}) \tag{B.48}$$

Using Eq. (B.48), we can find the voltage $v_O(t)$ across the inductance L:

$$v_O(t) = v_L(t) = L\frac{di}{dt} = V_S e^{-t/\tau} \tag{B.49}$$

In the steady state (at $t = \infty$),

$$v_O(t) = 0 \qquad \text{[from Eq. (B.49)]}$$

$$i(t) = \frac{V_S}{R} \qquad \text{[from Eq. (B.48)]}$$

If the output is taken across resistance R, the output voltage $v_O(t)$ becomes

$$v_O(t) = v_R(t) = Ri(t) = V_S(1 - e^{-t/\tau}) \tag{B.50}$$

In the steady state (at $t = \infty$), $v_R(t) = V_S$ and $i(t) = V_S/R$.

▶ **NOTE** Under steady-state conditions, the current through the inductor is V_S/R. If the input voltage v_S is turned off with a sharp edge, a very high voltage will be induced by the inductor in order to oppose this change of current. This voltage could be destructive. A series RL circuit is not operated with a pulse (or step) signal input unless there is a protection circuit to suppress the voltage transient caused by the inductor. If the input voltage v_S switches between $+V_S$ and $-V_S$, the inductor current can rise and fall, resulting in positive and negative voltages.

B.9 Resonant Circuits

The effective impedance of an RLC circuit is a function of the frequency, and the voltage or current becomes maximum at a frequency f_n, known as the *resonant* (or *natural*) *frequency*. At resonance, the energy absorbed at any instant by one reactive element (say, inductor L) is exactly equal to that released by another element (say, capacitor C). The energy pulsates from one reactive element to the other, and a circuit with no resistive element requires no further reactive power from the input source. The average input power, which is the power dissipated in the resistive element, becomes maximum at resonance. Resonant circuits are of two types: series resonant circuits and parallel resonant circuits.

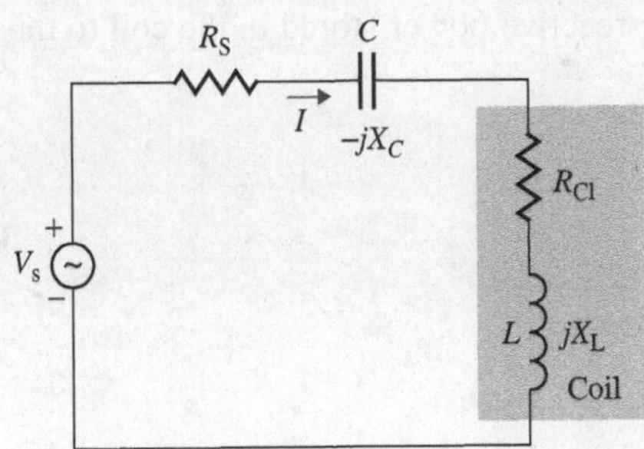

FIGURE B.31 Series resonant RLC circuit

B.9.1 Series Resonant Circuits

A series resonant *RLC* circuit is shown in Fig. B.31, where R_{Cl} is the internal resistance of the coil and R_S is the source resistance. If we define $R = R_{Cl} + R_S$, the total series impedance Z of the circuit is given by

$$Z = R + j(X_L - X_C) \tag{B.51}$$

The series resonance will occur at $f = f_n$ when

$$X_L = X_C \tag{B.52}$$

Equation (B.52) can also be stated as

$$\omega L = \frac{1}{\omega C} \quad \text{or} \quad 2\pi f_n L = \frac{1}{(2\pi f_n C)}$$

which gives the series resonant frequency f_n as

$$f_n = \frac{1}{2\pi\sqrt{LC}} \tag{B.53}$$

The impedance Z_n at resonance becomes

$$Z_n = Z = R \tag{B.54}$$

A series resonant circuit is normally defined by a *quality factor* Q_s, which is defined as the ratio of the reactive power stored in either the inductor or the capacitor to the average power dissipated in the resistor at resonance; that is,

$$Q_s = \frac{\text{Reactive power}}{\text{Average power}}$$

$$= \begin{cases} \dfrac{I^2 X_L}{I^2 R} = \dfrac{X_L}{R} = \dfrac{2\pi f_n L}{R} & \text{(for an inductive reactance)} \quad \text{(B.55)} \\[2ex] \dfrac{I^2 X_C}{I^2 R} = \dfrac{X_C}{R} = \dfrac{1}{2\pi f_n CR} & \text{(for a capacitive reactance)} \quad \text{(B.56)} \end{cases}$$

The quality factor Q_{C1} of a coil is defined as the ratio of the reactive power stored in the coil to the power dissipated in the coil resistance R_{C1}; that is,

$$Q_{C1} = \frac{\text{Reactive power}}{\text{Power dissipated}} = \frac{X_L}{R_{C1}}$$

The rms voltage V_L across inductor L at resonance can be found from

$$V_L = \frac{X_L V_s}{Z_n} = \frac{X_L V_s}{R} = Q_s V_s \tag{B.57}$$

The rms voltage V_C across capacitor C at resonance can be found from

$$V_C = \frac{X_C V_s}{Z_n} = \frac{X_L V_s}{R} = Q_s V_s \tag{B.58}$$

In many electronic circuits, the quality factor Q_s is high, in the range of 80 to 400. If $V_s = 30$ V and $Q_s = 80$, for example, then $V_C = V_L = 80 \times 30 = 2400$ V, and all electronic devices in the circuit will be subjected to this high voltage. Thus, a designer must be careful to protect the circuit from a high voltage across the inductor or the capacitor of a resonant circuit.

B.9.2 Parallel Resonant Circuits

A parallel resonant *RLC* circuit is shown in Fig. B.32(a). This circuit is also known as a *tank circuit*. The input signal to a tank circuit is usually a current source. This type of circuit is frequently used with active devices such as transistors, which have the characteristic of a constant current source. By replacing the series *RL* combination of the inductor with a parallel combination, we can obtain the circuit in Fig. B.32(b), for which the admittance Y_{RL} is

$$Y_{RL} = \frac{1}{R_{C1} + jX_L} = \frac{R_{C1}}{R_{C1}^2 + X_L^2} - j\frac{X_L}{R_{C1}^2 + X_L^2} = \frac{1}{R_p} - j\frac{1}{X_p}$$

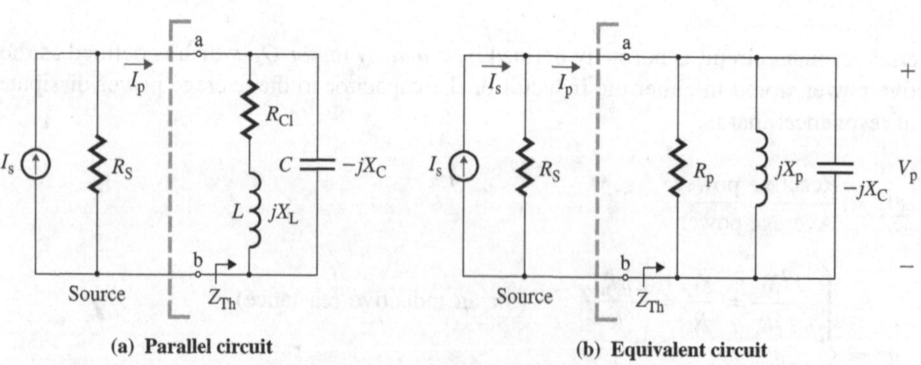

FIGURE B.32 Parallel resonant *RLC* circuit

$$\text{where} \quad R_p = \frac{R_{C1}^2 + X_L^2}{R_{C1}} \tag{B.59}$$

$$X_p = \frac{R_{C1}^2 + X_L^2}{X_L} \tag{B.60}$$

For the resonant condition,

$$X_p = X_C$$

Substituting X_p from Eq. (B.60) into the preceding equation, we get

$$\frac{R_{C1}^2 + X_L^2}{X_L} = X_C$$

$$R_{C1}^2 + X_L^2 = X_C X_L$$

$$X_L^2 = X_C X_L - R_{C1}^2$$

or

$$X_L^2 = \frac{L}{C} - R_{C1}^2$$

$$X_L = \left[\frac{L}{C} - R_{C1}^2\right]^{1/2} \tag{B.61}$$

which gives the parallel resonant frequency f_p as

$$f_p = \frac{1}{2\pi L}\left[\frac{L}{C} - R_{C1}^2\right]^{1/2} = \frac{1}{2\pi\sqrt{LC}}\left[1 - \frac{CR_{C1}^2}{L}\right]^{1/2} \tag{B.62}$$

$$= f_n\left[1 - \frac{CR_{C1}^2}{L}\right]^{1/2} \tag{B.63}$$

Thus, the parallel resonant frequency f_p, which is dependent on the coil resistance R_{C1}, is less than the series resonant frequency f_n. For the conditions $(CR_{C1}^2/L) << 1$ or $R_{C1} << \sqrt{L/C}$, and for $R_{C1} = 0$, Eq. (B.63) gives

$$f_p = f_n$$

The quality factor Q_p of the parallel resonant *RLC* circuit can be determined by the ratio of reactive power to the real power at resonance; that is,

$$Q_p = \frac{V_p^2/X_p}{V_p^2/(R_S \parallel R_p)} = \frac{R_S \parallel R_p}{X_p} \tag{B.64}$$

where V_p is the voltage across the parallel branches.

EXAMPLE B.13

Finding the parallel resonant frequency The parameters of the parallel resonant *RLC* circuit in Fig. B.32(a) are $R_{\text{Cl}} = 47\ \Omega$, $L = 5$ mH, $C = 50$ pF, $R_S = 20$ kΩ, and the current source $I_s = 6$ mA. Calculate **(a)** the parallel resonant frequency f_p, **(b)** the voltage V_p across the resonant circuit at resonance, **(c)** the quality factor Q_{Cl} of the coil, and **(d)** the quality factor Q_p of the resonant circuit.

SOLUTION

$R_{\text{Cl}} = 47\ \Omega$, $L = 5$ mH, $C = 50$ pF, $R_S = 20$ kΩ, and $I_s = 6$ mA.

(a) From Eq. (B.53),

$$f_n = \frac{1}{2\pi \times \sqrt{5 \times 10^{-3} \times 50 \times 10^{-12}}} = 318.3 \text{ kHz}$$

From Eq. (B.63),

$$f_p = 318.3 \times 10^3 \times \left[1 - 50 \times 10^{-12} \times \frac{47^2}{5 \times 10^{-3}}\right]^{1/2} \approx 318.3 \text{ kHz}$$

(b) We know that

$$X_L = 2\pi f_p L = 2\pi \times 318.3 \times 10^3 \times 5 \times 10^{-3} = 9999.7\ \Omega$$

From Eq. (B.59), the effective resistance R_p of the parallel circuit is

$$R_p = \frac{47^2 + 9999.7^2}{47} = 2127.6 \text{ k}\Omega$$

Using the current divider rule, the rms current I_p through the parallel circuit is

$$I_p = \frac{R_S}{R_S + R_p} I_s = \frac{20 \text{ k}\Omega \times 6 \text{ mA}}{20 \text{ k}\Omega + 2127.6 \text{ k}\Omega} = 55.876\ \mu\text{A}$$

$$V_p = I_p R_p = 55.876\ \mu\text{A} \times 2127.6 \text{ k}\Omega = 118.88 \text{ V}$$

(c) We have

$$Q_{\text{Cl}} = \frac{X_L}{R_{\text{Cl}}} = \frac{9999.7}{47} = 212.8$$

(d) From Eq. (B.60), the effective inductive reactance X_p of the parallel circuit is

$$X_p = \frac{47^2 + 9999.7^2}{9999.7} = 9999.92\ \Omega$$

$$R_S \| R_p = \frac{20 \times 2127.6}{20 + 2127.6} = 19.81 \text{ k}\Omega$$

From Eq. (B.64), $Q_p = 19{,}810/9999.92 = 1.98$.

B.10 Frequency Response of First- and Second-Order Circuits

A sine wave (or sinusoid) is generally used to characterize electronic circuits, such as amplifiers and filters. The frequency response refers to the output characteristic for a sine-wave input. If a sine-wave input voltage

$$v_s(t) = V_m \sin \omega t \tag{B.65}$$

where V_m is the peak input voltage and ω is the frequency of the input voltage in radians per second, is applied to a circuit, the output voltage $v_o(t)$ can have a different amplitude and phase than the input voltage. The output voltage $v_o(t)$ will be of the form

$$v_o(t) = V_p \sin(\omega t - \phi) \tag{B.66}$$

where V_p is the peak output voltage. If f is the input frequency in hertz,

$$\omega = 2\pi f$$

A typical relationship between the input and output voltages of an amplifier is shown in Fig. B.33. An amplifier has in general a voltage gain of $A_v = V_p/V_m > 1$. Thus, the peak output V_p has a higher value than the input voltage V_m.

If $V_s(j\omega)$ and $V_o(j\omega)$ denote the rms values of input voltage and output voltage, respectively, as a function of frequency, the voltage gain $G(j\omega)$ is defined as

$$G(j\omega) = \frac{V_o(j\omega)}{V_s(j\omega)} \quad \text{(dimensionless)} \tag{B.67}$$

$G(j\omega)$ is a complex function with a magnitude and a phase. The magnitude $|G(j\omega)|$ gives the magnitude response, and the phase of $G(j\omega)$ gives the phase response. The magnitude and the phase are generally plotted against the frequency, with a logarithmic scale used for the frequency. The magnitude $|G(j\omega)|$ is normally expressed in decibels (dB):

$$\text{Magnitude in dB} = 20 \log_{10} |G(j\omega)|$$

We will review the frequency responses of the following circuits: first-order low-pass *RC* circuits, first-order high-pass *CR* circuits, second-order series *RLC* circuits, and second-order parallel *RLC* circuits.

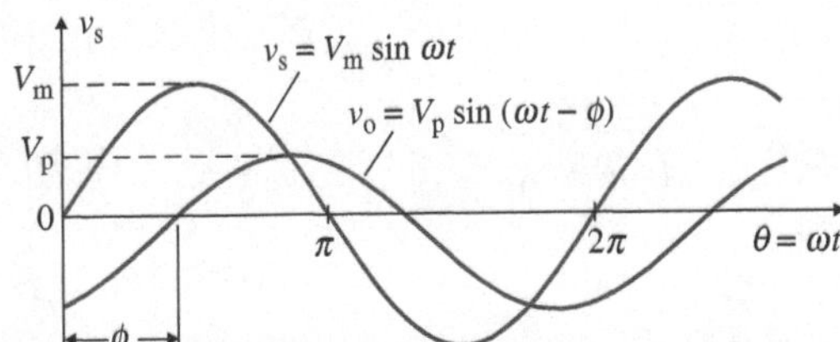

FIGURE B.33 Typical sinusoidal input and output voltages

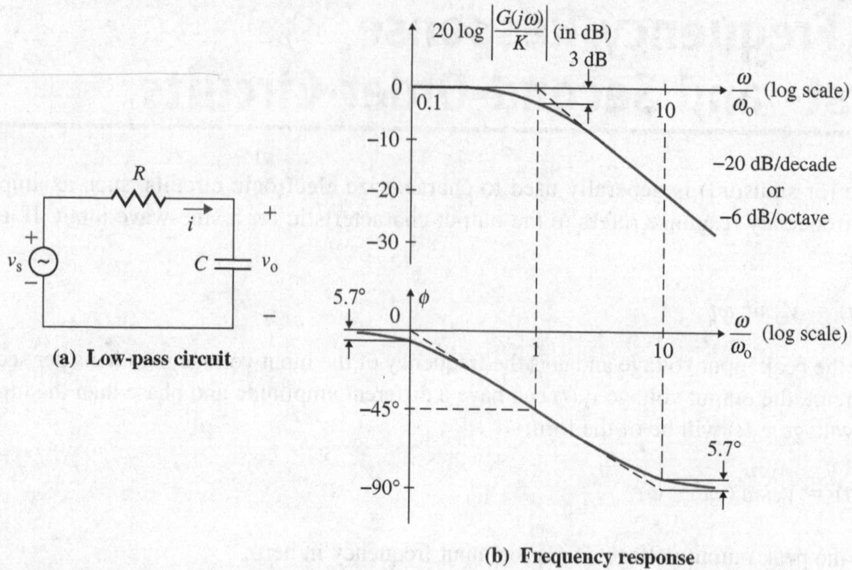

FIGURE B.34 First-order low-pass RC circuit

B.10.1 First-Order Low-Pass *RC* Circuits

A typical low-pass RC circuit is shown in Fig. B.34(a). The output voltage v_o is taken across capacitance C. The capacitor impedance in Laplace's domain is $1/Cs$. Using the voltage divider rule, we can find the voltage gain $G(s)$:

$$G(s) = \frac{V_o(s)}{V_s(s)} = \frac{1/Cs}{R + (1/Cs)} = \frac{1}{1 + sRC} \quad \text{(dimensionless)}$$

In the frequency domain, $s = j\omega$, and

$$G(j\omega) = \frac{1}{1 + j\omega RC} = \frac{1}{1 + j\omega\tau}$$

where $\tau = RC$. Thus, the magnitude $|G(j\omega)|$ of the voltage gain can be found from

$$|G(j\omega)| = \frac{1}{[1 + (\omega\tau)^2]^{1/2}} = \frac{1}{[1 + (\omega/\omega_o)^2]^{1/2}} \tag{B.68}$$

and the phase angle ϕ of $G(j\omega)$ is given by

$$\phi = -\tan^{-1}(\omega\tau) = -\tan^{-1}\left(\frac{\omega}{\omega_o}\right) \tag{B.69}$$

where $\omega_o = 1/RC = 1/\tau$.

For $\omega << \omega_o$,

$$|G(j\omega)| \approx 1$$

$$20 \log_{10}|G(j\omega)| = 0$$

$$\phi = 0$$

Therefore, at low frequency, the magnitude plot is a straight horizontal line at 0 dB. For $\omega >> \omega_o$,

$$|G(j\omega)| \approx \omega_o/\omega$$

$$20 \log_{10}|G(j\omega)| = 20 \log_{10}(\omega_o/\omega)$$

$$\phi = \pi/2$$

For $\omega = \omega_o$,

$$|G(j\omega)| = 1/\sqrt{2}$$

$$20 \log_{10}|G(j\omega)| = 20 \log_{10}(1/\sqrt{2}) = -3 \text{ dB}$$

$$\phi = \pi/4$$

Let us consider a high frequency such that $\omega_1 >> \omega_o$. At $\omega = \omega_1$, the magnitude is $20 \log_{10}(\omega_o/\omega_1)$. At $\omega = 10\omega_1$, the magnitude is $20 \log_{10}(\omega_o/10\omega_1)$. The change in magnitude between $\omega = \omega_1$ and $\omega = 10\omega_1$ becomes

$$20 \log_{10}(\omega_o/10\omega_1) - 20 \log_{10}(\omega_o/\omega_1) = 20 \log_{10}(1/10) = -20 \text{ dB}$$

If the frequency is doubled so that $\omega = 2\omega_1$, the change in magnitude becomes

$$20 \log_{10}(\omega_o/2\omega_1) - 20 \log_{10}(\omega_o/\omega_1) = 20 \log_{10}(1/2) = -6 \text{ dB}$$

The frequency response is shown in Fig. B.34(b). If the frequency doubled, the interval between the two frequencies is called an *octave* on the frequency axis. If the frequency is increased by a factor of 10, the interval between the two frequencies is called a *decade*. Thus, for a decade increase in frequency, the magnitude changes by −20 dB. The magnitude plot is a straight line with a slope of −20 dB/decade or −6 dB/octave. The magnitude curve is therefore defined by two straight-line asymptotes, which meet at the *corner frequency* (or *break frequency*) ω_o. The difference between the actual magnitude curve and the asymptotic curve is largest at the break frequency. The error can be found by finding the gain at $\omega = \omega_o$; that is, $|G(j\omega)| = 1/\sqrt{2}$, and $20 \log_{10}(1/\sqrt{2}) = -3$ dB. This error is symmetrical with respect to the break frequency. The break frequency is also known as the *3-dB frequency*.

The circuit in Fig. B.34(a) passes only the low-frequency signal, and the amplitude falls at higher frequencies. A circuit with this type of response is known as a *low-pass circuit*. The gain function (commonly known as the transfer function) of a low-pass circuit has the general form

$$G(s) = \frac{K}{1 + (s/\omega_o)} \tag{B.70}$$

where K is the magnitude of the gain function at $\omega = 0$ (or the *low-pass gain*). A low-pass circuit exhibits the characteristics of (a) finite output at a very low frequency, tending to zero, and (b) zero output at a very high frequency, tending to infinity.

EXAMPLE B.14

Using PSpice/SPICE to plot the frequency response of a low-pass *RC* circuit Use PSpice/SPICE to plot the frequency response of the low-pass RC circuit in Fig. B.34(a). Assume $V_m = 1$ V (peak AC), $R = 10$ kΩ, and $C = 0.1$ μF. The frequency f varies from 1 Hz to 100 kHz.

SOLUTION

The low-pass RC circuit for PSpice simulation is shown in Fig. B.35. The PSpice plot of the magnitude and phase angle are shown in Fig. B.36, which gives $f_o = 161$ Hz at -3 dB.

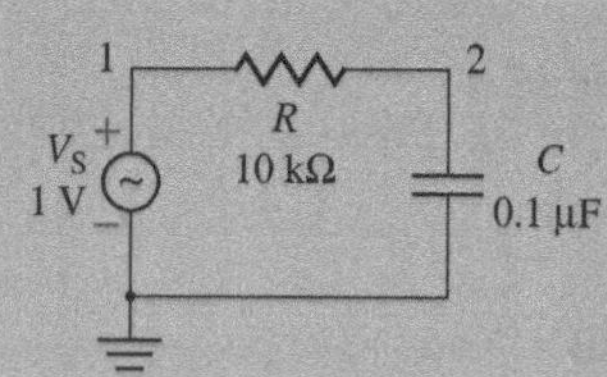

FIGURE B.35 Low-pass *RC* circuit for PSpice simulation

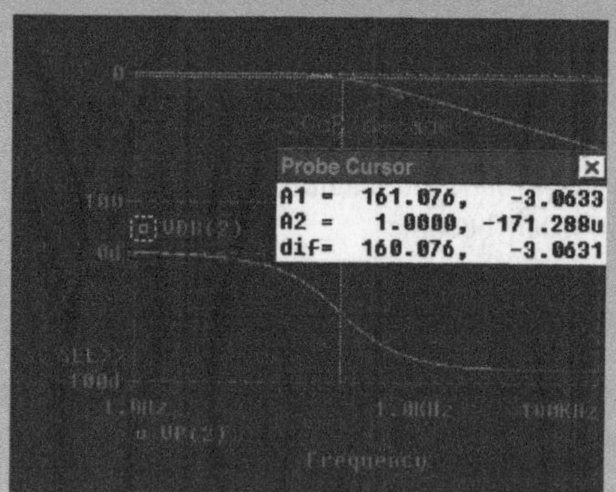

FIGURE B.36 Frequency response plots for Example B.14

B.10.2 First-Order High-Pass *CR* Circuits

A high-pass CR circuit is shown in Fig. B.37(a). The output voltage v_o is taken across resistance R. Using the voltage divider rule, we can find the voltage gain $G(s)$ in Laplace's domain:

$$G(s) = \frac{V_o(s)}{V_s(s)} = \frac{R}{R + (1/Cs)} = \frac{sRC}{1 + sRC} \quad \text{(dimensionless)}$$

In the frequency domain, $s = j\omega$, and

$$G(j\omega) = \frac{j\omega RC}{1 + j\omega RC} = \frac{j\omega\tau}{1 + j\omega\tau}$$

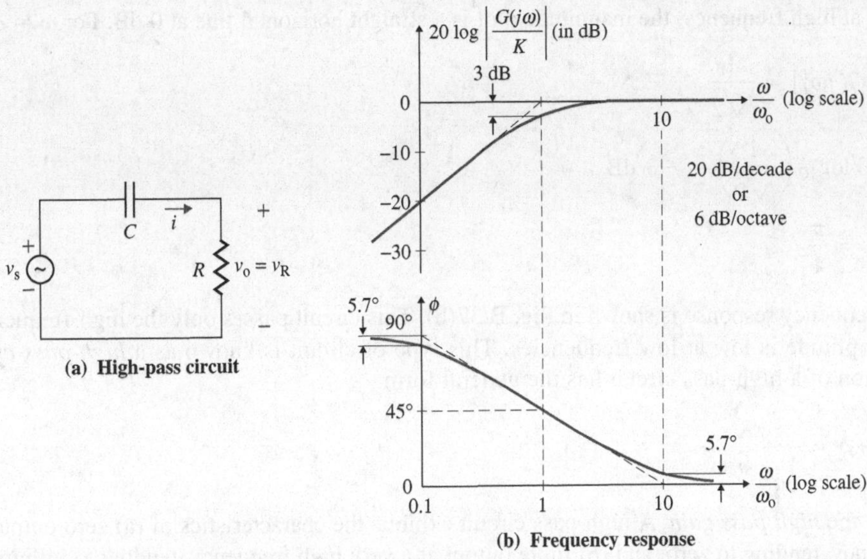

FIGURE B.37 First-order high-pass *CR* circuit

where $\tau = RC$. Thus, the magnitude $|G(j\omega)|$ of the voltage gain can be found from

$$|G(j\omega)| = \frac{\omega\tau}{[1 + (\omega\tau)^2]^{1/2}} = \frac{\omega/\omega_o}{[1 + (\omega/\omega_o)^2]^{1/2}} \tag{B.71}$$

and the phase angle ϕ of $G(j\omega)$ is given by

$$\phi = \frac{\pi}{2} - \tan^{-1}\left(\frac{\omega}{\omega_o}\right) \tag{B.72}$$

where $\omega_o = 1/RC = 1/\tau$.

For $\omega << \omega_o$,

$$|G(j\omega)| \approx \frac{\omega}{\omega_o}$$

$$20 \log_{10}|G(j\omega)| = 20 \log_{10}\left(\frac{\omega}{\omega_o}\right)$$

$$\phi = \frac{\pi}{2}$$

Therefore, for a decade increase in frequency, the magnitude changes by 20 dB. The magnitude plot is a straight line with a slope of +20 dB/decade or +6 dB/octave.

For $\omega >> \omega_o$,

$$|G(j\omega)| = 1$$

$$20 \log_{10}|G(j\omega)| = 0$$

$$\phi \approx 0$$

Therefore, at high frequency, the magnitude plot is a straight horizontal line at 0 dB. For $\omega = \omega_o$,

$$|G(j\omega)| = \frac{1}{\sqrt{2}}$$

$$20 \log_{10}\left(\frac{1}{\sqrt{2}}\right) = -3 \text{ dB}$$

$$\phi = \frac{\pi}{4}$$

The frequency response is shown in Fig. B.37(b). This circuit passes only the high-frequency signal, and the amplitude is low at low frequencies. This type of circuit is known as a *high-pass circuit*. The gain function of a high-pass circuit has the general form

$$G(s) = \frac{sK}{1 + s/\omega_o} \qquad \text{(B.73)}$$

where K is the *high-pass gain*. A high-pass circuit exhibits the characteristics of (a) zero output at a very low frequency, tending to zero, and (b) finite output at a very high frequency, tending to infinity.

EXAMPLE B.15

Using PSpice/SPICE to plot the frequency response of a high-pass *CR* circuit Use PSpice/SPICE to plot the frequency response of the high-pass *CR* circuit in Fig. B.37(a). Assume $V_m = 1$ V (peak AC), $R = 10$ kΩ, and $C = 0.1$ μF. The frequency f varies from 1 Hz to 100 kHz.

SOLUTION

The high-pass *CR* circuit for PSpice simulation is shown in Fig. B.38. The PSpice plots of the magnitude and phase angle are shown in Fig. B.39, which gives $f_o = 157$ Hz at −3 dB.

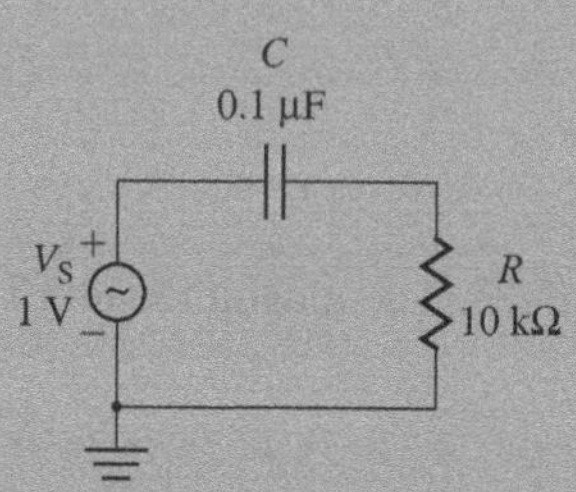

FIGURE B.38 High-pass *CR* circuit for PSpice simulation

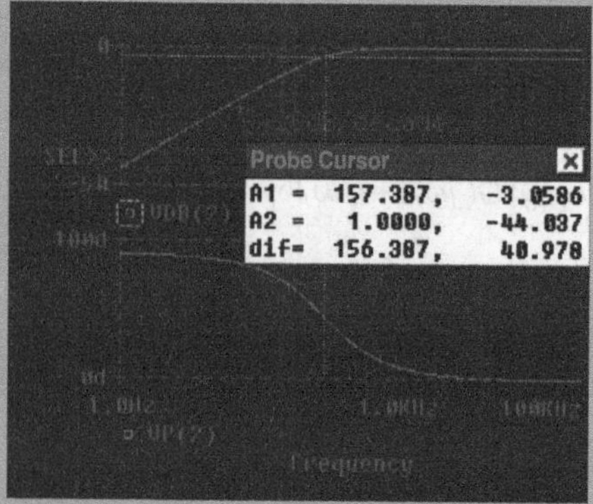

FIGURE B.39 Frequency response plots for Example B.15

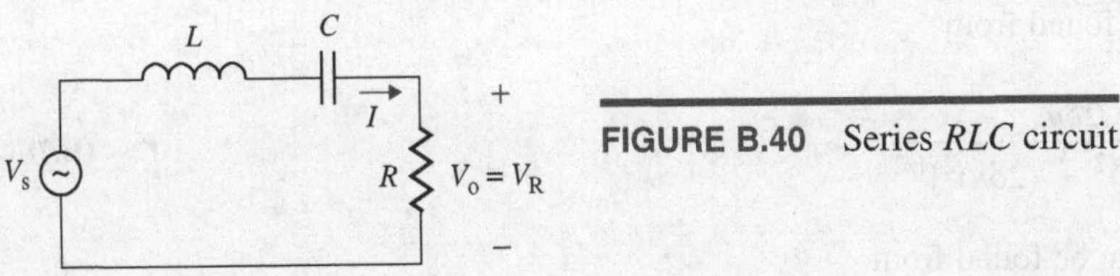

FIGURE B.40 Series *RLC* circuit

B.10.3 Second-Order Series *RLC* Circuits

A series *RLC* circuit is shown in Fig. B.40. The output voltage v_o is taken across resistance *R*. Using the voltage divider rule, we can find the voltage gain (or the transfer function) in Laplace's domain of *s*:

$$G(s) = \frac{V_o(s)}{V_s(s)} = \frac{R}{R + sL + (1/Cs)} = \frac{sR/L}{s^2 + (sR/L) + (1/LC)} \quad \text{(dimensionless)} \tag{B.74}$$

Defining $\omega_n = 1/\sqrt{LC}$ as the *natural frequency* in rad/s and $\alpha = R/(2L)$ as the *damping factor*, we can write Eq. (B.74) as

$$G(s) = \frac{2\alpha s}{s^2 + 2\alpha s + \omega_n^2} \tag{B.75}$$

Let us define

$$\delta = \frac{\alpha}{\omega_n} = \frac{R}{2L}\sqrt{LC} = \frac{R}{2}\sqrt{\frac{C}{L}}$$

as the *damping ratio*. Then Eq. (B.75) becomes

$$G(s) = \frac{2\delta\omega_n s}{s^2 + 2\delta\omega_n s + \omega_n^2} \quad \text{(dimensionless)} \tag{B.76}$$

where $\delta < 1$. (Note that δ is not necessarily less than 1, but has been assumed at less than 1 for this discussion.) In the frequency domain, $s = j\omega$. Thus,

$$G(j\omega) = \frac{2\delta\omega_n j\omega}{(j\omega)^2 + 2\delta\omega_n(j\omega) + \omega_n^2} = \frac{j2\delta\omega/\omega_n}{-(\omega/\omega_n)^2 + (j2\delta\omega/\omega_n) + 1}$$
$$= \frac{j2\delta\omega/\omega_n}{1 + (j2\delta\omega/\omega_n) - (\omega/\omega_n)^2} \tag{B.77}$$

Let us define $u = \omega/\omega_n$ as the *frequency ratio* or the *normalized frequency*. Then Eq. (B.77) can be simplified to

$$G(j\omega) = \frac{j2\delta u}{1 + j2\delta u - u^2}$$

The magnitude $|G(j\omega)|$ can be found from

$$|G(j\omega)| = \frac{2\delta u}{[(1-u^2)^2 + (2\delta u)^2]^{1/2}} \tag{B.78}$$

The phase angle ϕ of $G(j\omega)$ can be found from

$$\phi = \frac{\pi}{2} - \tan^{-1}\left(\frac{2\delta u}{1-u^2}\right) \tag{B.79}$$

For low frequencies $u << 1$,

$$|G(j\omega)| \approx 2\delta u$$

$$20 \log_{10}|G(j\omega)| \approx 20 \log_{10} (2\delta u)$$

$$\phi \approx \frac{\pi}{2}$$

Therefore, at low frequencies, the magnitude plot is a straight line with a slope of +20 dB/decade or +6 dB/octave. For $u = 1$, $|G(j\omega)| \approx 1$ only if

$$\delta \approx 1$$

$$20 \log_{10}|G(j\omega)| = 0 \text{ dB}$$

$$\phi = 0$$

For $u >> 1$,

$$|G(j\omega)| \approx \frac{2\delta u}{u^2} = \frac{2\delta}{u}$$

$$20 \log_{10}|G(j\omega)| \approx 20 \log_{10} (2\delta) - 20 \log_{10} (u) \approx -20 \log_{10} (u)$$

$$\phi \approx -\frac{\pi}{2}$$

Therefore, at high frequencies, the magnitude plot is a straight line with a slope of −20 dB/decade or −6 dB/octave. The actual characteristic will differ considerably from the asymptotic lines, and the error will depend on the damping ratio δ. The magnitude and frequency plots of the series *RLC* circuit are shown in Fig. B.41. The magnitude peaks at $u = 1$.

If the output voltage of the series *RLC* circuit falls below 70% of its maximum value, the output is not considered the significant value. The *cutoff frequency* is defined as that value of frequency for which the magnitude of the gain drops to 70.7% of its maximum value $|G(j\omega)|_{max} = 1$. Thus, at cutoff frequencies, Eq. (B.78) gives

$$|G(j\omega)| = \frac{2\delta u}{[(1-u^2)^2 + (2\delta u)^2]^{1/2}} = 0.707 = \frac{1}{\sqrt{2}}$$

or

$$\sqrt{2}(2\delta u) = [(1-u^2)^2 + (2\delta u)^2]^{1/2}$$

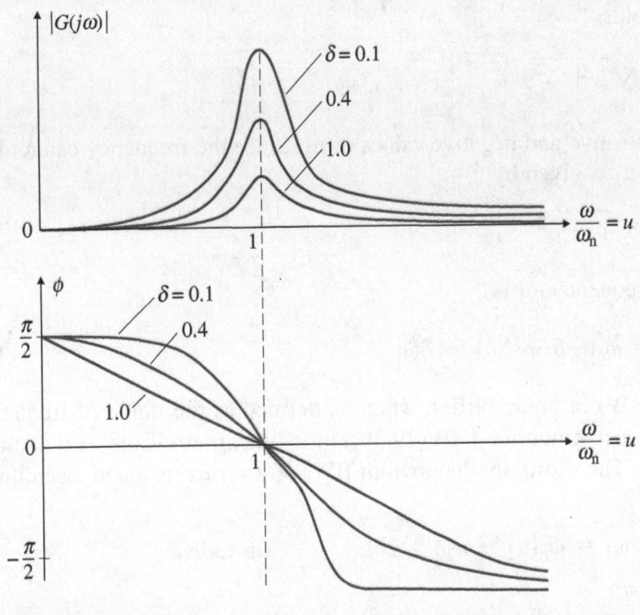

FIGURE B.41 Frequency response of a series *RLC* circuit

Squaring both sides yields

$$2(2\delta u)^2 = (1 - u^2)^2 + (2\delta u)^2$$

or $$(2\delta u)^2 = (1 - u^2)^2 \tag{B.80}$$

The possible solutions of Eq. (B.80) are

$$2\delta u_1 = 1 - u_1^2$$

$$u_1^2 + 2\delta u_1 - 1 = 0 \tag{B.81}$$

$$2\delta u_2 = -(1 - u_2^2) = u_2^2 - 1$$

$$u_2^2 - 2\delta u_2 - 1 = 0 \tag{B.82}$$

Solving Eq. (B.82) yields

$$u_2 = \delta \pm \sqrt{1 + \delta^2}$$

Since the frequency cannot be negative, the upper cutoff frequency ratio u_2 is given by

$$u_2 = \delta + \sqrt{1 + \delta^2} \tag{B.83}$$

and the upper cutoff frequency ω_2 is

$$\omega_2 = u_2\omega_n = \omega_n(\delta + \sqrt{1 + \delta^2}) \tag{B.84}$$

Solving Eq. (B.81) yields

$$u_1 = -\delta \pm \sqrt{1 + \delta^2}$$

which will give both positive and negative values of u_1. Since the frequency cannot be negative, the lower cutoff frequency ratio u_1 is given by

$$u_1 = -\delta + \sqrt{1 + \delta^2} \tag{B.85}$$

and the lower cutoff frequency ω_1 is

$$\omega_1 = u_1\omega_n = \omega_n(-\delta + \sqrt{1 + \delta^2}) \tag{B.86}$$

The *bandwidth* (BW) of an amplifier, which is defined as the range of frequencies over which the gain remains almost constant within 3 dB (29.3%) of its maximum value, is thus the difference between the cutoff frequencies. Therefore, the bandwidth $\mathrm{BW_s}$ of a series resonant circuit can be found from

$$\mathrm{BW_s} = \omega_2 - \omega_1 = \omega_\mathrm{n}(u_2 - u_1) = 2\delta\omega_\mathrm{n} = \frac{R}{L} \quad \text{(in rad/s)} \tag{B.87}$$

$$\mathrm{BW_s} = f_2 - f_1 = \frac{1}{2\pi}\frac{R}{L} \quad \text{(in Hz)} \tag{B.88}$$

From Eq. (B.55), $R/L = 2\pi f_\mathrm{n}/Q_\mathrm{s}$. Thus, Eq. (B.88) can be rewritten as

$$\mathrm{BW_s} = \frac{1}{2\pi}\frac{R}{L} = \frac{1}{2\pi}\frac{2\pi f_\mathrm{n}}{Q_\mathrm{s}} = \frac{f_\mathrm{n}}{Q_\mathrm{s}} \tag{B.89}$$

which shows that the larger the value of Q_s, the smaller the value of bandwidth $\mathrm{BW_s}$, and vice versa. It can be shown that Eq. (B.89) can be also applied to calculate the bandwidth $\mathrm{BW_p}$ of a parallel resonant circuit; that is,

$$\mathrm{BW_p} = \frac{f_\mathrm{p}}{Q_\mathrm{p}} \tag{B.90}$$

where f_p is the parallel resonant frequency in Eq. (B.63) and Q_p is the quality factor of a parallel resonant circuit in Eq. (B.64).

EXAMPLE B.16

Finding the frequency response of a series *RLC* circuit The series *RLC* circuit in Fig. B.40 has $R = 50\ \Omega$, $L = 4$ mH, and $C = 0.15\ \mu$F.

(a) Determine the series resonant frequency f_n, the damping ratio δ, the quality factor Q_s, the cutoff frequencies, and the bandwidth $\mathrm{BW_s}$.

(b) Use PSpice/SPICE to plot the magnitude and phase angle of the output voltage for $R = 50\ \Omega$, $100\ \Omega$, and $200\ \Omega$. The frequency f varies from 100 Hz to 1 MHz. Assume $V_\mathrm{m} = 1$ V peak AC.

SOLUTION

(a) $R = 50\ \Omega$, $L = 4$ mH, and $C = 0.15\ \mu$F, so

$$\omega_n = \frac{1}{\sqrt{LC}} = \frac{10^5}{\sqrt{4 \times 1.5}} = 40{,}825 \text{ rad/s}$$

The series resonant frequency is

$$f_n = \frac{\omega_n}{2\pi} = \frac{40{,}825}{2\pi} = 6497.5 \text{ Hz}$$

Since $\alpha = R/(2L) = 50/(2 \times 4 \times 10^{-3}) = 6250$, the damping ratio is

$$\delta = \frac{\alpha}{\omega_n} = \frac{6250}{40{,}825} = 0.1531$$

From Eq. (B.55),

$$Q_s = \omega_n \frac{L}{R} = 40{,}825 \times 4 \times \frac{10^{-3}}{50} = 3.266$$

For the lower cutoff frequency, Eqs. (B.85) and (B.86) give

$$u_1 = -\delta + \sqrt{1 + \delta^2} = -0.1531 + \sqrt{1 + 0.1531^2} = 0.85855$$

$$\omega_1 = u_1\omega_n = 0.85855 \times 40{,}825 = 35{,}050.4 \text{ rad/s}$$

Thus, $f_1 = 35{,}050.4/2\pi = 5578$ Hz. For the upper cutoff frequency, Eqs. (B.83) and (B.84) give

$$u_2 = \delta + \sqrt{1 + \delta^2} = 0.1531 + \sqrt{1 + 0.1531^2} = 1.16475$$

$$\omega_2 = u_2\omega_n = 1.16475 \times 40{,}825 = 47{,}551 \text{ rad/s}$$

Thus, $f_2 = 47{,}551/2\pi = 7568$ Hz. From Eq. (B.89), the bandwidth is

$$BW_s = f_2 - f_1 = \frac{f_n}{Q_s} = \frac{6497.5}{3.266} = 1989.4 \text{ Hz}$$

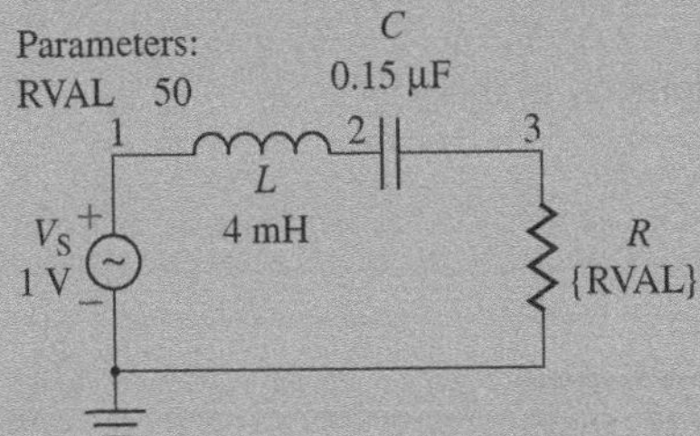

FIGURE B.42 Series *RLC* circuit for PSpice simulation

(b) The series *RLC* circuit for PSpice simulation is shown in Fig. B.42. The list of the circuit file is as follows.

```
Example B.16  Frequency Response of Series RLC Circuit
.PARAM RVAL = 50
.STEP PARAM RVAL LIST  50  100  200
Vm  1  0  AC  1V                  ; AC input of 1 V peak
L  1  2  4MH
C  2  3  0.15UF
R  3  0  {RVAL}
.AC DEC 100 100HZ 1MEGHz          ; AC analysis from f = 100 Hz to 1 MHz
                                  ; with a decade change and 100 points per decade
.PROBE
.END
```

The PSpice plots of the magnitude and phase angle (using EXB-16.SCH) are shown in Fig. B.43. The plot for $R = 50\ \Omega$ gives $f_1 = 5578$ Hz, $f_2 = 7568$ Hz, $f_n = 6457$ Hz, and $BW_s = f_2 - f_1 = 1990$ Hz.

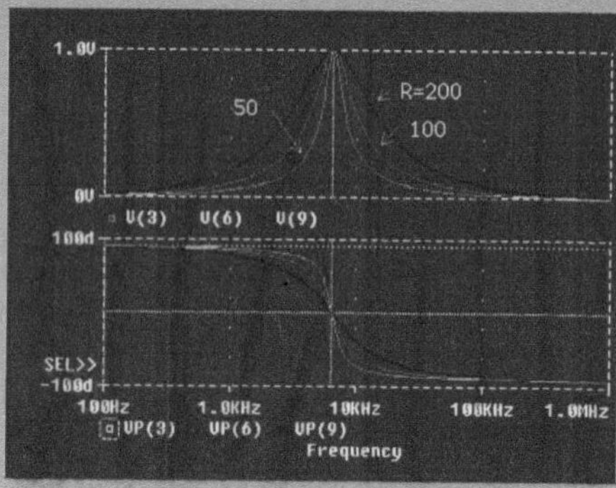

FIGURE B.43 Frequency response plots for Example B.16

B.10.4 Second-Order Parallel *RLC* Circuits

A parallel *RLC* circuit is shown in Fig. B.44. The output voltage v_o is taken across the parallel combination of *R*, *L*, and *C*. The transfer function $G(s) = V_o(s)/I_s(s)$ in Laplace's domain of *s* is the equivalent impedance *Z*(*s*).

The function

$$\frac{1}{Z(s)} = \frac{1}{R} + \frac{1}{sL} + sC = \frac{sL + R + s^2LCR}{sRL}$$

$$= \frac{s^2 + (s/RC) + (1/LC)}{s/C} \quad \text{(in siemens, or mhos)}$$

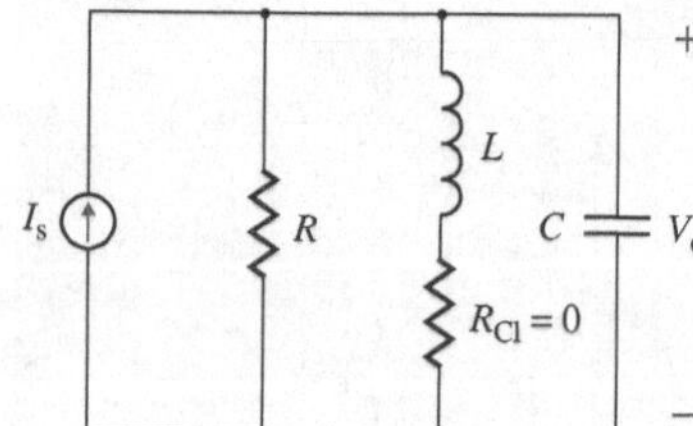

FIGURE B.44 Parallel *RLC* circuit

gives the transfer function $G(s)$ as

$$G(s) = \frac{V_o(s)}{I_s(s)} = Z(s) = \frac{s/C}{s^2 + (s/RC) + (1/LC)}$$

$$= R\frac{s/RC}{s^2 + (s/RC) + (1/LC)} \quad \text{(in ohms)} \tag{B.91}$$

Defining $\omega_n = 1/\sqrt{LC}$ as the *resonant frequency* in rad/s and $\alpha = 1/(2RC)$ as the *damping factor*, we can write Eq. (B.91) as

$$G(s) = R\frac{2\alpha s}{s^2 + 2\alpha s + \omega_n^2} \quad \text{(in ohms)} \tag{B.92}$$

Let us define

$$\delta = \frac{\alpha}{\omega_n} = \frac{1}{2RC} \times \sqrt{LC} = \frac{1}{2R}\sqrt{\frac{L}{C}} \tag{B.93}$$

as the *damping ratio*. Then Eq. (B.92) becomes

$$G(s) = R\frac{\delta\omega_n s}{s^2 + 2\delta\omega_n s + \omega_n^2} \quad \text{(in ohms)} \tag{B.94}$$

where $\delta < 1$. (Note that δ is not necessarily less than 1, but has been assumed less than 1 for this discussion.) The right-hand side of Eq. (B.94) is $R/2$ multiplied by Eq. (B.76). Following the development of Eqs. (B.78) and (B.79), we find the magnitude $|G(j\omega)|$ as

$$|G(j\omega)| = \frac{2\delta uR}{[(1 - u^2)^2 + (2\delta u)^2]^{1/2}} \quad \text{(in ohms)} \tag{B.95}$$

and the phase angle ϕ of $G(j\omega)$ as

$$\phi = \frac{\pi}{2} - \tan^{-1}\left(\frac{2\delta u}{1 - u^2}\right) \tag{B.96}$$

The magnitude and frequency plots of a parallel *RLC* circuit are shown in Fig. B.45. The maximum value $|G(j\omega)|_{max} = |Z(j\omega)|_{max} = 1$. At the cutoff frequencies, the magnitude of the gain drops to 70.7% of its maximum value R. Thus, Eq. (B.95) gives

$$|G(j\omega)| = \frac{2\delta uR}{[(1 - u^2)^2 + (2\delta u)^2]^{1/2}} = 0.707R = \frac{R}{\sqrt{2}}$$

or $$\sqrt{2}(2\delta u) = [(1 - u^2)^2 + (2\delta u)^2]^{1/2}$$

Squaring both sides yields

$$2(2\delta u)^2 = (1 - u^2)^2 + (2\delta u)^2$$

or $$(2\delta u)^2 = (1 - u^2)^2 \tag{B.97}$$

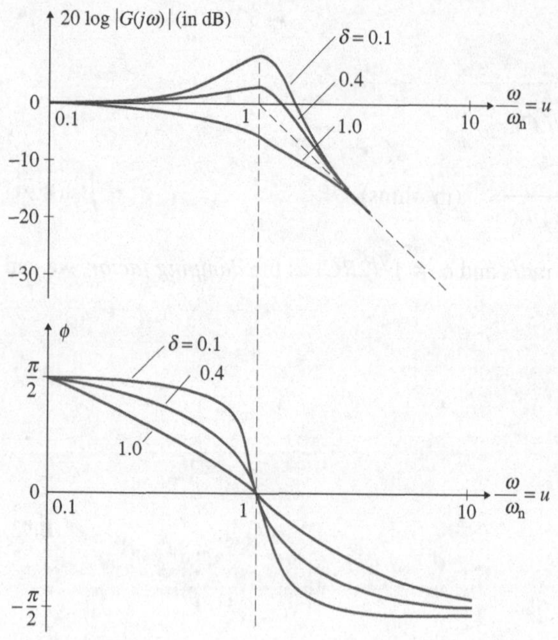

FIGURE B.45 Frequency response of parallel *RLC* circuit

which is the same as Eq. (B.80). Equations (B.81) to (B.86) can be applied to find ω_1 and ω_2. Then the bandwidth BW_p of the parallel resonant circuit can be found from

$$BW_p = \omega_2 - \omega_1 = \omega_n(u_2 - u_1) = 2\delta\omega_n$$

$$= 2\frac{1}{2R}\sqrt{\frac{L}{C}} \times \frac{1}{\sqrt{LC}} = \frac{1}{RC} \quad \text{(in rad/s)} \tag{B.98}$$

▶ **NOTE** For a parallel circuit, $BW_p = 1/RC$ only; for a series circuit, $BW_s = R/L$.

EXAMPLE B.17

Finding the frequency response of a parallel *RLC* circuit The parallel *RLC* circuit in Fig. B.44 has $R = 50\ \Omega$, $L = 4$ mH, and $C = 0.15\ \mu$F.

(a) Determine the parallel resonant frequency f_p, the damping ratio δ, the cutoff frequencies, the bandwidth BW_p, and the quality factor Q_p of the circuit.

(b) Use PSpice/SPICE to plot the magnitude and phase angle of the output voltage for $R = 50\ \Omega$, $100\ \Omega$, and $200\ \Omega$. The frequency f varies from 100 Hz to 100 kHz. Assume $I_m = 1$ A peak AC.

SOLUTION

(a) $R = 50\ \Omega$, $L = 4$ mH, $C = 0.15\ \mu$F, and $I_m = 1$ A peak AC, so

$$\omega_n = \frac{1}{\sqrt{LC}} = \frac{10^5}{\sqrt{4 \times 1.5}} = 40{,}825 \text{ rad/s}$$

The parallel resonant frequency is

$$f_p = \frac{\omega_n}{2\pi} = \frac{40{,}825}{2\pi} = 6497.5 \text{ Hz}$$

Since $\alpha = 1/(2RC) = 1/(2 \times 50 \times 0.15 \times 10^{-6}) = 66.667 \times 10^3$, the damping ratio is

$$\delta = \frac{\alpha}{\omega_n} = 66.667 \times \frac{10^3}{40{,}825} = 1.633$$

For the lower cutoff frequency, Eqs. (B.85) and (B.86) give

$$u_1 = -\delta + \sqrt{1 + \delta^2} = -1.633 + \sqrt{1 + 1.633^2} = 0.28186$$

$$\omega_1 = u_1\omega_n = 0.28186 \times 40{,}825 = 11{,}507 \text{ rad/s}$$

Thus, $f_1 = 11{,}507/2\pi = 1831$ Hz. For the upper cutoff frequency, Eqs. (B.83) and (B.84) give

$$u_2 = \delta + \sqrt{1 + \delta^2} = 1.633 + \sqrt{1 + 1.633^2} = 3.54786$$

$$\omega_2 = u_2\omega_n = 3.54786 \times 40{,}825 = 144{,}841 \text{ rad/s}$$

Thus, $f_2 = 144{,}841/2\pi = 23{,}052$ Hz. From Eq. (B.98), the bandwidth is

$$\text{BW}_p = f_2 - f_1 = \frac{1}{RC} = \frac{1}{50 \times 0.15 \times 10^{-6}} = 133{,}333.3 \text{ rad/s, or } 21{,}220 \text{ Hz}$$

Using Eq. (B.90), we can find the quality factor:

$$Q_p = \frac{f_p}{BW_p} = \frac{6497.5}{21{,}220} = 0.3062$$

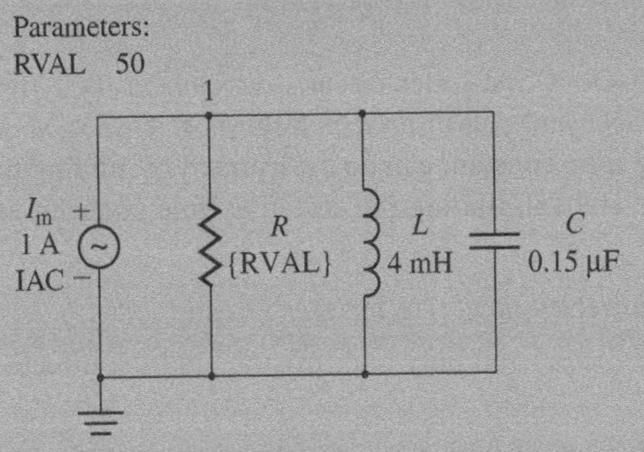

Parameters:
RVAL 50

1

I_m + 1 A IAC −

R {RVAL}

L 4 mH

C 0.15 μF

FIGURE B.46 Parallel *RLC* circuit for PSpice simulation

(b) The parallel *RLC* circuit for PSpice simulation is shown in Fig. B.46. The list of the circuit file is as follows.

```
Example B.17 Frequency Response of a Parallel RLC Circuit
.PARAM RVAL = 50
.STEP PARAM RVAL LIST  50  100  200
IM  0  1  AC  1A        ; ac input of 1 V peak
L   1  0  4MH
C   1  0  0.15UF
R   1  0  {RVAL}
.AC  DEC  100  100HZ  1MEGHZ
.PROBE
.END
```

The PSpice plots of the magnitude and phase angle (using EXB-17.SCH) are shown in Fig. B.47. The plot for $R = 50\ \Omega$ gives $f_1 = 1834$ Hz, $f_2 = 22.56$ kHz, $f_p = 6457$ Hz, and $BW_p = f_2 - f_1 = 20{,}726$ Hz.

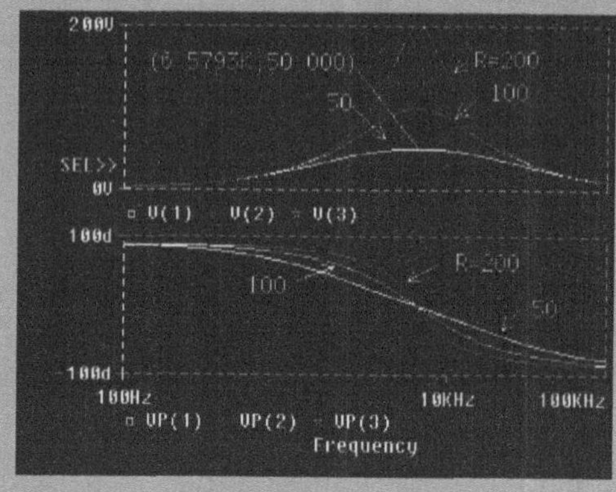

FIGURE B.47 Frequency response plots for Example B.17

B.11 Time Constants of First-Order Circuits

We have seen that the transient and frequency responses of first-order circuits depend on their time constants. The time constant of an *RC* network is $\tau = RC$, and that of an *RL* circuit is $\tau = L/R$. Many circuits have more than two components. The effective time constant can be determined by finding the effective resistance and capacitance of the circuit. The steps in finding the effective time constant are as follows:

Step 1. Set the voltage source(s) to zero and the current source(s) as an open circuit.

Step 2. If there is more than one capacitor (or inductor) and only one resistor, find the effective capacitance or inductance seen by the resistor.

Step 3. If there is more than one resistor and only one capacitive (or inductive) element, find the effective resistance seen by the capacitor (or inductor).

EXAMPLE B.18

Finding the effective time constant The circuit in Fig. B.48 has $R_1 = R_2 = R_3 = 6\ \text{k}\Omega$ and $C = 0.1\ \mu\text{F}$. Determine **(a)** the effective time constant τ, **(b)** the cutoff frequency ω_o, and **(c)** the bandwidth BW.

SOLUTION

If the source is shorted, the effective resistance seen by capacitor C is the parallel combination of R_1, R_2, and R_3. The effective resistance R is given by

$$\frac{1}{R} = \frac{1}{R_1} + \frac{1}{R_2} + \frac{1}{R_3} \quad \text{or} \quad R = \frac{R_1}{3} = \frac{6\ \text{k}}{3} = 2\ \text{k}\Omega$$

(a) The effective time constant is

$$\tau = CR = 2\ \text{k}\Omega \times 0.1\ \mu\text{F} = 0.2\ \text{ms}$$

(b) The cutoff frequency is

$$\omega_o = \frac{1}{\tau} = \frac{1}{0.2\ \text{ms}} = 5000\ \text{rad/s, or } 795.8\ \text{Hz}$$

(c) At $\omega = 0$, capacitor C is open-circuited, and the output voltage has a finite value. At a high frequency, tending to infinity ($\omega = \infty$), capacitor C is short-circuited, and the output voltage becomes zero. This is a low-pass circuit with $f_1 = 0$ and $f_2 = f_o = 795.8$ Hz. Thus, the bandwidth is

$$\text{BW} = f_2 - f_1 = 795.8\ \text{Hz}$$

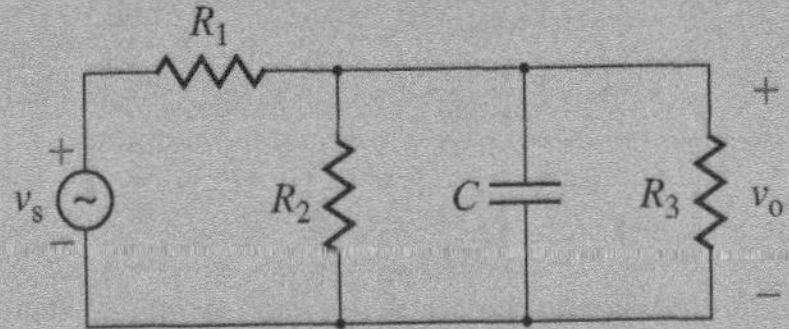

FIGURE B.48 Circuit for Example B.18

EXAMPLE B.19

Finding the effective time constant The circuit in Fig. B.49 has $R_1 = R_2 = R_3 = 10\ \text{k}\Omega$ and $C_1 = 0.1\ \mu\text{F}$. Determine **(a)** the effective time constant τ and **(b)** the cutoff frequency ω_o.

SOLUTION

If the source is shorted, the effective resistance is the sum of R_1 and $(R_2 \| R_3)$; that is,

$$R = R_1 + (R_2 \| R_3) = 10\ \text{k} + 10\ \text{k} \| 10\ \text{k} = 15\ \text{k}\Omega$$

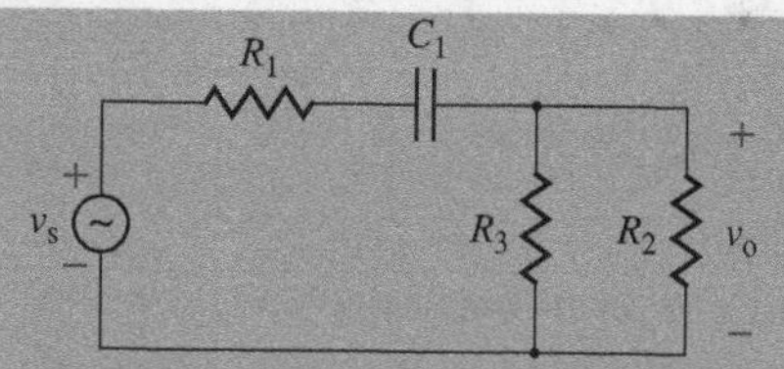

FIGURE B.49 Circuit for Example B.19

(a) The effective time constant is

$$\tau = CR = 15\ \text{k}\Omega \times 0.1\ \mu\text{F} = 1.5\ \text{ms}$$

(b) The cutoff frequency is

$$\omega_o = \frac{1}{\tau} = \frac{1}{1.5\ \text{ms}} = 667\ \text{rad/s, or } 106\ \text{Hz}$$

APPENDIX C

LOW-FREQUENCY HYBRID BJT MODEL

A bipolar transistor can be represented by hybrid (h) parameters. If i_b, v_{be}, i_c, and v_{ce} are the small-signal variables of a transistor, as shown in Fig. C.1(a), they are related to the hybrid parameters as follows:

$$v_{bc} = h_{ie} i_b + h_{re} v_{ce} \tag{C.1}$$

$$i_c = h_{fe} i_b + h_{oe} v_{ce} \tag{C.2}$$

where h_{ie} is the *short-circuit input resistance* (or simply the *input resistance*), defined by

$$h_{ie} = \left.\frac{v_{be}}{i_b}\right|_{v_{ce}=0} \quad \text{(in ohms)} \tag{C.3}$$

h_{re} is the *open-circuit reverse voltage ratio* (or the *voltage-feedback ratio*), defined by

$$h_{re} = \left.\frac{v_{be}}{v_{ce}}\right|_{i_b=0} \quad \text{(dimensionless)} \tag{C.4}$$

h_{fe} is the *short-circuit forward-transfer current ratio* (or the *small-signal current gain*), defined by

$$h_{fe} = \left.\frac{i_c}{i_b}\right|_{v_{ce}=0} \quad \text{(dimensionless)} \tag{C.5}$$

and h_{oe} is the *open-circuit output admittance* (or simply the *output admittance*), defined by

$$h_{oe} = \left.\frac{i_c}{v_{ce}}\right|_{i_b=0} \quad \text{(in siemens)} \tag{C.6}$$

The low-frequency hybrid model is shown in Fig. C.1(b). The input has a voltage-controlled voltage source in which the controlling voltage is the output voltage. The output circuit has a current-controlled current source in which the controlling current is the input current. The subscript e on the h parameters indicates that these hybrid parameters are derived for a common-emitter configuration. It is also possible to have common-base and common-collector configurations.

BJT manufacturers specify the common-emitter hybrid parameters. The parameter h_{re}, which takes into account the effect of v_{CE} on i_B, is very small, with a typical value of 0.5×10^{-4}. The parameter h_{oe}, which represents the admittance of the CE junction, is also very small; its value is typically 10^{-6} S. The parameters h_{re} and h_{oe} can often be omitted from the circuit model without significant loss of accuracy, especially for hand calculations.

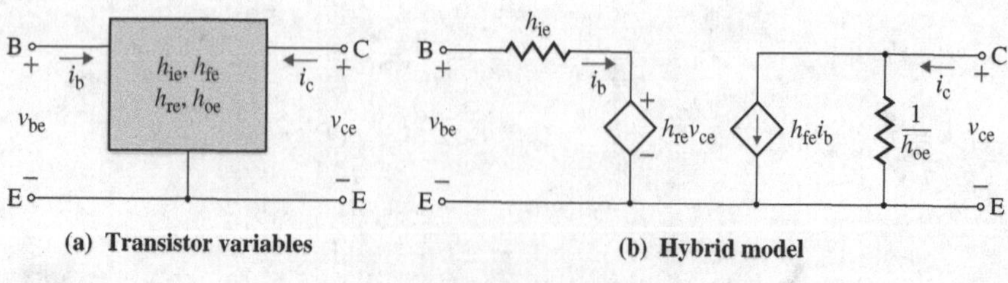

FIGURE C.1 Low-frequency hybrid model

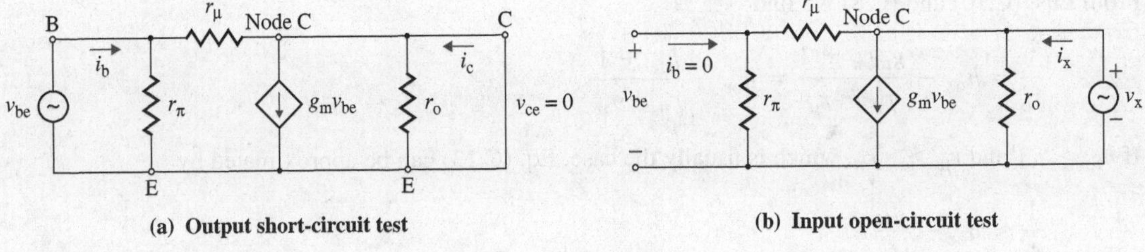

FIGURE C.2 Equivalent circuits for deriving h parameters

The π-model parameters of Fig. 8.12(a) can be related to the h parameters by applying short-circuit and open-circuit tests. The parameters h_{ie} and h_{fe}, which are short-circuit parameters, can be derived in terms of π-model parameters from Fig. C.2(a) by shorting the collector-to-emitter terminals.

$$h_{ie} = \left.\frac{v_{be}}{i_b}\right|_{v_{ce}=0} = r_\pi \| r_\mu = \frac{r_\pi r_\mu}{r_\pi + r_\mu} \tag{C.7}$$

$$h_{fe} = \left.\frac{i_c}{i_b}\right|_{v_{ce}=0} = \frac{g_m v_{be}}{v_{be}/r_\pi} = g_m r_\pi \tag{C.8}$$

As shown in Fig. C.2(b), h_{oe} and h_{re} can be determined by open-circuiting the base-to-emitter terminals and then applying the voltage divider rule.

$$h_{re} = \left.\frac{v_{be}}{v_x}\right|_{i_b=0} = \frac{r_\pi}{r_\pi + r_\mu} \tag{C.9}$$

Summing the currents at the collector node in Fig. C.2(b), we get

$$i_x = \frac{v_x}{r_o} + g_m \frac{r_\pi v_x}{r_\pi + r_\mu} + \frac{v_x}{r_\pi + r_\mu} = v_x\left[\frac{1}{r_o} + \frac{g_m r_\pi + 1}{r_\pi + r_\mu}\right]$$

which gives

$$h_{oe} = \frac{i_x}{v_x} = \frac{1}{r_o} + \frac{g_m r_\pi + 1}{r_\pi + r_\mu} \tag{C.10}$$

Using Eq. (C.8), we get

$$r_\pi = \frac{h_{fe}}{g_m} \tag{C.11}$$

Using Eq. (C.7) and $h_{fe} = g_m r_\pi$ from Eq. (C.8), we get

$$r_\mu = \frac{r_\pi + r_\mu}{r_\pi} h_{ie} = \frac{h_{ie}}{h_{re}} \tag{C.12}$$

From Eqs. (C.10) and (C.8) we find

$$\frac{1}{r_o} = h_{oe} - \frac{g_m r_\pi + 1}{r_\pi + r_\mu} = h_{oe} - \frac{h_{fe} + 1}{r_\pi + r_\mu} \tag{C.13}$$

If $h_{fe} >> 1$ and $r_\mu >> r_\pi$, which is usually the case, Eq. (C.13) can be approximated by

$$\frac{1}{r_o} \approx h_{oe} - \frac{h_{fe}}{r_\mu}$$

which gives

$$r_o = \left[h_{oe} - \frac{h_{fe}}{r_\mu} \right]^{-1} \tag{C.14}$$

$$= \frac{V_A}{I_C}$$

r_μ is very large, and h_{re} is negligibly small. The π model in Fig. 8.12(b) becomes similar to the h model in Fig. C.2(b) such that $r_\pi \equiv h_{ie}$, $\beta_f \equiv h_{fe}$, and $r_o \equiv 1/h_{oe}$. Being open-circuited, r_μ has a high value and can often be neglected.

EXAMPLE C.1

Converting hybrid parameters to π-model parameters The h parameters of a transistor are $h_{ie} = 1\ \text{k}\Omega$, $h_{fe} = 100$, $h_{re} = 0.5 \times 10^{-4}$, and $h_{oe} = 10 \times 10^{-6}$ S. The collector biasing current is $I_C = 11.62$ mA. Calculate the small-signal π-model parameters of the transistor. Assume $V_T = 25.8$ mV at 25°C.

SOLUTION

From Eq. (8.42),

$$g_m = \frac{I_C}{V_T} = \frac{11.62\text{ mA}}{25.8\text{ mV}} = 0.4504\text{ A/V}$$

From Eq. (C.11),

$$r_\pi = \frac{h_{fe}}{g_m} = \frac{100}{0.4504} = 222\ \Omega$$

From Eq. (C.12),

$$r_\mu = \frac{h_{ie}}{h_{re}} = \frac{1\ \text{k}\Omega}{0.5 \times 10^{-4}} = 20\ \text{M}\Omega$$

From Eq. (C.14),

$$r_o = \left[h_{oe} - \frac{h_{fe}}{r_\mu} \right]^{-1} = \left[10 \times 10^{-6} - \frac{100}{20 \times 10^6} \right]^{-1} = 200\ \text{k}\Omega$$

EXAMPLE C.2

Finding the hybrid-model parameters of the circuit shown in Fig. C.3 The circuit values are $R_1 = 4\ \text{k}\Omega$, $R_2 = 10\ \text{k}\Omega$, and $R_f = 16\ \text{k}\Omega$. Determine the hybrid parameters h_{ie}, h_{re}, h_{fe}, and h_{oe}.

SOLUTION

$R_1 = 4\ \text{k}\Omega$, $R_2 = 6\ \text{k}\Omega$, and $R_f = 20\ \text{k}\Omega$. Using Eq. (C.3) and the equivalent circuit as shown in Fig. C.4(a), h_{ie} can be determined from

$$h_{ie} = \left.\frac{v_S}{i_S}\right|_{v_O=0} = R_1 \| R_f = 4\text{ k} \| 16\text{ k} = 3.2\ \text{k}\Omega$$

Using Eq. (C.4) and the equivalent circuit shown in Fig. C.4(b), h_{re} can be determined from

$$h_{re} = \left.\frac{v_S}{v_O}\right|_{i_S=0} = \frac{R_1}{R_1 + R_f} = \frac{4\text{ k}}{4\text{ k} + 16\text{ k}} = 0.2\text{ V/V}$$

Using Eq. (C.5) and the equivalent circuit shown in Fig. C.4(c), h_{fe} can be determined from

$$h_{re} = \left.\frac{i_O}{i_S}\right|_{v_O=0} = \frac{-R_1}{R_1 + R_f} = \frac{-4\text{ k}}{4\text{ k} + 16\text{ k}} = -0.2\text{ A/A}$$

Using Eq. (C.6) and the equivalent circuit shown in Fig. C.4(d), h_{oe} can be determined from

$$h_{oe} = \left.\frac{i_O}{v_O}\right|_{i_S=0} = \frac{1}{R_1 \| (R_1 + R_f)} = \frac{1}{4\text{ k} \| (4\text{ k} + 16\text{ k})} = 1.5 \times 10^{-4}\ \mho$$

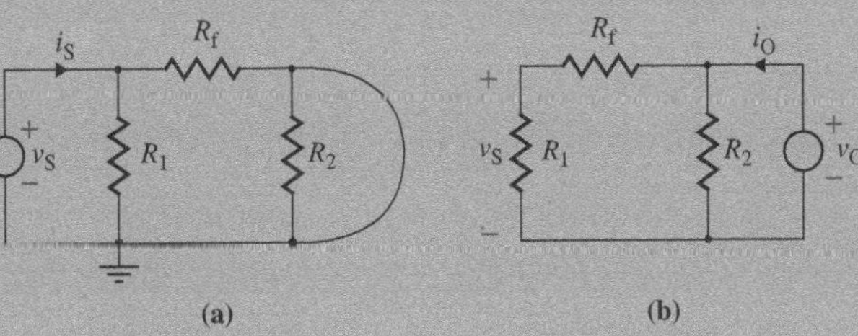

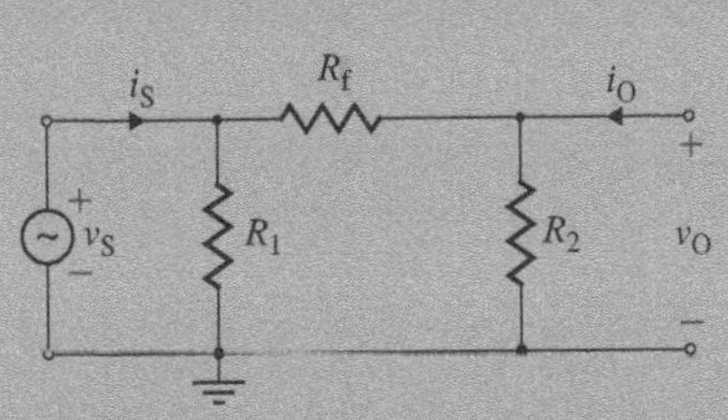

FIGURE C.3 Circuit for Example C.2

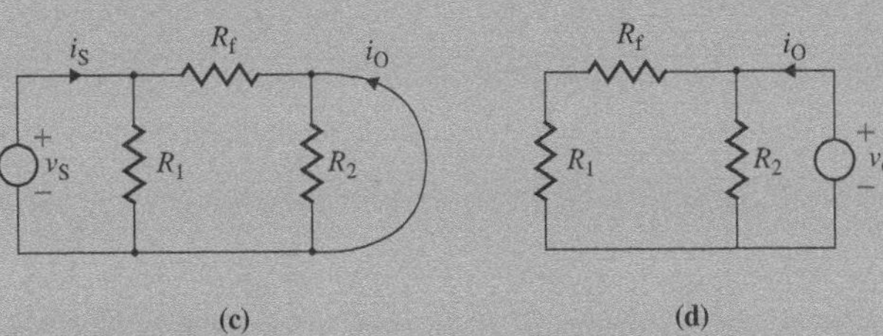

FIGURE C.4 Equivalent circuits for determining hybrid parameters for Example C.2

APPENDIX D

EBERS–MOLL MODEL OF BIPOLAR JUNCTION TRANSISTORS

A transistor is a nonlinear device that can be modeled using the nonlinear characteristics of diodes. The Ebers–Moll model is a large-signal model commonly used for modeling BJTs. One version of the model is based on the assumption of one forward-biased diode and one reverse-biased diode. This arrangement is shown in Fig. D.1 for an *npn* transistor. This model, referred to as the *injection version* of the Ebers–Moll model, is valid for active, saturation, and cutoff regions. Under normal operation in the active region, one junction of the BJT is forward biased and the other is reverse biased.

The emitter–base and collector–base diodes can be described using the Shockley diode characteristic of Eq. (4.1):

$$I_F = I_{ES}\left(\exp\frac{V_{BE}}{V_T} - 1\right) \tag{D.1}$$

$$I_R = I_{CS}\left(\exp\frac{V_{BC}}{V_T} - 1\right) \tag{D.2}$$

where $V_T = kT/q = 25.8$ mV at 25°C
I_{ES} = reverse saturation current of base–emitter diode
I_{CS} = reverse saturation current of base–collector diode

Both I_{ES} and I_{CS} are temperature dependent. If $V_{BE} > 0$, diode D_F is forward biased and its current I_F causes a corresponding current $\alpha_F I_F$. If $V_{BC} > 0$, diode D_R is reverse biased. The subscripts F and R are used to designate forward and reverse conditions, respectively. Using Kirchhoff's current law (KCL) at the emitter and collector terminals, we can write the emitter current I_E as

$$\begin{aligned} I_E &= -I_F + \alpha_R I_R \\ &= -I_{ES}\left(\exp\frac{V_{BE}}{V_T} - 1\right) + \alpha_R I_{CS}\left(\exp\frac{V_{BC}}{V_T} - 1\right) \end{aligned} \tag{D.3}$$

and the collector current I_C as

$$\begin{aligned} I_C &= \alpha_F I_F - I_R \\ &= \alpha_F I_{ES}\left(\exp\frac{V_{BE}}{V_T} - 1\right) - I_{CS}\left(\exp\frac{V_{BC}}{V_T} - 1\right) \end{aligned} \tag{D.4}$$

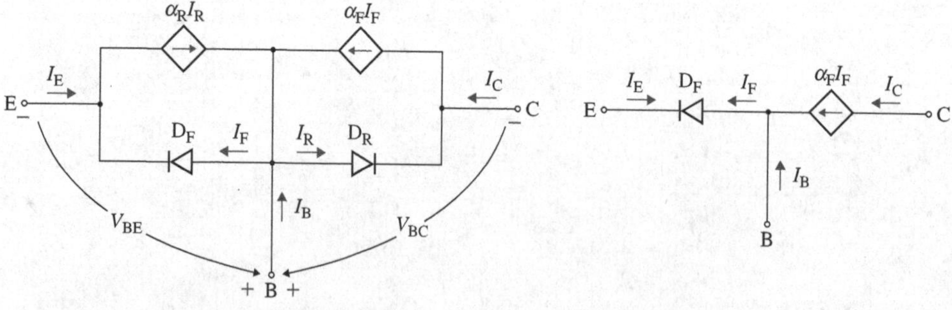

(a) Injection version model

(b) Approximate injection version model

FIGURE D.1 Ebers–Moll injection version model for *npn* transistor

If $V_{BE} = 0$, $\alpha_R I_{CS} = I_S$ represents the reverse saturation leakage current of diode D_R. Similarly, if $V_{BE} = 0$, $\alpha_F I_{ES} = I_S$ represents the reverse saturation leakage current of diode D_F. If we assume ideal diodes, the forward and reverse saturation leakage currents are related by

$$\alpha_R I_{CS} = \alpha_F I_{ES} = I_S \tag{D.5}$$

where I_S is known as the *transistor saturation current.*

The current from the collector to the base with the emitter open-circuited can be found by letting $I_C = I_{CBO}$ and $I_E = 0$. Since the collector–base junction is normally reverse biased, $V_{BC} < 0$ and $|V_{BC}| >> V_T$, $\exp(V_{BC}/V_T) << 1$. With these conditions in Eqs. (D.3) and (D.4), we get

$$0 = -I_{ES}\left(\exp\frac{V_{BE}}{V_T} - 1\right) - \alpha_R I_{CS}$$

$$I_{CBO} = \alpha_F I_{ES}\left(\exp\frac{V_{BE}}{V_T} - 1\right) + I_{CS}$$

Solving these two equations for I_{CBO} yields

$$I_{CBO} = -\alpha_F\alpha_R I_{CS} + I_{CS} = (1 - \alpha_R\alpha_F)I_{CS} = I_{CS} - \alpha_F I_S \tag{D.6}$$

The current from the emitter to the base with the collector open-circuited can be found by letting $I_E = I_{EBO}$ and $I_C = 0$. Since the emitter junction is reverse biased, $V_{BE} < 0$ (i.e., $V_{EB} > 0$) and $|V_{BE}| >> V_T$, $\exp(V_{BE}/V_T) << 1$. Equations (D.3) and (D.4) give

$$I_{EBO} = I_{ES} + \alpha_R I_{CS}\left(\exp\frac{V_{BC}}{V_T} - 1\right)$$

$$0 = -\alpha_F I_{ES} - I_{CS}\left(\exp\frac{V_{BC}}{V_T} - 1\right)$$

Solving these two equations for I_{EBO} gives

$$I_{EBO} = I_{ES} - \alpha_R\alpha_F I_{ES} = (1 - \alpha_R\alpha_F)I_{ES} \tag{D.7}$$

From Eqs. (D.5), (D.6), and (D.7), we get

$$\alpha_F I_{EBO} = I_{ES}(1 - \alpha_R\alpha_F)\alpha_F = \alpha_R I_{CS}(1 - \alpha_R\alpha_F) = \alpha_R I_{CBO} \tag{D.8}$$

Since diode D_F is forward biased and diode D_R is reverse biased, $V_{EB} < V_{BC}$. Thus, I_{EBO} is less than I_{CBO}, and α_F is greater than α_R. In the active region, diode D_R is reverse biased, and $I_R \approx 0$; that is, $I_E = -I_F$, and $I_C = \alpha_F I_F = -\alpha_F I_E$. Thus, Fig. D.1(a) can be approximated by Fig. D.1(b).

The circuit model of Fig. D.1(a) relates the dependent sources to the diode currents. In circuit analysis, it is convenient to express the current source in a form that is controlled by the terminal currents. Eliminating $[\exp(V_{BE}/V_T) - 1]$ from Eqs. (D.3) and (D.4) and then using Eq. (D.6), we get

$$I_C = -\alpha_F I_E - (1 - \alpha_F\alpha_R)I_{CS}\left(\exp\frac{V_{BC}}{V_T} - 1\right)$$

$$= -\alpha_F I_E - I_{CBO}\left(\exp\frac{V_{BC}}{V_T} - 1\right) \tag{D.9}$$

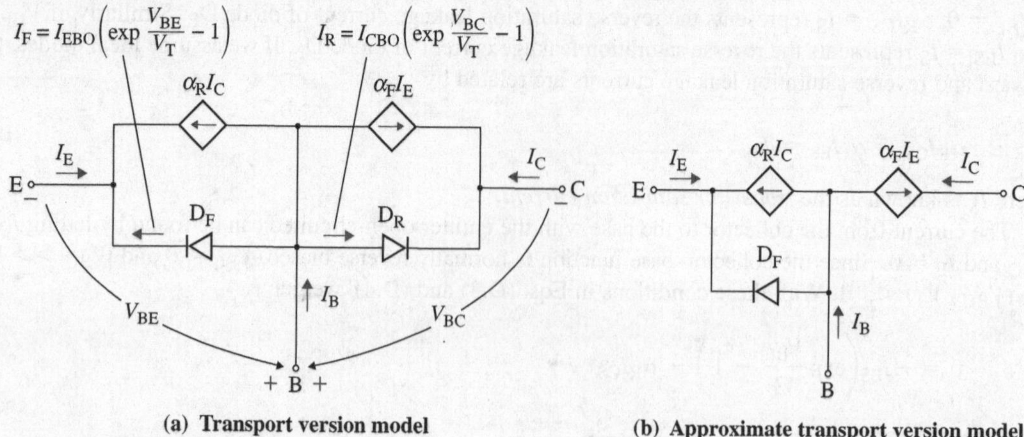

FIGURE D.2 Ebers–Moll transport version model

Similarly, eliminating $[\exp(V_{BC}/V_T) - 1]$ from Eqs. (D.3) and (D.4) and then using Eq. (D.7) gives

$$I_E = -\alpha_R I_C - (1 - \alpha_R\alpha_F)I_{ES}\left(\exp\frac{V_{BE}}{V_T} - 1\right)$$

$$= -\alpha_R I_C - I_{EBO}\left(\exp\frac{V_{BE}}{V_T} - 1\right) \tag{D.10}$$

The circuit model corresponding to Eqs. (D.9) and (D.10) is shown in Fig. D.2(a). The current sources are controlled by collector current I_C and emitter current I_E. This model, referred to as the *transport version* of the Ebers–Moll model, is normally used in computer simulations with PSpice/SPICE. In fact, the linear models of Fig. 8.12 are the approximate versions of the Ebers–Moll model in Fig. D.2(a).

Assuming that $V_{BC} > 0$ and $I_R \approx 0$, $I_C = -\alpha_F I_E$ and Fig. D.2(a) can be approximated by Fig. D.2(b). If we substitute $I_C = -\alpha_F I_E$, Eq. (D.10) becomes

$$I_E = \alpha_R\alpha_F I_E - I_{EBO}\left(\exp\frac{V_{BE}}{V_T} - 1\right)$$

which relates I_E to α_R, α_F, and V_{BE} by

$$I_E = \frac{I_{EBO}\exp(V_{BE}/V_T - 1)}{1 - \alpha_R\alpha_F} \tag{D.11}$$

I_{ES} and I_{CS} are also known as the *short-circuit saturation currents*, whereas I_{CBO} and I_{EBO} are known as the *open-circuit saturation currents*. Note that the injection and transport versions of the Ebers–Moll model are interchangeable. Once the parameters of one version are known, the parameters of the other version can be found.

EXAMPLE D.1

Finding the currents for the Ebers–Moll model of an *npn* transistor An *npn* transistor is biased so that $V_{BE} = 0.3$ V and $V_{CE} = 6$ V. If $\alpha_F = 0.99$, $\alpha_R = 0.90$, and $I_{CBO} = 5$ nA, find all the currents for the injection version of the Ebers–Moll model of Fig. D.1(a). Assume thermal voltage $V_T = 25.8$ mV.

SOLUTION

$V_T = 25.8 \text{ mV} = 0.0258$ V. Since V_{BE} is positive, the base–emitter junction is forward biased. The voltage at the collector–base junction is

$$V_{CB} = V_{CE} + V_{EB} = V_{CE} - V_{BE} = 6 - 0.3 = 5.7 \text{ V}$$

or $\quad V_{BC} = -5.7$ V

Since V_{CB} is positive, the collector–base junction is reverse biased, and the transistor operates in the active region. From Eq. (D.8),

$$I_{EBO} = \frac{\alpha_R I_{CBO}}{\alpha_F} = 0.9 \times \frac{5 \text{ nA}}{0.99} = 4.545 \text{ nA}$$

From Eq. (D.6),

$$I_{CS} = \frac{I_{CBO}}{1 - \alpha_R \alpha_F} = \frac{5 \text{ nA}}{1 - 0.9 \times 0.99} = 45.87 \text{ nA}$$

From Eq. (D.7),

$$I_{ES} = \frac{I_{EBO}}{1 - \alpha_R \alpha_F} = \frac{4.545 \text{ nA}}{1 - 0.9 \times 0.99} = 41.697 \text{ nA}$$

From Eq. (D.1), the forward diode current is

$$I_F = 41.697 \times 10^{-9} \times \left(\exp\frac{0.3}{0.0258} - 1\right) \approx 4.678 \text{ mA}$$

From Eq. (D.2), the reverse diode current is

$$I_R = 45.87 \times 10^{-9} \times \left(\exp\frac{-5.7}{0.0258} - 1\right) \approx -45.87 \text{ nA}$$

and $\quad \alpha_R I_R \approx -0.9 \times 45.87 \text{ nA} = -41.28 \text{ nA}$

$$\alpha_F I_F \approx 0.99 \times 4.678 \text{ mA} = 4.63 \text{ mA}$$

$$I_E = -I_F + \alpha_R I_R = -4.678 \times 10^{-3} - 41.28 \times 10^{-6} \approx -4.719 \text{ mA}$$

$$I_C = \alpha_F I_F - I_R = 4.63 \times 10^{-3} + 45.87 \times 10^{-9} \approx 4.63 \text{ mA}$$

$$I_B = -(I_E + I_C) = -(-4.719 \times 10^{-3} + 4.63 \times 10^{-3}) = 89 \text{ } \mu\text{A}$$

EXAMPLE D.2

Finding the currents for the Ebers–Moll model of a *pnp* transistor The model parameters of the *pnp* transistor in Fig. 8.8(b) are $\alpha_R = 0.9$, $\alpha_F = 0.99$, $I_{ES} \approx I_{CS} = 45$ nA, $V_{EB} = 0.4$ V, and $V_{CB} = 0.3$ V. Determine **(a)** the currents for the transport version of the Ebers–Moll model in Fig. D.2(a) and **(b)** $\beta_F = \beta_{forced}$. Assume thermal voltage $V_T = 25.8$ mV.

SOLUTION

For $V_{EB} = 0.4$ V and $V_{CB} = 0.3$ V, the collector–base and emitter–base junctions are forward biased. Therefore, the transistor is operated in the saturation region.

(a) For a *pnp* transistor, all the polarities of voltages and currents are reversed; thus, Eqs. (D.3) and (D.4) become

$$I_E = I_{ES}\left(\exp\frac{V_{EB}}{V_T} - 1\right) - \alpha_R I_{CS}\left(\exp\frac{V_{CB}}{V_T} - 1\right) \tag{D.12}$$

$$= 45 \times 10^{-9}\left(\exp\frac{0.4}{25.8\text{ m}} - 1\right) - 0.9 \times 45 \times 10^{-9}\left(\exp\frac{0.3}{25.8\text{ m}} - 1\right)$$

$$= 243.48\text{ mA} - 4.54\text{ mA} = 238.94\text{ mA}$$

and

$$I_C = -\alpha_F I_{ES}\left(\exp\frac{V_{EB}}{V_T} - 1\right) + I_{CS}\left(\exp\frac{V_{CB}}{V_T} - 1\right) \tag{D.13}$$

$$= -0.99 \times 45 \times 10^{-9}\left(\exp\frac{0.4}{25.8\text{ m}} - 1\right) + 45 \times 10^{-9}\left(\exp\frac{0.3}{25.8\text{ m}} - 1\right)$$

$$= -241.04\text{ mA} + 5.05\text{ mA} = -235.99\text{ mA}$$

Thus, $I_B = -(I_E + I_C) = -(238.94 - 235.99) = -2.95$ mA

(b) Since the transistor is operated in the saturation region, the value of the forward current gain becomes less than that of the active region. The forward current gain is known as the *forced current gain* β_{forced}:

$$\beta_{forced} = \frac{I_C}{I_B} = \frac{235.99\text{ mA}}{2.95\text{ mA}} = 80$$

EXAMPLE D.3

Finding the collector–emitter saturation voltage of a BJT An *npn* transistor is operated in the saturation region, and its parameters are $\alpha_R = 0.9$, $\alpha_F = 0.989$, $\beta_F = 89.91$, and $V_T = 25.8$ mV. Calculate the collector–emitter saturation voltage $V_{CE(sat)}$.

SOLUTION

From Eq. (8.6),

$$\beta_F = \frac{\alpha_F}{1 - \alpha_F} = \frac{0.989}{1 - 0.989} = 89.91$$

We know that

$$V_{BC} = V_{BE} + V_{EC} = V_{BE} - V_{CE}$$

In the saturation region, $V_{BE} > 0$ and $V_{BC} > 0$; that is,

$$\exp\left(\frac{V_{BE}}{V_T}\right) >> 1 \quad \text{and} \quad \exp\left(\frac{V_{BC}}{V_T}\right) >> 1$$

Using Eqs. (D.3) and (D.5), we get

$$I_E = -I_{ES}e^{V_{BE}/V_T} + a_F I_{ES}e^{V_{BC}/V_T}$$

$$\approx -I_{ES}e^{V_{BE}/V_T} + \alpha_F I_{ES}e^{(V_{BE}-V_{CE})/V_T}$$

$$\approx -I_{ES}e^{V_{BE}/V_T}[1 - \alpha_F e^{-V_{CE}/V_T}]$$

But $I_E + I_C + I_B = 0$, so $-I_E = (I_C + I_B)$; that is,

$$I_C + I_B = I_{ES}e^{V_{BE}/V_T}[1 - \alpha_F e^{-V_{CE}/V_T}] \tag{D.14}$$

Similarly, using Eqs. (D.4) and (D.5), we get

$$I_C = \alpha_F I_{ES}\left(\exp\frac{V_{BE}}{V_T} - 1\right) - I_{CS}\left(\exp\frac{V_{BC}}{V_T} - 1\right)$$

$$= \alpha_F I_{ES}e^{V_{BE}/V_T}\left[1 - \frac{1}{\alpha_R}e^{-V_{CE}/V_T}\right] \tag{D.15}$$

Dividing Eq. (D.14) by Eq. (D.15) yields

$$\frac{I_C + I_B}{I_C} = 1 + \frac{I_B}{I_C} = \frac{1 - \alpha_F e^{-V_{CE}/V_T}}{\alpha_F[1 - (e^{V_{CE}/V_T})\alpha_R]}$$

which, after simplification, becomes

$$V_{CE} = V_T \ln\left\{\frac{\alpha_F[1 + I_C/I_B\,(1 - \alpha_R)]}{\alpha_R[\alpha_F + I_C/I_B\,(\alpha_F - 1)]}\right\} \tag{D.16}$$

In the active region, the base current I_B is related to the collector current I_C by $I_B = I_C/\beta_F$. The saturation region is considered to begin at the point where the forward current gain β_F is 90% of the value in the active region. That is, $\beta_{sat} = \beta_{forced} = 0.9\beta_F$ and $I_C = 0.9 I_B \beta_F$. Equation (D.16) gives the collector–emitter saturation voltage as

$$V_{CE(sat)} = V_T \ln\left\{\frac{\alpha_F[1 + 0.9\beta_F(1 - \alpha_R)]}{\alpha_R[\alpha_F + 0.9\beta_F(\alpha_F - 1)]}\right\}$$

Substituting $\beta_F = \alpha_F/(1 - \alpha_F)$ in the denominator gives

$$V_{CE(sat)} = V_T \ln\left[\frac{1 + 0.9\beta_F(1 - \alpha_R)}{\alpha_R(1 - 0.9)}\right] \tag{D.17}$$

For $V_T = 0.0258$ V, $\alpha_R = 0.9$, and $\beta_F = 89.91$, we get

$$V_{CE(sat)} = 0.0258 \times \ln\left[\frac{1 + 0.9 \times 89.91 \times (1 - 0.9)}{0.9 \times (1 - 0.9)}\right] = 0.119 \text{ V}$$

APPENDIX E

PASSIVE COMPONENTS

E.1 Resistors

TABLE E.1 Standard values (in ohms) for metal film resistors (tolerance ±1%) and carbon resistors (tolerance ±5% and ±10%)

1%	5%	10%	1%	5%	10%	1%	5%	10%	1%	5%	10%
10.0	10	10	20.0	20		40.2			61.9	62	
10.2			20.5			41.2			63.4		
10.5			21.0			42.2			64.9		
10.7			21.5			43.2	43		66.5		
11.0	11		22.1	22	22	44.2			68.1	68	68
11.3			22.6			45.3			69.8		
11.5			23.2			46.4			71.5		
11.8	12	12	23.7			47.5	47	47	73.2		
12.1			24.3	24		48.7			75.0	75	
12.4			24.9			49.9			76.8		
12.7			25.5			51.1	51		78.7		
13.0	13		26.1			52.3			80.6		
13.3			26.7	27	27	53.6			82.5	82	82
13.7			27.4			54.9			84.5		
14.0			28.0			56.2	56	56	86.6		
14.3			28.7			57.6			88.7		
14.7			29.4			59.0			90.9	91	
15.0	15	15	30.1	30		60.4			93.1		
15.4			30.9						95.3		
15.8			31.6						97.6		
16.2	16		32.4	33	33						
16.5			33.2								
16.9			34.0								
17.4			34.8								
17.8			35.7	36							
18.2	18	18	36.5								
18.7			37.4								
19.1			38.3								
19.6			39.2	39	39						

▶ **NOTE** The available values for carbon are $1\ \Omega \le R \le 100\ \text{M}\Omega$ and for metal film are $10\ \Omega \le R \le 10\ \text{M}\Omega$. The available values are obtained by multiplying the sequence number by a power of 10 (i.e., 10^{-1}, 10^{0}, 10^{1}, 10^{2}, 10^{3}, and so on).

TABLE E.2 Standard values (in ohms) for wire-wound resistors (tolerance ±5%)

0.008	0.75	7.5	27	62	180	450	1 k	4 k
0.01	1.0	8	30	70	200	470	1.2 k	5 k
0.02	1.5	10	33	75	220	500	1.3 k	10 k
0.03	2.0	12	35	80	250	560	1.5 k	15 k
0.05	2.5	15	40	82	270	600	1.8 k	20 k
0.1	3.0	16	45	100	300	680	2 k	25 k
0.15	3.3	20	47	110	330	700	2.2 k	40 k
0.2	4.0	22	50	120	390	750	2.5 k	50 k
0.26	5.0	22.5	56	150	400	910	3 k	100 k
0.3	6	25	60	160	430		3.5 k	150 k
0.5	7							

TABLE E.3 Power ratings for resistors

Type	Tolerance	Power Rating
Carbon resistors	5% and 10%	$\frac{1}{8}$ W $\frac{1}{4}$ W $\frac{1}{2}$ W 1 W 2 W
Metal film resistors	1%	$\frac{1}{8}$ W $\frac{1}{4}$ W $\frac{1}{2}$ W
Wire-wound resistors	5%	5 W 12 W 25 W 50 W 100 W 225 W

TABLE E.4 Color code for resistors

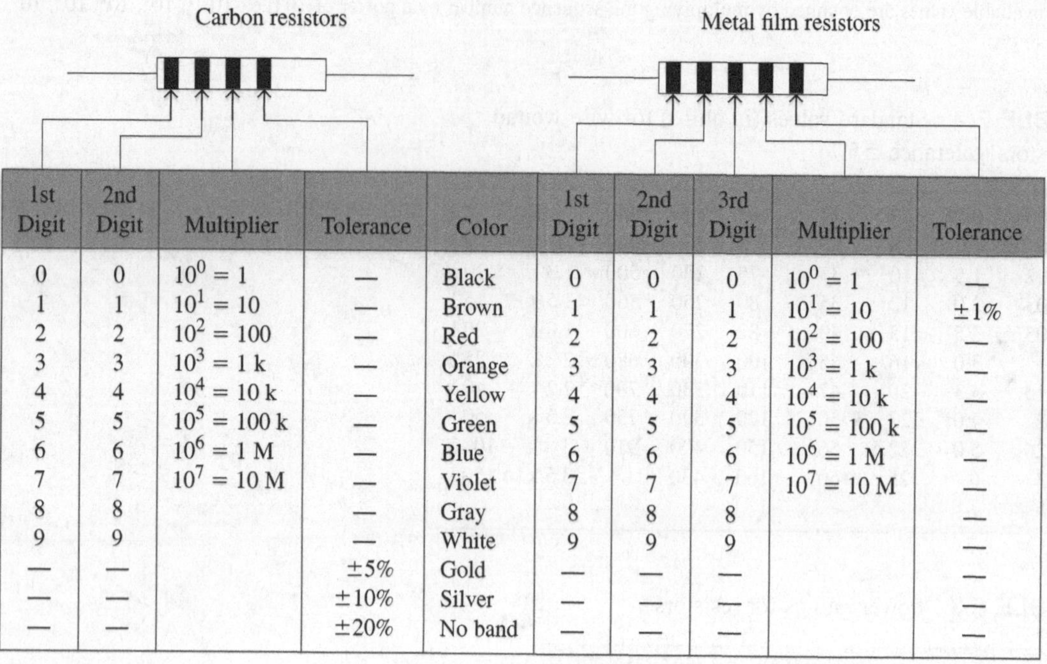

1st Digit	2nd Digit	Multiplier	Tolerance	Color	1st Digit	2nd Digit	3rd Digit	Multiplier	Tolerance
0	0	$10^0 = 1$	—	Black	0	0	0	$10^0 = 1$	—
1	1	$10^1 = 10$	—	Brown	1	1	1	$10^1 = 10$	±1%
2	2	$10^2 = 100$	—	Red	2	2	2	$10^2 = 100$	—
3	3	$10^3 = 1$ k	—	Orange	3	3	3	$10^3 = 1$ k	—
4	4	$10^4 = 10$ k	—	Yellow	4	4	4	$10^4 = 10$ k	—
5	5	$10^5 = 100$ k	—	Green	5	5	5	$10^5 = 100$ k	—
6	6	$10^6 = 1$ M	—	Blue	6	6	6	$10^6 = 1$ M	—
7	7	$10^7 = 10$ M	—	Violet	7	7	7	$10^7 = 10$ M	—
8	8		—	Gray	8	8	8		—
9	9		—	White	9	9	9		—
—	—		±5%	Gold	—	—	—		—
—	—		±10%	Silver	—	—	—		—
—	—		±20%	No band	—	—	—		—

E.2 Potentiometers

TABLE E.5 Standard values (in ohms) for carbon composition, linear taper potentiometers (tolerance ±10%; power rating 2.25 W)

50	1 k	10 k	100 k	1 M
150	1.5 k	15 k	150 k	1.5 M
200	2 k	20 k	200 k	2 M
250	2.5 k	25 k	250 k	2.5 M
350	3.5 k	35 k	350 k	3.5 M
500	5 k	50 k	500 k	5 M
750	7.5 k	75 k	750 k	

TABLE E.6 Standard values (in ohms) for conductive plastic potentiometers (tolerance ±10%; power rating $\frac{1}{2}$ W)

250			
1 k	10 k	100 k	1 M
2.5 k	25 k	250 k	2.5 M
5 k	50 k	500 k	5 M

TABLE E.7 Standard values (in ohms) for CERMET potentiometers (tolerance ±10%; power rating $\frac{1}{2}$ W)

50				
100	1 k	10 k	100 k	1 M
200	2 k	20 k	200 k	2 M
500	5 k	50 k	500 k	

E.3 Capacitors

TABLE E.8 Standard values for polarized aluminum electrolytic capacitors (tolerance −10% to +50%)

Voltage (V)	Capacitance (μF)	Voltage (V)	Capacitance (μF)	Voltage (V)	Capacitance (μF)
10	22	25	10	50	0.1
	33		22		0.22
	47		33		0.33
	100		47		0.47
	220		100		1.0
	330		220		2.2
	470		330		3.3
	1000		470		4.7
	2200		1000		10
	3300		2200		22
	4700		3300		33
	6800		4700		47
	10,000				100
					220
					330
					470
					1000
					2200

TABLE E.9 Standard values for ceramic disc capacitors (tolerance ±10%)

Voltage (V)	Capacitance (pF)
200	10
	15
	22
	33
	47
	68
	100
	150
	220
	330
	470
	680
	1000
	1500
	2200
	3300
	4700
	6800
	10,000
	15,000

TABLE E.10 Standard values for mylar polyester capacitors (tolerance ±10%)

Voltage (V)	Capacitance (μF)
100	0.001
	0.0015
	0.0022
	0.0033
	0.0047
	0.0068
	0.0082
	0.01
	0.015
	0.022
	0.027
	0.033
	0.039
	0.047
	0.056
	0.068
	0.082
	0.1
	0.12
	0.15
	0.18
	0.22
	0.27
	0.33
	0.39
	0.47
	0.56
	0.68
	0.82
	1

TABLE E.11 Standard values for ceramic variable capacitors (tolerance ±10%)

	Capacitance (pF)	
Voltage (V)	Min	Max
250	1	4.5
	2.5	10
	4	18
	6	35
	7	40
	8	50

APPENDIX F

DESIGN PROBLEMS

Mini Design Projects

F.1 Design a circuit that will sum the sensing signals generated by the probes of an electrocardiogram. The output can be expressed in terms of the input signals (V_A, V_B, V_C, and V_D) as follows:

$$V_O = 1.4V_A + 5V_B + 3V_C + 0.6V_D$$

The accuracy should be better than 2%. The DC power supplies are ± 15 V.

F.2 Design a BJT buffer amplifier to give a midfrequency voltage gain of $|A_v| = v_L/v_s \approx 1$ and an input resistance of $R_{in} = v_s/i_s \geq 50$ kΩ. The load resistance is $R_L = 10$ kΩ. Assume a source resistance of $R_S = 500$ Ω and $V_{CC} = 15$ V.

F.3 Design an NMOS buffer amplifier to give a midfrequency voltage gain of $|A_v| = v_L/v_s \approx 1$ and an input resistance of $R_{in} = v_s/i_s \geq 500$ kΩ. The load resistance is $R_L = 10$ kΩ. Assume a source resistance of $R_S = 500$ Ω and $V_{DD} = 15$ V.

F.4 Design a BJT buffer amplifier with an active load to give a midfrequency voltage gain of $|A_v| = v_L/v_s \approx 1$ and an output resistance of $R_{in} = v_s/i_s \geq 50$ kΩ. The load resistance is $R_L = 10$ kΩ. Assume a source resistance of $R_S = 500$ Ω and $V_{CC} = V_{EE} = 15$ V.

Medium Design Projects

F.5 Design a DC power supply to electronic equipment from an AC supply of 120 V (rms) $\pm$ 10%, 60 Hz. The load requires ± 12 V $\pm$ 5% at 0.5 A. Use discrete devices only and design for minimum expense.

F.6 **a.** Design a BJT amplifier to give a midfrequency voltage gain of $|A_v| = v_L/v_s = 20 \pm 5\%$ at a load resistance of $R_L = 10$ kΩ. Assume a source resistance of $R_S = 500$ Ω and $V_{DD} = 15$ V.

b. Modify the design so that the amplifier operates in the frequency range from 10 Hz to 100 kHz.

F.7 **a.** Design a depletion NMOS amplifier to give a midfrequency voltage gain of $|A_v| = v_L/v_s = 20 \pm 5\%$ at a load resistance of $R_L = 10$ kΩ. Assume a source resistance of $R_S = 1.5$ kΩ and $V_{DD} = 15$ V.

b. Modify the design so that the amplifier operates in the frequency range from 5 Hz to 50 kHz.

F.8 **a.** Design a MOSFET amplifier to give a midfrequency voltage gain of $|A_v| = v_L/v_s = 10 \pm 5\%$ at a load resistance of $R_L = 10$ kΩ. Assume a source resistance of $R_S = 1.5$ kΩ and $V_{DD} = 12$ V.

b. Modify the design so that the amplifier operates in the frequency range from 10 Hz to 50 kHz.

F.9 **a.** Design an NMOS amplifier with an active load to give a midfrequency voltage gain of $|A_v| = v_L/v_s \geq 250 \pm 5\%$ at a load resistance of $R_L = 20$ kΩ. Assume a source resistance of $R_S = 1.5$ kΩ and $V_{DD} = 15$ V.

b. Modify the design so that the amplifier operates in the frequency range from 10 Hz to 50 kHz.

Large Design Projects

F.10 The input signal to an amplifier is $v_s = 2$ mV, and the amplifier has a source resistance of $R_S = 1$ kΩ.

a. Design a BJT amplifier to give a midfrequency voltage gain A_v ($= v_L/v_s$ with a load resistance $R_L = 10$ kΩ) of greater than 650. The input resistance R_{in} of the amplifier should be greater than 70 kΩ, and the output resistance R_{out} should be less than 250 Ω. The DC supply voltage is $V_{CC} = 12$ V.

b. Modify the design so that the amplifier operates in the frequency range from 10 kHz to 80 kHz.

c. Apply feedback and modify the design. The input resistance must be increased by 20 (i.e., $R_{if} \geq 20\ R_{in}$), and the output resistance must be decreased by 20 (i.e., $R_{of} = R_o/20$).

d. Apply feedback and modify the design so that the amplifier oscillates at a frequency of $f_o = 20$ kHz.

F.11 The input signal to an amplifier is $v_s = 2$ mV, and the amplifier has a source resistance of $R_S = 1$ kΩ.

a. Design an NMOS amplifier to give a midfrequency voltage gain A_v ($=v_L/v_s$ with a load resistance $R_L = 10$ kΩ) of greater than 450. The input resistance R_{in} of the amplifier should be greater than 500 kΩ, and the output resistance R_{out} should be less than 250 Ω. The DC supply voltage is $V_{DD} = 12$ V.

b. Modify the design so that the amplifier operates in the frequency range from 20 kHz to 60 kHz.

c. Apply feedback and modify the design. The input resistance must be increased by 10 (i.e., $R_{if} \geq 10\,R_{in}$), and the output resistance must be decreased by 10 (i.e., $R_{of} \leq R_o/10$).

d. Apply feedback and modify the design so that the amplifier oscillates at a frequency of $f_o = 20$ kHz.

F.12 Design a BJT operational amplifier to give a large-signal differential gain of $10^3 \pm 10\%$. The input resistance should be higher than 100 kΩ ± 5%, and the output resistance should be less than 210 Ω ± 5%. The op-amp should have a common-mode rejection ratio of CMRR = $10^4 \pm 10\%$. The unity-gain bandwidth must be better than $10^4 \pm 10\%$.

F.13 Design a multistage BJT amplifier to meet the following specifications: voltage gain, $|A_v| = v_L/v_s = 600 \pm 5\%$ (with load); input resistance, $R_i = v_s/i_s \geq 25$ kΩ; output resistance, $R_o \leq 300$ Ω; load resistance, $R_L = 25$ kΩ; source resistance, $R_s = 1$ kΩ; DC supply, $V_{CC} = 15$ V; input signal, $v_s = 1$ mV to 2 mV (peak sinusoidal), 1 kHz. The amplifier should operate in the frequency range from 10 Hz to 30 kHz.

F.14 Design a multistage MOS amplifier to meet the following specifications: voltage gain, $|A_v| = v_L/v_s = 400 \pm 5\%$ (with load); input resistance, $R_i = v_s/i_s \geq 25$ kΩ; output resistance, $R_o \leq 300$ Ω; load resistance, $R_L = 25$ kΩ; source resistance, $R_s = 1$ kΩ; DC supply, $V_{DD} = 15$ V; input signal, $v_s = 1$ mV to 2 mV (peak sinusoidal), 1 kHz. The amplifier should operate in the frequency range from 100 Hz to 25 kHz.

F.15 Design a differential amplifier to satisfy the following conditions: a differential voltage gain of $A_o \geq 2500$; common-mode rejection ratio of CMRR ≥ 4500; load of $R_L = 50$ kΩ; coupling capacitor of $C_L = \infty$. Use all BJTs and only one resistance. Use PSpice to verify your design with $v_d = 5$ μV and $v_c = 10$ mV. Supply voltages are $V_{CC} = V_{EE} = 15$ V. The amplifier should operate in the frequency range up to 25 kHz.

F.16 Design a differential amplifier to satisfy the following conditions: a differential voltage gain of $A_o \geq 2500$; common-mode rejection ratio of CMRR ≥ 4500; load of $R_L = 50$ kΩ; coupling capacitor of $C_L = \infty$. Use all MOSFETs. Use PSpice to verify your design with $v_d = 5$ μV and $v_c = 10$ mV. Supply voltages are $V_{DD} = V_{SS} = 15$ V. The amplifier should operate in the frequency range up to 25 kHz.

ANSWERS TO SELECTED PROBLEMS

Chapter 1

1.5 80, 60°

1.7 6.66 mS

1.9 **(a)** $v_{CE} = (6 + 0.1 \sin 2000\pi t)$ V, $v_{BE} = (700 + 1 \sin 2000\pi t)$ mV
(b) 100

1.10 **(a)** $v_{DS} = (6 + 0.05 \sin 1000\pi t)$ V, $v_{GS} = (3 + 0.002 \sin 1000\pi t)$ V
(b) 25

Chapter 2

2.1 **(a)** 42.28 dB, 62.28 dB, 104.56 dB, 50 kΩ
(b) 0.94% **(c)** 115.4 mV

2.4 **(a)** 133.6, 56.84, 7594

2.6 225 Ω

2.13 **(a)** 191.7, 3333, 639×10^3

2.14 2.33 A

2.16 $R_o = 47.5$ kΩ, $R_i = 112$ Ω

2.21 12.25 kΩ

2.34 **(a)** 62.72 dB, 100.89 dB, 163.6 dB

2.35 **(a)** 94.77 dB, 97.62 dB, 192.39 dB

2.37 **(a)** −3.96

2.38 **(a)** $R_i = 1$ MΩ, $C_1 = 15.9$ pF, $C_2 = 637$ pF

2.43 **(a)** $\omega_1 = 20.48$ rad/s, $\omega_2 = 1.02 \times 10^6$ rad/s

(b) $-\dfrac{956.48}{(1 + j\omega/20.48)[1 + j\omega/(1.02 \times 10^6)]}$

2.44 **(a)** 20.1 μF **(b)** 0.2 μF

2.45 **(a)** $A_v = -19.63$

Chapter 3

3.1 **(a)** ±75 μV **(b)** ±37.5 pA

3.4 −14 V

3.5 +14 V

3.6 **(a)** 40 μV **(b)** 80 μV **(c)** 6.32
(d) $(8 \pm 50 \times 10^{-6})$ V

3.10 **(a)** 250×10^{12} Ω, 0.2 Ω
(b) 5×10^{15} Ω, 0.01 Ω

3.15 10^{12} Ω, 1.7×10^{-4} Ω

3.17 2×10^6 Hz, 1.99 ∠ −5.7°, 1 MHz

3.21 −14 V

3.23 **(a)** 2 MHz, −3.257 ∠ −12.22°, 461.53 kHz

3.28 **(a)** 4.95 V **(b)** 5 V

3.30 −0.2 V

3.39 3.18×10^{-3} Hz

3.42 **(a)** 20 μs **(b)** 7958 Hz **(c)** 5

3.45 −3 V

3.51 20 μA

Chapter 4

4.12 **(b)** 2.14×10^{-14} A

4.16 3.56 mA

4.22 **(a)** 3.58 V **(b)** 6.42 mA

4.25 **(a)** 0.613 V, 18.8 mA **(b)** 16.3 Ω
(c) 0.3 V **(d)** 2.36 ∠ 79.2° mV rms

4.29 **(a)** 6.73 V **(b)** 6.715 V, 6.707 V
(c) 2.5 V

4.31 **(a)** 1317 Ω, 131.7 mW

4.34 **(a)** $R_1 = 1398$ kΩ, $R_2 = 66.7$ kΩ

4.37 **(a)** $v_{D2} = V_Z$, $v_{D1} = V_S - v_{D2}$
(b) $V_{D1} = 0.8$ V, $V_{D2} = 6.7$ V
(c) 115.3 mA

Chapter 5

5.2 $$v_O(t) = \frac{169.7}{\pi} + \frac{169.7}{2}\sin 314t - \frac{2 \times 169.7}{\pi}\cos 62t + \frac{2 \times 169.7}{15\pi}\cos 1256t - \frac{2 \times 169.7}{35\pi}\cos 1884t + \cdots$$

5.4 **(a)** 5/4 V **(b)** 2.04 V **(c)** 0.577

5.6 (a) $v_O(t) = \frac{2V_m}{\pi} \times \left[1 + 2\sum_{n=1}^{\infty} \frac{1}{(1-4n^2)} \cos n\omega_o t\right]$,
$f_o = 50$ Hz, $V_m = 31.1$ V

5.8 (a) 566.5 V (b) 1.0396

5.9 (a) $R_s = 3485\ \Omega$

5.13 (a) 14.9 H

5.15 (a) 64 μF

5.16 (a) 126 μF (b) $V_{o(av)} = 169.7$ V, $V_{o(\text{no load})} = 158.7$ V

5.19 (a) 6.47 H

5.22 (a) 150 V, 0.4714

5.31 (a) $R_s = 70\ \Omega$, 100 kΩ, 0.1 μF

5.32 (a) $R_s = 50\ \Omega$, 100 kΩ, 0.1 μF

Chapter 7

7.16 1.94 V

7.18 (a) 3.4 mA (b) −5.5 V (c) −2.0 V (d) 3.33

7.21 (a) 2.9 mA, 3.98 V, −3.34 V (b) 1447 Ω

7.24 (a) −0.67 V (b) 89.3 Ω, 1244 Ω

7.28 11.04 mA, 6.96 V, 0.52 V

7.31 9.05 V, 0.905 mA, 2.95 V

7.34 (a) 100.5 kΩ (b) −4.9 (c) 4.63 kΩ (d) −3.35

7.37 (a) 10 MΩ (b) −68.3 (c) 3.13 kΩ (d) −42

7.44 (a) 10 MΩ (b) 0.78 (d) 0.64

7.47 (a) 4.58 mA/V, 58.36 kΩ (b) 60 MΩ, 0.813, 0.714

7.63 (a) $C_{gs} = 2.3$ pF, $C_{gd} = 2.87$ pF (b) 9.6×10^{10} rad/s

7.65 (a) $C_{sb} = 0.3$ pF, $C_{gd} = 1.5$ pF (b) 0.91×10^{-9} rad/s

Chapter 8

8.12 (a) 150.5 (b) 3.76 mA (c) 3.787 mA

8.21 (a) $\beta = 50$: 158.7 μA, 7.935 mA, 8.09 mA, 2.29 V
$\beta = 250$: 36.3 μA, 9.11 mA, 9.146 mA, 2.53 V
(b) $\beta = 50$: 185.2 μA, 9.26 mA, 9.445 mA, 2.55 V
$\beta = 250$: 43.29 μA, 10.82 mA, 10.86 mA, 1.13 V

8.23 5.66 kΩ, 3.19 V

8.26 (a) 22.9 μA, 2.29 mA, 1.22 V
(b) 88.76 mA/V, 1131 Ω, 87.3 kΩ
(c) 1090 Ω, −4.95, −86.5, −22.33

8.32 (a) 2.26 mA, 88.5 kΩ
(b) 101.1 kΩ, 11.3 Ω

8.50 (a) $C_{je} = 16$ pF, $C_\mu = 2.16$ pF, $C_\pi = 415$ pF (b) 507 ps

8.52 (a) $C_\mu = 2.29$ pF, $C_\pi = 1538$ pF (b) 786 ps

8.60 0.22 Hz, 1.59 MHz

Chapter 9

9.13 $W/L = 39$, $V_t = 5.06$ V

9.19 (a) $I_{D1,2} = 0.168$ mA
(b) $A_d = -1.832$, $A_c = 9.945 \times 10^{-3}$, CMRR = 184.2, $v_O = -54.96$ mV

9.21 (a) 96.46 μA, 103.54 μA

9.22 (a) 0.715 mA (b) 139.8 kΩ (c) 100 V (d) 1.2909

9.24 (a) 1.36 mA (b) 73.53 kΩ (c) 100 V (d) 1.282

9.26 (a) 4.152 mA (b) 24 kΩ (c) 100 V (d) 1.142

9.28 $R_2 = 11.88$ kΩ, $R_o = 55.69$ MΩ

9.30 (a) $R_1 = 2.86$ MΩ (b) $R_o = 760$ MΩ

9.31 1.231 mA, 9.26 MΩ

9.35 For $V_A = 100$ and $\beta_F = 100$, $r_o = 100$ kΩ; for $V_{CC} = 15$ V, $R_1 = 13.6$ kΩ

9.37 (a) 2.955 mA, 2.011 mA (b) −961.5, −0.25, 38.44×10^6, 20.8 Ω, 6.04 MΩ

9.41 769.2, 2.08 MΩ, 8 MΩ, 587.9

9.45 −28.57, 0, 0

9.47 $-60.14, 0, 0$

9.49 $A_{do} = -961$, $\omega_H = 51.1$ krad/s

Chapter 10

10.1 **(a)** -9.5×10^{-3}
(b) -101.26

10.4 **(a)** 22.173, 10.4% **(b)** 9.99, 4%

10.5 9

10.6 **(a)** 2 **(b)** 10^6 Hz **(c)** 2×10^6

10.9 **(a)** 3.17×10^{11} Ω **(b)** 0.475 mΩ
(c) 1.25

10.11 **(a)** 804 kΩ **(b)** 7.29 Ω
(c) 1.31

10.17 5,355,006 Ω, 0.3 Ω, 7.43

10.20 **(a)** 14.06 kΩ **(b)** 30.2 Ω
(c) 1427.7

10.24 **(a)** 28.5 kΩ **(c)** 3.96×10^{-4} A/V

10.25 **(a)** 3.15 MΩ **(b)** 6.3 MΩ
(c) 0.4×10^{-3} A/V

10.28 $R_{if} = 1{,}321{,}326$ Ω, $A_f = 93.6$ mA/V

10.32 **(a)** 3.5 kΩ **(b)** 49 Ω
(c) -97.67 Ω

10.35 26.26 Ω, 13.72 Ω, -3973 V/A

10.38 **(a)** 30.5 Ω **(b)** 342 Ω
(c) 225,077 V/A

10.45 **(a)** 17,029 Ω **(b)** 37,072 Ω
(c) 1.88

10.47 **(a)** 12.86 Ω **(b)** 5 kΩ
(c) 19.61

10.51 $-\frac{1}{4} + \frac{3}{20}e^{-4t} + \frac{3}{30}e^{6t}$

10.52 $e^{0.5t}[2 \sin 10t]$

10.56 $50/(0.234 - j0.643)$

10.58 44.44 Hz

10.60 **(a)** 401.9 kHz **(b)** 127 pF

10.61 10.47 μF

10.64 **(a)** 3.32 nF, 4.8 kΩ
(b) 79.7 pF, 199.8 kΩ

Chapter 11

11.2 $R_L = 1.43$ kΩ, 29 V

11.6 **(a)** $v_O = 14.5$ V, $R_1 = 493$ Ω
(b) 525.5 mW **(c)** 60.4%

11.8 **(a)** $I_C = 3.75$ A **(b)** 28.125 W
(c) 50%

11.11 $V_p = 14.8$ V, $P_L = 10.952$ W, 77.5%, $R_1 = 47$ Ω

11.14 **(a)** 7.5 A, $V_{p(max)} = 9.55$ V
(b) $P_{L(max)} = 56.25$ W **(c)** 78.55%

11.16 **(a)** $R_{ref} = 255$ Ω

11.32 76.48 W

Chapter 13

13.1 14.29, 135°

13.5 27 Ω

13.7 318 Ω

13.10 For $C = 0.1$ μF, 318 Ω, 20 kΩ

13.14 58.1 kHz, 2.5 Ω

13.16 For $C = 0.01$ μF, 25.3 mH, 133 Ω

13.19 435.8 kHz, $R_L/2R_1$

13.24 0.046 pF

Chapter 14

14.3 26 mV

14.5 1 pV

14.9 **(a)** ±2.7 mV **(b)** -0.5 V ± 2.7 mV

14.12 **(a)** 0.6 mV **(b)** 2.6 mV

14.17 **(a)** 835.96 kΩ **(b)** 7.75 mA/V
(c) 6480 **(d)** 125 kΩ

14.20 **(a)** 0.2 μA **(b)** 258 kΩ
(c) 6.16 MHz

14.21 **(a)** 0.1 μA **(b)** 516 kΩ

14.23 521,963

14.24 347,104

14.28 **(a)** ±66.66 V/μs **(b)** 205.6 MHz

Chapter 15

15.1 **(a)** 2 kΩ, 3.28 kΩ **(b)** 94

15.2 **(a)** 89.25 ps **(b)** 4.635 mW

15.4 27.38 mA/V^2

15.7 **(a)** 3 V **(b)** 0.4 V, 3.04 mA

15.9 **(a)** 3.5 V, 0.39 V **(b)** 1.5 V, 3.37 mA, 8.438 mW, 0 W

15.11 4.2 V

15.16 **(a)** 2.5 V, 4 V, 1 V **(b)** 0

15.17 **(a)** 4.914 V, 1.914 V

15.18 **(a)** 2.5 V, 3.5 V, 1.5 V

15.19 0.464

15.22 **(a)** 2.5 V, 3.5 V, 1.5 V **(b)** 4.95 V, 0.25 mA, 0.05 V, 0.25 mA

15.23 **(a)** 1/1.5 **(b)** 2.1 V, 2.276 V **(c)** 1.02 ns **(d)** 0.51 pJ

15.28 **(a)** 17.55 kΩ **(b)** 2 **(c)** 0.049 mW

15.30 **(a)** 1175 Ω, 3384 Ω **(b)** 1.006, 1.63 V **(c)** 42

15.32 **(a)** 4.8 kΩ, 14.4 kΩ **(b)** $N = 1$

15.36 0.2 V, $I_B = 2.102$ mA, $I_3 = 1.92$ mA, $N = 58$

15.37 $I_1 = 0.58$ mA, $I_2 = 1.64$ mA, $I_{B3} = 2.102$ mA, $N = 64$

15.38 $R_1 = 4$ kΩ, $I_{B1} = 725$ μA, $I_{LO} = 410$ μA

15.41 21.15 mA, 1.15 mA, 0.846 mA

15.43 **(a)** $v_{O1} = -0.8$ V, $R_{C1} = 400$ Ω, $R_4 = 4.5$ kΩ

15.44 **(a)** $v_A = 2.1$ V, $R_E = 2.3$ kΩ, $v_{O1} = 2.2$ V, $R_4 = 3.7$ kΩ, $V_{B6} = 0.2$ V

15.45 $R_3 = 1.75$ kΩ, $R_E = 1.75$ kΩ, $R_4 = 1250$ Ω

15.47 **(a)** 0.106 ns

Chapter 16

16.4 **(a)** 3.32 V, 0.57 V **(b)** −2.66 V, 0.456 V **(c)** −3.1 V

16.10 For $R_1 = 10$ kΩ, $R_F = 14$ kΩ, $R_x = 5833$ Ω

16.14 For $R_1 = 10$ kΩ, $R_F = 50$ kΩ, $V_{ref} = 4.8$ V

16.18 2127.6 Hz

16.21 **(a)** 16 kHz **(b)** 12.8 kHz

16.26 **(c)** 196 Hz

16.31 **(a)** 2.5 kHz **(b)** ±200/3 kHz **(c)** 543 Hz

16.37 0.747 V

16.38 8

INDEX